Basic Laboratory Methods
for Biotechnology

Basic Laboratory Methods for Biotechnology
Textbook and Laboratory Reference

Third Edition

Lisa A. Seidman, Cynthia J. Moore, and Jeanette Mowery

CRC Press
Taylor & Francis Group
Boca Raton London New York

CRC Press is an imprint of the
Taylor & Francis Group, an **informa** business

Third edition published 2022
by CRC Press
6000 Broken Sound Parkway NW, Suite 300, Boca Raton, FL 33487-2742

and by CRC Press
4 Park Square, Milton Park, Abingdon, Oxon, OX14 4RN

© 2022 Taylor & Francis Group, LLC
Second edition published by Pearson 2009

CRC Press is an imprint of Taylor & Francis Group, LLC

Library of Congress Cataloging-in-Publication Data
A catalog record for this title has been requested

ISBN: 978-0-367-24490-3 (hbk)
ISBN: 978-0-367-24488-0 (pbk)
ISBN: 978-0-429-28279-9 (ebk)

DOI: 10.1201/9780429282799

Typeset in Times
by codeMantra

Contents

UNIT I Biotechnology Is the Transformation of Knowledge into Useful Products

UNIT II   *Introduction to Quality in Biotechnology Workplaces*

UNIT III   *Safety in the Laboratory*

UNIT IV   *Math in the Biotechnology Laboratory: An Overview*

UNIT V   Obtaining Reproducible Laboratory Measurements

UNIT VI   Laboratory Solutions

UNIT VII   Quality Assays and Tests

UNIT VIII   Cell Culture and Reproducibility

UNIT IX *Basic Separation Methods*

UNIT X Biotechnology and Regulatory Affairs

Preface

Consider these headlines from respected scientific journals:

On rigor and replication: "A journalist shines a harsh light on biomedicine's reproducibility crisis" [1].

Reliability of 'new drug target' claims called into question: "Bayer halts nearly two-thirds of its target-validation projects because in-house experimental findings fail to match up with published literature claims…" [2].

Irreproducible biology research costs estimated at $28 billion per year: "Study calculates cost of flawed biomedical research in the United States" [3].

In the preface to the first edition we wrote: "We are aware that basic methods, such as how to mix a solution or weigh a sample, are less glamorous than learning how to manipulate DNA or clone a sheep. However, we also know that, in practice, the most sophisticated and remarkable accomplishments of biotechnology are possible only when the most basic laboratory work is done properly." That was written nearly twenty years ago. Since then, a series of significant publications have sounded an alarm about irreproducible research results. Issues relating to good laboratory practice are on scientist's minds, discussed in journals, and subjected to scrutiny in conferences. Granting agencies and journals have revised their requirements to encourage more robust research practices. Both of the earlier editions of this textbook addressed key practices required to achieve reproducible results. This edition even more explicitly discusses the relationship between proper fundamental practices and reproducibility. A solid grounding in basic, quality practices is essential for success in a biotechnology career, wherever that career pathway leads.

New biotechnology discoveries – such as CRISPR gene editing, RNAi, and stem cell technologies – have changed the landscape of biotechnology. And, an unforeseen event, the intrusion of a pandemic beginning in 2019 and extending through the writing of this book has focused even more attention on the field. It is hard to imagine anything that could more dramatically demonstrate the importance of biotechnology than this pandemic. It is only because of the work of biotechnologists that we have a pathway out of the pandemic that does not involve millions more deaths. Every student of biotechnology should be proud to be entering this critically important field.

Yet, even as biotechnology advances dominate the general and scientific news, the fundamental practices of quality laboratory work have not changed. As authors, we have tasked ourselves with identifying that which is fundamental and will enable students and early-stage professionals to be successful in a changing world. At the same time, we want to highlight biotechnology's stunning achievements in recent years. We have therefore integrated new discoveries into this edition largely via case studies. Case studies have the virtues of being engaging and human. Case studies allow us to introduce both scientific discoveries and the complex and important social and ethical issues presented by these advances.

Another change over the years since the first edition of this text is an increase in the number and quality of educational resources freely or inexpensively available online. These resources include animations, videos, and interactive textbooks. Given the availability of such resources, students might wonder whether there is value in a published textbook. As educators, we think that a textbook, such as this, plays a vital role by presenting and organizing information in a cohesive way such that students can develop a comprehensive framework. Relying on online resources leaves gaps in student understanding. We do, however, value online resources as they can complement and enhance this textbook. We have provided references to some excellent online materials throughout the text.

We also note that this text includes introductory sections on regulatory affairs and other content that is relevant in biotechnology production settings, as contrasted with laboratory workplaces. This is because the term "biotechnology" is not a synonym for "molecular biology." We do not have "biotechnology" until scientific knowledge is transformed into useful products. This transformation involves commercial interests, government regulations, societal and ethical concerns, biomanufacturing technologies, and more. Thus, while this is not a textbook on biomanufacturing, we do think it is important for students of biotechnology to have a mental map of biotechnology as a whole, beginning in the research laboratory and ending up as products that profoundly affect individuals and society.

As with the previous editions, this textbook is written for a broad audience, but is particularly targeted toward individuals preparing for a career in biotechnology, their instructors, and early-stage professionals.

We have endeavored to make the language and discussions accessible – and interesting – to students with some, but not extensive, science backgrounds. We expect that more experienced professionals will also find this to be a useful reference. Achieving reproducible results requires attention to fundamental principles – for example, the metrology concepts that lead to consistent and accurate measurements – and it is difficult to find a comprehensible, thorough introduction to these basic concepts. This text is designed to fill that void in a manner useful to those in the biological sciences, both those entering the field and those with more experience.

This third edition updates references. It provides new illustrations that are so valuable in enhancing learning. We have further included additional practice problems to support educators and students.

Changes in various areas (for example, patent law, regulatory initiatives, and safety regulations) have been updated. New sections have been added on medical devices and cell and tissue manufacturing because of ever-expanding biotechnology applications in diagnostics and regenerative medicine. Working with cells has always been a fundamental topic in biotechnology and was addressed in various places in earlier editions of this textbook. Looking not very far into the future, we see regenerative medicine – which requires growing cells in new and complex ways – as being a "game-changer." Therefore, additional information about reproducibility and cell culture has been added. Properly performing assays, including but not limited to immunoassays, is of fundamental importance in any biotechnology setting. We have therefore expanded the assay unit to include immunoassays and more about PCR assays. PCR, which was already a key technique in molecular biology, became even more important during the COVID-19 pandemic. Problems with antibodies and immunoassays are frequently cited as contributors to irreproducibility in biological research. We therefore added a chapter that addresses immunoassays. As always, in the new material, we have tried to emphasize fundamental principles and practices that lead to quality results.

1. Kaiser, Jocelyn. "Mixed Results from Cancer Replications Unsettle Field." *Science*, vol. 355, no. 6322, 2017, pp. 234–235. doi:10.1126/science.355.6322.234.
2. Mullard, Asher. "Reliability of 'New Drug Target' Claims Called into Question." *Nature Reviews Drug Discovery*, vol. 10, no. 9, 2011, pp. 643–644. doi:10.1038/nrd3545.
3. Baker, Moyna, "Irreproducible Biology Research Costs Put at $28 Billion per Year." *Nature News*, June 9, 2015. doi:10.1038/nature.2015.17711.

Ancillary Materials

LABORATORY MANUAL FOR BIOTECHNOLOGY AND LABORATORY SCIENCE

Learning by doing is clearly vital, and so an online laboratory manual is freely available to students and educators using this textbook: *Laboratory Manual for Biotechnology and Laboratory Science*. Early in our teaching careers, we found that, with beginning students, the best way to teach a fundamental principle is to focus an exercise specifically on that principle. For example, our early attempts to teach students about spectrophotometry included a conceptually complex laboratory exercise on photosynthesis. We found that as students focused on understanding photosynthesis, they used the spectrophotometers in a "cookbook"

fashion, with little understanding of the requirements for making accurate and precise light absorbance measurements. Students learned much more about quality spectrophotometric measurements when we gave them food coloring, and focused their attention on the principles of spectrophotometry. Therefore, the exercises in the online laboratory manual focus on quality principles, such as achieving accuracy and precision in measurements and assays, preparing laboratory solutions, and, importantly, evaluating the consistency and quality of results. The manual does venture into basic cell culture methods and agarose gel electrophoresis, but it is not a molecular biology or cell culture manual. Table 1 below shows the alignment of the laboratory manual with this *Basic Laboratory Methods* textbook.

TABLE 1

Alignment of *Basic Laboratory Methods for Biotechnology* with the Online Laboratory Manual

Textbook Chapter	Laboratory Manual: Classroom Activities	Laboratory Manual: Laboratory Exercises
Topic: Safety in the Biotechnology Laboratory		
Chapters 7, 8, and 9: Introduction, Physical and Chemical Hazards	• Understanding Chemicals with Which You Work • Personal Protection • Analyzing Safety Issues in a Laboratory Procedure	• Tracking the Spread of Chemical Contamination
Chapter 10: Working Safely with Biological Materials		• Production of Bioaerosols and Factors Affecting Bioaerosol Production
Topic: Documentation		
Chapter 6: Documentation; the Foundation of Quality	• Being an Auditor • Writing and Following an SOP	• Keeping a Laboratory Notebook
Topic: Metrology		
Chapter 15: Introduction to Quality Lab Measurements	• Recording Measurements with the Correct Number of Significant Figures	
Chapter 17: The Measurement of Weight **Chapter 18:** The Measurement of Volume	• Constructing a Simple Balance	• Weight Measurements 1: Good Weighing Practices • Weight Measurements 2: Performance Verification • Volume Measurements 1: Proper Use of Volume-Measuring Devices • Volume Measurements 2: Performance Verification of a Micropipette
Chapter 20: The Measurement of pH		• Measuring pH with Accuracy and Precision

(Continued)

TABLE 1 (*Continued*)

Alignment of *Basic Laboratory Methods for Biotechnology* with the Online Laboratory Manual

Textbook Chapter	Laboratory Manual: Classroom Activities	Laboratory Manual: Laboratory Exercises
Chapter 21: Measurements Involving Light – Part A: Basic Principles and Instrumentation	• Beer's Law and Calculating an Absorptivity Constant	• Color and the Absorbance of Light • Concentration, Absorbance, and Transmittance • Preparing a Standard Curve with Food Coloring and Using it for Quantitation • Determination of the Absorptivity Constant for ONP (*o*-nitrophenol, used to assay β-galactosidase activity)

Topic: Laboratory Solutions

Chapter 22: Preparation of Laboratory Solutions – Part A: Concentration Expressions and Calculations	• Getting Ready to Prepare Solutions with One Solute: Calculations • Getting Ready to Prepare Solutions with One Solute: Ordering Chemicals	• Preparing Solutions with One Solute • Preparing Solutions to the Correct Concentration
Chapter 23: Preparation of Laboratory Solutions – Part B: Basic Procedures and Buffers		• Working with Buffers • Preparing Breaking Buffer • Preparing TE Buffer • More Practice Making a Buffer • Making a Quality Product in a Simulated Company

Topic: Assays

Chapter 26: Introduction to Quality Laboratory Assays and Tests		• Two Qualitative Assays
Chapter 28: Reproducible Assays Involving Light		• UV Spectrophotometric Assay of DNA: Quantitative Applications • UV Spectrophotometric Assay of DNA and Proteins: Qualitative Applications • The Bradford Protein Assay: Learning the Assay • The Bradford Protein Assay: Exploring Assay Verification • The Beta-Galactosidase Enzyme Assay • Comparing the Specific Activity of Two Preparations of Beta-Galactosidase • Using Spectrophotometry for Quality Control: Niacin

Topic: Biological Separation Methods

Chapter 33: Introduction to Centrifugation	• Planning for Separating Materials Using a Centrifuge	

(*Continued*)

TABLE 1 (*Continued*)

Alignment of *Basic Laboratory Methods for Biotechnology* with the Online Laboratory Manual

Textbook Chapter	Laboratory Manual: Classroom Activities	Laboratory Manual: Laboratory Exercises
Chapter 34: Introduction to Bioseparations		• Separation of Two Substances Based on Their Differential Affinity for Two Phases • Separation and Identification of Dyes Using Paper Chromatography • Separating Molecules by Agarose Gel Electrophoresis • Using Agarose Gel Electrophoresis to Perform an Assay • Optimizing Agarose Gel Electrophoresis • Quantification of DNA by Agarose Gel Electrophoresis • Introduction to Ion Exchange Chromatography
	Topic: Growing Cells	
Chapter 30: Introduction to Quality Practices for Cell Culture		• Using a Compound Light Microscope • Aseptic Technique on an Open Lab Bench • Working with Bacteria on an Agar Substrate: Isolating Individual Colonies • Gram Staining • The Aerobic Spread-Plate Method of Enumerating Colony-Forming Units • Preparing a Growth Curve for *E. coli* • Aseptic Technique in a Biological Safety Cabinet • Examining, Photographing, and Feeding CHO Cells • Counting Cells Using a Hemacytometer • Subculturing CHO Cells • Preparing a Growth Curve for CHO Cells
Chapter 31: Culture Media for Intact Cells		• Preparing Phosphate-Buffered Saline • Making Ham's F-12 Medium from Dehydrated Powder

BASIC LABORATORY CALCULATIONS FOR BIOTECHNOLOGY

This *Basic Laboratory Methods in Biotechnology* textbook includes a unit that briefly reviews math tools commonly used in the biotechnology laboratory. However, we have found over the years that many students struggle to perform the math calculations required in a typical biotechnology setting, and this brief review was insufficient for their needs. We do not believe that this is because students do not know how to "do" math. In fact, we have found that most of our students possess sufficient knowledge of basic math and introductory algebra to perform routine calculations in the laboratory. Rather, we find that many students have not learned how to apply the abstract math skills that they learned in math classes to the contextualized problems they encounter in the real world. Therefore, we created a separate math calculations textbook to help students systematically practice, and become confident, solving practical, common, laboratory problems. This separate math-centered textbook, *Basic Laboratory Calculations for Biotechnology, (CRC Press, 2022),* is closely aligned with the *Basic Laboratory Methods for Biotechnology* text, but it includes additional practice problems and additional discussion of how math is used as a tool in day-to-day operations in the biotechnology laboratory.

Acknowledgements

As in the first two editions, many people have contributed to this book and we appreciate all their help. We thank our students who provided feedback and purpose over the years. We thank our many talented colleagues who used the first two editions in their classes, and who provided support, ideas, editing, encouragement, and feedback. Thank you to Diana Brandner, Elaine Johnson, Jim De Kloe, Rebecca Dunn, Linnea Fletcher, Mary Ellen Kraus, Jessie Bathe, and Joseph Lowndes for their much-appreciated support. Previous editions were made possible by the skilled staff at Benjamin Cummings including Gary Carlson, Kaci Smith, and Shannon Tozier. This edition was supported by the excellent staff at CRC Press, including our highly supportive editor, Barbara Knott; Danielle Zarfati, editorial assistant; Glenon C. Butler, Jr, production editor, Christian Munoz, cover designer and Karthik Orukaimani, project manager at CodeMantra and the production team at Code Mantra.

Many thanks to the expert reviewers who helped make this a much better book: Cynthia A. Blank, Hagerstown Community College; David Blum, University of Georgia; Jane Breun, Madison Area Technical College; Craig Caldwell, Salt Lake Community College; Michael Fino, MiraCosta College; Todd Freeman, Illinois State University; Collins Jones, Montgomery College; Nick Kapp, Skyline College; Bridgette Kirkpatrick, Collin College; Mary Ellen Kraus, Madison Area Technical College; Melanie Lenahan, Raritan Valley Community College; Ying-Tsu Loh, City College of San Francisco; Nancy Magill, Indiana University; Oana Martin, Madison Area Technical College; Becky A. Mercer, Palm Beach State College; Charlotte Mulvihill, Oklahoma City Community College; Traci Nanni-Dimmey, Berdan Institute; Virginia Naumann, St Louis Community College; Jack O'Grady, Austin Community College; Beverly Owens, Novo Nordisk; Trish Phelps, Austin Community College; Sandra Porter, Digital World Biology, LLC; David Shaw, Madison Area Technical College; Rebecca Siepelt, Middle Tennessee State University; Salvatore Sparace, Clemson University; Thomas Tubon, Madison Area Technical College; Jason Tucker, North Central State College, Laura Vogel, Illinois State University; Duncan Walker, Array BioPharma; Luanne Wolfgram, Johnson County Community College; and Dwayne Zeiler, Bradley University.

We thank Andres Vidal-Gadea and Tom Hammond, Illinois State University, for their expert assistance with specialized molecular biology techniques. We also thank Noreen Warren, who made major contributions to the development of the first edition of this book; Jennifer Leny, Independent Consultant, for her extensive and expert assistance with the chapters on regulatory affairs; Tenneille Ludwig, Director of the WiCell Stem Cell Bank, for assistance with images; and Chiharu Johnston for technical assistance. David Casimir graciously assisted with sections on business and biotechnology.

While we do not want to endorse the products of any company over any other, we do want to acknowledge and thank the many companies that so graciously provided us with illustrations and technical information.

This material is based, in part, on work supported by the National Science Foundation Advanced Technology Education Initiative, under grant number 0501520. Any opinions, findings, conclusions, or recommendations expressed in this material are those of the authors and do not necessarily reflect the views of the National Science Foundation.

Authors

Lisa Seidman obtained her Ph.D. from the University of Wisconsin and has taught for more than 30 years in the Biotechnology Laboratory Technician Program at Madison Area Technical College. She is presently serving as Emeritus Faculty at the college.

Cynthia Moore received her Ph.D. in Microbiology from Temple University School of Medicine. She has taught for more than 20 years in the School of Biological Sciences at Illinois State University, where she currently serves as Emeritus Faculty.

Jeanette Mowery obtained her Ph.D. in Biomedical Science from the University of Texas Health Science Center at Houston. She has taught for more than 20 years in the Biotechnology Laboratory Technician Program at Madison Area Technical College and is currently serving as Emeritus Faculty at the college.

UNIT I

Biotechnology Is the Transformation of Knowledge into Useful Products

---

Chapters in This Unit

- ◆ Chapter 1: Techniques to Manipulate DNA: The Root of the Biotechnology Industry
- ◆ Chapter 2: The Biotechnology Industry Branches Out
- ◆ Chapter 3: The Business of Biotechnology: The Transformation of Knowledge into Products

---

If you are reading this text, you must have an interest in the fascinating field of biotechnology. Biotechnology is a broad term that is used to refer to many things – scientific discoveries, laboratory techniques, commercial enterprises, and more. Biotechnology is ancient, and yet is as cutting edge as today's latest scientific discoveries. Biotechnology is solidly rooted in the ever-expanding and deepening discoveries of basic biological research. *Biotechnology*, however, is about *the transformation of knowledge into products that are valued by people*. This unit will discuss how scientists build on knowledge to create these products.

DOI: 10.1201/9780429282799-1

Case Study: Fast-Tracking a Vaccine

Vaccine development is usually a long process; 12 years from start to public distribution is average. Prior to 2020, the record holder for fastest vaccine development was 4 years for the mumps vaccine in 1967. Yet scientists were able to develop three vaccines for COVID-19 and have them approved for emergency use in the United States in less than a year. How was this possible?

SARS-CoV-2, the agent that causes COVID-19 (which stands for <u>corona</u>virus <u>disease 2019</u>), is a member of the coronavirus family, a group of viruses that has been studied for more than 60 years (Figure 1). This means that scientists already had basic information about the virus lifecycle, as well as its structure and genome. Due to modern advances in nucleic acid sequencing, the genome sequence of this specific virus was determined 10 days after the first case of COVID-19 pneumonia was identified. The laboratory work on potential vaccines was started a few days after the sequence was determined. Online venues for publishing preliminary research results were established to allow scientists worldwide to work with the latest information.

Testing a vaccine for safety and effectiveness in animals and humans is the most time-consuming phase of vaccine development. Due to the pandemic emergency, companies developing vaccines were allowed to run animal testing and early human safety trials simultaneously, shortening the testing timeline significantly. Substantial public and private investments allowed companies to build their vaccine manufacturing facilities, and actually begin manufacturing millions of doses, before testing was completed. Normally, companies would never risk putting such substantial resources into manufacturing until test results proved that their vaccine was effective and safe.

The first two vaccines approved for use in the United States, from Pfizer and Moderna, are the first messenger RNA (mRNA)-based vaccines to be widely distributed for human use. Traditionally, vaccines are virus-based, but in this case, only a small portion of the viral genome, the part containing the code for the S protein, is injected into people. While this is a new technology, it is backed up with more than 10 years of basic research.

As of January 2021, there were nearly 240 more COVID-19 vaccines in development, using a variety of mechanisms for instilling immunity to SARS-CoV-2. Some of these may prove to be easier to store and cheaper to manufacture, making worldwide distribution more efficient than is possible with the mRNA vaccines described above. The development of COVID-19 vaccines is a compelling example of how biotechnology takes basic science and applies it to practical purposes with new products.

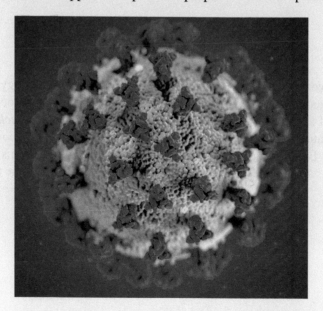

FIGURE 1   The SARS-CoV-2 virus. The focus of vaccine research was the virus spike (S) protein (shown in red), which is responsible for viral attachment and infection of respiratory cells. (Image credit: U.S. Centers for Disease Control.)

This unit introduces the complex biotechnology community, beginning with an introduction to the science of biotechnology and some of the many products that emerge from that science.

Chapter 1 is an overview of the principles that are termed "recombinant DNA technology" and the products that are created with recombinant DNA techniques. The "modern" biotechnology industry emerged from discoveries about how to manipulate and recombine DNA from disparate organisms.

Chapter 2 introduces some of the other products that now fall under the umbrella of modern biotechnology, particularly those that are termed "regenerative medicine" and those that relate to genomics. Biotechnology advances at a breathtaking pace, and so these two chapters are not intended to be comprehensive. Rather they explore the landmark discoveries that led to the "biotechnology revolution" and show how scientific knowledge is continuously being transformed into new products that benefit people.

Chapter 3 introduces the processes that transform knowledge into commercial products. The business side of biotechnology is introduced in this chapter.

BIBLIOGRAPHY FOR UNIT I

There are many good books, journals, and websites on biotechnology. A few examples that discuss topics covered in this unit are listed here, but this list is by no means comprehensive and new resources are appearing constantly. Specific quotes and article references are directly cited in the text.

Books

Clark, David, and Russell, Lonnie. *Molecular Biology Made Simple and Fun*. 4th ed. Cache River Press, 2010. "(An easy-to-read basic introduction to general principles of molecular biology.)

Micklos, David A., Freyer, Greg A. with Crotty, David A. *DNA Science, a First Course*. 2nd ed. Cold Spring Harbor Laboratory Press, 2003.

Renneberg, Reinhard, Viola Berkling, and Vanya Loroch. *Biotechnology for Beginners*. 2nd ed. Academic Press, 2017. (A beautifully illustrated and interesting to read introductory text.)

Thieman, William J., and Palladino, Michael A. *Introduction to Biotechnology*. 4th ed. Pearson, 2019.

Articles

Bell, Jordana T, and Spector, Tim D. "A Twin Approach to Unraveling Epigenetics." *Trends in Genetics: TIG*, vol. 27, no. 3, 2011, pp. 116–25. doi:10.1016/j.tig.2010.12.005.

Chapman, Kenneth, Fields, Timothy, and Smith, Barbara. "The Case of the Q.C. Unit." *Pharmaceutical Technology*, January 1996, pp. 74–9.

Collins, Francis. "A CRISPR Approach to Treating Sickle Cell." National Institutes of Health, U.S. Department of Health and Human Services, 4 April. 2019, directorsblog.nih.gov/2019/04/02/a-crispr-approach-to-treating-sickle-cell/.

Cruz, Martin Paspe. "Conestat Alfa (Ruconest): First Recombinant C1 Esterase Inhibitor for the Treatment of Acute Attacks in Patients With Hereditary Angioedema." *P & T : A Peer-Reviewed Journal for Formulary Management*, vol. 40, no. 2, 2015, pp. 109–14.

Daley, Jim. "Gene Therapy Arrives." *Scientific American*, vol. 322, no. 1, 2020. www.scientificamerican.com/article/gene-therapy-arrives/.

Das, Rathin and Morrow, K. John Jr. "Angiopoietins: Novel Targets for Anti-Angiogenesis Therapy." *BioPharm International*, October 2013, pp. 28–32.

Flamm, Eric L. "How FDA Approved Chymosin: A Case History." *Nature Biotechnology*, vol. 9, no. 4, 1991, pp. 349–51. doi:10.1038/nbt0491–349.

Kaiser, Jocelyn. "Gut Microbes Shape Response to Cancer Immunotherapy." *Science*, vol. 35, no. 6363, 2017, p. 573. doi:10.1126/science.358.6363.573

Keefer, Carol L. "Artificial Cloning of Domestic Animals." *Proceedings of the National Academy of Sciences*, vol. 112, no. 29, 2015, pp. 8874–78. doi:10.1073/pnas.1501718112.

Mullard, Asher. "FDA Approves Landmark RNAi Drug." *Nature Reviews Drug Discovery*, vol. 17, no. 9, 2018, p. 613. doi:10.1038/nrd.2018.152.

Perica, Karlo, et al. "Adoptive T Cell Immunotherapy for Cancer." *Rambam Maimonides Medical Journal*, vol. 6, no. 1, 2015, p. e0004. doi:10.5041/rmmj.10179. (A readable, general introduction to various cell-based immunotherapies for cancer.)

Relling, Mary V., and Evans, William E. "Pharmacogenomics in the Clinic." *Nature*, vol. 526, no. 7573, 2015, pp. 343–50. doi:10.1038/nature15817.

Scudellari, Megan. "Tumor Snipers." *The Scientist*, November 2012, pp. 65–67.

Stein, Rob. "News: First U.S. Patients Treated With CRISPR as Human Gene-Editing Trials Get Underway (NPR News) - Behind the Headlines - NLM." April 16, 2019. www.ncbi.nlm.nih.gov/search/research-news/828.

Stix, Gary. "Hitting the Genetic Off Switch." *Scientific American*, vol. 291, no. 4, 2004, pp. 98–101. doi:10.1038/scientificamerican1004–98. (An article about RNAi.)

Stix, Gary. "The Land of Milk & Money." *Scientific American*, vol. 293, no. 5, 2005, pp. 102–05. doi:10.1038/scientificamerican1105-102. (This article is about the use of animals to produce biopharmaceuticals.)

Usmani, S.S. et al. "THPdb: Database of FDA-approved Peptide and Protein Therapeutics." *PLoS One*, vol. 12, no. 7, 2017, p. e0181748. doi.org/10.1371/journal.pone.0181748

Wilmut, Ian, et al. "Somatic Cell Nuclear Transfer: Origins, the Present Position and Future Opportunities." *Philosophical Transactions of the Royal Society B: Biological Sciences*, vol. 370, no. 1680, 2015, p. 20140366. doi:10.1098/rstb.2014.0366. (A comprehensive review of cloning methods historically and until 2015. The first author is the scientist responsible for cloning Dolly, the sheep.)

Yao, Jian et al. "Plants as Factories for Human Pharmaceuticals: Applications and Challenges." *International Journal of Molecular Sciences*, vol. 16, no. 12, 2015, pp. 28549–65. doi:10.3390/ijms161226122.

Zakrzewski, Wojciech, et al. "Stem Cells: Past, Present, and Future." *Stem Cell Research & Therapy*, vol. 10, 2019, p. 68. doi:10.1186/s13287-019-1165–5.

WEBSITES

The Dolan DNA Learning Center. Award-winning website with information, videos, animations, and interactive activities covering molecular biology and biotechnology. This extensive website can be used as an online textbook for high school and college courses. http://www.dnalc.org.

InnovATEBIO National Biotechnology Education Center. Funded by the National Science Foundation, this extensive website provides multiple resources for teachers and students and links to a comprehensive biotechnology career-exploration website. https://innovatebio.org/.

ISAAA. This is a not-for-profit international organization that advocates for genetically modified food and provides information about biotechnology and agriculture. While it is clearly an advocacy organization, it provides specific and helpful information, statistics, and descriptions of modified crops and methods. https://www.isaaa.org/inbrief/default.asp.

Mayo Clinic. The Mayo Foundation for Medical Education and Research has a good website to address how monoclonal antibodies work to treat cancer. https://www.mayoclinic.org/diseases-conditions/cancer/in-depth/monoclonal-antibody/art-20047808.

National Cancer Institute, NCI. This U.S. federal agency has an informative website with information about cancer and its treatment, including gene therapies. http://www.cancer.gov.

Rising Tide Biology. A personal blog website by Kevin Curran PhD. Normally we would not include a personal blog, but Curran does a nice job summarizing scientific topics. His March 2020 summary of stem cells is a good resource for students. https://www.risingtidebio.com/.

United States Patent and Trademark Office. This U.S. federal agency has an informative website with information about patenting. https://www.uspto.gov/learning-and-resources/inventors-entrepreneurs-resources.

1 Techniques to Manipulate DNA
The Root of the Biotechnology Industry

1.1 SCIENTISTS DISCOVER TECHNIQUES TO MANIPULATE DNA

1.1.1 INTRODUCTION

If you do a Google® image search on the word "biotechnology," you will get a number of interesting graphics, but by far the most popular imagery involves the DNA double helix. Indeed, the twisting form of DNA has become an icon for all of biological science. DNA is a reasonable logo – it is recognizable, is eye-catching, and feels suitably technical. It is also true that what we are calling "modern biotechnology" did originate with discoveries that involved the manipulation of DNA. However, biotechnology now involves much more than DNA. It can be said that "DNA is the flash – but proteins are the cash.[1]" It is the products of biotechnology – ranging from protein-based cancer therapeutics to lens cleaning solutions – that have fueled the explosive growth of this field.

As we will see in this chapter, modern biotechnology is deeply rooted in research from the biology laboratory. The research laboratory is a relatively recent manifestation of human curiosity, a place invented to study nature. In our time, significant discoveries about the intricacies of living systems have emerged from laboratory studies and have been transformed into the products of "biotechnology." Over the years, researchers around the world have explored fundamental questions in biology such as the complexities of cellular function; the mechanisms by which information is passed from generation to generation; how individuals develop from a single, fertilized egg cell; and how the complex immune system is coordinated. This is **basic research**, *research that is performed in order to understand nature.* The modern biotechnology industry emerged as knowledge from basic biological research has been transformed into products. **Biotechnology** *is thus the transformation of biological knowledge and discovery into useful products.*

It is important for individuals working in the biotechnology industry to have a sense of its breadth and scope. This chapter and the next chapter therefore provide a brief (and selective) overview of the origins of the modern biotechnology industry, the varied products of the biotechnology industry, and the processes by which biological knowledge is transformed into a host of products. We begin this overview with an introduction to three key biological molecules – DNA, RNA, and proteins – because it is not possible to understand the science of biotechnology without some understanding of these molecules.

---

[1] Our thanks to Ellen Daugherty for coining this memorable phrase.

DOI: 10.1201/9780429282799-2

1.1.2 A Brief Overview of Molecular Biology

1.1.2.1 DNA Tells the Cell How to Make Proteins

Biotechnology utilizes powerful tools to manipulate DNA (deoxyribonucleic acid); in fact, the term "biotechnology" is sometimes used to refer to the manipulation of DNA. The image of the DNA double helix has become so pervasive in our culture that it is easy to forget that biologists did not understand the structure and function of DNA until the middle of the twentieth century. At that time, the work of many researchers led to the realization that DNA is the chemical of inheritance. **DNA** *is the substance by which parents pass information to their offspring.* DNA tells the offspring how to grow and develop to form an organism with that individual's unique traits.

When we consider the complexity of a living organism, such as a human, it is difficult to imagine how molecules of DNA can contain and transmit all the information necessary to "construct" and "operate" an individual. The discovery of the molecular structure of DNA was an important milestone in the search to understand heredity, that is, how information is passed from parent to offspring. The structure of DNA was elucidated in the 1950s by the research of various scientists including Erwin Chargaff, Rosalind Franklin, Maurice Wilkins, James Watson, and Francis Crick. The work of these scientists culminated in a series of scientific papers, the most famous of which, by Watson and Crick, was published in 1953 in the prestigious journal, *Nature.* (Watson, J. D., and F. H. C. Crick. "Molecular Structure of Nucleic Acids: A Structure for Deoxyribose Nucleic Acid." Nature, vol. 171, no. 4356, 1953, pp. 737–38. doi:10.1038/171737a0.) Watson, Crick, and Wilkins won a Nobel Prize in 1962 for their contributions.

DNA is a linear molecule consisting of four types of *molecular subunits,* called **nucleotides,** connected one after another into long strands. The four types of nucleotide are distinguished from one another because they each contain a different **base**. There are four bases in DNA: adenine, guanine, thymine, and cytosine. In most situations, DNA is double-stranded, meaning that two linear strands of DNA associate with one another. Double-stranded DNA twists to form the famous "double helix" (Figure 1.1).

It is the sequence (order) of the four types of nucleotides making up a strand of DNA that contains information. A particular sequence of nucleotides "tells" the cell how to build a particular protein. Therefore, a **gene** *can be considered to be an ordered sequence of nucleotides (thus a stretch of DNA) that contains information that "tells" the cell how to make a particular*

protein. The proteins provide structure to the cell and do the work of the cell; hence, by encoding proteins, the DNA constructs and operates the cell.

Cells contain many *genes that are organized into long DNA macromolecules called* **chromosomes**. In bacteria, *the entire collection of genes,* the **genome**, lies on a single chromosome. Organisms that are more complex have a genome with many more genes that are arranged onto multiple chromosomes. The chromosomes are sequestered in *a specialized membrane-bound area of the cell, the* **nucleus**. Bacterial cells do not have a nucleus. Bacteria are termed **prokaryotic**, *meaning they lack a nucleus. Cells with a nucleus are termed* **eukaryotic**. The cells of plants, animals, and yeast are eukaryotic.

Human cells each (with a few exceptions) contain two copies of an individual's genome, one copy from the mother and another from the father. The human genome consists of about three billion base pairs of DNA where a *base pair* is two nucleotides across from each other, one on each of the two opposing DNA strands (Figure 1.1c). These three billion base pairs are arranged onto 23 chromosomes. Since most human cells have two copies of the genome, they contain 46 chromosomes, 23 from each parent. These chromosomes, which are inherited from an individual's parents, contain the information required to create that individual.

Every cell in a particular individual (with a few exceptions) contains the same genes. Clearly, however, a nerve cell is different from a muscle cell or a skin cell. Different cells have different characteristics because only some of the genes present in that cell are "turned on," that is, make the protein for which they code. *When a gene is "turned on," the cell makes the protein that the gene encodes, and we say that the gene coding for that protein is* **expressed**. In muscle cells, for example, genes are expressed that code for proteins required for muscle contraction. Other genes are expressed in blood cells, yet others in nerve cells, still others in skin cells, and yet others in retinal cells. **Differentiation** *is the process in which a cell matures into a particular cell type with a specialized structure and function (e.g., muscle, nerve, skin, and retina)* (Figure 1.2). **Stem cells** *are undifferentiated cells that have the potential for indefinite self-renewal and the potential to develop into multiple mature cell types.* **Regulatory DNA sequences** *act as "switches" to control which genes are turned on at a given time in a given cell.* DNA thus plays its vital role by directing each cell to make the correct proteins required to do the work of that cell.

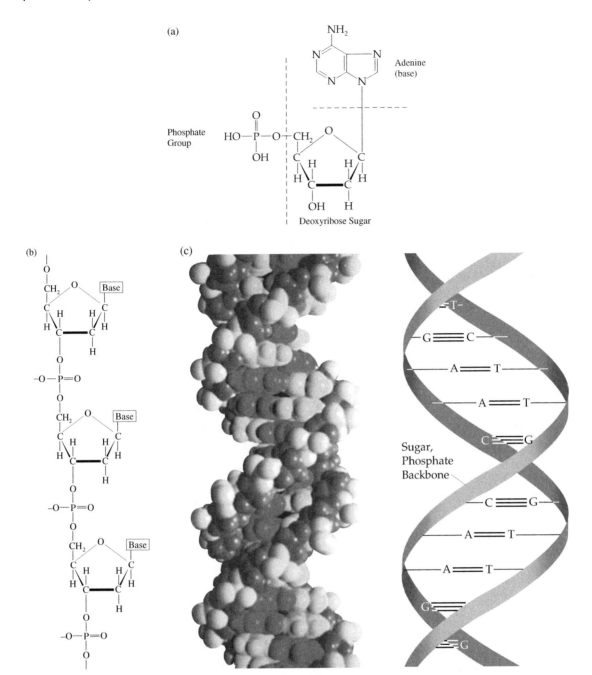

FIGURE 1.1 DNA structure. (a) The basic subunit of DNA is the nucleotide, which consists of a sugar, a phosphate group, and a nitrogenous base, in this case, adenine. Each of the four types of DNA nucleotide has a different base, but the rest of the nucleotide is the same. (b) The subunits of DNA are connected by covalent phosphodiester bonds between the phosphate group on one nucleotide and the sugar on the next. The bases extend out to the side. In solution, the phosphate groups lose a H+ and so have a negative charge. Three nucleotides linked together are shown here. (c) Chromosomal DNA consists of two strands of DNA held together by hydrogen bonds between complementary bases. A guanine will always pair with a cytosine, and an adenine with a thymine. Observe that three hydrogen bonds stabilize each G–C linkage, but only two hydrogen bonds stabilize each A–T linkage. The order in which the four types of nucleotides are arranged along the strands encodes genetic information.

1.1.2.2 Proteins Perform the Work of Cells

Every cell in an organism has many **proteins**. It is these proteins that give the cell its structure and that perform the work of that cell. **Proteins** *are diverse* *molecules composed of chains of varying numbers of amino acid building blocks that link together, like beads on a chain, and then fold into complex three-dimensional structures. Each protein has its own*

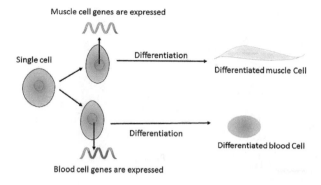

FIGURE 1.2 Examples of differentiation and expression. All cells in an individual have the same genes (with a few exceptions). Differentiation is the process in which cells become specialized to perform their particular function, for example, motility (muscle cells) or transporting oxygen (red blood cells). Differentiation occurs because different genes are expressed in different cell types.

structure that results from the amino acids that comprise it; different amino acids arranged in different orders give rise to differently structured proteins. The varied structures of proteins allow them to perform many tasks. Hemoglobin, for example, is a protein found in red blood cells; it has a unique structure that enables it to transport oxygen. Antibodies are proteins

that recognize and help the body neutralize foreign invaders (e.g., bacterial or viral pathogens).

1.1.2.3 The Assembly of Proteins

Proteins are assembled from amino acid building blocks by a specialized cellular component, the **ribosome**. DNA conveys information to the ribosome, directing it to stitch together specific amino acids in a specific order to make a particular protein. *Information moves from DNA to ribosomes via an intermediary molecule, called* **messenger ribonucleic acid, mRNA**. When a protein is to be made by the cell, the DNA that encodes the information for how to manufacture that protein is used to **transcribe**, or *synthesize*, mRNA. The mRNA molecules travel to the ribosomes where they direct the assembly of a protein molecule from amino acid subunits. Information thus flows as follows:

$$DNA \rightarrow mRNA \rightarrow protein$$

This *pathway by which information flows from DNA via mRNA* to code for protein is sometimes termed **the Central Dogma** of biology. *The process in which a protein is manufactured is called* **translation** (Figure 1.3).

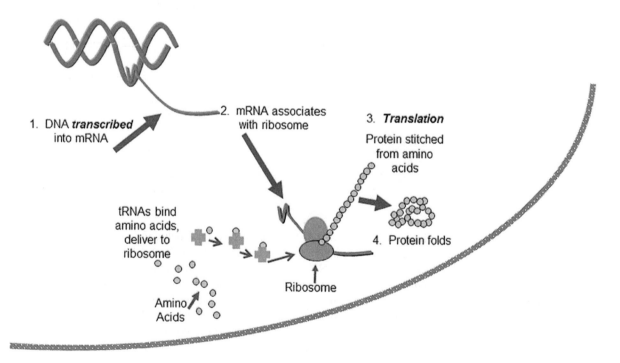

FIGURE 1.3 DNA codes for proteins. (1) The first step in the production of a protein involves the unwinding of a specific section of DNA so that the information in that section can be *transcribed* into mRNA. In eukaryotic cells, transcription occurs in the membrane-bound nucleus and the resulting mRNA exits the nucleus. (2) The mRNA associates with a ribosome. (3) Amino acid subunits are picked up by transporter molecules, called tRNA, and are taken to the ribosome where they are added to the growing protein chain in the order directed by the mRNA. This is called *translation*. (4) Proteins fold into specific three-dimensional shapes based on their amino acid sequences.

1.1.3 INTRODUCTION TO RECOMBINANT DNA TECHNIQUES

1.1.3.1 The Tools of Biotechnologists

As scientists have come to understand the structures and functions of DNA, RNA, and proteins, they have also devised tools to manipulate these biological molecules. Tools to manipulate DNA include the following:

- *Enzymes that cut DNA at specific sites.*
- *Enzymes that ligate (join) DNA strands together.*
- *Techniques to visualize DNA.*
- *Techniques to separate DNA fragments from one another.*
- *Techniques to identify fragments of DNA with specific sequences.*
- *Enzyme-based techniques that amplify DNA (generate many copies of a specific gene fragment).*
- *Techniques to determine the nucleotide sequence of a piece of DNA.*
- *Techniques to synthesize DNA.*
- *Techniques to edit (alter) specific bases in DNA.*
- *Techniques to turn specific genes off or on, or lower the expression of specific genes.*

These tools are discussed in various places throughout this text. For example, the genetic modification of cells, as shown in Figure 1.4, uses enzymes that cut a gene of interest out of one stretch of DNA, and other enzymes that ligate the gene of interest into another stretch of DNA.

The process of studying DNA and other biological molecules is ongoing. As scientists develop ever more sophisticated methods of manipulating biological molecules, they use these methods to probe more deeply into the workings of biological systems. As the functions of biological molecules are better understood, researchers are better able to manipulate these molecules, making new discoveries about the intricacies of life.

The "modern" biotechnology industry emerged from the development of tools to manipulate DNA in such a way that genetic information (DNA) from one organism is transferred to another. *When a biologist causes a cell or organism to take up a gene from another organism, we say the cell or organism is* **genetically modified** *or* **genetically engineered** (Figure 1.4). The related term **recombinant DNA (rDNA)** *refers to DNA that contains sequences of DNA from different sources that were brought together (i.e.,*

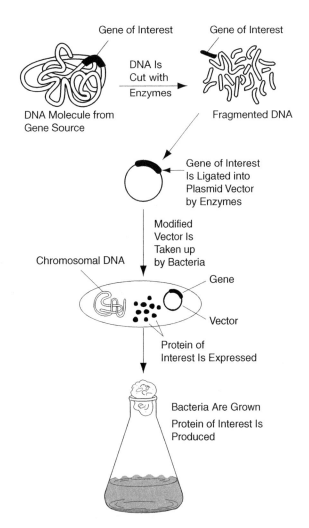

FIGURE 1.4 The genetic modification of bacteria. A gene of interest is isolated and inserted into a vector. The vector is taken up by bacteria, which are now said to be *transformed*, or genetically modified. The transformed bacteria can be grown in flasks or larger containers. Under the proper conditions, the bacteria express the protein encoded by the foreign DNA. This process laid the foundation for the explosive growth of the biotechnology industry.

recombined) using the tools of biotechnology. Under the proper conditions, a genetically modified cell will express (produce) a protein encoded by an introduced gene. The ability of biologists to create genetically modified organisms that make proteins they would not ordinarily produce is so powerful that the term "revolutionary" is often applied to it.

DNA from one organism can be transferred into another in various ways. **Plasmids** are small circular molecules of DNA that occur naturally in many types of bacteria and yeast and that exist separately from their chromosomes. Scientists often ligate a gene of interest into a plasmid using enzymes that are able to stitch together two pieces of DNA. Under proper conditions, plasmids are readily taken up by bacterial cells (and other types of cells as well) and can carry with them a gene of interest. Plasmids are called **vectors** *when they carry a desired gene into recipient cells.* A gene of interest can also be introduced into the DNA of a virus so that the virus acts as a vector when it infects a host cell. In some cases, DNA can be directly injected into a recipient cell. Electrical current can also be used to induce cells to take up foreign DNA. *When a bacterial cell takes up foreign DNA (e.g., takes up a plasmid vector), it is said to be* **transformed**. *When a eukaryotic cell takes up foreign DNA, it is said to be* **transfected**.

1.1.3.2 Using Genetically Modified Cells

By the early 1970s, scientists and entrepreneurs realized that the powerful techniques of manipulating DNA could be utilized to make products of commercial importance; the modern biotechnology industry emerged from this vision. The scientific community, however, was also concerned that this new, powerful technology might create unknown and potentially significant safety issues. Prominent scientists called for a temporary halt in research – a request without precedent in the scientific community – in order to assess the safety of recombinant DNA methods. A Recombinant DNA Advisory Committee (RAC) was established by the National Institutes of Health (NIH) to study the issues. A larger gathering, the Asilomar Conference, was convened in February 1975, to discuss the safety of recombinant DNA technology. The conclusion that emerged from this scientific conference was that most rDNA work should continue, but appropriate safeguards in the form of physical and biological containment procedures should be put in place.

Shortly after the Asilomar Conference, the company, Genentech, was founded; Genentech and Cetus Corporation are generally acknowledged to be the first modern biotechnology companies. According to the Genentech website (http://www.gene.com), the company was formed in 1976 by business investor Robert Swanson and scientist Herbert Boyer who saw the commercial potential of the new methods of manipulating DNA. By 1978, Genentech scientists were able to transfer the gene coding for human insulin into bacteria. Insulin is a protein hormone required for proper regulation of sugar levels in blood and cells, and it is used to treat diabetes. **Type 1 diabetes** *is a serious, relatively common disease that occurs when pancreatic cells that normally produce insulin are destroyed.* The researchers were also able to induce the bacteria to express the human gene, and

thus produce human insulin at levels that allowed commercial production.

Bacteria that contain an introduced gene, such as insulin, can be grown in large quantities in *special vats called* **fermenters**. *The large-scale cultivation of bacteria to produce a product is called* **fermentation**. The bacteria produce the product of interest, which can then be isolated and purified using protein separation techniques. The bacteria thus become a "factory" to manufacture the protein product of interest. (Refer to Figure 1.4.) This technology, which takes advantage of a cellular process that nature has optimized over eons, lies at the heart of modern biotechnology.

The production of insulin in bacteria was a scientific accomplishment; bacteria have absolutely no use for insulin and would never produce it without human intervention. Insulin made by recombinant DNA methods was also a medical achievement. Prior to the 1980s, insulin to treat diabetics was purified from the pancreas of animals slaughtered for human consumption. To make one pound of insulin, 8,000 pounds of pancreas glands from 23,500 animals are required. (Genentech. "Cloning Insulin." *Genentech: Breakthrough Science. One Moment, One Day, One Person at a Time,* www.gene.com/stories/cloning-insulin.) Animal insulin is similar, but not identical to human insulin; therefore, some diabetic patients developed allergies to the drug. In 1982, human insulin made by recombinant DNA technology was approved for use by patients. The production of insulin by bacteria is a classic example of how basic scientific knowledge was transformed into a product that helps millions of diabetic patients – and also confirmed the significant commercial potential of recombinant DNA methods.

Bacteria were the first genetically engineered cells used to make a commercial product. Researchers, however, soon learned to genetically modify other types of cells as well, including those from mammals, yeasts, insects, and plants. **Cultured cells** *are those grown in flasks, dishes, vats, or other containers outside a living organism.* (The term "cell culture" most commonly refers to eukaryotic cells, although prokaryotic cells are also grown under "culture" conditions.) *Some cells can be induced to divide indefinitely in culture using particular procedures, and so immortal* **cell lines** have been established. Certain immortal cell lines are commonly used for production purposes. The techniques used to grow eukaryotic cells for production are analogous to those used in bacterial fermentation, although the growth conditions microbes require are somewhat different than those of eukaryotic cells. *The specialized growth chambers used for eukaryotic cells*

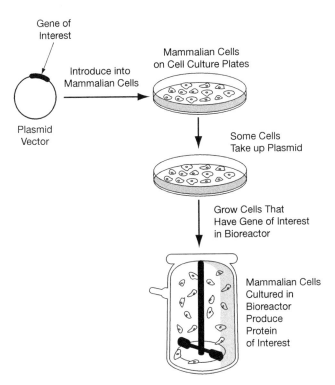

FIGURE 1.6 The first transgenic animal. The gene for rat growth hormone was microinjected into a fertilized mouse egg, which was then implanted into a surrogate mother. The resulting transgenic mouse on the left expressed the gene, resulting in an exceptionally large animal. The mouse on the right is of normal size. (Photo courtesy of R. L. Brinster/ Peter Arnold.)

FIGURE 1.5 The use of genetically modified cultured mammalian cells to produce a protein of interest. Once cells have been transfected with a gene of interest, they can be grown in large quantities in a bioreactor and the protein product can be isolated.

are usually called **bioreactors** (instead of *fermenters*). Figure 1.5 illustrates the use of genetically modified mammalian cells to produce a protein product.

We have so far talked about introducing foreign DNA into cells and then growing the cells to high densities in fermenters or bioreactors. It is also possible to introduce a gene of interest into whole plants and animals, although this is more complex than manipulating cultured cells. *A plant or animal whose cells are genetically modified is called* **transgenic**. Figure 1.6 shows one of the first transgenic animals. The two animals in this photo are littermates. The mouse on the left, however, was genetically modified by the introduction of the gene for rat growth hormone. This was accomplished by microinjecting the growth hormone gene into a fertilized mouse egg; thus, all the cells in the resulting mouse contained the foreign gene. The rat growth hormone gene was expressed, causing the transgenic animal to be unusually large.

Biotechnologists can produce genetically modified microorganisms and cultured cells, transgenic animals, and transgenic plants. There are many commercial applications that involve genetically modified organisms. One application is to use cultured cells to produce proteins of value, such as insulin. It is also

possible to create transgenic plants and animals that have desirable characteristics. For example, transgenic crop plants may have enhanced resistance to disease. Figure 1.7 summarizes some of these points regarding the genetic modifications of organisms.

1.2 APPLICATIONS OF RECOMBINANT DNA TECHNOLOGY

1.2.1 Biopharmaceuticals Overview

The use of genetically modified cells as "factories" to manufacture therapeutic proteins, such as human insulin, was the first and is still the most commercially important modern biotechnology application. **Biopharmaceuticals** *are defined here as therapeutic products, such as insulin, that are manufactured using genetically modified organisms as production systems.* A few examples of biopharmaceutical products are shown in Table 1.1; there are many more in addition to these. We will devote much of this chapter to exploring biopharmaceuticals because of their value in alleviating illness, their significance in the development of the biotechnology industry, and their commercial importance.

"Traditional" drugs manufactured by pharmaceutical companies are small molecules that are most often produced by chemical synthesis, for example, aspirin. Chemically synthesized drugs have been manufactured for a long time and are well characterized, and their manufacture has been optimized by pharmaceutical companies. But some disorders cannot be

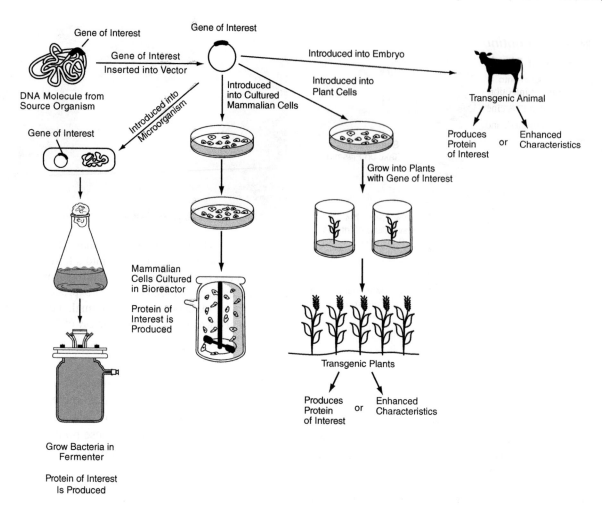

FIGURE 1.7 The genetic modification of organisms using biotechnology. The tools of biotechnology allow biotechnologists to insert recombinant DNA into various cell types including those of bacteria, cultured mammalian cells, whole plants, and animals.

TABLE 1.1

Examples of Biopharmaceutical Products

Recombinant DNA Products Produced in Bacteria (*E. coli*)

Human Insulin. Approved 1982, First Produced by Genentech

Used by about 7.4 million people to treat diabetes. Before insulin from genetically modified bacteria was available, it was purified from the pancreas of animals. Insulin derived from genetically modified bacteria provides a more reliable source than animals and is less likely to cause allergic reactions in patients.

Human Growth Hormone. Approved 1985, Genentech

Used to treat dwarfism. Prior to the introduction of recombinant DNA-derived human growth hormone, dwarfism was treated with a hormone purified from the pituitary glands collected from human cadavers. Some children who received hormone isolated from human sources eventually died of the neurodegenerative disease Creutzfeldt–Jakob disease.

Interferon-2b. Approved 1986 (for the Treatment of Leukemia), Hoffmann-La Roche

Various types of interferons are used to treat a variety of cancers and viral diseases. The potential of interferons in the treatment of disease was recognized in 1957, but it was not available in the amounts and purity required for experimentation and clinical trials until the advent of biotechnology.

Recombinant DNA Products Produced in Yeast Cells

Hepatitis B Vaccine. Approved 1986, Merck & Co., Inc.

(Continued)

TABLE 1.1 (*Continued*)
Examples of Biopharmaceutical Products

Used to prevent infection with hepatitis B. Hepatitis B is a common and serious viral illness for which there is no known cure. Before the recombinant vaccine was developed, a vaccine was prepared from the plasma of hepatitis-infected humans. This source of vaccine was limited, and there were concerns about its purity. The protein required to make the hepatitis B vaccine is now produced by genetically modified yeast, which makes the vaccine more available and reduces the possibility of contamination.

Human Albumin. Approved 2005, Delta Biotechnology Ltd.

Used as a stabilizer in vaccine manufacture. Human albumin, a blood protein, has been shown to help prevent fever in vaccinated children. There are concerns that blood proteins isolated from natural sources (humans or animals) might be contaminated with viruses or other pathogens. Human albumin made using recombinant DNA technology alleviates this concern.

Human Papillomavirus (HPV) Vaccine. Approved 2006, Merck & Co., Inc.

Used to prevent infection with several types of HPV. HPV is a virus that is sexually transmitted and that causes genital warts and, in some cases, cervical cancer. This vaccine is therefore considered to be an anticancer vaccine.

Recombinant DNA Products Produced in Cultured Mammalian Cells

Erythropoietin. Approved 1989, Amgen

Used to treat anemia due to renal failure, chemotherapy (in cancer patients), and AZT treatment (used for AIDS patients). Erythropoietin (EPO) is a glycoprotein, produced in the kidney, which stimulates the production and maturation of red blood cells. EPO occurs naturally in very small quantities in human urine and therefore, prior to its production by recombinant DNA methods, had never been available in sufficient quantity for clinical testing. The conventional treatment for anemia due to renal failure was blood transfusion. Recombinant EPO provided a new approach to the treatment of anemia.

Factor VIII. Approved 1992, Baxter

Used to treat hemophilia in more than 1 million people worldwide. Before the introduction of recombinant Factor VIII, many hemophilia patients contracted AIDS from Factor VIII derived from infected human plasma.

Tumor Necrosis Factor Blocker. Approved 1998 (for Rheumatoid Arthritis), Immunex

Used to treat arthritis and psoriasis. People with an inflammatory disease (e.g., rheumatoid arthritis, ankylosing spondylitis, Crohn's disease, and psoriasis) have too much tumor necrosis factor (TNF) in their bodies, which stimulates an inflammatory response. The recombinant DNA drug, Enbrel, reduces the amount of TNF to normal levels.

Monoclonal Antibody Products

Murine Monoclonal Antibody to CD3. Approved 1986, Ortho Biotech

Used to suppress organ rejection by patients receiving kidney transplants. This product is a highly purified antibody that attacks the T cells of patients. T cells are involved in transplant rejection. This antibody has been effective in treating patients who do not respond to conventional anti-rejection treatments.

See Table 1.2 for more examples of monoclonal antibody products.

Recombinant DNA Products Produced in Transgenic Animals

Antithrombin. Approved 2009, GTC Biotherapeutics

Used to prevent blood coagulation. Antithrombin prevents life-threatening blood clots during surgery and childbirth in patients who have a disease that causes a deficiency in this protein.

Ruconest®. Approved 2014, Pharming Group and Salix Pharmaceuticals

Used to treat a rare genetic disease (hereditary angioedema) that causes painful and sometimes life-threatening swelling. The disease results from a deficiency of the protein C1 esterase inhibitor, which has a normal role in regulating inflammatory pathways in the body. Shortage of the protein leads to leakage of fluids from blood vessels, resulting in swelling and severe pain. Ruconest ® supplies the missing protein at the onset of symptoms. It is purified from the milk of transgenic rabbits.

Kanuma®. Approved 2015, Alexion Pharmaceuticals

Used to treat a rare and often fatal genetic disease (lysosomal acid lipase deficiency) caused by the absence of an enzyme that normally breaks down fats in lysosomes (a membrane-bound compartment inside of cells). As a result, fat accumulates in the liver, spleen, and blood vessels of affected children. Kanuma® is an enzyme that is internalized into the lysosomes of treated patients where it catalyzes the breakdown of fats. Kanuma® is purified from the eggs of transgenic chickens.

Recombinant DNA Product Produced in Transgenic Plant Cells

Elelyso®. Approved 2012, Protalix Biotherapeutics

Used to treat Gaucher disease, which is caused by mutations in the gene that encodes the enzyme beta-glucocerebrosidase. In affected patients, the fatty compound, glucocerebroside, accumulates in the body, causing an enlarged spleen and liver, blood abnormalities, and brittle bones that are prone to fractures. Elelyso® is used for enzyme replacement therapy.

treated with small-molecule drugs. Type 1 diabetics, for example, require regular administration of insulin, which is a protein. Proteins are large complex molecules that cannot easily be chemically synthesized on the scale required for their use as a therapeutic. Before the advent of recombinant DNA technology, proteins and other complex biopharmaceuticals had to be purified from natural sources (often plants or animals). For instance, growth hormone, used to treat a form of dwarfism in children, was isolated from the pituitary glands of human cadavers. Isolating natural products from animals and plants has drawbacks. Some children who received growth hormone from human cadavers tragically contracted an otherwise rare, fatal neurological disease, Creutzfeldt–Jakob disease, caused by contaminants in the brain tissue from which the hormone was isolated. In the 1980s, thousands of individuals with hemophilia contracted the life-threatening disease, AIDS, because the blood factors they require to control bleeding were isolated from human blood contaminated with the virus that causes AIDS. Contamination is always a concern with products isolated from natural sources.

Another problem with isolating drugs from plants and animals is that the drug might not be available in large quantities from its natural source. Interferon, discovered in 1957, is a protein made by animal cells as a defense against viral infection. Interferon is released into the blood and causes the body to mount a response against the attacking viruses. Scientists quickly realized that interferon had the potential to be a therapeutic agent against infectious agents, but only tiny quantities could be isolated from blood, not nearly enough to test or to use clinically. It was not until recombinant DNA technology was developed in the early 1980s that researchers could obtain enough interferon to test the compound in animals and humans. Interferon drugs did not turn out to be as useful in fighting infectious agents as scientists predicted, but interferon has proven efficacious in treating patients with certain cancers and other conditions.

There are so many advantages to producing protein therapeutics using recombinant DNA technology, and the market for these products is so large that pharmaceutical companies rapidly created biopharmaceutical production facilities and new biotechnology companies sprang up to develop new products. There are now many types of biopharmaceuticals on the market including enzymes, vaccines, antibodies, thrombolytics (drugs that dissolve blood clots), blood clotting factors (drugs that help blood to clot), and cytokines (small proteins that act as chemical messengers). A financial report stated that more than $219 billion was spent on biopharmaceuticals in the year 2018. Drugs that use antibody technology (which will be discussed below) comprised about half of the total sales. In 2018, the biopharmaceutical industry was estimated to provide 4 million jobs in the United States and that number is almost certainly growing. (Hardman & Co. "Global Pharmaceuticals: 2018 Industry Statistics." *Hardman & Co.*, www.hardmanandco.com/research/corporate-research/global-pharmaceuticals-2018-industry-statistics.)

Insulin and growth hormone are natural products that existed before recombinant DNA technology was invented. Other biopharmaceutical products are entirely new and could not exist without recombinant DNA technology; Herceptin® and other monoclonal antibody cancer drugs, described later in this chapter, are examples. Small chemically synthesized drugs are still very much in use, and pharmaceutical companies continue to develop new ones, but the existence of biopharmaceuticals greatly expands the possibilities for therapeutic agents.

1.2.2 PRODUCTION SYSTEMS FOR BIOPHARMACEUTICALS

1.2.2.1 Cultured Cells

Genetically modified bacterial cells were the first type of host cells used to manufacture biopharmaceuticals. Bacterial cells are often still used because they require relatively inexpensive growth medium and are easily grown in culture. Yeast cells are also used sometimes to manufacture products and are also relatively inexpensive and simple to grow. But bacteria and yeast are not able to produce all human proteins in an active form. This is because many human proteins are modified by cells after they are assembled from their amino acid subunits, and bacteria and yeast often do not perform these modifications the same way that human cells do. **Glycosylation** *is a common, important type of modification in which complex, specific, branched carbohydrates are attached to a protein* (Figure 1.8). These branched structures affect how the protein functions when it is administered to a patient. When proteins must be modified in specific ways in order to be functional, bacteria and yeast are generally not used as host cells and mammalian cells are commonly used instead.

Mammalian cells are relatively fragile, and they grow more slowly than bacterial cells. This means mammalian cells are more expensive to grow than bacteria. Mammalian cells require a more complex growth medium than bacteria or yeast. This means

Sugar Chains

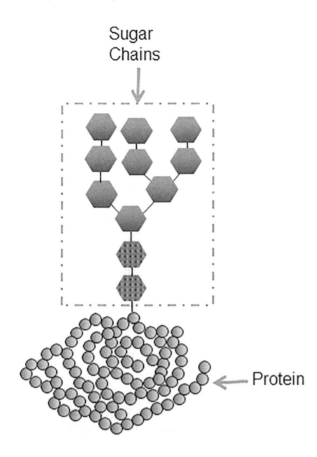

← Protein

FIGURE 1.8 Glycoproteins are proteins with attached sugar chains. (Not drawn to scale.)

that more impurities from the medium must be removed from products when mammalian cells are used in manufacturing than is the case when bacteria or yeast are used. Mammalian cells have the further disadvantage that they are more likely than bacteria or yeast to harbor contaminants that are pathogenic to humans; these contaminants must be removed during processing. Despite all these disadvantages, mammalian cells are the most common type used for biopharmaceutical production. There are various mammalian cell lines used for biopharmaceutical production, the most common of which is currently the CHO cell line. CHO cells are the descendants of cells originally isolated from a Chinese hamster ovary, hence, their acronym. (Dumont, Jennifer, et al. "Human Cell Lines for Biopharmaceutical Manufacturing: History, Status, and Future Perspectives." *Critical Reviews in Biotechnology*, vol. 36, no. 6, 2015, pp. 1110–22. doi: 10.3109/07388551.2015.1084266.) CHO cells were first described by Theodore Puck and his colleagues in a 1958 paper. (Puck, Theodore T., et al. "Genetics of Somatic Mammalian Cells." *Journal of Experimental Medicine*, vol. 108, no. 6, 1958, pp. 945–56. doi:10.1084/ jem.108.6.945.) Puck and his colleagues not only

described their work, but they also deposited some of their CHO cells with ATCC. **ATCC** is *a global, nonprofit organization that stores and distributes biological resources, particularly cells and tissues.* Puck's CHO cell line and other cell lines derived from it can be easily purchased from the ATCC.

Puck and his colleagues were probably not thinking about creating a production system for the then nonexistent biopharmaceutical industry when they isolated CHO cells in the 1950s. The purpose of their work was to find improved methods for culturing cells in the laboratory, which, at the time, was difficult and often unsuccessful. It is interesting to note that in order to promote healthy cell growth, Puck's group tried adding serum isolated from the blood of fetal calves to the growth medium. They found that fetal calf serum was very helpful in coaxing cells to grow. Indeed, fetal calf serum is so helpful to cultured cells that its addition became standard in mammalian cell culture. Unfortunately, a few people in Great Britain have died of variant Creutzfeldt–Jakob disease thought to be caused by a pathogen found in cows. This means the pathogen could also be present in fetal calf serum. This concern, and other similar problems relating to animal products, has led the biotechnology industry to make a concerted (and still ongoing) effort to find alternatives to the use of fetal calf serum and other animal-derived materials in cell culture (see also Chapter 31).

1.2.2.2 Animals

At this time, the usual production systems for protein products are cultured bacteria, yeast, mammalian, or sometimes insect cells. However, transgenic plants and farm animals are also being developed for biopharmaceutical production. Since the 1980s, biotechnologists have speculated that farm animals, such as goats, sheep, chickens, and cows, might be genetically modified so that they would produce a therapeutic protein in their milk or eggs. When transgenic mammals are used as production systems, the gene for the therapeutic protein is joined to the "on-switch" for a milk production gene. The resulting DNA is called the *genetic construct.* The genetic construct is sometimes painstakingly injected under a microscope into the nucleus of a fertilized egg from a sheep or other host species. Each embryo that results from a successful microinjection is transferred into a sheep surrogate mother (assuming sheep are the animal of choice) who gives birth to transgenic lambs. Some of these transgenic females will produce the therapeutic protein in their milk. Any transgenic sheep can then be bred in a

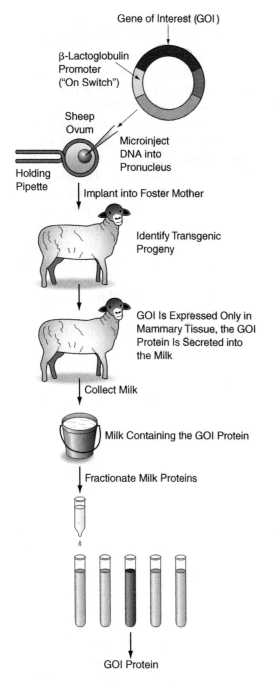

FIGURE 1.9 Production of a protein product using a transgenic farm animal. In this example, the gene of interest (GOI) is a therapeutic protein that is to be harvested from the milk of a transgenic farm animal.

normal fashion to form a herd of animals that produce the therapeutic protein in their milk.

The therapeutic protein is harvested from the milk of transgenic sheep (or some other animals) using standard dairy methods. Then, the protein goes through a series of purification steps to ensure that it is pure and uncontaminated by any material from the host animal (Figure 1.9).

Animals have advantages as production systems for proteins. Transgenic animals can produce proteins that are folded into their proper shapes and are properly glycosylated. CHO cells will accomplish these tasks, but animals can, in principle, provide much larger amounts of high-quality protein that is noninvasively harvested from their milk. Transgenic animal technology has been slow to develop despite its advantages, due to technical difficulties, safety concerns, and societal concerns about this use of animals. At the time of writing, only three drugs made in transgenic animals have been approved for marketing; see Table 1.1.

1.2.2.3 Plants

Materials derived from plants have been used as medicines for thousands of years, and some modern drugs are still derived from plants. However, the use of genetically modified plants to manufacture drugs is new. Plants that have been transfected with a genetic construct can be used to produce a protein of interest in their cultured cells, stems, leaves, shoots, or roots. At the time of writing, only two plant-based biopharmaceutical have entered commercial production (Elelyso®, produced by cultured carrot cells, and Palforzia®, produced by peanut plants), but many companies are testing them. Transgenic plants, like transgenic animals, offer the potential of high yields while avoiding controversy relating to the use of animals. Also, plants are unlikely to harbor pathogens that are dangerous to human patients. At this time, a major concern with whole plant-based biopharmaceuticals is the possibility that altered genetic material contained in pollen grains might escape and fertilize nearby crops or wild plants. This would allow the altered genetic material to spread where it should not, and might expose people to unforeseen drug products in foods. Another concern relates to difficulties in purifying the desired product away from plant tissues. It is difficult to predict whether plant-based biopharmaceuticals will eventually become commercially significant.

1.2.2.4 Other Production Systems

Scientists are working on new strategies as all these current issues relating to cell, plant, and animal production systems are being resolved. Some scientists, for example, are attempting to engineer bacterial and yeast cells so they can fold and glycosylate proteins in a manner like human cells. Cultured insect cells are also in occasional use. Scientists in some companies are genetically engineering the tiny aquatic duckweed plant, *Lemna*, which can be grown in a sealed vessel inside a production facility. This system reduces the potential for the escape of genetic information to other plants. Other scientists are working on synthesizing proteins by chemical reactions – without using cells at all. It is hoped that eventually cheaper and more efficient drug production systems will be developed that have no environmental risks and do not raise ethical concerns.

1.2.3 MONOCLONAL ANTIBODIES

Antibodies are familiar to us as agents inside our bodies that provide protection from invading bacterial and viral pathogens. However, antibodies have a much broader role in the biotechnology arena. Over the past decade or so, an ever-increasing number of antibody-based drugs have been created. Indeed, at the time of writing, antibody therapeutics are the most commercially important biotechnology products. This comes as a bit of a surprise because, as we will see later, for many years, research into antibody therapeutics languished. In addition to their role as therapeutic agents, antibodies have been a vital tool in basic research laboratories for decades. This is because antibodies can be used to find and label molecules within biological samples. However, as we will see in Chapters 5 and 29, the use of antibodies in research has led to dramatic advances and equally frustrating confusion. Yet another application of antibodies is in medical diagnosis. Home pregnancy test kits, for example, are based on antibodies that specifically bind to a hormone that increases during pregnancy. Most recently, antibodies have occupied a prominent role in the news because of their roles in treating the pandemic disease, COVID-19, and preventing disease through vaccination. The antibody-based strategies that are being developed at the time of writing to overcome COVID-19 are likely to be important in the future for treating other diseases. For all these reasons, we now turn to a discussion of antibodies.

Monoclonal antibodies (mAbs) are biopharmaceuticals made in a somewhat different way than has been discussed so far. **Antibodies** *are proteins made by the immune system that recognize and bind to substances invading the body* – such as bacteria, viruses, and foreign proteins – thus aiding in their destruction. *Substances that trigger the production of antibodies are called* **antigens**. Antibodies are produced by B cells, a type of white blood cell. A particular antibody binds only a particular target antigen, somewhat like a specific key only fitting into a particular lock (Figure 1.10).

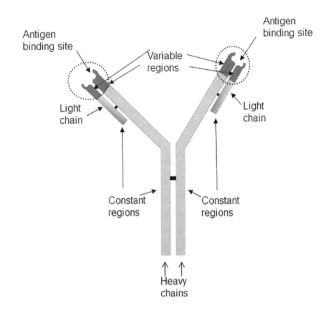

FIGURE 1.10 Antibody structure. Antibodies are large proteins that recognize and help neutralize foreign substances. Antibodies have a characteristic Y-shaped structure. They are composed of subunits: two identical heavy protein chains plus two identical light chains that are linked to each other. Antibodies are all structurally similar except for a variable region at the end of each Y arm. An individual has millions of different antibody populations, each of which has a structurally distinct variable region. *The part of an antigen that an antibody recognizes and binds is called an* **epitope**. Each antibody is able to recognize a specific antigen from among millions of different molecules because each epitope interacts in a highly specific way with the structure of a particular matching variable region.

Before the 1970s, antibodies could only be manufactured by exposing an animal, such as a mouse or a rabbit, to the target material, the antigen, of interest. *The animal would mount an immune response against the antigen, producing many antibodies,* **polyclonal antibodies**, *which could then be isolated from the blood serum of the animal.* Polyclonal antibodies are valuable and are widely used by research scientists, but they have two problems. First, they are not homogeneous molecules because an animal will usually produce diverse antibodies directed against various epitopes of the antigen. Second, polyclonal antibodies cannot be indefinitely obtained; a certain population of antibodies can be harvested only from the animal in which it is produced. When the animal dies, the source of antibodies is gone. These limitations have kept polyclonal antibodies from being widely used in medicine. **Monoclonal antibodies** do not have these two limitations; *they are exceptionally homogenous* *populations of antibodies directed against a specific target, and they can be produced indefinitely in culture.*

Monoclonal antibodies were first described in 1975 by George Köhler and Cesar Milstein. (Köhler, G., and Milstein, C. "Continuous Cultures of Fused Cells Secreting Antibody of Predefined Specificity." *Nature*, vol. 256, no. 5517, 1975, pp. 495–97. doi:10.1038/256495a0.) Köhler and Milstein produced monoclonal antibodies by fusing together two cells: an antibody-producing cell from a mouse, and a tumor cell. Suppose, for example, that one wants to make a monoclonal antibody against a specific protein that is thought to be important in making cancer cells divide. One obtains the protein of interest and injects it into a laboratory mouse. The mouse mounts an immune response against the protein, making different types of antibodies that recognize different parts of the antigen. The different types of antibodies are made by different B cells. *Each B cell divides many times to form a* **clone**, and each clone makes only one type of antibody. The mouse is later sacrificed, and its B cells are isolated from its spleen. *B cells are then fused in culture with mouse myeloma cells to form* **hybridoma cells**. The myeloma cells are immortal; that is, they will divide indefinitely in culture. Each fused hybridoma cell thus has two important qualities: It produces a single, identical type of antibody molecule, as does each clone of B cells, and it will divide indefinitely like the myeloma cells. It is necessary to fuse the B cells with the myeloma cells because the B cells are not immortal; they will eventually stop dividing. *The fused cells are diluted in culture to isolate individual hybridoma cells that divide repeatedly to form a uniform clone of cells, all of which make the same antibody*, hence, **monoclonal antibodies (mAbs)**.

Monoclonal antibodies were rapidly adopted for use in research laboratories to detect and visualize specific proteins in cells and tissues. Their potential use in the clinic was also quickly imagined; people reasoned that monoclonal antibodies could be used to search out and destroy specific pathogens and damaged cells. Scientists further thought that monoclonal antibodies could be attached to a cytotoxic agent that would kill targeted cells – such as infected cells or cancer cells – while sparing healthy cells. Monoclonal antibodies were therefore optimistically termed "magic bullets" because they could, in theory, find, bind, and possibly destroy a very specific target. Monoclonal antibodies, however, were not easily adapted to clinical uses. The first monoclonal antibody was approved for medical use in the United States in 1986 (see Table 1.1), but further development into useful products was slow, and many scientists concluded that they would never find use as therapeutics. One scientist remarked that in the 1990s, "We wanted to present posters at meetings, but if we had the word 'antibody' in the title, no one would look at it." (John Lambert, Immunogen's chief scientific officer, as quoted in Megan Scudellari, "Tumor Snipers," *The Scientist* (November 2012): 65–67.) One of the biggest obstacles to using monoclonal antibodies to treat humans was that the standard procedure of producing them yields mouse antibodies. Even though mouse antibodies are similar to human ones, mouse antibodies are detected as being non-self by the human immune system and are destroyed. The result was that nearly 20 years after Köhler and Milstein's discovery, only one monoclonal antibody drug had been approved for clinical use. Over time, however, various approaches to "humanize" monoclonal antibodies were devised as scientists developed ever more sophisticated recombinant DNA methods. Scientists furthermore learned to efficiently manufacture large amounts of humanized mAbs using mammalian cells growing in bioreactors, which spared animals, and also resulted in higher and more consistent antibody yields. Figure 1.11 illustrates one such approach to antibody production.

Improvements in the technologies for humanizing and manufacturing monoclonal antibodies eventually resulted in successful mAb therapeutics. By 2013, the monoclonal antibody market was estimated to be worth $75 billion, and by 2015, monoclonal antibody development dominated the biopharmaceutical industry. In 2015, analysts predicted that by 2020 there would be 70 monoclonal antibody products. In fact, by 2020 there were roughly 100 mAb drugs that had been approved for human use, with a value sometimes estimated to be more than $100 billion.

One of the reasons that antibody drugs are so valuable is that they can be used to treat cancer with fewer side effects than conventional chemotherapy drugs. Conventional cancer chemotherapy drugs attack normal as well as cancerous cells; monoclonal antibodies can be targeted to attack cancer cells more specifically. More than half of the monoclonal antibodies that have been approved for clinical use are cancer therapeutics. While cancer is still one of the most significant causes of death, many lives, including those of

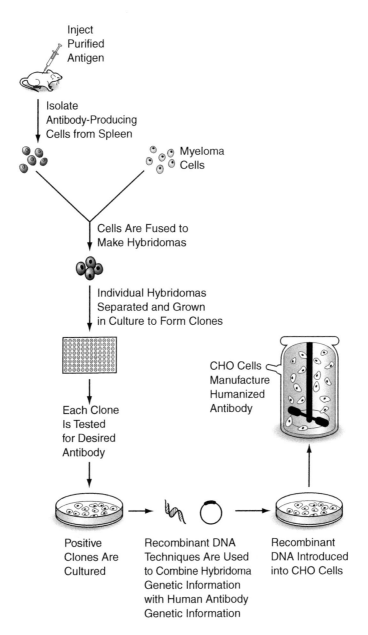

FIGURE 1.11 A method of manufacturing humanized monoclonal antibodies. Various methods exist to manufacture humanized monoclonal antibodies. Here, hybridomas are first produced. A mouse is inoculated with the antigen of interest, which stimulates the proliferation of cells that produce antibodies against the antigen. These cells are harvested from the mouse's spleen and are then fused with myeloma cells to produce hybridomas. Individual hybridoma cells are transferred to separate wells in a 96-well plate where they are cultured for several days. The original hybridomas divide to form clones of cells. Each clone is tested to see if it is producing the desired antibody. Positive clones are maintained in culture. Using the tools of genetic engineering, the genetic information from a clone that is producing the desired antibody is combined with the genetic information for producing a human antibody. CHO cells are transfected with the recombined DNA. The CHO cells are placed in a bioreactor where they manufacture an antibody that recognizes the antigen of interest (as did the hybridoma cell) yet has most of the structural features of a human antibody.

children with cancer, have been saved by monoclonal antibody drugs.

Antibody drugs are also effective therapeutics for other difficult-to-treat disorders including inflammatory and autoimmune diseases, severe arthritis (see the case study "A Personal Story from Chris B"), transplant rejection, viral disease, and macular degeneration (a cause of blindness). Several mAbs have received Emergency Use Authorization for the treatment of COVID-19 patients at high risk of developing severe disease.

Case Study: A Personal Story from Chris B.

Chris B. is a long-time biotechnologist with experience in developing and manufacturing various important biopharmaceuticals. This is an excerpt from an interview with him:

I went to school to be a plumber...I was working in the trades in the early 90s when the industry really crashed in the economic downturn. I needed to make a change and the first thing I looked at about the biotech industry was it was great paying, good job stability. At a time when the economy's bad, they're still hiring. I get into it and started working hard, working my way up. So, I'm into my third organization in a management role and we were working on a [monoclonal antibody] product that was going to be launched for public use and we were doing our clinical phase 3 studies. We were working on a product for rheumatoid arthritis and we had people who are in our study come in and they actually met with all of us. Before they came in, they showed all of us who were working on the product these videos showing these people struggling with these little things that all of us take for granted, tying our shoes, holding a fork, feeding ourselves. After we watched the videos, the people in the study came in and say... fifty percent of these people that we had watched struggle came in doing all these different things, juggling, tying their shoes, etc. At that point, it really opened my eyes as to what we were doing - you think we're manufacturing a product but you're really manufacturing quality of life for these people. That was the biggest eye-opener for me. It really makes you understand that it's somebody's life that you're affecting and how can I improve that quality of life by my work.

(To learn more about Chris B.'s career and hear his story in his own voice, go to the YouTube video at https://www.franklinbiologics.org/for-educators/. Scroll down the page to find the interview with Chris.)

Developing techniques to "humanize" monoclonal antibodies was a critical technological step in the development of monoclonal antibody treatments, but technology advances alone did not enable them to be used as drugs. Knowledge is another essential requirement for developing monoclonal antibody therapeutics. To create a monoclonal antibody treatment, it is necessary to understand, at least to some degree, the molecular mechanisms of a disease. This understanding allows scientists to find **targets** for antibodies. A **target** *is a protein involved in a disorder to which the antibody can bind and exert an effect*. Ideally, targets are proteins that are only present when cells are diseased. In practice, many targets are proteins that are more abundant or are overexpressed when disease occurs, but are also present in normal cells. Finding targets for a particular disorder requires basic scientific research into the mechanism of that disorder.

Let's consider as an example the development of the monoclonal antibody drug Herceptin®. Herceptin® demonstrates how scientific knowledge of a disease process can be used to rationally design a specific monoclonal antibody treatment. It is an excellent example of what defines biotechnology, that is the transformation of knowledge into a product of value to humans.

Cancer researchers in the 1970s and 1980s learned that cancer cells, which divide over and over again malignantly, often have genetic alterations that drive their abnormal growth. Dr. Dennis Slamon and colleagues at the University of California found that there is a genetic alteration in a specific gene called Her2 in about 25% of women with breast cancer. The Her2 gene codes for the Her2 protein. The job of the Her2 protein is to reside on the surface of cells and act as a receptor to accept signals from growth factors – chemicals from outside the cell that carry growth-regulating orders. In a normal cell, two copies of the Her2 gene are present and the cell makes modest amounts of Her2 protein. Sometimes a mutation (change) occurs in a cell so that the Her2 gene is amplified, resulting in more than two copies of the gene. This amplification results in the production of too much receptor protein. When too much receptor protein is present, the cell binds too much growth factor and divides and multiplies more actively than normal. This mutation is associated with an aggressive form of breast cancer. The Her2 gene is thus a normal gene that causes cancer when it becomes overexpressed.

Scientists reasoned that if they designed a drug that would locate and bind to the Her2 receptor protein, it would block the receptor and prevent the cell from receiving growth signals and dividing malignantly. Monoclonal antibodies can target and bind to a specific protein, in this case the Her2 receptor, so scientists decided to create a monoclonal antibody as the blocking agent. Scientists successfully created a monoclonal antibody, named Herceptin®, which recognizes and binds to the Her2 receptor protein (Figure 1.12). Experiments in human volunteers showed that treatment with Herceptin® improves survival in women with Her2-positive breast cancer. Herceptin® thus became one of the first novel cancer treatments to emerge from basic research into the fundamental mechanisms of cancer cell growth. Herceptin® does not target all dividing cells, like standard chemotherapy treatments for cancer, but rather is specific for those expressing the Her2 receptor on their surface. Herceptin® is manufactured and marketed by Genentech Corporation.

The Herceptin® story illustrates some important principles about monoclonal antibody therapeutics. Antibody drugs are designed to find and bind a particular protein target. The key to their development is finding a suitable target, in this case, the Her2 receptor. Sometimes the binding of the monoclonal antibody blocks a receptor and therefore interferes with a cellular pathway that is associated with a disease condition. This is the case with Herceptin®.

Monoclonal antibody drugs treat disease in a variety of ways, in addition to blocking receptors. Sometimes monoclonal antibody drugs act by delivering a cytotoxic agent that destroys unhealthy cells. *Monoclonal antibodies that deliver cytotoxic agents are called* **antibody drug conjugates (ADCs).** For example, Adcetris®, a drug produced by Seattle Genetics, Inc., is used to treat Hodgkin lymphoma. Hodgkin lymphoma is a type of cancer that affects white blood cells. A protein receptor called CD30 is highly expressed on the surface of these cancer cells and is less commonly expressed on normal cells. Adcetris® consists of a monoclonal antibody against CD30 attached to the drug monomethyl auristatin E. Monomethyl auristatin E is an agent that disrupts microtubules inside cells. **Microtubules** *are hollow cylinders that are part of the "skeleton" inside of cells.* Microtubules have various critical cellular functions: They help maintain the structure of cells, are involved in cell division, assist in the transport of materials within the cell, and assist in cellular motility. When Adcetris® is administered to patients, the monoclonal antibody seeks out and binds to cancer cells displaying the CD30 protein. The monomethyl auristatin E disrupts the microtubules of the targeted cancer cells, which causes them to self-destruct. Adcetris® was approved in 2011 for use in human patients and was the first ADC entering the market. It was approved after only phase II trials in human patients (normally three phases of testing are required) because it was so effective against lymphoma. Adcetris® is much like that "magic bullet" imagined by scientists years ago; it seeks out and finds diseased cells, and delivers a killing agent.

Adcetris® was the first approved ADC, but others are now being used successfully. For example, the type of antibody used in Herceptin® can be conjugated to a cytotoxic drug to make an ADC. Kadcyla® is an ADC that targets the Her2 receptor in Her2-positive breast cancer patients and augments the arsenal of drugs against that disease. The case study below,

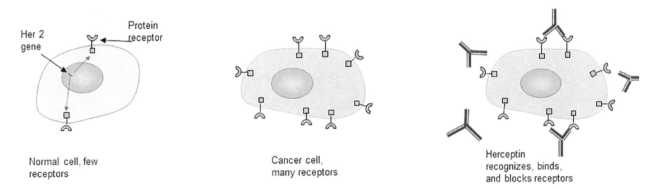

FIGURE 1.12 Herceptin® blocks receptor protein in cancerous cells and slows disease progression. In normal breast tissue, the Her2 gene causes a receptor protein to be made that accepts growth-regulating signals. Some breast cancer cells have more copies of the Her2 gene than normal and therefore produce too much receptor protein, leading to an accelerated rate of cell division and proliferation. Herceptin® recognizes the receptor protein, binds to it, and thus interrupts the growth signals.

"Checkpoint Inhibitors as Game-Changers," describes another important class of monoclonal antibodies used in cancer treatment.

While many monoclonal antibody drugs target cancer, monoclonal antibodies have value beyond the treatment of cancer. We will consider one more class of monoclonal antibody drug to illustrate another mechanism of action in a non-cancer disease. Age-related macular degeneration (wet AMD) is an eye disorder that causes loss of vision. It occurs when abnormal blood vessels grow under the macula, which is the part of the eye that is involved in central vision and seeing detail. These abnormal blood vessels are weak, and they leak blood and fluid into the back of the eye. The blood and fluid damages the macula causing blurred vision, wavy lines, dull colors, and blind spots (Figure 1.13). Ten or twenty years ago, wet AMD was considered to be untreatable and two-thirds of people with this disorder could expect to be legally blind within 2 years of developing the disease. Now, monoclonal antibody drugs are available that can protect the vision of many patients. What is the target of wet AMD drugs? Scientists discovered that targeting a protein called vascular endothelial growth factor (VEGF) is effective in treating wet AMD. **VEGF** *is a signaling protein that sends the message to the body that new blood vessels are required.* Blood vessels supply nutrients and oxygen to cells. For example, after an injury, the normal healing process requires the formation of new blood vessels. VEGF is involved in signaling the body to create these new vessels. VEGF is also normally active during embryonic development when the body's vascular system of blood vessels is formed. It is also secreted normally in muscles following exercise. VEGF is therefore important for the normal function of the body. However, in wet AMD, VEGF is overexpressed, and this causes the abnormal blood vessels to

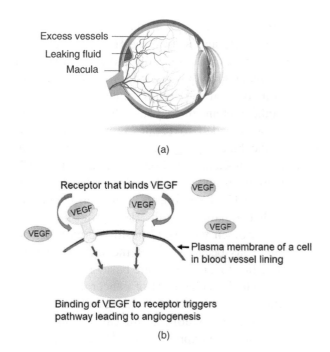

(a)

(b)

FIGURE 1.13 Wet AMD is driven by abnormal blood vessel growth. (a) Eye affected by wet AMD with excessive blood vessel growth causing leaking of fluid and blood. (b) Excessive VEGF binds to receptors located on the plasma membrane of cells that line blood vessels. This binding triggers a pathway in the cells that ultimately causes angiogenesis. (Image 1.13a © 2019 Novartis Pharmaceutical Corporation, used with permission.)

develop where they should not. The drugs Lucentis™, Avastin™, Eylea™, and Beovu™ all target VEGF, bind to it, and therefore block VEGF from signaling the eye to grow more blood vessels (Figure 1.14). These drugs are injected directly into the eye. While an injection into the eye sounds quite unpleasant, most people report little pain or distress. Directing the drug directly into the eye helps ensure that it reaches the

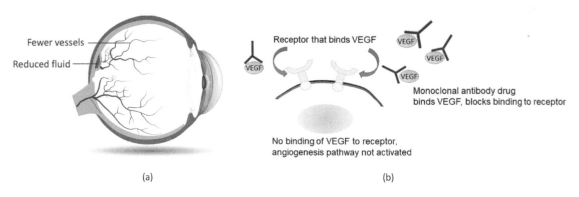

(a)   (b)

FIGURE 1.14 Wet AMD is treated by antibody binding to VEGF. (a) Eye after antibody treatment; fewer abnormal blood vessels are present. (b) Antibody drugs bind VEGF molecules and prevent them from binding to the cellular receptor. Angiogenesis is not induced. (Image 1.14a © 2019 Novartis Pharmaceutical Corporation, used with permission.)

location where it is needed and reduces the chance that it will have harmful side effects elsewhere in the body. It is important to note in this regard that monoclonal antibody drugs are powerful agents. Although they are directed toward diseased cells and tissues, they can still have intense and undesired side effects. Therefore, being able to deliver monoclonal antibodies directly to a specific site in the body is advantageous.

Case Study: Checkpoint Inhibitors as Game-Changers

In 1891, a physician named William Coley began experiments on a controversial cancer treatment – injecting pathogenic bacteria into tumors with the hopes of initiating a severe infection. Coley had discovered through researching medical cases that sometimes cancer patients who contract and recover from a major infection also experience regression of their cancer. Coley reasoned that this is because the patient's immune system was activated by the infection, and that activation also triggered the immune system to attack the cancer cells. While Coley's method of injecting pathogenic bacteria into cancer patients never achieved widespread acceptance, his basic concept, that the patient's own immune system could be harnessed to fight cancer, has turned out to be a critical step in the advancement of cancer treatments. More than 100 years after Coley's early work, editors of the prestigious journal, *Science*, named cancer immunotherapy (treatments based on harnessing a patient's immune system) to be the "2013 Breakthrough of the Year."

James P. Allison and Tasuku Honjo are contemporary scientists who performed pivotal work in immunotherapy. They were awarded the Nobel Prize for their discoveries in 2018. The basic premise of their research is that the immune system is capable of recognizing and destroying cancer cells. The problem in cancer is that something keeps the immune system from being successful. Researchers initially believed that the reason the immune system does not destroy cancer cells is because cancer cells are too much like normal cells, that is, the immune system does not recognize cancer cells as being foreign or defective. Instead, decades of basic immunology research demonstrated that immune cells could recognize cancer cells, but there are "brakes" that hold back immune cells, preventing them from aggressively destroying cancer cells. The existence of brakes in the immune system is necessary to keep immune cells from attacking one's own body. The immune system requires a careful balance that allows it to attack invading pathogens while not attacking the body's own cells. Allison and other researchers discovered that cancer cells "hijack" a brake that holds back T cells. **T cells** *are a type of white blood cell that search out and destroy invaders in the body.* Allison and other researchers discovered that a protein, cytotoxic T-lymphocyte-associated protein 4 (CTLA-4), plays a pivotal role as a brake that inhibits T cell activity. Allison theorized that if CTLA-4 could be temporarily blocked with an antibody, then T cells might proliferate and be more active than would normally occur. These active T cells might be able to recognize and attack cancer cells. Allison and colleagues showed that this is the case in animal models where blocking CTLA-4 caused tumor regression. Success in animal studies led to clinical trials in human patients. Two humanized monoclonal antibodies that blocked CTLA-4 entered clinical trials in 2000, ipilimumab and tremelimumab. While these drugs did not cure every patient, they did cure some, including some patients with very advanced cancers. The drugs were particularly effective for treating advanced metastatic melanoma, for which few effective treatments existed. In 2011, ipilimumab was approved for the treatment of metastatic melanoma, marking a medical breakthrough and confirmation of Coley's hypothesis that the immune system can be harnessed to treat cancer.

CTLA-4 blockade has significant side effects in many patients, and so other drugs with a similar mechanism of action and fewer side effects have subsequently been introduced. PD-1 is a protein on the surface of T cells that also acts as a brake on the immune system. It was discovered that some cancer cells have a complementary structure on their surface, called PD-L1. Researchers found that when PD-L1 on cancer cells binds to PD-1 on T cells, the T cells do not attack the tumor cells. This binding seemed to be another brake, a way in which cancer cells deceive the immune system. Scientists reasoned that if they could prevent PD-1 from binding to PD-L1, it would allow the T cells to recognize and attack cancer cells. Scientists therefore created monoclonal antibodies that bind and block either PD-1 or PD-L1. This strategy has

(Continued)

Case Study (*Continued*): Checkpoint Inhibitors as Game-Changers

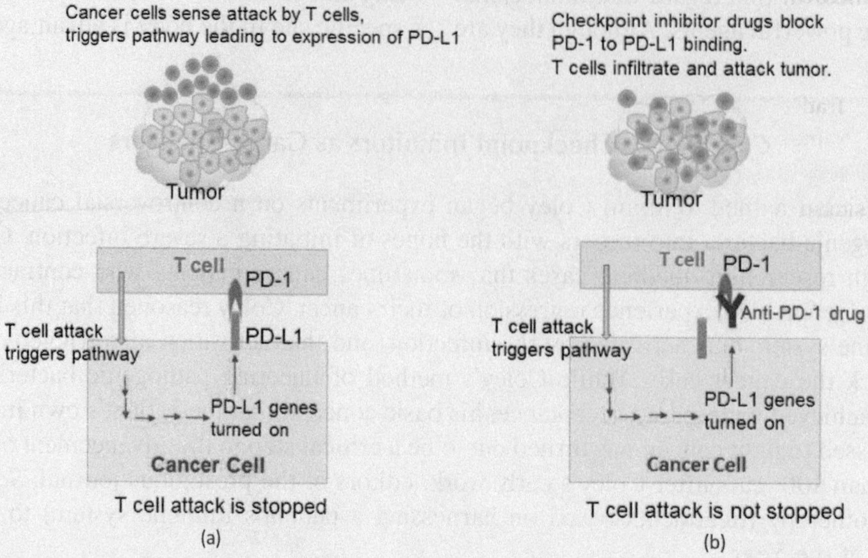

FIGURE 1.15 Checkpoint inhibitors. (a) Top panel: Patients produce T cells that recognize cancer cells as being defective and begin to attack them. Bottom panel: The attack of T cells on cancer cells triggers a pathway inside the cancer cells that ultimately leads to turning on the gene for the PD-L1 protein. The cancer cell makes PD-L1, which migrates to the cellular membrane where it binds to PD-1, a protein on the surface of the T cells. The binding of PD-1 to PD-L1 turns off the attack of the T cells, which remain on the margins of the tumor. (b) Checkpoint inhibitor drugs bind to either PD-1 or PD-L1 and prevent them from binding to one another. In this situation, the T cells remain active and infiltrate the tumor (top panel), resulting in tumor regression. (This figure is based on the information from Ribas, Antoni, and Jedd D. Wolchok, 2018, as cited in the text.)

worked, and by 2018, five antibody drugs targeting either PD-1 or PD-L1 had been approved (Figure 1.15). *The antibodies that bind CTLA-4, PD-1, and PD-L1 are termed* **checkpoint inhibitors**. According to some experts "The…durable response rates in patients with multiple types of cancer indicate that therapeutic blockade of the PD-1 pathway is arguably one of the most important advances in the history of cancer treatment." (Ribas, Antoni, and Jedd D. Wolchok. "Cancer Immunotherapy Using Checkpoint Blockade." *Science*, vol. 359, no. 6382, 2018, pp. 1350–55. doi:10.1126/science.aar4060.)

Checkpoint inhibitors are an example of how knowledge from basic biological research conducted over years by many investigators combined with a powerful technology – monoclonal antibodies – came together to help critically ill patients. These drugs do not cure every patient, nor are they effective against all types of cancer. They can, in some cases, elicit serious side effects. But these drugs have proven that Coley was right and that patients have powerful immune cells in their bodies that sometimes, with help from drugs, can cure their cancer.

Table 1.2 summarizes the monoclonal antibody market by providing varied examples of approved monoclonal antibody drugs. This is not a comprehensive list, but it provides a sense of the importance of this class of biopharmaceuticals for treating disease. Each of these monoclonal antibodies works by selectively finding, binding, and altering a specific cellular protein target that is associated with diseased cells. Observe that there can be more than one antibody product based on a particular target. One of the major goals of biomedical research is to identify more targets

for antibody drug interventions; identifying suitable targets is the limiting factor in monoclonal antibody drug development.

1.2.4 Vaccines

Vaccines *are agents that are used to enhance the immune system*, in the most familiar case to protect against infection by a specific pathogen, such as polio or diphtheria. Vaccines can therefore be considered a type of immunotherapy, a treatment that helps the body's own immune system fight disease. Vaccines are

TABLE 1.2

Monoclonal Antibody Therapeutics

mAb Name[a]	Trade Name[a]	Disease Treated	Year of Approval for Use in Patients	Target Protein	Drug's Mode of Action
Trastuzumab	Herceptin®	Breast cancer	1998	Her2	See text.
Isatuximab-irfc	Sarclisa®	Multiple myeloma	2020	CD38	Myeloma is a type of cancer that forms in a type of white blood cells called plasma cells. CD38 is highly expressed on cancerous myeloma cells. Binding of the antibody drug to myeloma cells triggers destruction of CD38-expressing cells.
Gemtuzumab ozogamicin	Mylotarg®	Acute myelogenous leukemia (AML)	2000	CD33	CD33 is a cell surface molecule expressed by cancerous blood cells, but not found on normal stem cells needed to repopulate the bone marrow. Mylotarg® is an ADC with a cytotoxic agent that causes death of CD33-expressing cells.
Cetuximab	Erbitux®	Colorectal cancer Head and neck cancers	2004 2006	Epidermal growth factor receptor (EGFR)	EGFR is a cell surface protein that receives growth-promoting messages associated with abnormal growth in these cancers. When EGFR is bound to Erbitux®, it cannot receive growth-promoting signals.
Bevacizumab	Avastin®	Colorectal cancer	2004	Vascular endothelial growth factor (VEGF)	VEGF stimulates new blood vessel formation that, in cancer, is associated with the growth of tumors. The antibody blocks VEGF. Avastin® is sometimes also used to treat wet AMD, as described in the text.
Ado-trastuzumab emtansine	Kadcyla®	Early Her2-positive breast cancer Late-stage breast cancer	2019 2013	Her2	See text.
Pembrolizumab	Keytruda®	Melanoma Non-small cell lung cancer Head and neck squamous cell cancer Hodgkin lymphoma Urothelial cancer Cervical cancer Gastric cancer Merkel cell carcinoma Hepatocellular carcinoma	2014 and later (different dates of approval for different cancers)	PD-1	See text.

(Continued)

TABLE 1.2 (Continued)
Monoclonal Antibody Therapeutics

mAb Name[a]	Trade Name[a]	Disease Treated	Year of Approval for Use in Patients	Target Protein	Drug's Mode of Action
Palivizumab	Synagis®	Human respiratory syncytial virus, RSV (virus associated with infant mortality)	1998	**RSV fusion protein**	The target is a protein found on the surface of RSV that is involved in viral insertion into lung cells. The drug binds this protein and thereby prevents virus from inserting into infant's cells.
Infliximab	Remicade®	Rheumatoid arthritis and Crohn's disease (autoimmune diseases)	1999	**Tumor necrosis factor (TNF)**	TNF is a protein associated with immune responses. Remicade® binds TNF and prevents it from initiating inflammation, which reduces the severity of several autoimmune diseases.
Adalimumab	Humira®	Rheumatoid arthritis, plaque psoriasis, Crohn's disease, ulcerative colitis	2002	**Tumor necrosis factor (TNF)**	See Remicade® mode of action.
Omalizumab	Xolair®	Asthma	2003	**Immunoglobulin E**	Xolair® inhibits the binding of immunoglobulin E to receptors on specific types of white blood cells. When this binding is reduced, the cells do not release certain chemical agents that are associated with an allergic response.
Brolucizumab	Beovu®	Age-related macular degeneration	2019	**Vascular epithelial growth factor (VEGF)**	See text.
Crizanlizumab	Adakveo®	Sickle cell anemia	2019	**P-selectin**	Sickle cell anemia is a genetic disease in which red blood cells are abnormally shaped and get stuck in small blood vessels where they block blood flow and oxygen transport, resulting in painful and debilitating episodes. Adakveo® blocks interactions between cells circulating in blood and reduces the severity of sickle cell episodes.

[a] Monoclonal antibody drugs have two names. The first is a generic name that is assigned to the antibody using naming conventions. For example, all monoclonal antibodies end in -mab, while the letters -xi- and -zu- provide information about how the antibody was humanized. The second name is a simpler one created by the manufacturer to identify their brand.

sometimes, though not always, made using recombinant DNA technology.

The eighteenth-century English physician, Edward Jenner, is usually credited with devising the first vaccine. Jenner realized that milkmaids who had been infected with cowpox did not get smallpox. Since cowpox infection is mild and smallpox is often fatal, Jenner got the idea of intentionally infecting people with the milder disease to prevent the more severe one. Jenner experimented by injecting a healthy boy, who had never had cowpox or smallpox, with some fluid from the cowpox sore of a milkmaid. When the boy recovered from cowpox, Jenner exposed him to smallpox. The boy did not get sick, and Jenner termed the procedure "vaccination," which is Latin for "pertaining to cows." See also the case study "Transporting a Vaccine to the New World."

Case Study: Transporting a Vaccine to the New World

An interesting side note to the smallpox vaccine story relates to the storage, stability, administration, and transport of the vaccine. An important part of the development of any new therapeutic product is determining how best to package and store it, how to administer it, and how long it is stable. These problems are particularly acute for drugs to be used in countries where there is a scarcity of refrigeration, sterile water for reconstituting drugs, and sterile syringes. There were no refrigerators or pharmaceutical manufacturers in Edward Jenner's time. The Spanish physician, Francisco Xavier Balmis, was ordered by the Spanish King, Charles IV, to sail across the ocean to bring vaccination to the Spanish colonies in the Americas. Because the cowpox fluid was not stable over long periods, Balmis used a chain of orphan boys on the voyages. He would infect one boy with cowpox, wait a week or so, and then infect the next, and so on, as they sailed all the way across the sea. On reaching land, the boys were used to vaccinate other people and then were fostered out to families. Due to the efforts of Balmis and other physicians, vaccination against smallpox spread rapidly across Europe and the Americas.

The story of smallpox vaccination is still relevant today for several reasons. It is interesting from the point of view of understanding the immune system, the nature of vaccines, and issues with storage, administration, and transport. The story also raises interesting ethical questions. Jenner's experiments, intentionally exposing a boy to smallpox, and the use of children to store and produce a vaccine seriously violate the ethical principles applied today, but were within the accepted ethical boundaries of the 1700s and 1800s. We will further note that the first vaccines that were developed to combat the much more recent COVID-19 pandemic require stringent conditions of cold during storage and transport. While these conditions can be achieved in high-income countries using modern technology, the logistics of storing and transporting the vaccines, and making them widely available to everyone, remain a challenge.

Many other vaccines have been developed since Jenner's time, for example, to prevent polio, diphtheria, and tetanus. Prior to the advent of recombinant DNA technology, most vaccines were made by growing the virus or bacterium that caused the disease and then either killing the pathogen or damaging it in some way so that it could not cause full-blown illness. The dead or attenuated (weakened) pathogen was then administered to people, eliciting an immune system response that protected the person if they were later exposed to the pathogen. The flu vaccine is still often made this way; flu virus is grown each year in millions of fertilized chicken eggs. Vaccines made in the traditional fashion are usually fine, but they may, in a few cases, cause the disease they are intended to prevent.

Recombinant DNA technology provides a new method of vaccine production that avoids the risk of causing the disease it is meant to prevent. The vaccine against hepatitis B, for example, is made by taking a gene that codes for a protein found on the surface of the virus and inserting the gene into host yeast cells. The yeast then manufacture large amounts of the viral protein. The resulting protein is purified and formulated into an injectable vaccine that, when administered, triggers an immune response against the hepatitis B virus. There is no risk that the vaccine

will cause the disease it is intended to prevent because only a single viral protein is used to make the vaccine. More and more vaccines are being made in this way. There is, for example, currently one recombinant DNA flu vaccine approved for commercial use in the United States; almost certainly, more will be available in future years.

Yet another type of biotechnology-based vaccine, DNA/mRNA vaccines, may become common in the future. **DNA vaccines** are made from vectors that have been genetically engineered to include the DNA coding for one or two specific proteins from the infectious agent. mRNA vaccines are similar, but inject the mRNA for a protein, instead of the DNA that codes for it. When the DNA or mRNA is injected into the cells of a person (or some other animal), the cells synthesize a protein associated with the infectious agent (Figure 1.16). The recipient's immune system responds to the new protein(s) by mounting a protective immune response. mRNA vaccines bypass the use of cells in a fermenter or bioreactor to manufacture a vaccine. Threats from a number of emerging diseases in the past decades have led to increased interest in DNA and mRNA-based vaccines, which allow faster and more efficient vaccine production than previously used methods. For example, the U.S. National

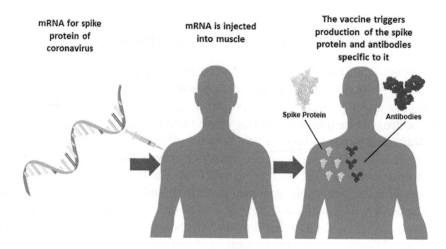

FIGURE 1.16   A strategy to make an mRNA vaccine to prevent COVID-19. The virus's spike protein is a part of the viral particle that mediates the entry of the virus into cells. Manufacturers synthesize viral spike protein mRNA and make it into a vaccine that is injected into the cells of a person (or some other animal). The recipient's cells then synthesize spike protein, and his or her immune system responds to the spike protein with the production of antibodies. (Image credit: U.S. National Institutes of Health.)

Institute of Allergy and Infectious Diseases Vaccine Research Center has developed candidate DNA vaccines to address several viral disease threats, including SARS coronavirus (SARS-CoV-1) in 2003, H5N1 avian influenza in 2005, H1N1 pandemic influenza in 2009, and Zika virus in 2016. In 2019, the first DNA vaccine was approved in the United States. It is called Ervebo® and is credited with stopping an outbreak of the deadly disease, Ebola, in Guinea.

At the time of writing, several vaccines have been developed to prevent COVID-19, a disease caused by a virus, SARS-CoV-2, that is spreading very quickly around the world. The first vaccines to enter clinical trials to protect against COVID-19 were mRNA-based vaccines. These vaccines have been dramatically effective in preventing serious COVID-19 illness and death. The time from selection of the viral genes to be included in the vaccine to initiation of clinical studies keeps getting shorter and shorter, which is a critical advantage when pandemics strike. In the case of COVID-19, just 42 days elapsed between the announcement of the sequence of the genome of SARS-CoV-2 and a candidate vaccine – this is unprecedented in the history of medicine and could only have occurred because of the years of research and development in the biotechnology arena and international cooperation efforts.

Another area of active research is in therapeutic cancer vaccines. Unlike familiar prophylactic vaccines that prevent illness, such as the flu or polio vaccines, therapeutic vaccines are intended to treat patients who already have cancer. These vaccines might help patients fight their cancer by mobilizing their immune system. Hence, they might become another weapon in the immunotherapy arsenal, alongside the checkpoint inhibitors described above. The strategy to create a cancer vaccine is to administer to patients a preparation containing many copies of an antigen molecule that is found on the surface of their cancer cells, but not on their normal cells. Administering these cancer-specific antigens is intended to stimulate the patient's immune system, causing it to produce antibodies and immune cells that recognize and destroy cancer cells. One FDA-approved vaccine for cancer is sipuleucel-T (Provenge®), which is used for prostate cancer that has metastasized (spread). Provenge rallies the immune system's disease-fighting forces in men who already have prostate cancer. Provenge is created by removing some immune cells, exposing them to a molecule from prostate cancer cells, and then infusing them back into the body. Provenge has been shown to extend survival in men with metastatic prostate cancer.

It remains to be seen if cancer vaccines will become a common strategy for cancer treatment.

1.2.5   Genetic Engineering and Food Production

1.2.5.1   GMO Crops

Recombinant DNA technology has not only impacted medicine, but also agriculture. Conventional methods of plant and animal breeding have long been used to enhance the characteristics of crops and livestock. In conventional breeding, plants or animals with desirable qualities, for example, plants with large fruit or resistance to insects, have been selectively mated with other plants with desired traits. It is possible, however, to use recombinant DNA technologies to genetically modify plants and animals more quickly, and with better control, than is possible using traditional methods. Conventional breeding does not allow genes from unrelated plant or animal families to combine because the plants or animals do not breed with one another. Also, in conventional breeding, there is an uncontrolled mixing of all the genes from both parents. Recombinant DNA methods allow the introduction of only a specific gene of interest into the offspring.

Genetically modified (GMO) varieties of soybeans, corn, and cotton were introduced into commercial production beginning in 1996. The first trait that was introduced into crop plants using recombinant DNA methods was a feature that makes the plants easier for farmers to grow, not a trait with obvious appeal to consumers. These plants contain an introduced gene that makes them resistant to the herbicide glyphosate (trade name: Roundup®). Glyphosate kills plants and has been thought to be nontoxic to animals (although there is ongoing controversy about whether or not Roundup® is toxic to humans). The use of glyphosate in the past was limited because it kills all plants – including the crop. Farmers now plant genetically modified, herbicide-resistant plants so they can use glyphosate to kill weeds without harming their crops.

A second widely adopted genetic modification of plants was the introduction of a gene from the soil bacterium *Bacillus thuringiensis* (Bt). Plants with the Bt gene produce a protein that is toxic to insects that destroy crops, thus helping protect genetically modified plants from insect damage. This method has the environmental advantage of protecting crops while reducing the use of pesticides.

Genetically modified crop varieties rapidly gained acceptance with farmers in the United States. Survey data from the U.S. Department of Agriculture (USDA) in 2005 indicated that more than 50% of all corn, 79% of all cotton, and 87% of all soybean crops in the United States were genetically modified with an herbicide resistance gene, a Bt gene, or both. By 2019, the USDA stated that over 90% of US corn, upland cotton, soybeans, canola, and sugar beets are produced using genetically engineered varieties (http://www.ers.usda.gov). Other crops that are sometimes genetically modified include alfalfa, papaya, squash, eggplant, potatoes,

and apples. While the majority of genetically modified crops now have herbicide or insect resistance, other genetically modified crop varieties are being field-tested and are beginning to enter the marketplace. These new genetically modified crops are expected to provide varied benefits such as resistance to fungal diseases, tolerance to drought and cold, increased nutritional value, and changes that allow crops to be grown in marginal soils and climates. At the time of writing, many of these crops are not yet approved for commercial use, but there are already commercial corn and soybean crops that are modified with genes that confer drought resistance; GMO apples that resist browning when sliced; corn high in the amino acid lysine (which is beneficial for livestock fed the corn); potatoes that resist browning, have reduced acrylamide (a toxic agent that can occur in potatoes), and are blight-resistant; and corn that is enhanced for bioethanol production.

1.2.5.2 GMO Animals Used for Food

Biotechnology methods are also being introduced to modify animals used for food. A recent entry into the GMO food marketplace is a type of genetically modified salmon that contains genes from Chinook salmon and an eel-like creature called ocean pout. The pout genes allow the salmon to grow twice as fast on less food than normal Atlantic salmon. This fish was approved for sale in the United States in 2019.

1.2.5.3 Controversy Surrounding GMO Food

While genetic engineering is increasingly being applied to agriculture, particularly in the United States, the use of these technologies in food production is among the more controversial of its applications. One issue relates specifically to the use of Bt crops. Organic farmers have a long tradition of applying the bacterium, *Bacillus thuringiensis*, to their crops. Organic farmers fear that the widespread use of genetically modified Bt crops will result in target pests that develop resistance to the Bt toxin. Bt seed producers and farmers planting genetically modified Bt crops therefore use resistance management strategies to avoid, or at least postpone, the appearance of Bt-resistant pests. Seed producers genetically modify their seeds in such a way that the plants will produce high levels of Bt toxin sufficient to kill all target insects except for a few very rare resistant individuals. Farmers plant small refuges of non-genetically modified crops within their Bt crop fields. The logic behind the refuges is that susceptible, nonresistant insects that come from the refuges will mate with the few Bt-resistant individuals. Their resulting offspring will still be susceptible to the Bt toxin. This combined strategy appears to be successful to date, although scientists are working on new methods to protect crops in anticipation of the appearance of Bt-resistant pests.

Another concern associated with Bt crops is that desirable insects may be harmed along with target pests. Early research, for example, suggested that monarch butterflies might be harmed by eating pollen from Bt corn plants. Later research indicated that monarch butterflies are not at risk, although the general concern remains a topic of investigation.

Other concerns relating to genetically modified foods are that food produced using genetic engineering methods might not be as safe as conventional foods, that animals may be stressed by the introduction of foreign genes, that genes that make plants hardier may unintentionally be transferred to weeds, and that the introduction of genetically modified organisms into the environment may have unforeseen, adverse effects. There are also economic and political issues that are of concern. These various issues will presumably be resolved as the agricultural biotechnology industry matures. In the meantime, the trend has been toward a gradual introduction of more genetically modified crops in the United States with slower acceptance in Europe and other parts of the world.

1.2.5.4 Cloning

The term "clone" means a copy. When used as a verb, the term *clone* means to make a copy. In biotechnology, a clone can refer to an exact copy of a DNA segment produced using recombinant DNA technology, as shown in Figure 1.3. A clone can also be one or more cells derived from a single ancestral cell (as discussed previously in the section on monoclonal antibodies). The term "clone" can also refer to one or more organisms derived by asexual reproduction that are genetically identical to a parent. In this section, we discuss the last of these definitions, particularly as it relates to agriculture. **Whole animal cloning** *is the production of one or more adult animals that have the same genetic information.*

Cloning occurs naturally in people and in other animals when an early embryo splits, giving rise to identical twins. Scientists working with laboratory and farm animals in the 1980s intentionally created twins by splitting embryos in the laboratory into parts, each of which was then implanted into a surrogate mother. Using this technique, scientists cloned embryos that resulted from a normal mating of two animals, so the clones were identical to one another, but they were not identical to either parent, and their genetic traits could not be fully predicted.

While these early cloning activities proceeded with little fanfare, Dolly, the cloned sheep, became an international celebrity. Dolly was not produced by splitting an embryo. *She was made by taking an adult sheep's mammary cell nucleus and inserting it into another sheep's egg whose nucleus had been removed; this is called* **somatic cell nuclear transfer** (Figure 1.17). The resulting egg thus contained only the genetic information

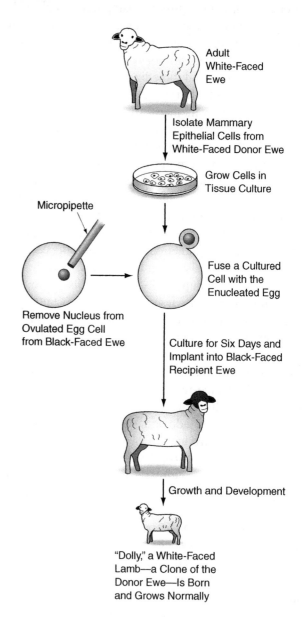

FIGURE 1.17 Somatic cell nuclear transfer was used to create Dolly. A somatic cell *is any cell in an organism other than eggs or sperm cells.* The isolated mammary epithelial cells diagrammed here are somatic cells. The genetic information from a somatic cell was transferred to an enucleated egg, which was subsequently incubated in a surrogate mother ewe. The resulting lamb, Dolly, had only the genetic information of the parent from whom mammary cells were removed.

from the mammary cell. The egg was placed inside a surrogate mother sheep where it developed into a baby sheep, named Dolly. Dolly did not arise from any mating between two animals (so she was the product of asexual reproduction), and she was genetically equivalent to the adult sheep from whom the mammary cell had been removed. Dolly was born in July 1996, at the Roslin Institute in Scotland under the direction of Ian Wilmut. She was euthanized in 2003 because she suffered from various health problems, including serious lung disease. Her illness was not believed to be cloning-related; rather, it was due to a viral illness that affects sheep.

Dolly was a public service success, becoming a celebrity and "spokesperson" for biotechnology – at least in the late 1990s. From a basic research point of view, the cloning of Dolly was a monumental advance. She proved that an adult cell can be "reprogrammed" to act like an embryonic cell. Prior to Dolly, many scientists thought this reprogramming was impossible. Cloning continues to be a research tool for studying development and for medical research. Researchers have cloned diverse animals including mice, rats, rabbits, cats, mules, fish, horses, dogs, and, in 2018, macaque monkeys. The monkeys are the first primate clones. No human has ever been artificially cloned, despite many popular fictional stories featuring emotionally troubled cloned humans. Most scientists are ethically opposed to conducting research on human cloning, and many think it would be extremely difficult or impossible to accomplish.

From the point of view of generating biotechnology products, cloning is a more nuanced achievement. The major application for cloning was thought to be, and still is, in livestock production. For example, consider a farmer with a goat who produces abundant milk that yields prize-winning goat cheese. Eventually, this goat will grow old and will die. If, however, the farmer could clone her, then the farmer could obtain another prize-winning goat. In fact, since Dolly, livestock species including cattle, swine, sheep, and goats have been cloned. Dolly herself was cloned to create Dianna, Daisy, Denise, and Debbie. In 2008, the U.S. Department of Agriculture estimated that there were about 600 cloned farm animals in the United States alone and there are probably thousands now. The purpose of these cloned animals is not actually to produce milk or meat; they are used for breeding. Their offspring might be used for milk or meat production, but the cloned animals are not. The reason that livestock cloning has not become widespread is that it is complex, somewhat unpredictable, and expensive.

Cloning is technically challenging, and it requires many attempts to produce a viable animal. In 2020, one company was charging $20,000 to clone a single cow. A website that promises to clone a beloved pet dog or cat charges $50,000 for a dog and $25,000 for a cat.

An additional problem with cloning is that, even if cloning technology improves and prices go down, clones are not identical to their parent, and their differences are unpredictable. These differences between parent and clones are because of a phenomenon called **epigenetics** (Figure 1.18). According to Dr. Randy Jirtle, a prominent scientist who studies epigenetics, "Epigenetics literally translates into…'above the genome.' I think of the DNA as being like the hardware of your computer. So, then the epigenome is the software that tells the genes when, where, and how to work." (As quoted in a 2018 interview with Dr. Kara Fitzgerald, "Episode 40: Heal Your Gut, Boost Your Brain and Live Longer with Polyphenols with Dr. Randy Jirtle." 12 April 2018, www.drkarafitzgerald. com/2018/04/12/epigenetics-the-agouti-mice-study-with-dr-randy-jirtle/.) One mechanism of epigenetics is the addition of methyl tags to DNA bases. A methyl group, CH_3, is a carbon atom bonded to three hydrogen atoms. When a methyl group is attached to certain bases in DNA, it prevents the expression of specific genes. Another epigenetic mechanism involves histones. Chromosomes are made of DNA strands that are tightly wrapped around **histones**, *proteins that alter the structure of DNA so that the entire genome can fit inside the nucleus*. Like methyl tags, histone modifications can turn off specific genes. Epigenetic changes are critical during development because they allow cells in different parts of an organism to express different genes. For example, muscle cells and nerve cells in an individual have the same genes, but different ones are turned on in each type of tissue. Epigenetic markers are one of the mechanisms by which genes are turned off and on in specific cells during development. Epigenetics is also important after an organism is born and throughout its life. Scientists have found that the environment, food choices and availability, exercise, and other influences can alter the epigenetics of an organism. It has been found, for example, that identical twins have slightly different characteristics when they are born, even though they have the same genes. As twins age, they have different life experiences and their "epigenomes" become progressively more distinct from one another. Epigenetics is relevant to cloning because clones are genetically identical to their parent, yet, because of epigenetics, they

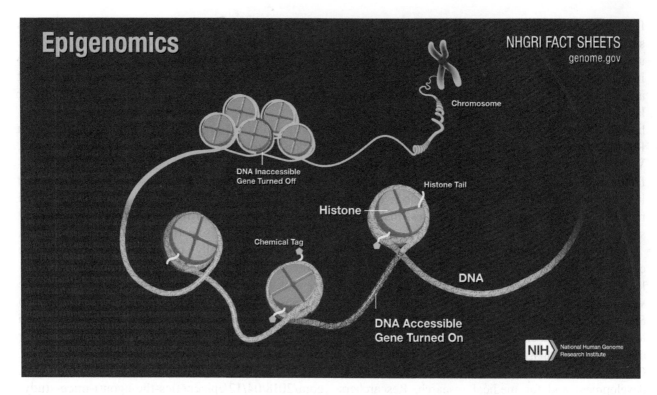

FIGURE 1.18   Epigenetics affects which genes are expressed. This illustration shows how modifying histones can make DNA more accessible for transcription. (Image credit: U.S. National Human Genome Research Institute.)

can differ in significant and not totally predictable ways. Because of epigenetics, it might be possible to clone a prize-winning racehorse, but that clone may or may not cross the finish line anywhere near the front of the pack.

1.2.5.5 Tissue Engineering to Produce Meat

In Chapter 2, we will discuss tissue engineering, a rapidly growing area of biotechnology that is being developed to produce therapeutic tissue, such as skin for burn victims, and organs for transplantation. Here, we will mention that the same techniques that are used to grow human tissues in artificial environments can be used to grow meat for human consumption. Cultured beef is created by harvesting muscle stem cells from a living cow. Scientists then feed and nurture the cells in culture dishes and bioreactors so that the cells multiply and generate strands of muscle tissue, which is the main component of the beef that we eat. Culturing meat in this way has many advantages. First and foremost, no animals are killed. In terms of sustainability, proponents say that cultured meat will save vast areas of land that can be used to grow crops and feed more people than can livestock production. Commercial

livestock production now often involves the use of large amounts of antibiotics. This use of antibiotics may be contributing to the worrisome increase in antibiotic-resistant bacterial infections in humans. Cultured meat alleviates concerns about antibiotic use. According to the Mosa Meat company website, a company working on this product, "From one sample from a cow, we can produce 800 million strands of muscle tissue (enough to make 80,000 quarter pounders)." The technology therefore has the potential to be scaled up to become a significant source of food for a rapidly growing global population. At the time of writing, cultured meat has entered the marketplace and restaurants in your neighborhood may be touting their eco- and bovine-friendly burgers.

1.2.6 OTHER PRODUCTS OF RECOMBINANT DNA TECHNOLOGY

Although pharmaceuticals and modified crops are the most well-known applications of genetic engineering, there are other applications as well. Many industries have been quietly incorporating recombinant DNA technology into their manufacturing for years.

For example, rennin is an enzyme used in the manufacture of cheese. Before the advent of genetically modified organisms, rennin was isolated from the fourth stomach of unweaned calves. Rennin is now produced by genetically modified bacteria.

Almost all modern laundry detergents have enzymes to break down and remove substances that soil clothing. Lipases break down oils, proteases break down proteins, and amylases break down starches. These enzymes are typically manufactured by genetically modified microorganisms. Subtilisin is an example of a protease manufactured by genetically modified bacteria. At the time of writing, one listing of household products shows 304 common brands containing subtilisin, ranging from laundry detergents to contact lens cleaner (https://www.ewg.org/guides/substances/5852-SUBTILISIN/?page=3).

There are other industrial processes that have been enhanced by biotechnology. Traditional methods of paper manufacturing, for example, use harsh chemicals that are released into the environment. Paper manufacturers now can substitute enzymes manufactured by genetically modified bacteria for these chemicals. Another example is genetically modified bacteria that can break down pollutants in contaminated soil and water. Another promising area is the introduction of "bioplastics" that can potentially replace conventional plastics. In recent years, environmental pollution caused by the ubiquitous manufacture and disposal of plastic items has become increasingly severe. Using renewable feedstocks such as plants, industrial and food waste, and agricultural residues, it is possible to create renewable chemicals to replace fossil fuel-derived ingredients used in conventional plastics. Bioplastics are expected to have many advantages in reducing environmental pollution and allowing the recycling of waste.

This chapter has introduced the basic science and applications that are the foundation of the "modern" biotechnology industry. This industry has developed in many new and exciting directions, some of which will be introduced in Chapter 2.

Practice Problems

1. The process illustrated in Figure 1.4 is sometimes called *gene cloning*. Why is this process called *cloning*?
2. The following words are sometimes confused with one another. Define them:
   a. Transfection
   b. Transformation
   c. Translation
   d. Transcription
3. Which of the following biotechnology applications involve transfection of cells, which involve transformation of cells, and which involve neither transfection nor transformation?
   a. Producing crop plants that are resistant to the herbicide glyphosate.
   b. Growing cultured human skin cells to use for tissue engineering (e.g., to treat burn victims).
   c. Making wine.
   d. Manufacturing human epidermal growth factor (a protein) in bacteria.
   e. Harvesting stem cells from an embryo.
4.
   a. What is the role of the antibodies that a person, or some other animal, normally produces? (If you are unsure, consult a biology reference.)
   b. Compare and contrast monoclonal antibody therapeutic agents to the antibodies normally produced in the body. Refer to Table 1.2 in your answer.
5. The monoclonal antibody Synagis® is used to fight infection by respiratory syncytial virus (RSV). Use an Internet search engine to find information about this drug.
   a. The general mechanism of action of monoclonal antibody drugs is to recognize a specific protein target, bind to this target, and block the target's action with therapeutic effect. Find the mode of action (mechanism of action) of the drug Synagis®, and explain its mode of action.

b. Synagis® protects babies against a viral agent, but it is not a vaccine, at least not in a conventional sense. What is the difference between Synagis® and a conventional vaccine (like the one Jenner invented against smallpox)?

6. Insulin and other biopharmaceuticals are seldom administered orally. What might happen to these drugs if they were swallowed?

7. The first vaccine candidate to go into clinical trials against the virus that causes COVID-19 was based on mRNA. This type of vaccine has a major advantage in a situation where billions of doses are required quickly – as is the case when a global pandemic is raging. What is this advantage?

8. The following text is from a package insert (from GlaxoSmithKline Biologicals) that accompanies a vaccine to prevent the disease shingles. **Shingles** is *a viral infection that causes a rash that is frequently very painful and disabling*. Shingles is caused by the varicella-zoster virus – the same virus that causes chickenpox.

a. What type of vaccine is this, one based on mRNA, or DNA, or protein, or inactivated virus?

b. Draw a sketch of how this vaccine is manufactured, based on the information provided by the manufacturer (Figure 1.19).

11   DESCRIPTION

SHINGRIX (Zoster Vaccine Recombinant, Adjuvanted) is a sterile suspension for intramuscular injection. The vaccine is supplied as a vial of lyophilized recombinant varicella zoster virus surface glycoprotein E (gE) antigen component, which must be reconstituted at the time of use with the accompanying vial of AS01$_B$ adjuvant suspension component. The lyophilized gE antigen component is presented in the form of a sterile white powder. The AS01$_B$ adjuvant suspension component is an opalescent, colorless to pale brownish liquid supplied in vials.

The gE antigen is obtained by culturing genetically engineered Chinese Hamster Ovary cells, which carry a truncated gE gene, in media containing amino acids, with no albumin, antibiotics, or animal-derived proteins. The gE protein is purified by several chromatographic steps, formulated with excipients, filled into vials, and lyophilized.

The adjuvant suspension component is AS01$_B$ which is composed of 3-*O*-desacyl-4'-monophosphoryl lipid A (MPL) from *Salmonella minnesota* and QS-21, a saponin purified from plant extract *Quillaja saponaria* Molina, combined in a liposomal formulation. The liposomes are composed of dioleoyl phosphatidylcholine (DOPC) and cholesterol in phosphate-buffered saline solution containing disodium phosphate anhydrous, potassium dihydrogen phosphate, sodium chloride, and water for injection.

After reconstitution, each 0.5-mL dose is formulated to contain 50 mcg of the recombinant gE antigen, 50 mcg of MPL, and 50 mcg of QS-21. Each dose also contains 20 mg of sucrose (as stabilizer), 4.385 mg of sodium chloride, 1 mg of DOPC, 0.54 mg of potassium dihydrogen phosphate, 0.25 mg of cholesterol, 0.160 mg of sodium dihydrogen phosphate dihydrate, 0.15 mg of disodium phosphate anhydrous, 0.116 mg of dipotassium phosphate, and 0.08 mg of polysorbate 80. After reconstitution, SHINGRIX is a sterile, opalescent, colorless to pale brownish liquid.

SHINGRIX does not contain preservatives. Each dose may also contain residual amounts of host cell proteins (≤3.0%) and DNA (≤2.1 picograms) from the manufacturing process.

The vial stoppers are not made with natural rubber latex.

FIGURE 1.19   Description of the Shingrix vaccine.

2 The Biotechnology Industry Branches Out

2.1 REGENERATIVE MEDICINE: AN INTRODUCTION

2.1.1 OVERVIEW

The technologies that fall under the umbrella of "biotechnology" are constantly evolving and expanding. In this chapter, we consider two of these active areas: regenerative medicine and genomics.

Regenerative medicine *is a broad area of biotechnology that encompasses strategies to restore normal function to tissues and organs that have been damaged due to injury, genetic problems, aging, or disease.* The strategies included in the term "regenerative medicine" vary somewhat depending on the source. We will consider the following to be part of regenerative medicine (Figure 2.1):

- *Gene therapies.*
- *Immunotherapies*, including checkpoint inhibitors, as described in Chapter 1, and cell-based immunotherapies, which will be discussed in this chapter.
- *Administration of stem cells, which are undifferentiated cells that can differentiate into various types of cell in the body.*
- *Tissue engineering, which typically involves the implantation of materials that combine*

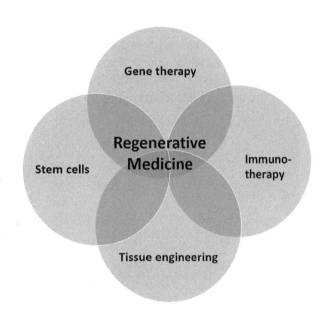

FIGURE 2.1   Areas of regenerative medicine.

living cells with engineered structural materials to restore, maintain, or improve damaged tissues or whole organs.

There is a process in each country by which a new medical intervention becomes approved for commercial sale and distribution. Clinical trials are generally an essential part of the process. A **clinical trial** *is a*

DOI: 10.1201/9780429282799-3

research study performed on human volunteers to evaluate the safety and effectiveness of a new medical treatment. In the United States, if a treatment proves to be reasonably safe and effective in clinical trials, then it can be approved by the Food and Drug Administration (FDA) for widespread use. If it is not successful, then the intervention cannot be approved and used commercially. Other countries have comparable agencies that approve new medical treatments. (More information about the important role of clinical trials and the FDA is provided in Chapter 35.)

Clinicaltrials.gov *is a web-based resource that provides the public with information about clinical studies being performed in a number of countries, dealing with treatments for a wide range of diseases and conditions.* When there are a lot of clinical trials in some area, such as stem cell treatments, it indicates that this is an active and promising area of medical research.

2.1.2 GENE THERAPIES

2.1.2.1 Introducing a "Good" Gene to Patients

Biopharmaceuticals, as discussed in Chapter 1, are valuable to patients and to the companies that produce them. But biopharmaceuticals and other drugs have limitations. Diseases such as diabetes and hemophilia can be managed by repeated administration of drugs, but the disease is not cured in this way. Some disorders are not treatable at all with drugs. Therefore,

scientists have been actively exploring the possibility of genetically modifying the cells of patients to treat, and ideally cure, various disorders. **Gene therapy** *involves correcting the function of a faulty gene, in order to treat or cure an illness.* Gene therapy utilizes the tools of recombinant DNA technology discussed in Chapter 1. However, in the case of gene therapy, the cells that are altered are not bacterial or cultured mammalian cells, but are instead the cells of a patient.

The most obvious application for gene therapy is to treat diseases that are caused when an individual inherits a genetic defect in a single gene. This includes such disorders as sickle cell anemia and hemophilia. It is comparatively straightforward in such cases to imagine treating the patient by administering a properly functioning gene to replace or augment the defective one. The first gene therapy trials thus treated genetic disorders caused by single gene defects.

Figure 2.2 illustrates two general strategies of gene therapy. In **ex vivo gene therapy**, *stem or progenitor cells are removed from the patient, genetically modified, and then returned to the patient.* (Stem and progenitor cells will be described in more detail later in this chapter.) In **in vivo gene therapy**, *genes are introduced directly into the body of the patient using a vector.* The vectors are often viruses, because viruses have the ability to insert themselves into cells and deliver genetic information. The virus is genetically modified to carry the gene of interest into the body.

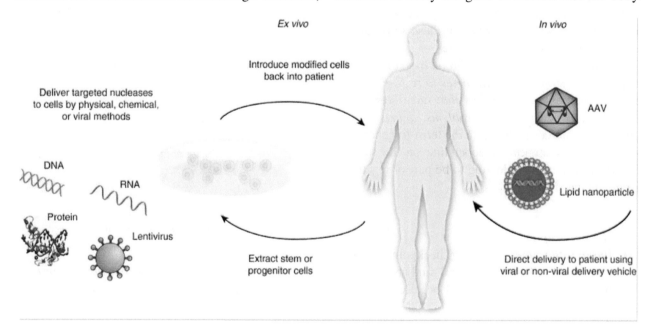

FIGURE 2.2   Two general strategies of gene therapy. See text for more information. (From: Maeder, Morgan L., and Gersbach, Charles A. "Genome-Editing Technologies for Gene and Cell Therapy." *Molecular Therapy*, vol. 24, no. 3, 2016, pp. 430–46. doi:10.1038/mt.2016.10. Copyright © 2016 Official journal of the American Society of Gene & Cell Therapy. Used under Creative Commons Attribution 4.0 International License.)

Other vectors are being used experimentally, such as very small lipid particles.

The first successful gene therapy strategy, an ex vivo gene therapy, involved genetically modifying the patient's cells in the lab and then returning them to the patient (Figure 2.3). In 1990, Ashanti De Silva became the first patient to receive such gene therapy in a procedure approved by government authorities. She was 4 years old and suffering from severe combined immunodeficiency (SCID), an inherited condition that caused her to lack a functional gene for an enzyme, adenosine deaminase, which is needed for proper functioning of the immune system. Ashanti was therefore highly vulnerable to infection. Her condition improved after the therapy illustrated in Figure 2.3, but she continued to receive repeated gene therapy treatments because, in her case, the treated blood cells do not survive indefinitely. She also takes the enzyme adenosine deaminase as a drug; therefore, no one is certain how well her gene therapy treatment works by itself. As an adult, De Silva has become a genetic counselor and patient advocate and works to help others struggling with disease.

In 2000, French physicians used ex vivo gene therapy to treat 10 boys with a different form of immunodeficiency disease, X-SCID. Children with this disorder completely lack certain cells that are part of a normal immune system, so they suffer from recurrent and often fatal infections. Unlike the immunodeficiency disease afflicting Ashanti, X-SCID cannot be treated by administering a missing enzyme as a drug. Infants with this disorder are sometimes treated successfully with bone marrow transplants that provide them with normally functioning immune cells. Other times, however, no matching bone marrow donor can be found, as was the case for the children treated in this gene therapy trial.

The X-SCID gene therapy trial involved administering the therapeutic gene into the children's bone marrow stem cells using a nonpathogenic viral vector. Unfortunately, four of the ten children developed leukemia and one child died. In a similar study performed soon after, another child developed leukemia. Researchers found that leukemia occurred because the viral vector inserted itself into the children's chromosomes near a gene that codes for a protein that causes cancer. Researchers at the Salk Institute for Biological Studies treated mice with the same gene therapy used in the X-SCID trial and found that one-third of the animals developed a lymph node cancer later in their lives. (Woods, Niels-Bjarne, et al. "Therapeutic Gene Causing Lymphoma." *Nature*, vol. 440, no. 7088, 2006, p. 1123. doi:10.1038/4401123a.) A follow-up report published in 2016 indicated that four of the five children treated for leukemia survived and are in remission. The report further stated that "Taken as a whole, these data demonstrate that genetic correction of T cell immunity restored the patients' general health status and enabled them to lead a normal life with long-lasting beneficial effects (median follow-up, 13 years)." (Cavazzana, Marina, et al. "Gene Therapy for X-Linked Severe Combined Immunodeficiency: Where Do We Stand?" *Human Gene Therapy*, vol. 27, no. 2, 2016, pp. 108–16. doi:10.1089/hum.2015.137.)

The X-SCID gene therapy trial was an indication that inserting a missing gene into the cells of patients could provide clinical benefit, but the initiation of cancer in a substantial proportion of patients caused serious concern. In 1999, 18-year-old Jesse Gelsinger died after receiving an in vivo experimental gene therapy treatment for a metabolic deficiency. His death was apparently due to a massive immune response triggered by direct exposure to a viral vector. Thus, safely delivering and targeting a "good" gene to the right cells is a complex problem. As a result of these setbacks, researchers were cautious, and gene therapy slowly evolved over the next 15 or so years. During that time, substantial investment went into developing viral and nonviral vectors with better safety profiles than the ones used in early gene therapy trials.

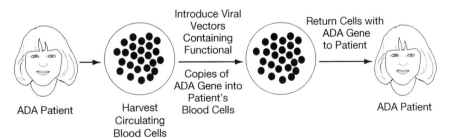

FIGURE 2.3 Ex vivo gene therapy to treat ADA deficiency. Blood cells were removed from the patient. Copies of the ADA gene were inserted into the cells, which were then returned to her. The genetically modified cells produce the protein that is missing in the affected child, thereby improving the function of her immune system.

By 2021, around ten gene therapy treatments had been approved in the United States and/or Europe. (The exact number depends on what one considers to be "gene therapy.") These include the following:

- Another SCID therapy that uses an approach similar to the one described in the French study (Strimvelis®).
- A gene therapy to treat an inherited form of blindness (Luxturna®).
- A gene therapy to treat the inherited disease, beta-thalassemia, that interferes with a patient's ability to produce hemoglobin. This therapy uses a virus to introduce healthy copies of the beta-globin gene into stem cells taken from the patient. The cells are then infused back into the patient where they migrate to the bone marrow and produce healthy precursor blood cells that manufacture appropriate amounts of hemoglobin (Zynteglo®).
- A gene therapy for children with spinal muscular atrophy, a neuromuscular disorder that is one of the leading genetic causes of infant mortality. The treatment delivers a healthy copy of the human SMN gene to a patient's motor neurons. The SMN gene is necessary for motor neuron function (Zolgensma®). This is an example of in vivo gene therapy.

While the early gene therapy trials involved children who had inherited defects in a single gene, researchers imagined treating more complex diseases with gene therapy. In 2017, two such gene therapies were approved by the FDA to treat leukemia and lymphoma. These two therapies (Kymriah® and Yescarta®) will be discussed later in this chapter.

Gene therapies have thus evolved in more than one direction. Some gene therapies target inherited diseases caused by a single gene defect, while others target cancer, a more complex disease. Some gene therapies involve treating patients with whole cells, as will be discussed in more detail below, while others involve delivering a genetically modified viral vector to a site of action. At the time of writing, hundreds of gene therapies are being researched and some will almost certainly be successful. Gene therapy has thus been challenging and slow to mature, but is expected to become an important part of medical practice in coming years.

2.1.2.2 RNAi, Another Kind of Gene Therapy

Biotechnology continues to branch out, sometimes in surprising ways, as biologists make new discoveries about the workings of living systems. An example of this is the discovery of RNA (ribonucleic acid) interference, RNAi. **RNAi** *is a phenomenon in which short RNA molecules inside cells prevent or reduce the synthesis of a specific gene product at the messenger RNA level* (Figure 2.4). The existence of RNAi came as a surprise to the scientific community when it was first discovered in plants and small nematode worms; the existence of such a system of turning off gene expression was never suspected. While the novelty of the discovery created something of a stir in the 1990s, the phenomenon achieved blockbuster status in the scientific community in 2001 when scientists demonstrated that the RNAi phenomenon occurs in mammalian cells. In 2006, two American researchers, Andrew Z. Fire and Craig C. Mello, won the Nobel Prize in Physiology or Medicine for their pioneering work on RNAi in nematode worms.

Biologists were quick to see many practical applications of RNAi. Researchers immediately began using RNAi as a tool to investigate the role of genes and the inner workings of cells. Scientists can, for example, add carefully designed RNA fragments to cells in order to "turn down" the expression of a specific gene and then observe the effect that the loss of a gene product has on the cell. This allows the researchers to understand the normal role of that gene without changing the gene itself.

By 2004, just 3 years after RNA interference was demonstrated to occur in human cells, many companies were beginning to experiment with RNAi as a method of gene therapy (Figure 2.5). The first RNAi experiments began in humans in 2004. These experiments involved treating macular degeneration. Early results were promising, but later studies showed that the RNAi treatment was ineffective and might actually cause another form of blindness in certain patients. Other experiments were conducted on a wide variety of ailments including AIDS, respiratory diseases caused by pathogens, and various inherited diseases. However, by 2010 no RNAi treatments had been successful and the pharmaceutical industry largely

RNAi molecule   AGCGUUAC
                                                  ||||||||
mRNA molecule   ---GUAAGAUCGCAAUGCAUGUCGC---

FIGURE 2.4 RNAi molecules are complementary to short sequences within larger mRNA molecules. They prevent the mRNA molecules from being translated into proteins.

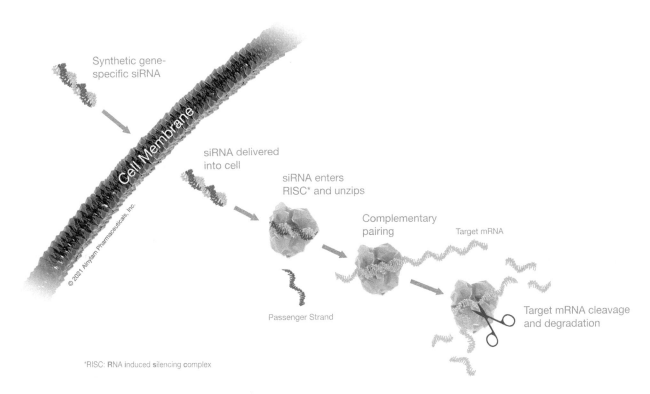

Synthetic gene-
specific siRNA

Cell Membrane

© 2021 Alnylam Pharmaceuticals, Inc.

siRNA delivered
into cell

siRNA enters
RISC* and unzips

Complementary
pairing

Target mRNA

Passenger Strand

Target mRNA cleavage
and degradation

*RISC: RNA induced silencing complex

FIGURE 2.5 This is the mechanism of action of the approved RNAi drugs that are made by Alnylam Pharmaceuticals. The goal of their treatments is to turn off, or decrease, the expression of a disease-causing gene using the process of RNA interference that occurs naturally in cells. An RNAi drug made by Alnylam Pharmaceuticals is a synthetic piece of RNA, called siRNA (small interfering), that is administered to patients. Once inside the cell, the siRNA is incorporated into a protein complex called RISC (RNA-induced silencing complex), which cleaves and discards one strand of the siRNA. The remaining strand guides the complex to bind with its complementary target mRNA, which codes for the protein whose expression is being turned down or off. The captured target mRNA is then cleaved and degraded. (Image courtesy of Alnylam Pharmaceuticals.)

lost interest in funding this work. Nonetheless, steady research by some scientists and companies over the next ten or so years (along with around $1 billion in investments) has started to pay off with three approved RNAi drugs by 2020 and a number of others in clinical trials. Givlaari® (Alnylam Pharmaceuticals) treats acute hepatic porphyria, a rare, life-threatening disease that causes severe abdominal pain, kidney disease, and liver disease. Onpattro® (Alnylam Pharmaceuticals) treats hereditary transthyretin-mediated amyloidosis, a rare and progressive genetic disease that affects nerves, the heart, and the kidneys. A third Alnylam Pharmaceuticals RNAi drug, Oxlumo®, has been approved for treatment of the rare genetic disorder primary hyperoxaluria type 1, which causes kidney stones and severe kidney disease. RNAi thus seems to be on a path similar to that of monoclonal antibodies – initial enthusiasm and investment, followed by a period of disillusionment, and finally, due to the perseverance of a small group of researchers, eventual success.

We note also that RNAi is following a similar pathway in another arena, that is, in agriculture. In 2017, the Environmental Protection Agency approved Monsanto's new genetic engineering technology that uses RNAi to kill insect pests in corn. In this case, RNAi is used to silence the activity of a gene critical to the survival of an insect pest called corn rootworm. Corn rootworm causes serious damage to corn crops by eating the plants' roots. Scientists genetically modified the corn plant so that when the corn rootworm feeds on the plant, RNAi disrupts a critical rootworm gene, leading to death of the insect. This RNAi treatment is only harmful to rootworms. Chemical pesticides, in contrast, are often toxic to humans and other organisms. Other approved RNAi-modified foods include potatoes and apples that do not bruise or brown when sliced.

RNAi is a good example of how a basic research discovery, originally made in plants and small worms, turns out to have practical applications and, with effort, can be transformed into commercial products.

2.1.2.3 CRISPR and Gene Therapy

The methods of manipulating DNA described in Chapter 1 enabled many scientific advances and facilitated the development of a robust biotechnology industry. As we have discussed so far, genetic engineering allows the transfer of existing genes or DNA sequences from one organism to another. Thus, bacteria or cultured mammalian cells can become "factories" to make human proteins, healthy genes can be transferred to patients in gene therapy, and so on. But scientists have also had the additional goal of being able to *edit* specific DNA nucleotides, that is, to change the nucleic acid sequence of a specific gene or section of DNA, in order to modify the gene's function. In 2013, a new genetic editing technology, CRISPR (pronounced "crisper"), was developed with the promise of doing exactly that. As is often the case, scientists who discovered this technique were not looking for it; they were trying to understand how the immune system of bacteria works. The CRISPR system is a defense mechanism in bacteria that confers immunity from phage infection. (Phages are viruses that infect bacteria.) Scientists learned how to adapt the bacterial strategy to edit a sequence of DNA in a precise, targeted way.

Although gene editing is not new, previous methods were cumbersome, expensive, and inefficient, with unpredictable results. CRISPR technology is easier to use and permits precise changes to a gene. A common analogy is that CRISPR is like a "cut and paste" application in a word processor: A sequence of DNA "letters" (nucleotides) can be cut out, and a new set of precisely the right nucleotides can be pasted into the space. CRISPR was quickly adopted in biological research laboratories. In addition, the potential to use CRISPR for a variety of applications, ranging from medicine to agriculture, was immediately evident (Figure 2.6). In fact, the method was so obviously powerful that Emmanuelle Charpentier and Jennifer A. Doudna were awarded the Nobel Prize in

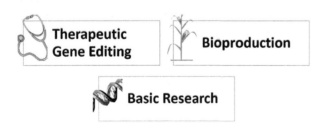

FIGURE 2.6 Additional applications of gene editing. In addition to medical applications, gene editing can be used by biotechnologists to create novel organisms for bioproduction of useful molecules such as biofuels or proteins. Gene editing is an invaluable tool for basic research as well.

Chemistry for its discovery in 2020. It generally takes longer for the importance of a discovery to be recognized and lead to a Nobel Prize. This award was also notable in that it was the first time that a Nobel Prize was awarded to an all-female team.

By 2019, human gene therapy trials using CRISPR had been initiated, and by 2021 successful results had been reported. One such trial involves removing immune cells from patients, genetically modifying the cells in the laboratory using CRISPR so that they are better able to destroy cancer cells, and then infusing the modified cells back into the patients. Other human trials aim to repair disorders caused by a single-gene defect. One such trial is directed at sickle cell disease, and another at beta-thalassemia. Both sickle cell disease and beta-thalassemia are caused by mutations that affect hemoglobin, the protein in red blood cells that allows the cells to carry oxygen through the body.

CRISPR gene therapy avoids the use of viruses to carry a new gene into cells. As we have seen, viral vectors have safety issues. Also, viral vectors deliver new genes into cells, but they do not always place the new genes in a desirable chromosomal location. Delivering genes to the wrong location caused leukemia in the X-SCID gene therapy trials described above. CRISPR is different in that it is much more likely to place a new gene sequence in the proper place in the cell's DNA.

2.1.3 CELL-BASED CANCER IMMUNOTHERAPY

Immunotherapy harnesses the patient's own immune system to fight disease. In Chapter 1 we talked about monoclonal antibody checkpoint inhibitors, one form of cancer immunotherapy, in which a *therapeutic protein* (the monoclonal antibody) is the treatment. In this chapter we discuss immunotherapies in which the treatment involves *whole, living cells* that are infused into patients.

The type of cell therapy we will discuss in this section involves yet another – and very sophisticated – application of the tools of recombinant DNA technology that were introduced in Chapter 1. In this application, these tools are used to genetically modify a patient's immune cells, which are then administered to the patient to treat cancer.

Adoptive cell therapy (ACT) *is the term used when whole cells are administered to patients.* Another (and more understandable) term for treatments using whole cells is **cell transfer therapy**. There are various types of ACT, but the one that is currently FDA-approved for use in the clinic is CAR-T cell therapy. In 2017, two CAR-T cell therapies were approved: one (named

Kymriah®) to treat children with acute lymphoblastic leukemia (ALL) and the other (Yescarta®) to treat adults with advanced lymphomas.

Leukemias and lymphomas *are cancers of white blood cells.* B cells are the type of white blood cell that makes antibodies. In leukemia and lymphoma, malignant white blood cells proliferate uncontrollably and crowd out normal cells in the bone marrow and/or the lymph nodes. The main difference between leukemias and lymphomas is that in leukemia, the cancer cells arise in the bone marrow and move to the bloodstream, while in lymphoma, the cancer cells tend to be in lymph nodes and other tissues. CAR-T cell therapy has led to remarkable responses in some leukemia and lymphoma patients for whom other treatments had stopped working. For example, remission rates up to 90% have been observed in children with ALL.

To begin to explore this form of cancer treatment, let's consider the acronym "CAR-T cell" therapy.

- *T cells* are part of the immune system, as are B cells. But T cells do not make antibodies; rather, they recognize infected and damaged cells (like those involved in cancer) and destroy them or assist in their destruction. This cell therapy utilizes T cells.
- The letter *"C"* stands for "chimeric." A chimera is a mythical animal formed from parts of more than one animal, for example, the body of a horse and the head of a man.
- The letters *"AR"* stand for "antigen receptor."

What is an "antigen receptor?" In this context, a **receptor** *is a protein that resides on the surface of a cell and "communicates" with things outside the cell, such as other cells or chemical messengers.* In CAR-T therapy, the words **chimeric antigen receptor** (CAR) *refer to a receptor protein that is genetically engineered to have parts derived from more than one protein.* One part of the chimeric protein is a receptor that is normally found on the surface of the T cell. Another part of the chimera is the recognition region of a monoclonal antibody. A protein called "CD19" is the antigen recognized by the monoclonal antibody (in the approved drugs). **CD19** *is a protein that is overexpressed on cancer cells in leukemia and lymphoma.* The recognition region of the chimeric antigen receptor thus improves the T cell's ability to recognize and bind to malignant B cells. The CAR also has sections that help control the activity of the T cell, which makes it better able to attack cancer cells.

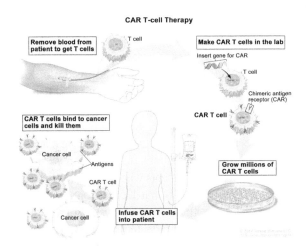

FIGURE 2.7 CAR-T cell therapy. See text for explanation. https://www.cancer.gov/about-cancer/treatment/types/immunotherapy/t-cell-transfer-therapy. (Image © 2017 Terese Winslow LLC; U.S. Govt. has certain rights.)

CAR-T cell therapy is illustrated in Figure 2.7. Therapy begins with the removal of T cells from the blood of the cancer patient. The T cells are sent to a manufacturing facility where they are genetically modified by inserting a gene that codes for the chimeric antigen receptor protein. As is shown in Figure 2.7, the genetically modified T cells express (make) the chimeric antigen protein, which moves to the cell's plasma membrane. The genetically modified CAR-T cells are expanded (grown) in culture until there are millions of them present. When there are a sufficient number of genetically modified T cells, they are sent to a hospital where they are infused back into the patient. The CAR-T cells move through the patient's blood where they recognize malignant cells, bind to them, and destroy them.

Using T cells as a therapeutic has advantages as compared to monoclonal antibodies and other drugs. T cells can proliferate after they are administered to patients. Thus, in theory, T cells only need to be administered once, although this will need to be confirmed as more patients are treated. T cells can travel throughout the body, which means that they can potentially travel to sites of cancer metastases throughout the body. T cells have "memory" and can theoretically persist and exert effects for years after administration. However, there are some problems that must be resolved to make these cell therapies more widely useful. First, these therapies can have serious side effects. While the effects can generally be managed, they do require hospitalization and sometimes intensive treatment. Another concern is that, at this time, approved CAR-T cell therapy is **autologous**, *which means that the cells that are used come from the patient.* Each patient's cells must be taken through the laborious process of genetic engineering and expansion. This process is expensive and time-consuming. As a result, current CAR-T cell therapy can cost more than $100,000 per dose – clearly an impediment to widespread use. Many scientists are therefore working to find ways to use **allogeneic cells**, *which are cells that come from a single donor person and are used to treat many patients.* Using allogeneic cells would be less time-consuming and costly because they could be manufactured in advance, stored, and then distributed to patients who need them. The problem with allogeneic cells is that T cells from a donor may not be compatible with the immune system of a recipient. Scientists in many research centers are working to create allogeneic treatments in such a way that the donor cells are compatible with the immune system of any recipient. At the time of writing, this has not been consistently achieved, but hopefully will be in the near future.

CAR-T cell therapy, so far, has only proven to be effective for blood cell cancers, not solid tumors such as those of the breast or lung. Finding ways to make ACT methods that work against multiple kinds of cancer is another area of active research.

2.1.4 STEM CELLS

The human body is composed of trillions of cells organized into more than 200 different types. Each cell type has structural and functional features that match its role in the body. For example, there are nerve cells that are long and thin and function to transmit signals from one part of the body to another. Liver cells have a different morphology, and they have specialized enzymes to detoxify hazardous substances.

As mentioned in Chapter 1, **stem cells** *are a type of cell that have the potential for indefinite self-renewal and also have the potential to develop into multiple mature cell types.* "Self-renewal" means they can divide to make more daughter cells that are like the parent cells. "[D]evelop into multiple mature cell types" means that stem cells can mature into more than one type of cell. *The process by which a stem cell matures into a particular cell type is called* **differentiation**. The term **pluripotent** *is used to describe stem cells and means they can differentiate in more than one type of cell.* Figure 2.8 diagrams the features of stem cells.

Prior to 1998, stem cells could not be cultured in the laboratory. In 1998, Dr. James Thomson and colleagues at the University of Wisconsin published the first report describing the identification and isolation of stem cells from human embryos. These embryos were obtained from fertility clinics where

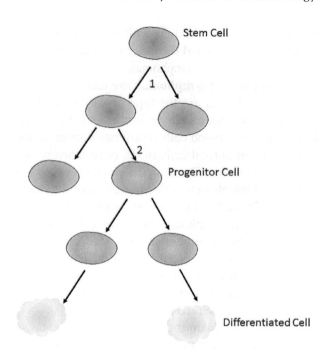

FIGURE 2.8 Stem cell development: Stem cells can self-renew or differentiate. 1. A stem cell has the capability of indefinitely dividing and self-renewing, thus creating more undifferentiated stem cells. 2. Signals from the body can cause a stem cell to begin to develop into a particular cell type. Once a stem cell begins to differentiate, it is said to have entered a developmental pathway. Such a cell is called a progenitor cell. Each developmental pathway results in a different type of mature cell. For example, cells that enter the hematopoietic pathway are destined to become blood cells. Other pathways form nerve cells, or skin cells, and so on.

they were originally produced to treat infertility, and were donated for research with the informed consent of donor couples who no longer wanted them. Once established, an embryonic stem cell line can be maintained indefinitely in culture, or the cells can be frozen for later use.

Thomson's report created much excitement. To understand the interest in stem cells, consider that every human begins as a fertilized egg. This single egg cell must divide many times to form a complete individual. If the cells of the embryo, however, simply divided over and over again, no person would form because the cells also need to differentiate into muscles, nerves, skin, bones, arms, legs, brains, eyes, and all the other structures of the body. **Embryonic stem cells** *are cells from an embryo that have the potential to differentiate into any cell type in the body* (Figure 2.9). The ability of stem cells to become any cell type means they have enormous potential as therapeutics. Stem cells, for example, might someday be used to replace pancreatic

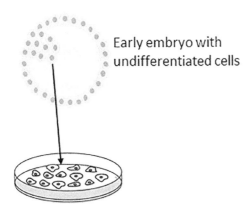

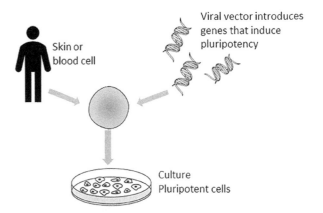

FIGURE 2.9 Embryonic stem cells are taken from early-stage embryos.

FIGURE 2.10 Induced pluripotent stem cells are produced by transfecting adult cells with a few genes.

islet cells in people with type I diabetes so their bodies would again produce insulin. Stem cells are being explored to treat patients with Parkinson's disease who experience a loss of motor control when brain cells that are supposed to produce dopamine fail to do so. Stem cells are being studied to see if they can repair heart muscle cells damaged in a heart attack, nerve cells damaged in spinal cord injuries, cells damaged by arthritis, and a vast number of other conditions.

Thomson's report of culturing embryonic stem cells created controversy as well as excitement. The controversy over the use of human stem cells centers on the fact that they were originally, and sometimes still are, isolated from human embryos. Critics voiced concerns about the ethics of destroying human embryos. To a large extent, the controversy has subsided because sources of stem cells other than embryos have been discovered. Two reports published in 2007 significantly changed the direction of stem cell work away from embryonic stem cells. The first report was from Shinya Yamanaka's laboratory in Japan, and the other was from James Thomson's laboratory. *These two groups were able to transfect human skin cells with genes that caused them to look and act like human embryonic stem cells; these cells are called* **induced pluripotent stem cells (iPS cells)** (Figure 2.10). iPS cells quickly found application in research laboratories around the world, and Dr. Yamanaka was awarded a Nobel Prize for this discovery in 2012. It is interesting to note that Dolly the sheep had a role in this advance. Shinya Yamanaka is reported to have said "Dolly the sheep told me that nuclear reprogramming is possible even in mammalian cells and encouraged me to start my own project." (As quoted by Weintraub, Karen. "20 Years after Dolly the Sheep Led the Way—Where Is Cloning Now?" *Scientific American.* July 5, 2016. https://www.scientificamerican.com/article/20-years-after-

dolly-the-sheep-led-the-way-where-is-cloning-now/.) Prior to the cloning of Dolly, scientists believed that differentiation of a cell into a particular mature type is irreversible. Dolly showed that a fully differentiated cell – a mammary cell in Dolly's case – can be reprogrammed back to a pluripotent state.

In the clinic, iPS cells have a significant advantage over embryonic stem cells, in addition to the fact that they do not involve embryos. iPS cells can be derived from an individual patient and therefore can be used to treat that individual without the risk of immune incompatibility. Despite these advantages, iPS cells have not yet been widely adapted for use in the clinic. Concerns about their safety, their potential to cause cancerous transformation, and technical challenges are still being resolved. It was not until 2019 that the first clinical trial of iPS cells in human patients was reported. Researchers at the National Eye Institute (NEI) launched a clinical trial with 12 patients to test the safety of an iPS therapy to treat geographic atrophy, the advanced "dry" form of age-related macular degeneration (AMD). This disease is a leading cause of vision loss among people aged 65 and older. In geographic atrophy, there are regions of the retina where retinal pigment epithelial (RPE) cells waste away and die (atrophy). RPE cells nurture photoreceptors, the light-sensing cells in the retina. In geographic atrophy, the RPE cells die and then the photoreceptors eventually also die, resulting in blindness.

The experimental iPS therapy involves removing a patient's blood cells and, in a lab, converting them into iPS cells. Next, the iPS cells are programmed to become RPE cells (Figure 2.11). The iPS RPE cells are grown in sheets one cell thick (monolayers) to replicate their normal structure in the eye. The monolayer is supported on a biodegradable polymer scaffold that allows the cells to be surgically implanted into the eye.

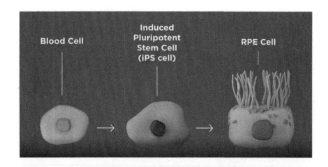

FIGURE 2.11 First iPS cell trial in 2019. Blood cells from patients with advanced dry form macular degeneration were removed and converted into iPS cells. They were then differentiated into RPE cells and grown on a scaffold (not shown) to be surgically implanted into the patients. (Image credit: U.S. National Eye Institute.)

The use of engineered scaffolds for supporting cells is common in regenerative medicine products (see Tissue Engineering, Section 2.1.5).

There is yet another source of stem cells that is currently under active investigation, that is adult stem cells. **Adult stem cells (ASCs)** *are precursor cells that are found in adult tissues and have some, but limited, self-renewal and differentiation capacity.* These are cells that have started down a developmental pathway as illustrated in Figure 2.8. ASCs are also called "tissue-specific stem cells" or "somatic stem cells." ASCs are thought to be present in all tissues and function to replace cells that die in that tissue. For example, *blood-forming*, **hematopoietic stem cells** can differentiate into red or white blood cells, or platelets (important in blood clotting), but they cannot become other sorts of cells, such as liver cells. Adult stem cells in a human are relatively rare and therefore they are difficult to isolate, but they can sometimes be identified because of protein markers on their surface. For example, human hematopoietic stem cells have certain proteins on their plasma membrane surface called CD34 and CD133. Cell sorting instruments can be used to separate cells that have a particular surface marker from those that do not.

Multipotent hematopoietic stem cell (HSC) transplantation is currently the most widely used, FDA-approved stem cell therapy. This treatment, also called bone marrow transplant, has been used for more than 50 years to treat cancers of the blood and other blood disorders. This form of stem cell therapy predates the discovery of methods to culture embryonic stem cells. In HSC transplantation, stem cells are obtained from bone marrow, peripheral blood, or umbilical cord blood. Sometimes autologous cells are used, and other times allogeneic cells. In the latter case, it is important to find a donor whose tissue matches the patient as closely as possible, and a sister or brother is often the donor. When this transplantation treatment is used, the patients' diseased hematopoietic stem cells are usually destroyed by radiation or chemotherapy and then the new hematopoietic stem cells are infused. HSC transplantation has successfully treated tens of thousands of patients over the years since it was introduced. (A dramatic example of HSC therapy will be discussed later in this chapter in the Nicholas Volker case study (pp. 53–54).)

Mesenchymal stem cells are a type of adult stem cells that are currently under intense investigation for use in the clinic. **Mesenchymal (stromal) stem cells (MSCs)** *are a type of adult stem cells that are isolated from the connective tissue that surrounds tissues and organs.* MSCs are relatively easy to isolate and have been harvested from adipose tissue, amniotic fluid, endometrium, dental tissues, and umbilical cord. They have the capacity to differentiate into multiple cell types (Figure 2.12). MSCs play a normal role in injury healing. Additionally, this type of stem cell is thought to produce chemicals that might have a beneficial effect on the immune system and might reduce inflammation. MSCs are being used in clinical studies to treat heart disease, diabetes, spinal cord injury, Crohn's disease, aplastic anemia, rheumatoid arthritis, brain injury, liver cirrhosis, osteoarthritis, multiple sclerosis, lupus, Parkinson's disease, and other less common conditions. In 2020, they are being used experimentally to treat life-threatening complications of COVID-19.

Although stem cells are of great interest, there have been challenges in adapting them to clinical use. For

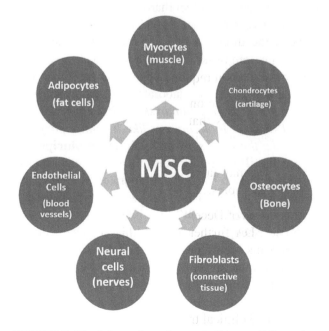

FIGURE 2.12 Mesenchymal stem cells can differentiate into a variety of cell types.

example, stem cells that are used to treat a disorder must differentiate into the right type of cell within a patient. Presumably, the cells will take cues from the environment in which they are placed, but this is not yet proven. Also, stem cells have the potential to divide indefinitely – a hallmark of cancer cells and a potential obstacle to their use in patients. Although stem cells are being used in thousands of clinical trials and research studies, at the end of the year 2020, there were few stem cell treatments approved by the FDA or regulatory bodies in other countries. This, however, has not stopped some clinics from providing stem cell treatments as described in the case study "Stem Cell Therapies and the Regulatory Process: A Conundrum."

Case Study: Stem Cell Therapies and the Regulatory Process: A Conundrum

The announcement of the successful isolation and culture of human embryonic stem cells in 1998 was greeted with much fanfare. Within a short time, the possibility of using stem cells to cure intractable disease had been discussed by scientists and widely reported to the public. Nearly 10 years later, iPS cells were greeted with similar enthusiasm. By 2020, there were thousands of clinical trials involving stem cells derived from varied sources, yet there were almost no novel FDA-approved stem cell treatments. Thus, it is not surprising that the public grew impatient. It is also not surprising that businesses emerged to supply the demand for stem cell treatments. The Alliance for Regenerative Medicine stated that in 2017 there were more than 700 clinics in the United States (and many more in other countries) anxious to provide stem cell treatments of various sorts. An Internet search easily finds these clinics with ample testimonials from presumably cured customers.

One of the authors received a letter in the mail, much like the one in Figure 2.13, inviting her to attend a free dinner at a local restaurant in order to learn about how stem cells could relieve all sorts of aches and pains. Intrigued, she attended the seminar, had a pleasant dinner, and then, for 2 hours heard that an injection of cord blood cells could relieve all sorts of arthritic pains and musculoskeletal injuries. Attendees were assured that that cord blood stem cell therapy is totally safe, helps nearly 100% of patients, and costs only $5,000 per treatment. When asked if the treatment is FDA-approved, we were assured that this particular form of stem cell therapy does not require FDA approval. The opportunity to have these stem cells sounded too good to pass up, and nearly all of the seminar attendees signed up for individual conferences in order to discuss the next step in obtaining a stem cell treatment. After all, common aging ailments, particularly arthritic pain, and disability caused by old injuries, are not well managed by traditional medicine. So, many of the seminar attendees, along with tens of thousands of other Americans, decided to give stem cells a try.

But there are some problems here. First, in the seminar, participants were told that the treatment being offered does not require FDA approval. The FDA begs to differ. The FDA states:

> Currently, the only stem cell treatments approved by the Food and Drug Administration (FDA) are products that treat certain cancers and disorders of the blood and immune system. If the products are being used for arthritis, injury-related pain, chronic joint pain, anti-aging or other health issues, they have not been approved by FDA and are being marketed illegally…Some clinics may falsely advertise that it is not necessary for FDA to review and approve their stem cell therapies… These claims are false.

(Current as of December 2019. https://www.cdc.gov/hai/outbreaks/stem-cell-products.html.)

The FDA further points out that stem cell treatments are not without risk. For example, stem cell treatments in unregulated clinics have led to severe infections, blindness (in a case where the cells were injected into the eye), and other problems. Moreover, stem cells share certain traits with cancer cells and there is a concern that they might, in some cases, promote cancer. The cells might also move from the placement site in the body and change into an inappropriate cell type. Meanwhile, the thousands of controlled clinical trials, in which the effects of stem cells are compared to conventional treatments, have so far, resulted in few approved therapies. This appears to be because the experimental stem cell treatments

(Continued)

Case Study (*Continued*): Stem Cell Therapies and the Regulatory Process: A Conundrum

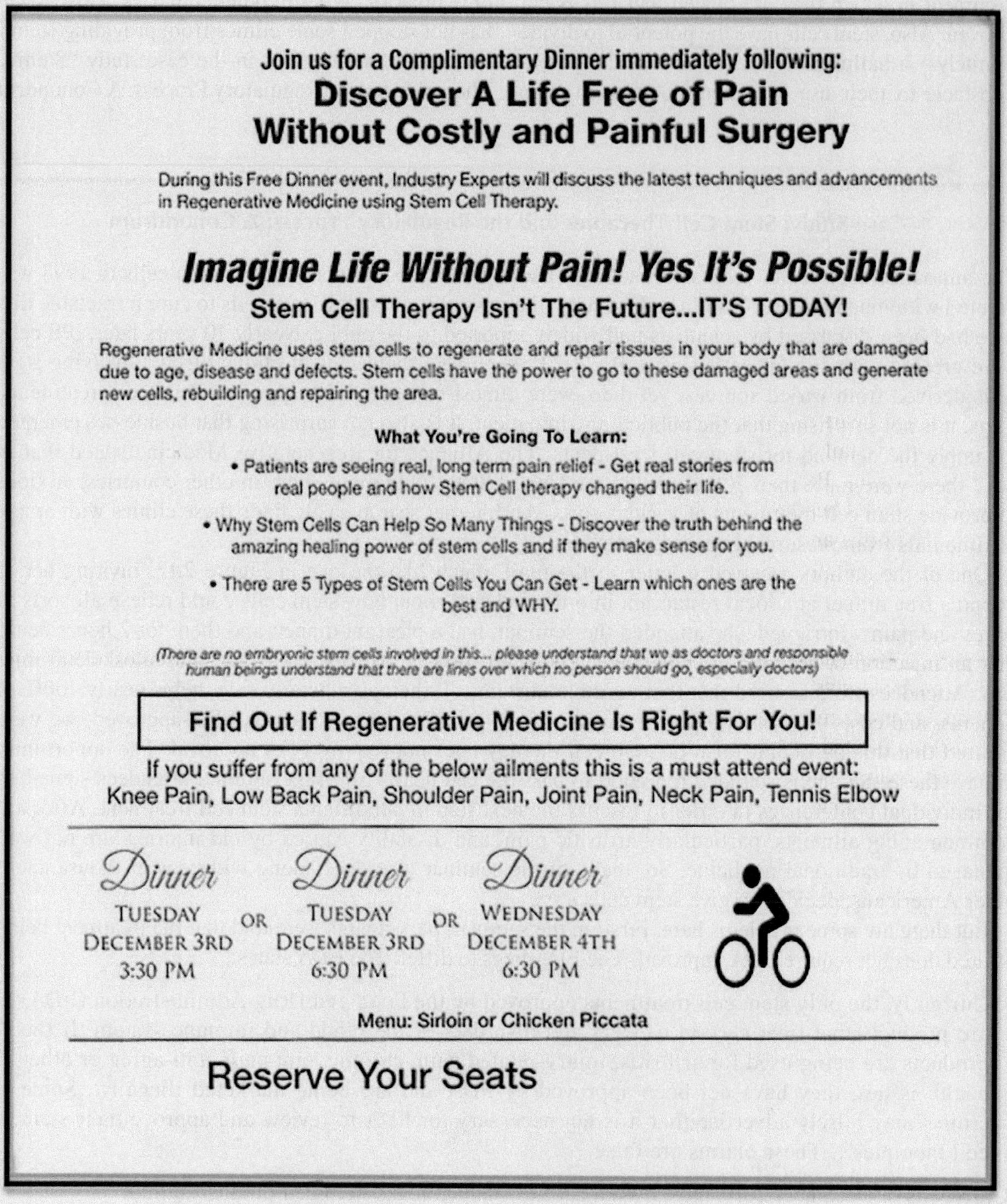

FIGURE 2.13   An invitation to a seminar promoting a stem cell clinic.

have not resulted in clear benefits to patients. Scientists are optimistic that stem cells will play a therapeutic role in the future and, by the time you read this, they may be fulfilling that promise. In the meantime, stem cell clinics are treating thousands of patients without knowing exactly how their treatments might or might not work, nor with good information about the long-term safety of these treatments.

Although, at the time of writing, stem cells have not lived up to their expectations in the clinic, they have been invaluable tools for research and testing. The discovery of methods to induce pluripotent stem cells was a major basic research advance that changed the way biologists understand genes and development. Stem cells are used every day to investigate development, aging, and the genetics of various diseases. Stem cell lines have been created that have genetic alterations associated with various disorders, such as fragile X syndrome, in order to better study and eventually treat these disorders. Stem cells are used in drug discovery to identify drugs that might be useful for a particular application, and to test whether drugs have adverse effects on cells. One way this can be done is to take human pluripotent stem cells and coax them into differentiating into a particular cell type, such as liver or heart. Liver and heart are critical organs, and many prospective drugs fail because they damage one or both of these organs. Prospective drugs can be tested in cultured liver and heart cells, thus avoiding animal studies for drugs that are doomed to failure.

2.1.5 Tissue Engineering

The human body can naturally repair itself to a limited extent; for example, cuts, scrapes, and broken bones can heal. More extensive medical repair of damaged body parts is usually limited to transplantation of tissue and organs from donors, but there are far more people in need of transplants than there are donors. Tissue engineering has the potential to greatly expand the options for repairing diseased and damaged tissues and organs or creating entire replacement organs. (See the case study "Engineering New Tissues and Organs Using Our Own Living Cells"). **Tissue engineering** *refers to strategies to create functional human tissues using cells grown in the laboratory.*

Tissue engineering requires four components:

- **Cells that can perform the role of the target tissue or organ**. Various sources of cells are being explored, some of which are autologous and others allogeneic.
- **A supportive scaffold that guides tissue growth**. Artificial polymers and gels, and substances derived from organisms are all being explored (Figure 2.14).
- **Biomolecules, such as growth factors, that cause the cells to grow and function**.
- **Physical and mechanical forces that influence the growth of the tissue or organ**.

The first tissue-engineered products to be approved for commercial use are those that treat chronic wounds, burns, and other skin disorders. For example, Apligraf® (Organogenesis, Canton, MA) is an FDA-approved product used to treat severe wounds that do not heal. Such wounds sometimes occur in people with diabetes and, in severe cases, can require amputation of an affected limb. The manufacture of Apligraf® occurs in a specialized production facility where fibroblast cells are isolated from human foreskin that is donated after circumcisions. **Fibroblasts** *are cells that provide a structure for tissues and also have a role in wound healing.* The cells are placed over a matrix composed of the protein, collagen, which has been isolated from cows. Over time, the fibroblasts form a layer, like skin, over the collagen matrix. A second layer of cells derived from foreskin are layered on top of the first. When the cellular layers have grown sufficiently, the product is packaged and delivered to a physician who applies it to a patient's wound. Applying Apligraf® (and other products like it) often promotes healing when other conventional treatments have failed. The cells that comprise Apligraf® do not appear to persist more than a month or so. Therefore, it is assumed that the treatment works because the cells secrete chemical messengers, including growth factors, that tell the patient's own cells to repair the wound. Interestingly, although this skin product contains cells that are foreign to the patient, it does not appear to elicit an immune response by the patient. (Zaulyanov, Larissa, and Robert S. Kirsner.

FIGURE 2.14 Using a scaffold to regenerate a human ear. Cells are adhered to the scaffold and are fed with medium (the pink solution) containing nutrients and growth factors until they cover the scaffold. (Originally posted to FLICKR by Army Medicine at https://flickr.com/photos39582141@ N06/6127848729, CC-BY-2.0.)

"A Review of a Bi-Layered Living Cell Treatment (Apligraf) in the Treatment of Venous Leg Ulcers and Diabetic Foot Ulcers." *Clinical Interventions in Aging*, vol. 2, no. 1, 2007, pp. 93–98. doi:10.2147/ciia.2007.2.1.93.) Apligraf® is not the only living skin therapeutic on the market, and new ones are moving through clinical trials.

As is the case for stem cells, tissue engineering is playing an important role in research and testing.

For example, **organs on a chip** are devices that are about the size of a flash drive. They contain tiny tubes that are lined with human cells from an organ of interest. Nutrients, blood, and test compounds such as experimental drugs, can be pumped through the tubes to simulate the organ of interest and to look for effects on the cells. Such organ microchips are being used for basic research and to test prospective drugs.

Case Study: Engineering New Tissues and Organs Using Our Own Living Cells

In 2004, 10-year-old Luke Massella learned that his malfunctioning bladder was leading to kidney failure. "I was kind of facing the possibility I might have to do dialysis (blood purification via machine) for the rest of my life," he says. "I wouldn't be able to play sports, and have the normal kid life with my brother." But Dr. Anthony Atala, then a surgeon at Boston Children's Hospital, provided an alternative future. He took a small piece of Luke's bladder and over 2 months grew a new one in the lab; then, after a 12-hour surgical procedure, he replaced the defective bladder with this new one. Luke went on to become a wrestling coach in the Connecticut public school system. Since that time, Dr. Atala, currently at Wake Forest Institute for Regenerative Medicine, has led a team that has developed "eight cell-based tissues we put into patients," including engineered skin, urethras, and cartilage, all grown in the lab. Dr. Atala's research work involves bioprinting, using modified 3D inkjet machines to produce human tissues. Engineering tissue and its fabrication requires a source of cells, usually from a tissue biopsy, that is then expanded and paired with sophisticated structural materials that are produced using the bioprinting process to provide "connective tissue" that holds the cells together.

According to Dr. Atala, flat structures such as skin are easiest to print, whereas tubular structures like blood vessels and hollow non-tubular organs such as bladders are more complex. Solid organs such as hearts, lungs, and kidneys are the most difficult to bioprint as they have more cells per centimeter, although some researchers have had small successes in this field.

As with many exciting areas of biotechnology, early reports of this work were prone to overstating progress and raised hopes that these 3D-printed organs would make organ transplants unnecessary in the immediate future. In fact, clinical application of this technology is extremely challenging with widespread implementation likely many years in the future. However, in the meantime, application of bioprinting technology in the field of regenerative medicine has enabled innovation in several important areas. For example, because it allows construction of accurate 3D physiological structures, bioprinted tissues can be interconnected to form a "body on a chip" that can be used for initial testing before costly clinical trials are needed in the drug development process.

2.2 GENOMICS

2.2.1 INTRODUCTION

2.2.1.1 The Human Genome Project

In the 1980s, prominent scientists introduced the idea of a project to determine the sequence of all the nucleotides in the human genome. The main goals of the project, as later stated on the project's website, were to "determine the complete sequence of the 3 billion DNA subunits (bases) [that comprise the human genome], identify all human genes, and make them accessible for further biological study." At the time the idea of sequencing the human genome was introduced, only a handful of genes had been identified and isolated, each one of which was painstakingly studied by researchers over a period of many years. Given the laborious methods that were then available, the idea to sequence the entire genome was ambitious and costly, with an estimated price tag in the billions of dollars range. In 1987, Robert Weinberg wrote an editorial for *The Scientist Magazine* in which he outlined serious criticisms of the proposed project. In that editorial he said: "Those who fret about the sanctity of the human genome needn't worry—we may understand less about ourselves at the end of this project than when we began." Weinberg was not the only critic, there were others, many of whom were concerned that the project would compete with other important tasks, such as finding a cure for AIDS. Despite the criticisms and the high price tag, the international sequencing project, named the Human Genome Project (HGP) began formally in 1990 and ended officially in 2003 with the announcement of a reasonably complete draft of the human genome. During the course of the project, sequencing methods improved dramatically. By the end of the project, automated sequencing instruments had dramatically reduced the time and the cost required to sequence a gene (Figure 2.15). By 2020, while it was still not common practice, it became financially feasible to sequence an individual's entire genome – a feat that was almost unimaginable in 1987.

2.2.1.2 HGP Benefits

With the luxury of hindsight, it is clear that the HGP had numerous, unforeseen benefits. Some of these benefits relate to the information acquired about the human genome, but others relate to the methods for rapid DNA sequencing that were developed during the project. For example, in 1987 it was hard to foresee that in December 2019 a novel coronavirus would sweep

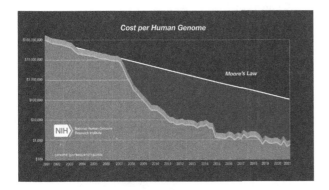

FIGURE 2.15 The dramatic reduction in DNA sequencing costs. Moore's law is a principle that predicts that the speed and capability of computers can be expected to double every 2 years. Scientists had initially hoped that sequencing technologies would improve in the same way as computers. In fact, DNA sequencing methods improved even more rapidly. Note the logarithmic scale on the graph. The cost of sequencing a single person's genome went from $100,000,000 to $1,000 between 2001 and 2019. (Image credit: U.S. National Human Genome Research Institute.)

out of China, sickening millions of people, causing hundreds of thousands of deaths, and leading to major social and economic disruptions worldwide. Less than 1 month after the virus was identified, Chinese scientists were able to sequence the virus's genome using the rapid methods that had been previously developed. Within a few days of viral genome sequencing, a diagnostic testing kit was developed, based on that gene sequence. Having a diagnostic kit is of vital importance because it distinguishes individuals infected with the new virus from those with an ordinary cold or flu. In a pandemic, knowing who is infected allows proper quarantine and treatment of affected individuals and guides strategies to slow the spread of the disease. Only a few months after the viral genome was sequenced, at least 43 vaccines were in, or approaching, clinical trials. Many of these vaccines were based on synthesizing DNA or RNA sequences that are part of the viral genome. (See Section 1.2.4 for a discussion of DNA/RNA vaccines.)

But it did not require a pandemic to prove that sequencing genomes has value. The following sections provide "snapshots" of a few of the applications of **genomics**, *the study of genomes, their functions, and their regulation.* Observe that the applications that are described below are biotechnology "products," but they are different from the products of biotechnology that we have described so far in Chapters 1 and 2. Genomics is about information and knowledge, not products that you can hold in your hand, like a vial of a drug, or touch, like a transgenic animal. Knowledge is the beginning and the basis of all biotechnology products. Biotechnology, more than almost any other industry, is one that emerges from human thought and intellect.

2.2.2 GENOMICS AND MEDICINE

2.2.2.1 Precision Medicine

Medicine is an area where the applications of genetic knowledge are evident. Analysis of the human genome led to information about dozens of genetic conditions, such as muscular dystrophies, inherited colon cancer, Alzheimer's disease, and familial breast cancer (Figure 2.16).

Precision medicine *refers to strategies that tailor a medical treatment to the characteristics, particularly the genetic traits, of an individual.* Precision medicine does not mean the creation of drugs or medical devices that are unique to that person. Rather, it refers to the classification of individuals into subpopulations that differ in their susceptibility to a particular disease, or their response to a particular treatment. This is not a

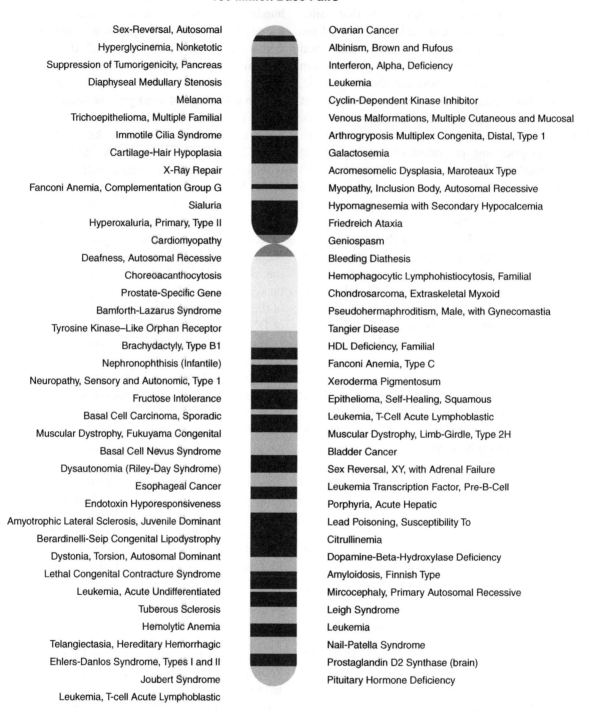

136 Million Base Pairs

Sex-Reversal, Autosomal	Ovarian Cancer
Hyperglycinemia, Nonketotic	Albinism, Brown and Rufous
Suppression of Tumorigenicity, Pancreas	Interferon, Alpha, Deficiency
Diaphyseal Medullary Stenosis	Leukemia
Melanoma	Cyclin-Dependent Kinase Inhibitor
Trichoepithelioma, Multiple Familial	Venous Malformations, Multiple Cutaneous and Mucosal
Immotile Cilia Syndrome	Arthrogryposis Multiplex Congenita, Distal, Type 1
Cartilage-Hair Hypoplasia	Galactosemia
X-Ray Repair	Acromesomelic Dysplasia, Maroteaux Type
Fanconi Anemia, Complementation Group G	Myopathy, Inclusion Body, Autosomal Recessive
Sialuria	Hypomagnesemia with Secondary Hypocalcemia
Hyperoxaluria, Primary, Type II	Friedreich Ataxia
Cardiomyopathy	Geniospasm
Deafness, Autosomal Recessive	Bleeding Diathesis
Choreoacanthocytosis	Hemophagocytic Lymphohistiocytosis, Familial
Prostate-Specific Gene	Chondrosarcoma, Extraskeletal Myxoid
Bamforth-Lazarus Syndrome	Pseudohermaphroditism, Male, with Gynecomastia
Tyrosine Kinase–Like Orphan Receptor	Tangier Disease
Brachydactyly, Type B1	HDL Deficiency, Familial
Nephronophthisis (Infantile)	Fanconi Anemia, Type C
Neuropathy, Sensory and Autonomic, Type 1	Xeroderma Pigmentosum
Fructose Intolerance	Epithelioma, Self-Healing, Squamous
Basal Cell Carcinoma, Sporadic	Leukemia, T-Cell Acute Lymphoblastic
Muscular Dystrophy, Fukuyama Congenital	Muscular Dystrophy, Limb-Girdle, Type 2H
Basal Cell Nevus Syndrome	Bladder Cancer
Dysautonomia (Riley-Day Syndrome)	Sex Reversal, XY, with Adrenal Failure
Esophageal Cancer	Leukemia Transcription Factor, Pre-B-Cell
Endotoxin Hyporesponsiveness	Porphyria, Acute Hepatic
Amyotrophic Lateral Sclerosis, Juvenile Dominant	Lead Poisoning, Susceptibility To
Berardinelli-Seip Congenital Lipodystrophy	Citrullinemia
Dystonia, Torsion, Autosomal Dominant	Dopamine-Beta-Hydroxylase Deficiency
Lethal Congenital Contracture Syndrome	Amyloidosis, Finnish Type
Leukemia, Acute Undifferentiated	Mircocephaly, Primary Autosomal Recessive
Tuberous Sclerosis	Leigh Syndrome
Hemolytic Anemia	Leukemia
Telangiectasia, Hereditary Hemorrhagic	Nail-Patella Syndrome
Ehlers-Danlos Syndrome, Types I and II	Prostaglandin D2 Synthase (brain)
Joubert Syndrome	Pituitary Hormone Deficiency
Leukemia, T-cell Acute Lymphoblastic	

FIGURE 2.16 Illustration of chromosome 9. This is a graphical representation of one human chromosome, chromosome 9, as elucidated in the Human Genome Project. It shows a number of traits and disorders that are now known to be controlled or partially controlled by genes located along chromosome 9. (Image credit: Genome Management Information System, Oak Ridge National Laboratory, http://genomics.energy.gov.)

new concept; physicians have attempted to match their treatments to their patients for thousands of years. But, with the advent of genetic information, it is possible to tailor treatments with far more accuracy than was possible before.

Precision medicine typically involves a diagnostic test that detects one or more traits in an individual. This information can be used in various ways. Genetic information might be used to detect a predisposition to a particular disease. For example, there are genetic panels that look

---

ColoNext: Analyses of 14 Genes Associated with Hereditary Colon Cancer

Panel Results

CHEK2	Pathogenic Mutation	p.S428F

POSITIVE: Pathogenic Mutation Detected

- This individual is heterozygous for the p.S428F pathogenic mutation in the CHEK2 gene
- Cancer Risk estimate: up to a 2 fold increased risk of breast and colon cancer.
- The expression and severity for this individual cannot be predicted.
- Genetic counseling is a recommended option for all individuals undergoing genetic testing.

No pathogenic mutations, variants of unknown significance, or gross deletions or duplications were detected in the other genes analyzed. In total, 14 genes were analyzed as part of this panel: APC, BMPR1S, CDH1, CHEK2, EPCAM, MLH1, MSH2, MSH6, MUTYH, PMS2, PTEN, SMAD4, STK11, and TP53.

The p.S428F variant (also known as c.1283 C>T) is located in coding exon 11 of the CHEK2 gene. This mutation results from a C to T substitution at nucleotide position 1283. The serine at codon 428 is replaced by phenylalanine, an amino acid with highly dissimilar properties. The p.S428F mutation...

FIGURE 2.17 Excerpt from a genetic panel that reveals a mutation associated with an elevated risk of breast and colon cancer.

for mutations that have been associated with an increased risk of colon and/or breast cancer (Figure 2.17). A woman who discovers that she is at a higher risk of breast cancer might have more frequent examinations and mammograms than normal. Women with mutations in particular genes, BRCA1 and BRCA2, might opt for a prophylactic (preventive) mastectomy and ovariectomy because mutations in these genes are associated with very high rates of breast and ovarian cancer.

Pharmacogenomics/pharmacogenetics is a part of precision medicine. *These terms refer to the use of genetic tests to decide if a certain drug will be safe and effective for a particular person.* Individuals often vary in the ways they absorb, transport, process, or metabolize drugs. These variations are due to differences in genes that code for proteins that interact with drugs, such as enzymes that metabolize drugs or protein receptors that recognize drugs. Thus, a drug and dosage that is useful and safe for one person might be ineffective or even harmful in another. Adverse drug reactions are unfortunately common, leading to an estimated 2 million hospitalizations and 100,000 deaths in the United States alone. (Shastry, B. S.

"Pharmacogenetics and the Concept of Individualized Medicine." *The Pharmacogenomics Journal*, vol. 6, no. 1, 2005, pp. 16–21. doi:10.1038/sj.tpj.6500338.) An example is illustrated in Figure 2.18. Statins are a type of drug used to lower cholesterol. In order to work properly, statins must be taken up by the liver. They are transported into the liver by a protein encoded by the SLCO1B1 gene. Some people have a mutation in this gene that causes less of a statin drug, simvastatin, to be taken up by the liver. In these individuals, simvastatin can build up in the blood, causing muscle weakness and pain. Therefore, a physician might recommend a genetic test to see if simvastatin is a suitable drug for an individual, and what dose will work best.

We saw another example of pharmacogenetics in Chapter 1 where the drug Herceptin® was introduced. There is a genetic alteration in a specific gene called Her2 in about 25% of women with breast cancer, and this alteration is associated with an aggressive form of breast cancer. Herceptin® is a drug that benefits breast cancer patients with this Her2 mutation, but not other patients. Kadcyla® is a combination of Herceptin conjugated to a toxic agent. A diagnostic test is used

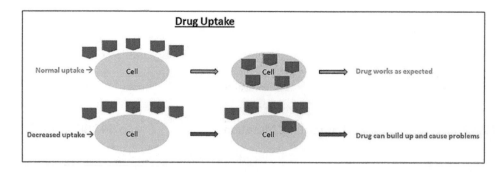

FIGURE 2.18 Genes can affect the uptake, efficacy, and safety of certain drugs. Statins must be taken up by the liver in order to function properly. In people with a certain mutation in the SLCO1B1 gene, this uptake does not occur correctly, and the drug can build up in the blood with harmful effects. (Image credit: U.S. Centers for Disease Control.)

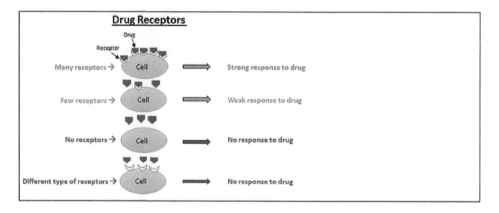

FIGURE 2.19 Genes and a targeted breast cancer drug therapy. Breast cancers that make too much Her2 receptor can be successfully treated with drugs such as Herceptin® and Kadcyla®. A test can be performed to determine if a patient's cancer cells overexpress the Her2 receptor protein and are likely to respond to these drugs. (Image credit: U.S. Centers for Disease Control.)

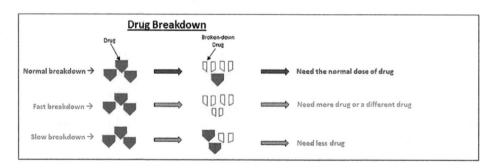

FIGURE 2.20 Genes affect how people break down drugs, and therefore how drugs function in their bodies. The breakdown of the antidepressant drug, amitriptyline, is influenced by the genes CYP2D6 and CYP2C19. It is therefore useful to perform genetic tests for these genes before prescribing this drug in order to determine the right dose. If a person breaks down the drug very slowly, then a lower dose is needed to avoid an adverse reaction. Conversely, a higher dose is required if a person breaks down the drug quickly. (Image credit: U.S. Centers for Disease Control.)

to determine if a patient should receive these drugs (Figure 2.19).

As another example, consider the antidepressant drug amitriptyline (Figure 2.20). The effect of this drug on an individual is influenced by two proteins encoded by the genes CYP2D6 and CYP2C19. These two proteins affect how quickly drugs break down in the body. People have genetic differences in these two proteins that affect their breakdown of various drugs.

Codeine is another example of a drug whose effects are influenced by a person's genes. In most people, an enzyme called CYP2D6 (which is encoded by the CYP2D6 gene) breaks down the drug into its active ingredient, morphine, which provides pain relief. But, as many as 10% of the Caucasian population have genetic variants and do not produce enough of this enzyme to convert codeine to morphine. In these people, codeine is ineffective. About 2% of the population have the opposite problem and have too many copies of the gene that produces the enzyme, leading to overproduction of morphine. In these people, a little codeine can turn into too much morphine, which can lead to a fatal overdose. Codeine-related deaths have occurred in children.

Researchers in 2015 identified approximately 20 genes known to affect an individual's response to 80 different drugs. (This identification process is described on pp. 57–59.) (Relling MV, Evans WE. "Pharmacogenomics in the Clinic." *Nature*, vol. 526, no. 7573, 2015, pp. 343–50. doi:10.1038/nature15817.) However, for a variety of reasons (including reluctance of insurance companies to pay for them), it is presently uncommon for patients to receive genetic testing for their responses to drugs. (Wu, Ann, et al. "The Implementation Process for Pharmacogenomic Testing for Cancer-Targeted Therapies." *Journal of Personalized Medicine*, vol. 8, no. 4, 2018, p. 32. doi:10.3390/jpm8040032.) Pharmacogenomics is an evolving field, and it is expected that, in the future, matching of individuals and drugs will become common and perhaps, routine.

2.2.2.2 Whole Exome Sequencing

In addition to sequencing individual genes, or panels of disease susceptibility genes, or the genes that are responsible for drug metabolism, it is now possible to sequence a patient's entire genome or the *sections of the genome that code for proteins*, known as the **exome**. Although there are many barriers to widespread implementation of this method, it is increasingly being used in the clinic to diagnose rare diseases when other diagnostic approaches have failed. Results from some clinics where whole exome sequencing is performed indicate that a disease gene can be identified in about 25%–30% of cases where sequencing is used to solve a diagnostic dilemma. This is particularly important for neonatal conditions where time is limited to produce a diagnosis and establish a treatment plan. ("Clinical Whole Genome Sequencing." *Duke Center for Applied Genomics and Precision Medicine*, 12 December 2020, precisionmedicine. duke.edu/researchers/precision-medicine-programs/ clinical-whole-genome-sequencing.) One of the first examples of using whole genome sequencing to solve a diagnostic conundrum is the compelling story of Nicholas Volker, often said to be the first person saved by whole genome sequencing. (See the case study below, "Nicholas Volker: Genome Sequencing to Uncover Disease.")

Case Study: Nicholas Volker: Genome Sequencing to Uncover Disease

In 2009, 4-year-old Nicholas (Nic) Volker was dying from a mysterious and devastating bowel disease of unknown origin that began when he was a toddler, necessitating over 100 surgeries including removal of his colon. While Nic's symptoms were similar to Crohn's disease, the usual treatments were ineffective, and he had been taken off solid food. Doctors and researchers on his treatment team at Children's Hospital and the Medical College of Wisconsin had spent many months trying to figure out what was wrong by testing individual genes, scouring the medical literature, and performing a host of diagnostic tests, but they were running out of options. Because it appeared that immune cells were attacking the intestinal cells, some doctors proposed a bone marrow transplant, but, without knowing the exact cause, others were afraid to subject him to such a risky procedure.

As a last resort, after an urgent letter from the pediatrician who was treating Nic, a team was assembled to decide whether to use genome sequencing to search for clues to his disease. Getting agreement to apply this new technology for patient use was complicated. At that time, there had been no published reports of any patient diagnosed using the new, "next-gen" (next-generation) sequencing technology. Although sequencing DNA was relatively straightforward in 2009, analyzing the data was not. There was considerable skepticism about the value of genetic sequencing because sorting through genetic sequences to find the

(*Continued*)

Case Study (*Continued*): Nicholas Volker: Genome Sequencing to Uncover Disease

cause of Nic's illness was akin to searching for a needle in a haystack. Scientists predicted that Nic could have as many as 20,000 nucleotide variations (in fact, Nic would have 16,124) from a "normal" (reference) human sequence. The scientists would need to sift through all the thousands of variants to, hopefully, find one that could explain the damage in Nic's body. Despite the uncertainties about this method, the scientists decided to go ahead with genetic sequencing although they did not sequence Nic's entire genome. Rather they focused on the **exons**, *the portion of the genome that codes for proteins*, since it is the failure to make proteins correctly that causes many diseases. Although less expensive than sequencing the entire genome, it would still cost around $750,000 at that time, so donations were needed.

For the analysis of Nic's exome sequence, new software had to be developed that could eliminate those variations that do not disrupt normal function and could detect variations that could potentially be the root cause of the disease. The new software was named Carpe Novo, Latin for "seize the new." Two assumptions guided the computer analysis: The critical difference in Nicholas's DNA must affect a vital process in the body and must be previously undiscovered, since his disease had not appeared in the medical literature. Early research in the literature had identified a list of 2,000 genes that could possibly be involved. After results from the first sequencing run, 32 genes seemed promising. As analysis continued, all except one of those genes was eliminated because the variants found in Nicholas were also identified in healthy people. The one candidate that remained was the XIAP gene, a gene on the X chromosome that functions in the inflammation pathway. XIAP codes for a protein that has two jobs: It blocks a process that makes cells die (apoptosis), and it regulates the immune system's inflammatory response. Nic has a single nucleotide change from a G (guanine) to A (adenine) in his DNA that resulted in one amino acid change, from cysteine to tyrosine, affecting just one of the 500+ amino acids of the XIAP protein. All species from humans down to the fruit fly have a cysteine at this site.

Finally, Nic's disease started to make sense. Because his XIAP protein was made incorrectly, his immune system was attacking his intestine. Once the cause of his disease was known, the bone marrow transplant option became justifiable because it would give him a new immune system. He was given this transplant, and 6 months later, he was able to eat solid food for the first time in 9 months, including his favorite meal, steak with A-1 sauce.

Primary Sources

Johnson, Mark, and Kathleen Gallagher. *One in a Billion: The Story of Nic Volker and the Dawn of Genomic Medicine*. Reprint, Simon & Schuster, 2017.

Johnson, Mark, and Gallagher, Kathleen. *One in a Billion*: *A Boy's Life, a Medical Mystery*, *Journal Sentinel* (Milwaukee, WI), December 27, 2010. https://www.jsonline.com/story/news/health/2016/04/09/one-in-a-billion-update-although-still-in-infancy-dna-sequencing-rapidly-evolving/84957846/.

2.2.3 BIOINFORMATICS

2.2.3.1 Introduction

As a result of the Human Genome Project, fast and cost-effective DNA sequencing is available to laboratories all over the world. However, the results of sequencing are only the beginning, as illustrated in the Nicholas Volker case study, where finding meaning in thousands of genetic sequences required the creation of a new, sophisticated software program (Figure 2.21).

The field of **bioinformatics** acts as an intersection for biologists, mathematicians, and computer scientists and *allows investigators to make sense of sequence data*. Raw sequence data must be analyzed in order

---AGTTCGCGATAAGATCCATGGCGTTAAATGTGCC---

FIGURE 2.21 DNA sequencing provides just a string of nucleotides that are meaningless without further analysis. Computer programs are used to bring meaning to sequence information.

to provide useful knowledge. To quote the National Human Genome Research Institute:

> Bioinformatics is the branch of biology that is concerned with the acquisition, storage, and analysis of the information found in nucleic acid and protein sequence data. Computers and bioinformatics software are the tools of the trade.

The dramatic results described in the Nicholas Volker case study provide an example of output from the field of bioinformatics, which has three essential components, which we will discuss below:

- **Establishment of databases** to store large quantities of molecular data (for example, human genome databases for comparison with Nic's DNA sequence).
- **Creation of software for data analysis** (for example, development of Carpe Novo software that was used to compare Nic's sequence to the database).
- **Application of the software and databases to solve specific problems**, also called "data mining" (for example, application of the software to discover Nic's mutation).

2.2.3.2 Databases

A **database** is *an organized collection of data that is accessed through database management software.* The stored data could include anything; for example, a retailer will have a database of its product inventory, prices, and where the items are stored. **Database management systems** *allow the user to search for, sort, look for patterns, and report selected data within a database. When examining databases with genomic information, the database management system is called a* **genome browser**.

There are many nucleic acid and protein database resources available to scientists, a few of which are described in Table 2.1. One well-known example is GenBank, which contains virtually all reported DNA sequences. GenBank is maintained by the **National Center for Biotechnology Information (NCBI)**, *a subdivision of the National Institutes of Health that acts as a public resource for molecular data and other biomedical information.* NCBI acts as a portal to dozens of additional molecular databases and is coordinated with multiple international databases, allowing it to provide a comprehensive data resource. It is also a clearinghouse for many of the most essential software tools available in genomics and bioinformatics. NCBI's main goals are as follows:

- **Establishing public databases to store biological data**.
- **Performing research in computational biology** (*developing programs and models to interpret biological data*).
- **Developing software tools for sequence analysis**.
- **Disseminating biomedical information**.

GenBank was established as a public resource in 1982, and the number of sequences grew exponentially in the subsequent 40 years. **Sequence annotation**, *notes that include structural and functional information about a nucleotide or amino acid sequence*, continues to be a major undertaking at NCBI.

2.2.3.3 Software and Data Mining

In the Nicholas Volker case, a personal DNA sequence was compared to a database containing the sequence for the human genome. In order to do the comparison, new analytical software was needed. Bioinformaticians are frequently involved in computer programming, or creating bioinformatics databases, algorithms, and software. **Computer algorithms** are *a set of specific, stepwise instructions in a computer language used to solve a designated problem*. They are the basis for computer software.

Much of bioinformatics software is designed for the purpose of **data mining**, *the process of discovering*

TABLE 2.1

Examples of Bioinformatics Databases

Sample Question	Database	Contents	Location
Has the DNA sequence I discovered been described previously?	GenBank	DNA sequences	https://www.ncbi.nlm.nih.gov/genbank/
How similar are the genomes of humans and gorillas?	Ensembl	Genomes	https://www.ensembl.org/
How do the amino acid sequences of these two proteins differ?	UniProt	Proteins	https://www.uniprot.org/
What gene is associated with this genetic disorder? What is known about this genetic disorder?	OMIM (Online Mendelian Inheritance in Man)	Genes and phenotypes associated with genetic disorders	https://www.ncbi.nlm.nih.gov/omim

patterns in the large quantities of data within a database, which has become an essential tool for biological research. Data mining is not a new idea, but advances in computer technology have greatly enhanced the ability of researchers to separate small subsets of relevant data from massive data sets, which would be impossible to analyze by eye. For example, current models of global climate change are based on data mining of historical weather records. Many businesses develop marketing strategies based on data mining of consumer practices. The overall process of data mining is summarized in Figure 2.22.

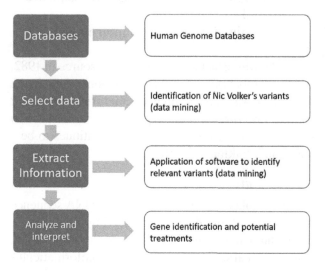

FIGURE 2.22   The data mining process requires multiple steps, shown on left. The application of these steps to the Nic Volker case is shown on the right.

Any large repository of information can be data mined. One familiar application is text mining, searching databases for topics of interest using keywords. Google Search is a good example of how everyday users perform data mining operations. The size and complexity of the data repositories involved is what separates data mining from a simple search. Ordinary database management software is not suited for simultaneously sorting through vast amounts of information in multiple databases on non-centralized computers.

For both Google Search and bioinformatics software, the user retrieves information using a **query** – *a text-based set of criteria used to search for and extract a desired subset of data from a database and then present it in a specific format*. A query is essentially a database filter, which shows (or hides) data records with designated characteristics. Most popular search tools (including search engines such as Google as well as the data mining programs discussed below) offer user-friendly query forms that automatically translate user requests into specialized retrieval languages (Figure 2.23). However, many serious bioinformatics researchers prefer to write their own queries.

Another common form of text mining in the laboratory is using a computer to look for references when writing a paper. **PubMed** (https://pubmed. ncbi.nlm.nih.gov/) *is a database of more than thirty million scientific reference articles* and is an exceptional resource for researching biomedical topics. One adjunct is PubMed Central, a free online archive of medical and biological journal contents. PubMed and PubMed Central are maintained by NCBI. Note that

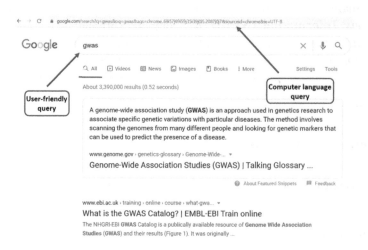

FIGURE 2.23   Google translates a simple search term into a computer language query. The user simply types in the desired search term, and the search engine translates that into computer query language (the address bar). It then reports the search results in a specific, programmed format. The simple search term in this example is "gwas," which is an abbreviation for "genome-wide association studies." (See the case study "Genome-Wide Association Studies" below for an explanation of GWAS.)

Google Search has an adjunct tool, Google Scholar, which also searches scientific and other academic articles. However, this tool is not as robust as PubMed for specialized biological topics.

Biomolecular data mining concentrates on many types of knowledge discovery relating to genes, proteins, and other biological molecules. This type of data mining is used for such purposes as determining the function of newly reported gene sequences, discovering which genes are similar across species, and which sets of genes are expressed together in various organisms.

Commercial data mining programs are available, but a variety of free biological data mining software is also accessible, usually associated with database websites. NCBI alone offers a wide range of databases and analytical tools for investigative purposes. It would take hours to scour the NCBI website and examine the variety of analytical tools available, and new tools are constantly developed.

Probably the most frequently used analytical tool is **BLAST** (<u>B</u>asic <u>L</u>ocal <u>A</u>lignment <u>S</u>earch <u>T</u>ool; https://blast.ncbi.nlm.nih.gov/Blast.cgi), *a pattern recognition tool used to search for similarities between nucleotide or amino acid sequences*. This program is one of the foundations of bioinformatics and is routinely used by those who work with nucleic acid or protein sequences. It is easy to do a simple BLAST search that compares a nucleotide or amino acid sequence of interest to a wide range of sequence databases. BLAST compares the entered nucleotide or protein sequences to database sequences and finds matches or partial matches. BLAST then displays which base pairs or amino acids match in the two sequences. This type of search can be used to help determine the potential function of an unknown nucleotide or amino acid sequence by finding similar sequences of known function. It also can be used to suggest molecular relationships among groups of genes or proteins, or evolutionary relationships among organisms.

Bioinformatics tools allow scientists to perform an ever increasing array of complex analytical tasks using molecular data. Nic Volker's case was only the beginning of our ever-increasing use of genomic knowledge to solve human medical problems, as suggested by the case study "Genome-Wide Association Studies."

Case Study: Genome-Wide Association Studies

So far, we have discussed the case of Nic Volker and how bioinformatics was able to identify his rare disorder. However, what about common disorders that affect large numbers of people? One widely used process used in identifying genes associated with human disorders in populations is **genome-wide association studies (GWAS)**. GWAS are *studies that look at genomic markers in large numbers of people in order to identify variations associated with complex disorders*. Some scientists have rather accurately described them as "fishing expeditions," because the entire genome is scanned for unknown associations between a disorder and millions of variant nucleotide markers. These studies are greatly complicated by the fact that complex human disorders, such as heart disease and cancer, have multiple genetic components, so multiple genes are involved, unlike in Nic Volker's case.

The variants examined in these studies are previously identified human <u>s</u>ingle <u>n</u>ucleotide <u>p</u>olymorphisms (SNPs, pronounced "snips"), that is, *single nucleotide locations where 1% or more of the population exhibits a specific variant*. Because these variants are relatively common, large populations of patients are compared to large populations of unaffected individuals. There are millions of SNPs found in the human genome. Scientists use statistics to look for variants that occur more frequently in a population of patients than in the unaffected population. The general GWAS screening process is shown in Figure 2.24.

Because many people will have each version of the variant SNPs, it is unlikely that a specific variant will be found always and only in patient populations. If that were the case, this would indicate that there is a single mutation in one gene that is involved in the disorder. In reality, complex disorders (by definition) involve multiple genes, with different subsets of variants in different patients. Therefore, complex statistical analysis is required to determine which differences are potentially meaningful. These statistical results can only be applied to populations, not to individuals, who may or may not have any specific variant. An example of how these studies are applied is shown in Figure 2.25.

Despite difficulties in analysis, GWAS have been successful in identifying gene variants strongly associated with some complex disorders. For example, GWAS have identified more than 80 genetic variants

(Continued)

Case Study (*Continued*): Genome-Wide Association Studies

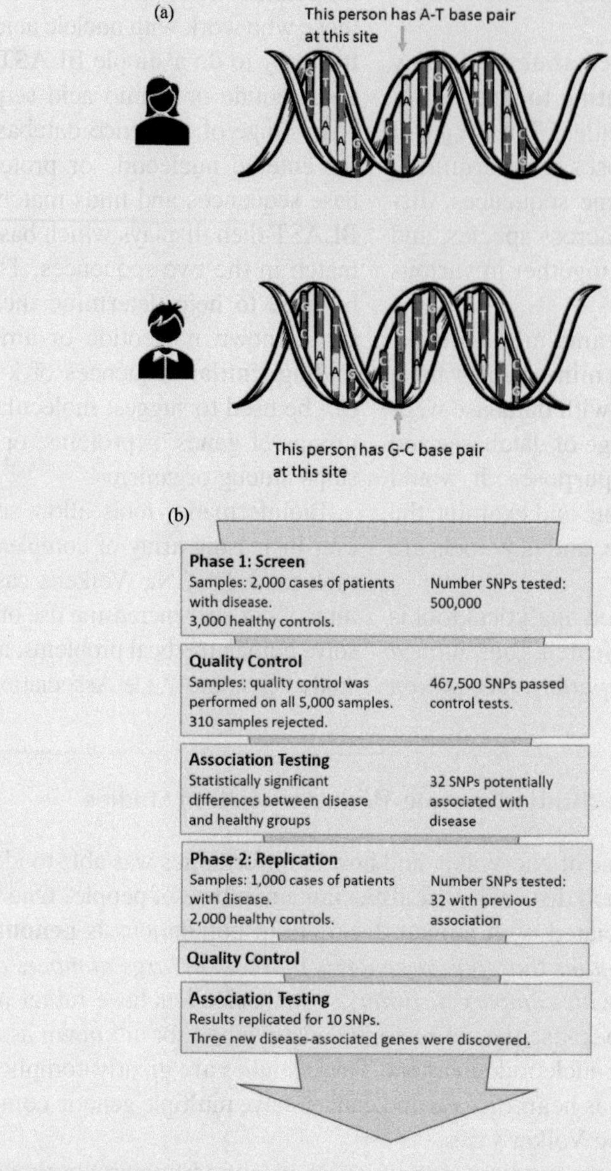

FIGURE 2.24 SNPs and the general stepwise process for a GWAS. (a) SNPs are found at the same specific locations in everyone's genome. There are usually two common variants (in this illustration, A or G) found within a population. (b) For Phase 1 of a GWAS, large groups of people with and without a disease are screened for SNPs. After quality control, statistics are applied to determine which SNPs potentially correlate with the disease. For Phase 2, new groups of disease and control patients are selected and tested for the SNPs identified in Phase 1. After quality control, the SNPs that are still associated with the disease in the Phase 2 group are chosen for further investigation, to see if the SNPs are associated with a gene that may contribute to the disease. The function of these genes will then be analyzed.

associated with type 2 diabetes. This suggests that although this disorder is genetically heterogeneous in a population, there are certain genes that can be identified and that do tend to be associated with type 2 diabetes.

GWAS have helped physicians find effective drugs for at least some patients with complex disorders such as type 2 diabetes, psoriasis, schizophrenia, and osteoporosis, as introduced in Section 2.2.2.1. It is possible to use GWAS to compare large groups of people who have a good response to a particular drug versus those who do not. Based on statistical differences in the patterns of SNPs between the two

(*Continued*)

Case Study (*Continued*): Genome-Wide Association Studies

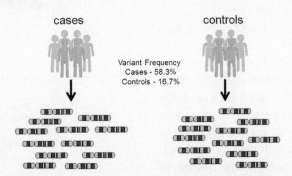

FIGURE 2.25 Results for a single nucleotide variant from a GWAS. Large groups of people with the disorder, and large groups of people without the disorder, are tested for a single nucleotide marker, found at the same location on everyone's chromosomes. One specific variant, such as cytosine, shown as red bands on the same chromosome from many individuals, is more common in the population with heart disease, found in 58.3% of cases, while only found in 16.7% of controls. If this proves to be a statistically significant difference, then that genomic location may be associated with a gene involved in the disorder, at least in some patients. See text for further explanation. (Image credit: EMBL-EBI Training, https://www.ebi.ac.uk/training/online/course/gwas-catalog-exploring-snp-trait-associations-2019/what-gwas-catalog/what-are-genome-wide, CC-SA 4.0.)

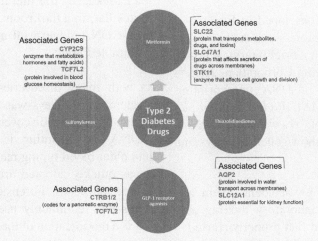

FIGURE 2.26 Application of pharmacogenetics to treatment of type 2 diabetes (T2D). The colored circles represent examples of families of drugs that are commonly used to treat the symptoms of T2D. Beside each circle are the names of genes that may be associated with individual drug response in patients, with information about their main function. GWAS is used to better understand these genes' effects on drug responses in order to predict for an individual patient which drug will be most effective. (Based on information in Brunetti, Antonio, et al. "Pharmacogenetics in Type 2 Diabetes: Still a Conundrum in Clinical Practice." *Expert Review of Endocrinology & Metabolism*, vol. 12, no. 3, 2017, pp. 155–58. doi:10.1080/17446651.2017.1316192.)

groups, it is sometimes possible to identify SNPs that are associated with a favorable drug response and that make sense in terms of what is known about certain drugs. Figure 2.26 summarizes such work for drugs used to treat type 2 diabetes. For example, metformin is a drug that is used to treat the symptoms of type 2 diabetes. For many patients, metformin is highly effective. But some patients do not respond well to metformin. GWAS are being used in an effort to predict who will respond to metformin and to understand why some people respond to this drug better than others. This is an application where GWAS comes together with, and drives, pharmacogenetics.

The pharmacogenetics of complex diseases is still in its infancy, but there have already been discoveries that have helped some patients. We can expect to see better tailored treatments in the future as our knowledge of the genetics of complex disorders increases.

2.2.4 Genetic Identity Testing and Forensics

DNA identity testing, also known as DNA fingerprinting or DNA profiling, *is a method of identifying individuals based on differences in their DNA*. Identity testing uses DNA extracted from tissue (hair, bone, blood, etc.) to distinguish individuals. About 99.9% of every person's DNA is identical, but in a few stretches of DNA, there are variations between humans that make each of us distinct from one another (except for identical twins, whose DNA is the same). These variable regions are used to make a DNA "fingerprint" that is unique for each person. (Keep in mind that the human genome consists of about 3 billion base pairs. A 0.1% variation between individuals could represent 3 million base pair differences in sequences.) The most famous use of this technology is to match DNA from a crime scene with that of a suspect. If the DNA matches, it provides evidence that the individual was at the scene. The idea of DNA fingerprinting is well known, thanks to the high-profile murder trial of O.J. Simpson in 1995 and some very successful television crime dramas.

DNA fingerprinting was developed in England in 1984 by Sir Alec Jeffreys, a professor at the University of Leicester (Figure 2.27). In a brief interview posted by the Australian Broadcasting Corporation, Jeffreys talks about his discovery of DNA fingerprinting. ("Genetic Fingerprinting." *ABC Radio National*, 24 September 2002. Transcript. www.abc.net.au/radio-national/programs/scienceshow/genetic-fingerprinting/3520000.) He says,

> It was invented purely by accident...from a different project in my laboratory looking at how genes evolve, and, in fact properly started with a lump of...seal meat which we used to get at a gene which we were interested in studying its evolution...We got that gene, looked at its human counterpart and purely by chance inside that gene was a bit of DNA which was a key to unlock the door on genetic fingerprinting...

FIGURE 2.27   Sir Alec Jeffreys showing an example of a DNA fingerprint. (Photo courtesy of Homer Sykes/Alamy.)

> It was absolutely blindingly obvious that what we had was a technology that could be used for identification, establishing family relationships and the like. So for me, it was very much a Eureka, my life literally changed in five minutes flat, in a darkroom, when I pulled out that first DNA fingerprint and saw just what we'd stumbled upon.

As with many discoveries, this one was made by accident, but Jeffreys was quick to see the profound implications of his discovery.

DNA fingerprinting is so much more accurate than older blood typing methods of identification that it was quickly adopted around the world for identification purposes. Forensic scientists in the United States now commonly use a fingerprinting method that is a modification of the one pioneered by Jeffreys. (Forensic identity testing is also discussed in the case study below).

DNA identity testing, although it is probably most famous for its use in criminal investigation, is also used to reliably determine paternity and other family relationships. It is used to identify bodies in the

---

Case Study: DNA Fingerprinting

Sir Alec Jeffreys's professional life did indeed change with his discovery of DNA fingerprinting. Almost immediately after Jeffreys published his fingerprinting method, he was called upon to help a family embroiled in an immigration dispute. The family, who were citizens of the United Kingdom, claimed that their son had visited Ghana and was returning home to his family after the trip. The immigration authorities would not allow the boy to return to his family, claiming that his passport had been altered and that he was an unrelated imposter. Jeffreys used his DNA fingerprinting technique to conclusively show that the boy was indeed the son of the family in question and, in a happy ending, the boy was reunited with

(Continued)

Case Study (*Continued*): DNA Fingerprinting

his family. (*"Pioneering DNA Forensics" All Things Considered,* interview by Michelle Norris, National Public Radio, 13 April 2015, Transcript. https://www.npr.org/templates/story/story.php?storyId=4756341.)

Soon after, Jeffreys was called on to use his method in a tragic crime investigation involving two teenage girls who had been raped and murdered. A man had confessed to the rape and murder of one of the victims, but not the other. The police were certain the murders were related and asked Jeffreys to use his method to analyze semen isolated from the victims and the suspect. The DNA analysis showed that the semen isolated from both girls was from the same man. DNA fingerprinting, moreover, showed that the man who had confessed was not the perpetrator of either crime. Thus, the very first time DNA fingerprinting was used in a crime scene investigation, it established a suspect's innocence, not his guilt. The police went on to mount a dramatic investigation in which they asked all the men in the local community to submit their blood for DNA testing. More than 5,000 men complied, but none were a match for the crime scene DNA. The case would have ended there were it not that someone overheard a man telling his friends that he had been asked by the local baker to give his blood in place of the baker. This conversation was reported to investigators who took a DNA sample from the baker. The baker's DNA was a perfect match for the semen taken from the victims and the baker confessed to both crimes. DNA fingerprinting thus quickly proved its value as a forensics tool.

Dr. Mary-Claire King, a scientist recognized for her pioneering work studying genes that cause familial breast cancer, used DNA fingerprinting in another significant way. Between 1976 and 1983, there was a brutal dictatorship in Argentina. Thousands of men, women, and children were tortured, killed, and "disappeared" during this reign of terror. Some of the women who disappeared were pregnant at the time of their capture and gave birth before their death. Also, some young children were seized before their parents were killed. The babies were sold or given to military families who illegally adopted and raised them. When the military regime was deposed, grandmothers of these kidnapped children began to search for them. The grandmothers followed leads from school registrars, who saw children arrive with forged papers, and from others who had noticed babies suddenly appearing in military families. As the grandmothers found individuals who they thought were their grandchildren, they appealed to scientists to use DNA fingerprinting to definitively identify their family members. Mary-Claire King, then a professor at the University of California, went to Argentina and obtained court orders that enabled her to test the DNA of individuals who were thought to have been kidnapped as babies. In this way, about 50 children were eventually matched with their grandmothers and reunited with their birth families.

case of a disaster, such as the 9/11 terrorist attack on the World Trade Center in New York. Analysts use DNA fingerprinting in wildlife investigations, for example, to identify whether an item was illegally poached from an endangered species. DNA fingerprinting can be used to detect pathogens in food, air, and water. It can be used to determine the pedigree of a valuable animal, such as a racehorse. DNA fingerprinting methods are therefore powerful and widely applicable to a variety of circumstances. The case study below, "DNA Ancestry Sites," provides yet more examples of how genetic identity information is being used. Yet another example of genetic identity testing is provided in Chapter 30 where we consider how cultured cells in the laboratory are tested to determine their origin.

Case Study: DNA Ancestry Sites

Tens of millions of people perished as a result of World War II, including, to his knowledge, all of David Green's[1] family. David escaped from Nazi Germany as a teenager, was taken in by a British family, and later emigrated to the United States. He married a woman who had also escaped from Europe, and they raised two sons, Carl and Nathan. When their parents died, the two men were unaware of any living relatives – that is, until Carl provided a saliva sample to a genetic testing service. When he logged onto the

(*Continued*)

Case Study (*Continued*): DNA Ancestry Sites

service's site, Carl found he had a cousin living in California. They corresponded and eventually met in person. The cousin turned out to be a fascinating celebrity, and she had traced one branch of their shared family tree back to the 1500s. Success stories such as this are common on the genetic testing sites. Adoptees have found their biological families, and people have discovered relatives and their ancestral roots. These successes – together with effective marketing – have led millions of people to provide a sample of their saliva sample for DNA testing.

While many people have experienced heartwarming reunions, using a genetic testing service can have unforeseen and sometimes undesired results. It is not uncommon for people to discover that the man who raised them was not their biological father. There are cases where individuals who were conceived in fertility clinics by artificial insemination discover that they have as many as thirty or more half-siblings. Sometimes this is good news, and the siblings establish Facebook groups, host reunions, and warmly welcome new siblings, each time one is discovered on a genetic testing site. In other cases, it is an unwelcome surprise to discover that one was conceived using an anonymous sperm donor – and so were thirty or more strangers.

Most people who sign up for genetic testing services assume that their genetic information will be confidential, as is guaranteed by the services, unless they voluntarily choose to share information. But what has come as a surprise – even to some geneticists – is that soon the information on these sites can be used to find nearly anyone of European descent in the United States. Yaniv Erlich is the chief science officer of MyHeritage, a genetic testing service. Erlich and colleagues published an analysis that showed that about 60% of searches for individuals of European descent will result in a third cousin or closer match. What is the significance of this? Suppose, for example, that you participate in a research study that collects DNA data. When signing up for the study, you were guaranteed anonymity; that is, your genetic information could be made available in research reports, but your name would be withheld. Perhaps the study reveals that you have a predisposition to a certain disease. Although there are laws protecting you from discrimination based on this predisposition, you still do not want this medical information to be available. What Erlich and his colleagues demonstrated is that it is possible to take DNA information that was posted anonymously, put it into a genetic testing services database, and find some of your relatives. Once your relatives are identified, it is possible to use conventional genealogical tools (such as family trees, birth records, and social media profiles) to link your name with your DNA data. Thus, you (and probably also the researchers) thought your DNA information was anonymous, but, in reality, it was not.

DNA fingerprinting, as pioneered by Sir Alec Jeffreys, turned out to be a "game-changer" in forensics. But these DNA fingerprinting methods require that a suspect has been identified. The suspect's DNA is then compared to DNA from a crime scene. What if there is crime scene DNA and no suspect? This was the case in California where a series of rapes and murders by the "Golden State Killer" terrorized people in the 1970s and 1980s. DNA was available from the crime scenes, but no suspect was found. In 2017, a detective, Paul Holes, recruited Barbara Rae-Venter, to work on the case. Rae-Venter is a retired patent lawyer with a Ph.D. in biology. As a patent lawyer, she specialized in biotechnology inventions, including the Flavr Savr tomato, the first genetically engineered fruit licensed by the FDA. Rae-Venter's ex-husband is J. Craig Venter, a geneticist who was instrumental in the sequencing of the first human genome. Rae-Venter became interested in genetic genealogy after using a genetic testing service and discovering a cousin. The cousin had learned through the testing service that the man who raised him was not his biological father, and he wanted to find out who was. Rae-Venter was not a genetic genealogist, but she had the molecular biology background to understand the field. She studied genetic genealogy in order to help her cousin and subsequently began teaching classes for those interested in finding their biological parents. Detective Holes thought that Rae-Venter's expertise in finding people using genetic service databases could help find the California killer. It is not possible to search the databases of consumer genetic testing sites, such as 23andMe or Ancestry.com, without a court order. However, another site, GEDMatch, had a looser customer agreement. Rae-Venter took crime scene DNA provided by Detective Holes, converted it to a format compatible with GEDMatch, and used this database to identify individuals who were distant cousins of the Golden State perpetrator. (This is

(Continued)

Case Study (*Continued*): DNA Ancestry Sites

an example of data mining.) She then painstakingly used conventional genealogical methods that provided clues eventually leading to the arrest of Joseph James DeAngelo, a former police officer. Once this suspect was identified, conventional DNA fingerprinting matched his DNA to that at the crime scenes.

The DeAngelo case generated a lot of interest, and other suspects have been apprehended using similar methods. In recent years, several companies have formed that specialize in forensic genetic genealogy. While most people agree that the apprehension of serial killers is desirable, ethicists are concerned about the rapidly increasing use of genetic genealogy methods. Ethicists point out that, for example, unscrupulous people could use these methods to find individuals whose political views do not agree with those in power. They further note that genetic forensics is complex and requires deep understanding of genetic testing, genealogical methods, and the limitations of these methods. It would be easy for errors to lead to the wrong person being accused of a crime. As more and more people, many without a great deal of training, enter this forensic field, the risk of error increases. There have already been reports of adoptees who have erroneously been told that someone is their parent, who is not. With all genetic information comes the concern that people with certain medical conditions will face discrimination, despite the existence of protective laws. Yaniv Erlich and colleagues warn that care is required and conclude with the following statement: "Overall, we believe that technical measures, clear policies for law enforcement..., and respecting the autonomy of participants in genetic studies are necessary components for long-term sustainability of the genomics ecosystem."[1]

Primary Sources

Murphy, Heather. "She Helped Crack the Golden State Killer Case. Here's What She's Going to Do Next." *The New York Times*, 4 September 2018, www.nytimes.com/2018/08/29/science/barbara-rae-venter-gsk.html.

Molteni, Megan. "A New Type of DNA Testing Is Entering Crime Investigations." *Wired*, 28 December 2018, www.wired.com/story/the-future-of-crime-fighting-is-family-tree-forensics.

Erlich, Yaniv, et al. "Identity Inference of Genomic Data Using Long-Range Familial Searches." *Science*, vol. 362, no. 6415, 2018, pp. 690–94. doi:10.1126/science.aau4832.

---

[1] Names have been changed, but the story is real.

2.3 SUMMARY

Biotechnology can be understood as the transformation of knowledge about living systems into useful applications and products. Biotechnology is not a new phenomenon because for thousands of years curious humans have made discoveries that have led to useful products. But in the past 50 or so years, the pace of discovery has exploded, and applications of this knowledge are expanding every day. Often, as was the case with DNA identity testing, new knowledge is quickly transformed into widespread application. In the 1980s, DNA fingerprinting was discovered; in the 1990s, it was developed for use around the world; and by the beginning of the next decade, DNA fingerprinting was a fixture of popular culture. Other applications are slower to mature, but are likely to do so in the future.

Modern biotechnology is based on the discovery of methods to manipulate DNA, but the industry clearly incorporates many other methodologies as well. Protein scientists isolate, model, purify, and study proteins. Cell culture biologists and fermentation specialists develop and refine methods for growing cells in culture. Engineers work together with scientists to develop manufacturing and protein purification processes. There are a host of analytical methods that are used in the laboratory for testing materials at all stages of product discovery and development. Pharmacology, medicine, forensics, agricultural science, environmental science, and manufacturing technology are key parts of the biotechnology industry. Combine all these technical fields with business, ethics, and more, and this becomes the modern biotechnology industry.

As we will see in Chapter 3, biotechnology is a commercial enterprise, and as such, it is driven by profit and finances, patents, and business concerns. Yet, it is important to remember that biotechnology is also driven by people who study science out of curiosity, and by many people who hope their work will help others in ways unrelated to monetary profit.

Practice Problems

Note: You will likely need to use the Internet to find information required to answer some of these questions.

1. The gene therapy treatment used for Ashanti De Silva must be periodically repeated, whereas the surviving children who received gene therapy for X-SCID are presumed to be cured. What is the difference between the two treatments?
2. Biotechnologists often use cells as "factories" to make products. In this type of manufacturing, once the product is made, the cells are discarded. In other cases, biotechnologists grow cells that are used therapeutically; the cells *are* the product. For each of the following products of biotechnology, are cells being used as "factories" or are the cells themselves the product?
    a. Humulin®
    b. Hepatitis B vaccine
    c. Apligraf®
    d. Herceptin®
3.
    a. What are CD33, CD34, and CD133?
    b. Which of these is a target for a monoclonal antibody drug?
    c. Which of these is used to identify hematopoietic stem cells?
4. Bone marrow transplants take hematopoietic stem cells from a donor and deliver them to a recipient patient. These stem cells cannot differentiate into all the types of cells in a human body, but can differentiate into more than one cell type. What types of cells can hematopoietic stem cells become?
5. Stem cells are being researched for treating the diseases listed below. For each disease, the stem cells must differentiate into a particular cell type in order to treat the disease. What cell type is required to cure or reduce the symptoms of these diseases?
    a. Parkinson's disease.
    b. Diabetes.
    c. Amyotrophic lateral sclerosis
6. The bioinformatics tool, BLAST, is being used to study mutations in the genetic sequence of the SARS-CoV-2 virus. Can you think of any reasons why scientists would be interested in mutations in this viral genome?
7. Use the Internet to find out what therapeutic cloning is. (Another name for this process is somatic cell nuclear transfer.) Compare and contrast *therapeutic cloning* with the form of cloning used to create the sheep Dolly.

3 The Business of Biotechnology
The Transformation of Knowledge into Products

3.1 PRODUCT LIFECYCLES IN BIOTECHNOLOGY

3.1.1 INTRODUCTION

In 1918, an 11-year-old girl, Elizabeth Hughes, was diagnosed with type 1 diabetes. This is a relatively common illness that occurs when the pancreas loses the ability to make insulin, a hormone necessary for life. In 1918, a diagnosis of type 1 diabetes was a death sentence; afflicted individuals seldom lived more than a couple of years after diagnosis. But Elizabeth was different; as a result of pioneering work by scientists and physicians, Elizabeth was able to receive insulin that had been isolated from the pancreas of animals. She went from the edge of death to survival, growing up to become a civic leader and parent, dying at the age of 73.

Insulin is notable as a drug that has saved the lives of tens of millions of people since Elizabeth Hughes was given her first injection. As noted in Chapter 1, insulin is also noteworthy as being the first major product of "modern" biotechnology. But good science alone does not bring products to people's lives. This chapter introduces the business processes that transform knowledge into commercial products.

Biotechnology is fueled by commerce, and the companies that work in the biotechnology arena are essential participants. This means that social issues are also part of the biotechnology landscape. Even insulin, a product that has clear benefit to millions of people, can raise societal issues. In the United States, commercial interests have priced insulin out of the range of some patients who need it, and on an international scale, the products of biotechnology are often not available

DOI: 10.1201/9780429282799-4

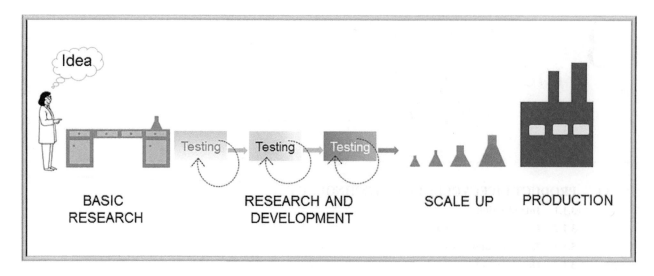

FIGURE 3.1 The big picture. Biotechnology is the transformation of knowledge into products. Typically, an idea for a biotechnology product emerges from work done in a basic research laboratory. The idea must go through a period of extensive testing, evaluation, modification, further testing, and development into an actual product. If all goes well, and the concept proves to be viable, then scientists must learn how to make the product in larger and larger quantities; this is called scale-up. The final phase is production where the manufactured product is available to consumers.

to everyone. Finding ways to provide broad access to biotechnology's advances – insulin is just one example – is a critically important topic of discussion.* It is beyond the scope of this text to explore all the complex social issues relating to biotechnology, but it is important for biotechnologists to be aware that it is an interplay between science and business that transforms scientific advances into tangible products.

3.1.2 OVERVIEW OF PRODUCT LIFECYCLES

The transformation of an idea into a product does not happen all at once; rather, it is a process that usually requires years of effort. It is possible to describe this process as a product's *lifecycle*. The typical biotechnology product is "born" in the research laboratory when a discovery results in an idea for a product. The potential product must then go through a period of **development** *during which the idea is transformed into an actual, workable product*. If the development period is successful, then the product enters **production** – *the final stage in a successful lifecycle including sales and expansion and growth of the product* (Figure 3.1). As we will see later in this chapter, this product lifecycle is roughly mirrored in the organization of biotechnology companies.

3.1.3 RESEARCH AND DISCOVERY

The biotechnology industry is rooted in scientific discoveries that are usually made by scientists and their students in universities, colleges, research institutes, and medical centers. Sometimes research scientists, who are working to understand nature, make a discovery that turns out to have commercial application. They, or other colleagues, might decide to explore that potential. The initial exploration of the commercial applications of a discovery is conducted in a research laboratory, often still in an academic institution. Early experiments may involve, for example, testing a treatment in cultured cells, testing a compound in animals, using biochemical assays to explore a molecule of significance, using computer simulations of molecules and their interactions, and growing plants in controlled environments. This is also the period when intellectual property is secured (discussed later in this chapter) for the idea, the product, its potential uses, and its processes of manufacturing.

If early studies yield promising results, then researchers may decide to form a company to develop the commercial applications of the idea. Many biotechnology companies are thus initially founded by research scientists from universities and other research institutions. Sometimes the discovery and early research that results in a product is conducted by scientists in a company that already exists. Most biotechnology companies have research scientists and technicians whose job is to explore new ideas for products.

---

[1] See, for example, Rajkumar, S. Vincent. "The High Cost of Insulin in the United States: An Urgent Call to Action." *Mayo Clinic Proceedings*, vol. 95, no. 1, 2020, pp. 22–28. doi.org/10.1016/j.mayocp.2019.11.013.

3.1.4 DEVELOPMENT

Product development requires a cycle of rigorous testing, modification, and continued testing of a potential product to optimize its utility for its intended purpose. The development phase of a product's lifecycle is the transition between its discovery in a research laboratory, and its use as a commercial product.

The development stage includes determining the properties of the product, specifying the properties the product must have to be effective, and describing how to make the product. The development phase is a time of transition, evolution, and evaluation. As development progresses, the characteristics of the product become established, methods of production become increasingly consistent and systematized, and the specifications for raw materials are codified. It is during this evolutionary period that quality is designed and built into the product.

A manufactured product is initially made in small quantities in the laboratory. A major part of development is the **scale-up** *of the processes used in the laboratory so that larger amounts of the product can be made in a consistent, reproducible, and economically efficient manner.* Depending on the quantities of the product to be sold, scale-up may be a fairly simple process of moving from one size flask to another. In other situations, scale-up involves initially making the product in a medium-sized **pilot plant** *where further development occurs*, and then eventually in full-scale production facilities. Scale-up of an agricultural product may mean moving from the greenhouse to a small experimental plot, and finally to large fields.

Sometimes ideas are transformed into commercial products in an academic institution or a research institute, but it is often in biotechnology companies that this transformation occurs. Biotechnology companies have organizational units, generally called **research and development (R&D)**, *which find ideas for products, perform research and testing to see if the ideas are feasible, and develop promising ideas into actual products.* Larger companies often separate the *research* part of a product's lifecycle from its *development* and have separate functional units for each. The development of pharmaceutical/biopharmaceutical products is so extensive that multiple teams and more than one company often participate in their development. The general responsibilities of the R&D functional unit(s) are summarized in Table 3.1.

TABLE 3.1

The Responsibilities of the Research and Development Unit(s) in a Biotechnology Company

- *Discovering a potential product with commercial value*
  - performing scientific research.
- *Genetically modifying cells, if this is required to make the product*
- *Describing and documenting the features of any cells or organisms required to make the product*
- *Characterizing and documenting the properties of the product, such as:*
  - composition, physical, and chemical properties
  - strength, potency, or effect of the product
  - purity of the product required, and steps required to avoid contamination
  - applications of the product
  - safety concerns in the use of the product.
- *Establishing and documenting product specifications* (descriptions of properties that every batch of the final product must have to be released for sale)
- *Developing and documenting methods to test the product to be sure it meets its specifications*
- *Developing and documenting processes to make the product*
- *Determining and documenting the raw materials required to make the product and establishing specifications to characterize those materials*
- *Describing and documenting equipment and facilities required to make the product*
- *Determining and documenting stability and shelf life of the product*
- *Scaling up production.*

Case Study: Crime Scene Investigation: A Behind-the-Scenes Story

There is a gritty crime scene; enter a freshly coiffured investigator who discovers a bit of blood, rushes it back to the laboratory, and translates the blood stain into the evidence that unlocks the case. Behind the scenes of the popular television crime scene investigation dramas are real scientists who create laboratory tools that are used in actual forensic investigations. DNA "fingerprinting," one of the most famous of the modern forensic technologies, was developed in 1984 by the scientist Sir Alec Jeffreys. Although about 99.9% of every person's DNA is identical, there are a few differences between humans. These *genetic variations*, called **polymorphisms**, are the basis for DNA typing. Jeffreys found that certain regions of human DNA contain specific sequences of nucleotides that are repeated over and over again, one after the other. He also discovered that the number of times a sequence repeats is polymorphic (has many forms) and so can be used to distinguish the DNA from individuals. Jeffreys used a method called restriction fragment length polymorphism (RFLP) analysis to detect the differences in the numbers of repeats in the DNA from different people. Although RFLP analysis was successfully used in criminal investigations in the 1980s and 1990s, it has limitations. RFLP analysis requires relatively large amounts of undegraded DNA, more than is often available from a crime scene. RFLP analysis is used to examine only one sequence of DNA at a time, is not easily automated, and is slow. Many researchers in academia and in biotechnology companies sought to improve DNA typing technology after Jeffreys's initial discovery. Scientists quickly adapted other types of polymorphisms in addition to the repeated sequences Jeffreys used, including SNPs and STRs. **Single nucleotide polymorphisms (SNPs)** *are places in the genome where a single nucleotide differs from person to person.* A **short tandem repeat (STR)** *is a type of polymorphism, similar to that used by Jeffreys, but smaller.*

FIGURE 3.2   Dawn Rabbach preparing laboratory samples for analysis.

(Continued)

Case Study (*Continued*): Crime Scene Investigation: A Behind-the-Scenes Story

Scientists also learned to use the **polymerase chain reaction (PCR)**, *a powerful method for amplifying DNA*, to deal with the problem that DNA is often limited in quantity and poor in quality. Furthermore, scientists reasoned that if they could look at many polymorphic DNA locations (loci) at one time, they could quickly and reliably distinguish one individual from another with little chance of two people sharing the same "fingerprint."

Research scientists at Promega Corporation were among the many creative people studying DNA typing. Their challenge was to develop a method to distinguish individuals from one another that is fast, is reliable, has virtually no chance of error, and could be readily used by forensic analysts with different levels of training and possibly different equipment. At the time Promega was beginning their DNA typing project, Dawn Rabbach (Figure 3.2) was a student in a 2-year associate degree biotechnology program and a part-time entry-level technician at Promega whose job was to prepare laboratory reagents. Dawn was selected to work on the DNA typing team because of her hands-on skills at the lab bench and because she was known to pay careful attention to detail. Dawn and her colleagues worked feverishly to be one of the first companies to develop a commercial method of DNA typing that would meet the stringent requirements of forensics laboratories.

During this fast-paced period of research and development, Ms. Rabbach performed many studies, carefully modifying one factor at a time to help determine the best methods of typing. She tested STRs to see if this type of polymorphism would work in an automated system. She tested different methods of labeling DNA and experimented to determine how much DNA should be analyzed in each test to obtain the most accurate results. The efforts of the Promega R&D team led to a patented, automated DNA typing system, which provides a fingerprint from 16 polymorphic loci simultaneously (Figure 3.3). The team packaged the method into a kit that was tested and accepted by the forensics community. The team's efforts also contributed to the development of the **Combined DNA Index System (CODIS)**, *a national database of DNA fingerprints from convicted offenders*. In the late 1990s, the FBI was creating their database and they needed to select a standard set of polymorphisms that every crime lab in the United States would analyze. They selected for their database a set of 13 loci, many of which had been developed by the Promega scientific team.

Ms. Rabbach, who is now a Senior Research and Development Scientist at Promega Corporation, continues to transform the ideas of research scientists into commercial products. Her job responsibilities are typical of those in biotechnology company R&D laboratories and include the following:

* performing laboratory experimentation necessary to develop new molecular biology products.
* documenting her work so that her discoveries can be patented and her results are repeatable.
* writing **standard operating procedures (SOPs)** *that are used by manufacturing personnel when her team's products are ready for commercial production*.
* developing assays that are used by quality-control analysts when they test the products her team developed.
* writing standard operating procedures that are used by quality-control analysts.
* troubleshooting problems that arise as customers with different equipment and processes adopt Promega products.
* anticipating issues that arise as software and instrumentation change.
* traveling to customer laboratories to test methods and equipment.
* speaking at forensic conferences.

(Continued)

Case Study (*Continued*): Crime Scene Investigation: A Behind-the-Scenes Story

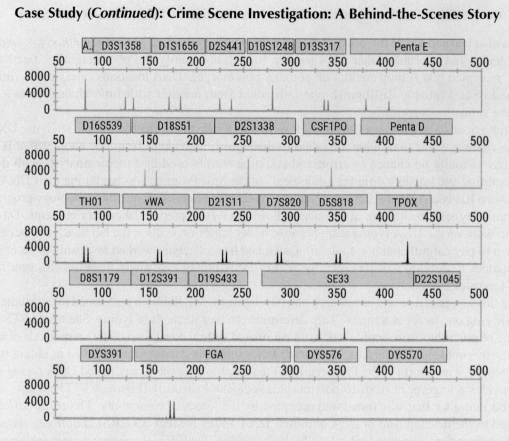

FIGURE 3.3 Example of a DNA fingerprint using STRs. This is an STR fingerprint of an individual. Each of the colored peaks represents one site on this person's chromosomes where there is a certain repeated DNA sequence. Twenty-seven different STRs were tested to produce this fingerprint. The lengths of each STR can be determined from this picture and assembled to create a "fingerprint" for this individual. The probability that two people selected at random will have identical fingerprints using this system is 2.3×10^{-32}; in other words, *extremely* unlikely. (Image reproduced with permission from Promega Corporation.)

3.1.5 PRODUCTION

If a product successfully makes it through the development phase, then it enters the production phase and is produced, marketed, sold, and used. In most biotechnology companies, a distinct production unit is responsible for these tasks.

The systems used for production in biotechnology companies are diverse. Manufacturing a product may involve growing bacteria in laboratory-sized flasks and isolating product from the cultures. In other situations, manufacturing involves growing bacteria or other types of cells in fermenters or bioreactors that can be several stories tall and using industrial-scale equipment to purify products from the cultures (Figure 3.4). Production may involve isolating biological molecules, growing plant cells in plastic dishes, cultivating crops in a field, maintaining laboratory animals, or even keeping farm animals. The

details of production thus vary greatly from company to company. Certain functions, however, are generally the responsibility of the production team, regardless of the nature of the product and company (Table 3.2).

3.1.6 QUALITY CONTROL AND QUALITY ASSURANCE

As a product moves through its lifecycle from development and into production, the tasks of the quality-control (QC) and quality-assurance (QA) units mature along with it. These are functional units in a facility that are responsible for assuring the quality of the product. You may see the terms *quality control* and *quality assurance* used differently in various sources. Here, we define **quality control (QC)** *as the unit that is responsible for monitoring processes and performing laboratory testing.* A quality-control technician might, for example, test a product in the laboratory

FIGURE 3.4   Industrial bioreactors. The bioreactor in this photo contains a cell broth that produces proteins that are purified to make biopharmaceuticals. (Photo courtesy of the National Research Council of Canada.)

TABLE 3.2

The Responsibilities of the Production Unit in a Biotechnology Company

- *Making the product*
- *Working with large-scale equipment and/or large-volume reactions* (not applicable to all biotechnology companies or products)
- *Routine monitoring and control of the environment as required for the product* (e.g., maintaining the proper temperature or sterility requirements)
- *Working with computer-controlled instruments and equipment*
- *Routine cleaning, calibration, and maintenance of equipment*
- *Following written procedures and performing tasks associated with producing the product*
- *Monitoring processes associated with making the product*
- *Initiating corrective actions if problems arise*
- *Completing forms, entering information into a computer, labeling, maintaining logs, and producing other required documents.*

---

to make sure it has the proper attributes before it is released for sale to customers. The quality-control laboratory is discussed in more detail in the last section of this chapter.

Quality assurance (QA) *refers to all the activities, people, and systems that ensure the final quality of products.* A person who works in quality assurance might, for example, check the documentation (defined and discussed in Chapter 6) associated with a product to be certain that it is completed properly, stored in the right place, and accessible if needed. Another example of a quality-assurance task would be participating in the investigation and correction of a problem, as will be introduced

in Chapter 4. The goal of the quality-assurance team is to ensure that all the processes in the company come together to produce a quality product. The responsibilities of the quality units are summarized in Table 3.3.

3.1.7 REGULATORY AFFAIRS

When products are regulated by the government (e.g., as are biopharmaceuticals) a **regulatory affairs staff** is needed *to interpret the rules and guidelines of regulatory agencies and to ensure that the company complies with these requirements.* Regulatory affairs specialists work with their counterparts in

TABLE 3.3

The Responsibilities of the Quality-Control and Quality-Assurance Units in a Biotechnology Company

- *Developing and managing systems and processes that help ensure product quality*
- *Monitoring equipment, facilities, environment, personnel, and product*
- *Testing samples of the product and the materials that go into making the product to determine whether they are acceptable*
- *Comparing data to established standards*
- *Auditing records and operations*
- *Investigating and correcting problems; developing systems to avoid future problems*
- *Managing changes to procedures, documents, and processes*
- *Ensuring that all documents are accurate, complete, secure, and available when needed*
- *Deciding whether or not to approve each batch of product for release to consumers*
- *Reviewing customer complaints.*

the regulatory agencies (e.g., the Food and Drug Administration and the Environmental Protection Agency) to make sure that the data, documentation, and forms that the company produces are sufficient to support their product (e.g., pharmaceuticals and genetically modified crops). Regulatory affairs personnel must work with the quality-assurance and quality-control units to ensure that the company's quality systems are compliant with regulatory requirements.

3.1.8 The Lifecycle of a Company

Biotechnology companies usually undergo a maturation process that reflects the lifecycle of their products. **Start-up companies** *are those that, like their product(s), are at the beginning of their lifecycle.* The emphasis in a start-up company is on research and development, and nearly everyone in the company is likely to be working in this area. Start-up companies typically have only a few employees, and a single individual may play a number of roles, perhaps doing research and making business connections. Depending on the nature of the product, the same facilities may be used to perform research, development, and manufacturing of the first batches of product for sale.

The start-up of a company or the beginning of a new product's lifecycle is a time of excitement, change, and unpredictability. Research and development scientists often do not know which ideas will result in a product, which methods will work best, and how much success they will ultimately have in their endeavors. Creativity and a willingness to work with uncertainty characterize most investigators in R&D and in academic research laboratories.

If a product is successful, then the company must evolve to manufacture it. During development,

the company likely will expand, more individuals might be hired, and staff will start to specialize in certain functions. Mature companies have a separate production team that works in a relatively predictable environment.

As a company grows along with its product(s), a variety of job opportunities open (Figure 3.5) and new functional areas are established to support the development, production, marketing, and sales of the product(s) including:

- *Business development*, which identifies, explores from an economic perspective, analyzes, and brings to market new products or collaborations.
- *Marketing and sales*, which is responsible for interacting with customers.
- *Business divisions*, such as accounting and human relations, which keep the organization running.
- *Dispensing*, which puts products that are produced in bulk into individual containers for customer use.
- *Metrology*, which ensures that instruments (such as those used in laboratories and those used to monitor production conditions) operate properly.
- *Clinical research management* (for companies whose products are therapeutic agents), which monitors experiments with human subjects to ensure compliance with regulations, guidelines, and company procedures.

The culture of a successful company is likely to change as it makes the transition from being a small start-up with few employees who fill many roles, to becoming

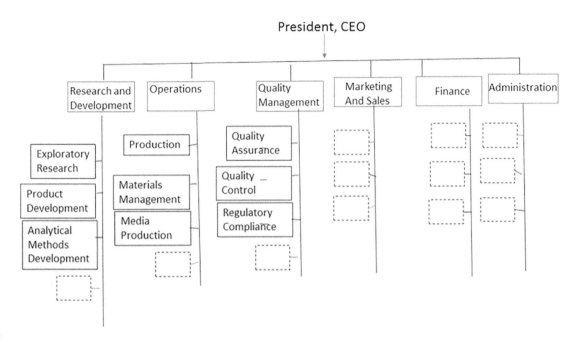

FIGURE 3.5 The organization of a mature biotechnology company. There are various ways to organize all functional units in a company; this is an example of an arrangement. This textbook is primarily concerned with the three areas of research and development, operations (which can encompass production and associated areas), and quality management. The dashed boxes indicate that there are other functional units in companies that are outside the scope of this book.

a larger company with specialized staff. Everyone in a small company knows everyone else; this is not true of larger companies. Communication becomes less direct as the company grows physically and in numbers of staff, so processes need to be established to ensure good communication. Small companies, if they have a promising product, are often purchased by larger ones. When a company is bought, its staff suddenly acquires new colleagues they have never met, who work in different facilities – perhaps on different continents. As a company grows, and is perhaps bought, decisions will be made by different people and individuals may have less input. Changes that affect individual workers are common as the products of a company move through their lifecycle.

3.2 INTELLECTUAL PROPERTY AND THE BIOTECHNOLOGY INDUSTRY

3.2.1 INTELLECTUAL PROPERTY AND PATENTS: OVERVIEW

The biotechnology industry is fueled by scientific discoveries that are transformed into valuable products. The economic value of a company therefore depends not only on its physical assets (e.g., buildings, laboratories, and production facilities), but also on its scientific knowledge. The knowledge base of a company,

individual, or other entity is a kind of property, called intellectual property. **Intellectual property (IP)** *embodies creations of the mind and intellect*. The success of most biotechnology organizations depends on their intellectual property. It is not enough, however, for an organization to have intellectual property – it must also protect that property. This section introduces the use of patents and trade secrets to protect IP.

IP includes such familiar items as literary and artistic creations, that are protected by copyright, and product branding, that are protected by trademarks. Intellectual property also encompasses *inventions*, which is the category of most interest to us as biotechnologists. Machines, such as digital cameras or computers, often come to mind when we think of inventions, but the term *invention* can broadly refer to many types of innovations. New therapeutic compounds, diagnostic kits, genetically modified bacteria, methods to genetically transform bacteria, and industrial enzymes are examples of biotechnology *inventions*.

Intellectual property is different from more tangible forms of property, like a car or a piece of land. One of the differences is that it is relatively easy to steal IP; one need only copy it. A person might expend considerable efforts devising something new only to have another person copy and profit from it. Governments therefore provide protections for creators of intellectual property. These protections include patents, trade secrets, and copyrights. A **patent** *is a type of intellectual property protection that is an agreement between the government, represented by the Patent Office, and an "inventor" whereby the government gives the inventor the right to exclude others from using an "invention" without permission from the inventor.* In exchange for this right, the inventor must fully disclose the invention to the public. Governments protect intellectual property not only to reward inventors with a means to profit from their work, but also to promote innovation. To obtain a patent, an inventor must, in effect, teach everyone to make and use the invention. In this way, new innovations spread quickly.

Useful chemicals, enzymes, and drugs are considered inventions for patent purposes. Processes or methods of accomplishing a task can also be patented as inventions. A process to manufacture a biopharmaceutical agent, a method to manufacture an industrial enzyme, a method for DNA identity testing, a method to insert genetic information into mammalian cells, a method of purifying a biological compound, and a method of administering a therapeutic compound are all examples of inventions that are processes. Manufactured items that are parts of other devices, and improvements of previous inventions, are also inventions.

One cannot patent a human being, a naturally occurring organism, a law of nature, natural or physical phenomena, or abstract ideas. It is not possible, for example, to patent a mathematical theorem, nor the discovery of the structure of DNA. One cannot patent an idea or discovery alone. It is the practical application of an idea or a discovery in order to solve a problem or supply a need that makes it patentable.

In the United States, patents are granted by the U.S. Patent and Trademark Office (USPTO). Other countries have similar offices. Patents only apply in the country in which they are granted. It is therefore necessary to receive a patent in every country where one wants protection.

The term of a new patent is commonly 20 years from the date on which the application for the patent was filed. Anyone can use the invention when the patent expires. Generic drugs, for example, are compounds that were originally patented by a pharmaceutical company, but their patent protection has expired.

If you own property such as land or a car, you have the right to use or sell it. A patent is different. A patent grants the inventor the right to *stop others* from selling, using, or exporting the invention. A patent can, however, be *licensed*. A **license** *is an agreement by the patent holder that it will not enforce the right of exclusion against the licensee (the party wishing to use the patented invention).* The licensee pays the patent holder fees and/or a share of revenues. There are many situations where a company wants to manufacture a product that is partly protected by one or more other patents. The company may then license the technologies it needs in order to develop and produce its own product.

If one party holds a patent and another party uses that invention without permission, it is called **infringement**. In this case, the patent holder will normally send the other party a letter telling them to stop. The patent holder may offer the offending party a license.

There are cases where two parties disagree as to whether one is infringing on the other's patent, in which case the dispute may be decided in federal civil court. If the court agrees with the patent holder, then the court might compel the other party to pay damages to the patent holder and may order the other party to stop making, using, selling, or importing the invention. This situation is not uncommon in the biotechnology industry, and the economic stakes are often high in these disputes.

3.2.2 INVENTION AND BIOTECHNOLOGY

The Boyer–Cohen patent, awarded to Stanford University in 1980, was a landmark in the history of modern biotechnology. This patent covered the discovery by Stanley N. Cohen and Herbert W. Boyer of a method to transform one organism with the DNA from another organism (see Chapter 1 for an explanation). Stanford required that any commercial party (business) wanting to use this method of genetic transformation had to obtain a license and pay fees, plus a royalty on the sales of any product developed using the method. Stanford allowed multiple businesses to license the Boyer–Cohen invention, and so the method of "genetic engineering" quickly spread around the world.

Biotechnology involves living things and so leads to the issue of whether it is possible to patent an organism. This is not a new question, although it remains a topic of much discussion. In 1873 Louis Pasteur patented a yeast "free from... germs or disease." More recently, Ananda Chakrabarty genetically modified a bacterium so that it could degrade crude oil. His attempt to patent this bacterium was originally turned down by a patent examiner, who argued that the law dictated that living things are not patentable. Ultimately however, in 1980, the Supreme Court, in a close 5 to 4 decision, ruled that this bacterium was patentable because it was not found in nature. The Supreme Court's interpretation of US patent law opened the door to patents for other genetically modified organisms and cells including plants, non-human animals, hybridoma cells, viruses, cell lines, and embryonic stem cells (Figure 3.6).

FIGURE 3.6 The first patented transgenic animal. The OncoMouse, a genetically altered mouse that gets cancer very easily, was developed by researchers at Harvard University to serve as a model organism in cancer research. In 1988, the OncoMouse became the first transgenic animal to be patented in the United States. The patenting of an animal raises ethical and societal concerns, and OncoMouse patents have repeatedly been challenged by activists, particularly in Europe and Canada. (Photo courtesy of the Science Museum/Science & Society Picture Library.)

Humans cannot be patented, but rulings allow the patent of products isolated from humans, for example, proteins and cell lines. A scientist, for example, who purifies a type of interferon from blood can patent that molecule. Interferon existing in a person, however, is not patentable.

3.2.3 GETTING A PATENT

Intellectual property is the foundation for biotechnology businesses. It is therefore essential that biotechnologists, particularly those who work in a research or development environment, are familiar with the requirements for obtaining a patent so that their innovations can be protected.

Obtaining a patent requires submission of a patent application and fees (that can run from hundreds to thousands of dollars). If an employee invents something in the course of their work for an employer, then the employer (e.g., a biotechnology company or university) will generally submit the application and will own that patent if it is awarded. However, inventors have the right to be recognized for their inventions, so even if the patent is owned by someone else, the inventor's name(s) is required on the patent application. If an employee creates an invention outside of their scope of work for the company, then the employee would normally submit a patent application on her or his own behalf.

The patent application must describe the innovation in sufficient detail that another skilled person could use the invention described using the information in the patent application and existing knowledge in the field. Patent applications submitted in the United States must contain one or more *claims*. The invention is described in the claims, and the scope of protection requested by the applicant is defined by the claims. Like a deed to land establishes the legal boundaries of physical property, the claims establish the legal boundaries of the intellectual property of the patent.

A patent applicant must sign an oath claiming inventorship and must prove the following in the patent application:

- *the innovation is new*
- *the invention is not obvious*
- *the invention is useful*.

To prove that an invention is new and that it is not obvious, the invention must not have been previously described or discovered by another before the patent application was filed. The innovation must not be obvious from other people's publicly known work. For a protein, for example, this usually means the inventor must learn the chemical structure of the protein. If the structure was already known, then this requirement cannot be met. This requirement means that it is often important to conduct a patent search to see what other people have already patented. It is now possible to search online, full-text patent databases that make it vastly more convenient to conduct a patent search than was possible when only paper copies were available.

The patent application must clearly specify the invention's potential usefulness. A newly discovered protein, for example, might be useful as a drug to treat a particular illness or might have application in an industrial process.

There are situations where more than one party applies for a patent on the same invention. Only one party will be awarded a patent on this invention. Patent laws assign the patent to the party that submits a patent application first. In the past, this was not true in the United States, where the date of invention was used to establish priority. However, in 2013, the United States switched to the "first-to-file" system that is used in other parts of the world.

Before one can file a patent application, two steps must be completed. First is *conception*, and the other is *reduction to practice*. **Conception** *is defined as the formation, in the mind of the inventor, of the complete invention (as defined in the patent claim)*. Conception requires that the inventor knows how to make the invention and how to use it in a practical way. If an invention, for example, is a new compound isolated from blood, then the inventor often must know its structure, how to isolate or synthesize it, and how to use it for a practical purpose. Conception occurs in the mind of the inventor. **Reduction to practice** *is constructing a prototype of the invention or performing a method or process* (as described in the patent claim). Reduction to practice requires establishing that the invention works for its intended purpose. If the invention can be described in writing in a way that allows those in the field to make and use the invention, no physical prototype may be required and the writings in the patent application provide the reduction to practice (called "constructive reduction to practice"). If, on the other hand, the nature of the invention is unpredictable and requires experimental testing to demonstrate that it works, a physical prototype is required (called "actual reduction to practice").

3.2.4 Trade Secrets

Science advances due to the open exchange of knowledge; publication of scientific discoveries is critical to the success of academic research scientists. The patent system similarly promotes exchange of information and knowledge. There are, however, situations where biotechnology companies do not want to share information or materials so they can protect their competitive advantage. **Trade secrets** *are private information or physical materials that give a competitive advantage to the owner.* To qualify as a trade secret, information must be valuable, it must be secret, and it must give the holder a competitive advantage. The formula for making Coca Cola is a commonly cited example of a trade secret. A biotechnology trade secret might be, for example, a cell line to manufacture a product, or information about how to grow a valuable cell line. Trade secrets are not registered with the government, as are patents. Rather they are actively protected (e.g., with encryption, locks on file cabinets, and limited access to the facility).

One of the most important ways to protect trade secrets is by the use of nondisclosure (confidentiality) and non-compete agreements. A **nondisclosure agreement** *is a contract in which the parties promise to protect the confidentiality of secret information that is disclosed during employment or another type of business transaction.* A **non-compete agreement** *is a contract that the party will not leave the company and compete with the business for a fixed period of time.* A person who violates agreements can be taken to court and sued for damages. It is standard practice for employers to require employees, interns, consultants,

vendors, and anyone else involved in a company to sign a nondisclosure agreement. The knowledge of a biotechnology company is critical to its success. It is essential that employees respect confidentiality agreements and not disclose secret information intentionally or through carelessness.

3.2.5 Patent Issues

The use of patents to protect inventions is generally viewed as being good both for individuals and for societies because the system protects inventors and promotes innovation. The system, however, provides many opportunities for discussion and disagreement.

It is not uncommon for two parties to disagree as to whether an innovation is new, or whether it is obvious. One party may have a patent that it believes covers an invention that another party is claiming is new. One party may believe that another party's patent claims are too broad and therefore stifle their innovations. These sorts of disputes often must be resolved by the court system and can be very expensive for both parties. Winning or losing these court battles can have a profound effect on a company's fortunes and future.

Society as a whole has a stake in patent disputes because society wants patents to be broad enough to reward innovation, but not so broad as to prevent others from building on previous work. It is often difficult for a patent examiner to have sufficient knowledge in the complex, technical arena of biotechnology to evaluate whether a patent's claims are too broad.

Some people believe that it is unethical to patent a living organism, a gene, or a protein and hence object to many biotechnology patents. There is also the concern that patenting of genes prevents the use and development of important diagnostic tools and treatments if the patent holder chooses not to develop them or prices them out of the reach of the needy; see the case study "Breast Cancer Testing Goes to the Supreme Court."

Some people object to patents on plants because food is an absolute necessity for life, and patents might cause certain foods to be too expensive for the poor. In the past, traditional breeding of varieties of crop staples that are hardy and high-producing was largely the concern of researchers in public, government institutions and universities. The results of their work were freely available. Biotechnology, however, has introduced new commercial interests to agriculture.

These are only a few of the issues relating to IP, a complex, sometimes contentious area that is of vital importance to the biotechnology industry. Every biotechnologist should know something about intellectual property, and many find it provides a stimulating career.

Case Study: Breast Cancer Genetic Testing Goes to the Supreme Court

Biotechnology generates numerous, diverse products that provide wide-ranging benefits to society. These products also provide major financial benefits to some companies and individuals. Biotechnology is a high-stakes industry that, not surprisingly, has created perplexing societal conflicts, many of which play out in the arena of patent law. A particularly important patent case involving the biotechnology company, Myriad Genetics, found its way to the United States Supreme Court.

The subject of this patent case was a pair of genes called BRCA1 and BRCA2. In 1990, researchers at the University of California, Berkeley, announced that they had located a gene on chromosome 17, called BRCA1, that, when mutated, dramatically increases a woman's chance of contracting breast or ovarian cancer. The average woman has a 12%–13% chance of contracting breast cancer. That chance can rise as high as 80% if a woman inherits certain mutations of the BRCA genes. Shortly after the discovery of BRCA was announced, a small biotechnology company, Myriad Genetics, was created with backing from the pharmaceutical giant, Eli Lilly. Myriad sequenced BRCA and, in 1994, obtained patents covering the sequenced gene, more than 40 mutations of the gene, and numerous diagnostic tests and methods for identifying mutations of the gene. Myriad then launched an aggressive campaign to benefit from its patent. They sent letters to researchers whose work involved isolating BRCA genes telling them to cease work. They filed patent infringement suits against parties engaging in BRCA testing. Many organizations, patients, patient advocacy groups, and healthcare workers believed that Myriad was endangering patient health by making BRCA testing unduly expensive (Myriad's list price for this test was $3,340) and not widely available. Eventually, several organizations, including the American Civil Liberties Union (ACLU), filed legal claims against Myriad arguing that their BRCA patents were invalid because DNA segments are not separate from nature. The case worked its way through several lower courts and eventually made its way up to the United States Supreme Court. The question, simply stated, was whether or not human genes are patentable, a question that is of great interest to the biotechnology community. In 2013 the Supreme Court unanimously decided that human genes are not patentable. The Supreme Court decision was a cause for celebration among many groups, and various organizations quickly stepped up to provide BRCA testing. At the same time, analysts in the biotechnology community warned that limiting patent protection for biotechnology products will undermine investor confidence in biotechnology and ultimately lead to fewer biotechnology products and fewer societal benefits. Today, years later, there are still questions about what biotechnology products are or are not patentable, and also about how patent law does or does not serve the greater good of society.

In the 1950s, a young school child, Mary-Claire, watched baseball games with her father who used the opportunity to teach her how to calculate batting averages. With this early introduction to math, Mary-Claire went on to study statistics in college, but after taking a genetics course, found her passion in biology. During the turbulent 1960s, Mary-Claire was involved in advocating for civil rights, opposing the Viet Nam War, and aiding the consumer rights activist, Ralph Nader. Eventually, she obtained a Ph.D., and after a stint working in Chile, she was hired as a professor at the University of California, Berkeley.

One of Dr. King's most important scientific contributions was her work in cancer. Into the 1970s and 1980s, cancer was not understood to be a genetic disease. Cancer did sometimes seem to run in families, but no particular genes had been identified that caused cancer. When Dr. King and her colleagues announced the discovery of the BRCA1 gene in 1990, it became clear that cancer is, indeed, a genetic disease. It was Dr. King's insights in math and genetics that enabled her to lead the research team that discovered the BRCA genes.

In 2013, when the Myriad Supreme Court decision was announced, Dr. Mary-Claire King was not much in the news. But, in an interview with Ushma Neill, Dr. King reports:

Nina Totenberg [correspondent for National Public Radio] called after the decision. She said, "The Supreme Court decision has just come out: it's 9 to 0." I said, "Which way?" She said, "Oh, in favor of the ACLU position." I said, "Wow." She said, "Surely you have a longer comment than that." I said, "I'm as high as the flag on the fourth of July." She said, "That'll do."

(Continued)

Case Study (*Continued*): Breast Cancer Genetic Testing Goes to the Supreme Court

Primary Sources:

Wales, Michele, and Eddie Cartier. "The Impact of Myriad on the Future Development and Commercialization of DNA-Based Therapies and Diagnostics." *Cold Spring Harbor Perspectives in Medicine*, 2015, p. a020925. doi:10.1101/cshperspect.a020925.

Neill, Ushma S. "A Conversation with Mary-Claire King." *Journal of Clinical Investigation*, vol. 129, no. 1, 2019, pp. 1–3. doi:10.1172/jci126050.

"Supreme Court to Myriad Genetics: Synthetic DNA Is Patentable but Isolated Genes Are Not." *AMA Journal of Ethics*, vol. 17, no. 9, 2015, pp. 849–53. doi:10.1001/journalofethics.2015.17.9.hlaw1-1509.

3.3 THE MANY ROLES OF THE LABORATORY IN THE BIOTECHNOLOGY INDUSTRY

3.3.1 WHAT IS A LABORATORY?

We have talked about the lifecycle of products and how the organization of biotechnology companies reflects those lifecycles. Laboratory scientists, technicians, and analysts play key roles at all stages of a biotechnology product's lifecycle, from its discovery, through its development, and into the production stage. This section explores the laboratory in more detail.

Research laboratories, which have existed for hundreds of years, *are spaces set aside to study the complexities of nature in a controlled manner.* Observations can be made, and experiments can be performed in the laboratory in which the researcher controls the factors of interest. For example, if researchers are interested in the effect of light on plant growth, they can carefully control the light that plants receive in a laboratory. Outside the laboratory, the researcher has little control over light exposure or many other important factors – such as rain, temperature, and insects – that may affect the plants' growth. The research laboratory is the site of discoveries that root biotechnology; without research, modern biotechnology would not exist.

Many research laboratories are located in academic institutions and research institutes. Biological research is also conducted in medical centers. Some biotechnology researchers work in laboratories associated with biotechnology companies. In all of these settings, the tasks of biological research, that is, making observations and performing experiments, are similar. The purpose, however, of research in a medical center or a company is usually to find and develop practical applications of knowledge, whereas university research

may be "basic" – that is, the pursuit of knowledge for its own sake.

There is another important category of laboratory that can be distinguished from the research laboratory. This is the **testing laboratory**, *a place where analysts test samples.* The product of a testing laboratory is a test result, such as a measurement of the blood glucose level in a sample, a DNA "fingerprint," or a report on pollutants in a lake. Clinical laboratories are a familiar type of testing lab where blood and other samples from patients are tested. Forensics laboratories are testing labs where samples from crime scenes are tested. Samples from the environment are tested in environmental laboratories. A quality-control laboratory in a company is a type of testing laboratory where samples of products and raw materials are tested.

What then, makes a place a laboratory? We can say that a **laboratory** *is a workplace whose product is data, information, or knowledge.* This is a reasonable definition with a couple of caveats. First, people in laboratories do produce tangible items such as photographs, antibodies, purified proteins, and printouts. These materials, however, are produced with the purpose of learning more about a system or a sample, answering a research question, or documenting what has been discovered. The tangible materials that emerge from a laboratory are not produced for commercial sale. Another caveat is that biotechnology companies often produce small amounts of products for sale in facilities that are also used for research and development, or that are similar to research laboratories. However, we consider a laboratory-like facility used for producing a commercial product to actually be a small-scale production facility, rather than a laboratory.

There is one more type of laboratory that is perhaps the most familiar one – the teaching laboratory. A **teaching laboratory** *is a space set aside in which students learn about nature.* Often students learn things that other people (such as their teachers) already know. Students also practice using the methods that research scientists use when performing experiments. The product of a teaching laboratory is knowledgeable students.

3.3.2 LABORATORIES AND THE LIFECYCLE OF A BIOTECHNOLOGY PRODUCT

3.3.2.1 Research and Development

The discoveries that lead to biotechnology products occur in research laboratories. The efforts of many scientists from many laboratories often interconnect in the discovery of a single product. Chapter 1 discussed, for example, how the scientific discoveries of various scientists over the years led to genetic engineering and how research into the mechanisms of cancer led to the drug Herceptin. Another example is provided in the Gleevec case study in Section 36.2.2.

Development, the transformation of discoveries into products and applications, is primarily a laboratory function. For example, development of a new product made by genetically modified cells might involve experimenting with different host cells to see which is best, optimizing their culture medium to maximize protein expression, and developing optimized purification methods. All these tasks require scientifically skilled laboratory personnel.

3.3.2.2 Quality Control

The quality-control laboratory is a type of testing laboratory that is essential in helping ensure the quality of products through development and during production. The quality-control tests performed in biotechnology companies are often sophisticated and diverse. Quality-control analysts perform:

- *environmental monitoring*
- *tests of raw materials*
- *tests of in-process samples (samples obtained at intermediate stages of production)*
- *tests of final product.*

Environmental monitoring in this context refers to the air, water, surfaces, and equipment in a facility. QC technicians might, for example, take swabs from a piece of production equipment after it has been cleaned to be certain that no remnants of the last product run remain.

Incoming raw materials are evaluated by quality-control analysts to be sure they meet the requirements of production. In a pharmaceutical company,

regulations require that every incoming raw material be quarantined and tested to assure its quality before it is released to production. This means that even if the manufacturer of the raw material has tested the material and provided documentation to that effect, the pharmaceutical company still needs to confirm, often with laboratory testing, that the raw material is acceptable. In a company that does not make pharmaceuticals, the requirements for testing raw materials may be less stringent, but a process still needs to exist to confirm that raw materials are acceptable.

Quality-control analysts may perform *in-process* testing, tests done while a product is in the process of being made, to be sure everything is proceeding normally. For example, analysts might check the purity of a material at an intermediate stage in a purification process.

Quality-control analysts run a series of tests on samples from each batch of final product in order to see if that batch is good enough to be released for sale. They compare the results of the tests to **specifications** that are *numerical limits, ranges, or other results that the tests must meet if the product is good.* For example, if the product is a chemical entity, QC analysts might perform a battery of chemical tests to confirm that the batch contains the right compound, and that it is pure.

For pharmaceutical products, final product testing is tightly controlled, and QC laboratory personnel have a key role in ensuring the quality of the final product. Regulations require that *samples from every batch manufactured for clinical trials or for sale are tested in a variety of ways to be certain that they meet all specifications for the product;* this is called **lot release testing**. The drug will only be released if it meets all the specifications; if it does not, the drug is rejected. The criteria on which to base acceptance or rejection of a drug product are established and justified based on data obtained from material used during development of the product. The R&D unit must optimize every test and prove that all the assays used in QC are effective and contribute to evaluating the quality of the product.

The various tests that are used for testing a product are sometimes categorized as follows.

1. **Tests of general characteristics**. Examples of such tests include evaluating appearance, color, and clarity; and measuring pH, particulate, and moisture content.
2. **Tests of identity**. In this context, **identity** refers to *whether a particular substance or*

substances *(e.g., the active ingredient in a drug) is present*. For drug products, identity tests should be highly specific for the drug substance and should be based on unique aspects of its molecular structure or other specific properties. (*Identity testing* may also refer to the identification of individual people or other organisms. In the present context, identity testing means to identify a particular molecular entity.)

3. **Tests of purity**. *Purity is the relative absence of undesired, extraneous matter in a product*. For drug products, the absence of contaminants is of utmost importance. There are various methods to test for purity or impurity. Some tests detect specific contaminants (e.g., the HIV virus), and other tests detect classes of contaminants (e.g., bacteria). In the pharmaceutical industry, a battery of different tests for many potential contaminants is required.

4. **Quantitation/concentration tests**. These assays test the amount or concentration of a substance. For example, many biotechnology products are proteins. The amount of protein in a product, or at an intermediate stage of processing, is frequently determined.

5. **Potency/activity tests**. *Potency or activity is the specific ability of the product to produce a desired result*. For a drug, potency refers to the drug's ability to have its desired therapeutic effect on a human or another animal. The same principle applies to many other products; activity refers to the product's ability to perform as desired. For example, a restriction enzyme is intended to cut DNA at specific sites. The activity of a restriction enzyme therefore relates to how much DNA it can specifically cut in a set amount of time.

Example 3.1

A company develops a recombinant DNA production system in the bacterium, *Escherichia coli*, to produce a therapeutic protein. They establish final product specifications and assay methods to ensure that the product meets the specifications. Before a drug lot can be released for sale, it must be tested to be sure that it meets all these specifications. A number of tests would likely be performed and might include those shown in Table 3.4.

3.4 THEMES

Unit I discusses how the biotechnology industry transforms scientific knowledge and discovery into a wide variety of useful products. Chapters 1 and 2 provide a broad overview of the science and the products of biotechnology. We saw that these products are numerous and include items as diverse as biopharmaceuticals and genetically modified salmon. Chapter 3 introduces the business side of biotechnology, including the organization of companies and the role of intellectual property in biotechnology.

One of the themes that recurs in this unit is the importance of the laboratory and laboratory personnel in all aspects of biotechnology. Laboratory scientists, technicians, and analysts perform experiments and make discoveries that are the basis for biotechnology. Laboratory personnel transform knowledge from the research laboratory into effective products and perform a wide variety of important laboratory analyses.

Another theme that emerges is that biotechnology fosters interconnections. Research discoveries are

---

TABLE 3.4

Release Specifications for Protein Therapeutic XYZ

Characteristic	Specification	Assay Method
Appearance	Clear, Colorless Solution	Visual Inspection
E. coli DNA	< 0.01 μg/μg product	PCR
E. coli protein	Undetectable	Specific immunoassays
E. coli RNA	Undetectable	Agarose gel electrophoresis
Residual ethanol	< 250 ppm	Gas chromatography
Endotoxin	< 0.1 EU/μg	LAL assay
Sterility	No growth in 14 days	Assay described in U.S. Pharmacopeia
Retrovirus	Undetectable	Infectivity assay

almost always the result of the efforts of many people who are connected through publications and personal communications. The staff of multiple research laboratories and companies often share their expertise as new ideas are brought through development, testing, and into production. Biotechnology companies are further interconnected with one another in business partnerships. Many biotechnology products take advantage of sophisticated scientific processes that involve patents from multiple parties leading to business partnerships. Moreover, the development and production of many products is so complex that companies must partner with one another to be successful.

A third theme that is introduced in this unit is that of quality. Every biotechnology product must be of high quality; that is, it must be suitable for its intended use. The word "quality" occurs repeatedly in the world of biotechnology. There is a quality-control/quality-assurance unit in almost every company. Regulatory authorities, such as the Food and Drug Administration in the United States, are charged with monitoring the quality of medical products. Scientists strive to produce quality results. This theme will be explored in detail throughout this text because understanding quality is essential for a successful career in biotechnology.

Practice Problems

1. A new, effective anticancer compound is discovered. The compound is unfortunately found in the stems of a rare, slow-growing plant; therefore, very little of the compound is available. A small start-up biotechnology company is formed by a team of research scientists who plan to isolate the gene for this anticancer agent. After several years of dedicated, difficult work, they obtain the gene that codes for the anticancer agent. They are then able to insert the gene into bacteria and are elated to find that the bacteria make the anticancer agent. They develop methods to grow and harvest large quantities of these bacteria and to isolate the product from them. At this point, the company is purchased (for a lot of money) by a pharmaceutical company that takes over the tasks of testing the compound in animals and humans. Eventually, the drug is approved and is sold for patients. Over a number of years, the following tasks were performed by staff in the biotechnology and pharmaceutical companies. Label each task as being primarily the job of research and development personnel, production personnel, quality-control technicians, or quality-assurance personnel. The tasks are not necessarily listed in the order in which they would be performed. Note also that there is some overlap in tasks so that in some cases more than one answer may be correct, depending on the organization of a company.
    a. Identify the gene that codes for the anticancer agent.
    b. Design and develop a new assay to check whether the anticancer compound is present in a sample at a certain level.
    c. Devise methods to grow large amounts of the bacteria in such a way that the compound is consistently produced.
    d. Develop a system to keep track of all documents.
    e. Sterilize the equipment used to produce the drug product.
    f. Devise a method to insert the gene of interest into bacterial cells so that the cells make the desired anticancer agent.
    g. Perform laboratory tests of the anticancer agent to see that it meets its specifications before it is released for sale to patients.
    h. Design purification methods to purify the compound from bacteria.
    i. Monitor pH and temperature levels in the fermenters during production runs.
    j. Purify the anticancer product that is to be used by patients.
    k. Determine the chemical composition of the anticancer agent.
    l. Review all documents associated with a batch of product to be sure they are correct and complete.
    m. Perform laboratory tests of incoming raw materials to be sure they are suitable for use.
2. Suppose you are beginning a small start-up biotechnology company. You have a strain of bacteria that is good at degrading certain industrial by-products, and you envision that your bacteria can be used to remediate (clean) contaminated soil. You need to rent a space for your company. What features would you be looking for in your first space? There are many answers to this; you can be creative.

3. An excerpt from a job posting is shown below.
   a. Would you classify this job as primarily involving R&D, production, or QC/QA?
   b. This company says that it is "dedicated to helping find a cure for diabetes." Based on this job posting, what is the company's role in helping find a cure?
   c. Does the company appear to be involved in performing clinical trials of stem cells to treat diabetic patients?

   Job Posting

   Regenerative Medical Solutions (RMS) is dedicated to helping find a cure for diabetes. RMS' extensive research has fostered human pluripotent stem cell derived pancreatic lineage cells for use in drug discovery, toxicity testing, assay models and therapeutic solutions.

   Job Purpose

   The Research Specialist will provide technical services including performing human induced pluripotent stem cell (iPSC) culture, various assays, and other related responsibilities.

   Primary Responsibilities

   - Thaw, expand, maintain, and cryopreserve human pluripotent stem cells and differentiated derivatives.
   - Under some supervision, conduct independent research projects involving human pluripotent stem cells and related materials.
   - Maintain accurate records of experiments and analysis.
   - Carry out quality assurance and quality control programs on cell lines and related materials and equipment.
   - Carry out laboratory maintenance, record-keeping, equipment maintenance, and order supplies.
   - Follow all safety guidelines, facility use guidelines, standard operating procedures, and other relevant regulatory requirements.

4. Using *Google Patent Search*, look up the following three patent applications:

   EP3472352A1
   US5972346A
   US9284371B2.

   Answer the following questions:
   a. Who is/are the inventor(s) on each of these three patent applications?
   b. Who is the assignee? What is the difference between an assignee and inventor?
   c. Based on the abstracts for each patent (found near the top of the application) what is the goal of each invention? You may want to also look at the Descriptions and Claims to better understand each invention.
   d. How does the invention described in EP3472352A1 relate to Figure 2.17?
   e. Patent US9284371B2 relates to the product adalimumab. What is this product? What does it do? (Note that this invention relates to improving the methods of mammalian cell culture. Concepts outlined in this patent application are discussed in this textbook in Chapters 30 and 31.)
   f. Patent US5972346A relates to a hepatitis B vaccine. How does a vaccine like this one work in the body? How might this concept be applied to make a vaccine to prevent COVID-19, the disease caused by the virus SARS-CoV-2?

UNIT II

Introduction to Quality in Biotechnology Workplaces

Chapters in This Unit
◆ Chapter 4: An Overview of Quality Principles in Biotechnology
◆ Chapter 5: Quality in Research Laboratories
◆ Chapter 6: Documentation: The Foundation of Quality

As we saw in Unit I, biotechnology begins with knowledge about the natural world that emerges from the work of people in biological research settings. That knowledge is then transformed into tangible products, such as drugs, transgenic livestock, modified plants, and enzymes. A biotechnology product can also be a test result, such as the sequence of a gene, or a measure of the purity of a product. Biotechnology products are thus diverse, but one thing is true of all of them – they must be of high quality. Producing a quality product does not happen by accident – quite the contrary, accidents usually result in an inferior product.

Quality is equally important in research laboratories and in production settings, but as we will see in this unit, there are differences in how quality is achieved and, in fact, how quality is understood. Research, by definition, values experimentation and requires ongoing change to create new knowledge. As we will see in Chapter 5, in a research setting, quality is associated with results that are reproducible by the investigator and by others. In a production setting, quality is associated with reduced product variability and with controlling change so that consistency is maintained. To achieve quality in a production setting, quality systems are implemented that are broad in scope, and rely on coordinated efforts of all the individuals in the organization.

DOI: 10.1201/9780429282799-5

Case Study: Achieving Quality Requires Attention

A laboratory technician was running enzyme assays (tests of how an enzyme catalyzes a reaction) on cell samples in one of the authors' research programs. It is always customary to include a negative control, a sample that contains no enzyme, to determine the background levels of enzyme product that might be present in the samples. This background, which should be low, can then be subtracted from the enzyme levels measured in the treated samples. The technician had performed these assays several times a week for several weeks, so he was quite familiar with the procedure and typical results. One day, the background level for the negative controls in the assay was five times higher than the typical value. It could have been tempting, given time constraints, to follow the procedure and simply subtract the higher background levels from the sample levels and move on to the next task. However, the laboratory technician was attentive enough to question the result, reasoning that the control values should not have been so high. Quality results, in this case, required repeating the experiment with freshly made reagents. When the assay was repeated, the control values were again at typical levels and the sample data were considered to be valid. Because of this occurrence, the technician began to prepare fresh reagents on a more frequent schedule than he had previously. The technician's care in performing this assay was important in achieving trustworthy experimental results.

We will note here that in a company that complies with government regulations and/or with quality standards, this same situation (unexpected assay results) would have triggered a specific investigatory process. The investigation would have involved not only the technician, but also other team members. Once the problem was identified and fresh reagents had been prepared, the team would establish a formal system to avoid future problems with degraded reagents.

This unit is a broad introduction to the concepts of quality, the means by which people ensure that the products of their work are of good quality. The consistent production of quality products requires resources, planning, and commitment. This unit explores how a company, organization, or laboratory translates a commitment to quality into practice. The idea of quality is a major theme throughout this textbook.

Chapter 4 is a broad overview of the issues and vocabulary relating to product quality systems.

Chapter 5 introduces the idea of "reproducibility," a key issue relating to quality in research laboratories.

Chapter 6 discusses documentation, one of the most important aspects of any quality system.

BIBLIOGRAPHY FOR UNIT II

There is a vast, fluid literature in the area of product quality, and the easiest way to enter the literature is through the Internet. Some useful sites are included in these references. Specific quotes and article references are directly cited in the unit text. For more references relating to quality and regulatory affairs, see the Bibliography in the Introduction to Unit X.

BOOK

DeSain, Carol, and Sutton, Charmaine Vercimak. *Documentation Practices.* Advantstar Communications, 1996. (Despite being a somewhat older resource, this remains a good overview of documentation principles with many useful examples.)

ARTICLES

Baker, Monya. "Reproducibility Crisis: Blame It on the Antibodies." *Nature*, vol. 521, no. 7552, 2015, pp. 274–76. doi:10.1038/521274a.

Freedman, Leonard P., et al. "The Economics of Reproducibility in Preclinical Research." *PLoS Biology*, vol. 13, no. 6, 2015, p. e1002165. doi:10.1371/journal.pbio.1002165.

Huber, L. "Implementing 21 CFR Part 11 in Analytical Laboratories: Part 1, Overview and Requirements." *BioPharm International,* November 1999, pp. 28–34.

Huber, L., and Winter, W. "Implementing 21 CFR Part 11 in Analytical Laboratories: Part 4, Data Migration and Long-Term Archiving for Ready Retrieval." *BioPharm International,* June 2000, pp. 58–64.

Ioannidis, John P. A. "Why Most Published Research Findings Are False." *PLoS Medicine*, vol. 2, no. 8, 2005, p. e124. doi:10.1371/journal.pmed.0020124.

Palovich, Tracy U. "Electronic Notebooks in the Post-America Invents Act World." *ACS Medicinal Chemistry Letters*, vol. 5, no. 12, 2014, pp. 1266–67. doi: 10.1021/ml500442p.

Prinz, Florian, et al. "Believe It or Not: How Much Can We Rely on Published Data on Potential Drug Targets?" *Nature Reviews Drug Discovery*, vol. 10, no. 9, 2011, p. 712. doi:10.1038/nrd3439–c1.

Sandle, Tim. "Important cGMP Considerations for Implementing Electronic Batch Records." *Pharmaceutical Online,* February 26, 2021.
https://www.outsourcedpharma.com/doc/important-cgmp-considerations-for-implementing-electronic-batch-records-0001

Vines, Timothy H., et al. "The Availability of Research Data Declines Rapidly with Article Age." *Current Biology*, vol. 24, no. 1, 2014, pp. 94–97. doi: 10.1016/j.cub.2013.11.014.

Winter, W., and Huber, L. "Implementing 21 CFR Part 11 in Analytical Laboratories: Part 2, Security Aspects for Systems and Applications." *BioPharm International,* January 2000, pp. 44–50.

Winter, W., and Huber, L. "Implementing 21 CFR Part 11 in Analytical Laboratories: Part 3, Ensuring Data Integrity in Electronic Records." *BioPharm International,* March 2000, pp. 45–9.

Winter, W., and Huber, L. "Implementing 21 CFR Part 11 in Analytical Laboratories: Part 5, The Importance of Instrument Control and Data Acquisition." *BioPharm International,* September 2000, pp. 52–6.

WEBSITES

ANAB. "ISO/IEC 17025 | Laboratory Accreditation Documents."*ANAB,*2020,anab.ansi.org/en/laboratory-accreditation/iso-iec-17025-docs.

Center for Drug Evaluation and Research. "Data Integrity and Compliance With Drug CGMP Questions and Answers Guidance for Industry." *U.S. Food and Drug Administration*, 13 December 2018, www.fda.gov/regulatory-information/search-fda-guidance-documents/data-integrity-and-compliance-drug-cgmp-questions-and-answers-guidance-industry.

Office of the Commissioner. "Part 11, Electronic Records; Electronic Signatures - Scope and Application." *U.S. Food and Drug Administration*, 24 August 2018, www.fda.gov/regulatory-information/search-fda-guidance-documents/part-11-electronic-records-electronic-signatures-scope-and-application.

"Principles on Conduct of Clinical Trials." *PhRMA*, 14 October 2020, www.phrma.org/en/Codes-and-guidelines/PhRMA-Principles-on-Conduct-of-Clinical-Trials. (The Pharmaceutical Research and Manufacturers of America (PhRMA) represents the country's leading pharmaceutical research and biotechnology companies. This is a reference to a handbook they distribute.)

4 An Overview of Quality Principles in Biotechnology

4.1 INTRODUCTION

4.1.1 WHAT IS QUALITY?

As consumers of products and services, we are all familiar with the concept of product quality. At the supermarket, we may find certain brands to be superior. Before purchasing a major appliance, we might consult a consumer guide to learn about the performance of various models. This unit introduces basic principles of quality, but not from the perspective of being a consumer, rather from the perspective of being the producer of a biotechnology product. Quality, and how it is achieved in biotechnology workplaces, is a theme that flows through and integrates this entire text.

There are many definitions of the word "quality," but for our purposes, we will consider a **quality product** *to be one that is suitable for its intended purpose.* For example, consider salt, NaCl. NaCl is a chemical that is commonly found on the dining table to be used as a flavor enhancer. NaCl is also often used in biotechnology laboratories for a variety of purposes, some of which will be discussed in Chapter 25. Yet another common application of salt is for melting winter's ice on roadways and sidewalks. All three applications require salt, but the features that make a quality product are different. Purity is of utmost importance in the laboratory. But we would not want to pay for salt of laboratory purity when applying tons of it to roadways. Quality road salt must have certain additives and must be relatively coarse, in contrast to laboratory salt. Table salt need not be as pure as laboratory salt, but it must not contain additives or contaminants that are harmful when ingested. We want all three products to be of good quality, but what constitutes good quality depends on the purpose of the product. Therefore, determining what constitutes quality for a given product requires understanding how the product will be used.

As we saw in Chapter 1, the products of biotechnology are varied. There are tangible products such as drugs, transgenic livestock, modified plants, and enzymes for research. There are laboratories whose products are test results, such as measurements of a product's activity, the sequence of a gene, or the level

DOI: 10.1201/9780429282799-6

of a substance in the blood. There are research laboratories whose product is knowledge. The features that constitute a "suitable," quality product are different for these varied products, as will be discussed more fully in this chapter and in Chapter 5, where quality in research laboratories is introduced.

4.1.2 INTRODUCTION TO QUALITY SYSTEMS

How is a quality product made? There is no single or simple answer to this question. Producing a quality product requires many coordinated elements such as skilled and knowledgeable personnel, a well-designed and managed facility, and ready access to raw materials. All of these coordinated elements together are called a **quality system**. A **quality system** *is the organizational structure, responsibilities, procedures, processes, and resources that together ensure the quality of a product or service.* A product quality system has the goal of ensuring a quality product. There are different quality systems for each company or institution. For example, one would expect a company that produces a pharmaceutical-grade product under stringent regulations to require a more encompassing quality system than a company that produces coffee mugs.

This definition of a quality system may seem vague, so let's think about quality for a familiar product. Suppose a team of entrepreneurs want to begin a company to manufacture and sell chocolate chip cookies. The team plans to use a beloved family recipe when making their cookies. Before the team bakes its first batch, there will be many steps they must complete to ensure a consistent quality cookie product.

First and foremost, in order to produce a quality product, the team will need to know what features their product must have. They may know that their grandmother's cookies were remarkable, but what features made them so good? Did her cookies have a particular flavor? How much can flavor vary from batch to batch and still be "good enough?" Are there requirements for cookie texture, stability (how long the cookies can be stored), or color? In this early planning phase, they will need to identify the features of their product that will guide their decisions about how to manufacture their cookies.

Many other tasks will follow. The team will need to develop methods that enable them to test for the features the cookies must have. For example, they must develop methods to test for flavor, texture, color, and whatever else they decide are the essential features of their product. The team will need to formalize a process to turn their family recipe into finished cookies.

The process will consist of steps including purchasing ingredients (raw materials), measuring out the ingredients, mixing, baking, cooling, and packaging. At some point, the team will need to obtain facilities and equipment, including a space with ovens, counters, mixers, and other physical items. For a moment, consider just one item they will need, that is, an oven. Before the team can decide what kind of oven to buy, they must determine the oven features they require. For example, they may determine that the baking temperature is critical in baking perfect cookies. If the temperature is too high, the cookies might be too crisp or brown. If the temperature is too low, the cookies may not fully bake or develop the proper color. If they decide that a consistent oven temperature is critical, certain models of oven might work better than others. They will also need to think about how often to test the temperature in the oven and what to do if the temperature varies from what it is supposed to be. The team will need to write documents to guide all the aspects of their operation and will need to develop a documentation system to track the flow of raw materials into products and out to customers. Once the cookie company is in operation, they will need procedures to ensure that their activities always produce quality products.

We could devote many pages to this cookie example, but we will stop here and summarize some key points about quality systems. First, quality systems are complex and require thought, planning, and investment. Therefore, establishing and maintaining a quality system requires commitment by everyone in the organization, beginning with the highest level of management. Another guiding principle is that the systems a company develops to make a product should be designed based on an understanding of their product's requirements. Thus, the cookie manufacturers must design their manufacturing operations to produce a cookie product that meets predetermined requirements.

A few more points about developing a quality system that were introduced in this example are as follows:

- The team must develop methods **to evaluate the essential features** of the product; otherwise, how will they know if their cookies are suitably good?
- The team must **establish a process** that turns their family recipe into actual cookies.
- The team must **obtain suitable resources**, such as facilities, equipment, and personnel.
- The team must **establish methods to monitor the performance of their process** (such as testing the oven temperature).

- The team must **determine the raw materials required** to make the product (such as flour and chocolate chips) and must determine the required features of those materials (e.g., does it matter which kind of chocolate is used for the chips?).
- The team must **create documentation** to guide and keep track of operations.

All these bulleted factors (and more) are part of establishing a quality system for a biotechnology product, regardless of what the actual product is.

Typically, a quality system is established and managed by a quality department in a company, although producing quality products is everyone's responsibility. There are existing formal quality systems to guide people as they develop and maintain their own quality system. Since products differ, there is more than one formal quality system that is relevant to biotechnology companies. The general features of the formalized quality systems that are important in biotechnology workplaces are introduced in this chapter, and we will refer to them in various places throughout this text. We will also revisit quality systems and the regulations relating to quality systems in more detail in Unit 10.

4.2 QUALITY SYSTEMS IN COMPANIES THAT PRODUCE MEDICAL PRODUCTS AND FOODS

4.2.1 MEDICAL PRODUCTS

As we saw in Chapters 1–3, many significant biotechnology products are medical in nature. There are also various food/agricultural products. Poor-quality drugs and contaminated foods have led directly to human injuries and deaths. Because the consequences of a poor-quality product are extreme, the governments in most countries are involved in the quality systems of companies that make medical and food products. Governments provide regulations (laws) that define quality for drugs, biopharmaceuticals, other medical products, and foods. In the United States, a quality drug or food must be *safe, effective, reliable*, and *nutritious* (for foods).

There are a number of regulations that relate to quality systems for medical products and foods, but the core principle of these regulations is that quality, safety, and effectiveness are designed and built into a product. In the past, companies sometimes relied on testing their final products to determine if they were good enough for sale. That approach is no longer considered to be sufficient, because testing samples of a product may not reveal every defect. If we consider our cookie manufacturers, they will almost certainly test some of their finished cookies from every batch. But they cannot test every single cookie – if they tried, they would have no cookies left to sell. Instead, they must rely on designing systems (such as carefully controlling the oven temperature and purchasing proper ingredients) that ensure a quality product.

The system of regulations that relate to quality in companies that produce drugs and biopharmaceuticals is called **Current Good Manufacturing Practices (CGMPs)**. The CGMPs have evolved within an ongoing relationship between the government (which seeks to protect the consumer) and the medical products industry. The CGMPs constitute a quality system that has a sweeping effect in every company that produces regulated medical products. The requirements of CGMPs also radiate out to impact those who package, distribute, market, sell, and use these items. *The CGMPs are enforced in the United States by the federal agency,* **the Food and Drug Administration (FDA)**. Other countries have analogous regulations for drugs and pharmaceuticals, and analogous government agencies to enforce the regulations.

The Current Good Manufacturing Practices are quality principles formalized into regulations. **Regulations** *are requirements that government-sanctioned agencies, such as the Food and Drug Administration or the Environmental Protection Agency, impose on an industry and on companies within that industry.* Compliance with regulations is required by law.

Regulations are objective and generally focus on safety (or reducing risk), efficacy, and honesty (for example, in labeling). Regulations tend not to cover subjective areas of product quality, such as the flavor of cookies. Thus, local government agencies will be interested in how our hypothetical chocolate chip cookie manufacturers clean their equipment – a safety consideration – but the government will not care about the flavor of the cookies. A pharmaceutical company might add grape flavoring to cough syrup to enhance its appeal. The individual cold sufferer is capable of evaluating whether the flavoring improves the quality of the product, and this feature is not a safety concern. Therefore, the flavoring aspect of product quality is not subject to government regulation. In contrast, the safety of the cough syrup is subject to regulation because the consumer cannot perceive whether a medication is safe, and depends on regulations for protection.

Because every company, laboratory, and organization is different, quality regulations must be written in a general fashion. For example, the CGMP regulations contain the statement:

21 CFR 211.63 *Equipment used in the manufacture, processing, packing, or holding of a drug product shall be of appropriate design, adequate size, and suitably located to facilitate operations for its intended use …*

This regulation does not specify the particular equipment of concern. The terms *appropriate*, *adequate*, and *suitably* are not clarified. This statement provides no concrete guidance as to how a company might comply with it. In the case of CGMPs, such generic regulatory statements are interpreted and enforced by the Food and Drug Administration. FDA publishes their interpretation of the requirements of CGMPs in the form of guidance documents called "Guidelines" and "Points to Consider." Guidance documents are not laws and are intended to help companies apply the general principles of CGMPs to their own situation. Each company is different, and each CGMP-regulated company must ultimately develop its own quality system to implement the quality requirements of CGMPs.

The CGMP regulations outline a quality system for manufacturing pharmaceutical and biopharmaceutical products. *There are also US government regulations that relate to performing laboratory studies of pharmaceutical and biopharmaceutical products; these are called* **"Good Laboratory Practices."** *Yet other regulations relate to how studies involving human subjects must be performed; these are called* **"Good Clinical Practices."** There is another set of quality regulations for companies that make medical devices. **Medical devices** *are a category of medical products that includes such items as tongue depressors and pacemakers.* Many medical devices fall outside the scope of what we consider to be "biotechnology." However, the category of medical devices includes in vitro (performed outside a living organism) diagnostic products, such as pregnancy testing kits and rapid strep tests. In vitro testing products often take advantage of biotechnology discoveries and methods. (Some of these tests are discussed in Chapter 28.) *The system of regulations that relate to quality for medical devices is called* **Quality System Regulations (QSR)**. Table 4.1 summarizes these US government quality systems for products with medical application. Individuals who work in biotechnology companies that are involved in health-related products must have at least some familiarity with these regulations.

TABLE 4.1

Regulatory Quality Systems that are Important in Biotechnology

1. **Current Good Manufacturing Practices (CGMPs)**

   The purpose of CGMPs is to ensure the safety, efficacy, and reliability of drugs and other products that treat disease and injury. These regulations are written in a government document, the Code of Federal Regulations (CFR). (More specifically, they are found in 21 CFR 210 *Current Good Manufacturing Practice in Manufacturing, Processing, Packaging or Holding of Drugs* and 21 CFR 211 *Current Good Manufacturing Practice for Finished Pharmaceuticals*.)

2. **Good Laboratory Practices for Nonclinical Laboratory Studies (GLPs)**

   Before any drug or medical product is tested in humans or manufactured in quantity, it undergoes a number of tests to determine its safety and efficacy. Many, though not all, of these studies involve animal subjects. GLPs aim to ensure quality data from these studies. (The GLPs are found in 21 CFR 58 *Current Good Laboratory Practices*.)

3. **Good Clinical Practices (GCPs) and Clinical Trials**

   These regulations concern the design, conduct, monitoring, and reporting of studies that test medical products using human volunteers as the test subjects. Much of the purpose of these regulations is to provide protection for the people who volunteer to participate in these studies. These regulations also aim to ensure that quality data are obtained. The GCPs are found in 21 CFR 50 *Protection of Human Subjects*.

4. **Quality Systems Regulation (QSR)**

   QSR is a quality system outlining the methods used in the design, manufacturing, and servicing of medical devices. This quality system is slightly different from the CGMPs because medical devices are often designed by engineers who create equipment and instruments. Engineering design is not exactly the same as designing a drug or protein biopharmaceutical. The medical device category includes diagnostic medical tests that are of interest to us as biotechnologists. (The QSRs can be found in 21 CFR 820 *Quality System Regulation*.)

4.2.2 Food Products

In the late 1950s and 1960s, the United States was actively competing in the "space race," with the goal of sending an astronaut to the moon. Pillsbury, a major food company, was recruited to develop food for the NASA space program. It was essential that the food be completely pathogen-free to ensure the safety of the astronauts. Scientists considered testing the food for pathogens. Testing final products was, at that time, a widely accepted method of ensuring quality. If a certain percent of tested products was acceptable, it was assumed that the entire batch was of acceptable quality. Analysis quickly convinced the Pillsbury scientists that simply testing some final food products would not ensure their purity. This is because not every food item can be tested; most testing procedures destroy the food or its packaging. Even if nondestructive methods exist, it would likely be too time-consuming and costly to test every item. The Pillsbury scientists did an analysis in which they supposed that there was one pathogenic *Salmonella* organism per 1,000 units of food. If 20 of the 1,000 units were to be tested, then there would be a 98% chance of missing the single pathogen and accepting the defective lot. The Pillsbury staff concluded that testing samples of the final food products was not an adequate method of ensuring their quality and safety. They decided instead to make sure that the production of the foods was so tightly controlled that no contaminants could enter the process. Pillsbury adopted quality methods that were already used in engineering and in the military and devised a *quality system for food manufacture that is now called* **Hazard Analysis and Critical Control Points (HACCP)**. The basic idea of HACCP is that biological, chemical, physical, and radiological risks to food safety must be identified and then controlled.

In 2011, the Food Safety Modernization Act (FSMA) was signed into law, giving the FDA greater power to prevent food safety problems, instead of focusing on reacting to problems. In contrast to HACCP, FSMA requires all food facilities to have a written food safety plan (called a HARPC; Hazard Analysis and Risk-Based Preventive Controls) based on an analysis of all hazards, and designating prevention or minimization methods for each of these hazards. To ensure appropriate compliance to the new standard, FDA requires the food safety plan be created and implemented by a preventive controls-qualified individual (defined by FSMA). Particular emphasis is placed on monitoring supply chains. These activities begin with farmers and other suppliers and extend through all activities involved in bringing food to the consumer.

FSMA requires identifying the risks associated with a product, assessing the importance of each risk, and controlling the risks. The idea that one must understand and control risk is an important concept that has influenced how FDA regulates not just the food industry, but also all of the medical products industries.

4.3 STANDARDS

4.3.1 Introduction

Not all biotechnology companies make products whose quality is regulated by the government. For example, there are biotechnology companies that produce enzymes and other molecular biology products that are used in research laboratories. These products are not food or medical products used in humans, and the quality of these materials is not regulated by a government agency. There are, however, **quality standards** that may apply to the production of such products. A **standard**, *broadly defined, is any concept, method, or procedure that is established by some authority or by general agreement.* Quality standards are established by various organizations, agencies, and other entities. Like regulations, quality standards are written and are followed consistently in workplaces. Unlike regulations, standards are not imposed by the government, and compliance with standards (as we are defining them here) is voluntary.

There are many types of standards and many organizations that develop and disseminate standards. Standards can be narrow in focus (e.g., there are standards that detail how to correctly place the markings on a laboratory flask). Standards can also be broad and cover many aspects of quality in an organization. Standards are valuable because they assemble the thinking of many experts into a single, concise, available document. Standards help to ensure consistency in practices among individuals and nations.

There are standards that relate to measurement practices, performing chemical analyses, and other common laboratory activities. We will encounter such standards in various sections of this text.

Companies and organizations comply with standards to improve the quality of their products and to be more competitive in the international marketplace. For example, our hypothetical chocolate chip cookie bakers might voluntarily join an association, such as the (fictitious) "International Association of Chocolate Chip Cookie Bakers." The association might have standards with which all members comply, such as a requirement that members use only real chocolate chips and not artificially flavored ones.

4.3.2 ISO 9000 Standards

ISO 9000 *is a family of quality standards published by the International Organization for Standardization (ISO).* **ISO** *is a non-governmental organization, head-quartered in Geneva, Switzerland, that oversees voluntary, international quality standards.* Companies, including many biotechnology companies, comply with the requirements of the ISO 9000 series to:

- improve the quality of their products;
- make their processes more cost-effective;
- demonstrate to potential customers that their products are well made; and
- increase the profitability of their company.

Observe that CGMP regulations address the safety and efficacy of products, but not the efficiency with which products are made, or other issues that impact the competitiveness of a company. Therefore, companies that comply with a regulatory quality system may also adopt ISO 9000 standards for business reasons.

ISO 9000 standards are written in a generic, streamlined way so as to be applicable to any company that makes a product. ISO 9000 standards can also be applied to companies that provide a service, for example, disposing of hazardous waste, or repairing equipment.

There is no government agency that enforces and monitors compliance with ISO 9000. If a company or organization decides to voluntarily comply with ISO 9000, then they develop and implement their own ISO-compliant quality system. This quality system is formalized and documented. The company may then hire a third-party certified auditor to evaluate whether they are meeting their commitments based on their own plan. If the auditor determines that the company is in compliance with its plan and if the plan includes all the required parts of a quality program, then the company achieves ISO 9000 certification. The company must periodically hire a third-party auditor to conduct inspections to assure that the company remains in compliance with its plan if the company wants to remain certified. Although being ISO 9000-certified does not actually guarantee that a company is producing a high-quality product, it does show that the company has systems in place that support quality. Further information about ISO 9000 is summarized in Table 4.2, and ISO 9000 is contrasted with CGMPs in Table 4.3.

TABLE 4.2

Information about the ISO 9000 Series of Standards

1. *The ISO 9000 quality standards were first issued in 1987 by the International Organization for Standardization, based in Geneva, Switzerland.* ISO is a worldwide federation of national standards groups from more than 160 countries.

2. *ISO 9000 standards address quality management and promote international trade and cooperation.* The standards aim to:
   a. Enhance the quality of goods and services.
   b. Promote standardization of goods and services internationally.
   c. Promote safety practices and environmental protection.
   d. Assure compatibility between goods and services from various nations.
   e. Increase efficiency and decrease costs.

3. *The term "ISO 9000" actually refers to a family of standards that includes these documents:*
   a. **ISO 9000:2015, Quality management systems – Fundamentals and vocabulary** *describes fundamentals of quality management systems, which form the subject of the ISO 9000 family, and defines related terms.*
   b. **ISO 9001:2015, Quality management systems – Requirements** *provides a number of requirements that an organization needs to fulfill if it is to achieve consistent products and services that meet customer expectations.*
   c. **ISO 9004:2018, Quality management – Quality of an Organization – Guidance to Achieve Sustained Success (continuous improvement)** *provides guidelines beyond the requirements in ISO 9001 in order to consider both the effectiveness and efficiency of a quality management system, and consequently the potential for improvement of the performance of an organization.*

4. *ISO 9000 is based on seven quality management principles:*
   Principle 1: Customer focus *states that organizations depend on their customers and therefore should understand current and future customer needs, meet customer requirements, and strive to exceed customer expectations.*

(Continued)

TABLE 4.2 (*Continued*)

Information about the ISO 9000 Series of Standards

> **Principle 2: Leadership** *states that leaders establish unity of purpose and direction of the organization and they should create and maintain the internal environment in which people can become fully involved in achieving the organization's objectives.*
>
> **Principle 3: Engagement of people** *states that people at all levels are the essence of an organization and their full involvement benefits the organization.*
>
> **Principle 4: Process approach** *states that a desired result is achieved more efficiently when activities and related resources are managed as a process.*
>
> **Principle 5: Improvement** *states that continual improvement of the organization's overall performance should be a permanent objective of the organization.*
>
> **Principle 6: Evidence-based decision making** *states that effective decisions are based on the analysis of data and information.*
>
> **Principle 7: Relationship management** *states that an organization and its suppliers are interdependent, and a mutually beneficial relationship enhances the ability of both to create value.*

TABLE 4.3

Differences Between the ISO 9000 and CGMP Quality Systems

ISO 9000	CGMPs
Compliance is voluntary.	Compliance is required by law for companies making regulated products.
Compliance is monitored by outside, third-party auditors who are paid by the company. The company complies voluntarily with the auditors' suggestions to improve their product quality.	CGMP regulations are monitored by FDA inspectors who have enforcement authority.
Standards are generic and can be applied to any manufacturing or service industry.	CGMP regulations are specific to the pharmaceutical/medical products industry.
Originated in Europe.	Originated in the United States.

The ISO 9000 family is the most familiar of the ISO products, but it is not the only one; there are more than 20,000 ISO standards. A few examples of other ISO standards that are particularly relevant to biotechnology organizations are shown in Table 4.4.

Technology, science, and social milieu change with time. Similarly, standards (and regulations) are not static entities, but rather evolve over time to reflect new conditions. New standards are constantly being developed by teams of experts, and existing standards undergo periodic revisions. This evolution will be explored in more depth in Chapter 35.

TABLE 4.4

Examples of ISO Standards Relevant to Biotechnology Organizations in Addition to ISO 9000

Standard	Purpose
ISO 18385	Describes how an organization minimizes the risk of human DNA contamination in products used to collect, store, and analyze biological material for forensic purposes.
ISO 13485	Provides a framework for adopting a risk-based approach to product development and the quality management system for medical devices.
ISO 17025	Enables testing laboratories to demonstrate that they operate competently and generate valid results, thereby promoting confidence in their work both nationally and around the world.
ISO 23033	"General guidelines for the characterization and testing of cellular therapeutic products." This is just one of a group of new standards being developed to meet the needs of the rapidly growing regenerative medicine area.

Example Problem 4.1

Which of the following is likely to be subject to regulatory oversight; which is likely to be the subject of standards; which is likely to be neither?

a. The smoothness of peanut butter.
b. The microbial content of peanut butter.
c. The numbers of peanut chunks in peanut butter.
d. The levels of aflatoxins in peanut butter (aflatoxins are potent carcinogens formed by certain molds).
e. The electrical wiring of a food blender.
f. The color of a food blender.
g. The activity of a DNA-cutting enzyme.
h. The activity of a substance used to treat asthma.

Answer

a. and c. The smoothness and chunkiness of peanut butter are subjective features that are not subject to regulation. However, they are addressed in a standard put out by the United States Department of Agriculture.
b. and d. Microbial and aflatoxin contamination relate to food safety and therefore are subject to regulatory oversight. In this case, FDA is involved in regulating these features of the food product.
e. The wiring of devices is covered by electrical codes (regulations) because it is a safety concern.
f. Color is not subject to regulations or standards.
g. Enzymes used in research or teaching laboratories are not subject to regulations or standards. (A company that makes enzymes, however, is likely to voluntarily conform to a quality system like ISO 9000 or to its own internal quality system.)
h. Substances used in the treatment of illness are subject to FDA regulation.

4.3.3 Some Common Elements

Although various quality systems differ from one another (as shown, for example, in Table 4.3), there are certain elements they tend to have in common. Of these common elements, documentation is among the most important. **Documentation** *consists of written records that guide activities and substantiate and prove what occurred.* For example, written procedures guide the work of individuals, thus ensuring consistency throughout a facility. Written records show what was done, by whom, and when, thus providing accountability. There is a common saying about the importance of documentation:
If it isn't written down, it wasn't done.

This saying emphasizes the importance of keeping good records and is applicable in any biotechnology workplace, whether it is a laboratory or production facility. Even though many records are not actually "written down," but rather are recorded on a computer, the principle still applies. Another common saying is:
Do what you say and say what you do.

Documentation is part of the job of every bench scientist, technician, and operator in every biotechnology workplace. Because this element of quality is so important, we will devote all of Chapter 6 to it.

Another significant common element relates to resources. Every company, laboratory, and facility needs resources to produce quality products. If any of these resources is missing or is unsuitable for its purpose, then the final product is likely to be inadequate. Skilled personnel are one of the most important resources in any company or organization. The employer is responsible for ensuring that all employees have the education and training needed to perform consistently under ideal and unusual circumstances; employees are familiar with all quality requirements pertinent to their work; employees are well supervised; and employees receive the information, equipment, and tools to do their jobs properly. In a quality environment, there are usually records for each individual to show their qualifications and to keep track of their ongoing training, education, and acquisition of skills. In almost all companies, employees participate in some sort of safety training so that they know how to deal effectively with hazards.

Employees are responsible for the accuracy and completeness of their work. Employees are required to follow instructions, document their work, observe problems and report them as appropriate, understand the impact and consequences of their actions, and undergo continuous training. The language with which GLP regulations describe personnel is:

21CFR58.29 "Each individual engaged in the conduct of or responsible for the supervision of a nonclinical laboratory study shall have education, training, and experience, or combination thereof, to enable that individual to perform the assigned functions ..."

Depending on where an individual is employed, one or another quality system may impact their everyday work. For example, the work of a production operator in a biopharmaceutical production environment is stringently monitored and controlled in order to assure the safety and efficacy of the products she helps to produce. This operator will have to follow written procedures meticulously and will have to record each activity in a specific way. An FDA inspector can come in at any time and review her records. A technician working in a testing laboratory may devote considerable attention to recording and tracking incoming samples. He will verify and document that analytical instruments are functioning properly. His activities are directed at ensuring consistent, reliable test results. Thus, individuals who work in biotechnology must all be committed to consistently producing quality products.

Facilities, equipment, instruments, and raw materials are other resources that are necessary to make a product. There are various types of biotechnology facilities, including laboratories where research and development occur, laboratories where quality-control testing is performed, greenhouses, animal facilities, fermentation plants, and purification facilities. To produce a quality product, each area must efficiently accommodate the activities that occur there, be suitably heated, cooled, and otherwise have an appropriate environment, and so on. Within the facility, there must be properly maintained equipment, instruments, and materials.

4.4 QUALITY SYSTEMS AND TESTING LABORATORIES

In 1989, a mother, Patricia Stallings, was accused of murdering her baby by feeding him antifreeze in his bottle. The child's blood and his bottle were sent for analysis to a commercial testing laboratory and the hospital laboratory. Workers at both laboratories confirmed that there was ethylene glycol (antifreeze) in the blood and bottle. Stallings was convicted of first-degree murder and sentenced to life in prison. Two scientists fortunately became interested in the case after hearing about it on a television broadcast. The scientists proved that the child had not died of poisoning, but rather because he had a rare metabolic disorder. The mother was exonerated (after serving time

in prison). When the scientists investigating the case obtained the original laboratory reports, what they saw was, in their words, "scary." One laboratory said that the child's blood contained ethylene glycol, even though the sample did not match the profile of a known ethylene glycol standard. The second laboratory found an abnormal component in the child's blood and "just assumed it was ethylene glycol." In fact, samples from the bottle had not shown evidence of ethylene glycol, yet the laboratory report claimed it did. In this case, an innocent person was convicted of murder based on the erroneous statements of laboratory workers. (Shoemaker, James D., et al. "Misidentification of Propionic Acid as Ethylene Glycol in a Patient with Methylmalonic Acidemia." *The Journal of Pediatrics*, vol. 120, no. 3, 1992, pp. 417–21. doi:10.1016/s0022-3476(05)80909-6.) The staff of these laboratories were apparently not committed to the principles of quality.

In any testing laboratory, a quality product is a test result that can be trusted when making a decision. A laboratory quality system is intended to help laboratory analysts obtain quality results. In the Stallings case, if effective quality systems had been in place in the laboratories involved, Stallings would never have been prosecuted.

There are different types of testing laboratories, all of which need to produce trustworthy information. For example:

- **Forensics laboratories** *provide results that must be trusted when investigating criminal cases.*
- **Clinical laboratories** *provide test results that must be trusted when deciding how to treat a patient.*
- **Environmental testing laboratories** *provide results that might be used to guide an environmental remediation project, to determine if a pesticide is safe, or to determine whether a location is suitable for a new building.*
- **Quality-control laboratories** *provide information that guides decisions in a company that makes a product.* For example, quality-control laboratories perform tests of raw materials to determine if they should be used in making products. Quality-control laboratories also run tests of final products to see if they have whatever features are required and should be released to consumers.

Some types of testing laboratories are regulated. For example, the Environmental Protection Agency has a regulatory quality system, also called **Good**

Laboratory Practices, *that guides laboratories involved in the testing of agrochemical products.* **The Clinical Laboratory Improvement Amendments (CLIA)** of 1988 *is a system of regulations intended to ensure the quality of results generated by laboratories that perform tests on human specimens.* The CLIA regulations play an important role in ensuring that medical test results can be trusted when making a decision about how a patient should be treated.

ISO provides a voluntary quality system that is applicable to any laboratory. **ISO 17025** *is a quality system that is related to ISO 9000, but includes technical issues specific to laboratories.* A laboratory that meets the requirements of ISO 17025 will be inspected by a national organization that accredits laboratories. Like ISO 9000, ISO 17025 accreditation requires developing and complying with a quality system. For example, managing a quality laboratory requires appropriate resources. Technically skilled personnel are one of the most important resources in the laboratory. The laboratory should be managed so as to ensure that there are personnel with the required skills, that each individual's responsibilities are established, and that there is adequate supervision. Each laboratory must establish procedures to train new analysts, and that training should be documented. Experienced analysts require ongoing training to refresh their skills and to learn new technologies as they are developed. An important management issue in testing laboratories is that laboratory technicians and managers should never be placed in a situation where they feel obligated or pressured to report a particular result. Whatever test result is obtained should be exactly recorded and reported (after proper verification). Other resources include a laboratory facility with appropriate environmental controls, suitable reagents, and equipment that is properly maintained and calibrated. (Calibration is discussed in Chapter 15.)

Quality systems in laboratories include regular reviews of practices, progress, and results. This includes both internal and external reviews. Internal review involves an audit of records, data, and practices by someone within the laboratory or within the company. External review involves auditors from the accrediting agency who audit the laboratory to see that it operates in compliance with quality practices.

So far, the requirements described for ISO 17025 accreditation are analogous to those of the ISO 9000 series. But laboratory accreditation also requires proving technical competence. For example, if a laboratory wants to be accredited as a cannabis testing laboratory according to ISO 17025, they must demonstrate that they can accurately measure such things

as THC content (a chemical component of cannabis plants), pesticide residues, microbial contaminants, heavy metals, and other items specific to their mission. A laboratory that wants to be accredited for another purpose will need to demonstrate competence in other testing areas.

4.5 MANAGING CHANGE, VARIABILITY, AND PROBLEMS

4.5.1 CONTROLLING CHANGE

Change is an important topic when discussing product quality systems. In a manufacturing environment, once a product is proven safe, effective, reliable, or otherwise acceptable, it must be produced consistently. Changing various aspects of manufacturing can be complex because the final product quality must not be adversely impacted. However, change is inevitable. Better processes may be developed, raw materials may change, equipment may be upgraded, computer methods will become more powerful, regulatory requirements change, and so on. It is necessary to respond to changing circumstances and, at the same time, maintain control over processes, materials, and documentation so that product quality does not suffer. Change management is therefore a critical part of any quality system, particularly where regulated products are made.

Effectively controlling change requires that changes be reviewed, evaluated, and approved before and after they are made. Changes require assessment and evaluation. This is because there is the potential that a simple, seemingly unimportant change might alter the way a process runs or the characteristics of a finished product. (See, for example, the case study below, "Example Relating to Uncontrolled Change.") In regulated companies, some changes require approval by the proper regulatory agency.

The need for controlling change requires a way of thinking that is alien to most scientist researchers who are continuously looking for new and better ways to accomplish a task. There can be tension between the need for flexibility during research, and the need to control change in a product. This tension can be evident when a biotechnology company is making the transition from being involved primarily in research and development, to being focused on production. During this transition, companies must institute efficient, thorough procedures to manage change, and they must educate personnel in how to use the procedures.

Some points regarding the control of planned changes are shown in Table 4.5.

TABLE 4.5

Managing Change

1. *Each laboratory, facility, or organization should have a procedure for making changes.* In a non-research environment, this usually involves having a standard procedure that outlines the process for making a change.
2. *Change should always be justified.*
3. *In a manufacturing facility, proposed changes usually should be evaluated and preapproved by the research and development (R&D) department.*
4. *A technical review of the proposed change should be performed to assess its value, and to address risks associated with the change.*
5. *In a regulated company, the quality-assurance unit must review the change to see if it requires approval by regulatory agencies.*
6. *If necessary, the change should be evaluated for financial implications.*
7. *After the change is made, its effects should be investigated and documented.*
8. *In a CGMP-compliant company, changes may need to be validated (as discussed in Chapter 38) to ensure that the quality of the product or result is unaffected.*

Example Problem 4.2 (A Fictitious Scenario)

A new quality-control technician is hired by a biopharmaceutical company. She is assigned the task of routinely checking samples of the company's product for the presence of the deadly (fictitious) pathogen, *Badbeastea miserabilis*. The test consists of applying 1 mL of each sample to a petri dish containing a special nutrient medium, allowing the plate to incubate at a specific temperature for 2 days, and then checking the plate for the characteristic colonies of *B. miserabilis*. The technician (who is eager to make a positive contribution in her new job) reads about a new nutrient medium that supports the growth of *B. miserabilis* better than the old medium, is less costly, and allows the assay to be performed in only 1 day. Elated, the technician orders the new nutrient medium and begins checking samples of the product with the new medium as soon as it arrives. She then tells her supervisor how well this new medium has worked. What do you think about this (imaginary) scenario?

DISCUSSION

Even though the enthusiasm of this new technician is laudable, her new job has not begun auspiciously. The issue in this scenario is change and how change is controlled. The assay for *B. miserabilis* would have been developed and tested by the R&D unit. During that testing period,

R&D scientists would have optimized the assay using the original nutrient medium, checked for potential problems, and written documentation for the assay. Switching to a new nutrient medium could result in an unforeseen difference in the results of the test. For example, certain strains of the pathogen might not grow on the new medium. Before switching to a new nutrient medium, there would need to be testing to see that results with the original and new media are comparable. These tests might require testing samples on both media side by side to see if the results were the same. Changing the nutrient medium would also require that proper forms be completed, that approval from a supervisor(s) and from the quality-assurance unit be obtained, and that new directions for performing the test be written. The new technician failed to complete the proper steps.

Note that the new technician made a serious mistake that suggests that her introduction to her job and the way in which she was supervised were inadequate. The company should therefore look at its methods of training and supervising its employees so that such problems do not reoccur.

Also note that with the appropriate tests and controls, the company may decide to switch to the new medium because of its advantages. The mistake the technician made was to prematurely use the new medium in QC of products, thereby putting a production run at risk.

Case Study: Example Relating to Uncontrolled Change

A biotechnology company that makes products for research was producing an enzyme for use in molecular biology laboratories. Several customers observed lot-to-lot variations in the performance of the enzyme. The company was concerned about these customer comments and began an investigation of the product. The production records showed that there had been a large increase in demand for the enzyme in recent months and so the company had combined scale-up with production. During this period, production scientists had made substitutions and changes without review. This made it very difficult to identify specific sources of the lot-to-lot variability. Scientists performed a number of analyses on the enzyme. They eventually observed that the enzyme's storage solution, which contained Mg^{+2}, was losing the Mg^{+2} over time. Mg^{+2} is a cofactor required by the enzyme for activity. The loss of magnesium was eventually traced to a change in the plastic tubes in which the enzyme was stored. The tube manufacturer had begun adding an antioxidant to the plastic, and the antioxidant reduced the magnesium level in the enzyme solution. In this case, a seemingly trivial change in a raw material (the tubes) caused a problem. Finding the cause of the problem required a costly investigation. Careful control of change aims at preventing this type of predicament, or making it easier to trace the cause of difficulties should they occur.

4.5.2 REDUCING VARIABILITY; CONTROLLING PROCESSES

Quality and productivity improvement share a common element—reduction in variability through process understanding (e.g., application of knowledge throughout the product life-cycle).

(U.S. Food and Drug Administration. "Pharmaceutical CGMPs for the 21st Century—A Risk-Based Approach, Final Report." September 2004)

In the context of product quality, **variability** *is when some characteristic of a product fluctuates.* Product variability is generally undesirable, although a small amount of variability is usually inevitable. Consider, for example, a drug product; it is imperative that its potency is known and does not vary from dose to dose. All quality systems aim to reduce variability in products.

There are multiple sources of variability. Perhaps one person performs a task differently than others do. Perhaps one product is different than another due to overall processing differences (e.g., temperature and pressure). Perhaps a piece of equipment responds differently to temperature changes in a facility. Perhaps an incoming raw material is different than in previous shipments. Each source of variability contributes to the overall variability in the final product (Figure 4.1).

To protect product quality, people attempt to identify and understand all the sources of variability, and to eliminate them whenever possible. The more completely a process is understood, the more the sources of variability can be controlled. The chocolate chip cookie bakers,

for example, must identify the qualities that their ingredients must have in order to make a consistent product. The company then would put in place a system to test all incoming raw materials to be sure they have those qualities. In a **controlled process**, *the sources of variability are understood and, as much as possible, eliminated.*

Reducing variability is obviously important when tangible products are made, such as drugs, but the principle is also relevant in laboratories. For example, many laboratories perform tests that involve an instrument. If the instrument is not well maintained, its response might vary, resulting in untrustworthy test results. The same principle also applies in research laboratories. Suppose, for example, a researcher is trying

Sources of Variability

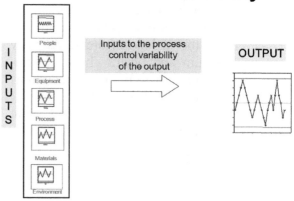

FIGURE 4.1 Sources of variability. Variability in how people perform tasks, equipment, materials, processes, and environment (e.g., facility temperature) all contribute to the overall variability of a process. Understanding all of these inputs is necessary to reduce variability. (Illustration concept courtesy of Michael Fino.)

to investigate changes in gene expression at various stages of development. If the researcher uses reagents that are not made and stored consistently, then none of the experimental data will be reliable. If there is variation in gene expression, it might be related to the stage of development of the test subjects – but it also might be attributable to variation in the reagents.

4.5.3 When Something Goes Wrong; CAPA

Things go wrong – even in the best-run facilities. Every quality system must therefore include a mechanism to deal with problems. In companies that are regulated or audited, inspectors will look to see that problems are investigated, mistakes are corrected, and a system is in place to prevent recurrences.

FDA uses the term **CAPA**, *Corrective and Preventive Actions, to refer to the processes by which a company responds to problems and failures.* **Corrective action** *means to fix problems that have already occurred and may happen again.* **Preventive action** *involves looking for problems that have not yet occurred and preventing them.* A general strategy to deal with problems is to:

- Describe the problem and assess the level of risk it poses.
- Take short-term action as required to prevent further problems or correct the error.
- Conduct an investigation to determine the root cause.
- Develop and implement a long-term solution to prevent the recurrence of the problem.
- Follow up to ensure that the solution was properly implemented and was effective.

Part of the process of evaluating problems is evaluating the risk they pose. In a facility that makes medical products or performs tests of patient samples, a problem might pose the serious risk that patients will be harmed. The more severe the risk posed by the problem, the more rapidly an investigation must occur, and the more resources that should go into fixing the problem. In a research laboratory, problems that are likely to seriously compromise the experimental results of the laboratory are dealt with aggressively.

Root causes *are the "real" or underlying cause(s) of a problem.* If the root cause is not corrected, then the problem might happen again. There may be one or more than one root cause for a problem, and the root cause(s) may be hidden. A root cause might involve, for example, a human error, a malfunctioning instrument, a method or a process that is not effective, a fault in a material, or a problem in the environment (e.g., air, water supply, and dirty surfaces). The investigation to find the root cause(s) might include looking at historical data to see if the problem is recurring, reviewing written records from the affected batch or test, and interviewing associated personnel. Once the root cause of a problem is found, it must be fixed. Other potential problems might be uncovered during an investigation that should also be fixed; this is preventive action.

The following two case studies relate to CAPA systems. The first case study, "Analyst Errors," is an example of CAPA analysis in a quality-control laboratory. This case study demonstrates the importance of identifying and correcting the root cause of a problem. To understand this case study, recall that quality-control analysts perform tests of final products to see if the products are of suitable quality to release to customers. An **out-of-specification (OOS) result** *is one in which a product fails to meet its requirements.* The cause of an OOS result may be a poorly made product – but it may also be due to an error made in laboratory testing. An OOS result that is invalidated is presumably due to a laboratory error. The laboratory in this case study is having problems with OOS results that were due to errors made in the laboratory. (This case study is further discussed in Practice Problem 4 at the end of this chapter.)

The second case study, "Warning Letter from the FDA Relating to CAPA Violations," relates to a laboratory that was visited by an FDA inspector and was found to have an inadequate CAPA system that was not in compliance with the FDA's requirements. (This case study is further discussed in Practice Problem 5 at the end of this chapter.)

Case Study: Analyst Errors

As part of efficiency analysis, several QC work units were asked to report metrics on the frequency of out-of-specification (OOS) results and the root causes found. One work unit reported a high rate of OOS results that were not confirmed and resulted in invalidation of test data. Because 70% of the observed OOS results were invalidated, the laboratory performed a great deal of retesting and investigation and its capacity plummeted while cycle time became unpredictable.

(Continued)

Case Study (*Continued*): Analyst Errors

TABLE 4.6

Reasons for Invalidation of OOS QC Results

Cause Listed	Corrective Action
Incubation done at 35°C instead of 37°C	Retrained analyst
ELISA plate stored at 4°C for 30 minutes before reading	Retrained analyst
Four instead of five replicates were tested	Retrained analyst
Step four performed before step three	Retrained analyst
Test method run outside of validated range	Retrained analyst
Over-incubated ELISA plates	Retrained analyst and updated standard operating procedure
Dilution error	Retrained analyst
Expired standard was used	Retrained analyst

Table 4.6 lists brief summaries of the causes ascribed to several laboratory failures. The pattern was initially missed because it did not correlate with a single analyst, a shift, a test method, or a production sample. When the work unit was compared with others in QC, the invalidation rate stood out, and the causes listed clearly indicated a single problem: failure to follow the procedures exactly. What then were the root causes? Several hypotheses were examined:

- Analysts not trained.
- Procedures written poorly.
- Lack of supervision.
- Lack of resources causing analysts to rush.

The corrective action taken after each event – retrain the analyst on the specific method – clearly wasn't changing the overall metric. What did affect the OOS rate was additional training and a QA person in place to support the supervisor, who was spread too thin in overseeing a large number of employees. Many analysts had less than six months on the job and were in various stages of training. Some …[cultural change] was needed to convince the staff that innovation was not acceptable [without suitable change control], and group retraining was found to be more effective in changing work habits.

Source: This case study is reprinted verbatim by permission from: Shadle, Paule, J. "Navigating CAPA." *BioProcess International*, October 2004, p. 16.

Case Study: Warning Letter from the FDA Relating to CAPA Violations

The Food and Drug Administration has inspectors who periodically inspect companies that make medical products. If the inspectors observe violations, they note them on forms, called "483s," and in official warning letters sent to the company. These letters are posted on the FDA's website. Excerpts from a real warning letter are reprinted below. These excerpts relate to the company's failures to conduct thorough CAPA investigations, initiate corrective actions, and institute a preventive action plan when employees made serious mistakes. As you read this warning letter, consider how the company might improve its CAPA program.

Warning Letter

Dear Mr. T… :

We are writing to you because on March 28 through May 20, 2005, the Food and Drug Administration (FDA) conducted an inspection of your … facility which revealed serious regulatory problems involving your medical devices, including the implantable … Infusion Ports, … drug eluting stents, and … balloon dilatation catheters … [A list of 6 violations follows. Only those violations relating to CAPA are excerpted here.]

(Continued)

Case Study (*Continued*): Warning Letter from the FDA Relating to CAPA Violations

4. Failure to document all activities performed in regards to your corrective and preventive activities, including the investigations of causes of nonconformities, and the actions needed to correct or prevent recurrence of nonconforming product and other quality problems, as required by 21 CFR 820.100(b). During our inspection, we reviewed several ... CAPAs [investigative reports]. This review indicated that your CAPAs fail to include all the necessary information to describe the incident and/or the nonconforming condition.

For example, [report] CAR-05-004 ... involved the shipment of ... units of failed ... [products] to 5 separate hospitals. This CAPA only states, "... product part number H7493897012250, Batch ... was removed by an operator from a QA quarantined location for shipping. The skid containing these units was labeled 'Pending KDR Test Results' and was also 'S' blocked in SAP. A second operator performed an SAP transaction, removing the 'S' block status. This resulted in the units to ship to customers." The CAPA did not include the dates of these serious occurrences, the employees involved, or the number of instances that product was actually either removed from quarantine or overridden in the computer system (SAP). The CAPA also did not list the number of units that were actually shipped, or the number of hospitals that actually received nonconforming product. We learned through interviews with employees, that there were actually 5 separate removal actions of ... product from the quarantined area. On January 12, 2005 ... separate batches of ... product were removed from the quarantined area. These were caught by an employee; however, a CAPA was not generated for the incident. Actual shipment of the ... units occurred on January 20 and 21, 2005 when it was realized there was a 5th removal of [product] from quarantine. The CAPA for this instance was initiated on January 27, 2005.

A serious event such as the one described above, requires a thorough investigation into the activities that precipitated the actual shipment of adulterated ... product. Without a thorough investigation into these events, it is difficult to implement an adequate corrective or preventive action that is required by our regulations.

We note in your response letter dated June 20, 2005, that you have supplemented information to the above CAPA and consider the remedial action to be complete and closed. We are concerned that you have not taken adequate action to prevent this type of serious failure from recurring.

5. Failure to establish and maintain an adequate corrective and preventive action procedure which ensures identification of actions needed to correct and prevent the recurrence of nonconforming product and other quality problems, as required by 21 CFR 820.1 00(a)(3). Your CAPA system has failed to identify the necessary actions to correct and prevent the continued distribution of nonconforming product.

For example, we noted that CAPA 04-125 was initiated on November 2, 2004 to address the release of ... batches of ... balloon dilatation catheters into finished goods inventory. The only preventive action taken as the result of this CAPA was to "Update procedure S801280-00." This CAPA was closed on February 15, 2005 with a notation that the CAPA plan was effective. This CAPA did not identify any corrective action for the ... units that were released into finished goods without proper authorization. During our inspection, you had no information or documents to establish that a final release from your ... manufacturing facility was provided for a final disposition of these ... batches that had already been distributed to customers ...

The specific violations noted in this letter and in the Form FDA-483 issued at the conclusion of the inspection may be symptomatic of serious underlying problems in your establishment's quality system. You are responsible for investigating and determining the causes of the violations identified by the FDA. You also must promptly initiate permanent corrective and preventive action on your Quality System ... You should know that these serious violations of the law may result in FDA taking regulatory action without further notice to you. These actions include, but are not limited to, seizing your product inventory, obtaining a court injunction against further marketing of the product, or assessing civil money penalties.

Sincerely yours,

4.6 SUMMARY

This chapter introduces Unit 2, An Introduction to Quality in Biotechnology Workplaces, by discussing broad concepts relating to product quality and product quality systems. In every workplace, producing a quality product requires the commitment of everyone who works there, beginning with senior personnel and extending to everyone in the organization.

A quality system is the organizational structure, responsibilities, procedures, processes, and resources that work together to provide a quality product or service. Different biotechnology workplaces produce different types of products, and therefore adhere to somewhat different quality systems. The government steps in to enforce quality practices when the consequences of a poor-quality product are life-threatening, as is the case for food and drugs. For example, the Current Good Manufacturing Practices, CGMPs, are regulations that outline a quality system that pharmaceutical manufacturers are required by law to follow. Other examples of quality systems enforced by the government include the FDA's Good Laboratory Practices, Good Clinical Practices, and Quality System Regulations, and the EPA's Good Laboratory Practices.

Many companies and organizations are not required to comply with quality regulations, but voluntarily adopt quality standards. Companies comply with these voluntary quality systems to improve the quality of their product and therefore, presumably, to be more successful. The ISO 9000 series comprise a voluntary quality system that is followed by many organizations around the world.

Although quality systems vary in their details, they share common themes. Quality systems help organizations design and build quality into their operations. Methods to reduce variability, control change, and respond to problems are common issues addressed by quality systems. Another commonality is the fundamental role of documentation as a means to guide and record the work of an organization.

We have not yet focused on the issues of quality in academic research environments and will do so in Chapter 5. Quality is just as important in a research laboratory as it is in any other biotechnology environment, but researchers face somewhat different challenges and must use somewhat different strategies than those who work in other biotechnology settings.

Documentation is such an important part of a quality system in any biotechnology environment that all of Chapter 6 is devoted to it. Documentation ends Unit 2 and provides a transition to the basic methods used to obtain good quality results in the laboratory. In the last unit of this text, we will return to regulatory and quality issues in biotechnology production settings.

Practice Problems

1. Why do companies voluntarily subject themselves to the difficult, time-consuming, and costly process of complying with the ISO 9000 standards?
2. This chapter discussed issues relating to change. A significant change in a biopharmaceutical manufacturing process (for example, a change in cell culture medium or a change in the genetic construct inside the host cells) can be extremely costly and might have significant consequences. Discuss why change in a biopharmaceutical manufacturing environment poses special challenges.
3. Personnel who work in research laboratories are not required by law to follow standard procedures for performing most routine tasks (such as preparing laboratory reagents). It is, however, common for research laboratory staff to write and follow standard procedures for various tasks. What advantage is there to having standard procedures in a research environment?
4. Consider the problem and CAPA investigation in the case study "Analyst Errors," pp. 99–100.
   a. Explain the problem here. As part of your explanation, consider: What is an OOS result? What are the general causes of an OOS result? What do OOS results have to do with QC work units?
   b. Examine the table showing reasons for the erroneous results. What was happening here? What did the laboratory staff do initially to try and fix the problem?
   c. An investigation of the laboratory was performed to look for the root cause. What does "root cause" mean in this situation? What hypotheses were explored in this investigation?
   d. What did the root cause investigation discover? Were any of their hypotheses shown to be true?
   e. The P in CAPA stands for prevention. What did the company do to prevent future problems?

5. Consider the case study "Warning Letter from the FDA Relating to CAPA Violations," pp. 100–101.
   a. Carefully read the description of events in item number 4, relating to the product number H7493897012250. Explain in your own words what went wrong here and why this is a problem of concern.
   b. The FDA says it is looking for some sort of preventive action. Speculate as to how the company might act to prevent this type of mishap. What would you consider to be an adequate preventive action plan?
   c. What consequences has the company experienced due to these errors? If the company does not adequately respond to the FDA's concerns about its CAPA plan, what might be the consequences in the future?
6. Match the word with the definition:
   a. Federal agency that enforces CGMPs.
   b. Provides a framework for adopting a risk-based approach to product development and the quality management system for medical devices.
   c. Organizational structure, responsibilities, and resources that together ensure the quality of a product or service.
   d. The regulatory system used in companies that produce drugs and biopharmaceuticals.
      1. Current Good Manufacturing Practices (CGMPs)
      2. Quality system
      3. ISO 13485
      4. The Food and Drug Administration (FDA)
7. What is a root cause? Why is it important to address the root causes of problems?
8. What is the difference between corrective action and preventive action?
9. Identify at least two potential causes of variability in a production environment.

5 Quality in Research Laboratories

5.1 INTRODUCTION

An article by Sandra Andersson and colleagues published in 2017 illustrates a vexing problem in the biological research community. The authors describe the excitement, followed by confusion over inconsistent results in studies relating to a receptor for the hormone, estrogen. (Andersson, Sandra, et al. "Insufficient Antibody Validation Challenges Oestrogen Receptor Beta Research." *Nature Communications*, vol. 8, no. 1, 2017, pp. 1–12. doi:10.1038/ncomms15840.) Estrogen has many normal roles in the body, but it can also drive the pathogenic growth of breast cancer cells. Inhibiting the effects of estrogen is therefore the goal of certain breast cancer therapies. Estrogen exerts its effects via interaction with specific protein receptors found in cells. The first such receptor, estrogen receptor alpha, ERα, was discovered in 1986. About 70% of breast cancer patients overexpress ERα, and the interaction of this receptor with estrogen drives their cancers' growth. Drugs that inhibit ERα, such as tamoxifen and raloxifene, are effective treatments for many patients. Unfortunately, some breast cancer patients do not respond to drugs that target ERα, or they stop responding after time. Therefore, scientists were excited by the discovery of a second receptor, estrogen receptor beta, ERβ, which was hoped to be a new target for breast cancer drug therapy.

Initial studies using immunohistochemistry (IHC) suggested that ERβ is found in breast tissue and so might, indeed, be a target for treatments. IHC, which will be described in more detail in Chapter 29, is a method to find the cellular location of proteins. Antibodies are the critical reagent in IHC research; they are used to locate and bind the specific protein

of interest. Disappointingly, after 20 years of intense research, ERβ has become a source of confusion, not treatments. Andersson and colleagues attribute this confusion to poorly functioning antibodies. They state that, in multiple studies, scientists used antibodies that they thought were localizing ERβ, when, in fact, they were not. Of the 13 supposedly ERβ antibodies Andersson and colleagues studied, only one actually recognized ERβ. Moreover, it turned out that the one antibody that truly recognized ERβ did not find it in breast tissue. Thus, attempts to use ERβ in breast cancer therapy were misguided. The authors state: "We do not find evidence of expression [of ERβ] in normal or cancerous human breast… This [finding]… contradicts a multitude of studies… Our study highlights how inadequately validated antibodies can lead an exciting field astray."

5.2 QUALITY IN ACADEMIC RESEARCH: "GOOD SCIENCE" IS REPRODUCIBLE

The ERβ antibody study is not an isolated case; rather, it is an example of a perplexing problem that occurs in basic scientific research. Basic research is the root from which all the products of biotechnology emerge. We have seen in previous chapters how innovative products have resulted from a deeper understanding of the fundamental processes of nature. It is therefore vital that biological research be conducted in such a way as to provide trustworthy, "quality" results. The ERβ studies exemplify the confusion that arises when research results do not meet this standard.

Scientists who work in academic research laboratories, in contrast to those in testing laboratories, are unlikely to use the term "quality system." One would be hard-pressed to find a copy of ISO 17025 (*General Requirements for the Competence of Testing and Calibration Laboratories*) or 21CFR211 (*Current Good Manufacturing Practice for Finished Pharmaceuticals*) in a university research laboratory. But research scientists have always pursued "good science" and strived to produce trustworthy results. "Good science" is indeed a quality system, though not one that is encoded in any single regulatory document or standard. What then, is "good science"?

In testing laboratories, it is generally possible to obtain standards with known properties. For example, analysts in an environmental testing laboratory who are testing levels of heavy metals can obtain standards with known concentrations of purified metals. The analysts can use these standards to evaluate their methods; these standards give them a "true" value as a basis for comparison. In contrast, scientists working in research laboratories do not have straightforward ways to be sure a research discovery is "right." When venturing into uncharted territory, "truth" is difficult to define. Therefore, research scientists use a marker for "truth." Scientists speak of **reproducibility**, *the ability of other researchers to repeat an experiment and get the same result.*[1] Reproducibility is used as the indicator of the validity, and hence, quality of a research result and its interpretation. "**Good science**," *by this definition, produces reproducible results.* As Andersson and colleagues demonstrated, irreproducibility can lead to disappointment and wasted time and resources. In the situation they describe, many years went by before the cause of irreproducibility, that is, poorly characterized antibodies, was uncovered.

Concern over reproducibility in research is not new, but a paper published in 2012 galvanized the scientific community into a widespread discussion of the topic. In this 2012 paper, C. Glenn Begley and Lee M. Ellis reported that scientists at Amgen had tried to reproduce the most important findings in 53 landmark papers in cancer biology. Scientists at biopharmaceutical companies, such as Amgen, routinely scour scientific literature looking for leads that might result in a new drug or treatment. If they find a promising research paper, Amgen scientists try to replicate the study and confirm its results. Unfortunately, in most cases, they

were unsuccessful. Begley decided to study this phenomenon formally. His team selected 53 papers with results that might lead to ground-breaking drugs and tried to reproduce those studies in the Amgen laboratories. Begley and Ellis reported that of these 53 papers, the scientific findings were confirmed in only 6, which is 11% of the studies. Begley and Ellis further showed that the papers whose findings they failed to reproduce often were cited hundreds of times in other researchers' papers. This suggests that the work that was not reproducible was leading other researchers astray. Begley and Ellis stated: "Even knowing the limitations of preclinical research, this was a shocking result." (Begley, C. Glenn, and Ellis, Lee M. "Raise Standards for Preclinical Cancer Research." *Nature*, vol. 483, no. 7391, 2012, pp. 531–533. doi:10.1038/483531a.)

At around the same time as the Begley and Ellis paper was published, scientists at Bayer reported that they had halted nearly two-thirds of their drug target projects because their in-house experimental findings did not reproduce the claims in the original scientific papers. (Mullard, Asher. "Reliability of 'New Drug Target' Claims Called into Question." *Nature Reviews Drug Discovery*, vol. 10, no. 9, 2011, pp. 643–644. doi:10.1038/nrd3545.) Other analyses of irreproducibility quickly followed with estimates that anywhere from about 50% to 89% of published preclinical studies were not reproducible (Figure 5.1). One estimate put the economic value of irreproducible work at 28 billion dollars. (Freedman, Leonard P., et al. "The Economics of Reproducibility in Preclinical Research." *PLOS Biology*, vol. 13, no. 6, 2015, p. e1002165. doi:10.1371/journal.pbio.1002165.)

The phrase "reproducibility crisis" began to reverberate in the biological community as these reports were published. Irreproducibility strikes at the heart of the scientific process: Science progresses as researchers build on previous work – it is not possible to build a strong structure on a shaky foundation. Irreproducible scientific work can have serious adverse consequences, including the following:

- The public's trust in scientists' work is diminished.
- Scientists waste time and effort pursuing false leads.
- Research funds, a very limited commodity, are wasted.
- Patients may be subjected to clinical trials of treatments based on flawed assumptions.
- Patient advocacy groups become disillusioned as new treatments fail.

---

[1] In this chapter, we use the terms "replication" and "reproducibility" interchangeably, as is the practice in much of the literature about this topic. Note, however, that some sources distinguish the two terms.

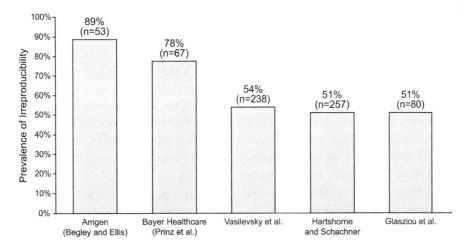

FIGURE 5.1 Studies reporting the prevalence of irreproducibility. (From "The Economics of Reproducibility in Preclinical Research," as cited in text. CC BY license.)

The Amgen and Bayer reports were influential, but had one serious flaw: They were confidential. The authors did not report which landmark studies failed to be reproduced, nor did they provide any details as to what, exactly, they meant by replication. Without such information, their reports could not be evaluated by the scientific community. In an effort to overcome this flaw, *The Reproducibility Project: Cancer Biology (abbreviated hereafter as "Reproducibility Project")* was initiated in 2013. The project's goal was to test the reproducibility of 50 important cancer research studies, this time, using a transparent process. Before each replication study could begin, the team would openly publish the proposed experimental design and protocol. These open reports would be reviewed and could be improved upon by other scientists. The scientists whose work was being replicated were invited to

provide as many details and reagents as possible to best ensure that the replication effort matched their original study. The project was led by Science Exchange, a company based in Palo Alto, California, that found contract labs to reproduce a few key experiments from each paper. Funding included a $1.3 million grant from the Laura and John Arnold Foundation, enough for about $25,000 per study. Experiments were expected to take 1 year – a timeline that was overly optimistic.

In 2017, the first results from the *Reproducibility Project* were published. Two of the replication studies successfully reproduced important parts of the original papers and one did not (Figure 5.2). The other two replication studies were uninterpretable because samples in the control group, that is, tumors that did not experience the experimental intervention, grew too quickly, grew too slowly, or regressed. Therefore, researchers

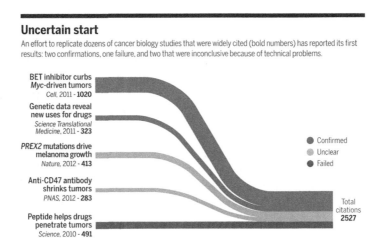

FIGURE 5.2 Results of first replication efforts. These replication studies had mixed results, highlighting the challenges of reproducing others' work. (From Kaiser, Jocelyn. "Mixed results from cancer replications unsettle field." *Science*, vol. 355, no. 6322, 2017, p. 234. Reprinted with permission.)

could not determine whether the experimental intervention had the predicted effect. By 2018, the results of 10 studies were published. In summary, five studies were mostly repeatable, but some aspects of the studies were not replicated; three were inconclusive; and two were not replicated, but their original findings were confirmed in other laboratories. The cost and complexity of each replication was much greater than expected, and the project team significantly scaled back their plans. (For more information about this project, see their website at "Reproducibility Project: Cancer Biology." *ELife*, 10 Dec. 2014, elifesciences.org/collections/9b1e83d1/reproducibility-project-cancer-biology.)

Although the results of the Reproducibility Project are not clear-cut, they did demonstrate conclusively that reproducibility is difficult to achieve, and performing reproducible research requires effort and attention. The project also provided information about what reproducibility means, why it is difficult to achieve, and practices that can promote reproducibility in basic research. Project team members reported: "One of the effort's lessons: Disclosing more protocol details and making materials freely available directly from the original lab or through services…would speed scientists' ability to build on the work of others…Communication and sharing are low-hanging fruit that we can work on to improve…Another problem…is that academic labs rarely validate their assays, making it difficult to know whether a positive result is real or 'just noise'." [Assay validation will be discussed in detail in Chapter 26 of this text.] (Kaiser, Jocelyn. "Plan to Replicate 50 High-Impact Cancer Papers Shrinks to Just 18." *Science*, July 31, 2018. doi:10.1126/science.aau9619.)

The word **transparency** is often used in the irreproducibility literature. **Transparency** *is used to refer to the sharing of data and information describing data collection methods, technical details, the data underlying figures/graphs/conclusions, and research analysis methods (e.g., statistical tools).* The trend in biological journals and grant applications is toward ever increasing transparency in order to improve reproducibility and thus improve research quality.

While there is much concern about irreproducibility, it is worth noting that some prominent scientists have argued that irreproducibility is not really a problem at all, and the idea that there is a crisis is simply "hype." They argue that the very nature of cutting-edge research – which is difficult and leads into the unknown – means that there will be errors and uncertainties. Indeed, errors are the cost of high-impact science. Moreover, scientists learn from these errors so as to make further advances. Other prominent scientists

argue that while it is true that some errors are inevitable, it is the scientific community's obligation to greatly reduce the impact of errors through better research practices. In any event, the discussion about reproducibility is causing shifts in the scientific community that are expected to improve scientists' ability to "do good science." The scientific community is using this crisis – real or "over-hyped" – as an opportunity to improve how science is done (Figure 5.3).

5.3 UNDERSTANDING AND MINIMIZING IRREPRODUCIBILITY

5.3.1 Issues Particularly Relevant to the Academic Research Community

The first step in avoiding irreproducibility is to understand its root causes. Solutions can emerge as these causes are uncovered. Let's consider some of the problems that are being discussed as people converse about irreproducibility and potential solutions:

- **Statistical analysis of results**.[2] The proper use of statistics to analyze research data is an area of active discussion. This application of statistics is beyond the scope of this book. Suffice it to say that many areas of biological research rely heavily on statistical methods to reach conclusions. If those methods are applied incorrectly, then the ways scientists interpret their results – or initially design their experiments – will be flawed.

- **Creating a hypothesis *after* an experiment is conducted**. Many scientists have fallen into a trap called "HARKing," that is, "hypothesizing after the results are known." This is where a research study begins with a specific hypothesis to be tested. Then, when the data are analyzed, researchers might see that the data support a completely different hypothesis that was not the intent of the study. They then publish the results with this unintended hypothesis as though that were the point all along. This practice is wrong; one cannot run an experiment first and come up with a hypothesis that fits the data later. This is because, in general, if one looks at any set of data long enough and in enough different ways, *something* will seem

---

[2] Statistical tools are also used in testing laboratories and production facilities, but there are particular statistical tools used for hypothesis testing in basic research that are related to the reproducibility discussion.

FIGURE 5.3 Scientists, scientific societies, and prominent journals discuss issues in scientific research. This cover from the *Journal of the American Association for the Advancement of Science* (September 2018) illustrates efforts to understand causes and cures of irreproducibility. (From *Science*, vol. 361, no. 6408, 2018. Reprinted with permission from AAAS.)

to be significant. The key to avoiding this error is to confirm the unexpected finding with new studies that are based on the new hypothesis.

The *Reproducibility Project* addressed this concern about HARKing by publishing their study design and hypotheses for each replication project before conducting any work. They could not later change their hypothesis. Similarly, the Food and Drug Administration Modernization Act of 1997 requires scientists running clinical trials to publish their hypotheses in advance at ClinicalTrials.gov. This move by FDA helps to avoid HARKing and also ensures that negative results are published so others do not repeat tests on treatments that fail. Journals and funding organizations have introduced the idea of **preregistration** *where researchers publish a document outlining their hypothesis and methods for a planned experiment before beginning that experimental work*. Some journals have gone even further by providing a mechanism for reviewing these preregistration documents. If the planned work is deemed to be strong and important, then the journal guarantees it will publish the final work, whatever the results.

- **Study design**. Designing experiments is difficult, and many factors must be considered. For example, blinding is a method intended to eliminate unconscious bias, that is, where a researcher sees a particular result because he or she wants, or expects, that result. When a study is **blinded**, *the experimenters do not know which subjects have received an experimental treatment and which subjects are the controls that have not been given the treatment*. Research protocols are often not blinded due to expense or complex logistics, allowing bias to affect the results. Other aspects of study design include the use of proper control groups to detect unintended interactions and sufficient numbers of subjects. Scientists and those who educate scientists have always talked about good study design – this is not new – but the conversation about reproducibility has intensified the discussion. Part of the solution is more aggressively educating new (and established) researchers about study design practices that contribute to quality results.

- **Culture**. Success in academia depends on obtaining grant funding, which is limited. It also depends on publishing papers quickly and in high-prestige journals. Given this high-stress culture, it can be difficult for researchers to take the time and devote the resources to practices that might ensure better quality. This is a thorny problem that can only be resolved if funding agencies, institutions, professional societies, scientists, private industry, and educators work together to improve the culture of science. In fact, all of society must consider this issue because society drives funding priorities and the resources available to scientists.
- **Education**. Changing the culture of science goes hand-in-hand with education. Policies and practices will change as scientists develop an understanding and appreciation for the causes of irreproducibility and methods to reduce it.
- **Publishing**. Academic scientists must publish their work. Those involved in scientific journals (e.g., editors, reviewers, publishers, and scientific societies) are therefore important members of the scientific community, and they have been looking at ways to improve publication practices. For example, to reproduce someone else's work, it is essential to know the details of that work. Journals and societies are now asking scientists to publish these methodological details online and/or in print. New platforms for sharing details have become available.

Additionally, journals have started to demand (or strongly encourage) authors to submit the data that support their figures, graphs, and conclusions. A figure, graph, or conclusion is usually a synopsis of a number of data points. In order for reviewers and other scientists to evaluate the figure/graph/conclusion, it is often necessary to see the underlying data. In the past, original or source data were not usually included in a research paper.

5.3.2 Irreproducibility and Variability – A Fundamental Issue in Quality throughout the Biotechnology Community

5.3.2.1 Overview

Let's consider irreproducibility and its relationship to variability. We have seen in Chapter 4 that controlling variability is fundamental to every quality system.

Hence, as people discuss solutions to irreproducibility in research, the causes of variability must be considered. Some of these causes will be introduced here and discussed in far more detail throughout this text.

5.3.2.2 Reagents Must Be Prepared Consistently to Reduce Variability[3]

Reagents, broadly defined, are one of the key areas of concern in achieving reproducibility. In fact, at least one analyst considers reagents to be the most important factor when considering reproducibility. (Freedman, Leonard P., et al. "The Economics of Reproducibility in Preclinical Research." *PLOS Biology*, vol. 13, no. 6, 2015, p. e1002165. doi:10.1371/journal.pbio.1002165.) Reagents are the beginning of every research study, every test in a testing laboratory, and every product in a production environment. For example, at the beginning of this chapter, we saw how poorly characterized antibodies (a reagent) against ERβ derailed 20 years of research.

Cultured cells are another key reagent that have caused great difficulties. **Cultured cells** *are cells isolated from an organism and grown in vessels, such as dishes or vats*. Thousands of research studies have been performed using cells that turned out to be of a different type than they were thought to be. This problem has been recognized and discussed for at least 50 years. But now, misidentified cells are recognized as being an important part of the broader problem of irreproducibility. The National Institutes of Health (NIH) is a government agency that is responsible for conducting and funding biological research. In 2016,

---

[3] In a broad sense, laboratory animals may be considered to be "reagents," similar to living cells that are used in assays. Animals have a particularly important role in research, but the issues relating to animals are outside the scope of this text. Here we will simply note that many animal studies result in valid conclusions; after all, there is a considerable commonality among organisms. The genetic code, for example, is virtually the same across plants, animals, and microbes. Research on drugs and medical treatments involves extensive testing in animals before the treatment is attempted in humans, as described in Chapter 35. This testing provides critical insights into the activity of a treatment in the body. However, it is also true, for a variety of reasons, that the effects of a treatment in animals do not always predict the effects in humans. The use of animal models can therefore be problematic in biomedical research. Some scientists argue that the standard practice of using genetically homogenous animals for testing drugs is flawed and that using more heterogeneous animal models would provide better predictions. Others are looking at improving the ways animals are handled. There is also ongoing work looking for alternatives to animal testing, such as more use of cultured cells and more sophisticated computer modeling.

the NIH amended its grant application instructions to require the authentication of key biological and chemical resources; cultured cells and antibodies are such key resources. Dr. Michael Lauer, Deputy Director for Extramural Research, stated: "research performed with unreliable or misidentified resources can negate years of hard work…it is imperative that researchers regularly authenticate key resources used in their research." (Lauer, Mike. "Authentication of Key Biological and/or Chemical Resources in NIH Grant Applications." *NIH Extramural Nexus*, January 29, 2016. https://nexus.od.nih.gov/all/2016/01/29/authentication-of-key-biological-andor-chemical-resources-in-nih-grant-applications/.) Similarly, many journals have also begun to require authentication of key resources. One of the first journals to do so was the *International Journal of Cancer*, in 2012. This makes sense because antibodies and cultured cells are of critical importance in cancer research.

Problems with antibodies and cultured cell lines have received a great deal of attention as contributors to irreproducibility. But, in later chapters we will see that small changes in reagents that are so basic that their preparation is thought to be "easy" (such as buffers) can also profoundly affect the results of experiments and assays. Raw materials (such as chemicals and plastic tubes) can introduce variability; their control requires constant vigilance. It is essential to understand these unassuming reagents and establish procedures to ensure their quality.

Because reagents are so fundamental to successful practices throughout biotechnology, a number of units in this textbook address them:

- Unit IV includes common **math calculations** that relate to making reagents.
- Unit VI addresses issues relating to **basic solutions**, such as buffers.
- Unit VII includes discussion of the reagents involved in **assays**, including antibodies. Unit VII also discusses assay validation, an issue identified to be of importance when thinking about reproducibility.
- Unit VIII addresses issues relating to **cell culture**, including the problem of misidentified cells.

The case study below, "The Critical Importance of Reagents," provides another example of how problems with reagents can have profound effects.

Case Study: The Critical Importance of Reagents

In the spring of 2020, the population of the United States received an unpleasant introduction to the critical role of reagents in science. At that time, the pandemic caused by the virus, SARS-CoV-2, was strengthening, and new infections were occurring at an alarming rate. In some regions around the world, emergency rooms, hospitals, and morgues were becoming overwhelmed. A key question for treatment of individuals and for understanding the broadening of the pandemic was "who is infected?" Around the world, scientists rapidly developed diagnostic test kits to answer this question. Unfortunately, for a variety of regulatory and political reasons, in the United States, only the Centers for Disease Control (CDC) was allowed to provide diagnostic test kits for the new virus. Weeks passed without a test kit in the United States as the need to know who was infected became increasingly urgent. After several anxious weeks in which the public was assured that a test kit was imminent, the CDC finally released kits to public health laboratories around the country (Figure 5.4). But something went wrong; analysts at various testing sites reported that the kits did not work properly. The problem was traced to a faulty reagent, reportedly one that was involved in the negative control. In a properly performing assay (test), the negative control should not indicate the presence of virus. Yet, in the CDC kits, sometimes the negative control seemed to detect virus where there was no virus. Any assay where the negative control produces a positive result is suspect, and generally the results of such a test are considered to be invalid. Determining the cause of this problem with the CDC test kits took yet more time and set back the testing efforts in the United States even further. Because testing was delayed in the United States, it is likely that infected people who should have been quarantined were not, because they were not identified. It is possible that the epidemic spread more rapidly in some areas because authorities did not know how it was spreading and where it was clustering. The issue here is not that the CDC staff were at fault – they were working under tremendous pressure and were developing a new test, a

(*Continued*)

Case Study (*Continued*): The Critical Importance of Reagents

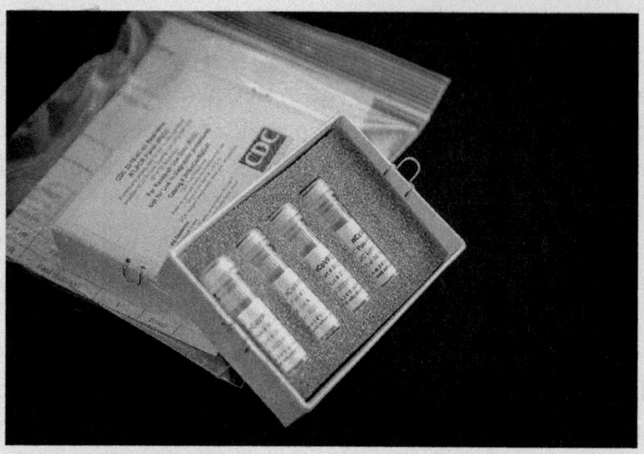

FIGURE 5.4   The CDC COVID-19 Test Kit. (Image credit: U.S. Centers for Disease Control.)

task that normally requires months or even years, in a matter of weeks. Rather, the point is that the quality of reagents in any scientific endeavor is absolutely critical. In fact, in some cases, as the public sadly learned, reagent quality is a matter of life and death.[4]

---

[4] It is worth noting that the negative control in the faulty test kit did what it was intended to do – it alerted analysts to a problem with the kit. The particular problem, contamination of reagents, is a concern with this type of test, and a negative control is essential to detect contamination. The use of controls in assays is always of vital importance and is discussed in more detail in Chapter 26.

5.3.2.3 Measurements and Assays Must Be Performed Properly to Reduce Variability

Measurements include determinations of such things as weight, volume, and pH; properties that are daily assessed in almost every laboratory. **Assays** *are tests run on samples*. Research requires collecting information about samples, and so assays are the basis for experimental results. Assays are also the core activity of testing laboratories, and they are used for control and monitoring in production facilities. Assays must be demonstrated to provide meaningful data. Because of their fundamental importance, measurements and assays are explicitly explored in this textbook:

- Unit V explores basic **measurement science** and how to control variability in fundamental measurements.
- Unit VII discusses methods that help assure trustworthy **assay** results, including the validation of assays.

5.3.2.4 Documentation Provides Transparency

Good documentation is essential in all biotechnology workplaces. It is impossible to be transparent

without high-quality documentation. Being able to reproduce an experiment requires knowing how that experiment was performed – in detail. For example, all raw materials must be specified completely (e.g., by exact name, manufacturer, catalog number, and lot number). If the raw materials in a study are not detailed at the time of use, it may be impossible to reconstruct exactly what was done in an experiment. Documentation is addressed throughout this text, and Chapter 6 provides a basis for understanding this critical issue. See the Case Study below, "Mix-Up in the Mad Cow Freezer," for an example of how a seemingly minor documentation error can cost years of work.

In summary, striving for quality is important throughout the biotechnology enterprise, beginning with basic research and continuing as the results of that research are transformed into a myriad of valuable products. In a biological research setting, striving for quality involves addressing the problems of irreproducibility. The fundamental principles that are the central part of this textbook are the basis for quality results, wherever one works in biotechnology.

Case Study: Mix-Up in the Mad Cow Freezer

According to various reports (Butler, Declan. "Brain Mix-up Leaves BSE Research in Turmoil." *Nature*, vol. 413, no. 6858, 2001, p. 760. doi:10.1038/35101729), a labeling mix-up caused 3 years of important research on BSE (Bovine Spongiform Encephalopathy), known as "mad cow disease," to be scrapped. Researchers at the Institute for Animal Health in Edinburgh were conducting pivotal studies to see whether British sheep had become infected with the BSE pathogen. The researchers were studying what they thought was tissue from 3,000 sheep collected in the early 1990s. Initial findings from the laboratory suggested that sheep brains might be infected with BSE, leading to fears that entire herds of British sheep would need to be destroyed. It was subsequently determined that the experimenters were actually studying tissue from cow brains. An audit carried out by the United Kingdom Accreditation Service reported that there was "no formal documented quality system" covering this work and that record-keeping was "inadequate." The director of the research institute disputed the findings, but, in any event, everyone agreed that the results of the research were not salvageable.

Practice Problems

1. Suppose you make a calculation or measurement mistake and accidentally add ten times too much salt when baking a cake.
   a. Will this mistake be evident?
   b. Will your cake be edible?
2. Suppose you make a calculation or measurement mistake and accidentally add ten times too much of a critical component into a solution that you will use in an experiment. (Note that most biological solutions are mixtures of various components, for example, salts and buffers, dissolved in water. Once the components are dissolved, these solutions generally look completely clear, like plain water.)
   a. Will you know that you made this mistake?
   b. How might this error relate to reproducibility?
   c. Suggest some strategies to avoid making this type of error.
3. Suppose you are preparing the laboratory solution mentioned in problem 2. This time, your calculation is correct, but, unbeknownst to you, the scale that you are using is not operating properly. As a result, you add 5% too little of every component into your mixture.
   a. Will you know that you made this mistake?
   b. How might this error relate to reproducibility?
   c. Suggest some strategies to avoid making this type of error.

6 Documentation
The Foundation of Quality

6.1 INTRODUCTION: THE IMPORTANCE OF DOCUMENTATION

6.2 DATA INTEGRITY, QUALITY, AND GOOD DOCUMENTATION PRACTICES

6.3 TYPES OF DOCUMENTS

    6.3.1 Overview

    6.3.2 Controlled Documents

    6.3.3 Documents in Various Workplaces

    6.3.4 Other Documents that Are Common in Laboratories

    6.3.5 Documents That Are Specific to Production Facilities

6.4 ELECTRONIC DOCUMENTATION

6.1 INTRODUCTION: THE IMPORTANCE OF DOCUMENTATION

Everyone who has taken a science class is familiar with laboratory notebooks. Not every student, however, realizes their importance in the workplace, nor that laboratory notebooks are just one piece of a broad system of documentation.

Documentation is a difficult word to define. Documentation includes a variety of tangible items (e.g., forms, information recorded on paper, and written instructions). Documentation may also be electronic, that is, recorded using a computer. The term "documentation" also refers to the methods of verifying that tasks were performed in a certain way, and that results and products are of good quality. As a simple definition, we will say that **documentation** *is a system of records*, where a **record** *is anything that provides permanent evidence of, or information about, past events.*

Documents have many important functions. They provide a record of what was done, by whom, when, how, and why. They chronicle a research study in detail. They provide objective evidence that a product was made properly, that all personnel followed proper procedures, and that all equipment was operating correctly. They provide evidence of invention that can be important in patent disputes. Some of the many functions of documentation are summarized in Table 6.1.

FDA's policy regarding documentation is "if it isn't documented, it wasn't done." If the documentation relating to a particular batch of a regulated product (such as a drug) is lost, accidentally destroyed, or is badly prepared, then that batch of product cannot be sold. Such an error could cost the company millions of dollars. Regulated companies, therefore, have extensive systems in place to ensure that the work is recorded, that the documents associated with every product are completed, that all documents are securely stored and can be retrieved from storage, and that documents are protected, just as is the product itself.

Documentation is equally important in a research setting. If a researcher cannot show evidence of experiments and results, then those results are not credible. Reliable documentation is essential to achieve reproducibility. If a researcher cannot determine how an experiment was performed initially, it is impossible to reproduce it. Missing information that makes replication difficult includes, but is not limited to, complete descriptions of reagents used, suppliers, assay conditions, instrument settings, and original data. The scientists involved in the Reproducibility Project, discussed in Chapter 5, frequently reported inadequate documentation, leading to problems ascertaining how an original research study was performed.

6.2 DATA INTEGRITY, QUALITY, AND GOOD DOCUMENTATION PRACTICES

In our discussion of documentation, it is important to consider the term "data integrity." A **datum** (singular of data) *is a piece of information*. Many pieces of

DOI: 10.1201/9780429282799-8

TABLE 6.1

The Functions of Documentation

1. *Record what an individual has done and observed.*
2. *Establish a record of experiments, data, and conclusions.*
3. *Record operating parameters of a laboratory instrument or manufacturing vessel.*
4. *Establish ownership for patent purposes.*
5. *Demonstrate that a procedure was performed correctly.*
6. *Tell personnel how to perform particular tasks.*
7. *Establish the specifications by which to evaluate a process or product.*
8. *Demonstrate by an evidence "trail" that a product meets its requirements.*
9. *Ensure traceability.* Here we define **traceability**, as does the International Organization for Standardization (ISO), *as the ability to trace the history, applications, and location of a product and to trace the components of a product.* Traceability helps to ensure that if problems arise in a product, then the origin of the problem can be traced to its components, and the product itself can be found and recalled if necessary. Traceability depends on an organized, well-designed system of documentation.
10. *Establish a contract between a company and consumers.* The written specifications, labels, and other documents associated with a product establish that contract.
11. *Establish a contract between a company and regulatory agencies.*

---

information are recorded (documented) daily in biotechnology facilities. For example, analysts record operating parameters for instruments, results of tests on raw materials and products, and names of people who performed certain tests. Production operators record information about raw materials, fermenters, processes, and much more. Researchers record information about how experiments were performed, the samples involved in the experiments, and the experimental results. All these many records constitute the data of the organization or laboratory. According to FDA, "**data integrity** *refers to the completeness, consistency, and accuracy of data.*" The term "data quality" may be used as a synonym for data integrity, as described in this FDA definition.

The FDA is very concerned with data integrity. This is because a pharmaceutical/biopharmaceutical company must prove that its products are manufactured in such a way as to be safe, effective, and reliable. Between 2013 and 2015, the FDA sent at least 15 letters to pharmaceutical manufacturers accusing them of not ensuring data integrity. In these cases, the manufacturers had deleted, altered, or omitted test data that indicated a drug product did not meet its requirements. After altering the test data, the companies distributed the drugs in question to consumers. This practice of manipulating and altering test data is clearly unacceptable.

In the cases of the pharmaceutical manufacturers mentioned above, the FDA was concerned about companies providing intentionally fraudulent test data. But

data need not be intentionally altered or omitted to be of poor quality. It is easy to imagine situations where unintentional mistakes or omissions can cause difficulties. For example, a scientist might forget to document details about a critical reagent used for an experiment and then be unable to later duplicate the experimental results. An analyst may not be able to troubleshoot a problem because of insufficient records. Every biotechnologist must be concerned with the quality of their records, and organizations must develop systems that help personnel minimize mistakes and omissions.

The FDA uses the acronym **ALCOA** *to summarize the requirements for data quality.* These letters stand for:

A = Attributable   **L** = Legible   **C** = Contemporaneous

O = Original   **A** = Accurate

Attributable *means that each piece of data recorded must be associated with a specific person.* It is not enough to record a piece of information; the person who makes the record must also be noted. This can be achieved manually by signing or initialing a paper document. When written signatures and initials are used, there must be a signature log that connects signatures and initials to specific persons. A person can also be identified when using a computer by using a secure log-in system.

Legible *means that all records must be readable and permanent.* For paper records, this means that indelible ink must be used and the person making the record must write so that others can read it. For electronic records, there must be a way to store the data so that they can be read in the future, even when methods of storing and retrieving electronic data change.

Contemporaneous *means that records should be made when an event happens, not later on.* Also, the date and time of each record must be noted accurately. This means, for example, that if one is using a logbook or a laboratory notebook, it is not allowable to skip pages with the intent of filling them in later. Each page must be completed in sequence, and each page must be dated. With an electronic system, automatic date and time stamps must be used and security systems must ensure that the date and time stamps are never altered.

Original data (sometimes called **source data** or **raw data**) *refer to the first time that data are recorded.* If, for example, an analyst was to record a piece of information on a paper towel (which should never happen, but does), then the original data are on the paper towel and that paper towel must be securely preserved.

More commonly, original data are a printout from a device, an observation recorded with a pen on paper, or a record in a computer. It is important for an organization to provide effective documentation systems that capture and archive data at the time they are created.

Accurate *means the records must be free from errors, complete, and truthful.* Obviously, falsifying records results in inaccurate data and should never happen. It is likely, however, that most inaccurate records result from unintentional mistakes or unforeseen problems.

Good documentation practices (GDocP or GDP) *are the ways that people ensure data quality.* When good documentation practices are followed, it is likely that the data of the organization will be attributable, legible, contemporaneous, original, and accurate. Good documentation practices are important in any biotechnology setting, regardless of whether or not the organization is regulated or complies with a formal quality system. These documentation practices are summarized in Table 6.2.

TABLE 6.2

Good Documentation Practices

Rule 1: Records must be accurate, legible, and understandable. To accomplish this:
  a. Write legibly.
  b. Define abbreviations.
  c. Paginate all forms and documents with the page number and the total number of pages (e.g., page 2 of 10). This helps ensure that no pages get lost or omitted over time.
  d. Avoid ditto (") marks or arrows when recording repetitive information.

Rule 2: Records must record events with clear and verifiable dates. To accomplish this:
  a. Date every record (e.g., every label, every signature, and every page of a laboratory notebook).
  b. Never "backdate"; show the actual date.
  c. Use the format that is required in your organization, e.g., April 15, 2021. Note that the conventions for writing dates vary in different countries.

Rule 3: Documents must be secure and safe from natural disaster, theft, and access by unauthorized individuals. To accomplish this:
  a. Use filing cabinets and locks to protect paper documents.
  b. Use software security to safeguard electronic documents.

Rule 4: Records must be attributable to a particular individual. To accomplish this:
  a. Sign every record to identify the person making the record and to attest to the truth of the data recorded.
    i. A traditional signature is used with paper documents.
    ii. An electronic signature is used for computer records. This typically requires entering a unique user ID and password. Everything recorded onto the computer while that individual is logged on is attributed to that person.
  b. Never sign a document for another person or log in as another person.
  c. Never sign before you have completed a task.

Rule 5. Documents must provide information for traceability, so that, for example, all the materials used in an experiment or analysis, or used in making a product, and all the associated documents can be identified. To accomplish this:

(Continued)

TABLE 6.2 (*Continued*)

Good Documentation Practices

   a. Identify all chemicals, equipment, documents, samples, and other items with unique identification numbers, tags, or labels.

   b. Always record complete information about all chemicals, equipment, documents, samples, and so on when the item is used.

Rule 6. Records should not be capable of being altered, either accidentally or intentionally. To accomplish this:

   a. Enter data directly onto the correct form or into your laboratory notebook, never onto a piece of scrap paper or the back of your hand.

   b. Always use permanent ink.

   c. Cross out mistakes with a single line so that the original recording is visible. Never use whiteout or write over data to obscure it. In many organizations, corrections must be initialed, dated, and briefly explained.

   d. Draw a blank line through any unused space in a laboratory notebook so that no one can later add anything. If a whole page is supposed to be blank, label the top of the page as "intentionally left blank."

   e. If you are filling out a form and a blank does not apply, write NA, "not applicable." Do not leave any fields empty as this can be interpreted as missing data or accidentally omitted data.

   f. Electronic documents require a software method of tracking modifications to data to ensure that original recordings are not erased. If someone legitimately tries to add information to a record, the computer must be programmed to "know" that this act is legitimate and to show both the original record and the revision.

Case Study: Data Integrity and Research Scientists

With the advent of the Internet and the ability of scientists to closely analyze each other's data, it is not surprising that concerns about data integrity and interpretation have become more frequent. Many of these concerns are cleared up amicably with the sharing of raw data and detailed methodology that might not have been published with the analyzed data. The issue of data integrity becomes more problematic when universities, grant agencies, and other institutions launch investigations of individual scientists. For example, in 2020, a prominent and well-funded spider researcher was accused of fabricating data when reporting his studies of spider social behaviors. Other researchers spent hours poring over his data and found anomalies that might indicate that his data were fabricated. The researcher retracted several papers as a result of these investigations. (Retraction of an article means that the article should not have been published and warns other researchers not to consider it as a basis for their work.) The researcher's explanation is that the data anomalies are the result of honest mistakes, not intentional fraud. In any event, whether the spider data were intentionally fabricated or were simply not recorded and archived properly, the career of this researcher (and all of his colleagues who co-authored papers with him) is seriously damaged.

A related issue that appears frequently in investigations of research data is that the scientist in question cannot produce all of the raw data supporting their publications. This sometimes happens when multiple scientists collaborate and only provide their analyzed data to their colleagues, or when graduate students and post-doctoral fellows change laboratories and do not leave their laboratory notebooks behind. It is very difficult for a scientist to clear their name and prevent retraction of their research articles when they have no documentation that the reported experiments even occurred. For example, in 2020, an article in the prestigious journal, Science, was retracted because "careful examination of the first author's lab note-book…revealed *missing contemporaneous entries* and *raw data* for key experiments." [*Emphasis added* to highlight ALCOA principles.]

Clearly, the quality principles of documentation apply to researchers. Scientists and their institutions must take responsibility for instituting robust practices that help reduce errors, safely archive data, and promote data quality.

6.3 TYPES OF DOCUMENTS

6.3.1 OVERVIEW

There are various types of documents, and a single company or organization is likely to have many documents, each with a particular purpose. It may be helpful to broadly classify documents into three categories to better understand their roles.

Directive documents *tell personnel how to do something.* Standard operating procedures and protocols, discussed later in this chapter, are examples of directive documents. Information is not added to these documents when work is performed.

Data collection documents *facilitate the recording of data and provide evidence that a directive document has been properly followed.* Information is added to data collection documents during operations. Laboratory notebooks, reports, forms, and logbooks, all discussed later in this chapter, are examples.

Commitment documents *lay out the organization's goals, standards, and commitments.* Documents submitted to the FDA (e.g., a New Drug Application, as will be described in Chapter 35) are examples.

6.3.2 CONTROLLED DOCUMENTS

In a company or a regulated workplace, most documents are likely to be **controlled**; *that is, they are prepared and distributed using a formal process to ensure that each document is identified and accounted for.* Consider, for example, a laboratory where an analyst writes a procedure that describes how to perform a laboratory test in a step-by-step format. The analyst's draft procedure is reviewed and possibly revised by a supervisor, other analysts, and a representative from the quality assurance department (the department charged with overseeing quality processes and documentation). Once the procedure is completed and approved, the original (master) procedure is signed, dated, and stored in a secure location. An analyst who needs to perform this laboratory test is issued a working copy of the approved procedure; receipt of the copy is recorded. If the procedure is revised, the new version is reviewed, approved, signed, and dated. The revision number is noted on the new version. The older version is removed from active use so that only copies of the most recent version are available to analysts. A controlled system ensures that everyone in the laboratory always uses the correct, most up-to-date version of every document. Controlling documents also helps

to prevent unauthorized individuals from obtaining access to confidential information.

In an academic research setting, documents are usually not formally controlled, as they are in a company or regulated environment. In an academic environment, documents are typically maintained by individuals who generally share them as they, or their laboratory director, see fit.

6.3.3 DOCUMENTS IN VARIOUS WORKPLACES

Although documentation is essential in all biotechnology work environments, the specific types of documents and the systems for documentation vary in different workplaces. In an academic research laboratory, the major documentation requirements are that investigators or colleagues can reconstruct their work based on their records, solve problems, detect mistakes, prove to the scientific community that their results were properly obtained and were accurately reported, and provide a trustworthy chronological record of their work. The laboratory notebook is the primary document in a research laboratory and will become a matter of public record in patent applications, in patent disputes, and if there ever should be questions about the correctness or authenticity of reported findings. (See, for example, the case study above, "Data Integrity and Research Scientists.")

Individuals in research and development (R&D) laboratories likewise rely heavily on laboratory notebooks to document findings, especially when applying for patents. R&D workers also prepare other documents that describe how to manufacture the product and that detail properties of a product.

People in production facilities use documents other than laboratory notebooks. For example, production operators use **batch records** *that describe how to make a product and also record details of product manufacture.* Batch records are described in more detail later in this chapter.

Table 6.3 summarizes common types of documents. The first part of the table focuses on documents that are typically found in laboratory environments. This includes academic research laboratories, R&D laboratories, testing laboratories, and quality-control laboratories associated with production facilities. With a few exceptions, the types of documents in the first part of Table 6.3 are also used in production environments. The second part of the table gives examples of documents that are specific to production facilities and are seldom, if ever, used in laboratories.

TABLE 6.3

Examples of Documents That Are Common in Testing and Research Laboratories

Directive Documents

1. **Standard operating procedures (SOPs)** or simply "procedures" *detail what is to be done to complete a specific task and how to document that the task was done correctly.*
2. **Protocols** are similar to SOPs in that they *explain how to do a task.* The term *protocol*, however, is often reserved for situations where a question or hypothesis is to be investigated (an experiment will be performed) or the procedure is going to be performed only once.
3. **Labels** *are attached to solutions, products, or items to identify them.*

Data Collection Documents

4. **Laboratory notebooks** *are a chronological log of everything that an individual does in a laboratory.*
5. **Forms** *contain blanks that are filled out by an analyst to record information. Forms are typically associated with SOPs or other documents.*
6. **Reports** *are documents generated as a result of performing a protocol.* Reports summarize and interpret data that were previously collected as a result of following a protocol.
7. **Equipment/instrument logbooks** *keep track of maintenance, calibration, and problems for a given instrument or piece of equipment.*
8. **Analytical laboratory documents** *record information regarding the testing of a sample.*
9. **Recordings from instruments.**
10. **Chain of custody forms** *are used to trace the movement of a sample throughout a facility and to keep samples and sample test results from being confused with one another.*
11. **Training reports** *document that individuals were properly trained to perform particular tasks.*

Examples of Documents That Are Specific to Production Facilities

1. **Batch records** *are collections of documents associated with a particular batch of a product. (A batch record is both a directive and data collection document.)*
2. **Regulatory submissions** *are forms filled out and sent to regulatory agencies to inform them of what a company is doing and/or to ask permission to test or sell a product. (These are a type of commitment document.)*
3. **Release of final product record** *is filled out when a product has been approved for sale. (This is a type of data collection document.)*

Example 6.1

Consider the documentation that might be required when performing a routine laboratory task in a regulated industry, such as mixing a 1 M solution of NaCl.

1. The technician will follow an **approved standard operating procedure** for mixing the solution.
2. The raw materials – clean glassware, NaCl, and purified water – will all have documents associated with them that show they were tested and found satisfactory prior to being released for use by a technician.

3. The instruments used to weigh the NaCl and measure the water will have **logbooks** or other documents showing they were properly maintained.
4. The technician will record information about the solution on a **form.** The form identifies the person making the solution, the date, the procedure followed, quantities and source of raw materials used, the storage location and conditions of the solution, the amount made, and the ID number of the solution.
5. The resulting solution will need an **identifying number**, and there will be documentation associated with assigning this number.

6. The solution will need a **distinguishing label**.
7. If the solution is split into more than one container, **documented ID numbers** will need to be assigned to each container referring back to the original solution. Each container will need a **distinguishing label**.

6.3.3.1 Laboratory Notebooks: Functions and Requirements

Laboratory notebooks *are assigned to individuals and are a chronological log of everything that individual does and observes in the laboratory.* Of all the documents that are described in this Chapter, the laboratory notebook is the most important to those who work in research laboratories. Laboratory notebooks are a researcher's primary data collection document. They may also be used in testing laboratories for nonroutine tasks, such as investigations and method development.

The primary user of a laboratory notebook is the researcher who uses the notebook to track the progress of a project, archive the data generated in experiments, record observations, and record all the details that must be remembered. Researchers use their laboratory notebooks when they analyze their results, write reports, plan new investigations, and troubleshoot problems.

Laboratory notebooks can be either paper or electronic, and both types are widely used. While paper laboratory notebooks have been used for hundreds of years and continue to be preferred by many investigators, there are also a number of electronic laboratory notebooks that are commercially available. The general principles of documentation are the same, regardless of whether a pen or electrons are the recording implement. There are, however, some issues that arise with electronic documentation that will be discussed later in this chapter.

A paper laboratory notebook is a permanently bound, rigidly constructed book that consists of mostly blank pages. It is typical for each page in the notebook to have a page number, a blank line for the title of the page or project, a blank line for a dated signature at the bottom of the page, and a blank for a witness to sign and date the page. Otherwise, the page does not provide guidance as to what to record. Blank pages provide the researcher with total flexibility to record thoughts, sketches, tables, and text, all in any format. This freedom can also make it easy to forget to record important information, such as lot numbers for raw materials, or details about samples.

As we will see later in this chapter, all other types of documents, such as forms and batch records, are more directive and provide blanks that are filled in with specific pieces of information.

Laboratory notebooks in a biotechnology company are generally distributed to individual investigators by a company representative. The notebook, and the ideas and information recorded in it are intellectual property that belongs to the company. The ownership rules in academic research institutions vary.

A laboratory notebook is an important legal document that may be viewed by people in addition to the researcher. A laboratory notebook may provide evidence that is used by the scientific community to assign credit for a research discovery. The notebook documents the honesty and integrity of data that are published in research journals and used in grant applications. Laboratory notebooks can be subpoenaed in litigations, and they can be examined by auditors from the FDA, EPA, and other regulatory agencies.

Laboratory notebooks are of importance in patent law to prove ownership of an invention (see Chapter 3). Even academic scientists who do not seek commercial gain from their research may find that they want to patent an invention to protect their rights to it, or to ensure that their invention will remain in the public domain. It is therefore essential that all researchers maintain proper laboratory notebooks and that these notebooks unequivocally document the dates at which ideas were conceived and experiments were performed.

Table 6.4 summarizes good documentation guidelines for keeping a paper laboratory notebook that should always be followed and that allow the notebook to be used for all the purposes described above. Electronic laboratory notebooks have analogous requirements that will be discussed later in this chapter.

6.3.3.2 The Content of Laboratory Notebooks

Table 6.5 summarizes the items that are generally recorded in a laboratory notebook. All notebooks include basic information, such as the name of the person to whom the notebook was assigned and a page number and date on every page.

A laboratory notebook must be complete enough that the researcher or another individual could exactly repeat the work described based on the information recorded. It is essential that laboratory notebooks include raw data. **Raw data** *are the first records of an original observation.* Depending on the situation, raw data may be written into the notebook with a pen by the operator, may be a paper output from an instrument, or, increasingly, may be recorded into a computer medium

---

TABLE 6.4

Guidelines for Keeping a Paper Laboratory Notebook

1. *Use only a bound notebook, not a spiral or loose-leaf notebook from which pages can be removed or into which pages can be inserted.* This helps to ensure that the dates are correct, and that the notebook honestly records what happened and when it happened.
2. *Make sure every page is numbered consecutively before using the notebook.*
3. *Never rip out a page.*
4. *Keep the laboratory notebook in chronological order.* Never skip a page to insert information later.
5. *Blank lines or unused portions of the page should be crossed out with a diagonal line so nothing may be added to the page at a later date.*
6. *Make all entries with indelible ink to ensure the integrity of the data.*
7. *Be legible, clear, and complete in your entries.* Remember that you, supervisors, colleagues, patent attorneys, and regulatory agency inspectors may review your entries.
8. *Enter all observations and data directly into the notebook – not onto a paper towel or the back of your hand.*
9. *Cross out all errors with a single line so that the underlying text is still clearly legible.* Date when the cross-out was made, explain it briefly, and initial or sign it. In some settings, cross-outs must also be witnessed. This helps ensure that entries are not obscured, altered, or changed at a later date.
10. *Note all problems; never try to obscure, erase, or ignore a mistake; be honest.* Be objective. Avoid derogatory statements about your ideas.
11. *Date and sign each page. In many laboratories, a corroborating witness should also sign and date the page.* The scientist and, where relevant, the witness should verify that there are no blank spaces on the pages, that all tables are complete, and that the page is complete. If corrections are later made to an entry, the corrections should be signed and dated by both the scientist and witness.
12. *Be certain that the laboratory notebook is stored in a secure location.*

(discussed later in this chapter). Paper printouts from instruments, photos, and other forms of raw data can be taped securely in the laboratory notebook, as described in Table 6.5. Systems to deal with data existing on a computer (such as those collected by an instrument) vary. In some cases, a printed version of the data is taped into the notebook or filed and referred to in the notebook. In any event, the researcher must save all raw data, ensure that these data are never altered or edited, and ensure that they are retrievable.

Laboratory notebook entries should include ideas as well as experiments. Researchers should explain why each experiment is performed and, at the end, summarize what the results show, avoiding derogatory comments about their work or ideas (even if something did not go as planned).

One of the challenges in keeping a good laboratory notebook is that it must record clearly what actually happened in the laboratory. It is common to outline in a laboratory notebook what one *intends* to do; these plans should be written in the present or future tense. What *actually* occurred must be clearly recorded using the past tense.

Another challenge is that a laboratory notebook must be chronological – it must keep moving forward in time. A researcher might be working on more than one project, in which case it is often simplest to keep a different laboratory notebook for each project. If only one laboratory notebook is used, it is not correct to leave empty pages or spaces on a page in order to fill them in later. Rather, if a researcher returns to a project after recording information on something else, it is correct to state at the top of the page that the work is "continued from page___".

In many situations, such as when research might lead to a patent or regulated product, the laboratory notebook must be witnessed. The witness, who is not one of the inventors, reads, signs, and dates each entry. The witness should be sure that he/she understands the entries because the witness is corroborating that the work described really happened. It is preferable that the witness is someone who actually observed the experiments, although this may not be possible in practice. The witness must carefully look for mistakes or omissions, such as an erroneous calculation or a missing date. These mistakes are then corrected, briefly explained, signed by both the researcher and witness, and dated. After a notebook is witnessed, no changes should be made to that page. Ideally, notebooks should be witnessed every day. Figure 6.1 shows an annotated laboratory notebook that illustrates some of these ideas.

TABLE 6.5

Typical Components of a Laboratory Notebook

1. *In the front of the notebook: the person to whom the book is assigned, the project, the date of assignment, the company/institution, and any other identifying information.*

2. *A table of contents on the first pages.* The table of contents should include page numbers and descriptions with sufficient detail to allow easy searching of the notebook's contents.

3. *For each project, a listing of the results of any literature search and any experimental information collected from colleagues.*

4. *A page number on every page in consecutive order.* It is essential that every page is numbered before the notebook is used. This is to ensure that no pages are torn out.

5. *Dates, titles, and descriptions.* Begin the record of each day's work with the date, title, and description of the objectives for the work. It is common to begin each day on a new page with a diagonal line drawn across the unused part of the previous day's page.

6. *The rationale for each activity performed.* Documenting ideas is important.

7. *Any relevant equations or calculations.*

8. *Complete descriptions of all instrumentation (including models and serial numbers), chemicals used (including manufacturers, catalog and lot numbers, and expiration dates), reagents used (including recipes or references to SOPs), supplies used, samples assayed, standards or reference materials used.*

9. *Procedural details.* If an SOP or protocol from the researcher's institution or company is followed, it should be referenced in a unique fashion (e.g., by title and revision date, and/or by ID number). It is usually not necessary to copy an SOP or protocol into the notebook, but any deviations and their justification should be noted. If a procedure comes from a compendium, journal, book, or manual, the complete reference for the procedure should be cited, and it may be necessary to record procedural details in the laboratory notebook.

10. *Sample information.* When samples are tested, information should be provided regarding their source, storage, identifying information, disposal, and so on.

11. *Data.* Data take many forms, for example, values read from instruments, color changes observed, photos, and instrument printouts. Printouts from instruments, photos, and other paper data are generally signed, dated, and securely taped into the laboratory notebook with a permanent adhesive. It is common to sign or write across both the inserted paper and the page on which it is taped in order to authenticate the paper's placement on the page. In such cases, the laboratory notebook page, signature, and date should be noted on the printout so they can be replaced in the proper place if they become detached. It is common to record on a page how many documents are taped to that page so they can be found if necessary. If printouts cannot be affixed in the notebook, they may be titled, signed, explained, dated, and filed securely. The data and its storage location should be referenced in the laboratory notebook.

12. *Observations.* Observations might include, for example, changes in pH or temperature, humidity readings, and instrument operational parameters.

13. *A brief summary of the work completed.*

14. *A conclusion and brief interpretation of data collected is usually appropriate.* For example, if a particular line of investigation is pursued on the basis of preliminary results, note this in the laboratory notebook.

6.3.4 OTHER DOCUMENTS THAT ARE COMMON IN LABORATORIES

6.3.4.1 Standard Operating Procedures

Most production facilities and many laboratories use procedures to instruct personnel in how to perform particular tasks. A **procedure** *is a written document that provides a step-by-step outline of how a task is to be performed.* Such documents are often called standard operating procedures (SOPs). Everyone follows the same procedures to assure that tasks are performed consistently and correctly. Standard operating procedures describe what is required to perform a task, who is qualified or responsible for the work, what problems may arise and how to deal with them, and how to document that the task was performed properly. SOPs must be written so they are clear, are easy to follow, and can accommodate minor changes in instrumentation.

Many organizations use templates so that within the organization, every procedure has the same format and includes the same sections and types of information. There can be slight differences between organizations in how SOPs are formatted and what information they include, but they all usually have the components outlined in Table 6.6.

434   ①   PROJECT: TAT3 Blockers   Page 4
Continued from page 3   Date: Jan 3 '08

Title: Inhibition of MabTAT3 binding by test compounds ANTAT3-1 and ANTAT3-10

②   ③
Purpose: To test compounds ANTAT3-1 and ANTAT3-10 for their ability to competitively block
antibody binding to TAT3. As described (nbk. 425, p.1). We have isolated a cell surface molecule,
TAT3, involved in spermatozoa maturation and development. Monoclonal antibody MabTAT3 binds
TAT3 with high affinity. Testing and preparation is described in nbk. 426, pp. 67-92. TAT3 is
characterized in nbk. 425, pp. 5-32. The preparation and initial screening of ANTAT3-1 and
ANTAT3-10 is described in nbk. 432, pp. 1-22.   ⑤

                 ⑥   ④
Methods: ATCC cell line 7456A was cultured and plated (10⁴ cells/well) in 35 mm² wells and grown to
confluency. Culturing conditions are described in nbk. 425, p. 3. ANTAT3-1
and ANTAT3-10 were diluted in HBSS (GIBCO) in 10-fold dilutions ranging from 4×10^{-3} ug/ml to 40
ug/ml. MabTAT3 was diluted from a stock solution of 25 ug/ml and used at between 0.05 ug/ml
and ~~10 ug/ml 15 ug/ml~~ obtained from MLP. ⑮

        ⑧ 10 ug/ml MabTAT3   JCL 1/4/08
                        ⑦   Ran out of reagent ⑨

⑩
Samples were distributed in the cell samples as diagrammed on page 5. Cells were incubated
for 2 hrs, 37 C and harvested and lysed as described (nbk. 426, p. 140).
MabTAT3 bound to 7456A cells was quantitated by modified ELISA described in nbk. 432, p. 80.
Raw data are provided on p. 6. ⑪

⑫
Results: The experiment failed, no ANTAT3-1 or ANTAT3-10 binding was detected in any of the
wells except at the highest doses, but I think this is an artifact.

                                                        ⑬
⑭   ⑯
1/3/08 Title: Abiloty of ANTAT3-1 and ANTAT3-10 to inhibit spermatozoa maturation in mice.
Method: BalbC mice were injected i.p. with 0.5 ml of a 1 mg/ul solution of

                                            Work continued to page 5

Signature: Julie Lawson   Date: January 3, 2008 ⑰
Read and Understood by: ⑱   Date: March 10, 2008   ⑲
Read and Undersood by:

                Confidential Property of Receptor Blockers, Inc. ⑳

(a)

FIGURE 6.1   An annotated laboratory notebook. (a) A notebook page. (b) Legend. (Reprinted with permission from Merchant & Gould Law Firm.)

(Continued)

Legend for Notebook Diagram

1. Top of the page identifies the notebook number, the page the work continued from, the page number, date, and project.

2. The entries are organized and legible.

3. Laboratory abbreviations and designations are defined and referenced.

4. Methods are provided in sufficient detail so that a third party could repeat the experiment using only the references and materials supplied in the notebook.

5. Methods are referenced to earlier notebooks. The term "nbk" should be spelled out in a designated place in each notebook.

6. The cell line is sufficiently identified by its supplier.

7. Well-known abbreviations do not need to be defined further.

8. An initial correction is made with a single line. The corrected text is placed in line, next to the error.

9. A later correction is properly initialed, dated, and explained.

10. The entry is in a single permanent ink.

11. Raw data are identified and entered into the notebook.

12. If the experiment did fail, a simple statement that the experiment will be repeated is sufficient. Here it appears that the results may show that the blockers did work at higher doses. Results should be stated positively and repeated as necessary.

13. Blank regions are blocked out in pen.

14. New entries are re-dated.

15. We do not know who or what MLP is. If it is a supplier, it should be spelled out or referenced to an earlier page. If it is a person, the name should be spelled out or the initials provided in the abbreviations index.

16. "i.p." is a well-known scientific term of art for individuals in this field and need not be further identified. The test is whether an abbreviation could be reasonably interpreted by someone similarly skilled in the art.

17. This notebook page was timely signed and dated.

18. The signature is illegible. Where this is a problem, the name should be printed at least once, beneath the similar signature. Also, if MLP is Mark Peterson, then a question arises as to whether Mark is an inventor. Inventors must not witness notebooks reducing their invention to practice.

19. The witness date is much too late. Preferably, the witnessing signature is provided within the same week or two week period.

20. Each page is labeled as confidential and the property of the particular research organization.

(b)

FIGURE 6.1 (*Continued*) An annotated laboratory notebook. (a) A notebook page. (b) Legend. (Reprinted with permission from Merchant & Gould Law Firm.)

TABLE 6.6

Typical Components of an SOP

1. *Title.*
2. *Any safety concerns relating to the procedure.*
3. *ID number, revision number, date of revision.* Most SOPs are revised periodically. It is essential that all staff use only the most up-to-date revision.
4. *Statement of purpose* (may restate the title with a little more detail).
5. *Scope* describes when the procedure is relevant. For example, if the procedure is for verifying the performance of a balance, then it might be only for a particular brand, or a particular model.
6. *A statement of responsibility,* who does this task.
7. *Materials required, including manufacturers, and identifying information.*
8. *Calculations required, preferably with an example.*
9. *The procedure itself, written as a series of steps.* The actions required, how they are performed, and the endpoints of the steps should be included.
10. *References to other documents, as required.*
11. *How to document that the procedure was performed, references to any associated forms.* An SOP is a directive document, but it may be associated with a form that is used to collect data.

Example 6.2

Discuss the considerations in writing an SOP to test the oven that is used in a company that makes chocolate chip cookies.

1. **Preliminary steps**. A first step is to identify the critical features of the oven. Temperature is obvious and will need to be described as a range of acceptable temperatures. Consistency of temperature and oven cleanliness are other features that might be important and must be described clearly.
2. **Raw materials/equipment**. What equipment is necessary to complete this task? Obviously, a thermometer is required, but what type and how is the thermometer's accuracy ensured? (See Chapter 19 for more details about temperature measurement.) Safety equipment, such as gloves for thermal protection, should also be identified.
3. **Writing the SOP**. The SOP will include all the components listed in Table 6.6.
4. **A form** or other data collection document should be associated with the SOP in which the baker will record information about the condition of the oven.
5. **Integration into processes of the company**. Testing the oven is just one part of the cookie manufacturing process. The results of the oven test must be integrated into the whole process. When will this testing occur? Who will do it? How can the company be certain that the oven was tested? If the oven is not performing properly, what should the baker do about it? All these questions need to be answered to ensure quality in the cookie products.

Example 6.3

Consider briefly the language used as part of an SOP to prepare a chemical solution.

a. Mix the components of the solution for 10 minutes at room temperature.
b. Mix the components on a magnetic stir plate for 10 minutes at room temperature.
c. Mix the components on a magnetic stir plate until they are well mixed.

    d. Mix the components on a magnetic stir plate until they are well mixed. Ten minutes at room temperature is usually sufficient, but if particulates are observed, continue mixing until they disappear.

Statement **a** is vague and does not direct the technician in how to accomplish the task or its endpoint. Statements **b and c** tell the technician to use a magnetic stir plate, which is helpful, but might not be sufficient information for an inexperienced person. Statement **d** has the clearest description of an endpoint and is reasonably flexible. Note that if stirring too long at room temperature is problematic, this should be noted.

Example 6.4

Carol DeSain (DeSain, Carol, and Sutton, Charmaine Vercimak. *Documentation Practices*, Advanstar Communications, 1996) cites an example in which the language of a procedure is clear, but the intent is not to a new employee. The statement in the SOP is: "Wash the filter press in mild detergent and rinse with WFI [purified water]. Make sure that the filter support grid is completely dry before placing a new filter on the grid." The writer expects the grid to be air- or oven-dried. The technician, on the other hand, focuses on the need for the grid to be <u>dry</u>. She therefore retrieves a box of laboratory wipes and wipes the grid dry. This action is not in violation of the SOP, but it results in the contamination of several batches of product with fibers from the laboratory wipes.

This example highlights the importance of knowing the audience. The degree of detail in an SOP and the particular points that are noted will, to some extent, depend on the training and background of the people who will be performing the procedure. It is important when writing an SOP to be conscious of the needs of the reader and when following an SOP to be certain that you understand its intent.

In a company, every SOP must be reviewed and accepted before it is used. The person who writes the SOP has responsibility for it and must sign it. A second individual who is knowledgeable about the work also normally shows approval with a signature and, in a company, an individual from the quality assurance unit also signs it.

SOPs periodically require changes. It is important that the old SOPs are destroyed or made unavailable when changes are made, except for a copy(ies) kept in a historical file. Only the latest revision of the SOP should be available to processing or laboratory technicians. In a regulated setting, depending on the magnitude of the change, revisions may have to be checked and approved by several levels of responsible individuals, and by the quality-assurance unit before the change can be implemented. Each revision of an SOP needs a date and revision number so that it can be uniquely identified.

An example of an SOP is shown in Figure 6.2, and some of the potential problems with SOPs that must be avoided are summarized in Table 6.7.

6.3.4.2 Forms

An SOP is often associated with a **form** that is filled in as the procedure is being performed. Filling in the blanks requires the individual performing the task to monitor the processes as they go along, thus ensuring that everything is going smoothly. In addition, the form will remind the technician to record information about lot numbers, raw materials, times, temperatures, and other relevant information that is easy to forget to record. In some production laboratories, a witness must sign key steps. Figure 6.3 shows an example of a form.

6.3.4.3 Protocols

The terms "protocol," "procedure," and "SOP," are similar, and in many research laboratories, the terms "protocol" and "procedure" are used synonymously. They can, however, be distinguished. The term **protocol** *is used in some industries to refer to a procedure that tells an operator how to perform a task or an experiment that is intended to answer a question or test a hypothesis.* The term *protocol* may also be used for a procedure that will only be performed one time. In contrast, the term "procedure" or "SOP," using this terminology, refers to procedures that do not lead to the answer to a question. For example, one follows a *procedure* or an *SOP* to clean a laminar flow hood, but one follows a *protocol* to investigate the effectiveness of cleaning a laminar flow hood with different cleaning agents.

Clean Gene, Inc.

STANDARD OPERATING PROCEDURE

SUBJECT: Identification Method for Calcium Sulfate   Prepared By *N. Warren*

Effective Date: 7/1/20

SOP # 2648   Approved By *J. McMillan*   Date *6/24/20*

Revision 01

Page 1 of 3   Approved By *J. Lownd*   Date *6/26/20*

1. Scope

    Provides the method for assessment of the raw material calcium sulfate.

2. Definitions—NA

3. References:

    3.1 USP, current—General Identification Tests

    3.2 Incoming Raw Material/Component specification: $CaSO_4$, Grade I

4. Reagents

    4.1   3N Hydrochloric Acid, HCl

    4.2   6N Acetic Acid

    4.3   Methyl Red

    4.4   Ammonium Oxalate

    4.5   Ammonia

    4.6   Barium Chloride

5. Responsibility

    5.1 This test is to be performed by an appropriately trained analyst. Analyst training and documentation of training will be conducted per SOP 5688. Approval of analyst data is the responsibility of the Quality Control Manager.

    5.2 The results of the test will be verified by a second qualified analyst.

6. Hazard Communication

    6.1   *3N Hydrochloric Acid, HCl*

    DANGER: Corrosive. Avoid contact with skin and eyes. Avoid inhalation of fumes and mist. Do not mix with caustics or other reactives.

    6.2   *6N Acetic Acid*

    DANGER: Corrosive. Avoid contact with skin and eyes. Avoid inhalation of fumes and mist. Do not mix with caustics or other reactives.

    6.3   *Methyl Red*   **CAUTION: Irritant. Avoid contact with skin and eyes.**

    6.4   *Ammonium Oxalate*

    CAUTION: Irritant. Avoid contact with skin and eyes.

    6.5   *Ammonia*   **CAUTION: Irritant. Avoid contact with skin and eyes.**

    6.6   *Barium Chloride*

    CAUTION: Irritant. Avoid contact with skin and eyes.

7. ATTACHMENTS

    7.1 Attachment I—Form No. 687

8. PROCEDURES

    8.1 Sample Preparation

    8.1.1   In a clean beaker or test tube, add approximately 200 mg of the incoming Calcium Sulfate to be tested to 4 mL of 3N Hydrochloric Acid and 16 mL reagent water.

    8.1.2   Gently warm with low stirring on a hot plate to aid dissolution.

    8.1.3   Portions of the solution should respond to the tests for calcium and sulfate.

    8.2 Calcium Identification

    8.2.1   Place about 10 mL of the solution from 8.1 into a clean test tube.

    8.2.2   Place 2 drops of methyl red into this solution followed by enough ammonia to turn the solution YELLOW.

    [and so on . . . the entire SOP is not shown here]

FIGURE 6.2   A portion of an SOP. This SOP is used to test an incoming raw material to ensure it is calcium sulfate.

A protocol must include information on what data are to be collected, how the data are to be gathered, what outcome proves or disproves the hypothesis, and any statistical methods that need to be used. A protocol may refer to SOPs. Both research and production facilities use protocols. In research laboratories people obviously investigate questions all the time. In production facilities people investigate whether

TABLE 6.7
Problems to Avoid Relating to SOPs

1. *The SOP says what to do, but not how to do it.*
2. *The procedure was written by someone who does not have experience doing the work.*
3. *The SOP has too much detail or too little detail.*
4. *The procedure is not written in the order in which the tasks are actually performed.*
5. *The SOP is not updated as needed.*
6. *Employees cannot find the right SOP or use an older version.*

TABLE 6.8
Typical Components of a Protocol

1. *The hypothesis or question the study is designed to answer.*
2. *A description of the study.*
3. *A plan for how the study is to be conducted.*
4. *Information about how the samples are to be collected, processed, and identified.*
5. *The methods that will be used to test the hypothesis.*
6. *The schedule of testing.*
7. *The way the study results and conclusions will be reported.*
8. *The criteria that will be used to reach conclusions.*

a product performs as expected, the qualities of the product under certain conditions (such as long storage), the effects of the product in a test population, and so on. Example statements or hypotheses that might be seen in an industrial protocol are as follows:

> This study is designed to demonstrate that the cleaning process for laminar flow hoods does not leave detectable detergent residues on the surface of the hood.
>
> This study is designed to demonstrate that the antitumor drug XYZ has no adverse medical effects on the livers of test subjects.

The components of a protocol are somewhat different from a procedure and are summarized in Table 6.8. An example of a portion of a protocol is shown in Figure 6.4.

6.3.4.4 Reports

A **report** *is a document that describes the results of an executed protocol*. The report summarizes what was done, by whom, why, the data, and the conclusions. A report is written in narrative format. Reports from basic scientific research are published in scientific journals. Reports from investigations performed in a company may or may not be published, but must be available for inspection.

Note that the documentation provided by a laboratory *notebook* is quite different from a *report*. A scientific research report is based on the information that is recorded in a laboratory notebook, but it is not located on the pages of the notebook. Rather, the report is prepared using a computer word processor. A report is a formal discussion of laboratory work

---

Clean Gene, Inc.

Page 1 of 1

AGAROSE GEL PREPARATION FORM, REVISION 01

1/31/18 Form # 992

Technician Name _____ Date _____

1. *Add between 1.8 and 2.0 grams of agarose to 100 mL of buffer.*
   Record the ID/lot number for the agarose _____
   Weight of agarose added _____
   Source of buffer used _____
2. *Place the mixture in a microwave oven on its highest setting and heat until the mixture just begins to boil.*
   Time to boiling _____
3. *Let the mixture cool until it is between 58 and 62°C and then pour gel.*
   Temperature when pouring gel _____

FIGURE 6.3   A portion of a form associated with preparing an agarose solution.

Clean Gene, Inc.

CLEANING PROTOCOL CL 898; REVISION 01

Prepared by *Tania Seid*   Date *3/6/20*

Approved by *J. Mavey*   Date *5/15/20*   Approved by *Jon Reidman*   Date *5/16/20*

PROTOCOL

A TEST OF THE EFFICACY OF ALCOHOL FOR DISINFECTING THE SURFACE OF BIOLOGICAL HOODS

1.0   Study Question

This study is designed to determine the efficacy of alcohol for disinfecting the surface of biological hoods, BH655.

2.0   Objectives

    —To determine whether surface cleaning of a contaminated biological hood with ethanol is an effective method of eliminating bacteria.

    —If ethanol is effective, to establish the optimal ethanol concentration.

3.0   Plan

    —Contaminate the surface of the biological hood with 10^9 cfu of *E. coli* strain 56298.

    —Treat the surface with ethanol of different concentrations

    —Determine the success of disinfection by using the Standard Wipe Method of detection of bacteria, SOP 87-68.

4.0   Equipment required

    —Biological cabinet # 4

    —Spectrophotometer # 7

    —Overnight culture of *E. coli* strain 56298

    —Ethanol at a concentration of: 60, 70, 80, 95%.

    —Sterile bacteria wipes (ID No. 932-8)

    —Microbial test agar (ID No. 56-84)

    —Sterile forceps (ID No. 6-1-97)

    —Nutrient agar plates (prepared according to SOP 87-75)

5.0   Procedure

    5.1.   Grow overnight culture of *E. coli* (according to SOP 95-17).

    5.2   Measure the absorbance of culture at wavelength 520 nm.

    5.3.   Calculate the number of cfu/mL.

         1 AU $= 10^9$ cfu/mL

    5.4.   Dilute a portion of the culture to get a concentration of 10^9 cfu/mL

    5.5.   Spread 1 mL of diluted culture over a 2 square inch section of the hood.

    5.6.   Wait 5 minutes.

    5.7.   Spread 10 mL of ethanol across the surface of the contaminated hood.

    5.8.   Wait 5 minutes.

    5.9.   Using a clean, sterile absorbent wiper, wipe up the ethanol.

    5.10.   Allow the area to dry for 5 minutes.

    5.11.   As directed in SOP 87-68, using the sterile forceps and wipes, wipe the entire 2 square inch section of the "spill."

    5.12.   Place the bacterial test swipe on nutrient medium.

    5.13.   Repeat steps 5.5 through 5.12, three times for each concentration of ethanol, for a total of 12 plates.

    5.14.   After 36 hours record the results, as directed in SOP 87-68.

         [. . . and so on]

FIGURE 6.4   A portion of a protocol. The protocol directs the activities of the technician in conducting a study.

that is written in a particular format. A conventional report format begins with an *Introduction*, followed by *Materials and Methods*, then *Results*, and finally a *Discussion*. In contrast, a laboratory notebook is written as a chronological account of what an individual did, in the order that it was done. The key issue in a laboratory notebook is to make sure it is chronological. While a laboratory notebook should include all the materials, methods, and results that will be included in the formal report, the laboratory notebook seldom includes the formal discussion required for a report. A report is written in paragraph form with correct grammar and careful attention to spelling. A laboratory notebook may or may not include text written in paragraph form. A laboratory notebook must be clear and legible, but it is not supposed to be a "thing of beauty."

6.3.4.5 Logbooks

Logbooks *are used to record information chronologically about the status and maintenance of equipment or instruments.* When an item is used, calibrated, maintained, or repaired, this is indicated in the logbook. Logbooks are conventionally bound and labeled paper notebooks that are associated with a specific instrument, area, or piece of equipment. However, modern equipment commonly incorporates automatic logging software that keeps track of its operation, maintenance, and so on, thus avoiding the use of paper notebooks that can be damaged or misplaced.

6.3.4.6 Recordings from Instruments

Many analytical instruments generate printouts of results. Printouts are generally considered to be raw data. If the results from the instrument belong in the laboratory notebook, then it is usually acceptable to tape them there. Alternatively, instrument paper printouts can be filed and referred to in the notebook or on a form using some clear and organized tracking method.

Some instruments monitor themselves and record their operating parameters as they operate. For example, modern autoclaves continuously record the date, time, temperature, and pressure throughout their cycle. This information is important in demonstrating that the instrument performed properly. Files may be kept of such recordings, or they may be associated with the production records for a particular batch of product. (Batch records will be discussed later.)

Printouts from any instrument must be thoroughly identified (including the date, product name, batch number, and equipment number) and be signed and dated by the technician.

6.3.4.7 Analytical Laboratory Documents

Analytical tests are those that measure a property(ies) of a sample. For example, in an environmental testing laboratory, analysts might test a sample of lake water to determine the level of cadmium present. In a clinical laboratory, analysts might perform a drug screen of a patient's blood. The product is the test result(s). Documentation is required that provides information about the method used, the sample being tested, and the results. Specific types of information that must be documented are summarized in Table 6.9.

6.3.4.8 Identification Numbers

Identification numbers uniquely identify items. There are many types of items that require identification, including raw materials, documents, equipment, parts, batches of product, chemicals, solutions, and laboratory samples. Identification numbers convey two pieces of information: what the item is, and which one it is.

TABLE 6.9

Essential Information to Document in Analytical Laboratories

1. *Information regarding each assay method used*:
   - The SOP detailing the test
   - The purpose of the test
   - The limits of the test (e.g., what is the lowest level of the material of interest that the test can detect)
   - The origin of the test method (e.g., whether it came from a compendium of commonly accepted methods)
   - Method validation information (method validation is discussed in Chapter 26)
   - The suitability of the test for a given purpose
2. *Information regarding each sample*:
   - The sample ID number
   - How the sample was collected, by whom, and on what date
   - Where the sample is stored and the conditions of storage
   - How and when the sample is to be discarded
3. *Information regarding each assay*:
   - The sample tested
   - The date of the test
   - Analyst who performed the test
   - Reagents and materials used
   - The method used to test the sample
   - The raw data collected during testing
   - Calculations for the sample results
   - Reported conclusions based on the test

For example, the first part of an ID number might logically tell whether the item is a particular type of instrument or a particular type of solution. The next part of the ID number might tell which particular one of those instruments it is, or which batch of solution it is.

6.3.4.9 Labels

Labels identify equipment, raw materials, products, and other items. Information that may be found on labels is summarized in Table 6.10. An example of a label is shown in Figure 6.5.

6.3.4.10 Chain of Custody Documentation

In laboratories that handle many samples that come from diverse subjects or sites, it is critical that information and results are always associated with the correct sample. **Chain of custody documentation** *provides a chronological history, or "paper trail," for samples.* For example, in a clinical testing laboratory patient samples must not be confused with one another. An environmental testing laboratory tests samples from many sites and must keep track of them. Forensics laboratories must scrupulously keep evidence from various cases in order; otherwise, their test results will be invalid in court.

Chain of custody documents are a method of organizing information about samples. Each sample must be assigned a unique ID number. Records for each sample must show the source of the sample, who collected the sample, who transported the sample, its condition upon receipt, the date of receipt, how the sample was processed and tested in the laboratory and by whom, how it was stored, and how it was disposed of, if relevant. The sample is logged in and out as it is moved and processed. The format and the exact nature of these records are variable as each organization has its own requirements.

6.3.4.11 Training Reports

Training reports are associated with individuals working in a facility. The training report shows the training the person has completed, the dates, the purpose, and so on. Training reports help to show that individuals are competent to perform their work. In research laboratories, for example, training records are used when employees have been trained in the use of radioisotopes. An example of a training report is shown in Figure 6.6.

TABLE 6.10
Typical Components of a Label

1. *The ID number of the item*
2. *The person responsible for the item*
3. *Date the item was prepared*
4. *The lot number*
5. *The identity, name, or composition of the item*
6. *Safety information (see Section 9.4.4.2)*
7. *The name of the company or institution*
8. *Storage and stability information*

Clean Gene, Inc.

EMPLOYEE SOP TRAINING RECORD

Form # 875
Revision 01, 06/07/18

EMPLOYEE'S NAME

Rachel Allen

SOP TITLE	TRAINING DATE	EMPLOYEE'S INITIALS	TRAINER'S SIGNATURE
Cleaning Hoods	01/19/20	RA	JES
Formulation 10	02/12/20	RA	LW
$CaSO_4$ Analysis	04/08/20	RA	JMM

FIGURE 6.6 Portion of an employee SOP training record.

Clean Gene, Inc.

Solution Name: ____________

Concentration: ____________

Date Prepared: ____________ By: ____________ ID No.: ____________

Expiration Date: ____________

Reagent Source: ____________ Lot No: ____________

Storage Conditions: ____________ Special Precautions: ____________

FIGURE 6.5 An example of a label for a reagent.

6.3.5 Documents That Are Specific to Production Facilities

6.3.5.1 Batch Records

There are various types of documents found in production facilities; a few of these will be explained here.

Batch records accompany a particular batch of product. A **batch record** *includes step-by-step instructions that detail how to formulate or produce a product, including raw materials required, processing steps, controls, and required testing.* In this sense, it is a directive document, like an SOP. *A batch record also provides blanks in which the operator(s) records information and documents activities as they are performed.* In this sense, the batch record is a data collection document.

A **master batch record** *is the original signed batch record.* Each time a new production run begins, the master batch record is copied to generate a batch record for that batch. The batch record is officially issued to the production crew by the quality department. It is essential that the batch record that is issued is complete, readable, and correct.

Table 6.11 describes some of the major components of a batch record. Figure 6.7 shows portions of a batch record.

6.3.5.2 Regulatory Submissions

Regulatory submissions *are documents completed to meet the requirements of an outside regulatory agency.* For example, before testing an experimental drug in humans, a company must submit an application to FDA showing its preliminary research on the drug, its plan for human studies, and other relevant information. (This document is known as an IND, or Investigational New Drug Application.)

6.3.5.3 Release of Final Product Records

Companies must complete product release documents when a product has been manufactured and tested. The release document certifies the product, shows its specifications and any testing performed, establishes that the product documentation has been reviewed and approved, and states that the product is ready to be sold.

6.4 ELECTRONIC DOCUMENTATION

The entire system of good documentation practices in the biotechnology industry was created in a world of paper records. But this assumption does not meet the reality of our "electronic age." Computers are widely used to obtain, analyze, and store information. Most modern laboratory instruments are connected to computers and/or are controlled by internal microprocessors. Raw materials, products, and samples are routinely bar-coded, and their movements and disposition tracked by computer.

It is easy to imagine the advantages to replacing paper with electronic documentation. For example, manufacturing technicians traditionally record critical information (e.g., material lot numbers, times tasks are performed, temperatures, test results, calculations, and equipment identifications) on paper batch records. An electronic batch record system has the potential to use the computer to detect errors such as an omitted lot number, a temperature that is not in the correct range, and a test result that is improperly transcribed. For example, Practice Problem 1 at the end of this chapter shows a batch record with errors. A computerized system can help avoid these errors by requiring properly formatted data in every blank (Figure 6.8). The computer's ability to instantly recognize errors can prevent costly mistakes.

TABLE 6.11

Essential Components of a Batch Record

1. *Product identification*
2. *Document identification*
3. *Company name*
4. *Dates of manufacturing*
5. *A step-by-step account of the processing and testing to be done*
6. *The monitoring specifications – how will the operators know if the process is proceeding properly?*
7. *Raw data that must be collected and blanks to fill in to record it*
8. *Bill of materials and equipment.* This is a list of the reagents, equipment, and other materials required to make the batch. Each item is listed in the master batch record, and information about each item (for example, lot numbers) is filled in when the product is manufactured.
9. *Required signatures*

(a)

Effective 6/10/2020 Batch #__________ page 1 of 2

BATCH RECORD
LYSIS SOLUTION – VP SP-0207-00

Batch Number:______________________ Exp. Date:______________
(6 months from date of mfg.)

Date of Manufacture:____________________

1.0 BILL OF MATERIALS:

REAGENTS	PART NO.	LOT NO.	EXP. DATE	*STD. QTY	MULT BY	QTY USED
Tris base	RR-0111-00			11 g		
EDTA	RR-0040-00			3.7 g		
N-lauryl-sarcosine	RR-0088-00			50.0 g		
Sodium dodecyl sulfate (SDS)	RR-0095-00			5 g		
ProClin 150	RR-0176-00			1 ml		
HCl, 1 N	RR-0171-00			~50 ml		
NaOH, 1 N	RR-0196-00			as req.		
Water, deionized	RR-0116-00	N/A	N/A	qs to 1 L		

***Standard formula is for 1000 ml.**

1.1 ACCOUNTABILITY.

Amount Requested		Amount to QC (10 ml)	
Amount Manufactured		Amount to Retention (10 ml)	
Amount Lost/Discard		Other__________	
		Yield	

1.2 COMMENTS:

FIGURE 6.7 A portion of a batch record.

(Continued)

Computers similarly have advantages in a research environment. Researchers are beginning to use **electronic laboratory notebooks** (ELNs) *which are computers with software designed to perform the roles of a traditional laboratory notebook.* Electronic laboratory notebooks can easily store huge amounts of data, search the data, and enable researchers to readily communicate with others via computer. Researchers often have difficulty in retrieving their raw data when they were recorded in paper laboratory notebooks. If

(b)

Part No SP-0207-00D Rev. Level: B DRF # 989 Superseded 11/7/2019
Effective 6/10/2020 Batch # ___________ page 2 of 2

2.0 MATERIALS\EQUIPMENT\REFERENCE DOCUMENTATION:

3.0 PROCEDURE:

NOTE: The solution may be allowed to stir overnight in a sealed container to ensure that all reagents are completely dissolved.

3.1____ Add approximately 3/4 the volume of water and a stir bar to a glass container. Start stirring.

3.2____ Weigh and add Tris base.

Witness to Tris addition: ______________________ Date: __________

3.3____ Weigh and add EDTA. Stir to dissolve.

Witness to EDTA addition: ______________________ Date: __________

3.4____ Weigh and slowly add N-lauryl sarcosine.

Witness to sarcosine addition: ____________________ Date: __________

3.5____ Weigh and add SDS. Stir until dissolved.

Witness to SDS addition: ______________________ Date: __________

3.6____ Measure and add ProClin. Mix well.

Witness to ProClin addition: _____________________ Date: __________

3.7____Adjust the pH to 7.9 - 8.1 at 20-25 C with 1 N HCl, and 1 N NaOH as required.

Starting pH__________________ Temp_________C

Vol. HCl added_____________________________

Vol. NaOH added_____________________________

Final pH____________________ Temp_________C

3.8____ Adjust to final volume with water.

FIGURE 6.7 ***(Continued)*** A portion of a batch record.

pH METER #7897 CALIBRATION RESULTS			
Standard	Standard ID Number	User	Date
pH 4	ID number missing	*MFS*	*05/27/20*
pH 7	ID number missing		
pH 10	ID number missing		

FIGURE 6.8 Computers can be programmed to recognize missing data or data that are out of range.

a researcher leaves a laboratory and takes their notebook, it can be hard to obtain their raw data later on. Also, searching through years of paper records looking for a certain table or a bit of information can be a formidable task. An electronic notebook can make searching much easier and can archive information in an organized way. An ELN also allows multiple people to be granted secure access to data, even if they are not located in the same geographic region.

However, while the advantages to using computers for data management are substantial, there are problems associated with ensuring the quality of data captured and stored electronically. In March 1997, the FDA issued regulation 21 CFR Part 11: Electronic Records; Electronic Signatures; Final Rule, to address the role of computers in documentation in the pharmaceutical industry. The purpose of the Part 11 regulations is to "provide criteria for acceptance by FDA, under certain circumstances, of electronic [computer] records, electronic signatures, and handwritten signatures executed to electronic records as equivalent to paper records and handwritten signatures executed on paper. These regulations, which apply to all FDA program areas, were intended to permit the widest possible use of electronic technology, compatible with FDA's responsibility to protect the public health." (Food and Drug Administration. "Guidance for Industry Part 11, Electronic Records; Electronic Signatures—Scope and Application." August, 2003. https://www.fda.gov/media/75414/download) Table 6.12 provides some terminology relating to electronic documentation and 21 CFR Part 11.

The Part 11 regulations are intended to encourage pharmaceutical companies to adopt modern electronic documentation methods. At the same time, they require that companies validate new electronic documentation methods to prove they are as secure, reliable, and searchable as paper systems. Creating a computer

documentation system that is compliant with Part 11 requires a sophisticated understanding of computer software and hardware – on the part of both computer vendors and users. The pharmaceutical industry has spent considerable effort and resources developing this understanding, and the lessons they have learned are of general interest.

To understand the challenges in using computers for documentation, consider the ALCOA (attributable, legible, contemporaneous, original, and accurate) requirements. As we have seen previously in this chapter, a variety of good documentation practices have evolved for paper documentation to meet these requirements. Computer systems must have different methods to provide the same controls as paper documents. Consider, for example, signatures, a key element in a paper documentation system. A signature identifies the signer and generally means that a person consents to something. This is the case, for example, when you sign a credit card authorization. The signature is in ink that permanently binds the signature to the paper so that it is difficult to remove without leaving a trace. Your signature on the form means that you are the proper holder of the credit card and that you agree to the charge. In a paper laboratory notebook, an individual's signature identifies that person and is a method of attesting to the truth of the data recorded. A commonly accepted electronic signature is a log-on procedure in which individuals must enter a unique user ID and secret password. Everything recorded onto the computer while that individual is logged on is attributed to that person. If a person logs on with another's password, it is comparable to forging another person's signature. In situations where more assurance of an individual's identity is required, sophisticated methods of authentication, such as voice recognition and retinal scans, can be used.

There are guidelines (as summarized in Table 6.4) to assure the chronology of events recorded in a paper laboratory notebook. Chronology in electronic systems is usually handled with a "time stamp" that is automatically added by the computer. Time stamps require software that "knows" the correct time, and that can detect if someone attempts to alter a time stamp.

It is important that records are safe and that only those with the authority to see them have access. Paper laboratory notebooks can be stored in a locked file cabinet or secure storage facility to meet these requirements. With a computer system, passwords are a primary method of maintaining security and much effort has gone into designing systems to prevent

TABLE 6.12

Some Vocabulary Often Used with Reference to 21 CFR Part 11 and Electronic Documentation

Audit trail. A secure, computer-generated, time- and date-stamped record that allows the reconstruction of a course of events relating to the creation, modification, and deletion of an electronic record.

Biometrics. A method of verifying an individual's identity based on the measurement of physical features or repeatable actions that are unique to that person. For example, fingerprint or retinal scans can be used to identify individuals. A handwritten signature can be considered to be a biometric method.

Closed system. A computer system in which access is controlled by the people who are responsible for the content of the system's records. For example, a system of computers that is only accessible to the individuals who work in a company is a closed system.

Electronic Laboratory Notebook, ELN. Any of a wide variety of software programs/computer systems designed to fulfill the functions of traditional paper laboratory notebooks. ELNs provide the advantages of computers: They can be electronically searched for specific information, they can archive very large data files, they can hold and display graphics, and they can be used to share information locally and remotely.

Electronic records. Text, graphics, data, audio, or pictorial information that is created, modified, maintained, archived, retrieved, or distributed by a computer system.

Electronic signature. A computer equivalent to a handwritten signature. In its simplest form, it can be a combination of a user ID plus password. It may also include identification based on biometric characteristics.

Encryption software. Software that translates information into a secret code in order to provide security. To read an encrypted file, one must have access to a secret key or password.

Hybrid systems. Systems that use both electronic and paper records. For example, a laboratory instrument might be attached to a computer that retrieves and processes data from the instrument and then prints out a result on paper that is signed and dated.

LIMS, Laboratory Information Management Systems. Computer-based laboratory management systems that automate such activities as tracking work requests, tracking samples, printing analytical worksheets, storing data, analyzing data, performing calculations, providing financial statistics, and tracking client requests.

Metadata. Information that describes the content and context of the data. Metadata helps to reconstruct the original raw data. For example, a digital camera produces both a picture and also metadata that includes the camera's shutter speed, f-stop, and other camera settings when the photo was taken.

Open system. A computer system that is not controlled by the persons who are responsible for the content of the system. For example, if a contract laboratory sends data to a company via the Internet, the system is open. Additional security must be in place for open systems as compared to closed systems.

Predicate rules. The CGMP, GCP, GLP, and other regulations (as contrasted with the 21 CFR Part 11 regulations).

Source: Many of these definitions are modified from: Huber, Ludwig. "21 CFR Part 11: Overview of the Final Document and its New Scope." In *21 CFR Part 11; A Technology Primer, Supplement to Pharmaceutical Technology.* Advanstar Communications, 2005.

unauthorized access. Data that are sent through an open system (such as the Internet) pose a particular security problem that is usually solved with encryption software.

With paper documentation systems, operators record entries in permanent ink to prevent their change. With a computer, a software method of tracking modifications to data must be in place to ensure that original recordings are not erased. If someone legitimately tries to add information to a record, the computer must be programmed to "know" that this act is legitimate and to show both the original record and the revision.

The rules in Part 11 require that pharmaceutical companies prove that the various software controls (such as those mentioned in the above paragraphs) are effective in protecting the integrity of their documents. For example, companies (and software providers) must demonstrate that if someone intentionally tries to change a time stamp or penetrate a password-protected site, the software is capable of resisting these incursions and recording the attempts. Companies must show that the integrity of data being recorded at the moment a computer crashes can be guaranteed. Computer software and storage devices rapidly become obsolete. This is a major challenge to companies since they must guarantee that their records will be accessible in the future, even when technology changes.

Resolving these technical issues has slowed the adoption of computer-based documentation in regulated industries. In some cases, the FDA has decided to be flexible to encourage the adoption of electronic documentation. For example, the FDA has decided that companies may archive records in a copied form, such as a PDF file, or a paper record, to avoid problems if an electronic storage medium becomes obsolete. Guidance from the FDA in 2003 indicated that the agency would adopt a risk-based approach to

enforcing the requirements of Part 11. This means that documentation that could be expected to impact patient health must be fully validated and compliant with all the requirements of Part 11, but systems of low risk to public safety and health might be less rigorously secured and validated. A critical record would be, for example, the results of quality-control laboratory tests on a final product that are used to decide whether or not the product should be released for sale. These records are critical because the tests help ensure that the product is of high quality. A lower-risk record would be, for example, a schedule for employee CGMP training sessions.

Case Study: Examples Relating to Documentation

The Federal Food and Drug Administration has inspectors who periodically inspect pharmaceutical facilities and biotechnology companies that make regulated products. If the inspectors observe violations of Good Manufacturing Practices, they note the violations on forms, called "483s," and in official warning letters sent to the company. If companies fail to correct their deficiencies, then FDA can cause products to be seized and destroyed, fines to be levied, and, in the most extreme cases, individuals in the company may be charged as criminals and may be imprisoned if convicted. The following are excerpts from actual warning letters sent by FDA to various companies. Improper documentation is frequently cited in warning letters. Observe in these warning letters how carefully inspectors checked for proper documentation and the details of their findings.

Warning Letter, Example 1

Dear ...:

From March 12, to March 26, 2018, the U.S. Food and Drug Administration (FDA or we) conducted an inspection of your facility ...Based on the inspection and the samples collected during the inspection, we have identified serious violations of the Federal Food, Drug, and Cosmetic Act (the Act) and applicable regulations.

[A description of multiple violations follows. Only some of those violations relating to documentation are excerpted here.]

1. You failed to establish and follow written procedures for the responsibilities of the quality control operations, including written procedures for conducting a material review and making a disposition decision, and for approving or rejecting any reprocessing, as required by 21 CFR 111.103.
2. You failed to prepare and follow a written master manufacturing record (MMR) for each unique formulation of dietary supplement that you manufacture, and for each batch size, to ensure uniformity in the finished batch from batch to batch as required by 21 CFR 111.205(a). During the inspection, you told the investigator that you were not aware of the requirement to prepare an MMR for each unique dietary supplement formulation.
3. You failed to include the complete information relating to the production and control of each batch in the batch production record (BPR) as required by 21 CFR 111.255(b). Specifically, the BPR's for (b)(4) failed to include all the required elements specified under 21 CFR 111.260.

(Continued)

Case Study (*Continued*): Examples Relating to Documentation

Please respond to this office in writing within 15 working days of the receipt of this letter as to the specific steps you are taking to correct the stated violations, including an explanation of each step to identify violations and make corrections to ensure that similar violations will not occur. In your response, you should include documentation, including revised procedures, photographs, results of tests you have conducted, and any other useful information that would assist us in evaluating your corrections.

Warning Letter, Example 2

Dear . . .

Inspection of your unlicensed hospital blood bank . . . revealed serious violations . . .

Inspection revealed that [prior] blood product disposition records . . . are not available. According to your blood bank supervisor, the missing disposition records were transferred to a computer system and were subsequently "lost" by that system. Your supervisor stated that the computer system has not been validated and is being used only for "practice." Your supervisor also stated the blood bank had written back-up records for the data in the computer system. However, written disposition records could not be produced during the inspection for review . . .

Sincerely . . .

[Signed by the district director]

The general trend in the industry is toward increasing use of electronic documentation in all areas, from researcher's laboratory notebooks to sample labeling. Computer technology continues to improve, and methods to ensure data integrity are being developed. Remember, however, that the core principles of documentation are the same, whether paper or computers are used. The principles of keeping a chronological, honest, complete laboratory notebook; writing clear SOPs and forms; ensuring chain of custody for samples; and so on will remain relevant whether the professional wields a pen or a mouse.

Practice Problems

1. Figure 6.9 is a brief portion of a batch record that covers the formulation of a particular product. Note that it consists of a series of steps to be performed by the operator and a series of blanks that the operator fills out as he or she performs each step. The operator initials each step as it is performed, and another person verifies that the procedure was properly executed. There are a number of errors in how this batch record was completed. Circle the errors. (Note that it is not necessary to understand the actual procedure in order to detect the errors in how the form is completed.)

2. People commonly explain documentation with the slogan "Do what you say, say what you do." Explain this slogan with reference to the types of documents described in this chapter.

3. Imagine that a team of entrepreneurs, as introduced in Chapter 4, opens a new chocolate chip cookie bakery. Discuss the documentation requirements for this new operation.

4. Discuss the warning letters from the FDA in the case study Examples Relating to Documentation (pp. 138–139). What had these companies failed to document? What are the potential adverse consequences of these problems with documentation?

FORMULATION OF XYZ COMPOUND

Clean Gene, Inc.

3550 Anderson St.
MADISON, WI 54909
Revision 01
Batch Record # 133
Approved by _____Aaron Reid_____   _____Anna Gold_____   _____Sam Rothstein_____
           Date 2/14/18.   Date 2/14/18   Date 2/16/18

Issued by: _Erin Jane_   Date _12/3/18_   Lot # _15.987_
Product Name: **Very Good Product**
Strength: **10 Units/mL**   Vial Size **10 mL**   Batch quantity: **350 L**

Reference: **Refer to separate Formulation SOP Q75 for quantities of each component.**
          Refer to separate instrument/equipment SOP 76 for ID information

---

NOTE: PRODUCT IS TO BE STIRRED CONTINUOUSLY DURING COMPOUNDING AND FILLING

A.   COLLECTION OF WATER FOR INJECTION (WFI), USP.

A1. Collect approximately 370 L of WFI, USP, in a clean, calibrated vessel and cool to 24°C–28°C.
Vessel # ___7___ Amount collected ___365 ℓ___ Initial Temperature ___23°C___ Time ___8:00___ (am)/pm
Final Temperature ___26°C___ Time ___08:15___ am/pm

                                                            ℓₗ / ℊℳ
                                                        12/10/18 12/10/18

A2. Close the water for injection valves.   ℓₗ / ℊℳ
                                                        12/10/18 12/10/18

A3. Remove about 20 L of the cooled WFI from step 1 and place in a clean, calibrated vessel.
     (This water will be used to bring the final formulation to the proper volume.)

Vessel # _____2_____   ℓₗ / ℊℳ
                                                        12/10/18 12/10/18

A4. Remove about 5 L of the cooled WFI from step 1 into a clean, calibrated vessel. (This water will be used to
     prepare the solutions used to adjust the pH.)

B.   PREPARATION OF pH ADJUSTING SOLUTIONS

B1. Collect 750 mL cool WFI from the vessel in step A4 and place into a 1000 mL volumetric flask and add
     100 g of NaOH (Sodium Hydroxide # 875) USP and dissolve.

Amount of WFI collected _750 mℓ_   Amount NaOH added _115 g_   Lot # _____

Time step completed ___09:15___ am/pm   _____ / ℊℳ
                                                               12/10/18

B2. Using a water bath containing cold WFI, cool the solution prepared in B1 to 25°C ± 5°C.

Final Temperature ___31°C___   ℓₗ / ℊℳ
                                                        12/10/18 12/10/18

B3. Bring the solution to 1000 mL with cool WFI from the vessel in step A4.

Approximate volume of WFI added _300 mℓ_   Time completed ___09:00___ (am)/pm

                                                            ℓₗ / ℊℳ
                                                        12/10/18 12/10/18

          and so on

FIGURE 6.9   Brief portion of a batch record for Practice Problem 1.

UNIT III

Safety in the Laboratory

Everyone wants to work in a safe environment. It is essential in a safe workplace to recognize hazards and reduce risks to the personnel. **Hazards** *are the equipment, chemicals, and conditions that have a potential to cause harm*, and **risk** *is the probability that a hazard will cause harm*. For example, even though toxic chemicals are hazardous, the risk of working with them is reduced by using smaller working volumes, proper ventilation, shorter working times, and good experimental technique.

Laboratory hazards generally fall into several categories:

• physical hazards
• chemical hazards
• biological hazards.

This unit discusses examples of these classes of hazards and specific approaches to risk reduction.

DOI: 10.1201/9780429282799-9

Case Study: It's Important to Know Your Chemicals

A post-doctoral researcher at the University of California narrowly escaped serious injury from a chemical explosion in the fume hood where he was working. The event occurred after he improperly added nitric acid to a waste container of organic solvents. Nitric acid is a strong oxidizing agent that caused a violent reaction with the organic chemicals. Luckily, the worker had stepped away from the hood when the explosion occurred. The hood and its ductwork were severely damaged, and the shock wave carried out into the adjacent hallway, where the ceiling panels collapsed. The forceful reaction was both predictable and avoidable, based on the known properties of the chemicals involved. This unit will discuss how to access and read safety information, so that you will be able to work with chemicals safely.

Safety information is critical in the laboratory so that hazard exposure can be reduced or eliminated by using good laboratory practices. In addition, in the event of an accident, a quick and appropriate response usually results in less harm to people and property. Knowing about the potential for laboratory injuries can help you anticipate the types of emergencies that are most likely to occur and to plan how you would react.

This unit is intended to provide practical advice that stems from a variety of general information sources, as well as the personal experience of the authors. It is not a substitute for a safety manual, which is specific for an institution or facility.

Unit III is organized as follows:

Chapter 7 discusses general regulatory requirements for laboratory safety management.

Chapter 8 surveys general risk reduction strategies, personal protective equipment, and the most common physical hazards found in laboratories.

Chapter 9 discusses safe handling of chemicals, with special emphasis on those most likely to be found in the biotechnology laboratory.

Chapter 10 gives an overview of biosafety issues, including universal precautions, containment and sterilization strategies, animal handling, and recombinant DNA guidelines.

BIBLIOGRAPHY FOR UNIT III

There are many excellent books available that can serve as safety references in the laboratory. It is a good practice for every laboratory to have at least one comprehensive reference book covering the general types of hazards (i.e., biological, chemical, etc.) found in that setting, as well as references that address specific

hazards present in your laboratory. This bibliography includes basic references and some books dealing with specific safety topics.

GENERAL SAFETY REFERENCES

Cold Spring Harbor Laboratory Press. *Safety Sense: A Laboratory Guide: 2nd Edition.* Cold Spring Harbor Laboratory Press, 2007. (This is a short basic guide for students and individuals new to lab safety.)

Furr, A. Keith. *CRC Handbook of Laboratory Safety.* 5th ed. CRC, 2000. (This is a comprehensive reference for laboratory safety issues.)

Stricoff, R. Scott, Walters, Douglas B., and DiBerardinis, Louis J. *Handbook of Laboratory Health and Safety.* 3rd revised ed. Wiley-Interscience, 2017.

WORKPLACE SAFETY REFERENCES

Goetsch, David L. *Occupational Safety and Health for Technologists, Engineers, and Managers.* 9th ed. Pearson, 2019. (A comprehensive look at workplace safety and health issues and practices.)

National Safety Council. *Supervisors' Safety Manual.* 11th ed. National Safety Council, 2018. (This book discusses the obligations of work supervisors to inform and monitor their employees concerning safety issues.)

Plog, Barbara A, and Quinlan, Patricia J. eds. *Fundamentals of Industrial Hygiene.* 6th ed. National Safety Council, 2012. (This book provides an overview of workplace safety concerns and regulations.)

ERGONOMIC SAFETY REFERENCES

Kroemer, Karl H.E. *Fitting the Human: introduction to Ergonomics/Human Factors Engineering.* 7th ed. CRC Press, 2017.

National Safety Council. *Ergonomics: A Practical Guide.* 2nd ed. National Safety Council, 1993.

CHEMICAL SAFETY REFERENCES

Alaimo, Robert J., ed. *Handbook of Chemical Health and Safety* (ACS Handbooks). American Chemical Society, 2001.

Ballinger, Jack T. and Shugar, Gershon J., *Chemical Technicians' Ready Reference Handbook*. 5th ed. McGraw-Hill Education, 2011.

Flinn Scientific. *The Flinn Chemical and Biological Catalog Reference Manual*. This annual publication contains extremely valuable information on safety issues, particularly chemical safety. They provide accurate information about chemical storage and disposal, proper use of safety equipment, and lab design, among other topics. The free catalog can be obtained from the company at http://www.flinnsci.com/.

National Research Council. *Prudent Practices in the Laboratory. Handling and Disposal of Chemicals*. Updated ed. National Academy Press, 2011. doi:10.17226/12654. (A free copy of the 1995 edition can be found online at https://www.nap.edu/read/4911/chapter/1.)

BIOLOGICAL SAFETY REFERENCES

The Centers for Disease Control, Office of Health and Safety and National Institutes of Health, Division of Safety. *Primary Containment for Biohazards: Selection, Installation and Use of Biological Safety Cabinets*. 3rd ed. CDC, 2007.

Clinical and Laboratory Standards Institute. M29-A4: *Protection of Laboratory Workers from Occupationally Acquired Infections; Approved Guideline*. 4th ed. Clinical and Laboratory Standards Institute, 2014.

Cox, C.S., and Wathes, C.M. *Bioaerosols Handbook*. CRC Press, 1995.

National Institutes of Health. *NIH Guidelines for Research Involving Recombinant or Synthetic Nucleic Acid Molecules*. NIH, 2019. Available at: https://osp.od.nih.gov/wp-content/uploads/NIH_Guidelines.pdf.

U.S. Department of Health and Human Services, Public Health Service, Centers for Disease Control and Prevention, and National Institutes of Health. *Biosafety in Microbiological and Biomedical Laboratories*. 6th ed. U.S. Government Printing Office, Revised 2020. (Contains tables of biosafety level requirements and pathogenicity levels of specific organisms.) Available at: https://www.cdc.gov/labs/pdf/SF__19_308133-A_BMBL6_00-BOOK-WEB-final-3.pdf.

U.S. Department of Labor, Occupational Safety and Health Administration. *Bloodborne Pathogens (29 CFR 1910.1030)*. OSHA, 2012. Available at: https://www.osha.gov/pls/oshaweb/owadisp.show_document?p_id=10051&p_table=STANDARDS

Wooley, Dawn P., and Byers, Karen B., eds. *Biological Safety: Principles and Practices*. 5th ed. ASM Press, 2017. (An excellent general reference on biological safety.)

World Health Organization. *Laboratory Biosafety Manual*. 4th ed. World Health Organization, 2020. Available at: https://www.who.int/publications/i/item/9789240011311.

ANIMAL CARE AND SAFETY REFERENCES

Institute for Laboratory Animal Research, National Research Council. *Guide for the Care and Use of Laboratory Animals*. 8th ed. National Academy Press, 2011. Available at: https://grants.nih.gov/grants/olaw/guide-for-the-care-and-use-of-laboratory-animals.pdf.

Office of Laboratory Animal Welfare. *Public Health Service Policy on the Humane Care and Use of Laboratory Animals*. National Institutes of Health, Revised 2015. Available at: https://grants.nih.gov/grants/olaw/references/phspolicylabanimals.pdf.

Suckow, Mark A., Douglas, Fred A., and Weichbrod, Robert H., eds. *Management of Laboratory Animal Care and Use Programs in Research, Education, and Testing*. 2nd ed. CRC Press, 2018. Available at: https://www.ncbi.nlm.nih.gov/books/NBK500419/.

LABORATORY SAFETY WEBSITES

These are current websites that provide helpful information and links to other sites about lab safety issues.

ABSA International: The Association for Biosafety and Biosecurity: http://www.absa.org/.

Centers for Disease Control: http://www.cdc.gov/.

SDS data can be found on the Internet at a variety of sites. For example: https://chemicalsafety.com/sds-search/

https://www.fishersci.com/us/en/catalog/search/sdshome.html

https://www.flinnsci.com/sds/

7 Introduction to a Safe Workplace

7.1 A BIT OF HISTORY TO PUT THINGS IN PERSPECTIVE

If you talk to someone who worked in a chemistry or biology laboratory in the mid- or late twentieth century, they can likely tell you stories of unsafe techniques they and their colleagues practiced. For example, consider mouth pipetting. A pipette is a device used to dispense a set amount of liquid. It is a glass or plastic tube with markings to indicate volume. To use a pipette, one draws liquid up the tube and then allows the desired volume to flow out of the tube into a vessel. In the "bad old days," people who worked in laboratories would routinely measure volume by sucking liquids into the pipette with their mouth, like sucking on a straw, and then quickly placing a finger over the top of the pipette. By slowly releasing their finger, the person could control the volume of liquid dispensed out of the pipette. This practice is obviously unsafe. Mouth pipetting exposes the person to any chemical fumes or airborne pathogens in the material being pipetted. It was easy to accidentally ingest the liquid, which might be a hazardous chemical, radioactive material, sample with pathogens, or other nauseating substance. No one today would ever consider mouth pipetting, and convenient mechanical and electronic devices are used to safely draw liquids up into pipettes. It is surprising that such a practice was

allowed not that long ago. It is shocking that such a practice was not just allowed, it was required. One of the authors had to demonstrate skill at mouth pipetting before being allowed to take a required chemistry laboratory course. Another author routinely used large amounts of uranyl acetate in the laboratory without realizing that, as a uranium compound, it is radioactive and toxic. Uranyl acetate should be handled with special precautions. Fortunately, those of you beginning a laboratory career will enter a much safer work environment. Things have changed, partly because the federal government (we will talk mostly about the United States) passed laws to protect workers in a variety of workplaces, including laboratories. Things also changed because laboratory culture changed, and science professionals today expect a safer workplace. Possibly the laws changed because the culture changed; possibly the culture changed in response to regulations – probably both occurred together. This chapter will introduce you to the regulations, agencies, and organizational processes relating to laboratory safety that have been established over the past few decades. For example, it is now legally required for every laboratory to have readily available information about every chemical used in that laboratory, and precautions for handling it. Had that law been in place, the author who worked with uranyl acetate would have handled it much more carefully than

DOI: 10.1201/9780429282799-10

she did. It is important to also note that information alone is not sufficient – you, the laboratory professional, need to use information to protect yourself, your colleagues, and the environment.

We do not want to leave you with the idea that laboratory professionals in the past did not care about safety – of course they did. Today, however, there are more organizational structures in place to enhance the safety of laboratory workplaces. This chapter introduces the general processes by which organizations (such as a university, company, or research institute) promote safety, and the ways that individuals in those organizations maintain a safe environment. Later chapters in this unit continue our discussion by looking at specific safety issues:

Chapter 8 focuses primarily on physical hazards (such as fire and cold).
Chapter 9 addresses chemical hazards.
Chapter 10 discusses biological hazards.

7.2 BASIC TERMINOLOGY

I think good lab practice, consideration for other people, and safety are three totally related issues.

-David H. Beach, Ph.D.
Cold Spring Harbor Laboratory

What do we mean when we talk about "safety?" **Safety** *is defined as the elimination of potential threats to human health and well-being.* While this is an essential goal in every profession, complete safety can never be achieved. All workplaces have the potential for **accidents**, *unexpected and usually sudden events that cause harm.*

Although laboratories may present special safety challenges, those of us who have worked in laboratories for many years can attest to the fact that most work-related accidents are mundane, and fortunately they are usually minor. They include:

* tripping on unexpected items left on the floor;
* falls on slippery floors (especially around sinks and ice machines);
* slamming fingers in cabinet doors;
* hallway collisions with co-workers; and
* minor cuts while picking up pieces of broken glass.

It is estimated that 30% of all workplace accidents involve trips, slips, and falls, mostly on flat surfaces. These incidents can usually be prevented by using care and common sense, which are the best approaches to avoiding all accidents.

Although most laboratory accidents relate to common problems such as falls and cuts, there are some particular concerns in laboratories. The first step in improving the safety of any workplace is understanding the hazards that are present. Laboratories, by their nature, contain hazards. **Hazards** *are the equipment, chemicals, and conditions that have a potential to cause harm.* Heavy equipment, chemicals, electricity, animals, and infectious agents are examples of hazards that are frequently present in biotechnology laboratories.

Because it is impossible to remove all hazards from the biotechnology workplace, the most useful measure of safety in a laboratory is risk. **Risk** *is the probability that a hazard will cause harm.* **Risk assessment** *is a process where people:*

* *Identify hazards that have the potential to cause harm.*
* *Analyze and evaluate the risk associated with that hazard.*
* *Determine appropriate ways to eliminate the hazard, or control the risk when the hazard cannot be eliminated.*

By assessing and understanding the risk of various laboratory hazards, it is possible to institute practices that prevent harm from occurring. **Safety guidelines** and standards are *procedures that are designed to reduce the risk of hazards in situations where the hazards cannot be eliminated entirely.* When you begin working in a laboratory workplace, you will probably find that this process of identifying risks and establishing safety guidelines has already been performed for various hazards. By following the guidelines you will be able to protect yourself and others from the hazards in the laboratory. Over time you may also take on responsibilities for assessing risk and establishing safety guidelines as new processes and equipment are introduced into the laboratory.

Emergencies occur, even in organizations with good safety practices. An **emergency** *is a situation requiring immediate action to prevent an accumulation of harm or damage to people or property.* While it is difficult, establishing a safe work environment requires preparing for the unexpected. Consider the case study "Fire in the Workplace," which describes an emergency that occurred in a laboratory where one of the authors worked.

Case Study: Fire in the Workplace

One evening, three laboratory co-workers were finishing experiments when one commented that he smelled smoke. Within 1 minute, the laboratory filled with black, acrid smoke so thick that the workers could not see their hands 12 inches from their faces. Two workers were together and quickly located the third, who was trying to gather her research notebooks and experimental materials. They convinced her to leave behind her notebooks and leave quickly. At this point, the smoke was so thick that conversation was impossible. The workers bent over because they had been trained that smoke rises and is least concentrated near the floor. The trio moved to the nearest stairwell, evacuating the building. The fire department quickly responded and extinguished the fire, which was traced to electrical wires in an interstitial space. Luckily, there were no injuries. Even though no flames reached the laboratory, smoke damage to equipment and materials was extensive. The employees could not have anticipated or prevented the fire, which started in an inaccessible area. In this emergency situation, these colleagues responded quickly, took care of one another, and remembered their safety training, thus avoiding serious injuries from smoke inhalation.

7.3 WHO IS RESPONSIBLE FOR WORKPLACE SAFETY?

Case Study: Everyone Is Responsible for Laboratory Safety

In 2010, two graduate students working in a laboratory that manufactured high-energy metal compounds decided to synthesize a large batch of a test chemical. They followed the standard safety procedures designed for smaller chemical batches. Unfortunately, when one student later approached the hood, the compound exploded, leading to severe injuries for the student, including eye damage and the loss of several fingers. There was also massive damage to the laboratory (Figure 7.1).

A thorough investigation by the U.S. Chemical Safety and Hazard Investigation Board identified multiple contributing factors to the incident, including a lack of standard operating procedures, a lack of specific hazard training for individuals at the laboratory level, insufficient institutional oversight of lab safety, and lack of safety oversight from granting agencies that provided financial support for the research performed in the laboratory. In addition to changes in laboratory policies, the university made significant changes to its organizational structure in response to the investigation, to provide increased and more effective oversight of laboratory safety on campus. This incident highlights an important point about safety. Establishing a safe workplace is complex and requires efforts at all levels of an organization.

FIGURE 7.1   Damage to lab bench after chemical explosion. (Photo courtesy of the U.S. Chemical Safety Board.)

(Continued)

Case Study (*Continued*): Everyone Is Responsible for Laboratory Safety

Individuals certainly have a major role in protecting themselves and others, but, as we can see in this situation, the graduate students involved thought they were acting appropriately; they did not have the background and knowledge to avoid this serious accident. So, if we ask who is responsible for laboratory safety, the answer is that it requires efforts at all levels of the organization. In fact, as we pointed out at the beginning of this chapter, safety practices extend outside the organization because they are built on a foundation of laws passed by the government (Figure 7.2).

Safety is everyone's business. Federal agencies (particularly OSHA which will be described below) and other outside organizations are responsible for creating regulations and codes for safe workplaces. The institution (employer) has the responsibility to provide a safe work environment and a general institutional attitude of "Safety First," to train employees to work responsibly, and to develop an emergency response plan. Laboratory personnel must establish practices based on understanding the hazards present in that specific laboratory. Individual employees have the right to work in a safe environment, and to be well trained and informed about workplace hazards. It is the responsibility of the employee to apply their training, and to implement the safety plans of the institution and their specific laboratory. The following sections of this chapter talk more about the responsibilities of these different levels of organization.

7.4 SAFETY RESPONSIBILITIES AT VARIOUS LEVELS OF ORGANIZATION

7.4.1 REGULATORY AGENCIES AND OTHER OUTSIDE ORGANIZATIONS

There is a vast number of state, federal, and local regulations, as well as industry standards that affect

biotechnology companies and academic biotechnology laboratories. **Regulations** *are operating principles that are required by law.* **Standards** *are operating principles or requirements that are often voluntary.* Note, however, that the term "standard" in the safety literature is sometimes used for practices that are required. The many regulations and standards related to safety can be arranged into categories:

- **Worker safety**. For example, there are regulations that require laboratory chemicals to be labeled and that require employees to be informed about hazards (discussed in this chapter).
- **Environmental protection**. For example, the disposal of hazardous laboratory chemicals is regulated in order to minimize the impact to the environment (discussed in this chapter).
- **The use and handling of animals**. For example, there are regulations regarding the cages used to house laboratory animals and regulations aimed at preventing the spread of contagious disease. These regulations protect animals from inhumane treatment, prevent faulty experimental results due to sick animals or inconsistent treatment of animal subjects, and protect the environment from disease spread. These regulations will be discussed in Chapter 10.
- **Regulation of radioisotopes**. For example, these regulations cover such issues as how radioisotopes should be handled and stored, who has access to such compounds, and what documentation is required when radioisotopes are used. We do not cover radioisotopes in this text because they are no longer routinely used in the majority of biotechnology workplaces. If you do eventually work with radioisotopes, you will receive specialized training; there are regulations that require employers to provide this training.

FIGURE 7.2 Who is responsible for a safe workplace?

TABLE 7.1

Examples of Federal Agencies That Regulate Safety and Environmental Protection in Biotechnology Organizations

The Occupational Safety and Health Administration (OSHA)

OSHA is the federal agency charged with ensuring worker safety. The Occupational Exposure to Hazardous Chemicals in Laboratories Standards (29 CFR 1910.1450; revised 2012) applies to non-production laboratories. Additional OSHA standards provide rules that protect workers in all laboratories from physical, chemical, and biological safety hazards. Central requirements of the laboratory standards are that the employer develop, document, and implement a plan that protects workers from hazards.

The Environmental Protection Agency (EPA)

EPA is responsible for protecting the environment. EPA regulations affect how laboratories and companies handle and dispose of waste, what substances can be emitted into the air and water, the movement, storage, and disposal of hazardous substances, and records relating to chemicals. EPA also regulates certain types of biotechnology field work that involve releasing genetically modified organisms into the environment.

The Department of Transportation (DOT)

DOT regulates the transportation of hazardous materials, such as chemicals, compressed gas cylinders, and hazardous wastes. The regulations cover packaging, labeling, transport, and reporting procedures.

The Nuclear Regulatory Commission (NRC)

NRC is responsible for the safe use of radioactivity. Facilities that use radioactive substances for research purposes and medical applications, or in products must comply with NRC regulations, including those related to worker safety, waste disposal, and record-keeping.

Table 7.1 summarizes the roles of US government agencies whose regulations directly affect biotechnology companies and research laboratories. Even though this book cannot discuss all of the statutes related to laboratory safety, we will discuss some of the most frequently encountered agencies and their regulations. For a more detailed and comprehensive discussion, we recommend starting with the National Research Council's *Prudent Practices in the Laboratory: Handling and Disposal of Chemicals* (see the Bibliography in the Introduction to this unit). Significant amounts of information are also available online from the Occupational Safety and Health Administration (http://www.osha.gov) and the Environmental Protection Agency (http://www.epa.gov).

Although occupational safety is regulated primarily by federal agencies, there are many other organizations concerned with safety in the workplace. Many of these organizations establish standards or develop **codes**, *which are sets of standards centered on a specific topic.* A well-known example is the **Underwriters Laboratories (UL),** *an organization that has developed codes for safe electrical devices.* As with many of these organizations, UL has no enforcement powers. Table 7.2 provides a list of some of the organizations that have developed standards or codes related to worker safety. Even though the standards or codes developed by these organizations are recommendations, they are frequently the basis for federal regulations. For example, OSHA requires that protective eyewear meet

American National Standards Institute (ANSI) standards for impact and penetration resistance.

7.4.2 OSHA WORKER SAFETY REGULATIONS

As part of the U.S. Department of Labor, the **Occupational Safety and Health Administration (OSHA)** *is the main federal agency responsible for monitoring workplace safety.* Its mission:

> With the Occupational Safety and Health Act of 1970, Congress created the Occupational Safety and Health Administration (OSHA) to ensure safe and healthful working conditions for working men and women by setting and enforcing standards and by providing training, outreach, education, and assistance.

> *From the OSHA website: https:// www.osha.gov/aboutosha*

Since the passage of the Occupational Safety and Health Act of 1970, OSHA has both developed and enforced safety regulations that encourage employers to reduce hazards in the workplace. In 1983 OSHA created the **Federal Hazard Communication Standard (HCS or HazCom),** *which regulates the use of hazardous materials in industrial workplaces. It focuses on the availability of information concerning*

TABLE 7.2

Professional Organizations Concerned with Workplace Safety

American Board of Industrial Hygiene (ABIH)

American College of Occupational and Environmental Medicine (ACOEM)

American Conference of Governmental Industrial Hygienists (ACGIH®)

American Industrial Hygiene Association (AIHA)

American National Standards Institute (ANSI)

Institution of Occupational Safety and Health (IOSH)

The National Association of Safety Professionals (NASP)

National Safety Council (NSC)

---

employee hazard exposure and applicable safety measures. In 2012, the HCS was aligned with the Globally Harmonized System of Classification and Labelling of Chemicals (GHS), providing a standardized approach to hazard identification, chemical labeling, and Safety Data Sheets (formerly known as Material Safety Data Sheets, MSDS). Safety Data Sheets will be discussed in detail later in this chapter.

The HCS mandates that employers fulfill specific requirements for employee safety and knowledge. Employers must provide:

• a workplace hazard identification system
• a written hazard communication plan
• files of Safety Data Sheets for all hazardous chemicals
• clear labeling of all chemicals according to international standards
• worker training for the safe use of all chemicals.

The original HCS applied mainly to manufacturing employers until 1987, when it was more broadly applied to all industries where workers are exposed to hazardous chemicals.

After years of development, OSHA provided a set of general safety regulations specifically aimed at laboratories. The 1990 **Occupational Exposure to Hazardous Chemicals in Laboratories Standards (29 CFR Part 1910)** *adapts and expands the HCS to apply to academic, industrial, and clinical laboratories.* The main requirement of these standards is the Chemical Hygiene Plan that each institution must develop for every laboratory. The **Chemical Hygiene Plan (CHP)** *is a written manual that outlines the specific information and procedures necessary to protect workers from hazardous chemicals.*

Although institutions have considerable latitude in developing their CHP, certain issues must be addressed, as outlined in Table 7.3. Another important provision for laboratories is that all work-related injuries and health problems must be reported to OSHA.

TABLE 7.3

Required Elements of a Chemical Hygiene Plan

A CHP must provide institutional policies or procedures to address each of the following issues:

• general chemical safety rules and procedures
• purchase, distribution, and storage of chemicals
• environmental monitoring
• availability of medical programs
• maintenance, housekeeping, and inspection procedures
• availability of protective devices and clothing
• record-keeping policies
• training and employee information programs
• chemical labeling requirements
• accident and spill policies
• waste disposal programs.

A CHP also generally provides information about emergency response plans, as well as the designation of institutional safety officers. The CHP must be reviewed at least annually to ensure that it is effective.

As we introduced at the beginning of this chapter, these relatively recent regulations reflect an important and positive shift in attitudes about laboratory safety. OSHA regulations demonstrate that laboratory safety is of sufficient concern to warrant the involvement of the federal government. The regulations require that institutions provide resources that help individuals to understand hazards and work more safely. They encourage employees and their supervisors to take extra time, if necessary, to perform safety-related tasks. They require institutions (which may or may not have actively promoted safety in the past) to invest in safety equipment, training, and protective clothing.

7.4.3 Responsibilities at the Institution Level of Organization

Much of the responsibility for safety lies at the institutional level of organization. An institution, such as a major university, is likely to have an entire department with specialized personnel who oversee safety practices, set up policies, inspect individual laboratories, provide training, and make sure the institution complies with all safety regulations. Smaller organizations are likely to have one or more people who take on these tasks, and develop specialized expertise.

Safety training is an important institutional responsibility. It is well documented that most injuries occur to employees with less than 2 years of experience on the job. This points to the need for safety training programs aimed at all new employees. Training program requirements originate with government agencies and are then instituted by institutions. The key aims of any safety training program are to allow employees to:

- understand the risks inherent in their jobs
- recognize their personal susceptibility to accidents
- learn about preventive measures that reduce the risk of accidents
- accept personal responsibility for accident prevention.

Safety training is not just for new employees. Refresher courses should be available for more experienced personnel, allowing them to learn about new policies and resources, and practice safety skills that have not previously been needed. Table 7.4 lists some of the elements that should be covered in a safety training program.

In addition to safety training, institutions are required to develop manuals that explain hazards, and practices to minimize the risk of the hazards. You can find examples of such manuals from various universities using a

---

TABLE 7.4

Elements of a Safety Training Program

Safety training programs vary widely among institutions. Employees are required by federal law to attend these programs, and most institutions require employees to sign a statement indicating that they have received training. Some of the training elements required by OSHA regulation 29 CFR 1910 most relevant to laboratories are as follows:

- Emergency planning
- Hazardous materials
- Personal protective equipment
- Medical services and first aid
- Fire protection
- Toxic and hazardous substances

In addition, every laboratory safety program should include the following topics:

- Institutional policies – hazard information, inspections, reporting systems, waste disposal
- Safety rules – practices, manual, signs, labels
- Protective equipment – location and use
- Emergency procedures – alarms, injuries, medical assistance
- Chemical hazard awareness, including:
  o location of Safety Data Sheet reference materials
  o symptoms of chemical exposure
  o detection methods for chemical exposure
  o protective mechanisms
  o emergency procedures.

In addition, there are many laboratory-specific safety issues that may need to be addressed, such as biological safety.

web browser. Biotechnology companies will similarly develop safety manuals specific to their organization. The Chemical Hygiene Plan (CHP), as previously introduced in this chapter, can be incorporated into the manual, but there are likely safety issues in a biotechnology setting in addition to those in the CHP. Sometimes people complain about "ten-pound" institutional safety manuals that attempt to regulate virtually every job process, including breathing rates. These manuals do not tend to be effective in motivating worker compliance. In this situation, individual laboratories may need to develop a more focused safety manual that directly addresses the safety issues in that laboratory. Rules should be written in simple, clear language that can be understood by a wide range of laboratory personnel. The manual should provide a mechanism to quickly find guidance that pertains to specific situations.

7.4.4 Responsibilities at the Laboratory Level of Organization

7.4.4.1 Overview

Although institutions develop policies to promote employee safety, these policies are carried out at the level of the individual laboratory. For example, even though there is usually a CHP for the institution in general, each laboratory is required to have its own additions to the CHP to describe hazards and safety measures unique to that laboratory. The supervisor or mentor is responsible for setting the tone of the daily operations, as well as for modeling safety and good laboratory practices. Every laboratory with more than three people should have a designated safety officer, who is responsible for monitoring safety practices. In addition, this individual (or committee, in larger groups) will also:

- serve as a safety advisor to the laboratory;
- ensure that safety procedures are documented and understood;
- act as a liaison with the institution's safety officers;
- communicate policy changes to co-workers;
- coordinate internal safety inspections;
- ensure that equipment is properly maintained; and
- keep records of hazards and problems within the laboratory.

7.4.4.2 Labeling and Documentation

Standardized labeling of hazardous chemicals is required under the HCS, as well as by common sense. This means that all containers of potentially hazardous chemicals must be labeled to an extent that makes them readily identifiable to new employees or to outsiders in case of a spill or emergency. Lack of proper labeling is one of the most common OSHA citations against laboratories; this topic of chemical labeling is discussed in more detail in Chapter 9. Of course, proper labeling of nonhazardous chemicals is good laboratory practice as well.

7.4.4.3 Safety Data Sheets

A Safety Data Sheet (SDS) for every chemical used in a given laboratory must be readily available to all personnel. The **Safety Data Sheet (SDS)** *is a legally required technical document provided by chemical suppliers that describes the specific properties of a chemical.* The HCS specifies information that must be included in an **SDS**, as shown in Table 7.5.

A portion of an SDS (some are several pages long) for the chemical benzene is shown in Figure 7.3. Benzene is widely used as an industrial solvent. It is a powerful carcinogen (among other hazards), but it can be used safely with proper precautions, which are indicated in the **SDS**.

7.4.4.4 Labeling of Work Areas

In addition to labeling chemicals, laboratory rooms and work areas must also be labeled with signs that indicate hazards. These must provide enough information to alert visitors to take appropriate precautions. Any area that is unsafe for visitors without training or specific precautions should be labeled with a "Do Not Enter" sign.

7.4.4.5 Job Safety Analysis

One task that is extensively used in industry to provide both safety guidelines for personnel and compliance with OSHA regulations is the preparation of a Job Safety Analysis. A **Job Safety Analysis (JSA)** *is a detailed analysis of each step in a procedure, identifying hazards and outlining accident prevention strategies.* An example is provided in Figure 7.4. (This example relates to centrifugation, a common laboratory procedure that will be discussed in detail in Chapter 33.) An effective **JSA** is usually prepared jointly by safety officers and individuals who perform the procedures, and can be used for both training and documentation of laboratory safety measures.

7.4.4.6 Housekeeping

Many hazards can be eliminated or reduced by the simple policy of good housekeeping. The majority of routine maintenance and cleaning in laboratories

TABLE 7.5

Contents of a Safety Data Sheet

An **SDS** is required to provide the following content by the Hazard Communication Standard:

Section 1. Identification

Section 2. Hazard(s) identification

Section 3. Composition/information on ingredients

Section 4. First aid measures

Section 5. Fire-fighting measures

Section 6. Accidental release measures

Section 7. Handling and storage

Section 8. Exposure controls/personal protection

Section 9. Physical and chemical properties

Section 10. Stability and reactivity

Section 11. Toxicological information

Section 12. Ecological information[a]

Section 13. Disposal considerations[a]

Section 14. Transport information[a]

Section 15. Regulatory information[a]

[a] These SDS sections are regulated by agencies other than OSHA.

must be performed by the personnel who are familiar with the hazards present. In most institutions, outside staff do not clean benchtops or equipment. This prevents accidental exposure of staff to hazards that are unknown to them. It protects experimental materials from inadvertent contamination or disposal. It also means, however, that the laboratory workers themselves are responsible for maintaining a clean, orderly work space.

Safety, as well as good laboratory practice, requires that clutter on benchtops and shelves be kept to a minimum (Figure 7.5). This lessens the risk of reagent mixups and potential degradation of old chemicals. Fewer objects in a work area provide fewer opportunities to accidentally contaminate equipment or containers that are not part of the current experiment. Developing a habit of regularly cleaning your work area will avoid situations where a massive effort must be undertaken to clear working space.

Laboratory benchtops should be routinely cleaned both before and after work sessions. This will guard against any residual chemical or biological contamination that may be present from a previous user, as well as prepare the space for the next user. Cleaning techniques should be based on a worst-case scenario of the hazardous contaminants that may be present. Specific cleaning methods are discussed in Chapters 9 and 10. Cleanup should always include decontaminating and rinsing glassware in preparation for dishwashing procedures. In most institutions, dishwashing facilities will not accept laboratory items that still contain obvious residues of experimental materials.

Even though institutional regulations will require periodic laboratory inspections by external safety officers, a laboratory should not wait for these inspections to identify problems. Regular internal inspections noting housekeeping problems and potential risks can provide timely information about unsuspected hazards as well as the current state of compliance with safety regulations. It is particularly important to perform regular inspections of specific types of hazards, such as gas cylinders and chemical storage and labeling. In this way, small problems can be remedied before accidents occur.

Every laboratory needs to have a system of waste collection and disposal for specific hazardous materials. These may include broken glass and other sharp objects, solid and liquid radioactive waste, chemical waste, and biologically contaminated materials. Institutions must comply with a variety of regulations concerning waste disposal, including detailed labeling of hazardous contents. Proper waste disposal can be difficult and extremely expensive for unidentified materials; therefore, every laboratory should have a system for labeling waste at its source. Environmental concerns, as well as common sense, dictate that laboratory waste should be minimized when possible. It is also important that hazardous waste not be mixed with regular trash both to reduce the volume of hazardous waste and to avoid serious injuries for housekeeping staff.

FLINN SCIENTIFIC, INC.
Safety Data Sheet (SDS)

SDS #:   105.00

Revision Date: March 21, 2014

SECTION 1 — CHEMICAL PRODUCT AND COMPANY IDENTIFICATION

Benzene

Flinn Scientific, Inc. P.O. Box 219, Batavia, IL 60510 (800) 452-1261
CHEMTREC Emergency Phone Number: (800) 424-9300   **Signal Word**   DANGER   **Pictograms**

SECTION 2 — HAZARDS IDENTIFICATION

Hazard class: Flammable liquids (Category 2). Highly flammable liquid and vapor (H225). Keep away from heat, sparks, open flames, and hot surfaces (P210).

Hazard class: Acute toxicity, oral (Category 4). Harmful if swallowed (H302). Do not eat, drink or smoke when using this product (P270).

Hazard class: Skin and serious eye damage, corrosion or irritation (Category 2, 2A). Causes skin and serious eye irritation (H315+H319).

Hazard class: Acute toxicity, inhalation (Category 5). May be harmful if inhaled (H333).

Hazard class: Germ cell mutagenicity and carcinogenicity (Category 1). May cause genetic defects (H340). May cause cancer (H350). Obtain special instructions before use (P201). Do not handle until all safety precautions have been read and understood (P202). Use personal protective equipment as required (P281). Known human carcinogen (IARC-1).

SECTION 3 — COMPOSITION, INFORMATION ON INGREDIENTS

Component Name	CAS Number	Formula	Formula Weight	Concentration
Benzene	71-43-2	C_6H_6	78.11	

Synonym: Benzol

SECTION 4 — FIRST AID MEASURES

If exposed or concerned: Get medical advice or attention (P308+P313).
If inhaled: Remove victim to fresh air and keep at rest in a position comfortable for breathing. If in eyes: Rinse cautiously with water for several minutes. Remove contact lenses if present and easy to do so. Continue rinsing (P305+P351+P338). If eye irritation persists: Get medical advice or attention (P337+P313). If on skin: Immediately remove all contaminated clothing. Rinse skin with water (P303+P361+P353).
If swallowed: Immediately call POISON CENTER or physician. Do NOT induce vomiting (P301+P310+P331).

SECTION 5 — FIRE FIGHTING MEASURES

Class IB flammable liquid.
Flash point: -11 °C Flammable limits: Lower: 1.3% Upper: 7.8% Autoignition Temperature: 498 °C
When heated to decomposition, may emit toxic fumes.
In case of fire: Use a tri-class dry chemical fire extinguisher.

NFPA Code
H-1
F-3
R-0

SECTION 6 — ACCIDENTAL RELEASE MEASURES

Remove all ignition sources and ventilate area. Contain the spill with sand or other inert absorbent material and deposit in a sealed bag or container. See Sections 8 and 13 for further information.

© 2014 Flinn Scientific, Inc. All Rights Reserved.   Page 1 of 2

FIGURE 7.3   Partial Safety Data Sheet for benzene. (Image © Flinn Scientific, Batavia, IL, the USA. Used with permission.)

7.4.4.7 Emergency Response

Although all institutions have emergency plans, individual laboratories also need to prepare for potential accidents and emergencies. Some of the basic preparations needed are shown in Table 7.6. Everyone in the laboratory should ideally be trained in basic first aid and CPR. The laboratory or a nearby area must be equipped with basic safety items such as fire extinguishers and fire alarms, first aid kits, chemical spill kits, a safety shower, and an eye wash station. Each employee needs to know the evacuation plan for the laboratory and the location of emergency telephone numbers and procedures. Remember that most accidents happen very quickly; there is usually no time for carefully planning a response after the accident occurs. (Recall the case study "Fire in the Laboratory" earlier in this chapter.) This is the reason

that emergency procedures need to be understood and practiced in advance.

Laboratory accidents, no matter how minor, must be reported as soon as possible after the occurrence to the appropriate person, according to institutional policy. There is a natural tendency for many people to cover up minor accidents, such as cuts or spills, either from embarrassment, from a dread of paperwork, or from concerns about being blamed for the incident; however, these are not adequate excuses for secrecy. Accident reporting is a legal requirement and is also essential in order to prevent repeated problems. There are numerous studies that show that people learn to avoid accidents by being informed about other accidents, as well as near misses.

If you are injured on the job and require medical treatment, you need to fill out a workers' compensation

Genes-R-Us Technology, Inc **JOB SAFETY ANALYSIS**	JSA No. and Title **#3 – Loading centrifuge**		Date **12/09/21**
	Job Title: **Lab technician**	Job Analysis Performed By: **Cynthia Moore**	
Sequence of Basic Job Steps	**Potential Hazards**	**Recommended Actions or Procedures**	
Match adapters to centrifuge tubes	Incorrect size of adapter could create a catastrophic rotor imbalance Damaged adapter could fail and create a catastrophic rotor imbalance	1. Visually match adapters to tube and rotor size. 2. Examine adapters for any signs of wear or chemical contamination. 3. Assemble tubes, adapters, and rotor to determine if the fit is appropriate.	
Balance tubes before final loading into rotors	Improperly balanced samples could create a catastrophic rotor failure	1. Confirm that tubes are comparably filled with liquid of the same density, especially for ultracentrifugation. 2. Weigh tubes and match pairs with the same weight to be placed on opposite sides of the rotor. 3. Be sure weight does not exceed manufacturer recommendations for the rotor. 4. Distribute tubes symmetrically in the rotor.	
Install rotor in the centrifuge	Improperly installed rotor could cause the rotor to come loose during the spin with catastrophic results	(Etc.)	

FIGURE 7.4 Sample Job Safety Analysis form.

form. **Workers' compensation** *is a no-fault state insurance system designed to pay for the medical expenses of workers who are injured on the job, or develop work-related medical problems.* It provides a mechanism for your employer to pay for your job-related medical treatments without the necessity of a lawsuit.

7.4.5 WHAT DOES THIS MEAN FOR YOU, THE INDIVIDUAL?

7.4.5.1 Laboratory Safety and Common Courtesy

Training courses, manuals, Safety Data Sheets, and institutional plans all help provide employees with the

FIGURE 7.5   I know I left my experiment here somewhere.

resources needed to create a healthy, safe workplace. Ultimately, each person must use these resources to protect themselves and others from workplace hazards. This section provides some basic, general practices to maintain a safe workplace.

Good laboratory technique and consideration for co-workers are essential parts of laboratory safety. Proper labeling of all materials notifies co-workers of hazardous chemicals. Prompt cleanup of spills, reduction in clutter, and the return to storage of unused chemicals and equipment free up working space and reduce accidental spills and mistakes. Although you may deliberately choose to expose yourself to a known risk under some circumstances, it is inexcusable to expose others who have not made that choice.

It is important that you and each of your co-workers know how to respond to an emergency. Be familiar with the emergency procedures established for the laboratory and know what to do if you or a co-worker needs medical assistance. Acquaint yourself with basic first aid and the location of the first aid materials. Know where to find and how (and when) to use the fire extinguishers, and know the evacuation plan for the building in case of fire or other physical emergency (see the case study on p. 147).

One of the primary guidelines for behavior in the laboratory is often referred to as the "rule of reason," where an individual asks themselves how reasonable an action is in the current situation. Considering whether an action makes sense under the circumstances frequently suggests an appropriate course of action. When tempted to cut corners, think about what your opinion would be of a co-worker who did the same thing.

7.4.5.2 Personal Hygiene

This does not refer to bathing (although that is a good idea, too); rather, it refers to personal habits that may increase your risk of hazard exposure. Never eat, drink, smoke, chew gum, or apply cosmetics in a laboratory. Even though you are "not working with anything dangerous," you can never be sure about other hazards in the immediate area. Never use the laboratory or your pockets to store food, beverages, or anything that will be consumed. If you do smoke (outside the laboratory, of course), be aware that cigarettes in open packs can absorb vapors from volatile chemicals. Develop the habit of washing your hands every time you move away from your lab bench, and at regular intervals while you are working.

Drink only from hall fountains. Water from laboratory faucets may not be **potable** *(suitable for human consumption)*, due either to water quality or to contamination of the faucet with hazardous materials. For the same reasons, do not consume ice taken

---

TABLE 7.6

Basic Emergency Precautions for the Laboratory

The following list provides some of the fundamental preparations that every laboratory should have in place for emergency situations.

- Everyone in the laboratory should be aware of basic emergency procedures.
- There should be at least one person trained in first aid and CPR present at all times.
- The first aid kit must be readily accessible and fully stocked.
- All required protective devices, such as fire extinguishers and eyewash stations, must be well marked and easily accessible.
- Emergency telephone numbers and instructions should be prominently posted by every telephone.
- Evacuation routes should be kept clear of boxes or clutter.

from laboratory machines. Even if you are certain that the ice is made from potable water, it is impossible to know exactly what types of containers may have come into contact with the ice. Avoid storing personal items such as coats in the laboratory, where they may become contaminated with hazardous materials.

Dress appropriately for the laboratory. That means covering as much bare skin as possible and wearing closed-toe shoes. See Figure 8.1 for more details. Long hair must be tied back in the laboratory to prevent a variety of problems. Hair has an unfortunate tendency to fall forward, which can lead to chemical or biological contamination of the hair, or, even worse, ignition due to contact with an open flame. A greater danger that is not often considered is that most people with long hair frequently push the hair back from their faces without conscious thought, leading to potential skin and hair contamination with hazardous substances from a gloved hand. A related issue is the wearing of beards in laboratories where biological agents are used. Many microbiological standards warn against beards because of the possibility of experimental and personal contamination, and laboratory situations that require the use of fitted respirators may preclude beards entirely. This is an issue that is best judged in the individual situation.

Another variable issue is the wearing of contact lenses in the laboratory. The American Chemical Society Committee on Chemical Safety has concluded that contact lenses are acceptable in most laboratory situations, as long as standard eye protection is used, and may actually provide some protection against injuries. ("American Chemical Society: Chemical & Engineering Safety Letters." *Chemical and Engineering News*, June 1, 1998, pubsapp.acs.org/cen/safety/19980601.html.) The ACS recommendation has been accepted by OSHA. However, some laboratories have a policy that contact lenses should never be worn in the laboratory because they might exacerbate injury in the case of exposure to certain chemical fumes, or splashes to the eye. When wearing contact lenses in the lab, always wear eye protection gear, such as safety glasses or goggles. Never insert or remove contact lenses from your eyes in or near the laboratory. Do not adjust an uncomfortable lens with a possibly contaminated finger. Be sure that your supervisor and co-workers are aware that you wear contact lenses, in case of an emergency. In case of eye contamination, lenses should be left in place unless they are easily flushed out in the eyewash; emergency workers should be notified of their potential presence.

7.4.5.3 Work Habits

Many laboratory accidents are the result of simple carelessness, coupled with the fatigue and distractions that everyone experiences. The obvious prevention measure is to avoid working when tired or distracted, but this of course is not always practical advice. You can address this problem, however, by personally acknowledging your temporarily diminished state of attention, slowing down, and taking extra precautions. Take a few additional moments to plan what you are doing and anticipate any potential problems before they occur.

Do not work alone in the laboratory. Let your co-workers know if you will be in an isolated part of the building, such as a cold room, for extended periods of time. It is a frequent temptation to finish experiments in the evening or on weekends when the lab is quiet, but this can be a dangerous strategy in the event of a serious accident (see the case study "The Dangers of Working Alone").

Case Study: The Dangers of Working Alone

In 1994, a senior researcher at a major university nearly died while working alone at night. While performing a familiar procedure involving small solvent volumes, a distillation flask exploded, starting a fire. Because he was not wearing a lab coat, his shirt ignited. Even worse, a flying piece of glass severed a major artery in his arm. He collapsed from loss of blood and shock before reaching the emergency phone. His life was saved because a colleague in a nearby office heard the fire alarm, called 911, and then performed first aid until the ambulance arrived. This university officially prohibited working alone in laboratories, but the prohibition was widely ignored. Following this incident, an enforced policy of using a "buddy system" for work at odd hours was developed. Each worker must have at least one co-worker in the immediate vicinity, in case of emergencies.

Another time for concern about hazards is when passing from the laboratory into public areas and back. Do not wear lab apparel from work, especially lab coats and gloves, into public areas. You may know that your hands and coat are uncontaminated by hazardous materials, but the people you encounter will not. If you do feel a need to wear a lab coat in public, keep a clean coat for this purpose (although this will not reassure the people who see you in the lab coat). Never wear safety gloves outside the laboratory, and never handle common use items such as phones, radios, or light switches while wearing gloves. This is an easy way to spread chemical or biological contamination. In labs where radioactive chemicals are used, radiation inspections specifically include checking door knobs, equipment controls, and laboratory desks for radioactive contamination.

If you must transport hazardous materials through public areas, handle them with one gloved hand, leaving the other hand ungloved for opening doors. Samples should be carried in sealed double containers, to prevent spills. Try not to hurry through halls and around corners, to avoid collisions with co-workers. Many experiments have ended up on the floor (or worse, on a person) because of sudden hallway encounters.

Table 7.7 provides a set of general guidelines for safe laboratory practices for the biotechnologist.

7.5 THE EPA AND ENVIRONMENTAL PROTECTION

As a final note in this chapter, it is important to consider the safety of the environment outside the laboratory or institution. Generation and disposal of toxic biological and chemical wastes is a significant factor in laboratory management. The **Environmental Protection Agency (EPA)** *has primary responsibility for enforcement of laws to prevent environmental contamination with hazardous materials.* Some of the legislative regulations enforced by the EPA are the Clean Water Act, the Safe Drinking Water Act, the Clean Air Act, and the Toxic Substances Control Act. The **Toxic Substances Control Act (TSCA)** *was designed to regulate chemicals that pose health or environmental risks.* It has a major impact on the chemical industry and associated laboratories. TSCA establishes chemical inventory and record-keeping requirements, allows the EPA to control or ban hazardous chemicals in commerce, and requires companies to notify the EPA of their intentions to manufacture new chemicals. TSCA was updated in 2016 by the Frank R. Lautenberg Chemical Safety for the 21st Century Act to broaden EPA authority, mandate risk-based chemical evaluations, and increase public transparency about chemical information.

Laboratories are also affected by the requirements of the **Resource Conservation and Recovery**

TABLE 7.7

Personal Laboratory Safety Practices

- Be sure that you are informed about the hazards that you encounter in the laboratory.
- Be aware of emergency protocols.
- When in doubt about a hazardous material or a procedure, ask.
- Use personal protective wear such as lab coats and safety glasses at all times.
- Do not eat, drink, chew gum, or smoke in the laboratory.
- Avoid practical jokes or horseplay, which can unintentionally create a hazard.
- Use gloves whenever in doubt (see Chapter 8 for guidelines on proper use of gloves).
- Wash your hands regularly, regardless of whether your work requires gloves.
- Always wash your hands thoroughly before leaving the laboratory.
- Read the labels of chemicals and reagents carefully.
- Read procedures before performing them and visualize hazardous steps.
- Minimize the use of sharp objects and be sure that you properly dispose of them.
- Clean up spills and pick up any dropped items promptly.
- Label everything clearly.
- Use a fume hood for any chemical or solvent that you can smell, that has known toxic properties, or that is unfamiliar to you. Never assume that an odorless chemical is safe outside a fume hood.
- Record everything in your lab notebook.
- Always report accidents, however minor, immediately.

Act (RCRA) of 1976, *which provides a system for tracking hazardous waste, including poisonous or reactive chemicals, from creation to disposal.* The RCRA provides the EPA with authority for regulating transport, storage, emergency procedures, and waste management plans for toxic materials. In 1984 the Federal Hazardous and Solid Waste Amendments (HSWA) were added to RCRA, focusing on minimizing waste, strengthening hazardous waste management standards, and phasing out land

disposal of hazardous waste, among other facets of waste management.

Most institutions have specialized staff who set up systems to comply with these environmental requirements. It is the responsibility of laboratory personnel to follow these systems. For example, there may be specific containers designated in the laboratory for disposal of specific types of waste. Be sure that you understand what goes into what container. When in doubt, ask!

Practice Problems

1. Consider a laboratory situation where large amounts of a toxic and flammable solvent are required for experimental work. How would you approach general risk reduction for this hazard?
2. Analyze the case study "Fire in the Workplace" on p. 147, and list the emergency procedures that would apply to this situation.
3. Analyze the case study "Everyone is Responsible for Laboratory Safety" on pp. 147–148. What levels of responsibility (student, supervisor, etc.) were not adequate in avoiding this incident?
4. Analyze the case study "The Dangers of Working Alone" on p. 157. What basic safety precautions were ignored by the researcher and institution involved?

Questions for Discussion

1. Think of an accident that you were involved in or witnessed. What safety standards (if they had been followed) might have prevented this accident?

8 Working Safely in the Laboratory
General Considerations and Physical Hazards

8.1 RISK REDUCTION IN THE LABORATORY

Chapter 7 introduced the concept of evaluating and reducing the risk of hazards in the laboratory. This chapter will delve further into risk reduction, particularly related to physical hazards (e.g., fire and extreme cold).

There are four general approaches to risk reduction in the laboratory that apply to all categories of hazards:

- Reduce the presence of hazards.
- Reduce the risk of inevitable hazards with good laboratory design.
- Establish good laboratory practices for handling hazards.
- Use personal protective equipment.

The presence of hazards should be reduced as much as possible. For example, amounts of flammable solvents and other hazardous substances on the premises should be limited. It may be possible to eliminate hazardous substances or replace them with safer substitutes. Next, the risk of those hazards that cannot be eliminated should be reduced by good laboratory engineering. This means, for example, the installation of properly functioning fume hoods, protective shielding, and fire-resistant chemical storage facilities. Appropriate lab facilities should provide safe separation between personnel and hazards whenever possible. The third approach is to establish good laboratory practices in ways that reduce risk. All personnel must take advantage of the engineered solutions such as fume hoods, be aware of proper procedures for performing hazardous operations, and exercise caution in their work behavior.

DOI: 10.1201/9780429282799-11

Adequate personnel training and maintenance of good housekeeping practices are essential. Finally, the provision of personal protective equipment (PPE), such as safety goggles, is essential to create a barrier between the worker and hazards, to reduce residual hazards and to guard against unexpected events.

8.2 PERSONAL PROTECTION IN THE LABORATORY

8.2.1 Clothing

8.2.1.1 General Dress

Proper clothing is required whenever entering a laboratory. Even though a lab coat will protect you and your clothing from some hazards, what you wear under a lab coat can be just as important (Figure 8.1). It is important that clothes cover all parts of the body, including legs. For this reason, pants or long skirts are appropriate. Avoid dangling jewelry or ties and long loose hair that can fall into your experiment or get caught in moving equipment. It is also a good idea to refrain from wearing rings, bracelets, or watches in the laboratory. It is easy for chemicals to seep under these items. Any clothing worn in the laboratory should be fire-resistant and easily removable in case of chemical or biological contamination. Many experienced laboratory workers keep a spare change of clothing handy in case of spills, or for wearing after work.

PROPER PERSONAL ATTIRE*

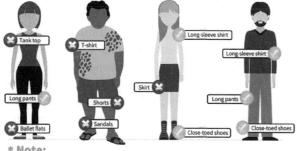

* Note:

These examples do NOT represent a list of all (im)proper forms of personal attire in the laboratory

Please review posted signage in each facility for additional Proper Personal Protective Equipment (PPE) room requirements

FIGURE 8.1 Examples of appropriate (green) and inappropriate (red) personal dress in the laboratory. It is essential to wear proper attire even if you are wearing a lab coat. (Image courtesy of Amy Puffenberger, University of Michigan Animal Care & Use Program. animalcare.umich.edu/research-role-models.)

Example Problem 8.1

Look carefully at Figure 8.1. Not all appropriate attire is shown in the figure. What important items are missing from the illustration?

Answer

The most important item not shown is safety glasses. These should be covering the eyes, not pushed up on top of the head. The woman with long hair should have it tied back. An ankle-length skirt would be acceptable.

8.2.1.2 Lab Coats

Lab coats should be worn at all times in the laboratory. Even when you are not using hazardous materials yourself, other people's activities in the laboratory might present unexpected hazards. Lab coats provide a barrier against harmful agents and prevent contamination of street clothes. By soaking up spills, they allow more time to recognize contamination problems and protect yourself. They also protect experiments from contaminants outside the laboratory that might be carried in on clothing.

Many types of lab coats are available, and selection should be based on the hazards that are of most concern. For example, front-buttoning coats are desirable for protection against chemical spills because they can be removed quickly. All lab coats should be flame-resistant, with cotton frequently providing the best resistance to both chemicals and heat in a comfortable garment (Table 8.1). For specialized work, such as pouring large quantities of corrosive chemicals, an impermeable apron may be most appropriate.

To be effective, lab coats must fit properly and remain buttoned at all times in the laboratory. Sleeves must be long enough to provide arm protection and should fit the arm fairly snugly to avoid flapping. Rolling up the sleeves provides a holding area for chemical and biological contaminants and is not recommended. This practice also leaves the wrists and lower arms without protection. Some lab coats have knit cuffs that provide snug but comfortable wrist protection, especially for individuals with shorter arms.

Lab coats should be laundered regularly at your institution, even in the absence of any known contamination. Never take laboratory clothing home for washing. In case of known contamination, the coat can be decontaminated in the laboratory before washing, or discarded. Do not wear lab coats used in the laboratory into common areas such as lunchrooms or lavatories.

TABLE 8.1
Protective Clothing Materials

Material	Use	Properties
Cotton	Lab coats	Lightweight, degraded by acids
Cotton/polyester blend	Lab coats	Lightweight, neat appearance
Modacrylic	Lab coats	Nonflammable, resistant to most chemicals, easy to clean, low static
Nomex® IIIA	Lab coats	Nonflammable, liquid-resistant, expensive
Nylon	Lab coats, hair nets	Lightweight, strong, water-resistant, highly flammable unless treated
Neoprene	Aprons	Excellent chemical resistance, inflexible
Rubber	Aprons, long gloves	Very good chemical resistance
Vinyl	Aprons, sleeves, shoe covers	Lightweight, prone to develop static charge
Polypropylene	Aprons, full-body suits, caps, shoe covers	Chemical resistance, strong, lightweight, water-repellent
Tyvek (high-density polyethylene)	Full-body suits, shoe covers, disposable lab aprons, caps, sleeves	Strong, lightweight, excellent barrier protection for user, protects lab materials from human contamination, recyclable material

8.2.1.3 Shoes

Proper footwear for the laboratory includes shoes with covered toes and nonslip soles, which will protect the feet from broken glass and hazardous spills. Sandals, sneakers, or woven shoes provide little protection. Low heels are the most comfortable while standing at the lab bench and also protect against falls.

Laboratory workers may want to consider keeping a special pair of shoes to wear only in the lab. Changing to and from street shoes prevents the tracking of hazardous materials from the lab into the outside environment and also prevents the introduction of potential contaminants into a cleanroom. (A **cleanroom** *is a special laboratory facility where all contaminating materials and any particulate matter in the air must be limited.*) Numerous studies have shown that shoes worn in a bacterially contaminated environment may carry higher concentrations of bacteria on their soles than the floor itself.

One method to prevent contamination of shoes is the use of disposable shoe covers, which are routinely used for animal surgery and cleanroom operations, and which are removed before exiting the laboratory. Shoe covers will also prevent your shoes from carrying bacterial contamination into public areas and your home.

8.2.2 GLOVES

8.2.2.1 Choice of Gloves

The proper use of gloves in the laboratory provides a significant measure of protection against many types of hazards. One of the most obvious benefits is the creation of a barrier between your skin and chemical or biological contamination.

If you look in a laboratory supply catalog, you will find a bewildering variety of gloves available, in many materials and styles. It is important to remember that although every type of glove provides a barrier, none can protect against all types of hazards. Each glove type will provide at least some protection against one or more of the following:

- corrosive or toxic chemicals
- biological contaminants
- sharps
- extreme temperatures.

Because no glove can provide all the necessary protective features, most laboratories have several glove types available, including:

- thin-walled gloves for dexterity
- heavy rubber gloves for dishwashing
- insulated gloves for handling hot and cold materials
- puncture-resistant gloves for handling animals.

The first step in choosing the right glove for a job is deciding what protection is required. Are you trying to protect yourself or your work materials from contamination? Do you need maximum protection from a

TABLE 8.2

Protective Glove Materials

Type	Advantages	Disadvantages	Recommended for Protection from:
Latex	Low cost, good flexibility, comfortable	Poor protection from oils, grease, organic solvents; may trigger allergic reactions	Bases, alcohols, bloodborne pathogens
Vinyl	Low cost, medium chemical resistance	Protection against some chemicals less than latex	Strong acids and bases, salts, other aqueous solutions, alcohols
Neoprene	Medium cost, medium chemical resistance, abrasion-resistant	Not as flexible as rubber; can give poor grip	Oxidizing acids, phenol, glycol ethers
Nitrile	Puncture- and abrasion-resistant, dexterity, comfortable for longer wear, hypoallergenic, good chemical resistance	Poor protection from benzene, methylene chloride, ethylene, many ketones	Oils, aliphatic chemicals, bloodborne pathogens
Butyl	Specialty glove, resistant to polar organics	Expensive, poor protection from hydrocarbons and chlorinated solvents	Gases, aldehydes, glycol ethers, ketones, esters
Polyvinyl chloride (PVC)	Specialty glove, good abrasion resistance	Poor protection from most organic solvents	Oils, acids, bases, peroxides
Polyvinyl alcohol (PVA)	Specialty glove, resists a very broad range of organic solvents	Very expensive, water-sensitive, poor protection from light alcohols	Aliphatics, aromatics, chlorinated solvents, ketones (except acetone), esters, ethers
Fluoroelastomer (Viton)	Specialty glove, resistant to organic solvents	Extremely expensive, poor physical properties, poor protection from some ketones, esters, amines	Carcinogens, aromatic and chlorinated solvents
Norfoil (silver shield)	Specialty glove, excellent chemical resistance, lightweight, flexible	Poor fit, easily punctured, poor grip	Use as glove liner, good for emergency use in chemical spills

highly toxic chemical? Table 8.2 provides an overview of the most common glove materials and their advantages and disadvantages.

When choosing gloves for protection against chemicals, always consult the specific glove manufacturer's chemical resistance chart, which is usually supplied with the gloves or found in the supplier's catalog. This will provide information about the properties of specific glove materials. The information provided generally includes the following:

- **degradation rate**, *which indicates the tendency of a chemical to physically change the properties of a glove on contact;*
- **permeation rate**, *which measures the tendency of a chemical to penetrate the glove material; and*
- **breakthrough rate**, *which indicates the time required for a chemical that is spilled on the outside of a glove to be detected on the inside of the glove.*

Example: Choosing a Glove for Chemical Resistance

Assume that you are performing an experiment and are concerned about the possibility of acetone spills. After taking precautions to minimize the risk of skin exposure to acetone, you will still want to wear gloves. The two types of gloves you have available are made of either PVC or butyl. You then check the chemical resistance guide from the glove manufacturer and find the following information for acetone:

PVC gloves: Degradation rate >25% in 30 minutes
Permeation and breakthrough rate <1 minute

> Butyl gloves: Degradation rate – no effect
>
> Permeation and breakthrough rate > 17 hours
>
> This indicates that these PVC gloves are susceptible to chemical breakdown by acetone, and that any acetone spilled on the gloves will contact your skin in less than 1 minute. Butyl, on the other hand, appears to be highly resistant to acetone, the better choice for this situation.

The best option when dealing with highly toxic agents is sometimes to double glove using two different types of gloves (e.g., using chemically resistant gloves under puncture-resistant gloves). This provides the benefits of two glove types.

In addition to choosing the proper glove material, you may also have choices in the thickness of glove material. Glove thickness is usually measured in **mils**, *a unit where 1 mil = 0.001 in.* Thinner gloves generally provide more flexibility but less protection. Therefore, use the thickest glove that does not decrease the necessary dexterity for the task.

Do not allow the wearing of gloves to provide a false sense of security when working with highly toxic materials. Because all gloves are permeable to some extent, assume that your gloves may leak and do not rely on them to protect you when a spill occurs. If you are working with potentially hazardous materials that are unfamiliar to you, do not proceed with your experiments until you confirm that you have proper protection; see the case study below.

Case Study: Proper Gloves Could Save Your Life

The research world was horrified in June 1997 when Dartmouth Professor Karen Wetterhahn died of mercury poisoning, 10 months after what seemed at the time to be a minor incident. Dr. Wetterhahn, who was considered a careful laboratory worker by her colleagues, was following standard precautions and wearing the recommended latex gloves when she spilled one or two drops of highly toxic dimethylmercury on her gloved hand. About 3 months later, she began to develop symptoms of mercury poisoning, starting with nausea and proceeding to neurological problems. Tests revealed that she had been exposed to a single dose of mercury far above toxic levels. Later testing showed that latex disposable gloves offered virtually no protection against dimethylmercury, with a breakthrough rate of 15 seconds or less. The incident has led to improved safety information provided to the users of the chemical, along with a recommendation for double gloving with a silver laminate glove under a heavy-duty neoprene or nitrile glove. This knowledge certainly came at a very high price.

8.2.2.2 Proper Use of Gloves

Glove choice is only the first step in ensuring safety. Gloves must be used properly in order to provide full protection. In fact, improper use of gloves can actually increase the risk of hazard exposure in the laboratory by spreading contamination or sealing it against the skin of the user.

To provide maximum user protection, before using a pair of gloves, check them for any holes or openings. Disposable gloves are usually mass-produced, which means that a certain percent (depending on the manufacturer) may be defective. These should be discarded immediately. Any cuts or abrasions on the hands should be bandaged or covered before donning gloves because these are possible entry sites for contamination. Long or ragged fingernails can easily tear many disposable glove materials. Gloves must be long enough to provide wrist protection; if not, use arm protectors for hazardous work. It is best not to wear wrist watches or bracelets that can trap contamination against the skin.

When working with hazardous materials, it is important to change gloves regularly. Always have a plentiful supply of disposable gloves nearby for quick changes. Change gloves immediately if you think they might have come into contact with hazardous material, and also change them regularly even if you think that they are clean. Remember that all gloves are permeable to some extent, and the longer you wear them, the more likely they are to develop small holes or tears. When removing fitted disposable gloves, use a removal technique that will not spread contamination from the

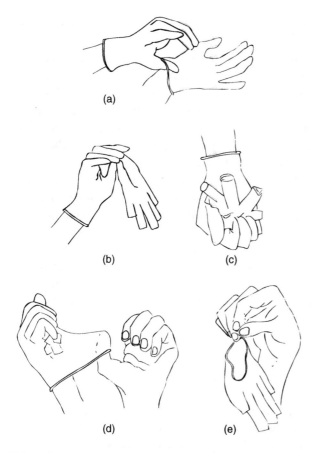

FIGURE 8.2 Proper glove removal technique. (a) Hook a finger on the cuff of one glove, being careful not to contact the skin of the wrist. (b) Pull the glove from the hand inside out. (c) Roll up the removed glove in the palm of the remaining gloved hand. (d) Use an ungloved finger to hook the inside of the glove cuff. (e) Pull off the glove inside out and discard properly. (Artist: Dana Benedicktus.)

outside of the glove to the skin. A useful technique is illustrated in Figure 8.2. Never remove thin gloves by pulling on the fingertips; these are likely to tear. Wash your hands thoroughly after glove use and in between glove changes if any contamination is suspected.

In addition to user risk, improper glove use can extend hazards to other members of the laboratory. Once the outside of a glove is contaminated with hazardous material, that contamination will be spread to any surface touched by the glove. Always remove at least one glove when opening refrigerator doors, using laboratory equipment with adjustable controls, touching doorknobs or light switches, answering telephones, or any time you leave the laboratory and pass through common areas. Be careful when writing in lab notebooks while using gloves. The pen used should be kept at the lab bench and considered contaminated. If you need to remove gloves to answer the telephone, for example, do not reuse disposable gloves; get a fresh pair to avoid the risk of a tear or skin contamination.

8.2.2.3 Potential Health Risks from Disposable Gloves

Many laboratory workers report minor problems from working with disposable gloves. This frequently takes the form of general irritation from the gloves. This can be alleviated by changing gloves often and allowing your hands to dry between changes. Use larger gloves to allow more air circulation around the fingers. It may help to apply a barrier cream under your gloves. These are hand lotions (available in scientific supply catalogs) that are designed to prevent irritation from glove materials and prevent drying of hands. Cotton glove liners that absorb perspiration are also available, although these may be unsuitable for highly hazardous work.

A much more serious problem arises when laboratory workers develop allergies to rubber **latex**, *a natural product that is commonly found not only in gloves, but in many pieces of laboratory equipment and household items*. **Allergies** *are reactions by the body's immune system to exposure to specific chemicals*, in this case, proteins that are found in latex gloves. Although this problem is less common in the general population, as many as 8%–12% of healthcare workers have adverse reactions to latex exposure. If you develop a rash on your hands, persistent chapping, or other annoying skin symptoms, you should stop wearing latex gloves altogether, even if the symptoms don't seem too serious. These allergies tend to become more severe over time and exposure.

Latex allergies apparently originate in sensitization to proteins found in natural rubber latex and appear to be more likely to develop in users of powdered gloves. Glove powder, which is generally USP-approved cornstarch, is believed to make skin adherence or inhalation of latex proteins more likely, thereby acting as a sensitizer. A **sensitizer** *is an agent that can trigger allergies by itself, or cause an individual to develop an allergic reaction to an accompanying chemical*. The resulting reactions can range from minor skin irritation to respiratory shock and even death after exposure.

Current recommendations to avoid the development of latex allergies (assuming that your work includes the use of latex gloves) include the following:

- choosing gloves marked as **hypoallergenic**, *which indicates that the gloves are less likely to trigger allergic reactions than similar gloves*

- choosing powder-free gloves
- checking the glove manufacturer's information sheets for data about the levels of latex protein present in the gloves, and choosing gloves with the lowest levels of latex proteins.

Individuals who exhibit any symptoms of latex allergies should avoid latex gloves and all other items containing latex. Other glove materials should be substituted.

8.2.3 Eye Protection

According to OSHA, an estimated 2,000 eye injuries per day occur in US workplaces. These injuries cost businesses and individuals more than $300 million per year in lost time and money. Of the injured individuals, approximately 60% were not wearing any eye protection, and the remainder were wearing inappropriate devices. The most common injury type was the result of small flying particles, and about one in five injuries were caused by chemicals. Laboratory workers and visitors must be provided with adequate means of preventing eye injuries. OSHA regulations require that workplaces provide suitable eye protection gear that:

- protects against the hazards found in that workplace;
- fits securely and is reasonably comfortable; and
- is clean and in good repair.

Virtually all protective eyewear sold in scientific catalogs meet strict **ANSI** standards for impact resistance, but there can be many other types of hazards found in biotechnology laboratories (Table 8.3).

There are three general types of eye and face protection devices that should be available in all laboratories (Figure 8.3). First, safety glasses are a minimum precaution against small splashes and minor hazards. Safety glasses must have side protection in order to provide splash protection. For this reason, regular eyeglasses are not considered adequate eye protection in the laboratory. The next step up in eye protection is goggles, which are sealed around the eyes, providing good protection against large splashes or caustic agents. As with gloves, the laboratory may stock different types of goggles. Goggle materials may provide protection against chemicals or against ultraviolet radiation, not necessarily both.

Finally, full-face shields should be used when working with materials under vacuum or where there is any threat of explosion. It is essential to wear additional eye protection under a face shield, which is not sealed. Shields made of polycarbonate or other synthetic materials, coupled with appropriate eye wear, can protect against UV radiation, liquid nitrogen, or chemical splashes.

Individuals who require vision correction should wear goggles over their regular glasses, or request prescription safety glasses. Contact lenses are generally considered safe to wear in the lab, but if they are worn, they must not be considered eye protection. If you do wear contact lenses in the laboratory, be sure that co-workers are aware of this, so that in case of an accident they can inform emergency personnel.

TABLE 8.3

Why Do You Need Eye Protection?

Eye protection should be worn at all times in the laboratory. This is also true for visitors, who are not aware of hazards. Chemical goggles may be uncomfortable, but modern safety glasses are often sufficient and are designed to be worn comfortably. Every worksite will have its own requirements because laboratories present many hazards to vision, such as:

- danger of explosion or flying particles;
- liquids that may splash into the eyes;
- glassware under vacuum;
- corrosive liquids such as acids or bases;
- cryogenic materials;
- compressed gases;
- blood and other fluids containing infectious materials, that may splash or form aerosols;
- radioactive materials; and
- ultraviolet light and other radiation.

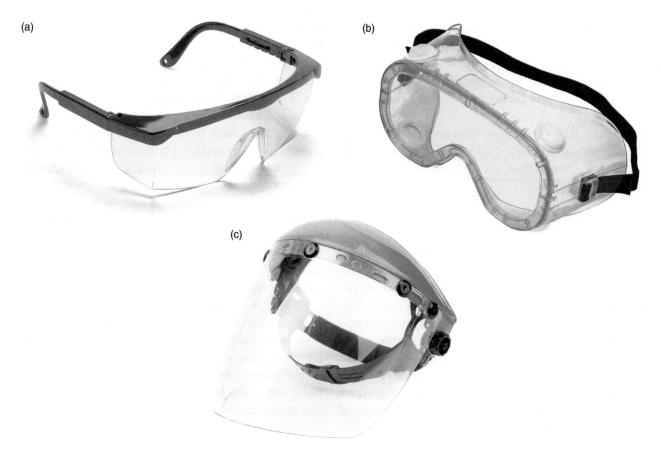

FIGURE 8.3 Eye and face protection. (a) Safety glasses with side protection. (Credit:Photo Melon/Shutterstock.com)
(b) Goggles. (Credit:Oleksandr Kostiuchenko/Shutterstock.com) (c) Face shield. (Credit:indigolotos/Shutterstock.com)

Everyone in the laboratory should be aware of the appropriate procedures to follow in case of an accident involving the eyes or face (Table 8.4). Many studies have shown that a quick emergency response can significantly reduce the possibility of permanent damage to vision. For this reason, ANSI standards recommend that emergency eye wash stations (Figure 8.4) be installed within "ten-second" access from all points in a lab. This may translate to anywhere from 10 to 50 ft from lab benches, depending on possible obstructions. Be certain that these stations are easy to locate, that each worker knows how to use them properly, and that they are tested on a weekly basis, both for function and to wash out any contamination that may have accumulated in the water.

8.2.4 HEARING PROTECTION

Noise-emitting laboratory equipment, such as centrifuges, often produce sound that is uncomfortable or hazardous. Long-term exposure to high noise levels may cause loss of hearing sensitivity. Disposable or personal ear plugs should be available to reduce ambient noise when necessary. **Sonication devices**, *which are used to disrupt cells with high-frequency sound waves,* produce particularly high levels of noise and should only be used with ear protection for everyone in the area, unless the device is contained in a noise-reducing enclosure. The manufacturer may suggest proper ear protection; otherwise, standard hearing protection devices should be worn, either in or over the ears. Several designs are effective at reducing noise levels, and comfort is a prime consideration when choosing ear protection.

8.2.5 MASKS AND RESPIRATORS

Laboratories generally provide access to at least basic respiratory protective equipment. These items may include the following:

- masks, which filter dirt and large particles from the air, and provide splash protection
- air purification or filtration respirators
- self-contained breathing systems (in specialized situations).

TABLE 8.4

Emergency Eye Wash Procedures

An important line of defense against eye injuries in the laboratory (or other workplace) is the use of the emergency eye wash station (Figure 8.4).

- Know the location of the eye wash station and how to operate it.
- If you have an accident involving the eyes or face, yell for help and immediately move to the nearest eye wash station if you can.
- Be prepared to help an accident victim to the eye wash station – seconds count! Never assume that victims can take care of the problem themselves.
- Turn on the water at the station and hold the victim's eyes into the double streams of water. Do not worry about wet clothing.
- Most chemical splashes to the eye and many other injuries involve both eyes, so be sure each eye is properly flushed.
- The eyes should receive a constant stream of warm water for at least 15 minutes. (For this reason, personal eye wash bottles are not adequate for an emergency.)
- Help the involved person hold their eyelids open and roll their eyes around to aid in proper flushing.
- Do not attempt to remove any particulate matter from the eye by hand. (This includes contact lenses.)
- In case of a chemical injury, have someone determine the chemical involved and call for medical help.
- All potentially injured eyes should be examined by a medical professional after the flushing period, even if the person does not feel pain.

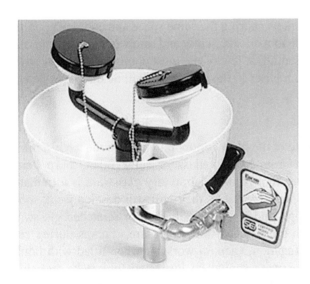

FIGURE 8.4 An emergency eye wash station. There are two nozzles to direct streams of water at each eye simultaneously. The hand lever should allow the water to stay on with a single push. (Image © Flinn Scientific, Batavia, IL, USA. Used with permission.)

These items are illustrated in Figure 8.5. The majority of lab personnel never use more respiratory protection than a surgical-type mask made of cloth-like materials. These masks are usually disposable. These mainly filter dust and larger aerosols from the air and shield the lower face from minor splashes. These are ideal for animal work because they can remove allergens from the air. It is important to read the manufacturer's description for the masks available in your laboratory. Some types offer little protection against airborne infectious agents, although they may be helpful in preventing lab personnel from touching their noses or faces while gloved. Cloth or paper masks offer no protection from toxic gases.

Air purification and self-contained breathing systems are examples of respirators, which are devices that improve air quality for the user. **Respirators,** *breathing devices designed to reduce airborne hazards by manipulating the quality of the air supply,* should not be used by untrained personnel. They are required by OSHA under circumstances where toxic fumes or hazardous air contaminants cannot be removed from the environment by other means.

Air purification respirators *work by filtering the room air through purification filters and canisters of various adsorbent materials that remove specific contaminants from the air.* When used properly, they can significantly reduce, but not eliminate, airborne contaminants. The most common use for these respirators is the removal of toxic chemicals from the air. There are more than a dozen different types of canisters (shown in Figure 8.5b) that can be attached to the respirator, each suitable for various chemical concerns. There is no "universal" canister that can protect

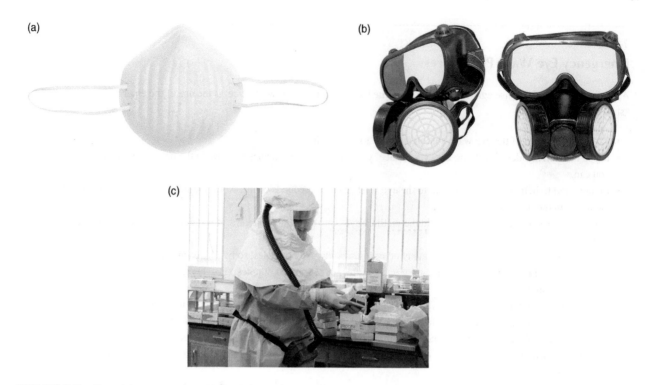

FIGURE 8.5 Breathing protection. (a) Particle mask. (Credit:Thananya Apiromyanon/Shutterstock.com) (b) Air-purifying mask. (Credit:breakermaximus/Shutterstock.com) (c) Air-purifying hood. (Credit:Edgloris Marys/Shutterstock.com)

against all types of chemicals. Different types of particulate filters can also be added to the system.

These respirators require an excellent fit for effectiveness, especially the half-mask models. They are not suitable for workers with facial hair. When fit is a problem, a **full-face air-purifying hood** should be used (Figure 8.5c). These powered hoods *force clean air through installed filters and task-appropriate canisters.* For all air-purifying devices, filters and absorbent canisters must be replaced on a regular basis. Respirators are ineffective if not used properly, so only trained personnel should be placed in situations where respirators are required.

A **self-contained breathing apparatus** *contains its own air supply and is required in situations where the user is exposed to highly toxic gases.* These systems, the same type worn by firefighters, are heavy, uncomfortable, and limited to the short period of protection provided by the attached air tank. They are not routinely used in the laboratory and are not appropriate for use by untrained personnel.

8.3 PHYSICAL HAZARDS IN THE LABORATORY

8.3.1 INTRODUCTION

There are numerous physical hazards that are encountered in laboratories, which are busy places with many workers sharing the same space and equipment. Being able to work efficiently and safely in crowded spaces is essential. A few of the most common causes of physical injuries in the laboratory will be discussed here.

8.3.2 GLASSWARE AND OTHER SHARP OBJECTS

One of the most common injuries in the lab is a minor cut from broken glass or some other sharp items. Even though most laboratory glassware is formulated to resist breakage, they will still develop weak spots, chips, and scratches. Chipped or scratched glassware can break when under pressure, such as during centrifugation, vacuum work, or when filled with liquid. Glassware should be inspected for cracks and chips before being washed and again before laboratory use. Damaged glassware should be discarded or repaired, because it is especially fragile and can easily shatter. Cut glass plates or glass tubing and rods should be sanded or fire polished to dull the sharp edges.

A significant number of injuries from broken glass occur during the cleaning process. Any sink used to collect dirty glassware should be equipped with a soft mat to prevent breakage. Discard any cracked or chipped glassware after your experiments because these items are especially dangerous for dishwashing personnel. Do not leave broken glass in a sink, where someone else may reach in and be cut. For this and other reasons, never reach into a laboratory sink without hand protection.

Sharps *is a term that describes laboratory items, such as razor blades and needles, that can cause cuts and lacerations.* Razor and scalpel blades should be handled with care, in blade holders whenever possible. Careful covering of the sharp edge with tape when not in use or before discarding can reduce accidental cuts. Never leave these items sitting on lab benches. If using a razor blade to scrape a label from a bottle, be sure to scrape away from yourself, bracing the bottle on a solid surface.

Needles should be handled with caution in the laboratory, especially if they are used with biological materials. It is safest not to reuse needles since many needle punctures are the result of attempts to recap a needle. Gloves will not provide adequate protection against a puncture by a small gauge needle. If it is absolutely necessary to reuse a needle, then recap it by placing the cap on a flat surface and then placing the needle in the cap. Do not hold the needle cap in your hand. All needles, broken glassware, and other sharps should be disposed of in a properly labeled container and not in the general trash (Figure 8.6).

8.3.3 COMPRESSED GASES

Cylinders of compressed gas are commonly found in laboratories because certain laboratory instruments require a supply of specific gases. For example, incubators used to contain cultured mammalian cells require a steady flow of carbon dioxide gas. Gas chromatography instruments require gases such as helium and nitrogen. Gases for laboratory use are stored under high pressure in metal tanks, thus allowing a large amount of gas to be stored in a relatively small volume. Although most gases used in biotechnology laboratories are nontoxic and nonflammable, the gas cylinders themselves can be dangerous because they are under high pressure. Accidents involving the rupture of gas cylinders are rare, but quite dramatic (Figure 8.7).

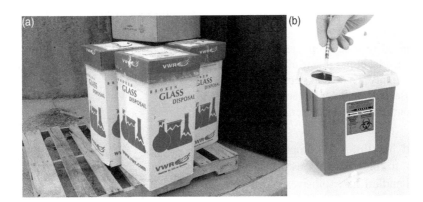

FIGURE 8.6 Sharps disposal. (a) Broken glass should be collected in a rigid, clearly marked container. (Credit:Red Herring/Shutterstock.com) (b) Proper needle disposal in a marked receptacle. (Credit:Thom Hanssen Images/Shutterstock.com)

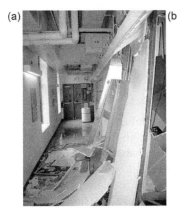

FIGURE 8.7 Compressed gas cylinders are under extreme pressure. This cylinder exploded during storage due to blocking of the pressure release valve and was thrown through the roof of the building, landing 330 ft from the explosion site. (a) Hallway next to laboratory; (b) laboratory after explosion; (c) hole in ceiling where cylinder was thrown. (Images courtesy of Texas State Fire Marshal's Office, Fire Safety Inspection Division.)

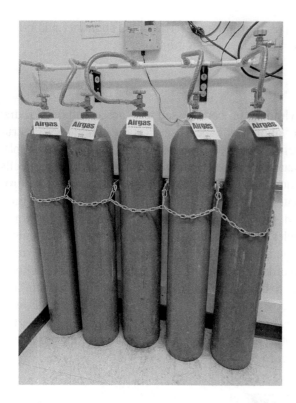

FIGURE 8.8   Proper storage of gas cylinders. Note that the tanks are safely chained to an appropriate rack, with labels facing outward. Each label has a tear off portion to indicate when the tank is empty. These tanks are hooked up to a manifold; if not in use, they would have the metal caps over the valve. (Photo courtesy of Dan Felkner, WiCell.)

Proper storage and handling of compressed gas cylinders is essential to prevent serious accidents. Cylinders should be handled as explosives, with the following guidelines:

- Wear eye protection whenever handling tanks, whether they are full or empty.
- Cylinders should be stored upright, attached to a wall or other solid surface with a strong canvas strap or chain (Figure 8.8).
- Extra or unneeded gases should not be stored in the laboratory.
- All cylinders should be delivered with a safety cap over the delivery valve. This cap should be in place at all times when the tank is not in use. It protects the cylinder valve from damage, and it also prevents accidental opening of the release valve.
- Any flammable gas container (not common in biotechnology laboratories) must be grounded with a wire cable to avoid electrostatic discharges. This should be done with the guidance of an electrician.

- Fuel and oxygen cylinders must be stored at least 20 ft apart, or separated by a fireproof wall.

When selecting a gas cylinder for use, always read the label. Do not rely on gas-indicating color-coding, which is not uniform among manufacturers. Cylinders should be transported carefully one at a time, using a cart that allows the cylinder to be secured with a strap. Never roll a cylinder on end. These tanks are heavy, and if the cylinder falls, one of the least destructive results would be a broken foot. Cylinders can become high-energy missiles if the gas valve is accidentally knocked off on impact, so be sure that the safety cap is securely fastened over the valve when moving tanks.

Compressed gases are dispensed with gas pressure **regulators,** *which decrease and modulate the pressure of the gas leaving the cylinder* (Figure 8.9). A gas storage tank cannot be directly connected to a laboratory incubator or instrument because of the high pressure of the gas. The gas pressure regulator:

- reduces and controls the pressure of the gas flowing from the storage tank to the instrument; and
- displays the pressure in the storage tank and the pressure of gas flowing to the instrument.

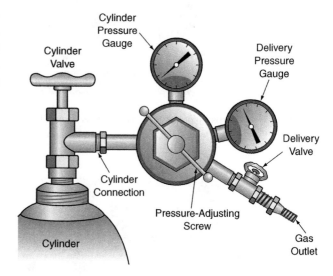

FIGURE 8.9   Gas cylinder valve and regulator. The cylinder valve at the top of the high-pressure tank is opened and closed by turning the handwheel on the cylinder valve. When the valve is open, gas can flow out of the cylinder. The cylinder connection is where the regulator is attached to the tank. The cylinder pressure gauge measures the pressure in the storage tank. The delivery pressure gauge measures the pressure of the gas being delivered to the instrument. Box 8.1 describes the procedure for opening and closing the gas cylinder.

It is the responsibility of the instrument operator to adjust the flow of gas to the instrument, to check that the tank is not depleted, and to replace an empty tank when necessary.

Regulators are threaded to fit the valve outlets of cylinders designed for specific gas types. The gas supplier can help you make sure that you have the correct type of regulator. Never attempt to adapt a regulator to an outlet that does not fit. Do not use grease on the valve, washers or O-rings, or regulator fittings. Use a nonadjustable, fitted wrench to attach the regulator to the valve. Pliers or other inappropriate devices may damage the cylinder or regulator fittings.

Although there are various types of regulators, the most common type in biology laboratories is the **two-stage gas regulator.** The first stage of a two-stage regulator greatly reduces the pressure of the gas leaving the cylinder. The second stage is used to "fine-tune" the pressure reaching the instrument. The regulator thus controls the pressure reaching the instrument and protects it from a "blast" of pressurized gas. The parts of a regulator are illustrated in Figure 8.9.

Before using any gas, learn about the properties and potential hazards of that gas. Because of the high pressure in gas cylinders, never direct a stream of any gas at another person or yourself. Cylinder valves and their regulator connections can be checked for leaks with a dilute solution of soap and water applied to the fittings. Leaks will appear as bubbles. Special instruments that detect leaks of certain gases are also available and are recommended by instrument manufacturers under some circumstances.

There is a proper procedure to shut off the flow of gas after use that will decrease the possibility of accidents or damage to regulators (see also Box 8.1). Whenever the gas is not in use, close the cylinder valve completely. Leave the regulator open to empty the line to the gas outlet, and then close the regulator. This prevents any residual pressure on the regulator that may cause leakage.

When a tank is almost empty, shut off the gas as described earlier, remove the regulator, and replace the safety cap over the valve. It is best not to drain gas cylinders completely, to avoid contamination of the

---

BOX 8.1   STANDARD PROCEDURE FOR USING A COMPRESSED GAS CYLINDER

See Figure 8.9 as a reference for the parts of a gas regulator.
To deliver gas to an instrument or other outlet:

1. Use a hand cart specifically designed for this purpose to move the gas cylinder into position for the instrument. The gas cylinder should be secured and supervised at all times.
2. Securely fasten the gas cylinder to its new location.
3. Remove the protective cap from the top of the cylinder.
4. Attach the appropriate gas regulator to the gas cylinder through the side cylinder connection valve. Be sure that the pressure gauges are positioned at the top of the regulator and that the delivery valve is closed.
5. The regulator should be tightened to the connection using an appropriately sized, designated wrench kept handy for this exact purpose.
6. Slowly open the cylinder valve using hand pressure. In general, you will not open the valve all the way.
7. Never force a valve open. If hand pressure is insufficient to open a valve, it may be damaged and dangerous.
8. Wait until the cylinder pressure valve indicates a steady pressure from the cylinder.
9. Next use the pressure-adjusting screw to create an appropriate level of gas flow to the delivery pressure gauge of the regulator. Wait until the gauge shows a steady pressure value.
10. Connect the instrument to the tubing from the gas outlet.
11. Slowly open the delivery valve to allow gas flow from the regulator to the instrument; the gas outlet is where the regulator is attached to the instrument or other required outlet.
12. Be sure that the delivered gas pressure is appropriate for your needs. If not, adjust with the delivery valve.

(Continued)

BOX 8.1 (*Continued*) STANDARD PROCEDURE FOR USING A COMPRESSED GAS CYLINDER

When finished using a gas cylinder:

1. Close the cylinder valve first. This allows drainage of the gas from the regulator.
2. When the gas flow stops and both pressure gauges read zero, close the delivery valve and the pressure-adjusting screw.
3. Using the fitted wrench, unscrew the regulator from the gas cylinder. Be sure to keep a good grip on the regulator so that it does not fall. Return the regulator to the appropriate location.
4. Replace the protective cap on top of the gas cylinder.
5. Secure the cylinder to the gas cart and carefully return the cylinder to its original location.
6. Secure the cylinder at the storage site. It is polite to make sure that the gas label is facing outward so that it can be easily read.

empty cylinder with air. Mark empty cylinders with the symbol MT (empty) to prevent others from trying to guess whether a tank is full or not. Use tape or other large markings. Do not check the fullness of the cylinder by banging on it. Regardless of labels, treat all gas cylinders as though they were full. Table 8.5 provides a summary list of guidelines for safely handling compressed gas cylinders.

8.3.4 Heat

Laboratory burns due to hot plates, Bunsen burners, and autoclaves are not as unusual as they should be. Most laboratory heat burns are minor. In the case of first-degree burns, which involve reddening of skin and pain, flood the burned area immediately with cold water for 10–15 minutes to reduce the pain and spread of tissue damage. Blistering indicates a more serious second-degree burn. If any blistering or blackening occurs, or any chemical contamination is involved, the burn should receive prompt medical attention. Note that all electrical burns should receive immediate medical attention, because tissue and organ damage can occur that is not externally obvious.

Any heat-producing equipment can cause burns or a fire and should be treated with respect. Never leave an

TABLE 8.5
General Guidelines for Handling Compressed Gas Cylinders

- Be sure all cylinders are clearly labeled.
- Secure cylinders in an upright position at all times, using a strong strap or chain.
- Store only necessary amounts of gases.
- Store gas tanks in a well-ventilated area.
- Never store gas tanks at temperatures above 125°F.
- Do not drop or strike cylinders.
- Use a proper cart with a strap for moving cylinders – do not roll them.
- Do not move a gas tank while the regulator is attached.
- Keep the protective cap on the cylinder head when not in use.
- Use a proper gas pressure regulator, appropriate for the gas in the tank.
- Open valves slowly, and do not attempt to force a valve open.
- When changing tanks, close the valve between the tank and regulator, and be certain that there is no residual pressure in the gas lines.
- Keep the valves from the tanks closed when the tanks are not in use.
- Never try to repair a cylinder or valve.
- Check cylinders of toxic or flammable gas for leaks on a regular basis.
- Always use safety glasses when handling compressed gases, especially while connecting or disconnecting regulators or supply lines.

uncontrolled heat source, such as a Bunsen burner, unattended. Keep the gas regulated on open flames so that they are visible to any observer. Do not place a device with an open flame in a position on the lab bench or hood where you or anyone else might need to reach past it.

Another source of laboratory burns are hot plates that are left on unintentionally, or which have just been turned off. Most people would not touch a hot-plate surface that contained a boiling container, but they might contact a hot surface when it is empty. When you turn off a hot plate, it is courteous to leave a note indicating that the equipment is still hot (and remember to remove the note when the item has cooled).

Boiling or heated liquids are hazards if not handled properly. Use insulated gloves or tongs to handle hot beakers and flasks. Never heat a sealed container of liquid, because this creates a risk of an explosion and the violent splashing of hot liquid. Be cautious around liquids that may have **superheated**, *which means that they have been heated past their boiling point without the release of the gaseous phase*. Superheated liquids may boil over, sometimes violently, when jarred. This happens regularly with agarose solutions heated in microwave ovens, as well as with liquids that have been autoclaved. It is best to avoid superheating liquids, for example, by heating agarose solutions only enough to melt the gel properly. If you suspect the possibility of superheating, approach the container cautiously, and allow some cooling time before moving the solution container.

8.3.5 Fire

Fire is a chemical chain reaction between fuel and oxygen that requires heat or other ignition source (Figure 8.10). Fuels include any flammable materials, such as paper or solvents. A **flammable** substance *is one that will ignite and burn readily in air.*

The most common source of laboratory fires is the ignition of flammable organic liquids and vapors. This hazard will be discussed more fully in the next chapter on chemical safety. The best overall fire prevention strategy for labs is to limit sources of flammable materials. This and other general strategies for fire prevention are provided in Table 8.6.

In the event of a fire, it is critical to know what to do. Attempting to extinguish a fire (which may not be the best strategy) requires an understanding of fires and how they spread. The fire triangle is important to remember here, because all fires require fuel, heat, and oxygen. If any of these factors are removed, the fire will be extinguished. The use of appropriate fire extinguishers can control small fires if used properly (Figure 8.11).

TABLE 8.6

Fire Prevention in the Laboratory

- Store only minimum amounts of flammable materials.
- Keep open solvent containers far away from heat sources.
- Store flammable solvents in appropriate containers.
- Use water baths or hot plates in preference to Bunsen burners.
- Limit other ignition sources, such as sparks or static electricity.
- Never leave an open flame unattended.
- Mark heated hot plates.
- Reduce electrical hazards in the laboratory (Table 8.7).

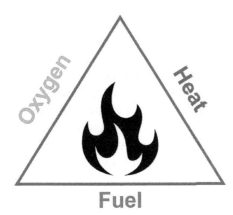

FIGURE 8.10 The fire triangle. Fuel, heat, and oxygen must each be present for fire to occur.

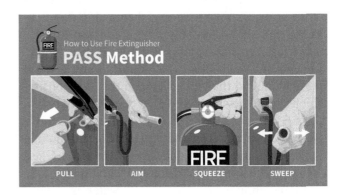

FIGURE 8.11 Proper use of a fire extinguisher. Use the PASS technique: pull, aim, squeeze, and sweep. (Credit:Toywork/Shutterstock.com)

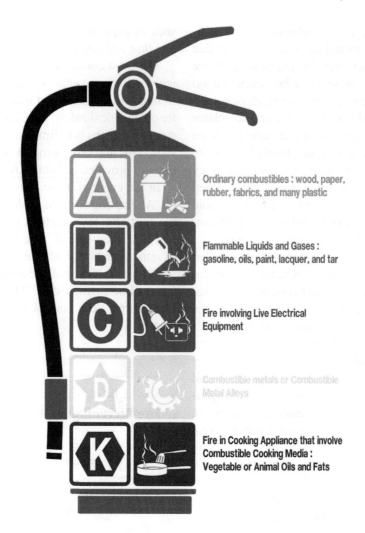

Ordinary combustibles : wood, paper, rubber, fabrics, and many plastic

Flammable Liquids and Gases : gasoline, oils, paint, lacquer, and tar

Fire involving Live Electrical Equipment

Combustible metals or Combustible Metal Alloys

Fire in Cooking Appliance that involve Combustible Cooking Media : Vegetable or Animal Oils and Fats

FIGURE 8.12   Classification of fires and appropriate portable fire extinguishers. (Credit:Heavypong/Shutterstock.com)

There are five types of fires that might occur in the laboratory, summarized in Figure 8.12. Most fires in homes involve Class A combustibles such as paper, cloth, or wood. Class A fires can be extinguished by water or multipurpose dry chemical extinguishers. In the laboratory Class B fires are more likely than Class A. Class B fires usually involve organic solvents and other flammable liquids. Water is more likely to spread than to extinguish this type of fire, which should be smothered with chemical foam or a Class B carbon dioxide fire extinguisher to remove the oxygen supply from the flames. (Carbon dioxide is heavier than oxygen and covers the fire.) Class C fires, which involve electrical equipment, also occur in laboratory settings, and water is a poor choice for fighting these fires. Water may increase the possibility of serious electric shock to individuals in the area. Use a carbon dioxide extinguisher instead. Many laboratories have multipurpose ABC extinguishers available. Class D fires, which involve combustible metals, are seldom a concern in biotechnology laboratories. There is also a fifth type of fire, Class K, involving oils and fats used in restaurants, again uncommon in laboratories.

All laboratories should be equipped with a multiclass fire extinguisher. Dry chemical fire extinguishers are highly effective against Class B and C (solvent and electrical) fires. Carbon dioxide extinguishers can be used on the same types of fire, although they tend to have a limited distance of effectiveness. Multipurpose chemical fire extinguishers that can be used for Class A+B+C fires are convenient, but most types leave residue behind that requires significant cleanup. (This is the type most likely found in homes.) Residue is not a problem with halon A+B+C fire extinguishers, although halon is designated as an ozone-depleting chemical. More recently, environmentally friendly extinguishers, containing halocarbon agents, have become increasingly available. These are effective against Class A+B+C fires and do not leave residue behind for cleanup.

Every laboratory must be equipped with fire extinguishers that are installed close to exits, are easily accessible, and are regularly checked for pressure. Some institutions have safety plans that forbid general employees from using the fire extinguishers. However, even if there is no policy against extinguisher use, do not attempt to use these devices unless you have been trained to operate them (some standard extinguishers weigh up to 40 pounds) and you have practiced recently. Your institutional safety office should be able to provide this training. If you are certain that the fire is contained and minor, and you are qualified to extinguish it, have someone call the fire department before you start. Be sure that all people are out of danger and that they are aware of the fire. Many laboratory flammables will create toxic fumes and heavy smoke as they burn, which makes them difficult to control. Be alert for this possibility. Be sure that you have a clear route of retreat; this is why fire extinguishers are usually installed near exits.

When a fire occurs, it is frequently best to evacuate. Everyone working in a laboratory should be familiar with evacuation routes and procedures, and it is particularly important that no one stay behind after an evacuation is ordered. Close all fire doors and windows when leaving, if it is safe to do so. Evacuation procedures should include a designated meeting place away from the building. Do not attempt to reenter the building until the appropriate emergency director indicates that it is safe to do so.

8.3.6 COLD

The most common cold hazards encountered in biotechnology laboratories are low-temperature freezers, dry ice baths, and liquid nitrogen. Skin contact with **cryogenic,** or *extremely cold (usually defined as temperatures below − 78°C)* substances, or their cooled containers, can result in skin damage similar to heat burns. When handling cryogenic materials, proper hand and eye protection is essential. Gloves should be insulated and loose-fitting for quick removal.

Cryogenic liquids are not compatible with ordinary lab glassware or, in particular, plasticware. They are usually contained in **Dewar flasks**, *which are heavy multi-walled evacuated metal or glass containers.* While Dewar flasks are similar in appearance to Thermos bottles, they are not the same. Thermos bottles are thin-walled and will shatter if exposed to cryogenic temperatures.

Dry ice, *which is frozen carbon dioxide in solid form*, is frequently used to prepare coolant baths for samples. Most of us have handled dry ice casually at Halloween parties, where it is used to create fuming cauldrons of punch; however, dry ice, which can readily burn wet skin, should always be handled with insulated gloves. Most laboratories that use dry ice keep a supply in a special dry ice chest. If the dry ice is provided as chips, do not use a fragile Dewar flask to scoop up the chips. Use a metal scoop to place the chips in an ice bucket. If the dry ice is furnished as a large block, you will need to make chips of the appropriate size. Do not chip dry ice without goggles for eye protection. A less obvious hazard is the danger of induced hyperventilation (*increased respiration rate*) or even asphyxiation (*interruption of normal breathing*) from breathing high concentrations of carbon dioxide. Do not lean into the dry ice chest for extended periods of time.

To make a coolant bath, a Dewar flask is usually filled about one-third full with an appropriate solvent, and then small pieces of dry ice are added to cool the solvent. Acetone was used historically as the solvent, but this is not recommended for safety reasons; see the case study below. Isopropyl alcohol, ethanol, or various solvent mixtures that are relatively nontoxic and nonflammable are recommended. (See National Research Council's "Prudent Practices in the Laboratory," as referenced in the Bibliography in the Introduction to this unit, or the 1995 edition free online at https://www.nap.edu/read/4911/chapter/1; chemical properties have not changed.) It is essential to wear gloves and use a face shield and eye protection while preparing the coolant bath. The dry ice chips must be added slowly, waiting until the solvent stops bubbling before adding additional dry ice. Once the solvent is suitably cooled, the sample can be introduced to the bath. Always add the item to be cooled slowly. A sudden change in bath temperature can cause the solvent to splatter or overflow the container.

Case Study: Hazards of an Acetone–Dry Ice Bath

A graduate student was making a dry ice and acetone cooling bath on a general-use lab bench. Being in a hurry, he added the dry ice chips rapidly and ignored the rapid bubbling of the acetone. He did not notice that the acetone was spilling onto the smooth lab bench and spreading flammable vapors along the surface. When the acetone vapors contacted the motor of a working water bath several feet away, the solvent vapors ignited, causing a fire that completely destroyed the water bath and surrounding materials. Of more immediate concern to the worker himself was the fact that the fire, which started at the water bath, traveled along the stream of solvent vapors back to the dry ice bath. The vapors there ignited as well. The student could have been seriously injured if he had not been standing away from the flask and wearing proper eye protection.

Another cryogenic hazard frequently encountered in laboratories is **liquid nitrogen,** *a liquid form of nitrogen that is supplied in compressed gas cylinders at* $-198°C$. This substance is cold enough to change the physical properties of many materials, and it will cause the equivalent of **third-degree** (*full thickness; through all skin layers*) **burns** if it contacts skin. It is essential to handle liquid nitrogen with great care, and the involved personnel should receive specific training for this task. When drawing liquid nitrogen from the original cylinder, always wear a face shield and goggles, as well as protective clothing, including heavy leather gloves and leg and foot protection. Because of the high pressure, be certain that you have a firm grip on the metal or insulated dispensing hose, and that the container you are dispensing into is firmly secured in place.

Liquid nitrogen poses both a cryogenic hazard and a hazard from high-pressure gas release. Liquid nitrogen is formed and stored under pressure and can convert to the gaseous form quickly, with an expansion of about 700 volumes of nitrogen gas from 1 volume of liquid nitrogen. Special tanks for sample storage have loose-fitting caps with vents. Never place liquid nitrogen into a tightly sealed container. Always wear a face shield and eye protection when removing storage vials from liquid nitrogen. If liquid nitrogen has leaked into a vial during storage, the pressure buildup from nitrogen gas expansion can cause the vial to explode if warmed quickly.

8.3.7 ELECTRICITY

Laboratories are full of electrical equipment; therefore, they contain electrical hazards. These issues will be discussed in Chapter 16. This is an appropriate place, however, to remind the reader that even small levels of current flowing through the human body can cause harm or even death. An **electrical shock** *is the*

sudden stimulation of the body by electricity, when the body becomes part of an electrical circuit. Make sure that all electrical equipment is in good working condition and properly grounded. When appropriate, GFI circuits should be installed. **GFI,** or **ground fault interrupt circuits,** *are safety circuits designed to shut off electrical flow into the circuit if an unintentional grounding is detected.* These are usually installed around sinks and other water sources. All electrical burns and shocks should receive prompt medical attention to rule out the possibility of internal tissue damage.

High-voltage power supplies and electrophoresis equipment should be used with caution. Handle power leads one at a time. Table 8.7 provides some general suggestions for avoiding electrical hazards in the laboratory.

As suggested by the case study above, electrostatic discharges can also have explosive effects when the equipment used around flammable materials is not properly grounded. Prevent the personal buildup of static electricity by avoiding wool and nylon clothing and by touching a safe metal object, such as a doorknob, periodically. Unfortunately, one excellent preventive measure, increasing room humidity, is frequently not appropriate in the laboratory.

8.3.8 ULTRAVIOLET LIGHT

Ultraviolet (UV) light *is a form of nonionizing radiation that makes up the light spectrum between visible light and X-rays.* (The light spectrum is discussed in more detail in Chapter 21.) UV radiation is generally divided into three classes:

- UV-A – 315–400 nm wavelength
- UV-B – 280–315 nm wavelength
- UV-C – 180–280 nm wavelength

TABLE 8.7

Reducing Risk from Electrical Devices

- Use only Underwriters Laboratories (UL)-approved electrical equipment.
- Be sure that all equipment is properly grounded with a three-pronged plug (see Chapter 16).
- Use GFI circuits whenever appropriate.
- Check electrical cords regularly to be sure they are in good condition.
- Keep your hands dry when handling electrical equipment.
- Never use electrical equipment with puddles of liquid underneath.
- Avoid using extension cords when possible. If one is necessary, use only extension cords marked as "heavy duty," to accommodate higher electrical loads from equipment such as hot plates. Be sure to avoid possible tripping hazards.
- Do not perform repairs inside equipment unless you are qualified to do so.
- Keep equipment unplugged when not in use.
- Plug in fume hood equipment outside the hood to prevent electrical sparks within the hood.

TABLE 8.8

Guidelines for Minimizing Exposure to Ultraviolet Radiation in the Laboratory

- Wear UV-absorbing eye protection in the presence of UV radiation.
- Protect skin surfaces from UV exposure. Remember your wrists.
- Turn on germicidal lamps in hoods or rooms only when the area is not in use.
- Be careful to avoid reflection from handheld lamps into your eyes.
- Make sure UV light is not directed toward an unsuspecting co-worker.
- Always use a UV-absorbing face shield for UV transilluminator work.

UV-A, which is also called "black light," is used in tanning booths and is not usually generated in laboratories. The major sources of UV in labs are germicidal lamps and transilluminators for visualizing stained DNA bands in electrophoresis gels, which operate in the UV-B and UV-C ranges. Handheld UV lamps for various purposes usually operate around 254 nm. UV-B and UV-C wavelengths damage skin and eyes, and exposure to radiation shorter than 250 nm is considered dangerous.

UV light sources used in the laboratory can cause burns and severe damage to the cornea and conjunctiva of the eyes; see the case study "Hurrying Can Slow You Down." These tissues absorb the energies of UV light, with eye irritation developing 3–9 hours after exposure. Skin can also be severely burned by UV radiation, as in a bad sunburn. UV radiation can easily be blocked with certain types of glass and plastic. Eye and skin protection is essential when UV exposure is a hazard, so check that the materials in the protection devices used will guard against UV. Table 8.8 suggests strategies for minimizing laboratory exposure to UV radiation.

Case Study: Hurrying Can Slow You Down

One late night in the laboratory, a graduate student of one of the authors was in a hurry and forgot to use the appropriate face shield while visualizing DNA bands in a gel using a UV transilluminator. He was not wearing the required safety glasses either. There were no immediate effects, and he did not notice any consequences of the incident. However, when he woke the next morning, he was unable to open his eyes more than a slit. A trip to the emergency room revealed that he had burned the corneas of both eyes as a result of the exposure to UV radiation. His face was also reddened, comparable to a significant sunburn. Luckily, the corneal burns were superficial and did not result in permanent vision damage. However, he was required to wear sunglasses indoors for 2 weeks, thus providing a cautionary example to other students who might be tempted to cut corners.

TABLE 8.9

Safe Use of Autoclaves

- Learn correct loading procedures to prevent broken glassware.
- Make sure glassware placed in an autoclave is not cracked or chipped.
- Be sure that plastics are autoclavable. (Teflon and polypropylene are suitable for this purpose.)
- Never autoclave a sealed container.
- Check that the inside is at room pressure before opening the autoclave door.
- Use protection for eyes and hands when opening an autoclave.
- Stand back while opening the autoclave door.
- Release autoclave pressure slowly for liquids to prevent boilover.
- Allow autoclaved liquids to rest for at least 10 minutes before moving the containers to prevent boilover from superheating.

8.3.9 PRESSURE HAZARDS

Many common laboratory operations, such as filtration, are carried out using a vacuum. All low-pressure work should be carried out inside a shielded enclosure, such as a fume hood, that will provide a partial safety jacket in case of an implosion. An **implosion** *is the collapsing of a vessel under low pressure compared with the outside atmosphere.* To reduce the risk of implosion, only suitably heavy-walled glassware should be used for vacuum work. All vacuum glassware should be rigorously checked for chips or cracks before use. Although the fume hood sash can be used as a partial body shield, you should still use a face shield and goggles when working with systems under strong (high) vacuum.

High vacuums require a vacuum pump, but moderate vacuums suitable for filtration and other purposes can be provided by water aspirators. A **water aspirator** *creates a vacuum through a side arm to a faucet with flowing water.*

All vacuum lines should contain a trap or in-line filter for any liquids or vapors that may be pulled into the vacuum source; refer to Figure 32.1b for an example.

A common high-pressure application in biotechnology laboratories is the use of autoclaves for sterilization of glassware and solutions. All modern autoclaves are equipped with safety features to reduce risks, but careful operation is still required. Table 8.9 provides some suggestions for the safe use of autoclaves. (More information about autoclaves can be found in section 24.3.)

8.4 ERGONOMIC SAFETY IN THE LABORATORY

While not as dramatic as some safety issues, musculoskeletal disorders – and in particular, repetitive stress injuries (RSIs) – have become the most commonly reported health problems in the laboratory workplace. RSIs, such as carpal tunnel syndrome, are caused by **repetitive stress**, *where the same movements and actions are repeated until physical fatigue results.* RSIs are chronic conditions that increase in severity over time. They occur due to muscle and joint stress, inflammation of tendons and associated tissues, and constricted nerves and blood flow to the affected regions.

OSHA data indicate that RSIs account for more than 50% of reported workplace injuries. It is estimated that workers' compensation claims from RSIs cost at least $20 billion each year in lost work time. Employers and employees alike have a strong interest in minimizing these injuries. This is generally achieved through ergonomic safety measures in the workplace. **Ergonomics** *is the study of the effects of environmental factors on worker health and comfort, and the design of environments to increase worker health and productivity.* Ergonomic analysis is used to find the body positions, lab furniture, and equipment that minimize stress and strain on workers. Ergonomic laboratory design can prevent or minimize many health concerns related to the lab.

The two most common musculoskeletal hazards in the laboratory are neck and back injury due to poor posture, and hand and wrist injury due to repetitive hand movements. Neck and back pain can result from the awkward postures dictated by laboratory design. It is impossible for a fixed lab bench to be the ergonomically appropriate height for all workers, given the differences in size among individuals. Likewise, biological and fume hoods are generally fixed in height and require unnatural arm movements to work inside the enclosure. Any task that encourages rounding the shoulders and working with your head and arms in front of the body encourages muscle and joint

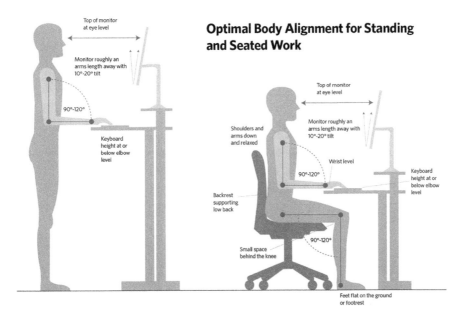

FIGURE 8.13   Optimal body alignment for standing and seated work. The same parameters for neutral position apply to both computer and laboratory bench work. (Artwork provided with permission courtesy of Cal State East Bay.)

pain. Considering how many laboratory tasks require leaning forward, such as looking into a microscope, loading samples into electrophoresis gels, performing dissections, and peering at computer screens, it is clear why laboratory workers may experience back and neck pain.

Back and neck problems can frequently be avoided by the use of ergonomically designed chairs, deliberate good posture, and frequent breaks. Any office furniture catalog will offer a variety of well-designed chairs, which should at minimum provide height and tilt adjustment, along with good back support. The best posture to avoid problems is what is called **neutral position**, *the body alignment that uses the least muscular energy and provides the best possible blood circulation* (Figure 8.13).

The repetitive hand movements of pipetting and computer keyboarding frequently contribute to the development of RSIs. These are two of the most common tasks in a biotechnology laboratory. An experienced technician may dispense as many as 60 or more samples a minute. This task causes repetitive stress and also requires the use of thumb force and awkward bending of the arm and wrist. If a computer keyboard is not located at the optimal height and configuration, the wrists can be held at inappropriate angles. Repetitive pipetting and keyboarding can lead to carpal tunnel syndrome and other problems. **Carpal tunnel syndrome** *is a chronic irritation and swelling of tendons and associated membranes in the wrist.* The irritation and swelling create pressure on the median nerve, resulting in pain and numbness in the thumb and fingers, and loss of manual dexterity. An estimated 4%–5% of the US population has experienced carpal tunnel syndrome. Another common problem is de Quervain's tendinitis, which is a painful inflammation of the tendons involved in thumb movement.

Wrist and hand problems may be avoided by purchasing specially designed, ergonomic pipettes and taking frequent breaks. It is good practice to keep your wrists straight and relaxed while using the keyboard and mouse. It may be useful to purchase an inexpensive, soft wrist rest to keep your wrists elevated while typing and to wear a wrist protection device. Most importantly, do not ignore wrist or arm pain when it starts; take immediate action to prevent the problem from getting worse.

Ergonomic risks are to some extent unavoidable in the laboratory, but you can take steps to protect yourself from health problems. Table 8.10 summarizes general recommendations to avoid musculoskeletal problems. These practices are applicable to activities outside of work as well. Table 8.11 provides additional strategies to deal with individual laboratory tasks, indicating specific stressors and suggesting solutions to minimize the associated physical stress. By developing careful technique, investing in ergonomically designed furniture and equipment, and paying attention to your body's signals, you can minimize your risk of developing a potentially debilitating RSI.

TABLE 8.10

General Ergonomic Recommendations in the Workplace

- Maintain good posture at all times.
- Frequently take short breaks from extended activities.
- Change tasks every 20–30 minutes if possible.
- Perform gentle stretches and exercises for hands, neck, and other stressed body parts on a regular basis throughout the day.
- Request that your employer provide ergonomically designed chairs for working at your desk, at the laboratory bench, and in a hood.
- Purchase and use ergonomically designed devices for repetitive tasks.

TABLE 8.11

Examples of Physical Stress Sources in the Laboratory

Equipment and Activities	Potential Stressors	Possible Solutions
Lab benches	Inappropriate height, hard edges	Use height-adjustable chair with back support; install padding on laboratory bench edges; use elbow pads.
Biological and chemical fume hoods	Inappropriate height; face guard requires awkward movements	Use height-adjustable chair with back support; keep items in hood as close to front as feasible. However, when using a chemical fume hood, be sure your face is protected by the sash.
Centrifuges	Lifting and carrying heavy rotors, bending	Use cart for transporting rotors; use proper lifting techniques.
Microscopes	Eyepiece placement too low, awkward placement of hand controls	Purchase ergonomically designed eyepieces that do not require bending; use computer monitor for viewing.
Pipetting	Repetitive hand movements, application of force, awkward arm angle	Purchase ergonomically designed pipettors; use automation when possible; experiment with alternate grips; use shorter pipettes.
Standing	Back pressure, constriction of blood flow in legs	Avoid standing for more than 20 minutes without walking around; wear comfortable shoes with arch support; install anti-fatigue floor matting.
Keyboarding	Repetitive hand movements, application of force, inappropriate height, inadequate wrist support	Install keyboard holder at appropriate height; use wrist support devices to maintain the proper wrist angle; use a good chair with arm rests; use an ergonomically designed keyboard and mouse.
Vortex mixing	Vibration	Use minimum speeds; wear elbow pads; utilize closed containers to avoid the need for finger force.
Handling vials	Repetitive hand movements, application of force, twisting	Use easy-open vials; use mechanical vial openers.

8.5 FINAL NOTES

It is essential for everyone who works in a laboratory to be familiar with the specific physical hazards in that workplace. Risk of injury decreases with the knowledge of how to work safely with these potentially harmful elements. It is also important for you to learn about proper use of PPE and the operation of safety equipment. Good laboratory practices in handling hazards, described in this chapter, are key for preventing accidents and injuries.

Practice Problems

1. You are attempting to open the valve on a cylinder of carbon dioxide gas. The valve will not turn despite your best efforts. What should you do?
2. Why are canvas shoes, such as sneakers, not recommended in the laboratory?
3. Are disposable latex gloves suitable for handling dry ice?
4. How long should the eyes be flushed with water after a chemical splash?

Questions for Discussion

1. Think of common situations in the laboratory when gloves are worn. Make a list of possible actions you might take during a procedure and the appropriate times to remove or change your gloves.
2. Consider how you would handle a fire at work or at school. What safety equipment is available? Where are the exits? Are emergency telephone numbers accessible?
3. Identify the locations of all safety equipment intended for the laboratory, including:
   - Fire extinguishers
   - Eyewash stations
   - First aid kit
   - Automated electrical defibrillator (AED) if available
   - Instructions on how to report safety issues and injuries.

9 Working Safely with Chemicals

9.1 INTRODUCTION TO CHEMICAL SAFETY

Biotechnology laboratories contain a wide variety of chemicals with different health and environmental hazards. In this chapter, we discuss numerous examples. Some of these chemical names may be unfamiliar to you, but they were chosen to illustrate certain types of hazards and where to find information about these hazards.

It would be an insurmountable task for an individual to memorize all the properties of every compound in the laboratory. Therefore, the Federal Hazard Communication Standard (HazCom 2012) regulates the use of hazardous materials in industrial workplaces, and ensures that chemical hazards in the workplace are identified and that this information is communicated to employees. This law requires the employer to:

- identify all chemicals used in the workplace;
- have a Safety Data Sheet, SDS (see Section 7.4.4.3), available for each chemical;
- properly label all chemicals;
- provide a written program for handling the chemicals;
- train employees on the proper use of all chemicals they will encounter; and
- provide complete information to healthcare professionals in emergencies.

All chemicals can be hazardous (e.g., common table salt can be related to high blood pressure in some cases; sugar can be hazardous to an individual with

DOI: 10.1201/9780429282799-12

TABLE 9.1

Good Practices for Working with All Laboratory Chemicals

- Learn about the physical and toxic properties of a chemical before you start working (e.g., by reading the chemical label and SDS).
- Minimize the amount of chemical used.
- Handle, store, and dispose of the chemical according to recommended procedures.
- Work only in well-ventilated areas.
- Label all containers with the chemical name and hazard warnings.
- Wear appropriate personal protective clothing.

diabetes). When handling any chemicals, even those with no known hazards, the best strategy is to use good general laboratory practices (Table 9.1).

One excellent resource for laboratories is National Research Council's *Prudent Practices in the Laboratory: Handling and Management of Chemical Hazards*, Updated Version, National Academies Press, 2011, doi:10.17226/12654, which provides an overview of the issues related to chemical hazards, provides general information about many laboratory chemicals, and also suggests appropriate reference materials for more detailed information.

9.2 CHEMICAL HAZARDS

9.2.1 INTRODUCTION TO HAZARDOUS CHEMICALS

Many chemicals found in laboratories can present risks to users who do not practice good laboratory techniques. A chemical is defined as *hazardous* if:

- it has been shown to cause harmful biological effects;
- it is flammable, explosive, or highly reactive (participates in reactions that quickly release a great deal of energy); or
- it generates potentially harmful vapors or dust.

The preceding categories are not mutually exclusive, and many chemicals exhibit more than one of these properties. In addition, many chemicals are considered potential hazards because of their structural similarities to known hazards, even though no data are available about the chemical itself.

Because no one can be expected to remember all of the hazards of every chemical used in a laboratory, it is essential to have information about the hazards of laboratory chemicals on hand at all times. As previously

noted, HazCom 2012 requires that a set of SDSs be available for all chemicals used in a particular workplace. Many laboratories keep their own set in a notebook. SDS information is also readily available on the Internet for download. However, a good overview of the basic hazards is available by reading the chemical label carefully.

There are a number of labeling systems that chemical suppliers use to indicate hazards on chemical containers. When OSHA revised the Hazard Communication Standard (HCS, see Section 7.4.2) in 2012 to align with the Globally Harmonized System of Classification and Labelling of Chemicals (GHS), they designated six required elements for the labels of all shipped chemical containers (Figure 9.1). The standard GHS pictograms are illustrated in Figure 9.2

There are two additional label elements that also may be found on chemical containers Historically, one of the most widespread has been the hazard diamond system developed by the National Fire Protection Association (NFPA). The **hazard diamond system** *rates chemicals according to their fire, reactivity, and general health hazards* (Figure 9.3). This system provides easy-to-read, color-coded information with simple numerical scales. Special hazards, such as radioactivity, are indicated by standard symbols. While the NFPA hazard diamond is not included in the HCS, its use is still commonplace as an addition to the required labeling. It is used by fire fighters and other emergency personnel for a quick assessment of acute health hazards in a laboratory incident.

There is another optional system you might see, called the Hazardous Materials Identification System (HMIS). It was developed by the American Coatings Association to indicate major hazards according to HazCom 2012 standards (Figure 9.4).

The NFPA hazard diamond and the HMIS color-coding system are both voluntary and seldom are found

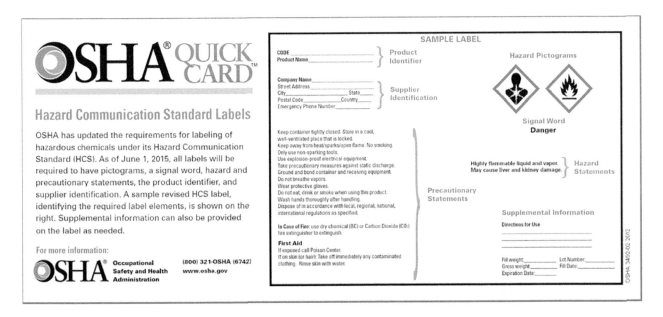

FIGURE 9.1 Required elements for hazardous chemical labels. These elements are indicated in blue in the figure. (Supplemental Information at the bottom right is optional.) The Signal Word is either "Danger," for the most severe hazards, or "Warning," for lesser hazards. This label indicates a health and fire hazard. The standard GHS pictograms are illustrated in Figure 9.2. (Image credit: U.S. Occupational Safety and Health Administration.)

together on the same chemical. Despite their similarity, the NFPA and HMIS systems serve different purposes. The NFPA diamond is intended for firefighters and other first responders, and only indicates acute hazards. The HMIS is intended for personnel in the facility and indicates both acute and chronic toxicities.

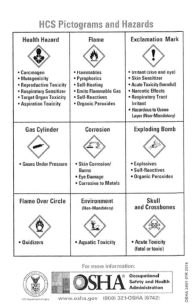

FIGURE 9.2 GHS pictograms for chemical labeling. Standard symbols are used to indicate a variety of hazards. (Image credit: U.S. Occupational Safety and Health Administration.)

9.2.2 Flammable Chemicals

As discussed in the previous chapter, laboratory fires frequently involve flammable chemicals (Class B fires). **Flammable** *refers to materials that are relatively easy to ignite and burn.* Note that the term *inflammable* also *refers to flammable materials.* (The term **nonflammable** *is sometimes applied to chemicals that do not readily ignite and burn.*) The most common flammable chemicals found in laboratories are liquid organic **solvents,** *which are chemicals that dissolve other substances.* Many of these liquids are **volatile,** *evaporating quickly at room temperature.* It is the vapor phase of flammable liquids that burns, not the liquid itself.

Flammability is a relative term because all materials can be ignited in the presence of adequate heat. **Flash point** *is the temperature where a chemical produces enough vapor to burn in the presence of an ignition source.* The lower the flash point, the more flammable the compound. Many common laboratory chemicals, such as acetone and hexane, have flash points well below 0°C; see Example 9.1. This means that flammability is always a concern when working with these materials. Once ignition of a flammable solvent occurs, the increasing temperature increases the rate of vaporization, which, in turn, provides more fuel to burn. Solvent fires can spread out of control very quickly if there is no prompt corrective action. Another concern with flammable liquids is that solvent vapors can diffuse along

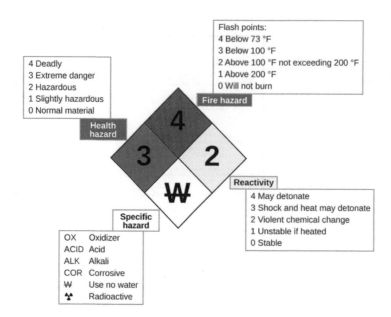

FIGURE 9.3 National Fire Protection Association system for hazard classification. The hazard diamond has four hazard codes to indicate health hazards (blue), flammability (red), reactivity (yellow), and any special hazards (white). The numerical rating system uses the numbers 0–4, with 0 indicating a nonhazard and 4 denoting the highest hazard level of that type. Special hazards are indicated with symbols as shown in the figure. (Image credit: OpenStax, CC BY 4.0 <https://creativecommons.org/licenses/by/4.0>, via Wikimedia Commons.)

lab benches or floors, mixing with air and eventually contacting an ignition source. The flames can then follow the invisible vapor trail back to the original container, igniting the larger source of fuel.

Example 9.1 Acetone

Acetone is an example of a relatively non-toxic, but highly flammable solvent that is commonly found in biotechnology laboratories. It has a flash point of –18°C (NFPA Fire Hazard Rating 3), and it is susceptible to "flash back" because of its volatility (remember the case study in Chapter 8, p. 178). The odor of acetone is detectable at concentrations in air well below toxic levels. Acetone can act as an eye and nasal irritant, but these effects are generally short-lived.

There are many systems available for rating the flammability of laboratory chemicals, usually based on their flash points. The NFPA, for example, rates chemicals on a scale of 0–4, with 0 being nonflammable and 4 being extremely flammable. Any chemical with a rating of 3 or 4, which includes those with flash points at room temperature or below, should be treated as a fire hazard. Table 9.2 provides a set of guidelines for handling flammable chemicals to reduce the risk of fire.

FIGURE 9.4 Hazardous Materials Identification System (HMIS) table. This hazard notification system may be optionally included on a hazardous chemical label. The color-coded categories are the same as for the NFPA diamond and similarly numbered 0–4 from least to worst hazard in that category. The white Personal Protection box follows a letter code corresponding to the necessary PPE items.

Handling chemicals safely depends on knowledge of the properties of individual chemicals. Some powdered metals, for example, are **pyrophoric**, *which means that they will ignite on contact with air.* Other chemicals (e.g., elemental sodium) react violently on contact with water. These types of chemicals must be manipulated only within a controlled environment.

Possibly the most dangerous flammable liquid handled regularly in biotechnology laboratories is diethyl ether, with a flash point of –45°C (NFPA Fire Hazard Rating 4). This chemical should never be placed in

---

TABLE 9.2

Safe Handling of Flammable Chemicals

Standard safety practices as described in Table 9.1, plus:
- Keep all flammable substances away from ignition sources.
- Never heat a flammable chemical with an open flame.
- Remember that solvent vapors can mix with air and diffuse to distant ignition sources.
- Know the appropriate fire prevention methods for the chemical.
- Keep containers tightly closed at all times when not in use.
- Keep flammable chemicals away from reactive chemicals.
- Work only in fume hoods or other well-ventilated areas.
- Avoid static electricity discharges when working with flammable substances.
- Never pour flammable substances down a drain or into the trash.
- In case of a spill, deal with any skin contamination before beginning the laboratory decontamination process.

---

proximity to electrical equipment that might produce sparks (e.g., centrifuges or regular refrigerators). It should be handled in fume hoods that will prevent a buildup of flammable vapors.

Fire codes require that 10 gallons or more of flammable liquids be stored in safety cabinets that are designed to minimize the risk of fire or explosion (Figure 9.5). These cabinets do not require venting in the absence of toxic fumes, and the lack of an outside air supply will act as a limiting agent for any fire within the cabinet. If venting is introduced, the cabinet system must be constructed with proper ducts and blowers to vent fumes from the building.

FIGURE 9.5 Storage cabinet for flammable chemicals. (Image © Flinn Scientific, Batavia, IL, USA. Used with permission.)

9.2.3 Reactive Chemicals

Most laboratory chemicals are reactive to some extent, but those that pose the greatest hazard to laboratory workers are those that undergo violent chemical reactions. Reactive chemicals are those that *can undergo chemical reactions that release relatively large quantities of energy within a short period of time.* They can be categorized as follows:

- those that participate in **exothermic** (*heat-emitting*) or gas-generating reactions;
- unstable chemicals that break down over time to become reactive chemicals;
- **oxidizing agents** (*compounds that gain electrons through chemical reactions*); and
- incompatible chemical mixtures.

Some reactive chemicals may produce an **explosion**, *a sudden release of large amounts of energy and gas within a confined area.* A fire may result, depending on the elements involved. Table 9.3 describes specific types of reactive chemicals.

General guidelines for handling reactive chemicals are provided in Table 9.4.

In a laboratory, explosions are likely to occur as a result of combining reactive chemicals in a sealed container. Sealed containers can also explode if used for any **exothermic** (*heat-producing*) or gas-forming chemical reaction; see the case study "Mixed Waste Containers Are a Bad Idea" below. This happens with some regularity in mixed chemical waste containers (which are not recommended). Glass bottles should never be used to contain potentially reactive chemical mixtures.

TABLE 9.3

Types of Reactive Chemicals

Chemical Type	Examples
Inorganic oxidizers, react with organic compounds	Nitric acid, chlorates
Shock-sensitive, explosive when bottle is moved	Dry picric acid, certain nitroso compounds
Organic peroxide, fire and explosion hazard	Benzoyl peroxide
Peroxide-forming, when allowed to evaporate	Ethers, aldehydes
Polymerizable, small molecules link while producing heat	Methacrylates
Water-reactive, explodes on contact	Sodium, certain anhydrides
Air-reactive, explodes on contact	Some organometallic compounds

TABLE 9.4

Safe Handling of Reactive Chemicals

Standard safety practices as described in Table 9.1, plus:

- Know the reactive properties of the chemical (read the SDS).
- Never mix unknown chemicals together, especially in closed waste containers.
- Label containers of reactive chemicals carefully.
- Store only compatible chemicals in the same area.
- Store oxidizing chemicals away from flammable materials.

Case Study: Mixed Waste Containers Are a Bad Idea

A graduate student in one of the author's labs was working late and in a hurry. He needed to dispose of some liquid waste and poured it into a glass bottle, designated for organic wastes, inside a fume hood. Noticing that the bottle was full, he replaced the cap, pulled the hood sash 80% closed, and left the room. Within three minutes, the bottle, which contained acid as well as organic waste, exploded from a gas-producing chemical reaction. The sash of the fume hood, which was constructed of shatterproof material, contained much of the flying glass and liquid within the hood, but was sufficiently damaged to require replacement. Several large pieces of chemically contaminated glass flew under the hood sash, through the open lab door, across the hallway, and into the facing lab – a distance of almost 50 feet. It was fortunate that no one was present in the path of the debris. Cleanup and decontamination took several hours. After the incident, only appropriate plastic bottles, with the caps removed from the hood, were used as waste containers.

Some chemicals are unstable and are susceptible to chemical breakdown with time; see Example 9.2. This is one reason why all chemicals should be labeled with the date of receipt. In some cases, the breakdown products are shock-sensitive.

A peroxide former *is a chemical that produces peroxides with age or air contact.* **Peroxides** are *chemicals that contain an oxygen–oxygen bond.* The most commonly encountered examples are picric acid, dinitrophenol, and compounds that break down into organic peroxides. These chemicals, which include a variety of aldehydes, ethers, and ketones, are highly hazardous. They are flammable and may explode on exposure to heat or shock (sometimes even a slight movement). It is important to refer to the SDS and labels for information about the hazards of specific chemicals.

Example 9.2 DEPC (Diethyl Pyrocarbonate)

Diethyl pyrocarbonate (DEPC) is a toxic chemical sometimes found in biotechnology laboratories. It is used to treat solutions and glassware when isolating RNA because it is an effective agent for inactivating RNA-digesting enzymes. It is also a suspected carcinogen and should be handled only with gloves. In addition, DEPC breaks down to carbon dioxide gas and ethanol when exposed to moist air. If this decomposition takes place in a sealed bottle, pressure can build to explosive levels. DEPC should be stored under dry conditions in a refrigerator, within a desiccator. If possible, store the bottle in the original metal container to act as an explosion barrier. Always allow refrigerated DEPC to equilibrate to room temperature before opening the bottle. Because DEPC is an explosion hazard, always use goggles and a face shield when handling a stock bottle.

Peroxide formation is generally limited to liquids that have evaporated and undergone autoxidation (*spontaneous oxidation in the presence of oxygen*). Diethyl ether and tetrahydrofuran (THF) are the most likely peroxide formers to be found in biotechnology laboratories. These chemicals should be stored away from light and heat in carefully sealed containers. Any containers that show signs of evaporation, especially older containers of ether, should not be handled. Contact your institutional safety office for proper disposal instructions.

In addition to the preceding hazards, certain combinations of chemicals can undergo violent reactions that result in explosions or release of highly toxic gases or other products. The best protection against this phenomenon is to follow standard laboratory procedures when mixing chemicals. Never combine chemicals without an established set of instructions, or without researching the reactive properties of the chemicals involved. Know the hazards associated with the chemicals with which you work. Figure 9.6 provides a few examples of incompatible chemical mixtures.

Example	Acids, inorganic	Acids, organic	Acids, oxidizing	Bases, alkali	Organic solvents	Oxidizers
Hydrochloric acid	OK	No	OK	No	No	OK
Acetic acid	No	OK	No	No	OK	No
Nitric acid	OK	No	OK	No	No	OK
Sodium hydroxide	No	No	No	OK	No	OK
Acetone	No	OK	No	No	OK	No
Hydrogen peroxide	OK	No	OK	OK	No	OK

FIGURE 9.6 Examples of incompatible chemical mixtures. "OK" refers to chemicals that are generally safe to mix; "No" indicates incompatible chemicals.

9.2.4 CORROSIVE CHEMICALS

Corrosive chemicals *are those that can destroy tissue and equipment on contact.* Acids and bases are the most common corrosives found in biotechnology laboratories. Both will cause chemical burns and tissue damage on contact with skin or eyes. An even greater danger is that of inhalation of corrosive vapors, which can irritate or burn mucous membranes and potentially cause serious lung damage. Because of this inhalation hazard, strong solutions of corrosives should always be used in a fume hood. The extent of potential damage caused by a corrosive chemical will depend upon the nature of the chemical and the amount and length of exposure. Table 9.5 offers some guidelines for safe handling of corrosives.

9.2.5 TOXIC CHEMICALS

9.2.5.1 Acute versus Chronic Toxicity

What thing is not a poison? All things are poison and nothing is without poison. It is the dose only that makes a thing not a poison.

Paracelsus, 16th century

Toxicity *is the term used to describe the capacity of a chemical to act as a poison, creating biological harm to an organism. A toxic material may alter the function of essential organs of the body, the heart, lungs, liver, nervous system, or kidneys. The level of toxic*

TABLE 9.5

Safe Handling of Corrosive Chemicals

Standard safety practices as described in Table 9.1, plus:
- Work with corrosive chemicals in a fume hood to avoid respiratory irritation.
- Add acid to water, not the reverse; mixing acid and water produces heat, which may cause splashing if water is added to concentrated acid.
- Perform neutralization reactions between acids and bases slowly to minimize gas and heat generation.
- Be sure that protective gear is appropriate to the chemical in use.
- Store acids and bases in separate areas.
- Do not work with hydrofluoric acid without specific training and precautions.

TABLE 9.6

Classes of Relative Toxicity

Toxicity Level	Chemical Example	Approximate Human Oral Lethal Dose (g/70 kg adult)
Almost nontoxic	Sucrose	5,400
Slightly toxic	Ethanol	918
Toxic	Sodium chloride	258
Highly toxic	Sodium cyanide	7
Extremely toxic	Strychnine	0.05

hazard is dependent on the nature of the chemical itself, the concentration and length of exposure, the health of the individual, and the speed and success of corrective measures. As recognized by Paracelsus, all chemicals can be toxic at higher dose levels in some individuals (Table 9.6). Many laboratory chemicals, however, have well-documented toxic effects at low doses. The risk of harmful effects from these substances can be reduced or eliminated by proper laboratory technique and simple precautions.

Toxic materials can act in a variety of ways. An **acute toxic agent** *causes damage in a short period of time, and a single exposure may be adequate to cause harmful effects.* A fast-acting poison like hydrogen cyanide is an example of an acute toxic material. **Chronic toxic agents** *have cumulative effects or may accumulate in the body with multiple small exposures.* These small doses of the toxic material may not produce an immediate effect, but instead produce injury over time. Lead poisoning in children is an example where cumulative exposure to very low levels of a toxic material can have long-term, serious health effects. Many chemicals can act as both acute and chronic toxic materials; see Example 9.3. In some cases, exposure to mixtures of chemicals may increase their toxic effects.

Example 9.3 Toxicity of Organic Solvents

Non-chlorinated flammable organic solvents, such as ethyl acetate, toluene, and xylene, can exert a wide range of toxic effects. As acutely toxic materials, they can cause headaches and dizziness after inhalation. They are respiratory, skin, and eye irritants. Because of their solvent properties, contact with skin will cause drying, leading to **dermatitis**, *a painful reddening and inflammation of the skin.* Some organic solvents, such as phenol, can act as corrosives as well. Organic solvents may also cause chronic toxicity, with long-term exposure leading to liver, lung, and kidney damage. Chronic exposure to benzene is linked to human leukemia. Virtually all of these chemicals have characteristic odors, and some lab workers mistakenly use these to indicate exposure. The vapor levels necessary for detection by smell, however, have no relationship to toxicity levels in most cases, so this is not an adequate warning system. When working with organic solvents, always use a fume hood and refer to the SDS for hazard information.

TABLE 9.7

Safe Handling of Toxic Chemicals

Standard safety practices as described in Table 9.1, plus:
- Be aware of both the acute and chronic effects of a known toxic chemical.
- Treat all chemicals as toxic unless otherwise informed.
- Be certain that the gloves you are wearing provide appropriate protection from the chemicals being used.
- Do not rely on odors to warn you of exposure to chemicals.
- Minimize exposures to any toxic substance that can accumulate in the body.
- Maximize precautions when working with any chemical known to be mutagenic or carcinogenic.
- Be alert to symptoms of toxicity or sensitization to chemicals.
- Both men and women should consider reproductive hazards in the workplace. Handle toxic chemicals carefully if you plan on starting or increasing a family, or suspect that you might be pregnant.

Exposure to toxic chemicals can result in a wide variety of human health problems. The next few sections will discuss some of the types of toxic materials likely to be encountered in the laboratory. Table 9.7 provides some general suggestions for safe handling of toxic chemicals.

9.2.5.2 Irritants, Allergens, and Sensitizers

Many laboratory chemicals are **irritants**, *which produce unpleasant or painful reactions when they contact the human body*. Skin irritation frequently takes the form of redness, itching, or dermatitis. Chemical irritation can be more serious when the point of contact is the eyes or respiratory tract. It is always a good idea to minimize body contact with all chemicals, even those that seem innocuous. Wear proper protective clothing, including gloves and safety glasses, whenever handling chemicals. Some chemicals with acute irritating effects can also exert chronic effects; see Example 9.4.

Allergies *are reactions by the body's immune system to exposure to a specific chemical*, or **allergen**. Allergies can manifest themselves as skin or respiratory reactions, depending on the route of exposure to the allergen. Just as individuals will have a wide range of sensitivities to environmental allergens, such as dust and pollens, they will also have a wide range of sensitivities and symptoms to chemical allergens and a range of symptoms associated with an allergic response. Reactions can range from a mild rash, to nasal congestion, to **anaphylactic shock**, *a sudden life-threatening reaction to allergen exposure*.

Before allergies occur, an individual must be sensitized to the allergen. Some chemicals can act as

Example 9.4 Formaldehyde

Everyone who took an anatomy class in years past has probably encountered formaldehyde. Formaldehyde gas is a toxic and highly flammable substance. Diluted formalin, a 37% solution of formaldehyde, has historically been used to preserve tissue specimens for dissection. Formalin contains 7%–15% methanol to stabilize the formaldehyde. Most suppliers of preserved specimens are now providing propylene glycol-based or other types of tissue preservatives because of the formidable list of the toxic properties documented for formaldehyde, including:

- respiratory and eye irritation
- slow-developing burns to eyes and skin
- skin sensitization
- potential carcinogenicity.

sensitizers, *and may trigger allergies themselves, or cause an individual to develop an allergic reaction to an accompanying chemical*. Dimethyl sulfoxide (DMSO) is an example of a sensitizing agent that penetrates skin and carries other chemicals with it, as in the case study "Chemical Sensitization" below. Limiting contact with potential allergens and sensitizers, working under well-ventilated conditions, and wearing proper protective clothing are wise precautions to reduce exposure to these agents.

Case Study: Chemical Sensitization

As a beginning graduate student, one of the authors did extensive tissue culture work. This work was performed in an appropriate biological safety cabinet while wearing gloves. After about 4 months, she developed a rash on her left wrist, exactly corresponding to the shape and location of her wristwatch. Because she had worn this same watch almost continuously for more than 6 years, it seemed unlikely that she had suddenly developed an allergy to it. After examining the rash, a dermatologist questioned the author about her work. As soon as tissue culture was mentioned, the doctor asked if she worked with cells treated with DMSO (dimethyl sulfoxide). When confirmed, he pointed out that DMSO is a powerful penetrating and sensitizing agent, which probably had triggered a reaction to nickel found in the watchband. Even though the author had never spilled DMSO on her wrist, enough DMSO vapors had reached above the glove line to produce the sensitization reaction. After this, the author wore longer gloves while handling DMSO and did not wear a watch in the laboratory. The problem disappeared.

9.2.5.3 Neurotoxins

Neurotoxins *are compounds that can cause damage to the central nervous system.* In many cases, the neurological effects, which may include loss of coordination and slurred speech, may develop slowly after long-term exposure to small doses of the toxic agent. Organometallic compounds, like methylmercury, act as potent neurotoxins. Acrylamide, which is routinely used to prepare gels for protein separations, is a common neurotoxin in many biotechnology laboratories, as shown in Example 9.5.

9.2.5.4 Mutagens and Carcinogens

Mutagens *are compounds that affect the genetic material of a cell.* They cause alterations in DNA that will be inherited by offspring cells. Mutagens are considered chronic toxic agents because the induced damage may not be apparent for years. Mutational damage is cumulative. Because mutagens can affect the genetic material of cells, they may also act as cancer-causing agents. Although the relationship between carcinogenicity and mutagenicity has not been clearly demonstrated for many chemicals, it is prudent to assume that any mutagen is potentially carcinogenic as well. Mutagenicity data are generally derived from animal studies, so the exact human risk is difficult to determine. The SDS can provide a summary of the available data for individual chemicals. Some common laboratory chemicals that are known to be animal mutagens are shown in Table 9.8.

One of the most commonly used chemicals in many biotechnology laboratories is ethidium bromide. **Ethidium bromide (EtBr)** *is a fluorescent dye used to visualize nucleic acids in agarose gels.* It acts

Example 9.5 Acrylamide

Acrylamide is used in the preparation of polyacrylamide gels for protein separation. In its polymerized form, it is considered harmless and is sold in gardening stores as a water-absorbing agent to be mixed with potting soil for plants. Acrylamide in its unpolymerized form, however, is a potent neurotoxin. It can have both acute and chronic effects. Direct contact can result in eye burns and skin rashes. Symptoms of chronic overexposure include dizziness, slurred speech, and numbness of extremities. Acrylamide is also a suspected human carcinogen. This is a chemical that should always be handled with great respect for its toxic properties. Always wear a lab coat, gloves, and dust mask when handling the solid form. Acrylamide should only be weighed on a designated balance within a fume hood. Given the documented risk from respiration of acrylamide powder, laboratories should consider the extra expense of purchasing premade acrylamide solutions. Although polymerized polyacrylamide gels are nontoxic, do not handle these gels without gloves. There may be residual unpolymerized acrylamide present that can be absorbed through the skin.

by inserting itself into DNA molecules (Figure 9.7). This provides the mechanism for the strong mutagenicity EtBr demonstrates in animal models; see Example 9.6.

TABLE 9.8

Examples of Known Animal Mutagens

The following are examples of common laboratory chemicals that have been shown to be mutagenic in animal models and are presumed to be human mutagens as well.

Acridine orange	Potassium permanganate
Colchicine	Silver nitrate
Ethidium bromide	Sodium azide
Formaldehyde	Sodium nitrate
Hydroquinone	Sodium nitrite
Osmium tetroxide	Toluene

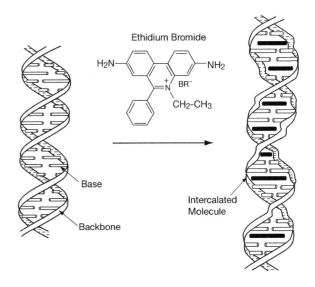

FIGURE 9.7 Interaction of ethidium bromide with DNA. The flat ethidium bromide molecule inserts itself into the DNA double helix, causing conformation changes.

Example 9.6 Ethidium Bromide

Ethidium bromide (EtBr) is a known mutagen that is commonly used in biotechnology laboratories to visualize nucleic acids under ultraviolet light (Figure 9.7). Most laboratories treat solid EtBr and concentrated stock solutions with great respect. However, workers commonly add low concentrations of EtBr to agarose gels and electrophoresis running buffers and these dilute solutions should still be considered potential health hazards. Decontaminating spills or solutions containing EtBr with bleach has been demonstrated to be ineffective and capable of creating breakdown products that may be more harmful than EtBr itself. (Lunn, George, and Eric Sansone. *Destruction of Hazardous Chemicals in the Laboratory*, Wiley, 1994, p. 185.) Several studies indicate that laboratories should follow more effective procedures using special detoxification resins, or a deamination procedure with sodium nitrite and hypophosphorous acid. Complete details of these procedures can currently be found at multiple sites on the Internet. The decontamination process can be monitored with a handheld ultraviolet light. All items that come into contact with EtBr, such as gloves, spatulas, or paper towels, should be properly decontaminated or treated as hazardous waste.

Carcinogens, or **cancer-causing agents**, *can initiate and promote the development of malignant growth in tissues.* They act as chronic toxic materials, and exposure to these compounds is often unrecognized for years.

Known and suspected laboratory carcinogens are identified and listed by OSHA and other agencies, including the **International Agency for Research on Cancer (IARC)**, *an agency that determines the relative cancer hazard of materials.* Carcinogenic substances need to be handled with great care. A list of types of chemicals known to have carcinogenic effects in animals is shown in Table 9.9. Information about carcinogenicity is included on the labels and in the SDS for specific chemicals.

TABLE 9.9

Examples of Known Types of Animal Carcinogens

Carcinogen Type	Example
Acylating agents (see label)	β-Propiolactone
Alkylating agents (see label)	Acrolein, ethylene oxide, ethyl methanesulfonate
Aromatic amines	Benzidine
Aromatic hydrocarbons	Benzene, benzo[a]pyrene
Hydrazines	Hydrazine
Miscellaneous inorganic compounds	Arsenic and certain arsenic compounds
Miscellaneous organic compounds	Formaldehyde (gas)
Natural products	Aflatoxins
N-Nitroso compounds	N-nitroso-N-alkylureas, N-nitrosodimethylamine
Organohalogen compounds	Carbon tetrachloride, vinyl chloride

9.2.5.5 Embryotoxins and Teratogens

Embryotoxins *are compounds known to be especially toxic to the developing fetus.* For example, organic mercury compounds, some lead compounds, and formamide are known to be embryotoxins. **Teratogens** *are a type of embryotoxin that causes fetal malformation.* They are known to interfere with normal embryonic development, but they do not necessarily cause direct harm to the mother. The greatest susceptibility of the fetus to these compounds is generally in the first 12 weeks of pregnancy, sometimes when the woman is unaware of the pregnancy.

Pregnant women and women of child-bearing age should be especially cautious when working with toxic chemicals. Substances that enter the bloodstream of the mother may be able to pass the placental barrier to the fetus. Always discuss an intended or actual pregnancy with your laboratory supervisor. It may be safest to request alternate duties during at least the first trimester of pregnancy.

Table 9.10 provides a summary of the hazards of some common biotechnology laboratory chemicals.

9.3 ROUTES OF CHEMICAL EXPOSURE

9.3.1 INTRODUCTION TO TOXICITY MEASUREMENTS

Toxic chemicals can enter the body by four main routes of exposure:

- inhalation
- skin and eye contact
- ingestion
- injection.

Types and levels of toxic injuries depend on the exposure route. Some chemicals are especially dangerous when inhaled; others only when ingested. Of the four routes of toxicity, inhalation and skin absorption are most likely in a laboratory workplace.

Every chemical has its own level of toxicity; some are more poisonous than others (Table 9.6). Many chemicals are found in the foods we eat and are considered relatively nontoxic. Some substances, such as hand creams, are good for the skin, but should not be ingested. Many compounds, such as over-the-counter pain relievers, are considered safe at one level, but toxic when the dosage is increased. Other chemicals will cause quick death even in small doses.

Example Problem 9.1

LD_{50} studies can be performed in a variety of animal species, with the results suggesting relative toxicity levels in humans. For example, consider a chemical that is tested for toxicity in rats and mice. The dose of chemical that kills 50% of the rats tested is 480 mg. The dose of chemical that kills 50% of the mice tested is 50 mg. How would you convert these data to LD_{50} values, and how might you extrapolate the data to humans?

Answer

The key additional information needed for calculating LD_{50} values is the mean body weights of the rats and mice tested. In this scenario, assume that the mean body weight of the rats is about 400 g and the mice, about 39 g. The LD_{50} values would be calculated as follows:

For rats — 480 mg of chemical divided by 400 g per rat

= 1.2 mg chemical per gram body weight or

  1,200 mg / kg body weight

= the LD_{50} in rats

For mice — 5 mg of chemical divided by 39 g per rat

≈ 1.3 mg chemical per gram body weight or

  1,300 mg / kg body weight

= the LD_{50} in mice

In order to reach a solid conclusion about human toxicity, you would need to have human data. In its absence, a hypothetical calculation can be made. Because the preceding LD_{50} values are similar, it would be reasonable to predict that the LD_{50} value for humans could be comparable. Based on this assumption, for a 70 kg (150 lb) human, a lethal dose of the chemical is in the range of 84–91 grams. This would be considered a mildly toxic chemical, unlikely to be consumed in toxic amounts by humans.

TABLE 9.10

Summary of Common Chemical Hazards in Biotechnology Laboratories[a]

Chemical	Flammable	Corrosive	Reactive	Low-Dose Toxicity	Major Route of Exposure
Acetic acid		Yes	Yes	Acute	Any
Acetone	Yes		Yes	Chronic	Any
Acetonitrile	Yes		Yes	Acute	Any
Acrylamide				Acute, chronic, neurotoxin	Skin, inhalation
Ammonium hydroxide		Yes	Yes	Acute	Skin, inhalation
Benzene	Yes		Yes	Chronic, carcinogen	Inhalation, skin
Chloroform		Yes	Yes	Possible carcinogen, teratogen	Inhalation
Diethyl ether	Yes		Yes	Acute	Inhalation
Dimethyl sulfoxide			Yes	Sensitizer	Skin contact with vapors
Ethanol	Yes		Yes	Acute, chronic	Ingestion
Ethidium bromide				Mutagen	Skin
Ethyl acetate	Yes		Yes	Acute, chronic	Any
Formaldehyde	Yes		Yes	Acute, sensitizer, carcinogen	Any
Hexane	Yes		Yes	Chronic, neurotoxin	Inhalation
Hydrochloric acid		Yes	Yes	Acute, corrosive	Any
Hydrogen peroxide			Yes	Irritant	Skin
Mercury				Acute, chronic, neurotoxin	Inhalation, skin
Methanol	Yes		Yes	Acute	Ingestion
Nitric acid		Yes	Yes	Acute, corrosive	Skin
Phenol		Yes	Yes	Acute, chronic, neurotoxin, hepatotoxin	Skin, inhalation
Pyridine	Yes			Chronic, hepatotoxin	Inhalation, skin
Sodium hydroxide		Yes	Yes	Acute, corrosive	Skin, inhalation
Sulfuric acid		Yes	Yes	Acute, corrosive	Skin, inhalation
Toluene	Yes		Yes	Chronic, irritant	Skin, inhalation

This table was prepared from the information provided in *Prudent Practices in the Laboratory*, as referenced in the Bibliography in the Introduction to this unit.

[a] This list provides a summary of the most significant hazards of some of the chemicals commonly encountered in biotechnology laboratories. Many have additional hazardous properties that are not listed here. Always consult the label, SDS, and other reference sources when working with an unfamiliar chemical.

One method scientists use to assess the toxic effects of different chemicals is to measure the amount of the compound that will cause a reaction or death in animals. Test animals receive known doses of the chemical, and the results are measured. One common measure of chemical toxicity is the LD_{50} level. **LD_{50}** (**L**ethal **D**ose, 50%) *is the amount of a chemical that will cause death in 50% of test animals.* Animals of different species are used. In order to compare these doses, the amount needed to kill 50% of the animals is recorded in amount (grams or milligrams) per kilogram of the animal's body weight. A larger absolute dose is generally necessary to kill a larger animal, although relative values may be similar (see Example Problem 9.1). LD_{50} studies obviously cannot be performed with humans, but the information from animal studies can be used to indicate the relative toxicity of certain compounds. Animal toxicity is usually, but not always, a good predictor of human toxicity; see the case study "Reducing the Use of Animals for Testing."

Case Study: Reducing the Use of Animals for Testing

LD_{50} testing originated in the 1920s and was long considered the gold standard for collecting toxicity data to apply to humans. As originally conceived, LD_{50} assays involved dozens to more than one hundred animals in order to gather reliable data. The LD_{50} data used in this textbook are based on historical assays. Given modern attitudes about animal welfare, these tests are no longer considered to be ethical or desirable by most of the scientific community. In more recent times, new statistical methods have been developed to allow effective data collection with a minimal number of animals (in some cases, fewer than ten). Animals are generally humanely euthanized before inevitable lethality. However, these animal tests are based on the premise that animal toxicity is an excellent predictor of human toxicity; this is not always the case.

In a similar vein, scientists have worked to develop *in vitro* tests for toxicity testing. For example, Botox is a prescription product that is most commonly injected into the face to reduce the appearance of wrinkles. It is also used as a medical treatment for migraine headaches. Botox is a protein derived from botulinum toxin, which is a paralyzing agent that is responsible for botulism. Every batch must be individually tested for potency, requiring hundreds of thousands of mice worldwide every year. In 2017, Allergan, Inc., a major producer of Botox, won FDA approval for an *in vitro*, cell-based assay for batch potency and stability. The European producer Ipsen has recently developed another *in vitro* method as well. Given the number of animals previously required to be tested, this is a major development. Animal testing of cosmetics is banned in the EU and highly frowned upon in the United States. However, even though the majority of Botox on the market is used for cosmetic purposes, Botox is considered an injectable medical substance rather than a "cosmetic," because it is injected rather than applied to the skin. Cell-based testing of potential new drugs is a welcome strategy for testing toxicity.

LD_{50} values depend on the route of chemical exposure. Ingestion, or oral, LD_{50} values may differ significantly from skin exposure LD_{50} values. Although toxicity is a relative term, certain LD_{50} values are generally accepted as indications of high toxicity. These are as follows:

- LD_{50} <500 mg/kg body weight by ingestion or injection
- LD_{50} <1,000 mg/kg body weight by skin contact.

Toxicity for inhaled chemicals is measured in a similar manner, although the results are expressed differently. With toxic chemicals that are likely to be inhaled, the **LC_{50} (Lethal Concentration, 50%)** is usually provided. The **LC_{50}** *is the chemical concentration in air that will kill 50% of exposed animals.* LC_{50} is usually expressed as **ppm** *(parts per million in air)* or **mg/m³** *(milligrams per cubic meter of air).* The approximate level indicating high toxicity is an LC_{50}<2,000 ppm inhalation.

9.3.2 INHALATION

Toxic vapors, gases, and dusts can all enter the body through inhalation. The respiratory system is wonderfully designed with a large surface area to deliver oxygen and other airborne chemicals to the blood. This unfortunately means that hazardous chemicals in the air can easily penetrate the mucous membranes of the nose, throat, and lungs and be delivered to the bloodstream for distribution to all the tissues of the body. Reactions to inhaled chemicals can range from mild discomfort to burning sensations in the throat to asphyxiation.

For example, exposure to small amounts of the odorless gas carbon monoxide can be lethal. Exposure can cause headache, nausea, and eventually unconsciousness. Only 0.03% in air (300 ppm) can be fatal within 30 minutes. OSHA has set a maximum allowable limit of 50 ppm of carbon monoxide in air for an 8-hour period (time-weighted average, as discussed later in this chapter).

For most volatile chemicals, inhalation exposure levels cannot be estimated by odor or other obvious characteristics. It is important, therefore, to limit exposure to the vapors of chemicals through the use of fume hoods and other methods.

OSHA and other agencies have created regulations and guidelines to indicate safe exposure levels to many, although not all, hazardous chemicals. The **American Conference of Governmental Industrial Hygienists (ACGIH)** *is an organization of governmental, academic, and industrial professionals who develop and publish recommended exposure standards for chemical and physical agents.* OSHA has used these standards as the basis for many of their regulations for chemical exposure levels.

ACGIH has established several types of chemical exposure guidelines of importance to laboratory personnel. The **threshold limit value (TLV)** *is the airborne concentration for a chemical that most healthy employees can safely be exposed to for 8 hours per day, repeatedly, with no adverse effects.* This is considered the highest safe level for the work environment. Because chemical exposures vary during a day, recommended exposure levels are frequently designated as **TLV-TWA (time-weighted average)** values.

For those chemicals that are extremely toxic at certain levels, ACGIH suggests upper limits to short-term exposures. The **TLV-STEL (short-term exposure limit)** *indicates the air concentration at which only 15 continuous minutes of exposure is allowed, up to four times during an 8-hour day.* Some rapidly acting toxic materials also have **TLV-C (ceiling)** values, *indicating the highest concentration allowable.* During the day, the airborne exposure to the chemical must never exceed this value. Exposures to ceiling levels of chemicals may have biological consequences.

The TLVs established by the ACGIH are merely advisory. OSHA used these values and additional data to set PEL values. **Permissible exposure limit (PEL)** *values are the legal limits set by OSHA for worker exposure to chemicals.* Employers have the legal responsibility not to allow employee exposure to exceed hazardous levels, and to provide safety equipment and training to protect employees. PEL values are often (but not always) the same as or similar to TLVs. Table 9.11 summarizes the main types of exposure limit information that are available to laboratory personnel.

9.3.3 SKIN AND EYE CONTACT

Direct skin or eye contact is another frequent route of laboratory injury by toxic chemicals. When working with chemicals, especially in large volumes, it is sometimes difficult to avoid all contact with the chemical. When potentially toxic chemicals contact the skin, there are four possible outcomes:

- The skin will act as a protective barrier – no harm will occur.
- The skin surface will react with the chemical – rashes or burns may result.
- The chemical will penetrate the skin – allergic sensitization may occur.
- The chemical will penetrate the skin and enter the bloodstream – systemic toxicity may result.

TABLE 9.11

Summary of Guidelines for the Amounts of Chemicals Allowed in the Air

Abbreviation	Name	Definition
TLV	Threshold limit value	Safe airborne concentration of a chemical that healthy employees can be exposed to on a daily basis.
TLV-TWA	Time-weighted average	The average allowable airborne concentration of a chemical within an 8-hour day.
TLV-STEL	Short-term exposure limit	The air concentration allowed for only a short period of time (15 minutes). Only four 15-minute periods of exposure to this concentration are allowed in a day.
TLV-C	Ceiling exposure limit	During the day, the exposure to airborne chemicals must never exceed this value.
PEL	Permissible exposure limit (OSHA)	Employers have the legal responsibility to keep employee exposure below this level.

Skin is designed to provide a natural protective barrier for the rest of the body, guarding against both chemical and biological invasion. However, there are natural entry sites for hazards at hair follicles and sweat glands, and any cuts or abrasions of the skin will also provide entry for toxic materials.

Most laboratory workers are naturally cautious when working with known corrosive agents such as acids and bases, which can cause extensive local tissue damage. It is important to remember, however, that many other types of laboratory chemicals can cause skin irritation or burns, and virtually all chemicals will cause problems if splashed into the eyes. Organic solvents can cause direct skin irritation because of their ability to strip natural oils,

diminishing the skin's natural barrier properties. This can also lead to increased skin irritation from other chemicals. It is essential to limit chemical contact by wearing proper laboratory attire, including gloves and eye protection.

In many cases, the most serious consequence of skin exposure to chemicals is the absorption of toxic substances into the bloodstream. Potent neurotoxic materials, such as mercury (see the case study on p. 165) or acrylamide, can easily be absorbed through skin. Phenol, which is found in biotechnology laboratories, can burn the skin directly and also be absorbed systemically. Exposure to phenol is extremely dangerous in large doses; see Example 9.7.

9.3.4 INGESTION

It is common sense that laboratory chemicals taken into the mouth and swallowed can cause harm. Because no one would deliberately consume laboratory chemicals, this is not a common path for biotechnology workplace poisonings. However, chemicals can inadvertently be transferred to the mouth in toxic quantities by contaminated fingers or pencil ends. Develop the habit of keeping your hands and other objects away from your face, mouth, and eyes while in the laboratory. Separate refrigerators must be designated for food storage, and all food and drinks should be consumed in designated areas where laboratory chemicals are not allowed. Maintaining complete separation of work areas and food areas helps prevent accidental ingestion of chemicals.

Oral toxicity levels for chemicals are determined with animal studies that measure LD_{50} values. Another indicator of chemical ingestion toxicity in humans is the **LD_{Lo}** (*Lethal Dose – Low*), *the lowest dose of a compound that has been reported to cause a human death.*

9.3.5 INJECTION

Direct injection of toxic chemicals is a relatively unlikely route of poisoning in the biotechnology laboratory, although it does play a more common role in the spread of biological hazards (see Chapter 10). The most likely method of chemical injection is through contact with chemically contaminated broken glass or other sharps. The injection of toxic chemicals that might not otherwise be able to pass the skin or respiratory barrier can be serious, and requires medical attention.

Table 9.12 provides information on the relative toxicities of some of the common chemicals found in biotechnology laboratories.

Example 9.7 Phenol

Phenol is used in biotechnology laboratories to aid in the isolation of nucleic acids. It is corrosive and can cause severe chemical burns. It damages cells by denaturing and precipitating protein. Phenol burns have a characteristic white appearance when they first appear. Because phenol also has anesthetic properties, these injuries may be unnoticed at first. Phenol splashes to the eye are likely to lead to serious burns and possible blindness, so phenol must never be handled without proper eye protection and a readily available eyewash station. Phenol is a respiratory irritant and readily penetrates skin. Whether exposure is by inhalation or skin contact, phenol can enter the bloodstream in significant amounts. From there, it can damage the nervous system, liver, and kidneys. Long-term exposure to phenol vapors can result in chronic symptoms, such as headache, nausea and vomiting, diarrhea, and loss of appetite. Lethal doses of phenol can be inhaled or absorbed through skin relatively easily, so anyone who has significant contact with phenol through a spill or other mishaps should seek medical attention. Some symptoms may be delayed in appearance. Phenol fortunately has a distinctive odor that is detectable well below toxic concentrations. This chemical should always be handled in a fume hood to avoid any air contamination of the laboratory.

TABLE 9.12
Relative Toxicities of Common Laboratory Chemicals

Chemical	LD$_{50}$, oral, rat (mg/kg)	TLV (ppm)	TLV (mg/m^3)	PEL (ppm)	PEL (mg/m^3)
Acetic acid	3,310	10	25	10	25
Acetone	5,800	750	1,780	1,000	2,400
Acetonitrile	2,730	40	70	40	70
Acrylamide	124	-	0.03 skin	-	0.3 skin
Ammonium hydroxide	350	25	17	35	27
Benzene	4,894	10	32	1	3
Chloroform	908	10	49	50	240
Diethyl ether	1,215	400	-	400	-
Ethanol	7,060	1,000	1,880	1,000	1,900
Ethyl acetate	6,100	400	1,440	400	1,200
Formaldehyde	500	0.3	0.37	0.75	1.5
Hexane	28,710	50	176	500	1,800
Hydrochloric acid	-	5	7.5	5	7
Hydrogen peroxide	75	1	1.4	1	1.4
Mercury	-	-	0.025	-	0.1
Methanol	5,628	200	262	200	260
Nitric acid	-	2	5.2	2	5
Phenol	384	5	19	5	19
Sodium hydroxide	140	-	2	-	2
Sulfuric acid	2,140	-	1	-	1
Toluene	2,650	50	188	200	750

9.4 STRATEGIES FOR MINIMIZING CHEMICAL HAZARDS

9.4.1 PREPARING A WORK AREA

An important part of good laboratory practice is planning each task, anticipating hazards, and preparing an appropriate work area. For any task, you should have all supplies and small equipment nearby to minimize wasted time and movement. Always wear proper PPE. Read the SDS and follow the prescribed safety recommendations. Prepare appropriate safety materials before the work begins. For example, when working with hazardous liquids, cover work surfaces with absorbent plastic-backed paper that can absorb small spills and be disposed of after your work is finished. Another containment strategy is to perform the work within a tray that can be easily cleaned. The work area should be marked with an appropriate hazard sign while in use.

9.4.2 USING CHEMICAL FUME HOODS

9.4.2.1 Structure and Function of Fume Hoods

Many tasks involving toxic chemicals are best performed in a chemical fume hood. A **chemical fume hood** *is a well-ventilated, enclosed, chemical- and fire-resistant work area that provides the user access from one side.* The hood isolates and removes toxic or noxious vapors from a working area and protects the user. A fume hood should be used whenever your work involves chemicals with the following characteristics:

- volatility
- unpleasant smells
- a TLV lower than 50 ppm in air.

The typical features of a laboratory fume hood are illustrated in Figure 9.8. The hood usually sits on top of a storage cabinet. The opening is covered by a transparent **sash**, *which is a window constructed of impact-resistant material, that can be raised and lowered by the user.* Gas and electrical outlets should be located on the outside of the hood. Inside is a smooth work surface tray, which is designed to contain spills and provide easy cleaning. Most modern fume hoods include interior **baffles**, *which direct the airflow within the hood.* **Airfoils** *located at the bottom and sides of the sash help to reduce air turbulence at the face of the hood.*

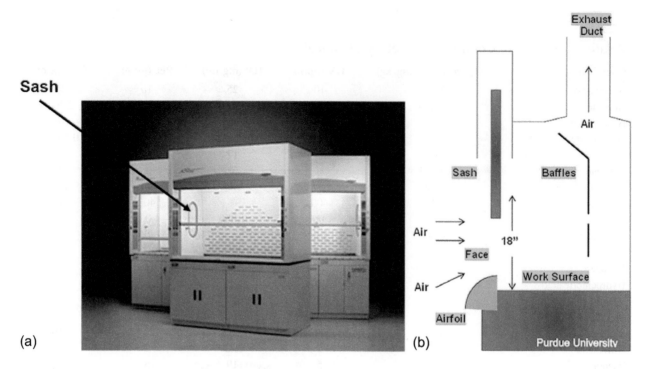

FIGURE 9.8   Features of a standard benchtop fume hood. (a) Freestanding fume hoods, with chemical storage underneath the cabinet, are the most common type found in laboratories. (Image courtesy of Labconco Corporation.) (b) Side view of the main components of a conventional fume hood. The shatter-resistant sash can be raised or lowered. The baffles in the back of the hood serve to redirect airflow to minimize turbulence at the work surface. (Image courtesy of Purdue University, Radiological and Environmental Management.)

There are several types of effective fume hoods available for biotechnology laboratories. The simplest is the **constant air volume (CAV) design**, *where a constant airflow is pulled through the exhaust duct.* In these hoods, raising and lowering the sash changes the face velocity at the sash opening. **Face velocity** *is the rate of airflow into the entrance of the hood, measured in* **linear feet per minute (fpm)**. In this conventional hood design, air enters the hood only at the bottom of the sash. When the hood sash is lowered, the face velocity of a CAV hood can increase dramatically (Figure 9.9). This can have the effect of blowing around paper and small items within the hood. The sash cannot be completely closed without disrupting airflow within the cabinet.

One partial solution to the problem of excessive face velocity is the use of a **bypass hood**, *which has an opening at the top of the hood behind the sash. This allows air to enter the hood and bypass the working face, restricting face velocity when the sash is lowered.* Face velocity will still increase, but not as dramatically. The sash can be completely closed, and the hood will still maintain a suitable

airflow. There are also **variable air volume (VAV) fume hoods** available, *which maintain a relatively constant face velocity by changing the amount of air exhausted from the hood.* VAV hoods are frequently designed with an air bypass that adjusts to maintain constant airflow within the hood based on the sash position. This is a preferred design in most laboratories.

9.4.2.2 Placement of Fume Hoods

The location of fume hoods within a laboratory can influence their effectiveness. Modern fume hood design has eliminated much, but not all, of the sensitivity of these hoods to air currents in the room. The movement of materials as well as hands and arms into and from the hood creates air drafts that can pull toxic vapors from the hood into the laboratory. The passage of other workers directly in front of the hood opening can similarly pull vapors through the face opening. Airfoils are designed to minimize this problem, but fume hoods should be installed in locations where they are isolated from foot traffic and drafts from doors and ventilation fans (Figure 9.10). Fume hoods should not

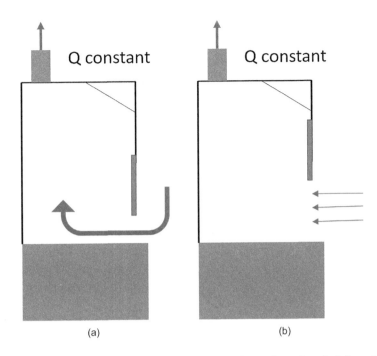

FIGURE 9.9 Effect of sash position on conventional constant air volume fume hood airflow. The arrows represent face velocity, and Q is the discharge volume flow rate. (a) When the sash (shown in blue) is lowered, the face velocity increases to allow a constant Q. This can create turbulence within the fume hood. (b) Raising the sash allows a gentler air influx. Note that the sash of variable air volume hoods (described in text) should be kept closed when not in use.

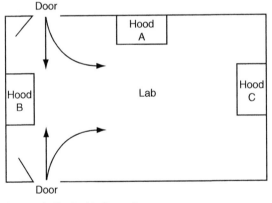

Arrows Indicate Air Currents

FIGURE 9.10 Placement of fume hoods and the effects of traffic air currents. Hoods B and A are poorly placed to avoid air currents. Hood C is in the better location for this laboratory.

be located in the corners of rooms, where air currents collide and create turbulence.

9.4.2.3 Testing Procedures

Laboratories must test the effectiveness of their fume hoods at least annually, according to OSHA

regulations. A complete performance check evaluates three parameters:

- general function and airflow patterns within the hood
- face velocity
- the uniformity of the face velocity.

The third measurement is usually performed when a fume hood is first installed, and later estimated using a smoke generator.

Smoke generators are small tubes of chemicals, frequently including titanium tetrachloride, that generate highly visible white smoke from a chemical reaction (Figure 9.11). They are used to check the general function of the hood. The tube is ignited and slowly moved across the front of the hood and then inside at various locations, to check the direction of airflow. This will indicate whether the hood is drawing air from all parts of the work surface, and whether fumes from inside the hood are entering the laboratory. A smoke generator is also useful for checking the effects of sudden hand motions and room traffic on the effectiveness of the hood.

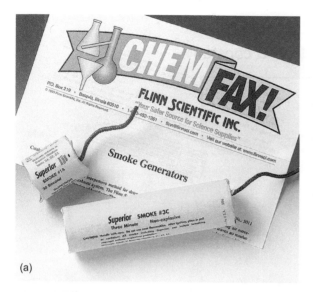

(a)

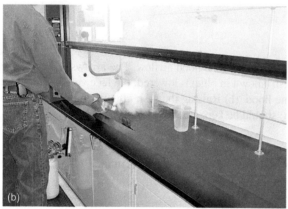

(b)

FIGURE 9.11 Using a smoke generator to test fume hood function. (a) Small smoke generators for informal tests. (Image © Flinn Scientific, Batavia, IL, USA. Used with permission.) (b) Using a smoke generator to test airflow near the rear baffle of a fume hood. (Courtesy of PSA Laboratory Furniture, LLC.)

Fume hoods are usually professionally tested yearly, usually with smoke-generating machines, but it is good practice to hand test between formal inspections. Smoke generators should be used to determine the most effective locations within the hood for drawing out toxic fumes. For example, in most hoods, volatile chemicals should be handled 5–6 inches back from the sash. Exhaust levels tend to increase toward the back of the hood and decrease somewhat at the edges. During the smoke test, it is helpful to mark the effective parts of the fume hood with tape, indicating to users the optimum placement for experimental materials. This is also a good time to check that large items located in the hood do not obstruct airflow. If they do, they can be relocated to optimize hood function. All equipment within the hood ideally should be elevated at least 2 inches above the work surface to aid in air circulation.

A **velometer (velocity meter)** *is an instrument used to measure the face velocity of the fume hood.* The recommended face velocity for most fume hoods is 100 linear feet per minute (fpm). Face velocities both lower and higher than this value are less effective in containing fumes. Higher velocities create internal turbulence that may be counterproductive, allowing small papers and other items to be drawn into the exhaust vent. This decreases airflow and therefore the effectiveness of the hood. For a standard CAV fume hood, the tester raises or lowers the hood sash to achieve the proper face velocity of 100 fpm. This sash location should be marked with a sign or tape on the hood frame to indicate the optimum sash height.

9.4.2.4 Optimal Use of a Fume Hood

Fume hoods cannot protect the user and others in the laboratory unless they are used properly. Before working in the hood, be certain that it is functioning properly. The exhaust fan in the hood should always be running, and in many models, the fan cannot be turned off. If yours is adjustable, check that it is turned on and operational. A quick test is to dangle a piece of tissue paper at the hood opening and see that it is drawn toward the hood. All supplies and equipment you will need should be loaded into the hood at the beginning of your work session, so that the sash will not have to be raised again while chemicals are in use. Consider chemical compatibility when placing materials within the hood. Never lower your head to peer under the hood sash. Table 9.13 summarizes guidelines for proper hood maintenance and use.

It is essential to distinguish between fume hoods and biological safety cabinets. **Biological safety cabinets** *are enclosures designed for the containment of biological hazards* (see Section 10.2.2). They are equipped with special filters that remove potentially dangerous particles from the air within the cabinet, and they provide varying degrees of safety and sterility for users and cabinet contents. Fume hoods are not equipped to filter biological hazards from the exhausted air, and do not protect the environment from these agents. Biological safety cabinets can function as fume hoods if they are vented to the outside of the building. However, many of the volatile and toxic chemicals appropriate for fume hoods will destroy the filters in biological safety cabinets, reducing or eliminating the removal of hazardous particles from the air. Biological filters will not detoxify chemically contaminated air and may not adequately protect the user from toxic chemicals.

TABLE 9.13

Proper Use of a Fume Hood

Hood maintenance:

- Regularly test the face velocity and general function of the hood.
- Mark the optimum sash position to attain a face velocity of 100 linear fpm.
- Mark the safe interior working area with tape.
- Keep the sash closed when not in use.
- Do not use the hood work surface for chemical or other long-term storage. Note that most hoods are constructed with storage cabinets underneath.
- Keep the exhaust fan on at all times.
- Do not adjust the interior baffles unless you are trained to do so.

Hood use:

- Be sure that the hood is functioning and drawing air.
- Open the sash and load all necessary items.
- Do not overload the hood, blocking airflow.
- Be sure that all chemicals in the hood are compatible.
- Elevate all equipment at least 2 inches above the work surface.
- Once all items are loaded, move the sash slowly to the optimum position.
- Keep your face outside the hood and behind the sash at all times.
- Use appropriate face and eye protection.
- Move your arms slowly while using the hood.
- Never stack materials on the bottom air foil.
- Work within the marked safe interior working area.
- Do not allow paper or other objects to enter the exhaust ducts.
- Decontaminate the working surface properly every day after use.
- Never use infectious materials within a fume hood. (This is the job of a biological safety cabinet.)

In case of a spill or fire within the hood:

- Immediately close the sash completely if you can safely.
- Do not turn off the exhaust fan.
- Unplug all equipment within the hood. (This assumes that the equipment is plugged into outlets outside the hood.)
- Warn other personnel and evacuate the area.
- Call 911 or other designated emergency number.

9.4.3 LIMITING SKIN EXPOSURE

As discussed earlier in the chapter, it is essential to avoid skin contact with laboratory chemicals. To protect yourself, always wear a lab coat to protect the body. Wear gloves that are appropriate for the chemicals being handled. Remember that gloves differ in materials and in thickness, which can directly affect their ability to protect the hands. Table 9.14 provides data on the chemical compatibility of various types of gloves, measured by breakthrough time (BT) in minutes. These data were compiled from several different manufacturers, as indicated by the wide range of BT values for certain glove–chemical combinations.

Always check the specific manufacturer's specifications when choosing a glove for toxic chemical work. Another factor to keep in mind is the length of the gloves (remember the case study on p. 194). The skin of the hands is generally thicker and withstands chemical penetration better than the thinner, more sensitive skin of the wrists or forearms. Gloves ideally should be pulled over the cuffs of your lab coat to provide complete protection for your arms (Figure 9.12). If the available gloves are too short, disposable arm protectors are desirable when working with highly toxic chemicals (Figure 9.12). Always wash your hands and wrists thoroughly before leaving the laboratory.

TABLE 9.14

Chemical Compatibility with Glove Materials

Chemical	Breakthrough Time (minutes)					
	Rubber	Neoprene	Nitrile	PVC	Butyl	Viton
Acetic acid	31–ND	ND	240–ND	47–300	ND	ND
Acetone	0	12–35	0	0	ND	0
Acetonitrile	0–16	40–65	0	0–24	ND	0
Ammonium hydroxide	58–120	ND	240–ND	60–ND	ND	ND
Benzene	0	15–16	16–27	2–13	30–34	ND
Chloroform	0	14–23	0	0	21	ND
Diethyl ether	0	12–18	33–64	0–14	8–19	12–29
Dimethyl sulfoxide	240	0	0	60	ND	90
Ethanol	ND	ND	225–ND	20–66	ND	ND
Ethyl acetate	0–72	24–34	0–30	0	212–ND	0
Formaldehyde	ND	ND	ND	ND	ND	ND
Hexane	0–21	39–173	234–ND	0–29	0–13	ND
Hydrochloric acid	211–ND	ND	ND	ND	ND	ND
Methanol	60–82	60–226	28–118	3–39	ND	ND
Nitric acid	233–ND	ND	0–72	114–240	ND	ND
Phenol	ND	ND	ND	32	ND	ND
Sodium hydroxide	ND	ND	ND	ND	ND	ND
Sulfuric acid	ND	ND	180–ND	210–ND	ND	ND
Toluene	0	14–25	26–28	3–19	21–22	ND

Note: These data are compilations from several glove manufacturers (resulting in ranges of values) and do not constitute specific recommendations. Always check the data for the specific brand and thickness of glove for intended use.

ND = none detected within 5 hours; 0 = not recommended, less than 10 minutes BT.

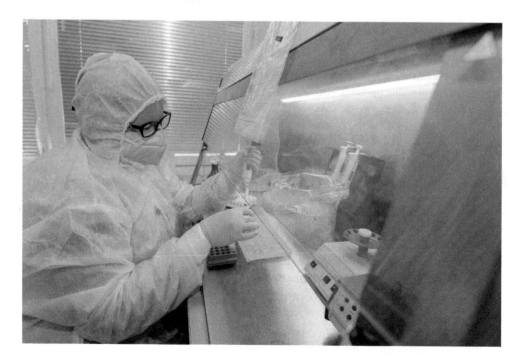

FIGURE 9.12 Wrist and forearm protection using gloves and arm protectors. Arm protectors (blue) are cuffed at the wrist and usually extend to the elbow. (Credit:Vital Hil/Shutterstock.com)

9.4.4 STORING CHEMICALS PROPERLY

9.4.4.1 Storage Facilities

There are numerous state and federal regulations that cover the storage of laboratory chemicals. These vary by location, but certain issues are invariably regulated. Examples include the following:

- storage of chemicals with fire and explosion risk in approved safety cabinets
- separation of stored chemicals by compatibility
- identification of all chemicals, with appropriate hazard labeling
- limiting access to radioactive materials, explosives, controlled substances, and other special hazards.

Laboratory chemicals should be stored in safe locations as close as possible to the point of use. This will minimize the risk of a spill during transport, which is a frequent source of accidents. Flammable and corrosive chemicals should be transported in secondary containers, such as buckets or trays, for protection from spills or breakage. Chemical storage areas must have spill containment materials immediately available that can handle the contents of the largest container of chemical present.

Every laboratory should maintain a current inventory of chemicals, indicating their date of receipt, user, and location. Every chemical should have a specific storage location, where it can easily be found in case of fire (see the case study "Organic Solvent Fire"). This information should be stored in a specific location outside the laboratory. Chemical containers removed for use should always be returned to their proper location, with only the amounts of chemical in immediate use kept at the lab bench or other work areas. Chemical storage areas should be inspected regularly to be sure no leaking containers or other signs of container damage are present. Laboratories should dispose of old chemicals or those that are no longer needed regularly. This is especially important when personnel change in the laboratory, leaving behind "orphan" chemicals that are unwanted or unidentifiable. Additional guidelines are provided in Table 9.15.

TABLE 9.15

Guidelines for Chemical Storage

- Date all chemicals on receipt.
- Be sure all stored chemicals are labeled by contents and general hazards.
- Purchase and store only minimum amounts of chemicals.
- Keep only small quantities of chemicals at your work station for immediate use.
- Use proper storage containers and cabinets.
- Do not store flammable chemicals in a standard refrigerator; use a certified spark-free refrigerator or freezer.
- Sort chemicals by hazard and store each type separately (Figures 9.13 and 9.14). Some of the suggested chemical categories are:
  - acids
  - bases
  - organic oxidizers
  - inorganic oxidizers
  - flammable liquids
  - flammable solids
  - acute poisons
  - water-reactive chemicals.
- Maintain a current inventory of chemicals for the laboratory.
- Store every chemical in a specific location.
- Never store chemicals on the floor or above eye level.
- Inspect containers weekly for signs of leakage or deterioration.
- Dispose of old chemicals properly and promptly.
- Be sure that all appropriate spill kits are easily available and fully stocked.

Case Study: Organic Solvent Fire

In 2005 at Ohio State University, a chemistry laboratory was completely destroyed by an explosion and fire, due to spilled organic solvent. A graduate student was loading twelve bottles of hexane onto the top shelf of a solvent storage cabinet when the shelf collapsed. A substantial amount of hexane spilled, and sensibly, all of the graduate students in the laboratory decided to leave due to the hexane fumes. An explosion occurred a few minutes later, and the laboratory was engulfed in flames. The students called 911, and several dozen firefighters were called in to ensure that most of the damage was confined to the room where the spill occurred. Luckily, the professor was able to return to the scene at the same time as the first responders arrived, and he was able to inform them as to the solvent involved and the laboratory's chemical inventory. CDF Battalion Chief Kevin M. O'Connor, one of the first responders, was quoted as saying, "… we were tremendously lucky that Dr. Coleman was on the scene," crediting the professor for his knowledge of what was in the lab so that firefighters knew they could safely go into the building and extinguish the blaze. Otherwise, he says, "we would have left the fire to burn." (Quote taken from Schulz, William G. "Fighting Lab Fires." *Chemical & Engineering News*, vol. 83, no. 21, 2005, pp. 34–35. doi:10.1021/cen-v083n021.p034.)

It is essential that chemicals are stored by compatibility and not in alphabetical order. As demonstrated in Figures 9.13 and 9.14, incompatible chemicals must not be stored next to one another. Figure 9.13 shows

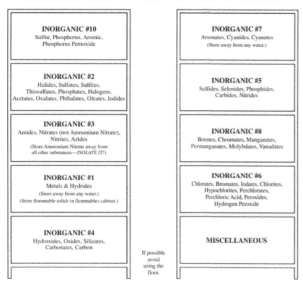

SUGGESTED SHELF STORAGE PATTERN—INORGANIC

INORGANIC #10 Sulfur, Phosphorus, Arsenic, Phosphorus Pentoxide	**INORGANIC #7** Arsenates, Cyanides, Cyanates (Store away from any water.)
INORGANIC #2 Halides, Sulfates, Sulfites, Thiosulfates, Phosphates, Halogens, Acetates, Oxalates, Phthalates, Oleates, Iodides	**INORGANIC #5** Sulfides, Selenides, Phosphides, Carbides, Nitrides
INORGANIC #3 Amides, Nitrates (not Ammonium Nitrate), Nitrites, Azides (Store Ammonium Nitrate away from all other substances—*ISOLATE IT!*)	**INORGANIC #8** Borates, Chromates, Manganates, Permanganates, Molybdates, Vanadates
INORGANIC #1 Metals & Hydrides (Store away from any water.) (Store flammable solids in flammables cabinet.)	**INORGANIC #6** Chlorates, Bromates, Iodates, Chlorites, Hypochlorites, Perchlorates, Perchloric Acid, Peroxides, Hydrogen Peroxide
INORGANIC #4 Hydroxides, Oxides, Silicates, Carbonates, Carbon	**MISCELLANEOUS**

If possible avoid using the floor.

FIGURE 9.13 One recommended scheme for storing inorganic chemicals by compatibility. The numbers in the layout refer to compatibility groups, indicated in the figure. Inorganic #9 includes acids that can be stored together and should be stored in a dedicated acid storage cabinet. Acute poisons should be kept in a separate, locked cabinet. Within a given compatibility group, chemicals are normally stored in alphabetical order. (Image © Flinn Scientific, Batavia, IL, USA. Used with permission.)

ten categories of inorganic chemicals. Those in the same category can be stored together in alphabetical order, but those in different categories must be separated from one another. Figure 9.14a similarly shows ten categories of organic chemicals. Observe that storing chemicals by compatibility requires categorizing them. While it is easy to identify sulfuric acid as being an acid, many chemicals are not as easy to classify based on their name. Most of us do not have the extensive chemistry background required to categorize a chemical as being an "organic oxidizer" or a "phthalate." Fortunately, chemical manufacturers categorize and provide storage information for the chemicals they sell. For example, the catalog for Flinn Scientific, Inc., Batavia, IL, has clear information on how to set up a chemical storage system, and how to properly categorize most of the chemicals you would encounter in a typical biotechnology laboratory. Other chemical manufacturers similarly provide storage information on labels. Storage compatibility information is always available in the Handling and Storage section (Section 7) of the SDS provided with each chemical. The SDS for individual chemicals can be found on the Internet at many locations in the event it is not immediately available in the laboratory. Virtually all chemical supply companies provide an online SDS for their products in addition to their printed sheets.

Note that chemical compatibilities are sometimes not what we might predict. For example, not all acids can be stored together. If you look at Figure 9.6, which indicates incompatible chemical pairings, you can see

SUGGESTED SHELF STORAGE PATTERN—ORGANIC

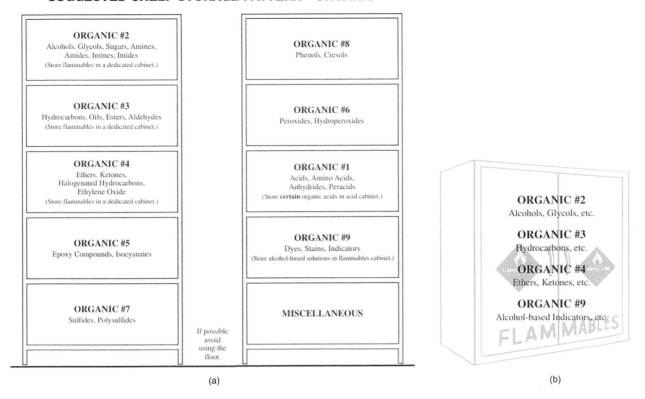

(a)

(b)

FIGURE 9.14   One recommended scheme for storing organic chemicals by compatibility. The numbers in the layout refer to compatibility groups, indicated in the figure. (a) Suggested compatibility arrangement. (b) Solvent groups should be stored in a dedicated flammables storage cabinet, arranged in compatibility groups. (Images © Flinn Scientific, Batavia, IL, USA. Used with permission.)

that there are several combinations of acids that are dangerous. Perchloric acid, for example, should not be stored in proximity to acetic or sulfuric acid. Again, it is important to use catalog information, the SDS, or the label for each chemical to determine optimal storage locations.

9.4.4.2 Labeling

All **primary chemical containers** (i.e., *those supplied by the manufacturer*) should immediately be labeled with a date and user name when they arrive in the laboratory. Stock chemicals should not be repackaged in other containers unless the original container is damaged. **Secondary containers** (*those supplied by the user*) that will not be emptied by the end of the workday should be clearly labeled with the chemical name and concentration, date, user, and appropriate GHS hazard information. Chemical name abbreviations are not considered adequate under OSHA regulations. Handwritten labels should be legible and written in indelible ink; organic solvents are notorious for dissolving writing in nonpermanent ink. Hazard labeling

should provide information about the greatest risk from the chemical. For example, strong acids should carry a GHS pictogram for "Corrosives" (Figure 9.15). Laboratories can purchase sets of standard hazard pictogram stickers that provide effective warnings for handlers who are unfamiliar with the specific chemical. Experimental materials should also be labeled by hazard.

FIGURE 9.15   Appropriate labeling for a secondary container of hydrochloric acid. Text can be used instead of the hazard pictogram.

9.4.5 HANDLING WASTE MATERIALS

Waste *refers to any laboratory material that has fulfilled its original purpose and is being disposed of permanently.* While OSHA provides guidelines for labeling chemicals within the laboratory, the U.S. Department of Transportation (DOT) and EPA regulate the handling and disposal of chemicals designated as hazardous waste. Within DOT, the Pipeline and Hazardous Materials Safety Administration (PHMSA) provides oversight for labeling, handling, and packaging hazardous materials for transport within the United States. This includes almost one million shipments each day. The regulations for hazardous waste disposal are complex, and it is essential that laboratories work with the safety office at their institution to ensure that waste is handled, packaged, and labeled appropriately.

Waste disposal has become increasingly expensive, and improper labeling or packaging of various materials can raise the disposal cost significantly. To minimize hazardous waste volume, do not dispose of nonhazardous materials in hazardous waste containers. Receptacles for nonhazardous materials, such as old notes or relatively clean paper towels, should be available away from immediate work areas. No sharps or any hazardous materials should ever be placed in these containers. In most institutions, this nonhazardous waste is handled by the custodial staff. Biotechnology laboratories must maintain several types of waste containers, to sort materials by hazard (e.g., separating biological waste from chemical waste).

Chemical wastes are generated as the end products of procedures, or as chemical stocks that are no longer needed in the laboratory. These substances should similarly be collected in a manner that keeps hazard groups separate. The EPA classifies chemical wastes by specific characteristics, such as flammability, corrosive properties, and reactive potential. Local waste disposal facilities may also designate chemical compatibility and hazard groups. The safety office at your institution will tell you how to sort waste materials and will place labels on each container for disposal. Always pay attention to these labels. Never put a chemical in a container if you are uncertain that it belongs there (see the case study on p. 190).

Chemical wastes that contain mixtures of compatible chemicals should be labeled with identification of each chemical and its approximate percentage in the mixture. Solid chemical wastes should be divided into those that can be incinerated (e.g., lightly contaminated paper towels) and those that cannot. Dry wastes should be double-bagged for safety and clearly labeled by hazard. Liquid wastes are collected in labeled glass (for unmixed chemicals only) or plastic containers, as appropriate. Never pour chemicals down a sink drain unless they have been neutralized or deactivated in accordance with local regulations for sewer disposal. Never use the drain for flammable solvents or potentially reactive chemicals. Even at very low concentrations, some chemicals, such as sodium azide, can form explosive mixtures within drain pipes, with predictably unpleasant results. Table 9.16 provides some guidelines for determining when stock chemicals should be discarded.

Because of the high expense of toxic chemical waste disposal, it is important to minimize the amounts of waste

---

TABLE 9.16

Disposal of Stock Chemicals

Most laboratories have a tendency to keep and store chemical stocks "just in case" someone needs them in the future. Although this is a reasonable approach with some chemicals, such as sodium chloride, which is stable and used on a regular basis, it can be a hazardous practice in other cases. Dispose of chemical stocks if any of the following criteria are observed:

- The chemical is more than 1 year old and not in current use.
- A formerly clear liquid has turned cloudy.
- Solids have clumped or show other signs of water absorption.
- Chemicals have changed color.
- Containers show signs of damage.
- There is suspicion of pressure buildup in a container.
- The chemical's identity is unclear.

These stocks should be disposed of in their original containers when possible, with appropriate identity and hazard labeling. Consult with your institutional safety office for proper disposal procedures.

TABLE 9.17

Minimization of Chemical Waste

- Maintain a current chemical inventory and avoid duplication of stocks.
- Order as little of a chemical as needed for planned work.
- Date all chemicals on arrival, to eliminate doubts about age.
- Use older chemicals first.
- Be sure all chemical containers are labeled by content.
- Keep all chemicals tightly sealed to prevent deterioration from air exposure.
- Do not contaminate chemical stocks by returning materials to the container.

---

generated by the laboratory. Table 9.17 contains some suggestions for reducing the amounts of chemicals that will require disposal. Unknown or unlabeled wastes are by far the most expensive type to dispose of, so all laboratory personnel should be diligent about chemical labeling.

9.5 RESPONSE TO CHEMICAL HAZARDS

9.5.1 CHEMICAL EMERGENCY RESPONSE

Every laboratory is required by OSHA to have an emergency response plan for accidents involving hazardous chemicals. These plans should be explained as part of laboratory safety training, and it is the responsibility of each person to be familiar with the appropriate actions. Emergency response needs to be an automatic reaction because accidents rarely provide time for detailed analysis and research into procedures. A case study later in this chapter, "Emergency Response to a Major Chemical Spill," provides an example where fast response by laboratory personnel prevented serious outcomes from a dangerous accident. It is important to be able to distinguish between a truly minor incident and an emergency. Table 9.18 provides some guidelines that should be considered before they are needed.

TABLE 9.18

When Is a "Problem" Involving Chemicals an Emergency?

Most of us are reluctant to appear to overreact to a potentially dangerous laboratory situation. The following list includes some of the circumstances where the best course of action is to summon help immediately. You probably require assistance from a professional safety officer or emergency response team when a chemical problem:

- causes a serious injury;
- involves a public area;
- creates a fire hazard;
- cannot be isolated or contained by those present;
- creates toxic vapors that could spread through the building;
- causes property damage;
- requires a prolonged cleanup;
- involves mercury compounds;
- involves hydrofluoric acid; and/or
- involves an unknown hazard.

Your laboratory should post the telephone numbers of the designated sources of local assistance near all telephones. Some of the numbers that might be useful in an emergency:

- Institutional safety office, in case of spills or questions about safety procedures.
- Fire department, in case of fire.
- Emergency medical assistance, in case of injury.
- Poison control, in case of exposure to toxic chemicals.

In cases where the exposure of a person to chemicals is obvious, immediately call a poison control center or a designated physician. These telephone numbers should be posted by every telephone in the laboratory area. In the case of a skin or body splash, the victim should move to the emergency shower with the aid of a nearby colleague, remove any contaminated clothing, and drench the affected area for at least 15 minutes. Eye exposure requires flushing of both eyes for at least 15 minutes as well. If chemicals have been inhaled, the person should be moved to an area with fresh air and remain quiet until medical help arrives.

Laboratory personnel should be familiar with the symptoms of toxic chemical exposure, which are summarized in Table 9.19. Specific information about individual chemicals can be found in the SDS, which should always be consulted before toxic chemical use (indicated on the chemical's label).

It is certainly better to prevent accidents or chemical exposures before they happen. Immediate preventive measures or notification of a supervisor is appropriate

under any circumstances where you have concerns about potential problems. Table 9.20 provides a few examples.

9.5.2 CHEMICAL SPILLS

9.5.2.1 Preventing Chemical Spills

In laboratory work, chemical spills occur. Most of these will be small, and many can be prevented by good laboratory practices. Proper advance preparation for spills ensures that most of these incidents will remain relatively minor in terms of scope and consequences. Many spills are caused by simply knocking over open containers of chemicals. These spills can be avoided by properly arranging experimental materials so that routine arm movements will not endanger chemical containers. Containers should be closed when possible, or anchored in a rack or holder for added stability.

Only small amounts of chemicals, in unbreakable containers when possible, should be kept at lab benches. Any toxic chemicals should have secondary

---

TABLE 9.19

Symptoms of Chemical Exposure

Any time you become aware of direct skin or eye contact with a laboratory chemical, you should take immediate action to remove any traces of the chemical and seek medical attention as needed. If you or your co-workers notice a chemical odor, especially for low-TLV substances, immediate action may be required. In the absence of known contact or inhalation, the following are some of the potential signs of toxic chemical exposure. The SDS for a chemical will provide specific details.

Acute exposure:

- headache or dizziness
- sudden nausea or vomiting
- coughing spasms
- eye, nose, or throat irritation.

Especially if:

- the preceding symptoms disappear with fresh air
- the symptoms reappear when work is resumed
- more than one person in the laboratory is affected.

Chronic exposure:

- persistent dermatitis
- unusual body or breath odor
- a strange taste in the mouth
- discolored urine or skin
- numbness or tremors.

TABLE 9.20
When to Take Special Precautions

There are many circumstances where a worker should take steps to prevent a problem directly, or to notify an appropriate supervisor. Some examples:

- Chemical leaks or spills are noticed or anticipated.
- Possible symptoms of chemical poisoning are noticed.
- Laboratory workers notice odd smells.
- A fume hood or other safety equipment fails to operate properly.
- A procedure creates chemical exposure that may exceed toxic thresholds.
- A new or altered procedure requires chemicals with unknown properties.

---

containment, such as a spill tray. Benches or fume hood work surfaces should be covered with absorbent, plastic-backed paper for liquid chemical work. This makes cleanup quick and simple.

Many serious spills occur when transporting chemicals through the laboratory. Large reagent containers should be handled one at a time, preferably in a secondary transport container such as a pan. Never carry glass reagent bottles by the cap or solely by the ring at the top. Support the bottle from underneath with one hand, and always wear gloves when transporting

chemicals. Be aware that the bottom seam of glass bottles is generally the weakest part of the container; see the case study below, "Emergency Response to a Major Chemical Spill."

One potential cause of major spills is the transfer of glass bottles from a warm-water bath or freezer directly to a countertop. Thermal shock can weaken the bottom seam and cause it to crack. When thawing reagents in a water bath, always place the bottle in a secondary container that will contain any leaked material.

Case Study: Emergency Response to a Major Chemical Spill

Major chemical spills require a quick response by all members of a laboratory. A biotechnology worker in one of the author's laboratories was thawing a 1 L stock bottle of frozen phenol in a water bath. Following proper procedures, she placed the bottle in a beaker for containment. When thawing was complete, she checked the bottle and beaker for any signs of leakage, and then picked up the phenol bottle at the side and began to transport it across the lab. Halfway across the room, the bottom of the bottle gave way, dumping the entire liter of phenol down her leg and onto the floor. Knowing that phenol is toxic through skin contact and is also corrosive, she immediately called for help and ran for the emergency shower. She was properly attired in a lab coat, gloves, long pants, and solid shoes, which limited her contact area. In the meantime, the lab supervisor arrived on the scene and determined that this was a major spill. She ordered that the room be evacuated, and both doors be closed, locked, and labeled with a "Danger: Phenol Spill" sign. Someone else called the building engineering hotline and ordered an immediate switch of the ventilation system for the floor to total exhaust to prevent the spread of toxic fumes within the building. Another worker was assisting the technician in the shower. The phenol had soaked through her pants fairly quickly, and she was convinced to remove them when someone brought a fire blanket to use as a modesty shield. Colleagues contacted the Institution Safety Office and requested an emergency spill team with respirators to clean up the phenol, and then escorted the injured person to the local emergency room. She suffered from nausea and headache for 2 days, but her chemical burns were relatively minor, and there were no apparent long-term effects. Aside from the inconvenience of being evacuated from the contaminated laboratory for several hours, no other personnel were affected. This would have been a very serious accident without the quick response and cooperation of many members of the laboratory team.

TABLE 9.21

Contents of a General Spill Kit

You can purchase general chemical spill kits or assemble your own, according to your needs. A general-use spill kit should contain:

- chemical-resistant, long-sleeved gloves
- chemical-resistant goggles
- chemical-absorbent materials for any anticipated chemical type
- spill pillows or sand for containment
- small whisk broom and dust pan
- disposable plastic bags for hazardous waste.

Most laboratories will require more than one type of spill kit because different absorbents and cleanup procedures are required for acids, bases, organic solvents, and other materials used in the laboratory. There are also special spill kits that should be purchased if your laboratory contains mercury or hydrofluoric acid hazards.

9.5.2.2 Chemical Spill Kits

Properly assembled and conveniently located spill kits can make the cleanup of minor spills significantly easier. **Chemical spill kits** *are preassembled materials for controlling and cleaning up small to medium size laboratory spills.* Every room where chemicals are used or stored should have a kit with sufficient materials to handle a spill of the largest container in the immediate area. A list of suggested contents for a general kit are provided in Table 9.21. Having all necessary materials assembled ahead of time, with all personnel trained to use the kits correctly, can make the difference between a minor inconvenience and a major problem.

Spill control kits may contain both loose absorbents and absorbent-filled pads and pillows. The absorbents should be chosen to reduce the vapor pressure from a liquid spill efficiently and to minimize any personal contact with the chemical. The type of absorbent material most applicable in biotechnology laboratories is labeled **universal absorbent**. *This is generally composed of polypropylene or expanded silicates and can absorb virtually any liquid, including some corrosives.* Check the manufacturer's recommendations, and be certain that kits are clearly and appropriately labeled. These absorbents control liquid spills, but do not reduce the toxic properties of absorbed liquids. Acids and bases, for example, may still require neutralization before disposal.

Loose absorbent works best for small spills. The absorbent should be poured around the edges of the spill to contain it and then sprinkled on the interior

(Figure 9.16). There are special neutralizer absorbents available for acid spills. For larger spills, absorbent pillows can be placed around and on top of the spill. Absorbent should usually be left undisturbed for a short period of time before final cleanup.

9.5.2.3 Handling Chemical Spills

When a chemical spill occurs, it is essential to assess the extent of the problem immediately. Even though a small spill is obviously easier to clean up than a

FIGURE 9.16 Proper cleanup for a chemical spill. The person cleaning up must don appropriate PPE. The spill is surrounded by absorbent to minimize the affected area. Then either loose absorbent or absorbent pads should be applied to the spill itself. Be sure to leave absorbent in place for enough time to compete the absorbance process. (Image courtesy of Creative Safety Supply, https://www.creativesafetysupply.com.)

larger mess, in many cases the volume of the spill is less significant than the toxicity of the spilled material. Whenever a highly toxic and volatile liquid is spilled, as in the previous case study, it is best to evacuate personnel and inform the safety experts at your institution. Your efforts may be best spent ensuring that others are warned of the hazard and that any injuries are promptly tended.

Small spills of relatively innocuous chemicals can be cleaned up without much concern. For hazardous chemical spills, however, a well-intentioned but inappropriate cleanup effort may cause more harm than good. For example, in the case of a volatile solvent, every liter spilled can produce up to $600\,L$ of flammable vapors. An attempt to wipe up a solvent spill with just paper towels will actually encourage vapor production by increasing the surface area of the spill. Liquid spills should be handled with appropriate chemical absorbents.

Dry chemicals should never be swept up dry, because this will create dust that can be inhaled. Wet mopping with damp paper towels is more appropriate in cases where absorbents are unnecessary.

Table 9.22 provides a summary of proper procedures for dealing with both minor and major chemical spills.

If hazardous chemicals are spilled on clothing, the clothing should be removed immediately and then rinsed. Any skin under the clothing should be flushed for 15 minutes in an emergency shower. When removing the clothes, be careful not to spread chemical to additional areas of the body. Never pull a contaminated shirt or sweater over the face; if the chemical is toxic, cut the clothing off with scissors. Do not attempt simply to rinse the chemicals out of the clothes while wearing them. A fire blanket or spare lab coat can be used to protect modesty if required.

9.6 FINAL NOTES

It is not uncommon in the laboratory to need to work with chemicals with which you are unfamiliar, and it is always bad practice to work with chemicals when you do not know their properties. While the names of these substances can be daunting, no one expects you to instantly know the characteristics and hazards of every

---

TABLE 9.22

Chemical Spill Procedures

It is essential that all laboratory personnel know what to do in advance in the event of a chemical spill. This topic should be discussed at periodic safety meetings, along with other routine safety procedures.

Minor Spills

- Take care of personal contamination first.
- Notify nearby workers and evacuate them as needed.
- Turn off heat or ignition sources if flammable chemicals are involved.
- Prevent the immediate spreading of the spill by layering with paper towels (do not wipe) or other absorbent materials.
- Avoid breathing any vapors from the spill.
- Dress properly and get the appropriate spill kit.
- First surround the spill with absorbent to contain the spill area.
- Slowly sprinkle absorbent over the spill, and follow any directions that accompany the spill kit.
- Collect the contaminated absorbent using a whisk and dust pan, and avoid creating dust or aerosols.
- Place the waste in a disposable bag, seal, and label; the whisk and pan should be thoroughly cleaned or discarded.
- Finish the cleanup by washing the area several times with detergent and water.

Major Spills

- Leave the spill site immediately.
- Warn others about the hazard.
- Tend to any injuries.
- Prevent others from approaching the spill site.
- Call for expert assistance.

chemical you encounter. It is important to understand that information about individual chemicals is readily available, on labels and in the SDS for each chemical, which is required to be available in your laboratory. There are also plentiful online sources of information.

Once you become familiar with the standard format of an SDS, it becomes simple to find the information that you require. When you know about the hazards of a chemical, it is simply a matter of good laboratory practice to handle the chemical in ways that minimize risk.

Practice Problems

1. As described in the text, small concentrations of the odorless gas, carbon monoxide, in air can be lethal. Exposure to low levels of carbon monoxide can cause headache, nausea, and eventually unconsciousness. OSHA has set a maximum allowable limit of 50 ppm in air (TWA) for an 8-hour period. What is the significance of the TWA designation?

2. A scientist in your laboratory is carrying a glass bottle of organic solvent across the laboratory when she slips on a wet spot on the floor, loses her grip, and drops the solvent bottle.
   a. What are several precautions that might have prevented this accident?
   b. What immediate steps should this person take to minimize any risk to herself and other laboratory workers?

3. Is the odor of organic solvents a reasonable indicator of safe exposure levels?

4. Safety Data Sheets (SDSs) are a major source of safety information, and it is essential to be able to use the information they contain. Excerpts from four Safety Data Sheets follow.

   Based on the information in these SDSs, answer the questions below. (*Note: These are only brief excerpts from the actual SDSs and are intended only for the purposes of study, not for use in the laboratory.*)

Safety Data Sheet General Information
   Item Name: **ACETONE, REAGENT**
   Date SDS Prepared: 01Jan19

Section 5: Fire-Fighting Measures
   Extinguishing Media: Water Spray, Dry Chemical, Carbon Dioxide, Alcohol-Resistant Foam
Section 8: Exposure Controls/Personal Protection
   OSHA PEL: 1000 PPM
   ACGIH TLV: 750 PPM/1000 STEL
Section 9: Physical/Chemical Properties
   Appearance and Odor: Clear, Colorless, Volatile Liquid with a Characteristic Sweetish Odor
   Boiling Point: 133°F, 56°C Melting Point: −139°F, −95°C
   Solubility in Water: Very Soluble
   Flash Point: −4°F, −20°C

Safety Data Sheet General Information
   Item Name: **ETHYL ALCOHOL ACS**
   Date SDS Prepared: 01Jan19
Section 5: Fire-Fighting Measures
   Extinguishing Media: dry chemical, carbon dioxide, water spray or alcohol-resistant foam.

Section 8: Exposure Controls/Personal Protection
 OSHA PEL: 1000 PPM
 ACGIH TLV: 1000 PPM
Section 9: Physical/Chemical Properties
 Appearance: Clear Colorless Liquid
 Flash Point: 55°F, 13°C

Safety Data Sheet General Information
Item Name: **PHENOL**, Liquid
Date SDS Prepared: 01Jan19

Section 5: Fire-Fighting Measures
 Extinguishing Media: Water Spray, CO_2, Dry Chemical or Foam
Section 8: Exposure Controls/Personal Protection
 OSHA PEL: 5 PPM
 ACGIH TLV: 5 PPM
Section 9: Physical/Chemical Properties
 Appearance and Odor: Colorless liquid, distinctive sweetish odor
 Flash Point: 174.9°F, 79.4°C

Safety Data Sheet General Information
 Item Name: **SODIUM CHLORIDE**
 Date SDS Prepared: 01Jan19

Section 5: Fire-Fighting Measures
 N/A
Section 8: Exposure Controls/Personal Protection
 N/A
Section 9: Physical/Chemical Properties
 Appearance and Odor: White Crystal
 Boiling Point: 1413°C
 Flash Point: N/A

 a. Rank each of the compounds in order from least toxic to most toxic based on their TLVs.
 b. For acetone, the SDS states "ACGIH TLV 750 ppm/1000 STEL." Explain the difference between the two numbers, 750 and 1000 ppm.
 c. Based on the flash points given in the SDSs, which of these compounds poses the most significant fire hazard? How should a fire involving that compound be extinguished?
5. Proper chemical storage is an important responsibility of any laboratory. It is critical to determine compatibilities and sort chemicals accordingly. Excerpts from the SDS of eleven common laboratory substances are shown in Table 9.23 below. Based on the information given, sort the chemicals into potential storage groups, that is, chemicals that can be stored together. You do not need to use the Flinn storage designations shown in Figures 9.13 and 9.14; simply note which chemicals can be stored together. (Note: *These are only brief excerpts from the actual SDSs and therefore are intended only for the purposes of study, not for use in the laboratory. These excerpts are in no way comprehensive.*)

TABLE 9.23
Chemical Storage Compatibilities

Chemical	Hazard Warnings	Incompatibilities
Sodium chlorate	Danger! Strong oxidizer	Aluminum, strong acids, strong reducing agents, organic matter.
Sodium hydroxide	Poison! Corrosive. Harmful if inhaled. Reacts with water, acids, and other materials.	Violent reactions with acids and organic halogens.
Sodium	Danger! Water-reactive. Flammable. Corrosive.	Water, oxygen, carbon dioxide, halogens, acetylene, metal halides, oxidizing agents, acids, alcohols, chlorinated organic compounds.
2-Propanol	Warning! Flammable liquid and vapor. Harmful if swallowed or inhaled.	Strong oxidizers, acids, acetaldehyde, chlorine, ethylene oxide, hypochlorous acid, aluminum.
D(+)-Galactose	Caution! May cause irritation to skin, eyes, and respiratory tract.	Strong oxidizers.
Nitric acid	Poison! Strong oxidizer. Contact with other materials may cause fire. Corrosive.	Incompatible with most substances, especially strong bases, metallic powders, and combustible organics.
Nicotinic acid	Avoid unnecessary exposure.	Strong oxidizers.
Ethyl acetate	Warning! Flammable liquid and vapor. Harmful if swallowed or inhaled. Affects central nervous system.	Contact with nitrates, strong oxidizers, and strong alkalis or acids may cause fire and explosions.
Agarose	Avoid unnecessary exposure.	Oxidizers.
Acrylamide	Warning! Neurotoxin. May cause cancer. Possible teratogen. Thermally unstable.	Acids, oxidizing agents, and bases. Spontaneously reacts with hydroxyl-, amino-, and sulfhydryl-containing compounds.
2-Amino-2-(hydroxymethyl)-1,3-propanediol	Warning! Harmful if swallowed or inhaled. Causes irritation to skin, eyes, and respiratory tract.	Copper, brass, aluminum, and oxidizing agents.

Question for Discussion

SDS information for most chemicals is available online. They can be found at manufacturer websites or currently at https://chemicalsafety.com/sds-search/.

Look around your house and identify three to five common household chemicals, such as cleansers or house paint. Look up the SDS for each of those chemicals and answer the following questions about each one:

a. Is the substance hazardous?
b. If the substance is hazardous, what is the nature of the hazard(s)?
c. What safety precautions should you use when working with each of these household substances?

10 Working Safely with Biological Materials

10.1 INTRODUCTION TO BIOLOGICAL SAFETY

Case Study: Biological Threats Apply Beyond the Laboratory and Enter the Public Consciousness

The global COVID-19 pandemic, caused by the virus SARS-CoV-2, has had profound effects around the world, causing illness, death, and major economic disruption. The pandemic has also introduced the general public to an issue that previously was of little general concern, that is, how to protect oneself in an environment that might contain highly infectious, pathogenic, and invisible particles. Perhaps it is not surprising that the public's response to this novel, alarming situation ranges from one extreme to the other. Some people have chosen to avoid all public contact, have not left their homes, and have carefully disinfected every food delivery package arriving on their doorstep – after totally avoiding the persons delivering them. Conversely, news media have shown photographs of individuals in close proximity earnestly conversing in crowded bars with no regard for their personal safety or the safety of others.

(Continued)

DOI: 10.1201/9780429282799-13

Case Study (*Continued*): Biological Threats Apply Beyond the Laboratory and Enter the Public Consciousness

FIGURE 10.1   COVID-19 has raised questions among the public about biological safety. (Art by: Luis "ULOANG" Angulo, 2020, Austin, TX.)

What is the "right" approach? There is not a single answer because each person differs in their tolerance of risk, their age, their health, and other significant factors. However, managing risk from a pathogenic agent is not a new problem and there is extensive literature on assessing and managing the risks of pathogenic agents. The fundamental concept is that you must understand the risk of the agent as clearly as possible and make thoughtful decisions based on that understanding. This chapter will introduce you to the knowledge that you need to protect yourself, whether it is from a pathogenic organism that may be lurking at the bus stop or one that is being carefully nurtured in a laboratory (Figure 10.1).

10.1.1 Biological Hazards

Biological safety is a major issue in biotechnology facilities. Biotechnology facilities may house a wide range of life forms, ranging from viruses and bacteria to plants, and even to farm animals. For example:

- Bacteria are used for many processes in the food industry, and as tools to explore and produce new protein and enzyme products.
- **Viruses** (*particles containing either DNA or RNA surrounded by a protein coat*) can be used to introduce DNA into cells of bacteria and higher organisms.
- Molds and fungi are sometimes used to produce complex metabolic products (such as penicillin).
- Yeasts are essential for making bread and wine. They can also be used as small eukaryotic factories for synthesis of complex proteins.
- Cells of higher organisms are grown in culture for product testing and producing recombinant proteins.

- Whole plants are studied to develop means of better food production for humans and animals, and as production systems for desirable proteins.
- Whole animals are used in research to test the products of biotechnology, and as factories for the production of antibodies or other complex proteins.

Any of these systems is potentially a **biohazard**, *which is a biological agent with the potential to produce harmful effects in humans.* **Pathogenicity** *is a term indicating the relative capability of an organism to cause disease in humans or other living organisms.* We commonly think of bacteria and viruses when we think of pathogens, but fungi, single-celled protozoa, and even multicellular organisms (such as nematodes and other worms) can also be pathogenic.

In order to cause a disease, an organism must be not only pathogenic, but also infectious. **Infectious** *refers to the ability of a pathogenic organism to invade a host organism. An organism that causes a specific*

TABLE 10.1

Risk Assessment for a Biohazardous Agent

It is essential for every laboratory worker to evaluate the health risks of specific biohazards in their workplace, addressing:

- Is this a known human or primate pathogen?
- What is the history of laboratory use of this organism or agent, and what are the recognized risks?
- Has this agent been associated with laboratory-acquired infections?
- If so, what are the health consequences of these infections?
- Is there an effective treatment or preventive vaccine?
- Does this agent frequently induce sensitivities or allergies in workers?
- What is my potential susceptibility to infection with this agent as a function of age, sex, or medical condition, as documented by previous cases?
- How can I limit my exposure to this agent?
- What are the recommended safety precautions for this agent, and are they being practiced in this laboratory?
- Is the estimated risk acceptable to me?

disease in an infected host is called an **etiological agent**. For example, the bacterium *Vibrio cholerae* is the etiological agent for cholera. Some organisms can infect a host without causing disease. In this case, the host is called a carrier. A **carrier** *is an infected individual who is capable of spreading the infecting agent to other hosts.*

In order to infect a host, an organism must be able to spread to the host and then penetrate its natural defenses. As with chemical hazards, the primary routes of exposure to biological hazards are inhalation, skin and eye contact, ingestion, and injection. Inhalation of airborne biological agents is the most likely route of laboratory exposure. Many routine laboratory procedures can produce aerosols. **Aerosols** *are very small particles or droplets suspended in the air.* A **bioaerosol** *is an aerosol that includes biologically active materials.* Strategies for the prevention of aerosols and aerosol exposure are discussed in detail later in this chapter.

Many of the same safety precautions that reduce risk from chemical hazards can also be applied to biological agents. The best health protection for a laboratory worker is to learn as much as possible about the biological materials present in their workplace. Table 10.1 provides criteria that can be used to evaluate the risk from a laboratory biohazard.

Note that the focus of this chapter is on biosafety concerns for laboratory personnel and how biosafety risks can be minimized. There are practices to protect the *product*, that is, whatever the person is working on, from contamination by microbes and other agents. This chapter only addresses the practices that protect personnel (although by protecting personnel we frequently also protect the product from contamination). Other chapters in this textbook, particularly

Chapters 30 and 38, address practices that are intended to protect products from contamination.

10.1.2 LABORATORY-ACQUIRED INFECTIONS

There are clear incentives for laboratories to use nonpathogenic organisms because of safety issues related to personnel, costs of specialized equipment and facilities, and product safety concerns. Therefore, production work (such as using cells to manufacture a drug) and the majority of the research work conducted worldwide involves the use of nonpathogenic organisms.

There are situations, however, where biotechnologists must work with pathogenic materials. Some laboratory pathogens have caused infections of laboratory workers, resulting in disease, illness, and even death; see, for example, the case study "Pathogen Issues 1." **Laboratory-acquired infections (LAIs)** are defined as *infections that can be traced directly to laboratory organisms handled by, or used in the vicinity of, the infected individuals.* OSHA estimates that as many as 12,000 cases of LAIs with hepatitis B (with 200 associated fatalities) occurred annually among US healthcare workers prior to the introduction of an effective vaccine. Figure 10.2 illustrates the origins of the majority of LAIs.

A 2016 study examining LAIs in BSL-3 and BSL-4 facilities (see Section 10.2.3 for an explanation of the terms BSL-3 and BSL-4) found that 42% of LAIs were associated with microbiology, 22% each with cell culture and microscopy, and 14% with animal care and experiments. (Wurtz, N., et al. "Survey of Laboratory-Acquired Infections around the World in Biosafety Level 3 and 4 Laboratories." *European Journal of Clinical Microbiology &*

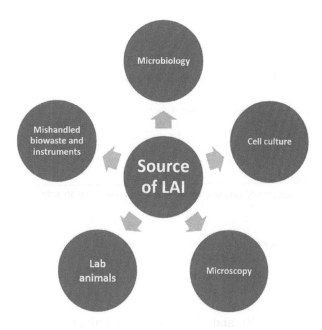

FIGURE 10.2   The common sources of laboratory-acquired infections.

Infectious Diseases, vol. 35, no. 8, 2016, pp. 1247–58. doi:10.1007/s10096-016-2657-1.) In this study, the most common individuals to become infected were those categorized as laboratory technicians (87% of the cases), followed by animal caretakers (7%) and researchers (6%). Note that there were many more laboratory technicians exposed to biological agents on a regular basis than there were individuals categorized as researchers in this study.

Case Study: Pathogen Issues 1

Pathogenic bacteria, viruses, and prions have made the headlines frequently in recent years. Infectious diseases that in the past would have afflicted only people in a single region, such as COVID-19, can now become global epidemics in a matter of a few months, due to widespread air travel. The rapid spread of SARS-CoV-2 has been a wake-up call to the international biomedical community. Other pathogens, such as West Nile virus, Ebola virus, and avian influenza (bird flu) are ongoing public health concerns as well. In the fall of 2001, anthrax bacterial spores that were intentionally sent through the US postal service sickened 22 people and killed 5. Governments and scientists have responded to these natural and terrorist-driven threats by expanding pathogen research programs.

Staff who knowingly handle biohazards are not the only individuals who contract LAIs. On average, one out of every four infections associated with a laboratory biohazard occurs among dishwashers, custodians, clerical staff, and maintenance personnel. (These cases were not included in the 2016 study described previously.) Families and friends of laboratory workers are also at potential risk. You should therefore be familiar with the basic safety issues, not only to protect yourself, but also to ensure that others are not exposed to biological hazards. Some precautions are as simple as marking rooms containing these hazards with biohazard warning signs to indicate that only trained personnel should enter the facilities (Figure 10.3). These signs can also designate the specific biohazards present. Proper waste disposal and decontamination procedures must also be followed to protect other people (see Sections 10.4 and 10.5)

10.1.3 REGULATIONS AND GUIDELINES FOR HANDLING BIOHAZARDS

There is a wide variety of guidelines, standards, and regulations that govern the use of biological materials in the laboratory. A list of examples is provided in Table 10.2. The government agencies whose biohazard guidelines are most often cited are as follows:

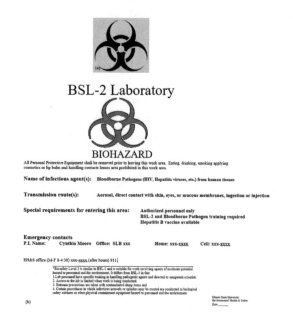

FIGURE 10.3   Standard biohazard warning signs. (a) The universal biohazard symbol. Either the symbol or the background is solid orange or red. (b) Example of a sign marking entrances to laboratory and storage areas where biohazards are handled, and access must be limited. BSL-2 (Biosafety Level 2) precautions are described in Section 10.2.3.

TABLE 10.2

Examples of Guidelines and Regulations That Cover Biological Hazards in the Laboratory

Biosafety in Microbiological and Biomedical Laboratories, HHS/CDC/NIH: This set of guidelines provides recommendations for facilities, operating procedures, hazard identification, and risk assessment for laboratories working with biological hazards. (See the Bibliography in the Introduction to this unit for full citation.)

OSHA standards for bloodborne pathogens (BBP) (detailed in the Code of Federal Regulations, 29 CFR 1910.1030) and personal protective equipment (PPE) (detailed in the Code of Federal Regulations, 29 CFR 1910 Subpart I): These OSHA standards describe the requirements for employers who must develop organizational plans for handling human blood and blood-related products, and also delineate the appropriate PPE measures for laboratories handling biological materials. These OSHA requirements are discussed later in the text. (Note that this is a situation where the term "standard" refers to legal requirements, not to practices that are voluntary.)

NIH Guidelines for Research Involving Recombinant or Synthetic Nucleic Acid Molecules, National Institutes of Health: These NIH Guidelines provide recommendations for facilities, operating procedures, hazard identification, and risk assessment for laboratories involved in research with recombinant DNA as well as **synthetic DNA**, which refers to *short DNA molecules that have been chemically synthesized, rather than having been derived from living organisms*. Synthetic DNA is used in applications such as CRISPR and RNA interference. (See Chapter 2 for a description of these techniques.)

Guide for the Care and Use of Laboratory Animals, National Research Council (NRC): This set of guidelines provides the basic procedures to be used for laboratory animal research and is considered a primary reference on animal care and use.

Animal Welfare Act and Animal Welfare Regulations, USDA Animal Care: This publication of the U.S. Department of Agriculture defines the requirements for licensing, registration, identification, records, facilities, health, and husbandry for all animals covered by the Animal Welfare Act (AWA). It consolidates AWA and other applicable regulations and standards.

The documents cited in this table are available on the Internet from the appropriate government agency. More references are included at the end of this unit's Introduction. In addition to these federal codes, there are also state, local, and institutional codes, guidelines, and design criteria that apply to biotechnology facilities.

- The **U.S. Department of Health and Human Services (HHS)**
- The **Centers for Disease Control and Prevention (CDC)**
- The **National Institutes of Health (NIH)**
- The **Occupational Safety and Health Administration (OSHA)**.

HHS *is the government agency responsible for public health and safety.* HHS is a key player in the **Science, Safety, Security project (S3)**, which is *a collaboration of US departments and agencies designed to share biorisk management information for major stakeholders.*

The **CDC** *is an agency of the federal Department of Health and Human Services whose mission is "to promote health and quality of life by preventing and controlling disease, injury, and disability"* (CDC Mission Statement). The **NIH** *is a federal health agency comprising 25 separate centers and institutions.* The agency performs and funds major biomedical research initiatives and provides guidelines for laboratory safety. The joint HHS/CDC/NIH guidelines form the basis for many of the standard biosafety precautions used in laboratories across the country. OSHA is also involved in biological safety as part of their overall mission to promote a safe workplace for all workers.

10.2 STRATEGIES FOR MINIMIZING THE RISKS OF BIOHAZARDS

Case Study: Pathogen Issues 2

In 2014, the CDC itself was forced to close two of its own laboratories and institute a system-wide review of biosafety practices. One case involved the transfer of a cell extract from anthrax bacteria between facilities, without proper sterilization to inactivate bacterial spores. More than 60 CDC staff were potentially exposed to these samples; luckily, no one became ill as a result. There was a specified biosafety procedure for sterilization to ensure that this could not happen, but the procedure was not followed.

(Continued)

Case Study (*Continued*): Pathogen Issues 2

A second case involved a number of violations in a CDC-associated laboratory. A sobering report in 2007 indicated that personnel in a secure biocontainment laboratory at Texas A&M University committed a number of safety violations. (Biocontainment will be described later in this chapter. At this point, note that pathogen research is conducted by trained personnel using specialized equipment and facilities.) Inspectors from the CDC, which provided direct oversight to this particular laboratory, cited violations including the loss of three vials of dangerous Brucella bacteria, an unreported case of an employee diagnosed with brucellosis caused by this type of bacteria, incidents where unauthorized employees worked with pathogenic agents, an instance where a faculty member performed an experiment with recombinant DNA without the necessary CDC approval, concerns about disposal of animals from experiments involving pathogenic agents, and three unreported cases of individuals exposed to the infectious bacterium that causes Q fever. The CDC reports that "There was no evidence that a coordinated response or biosafety assessment was performed as a result" of these exposures.

Chapter 4 discussed the CAPA (Corrective and Preventive Actions) process that institutions use to deal with issues such as those described above. While CAPA is related to FDA Good Manufacturing Practices, the same principles can be applied to assure that safety issues are addressed. The underlying causes of problems must be identified and corrected, and future problems must be prevented. Although the consequences of poor quality systems can be dramatic and severe in a facility that deals with deadly pathogens, the underlying cause of the problems is similar in every laboratory. According to Philip Hauck, a biosafety professional in New York City, "People get blasé, I hate to say it. After a while as microbiologists, you're like, 'This thing never bit me.'" In the absence of a robust quality system, people tend to become lax in their treatment of familiar hazards. Preventing these types of dangerous infractions requires a well-run, supervised system where activities are monitored and documented, and problems are investigated and fixed.

Primary Sources: Couzin, J. "Lapses in Biosafety Spark Concern." *Science*, vol. 317, no. 5844, 2007, p. 1487. doi:10.1126/science.317.5844.1487.

McCarthy, M. "Biosafety Lapses Prompt US CDC to Shut Labs and Launch Review." *BMJ*, vol. 349, 2014, p. g4615. doi:10.1136/bmj.g4615.

10.2.1 STANDARD PRACTICES AND CONTAINMENT

The hazards of pathogenic organisms must be reduced by strategies of good laboratory practices and containment. **Standard microbiological practices** *are the basic practices that should be used when working with all microbiological organisms.* These practices have two purposes: to separate the worker from the microorganism, and to maintain the purity of the microbial cultures. These recommended practices, which are similar to those used for chemical safety, are summarized in Table 10.3. Every biotechnologist should be familiar with the practices in this table. The same practices help maintain the integrity of whatever samples and materials with which one is working.

PPE is one of the most basic protections you must use when working with biohazards. Standard microbiological practices specify a minimum of lab coats and eye protection, but it is also good practice to wear

gloves anytime you are handling microorganisms. Remember, however, that gloves can spread contamination if you do not change them frequently. For example, do not touch items that should not be contaminated with a gloved hand. For example, do not touch a keyboard, drawer handle, door, or pen with a gloved hand unless the item is assumed by everyone to be contaminated. Even in this case, change your gloves before touching the item. Remember to change gloves any time you suspect your hands might have come in contact with microbes.

PPE in addition to a lab coat, safety glasses, and gloves may be needed in some situations. For example, work with human body fluids requires additional precautions (Figure 10.4 and Section 10.3.3). Wash your hands as soon as possible after the removal of gloves or other PPE, and immediately after any potential skin contact with biohazardous materials. All PPE must be removed before leaving the laboratory. If lab coats or other non-disposable items become contaminated,

TABLE 10.3

Standard Microbiological Practices

- Access to the laboratory should be limited to trained individuals, and a standard biohazard sign (Figure 10.3) must be posted on all entrances.
- Lab coats and eye protection should be worn at all times.
- Workers must wash their hands after any work with microorganisms and whenever they leave the lab.
- Eating, drinking, food storage, and smoking in the laboratory area are prohibited.
- Hand-to-mouth or hand-to-eye contact must be avoided.
- Appropriate pipetting devices must be used.
- Steps must be taken to minimize aerosol production.
- Strict policies regarding the use and disposal of sharps must be enforced.
- Work should be performed on a clean hard benchtop or other work surface with appropriate disinfectant readily available.
- Work surfaces should be decontaminated after any spill, and at the end of every work session.
- All biological materials must be properly decontaminated before disposal.

FIGURE 10.4   Protective personal apparel for routine work with body fluids. (a) Full PPE protection for biohazardous work includes a long gown, cap, eye shield, face mask, gloves, and shoe covers. Specialized PPE items are available for specific applications, such as (Credit:Poi NATTHAYA/Shutterstock.com) (b) shoe covers with anti-static strips to prevent static discharges. (Credit:Pongstorn Pixs/Shutterstock.com) (c) Appropriately attired COVID-19 laboratory worker using a biosafety cabinet, discussed below. (Credit:KYTan/Shutterstock.com)

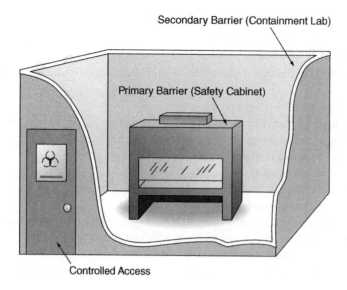

FIGURE 10.5 The relationship between primary and secondary barriers for hazard containment.

they should be disinfected and then laundered on site (not at home).

Containment *is the control of biohazards by isolation and separation of the organism from the worker.* There are many terms used in the literature to refer to containment practices. In this text, we use the term **primary containment** *to refer to equipment and practices that protect personnel and the immediate laboratory environment from hazardous exposure.* **Secondary containment** *refers to laboratory design features, equipment, and practices that protect the general environment* (Figure 10.5).

Personal containment *refers to standard practices, such as those outlined in Table 10.3, used to reduce the spread of microorganisms.* The procedures used for working with, and disposal of, the biohazardous material, along with the practices of proper laboratory hygiene, are examples of personal containment. **Physical containment** *includes laboratory design features and the physical barriers that workers use to isolate biohazards.* Special ventilation systems, PPE, gloves, and biological safety hoods are examples of physical containment; these are all discussed in various parts of this chapter.

10.2.2 Biological Safety Cabinets

10.2.2.1 What is a Biological Safety Cabinet?

Laboratory hoods *are enclosed spaces with separate air supplies or directed airflows.* There are many types of hoods, each designed for specific safety purposes. Chemical fume hoods, for example, are designed to separate volatile gases from the operator. This is

accomplished by drawing air from the room past the working space and venting the air outside the building (see Chapter 9).

A cabinet used to handle biohazards superficially resembles a chemical fume hood, but has a different purpose and construction. A **biological safety cabinet (BSC)** *provides primary containment for aerosols and separates the work material from the operator and the laboratory, while providing clean air within an enclosed area.* BSCs are intended to protect both the person using the cabinet, and the product with which the person is working. BSCs are designed to filter all air that passes through the cabinet using HEPA filters. A **HEPA (High-Efficiency Particulate Air) filter** *is a specially constructed filter made of highly pleated glass and paper fibers* (Figure 10.6). HEPA filters are designed to have a 99.97% efficiency in removing particles 0.3 μm in diameter or larger. This filter size will exclude most bacteria and microorganisms. Viruses

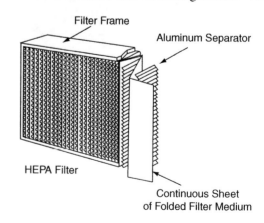

FIGURE 10.6 Structure of a HEPA filter.

found in aerosols are usually associated with particles larger than 0.3 µm, and so HEPA filters provide protection against most airborne virus particles. It is important to remember that HEPA filters are designed to remove particulate matter from the air. They are not effective for removing chemical vapors from air.

BSCs contain fans that cause the HEPA-filtered, sterile air to flow as a nonturbulent curtain into the cabinet. The air is directed over the work surface at a velocity that minimizes disturbances in the interior of the cabinet. This type of airflow is called laminar flow, and therefore, a **laminar flow cabinet** *is one that has a sterile, directed airflow.* All BSCs have laminar flow designs; however, not all laminar flow cabinets are BSCs. This is because some laminar flow hoods, such as a **clean bench**, or **horizontal laminar flow hood**, *are designed to provide a sterile work surface, but not worker protection* (Figure 10.7). Air passes through the HEPA filter onto the work surface and is vented directly into the room toward the operator. This type of laminar flow cabinet is useful for media preparation, or sterile work with non-harmful organisms. It is not a BSC because it does not provide protection for the operator.

10.2.2.2 Classification of Biological Safety Cabinets

There are three general classes of BSCs that provide different levels of protection against biological hazards. A **Class I cabinet** *is designed to protect the operator from airborne material generated at the work surface. The cabinet draws air directly from the room. The air then flows over the working area and is vented back into the room after being filtered.* This cabinet will filter aerosols released from the work materials and so can be used to reduce aerosol escape to the environment. Because the air flowing over the work surface is unfiltered room air, this type of hood is not designed to maintain sterility of the work surface or product integrity. A Class I cabinet is used for routine operations that might generate harmful aerosols, such as vortexing, pouring, or centrifuging solutions containing microorganisms.

A Class II BSC, which is the most common type found in biotechnology laboratories, provides protection for both the operator and the work product (Figure 10.8). It is designed to draw HEPA-filtered

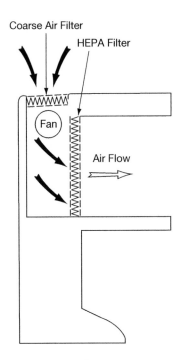

FIGURE 10.7 Design of a clean bench. Nonsterile air is shown as solid arrows, and sterile air as open arrows. Observe that the air flowing over the work surface has come through a HEPA filter located in the back portion of the cabinet. This sterile air then flows over the product and out the front of the cabinet, potentially exposing the user to whatever was on the work surface of the cabinet.

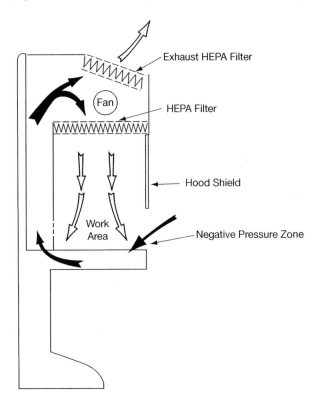

FIGURE 10.8 Design of a Class IIA BSC. A Class II, Type A1 BSC design. Observe that the air entering the cabinet is filtered to provide sterile air to protect the product. Air that has passed over the product is directed through the back of the cabinet, again through the HEPA filter, so that air returning to the laboratory is sterile. The operator sitting at the front of the cabinet is not exposed to potential pathogens inside the cabinet.

TABLE 10.4

Comparison of Types of Biological Safety Cabinets

Type	User Protection	Work Surface Protection	Face Velocity (in Linear Feet per Minute)	Suitable for Use with Toxic Chemicals	Airflow Pattern
Clean bench (not a BSC, does not protect user)	No	Yes	Varies	No	Air in at top, exhausted at front toward operator
Class I	Yes	No	75	No	Air in at front, exhaust duct within lab
Class IIA1	Yes	Yes	75	No	Air in at front, exhaust duct within lab or external, ~70% air recirculation within cabinet
Class IIA2	Yes	Yes	100	Low levels	Air in at front, exhaust duct within lab or external, ~50% air recirculation within cabinet
Class IIB1	Yes	Yes	100	Low levels	Air in at front, external exhaust duct, ~30% air recirculation within cabinet
Class IIB2 (100% exhaust cabinet)	Yes	Yes	100	Yes	Air in at front, external exhaust duct, no air recirculation within cabinet
Class IIC1	Yes	Yes	100	Yes	Air in at front, external exhaust duct with fan but can be used without exhaust, ~40% recirculation with exhaust
Class III	Yes	Yes	None	Yes	Filtered air ducts for supply and exhaust, gas-tight

air across the work surface and re-filter the air before venting. Because the air is filtered before it passes over the work surface, this cabinet type will maintain a sterile work area. A Class II cabinet is suitable for microbiology and tissue culture procedures.

Class II cabinets are subdivitded into several categories, Types A, B, and C, and then subcategories. These have different airflow patterns and uses. For example, some categories are suitable for use with low levels of toxic agents. A detailed description of these differences is provided in Table 10.4.

Class III cabinets, also called **glove boxes**, are gas-tight cabinets designed to provide total containment for extremely hazardous biological agents. They provide the highest level of protection possible for both the user and environment and are required in a BSL-4 facility, described below. Class III cabinets are designed to isolate the work material completely within a closed system. A Class III cabinet is accessed with air-tight gloves (Figure 10.9). Materials in these cabinets usually require decontamination before removal from the cabinet. Class III cabinets often have incubators, autoclaves, and air locks directly attached. A Class III cabinet would only be found when working with the most hazardous agents, and special rooms would also be required for proper use.

Table 10.4 summarizes the basic characteristics of the different classes of biological safety cabinets.

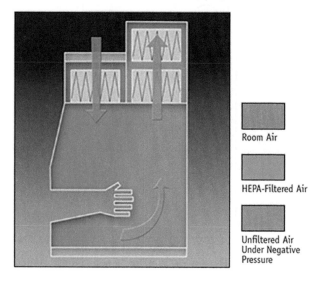

Room Air

HEPA-Filtered Air

Unfiltered Air Under Negative Pressure

FIGURE 10.9 A Class III BSC. These cabinets are completely enclosed from the front, and the operator accesses the materials using built-in gloves. The HEPA filters are shown as zigzag lines at the top of the cabinet. (Image courtesy of Labconco Corporation.)

Example Problem 10.1

Classify each of the following practices as (1) Protecting yourself from potential infection or harm from an agent with which you are working. (2) Protecting the product (the material with which you are working) from contamination. (3) Protecting both yourself and the product.

a. Wearing safety glasses.
b. Wearing gloves.
c. Wearing a laboratory coat.
d. Not talking while working in a biological safety cabinet.
e. Working at a clean bench.
f. Working in a biological safety cabinet.

Answer

a. Protects you.
b. Protects you and the product – when used properly.
c. Protects you and the product.
d. and e. Protects the product.
f. Protects both you and the product.

10.2.3 RECOMMENDED BIOSAFETY LEVELS

The NIH classifies organisms into Risk Groups (RGs) according to their health threats to laboratory personnel and the community at large. Risk is usually determined based on the pathogenicity of the organism, the infection rate after exposure, and the availability of effective treatment if infection occurs. Table 10.5 provides a summary of the RGs and examples for each group. Depending on the level of risk posed by a particular agent, it is assigned to one of four levels of biosafety practices, BSL-1, BSL-2, BSL-3, or BSL-4. **Biosafety levels (BSLs)** *are defined by the NIH as the combinations of laboratory facilities, equipment, and practices that protect the laboratory, the public, and the environment from potentially hazardous organisms.* The higher the number, the more stringent the practices required when working with that organism. A pathogen's biosafety level (BSL) can differ depending on how the organism is used. For example, current guidelines for handling SARS-CoV-2 call for BSL-3 conditions when working with concentrated samples, but BSL-2 is considered adequate for dilute samples. (These guidelines may change as we learn more about the virus.)

Biosafety Level 1 (BSL-1) *is used for well-characterized strains of living microorganisms that are not known to cause disease in healthy adult humans.* BSL-1 organisms have been used safely in many laboratories. High school and college students commonly work with BSL-1 organisms. Examples of organisms in this category are nonpathogenic strains of *E. coli*, yeasts, and most plants. The fact that BSL-1 organisms are generally nonpathogenic does not mean that they can be handled without caution. Standard microbiological practices must be applied to handling these organisms, but no special equipment is required.

Biosafety Level 2 (BSL-2) *is designated when working with agents that may cause human disease and therefore pose a risk to personnel.* Diseases caused by BSL-2 organisms are usually treatable or have preventive vaccines available. These organisms are not likely to spread to the external environment and create a general health threat. Typical organisms in the BSL-2 category are pathogenic bacteria such as *Salmonella* and *Clostridium*, molds such as *Penicillium*, and viruses, such as rabies virus, which are not generally spread by aerosols. Human blood and tissue products are handled in a BSL-2 environment.

The health risks associated with BSL-2 organisms require a higher level of containment than with BSL-1. Written procedures and special worker training are required. Physical containment procedures require that all work potentially generating aerosols or splashes be performed in a biological safety cabinet to minimize the release of microorganisms, especially in the form of aerosols. Waste must be decontaminated before disposal. For some infectious agents, the health of the worker may be monitored, and the worker may be required to be immunized against the infectious agent. Production of blood antibodies against the microorganism may be checked to determine whether an unvaccinated worker has been exposed to a specific organism.

The majority of laboratories in the United States are designated BSL-1 or BSL-2. High-containment **Biosafety Levels 3 (BSL-3)** and **4 (BSL-4)** *are generally associated with more dangerous agents that are highly infectious.* These hazardous agents require significantly higher containment barriers in the form of special facilities. The airflow into and out of the room where such organisms are handled is closely monitored. Examples of organisms requiring BSL-3 facilities are certain arboviruses (arthropod-borne viruses associated with several human

TABLE 10.5

Classification of Microorganisms by Risk Group

Risk Group (RG) (Required BSL)	Description	Typical Environment	Examples
RG1 (usually BSL-1)	Agents that are not associated with disease in healthy adult humans. No individual or community risk.	Open bench	*E. coli* strain K12, *Lactobacillus* sp., Many non-spore-forming *Bacillus* sp., Adeno-associated virus (AAV) types 1–4
RG2 (BSL-2)	Agents that are associated with human disease that is rarely serious and for which preventive or therapeutic interventions are *often* available, or which are not highly transmissible. Moderate individual and low community risk.	Any Class I or II BSC	*Salmonella* sp., *Streptococcus* sp., Adenoviruses, HIV (most protocols), Most poxviruses
RG3 (BSL-3)	Agents that are associated with serious or lethal human disease for which preventive or therapeutic interventions *may be* available. High individual risk and moderate community risk.	Usually Class II cabinets (possibly Class III)	SARS-CoV-2 (concentrated samples) *M. tuberculosis* *Yersinia pestis* HIV (large-scale protocols) Retroviruses
RG4 (BSL-4)[a]	Agents that are likely to cause serious or lethal human disease for which preventive or therapeutic interventions are *not usually* available. High individual and community risk.	Class III biosafety cabinets only Full isolation precautions	Ebola virus Lassa virus Marburg virus Other viruses

Source: Adapted from: *The National Institutes of Health, Office of Biotechnology Activities.* "NIH Guidelines for Research Involving Recombinant or Synthetic Nucleic Acid Molecules." NIH, 2019. https://osp.od.nih.gov/biotechnology/biosafety-and-recombinant-dna-activities/.

World Health Organization. "Laboratory Biosafety Manual." World Health Organization, 2020. https://www.who.int/publications/i/item/9789240011311.

[a] Only a few laboratories around the world work with BSL-4 conditions.

diseases such as West Nile virus) and large cultures of *Mycobacterium tuberculosis*, the etiological agent for human tuberculosis. Very dangerous BSL-4 agents require closed glove boxes, isolated air supplies, and an enclosed breathing apparatus for the operators. BSL-4 facilities are found in only a few places in the world. They are required for handling Ebola virus and related organisms that cause rapidly fatal human disease.

Table 10.6 provides an overview of the containment practices and equipment required at each of the four biosafety levels. Many of the requirements shown in this table are discussed later in this chapter. There are additional requirements for large-scale (over 10 liters of volume) laboratory operations that use live organisms. For details, see the NIH Guidelines for Research Involving Recombinant or Synthetic DNA Molecules, Appendix K. (Full reference is found in the Unit III Introduction.)

10.2.4 Use of Disinfectants and Sterilization

Bioaerosols or work-related splashes can leave contamination on laboratory work surfaces and equipment. An effective practice for preventing the spread of biohazards is therefore the use of proper decontamination procedures. Decontamination can be accomplished with the use of germicidal agents. Some of the most common agents are heat (autoclaving), gas exposure (using ethylene oxide, for example), and the use of liquid chemical

TABLE 10.6

Summary of Biosafety Levels for Infectious Agents

Level	Special Safety Practices	Special Equipment	Facilities
BSL-1	Standard microbiological practices	None required	Handwashing sink required
BSL-2	BSL-1 practices plus: • Biohazard warning signs • Biosafety manual specific to the laboratory provided • Potentially infectious materials kept in a sealed container • Medical surveillance available	Class I or II BSCs to prevent escape of aerosols	BSL-1 requirements plus: • Self-closing, locked doors • Autoclave available • Elimination of all porous furniture materials
BSL-3	BSL-2 practices plus: • Decontamination of all waste • Decontamination of all laboratory clothing before laundering • Baseline blood serum testing for workers	BSL-2 equipment plus: • Physical containment within a BSC for all manipulations • Additional protective clothing and respiratory protection as needed	BSL-2 requirements plus: • Physical separation from access corridors • Self-closing, double-door system • Handwashing sink must operate hands-free • Walls and ceiling must have a smooth, sealed finish for easy decontamination • Exhausted air not recirculated • Negative airflow[a] into laboratory • Facility and procedures must be inspected and documented before use
BSL-4	BSL-3 practices plus: • Specific and rigorous training for all personnel • A logbook of all entries and exits must be maintained • Special transport protocols for all biological materials • Clothing change before entering • Shower on exit • All materials decontaminated on exit from facility	All procedures are conducted in Class III BSCs or in Class I or II BSCs in combination with a full-body, air-supplied, positive pressure personnel suit	BSL-3 requirements plus: • Separate building or isolated zone • Dedicated air supply and exhaust, vacuum, and decontamination systems • Emergency power supply • All plumbing protected from backflow and all liquids decontaminated before exit from lab

Source: Adapted from: *Centers for Disease Control and Prevention (U.S.)*, et al. "Biosafety in Microbiological and Biomedical Laboratories." 6th ed, U.S. Dept. of Health and Human Services, Revised 2020.

BSC, biological safety cabinet; PPE, personal protective equipment.

[a] Air can flow into the laboratory from external spaces such as hallways, but not in the opposite direction.

disinfectants. **Disinfectants** *are chemicals that kill pathogenic microorganisms and other biohazardous particles.* There are many factors that determine the effectiveness of germicidal agents, such as:

- the type of contaminating organism and its susceptibility to disinfecting agents
- the level of contamination
- the chemical composition and concentration of the decontaminating agent
- the length of exposure of the organism to the decontaminant

- the shape and texture of the surface being decontaminated.

Biohazardous agents differ in their susceptibility to disinfection procedures (Table 10.7). For example, most growing bacteria are relatively easy to kill, but bacterial spores are difficult to eradicate. It is essential to know the proper disinfection procedure for the organism you are handling, as discussed in the case studies "The Right Disinfectant Choice Can Prevent LAIs" (below) and "Isolator Technology Is Only as Effective as the Operator," p. 1014.

TABLE 10.7

Disinfection Resistance of Microorganisms

Resistance to Disinfection	Type of Organism	Examples
Least resistant	Hydrophobic and/or medium-sized viruses	Human immunodeficiency virus
		Herpes simplex virus
		Hepatitis B virus
Low resistance	Bacteria	*Escherichia coli*
		Staphylococcus aureus
	Fungi	*Candida* sp.
		Cryptococcus sp.
Moderate resistance	Hydrophilic and/or small viruses	Rhinovirus
		Poliovirus
High resistance	Mycobacteria	*Mycobacterium tuberculosis*
Most resistant	Bacterial spores	*Bacillus subtilis* spores
		Clostridium sp. spores

Note: SARS-CoV-2 is a large virus and is currently considered relatively easy to disinfect from surfaces.

Case Study: The Right Disinfectant Choice Can Prevent LAIs

While LAIs can result from unforeseen circumstances, many are the result of poor laboratory practices. For example, in 2002 a lab worker in Texas developed cutaneous anthrax after working with cultures of *Bacillus anthracis*, a spore-forming bacterium. The infection was traced to contaminated freezer vials prepared by the technician, who wiped the outside of the vials with 70% isopropyl alcohol and then handled the containers with bare hands. The lab SOP for decontaminating surfaces required the use of bleach for spore inactivation, but the technician apparently substituted alcohol to avoid removing the vial labels. This LAI, which was treatable and relatively noncontagious, could have been prevented by using the appropriate disinfectant and wearing gloves at all times while handling cultures.

There are distinctions between sanitizing, disinfecting, and sterilizing a surface or object. **Sanitization** *is the reduction of the number of microorganisms on a surface.* Sanitization is generally the purpose of antiseptics. An **antiseptic** *is a type of mild disinfectant gentle enough to be applied to skin.* Chemical solutions that are suitable for use as antiseptics and hand sanitizers, such as 3% hydrogen peroxide or 70% isopropyl (rubbing) alcohol, may not provide enough disinfectant strength for work surfaces. Note that proper handwashing with soap is significantly more effective than hand sanitizer for skin cleaning. **Disinfection** *is the removal of all or almost all pathogenic organisms from a surface.* Disinfectants can be classified as low, intermediate, and high level, according to their capacity to destroy microorganisms. **Sterilization** *is the highest level of disinfection, the killing of all living organisms on a surface.* An item or surface is either sterile or not sterile. High-level disinfectants achieve or approach sterilization, whereas low-level disinfectants are of variable effectiveness, depending on the nature of the contaminating organisms. High-level disinfectants are appropriate for use on medical devices. Low-level disinfectants are frequently called sanitizers to denote their main purpose.

Table 10.8 provides a summary of the sterilization and disinfection capabilities of some commonly used agents. Note that most chemical agents are not strong enough to destroy every last microbe and its spores; that is, most chemical disinfectants are not sterilizing agents. Note also that products that are marketed as hand sanitizers are not appropriate for cleaning work surfaces.

A major consideration when choosing a disinfecting agent is the nature of the biohazardous organism

TABLE 10.8

Strength of Common Germicidal Ingredients

Ingredient	Concentration for Sterilization	Disinfection Concentration	Disinfection Activity Level	Example of Commercial Product
Alcohols	NA	70%	Low to intermediate	Purell Advanced Hand Sanitizer
Chlorine mixtures	NA	500–5,000 ppm chlorine (1%–10% bleach solution)	Intermediate	Clorox
Glutaraldehyde	2% at high pH	2%–3%	High	Cidex
Hydrogen peroxide	6%–30%	3%–6%	High	
Iodophor mixtures	NA	40–50 mg free iodine/liter	Intermediate	Betadine
Phenolic mixtures	NA	0.5%–3%	Intermediate	Lysol (some types)
Quaternary ammonium mixtures	NA	0.1%–0.2%	Low	Roccal

NA, not applicable.

TABLE 10.9

Antimicrobial Activity of Commonly Used Chemical Disinfectants

Antimicrobial Agent	Range of Effectiveness				
	Bacteria	*Mycobacterium tuberculosis*	Bacterial Spores	Fungal Spores	Viruses
Ethylene oxide gas (500–800 mg/L)	4	4	4	4	4
Ethyl alcohol (70%)	2	2	0	2	1
Sodium hypochlorite (5% bleach)	3	2	0	2	2
Glutaraldehyde (2%)	2	2	3	2	2
Iodophors (1%)	2	2	1	1	2
Phenolic derivatives (1%–3%)	2	2	0	2	1
Quaternary ammonium compounds (3%)	2	0	0	2	1

4: Superior, 3: very good, 2: good, 1: fair, 0: no activity.

that must be removed from a surface. For human safety reasons, it is best to use the least toxic agent that can get the job done. Table 10.9 shows the relative effectiveness of a selection of chemical disinfecting agents against specific types of biohazardous organisms.

Ethylene oxide sterilization is routinely used for medical devices such as catheters and electrical equipment. Because ethylene oxide vapors can be explosive when mixed with even small amounts of air, this treatment process is performed under vacuum and requires specialized chambers.

All disinfection and sterilization methods present certain safety hazards to the operator as well as to the targeted microorganisms. Table 10.10 provides examples of some of the human hazards associated with germicidal laboratory techniques.

10.2.5 Bioaerosol Prevention

Airborne materials pose a major concern in any laboratory or facility where potentially infectious agents or hazardous materials are used. In a 1978 study of LAIs, fewer than 20% could be linked to a specific spill or other accidents. Researchers believe that about 70% of all LAIs result from inhalation of infectious particles (bioaerosols). Airborne particles (carried by air currents) are highly effective in transmitting certain types

TABLE 10.10

Human Hazards Associated with Selected Antimicrobial Agents

Antimicrobial Agent	Potential Hazards
UV light	Eye and skin burns; mutagen; carcinogen
Steam (autoclave)	Burns; aerosols and chemical vapors; cuts from imploding glassware
Ethylene oxide	Severe respiratory and eye irritant; sensitizer; mutagen; suspected human carcinogen
Isopropyl alcohol	Flammable; toxic; severe eye irritant
Chlorine (bleach)	Gaseous form highly toxic; oxidizing agent
Formalin (37% formaldehyde)	Strong eye and skin irritant; toxic; sensitizer; mutagen; suspected human carcinogen
Glutaraldehyde	Toxic; skin irritant
Hydrogen peroxide	Oxidizing agent; corrosive; skin irritant
Iodine (iodophors)	Skin irritant; toxic vapors; sensitizer
Phenols	Chemical burns; toxic vapors; strong eye and skin irritant
Quaternary ammonium compounds	Skin irritant

of infections; see the case study "Poor Ventilation and Prolonged Contact Can Result in the Spread of COVID-19." The common cold, for example, can be spread by sneeze-generated aerosols that are inhaled by the next cold sufferer. Airborne organisms can remain suspended for significant periods of time and can be carried through a laboratory by air currents and even throughout buildings by way of the ventilation system. If inhaled, aerosols can carry hazardous materials into the lungs. Aerosols will also eventually settle on laboratory surfaces, where they can deposit hazardous substances.

Case Study: Poor Ventilation and Prolonged Contact Can Result in the Spread of COVID-19

On January 24, 2020, a New Year's meal resulted in a COVID-19 outbreak among three families in the Chinese city of Guangzhou. One family had just traveled from Wuhan and was seated at one end of a restaurant with no outside ventilation (Table A in Figure 10.10). Although everyone felt fine during the meal, one family member developed symptoms later that day. By the first week of February, a total of

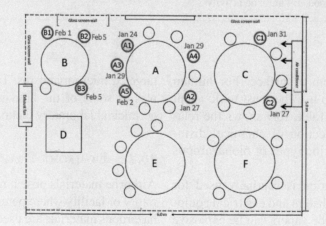

FIGURE 10.10   Spread of COVID-19 infection among restaurant tables. Patient zero, *the first individual to be diagnosed*, sat at Table A and is indicated in yellow (A1); all of the other nine cases came from three tables, including the family of patient zero. Note that the infected individuals at Table C were distanced ~12 feet from patient zero. (From Lu, Jianyun, et al. "COVID-19 Outbreak Associated with Air Conditioning in Restaurant, Guangzhou, China, 2020." *Emerging Infectious Diseases*, vol. 26, no. 7, 2020, pp. 1628–31. doi:10.3201/eid2607.200764.)

(Continued)

Case Study (*Continued*): Poor Ventilation and Prolonged Contact Can Result in the Spread of COVID-19

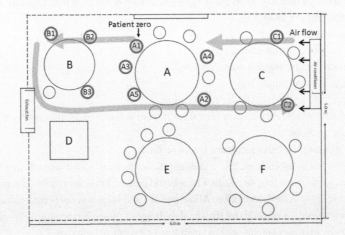

FIGURE 10.11   Recycled air moved in a closed circuit in the area of the restaurant where the outbreak occurred. No one at other tables in the restaurant was infected. (Adapted from Lu, Jianyun, et al. "COVID-19 Outbreak Associated with Air Conditioning in Restaurant, Guangzhou, China, 2020." *Emerging Infectious Diseases*, vol. 26, no. 7, 2020, pp. 1628–31. doi:10.3201/eid2607.200764.)

10 individuals from three families who ate at the restaurant in the same time frame were diagnosed with COVID-19.

Investigation suggested two crucial factors in this outbreak. First, the closed ventilation system of the restaurant led to recycling of the air supply through that portion of the dining room, suggesting that aerosol transmission was the most likely route of infection (Figure 10.11).

A second important factor seemed to be the length of time in proximity. The occupants of Tables A–C were in proximity for more than 50 minutes. In contrast, occupants of nearby Table D, who were not affected, were in proximity to Family A for less than 15 minutes (and are therefore not shown in the figures). This case indicates the importance of remaining in well-ventilated areas and limiting exposure time with large groups of people during a pandemic.

One of the most important steps that you can take to minimize your risk of LAI is to reduce the presence of aerosols in your working environment. Table 10.11 lists examples of common laboratory operations that generate airborne particles. Any activity that mechanically disturbs a liquid or a dry powder has the potential to release airborne substances. Care must therefore be used whenever handling any culture containing microorganisms, cells, or infectious agents, and when manipulating solutions containing hazardous substances. Because aerosols are not visible, and because they are created during routine laboratory operations, laboratory workers are often unaware of their presence. Always be careful and deliberate when handling potentially hazardous liquid cultures, and use containment devices as appropriate.

Complete prevention of aerosol formation is not possible. By using careful technique, however, it is possible to greatly reduce their formation rate. For example, it is good practice to wait a few minutes after centrifugation before opening centrifuge tubes and rotors. This allows time for aerosols to settle. Table 10.12 includes examples of other practices that reduce aerosol formation. Even when careful practices are used, some airborne particles are still generated. Most laboratories therefore provide various containment barriers to control unavoidable aerosols (Figure 10.12). For example, BSCs can be used for operations that are likely to generate aerosols, such as pouring fluids, blending tissue samples, and handling cultures. It is also possible to purchase accessories for centrifugation that safely enclose potentially hazardous materials.

TABLE 10.11

Examples of Routine Laboratory Procedures that Generate Aerosols

- **Shaking and mixing liquids**, particularly when using vigorous mixing devices such as blenders and vortex mixers.
- **Pouring liquids** from one container to another.
- **Pipetting liquids** generates aerosols when:
  - the last drop of liquid is ejected from a pipette tip;
  - a micropipette tip is ejected into a disposal container;
  - liquids splash upon entering a container; and
  - a used pipette is dropped into a vertical cylinder for cleaning.
- **Removing the cap from a tube** can cause aerosol formation if there is liquid under the cap; "snap cap" tubes are particularly likely to cause aerosols.
- **Removing a stopper or cotton plug from a culture bottle or flask.**
- **Inserting a hot loop into a bacterial culture.** It is common practice to flame metal loops or needles (to kill any microorganisms) and then to place the loop or needle into a liquid culture. If the loop is hot when inserted into the culture, splattering can occur that leads to aerosol formation. Allow needles and loops to cool before placing them in a culture.
- **Opening a sealed tube containing a lyophilized (freeze-dried) agent.** When reconstituting lyophilized materials, add liquid to the contents slowly. Mix the contents without bubbling or excessive agitation.
- **Breaking cells open by sonication.**
- **Grinding cells or tissues** with a mortar and pestle or homogenizer.
- **Bubbling air into a liquid (aeration).**
- **Centrifuging samples** is particularly problematic. Note the following:
  - Centrifugation in uncapped tubes causes aerosols to be released; never centrifuge hazardous materials in uncapped tubes.
  - A centrifuge tube that breaks during spinning will spew its contents into the air.
  - Even if a tube is unbroken and capped, it is possible for the cap to deform under the force of centrifugation, allowing liquid to leak out and form aerosols during spinning.

TABLE 10.12

Good Practices that Reduce Exposure to Aerosols

- Use a BSC when performing any operation that may generate hazardous aerosols.
- Before opening containers whose contents have been shaken, blended, or centrifuged, wait a few minutes to allow aerosols to settle.
- Always change gloves after any actions that potentially create an aerosol.
- Pipetting practices that reduce aerosol formation and protect the operator include:
  - Plug the tops of pipettes with cotton.
  - Never mix a liquid that contains hazardous material by pipetting up and down.
  - Use pipettes that do not require expulsion of the last drop of liquid.
  - Do not discharge biohazardous material from a pipette in a manner that induces splashing; when possible, allow the discharge to run down the container wall.
  - Place contaminated pipettes horizontally in a pan containing enough decontaminant to allow complete immersion.
- Centrifuge practices that reduce aerosol formation and/or protect the operator include:
  - Use centrifugation rotors and tubes designed to contain hazardous materials.
  - Fill and open centrifuge rotors and tubes in a BSC (Figure 10.12).
  - Carefully disinfect rotors and tubes used for hazardous materials according to the manufacturer's directions.
  - Discard chipped or cracked centrifuge tubes.
  - Routinely allow the contents of rotors and tubes to settle for a few minutes before opening.
  - Avoid pouring liquids or decanting supernatants if possible; it is preferable to use a vacuum system with appropriate in-line safety reservoirs and filters to remove the liquid from a centrifuge tube or other vessels.

FIGURE 10.12 Avoid the spread of centrifuge-related aerosols by using a safety cabinet. (Image credit: U.S. Army Medical Research Institute of Infectious Diseases.)

Example Problem 10.2

Rabies has the dubious distinction of having the highest fatality rate of any infectious disease. It is estimated that as many as 55,000 people die of rabies every year, and children are most often the victims. (Nandi S, and Kumar M. "Development in Immunoprophylaxis against Rabies for Animals and Humans." *Avicenna J Med Biotechnol.* vol. 2, no. 1, 2010, pp. 3–21.) There is thus a need for an inexpensive and effective vaccine against rabies. Suppose you are working on vaccine experiments involving rabies virus in a BSL-2 facility. At the moment, you have a vial containing concentrated virus that is in a rack inside the biosafety cabinet. You need to open the vial. How many pairs of gloves will this task require? Explain.

Answer

At least two pairs, possibly three. You will have a pair on when you open the vial. You might also double glove, meaning that you wear a pair of surgical (high-quality) gloves as a base layer and a pair of disposable gloves on top. You open the vial and then immediately remove the disposable gloves and put on a new pair. The disposable gloves are quite likely to be contaminated by aerosols when you opened the vial, and possibly by a droplet on the lip of the vial. Remember to remove the gloves with proper technique (Figure 8.2), and place the used gloves in the proper receptacle for later disinfection.

10.3 SPECIFIC LABORATORY BIOHAZARDS

10.3.1 VIRUSES

Viruses *are particles that contain nucleic acid surrounded by a protein coat of varying complexity.* They require a cellular host for replication. Viruses can contain either DNA or RNA as their genetic material. Both RNA- and DNA-containing viruses have the ability at times to integrate their nucleic acid sequences into an infected cell's genome. In some cases, this viral DNA can direct the cell to make new virus particles, which will then be released as infectious viruses.

Many types of viruses have the ability to remain dormant for long periods of time, and in some cases, to resist disinfection and sterilization procedures (described in Section 10.2.4). Because virus particles are much smaller than other microorganisms (Figure 10.13), they can aerosolize readily and are relatively difficult to filter from the air at high concentrations (although HEPA filters are effective at filtering virus particles associated with aerosol particles). It is essential to use standard microbiological practices when handling any viruses and virally infected cells, but these practices are insufficient when working with highly infectious, pathogenic viruses, which may require BSL-2, BSL-3, or in rare cases, BSL-4 practices.

Viruses pose risks of acute and sometimes chronic infection. Acute infection can be transitory and harmless, or it can result in disease. Viruses are classified into the same risk levels as other microorganisms. Always know the properties of and any potential risks posed by viruses that are used in your workplace. The case study below discusses the safety precautions for SARS-CoV-2.

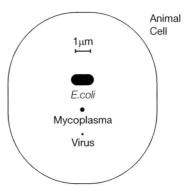

FIGURE 10.13 Relative sizes of biological organisms used in laboratories. These include animal cells, bacteria, and viruses.

Case Study: SARS-CoV-2 in the Laboratory and in Public

The guidelines for handling concentrated amounts of SARS-CoV-2 have been evolving as scientists learn more about the organism and its patterns of transmission. Because this virus causes a potentially lethal disease, the interim designation is RG3 for concentrated samples, meaning that BSL-3 precautions are advised. Procedures without concentrated virus but likely to generate aerosols should be performed in a Class II biosafety cabinet in a BSL-2 laboratory. This includes handling diagnostic samples from patients.

SARS-CoV-2 is mainly transmitted person to person and apparently spreads more efficiently than influenza virus. While public health officials were at first concerned about virus spread on surfaces, it has become clear that the main transmission route is through droplets and aerosols generated by infected individuals, including those who are not yet symptomatic and are unaware of being infected. There is growing evidence that airborne virus is a significant source of infection in locations with poor ventilation, as described in the case study on pp. 234–235.

One line of defense against the spread of COVID-19 is the wearing of masks in public locations and social situations. N95 respirators and surgical masks are generally limited to medical personnel, although they are also used in research laboratories, especially when handling animals. Most individuals wear some type of cloth mask in public when social distancing or good ventilation are not available. The effectiveness of cloth masks varies significantly by fabric choice, the number of layers, and fit around the nose and mouth. Cloth masks do not filter out all airborne virus particles, although they do block the passage of larger aerosols and provide some degree of protection for the wearer. They do prevent the spreading of the droplets and some larger aerosols generated by the wearer. They are also effective in reminding people not to touch their mouth or nose with their hands.

It is possible that transmission can occur by hand-to-mouth or hand-to-eye contact after touching a contaminated surface, but this does not appear to be a major route of infection. The CDC has published a long list of disinfectants that should be effective against SARS-CoV-2, and the agency has recently approved two Lysol products that were specifically tested against SARS-CoV-2 and found to be 99.9% effective on hard surfaces when used according to directions.

The greatest concern in laboratories that use human tissue or blood products is unknown contamination of materials with HBV, HIV, or other unknown viruses. **HBV**, or **hepatitis B virus**, *is the etiological agent for human hepatitis B.* **HIV**, or **human immunodeficiency virus**, *is the etiological agent for acquired immunodeficiency syndrome (AIDS).* Most of us are more aware of and concerned about HIV infection, but in reality, the rate of transmission of HIV is much lower than that of HBV. For example, the risk of infection with HIV after skin puncture with an HIV-contaminated needle is approximately 0.3%, based on studies of hospital exposures. The risk of infection with HBV is approximately 30%–35% after skin puncture with an HBV-contaminated needle (in unvaccinated individuals).

Hepatitis B *is a bloodborne infection that attacks the liver and causes inflammation.* Most people who contract hepatitis B exhibit moderate to severe flu symptoms. Up to 5% of these individuals will develop

chronic disease, which can eventually prove fatal. Another 5%–10% of infected individuals become chronic carriers of HBV. Fortunately, there is an effective (>90%) vaccine available that should be administered to all persons who contact human tissue, blood, or blood products. This vaccine requires a series of three injections for maximum efficacy. Federal guidelines mandate that this vaccine be offered free of charge to any individual who works with, or may be exposed to human tissue or blood products in the workplace. This availability must be included in laboratory signage (Figure 10.3b).

HIV is an example of a type of RNA virus called a retrovirus. A **retrovirus** *is an RNA-containing virus that also contains **reverse transcriptase**, an enzyme that transcribes RNA into DNA.* This DNA copy of the viral genetic material can then integrate into the DNA of a host cell. With some retroviruses, this alteration may be undetectable except by molecular analysis and does not lead to further problems. In other cases, such as HIV infection, adverse effects become apparent with time.

Without treatment, HIV infection eventually leads to the development of AIDS. **AIDS**, or **acquired immunodeficiency syndrome**, *is a condition in which the T cells of the immune system are destroyed, devastating the body's immune defenses and allowing other infectious agents to cause disease and frequently death*. Individuals with AIDS, or other conditions that weaken the immune system, such as chemotherapy, are especially susceptible to infection with microbiological agents that are considered to be relatively nonpathogenic in healthy individuals.

Although there is no vaccine currently available for HIV, there are treatments that help to delay the deterioration of the immune system and therefore the manifestation of immunodeficiency symptoms in HIV-infected individuals. These treatments can slow the development of AIDS and increase both life expectancy and quality of life for infected individuals. Fortunately, HIV is not highly transmissible under laboratory conditions, and the virus is susceptible to inactivation with desiccation or disinfection procedures.

Another potential consequence of exposure to certain types of viruses is the development of cancer.

Cancer *includes a variety of diseases characterized by the uncontrolled growth of cells and the ability of these cells to spread to other parts of the body* (**metastasis**). **Oncogenic viruses** *are agents that can induce cancer after infecting cells*. This occurs when the viral infection either interferes with normal cellular growth control, or alters the cell's genetic material in a manner that affects growth control.

Potentially oncogenic viruses are classified as being of low, moderate, or high risk to humans. A low-risk virus is one that has the theoretical capability of being oncogenic, but has not yet shown any evidence of a threat. Many viruses that cause cancer in animals fall into this category. A high-risk oncogenic virus would be one that has been shown to cause cancer in humans. A list of viruses that are currently believed to cause human cancer is shown in Table 10.13. It is important to note that many individuals become infected with these viruses without developing cancer as a result. For example, infection with Epstein–Barr virus most commonly results in mononucleosis, not cancer, in the United States. The criteria for designating oncogenic viruses as being of moderate risk, that is suspected, but not proven to be carcinogenic in humans, are shown in Table 10.14.

TABLE 10.13

Viruses Considered to be Causative Agents in Human Cancers (High Risk)

Virus	Associated Human Cancers
Human papillomavirus (HPV)	Cervical cancer (vaccine available)
Epstein–Barr virus (EBV)	Burkitt's lymphoma, nasopharyngeal carcinoma
Hepatitis B and C (HBV and HCV)	Liver cancer (HBV vaccine available)
Human herpes virus (HHV-8)	Kaposi's sarcoma, lymphoma
Human T cell lymphotropic virus type 1 (HTLV-1)	Adult T cell leukemia/lymphoma

---

TABLE 10.14

Criteria for Classification as a Moderate-Risk Oncogenic Virus

- The virus has been isolated from human cancers.
- The virus causes cancer in non-human primates without experimental manipulation of the host.
- The virus can produce tumors in healthy non-primate mammals.
- The virus can transform human cells in vitro.
- Any virus or genetic material that is a recombinant of an oncogenic animal virus and a microorganism infectious for humans is classified as a moderate oncogenic risk until its human oncogenic potential has been determined.
- Any large-scale or concentrated preparation of infectious virus or viral nucleic acid is classified as a moderate oncogenic risk until its human oncogenic potential has been determined.

10.3.2 Molds and Fungi

Although we have discussed safe practices for working with bacteria (Table 10.3), we have not so far discussed molds and fungi, which are also important. Molds and fungi are ubiquitous airborne organisms, with millions of fungal spores in an average cubic meter of unfiltered air. **Fungi** *are primitive eukaryotic microorganisms that have complex lifecycles and can form spores.* There is an incredible diversity of fungal species. **Molds** *are filamentous forms of fungi, presenting a "fuzzy" appearance when they grow* (Figure 10.14). The genus *Penicillium* is an example of a mold that produces a useful product, in this case, the antibiotic penicillin. **Yeasts** *are single-celled forms of fungi.* The yeast we are most familiar with is *Saccharomyces cerevisiae*, which is common baker's yeast. *S. cerevisiae* is often used in laboratories as a research organism and is placed in Risk Group 1.

Fungi can sometimes be dangerous to humans through several mechanisms. A **mycosis**, or **mycotic disease**, *is a disease caused by infection with a pathogenic fungus.* Some skin mycoses, such as "athlete's foot," are relatively common and do not pose a serious threat to healthy individuals. Systemic mycoses occur occasionally, especially in individuals with compromised immune systems, although the spread of systemic fungal disease between individuals is rare.

Many fungi produce **mycotoxins**, *which are toxic fungal products that may render the fungus poisonous, or contaminate the environment of the fungus.* The common mold *Aspergillus flavus*, for example,

produces aflatoxin, a highly potent carcinogen. Special precautions should be taken to avoid personal contamination when working with mycotoxins or any fungal species known to produce mycotoxins.

One of the most likely problems to arise from exposure to fungi in the laboratory is the development of allergies to the fungal spores. Fungal allergies are common and are easily triggered in the presence of significant numbers of airborne spores. Because biotechnology facilities are increasingly using fungi to produce enzymes or food products, occupational exposure becomes more likely. Proper containment and reduction of airborne spores in the immediate environment are essential measures to prevent allergies and environmental contamination.

10.3.3 Biological Materials from Humans

Biological materials from humans are a special problem for laboratory workers. Table 10.15 provides a list of potentially infectious materials that are considered by OSHA to constitute hazardous human products requiring special precautions. Sometimes commercially purchased human serum-derived antibodies, cells derived from humans, and blood factors are used in research. The supplier will provide a description of pathogen testing for the product, but the absence of all pathogens cannot be guaranteed. These products should therefore be handled with the same precautions as other human materials.

There are no test methods available that can guarantee the absence of all infectious agents in biological materials, and infectious agents that are present are likely to be targeted to humans if working with human or primate materials. OSHA therefore requires that all human blood and blood products be handled as though they were contaminated. This involves the routine use of **universal precautions**, *which are standard safety practices for handling human blood products and other human biological materials* (Table 10.16). In general, BSL-2 procedures and a Class II BSC should be used. All tissue and waste must be treated as pathogenic and be properly decontaminated.

Institutions and organizations where occupational exposure occurs must develop an exposure control plan to comply with the OSHA Bloodborne Pathogens Standard 29 CFR Part 1910.1030. **Occupational exposure** *is defined as reasonably anticipated skin, eye, mucous membrane, or other contact with blood or other potentially infectious materials that may result from the performance of an employee's duties* (U.S. Department of Labor. "Model Plans and Programs

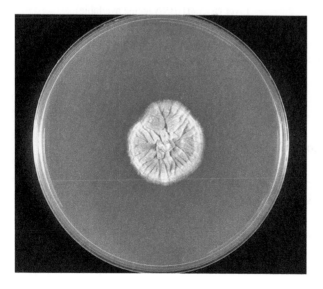

FIGURE 10.14   Example of a mold colony. Note the irregular edges and fuzzy appearance.

TABLE 10.15

Potentially Infectious Human Products

The following materials are potential sources of human bloodborne pathogens:

- Human blood and blood components
- Products made from human blood
- The following human body fluids:
  - semen
  - vaginal secretions
  - cerebrospinal fluid (fluid from the nervous system)
  - synovial fluid (fluid from the joints)
  - pleural fluid (fluid from the lungs)
  - pericardial fluid (fluid from the heart)
  - peritoneal fluid (fluid from the abdomen)
  - amniotic fluid
  - saliva in dental procedures
  - any other body fluid that is visibly contaminated with blood, such as saliva or vomitus
  - all body fluids in situations where it is difficult or impossible to differentiate between body fluids
- Any unfixed tissue or organ (other than intact skin) from a human (living or dead)
- HIV-containing cell or tissue cultures, organ cultures, and HIV- or HBV-containing culture medium or other solutions
- Blood, organs, or other tissues from experimental animals infected with HIV or HBV.

Source: Taken from the OSHA Bloodborne Pathogen Standard 29 CFR Part 1910.1030.

TABLE 10.16

Summary of Universal Precautions

This is a summary of OSHA standards describing safe handling procedures for the materials described in Table 10.15. Notice the similarities between many of these items and those incorporated into standard microbiological practices (Table 10.3).

- Universal precautions must be followed at all times when blood, other body fluids (as described in Table 10.15), and other potentially infectious materials are handled.
- Workers must wash their hands after handling potentially infectious material, whenever changing gloves, and whenever leaving the laboratory.
- Use of needles and other sharps must be avoided whenever possible; if use is necessary, needles must not be recapped, bent, or otherwise manipulated by hand; used needles and other sharps must be disposed of in nearby puncture-resistant containers.
- Gloves and other protective clothing must be worn when there is potential for contact with blood, or other potentially infectious materials; gloves must be replaced when visibly soiled or punctured; fluid-resistant clothing, surgical caps or hoods, face masks, and shoe covers should be worn if there is a potential for splashing or spraying of blood or other potentially infectious materials.
- Blood or other potentially infectious materials must be placed in leakproof, properly labeled containers during storage or transport.
- Appropriate pipetting devices must be used.
- Work surfaces must be clean and decontaminated with an appropriate disinfectant after any known contamination, after completion of planned work, and at the end of the day.
- All contaminated waste materials must be decontaminated prior to disposal, following proper institutional procedures.
- Laboratory equipment that has been in contact with blood or other potentially infected materials must be decontaminated by laboratory personnel before any servicing or shipping.
- Biohazard warning signs must be prominently placed in facilities where blood or other potentially infected materials are present.

for the OSHA Bloodborne Pathogens and Hazard Communications Standards." OSHA 3186-06R, 2003 [revised]. https://www.osha.gov/sites/default/files/publications/osha3186.pdf). In addition to specifying the use of universal precautions, the exposure control plan must also include detailed information about:

- exposure determination
- PPE and first aid materials available for employees
- provision of hepatitis B vaccination
- procedures for any post-exposure follow-up
- documentation requirements
- employee training.

Employee training should include information about the OSHA standard, explanations of bloodborne pathogens and universal precautions, use of PPE and other equipment, availability of HBV vaccination, and emergency procedures.

It is advisable to be cautious when assisting in a medical emergency that may occur in the laboratory. In an emergency where a loss of blood occurs, after assisting the victim, take time to protect yourself by proper disinfection of your skin and clothing and the contaminated areas. If appropriate, consult a healthcare professional.

10.3.4 TISSUE CULTURE

Mammalian cell culture is routinely performed in many biotechnology settings and poses particular biosafety concerns. These cell cultures generally involve growing a monolayer of cells, covered with a layer of appropriate growth-sustaining liquid medium, in glass or plastic dishes (see Chapter 30 for more details). Cultured cells have a variety of biotechnology applications, some of which are listed in Table 10.17.

Many healthy animal cells are unable to divide and grow in vitro (outside the body). Cells that do propagate in culture tend to lose some of the normal cell properties exhibited in their natural environments. However, early cell researchers sometimes found that tumor cells are easier to establish in culture. Moreover, many chemical and viral agents can transform normal animal cells so that they have the characteristics of tumor cells. In this context, **transform** *means to cause cells to become competent to grow in long-term culture and develop some properties similar to tumor cells.*

Primary cell cultures *are newly isolated cells from tissue or blood, growing outside the body for the first time.* After removal from the body's tissue, the cells are potential carriers for any infectious agents that were present in the organism. Primary cell cultures from humans may therefore contain pathogenic agents that can infect culturists. The most likely viral hazards of primary human cell lines are HIV and hepatitis viruses. Non-human cells, especially from monkeys and other primates, can also carry agents that are serious health hazards.

Cell lines, *which are cells that are established in long-term culture,* are less likely to pose an unknown health threat. Many cell lines used in laboratories have been analyzed and any potential hazards documented. Even these cells, however, may harbor hidden virus particles, which may be released upon unusual circumstances. This is illustrated in the case study "Culture Cross-Contamination Can Be Dangerous" (p. 243). There are general guidelines for the relative levels of human health risk when handling various types of animal cell lines, shown in Table 10.18. High-risk cell cultures should only be handled by trained personnel using at least a Class II biological safety cabinet (BSC). Note that all human- and primate-derived cell lines are considered high risk until they are fully characterized. Any procedures that infect cells with viruses, or have the potential to release endogenous viruses should be carried out at BSL-3.

Some established cell lines have the ability to induce tumor development when injected into laboratory animals. There have been several reported cases where laboratory workers have developed localized tumors after accidental injuries with sharps contaminated with cancer cell lines. Because one of the most common injuries among tissue culture workers is skin

---

TABLE 10.17

Uses for Cultured Cells in Biotechnology Laboratories

Cultured cells, especially from mammalian sources, have many uses in biotechnology. These include:

- Tools for biomedical research (e.g., for the study of mechanisms of cell division)
- Sources of nucleic acids for genomic studies
- Sources of species-specific proteins
- Propagation of viruses (e.g., for vaccine production)
- Production of recombinant eukaryotic proteins that cannot be expressed in bacteria (e.g., proteins with extensive post-translational modifications).

TABLE 10.18

General Risk Levels for Cultured Cell Lines

Relative Health Risk	Types of Cell Lines
Low risk	Most well-characterized human, primate, and other non-human permanent cell lines
Medium risk	Poorly characterized mammalian cell lines
High risk	Cells derived directly from human/primates
	Cell lines with endogenous pathogens (relative risk determined by the nature of the pathogen)
	Cell lines after experimental infection with a pathogen (relative risk determined by the nature of the pathogen)

punctures from broken glass or needles, it is essential to handle sharps with care.

When performing tissue culture, a sterile work surface and sterile materials are essential to avoid contaminating the cells with extraneous microorganisms. Mammalian cell culture is usually performed in a Class II BSC that protects both the cells and the culturist from exposure to contaminants. All pipettes, culture dishes, dissection instruments, and other tools must be sterilized before and after every use to protect both scientist and culture. When working with biohazardous cell cultures, it is important to keep an autoclavable pan filled with disinfectant to one side of the BSC work surface. This can be used to collect all contaminated reusable items, such as pipettes. The pan is covered before removal from the cabinet and autoclaved before reopening. Contaminated liquid culture media for disposal should be treated with a disinfectant such as bleach before discarding. As discussed in Section 10.2.4, the disinfectant must be chosen according to the organisms present.

An important concern with handling cell cultures is avoiding cross-contamination. **Cross-contamination** *is a type of contamination where cells from one culture accidentally enter another culture.* While cross-contamination of laboratory cultures usually results in loss of the cultures and invalidated research results or products, it can also create unknown hazards for lab workers who are unaware of their exposure to contaminating organisms. (See the case study "Culture Cross-Contamination Can Be Dangerous" below.) When working with multiple cell cultures, it is important to only work with one at a time. Also, separate bottles of media and reagents should always be used for different cell types or cultures, to avoid the risk of cross-contamination.

Case Study: Culture Cross-Contamination Can Be Dangerous: Two Examples

In 2002 and 2003, a major epidemic of the original severe acute respiratory syndrome (SARS) swept through much of East Asia. SARS is a life-threatening illness caused by a highly infectious and then little-known coronavirus. By June 2003, the World Health Organization documented 8,450 SARS cases in the epidemic, with 810 deaths. Because of the severity of the crisis, many laboratories were hastily converted to BSL-3 facilities to study the SARS virus and develop treatments. In one such lab in Singapore, a lab worker contracted SARS even though she was not working with the SARS virus personally. Although the worker recovered, the infection was spread to others outside the lab, with one resulting death in her family. Investigators discovered that the West Nile virus cultures the worker handled in the laboratory were cross-contaminated with SARS virus from other cultures. Conditions within the lab did not meet all safety standards for a designated BSL-3 facility, which was recommended for all work with these two organisms.

In the spring of 2020, a COVID-19 test kit sent out by the CDC was found to have a high false-positive rate. This turned out to be due to cross-contamination of the kits when they were assembled in the same facility that handled synthetic coronavirus materials, in a breach of standard procedures.

10.3.5 Recombinant and Synthetic DNA

Many of the benefits of biotechnology are a result of recombinant and synthetic DNA methods. **Recombinant DNA (rDNA)** *is a DNA molecule constructed outside of a living cell, joining DNA from more than one source, and capable of replication in a host cell.* Recombinant DNA methods are being used around the world in more than 40,000 laboratories. Since 1979, rDNA technology has led to many advances, particularly in the fields of medicine and agriculture.

When the possibilities of rDNA technology were first conceived, many scientists were concerned with the safety aspects of this type of experimentation and the possible introduction of new and unpredictable life forms. In 1975, leading scientists in the field assembled in Asilomar, California, to discuss the safety implications of rDNA methods. This was one of the few times that industry assembled to discuss the safety of a new technology before any major safety incident occurred. The consensus recommendations formulated at this meeting became the basis for NIH Guidelines for Research Involving Recombinant DNA Molecules (NIH Guidelines), which were officially established in 1976. The NIH Guidelines have substantially been revised several times afterward, most recently in 2019 to include synthetic DNA molecules. The NIH Guidelines are currently considered the general authority on rDNA safety procedures. While the NIH Guidelines do not have the force of federal regulations, any institution receiving NIH funding must comply with their recommendations. The NIH Guidelines are updated as new information becomes available.

The safety aspects of rDNA have vigorously been debated in both the popular literature and the scientific press. The safety model that is currently applied worldwide assumes that the safety of specific rDNA protocols depends on the DNA fragment used, the properties of the chosen host organism, and the properties of the vector used for delivering the DNA to the host cells. Some of the criteria used for evaluation of rDNA safety are shown in Table 10.19.

All US laboratories performing rDNA and gene editing research must register with their **Institutional Biosafety Committee** (IBC). An IBC is *an on-site committee that reviews and provides oversight for almost all experiments involving recombinant or synthetic DNA molecules,* and which then recommends the appropriate level of containment. In addition to applying NIH Guidelines, IBCs implement any local policies regarding rDNA and synthetic DNA work as well.

In recent years, the use of synthetic DNA molecules that can base-pair with naturally occurring nucleic acids has become commonplace. In addition to the short PCR primer molecules used in most biotechnology laboratories (see Chapter 27), the DNA synthesized for use in CRISPR gene editing (see Section 2.1.2.3) is increasingly common in laboratories. The potential hazards of these molecules are regulated in the same manner as recombinant nucleic acid systems.

TABLE 10.19

Some Factors Related to the Safety of Recombinant DNA Experiments

There are many factors that enter into a risk assessment for planned rDNA work. These factors center on the properties of the host organism whose DNA is modified, and the vector that carries DNA from one organism into another. The nature of the DNA sequence being transferred also matters. Examples of these concerns include:

Host Organism and/or Vector
- Pathogenicity or infectious properties of the host organism or vector
- Potential routes of human infection (aerosols, direct or indirect contact, etc.)
- Ability of an altered host cell or vector to survive outside the biotechnology facility
- Ability of an altered organism to reproduce
- Capacity of host cells to transfer genes into other, unintended organisms

Nature of the DNA Sequence
- Level of product expression
- Biological stability of the product
- Coding for:
  - Production of toxins
  - Antibiotic resistance
  - Protein products that trigger allergies
- Genetic elements that could increase invasive properties of the host.

In general, gene editing protocols must undergo a risk assessment by the appropriate IBC. The majority of gene editing protocols do not require specialized safety measures beyond standard precautions.

The NIH Guidelines for working with recombinant or synthetic DNA are similar to the regulations for work with infectious agents. The recommended way to protect personnel and the environment is with containment barriers. As risks become greater, the procedures and facilities are designed to have a greater level of containment. The NIH Guidelines classify rDNA and synthetic DNA hazards into four risk groups related to their potential pathogenicity for humans, based on the factors listed in Table 10.19. The Guidelines also designate the four levels of physical biosafety containment, BSL-1 through BSL-4. These levels are very similar in concept to those assigned to microorganisms.

The NIH Guidelines provide an "exempt" category, to designate DNA experiments that do not require containment measures beyond those normally used for the organisms involved. Section III-F-8 of the Guidelines states that exempt experiments are "those that do not present a significant risk to health or the environment" when usual containment procedures are followed. A substantial percentage of all rDNA experiments are in the exempt category. Much of this exempt work is conducted using *E. coli* strain K12, *Saccharomyces cerevisiae*, and non-spore-forming *Bacillus subtilis* host cell systems.

E. coli strain K12 is a particularly well-characterized bacterium. Despite the sensational news stories about *E. coli*-induced food poisonings, the vast majority of *E. coli* strains have been characterized as nonpathogenic. Strain K12 is a particularly safe variety of *E. coli* that is used in high school and college laboratories, because studies indicate that strain K12 cannot colonize in humans, even after deliberate inoculation.

When host cells that require stringent growth conditions and that are generally not infectious (such as *E. coli* K12) are coupled with vectors that only infect this host cell type, biological containment of rDNA is relatively complete. Of course, not every laboratory can conduct their work in these systems, so it is essential to learn about the host cell and vector system used in your laboratory, to better understand any possible biohazards involved.

10.3.6 Laboratory Animals

10.3.6.1 Humane Handling

Animals are often used in biological research and in drug-testing programs. The decision to use animals is a difficult one, because of ethical concerns relating to their welfare, and also because they are expensive, in terms of maintenance and personnel costs. However, there are situations where no other method of obtaining information is adequate. In these cases, there must be a thoughtful evaluation of health and safety issues for both animals and personnel.

There is a variety of regulations and guidelines that address both humane treatment of animals and human safety. The Animal Welfare Act, originally enacted in 1966 and amended several times, is the main federal statute governing the handling and use of animals in the United States. This act applies to all institutions performing animal research. Animal welfare regulations are published each year in *Code of Federal Regulations, Title 9, Chapter 1, Subchapter A—Animal Welfare*, known as 9 CFR. The *Public Health Service Policy on Humane Care and Use of Laboratory Animals* (PHS Policy) (found at http://grants1.nih.gov/grants/olaw/references/PHSPolicyLabAnimals.pdf) outlines the basic 9 CFR mandatory principles and goals for the care and use of vertebrate animals. This policy is regulated by the Office of Laboratory Animal Welfare (OLAW; part of the NIH) and applies to all PHS-supported agencies, research, and animal-related activities. The PHS includes the NIH, FDA, CDC, and many other federal agencies. Current guidelines describing specific practices used to achieve compliance with PHS Policy for use of animals in the laboratory under these regulations are contained in the *Guide for the Care and Use of Laboratory Animals*. (Committee for the Update of the Guide for the Care and Use of Laboratory Animals, et al. *Guide for the Care and Use of Laboratory Animals*. 8th ed., National Academies Press, 2010.) Some of the basic principles for the care and use of vertebrate animals as outlined by PHS Policy are shown in Table 10.20.

All institutions and organizations subject to the Animal Welfare Act, including biotechnology companies that perform studies using animal subjects, must develop an animal care and use program that includes (among other requirements):

- An Institutional Animal Care and Use Committee (IACUC)
- Appropriate animal housing, maintenance, and support facilities
- A veterinary care program
- Training for all personnel with animal-related duties
- An appropriate program for monitoring activities, record-keeping, and reporting.

TABLE 10.20

Principles for the Care and Use of Vertebrate Animals in the United States

These are examples of the principles that govern the 9 CFR mandates for treatment of animals related to testing, research, or training procedures. Refer to the *Guide for the Care and Use of Laboratory Animals* for further information about these principles and their implementation.

- Transportation, care, and use of animals must conform to the Animal Welfare Act and other applicable federal laws, guidelines, and policies.
- Procedures should be designed and performed with due consideration of their relevance and contribution to scientific knowledge.
- Animals should be an appropriate species and include the minimum number required for valid results. Alternatives to animal use, such as computer modeling and in vitro cell systems, should be considered.
- Animal discomfort, distress, and pain should be avoided or minimized whenever possible.
- Any procedures involving more than slight pain or distress should be performed under anesthesia or other means to minimize these factors.
- Animals in severe or chronic pain or distress that cannot be relieved should be painlessly euthanized when appropriate.
- Animals should be housed under healthy and comfortable conditions appropriate for their species. Veterinary care and consultation must be available.
- All investigators and personnel who handle animals must receive proper training.

Source: Adapted from: Office of Laboratory Animal Welfare. *Public Health Service Policy on the Humane Care and Use of Laboratory Animals.* National Institutes of Health, Revised 2015. Available at: https://grants.nih.gov/grants/olaw/references/phspolicylabanimals.pdf.

An **Institutional Animal Care and Use Committee (IACUC)** *reviews the institution's programs for humane care and use of animals, inspects institutional animal facilities, and reviews and approves all protocols using live vertebrate animals, among other duties.* An IACUC is required to have at least five members, including a scientist familiar with animal research, a veterinarian, a non-scientist such as an ethicist or lawyer, and a member external to the institution.

There are two categories of institutions that perform animal research. Category 2 institutions are evaluated internally by their IACUC, which ensures that the facilities and programs involving animals meet PHS Policy. Category 1 institutions, in addition to meeting all Category 2 criteria, are also voluntarily accredited by the **American Association for Accreditation of Laboratory Animal Care (AAALAC)**. AAALAC *is an independent peer-review organization that ensures that companies, universities, hospitals, government agencies, and other research institutions surpass minimal animal care standards.* Currently, more than 1,000 institutions in 47 countries are AAALAC-accredited.

One significant requirement of the agencies that oversee animal care is that all personnel who work with animals receive appropriate training. In addition to the considerations of humane treatment, animals must be handled in ways that minimize health threats to both the animals and handlers. Sick or mistreated

animals will not provide useful information in any study where they are used. All personnel who perform animal surgery (which must be performed under aseptic conditions) or any other experimental manipulations must receive training in the performance of these techniques.

10.3.6.2 Working Safely with Animals

Animal work introduces specific safety concerns in biotechnology facilities. Staff who come into contact with laboratory animals can be at risk for allergies, animal bites and scratches, or exposure to infectious diseases. As shown in Figure 10.15, animal research frequently uses large numbers of animals housed in concentrated areas, which exposes workers to animal-generated bioaerosols. As a consequence, approximately 20% of workers who frequently contact or work in the vicinity of rodents will develop animal-related allergies. These allergies may take months to develop and can be triggered by direct contact with the animals, or by inhalation of the allergens. It is believed that the main causes of animal-related allergies are proteins in the urine and/or saliva of the animals (rather than the fur or dander as previously thought). Worker exposure to these allergens can be reduced (but not entirely eliminated) through the use of cage-top filters that prevent the escape of airborne particles (Figure 10.15b).

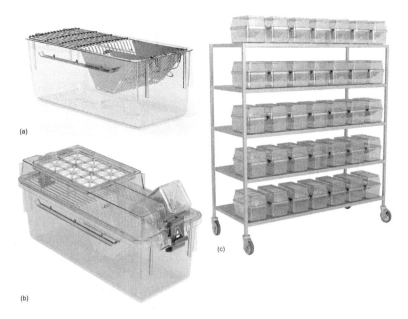

FIGURE 10.15   Rodent housing in an animal facility. (a) Cages typically have wire covers to prevent rodents from chewing through the enclosure. The number of animals per cage depends on cage size and experimental design. (b) Filter tops cover the wire cage top and minimize the spread of bioaerosols and bedding particles. (c) A rolling rack for multiple animal cages. This arrangement facilitates animal care and can reduce the risk of confusing the cages. (Photos courtesy of Allentown, Inc.)

Most individuals with animal-related allergies will experience only symptoms of rhinitis (hay fever), but a small percentage will progress to animal-related asthma, a severe and potentially debilitating condition. For this reason, animal caretakers who develop animal-related allergies generally should request a reassignment of duties to minimize animal contact, because allergies tend to worsen with continuing exposure. If reassignment is not an option, the caretaker must use full-body PPEs, including gloves and respiratory protection. Once the symptoms of an allergy occur, it is best to contact a physician with experience in animal-triggered allergies. The physician can advise you about the dangers of continuing to work with animals and any treatments that may be available.

Bites and scratches are another problem that can occur when handling animals. All animal bites must be reported to the safety authorities at your institution, who are required to record all incidents. Bite and scratch areas should be washed immediately and thoroughly. If the injury is deep or extremely painful, further medical attention should be obtained. The biggest health concerns from animal bites and scratches are localized wound infection and tetanus. The probability of infection can be minimized by proper cleansing and bandaging of the wound. Anyone bitten by a laboratory animal should receive a booster injection of tetanus vaccine unless they have a current vaccination. Always keep a record of any immunizations you have received.

Whenever handling animals, it is important to remember that they are a powerful source of aerosols and potential biological contaminants. **Zoonotic diseases (zoonoses)** *are those that can be passed between different species.* Luckily, zoonotic infections are rare in laboratories not involved in infectious disease research. Most of the laboratory animals you are likely to encounter, such as rats or mice, have been bred and raised in clean environments, and rodents carry few zoonotic disease agents that can be passed to humans. There are some exceptions, such as hantaviruses, which are of concern when handling wild-caught animals. Laboratory primates, however, may harbor organisms that can easily infect humans, and any worker who comes into contact with these animals must receive special training that will include information about possible health hazards from zoonotic diseases.

Special precautions are required whenever laboratory animals are treated with biohazardous materials or infectious agents. In these cases, the worker must consider both animal welfare and biohazard containment. Only personnel who have specific training in handling pathogenic agents and animals should be allowed in the areas where this work is conducted.

In the case of highly pathogenic agents, special animal facilities must be used, with provisions for safe handling of animals and disposal of infectious materials. In these situations, there are defined Animal Biosafety Levels for infected animals, similar to the BSLs described earlier in this chapter in Table 10.6. Animal BSLs are described in detail in *Biosafety in Microbiological and Biomedical Laboratories* (see the Bibliography in the Introduction to this unit for full citation).

The generation of bioaerosols and biohazardous waste materials by the animals is a major concern. All bedding, food, excrement, and other materials that contact the infected animals must be treated as biohazardous waste. Animal cages must be disposable or decontaminated before washing. It is essential that appropriate warning signs be placed on the doors of any containment facilities used for infected animals.

10.4 HANDLING BIOHAZARDOUS WASTE MATERIALS

Biohazardous waste *consists of biological and biologically contaminated materials that are no longer needed in the laboratory*. These materials include the following:

- Discarded cultures of bacteria and other cells.
- Outdated stocks of organisms.
- Human and animal waste, including blood and other body fluids.
- Used culture dishes and tubes.
- Biologically contaminated sharps.

All biohazardous wastes should be placed in labeled, closable, leakproof containers or bags. These waste containers should usually be placed inside secondary containers in case of accidental punctures. As with chemical wastes, every institution will have specific procedures for sorting and labeling biological wastes.

In most situations, heat sterilization by autoclaving is considered an appropriate decontamination procedure for biohazardous waste. If this method is used, the resulting materials can be discarded as regular laboratory waste (in accordance with local regulations). Figure 10.16 illustrates materials used for routine biohazard waste disposal. Table 10.21 provides guidelines for effective sterilization of biological waste using an autoclave.

FIGURE 10.16 Proper biohazard waste disposal materials. These include bags for solid waste disposal, appropriate containers for liquid waste, and biohazard label stickers for all containers. (Image courtesy of US Bio-Clean, https://usbioclean.com/.)

10.5 RESPONSE TO BIOHAZARD SPILLS

Despite precautions, spills of biological materials occur. All laboratories that work with biohazards must have an emergency plan available in case of accidents. This plan should be written and posted at convenient locations for easy access. It is essential that all personnel be familiar with these plans before they are needed. Many of the same precautions that apply to chemical spills are also appropriate for biological spills. Always consider the nature of the hazard before attempting to handle the situation on your own. If a highly pathogenic agent is involved, evacuate the premises, and summon help.

In cases where a spill is small, involves a relatively nonpathogenic organism (BSL-1 or BSL-2), and a biological spill kit is available, the worker can deal with the spill directly. Spill kits are commercially available, or they can be assembled ahead of time in the laboratory (Figure 10.17). A biological spill kit should include the following:

- personal protective equipment (e.g., gloves, eye, and face protection)
- absorbent materials (e.g., chemical absorbent or paper towels)
- cleanup tools (e.g., a whisk and dustpan)
- waste container (e.g., biohazard bags)
- disinfectant (e.g., fresh 10% bleach solution).

The bleach solution must be made fresh on at least a weekly basis.

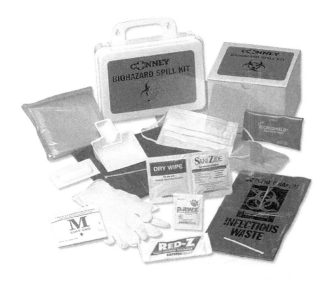

FIGURE 10.17 Biohazard emergency response kit. These kits typically contain PPE, absorbents with disinfectant, biohazard disposal bags, and instructions for general biohazard cleanup. (Reproduced with permission from Merck KGaA, Darmstadt, Germany, and/or its affiliates.)

When cleaning up a spill, remember that there are three hazards that must be addressed. The main spill area may be obvious, but there is also a surrounding splash area. This consists of small droplets of liquid that have splashed from the main spill site. It is essential that the cleanup effort does not spread contamination from these splash spots. Perhaps the most significant consideration at the time of the spill is the third element: the generation of aerosols. When the spill is large, or involves a pathogenic organism,

it is best to leave the room, close the doors, and wait at least 10 minutes for the aerosols to settle. Keep in mind that every surface where the aerosols settle is also contaminated. Table 10.22 provides a general set of guidelines for cleaning up minor biohazardous spills.

In cases of personal contamination with biohazardous material, it is essential to take immediate steps for decontamination. Remove all contaminated clothing and soak lab coats in a bleach solution to disinfect before washing. Wash any potentially contaminated skin areas vigorously for at least 10 minutes. Use an antiseptic such as Betadine if available. In case of eye contamination, use the eye wash station and flush for at least 15 minutes. Your supervisor should be informed immediately of the accident. Depending on the biohazard involved, it may be necessary to seek medical attention and notify proper authorities as designated by your institution's biohazard plan.

10.6 FINAL NOTES

Like other aspects of laboratory work, biosafety is a matter of attention to detail and adherence to quality practices. In this way you can avoid the majority of hazards inherent in working with biological samples and organisms. Appropriate use of a biological safety cabinet is an important skill for handling microorganisms, cells, and blood products safely, both for you and for the products of your labor. It is essential to familiarize yourself with the biohazards present in the laboratory and best practices for avoiding the risk of exposure to these hazards.

TABLE 10.21

Guidelines for Safe Decontamination of Biohazardous Waste by Autoclaving

- Separate chemical and biological wastes for autoclaving.
- All biohazardous waste must be placed in orange/red biohazard bags with a heat-sensitive sterilization indicator.
- Before autoclaving, biohazard waste bags should be kept closed to prevent airborne contamination and odors; while autoclaving, however, the bag must be open to allow the steam to penetrate. Be sure to wear appropriate PPE for all waste handling steps prior to autoclaving.
- Add at least 100 mL water to each biohazard bag before autoclaving for effective steam generation.
- Autoclave biohazardous materials for at least 45 minutes at the standard 121°C and 15 psi for a single bag, and at least 60 minutes for a run with multiple bags.
- After autoclaving, the bag should be closed and disposed of in an opaque regular waste bag.
- All decontamination autoclaves should be tested regularly for sterilization effectiveness (at least annually).

TABLE 10.22

General Guidelines for Handling Minor Biohazard Spills

BSL-1 Organism

- Always wear appropriate PPE, including lab coat, gloves, and eye protection.
- Soak absorbent materials such as towels in an appropriate disinfectant (usually a fresh 10% solution of bleach).
- Use these towels to soak up the spill; do not wipe the spill with dry towels.
- Dispose of the towels as biohazardous waste.
- Clean the spill area with fresh disinfectant and towels.

BSL-2 Organism

- Notify all personnel in the area and evacuate for at least 10 minutes to allow aerosols to settle.
- In the meantime, remove any contaminated clothing and decontaminate it.
- Be sure to wear appropriate PPE, including lab coat, gloves, mask, and eye protection.
- Soak absorbent materials such as towels in an appropriate disinfectant (usually a fresh 10% solution of bleach).
- Cover the main spill area with the towels or other absorbent materials that contain disinfectant.
- Carefully (to avoid creating aerosols) flood the immediately surrounding area with disinfectant.
- Cover all items in the spill area with disinfectant and then remove them from the spill area.
- Remove any broken glassware with forceps and place in a sharps container; never pick up contaminated sharps with your hands.
- Remove the contaminated towels and any other absorbent material and place in a biohazard bag.
- Gently apply additional disinfectant to the spill area and allow at least 20 minutes for decontamination.
- Remove the disinfectant by applying paper towels or other absorbent materials, which are then discarded in a biohazard bag.
- Wipe off any residual spilled material and reapply disinfectant for final cleanup.
- Collect all materials in the biohazard bag and seal for autoclaving.
- Reopen the area to general use only after spill cleanup and decontamination is complete.

Spills involving BSL-3 and BSL-4 organisms are not considered minor incidents; they require special considerations that should be part of institutional policy.

Practice Problems

1. Many germicidal agents cannot be used on the skin. Fortunately, thoroughly washing your hands with soap and water is the standard way to protect yourself from pathogens, such as the SARS-CoV-2 virus, that might have gotten onto your hands from a contaminated surface. Use an Internet browser to research how soap and detergents act to remove dirt. Then look at the diagram of the SARS-CoV-2 virus on p. 2. Explain why this virus is readily removed from your skin by washing with soap and water.
2. A laboratory scientist drops a glass flask containing pathogenic bacteria. The glass shatters and splatters the scientist with liquid.
   a. What are several precautions the scientist should have taken ahead of time to reduce the risk of this incident?
   b. What should this person do immediately?

3. HEPA filters are designed to remove 99.97% of all particulate matter above the size of 0.3 µm from air. If a cubic meter of air containing 2,880,000 particles of about 0.5 µm passes through the filter, how many particles will remain in the air?

4. What are the similarities and differences between standard microbiological practices (Table 10.3) and universal precautions (Table 10.16)?

5. Classify the following experiments as requiring precautions at BSL-1, or BSL-2, or BSL-3. (Hint: ATCC is a resource that distributes a variety of biological products, including cultured cells. Their website has information about the biosafety level of cell lines and bacterial strains.)

   a. Transforming *E. coli* strain K12 with a gene that makes the bacteria glow under a UV light.

   b. Using *Lactobacillus bulgaricus* to make yogurt.

   c. Culturing the cell line HeLa and using the cultured cells to see the effects of various nutrient medium supplements.

   d. Culturing the cell line ZFL (zebrafish liver) and using the cultured cells to see the effects of various nutrient medium supplements.

   e. Growing large amounts of SARS-CoV-2 virus to use in experiments of different drug agents.

   f. A colleague sends you DNA samples extracted from spinach. The spinach was thought to have been contaminated with pathogenic *Salmonella sp.* bacteria. Your job is to sequence the DNA.

   g. Determining the blood types of the students in a class.

6. Fill in the door sign for the rabies laboratory shown in Figure 10.18. Note that ingestion is a possible route of exposure for hazardous substances in a laboratory. Obviously, no one would intentionally ingest hazardous materials, but, if they get onto one's hands, then they can find their way into one's mouth.

7. A portion of a procedure that transfers DNA into bacterial cells is shown below. Identify all the steps during this procedure when aerosols are likely to be generated.

   Step 1. Take a vial of bacterial cells out of the −80°C freezer and thaw on ice (approximately 20–30 minutes).

   Step 2. Remove agar plates from storage at 4°C and let them warm up to room temperature.

   Step 3. Mix 1–5 µL of DNA into 20–50 µL of bacterial cells in a small tube. Gently mix by flicking the bottom of the tube with your finger a few times.

   Step 4. Centrifuge for 10 seconds to bring the mixture to the bottom of the tube.

   Step 5. Incubate the bacterial cell/DNA mixture on ice for 20–30 minutes.

   Step 6. Heat shock each transformation tube by placing the bottom 1/2–2/3 of the tube into a 42°C water bath for 30–60 seconds.

   Step 7. Put the tubes back on ice for 2 minutes.

   Step 8. Add 250–1,000 µL nutrient medium (without antibiotic) to the bacteria and grow in a 37°C shaking incubator for 45 min.

   Step 9. Apply 50 µL of the suspension to an agar plate.

   Step 10. Sterilize a spreader by passing it through a flame.

   Step 11. Use the spreader to spread the liquid evenly across the agar plate.

   Step 12. Repeat Steps 9–11 with remaining liquid.

   Step 13. Incubate the plates overnight at 37°C.

8. Consider the case study in this chapter relating to pathogen research at Texas A&M University. Discuss how adherence to quality practices (such as described in Chapter 4) might have prevented the problems identified at this university.

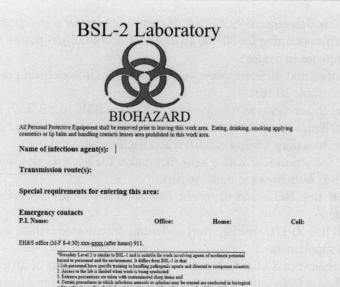

FIGURE 10.18   Biohazard door sign template (for Practice Problem 6.).

Questions for Discussion

1. One part of your laboratory responsibilities is to safely unpack and process biological samples. Demonstrate everything you would do to accomplish this.
2. While unpacking samples one morning, you notice that one of the samples is leaking from the container. What should you do?
3. After inoculation of a bioreactor with a microorganism, a co-worker points out that the bioreactor exit air filter cartridge is not installed. This means that there is no filter between the recombinant cells in the bioreactor and the outside environment. Indicate how you would handle this.

UNIT IV

Math in the Biotechnology Laboratory

An Overview

Chapters in This Unit

Mathematical descriptions and operations permeate daily life in the laboratory workplace. Math is an important tool for scientists and technicians, just as a hammer is a tool for a carpenter. To be a skillful laboratory biologist you do not need to "like" math any more than a carpenter needs to "like" a hammer.

Most carpenters are probably neither happy nor anxious when they encounter a hammer (unless they have recently smashed their finger with one). Students in the sciences, however, sometimes do have to overcome negative feelings about math. If you are anxious about math, remember that with practice and time

DOI: 10.1201/9780429282799-14

people learn to perform the math required in their profession.[1]

The purpose of this unit is to provide a brief review of common math manipulations used in bioscience laboratories and to introduce applications of math relevant to everyday laboratory work. This unit is not intended to replace a math textbook. Rather, these chapters discuss topics that relate to problems commonly encountered in the laboratory. Although these chapters review various basic math concepts, it is assumed that readers can manipulate fractions and decimals, prepare simple graphs, solve an equation with one unknown, and use a scientific calculator.

Readers who are comfortable with the mathematical operations reviewed may still want to work the "Application Problems" in each section. These problems demonstrate the ways in which familiar mathematical tools are applied in the laboratory. Readers who are less comfortable with mathematical calculations will find it beneficial to complete the "Manipulation Problems." Solving the "Manipulation Problems" will increase your comfort and efficiency when performing routine laboratory calculations.

There is often more than one strategy to solve a math problem. In some instances in this unit, we demonstrate two ways to approach a problem. For example, we show how to convert between units using both proportions and a unit canceling strategy. Both strategies, when applied correctly, will provide the correct answer. Individuals may prefer one strategy over another. (Of course, in a classroom situation students are encouraged to use strategies as directed by their instructor.) More experienced scientists and technicians often move fluidly from one problem-solving strategy to another, using whatever approach is most efficient for solving a particular problem. Experienced workers also are likely to check their answers by first using one approach to solve a problem and then testing another approach to see if it gives the same answer.

A few tips for math problem solving include the following:

1. *Keep track of all information.* Begin with a sheet of paper with plenty of room for each problem. Write down relevant information and record the results of each step.
2. *Use simple sketches, arrows, or other visual aids to help define problems.*
3. *Check that each answer makes sense.*
4. *Keep track of units.* It is tempting to ignore units when performing calculations, but this practice has been the cause of much grief. (We will talk more about units later.)
5. *State the answer clearly; remember to indicate the units.*

Chapter 11 is a brief review of basic mathematical tools used in the biotechnology laboratory, including exponents, scientific notation, logarithms, units of measurement, and equations.

Chapter 12 introduces the use of proportional relationships to solve a variety of practical laboratory calculation problems. Proportions can be used, for example, to convert from one unit of measurement to another or to determine how much solute is required to prepare a solution of a given concentration. The unit canceling method (dimensional analysis), which is commonly taught in chemistry classes, is introduced as another strategy to solve these types of problems.

Chapter 13 introduces graphical methods of data analysis, focusing in particular on linear relationships and also briefly discussing exponential relationships.

Chapter 14 introduces the basic vocabulary, concepts, and mathematical tools of descriptive statistics and how they are used to organize, display, and interpret data in the laboratory. The use of descriptive statistics in a quality-control setting is also introduced.

---

[1] Many people believe that they are not good at math and/or have negative feelings about math. Sheila Tobias wrote a book about "math anxiety" (Tobias, Sheila. *Overcoming Math Anxiety.* WW Norton and Co., 1995). Tobias points out that people in Japan and Taiwan believe that hard work leads to good performance in math. In contrast, people in the United States are apt to believe that math ability leads to good performance, that one is either born with this ability or not, and that no amount of hard work can make up for the lack of math ability. Because of these beliefs, Americans may give up on math and try less hard in math classes. In reality, the ability to use math is not a genetic gift, but rather is learned with practice.

BIBLIOGRAPHY

Adams, Dany Spencer. *Lab Math: A Handbook of Measurements, Calculations, and Other Quantitative Skills for Use at the Bench.* 2nd ed. Cold Spring Harbor Laboratory Press, 2013.

Campbell, June, and Campbell, Joe. *Laboratory Mathematics: Medical and Biological Applications.* 5th ed. Mosby Co., 1997.

Johnson, Catherine W., Timmons, Daniel L., and Hall, Pamela E. *Essential Laboratory Mathematics.* Delmar Learning, 2003.

Seidman, Lisa A. *Basic Laboratory Calculations for Biotechnology*. CRC Press, 2022. (A friendly guide and workbook with multiple solved problems.)

Stephenson, Frank. *Calculations for Molecular Biology and Biotechnology: A Guide to Mathematics in the Laboratory*. 3rd ed. Academic Press, 2016. (Excellent reference for calculations involving some more advanced techniques in molecular biology.)

Zatz, Joel L., and Teixeira, Maria Glaucia. *Pharmaceutical Calculations*. 5th ed. John Wiley and Sons, 2017.

BASIC STATISTICS REFERENCES

Ambrose, Harrison W., and Ambrose, Katharine Peckham. *A Handbook of Biological Investigation*. 7th ed. Hunter Textbooks, 2007. (This short handbook includes practical statistical information in an easy to understand format.)

Gunter, Bert. "Fundamental Issues in Experimental Design." *Quality Progress,* June 1996, pp. 105–13. (A short, well-written article that addresses some basic principles of statistical thinking.)

Haaland, Perry. *Experimental Design in Biotechnology*. Marcel Dekker, 1989.

Hampton, Raymond E., and Havel, John E. *Introductory Biological Statistics*. 2nd ed. Waveland Press, 2014.

Miller, James and Miller, Jane. *Statistics and Chemometrics for Analytical Chemistry*. 7th New ed. Coronet Books Inc., 2018. (Contains good explanations of how statistics are applied in the laboratory.)

11 Basic Math Techniques

11.1 EXPONENTS AND SCIENTIFIC NOTATION

Example Application 11.1: Exponents and Scientific Notation

a. The human genome consists of approximately 3×10^9 base pairs of DNA. If the sequence of the human genome base pairs was written down, then it would occupy 200 books of 10^3 pages each. How many base pairs would be on each page?

b. In the early days of gene sequencing, it cost \$0.50 to determine the location (sequence) of a single base pair. At this price, how much would it cost to determine the location of all 3×10^9 base pairs?

Answers on p. 264.

11.1.1 EXPONENTS

11.1.1.1 The Meaning of Exponents

We begin this chapter by discussing exponents and scientific notation. Exponents and scientific notation are routinely used in the laboratory, and so it is important to be able to manipulate them easily and quickly.

An **exponent** *is used to show that a number is to be multiplied by itself a certain number of times.* For example, 2^4 is read "two raised to the fourth power" and means that 2 is multiplied by itself 4 times:

$$2 \times 2 \times 2 \times 2 = 16$$

Similarly:

$$10^2 \text{ means} : 10 \times 10, \text{ or } 100$$

$$4^5 \text{ means}: 4 \times 4 \times 4 \times 4 \times 4 \text{ or } 1024$$

DOI: 10.1201/9780429282799-15

The number that is multiplied is called the **base** *and the power to which the base is raised is the* **exponent**. In the expression 10^3, the base is 10 and the exponent is 3.

A negative exponent indicates that the reciprocal of the number should be multiplied times itself. For example:

$$10^{-3} = \frac{1}{10} \times \frac{1}{10} \times \frac{1}{10} = \frac{1}{1000} = 0.001$$

Rules that govern the manipulation of exponents in calculations are summarized in Box 11.1.

BOX 11.1   CALCULATIONS INVOLVING EXPONENTS

1. To multiply two numbers with exponents where the numbers have the same base, add the exponents:

$$a^m \times a^n = a^{m+n}$$

$$\text{Two examples}: \quad 5^3 \times 5^6 = 5^9$$

$$10^{-3} \times 10^4 = 10^1$$

To convince yourself that this rule makes sense, consider the following example:

$$2^3 \times 2^2 = 2^5 = (2 \times 2 \times 2) \times (2 \times 2)$$

$$= 2 \text{ multiplied 5 times} = 2^5 = 32$$

2. To divide two numbers with exponents where the numbers have the same base, subtract the exponents:

$$\frac{a^m}{a^n} = a^{m-n}$$

$$\text{Two examples}: \quad 5^3/5^6 = 5^{3-6} = 5^{-3}$$

$$2^{-3}/2^{-4} = 2^{(-3)-(-4)} = 2^1 = 2$$

You can convince yourself that this rule makes sense by rewriting an example this way:

$$\frac{5^3}{5^6} = \frac{5 \times 5 \times 5}{5 \times 5 \times 5 \times 5 \times 5 \times 5} = \frac{1}{5 \times 5 \times 5} = \frac{1}{125} = 5^{-3}$$

3. To raise an exponential number to a higher power, multiply the two exponents.

$$\left(a^m\right)^n = a^{m \times n}$$

$$\text{Two examples}: \quad \left(2^3\right)^2 = 2^6$$

$$\left(10^3\right)^{-4} = 10^{-12}$$

(Continued)

To convince yourself that this rule makes sense, examine this example:

$$\left(2^3\right)^2 = 2^3 \times 2^3 = (2 \times 2 \times 2) \times (2 \times 2 \times 2) = 2^6$$

4. To multiply or divide numbers with exponents that have different bases, convert the numbers with exponents to their corresponding values without exponents. Then, multiply or divide.

$$\text{Two examples:} \quad \text{Multiply}: 3^2 \times 2^4$$

$$3^2 = 9 \quad \text{and} \quad 2^4 = 16,$$

$$\text{so } 9 \times 16 = 144$$

$$\text{Divide}: 4^{-3} / 2^3$$

$$4^{-3} = \frac{1}{4} \times \frac{1}{4} \times \frac{1}{4} = \frac{1}{64} = 0.015625$$

$$\text{and } 2^3 = 8$$

$$\text{so } \frac{0.015625}{8} \approx 0.00195$$

5. To add or subtract numbers with exponents (whether their bases are the same or not), convert the numbers with exponents to their corresponding values without exponents.

$$\text{For example:} \quad 4^3 + 2^3 = 64 + 8 = 72$$

6. By definition, any number raised to the 0 power is 1.

$$\text{For example:} \quad 85^0 = 1$$

Example Problem 11.1

Perform the operations indicated:

$$\frac{10}{43^2 + 13^3}$$

Answer

The denominator involves addition of numbers with exponents. Convert the numbers with exponents to standard notation. Then perform the calculations.

$$\frac{10}{(43)^2 + (13)^3} = \frac{10}{1849 + 2197} = \frac{10}{4046} \approx \mathbf{0.0025}$$

11.1.1.2 Exponents Where the Base Is 10

In preparation for discussing scientific notation, consider the particular case of exponents where the base is 10. Observe the following rules as illustrated in Figure 11.1:

1. *For numbers greater than 1*:
   - *the exponent represents the number of places after the number (and before the decimal point)*
   - *the exponent is positive*
   - *the larger the positive exponent, the larger the number.*
2. *For numbers less than 1*:
   - *the exponent represents the number of places to the right of the decimal point including the first nonzero digit*

- *the exponent is negative*
- *the larger the negative exponent, the smaller the number.*

People commonly use the phrase **orders of magnitude** where *one order of magnitude is* 10^1. Using this terminology, 10^2 is said to be two orders of magnitude less than 10^4. Similarly, 10^8 is three orders of magnitude greater than 10^5.

11.1.2 Scientific Notation

11.1.2.1 Expressing Numbers in Scientific Notation

Scientific notation is a tool that uses exponents to simplify handling numbers that are very big or very small. Consider the number:

$$0.0000000000000000000000602$$

In scientific notation, this lengthy number is compactly expressed as:

$$6.02 \times 10^{-23}$$

A value in scientific notation is customarily written as a number between 1 and 10 multiplied by 10 raised to a power. For example:

$$100\,(\text{Standard Notation}) = 10^2$$

$$= 1 \times 10^2\,(\text{Scientific Notation})$$

$$1 \times 10^2 = 1 \times 10 \times 10 = 100$$

$$300\,(\text{Standard Notation}) = 3 \times 10^2\,(\text{Scientific Notation})$$

$$3 \times 10^2 = 3 \times 10 \times 10 = 300$$

The number of bacterial cells in 1 liter of culture might be

$$100,000,000,000\,(\text{Standard Notation})$$

$$= 1 \times 10^{11}\,(\text{Scientific Notation})$$

A number in scientific notation has two parts. *The first part is sometimes called the* **coefficient**. The second part is 10 raised to some power, the **exponential term**. For example:

As shown in Figure 11.1, a negative exponent is used for a number less than 1. Three examples:

$$1 \times 10^{-5} = \frac{1}{10} \times \frac{1}{10} \times \frac{1}{10} \times \frac{1}{10} \times \frac{1}{10} = \frac{1}{100,000} = 0.00001$$

$$0.000135 = 1.35 \times 10^{-4}$$

A bacterial cell wall is about $0.00000001\ \text{m} = 1 \times 10^{-8}$ m thick

A procedure to convert a number from standard notation to scientific notation is shown in Box 11.2.

So far, we have shown the customary manner of writing numbers in scientific notation, that is, with the coefficient written as a number between 1 and 10. It is not necessary, however, always to write numbers in

	First Part (*Coefficient*)	**Second Part** (*Exponential Term*)
1000	$= 1$	$\times 10^3$
235	$= 2.35$	$\times 10^2$

scientific notation in this way. For example, the value 205 may be expressed as:

$$205. = 0.205 \times 10^3$$
$$205. = 2.05 \times 10^2$$
$$205. = 20.5 \times 10^1$$
$$205. = 2050 \times 10^{-1}$$
$$205. = 20500 \times 10^{-2}$$

$1,000,000 = 10^6$ = one million	(6 places before decimal point)
$100,000 = 10^5$ = one hundred thousand	(5 places before decimal point)
$10,000 = 10^4$ = ten thousand	(4 places before decimal point)
$1,000 = 10^3$ = one thousand	(3 places before decimal point)
$100 = 10^2$ = one hundred	(2 places before decimal point)
$10 = 10^1$ = ten	(1 place before decimal point)
$1 = 10^0$ = one	(no places before decimal point)
$0.1 = 10^{-1}$ = one tenth	(1 place right of the decimal point)
$0.01 = 10^{-2}$ = one hundredth	(2 places right of the decimal point)
$0.001 = 10^{-3}$ = one thousandth	(3 places right of the decimal point)
$0.0001 = 10^{-4}$ = one ten thousandth	(4 places right of the decimal point)
$0.00001 = 10^{-5}$ = one hundred thousandth	(5 places right of the decimal point)
$0.000001 = 10^{-6}$ = one millionth	(6 places right of the decimal point)

FIGURE 11.1 Using exponents where the base is 10. (When the decimal point is not written, it is assumed to be to the right of the final digit in the number.)

BOX 11.2   A PROCEDURE TO CONVERT A NUMBER FROM STANDARD NOTATION TO SCIENTIFIC NOTATION

Step 1. a. If the number in standard notation is greater than 10, then move the decimal point to the left so that there is one nonzero digit to the left of the decimal point. This gives the first part of the notation.

   b. If the number in standard notation is less than 1, then move the decimal point to the right so that there is one nonzero digit to the left of the decimal point. This gives the first part of the notation.

   c. If the number in standard notation is between 1 and 10, then scientific notation is seldom used.

Step 2. Count how many places the decimal was moved in Step 1.

Step 3. a. If the decimal was moved to the left, then the number of places it was moved gives the exponent in the second part of the notation.

   b. If the decimal point was moved to the right, then place a − sign in front of the value. This is the exponent for the second part of the notation.

EXAMPLE

Express the number 5467 in scientific notation.

Step 1. This number is greater than 10. Therefore, move the decimal point to the left so that there is only one nonzero digit to the left of the decimal point: = **5467**. *Move decimal point 3 places left* → **5.467**

Step 2. The decimal point was moved three places to the left.

Step 3. The exponent for the second part of the notation is therefore 3. This means the number in scientific notation is: 5.467×10^3.

Example Problem 11.2

Convert the number **0.000348** to scientific notation.

Answer

   Step 1. This number is less than 1. Move the decimal point to the right: **0.000348** → **3.48**.

   Step 2. The decimal point was moved four places to the right.

   Step 3. The exponent is −4, so the answer in scientific notation is: 3.48×10^{-4}.

Similarly:

$$1.00 \times 10^4 = 10.0 \times 10^3 = 100. \times 10^2$$

$$3.45 \times 10^{23} = 0.0345 \times 10^{25} = 345 \times 10^{21}$$

There are situations where it is useful to manipulate coefficients and exponents without changing the value of the numbers. This is the case in addition and subtraction of numbers expressed in scientific notation.

Example Problem 11.3

Fill in the blank so that the numbers on both sides of the = sign are equal. (For example: $2.58 \times 10^{-2} = 25.8 \times 10^{-3}$)

$$0.0055 \times 10^4 = 5500 \times 10 —$$

Answer

One way to think about this problem:

1. Convert the expression on the left to standard notation; it equals **55**.
2. The number on the right-hand side of the expression, that is, 5500, is larger than the number on the left, that is, 55. The exponent that fills in the blank, therefore, will need to be a negative number (to make 5500 smaller).
3. 5500 times 10^{-2} equals 55. The answer is therefore −2.

11.1.2.2 Calculations with Scientific Notation

Box 11.3 summarizes methods of performing calculations with scientific notation.

BOX 11.3   CALCULATIONS INVOLVING NUMBERS IN SCIENTIFIC NOTATION

1. To multiply numbers in scientific notation use two steps:
   Step 1. Multiply the coefficients together.
   Step 2. Add the exponents to which 10 is raised.
   For example:

(multiply the coefficients) (add the exponents)

$$\left(2.34 \times 10^2\right)\left(3.50 \times 10^3\right) = \ (2.34 \times 3.50) \ \times \ 10^{2+3} \ = 8.19 \times 10^5$$

2. To divide numbers in scientific notation, use two steps:
   Step 1. Divide the coefficients.
   Step 2. Subtract the exponents to which 10 is raised.
   For example:

(divide the coefficients) (subtract the exponents)

$$\frac{\left(4.5 \times 10^5\right)}{\left(2.1 \times 10^3\right)} = \ \left(\frac{4.5}{2.1}\right) \ \times \ 10^{5-3} \ \approx 2.1 \times 10^2$$

3. To add or subtract numbers in scientific notation:
   a. If the numbers being added or subtracted all have 10 raised to the same exponent, then the numbers can be simply added or subtracted as shown in these examples:

$$\left(3.0 \times 10^4\right) + \left(2.5 \times 10^4\right) = ? \quad \left(7.56 \times 10^{21}\right) - \left(6.53 \times 10^{21}\right) = ?$$

$$
\begin{array}{r}
3.0 \times 10^4 \\
+2.5 \times 10^4 \\
\hline
5.5 \times 10^4
\end{array}
\qquad
\begin{array}{r}
7.56 \times 10^{21} \\
-6.53 \times 10^{21} \\
\hline
1.03 \times 10^{21}
\end{array}
$$

   b. If the numbers being added or subtracted do not all have 10 raised to the same exponent, then there are two strategies for adding and subtracting numbers.
      Strategy 1: Convert the numbers to standard notation and then do the addition or subtraction:

For example : $\left(2.05 \times 10^2\right) - \left(9.05 \times 10^{-1}\right) = ?$

Convert both numbers to standard notation:

$$2.05 \times 10^2 = 205$$

$$9.05 \times 10^{-1} = 0.905$$

$$
\begin{array}{r}
\text{Perform the calculation : } 205 \\
- 0.905 \\
\hline
204.095
\end{array}
$$

(Continued)

Strategy 2: Rewrite the values so they all have 10 raised to the same power:

For example: $(2.05 \times 10^2) - (9.05 \times 10^{-1}) =?$

To convert both numbers to a form such that they both have 10 raised to the same power:

Step 1. Decide what the common exponent will be. It should be either 2 or −1. Suppose we choose 2.

Step 2. Convert 9.05×10^{-1} to a number in scientific notation with the exponent of 2:

$$9.05 \times 10^{-1} = 0.00905 \times 10^2.$$

Step 3. Perform the subtraction

$$
\begin{array}{r}
2.05 \quad \times 10^2 \\
-0.00905 \times 10^2 \\
\hline
2.04095 \times 10^2
\end{array}
$$

Example Problem 11.4

$$\left(3.45 \times 10^{23}\right) + \left(4.56 \times 10^{25}\right) = ?$$

Answer

The two numbers in this example would require many zeros if written in standard notation. The strategy of expressing both values in scientific notation with the same exponent, therefore, is preferred over converting both numbers to standard notation.

Step 1. Decide on a common exponent. Suppose we choose 25.

Step 2. Express both the numbers in a form with the same exponent.

$$3.45 \times 10^{23} = 0.0345 \times 10^{25}$$

Step 3. Perform the addition.

$$
\begin{array}{r}
0.0345 \times 10^{25} \\
+ 4.56 \times 10^{25} \\
\hline
4.5945 \times 10^{25}
\end{array}
$$

A calculator will hold a limited number of places and will not accept very large or very small numbers if you try to key in all the zeros. A scientific calculator, however, works easily with large and small numbers expressed in scientific notation. On my calculator, to key in the number 1×10^3, I push the following keys:

Note that I do not key in the number 10, the base in scientific notation. "exp" tells my calculator that the 10 is present. To key in the number 3×10^{-4} on my calculator, I press:

The +/− key tells the calculator I want a negative exponent. "EE" is sometimes used to indicate that 10 is being raised to a certain power. Consult your instruction manual to see how to key in a number in scientific notation.

Example Application 11.1
Answer (from p. 257)

a. The total number of pages needed to record the genome would be

$$200 \times 10^3 = 2 \times 10^5 = 200,000$$

Then, the number of base pairs on each page would be:

$$\frac{3 \times 10^9}{2 \times 10^5} = 1.5 \times 10^4 = 15,000$$

b. $3 \times 10^9 \times \$0.50 = 1.5 \times 10^9 =$ 1.5 billion dollars! (However, by 2020, companies were advertising complete genome sequencing for \$1,000, or even less.)

Manipulation Practice Problems: Exponents and Scientific Notation

1. Give the whole number or fraction that corresponds to these exponential expressions.
   a. 2^2
   b. 3^3
   c. 2^{-2}
   d. 3^{-3}
   e. 10^2
   f. 10^4
   g. 10^{-2}
   h. 10^{-4}
   i. 5^0

2. Perform the operations indicated:
   a. $2^2 \times 3^3$
   b. $(14^3)(3^6)$
   c. $5^5 - 2^3$
   d. $5^7/8^4$
   e. $(6^{-2})(3^2)$
   f. $(-0.4)^3 + 9.6^2$
   g. $a^2 \times a^3$
   h. c^3/c^{-6}
   i. $(3^4)^2$
   j. $(c^{-3})^{-5}$
   k. $\dfrac{13}{43^2 + 13^3}$
   l. $10^2/10^3$

3. Underline the larger number in each pair. Underline both if their values are equal.
   a. 5×10^{-3} cm, 500×10^{-1} cm
   b. 300×10^{-3} μL, 3000×10^{-2} μL
   c. 3.200×10^{-6} m, 3200×10^{-4} m
   d. 0.001×10^1 cm, 1×10^{-3} cm
   e. 0.008×10^{-3} L, 0.0008×10^{-4} L

4. Convert the following numbers to scientific notation.
   a. 54.0   b. 4567
   c. 0.345000   d. 10,000,000
   d. 0.009078   f. 540
   g. 0.003040   h. 200,567,987

5. Convert the following numbers to standard notation.
   a. 12.3×10^3 b. 4.56×10^4 c. 4.456×10^{-5}
   d. 2.300×10^{-3} e. 0.56×10^6 f. 0.45×10^{-2}

6. Perform the following calculations.
   a. $\dfrac{(4.725 \times 10^8)(0.0200)}{(3700)(0.770)}$

   b. $\dfrac{(1.93 \times 10^3)(4.22 \times 10^{-2})}{(8.8 \times 10^8)(6.0 \times 10^{-6})}$

   c. $(4.5 \times 10^3) + (2.7 \times 10^{-2})$
   d. $(35.6 \times 10^4) - (54.6 \times 10^6)$
   e. $(5.4 \times 10^{24}) + (3.4 \times 10^{26})$
   f. $(5.7 \times 10^{-3}) - (3.4 \times 10^{-6})$

7. Fill in the blanks so that the numbers on both sides of the = sign are equal. For example: $2.58 \times 10^{-2} = 25.8 \times 10^{-3}$.
   a. $0.0050 \times 10^{-4} = 0.050 \times 10^{—} = 0.50 \times 10^{—} = 5.0 \times 10^{—}$
   b. $15.0 \times 10^{-3} = \underline{} \times 10^{-2} = \underline{} \times 10^{-1} = \underline{} \times 10^{1}$
   c. $5.45 \times 10^{-3} = 54.5 \times 10^{—}$
   d. $100.00 \times 10^{1} = 1.0000 \times 10^{—}$
   e. $6.78 \times 10^{2} = 0.678 \times 10^{—}$
   f. $54.6 \times 10^{2} = \underline{} \times 10^{6}$
   g. $45.6 \times 10^{8} = \underline{} \times 10^{6}$
   h. $4.5 \times 10^{-3} = \underline{} \times 10^{-5}$
   i. $356.98 \times 10^{-3} = \underline{} \times 10^{1}$
   j. $0.0098 \times 10^{-2} = 0.98 \times 10^{—}$

11.2 LOGARITHMS

Example Application 11.2: Logarithms

The concentration of hydrochloric acid, HCl, secreted by the stomach after a meal is about 1.2×10^{-3} M. What is the pH of stomach acid?

Answer on p. 268

11.2.1 COMMON LOGARITHMS

Common logarithms (also called **logs** or **log₁₀**) are closely related to scientific notation. The **common log** *of a number is the power to which 10 must be raised to give that number* (Figure 11.2).

$$100 = 10^2$$

The log of 100 is 2 because 10 raised to the second power is 100.

$$\log 10^2 = 2$$

$1{,}000{,}000 = 10^6 = $ one million	$\log 10^6 = 6$
$100{,}000 = 10^5 = $ one hundred thousand	$\log 10^5 = 5$
$10{,}000 = 10^4 = $ ten thousand	$\log 10^4 = 4$
$1{,}000 = 10^3 = $ one thousand	$\log 10^3 = 3$
$100 = 10^2 = $ one hundred	$\log 10^2 = 2$
$10 = 10^1 = $ ten	$\log 10^1 = 1$
$1 = 10^0 = $ one	$\log 10^0 = 0$
$0.1 = 10^{-1} = $ one tenth	$\log 10^{-1} = -1$
$0.01 = 10^{-2} = $ one hundredth	$\log 10^{-2} = -2$
$0.001 = 10^{-3} = $ one thousandth	$\log 10^{-3} = -3$
$0.0001 = 10^{-4} = $ one ten thousandth	$\log 10^{-4} = -4$
$0.00001 = 10^{-5} = $ one hundred thousandth	$\log 10^{-5} = -5$
$0.000001 = 10^{-6} = $ one millionth	$\log 10^{-6} = -6$

FIGURE 11.2 Common logarithms of powers of ten.

$$1,000,000 = 10^6$$

The log of 1,000,000 is 6 because 10 raised to the sixth power is 1,000,000.

$$\log 10^6 = 6$$

$$0.001 = 10^{-3}$$

The log of 0.001 is −3 because 10 raised to the −3 power is 0.001.

$$\log 10^{-3} = -3$$

$$1 = 10^0$$

The log of 1 is 0 because 10 raised to the zero power is 1 (by definition).

$$\log 10^0 = 0$$

So, by definition, $\log\left(10^X\right) = X$

To what power must 10 be raised to equal the number 5? This is not obvious. We can reason that the number 5 is between 1 and 10 so the log of 5 must be greater than the log of 1 (which is 0) and less than the log of 10 (which is 1). The same is true for the numbers 2, 3, 4, 6, 7, 8, and 9; their logs are decimals between 0 and 1. There is no intuitive way to know the exact log of 5, although we know it is between 0 and 1. Rather, it is necessary to look up the log in a table of logarithms, or to find it using a scientific calculator. The log of 5 is approximately 0.699, which means that $10^{0.699} \approx 5$. Thus:

$$5 = 10^X$$

Take the log of both sides:

$$\log 5 = X$$

From a calculator we find that:

$$X \approx 0.699$$

which means that:

$$5 \approx 10^{0.699}$$

The same logic can be applied to numbers between 10 and 100. For example, what is the log of 48? The log of 10 is 1; the log of 100 is 2. Therefore, the log of the number 48 must fall between 1 and 2. It is, in fact, approximately 1.681.

$$\log 48 = X$$

From a calculator we find that:

$$X \approx 1.681$$

which means that:

$$48 \approx 10^{1.681}$$

All numbers between 10 and 100 have a log between 1 and 2. We can continue and apply the same logic to all positive numbers. For example, the log of 4,987,000 must fall between 6 and 7 because 4,987,000 is between 1 million (10^6) and 10 million (10^7).

$$\log 4,987,000 = X$$

$$X \approx 6.698$$

which means that:

$$10^{6.698} \approx 4,987,000$$

What about numbers between 0 and 1? Observe in Figure 11.2 that the log of 1 is 0 and that the logs of numbers between 0 and 1 are negative numbers. Therefore, all numbers between 0 and 1 have a negative log. For example:

The log of 0.130 is approximately −0.886, which means that: $10^{-0.886} \approx 0.130$.
The log of 0.00891 is approximately −2.05, which means that: $10^{-2.05} \approx 0.00891$.

Note that 0 and numbers less than 0 do not have logs at all because there is no exponent we can use that will make 10 to that number equal 0 or a negative. Formally, we say that logarithms of 0 or negative numbers are undefined.

11.2.2 ANTILOGARITHMS

An **antilogarithm (antilog)** is the number corresponding to a given logarithm. For example:

The log of 100 is 2; therefore, 100 is the antilog of 2.
The log of 5 is approximately 0.699; therefore, 5 is approximately the antilog of 0.699.

What is the antilog of 3? Remember that logs are exponents; therefore, to find the antilog of a number, n, use n as an exponent on a base of 10. If n = 3, then antilog $3 = 10^3 = 1000$. 1000 is the antilog of 3.

What is the antilog of 2.5?
Antilog $2.5 = 10^{2.5}$.

While it is not obvious what $10^{2.5}$ equals, we can reason that the antilog of 2.5 must be a number between 100 and 1000. (If this is not clear, think about the "2" in 2.5.) A scientific calculator is the simplest way to find antilogs. On many calculators the "antilog key" is the second function of the "log key." (Consult your calculator directions to determine how to perform this function.) Using my calculator I can find the antilog of 2.5 in two ways:

$$\boxed{2.5}$$

$$\boxed{2^{nd}}$$

$$\boxed{log}$$

The answer **316.227766** appears.

Alternatively I can press:

$$\boxed{10}$$

$$\boxed{y^x}$$

$$\boxed{2.5}$$

$$\boxed{=}$$

The answer **316.227766** appears.

Your calculator might work in a slightly different order; try it out.

Example Problem 11.5

What is the antilog of 3.58?

Answer

We can reason that the answer is between 1000 and 10,000 because of the 3.

$$\text{Antilog}(3.58) = 10^{3.58} \approx \mathbf{3801.89}$$

Example Problem 11.6

What is the antilog of −0.780?

Answer

The antilog of a negative number must be a value between 0 and 1.

$$\text{Antilog}(-0.780) = 10^{-0.780} \approx \mathbf{0.166}$$

11.2.3 NATURAL LOGARITHMS

Common logarithms have a base of 10. *There are also logarithms whose base is ≈ 2.7183, called* **e**. *When e is the base, the log is called a* **natural log**, *abbreviated* **ln**. There are tables of natural logs, and there are keys on scientific calculators to find the natural log. Note that the terms *log* and *ln* are not synonyms and should never be used interchangeably. Natural logs are less commonly used in biotechnology laboratories than logs with a base of 10.

11.2.4 AN APPLICATION OF LOGARITHMS: pH

The proper function of aqueous biological systems depends on the medium having the correct concentration of hydrogen ions. pH is a convenient means to express the concentration of hydrogen ions in a solution.

The concentration of H^+ ions in aqueous solutions normally varies between 1.0 M and 0.00000000000001 M (where M is a unit of concentration). The number 0.00000000000001 can also be written as 1×10^{-14}. Søren Sørenson devised a scale of pH units in which the wide range of hydrogen ion concentrations can be written as a number between 0 and 14. If the hydrogen ion concentration in a solution is 1×10^{-14} M, then using Sørenson's method we say its pH is 14. If the hydrogen ion concentration of a solution is 0.0000001 M, or 1×10^{-7} M, then we say its pH is 7. If the hydrogen ion concentration of a solution is 1 M, then we say its pH is 0. Thus, to find the pH of a solution:

Step 1. Take the log of the hydrogen ion concentration (expressed as molarity).
Step 2. Take the negative of the log.

For example:

What is the pH of a solution with a H^+ concentration of 0.000234 M?

Step 1. The log of 0.000234 is approximately −3.63.
Step 2. The negative of − 3.63 is 3.63. So, 3.63 is the pH of this solution.

We can express the relationship between pH and hydrogen ion concentration more formally using the following expression:

$$pH = -\log\left[H^+\right]$$

The symbol [ ] means "concentration." In words, this expression means "pH is equal to the negative log of the hydrogen ion concentration."

If we know the pH and want to find the [H⁺] in a solution, then we apply antilogs. To do this:

Step 1. Place a negative sign in front of the pH value.
Step 2. Find the antilog of the resulting number.

For example:

What is the concentration of hydrogen ions in a solution whose pH is 5.60?

Step 1. Place a – sign in front of the pH: –5.60.
Step 2. The antilog of –5.60 is ≈ 2.51×10^{-6}. The concentration of hydrogen ions in this solution, therefore, is 2.51×10^{-6} M.

The relationship between hydrogen ion concentration and pH is expressed more formally using the following expression:

$$\left[H^+\right] = antilog(-pH)$$

Example Application 11.2
Answer (from p. 265)

The HCl dissociates to release 1.2×10^{-3} M of hydrogen ions. The log of 1.2×10^{-3} is ≈ –2.9. The negative of this log value is 2.9. So the pH of the stomach after a meal is about 2.9.

Manipulation Practice Problems: Logarithms

1. Answer the following without using a calculator:
   a. Is the log of 445 closer to 2 or 4?
   b. Is the log of 1876 closer to 3, 9, or 10?

2. Give the common log of the following numbers.
   a. 100   **b.** 10,000   **c.** 1,000,000   **d.** 0.0001   **e.** 0.001

3. The log of 567 must lie between 2.0 and 3.0. For each of the following numbers, state the values the log must lie between.
   a. 7   **b.** 65.9   **c.** 89.0   **d.** 0.45   **e.** 0.0078

4. Use a scientific calculator to find the log of the following numbers.
   a. 1.50×10^4 **b.** 345 **c.** 0.0098   **d.** 2.98×10^{-5} **e.** 1209   **f.** 0.345

5. Use a scientific calculator to find the antilog of the following numbers.
   a. 4.8990   **b.** 3.9900   **c.** –0.5600   **d.** –0.0089   **e.** 9.8999   **f.** 1.0000
   g. 8.9000

6. Use a scientific calculator to convert each of the following [H⁺] to a pH value.
   a. [0.45 M]   **b.** [0.045 M]   **c.** [0.0045 M]   **d.** [0.00000032 M]

7. Use a scientific calculator to determine the hydrogen ion concentration of the solutions with the following pH values.
   a. 4.56   **b.** 5.67   **c.** 7.00   **d.** 1.09   **e.** 10.1

11.3 UNITS OF MEASUREMENT

Scientists and technicians often measure things such as the length of insects, the concentration of pollutants in an air sample, the number of viral particles in a sample, or the amount of air in a patient's lungs. **Measurement** *is basically a process of counting and comparison.* For example, consider an insect that is 1.55 cm in length. Length is the property that is measured, 1.55 is the **value** of the measurement, and centimeter (cm) is the **unit** of measurement. Another way to write this measurement is: 1.55×1 cm. This latter statement tells us that the length of the insect was determined by *comparison* with a standard of 1 cm in length and was found to be 1.55 times as long as the standard. Measuring length is therefore comparable to *counting* how many times the standard must be placed end-to-end to equal the length of the object. Another

Example Application 11.3: Units of Measurement

The bacterial toxin botulinum is extremely toxic and acts by paralyzing muscles. Exposure to a few micrograms of toxin causes death. The toxin, however, can be used to treat patients with a rare disorder, blepharo-spasm. These patients are blind because their eyes are squeezed shut. Minute doses of the botulinum toxin allow patients to open their eyes. If a patient is administered 100 pg of toxin, will the patient be able to see or will the patient die?

Answer on p. 271.

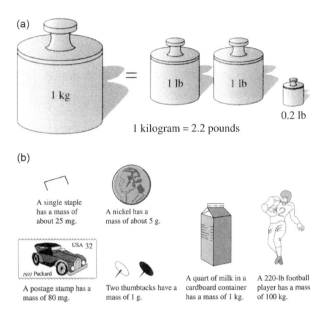

FIGURE 11.3 Units of measurement. (a) A comparison of the metric system kilogram and the USCS pound. (b) The approximate weights of everyday objects in the metric system.

example is the characteristic of time that can be measured by counting how many times the sun "rises" and "sets." The standard in this case is the apparent motion of the sun, and the unit is day.

All measurements require the selection of a **standard** and then counting the number of times the standard is contained in the material to be measured. The terms **standard** and **unit** are related, but they can be distinguished from one another. A **unit of measure** *is a precisely defined amount of a property, such as length or mass.* A **standard** *is a physical embodiment of a unit.* Centimeter and millimeter are examples of units of measure; a ruler is an example of a standard. Because units are not physical entities, they are unaffected by environmental conditions such as temperature or humidity. In contrast, standards are affected by the environment. A strip of metal used to measure meters will differ in length with temperature changes and will therefore only be correct at a particular temperature and in a particular environment.

A group of units together is a **measurement system***. The measurement system common in the United States, which includes miles, pounds, gallons, inches, and feet, is called* **the United States Customary System (USCS)***.* In most laboratories and in much of the world, the **metric system**, *with units including meters, grams, and liters, is used* (Figure 11.3).

The units in the US system are not related to one another in a systematic fashion. For example, for length, 12 in = 1 ft; 3 ft = 1 yd. There is no pattern to these relationships. In contrast, in the metric system, there is only one basic unit to measure length, the meter. The unit of a meter is modified systematically by the addition of prefixes. For example, the prefix **centi** *means*

1/100, so a **centimeter** *is a meter/100.* **Kilo** *means 1000,* so a **kilometer** *is a meter × 1000.* The same prefixes can be used to modify other basic units, such as grams or liters. The prefixes that modify a basic unit always represent an exponent of 10 (Table 11.1).

The most important basic metric units for biologists are **meter (m)**, *for length*; **gram (g)**, *for weight*; and **liter (L)**, *for volume.* Table 11.2 shows how prefixes are used to modify the basic units. Table 11.3 shows some commonly used conversion factors that relate the USCS to the metric system. In the next chapter, conversions between the USCS and the metric system are discussed.

It is always correct to convert between units that measure the same basic property, such as length. Length can be expressed in many units, including mile, meter, centimeter, and nanometer, and these units may be converted from one to the other. In general, it is not correct to convert units from one basic property to another. For example, time is a basic unit that is measured in second, minute, and so on. It is unreasonable to try to convert a number of seconds to centimeters because these units measure different characteristics.

There are a few terms used by biologists that are not part of the "regular" metric system. A microliter is 1/1,000,000 liters and is normally abbreviated μL, but biologists sometimes refer to 1 μL as 1 "lambda" (λ). 10^{-10} meters may be called an "angstrom" (Å). A micrometer is often called a "micron," although micrometer is the preferred term. The terms "cc" and "cm³" (in reference to volume) are both the same as an mL.

TABLE 11.1
Prefixes in the Metric System

Decimal	Prefix	Symbol	Power of 10
1,000,000,000,000,000,000	exa-	E	10^{18}
1,000,000,000,000,000	peta-	P	10^{15}
1,000,000,000,000	tera-	T	10^{12}
1,000,000,000	giga-	G	10^{9}
1,000,000	mega-	M	10^{6}
1,000	kilo-	K	10^{3}
100	hecto-	h	10^{2}
10	deca-	da	10^{1}
1	Basic unit, no prefix		10^{0}
0.1	deci-	d	10^{-1}
0.01	centi-	c	10^{-2}
0.001	milli-	m	10^{-3}
0.000001	micro-	μ	10^{-6}
0.000000001	nano-	n	10^{-9}
0.000000000001	pico-	p	10^{-12}
0.000000000000001	femto-	f	10^{-15}
0.000000000000000001	atto-	a	10^{-18}

The prefixes used commonly in biology laboratories are underlined.

TABLE 11.2
Common Metric Units in Biology

Multiple of Ten	Mass	Abbreviation	Volume	Abbreviation	Length	Abbreviation
Basic unit	gram	g	liter	L	meter	m
$\times 10^{3}$	Kilogram	kg	Kiloliter	kL	Kilometer	km
$\times 10^{-1}$	Decigram	dg	Deciliter	dL	Decimeter	dm
$\times 10^{-2}$	Centigram	cg	Centiliter	cL	Centimeter	cm
$\times 10^{-3}$	Milligram	mg	Milliliter	mL	Millimeter	mm
$\times 10^{-6}$	Microgram	μg	Microliter (formerly, λ[a])	μL	Micrometer (formerly, micron[a])	μm
$\times 10^{-9}$	Nanogram	ng	Nanoliter	nL	Nanometer	nm
$\times 10^{-10}$					Angstrom[a]	Å
$\times 10^{-12}$	Picogram	pg	Picoliter	pL	Picometer	pm

[a] Older term, not part of SI system.

Note: There is inconsistency in capitalization. For example, you may see "Km" or "km" and "mL" or "ml."

The **SI measurement system** (*Système International d'Unités*) *is an updated version of the metric system that is intended to standardize measurement in all countries.* The SI system was agreed upon in 1960 by a number of international organizations at the *Conférence Générale des Poids et Mesures.* The International Organization for Standardization (ISO) has also endorsed the SI system.

There are seven **basic properties** in the SI system: *length, mass, time, electrical current, thermodynamic temperature, luminous intensity,* and *amount of substance.* The units for these basic properties are shown in Table 11.4. To stay strictly within the SI system, the only units that can be used are those in Table 11.4 or those that are a combination of two or more units shown in this table. For example, the SI unit for **volume**, *the amount of space a substance occupies,* is: *length* × *length* × *length in meters, or* **meters**^{3}. The unit of liter is not part of the SI system, although liter remains the conventional metric unit for volume measurements in biology laboratories (one liter = 1 dm^3).

TABLE 11.3
Commonly Used Conversion Factors for Units of Measurement

Length

1 mm = 0.001 m = 0.039 in
1 cm = 0.01 m ≈ 0.3937 in ≈ 0.0328 ft
1 m ≈ 39.37 in ≈ 3.281 ft ≈ 1.094 yd
1 km = 1000 m ≈ 0.6214 mi

1 in = 2.540 cm (exactly)
1 ft = 12 in = 0.3048 m (exactly)
1 mi = 5280 ft ≈ 1.609 km

Mass

1 g ≈ 0.0353 oz ≈ 0.0022 lb
1 kg = 1000 g ≈ 35.27 oz ≈ 2.205 lb

1 oz ≈ 28.35 g
1 lb = 16 oz = 453.59237 g (exactly)
1 ton = 2000 lb

Volume

1 mL = 0.001 L ≈ 0.03381 fl oz
1 L ≈ 2.113 pt ≈ 1.057 qt ≈ 0.2642 gal (U.S.)

1 fl oz ≈ 0.0313 qt ≈ 0.02957 L
1 pt ≈ 0.4732 L
1 qt = 2 pt ≈ 0.9463 L
1 gal = 8 pt ≈ 3.785 L

Temperature

$°C = (°F − 32)\,0.556$
$K = °C + 273$

$°F = (1.8 × °C) + 32$
$°R ≈ °F + 460$
(R is Rankine)

USCS abbreviations: in = inch; ft = foot; yd = yard; mi = mile; lb = pound; oz = ounce; fl = fluid; pt = pint; qt = quart; gal = gallon.

TABLE 11.4
The SI System

Basic Property Measured	Unit of Measurement	Abbreviation
Length	meter	m
Mass	gram	g
Time	second	s
Electrical current	ampere	A
Temperature	kelvin	K
Luminous intensity	candela	cd
Amount of substance	mole	mol

Example Application 11.3
Answer (from p. 269)

A pg = $1 × 10^{-12}$ g and

100 pg = $1 × 10^{-10}$ g.

A few micrograms would be about $3 × 10^{-6}$ g; therefore, 100 picograms is thousands of times less than a few micrograms and this dose of botulinum toxin is likely to be safe.

$$\left(100\ \text{pg} = 0.0001\ \mu g\right)$$

Manipulation Practice Problems: Measurements

1. For each pair, underline the larger value. If two values are the same, underline both.
   a. 1 µm, 1 nm
   b. 1 cm, 1 mm
   c. 10 cm, 1000 mm
   d. 10,000 µm, 10 m
   e. 1000 g, 1 kg
   f. 100 nm, 1000 µm
   g. 100 µg, 1 mg
   h. 1 nm, 10 pm
   i. 10 nm, 1 Å
   j. 100 mg, 1 g
   k. 1000 µL, 1 mL
   l. 100,000 µm, 1 m
   m. 1 L, 1000 µL
   n. 1 m, 500 cm
   o. 1 pL, 0.001 nL
   p. 10 nm, 0.1 µm
   q. 100 cm, 0.1 m
   r. 0.0001 m, 10 µm
2. If a sample weighs 100 g, how much does it weigh in kg?
3. What is the approximate volume of the common soda can in mL?

11.4 INTRODUCTION TO THE USE OF EQUATIONS TO DESCRIBE A RELATIONSHIP

11.4.1 EQUATIONS

Much of scientific inquiry is about determining the relationship between two or more entities. For example, scientists might study how wolves affect deer populations, how size affects the rate of migration of DNA fragments in electrophoresis, or the relationship between resistance and current in an electrical circuit. Math provides various tools for describing such relationships, including equations, graphs, and statistics.

Equations *are a way to describe relationships using mathematical symbols.* An equation is about a relationship in which two or more things are equal. Letters in a scientific equation represent the items involved in the relationship. Scientific equations may seem intimidating if they use unfamiliar symbols. However, even complicated equations with unfamiliar symbols can tell you about relationships in a system. There are many equations sprinkled throughout this book, all of which convey information about relationships.

For example, consider the equation $V = I\ R$. This equation (to be discussed in Chapter 16) describes the relationship between voltage (V), current (I), and resistance (R) in an electrical circuit. Even if you have little understanding of electricity, you can learn from this equation that:

1. *Voltage, current, and resistance in an electrical circuit are related to each other in a specific way.*

2. *If resistance increases, either the voltage must also increase, or the current must decrease. Similarly, if current increases, then either voltage must increase or resistance must decrease.*

3. *If voltage goes down, then current and/or resistance must also go down.*

For example:
Suppose that initially the voltage = 12, resistance = 4, and current = 3.

(Voltage, current, and resistance have units, but we will momentarily ignore them to simplify the example.)

$$V = IR$$

$$12 = 3(4)$$

What happens if the resistance increases to 5? Perhaps voltage increases and becomes 15. Or, perhaps current decreases and becomes 2.4:

$V = I\ R$	$V = I\ R$
$15 = 3\ (5)$	$12 = (2.4)(5)$
In this case, the current did not change, so voltage must increase.	In this case, voltage remained constant.

You will encounter the $V = I\ R$ equation when performing electrophoresis, a technique that uses electricity to separate various proteins and nucleic acids from one another. During electrophoresis, the resistance of the system increases as the separation of molecules occurs. The current, therefore, goes down – unless the operator increases the voltage.

Consider another equation:

$$°F = 1.8(\text{temperature in } °C) + 32$$

where:
°F is degrees in the Fahrenheit scale
°C is degrees in the Celsius scale.

Celsius and Fahrenheit are scales used to measure temperature. The size of a degree is different in the Celsius and Fahrenheit scales.

1. The numbers 1.8 and 32 in this equation are called **constants** *because they are always present in the equation and always have the same value.* The constant 1.8 is needed in the equation because a Celsius degree is 1.8 times larger than a Fahrenheit degree. The number 32 is necessary because the two scales put the zero point in a different place. Celsius places zero degrees at the temperature of frozen water. Zero degrees Fahrenheit is the lowest temperature that can be obtained by mixing salt and ice.
2. °F and °C are called **variables** *because their value can vary.* (In the previous example, V, I, and R are variables.)
3. Given either of the variables in the equation, it is possible to calculate the other variable. Thus, if we know the temperature in either Celsius or Fahrenheit, the temperature can be converted to the other scale.

11.4.2 UNITS AND MATHEMATICAL OPERATIONS

Units can be manipulated in mathematical operations. Like numbers, units can be multiplied by themselves and can have exponents. For example, the area of a rectangle area is defined as "length times width." In equation form this is:

$$A = (L)(W)$$

where A is area, L is length, and W is width. For example, if the length of a room is 15 feet and its width is 10 feet, then its area is:

$$(L)(W) = A$$

$$(10 \text{ ft})(15 \text{ ft}) = 150 \text{ ft}^2$$

The rules for the manipulation of exponents, as shown in Box 11.1, apply to units with exponents. Exponents are added for multiplication and subtracted in division. Two examples:

$$\frac{40 \text{ cm}^3}{20 \text{ cm}^2} = \frac{(40) \text{ cm}^{(3-2)}}{(20)} = 2 \text{ cm}^1$$

$$(2.1 \text{ mL})(3.4 \text{ mL}) = (2.1 \times 3.4)\left(\text{mL}^{1+1}\right) = 7.1 \text{ mL}^2$$

When working with equations, units can cancel and can be multiplied and divided. For example:

$$2.00 \frac{\text{mg}}{\text{mL}} \times 13.0 \text{ mL} \times 4.00 = 104 \text{ mg}$$

Another example, which comes from a topic in statistics, is:

$$m = \left[\frac{(4716.20)(\text{mg})(\text{g}) - \left(\dfrac{170 \text{ mg}}{7}\right)(161.03 \text{ g})}{\left(5750 \text{ mg}^2\right) - \left(\dfrac{170^2}{7}\right)(\text{mg})^2} \right]$$

At this point, the relationship expressed by this equation is not important, but watch the units as this equation is solved algebraically:

$$m = \frac{(4716.20)(\text{mg})(\text{g}) - 3910.728571(\text{mg})(\text{g})}{5750 \text{ mg}^2 - 4128.571429 \text{ mg}^2} =$$

$$\frac{805.4714286(\text{mg})(\text{g})}{1621.428571 \text{ mg}^2} \approx 0.497 \frac{\text{g}}{\text{mg}}$$

Example Problem 11.7

A box has a length of 3.45 cm, a width of 2.98 cm, and a height of 3.00 cm. What is its volume?

Answer

$$V = (L)(W)(H)$$

$$V = 3.45 \text{ cm} \times 2.98 \text{ cm} \times 3.00 \text{ cm} =$$

$$(3.45 \times 2.98 \times 3.00)(\text{cm} \times \text{cm} \times \text{cm}) \approx \textbf{30.8 cm}^3$$

Example Problem 11.8

Perform the following additions:

    a. 23 cm + 56 cm = ?
    b. 2 cm + 4 s = ?

Answer

    a. The correct answer is **79** cm – not just 79.
    b. The correct answer is: **2** cm + 4 s. (This is because s and cm are different units and cannot be added together.)

This is a good time to emphasize that a unit of measurement is not the same thing as a variable. Consider the equation, V = IR. In this equation, there are three *variables*, voltage (V), current (I), and resistance (R). Voltage is measured in *units* of volts; volts are also abbreviated with the letter "V". Current is measured in units of amps; amps are abbreviated with the letter "A". Resistance is measured in units of ohms; ohms are abbreviated with the Greek letter omega, "Ω". Remember that a variable and a unit of measurement are quite different from one another – even in those cases (like volts) where the abbreviation is the same.

Manipulation Practice Problems: Equations

1. Explain in words what the following relationships mean:
     a. A = 2C   **b.** C = A/D   c. Y = 2 (X) + 1
2. An important equation in centrifugation is:
       $RCF = 11.18 \times r(RPM / 1000)^2$
       It is not necessary to understand the abbreviations or to be familiar with the term "RCF" to answer the following questions:
     a. What happens to RCF if RPM increases?
     b. What happens to RCF if r is decreased?
     c. There are two constants in this equation. What are they?
3. Insert numbers into each equation in Problem 1 that will satisfy the equation. For example, if the equation is A = 2 C, then 710 = 2 (355) will satisfy the equation. (So will many other answers.)
4. Solve the following equations; that is, find the value of X. Pay attention to the units, if present.
     a. $3X = 15$
     b. $X = 25(5 - 4)$
     c. $-X = 3X - 1\,mg$
     d. $5X = 3X - 5\ mL + 34\,mL$
     e. $\dfrac{X}{2} = 25(3)\ cm^2$
     f. $X = 3.0\,cm\ (2.0\,mg/mL)\ (2.0)$
     g. $X = \dfrac{25\ mg}{mL}\ (40\ mL)\dfrac{3.0\ oz}{mg}$
5. Perform the following calculations.
     a. 23.4 pounds × 34.1 pounds = ?
     b. 15.2 g/3.1 g = ?
     c. $\dfrac{25.2\ cm \times 34.5\ cm}{3.00}$
     d. $\dfrac{5\ mL\ \times\ 3\ mL \times 2\ cm}{2\ cm}$

12 Proportional Relationships

12.1 INTRODUCTION TO RATIOS AND PROPORTIONS

Example Application 12.1: Proportional Relationships

a. A transgenic animal is one that produces a protein or has a trait from another species as a result of incorporating a foreign gene(s) into its genome. In 1993, transgenic sheep were born that secrete a human protein, AAT, into their milk. AAT is valuable in the treatment of emphysema. A sheep can produce 400 L of milk each year. If a transgenic sheep secretes 15 g of AAT into each liter of milk she produces, how many grams of AAT can this transgenic sheep produce in a year?

b. If AAT is worth $110/g, what is the value of a year's production of AAT from this sheep?

c. In 2014 a drug (Ruconest®) was approved that is purified from the milk of transgenic rabbits. The transgenic rabbits can produce as much as 12 g of drug per liter of milk. One rabbit can produce up to 10 L of milk per year. How much drug can one rabbit produce in a year?

Information from:

Amato, Ivan. "A Biotech Bonanza on the Hoof?" *Science*, vol. 259, March 19, 1993, p. 1698.

Megget, Katrina. "Milking Transgenic Rabbits Gets Approval." *Outsourcing-Pharma.com*. July 15, 2007. www.outsourcing-pharma.com/Article/2007/07/16/Milking-transgenic-rabbits-gets-approval

Answers on p. 277.

A **ratio** *is the relationship between two quantities using division. For example, we might say a car "gets*

DOI: 10.1201/9780429282799-16

30 miles per gallon." This statement describes the relationship between gas consumption and miles traveled. The word "per" means "for every." The preparation of a cake might require 10 oz of chocolate. The relationship between the cake and the amount of chocolate required is a ratio. Other commonplace examples of ratios are "revolutions per minute" (RPM) and "cost per pound."

A ratio can be expressed with a numerator and denominator, like a fraction:

$$\frac{30 \text{ mi}}{1 \text{ gal}} \text{ or, it is also correct to say } \frac{1 \text{ gal}}{30 \text{ mi}}$$

$$\frac{1 \text{ cake}}{10 \text{ oz chocolate}} \text{ or } \frac{10 \text{ oz chocolate}}{1 \text{ cake}}$$

Observe that a ratio is not a type of equation because there is no = sign. Also, even though ratios have a numerator and denominator, like a fraction, they are not the same as fractions. Fractions, for example, can be added together and ratios cannot.

Example Problem 12.1

A laboratory solution contains 58.5 g of NaCl per L. Express this ratio as a fraction.

Answer

The relationship can be expressed as:

$$\frac{58.5 \text{ g}}{1 \text{ L}}$$

A *proportion is a statement that two ratios are equal.* For example, given that a single cake requires 10 oz of chocolate, 20 oz of chocolate is needed to bake two cakes. This can be expressed as a proportion equation showing that the two ratios are equal:

$$\frac{1 \text{ cake}}{10 \text{ oz chocolate}} = \frac{2 \text{ cakes}}{20 \text{ oz chocolate}}$$

Read as "1 cake is to 10 ounces of chocolate as 2 cakes are to 20 ounces of chocolate."

A proportion statement is an equation because there is an = sign.

Many situations in the laboratory require the same reasoning as this "chocolate cake" example. For example, if 10 g of glucose is required to make 1 L of a nutrient solution, how many grams of glucose are needed to make 3 L of nutrient solution? Of course, 30 g is the answer.

Even if you are not a great baker, you probably "knew" that if 1 cake requires 10 oz of chocolate, then 2 cakes require 20 oz, but suppose that 16 oz of chocolate is needed for 3 cakes and you want to make 5 cakes. It may not be obvious in this case how much chocolate is necessary. It is possible to use an equation as a helpful tool to solve this problem:

1. The ratio "there are 16 ounces of chocolate per 3 cakes" can be written as:

$$\frac{16 \text{ oz chocolate}}{3 \text{ cakes}}$$

2. The unknown, ?, is how much chocolate is required for 5 cakes.
3. If 3 cakes require 16 oz of chocolate, then how many ounces of chocolate are required for five cakes? This proportional relationship is written as:

$$\frac{16 \text{ oz chocolate}}{3 \text{ cakes}} = \frac{?}{5 \text{ cakes}}$$

(Note that the units must be in the equation. The units in both numerators are the same, and the units in both denominators are the same.)

4. To solve for the unknown, cross-multiply and divide:

Cross Multiply

$$\frac{16 \text{ oz chocolate}}{3 \text{ cakes}} \times \frac{?}{5 \text{ cakes}}$$

$$(16 \text{ oz chocolate})(5 \text{ cakes}) = (?)(3 \text{ cakes})$$

Divide

$$\frac{(16 \text{ oz Chocolate})(5 \text{ cakes})}{3 \text{ cakes}} = ?$$

The unit of cakes cancel:

$$\frac{(16 \text{ oz chocolate})(5)}{3} = ?$$

Simplifying further: $? \approx 27$ oz chocolate = how much chocolate 5 cakes require.

5. Thus:

$$\frac{16 \text{ oz chocolate}}{3 \text{ cakes}} \approx \frac{27 \text{ oz chocolate}}{5 \text{ cakes}}$$

(The fractions 16/3 and 27/5 are about equal, although there is a slight difference due to rounding.)

Some important points about proportional relationships include the following:

1. *There are many problems in the laboratory that can be easily solved using the same reasoning as the "chocolate cake" example, even though the units may be less familiar, and the numbers may be more complex.*
2. *It is necessary to keep track of units.* The units in the chocolate cake example are "ounces" (oz) (of chocolate) and "cake" (or "cakes"). In a proportion equation, the units in both denominators must be the same and the units in both numerators must be the same. For example, if one cake requires 10 oz of chocolate, how much chocolate is needed for two cakes? The **wrong** way to set this up is:

$$\frac{1 \text{ cake}}{10 \text{ oz}} = \frac{?}{2 \text{ cakes}}$$

Cross-multiply and divide: $? = 0.2$ cakes^2/oz.

Clearly, these units are absurd and the answer is wrong.

It is, however, correct to either write:

$$\frac{1 \text{ cake}}{10 \text{ oz}} = \frac{2 \text{ cakes}}{?} \quad \text{or} \quad \frac{10 \text{ oz}}{1 \text{ cake}} = \frac{?}{2 \text{ cakes}}$$

Example Problem 12.2

If there are about 100 paramecia in a 20 mL water sample, then about how many paramecia would be found in 10^3 mL of this water?

Answer

A couple of ways to think about this problem are as follows:

1. Strategy 1: Use a proportion equation as follows:

$$\frac{100 \text{ paramecia}}{20 \text{ mL}} = \frac{?}{1,000 \text{ mL}}$$

Cross-multiply and divide:

$$(100 \text{ paramecia})(1,000 \text{ mL}) = (20 \text{ mL})(?)$$

$$? = \frac{(100 \text{ paramecia})(1,000 \text{ mL})}{(20 \text{ mL})}$$

$$? = 5,000 \text{ paramecia}$$

2. Strategy 2: Use a proportion equation to calculate how many paramecia are in each milliliter:

$$\frac{100 \text{ paramecia}}{20 \text{ mL}} = \frac{?}{1 \text{ mL}}$$

$$? = 5 \text{ paramecia}$$

There are 5 paramecia in each milliliter; therefore, multiply 5 by 1,000 to calculate how many paramecia there are altogether. The answer is 5,000 paramecia.

Example Application 12.1
Answer (from p. 275)

a. This is a proportional relationship: If the sheep secretes 15 g of AAT per L of milk, how much will she secrete into 400 L? Using a proportion equation:

$$\frac{15 \text{ g}}{1.0 \text{ L}} = \frac{?}{400 \text{ L}}$$

$$? = 6,000 \text{ g}$$

This is the amount of AAT a single sheep might produce in a year.

b. This is also a proportional relationship:

$$\frac{\$110}{1.0 \text{ g}} = \frac{?}{6,000 \text{ g}}$$

$$? = \$660,000$$

This is the potential value of the AAT from one sheep.

c. If a rabbit secretes 12 g of drug per L of milk, and she can produce 10 L of milk per year, how much drug can one rabbit produce in a year?

Using a proportion equation:

$$\frac{12\ g}{1.0\ L} = \frac{?}{10\ L}$$

$$? = 120\ g$$

Manipulation Practice Problems: Proportions

1. Solve for ?.

a. $\dfrac{?}{5} = \dfrac{2}{10}$

b. $\dfrac{?}{1\ mL} = \dfrac{10\ cm}{5\ mL}$

c. $\dfrac{0.5\ mg}{10\ mL} = \dfrac{30\ mg}{?}$

d. $\dfrac{50}{?} = \dfrac{100}{100}$

e. $\dfrac{?}{30\ in^2} = \dfrac{15\ lb}{100\ in^2}$

f. ? is to 15, as 30 is to 90

g. 100 is to 10, as 50 is to ?

2. In the following problems, first state the unknown, set up the proportion equation with the units, and then solve for the unknown.

a. If it requires 50 minutes to drive to Denver, then how long does it take to drive to Denver and back?

b. If it requires 1 teaspoon of baking soda to make one loaf of bread, then how many teaspoons of baking soda are required for 33 loaves?

c. If it costs $1.50 to buy one magazine, then how much do 100 magazines cost?

d. If a recipe requires 1/4 cup of margarine for one batch, then how much margarine is required to make 5 batches?

e. If a recipe requires 1/8 teaspoon of baking powder to make one batch, then how much baking powder is required to make a half batch?

f. If one bag of snack chips contains 3 oz, then how many ounces are contained in 4.5 bags of chips?

Application Practice Problems: Proportions

The following are all proportion-type problems, like the chocolate cake problem.

1. If there are about 1×10^2 blood cells in a 1.0×10^{-2} mL sample, then about how many blood cells would be in 1.0 mL of this sample?

2. Ten milliliters of buffer is needed to fill a particular size test tube. How many milliliters are required to fill 37 of these test tubes?

3. If there are about 5×10^1 paramecia in a 20 mL water sample, then about how many paramecia would be in 10^5 mL of this water?

4. If there are about 315 insect larvae in 1×10^{-1} kg of river sediment, then about how many larvae would be in 5.0×10^4 g of this sediment?

5. A fermenter is a vat in which microorganisms are grown. The microorganisms produce a material, such as an enzyme, that has a commercial value. Suppose a certain fermenter holds 1,000 L of broth.

There are about 1×10^9 bacteria in each milliliter of the broth. About how many bacteria are present in the entire fermenter?

6. Thirty grams of NaCl is required to make 1 L of a salt solution. How much NaCl is needed to make 250 mL of this solution?

7. A particular solution requires 100 mL of ethanol per L final volume. How much ethanol is needed to make 10^5 mL of this solution?

8. A particular solution contains 50 mg/mL of enzyme. How much enzyme is needed to make 10^4 μL of solution?

9. A particular enzyme breaks down proteins by removing one amino acid at a time from the proteins. The enzyme removes 60 amino acids per minute from a protein. If a protein is initially 1,000 amino acids long and the enzyme is added to it, how long will the protein chain be after 10 minutes?

10. The components required to make 100 mL of a solution are listed as follows:

Solution Q	
Component	**Grams**
NaCl	20.0
Na azide	0.001
Mg sulfate	1.0
Tris	15.0

(Do not worry at this time about what these components are.) Prepare a table that shows how to prepare 1.5 L of Solution Q.

11. One mole of carbon has a mass of 12 g. (Although the definition of a "mole" is not important for solving this problem, you can see Chapter 22 for a brief explanation.)
    a. What is the mass of 2 moles of carbon?
    b. What is the mass of 0.5 moles of carbon?

12. One mole of table salt (NaCl) has a mass of 58.5 g.
    a. What is the mass of 1.5 moles of table salt?
    b. What is the mass of 0.75 moles of table salt?

12.2 PERCENTS

Example Application 12.2: Percents

a. There are about 3×10^9 DNA base pairs in the human genome. Human chromosome 21 is the smallest chromosome (besides the Y chromosome) and contains about 2% of the human genome. About how many base pairs comprise chromosome 21?

b. The goal of the Human Genome Project was to find the base pair sequence of the entire human genome. If chromosome 21 is sequenced, and if the sequencing techniques used are correct 99.9% of the time, then how many of the base pairs determined will be incorrect?

c. Approximately 1 out of every 700 babies born has an extra copy of chromosome 21. Such children have Down syndrome, which is associated with mental retardation and heart disease. About what percent of all children are born with Down syndrome?

Answers on p. 282.

12.2.1 BASIC MANIPULATIONS INVOLVING PERCENTS

Percents are a familiar type of ratio. The % sign symbolizes a numerator over a denominator, and the word **percent** *means "of every hundred."* For example:

10% means 10/100 or "ten out of every hundred."
0.1% means 0.1/100 or "one tenth out of every hundred."

Suppose that 300 students are surveyed to determine their favorite computer game company. Of those, 225 students prefer Brand Z. This information can be expressed as a ratio, that is:

$$\frac{225 \text{ students}}{300 \text{ students}}$$

To convert this information to a percent, it is necessary to remember that percent means "out of every hundred." The ratio 225/300 can be converted to a percent using the logic of proportions:

$$\frac{225 \text{ students}}{300 \text{ students}} = \frac{?}{100}$$

$$? = 75$$

This means "75 out of every hundred" or 75% of students preferred Brand Z.

We can generalize from this example and say that the percent of individuals with a particular characteristic is:

$$\frac{\text{number with the characteristic}}{\text{total number}} \times 100\%$$

$$= \% \text{ with characterisitic}$$

Thus, the percent of students who prefer Brand Z computer game is:

$$\frac{225 \text{ students}}{300 \text{ students}} \times 100\% = 75\%$$

Observe that the numerator and denominator must have the same units.

Rules for manipulating percents are summarized in Box 12.1.

BOX 12.1   RULES FOR MANIPULATING PERCENTS

1. *To convert a percent to its decimal equivalent, move the decimal point two places to the left.* For example:

$$10\% = 0.10$$
$$15\% = 0.15$$
$$0.1\% = 0.001$$

2. *To change a decimal to a percent, move the decimal point two places to the right and add a % sign.* For example:

$$0.10 = 10\%$$
$$0.87 = 87\%$$
$$1 = 100\%$$

3. *To change a percent to a fraction, write the percent as a fraction with a denominator of 100.* For example:

$$30\% = 30/100 = 3/10$$
$$1.5\% = 1.5/100$$
$$100\% = 100/100 = 1$$

(Continued)

4. *To change a fraction to a percent:*
   Strategy 1: Use the logic of proportions.
   For example: Change 50/75 to a percent.
   a. Set the problem up as a proportion equation

$$\frac{50}{75} = \frac{?}{100}$$

   b. Cross-multiply and divide
       $? = 66.66\ldots$ So, $50/75 \approx 66.67\%$

Strategy 2: Convert the fraction to its decimal equivalent and then move the decimal two places to the right and add a % sign:

   Change 50/75 to a percent.
   a. Change the fraction to a decimal: $50/75 \approx 0.6667$
   b. Change the decimal to a percent: $0.6667 \approx 66.67\%$

5. *To find a percentage of a particular number convert the percent to its decimal equivalent and multiply the decimal times the number of interest.* For example:
   To find 45% of 900,
   a. Convert the percent to a decimal: $45\% = 0.45$
   b. Multiply the decimal times the number of interest: $0.45 \times 900 = 405$

Example Problem 12.3

Show two strategies to convert the fraction 34/67 into a percent.

Answer

Strategy 1: Use proportions.

$$\frac{34}{67} = \frac{?}{100}$$

Cross-multiply and divide

$$? \approx \mathbf{50.7}$$

Therefore, the answer is 50.7%.
   Strategy 2: Convert the fraction to a decimal and then move the decimal point two places to the right and add a percent sign:

$$\frac{34}{67} \approx 0.507$$

$$\rightarrow \mathbf{50.7\%}$$

12.2.2 An Application of Percents: Laboratory Solutions

Percents have many applications in the laboratory, one of which is in expressing the components of a solution. We address "percent solutions" in detail in Chapter 22. In this section, we will introduce some simple example problems relating to laboratory solutions.

Example Problem 12.4

A laboratory solution is composed of water and ethylene glycol. How could you prepare 500 mL of a 30% ethylene glycol solution?

Answer

Strategy 1: We can use the logic of proportions to calculate how much ethylene glycol is required for 500 mL:

$$\frac{30}{100} = \frac{?}{500 \text{ mL}}$$

Cross-multiply and divide.

$$? = 150 \, mL$$

This is how much ethylene glycol is needed.

The remainder of the solution will be water. The best way to prepare this solution is to place the 150 mL of ethylene glycol in a piece of glassware that is marked at 500 mL and then to fill it to the 500 mL mark with water.

Strategy 2: Since the total volume of the solution will be 500 mL and 30% of it will be ethylene glycol:

$$500 \, mL \times 0.30 = 150 \, mL$$

This is how much ethylene glycol is needed.

The remainder of the solution will be water. The best way to prepare this solution is to place the 150 mL of ethylene glycol in a piece of glassware that is marked at 500 mL and then to fill it to the 500 mL mark with water.

Example Application 12.2
Answers (from p. 279)

a. 2% = 0.02

$3 \times 10^9 \times 0.02 = 6 \times 10^7$ = the number of base pairs in chromosome 21.

b. If 99.9% of the base pairs are correct, then 0.1% will be incorrect.

0.1% = 0.001

$6 \times 10^7 \times 0.001 = 6 \times 10^4$, which is the number of base pairs that will be incorrect.

c. 1 out of 700 equals 1/700 ≈ 0.0014 = 0.14% = percent of children born with Down syndrome.

Manipulation Practice Problems: Percents

1. Without using a calculator, give the approximate percent of each of the following. For example, 35 out of 354 is about 10% because it is close to 35 out of 350 which is close to 10/100 which is 10%.
   a. 98 out of 100
   b. 110 out of 10,004
   c. 3 out of 15
   d. 45 out of 45,002
2. Express the following percents as fractions and as decimals.
   a. 34%
   b. 89.5%
   c. 100%
   d. 250%
   e. 0.45%
   f. 0.001%
3. Calculate the following.
   a. 15% of 450
   b. 25% of 700
   c. 0.01% of 1,000
   d. 10% of 100
   e. 12% of 500
   f. 150% of 1,000
4. Express the following fractions or decimals as percents.
   a. 15/45
   b. 2/2
   c. 10/100

    d. 1/100
    e. 1/1,000
    f. 6/40
    g. 0.1/0.5
    h. 0.003/89
    i. 5/10
    j. 0.05
    k. 0.0034
    l. 0.25
    m. 0.01
    n. 0.10
    o. 0.0001
    p. 0.0078
    q. 0.50

5. Calculate the following without using a calculator.
    a. 20% of 100
    b. 20% of 1,000
    c. 20% of 10,000
    d. 10% of 567
    e. 15% of 1,000
    f. 50% of 950
    g. 5% of 100
    h. 1% of 876
    i. 30% of 900

6. State the following as percents:
    a. 1 part in 100 total parts
    b. 3 parts in 50 total parts
    c. 15 parts in (15 parts+45 parts)
    d. 0.05 parts in 1 part total
    e. 1 part in 25 parts total
    f. 2.35 parts in (2.35 parts+6.50 parts)

Application Practice Problems: Percents

1. There are 55 students in a class: 25 of them work 20 hours a week or more at jobs outside school; 14 of them work 10–19 hours per week at jobs. The rest work 0–10 hours per week at outside jobs. What percent of the students work 20 hours or more per week at jobs? What percent work 19 hours or less at outside jobs?
2. How much ethanol is present in 100 mL of a 50% solution of ethanol?
3. A laboratory solution is made of water and ethylene glycol. How could you prepare 250 mL of a 30% ethylene glycol solution?
4. A laboratory solution is required that is 10% acetonitrile, 25% methanol, and the rest water. How could you prepare this solution? (Note that the total volume is unspecified, so you can prepare any volume you want. Also, these components come as liquids.)
5. A particular laboratory solution contains 5 mL of propanol per 100 mL solution. Express this as a percent propanol solution.
6. A solution contains 15 mL of ethanol per 700 mL total volume. Express this as a percent ethanol solution.
7. A solution contains 10 μL of methanol in 1 mL of water. Express this as a percent methanol solution.
8. A solution contains 15 mL of acetone/L. Express this as a percent solution.

9. Suppose there is a population of insect in which 95% die over the winter, but the rest survive. Each surviving insect produces 100 offspring in the spring, and then dies. Assuming there is no mortality during the summer and the population has 1,000 insects at the beginning of the first winter, how many insects will there be after two winters?

10. Osteoporosis is a disease that leads to brittle bones and is partially caused by a lack of calcium. Each year, 250,000 people suffer osteoporosis-related hip fractures, about 15% of whom die after sustaining such fractures. How many people die each year after fracturing their hip?

11. Double-stranded DNA consists of two long strands. An adenine on one strand is always paired with a thymine on the opposite strand, and a guanine on one strand is always paired with a cytosine on the opposite strand. If a purified sample of DNA contains 24% thymine, what are the percentages of the other bases?

12. Olestra is a fat substitute for food products that has been associated with gastrointestinal problems. (Olestra has therefore been largely phased out, but is still present in a few products under the name Olean.) A study was performed in which 1,000 subjects were given a free soda and a movie pass. Half the subjects received an unlabeled bag of potato chips made with olestra; half received chips without olestra. The incidence of intestinal problems was monitored for 10 days. Of the group fed regular potato chips 17.6% experienced intestinal problems versus 15.8% of the group given olestra potato chips. How many individuals in each group experienced gastrointestinal problems?

Source: Data from "Olestra at the Movies," *Harvard Heart Letter*, vol. 8, no. 9, 1998, pp. 7–8.

12.3 DENSITY

Density, d, *is the ratio between the mass and volume of a material.* Thus:

$$density = \frac{mass}{volume}$$

For example, the density of benzene is

$$\frac{0.880\ g}{mL}$$

Various materials have different densities. For example, balsa wood is less dense than lead (Figure 12.1). A lead brick, therefore, weighs more than a piece of balsa wood that occupies the same volume.

Density can be expressed in various units, so it is important to record the units. In addition, the density of a material changes with temperature: Most materials expand when heated and contract when cooled. It is therefore conventional to report the density of a material at a particular temperature. For example:

$$benzene \quad d^{20°} = 0.880\,g\,/\,mL$$

This means that the density of benzene is 0.880 g/mL at 20°C.

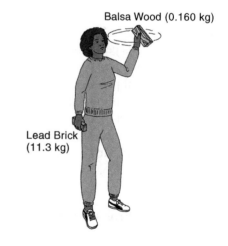

Balsa Wood (0.160 kg)

Lead Brick (11.3 kg)

FIGURE 12.1 Density. A piece of balsa wood is lighter than a lead brick of the same volume.

The densities of solids and liquids are often compared with the density of water. If the density of a material is less than water, then that material will float on water. If the material is denser than water, then it will sink. Similarly, the densities of gases are often compared with that of air. Materials that are less dense than air (such as balloons filled with helium) rise; materials that are denser than air do not rise.

Consider different liquids that do not dissolve in one another. If these liquids are mixed in a test tube, then they will separate according to their densities, with the densest liquid on the bottom (Figure 12.2).

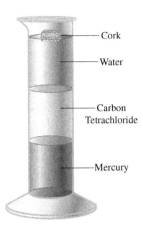
— Cork

— Water

— Carbon Tetrachloride

— Mercury

FIGURE 12.2 Materials of different densities. The density (at 20°C) of mercury is 13.5 g/mL, of carbon tetrachloride is 1.59 g/mL, and of water is 0.998 g/mL. Because mercury is the densest, it is found on the bottom of the cylinder. Cork is the least dense and so floats on water.

Example Problem 12.5

The density of water is:

$$d^{4°} = 1.000\,\text{g}\,/\,\text{mL}$$

$$d^{0°} = 0.917\,\text{g}\,/\,\text{mL}$$

a. What is the volume occupied by 5.6 g of water at 4 °C?
b. Why does ice float?

Answer

a. This can easily be solved intuitively. The answer is simply 5.6 mL. This can also be solved using a proportion equation:

$$\frac{1.000\ \text{g}}{1\ \text{mL}} = \frac{5.6\ \text{g}}{?}$$

Cross-multiply and divide:

$$? = \mathbf{5.6\,mL}$$

b. Water is less dense at 0°C than it is at 4°C; therefore, ice (which is 0 °C) floats on more dense, unfrozen water.

Example Problem 12.6

What is the mass of 25.0 mL of benzene at 20°C? (Density = 0.880 g/mL.)

Answer

Strategy 1: Use a proportion equation:

$$\frac{0.880\ \text{g}}{1\ \text{mL}} = \frac{?}{25.0\ \text{g}}$$

$$? = \mathbf{22.0\ g}$$

Strategy 2: Since we know that 1 mL has a mass of 0.880 g, we can multiply by 25.0 to calculate the mass of 25.0 mL of benzene:

$$\frac{0.880\ \text{g}}{1\ \text{mL}} \times 25.0\ \text{mL} = \mathbf{22.0\ g}$$

Manipulation Practice Problems: Density

1. At 25°C, the density of olive oil is 0.92 g/mL and water is 0.997 g/mL. Vinegar has a density similar to that of water. Which is the top layer in vinegar and oil dressing?
2. The mass of a gold bar that is 3.00 cm³ is 57.9 g. What is the density of gold?
3. The density of glycerol at 20°C is 1.26 g/mL. What is the volume of 20.0 g of glycerol?

12.4 UNIT CONVERSIONS

Example Application 12.3: Unit Conversions

a. Phthalates are compounds that make plastics flexible and are used in many products, including food wrap. Studies suggest that phthalates may activate receptors for estrogen, the primary female sex hormone. There is speculation that exposure to such estrogenic compounds may increase breast cancer incidence in women, reduce fertility in men, and adversely affect wildlife. One study suggested that margarine may pick up as much as 45 mg/kg of phthalates from plastic wrap. If you use a half pound of margarine that has absorbed 45 mg/kg of phthalates to bake three dozen cookies, how much phthalates will three cookies contain?

b. The pesticide DDT also has estrogenic effects. Although DDT is no longer used in the United States, it is still widely used worldwide. Humans have been shown to accumulate as much as 4 μg/g of body weight of DDT in their tissues. If a woman weighs 148 pounds, how much DDT might her body contain? (Let us simplify the situation by assuming that all tissues in her body accumulate the same maximum amount of DDT. This is probably not correct.)

Information from:

"Newest Estrogen Mimics the Commonest?" *Science News*, vol. 148, p. 47, July 15, 1995.

Raloff, Janet. "Beyond Estrogens: Why Unmasking Hormone-Mimicking Pollutants Proves So Challenging," *Science News*, vol. 148, July 15 1995, pp. 44–46.

Guillette, L. J., Jr. "Endocrine-Disrupting Environmental Contaminant and Reproduction: Lessons from the Study of Wildlife." *Women's Health Today: Perspectives on Current Research and Clinical Practice*, 1994, pp. 201–207.

For a more recent review, see: Paumgartten, Francisco José Roma. "Commentary: 'Estrogenic and Anti-Androgenic Endocrine Disrupting Chemicals and Their Impact on the Male Reproductive System'." *Frontiers in Public Health*, vol. 3, 2015. doi:10.3389/fpubh.2015.00165.

Answers on p. 288.

12.4.1 OVERVIEW

A given quantity, amount, or length can be measured in different units. For example, a candy bar that costs $1.00 also costs 100 cents. A dollar is a larger unit of currency than a cent; therefore, it takes many cents (100) to equal the value of a single dollar. A snake that is 1 foot long is also 12 in or 30.48 cm long. A foot is a larger unit of length than a centimeter. It takes more than 30 cm to equal the length of a single foot.

It is often necessary to convert numbers that have units in the United States Customary System (USCS) to numbers with metric units (e.g., from pounds to kilograms). It is also often necessary to convert from one USCS unit to another USCS unit (e.g., from feet to inches) or from one metric unit to another metric unit (e.g., from centimeters to meters). This section discusses such conversions.

There are two common strategies for performing unit conversions. The first approach is to think of conversion problems in terms of proportions and to use a proportion equation to solve them. The second approach is the use of conversion factors (also called the "unit canceling method" or "dimensional analysis"). Some people prefer one strategy to perform conversions; other people prefer the other. Any strategy is correct as long as it consistently yields the right answers.

12.4.2 PROPORTION METHOD OF UNIT CONVERSION

Let us illustrate the proportion approach with an example of a conversion from the USCS to metric units:

If a student weighs 150 pounds, how much does he weigh in kilograms?

$$(1 \, kg \approx 2.2 \, lb)$$

Using proportions, if 1 kg = 2.2 lb, then how many kilograms are 150 lb?

$$\frac{2.2 \, lb}{1 \, kg} = \frac{150 \, lb}{?}$$

$? \approx 68.2 \, kg =$ the student's weight in kilograms

It is a good idea to examine the answer to see if it makes sense. A kilogram is a larger unit than a pound (just as a dollar is a larger unit than a cent). When the student's weight is converted from pounds to kilograms, therefore, the number should be lower – this answer makes sense.

Another example:

A student is 6.00 ft tall. How tall is she in centimeters? (1 in = 2.54 cm; 1 ft = 12 in.)

This is a proportion problem that, with the information given, needs to be solved in two steps:

Step 1. Convert the height in feet to inches.

$$\frac{1 \text{ ft}}{12 \text{ in}} = \frac{6.00 \text{ft}}{?}$$

$$? = 72.0 \text{ in} = \text{her height in inches}$$

Step 2. Convert the height in inches to height in centimeters.

$$\frac{1 \text{ in}}{2.54 \text{ cm}} = \frac{72.0 \text{ in}}{?}$$

$$? = 183 \text{ cm} = \text{her height in centimeters}$$

Again, it is a good idea to examine the answer to see if it makes sense. A centimeter is a smaller unit than a foot; therefore, when the student's height is converted from feet to centimeters, the number will have to be larger.

Proportions can also be used to convert from one metric unit to another:

How many meters are 345 cm?

There are 100 cm in a meter. So:

$$\frac{1 \text{ m}}{100 \text{ cm}} = \frac{?}{345 \text{ cm}}$$

$$? = 3.45 \text{ m}$$

This answer makes sense. A meter is a larger unit than a centimeter; therefore, when centimeters are converted to meters, the value is a smaller number.

Another example:

Convert 105 cm to nm.
($1 \text{ m} = 10^2 \text{cm}$, and $1 \text{ m} = 10^9 \text{nm}$.)
This problem can be solved using two steps:

$$\text{Step 1.} \quad \frac{1 \text{m}}{10^2 \text{ cm}} = \frac{?}{105 \text{cm}}$$

$$? = 1.05 \text{m}$$

$$\text{Step 2.} \quad \frac{1 \text{ m}}{10^9 \text{nm}} = \frac{1.05 \text{ m}}{?}$$

$$? = 1.05 \times 10^9 \text{nm}$$

This answer is reasonable. A nanometer is a very small unit of length; therefore, it will take many nanometers to equal the length of 105 cm. The large answer obtained, $1.05 \times 10^9 \text{nm}$, makes sense.

12.4.3 CONVERSION FACTOR METHOD OF UNIT CONVERSION (UNIT CANCELING METHOD)

A second strategy for doing conversion problems is to multiply the number to be converted times the proper conversion factor. For example:

Convert 2.80 kg to pounds.

A pound is 0.454 kg. The conversion factor is expressed as a ratio:

$$\frac{1 \text{ lb}}{4.454 \text{ kg}}$$

Observe that this ratio equals 1. All conversion factors equal 1.

Multiply 2.80 kg by the conversion factor:

$$2.80 \text{ kg} \times \frac{1 \text{ lb}}{0.454 \text{ kg}} = 6.17 \text{ lb}$$

The units of kilograms cancel, and the answer comes out with the correct unit, pounds.

As we demonstrated earlier with proportions, it is good practice to examine the answer to see that it makes sense.

Suppose you had another reference that told you not that 1 lb is 0.454 kg, but rather that 1 kg is 2.205 pounds. If you use this factor directly, you will get the wrong answer:

$$2.80 \text{ kg} \times \frac{1 \text{ kg}}{2.205 \text{ lb}} \approx 1.27 \text{ kg}^2/\text{lb}$$

You can tell that this is the wrong answer because the units are wrong. If you "flip over" the conversion factor, however, the kilogram units cancel, resulting in the correct unit at the end:

$$2.80 \text{ kg} \times \frac{2.205 \text{ lb}}{1 \text{ kg}} \approx 6.17 \text{ lb}$$

When using conversion factors, the units guide you in setting up equations. The units must cancel so that the result has the correct units. The term **unit canceling method** is thus a good description for this strategy.

For example:

A student is 6.00 ft tall. How tall is she in centimeters? The conversion factors are:

$$\frac{1\,\text{in}}{2.54\,\text{cm}} \quad \text{and} \quad \frac{1\,\text{ft}}{12\,\text{in}}$$

The inches will have to cancel, the feet will have to cancel, and centimeters must remain. Let us try a couple of ways of setting this up:

$$6.00\,\text{ft} \times \frac{1\,\text{ft}}{12\,\text{in}} \times \frac{1\,\text{in}}{2.54\,\text{cm}} = ?$$

The units of feet do not cancel if the equation is set up this way, and cm is in the denominator instead of the numerator; however, if the equation is set up:

$$6.00\,\text{ft} \times \frac{12\,\text{in}}{1\,\text{ft}} \times \frac{2.54\,\text{cm}}{1\,\text{in}} = ?$$

the units of inches and feet cancel leaving the answer with the unit of centimeters. The answer is 183 cm, which is the same as the answer obtained previously using proportions. Remember to examine the answer to see that it makes sense.

With the unit cancellation method, you can string together as many conversion factors as you want into one long equation. If the equation is correct, then the units will cancel, leaving the answer in the desired unit.

Consider the conversion of metric units from one another. For example:

How many meters are 345 cm?
The conversion factor is:

$$\frac{1\,\text{m}}{100\,\text{cm}}$$

$$\text{Multiply } 345\,\text{cm} \times \frac{1\,\text{m}}{100\,\text{cm}} = 3.45\,\text{m}$$

The units of cm cancel, leaving the answer in the correct unit.

The answer makes sense.
Another example:

Convert 105 cm to nm.
There are 10^9 nm in 1 m, and there are 10^2 cm in 1 m. These are the conversion factors. Then:

$$105\,\text{cm} \times \frac{1\,\text{m}}{10^2\,\text{cm}} \times \frac{10^9\,\text{nm}}{1\,\text{m}} = ?$$

The centimeters and meters cancel, leaving the answer in nanometers:

$$105\,\text{cm} = 1.05 \times 10^9\,\text{nm}.$$

This answer makes sense.

Both the proportion and the conversion factor strategies are effective ways to convert numbers from one unit to another. Both strategies require paying attention to the units. Observe that when using proportions to do conversion problems, there is always an equals sign (=) between two ratios. In addition, when using proportions, the units must be the same in both the denominators and the same in both the numerators. In contrast, with the conversion factor (unit canceling) method, there is not an equals sign between two ratios. There is a multiplication sign (×) between the number to be converted and the conversion factor(s).

Example Application 12.3
Answers (from p. 286)

a. Let us solve the first part of the question using the proportion method. First, convert 0.50 lb of margarine to kg:

$$\frac{1\,\text{lb}}{0.454\,\text{kg}} = \frac{0.50\,\text{lb}}{?}$$

$? \approx 0.227$ kg = weight of margarine in kilogram units.

Margarine can absorb 45 mg of phthalates per kilogram, so 0.227 kg of margarine can take up:

$$\frac{45\,\text{mg}}{1\,\text{kg}} = \frac{?}{0.227\,\text{kg}}$$

? ≈ 10.22 mg = milligrams of phthalates that 0.227 kg of margarine can absorb.

There is therefore potentially 10.22 mg of phthalates in 36 cookies. To calculate the milligrams of phthalates in three cookies:

$$\frac{10.22 \text{ mg}}{36 \text{ cookies}} = \frac{?}{3 \text{ cookies}}$$

? ≈ 0.85 mg = mg of phthalates in three cookies.

b. Let us use the conversion factor method to solve part b of the problem. Conversion factors can be strung together into one long equation. First, it is necessary to convert the woman's weight from pounds to grams. A factor must then be included to account for the fact that the woman accumulated 4 µg/g of DDT in all her tissues. Finally, it would be helpful to covert the answer from micrograms to grams. The resulting single equation is:

$$148 \text{ } \cancel{\text{lb}} \times \frac{454 \text{ } \cancel{g}}{1 \text{ } \cancel{\text{lb}}} \times \frac{4 \text{ } \cancel{\mu g}}{\cancel{g}} \times \frac{1 \text{ g}}{10^6 \text{ } \cancel{\mu g}}$$

$$\approx 0.269 \text{ g} = \text{accumulated DDT}$$

Manipulation Practice Problems: Unit Conversions

(Use either the proportion method or the conversion factor (unit canceling) method to solve these problems. Conversion factors are shown in Table 11.3.)

1. Convert:
   a. 3.00 ft to cm
   b. 100 mg to g
   c. 12.0 in to miles (use scientific notation for your answer)
   d. 100 in to km
   e. 10.0555 pounds to ounces
   f. 18.989 pounds to g
   g. 13 miles to km
   h. 150 mL to L
   i. 56.7009 cm to nm
   j. 500 nm to µm
   k. 10.0 nm to in
2. How far is a 10 km race in miles?
3. A marathon is 26.2 miles. Express this in kilometers.
4. How tall is a person who is 5 ft 4 inches in m?
5. In kilometers, how far is a town that is 45 miles away?
6. A car is going 55 mph. How fast is it moving in kilometers per hour?
7. How much does a 3.0-ton elephant weigh in the metric system?
8. Which jar of hot fudge is the best value?
   a. $2.50 for 12 oz
   b. $3.67 for 250 g
   c. $4.50 for 0.300 kg
   d. $2.35 for 0.75 lb
   <u>Relationships</u>
   52 weeks = 1 year
   1 week = 7 days
   1 day = 24 hours
   1 hour = 60 minutes
   1 minute = 60 seconds

Fill in the blanks:

9. 1 week = ___ days = ___ hours = ___ minutes = ___ seconds
10. ___ year = 1 week = ___ hours = ___ minutes = ___ seconds
11. 1 year = ___ weeks = ___ days = ___ minutes = ___ seconds

Relationships
1,760 yds = 1 mi
1 yd = 3 ft
1 ft = 12 in

12. ___ mi = ___ yds = 1 ft = ___ in

Relationships
1 gal = 4 qts
1 qt = 2 pts
1 pt = 16 oz

13. ___ gal = ___ pt = 2 oz

Relationships
1 km = 1,000 m = 10^3 m
1 m = 100 cm = 10^2 cm
1 cm = 10 mm
1 mm = 1,000 μm = 10^3 μm
1 μm = 1,000 nm = 10^3 nm
2.5 cm ≈ 1 in

14. 1 km = ___ m = ___ cm = ___ mm = ___ μm = ___ nm
15. ___ km = ___ m = ___ cm = ___ mm = ___ μm = 1 nm
16. ___ km = ___ m = 2.5 cm = ___ mm = ___ μm = ___ nm
17. a. 6.25 mm = ___ μm
    b. 0.00896 m = ___ mm
    c. 9876000 nm = ___ mm
18. ___ km = 3.0 m = ___ cm = ___ in

Relationships
1 g = 10^{-3} kg
1 mg = 10^{-3} g
1 μg = 10^{-3} mg

19. 5 kg = ___ g = ___ mg = ___ μg
20. ___ kg = 0.0089 g = ___ mg = ___ μg
21. ___ kg = ___ g = ___ mg = 2×10^{-8} μg
22. a. 0.8657 g = ___ mg
    b. 526 kg = ___ mg
    c. 63 g = ___ μg
    d. 2.63×10^{-6} μg = ___ kg

Relationships
(These units relate to the decay of radioactivity.) (Ci is a unit called "Curie.")
1 Ci = 3.7×10^{10} dps (disintegrations per second)
1 Ci = 1,000 mCi = 10^3 mCi
1 mCi = 1,000 μCi = 10^3 μCi
1 Bq (becquerel) = 1 dps

23. 1 Ci = ___ dps = ___ dpm (disintegrations per minute)
24. 1 Ci = ___ mCi = ___ μCi = ___ dps
25. ___ Ci = ___ mCi = ___ μCi = 10^5 dps
26. ___ Ci = ___ mCi = 100 μCi = ___ dps = ___ dpm
27. 1 Ci = ___ mCi = ___ μCi = ___ dps = ___ Bq
28. ___ Ci = ___ mCi = 250 μCi = ___ dps = ___ Bq

Application Practice Problems: Unit Conversions

(Use either the proportion method or the conversion factor method to solve these problems. Conversion factors are shown in Table 11.3.)

1. Suppose bacteria are growing in a flask. The growth medium for the bacteria requires 5 g of glucose per liter. A technician has prepared some medium and added 0.24 lb of glucose to 25 L. Did the technician make the broth correctly?
2. A recipe to make 1 L of a laboratory solution is shown as follows:

Solution X	
Component	**Grams**
NaCl	20.00
Na azide	0.001
Mg sulfate	1.000
Tris	15.00

Prepare a table that shows how to prepare 1 mL of the same solution. Express the amounts of each component needed in mg.
3. Suppose a particular enzyme must be added to a nutrient solution used to grow bacteria. The enzyme comes as a freeze-dried powder. The manufacturer of the enzyme states that every gram of enzyme powder actually contains only 680 mg of enzyme and the rest is an inert filler that has no effect. If a recipe calls for 10.0 oz of this enzyme for every 100.0 L of broth, and if you prepare 500.0 L of broth, how much of the enzyme powder will you need to add? Remember to compensate for the inert filler.

12.5 CONCENTRATION AND DILUTION

12.5.1 CONCENTRATION

Concentration *is the amount of a particular substance in a stated volume (or sometimes mass) of a solution or mixture.* Concentration is a ratio where the numerator is the amount of the material of interest and the denominator is usually the volume (or sometimes mass) of the entire mixture. For example:

$$\frac{2 \text{ g NaCl}}{1 \text{ L Water}}$$

means that 2 g of NaCl is dissolved in enough water so that the total volume of the solution is 1 L.

The substance that is dissolved is called the **solute**. *The liquid in which the solute is dissolved is called the* **solvent**. In this example, NaCl is the solute and water is the solvent.

Note that the words "concentration" and "amount" are not synonyms. **Amount** *is how much of a* *substance is present* (e.g., 2 g, 4 cups, or one teaspoon). In contrast, concentration is a ratio with a numerator (amount) and a denominator (usually volume).

Because concentration is a ratio, problems involving concentrations use the same reasoning as other proportion problems. For example, a concentration of 1 mg NaCl in 10 mL of solution is the same as a concentration of 10 mg NaCl in 100 mL of solution. Similarly:

How could you make 300 mL of a solution that has a concentration of 10 g of NaCl in 100 mL total solution?

This is a proportion problem:

$$\frac{10 \text{ g}}{100 \text{ mL total}} = \frac{?}{300 \text{ mL total}}$$

$$? = 30 \text{ g}$$

Thirty grams of NaCl in 300 mL is the same concentration as 10 g of NaCl in 100 mL.

Example Problem 12.7

Which is purer: a chemical that contains 0.025 g of contaminating material in 10^4 kg or a chemical that contains 10^2 mg of contaminant in 10^4 kg?

Answer

There is more than one way to solve this problem. One approach, shown here, is to convert the concentrations of both the chemicals into the same unit so they can be compared more easily.

Chemical 1: $\dfrac{0.025 \text{ g}}{10^4 \text{ kg}}$ of contaminant

Chemical 2: Convert 10^2 mg to g.

$$10^2 \text{ mg} = 10^{-1} \text{ g}$$

Thus, chemical 2 has a contaminant concentration of

$$\frac{10^{-1} \text{ g}}{10^4 \text{ kg}} = \frac{0.1 \text{ g}}{10^4 \text{ kg}}$$

Chemical 1 is purer because it has less contaminant per 10^4 kg than chemical 2.

Manipulation/Application Practice Problems: Concentration

1. If a solution requires a concentration of 3 g of NaCl in 250 mL total volume, how much NaCl is required to make 1,000 mL?
2. If the concentration of magnesium sulfate in a solution is 25 g/L, how much magnesium sulfate is present in 100 mL of this solution?
3. If the concentration of magnesium chloride in a solution needs to be 1 mg/mL, how much magnesium chloride is required to make 15 L of this solution?
4. If a solution requires 0.005 g of Tris base per liter, how much Tris base is required to make 10^{-3} L of this solution?
5. If there is 300 ng of dioxin in 100 g of baby diapers, how much dioxin is there in 1 kg of diapers?
6. A *mole* is an expression of amount (which is discussed in Chapter 22). If the concentration of a solute in a solution is 5.00×10^{-3} moles/L, how many moles are there in 5 mL of this solution?
7. If a solute has a concentration of 0.1 moles/L, how much solute is present in 1 μL of solution?
8. If a solute has a concentration of 10^{-2} g/L, how much solute is present in 78 mL of solution?
9. Enzymes are proteins that catalyze chemical reactions in biological systems. Enzymes are rated on the basis of how active they are or how quickly they can catalyze reactions. Every preparation of enzyme has a certain activity expressed in units of activity. Suppose you buy a vial of an enzyme called beta-galactosidase. The vial is labeled: 3 mg solid, 500 units/mg. How many units are present in the vial?
10. A vial of the enzyme horseradish peroxidase is labeled: 5 mg solid, 2,500 units/mg. How many units are present altogether in this vial?
11. If a solution contains 3 g/mL of compound A, how much compound A is present in 1 L of this solution?
12. A solution has 5 μg/L of enzyme Q. How much enzyme Q is present in:
    a. 50 mL of solution
    b. 500 mL of solution
    c. 100 mL of solution
    d. 100 μL of solution
13. A solution has 0.5 mg/mL of the enzyme lysozyme. How much lysozyme is present in:
    a. 5 mL of solution
    b. 0.5 mL of solution
    c. 100 μL of solution
    d. 1,000 μL of solution

14. There are analytical instruments in the laboratory that are capable of detecting extremely small amounts of specific chemicals. Suppose a particular instrument can detect as little as a single molecule of benzo(a)pyrene (a carcinogen) out of 10^6 molecules of various compounds. Is the instrument sensitive enough to detect 100 molecules of benzo(a)pyrene out of 10^9 molecules?

15. Which is purer: a chemical that contains 1 g of contaminating material in 10^6 kg, or a chemical that contains 10^{-2} mg of contaminant in 10^{-3} kg?

Example Application 12.4: Dilutions

Plasmids are circular DNA molecules that can transport a gene from one bacterium into another bacterium. In nature, plasmids may carry genes that make a bacterium resistant to an antibiotic. As bacteria exchange plasmids carrying resistance genes, resistance to antibiotics spreads among bacterial populations. Plasmids can also be used in the laboratory to transport useful genes into bacteria. Suppose that in a particular experiment, it is necessary to add 0.01 μg of plasmid to a tube. Suppose further that 1 mL of plasmid in solution at a concentration of 1 mg plasmid/mL is available. How might the addition of only 0.01 μg of plasmid be accomplished? Assume that it is not possible to accurately measure a volume less than 1 μL.

Answer on p. 302.

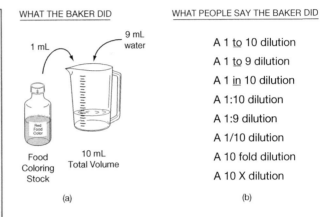

FIGURE 12.3 Dilutions and terminology. (a) What the baker did. (b) Various terminologies to describe the same dilution.

12.5.2 INTRODUCTION TO DILUTIONS: TERMINOLOGY

There are many situations in the laboratory that require dilutions, and we will see examples in various chapters in this textbook. A **dilution** *is when one substance (often but not always water) is added to another to reduce the concentration of the first substance. The original substance being diluted is generally called the* **stock solution**.

There are various ways to speak about dilutions; unfortunately, this variation in terminology can lead to confusion. Let us illustrate dilution terminology with an example. A baker buys a bottle of food coloring and dilutes it to decorate cookies. The baker takes 1 mL of the concentrated food coloring and adds 9 mL of water so that the total volume of the diluted food coloring solution is 10 mL (Figure 12.3a). Various people might

refer to this same dilution using different terminology (Figure 12.3b).

Observe that the word **to** and the symbol **:** are used inconsistently. The word **to** or the symbol **:** is sometimes used before the volume of the *diluting substance*, the **diluent**. (In this example, the diluent is 9 mL of water.) Other times, the word **to** or the symbol **:** is used before the total volume of the final mixture. (In this example, the total volume of the final mixture is 10 mL.) The key to avoiding confusion is to keep track of whether you are talking about the *total volume of the final mixture or about the amount of diluting substance*. When reading what other people have written, try to determine what they mean.

In this book, the dilution terminology conforms to that suggested by the American Society for Microbiology (ASM) Style Manual (*ASM Style Manual for Journals and Books*, American Society for Microbiology, 1991). This terminology is logical and consistent, even when three or more substances are combined. This terminology is summarized in Box 12.2.

BOX 12.2   DILUTION TERMINOLOGY BASED ON ASM RECOMMENDATIONS

1. *1 part food coloring combined with 9 parts water means the food coloring is 1 part in 10 mL total volume or 1/10 food coloring.* The denominator in an expression with a slash (/) is the total volume of the solution, never the amount of the diluting substance.
2. *An undiluted substance, by definition, is called 1/1.*
3. *When talking about dilutions, the symbol ":" means parts.* If 1 mL of food coloring is combined with 9 mL of water, that is 1 part food coloring plus 9 parts water or 1:9 food coloring **to** water. In this text, a dilution of 1 part plus 9 parts diluent is *not* referred to as a 1:10 dilution. However, many other authors *do* say that 1 part plus 9 parts diluent is a 1:10 dilution; hence, there is confusion in the terminology.

   For example:

   A **1:2 dilution** *means there are three parts total volume.*

   A $\dfrac{1}{2}$ **dilution** *means there are two parts total volume.*

   Therefore:

   1/2 is the same as 1:1.
   1:2 is the same as 1/3.
   1:9 is the same as 1/10.
   1:3:5 A:B:C means that 1 part A, 3 parts B, and 5 parts C are combined for a total of 9 parts.

(The parts can be any unit. For example, this might mean 1 mL of A, 3 mL of B, and 5 mL of C. Or, it might mean 1 g of A, 3 g of B, and 5 g of C.)

12.5.3 DILUTIONS AND PROPORTIONAL RELATIONSHIPS

The concepts of dilution and proportion are related.

For example:

One milliliter of food coloring mixed with 9 mL of water is the same dilution as 10 mL of food coloring mixed with 90 mL of water.

$$\frac{1 \text{ mL}}{10 \text{ mL}} = \frac{10 \text{ mL}}{100 \text{ mL}} = \frac{1}{10}$$

1 mL in 10 mL total = 10 mL in 100 mL total = 1/10 dilution.

Note that the units in the numerator and the denominator are the same and cancel.

All of the following are the same dilution, **1 in 10 total**:

2 mL food coloring

+18 mL water

_____

20 mL total volume

$$\frac{2 \text{ mL}}{20 \text{ mL}} = \frac{1}{10}$$

100 μL enzyme solution

+ 900 μL buffer solution

_____

1,000 μL total volume

$$\frac{100 \text{ μL}}{1,000 \text{ μL}} = \frac{1}{10}$$

17 fluid oz juice

+153 fluid oz

_____

170 fluid oz total

$$\frac{17 \text{ oz}}{170 \text{ oz}} = \frac{1}{10}$$

Manipulation Practice Problems: Dilutions (Part A)

(Follow ASM recommendations whenever applicable.)

1. Suppose you dilute 1 oz of orange juice concentrate with 3 oz of water.
   a. Express this dilution using the word *in*.
   b. Express this dilution using the word *to*.
   c. Express this dilution with a : and then with a /.
2. Express each of the following as a dilution (using a /):
   a. 1 mL of original sample+9 mL of water.
   b. 1 mL sample+10 mL water.
   c. 3 mL sample in a total volume of 30 mL.
   d. 3 mL sample+27 mL water.
   e. 0.5 mL sample+11.0 mL water.
3. Express each of the following ratios as a dilution (using a /).
   a. 1 part sample:9 parts diluent
   b. 1 part sample:10 parts diluent
   c. 1:3
   d. 1:1
   e. 1:4
4. Express each of the mixtures in Problem 2 as a 1:___ ratio.
5. If you take a 0.5 mL sample of blood and add 1.0 mL of water and 3.0 mL of reagents, what is the final dilution of the blood?

12.5.4 CALCULATIONS FOR PREPARING ONE DILUTION

Let's consider the calculations for making a single dilution. There will be two components in the dilution, the solute and the solvent. We will assume in all our examples that the solvent is water, as is true in the majority of biological solutions. Proportions can be used as a tool to calculate how to prepare a particular dilution.

For example, how could you prepare 10 mL of a 1/10 dilution of food coloring. The solute is food coloring and the solvent is water. This is a simple dilution and is easily accomplished as illustrated in Figure 12.4:

Combine 1 mL of the food coloring with 9 mL of water. The total volume will be 10 mL. Thus, the dilution will be 1/10.

Proportions can be used to determine how to make this dilution. The desired dilution is 1/10 and the total volume is 10 mL. Using a proportion equation:

$$\frac{1}{10} = \frac{?}{10 \text{ mL}}$$

? = 1 mL, which means that we need 1 mL of food coloring. Then, in order to get the desired volume, we will need to add 9 mL of water, as illustrated.

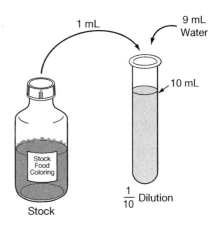

FIGURE 12.4 Ten milliliters of a 1/10 dilution of food coloring.

Let's try another example. How could you make 10 mL of a 1/5 dilution of food coloring?

Using a proportion equation:

$$\frac{1}{5} = \frac{?}{10 \text{ mL}}$$

? = 2 mL, so this means you would combine 2 mL of food coloring with 8 mL of water. The total volume will be 10 mL. Thus, the dilution will be 2/10 = 1/5. This is illustrated in Figure 12.5.

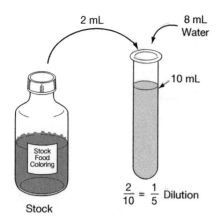

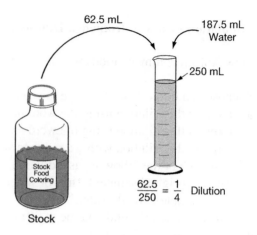

FIGURE 12.5 Ten milliliters of a 1/5 dilution of food coloring.

Another example: How could you make 100 mL of a 1/10 dilution of food coloring?

$$\frac{1}{10} = \frac{?}{100 \text{ mL}}$$

? = 10 mL, so combine 10 mL of food coloring with 90 mL of water. The total volume will be 100 mL. Thus, the dilution will be 10/100 = 1/10. This is illustrated in Figure 12.6.

The dilutions illustrated in Figures 12.4–12.6 are relatively simple, and with practice, the amount of water and food coloring required are likely to be obvious to you. Sometimes figuring out how to prepare a dilution is less intuitive. For example, how could you make 250 mL of a 1/4 dilution of food coloring? This may be less obvious than the three previous examples, but it can be solved easily using the logic of proportions:

$$\frac{1}{4} = \frac{?}{250 \text{ mL}}$$

$$? = 62.5 \text{ mL}$$

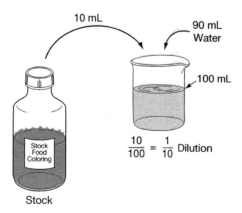

FIGURE 12.6 One hundred milliliters of a 1/10 dilution of food coloring.

FIGURE 12.7 Two hundred fifty milliliters of a 1/4 dilution of food coloring.

? = 62.5 mL, so take 62.5 mL of food coloring and add enough water to get 250 mL total volume (187.5 mL). The food coloring dilution will be:

$$\frac{62.5 \text{ mL}}{250 \text{ mL}} = \frac{1}{4}$$

Figure 12.7 illustrates this dilution.

Example Problem 12.8

Consider the dilutions illustrated in Figures 12.4–12.7. If you actually prepared these dilutions, which dilution would have the most intense color? (As the food coloring becomes progressively more dilute, the color intensity is reduced.) If you prepared these dilutions, which one would have the least intense color?

Answer

The denominator in the dilution is an indication of how much the solute has been diluted. For example, compare Figures 12.4 and 12.5. The denominator in Figure 12.4 is larger than in Figure 12.5, and there is less water relative to solute in Figure 12.5. Thus, the color will be more intense in the 1/5 dilution than in the 1/10 dilution. The most dilute of all the examples is 1/10, and the least dilute is ¼. This means that the color will be most intense in the ¼ dilution and least intense in the 1/10 dilution.

Manipulation Practice Problems: Dilutions (Part B)

1. How would you prepare 10 mL of a 1/10 dilution of blood?
2. How would you prepare 250 mL of a 1/300 dilution of blood?
3. How would you prepare 1 mL of a 1/50 dilution of blood?
4. How would you prepare 1,000 μL of a 1/100 dilution of food coloring?
5. How would you prepare 23 mL of a 3/5 dilution of Solution Q?
6. Suppose you have a stock of a buffer solution that is used in experiments. The stock is ten times more concentrated than it is used (like frozen orange juice that is sold in a concentrated form). How would you prepare 10 mL of the buffer solution at the right concentration?
7. Suppose you have a stock of buffer that is five times more concentrated than it is used. How would you prepare 15 mL at the correct concentration?
8. Suppose you need 10^3 μL of a solution. The solution is stored at a concentration that is 100 times the concentration at which it is normally used. How would you dilute the solution?
9. How would you prepare 50 mL of a 0.01 dilution of buffer?

12.5.5 DILUTION AND CONCENTRATION

When dilutions are prepared, it is important to keep track of the concentration of the solute in the diluted tubes. Let's look at some examples of the relationship between concentration and dilution. *This example is illustrated in Figure 12.8:*

A stock solution contains 10 mg/mL of a particular enzyme. This means that every mL of stock contains 10 mg of enzyme.

The stock solution is diluted by taking 1 mL of the solution (which contains 10 mg of enzyme) and mixing it with 4 mL of water. What is the concentration of enzyme in the diluted solution?

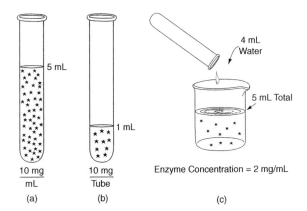

FIGURE 12.8 Dilution of an enzyme solution. (a) The original solution has a concentration of 10 mg enzyme/mL. Each * represents a mg of enzyme. (b) One milliliter of solution is removed containing 10 mg of enzyme. (c) The 1 mL of solution is diluted with 4 mL of water. The resulting diluted solution contains 10 mg of enzyme in 5 mL. The concentration of enzyme is therefore 2 mg/mL.

Answer:
The resulting diluted solution has a total volume of 5 mL and contains 10 mg of enzyme. The concentration of enzyme is:

$$\frac{10 \text{ mg enzyme}}{5 \text{ mL}} = \frac{2 \text{ mg enzyme}}{1 \text{ mL}}$$

Thus, the stock solution had an enzyme concentration of 10 mg/mL. The diluted solution has an enzyme concentration of 2 mg/mL.

Another example:
A stock solution initially has a concentration of 20 mg of solute per L.
A diluted solution is prepared by removing 1 mL of stock solution and adding 14 mL of water.
What is the concentration of solute in the diluted solution?
We can figure this out using a series of steps. The concentration in the stock solution is:

$$\frac{20 \text{ mg}}{1 \text{ L}} = \frac{20 \text{ mg}}{1,000 \text{ mL}} = \frac{0.020 \text{ mg}}{1 \text{ mL}}$$

One milliliter was removed to make the dilution, so 0.020 mg was removed from the stock. The total volume of the dilution was 15 mL. So, the concentration of solute in the diluted solution is:

$$\frac{0.020 \text{ mg}}{15 \text{ mL}} \approx \frac{0.0013 \text{ mg}}{\text{mL}} = \frac{1.3 \text{ mg}}{\text{L}}$$

It works to think about dilutions as we have explained so far. But there is a simpler rule that relates to this type of problem:

> **The concentration of a diluted solution is determined by multiplying the concentration of the original solution times the dilution (expressed as a fraction).**

Thus, for the example where the stock concentration is 20 mg/L and the dilution is 1/15, applying the rule:

Original solute concentration × dilution =

Solute concentration in dilution

$$\frac{20 \text{ mg}}{1 \text{ L}} \times \frac{1}{15} \approx \frac{1.3 \text{ mg}}{1 \text{ L}}$$

Another example:

A solution of 100% ethanol is diluted 1/10. What is the concentration of ethanol in the diluted solution?

$$100\,\% \text{ ethanol} \times \frac{1}{10} = 10\% \text{ ethanol}$$

The diluted solution has a concentration of 10% ethanol.

Another example:

A solution has an original concentration of 10 mg/mL of an enzyme.

The solution is diluted 1/5. The concentration of enzyme in the diluted solution is:

$$\frac{10 \text{ mg}}{1 \text{ mL}} \times \frac{1}{5} = \frac{2 \text{ mg}}{1 \text{ mL}}$$

Note that this is the same example as in Figure 12.8, simply described in a different way.

Example Problem 12.9

A stock solution of enzyme contains 10 mg/mL of enzyme. One hundred milliliters of this stock is diluted with 400 mL of buffer.

a. What is the concentration of enzyme in the resulting diluted solution?
b. How much enzyme will be present in 300 μL of the resulting dilution?

Answer

a. When 100 mL of stock solution is mixed with 400 mL of buffer, the dilution can be expressed as:

$$\frac{100 \text{ mL}}{500 \text{ mL}} = \frac{1}{5}$$

The concentration of enzyme in the resulting dilution, therefore, is:

$$\frac{10 \text{ mg}}{1 \text{ mL}} \times \frac{1}{5} = \frac{2 \text{ mg}}{1 \text{ mL}}$$

b. 300 μL = 0.300 mL. Because there is 2 mg/mL enzyme in the diluted solution, in 0.300 mL there is:

$$\frac{2 \text{ mg}}{\text{mL}} \times 0.300 \text{ mL} = 0.6 \text{ mg}$$

It is also possible to use the logic of proportions:

$$\frac{2 \text{ mg}}{1 \text{ mL}} = \frac{?}{0.300 \text{ mL}}$$

$$? = 0.6 \text{ mg}$$

Suppose you have a concentrated solution and a dilution made from the concentrated stock, and you know the concentration of solute in the *diluted solution* but not in the original stock. This is the inverse of the situation we have discussed so far. To calculate the concentration of material in the original stock solution, apply this rule:

> **When you know the concentration of a substance in a diluted solution, to calculate the concentration of a substance in an original stock solution, multiply the concentration of the substance in the diluted sample times the reciprocal of the dilution (expressed as a fraction).**

Example Problem 12.10

A solution of ethanol is diluted 1/10. The concentration of ethanol in the diluted solution is 10%. What was the concentration in the original stock solution?

Answer

Using the rule, the concentration of solute in the original solution is:

Solute concentration in dilution × reciprocal of dilution = Solute concentration in stock

$$10\% \times \frac{10}{1} = 100\%$$

The original solution had a concentration of 100% ethanol.

A note about terminology: You may come across the term "**dilution factor.**" Sometimes this term is used to mean a dilution, expressed as a fraction, but other times this term is used to refer to the *reciprocal* of the dilution. For example, if 1 mL of stock is diluted with 9 mL of diluent, some people might say the "dilution factor is 1/10," but other people might say the "dilution factor is 10." We avoid the term "dilution factor" in this book because it is not used consistently in the literature.

Example Problem 12.11

A stock solution has a total volume of 100 mL. It is diluted by removing 1 mL of stock solution and adding 24 mL of water. The concentration of solute in the dilution is 1.5 g/L.

a. What is the concentration of solute in the stock solution?
b. How much solute is present in the total 100 mL of stock?

Answer

a. *Solute concentration in dilution × reciprocal of dilution = Solute concentration in stock*

$$1.5 \frac{g}{L} \times \frac{25}{1} = 37.5 \text{ g}/\text{L}$$

b. This can be solved using the logic of proportions:

$$\frac{37.5 \text{ g}}{1,000 \text{ mL}} = \frac{?}{100 \text{ mL}}$$

$$? = \textbf{3.75 g}$$

This is the amount of solute present in 100 mL of the original solution.

Manipulation Practice Problems: Dilutions (Part C)

1. If you prepare a 1/40 dilution of a 50% solution, what is the final concentration of the solution?
2. If you prepare a 1/10 dilution of a 10 mg/mL solution, what is the final concentration of the solution?
3. If you prepare a 1:1 dilution of a 10 mg/mL solution, what is the final concentration of the solution?
4. How much 1/5 diluted solution can be made if you have 1 mL of original solution?
5. To prepare 1,000 mL of food coloring at a 1/100 dilution, how much of the original food coloring stock solution is required?

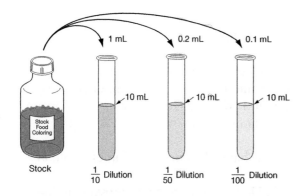

FIGURE 12.9 A dilution series where each dilution is independent of the others.

12.5.6 DILUTION SERIES

A **dilution series** *is a group of solutions that have the same components but at different concentrations.* One way to prepare a dilution series is to make each diluted solution independently of the others beginning with the initial concentrated stock solution. An example is explained next and is illustrated in Figure 12.9.

12.5.6.1 An Independent Dilution Series with Three Dilutions

How could you make 1/10, 1/50, and 1/100 dilutions of food coloring from an original bottle of concentrated food coloring?

First, decide how much of each dilution to prepare, for example, 10 mL.

a. To make 10 mL of a 1/10 dilution:
    Take 1 mL of the stock and bring it to a volume of 10 mL using water. This is the first diluted solution.
b. To make 10 mL of a 1/50 dilution:

$$\frac{1}{50} = \frac{?}{10\ mL}$$

$$? = 0.2\ mL$$

Thus, take 0.2 mL of the original stock and dilute to a volume of 10 mL. This is the second diluted solution.
c. To make 10 mL of a 1/100 dilution:

$$\frac{1}{100} = \frac{?}{10\ mL}$$

$$? = 0.1\ mL$$

Thus, take 0.1 mL of the original stock and dilute to a volume of 10 mL. This is the third diluted solution.

Thus, this strategy requires removing some of the original stock solution three times, once to make each dilution. Each of the three dilutions, therefore, is independent of the others. This strategy will work effectively in some situations.

12.5.6.2 Dilution Series Where the Dilutions Are Not Independent of One Another

Let us consider a situation where the strategy of preparing independent dilutions is not effective. Suppose you are working with bacteria and there are so many microorganisms in 1 mL of medium that the broth must be diluted 100,000 times to get a reasonable number of bacteria for counting. Thus, you want a 1/100,000 dilution of the original stock solution of bacteria. Assuming 10 mL is the final volume needed, the proportion is:

$$\frac{1\ mL}{100,000\ mL} = \frac{?}{10\ mL}$$

$$? = 0.0001\ mL$$

To dilute the bacterial broth in one step would require taking 0.0001 mL of bacterial broth and diluting it to 10 mL. It is very difficult, however, to measure 0.0001 mL accurately. The common strategy to prepare such a dilution, therefore, is to use two or more steps. First, a dilution is prepared and then some of this first dilution is removed and used to make a second dilution. Some of the second dilution is used to make a third dilution and so on until the solution is dilute enough. The following example shows how one might prepare 10 mL of a 1/100,000 dilution of bacterial cells in three steps (Figure 12.10).

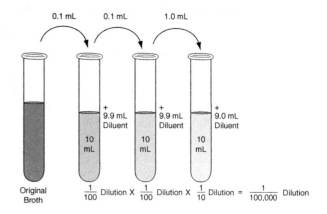

FIGURE 12.10 A dilution series used to dilute a bacterial stock solution 100,000×.

1. Mix 0.1 mL of the original broth with 9.9 mL of diluent. This will give a dilution of 0.1 mL/10 mL = 1/100.

↓

2. After thorough mixing, remove 0.1 mL from the diluted bacterial broth prepared in Step 1 and add 9.9 mL of diluent; the total volume in the second dilution is 10 mL. The second dilution is 1/100, as is the first dilution.

↓

3. Remove 1 mL from the second dilution tube prepared in Step 2, and add 9 mL of diluent; the total volume in the third dilution tube is 10 mL. The third dilution is 1/10.

The bacteria were therefore diluted 1/100 in the first tube, 1/100 in the second tube, and 1/10 in the third tube. The final, total dilution is:

$$1/100 \times 1/100 \times 1/10 = 1/100,000.$$

The third tube has the desired dilution and volume.

Note in the preceding example that to determine the dilution of solute in the final dilution tube, the dilutions in each intermediate tube were multiplied by one another. Thus, the dilution of solute in the final dilution tube was 1/100 × 1/100 × 1/10 = 1/100,000. The procedure to find the concentration of solute in the final tube can be generalized into the following rule:

The concentration of a diluted solution in the final tube is determined by multiplying the concentration of the original solution times the dilution in the first tube, times the dilution in the second tube and so on until reaching the last tube.

Continuing with this bacteria example, suppose that the original broth contained 1×10^9 bacteria per milliliter. What is the concentration of bacteria in the third (final) dilution?

$$\frac{1 \times 10^9 \text{ bacteria}}{1 \text{ mL}} \times \frac{1}{100,000} = \frac{1 \times 10^4 \text{ bacteria}}{1 \text{ mL}}$$

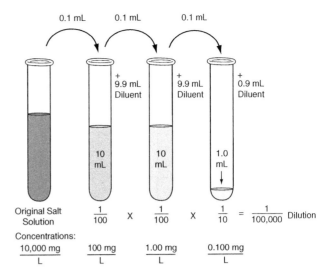

FIGURE 12.11   Diluting a salt solution.

12.5.6.3 Another Dilution Series Where the Dilutions Are Not Independent of One Another

Consider another example of a dilution series where the dilutions are not independent of one another (Figure 12.11).

A solution contains 10 g salt/L. How could this solution be diluted to obtain 1 mL of solution with a salt concentration of 0.100 mg/L?

There are various strategies that will work. One example is:

1. Remove 0.1 mL of the original salt solution, and add 9.9 mL of water. This is a 1/100 dilution.

The concentration of salt in this dilution is 100 mg/L because:

original concentration		dilution		concentration first dilution
$\dfrac{10,000 \text{ mg}}{1 \text{L}}$	×	$\dfrac{1}{100}$	=	$\dfrac{100 \text{ mg}}{1 \text{L}}$

⇓

2. Remove 0.1 mL from the first dilution, and place it in 9.9 mL of water. This is also a 1/100 dilution.

The concentration of salt in this dilution tube is 1.00 mg/L because:

concentration after 1st dilution		dilution		concentration after second dilution
$\dfrac{100 \text{ mg}}{1 \text{L}}$	×	$\dfrac{1}{100}$	=	$\dfrac{1.00 \text{ mg}}{1 \text{L}}$

⇓

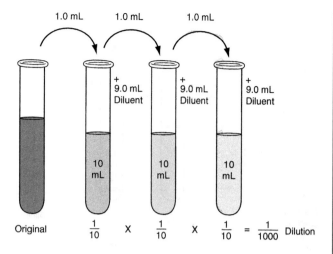

FIGURE 12.12   A serial dilution.

3. Remove 0.1 mL from the second dilution, and place it in 0.9 mL of water. This is a 1/10 dilution. The volume in this tube is 1 mL, as desired.

The concentration of salt in this tube is 0.100 mg salt/L because:

concentration after 2nd dilution	dilution	concentration after final dilution
$\dfrac{1.00\ mg}{1L}\ \times$	$\dfrac{1}{10}\ =$	$\dfrac{0.100\ mg}{1L}$

This is the desired final concentration and volume.

A **serial dilution** *is a series of dilutions that all have the same dilution (for example, all are 1/10 dilutions, or all are 1/2 dilutions).* (Note that when people use the phrase *dilution series*, the dilution may vary.) Figure 12.12 shows an example of a 1/10 serial dilution.

Example Application 12.4 Answers (from p. 293)

The concentration of plasmid in the stock solution is 1 mg/mL, which is equal to 1 µg/µL. If the plasmid is drawn directly from the stock tube, only 0.01 µL is needed, which is a volume too small to measure accurately. Therefore, the stock must be diluted. If the stock is diluted 1,000×, its concentration will be:

$$\frac{1\ \mu g}{1\ \mu L} \times \frac{1}{1,000} = \frac{0.001\ \mu g}{1\ \mu L}$$

Then, 10 µL of the diluted plasmid solution will contain 0.01 µg of plasmid.

There are pipettes available that can accurately measure 10 µL volumes.

There are many strategies to dilute the stock plasmid solution 1,000 times. One strategy is:

1. Remove 10 µL of the original plasmid stock, and add 990 µL of buffer or water resulting in a 1/100 dilution.
2. Remove 10 µL from the dilution in Step 1, and add 90 µL of buffer or water resulting in a 1/10 dilution. The total dilution in the second dilution tube is therefore:

$$1/100 \times 1/10 = 1/1,000$$

The concentration of plasmid in this tube is 0.001 µg/µL, which is the desired concentration.

This two-step strategy involves volumes that can be measured with reasonable accuracy. This strategy also uses only a small amount of the stock solution, which might be an advantage if the stock is being saved for other experiments.

Manipulation Practice Problems: Dilutions (Part D)

1. Dilution series: Explain how to prepare a 1/10 dilution of food coloring. Then, use the 1/10 dilution to prepare a 1/250 dilution of the original stock of food coloring. Use the 1/250 dilution to prepare a 1/1,000 dilution of the stock of food coloring.
2. **a.** Explain how a dilution series could be used to prepare a $1/10^6$ dilution of bacterial cells.
   b. Explain how a 1/10 serial dilution could be used to prepare the same dilution of bacterial cells.
3. Explain how 1/5, 1/50, and 1/250 dilutions of a blood sample could be prepared.

Application Practice Problems: Dilutions

1. Suppose you have $20\,\mu L$ of an expensive enzyme and you cannot afford to purchase more. The enzyme has a concentration of 1,000 units/mL. You are going to do an experiment that requires tubes with a concentration of 1 unit/mL of enzyme, and each tube will have 5 mL total volume. How much enzyme does each tube require? ____ How many tubes can you prepare before you run out of enzyme? ____

2. Suppose you have $20\,\mu L$ of an expensive enzyme that has 1,000 units/mL. You are going to use the enzyme in an experiment that requires tubes with 0.01 units/mL of enzyme, and each tube will have 5 mL total volume. How much enzyme will each tube require? ____ Will you be able to directly measure this amount of enzyme accurately? ____ Show how you can dilute $10\,\mu L$ of the original $20\,\mu L$ of enzyme so that you can use it in your experiment. Use a diagram and words to demonstrate your strategy for preparing the enzyme.

3. Antibodies are frequently very concentrated relative to how much is necessary in an experiment. Show how you would use a dilution series to dilute an antibody solution 500,000×. Assume you have 1 mL of antibody to begin with, but you want to save at least 0.5 mL of the antibody for future experiments.

4. Counting seems like a simple mathematical process; however, counting microorganisms is not so simple. For one thing, microorganisms, such as bacteria, are not visible to the eye. Another problem is that bacteria may be present in extremely large numbers in a sample. For example, in nutrient broth, there might be 1×10^9 bacterial cells per milliliter. Therefore, microbiologists have devised various methods to count bacterial cells. One such method is called viable cell counting. To perform a viable cell count, a sample of bacterial cells is first diluted in series. Then, 0.1 mL of diluted cells is spread on a petri dish that contains nutrient agar. It is assumed that every living cell in the 0.1 mL placed on the agar divides to form a colony of bacterial cells. A colony contains so many individual cells that it is visible to the eye. It is also assumed that each colony originates from a single cell. It is possible to count the colonies and therefore to estimate the number of bacteria in the 0.1 mL of broth. Assume you begin with a culture of bacteria that contains 1×10^9 bacteria/mL. Show how you could dilute the culture so that the final tube has a concentration of 200 bacteria/0.1 mL.

5. You are performing a viable cell count of bacteria. The original broth has an unknown concentration of bacteria. You dilute the culture as shown in Figure 12.13 and plate 0.1 mL of the last three dilutions onto three petri dishes with nutrient agar. The following day you count the number of colonies on the plates with the results shown in Figure 12.13. What was the concentration of bacteria in the original tube?

6. You are performing a viable cell count of bacteria. The original broth has an unknown concentration of bacteria. You dilute the culture as shown in Figure 12.14 and plate 0.1 mL of the last three dilutions on three nutrient agar plates. The following day you count the number of colonies on the plates with the results shown in Figure 12.14. What was the concentration of bacteria in the original tube?

7. Suppose you have a bacterial culture with 10^7 cells/mL. How would you dilute this culture so that if you plate 0.1 mL of the last dilution, you will get (in theory) 100 colonies?

8. Suppose you do an assay (test) to determine how much protein is present in a sample. To perform the assay, it is necessary to dilute the original sample 1/100. If the assay shows that the concentration of protein in the diluted sample was 50 mg/mL, what was the concentration of protein in the undiluted sample? How much protein was present in 100 mL of the original, undiluted sample?

9. In a protein assay, the amount of protein in 1 mL of diluted sample was 87 mg. If the original sample was diluted 1/50, what was the concentration of protein in the original sample?

10. In a protein assay, the original sample was first diluted 1/5. Then, 5 mL of the diluted sample was mixed with 20 mL of reactants to give 25 mL total. Five milliliters was removed from the 25 mL mixture. The 5 mL contained 3 mg of protein. What was the concentration of protein in the original sample?

11. In a protein assay, the original sample was diluted 1/4. Then, 10 mL of the diluted sample was added to 20 mL of reactants to give 30 mL total. Five milliliters was removed from the 30 mL mixture. The 5 mL contained 10 mg of protein. What was the concentration of protein in the original sample?

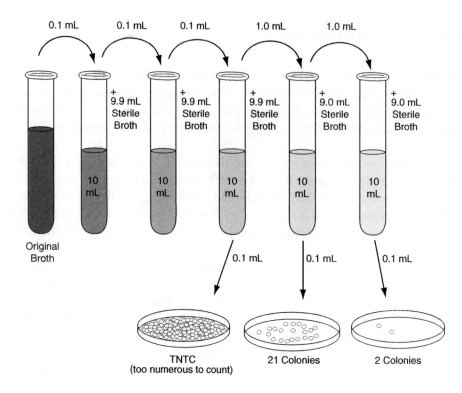

FIGURE 12.13   Diagram for Application Practice Problem 5.

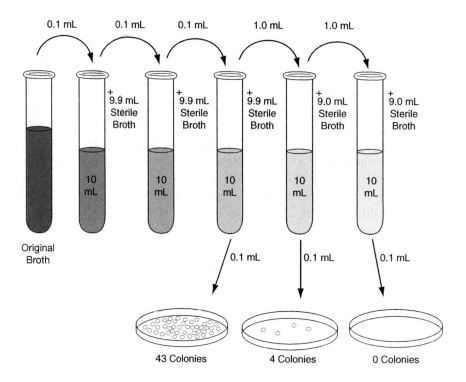

FIGURE 12.14   Diagram for Application Practice Problem 6.

13 Relationships and Graphing

13.1 GRAPHS AND LINEAR RELATIONSHIPS

13.1.1 A BRIEF REVIEW OF BASIC TECHNIQUES OF GRAPHING

Graphs, like equations, are a tool for working with relationships between two (or sometimes more) variables. The general rules regarding graphing will briefly be reviewed in the first part of this chapter. The use of graphing as a tool to perform various tasks in a laboratory will be demonstrated in later sections. Note that computer programs (such as Excel and specialized graphing programs) are now used to prepare graphs in most workplaces and academic settings. However, even though computers make the task of drawing graphs easier, it is still important to understand how graphs illustrate relationships, and how they are used to summarize and interpret data.

A simple two-dimensional graph, as shown in Figure 13.1, consists of a vertical and a horizontal line that intersect at a point called the **origin.** *The horizontal line is the* **X-axis***; the vertical line is the* **Y-axis.** The X- and Y-axes are each divided into evenly spaced subdivisions that are assigned numerical values. To the right of the origin on the X-axis, the X values are positive numbers; to the left of the origin, the X values are negative. On the Y-axis, the values above the origin are positive; below the origin, they are negative.

A basic two-dimensional graph shows the relationship between two variables, one of which is assigned to the X-axis and the other to the Y-axis. *For any value of a variable on the X-axis, there is a corresponding value of a variable on the Y-axis.* The two values are

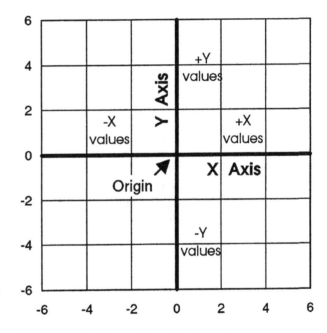

FIGURE 13.1 A two-dimensional graph.

called **coordinates** because they are coordinated, or associated, with one another. A pair of coordinates can be plotted on the graph as shown in Figure 13.2a. *The distance of a point along the X-axis is sometimes called the* **abscissa***, and the distance of a point along the Y-axis is sometimes called the* **ordinate.** The first point shown on the graph has an X-coordinate of 2 because it is above the 2 on the X-axis, and a Y-coordinate of 3, because it is across from the 3 on the Y-axis. This point can be written as X = 2, Y = 3, or as (2,3). The second point on this graph has the coordinates (4,6). The point

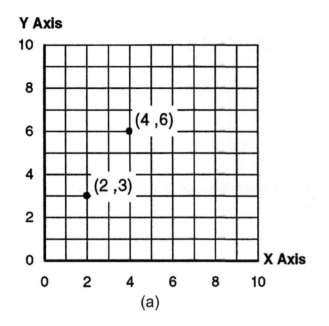
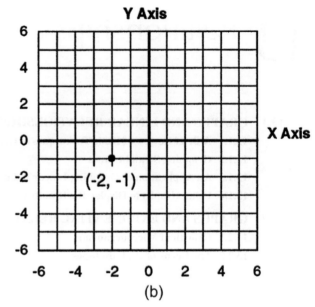

FIGURE 13.2 Coordinates and graphs. (a) The points (2,3) and (4,6). (b) The point (−2,−1).

in Figure 13.2b has coordinates X = −2, Y = −1. To prepare a graph, one marks the locations of a series of points by using their coordinates.

13.1.2 GRAPHING STRAIGHT LINES

Two variables may be related to one another in such a way that when plotted on a graph, the points form a straight line. The two variables are then said to have a **linear relationship.** There are many applications in the laboratory that involve linear relationships. Consider the simple equation:

$$Y = 2X$$

It is possible to find numbers that satisfy this equation. For example:

If X = 3, then Y = 6

If X = 4, then Y = 8

and so on

These solutions and more are summarized in both tabular and graphical form in Figure 13.3. Each pair of X and Y values is the coordinates of a point on the graph. The points form a straight line when connected.

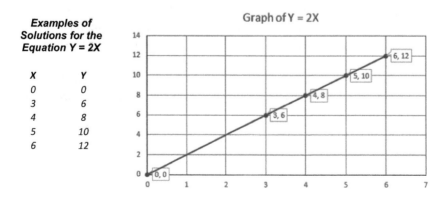

FIGURE 13.3 Graphing the equation Y = 2X.

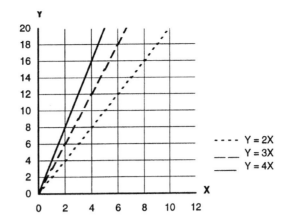

Y = 3X		Y = 4X	
X	Y	X	Y
0	0	0	0
1	3	1	4
2	6	2	8
3	9	3	12
4	12	4	16
5	15	5	20

FIGURE 13.4   Slope. The greater the value for the slope of a straight line, the steeper the line.

In the linear equation $Y = 2X$, the value 2 is called the **slope.** In the equation $Y = 3X$ the slope is 3, and in the equation $Y = 4X$ the slope is 4. The equations $Y = 2X$, $Y = 3X$, and $Y = 4X$ are plotted on the same graph in Figure 13.4. Each equation is linear; the difference between the three lines is their steepness. Just as a hill may be more or less steep, so a line has a slope that is more or less steep. The equation $Y = 4X$ defines a steeper line than the other two equations because its slope, 4, is the largest.

Given a straight line plotted on a graph, the slope of the line is calculated by determining how steeply the line rises. For example, consider the line for $Y = 2X$ in Figure 13.5. From point a to point b, X increases by +1 and Y increases by +2. *The amount by which the X-coordinate increases is called the* **run;** *the amount*

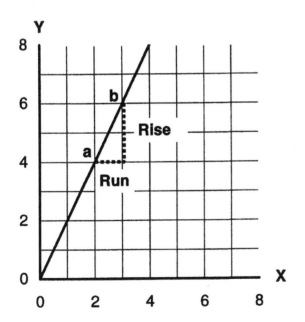

FIGURE 13.5   Determining the slope of a line. The slope can be calculated based on the coordinates of any two points on the line. Slope=rise/run, which in this case=2/1=2.

by which the Y-coordinate increases is called the **rise.** The rise divided by the run is a numerical measure of the steepness of the slope. To calculate the slope of any straight line, choose any two points on the line. The coordinates for the first point are (X_1, Y_1) and for the second point are (X_2, Y_2). Then:

$$\text{slope} = \frac{\text{rise}}{\text{run}} = \frac{\text{change in Y}}{\text{change in X}} = \frac{Y_2 - Y_1}{X_2 - X_1}$$

For the line $Y = 2X$:

$$\text{slope} = \frac{\text{rise}}{\text{run}} = \frac{2}{1} = 2$$

Choosing any two points on the line $Y = 2X$ will give a slope of 2. Choosing any two points on the line $Y = 3X$ will give a slope of 3 and any two points on the line $Y = 4X$ will yield a slope of 4.

Figure 13.6 shows the graph of a line that goes "downhill" from left to right. For this line, as the X values increase, the Y values decrease. The slope is therefore negative.

Figure 13.7 shows the plot for the equation $A = 2C + 4$. (Substitution of other symbols for X and Y does not change the basic meaning of the equation.) The difference between the plots of $Y = 2X$ and $A = 2C + 4$ is that the latter line is higher and does not pass through the origin. The line $A = 2C + 4$ intercepts, or passes through, the Y-axis at 4. *The point at which a line passes through the Y-axis, that is, where X = 0, is termed the* **Y-intercept.** If the line passes through the Y-axis at the point where both X and Y are zero, the intercept is often not written. Thus, the Y-intercept of the line $Y = 2X$ is 0.

If an equation gives a straight line when it is graphed, then it will have the form:

$$Y = \text{slope}(X) + \text{Y-intercept}$$

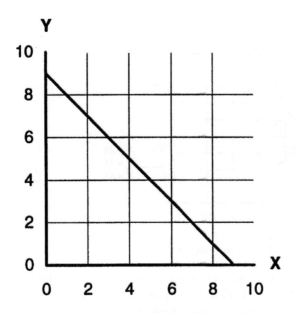

FIGURE 13.6   A line with a negative slope. All lines that go "downhill" from left to right have a negative slope.

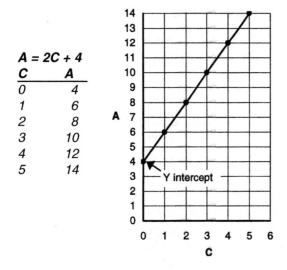

FIGURE 13.7   A graph with a nonzero Y-intercept.

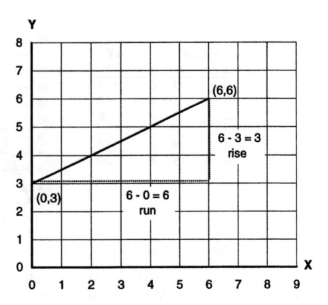

FIGURE 13.8   Determining the equation for a line. Slope = $(6-3)/(6-0) = 0.5$. Y-intercept = 3. Equation is $Y = 0.5X + 3$.

This general equation for a straight line is sometimes written:

General Equation for a Straight Line

$$Y = mX + a$$

where:

m = the slope and
a = the Y-intercept.

It is possible to determine the equation for a straight line from its graph. For example, the line drawn in Figure 13.8 has a Y-intercept of 3 and a slope of 0.5; therefore, the equation for this line is: **$Y = 0.5 (X) + 3$.** The general procedure to find the equation for any graphed straight line is shown in Box 13.1.

BOX 13.1   PROCEDURE TO FIND THE EQUATION FOR A STRAIGHT LINE ON A GRAPH

Straight lines can be described by an equation in the form:

$$Y = mX + a$$

where:
m is the slope and
a is the Y-intercept.

1. Find the Y-intercept, that is, the value of Y when X = 0. The intercept may be positive or negative.
2. Find the slope by picking any two points and calculating $(Y_2 - Y_1)/(X_2 - X_1)$. The slope may be positive, negative, or zero. (A horizontal line has a slope of zero.)
3. Put the slope and intercept into the proper form by filling in the blanks for slope and Y-intercept:

$$Y = \underline{slope}(X) + \underline{Y\text{-}intercept}$$

(Continued)

Notes: The axes of a graph can have units in which case the slope and the intercept will also have units. Include the units when writing the equation for a line.

The equation for a horizontal line is $Y = 0\ (X) + \text{Y-intercept}$, or just $Y = \text{Y-intercept}$. For example, $Y = 3$ is the equation for a horizontal line.

This pattern does not fit vertical lines, such as $X = 2$.

Example Problem 13.1

Which of the following equations describe a straight line?

a. $C = 2B$ **b.** $Q = 25.4T - 5$
c. $Y = X^2 + 3$ **d.** $C = -V - 34$

Answer

All except c are equations for a straight line. The exponent in equation c means that it does not describe a straight line. The rest of the equations have two variables, a slope, and a Y-intercept, which may or may not be zero. Equation b fits the linear equation as shown here:

$$Y = mX + a$$

$$Q = 25.4\,T + (-5)$$

Equation d is also the equation for a line:

$$Y = mX + a$$

$$C = (-1)\,V + (-34)$$

Manipulation Practice Problems: Graphing

1. For each of the following graphs, describe in words the relationship that is displayed.

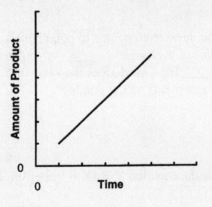

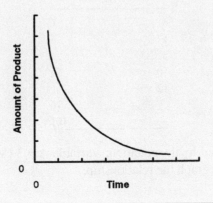

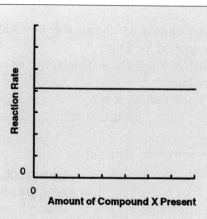

2. The following figure shows a graph with four points, labeled a, b, c, and d. Write the coordinates for each of the four points.

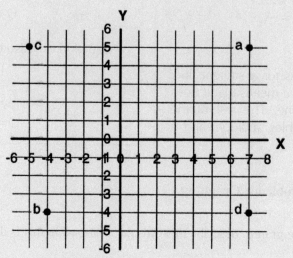

3. In the graph in Problem 2, as you move from point a to point b, how much does the value of X change? _____ How much does the value of Y change? _____ As you go from point b to point c, how much does the value of X change? _____ How much does the value of Y change? _____

4. Draw a graph and plot each of these points on the graph:
   X= 4, Y = 6
   X= 5, Y =–2
   (–4,3)
   (1,1)

5. A table showing three values for the equation Y = 5X + 1 is given. Fill in the blanks in the table, and then plot the equation.

X	Y
1	6
5	26
10	51
12	—
—	76
—	101

6. Suppose two variables are related by the equation: Variable A = 3 (Variable Q) – 4. Prepare a table to show this relationship, and then graph the relationship.

7. What is the slope and the Y-intercept for each of these equations?
   a. $Y = 3X + 2$
   b. $C = 0.2X - 1$
   c. $Y = 0.005X$
8. What is the slope and Y-intercept for lines a–d?

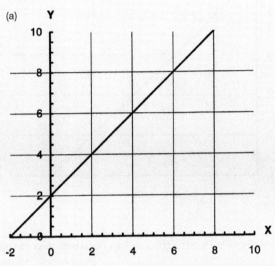

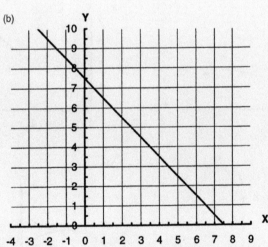

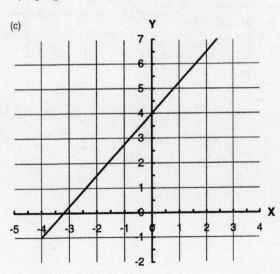

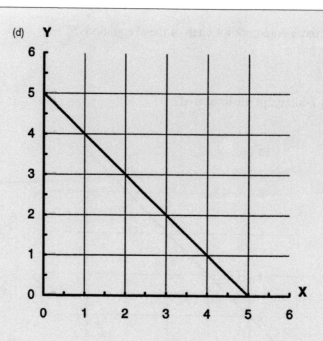

(d)

9. Which of these graphs below show a linear relationship between two variables?

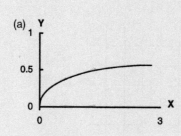

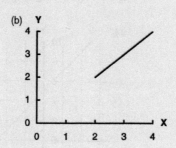

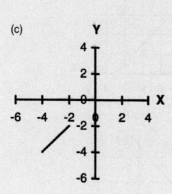

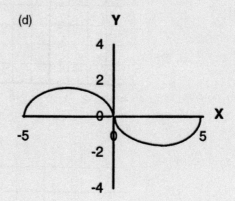

10. Which of these equations will form a straight line when plotted on a graph?
    a. $Y = 45X - 1$ b. $C = 34D + 17$
    c. $34 + 2 - 4D = E$ d. $Q = R$
11. a. For each of the following equations, what is the slope?
    b. For each of the following equations, what is the Y-intercept?
    c. Graph the equations.
        i. $Y = (10 \text{ cm/min}) X + 1 \text{ cm}$
        ii. $3 - X = Y$
        iii. $(12 \text{ mg}) + (7 \text{ mg/cm}) X = Y$
12. a. Draw the line that has a Y-intercept of 2, that is, (0, 2), and a slope of 0.25.
    b. Draw the line that goes through the point (−4, 3) and has a slope of − 2.
13. Find the equations for the lines graphed in each of the following examples.

(a)

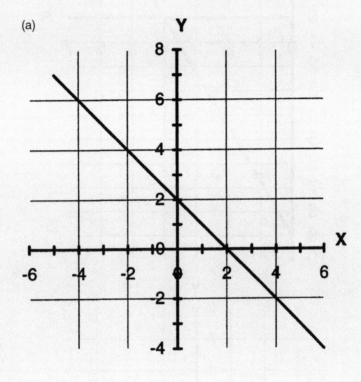

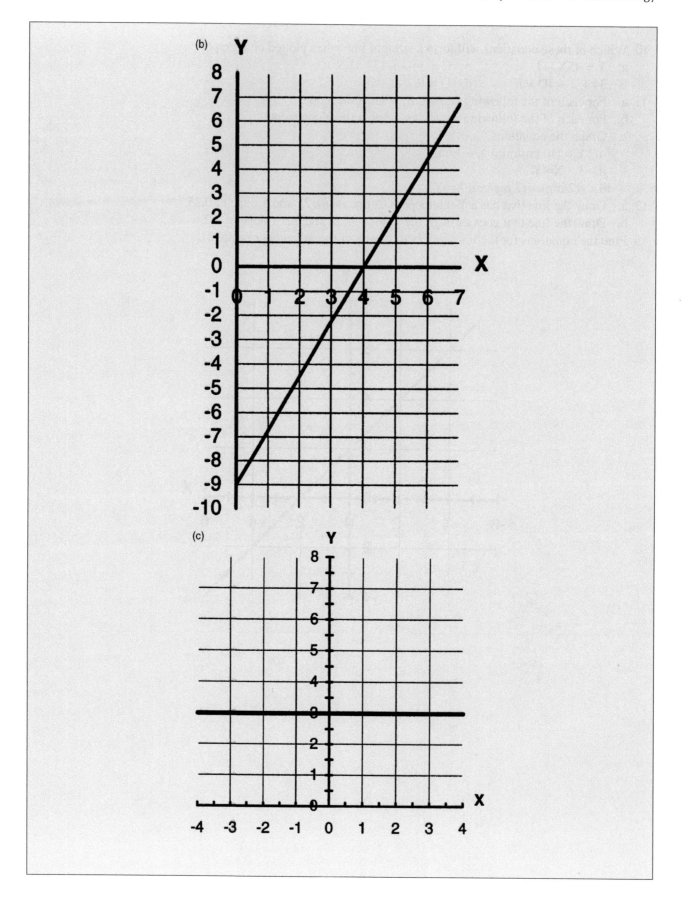

(d)

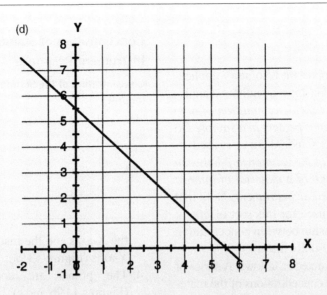

14. The following figure has three lines. Match each line with its equation:

$$Y = \frac{1}{2}X + 2 \quad Y = \frac{3}{4}X \quad Y = 2X + 2$$

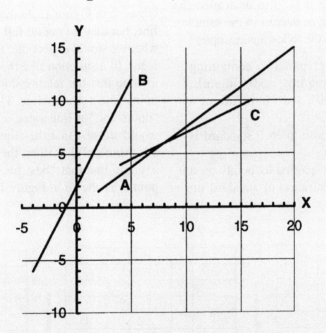

15. Plot the line that contains the following points: (1,1), (6,6), (9,9). What is the equation for this line?

16. The relationship between the temperature in degrees Fahrenheit and degrees Celsius is given by the equation: $^\circ F = 9/5(^\circ C) + 32$.

    a. There are two constants in this equation. What are they?

    b. Which of the two constants is the Y-intercept?

    c. Which of the two constants is the slope of the line?

    d. Graph this relationship.

13.1.3 AN APPLICATION OF GRAPHING LINEAR RELATIONSHIPS: STANDARD CURVES AND QUANTITATIVE ANALYSIS

This section discusses an important laboratory application of graphs of linear relationships: quantitative analysis. **Quantitative analysis** *is the determination of how much of a particular material is present in a sample.* To make such determinations, a standard curve is used. A **standard curve** *is a graph of the relationship between the concentration (or amount) of a material of interest and the response of a particular instrument.* Note that the term *standard curve* is used for this sort of graph; however, the desired relationship between concentration and instrument response is usually linear.

A standard curve is constructed as follows: A series of standards containing known concentrations of the material of interest are prepared. The response of an instrument to each standard is measured. A standard curve is plotted with the response of the instrument on the Y-axis and the concentration of standard on the X-axis. Once graphed, the standard curve is used to determine the concentration of the material of interest in the samples. This process is illustrated in the following example:

1. Standards are prepared containing 1.0 mg/mL, 2.0 mg/mL, 3.0 mg/mL, 5.0 mg/mL, 6.0 mg/mL, and 7.0 mg/mL of a compound of interest.
2. An instrument's response to each standard is measured (Table 13.1).
3. The resulting data are plotted as points on a graph with the concentration of standard on

TABLE 13.1

Concentration of Standards versus Instrument Response

Concentration of the Standard (mg/mL)	Instrument Response
1.0	0.24
2.0	0.53
3.0	0.72
5.0	1.28
6.0	1.51
7.0	1.79

the X-axis and the instrument response on the Y-axis (Figure 13.9a).
4. The points are connected into a line (Figures 13.9b and c).
5. The standard curve is used to determine the concentration of the material of interest in samples (Figure 13.10).

Observe in Figure 13.9a that the points approximate a line, but they do not all fall exactly on the line. This is what we would expect due to small errors in measurement. In a situation like this, where it is reasonable to assume that the relationship is linear, we connect the points into a straight line. The points are not connected "dot to dot," as illustrated in Figure 13.9b, because this would suggest that the slight variations from the line are meaningful. Rather, the points are connected into a single line that "best fits," or best averages, all the points, as shown in Figure 13.9c.

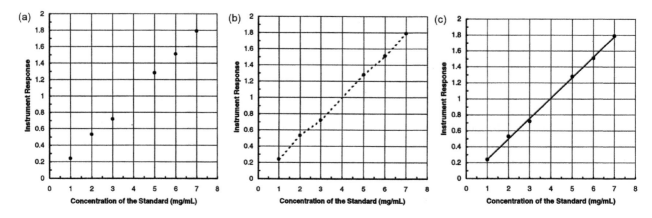

FIGURE 13.9 A standard curve. (a) Data from Table 13.1 are graphed. The concentration of compound in the standards is on the X-axis, and the instrument response is on the Y-axis. (b) The points are not connected "dot to dot," as shown here, because this would suggest that slight variations from a straight line are meaningful. (c) The points are connected into a "best-fit line," which best averages all the points.

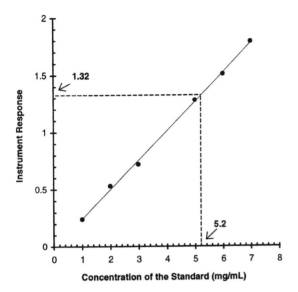

FIGURE 13.10 Using a standard curve to determine the concentration of a material in a sample. In this example a sample gives an instrument reading of 1.32, which corresponds to a concentration of 5.2 mg/mL.

There are two ways to get a "best-fit" line for the points on a graph. One method is to place a ruler over the graph and draw a straight line that appears "by eye" to be closest to all the points. Most people are able to draw a reasonable "best-fit" line "by eye," although two people will seldom draw a line with exactly the same slope and Y-intercept. *A more accurate method to draw a line of best fit is the statistical technique,* **the least squares method.** (This statistical method is explained in the Appendix to Chapter 28.)

Regardless of whether the points are connected into a line "by eye" or using the method of least squares, the standard curve can be used to determine the level of compound in each unknown sample. In the example illustrated in Figure 13.10, a sample gives an instrument reading of 1.32. From the standard curve, one can see that a reading of 1.32 corresponds to a concentration of 5.2 mg/mL.

Example Problem 13.2

Biologists commonly use a protein assay to determine the concentration of protein in a solution. Various protein assays are available, many of which measure the color change in a protein solution when it reacts with various dyes. The amount of color appearing is generally proportional to the amount of protein present.

The following graph shows the relationship between the concentration of protein in a series of standards and the amount of color after the standards are reacted with dye. The amount of color is measured in terms of the amount of light absorbed by the dye expressed as "absorbance."

a. In what range of protein concentration does the assay give linear results? What happens at higher concentrations of protein?

b. Suppose you have a sample containing an unknown amount of protein. The sample is reacted with the dye and has an absorbance of 0.70. Based on the standard curve, what is the concentration of protein in the unknown?

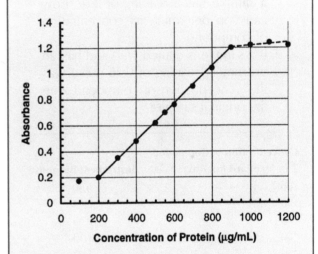

Answer

a. The assay is linear in the middle range; above about 900 μg/mL and below 200 μg/mL, the assay is not useful because the absorbance does not change, even if the concentration of protein changes.

b. The concentration of the unknown is about 550 μg/mL.

Example Problem 13.3

The concentration of compound Z in samples needs to be determined. The response of an instrument to compound Z is related to its concentration. A series of standards are prepared, and the instrument's response is measured. The resulting data are shown in the following table.

a. Plot a standard curve for compound Z based on the data in the table. Draw the line that best fits the points.

b. A sample with an unknown concentration of compound Z gives an instrument response of 1.35. Based on the standard curve, what is the concentration of compound Z in the sample?

c. Suppose that the response of the instrument used in this example is known to be inaccurate at readings above 2. If a sample has a reading of 2.67, how can you determine its concentration of compound Z?

d. If a sample is diluted 1/20 and has an instrument reading of 0.45, what was the concentration of compound Z in the original sample?

Concentration of Compound Z in Standard (mg/mL)	Instrument Response
100	0.30
150	0.44
250	0.68
400	1.13
550	1.48
650	1.82
850	2.10

Answer

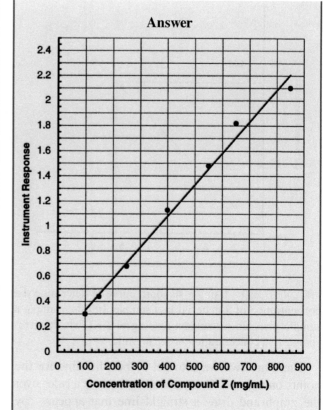

b. Based on the standard curve, the concentration of compound Z in the sample is 500 mg/mL.

c. It is necessary to dilute the sample before reading its absorbance.

d. A value of 0.45 corresponds to a concentration of 150 mg/mL; however, the sample was diluted and this dilution needs to be taken into consideration. We therefore multiply 150 mg/mL times the reciprocal of the dilution.

$$150\,\frac{mg}{mL} \times \frac{20}{I} = 3,000 \text{ mg/mL}$$

The concentration of compound Z in the original undiluted sample was 3,000 mg/mL.

Manipulation and Application Practice Problems: Quantitative Analysis

1. a. Plot a standard curve based on the data in the following table. Draw the line that best fits the values.

   b. Suppose you have a sample that gives an instrument response of 2.35. This value is higher than the reading of the highest standard; therefore, you dilute the sample 1/100. The instrument reading of the diluted sample is 0.78. What amount of the material of interest was present in the original solution?

Amount of Standard (in grams)	Instrument Response
0	0.00
10	0.14
20	0.28
30	0.41
40	0.52
50	0.71
65	0.84
75	1.04
80	1.15

2. A stock solution of copper sulfate has a concentration of 100 mg/mL. How would you dilute the stock solution to prepare each of the following standards?

1 mg/mL

5 mg/mL

15 mg/mL

25 mg/mL

50 mg/mL

75 mg/mL

100 mg/mL

13.1.4 USING GRAPHS TO DISPLAY THE RESULTS OF AN EXPERIMENT

Many experiments involve manipulating one variable and measuring the result of that manipulation on a second variable. For example, an investigator interested in the effect of light intensity on the rate of seedling growth could expose different groups of seedlings to different light intensities and measure their growth rates. The two experimental variables – light intensity and seedling growth rate – can be plotted on a two-dimensional graph.

When plotting data from experiments, one distinguishes between the **dependent variable** and the **independent variable.** In the preceding example, the investigator is looking at whether seedling growth rate is dependent on the light intensity. Growth rate is therefore called the dependent variable. *The variable the investigator controls is the* **independent variable.** *The variable that changes in response to the independent variable is called the* **dependent variable.** It is conventional to plot the dependent variable on the Y-axis and the independent variable on the X-axis. In this example, light intensity is plotted on the X-axis and seedling growth rate on the Y-axis.

Let us consider how the results of a hypothetical experiment might be displayed graphically. Suppose investigators are interested in the effects of a plant hormone on the number of fruits produced by a certain plant. Investigators perform an experiment to determine the relationship, if any, between this hormone and fruit production. The investigators divide plants into 11 groups, each of which is treated identically except for the application of differing levels of the hormone. The investigators count the fruits produced by each plant. The results of this hypothetical experiment are shown in tabular form in Table 13.2 and graphically in Figure 13.11. Fruit production is the dependent variable and is plotted on the Y-axis; applied hormone concentration is on the X-axis. These data strongly suggest that the hormone boosts fruit production.

There are several important concepts illustrated by this graph:

1. *Thresholds.* In the central portion of the graph the points appear to form a straight line; that is, there appears to be a linear relationship between hormone level and fruit production. Below about 4.0 mg/L of hormone and above about 48.0 mg/L of hormone the relationship changes; that is, the data points are no longer approximated by the best-fit line for the points from 4.0 mg/mL to 48.0 mg/mL. 4.0 mg/L and 48.0 mg/L are threshold values. A **threshold** *is a point on a graph where there is a change in the relationship.* Thresholds at low and high values, as illustrated in Figure 13.11a, are common when

TABLE 13.2

Experimental Data

Hormone Level (mg/L) (Independent Variable)	Average Fruit Production Per Plant (Dependent Variable)
0.0	3.2
3.0	3.7
5.0	3.8
10.0	6.5
15.0	12.7
20.0	15.2
25.0	17.0
30.0	24.8
40.0	32.3
50.0	36.0
55.0	36.2
60.0	36.5

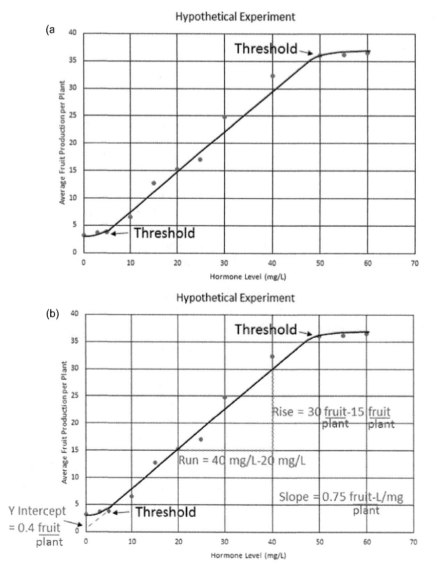

FIGURE 13.11   Hypothetical experiment. (a) The effect of hormone application on fruit production. (b) Determining the equation for the line.

working with biological data. Above 48.0 mg/mL and below 4.0 mg/mL the line is "flat," which means that there is no relationship between fruit production and hormone level.

2. ***Best-fit line.*** The middle points on the graph (those between the threshold points) are close to forming a line, and it is reasonable to conclude that the relationship between fruit production and hormone level is linear between 4.0 and 48.0 mg/L of applied hormone. Therefore, it is acceptable to represent these middle points with a straight line. A line may be drawn either "by eye" or using the statistical method of least squares.

Once the points between the threshold points are represented by a line, it is possible to determine the equation for the best-fit line. The slope has units since both the X- and Y-axes have units. The Y-axis has the unit of "average fruit production per plant," or, more simply, "fruit/plant"; the X-axis has the unit of mg/L. The slope of the line, shown in Figure 13.11b, is:

$$\frac{0.75(\text{fruit})(\text{L})}{(\text{plant})(\text{mg})}$$

To determine the Y-intercept of the line, it is necessary to determine where the intercept would be <u>if</u> there were no lower threshold. By using a ruler to extend the line to the Y-axis, the Y-intercept can be determined to be 0.4 fruit (Figure 13.11b). The equation for this line in Figure 13.11 is therefore:

$$Y = \frac{0.75(\text{fruit})(\text{L})}{(\text{plant})(\text{mg})} X + 0.40 \text{ fruit / plant}$$

The slope is the amount by which average production of fruit per plant increases with 1 mg/L of increased hormone. The Y-intercept is the amount of fruit production per plant expected if there were no hormone added and if the relationship was linear for all values of hormone (which is not the case).

3. ***Prediction.*** Equations and graphs are tools that can be used to make predictions. For example, the experiment did not involve testing the effect of 13.0 mg/L of hormone on fruit production. Yet, we can infer from the graph that if 13.0 mg/L of hormone were to be applied, the average fruit production per plant would be about 10. It is also possible to use the equation to predict the amount of fruit with 13.0 mg/L of hormone:

$$Y = \frac{0.75(\text{fruit})(\text{L})}{(\text{plant})(\text{mg})} \left(\frac{13 \text{ mg}}{\text{L}} \right) + \frac{0.40 \text{ fruit}}{\text{plant}}$$

$$\approx \textbf{10.2 fruit/plant}$$

The use of equations and graphs for prediction is powerful. However, when studying natural systems there are often thresholds. We cannot predict how much fruit production there will be with 100 mg/L of hormone because this is outside the linear range of our data. Even if there were no threshold at 48.0 mg/L, we still could not predict fruit production with 100 mg/L of hormone because we would not know if there was a threshold before reaching 100 mg/L.

4. ***Using graphs to summarize data.*** The graph of the hypothetical experiment contains the same information as Table 13.2. However, it is usually easier to see a relationship between two variables when the data are displayed on a graph than when they are listed in a table.

So far, we have looked at data where two variables are indeed related to one another. What would a graph look like if the two variables studied are not related to one another? One possibility is shown in Figure 13.12a, where the points appear to be scattered without pattern on the graph. Another possibility is shown in Figure 13.12b, where the graph is "flat." The value of the Y variable is constant regardless of the value for the variable on the X-axis. Observe in the hypothetical fruit and hormone experiment (Figure 13.11) the graph of the data is "flat" at high and low hormone levels. We can, therefore, reasonably conclude that at low and high levels of hormone, the production of fruit is controlled by factors other than the level of this hormone.

Figures 13.10 and 13.11 both illustrate situations where the variable on the Y-axis clearly appears to be related to the variable on the X-axis. The graphs in Figure 13.12 illustrate situations where two variables clearly appear to be unrelated to one another. In practice, it is sometimes ambiguous whether or not there is a relationship between two variables. For example, in Figure 13.13 there appears to be a weak relationship between the adult height of a daughter and the height of her mother. The relationship is inconsistent, however, because maternal height is not the only factor affecting a woman's adult height. Paternal genes and nutrition also affect height. The plot of daughter versus mother's height, therefore, does not form a very "good" line.

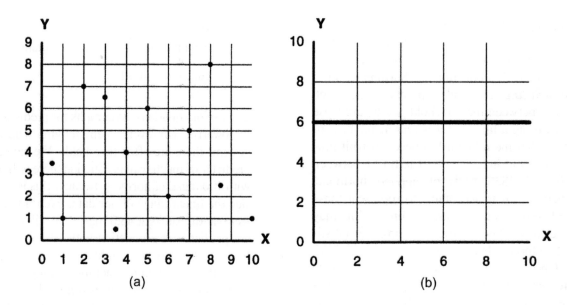

FIGURE 13.12   Graphs of variables that are not related.

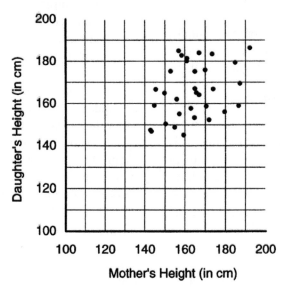

FIGURE 13.13   Two variables whose relationship to one another is ambiguous.

Application Practice Problems: Graphing

1. It is important to know whether exposure to agents such as low-level radiation, pesticides, or asbestos is likely to cause cancer. Two different predictions of cancer risk due to exposure to small amounts of cancer-causing materials are shown in the graphs.

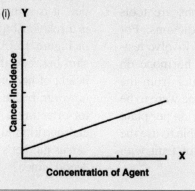

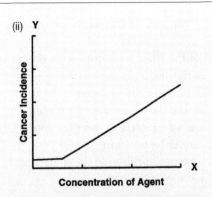

(ii)

Cancer Incidence

Concentration of Agent

a. If graph *i* is correct (no threshold) and a population of humans is exposed to small amounts of potentially carcinogenic materials, will the incidence of cancer in the population increase? Explain.

b. If graph *ii* is correct, will exposure to small amounts of potentially carcinogenic materials result in an increase in the incidence of cancer?

c. Speculate as to why exposures to small amounts of carcinogenic materials may not increase the likelihood of cancer while exposure to large amounts does increase the probability of cancer.

d. Explain why neither graph *i* nor *ii* intersect the Y-axis at zero.

e. Why is it important to know whether graph *i* or *ii* more accurately reflects the truth about a given material? (To learn more about this issue, see, for example, Goldman, M. "Cancer Risk of Low-Level Exposure." Science, vol. 271, no. 5257, 1996, pp. 1821–22. doi:10.1126/science.271.5257.1821. More recently, Nohmi, Takehiko. "Thresholds of Genotoxic and Non-Genotoxic Carcinogens." Toxicological Research, vol. 34, no. 4, 2018, pp. 281–90. doi:10.5487/tr.2018.34.4.281.)

2. Suppose an investigator is studying the genetics of plant productivity. The investigator determines the mass of 500 parent plants and 1,000 of their offspring plants. The investigator plots parent plant mass versus the average of offspring mass.

a. Which of the following graphs (i or ii) would indicate that there is a relationship between the mass of the parent plant and the mass of the offspring? Explain.

b. Suppose the data suggest that there is no relationship between the mass of the parent plant and the mass of its offspring. Suggest a hypothesis to explain that observation, and suggest an experiment to test your hypothesis.

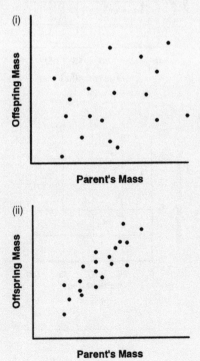

(i)

Offspring Mass

Parent's Mass

(ii)

Offspring Mass

Parent's Mass

3. a. Draw the best-fit line "by eye" for each of the following graphs. Be careful not to extrapolate (extend the line) past the data.
   b. Calculate the slope for each of the lines – do not forget the units.
   c. What is the Y-intercept for each line?
   d. What is the equation for each of these lines? Include the units.
   e. Examine graph i to determine the mosquito density if there are 5 in of rain.
   f. Use the equation for the line in graph i to predict mosquito density if there are 5 in of rain. (The answers for 3e and 3f should be the same.)
   g. From graph ii determine the average shrub's height at 20 months of age. Confirm your determination using the equation for the line in graph ii.
   h. From graph iii determine average seedling height with 50 mg of nutrient. Confirm your determination using the equation for the line in graph iii.

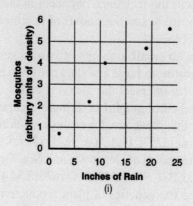

(i)

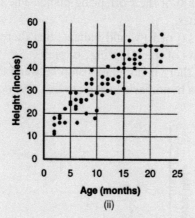

(ii)

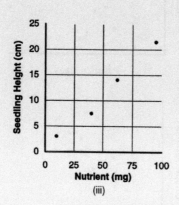

(iii)

4. Ten students took a midterm and final in a course. The scores for each student are plotted in the next graph.
   a. What was the approximate average score for the midterm: 20, 40, 60, or 80?
   b. What was the approximate average score for the final: 20, 40, 60, or 80?
   c. Which exam had lower scores?
   d. Was there a clear relationship between a student's midterm grade and their final grade? If the same class and the same exams were given the following year, could the teacher predict the final score of a student based on their midterm? Explain.

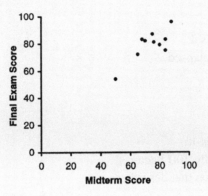

13.2 GRAPHS AND EXPONENTIAL RELATIONSHIPS

13.2.1 GROWTH OF MICROORGANISMS

13.2.1.1 The Nature of Exponential Relationships

Previous sections of this chapter focused on relationships that form a straight line when graphed. Although linear relationships are extremely important, not all relationships in the laboratory are linear. This section discusses two important examples of nonlinear relationships: (1) the relationship between the number of bacteria present and time elapsed; (2) the relationship between the amount of radioactivity present and time elapsed.

Suppose there is a single bacterial cell that divides to form two cells. The two cells each divide to form four cells, which divide into eight cells and so on. Suppose that the cells divide every hour and that the number of bacterial cells therefore doubles every hour. We say that these bacteria have a generation time of 1 hour.

The relationship between time elapsed and the number of bacteria in this example is summarized in Table 13.3 and graphed in Figure 13.14. Observe that this relationship does not form a straight line.

How could we write an equation that describes the relationship graphed in Figure 13.14? There are two variables: the number of generations that have occurred and the number of bacteria. The equation needs to include both variables to show that the population doubles at a regular interval. The equation that describes this relationship is:

$$Y = 2^t$$

TABLE 13.3

Bacterial Growth Where Generation Time Is 1 Hour

Time Elapsed (Hours)	Number of Bacterial Cells Present (N)
0	1
1	2
2	4
3	8
4	16
5	32
6	64
7	128

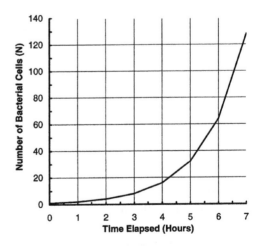

FIGURE 13.14 Bacterial growth. A plot where there is initially only one bacterial cell present and the bacteria population doubles every hour. A graph of this form is called *exponential*.

where:

t = the number of generations that have elapsed
Y = the number of bacterial cells present.

For example:

When t = 2, two generations have elapsed.

$$Y = 2^2 = 4$$

The number of bacteria cells present is 4.

When t = 4, four generations have elapsed.

$$Y = 2^4 = 16$$

The number of bacteria cells present is 16.

Now, suppose that there are initially 100 bacterial cells that double as before. The equation that describes these data is:

$$Y = 2^t(100)$$

This example is illustrated in tabular form in Table 13.4 and graphically in Figure 13.15. The graph where there are initially 100 cells present is much like the graph where there is initially only one bacterium, but with a Y-intercept of 100.

Being able to calculate bacterial numbers is important in the laboratory in situations where we want to predict approximately how many bacteria will be present in a culture after a certain time, or want to know how long a culture must be incubated to get a certain density of cells. It is possible to write a general growth equation that applies to any bacterial population, regardless of how many cells there are initially and regardless of how long it takes the population to double:

General Equation for Bacterial Population Growth

$$N = 2^t (N_o)$$

where:

N_o = the number of bacteria initially
N = the number of cells after t generations
t = the number of generations elapsed. (For example, if the bacteria double every 2 hours, then, after 6 hours three generations have elapsed.)

TABLE 13.4

Bacterial Growth Beginning with 100 Cells and Generation Time of 1 Hour

Time Elapsed (Hours)	Number of Bacterial Cells Present (N)
0	100
1	200
2	400
3	800
4	1,600
5	3,200
6	6,400
7	12,800

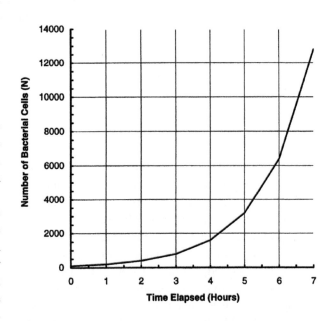

FIGURE 13.15 Bacterial growth. A plot where there are initially 100 bacterial cells present and the bacteria population doubles every hour.

Example Problem 13.4

A type of bacterium has a doubling time of 20 minutes. There are initially 10,000 bacteria present in a flask. How many bacteria will there be in one hour?

Answer

Strategy 1: Using a Table

Time Elapsed	Number of Generations Elapsed (t)	Number of Bacteria Present (N)
0	0	10,000
20 min	1	20,000
40 min	2	40,000
60 min	3	80,000

Thus, after 60 minutes we predict there will be 80,000 bacteria in the flask.

Strategy 2: Using the Equation for Bacterial Population Growth

Three generations have elapsed (the first at 20 minutes, the second at 40 minutes, and the third at 60 minutes). Substituting the number of generations elapsed and the number of cells originally present into the equation gives:

$$N = 2^t (N_0)$$

$$N = 2^3 (10,000)$$

$$N = 8(10,000)$$

$$N = 80,000$$

The equation for population growth and the "intuitive" approach both give the same answer.

These relationships between the number of bacteria present and the time elapsed (or the number of generations) are called **exponential** *because there is an exponent in their equation.* As you can see in Figures 13.14 and 13.15, an exponential relationship is not linear if it is plotted on a regular two-dimensional graph. Note that the relationship between population number and the number of generations elapsed is the same for other types of cultured cells as well.

13.2.1.2 Semi-log Graphs

It is more convenient to work with graphs that are linear in form than with those that are not. It is often desirable, therefore, to convert an exponential relationship to a form that is linear when plotted. The way this is accomplished involves logarithms. If, instead of plotting the number of bacterial cells on the Y-axis, we plot the log of the number of bacterial cells, then the plot is linear. This approach is illustrated in tabular form in Table 13.5 and graphically in Figure 13.16.

There is an alternative method to graph bacterial growth so that the plot is linear in form. This method involves the use of a **semi-log plot**, *a type of graph that substitutes for calculating logs.* A semi-log graph has normal, linear subdivisions on the X-axis; see Figure 13.17. The divisions on the Y-axis, however, are not even; rather, they are initially widely

TABLE 13.5

Bacterial Growth – Log of the Number of Cells Where Generation Time Is 1 Hour

Time Elapsed (Hours)	The Number of Cells Present (N)	Log of the Number of Cells
0	100	2.00
1	200	2.30
2	400	2.60
3	800	2.90
4	1,600	3.20
5	3,200	3.51
6	6,400	3.81
7	12,800	4.11

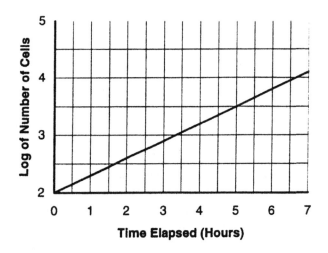

FIGURE 13.16 A log plot of bacterial growth. When the log of bacterial cell number is plotted versus time, a straight line is formed.

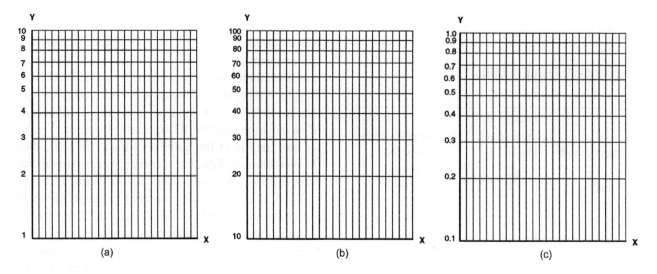

FIGURE 13.17 Labeling semi-log graphs. (a) Labeling the Y-axis when the first division is 1. (b) Labeling the Y-axis when the first division is 10. (c) Labeling the Y-axis when the first division is 0.1.

spaced and then become narrower toward the top of the graph. This spacing takes the place of calculating logs. The graphs shown in Figure 13.17 are called "semi-log" because only the Y-axis has logarithmic spacing.

Observe in Figure 13.17a that the Y-axis has 10 major divisions. These ten divisions together are called a "cycle." The first division in a cycle never begins with 0 because there is no log of 0. Rather, the first division is a power of 10, for example, 0.1, 1, 10, or 100. The second division in each cycle is twice the first, the third division is three times the first, and so on. For example, the Y-axis in Figure 13.17a begins at the bottom with 1. The next division, therefore, is 2, the next 3, and so on to 10. The Y-axis in Figure 13.17b begins with 10. The next division, therefore, is 20, then 30, and so on to 100. The Y-axis in Figure 13.17c begins with 0.1. The next division, therefore, is 0.2, then 0.3, and so on to 1.0.

Figure 13.17 illustrates one-cycle semi-log graphs. The graph in Figure 13.18a is called a *two-cycle semi-log graph* because it repeats the same pattern twice. A three-cycle semi-log graph, shown in Figure 13.18b, repeats the pattern three times. The beginning of each cycle is 10 times greater than the beginning of the previous cycle. For example, if the first cycle goes from 1 to 10, then the second cycle must range from 10 to 100, the third cycle from 100 to 1,000, and so on.

It is possible to prepare a semi-log graph with various numbers of cycles. To determine how many cycles you need, examine the data to be plotted. For example, in Tables 13.4 and 13.5 the Y values vary between 100 and 12,800. In this case, there is no need for a cycle from 1 to 10 or from 10 to 100, so the bottom cycle

can begin at 100. The first cycle required to plot these data runs from 100 to 1,000. The second cycle ranges from 1,000 to 10,000, and the third cycle, from 10,000 to 100,000, is needed to plot the value "12,800." Thus, a three-cycle semi-log graph is required to plot these data. Figure 13.19 shows the semi-log plot of the bacteria data from Table 13.4. Observe how the Y-axis is labeled in this figure. Also, compare the plots in Figures 13.16 and 13.19. Both graphs are linear in form, and either is an acceptable way to plot these data.

In principle, all biological populations have the potential to grow exponentially. If any population continued to grow exponentially for a long enough period of time, however, it would cover the earth and eventually the universe. In reality, population growth is limited by space, nutrients, waste buildup, and other factors. Bacterial growth normally looks like the plot in Figure 13.20. When a few bacteria that are not reproducing rapidly are placed in fresh media, there is initially a lag period as they utilize nutrients and prepare to divide. A period of exponential growth follows the lag period. The bacteria eventually deplete the nutrients in the broth and generate toxic waste products. Reproduction slows and stops, and the population declines.

13.2.2 THE DECAY OF RADIOISOTOPES

Radioactive substances decay over time so that there is progressively less and less radioactivity present. The **half-life** *of a radioactive substance is the time it takes for the amount of radioactivity to decay to half its original level.* Radioactivity can be measured in various units including disintegrations/minute, curie (Ci), and microcurie (μCi).

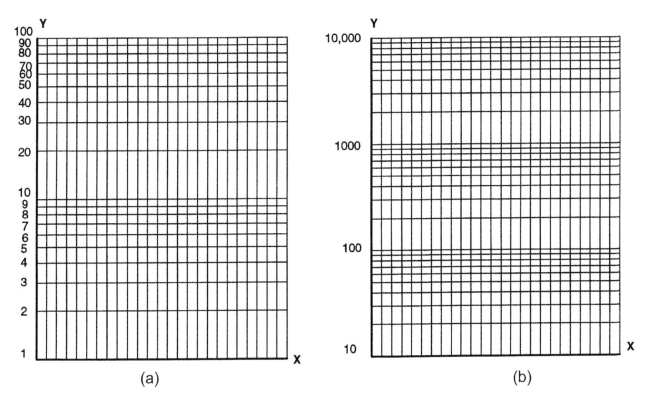

FIGURE 13.18   Cycles on semi-log graphs. (a) A two-cycle semi-log graph. (b) A three-cycle semi-log graph.

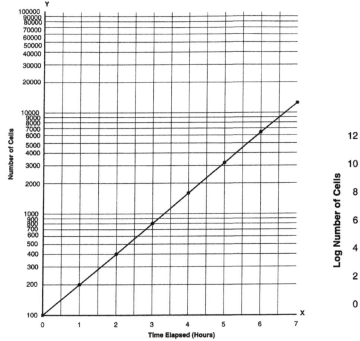

FIGURE 13.19   A semi-log plot of bacterial growth.

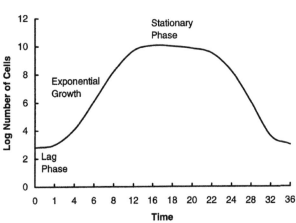

FIGURE 13.20   The limits to bacterial growth.

TABLE 13.6
Radioactive Decay (Half-Life Is 1 Hour)

Time Elapsed (Hours)	Half-Lives Elapsed	Radioactive Substance Remaining (Disintegrations/Minute)
0	0	400
1	1	200
2	2	100
3	3	50
4	4	25

For example, suppose a solution containing a radioactive substance initially undergoes 400 disintegrations/minute and that the half-life of the substance is 1 hour. Then, after 1 hour, there will be 200 disintegrations/minute of radioactivity remaining in the solution. After 2 hours there will be 100 disintegrations per minute, and so on. The relationship between time elapsed and radioactivity for this example is shown in tabular form in Table 13.6 and is graphed in Figure 13.21.

An exponential equation can be used to calculate the amount of radioactivity remaining after a certain number of half-lives have elapsed:

General Equation for Radioactive Decay

$$N = (1/2)^t\, N_0$$

where:

N = the amount of radioactivity remaining
t = the number of half-lives elapsed

N_0 = the amount of radioactivity initially.

For example, let us return to the example of the radioactive solution that has a half-life of 1 hour and an activity of 400 disintegrations/minute initially. How much radioactivity will remain after 3 hours? We can see from Table 13.6 that there are 50 disintegrations/minute after 3 hours. It is possible to get the same answer by substituting into the general equation for radioactive decay as follows (3 hours is the same as three half-lives for this substance):

$$N = \left(\frac{1}{2}\right)^3 \left(\frac{400 \text{ disintegrations}}{1 \text{ min}}\right)$$

$$= 50 \text{ disintegrations / min}$$

There is similarity between the equations for the decay of radioactivity and the growth of microorganisms. You can see this similarity by comparing Figures 13.15 and 13.21a. Note also the similarity in their equations, which can be expressed as shown on page 331.

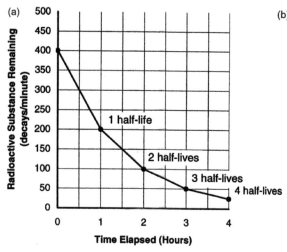

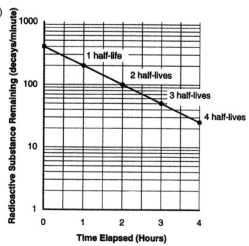

FIGURE 13.21   The relationship between time elapsed and radioactivity remaining. Half-life is 1 hour. (a) The relationship on a normal graph. (b) A semi-log plot of the same relationship.

Example Problem 13.5

^{131}Iodine has a half-life of 8 days. If you start with 500 μCi of ^{131}I, how much will remain after 80 days?

Answer

Strategy 1: Using a Table

Time Elapsed (days)	Half-Lives Elapsed	Radioactive Substance Remaining (μ*Ci*)
0	0	500
8	1	250
16	2	125
24	3	62.5
32	4	31.25
40	5	15.625
48	6	7.8135
56	7	3.9063
64	8	1.9531
72	9	0.9766
80	10	0.4883

The answer is 0.4883 μCi, or about 0.5 μCi remains.

Strategy 2: Using the General Equation for Radioactive Decay

1. Calculate how many half-lives have passed. The half-life of this isotope is 8 days, so the number of half-lives that have passed is: 80/8 = 10.
2. Substitute into the equation:

$$N = \left(\frac{1}{2}\right)^t (N_0)$$

$$N = \left(\frac{1}{2}\right)^{10} (500 \ \mu Ci)$$

$$N \approx 0.4883 \ \mu Ci$$

The answer is that only about 0.5 μCi remains. Eight days is a relatively short half-life. Isotopes with short half-lives rapidly disappear.

Example Problem 13.6

The half-life for the radioisotope ^{32}P is 14 days. You receive 200 μCi to perform an experiment on March 3. On April 30 you must complete paperwork showing how much ^{32}P activity remains. How much is present?

Answer

1. Calculate how many half-lives have passed. There are 58 days between March 3 and April 30. The half-live of this isotope is 14 days, so the number of half-lives that have passed is: 58/14 = 4.1.
2. Substitute into the equation:

$$N = \left(\frac{1}{2}\right)^t (N_0)$$

$$N = \left(\frac{1}{2}\right)^{4.1} (200 \ \mu Ci)$$

$$N \approx 11.7 \ \mu Ci$$

The answer to be recorded is 11.7 μCi remains (assuming you have not used any for an experiment).

A Comparison of the Equations for Bacterial Growth and Radioactive Decay

The growth of microorganisms: $N = 2^t (N_0)$

The decline of radioactivity $N = \left(\frac{1}{2}\right)^t (N_0)$

where:
N = the amount (of bacteria or radioactivity) present after a certain period of time
t = the time elapsed (expressed as the number of generations or the number of half-lives)
N_0 = the initial amount (of bacteria or radioactivity).

Application Practice Problems: Exponential Equations

1. There is an insect population in which each adult leaves 3 offspring and then dies. The generation time is 10 months; that is, the offspring grow up and reproduce every 10 months. In this type of organism, the adults die after they reproduce. The population begins with 20 adults and reproduces for four generations. The resulting population growth is shown in this table:

Generation	Time (months)	Number
0	0	20
1	10	60
2	20	180
3	30	540
4	40	1,620

    a. Graph this relationship.
    b. Determine the equation that describes this relationship.

2. The half-life of the radioisotope ^{32}P is 14 days. If there are initially 300 µCi of this radioisotope:
    a. How much will be left after 28 days?
    b. How much will be left after 140 days?
    c. How much will be left after 200 days?
    d. How much will be left after one year?

3. The half-life of the radioisotope ^{22}Na is 2.6 years. If there is initially 600 µCi of this radioisotope, how much will be left after:
    a. 1 year
    b. 2 years
    c. 3 years.

4. It is difficult to kill all microorganisms in a material. A solution or material to be sterilized is typically heated under pressure. The following graph shows the effect of time of exposure to heat and bacterial death. (Based on information from Perkins, John J. *Principles and Methods of Sterilization in the Health Sciences.* 2nd ed. Charles C. Thomas, 1983.)
    a. About how many bacteria were there at the beginning of the experiment, before exposure to heat?
    b. About how many bacteria were present after 2 minutes of treatment?
    c. Why did the investigators show their data on a semi-log plot?

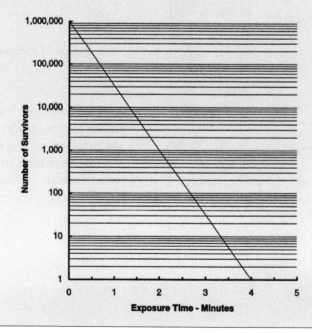

5. Label the Y-axis of this semi-log graph.

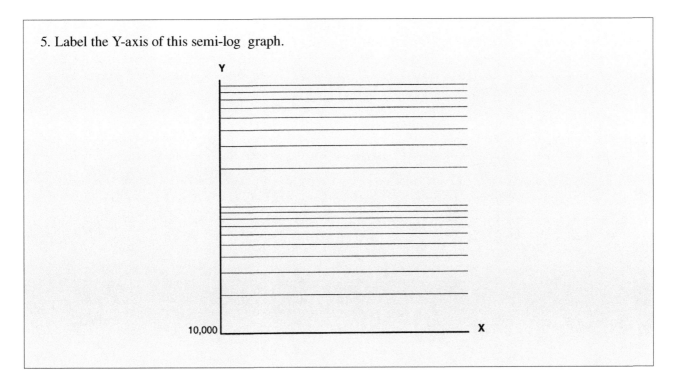

14 Descriptions of Data
Descriptive Statistics

14.1 INTRODUCTION

14.1.1 POPULATIONS, VARIABLES, AND SAMPLES

The natural world is filled with variability: Organisms differ from one another; the weather changes from day to day; the earth is covered by a mosaic of different types of habitats. The natural world is also characterized by uncertainty and chance. The outcome of a card game, the weather, and the personality of a newborn child are all uncertain. Scientists strive to observe and to understand this world, which is characterized by variability and uncertainty.

Statistics is a branch of mathematics that has developed over many years to deal with variability and uncertainty. Statistics provides methods to summarize, analyze, and interpret observations of the natural world. Statistical methods are also used to reach conclusions, based on data, with a certain probability of being right – and a certain probability of being wrong. In this chapter, we will explore a small area of statistics, and its use in summarizing, displaying, and organizing numerical data. We will begin by introducing several fundamental statistical terms.

A **population** *is an entire group of events, objects, results, or individuals, all of whom share some unifying characteristic(s).* Examples of populations include all of a person's red blood cells, all of the enzyme molecules in a test tube, and all of the college students in the United States.

Characteristics of a population often can be observed or measured, such as blood hemoglobin levels, the activity of enzymes, or the test scores of students. *Characteristics of a population that can be measured are called* **variables.** They are called "variables" because there is variation among individuals in the characteristic. A population can have numerous variables that can be studied. For example, the same population of 6-year-old children can be measured for height, shoe size, reading level, or any of a number of other characteristics. *Observations of a variable are called* **data** (singular "datum").

Observations may or may not be numerical. For example, the lengths of insects (in centimeters) are numerical data. In contrast, electron micrographs of mouse kidney cells are non-numerical observations. The statistical methods discussed in this unit are used to interpret numerical data.

DOI: 10.1201/9780429282799-18

Population—All 20-year-old men in the United States

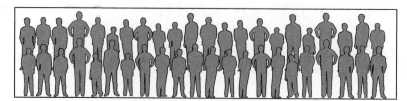

Variables—Many variables could be measured for this population such as:
height, hair color, college entrance exam score, and birthplace

Sample—A sample of the population of all 20-year-old men is drawn

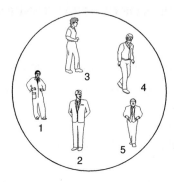

Data—Observations are made of the individuals in the sample

Individual	Height (in cm)	Hair Color	College Exam Score	Birthplace
1	172	brown	1500	WI
2	173	blond	none	IL
3	169	black	1800	NY
4	177	brown	none	WY
5	175	blond	1310	MO

FIGURE 14.1 Populations, samples, variables, and data. A sample is drawn from the population to represent that population. In this illustration, data for four variables are collected for each individual in the sample.

It is seldom possible to determine the value for a given variable for every member of a population. For example, it is virtually impossible to measure the level of hemoglobin in every red blood cell of a patient. Rather, a **sample** of the patient's blood is drawn and the hemoglobin level is measured and evaluated. Investigators draw conclusions about populations based on samples (Figure 14.1).

Statistical methods are generally based on the assumption that a sample is **representative of its population** – that is, *the sample truly reflects the variability in the original population.* In order for a sample to be representative of its population, it must meet two requirements: (1) *All members of the population must have an equal chance of being chosen.* If this is the case, the sample is called **random.** (2) *The choice of one member of the sample should not influence the choice of another.* If this is the case, the sample is said to have been drawn **independently.**

Note that the statistical meaning of "random" is different from the meaning in common English. In popular usage random means "haphazard," "unplanned," or "without pattern"; this is not its meaning in statistics. Randomness refers to how a sample is taken. For example, a lottery is random if every ticket has an equal chance of being drawn. If a deck of cards is properly shuffled and if a card is picked by chance from that deck, then the choice of the card is random.

To understand independence, consider a coin that is flipped twice. The second toss has a 50% probability of being heads regardless of whether or not the first toss was heads. This is because the outcomes of the two tosses are independent of one another. In contrast, suppose we begin with a full deck of cards and randomly draw and set aside one card, which turns out to be the five of hearts. A second draw from this deck is not independent because there is now no chance of drawing the five of hearts.

The individuals that comprise a population vary from one another; therefore, even when a sample is drawn randomly and independently, there is uncertainty when a sample is used to represent a whole population. The sample could truly reflect the features of the population from which it was drawn, but perhaps it does not. As you might expect, if a sample is properly drawn, then the larger the sample, the more likely it is to accurately reflect the entire population.

Example Problem 14.1

a. *In a quality-control setting,* 15 vials of product from a batch were tested. What is the sample? What is the population?
b. *In an experiment,* the effect of a carcinogenic compound was tested on 2000 laboratory rats. What is the sample? What is the population?
c. *A political poll* of 1000 voters was conducted. What is the sample? What is the population?
d. *A clinical study* of the effect of a new drug was tested on 50 patients. What is the sample? What is the population?

Answer

a. The sample is the 15 vials tested. The population is all the vials in the batch.
b. The sample is the 2000 rats. The population is presumably all laboratory rats. However, if the rats in the study were all of the same strain, then the population is all laboratory rats of that strain.
c. The sample is the 1000 voters polled. The population is all voters.
d. The sample is the 50 patients tested. The population is all similar patients.

Example Problem 14.2

What is meant by the statement "two out of three doctors surveyed recommend Brand X..."? What is the population of interest? What is the sample? Is this sample representative of the population? Does this statement ensure that the product being endorsed is better than any other?

Answer

Many abuses of statistics relate to poor sampling. A classic type of abuse is statements such as "two out of three doctors surveyed recommend Brand X...". The population of interest is all doctors. This statement suggests that two-thirds of all doctors prefer Brand X to other brands; however, we have no way to know which doctors were sampled. For example, perhaps the survey only looked at physicians who prescribe Brand X, or only physicians in a certain region. It is possible that the survey sampled doctors who were not representative of all doctors; therefore, the statement does not ensure that most doctors prefer Brand X. In addition, even if two-thirds of all doctors really do recommend Brand X, it may or may not be the best brand – it may simply be the most accessible, least expensive, most advertised, and so on.

14.1.2 Describing Data Sets: Overview

When a sample is drawn from a population, and the value for a particular variable is measured for each individual in the sample, the resulting values constitute a data set. Because individuals differ from one another, the data set consists of a group of varying values.

A set of data without organization is something like letters that are not arranged into words. Like letters of the alphabet, numerical data can be arranged in ways that are meaningful. (Numerical data, like letters of the alphabet, can also be organized in ways that are confusing, deceptive, or irrelevant. Delving into the many ways that statistics are misused, however, is beyond the scope of this book.) **Descriptive statistics** *is a branch of statistics that provides methods to appropriately organize, summarize, and describe data.*[1]

Consider, for example, the exam scores for a class of students. The variable that is measured is the exam score for each individual. A list of the scores of all the students constitutes the data set. In order to summarize and describe the performance of the class as a whole,

---

[1] There is another branch of statistics, called *inferential statistics,* that provides methods to make predictions about a population based on a sample. Inferential statistics is beyond the scope of this book.

the instructor might calculate the class's average score. The average is an example of a measure that summarizes the data. A measure that describes a sample, such as the average, is sometimes referred to as a "statistic."

The average gives information about the "center" of a data set. There are other measures, such as the median and the mode (described below), which also describe data in terms of its center. *Such measures, which describe the center of a data set, are called* **measures of central tendency.**

Data sets A and B both have the same average:

A : 4 5 5 5 6 6

B : 1 2 4 7 8 9

Inspection of these values, however, reveals that the two data sets have a different pattern. The data in A are more clumped about the central value than the data in B. We say that the data in B are more **dispersed,** that is, *spread out.*

Measures of central tendency, such as the average, do not describe the dispersion of a set of observations; therefore, there are statistical **measures of dispersion** *which describe how much the values in a data set vary from one another.* Common measures of dispersion are range, variance, standard deviation, and coefficient of variation, all of which will be discussed below.

Measures of central tendency and of dispersion are determined by calculation. The next section discusses the calculation of these measures. It is also possible to organize and effectively display data using graphical techniques. Graphical techniques and their interpretation are discussed in Section 14.3 of this chapter. Section 14.4 applies these ideas to the area of controlling product quality in a company.

14.2 DESCRIBING DATA: MEASURES OF CENTRAL TENDENCY AND DISPERSION

14.2.1 Measures of Central Tendency

Consider a hypothetical data set consisting of these values:

2, 5, 6, 7, 8, 3, 9, 3,

10, 4, 7, 4, 6, 11, 9

The simplest way to organize these values is to order them as follows:

2, 3, 3, 4, 4, 5, 6, 6, 7, 7, 8, 9, 9, 10, 11

Inspection of the ordered list of numbers reveals that they center somewhere around 6 or 7. The center can be calculated more exactly by determining the average or mean. The **mean** *is the sum of all the values divided by the number of values.* In this example the mean is calculated as:

$$2+3+3+4+4+5+6+6+7$$

$$+7+8+9+9+10+11 = 94$$

The average or the mean is 94 / 15 = 6.3

In algebraic notation, the observations are called X_1, X_2, X_3, and so on. In this example there are 15 observations, so the last value is X_{15}. The method to calculate the mean can be written compactly as:

$$mean = \frac{\sum X}{n}$$

where:
n = the number of values.
Σ (the Greek letter, sigma) is short for "add up." X means that what is to be added is all the values for the variable, X. Thus:

$$mean = \frac{\sum X}{n} = \frac{X_1 + X_2 + X_3 + \cdots + X_n}{n}$$

Statisticians distinguish between the true mean of an entire population and the mean of a sample from that population (Figure 14.2). The true mean of a population is represented by μ (the Greek letter, mu). The mean of a sample is represented by $\bar{x}$ (read as "X bar"). In practice, it is rare to know the true mean for an entire population; therefore, the sample mean, or $\bar{x}$ must be used to represent the true population mean.

The median is another measure of central tendency. The **median** is the middle value of a data set, or the number that is halfway between the highest and lowest value, when the values are arranged in ascending order. The median for the following values is 6:

2, 3, 3, 4, 4, 5, 6, <u>6,</u> 7, 7, 8, 9, 9, 10, 11

When there is an even number of values, the median is the value midway between the two central numbers. The median for these values is 7.5 or the average of the numbers 7 and 8:

3, 5, <u>7, 8,</u> 9, 11

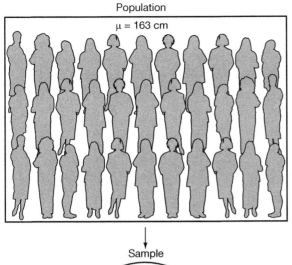

Population

$\mu = 163\ cm$

Sample

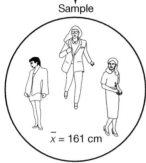

$\bar{x} = 161\ cm$

FIGURE 14.2 The true mean of a population is distinguished from the mean of a sample. The population of interest is all women in the United States; the variable of interest is height. The mean height, μ, for all women is 163 cm. A sample of women is drawn, and their heights are measured. The mean height in the sample, $\bar{x}$, is 161 cm.

Observe that in both of the following data sets the median is 5 even though their means are quite different:

2, 3, 4, 5, 6, 7, 8	$\bar{x} = 5$
2, 3, 4, 5, 6, 100, 1000	$\bar{x} = 160$

The **mode** *is the value of the variable that occurs most frequently.* For example, the mode in the following data set is "3" since it appears more often than the other numbers:

$$2,\ 3,\ 3,\ 3,\ 4,\ 5,\ 6,\ 7,\ 7,\ 8$$

It is possible for a set of data to have more than one mode. *If there are two modes, the data are* **bimodal.** *If there are more than two modes, the data are* **polymodal.**

Box 14.1 summarizes calculations of these three measures of central tendency – the mean, median, and mode.

Example Problem 14.3

Suppose you are investigating the height of college students. You measure the height of every student in your classes and obtain the data shown in the following table.

a. What is the average height of the men in the class?
b. What is the average height of the women in the class?
c. What is the average height of all students in the class?

Heights of Students in the Class (in cm)					
Men			**Women**		
175.3	176.5	177.8	162.5	166.7	155.6
168.5	165.2	160.2	159.7	163.4	164.2
180.9	188.9	171.6	160.1	168.6	162.6
175.9	174.8	179.2	162.8	158.4	174.9
175.9	175.8	174.8	161.3	166.2	166.7
$n_{men} = 15$ (# of men)			$n_{women} = 15$ (# of women)		

Answer

a. $\bar{x} = \dfrac{2621.3\ cm}{15} \approx 174.75\ cm\ (men)$

b. $\bar{x} = \dfrac{2453.7\ cm}{15} \approx 163.58\ cm\ (women)$

c. $\bar{x} = \dfrac{5075.0\ cm}{30} \approx 169.17\ cm\ (men + women)$

14.2.2 Measures of Dispersion

14.2.2.1 The Range

The **range** is the simplest measure of dispersion. The **range** is the difference between the lowest value and the highest value in a set of data:

$$2\ 3\ 3\ 3\ 4\ 4\ 5\ 6\ 7\ 7\ 8\ 9\ 9\ 10\ 11$$

The range is 2 to $11 = 9$

The range is not a particularly informative measure because it is based on only two values from the data set. A single extreme value will have a major effect on the range.

BOX 14.1 MEASURES OF CENTRAL TENDENCY: CALCULATING THE MEAN, MEDIAN, AND MODE

1. *To calculate the mean, add all the values and divide by the number of values present.*

$$\text{mean} = \frac{\sum X}{n}$$

2. *To find the median:*
   Order the data values.
   If there is an odd number of values, the median is the middle value in the sequence.
   If there is an even number of values, the median is the average of the two middle values.
3. *Mode:* Order the data values. The mode is the most frequent observation. There may be more than one mode.

EXAMPLE

Given the data set: 12, 15, 13, 12, 13, 14, 16, 19, 21, 13, 15, 14
  1. Order the data 12, 12, 13, 13, 13, 14, 14, 15, 15, 16, 19, 21
  2. The mean is $\dfrac{12+12+13+13+13+14+14+15+15+16+19+21}{12} = 14.75$
  3. The median is $\dfrac{14+14}{2} = 14$
  4. The mode is 13.

14.2.2.2 Calculating the Variance and the Standard Deviation

The variance and the standard deviation are more informative measures of dispersion than the range because they summarize information about dispersion based on all the values in the data set. Variance and standard deviation are best explained with a simple example. Assume that the following data are lengths (in millimeters) of eight insects:

$$4, 5, 6, 7, 7, 7, 9, 11$$

The mean for these data is 7 mm. An intuitive way to calculate how much the data are dispersed is to take each data point, one at a time, and see how far it is from the mean (Table 14.1). *The distance of a data point from the mean is its* **deviation.** There is a deviation for each data point: The deviation is sometimes positive, sometimes negative, and sometimes zero.

It is useful to summarize all the deviations with a single value. An obvious approach would be to calculate the average of the deviations, but, for any data set, the sum of the deviations from the mean is always

TABLE 14.1

Calculation of Deviation from the Mean

(Value – Mean)	Deviation
First value – mean =	(4−7) = −3 mm
Second value – mean =	(5−7) = −2 mm
Third value – mean =	(6−7) = −1 mm
Fourth value – mean =	(7−7) = 0 mm
Fifth value – mean =	(7−7) = 0 mm
Sixth value – mean =	(7−7) = 0 mm
Seventh value – mean =	(9−7) = +2 mm
Eighth value – mean =	(11−7) = +4 mm
	0
	Sum of all the deviations

zero. Mathematicians therefore devised the approach of squaring each individual deviation value to result in all positive numbers. *The squared deviations can be added together to get the* **total squared deviation,** also called the **sum of squares.** In this example, the sum of squares is 34 mm² (Table 14.2). The sum of squares is always a positive number that indicates how much the values in a set of data deviate from the mean.

TABLE 14.2

Calculation of the Sum of Squares

(Value – Mean) (mm)	(Deviation) (mm)	(Deviation Squared)
(4−7)	−3	$9\,mm^2$
(5−7)	−2	$4\,mm^2$
(6−7)	−1	$1\,mm^2$
(7−7)	0	$0\,mm^2$
(7−7)	0	$0\,mm^2$
(7−7)	0	$0\,mm^2$
(9−7)	+2	$4\,mm^2$
(11−7)	+4	$\underline{16\,mm^2}$
		$34\,mm^2$
		Total squared deviation=the sum of squares

The **variance** *is the total squared deviation divided by the number of measurements*:

variance (of a population)

$$= \frac{\text{total squared deviations from the mean}}{n}$$

$$= \frac{\sum (X_i - \text{mean})^2}{n}$$

where:

Σ=the sum of

X_i=individual value

n=the number of values.

In this example n=8 and the variance is:

$$\text{variance} = \frac{34\ mm^2}{8} = 4.25\ mm^2$$

The standard deviation is calculated by taking the square root of the variance:

$$\text{standard deviation} = \sqrt{\text{variance}}$$

In this example:

$$\text{standard deviation} = \sqrt{4.25\ mm^2} = 2.06\ mm$$

Note that the standard deviation has the same units as the data.

The standard deviation and the variance are values that summarize the dispersion of a set of data around the mean. The larger the variance and the standard deviation, the more dispersed are the data.

14.2.2.3 Distinguishing between the Variance and Standard Deviation of a Population and a Sample

In the preceding discussion we skipped over a significant detail regarding the calculation of the variance and standard deviation. Statisticians distinguish between the variance and standard deviation of a population and the variance and standard deviation of a sample (Figure 14.3). *The variance of a population is called* σ^2 *(read as "sigma squared"). The variance of a sample is called* **S²**. *The standard deviation of a population is called* **σ** *(sigma). The standard deviation of a sample is sometimes abbreviated* **S** *and sometimes* **SD**. Terminology relating to populations and samples is summarized in Table 14.3.

The formulas used to calculate the variance and standard deviation of a sample are slightly different than those for a population. In the preceding discussion we showed how to calculate the variance and standard deviation of a population. For a sample, the denominator is $n - 1$ rather than n. The formulas for variance and standard deviation of a sample are thus:

variance (of a sample) $= S^2$

$$= \frac{\text{sum of squared deviations from the mean}}{n-1}$$

$$= \frac{\sum (X_i - \bar{x})^2}{n-1}$$

TABLE 14.3

Terminology Relating to Populations and Samples

Population	Sample
An entire group of events, objects, results, or individuals where each member of the group has some unifying characteristic(s).	A subset of a population that represents the population.
Parameters	**Statistics**
Measures that describe a population.	Measures that describe a sample.
Mean μ	Mean $\bar{x}$
Variance σ^2	Variance S^2
Standard deviation σ	Standard deviation S or SD.

Example Problem 14.4

The heights for a sample of seven college women are shown in the following table. Calculate the mean ($\bar{x}$), the deviation (d) of each value from the mean, the squared deviation (d^2), the variance (S^2), and the standard deviation (SD).

Measured Value (cm)	d (cm)	d^2 (cm²)
162.5		
166.7		
155.6		
159.7		
163.4		
164.2		
160.1		

Answer

$\bar{x} \approx 161.74 \text{ cm}$

Measured Value (cm)	d (cm)	d^2 (cm²)
162.5	0.757	0.573
166.7	4.96	24.6
155.6	−6.14	37.7
159.7	−2.04	4.16
163.4	1.66	2.76
164.2	2.46	6.05
160.1	−1.64	2.69
	Summation	≈78.53 cm²

$n - 1 = 6$

$$S^2 = \frac{78.53 \text{ cm}^2}{6} \approx 13.09 \text{ cm}^2$$

$$SD = \sqrt{13.09 \text{ cm}^2} \approx 3.618 \text{ cm}$$

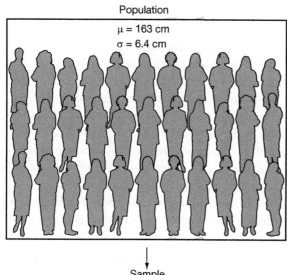

Population

$\mu = 163$ cm
$\sigma = 6.4$ cm

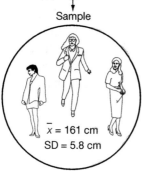

Sample

$\bar{x} = 161$ cm
SD = 5.8 cm

FIGURE 14.3 The mean and standard deviation of a population versus a sample. The population has a true mean and standard deviation. Every sample drawn from a population has its own mean and standard deviation, which are likely to differ from the true mean and standard deviation of the population.

standard deviation (of a sample) $= \text{SD} = \sqrt{S^2}$

$$\text{or SD} = \sqrt{\frac{\sum (X_i - \bar{x})^2}{n-1}}$$

where:

$\Sigma =$ the sum of

$X_i =$ each value

$n-1 =$ the number of values minus 1.

In this text we are generally looking at data from a sample and so routinely use $n-1$ in the denominator of the equation.

The standard deviation is a widely used statistic, so being able to calculate it for a given set of data is an important skill. Many scientific calculators are set up to calculate the standard deviation easily. The operator can often enter the data and then press one key to get the standard deviation for a population and a different key to get the standard deviation for a sample. (Refer to the calculator instructions.) There is an alternative equation for standard deviation that is useful if your calculator does not have the standard deviation function built in, which is given in Box 14.2. There are also statistical computer programs that are useful for analyzing large sets of data.

14.2.2.4 The Coefficient of Variation (Relative Standard Deviation)

Table 14.4 shows the weights of laboratory mice and rabbits. Observe that the standard deviation for the

rabbits' weights has a higher value than that for the mice, but, of course, rabbits weigh more than mice.

The dispersion of a data set can be expressed as the standard deviation divided by the mean. This is called the **coefficient of variation (CV)** or the **relative standard deviation (RSD)**. The formula for the CV is:

$$\text{CV} = \frac{\text{standard deviation} (100\%)}{\text{mean}}$$

When the dispersions of the weights of mice and rabbits are expressed in terms of the coefficient of variation, the value is slightly higher for the mouse data than it is for the rabbit data.

Note that the mean and the standard deviation have units, but the units cancel when the CV is calculated. We will come across both the standard deviation and the coefficient of variation in Unit V, where we talk about measurements.

Methods of calculating various measures of dispersion are summarized in Box 14.2.

TABLE 14.4
Weights of Laboratory Mice and Rabbits

Weights of Laboratory Mice (g)	Weights of Laboratory Rabbits (g)
32	3178
34	3500
24	3428
33	2908
36	2757
30	3100
36	2876
36	3369
30	3682
34	2808
$\bar{x} = 32.5\,\text{g}$	$\bar{x} \approx 3161\,\text{g}$
$\text{SD} \approx 3.75\,\text{g}$	$\text{SD} \approx 322.8\,\text{g}$
$\text{CV} \approx 11.5\%$	$\text{CV} \approx 10.2\%$

Example Problem 14.5

A biotechnology company sells cultures of a particular strain of the bacterium *E. coli*. The bacteria are grown in batches that are freeze-dried and packaged into vials. Each vial is expected to contain 200 mg of bacteria. A quality-control technician tests a sample of vials from each batch and reports the mean weight and the standard deviation.

Batch Q-21 has a mean weight of 200 mg and a standard deviation of 12 mg. Batch P-34 has a mean weight of 200 mg and a standard deviation of 4 mg. Which lot appears to have been packaged in a more controlled (consistent) fashion?

Answer

The standard deviation can be interpreted as an indication of consistency. The SD of the weights in Batch P-34 is lower than in Batch Q-21. The values for Batch P-34, therefore, are less dispersed than those for Batch Q-21. The packaging of Batch P-34 appears to have been better controlled.

BOX 14.2   MEASURES OF DISPERSION: THE VARIANCE, THE STANDARD DEVIATION, AND THE RELATIVE STANDARD DEVIATION

1. To calculate the variance (S^2) for a sample, use the equation:

$$S^2 = \frac{\sum (X - \bar{x})^2}{n-1}$$

2. To calculate the standard deviation (SD) for a sample:

$$SD = \sqrt{\frac{\sum (X - \bar{x})^2}{n-1}}$$

There is an alternative equation that can be used to calculate SD:

$$SD = \sqrt{\frac{\sum X^2 - \frac{\left(\sum X\right)^2}{n}}{n-1}}$$

3. To calculate the coefficient of variation (CV):

$$CV = \frac{\text{Standard deviation}(100\%)}{\text{mean}}$$

EXAMPLE

Given these data from a sample:

1.00 mm, 2.00 mm, 2.00 mm, 3.00 mm, 3.00 mm, 4.00 mm, 4.00 mm, 5.00 mm

$\bar{x} = 3.00$ mm

1. $S^2 = \dfrac{((1-3)\text{mm})^2 + ((2-3)\text{mm})^2 + ((2-3)\text{mm})^2 + ((3-3)\text{mm})^2 + ((3-3)\text{mm})^2 + ((4-3)\text{mm})^2 + ((4-3)\text{mm})^2 + ((5-3)\text{mm})^2}{8-1}$

$= \dfrac{12.00 \text{ mm}^2}{7} \approx 1.71 \text{ mm}^2$

2. The standard deviation is the square root of the variance ≈ 1.31 mm.

   Note that the standard deviation and the variance have units if the data have units.

   Note that using the alternate formula shown in # 2 gives the same answer:

$$\sum X^2 = 1 + 4 + 4 + 9 + 9 + 16 + 16 + 25 = 84 \text{ mm}^2$$

$$\frac{\left(\sum X\right)^2}{n} = 1 + 2 + 2 + 3 + 3 + 4 + 4 + 5 = \frac{(24)^2}{8} = 72 \text{ mm}^2$$

$$SD = \sqrt{\frac{84 \text{ mm}^2 - 72 \text{ mm}^2}{7}} \approx 1.31 \text{ mm}$$

3. The coefficient of variation is:

$$CV = \frac{1.31 \text{ mm} \times 100\%}{3.00 \text{ mm}} \approx 43.67\%$$

14.2.3 Summarizing Data by the Mean and a Measure of Dispersion

It is common in a scientific paper or in other technical literature to see a set of data summarized using the mean followed by a measure of dispersion. The smaller the measure of dispersion, the closer the data points are to one another. Thus, you might encounter a mean reported in any of the following styles:

1. A mean may be reported as the mean value ± the standard deviation (Figure 14.4a). For example, 6.02 ± 0.23 (SD) means the sample

mean is 6.02 and the standard deviation of the sample is 0.23.

2. **You may see a mean reported in a scientific paper as the mean value ± standard error of the mean** (Figure 14.4b). The standard error of the mean (SEM) is not actually a measure of error; rather, it is a measure of dispersion relating to the mean. It is an indication of how much discrepancy there is likely to be in a sample's mean compared to the true mean of the population. The complete meaning of the SEM is outside the scope of this text, but, when encountered in graphs and tables,

---

EFFECT OF EXERCISE ON BLOOD PRESSURE: BLOOD PRESSURE BEFORE STARTING AN EXERCISE PROGRAM AND AFTER 16 WEEKS OF SOFT TRAINING

Variable	Baseline mean ± SD n = 43	16 Weeks mean ± SD n = 43
Maximal heart rate	153 ± 15	153 ± 11
Weight (kg)	97 ± 16	97 ± 17
Peak oxygen uptake (mL/kg/min)	21 ± 4	23 ± 4
Submaximal systolic blood pressure (mm Hg)	218 ± 23	187 ± 30
Submaximal diastolic blood pressure (mm Hg)	107 ± 10	94 ± 9

(a)

PHYSICAL TRAITS OF FOUR DIFFERENT SUBSPECIES OF THE WHEAT *TRITICUM MONOCOCCUM*

		Mean Value (± SEM)	
Subspecies	Number Studied	Seed Weight (mg)	Awns per Spikelet
T.m. a	250	21.1 ± 0.2	2.50 ± 0.05
T.m. b	68	30.2 ± 0.7	1.00 ± 0.04
T.m. c	9	22.9 ± 0.7	1.10 ± 0.14
T.m. d	11	21.5 ± 0.9	2.33 ± 0.14

(b)

TOXICITY OF INSECTICIDE PRODUCED BY *E. COLI* RECOMBINANT CELLS

Insect	LC_{50} (mg /mL)*
Cutworm	10.9 (7.0–17.0)
Aphids	17.4 (9.7–29.2)
Beetles	147.8 (79.2–285)

*95% confidence intervals are shown in parentheses

(c)

FIGURE 14.4 Examples of how the mean is reported in the scientific literature. (a) A study of the effect of exercise on individuals with high blood pressure. Values are reported as the mean ± the standard deviation. (b) A study of wheat in which physical traits are reported as the mean ± the SEM. (c) A study of insecticide toxicity on three species of insects. Results are reported as the mean and the 95% confidence interval.

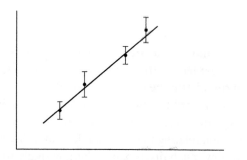

FIGURE 14.5 The graphical representation of sample means; error bars. Each point on the line in this graph represents the mean of the observations of 20 individuals. The error bar represents ± 1 SD.

the simplest way to understand the SEM is to interpret it like the standard deviation: A larger value means there is more variability in the population than a smaller value.

3. **You may see a mean reported in a scientific paper as the mean value and the 95% (or 90% or 99%) confidence interval** (Figure 14.4c). A 95% confidence interval is a range of values that is constructed in such a manner that if 100 random samples were drawn from the population and a confidence interval were computed for each sample, then we would expect 95 of the 100 confidence intervals to include the true mean of the population. It can also be said that "We are 95% confident that μ is within this range of values." The variability of the data and the sample size are considered when a confidence interval is calculated: the more the variability in the data, the wider the range of the interval.

4. **You may see a mean reported graphically.** The points on the graph in Figure 14.5 represent sample means. The vertical lines are called *error bars*. The error bars may represent ± one standard deviation. Error bars may also be used to represent ± the standard error of the mean, or they may show the confidence interval. The interpretation of error bars is customarily explained in the figure caption.

14.3 DESCRIBING DATA: FREQUENCY DISTRIBUTIONS AND GRAPHICAL METHODS

14.3.1 ORGANIZING AND DISPLAYING DATA

Consider a set of data consisting of the times it took 10 students to run a race (in minutes):

9.6,   10.3,   9.1,   7.5,   10.9,   7.2,   6.9,   9.5,   8.7,   8.1

These values are displayed in a **dot diagram** (Figure 14.6).

A dot diagram is a simple graphical technique used to illustrate small data sets. In this technique, each data value is represented as a dot. The dot is situated above the corresponding value on an axis. With a dot diagram it is easy to see the general location and center of the data. The dispersion of data is also evident in such diagrams (Figure 14.7).

A dot diagram is difficult to construct and interpret when a data set is large, and there are other graphical methods that are more useful for organizing and displaying large sets of data. For example, suppose a researcher collects data for the weights of a number of field mice (Table 14.5). An informative way to tabulate these data is to list them by frequency where the **frequency** *is the number of times a particular value is observed*. Table 14.6 shows the values for the mouse weights and the frequency of each value.

Observe in Table 14.6 that the values for mouse weights have a pattern. Most of the mice have weights in the middle of the range, a few are heavier than average, and a few are lighter (Figure 14.8). The term **distribution** *refers to the pattern of variation for a given variable*. It is very important to be aware of the patterns, or distributions, which emerge when data are organized by frequency.

The frequency distribution in Table 14.6 can be illustrated graphically as a **frequency histogram** (Figure 14.9). A frequency histogram is a graph where the X-axis is the relevant unit of measurement. In this example, it is weight in grams. The Y-axis is the frequency of occurrence of a particular value. For example, 11 mice are recorded as weighing 19 g. The values for these 11 mice are illustrated as a bar. Note that when the mouse data were collected, a mouse recorded as

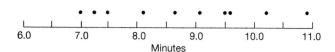

FIGURE 14.6 A data set displayed as a dot diagram.

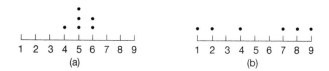

FIGURE 14.7 Dot diagrams illustrate dispersion. (a) Less dispersed data. (b) More dispersed data.

TABLE 14.5
The Weights of 175 Field Mice (in Grams)

21	23	22	19	22	20	24	22	19	24	27	20	21
22	20	22	24	24	21	25	19	21	20	23	25	22
19	17	20	20	21	25	21	22	27	22	19	22	23
22	25	22	24	23	20	21	22	23	21	24	19	21
22	22	25	22	23	20	23	22	22	26	21	24	23
21	25	20	23	20	21	24	23	18	20	23	21	22
22	25	21	23	22	24	20	21	23	21	19	21	24
20	22	23	20	22	19	22	24	20	25	21	22	22
24	21	22	23	25	21	19	19	21	23	22	22	24
21	23	22	23	28	20	23	26	21	22	24	20	21
23	20	22	23	21	19	20	26	22	20	21	22	23
24	20	21	23	22	24	21	23	22	24	21	22	24
20	22	21	23	26	21	22	23	24	21	23	20	20
21	25	22	20	22	21							

TABLE 14.6
Frequency Distribution Table of the Weights of Field Mice

Weight (g)	Frequency
17	1
18	1
19	11
20	25
21	34
22	40
23	27
24	19
25	10
26	4
27	2
28	1

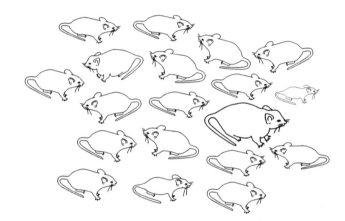

FIGURE 14.8 Distribution of mouse weights. Most mice are of about the average weight: Some are a bit heavier; some a bit lighter than others. A few mice are substantially heavier, and a few mice are substantially lighter than average.

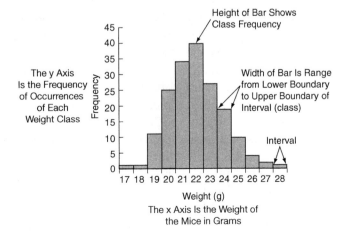

FIGURE 14.9 Frequency histogram for mouse data.

BOX 14.3 CONSTRUCTING A FREQUENCY HISTOGRAM

1. ***Divide the range of the data into intervals, or classes.*** It is simplest to make each interval the same width. There is no set rule as to how many intervals should be chosen; this will vary depending on the data. (For example, length data could be divided into intervals of 0–9.9 cm, 10.0–19.9 cm, 20.0–29.9 cm, and so on.)
2. ***Count the number of observations that are in each interval.*** These counts are the frequencies.
3. ***Prepare a frequency table showing each interval and the frequency with which a value fell into that interval.***
4. ***Label the axes of a graph with the intervals on the X-axis and the frequency on the Y-axis.***
5. ***Draw in bars where the height of a bar corresponds to the frequency with which a value occurred.*** Center the bars above the midpoint of the class interval. For example, if an interval is from 0 to 9 cm, then the bar should be centered at 4.5 cm.

Note that **continuous data** *can take any value.* For example, height and time are continuous. A prairie flower might be 100 cm, 101 cm, or any height in between. **Discrete data**, in contrast, *can only take on certain values.* For example, there might be 20 students in a class or 21, but not any value in between. When data are continuous, the bars on a histogram touch each other, as shown in Figure 14.9. When data are discrete, the bars should be drawn so they do not touch each other.

19 g might actually have weighed anywhere between 18.5 and 19.4 g. The bar in the histogram, therefore, spans an interval. In this graph the intervals are 1 g in width. You can think of the bars on the histogram as representing individual mice piled up on the number line with each individual sitting above its score.

It is possible to construct a frequency histogram for any numerical data set consisting of the values for a single variable. The procedure for doing so is given in Box 14.3.

14.3.2 The Normal Frequency Distribution

The same data that are shown in Figure 14.9 are graphed in a slightly different form, as a **frequency polygon,** in Figure 14.10a. A frequency polygon does not use bars; rather, each class is represented as a single point. The point is placed halfway between the smaller and the larger limit for that class, and the points are connected with lines.

Figure 14.10a shows the distribution of weights for a sample of only 175 mice. If the weights of a great many laboratory mice were measured, then it is likely that the frequency distribution would approximate a bell shape, or normal curve (Figure 14.10b). *The frequency distribution pattern that has a bell shape when graphed is called a* **normal distribution.**

There are many situations in nature and in the laboratory where the frequency distribution for a set of data approximates a normal curve (Figure 14.11). For example, the height of humans, the weight of animals, and measurements of the same object all tend to be normally distributed.

At this point, it is possible to relate the first part of this chapter, which discussed calculated measures (such as the mean and standard deviation), to the graphical techniques in this section. The center of the peak of a perfect normal curve is the mean, median, and mode. Values are equally spread out on either side of that central high point. Moreover, the width of the normal curve is related to the standard deviation (Figure 14.12). The more dispersed the data, the higher the value for the standard deviation and the wider the normal curve.

If we know that the pattern of variation for a variable is normally distributed, then we can make certain predictions about individuals. For example, consider a 20-year-old man, Robert Smith, who is from a large population of all 20-year-old men. Height in adults is a variable that is normally distributed. This allows us to predict that Robert is most likely about average height. He is equally likely to be either a bit taller or a bit shorter than average. He is unlikely to be substantially taller or shorter than average.

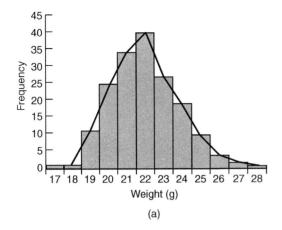

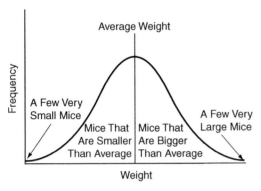

FIGURE 14.10 The frequency distribution for mouse weights. (a) The weights of 175 mice are graphed both as a frequency histogram and as a frequency polygon. (b) If a frequency polygon were prepared for a great many mice, then the shape of the plot would approach a bell shape. The peak in the center is the average weight. Most of the mice have weights somewhere in the middle of the plot, that is, around the average weight, while a few mice are substantially heavier or lighter than the average. This distribution pattern is called a *normal distribution*.

Example Problem 14.6

A student weighs himself one morning 25 times on his bathroom scale. The resulting weight values (in pounds) are shown.

a. Show these data in a frequency table and graphically.
b. Do these data appear to be approximately normally distributed?

159 159 158 161 160 158 159 157 158 160 159 158 160
159 158 161 160 159 157 159 162 161 159 158 157

Answer

These data can be readily summarized using six classes:

a.

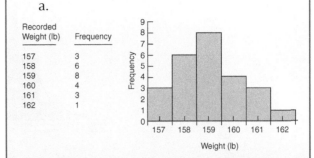

Recorded Weight (lb)	Frequency
157	3
158	6
159	8
160	4
161	3
162	1

b. These data suggest a normal distribution. In fact, it has been found that if a great many measurements of the same item are made, and if those measurements are made properly with a sensitive measuring device, then there is variation in the measurements which tends to be normally distributed.

Example Problem 14.7

Multiple choice:
What frequency distribution is illustrated in this histogram?

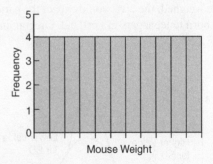

a. All the mice are of the same weight.
b. There is the same number of mice in each weight class.
c. Neither of the above.

Answer

Option **b** is correct. The histogram illustrates a situation where there are four mice in each weight class.

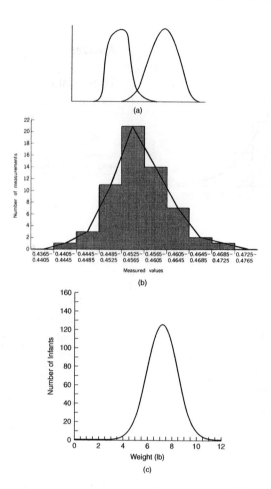

FIGURE 14.11   Examples of relationships that approximate a normal curve. (a) An example from chromatography, an analytical technique used to separate and detect various components in a mixture of compounds. Under ideal conditions each compound appears on the chromatographic printout as an approximately normal curve. (b) The weights of a number of vials from the same batch. The vials vary in weight because of both variability in the manufacturing process and variability that is inherent in repeated measurements. The pattern of variation approximates a normal distribution. If more vials were weighed, then we would expect the pattern to be even closer to a normal curve. (c) Distribution of birth weight of babies born to teenagers in Portland, Oregon, in 1992.

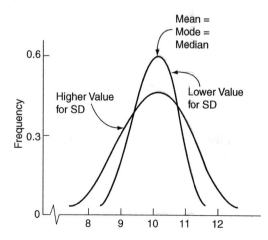

FIGURE 14.12   The normal curve. When data are perfectly normally distributed, their mean, median, and mode all lie at the peak of the distribution plot and the standard deviation determines the width of the curve.

Although data are commonly normally distributed, not all data have a normal distribution. For example, the frequency distribution of a set of data may be **skewed,** *in which case the values tend to be clustered either above or below the mean* (Figure 14.13). The frequency distribution in Figure 14.14 is **bimodal,** *which means that it has two peaks.* Note that when a distribution is not normal, the mean, the median, and the mode do not coincide.

14.3.3   THE RELATIONSHIP BETWEEN THE NORMAL DISTRIBUTION AND THE STANDARD DEVIATION

The normal distribution is extremely important in interpreting and understanding data; therefore, we will now consider some of the features of this distribution in more detail.

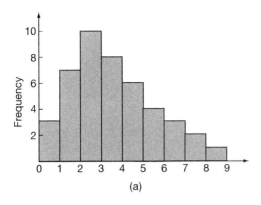

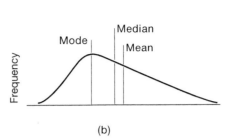

FIGURE 14.13 Examples of distributions that are skewed.

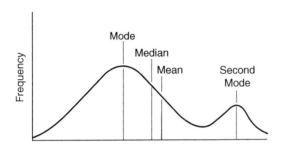

FIGURE 14.14 A bimodal distribution.

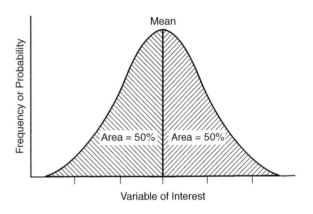

FIGURE 14.15 The normal curve. The center of the peak is the mean. The variable of interest is plotted on the X-axis; either probability or frequency is on the Y-axis. The normal curve is symmetrical; half its area lies to the left of the mean and half lies to the right.

The area under any normal curve, by definition, is equal to 100% or 1.0. The center of the peak of a normal curve is the mean. The normal curve is symmetrical; half of the area under the normal curve lies to the right of the mean, and half lies to the left of the mean (Figure 14.15).

Figure 14.16 shows four normal probability distributions. The overall pattern of the four curves is the same; each has a peak centered at the mean. Each curve is symmetrical so that half its area lies to the left of the mean and half to the right; however, the four curves are not identical. Three of the four distributions

have different means; therefore, the curves lie at different locations along the X- axis. The curves also differ in width. Those with smaller standard deviations are narrower than those with larger standard deviations. Every normal curve is thus described by two characteristics:

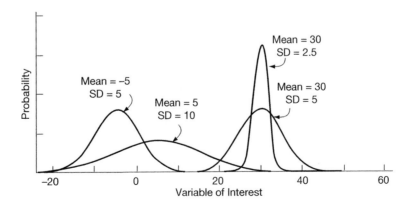

FIGURE 14.16 The patterns of four normal distributions having different means and different standard deviations. The mean locates the center of the curve along the X-axis. The standard deviation determines how wide or narrow the curve is.

1. The mean, which locates the center of the curve along the X-axis.
2. The standard deviation, which determines how "fat" or "thin" the curve appears.

The normal curve in Figure 14.17 shows the distribution of heights of men in the United States. Observe that this plot is "marked off" in terms of the standard deviation. The mean height for the population is about 175 cm, and the standard deviation is about 7.6 cm. The mean +1 SD equals 175 cm +7.6 cm = 182.6 cm. The mean − 1 SD is 167.4 cm. Observe the location of these two points on the graph. The mean height + 2 SD is 190.2 cm, and the mean −2 SD is 159.8 cm. Observe these values on the graph and also the location of the mean ± 3 SD.

A graph labeled in terms of standard deviations from the mean can be constructed for any set of data that is approximately normally distributed. There is a certain pattern that is always true of such plots (Figure 14.18). The pattern is such that about 68% of the area under the normal curve lies in the section between the mean +1 SD and the mean −1 SD. About 95.4% of the total area under the curve lies in the section between the mean + 2 SD and the mean − 2 SD. About 99.7% of the total area under the curve lies in the section between the mean + 3 SD and the mean − 3 SD. The pattern shown in Figure 14.18 is consistent for any normal curve, regardless of how spread out the curve is, or the value for the mean.

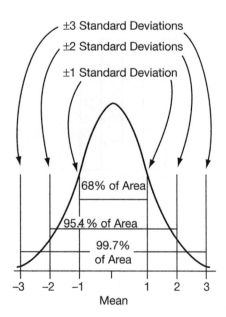

FIGURE 14.18 The relationship between the standard deviation and the area under the normal curve. For any normal distribution, about 68% of the area under the curve lies within ± 1 standard deviation from the mean; about 95.4% of the area under the curve lies within ± 2 standard deviations from the mean; about 99.7% of the area lies within ± 3 standard deviations from the mean.

The relationship between the area under a normal curve and the approximate number of standard deviations from the mean is summarized in tabular form in Table 14.7.

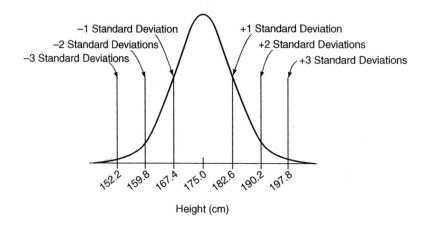

FIGURE 14.17 The normal curve subdivided in terms of the standard deviation. The frequency distribution for the height of men in the United States in terms of the mean and the standard deviation. (The mean = 175.0 cm; SD = 7.6 cm.)

TABLE 14.7

The Relationship between the Area under a Normal Curve and the Approximate Number of Standard Deviations from the Mean

Standard Deviation	Approximate Area under the Curve
± 1.0	68%
± 2.0	95%
± 2.6	99%
± 3.0	99.7%

So far, the significance of this discussion of standard deviation and the normal curve may seem obscure; however, it is actually of great practical significance.

1. *When a variable is normally distributed, the standard deviation tells us the percent of individuals whose measurements are within a certain range.* For example, height in humans is a normally distributed variable; therefore, we know that about 68% of all values measured for height will fall within 1 standard deviation on either side of the mean (Figure 14.19).

2. *When a variable is normally distributed, the standard deviation tells us the probability of obtaining a particular value if a single member of the population is randomly chosen.* For example, if a man is picked randomly there is a 68% probability that his height will be in the range of the mean ± one standard deviation.

We previously predicted that a 20-year-old man selected at random, Robert Smith, is probably about average height, might be a bit taller or shorter than average, but is unlikely to be substantially taller or shorter than average. We can now be more specific in our predictions. If Robert's name was chosen randomly, then there is about a 95.4% probability that his height is within two standard deviations on either side of the mean. The mean is 175 cm, so there is a 95.4% probability that he is between 159.8 cm and 190.2 cm. If we predict the height of a randomly chosen man to be between 159.8 cm and 190.2 cm, then, about 4.6% of the time, we expect to be wrong, but about 95.4% of the time we expect to be right. Thus, when data are normally distributed, knowing the standard deviation enables us to make predictions and assign them a certain probability of being correct. We can generalize to say that:

Statistics provides tools that enable us to make numerical statements and predictions in the presence of chance and variability. Statistics gives answers that are not expressed as "right" or "wrong"; rather, they are expressed in terms of probability.

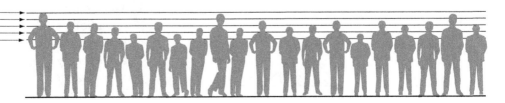

$\bar{x}$ +2 SD = 190.2 cm
$\bar{x}$ +1 SD = 182.6 cm
$\bar{x}$ = 175.0 cm
$\bar{x}$ −1 SD = 167.4 cm
$\bar{x}$ −2 SD = 159.8 cm

FIGURE 14.19 Men's height and the normal distribution. About 68% of all men are of a height that falls within one standard deviation on either side of the mean. About 95.4% of men are of a height that falls within 2 SD on either side of the mean.

Example Problem 14.8

For the example and practice problems, use the values in Table 14.7.

a. What percent of the area under a perfect normal curve lies between the mean and the mean+1 standard deviation, as shown in the shaded area?

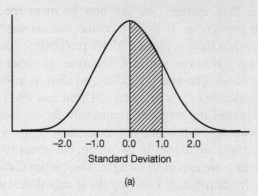

(a)

b. What percent of the area under a perfect normal curve lies in the shaded section in this graph?

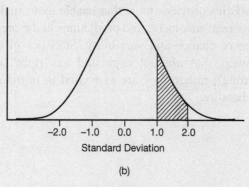

(b)

c. What percent of the area under a perfect normal curve lies in the section shown in this graph?

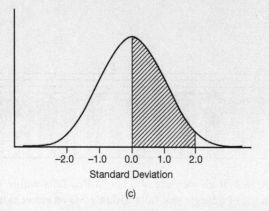

(c)

Answer

Using the values from Table 14.7:

a. About 34% of the area lies between the mean and +1 standard deviation.
b. About 13.5% of the area lies between +1 and +2 standard deviations.
c. The area under the curve that lies between the mean and 2 standard deviations is about 47.5%.

Example Problem 14.9

The average for US women's height is about 163 cm, and the standard deviation is about 6.4 cm. Assuming that women's height is normally distributed:

a. What is the probability that a woman selected at random will have a height between 156.6 cm and 169.4 cm?
b. What is the probability that a woman selected at random will have a height between 150.2 cm and 175.8 cm?
c. What percent of women have heights between 143.8 cm and 182.2 cm?
d. Show the distribution of women's heights graphically.

Answer

a. This is the range of ± one standard deviation (163 cm ± 6.4 cm). We know that about 68% of all values can be expected to fall within this range, so there is a 68% chance that a woman selected at random will be in this range.
b. This is the range of ± two standard deviations (163 cm ± 12.8 cm). There is therefore about a 95% chance that a randomly selected woman will fall in this range.
c. This is the range of ± three standard deviations (163 cm ± 19.2 cm). About 99.7% of women will have heights in this range.

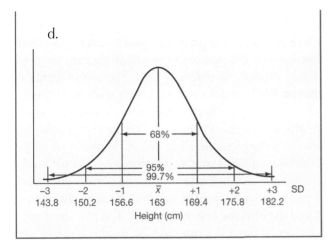

d.

68%

95%
99.7%

| -3 | -2 | -1 | x̄ | +1 | +2 | +3 | SD |
| 143.8 | 150.2 | 156.6 | 163 | 169.4 | 175.8 | 182.2 | |

Height (cm)

Example Problem 14.10

The height of adult males is normally distributed with a mean of 175 cm and a standard deviation of 7.6 cm.

    a. About what percent of men are taller than 190.2 cm?
    b. About what percent of men are shorter than 159.8 cm?

Answer

Roughly 95% of adult men have heights within two standard deviations of the mean. This means that a total of about 5% of men are either taller than 190.2 cm or shorter than 159.8 cm. We expect, therefore, that about 2.5% of men will be taller than 190.2 cm and about 2.5% of men will be shorter than 159.8 cm.

14.3.4 A BRIEF SUMMARY

There is variability in nature, and so there is variability in the data collected when we observe the natural world. Statistics provides methods for organizing, summarizing, and displaying such data. Measures of central tendency and measures of dispersion are calculated values that describe and summarize data. Frequency histograms and frequency polygons display variation graphically.

Patterns emerge when data values are organized by frequency. The normal distribution is a frequency distribution that is often approximated in the natural world. When values are normally distributed, they center around the average. Most individuals have values that are about the average, some have values that are a bit higher or lower than the average, and relatively few individuals have values that deviate greatly from the average.

Statistics provides tools to organize and help us interpret variability in our observations of the natural world.

14.4 AN APPLICATION: CONTROLLING PRODUCT QUALITY

14.4.1 VARIABILITY

There are many applications of the ideas discussed so far in this chapter. This section focuses on the application of these principles to the area of ensuring product quality.

It might seem that products made by the same process – or samples of product from the same batch, or measurements of the same sample performed in the same way – ought to be identical. In fact, the goal of product quality systems [including ISO 9000 and the Good Manufacturing Practices (GMPs), which govern pharmaceutical production] is to reduce variability so that products are consistently of high quality. (See also Chapter 4.) Variability is reduced through such means as the consistent use of standard operating procedures and instrument maintenance programs. In practice, however, even when product quality systems are implemented there is always some variability in products. This is partly because there is variability inherent in any production process. In addition, measurements vary, even repeated measurements of a single characteristic of a single item. In fact, if there is no variability in a set of measurements relating to a product, then there is likely to be a malfunction or a lack of sensitivity in the system. For example, if the activity of an enzyme product is measured in 10 different vials from a single batch, or if the amount of a drug is measured in 25 tablets, then there will inevitably be some variability in the results. Product variability is not desirable; it is, however, always present.

Statistics provides important tools that can help to maintain the quality of products in the face of this inevitable variability. Statistical tools can be used to:

- determine how much variability is naturally present in a process
- determine whether there is likely a malfunction or problem in a process
- determine how certain we can be of a measurement result in the presence of variability.

Consider Example Problem 14.11.

Example Problem 14.11

A technician is responsible for purifying an enzyme product made by fermentation. During the purification process she checks the amount of protein present. The results of six such measurements made during typical runs were (in grams):

23.5, 31.0, 24.9, 26.7, 25.7, 21.0

One day the technician obtains a value of 15.3 g. Should she be concerned by this value? Another day the technician obtains a value of 32.1 g. Is this value a cause for concern?

Answer

Let us begin by looking at these values in an intuitive way. The six typical values appear to hover around about 25 g ± about 5 g. A value of 15.3 g is quite a bit lower than the other values; it is around 10 g below the approximate mean. In contrast, the value 32.1 is not as far from the mean and may therefore be less of a cause for concern.

Let us now apply statistical reasoning to this problem. The six measurements that were made during normal runs are assumed to represent values obtained when the purification process was working properly. The mean for these six values is 25.5 g, and their standard deviation is 3.36 g. Based on the normal distribution, we expect 95.4% of the values to fall by chance within 2 SD of the mean. The range that includes two standard deviations is 18.8 g–32.2 g. Thus, based on the normal distribution, the value of 15.3 g is clearly unexpected on the basis of chance alone and is therefore a cause for concern. The value of 32.1 g does not immediately suggest there is a problem (although, if it signals the beginning of an upward trend, there might in fact be a problem in the process).

Note that in this example the technician is applying statistical thinking to her everyday work. She is taking advantage of knowledge about normal distributions and about variability to identify potential problems.

14.4.2 CONTROL CHARTS

There is a common graphical quality-control method that extends the reasoning demonstrated in the preceding example problem. This is the use of **control charts.** We introduce control charts with an example. A biotechnology company uses a fermentation process in which bacteria produce an enzyme that is sold for use in food processing. The pH of the bacterial broth must be controlled over time; otherwise, the pH tends to drop and the production of the enzyme diminishes. A preliminary fermentation run that successfully produced the enzyme was performed. The pH was monitored throughout the run, and the resulting values are shown in Table 14.8.

Let us assume that this fermentation process is tested more times and that the pattern of variability illustrated in Table 14.8 is characteristic of successful runs. The mean hovers around pH 7.05, and the SD is 0.17 pH units. When the company begins to produce this enzyme for sale, the pattern of pH variability should be similar to that observed in the preliminary, successful tests. In a production setting, *a variable that has the same distribution over time is said to be* **in statistical control,** *or, is simply* **in control.** Production processes should be consistent; therefore, having "in-control" processes is critical.

A control chart is a graphical representation of the variability in a process over time. A control chart for the data in Table 14.8 is shown in Figure 14.20. When a control chart is constructed, a line is drawn across the chart at the mean (in this case, at pH 7.05). A second

TABLE 14.8

The pH during Fermentation in Preliminary Studies

Time	pH
0800	7.22
0830	7.13
0900	6.84
0930	6.93
1000	7.12
1030	7.33
1100	7.04
1130	6.79
1200	6.94
1230	7.03
1300	7.22

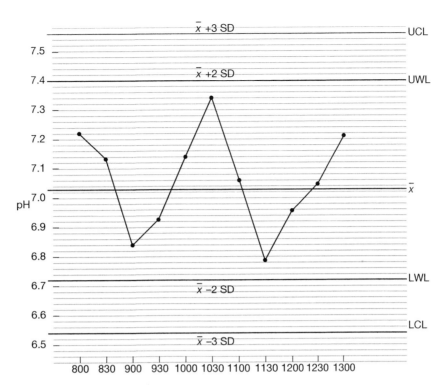

FIGURE 14.20 A control chart for pH values from a fermentation run.

line is drawn at $\bar{x}$ +2 SD (in this example, at 7.39). A third line is drawn on the graph at $\bar{x}$ −2 SD (in this example at 6.71). Two more lines are drawn at $\bar{x}$ ± 3 SD (pH 7.56 and 6.54, respectively).

The lines drawn across the control chart have names. The line at $\bar{x}$ −2 SD is called the **lower warning limit, LWL.** The line at $\bar{x}$ +2 SD is called the **upper warning limit, UWL.** The **lower control limit, LCL,** is $\bar{x}$ −3 SD, and the **upper control limit, UCL,** is $\bar{x}$ +3 SD.

The use of control charts is based on the important assumption that the inherent variability in production processes is approximately normally distributed. Given that this is the case, then 95.4% of all observations are expected to fall within ± 2 SD of the mean; therefore, the warning limits define the range in which about 95% of all values are predicted to lie. One pH observation falling outside the warning limits is not a cause for concern because about 5% of the time values may fall outside these limits simply because of chance variability. Two or more points outside the warning limits, however, may indicate a problem in the process. The control limits define the range in which 99.7% of all points should lie. The probability that an observation, by chance, would fall outside the control lines is only 3 in 1000. It is important, therefore, to immediately find the problem if a value outside these action lines is obtained.

When a process is in control, its control chart has features illustrated in Figure 14.20. All the points are between the upper and lower warning limits. About half the points are above the mean and half below. It is also important that there is no trend where the points seem to be increasing in pH over time or decreasing in pH over time.

A control chart helps operators distinguish between variation that is inherent in a process and variation that is due to a problem and should be investigated. Figure 14.21 is a control chart that illustrates several problems. On days 7–10 the process is showing an upward trend. At day 10 there is a high point that is outside the control limits. At day 20 there is a point that exceeds the lower control limit. When these types of problems are observed, the process would probably be stopped and the cause(s) of the deviations would be investigated.

A point that has a value much higher or much lower than the rest of the data is called an **outlier.** For example, an aberrant pH value in the fermentation process might indicate that the bacteria are not growing properly and that enzyme production is likely to be adversely affected. In the real world, outliers may

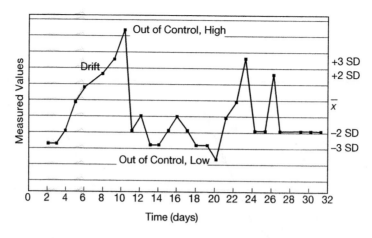

FIGURE 14.21   A control chart illustrating possible problems.

or may not signal that a process is out of control. Outliers are sometimes simply errors in how a test was performed. Errors might include, for example, improperly removing and preparing the sample, using a malfunctioning instrument, or inattentiveness on the part of the analyst. An aberrant pH reading might also be due to a malfunction in the pH meter – a testing error. In production settings the initial investigation of an outlier looks at whether it is simply due to testing error.[2]

Control charts are a statistical tool based on an understanding of variability and the normal distribution. Control charts allow operators to quickly distinguish between inherent, expected variability in measurement values and variation that is caused by a problem in a process. As long as the values remain within the limits of expected variability, the process is considered to be in control and variation

---

[2] When an aberrant or unexpected test result of a regulated product (such as a pharmaceutical product) is obtained, the company must investigate and respond. How the company must conduct its investigation and the company's response to an unexpected test result has sometimes been a cause of contention between the company and the FDA. This is because an unexpected test result may indicate that a product is defective. An unexpected test value may also indicate that an analyst made a mistake, that a piece of test equipment was malfunctioning, that a value was recorded inaccurately, or other such problems that do not relate to the quality of the product. Moreover, very high or very low measurement results occasionally occur simply by chance and are the result of normal variability in the measurement system. Companies therefore sometimes attribute unexpected test results to laboratory error or to chance while regulators are concerned that the test result might reflect a problem with a product. Disagreements between a company and the FDA in how unexpected test values should be handled have been contested in the courtroom. (See the discussion of the *Barr* decision in Chapter 38.)

Example Problem 14.12

The company that produces an enzyme by fermentation performs a production run. The following pH data are obtained. Prepare a control chart for these data. Is the process in control during this run? (Note that the expected pattern of variability in this process was determined in preliminary studies; therefore, use the same mean, standard deviation, UWL, LWL, UCL, and LCL as in Figure 14.20.)

Time	pH
0800	7.34
0830	7.22
0900	7.27
0930	6.88
1000	7.02
1030	7.01
1100	6.78
1130	6.88
1200	6.76
1230	6.57
1300	6.54

Answer

The points for this run are plotted on a control chart. Observe that the data show a downward trend. Moreover, the final point is close to the control limit. These data strongly suggest that the process is not in control.

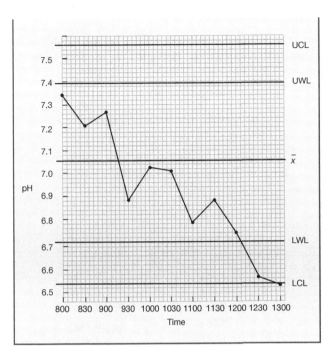

is assumed to be due to chance alone. If a value falls outside the range predicted by chance, or if the values begin to be consistently high or low, then there may be a problem in the production process that needs to be examined. There are many types and forms of control charts other than the one illustrated here. The basic purpose of a control chart is the same, however, regardless of the details of its construction.

Practice Problems

1. a. Ten vials of enzyme were chosen from a batch for quality-control testing. What was the sample? What was the population?
   b. The glucose level in a blood sample was measured. What was the sample? What was the population?
   c. The reading ability of all first graders in Franklin Elementary School was measured. What was the sample? What was the population?

2. a. Ten measurements of a 10 g standard weight were made. Calculate the mean, standard deviation, and coefficient of variation.

10.001 g	10.000 g	10.001 g	9.999 g	9.998 g
10.000 g	10.002 g	9.999 g	10.000 g	10.001 g

   b. Calculate the mean, standard deviation, and coefficient of variation for these test scores.

75%	71%	88%	99%	76%	83%	89%	91%

   c. The rate of an enzyme reaction was timed in six reaction mixtures. Calculate the mean, standard deviation, and coefficient of variation.

235 sec	287 sec	198 sec	255 sec	234 sec	201 sec

   d. The analyst who performed the enzyme reactions in problem c tried an alternative assay method and obtained the following results. Calculate the mean and the standard deviation for the alternative method. Compare the two methods.

245 sec	235 sec	256 sec	228 sec
237 sec	244 sec	234 sec	215 sec

3. (Optional: These calculations are best performed using a calculator that automatically performs statistical functions.) Calculate the mean, standard deviation, and relative standard deviation for the following data sets:

a. The weights of 33 vials from a lot were sampled. The results in milligrams are:

115	133	125	124	87	95	136	146	114	133	139
177	193	136	177	193	136	123	171	147	153	149
110	111	121	71	178	173	201	149	94	100	103

b. The weights of 25 samples were measured. The results are:

9.27 g	9.88 g	6.19 g	8.38 g	9.62 g
5.48 g	9.71 g	9.54 g	7.41 g	6.76 g
9.33 g	8.64 g	6.43 g	7.66 g	9.60 g
7.50 g	9.89 g	7.73 g	5.68 g	6.74 g
8.04 g	7.23 g	7.35 g	10.82 g	8.94 g

c. The volumes of 19 samples were measured. The results are:

1022 mL	1045 mL	1103 mL	1200 mL
1189 mL	1234 mL	1043 mL	1089 mL
1103 mL	1020 mL	1058 mL	1197 mL
1201 mL	1189 mL	1098 mL	1155 mL
1056 mL	1023 mL	1109 mL	

4. A biotechnology company is developing a new variety of fruit tree, and researchers are interested in the number of blossoms produced. The following are the total number of blossoms from 20 trees. Find the mean, median, range, and standard deviation for these data:

| 34 | 36 | 29 | 27 | 30 | 35 | 32 | 31 | 39 | 30 |
| 44 | 30 | 33 | 43 | 32 | 21 | 35 | 33 | 40 | 27 |

5. A company plans to grow and market shiitake mushrooms. The yields of mushrooms obtained in preliminary experiments are shown, expressed in kilograms. Calculate the mean, median, range, and standard deviation.

| 38.2 | 31.7 | 28.1 | 29.1 | 33.4 | 25.2 |
| 18.2 | 19.7 | 41.7 | 26.2 | 21.0 | 52.2 |

6. The special average test score for a class of thirty students is 75%. One additional student takes the exam and scores 100%. What is the new average for the class?

7. For each of the following three distributions, estimate the mean and mode.

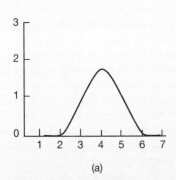

(a)

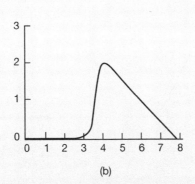

(b)

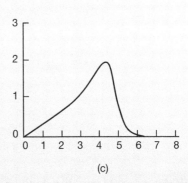

(c)

8. Which of the following frequency histograms most closely approximate a normal distribution? Which appear to be bimodal? Which appear skewed?

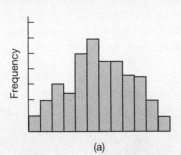

(a)

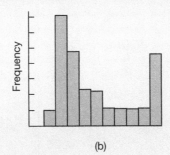

(b)

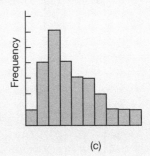

(c)

9. Which of these two distributions is less dispersed?

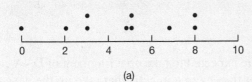

(a)

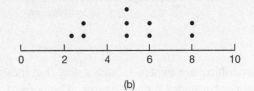

(b)

10. Which of these two distributions is less dispersed?

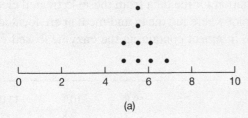

(a)

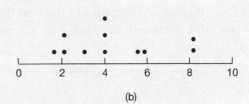

(b)

11. Which of these two distributions is less dispersed?

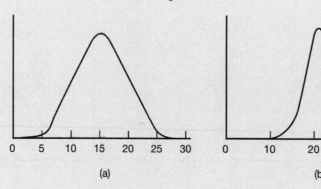

(a)   (b)

12. Which of these two distributions is less dispersed?

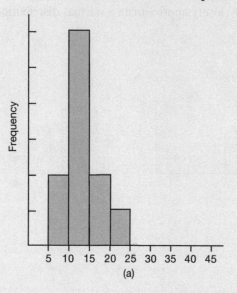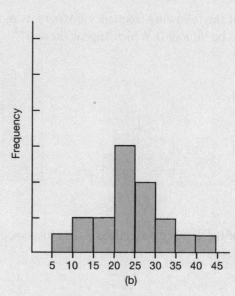

(a)   (b)

13. Cells in culture are treated in such a way that they are expected to take up a fragment of DNA containing a gene that codes for an enzyme. The activity of the enzyme in cells that take up the gene can be assayed. The more active the enzyme, the better. Suppose a researcher isolates 45 clones of treated cells and measures the enzyme activity in each clone. The results are shown in activity units below.
   a. Find the range, median, mean, and standard deviation for the data from these 45 treated clones.
   b. Plot these data on a histogram and show on the plot where the mean and median are located.
   c. Do you think these cells have taken up the gene fragment containing the enzyme, based on these data? Explain.

10.4	12.2	12.0	9.1	5.8	3.2	9.8	10.1	13.0
2.1	9.8	10.1	1.5	7.8	5.6	2.3	9.8	10.2
9.1	12.3	10.1	12.3	14.2	15.1	13.6	12.1	10.8
9.2	8.9	12.4	11.0	13.1	8.9	9.4	10.2	11.3
12.8	9.0	8.6	12.3	12.0	2.1	0.4	10.6	13.0

14. Suppose that the assay for enzyme activity (discussed in question #13) is performed on cultured cells that have not been treated with the DNA fragment and are not expected to have any enzyme activity at all. In this case, one would expect to see little or no activity when performing the assay. You test this

assumption on 20 untreated cell clones that are not expected to have the enzyme and obtain the following results in activity units:

| 0.0 | 1.0 | 0.3 | 3.0 | 2.8 | 1.2 | 2.3 | 1.1 | 0.9 | 2.5 | 1.9 | 0.9 | 2.2 | 1.2 | 0.2 | 0.1 | 0.2 | 0.9 | 1.0 | 2.0 |

a. Calculate the mean and the standard deviation for these results.
b. Prepare a histogram for these results.
c. Discuss the results shown in question #13 in light of these data in question #14.

15. The following are heights in centimeters of a common prairie aster measured from one field.
a. Determine the range and mean for these data.
b. Plot these data as a frequency histogram. (You will need to divide the data into groups. After finding the range, make 10–12 divisions of equal size and assign each value to the proper class.)

151	182	182	162	177	166	197	144
174	160	131	156	125	170	153	172
146	127	156	140	159	155	158	165
155	141	180	145	145	150	145	135
105	122	180	152	161	170	156	150
122	140	133	145	190	165	176	170
144	160	162	157	155	146	155	143
154	157	141	150	142	139	138	156
130	126	189	138	103	163	135	158
151	129	147	153	150	140	146	138
154	138	179	165	142	140	141	160
125	156	145	159	147	155	189	195
140	141	160	156				

16. Show the following data in the form of a frequency distribution and a frequency histogram and polygon.

23 mg	28 mg	22 mg	23 mg	20 mg	19 mg
22 mg	24 mg	26 mg	23 mg	23 mg	24 mg
21 mg	25 mg	24 mg	25 mg	21 mg	

Where relevant, use the values in Table 14.7 to solve the following problems.

17. As a trouser manufacturer, you are interested in the average height of adult males in your town. You take a random sample of five male customers and find the mean to be 5 feet 10 inches and the standard deviation to be 3 in. How certain are you of the following statements and why?
a. The average height of the population is between 0 and 50 ft.
b. The average height of the population is between 4 ft and 7.5 ft.
c. The next new customer will be between 5 ft 7 in and 6 ft 1 in.
d. The average height of the entire male population is 5 ft 10 in.

18. Refer to example problem 14.9 regarding women's heights.
a. What is the probability that a woman selected at random will be shorter than 150.2 cm?
b. What is the probability that a woman selected at random will be taller than 169.4 cm?

19. A pharmaceutical company finds that the average concentration of drug in each vial is 110 µg/vial with a standard deviation of 6.1 µg.
    a. About what percent of all vials can be expected to have between 94.1 µg and 125.9 µg of drug?
    b. What is the likelihood that a vial selected at random will have more than 122.2 µg of drug?
    c. What percent of all the vials are expected to have a value between 103.9 µg and 116.1 µg?
    d. Suppose 100 vials are checked. About how many of them would be predicted to have more than 125.9 µg?

20. A technician customarily performs a certain assay. The results of 8 typical assays are:

    | 32.0 mg | 28.9 mg | 23.4 mg | 30.7 mg | 23.6 mg | 21.5 mg | 29.8 mg | 27.4 mg |

    a. If the technician obtains a value of 18.1 mg, should he be concerned? Base your answer on estimation without performing actual calculations.
    b. Perform statistical calculations to determine whether the value 18.1 mg is out of the range of two standard deviations.

21. A technician customarily counts the number of leaves on cloned plants. The results of nine such counts in successful experiments are:

    | 75 | 54 | 55 | 61 | 71 | 67 | 51 | 77 | 71 |

    a. If the technician obtains a count of 79, is this a cause for concern? Base your answer on estimation without performing actual calculations.
    b. Perform statistical calculations to determine whether 79 leaves is out of the range of two standard deviations.

22. Examine this control chart. Discuss the points on March 22, April 15, May 31, and June 23.

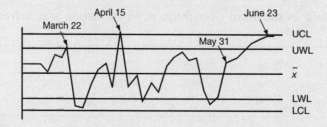

23. Suppose a biotechnology company is using bacteria to produce an antibiotic. During the production of the drug, the pH of the bacterial broth must be adjusted so that it remains optimal; otherwise, the production of the antibiotic diminishes. Preliminary studies were performed to determine the optimal pH. The results of a successful preliminary study are shown in the following table.

Time	pH
0800	6.12
0830	5.13
0900	5.84
0930	6.53
1000	6.12
1030	6.23
1100	6.04
1130	5.79
1200	5.94
1230	6.03
1300	6.12

a. Calculate the mean pH and the SD.
b. Construct a control chart with a central line, UWL, LWL, UCL, and LCL.
   The enzyme goes into production, and periodic samples of the broth are assayed for pH. The results are shown in the following table.
c. By simply observing these points, what can you say about the process?
d. Plot these data on the control chart.
e. Comment on the process; did it ever reach the warning level or action-required levels?

Time	pH
0800	6.54
0830	6.12
0900	5.87
0930	5.18
1000	4.95
1030	4.89
1100	5.03
1130	5.43
1200	5.34
1230	5.37
1300	5.38

UNIT V

Obtaining Reproducible Laboratory Measurements

Chapters in This Unit

- ✦ Chapter 15: Introduction to Quality Laboratory Measurements
- ✦ Chapter 16: Introduction to Instrumental Methods and Electricity
- ✦ Chapter 17: The Measurement of Weight
- ✦ Chapter 18: The Measurement of Volume
- ✦ Chapter 19: The Measurement of Temperature
- ✦ Chapter 20: The Measurement of pH, Selected Ions, and Conductivity
- ✦ Chapter 21: Measurements Involving Light – Part A: Basic Principles and Instrumentation

---

Measurements are *quantitative observations, or numerical descriptions.* Examples of measurements include the weight of an object, the amount of light passing through a solution, and the time required to run a race. Measuring properties of samples is an integral part of everyday work in any biotechnology laboratory or production setting. For example, solutions are required to support the activity of cells, enzymes, and other biological materials. Preparing solutions involves measuring the weights and volumes of the components. Estimating the quantity of DNA in a test tube may involve measuring how much light passes through the solution. The pH of the media in which bacteria grow during fermentation must be monitored continuously. There are countless measurements made in most biotechnology settings, each of which must be a "good" measurement. Obtaining reproducible research results or consistent product quality begins with making "good" measurements.

DOI: 10.1201/9780429282799-19

Case Study: Accurate Measurements Make a Difference

Everyone has experience in making measurements in everyday life. For example, the COVID-19 crisis made everyone more aware of the importance of temperature measurements. In response to the pandemic, many venues instituted a requirement for a temperature check before entering. Any human temperature reading over 100.4° is considered indicative of a fever, according to Centers for Disease Control and Prevention (CDC) guidelines, and therefore a possible sign of illness.

Baking is a familiar area where careful measurements can have a major impact on one's success. While, unlike in the biotechnology laboratory, many recipes allow a cook to be creative with quantities of ingredients, this is not the case if you want to bake a fluffy, moist cake from scratch. In this case, a key factor is the ratio of flour to liquid and other ingredients. In the United States, home cooks usually measure flour in cups, which is a measure of volume. However, the volume of flour can vary dramatically if it is scooped into a measuring cup from the bag, or sifted and gently sprinkled into the cup. The difference in the actual amount of flour measured as one cup can be as much as 25%! That is more than enough variation to make the difference between a great cake and one that is heavy or dry.

For this reason, experienced cooks and pastry chefs weigh their flour when they bake. No matter how fluffy or packed the flour is, the weight will always correspond to the same amount of flour. If you have a recipe that calls for 270 grams of flour, that is the equivalent of ≈ 2.16 cups of carefully measured flour, an amount that would be difficult for a home cook to recreate accurately. If you have ever read a European cookbook, you will see that the ingredients are weighed out in gram units, rather than measured by volume.

Metric measurements are always used in laboratories to avoid confusion between units of volume and weight. For example, ounces can refer to weight, or to fluid ounces, a measure of volume. For this reason, a good cook would never measure a cup of flour with a liquid measuring cup. Sixteen ounces of flour by weight would measure about twenty-eight fluid ounces in volume. This would not qualify as a good measurement if you use the wrong instrument.

Although it seems obvious that laboratory measurements should be "good," it is surprisingly difficult to define a "good" measurement. One definition is that a "good" measurement is correct; however, this leads to the question of what is "correct?" Suppose a man weighs himself in the morning on a bathroom scale that reads 165 pounds. Shortly after, he weighs himself at a fitness center where the scale reads 166 pounds. At this point, the man might be somewhat uncertain as to his exact, correct weight; perhaps he weighs 165 pounds, or perhaps 166. Perhaps his weight is somewhere between 166 and 165 pounds. It is also possible that both scales are wrong. He is likely to conclude, however, that he weighs about 165 pounds and leave it at that.

Uncertainty of a pound or so in an adult's weight is seldom of great concern. A bathroom scale that gives a weight value within 1 pound of the true weight is a reasonably "correct" instrument. In other situations, however, a measurement must be much more correct. For example, a 1-pound difference in the weight of an infant could mean the difference between a healthy baby and one that is severely dehydrated. In

the laboratory, an error of 1 g in a measurement could mean the difference between a successful experiment and a disastrous failure. In a drug product, a 1 mg error in measurement could endanger a patient. In each of these situations, a "good" measurement is one that can be trusted when making a decision. A good measurement can be trusted by a physician selecting a treatment for a patient, by a research team drawing conclusions from a study, and by a pharmaceutical company deciding whether to release a drug product to the public.

Laboratory workers play a key role in performing measurements of properties of samples. To make good, trustworthy measurements, laboratory workers must understand the principles of measurement, know how to maintain and operate instruments properly, and be aware of, and avoid, potential pitfalls in measurement. It takes knowledge and careful technique to produce measurements that can be trusted in a particular situation (Figure 1).

This unit discusses methods of making "good," trustworthy measurements in the laboratory.

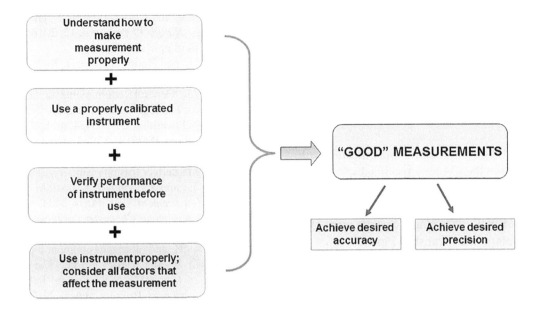

FIGURE 1   Making "good" measurements.

- **Chapter 15** introduces basic principles underlying measurements and terminology relating to metrology (the study of measurements).
- **Chapter 16** introduces basic principles of electricity and electronics, and instrumental methods of measurement. This chapter serves as a transition to the next five chapters, each of which discusses a specific type of measurement.
- **Chapter 17** introduces principles and practices relating to weight measurements in the biotechnology laboratory.
- **Chapter 18** introduces principles and practices relating to volume measurements in the biotechnology laboratory.
- **Chapter 19** introduces principles and practices relating to temperature measurements in the biotechnology laboratory.
- **Chapter 20** introduces principles and practices relating to the measurement of pH, selected ions, and conductivity.
- **Chapter 21** introduces principles and practices relating to the measurement of light transmittance and absorbance.

BIBLIOGRAPHY FOR UNIT V

Manufacturers are an excellent source of up-to-date information on instrument design, operation, and performance verification. Many manufacturer guides provide clearly written technical information that is not specific to any brand of instrumentation. Checking manufacturer's websites is often a useful way to get technical information on measurement theory and practice. Some of the manufacturer resources that we find particularly useful are listed below.

NIST publications are an excellent source of information on measurement topics and are generally available for free through NIST at: "National Institute of Standards and Technology." *NIST*, www.nist.gov. Specific examples of these resources are included in the references below.

ASTM International and ISO standards provide specific, technical information. They are available in some libraries and for purchase at:

"ASTM International - Standards Worldwide."
ASTM International, www.ASTM.org. and
"International Organization for Standardization."
ISO, www.iso.org/home.html.

ASTM and ISO Standards Relevant to this Unit (see the ASTM and ISO websites for periodic updates):

- ASTM E456, "Standard Terminology Relating to Quality and Statistics."
- ASTM E617-18, "Standard Specification for Laboratory Weights and Precision Mass Standards."
- ASTM E1044, "Standard Specification for Glass Serological Pipets (General Purpose and Kahn)."

- ASTM E1154, "Standard Specification for Piston or Plunger Operated Volumetric Apparatus." (Recommended for individuals responsible for checking the calibration of micropipetting devices.)
- ASTM E1878, "Standard Specification for Laboratory Glass Volumetric Flasks, Special Use."
- ASTM E288, "Standard Specification for Laboratory Glass Volumetric Flasks."
- ASTM E542, "Standard Practice for Calibration of Laboratory Volumetric Apparatus."
- ASTM E969, "Standard Specification for Glass Volumetric (Transfer) Pipets."
- ASTM E1, "Standard Specification for ASTM Liquid in-Glass Thermometers."
- ASTM E344, "Terminology Relating to Thermometry and Hydrometry."
- ASTM E77, "Standard Test Method for Inspection and Verification of Thermometers."
- ISO/IEC 17025, "General Requirements for the Competence of Testing and Calibration Laboratories."
- ISO/IEC Guide 99:2007, "International vocabulary of metrology—Basic and general concepts and associated terms (VIM)." *ISO,* www.iso.org/standard/45324.html. (2007 version is current as of spring 2020.)

GENERAL STATISTICS

General statistics books have information about standard deviation, accuracy, precision, and other important concepts in metrology. Examples include the following:

Freedman, David, et al. *Statistics.* 4th ed., W. W. Norton & Company, 2007.

Miller, James and Miller, Jane. *Statistics and Chemometrics for Analytical Chemistry.* 7th New edition, Coronet Books Inc., 2018.

GENERAL MEASUREMENTS

Benham, Elizabeth. "Busting Myths about the Metric System." *NIST,* 6 October 2020, https://www.nist.gov/blogs/taking-measure/busting-myths-about-metric-system (An interesting blog posting about the relationship between the United States and the metric system.)

Fields, Lawrence D., and Hawkes, Stephen J. "Minimizing Significant Figure Fuzziness." *Journal of College Science Teaching,* vol. 16, no. 1, Sept/Oct 1986, pp. 30–34.

Fischer, Joachim, and Ullrich, Joachim. "The New System of Units." *Nature Physics,* vol. 12, 2016, pp. 4–7. doi.org/10.1038/nphys3612.

Grinberg, Nelu, and Rodriguez, Sonia. *Ewing's Analytical Instrumentation Handbook.* 4th ed., CRC Press, 2019.

ISO/IEC 17025, "General Requirements for the Competence of Testing and Calibration Laboratories," *ISO,* www.iso.org/publication/PUB100424.html. (A free publication about ISO/IEC 17025, and how it can help testing and calibration laboratories demonstrate their capacity to deliver trusted results.)

Jones, Frank E., and Schoonover, Randall M. *Handbook of Mass Measurement.* CRC Press, 2002.

Kenkel, John. *Analytical Chemistry for Technicians.* 4th ed. CRC Press, 2014. (Good explanation of instrumental analysis, pH, and other measurement topics.)

Newell, David B. and Tiesinga, Eite. *The International System of Units (SI).* NIST Special Publication 330. *U.S. Government Printing Office,* 2019. (A comprehensive guide to the SI system. Check the NIST website for periodic updates to this definitive publication.)

NIST. "A Turning Point for Humanity: Redefining The World's Measurement System." *NIST,* https://www.nist.gov/si-redefinition/turning-point-humanity-redefining-worlds-measurement-system.

Payne, Graeme, C. "Calibration: What is It?" *Quality Progress: The Official Publication of ASQ,* May 2005, asq.org/quality-progress/2005/05/measure-for-measure/calibration-what-is-it.html

Pinkerton Richard C., and Gleit, Chester E. "The Significance of Significant Figures." *Journal of Chemical Education,* vol. 44, no. 4, 1967, pp. 232–234.

Schoonover, Randall M., and Jones, Frank E. "Air Buoyancy Correction in High-Accuracy Weighing on Analytical Balances." *Analytical Chemistry,* vol. 53, no. 6, 1981, pp. 900–02. doi:10.1021/ac00229a036.

Taylor, Barry N., and Kuyatt, Chris E. "Guidelines for Evaluating and Expressing the Uncertainty of NIST Measurement Results." NIST Technical Note 1297. *U.S. Government Printing Office,* 1994. (Check the NIST website for periodic updates to this definitive publication on the NIST policy on expressing measurement uncertainty.)

Taylor, John K. "Standard Reference Materials Handbook for SRM Users." NIST Special Publication 260-100. *U.S. Government Printing Office,* 1993.

MEASUREMENTS OF WEIGHT

Kupper, Walter E. "Honest Weight and True Mass— (They are Not the Same)." *American Laboratory.* December 1990, pp. 8–9.

Levit, Larry B. "Dealing with Electrostatic Charge in Analytical Balances." *NRD Static Control, LLC.* https://nrdstaticcontrol.com/images/whitepapers/StaticInAnAnalyticalBalance.pdf, 2020.

Mettler-Toledo has extensive, well-written resources on weight measurements freely available on their website, www.mt.com/us/en/home.html, in their expertise library. These resources are periodically updated. Their resources include the following:

- *Weighing the Right Way:– Guidelines for Accurate Results and Better Weighing Techniques*
  (A nicely written booklet on basic principles of weighing.)
- *Understanding Weighing*
- *Electrostatic Changes During Weighing*
- *The Unknown Sources of Error in the Weighing Process*
- *Correct Weight Handling 12 Practical Tips*
- *SOP for Periodic Sensitivity Tests (Routine Tests)*
- *Calibration: What is It? Separate from Adjustment*
- *Eliminating Weighing Uncertainty Adds Confidence to Manufacturing Process*

MEASUREMENTS OF VOLUME

Gilson, Inc. "Gilson Guide to Pipetting." *Gilson, Inc.*, 2019. (An excellent primer on micropipettes. Available at www.gilson.com/default/guide-to-pipetting)

From Mettler-Toledo's expertise library (www.mt.com/us/en/home.html):

- *The Pipetting Handbook*
- *An Introduction to Good Pipetting Practice*
- *Pipette Challenging Liquids*
- *Rainin Technique Posters*
- *How to Clean a Pipette*

MEASUREMENTS OF TEMPERATURE

Nicholas, J., and D. White. *Traceable Temperatures: An Introduction to Temperature Measurement and Calibration*. 2nd ed., Wiley, 2001. (A good introduction to both temperature measurement and metrology in general.)

MEASUREMENTS OF pH, CONDUCTIVITY, AND SPECTROPHOTOMETRIC MEASUREMENTS

Frant, Martin S. "The Effect of Temperature on pH Measurements." *American Laboratory*, July 1995, pp. 18–23.

Hach. "How Can mV Results for pH Be Used for Troubleshooting?" *Hach*. 2019. https://support.hach.com/app/answers/answer_view/a_id/1019197/~/ how-can-mv-results-for-ph-be-used-for-troubleshooting. (Contains information about using mV readings during calibration to troubleshoot problems.)

From Mettler-Toledo's expertise library (www.mt.com/us/en/home.html):

- *Conductivity Guide: Theory and Practice of Conductivity Applications*
- *Perform Your Next pH Measurement In compliance with USP <791>*
- *pH Measurement Theory Guide*

Radiometer Analytical. "Conductivity Theory and Practice." *Radiometer Analytic*. https://support.hach.com/ci/okcs Fattach/get/1002532_4

Thermo Scientific, "Thermo Scientific pH Electrode Selection Handbook." *Thermo Scientific*. 2013. www.fondriest.com/pdf/thermo_ph_electrode_hb.pdf.

YSI Instruments. "What Is the pH of Water and How Is pH Measured? What Does pH Measure?" www.ysi.com/parameters/ph.

SPECTROPHOTOMETRY

Gore, Michael G., ed. *Spectrophotometry and Spectrofluorimetry: A Practical Approach*. Oxford University Press, 2000.

Hammond, John, Irish, Doug, and Hartwell, Steve. "Calibration Science for UV/Visible Spectrometry, Part III in a Series on Quality Issues in Spectrophotometry." *Spectroscopy*, vol., 13, no. 2, 1998, pp. 64–71.

Levy, Gabor B. "The Editor's Page: A Literature Search." *American Laboratory*, October 1992, p.10. (A discussion of who really discovered "Beer's" law for those with historical interests.)

Mavrodineanu, Radu, Burke, R. W. et al. "Glass Filters as a Standard Reference Material for Spectrophotometry—Selection, Preparation, Certification and Use of SRM 930 and SRM 1930." NIST Special Publication 260-116. *Department of Commerce/Technology Administration*, March 1994. (A good detailed source for those who are verifying the performance of spectrophotometers.)

From Mettler-Toledo's expertise library (www.mt.com/us/en/home.html):

- *UV/VIS Spectrophotometry Applications and Fundamentals*
- *Wavelength Accuracy in UV/Vis Spectrophotometry*
- *Fundamentals of Color Measurement at a Glance*

15 Introduction to Quality Laboratory Measurements

15.1 MEASUREMENTS AND EXTERNAL AUTHORITY: STANDARDS, CALIBRATION, AND TRACEABILITY

15.1.1 OVERVIEW

This chapter discusses the general terminology and concepts relating to making "good" measurements. Recall that measurements are numerical descriptions. For example, if an object is said to be "15 centimeters," then length is the property that is described, 15 is the value of that property, and centimeter is the measurement **unit**. A **unit of measure** *is an exactly defined amount of a property.*

We are accustomed to using units of measure, such as grams, pounds, inches, and centimeters. But what, exactly, is the mass of 1 g? This may seem like a trivial question; obviously, the mass of a gram is a gram. However, exactly defining the meaning of a "gram," or of any other unit of measurement, is anything but trivial. A unit must be defined in some clear way, and everyone who uses the unit must agree on that definition. Establishing the meaning of a unit of measure, therefore, requires international agreement.

Metrology *is the science and practice of measurements.* **Metrologists,** *people who study measurements,* devote much effort toward ensuring international consistency in measurement. One result of this effort is the **SI (Le Système International d'Unités) measurement system,** *which defines units of measurement* (see Section 11.3 for more detail on the SI system). The SI definitions of units are an authority to which people in many nations refer when making measurements.

Measurements are always made by comparison to an external authority. As a simple example, we commonly measure length using a ruler. The ruler is our external authority when measuring length. The ruler was marked by the manufacturer so that its lines are correct according to an internationally accepted definition of a "meter."

There is a distinction between a unit of measurement and a standard, such as a ruler. A **standard** *is a physical embodiment of a unit.* Units are not physical entities and are unaffected by environmental conditions, but standards are affected by the environment. For example, units of centimeters are unaffected by corrosion, but a metal ruler, a physical embodiment of centimeters, may become corroded.

DOI: 10.1201/9780429282799-20

International efforts to promote "good" measurement practices are intimately associated with the ever-increasing importance of quality systems, such as ISO 9000, ISO Guide 17025, and CGMPs. (See Chapter 4 for explanation of these terms.) Scientists have always been aware of the importance of "good" measurements. However, as people implement quality systems, they become even more concerned with establishing methods of measurement that are consistent, that are widely accepted, and whose accuracy they can document.

The next sections of this chapter explore three key interrelated words in measurement: **standard, calibration**, and **traceability**. Later sections explore four more interrelated terms: **error, precision, accuracy**, and **uncertainty**.

15.1.2 STANDARDS

The term *standard* was defined earlier as a physical object that embodies a unit. The most famous physical standard is probably the kilogram standard.

Until May 2019, the unit of a kilogram was defined by international treaty to have as much mass as a special platinum–iridium bar located at the International Bureau of Weights and Measures near Paris. All other mass standards were defined by comparison with this special metal bar. Every country that signed the treaty received one or more national kilogram prototypes whose mass was determined by comparison with the standard in France. The US standards are still housed at the National Institute of Standards and Technology (NIST). However, there are disadvantages to using a physical object, like the platinum–iridium bar, as the ultimate definition of a unit. The bar could potentially be damaged or destroyed. Its mass changes slightly with dust and with cleaning. It requires a secure and environmentally stable storage site. Therefore, in 2019, the definition of a kilogram changed so that it is no longer defined by a physical object. Rather, the kilogram is now defined in terms of Planck's constant, as discussed in Box 15.1.

BOX 15.1   DEFINING THE METER AND THE KILOGRAM

Thousands of years ago in Egypt, length was defined using the unit of a cubit, which was based on the length of the pharaoh's forearm. This unit of measure must have been adequate to construct the pyramids, which remain architectural masterpieces. However, defining a unit of measure in this way obviously suffers from the problem that with each new pharaoh, the definition changes. By the time of the French Revolution, there were many measurement systems in the world, all of which were plagued by inconsistencies. People were anxious to define measurements using consistent, unchanging definitions – they wanted a measurement system "for all times, for all peoples." The SI system of units eventually emerged from this goal.

Two important standards were created in the late 1800s as part of the SI system. Both were metal bars constructed from a platinum–iridium alloy. One bar defined the unit of a "meter." The other defined the unit of a "kilogram" (Figure 15.1). These two objects were deposited in a secure location at the International Bureau of Weights and Measures near Paris. They are called "international prototypes." Representatives from a number of countries agreed that these, and only these two metal prototypes would define the units of length and mass.

(a)   (b)

FIGURE 15.1   International metal prototypes for length and mass. (a) Length was defined based on a platinum–iridium bar. (b) The round cylinders in this image are four of the US national prototype metal mass standards made by comparison with the prototype in France. The one in front is named K20. They are housed inside a bell jar that holds a vacuum. (Image credit: U.S. National Institute of Standards and Technology.)

(Continued)

These metal prototypes served admirably over many years to unify length and mass measurements internationally, but they were not a perfect solution. The metal bars could potentially be damaged or destroyed. Their masses changed slightly over time. Scientists from every country had to travel each time they wanted to compare their physical standards with the prototypes in France. Measurement scientists therefore spent years looking for ways to avoid using physical objects to define units. They have now created methods of defining units that depend on universal physical constants. For length, the speed of light is the physical constant used. Scientists took the prototype that previously defined a meter and measured its length in terms of the speed of light. They found that the length of this bar is the distance light travels in a vacuum in 1/299,792,458 of a second. Therefore, in 1983, the meter was redefined as the distance light travels in 1/299,792,458 of a second. While not every individual has the ability to measure the speed of light, most countries have scientists who can make such measurements. Thus, scientists in each country can create metal length standards based on the speed of light. If a standard gets old or degraded, the scientists can make a new one and no one needs to travel to France to compare their standard to the prototype deposited there.

The kilogram proved to be more difficult than the meter to redefine based on a universal constant. Until May 2019, the unit of a kilogram continued to be defined by international treaty to have as much mass as the prototype deposited in France more than 100 years earlier. Finally, in 2019, the kilogram was redefined based on a physical constant, that is, Planck's constant, which relates the energy carried by a photon to its frequency. Laboratories, such as those at NIST, have instruments, called "Kibble balances," that can measure the mass of objects in terms of Planck's constant. Thus, each country can now transform the definition of a kilogram, based on Planck's constant, into a mass standard that is the authority for that country.

15.1.3 CALIBRATION

15.1.3.1 Calibration as Adjustment to an Instrument or Measuring Device

In common usage, **calibration** *adjusts a measuring system so that the values it gives are in accordance with an external standard(s)*. For example, calibration of an instrument might involve placing a standard in or on the instrument and pressing keys until the instrument displays the value of the standard. After calibration, the response of the instrument is in accordance with the standard. The result of calibration (according to this definition) is that the instrument or measuring device is adjusted and it gives more correct values after calibration than before.

Calibration as an adjustment to a measuring system applies to the way a manufacturer makes a measuring device. For example, glassware that is used to measure the volume of liquids is marked with lines that indicate volume. The manufacturer "calibrates" the glassware so that the lines are in the proper place.

Once instruments enter the laboratory, they are subject to the effects of aging and their response may be altered by changes in their environment. As a result, the response of instruments in the laboratory drifts. To correct for drift, laboratory equipment must be periodically recalibrated by the user or a service technician. For example, pH meters are very sensitive to the effects of aging and therefore require frequent calibration. Calibration involves adjusting the readings of the pH meter according to the pH of standard solutions of known pH. The calibration of pH meters and other measuring devices will be discussed in more detail in later chapters in this unit.

If any instrument or piece of apparatus is not properly calibrated, its measurements will deviate unacceptably from their correct values. Improper calibration is a common cause of laboratory error; therefore, maintaining instruments "in calibration" is critical.

There is always some error when items are calibrated. The error can be reduced by using more expensive, exacting procedures, but it cannot be eliminated. **Tolerance** *is the amount of error that can be tolerated in the calibration of a particular item*. For example, the tolerance for a "500 g" Class 1 mass standard is 1.2 mg, which means the standard must have a true mass between 500.0012 and 499.9988 g. The tolerances for a standard, a measuring device, or an instrument vary depending on the purpose for which the item is being used.

15.1.3.2 Calibration as a Formal Assessment of a Measuring Instrument

In formal usage by metrologists, calibration and adjustment are two separate actions. You *calibrate* a measuring device to understand how its response relates to standards. You may or may not *adjust* the measuring device after calibration. During the assessment, the measuring system being calibrated is compared with a trustworthy standard. Calibration quantifies the relationship between the values read by the instrument or device (e.g., a balance or thermometer) and the relevant units in the SI system. This relationship is established following a specific procedure under controlled conditions. Errors in the readings of the item being calibrated are determined and documented. The result of this calibration is a *document* that certifies that the measuring item was functioning in a particular way, what corrections, if any, need to be made to its readings when it is used, and how certain the user can be of its readings. After calibration, in this sense of the word, the item being calibrated does not perform any better than it did before. However, the calibration document may allow the operator to use the item with better accuracy by applying correction factors to its readings. For example, it may be established by calibration that the readings of a thermometer are consistently 0.1° too high. Then, every time the thermometer is used, 0.1° must be subtracted from its readings.

The more common usage of the term "calibration," at least in biotechnology facilities, includes adjustment to the instrument so that its readings are in accordance with the external standard. In this textbook, we use the term calibration to include adjustment.

15.1.3.3 Verification

It is good laboratory practice to check the performance of instruments regularly to make sure they are functioning properly. **Verification** or **performance verification** *means to check the performance of an instrument or system*. Verification includes checking whether the instrument is properly calibrated. The checking done during verification is usually a simpler, less rigorous evaluation of the performance of a measuring item than is calibration. Verification is performed in the laboratory of the user and is typically documented, possibly in a paper logbook or on an electronic form. For example, in many laboratories balance accuracy is checked regularly by weighing a standard and recording its weight.[1] If the value for the standard's weight does not fall within a particular range, as specified by a quality-control procedure, then the balance is repaired. This check of the balance's performance verifies that the balance is properly calibrated, but it should not be mistakenly called "calibrating the balance."

---

[1] Weight and mass are not synonyms, but in common practice, the terms are used interchangeably, as we do in this text. The distinction between mass and weight will be explored in Chapter 17.

15.1.3.4 Calibration (Standard) Curves

The term "calibration curve" relates to chemical and biological assays (tests) where the response of an instrument to an analyte is determined. For example, it is common to evaluate the amount of protein present in a solution using a spectrophotometer. It is necessary to "calibrate" the response of the spectrophotometer to the amount of protein present. To perform this "calibration," the response of the instrument to solutions containing known amounts of protein is plotted on a graph. The solutions with known amounts of protein are *standards*. The resulting graph is a *standard curve* or a *calibration curve* (see Section 28.3 for more details).

15.1.4 TRACEABILITY

15.1.4.1 The Meaning of Traceability (in the Context of Measurements)

Suppose you use a balance to weigh an object. You obtain and record a particular weight. How can you be certain, and demonstrate to others, that the balance you used was indeed properly calibrated based on international standards? This is a quality-assurance problem and leads to the concept of **traceability**. **Measurement traceability** *describes the chain of calibrations that establishes the value of a standard or of a measurement*.

The concept of traceability dates at least back to the ancient Egyptians, who used the pharaoh's arm as the national standard of length. Because the pharaoh could not always be present when measurements were made, the length of his arm was reproduced using a granite rod. The rod, in turn, was duplicated to make wooden measuring sticks that were used by workers building the pyramids. This system was an early example of using an external standard for measurement, and traceability to this standard.

Mass standards provide another example of traceability. A mass standard used in an individual laboratory would have been calibrated according to a standard that was, in turn, calibrated by comparison with a standard at a national standards laboratory, such as NIST. Thus, there is a "genealogy" for the standard that is used in a particular laboratory (Figure 15.2). For traceability purposes, the genealogy for a standard must be documented in a formal certificate.

The purpose of tracing the genealogy of a standard is to ensure that measurements made with that standard, or calibrations performed using that standard,

FIGURE 15.2 Traceability of mass standards showing their "genealogy." Planck's constant is transformed into a national standard that is a physical object held at NIST. This standard is the basis to which all mass measurements in the United States are traced. Calibration laboratories have standards that were calibrated (compared) with the NIST standard and are used to make working standards. Working standards are sold and used in individual workplaces. Thus, the measured values obtained for weight in an individual laboratory have a "genealogy" that is traceable back to a national standard at NIST.

are trustworthy. A statement of traceability is a quality statement. Manufacturers use the term *traceable to NIST* in their catalogs when they have documented the genealogy of standards used in manufacturing their product.

Table 15.1 summarizes terminology relating to standards, calibration, and traceability.

15.1.4.2 Summary: The Relationship between Standards, Calibration, and Traceability

To summarize the relationship between standards, calibration, and traceability, consider the example of making a measurement of weight using a balance. The balance was *calibrated* by a technician so that its readings are correct according to an internationally accepted definition of a gram. To calibrate the balance, the technician used working *standards*. The technician knows that the standards are correct embodiments of the unit called a "gram" because the comparisons of those standards to the national kilogram standard were properly performed and documented. The working standards are therefore *traceable* to NIST.

15.1.4.3 A Note about the United States and the Metric System

In the United States, we routinely see units of miles, inches, feet, pounds, gallons, ounces, and Fahrenheit – none of which are metric units. For this reason, people believe that the United States is one of few countries in the world that does not use the metric system. In fact, the United States has adopted the metric system. All the US measurement units (feet, pounds, gallons, Fahrenheit, etc.) are defined in terms of the SI – and mass, length, and volume have been defined in metric units since 1893. Measurements in the United States are defined in terms of the SI system through an unbroken chain of traceable measurements, as described in this chapter. Even if a commercial package is labeled as containing "3 *ounces* net weight," the measuring device used to determine its weight was calibrated using standards traceable to a NIST kilogram standard. Thus, although it appears on the surface that the United States does not use the metric system, in fact, below the surface, measurements in the United States are traceable to the metric system (Figure 15.3).

15.2 MEASUREMENT ERROR

15.2.1 VARIABILITY AND ERROR

If you were to weigh the same standard ten times, would you get the same value every time? It seems reasonable to expect the ten values to be identical if you follow the same procedure every time under uniform conditions. In reality, if you were to meticulously weigh the same object repeatedly with a high-quality balance, then the results would vary slightly each time. Twenty such weight measurements are shown in Table 15.2. Note that the first four digits in the weight values for the standard were always the same, but the last three digits varied.

It turns out that there is variability inherent in all observations of nature, including measurements. There is also uncertainty in all measurement values. What does this standard really weigh – exactly 10, 9.999591, 9.999594 g? Because of the variability in the measurements, we do not know the exact, true value for this standard. In fact, in principle, we can never be certain of the exact "true" value for any measurement, although we can approach that true value.

Error *is responsible for the difference between a measured value for a property and the "true" value for that property.* Statisticians classify measurement errors into types. One such classification scheme is the following:

TABLE 15.1

Terminology Relating to Measurement Standards, Calibration, and Traceability

Standards

There are 13 definitions for the word *standard* in the *Merriam-Webster Collegiate Dictionary* (G. & C. Merriam Co., MA, 1977). The broadest of these definitions is "something established by authority, custom, or general consent…" This meaning encompasses the others described below.

1. *A standard is a physical object, the properties of which are known with sufficient accuracy to be used to evaluate another item, a physical embodiment of a unit.* For example, a metal object whose mass is accurately known can be used to determine the response of a balance.

2. *In chemical or biological measurements, a standard often describes a substance or a solution that is used to establish the response of an instrument or an assay method to the material being studied.* This definition includes the following:

   a. *Standard Reference Material (SRM).* A substance issued from NIST that is accompanied by documentation that shows how its composition was determined and how certain NIST is of the given values.

   b. *Certified Reference Material.* A reference material from any source that is issued with documentation.

   c. *Reference Material.* Any substance for which one or more properties are established sufficiently well to allow its use in evaluating a measurement process or an assay.

   d. *Standard.* In common usage, any substance used to determine the response of an instrument or a method to the analyte of interest. The information obtained from a standard solution is often portrayed graphically in a **standard curve** (also called a **calibration curve**).

3. *A standard is a document established by consensus and approved by a recognized body that establishes rules or guidelines to make a procedure consistent among various people.* For example, ASTM International specifies standard methods to calibrate volumetric glassware. The U.S. Pharmacopeia specifies methods to perform tests of pharmaceutical products.

Calibration

1. *In common usage, calibration brings a measuring system into accordance with an external standard(s).* Calibration commonly includes adjusting an instrument so that its readings are consistent with the external standard.

2. *Calibration is formally defined as an assessment that establishes, under specified conditions, the relationship between values indicated by a measuring instrument or measuring system and known values based on a trustworthy standard.* The result of such calibration is a document. The instrument is not necessarily adjusted according to this definition of calibration.

3. *Tolerance is the amount of error that can be allowed in the calibration of a particular item.* National and international organizations (including ASTM International and NIST) specify the tolerances allowed for particular classes of glassware, weight standards, and other measurement items.

4. *Verification is a check of the performance of an instrument or system.*

5. *Instrumental quantitative analysis frequently involves the construction of a calibration (standard) curve that shows graphically the relationship between the response of the instrument and the amount of reference standard present.*

Traceability

1. *Traceability describes the chain of calibrations ("genealogy") that establishes the value of a standard or a measurement.*

2. *In the United States, traceability for physical and some chemical standards generally leads back to NIST, since NIST maintains national standards.*

3. *For the purpose of traceability, measurement values reported must include uncertainties* (discussed later in this chapter).

1. **Gross errors** *are caused by blunders.* For example, if a technician were to drop a mass standard (or drop a balance), these would be gross errors. Gross errors are recognizable and, of course, should be avoided.

2. **Systematic errors** are normally more subtle and harder to detect than gross errors. There are a vast number of causes of systematic errors, such as a contaminated solution, a malfunctioning instrument, and an environmental inconsistency (such as a change in humidity). An important feature of systematic error is that it results in **bias**, *measurements that are consistently either too high or too low.* For example, if a flask used to measure 500 mL has its 500 mL mark set slightly too low, then it will consistently deliver

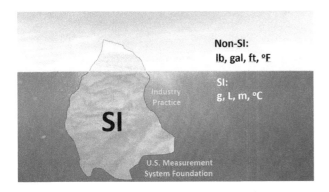

FIGURE 15.3 Measurements in the United States on the surface appear to be incompatible with the SI system, but are, in fact, based on SI definitions. (Image credit: E. Benham/NIST.)

volumes that are a little less than they should be. An experiment or series of measurements may involve more than one systematic error and these errors may nullify one another, but each alters the true value in one direction only.

3. **Random errors** *are extremely difficult or impossible to find and eliminate. Random errors cause measurement values that are sometimes too high, and sometimes too low.* Although random error is called "error," in the presence of only random error (and not gross or systematic error), measurements will average the correct "true" value.

What type of error is likely responsible for the variability in weight measurements shown in Table 15.2? If the technician making these measurements blundered or missed subtle factors, then the variability would be due to gross or systematic error. If, however, the operator was very skilled and the method of measurement was carefully planned (as we will assume was the case), then the variability in weight measurements was due to random error, which is difficult or impossible to eliminate. Thus, there is error every time a measurement is made. Even if one uses such an excellent technique that all gross and systematic errors are eliminated, there will still be random errors.

If we assume that there is no systematic error or bias in the measurements, then we must also conclude that the standard in Table 15.2 weighs slightly less than 10 g. If it was intended to be 10 g, then there was an error made at the time of its manufacture. (Note that if this standard is used to calibrate a balance, in the sense of adjusting the instrument, then all subsequent readings of this balance will be a bit too high and systematic error will be introduced.)

Observe also in Table 15.2 that if a less sensitive balance were used, for example, one that only weighed to the nearest 0.01 g, then all the observed values would be 10.00. The random variability in measurements would not be detected with this less sensitive balance. Observing random variability requires a sensitive measuring system.

The word *error* has another, somewhat different, usage than we have given so far. Error is sometimes expressed as:

$$\text{Absolute error} = \text{True value} - \text{Measured value}$$

Although, in principle, we can never be certain of the exact "true" value for a measurement, in practice, this expression of absolute error has practical application and will be further explored later in this chapter.

15.2.2 Accuracy and Precision

At this point, we introduce two important words: **accuracy** and **precision**. Measurement errors lead to a loss of accuracy and precision. These words may be defined as follows:

Precision *is the consistency of a series of measurements or tests.*
Accuracy *is how close a measurement value is to the true or accepted value.*

It is common to talk about the precision and accuracy of instruments, tests, assays, and methods. For example, we can speak of how consistent (precise) a

TABLE 15.2
Values of Repeated Weight Measurements of a Standard Object

9.999600 g	9.999597 g	9.999601 g	9.999601 g	9.999600 g
9.999596 g	9.999600 g	9.999603 g	9.999595 g	9.999592 g
9.999600 g	9.999592 g	9.999604 g	9.999593 g	9.999599 g
9.999595 g	9.999599 g	9.999598 g	9.999597 g	9.999590 g

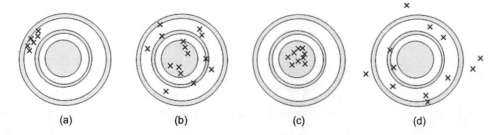

FIGURE 15.4 Accuracy and precision. Precision and accuracy are illustrated using the analogy of a target where the bull's eye is the correct value. (a) Archer A is precise but not accurate. (b) Archer B is inconsistent. Archer B's results approximately average the correct value, although they are not precise. (c) Archer C is expert and is both precise and accurate. (d) Archer D is not very skilled; the results are neither accurate nor precise.

balance's readings are, or the "correctness" (accuracy) of a test for blood glucose.

Accuracy and precision are not synonyms to scientists, although they are often used interchangeably in nonscientific English. It is possible for a series of measurements to be precise (consistent), but not accurate (Figure 15.4a). A series of measurements may also average the correct answer, yet lack precision (Figure 15.4b). A series of measurements may be both accurate and precise (Figure 15.4c), or may be neither accurate nor precise (Figure 15.4d). One definition of a "good" measurement is that it is both accurate and precise.

Consider precision in more detail. Laboratory workers are aware that measurements repeated in succession on the same day tend to be relatively consistent. In contrast, measurements performed on different days, by different people, and using different materials and equipment tend to be more variable. There are different words for precision to make this distinction clear. **Repeatability** *is the precision of measurements made under uniform conditions.* **Reproducibility** *is the precision of measurements made under nonuniform conditions, such as in two different laboratories.* Repeatability and reproducibility are therefore two practical extremes of precision. It is challenging, but important, to work to develop procedures that give results that are as reproducible as possible when performed on different days, by different people, and so on.

Example Problem 15.1

Four students each completed a laboratory exercise in which they weighed a standard known to be 5.0000 g. The values they obtained (in grams) are shown in the following table. Comment on the accuracy and precision of each student's results.

Juan	Chris	Ilana	Mel
4.9986	5.0021	5.0001	5.1021
5.0020	5.0020	4.9998	4.9987
5.0007	5.0021	4.9999	5.0003
4.9995	5.0022	4.9998	4.9977

Answer

Juan's data have a mean value that is reasonably accurate; however, their precision is not as good as the data of other students.

Chris has precise but inaccurate values. It is likely that there is a systematic error in Chris's work that causes all the values to be too high (to be biased).

Ilana's values are both accurate and precise with some random fluctuations affecting only the last figure (the fourth place to the right of the decimal point). These data are considered to be "good"; we expect *some* variability in measurements made with a sensitive instrument.

Mel's values are neither accurate nor precise. We may suspect that Mel had problems with the equipment.

15.2.3 THE RELATIONSHIP BETWEEN ERROR, ACCURACY, AND PRECISION

15.2.3.1 Random Error and Loss of Precision

Random error leads to a loss of precision because it leads to inconsistency in measurements; due to random error, values are sometimes too high and sometimes too low. We saw the effects of random error on the precision of weight measurements in the example in Table 15.2.

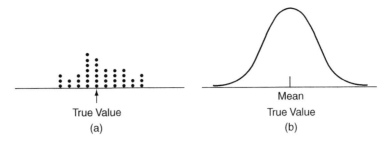

FIGURE 15.5 Random error. (a) Hypothetical results of repeated measurements using a flask that is correctly marked at 500.0 mL. Each point represents one measurement. (b) An idealized frequency distribution assuming the same volume measurement was made a great many times with a flask perfectly marked at 500.0 mL. The mean value and the true value are the same.

We can also consider a flask that is used to measure a volume of 500.0 mL. Suppose the flask is perfectly marked with a line at exactly 500.0 mL. Although the flask is marked correctly, every time it is used there will be a tiny variation in how much liquid is delivered due to imperceptible variations in the environment and the person using the flask. This slight variability is random error and will result in volumes being delivered that are sometimes a bit more than 500.0 mL and sometimes a bit less (Figure 15.5a). In this example, because the flask is perfectly marked at exactly 500.0 mL, if the flask is used many times, then the average volume delivered will equal the true value, 500.0 mL (Figure 15.5b).

It is good laboratory practice to repeat critical measurements, assays, and tests to determine the impact of random error. As measurements are repeated, you can see their variability. In addition, blunders often are detected when repeating measurements. For example, one might accidentally misread a meter or remove the wrong amount of a sample, but notice the error when the measurements are repeated.

15.2.3.2 Errors and Loss of Accuracy

Gross, systematic, and random errors all lead to a loss of accuracy. It is obvious how a gross error could cause a value to be incorrect. A systematic error will also affect the accuracy of a measurement. For example, consider a systematic error where a flask intended to measure out exactly 500.0 mL has its line drawn slightly too low on the flask. Every time this flask is used, it will tend to deliver slightly less than 500.0 mL. The volume it delivers is obviously inaccurate. When there is systematic error, the system is **biased**, *and the mean (average) of the measurements will be too high or too low* (Figure 15.6).

Systematic error can result from many causes. For example, equipment that is improperly calibrated, is not well maintained, or is used improperly; solutions that have degraded; and environmental fluctuations all cause systematic error. Unlike random error, the

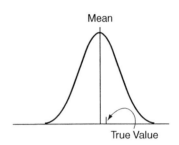

FIGURE 15.6 Measurements from a biased system. In the presence of bias, the mean is not the true value for the measurement.

impact of systematic error is not detected by repeating measurements. If a systematic error is present, then every time the measurement is repeated it will incorporate the same error and tend to be too high or too low. To minimize systematic errors in measurement, it is necessary to be knowledgeable in the methods used, to be attentive to potential problems, and to properly maintain and calibrate equipment. Chapters 17–21 in this text explore methods of avoiding systematic error when using common laboratory instruments.

We know that random error causes a loss of precision. Does random error also cause a loss of accuracy? This question is somewhat more complex. Consider again Figure 15.4b. On average, the accuracy of this archer is fairly good; however, if we look only at a single measurement, it may be way off center. Thus, when considering an individual measurement, or only a few measurements, if a system is relatively imprecise, then its accuracy will also be poor. The same idea is illustrated in Figure 15.5a and b. If a single measurement from Figure 15.5a is selected, that measurement may or may not be close to the correct reading of 500 mL even though the average, as shown in Figure 15.5b, is accurate. Random error can thus lead to both a loss of precision and a loss of accuracy.

In practice, there are both random and systematic errors present when measurements are made in the laboratory. It is often difficult to tease apart the effects of random errors and systematic errors.

Terms relating to accuracy, precision, and error are summarized in Table 15.3.

15.2.4 Evaluating the Precision of a Measurement System

Variability in a series of measurements can be evaluated using statistical tools. For example, consider hypothetical values obtained by a technician analyzing lead levels in paint samples (Table 15.4). The analyst observes on February 20 that the values from two houses seem to fluctuate more than normal. This observation alerts her to the possibility that there might be a problem in the measurement system. She can analyze the variability in the data using various statistical tools including the **range**, the **standard deviation (SD)**, and the **coefficient of variation (CV)**.

The **range** *is the difference between the highest and the lowest values of a set of measurements.* Range is a simple indicator of precision: If the range is narrow, then the data are less variable than if the range is wide. Table 15.5 shows the range for the lead measurements.

TABLE 15.3

Error, Accuracy, and Precision

1. *Error is sometimes defined as the difference between a measured value and the true value.*
2. *Error sometimes refers to the cause of variability and inaccuracy. Errors can be classified into three types: gross, systematic, and random.*
   a. *Gross errors. Human blunders.*
   b. *Systematic errors. Errors that cause results to generally be either too high or too low. Repeating measurements is not a useful way to detect systematic errors.*
   c. *Random errors. Errors that cause results to be sometimes too high and sometimes too low and which are difficult or impossible to eliminate completely. Repeating measurements is a useful way to determine the magnitude of random error.*
3. *Errors occur whenever laboratory measurements are performed.*
4. *Errors cause uncertainty in measurements.*
5. *Bias occurs when there is a systematic error(s) that causes measurement values to tend to be too high or too low.* In a biased system, the mean differs from the true value.
6. *Precision is the consistency of a series of measurements.* Precision is affected by random error.
   a. **Repeatability** *is the precision of measurements repeated in succession.*
   b. **Reproducibility** *is the precision of measurements performed under varying conditions.*
7. *Accuracy is sometimes defined as how close a measurement is to the true or accepted value.*
8. *Uncertainty is an estimate of the inaccuracy of a measurement that includes both the random and bias components.*
9. *Uncertainty is also defined as an estimate of the range of values in which the "true" value lies.*

TABLE 15.4

Lead Levels in Paint from Houses

Feb. 4 House 1 Window (mg/cm²)	Feb. 4 House 2 Window (mg/cm²)	Feb. 8 House 3 Window (mg/cm²)	Feb. 8 House 4 Window (mg/cm²)	Feb. 8 House 5 Window (mg/cm²)	Feb. 10 House 6 Window (mg/cm²)	Feb. 20 House 7 Window (mg/cm²)	Feb. 20 House 8 Window (mg/cm²)
1.21	0.43	0.92	0.78	1.51	2.12	0.98	1.23
1.18	0.40	0.93	0.67	1.43	1.99	0.78	0.21
1.31	0.34	0.79	0.71	1.34	2.13	0.21	0.11
1.23	0.41	0.93	0.65	1.47	1.98	1.24	0.89

Some information for this example was taken from Driscoll, J.N., Laliberte, R., and Wood, C. "Field Detection of Lead in Paint and Soil by High-Resolution XRF." *American Laboratory*, March 1995, p. 34H.

TABLE 15.5

Statistical Analysis of Paint Lead Levels Data

Feb. 4 House 1 Window (mg/cm²)	Feb. 4 House 2 Window (mg/cm²)	Feb. 8 House 3 Window (mg/cm²)	Feb. 8 House 4 Window (mg/cm²)	Feb. 8 House 5 Window (mg/cm²)	Feb. 10 House 6 Window (mg/cm²)	Feb. 20 House 7 Window (mg/cm²)	Feb. 20 House 8 Window (mg/cm²)
1.21 1.18	0.43 0.40	0.92 0.93	0.78 0.67	1.51 1.43	2.12 1.99	0.98 0.78	1.23 0.21
1.31 1.23	0.34 0.41	0.79 0.93	0.71 0.65	1.34 1.47	2.13 1.98	0.21 1.24	0.11 0.89
Range=1.31– 1.18=0.13	Range=0.09	Range=0.14	Range=0.13	Range=0.17	Range=0.15	Range=1.03	Range=1.12
Mean ≈ 1.23	Mean ≈ 0.40	Mean ≈ 0.89	Mean ≈ 0.70	Mean ≈ 1.44	Mean ≈ 2.06	Mean ≈ 0.80	Mean ≈ 0.61
SD ≈ 0.06	SD ≈ 0.04	SD ≈ 0.07	SD ≈ 0.06	SD ≈ 0.07	SD ≈ 0.08	SD ≈ 0.44	SD ≈ 0.54
CV ≈ 4.5%	CV ≈ 9.8%	CV ≈ 7.7%	CV ≈ 8.2%	CV ≈ 5.1%	CV ≈ 3.9%	CV ≈ 55%	CV ≈ 88%

Note that the range of values for February 20 is greater than for the other days.

Standard deviation (SD) is commonly used to evaluate the variability of a group of measurements. (See Chapter 14 for a discussion of SD and CV.) A series of measurements with a lower SD has better precision than a series of measurements from a similar system with a higher SD. The **coefficient of variation (CV)** expresses the standard deviation of a series of measurements in terms of the mean. The standard deviation and the coefficient of variation for the measurement values on February 20 were higher than on the other days (Table 15.5).

These results suggest that for some reason the data obtained on February 20 were more variable than on the other dates. There are many possible explanations for this variability. The variability might be due to a characteristic of the houses themselves. The windows from which the paint samples were taken may have had many layers of different paints applied, in which case the variability resulted from lack of homogeneity in the samples. An instrument could have been malfunctioning on February 20 and required repair. A reagent involved in the measurement procedure may have degraded. The technician at this point will want to identify the source of variability and correct any problems that may be present.

Another example of the use of standard deviation to express the precision of a measurement system is illustrated in Figure 15.7. Catalog descriptions of instruments may report the precision of the instrument in terms of the standard deviation of a series of measurements made with the instrument. The more consistent the instrument, the lower its SD, and, presumably, the better its quality.

Catalog Number	pH 20 007
Type	pH
Range	2.00 to 12.00
Resolution	0.01 pH unit
Repeatability	± 0.10 pH units

FIGURE 15.7 Catalog description of a measurement instrument using standard deviation to indicate precision. A pH meter description. Repeatability (± 0.1 pH unit) is the standard deviation around an undisclosed mean. Note that the term *range* is not an indicator of precision in this case; rather, it indicates the span from the lowest to the highest pH that the meter can read.

Example Problem 15.2

Which result from a series of measurements is more precise: 15.0 ± 0.3 g (mean ± SD) or 15.00 ± 0.03 g (mean ± SD)?

Answer

The SD "0.03" is smaller, or more precise, than 0.3 g; therefore, 15.00 ± 0.03 g is more precise.

Example Problem 15.3

A biotechnology company manufactures a particular enzyme that is used to cut DNA strands. The enzyme's activity can be assayed and is reported in terms of "units/mg." Each batch of enzyme is tested before it is sold. The results of repeated tests on four batches of enzyme are shown in the table below.

a. What is the mean activity of the enzyme for each batch?
b. What is the SD for each batch?
c. What is the mean activity for all batch values combined?
d. What is the SD for all batch values combined?

ENZYME ACTIVITY (UNITS/MG)

Batch 1	Batch 2	Batch 3	Batch 4
100,900	100,800	110,000	123,000
102,000	101,000	108,000	121,000
104,000	100,100	107,000	119,000
104,100	100,800	109,100	121,000

Answer

Mean batch 1 = 102,750 units/mg;
SD = 1,567 units/mg.
Mean batch 2 = 100,675 units/mg;
SD = 395 units/mg.
Mean batch 3 = 108,525 units/mg;
SD = 1,305 units/mg.
Mean batch 4 = 121,000 units/mg;
SD = 1,633 units/mg.
Mean all batches combined = 108,238 units/mg.
SD all batches combined = 8,254 units/mg.

Note that there is, as we might expect, more variability *between* the batches than there is *within* one batch. This is reflected in the fact that the standard deviation for all the batches combined is higher than the standard deviation of any one batch.

15.2.5 EVALUATING THE ACCURACY OF A MEASUREMENT SYSTEM

We saw in the last section how the precision of an instrument or measuring system is evaluated by performing a series of measurements and applying statistical tests to the results. Next, consider the evaluation of accuracy. Accuracy is the closeness of agreement between a measurement or test result and the true value, or the accepted reference value, for that measurement or test.

We generally do not know the "true" value for a measurement. For example, suppose an analyst is testing the level of glucose in a sample of blood. The analyst does not know the *true* blood glucose value for that sample – if the value were known, there would be no point in doing the test. The analyst, therefore, cannot calculate the accuracy of the measurement for that sample. The analyst can, however, evaluate the accuracy of the method itself.

The most obvious way to assess the accuracy of a method or an instrument is to use a standard. For example, the assay method for glucose can be evaluated by testing a blood sample with a known amount of glucose. This sample with a known amount of glucose is a quality-control (or simply control) sample. The analyst assays the level of glucose in the control sample using the same assay method that is used for patient blood samples. The result obtained for the control is compared with the expected result to determine the accuracy of the test procedure.

Two simple ways to mathematically express the accuracy are as follows:

Expressing Accuracy as "Absolute Error"

$$\text{Absolute Error} = \text{True Value} - \text{Average Measured Value}$$

where "error" is an expression of accuracy and the true value may be the value of an accepted reference material.

Expressing Accuracy as "Percent Error"

$$\% \text{ Error} = \frac{\text{True Value} - \text{Average Measured Value}}{\text{True Value}} \times 100\%$$

where "% error" is an expression of accuracy and the true value may be the value of an accepted reference material.

It is good practice to check the performance of a test using one or more quality-control samples on a regular basis. This practice of checking methods and

Example Problem 15.4

Suppose the glucose level in a control sample is stated to be 1,000 mg/L. An analyst performs a glucose test ten times on this control sample and gets the following values:

996, 1,009, 1,008, 998, 1,001, 999, 997, 1,000, 1,008, and 1,010 mg/L.

What are the absolute error and percent error based on these data?

Answer

The average value from the ten tests is 1002.6 mg/L. The absolute error, based on this average, is:

1,000 mg/L −1,002.6 mg/L=−2.6 mg/L.

The percent error, based on this average, is:

$$\frac{1,000\frac{mg}{L}-1,002.6\frac{mg}{L}}{1,000\ mg/L}\times100\%=\ -0.26\%$$

instruments with controls is often a documented part of a quality-control program. Based on experience, repetition of the test, and knowledge of the system, it is possible to establish a range within which the values for the test should fall. If the results of a test lie outside of this range, then it is necessary to look for problems.

Example Problem 15.5

a. Suppose you want to evaluate the precision of a balance. Do you need to use a standard whose mass value is traceable to NIST?

b. Suppose you want to check whether a balance is giving measurement values that are accurate. Do you need to use a standard whose mass value is traceable to NIST?

Answer

a. The evaluation of precision does not require a standard whose properties are known. Any object can be used. For example, you could determine the precision of the balance by checking the mass of the same pencil 20 times

and calculating the standard deviation for the values.

b. In contrast, the evaluation of accuracy requires a standard against which the readings of the balance are compared. This is where traceability comes in: Traceability ensures that the mass of the standard used to check the balance is known.

Example Problem 15.6

An environmental testing laboratory is about to begin testing water for chromium. The company's scientists learn to perform the chromium assay and test their skills by assaying commercially prepared standards with known chromium concentrations. They use four standards and obtain the results as below. The scientists want to be able to guarantee their customers that they will be able to analyze chromium in samples with less than 2% error. Fill in the table to show the relative percent error for each measurement. Have the scientists met their goal for accuracy based on these results?

Actual Concentration of Chromium in the Standard (µg/L)	Assayed Concentration of Chromium Obtained in the Laboratory (µg/L)	% Error
1.00	0.91	____
5.00	4.78	____
10.00	9.89	____
15.00	15.08	____

Answer

The percent errors are, in order: 9.00%, 4.40%, 1.10%, and −0.53%.

These data suggest that at low concentrations they have not achieved the desired accuracy, although they have at higher concentrations. Refinements in their technique are required.

Example Problem 15.7

An assay used to measure the amount of protein present in samples is giving erroneous results that are low by about 5 mg. What will be the percent error due to this problem if the actual protein present in a sample is:

i. 30 mg **ii.** 50 mg **iii.** 100 mg **iv.** 550 mg

Answer

i. (true – measured value)/true value

$$= \frac{30\,\text{mg} - 25\,\text{mg}}{30\,\text{mg}} \times 100\% \approx 16.7\%.$$

ii. 10.0%; iii. 5.0%; iv. 0.91%.

Note that the impact of this error is more pronounced when protein is present in low amounts than when protein is present at higher levels; the 5 mg error is a larger percent of 30 mg than it is of 550 mg.

Example Problem 15.8

A new balance must be purchased for a teaching laboratory. An instructor compares three competing brands by measuring a standard of 1.0000 g five times on each balance.

a. Which balance is most accurate? Show the percent error for each balance.
b. Calculate the SD for the measurements from each balance. Which balance is most precise?
c. What is the CV for each of the balances?
d. Report the mean value for the standard from each balance ± the SD.
e. Which balance would you buy?

Brand A (g)	Brand B (g)	Brand C (g)
1.0004	0.9997	1.0003
1.0005	0.9996	0.9996
1.0004	1.0003	1.0002
1.0003	1.0002	0.9995
1.0005	0.9999	1.0004

Answer

The percent error is an indication of the accuracy of the balances. Precision is shown by

Balance A	Balance B	Balance C
Mean ≈ 1.0004 g	Mean ≈ 0.9999 g	Mean = 1.0000 g
% Error ≈ −0.04%	% Error ≈ 0.01%	% Error = 0.00%
SD ≈ 0.0001 g	SD ≈ 0.0003 g	SD ≈ 0.0004 g
CV ≈ 0.01%	CV ≈ 0.03%	CV ≈ 0.04%
Mean ± SD ≈ 1.0004 ± 0.0001 g	Mean ± SD ≈ 0.9999 ± 0.0003 g	Mean ± SD ≈ 1.000 ± 0.0004 g

the SD and CV. Although the mean value of Balance C is correct, its precision is poor compared with the other balances, and any single measurement made on Balance C is likely to be inaccurate. Balance A has the best precision, but it is biased – all the readings are a bit high, which suggests a systematic problem. If Balance A can be recalibrated or otherwise be made accurate, it might be a good choice. Balance B might be an acceptable compromise, particularly if it has other qualities that are desirable, such as ruggedness and ease of use.

15.3 INTRODUCTION TO UNCERTAINTY ANALYSIS

15.3.1 What Is Uncertainty?

A "good" measurement is one that can be trusted when making a decision in a given situation. "Good" measurements have these characteristics:

- "Good" measurements are traceable to international standards.
- "Good" measurements achieve a required level of accuracy and precision.
- "Good" measurements include an indication of their uncertainty.

We have discussed the first two bulleted requirements, traceability to international standards and accuracy and precision. What is uncertainty? The term *uncertainty* is used often in the literature of metrology. To some extent, the meaning of *uncertainty* is familiar. For example, consider a statement in the newspaper that "the distance from the earth to the sun is 156,300,000 km." Next, consider a statement that "there are 10 microscopes in our laboratory." You would know from experience that the figure given for the distance to the sun is not likely to be exactly correct; in fact, it could be off by many kilometers. In contrast, the statement that "there are 10 microscopes in the laboratory" is likely to be exactly correct. There is uncertainty in the value reported for the distance to the sun, while there is little, if any, uncertainty in a count of ten microscopes.

As there is uncertainty in the figure reported for the distance to the sun, so there is uncertainty in the measurements we make in the laboratory. The reason there is uncertainty in laboratory measurements is that error exists. Even when people are careful, random errors and sometimes systematic errors persist.

Metrologists try to estimate the effect of errors so they can know how much confidence to place in a measurement. Metrologists state that a "good" measurement must include not only the value for the measurement, but also a reasonable estimate of the uncertainty associated with the value. Calibration and traceability documents include statements of uncertainty. Thus, a measurement should properly be of the form: measured value ± an estimate of uncertainty.

Consider, for example, the measurements of the standard weight as shown in Table 15.2. We can be fairly certain that the standard weighs a little less than 10 g although we cannot be sure of its exact true weight. The best estimate of the true weight of the standard is the mean of a large number of measurements. The mean weight of the standard based on the 20 measurements in Table 15.2 is 9.999598 g. Therefore, to the best of our knowledge, the standard's true weight is 9.999598 g – "give-or-take" something. The "give-or-take" something is the uncertainty in the measurement.[2] **Uncertainty** *is an estimate of the inaccuracy of a measurement due to all the errors present.*

Estimating uncertainty is complex. It requires identifying, to the best of one's knowledge, all sources of error, estimating the magnitude of each error, and combining all the effects of error into a single value.

Not surprisingly, it is difficult to state, with certainty, how much uncertainty there is in a measurement.

15.3.2 Using Precision as an Estimate of Uncertainty

Although it is difficult to perform an in-depth uncertainty analysis, it is reasonably straightforward to evaluate the uncertainty due to random error. It is common for analysts to repeat a particular measurement and to summarize the uncertainty due to random error using the standard deviation of the repeated measurements.[3]

Random error is not the only type of error that may be present when a measurement is made. However, it is more difficult to estimate the uncertainty due to other types of errors. In situations where a measurement process is well understood, the analyst may make estimates of the uncertainty caused by a variety of errors. For example, a measuring system may be known to be affected by changes in barometric pressure. The analyst may therefore add in a factor for uncertainty caused by fluctuations in barometric pressure. There are statistical methods that have been developed to deal with these types of uncertainty. (For an explanation of these methods and of current practice in uncertainty measurement, consult, for example, Nicholas and White, or the NIST publication "Guidelines for Evaluating and Expressing the Uncertainty of NIST Measurement Results"; both are referenced at the beginning of this unit.)

15.3.3 Using Significant Figures as an Indicator of Uncertainty

15.3.3.1 The Meaning of Significant Figures

Measurements are properly displayed as a value ± an estimate of uncertainty. We previously discussed methods to estimate the figure that goes after the ± sign. In common practice, biologists routinely report measurement values without explicitly adding any uncertainty estimate to the value. There are, however, conventional practices used to report measurement values that roughly indicate the certainty of the measurement. These practices are covered under the heading of "significant figures."

---

[2] In this particular example, if all the uncertainty is due to random error, then it is possible to give a value to the "give-or-take something." In this case, the standard error of the mean is used to estimate uncertainty. See, for example, *Statistics* by Freedman et al., which is referenced at the beginning of this unit.

[3] The standard error of the mean (SEM) is sometimes used in a similar fashion, as is a confidence interval for the mean. The calculation of both the SEM and a confidence interval begins with determining the mean and the standard deviation for a series of repeated measurements.

Significant figures are the most basic way that scientists show the certainty in a measurement value. A **significant figure** *is a digit within a number that is a reliable indicator of value.* It is easiest to understand significant figures by looking at practical examples from the laboratory and everyday life (Examples 15.1–15.3).

Example 15.1 Rulers

Consider the length of the arrow drawn in Figure 15.8a and b. In Figure 15.8a, the ruler's gradations divide each centimeter in half. We know the arrow is somewhat longer than 4 cm and it is reasonable to estimate the tenths place and report the arrow's length as "4.3 cm." Information will be lost if we report the length as simply "4 cm" because it is possible to tell that the arrow is somewhat longer than 4 cm. It would be unreasonable to report that the arrow is "4.35 cm" because the ruler gradations give no way to read the hundredths place. When measurements are recorded, it is customary to record all the digits of which we are certain, plus one that is estimated. In this example, the 4 is certain and the 0.3 is estimated, so the measurement is said to have two significant figures. Thus, by reporting the measurement to be 4.3 cm, we are telling the reader something about how certain we are in the length. We are sure about the 4 and not as sure about the 0.3.

The subdivisions in the second ruler (Figure 15.8b) are finer and divide each centimeter into tenths. With the second ruler, we can reliably say the arrow is "4.3 cm," and, in

fact, it is reasonable to estimate that the arrow is about "4.35 cm." The 4.3 is certain; the 0.05 is estimated and so there are three significant figures in the measurement. Thus, the arrow is the same length in both Figure 15.8a and b, yet the length recorded is different because the rulers are subdivided differently. The fineness of the measuring instruments, in this case, the rulers, determines our certainty of the length of the arrow. The measurement certainty is reflected by the number of significant figures used in recording the measurement.

A Note about Terminology

- Some people would say the second ruler is more accurate than the first, because it gives values with more significant figures (i.e., with more certainty).
- Some would say the second ruler is more precise than the first because it allows us to read further past the decimal point. The word *precise* thus has two meanings. Precision may refer to the fineness of increments of a measuring device; the smaller the increments, the better the precision. Precision is also used, as explained previously in this chapter, to refer to the consistency of values: the more consistent a series of measurements, the more precise.
- Some people might say the second ruler has more resolution because it allows us to discriminate a smaller length change than the first ruler.

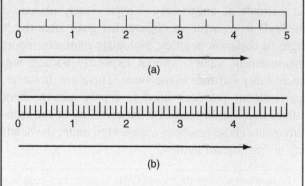

(a)

(b)

FIGURE 15.8 Measurements of length with two rulers. (a) A ruler where each centimeter is subdivided in half. (b) A ruler where each centimeter is subdivided into tenths. (Not drawn to scale.)

Example 15.2 Balances

Suppose a particular balance can reliably weigh an object as light as 0.00001 g. On this balance, sample Q is found to weigh "0.12300 g." It would not be correct to report that sample Q weighs "0.123 g" because information about the certainty of the measurement is lost if the last zeros are discarded. In contrast, if sample Q is weighed on a balance that only reads three places past the decimal point, it is correct to report its weight as "0.123 g." It would not be correct to record the weight as "0.1230 g" because the balance could not read the fourth place past the decimal point.

The first balance gives more certainty about the sample weight. The difference in measurement certainty between the two balances is shown by the number of significant figures used: "0.12300 g" has five significant figures, whereas "0.123 g" has only three.

Example 15.3 Weights of a Standard

Suppose that the technicians who made the weight measurements of the standard in Table 15.2 used a balance that could only weigh objects to the nearest 0.1 g rather than six places to the right of the decimal point. Then, every time they weighed the standard, they would have recorded its weight as "10.0 g," a value with three significant figures. Their results would be consistent, but they would never realize that the standard truly weighs a little less than 10 g. Although consistent, there is less certainty and fewer significant figures in the value "10.0 g" than in the values shown in Table 15.2.

Examples 15.1–15.3 show that a basic principle in recording measurements from instruments is to report as much information as is reliable plus one last figure that is estimated and might vary if the measurement were repeated. The number of figures reported by following this principle is the number of significant figures for the measurement. The number of significant figures reported is a rough estimate of the certainty of a measurement.

Note that most modern electronic instruments show results with a digital display. There is no meter to read. With any instrument having a digital display, the last place is assumed to have been estimated by the instrument.

As a final example, let's return to the value for the distance to the sun discussed at the beginning of this section.

Example 15.4 Zeros and Significance

Suppose that one source reports the distance to the sun to be 156,000,000 km, but another reports it as 156,155,300 km. Let's assume that both sources are correct, but the first rounded the number to make it easier to read. The second number, 156,155,300 km, is a more exact figure for the distance, which allows the reader to be more certain about the actual distance than does the first number. We say that the number 156,155,300 has more significant figures (seven) than the number 156,000,000 (which has three significant figures).

This example can be used to illustrate an important point regarding zeros. The zeros in the reported values are essential; without them, the distance would be reported as a paltry 156 km or as 1,561,553 km. The zeros are "placeholders" that tell us we are talking about a very large distance, but they are not correct indicators of value. Perhaps the exact distance is really 156,155,333 km or 156,155,329 km, or any of a number of other possibilities. In contrast, when I report that "there are 10 microscopes in my laboratory" the zero is a reliable indicator of value. The zero shows that there are 10, not 9 or 11, microscopes. Zeros that are placeholders are not called significant figures, whereas zeros that indicate value are significant.

Suppose the number given in a report is 45,000. The three zeros in this number each may be placeholders, or they may be indicators of value. There are various ways to tell the reader whether the zeros at the end of a number are significant. One method is to use scientific (exponential) notation. For example, there are three zeros at the end of the number 45,000. If none of the zeros are significant, then the number could be reported as 4.5×10^4: two significant figures. If there are three significant figures, the number could be reported as 4.50×10^4, and so on. If all the zeros in the number 45,000 are significant, this can be shown by placing a decimal point after the number. The number 45,000. and 4.5000×10^4 both have five significant figures.

Table 15.6 summarizes rules regarding how to record measurements with the accepted number of significant figures. As a laboratory professional, it is essential that you record all measured values so that they properly report the number of significant figures provided by the measuring instrument. If you fail to do so, important information may be lost. These rules also show how to decide when a zero is significant and when it is a placeholder.

TABLE 15.6

Rules to Record Measurements with the Correct Number of Significant Figures

1. *The number of significant figures is related to the certainty of a measurement.* (Note that counted values, such as "there are 10 microscopes in the laboratory," or "there are 30 students in the class," are not measurements, but rather are considered to be "exact." The rules of significant figures do not apply to counted values.)

2. *When reporting a measurement, record as many digits as are certain plus one digit that is estimated.* When reading a meter or ruler, estimate the last place. When reading an electronic digital display, assume the instrument estimated the last place.

3. *All nonzero digits in a number are significant.* For example, all the digits in the number 98.34 are significant; this number has four significant figures. A reader will assume that the 98.3 is certain and the 4 is estimated.

4. *All zeros between two nonzero digits are significant.* For example, in the number 100.4, the zeros are reliable indicators of value and not just placeholders.

5. *Zero digits to the right of a nonzero digit but to the left of an assumed decimal point may or may not be significant.* Consider the number for the distance to the sun, 156,000,000 km. The decimal point is assumed to be after the last zero. In this case, the zeros are ambiguous and may or may not be reliable indicators of value. Methods of clarifying ambiguous zeros are discussed in the text.

6. *All zeros to the right of a decimal point and to the right of a nonzero digit before the decimal place are significant.* The following numbers all have five significant figures: 340.00, 0.34000 (the zero to the left of the decimal point only calls attention to the decimal point), and 3.4000.

7. *All the zeros to the left of a nonzero digit and to the right of a decimal point are not significant unless there is a significant digit to their left.* The number 0.0098 has two significant figures because the two zeros before the 98 are placeholders. On the other hand, the number 0.4098 has four significant figures.

15.3.3.2 Calculations and Significant Figures

Calculations, such as addition or multiplication, bring together numbers from separate measurements. Each value in the calculation has a particular number of significant figures. The result from the calculation is limited to the certainty (number of significant figures) of the starting number that is least certain.

The result displayed on a calculator is seldom what should be recorded when a calculation is performed because calculators do not give any indication of certainty. The calculator result must be rounded to the proper number of significant figures. When numbers are brought together in calculations, it is often not obvious how many significant figures there are. There are various rules to determine the correct number of significant figures (see the Bibliography to this unit for more information). Table 15.7 summarizes simple rules that are likely to be adequate for situations in biology laboratories. Remember that these rules are guides to decide where to round the result of a calculation(s).

Example Problem 15.9

Read the following meter and the digital display. How many significant figures does each measurement have? (Figure 15.9)

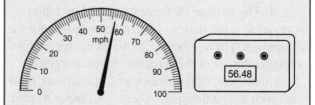

FIGURE 15.9 Meter and digital display for Example Problem 15.9.

Answer

The meter reads **"56.6," three significant figures**; the operator estimates the last place. The digital display reads **"56.48," four significant figures**; the instrument estimates the last place.

TABLE 15.7

Rules for Recording Values from Calculations with the "Correct" Number of Significant Figures

1. *It is assumed that the last digit of a result from a calculation is rounded.* For example, given the result "45.6," the .6 is assumed to have been rounded; therefore, the calculated value must have been between *45.55* and *45.64*.

2. When rounding:
   a. *If the digit to be dropped is less than 5, then the preceding digit remains the same.*
   b. *If the digit to be dropped is 5 or more, the preceding digit increases by 1.*
      For example:
      54.78 is rounded to 54.8
      54.83 is rounded to 54.8
      54.65 is rounded to 54.7
      (There are different approaches to rounding when the digit to be dropped is 5. Some people, for example, round a five to the nearest even number. Then, 9.65 is rounded to 9.6 and 4.75 is rounded to 4.8.)

3. *Round* **after** *performing a calculation.* If a problem requires a series of calculations, round after the *final* calculation and not after the intermediate calculations.

4. *The rule for expressing the answer after addition or subtraction is different than the rule for multiplication and division. The addition/subtraction rule focuses on the number of places to the right of the decimal point. The answer can retain no more numbers to the right of the decimal point than the number involved in the calculation having the least number of places past the decimal point.* For example, if adding the numbers 98.0008, 7.9878, and 56.2:

   98.0008
     7.9878
    56.2
   _____
   162.1886   Round to: 162.2

   The answer can be expressed only to the nearest tenth place because the value 56.2 has only that many places past the decimal point.

5. *In multiplication and division, keep as many significant figures as are found in the number with the least significant digits.* For example: $0.54678 \times 0.980 \times 7.899$. A calculator might display the answer as 4.232634916, but the answer should be reported as 4.23, three significant figures. Another example: 7987×12. The answer should have only two significant figures and is therefore not 95,844, but rather 96,000. (The zeros are placeholders.) (These examples assume that all the values given are measured values.)

6. *Constants are numbers whose value is exactly known.* Constants are assumed to have an infinite number of significant figures. For example, given that "12 inches equal a foot," 12 inches is a constant.

Example Problem 15.10

Which of the following are measured values and which are counted (exact) values?

a. The density at 25°C is 1.59 g/mL.
b. The distance from Chicago to Milwaukee is 75 miles.
c. Human body temperature is 37°C.
d. There are four regional campuses in the system.

Answer

Only d is counted. The other values are measured, and so the rules of significant figures apply to them.

Example Problem 15.11

A catalog advertises a particular product to be 99% pure. A competitor advertises their version of the same product to be 98.8% pure. Which one is purer?

Answer

First, assume that both competitors have faithfully followed the significant figures conventions. Second, assume that the value for purity is based on a calculation(s). Then the first product might reasonably be expected to be anywhere from 98.5% to 99.4% pure because anywhere in that range the value would be rounded to 99%. The second brand would be between 98.75% and 98.84% pure. It is unclear, therefore, which brand is actually the most pure. The second manufacturer, however, has presumably been able to ascertain the purity of their product with more certainty.

Example Problem 15.12

A biotechnology company specifies that the level of RNA impurities in a certain product must be less than or equal to 0.02%. If the level of RNA in a particular lot is 0.024%, does the lot meet the specification?

Answer

The specification is set at the hundredths decimal place; therefore, the result is also reported to that place. Rounded to the hundredths place, 0.024% is 0.02%. This lot therefore meets its specification.

Practice Problems

Statistical Formulas:

The variance for a sample is:

$$\text{Variance for a sample} = \frac{\Sigma(X - \bar{X})^2}{n-1}$$

The standard deviation for a sample is the square root of the variance:

$$\text{Variance for a sample} = \sqrt{\frac{\Sigma(X - \bar{X})^2}{n-1}}$$

The relative standard deviation is:

$$RSD = \frac{\text{Standard deviation} \times 100\%}{\text{mean}}$$

1. a. What is the purpose of a mass standard in a laboratory?
   b. Working mass standards need to be periodically recalibrated. What do you think is involved in recalibrating a mass standard? Why do you think calibration needs to be periodically repeated for a working mass standard?
2. If a balance is calibrated with a standard that is supposed to be 100.0000 g, but is actually 99.9960 g, how will this affect subsequent results from this balance?
3. If a balance is improperly calibrated (in the sense of being improperly adjusted or set):
   a. Do you expect this problem to affect the precision of the instrument? Explain.
   b. Will this problem affect the accuracy of the instrument? Explain.
4. If a flask that is used to measure volume is supposed to be marked at 10.00 mL and is actually marked at 9.98 mL, will this affect its accuracy? Will this affect its precision?
5. When a pH probe is placed in a sample, it requires a period of seconds to minutes to stabilize. If a technician using a pH meter sometimes allows the meter to stabilize for a minute or so and other times reads the meter immediately after placing the probe in the sample, will this affect the precision of the technician's results? Will it affect the accuracy of the results? Explain. Is this failure to allow the probe to stabilize an example of a systematic error?
6. Is SD a measure of precision or accuracy? Explain.

7. Suppose you prepare a solution that is intended to have 5.00 mg/mL of protein. You perform a protein assay on samples of this solution three times and obtain the following results:

5.12 mg/mL   4.86 mg/mL   5.13 mg/mL

If you perform a fourth assay of the solution, would you expect the result to be 5.00 mg/mL give or take:

0.03 g/mL or so   0.15 mg/mL or so   0.06 mg/mL or so

8. The graphs in Figure 15.10 represent the distributions of measurements from four methods that are being compared with one another. Which method is most accurate? Which method is most precise? (Assume the values on the X-axis are the same in all four cases.)

9. A standard preparation of human blood serum is prepared that has 38.0 mg/mL of albumin. Technicians from four different laboratories analyze the standard four times in 1 day and obtain the results (in mg/mL) as below. Comment on the various laboratories' accuracy and precision.

Laboratory 1	Laboratory 2	Laboratory 3	Laboratory 4
37.1	38.3	38.6	38.7
37.8	38.0	37.1	39.1
37.7	37.8	39.1	37.5
37.4	38.2	37.0	38.2

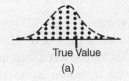

True Value
(a)

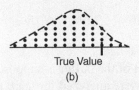

True Value
(b)

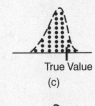

True Value
(c)

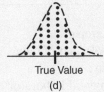

True Value
(d)

FIGURE 15.10   Graphs for Practice Problem 8.

10. Technicians from Laboratory 2 repeat the analyses, but this time do the analyses over a period of 4 months. Their results are:

    37.0 mg/mL   37.4 mg/mL   38.1 mg/mL   37.6 mg/mL

    Comment on these results in terms of their precision. Are the results more, less, or equally precise when the tests are spread over a period of months?

11. Refer to example problem 15.1. Calculate the precision of each student's results using the SD. (Remember the units.) Calculate the percent error for the average of each student's results.

12. Antibodies bind to antigens (such as proteins found on viruses and bacteria). Monoclonal antibodies are populations of antibodies produced in such a way that they are nearly identical. A population of monoclonal antibodies that bind to the HIV virus is prepared. The number of binding sites on the virus per antibody is investigated, and the results of five tests are:

    3.13   3.11   3.14   3.16   3.12

    What is the accuracy and precision of the test used to determine the number of viral binding sites?

13. The following table shows data from a study of blood calcium levels in several individuals:

Subject	Mean Calcium Level (mg/L)	Number of Observations	Deviation of Results from Mean Values
1	87.5	4	0.13, 0.19, 0.05, 0.11
2	97.6	4	0.18, 0.13, 0.10, 0.02
3	104.8	4	0.09, 0.04, 0.12, 0.06

    a. Calculate the SD for each subject's values.
    b. Pool the data and calculate the mean value for blood calcium level.

14. A company manufactures buffer solutions for use in calibrating pH meters. A new lot of pH 7.00 buffer was produced. The pH of this new lot was tested on an instrument known to be properly functioning. The results of seven measurements were:

    7.12   7.20   7.15   7.17   7.16   7.19   7.15

    a. Calculate the mean and SD for these data.
    b. Comment on these results if the pH of the buffer is supposed to be 7.00.

15. Samples of air in a particular factory were analyzed for lead. Each individual sample was tested three times and samples were taken on three occasions:

Sample	µg Pb/m³ Air
1	1.4, 1.3, 1.2
2	2.3, 2.3, 2.1
3	1.6, 1.5, 1.7

    a. Calculate the mean and SD for each sample.
    b. Calculate the mean and SD for the pooled set of data.
    c. Is the SD higher in a or b? Explain why.

16. A technician at LCJ Associates Environmental Laboratories evaluated a method to determine the levels of ammonia in water samples. To perform this assessment, the technician carefully prepared two control samples: one with 10 µg/L ammonia in water and the other with 100 µg/L ammonia in water.

LCJ Associates Environmental Laboratories

PRECISION AND ACCURACY FORM
QUALITY CONTROL DATA

Parameter   _Ammonia NH$_3$-N_   Date   _5-6-18_
Method   _Am-3_   SOP#   _27931_
Reference   _Water & Wastewater Manual_   Analyst   _N.W._

	Sample # 72807	Sample # 72808
Concentration of standard	10 µg/ℓ	100 µg/ℓ
n		
1	09.	103.
2	10.	103.
3	08.	104.
4	11.	97.
5	10.	98.
6	13.	102.
7	09.	98.

$\bar{x}$
SD
RSD
% error

FIGURE 15.11   Form for Practice Problem 16.

He tested each sample seven times using the method being evaluated and documented the results on the form above (Figure 15.11).
   a.   Calculate the mean, SD, RSD (CV), and percent error based on these data.
   b.   Why did the technician prepare two standards at two different concentrations?
   c.   Explain how these analyses and the completion of this form are part of a quality-control program.
   d.   How could the technician evaluate the accuracy of the method? What is the accuracy based on these data?
17. For each illustration (below), what measurement should be reported? How many significant figures does the measurement have? (Figure 15.12)
18. Put a line through each of the zeros in this problem that are placeholders. Place a ? above each zero that is ambiguous (i.e., may be a placeholder or may convey a value).
   a.   2,000
   b.   1,000,000
   c.   0.00677
   d.   134,908,098
19. How many significant figures does each of the following numbers have:
   a.   45.789
   b.   0.00650
   c.   10.009
   d.   0.000878

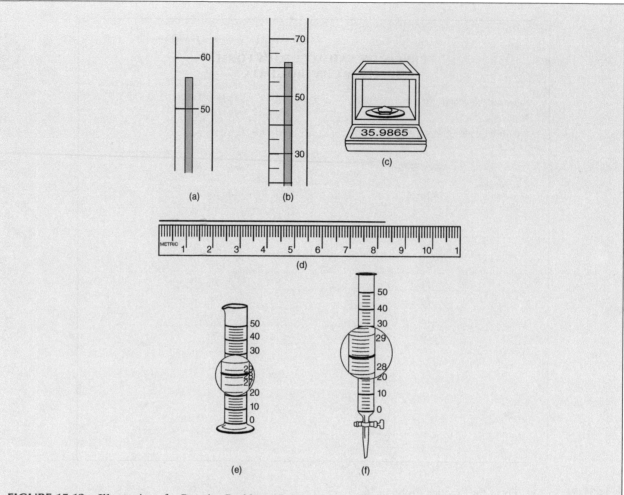

FIGURE 15.12   Illustrations for Practice Problem 17.

20. Round the following numbers to the nearest tenth decimal place.
    a. 0.0345
    b. 0.98
    c. 0.55
    d. 0.2245

21. A biotechnology company specifies that a product has at least 10 mg/vial. Do the following lots meet the specification?
    Lot a. 10.2 mg/vial
    Lot b. 9.899 mg/vial
    Lot c. 7.82 mg/vial
    Lot d. 9.400 mg/vial

22. A biotechnology company specifies that a product has ≤0.02% of impurities. Do the following lots meet the specification?
    Lot a. 0.025%
    Lot b. 0.015%
    Lot c. 0.027%
    Lot d. 0.024%

23. A spectrophotometer was used to determine the concentration of protein in a solution. The solution was analyzed six times, and the absorbance values were:

    0.956   0.948   0.958   0.991   0.963   0.957

According to the spectrophotometer manufacturer, the imprecision of the instrument should not exceed 1% relative standard deviation.

a. Do these results exceed 1% RSD?

b. If so, does this indicate that the spectrophotometer is malfunctioning or does not meet its specifications?

Questions for Discussion

1. If you work in a laboratory, find several examples of instruments or devices that are used to make measurements. Discuss these items in terms of calibration. How are they calibrated (in the sense of being adjusted according to a standard)? Who is responsible for their calibration? Are there standards used to calibrate the items that are traceable to national standards?

2. Examine a catalog of scientific supplies. What devices are specified to be "NIST traceable?"

3. Obtaining the utmost accuracy and precision in a measurement or assay is generally expensive in terms of money and/or time investment. Discuss the following situations. Is it more important for the test to be as accurate and precise as possible, or is low cost and/or speed a higher priority?

   a. Routine testing of the level of pollutant emissions from automobiles in a city where automobiles are required to have functioning antipollution devices.

   b. Determination of the level of drug in a blood sample after an overdose to enable physicians to make a rapid decision about treatment.

   c. Study of the stability of an enzyme with storage in a freezer to determine how long a product can be stored.

   d. Determination of glucose levels in the urine of pregnant women to detect pregnancy-related diabetes.

16 Introduction to Instrumental Methods and Electricity

16.1 USING INSTRUMENTS TO MAKE MEASUREMENTS

16.1.1 OVERVIEW

16.1.1.1 Introduction

The previous chapter discussed the fundamental principles of measurement. The five chapters following this one each focus on a specific type of measurement, such as the measurement of volume and of weight. This chapter is a transition that introduces instrumental methods of measurement, basic vocabulary, and also concepts relating to electricity and electronics. We include this transitional chapter because to best understand electronic laboratory instruments, it is necessary to have some knowledge of the principles and vocabulary of electricity and electronics. There are also electrical safety issues with which everyone should be familiar when operating instruments.

Electricity and electronics are relevant to anyone using laboratory equipment. In addition, there are many interesting applications of electronics that are relevant to biologists, as introduced in the Case Study "The Burgeoning Field of Bioelectronics" on pp. 402–403.

16.1.1.2 Mechanical Versus Electronic Measuring Instruments

People made measurements before measuring instruments were invented. For example, everyone uses the sense of "feeling" to estimate the air temperature or the weight of an object. Measurements made with senses alone, however, are not very accurate or reproducible; therefore, measuring instruments were invented, such as pan balances and mercury thermometers (Figure 16.1a and c). These early measuring devices did not require electricity to make a measurement and are sometimes referred to as "mechanical" instruments.

Mechanical instruments are still used in the laboratory since they are simple to understand, reliable, and usually inexpensive. Electronic measuring instruments, however, are now preferred for most applications. Electronic instruments convert a physical or chemical property of a sample into an electrical signal (Figure 16.1b and d). For example, an electronic balance converts a property relating to the object, that is, the force of gravity on the object, to an electrical signal. The balance then processes the electrical signal to convert it to a display of weight. Electronic instruments are generally more convenient, faster, and more

DOI: 10.1201/9780429282799-21

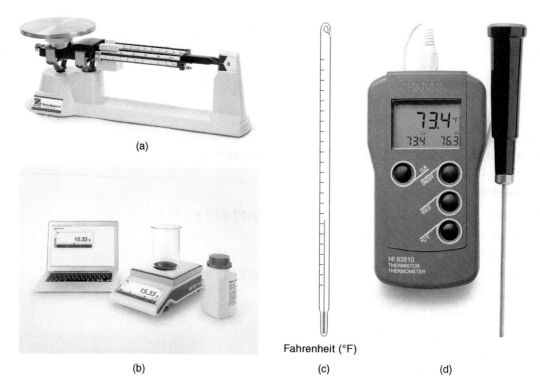

(a)

(b)   (c)   (d)

Fahrenheit (°F)

FIGURE 16.1   Mechanical versus electronic devices. (a) A mechanical balance. (Courtesy of OHAUS®.) (b) An electronic balance. (Courtesy of METTLER TOLEDO.) (c) A mechanical thermometer. (d) An electronic thermometer. (Courtesy of Hanna Instruments.)

consistent than mechanical ones. Electronic devices also can generally measure a smaller change in the property of interest than can mechanical instruments. For example, a high-quality mercury thermometer (a mechanical device) was able to give a temperature measurement to, at best, the closest tenth of a degree. In contrast, some electronic thermometers can detect a temperature change of a hundredth of a degree or better.

Early electronic instruments were characterized by control knobs, dials, and switches by which the analyst controlled the device. These instruments had meters in which a needle pointed to the value of the measurement. Modern electronic instruments are usually associated with computers, so they have keypads, display screens, and printers. The computer is sometimes an independent unit that is connected to the instrument via a cable; other times, *small computers, called* **microprocessors,** are integrated into the instrument itself. Computers and microprocessors can perform complex manipulations of data, store large amounts of information, control instruments so they operate consistently, and detect instrument malfunctions.

The computers associated with modern laboratory instruments are powerful, so these instruments

are often simple to operate and appear "intelligent." Nonetheless, it is important to remember that the human operator, although seemingly removed from the measurement process, is still a key part of making a "good" measurement. It remains the operator's responsibility to understand the measurement system, to be aware of problems, and to be skeptical of the results. Thus, it is essential for laboratory workers to have a basic understanding of how laboratory instruments work.

16.1.2 MEASUREMENT SYSTEMS

A mercury thermometer is a relatively simple measuring device. (Mercury thermometers are no longer in common use, but they still provide a good example of the principles of measurement.) The thermometer contains liquid mercury enclosed in a narrow glass tube or stem. If the thermometer is moved from a cooler to a warmer medium, the mercury expands and so moves up the stem. If the temperature decreases, the liquid contracts down the stem. There is a scale of "degrees" etched onto the glass stem. The operator reads the temperature of the medium by viewing the height of the mercury column relative to the scale of degrees.

In a mercury thermometer, the mercury "senses" one form of energy, the heat of the medium, and converts it to another form of energy, the mechanical expansion or contraction of mercury. There is an **interface** where the surface of the thermometer is in contact with the sample medium. The mercury acts as a **transducer,** *a device that senses one form of energy and converts it to another form.* The degree scale etched on the stem of the thermometer is the **display** or **readout.** We can generalize from this example and say that a mechanical measurement instrument has these components:

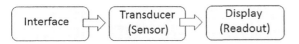

Consider a second measuring instrument, an electronic balance. The sample to be weighed is placed on the balance's weighing pan. The weighing pan is the interface. Within the balance an electronic component acts as the transducer and responds to the force of gravity on the sample. The transducer converts that force to an **electrical signal.** A **signal processor** *modifies the electrical signal so that it can be displayed as a weight value.* Thus, in an electronic instrument the components of the measuring system are:

There are many different transducers. Some respond to physical energies, such as pressure, temperature, and light, while others detect chemical properties such as pH.

For a measuring instrument to be useful, certain requirements must be met:

- *The instrument's response must have a consistent and predictable relationship with the property being measured.* For example, a certain temperature must predictably cause the mercury to rise to a certain height. A certain weight sample must generate a consistent electrical signal.
- *The instrument's response must be related to internationally accepted units of measurement.* For example, there must be a way to relate a particular height of mercury to an internationally accepted definition of the unit "degree." A certain electrical signal in a balance must be related to the unit of "gram."

Calibration *is the process by which the response of an instrument is related to internationally accepted measurement units.* (See Chapter 15 for a more complete explanation of calibration.) Every measuring instrument needs to be periodically calibrated and then needs to be checked regularly to be sure it is responding predictably and consistently. Calibration and performance verification of specific instruments are discussed in subsequent chapters as these instruments are introduced.

The relationship between the property being measured and the response of an electronic instrument is most often linear. If there is a linear response, adjustment at two points (e.g., zero and full-scale) calibrates the device because two points determine a straight line. If a device is nonlinear, additional points are needed to establish the relationship between instrument response and the property being measured. For example, a pH meter is calibrated using two standards of known pH. Two standards are sufficient because there is a linear relationship between the response of the instrument and the pH of the sample (Figure 16.2). **Two-point calibration** *refers to a situation where two standards are used to calibrate an instrument whose response is linear.*

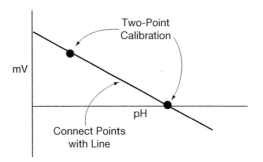

FIGURE 16.2 Calibration of a pH meter. If there is a linear relationship between instrument response and the property being measured, then two points define the relationship. In this graph, the instrument's response is an electrical signal that has the unit of millivolts (mV), and it is plotted on the Y-axis. The property being measured is pH, as plotted on the X-axis. Calibration involves determining the instrument's response at two points, shown by the arrows. To do this, two standards with known values of pH are used. For example, one standard might have a known pH of 7.0 and the other 4.0. The instrument's response (in mV) to each standard is measured, one at a time, and the points are plotted on the graph. The two points are connected by a line. The instrument uses this relationship when samples are tested. Each sample will generate an electrical signal (in mV). The calibrated instrument "knows" the relationship between pH and mV readings and is therefore able to display the pH of each sample.

Case Study: The Burgeoning Field of Bioelectronics

This chapter discusses electricity and electronics as these topics relate to laboratory instruments. But the bodies of humans and other animals also incorporate electrical systems. At every moment throughout the life of a human or other animal, information is passing back and forth between the brain and all parts of the body via the electrical conduits that are part of the nervous system. Consider the human eye. Light, which is a form of energy, reflects off of the objects around us and enters the eye. The light travels through the eye to the retina, where it strikes specialized photosensitive cells that are sensitive to light energy. The photosensitive cells convert the light energy into electrical impulses that travel through the optic nerve to the brain. In the brain the nerve impulses are converted into a picture of the environment. Thus, the eye is analogous to an electrical device as diagrammed on page 401:

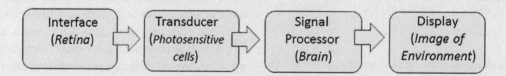

Suppose that a person is injured or experiences a disease such that there is a break in the pathway from light entering the eye to the formation of an image in the brain. This break in the pathway causes blindness. Similarly, suppose that, due to spinal cord injury or disease, the neural pathways that connect the arms and legs with the brain are broken. This person is paralyzed and can no longer detect sensation in the arms and legs nor move the affected limbs. In the past, such disorders of the body's information pathways were difficult or impossible to repair. But research is now rapidly advancing in the field of **brain–computer interfaces, BCIs.** BCIs are devices *that can detect the electrical signals from the brain, that is, "thoughts," analyze those signals, and translate them into commands that are relayed to output devices, such as artificial hands or limbs.* Sometimes BCIs are attached to the scalp, but often these computer interfaces are implanted directly into the brain. BCIs are being developed to restore function to people paralyzed by neuromuscular disorders, including spinal cord injury. In 2016, BCI researchers were able to help a man paralyzed from the chest down in a car accident regain a sense of touch through a robotic arm that he controls with his thoughts. The robot, in turn, sends electrical signals to this individual's brain that he interprets as touch. Experiments are in progress with BCIs that are connected to cameras that send information directly to the visual cortex of the brain, potentially restoring an image of the environment to the blind.

BCIs are only one area of research in the growing field of **bioelectronics,** *a discipline that brings together biology and electronics.* Bioelectronic devices have an interface where a biological entity is connected with an electronic device. Pacemakers, one of the first bioelectronic devices implanted in humans, control the beating of the heart by delivering electrical signals from a battery.

Biosensors are another type of bioelectronic device. Biosensors detect the presence or concentration of an analyte (substance), such as a biomolecule, a biological structure, or a pathogen. Binding of the analyte to the biosensor produces a response that is converted to an electrical signal that is sent to an output device. Biosensors have many applications. For example, there are biosensors that can be implanted in the body of a person with diabetes. The biosensor helps the individual control their disease by providing feedback about the glucose level in the blood. Other types of biosensors are used in settings where food is processed in order to detect pathogens. Biosensors are also applied in homeland security to detect harmful chemical and biological agents in the environment.

The field of bioelectronics offers career opportunities for individuals who are interested in biology, computers, and engineering. For those more interested in biology, understanding electricity and electronics can provide important insights into the function of the body. Everyone may be affected in the future by devices that translate thoughts into actions. Such devices may someday provide great benefits to humans who are now disabled. But, as with any powerful technology, such devices are likely to create ethical and social issues that will need to be understood and resolved by all citizens, scientists and non-scientists alike.

(Continued)

Case Study (*Continued*): The Burgeoning Field of Bioelectronics

Primary Sources:

Brown, Kristen, V. "DARPA is funding Brain-Computer Interfaces to Treat Blindness, Paralysis, and Speech Disorders." *Gizmodo*, October 7, 2017. gizmodo.com/darpa-is-funding-brain-computer-interfaces-to-cure-blin-1796779062.

Flesher, Sharlene N. et al. "Intracortical Microstimulation of Human Somatosensory Cortex." *Science Translational Medicine*, October 19, 2016. vol. 8, no. 361, pp. 361ra141. doi: 10.1126/scitranslmed.aaf8083

Shih, Jerry J., et al. "Brain-Computer Interfaces in Medicine." *Mayo Clinic Proceedings*. March 2012, vol. 87, no. 3, pp. 268–279. www.ncbi.nlm.nih.gov/pmc/articles/PMC3497935/

Wood, Max. "Neuroscience Researchers Receive $3.4 Million NIH Grant to Develop Brain-Controlled Prosthetic Limbs." *At the Forefront: UChicago Medicine*. October 16, 2018. https://www.uchicagomedicine.org/forefront/neurosciences-articles/2018/october/neuroscience-researchers-receive-grant-to-develop-brain-controlled-prosthetic-limbs.

16.2 BASIC TERMINOLOGY AND CONCEPTS OF ELECTRICITY

16.2.1 CURRENT, VOLTAGE, AND RESISTANCE

16.2.1.1 Current

Electricity is understood in terms of the atomic theory of matter. In short, according to this theory, matter consists of atoms that have electrons, protons, and neutrons. Protons have a positive charge; electrons have a negative charge. A material is said to be **negatively charged** *if electrons are in excess* and **positively charged** *if electrons are depleted.* Objects that have the same charge repel one another, whereas objects that have opposite charges attract one another. Therefore, if two objects are separated, they will be attracted if one is positive and the other is negative. On the other hand, if both are positively or negatively charged, the objects will repel one another.

Metals have a property that is important for electricity. In metals, the outer electrons of atoms are loosely held by the atoms, so these outer electrons move easily and randomly from one atom to another. Suppose there is a metal wire with an excess of electrons at one of its ends and a deficiency of electrons at the other. In this situation, the electrons in the wire will not move randomly. Rather, they will flow away from the end with an excess of electrons and toward the end with a deficiency (Figure 16.3). It may be helpful to imagine that the flow of electrons in a wire is like water flowing in a riverbed. *The flow of either water or electricity is called* **current.**

The flow of electrons, abbreviated I, is measured in the unit of **ampere,** abbreviated **amps** or **A.** If 6.25×10^{18} electrons pass a certain point every second, the current is said to be 1 amp. An ampere is a rather large unit. The current in many electronic instruments is on the order of **milliamperes (mA)** or even **microamperes (μA).** A **milliampere** *is one thousandth of an ampere*; a **microampere** *is one millionth of an ampere.*

There are two types of electrical current: **alternating current (AC)** and **direct current (DC).** In **direct current** *electrons flow in one direction through a wire.* A battery generates direct current (Figure 16.4). In **alternating current** *electrons change directions many times per second.* The current that comes from the power company to an outlet in a wall is alternating. In the United States, AC current cycles back and forth with a **frequency** of 60 times per second. In other countries AC current has a frequency of 50 times per second. Frequency is measured in the unit of **Hertz (Hz),** *where 1 Hz = 1 cycle per second.*

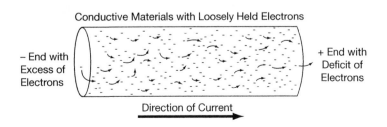

Conductive Materials with Loosely Held Electrons

– End with Excess of Electrons

+ End with Deficit of Electrons

Direction of Current

FIGURE 16.3 The flow of electricity in a wire. Electrons move from atom to atom.

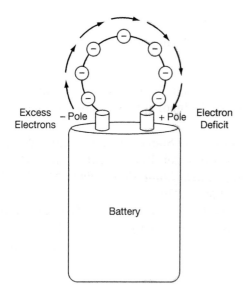

FIGURE 16.4 Direct current. In a battery, current flows directly from a pole with an excess of electrons to a pole with an electron deficit. (Credit: Audrius Merfeldas/Shutterstock.com)

Most laboratory instruments require direct current. Direct current is occasionally supplied to an instrument by batteries. It is more common, however, for a **power supply** *to convert the AC current of the power company to a DC current through a process called* **rectification.**

We are accustomed to electrical current that flows through wires; however, electrical current can also flow through liquids. Electrical current in a liquid is carried by positive and negative ions derived from salts added to the liquid. This type of current flow is important in the function of pH meters, as will be discussed in Chapter 20.

16.2.1.2 Voltage

Energy *is defined as the ability to do work;* **potential energy** *is stored energy.* For example, water poised at the top of a waterfall has **potential energy.** As the water plummets over the waterfall, its potential energy is converted into the mechanical energy of falling. Gravity is responsible for the potential energy of the water (Figure 16.5a).

Just as water can have potential energy, there is also **electrical potential energy. Electrical potential** is the potential energy of charges that are separated from one another, and attract or repulse one another because of their unlike or similar charges (Figure 16.5b and c). For example, in a battery there are a negative terminal with an excess of electrons and a positive pole with a deficit. The excess electrons at the negative terminal can be thought of as the force that "pushes" electrons toward the positive pole. Thus, just as gravity is the force that causes water to flow over a waterfall, so electrical potential is like a force that causes electrons to flow in a wire.

Electrical potential is also called **electromotive force (EMF or €)** or **voltage.** *The units of voltage are the* **volt (V)** *and the* **millivolt (mV).** The voltage that is supplied to a wall receptacle by the power company either is in the range from 110 to 120 V or is 220 V.

The development of voltage requires a method of separating charges from one another so that there is an excess of electrons at one site and a deficiency of electrons at another. There are many ways that voltage can be generated, all of which involve the conversion of some other form of energy into electrical potential energy. Fossil fuels, such as coal, gas, and oil, can be

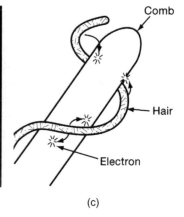

(a)   (b)   (c)

FIGURE 16.5 Potential energy. (a) The potential energy of water is converted to mechanical energy in a waterfall. (b) A familiar example of electrical potential energy. When two materials are rubbed together, electrons can be removed from one and deposited on the other causing them to have opposite charges and to attract one another. (c) Similarly, static electricity can develop between a comb and hair. (Static electricity is also discussed in the Safety Unit, Chapter 8, Section 8.3.7, and in Chapter 17 when weighing is discussed, Section 17.4.3.3.)

burned to create electrical potential. It is also possible to generate electrical potential by harnessing the mechanical energy of a waterfall or the wind, the solar power of the sun, the energy of nuclear reactions, and the energy of chemical reactions.

16.2.1.3 Resistance

Water in a river cannot flow if its path is blocked by a dam. Similarly, the flow of electricity is impeded if it encounters **resistance** in its path. **Resistance** *is the impedance to electron flow. The unit of resistance is* **ohm,** *abbreviated Ω. One* **ohm** *is the value of resistance through which one volt maintains a current of one amp.*

All materials (except the so-called superconductors) offer some resistance to current flow. The amount of resistance depends on the material. Electricity flows most readily in **conductors**. **Conductors** *are materials, such as metals, whose outer electrons are free to flow from one atom to another.* The best conductors are those materials whose outer electrons are most loosely bound. Silver is the best conductor, followed by copper and gold. In contrast, **insulators** *are materials in which the outer electrons of atoms are not free to move and so electricity does not flow readily.* Electricity usually does not flow in the air (lightning is an exception), nor does it flow in glass, plastic, or rubber.

When electricity encounters resistance, some or all of its energy is converted into heat energy. Because even copper wire offers some resistance to current flow, as electricity flows in a wire, some energy is lost as heat. The longer a wire, the more electrical energy is dissipated as heat. This is why the use of extension cords results in a loss of power. A burner on an electric stove is an application where we take advantage of the heat generated when electricity encounters resistance.

Semiconductors *offer intermediate resistance to electron flow and are used in the manufacture of transistors and other electronic devices.* Semiconductors are composed of crystalline silicon or germanium. Silicon and germanium do not readily conduct electricity; however, if impurities are added to their crystals, their structure changes so that electrons can move more easily. Semiconductors play a critical role in electronics because engineers can adjust their resistance and thereby control the flow of current through them.

16.2.2 Circuits

16.2.2.1 Simple Circuits

Electricity only flows if there is a complete conductive pathway in which the electrons can move. At one end of the pathway there must be a source providing an excess of electrons; at the other end there must be an electron deficit. *Such a pathway for electricity is called a* **circuit**. Figure 16.4 shows a battery with a negative pole, having an excess of electrons, and a positive pole, having a deficit. The wire connecting the positive and negative poles completes the pathway for electron flow.

In practice, one would not encounter a simple circuit such as the one in Figure 16.4 because the flow of electrons in the wire would produce no useful work. Electricity is useful when electrical energy flowing in a circuit is transformed into another form, such as heat, light, or mechanical work. Figure 16.6a shows a

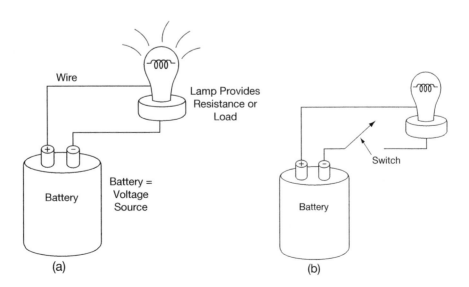

FIGURE 16.6 A simple circuit. (a) A simple circuit consisting of a voltage source, a wire, and a resistive element – in this case, a light bulb. (b) A switch is added to the simple circuit in (a).

diagram of a simple circuit that includes a light bulb. As electricity flows through the bulb, a thin wire heats up and emits light. The bulb thus converts electrical energy to light. Items such as burners on a stove, light bulbs, and motors convert electrical energy into useful heat, light, and motion. Motors, bulbs, and other such elements constitute resistance to electron flow. These elements, which provide resistance, are sometimes referred to as the "load" in a circuit. As electricity flows through such devices, work is performed, and heat is generated.

Every practical circuit at a minimum consists of conductors through which electrons can travel, a voltage source, and some type of resistance. In addition, insulation is usually added, serving to confine the current to the desired paths. In Figure 16.6b, a **switch** *to control electrical flow* has been added to the circuit. When the switch is closed, the circuit is complete and current flows; when the switch is open, the circuit is interrupted by air and current ceases to flow.

The circuit shown in Figure 16.6b is redrawn in Figure 16.7a using schematic symbols instead of pictures. There are a number of symbols used in electrical diagrams; examples are shown in Figure 16.7b.

16.2.2.2 Ohm's Law

Ohm's law (developed by George Simon Ohm) is an equation that relates voltage, current, and resistance in a circuit. Ohm's law states that the current in a circuit (expressed in unit of amps) is directly proportional to the applied voltage (expressed in unit of volts) and is inversely proportional to the resistance (expressed in unit of ohms). Ohm's law is written as:

$$\text{Voltage} = (\text{current})(\text{resistance})$$

$$V = IR \text{ or}$$

$$I = \frac{V}{R}$$

For example, consider a circuit in which there is a motor that provides 5 ohms of resistance. The voltage from the power company that supplies the circuit is 110V. Assuming there are no other devices in the circuit, it is possible to calculate the current in the circuit:

$$V = IR$$

$$110\,V = (?)(5\text{ ohms})$$

$$? = 22\,\text{amps}$$

Note that the proper units must be used to apply Ohm's law. For example, if the units in a problem are given as millivolts or microamperes, they must be converted to volts and amps.

Example Problem 16.1

According to Ohm's law, if the resistance in a circuit suddenly becomes very low, what will happen to the current in that circuit?

Answer

Assuming the voltage remains constant, the current will become very high.

Electrophoresis is a technique commonly used in the biotechnology laboratory that provides a good example of how Ohm's law is applied in practice. **Electrophoresis** *separates charged biological molecules from one another based on their rate of migration when placed in an electrical field.* In gel electrophoresis, the sample mixture is placed in a gel

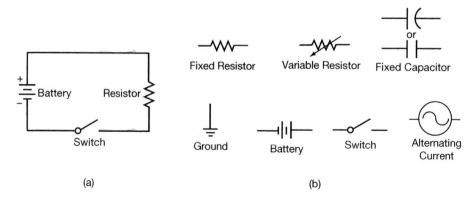

      (a)   (b)

FIGURE 16.7   An electrical diagram. (a) The same circuit as shown in Figure 16.6b, but using standard electrical symbols instead of pictures. (b) Examples of symbols used in electrical schematic diagrams. You might encounter such a diagram in the manual for a laboratory instrument.

matrix. (If you have never seen electrophoresis, imagine that the gel is like a rectangular slab of Jell-O.) The gel is positioned in a box and is covered with a thin layer of buffer. Positive and negative ions in the gel and the buffer are capable of conducting current. A power supply is used to provide voltage so that there is a positive pole at one end of the gel and a negative pole at the other end (Figure 16.8a). Negatively charged sample components migrate toward the positive pole; positively charged components move to the negative pole. Their speed of migration is determined by the magnitude of their charge. If two or more components have the same charge, then the larger ones will tend to move more slowly through the gel matrix than the smaller ones. Thus, different biological molecules migrate at different rates depending on their charge and size. Once the components are separated, the gel is removed from the box, a dye is used to stain the biological molecules, and the separated components appear as bands (Figure 16.8b).

In the early stages of an electrophoresis run, the resistance of the gel increases somewhat as ions are electrophoresed out of the gel. Therefore, if the voltage supplied to the gel remains constant, then the current decreases during the run, in accordance with Ohm's law. To maintain a constant current during electrophoresis, the voltage must be increased as the run progresses. Most power supplies designed for electrophoresis can be adjusted so that they provide either constant voltage or constant current, depending on the user's preference (Figure 16.8c). (Electrophoresis power supplies are also made that maintain constant power. Power will be discussed later in this chapter.)

Example Problem 16.2

a. At the beginning of an electrophoresis run, the voltage is 75 V and the current is 30 mA. What is the resistance of the gel?
b. The voltage is held constant, and the current decreases to 12 mA by the end of the separation. What is the resistance of the gel?

Answer

According to Ohm's law:

a. $V = IR$

$$75 \text{ V} = (0.030 \text{ A})(?)$$

$$? = 2.5 \text{ k}\Omega$$

b. $V = IR$

$$75 \text{ V} = (0.012 \text{ A})(?)$$

$$? = 6.25 \text{ k}\Omega$$

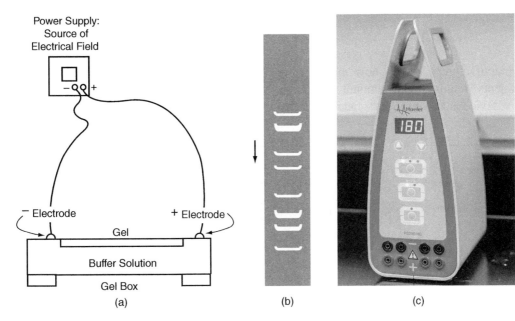

Power Supply:
Source of
Electrical Field

− Electrode   + Electrode
Gel
Buffer Solution
Gel Box
(a)

(b)

(c)

FIGURE 16.8 Electrophoresis. (a) An electrophoresis setup including a gel, a gel box, and a power supply. (b) A stained gel. The bands represent separated molecules. (c) An electrophoresis power supply. Constant current or constant voltage can be selected.

16.2.2.3 Series and Parallel Circuits

Most electrical circuits contain more than one resistive element. Circuits with multiple resistant elements (loads) can be arranged in two basic ways: in **series** or in **parallel.**

A circuit with three resistive elements is shown in Figure 16.9a. The three resistances are arranged one after the other; hence, they are said to be in series. In a **series circuit,** the *current flow has only one possible pathway.* The current passing through resistance 1 is the same as the current passing through resistances 2 and 3. A familiar example of a series circuit is a string of old-fashioned Christmas tree lights. If one bulb burns out, the pathway is no longer complete, and all the bulbs go out.

A **parallel circuit** with three resistive elements is shown in Figure 16.9b. The three paths split the total amperage, and so the total amperage in the entire circuit is equal to the sum of the currents in the three parallel pathways. In a parallel circuit, if one pathway is broken, current can still flow in the others. Circuits in most modern instruments are complex and consist of both parallel and series pathways.

16.2.2.4 Power, Work, and Circuits

Electricity is used to perform work, such as to generate light, or power a pH meter. As an electric device performs work, power is consumed. *The unit of power is* **watt,** abbreviated **W. Power** is defined as:

$$Power = (Voltage)(Current)$$

or

$$P = (V)(I)$$

Instrument manufacturers specify how much power their instruments require. The amount of power required by a device is related to how much heat that device will produce during operation. For example, a 100 W light bulb generates more light – and more heat – than a comparable 25 W light bulb. The more the power being used in a circuit, the more the heat that is generated in that circuit. Power ratings are useful to determine whether a circuit is being overloaded (i.e., too much heat is being generated) by devices.

Example Problem 16.3

A spectrophotometer requiring 300 W, an oven requiring 600 W, and a cell culture hood requiring 200 W are all plugged into the same laboratory circuit and are all turned on. Suppose further that the laboratory is served by a 120 V line with a current of 15 A. Will the load on the circuit exceed its capacity?

Answer

The total carrying capacity of the laboratory circuit is:

$$P = (V)(I)$$

$$= (15\ A)(120\ V) = 1{,}800\ W$$

The total power consumed by the three instruments is: 300 W + 600 W + 200 W = 1,100 W.

The circuit can therefore handle these three items. If, however, three more spectrophotometers at 300 W each are plugged into the same circuit and operated at the same time, then the circuit will be overloaded and the circuit breaker or fuse protecting that circuit will blow, as described in the next section.

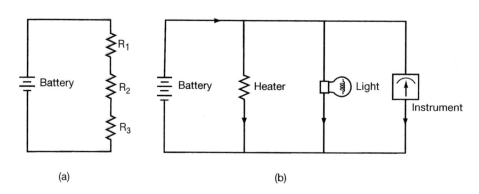

(a)   (b)

FIGURE 16.9   Series versus parallel circuits. (a) Three resistive elements arranged in series. Current passes through all three elements. (b) A parallel circuit. The current is split into three paths.

16.2.2.5 Grounding, Short Circuits, Fuses, and Circuit Breakers

Fuses or circuit breakers, and grounding wires are basic electrical safety devices that are incorporated into circuits. These devices are necessary to prevent electrical fires and to protect operators from serious shock.

If an unusually large amount of current flows in a wire, then the wire will become extremely hot – to the point where instruments can be damaged and/or a fire can be ignited. **Fuses** *are simple devices that protect a circuit from excess current.* A fuse contains a thin wire. Heat is generated whenever current flows through the wire; the more current, the more heat. The wire will break if too much current passes through it, thus breaking the circuit and preventing the further flow of current (Figure 16.10a and b). Fuses are rated according to amps and volts. For example, it is possible to purchase fuses for 10, 20, and 30 amps; 110 V. If a fuse is rated for 20 amps and more than 20 amps passes through it, it will "blow."

A circuit breaker performs the same function as a fuse; however, circuit breakers are more convenient than fuses. A blown fuse must be replaced, whereas a circuit breaker is simply switched back to its original position, after the original problem is solved.

Fuses or circuit breakers are placed in the circuits in houses and buildings. A typical house will receive from 50 to 220 amps from the power company, which is subdivided in the house into individual circuits. There is usually a main fuse or circuit breaker where the electric power line enters the building and another fuse or circuit breaker where each circuit branches from the main circuit (Figure 16.11). These branch fuses or circuit breakers are the ones with which you might be familiar, because you may have had to replace or reset one after plugging too many appliances into one circuit.

There are several common causes of excess current leading to blown fuses or tripped circuit breakers:

1. ***There may be too many devices plugged into the same circuit.*** (Anyone who has lived in an older apartment has probably experienced this problem.)

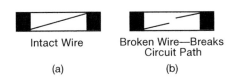

Intact Wire   Broken Wire—Breaks Circuit Path

(a)   (b)

FIGURE 16.10 Fuses and circuit breakers. (a) Good fuse. (b) "Blown" fuse.

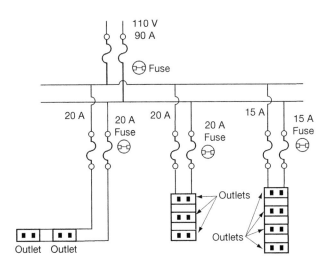

FIGURE 16.11 Fuses or circuit breakers protect the circuits in every home and building.

2. ***One device in the circuit may be using too much power.*** In this case, an electrician must be called to solve the problem.
3. ***There may be a short circuit in a device.*** In this case, the device should be unplugged, and a service technician called.

In addition to the fuses and circuit breakers that protect the circuits in a building, most laboratory instruments have their own fuses to protect delicate components. (One of the challenges of laboratory management is keeping track of the spare fuses that go with each instrument so that they are available when needed.) These fuses blow if for any reason too much current flows through the instrument. If an instrument's fuse blows, then it can indicate a problem that requires repair by a qualified person.

Figure 16.12a illustrates a properly functioning instrument that is plugged into a wall outlet. Voltage causes current to flow from the power company, through the instrument, and then back to the wall outlet, thus completing the circuit. In this example, the electricity powers a motor and so the motor provides resistance to current flow.

Figure 16.12b illustrates the same instrument, but this time the insulation around the power cord is frayed so that the metal wire touches the metal case of the instrument. No current flows through the metal frame because the circuit is not complete. The fuse does not blow, and the motor may be operative. Unfortunately, in this situation, if a person touches the metal casing of the instrument that person may provide a complete path for electrons to flow to the ground (Figure 16.12c). The person will be shocked; in some cases, such a shock

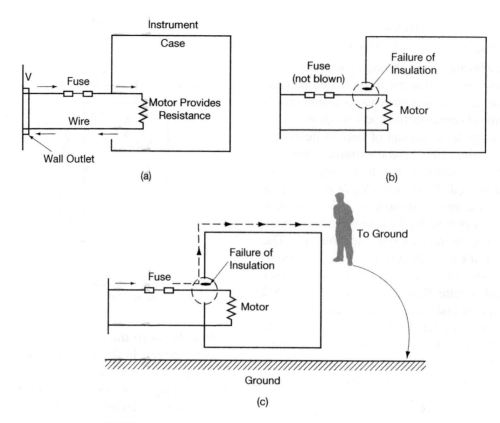

FIGURE 16.12 Short circuits and safety devices. (a) A properly functioning instrument. Voltage causes current to flow through the instrument and then back to the wall outlet. The electricity powers a motor, and so the motor provides resistance to current flow. (b) A short circuit. The insulation is frayed so that the metal wire touches the metal case of the instrument. The fuse does not blow, and the instrument may continue to be operative. (c) If a person touches the instrument in b, current will flow through the person to the ground causing dangerous shock.

can be fatal. This situation is called a **short circuit**, *a situation in which electrical current is able to flow without passing through the resistance, or load.*

Because the situation illustrated in Figure 16.12c is dangerous, modern electrical instruments are provided with a grounding wire (Figure 16.13). To understand this safety mechanism, it is necessary to know that the earth can gain or lose electrical charges and yet remain neutral. The earth, therefore, is considered to be at zero potential and can act as a charge neutralizer.

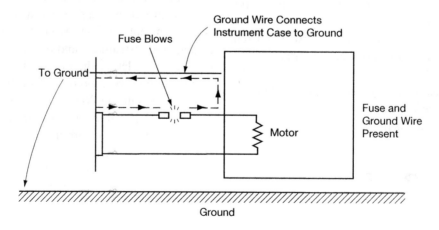

FIGURE 16.13 A properly protected instrument. The ground wire connects the instrument to the earth, thus harmlessly draining off the current and protecting the user from shock. A fuse is also present to prevent excess current from flowing in the circuit.

An **electrical ground** *is any conducting material connected to the earth.* A **ground wire** *is attached to the metal frame of the instrument and is connected to* *the earth via the third prong on the power cords of most modern instruments.* The third prong is a metal conductor that, when plugged into the wall, contacts another wire that is connected to a metal water pipe or another conductor. The water pipe, in turn, contacts the ground under the building. In the event of a short circuit, current flows from the instrument chassis to the third prong and is then discharged harmlessly to the ground. Note also that the fuse will blow in this situation because the current flowing to the ground will be great. The use of a three-prong cord to ground an instrument is an important safety feature and should not be bypassed by using adapters or other methods.

Example 16.1

The following figure shows the back panel of a representative instrument illustrating some of the ideas discussed so far.

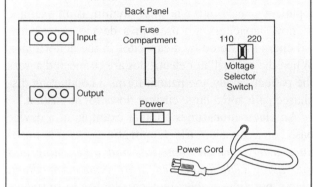

Voltage Selector. The voltage that comes to a particular receptacle from the power company may supply a voltage of either 110 or 220. An instrument can sometimes be used with either input voltage. The voltage selector is a switch that allows the user to select the proper voltage. Note that an instrument may require a fuse to be exchanged if the selected voltage is changed.

Fuse Compartment. Instrument fuses or circuit breakers are sometimes accessible to the user and are often located on the back panel of the instrument. Sometimes fuses are internal and require a service technician for replacement.

Output Connectors. Each electronic measuring instrument has some sort of output signal, such as a voltage or a current. This output can be connected via cables or cords to other devices, such as a computer.

Input Connector. Some instruments are able to receive electrical information (signal) from other instruments. For example, chromatography instruments (discussed in Chapter 34) may be controlled by computers, in which case connections are required from the computer to the instrument.

Power Connection. This is where the power cord attaches to the instrument.

16.3 BASIC TERMINOLOGY AND CONCEPTS OF ELECTRONICS

16.3.1 ELECTRONIC COMPONENTS

16.3.1.1 Overview

The term **electronic** *applies to electrical devices that are able to generate, amplify, and process electrical signals.* Most laboratory instruments – and also devices in the home – combine electrical and electronic circuitry. Both electrical and electronic circuits involve the flow of electrons to perform a useful service, but "electrical" and "electronic" are not synonyms. Let's consider the distinctions between these terms.

Electronic circuits (unlike electrical ones) can "make decisions" that control the flow of electrons. Consider a clothes dryer, a device that incorporates both electrical and electronic circuitry. The dryer has an *electrical* circuit that receives electrons from the power company, and includes an on/off switch, a heater, and a motor to rotate a drum. *Electrical* current in this circuit powers the heater and the motor that rotates the drum to dry the clothing. The user inputs settings to the dryer that are interpreted by *electronic* circuits. The electronic circuits control the electrical circuits in compliance with the user's directions regarding temperature, speed of drum rotation, length of drying time, buzzer signals, and so on.

Other distinctions between electrical and electronic circuits are the following:

- Electrical circuitry involves metals such as copper; electronic circuits use semiconductor materials.
- Electrical circuits rely on alternating current. Electronic circuits usually use direct current.
- Electrical devices consume more power than electronic circuits.

- Electrical circuits do not manipulate data. Electronic circuits manipulate data – think about computers as an example.
- Electrical circuits convert electrical energy into another form of energy – such as light. Electronic instruments perform a particular processing task, such as analyzing how light interacts with a laboratory sample.

In an electronic device, components that generate, amplify, and process electrical signals are combined and arranged in circuits. Depending on the components used and their arrangement, electronic circuits can perform a vast number of functions ranging from playing music, to word processing, to measuring weight in the laboratory.

Although the functions that electronic devices perform are numerous, the basic components of their circuits are relatively few. Some of the most common of these components are briefly described in this section, including **resistors, capacitors, diodes,** and **transistors.** The following section discusses how these components are assembled to make functional instruments for measurement.

16.3.1.2 Resistors, Capacitors, Diodes, and Transistors

Resistors are electronic components that are used to impede the flow of current in order to control the amount of current in a circuit. There are **fixed resistors** that have one value of resistance, and **variable resistors** that can be adjusted to provide different levels of resistance. Fixed resistors are manufactured to have a specific resistance and are color-coded to indicate what that resistance is.

There are two types of variable resistor: the **rheostat** and the **potentiometer**, abbreviated "pot." There are many situations where it is necessary to be able to vary the amount of current in a circuit. For example, most centrifuges have a speed control adjustment that is used to control how fast the centrifuge spins samples. This is accomplished by adjusting the voltage in a potentiometer. When you, as a user, adjust the speed of rotation of a centrifuge, you are really adjusting the potentiometer that raises and lowers the voltage supplied to the motor that spins the samples. Calibration is another example where a variable resistor is needed. Calibration of an electronic instrument usually involves changing the voltage or current in a particular circuit. This can be accomplished by varying the resistance in the circuit with a potentiometer.

A **capacitor** is a device that stores electrical charge and can be used to provide current in the absence of a battery or current from an outlet. Capacitors consist of two thin plates made of a conducting material and separated by a layer of insulating material. When current passes through a capacitor, one of the plates acquires a positive charge, and the other a negative charge. After current flow ceases through the capacitor, the charge can remain on the plates for days. A charged capacitor can function as if it were a DC voltage source.

A **defibrillator**, which is used medically to shock a patient's heart into a normal rhythm, is an example of how capacitors are used in a device. An electrical charge is stored by a capacitor in the defibrillator. When the defibrillator electrodes are connected across the patient's body, the patient forms a conducting discharge path and a large current flows for an instant.

An **electroporator** is another example of a device based on a capacitor. **Electroporation** *is used to introduce drugs, genetic material, and other molecules into living cells suspended in an aqueous medium.* Electroporation enables biotechnologists to introduce functional DNA into cells. In electroporation a short-duration, high-voltage electrical field is applied to a chamber containing cells. The electric field causes pores to be temporarily created in the cells' membranes through which the molecules enter. The simplest approach to creating a high-voltage pulse is to charge a capacitor using a high-voltage power supply and then to discharge the capacitor into the chamber with the cells.

Diodes *are components made of a crystalline material that is capable of conducting electricity in only one direction.* The primary function of **diodes** in instruments is to *convert alternating current to direct current, which is called* **rectification.**

There are also diodes, called **light-emitting diodes (LEDs)** that *produce light when electrical current flows through them.* LEDs are familiar to us as the displays on digital clocks and as a type of energy-efficient light bulb.

Transistors *are devices made from semiconductor materials that are used to amplify an electrical signal.* Transistors also function as switches to turn a circuit on or off, and also can be used to convert direct current to alternating current. Transistors have played a key role in the development of such devices as pocket calculators, home computers, and video games.

16.3.1.3 Integrated Circuits and Circuit Boards

Two manufacturing processes are used to make electronic circuits: conventional and integrated. Most

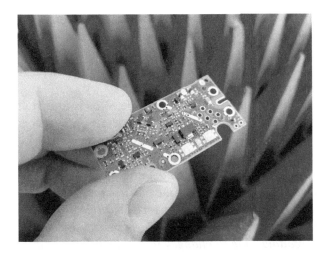

FIGURE 16.14 Printed circuit board.

laboratory instruments combine both circuit types. A conventional circuit contains various electronic components connected to one another by wires and attached to a base. The base is normally a **printed circuit board** *made of a thin piece of plastic or other insulating material.* Copper wires are "printed" onto the board using a special chemical process. Components can be connected to one another on the board in series, parallel, or in a combination of both (Figure 16.14).

An **integrated circuit (IC)** *is a very small electronic circuit that is assembled on a piece of semiconductor material called a* **chip.** The chip acts both as a base and as a part of the circuit. Integrated circuits contain components such as diodes, transistors, and capacitors linked together by tiny conducting wires. A single chip can contain millions of microscopic parts. The chips are packaged in protective carriers with electrical pin connectors that allow them to be plugged into a circuit board. Integrated circuits are found in almost all modern electronic laboratory instruments. ICs are often combined with one another and with other components onto a printed circuit board.

Electronic circuits are combinations of components connected to one another in various configurations to form **functional units.** Functional units perform the various tasks of electronic instruments. The next section discusses functional units as they relate to instruments that make measurements.

16.3.2 Functional Units

16.3.2.1 Transformers

Transformers *are used to vary the input voltage entering an instrument.* For example, a transformer can convert the 110 V AC voltage from the power company to 150 volts for an instrument. It may seem surprising that more voltage can be "created" than originally existed; however, the power supplied to the instrument remains constant. Power equals voltage multiplied by current; therefore, if the voltage supplied to an instrument is increased ("stepped up") by a transformer, then the current to that instrument is reduced. Similarly, if the voltage to the instrument is reduced ("stepped down"), then the current increases.

16.3.2.2 Power Supplies

Power supplies were previously mentioned as they relate to electrophoresis. In electrophoresis the power supply is conspicuous because the operator attaches it to the gel box and selects how much current or voltage is to run through the gel. Most laboratory instruments contain less conspicuous internal power supplies.

The **power supply** plays a variety of roles including:

- *Converting alternating current to direct current.* The power company sends out alternating current, but most instruments require direct current. A power supply is used to convert alternating current to direct current.
- *Acting as a transformer.*
- *Regulating the voltage in the event of fluctuations in voltage coming from the power company.* It is not unusual for the voltage to surge or change unexpectedly. Instruments are protected from these variations by the power supply.
- *Distributing voltage to multiple circuits within an instrument.* Instruments are composed of multiple circuits, each of which requires a source of voltage.

16.3.2.3 Detectors

Detectors are a key part of any laboratory measuring instrument. Different types of laboratory instruments have different types of detectors. Earlier in this chapter, we introduced a related term, "transducer." The term *transducer* tends to be used broadly and so would include simple devices, such as the mercury in a thermometer. The term **detector** *is used to refer to an electronic transducer that generates an electrical signal in response to a physical or chemical property of a sample.* The electrical output (signal) from a detector can be a change in voltage, current, or resistance. For example, pH meters (Chapter 20) respond to changes in the pH of a sample with a change in voltage. Thermistors (Chapter 19) respond to temperature changes with a change in resistance in a wire.

TABLE 16.1

Examples of Detectors

Sample Property that Is Measured	Transducer	Electrical Output
Concentration of H⁺ in a solution	pH electrode	Voltage
Temperature	Thermistor	Resistance
Temperature	Thermocouple	Voltage
Light intensity	Photomultiplier tube	Current

Table 16.1 lists the types of detectors that are described in later chapters in this Unit.

Because the detector generates the signal in a measuring instrument, its capabilities limit the instrument's overall performance. When evaluating detectors, it is common to talk about their **detection limit, sensitivity,** and **range.** (These terms are used not only to refer to detectors, but also to describe methods, as is discussed in Chapter 26. Note the context when these terms are used in order to understand the author's intention.)

The **detection limit** *of a detector is the minimum level of the material or property of interest that causes a detectable signal.* The detection limit is a useful figure when comparing different techniques for measuring an analyte. In practice, the term *detection limit* is sometimes used synonymously with the term **sensitivity,** although these two terms are different. The **sensitivity** *of a detector is its response per amount of sample.* The better the sensitivity, the lower the level of sample the instrument can reliably detect.

Figure 16.15 is part of an advertisement that contrasts a high-sensitivity and a lower-sensitivity detector. Only the high-sensitivity detector can generate a detectable signal when trace concentrations of certain materials are present.

Although the sensitivity of a detector is important in determining its lower limit of detection, a second factor is also involved. This factor is **electrical noise.** Electrical noise may have two components. **Short-term electrical noise** *may be defined as random, rapid "spikes" arising in the electronics of an instrument* (Figure 16.16a). **Long-term electrical noise,** or **drift,** *is a relatively long-term increase or decrease in readings due to changes in the instrument and the electronics* (Figure 16.16b). When there is no sample in the instrument, the detector should ideally provide a steady zero response, resulting in a flat baseline on a recorder. In practice, because of short-term noise and drift, the baseline may not be absolutely stable (Figure 16.16c). Some instruments must be turned on for a period of time before use; that is, they must be "warmed up." Allowing this warming up period may help reduce drift.

Electrical noise can arise spontaneously in an instrument's electrical circuitry. It can also be caused by other electrical devices, such as nearby power lines. In either event, noise is not related to the sample and can occur even if no sample is present. Electrical noise is a problem because it can interfere with our ability to see a small signal due to the sample (Figure 16.17).

There are various methods used to express how much electrical noise is present in an instrument. **Signal-to-noise ratio** *expresses the relationship*

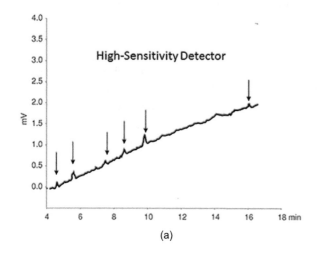

(a)

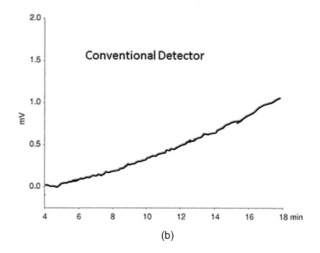

(b)

FIGURE 16.15   Limit of detection is an important factor in evaluating a detector. (a) A more sensitive detector generates signal, which appears here as "blips," in response to low levels of the material of interest. Each blip, indicated with an arrow, indicates a substance that is present in a sample mixture. (b) A less sensitive detector does not noticeably respond to low levels of the substances that comprise the sample.

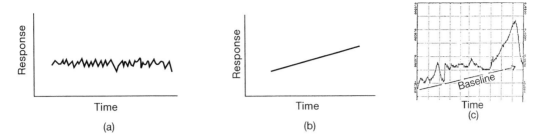

FIGURE 16.16   Electrical noise. (a) Electrical noise is short-term spikes that occur even when no sample is present. (b) Drift causes changes in the baseline over time, such as hours or days. (c) Both short-term noise and drift may be present in an instrument.

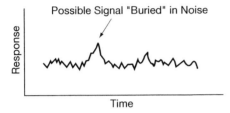

FIGURE 16.17   Electrical noise can interfere with our ability to see the signal arising from the detector's response to the sample.

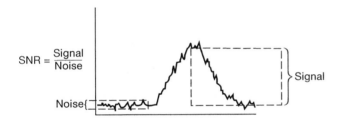

FIGURE 16.18   Signal-to-noise ratio. The relationship between signal and noise is illustrated by dividing the peak due to the signal by the average height of the spikes due to noise.

between signal and noise as the instrument response due to the sample, divided by the noise present in the system (Figure 16.18). The higher the signal-to-noise ratio, the better the performance of the instrument.

The term **root-mean-square noise (RMS)** is sometimes used to indicate how much noise is present in an instrument. The RMS value is based on a statistical calculation that "averages" the noise present over a period of time. The lower the value, the less noise is present in the system.

Thus, the limit of detection of a detector is affected both by its inherent sensitivity and by electrical noise present in the instrument. The **detection limit of a detector** *is therefore often defined in practice as the minimum level of sample that generates a signal at least twice the average noise level.*

The **dynamic range** *of a detector is the range of sample concentrations that can be accurately measured by the detector.* Some detectors have a narrow dynamic range, whereas others have a wide one. The detection limit and dynamic range of an instrument's detector determine what types of samples can be analyzed by that instrument, and how those samples must be prepared.

16.3.2.4 Signal Processing Units

a. *Overview*

The electrical signal that is generated by a detector must be processed before it can be displayed by a readout device. The processing that is required depends on the detector used and the final form of information required. For example, signal processing units may amplify the signal, count it, or convert it from one form to another (Table 16.2). A few of these processes will be discussed in more detail in this section.

b. *Amplification*

An **amplifier** *boosts the voltage or current from a detector in proportion to the size of the original signal.* Amplification is one of the most important types of signal processing because the current or voltage change produced in a detector is often very small. For example, suppose the output of a detector is only 3 μA. It is possible to amplify this current by a thousand times to 3 mA. In order to amplify a current, there must be a power supply, so an amplifier is associated with a power supply and with other electronic components.

Gain *is the degree to which a signal is increased or decreased.* The amplifier gain is the ratio of the output voltage from the amplifier to the input voltage arriving at the amplifier. For example:

TABLE 16.2

Examples of Electrical Signal Processing

Type of Transformation	Purpose
Amplification	Boosts voltage or current in proportion to the size of the original signal
Analog-to-digital conversion	Converts analog signal to digital signal
Filtering	Separates and removes unwanted noise from the signal generated by a detector
Attenuation	Reduces an amplified signal to make it best fit a readout device
Log to linear conversion	Converts a signal which has a logarithmic relationship to the property being measured to a signal with a linear relationship[a]
Integrator	Calculates the area under a peak (as in chromatography)
Counting	Keeps track of and counts signals from the detector

[a] Important in spectrophotometry (discussed in later chapters) where transmittance values are converted to absorbance values.

An input voltage of 1 mV is amplified to an output voltage of 100 mV.

$$\text{Gain} = \frac{100\,\text{mV}}{1\,\text{mV}} = 100$$

In some cases, the operator will need to adjust the gain (amplification) of an instrument. If the signal is amplified too little, then the signal will be difficult to see, and very small signals may be undetectable on the readout device. If the amplification is set too high, however, the signal may exceed the ability of the readout device to handle it.

Be aware that once a detector has generated an electrical signal, the signal-to-noise ratio cannot be changed by simple electronic amplification because the amplification of the signal is accompanied by a corresponding boosting of the noise. There are, however, electronic filtering devices and also software programs that extract the signal from noise. These devices are used to process the signal after it leaves the amplifier.

c. *Analog-to-Digital Converter*

There are two types of signal: **analog** and **digital**. A signal that can change values continuously is called **analog** (*i.e., "smoothly changing"*). A meter is often used to display an analog signal (Figure 16.19a). A meter has a needle that deflects from its zero position when a current is applied. The degree of deflection is related to the electrical signal reaching the meter. Note that the needle can point to any number and can also point

to a value anywhere in between two numbers. The speedometer of a car is a familiar example of such a meter. The signal generated by a laboratory instrument detector in response to a property of a sample is usually analog. For example, a pH probe develops voltage in response to the H^+ concentration of a solution. This voltage might be 50 mV or 51 mV or any value in between.

In contrast to an analog signal, a **digital signal** *represents information by a variable that can have only a limited number of discrete values.* The display from a digital clock radio is shown in Figure 16.19b. The clock radio can only display certain values. In this illustration the time reads 10:20. The next value it will display is 10:21. This digital clock radio cannot display any time in between 10:20 and 10:21.

Computers are digital instruments that represent all information using only two values: on and off (0 or 1). Each letter in the alphabet and every number processed by a computer is

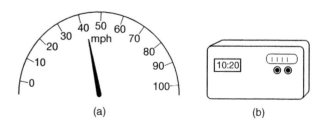

(a)   (b)

FIGURE 16.19 Displays of analog and digital signals. (a) A speedometer is an analog display. (b) A digital clock radio has a digital display.

converted into a distinct pattern of on and off signals.

A digital system is a collection of components that can store, process, and display information in a digital form. Digital devices, such as computers, cannot directly manipulate an analog signal, such as the signal produced by a detector. When the signal from a detector is to be displayed, stored, or manipulated by a digital device, the analog signal must be converted to a digital signal. An **analog-to-digital converter (A/D converter)** *is a type of signal processor that very rapidly converts an analog signal into a digital one*. There are also situations where a **digital-to-analog converter** converts a digitized signal back to an analog signal.

d. *Attenuator*

An amplified signal must sometimes be reduced in order to be best displayed by a readout device. An operator adjusts an **attenuator** *to reduce a signal*.

16.3.2.5 Readout Devices

Various devices, such as meters and computer screens, are used to display the information from the signal processing unit in a form that is interpretable to a human or to a computer. A display device may be either analog or digital (Figure 16.20). Table 16.3 lists various types of readout devices, a few of which are discussed in this section.

Digital devices commonly display individual measurement values with light-emitting diodes (LEDs) or with a **liquid crystal display (LCD)**. LEDs produce their own light and can therefore be seen in the dark. A digital clock radio is an example of a device with an LED display. LCDs play the same role as LEDs, but they are used where lower power consumption is required. An LCD has a thin layer of a liquid crystalline material sandwiched between two plates of glass. The sandwich normally reflects light; however, if a voltage signal is passed to an area of the LCD, that area darkens. The darkened portions form the letters and numbers of the display. LCDs are used, for example, on digital wrist watches.

Strip chart recorders *are analog display devices that record the output of a detector continuously over time*. These devices have become uncommon as digital recorders have taken their place. However, strip chart recorders elegantly illustrate the principle of converting an electrical signal to a readout that is interpretable by a human. A strip chart recorder has a pen that moves up and down over a strip of paper in response to the signal from a detector. The paper slowly moves along a track, so the recorder plots the level of the electrical signal versus time. Figure 16.21a shows an example of a strip chart recording generated

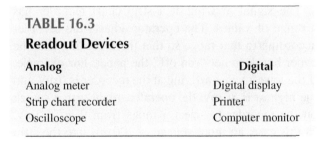

TABLE 16.3

Readout Devices

Analog	Digital
Analog meter	Digital display
Strip chart recorder	Printer
Oscilloscope	Computer monitor

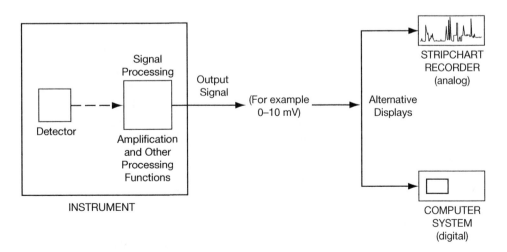

FIGURE 16.20 A chromatography system separates and detects the components of a sample mixture. The output can be displayed with an analog strip chart recorder or a computer (as is more common in modern instruments).

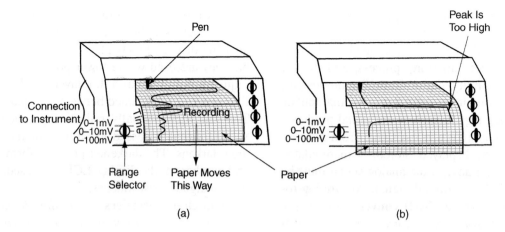

FIGURE 16.21 Strip chart recorder. (a) A strip chart recorder has a pen that responds to the signal from the detector, resulting in a continuous recording of instrument response over time. In this recording, the peaks represent biological compounds. The amount of compound present is related to the area under the peak. (b) The strip chart recorder range should be selected so that the recording fills the paper, but does not flatten out, as is shown by the arrow in this illustration.

as a sample flowed through a detector. The detector responded to biological compounds in the sample; the higher the signal, the higher the level of compound in the detector at that moment. In this type of recording, the area under the peak is a measure of the amount of the material being analyzed.

The signal arriving at a strip chart recorder has a range of values. The operator adjusts the recorder according to that range so that the recording fills the paper but does not "run off" the paper. For example, if the lowest value arriving at the recorder is 0 mV and the highest is 80 mV, the operator might turn a switch on the recorder to select a range from 0 to 100 mV. In this case, an input voltage of 100 mV into the strip chart recorder will move the pen all the way from the left edge of the paper to the right; 100 mV is said to be "full scale." If the recorder is set in this way, and a signal of 125 mV is unexpectedly produced, the pen will go "off the scale" (Figure 16.21b). The same concept applies to digital devices.

Most instruments are now attached to computers or incorporate microprocessors. Computers and microprocessors can take a signal and store it for later retrieval, calculate the concentration of a component in a sample, graph the signal in various ways, make corrections, produce a report, and perform many other functions. Printers are used to display the results after they have been processed. Because the signal has been processed by a microprocessor or computer, the X-axis of a graph need not represent time, as it does with a

strip chart recorder. The X-axis can be, for example, wavelength or concentration. The Y-axis similarly is not the direct signal from the detector, but is the result of various types of processing.

16.4 QUALITY AND SAFETY ISSUES

16.4.1 QUALITY ISSUES

It is important that laboratory instruments are properly maintained and repaired when malfunctions occur. **Preventive maintenance** *is a program of scheduled inspections of laboratory instruments and equipment that leads to minor adjustments or repairs and ensures that the instruments are functioning properly.*

Preventive maintenance programs are intended to:

- ensure correct results;
- identify components that require replacement;
- ensure that instruments are safe to use;
- ensure that instruments will not be shut down by a major problem; and
- lower the costs of repairs.

Recall that **calibration** in common usage *can be defined as the adjustment of an instrument so that its readings are in accordance with the values of internationally accepted standards.* Every measuring instrument must be calibrated, and this calibration may or may not be part of its routine maintenance. The frequency of calibration depends on the type of instrument and its use. The individuals who operate an instrument should ensure that it has been properly calibrated, although the operator may or may not be the one to actually perform the calibration.

Performance verification *is a process of checking that an instrument is performing properly.* One aspect of this process is to check that it is correctly calibrated. Note that we are distinguishing between *calibration* and *performance verification*. We use *calibration* here to mean that adjustments to the instrument are performed so that its readings are correct. *Performance verification* is used to mean that the instrument's performance is checked, but it is not adjusted. Performance verification is usually the responsibility of the individuals who use an instrument.

Instruments can also be **validated**. **Instrument validation** *is a comprehensive set of tests done before an instrument is put in service that demonstrate it will work, and the conditions under which it will function properly.* Instrument validation is a formal process that is required in regulated laboratories and in laboratories that meet certain voluntary standards.

An important part of a quality-control program in a laboratory is that all operations performed with an instrument are documented. It is common to have a logbook (either paper or electronic) associated with each instrument in which performance verification checks are recorded, preventive maintenance and repairs are noted, and routine operations are logged.

Table 16.4 lists some environmental factors that may affect the performance of most types of electronic instruments. These factors should be controlled whenever possible in the laboratory.

16.4.2 ELECTRICAL SAFETY

The fluids in a person's body can serve as a conducting pathway for current flow. Therefore, if a person touches two objects that differ in electrical potential, current can flow through the person – this is electrical shock. Electrical shock can result in severe internal and external burns, paralysis, muscle contractions leading to falls, and, in extreme cases, death. The more current that flows through a person's body, the more dangerous. As little as 1 mA is detectable, 15 mA can cause muscles to freeze, and about 75 mA is fatal. Take precautions, therefore, to avoid having current pass through your body.

Human skin is an excellent insulator when it is dry. The resistance of dry skin is in the megohm range. Wet skin, unfortunately, offers far less resistance to current flow. The resistance of wet skin is about 300 Ω. If a person with wet skin is exposed to a voltage of 120 V, the current passing through that person can be as high as 0.400 A (400 mA). Table 16.5 lists common (and common sense) safety rules. (See also Chapter 8, Section 8.3.7.)

Most people who operate laboratory instruments are not trained in instrument repair and should never attempt to repair a broken instrument, other than to replace accessible fuses, worn-out bulbs, or other simple tasks as directed by the manufacturer. Capacitors can hold a charge for weeks, so there is the risk of shock from an instrument, even when it is unplugged. Only a trained technician should bleed off the current from a capacitor. An untrained individual seeking to repair an electronic instrument can not only injure themselves, but can also damage an instrument. Table 16.6 lists simple things that can be safely checked by a user when an instrument appears to be malfunctioning.

TABLE 16.4

Environmental Factors that Commonly Affect the Performance of Electronic Instruments

1. *Temperature.* Changes in ambient temperature often affect instruments; therefore, avoid placing instruments in areas that are subject to direct sunlight or drafts, such as near a window or an open doorway.
2. *Contaminants.* Most instruments are sensitive to smoke, acid fumes, solvent fumes, dust, and dirt. Liquid spills on the top of instruments can also cause damage. Avoid environmental contaminants.
3. *Ventilation.* Most instruments have exhaust vents to dissipate heat generated by the instrument. Do not block the ventilation holes, and allow an adequate area around instruments to dissipate heat.
4. *Humidity.* The humidity in the room where instruments are used should ideally be about 40%–60%.

TABLE 16.5

Electrical Safety Rules

1. *Use three-prong power cords and receptacles to ensure proper grounding.* Do not attempt to bypass grounding provided on instruments.
2. *Avoid extension cords if possible.* When required, use only heavy-duty, grounded extension cords.
3. *Use properly insulated wires and connections.* Frayed wires or connectors should be repaired by a qualified individual.
4. *Do not handle connections with wet hands or while standing on a wet floor.*
5. *Avoid operating instruments that are placed on metal surfaces, such as metal carts.*
6. *Do not operate wet electrical equipment (unless it is intended to be used in this way) or devices with chemicals spilled on them.*
7. *Be certain that electronic equipment has adequate ventilation to avoid overheating.* Many instruments have fans. Be sure there is enough space around the vents for heat to dissipate.
8. *Never attempt to troubleshoot or repair the electronics of an instrument unless you are trained to do so.*
9. *Never touch the outside of a leaking electrophoresis gel box, or one with a puddle under it, if it is plugged into a power supply.* An electrophoresis gel box normally is closed when the box is connected to a power supply, thus protecting the user from current. If the box is cracked or broken, however, it can leak buffer and therefore also "leak" current.
10. *In case of an electrical fire:*
    a. *If an instrument is smoking or has a burning odor, turn it off and unplug it immediately.*
    b. *Use only a carbon dioxide-type fire extinguisher.* Water and foams conduct electricity and so increase the hazard.
    c. *Some electrical components (specifically selenium rectifiers) emit poisonous fumes when burning.* Avoid the fumes of burning instruments.

TABLE 16.6

Simple Checks of an Instrument that Appears to Be Malfunctioning

1. *Is the instrument plugged in?*
2. *Is the power on?*
3. *If there is a power strip, is it turned on?*
4. *Is the wall outlet functioning? An outlet can be checked by plugging a lamp into it.*
5. *Is the power cord frayed? If so, have it replaced.*
6. *Is there a blown fuse or tripped circuit breaker?*
7. *Is there an accessible bulb that needs replacement?*
8. *Does the instrument have a "reset" button? If so, try pressing it.*
9. *If there is a switch to select either 110 or 220V, is it properly set?*
10. *It sometimes helps to unplug an instrument and plug it back in or turn it off and on.*
11. *Check the instrument manual. There may be a series of troubleshooting steps that the user can easily and safely perform.*
12. *Call the instrument manufacturer and ask their advice.*

Practice Problems

1. In each of the following devices what is the voltage source? What is the resistance or load?

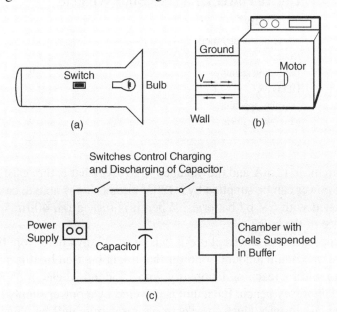

2. Most household appliances in the United States are designed so that they remain in a "standby" mode in which they continue to draw power, even when they are supposedly turned off. Manufacturers make appliances this way either because it is less expensive or because the device has a feature, such as a memory of previous settings, that requires constant power. About a quarter of all residential energy consumption is used on devices in idle power mode, according to a study of Northern California by the Natural Resources Defense Council. (Natural Resources Defense Council. "Home Idle Load: Devices Wasting Huge Amounts of Electricity When Not in Active Use." *NRDC Issue Paper.* May 2015. www. nrdc.org/sites/default/files/home-idle-load-IP.pdf.) That means that devices that are "off," or in standby, or in sleep mode can use up to the equivalent of 50 large power plants' worth of electricity and cost more than $19 billion in electricity bills every year. And there's an environmental cost: Overall electricity production represents about 37% of all carbon dioxide emissions in the United States, one of the main contributors to climate change. (Schlossberg, Tatiana. "Just How Much Power Do Your Electronics Use When They Are 'Off'?" *New York Times on-line.* www.nytimes.com/2016/05/08/science/just-how-much-power-do-your-electronics-use-when-they-are-off.html.)

    Suppose a home has the devices shown in the table below that are constantly drawing power, 24 hours per day, 7 days a week. (For simplicity, ignore the time that the devices are actually in use and draw more power than is listed.) Assume that power to this house costs $0.09 per kilowatt-hour. (A kilowatt-hour is equivalent to 1,000 W of power used for 1 hour.) How much would the owners of this home pay per year for standby current? How much would a city with 50,000 similar homes pay for standby current? (Information in this table is from various sources; individual brands may vary, and companies may improve energy efficiency of their products.)

TABLE 16.7

Device Power Drain Per Hour While in "Standby"

Alexa in standby mode	3.6 W
Router	4.2 W
Laptop on standby	1.6 W
HD DVR	31.0 W
Washing machine	4.0 W
Microwave oven	3.1 W

3. If the current in a circuit is 12 mA and the voltage is 120 mV, what is the resistance?

4. How many watts of power can be supplied by a 120 V circuit that is able to carry 20 A?

5. A flashlight is supplied with 6 V by batteries. When it is turned on, 400 mA flows through the bulb. What is the resistance of the bulb?

6. Suppose a power supply used in electrophoresis can be set at anywhere from 0 to 500 V and from 0 to 400 mA. What is the maximum amount of power that might be used by this power supply?

7. A professor wants to outfit a teaching laboratory with 15 identical electrophoresis units arranged one after the other on a laboratory bench. Each unit is powered by a power supply like the one described in #6. Suppose that the circuit into which the electrophoresis units will be plugged is served by a 120 V line with a current of 20 A. If students use all 15 devices at the same time, is it possible that the load on the circuit will exceed its power rating?

8. An instrument is rated at 500 W. Assuming that the voltage is 120 V, what is the current in this instrument? What is the resistance of this instrument?

Question for Discussion

If you work in a laboratory, list the types of electronic transducers found in the laboratory. List the types of readout devices in the laboratory.

17 The Measurement of Weight

17.1 BASIC PRINCIPLES OF WEIGHT MEASUREMENT

Weighing materials is an activity that occurs at the beginning of nearly any biotechnology experiment, assay (test of a sample), or production process. Chemicals are weighed out and combined to make necessary reagents; samples are weighed before they are tested. Any errors made during weighing will be propagated throughout the experiment or process and can adversely affect the result. In a research or testing laboratory, weighing errors can lead to irreproducible results. In a production setting, weighing errors can lead to a defective product. We therefore consider how to make accurate and precise weight measurements.

Weight *is the force of gravity on an object.* **Balances** *are instruments that measure this force.* The term *weight* is commonly used interchangeably with *mass*, as we do in this text, although these words are not synonyms. **Mass** *is the amount of matter in an object expressed in units of grams.* An object's mass does not change when it is moved to a new location,

but its weight will change if the force of gravity differs at the new site. For example, an astronaut in space is "weightless" because there is no gravity, but the astronaut's mass does not change with blast-off from the earth (Figure 17.1). The significance of the distinction between mass and weight in terms of measurements is explored in Section 17.5 of this chapter.

The process of weighing an object involves comparing the pull of gravity on the object with the pull of gravity on standard(s) of established mass. The weighing method illustrated in Figure 17.2a dates back to antiquity (Figure 17.2b). The object to be weighed and a standard are each placed on a pan hanging from opposite ends of a **lever** or **beam**. When the beam is exactly balanced, gravity is pulling equally on the sample and the standard. They are the same weight. Thus, instruments that weigh materials in the laboratory have come to be called *balances*. Note that the word *balance* is sometimes reserved for sensitive laboratory instruments, whereas the term *scale* often refers to less sensitive weighing instruments including those used in business, transportation, and the home. The two terms, however, are often used interchangeably.

DOI: 10.1201/9780429282799-22

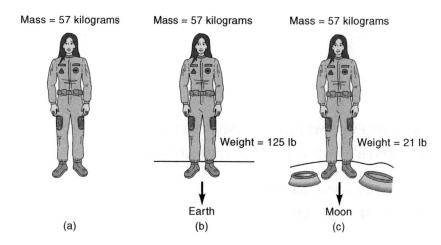

Mass = 57 kilograms   Mass = 57 kilograms   Mass = 57 kilograms

Weight = 125 lb   Weight = 21 lb

Earth   Moon

(a)   (b)   (c)

FIGURE 17.1 Mass versus weight. (a) An astronaut is composed of matter. Her mass is a measure of that matter. (b) The weight of the astronaut is a measure of the effect of gravity on her. (c) On the moon, the astronaut weighs less than on the earth because of the moon's weaker gravity, but her mass is unchanged.

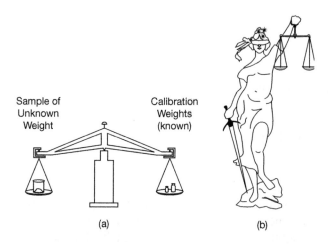

Sample of Unknown Weight   Calibration Weights (known)

(a)   (b)

FIGURE 17.2 Balances. (a) A simple "double-pan" balance with one beam and two pans where a sample on the left pan is balanced by objects of established weight on the right pan. (b) A double-pan balance has been used to symbolize justice since ancient times. In Greek mythology, Themis is the goddess of justice. She is usually represented wearing a crown of stars and holding a balance in her hand. (Double-pan balance sketch by Sandra Bayna, a biotechnology student.)

17.2 CHARACTERISTICS AND TYPES OF LABORATORY BALANCES

17.2.1 RANGE, CAPACITY, AND SENSITIVITY

There are many types and brands of balances with various designs and features. The fundamental characteristics of a balance, however, are its *range, capacity, and sensitivity (often expressed as readability):*

1. **Range and capacity**. Some laboratory balances are intended to weigh heavier objects, while others, lighter ones. The **range** *of a balance is the span from the lightest to the heaviest weight*

the balance is able to measure. **Capacity** *is the heaviest sample that the balance can weigh.*

2. **Sensitivity and readability. Balance sensitivity** *may be explained as the smallest value of weight that will cause a change in the response of the balance.* Sensitivity determines the number of places to the right of the decimal point that the balance can read accurately and reproducibly. Extremely sensitive laboratory balances can weigh samples accurately to the nearest 0.1 μg (or 0.0000001 g). A less sensitive laboratory balance might read weights to the nearest 0.1 g. In catalog specifications, manufacturers express the sensitivity of their balances by their **readability**. **Readability** *is the value of the smallest unit of weight that can be read; it is the smallest division of the scale dial or the digital readout.*

Range, capacity, and sensitivity are interrelated. A very sensitive balance will not be able to weigh samples in the kilogram range, but a balance intended for heavier samples will not detect a weight change of a microgram.

Analytical balances *are designed to optimize sensitivity and can weigh samples to at least the nearest tenth of a milligram (0.0001 g). The most sensitive analytical balances,* **microbalances** *and* **ultramicrobalances,** *can weigh samples to the nearest microgram (0.000001 g) and the nearest tenth of a microgram (0.0000001 g), respectively.*

Table 17.1 shows common names for laboratory balances with different readabilities. Figure 17.3 shows catalog descriptions of various balances. The descriptions show the readability, range, and capacity of each balance. The descriptions also show the repeatability

TABLE 17.1
Common Naming of Laboratory Balances Based on Their Readability

Ordinary Name	Number of Digits Past the Decimal Point That Are Displayed (g)
Technical balance	0–1 (e.g., 0.0)
Precision balance	1–3 (e.g., 0.00)
Analytical balance	4 (0.0000)
Semi-microbalance	5 (0.00000)
Microbalance	6 (0.000000)
Ultramicrobalance	7 (0.0000000)

(a)

(b)

FIGURE 17.4 Two types of balance. (a) A top-loading balance with a capacity of 4.2 kg and a readability of 0.1 g, (b) An analytical balance with a capacity of 320 g and a readability of 0.1 mg. A draft shield is present, and the sample is loaded from the side. (Photos courtesy of METTLER TOLEDO.)

of the instruments and their linearity (which will be discussed later in this chapter).

Most biotechnology laboratories have more than one type of balance. There is usually at least one balance with a higher capacity, for example, into the kilogram range. These higher-capacity balances tend to be "top-loaders," meaning the sample is placed directly on a weighing pan on top of the balance (Figure 17.4a). Laboratories usually have at least one analytical balance that is of lower capacity and can weigh samples in the milligram range with accuracy and precision. Analytical balances have draft shields to protect the samples from drafts, so samples are loaded from the side (Figure 17.4b).

The laboratory might also include other balances. For example, a double-pan balance (Figure 17.5a) might be used to ensure that test tubes are of the same weight before placing them in a centrifuge. The balance that should be selected for a particular application depends on the weight of the sample, the sensitivity required, and other characteristics of the sample. For example, it is not possible to weigh a person on a balance that has a readability of 0.0001 g, nor should one use an analytical balance for a laboratory sample that weighs in the kilogram range. An analytical balance should be

chosen, for example, to weigh milligram amounts of enzyme to use in a reaction mixture.

17.2.2 Mechanical versus Electronic Balances

17.2.2.1 Mechanical Balances

Laboratory balances can be broadly classified as either *mechanical* or *electronic*. **Laboratory mechanical balances** *typically have one or more beams; the*

MODEL NUMBER	CAPACITY (g)	READABILITY	REPEATABILITY (Standard Deviation)	LINEARITY
BA50	50	0.1 mg	± 0.1 mg	± 0.2 mg
BA300	300	0.1 mg	± 0.2 mg	± 0.5 mg
BA4000	4000	0.1 g	± 0.1 g	± 0.1 g

FIGURE 17.3 Catalog specifications for balances. Catalogs typically specify the capacity, readability, repeatability, and linearity for balances.

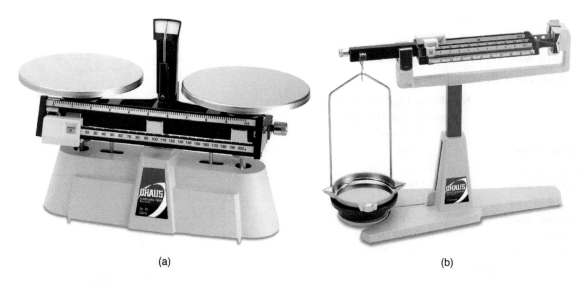

(a)   (b)

FIGURE 17.5 Mechanical balances. (a) In centrifugation, the tubes and their contents that are opposite one another in the centrifuge must be of equal weight. A single-beam, double-pan balance similar to this one can be used to easily compare the weights of centrifuge tubes. (b) Triple-beam hanging-pan balance. This balance has three beams. External standard weights are not needed because the weights (sometimes called "riders") slide to the right on the beams to balance the sample. These types of balances typically have readabilities of 0.1 g. (Photos courtesy of OHAUS®.)

object to be weighed is placed on a pan attached to a beam and is balanced against standards of known weight (Figures 17.2 and 17.5). Mechanical instruments do not generate an electronic signal. Balances or scales that consist of one beam with two pans are called **single-beam, double-pan devices**. Scales that employ the principle of balancing objects across a beam are still used in certain applications because they are relatively inexpensive and generally do not require electricity.

In the past, it was common to use mechanical analytical balances that could weigh small samples, down to 0.1 mg and even lower. These balances have a beam and a number of standard weights, each of which must be of the proper mass and in the proper orientation to ensure accurate weighing results. Although mechanical analytical balances can produce high-quality measurements, they are complicated to operate, are fragile, and must be maintained and calibrated by trained technicians. Mechanical analytical balances have therefore been replaced by electronic analytical balances, which are far easier to operate and maintain.

17.2.2.2 Electronic Balances

Electronic balances use an electromagnetic force rather than weights to counterbalance the sample; the balances shown in Figure 17.4 are electronic. An electronic balance produces an electrical signal when a sample is placed on the weighing pan, the magnitude of which is related to the sample's weight. The way an electronic balance works is as follows:

1. The weighing pan is depressed by a small amount when an object is placed on it.
2. The balance has a detector that senses the depression of the pan.
3. An electromagnetic force is generated to restore the pan to its original ("null") position. The amount of electromagnetic force required to counterbalance the object and move the pan back to its original position is proportional to the weight of the object.
4. This electromagnetic force is measured as an electrical signal that is converted to a digital display of weight value. Calibration, discussed in the next section, enables the balance to convert an electrical signal to a weight value.

Electronic balances are easy to use, automatically perform many functions, can be interfaced with computers and other instruments, permit automatic documentation, and are better able to compensate for environmental factors than mechanical balances. For these reasons, electronic balances are standard in most biotechnology settings.

17.3 CALIBRATION OF ELECTRONIC BALANCES

17.3.1 PROCEDURE

A balance must be properly calibrated in order to give accurate readings. We are using the term "calibration" to mean *adjustment* of the balance so that calibration

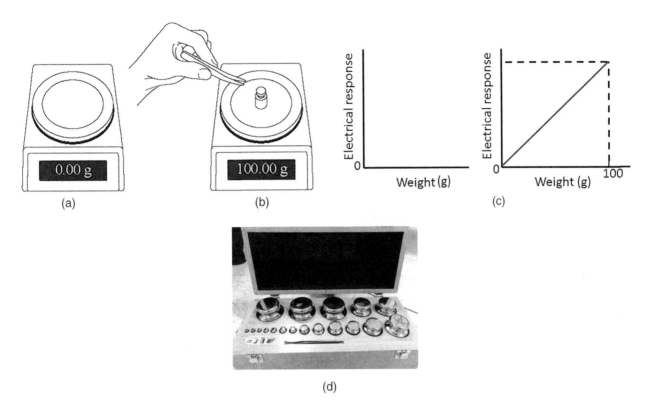

FIGURE 17.6 In the calibration mode of a balance, (a) the balance is set to zero when the weighing pan is clean and empty. There is no material on the balance, and so the electrical response is zero. (b) A calibration standard is placed on the pan; in this example, the standard is 100 g. The standard used for calibration is usually a weight that is close to the balance's capacity. The balance is adjusted to the calibration weight by pressing the appropriate button or key. (c) The relationship between weight and electrical response of a balance is linear. Once this relationship has been established using the standard, the balance can convert the electrical response generated by a sample to its weight. For example, if a 100 g standard results in an electrical response of 80, and a sample provides an electrical response of 40, then the balance "knows" that the sample weighs 50 g. (d) A box of calibration standards. (Credit:Talay and Pupha/Shutterstock.com)

brings its readings into accordance with the values of internationally accepted standards. (Recall that a formal definition of calibration does not involve adjustment of the instrument.)

Although balances are calibrated by the manufacturer when they are produced, calibration must be periodically checked in the laboratory of the user. This is because time and use can affect balance response. Calibration must also be checked when a balance is moved from one location to another. For measurements in the microgram range, the balance ideally is recalibrated each time the weather changes. Recalibration of the balance compensates for the effects of location and weather.

It is usually straightforward to calibrate an electronic balance using a mass standard and most manufacturers include directions for doing so in the instruction manual. Some electronic balances contain internal mass standards; others require that a standard be placed on the weighing pan during calibration. A basic calibration procedure for electronic balances

is illustrated in Figure 17.6 and outlined in Box 17.1. Observe that with electronic balances, the comparison between the sample and standards of known mass is made sequentially. The balance is calibrated to a standard at one time, and the samples are weighed on the balance afterward. This is in contrast to mechanical balances, such as those shown in Figures 17.2 and 17.5, where the samples and standards are placed on the balance at the same time.

17.3.2 STANDARDS

The accuracy of any weight determination is limited by the accuracy of the standards used for comparison. Weight (mass) standards[1] are metal objects whose

---

[1] A note about terminology: Standards used for balance calibration are often called "weight standards" because they are used in the process of weighing samples. Recall, however, that the weight of any object varies with its location, but its mass is constant. It is the *mass* of the standards that is established by traceability to NIST standards.

BOX 17.1   GENERAL TWO-POINT CALIBRATION OF AN ELECTRONIC BALANCE

1. *When the weighing pan is clean and empty, the balance is set to zero using the appropriate button, keypad, or other command method.* Zero is the first point in the calibration.
2. *The second calibration point is usually at the heavier end of the balance's range.* It is common to enter into a calibration mode that "tells" the balance to calibrate. A standard of established mass is placed on the weighing pan, and the balance is set to the value of the standard using an appropriate command method. Alternatively, some electronic balances have a button that when pressed causes a calibration standard inside the balance to be weighed and the balance to adjust itself. In fact, there are microprocessor-controlled balances that recalibrate automatically to compensate for environmental changes without any operator intervention.
3. *There is a linear relationship between the electronic response of a balance and the weight of an object.* Therefore, after calibration the balance can convert an electrical signal generated in response to a sample into a weight reading.

---

masses are known (within the limits of uncertainty) relative to international standards. Mass standards can be purchased with a certificate showing their traceability to NIST. This certificate documents that the standard's mass went through a series of comparisons that ultimately links it to a national primary mass standard. (The concept of standard traceability was introduced in Chapter 15.) The checking of standards against one another requires stringently applied methods and specific environmental conditions.

Standards for use in weighing can be purchased in different conformations of varying construction and with differing tolerances, where tolerance is the amount of error permitted in the manufacture of the standard. The user determines the type of standards required depending on the balance being used and the requirements of the application. In the United States, the specifications for labeling and classification of standards for laboratory balances have been established by ASTM International (formerly the American Society for Testing and Materials). (These specifications are reported in ASTM Standard E617–18, "The Standard Specification for Laboratory Weights and Precision Mass Standards.")

The finest standards, those with the smallest tolerance, are **Class 1**. A "500 g" Class 1 standard, for example, must have a mass within 1.2 mg of 500.0000 g (i.e., it must be between 500.0012 and 499.9988 g), whereas a **Class 4** standard must be within 10 mg of 500.0000 g (i.e., it must be between 500.0100 and 499.9900 g). In routine biology laboratory work, it is common to use **Class 2, 3, or 4** standard weights. Class 4 is recommended for student use. The calibration standards that manufacturers provide with electronic analytical balances are usually Class 2.

Table 17.2 has information about proper care of standards. Additional information regarding the classification and features of standards is excerpted in the Appendix to this chapter.

TABLE 17.2

Proper Handling of Weight Standards

1. *Store standard weights in their original packing.* Standards are usually supplied in padded cases that protect them from dust and scratches.
2. *Store standard weights near the balance so that they experience the same environmental conditions.* Alternatively, move standards close to the balance at least an hour before calibration.
3. *Handle standard weights carefully.* Standards can be picked up when wearing clean gloves, but it is better to lift them with special tweezers that are equipped with coated tips to avoid scratches. The oil from bare skin can permanently damage standards, so never touch them with bare hands.
4. *Ideally standard weights should be carefully used and stored and not require cleaning.* If necessary, it is possible to remove dust with brushes designed for this purpose or soft microfiber cloths. Do not clean standards with caustic or abrasive cleaners.
5. *Inspect standard weights before use to make sure they are not scratched or dusty.*
6. *Standards require periodic recalibration.* This is usually performed by specialized laboratories.
7. *To protect their standards, a laboratory might keep one set for routine use and a second set in storage.* Every 6 months or so, the standards for routine use can be checked against the set that is in storage.

17.4 FACTORS THAT AFFECT THE QUALITY OF WEIGHT MEASUREMENTS

17.4.1 OVERVIEW

Recall that accuracy relates to the correctness of measurements, and precision to the reproducibility of a series of measurements. You might assume that the accuracy and precision achievable with a balance are dependent on the quality designed and built into that balance. It is true that a balance must be properly manufactured to give accurate, reproducible results. In normal laboratory practice, however, the accuracy and reproducibility of weight measurements are greatly affected by the conditions in that laboratory, by the user's technique, and by the routine maintenance of the balance.

17.4.2 LEVELING THE BALANCE

Balances must be level when measurements are made; this is particularly critical with analytical balances. There is often a **leveling bubble** *to show whether the balance is level* and "feet" that can be adjusted up and down, if necessary, to even the balance (Figure 17.7).

17.4.3 ENVIRONMENTAL FACTORS

17.4.3.1 Drafts and Vibrations

Environmental factors can have a significant effect on the accuracy and precision of weight measurements. These factors are most relevant when working with analytical and microbalances, but they may be important in other situations as well.

Analytical balances are sensitive to movement caused by drafts and vibration. Analytical balances are therefore equipped with enclosures, called draft shields, that must be closed when a sample measurement is taken. In buildings where vibration is a problem, special counters that minimize the effects of vibration can be used to support the balances.

17.4.3.2 Temperature

Temperature can have an observable effect on weight measurements. If a sample and its surroundings are at different temperatures, then air currents are generated that cause the sample to appear heavier or lighter than it really is (Figure 17.8). You can test this effect of temperature on weight. Weigh a small flask or beaker on an analytical balance, then hold it in your hand to warm it for 1 minute and weigh it again. The flask should weigh less the second time. Perspiration and oils from your hands are of minor importance; otherwise, the flask would weigh more the second time. This temperature effect may appear as a weight reading that slowly increases or decreases. Microbalances and ultramicrobalances have thick enclosures to protect them from thermal disturbances outside the balance.

Balances are also affected by the heat from their own electronic components. When a balance is first turned on it goes through a "warming up" period when its temperature is increasing and its readings are unstable. Many balances are therefore left on in a permanent standby mode to avoid delay while the balance warms up.

17.4.3.3 Static Charge

Static charge can be a vexing problem when weighing small amounts of samples. The effect of static can be significantly more dramatic than the effects of temperature described above. Static charge is generated when different materials touch one another. One material acquires an excess of electrons, resulting in a negative

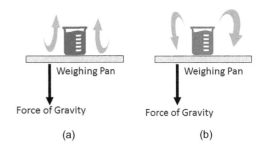

FIGURE 17.8 The effect of temperature on apparent weight. (a) When the sample is warmer than the environment, air currents flow upward generating a force that causes the sample to appear slightly lighter than it really is. (b) If a sample is colder than the environment, air currents cause it to appear slightly heavier than it is. This temperature effect is subtle, but can matter when high-accuracy results are required and small quantities of sample are being weighed.

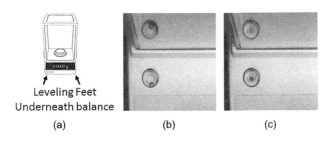

FIGURE 17.7 Leveling a balance. The balance is level when the bubble is centered. (a) "Feet" (underneath the balance) are moved up and down until the bubble is centered. (b) The bubble is not centered. (c) The bubble is centered.

charge, while the other loses electrons and acquires a positive charge. The magnitude of this effect depends on the affinity of each material for electrons, the environmental conditions, and other details. If the different materials rub back and forth against each other, friction results, which amplifies the effect. For example, drying a beaker by rubbing it with a cloth will create static charge on the beaker. The glass draft shields of analytical balances, plastic and glass weighing containers, and samples that are powders or granules are all prone to develop and retain static electrical charge.

If a sample or weighing vessel and its surroundings have the same charge, they repel each other; if they have opposite charges, they attract one another. This attraction or repulsion creates a field between the sample and the nonmoving parts of the balance. The force of this field can simulate a change in weight that may extend into the gram range (Figure 17.9a). You can recognize this effect by a weight display that drifts or fluctuates. In extreme cases, a charged sample can literally fly out of its container and adhere to the balance chamber walls (Figure 17.9b). It is sometimes possible to reduce static charge by using metal weigh boats instead of plastic or glass vessels. Sometimes analysts put a small petri dish with room temperature water in the weighing chamber to increase the humidity, which may help reduce static charge. Ionizing blowers are a more expensive but possibly more effective solution. These devices shoot out positive and negative ions that neutralize charge on the sample and the balance. Examples of commercial products to reduce static charge are illustrated in Figure 17.10.

Drafts, temperature, and static charge are only a few of the factors that a careful operator must consider when making weight measurements. A number of points relating to good weighing practices are summarized in Table 17.3.

FIGURE 17.10 Examples of commercial products to reduce the effects of static charge. (a) Ionizing blower "shoots out" ions that neutralize static charge on the item to be weighed. (Photo courtesy of Sartorius Corp.) (b) Antistatic brush removes electrostatic charge when balance cases and pans are brushed. The brush contains a small amount of radioactive polonium that creates an ionized atmosphere that neutralizes ions when surfaces are brushed. (Used with permission from Thermo Fisher Scientific, the copyright owner.)

(a)   (b)

FIGURE 17.9 Static charge causes problems in weighing. (a) In this situation, static charge creates a force field between the sample and the draft shield causing the sample to appear lighter than it really is. (b) In the presence of static charge, powders can become airborne, causing problems in weighing and possible safety hazards. This is not a subtle problem. (Photo courtesy of METTLER TOLEDO.)

TABLE 17.3

Factors That Affect Weight Measurements and Good Weighing Practices

1. *Analytical balances must be level.* Check the leveling bubble before use.
2. *Vibration will affect balance readings.* Use special tables designed to prevent vibration for microanalytical balances and in buildings where vibration is a problem. Locate all balances away from equipment that vibrates, such as refrigerators and pumps.
3. *Drafts will affect weight measurements.* Do not place balances near fans, doors, or windows. If a balance has a draft shield, keep its doors closed when taking a reading. Whenever possible, situate analytical balances in locations with little activity.
4. *Balances should not be jostled.* Avoid leaning on the table holding the balance. Avoid moving a balance unless it is stabilized. (Consult the owner's manual for information on how to move the balance.)
5. *Temperature changes can affect the apparent weight of a sample and can subtly affect the actual weight of a sample.*
   a. Keep the sample, balance, and surroundings at the same temperature, ordinarily, room temperature.
   b. Avoid touching samples and containers and placing your hands in the weighing chamber when using an analytical balance. Wearing gloves when handling samples is helpful (and is common practice), but using tongs provides better protection from temperature effects.
   c. Do not locate balances near windows, radiators, air conditioners, or equipment that produces heat.
   d. Allow a balance time to warm up before use, or leave it in standby mode to avoid warm-up time.
6. *Static charge will affect measurements.* Various sources recommend the following procedures to help reduce static charge:
   a. A variety of products are available commercially to help reduce static (Figure 17.10).
   b. Manufacturers make balances with weighing pans designed to remove static from samples.
   c. Static charge is more severe when the humidity is low. It is ideal to maintain the humidity between 40% and 60% in a weighing room.
   d. Metal does not tend to become electrostatically charged because it is conductive and charge readily dissipates from it. Metal weighing containers may therefore be the best choice if electrostatic charge is a problem.
   e. Cleaning the walls of the balance chamber with alcohol or a product intended for cleaning glass may reduce static charge.
   f. Do not place balances on tables with glass or plastic tops because these surfaces can build up static charge.
   g. Avoid wearing woolen or nylon clothing when weighing samples.
7. *Some samples gain moisture from the air and others lose volatile components to the air.* The weight of such samples will slowly increase or decrease during weighing.
   a. Work quickly but carefully to minimize such changes in the sample's composition.
   b. Maintain the humidity in a weighing room between 40% and 60%, if possible, because the amount of moisture loss or gain by a sample may be affected by the room's humidity.
8. *Samples that are magnetic may perturb the electromagnetic coil in an electronic balance, leading to incorrect results.* A magnetic sample may appear to have a different weight depending on its position on the weighing pan. It is helpful to distance magnetic samples from the weighing pan by placing a support under the sample.
9. *Electronic balances may give incorrect readings if the load is placed off-center.* Place the sample near the center of the pan.
10. *It is possible to damage electronic balances by placing a load that is too heavy on the weighing pan or by dropping items onto the pan.* Also, do not leave samples on the weighing pan for an extended period of time.
11. *Selection of the weighing container may affect the results.*
    a. Samples can be weighed in a wide variety of containers including beakers, glassine-coated **weighing paper**, and *plastic or metal containers designed for weighing*, called **weigh boats**.
    b. If a sample is likely to absorb or lose moisture, or contains volatile components, use a weighing vessel that is capped or has a narrow opening.
    c. Weigh samples in the smallest possible container. Be alert to sample particles that may adhere to the weighing vessel when poured out.
12. *Do not touch chemicals being weighed, magnetic stir bars, or the insides of beakers with your fingers.*

(Continued)

TABLE 17.3 (*Continued*)

Factors That Affect Weight Measurements and Good Weighing Practices

13. ***Do not return unused chemicals to their storage bottles.***
14. ***Balances and weighing rooms should be clean.*** Sloppiness can have detrimental effects. Some chemicals are toxic; chemicals left on balances can cause corrosion and damage; and materials from one person's work can contaminate the work of someone else. Remember that balances are expensive instruments and should be well maintained.
    a. Remove powder and dust first when cleaning a balance. Use a laboratory wipe, and do not blow air onto or into the balance.
    b. Use a damp, lint-free cloth and mild solvent such as 70% ethanol or isopropanol to remove sticky substances. Avoid abrasive materials.
    c. Do not spray or pour liquids directly onto the balance.
    d. Be careful to avoid damaging the cone opening that holds the weighing pan and to avoid allowing liquids to enter interior parts of the balance.
    e. The weighing pan should be removed for cleaning whenever it is easily removed. Also, the draft shield panes may be removable for cleaning. Only remove parts that can be removed without tools and whose removal is described in the operating instructions.
    f. Remember that your colleagues may have weighed hazardous substances, so always use PPE when cleaning a balance – and other laboratory equipment as well.

17.4.4 A Note about Uncertainty in Weight Measurements

As discussed in Section 15.3, all measurements are associated with some uncertainty and are affected by random and unknown systematic errors. For example, we discussed the tolerances of the weight standards used to calibrate balances. There is some uncertainty, even in the finest weight standards used for calibration, so there is uncertainty in every measurement made with a balance. The readability, sensitivity, and other factors associated with the design of the balance also affect measurement uncertainty. Additionally, the environment of the balance with respect to temperature fluctuations, vibrations, and drafts affects its performance. In practice, the uncertainty of weight measurements may be of little significance if one is weighing an object that is large relative to the sensitivity of the balance. But as one attempts to weigh items that are close to the minimum weight that can be read by the balance, the relative uncertainty can become significant. For example, suppose a balance in the laboratory is accurate to plus or minus 0.1 g. If you are weighing a 1000 g sample, this uncertainty is 0.01% of the sample's weight. Much of the time, such uncertainty is small enough that it will not affect the quality of your product or result, and most biologists would not be concerned. However, suppose that you are weighing a 1 g sample on this same balance. Now the uncertainty is 10% of the weight, which means that the actual sample weight may be 10% greater than or 10% less than what the balance reports. For this reason, it is important not to use the wrong laboratory balance. A top-loading balance that is intended to weigh samples into the kilogram range is not the correct balance to weigh out 100 mg of chemical. Rather, use an analytical balance that can weigh into the 0.1 mg range.

This concept applies to other measurement systems, in addition to balances. As measurements approach the limits of sensitivity of the system, relative uncertainty becomes more problematic.

17.4.5 Quality Programs and Balances

17.4.5.1 Overview

Balances must be operated correctly to ensure that they perform properly. They also must be regularly calibrated, cleaned, and repaired. The performance of a balance should be regularly checked to verify and document that it is giving accurate and precise readings. Laboratories that meet the requirements of a formal quality system, such as ISO 17025 or CGMPs, must have documented procedures that detail how to operate each balance and how to maintain, calibrate, and verify its performance. Even if a laboratory does not conform to the requirements of a formal quality system, it is still good practice to perform and document these tasks on a regular basis.

17.4.5.2 Verifying Balance Accuracy, Precision, and Linearity

Performance verification of a balance involves testing its accuracy, precision, and linearity. *Accuracy* is tested simply by weighing one or more mass standards. The resulting weight must be correct within a tolerance established by the laboratory's quality-control procedure. This test is simple enough that it can be performed each time a balance is used or each day. This type of test is sometimes called a test of the balance's *sensitivity*.

Precision is an important aspect of the quality of a balance's performance. Precision (repeatability) is measured by weighing a sample multiple times and calculating the standard deviation. Manufacturers test the repeatability of their balances under ideal conditions in their testing laboratory and report the results in their balance specifications (see Figure 17.3). To determine whether a balance is performing adequately, its repeatability in the laboratory can be compared with the manufacturer's specifications, or the laboratory's own requirements.

Linearity error *occurs when a balance is properly calibrated at zero and full scale (the top of its range), but the values obtained for weights in the middle of the scale are not exactly correct.* Linearity can be tested by weighing individual subsets of objects and comparing the sum of the subsets to the weight of the objects all together, as shown in Box 17.2. If a balance is discovered to have linearity error, it is necessary to have it repaired professionally. Manufacturers report the linearity error of their balances in the balance specifications, and this figure is an indication of the quality of the instrument.

BOX 17.2   CHECKING THE LINEARITY OF A BALANCE

NOTES

a. Linearity tests do not require standard calibration weights. Any four individual objects that weigh about one-fourth the capacity of the balance can be used. For example, if a balance has a capacity of 100 g, then four items with weights of about 25 g each are needed.

b. Each of the following weighings should be repeated, ideally ten times, and the results of the ten weighings averaged.

TEST 1: CHECK THE BALANCE RESPONSE AT ITS MIDPOINT

1. *Select four items whose total weight is about the capacity of the balance and label them A, B, C, and D.*
2. *Weigh all four pieces together.* Record this weighing as the "full-scale value."
3. *Weigh pieces A and B together and record their weights.*
4. *Weigh and record the weight of pieces C and D together.*
5. *Add the two values from step 3 and step 4 together.* This is the "weight sum."
6. If the balance is exactly linear at its midpoint, then the weight sum will equal the full-scale value. If the two are not equal, divide the difference by two because the two measurements were made at this point. The result is the linearity error at the midpoint.

TEST 2: CHECK THE BALANCE RESPONSE AT 25% OF CAPACITY

1. *Weigh all four pieces individually.* Add their weights together and call this the "weight sum."
2. *If the balance is linear at 25% of its capacity, then the weight sum should equal the full-scale value as determined earlier.* If the two differ, divide the difference by four. This is the linearity error at 25% of capacity.

TEST 3: CHECK THE BALANCE RESPONSE AT 75% OF CAPACITY

3. *Divide the pieces into the following groups:*
   Group 1: pieces A, B, C
   Group 2: pieces A, B, D
   Group 3: pieces A, C, D
   Group 4: pieces B, C, D

(Continued)

BOX 17.2 (*Continued*)   CHECKING THE LINEARITY OF A BALANCE

4. ***Weigh all four groups one at a time***. Add their weights together and call the result the "weight sum." Multiply the full-scale value by three and call it "75% full-scale value."
5. ***If the balance is linear at the 75% point, this weight sum will equal the 75% full-scale value***. If the two differ, divide the result by four. This is the linearity error at 75% of capacity.

Source: Based on information in Weil, Jerry. "Assuring Balance Accuracy."
Cahn Instruments Inc., Product note, November 1991.

Linearity testing as shown in Box 17.2 is not usually performed on a frequent basis; once or twice a year is generally sufficient. However, it is possible to perform a simple assessment of linearity by testing standard weights that are at roughly 5%, 50%, and 100% of the balance's capacity. In the absence of linearity error, there will be no difference in the accuracy of the results for the three standards. To perform this abbreviated linearity test, each standard should be weighed at least five times and the results averaged. The results can be compared to the laboratory's quality requirements.

Example Problem 17.1

Suppose a technician is verifying the performance of a laboratory balance. The technician weighs a 10 g standard six times and obtains these results:

10.002 g, 10.002 g, 10.001 g, 9.998 g,
9.999 g, 10.002 g

a. What is the standard deviation for the balance? (Remember the units.)
b. The balance specifications require that its precision (as measured by the standard deviation) be better than or equal to 0.001 g. Does it meet these specifications? If not, what should be done?

Answer

The standard deviation rounds to 0.002 g, which does not meet the specification. There may be a problem with the balance, although it is possible that the operator is not using good weighing practices. For example, the operator might forget to close the doors of the balance chamber or might place the standard in different locations on the weighing pan. The balance should be taken out of use, and the cause of the imprecision should be investigated.

17.4.5.3 Frequency of Testing

The frequency of balance performance verification and calibration varies among laboratories. In some laboratories, the weights of standards are checked each time the balance is used, and the results are evaluated and recorded. If a problem is discovered as a result of this check, an investigation ensues. In other laboratories, this check is performed less frequently. Some laboratories have the policy that analytical balances are calibrated every time they are used; in other laboratories, calibration is less frequent. Balances might similarly be checked at prescribed intervals for linearity, repeatability, and other features of importance.

Maintenance and quality checks should be documented each time they are performed. Newer balances may have a feature that enables them to automatically record the date, time, and results of calibration for documentation purposes. Consistently checking and recording the weights of standards documents that measurements were made on a properly functioning instrument and alerts the operator to malfunctions that do occur. Whatever process is used in a laboratory to ensure the quality of balance measurements, it is good practice to have a formal, written policy and standard operating procedures that detail the process.

Note that in this chapter we have been using the common meaning of the term "calibration"; that is, calibration is an *adjustment* to an instrument that brings its readings into accordance with standards. However, metrologists do not use the term in the same way. Calibration, as metrologists use the term, is a detailed *assessment* that determines the balance's performance under normal operating conditions. This assessment must quantify the uncertainty of the balance's readings. The assessment also involves an evaluation of the instrument's repeatability, linearity, and its **eccentricity**, *which determines if the balance responds differently when a sample is placed on the center of the pan versus when it is placed off-center.* Calibration, in the metrologist's formal sense of the word, states how a balance performs but does not change it. Adjustment is a separate action from calibration; adjustment changes how the balance performs. Calibration, in a formal sense, is generally performed by a trained technician, once or twice a year, depending on the quality program in a given facility. In contrast, users should much more frequently check the performance of their balance and also adjust it with standards, if necessary.

17.4.5.4 Operating Balances

Modern electronic balances are relatively easy to operate, as shown in Boxes 17.3 and 17.4, and so it may seem odd that laboratories have standard operating procedures (SOPs) to detail their operation. SOPs,

however, can play a valuable role in ensuring that operators use balances properly, that the performance of each balance is checked regularly, and that these checks are documented. They also remind operators of simple problems that can lead to systematic errors. For example, the SOP in Figure 17.11 specifies a warm-up period for the balance. Note also in the SOP that the balance is first calibrated and then the weight of a second standard is checked to verify that the balance is reading correctly.

BOX 17.3 GENERAL PROCEDURE FOR WEIGHING A SAMPLE WITH AN ELECTRONIC ANALYTICAL BALANCE

1. *Make sure the balance is level*.
2. *Adjust the balance to zero*. Make certain the weighing pan is clean and empty, and the chamber doors (if present) are closed to avoid air currents and dust.
3. *Tare the weighing container or weigh the empty container*. Most balances can be **tared**; that is, they *can be set to subtract the container weight from the total weight automatically*. If a balance does not have the tare feature, then it is necessary for the operator to weigh the empty container and subtract its weight from the total weight of the sample plus container.
4. *Place the sample in the tared container and read the value for the measurement*.
5. *Remove the sample; clean the balance and area around it*.

Clean Gene, Inc.

Standard Operating Procedure
SUBJECT: Operation of Balance: BRP 88
Balance Location: Media prep lab
Effective Date: *1/22/21*
Page 1 of 2

---

1. Scope: The routine operation and use of Balance model BRP 88
2. Resources: NIST traceable standard mass set (2000 g)
3. References: Manufacturer's bulletin No 2
4. Responsibility: Training level of "General Lab" personnel or greater is required for use of this equipment.
5. Procedure
5.1. <u>Pre-use calibration and verification of balance</u>
    Frequency: Daily before use
    Form: Balance Instrument record (QF 15.3.6.3)

5.1.1 Before calibration and verification, the balance must be switched on for at least 20 minutes to allow the internal components to come to working temperature.
5.1.2 Check that the leveling bubble is centered.
5.1.2.1 If the bubble is off center, adjust the leveling feet to bring back to center.
5.1.3 Check that the balance is clean.
5.1.4 Remove any load from the pan and press "ON" briefly. When zero "0.00" is displayed, the balance is ready for operation.
5.1.5 Press and hold "CAL" until "CAL" appears in the display, release key. The required standard value flashes in the display.
5.1.6 With tongs, place the traceable standard required from 5.1.5 in the center of the pan. The balance adjusts itself.
5.1.7 When zero "0.00" flashes, remove calibration standard.
5.1.7.1 If "0.00" does not flash, the balance is not verified and must be checked by the metrology department. Refer to SOP QI 15.3.20 for procedure to remove balance from use until metrology can evaluate it.
5.1.8 Choose a NIST traceable standard whose weight is close to that of the sample to be weighed. When "0.00" is displayed the balance is ready.

(Continued)

FIGURE 17.11 Example of a standard operating procedure for checking and operating a balance.

Clean Gene, Inc.

Standard Operating Procedure
SUBJECT: Operation of Balance: BRP 88
Balance Location: Media prep lab
Effective Date: *1/22/21*
Page 2 of 2

5.1.9   With tongs, place the traceable standard on the weighing pan.

5.1.10   Wait until stability detector "o" disappears.

5.1.11   Read results.

5.1.12   Record the daily verification on the form QF 15.3.6.3 with date, time, and operator's name.

5.1.13   The balance is now in the weighing mode and is ready for operation. Reverification with standards is not necessary more than once a day.

5.2. Operation

5.2.1   Remove any load from weighing pan and press "ON" briefly. When zero "0.00" is displayed the balance is ready for operation.

5.2.2   Place sample on the weighing pan.

5.2.3   Wait until the stability detector "o" disappears.

5.2.4   Read results.

5.2.5   Switch off by pressing OFF until "OFF" appears in the display. Release key. The balance may be left in this mode for the remainder of the day.

5.3. Taring weighing vessel

5.3.1   Place the empty weighing vessel on the balance, the weight of the container is displayed.

5.3.2   Press "TARE" briefly, the balance will re-zero.

5.3.3   Add sample to the container. The sample weight is displayed.

5.4. Maintenance

5.4.1   Cleaning

5.4.1.1   After working with the balance, the work surface must be checked for residual chemicals or soils. Solid materials must be removed and disposed of properly.

5.4.1.2   Each week, or more often if necessary, the unit should be wiped down with general purpose cleaning agents. Wipe down the weighing pan, baseplate, sides, and lid of the balance.

5.4.1.3   In the event of a chemical spill, assess the hazard. If necessary, call hazard control team (X 3762). If nonhazardous material, remove balance pan and clean thoroughly.

5.4.2   Semi-Annual Calibration

5.4.2.1   Initial and semi-annual calibration is performed by trained service department personnel.

5.4.2.2   If the balance is out of calibration according to 5.1.7.1 it must be re-calibrated immediately by trained personnel in the QC department.

5.4.2.3   Form SOP QI 15.3.20 is to be filled out and the QC department notified for repair or recalibration.

FIGURE 17.11 (*Continued*)   Example of a standard operating procedure for checking and operating a balance.

BOX 17.4 TWO STRATEGIES FOR WEIGHING OUT A CHEMICAL WITH AN ELECTRONIC ANALYTICAL BALANCE

Weighing out chemicals when preparing reagents is one of the most common uses of a balance in a biotechnology setting. Figures 17.12 and 17.13 show two strategies for weighing out a chemical with an analytical balance.

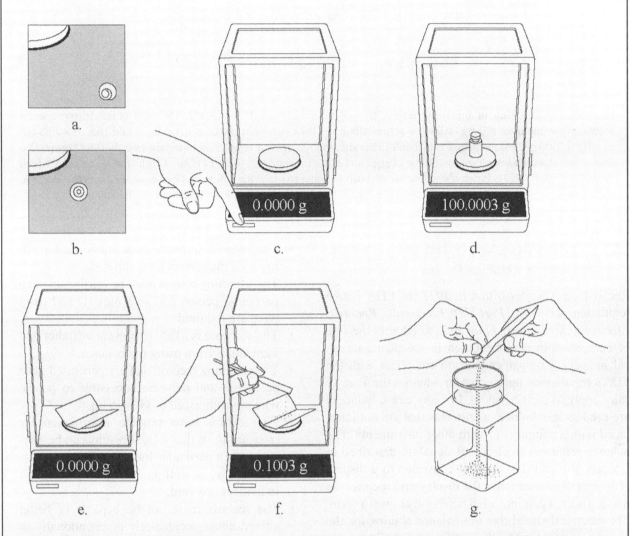

FIGURE 17.12 Using an analytical balance to weigh out a chemical: Strategy 1. In this example, the analyst needs to prepare a solution that contains 100 mg of NaCl and 100 mL of water. An analytical balance is required. (a and b) The balance is leveled so that the bubble is centered using the adjustable feet. (c) The balance is set to 0.0000 with the pan clean and empty. (d) The performance of the balance is verified and is found to be acceptable based on the laboratory's quality procedure. In this example, the deviation from zero in the last decimal place is acceptable. (e) The analyst folds a piece of weighing paper and places it on the balance's pan. (Alternatively, a small plastic weigh boat would work.) He closes the draft shield and tares the balance to subtract out the weight of the paper. (f) The analyst carefully adds NaCl until the display shows 0.1000 g (=100 mg). The figure in the last decimal place is likely to fluctuate, which is generally considered to be acceptable. (g) The analyst carefully pours the chemical into a container to which the water will be added.

(Continued)

BOX 17.4 (*Continued*)   TWO STRATEGIES FOR WEIGHING OUT A CHEMICAL WITH AN ELECTRONIC ANALYTICAL BALANCE

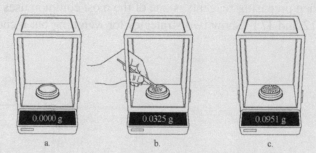

0.0000 g
a.

0.0325 g
b.

0.0951 g
c.

FIGURE 17.13   Using an analytical balance to weigh out a chemical: Strategy 2. The analyst needs to prepare a solution that contains 100 mg of bovine serum albumin (BSA) and 100 mL of water. She would like to weigh out exactly 0.1000 g. However, BSA is difficult to handle and the analyst is not able to obtain exactly 0.1000 g. (a) She uses a metal weigh boat to reduce static charge and tares the balance. (b and c) She weighs out some BSA which turns out to be 0.0951 g. Later, she adds only 95.1 mL of water to make her solution. This approach is correct because she adjusts the volume of water to compensate for the lesser amount of BSA weighed out on the balance.

17.5 COMPLIANCE WITH THE ELECTRONIC RECORDS REGULATIONS

Recall from Chapter 6 that in 1997 the FDA issued regulation *21 CFR Part 11: Electronic Records; Electronic Signatures; Final Rule* to address the role of computers in documentation in the pharmaceutical industry. Companies that are compliant with the FDA's regulations must consider whether the Part 11 rules apply to their balances. In many cases, balances are used as stand-alone instruments that are not interfaced with a computer or with other instruments. The balance generates an electrical signal (as described in Chapter 16), and this signal is converted to a display of weight. The user reads the display and records the reading with a pen into a lab notebook or onto a form. The electrical signal that the balance obtains for the sample is not stored by the instrument – it disappears when the sample is removed from the balance. The documentation with a stand-alone balance is therefore a conventional "paper" (and pen) record and is not subject to the requirements of Part 11.

Balances are, however, often interfaced with an electronic system that receives weight readings, stores the information (e.g., in Excel or in a database), perhaps analyzes the data, and provides an output. In these situations, if the company is compliant with FDA regulations and if the weight readings are critical to product safety, efficacy, and quality, then the rules of 21 CFR Part 11 do apply. This means that appropriate controls must be implemented for these weighing records. For example:

- The balance software must stamp each reading with the correct date and time.
- The weighing results must be attributable to a particular person and so a user ID and password are required.
- The software is likely to prevent unauthorized individuals from using the balance.
- The weighing records that are generated must be secure and must be accessible to people who are authorized to view them.
- The records must provide information for traceability so that weight results can be connected to a particular lot of product, or a particular task, as well as to the date, time, and individual involved.
- The records must not be capable of being altered either accidentally or intentionally in an uncontrolled manner.

Manufacturers have created newer models of laboratory balance that meet the electronic records requirements. You can find such balances by reading the literature provided by balance manufacturers.

17.6 MASS VERSUS WEIGHT

So far, we have used the terms *mass* and *weight* without much discussion of the distinction between them. In this section, we discuss these two words.

The value we read from a balance is the weight of an object, not its mass. This may seem surprising. After all, the object is directly compared with a

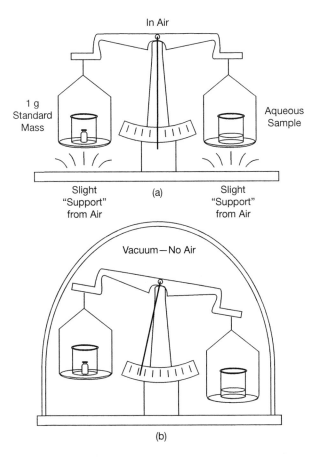

In Air

1 g
Standard
Mass

Aqueous
Sample

Slight
"Support"
from Air

(a)

Slight
"Support"
from Air

Vacuum—No Air

(b)

FIGURE 17.14 Mass versus weight. (a) A sample is exactly balanced against a 1 g mass standard in air. Although the sample and the standard have the same weight in air, they do not have the same mass. b. In a vacuum, the beam will tilt to the side with the sample.

standard whose mass is known. For example, consider the objects in Figure 17.14a. The standard is known to have a mass of 1 g and the sample exactly balances the standard. It seems logical, therefore, that the object should also have a mass of 1 g; however, there is another factor that affects this system – the air.

The major force measured in weighing is the force of gravity pulling down on an object. However, there is also a slight buoyant force from air. The air around a sample or standard gives it a bit of support and makes it appear lighter than it really is. This means that if the same object is weighed in air and in a vacuum, the object will be slightly heavier in the vacuum. This is

the **principle of buoyancy:** *Any object will experience a loss in weight equal to the weight of the medium it displaces.* Buoyancy is why ships float, helium balloons rise, and the weight of an object is different in a vacuum than it is in air (Figure 17.14b).

Balances are calibrated with metal mass standards. Metal has a relatively high density compared to powders, aqueous solutions, and other materials typically weighed in biotechnology settings. A 1 g metal mass standard, therefore, displaces less air than does a 1 g mass of water or a powdery chemical. Because the metal mass standard displaces less air than the water or chemical, the metal standard is buoyed less by the air. This difference in buoyancy of a mass standard and a sample explains why the standard and sample in Figure 17.14 can have the same weight yet be of different masses. The sample is buoyed by the air more than the standard. If a balance were calibrated with a mass standard whose density was identical to that of the sample, then the weight of the sample would equal its mass. Because we use metal mass standards and we weigh samples in air, the value we read in the laboratory for a sample is not its true mass.

This discrepancy between true mass and the value measured by the instrument is called the **buoyancy error**. The weight readings for aqueous solutions have a buoyancy error of roughly 1 part in 1,000. This buoyancy error is small enough to be of little concern in most applications. The distinction between mass and weight, therefore, is generally ignored in biotechnology laboratories, except when very high-accuracy measurements must be made. One situation where the distinction may be important is in verifying the performance of micropipettes that measure very small volumes of liquids, as will be discussed in Chapter 18. In other situations where buoyancy error must be considered, equations can be used to correct for the buoyancy effect. These corrections require knowing the barometric pressure and humidity at the time of measurement, and most biotechnology laboratories do not routinely make these measurements. (To obtain more information about this correction, see, for example, a guide from Mettler-Toledo, Inc. *Weighing the Right Way*. In 2021 this guide is available at https://www.mt.com/us/en/home/library/guides/laboratory-weighing/weighing-the-right-way.html)

Practice Problems

1. Suppose you need to prepare a solution with a concentration of 15 mg/mL of a particular enzyme. You try to weigh out 15 mg of the enzyme on the analytical balance, but find that it is extremely difficult to get exactly 15 mg. Suggest a strategy to get the correct concentration even if you cannot weigh exactly 15 mg.

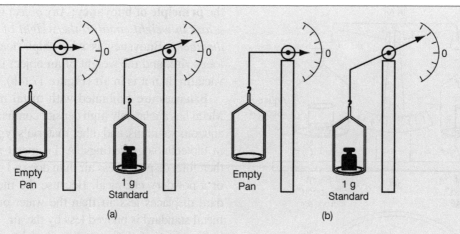

FIGURE 17.15 Diagrams for Practice Problem 4.

2. How much volume of solution at a concentration of 35 µg/mL can be made in each of these cases?
   a.   Solute weighs 0.003560 g
   b.   Solute weighs 0.0500 mg
   c.   Solute weighs 1.0897 g
   d.   Solute weighs 354.8 µg
3. A technician weighs a cell preparation on an analytical balance and observes that the initial weight is 0.0067 g. A while later the weight is 0.0061 g. What might be happening here?
4. Which scale shown in Figure 17.15 is more sensitive?
5. Suppose you are verifying the performance of a laboratory analytical balance. You first calibrate the balance, as directed by an SOP. You then repeatedly weigh a Class 1 standard weight that is 10 g. The results are as shown. According to the specifications for the balance, its standard deviation (precision) must be equal to or better than ± 0.0001 g and its accuracy must be better than ± 0.0001 g. Does the balance meet its performance specifications?

Weight Values Obtained
10.0000 g
10.0001 g
9.9999 g
10.0000 g
9.9998 g
10.0001 g

6. A standard operating procedure to check and operate a balance is shown in Figure 17.11. Read the SOP and answer the following.
   a.   How does a user determine and verify that the balance is operating acceptably?
   b.   How does a user document that the balance is operating acceptably?
7. Suppose a 100 g standard mass is accidentally dropped on the floor. As a result, its mass is slightly less than 100 g. If the standard is used to calibrate a balance, what will happen to subsequent readings from that balance? What type of error is this?
8. The linearity of a balance is being checked. The balance has a capacity of 500 g, and four weights of about 125 g each are used. The results follow. Use the procedure in Box 17.2 to calculate the linearity error (if any) at midpoint, 25%, and 75% capacity.

Midpoint Check
The four pieces all together weighed 500.001 g
A and B together weighed 250.003 g
C and D together weighed 250.000 g

Check at 25% of Capacity
A weighed 125.001 g
B weighed 125.001 g

C weighed 125.003 g

D weighed 124.996 g

Check at 75% of Capacity

Group 1: pieces A, B, C weighed 375.001 g

Group 2: pieces A, B, D weighed 374.998 g

Group 3: pieces A, C, D weighed 375.001 g

Group 4: pieces B, C, D weighed 374.996 g

9. (Optional) **a.** The density of air is about 1.2 mg/cm^3 (depending on the temperature and atmospheric pressure). 1,000 mL of water displaces 1,000 cm^3 of space (from the definition of a mL). What is the weight of air displaced by 1,000 mL of water?

b. Will the measured weight of a sample change if it is weighed first in air and then in a vacuum? Will its mass change?

Questions for Discussion

1. The ISO 9000 standards state that companies shall "define the process employed for the calibration of inspection, measuring, and test equipment, including details of equipment type, unique identification, location, frequency of checks, check method, acceptance criteria, and the action to be taken when results are unsatisfactory" (Section 4.11.2c). Explain how the SOP in Figure 17.11 accomplishes the requirements as stated by ISO 9000.

2. Consider weighing an object in the laboratory.
   a. List as many possible sources of bias as you can.
   b. What would be required to eliminate each source of bias you mentioned in part a?

3. If you work in a laboratory, write an SOP to operate a balance in your laboratory. Be sure to include steps to verify that the balance is responding properly.

CHAPTER APPENDIX: LABORATORY MASS STANDARDS

Standards are classified by Type and Class. **Type** *refers to the method of construction of the standard.* **Type I standards** *are of one-piece construction and are used when the most accurate and stable standards are required.* **Type II standards** *do not need to be constructed of a single piece of metal and may include additional adjusting material.*

Class indicates the permitted tolerance of the standard. There are two sources of classifications for laboratory mass standards that are commonly cited in the United States:

- ASTM E617–18 "Standard Specification for Laboratory Weights and Precision Mass Standards," with Classes 000, 00, and 0–7 where the lower the class number, the smaller the tolerance.
- OIML R 111 (from the International Organization of Legal Metrology), which identifies Classes E1, E2, F1, F2, M1, M1–2, M2, M2–3, and M3.

The different classes of standard each have different applications. For example:

ASTM Class 0: Used as primary reference standards for calibrating other reference standards and weights.

ASTM Class 1: Can be used as a reference standard in calibrating other weights and is appropriate for calibrating high-precision analytical balances with a readability as low as 0.1 mg to 0.01 mg.

ASTM Class 2: Appropriate for calibrating high-precision, top-loading balances with a readability as low as 0.01 g to 0.001 g.

ASTM Class 3: Appropriate for calibrating balances with moderate precision, with a readability as low as 0.1 g to 0.01 g.

ASTM Class 4: For calibration of semi-analytical balances and for student use.

ASTM Class 5: For student laboratory use.

ASTM Class 6: Student brass weights are typically calibrated to this class. (This class meets the requirements for OIML R 111 Class M2.)

ASTM Class 7: For rough weighing operations in physical and chemical laboratories.

OIML Class E1: Used as primary reference standards for calibrating other reference standards and weights.

OIML Class E2: Can be used as a reference standard in calibrating other weights and is appropriate for calibrating high-precision analytical balances with a readability as low as 0.1 mg to 0.01 mg.

OIML Class F1: Appropriate for calibrating high-precision, top-loading balances with a readability as low as 0.01 g to 0.001 g.

OIML Class F2: For calibration of semi-analytical balances and for student use.

OIML Class M1, M2, M3: Economical weights for general laboratory, industrial, commercial, technical, and educational use. Typically fabricated from cast iron or brass. Class M2 is commonly used for student brass weights.

TOLERANCE FOR MASS STANDARDS[A] ($\pm$ MG)

	Class 0	Class 1	Class 2	Class 3	Class 4	Class 5	Class 6	Class 7
1 kg	1.3	2.5	5.0	10	20	50	100	470
500 g	0.60	1.2	2.5	5.0	10	30	50	300
200 g	0.25	0.50	1.0	2.0	4.0	15	20	160
100 g	0.13	0.25	0.50	1.0	2.0	9	10	100
50 g	0.060	0.12	0.25	0.60	1.2	5.6	7	62
10 g	0.025	0.050	0.074	0.25	0.50	2.0	2	21
1 g	0.017	0.034	0.054	0.10	0.20	0.50	2.0	4.5
100 mg	0.005	0.010	0.025	0.050	0.10	0.20	1.0	1.2

[a] Excerpted from ASTM E617–18 "Standard Specification for Laboratory Weights and Precision Mass Standards."

Examples

When a balance is being calibrated and the utmost accuracy possible is required, it is necessary to choose the correct mass standard. It is recommended that the tolerance of the standard be four times more accurate than the readability of the balance. For example:

a. A balance has a readability of 1 mg and is to be calibrated with a 100 g standard weight. What class of standard should be chosen? To make this decision, divide 1 mg by 4; the result is 0.25 mg. This is the tolerance that is needed in the 100 g standard. Refer to the tolerance chart for weight standards. For a 100 g standard, the tolerance of a Class 1 standard is 0.25 mg.

A Class 1 standard, therefore, should be chosen.

b. A balance has a capacity of 60 g and a readability of 1 mg. It is to be calibrated with a 50 g standard. What class of standard is required? The readability is 1 mg; therefore, the tolerance of the standard should be ± 0.25 mg. From the chart, observe that a 50 g standard of Class 2 has a tolerance of 0.25 mg.

c. A balance has a capacity of 500 g and a readability of 0.1 g, and it is to be calibrated with a 500 g standard. What class of standard is required? The readability is 0.1 g; therefore, the tolerance of the standard should be ± 0.025 g or ± 25 mg. A Class 4 standard is adequate.

18 The Measurement of Volume

18.1 PRINCIPLES OF MEASURING THE VOLUME OF LIQUIDS

18.1.1 OVERVIEW

Biological systems (e.g., cells, proteins, DNA, and RNA) are active in an aqueous environment. Water-based liquids are therefore involved in almost every biotechnology procedure. For this reason, the measurement of liquid volumes is among the most fundamental activities performed in any biotechnology facility. As is true of weight measurements, it is essential that volume measurements are made accurately and precisely to avoid irreproducible research results and defective products.

Volume *is the amount of space a substance occupies.* The **liter** *is the basic unit of volume used in* *biology/biotechnology settings.*[1] Various devices are used to measure the volume of liquids, depending on the volume being measured and the accuracy required. For larger volumes, biologists use glass and plastic vessels, such as **graduated cylinders** and **volumetric flasks**. **Pipettes** are usually preferred for volumes in the 1–25 mL range. Various **micropipetting devices** are commonly used to measure volumes in the microliter range. These various methods of measuring volume are discussed in this chapter.

---

[1] The proper primary unit for volume in the SI system is not the liter; rather, it is the cubic meter, m^3. The cubic centimeter, cm^3, is also used in the SI system. The milliliter, mL, is a unit derived from the liter. One milliliter is equivalent to 1 cm^3. In this textbook we follow the common practice of measuring volume in units derived from the liter, rather than those derived from the meter.

DOI: 10.1201/9780429282799-23

18.1.2 BASIC PRINCIPLES OF GLASSWARE CALIBRATION

Manufacturers of glassware and plasticware for volume measurements (such as graduated cylinders, volumetric flasks, and pipettes) are responsible for the calibration of those items. **Capacity marks** and **graduations** *are lines marked on volume-measuring devices that indicate volume.* Calibration of glassware and plasticware involves placing capacity marks and graduations in such a way that they correctly indicate volume. ASTM International distributes standards that specify exacting procedures for how certain volume-measuring devices are to be calibrated, checked for accuracy, and labeled.

The meniscus must be considered when glassware is calibrated. The **meniscus** (Greek for "crescent moon") *is a curve formed by the surface of liquids confined in narrow spaces, such as in measuring devices* (Figure 18.1). Conventional practice and ASTM International standards specify that the lowest point of the meniscus is used as the point of reference in calibrating a volume-measuring device; therefore, the lowest part of the meniscus is also the reference when reading the volume of a measuring device. It may be helpful to place a dark background, such as a piece of black paper, behind the device to facilitate reading the meniscus accurately. It is important to hold your eyes level with the liquid surface when reading a meniscus.

There is inevitably a small amount of error in the volume calibration of glassware, plasticware, and pipettes, even when they are manufactured properly. The **tolerance,** *or how much error is allowed in the*

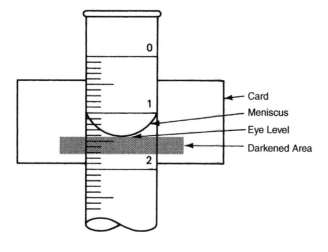

FIGURE 18.1 The meniscus. A dark card may be placed behind the meniscus to make it more visible. When reading volume, the eye should be level with the bottom of the meniscus.

calibration of a volume-measuring item, depends on the volume of the item and the purpose of the measurements to be made with it. *The most accurately calibrated glassware is termed* **volumetric;** volumetric glassware therefore has the narrowest tolerance allowed in its calibration. For example, Class A volumetric flasks are used where high accuracy is required, such as in the preparation of standard solutions. The tolerance of a 100 mL Class A volumetric flask that meets ASTM International standards is 0.08 mL (i.e., its true marked volume must be between 99.92 and 100.08 mL). A 100 mL Class B volumetric flask, which is used where slightly less accuracy is required, must be marked within 0.16 mL of 100 mL. The manufacturer must therefore calibrate a Class A flask more exactly than a Class B flask, and one might expect to pay more for it.

Glassware for measuring liquid volume is calibrated either **to contain (TC)** or **to deliver (TD)**. A device that is **TC** *will contain the specified amount when filled to the capacity mark.* It will not deliver that amount if the liquid is poured out because some of the liquid will adhere to the sides of the container. A **TD** *device is marked slightly differently so that it does deliver the specified amount, assuming the liquid is water at 20°C and it is poured out using a specific technique.* Consider the logic behind the two methods of calibration. A solution made to a specific concentration of solute can be prepared in a volumetric flask. The volume of water *in the flask* is critical in obtaining the right solute concentration, so a flask calibrated "to contain" is properly used. In contrast, if the volumetric flask is to be used to *pour* an exact volume of liquid to another container, then a flask calibrated "to deliver" should be used. (Plastics are considered to be nonwetting – that is, water does not stick to them – so there is no difference between TC and TD for plasticware.)

For high-accuracy volume measurements, it is important to consider two effects of temperature. First, the glass or plastic forming a volume-measuring device will expand and contract as the temperature changes, thus affecting the accuracy of its markings. Second, the volume of materials, including aqueous solutions, changes as the temperature changes. ASTM International standards specify that manufacturers calibrate devices at 20°C using water at the same temperature. In laboratory situations where the utmost accuracy is required, volume measurements should be made at 20°C or correction factors to account for temperature should be applied. (See ASTM Standard E 542, "Standard Practice for Calibration of Laboratory Volumetric Apparatus.")

To ensure accuracy when using volume-measuring devices, it is necessary to use the device in the same manner as it was calibrated. If devices calibrated with water are used to deliver other liquids, some accuracy may be lost unless the device is recalibrated. If labware is distorted by heating or is contaminated, then the calibration marks will not be accurate. Systematic error will occur if for any reason a piece of glassware does not read the correct volume. Repetition of the measurement will not reveal this error.

18.2 GLASSWARE AND PLASTIC LABWARE USED TO MEASURE VOLUME

18.2.1 BEAKERS, ERLENMEYER FLASKS, GRADUATED CYLINDERS, AND BURETTES

Beakers and Erlenmeyer flasks are containers intended primarily to hold liquids, not to measure volumes. These vessels are marked with graduation lines, but they are at best calibrated with a tolerance of ± 5%. This means, for example, that when the manufacturer marks a beaker at "100 mL," it may actually indicate anywhere between 95 and 105 mL.

Graduated cylinders *are cylindrical vessels calibrated with sufficient accuracy for most volume measurements in biology laboratories.* A moderately priced 100 mL graduated cylinder is calibrated with a tolerance of ± 0.6 mL.

Graduated cylinders are marked with a number of equal subdivisions. For example, a 100 mL graduated cylinder might be etched with 100 subdivisions, one for each mL, whereas a 500 mL cylinder might have a mark every 5 mL. This means that a single graduated cylinder can be used to measure various volumes; however, the correct cylinder should be chosen depending on the volume desired. For example, it is not possible to measure 83 mL accurately with a 500 mL graduated cylinder; a 100 mL graduated cylinder is more appropriate (Figure 18.2a–c). Note also that graduated cylinders are not designed as containers for mixing and for storing solutions because they are unstable and easily knocked over; they are expensive compared with storage bottles; and they often do not come with caps.

Burettes (also spelled *buret*) *are long graduated tubes with a stopcock at one end that are used to dispense volumes accurately* (Figure 18.2d). Burettes have a long tradition of use in chemistry laboratories for making accurate volume measurements.

18.2.2 VOLUMETRIC FLASKS

Volumetric flasks *are vessels used to measure specific volumes where more accuracy is required than is attainable from a graduated cylinder.* Each **volumetric flask** *is calibrated either "to contain" or "to deliver" a single volume* (Figure 18.3). Although volumetric flasks are calibrated with high accuracy, they have the disadvantages that they are relatively expensive and that each is calibrated for only one volume (e.g., 10, 50, or 1,000 mL). A volumetric flask cannot be used, for example, to measure 33 mL. Volumetric flasks, therefore, are typically used in biotechnology laboratories only in situations where high-accuracy volume measurements are required.

The ASTM International tolerances for volumetric flasks of different sizes are shown in Table 18.1. Observe that there is less error tolerated in the calibration of Class A flasks than in the calibration of Class B flasks.

Manufacturers sell special, serialized (numbered) volumetric flasks (Figure 18.4). These flasks are individually calibrated using equipment whose calibration is traceable to NIST, and they are accompanied by a certificate of traceability. The certificate documents that a given flask is accurate within the tolerance specified for its type and class. Certified, serialized glassware may be required in facilities meeting ISO 9000, CGMPs, or other formal quality requirements.

Volumetric flasks must be used correctly to ensure accurate volume measurements. Box 18.1 contains points relating to the proper use of a volumetric flask.

18.3 PIPETTES

18.3.1 PIPETTES AND PIPETTE AIDS

Pipettes (also spelled "pipet") *are hollow tubes that allow liquids to be drawn into one end and are generally used to measure volumes in the 0.1–25 mL range.* Pipettes may be made of glass or plastic and can be disposable or intended for multiple uses. Pre-sterilized disposable pipettes with cotton plugged tops are available for microbiology and tissue culture work. The cotton plug at the top helps prevent airborne contaminants from entering the pipette and also helps protect the user from exposure to the cells that are being manipulated.

Pipettes come in different styles and sizes. Manufacturers may place colored bands at the top of pipettes to indicate their capacity and graduation interval. These colored bands are a quick way to identify the pipette's capacity, and they facilitate both choosing the proper pipette and sorting pipettes after washing.

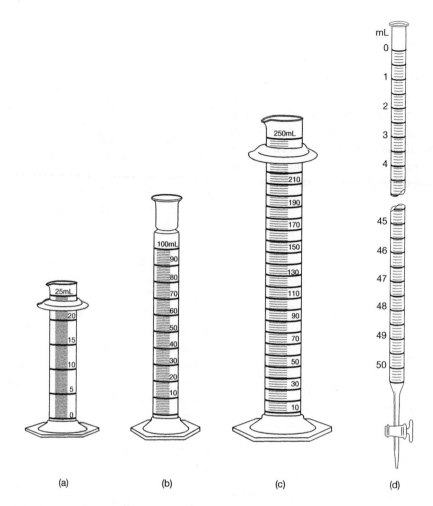

FIGURE 18.2 Graduated cylinders and burettes. (a) A 25 mL graduated cylinder has a graduation mark every 0.5 mL and is useful for volumes from 1 to 25 mL. (b) A 100 mL cylinder has a graduation mark every 1 mL and is most useful for volumes from 10 to 100 mL. (c) A 250 mL cylinder is marked every 2 mL and is most useful for volumes from 100 to 250 mL. For volumes less than or equal to 100 mL, a 100 mL graduated cylinder can be used with better accuracy. (d) A burette that is calibrated every 0.1 mL.

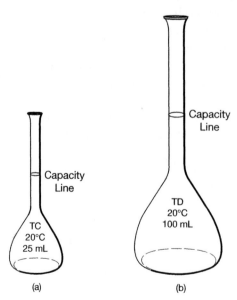

FIGURE 18.3 Volumetric flasks. (a) A 25 mL volumetric flask that is calibrated to contain 25 mL when filled with water at 20°C to exactly to its capacity line. (b) A volumetric flask that is calibrated to deliver 100 mL when it is filled exactly to its capacity line with water at 20°C and the liquid is then poured out.

TABLE 18.1

Excerpts from ASTM Standards for the Tolerance of Volumetric Flasks

Capacity (mL)	Class A Tolerance (± mL)	Class B Tolerance (± mL)
5	0.02	0.04
10	0.02	0.04
25	0.03	0.06
50	0.05	0.10
100	0.08	0.16
200	0.10	0.20
250	0.12	0.24
500	0.20	0.40
1,000	0.30	0.60
2,000	0.50	1.00

Source: From ASTM Standard E 288-10 (Reapproved 2017), "Standard Specification for Laboratory Glass Volumetric Flasks."

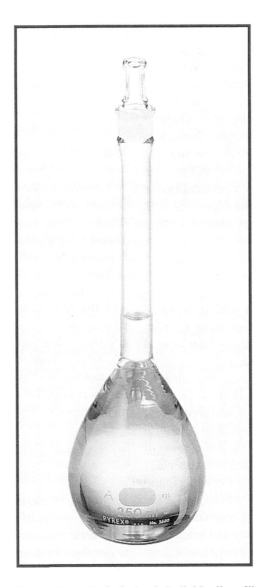

FIGURE 18.4 A serialized and certified volumetric flask that is individually calibrated according to ASTM standards. (Used with permission from Thermo Fisher Scientific, the copyright owner.)

BOX 18.1   PROPER USE OF VOLUMETRIC FLASKS

1. ***Choose the proper type of flask for the application: either "to contain" or "to deliver," either Class A or Class B, either serialized or not serialized.*** It may be prudent to check the calibration of new volumetric glassware before use. (Important Technical Guidance on Glassware, Tom Coleman and Georgia Harris, NIST, 2005. https://www.nist.gov/system/files/documents/2017/05/09/h-008.pdf.)
2. ***Be sure the flask is completely clean before use.***
3. ***Read the meniscus with your eyes even with the liquid surface.*** The bottom of the meniscus should exactly touch the capacity line.
4. ***If the flask is calibrated "to deliver," then pour as follows:***
   a. Incline the flask to pour the liquid; avoid splashing on the walls as much as possible.
   b. When the main drainage stream has ceased, the flask should be nearly vertical.
   c. Hold the flask in this vertical position for 30 seconds and touch off the drop of water adhering to the top of the flask by touching it to the receiving vessel.
5. ***Never expose volumetric glassware to high temperatures because heat causes expansion and contraction that can alter its calibration.*** For example, do not bake volumetric glassware to dry it.

Pipette aids *are devices used to draw liquid into and expel it from pipettes* (Figure 18.5). In the past, pipette aids were less common and people would suck liquid into pipettes, like sucking a drink into a straw. This procedure, called *mouth pipetting,* is no longer done because it was easy to accidentally inhale or swallow radioactive substances, pathogens, hazardous chemicals, or the like. Although mouth pipetting has been abolished, its terminology persists, and pipettes are often calibrated as "blow-out." A pipette calibrated as "blow-out" was previously literally blown out with one's mouth. Now, the last drop is ejected from a "blow-out" pipette using a pipette aid.

18.3.2 MEASURING PIPETTES

A measuring **pipette** *is calibrated with a series of graduation lines to allow the measurement of more than one volume.* Figure 18.6 illustrates how to read the meniscus on a measuring pipette.

Measuring pipettes are sometimes classified as being either **serological** or **Mohr.** Both these types are calibrated "to deliver." Pipettes termed **serological** *are usually calibrated so that the last drop in the tip needs to be "blown out" to deliver the full volume of the pipette* (Figure 18.7a and b). Manufacturers place one wide or two narrow bands at the top of a pipette to indicate that it is calibrated to be "blown out." This "blow-out" band(s) typically appears above any color-coding band that indicates the capacity and graduation interval for the pipette. The proper way to use a serological pipette is summarized in Box 18.2.

Mohr pipettes *are calibrated "to deliver," but, unlike the serological pipettes described earlier, the*

liquid in the tip is not part of the measurement and the pipette is not blown out (Figure 18.8a and b). Serialized Mohr pipettes that have a certificate of traceability to NIST can be purchased.

18.3.3 VOLUMETRIC (TRANSFER) PIPETTES

Volumetric (transfer) pipettes *are made of borosilicate glass and are calibrated "to deliver" a single volume when filled to their capacity line at 20°C* (Figure 18.9). Volumetric pipettes are the most accurately calibrated pipette type (Table 18.2). Serialized volumetric pipettes with a certificate of traceability to NIST can be purchased.

Volumetric pipettes are calibrated so that the "delivery of the contents into the receiving vessel is made with the tip in contact with the wall of the vessel and no after-drainage period is allowed." (ASTM Standard E 969-02 (Reapproved 2012), "Standard Specification for Glass Volumetric (Transfer) Pipets"). These pipettes are not "blown out." Remember to check that the tip is not cracked or chipped before using a volumetric pipette.

18.3.4 OTHER TYPES OF PIPETTES AND RELATED DEVICES

There are various types of pipettes in addition to those described above. For example, **Pasteur pipettes** *are used to transfer liquids from one place to another.* Pasteur pipettes are not actually volume-measuring devices because they have no capacity lines, but they are convenient for transferring liquids (Figure 18.10a). There are glass pipettes for measuring

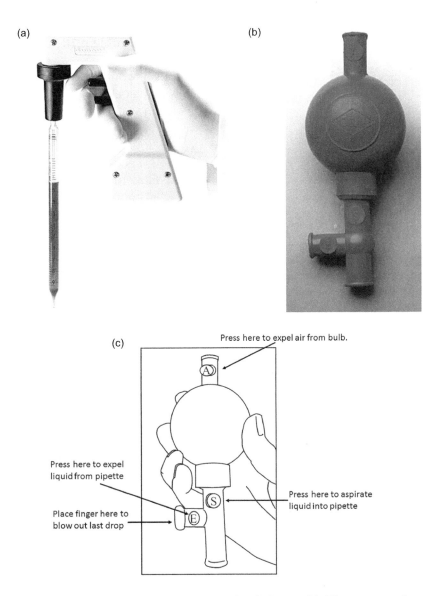

FIGURE 18.5 Pipette aids. (a) A convenient and common style of pipette aid. There are two buttons: one that, when pressed, aspirates liquid into the pipette, and another that is pressed to expel the liquid from the pipette. This device is motor-driven. With a little practice, this pipette aid facilitates accurate and ergonomic pipetting. (b) A rubber bulb style of pipette aid. Because these bulbs do not require batteries or electricity, they are relatively inexpensive and common. (c) Operating the rubber bulb shown in b. This style of bulb has three valves, labeled **A**, **S**, and **E**. To use this style of bulb, begin by forcing air out of the bulb by pressing button **A** and squeezing the bulb. Then insert a pipette into the end of the bulb. To aspirate the desired volume of liquid into the pipette, press button **S**. Dispense (expel) the liquid by pressing button **E**. Blow out the last drop by placing your finger on the side of the bulb, near to the letter **E**, and squeezing. (Photos used with permission from Thermo Fisher Scientific, the copyright owner.)

volumes less than 1 mL. These small-volume, reusable "micropipettes," however, are not widely used in biotechnology laboratories because their use has been supplanted by micropipetting instruments, as described in Section 18.4 of this chapter. Disposable pipettes for measuring less than 1 mL volumes are so convenient that they are sometimes used when high accuracy is unnecessary (Figure 18.10b). Inexpensive capillary glass pipettes that deliver in the microliter range are sometimes used (e.g., in teaching laboratories) (Figure 18.10c).

Manual dispensers for reagent bottles *are devices placed in a reagent bottle with a tube that extends to the bottom of the bottle. The dispenser has a plunger that is depressed to deliver a set volume of liquid* (Figure 18.10d). These devices effectively take the place of using a pipette or other device to remove liquids from reagent bottles.

FIGURE 18.6   Reading the meniscus on a measuring pipette. Liquid was drawn up to exactly the zero mark and was then dispensed. Reading the value at the bottom of the meniscus shows that about 3.19 or 3.20 mL of liquid was delivered.

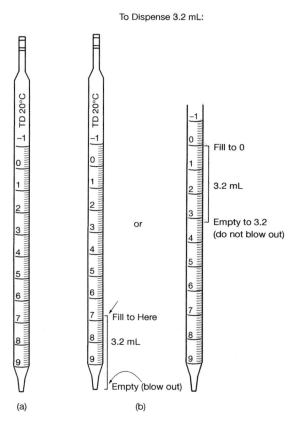

FIGURE 18.7   Using serological pipettes. (a) A serological pipette calibrated so that the tip includes the last milliliter. The two bands at the top indicate that this pipette is to be "blown out." Note that this pipette has a scale that extends above zero to expand the calibrated capacity of the pipette. (b) Two ways to use a serological pipette to dispense 3.2 mL.

Example Problem 18.1

Suppose you are going to prepare 100 mL of physiological saline solution, which is 0.9% NaCl in water.

a. How much solute is necessary? What is a good way to measure the solute?
b. How much solvent is necessary? What is a good way to measure it?
c. List some common measurement errors that must be avoided to ensure that the solution is made properly.

Answer

a. The solute is 0.9 g of NaCl. This would be measured on a balance that reads at least to the nearest 0.01 g. (Recall that the last digit on an instrument may be estimated.)
b. Close to 100 mL of water will be required to bring the solution to a volume of 100 mL. A graduated cylinder or volumetric flask would commonly be used to make this solution.
c. There are many potential sources of error in preparing this solution, such as improper reading of the meniscus, use of dirty glassware, use of an improperly maintained balance, and failure to pour all the salt from the weighing container into the mixing container. Care is required to avoid these major sources of error. There is also likely to be a small amount of error due to the fact that some uncertainty is tolerated in the calibration of a graduated cylinder or volumetric flask and the balance. This error, however, is very small compared with the potential error of the mistakes listed above. If a balance is properly maintained and used, the amount of error due to the balance is insignificant compared to these other factors.

BOX 18.2   USING A SEROLOGICAL PIPETTE

1. *Check that the pipette is calibrated to be "blown out" by looking for the band(s) at the top, or by looking at the final number at the bottom.* For example, a 10 mL pipette that only has numbers going up to 9 contains the last mL in the tip.
2. *Examine the pipette to be sure the tip is not cracked or chipped.*
3. *Attach the pipette to a pipette aid.*
4. *Fill the pipette about 10 mm above the capacity line desired and remove any water on the outside of the tip by a downward wipe with lint-free tissue.*
5. *Place the tip in contact with a waste beaker and slowly lower the meniscus to the capacity line.* Do not remove any water remaining on the tip at this time.
6. *Deliver the contents into the receiving vessel by placing the tip in contact with the wall of the vessel.*
7. *When the liquid ceases to flow, "blow out" the remaining liquid in the tip with one firm "puff" with the tip in contact with the vessel wall, if possible.* Because we no longer actually "puff" on pipettes, the last drop is ejected with the pipette aid.

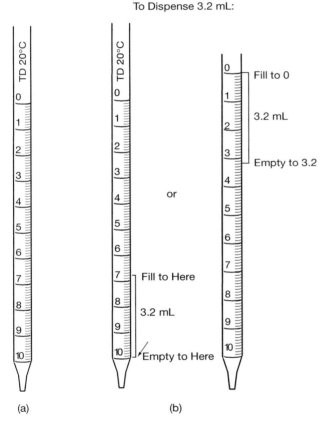

(a)   (b)

FIGURE 18.8 Mohr pipettes. (a) Mohr pipettes are calibrated so that the tip does not include the last milliliter. (b) Two ways to use a Mohr pipette to dispense 3.2 mL.

18.4 MICROPIPETTING DEVICES

18.4.1 POSITIVE DISPLACEMENT AND AIR DISPLACEMENT MICROPIPETTES

The measurement of small volumes of liquids, in the microliter range, is among the most common tasks performed in biotechnology settings. It is not unusual for laboratory personnel to measure out hundreds of small volumes every day when performing experiments and assays (tests of samples). Small inaccuracies in volume measurements can have profound effects on results. For example, in quantitative PCR, a minute amount of DNA is replicated thousands of times with the goal of quantifying the amount of a particular DNA sequence initially present in a sample. (Quantitative PCR is introduced in Chapter 27.) A small error in volume measurement at the beginning of the assay will be amplified, resulting in a serious error in the final result. Accurate and precise volume measurements are therefore necessary to ensure reproducible results. As another example, many common assays performed by biotechnologists require a standard curve that is created by diluting standards. Pipetting errors made during dilutions can invalidate assay results. (Assays requiring standard curves are discussed in Chapter 28.)

Glass and plastic pipettes, as discussed so far in this chapter, are typically used in biology

FIGURE 18.9   A volumetric pipette.

TABLE 18.2

Tolerances for Volumetric Class A and Class B Pipettes Compared with General-Purpose Serological Pipettes

Capacity (mL)	Volumetric Class A Tolerance	Volumetric Class B Tolerance	Glass Serological Tolerance
0.1	-	-	0.005
0.2	-	-	0.008
0.5	0.006	0.012	0.01
1.0	0.006	0.012	0.02
2.0	0.006	0.012	0.02
3.0	0.01	0.02	-
4.0	0.01	0.02	-
5.0	0.01	0.02	0.04
6.0	0.01	0.03	-
7.0	0.01	0.03	-
8.0	0.02	0.04	-
9.0	0.02	0.04	-
10.0	0.02	0.04	0.06
15.0	0.03	0.06	-
20.0	0.03	0.06	-
25.0	0.03	0.06	0.10
50.0	0.05	0.10	-
100.0	0.08	0.16	-

Source: Information from ASTM Standard E 969-02 (Reapproved 2012), "Standard Specification for Volumetric (Transfer) Pipets" and ASTM Standard E 1044-96 (Reapproved 2018), "Standard Specification for Glass Serological Pipets (General Purpose and Kahn)."

Tolerances are expressed as ± mL.

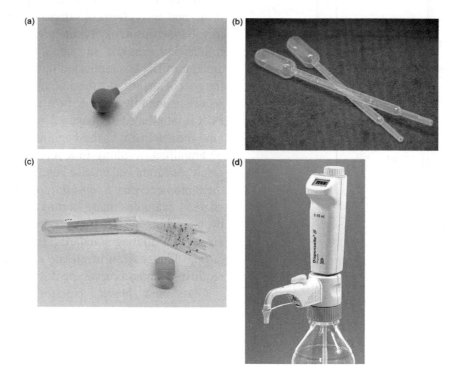

FIGURE 18.10 Examples of various types of pipette and liquid dispensing devices. (a) Pasteur pipettes. (Credit:BeataGFX/Shutterstock.com) (b) Disposable plastic pipettes to measure microliter volumes with an accuracy of about ±10%. (Credit: zairiazmal/Shutterstock.com) (c) Glass capillary pipettes for measuring microliter volumes. (Credit:Konokae/Shutterstock.com) (d) A manual dispenser for a reagent bottle. (Used with permission from Thermo Fisher Scientific, the copyright owner.)

laboratories to measure volumes as small as about 1 mL. **Micropipettes** *are devices commonly used to measure smaller volumes, in the 1–1,000 μL range.* These volume-delivering devices go by many names in addition to micropipette, such as **microliter pipette, piston- or plunger-operated pipette,** and, simply, **pipettor**.

There are two distinct designs of micropipette: **positive displacement** and **air displacement** (Figure 18.11). **Positive displacement micropipettes** *include syringes and other devices where the sample comes in contact with the plunger and the interior walls of the pipetting instrument.* Positive displacement devices are recommended for viscous and volatile samples.

Air displacement micropipettes *are designed so that there is an air cushion between the pipette and the sample such that the sample only comes in contact with a disposable tip and does not touch the micropipette itself.* Disposable tips reduce the chance that material from one sample will contaminate another, or that the operator will be exposed to a hazardous material. Air displacement micropipettes accurately measure the volume of aqueous samples and are among the most common instruments used in biotechnology laboratories.

The devices illustrated in Figure 18.11b and c are very common in biotechnology facilities. They are sometimes called **digital microliter pipettes.** They can be adjusted to deliver different volumes over a range, such as 1–10 μL, or 10–100 μL, or 100–1,000 μL. These devices are manually (by hand) controlled and deliver one volume at a time, for example 50 or 10 μL. There are other styles of micropipette as well (Figure 18.12). Some devices can deliver more than one liquid sample simultaneously (Figure 18.12a), some are electronically driven, and some are microprocessor controlled (Figure 18.12b).

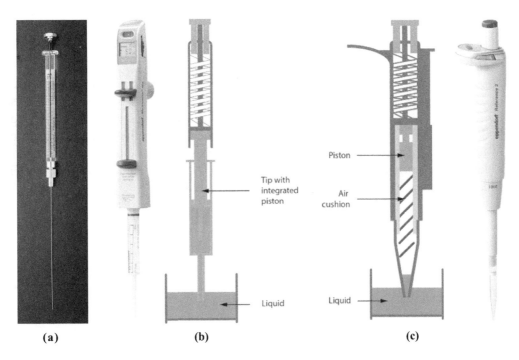

(a) **(b)** **(c)**

FIGURE 18.11 Comparison of positive and air displacement micropipettes. (a) A syringe is a type of positive displacement pipette. Liquids that are aspirated contact the interior of the syringe. Syringe-style micropipettes accurately dispense microliter volumes of nonaqueous liquids and are commonly used for injecting samples into chromatography instruments. (See Chapter 34 for information about chromatography.) (Used with permission from Thermo Fisher Scientific, the copyright owner.) (b) Another style of positive displacement micropipette. The liquid being aspirated is in direct contact with the piston, which therefore is contaminated with each liquid sample. The tip and piston are therefore both disposable, making this style of device expensive to use as compared to air displacement micropipettes. Positive displacement micropipettes are recommended for accurately dispensing microliter volumes of nonaqueous liquids, but they can also be used for aqueous pipetting. (c) An air displacement micropipette. There is a cushion of air that separates the liquid being aspirated from the piston, so the piston is not contaminated by each liquid sample. A new, disposable tip is used for each sample. This style of device is accurate when pipetting microliter volumes of aqueous samples, but is less accurate when pipetting nonaqueous samples. (Images b and c: Courtesy of Eppendorf AG.)

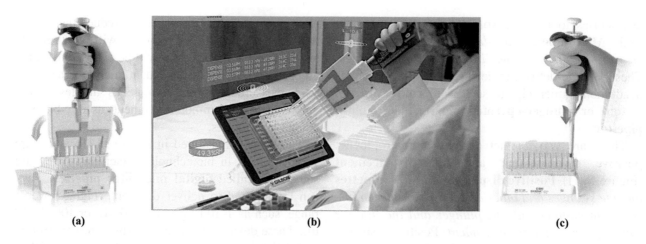

(a)   (b)   (c)

FIGURE 18.12 Various types of air displacement micropipettes and tips. (a) A multichannel micropipette used to simultaneously deliver the same volume to 12 wells of a plate that has 96 wells. (Each well is like a tiny test tube.) (b) An electronic, motorized, microprocessor-controlled micropipette. This device reduces operator fatigue and also allows the user to program the device, for example, to repeatedly dispense particular volumes. The micropipette's internal software allows the user to set a maintenance interval with an alarm that sounds when it is time for service. A computer can be used with this device to store and organize maintenance reports and data, thus assisting in adhering to quality system requirements. Repetitive pipetting can lead to injury, so, in laboratories where repetitive pipetting is performed, it is worthwhile to investigate motor-driven micropipettes designed to reduce operator fatigue. (c) Disposable tips are stored in racks inside autoclavable plastic boxes; racked tips do not need to be touched with one's fingers to attach them to a micropipette. (Photos provided courtesy of Gilson, Inc.)

18.4.2 OBTAINING ACCURATE MEASUREMENTS FROM AIR DISPLACEMENT MANUAL MICROPIPETTES

18.4.2.1 Procedure for Operation

It is important to operate micropipettes properly, or they will not deliver the correct volumes. Manual micropipettes have **plungers** *by which the operator controls the uptake and expulsion of liquids.* As the operator depresses the plunger, different "stop" levels can be felt. Although the detailed operating directions vary depending on the type and brand of micropipette, the general operation of most manual micropipettes is similar and is summarized in Figure 18.13 and in Box 18.3.

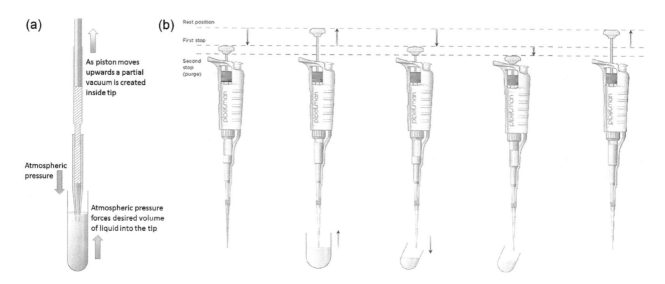

FIGURE 18.13 Using air displacement manual micropipettes. (a) The principle of air displacement. (b) Operation of a manual micropipette. See Box 18.3 for explanation. (Images courtesy of Gilson, Inc.)

BOX 18.3 PROCEDURE TO OPERATE A MANUAL MICROPIPETTE (IN THE USUAL MODE)

(Refer to Figure 18.13.)

1. ***Set the micropipette (if it is adjustable to different volumes) to the desired volume.***
2. ***Attach a disposable tip to the micropipette shaft.*** *Ensuring the insertion is vertical, press firmly to ensure an air-tight seal.*
3. ***Optional: Prewet the tip by aspirating and dispensing the solution to be measured into a waste container or back into the original solution.***

    Prewetting leaves a thin film of the liquid to be measured on the inside of the tip. Some operators prewet the tip; others do not. There is evidence that better accuracy is obtained with a prewetted tip. Since prewetting the tip will affect its performance, it is important that operators are consistent in whether they prewet the tip or not.
4. ***Hold the micropipette vertically.*** Manufacturers note that holding a micropipette at an angle greater than 20° can cause too much liquid to be drawn into the tip resulting in inaccurate measurements. Observe in Figure 18.13b that the micropipette is consistently held in a vertical position.
5. ***While observing the tip and the sample, depress the plunger to the first stop, and place the tip in the liquid.***

    For the utmost accuracy, it is recommended that the tip be immersed into the sample a specified distance during aspiration. This distance may be specified by the manufacturer. Otherwise, general guidelines are the following:
    * For volumes of 1–100 µL, immerse the tip 2–3 mm.
    * For volumes of 101–1,000 µL, immerse the tip 2–4 mm.
    * For volumes of 1001–10,000 µL, immerse the tip 3–6 mm.
6. ***Allow the plunger to slowly return to its undepressed position as the sample is drawn into the tip.***
    * Wait 1 second before removing the tip from the liquid when using micropipettes that measure less than 1,000 µL.
    * When using micropipettes that measure 1,000 µL, wait 2–4 seconds before removing the tip from the liquid.
    * When using larger-volume micropipettes, wait 4 or 5 seconds.
    * These waiting periods are necessary to allow the tip to fill to the proper height.

    Note 1: *Never allow the plunger to snap up!* (If the plunger "snaps," fluid can be aspirated into the interior of the micropipette. Also, the volume measured will be incorrect.)

    Note 2: If any liquid remains on the outside of the tip, remove it carefully by downward swiping with a lint-free tissue, taking care not to wick liquid from the tip orifice.

    Note 3: Pull the micropipette straight out of the container after aspirating the sample. Do not allow the tip to touch the side of the container.
7. ***Place the tip so that it touches the side of the container into which the sample will be expelled; depress the plunger to the second stop.*** Watch as the sample is expelled to the container. Wait about 2 seconds and be certain all the liquid is expelled. Remove the tip from the vessel carefully, with the plunger still fully depressed.

    Note: It is also correct to dispense the sample directly into a solution already in the tube and to rinse the tip with the solution. Discard the tip.
8. ***Eject the tip using the third stop, tip ejector button, or other mechanisms.***

    Note: The procedure outlined in this box is the most common method of pipetting, called *forward mode* pipetting. Forward mode pipetting is recommended for aqueous samples. A different mode of pipetting will be introduced in Table 18.3. The alternative mode is recommended for nonaqueous samples.

TABLE 18.3

Avoiding Error in the Operation of Air Displacement Digital Micropipettes

1. *Avoid damaging the micropipette:*
   - Never drop a micropipette.
   - Never rotate the volume adjuster of an adjustable micropipette beyond the upper or lower ranges of the instrument.
   - Never pass a micropipette through a flame.
   - Never use a micropipette without a tip, thus allowing liquid to contaminate the shaft assembly.
   - Never lay a filled micropipette on its side, thus allowing liquid to contaminate the shaft assembly.
   - Never immerse the barrel of a micropipette in liquid; only immerse the tip.
   - Never allow the plunger to snap up when liquid is being aspirated.
   - Never autoclave a micropipette unless the manufacturer says it is safe to do so.

2. *Use the proper disposable tip*:
   - Make sure the tip matches the micropipette. Tips come in various sizes for different sizes and brands of pipettes. A poorly fitting tip will leak and deliver inaccurate volumes.
   - Be sure the tip is firmly on the micropipette. If the tip is not firmly attached, the sample may leak and the volume dispensed will be too low.
   - Use an autoclaved (sterile) tip where appropriate.
   - Specialized tips are available for specific purposes, such as loading electrophoresis gels.
   - Replace the tip if liquid adheres inside it and does not drain easily.

3. *Move the plunger slowly and evenly.* Uncontrolled aspiration can lead to air bubbles in the tip, aerosol formation, and contamination of the pipette shaft.

4. *Wait a second or two after taking up a sample to allow complete filling.* (See Box 18.3 for details.)

5. *Hold the pipette as straight up and down as possible when taking up sample.* If the micropipette is tilted, too much liquid may be drawn in.

6. *Watch the sample as it comes into and leaves the tip.* The liquid sometimes does not successfully enter the tip or is not expelled completely. If an air bubble is observed, return the sample to its original vessel and try again, immersing the tip to the proper depth and pipetting more slowly. If an air bubble returns, try a new tip. Check for drops remaining in the tip after the liquid is expelled. It may be possible to rinse the tip with the liquid into which the sample is being expelled.

7. *Avoid cross-contamination (i.e., contaminating one sample with material from another).* Change tips every time a different material is pipetted. If the same material is being added to a number of tubes, it is possible, with care, to use the same tip repeatedly. To avoid cross-contamination, expel the drop onto the side of each tube and watch to be sure the tip does not touch the liquid already in the tube.

8. *Store micropipettes upright in racks designed for this purpose to avoid the possibility of liquid running back into the pipette shaft.*

9. *Micropipettes are calibrated with water and therefore are not accurate when used to measure liquids with densities that differ greatly from that of water.* It is possible to determine the actual volume of a liquid of any density delivered by a micropipette at a particular setting. This determination is done empirically by setting the pipette to a particular volume, dispensing the liquid, weighing it, and determining the actual volume of liquid dispensed at that pipettor setting. The weight of the liquid is converted to its volume based on its density. (See the Handbook of Chemistry and Physics, CRC Press, 2020, to determine the density of a liquid.) A table can then be prepared for the liquid of interest that shows the volume dispensed at each setting of the micropipette.

10. *Maintain the room humidity between 40 and 60%, if possible.* If the room humidity is too low, evaporation can occur quickly and may affect small-volume measurements.

11. *Maintain the micropipette and sample at room temperature, if possible.* If the micropipette and the sample are not at room temperature (e.g., if the measurements are made in a cold box), then the volumes delivered by the micropipette will be inaccurate. For the most accurate measurements, the micropipette, tips, and sample must all be at the same temperature. Do not pipette liquids with temperatures above 70°C or below 4°C.

12. *Avoid using a micropipette at the bottom of its range.* Manufacturers suggest that the working range of a micropipette is between 35 and 100% of the total volume of the device. For example, it is recommended that a 100 μL micropipette not be used to dispense volumes below 35 μL.

(Continued)

TABLE 18.3 (*Continued*)

Avoiding Error in the Operation of Air Displacement Digital Micropipettes

13. *Nonaqueous liquids can be difficult to pipette accurately and reproducibly. Viscous liquids, such as glycerol, do not flow easily into tips, and so the tip may not fill to the proper height. Liquids with a high vapor pressure may cause the tip to leak. Micropipettes are calibrated with water, and so liquids that have a density different than water will not be pipetted accurately.* Pipetting accuracy for these types of liquids may be improved by the following:

   • If available, use a positive displacement micropipette.
   • Prewet the tip when pipetting volatile liquids and viscous samples. Fill the tip and empty it once or twice before the desired volume is taken up and dispensed.
   • If an air displacement pipette must be used for a volatile or viscous liquid, a technique called "reverse mode" pipetting may be effective. This is accomplished by depressing the plunger to the *second stop* to take up the sample. Release the plunger after taking up the sample. The tip is then touched to the side of the receiving container, and the sample is expelled by depressing the plunger to the *first stop*. In this method, the liquid remaining in the tip is not "blown out," but rather is discarded. Reverse mode pipetting is shown in Figure 18.14.

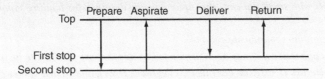

FIGURE 18.14   Reverse pipetting mode. (Diagram modified from one in ASTM Standard E 1154-14, "Standard Specification for Piston or Plunger Volumetric Apparatus.")

18.4.2.2 Factors That Affect the Accuracy of Manual Micropipettes

A variety of factors affect the accuracy of volume measurements by micropipettes:

1. **The operator's technique is the most important factor affecting the performance of a manual micropipette device.** If the operator does not operate the micropipette consistently, smoothly, and correctly, it will not accurately deliver the specified volume. Refer to Box 18.3 and Table 18.3.
2. The physical and chemical properties of the liquids being measured will affect the volumes delivered.
3. Measurements are affected by the environment in which they are made.
4. **The condition of a micropipette will affect its performance.** Basic micropipette maintenance will be discussed below.

18.4.2.3 Tips

Air displacement micropipettes require a disposable tip; the choice of tip can have a major impact on pipetting results. Some items to consider when choosing tips are summarized in Table 18.4.

18.4.3 CONTAMINATION AND MICROPIPETTES

Improper use of micropipettes can result in liquids being accidentally drawn into the micropipette assembly. Such liquids may contaminate later experiments, damage the micropipette, or put the operator at risk. **Aerosols** *are fine liquid droplets that remain suspended in the air.* Aerosols are readily produced during pipetting, and even proper use of a micropipette can result in aerosol contaminants being drawn into the instrument. Aerosols can be a serious problem if they contain pathogens, toxic materials, or radioactive substances. (Refer also to Section 10.2.5 in the Safety Unit.)

The polymerase chain reaction (PCR) is an extremely sensitive method of amplifying and detecting minute levels of DNA present in samples. If DNA from one PCR sample accidentally gets into the mixture for another sample, it is likely to adversely affect the results of the test. *DNA from one sample that contaminates another sample is called* **carryover**, and micropipettes are a major source of such contamination.

TABLE 18.4

Selecting the Proper Tips For Air Displacement Micropipettes

1. *There are different sizes of tips to match different models of micropipette; be sure to choose the right one.* Some manufacturers color-code their tips and micropipettes to avoid confusion.

2. *Micropipette manufacturers sell tips to match their devices.* It is possible to purchase less-expensive, generic tips in bulk, but these may not seal properly on every model of micropipette, may leak, and may not dispense liquids as accurately as tips recommended by the manufacturer. Generic tips are also more likely to be constructed using plastics that have contaminants that can leach from the plastic into the sample.

3. *Special wide-bore tips are useful to reduce shearing when pipetting DNA and other large molecules and intact cells.* These tips may, however, be susceptible to error due to changes in barometric pressure. Smaller-bore tips may be better for reaching inside small tubes or vials.

4. *Tips can be purchased loose in bags, or mounted in racks that sit inside plastic boxes.* (See Figure 18.12c.) Racked tips can be mounted on the end of a micropipette without touching them, which is desirable to avoid contamination.

5. *It is common practice to autoclave boxes of tips to sterilize them.* When removing an autoclaved tip from its box, minimize the time the box is left open to reduce the exposure of the remaining tips to the environment.

6. *Longer tips are available with small flat ends that are especially designed for loading samples into electrophoresis gels.*

7. *Some tips have calibration lines to indicate volume.*

8. *For molecular biology applications, it may be desirable to buy tips that are certified to be free of endotoxins (contaminants from bacteria), DNase (an enzyme that degrades DNA), and RNase (an enzyme that degrades RNA).* This is particularly important in pharmaceutical and biomedical research and testing facilities.

9. *Tips with filters are available to avoid contamination, as described in more detail below.*

To avoid micropipette contamination and carryover, consider the following:

1. *Manufacturers produce special tips that have a filter barrier that keeps aerosols and liquids from entering the micropipette assembly (Figure 18.15).* These tips are more expensive than normal tips, but they may be a good investment in certain circumstances.

2. *It is helpful to have micropipettes dedicated to a particular technique and used by only one operator.*

3. *Proper operation of the micropipette, as described in Box 18.3, is essential to avoid contamination problems.*

18.4.4 PIPETTING METHODICALLY

It is frequently necessary to pipette more than one liquid into a series of tubes. Each tube may receive the same ingredients, or one or more components may vary from tube to tube. Although this task seems simple, it is easy to make a mistake when doing repetitive work. Moreover, it is necessary to work quickly

if unstable ingredients are being added. There are electronic, computer-controlled pipettes for repeated pipetting that help ensure consistency in volume and reduce operator fatigue. The operator, however, must

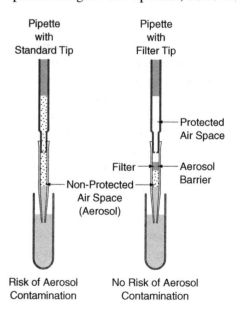

FIGURE 18.15 Avoiding carryover and contamination of a micropipette. A normal tip, shown on the left, is not designed to prevent aerosol contamination. The tip on the right has a barrier filter.

BOX 18.4   PIPETTING METHODICALLY AND ERGONOMICALLY

1. ***Plan exactly what will go into each tube.*** A table is a convenient way to display the ingredients of each tube.
2. ***For documentation, record the identity of the micropipette to be used.*** Information about its history (e.g., dates of service, repairs, and calibration checks) should be available.
3. ***Obtain all materials required and arrange them for easy access.*** Remove unnecessary clutter. A container for used tips should be easy to reach. Frequently used materials should be in front to avoid reaching.
4. ***Sit at a height such that you can rest your elbow on the work surface.***
5. ***Label all tubes carefully before the addition of liquids.***
6. ***Place tubes to be filled in the front of the rack; move each one back after material is added to it.***
7. ***Check off components that have been added.***
8. ***Change tips each time a new material is pipetted and whenever a tip touches components previously added to a tube.*** Discard any tip that does not drain completely.
9. ***After adding all the components to a tube, be sure the liquids are well mixed, are not adhering to the tube walls, and are in the bottom of the tubes.*** Mixing may be accomplished by inverting the tubes, tapping them with your finger, or using a vortex mixer (a simple mixing device that agitates tube contents). Note, however, that some materials, such as long strands of DNA, are damaged by vortexing and rough shaking. Fragile samples can be mixed by gently inverting the tubes or gently passing the samples up and down through a wide-bore micropipette tip. Brief centrifugation will bring the liquid to the bottom of the tube. Small centrifuges for this purpose are commonly found in laboratories.
10. ***Take time to relax.*** If possible, switch periodically to different types of work to avoid fatigue from repetitive pipetting. Let go of the pipette periodically to relieve stress on fingers and hands. Take short breaks when possible to change sitting position and to relax arms and shoulders. (See also Section 8.4.)
11. ***If using micropipettes frequently, consider investing in a motorized, ergonomically designed device.***

---

still be methodical to avoid errors. Box 18.4 includes suggestions on how to pipette multiple ingredients into many tubes methodically.

18.4.5 VERIFYING THAT MICROPIPETTES ARE PERFORMING ACCORDING TO THEIR SPECIFICATIONS

Micropipettes are calibrated by the manufacturer when they are made, but they can become less accurate as they are used. The performance of micropipettes, therefore, should be verified periodically. In some laboratories, users check their own micropipettes, some organizations regularly send their instruments back to the manufacturer to be inspected and repaired, and some facilities have in-house specialists for this purpose. Laboratories that are compliant with formal quality systems must have written policies for micropipette performance evaluation and routine maintenance.

There are no set rules for how often micropipette calibration should be checked. If a micropipette is dropped or known to be damaged, it should immediately be taken out of use and checked. In some laboratories, micropipettes are routinely checked and possibly calibrated every month, or even more regularly. In other laboratories service is performed every 3 or 6 months, and in some laboratories once a year. The ideal frequency depends on how the device is used, the applications for which it is used, the number of operators who use it, the nature of the liquids dispensed, and the recommendations of the manufacturer. It also depends on the risk associated with a poorly functioning device. Consider the consequences if a micropipette is found to be out of specification. This means that any work performed between performance tests might have been inaccurate. In this situation, will all work performed between performance tests need to be repeated? If so, more frequent performance checks are advisable. A survey of companies that calibrate micropipettes for clients revealed that, on average, 36% of the micropipettes sent in for routine calibration failed a test of accuracy and precision when they arrived. (Pavlis, Robert. "Surprising Statistics on Pipet Performance." *American Laboratory,* March 2004, pp. 8–10.) This means that these poorly performing micropipettes were used in the clients' laboratories for an indefinite period of time while users were unaware of the problem.

Micropipette performance is most often evaluated using a **gravimetric** (*which means "relating to weight"*) procedure. Gravimetric methods take advantage of the high accuracy and precision attainable with modern balances. This method is described in Box 18.5. Briefly, during gravimetric testing the micropipette is set to a desired volume. *This desired volume is termed the* **nominal volume.** The micropipette is then used to dispense purified water. The volume of water that was actually dispensed is determined by weighing the dispensed liquid on a high-quality, well-maintained balance. A calculation is performed to convert the weight of the dispensed water to a volume value. If the volume-measuring device is properly operating, then the nominal volume will be the same as the volume that was actually dispensed. Most laboratories have high-quality balances and can evaluate their micropipettes by a gravimetric procedure. (It is also possible to use spectrophotometric methods to evaluate micropipettes, but those methods are not covered here.)

Performance evaluation procedures test the accuracy and precision of the micropipette. These accuracy and precision values can be compared with the manufacturer's or ISO specifications (Table 18.5), or

TABLE 18.5
Micropipette Specifications

Model	Volume (µL)	Gilson Maximum Permissible Systematic Error (µL)	Gilson Maximum Permissible Random Error (µL)	ISO 8655 Maximum Permissible Systematic Error (µL)	ISO 8655 Maximum Permissible Random Error (µL)
P2	0.2	± 0.024	≤ 0.012	± 0.08	≤ 0.04
	0.5	± 0.025	≤ 0.012	± 0.08	≤ 0.04
	2	± 0.030	≤ 0.014	± 0.08	≤ 0.04
P10	1	± 0.025	≤ 0.012	± 0.12	≤ 0.08
	5	± 0.075	≤ 0.030	± 0.12	≤ 0.08
	10	± 0.100	≤ 0.040	± 0.12	≤ 0.08
P20	2	± 0.10	≤ 0.030	± 0.2	≤ 0.1
	5	± 0.10	≤ 0.040	± 0.2	≤ 0.1
	10	± 0.10	≤ 0.050	± 0.2	≤ 0.1
	20	± 0.20	≤ 0.060	± 0.2	≤ 0.1
P100	20	± 0.35	≤ 0.10	± 0.8	≤ 0.3
	50	± 0.40	≤ 0.12	± 0.8	≤ 0.3
	100	± 0.80	≤ 0.15	± 0.8	≤ 0.3
P200	50	± 0.50	≤ 0.20	± 1.6	≤ 0.6
	100	± 0.80	≤ 0.25	± 1.6	≤ 0.6
	200	± 1.60	≤ 0.30	± 1.6	≤ 0.6
P1000	100	± 3.0	≤ 0.6	± 8.0	≤ 3.0
	500	± 4.0	≤ 1.0	± 8.0	≤ 3.0
	1000	± 8.0	≤ 1.5	± 8.0	≤ 3.0

Notes: The first column is the model of the measuring device. The second column indicates the *volume that the device is set to dispense,* the **nominal volume** (i.e., the "true volume"). The third column is the permissible systematic error, which is a quantitative evaluation of the accuracy of the device. The systematic error is expressed here as the **absolute error,** *the difference between the dispensed volume and the nominal volume.* It is generally determined based on the average of 10 measurements. The fourth column is the **random error,** *which is a quantitative evaluation of the precision of the device.* The random error is expressed as the standard deviation of a series of measurements, usually 10. The fifth and sixth columns are from the ISO document, ISO 8655-2, which specifies general requirements for piston-operated volumetric apparatus (micropipettes). (Reproduced courtesy of Gilson, Inc. Specifications are subject to change without notice.)

with the requirements of the laboratory's quality procedure. If a micropipette is performing outside of its specifications for accuracy, then it may be possible to recalibrate it using directions from the manufacturer. In other cases, the instrument will need to be repaired by the manufacturer or an in-house specialist.

BOX 18.5   A GRAVIMETRIC PROCEDURE TO DETERMINE THE ACCURACY AND PRECISION OF A MICROPIPETTE

(Based primarily on ASTM Method E 1154, "Standard Specification for Piston or Plunger Operated Volumetric Apparatus.")

Summary of the method: This is a general procedure to verify that a micropipette is functioning properly by checking the accuracy and precision of the volumes it delivers. The procedure is based upon the determination of the weights of water samples delivered by the instrument. The weight of the water is converted to a volume based on the density of water.

NOTES

a. Clean the micropipette according to the manufacturer's instructions before checking its performance.

b. The analytical balance used to check the micropipette must be well maintained and properly calibrated. It must also be in a draft-free, vibration-free environment. It is recommended that the balance meet the minimum requirements shown in the table shown later in this Box.

c. The water used should be purified and degassed to remove air. Discard water after one use.

d. Document all relevant information including date, micropipette serial number, temperature when the check was performed, and name of person performing evaluation.

e. Allow at least 2 hours for the micropipette, tips, vials, and water to equilibrate together to room temperature.

f. This gravimetric procedure depends on converting a weight measurement to a volume value. To make this conversion with the utmost accuracy, it is necessary to correct for the following: the evaporation of water during the test procedure, the exact temperature of the water, the barometric pressure at the time of measurement, and the buoyancy effect. These corrections are found in the notes later in this Box. **The calculation for converting weight to volume that is shown in the body of this Box is a commonly used approximation that does not correct for all these factors.** If the performance verification is being performed to meet external quality requirements, it is likely to be necessary to apply the corrections.

g. Replicate number: For a more thorough check of the micropipette performance, 10 replicate measurements at each volume are recommended. For a quick check of the micropipette, 4 or 6 replicate measurements are sufficient.

h. Perform the following procedure as quickly as possible, but do not compromise the consistency of volume delivery and technique. The time required to make each measurement should be as consistent as possible because evaporation will affect the results.

To reduce the effects of evaporation it is recommended to:

  i. Keep the humidity of the room between 50% and 60%.

  ii. Surround the weighing vessel with a reservoir of water.

i. Computer programs are available that automatically perform the required calculations. Computer-based calibration systems that interface directly with the balance used for calibration are also commercially available.

PROCEDURE

1. *Set the micropipette to deliver a particular volume, referred to as the "nominal volume."*

2. *Optional: If corrections are to be applied (see Note f above), then measure and record the barometric pressure in the environment and the temperature of the water.*

3. *Place a small amount of water in a weighing vessel, such as a vial. The exact amount of water does not matter, but should be about 0.5 mL.*

(Continued)

**BOX 18.5 (*Continued*)   A GRAVIMETRIC PROCEDURE TO DETERMINE
THE ACCURACY AND PRECISION OF A MICROPIPETTE**

4. *Tare the balance to the vial.* (Handle the vial with tongs.)
5. *Optional: If you customarily prewet the tip, then prewet the tip by aspirating one volume of purified water and dispensing it into a waste container.*
6. *Pipette the selected nominal volume of water into the pre-tared vial using proper technique.*
7. *Weigh the vial and record the weight of the delivered water.*
8. *Repeat the measurement 4 or 10 times by repeating Steps 3–7.*
9. *Optional: If a correction for evaporation is to be applied, then perform a control blank by repeating Steps 3–4 exactly as in a normal weighing, but without actually measuring any liquid into the tared vial.* The evaporation check is recommended when testing small volume micropipettes. To determine evaporation, reweigh the blank after leaving it on the balance for a period of time equal to the elapsed time of the test run to see how much weight has been lost. Under normal conditions, this value is reported to be between –0.01 mg and –0.03 mg.
10. *For adjustable micropipettes that can be set to different volumes, it is good practice to check the micropipette at three volumes: the maximum capacity of the micropipette, 50% of maximum capacity, and 10% of maximum capacity.* Perform the entire procedure at each of the three volumes.

CALCULATIONS

1. *Calculate the mean weight of the water.*
2. *Convert the mean water weight to the mean volume measured.*

Assume that the density of water is 0.9982 g/mL (its density at 20°C), then 1 μL of water weighs 0.9982 mg. Then:

$$\text{Mean Volume} = \frac{\text{Mean Weight}}{\text{Density}} = \frac{\text{Mean Weight (in mg)}}{0.9982 \text{ mg} / \mu L}$$

Alternatively, for highest accuracy, apply the corrections in Step 5 below.

3. *Determine the accuracy of the micropipette as follows:*

$$\text{Absolute Error} = \text{Mean Volume Measured} - \text{Nominal Volume}$$

or

$$\text{Percent Error} = \frac{\text{Mean Volume Measured} - \text{Nominal Volume}}{\text{Nominal Volume}} \times (100\%)$$

4. *Determine the precision of the micropipette by calculating the standard deviation (SD) or the coefficient of variation (CV) as follows:*

$$SD = \pm \sqrt{\frac{\sum (X_i - \text{mean})^2}{n-1}}$$

where: n = the number of measurements, X_i = each volume measurement,

$$CV = \pm \frac{\text{standard deviation} (100\%)}{\text{mean}}$$

(Continued)

5. *Alternative to Step 2: Calculations to convert the mean water weight to the mean volume taking into consideration corrections for evaporation, barometric pressure, temperature, and buoyancy:*
   i. *Calculate the evaporation, e, based on the loss in weight of the blank.*
   ii. *Calculate the mean volume of the water as follows:*

$$\text{Mean volume} = (\text{mean weight} + e)(z)$$

where: mean weight = the mean weight of the 4 or 10 repeats at a given nominal volume
   e = the evaporation
   z = a conversion factor in microliters per milligram that incorporates the buoyancy correction for air at the test temperature and barometric pressure. The values for z are found in the following table.
   iii. *Continue with the calculations for accuracy and precision as shown earlier.*

Temperature °C	AIR PRESSURE HPA					
	800	853	907	960	1013	1067
15	1.0018	1.0018	1.0019	1.0019	1.0020	1.0020
15.5	1.0018	1.0019	1.0019	1.0020	1.0020	1.0021
16	1.0019	1.0020	1.0020	1.0021	1.0021	1.0022
16.5	1.0020	1.0020	1.0021	1.0022	1.0022	1.0023
17	1.0021	1.0021	1.0022	1.0022	1.0023	1.0023
17.5	1.0022	1.0022	1.0023	1.0023	1.0024	1.0024
18	1.0022	1.0023	1.0024	1.0024	1.0025	1.0025
18.5	1.0023	1.0024	1.0025	1.0025	1.0026	1.0026
19	1.0024	1.0025	1.0025	1.0026	1.0027	1.0027
19.5	1.0025	1.0026	1.0026	1.0027	1.0028	1.0028
20	1.0026	1.0027	1.0027	1.0028	1.0029	1.0029
20.5	1.0027	1.0028	1.0028	1.0029	1.0030	1.0030
21	1.0028	1.0029	1.0030	1.0030	1.0031	1.0031
21.5	1.0030	1.0030	1.0031	1.0031	1.0032	1.0032
22	1.0031	1.0031	1.0032	1.0032	1.0033	1.0033
22.5	1.0032	1.0032	1.0033	1.0033	1.0034	1.0035
23	1.0033	1.0033	1.0034	1.0035	1.0035	1.0036
23.5	1.0034	1.0035	1.0035	1.0036	1.0036	1.0037
24	1.0035	1.0036	1.0036	1.0037	1.0038	1.0038
24.5	1.0037	1.0037	1.0038	1.0038	1.0039	1.0039
25	1.0038	1.0038	1.0039	1.0039	1.0040	1.0041
25.5	1.0039	1.0040	1.0040	1.0041	1.0041	1.0042
26	1.0040	1.0041	1.0042	1.0042	1.0043	1.0043
26.5	1.0042	1.0042	1.0043	1.0043	1.0044	1.0045
27	1.0043	1.0044	1.0044	1.0045	1.0045	1.0046
27.5	1.0044	1.0045	1.0046	1.0046	1.0047	1.0047
28	1.0046	1.0046	1.0047	1.0048	1.0048	1.0049
28.5	1.0047	1.0048	1.0048	1.0049	1.0050	1.0050
29	1.0049	1.0049	1.0050	1.0050	1.0051	1.0052
29.5	1.0050	1.0051	1.0051	1.0052	1.0052	1.0053
30	1.0052	1.0052	1.0053	1.0053	1.0054	1.0055

(Continued)

BOX 18.5 (*Continued*)   A GRAVIMETRIC PROCEDURE TO DETERMINE THE ACCURACY AND PRECISION OF A MICROPIPETTE

Minimum Balance Requirements for Gravimetric Test of a Micropipette

Test Volume (µL)	Balance Readability (mg)	Standard Deviation (mg)
≥ 1	≤ 0.001	≤ 0.002
≥ 11	≤ 0.01	≤ 0.02
≥ 101	≤ 0.1	≤ 0.1
≥ 1,000	≤ 0.1	≤ 0.2

Example

A digital micropipette with a range of 100–1,000 µL is evaluated for performance.

The micropipette is set for 1,000 µL, and high-purity water at 20°C is dispensed four times. The dispensed water is weighed on an analytical balance. The four aliquots of water weigh:

$$0.9931 \text{ g} \quad 0.9948 \text{ g} \quad 0.9928 \text{ g} \quad 0.9953 \text{ g}$$

The micropipette is then set for 500 µL, and four aliquots of water are dispensed that weigh:

$$0.4960 \text{ g} \quad 0.4964 \text{ g} \quad 0.4970 \text{ g} \quad 0.4968 \text{ g}$$

The micropipette is then set at 100 µL, and four aliquots of water are dispensed that weigh:

$$0.1010 \text{ g} \quad 0.0901 \text{ g} \quad 0.0959 \text{ g} \quad 0.0947 \text{ g}$$

Determine the accuracy and the precision of this pipette (as defined in this box) at all three volumes as shown in this box under the heading "Calculations" on p. 462. Compare the accuracy and precision with the manufacturer's specifications in Table 18.5.

For a Nominal Volume of 1,000 µL:

Step 1. Calculate the mean weight of the water from the four aliquots

$$\frac{0.9931 \text{ g} + 0.9948 \text{ g} + 0.9928 \text{ g} + 0.9953 \text{ g}}{4} = 0.9940 \text{ g} = 994.0 \text{ mg}$$

Step 2. Calculate the mean volume of the water:

$$\text{Volume} = \frac{994.0 \text{ mg}}{0.9982 \dfrac{\text{mg}}{\text{µL}}} \approx 995.8 \text{ µL}$$

Step 3. Calculate the percent error:

$$\text{percent error} = \frac{995.8 \text{ µL} - 1,000 \text{ µL} \ (100\%)}{1,000 \text{ µL}} = -0.42\%$$

Step 4. Calculate the standard deviation. The four weight measurements converted to volumes are:

$$\frac{993.1 \text{ mg}}{0.9982 \dfrac{\text{mg}}{\text{µL}}} \approx 994.890 \text{ µL}$$

(*Continued*)

$$\frac{994.8 \text{ mg}}{0.9982 \frac{\text{mg}}{\mu L}} \approx 996.593 \; \mu L$$

$$\frac{992.8 \text{ mg}}{0.9982 \frac{\text{mg}}{\mu L}} \approx 994.590 \; \mu L$$

$$\frac{995.3 \text{ mg}}{0.9982 \frac{\text{mg}}{\mu L}} \approx 997.095 \; \mu L$$

$$SD \approx \sqrt{\frac{(994.9 - 995.8)^2 + (996.6 - 995.8)^2 + (994.6 - 995.8)^2 + (997.1 - 995.8)^2}{3}}$$

$$= \sqrt{\frac{0.81 + 0.64 + 1.44 + 1.69}{3}} \approx 1.24 \; \mu L$$

$$CV = 1.24 \, \mu L / 995.8 \, \mu L \times 100\% \approx \mathbf{0.12\%}$$

For a Nominal Volume of 500 µL:

Mean Volume ≈ **497.4 µL**   % Error ≈ **−0.51%**   SD ≈ **0.44 µL**   CV ≈ **0.09%**

For a Nominal Volume of 100 µL:

Mean Volume ≈ **95.6 µL**   % Error ≈ **−4.4%**   SD ≈ **4.49 µL**   CV ≈ **4.69%**

Summary of the Results for All Three Volumes

(µL)	Mean Error (Relative) (%)	Mean Error (Absolute[a]) (µL)	Precision (CV) (%)	Precision (SD) (µL)
1,000	−0.42	−4.2	0.12	1.24
500	−0.51	−2.6	0.09	0.44
100	−4.4	−4.4	4.69	4.49

[a] Mean error (absolute) = (mean volume − nominal volume)

Comparing these values with Table 18.5 shows that the results at 1,000 and 500 µL are within the manufacturer's specifications, but the results at 100 µL are not. How this problem is handled depends on the policies in the laboratory.

18.4.6 CLEANING AND MAINTAINING MICROPIPETTES

Undamaged micropipettes require periodic performance evaluation but minimal maintenance. Routine maintenance includes cleaning the exterior of the micropipette, and some manufacturers suggest replacing internal seals and O-rings periodically (Figure 18.16).

Micropipettes that are heavily used or are roughly handled may become damaged. Common symptoms of damage include the following:

- **The micropipette leaks or drips.**
- **Visual inspection shows that the tip does not take up the right volume of liquid.**
- **The plunger jams, sticks, or moves erratically.**

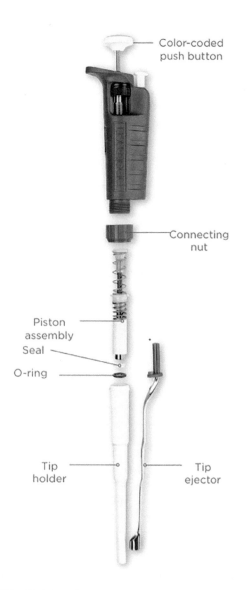

Color-coded
push button

Connecting
nut

Piston
assembly

Seal

O-ring

Tip
holder

Tip
ejector

FIGURE 18.16 The parts of a manual micropipette. Consult the manufacturer's guide for other models. (Photo courtesy of Gilson, Inc.)

Micropipettes should be frequently tested for leaks. A simple method to do so is to adjust the device to its highest volume setting and place a disposable tip on it. Aspirate purified water that is at the same temperature as the micropipette. Place the device vertically in a vibration-free stand. No droplets should appear on the tip before a period of 20 seconds has elapsed; if they do, there is a leak problem. Leaking is frequently caused simply by poorly fitting pipette tips. Replacing the tip will sometimes help; other times switching brands of tip will resolve this problem. A poorly fitting tip is also a common reason a micropipette aspirates the wrong volume.

Many users return damaged micropipettes to the manufacturer for repair. If you decide to service a micropipette yourself, determine whether the instrument is under warranty, and, if so, whether servicing

it will invalidate the warranty. Consult the manual for your micropipette if you decide to repair the device yourself. Table 18.6 has advice regarding cleaning and simple repairs that are applicable to various styles of micropipette. Table 18.7 has troubleshooting tips.

After a micropipette is serviced, it must be checked to see that it is performing according to its specifications. The gravimetric method described in Box 18.5 is suitable for this purpose.

18.5 AUTOMATED LIQUID HANDLING DEVICES

A typical biotechnology procedure might involve adding microliter volumes of multiple aqueous solutions to each sample – which is why biotechnologists must be able to properly operate a micropipette. However, when there are hundreds or thousands of samples, as is often the case, a human wielding a micropipette is not ideal. It is not easy for a person to track and accurately dispense liquids to thousands of samples, nor is it healthy for a person to perform repetitive micropipetting over a long period of time. Engineers therefore invented **liquid handling robots** *that can be programmed to accurately dispense multiple liquids into numerous samples.* Liquid handling robots are becoming increasingly common in testing laboratories, production settings, and research laboratories. Robots are routinely used in drug discovery laboratories, where thousands of drugs are tested on cells or other samples; in genomics laboratories where thousands of DNA samples are sequenced; in microarray production, where thousands of nanogram spots of specific DNA fragments are applied to small chips; and in many other biotechnology applications.

There are various types of liquid handling robots designed for different applications. There are, for example, robots with varying numbers of channels (where each channel has a pipette tip). Some robots allow each channel to be individually programmed to dispense different liquids or different volumes, while other robots do not have that flexibility. Some robots have a self-correcting feature that enables them to sense whether the correct volume has been aspirated. The robot shown in Figure 18.17 is analogous to an 8-channel micropipette. This robot is designed to add liquids to samples in 96-well plates. The robot is programmed with a computer to aspirate liquids from a reservoir and then move along a track dispensing a set volume of the liquid to specific wells on every plate.

TABLE 18.6
Notes on Cleaning and Simple Repair of Micropipettes

1. *Consult the manufacturer's manual before attempting to clean or repair a micropipette.* Follow the manufacturer's recommendations, as applicable.
2. *Use safety precautions when handling instruments that have been used to pipette hazardous materials.*
3. *Manufacturers recommend cleaning the exterior of micropipettes with purified water alone, or laboratory detergent solution, or 10% bleach solution, and/or with 60% isopropanol (depending on the type of contamination), followed by a rinse in purified water.* Dry with a lint-free cloth or tissue. Use care around the display window to avoid fogging or staining it. Do not allow liquids to enter the shaft.
4. *If you disassemble a micropipette, keep track of the order and orientation of the parts as you slowly remove each one.* There are springs inside the micropipette assembly, so when the micropipette is unscrewed, small parts may spring loose.
5. *If liquids are accidentally aspirated into the micropipette, it is most effective to clean the device immediately and not let liquids dry inside it.* The nose cone (tip holder) can usually be unscrewed, and its interior parts can be cleaned with distilled water. Dry the instrument after cleaning, and then reassemble it.
6. *Many types of micropipettes have internal seals and O-rings that can become worn and can be replaced by the user; see Figure 18.16.*
7. *Volatile organic compounds can cause the seals to swell, which in turn makes the piston move erratically.* Replace the seals and O-ring.
8. *Manufacturers may recommend that the piston, seals, and O-rings be greased after cleaning.* Check the manual.

TABLE 18.7
Manufacturers' Troubleshooting Tips

Problem	Possible Cause	Possible Solutions
Aspirated volume incorrect	• Incorrect operator technique	• Train operator
	• Tip holder is loose	• Tighten tip holder
	• Damaged tip holder	• Replace tip holder
	• Worn O-ring or seal	• Replace O-ring and seal
	• Damaged piston	• Clean or replace piston and replace seal and O-ring or return to manufacturer for repair
Liquid leaks from tip	• Unsuitable tip	• Use pipette manufacturer's tips
	• Tip not firmly seated	• Press tip tightly and evenly
	• Tip holder damaged	• Replace tip holder
	• Worn O-ring and seal	• Replace seal and O-ring
	• Piston contaminated	• Clean piston, replace O-ring and seal, or return to manufacturer for repair
	• Organic solvent is being dispensed	• Use a positive displacement device. Alternatively, with an air displacement device, saturate the air cushion of the pipette with solvent vapor by aspirating and dispensing solvent repeatedly. The leak may stop when the air space is saturated with vapor. Alternatively, try reverse mode pipetting. (See Table 18.3.)
Plunger jams, sticks, moves erratically	• Seal is swollen by reagent vapors	• Open pipette, ventilate, clean piston, replace seal and O-ring if necessary
	• Piston contaminated or damaged	• Clean or replace piston, seal, and O-ring or return to manufacturer for repair

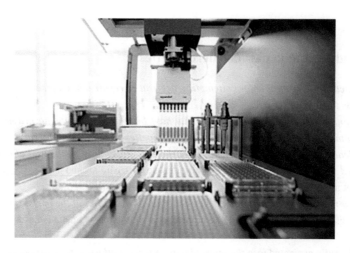

FIGURE 18.17 An automated liquid handling device. The moving head with eight pipette tips is positioned to automatically dispense liquid to multiwell plates underneath. A computer controls the movement of the moving head which aspirates and dispenses liquid. (Image courtesy of Eppendorf AG.)

Case Study: Analyzing Cancer at the Molecular Level

Cancer treatments tailored to an individual patient are an emerging application of modern genomic research. Genomic Health is a company that produces a diagnostic test kit to quantify the likelihood of breast cancer recurrence in women with newly diagnosed, early-stage breast cancer. This information is used by women and their physicians to help chart an individualized treatment strategy. Genomic Health's diagnostic test involves analyzing tissue from the patients for the expression of 21 genes that are associated with breast cancer (for example, Her2, a gene discussed in Chapter 1). Recall from Chapter 1 that when a gene is expressed, mRNA is transcribed by the cell using that gene as a template; the mRNA directs the synthesis of the protein encoded by that gene. The test method developed by Genomic Health analyzes the levels of mRNA for the 21 genes of interest in the patients' biopsy tissue. As is the case with many molecular biology procedures, their process involves a series of reactions; each reaction requires combining small volumes of liquids. The volumes of each liquid must be accurately dispensed in order to arrive at an accurate measurement of the expression levels of the genes. An inaccurate measurement of expression level can result in an erroneous diagnosis and treatment for a patient. The company is aware of the potentially severe consequences of error. They therefore use automated liquid handlers to help ensure consistency, and they regularly verify the precision and accuracy of the volumes dispensed by the robots. The company further verifies the performance of the system for each type and brand of tip that is used during dispensing. Attention to technical detail – such as the accuracy of dispensing – is critical when translating state-of-the-art research discoveries into practical clinical tools that help real patients.

Primary Sources

"When Analyzing Cancer at the Molecular Level, Microliters Matter," American Biotechnology Laboratory, November/December 2007. https://americanlaboratory.com/913-Technical-Articles/18985-When-Analyzing-Cancer-at-the-Molecular-Level-Microliters-Matter/

"American Society of Clinical Oncology 2007 Update of Recommendations for the Use of Tumor Markers in Breast Cancer." *Journal of Oncology Practice*, vol. 3, no. 6, 2007, pp. 336–339. doi:10.1200/jop.0768504.

Practice Problems

1. What is the volume of each of the liquids diagrammed in Figure 18.18?
2. What type of device(s) is used commonly in biology laboratories to measure a volume of:

a. 100 mL	**b.** 95 mL	**c.** 2 mL
d. 100 μL	**e.** 5 μL	**f.** 500 μL

3. List the common sources of error that might affect each of the measurements in question #2.
4. Each of the devices given as an answer in question #2 has a certain allowed error, or tolerance associated with it. Because of this tolerance there is some uncertainty in its measurements. Based on information given in this chapter, what is the tolerance of each device you gave as an answer to question #2?
5. Suppose you are planning to purchase a new micropipette to pipette volumes in the 100–200 μL range. You consult a catalog and find the following information for three brands of pipettor: Brand A, Brand B, and Brand C. Based on the catalog specifications, which micropipette is most accurate? Which micropipette is most precise? Which would you purchase?

	Volume Range (μL)	Accuracy (Expressed as % Error)	Precision (CV) (%)
Brand A	40–200	± 1%	0.5
Brand B	100–200	± 0.5%	0.3
Brand C	100–200	± 0.3%	0.4

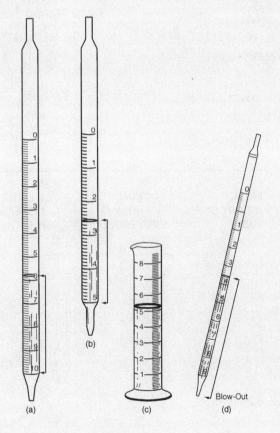

(a) (b) (c) (d) Blow-Out

FIGURE 18.18 Measuring devices for Practice Problem 1.

6. Suppose you are checking the accuracy of a micropipette. Working at 20°C you weigh 10 empty tubes.

   You set the micropipette to deliver 100 μL and carefully dispense that volume to each pre-weighed tube. You weigh the filled tubes on a calibrated balance and determine the weight of the water in each tube. The mean value for the weight of the water is 0.0994 g.
   a. The nominal volume is 100 μL. What do you expect the mean value for the weight of the water to be?
   b. What is the accuracy of your micropipette (expressed as percent error) based on this result?
   c. The manufacturer specifies that the micropipette should have an accuracy (expressed as % error) at least as good as ± 2%. Does this pipettor meet the specification for accuracy?

7. The accuracy and precision of a micropipette was checked using the procedure in Box 18.5. The micropipette was set to deliver 500 μL. A standard form was used to record the results of the evaluation. A portion of the form and the results are shown below.
   a. Fill in the rest of the form.
   b. Based on the information in Table 18.5, does this micropipette meet its specifications? Place the answer in the pass/fail blank on the form (Figure 18.19).

Clean Gene, Inc.

VERIFICATION OF PERFORMANCE OF MICROPIPETTOR REPORT
FORM 232

CALIBRATION TECHNICIAN ___*J.E.S.*___

CALIBRATION DATE ___*9/22/19*___   NEW DUE DATE FOR NEXT CALIBRATION ___*3/22/2020*___
MICROPIPETTOR ID NUMBER ___*2127*___   MANUFACTURER ___*Finestt Pipettes Inc.*___
MODEL NUMBER *micro B*   LOCATION ___*Lab #27*___   TEMPERATURE ___*20°C*___
RANGE ___*100–1000μL*___   PRIMARY USER ___*R.E.S.*___

SUMMARY:
PREVERIFICATION CLEANING AND ADJUSTMENTS ___*Changed Seals & O-rings*___
_____ PASS/FAIL _____
STATUS _____
Volume 1

NOMINAL VOLUME (μL)	TUBE NUMBER	INITIAL WEIGHT OF TUBE	WEIGHT AFTER H₂O DISPENSED	NET WEIGHT OF DISPENSED H₂O	H₂O VOLUME
500	1	1.9355 g	2.4352 g		
500	2	1.9877 g	2.4876 g		
500	3	1.9787 g	2.4789 g		
500	4	1.9850 g	2.4873 g		
500	5	1.9755 g	2.4763 g		
500	6	1.9387 g	2.4387 g		

Mean Water Volume ($\bar{x}$) _____ % INACCURACY _____ SD_____ CV _____

FIGURE 18.19   Form for Practice Problem 7.

19 The Measurement of Temperature

19.1 INTRODUCTION TO TEMPERATURE MEASUREMENT

19.1.1 THE IMPORTANCE AND DEFINITION OF TEMPERATURE

Biological systems in the laboratory are sensitive to temperature. For example, enzymes require a specific temperature for optimal activity. The polymerase chain reaction (PCR, an enzymatic process that is used to greatly amplify the quantity of specific sequences of DNA) requires that a reaction mixture containing the sample and enzymes is cycled between specific temperatures. Cells growing in culture must be incubated at a specific temperature. The calibration and operation of laboratory instruments requires temperature control. Temperature is one of the key process control variables that are closely monitored and controlled in the pharmaceutical industry. The laboratory contains many devices to control temperature, including ovens, heaters, incubators, water baths, refrigerators, freezers, freeze-dryers, and PCR thermocyclers. It is thus essential that there be convenient and accurate methods to measure temperature in the laboratory.

Although temperature is familiar to everyone and has been studied for hundreds of years, its physical basis is not obvious. Physicists now explain temperature as being related to the continuous, random motion of the molecules that make up substances. **Heat** *is a form of energy that is associated with this disordered molecular motion.* **Heating** *is the transfer of energy from an object with more random internal energy to an object with less.* For example, a burn results when a large amount of energy from a hot object is transferred to a person's skin.

Temperature *is a measure of the average energy of the randomly moving molecules in a substance. Temperature tells us which way heat will flow:* Objects with a higher temperature lose heat to objects with a lower temperature. *When two objects are at the same temperature, they are said to be in* **thermal equilibrium;** there is no net transfer of energy from one to the other. Temperature is measured by allowing a thermometer to come to thermal equilibrium with the material whose temperature is being determined. For example, taking a person's temperature involves placing a thermometer in contact with that person and waiting until the thermometer and person reach thermal equilibrium. The thermometer reading then indicates the person's temperature.

19.1.2 TEMPERATURE SCALES

19.1.2.1 A Bit of History

There are three temperature scales most commonly used in biology laboratories: Fahrenheit, Celsius (also previously called centigrade), and Kelvin. It is informative to discuss briefly how there came to be three scales in widespread use.

The Fahrenheit scale was invented in the early 1700s by Daniel Gabriel Fahrenheit. Fahrenheit made thermometers similar to a type of thermometer still used today. Fahrenheit placed mercury inside thin glass tubes. The liquid mercury moved up and down in the tube as it expanded and contracted in response to temperature changes. It is not entirely clear how Fahrenheit calibrated his thermometers, but it appears to have been more or less as follows. Fahrenheit chose two reference points. The first point was the coldest system he could produce, that is, a mixture of salt, water, and ice. He called the temperature of this mixture "zero." The second reference point was the body temperature of a person, which he called "96." (His methods were slightly in error because the temperature of a healthy person is 98.6 degrees on the Fahrenheit scale.) He positioned the thermometer in the salt, water, and ice mixture and placed a mark on the glass tube to indicate the height of the mercury. He then placed the assembly in contact with a person and similarly marked the glass tube to indicate the height to which the mercury rose. Next, he subdivided the length of the glass tube between the "zero" and "96" marks into 96 equal divisions, each of which he called a **degree.** The result was a thermometer marked in degrees according to the **Fahrenheit temperature scale.** Note that this method is based on the correct assumption that the relationship between mercury expansion and temperature is linear.

Anders Celsius also made mercury thermometers in the 1700s. Celsius chose as his fixed reference points the boiling point of water and the freezing point of water. Celsius divided the interval between his two reference points into 100 divisions, each called a degree. The result is a thermometer marked according to the **Celsius temperature scale.** It is interesting to note that Celsius originally assigned the boiling point of water "0" and the freezing point "100." This was later reversed so we now think of temperature as going down when it gets colder, rather than up.

Over the years, some people preferred and used Celsius thermometers, some used Fahrenheit, and some people invented other temperature scales based on other reference temperatures. For example, there were temperature scales based on the temperature at the first frost and the temperature of underground caverns.

The Kelvin scale was devised in the early 1800s by various scientists including William Thomson (later, Lord Kelvin). *The Kelvin scale uses a unit called a* **kelvin,** *which is the same size as a degree in the Celsius scale.* However, the Kelvin scale sets the zero point at **absolute zero,** *the temperature at which, theoretically, molecules stop their internal random motion.* Zero kelvin is −273.15°C (often rounded to −273°C); the zero point on the Celsius scale corresponds to 273.15 K. (The degree sign, "°", is not used when temperature is expressed in kelvins.) There are 100 K between the freezing and boiling points of water, just as there are 100°C between these two points.

As a result of these people's work in temperature measurement, there are now three temperature scales which you are likely to encounter, each of which assigns different temperature values to materials that are the same temperature. For example, the temperature at which water freezes is variously called 32°F, 0°C, and 273.15 K (Table 19.1 and Figure 19.1). In the

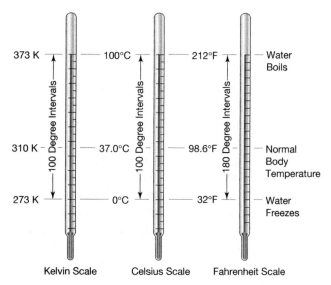

FIGURE 19.1 Comparison of Kelvin, Celsius, and Fahrenheit temperature scales.

TABLE 19.1

Comparison of the Fahrenheit, Celsius, and Kelvin Temperature Scales

Absolute zero	−460°F	−273°C	0 K
Freezing point of water	32°F	0°C	273 K
Average room temperature	68°F	20°C	293 K
Normal human temperature	98.6°F	37°C	310 K
Boiling point of water	212°F	100°C	373 K
Standard refrigerator temperature	≈ 39°F	≈ 4°C	≈ 277 K

laboratory or in a production setting, Celsius is generally the preferred scale.

19.1.2.2 Conversions from One Temperature Scale to Another

Although Celsius is the most common temperature scale used in professional settings, laboratory procedures may be written using any of the three temperature scales. It is therefore necessary to be able to convert temperatures from one scale to another. Simple equations can be used for such conversions, as shown in Box 19.1.

19.1.2.3 Fixed Reference Points and Thermometer Calibration

We have seen that temperature scales are based on two or more fixed reference points. **Fixed reference points** *are systems whose temperatures are determined by some physical process and hence are universal and repeatable.*

Like Anders Celsius, metrologists today use **phase transitions** as the references for calibration of laboratory thermometers and verification of their performance. A **phase transition** *is where a liquid turns to a solid or a vapor, or a vapor turns to liquid.* **Freezing point** *is the temperature at which a substance goes from the liquid*

BOX 19.1   CONVERSION FROM ONE TEMPERATURE SCALE TO ANOTHER

1. ***To convert from degrees Fahrenheit to degrees Celsius:***

$$°C = (°F - 32°)0.556$$

Subtract 32° from the Fahrenheit reading and then multiply the result by 0.556 or 5/9.
    Example: Convert human body temperature, 98.6°F, to degrees Celsius.

$$°C = (98.6° - 32°)0.556 \approx \mathbf{37.0°C}$$

2. ***To convert from degrees Celsius to degrees Fahrenheit:***

$$°F = (°C \times 1.8) + 32°$$

Multiply the Celsius reading by 1.8 or 9/5 and then add 32 degrees.
    Example: Convert human body temperature, 37.0°C, to degrees Fahrenheit.

$$°F = (37.0° \times 1.8) + 32° = \mathbf{98.6°F}$$

3. ***To convert from degrees Celsius to kelvin and from kelvin to degrees Celsius:***

$$°C = K - 273$$
$$kelvin = °C + 273$$

Example: Convert human body temperature, 37.0°C, to kelvin

$$kelvin = 37.0°C + 273 = \mathbf{310\,K}$$

4. ***To convert degrees Fahrenheit to kelvin:***
    Convert Fahrenheit to Celsius and then convert Celsius to kelvin.
    Example: Convert human body temperature, 98.6°F, to kelvin.

$$°C = (98.6° - 32°)0.556 \approx 37.0°C$$
$$kelvin = 37.0°C + 273 = \mathbf{310K}$$

phase to the solid phase. **Boiling point** *is the temperature at which a substance in the liquid phase transforms to the gaseous phase* (under specified conditions of pressure). People also use the **triple point** of various substances as references to define temperatures. The **triple point of water** *is a single temperature and pressure at which ice, water, and water vapor coexist with one another in a closed container.* The triple point temperature for water (at atmospheric pressure) is 0.01°C. At all other temperatures and pressures, only two phases can coexist (i.e., either water and ice or water and vapor).

The boiling and freezing points of water are adequate references for thermometers used for most biological applications. Scientists who study non-living systems use the freezing points, boiling points, and triple points not only of water but also of oxygen, hydrogen, and various metals to extend the range of temperatures at which there are fixed references. The reason that phase transitions are used for thermometer calibration is that given the proper apparatus and conditions, phase transitions always occur at a predictable temperature and pressure. It is possible, therefore, to consistently calibrate thermometers anywhere in the world.

19.2 THE PRINCIPLES AND METHODS OF TEMPERATURE MEASUREMENT

19.2.1 Overview

There are various transducers (sensors) used to measure temperature, all of which respond to the effects of temperature on a physical system. For example, liquid-in-glass thermometers measure the expansion or contraction of liquid inside a glass tube in response to temperature changes.

The types of temperature transducers used in the laboratory include the following:

- *Liquid expansion devices, also called liquid-in-glass thermometers*
- *Bimetallic expansion devices*
- *Change-of-state indicators*
- *Metallic resistance devices*
- *Thermistors*
- *Thermocouples.*

The first two types of sensors – liquid expansion and bimetallic devices – are similar in that they are both based on the fact that most materials expand as the temperature increases and contract when it decreases. These types of thermometers are relatively simple to use and inexpensive.

Liquid expansion thermometers, bimetallic devices, and change-of-state indicators are mechanical measuring instruments (i.e., they do not require electricity to make a measurement, nor do they generate an electrical signal). These devices are usually read by human eyes, so they cannot be directly interfaced with computers or electronic instrumentation. In contrast, resistance thermometers, thermistors, and thermocouples are electronic devices that generate an electrical signal and are readily interfaced with other equipment. Electronic thermometers are accurate, precise, versatile, and compact; can be used at a wide range of temperatures; and can be used for remote sensing. As these electronic thermometers have decreased in cost and become increasingly available, they are taking the place of liquid expansion thermometers in many applications.

Liquid expansion thermometers are traditionally the type most often handled in the biology laboratory, and we discuss them first. Bimetallic expansion thermometers and change-of-state indicators will then be described. The latter part of this section will outline the basic features of electronic temperature-sensing devices.

19.2.2 Liquid Expansion (Liquid-in-Glass) Thermometers

19.2.2.1 The Liquids in Liquid-in-Glass Thermometers

Liquid expansion thermometers *contain liquid that expands or contracts with temperature changes and moves up or down within a narrow glass capillary tube, or* **stem** (Figure 19.2). When the temperature of the surrounding medium increases, the liquid expands more than the glass, so the liquid moves up the stem. When the liquid and the substance whose temperature is being measured are in thermal equilibrium, the liquid stops rising. The temperature of the bulb is read by noting where the top of the liquid column coincides with a scale marked on the stem. If the temperature decreases, the liquid contracts down the stem until thermal equilibrium is reached. As was previously discussed, the scale of a liquid-in-glass thermometer is determined by marking the stem at two (or more) reference points and subdividing the interval between the marks.

Mercury or "spirit" (e.g., alcohol) were formerly the liquids used in liquid expansion thermometers. Mercury was used because it does not adhere to glass and its silvery appearance makes it easy to see. Mercury's expansion in response to temperature change is linear, and mercury thermometers could achieve a high level of accuracy and precision. Spirit-filled thermometers were (and sometimes still are) used as a safe

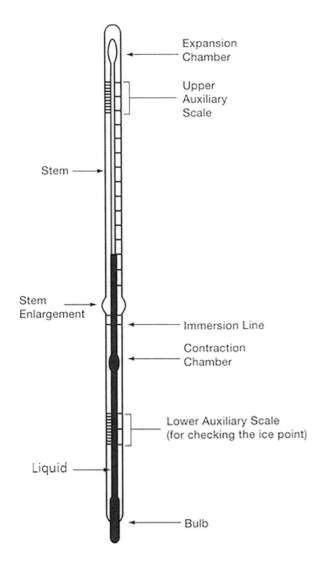

FIGURE 19.2   The parts of a liquid-in-glass thermometer.

calibration services for mercury thermometers, as part of an effort to phase out their use entirely. ASTM International followed NIST's lead by amending its standards that, in the past, required the use of mercury thermometers. ASTM International standards now permit the use of other types of liquid-in-glass devices and digital thermometers. Manufacturers also responded to the phase-out of mercury thermometers by developing proprietary liquid mixtures that provide high-accuracy measurements, similar to mercury, but with safer ingredients. As a result, it is becoming unlikely that you will work with mercury thermometers in the laboratory. You are more likely to see liquid expansion thermometers containing proprietary liquid mixtures, or digital devices.

19.2.2.2 The Construction of Liquid-in-Glass Thermometers

Liquid-in-glass thermometers have certain components; refer to Figure 19.2. These components are as follows:

- **Liquid**
- **Stem,** *a glass capillary tube through which the liquid moves as temperature changes*
- **Bulb,** *a thin glass container at the bottom of the thermometer that is a reservoir for the liquid*
- **Scale,** *marks that indicate degrees, fractions of degrees, or multiples of degrees*
- **Contraction chamber** *(not present in all thermometers), an enlargement of the capillary bore that takes up some of the volume of the liquid thereby allowing the overall length of the thermometer to be reduced*
- **Expansion chamber,** *an enlargement of the capillary bore at the top of the thermometer that prevents buildup of excessive pressure*
- **Stem enlargement** *(not present in all thermometers), a thickening of the stem that assists in the proper placement of the thermometer in a device (e.g., in an oven)*
- **Immersion line** *(not present on all thermometers), a line etched onto the stem to show how far the thermometer should be immersed into the material whose temperature is to be measured*
- **Upper and lower auxiliary scales** *(not present on all thermometers), extra scale markings at the zero degree and 100 degree areas to assist in calibration and verification of the performance of the thermometer*

alternative to mercury-containing devices. Spirit-filled thermometers contain organic liquids such as alcohol and toluene which are dyed red to make it easier to read the meniscus. Spirit-filled thermometers can be used at temperatures below −38.8°C, where mercury freezes. Spirit-filled thermometers, however, do not respond to temperature change as reliably as mercury, and these thermometers are not as accurate as mercury thermometers.

The liquids used to construct liquid expansion thermometers, which remained relatively unchanged through the careers of many generations of scientists, changed over the last ten years. Mercury is a dangerous neurotoxin, and its potentially adverse effects on individuals and the environment led many states to prohibit or limit its use. In 2011, NIST stopped providing

Liquid-in-glass thermometers come in many styles that differ in their scale divisions, range, length, accuracy, intended applications, and other qualities. For example, a thermometer for use in a refrigerator might have a range from −40°C to 25°C. A thermometer for monitoring the temperature in an autoclave might measure temperatures from 85°C to 135°C and may be specially designed to hold and display the maximum temperature reached during sterilization, even after the autoclave has cooled.

19.2.2.3 Immersion

The depth of immersion of a liquid thermometer into the material whose temperature is to be measured affects the accuracy of the measurement. This is because the liquid in the part of the stem that is not immersed is not at the same temperature as the liquid in the bulb. The manufacturer takes this difference into account when the thermometer is calibrated and indicates on the thermometer how far it is to be immersed. There are two common immersion types of

thermometer: partial immersion and total immersion. A third type, complete immersion, is less common.

A **partial immersion thermometer** *is manufactured to indicate temperature correctly when the bulb and a specified part of the stem are exposed to the temperature being measured* (Figure 19.3a). Partial immersion thermometers are easiest to read because their liquid column is not fully immersed in the medium. They are ideal for low-volume solutions. A partial immersion thermometer is marked with a line to indicate how far it should be immersed.

A **total immersion thermometer** *is designed to indicate temperature correctly when that portion of the thermometer containing the liquid is exposed to the medium whose temperature is being measured* (Figure 19.3b and c). Total immersion thermometers are more accurate than partial immersion thermometers and are suggested for applications where accuracy is critical or when a thermometer is used for calibrating other thermometers or equipment. If a total immersion thermometer is not immersed to the proper depth, there will be inaccuracies

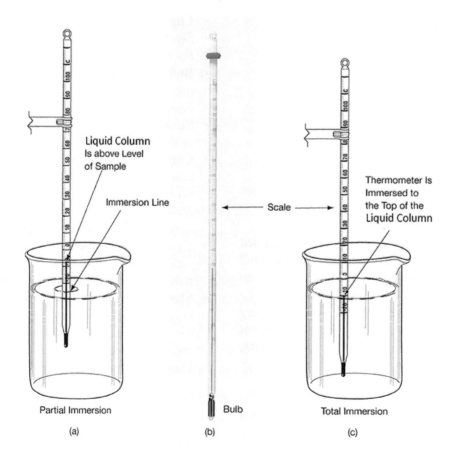

FIGURE 19.3 Partial and total immersion thermometers. (a) Partial immersion thermometer. A line on the stem indicates the required immersion depth. The thermometer is immersed only to the immersion line. (b) Total immersion thermometer. (c) A total immersion thermometer is positioned so that all the liquid within the thermometer is immersed in the material whose temperature is being measured. (Photo used with permission from Thermo Fisher Scientific, the copyright owner.)

in its readings. In general, at the temperatures used in a biology laboratory (e.g., between 0°C and 100°C), the correction is less than 1 degree. Nonetheless, it is good practice to always immerse a thermometer properly.

There are several ways to distinguish a partial from total immersion thermometer. A partial immersion thermometer has an immersion line marked on the stem; a total immersion thermometer does not. The immersion depth, in millimeters, is sometimes written on the stem of a partial immersion thermometer. Manufacturers specify the immersion type for each thermometer in their catalog (Figure 19.4).

A total immersion thermometer properly indicates temperature when that portion of the thermometer containing the liquid is exposed to the medium whose temperature is being measured. A total immersion thermometer is not intended to actually be placed inside an oven, or to be immersed *completely* in a hot liquid. Despite the name, a total immersion thermometer will be inaccurate if it is completely immersed, and there is the possibility that the thermometer will break at high temperatures if placed inside an oven or incubator. Most laboratory ovens have a slot in which the thermometer is inserted so that the proper length of the thermometer is inside the oven and the top part protrudes visibly outside the oven. There are **complete immersion thermometers** *that are designed so that the entire thermometer, from top to bottom, is exposed to the medium whose temperature is being measured.* Figure 19.5 shows a complete immersion thermometer specifically designed to be placed

inside a refrigerator. Observe that this thermometer is protected in case of breakage.

19.2.2.4 Limitations of Liquid Expansion Thermometers

Although liquid expansion thermometers are routinely used with acceptable results, they have inherent limitations. Liquid expansion thermometers have fairly wide tolerances (i.e., they may be in error by a fair amount). It is not unusual for manufacturers to specify that the tolerance for a general-purpose liquid expansion thermometer is ± 1°C or ± 5°C (± 2°F or ± 9°F). A thermometer may be in error by as much as its maximum tolerance. This means, for example, that if you are using a thermometer whose tolerance is ±5°C, and you measure the temperature of a solution as 37°C, the actual temperature of that solution may be anywhere from 32°C to 42°C.

ASTM International classifies thermometers and specifies how much error is tolerable for a particular class. Tolerances for partial immersion thermometers are greater than for total immersion thermometers. More expensive "precision thermometers" have narrower tolerances than other thermometers. To meet ASTM requirements, precision thermometers must have an error of less than 0.1°C–0.5°C, depending on their range. A precision thermometer whose range does not include 0°C will have a second **auxiliary scale** *that shows the zero point so that the performance of the thermometer can be verified with an ice point bath* (discussed later).

Other limitations of liquid expansion thermometers are described in Table 19.2.

Model	Temp Range (°C)	Divisions (°C)	Total Length (mm)	Immersion Depth (mm)
T24	24 to 38	0.05	305	76
TH01	-1 to 51	0.1	460	Total
TH101	-1 to 101	0.1	610	Total
TH20	-20 to 110	1	305	76
TH100	-100 to 50	1	305	76
TH201	-1 to 201	0.2	610	Total

FIGURE 19.4 Catalog descriptions specify how a thermometer should be immersed.

FIGURE 19.5 Fisherbrand™ Enviro-Safe™ liquid-in-glass verification thermometer. This thermometer can be used to verify the temperature in a refrigerator. Its calibration is certified to be traceable to NIST. The liquid-filled bottle mimics the temperature in the refrigerator. (Used with permission from Thermo Fisher Scientific, the copyright owner.)

19.2.2.5 Proper Use of Liquid Expansion Thermometers

Many of the problems that arise with liquid expansion thermometers are easy to diagnose. The most common problem is breakage. Another common problem is that the liquid column develops separations due to thermal or mechanical shocks during shipping or handling. If the liquid column is discontinuous, there will be substantial errors in the thermometer's readings. Thermometers should be inspected before use to be sure the liquid column is continuous. Box 19.2 describes several procedures that are used to reunite a liquid column.

Thermometers must be used properly to give correct measurements. Table 19.3 summarizes points regarding the proper use of liquid expansion thermometers.

19.2.3 BIMETALLIC EXPANSION THERMOMETERS

Materials generally expand as the temperature increases and contract when it decreases. Each material has a characteristic ***thermal expansion coefficient*** *that measures how much the material expands for a specific increase in temperature.* If two metals with different coefficients of expansion are bonded together and are then subjected to heat, the strips of metal will distort and bend. The amount of bending is related to the temperature.

A bimetallic dial thermometer *is constructed by fusing together two metal strips, typically one of brass and one of iron, which have different coefficients of expansion.* When the temperature changes, the two metals respond unequally, resulting in bending. One end of the

---

TABLE 19.2
Limitations of Liquid Expansion Thermometers

1. *The capillary bore within the thermometer in which the liquid moves must be very smooth so that the liquid is not impeded at any spot.* This is difficult to achieve, and less expensive thermometers may have slight imperfections leading to inaccuracy in their measurement.
2. *The scale must be very carefully etched onto the stem to ensure accuracy.* Not all thermometers are accurately subdivided into degrees.
3. *The working range of liquid thermometers is limited; see, for example, Figure 19.4.* The limited range of liquid expansion thermometers is not usually a problem for biologists because life exists only in a fairly narrow range of temperatures.
4. *Glass thermometers are fragile.* The glass comprising the bulb provides a thin interface between the liquid and the substance whose temperature is to be measured. This means the glass is thin and delicate and may easily be deformed or broken.
5. *Mercury, formerly used in liquid expansion thermometers, is hazardous to human health and the environment.* Mercury fumes and solids are toxic, and mercury from a broken thermometer can contaminate both the air and a water bath, a counter, or other sites.

BOX 19.2 REUNITING THE LIQUID COLUMN OF LIQUID EXPANSION THERMOMETERS

Be sure to wear personal protective equipment, particularly eye wear and gloves, while performing these steps!

1. *Lightly tap the thermometer by making a fist around the bulb with one hand and gently hitting your fist into the palm of the other hand.*
2. *Clamp the thermometer vertically. Slowly immerse the bulb of the thermometer in a freezing bath of salt, ice, and water.* The liquid should all slowly contract into the bulb, causing it to reunite. Be careful to cool the bulb only, not the stem. It may be helpful to remove the bulb periodically to slow the motion of the liquid. Remove and allow the thermometer to warm while it is supported in an upright position. Repeat several times if necessary.
3. *If 1 and 2 do not work, and if the separation is in the lower part of the column, then try cooling the thermometer, as described in 2, but use a mixture of dry ice and either acetone or alcohol.* Wear cold-resistant gloves. Cool only the bulb, not the stem. Remove the bulb from the cold. Do not touch the bulb until it has warmed to room temperature; it might shatter.
4. *As a last resort, particularly if the separation is in the upper part of the column, warm the thermometer.*
   a. Use this method only for thermometers with expansion chambers above the scale.
   b. Place the thermometer in a beaker of water or other nonflammable liquids, and heat it slowly.
   c. As the thermometer heats, the liquid should expand into the upper expansion chamber.
   d. When the liquid has moved into the expansion chamber, gently tap the thermometer with a gloved hand to reunite the column.
   e. Watch to be sure the expansion chamber does not overfill because if it does, the thermometer will break.
   f. Allow the thermometer to cool slowly.
   g. **Never use an open flame to heat the bulb.**
5. *If the column appears to be successfully reunited, check the ice point. If the ice point has shifted from its previous position, discard the thermometer.* (See Section 19.3. for a description of the ice point.)
6. *If the liquid column cannot be reunited, discard the thermometer.*

---

strips is fastened and cannot move, and the other moves along a scale to display the temperature (Figure 19.6).

Bimetallic expansion thermometers are simple and convenient, but they are generally not as accurate as high-quality liquid expansion thermometers. They do have various applications however, such as for checking the temperature of an oven or noting the air temperature outside the house. Note that bimetallic laboratory thermometers should be discarded if the stem is bent or deformed.

19.2.4 CHANGE-OF-STATE INDICATORS

Change-of-state indicators *are varied products, all of which change color or form when exposed to heat.* For example, items can be painted with lacquers that change appearance when heated, or marked with crayons that melt when a particular temperature is reached.

Labels are available that can be attached to items and change colors at specific temperatures (Figure 19.7a). There are encapsulated liquid crystals that change to various colors when exposed to particular temperatures. Crystals change colors reversibly and so can be reused (Figure 19.7b). Crystals have a fairly narrow temperature range, but can be useful at temperatures applicable to biological systems.

19.2.5 RESISTANCE THERMOMETRY: METALLIC RESISTANCE THERMOMETERS AND THERMISTORS

Metallic resistance thermometers and **thermistors** are two types of thermometers based on the principle that the electrical resistance of materials changes as their temperature changes. **Metallic resistance thermometers,** also called **resistance temperature detectors (RTDs),** *use metallic wires in which resistance*

TABLE 19.3

Proper Use of Liquid Expansion Thermometers

1. *Visually inspect a thermometer before use for breaks in the liquid column, bubbles, foreign materials in the stem, or distortions in the scale.* If the liquid is separated, reunite it as described in Box 19.2. If the thermometer contains foreign materials or has a distorted scale, it should be discarded.
2. *Avoid tapping a glass thermometer against a hard surface to prevent small fractures.*
3. *Support the thermometer by its stem; do not allow the bulb to rest on a surface.* The bulb is fragile.
4. *Do not stir solutions with thermometers.*
5. *Never subject a glass thermometer to rapid temperature changes.* For example, do not remove a thermometer from an ice bath and place it in a boiling water bath. This may cause the thermometer to break or the liquid filling solution to separate.
6. *Never place a thermometer in an environment above its maximum indicated temperature.*
7. *Immerse the thermometer to the depth indicated by the manufacturer.*
8. *Store thermometers as instructed by the manufacturer.*
9. *Place thermometers properly in equipment whose temperature is being measured.*
   • Do not allow thermometers to touch the sides of equipment.
   • Place the thermometer in a manner to avoid breakage.
   • Safely mounted thermometers for use in refrigerators, incubators, and other equipment are available commercially. Use these when possible.
   • In freezers and refrigerators, place the thermometer so that the stem does not touch any metal surface or ice. Air must be able to flow unobstructed around the stem of the thermometer.
   • For large freezers or refrigerators, it is good practice to check the temperature at more than one location.

(a)   (b)

FIGURE 19.6   Bimetallic expansion thermometers. (a) The unequal bending of two metals. (b) A bimetallic thermometer. (Photo copyright © Cole-Parmer. Used with permission.)

increases as the temperature increases. **Thermistors** *use semiconductor materials in which the resistance decreases as the temperature rises.*

Platinum wire resistance thermometers are considered to be the most accurate type of thermometer available. They are frequently used as references against which other thermometers are calibrated.

Thermistors are constructed of a hard ceramic-like material made up of a compressed mixture of metal oxides. The material can be molded into many shapes. Thermistors respond rapidly to temperature changes, can be miniaturized, function in a wide range of temperatures, and are very accurate. Thermistors are often used in a production setting as part of a temperature control device.

19.2.6 THERMOCOUPLES

A **thermocouple** *consists of two wires made of different metals that are joined together.* Two dissimilar wires joined as shown in Figure 19.8 generate a voltage. The magnitude of the voltage depends on the temperature at the junction between the wires. The voltage that develops in the wires, therefore, can be converted to a temperature measurement.

Thermocouples are rugged and versatile. They come in various types that are assigned a letter designation depending on the types of metals used in their construction. For example, a type T thermocouple has a copper wire and a wire made from a copper–nickel alloy. This type of thermocouple is used from 0°C to

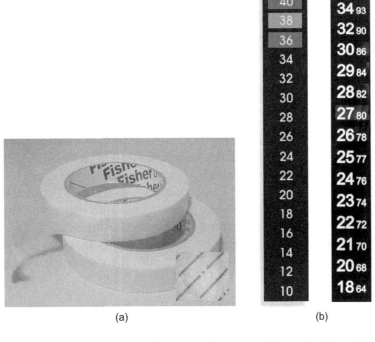

FIGURE 19.7 Change-of-state indicators. (a) Autoclave tape is commonly used to identify items that have been put through an autoclave cycle. It is affixed to materials before they are put into an autoclave, and it changes appearance after being subjected to sterilizing conditions. (Used with permission from Thermo Fisher Scientific, the copyright owner.) (b) Liquid crystal indicators that change color back and forth as the temperature changes. (Credit:Korke/Shutterstock.com)

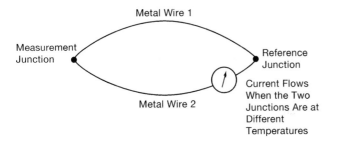

FIGURE 19.8 A simplified thermocouple.

350°C. Each type of thermocouple is useful in different applications and has a different temperature range.

Figure 19.9 shows a thermometer that has a thermocouple probe connected to a digital readout device.

19.3 VERIFYING THE PERFORMANCE OF LABORATORY THERMOMETERS

Thermometers have many routine uses in biology laboratories, such as checking the temperatures of solutions and monitoring the temperatures of water baths, incubators, ovens, refrigerators, and freezers. It is good laboratory practice to verify that the thermometers used to perform these tasks are performing

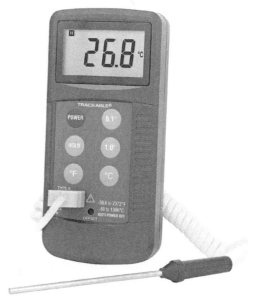

FIGURE 19.9 A thermocouple-based thermometer. This product includes NIST-traceable certification. According to the manufacturer, this means that "a 17025-accredited calibration test lab has confirmed that the product meets or exceeds all stated specifications. NIST-traceable calibration helps the user meet quality standards and regulatory compliance for ISO, FDA, cGMP, VFC, CAP, CLIA and Joint Commission requirements." (Photo copyright © Cole-Parmer. Used with permission.)

acceptably. Performance verification of a thermometer demonstrates that its readings are correct within its specified tolerance range.

For most thermometers, about 95% of all possible malfunctions will affect the ice point reading (Nicholas, J.V., and White, D.R. *Traceable Temperatures: An Introduction to Temperature Measurement and Calibration, 2nd edition,* Wiley, 2001). The freezing (ice) point of water is essentially independent of the environment, but the boiling point of water depends on the atmospheric pressure; therefore, the freezing point is more convenient for thermometer verification than the boiling point.

At the ice point, a Celsius thermometer should read at the "0" mark and a Fahrenheit thermometer should read at the "32" mark; otherwise, the thermometer is not performing adequately. Part A of Box 19.3

describes a relatively simple procedure for checking whether a liquid-in-glass thermometer is reading correctly at the ice point.

An ice point check is useful, but it only verifies the performance of a thermometer at one temperature. It is possible to check a thermometer more thoroughly by comparing its readings with those of a standard thermometer that is certified to be "NIST-traceable." A NIST-traceable thermometer is manufacturer-calibrated against a standard thermometer that, in turn, was calibrated at NIST. The readings of the thermometer that is to be checked are compared with the readings of the NIST-traceable thermometer at several temperatures spanning the thermometer's range. A procedure to check the performance of a liquid-in-glass thermometer by comparing it with a NIST-traceable thermometer is shown in part B of Box 19.3.

BOX 19.3   VERIFYING THE PERFORMANCE OF A LIQUID-IN-GLASS THERMOMETER

A. CHECKING THE ICE POINT OF THE THERMOMETER

1. *Visually inspect the thermometer whose performance is to be tested.* For a liquid expansion thermometer, check the column for separation or bubbles. Reunite the column and remove any bubbles present.
2. *Make certain the outside of the thermometer is clean.*
3. *Prepare an ice point apparatus:*
   a. *A Dewar flask (a vacuum-insulated flask for holding cold materials) is used to contain the ice point apparatus.*
   b. *Fill the flask one-third full of distilled water. Then add ice that has been shaved or smashed into tiny chips.* The ice should be made from purified water.
   c. *Compress the ice–water mixture into a tightly packed slush.* It is necessary to maintain tightly packed ice so that the thermometer is in close thermal contact with the ice–water mixture. Lack of good thermal contact is the major potential source of error in this procedure. Use only enough water to maintain good contact with the thermometer. The ice should not float. Remove excess water. It may be helpful to place a siphon tube that touches the bottom of the flask in order to drain off excess water that collects at the bottom of the Dewar. The ice should appear clear and not white. Avoid handling the ice to avoid contaminating it.
   d. *Wait 15–20 minutes for the mixture to reach equilibrium.*
   e. Periodically remove water from the bottom of the flask and add more ice.
4. *Immerse the thermometer in the apparatus to a depth approximately one scale division below the 0°C gradation.*

After at least 3 minutes have elapsed, tap the stem gently and observe the reading. Successive readings taken at least 1 minute apart should agree within one-tenth of a division. It may be necessary to repack the ice due to melting. The thermometer should read 0°C. It is recommended to use a 10X magnifying glass to read the scale display. Be certain your eyes are level with the display.

(Continued)

B. CHECKING A THERMOMETER AGAINST A CERTIFIED REFERENCE THERMOMETER AT TEMPERATURES OTHER THAN THE ICE POINT

If a certified reference thermometer is available, then the thermometer can be checked against the reference at various temperatures. The certified standard thermometer should have calibrated intervals ten times more precise than the working thermometer. (For example, to verify a thermometer with 0.1°C intervals, use a reference with intervals of 0.01°C.)

1. *Adjust a stable water heating bath to the temperature required for the analysis of interest.*
2. *Place the reference thermometer and the thermometer to be checked in the water.* The thermometers should be placed close to one another, but with sufficient space between them to ensure adequate circulation in the bath. Immerse the thermometers to their proper depth, and allow time for them to equilibrate to the temperature of the bath.
3. *After thermal equilibrium is reached (several minutes for liquid-in-glass thermometers), determine the temperature reading for both thermometers.* Note that reference thermometers come with a report of calibration that may include a correction factor. If there is a correction, then add or subtract that factor to or from the reading of the reference thermometer to obtain the actual temperature. It is good practice to take several readings of each thermometer and average them.
4. *Using a heated liquid bath, sequentially adjust the temperature of the bath to read each temperature for which the reference thermometer is calibrated.* For example, if the reference thermometer is calibrated at 25°C, 30°C, and 37°C, then sequentially adjust the water bath to these temperatures, as shown on the reference thermometer. Place the thermometer(s) whose performance is being checked into the water. The reading of the thermometer being checked should be the temperature of the water bath within the tolerance for that thermometer.
5. *Document the results of performance verification in a logbook or on appropriate forms.*

Information in this box is based on ASTM Standard E 77-14, "Standard Test Method for Inspection and Verification of Thermometers." Consult this standard for more details.

A thermometer should give the correct readings at the ice point and/or when compared to a standard thermometer. (Recall, however, that a given thermometer may have a fairly wide tolerance range.) Liquid expansion thermometers cannot be repaired or adjusted (except to reunite the liquid column). If the readings of a general-purpose liquid expansion thermometer fall outside its tolerance range, it is recommended that the thermometer be discarded.

Note that thermometers used to verify the performance of other thermometers may have correction factors. This means that the reading of the thermometer was tested at a specific temperature and was found to be slightly "off" by the amount of the correction factor. The correction factor is added to or subtracted from the reading of the standard thermometer to get the correct temperature. Thus, standard thermometers that have known corrections are not discarded even if they are marked slightly imperfectly.

Practice Problems

1. A thermocycler (an instrument used when DNA is amplified by the polymerase chain reaction method) is adjusted to cycle a reaction mixture between the temperatures shown. Convert these temperatures to °F.

94°C	(for 1 minute)
37°C	(for 1 minute)
72°C	(for 1 minute)

2. A thermophilic bacterium is found in nature at temperatures near 81°C. Convert this to °F.
3. A child is running a fever of 103.2°F. Convert this to °C.

Questions for Discussion

If you work in a laboratory, examine the laboratory thermometers and answer these questions:

1. For liquid expansion thermometers, what is the filling liquid?
2. What is the thermometer's range?
3. Identify each thermometer as either partial or total immersion. If a thermometer is a partial immersion type, what is its specified immersion depth?
4. Do any of your thermometers have an auxiliary scale?
5. Determine whether the laboratory thermometers have an expansion bulb.

20 The Measurement of pH, Selected Ions, and Conductivity

20.1 THE IMPORTANCE AND DEFINITION OF pH

20.1.1 OVERVIEW

The chemistry of life is based on water. Life evolved in the presence of water, and cells contain 80%–90% water. **pH,** *which is a measure of the acidity or alkalinity of an aqueous solution,* is intrinsically related to water chemistry. Maintaining the proper pH is essential for living systems. Plants do not grow if the soil pH is wrong; animals die if their blood pH deviates from normal; and microorganisms require their growth medium to be at a particular pH. Many industrial processes are likewise sensitive to pH, such as pharmaceutical production, food processing, sewage treatment, and water purification. Biological systems and many industrial processes, therefore, must be monitored and maintained at the proper pH.

Making accurate and reproducible pH measurements requires knowledge and proper equipment. While it is not recommended, it is usually possible to make acceptable measurements of temperature and weight using modern equipment with little understanding of thermometers or balances. pH measurement is different. There are a number of factors that influence pH, and it is easy to obtain readings on a pH meter

DOI: 10.1201/9780429282799-25

that look reasonable but are incorrect. For this reason, it is important to be knowledgeable about the factors that influence pH measurements. It is also critical to select the right equipment based on the requirements of the samples whose pH will be measured. Do not assume that because a particular pH meter system is available in your facility, it is the right one to use for your samples.

This chapter briefly defines pH and then discusses how it is measured with accuracy and precision.

20.1.2 A Brief Introduction to Water Chemistry and the Definition of pH

Pure water naturally dissociates to a limited extent to form **hydrogen (H^+)** and **hydroxide (OH^-)** ions:

$$H_2O \rightleftharpoons H^+ + OH^-$$

In pure water at 25°C, an equilibrium is established such that the concentration of H^+ ions is the same as the concentration of OH^- ions; it is 1×10^{-7} mole/L. (See Section 22.2.2 for an explanation of molarity.) *Pure water with a H^+ ion concentration of 1×10^{-7} mole/L* is called **neutral.**

When **acids** *are added to water, they release hydrogen ions into solution.* When **bases** *are added to water, they cause hydrogen ions to be removed from solution.* Thus, the addition of acids to water causes the concentration of hydrogen ions to be greater than 1×10^{-7} mole/L. The addition of bases to water causes the concentration of hydrogen ions to be less than 1×10^{-7} mole/L.

Strong acids and **strong bases** completely dissociate when dissolved in water and therefore have a strong effect on the hydrogen ion concentration. For example, hydrochloric acid (HCl) dissociates completely to release H^+ ions and Cl^- ions, thus increasing the H^+ ion concentration of the solution:

$$HCl \text{ in water} \rightarrow H^+ + Cl^-$$
(All is in this form.)

NaOH is an example of a strong base that dissociates completely in water as follows:

$$NaOH \text{ in water} \rightarrow Na^+ + OH^-$$
(All is in this form.)

The OH^- ions react with H^+ ions to form water, thereby lowering the concentration of hydrogen ions.

Weak acids and **weak bases** *do not completely dissociate in water* and therefore have a smaller effect on the concentration of hydrogen ions in the solution.

For example, acetic acid forms about one hydrogen ion for every 100 molecules:

$$CH_3COOH \text{ in water} \rightleftharpoons H^+ + CH_3COO^-$$
(Most of the acid is
in this form.)

The weak base, ammonia (NH_3), removes hydrogen ions from water with the formation of OH^- ions:

$$NH_3 + H_2O \rightleftharpoons NH_4^+ + OH^-$$
(Most of the base
is in this form.)

pH is a convenient way to express the hydrogen ion concentration[1] in a solution. *pH is the negative log of the H^+ concentration when concentration is expressed in moles per liter.*

To illustrate the definition of pH, consider the pH of pure water:

The H^+ concentration of pure water is
1×10^{-7} mole/L.
The log of 1×10^{-7} is -7.
The negative log of 10^{-7} is $-(-7) = 7$.
The pH of pure water is therefore 7.

Thus, the H^+ concentration of pure water is 10^{-7} mole/L and its pH is 7. By definition, the pH of any neutral solution is 7.

The pH scale is logarithmic. For example, a solution with a pH of 5.0 has 10 times more hydrogen ions than a solution with a pH of 6.0, and a solution of pH 7.00 is 100 times less acidic than a solution at pH 5.0. As hydrogen ion concentration, acidity, increases, the pH value decreases (Table 20.1).

The practical range of pH is between 0 and 14. (Very concentrated solutions of acid or base can fall outside this pH range and require special care when handling to protect both the technician and the equipment.) Solutions with a pH less than 7 are acidic, while those with a pH greater than 7 are basic.

The ultimate importance of H^+ concentration is that it profoundly affects the properties of aqueous solutions. For example, a strongly acidic solution, such as concentrated sulfuric acid, can dissolve iron nails.

---

[1] The term *pH* should properly be defined as the negative log of the H^+ *activity*, not its concentration. The activity of hydrogen ions is primarily related to their concentration, but it is also affected by other substances in the solution, by the solvent, and by the temperature of the solution. Defining pH in terms of the apparent concentration of hydrogen ions is an approximation that is very commonly used and is adequate for most purposes in a biotechnology laboratory.

TABLE 20.1

The Relationship between [H⁺] and pH

Hydrogen Ion Concentration (mole/L)		pH
Acidic		
10^0	1.0	0
10^{-1}	0.1	1
10^{-2}	0.01	2
10^{-3}	0.001	3
10^{-4}	0.0001	4
10^{-5}	0.00001	5
10^{-6}	0.000001	6
Neutral		
10^{-7}	0.0000001	7
Basic		
10^{-8}	0.00000001	8
10^{-9}	0.000000001	9
10^{-10}	0.0000000001	10
10^{-11}	0.00000000001	11
10^{-12}	0.000000000001	12
10^{-13}	0.0000000000001	13
10^{-14}	0.00000000000001	14

Example Problem 20.1

a. What is the pH of a solution with an H⁺ ion concentration of 10^{-4} mole/L?
b. What is the pH of a solution with an H⁺ ion concentration of 5.0×10^{-6} mole/L?

Answer

a. $pH = -\log [H^+] = -\log 10^{-4} = -(-4) = 4$
b. $pH = -\log [H^+] = -\log 5.0 \times 10^{-6} \approx -(-5.3)$
    $= 5.3$

Example Problem 20.2

What is the concentration of H⁺ ions in a solution with a pH of 9.0?

Answer

$$pH = -\log\left[H^+\right]$$

$$9.0 = -\log\left[H^+\right]$$

$$-9.0 = \log\left[H^+\right]$$

$$antilog(-9.0) = 1 \times 10^{-9}\, mole / L$$

(Strong acids and bases thus require safety precautions in the laboratory, as discussed in Section 9.2.4.) A weaker acid, citric acid, contributes to the taste of foods such as lemon and orange juice. The case study, "pH Matters," further illustrates the importance of pH.

Because maintaining the proper pH is critical to biological systems, methods are required to measure the concentration of hydrogen ions in solutions. Two common approaches to pH measurement in the laboratory are the use of **indicator dyes** and the use of **pH meter/electrode** measuring systems. We discuss first indicator dyes and then pH meter measuring systems.

20.2 pH INDICATORS

pH indicators *are dyes whose color is pH-dependent and that change colors at certain pH values.* Indicators can be directly dissolved in a solution or can be impregnated into strips of paper that are then dipped into the solution to be tested. **Phenolphthalein** is a well-known indicator dye (and formerly the main ingredient in some

---

Case Study: pH Matters

Flint, Michigan, was once known as a prominent auto manufacturing hub. In 2014, as a cost-cutting measure, the city changed its source of drinking water. Shortly thereafter, Flint entered the national consciousness as a place where children were being poisoned by lead in their drinking water that was present at levels hundreds of times above federally mandated limits. Even low levels of lead can impair the brain development of children, reducing their IQ and cognitive abilities. Lead exposure in children can also contribute to hearing problems, heart disease, and behavioral disorders. Lead exposure in adults has been linked to heart and kidney disease and reduced fertility. The high level of lead in the city drinking water was caused by corrosion in the lead and iron pipes that carry water to the city's homes. When the city changed the source of its drinking water, the chemistry of the water coursing through the hundreds of

(Continued)

Case Study (*Continued*): pH Matters

miles of pipes serving the city also changed, leading to the high lead levels. One of several issues relating to the lead contamination was the water's pH. Water utilities routinely treat municipal water to increase its pH. This is because a high pH decreases the solubility of hazardous lead carbonates and helps to create a layer of mineral deposits on pipes that prevents corroded metal ions from reaching the water. In Flint, Michigan, the pH of the water dropped as low as 7.3 in 2015. In contrast, Boston, another city with old lead-containing pipes, maintains a pH around 9.6 in their drinking water. Denver targets the pH of its water at 8.8 to reduce lead contamination. Thus, while it might seem reasonable to aim for a neutral pH in drinking water, in fact, in most municipalities, a pH closer to 9 is required.

Just as the pH of drinking water flowing into our homes is controlled to maintain our health, so too our bodies maintain tight control over the pH of the blood coursing through our circulatory system. The pH of blood is normally between 7.35 and 7.45. Blood levels above or below this range are a symptom of a health problem, for example, diabetes or severe asthma. The healthy human body has mechanisms in place to maintain the pH of our blood, regardless of what we eat or how active we are.

The pH of municipal water is maintained behind the scenes, and most of us seldom think about it. Similarly, unless we experience a health crisis, we do not think about the pH of our blood. But biotechnologists in laboratories and production facilities need to be conscious of pH because it has a profound effect on the growth and health of cultured cells; the rate and extent of enzymatic reactions; how molecules, such as DNA and proteins, interact with one another; and other biological processes. The first step in controlling pH is measuring it accurately and precisely, and so this chapter addresses pH measurement.

Primary Sources

CNN Editorial. "Flint Water Crisis Fast Facts." *CNN*, December 13, 2019. https://www.cnn.com/2016/03/04/us/flint-water-crisis-fast-facts/index.html

Torrice, Michael. "How Lead Ended Up in Flint's Tap Water: Without Effective Treatment Steps to Control Corrosion, Flint's Water Leached High Levels of Lead from the City's Pipes." *Chemical and Engineering News*, vol. 94, no. 7, February 11, 2016, pp. 26–29. cen.acs.org/articles/94/i7/Lead-Ended-Flints-Tap-Water.html.

laxatives). Phenolphthalein changes from colorless to red between pH 8 and 10. **Litmus,** which is extracted from lichens, is probably the oldest pH indicator. Litmus paper is pink in acidic solutions and blue in basic ones.

Indicators do not generally change colors sharply at a single pH; rather, they change over a range of one or two pH units. Thus, indicators are often used to broadly categorize a solution as acidic, basic, or neutral, or to determine roughly the pH of a solution. It is also possible to purchase pH indicators or pH paper that contain mixtures of dyes. These mixtures exhibit a range of colors and can be used to determine the pH of a sample to ± 0.3 pH units.

Indicators are a useful means of monitoring pH for some applications. For example, mammalian cells can be grown in the laboratory in a nutrient medium. Proper pH of the medium is critical for cell survival, but is easily altered by metabolites from the cells, contamination from microorganisms, and improper culture conditions. Therefore, a nontoxic, nonreactive pH indicator dye, phenol red, is frequently added to cell culture media. Changes in the indicator dye color provide immediate, visible information about the condition of the culture without opening the culture plates and exposing them to potential contaminants. When the medium is a certain shade of pink, the culture is healthy. A yellowish tinge indicates that the medium is becoming acidic and requires attention. Another example is the use of pH paper to measure the pH of toxic or radioactive solutions. The paper can be dipped in the hazardous solution and discarded, thus avoiding contamination of a non-disposable pH-measuring device.

pH indicators are inexpensive and simple to use; however, pH indicators show, at best, the pH of a solution within a range of about ± 0.3 pH units. In addition, indicators cannot be used in certain types of solutions. pH meter measuring systems should be used in situations where pH indicators provide inadequate information. Table 20.2 summarizes the potential sources of error when using pH indicators. Table 20.3 lists common pH indicators.

TABLE 20.2

Potential Sources of Error When Using pH Indicators

1. **Acid–base error.** Indicators are themselves acids or bases. When indicators are added to unbuffered or weakly buffered solutions, such as purified water or very dilute solutions of strong acids or bases, the indicator itself causes a change in pH of the sample solution.
2. **Salt error.** Salts at concentrations above about 0.2 M can affect the color of pH indicators, leading to inaccuracy.
3. **Protein error.** Proteins react with indicators and can have a significant effect on their color. Indicators, therefore, should not be used to measure the pH of protein solutions.
4. **Alcohol error.** The solvent in which a sample is dissolved can affect the color of an indicator; therefore, a sample dissolved in alcohol may be of a different color than if it is dissolved in an aqueous buffer.

TABLE 20.3

Examples of Common Indicators

Indicator	Color Change
Blue litmus	Changes from blue to red denoting a change from basic to acidic
Brilliant yellow	Changes from yellow at pH 6.7 to red at pH 7.9
Bromocresol green	Changes from yellow at pH 4.0 to blue at pH 5.4
Bromocresol purple	Changes from yellow at pH 5.2 to purple at pH 6.8
Congo red	Changes from red at pH 3.0 to blue at pH 5.0
Cresol red	Changes from yellow at pH 7.2 to red at pH 8.8
Methyl red	Changes from red at pH 4.2 to yellow at pH 6.2
Neutral litmus	Red in acid conditions and blue in basic conditions
Neutral red	Changes from red at pH 6.8 to orange at pH 8.0
Phenol red	Changes from yellow at pH 6.8 to red at pH 8–8.2
Phenolphthalein	Changes from colorless at pH 8.0 to red at pH 10.0
Red litmus	Changes from red to blue denoting a change from acidic to basic

Source: Pharmacopeia, U.S. *USP 33 NF 28 REISSUE 2010 (U.S. Pharmacopeia National Formulary) (USP 33 NF 28).* USP, 2021 (a good source of information about pH indicators, their chemical nature, solubility, and preparation, because indicators are used in testing drugs).

20.3 THE DESIGN OF pH METER/ELECTRODE MEASURING SYSTEMS

20.3.1 Overview

pH meter/electrode *systems are the most common method of measuring pH in the biotechnology laboratory.* Although more expensive and more complicated to use than chemical indicators, pH meters provide greater accuracy, sensitivity, and flexibility. When used properly, pH meters can measure the pH of a solution to the nearest 0.1 pH unit or better, and they can be used with a variety of samples.

Recall that we discussed electrical circuits in Chapter 16. We noted that **electrical current** *is a flow of charge* and this flow only occurs if there is a complete pathway for its movement. The *pathway is termed a* **circuit**. In Chapter 16, we were speaking of current that consists of electrons flowing in a wire. In this chapter, we will be talking about current that flows in a solution; in this case, it is ions that comprise the current, not electrons. A pH meter/electrode system involves a circuit in which current flows in the form of ions.

Figure 20.1 diagrams a pH meter/electrode circuit. Observe that it consists of (1) a **voltmeter** that measures voltage, (2) two **electrodes** connected to one another through the meter, and (3) the **sample** whose pH is being measured. When the two electrodes are

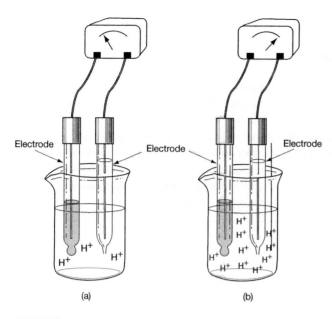

FIGURE 20.1 Overview of a pH meter measuring system. The system consists of two electrodes that are immersed in the sample and that are connected to one another through a meter. Note that most pH meter systems appear to have only one electrode, not two. That is because the two electrodes are combined into one housing. (a) The concentration of H^+ ions in the solution is relatively low. (b) A higher concentration of H^+ ions leads to a change in the voltage difference between the two electrodes.

immersed in a sample, they develop an electrical potential (voltage) that is measured by the voltmeter. (Appendix 20.1 at the end of this chapter has a brief explanation of how electrodes develop potential.) The magnitude of the measured voltage depends on the hydrogen ion concentration in the solution; therefore, the voltage reading can be converted by the instrument to a pH value. There is a complete pathway (circuit) for current to flow, from one electrode to the other, through the voltmeter and the sample.

Note that the term **pH meter** *refers to both the voltage meter and its accompanying electrodes.*

20.3.2 The Basic Design and Function of Electrodes

20.3.2.1 Electrodes and the Measurement of pH

The two electrodes are the heart of the pH-measuring system. One is a **pH-measuring electrode**, and the other a **reference electrode**. The **pH-measuring electrode** *has a thin, fragile glass bulb at its tip* and is therefore sometimes called a **glass electrode** (Figure 20.2a). The glass that composes the bulb is sensitive to the concentration of H^+ ions in the surrounding

medium. (Note that some sources refer to the glass bulb as the "membrane.") An electrical potential develops between the inner and outer surfaces of the glass bulb when the measuring electrode is immersed in a sample solution. It is this potential whose magnitude varies depending on the concentration of hydrogen ions in the solution. The measuring electrode's shaft contains a buffered chloride solution. This solution surrounds a metal wire that acts as a lead connecting the glass bulb to the voltmeter.

You might wonder why two electrodes are required when the glass measuring electrode by itself is sensitive to pH. In practice, it is not possible to measure the potential of a single electrode. The pH-measuring electrode must be paired with a second **reference electrode**, *which has a stable, constant voltage* (Figure 20.2b). The reference electrode has a strip of metal and an **electrolyte solution** (filling solution) enclosed in a glass or plastic shaft. **Electrolytes** *are substances, such as acids, bases, and salts, that release ions when dissolved in water.* The metal and electrolyte in the reference electrode react with one another to develop a constant voltage.

The two electrodes are placed in the sample whose pH is to be measured. The following events occur when electrodes are in the sample solution:

1. The electrical potential (voltage) of the glass electrode varies depending on the sample's hydrogen ion concentration.

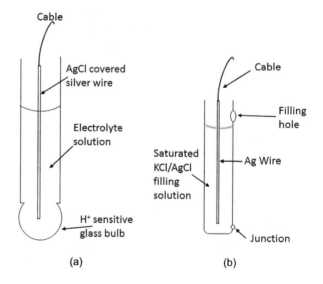

FIGURE 20.2 pH electrodes. (a) The measuring (glass) electrode. The bulb is composed of glass that develops an electrical potential when placed in solution. The magnitude of the potential depends on the concentration of hydrogen ions. (b) The reference electrode. More information about the parts of this electrode is provided in the text.

2. The electrical potential of the reference electrode is constant.

3. The two electrodes are connected to one another through the voltage meter.

4. The meter measures the voltage difference between the reference electrode and the measuring electrode. This voltage is an electrical signal.

5. The electrical signal is amplified and converted to a display of pH. As with other measurement instruments, calibration is required in order for the instrument to convert an electrical signal to a pH value. Calibration of pH meters is discussed later in this chapter.

20.3.2.2 Combination Electrodes and Compact Systems

The pH meters in biology laboratories usually appear to have only one electrode, not two. This is because they have convenient **combination electrodes. Combination electrodes** *are made by combining the reference and pH-measuring electrodes into one housing* (Figure 20.3). Combination electrodes work efficiently for most purposes although they are not suitable for measuring the pH of certain "difficult" samples. (Difficult samples will be discussed later in this chapter.) In this chapter, we assume you will be using a combination electrode because these are by far the most common in biotechnology settings. Note that the term "probe" is commonly used to refer to a combination electrode; you will see the terms "electrode," "electrodes," and "probe" used interchangeably in the literature and in this chapter.

pH meters can be made still more compact by combining the electrodes and the meter into a single housing. These compact pH meter systems are useful for field work, but are not typically used in biotechnology laboratories.

20.3.2.3 More about Measuring Electrodes

There are various types of measuring electrodes that vary in the type of glass used to make the bulb. A "general-purpose" measuring electrode is usually acceptable in a biotechnology setting. Most biotechnology procedures are buffered so as to maintain a pH that is suitable for life, for example, in the range of pH 6–9. However, if a procedure requires a pH above around 9 or 10, then *general-purpose measuring electrodes begin to respond to Na^+ and K^+, as well as H^+*; this response is termed **alkaline error.** There are specially designed measuring electrodes with a glass bulb that minimizes alkaline error.

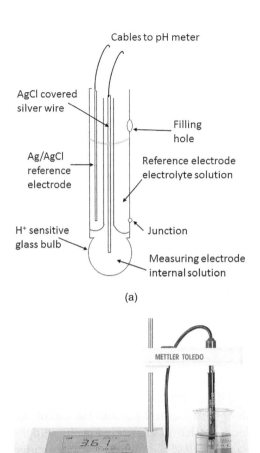

(a)

(b)

FIGURE 20.3 Combination electrodes. (a) Diagram of a combination electrode. (b) A combination electrode in use with a sample. (Photo courtesy of METTLER TOLEDO.)

Other situations that require specialized measuring electrodes include pH measurements at high and low temperatures and pH measurements of certain "difficult" samples. "Difficult" samples include high-purity water, solutions that are poorly conducting, and nonaqueous liquids. Manufacturers make specialized measuring electrodes for these purposes.

If your procedure requires the use of a specialized measuring electrode, then you may not be able to use a combination electrode and you might have to purchase the measuring and reference electrodes separately.

20.3.3 MORE ABOUT REFERENCE ELECTRODES

20.3.3.1 Overview

The reference electrode is the component of the pH-measuring system that is most likely to cause difficulties. It is therefore important to be familiar with reference electrode design and function.

There were formerly two common types of reference electrodes for pH meters: silver/silver chloride (abbreviated Ag/AgCl) and calomel. An **Ag/AgCl electrode** *contains a strip (wire) of silver coated with silver chloride.* The strip is immersed in an electrolyte solution that is normally saturated in both potassium chloride (KCl) and silver chloride (AgCl); refer to Figure 20.2b. A **calomel electrode** *contains a platinum wire coated with calomel, which is a paste consisting of mercury metal and mercurous chloride.* The calomel is in contact with a filling solution that is normally saturated KCl. Calomel electrodes, however, are being phased out to avoid mercury toxicity, just as are the mercury thermometers discussed in Chapter 19. The elimination of calomel electrodes has significance for biotechnologists as will be discussed in more detail later in this chapter.

Reference electrodes must maintain a constant electrical potential. This potential is sensitive to the concentration of Cl⁻ in the electrolyte filling solution. The concentration of Cl⁻ in the solution is usually held constant by using a filling solution that contains saturated KCl. Observe in Figure 20.2b that the end of the reference electrode that is submerged in the sample has a *porous plug, called a* **junction,** or **salt bridge.** The purpose of the junction is to allow the filling solution to slowly flow out of the electrode and into the sample. The slow flow of ions through the junction is necessary to maintain electrical contact between the reference electrode and the sample solution and complete the circuit for current flow.

Because reference electrode filling solution is lost through the junction, many electrodes have a **filling hole** *that is used to replenish the electrolyte*; refer to Figure 20.2b. The electrolyte in refillable electrodes should be topped up periodically. When the probe is not in use, the filling hole is plugged to avoid evaporation of the electrolyte. It is also possible to purchase reference electrodes that are manufactured with a gel-like filling solution that cannot be replenished. A gel-filled electrode must be discarded when the gel is depleted. Gel-filled electrodes are convenient, but they generally do not last as long as refillable electrodes and they are slower to respond to the pH of a sample than refillable electrodes. With any reference electrode, pH cannot be accurately measured if the electrolyte is depleted, or if the junction is blocked so that electrolyte does not contact the sample.

20.3.3.2 Types of Junctions

Different styles of junction are available that allow faster or slower flow of filling solution. (Flow rates vary from about 0.5 to 100 μL/hour.) Your choice of junction

should depend on the type of samples whose pH will be measured. We will consider three common types.

- A **ceramic junction** is a standard junction style that is used in general-purpose pH probes (Figure 20.4a). **Ceramic junctions** *consist of a small piece of ceramic material that is inserted through the glass shaft of the electrode.* Ceramic materials are porous, and so the filling solution flows slowly out of the small pores and into the sample.

- A **double junction** *is a more complex reference electrode configuration that is used when the sample is incompatible with the reference electrode filling solution.* As is shown in Figure 20.4b, the double junction separates the electrolyte solution that surrounds the wire element from the electrolyte that flows into the sample. This means that the electrolyte surrounding the element can have a different composition from the electrolyte that contacts the sample.

- A **sleeve junction** *is made by placing a hole in the side of the glass or plastic housing of the electrode and covering the hole with a moveable glass or plastic sleeve (Figure 20.4c).* The sleeve junction is the fastest flowing type, is the least likely to become clogged, and is the easiest to clean, since the sleeve can be moved to reveal the hole. Faster-flowing junctions have the disadvantage that the sample is more contaminated by filling solution than with slower-flowing junctions.

20.3.3.3 Measuring the pH of Samples in Tris Buffer and Protein-Containing Samples

This discussion of reference electrodes and their junctions may seem esoteric, but it is relevant to biotechnologists. Biotechnologists commonly work with samples in Tris buffers and with samples containing high concentrations of proteins. (See Chapter 23 for more information about Tris buffers.) Tris-buffered and protein-containing solutions are incompatible with the silver in Ag/AgCl electrodes. The electrolyte solution in Ag/AgCl electrodes typically consists of 3 mol/L KCl saturated with AgCl. The purpose of the AgCl is to prevent silver from leaching off the silver wire element inside the electrode. Tris and proteins can bind with silver ions to form solid precipitates. These precipitates form at the junction where the flowing silver ions meet the Tris or the proteins in the sample. The tiny pores of a standard ceramic junction are easily clogged by precipitates. When the

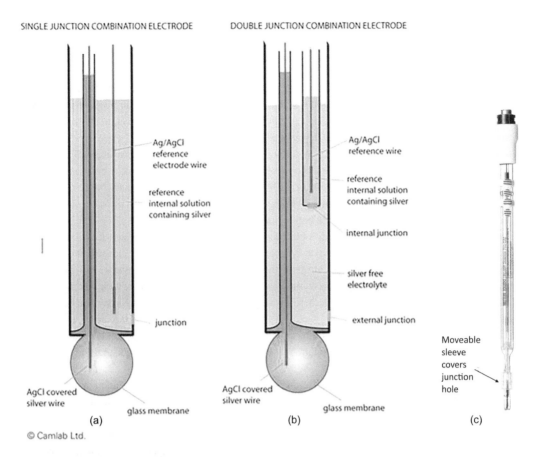

SINGLE JUNCTION COMBINATION ELECTRODE   DOUBLE JUNCTION COMBINATION ELECTRODE

© Camlab Ltd.

FIGURE 20.4 Reference electrode junctions. (a) A general-purpose ceramic-style junction is simply a small piece of a porous ceramic material inserted into the shaft of the electrode. (b) A double junction electrode separates the reference electrode electrolyte from the sample. (Images a–b courtesy of Camlab.) (c) A sleeve junction with a moveable sleeve allowing the junction to be easily cleaned. (Photo courtesy of METTLER TOLEDO.)

junction is clogged, the reference electrode no longer performs as it should, resulting in unstable and erroneous pH readings. With a clear glass electrode, the precipitates may be visible as a black coloration on the junction.

In the past, biotechnologists could use calomel electrodes that did not have silver in the electrolyte solution. Calomel electrodes provided more accurate measurements than Ag/AgCl electrodes when working with protein- and Tris-containing solutions. Calomel electrodes are no longer readily available due to their mercury content. Fortunately, manufacturers have developed alternative strategies for laboratories where the pH of Tris and proteins is commonly measured.

One strategy to make pH electrodes that are compatible with Tris and proteins (and with other materials that form precipitates in the presence of silver) is to use a double junction. The double junction prevents silver from reaching the sample; refer to Figure 20.4b. Another strategy is to use a relatively fast-flowing, easily cleaned junction, typically of the sleeve type. The sleeve junction is unlikely to clog and is easily cleaned if it does, but the fast flow can potentially contaminate the sample with electrolyte. Manufacturers have also developed new designs of electrode, such as an iodine/iodide reference system, that replace the Ag/AgCl reference system. Probes that are suitable for Tris and proteins are typically described in manufacturers' literature as "Tris-compatible."

In practice, what this means to you is that if the pH of Tris- and protein-containing samples is measured in your laboratory, then using a conventional, general-purpose pH probe may negatively affect the accuracy and reproducibility of your measurements. When you walk into a laboratory and see a pH meter, it is difficult to tell what type of probe is attached to it. Therefore, before using a pH meter, find out whether its probe is compatible with your samples. Consult with the manufacturer and be certain to select a pH probe that is suitable for your purposes.

Additional information about selecting and maintaining electrodes is provided in Table 20.4.

TABLE 20.4

Considerations in Selecting and Maintaining Electrodes

1. *Size and shape.* It is possible to purchase combination electrodes with various lengths and shapes to accommodate different samples. For example, common electrodes with a spherical tip provide a wide surface area that works well with aqueous samples in standard vessels. A conical tip can penetrate samples such as soils, sauces, and gels. A flat tip is useful for measuring the pH of samples with a flat surface.

2. *Glass or plastic body.* Glass is resistant to many chemicals, is easy to clean, and transfers heat readily to allow the sample and probe to come to thermal equilibrium. Glass also makes it easier to detect contamination on a junction or in the filling solution. The disadvantage to glass is that it is easily broken. Plastics are used to make break-resistant electrodes.

3. *Connections.* Electrodes are connected to pH meters with plug-in cables that must match the meter. Consult the meter manual to find out what kind of connector to use before purchasing electrodes.

4. *Solid-state (ISFET) electrodes.* ISFET pH probes incorporate a small silicon chip that is sensitive to H^+ ions into the head of a solid-state electrode. These electrodes do not require filling solution, are easily cleaned, and can measure the pH of very small-volume samples. The disadvantages of these electrodes are that they are more expensive than conventional probes; they do not offer the same stability and accuracy as glass electrodes; their readings are more likely to drift; and they require special pH meters. For these reasons, solid-state probes are not common in biotechnology laboratories.

5. *Filling solution.* Reference electrodes contain filling solution, which is generally saturated in KCl and AgCl.
   a. Gel-filled electrodes contain gelled filling solution that is never refilled, so the electrode must be discarded when the electrolyte is depleted.
   b. Refillable electrodes are periodically refilled through the filling hole. Use the filling solution recommended by the manufacturer for the electrode. Fill the electrode nearly to the top of the filling solution chamber to ensure proper flow out the junction.
   c. The filling hole must be open when a sample's pH is measured so that filling solution can flow from the junction properly. The filling hole must be closed during storage; if the filling hole is open, evaporation may occur leading to the formation of crystals in the electrode and the junction.

6. *Storage.* There is no single rule for how to store all types of electrodes. The easiest strategy is to purchase electrode storage solution from the electrode manufacturer. Some of the general considerations relating to storage are:
   a. Do not store electrodes in purified water; this affects the H^+ sensitive glass membrane. Electrodes should be stored in ion-containing, aqueous solutions.
   b. Manufacturers often recommend storing electrodes in pH 4.00 buffer, and some suggest adding 1/100 parts of KCl to the buffer. It may be helpful to add up to 4% sodium benzoate or sodium azide to the storage solution to discourage the growth of microorganisms.
   c. For refillable electrodes, be sure to close the filling hole for storage.
   d. New electrodes, an electrode stored dry, and electrodes that have been cleaned should be conditioned before use by soaking for at least 8 hours in buffer. Consult the electrode manufacturer to determine which buffer to use.
   e. For short-term storage and in between measurements, it is best to keep the electrode in a holder containing electrolyte solution (e.g., 3 mol/L KCl), or in a pH 4.00 or pH 7.00 buffer. Be sure that the level of solution in the beaker is below that of the filling solution in the electrode.

7. *Cleaning.*
   a. Between measurements, rinse the probe with purified water and gently blot it with lint-free laboratory tissue, but never wipe the probe to dry or clean it. Wiping can damage the fragile glass bulb and can also create static charge that results in inaccurate pH readings.
   b. For more thorough cleaning, manufacturers sell cleaning solutions that are compatible with their electrodes. Cleaning protocols for specific contaminants are in Appendix 20.2 "Troubleshooting pH Measurements."

8. *Fragility.* The measuring bulb of an electrode is fragile and cannot be repaired if it is broken or cracked. Keep the electrode bulb away from moving stir bars.

9. *Life expectancy.* Even properly maintained electrodes have a finite life span. Under typical usage conditions, a probe will last 1–3 years. Gel-filled electrodes may only last 6 months.

20.4 OPERATION OF A pH METER SYSTEM

20.4.1 CALIBRATION/STANDARDIZATION

20.4.1.1 Overview

In previous chapters, we discussed how calibration "tells" an electronic instrument the relationship between its electrical response and a parameter of the sample. In the case of balances, that parameter was weight; in the case of pH meters, it is pH. The electrical response of a pH meter is measured in units of millivolts, mV. Thus, **pH meter calibration,** also called **standardization,** *"tells" the meter how to translate the voltage difference between the measuring and reference electrodes into the unit of pH.* A note about language: Many analysts prefer the term "standardization" to describe the process of "telling" the meter the relationship between pH and millivolts. "Standardization" is repeated every day, every shift, or every time the pH meter is used because the response of electrodes changes over time. In this usage, the term "calibration" is reserved to describe tasks that are less frequently performed by metrology department personnel or third-party metrologists. In order to standardize/calibrate a pH meter, buffers with known values of pH are used. These buffers are termed pH calibration buffers, standards, or buffer standards.

There is a generally linear relationship between the voltage measured by the pH meter and the pH of the sample. Because the relationship is linear, two buffers of known pH can be used to standardize a pH meter (Figure 20.5). In normal practice, the first buffer is pH 7.00, although modern instruments are often designed so that buffer standards can be used in any order. (Also, modern meters may not require that one of the buffers is pH 7.00, but you may obtain more accurate results by always including pH 7.00 buffer in the standardization.) Using an appropriate key, button, or other command methods, the meter is set to display pH 7.00. The probe is rinsed and then placed in the second buffer standard. The second buffer is typically pH 4.00, 9.00, 10.00, or 12.00, although other buffer standards are available. If the sample whose pH is to be measured is acidic, an acidic buffer is used (e.g., pH 4.00), and if the sample is basic, a basic buffer is chosen (e.g., pH 9.00 or 10.00). Ideally, the second buffer's pH should be within 1.5 pH units of the sample's pH, if a suitable standard is available. However, it is possible to obtain acceptable results using the most common standards of pH 4.00, 7.00, and 10.00. The result of the standardization process is that the meter constructs a calibration line. The relationship provided by this calibration

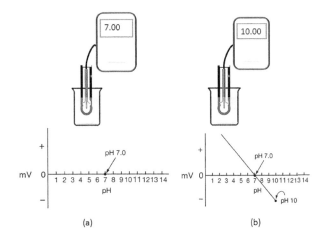

FIGURE 20.5 pH meter standardization. The process of standardization establishes the relationship between pH and mV so that the meter can convert any mV reading to a pH value. (a) The meter is adjusted to display pH 7.00 when the electrodes are immersed in pH 7.00 buffer standard. (b) The meter is adjusted to display the value of a second buffer standard, and the meter internally constructs the calibration line.

line can then be used to convert the voltage that arises when the probe is placed in a sample into a reading of the sample's pH.

The buffers used for standardization are critical in ensuring meaningful pH results. If they are contaminated or improperly prepared, then pH readings cannot be accurate. Information about buffer standards is provided in Table 20.5.

20.4.1.2 The Slope of the Calibration Line

To standardize a pH meter, one follows the procedure provided in the instrument's manual. This should be a straightforward process once you know which keys to press. However, obtaining accurate and reproducible pH measurements requires more than knowing how to operate the meter. It is also important to understand what is happening during standardization.

Consider the line that is created when the meter is standardized, called the calibration line (Figure 20.6a). This line shows the relationship between mV response and pH, where pH is on the X-axis and mV on the Y-axis. All straight lines have a slope. The slope of the pH meter calibration line is a measure of the response of the electrodes to pH. The steeper the slope, the more sensitive the electrodes are to hydrogen ions. Ideally, the slope of the calibration line of a new, perfectly performing Ag/AgCl probe should be −59.16 mV/pH unit when the temperature is 25°C. (See Appendix 20.1 section, "A Note about the Nernst Equation," for

TABLE 20.5

Buffers for Standardization

1. *While it is possible to mix buffer standards in your laboratory based on recipes, this is seldom recommended.* In practice, purchased, premixed buffer standards are convenient and are generally of higher and more dependable quality than those mixed in an individual laboratory. Many manufacturers provide buffers whose pH values are traceable to NIST, which means that the manufacturers compare the pH of their buffers to NIST standard buffers to be sure they are the same (with a small tolerance for error).

2. *Reaction with CO_2.* Some buffers react with CO_2 from the air, resulting in a change in their pH. This is more likely with basic than with acidic buffers; pH 10 and higher buffers are particularly susceptible. Buffer containers, therefore, should be kept closed. Remove a small amount of buffer each time an instrument is standardized, close the original container, and throw away the used portion after use.

3. *Expiration date.* Premixed buffers come with an expiration date. To ensure accuracy, do not use buffers after their expiration date.

4. *Opened bottles.* Manufacturers suggest discarding unused buffer standards 2–3 months after the bottle is first opened. pH 10 and higher buffers should be discarded after 1 month because they most readily react with CO_2 in the air. It is possible to purchase buffers in single-use packets to avoid discarding unused buffer.

5. *Temperature.*
   - The pH of a buffer will change as its temperature changes. For example, a common buffer has a pH of 7.00 at 25°C, but at 0°C, its pH is 7.12. When standardizing with this buffer at 25°C, adjust the meter to 7.00, but if the buffer is at 0°, then adjust the meter to 7.12. The pH listed on the container of a commercially purchased buffer is generally its pH at 25°C. Buffer manufacturers often place a table conveniently on the label of the buffer's container that lists its pH at various temperatures.
   - You may notice during standardization that you place the probes in a buffer that you think is a particular pH, but the meter automatically displays a slightly different pH. For example, you might be using a buffer standard that you think is pH 10.00, but during the standardization process, the meter displays a reading of 10.06. This is usually because the meter has been programmed to recognize common buffers, and it "knows" how to compensate for the effect of temperature. In this example, the meter "knows" that at the temperature of the buffer, the pH of that buffer is 10.06. Therefore, the meter uses a value of 10.06 when constructing the calibration line.
   - Allow sufficient time for the probe to equilibrate to the temperature of the buffer. If the difference in temperature between the probe and the buffer is extreme, or if the pH of the buffer is very high or low, equilibration can take as long as 3 or 4 minutes.
   - Store buffers at room temperature and out of direct sunlight.

6. *Avoid contamination.*
   - Never return a used buffer to its original bottle.
   - Pour out the needed amount of buffer for standardization; use it; discard after one use.
   - Buffers that are visibly contaminated by mold growth or that contain precipitates must be discarded.
   - Do not place electrodes directly into the original buffer bottle.

more information about the derivation of this value for the slope.) As electrodes age or become dirty, they are less able to generate a potential and the slope of the line declines (Figure 20.6b). Therefore, knowing the slope provides information about the condition of the electrodes.

The slope of the calibration line is often expressed as **efficiency**, *which is a percent of the ideal value.* For example, suppose the slope of the calibration line is −57.40 mV/pH unit. Ideally, the slope should be −59.16 mV/pH unit, assuming a temperature of 25°C. Then:

$$\text{Efficiency} = \frac{-57.40\ \dfrac{mV}{pH\ unit}}{-59.16\ \dfrac{mV}{pH\ unit}} \times 100\% \approx 97\%$$

Modern pH meters often display the calibration line's slope and/or efficiency each time the meter is calibrated. This value should be recorded, either by hand in a lab notebook or instrument log, or electronically. Some pH meters print information regarding the standardization, and this printout should be stored according to the policies of the organization.

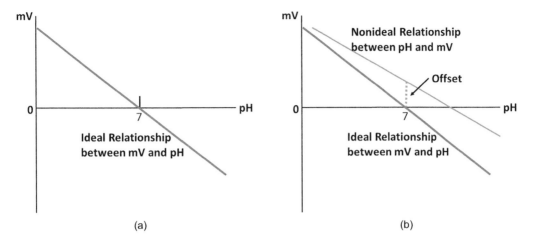

FIGURE 20.6 Relationship between millivolt response and pH. (a) With an ideal electrode (blue line), the slope of the calibration line is −59.16 mV/pH unit when the temperature is 25°C. At pH 7.00, the electrode response should be 0 mV. (b) The response of electrodes is not necessarily ideal (orange line). First, the slope may be less steep than −59.16 mV/pH unit. As electrodes age, the slope becomes less steep until the electrode is no longer usable. Additionally, the calibration line might not go through zero at pH 7.00. The distance from the calibration line to 0 mV at pH 7.00 (dashed line) is called the offset.

Your meter may or may not display the probe's efficiency, but you can calculate the slope and efficiency yourself, if your meter has a mode in which it displays mV readings. (Less expensive, simpler instruments may only display pH readings and not mV.) The method for calculating the slope of the calibration line and the efficiency of the probe is provided in Box 20.1.

20.4.1.3 The Offset of the Calibration Line

The calibration line of a new, ideally performing probe should read 0 mV at pH 7.00, as illustrated in Figure 20.6a. This is because the reference electrode's electrolyte has a pH of 7.00. If the measuring electrode's glass bulb is in a solution with the same pH (7.00), then ideally there should not be a difference in

---

BOX 20.1 DETERMINING THE EFFICIENCY AND OFFSET OF A pH METER SYSTEM

1. *Obtain two buffer standards, pH 7.00 and either pH 4.00 or 10.00.*
2. *Be sure the meter "knows" the temperature of the buffers.*
3. *Place the meter in mV mode so that it displays mV, not pH units.*
4. *Rinse the probe, blot it dry (do not wipe), and place it in the pH 7.00 buffer.* Allow the reading to stabilize, and record the mV reading.
5. *Rinse and blot the probe, and place it in the second buffer.* Allow the reading to stabilize, and record the mV reading.
6. *Calculate the slope as*:

$$\frac{|\text{Absolute change in Y}|}{|\text{Absolute change in X}|} = \frac{|\text{Absolute change in mV readings}|}{|\text{Absolute change in pH units}|}$$

Note: "Absolute" means the magnitude of a number without regard to whether the number is positive or negative. Brackets like this | | indicate an absolute value.

7. *Calculate the efficiency of the probe by dividing the electrode slope by the absolute theoretical maximum slope (59.16 mV/pH unit at 25 °C), and multiply by 100%*:

$$\text{Efficiency} = \frac{\text{Slope of probe}}{59.16 \frac{\text{mV}}{\text{pH unit}}} \times 100\%$$

(Continued)

BOX 20.1 (*Continued*)   DETERMINING THE EFFICIENCY AND OFFSET OF A pH METER SYSTEM

8. *Note the offset, which is simply the mV reading in the pH 7.00 standardization buffer.*
9. *Refer to Table 20.6 to interpret the values for efficiency and offset.*

   Example

   Suppose the measured mV value in pH 7.00 buffer is −5 mV and the measured value in pH 10.00 buffer is −172 mV.

   The absolute change in Y = −5 −(−172) = 167 mV.

   The absolute change in X = 3 pH units.

   The slope = 167 mV / 3 pH units ≈ 55.67 mV/pH unit.

$$\text{Efficiency} = \frac{55.67 \text{ mV} / \text{pH unit}}{59.16 \dfrac{\text{mV}}{\text{pH unit}}} \times 100\% \approx 94\%$$

   Offset = **−5 mV.**
   This electrode may need cleaning soon, as shown in Table 20.6.

TABLE 20.6

Recommendations for Evaluating the Performance of Your Probe

	Offset ± 0–20 mV	Offset ± 20–35 mV	Offset > 35 mV
Slope 95%–105%	Electrode is in good condition	Electrode requires cleaning soon	Electrode requires cleaning and/or regeneration
Slope 90%–95%	Electrode requires cleaning soon	Electrode requires cleaning soon	Electrode requires cleaning and/or regeneration
Slope 85%–90%	Electrode requires cleaning and/or regeneration	Electrode requires cleaning and/or regeneration	Electrode requires cleaning and/or regeneration
Slope < 85% or > 105%	Electrode is worn out and needs to be replaced	Electrode is worn out and needs to be replaced	Electrode is worn out and needs to be replaced

Source: Information is based on "Your pH Calibration." *Mettler Toledo*, 2014. www.mt.com/us/en/home.html.

---

potential between the measuring and reference electrodes. When the measuring and reference electrodes have the same potential, the meter should display 0 mV. In practice, there will be some deviation from zero as shown in Figure 20.6b. *The amount of this deviation* is the **offset**. As the electrodes age, or if they become dirty, then the offset increases. Therefore, like slope, the offset is an indication of the probe's performance. Modern pH meters may display the offset each time the meter is calibrated, and, like the slope, this value, if displayed, should be recorded. If your meter can display millivolts, then you can determine the offset yourself simply by checking the mV response when the probe is placed in a pH 7.00 standardization buffer.

Example Problem 20.3

a. If a probe has an ideal response, what should be the mV reading when a pH 7.00 buffer is checked? Assume the temperature is 25°C.

b. If a probe has an ideal response, what should be the mV reading when a pH 4.00 buffer is checked? Assume the temperature is 25°C.

c. If a probe has an ideal response, what should be the mV reading when a pH 10.00 buffer is checked? Assume the temperature is 25°C.

Answer

a. **0 mV**.

b. Observe in Figure 20.6a that at pH values less than 7, mV readings are positive. Therefore, the predicted mV reading of a pH 4.00 buffer should have a positive value. pH 4 is 3 pH units away from pH 7. The slope ideally is −59.16 mV/pH units, which means that for each pH unit of change, the slope changes by 59.16 millivolts. So:

3 pH units × 59.16 mV/pH unit = **177.48 mV** = the predicted value for pH 4.00 buffer under ideal conditions.

c. Observe in Figure 20.6a that at pH values greater than 7, mV readings are negative. pH 10 is three units away from pH 7. So:

3 pH units × 59.16 mV/pH unit → **−177.48 mV** = the predicted value for pH 10.00 buffer under ideal conditions.

20.4.1.4 Evaluating the Performance of Your Probe

Recommendations vary, but generally, it is recommended that the slope of a pH probe, expressed as its percent efficiency, should be in the range of 90%–105% and the offset should be <30 mV. Table 20.6 contains further recommendations to guide the interpretation of the efficiency and offset of a probe.

It is expected that over time and with use, the performance of a pH probe will decline. As the electrode response deteriorates, the system is eventually no longer able to adjust to the value of two buffers. Some pH meters will then display "calibration error." If this error is displayed or if the efficiency and offset of a probe are subpar, there are two possible explanations. Possibly the probe has aged to the point where it must be discarded and replaced. There is no set lifetime for a probe, and it is not unusual for electrodes to only last a year or two. It is also possible that the probe is contaminated in some way and it can be cleaned and restored to normal operation. Cleaning and rejuvenating probes is discussed in Appendix 20.2, "Troubleshooting pH Measurements."

Example Problem 20.4

Suppose the mV reading in pH 7.00 buffer is −15 mV and it is 160 mV in pH 4.00 buffer. a. What is the percent efficiency and the offset? b. What can you conclude about the performance of this probe?

Answer

a. Absolute change in mV = (−15 mV) − 160 mV = $|{-175}|$ mV = 175 mV.
Absolute change in pH = 3 pH units.
Slope = 175 mV/3 pH unit ≈ 58.33 mV/pH unit.
Efficiency = 58.33/59.16 × 100% ≈ **98.6% efficiency**.
Offset = **−15 mV**.

b. This probe is performing well, based on the information in Table 20.6.

Example Problem 20.5

If a meter displays an efficiency of 98.5%, what was the slope of the calibration line?

Answer

$$\frac{?}{59.16\dfrac{mV}{pH\ unit}} = 0.985$$

$$? = 0.985(59.16\,mV\,/\,pH\ unit)$$

$$\approx \mathbf{58.27\,mV\,/\,pH\ unit}$$

Example Problem 20.6

a. An electrode has a mV reading of +15 mV in pH 7.00 buffer and +160 mV in pH 4.00 buffer. What is the efficiency of this electrode?

b. What can you conclude about the performance of this electrode?

Answer

a. Absolute change in mV = 145 mV.
Absolute change in pH = 3 pH units.
Slope =
145 mV/3 ≈ 48.33 mV/pH unit.
48.33/59.16 × 100% ≈ **81.7 efficiency**.
Offset = **−15 mV**.

b. The offset of this electrode is satisfactory, but its slope is not. The recommendation in Table 20.6 is to replace this electrode, but you might try first changing the filling solution, cleaning the probe (as described later in this chapter), and recalibrating to see if the performance improves.

20.4.1.5 Multipoint Standardization

We have so far spoken of pH meters as if their calibration lines were strictly linear. In fact, that is not the case. If one calibrates a pH meter with pH 7.00 and 10.00 standardization buffers and then checks the pH of a pH 4.00 buffer, the measurement may be slightly "off." The deviation might not be large, perhaps less than 0.10 pH units, but some inaccuracy is to be expected. For example, after standardization we expect the meter to tell us that a pH 4.00 buffer standard has a pH of 4.00. But, because of nonlinearity, the meter might read the pH of this standard as 3.91. Because of the inherent nonlinearity of pH meters, the best way to calibrate a pH meter is with two buffers that bracket the pH of the samples. For example, if one expects all the samples to have pH values between 7.50 and 9.50, it would be appropriate to calibrate with pH 7.00 and 10.00 buffers. There are, however, situations where a number of samples are to be measured and their pH values range widely across the pH scale. Perhaps some of the samples are acidic, others basic, and others closer to neutral. In this situation, one strategy might be to calibrate the meter with pH 7.00 and pH 4.00 buffers, measure all the acidic samples, and then recalibrate with pH 7.00 and 10.00 before measuring the pH of the basic samples. However, if one does not know in advance which samples are acidic and which are basic, this strategy is cumbersome. Therefore, some pH meters allow standardization with three or more buffers, a "multipoint" standardization, rather than just the usual two-point standardization. This permits a wider standardization, perhaps with pH 4.00, pH 7.00, and pH 10.00 buffers.

If a meter is designed to allow a multipoint standardization, the standardization relationship might be determined in two ways. The meter might calculate the straight line that best fits all the points. (See the Appendix to Chapter 28 for a discussion of best-fit lines.) When this method is used, the calibration line may not exactly fit any of the points. For example, in Figure 20.7, three standardization buffers were used, pH 4.00, 7.00, and 10.00. The mV values for these standardization buffers were plotted as blue dots on the graph, and the line that best fits all three is shown in orange. This best-fit calibration line does not exactly fit any of the points; therefore, it will provide some error in pH readings. A better method is for the meter to calculate *two* calibration lines: one between pH 4.00 and 7.00 and the other between pH 7.00 and 10.00, as shown in green. When a meter provides more than one calibration line, it is called "segmented." In this example, the meter can automatically use one of the calibration lines for acidic samples and the other when the pH of basic samples is measured. This segmented standardization strategy is the most accurate strategy when samples with a wide range of pH values are measured.

In practice, what this means to you is that you might have a meter that allows a multipoint standardization. If, however, the meter simply creates one best-fit line for the multiple buffers, you might find that this multipoint standardization provides less accurate results than a two-point standardization. Experiment with your meter to determine how to get the best accuracy based on your meter's capabilities.

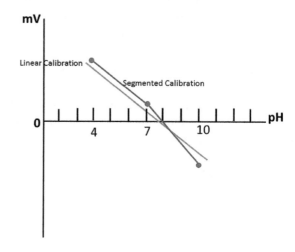

FIGURE 20.7 Multipoint standardization. Orange is the best-fit line; green is segmented into two lines. See text for further explanation. (Derived from a figure from: Mettler Toledo. "White Paper: Perform your Next pH Measurement in Compliance with USP <791>." *Mettler Toledo*, 2018. www.mt.com/us/en/home.html.)

20.4.2 EFFECT OF TEMPERATURE ON pH

Understanding the effects of temperature is essential for obtaining accurate and reproducible pH measurements. Temperature has two effects:

1. The **measuring electrode's response to pH** (the potential that develops) is affected by the temperature (Figure 20.8).
2. The **pH of the solution (or buffer standard)** that is being measured may increase or decrease as its temperature changes.

It is important to consider both temperature effects when operating a pH meter.

Most pH meters can compensate for the first factor (i.e., temperature-dependent changes in the electrode response). The meter needs to "know" the temperature of the solution to make this adjustment. It is possible to measure the temperature of a solution with a thermometer and, using the proper command method, "tell" the pH meter the solution's temperature. There are also *electronic temperature probes,* **automatic temperature compensating (ATC) probes,** that can be placed

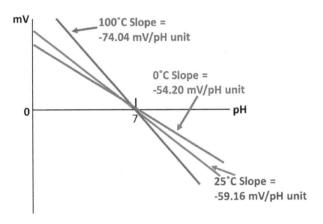

FIGURE 20.8 The effect of temperature on electrode response. Observe that the slope of the calibration line changes at different temperatures.

in the sample alongside the pH electrodes. The probes are connected to the pH meter and automatically measure the sample temperature and report it to the meter, which compensates accordingly. Electrodes may even have a temperature probe built into the electrode housing, so a three-in-one probe consists of the reference electrode, the measuring electrode, and an ATC probe.

The second temperature effect relates to the sample or buffer; the pH of some solutions changes as their temperature changes. For example, if Tris buffer is prepared to pH 8.00 at room temperature, its pH will be close to 7.60 when used at body temperature and 8.80 when placed on ice. It is best, therefore, to measure the pH of a solution when the solution is at the temperature at which it will be used.

When measuring the pH of samples that are not at room temperature, it is common practice to standardize the meter at room temperature using buffers at room temperature. Then, when it is time to measure the pH of the sample solution, the meter is adjusted to the temperature of the sample and the sample's pH is determined. It is also possible to standardize a pH meter with two buffers that are equilibrated to be the same temperature as the sample. The pH meter temperature setting is adjusted to this temperature before standardization (or an ATC probe is used). Because the pH of buffers varies with temperature, the meter must be adjusted to the pH of each buffer at the temperature of interest. Note that the sample and electrodes should be at the same temperature when a pH measurement is made. If they are not, the measurement value displayed may drift until the temperatures reach equilibrium.

20.4.3 SUMMARY OF THE STEPS IN OPERATING A pH METER

Although each brand and style of pH meter differs, there are certain steps that are usually performed when measuring the pH of a sample. These steps are summarized in Box 20.2.

BOX 20.2 GENERAL METHOD FOR MEASURING THE PH OF A SOLUTION

1. *Allow the meter to warm up as directed by the manufacturer.*
2. *Open the filling hole; be certain that the filling solution is nearly to the top* (refillable electrodes only).

(Continued)

BOX 20.2 (*Continued*)   GENERAL METHOD FOR MEASURING THE PH OF A SOLUTION

STANDARDIZATION

Standardize the system each day or before use:

1. *Obtain two standard buffers: commonly, one that has a pH of 7.00 at room temperature and the other that is acidic if the sample is acidic and basic if the sample is basic.*
   * For best accuracy, do not perform a two-point standardization with pH 4.00 and pH 10.00 buffers; choose pH 7.00 and either pH 4.00 or pH 10.00 buffer.
   * Pour sufficient volumes of each buffer into small beakers or vials that fit the pH probe.
   * Tightly seal the buffer bottles after pouring out the amount needed.
2. *Enter into the calibration mode on the meter.*
3. *Be sure the meter "knows" the correct buffer temperature if no automatic temperature probe is in use.*
4. *Rinse the probe with purified water, and blot dry.*
   * Do not wipe the probe as this may create a static charge leading to an erroneous reading.
   * Any crystals on the outside of the probe should be removed during rinsing.
5. *Immerse the probe in room temperature pH 7.00 buffer.*
   * Be certain that the junction is immersed and that the level of sample is below the level of the filling solution.
   * Allow the reading to stabilize.
   * Note the temperature of the buffer, and determine whether the pH of that buffer is exactly 7.00 at that temperature. There should be a label on the buffer bottle that lists the pH of the buffer at various temperatures. Make sure that the meter displays the correct pH for the buffer at the temperature in use. Most modern meters are programmed to "know" how temperature affects the pH of standard buffers and will automatically compensate for temperature effects, but older meters might not.
6. *Adjust the meter to read 7.00 (or the correct pH at the temperature in use) using the correct command method.*
7. *Remove the probe, rinse with purified water, and blot dry.* You can also rinse the probe with the next solution and do not blot dry.
8. *Place the probe in the second buffer, and allow the reading to stabilize.*
   * Adjust the meter to the pH of the second buffer with the proper method of adjustment.
   * As before, pay attention to the pH of the buffer at the temperature in use.
   * Remove, rinse, and blot the electrodes.
9. *Modern meters should display the slope and/or efficiency of the standardization and also possibly the offset; be sure these values are recorded.* Evaluate these values, as discussed in the text.
10. *Perform quality-control checks as follows:*
    a. *Linearity check, optional, may not be performed at each standardization.* To check the linearity of the measuring system, take the reading of a third buffer outside the range of the standardization. For example, if the meter was calibrated with pH 7.00 and pH 10.00 buffers, check the pH of a pH 4.00 buffer.
       - Immerse the probe in this third standard buffer.
       - Place the meter in the normal pH measurement mode (not calibration mode).
       - Allow the reading to stabilize; record the value. *Do not adjust the meter to this third buffer; the purpose of this third buffer is to check the system's linearity.*
       - If the reading is not within the proper range, as defined by your laboratory's quality-control procedures, the probe should be serviced as described in Appendix 20.2. Note that the response of pH meters is not expected to be perfectly linear throughout its entire range. A small amount of deviation in the linearity check is generally tolerated.

(Continued)

b. ***Accuracy ideally should be checked each time standardization is performed.*** Immerse the probe in a third standard buffer whose pH is bracketed by the buffers used for standardization. For example, if standardization used pH 7.00 and pH 10.00 buffers, you could perform this test with a pH 9.00 buffer.

- Place the meter in normal pH measurement mode (not calibration mode); allow the reading to stabilize; record the value. ***Do not adjust the meter to this third buffer; the purpose of this third buffer is to check the system's accuracy.***
- If the reading is not within the proper range, as defined by your laboratory's quality-control procedures, the probe should be serviced as described in Appendix 20.2. It is common practice in many laboratories to set the maximum allowable error of the control buffer to be 0.10 pH units.
- However, laboratories that are CGMP compliant will need a reading that is accurate ± 0.05 pH units, as described in the U.S. Pharmacopeia <791>.

If the results of standardization are satisfactory, proceed to measure the pH of the sample. If not, proceed to troubleshooting or consult with an instrument specialist.

MEASURING THE pH OF THE SAMPLE

1. ***Set the meter to the temperature of the sample or place an automatic temperature probe in the sample.***
2. ***Place the pH probe in the sample.***
3. ***Allow the pH reading to stabilize.***
   - Generally, the reading stabilizes within a minute or so. One of the causes of irreproducibility in pH measurements is not allowing the reading to stabilize.
   - If you wait too long, the pH of some samples will drift due to reaction with CO_2 in the air or chemical reactions in the sample.
   - Many new pH meters have an "autoread" feature that automatically determines when the reading has stabilized and locks on this reading. For example, the meter may lock on the reading if it changes less than 0.004 pH units over 10 seconds. The autoread feature can be helpful in reducing variability.
4. ***Record all relevant information.*** Be sure to note the temperature of the solution when its pH was measured.
5. ***Remove the probe from the sample, rinse, and store properly.*** Close the filling hole (refillable electrodes only).

20.4.4 Measuring the pH of "Difficult Samples"

The sample is a key part of the measurement system. The sample must be homogeneous to avoid drift in pH values. The sample must also be equilibrated to a particular temperature because if the temperature is changing, then the pH may drift. Some samples undergo chemical reactions that cause their pH to change over time. Some samples are inherently difficult to pH and may require special techniques or types of electrodes. Table 20.7 includes tips for handling difficult samples.

20.4.5 Quality Control and Performance Verification of a pH Meter

As with any instrument, pH meters require a quality-control plan. At the least, this includes keeping a logbook, performing regular standardization, and cleaning and replacing electrodes as needed. Although each laboratory must develop its own plan, Box 20.3 summarizes common, routine quality-control procedures for pH meter measuring systems.

BOX 20.3   QUALITY-CONTROL ACTIVITIES

Quality-Control Activity	Each Use	On Scheduled Basis
Standardize with two buffers	✓	
Perform linearity check as described in Box 20.2		✓
Read and record the pH of a control buffer whose pH is bracketed by the buffer standards	✓	
Check the level of filling solution (refillable electrodes only)	✓	
Clean and recondition electrodes; clear precipitates from the junction		✓
Remove all used filling solution and replace with new filling solution (refillable electrodes only)		✓ (for example, every month)

TABLE 20.7
Difficult Samples

1. **Nonaqueous solvents.**
   - The classical definition of pH relates to aqueous solutions, and so the interpretation of mV readings from nonaqueous samples is different.
   - Applications involving nonaqueous solvents commonly involve relative rather than absolute pH measurements. For example, there may be a jump in the mV reading of the meter when a reaction occurring in a sample goes to completion.
   - Nonaqueous solvents, such as acetone, can dry the glass electrode membrane, which must be hydrated to function properly. After measuring the pH of nonaqueous solvents, it may be necessary to soak the measuring electrode in an aqueous solution. Consult the manufacturer for further advice.
2. **Strongly acidic or basic solutions.** Take a reading of such solutions quickly, and then rinse the electrodes thoroughly.
3. **High-purity water.** High-purity water (e.g., acid rain, distilled, or boiler feedwater) is very difficult to pH because it does not readily conduct current and it absorbs CO_2 from the atmosphere, which affects its pH. Tips for measuring the pH of such samples include:
   - Avoid the absorption of CO_2 by taking a reading soon after immersing the probe deep within the sample.
   - Cover the sample container with a stopper that has a hole for the probe.
   - Blanket the surface of the sample with nitrogen gas.
   - Use special, low ionic strength buffers for standardization. (Such buffers are available commercially.)
   - Use commercially available additives to increase the ionic strength of the sample without affecting its pH.
   - Use a refillable reference electrode with a low resistance glass measuring bulb.
   - Add 0.30 mL of high-purity, saturated KCl to 100 mL of test solution, and measure the pH immediately without agitation.
4. **High-salt samples.** In samples with large amounts of salt, sample ions compete with the reference filling solution ions. This competition affects the voltage read by the meter. It may be possible to compensate for this effect by using buffer standards that also have a high salt concentration. An electrode with a double junction may also be helpful.
5. **Sample–electrode compatibility.** As previously introduced in Section 20.3.3.3, the sample must be compatible with the reference electrode. Tris buffer, proteins, sulfides, Br$^-$, and I$^-$ can precipitate or complex with the silver in Ag/AgCl electrodes, leading to a clogged junction. General-purpose electrodes are usually unsuitable when working with these materials. Use electrodes that are recommended by the manufacturer for these samples.
6. **Slurries, sludges, and viscous and colloidal samples.** These types of sample may require a fast-flowing junction, usually the sleeve type. These samples may also coat the electrode bulb. Cleaning procedures are shown later in this chapter.

Consult manufacturers for more information on specific electrodes to match particular types of samples.

20.4.6 COMPLIANCE WITH THE ELECTRONIC RECORDS REGULATIONS

Like balances and other laboratory instruments, pH meters are often operated in a stand-alone mode where the analyst reads and manually records a displayed pH value. Sometimes, however, pH meters are interfaced with computers that are able to store, analyze, and print data. Vendors who manufacture pH meters now provide software that is compatible with the requirements of the FDA's regulations in 21 CFR Part 11. Figure 20.9 shows an example of a printout from a pH meter that has the capability of storing and printing data, thus helping to ensure data integrity. (See Section 6.4 for an introduction to data integrity requirements when computers are involved.) This system should have these features:

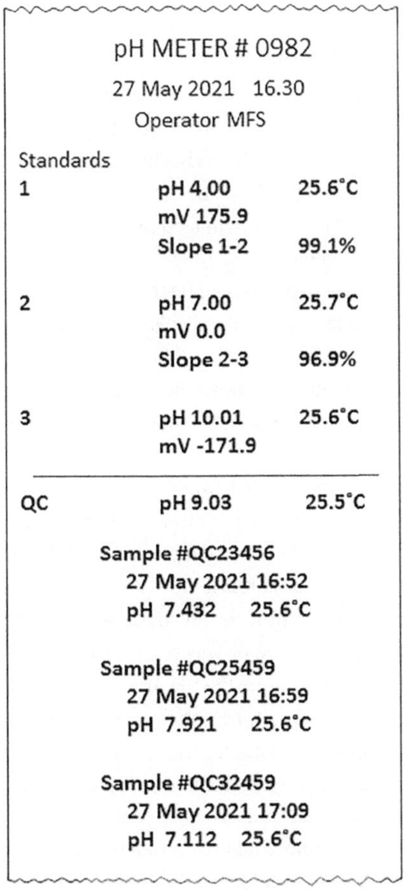

- *Date and time stamp* are provided.
- *The work is attributed to an individual*, and so it is likely that a user ID and password are required.
- The system *documents when the meter was last standardized/calibrated.*
- The *standardization information is linked to the sample data*.
- The system *documents standardization data*, including standards used, temperature, mV offset, and slope.
- The software can be programmed to *remind the user to recalibrate* the instrument after a specified time has elapsed.
- The *records that are generated must be secure* and must be accessible to people who are authorized to view them.
- The *records must provide information for traceability* so that pH results can be connected to a particular lot of product or a particular task as well as to the date, time, and individual involved.
- The *records must not be capable of being altered* either accidentally or intentionally in an uncontrolled manner.

```
        pH METER # 0982

       27 May 2021   16.30
          Operator MFS

Standards
1           pH 4.00        25.6°C
            mV 175.9
            Slope 1-2      99.1%

2           pH 7.00        25.7°C
            mV 0.0
            Slope 2-3      96.9%

3           pH 10.01       25.6°C
            mV -171.9

QC          pH 9.03        25.5°C

     Sample #QC23456
        27 May 2021 16:52
        pH 7.432    25.6°C

     Sample #QC25459
        27 May 2021 16:59
        pH 7.921    25.6°C

     Sample #QC32459
        27 May 2021 17:09
        pH 7.112    25.6°C
```

FIGURE 20.9 A printout from a pH meter that is interfaced with a computer. Observe that a multipoint, segmented standardization was performed with buffers of pH 4.00, 7.00, and 10.00. There are therefore two values for efficiency.

20.5 THE BASICS OF TROUBLESHOOTING pH MEASUREMENTS

A pH meter system must be able to respond to a range of hydrogen ion concentrations from 10^0 to 10^{-14} moles/L, must be responsive to very low concentrations of hydrogen ions and to very small changes in hydrogen ion concentration, and must be able to detect hydrogen ions selectively in the presence of other ionic species. Compared with many other measuring instruments, therefore, a pH meter must have a wide range of response and be very sensitive and specific. Given these stringent requirements, it is not surprising that problems sometimes arise when using a pH meter. This section discusses sources of common problems and general strategies for identifying and solving them.

The first step in troubleshooting is recognizing that there is a problem. Although this seems obvious, problems with pH meter systems may be subtle and are sometimes overlooked. Symptoms of pH system problems include the following:

1. *The reading drifts in one direction or takes an unusually long time to stabilize.*
2. *The reading is unstable and fluctuates.*

3. *The reading does not change at all, even when different samples are tested.*
4. *The meter cannot be adjusted to both buffer standards.*
5. *The apparent pH value for a buffer or sample seems to be wrong.* For example, the buffer that is checked as part of a quality-control procedure does not provide the correct pH value.
6. *There is a high value for the offset and/or a low percent efficiency for the slope after standardization.*
7. *Meter displays "calibration error."*
8. *There is no reading at all.*

Problems with a pH meter system can arise in any of its components: (1) the reference electrode, (2) the measuring electrode, (3) the meter and the cable connecting the electrode to the meter, (4) the buffer standards, or (5) the sample. It would be convenient if a particular symptom were always associated with a particular component of the pH meter system. Unfortunately, a particular symptom can have various causes. For example, a drifting or unstable reading might be due to temperature changes in the sample – but it might also be due to problems with the reference electrode. Similarly, difficulties in standardization might be due to electrode problems, but could also be caused by contaminated buffers. Therefore, the first step when a symptom is noted is to look for and correct the most common, simple mistakes and problems that relate to various parts of the measurement system (Box 20.4).

BOX 20.4   BASIC TROUBLESHOOTING STEPS FOR A pH METER SYSTEM

What to Check	More Information
Are the electrode measuring bulb and the reference electrode junction immersed in the sample?	Yes, it does happen that users do not immerse the probe in the sample, for example, when the sample is in an opaque or deep vessel.
Is the probe cracked; is the glass bulb scratched or broken?	This can be difficult to see with plastic-body electrodes, but you should be able to turn the electrode upside down and see obvious damage to the glass bulb. Discard a damaged probe.
Is the meter turned on and plugged in? Is the electrode cable connected to the meter? Has the meter had time to warm up?	Try turning the meter off and on. Try reattaching the electrode cable to the meter. Try gently bending the cable to see if the display changes; it is possible to have a faulty cable.
Is the reference electrode sufficiently filled with electrolyte (refillable electrodes only)?	The level of filling solution must be above the sample or storage solution level. Be sure that the electrode is filled with the proper filling solution.
Is the reference electrode filling hole open (refillable electrodes only)?	Unplug the filling hole before standardizing the system and reading the pH values of samples. Remember to close it when done.
Is the filling solution contaminated (refillable electrodes only)?	Look for precipitates and crystals inside the electrode (if the shaft of the electrode is glass and is transparent). Excessive crystallization can occur if the electrode was stored for a long time with the filling hole open or was stored in the cold. If the electrode was used to determine the pH of a sample when the filling solution was low, then the filling solution might have been contaminated by the entrance of sample. Try replacing the filling solution with fresh electrolyte (Box 20.5).
Is the sample well stirred and homogenous?	Placing the sample on a stir plate and stirring it while measuring pH is a common practice, but remember that the measuring electrode bulb is fragile and you do not want to hit it with a stir bar.

(Continued)

What to Check	More Information
Is the temperature of the sample changing?	Temperature changes in the sample can cause drift or fluctuation in readings.
	If the sample is on a stir plate, the stir plate may be warming, causing the sample to warm.
	Possibly a chemical reaction in the sample is causing its temperature to change. Before proceeding, be sure that the temperature is stabilized and that you are using an ATC probe or have manually adjusted the meter to the temperature of the sample.
Is the sample undergoing chemical reactions or reacting with CO_2 from the atmosphere?	Consider the chemistry of the sample. For example, high-purity water will react with CO_2 from the air.
	The autoread function on modern instruments might be helpful because it locks the reading as soon as it is stable for a certain length of time.
Are the electrodes and sample at the same temperature?	If not, allow them to equilibrate.
Is the sample inherently difficult to pH and does it require special techniques or specific types of electrodes?	See Table 20.7.
Are the buffer standards good?	Be sure buffer standards were freshly removed from their bottles right before use.
	Be sure the buffer standards' bottles had not been left open over a period of time.
	Be sure the buffer standards are not expired or contaminated.
	You might want to open new bottles of buffers to be sure that they are not the problem or use single-use buffer packets.

BOX 20.5   REPLACING ELECTROLYTE IN A REFERENCE ELECTRODE

1. *Wash the probe in warm purified water.*
2. *Shake gently until the crystals dissolve.*
3. *Remove the old filling solution.* It may be helpful to use a narrow plastic pipette, or a syringe attached to plastic tubing, to remove the solution. If a pipette or tubing is used, be careful not to disturb the internal elements of the electrode.
4. *Insert fresh solution of the proper type.*
5. *If you see bubbles in the electrolyte, try to remove them with gentle shaking because bubbles can cause unstable pH measurements.*
6. *It may be helpful to soak the electrode overnight in 1 M KCl or storage solution before using.*

If possible, a good way to check whether the probe is the source of problems is to swap out the probe and replace it with a new one, or with a probe that is known to be working. (This assumes another probe is available.) If you swap out the probe and the problem disappears, then you know the difficulty was with either the reference or measuring electrode. You can either discard the old probe or try to further troubleshoot the difficulty; see Appendix 20.2. Two of the more likely problems with a probe are a partially or totally occluded (plugged) junction (a problem with the reference electrode) or a problem with the measuring electrode bulb.

Problems with the reference electrode junction are fairly common. If the junction is dirty and partially occluded, then pH readings will not be correct. If the junction is completely clogged, the reading may never stabilize. The first indication that the junction is occluded is a long stabilization time where the reading drifts slowly toward the correct pH. Slow equilibration, however, can also be caused by changes in the temperature of the sample (such as is caused by a warm stir plate), by reactions occurring within the sample, by differences in the temperature of the electrodes and sample, or by incompatibility between the reference electrode and the sample. One way to test for an occluded junction, assuming you have a liquid-filled reference electrode, is to gently apply air pressure to the filling hole. A bead of electrolyte should appear readily on the junction. If not, the junction is likely fully or partially occluded. If the junction is occluded, you might be able to clean it as described in Box 20.7 in Appendix 20.2 to this chapter. Also, you might want to consult with the manufacturer of your electrodes for advice on cleaning the junction.

Problems with the measuring electrode bulb can take two forms. The measuring electrode bulb might be contaminated by materials from samples. Certain samples (e.g., those with high levels of protein) often leave deposits on the bulb that must be periodically cleaned, as described in Box 20.8 in Appendix 20.2. Another problem is that aqueous solutions slowly attack the glass bulb and form a gel layer on the bulb surface. As the gel layer thickens, the response of the electrode becomes sluggish and the slope of the calibration line decreases. It is sometimes possible to "rejuvenate" a measuring electrode bulb as described in Box 20.9 in Appendix 20.2. If you suspect a problem with the glass bulb, it might be advisable to consult with the electrode manufacturer.

An occluded junction or a difficulty with the measuring electrode glass bulb can often be remediated. However, note that pH-measuring electrodes age, even if they are properly maintained, and they have a finite lifespan after which time they must be replaced.

The meter is the least likely component of the system to cause problems. Complete lack of response (or sometimes, dramatically unexpected readings) is likely caused by problems with the meter itself or possibly the cable connecting the electrodes to the meter. If you have already tried the obvious solutions, such as turning the meter off and on and reconnecting the cable between the electrode and the meter, then consult the meter manufacturer.

More troubleshooting tips are provided in Appendix 20.2 along with manufacturer's advice on how to clean an occluded junction and clean and rejuvenate the measuring electrode bulb. Note, however, that the suggestions in Appendix 20.2 are compiled from various manufacturers. It is always best to consult the manufacturer of your own equipment and follow their recommendations.

20.6 OTHER TYPES OF SELECTIVE ELECTRODES

pH electrodes are only one type of indicator electrode. An **indicator electrode** *responds selectively to a specific molecule in solution.* For example, there are indicator electrodes that respond to dissolved CO_2 and others that respond to dissolved O_2. Cell culturists use dissolved O_2 electrodes to monitor the oxygen level in the broth during fermentation because the growth of most microorganisms is sensitive to oxygen levels.

Ion-selective electrodes (ISEs) *are a class of indicator electrodes, each of which responds to a specific ion.* A pH-measuring electrode is an example of an ion-selective electrode. Other ISEs include those that respond to Ca, Na, K, Li, F, and Cl. ISEs have many applications. For example, sodium levels in human blood are severely disturbed by cardiac failure and liver disease. Sodium levels in a patient's blood serum can be tested using a sodium ion-selective electrode. ISEs can be used to measure fluoride levels in drinking water, nitrates in soils, calcium in milk, and other applications. There are many methods published in official compendia that utilize ISEs. For example, ASTM International, the Environmental Protection Agency, and AOAC all have ISE methods for measuring levels of fluoride in drinking water.

An ion-selective electrode measuring system is diagrammed in Figure 20.10. An ISE has a membrane that, like the glass bulb of a pH electrode, develops an electrical potential in response to a particular ion. The ISE is connected to a reference electrode through a voltage meter. Although the electrodes are different, a standard pH meter can be used for these measurements, as long as it is able to display mV as well as pH.

Each type of indicator electrode is different; therefore, it is important to read the manual that comes with the electrode. Table 20.8 lists some general points of concern regarding indicator electrodes.

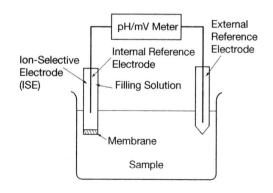

FIGURE 20.10 An ion-selective electrode measuring system.

TABLE 20.8

Practical Considerations for Indicator Electrodes

1. The sensing tips of the electrodes are fragile and should be protected from breakage and drying.
2. The response of the electrodes may be affected by the combined concentration of all the ions in the solution. The standards used for calibration and the samples should be roughly comparable in their total ionic concentration. It is possible to purchase ionic strength-adjusting buffers that are used to prepare standards and to dilute samples in order to achieve comparable ionic strength.
3. Temperature affects the response of ISEs, so standards must be at the same temperature as the samples are measured.
4. The standards should have the same pH as the samples.
5. Although indicator electrodes are usually specific for one substance, under certain conditions other substances may interfere with the measurement of interest. Be aware of contaminants that might adversely affect the measurement.
6. The sample must be well stirred.
7. Ion-selective electrodes are generally slower to stabilize than pH electrodes.
8. Consult the manufacturer's manual for storage directions.

20.7 CONDUCTIVITY

20.7.1 BACKGROUND

There are situations where it is useful to measure the combined concentration of all the ions in an aqueous solution (Table 20.9). The total ion concentration is typically determined by measuring the conductivity of the solution. **Conductivity** *is the ability of a material to conduct electrical current.* Metals are extremely conductive; electrons carry current through metals at almost the speed of light. Electrical current can also flow through solutions by the movement of ions, although this movement is not as rapid as current flow in metals. Thus, aqueous solutions containing ions are conductive. The higher the concentration of mobile ions in a solution, the more conductive the solution; therefore, conductivity is an indicator of total ion concentration.

TABLE 20.9

Some Uses of Conductivity Measurements

1. *If there is only one type of salt in a solution, it is possible to determine its concentration by comparing its conductivity with standards of known salt concentration.*
2. *Pure water does not readily conduct electricity, so a measurement of conductivity indicates the amount of impurities in the water.* Very low conductivity indicates high-purity water. There are a vast number of applications where the purity of water is important, such as in the production of pharmaceuticals, maintaining cells in culture, and ensuring the proper operation of industrial heating and cooling systems.
3. *Detectors that measure the conductivity of a solution can be attached to chromatography instruments and used to monitor the elution of sample components (Figure 20.11).*
4. *Conductivity can be used clinically to indicate electrolyte levels in body fluids.*
5. *Conductivity can be used to approximate "total dissolved solids" in a solution.* This measurement is frequently used, for example, in the pharmaceutical industry as an indicator of purity.
6. *Some enzyme-catalyzed reactions release ions, so changes in conductivity can be used to monitor these enzyme reactions.*
7. *Electrophoresis, a method commonly used to separate charged biological macromolecules, is dependent on using buffers with the proper conductivity.*
8. *In industrial processes, conductivity may be monitored as an indication of the concentration of an important chemical.*
9. *Conductivity may be monitored in fish tanks as an indicator of water quality.*
10. *Conductivity can be used as a routine quality-control check of laboratory reagents to see that they have been prepared properly.* The conductivity of a given reagent should be close to the same each time it is prepared. If it is not, then there was likely a mistake in the reagent's preparation or a problem with an ingredient.

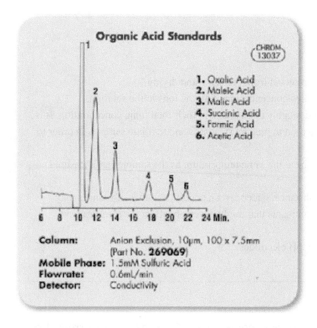

FIGURE 20.11 Display formed by a conductivity detector used to monitor the elution of sample components from a chromatography column. The Y-axis shows conductivity, and the X-axis is minutes. (Chromatogram courtesy of Dr. Maisch High Performance LC GmbH.)

We are accustomed to thinking of water as a good conductor of electricity (and so we would avoid being on a lake in a metal canoe during a lightning storm). However, because it is ions that conduct electricity in a solution, ultrapure water has a very low conductivity.

Some substances ionize more completely than others when dissolved in water, so their solutions are more conductive. **Electrolytes** *are substances that release ions when dissolved in water.* **Strong electrolytes** *ionize completely in solution.* For example, table salt completely dissolves in water to form Na^+ and Cl^- ions. A strong acid such as HCl likewise completely dissociates to form the ions H^+ and Cl^- in solution. **Weak electrolytes** *partially ionize.* **Nonelectrolytes** *(e.g., sugar, alcohols, and lipids) do not ionize at all when dissolved.* The magnitude of current that will flow through an aqueous solution depends on the temperature, the combined concentration of each electrolyte present, the degree to which each ionizes, and how readily the ions move through the solution.

20.7.2 The Measurement of Conductivity

20.7.2.1 The Design of Conductivity Instruments

The measurement of conductivity, like the measurement of pH, involves the use of an electrochemical system. A simple device to measure conductivity consists of two flat electrodes that are dipped into the sample solution

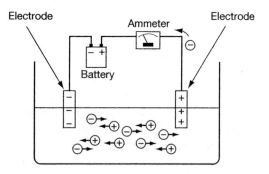

FIGURE 20.12 The measurement of conductivity. A simple meter consists of two flat electrodes, a battery, and a meter that measures current. Positive ions flow toward the negative electrode, and negative ions toward the positive electrode. Electrons flow in the wires between the electrodes and the battery and meter. This makes a complete circuit for current flow. The more ions present in the solution to carry current, the more current registered by the ammeter.

and are connected to a battery and to an **ammeter**, *a current-measuring instrument* (Figure 20.12). The battery generates electric current that can flow only if the sample solution is conductive and completes the circuit path. The electrode connected to the positive pole of the battery is positive, whereas the electrode connected to the negative pole is negative. Positive ions in the solution are attracted to the negative electrode; negative ions are attracted to the positive electrode. At the positive electrode negative ions give up electrons that flow in the wire to the positive pole of the battery. At the negative electrode positive ions take electrons. These processes result in continuing electron flow toward the negative electrode and continuing movement of ions through the solution. The more ions present in the solution, the more current is registered by the ammeter.

20.7.2.2 The Cell Constant, K

The amount of current flow that is measured by a conductivity instrument will vary depending on the distance between the two electrodes, d, and the surface area of the electrodes, A (Figure 20.13). The farther apart the electrodes, the less current is registered. Conversely, the larger the electrode area, the more current that is measured. For samples with low conductivity, it is best to configure the measuring device so that it is as sensitive as possible by having a small distance between the electrodes and/or a large electrode surface. If the sample is very conductive, however, then a wider spacing between electrodes and/or smaller electrodes are more effective.

Manufacturers make conductivity instruments with sample holders, or sample cells, having different

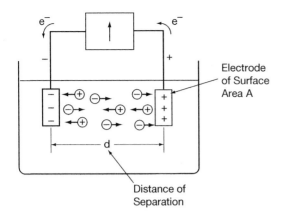

FIGURE 20.13 The configuration of a conductivity-measuring system. The surface area of the electrodes and the distance between them will affect the measurement.

configurations of distances and with different electrode surface areas. The configuration is described by the cell constant for the device. The **cell constant,** *abbreviated* K, *is defined as:*

$$K = d/A$$

where: d = the distance between the electrodes
   A = their area.
   The units of K are therefore $cm/cm^2 = 1/cm$.

The smaller the value for K, the shorter the distance between electrodes and/or the larger the electrodes. A device with a lower K value, therefore, is more sensitive than a device with a higher K value. For measuring the conductivity of high-purity water, a sensitive device with a K value of about 0.1/cm is preferred. For samples of moderate conductivity, a cell with a K value

of around 0.4/cm is useful, and for samples with very high conductivities, a cell with a K value of around 10/cm is chosen.

20.7.2.3 "Conductivity" and "Conductance"; Units Used for "Conductivity" Measurements

The terminology of conductivity measurements can be confusing, but as is true of all measurements, it is important to keep track of the units and always record them. There is a distinction between the terms *conductivity* and *conductance.* **Conductivity** is an inherent property of a material. **Conductance** *is a measured value that depends on the configuration of the conductivity-measuring device used.* Conductance is sometimes also called "measured conductivity." For example, the conductance of a sample of pure water will vary depending on the value for *K* in the measuring instrument used. The conductivity of the water sample is the same, regardless of the instrument used to measure it. The terms *resistivity* and *resistance* can similarly be distinguished from one another. Resistivity is an inherent property of the material, and resistance is a measured value.

Thus, there are four terms and four slightly different units used to describe a solution's ability to conduct current. These various units, all of which are used to describe the ability of a solution to carry current, are summarized in Box 20.6. The table may seem confusing, but the bottom line is that you must accurately record the units displayed by your conductivity instrument. Table 20.10 shows the resistivity and conductivity of some common types of solutions.

BOX 20.6   UNITS OF MEASUREMENT OF CONDUCTIVITY AND RESISTIVITY

1. ***Resistance.*** A measured value. The more readily a solution conducts current, the less resistance there is to current flow.
    Resistance is measured in ohms (abbreviated Ω).
    1 megohm (MΩ) = 1,000,000 Ω.
2. ***Conductance.*** A measured value. The conductance of a solution is the inverse of its resistance.
    Conductance = 1/resistance.
    1/ohm = 1 mho = a unit of conductance.
    1 mho = 1 siemens (abbreviated S) = another unit of conductance.
    1 micromho (μmho) = 1 microsiemens (μS) = 1 mho/1,000,000.

(Continued)

BOX 20.6 (*Continued*)   UNITS OF MEASUREMENT OF CONDUCTIVITY AND RESISTIVITY

3. ***Conductivity***. Conductance is the property that is measured, but to compare the results from various instruments, or from one laboratory, with another, it is necessary to convert conductance to conductivity. This is done using the following equation:

Conductivity = conductance (K).

The units of conductivity therefore are: (siemens) (1/cm) = siemens/cm.

In practice, a S/cm is too large, so most instruments display values as **µS/cm** or **mS/cm**. In catalogs, **µS/cm** and **mS/cm** may be abbreviated as **µS** and **mS**, respectively.

1 µS/cm = 0.001 mS/cm = 0.000001 S/cm = 1 µmho/cm.

4. ***Resistivity***. This term is:

1/conductivity.

The unit of resistivity, therefore, is ohm-cm.

Resistivity is often used in the measurement of ultrapure water due to its extremely low conductivity.

5. It is possible to convert between conductance and resistance.

For example:

Convert a conductance of 40 µS to resistance.

Conductance = 40 µS = 40×10^{-6} S = 4×10^{-5} S = 4×10^{-5} mho = $4 \times 10^{-5}/1$ ohm.

Resistance = 1/conductance.

Therefore, resistance = $(4 \times 10^{-5}/1 \text{ ohm})^{-1}$ = 1 ohm/4×10^{-5} = 25,000 ohm.

TABLE 20.10

The Conductivities and Resistivities of Common Types of Solutions

Solution	Conductivity	Resistivity
Ultrapure water	0.055 µS/cm	18,000,000 ohm-cm = 18 MΩ-cm
Distilled water	1 µS/cm	1,000,000 ohm-cm = 1 MΩ-cm
Deionized water	80 µS/cm	12,500 ohm-cm
0.05% NaCl	1000 µS/cm	1,000 ohm-cm
Seawater	50,000 µS/cm	20 ohm-cm
30% H_2SO_4	1,000,000 µS/cm	1 ohm-cm

Values shown are approximate.

20.7.2.4 Conductivity Measurements

As with any instrument, conductivity meters require standardization. Standards for conductivity measurements are commercially available.

Temperature has a large effect on conductivity. Conductivity standards will have a table to show their conductivity at various temperatures. Further notes about standardization and other practical comments about conductivity measurements are summarized in Table 20.11.

TABLE 20.11

Practical Considerations for Conductivity Measurements

1. **Standards.**
   a. Conductivity meters should be standardized before use.
   b. Conductivity standards have no buffering capacity and are easily contaminated. Always use fresh standard.
   c. Standardization is often performed with a single standard. Choose a standard whose conductivity is similar to that of the solution to be measured.
   d. Standards readily pick up contaminants from the air.
   e. Keep the standards' containers closed to avoid contamination and evaporation.
   f. A standard solution exposed to air will degrade within a day.
   g. Remember, too, that evaporation will increase the conductivity of a standard.
2. **Temperature.** Ions become excited at higher temperatures, and therefore raising the temperature increases the conductivity of a solution. The effect of temperature is different for every ion, but the conductivity typically increases by about 1–5% per °C. It is essential to use a consistent temperature when using conductivity to check the quality of solution. In industry, specific methods are used to compensate for the effects of temperature on conductivity. Note also that when reporting a conductivity value, temperature should also be reported.
3. **Depth of immersion.** The depth to which the electrodes are immersed can affect the current readings. Do not immerse to the bottom of the solution if there are particulates settled there. Some conductivity probes require deeper immersion than others; consult the manual.
4. **Air bubbles.** Make sure that air bubbles are not trapped in the probe when making measurements. Even tiny air bubbles can affect the conductivity measurement. To get rid of big air bubbles, try moving the probe up and down in the sample or standard. You can also gently tap the probe on the bottom of the beaker to dislodge remaining air bubbles.
5. **Electrode maintenance.** Probes should be rinsed with purified water after each series of measurements. Probes may be allowed to dry during storage, but they may require several minutes to stabilize when placed in a solution. Do not allow dried salts or particulates to build up on the probe.
6. **Probe configurations.** Conductivity probes come in different styles and with different cell constants. It is necessary to choose a conductivity probe that best suits the samples whose conductivities will be measured.
   a. There are four-pole probes that provide better accuracy for mid- to high-conductivity measurements than standard two-pole probes.
   b. A general-purpose probe might be of a two-pole style with a cell constant of $1.0\,cm^{-1}$. This probe is likely to work when measuring the conductivity of common laboratory reagents.
7. **Acids and bases.** Most acids and bases are considerably more conductive than their salts because H^+ and OH^- ions are extremely mobile. For example, NaOH is more conductive than NaCl.
8. **Particulates.** For greatest accuracy, make sure that there are no particulates suspended in the sample. Filter or settle out particulates.
9. **Linearity.** The relationship between conductivity and ion concentration is generally, but not always, linear. In highly concentrated solutions, the relationship between conductivity and concentration may be nonlinear due to interactions between ions.
10. **Total dissolved solids (TDS) measurement.** Conductivity is related to the concentration of all dissolved electrolytes in a solution. Conductivity, therefore, is sometimes used to indicate the concentration of "total dissolved solids" (often minerals such as calcium carbonate) in a solution. In these cases, the meter should be calibrated with a dissolved solids standard that contains the same type of solids as the sample to be tested. Otherwise, there will be serious errors in the measurement because each type of dissolved substance has a different effect on conductivity. Commercially available TDS standards include calcium carbonate (often used to test boiler water); sodium chloride (often used to test brines); potassium chloride; and a mixture of 40% sodium sulfate, 40% sodium bicarbonate, and 20% sodium chloride (often used to test water from lakes, streams, wells, and boilers).
11. **Cleaning the electrodes.** Clean probes with mild liquid detergent. Rinse several times with purified water and recalibrate before use.

Practice Problems

1. Calculate the hydrogen ion concentration of a solution whose pH is 9.0.
2. Calculate the hydrogen ion concentration of blood whose pH is 7.3.
3. The concentration of hydrochloric acid, HCl, secreted by the stomach after a meal is about 1.2×10^{-3} moles/liter. What is the pH of stomach acid?
4. Calculate the pH of 0.01 M HNO_3, nitric acid. Nitric acid is a strong acid that dissociates completely in dilute solutions. (Clue: Because the nitric acid dissociates completely, the concentration of H^+ ions will be 0.01 M.)
5. (Optional) What is the pH of 0.0001 M NaOH? (Sodium hydroxide is a strong base.) Clue: In aqueous solutions, $[H^+][OH^-] = 10^{-14}$.
6. Consider the response of a pH probe to a change in temperature as shown in the graph below (Figure 20.14).
    a. What is the slope of the line at 25°C?
    b. When the pH changes by one pH unit, for example, goes from 5 to 6, how much does the voltage change at 25°C?
    c. What is the slope of the line at 0°C?
    d. When the pH changes by one pH unit, for example, goes from 5 to 6, how much does the voltage change at 0°C?
       Note: We consider the slope to be negative. (See Chapter 13.) However, in this question, we are concerned with the magnitude of the slope and not its direction.
7. A recording from an industrial fermentation run is illustrated below (Figure 20.15). During this run, bacteria reproduced many times over and were harvested to be sold for use in commercial food production. A pH probe was present in the fermentation broth, and a continuous recording of the pH was made. When the microorganisms (the seed culture) were first added to the broth, the pH was about 6.8. The pH of the culture changed during the active growth phase. Answer the following questions about this recording.

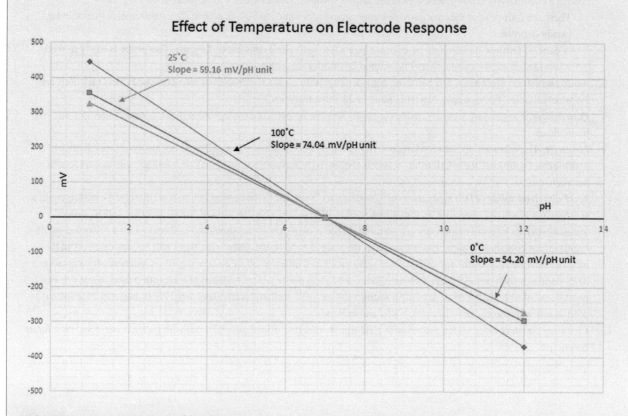

FIGURE 20.14 Graph for problem 6

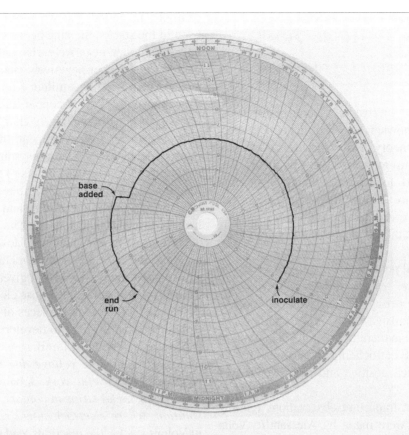

FIGURE 20.15   Chart for problem 7.

    a. What happened to the pH as the bacterial population grew?

    b. Speculate as to what might be the difficulties in designing pH probes for use in fermentation.

    c. Why do you think the company generates a continuous recording of pH during each fermentation run? (A digital recording would likely be made now, but this analog record nicely illustrates the change in pH over time.)

Questions for Discussion

1. Compare and contrast the standardization/calibration (in the sense of adjustment) of electrodes, pipettes, and balances. Based on this comparison, define *standardization or calibration* in your own words.
2. How does temperature affect the measurement of volume, weight, and pH?
3. Suggest some likely causes for the following pH measurement problems.
    a. The pH reading slowly drifts.
    b. The pH reading fluctuates.
    c. There is no reading at all.
4. Suggest a way to check each possibility you suggested in question 3.
5. If you work in a laboratory, write a mock standard operating procedure (SOP) to calibrate and operate a particular pH meter that is available in your laboratory.
6. a. Discuss how you would go about demonstrating that the pH meter is giving consistent, accurate results when measuring the pH of samples from a sewage treatment plant.
    b. Write a mock SOP to measure the pH of samples from a sewage treatment plant. Include safety precautions to protect the operator. Include precautions to avoid contamination of the meter and to avoid cross-contamination (where one sample contaminates another sample).

CHAPTER APPENDIX 20.1: A BRIEF INTRODUCTION TO ELECTROCHEMISTRY

HOW ELECTRODES GENERATE ELECTRICAL POTENTIAL (VOLTAGE)

Understanding how electrodes generate electrical potential requires knowledge of electrochemical reactions. Although the theory of electrochemistry is complex, it may be helpful to realize that this theory was devised by scientists to explain concrete, empirical observations of nature. This section briefly describes some of the early experimental observations in electrochemistry and the theory that arose to explain these observations.

Voltage (electrical potential) *is the potential energy of charges that are separated from one another.* Electrical potential arises because charged objects exert attractive or repulsive forces on one another. The development of voltage therefore requires a method of separating charges. Electrochemical reactions are one of the ways by which positive and negative charges become separated.

Some of the most important observations relating to electrochemistry were made by Alessandro Volta in the early 1800s (e.g., see Bordeau, Sanford P. *Volts to Hertz: The Rise of Electricity.* Burgess Publishing, 1982). Volta placed disks of two different metals adjacent to one another, but separated by a piece of cloth soaked in an electrolyte solution. Volta observed that one of the metal disks would acquire a positive charge and the other a negative charge. Volta assembled large stacks of disks separated by pieces of felt soaked in saltwater. He found that if he touched the top and bottom disks in the stack at the same time, he received an electrical shock.

Scientists explain the shock Volta received as being due to **electrochemical reactions** *in which metals either release electrons or take up electrons, becoming either positively or negatively charged.* Electrochemical reactions caused separation of positive and negative charges from one another, leading to the development of voltage. When Volta touched the top and bottom disks, the electrical potential caused current to flow between the disks through his body so that he was shocked.

Although Volta did not understand electricity as it is understood today, he was able to experimentally learn a great deal about electrochemical reactions. Volta made stacks with different metals and discovered that metals could be arranged in order of how likely they are to become positively charged. For example, he found that when zinc and copper were used in the stacks, the zinc became positively charged and the copper negatively charged. Iron and copper also left the copper negatively charged, but the shock Volta experienced was milder from stacks made with iron than with zinc. If copper and silver were used, the copper became positively charged. Scientists now interpret Volta's results to mean that metals differ in their tendency to lose electrons; some metals lose electrons more readily than others. The electrochemical reactions in Volta's stacks resulted in movement of electrons from the metal that had more tendency to lose electrons to other metals. The ordering of metals according to their tendency to lose electrons is now taught as the "activity series of metals." Scientists have investigated many metals and given numerical values to their relative tendency to lose electrons.

The idea of combining pieces of metal and an electrolyte solution in order to develop voltage has useful applications. The most familiar example is a **battery,** *which harnesses the voltage due to electrochemical reactions to do useful work.* A battery consists of two **electrodes,** *metal strips in contact with an electrolyte solution, where electrochemical reactions occur.* As in Volta's stacks, the reactions lead to the development of voltage. When the two electrodes are connected via a metal wire, current flows.

Ion-selective electrodes, such as pH electrodes, are another application that takes advantage of electrochemical reactions. Volta observed that the magnitude of the potential developed by his stacks was controlled by both the nature of the metals involved and the concentration of certain ions in the electrolyte solution. The higher the concentration of the ions, the higher the potential. This relationship between ion concentration and potential is the basis for ion-selective electrodes, such as pH-measuring electrodes, which develop a greater or lesser potential depending on the concentration of a particular ion in a solution. Thus, the design of pH meter measuring systems is based on the observations of Volta and other scientists dating back more than 100 years.

A NOTE ABOUT THE NERNST EQUATION

If you study pH measurements in more detail, you will encounter the Nernst equation. The discussion of standardization in this chapter is based on that equation. Recall that we described the relationship between pH and millivolt response for an ideal Ag/AgCl electrode. We noted that the relationship is linear with a slope of -59.16 mV/pH unit. The Nernst equation is the formal equation for this relationship. The Nernst equation, not surprisingly, was derived by a scientist named Walther Nernst, a theoretical chemist who studied energy transformations. This is the full form of the Nernst equation:

$$E = E_0 - \frac{2.3RT}{nF}(pH)$$

where: E is the measured potential from the electrode
E_0 is related to the potential of the reference electrode
R and F are constants that come from physics
n is the charge of the ion of interest, in this case, H^+, so n = 1
T is the temperature.

The Nernst equation is related to Figure 20.6.

- E is the response of the electrode in millivolts. If you look at Figure 20.6, you will see that the Y-axis of the calibration line we discussed is the millivolt response read on the meter.
- If you look at Figure 20.6, you will see that the X-axis of the calibration line we discussed is pH.
- The **slope** is the most intimidating part of the Nernst equation; it is (*2.3RT/nF*). It turns out that when the values for the constants 2.3, R, and F are substituted into the equation, when n = 1, and when the temperature is 25°C, then

the value for the slope is −59.16 mV/pH unit. As we discussed in the standardization section of this chapter, −59.16 mV/pH unit is the slope associated with an ideal electrode; the Nernst equation describes the performance of an ideal electrode.
- E_0 is the offset, which ideally is 0 at pH 7.0.

The derivation of the Nernst equation, and why it has the constants 2.3, F, and R, is beyond the scope of this book. Nonetheless, we have seen the practical significance of this equation. Using the Nernst equation, it is possible to evaluate the performance of a pH probe. As the slope and offset of the calibration line for a pH probe begin to deviate from the ideal relationship described by the Nernst equation, we know that the probe performance is declining. That is the basis for Table 20.6. Also, we can see in the Nernst equation that the slope of the calibration line is dependent on the temperature. Therefore, as we discussed, it is critical to consider temperature in order to obtain accurate and reproducible pH measurements.

CHAPTER APPENDIX 20.2: MORE ABOUT TROUBLESHOOTING pH MEASUREMENTS

Table 20.12 lists four common symptoms and the causes that might be associated with those symptoms.

Boxes 20.7–20.9 provide advice from manufacturers for solving common pH measurement problems.

TABLE 20.12
Common Symptoms and Causes of pH Measurement Problems

Possible Causes	Possible Solutions
SYMPTOM: Reading Drifts, Fluctuates, and Takes a Long Time to Stabilize	
1. Gel-filled electrodes are slower to respond than refillable electrodes, and this slow response may be perceived as drift.	1. Wait longer or consider purchasing a refillable electrode.
2. Unstable sample temperature.	2. Allow sample temperature to stabilize. Use ATC probe.
3. Nonhomogeneous sample.	3. Stir the sample.
4. The sample may be undergoing chemical reactions or may be reacting with CO_2 from the atmosphere.	4. Consider the sample chemistry. Use autoread function on meter.
5. The sample and electrodes may not be at thermal equilibrium.	5. Allow them to come to thermal equilibrium.

(Continued)

TABLE 20.12 (*Continued*)

Common Symptoms and Causes of pH Measurement Problems

Possible Causes	Possible Solutions
6. The sample may be inherently difficult to pH and may require special techniques or specific types of electrodes.	6. See Table 20.7.
7. The filling solution may be contaminated (refillable electrodes only).	7. Look for precipitates and crystals inside the electrode (if possible). • Excessive crystallization can occur if the electrode was stored for a long time with the filling hole open or was stored in the cold. • If the electrode was used to determine the pH of a sample when the filling solution was low, then the filling solution might have been contaminated by the entrance of sample. • Try replacing the filling solution with fresh electrolyte (see Box 20.5).
8. The reference electrode junction may be occluded by crystals or may be dirty. This is a common cause of sluggish or drifting response.	8. With a liquid-filled reference electrode, a clogged junction can be detected by applying air pressure to the filling hole. A bead of electrolyte should appear readily on the junction. If not, the junction is likely fully or partially occluded. If the junction is occluded, clean it as described in Box 20.7.
9. Long immersion in high-purity water can damage the junction.	9. The reference electrode (or the combination electrode) may need replacement.
10. The silver wire inside the probe is damaged.	10. This is less likely, but can occur due to long exposure to sulfides, proteins, heavy metals, and pure water. If this is the problem, replace the probe.
11. The measuring electrode bulb may be dirty.	11. To clean the bulb, see Box 20.8.
12. The response of the glass that constitutes the measuring bulb can decline due to prolonged use, excessive alkaline immersion, or high-temperature operation.	12. Try rejuvenating the glass electrode as shown in Box 20.9.

SYMPTOM: Difficulty in Standardization/Difficulty Standardizing to Two Buffers/"Calibration Error" Message Appears/Low Percent Efficiency or High Offset after Standardization/Poor Result of Linearity Check

1. Problem with buffer standards.	1. Use new bottles of buffers.
2. The filling solution may be depleted or contaminated.	2. Replace the electrolyte as described in Box 20.5 (refillable electrodes only).
3. The measuring electrode may be dirty.	3. Clean the bulb as described in Box 20.8.
4. The response of the glass that constitutes the measuring bulb can decline.	4. Rejuvenate the electrode as shown in Box 20.9.
5. Probe has aged and needs replacement.	5. See if a new or different probe resolves the problem.

SYMPTOM: The Value for a Buffer Standard or Sample Seems Wrong

1 The buffer standards have deteriorated.	1. This is the most likely problem; try fresh buffers.
2 The sample may not be homogenous.	2. Stir the sample.
3 The measuring electrode may be dirty.	3. Clean the bulb as described in Box 20.8.
4 The response of the glass that constitutes the measuring bulb has declined.	4. Rejuvenate the electrode as described in Box 20.9.

(Continued)

TABLE 20.12 (*Continued*)

Common Symptoms and Causes of pH Measurement Problems

Possible Causes	Possible Solutions

SYMPTOM: *There Is No Response at All or the Value Displayed on the Meter Does Not Change or Fluctuate in the Slightest*

1. The meter is malfunctioning. (A meter is the least likely component of the system to malfunction.)	1. Make sure the meter is plugged in and turned on. • Check the connections between the electrode and meter. • Connect the probe to another meter, if available, and see if this corrects the problem.
2. The cable connecting the meter and probe needs to be replaced.	2. Try another cable, if available. Try gently bending the connector cable to see if this changes its response. 3. Consult the meter manufacturer.

BOX 20.7 CLEANING THE JUNCTION OF A REFERENCE ELECTRODE

These methods are suitable for combination electrodes or individual reference electrodes. If you have an electrode with a sleeve junction, it is easily cleaned by moving the sleeve.

Refill the electrode filling solution after cleaning.

The first methods are least drastic and should be tried first. Continue to the next step only if the previous one has failed.

1. *If possible, inspect the reference cavity for excessive crystallization.* If there are excessive crystals, replace the old filling solution as discussed in Box 20.5 (refillable electrodes only).
2. *For gel-filled references, soak the probe's tip in warm water (about 60°C) for 5–10 minutes.* Do not heat the water above 60°C.
3. *Dissolve crystals from the end of the electrode by soaking it in a solution of 10% saturated KCl and 90% purified water for between 20 minutes and 3 hours.* Warm the solution to about 50°C. Immerse the electrode about 2 inches into the solution. Refillable electrodes can be soaked overnight in this solution.
4. *Soak the electrodes in a 0.1 N HCl or 0.1 M HNO_3 solution for 15 minutes.*
5. *Use a commercially available junction cleaner.*
6. *Remove any proteins that coat the outside of electrodes or penetrate into the junction.* Recommendations for removing protein deposits include rinsing in an enzyme detergent and using pepsin/HCl cleaners (5% pepsin in 0.1 M HCl). After removing the protein deposits, rinse the electrode and empty and replace the filling solution (refillable electrode only). Note that with combination electrodes, this method removes proteins that have deposited on the glass bulb at the same time as the junction is cleaned.
7. *Immersion in ammonium hydroxide may be helpful for refillable Ag/AgCl electrodes that are clogged with silver chloride.* Empty the filling solution, and immerse the electrode in 3 to 4 M NH_4OH for only 10 minutes. Ammonia can damage the internal element of the electrode if the electrode is soaked in it too long. (**NH_4OH is very caustic and the fumes are irritating. Use it in a fume hood.**) Rinse the inside and outside of the electrode thoroughly with purified water, and add fresh filling solution. If the probe is a combination electrode, soak it 10–15 minutes in pH 4.00 buffer.
8. *It may be possible to clear the junction by forcing filling solution through the junction. Do so by applying mild pressure to the filling hole or mild vacuum to the junction tip.*
9. *As a last resort, for ceramic junctions only, sand the junction with #600 emery paper.* For combination electrodes, be careful not to contact the glass bulb.

BOX 20.8   CLEANING THE MEASURING ELECTRODE BULB

Note that electrode cleaning methods generally involve soaking them in various chemicals and do not involve attempting to remove contaminants by rubbing with a cloth or tissue. However, it is sometimes possible to remove dirt and debris by gently cleaning with a cotton swab.

1. *Remove dirt with warm soapy water.* A mild laboratory dishwashing detergent can be used.
2. *Remove protein contaminants by washing in warm water with an enzyme detergent or 5% pepsin in 0.1 N HCl followed by rinsing in water.* If your application routinely involves many pH measurements of protein solutions, it is good practice to clean the bulb on a regular basis.
3. *Remove inorganic deposits by washing with EDTA, ammonia, or 0.1 N HCl for about 3 minutes.* Rinse the bulb with water, and soak it in dilute KCl.
4. *Remove grease and oil with methanol, acetone, or a laboratory soap solution.* Then rinse the bulb with water, and soak it in dilute KCl.
5. *To remove fingerprints, wipe the bulb gently with a 50/50 mixture of acetone and isopropyl alcohol.* Soak the electrode in a commercial soaking solution or pH 4.00 buffer for 1 hour.
6. *Alternatively, manufacturers provide convenient cleaning solutions for a variety of contaminants such as proteins, bacteria, and oils.* Follow their directions for these products.

BOX 20.9   "REJUVENATING" A MEASURING ELECTRODE BULB

Note that solutions used for soaking and conditioning a pH-measuring electrode are usually slightly acidic. This is because H^+ ions from the soaking solution replace contaminants in the glass of the bulb.

Some manufacturers recommend the following steps when the linearity of the system is poor, or the system response is sluggish.

The first methods are least drastic and should be tried first. Continue to the next step only if the previous one has failed.

1. *Soak the bulb in pH 4.00 buffer overnight.*
2. *Soak the bulb in 1 N HCl for 30 minutes.*
3. *Immerse the bulb in 0.1 N HCl for about 15 seconds, rinse with water, and then immerse in 0.1 M KOH (or 0.1 N NaOH) for 15 seconds.* Repeat several times and rinse. Soak the electrode in pH 4.00 buffer overnight.

Safety Note: Manufacturers sometimes recommend rejuvenating glass bulbs with ammonium bifluoride or with hydrofluoric acid. Ammonium bifluoride is hazardous and should be used with caution. Hydrofluoric acid is extremely hazardous! It will degrade glass containers. It can cause serious or fatal injuries even at low concentrations. We do not recommend its use except by those who are specially trained and have access to appropriate safety equipment.

21 Measurements Involving Light – Part A
Basic Principles and Instrumentation

21.1 LIGHT

21.1.1 INTRODUCTION

Previous chapters discussed the measurement of certain physical properties of a sample, including its weight, volume, and temperature. This chapter discusses the measurement of another physical characteristic of a sample: its ability to absorb light.

Light is essential for life. Almost every living creature on earth's surface is reliant on the energy from sunlight. For example, light enables humans to see, and light sustains plant growth. Light is able to play these roles because it interacts with molecules.

In the laboratory, we take advantage of the interaction of light with materials in samples to obtain information about those samples. **Spectrophotometers** *are instruments used to measure the effect of a sample on a beam of light.* This chapter, like other chapters in this unit, is about obtaining trustworthy measurements, in this case, measurements involving light. This chapter begins by exploring the nature of light. It then introduces spectrophotometers and how their design, operation, and performance enable us to obtain useful measurement results. Chapter 28 continues the discussion of how spectrophotometers are used to obtain information about biological systems. For example, we will see in Chapter 28 that spectrophotometers can be used to estimate how much DNA is present in an extract from cells, to estimate the activity of an enzyme, and to confirm the identity of an ingredient in a drug formulation.

21.1.2 ELECTROMAGNETIC RADIATION

21.1.2.1 The Electromagnetic Spectrum

Light is a type of **electromagnetic radiation,** that is, *energy that travels through space at high speeds.* There

DOI: 10.1201/9780429282799-26

are different categories of electromagnetic radiation, including gamma rays, X-rays, ultraviolet (UV) light, visible (Vis) light, infrared (IR) light, microwaves, and radio waves. *All the types of electromagnetic radiation together are termed the* **electromagnetic spectrum.** Visible, ultraviolet, and infrared radiation are all called "light" even though our eyes are only sensitive to visible light.

Each type of radiation within the electromagnetic spectrum has a characteristic way of interacting with matter. For example, X-rays, which have great energy, can easily penetrate matter. We take advantage of this characteristic when producing X-ray images for medical use. Cells in our eyes are able to interact with light in the visible region of the electromagnetic spectrum, leading to vision.

Scientists describe electromagnetic radiation as having a "dual nature." This is because sometimes electromagnetic radiation is best explained as consisting of waves moving through space with crests and troughs, much like waves in a pond. Other times electromagnetic radiation is portrayed as "packets of energy" moving through space. *Each "packet of energy" is called a* **photon**.

Figure 21.1 illustrates the wavelike nature of light. *The distance from the crest of one wave to the crest of the next is the* **wavelength** *(abbreviated λ) of the radiation.* The wavelength of electromagnetic radiation varies greatly depending on its type. For example, the wavelengths of X-rays are measured

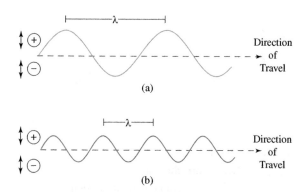

FIGURE 21.1 Electromagnetic radiation traveling as waves. Electromagnetic radiation is often visualized with peaks and troughs, similar to waves of water. Unlike water waves, however, electromagnetic radiation is not associated with the motion of matter. Rather, electromagnetic radiation is associated with regular changes in electric and magnetic fields. The electromagnetic radiation in (a) has a longer wavelength and carries less energy than the radiation in (b)

in nanometers, whereas radio waves can approach 10,000 meters (Figure 21.2).

Different types of electromagnetic radiation not only have different wavelengths, but also carry different amounts of energy. The shorter the wavelength of electromagnetic radiation, the more energy is carried by its photons. X-rays have very short wavelengths and consist of photons that carry a great deal of energy. In contrast, radio waves, which have long wavelengths, have photons with much less energy.

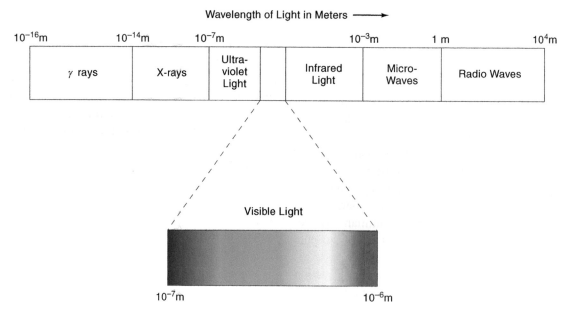

FIGURE 21.2 Types of electromagnetic radiation and their wavelengths. Visible light occupies a small portion of the electromagnetic spectrum. Within the visible portion of the spectrum, different wavelengths are associated with different colors.

21.1.2.2 The Visible Portion of the Electromagnetic Spectrum

Consider the portion of the electromagnetic spectrum to which the human eye is sensitive. Humans perceive different wavelengths of visible light as being different colors (Table 21.1). For example, if a light source emits light with wavelengths between about 505 and 555 nm, that light is perceived as green. "White" light emitted by the sun or an electric light bulb is **polychromatic;** *it is a mixture of many wavelengths.*

21.1.2.3 The UV Portion of the Electromagnetic Spectrum

Ultraviolet light is categorized into three types: UV-A (315–400 nm), UV-B (280–315 nm), and UV-C (180–280 nm) (Table 21.2). UV-A is the type used in tanning booths and "black lights." The illumination from a "black light" causes certain pigments to emit visible light; these materials are called **fluorescent**.

UV-B light has sufficient energy to damage biological tissues and is known to cause skin cancer. The atmosphere blocks most UV-B light; however, as the ozone layer thins, this form of radiation has become more of a health concern. UV-B light is routinely used in the biotechnology laboratory when visualizing DNA. The dye, ethidium bromide, intercalates into the grooves of DNA, forming a complex that fluoresces when exposed to UV-B radiation. The fluorescence makes the DNA visible to the eye and allows it to be photographed. There are other dyes that similarly fluoresce when exposed to UV light and can be attached to biological molecules. Fluorescence emitted by molecules allows investigators to visualize their location and will be discussed in more detail in Chapter 29.

UV-C, with high energy and short wavelengths, is rarely observed in nature because it is absorbed by oxygen in the air. UV-C is used in germicidal lamps to kill bacteria. Most hoods manufactured for culturing cells are equipped with UV-C lamps to help prevent bacterial contamination of the cultures. UV-C light damages DNA in cells and therefore can be used in research to intentionally induce mutations in bacteria and cultured cells.

Safety Note: There are sources of UV light in the laboratory, and it is important to avoid exposing your skin or eyes to this type of radiation. Safety glasses and goggles that specifically block UV radiation are available from scientific supply companies. Be careful to avoid exposing your hands or wrists to UV radiation.

TABLE 21.1

The Relationship between the Wavelength of Visible Light and the Color Perceived

λ of the Light (nm)	Color Our Eyes Perceive
380–430	Violet
430–475	Blue
475–495	Greenish-blue
495–505	Bluish-green
505–555	Green
555–575	Yellowish-green
575–600	Yellow
600–650	Orange
650–780	Red

Note: Because color is subjective, these wavelengths are approximations. The particular ranges here are based on various sources.

TABLE 21.2

Ultraviolet Light

Type	Wavelength (nm)	Examples of Applications
UV-A ("long wave")	315–400	Used in "black lights" and tanning booths.
UV-B ("medium wave")	280–315	Used in electrophoresis to visualize DNA tagged with ethidium bromide.
UV-C ("short wave")	180–280	Used in bactericidal lamps and for inducing mutations in cells.

21.1.3 INTERACTIONS OF LIGHT WITH MATTER

Various events can occur when light strikes matter (Figure 21.3). (1) The light can be **transmitted** (*i.e., pass without interaction through the material, as when light passes through glass*). (2) The light can be **reflected** (*i.e., change directions, as is light reflected by a mirror*). (3) The light can be **scattered**, *in which case the light is deflected into many different directions.* Scattering occurs when light strikes a substance composed of many individual, tiny particles. (4) The light can be **absorbed**, *in which case it gives up some or all of its energy to the material.* When light is absorbed, energy carried by photons is transferred to the electrons within a material. As this transfer occurs, light energy is converted to heat energy. Light absorption provides the basis for spectrophotometric measurements.

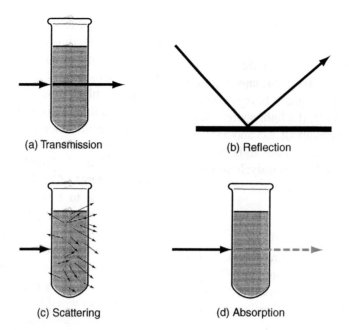

(a) Transmission

(b) Reflection

(c) Scattering

(d) Absorption

FIGURE 21.3   Events that can occur when light strikes matter. (a) The light may be transmitted through the object without interaction. (b) The light may be reflected by the surface of the object. (c) The light may be scattered by small individual particles. (d) The light may lose energy to the matter in which case it is absorbed.

Because absorption is the basis of spectrophotometry, we will consider it in more detail. Every material has a particular arrangement of electrons and of bonds involving electrons. For this reason, different materials absorb different wavelengths of light. The color of an object is determined by which wavelengths of light it absorbs.[1] If an object absorbs light of a particular color, then that color does not reach our eyes when we look at the object. If an object absorbs all colors of light, then that object appears to be black, as in Figure 21.4a. If an object reflects visible light of a particular color, we see that color when we look at the object. An object appears white if it reflects all colors of light (Figure 21.4b). The object in Figure 21.4c appears to be orange because it reflects orange light and absorbs all other colors.

The object in Figure 21.4d reflects all colors except blue. When blue light is removed from the mixture of wavelengths, our eyes perceive the object to be orange. Orange and blue are called *complementary colors*. Figure 21.5 shows a color wheel in which complementary colors are across from one another. Looking at the color wheel, we can determine, for example, that if green is removed from a beam of light, then the light will appear to be red.

In Figure 21.4, we illustrate solid objects that either absorb or reflect light. When we use a spectrophotometer, we are usually working with samples that are in liquid form. These liquids may either absorb or transmit light of a particular wavelength. The spectrophotometer enables us to quantify how much light is absorbed and how much is transmitted by a solution at a specific wavelength.

Table 21.3 is a guide that shows the general relationship between the color of light absorbed by a liquid solution and the color the solution appears to the eye. For example, if a solution absorbs blue light (light in the range from about 430 to 475 nm), then the solution will appear to be its complementary color, orange.

---

[1] *Color* is not truly a property of light, nor of objects that reflect light. It is actually a sensation that arises in the brain. Specialized cone cells in the eye transmit signals to the brain when stimulated by light of specific wavelengths.

Example Problem 21.1

What color will a solution appear to be if it absorbs primarily light that is 590 nm?

Answer

Light with a wavelength of 590 nm appears yellow to our eyes. From Table 21.3, you can see that a solution that absorbs primarily yellow light will appear to be violet. Yellow and violet are complementary colors.

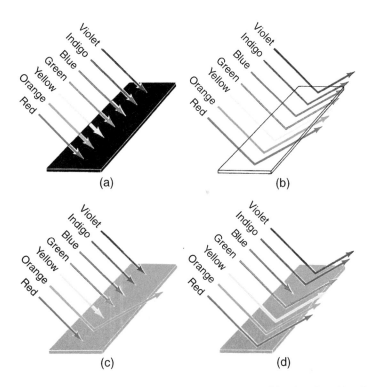

FIGURE 21.4 The colors of solid objects. (a) An object appears black if it absorbs all colors of light. (b) An object appears white if it reflects all colors. (c) An object appears orange if it reflects only this color and absorbs all others. (d) An object also appears orange if it reflects all colors except blue, which is complementary to orange.

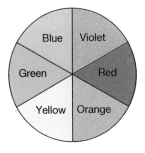

FIGURE 21.5 A color wheel showing complementary colors.

Consider the example illustrated in Figure 21.6. In Figure 21.6a, white light is shining on a test tube that contains a solution of red food coloring. The food coloring absorbs incoming light that is green. It also absorbs some light that is bluish-green and greenish-blue. Light of other colors is transmitted through the food coloring. Because blue and greenish light is removed from the white light shining on the red food coloring, our eyes perceive the solution to be orange-red. (See Table 21.3.) A spectrophotometer enables us to graphically display this pattern of light absorption by the red food coloring solution (Figure 21.6b). In this display, wavelength is on the X-axis. A measure of the amount of light absorbed by the food coloring is on the Y-axis. Observe on the graph that there is a peak between about 490 and 530 nm. This peak represents the absorption of light at those wavelengths. The graph in Figure 21.6b, *that shows the extent to which a particular material absorbs different wavelengths of light, is called an* **absorbance spectrum**.

TABLE 21.3

The Absorption of Light of Particular Wavelengths and the Color of Solutions

Wavelength (nm) of Light Absorbed by Liquid Solution	Color of Light Absorbed by Liquid Solution	Color Solution Appears to Be
380–430	Violet	Yellow
430–475	Blue	Orange
475–495	Greenish-blue	Red-orange
495–505	Bluish-green	Orange-red
505–555	Green	Red
555–575	Yellowish-green	Violet-red
575–600	Yellow	Violet
600–650	Orange	Blue
650–780	Red	Green

Note: Colors that appear across from one another on this table are complementary colors.

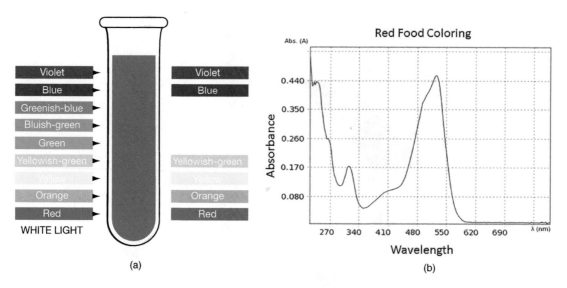

FIGURE 21.6   The absorption of light by a solution of red food coloring. (a) When white light is shined on a tube containing red food coloring, blue to green colored light is absorbed. (b) The absorbance spectrum for red food coloring, generated by a spectrophotometer. A peak of visible light absorption occurs between about 490 and 530 nm.

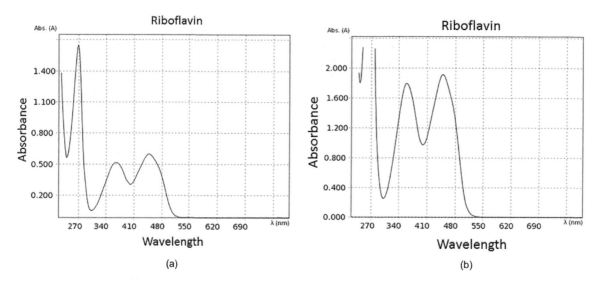

FIGURE 21.7   The absorbance spectrum for riboflavin. (a) The peaks and valleys of this absorbance spectrum are characteristic of riboflavin. (b) Increasing amounts of riboflavin have the same spectral pattern; however, the height of the peaks is greater if more compound is present.

Note in Figure 21.6b that red food coloring also absorbs light that has wavelengths shorter than 350 nm. This light, however, is in the ultraviolet range and therefore has no effect on the solution's apparent color.

Figure 21.7a shows the absorbance spectrum, as determined using a spectrophotometer, for the compound riboflavin. The absorbance spectrum of riboflavin has a pattern of "peaks" and "valleys" that is different than that of red food coloring. Every compound has its own characteristic absorbance spectrum. The absorbance spectrum of a particular compound always has the same pattern of peaks and valleys, regardless of how much of that compound is present (Figure 21.7b).

Example Problem 21.2

Solutions of proteins and of nucleic acids appear colorless. Does this mean they do not absorb any light?

Answer

It means they do not absorb any light in the *visible* range of the electromagnetic spectrum. Proteins and nucleic acids do, in fact, absorb light in the *ultraviolet* range of the spectrum.

Example Problem 21.3

What color is the solution whose absorbance spectrum is shown in the following graph?

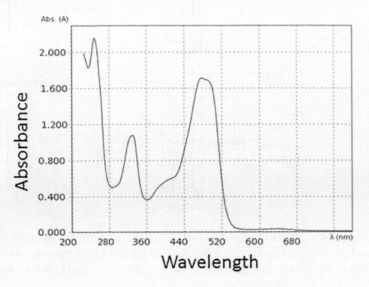

Abs. (A)

Absorbance

Wavelength

λ (nm)

Answer

This compound has an absorbance peak in the greenish-blue region of the spectrum. From Table 21.3 we would expect it to be orange, and it is, in fact, the dye Orange G.

Example Problem 21.4

Chlorophyll is one of the pigments responsible for the green color of leaves. Chlorophyll enables the leaves to absorb light from the sun and convert the energy of the light into food. Does chlorophyll capture all wavelengths of light from the sun?

Answer

No. Leaves are green because chlorophyll does not absorb much green light and those wavelengths are reflected.

Example Problem 21.5

What happens if red light is shined on a red solution?

Answer

The solution will not absorb the light; rather, it will transmit the light.

21.2 THE BASIC DESIGN OF A SPECTROPHOTOMETER

21.2.1 TRANSMITTANCE AND ABSORBANCE OF LIGHT

21.2.1.1 Transmittance

Figure 21.8 illustrates the basic principles underlying the operation of a spectrophotometer. In this illustration, two solutions are placed inside a spectrophotometer. The first is red food coloring dissolved in water. The analyte (the substance of interest) is red food coloring, and water is the solvent. The second solution is pure water (i.e., the solvent used to dissolve the red food coloring). The water is called the **blank**. (We consider the meaning of a blank in more detail later.) A beam of green light is incident on (shines on) both solutions. As we know from Figure 21.6, the red food coloring will transmit little or no green light. In contrast, the water will readily transmit green light. A spectrophotometer electronically compares the amount of light transmitted through the sample with that transmitted through the blank. *The ratio of the amount of light transmitted through a sample to that transmitted through the blank is called the* **transmittance**, abbreviated *t*:

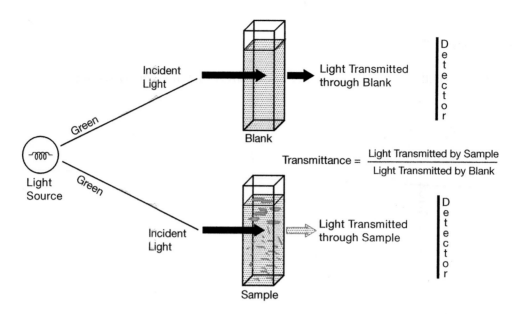

FIGURE 21.8 Transmittance. Two solutions are placed inside a spectrophotometer: red food coloring and a blank consisting of pure water. Green light is shined on the two solutions, and the amount of light transmitted through each is measured by a detector. The ratio of the amount of light transmitted through the sample to that transmitted through the blank is called the *transmittance*.

$$\frac{\text{light transmitted through the sample}}{\text{light transmitted through the blank}} = \text{transmittance} = t$$

It is also common to speak of **percent transmittance, %T**, *which is the transmittance times 100%*:

$$\%T = t \times 100\%$$

In the example illustrated in Figure 21.8, the transmittance is less than 1 because the amount of light transmitted through the sample is less than the amount transmitted through the blank. (The red food coloring in the sample absorbs some of the light.) In another situation, where both the sample and the blank transmit the same amount of light, the transmittance is 1, or %T is 100%. If the sample transmits no light at all, then the transmittance is 0. Thus, transmittance can range from 0 to 1, and percent transmittance can range from 0% (no light passed through the sample relative to the blank) to 100% (all light was transmitted).

21.2.1.2 Absorbance

Spectrophotometers measure transmittance (i.e., the ratio between the amount of light transmitted through a sample and the amount of light transmitted through a blank). Analysts, however, are generally interested in the amount of light *absorbed* by a sample, for example as illustrated in Figure 21.6 and Figure 21.7. It is customary, therefore, to convert transmittance to a measure of light absorption, called **absorbance.**

Absorbance *is calculated based on the transmittance*[2] *as*:

$$A = -\log_{10}(t)$$

Absorbance is also called **optical density, OD**.

Absorbance does not have units, but absorbance values are sometimes followed by the abbreviation A or AU. For example, an absorbance value might be written as "1.6," or "1.6 A," or "1.6 AU," or "OD 1.6." Transmittance and absorbance are measured at a particular wavelength, so it is customary to record both the absorbance and the wavelength. For example, "$A_{260} = 1.6$," means "an absorbance of 1.6 was measured at an analytical wavelength of 260 nm."

Figure 21.9a shows the relationship between how much light is transmitted through a solution and the concentration of analyte in that solution. Observe that the relationship between analyte concentration and transmittance is not linear. In contrast, there is a linear relationship between the absorbance of light by a sample and the concentration of analyte (Figure 21.9b). For this reason, absorbance values are more convenient than transmittance values.

---

[2] An alternative equation to calculate absorbance is $A = 2 - \log_{10}(\%T)$. Both this equation and the one in the body of the text will give the same answer.

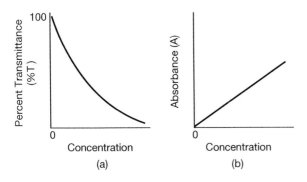

FIGURE 21.9 The relationship between the concentration of analyte in a solution and the % transmittance and absorbance values. (a) Concentration of analyte in a sample versus percent transmittance is not linear. (b) In contrast, the relationship between absorbance and analyte concentration is linear.

As one would expect, when the absorbance of light increases, the transmittance decreases, and vice versa (Table 21.4). For example, if the percent transmittance is 100%, then there is no light absorbance by the sample relative to the blank. Suppose in another case that a sample absorbs so much light that the percent transmittance is only 0.1%. In this example, the transmittance (t) is 0.001 and the absorbance is:

$$A = -\log(0.001) = -\log\left(10^{-3}\right) = -(-3) = \mathbf{3}$$

Most spectrophotometers can convert transmittance values to absorbances, so they can display either absorbance or transmittance. It is also possible to use a calculator to convert transmittance to absorbance, and vice versa.

TABLE 21.4

The Relationship between Absorbance and Transmittance

$A = -\log_{10}(t)$

t	%T	A
0.000	0.00	∞
0.001	0.10%	3
0.010	1.00%	2
0.100	10.0%	1
1.000	100%	0

Example Problem 21.6

Convert a transmittance value of 0.63 to absorbance.

Answer

$$A = -\log(t)$$

$$A = -\log(0.63)$$

$$A \approx -(-0.20) = \mathbf{0.20}$$

Example Problem 21.7

Convert an absorbance value of 0.380 to transmittance and % transmittance.

Answer

$$0.380 = -\log(t)$$

$$\text{antilog}(-0.380) = t$$

$$\mathbf{0.417} \approx \mathbf{t} \text{ or } \mathbf{41.7\%} \approx \mathbf{\%T}$$

Example Problem 21.8

A solution contains 2 mg/mL of a light-absorbing substance. The solution transmits 50% of incident light relative to a blank. (a) What is the absorbance in this case? (b) What is the expected absorbance of a solution containing 8 mg/mL of the same substance? (c) What is the expected transmittance of a solution containing 8 mg/mL of the same substance?

Answer

a. The absorbance of 2 mg/mL of this substance is equal to $-\log(0.50) \approx \mathbf{0.30}$.
b. The relationship between absorbance and concentration is linear (within a certain range, as discussed in a later chapter). Assuming linearity (and assuming that if there is no analyte in the sample, the absorbance is zero), if 2 mg/mL has an absorbance of 0.30, then 8 mg/mL should have an absorbance of **1.20**.

c. If the absorbance is 1.20, the transmittance = antilog (–1.20) ≈ **0.063 = 6.3%T**. Observe that the relationship between concentration and transmittance is not linear.

21.2.1.3 The Blank

A spectrophotometer is used to determine the absorbance of light by the analyte in a sample. In order to make this determination, a spectrophotometer always compares the interaction of light with a sample with its interaction with a blank, as illustrated in Figure 21.8. Let's consider in more detail the composition of the blank.

An analyte is seldom alone in a sample solution; rather, it is dissolved in a solvent and sometimes various reagents are added to the sample as well. When the sample is placed in the spectrophotometer, not only the analyte may absorb light, but the solvent and added reagents may also absorb light. Moreover, the sample is contained in a *holder, called a* **cuvette** (or **cell**), which may absorb a small amount of light. The **blank** *should contain no analyte, but it should contain the solvent and any reagents that are intentionally added to the sample.* The blank is held in a cuvette that is identical to that used for the sample, or the blank is alternately placed in the same cuvette as the sample. In the example in Figure 21.8, the blank is pure water because the analyte, red food coloring, was dissolved in water. In another situation, the blank may have a different composition.

21.2.2 Basic Components of a Visible/ UV Spectrophotometer

21.2.2.1 Overview

We can now consider the various components of a spectrophotometer. A basic spectrophotometer consists of (1) a source of light, (2) a wavelength selector, (3) a sample holder, (4) a detector, and (5) a display. There are various arrangements of these five components that we will discuss. The most basic of these arrangements is diagrammed in Figure 21.10, and we will begin by discussing this configuration.

21.2.2.2 Light Source

The **light source,** also called a **bulb** or **lamp,** *produces the light used to illuminate the sample.* There are various types of lamps available for spectrophotometry. All lamps produce polychromatic light, but each type emits a different range of wavelengths. The light used for visible spectrophotometry is usually produced by bulbs with a tungsten filament. *The filament is sealed into a transparent bulb that is filled with an inert gas and a small amount of a halogen, such as iodine or bromine.* This type of bulb is therefore called a halogen or **tungsten-halogen lamp.** A **deuterium arc lamp** *is commonly used to produce light in the UV range.* Spectrophotometers that can be used in both the visible and UV ranges have two light sources. Alternatively, a **xenon lamp,** *which contains xenon gas sealed in a bulb,* can be used to produce light in both the UV and visible ranges.

21.2.2.3 Wavelength Selector

Light emitted by a spectrophotometer bulb is polychromatic, but the absorption of light by a sample is wavelength specific. Spectrophotometry measures how much light of a particular wavelength is absorbed by a sample. Spectrophotometers, therefore, require a wavelength-selecting device. The wavelength-selecting component of a spectrophotometer isolates light of a particular wavelength (in practice, it may isolate a narrow range of wavelengths) from the polychromatic light emitted by the source lamp.

Some instruments, called **photometers,** use filters as an inexpensive wavelength selector. A **filter** *blocks all light above or below a certain cutoff* (Figure 21.11a and b). Two filters can be combined, as shown in Figure 21.11c, to make a filter that transmits light in a certain range.

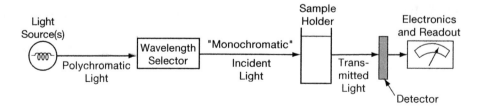

FIGURE 21.10 Simplified design of a spectrophotometer. Polychromatic light is emitted by the light source. A narrow band of light wavelengths is selected in the wavelength selector. The light passes through the sample (or the blank), where some or all of it may be absorbed. The light that is not absorbed falls on the detector.

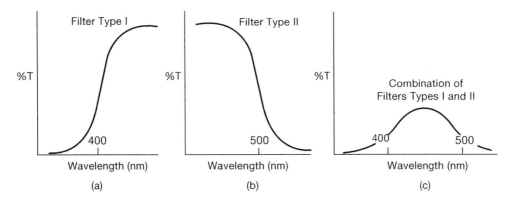

FIGURE 21.11   Filters. (a) A filter that blocks the transmission of light below 400 nm. (b) A filter that blocks the transmission of light above 500 nm. (c) Two filters can be combined to allow only light in a certain wavelength range to pass. In this example, light between 400 and 500 nm is transmitted through the filters.

Filters cannot produce **monochromatic light** *(i.e., light of a single wavelength)*. It is more effective to use a **monochromator** to produce light of a very narrow range of wavelengths. **Monochromators** *are devices that can disperse (separate) polychromatic light into its component wavelengths and that can select light of desired wavelengths.* A monochromator disperses light, much as light is dispersed into a rainbow by a prism (Figure 21.12).

A monochromator has three main components (Figure 21.13). These components are as follows:

1. **Entrance slit**. Light from the lamp is emitted in all directions. The entrance slit shapes the light entering the monochromator into a parallel beam that strikes the dispersing element.
2. **Dispersing element**. In earlier spectrophotometers, prisms were used to disperse light into its components. In modern instruments, **diffraction gratings** are the most common type of dispersing element. **Diffraction** *is the*

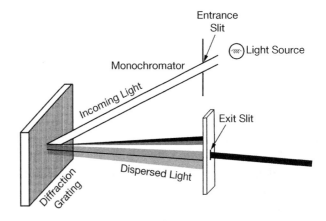

FIGURE 21.13   The components of a monochromator. The monochromator consists of an entrance slit, a dispersing element, and an exit slit. The dispersing element is a diffraction grating which consists of a series of microscopic grooves etched onto a mirrored surface.

bending of light that occurs when light passes an obstacle. **Diffraction gratings** *consist of a series of microscopic grooves etched onto a mirrored surface.* The grooves diffract light that strikes them in such a way that polychromatic light is dispersed into its component wavelengths.

3. **Exit slit**. The exit slit permits only selected light to exit the monochromator.

After light is dispersed by the diffraction grating, it is necessary to select light of the desired wavelength. This is accomplished by rotating the grating so that the light that reaches the exit slit is of the desired wavelength (Figure 21.14). The accuracy of wavelength selection in this type of spectrophotometer depends on the correct mechanical operation of the diffraction grating.

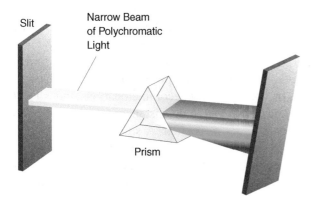

FIGURE 21.12   The dispersion of light into its component wavelengths by a prism.

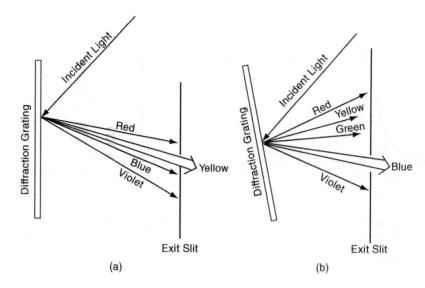

FIGURE 21.14   Rotating the grating of a monochromator so that light of a selected wavelength falls on the exit slit. (a) The monochromator selects for yellow light. (b) The monochromator selects for blue light.

If a perfect monochromator could be produced, and if the spectrophotometer were set to select a particular wavelength, then only light of that single wavelength would illuminate the sample (Figure 21.15a). In reality, the light that emerges from a monochromator normally consists of light spanning a narrow range of wavelengths, centered at the chosen wavelength (Figure 21.15b).

Modern spectrophotometers are well designed so that, in most situations, the light emerging from the monochromator has a sufficiently narrow span that it

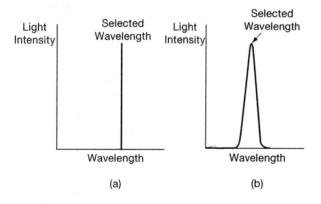

FIGURE 21.15 The light emerging from a monochromator. (a) If the light from a monochromator were truly monochromatic, then when the instrument was set for a certain wavelength, only that wavelength would be incident on the sample. (b) In reality, the light incident on a sample often consists of a narrow band of wavelengths, centered on the selected wavelength. (Note that this figure is not an illustration of a sample spectrum, as are Figures 21.6b and 21.7. Rather, it illustrates the wavelengths of light that are incident on the sample.)

can be treated as if it were truly monochromatic. There are situations, however, where the "monochromaticity" of the light incident on the sample affects an analysis, and more expensive instruments with higher-quality monochromators are preferred. The Appendix to this chapter discusses issues relating to the "monochromaticity" of light exiting the monochromator.

21.2.2.4 Sample Chamber

The sample (or the blank) is contained in a cuvette, which is placed inside a **sample chamber**. The **path length** *is the distance that the light travels through the sample or blank; it is determined by the interior dimension of the cuvette.* It is customary to use a 1 cm path length when working with sample volumes of 1–5 mL. Manufacturers make systems, including cuvettes and cuvette holders, that permit the analysis of small volumes, down to the microliter range. Being able to work with small volumes is useful in molecular biology applications where reaction mixes and samples are often less than 50 μL. Note that the path length in small-volume systems may not be 1 cm.

It is possible to replace the cuvette and sample holder with a fiber-optic probe. Fiber-optic cables are widely used in the telecommunications industry to transmit light over long distances without intensity losses or signal dispersion. A similar technology can be used in spectrophotometers to guide light from the source lamp through a fiber-optic cable to a probe. The end of the probe is dipped into the sample, which may be contained in almost any type of vessel; hence, these are called *dip probes*. Light interacts with the sample at the end of the dip probe tip. The path length of the dip probe may be 1 cm in length, or it may be narrower. Depending on the nature of the molecules in the sample, some of the light will be absorbed. The remaining light is then reflected back through a second fiber-optic cable that leads to a detector inside the spectrophotometer. The probe is connected to the spectrophotometer via a cable that may be a couple of meters long; this has the advantage that the sample can be located apart from the spectrophotometer.

21.2.2.5 Detector

The **detector** *senses the light coming through the sample (or the blank) and converts this information into an electrical signal*. The magnitude of the electrical signal is proportional to the amount of light reaching the detector. We are all familiar with a detector that converts light energy into an electrical signal; the human eye is such a detector. The eye converts light energy into a nerve signal that carries information to the brain. The human eye detection system is only sensitive to light in the range from about 400 to 700 nm. Similarly, detectors in spectrophotometers have a limited range in which they are sensitive, although that range is wider than the range of the human eye.

A **photomultiplier tube (PMT)** is a common detector in the type of spectrophotometer we are describing. A PMT contains a series of metal plates coated with a thin layer of a **photoemissive material** *that emits electrons when it is struck by photons*. Electrons are held loosely by the photoemissive material, so the energy of light is sufficient to "knock" electrons from the plates. Many electrons are released by each incident photon, and each electron can knock out even more electrons, resulting in a cascade that amplifies the original light signal. The stream of emitted electrons constitutes an electrical signal that is processed by the detector assembly and is converted to a reading of transmittance and/or absorbance. More recently, silicon photodiode detectors have been introduced as an alternative to PMTs. **Silicon photodiode detectors** *are semiconductor-based devices* that are less expensive than PMTs and do not require a dedicated power supply.

All detectors have limitations. First, detectors are not accurate at very high or very low levels of transmittance. It is best, therefore, to work with samples whose absorbance is in the mid-range of the instrument. Second, detectors are more responsive to some signals than others. In spectrophotometry, this means that the detector will respond more strongly to some wavelengths of light than others.

21.2.2.6 Display

There are many ways that the signal from the detector can be manipulated and displayed. The simplest method is to use a meter to display the resulting value. A meter has a needle that moves to indicate the value on a scale of either transmittance or absorbance. With modern instrumentation, the signal is converted into a digital display. Most of the instruments in use now have the capability to send the signal to a printer and/or to a computer for display, storage, and analysis.

21.3 DIFFERENT SPECTROPHOTOMETER DESIGNS

21.3.1 SCANNING SPECTROPHOTOMETERS

The spectrophotometer illustrated in Figure 21.10 is a **single-beam instrument** *in which the blank and sample are not simultaneously in the instrument*. Instruments with this basic design have successfully been used for years. In the earliest single-beam spectrophotometers, the operator had to reset the instrument to 100% transmittance (or zero absorbance) with the blank each time the wavelength was changed. This process of manually selecting each wavelength, inserting the blank, setting the instrument to 100% transmittance, removing the blank, and then inserting the sample each time the wavelength was changed is tedious when an absorbance spectrum is being prepared. Manufacturers, therefore, developed instruments called **scanning spectrophotometers** that are capable of rapidly scanning through a range of wavelengths and constructing an absorbance spectrum. (Recall that an absorbance spectrum is a plot of the absorbance of a sample at different wavelengths, as shown in Figures 21.6b and 21.7.) One approach to designing a scanning spectrophotometer takes advantage of microprocessor technology. In this type of instrument, the operator first places the blank in the sample compartment. The spectrophotometer scans the absorbance of the blank at all the wavelengths of interest, as a microprocessor inside the instrument "memorizes" the values. The blank is then removed from the sample compartment, and the sample is placed in the instrument and is similarly scanned across the wavelength range. The microprocessor compares the absorbance of the blank to the absorbance of the sample at each wavelength and generates a sample spectrum automatically.

Another approach that speeds up the process of creating an absorbance spectrum is to use a **double-beam scanning spectrophotometer**, as is illustrated in Figure 21.16. Double-beam instruments have two sample holders. Both the sample and the blank are

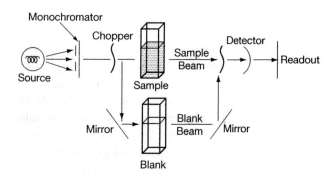

FIGURE 21.16 A double-beam scanning spectrophotometer. Light emitted by the source passes through a monochromator. After the monochromator, the "chopper" rapidly alternates the beam between the sample and the blank. The output of the detector is proportional to the ratio between the blank and the sample beams and so is a measure of transmittance.

placed in the instrument at the same time so that the absorbance of the sample can be continuously compared with that of the blank. The instrument alternates the light beam between the blank and the sample many times per second. Since the blank is always in the instrument, the operator does not need to manually reinsert it each time the wavelength is changed.

Scanning spectrophotometers often allow the operator to control the *rate at which the sample is scanned, the* **scan speed**. The scan speed is restricted to the rate at which the instrument can respond. A scan speed that is too fast creates an effect called *tracking error*. When **tracking error** *occurs, the absorbance peaks recorded for a sample are slightly shifted from their true locations.*

The instruments described so far, whether single beam or double beam, and whether or not they are microprocessor controlled, are termed "scanning spectrophotometers." In these instruments, light is dispersed by the monochromator *before* it reaches the sample. The monochromator moves to control the wavelength of light leaving the exit slit and reaching the sample, as illustrated in Figure 21.14. There is another style of spectrophotometer, termed **array spectrophotometer**, *in which the monochromator is located after the sample.* The original array spectrophotometers were expensive, high-end devices. In recent years, they have become more accessible and will be discussed in the next section.

21.3.2 ARRAY SPECTROPHOTOMETERS

Array spectrophotometers use a **photodiode array detector (PDA)** or a **charge-coupled device (CCD) detector**. CCDs are the same type of sensor that is used in digital cameras. **PDAs and CCDs** *both use a*

photosensitive semiconductor material to convert light into an electrical signal. We saw that PMT detectors can only measure the absorbance (or transmittance) of a sample one wavelength at a time and therefore acquire a spectrum over a period of seconds or minutes. In contrast, PDAs and CCDs *can determine the absorbance of a sample over the entire UV/Vis range virtually instantaneously.*

In an array spectrophotometer, all the light from the source is directed at the sample rather than being dispersed in a monochromator. There is a monochromator located *after* the sample chamber that disperses the light that has been transmitted through the sample or the blank (Figure 21.17). The dispersed, transmitted light strikes the PDA or CCD detector. For every wavelength in the spectrum, the computer compares the amount of light transmitted by the sample and the blank. The computer can convert this information to a complete absorbance spectrum for the sample. Generally, array spectrophotometers are single-beam instruments, which means that the there is one sample holder and the blank and sample are inserted into the instrument sequentially.

An array spectrophotometer has the advantage that it can produce an absorbance spectrum far more quickly than a scanning spectrophotometer. Also, the monochromator position is fixed; it does not need to rotate to select each wavelength. This makes array instruments more rugged and often smaller than scanning spectrophotometers.

21.3.3 MICROPROCESSORS AND SPECTROPHOTOMETERS

Microprocessors play a wide range of roles in modern spectrophotometers. Microprocessors control wavelength scanning, generate absorbance spectra, store and retrieve information, perform calculations and statistics

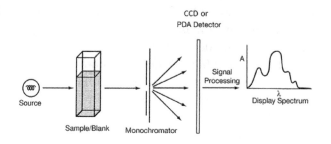

FIGURE 21.17 The optical design of an instrument with a PDA or CCD detector. The monochromator is located after the sample compartment. A PDA or CCD detector can analyze all the wavelengths of light simultaneously. A computer compares the amount of light transmitted by the sample and the blank at every wavelength and converts this information to a complete absorbance spectrum for the sample.

on data, and plot data on graphs. A variety of software is available that simplifies routine operations. Many of the spectrophotometry applications that will be described in Chapter 28 are now automated. As has been discussed in previous chapters on balances and pH meters, spectrophotometer manufacturers have modified their software so that their spectrophotometers can be operated in compliance with 21 CFR Part II requirements.

21.4 MAKING MEASUREMENTS WITH SPECTROPHOTOMETERS

21.4.1 MORE ABOUT THE BLANK

Recall that the transmission of light through a sample is always compared with the transmission of light through a blank. This is true whether the instrument is a scanning or an array-style spectrophotometer. "Blanking" the instrument compensates for the following:

1. There may be absorption of light by materials in the sample other than the analyte of interest, such as the solvent and reagents. This absorbance is subtracted using the blank.
2. The cuvette may absorb some light. Light absorbance by the cuvette is subtracted by placing the blank in the same cuvette as the sample, or in a cuvette that is nearly identical.
3. There may be variations in the light source due to age or power fluctuations. The sample and blank experience the same, or nearly the same conditions.
4. The intensity of light emitted by any light source varies with wavelength. For this reason, the sample and blank must be compared at every wavelength to compensate for this effect.
5. The detector is more sensitive to some wavelengths of light than others. The use of a blank at every wavelength compensates for this effect.

21.4.2 THE CUVETTE

Glass and certain types of plastic are transparent to light in the visible range, but they block UV light. These materials are therefore used to make cuvettes for work in the visible range (Figure 21.18a). Quartz glass is transparent to visible and UV light, so it is used to make cuvettes for both visible and ultraviolet work (Figure 21.18b). There are also disposable plastic cuvettes that may be used down to about 285 nm (Figure 21.18c). Note that the optical quality of plastic cuvettes may not be suitable for some applications.

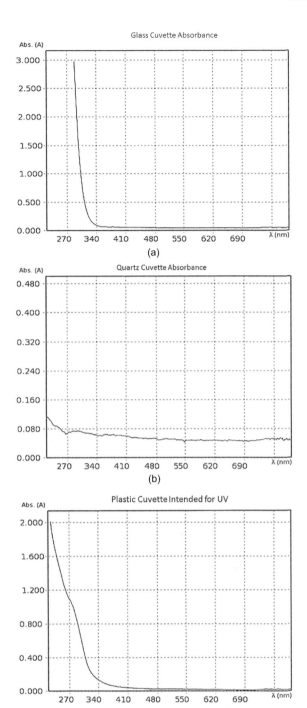

FIGURE 21.18 Absorbance of light by cuvettes made of various materials. (a) The absorbance spectrum of a glass cuvette intended for work in the visible range. This cuvette is unsuitable for work below about 330 nm. (b) The absorbance spectrum of a quartz cuvette that is suitable for work down to at least 200 nm. (c) The absorbance spectrum of a disposable polyacrylate plastic cuvette sold for work in the UV range. Note that this cuvette can work for some UV analyses, but it does absorb substantially more light in the UV range than a quartz cuvette. The advantage to plastic cuvettes is that they are much less expensive than quartz. (Air was used as the blank for all three spectra.)

TABLE 21.5

Proper Selection and Care of Cuvettes

1. *Use quartz cuvettes for UV work; glass, plastic, or quartz are acceptable for work in the visible range.* As shown in Figure 21.18, there are disposable plastic cuvettes that can work for some analyses in the UV range.
2. *Matched cuvettes are manufactured to absorb light identically so that one of the pair can be used for the sample and the other for the blank.*
3. *Make sure the cuvette is properly aligned in the spectrophotometer.*
4. *Do not touch the base of a cuvette or the sides through which light is directed.*
5. *Do not scratch cuvettes; do not store them in wire racks or clean with brushes or abrasives.*
6. *Do not allow samples to sit in a cuvette for a prolonged period of time.*

Even the highest quality cuvettes absorb a small amount of light. It is possible, therefore, to purchase paired cuvettes that are manufactured to be identical in their light-absorbing properties. One of the pair is used for the blank and the other for the sample.

Quartz and glass cuvettes are carefully manufactured, precision optical devices that must be properly maintained. Small scratches and dirt will absorb or scatter light and therefore damage the cuvette. Table 21.5 summarizes considerations relating to cuvettes. Box 21.1 outlines methods for determining whether two cuvettes match, for checking the cleanliness of cuvettes, and for cuvette cleaning.

BOX 21.1 CLEANING AND CHECKING CUVETTES

1. *Be certain to only use clean cuvettes.*
2. *Unless using disposable plastic cuvettes, wash cuvettes immediately after use. Recommended washing methods include the following*:
   - Rinse with distilled water immediately after use.
   - Wipe the outside of glass and quartz cuvettes with lens paper.
   - After an aqueous sample, wash in warm water, rinse with dilute detergent, and thoroughly rinse with purified water.
   - Avoid detergents with lotions. Be certain all detergent is removed after washing.
   - For organic samples, rinse thoroughly with a spectroscopy-grade solvent.
   - If simple rinses are not sufficient to remove all traces of samples, an acid cuvette washing solution may be used. For example, to make 1 liter of washing solution, combine:
     425 mL of distilled water
     525 mL of ethanol and
     50 mL of concentrated acetic acid.
   Do not allow acid to remain in the cuvette for more than 1 hour. Rinse thoroughly with distilled water.
   - An enzyme-containing detergent is sometimes recommended to remove dried proteins from cuvettes.
   - Avoid blowing air into cuvettes to dry them.
3. *For high-accuracy work, the cleanliness of cuvettes and matching response of two cuvettes can be checked as follows* (*from ASTM E 275-08 "Standard Practice for Describing and Measuring Performance of Ultraviolet, Visible, and Near-Infrared Spectrophotometers"*):
   a. Set the instrument to 100% T (or 0 A) with only air in the sample holder.
   b. Fill the cuvette with distilled water, and measure its absorbance at 240 nm (quartz cuvette) or 650 nm (glass). For scanning instruments, scan across the spectral region of interest. The absorbance should not be greater than 0.093 for 1 cm quartz cuvettes or 0.035 for glass.
   c. Rotate the cuvette 180° in its holder, and measure the absorbance again. Rotating the cuvette should not change the absorbance by more than 0.003.
   d. If an instrument is to be used with two cuvettes, one for the sample and one for the blank, check how well the two cuvettes are matched. Fill each with solvent, and measure their absorbances. The absorbance difference between the two cuvettes should be less than 0.01.

21.4.3 THE SAMPLE

Samples should be well mixed and homogenous without air bubbles. Particulates should be avoided (except in certain applications where turbid samples, such as bacterial suspensions, are assayed).

The solvent is an important part of the sample. Although solvents, such as water, buffers, or alcohol, appear transparent, they absorb light at certain wavelengths in the UV range. At wavelengths below a particular cutoff value, solvents absorb so much light that they interfere with the analysis of a sample. Table 21.6 is a general guide to typical solvents for UV/Vis work and the approximate cutoffs at which they begin to absorb light. The actual absorbance cutoffs for solvents vary depending on their grade and purity, the pH, the calibration of the spectrophotometer, and other factors. When working in the UV range, therefore, it may be necessary to prepare an absorbance spectrum of the solvent to check whether it absorbs appreciable amounts of light at the analytical wavelength(s). Solvents that are manufactured specifically for use in spectrophotometry and related applications are labeled "Spectroscopy" or "Spectro-Grade."

Figure 21.19a shows the absorbance spectrum for a commonly used biological buffer, Tris. Tris absorbs light in the lower UV region of the spectrum, but does not absorb much light above 230 nm. Figure 21.19b shows the absorbance for denatured alcohol (ethanol). Note the strong absorbance peak around 270 nm, a region of importance in the analysis of DNA, RNA, and proteins. Figure 21.19c is the spectrum for absolute (pure) ethanol, which has little absorbance above about 240 nm. Figure 21.19d is the spectrum for acetone.

TABLE 21.6

Approximate UV Cutoff Wavelengths for Commonly Used Solvents

Solvent	UV Cutoff Where Solvent Begins to Absorb Light (nm)
Acetone	320
Acetonitrile	190
Benzene	280
Carbon tetrachloride	260
Chloroform	240
Cyclohexane	195
95% Ethanol	205
Hexane	200
Methanol	205
Water	190
Xylene	280

Example Problem 21.9

You place a blank in a spectrophotometer, but the blank has so much absorbance that the instrument cannot be set to zero absorbance. How should you proceed?

Answer

Think of factors that might account for this absorbance, such as:

1. The blank might be cloudy or turbid. Check its appearance. Cloudy or turbid blanks (and samples) generally should be avoided.
2. If you are working in the ultraviolet range, make sure the cuvette is quartz or plastic suitable for UV use. If not, use a different cuvette.
3. Check that the cuvette is properly placed in the sample holder and is in the correct orientation. Many cuvettes have two sides that are transparent to light and two sides that are not transparent. The cuvette must be placed properly in the sample holder so that light can penetrate the cuvette.
4. Check the cuvette to make certain it is clean, without fingerprints or liquid droplets on the outside.
5. Some solvents appear transparent, but absorb light in the UV range. Be certain that the solvent does not absorb light at the wavelength being used; see, for example, Table 21.6.

21.5 QUALITY CONTROL AND PERFORMANCE VERIFICATION FOR A SPECTROPHOTOMETER

21.5.1 PERFORMANCE VERIFICATION

21.5.1.1 Overview

The performance of a spectrophotometer needs to be periodically checked, and the results documented. If necessary, the instrument must be repaired and/or calibrated to bring its readings in accordance with the values of accepted standards. To decide whether an instrument is performing satisfactorily, its performance is compared with specifications established by

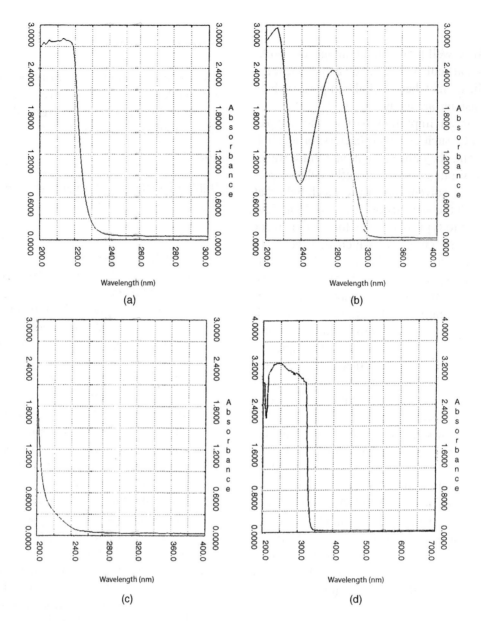

FIGURE 21.19 The absorbance spectra of various solvents. (a) 1 M Tris (Sigma) buffer absorbs strongly below about 230 nm. (b) Denatured ethanol absorbs strongly in the UV region. (Fisher Scientific denatured ethanol containing 1% ethyl acetate, 1% methyl isobutyl ketone, 1% hydrocarbon solvent, and methanol.) (c) The absorbance spectrum of pure ethanol (Fisher Scientific). (d) The absorbance spectrum for acetone (general-use grade, Caledon Laboratories Limited).

the manufacturer for each model. (In regulated facilities, spectrophotometers will also need to meet specifications established by regulatory authorities.) The methods that are used for verifying the performance of an instrument come from various sources, including manufacturers, the U.S. Pharmacopeia, ASTM International, and other standards and regulatory authorities.

The frequency with which the performance of a spectrophotometer is verified depends on the laboratory, the uses of the instrument, and whatever regulations or standards apply. Instruments that are handled roughly, or are used in environments that are dusty,

have chemical vapors, or vibration will require more frequent calibration and maintenance than other instruments.

Modern spectrophotometers are sufficiently complex that the individuals who operate the instruments are often not the ones who check their performance. Performance verification is often performed by trained service personnel, as is most repair and adjustment. Regardless of who performs the performance checks and repairs on a spectrophotometer, however, the operator should be aware of the characteristics of the instrument that affect its capabilities, range, and its ability to give accurate, reproducible results.

Understanding these characteristics is important, for example, in the following situations:

- *When purchasing a spectrophotometer, it is important to choose a model whose capabilities match the applications for which it will be used.*
- *When developing a spectrophotometry method for others to follow, the originator must describe the instrument characteristics required to duplicate the method.*
- *When following standard methods written by others, an analyst must ascertain that her/his spectrophotometer meets the requirements of the method.*
- *In situations where "difficult" samples are analyzed or high accuracy is required, it is important to have an instrument with the necessary features.* For example, if the absorbance of samples with very high or very low levels of analyte is to be measured, some models of spectrophotometer will perform better than others.

The performance characteristics that determine the accuracy, precision, and range of operation of a spectrophotometer include (but are not limited to) the following:

- **Calibration**, which has two components:
  - **Wavelength accuracy**
  - **Photometric (absorbance scale) accuracy**
- **Linearity of response**
- **Stray light**
- **Noise**
- **Baseline stability**
- **Resolution** (*discussed briefly here and in more detail in Appendix to this chapter*).

Note: For microprocessor-controlled instruments, the performance of the software should also be verified as described by the manufacturer.

This section explains the performance characteristics of a spectrophotometer. More information on calibration and performance verification is provided in the articles listed in this unit's Bibliography.

21.5.1.2 Calibration

Calibration of a spectrophotometer brings the readings of an individual instrument in accordance with nationally accepted values. Calibration is therefore a part of routine quality control/maintenance for a spectrophotometer. There are two parts to calibrating a spectrophotometer: **wavelength accuracy** and **photometric accuracy**. *Wavelength accuracy is the agreement between the wavelength the operator selects and the actual wavelength that exits the monochromator and shines on the sample.* **Photometric accuracy,** or **absorbance scale accuracy,** *is the extent to which a measured absorbance or transmittance value agrees with the value of an accepted reference standard.*

Consider wavelength accuracy. Recall that a monochromator provides a small range of wavelengths, not a single wavelength; see Figure 21.15. If an instrument is well calibrated with respect to wavelength accuracy, then when it is set to 550.0 nm, the center of the wavelength peak incident on the sample will be 550.0 nm plus or minus a certain tolerance.

Wavelength accuracy is determined using certified standard reference materials (SRMs) available through NIST, or standards that are traceable to NIST. An absorbance spectrum for the reference standard is prepared in the instrument whose performance is being checked. The absorbance peaks for reference standards are known, so the wavelengths of the peaks generated by the instrument can be checked for accuracy. The manufacturer specifies the wavelength accuracy of a given instrument. For example, a high-performance instrument may be specified to a wavelength accuracy with a tolerance of ± 0.5 nm. A less expensive instrument may be specified to have a wavelength accuracy with a tolerance of ± 3 nm. Figure 21.20 shows the spectrum for a commonly used reference standard, a **holmium oxide filter**.

Wavelength accuracy needs to be periodically checked. Wavelength accuracy can deteriorate if the source lamp, associated mirrors, and other parts are not properly aligned. This might occur, for example, after replacing a bulb. Problems leading to wavelength inaccuracy can also arise in the monochromator and in the display device.

Photometric accuracy ensures that if the absorbance of a given sample is measured in two different spectrophotometers at the identical wavelength and under the same conditions, the readings will be the same and will correspond to accepted values. This is a difficult objective to achieve, even with properly calibrated instruments, because spectrophotometers differ in their optics and designs.

Photometric accuracy is determined using SRMs available through NIST, or standards that are traceable to NIST. NIST establishes the transmittance of its standards using a special high-performance

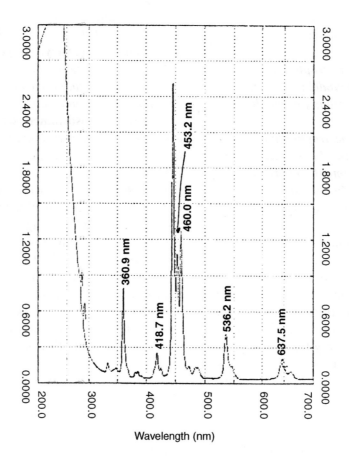

FIGURE 21.20 The spectrum of a holmium oxide filter used to test the wavelength accuracy of a spectrophotometer. The filter is placed in the spectrophotometer, and its spectrum is generated. If the instrument is properly calibrated with respect to wavelength, then the peaks in the spectrum will correspond to the specified wavelengths. (Based on wavelengths in ASTM E 275-08, "Standard Practice for Describing and Measuring Performance of Ultraviolet, Visible, and Near-Infrared Spectrophotometers.")

spectrophotometer. Thus, in the United States, the meaning of a transmittance value (and therefore of an absorbance value) is defined by NIST.

For many biological applications of spectrophotometry, photometric accuracy is not critical. As long as an individual instrument is internally consistent and its readings are linear and reproducible, it will work effectively to measure the concentration of analytes, as is discussed in detail in Chapter 28. There are, however, applications where a properly calibrated instrument is essential. This is the case, for example, where spectrophotometric measurements from various laboratories are being compared. Spectrophotometers used in regulated workplaces are likely to require absorbance scale (photometric accuracy) calibration with traceable standards.

21.5.1.3 Stray Light (Stray Radiant Energy)

Stray light *is radiation that reaches the detector without interacting with the sample.* This may occur

if light is scattered by various optical components of the monochromator or sample chamber. This scattered light does not pass through or interact with the sample, but some of it may fall on the detector, resulting in an erroneous transmittance value. Scratches, dust, and fingerprints on the surface of cuvettes can also deflect light from its proper path. Stray radiation is reduced by keeping surfaces and cuvettes clean and by avoiding fingerprints and scratches. Another source of unwanted light is leaks of room light into the sample chamber.

When the absorbance of a sample is very high, the amount of transmitted light reaching the detector becomes vanishingly small. If there is stray light present, that stray light will be erroneously considered as transmitted light. The presence of stray light, therefore, is a factor that limits the accuracy of a spectrophotometer at high absorbance values.

Figure 21.21 shows the effect of stray radiation on the relationship between concentration and absorbance.

That relationship should ideally be linear at all absorbance values. In Figure 21.21, *r* is a value that represents the level of stray light: The higher the value for *r*, the more stray light present. Given an instrument with high levels of stray light (i.e., where *r* = 10%), the relationship between absorbance and concentration begins to be nonlinear at absorbances of less than 1. A moderately priced spectrophotometer might be expected to have a stray light value on the order of 1%; therefore, it can be expected to be linear in response until the absorbance is between 1.5 and 2. More expensive spectrophotometers may have negligible levels of stray light and may respond in a linear fashion at absorbances of 3 or more.

21.5.1.4 Photometric Linearity

Linearity of detector response *is the ability of a spectrophotometer to yield a linear relationship between the intensity of light hitting the detector and the detector's response.* It is critical that the detector's response is proportional to the amount of light incident on it. A spectrophotometer may fail to respond in a linear fashion due to stray light, problems in the detector, the amplifier, the readout device, or a monochromator exit slit that is too wide.

21.5.1.5 Noise and Spectrophotometry

Recall from Chapter 16 that electrical noise includes background electrical signal arising from random, short-term "spikes" in the electronics of an instrument. These electrical "spikes" occur in the absence of a sample.

Signal in a spectrophotometer is electrical current arising in the photodetector that is related to the interaction of light with a sample (or with a blank). Signal is what we are interested in when using an instrument. At low concentrations of analyte, the electrical noise in a spectrophotometer can interfere with the measurement of absorbance, as illustrated in Figure 21.22. Figure 21.22a shows the absorbance spectrum for an undiluted sample of methylene blue. Methylene blue has three main peaks at about 240, 280, and 665 nm; the last peak has a "shoulder" at about 620 nm. This spectrum, with its peaks and valleys, is the "signal." Note that in Figure 21.22a the absorbance scale on the Y-axis runs from 0.0000 to 3.0000. In Figure 21.22a, electronic noise is not evident because there is ample signal from the sample. Figure 21.22b is the spectrum of a diluted solution of methylene blue that was prepared by taking 1 part methylene blue and adding 24 parts of water. The pattern of peaks and

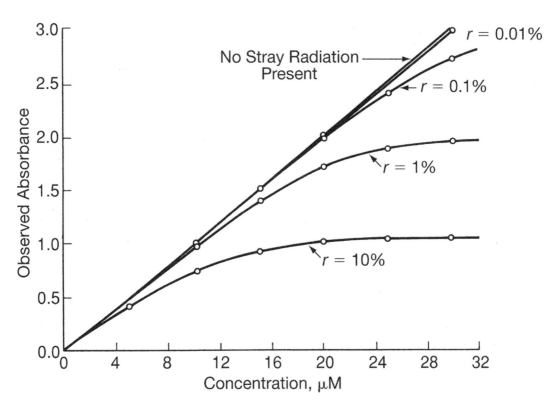

FIGURE 21.21 The effect of stray light. In the absence of stray light, the relationship between concentration and absorbance is perfectly linear. When stray light is present, it causes the relationship to deviate from linearity at higher absorbance levels. The more stray light (*r*) present (expressed here as a percent), the lower the absorbance level at which this deviation occurs.

Example Problem 21.10

A sample is expected to have an absorbance of 2 absorbance units. If a significant level of stray light is present (e.g., 1%), will the absorbance of the sample be greater than, less than, or equal to 2?

Answer

The absorbance will appear to be lower than 2 because stray light reaching the detector is interpreted as transmitted light.

valleys characteristic of methylene blue is difficult to distinguish. In order to better visualize the spectrum of this diluted sample, it is replotted in Figure 21.22c, with the Y-axis expanded to run from zero to 0.1320 absorbance. In the expanded plot in Figure 21.22c, the peaks and valleys characteristic of methylene blue are reasonably clear, but the "spikes" due to electrical noise also become barely visible. Figure 21.22d shows the spectrum of a very dilute sample of methylene blue that was prepared by taking 1 part of the original solution of methylene blue and adding 149 parts of water. The characteristic pattern of methylene blue is not detectable. If the Y-axis is expanded so that it runs from zero to 0.0100, then the signal of the methylene blue peaks can be barely discerned "buried" in the electrical noise (Figure 21.22e). Thus, at low levels of sample, electrical noise limits our ability to distinguish the pattern due to the sample.

21.5.1.6 Resolution

Figure 21.23 shows two absorbance spectra for benzene; however, the two spectra have a different appearance. In the first spectrum, the peaks can be clearly distinguished from one another. In the second spectrum, they largely have "run together." We say that the first spectrum has better **resolution,** meaning that *the individual peaks can be better distinguished from one another.*

Resolution (in spectrophotometry) is a property of an instrument. A more expensive, high-performance spectrophotometer is able to provide better resolution for a given sample than a less expensive instrument. Having an instrument with "good" resolution is important in qualitative analysis where distinctive peaks are used to identify a substance. "Good" resolution is also important in the

analysis of a sample containing more than one compound whose absorbance peaks are close together. The better the resolution of the instrument, the better the substances with close absorbance peaks can be distinguished.

The resolution of an instrument is primarily determined by how "monochromatic" the light exiting the monochromator is. The narrower the range of wavelengths incident on the sample, the better the resolution of the instrument. High-performance spectrophotometers are designed to minimize the wavelength range incident on the sample. In an instrument's specifications, the "monochromaticity" of the instrument and therefore its resolution may be expressed in terms of the "spectral bandwidth" or "spectral slit width." The smaller the value given, the more monochromatic the light and the better the resolution of the instrument. Resolution and spectral bandwidth are discussed in more detail in the Appendix to this chapter.

A practical way to check the resolution of an instrument is to prepare an absorbance spectrum of a reference material that has two or more peaks whose wavelengths are close together. The instrument's ability to distinguish these peaks is an indication of its resolution. For example, a holmium oxide filter has three peaks between 440 and 470 nm. If a holmium oxide filter is placed in the sample compartment and scanned, the three peaks should be distinct, given a properly functioning instrument with "good" resolution (Figure 21.20).

21.5.2 Performance Specifications

The choice of a spectrophotometer depends on the applications for which it will be used. For research and for qualitative work, a more expensive instrument with better resolution is desirable. A spectrophotometer that can scan an absorbance spectrum and that has both UV and visible capabilities is often required. In laboratories where a spectrophotometer is used only for routine colorimetric assays (some of which will be described in Chapter 28), less expensive models may be sufficient. The samples analyzed will also determine any special features required. Molecular biology experiments frequently involve very low volume samples, so microvolume cuvette systems become valuable. Another important issue in choosing a new spectrophotometer is its software capabilities. Ease of operation and availability of software to simplify biological applications are factors to consider. Table 21.7 contains the performance specifications for a high-quality, high-resolution spectrophotometer.

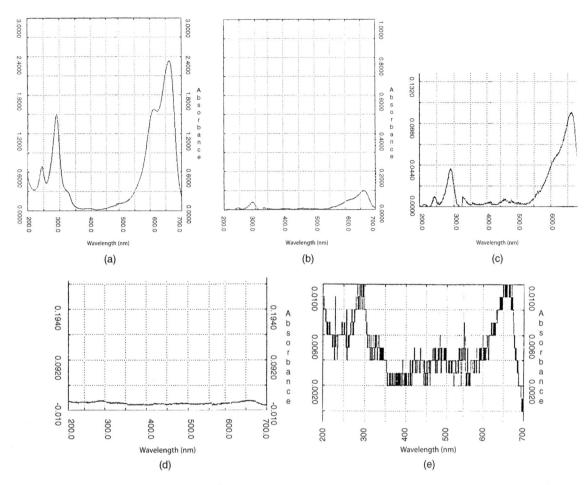

FIGURE 21.22 The relationship between signal and electrical noise illustrated with methylene blue. (a) The absorbance spectrum for methylene blue. Electrical noise is not evident. (b) The spectrum of methylene blue diluted 1/25. (c) The same plot as shown in (b), but with the Y-axis expanded to run from 0 to 0.1320. Noise becomes visible. (d) The spectrum of the methylene blue diluted 1/150. (e) The same plot as in (d), but the Y-axis is expanded. The signal of the methylene blue peaks can be barely discerned "buried" in the electrical noise. (All spectra were plotted on a Beckman DU 64 spectrophotometer.)

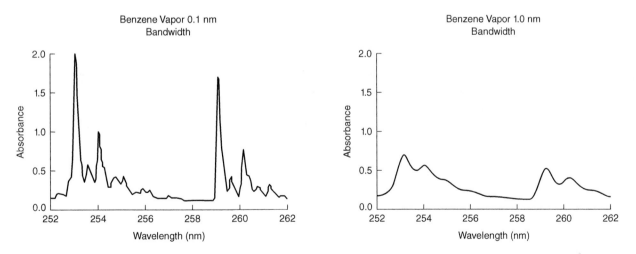

FIGURE 21.23 Resolution and spectral bandwidth. The better the resolution of an instrument, the better the peaks of an absorbance spectrum can be detected. These are absorbance spectra of benzene vapor in a sealed cuvette. In a higher-resolution spectrophotometer with a relatively narrow spectral bandwidth. left, benzene's peaks are clearly resolved. The peaks at 253.49 and 259.56 nm are visible if the spectral bandwidth is 0.2 nm or less. In an instrument with less resolution, right, the peaks are not resolved. (Reproduced courtesy of the Starna group of companies, www.starna.com.)

TABLE 21.7

Performance Specifications for a Spectrophotometer

Specification	Example
Capability (Minimally, states whether it has both UV and visible capability and scanning capability.)	Scanning UV/Vis with microcomputer electronics
Optics (Describes whether single or double beam. The monochromator type is also specified.)	Double-beam with concave holographic grating with 1,053 lines/nm
Wavelength range (Indicates the wavelengths at which the spectrophotometer can be used.)	190–325 (UV); 325–900 (Vis)
Sources (A UV/Vis spectrophotometer has a source to produce light in both spectral ranges.)	Pre-aligned deuterium and tungsten-halogen lamps
Wavelength repeatability (A measure of the ability of a spectrophotometer to return to the same spectral position.)	$\pm$ 0.1 nm
Wavelength accuracy (Agreement between displayed and actual wavelengths.)	$\pm$ 0.1 nm
Spectral slit width (An indication of the resolution attainable. The lower the value of spectral slit width, the better the resolution.)	$\pm$ 1 nm
Photometric accuracy (The ability of the detector to respond correctly to transmitted light.)	$\pm$ 0.003 A at 1.00 A measured with NIST 930 filters
Photometric stability (An indication of drift.)	$\pm$ 0.002 A/hr at 0.00 A at 500 nm
Noise (Random electrical signals. RMS is a statistical method of measuring noise. The lower the value, the less noise.)	0.00015 AU RMS
Stray light (The amount of light reaching the detector that was not transmitted through the sample.)	0.05% at 220 nm and 340 nm
Scanning speed (The rapidity at which the instrument changes from one wavelength to another during scanning.)	750 nm/min
Microprocessor capabilities: data handling programs (Specific programs to simplify operation or expand the data analysis capabilities of the instrument.)	Standard curves, linear and nonlinear; kinetics; spectral scanning
Accessories (Special features.)	Special sample holder for small volumes available Electrophoresis gel scanning accessories available Temperature-controlled sample chamber accessories available

Practice Problems: Spectrophotometry, Part A

1. Microwaves involve wavelengths in the range from 100 μm to 30 cm. Express the wavelengths of microwave radiation in terms of nm.
2. Which has the most energy: X-rays, green light, or ultraviolet light? Which has the least energy?
3. Are spectra A, B, and C probably from the same compound, or are they probably the spectra of different compounds? Explain.

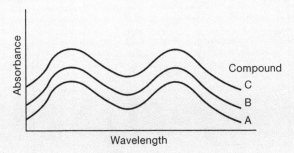

4. If blue light is shined on a solution of orange food coloring, will the light be absorbed?
5. What color are the dyes whose absorbance spectra are shown here? Explain.

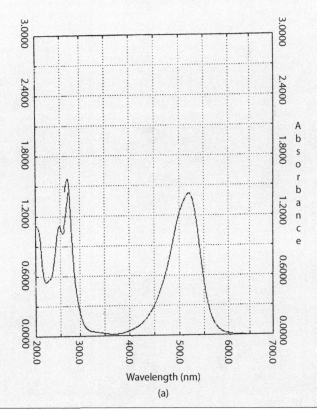

(a)

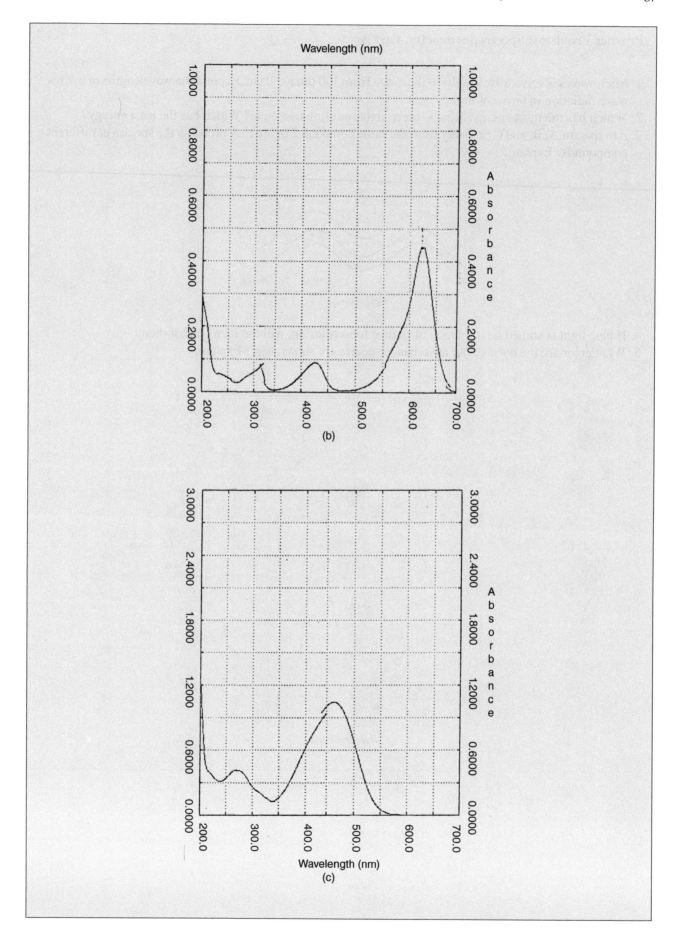

6. A spectrum for the plant pigment α-carotene is shown in the following graph. Like chlorophyll, this pigment absorbs light, allowing the plants to make carbohydrates. This molecule has attracted attention among those interested in nutraceuticals because there is some evidence that humans with high serum levels of α-carotene have a lower risk of death from all causes than humans with lower serum levels. α-Carotene is an abundant pigment in carrots, sweet potatoes, and pumpkins.

   a. What wavelengths of light does this pigment allow the plants to use?
   b. What color would a plant be if it contained only a-carotene?
      (The peaks for a-carotene are at 420, 440, and 470.)

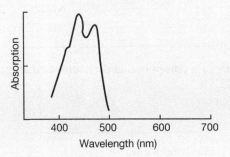

7. The absorbance spectrum for the plant pigment phycocyanin is shown in the following graph.

   a. What color light does this pigment absorb?
   b. What color would a type of algae be if it contained primarily phycocyanin?
   c. Do you think this pigment would be more common in red algae or in blue-green algae?

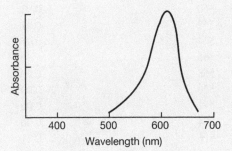

8.
   a. Plot an absorbance spectrum based on the following data.
   b. What color is this substance based on the absorbance spectrum?
   c. Optional: What is the natural bandwidth for this substance? (See the Chapter Appendix for information about natural bandwidth.)

Wavelength (nm)	Absorbance
380	0.01
382	0.00
384	0.01
386	0.02
388	0.07
390	0.10
392	0.15
394	0.38
396	0.46

(Continued)

Wavelength (nm)	Absorbance
398	0.54
400	0.55
402	0.52
404	0.42
406	0.37
408	0.26
410	0.11
412	0.08
414	0.03

9. The values for the transmittance of a compound at different wavelengths are given.
   a. Graph these data.
   b. What color is this compound?

Wavelength (nm)	t
550	0.01
552	1.06
554	1.02
556	1.01
558	1.08
560	0.98
562	0.91
564	0.45
566	0.36
568	0.20
570	0.09
572	0.06
574	0.06
576	0.21
578	0.35
580	0.46
582	0.98
584	1.04

10. Convert the following transmittance values to %T.

$$0.876 \quad 0.776 \quad 0.45 \quad 1.00$$

11. Convert the following values to absorbance values.

t=0.876	%T=25%
t=0.776	%T=15%
t=0.45	%T=95%
t = 1.00	%T=45%

12. Fill in the following table:

Absorbance	Transmittance (t)	%Transmittance (T)
0.01	_____	_____
_____	0.56	_____
_____	_____	1.0%

13. Convert the following absorbance values to transmittance.

1.24   0.95   1.10   2.25

14. The specifications for two spectrophotometers are given below. Answer the following questions about these spectrophotometers.
    a. Which spectrophotometer should be purchased to do qualitative analysis in the ultraviolet range?
    b. Which spectrophotometer would be expected to be best able to measure a sample with very high absorbance?
    c. Which spectrophotometer should be used to measure the density of bands in electrophoresis gels?
    d. Which would probably be better as a low-cost instrument for routine assays or for a teaching instrument?
    e. Which spectrophotometer would be expected to have the best resolution?
    f. Which spectrophotometer would be best able to distinguish the absorbance from compound A, which has a peak absorbance at 550 nm, and compound B, whose peak absorbance is at 557 nm?

	Instrument A
	Specification
Capability	Scanning Vis range with microcomputer electronics
Optics	Double beam with concave holographic grating with 1,053 lines/nm
Wavelength range	325–850 (Vis)
Sources	Pre-aligned tungsten-halogen lamps
Wavelength accuracy	± 0.5 nm
Wavelength repeatability	± 0.5 nm
Spectral slit width	± 12 nm
Photometric accuracy	± 0.0013 A at 1.00 A measured with NIST 930 filters
Photometric stability	± 0.006 A/hr at 0.00 A at 500 nm
Noise	0.00030 AU, RMS
Stray light	0.15% at 220 and 340 nm
Scanning speed	750 nm/minutes
Microprocessor capabilities: data handling programs	Standard curves, linear and nonlinear; kinetics; spectral scanning
Accessories	Small-volume holder available

(Continued)

	Instrument B
Capability	Scanning UV/Vis with microcomputer electronics
Optics	Double beam with concave holographic grating with 1,053 lines/nm
Wavelength range	170–325 (UV); 325–850 (Vis)
Sources	Pre-aligned deuterium and tungsten-halogen lamps
Wavelength accuracy	± 0.1 nm
Wavelength repeatability	± 0.1 nm
Spectral slit width	± 1 nm
Photometric accuracy	± 0.0005 A at 1.00 A measured with NIST 930 filters
Photometric stability	± 0.002 A/hr at 0.00 A at 500 nm
Noise	0.00014 AU, RMS
Stray light	0.05% at 220 and 340 nm
Scanning speed	550 nm/minutes
Microprocessor capabilities: data handling programs	Standard curves, linear and nonlinear; kinetics; spectral scanning
Accessories	Small-volume holder available
	Gel scanning
	Temperature control

15. Two laboratories are comparing the performance of their spectrophotometers. They take the identical sample and measure its absorbance. (Assume the temperature is the same in both laboratories.) One laboratory gets an average absorbance reading of 0.72 AU; the other laboratory gets 0.87 AU. What factors might account for the difference between the two laboratories?

16. On a certain spectrophotometer, 50 μg/mL of a certain compound is expected to have an absorbance of 1.95. If significant stray light is present, will the apparent absorbance be greater than, less than, or equal to 1.95?

Questions for Discussion

If you work in a laboratory with a spectrophotometer, find the answers to the following questions:

a. Does the spectrophotometer have the capability to measure absorbance in both the UV and visible range?

b. What is the smallest volume sample that can be measured using the cuvettes and sample holders available in your laboratory?

c. What is the instrument's specification for stray light?

d. Optional: What is the instrument's spectral slit width?

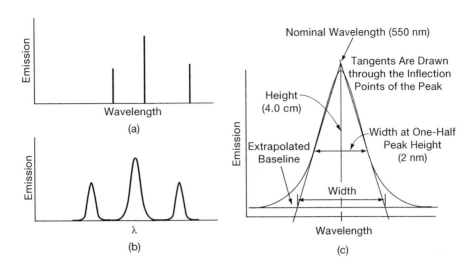

FIGURE 21.24 Spectral bandwidth. (a) The emission spectrum of a line source as it would appear in a "perfect" spectrophotometer. (b) The spectrum of a line source as it appears in practice. (c) The wavelength span is described by the spectral bandwidth that is measured at half the height of the peak.

CHAPTER APPENDIX: SPECTRAL BANDWIDTH AND RESOLUTION

Spectral Bandwidth and the Accuracy of a Spectrophotometer

Spectral bandwidth *is a measure of the range of wavelengths emerging from the monochromator when a particular wavelength is selected.* Recall that although a spectrophotometer may be set to a single, specific wavelength, in practice a narrow range of wavelengths emerges from the monochromator. Spectral bandwidth is a value that describes the ability of the spectrophotometer to isolate a portion of the electromagnetic spectrum.[3]

The spectral bandwidth is an instrument characteristic that affects the accuracy of the measurements and the resolution obtainable from a particular spectrophotometer. In general, the narrower the range of wavelengths emerging from the monochromator, the better the instrument.

Spectral bandwidth is determined by replacing the instrument's normal light source with a special line source (i.e., a lamp that emits light only at specific, individual wavelengths). (A method for measuring spectral bandwidth using a mercury lamp is detailed in ASTM Standard E 958, "Standard Practice for Measuring Spectral Bandwidth of Ultraviolet-Visible Spectrophotometers.") The intensity of light reaching the detector from the line source is plotted versus wavelength. The width of the emission peak at half the peak's height is measured. In the example illustrated in Figure 21.24, the wavelength at which the peak light intensity occurs is 550 nm. The peak height at this position is 4.0 cm above the baseline. Half the peak height is 2.0 cm. The width of the peak at half its height is 2 nm, so the spectral bandwidth is 2 nm.

Spectral bandwidth should not be confused with natural bandwidth. **Natural bandwidth** *describes the range of wavelengths absorbed by a particular substance.* For example, the natural bandwidth of DNA is around 45 nm, as shown in Figure 21.25. The natural bandwidth of a compound is an intrinsic property of that compound and is independent of the instrument's characteristics.

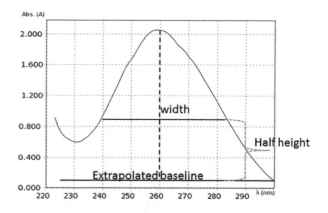

FIGURE 21.25 The natural bandwidth of DNA. The width at half the height is about 45 nm.

The fact that the light that is incident on a sample is not truly monochromatic leads to some inaccuracy in measurements of absorbance or transmittance. The amount of inaccuracy depends on both the spectral bandwidth and the natural bandwidth. The narrower the spectral bandwidth (i.e., the more monochromatic the light) the more accurate the instrument. If the natural bandwidth of a compound is relatively wide (as is, for example, the natural bandwidth of DNA), then the inaccuracy caused even by a wide spectral bandwidth is relatively insignificant. In contrast, if the natural bandwidth of a compound is comparatively narrow, then a narrow spectral bandwidth is required to get accurate measurements. A general rule is that for the most accurate measurements at an absorbance peak, the spectrophotometer must have a spectral bandwidth equal to or less than one-tenth of the natural bandwidth of the compound to be measured. (For DNA, this means the spectral bandwidth of the instrument should be less than or equal to 4.5 nm.)

Suppose the absorbance of the same sample is measured in two instruments. The first instrument has a very narrow spectral bandwidth; the second has a broader spectral bandwidth. It seems reasonable to suppose that the sample would have the same absorbance in both instruments. In fact, if all other factors are equal, the first instrument will give a higher absorbance reading for the sample than would the second. This is because the apparent absorptivity (absorptivity and absorptivity constants are discussed in Chapter 28) of an analyte approaches its maximum, "true," value as the spectral bandwidth decreases. The "true" absorptivity constant for a particular compound at a particular wavelength is a theoretical concept measurable only in the presence of truly monochromatic light.

---

[3] There are several terms used to describe the "monochromaticity" of light. These include *spectral bandwidth, bandwidth or bandwidth, effective bandwidth,* and *bandpass or band pass.* These terms are not always used in the same way by different authors. The definitions in this text are consistent with those of ASTM (ASTM Standard E 131, "Standard Definition of Terms and Symbols Relating to Molecular Spectroscopy").

THE FACTORS THAT DETERMINE THE SPECTRAL BANDWIDTH OF AN INSTRUMENT

The spectral bandwidth is controlled by the design of the monochromator. Some monochromators can provide more monochromatic light than others. There are two features of the monochromator that affect the "monochromaticity" of selected light. The first is the **slit width,** *which is the physical width of the entrance or exit slit of the monochromator.* The narrower the slit, the narrower the range of wavelengths selected.

The second feature of the monochromator that controls the range of wavelengths selected is the **linear dispersion of the monochromator. Linear dispersion** *is a measure of the dispersion of light by the diffraction grating.* The linear dispersion of the monochromator is not controlled by the user; rather, it depends on the manufacturing process used to produce the grating. Values for linear dispersion are obtained from the manufacturer of a spectrophotometer. Linear dispersion is generally expressed in terms of how many wavelengths (in nanometers) are included per millimeter at the exit slit.

Spectral slit width combines the value for the slit width of an instrument and its linear dispersion into a single value. **Spectral slit width** *is defined as the physical width of the monochromator exit slit, divided by the linear dispersion of the diffraction grating.* For example, consider the spectrophotometer in Figure 21.26. The width of the exit slit is 0.2 mm. The monochromator disperses light in such a way that at the plane of the exit slit there are 0.25 mm/nm. Thus, the spectral slit width is:

$$\frac{0.2\,\text{mm}}{0.25\,\text{mm}/\text{nm}} = 0.8\,\text{nm}$$

The term *spectral slit width* is not synonymous with *spectral bandwidth*. Both terms, however, relate to the "monochromaticity" of light emerging from the monochromator. Spectral bandwidth is difficult to measure so, in practice, the spectral slit width is used as an indicator of light "monochromaticity."

We have seen that the more monochromatic the light from the monochromator, the more accurate the readings of the spectrophotometer and the better its resolution. It seems reasonable, therefore, that manufacturers would always design instruments with the narrowest possible spectral slit width. In fact, it is true that more sophisticated spectrophotometers have

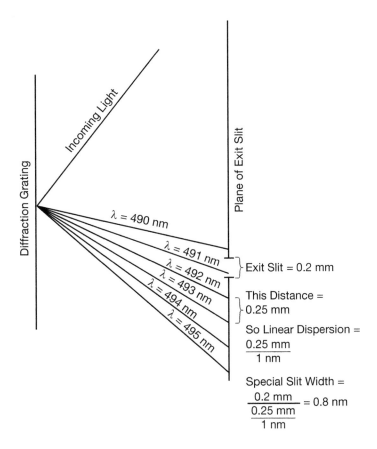

FIGURE 21.26 Spectral slit width. The spectral slit width in this example is 0.8 nm.

narrower spectral slit widths than do less sophisticated instruments; however, there are practical limitations to how narrow the spectral slit width can be.

One way to make an instrument with a narrow spectral slit width is to make the exit slit of the monochromator very narrow. Although it is possible to manufacture an extremely narrow exit slit, there are two problems with this approach. First, as the slit gets narrower and narrower, less and less light passes through it. Eventually, so little light is incident on the sample that electronic noise becomes significant relative to the amount of transmitted light. Noise then limits the ability of the spectrophotometer to make accurate measurements. Second, if the exit slit is too narrow, then its edges diffract light and the slit acts as a second unwanted diffraction grating.

RESOLUTION AND SPECTRAL BANDWIDTH

Recall that resolution affects our ability to distinguish peaks from one another if their wavelengths are close together. Spectral bandwidth affects resolution. A high-performance instrument has a narrow spectral bandwidth; therefore, it also has optimal resolution. Less expensive instruments tend to have wider spectral bandwidths and less ideal resolution.

When reading instrument specifications, the spectral slit width of the monochromator is commonly used as an indication of the resolution of a spectrophotometer. The lower the value for spectral slit width, the more monochromatic the light and the better the resolution (Figure 21.27).

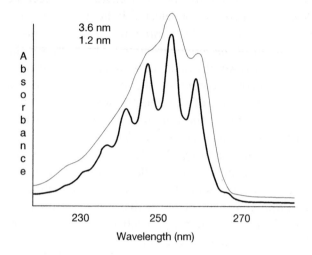

FIGURE 21.27 Spectral slit width and resolution. A narrower, 1.2 nm spectral slit width (black spectrum) has much better resolution than a 3.6 nm slit width (gray spectrum). (Courtesy Waters Corporation. Used with permission.)

UNIT VI

Laboratory Solutions

The preparation of laboratory solutions is an essential part of biotechnology. Biological systems evolved in an aqueous environment. Therefore, the first step in nearly every biological procedure, whether for research or production purposes, is the preparation of aqueous mixtures containing the proper components to support the biological system of interest. Chemists formally define a **solution** *as a homogeneous mixture in which one or more substances is (are) dissolved in another.* **Solutes** *are the substances that are dissolved in a solution. The substance in which the solutes are dissolved is called the* **solvent.** Although solutes and solvents may be gases, liquids, or solids, in biological applications, the solutes are solids or liquids and the solvent is a liquid – most often, water. Biologists tend to use the term "solution" more broadly than chemists to refer *to biological materials and the liquid environments in which they are contained.*

DOI: 10.1201/9780429282799-27

Case Study: Solutions Are Not Just Found in the Laboratory

We are surrounded by aqueous solutions in real life:

- Hand sanitizers are usually a solution of ethyl alcohol and water (at least 60% alcohol to be effective).
- White vinegar is a 5% solution of acetic acid in water.
- Colas are solutions of sugar, caffeine, and carbon dioxide in water.
- Tap water is a solution containing chlorine (less than 4 mg/L), ions, and other substances.
- Antifreeze for your car is usually a solution of either ethylene glycol or propylene glycol in water.

You may have experience in preparing a beverage such as Kool-Aid, which is an aqueous solution you make with a package of pre-measured powder (mostly citric acid and artificial color) and sugar. You may also know from experience that if you add too much water, or forget to add sugar, or don't mix the solution properly, it doesn't taste good. Similarly, laboratory solutions do not fulfill their purpose if they are not made properly. The good news is that laboratory solutions, like Kool-Aid, have recipes and instructions that, when followed, result in a successful product. Unlike Kool-Aid, some laboratory solutions have complex ingredients and procedures. But with knowledge, proper equipment, and some simple math tools, you will be able to prepare solutions that work for their application, even when solution recipes appear to be complex.

Biological solutions must be made properly to ensure reproducible results. There are a number of requirements for producing biological solutions consistently and correctly. These requirements include the following:

- Understanding various expressions of concentration, and correctly calculating the required amounts of solutes.
- Accurately weighing solutes, mixing them with water, and adjusting the pH of the resulting solution.
- Obtaining water of a proper purity for one's applications.
- Selecting chemical reagents that are suitable for a given application.
- Avoiding chemical and microbial contamination of biological solutions.
- Selecting appropriate glassware and plasticware.
- Understanding how changes in the components of a solution can alter the results of procedures.
- Properly storing biological solutions.
- Thoroughly documenting solution preparation, storage, use, and labeling.
- Performing quality-control tests to ascertain that biological solutions are acceptable.
- Ensuring the safety of personnel and the environment when working with chemicals.

This unit discusses the bulleted considerations relevant to the preparation and use of solutions in biotechnology laboratories.

Chapter 22 discusses how solute concentrations are expressed in "recipes" and introduces the calculations associated with solution preparation.

Chapter 23 introduces practical considerations relating to buffer solution preparation.

Chapter 24 discusses the purification of water for laboratory use; cleaning of glassware and plasticware; sterilization of solutions; and storage of solutions containing biological materials.

Chapter 25 explores the reasons why biological solutions for proteins and nucleic acids contain certain components.

BIBLIOGRAPHY

For general information about molarity, molality, normality, and percent solutions, consult any basic chemistry textbook.

Manufacturers' catalogs are a valuable source of information about chemicals. The Sigma-Aldrich Company catalog, for example, contains formula weights, compound names, and formulas for many chemicals (www. sigmaaldrich.com).

Alconox, Inc. "Guidance for Labware Washer Cleaning. Better, Safer Science Through Proper Detergent Selection and Cleaning." *Alconox, Inc.* 2017. alconox. com/resources/pdf/alconoxlabwarewasherguide.pdf.

Ballinger, Jack, and Shugar, Gershon. *Chemical Technicians' Ready Reference Handbook*. 5th ed., McGraw-Hill Education, 2011.

Chakravarti, Deb N., Chakravarti, Bulbul, and Mallik, Buddhadeb. "Reagent Preparation: Theoretical and Practical Discussions." *Current Protocols*. 28 October 2014. doi: 10.1002/9780470089941.et0301s9.

Deutscher, Murray P., ed. "Guide to Protein Purification," in *Methods in Enzymology*, vol. 182, Academic Press, 1990.

Favre, Nicolas, and Rudin, Werner. "Salt-Dependent Performance Variation of DNA Polymerases in Co-Amplification PCR." *BioTechniques*, vol. 21, no. 1, 1996, pp. 28–30. doi: 10.2144/96211bm04.

Goebel-Stengel, Miriam, et al. "The Importance of Using the Optimal Plasticware and Glassware in Studies Involving Peptides." *Analytical Biochemistry*, vol. 414, no. 1, 2011, pp. 38–46. doi: 10.1016/j.ab.2011.02.009.

Green, Michael, and Sambrook, Joseph. *Molecular Cloning: A Laboratory Manual*, Volume 1, 2, & 3. 4th ed., Cold Spring Harbor Laboratory Press, 2012.

Grzeskowiak, Rafal, and Gerke, Nils. *Leachables: Minimizing the Influence of Plastic Consumables on the Laboratory Workflows*. White Paper No. 26 I. September 2015. Eppendorf. www.eppendorf.com/product-media/doc/en/146279/Consumables_White-Paper_026_Consumables_Leachables-Minimizing-Influence-Plastic-Consumables-Laboratory-Workflows.pdf.

Ignatoski, Kathleen M. Woods, and Verderame, Michael F. "Lysis Buffer Composition Dramatically Affects Extraction of Phosphotyrosine-Containing Proteins." *BioTechniques*, vol. 20, no. 5, 1996, pp. 794–96. doi: 10.2144/96205bm13.

Invitrogen. "Working with RNA: The Basics. Technical Note." *Invitrogen*. 2019. https://www.thermofisher.com/us/en/home/references/ambion-tech-support/working-with-rna-the-basics.html.

Lewis, L. Kelvin, et al. "Interference with Spectrophotometric Analysis of Nucleic Acids and Proteins by Leaching of Chemicals from Plastic Tubes." *BioTechniques*, vol. 48, no. 4, 2010, pp. 297–302. doi: 10.2144/000113387.

Lundblad, Roger L., and Macdonald, Fiona, eds. *The Practical Handbook of Biochemistry and Molecular Biology*. Taylor and Francis, 2018. (Contains extensive information about biological materials including buffers for biological systems. This handbook is periodically updated and revised.)

McDonald, G. Reid, et al. "Bioactive Contaminants Leach from Disposable Laboratory Plasticware." *Science*, vol. 322, no. 5903, 2008, p. 917. doi:10.1126/science.1162395.

McLaughlin, Malcolm. "Stopping Residue Interference on Labware and Equipment." *American Laboratory News Edition*, June 1992.

Perkins, John. *Principles and Methods of Sterilization in Health Sciences*. 2nd ed., Charles C Thomas Pub Ltd, 2008. (A classic in-depth reference on sterilization methods.)

Scopes, Robert. *Protein Purification: Principles and Practice (Springer Advanced Texts in Chemistry)*. Softcover reprint of hardcover 3rd ed. 1994, Springer, 2010.

The Merck Index: An Encyclopedia of Chemicals, Drugs, and Biologicals. Merck & Co, 2021. (An encyclopedia that contains information about thousands of biologically relevant chemicals including their structures, densities, molecular weights, and solubilities. This index is periodically updated and revised.)

Thermo Fisher Scientific. "Technical Bulletin #159: Working with RNA." *Thermo Scientific*. 2019. www.thermofisher.com/us/en/home/references/ambion-tech-support/nuclease-enzymes/general-articles/working-with-rna.html.

Thermo Fisher Scientific. "Technical Bulletin #43: Protein Stability and Storage." *Thermo Scientific*, 2009. tools.thermofisher.com/content/sfs/brochures/TR0043-Protein-storage.pdf.

Thermo Scientific. "Nalgene Bottles and Carboys Technical Brochure." *Thermo Scientific*. 2012. tools.thermofisher.com/content/sfs/brochures/D01705.pdf.

Thermo Scientific. "Thermo Scientific Waterbook." *Thermo Scientific*. 2016. www.thermofisher.com/us/en/home/products-and-services/promotions/life-science/thermo-scientific-waterbook.html.

22 Preparation of Laboratory Solutions – Part A

Concentration Expressions and Calculations

22.1 OVERVIEW

Solution preparation involves various basic laboratory procedures, such as weighing compounds and measuring the volumes of liquids. Solution preparation also requires interpreting the "recipe" (procedure) for making the solution. The following two chapters discuss the basic procedures and the calculations that are the cornerstones of solution preparation.

Preparing laboratory solutions can be compared with baking. Suppose you decide to bake a batch of apple crisp. The first step is to find a recipe (Figure 22.1) that states:

1. The *components* of apple crisp.
2. *How much* of each component is needed.
3. *Advice for preparing* (mixing and baking) the components properly.

To make apple crisp successfully, you need to understand the terminology in the recipe and to measure, combine, and prepare the ingredients in the proper fashion.

When preparing a laboratory solution, you also need a procedure for preparing the solution. You follow the procedure and combine the right *amount(s)* of each *component* (solute) in the right volume of solvent to get the correct *concentration* of each solute at the end. You may also need to adjust the solution to the proper pH, sterilize it, or perform other manipulations. There are differences, however, between baking and preparing laboratory solutions. One important difference is that it is often appropriate to creatively modify a

KITCHEN RECIPE I APPLE CRISP	LABORATORY RECIPE I
Preheat oven to 375°F	Na_2HPO_4 6 g
12 oz can frozen lemonade	KH_2PO_4 3 g
3 lb sliced apples	NaCl 0.5 g
Reconstitute juice with water.	NH_4Cl 1 g
Cover apples with juice and refrigerate.	Dissolve in water Bring to a volume of 1 L
Blend together: 3 c rolled oats 1/2 c honey 1 c flour 3/4 c butter 1/2 tsp cinnamon	

Place apples in bottom of 9 x 13 inch baking pan. Crumble oat mixture over apples. Pour juice over apples. Bake 1 hour.

FIGURE 22.1 Comparison of a kitchen recipe and a laboratory recipe.

DOI: 10.1201/9780429282799-28

recipe when baking, but laboratory procedures should be strictly followed.

The recipe for apple crisp lists the *amount* of each component needed. A procedure to make a laboratory solution may similarly list the amount of each solute to use. For example, laboratory recipe I (Figure 22.1) lists the amount of each of the four solutes required: Na_2HPO_4, KH_2PO_4, NaCl, and NH_4Cl. Purified water is the solvent. The total solution has a volume of 1 L. In overview, the procedure for preparing laboratory recipe I is:

1. Weigh out the needed amount of each solute.
2. Dissolve the solutes in less than 1 L of water.
3. Add enough water so that the final volume is 1 L.

Laboratory recipe I is easy to interpret. Many recipes for laboratory solutions, however, do not show the *amount* of each solute required; rather, they show the *concentration* of each solute needed. **Amount** and **concentration** are not synonyms. **Amount** *refers to how much of a component is present.* For example, 10 g, 2 cups, and 30 mL are amounts. **Concentration** *is an amount per volume; it is a ratio.* The numerator is the amount of the solute of interest. The denominator is usually the volume of the entire solution, the solvent and the solute(s) together. For example, "2 g/L of NaCl" means there is 2 g of NaCl dissolved in enough liquid so that the total volume of the solution is 1 L.

Concentration is always a ratio, although concentrations are sometimes expressed in ways that do not look like fractions. For example, 2% milk is an expression of concentration that does not at first glance appear like a fraction. But, in fact, the % symbol represents a numerator and denominator separated by a line. "2%" milk has about 2 g fat/100 g of liquid. ("Percent" comes from Latin and means *per hundred*.) Molarity, molality, and normality, which are covered later in this chapter, are also expressions of concentration where a numerator and denominator are not explicitly shown.

There are various ways to express concentration, each of which is associated with certain mathematical calculations. The key question that is solved by calculation in this chapter is:

How much solute (an amount) is required to prepare a solution with a given concentration of this solute?

The next sections discuss these concentration expressions and calculations and give examples that biologists are likely to encounter.

22.2 TYPES OF CONCENTRATION EXPRESSIONS AND ASSOCIATED CALCULATIONS

22.2.1 WEIGHT PER VOLUME

The simplest way to express the concentration of a solution is as a fraction with the amount of solute, expressed as a weight in the numerator and the volume in the denominator. This type of concentration expression is often used for small amounts of chemicals and specialized biological reagents. For example, a solution of 2 mg/mL proteinase K (an enzyme) contains 2 mg of proteinase K for each milliliter of solution.

When concentrations are expressed as fractions, proportions can easily be used to determine how much solute is required to make the solution. For example:

How much proteinase K is needed to make 50 mL of proteinase K solution at a concentration of 2 mg/mL?

$$\frac{?}{50 \text{ mL}} = \frac{2 \text{ mg proteinase K}}{1 \text{ mL}}$$

? = **100 mg** = amount proteinase K needed.

Example Problem 22.1

How many micrograms of DNA are needed to make 100 μL of a 100 μg/mL solution?

Answer

The amount of DNA required is found by using a proportion, but first 100 μL must be converted to milliliters, or milliliters must be converted to microliters, so the units are consistent.

$$100 \text{ μL} = 0.1 \text{ mL}$$

$$\frac{?}{0.1 \text{ mL solution}} = \frac{100 \text{ μg DNA}}{1 \text{ mL solution}}$$

? = **10 μg** = amount of DNA needed.

TABLE 22.1

Practical Information Relating to Solution Preparation

1. *Solvent.* When the solvent is not stated, assume it is distilled or otherwise purified water. (Water purification is discussed in Chapter 24.)
2. *"Bringing a solution to volume."* In the proteinase K example above, the enzyme will take up some room in the solution. If you add the proteinase K to 50 mL of water, the volume will be slightly more than 50 mL. The most accurate way to prepare the solution, therefore, is to dissolve the proteinase K in *less* than 50 mL of water and then bring the solution to a final volume of exactly 50 mL using a volumetric flask or graduated cylinder.

   This procedure is called **bringing the solution to the desired final volume (BTV or "bring to volume").**

   Figure 22.2 illustrates a method of bringing a solution to volume.

   Note that, in practice, 100 mg of proteinase K takes up so little volume that it is sensible simply to add the enzyme to exactly 50 mL of water. In general, however, the correct way to make a solution is to "bring it to volume."

   Note also that some sources use the term ***QS*** (Latin for quantum sufficit) instead of BTV.

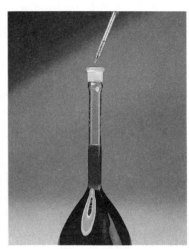

FIGURE 22.2 Bringing a solution to volume. (a) The desired amount of solute is weighed out on a balance. (b) The solute is dissolved with mixing in less than the total solution volume. (A beaker is often used when stirring the solution at this step.) (c) The solution is brought to its final volume by slowly filling a volumetric flask to its mark. In many situations, the accuracy provided by a graduated cylinder is suitable for bringing a solution to its final volume.

Practice Problems: Weight per Volume

1. How would you prepare 100 mL of an aqueous solution of $AgNO_3$ of strength 0.1 g $AgNO_3$/mL?
2. How many milligrams of NaCl are present in 50 mL of a solution that is 2 mg/mL NaCl?
3. How would you prepare 50 mL of proteinase K at a concentration of 100 μg/mL?

Table 22.1 summarizes some practical considerations regarding the preparation of solutions.

22.2.2 MOLARITY

Molarity *is a concentration expression that is equal to the number of moles of a solute that is dissolved per liter of solution.* Molarity is used to express concentration when the number of molecules in a solution is important. For example, in an enzyme-catalyzed reaction the numbers of molecules of each reactant and the enzyme are important. If there is too little enzyme relative to the number of molecules of reactants present, the reaction may be incomplete, whereas adding too much enzyme is costly and inefficient.

A brief review of terminology: A **mole** *of any element always contains* 6.02×10^{23} *(Avogadro's number) atoms.* Because some atoms are heavier than others, a mole of one element weighs a different amount than a mole of another element. *The weight of a mole of a given element is equal to its atomic weight in grams, or its* **gram atomic weight.**[1] Consult a periodic table of

---

[1] Weight and mass are not synonyms, and it is correct to speak of gram molecular mass in the context of molarity. It is common practice, however, to speak of "atomic weight," "molecular weight," and "formula weight," as we do in this book.

elements to find the atomic weight of an element. For example, one mole of element carbon weighs 12.0 g.

Compounds *are composed of atoms of two or more elements that are bonded together.* A mole of a compound contains 6.02×10^{23} molecules of that compound. The **gram formula weight (FW)** or **gram molecular weight (MW)** *of a compound is the weight in grams of 1 mole of the compound* (Figure 22.3). The FW is calculated by adding the atomic weights of the atoms that make up the compound. For example, the gram molecular weight of sodium sulfate (Na_2SO_4) is 142.04 g:

$$\text{2 sodium atoms } 2 \times 22.99 \text{ g} = 45.98 \text{ g}$$

$$\text{1 sulfur atom } 1 \times 32.06 \text{ g} = 32.06 \text{ g}$$

$$\text{4 oxygen atoms } 4 \times 16.00 \text{ g} = \underline{64.00 \text{ g}}$$

$$\text{Total} = \textbf{142.04 g}$$

By Definition: A 1 molar solution of a compound contains 1 *mole* of that compound dissolved in 1 L of total solution.

For example, a 1 molar solution of sodium sulfate contains 142.04 g of sodium sulfate in 1 liter of total solution (Figure 22.4). We also say that a "1 molar" solution has a "molarity of 1." Note that a "mole" is an expression of *amount* and "molarity" and "molar" are words referring to the *concentration* of a solution.

The word *molar* is abbreviated with an uppercase *M*. It is also common in biology to speak of "millimolar" (mM) and "micromolar" (µM) solutions. A **millimole** *is 1/1,000 of a mole;* a **micromole** *is 1/1,000,000 of a mole.* For example:

1 M NaCl = 1 mole or 58.44 g of NaCl in 1 L of solution.

1 mM NaCl = 1 mmole or 0.05844 g of NaCl in 1 L of solution.

1 µM NaCl = 1 µmole or 0.00005844 g of NaCl in 1 L of solution.

Note also that mol is sometimes used as an abbreviation for the word mole.

The following example problems illustrate calculations where concentration is expressed in terms of molarity.

Example Problem 22.2

How much solute is required to prepare 1 L of a 1 M solution of copper sulfate $(CuSO_4)$?

Answer

Calculate the FW or find it on the label of the chemical's container. By calculation:

$$\text{1 copper} = 1 \times 63.55 \text{ g} = 63.55 \text{ g}$$

$$\text{1 sulfur} = 1 \times 32.06 \text{ g} = 32.06 \text{ g}$$

$$\text{4 oxygens} = 4 \times 16.00 \text{ g} = \underline{64.00 \text{ g}}$$

$$\text{FW} = \textbf{159.61 g}$$

Therefore, **159.61 g** of $CuSO_4$ is required to make 1 L of 1 M $CuSO_4$.

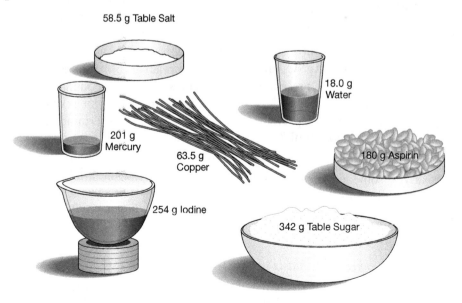

58.5 g Table Salt

18.0 g Water

201 g Mercury

63.5 g Copper

180 g Aspirin

254 g Iodine

342 g Table Sugar

FIGURE 22.3 One mole of various substances.

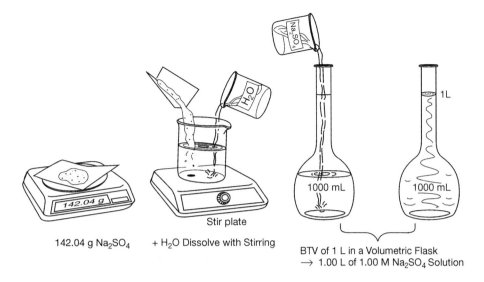

142.04 g Na_2SO_4 + H_2O Dissolve with Stirring

Stir plate

BTV of 1 L in a Volumetric Flask
→ 1.00 L of 1.00 M Na_2SO_4 Solution

FIGURE 22.4 A one molar solution of sodium sulfate (Na_2SO_4). The FW of Na_2SO_4 is 142.04. This amount is dissolved in water so that the total volume is 1 liter.

Example Problem 22.3

How much solute is required to prepare 1 L of a 1 mM solution of $CuSO_4$?

Answer

The formula weight of copper sulfate is 159.61; therefore,

$$1 \text{ mmole} = \frac{159.61 \text{ g}}{1{,}000} = 0.15961 \text{ g}$$

Therefore, 1 L of 1 mM $CuSO_4$ requires **0.15961 g**.

Molarity is a concentration expression; concentration expressions are ratios. We can determine how to prepare solutions of different molarities by using straightforward proportional relationships. For example:

How much solute is required to prepare 1 L of 0.25 M sodium chloride solution? Using the reasoning of proportions, if a 1 M solution of NaCl requires 58.44 g of solute, then:

$$\frac{?}{0.25 \text{ M}} = \frac{58.44 \text{ g}}{1 \text{ M}}$$

? = **14.61 g** = amount of solute to make 1 L of 0.25 M NaCl.

Example Problem 22.4

How many micromoles are there in 17.4 mg of NAD, FW 663.4? (NAD, nicotinamide adenine dinucleotide, is an organic compound that is important in metabolism.)

Answer

First, convert 17.4 mg to grams.
17.4 mg = 0.0174 g.
  Using a proportion:

$$\frac{1 \text{ mole}}{663.4 \text{ g}} = \frac{?}{0.0174 \text{ g}}$$

? = 2.62×10^{-5} moles = **26.2 micromoles**

Example Problem 22.5

How much solute is required to prepare a 1 L solution of NaCl of molarity 2.5?

Answer

If a 1 M solution of sodium chloride requires 58.44 g of solute, then:

$$\frac{58.44 \text{ g}}{1 \text{ M}} = \frac{?}{2.5 \text{ M}}$$

? = **146.1 g** = amount of solute required to make 1 L of 2.5 M NaCl.

It is sometimes necessary to make more or less than 1 L of a given solution. In these cases, proportions can again be used to determine how much solute is required. For example:

How much solute is required to make 500 mL of a 0.25 M solution of NaCl?

We know from the previous example that 14.61 g of solute is required to make 1 L of 0.25 M NaCl. Using the reasoning of proportions:

$$\frac{?}{500 \text{ mL}} = \frac{14.61 \text{ g}}{1,000 \text{ mL}}$$

$? = \mathbf{7.31 \text{ g}}$ NaCl = amount of solute to

make 500 mL of 0.25 M NaCl.

Thus, it is possible to use proportions to calculate how much solute is needed to make a solution with a molarity other than 1 M and a volume other than 1 L. In practice, however, most often people use an equation to calculate the amount of solute required to make a solution of a particular volume and a particular molarity. This can be written as shown in Equation 22.1.

Equation 22.1. Calculation of How Much Solute Is Required for a Solution of a Particular Molarity and Volume

Solute Required in grams

= (grams/1 mole)(Molarity)(Volume)

This equation can also be written as:

$$\text{Solute required in grams} = \left(\frac{\text{grams}}{1 \text{ mole}}\right)\left(\frac{\text{mole}}{\text{liter}}\right)\left(\frac{\text{liter}}{1}\right)$$

where:

Solute Required is the amount of solute needed, in grams
Grams/mole is the number of grams in 1 mole of solute
Molarity is the desired molarity of the solution expressed in moles per liter
Volume is the final volume of the solution, in liters.

Note: To use this equation, volume must be expressed in units of liters. To convert a volume expressed in milliliters to liters, simply divide the volume in milliliters by 1,000. For example:

To convert 100 mL to liters:

$$\frac{100}{1,000} = 0.1$$

Therefore, **100 mL = 0.1 L**.

Similarly, to convert 3,420 mL to liters:

$$\frac{3,420}{1,000} = 3.420$$

Therefore, **3,420 mL = 3.420 L**.

The following example problem illustrates both the proportion strategy and the use of Equation 22.1 to calculate how much solute is required to make a molar solution.

Example Problem 22.6

How much solute is required to prepare 300 mL of a 0.800 M solution of calcium chloride (FW = 111.0)?

Answer

Strategy 1: Proportions
First, determine how much solute is needed to make 1 L of 0.800 M.

We know that 111.0 g is required to make 1 L of a 1 M solution. So:

$$\frac{?}{0.800 \text{ M}} = \frac{111.0 \text{ g}}{1 \text{ M}}$$

$? = \mathbf{88.8 \text{ g}}$ = amount of solute to make

1 L of 0.800 M solution

Second, determine how much solute is needed to make 300 mL.

$$\frac{?}{300 \text{ mL}} = \frac{88.8 \text{ g}}{1,000 \text{ mL}}$$

$? = \mathbf{26.64 \text{ g}}$ = grams of solute required to make

300 mL of 0.800 M solution

Strategy 2: Using Equation 22.1
First, convert 300 mL to liters by dividing by 1,000:

$$\frac{300}{1,000} = 0.300$$

Therefore, 300 mL = 0.300 L.
   Plug values into Equation 22.1:

Solute required = (grams/mole)(molarity)(volume)

$$=\left(\frac{111.0\ g}{1\ \text{mole}}\right)\left(\frac{0.800\ \text{mole}}{1\ L}\right)(0.300\ L) = \textbf{26.64 g}$$

= grams of solute required

Observe that the units cancel, leaving the answer expressed in grams.

We have now shown how to calculate the amount of solute required to prepare a solution of a given molarity and volume using two different strategies. Either strategy when performed correctly will give the right answer. Box 22.1 outlines the procedure for making a solution of a particular volume and molarity.

Example Problem 22.7

How would you prepare 150 mL of a 10 mM solution of $Na_2SO_4 \cdot 10H_2O$ using the procedure outlined in Box 22.1?

Answer

1. Find the FW of the solute, preferably from the label on its container. The FW is 322.04.
2. Determine the molarity required. The molarity required is 10 mM, which is equal to 0.010 M.
3. Determine the volume required. The volume required is 150 mL, which is equal to 0.150 L.
4. Determine how much solute is necessary.

Strategy 1: Proportions
If it requires 322.04 g to make 1 L of a 1 M solution, then:

$$\frac{322.04\ g}{1.000\ M} = \frac{?}{0.010\ M}$$

$? = \textbf{3.2204 g}$ = solute needed to make 1 L of

10 mM (i.e., 0.010 M) solution.

To make 150 mL:

$$\frac{?}{150\ mL} = \frac{3.2204\ g}{1,000\ mL}$$

$? \approx \textbf{0.4831 g}$ = amount of solute required to

make 150 mL of 10 mM solution.

Strategy 2: Using Equation 22.1
Convert 150 mL to liters, 150/1,000 = 0.150 L. Substitute values into Equation 22.1:

Solute required = (grams / mole)(molarity)(volume)

$$=\left(\frac{322.04\ g}{1\ \text{mole}}\right)\left(\frac{0.010\ \text{mole}}{1\ L}\right)(0.150\ L) \approx \textbf{0.4831 g}$$

= grams of solute required

5. Weigh out the amount of solute required as calculated in Step 4, that is, 0.4831 g of $Na_2SO_4 \cdot 10H_2O$.
6. Dissolve the weighed out compound in less than the desired final volume of solvent.
7. Place the solution in a volumetric flask or graduated cylinder. Add solvent until exactly 150 mL is reached (BTV).

22.2.3 CONCENTRATION EXPRESSED AS A PERCENT

When concentration is expressed in terms of percent, the numerator is the amount of solute and the denominator is 100 units of total solution. There are three types of percent expressions that vary in their units.

Type I: Weight per Volume Percent

Grams of Solute per 100 mL of Solution
A **weight per volume** *expression is the weight of the solute (in grams) per 100 mL of total solution* (Figure 22.5). This is the most common way to express a percent concentration in biology applications. If a solution recipe uses the term % and does not specify type, assume it is a weight per volume percent. This type of expression is abbreviated as **w/v**. For example:

20 g of NaCl in 100 mL of total solution is a 20%, w/v, solution.

Box 22.2 shows the procedure for preparing a w/v solution.

BOX 22.1 PROCEDURE TO MAKE A SOLUTION OF A PARTICULAR VOLUME AND MOLARITY

1. *Find the FW of the solute.* While it is possible to calculate the FW of a compound by adding the atomic weights of its constituents, for compounds with complicated formulas or that come in more than one form, it is better to find the FW on the label of the chemical's container.
2. *Determine the molarity required.*
3. *Determine the volume required.*
4. *Determine how much solute is necessary by using either proportions or Equation 22.1.*
5. *Weigh out the amount of solute required as calculated in Step 4.*
6. *Dissolve the weighed out compound in less than the desired final volume of solvent.*
7. *Bring to volume (BTV).* Place the solution in a volumetric flask or graduated cylinder, and add solvent until the right volume is reached.

Notes

i. **Hydrates** *are compounds that contain chemically bound water.* The bound water does not make the compounds liquid; rather, they remain powders or granules. The weight of the bound water is included in the FW of hydrates. For example, calcium chloride can be purchased either as an anhydrous form with no bound water, or as a dihydrate. Anhydrous calcium chloride, $CaCl_2$, has a formula weight of 111.0. The dihydrate form, $CaCl_2 \cdot 2H_2O$, has a formula weight of 147.0 (111.0 plus the weight of two waters, 18.0 each). When hydrated compounds are dissolved, the water is released from the compound and becomes indistinguishable from the water that is added as solvent.

ii. Compounds that come as liquids can be weighed out, but it may be easier to measure their volume with a pipette or graduated cylinder. Convert the required weight to a volume based on the density of the compound. (See Section 12.3 for a discussion of density.)

Practice Problems: Molarity

Assume purified water is the solvent for all questions. If the formula weight is not given, use an internet browser to find it.

1. If you have 3 L of a solution of potassium chloride at a concentration of 2 M, what is the solute? _____ What is the solvent? _____ What is the volume of the solution? _____ Express 2 M as a fraction. _____
2. How much solute is required to prepare 250 mL of a 1 molar solution of KCl?
3. How would you prepare 10 L of 0.3 M KH_2PO_4?
4. How would you prepare 450 mL of a 100 mM solution of K_2HPO_4?
5. How much solute is required to make 600 mL of a 0.4 M solution of Tris buffer (FW of Tris base 121.10)?
6. Suppose you are preparing a solution that calls for 25 g of $FeCl_3 \cdot 6H_2O$. You look on the shelf in your laboratory and find anhydrous ferric chloride, $FeCl_3$. What should you do?
7. How many micromoles are there in 150 mg of D-ribose 5-phosphate, FW = 230.11?

Type II: Volume Percent

Milliliters of Solute per 100 mL of Solution

In a **percent by volume expression**, abbreviated **v/v**, *both the amount of solute and the total solution are expressed in volume units.* This type of percent expression may be used when two compounds that are liquid at room temperature are being combined. For example:

100 mL of methanol in 1,000 mL of total solution is a 10% by volume solution.

Box 22.3 shows the procedure for making a v/v solution.

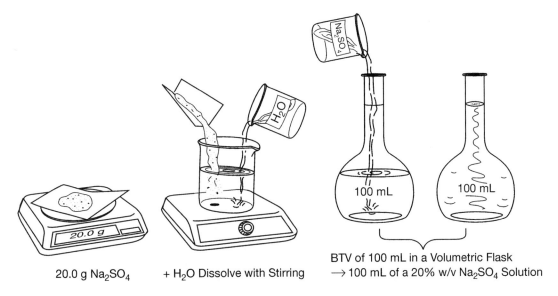

FIGURE 22.5 A 20% weight per volume percent solution of sodium sulfate (Na_2SO_4). Twenty grams of Na_2SO_4 is dissolved in water and brought to a volume of 100 mL.

BOX 22.2   PROCEDURE FOR PREPARING A WEIGHT PER VOLUME PERCENT SOLUTION

1. *Determine the percent strength (concentration) and volume of solution required.*
2. *Express the percent strength desired as a fraction (g/100 mL).*
3. *Multiply the total volume desired (Step 1) by the fraction in Step 2.*
4. *Dissolve the amount of material needed as calculated in Step 3.* (Assume purified water is the solvent if solvent is not specified.)
5. *Bring the solution to the desired final volume.*

Example Problem 22.8

How would you prepare 500 mL of a 5% (w/v) solution of NaCl?

Answer

1. Percent strength is 5% (w/v). Total volume required is 500 mL.

2. Expressed as a fraction, $5\% = \dfrac{5\,g}{100\,mL}$

3. $\dfrac{(500\ \cancel{mL})}{1}\left(\dfrac{5\,g}{100\ \cancel{mL}}\right) = \mathbf{25\ g}$ = amount of NaCl needed

4. Weigh out 25 g of NaCl. Dissolve it in less than 500 mL of water.

5. In a graduated cylinder or volumetric flask, bring the solution to 500 mL.

Type III: Weight Percent

Grams of Solute per 100 Grams of Solution
Weight (mass) percent, *w/w*, *is an expression of concentration in which the weight of solute is in the numerator and the weight of the total solution is in the denominator.* This type of expression is uncommon in biology manuals, but you may encounter it if you work with thick, viscous solvents whose volumes are difficult to measure. For example:

BOX 22.3   PROCEDURE FOR PREPARING A PERCENT BY VOLUME SOLUTION

1. *Determine the percent strength and volume required.*
2. *Express the percent desired as a fraction (mL/100 mL).*
3. *Multiply the fraction from Step 2 by the total volume desired (Step 1) to get the volume of solute needed.*
4. *Place the volume of the material desired in a graduated cylinder or volumetric flask.* Add solvent to BTV.

Example Problem 22.9

How would you make 100 mL of a 10% by volume solution of ethanol in water (v/v)?

Answer

1. Percent strength is 10% (v/v) and total volume wanted = 100 mL.

2. $10\% = \dfrac{10\ mL}{100\ mL}$

3. $\dfrac{(100\ \text{mL})}{1}\left(\dfrac{10\ mL}{100\ \text{mL}}\right) = 10\ mL =$

   amount of ethanol needed

4. Place 10 mL of ethanol in a 100 mL volumetric flask or graduated cylinder. BTV 100 mL.

Note: You cannot assume that 10 mL of ethanol+90 mL of water will give 100 mL total volume; their combined volume may be slightly less than 100 mL. The most accurate way to prepare this solution, therefore, is to bring it to volume.

5 g of NaCl plus 20 g of water is a 20% by mass solution because:
The weight of the NaCl is 5 g.
The total weight of the solution =
20 g+5 g = 25 g weight of solute.

$$\text{Weight percent} = \frac{\text{weight of solute}}{\text{total weight of solution}} \times 100$$

$$\frac{5\ g\ NaCl}{25\ g} \times 100 = 20\%$$

Box 22.4 shows the procedure for preparing a w/w solution.

BOX 22.4   PROCEDURE FOR PREPARING A PERCENT BY WEIGHT SOLUTION

1. *Determine the percent and weight of solution desired.*
2. *Change the percent to a fraction (g/100 g).*
3. *Multiply the total weight of the solution from Step 1 by the fraction in Step 2 to get the weight of the solute needed to make the solution.*
4. *Subtract the weight of the material obtained in Step 3 from the total weight of the solution (Step 1) to get the weight of the solvent needed to make the desired solution.*
5. *Dissolve the amount of the solute found in Step 3 in the amount of solvent from Step 4.*

Example Problem 22.10

How would you make 500 g of a 5% NaCl solution by weight (w/w)?

Answer

1. Percent strength is 5% (w/w), and total weight of solution wanted is 500 g.
2. $5\% = \dfrac{5\ g}{100\ g}$
3. $\dfrac{5\ g}{100\ g} \times 500\ g = 25\ g = $ NaCl needed
4. $500\ g - 25\ g = 475\ g = $ the amount of water needed.
5. Dissolve **25 g of NaCl in 475 g of water** to get a 5% w/w solution of NaCl.

Practice Problems: Percent Solutions

1. How would you prepare 35 mL of a 95% (v/v) solution of ethanol?
2. How would you prepare 200 g of a 75% (w/w) solution of resin in acetone?
3. How would you prepare 600 mL of a 15% (w/v) solution of NaCl?
4. Suppose you have 50 g of solute in 500 mL of solution.
   a. Express this solution as a %.
   b. Is this a w/w, w/v, or v/v solution?
5. Suppose you have 100 mg of solute in 1 L of solution.
   a. Express this as a %.
   b. Is this a w/w, w/v, or v/v solution?
6. What molarity is a 25% w/v solution of NaCl?

22.2.4 CONCENTRATION EXPRESSED AS PARTS

22.2.4.1 The Meaning of "Parts"

Parts solutions tell you how many parts of each component to mix together. The parts may have any units, but must be the same for all components of the mixture. For example:

A solution that is 3:2:1 ethylene:chloroform: isoamyl alcohol is:

3 parts ethylene, 2 parts chloroform, and 1 part isoamyl alcohol.

There are many ways to prepare this solution. Two examples are as follows:

Combine:		Combine:
3 L ethylene	or	3 mL ethylene
2 L chloroform		2 mL chloroform
1 L isoamyl alcohol		1 mL isoamyl alcohol

Example Problem 22.11

How could you prepare 50 mL of a solution that is 3:2:1 ethylene:chloroform:isoamyl alcohol? (Clue: The amount required of each component is calculated using a proportion equation.)

Answer

1. Add up all the parts required. In this case, there are $3 + 2 + 1$ parts = 6 parts.
2. Use a proportion to figure out each component.
   You need 3 parts of ethylene out of 6 parts total, so:
   Ethylene: $\dfrac{3}{6} = \dfrac{?}{50}$ $? = 25$ so you need **25 mL** of ethylene.
   You need 2 parts of chloroform out of 6 parts total, so:
   Chloroform: $\dfrac{2}{6} = \dfrac{?}{50}$ $? \approx 16.7$ so you need **16.7 mL** of chloroform.
   You need 1 part of isoamyl alcohol out of 6 parts total, so:
   Isoamyl alcohol: $\dfrac{1}{6} = \dfrac{?}{50}$ $? \approx 8.3$ so you need **8.3 mL** of isoamyl alcohol.
   Thus, this solution requires: 25.0 mL of ethylene + 16.7 mL of chloroform + 8.3 mL of isoamyl alcohol = 50 mL total.

Note that a recipe expressed as it is in this example is not brought to volume; rather, the components are added to one another, as shown.

22.2.4.2 Parts per Million and Parts per Billion

A variation on the theme of parts is the concentration expression "parts per million (ppm)". **Parts per million** *is the number of parts of solute per 1 million parts of total solution.* Any units may be used, but must be the same for the solute and total solution. **Parts per billion (ppb)** *is the number of parts of solute per billion parts of solution.* (Percent solutions are the same class of expression as ppm and ppb. Percent means "parts per hundred.") For example:

> 5 ppm chlorine might be:
> 5 g of chlorine in 1 million g of solution or
> 5 mg chlorine in 1 million mg of solution or
> 5 lb of chlorine in 1 million lb of solution, and so on.

22.2.4.3 Conversions

Concentration is most often expressed in terms of ppm (or ppb) in environmental applications. This expression of concentration is useful when a very small amount of something (such as a pollutant) is dissolved in a large volume of solvent (such as a lake). For example, an environmental scientist might speak of a pollutant in a lake as being present at a concentration of 5 ppm.

To prepare a 5 ppm solution in the laboratory, you must convert the term "5 ppm" to a simple fraction expression, such as grams per liter or milligrams per milliliter, to determine how much of the solute to weigh out. Grams per liter, however, has units of weight in the numerator and volume in the denominator, but ppm and ppb expressions have the same unit in the numerator and denominator.[2] To get around this problem, convert the weight of the water into milliliters based on the conversion factor that 1 mL of pure water at 20°C weighs 1 g. For example, suppose we are studying the pollutant polychlorinated biphenyls (PCBs) and we want a concentration of 5 ppm PCBs in water:

$$5 \text{ ppm PCB} = \frac{5 \text{ g PCB}}{1 \text{ million g water}} = \frac{5 \text{ g PCB}}{1 \text{ million mL water}}$$

$$= \frac{5 \text{ g PCB}}{1{,}000 \text{ L water}}$$

---

[2] Percent expressions are actually "parts per hundred." Thus, technically the units should be the same in the numerator and denominator. When we use % expressions with units of weight in the numerator and volume in the denominator, we are using the approximation that 1 mL of solvent weighs 1 g, which is reasonable if the solvent is water.

To make the expression simpler, it is possible to divide the numerator and denominator both by 1 million:

$$5 \text{ ppm PCB} = \frac{5 \times 10^{-6} \text{ g PCB}}{1 \text{ mL water}}$$

5×10^{-6} g is the same as 5 µg. So, 5 ppm PCB is $= \dfrac{5 \text{ µg PCB}}{\text{mL}}$

For any solute: 1 ppm in water $= \dfrac{1 \text{ µg}}{\text{mL}} = \dfrac{1 \text{ mg}}{\text{L}}$

Also, 1 ppb in water $= \dfrac{1 \text{ ng}}{\text{mL}} = \dfrac{1 \text{ µg}}{\text{L}}$

Example Problem 22.12

How could you prepare a 500 ppm solution of compound A?

Answer

Recall that: 1 ppm in water $= \dfrac{1 \text{ mg}}{\text{L}}$

So, 500 ppm must be equal to $\dfrac{500 \text{ mg}}{\text{L}}$

$$500 \text{ mg} = 0.500 \text{ g}$$

Thus, an acceptable method to prepare 500 ppm of compound A is to weigh out 0.500 g, dissolve it in water, and BTV 1 L.

Practice Problems: Parts

1. Suppose you have a recipe that calls for 1 part salt solution to 3 parts water. How would you mix 10 mL of this solution?
2. Suppose you have a recipe that calls for 1:3.5:0.6 chloroform:phenol:isoamyl alcohol. How would you prepare 200 mL of this solution?
3. How would you mix 45 µL of a solution that is 5:3:0.1 Solution A:Solution B:Solution C?

Assume water is the solvent in Problems 4–6.

4. Convert 3 ppm to:
   a. unit of milligrams per milliliter
   b. unit of milligrams per liter.
5. Convert 10 ppb to unit of milligrams per liter.
6. How would you prepare a solution that has 100 ppm cadmium?

22.2.5 MOLALITY AND NORMALITY

22.2.5.1 Molality

Molality and **normality** are two ways to express concentration that are used more commonly in chemistry than biology manuals. We will define them briefly here, but refer to a chemistry text for more information.

Molality, abbreviated with a lowercase *m, means the number of molecular weights of solute (in grams) per kilogram of solvent* (Figure 22.6). Molality has the unit of weight in both the numerator and denominator.

A 1 molal solution of a compound contains 1 mole of the compound dissolved in 1 kg of water.

For example:

58.44 g of sodium chloride (FW = 58.44) in 1 kg of water is a 1 m solution.

In practice, the difference between a 1 molar solution and a 1 molal solution is:

- When preparing a 1 molal (1 m) solution, add the solute(s) to 1 kg of water.
- When preparing a 1 molar (1 M) solution, add water to the solute(s) to bring to the final volume of 1 liter.

22.2.5.2 Normality

Normality, abbreviated *N*, is another way to express concentration. **Normality** *is the number of "equivalent weights" of solute per liter of solution* (Figure 22.7). Normality is used when it is important to know how many "reactive groups" are present in a solution rather than how many molecules.

The key to understanding normality expressions is knowing the terms *reactive groups* and *equivalent weight.* Because biologists are most likely to encounter normality in reference to acids and bases, consider these terms as they relate to acids and bases. Recall that acids dissociate in solution to release H^+ ions, and bases, such as NaOH, dissociate to release OH^- ions. The H^+ and OH^- ions can participate in various reactions; hence, they are "reactive groups." *For an acid,* **1 equivalent weight** *is equal to the number of grams of that acid that reacts to yield 1 mole of H^+ ions, that is, 1 mole of reactive groups. For a base,* **1 equivalent weight** *is equal to the number of grams of that base that supplies 1 mole of OH^-, that is, 1 mole of reactive groups.* For example:

$$NaOH \rightarrow Na^+ + OH^-$$

$$H_2SO_4 \rightarrow SO_4^{-2} + 2H^+$$

In the first reaction, 1 mole of NaOH dissociates to produce 1 mole of Na^+ and 1 mole of OH^-. Because 1 equivalent weight of NaOH is the number of grams that will produce 1 mole of OH^- ions, the equivalent weight of NaOH is the same as its molecular weight. The number of grams of sodium hydroxide that will produce 1 mole of OH is **40.0 g**. In contrast, 1 mole of sulfuric acid dissociates to form 2 moles of H^+ ions. It only requires 0.5 mole of H_2SO_4 to produce 1 mole of H^+ ions. Therefore, the FW of H_2SO_4 is 98.1, but its equivalent weight is $98.1 \times 0.5 = $ **49.1 g**.

A 1 normal solution of a compound contains 1 equivalent weight of a compound dissolved in enough water to get 1 L.

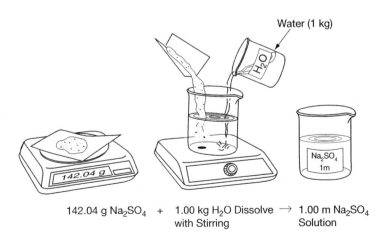

Water (1 kg)

142.04 g Na₂SO₄ + 1.00 kg H₂O Dissolve → 1.00 m Na₂SO₄
with Stirring   Solution

FIGURE 22.6 A 1 m (1 molal) solution of sodium sulfate (Na_2SO_4). 142.04 g of Na_2SO_4 is dissolved in 1 kg of water.

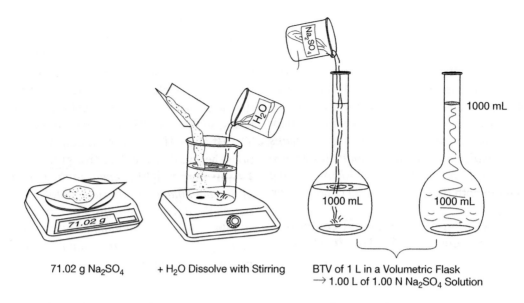

71.02 g Na$_2$SO$_4$ + H$_2$O Dissolve with Stirring   BTV of 1 L in a Volumetric Flask
$\rightarrow$ 1.00 L of 1.00 N Na$_2$SO$_4$ Solution

FIGURE 22.7 A 1 N solution of sodium sulfate (Na$_2$SO$_4$). When 1 mole of Na$_2$SO$_4$ dissolves, 2 moles of Na$^+$ ions are formed. The FW of Na$_2$SO$_4$ is 142.04; the equivalent weight is 71.02. (In this case, an equivalent is the number of grams of solute that will produce 1 mole of ionic charge in solution.)

Thus:

To make a 1 N solution of NaOH, dissolve 40.0 g of solute in enough water to make 1 L – the same as a 1 M solution.

To make a 1 N solution of H$_2$SO$_4$, however, dissolve 49.1 g of solute in water to make 1 L of solution – not the same as a 1 M solution.

Biology laboratory manuals seldom express concentration in terms of normality with the exceptions of HCl and NaOH solutions. Because HCl dissociates to produce one H$^+$ ion and NaOH dissociates to produce one OH$^-$ ion, their equivalent weight and FW are the same. For NaOH and HCl, 1 M = 1 N.

22.3 SUMMARY OF METHODS USED TO EXPRESS CONCENTRATION

Recipes in biology manuals sometimes list the *amounts* of each component required; other times, the *concentration* of each component required. There are various methods of expressing concentration. These methods are summarized in Tables 22.2 and 22.3. A couple of helpful rules are as follows:

1. ***When do you need to know the FW of a compound?***

    You need to know the FW when you have a recipe expressed in molarity, molality, or normality, or when you are converting to or from molarity, molality, or normality. With % solutions you do not need to know the FW of the compound (unless you are converting a % to molarity).

2. ***How do you convert molarity expressions to percent expressions (w/v), and vice versa?***

    There are several strategies to do these conversions. Some people find the following equation useful:

$$\text{Molarity} = \frac{\frac{w}{v}\% \times 10}{\text{FW}}$$

(This equation works because % is g/100 mL. Multiplying the percent by 10, therefore, gives g/1,000 mL or grams per liter.)

For example, convert 20% (w/v) NaCl to molarity.

w/v % is 20.
The FW is 58.44.
Substituting into the equation gives:

$$\text{Molarity} = \frac{20 \times 10}{58.44} \approx \textbf{3.42 M}$$

Another example, convert 3 M NaCl to a percent. The unknown, ?, is the percent (w/v).

$$3\text{ M} = \frac{? \times 10}{58.44} \qquad ? = \textbf{17.53\%}$$

TABLE 22.2

Summary of Methods Used to Express Concentration

A. **Weight per Volume**

A fraction with the weight of the solute in the numerator and the total volume of the solution in the denominator.

B. **Molarity**

The number of moles of solute per liter of solution.

C. **Percent**

i. WEIGHT PER VOLUME PERCENT (w/v %)

Grams of solute per 100 mL of solution.

ii. VOLUME PERCENT (v/v %)

Milliliters of solute per 100 mL of solution.

iii. WEIGHT PERCENT (w/w %)

Grams of solute per 100 g of solution.

D. **Parts**

i. Amounts of solutes are listed as "parts." The parts may have any units, but the units must be the same for all components of the mixture.

ii. PARTS PER MILLION AND PARTS PER BILLION

Parts per million: the number of parts of solute per 1 million parts of total solution. Parts per billion: the number of parts of solute per 1 billion parts of solution.

E. **Expressions of Concentration Not Common in Biology Manuals**

i. MOLALITY Moles of solute per kilogram of solvent.

ii. NORMALITY The number of gram-equivalents of solute per liter of solution. (Normality is reaction-dependent.)

TABLE 22.3

A Comparison of Methods of Expressing the Concentration of a Solute

Concentration of Solute (Na_2SO_4)	Amount of Solute	Amount of Water
1 M	142.04 g Na_2SO_4	BTV 1 L with water
1 m	142.04 g Na_2SO_4	Add 1.00 kg of water
1 N	71.02 g Na_2SO_4	BTV 1 L with water
1%	1 g Na_2SO_4	BTV 100 mL with water

CHAPTER APPENDIX: MOLARITY CALCULATIONS RELATING TO DNA AND RNA

DALTONS

The mass of one atom of an element is equal to the combined mass of its protons and neutrons. (The electrons' masses are so slight that they are usually ignored.) A periodic table provides the atomic mass for every element reported in units of atomic mass unit or dalton, abbreviated D or Da. A **dalton** is *defined to be 1/12th the mass of a ^{12}carbon nucleus, which is about the same as the mass of a hydrogen atom.* The mass in grams of one hydrogen atom is about 1.6605×10^{-24} g. So, there are $\approx 6.0222 \times 10^{23}$ D per gram.

The number 6.0222×10^{23} is **Avogadro's number,** *the number of molecules or atoms in a mole of any substance.* The mass of a *single* atom or molecule is its atomic mass or its formula mass in the unit of *dalton.*

The mass of a *mole* of any atom or compound has the same numerical value, but in the unit of *gram.*

THE MOLECULAR WEIGHTS OF NUCLEIC ACIDS

It is necessary to know the molecular or formula weight of a substance of interest in order to perform molarity calculations. The formula weight of a specific chemical compound is always the same, so it is usually straightforward to find the formula weights of chemicals by looking at the label on their containers. DNA is different, however, because its sequence and length vary depending on the source. A DNA molecule may be single-stranded (ss) or double-stranded (ds), and it may consist of anywhere from a few nucleotides to billions of base pairs. There is therefore no single MW for all DNA molecules. RNA molecules are similarly variable.

TABLE 22.4

Molecular Weights of Nucleotides Incorporated into Nucleic Acids

A in DNA	313.22 D
C in DNA	289.18 D
T in DNA	304.21 D
G in DNA	329.22 D
A+T	313.22+304.21 = 617.43
G+C	289.18+329.22 = 618.40
A in RNA	329.22 D
C in RNA	305.18 D
U in RNA	306.20 D
G in RNA	345.22 D

Source: Values from Roskams, Jane, and Rodgers, Linda., eds. *Lab Ref: A Handbook of Recipes, Reagents, and Other Reference Tools for Use at the Bench.* Cold Spring Harbor Laboratory Press, 2002.

Consider how to find the MW of DNA, first for oligonucleotides of known sequence and second for other types of DNA. Oligonucleotides are short segments of DNA that are often synthesized in a laboratory. To find the molecular weight of synthesized oligonucleotides, add the weights of all the nucleotides present. Table 22.4 shows the MW of individual nucleotides that have been incorporated into a nucleic acid. When nucleotides form phosphodiester bonds, water is lost so the MW of the incorporated bases is about 18 D less than that of solitary nucleotides. When oligonucleotides are synthesized, the last nucleotide in the chain is generally lacking a phosphate group; rather, it has a single H in that position. Synthesized oligonucleotides also usually have an OH group at the other end of the chain. The MW of the missing phosphate must be subtracted from the weight of the oligonucleotide, and the weight of the OH must be added. Therefore, 61.96 is subtracted from the total weight of the nucleotides. (Some sources subtract 61.) Equation 22.2 is used to determine the MW of a DNA oligonucleotide, based on the values in Table 22.4.

Sometimes the exact sequence of a DNA or RNA molecule is unknown, but its length in bases or base pairs is given. In these cases, conversion factors can be used to convert between length and MW. (These conversion factors provide estimates since the exact MW of a DNA or RNA molecule depends on its exact sequence.) A single nucleotide, on average, has a molecular weight of 330 D, and a base pair on average has a weight of 660 D. (These values are from Sambrook, J., Fritsch, E.F., and Maniatis, T. *Molecular Cloning: A Laboratory Manual.* Cold Springs Harbor, 1989, C1. Other references use slightly different average molecular weights, for example, 649 D for double-stranded and 325 D for single-stranded DNA.)

Equation 22.2. To Determine the MW of a DNA Oligonucleotide

$$MW = (N_c \times 289.18) + (N_a \times 313.22) + (N_t \times 304.21)$$
$$+ (N_g \times 329.22) - 61.96$$

where:

N_c = the number of cytosines
N_a = the number of adenines
N_t = the number of thymines
N_g = the number of guanines.

Example Problem 22.1A

What is the MW of a double-stranded DNA molecule that is 100 bp long?

Answer

$$100 \text{ bp} \times 660 \text{ D / bp} = 66,000 \text{ D}$$

Thus, one mole of this oligonucleotide would weigh about **66,000 g**. Obviously, this is a large number. In practice, we generally work with μmolar and pmolar concentrations of DNA.

CONVERTING BETWEEN MICROGRAMS AND PICOMOLES

There are helpful Equations, 22.3 and 22.4, to use in situations where it is necessary to convert an amount of DNA from micrograms into picomoles, or vice versa. Remember that for DNA:

dsDNA	ssDNA
$1\ mol \approx (660\,g)$ (# base pairs)	$1\ mol \approx (330\,g)$ (# bases)
$1\ pmol \approx (660\ pg)$ (# base pairs)	$1\ pmol \approx (330\ pg)$ (# bases)

Equation 22.3. To Convert from Micrograms to Picomoles DNA

$$\text{pmol of dsDNA} \approx \mu g(\text{of dsDNA})\left(\frac{1\ pmol}{660\ pg}\right)\left(\frac{1}{\#\ bp}\right)\left(\frac{10^6\ pg}{1\ \mu g}\right)$$

$$\text{pmol of ssDNA} \approx \mu g(\text{of ssDNA})\left(\frac{1\ pmol}{330\ pg}\right)\left(\frac{1}{\#\ bases}\right)\left(\frac{10^6\ pg}{1\ \mu g}\right)$$

Equation 22.4. To Convert from Picomoles to Micrograms DNA

$$\mu g \text{ of dsDNA} \approx \text{pmol}(\text{of dsDNA})\left(\frac{660\ pg}{pmol}\right)(\#\ bp)\left(\frac{1\ \mu g}{10^6\ pg}\right)$$

$$\mu g \text{ of ssDNA} \approx \text{pmol}(\text{of ssDNA})\left(\frac{330\ pg}{pmol}\right)(\#\ bases)\left(\frac{1\ \mu g}{10^6\ pg}\right)$$

Example Problem 22.2A

A vial contains 10 µg of pBR322 vector, which is 4,361 bp in length. How many picomoles are present?

Answer

By Equation 22.3:

$$10\ \mu g\left(\frac{1\ pmol}{660\ pg}\right)\left(\frac{1}{4,361}\right)\left(\frac{10^6\ pg}{1\ \mu g}\right) \approx \textbf{3.47 pmol}$$

Example Problem 22.3A

A vial contains 10 µg of a 1,000-base DNA fragment. How many picomoles are present?

Answer

By Equation 22.3:

$$10\ \mu g\left(\frac{1\ pmol}{330\ pg}\right)\left(\frac{1}{1,000}\right)\left(\frac{10^6\ pg}{1\ \mu g}\right) \approx \textbf{30.30 pmol}$$

Example Problem 22.4A

A vial contains 1 pmol of a 1,000 bp DNA fragment. How many μg are present?

Answer

By Equation 22.4:

$$1\,pmol \left(\frac{660\ pg}{pmol} \right) (1,000) \left(\frac{1\ \mu g}{10^6\ pg} \right) = \textbf{0.66}\ \boldsymbol{\mu}\textbf{g}$$

Equations relating to calculations for DNA and RNA solutions are summarized in Table 22.5.

TABLE 22.5

Commonly Used Equations Relating to DNA and RNA Solutions

Average Molecular Weights:

1. *Average MW of a deoxynucleotide ≈ 330 D.*
2. *Average MW of a DNA base pair ≈ 660 D.*
3. *Average MW of a ribonucleotide ≈ 340 D.*
4. *Average MW of dsDNA ≈ (number of bp) (660 D).*
5. *Average MW of ssDNA ≈ (number of bases) (330 D).*
6. *Average MW of ssRNA ≈ (number of bases) (340 D).*

To Calculate the Molecular Weight of a Synthesized Oligonucleotide of Known Sequence:

7. $\mathbf{MW = (N_c \times 289.18)\ (N_a \times 313.22)\ (N_t \times 304.21)\ (N_g \times 329.22) - 61.96}$

To Convert μg to pmol:

8. $pmol\ of\ dsDNA \approx \mu g\,(of\ dsDNA) \left(\dfrac{1\ pmol}{660\ pg} \right) \left(\dfrac{1}{\#\ bp} \right) \left(\dfrac{10^6\ pg}{1\ \mu g} \right)$

9. $pmol\ of\ ssDNA \approx \mu g\,(of\ ssDNA) \left(\dfrac{1\ pmol}{330\ pg} \right) \left(\dfrac{1}{\#\ bases} \right) \left(\dfrac{10^6\ pg}{1\ \mu g} \right)$

10. $pmol\ of\ ssRNA \approx \mu g\,(of\ ssRNA) \left(\dfrac{1\ pmol}{340\ pg} \right) \left(\dfrac{1}{\#\ bases} \right) \left(\dfrac{10^6\ pg}{1\ \mu g} \right)$

To Convert pmol to μg:

11. $\mu g\ of\ dsDNA \approx pmol\,(of\ dsDNA) \left(\dfrac{660\ pg}{pmol} \right) (\#\ bp) \left(\dfrac{1\ \mu g}{10^6\ pg} \right)$

12. $\mu g\ of\ ssDNA \approx pmol\,(of\ ssDNA) \left(\dfrac{330\ pg}{pmol} \right) (\#\ bases) \left(\dfrac{1\ \mu g}{10^6\ pg} \right)$

13. $\mu g\ of\ ssRNA \approx pmol\,(of\ ssRNA) \left(\dfrac{340\ pg}{pmol} \right) (\#\ bases) \left(\dfrac{1\ \mu g}{10^6\ pg} \right)$

23 Preparation of Laboratory Solutions – Part B

Basic Procedures and Buffers

23.1 PREPARING DILUTE SOLUTIONS FROM CONCENTRATED SOLUTIONS

23.1.1 THE $C_1V_1 = C_2V_2$ EQUATION

In the laboratory, we frequently use concentrated solutions that are diluted right before use. *The concentrated solutions are sometimes called* **stock solutions.** Frozen orange juice is a familiar example of a concentrated solution that is purchased as a concentrate in a can. The orange juice is prepared by diluting the concentrate with a specific amount of water. Similarly, in the laboratory, a 2 M stock solution of Tris buffer might be prepared that will later be used at concentrations of 1 M of less. The Tris stock is diluted to the proper concentration whenever needed.

There is an extremely helpful equation (Equation 23.1) that is used to determine how much stock solution and how much diluent (usually water) to combine to get a final solution of a desired concentration.

Equation 23.1 The $C_1V_1 = C_2V_2$ Equation

How to Make a Less Concentrated Solution from a More Concentrated Solution

$$(\text{Concentration}_{stock})(\text{Volume}_{stock}) = (\text{Concentration}_{final})$$
$$(\text{Volume}_{final})$$

This equation can be abbreviated : $C_1V_1 = C_2V_2$

Box 23.1 outlines the procedure to use the $C_1V_1 = C_2V_2$ equation, and the example problems illustrate its application.

A stock solution is sometimes written as an "X" solution where X means "times," *how many times more concentrated the stock is than normal.* A 10X solution is 10 times more concentrated than the solution

BOX 23.1   USING THE $C_1V_1 = C_2V_2$ EQUATION

1. *Determine the initial concentration of stock solution.* This is C_1.
2. *Determine the volume of stock solution required.* This is usually the unknown, ?.
3. *Determine the final concentration required.* This is C_2.
4. *Determine the final volume required.* This is V_2.
5. *Insert the values determined in Steps 1–4 into the equation, and solve for the unknown, ?.*

Note: The $C_1V_1 = C_2V_2$ equation is only used when calculating how to prepare a **less concentrated solution from a more concentrated solution.** Do not try to use this equation to calculate amounts of solute needed to prepare a solution of a given molarity or percent.

Example Problem 23.1

How would you prepare 100 mL of a 1 M solution of Tris buffer from a 2 M stock of Tris buffer?

Answer

Consider first, is this a situation where a less concentrated solution is being made from a more concentrated solution? Yes. It is appropriate, therefore, to apply the $C_1V_1 = C_2V_2$ equation:

1. Concentrated solution $= 2$ M $= C_1$.
2. The volume of concentrated stock necessary is ? (what you want to calculate) $= V_1$.
3. The concentration you want to prepare is 1 M $= C_2$.
4. The volume you want to prepare is 100 mL $= V_2$.

$$C_1V_1 = C_2V_2$$

$$2\,M(?) = 1\,M(100\,mL)$$

$$2M(?) = 100\,M\,(mL)$$

$$? = \frac{100\,\cancel{M}(mL)}{2\,\cancel{M}}$$

$$? = 50\,mL$$

5. Take 50 mL of the concentrated stock solution, and bring to volume (BTV) of 100 mL.

Example Problem 23.2

A recipe says to mix:

10X buffer Q	1 µL
Solution A	2 µL
Water	7 µL

What is the concentration of buffer Q in the final solution?

Answer

First consider whether this is a situation where a less concentrated solution is being made from a more concentrated solution. Yes. It is therefore appropriate to apply the $C_1V_1 = C_2V_2$ equation:

$C_1 = 10X$ $C_2 =$ unknown
$V_1 = 1$ µL $V_2 = 10$ µL $=$ total solution volume

$$C_1V_1 = C_2V_2$$

$$10X(1\,µL) = ?(10\,µL)$$

$$? = 1X$$

This means that 1 µL of a 10X stock in a final volume of 10 µL will give a final concentration of 1X. (Similarly, 1 mL of the 10X buffer in a solution with a final volume of 10 mL would give a final concentration of 1X.)

---

normally used. To work with "X" solutions, use the $C_1V_1 = C_2V_2$ equation.

In Figure 23.1, Recipe Version A begins with buffer that is 10 times more concentrated than it is usually used. In Recipe Version B, the buffer is 5 times more concentrated than its usual concentration. Either 5X or 10X buffer stock can be used, but how much is needed

Recipe Version A		Recipe Version B	
10 X buffer Q	1 µL	5 X buffer Q	2 µL
Solution A	2 µL	Solution A	2 µL
Water	7 µL	Water	6 µL
Total Volume	10 µL	Total Volume	10 µL

FIGURE 23.1 Two versions of the same solution recipe.

varies. If 10X buffer is used, then only 1 µL is required; however, if 5X buffer is used, then 2 µL is required. The volume of the total solution must not vary, so 7 µL of water is used in Recipe Version A. In Recipe Version B, only 6 µL of water is required.

23.1.2 DILUTIONS EXPRESSED AS FRACTIONS

Dilutions of concentrated stocks are sometimes expressed in terms of *fractions*. (Refer to Chapter 12 for a detailed discussion of such dilutions.) When a dilution is expressed as a fraction, most people do not use the $C_1V_1 = C_2V_2$ expression. For example:

a. A 95% solution is diluted 1/10. What is its final concentration?

(95%) (1/10) = 9.5%, so the final concentration is **9.5%.**

b. A 1 M solution is diluted 1/5. What is its final concentration?

(1 M) (1/5) = 0.2 M, so the final concentration is **0.2 M**.

c. Suppose you want to make 10 mL of a 1/50 dilution of food coloring from an original bottle of concentrated food coloring. Then, set up a proportion:

$$\frac{1}{50} = \frac{?}{10 \text{ mL}}$$

$$? = \mathbf{0.2 \text{ mL}}$$

So, take 0.2 mL of the original stock and dilute to a volume of 10 mL.

23.2 MAINTAINING A SOLUTION AT THE PROPER pH: BIOLOGICAL BUFFERS

23.2.1 PRACTICAL CONSIDERATIONS REGARDING BUFFERS FOR BIOLOGICAL SYSTEMS

23.2.1.1 The Useful pH Range

Biological systems require a particular pH. The interior of cells, blood, and other fluids are buffered in vivo (in the living organism). In the laboratory, comparable buffering

systems must be artificially reproduced. **Laboratory buffers** *are laboratory solutions that help maintain a biological system at the proper pH.* When a solution is buffered, it resists changes in pH, even if H^+ ions are added to, or lost from, the system.

There are many combinations of chemicals and some individual chemicals that are used as buffers. Each of these buffers has chemical characteristics that make it more or less useful for a particular purpose. Some of these chemical properties that distinguish buffers from one another will briefly be described.

The behavior of a buffer is displayed graphically in Figure 23.2. To prepare this graph, NaOH was slowly added to a buffer solution and the resulting pH was measured. Observe that the buffer tends to resist a change in pH at around pH 4.8. In other words, there is a pH at which OH^- or H^+ can be added to this solution with little change in its pH. The **pK_a** *of a buffer is the pH at which the buffer experiences little change in pH upon addition of acids or bases.* The buffer illustrated in Figure 23.2 is an acetate buffer that has a pK_a of 4.76. As shown in Table 23.1, different buffers have different values of pK_a. Buffers are effective in resisting a change in pH within a range of about 1 pH unit above and below their pK_a. For example, the pK_a for acetate buffer is about 4.8 and acetate buffer is effective in the range from about pH 3.8 to 5.8. Thus, different buffers are effective in different pH ranges. Note also that a buffer may have more than one pK_a. For example, phosphate buffer has a pK_a at 2.15, 7.20, and 12.33.

Every biological system has an optimal pH. For example, stomach enzymes work at around pH 2, whereas DNA is synthesized at about pH 6.9. In the

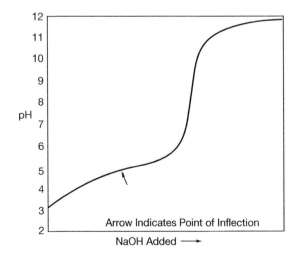

FIGURE 23.2 The effect of adding NaOH to acetate buffer. At around pH 4.8, the pK_a for acetate buffer, adding base has relatively little effect on the pH.

TABLE 23.1

The pK$_a$ Values for Some Common Buffers and the Effect of Temperature Change on the pH of These Buffers[a]

Trivial Name	Buffer Name	pK$_a$	dpK$_a$/dt
Phosphate (pK$_1$)	—	2.15	0.0044
Malate (pK$_1$)	—	3.40	—
Formate	—	3.75	0.0
Succinate (pK$_1$)	—	4.21	−0.0018
Citrate (pK$_2$)	—	4.76	−0.0016
Acetate	—	4.76	0.0002
Malate	—	5.13	—
Pyridine	—	5.23	−0.014
Succinate (pK$_2$)	—	5.64	0.0
MES	2-(N-Morpholino)ethanesulfonic acid	6.10	−0.011
Cacodylate	Dimethylarsinic acid	6.27	—
Dimethyl glutarate	3,3-Dimethylglutarate (pK$_2$)	6.34	0.0060
Carbonate (pK$_1$)	—	6.35	−0.0055
Citrate (pK$_3$)	—	6.40	0.0
Bis-Tris	[Bis(2-hydroxyethyl)imino]tris(hydroxymethyl)methane	6.46	0.0
ADA	N-2-Acetamidoiminodiacetic acid	6.59	−0.011
Pyrophosphate	—	6.60	—
EDPS (pK$_1$)	N,N′-Bis(3-sulfopropyl)ethylenediamine	6.65	—
Bis-Tris propane	1,3-Bis[tris(hydroxymethyl)methylamino]propane	6.80	—
PIPES	Piperazine-N,N′-bis(2-ethanesulfonic acid)	6.76	−0.0085
ACES	N-2-Acetamido-2-hydroxyethanesulfonic acid	6.78	−0.020
MOPSO	3-(N-Morpholino)-2-hydroxypropane-sulfonic acid	6.95	−0.015
Imidazole	—	6.95	−0.020
BES	N,N-Bis-(2-hydroxyethyl)-2-aminoethanesulfonic acid	7.09	−0.016
MOPS	3-(N-Morpholino)propanesulfonic acid	7.20	0.015
Phosphate (pK$_2$)	—	7.20	−0.0028
EMTA	3,6-Endomethylene-1,2,3,6-tetrahydrophthalic acid	7.23	—
HEPES	N-2-Hydroxyethylpiperazine-N′-2-ethanesulfonic acid	7.48	−0.014
TES	N-tris(hydroxymethyl)methyl-2-aminoethanesulfonic acid	7.40	−0.020
DIPSO	3-[N-Bis(hydroxyethyl)amino]-2-hydroxypropanesulfonic acid	7.60	−0.015
TEA	Triethanolamine	7.76	−0.020
POPSO	Piperazine-N,N′-bis(2-hydroxypropanesulfonic acid)	7.85	−0.013
EPPS, HEPPS	N-2-Hydroxyethylpiperazine-N′-3-propanesulfonic acid	8.00	—
Tris	Tris(hydroxymethyl)aminomethane	8.06	−0.028
Tricine	N-[Tris(hydroxymethyl)methyl]glycine	8.05	−0.021
Glycinamide	—	8.06	−0.029
PIPPS	1,4-Bis(3-sulfopropyl)piperazine	8.10	—
Glycylglycine	—	8.25	−0.025
Bicine	N,N-Bis(2-hydroxyethyl)glycine	8.26	−0.018
TAPS	3-{[Tris(hydroxymethyl)methyl]amino}propanesulfonic acid	8.40	0.018
Morpholine	—	8.49	—
PIBS	1,4-Bis(4-sulfobutyl)piperazine	8.60	—
AES	2-Aminoethylsulfonic acid, taurine	9.06	−0.022
Borate	—	9.23	−0.008
Ammonia	—	9.25	−0.031
Ethanolamine	—	9.50	−0.029
CHES	Cyclohexylaminoethanesulfonic acid	9.55	0.029
Glycine (pK$_2$)	—	9.78	−0.025

(Continued)

TABLE 23.1 (*Continued*)

The pK$_a$ Values for Some Common Buffers and the Effect of Temperature Change on the pH of These Buffers[a]

Trivial Name	Buffer Name	pK$_a$	dpK$_a$/dt
EDPS	N,N'-Bis(3-sulfopropyl)ethylenediamine	9.80	—
APS	3-Aminopropanesulfonic acid	9.89	—
Carbonate (pK$_2$)	—	10.33	−0.009
CAPS	3-(Cyclohexylamino)propanesulfonic acid	10.40	0.032
Piperidine	—	11.12	—
Phosphate (pK$_3$)	—	12.33	−0.026

[a] Values are at 25°C.

Source: Adapted from Burgess, Richard, and Deutscher, Murray. *Guide to Protein Purification (Methods in Enzymology, Volume 436).* 2nd ed., Academic Press, 2009.

laboratory, it is necessary to choose a buffer whose pK$_a$ matches that of the system of interest. For example, if one wants to work with a DNA solution at pH 6.9, acetate buffer should not be chosen.

23.2.1.2 Chemical Interactions with the System Being Studied

A laboratory buffer should be inert in the system being studied. For example, Tris buffer, which is widely used in molecular biology procedures involving nucleic acids, is unsuitable for some protein assays because it reacts with the assay components, giving erroneous results. Phosphate buffers contribute phosphate ions to a solution, which inhibit some types of enzyme reactions. An important biotechnology example is the use of Tris–borate–EDTA (TBE) buffer versus Tris–acetate–EDTA (TAE) buffer when DNA fragments are separated from one another by gel electrophoresis. TBE is most commonly used for electrophoresis of DNA because it gives good resolution of DNA fragments and has a good buffering capacity. It is difficult, however, to recover DNA from gels that were run with TBE buffer because the borate interacts with oligosaccharides in the gel. TAE, therefore, is the buffer of choice when DNA is to be extracted from gels after electrophoresis.

Some buffers chelate (bind) ions, such as Ca^{++} or Mg^{++}, making these ions inaccessible. If the biological system of interest requires such ions, then a buffer that binds them is inappropriate; in other cases, metal ion chelation is desirable.

23.2.1.3 Changes in pH with Changes in Concentration and Temperature

It is convenient to prepare buffers as concentrated stock solutions that are diluted before use. Some buffers, however, change pH when diluted. Tris buffer is fairly stable when diluted. Its pH decreases about 0.1 pH units for every tenfold dilution – which is acceptable in most applications. The pH value of a 100 mM sodium phosphate buffer increases from 6.7 to 6.9 with tenfold dilution. If the change with dilution of a buffer is not acceptable in an application, then the use of stock solutions should be avoided.

Temperature changes may markedly affect the pH of a buffer. For example, the pH of a Tris buffer will decrease approximately 0.028 pH units with every degree Celsius increase in temperature. Thus, Tris buffer prepared to pH 8.00 at room temperature (25°C) will be close to pH 7.66 when used at body temperature (37°C) and pH 8.70 when placed on ice. Buffers that are temperature-sensitive should generally be prepared at the temperature at which they will be used.

Some buffers are more sensitive than others to temperature. Table 23.1 gives values for various buffers that are used to calculate the change in pH with each degree change in temperature. The larger the numeric value for "dpK$_a$/dt," the more the buffer will change pH with each degree of temperature change. Negative values for dpK$_a$/dt mean that there is a decrease in pH with an increase in temperature, and vice versa.

Note that there are factors that affect the temperature of a solution while its pH is being measured. Magnetic stirring devices generate heat. Some chemical reactions are endothermic or exothermic (remove or generate heat), so, as chemicals are combined, their temperature may change. This has practical importance when preparing buffers because if you do not measure the temperature of the solution, you may think it is at room temperature (or another desired temperature) when it is not.

23.2.2 TRIS BUFFER

Tris (tris(hydroxymethyl)aminomethane) is probably the most common buffer used in biotechnology laboratories. Tris is used because it is very soluble in water; it does not interact with the components of many enzymatic systems; it is relatively inexpensive; and it has a high buffer capacity. The pK_a of Tris is 8.06 at 25°, which means that it is most effective between about pH 7.1 and 9.1 and therefore is not an ideal buffer for biological systems that require a pH of 7 or lower. Table 23.2 summarizes some considerations regarding Tris buffers.

23.2.3 ADJUSTING THE pH OF A BUFFER

23.2.3.1 Overview

Preparing buffers (and other laboratory solutions) properly and consistently is a critical step in ensuring reproducibility in research results and in minimizing variability in products. While it is not recommended, it is possible for a laboratory scientist or technician to follow established protocols without understanding why a particular buffer system was chosen. It is, however, the responsibility of the laboratory scientist or technician to prepare the buffers correctly and consistently.

To prepare a buffer, the correct amount of the buffer compound(s) is weighed out and dissolved in water to get the desired concentration. For example, Tris buffer comes as a crystalline solid. *Tris that is not conjugated to any other chemicals is called* **Tris base**. Tris base has a FW of 121.1 and a pK_a of 8.06. A 1 M solution of Tris can be prepared by dissolving 121.1 g of Tris base in 1 L of water. The pH of such a solution, however, will be greater than 10. This pH is far from the pK_a for Tris, so the solution will have little buffering capacity. Moreover,

few biological systems function at such a high pH. Before the Tris solution is useful as a biological buffer, its pH therefore must be lowered. Most buffering compounds are not at their pK_a when they are dissolved and must be adjusted to the desired pH. There are several strategies to obtain a buffer that has both the correct pH and the correct concentration. This section describes these strategies and illustrates them using Tris and phosphate buffers as examples.

23.2.3.2 Adjusting the pH of Tris Buffer

Two strategies to bring Tris to the correct pH will be described. The first is the most common strategy for preparing Tris buffers for molecular biology procedures and generally should be followed unless otherwise specified in a procedure.

Strategy 1 for Adjusting the pH of Tris (and other buffers).

A strong acid is added to the buffer until it is the desired pH. The amount of Tris required to get the desired concentration is dissolved in water, but it is not brought to its final volume. Tris base has a basic pH when dissolved, so it is necessary to add acid to bring the pH down to a value suitable for biological work. It is common for people to say that the buffer is *titrated* to bring it to the proper pH. If the acid required to bring Tris to the proper pH is unspecified, it is most common to use HCl, but other acids may be used as well. For example, boric acid, citric acid, and acetic acid can be used to prepare Tris–borate, Tris–citrate, and Tris–acetate buffers, respectively. Remember to use a Tris-compatible electrode when titrating the solution.

Some buffers when dissolved in water have a pH that is lower than the desired pH. For these buffers, a strong base (often NaOH) is added to raise the pH. Once the

---

TABLE 23.2

Considerations When Using Tris Buffers

1. *Tris buffers change pH significantly with changes in temperature.* Final pH adjustments to Tris buffers, therefore, should generally be made when the solution is at the temperature at which it will be used.

2. *Tris buffers require a compatible electrode for accurate pH determination.* As described in Chapter 20, Tris is not compatible with many "general-purpose" electrodes.

3. *Tris is subject to fungal contamination.* Contamination can be avoided by autoclaving the Tris solution or filter sterilizing it before long-term storage.

4. *Tris is minimally sensitive to concentration changes.* Tris is therefore often stored as a concentrated stock solution that is diluted before use.

desired pH is reached, the solution is brought to its final volume. It is common practice to use NaOH to titrate buffers that are too acidic when dissolved and HCl for the titration of buffers that are too basic when dissolved.

Box 23.2 outlines the steps involved in titrating a buffer by adding a strong acid or base until the proper pH is reached.

Strategy 2 for Adjusting the pH of Tris.

Combine Tris base with a conjugated form of Tris called Tris–HCl. Tris–HCl is purchased as a solid form of Tris that has been conjugated to HCl. If Tris–HCl is dissolved in water, its pH is less than 5. It is possible to combine Tris base with Tris–HCl to get a pH between 7 and 9. Depending on the pH desired, different amounts of the two forms of Tris are combined, as shown in Box 23.3. (Note that Figure 23.5b in Section 23.4.2.3 describes a procedure to prepare Tris buffer based on this strategy.)

23.2.3.3 How *Not* to Adjust the pH of Tris Buffer

It might seem that yet another strategy to prepare Tris buffer would be to begin by dissolving Tris–HCl, which has a pH of about 4.7, and bring it up to the

---

BOX 23.2 GENERAL PROCEDURE TO BRING A BUFFER TO THE CORRECT pH USING A STRONG ACID OR A STRONG BASE

1. *Determine the amount of solute(s) required to make the buffer at the correct concentration.*
2. *Mix the solute(s) with most, but not all, of the solvent required.* Do not bring the solution to volume yet. (To ensure consistency, a standard procedure may specify in how much water to dissolve the solute(s) initially.)
3. *Place the solution on a magnetic stir plate, add a clean stir bar, and stir.*
4. *Check the pH.* (Be careful not to crash the stir bar into the electrode.)
5. *Add a small amount of acid or base, whichever is needed to bring the solution toward the desired pH.* The recipe will often specify which acid or base to use; if the recipe does not specify, it is usually safe to add HCl or NaOH.
6. *Stir again and then check the pH.*
7. *Repeat Steps 5 and 6 until the pH is correct.* It is also useful to check the temperature of the buffer to make sure that it is correct for your application.
8. *Bring the solution to the proper volume once the pH is correct, and then recheck and record the pH.* (With many buffers, when you bring the solution to its final volume, the pH does not change significantly.)

NOTES ABOUT ADJUSTING THE pH OF A SOLUTION

i. *Temperature.* pH the solution when it is at the temperature at which you plan to use it. Note also that some solutions change temperature during mixing. For example, the addition of acid to Tris buffer may cause its temperature to change. Restore the buffer to the correct temperature before making the final adjustment of its pH.

ii. *Overshooting the pH.* If you accidentally overshoot the pH, the correct procedure is to remake the entire solution. It is not recommended to compensate by adding some extra acid or base. This is because the additional acid or base adds extra ions to the solution; these ions will change the composition of your solution and may adversely affect it. In addition, if you add more acid or base sometimes (because of a mistake), but not other times, there will be an inconsistency in the buffer that will be very difficult to diagnose if problems arise later.

iii. *Overshooting the volume.* If you accidentally overshoot the volume, begin again to avoid changing the concentration of solute(s) and to avoid introducing inconsistency. (Using more concentrated acid or base may help avoid this problem, but be careful not to overshoot the pH.)

iv. *Final adjustment of the pH after bringing to volume.* Most procedures do not involve readjusting the pH after bringing the solution to volume. However, in some facilities this might be done and should be explicitly described in a standard procedure to ensure consistency.

BOX 23.3   COMBINING TRIS BASE AND TRIS–HCL TO GET A 0.05 M BUFFER OF THE DESIRED PH

pH at 5°C	pH at 25°C	pH at 37°C	g/L Tris–HCl	g/L Tris Base
7.55	7.00	6.70	7.28	0.47
7.66	7.10	6.80	7.13	0.57
7.76	7.20	6.91	7.02	0.67
7.89	7.30	7.02	6.85	0.80
7.97	7.40	7.12	6.61	0.97
8.07	7.50	7.22	6.35	1.18
8.18	7.60	7.30	6.06	1.39
8.26	7.70	7.40	5.72	1.66
8.37	7.80	7.52	5.32	1.97
8.48	7.90	7.62	4.88	2.30
8.58	8.00	7.71	4.44	2.65
8.68	8.10	7.80	4.02	2.97
8.78	8.20	7.91	3.54	3.34
8.86	8.30	8.01	3.07	3.70
8.98	8.40	8.10	2.64	4.03
9.09	8.50	8.22	2.21	4.36

Source: "Sigma Catalog," *Sigma Company*. www.sigmaaldrich.com/content/dam/sigma-aldrich/docs/Sigma/Bulletin/1/106bbul. pdf. Note that Sigma company calls their Tris "Trizma."

To make 0.05 M Tris buffer: Dissolve the indicated amounts of Tris–HCl and Tris base together in water to a final volume of 1 L. Tris–HCl and Tris base are somewhat hygroscopic at high humidity. For precise work, desiccation before weighing is recommended.

Tris solution can be autoclaved.

It is possible to prepare other concentrations using these proportions of Tris base to Tris–HCl, but this may have a slight effect on the pH. See the SOP in Figure 23.5b for more information.

---

desired pH using NaOH or another base. However, this will NOT result in the same buffer as the two strategies described above. The difference is in the ionic strength. When Tris–HCl is combined with NaOH, the salt, NaCl, is formed, thus significantly increasing the ionic strength of the buffer. The importance of ionic strength will be considered in more detail in Chapter 25. For now, we will note that altering the method by which a solution is prepared can result in variability and unintended consequences.

Example Problem 23.3

Outline a procedure to make 300 mL of a 0.8 M solution of Tris buffer, pH 7.6 at 4°C. (FW of Tris base = 121.1.) Use the strategy in Box 23.2.

Answer

1. First, it is necessary to calculate the amount of solute (Tris base) required. This may be accomplished using Equation 22.1 (introduced in Section 22.2.2).

Solute Required =

(Grams/Mole)(Molarity)(Volume)

(121.1 g/mole)(0.80 mole/L)(0.300 L) = **29.064 g**

2. Dissolve 29.064 g of Tris base in about 200 mL of 4°C purified water. The pH will be above 9 or 10.

3. Add concentrated HCl (remember to use safety precautions) while stirring and reading the pH. (Use a pH meter equipped with a Tris-compatible electrode.)

4. Check the temperature of the solution. The temperature may change both during the initial dissolution of the Tris and during the addition of acid to the solution. Ensure that the temperature of the solution is 4°C before performing the final adjustment of its pH.

5. Continue adding HCl until the solution is pH 7.6.

6. Bring the solution to 300 mL with water. (Because the pH of Tris is relatively insensitive to dilution, the pH should still be close to 7.6 after the solution is brought to volume.)

Note: There is an alternate strategy that allows this buffer to be prepared at room temperature. From Table 23.1, we know that the change in pH with temperature for Tris is −0.028 pH units per degree Celsius. Suppose room temperature is 22°C. Then the difference between room temperature and 4°C is −18°C. Multiplying −18°C times −0.028 pH units per degree gives a change of +0.504 pH units. The solution, therefore, can be prepared at room temperature, but be brought to pH 7.096 (round to 7.1). When the buffer is used at 4°C, its pH should rise to be very close to pH 7.6. The advantage to this method is that it avoids the requirement of holding the temperature at 4°C as the solution is prepared.

Example Problem 23.4

Outline a procedure to make 300 mL of a 0.05 M solution of Tris buffer, pH 7.62 at 37°C. Use the information in Box 23.3.

Answer

1. Box 23.3 specifies that 1 L of 0.05 M buffer at pH 7.62 and at 37°C requires 4.88 g of Tris–HCl and 2.30 g of Tris base.

2. Because we do not want to make a liter of buffer, it is necessary to calculate how much of each chemical is required to make 300 mL. This can be easily accomplished using the logic of proportions:

Tris-HCl   Tris base

$$\frac{4.88 \text{ g}}{1{,}000 \text{ mL}} = \frac{?}{300 \text{ mL}} \qquad \frac{2.30 \text{ g}}{1{,}000 \text{ mL}} = \frac{?}{300 \text{ mL}}$$

$$? = \mathbf{1.464 \text{ g}} \quad ? = \mathbf{0.690 \text{ g}}$$

3. Combine 1.464 g of Tris–HCl and 0.690 g of Tris base in 250 mL of purified water that is at 37°C.

4. Ensure that the temperature of the solution is 37°C.

5. Check the pH of the solution with a pH meter having a Tris-compatible electrode.

6. If the pH is within 0.05 units of 7.62, then bring it to volume with water. (If the pH is not sufficiently close to 7.62, it may be acceptable to add small amounts of either acid or base to bring the pH to the exact pH before bringing to volume. This should be specified in the laboratory's standard procedure.)

23.2.3.4 Adjusting the pH of Phosphate Buffer Using a Strategy Involving Two Stock Solutions: One More Basic and the Other More Acidic

Many buffers are prepared by mixing two stock solutions in specific ratios to achieve the desired pH. We will illustrate this strategy with phosphate buffer, which is another buffer commonly used for biological systems. Phosphate buffer is made by combining solutions of sodium phosphate dibasic (Na_2HPO_4) and sodium phosphate monobasic (NaH_2PO_4) in the correct proportions to obtain a desired pH. At 25°C, a 5% solution of dibasic sodium phosphate has a pH of 9.1, whereas a 5% solution of the monobasic form has a pH of 4.2. Mixtures of these two solutions will make a buffer with a pH between the two extremes.

In this procedure, the stock solutions of the monobasic and dibasic compounds are prepared to the same molar concentration, for example, 0.2 M. When the two solutions are combined in the proportions required to achieve the proper pH, the concentration of the resulting buffer is determined by the molarity of the phosphate ions. This means that if 0.2 M sodium phosphate monobasic and 0.2 M sodium phosphate dibasic are combined, the resulting concentration will still be 0.2 M. Phosphate buffer prepared with a mixture of the monobasic and dibasic forms has a pK_a of 7.2, and so its buffering range is about pH 6 to pH 8. Box 23.4 shows a procedure to make phosphate buffer.

BOX 23.4   PREPARING PHOSPHATE BUFFER

To Prepare 0.1 M Sodium Phosphate Buffer of a chosen pH:

Step 1. **Prepare Solution A,** 0.2 M $NaH_2PO_4 \cdot H_2O$ (monobasic sodium phosphate monohydrate, FW = 138.0):

Dissolve **27.6 g $NaH_2PO_4 \cdot H_2O$** in purified water. Bring to volume of 1,000 mL.

Step 2. **Prepare Solution B,** 0.2 M Na_2HPO_4 (dibasic sodium phosphate):

Use either **28.4 g Na_2HPO_4** (anhydrous dibasic sodium phosphate, FW = 142.0) or **53.6 g $Na_2HPO_4 \cdot 7H_2O$** (dibasic sodium phosphate heptahydrate, FW = 268.1).

Dissolve in purified water, and bring to volume of 1,000 mL.
  Step 3. Combine Solution A + Solution B in the amounts listed in the following table.
  Step 4. Bring the mixture of Solution A + Solution B to a final volume of 200 mL.

Volume A (mL)	Volume B (mL)	pH
93.5	6.5	5.7
92.0	8.0	5.8
90.0	10.0	5.9
87.7	12.3	6.0
85.0	15.0	6.1
81.5	18.5	6.2
77.5	22.5	6.3
73.5	26.5	6.4
68.5	31.5	6.5
62.5	37.5	6.6
56.5	43.5	6.7
51.0	49.0	6.8
45.0	55.0	6.9
39.0	61.0	7.0
33.0	67.0	7.1
28.0	72.0	7.2
23.0	77.0	7.3
19.0	81.0	7.4
16.0	84.0	7.5
13.0	87.0	7.6
10.5	89.5	7.7
8.5	91.5	7.8
7.0	93.0	7.9
5.3	94.7	8.0

(Continued)

Tabular information from Chambers, J., et al. *LabFax Set: Biochemistry LabFax*. Academic Press, 1993, and Calbiochem Corporation. *Buffers: A Guide for the Preparation and Use of Buffers in Biological Systems*, 2003.

Additional recipes for sodium phosphate buffer are available from: *Cold Spring Harbor Protocols*. Cold Spring Harbor Laboratory Press. 2006. cshprotocols.cshlp.org/content/2006/1/pdb.rec8303.full, 2006.

Phosphate buffers can also be prepared with the potassium salt of phosphate and occasionally with a mixture of the sodium and potassium salts. Phosphate buffers are less sensitive than are Tris buffers to the effect of temperature, but are somewhat more sensitive to dilution. Phosphate buffers, therefore, are not normally prepared as concentrated stock solutions. They are instead prepared to the concentration at which they will be used, as described in Box 23.4.

Example Problem 23.5

Using Box 23.4, outline a procedure to make 0.1 M sodium phosphate buffer, pH 7.0.

Answer

1. Prepare the two stock solutions, Solution A and Solution B, as directed in Box 23.4. These stocks have a molarity of 0.2 M.
2. Mix the amounts indicated on the chart, that is, 39.0 mL of Solution A and 61.0 mL of Solution B. This solution has a concentration of 0.2 M.
3. Bring the solution to a volume of 200 mL as directed in the chart. Dilution of the buffer with water to 200 mL means that the final concentration of phosphate in the solution is 0.1 M. Check the pH.

23.3 PREPARING SOLUTIONS WITH MORE THAN ONE SOLUTE

23.3.1 INTRODUCTION

Chapter 22 discussed how to prepare a solution containing a single solute in such a way that the solute is present at the correct concentration. This chapter has so far added another issue to the preparation of a laboratory solution – ensuring that the solution is at the correct pH as well as the correct concentration.

In practice, very few biological solutions contain only one solute. More typically, solutions contain multiple solutes. Very often a solution contains a buffering substance with other chemicals added to it. Thus, in biology manuals, solutions that are called "buffers" may have functions in addition to maintaining the proper pH. For example, "buffers" may also contain salts, EDTA, β-mercaptoethanol, and so on. Each of these additional chemicals has a role to play in the biological system of interest. (These roles will be discussed in Chapter 25.) Biologists thus use the term *buffer solution* in a general way, as compared with a strict chemical definition.

Let's consider how to prepare solutions with more than one solute. Figure 23.3 shows three examples of recipes for solutions with more than one solute.

Recipe I is relatively simple to interpret. It states the *amounts* of four solutes to weigh out and combine in water.

Recipe II also states how much of each solute is required; however, the solutes are prepared as individual solutions before being combined. To prepare Recipe II, first make a 1 M solution of $MgCl_2$, a 0.4% solution of thymidine, and a 20% solution of glucose. Then, mix 5 mL of the $MgCl_2$ solution, 10 mL of the thymidine solution, and 25 mL of the glucose solution. The recipe does not call for additional water to be added to the mixture of the three solutes.

Recipe III is different in that it does not state the *amount* of each solute required. Rather, the recipe states the *final concentration* of each solute. This is the most common type of recipe in the scientific literature. If a recipe or procedure gives only the final

Recipe I		Recipe II		Recipe III
Na_2HPO_4	6 g	1 M $MgCl_2$	5 mL	0.1 M Tris
KH_2PO_4	3 g	0.4% thymidine	10 mL	0.01 M EDTA
NaCl	0.5 g	20% glucose	25 mL	1% SDS
NH_4Cl	1 g			
Dissolve in water.				
Bring to a volume				
of 1 liter.				

FIGURE 23.3 Examples of three recipes for laboratory solutions that contain more than one solute.

concentration(s) needed for the solute(s), you have to *calculate the amount(s)* you need of each. The reason authors often record concentrations instead of amounts is they do not know what volume you will need. Recipes such as Recipe III are discussed in more detail in this section.

23.3.2 A Recipe with Multiple Solutes: SM Buffer

Consider the following recipe:

SM Buffer	
0.2 M Tris, pH 7.5	1 mM $MgSO_4$
0.1 M NaCl	0.01% Gelatin

This is the entire recipe as it might appear in a technical manual. In a scientific paper, to save space, the recipe might be inserted in parentheses: SM buffer (0.2 M Tris, pH 7.5, 1 mM $MgSO_4$, 0.1 M NaCl, 0.01% gelatin).

The recipe lists the *final concentration* of each solute; Tris is present in the final solution at a concentration of 0.2 M, $MgSO_4$ at a final concentration of 1 mM, and so on. The first instinct people sometimes have when looking at a recipe like this is to prepare four separate solutions – 0.2 M Tris, 1 mM magnesium sulfate, 0.1 M NaCl, and 0.01% gelatin – and then mix them together. However, this instinct is wrong. If you combine any volume of 0.2 M Tris with any volume of 1 mM $MgSO_4$ or any of the other solutes, they will dilute one another, giving wrong final concentrations. Consider two correct strategies to prepare this solution.

Strategy 1: Preparing SM Buffer without Stock Solutions

In overview, first prepare 0.2 M Tris, pH 7.5, but do not BTV yet. Then, calculate how many grams of each of the other solutes are required. These solutes are weighed out and dissolved directly in the Tris buffer. Once all the solutes are dissolved, BTV. Note that when the temperature is not specified, assume a solution is to be at room temperature. The steps are as follows:

1. Decide how much volume to prepare, for example, 1 L.
2. One liter of 0.2 M Tris requires 24.2 g of Tris base (FW = 121.1). Dissolve the Tris in about 700 mL of water, and bring the pH to 7.5 by adding concentrated HCl. Do not bring the Tris to volume yet.

(The Tris buffer is brought to the proper pH before the other solutes are added because the recipe shows Tris to be a particular pH.)

3. One liter of 0.1 M NaCl requires 5.84 g of NaCl. Add this amount to the Tris buffer.
4. One liter of 1 mM $MgSO_4$ requires 1/1,000 of its FW. (Magnesium sulfate comes in more than one hydrated form. Read the container to determine the correct molecular weight.) Weigh out the correct amount, and add it to the Tris buffer. (For example, $MgSO_4 \cdot 7\ H_2O$ has a FW of 246.5. Add 0.2465 g of this form of $MgSO_4$.)
5. 0.01% Gelatin is 0.1 g in 1 L. Weigh out 0.1 g of gelatin, and add to the Tris buffer.
6. Stir the mixture until all the components are dissolved, and then BTV of 1 L with water.
7. It is good practice to check the pH at the end: It should be close to 7.5 because the Tris is fairly stable with dilution and the other components generally have little effect on the pH. Record the final pH, but do not readjust it (unless directed to do so by a procedure you are following). Note that this is a step where there might be subtle differences between researchers and facilities where some might perform a final pH readjustment and others do not. Such subtleties may need to be considered when trying to reproduce the work of others.

Strategy 2: Preparing SM Buffer with Stock Solutions

In this strategy, the four solutes are each prepared separately as *concentrated stock solutions*. Then, when the four stocks are combined, they dilute one another to the proper final concentrations.

1. Prepare a stock solution of Tris buffer at pH 7.5. There is no set rule as to what concentration a stock solution should be other than it should be significantly higher than the concentration of the diluted solution. In this case, a useful concentrated stock solution might be 1 M (i.e., 5X, that is, 5 times more concentrated than needed). To make 1 L of 1 M stock, dissolve 121.1 g of Tris base in about 900 mL of water. Bring the solution to pH 7.5 and then to a volume of 1 L.
2. Prepare a stock solution of magnesium sulfate (e.g., 1 M). To make 100 mL of this stock, dissolve 0.1 FW of $MgSO_4$ in water and bring to a volume of 100 mL. (For example,

$MgSO_4 \cdot 7 H_2O$ has a FW of 246.5. Use 24.65 g of this form of $MgSO_4$.)

3. Prepare a stock solution of NaCl, for example, 1 M. To make 100 mL of stock, dissolve 5.844 g in water and bring to a volume of 100 mL.

4. Prepare a stock solution of gelatin, for example, 1%, by dissolving 1 g in a final volume of 100 mL of water.

5. To make the final solution, it is necessary to combine the right amounts of each stock. Because this is a situation where concentrated solutions are being diluted, use the $C_1V_1 = C_2V_2$ equation four times, once for each solute. For example, to figure out how much Tris stock is required:

$$C_1V_1 = C_2V_2$$

$$(1 \text{ M})(?) = (0.2 \text{ M})(1,000 \text{ mL})$$

$$? = \textbf{200 mL}$$

To make 1,000 mL of SM buffer, therefore, combine:

Component	Final Concentration
200 mL of the 1 M Tris stock, pH 7.5	200 mM Tris (0.2 M)
1 mL of the 1 M magnesium sulfate stock	1 mM $MgSO_4$
100 mL of the 1 M NaCl stock	0.1 M NaCl
10 mL of the 1% gelatin stock	0.01% Gelatin

6. BTV 1,000 mL, and check the pH.

Both Strategy 1 and Strategy 2, preparing a solution with stocks or without, are correct. It is generally efficient to make stock solutions of solutes that are often used because weighing out chemicals is time-consuming. For example, many solutions require NaCl; therefore, it might be useful to keep a 1 M stock solution of NaCl on hand. Chemicals that are used infrequently should not be kept as stock solutions because microorganisms can grow in them and they may slowly degrade. (Sterilizing solutions can sometimes extend their shelf life.)

23.3.3 A RECIPE WITH TWO SOLUTES: TE BUFFER

Consider the preparation of TE buffer, a very common solution in molecular biology laboratories.

TE Buffer	
Component	Final Concentration
Tris, pH 7.6	10 mM
Na_2-EDTA	1 mM

This buffer contains Tris at a pH of 7.6 and a final concentration of 10 mM, and Na_2-EDTA at a final concentration of 1 mM. As was true for SM buffer, it is incorrect to prepare 10 mM Tris buffer and 1 mM Na_2-EDTA and then to mix them together because they will dilute one another. It is conventional to prepare TE buffer by combining stock solutions of Tris, at the desired final pH, and EDTA.

To make TE buffer:

Step 1. Decide how much volume to prepare, for example, 100 mL.

Step 2. Na_2-EDTA does not go into solution easily and is relatively insoluble in water if the pH is less than 8.0. Na_2-EDTA, therefore, is commonly prepared as a concentrated stock solution with a pH of 8.0. To make 100 mL of a 0.5 M stock of Na_2-EDTA, add 18.6 g of EDTA, disodium salt, dihydrate (FW = 372.2) to 70 mL of water. Adjust the pH to 8.0 slowly with stirring by adding pellets of NaOH. Switch to concentrated NaOH solution as the solution approaches pH 8.0. When the Na_2-EDTA is dissolved, bring the solution to volume. (Be patient; EDTA is slow to go into solution.)

Step 3. The Tris buffer needs to be dissolved and brought to the proper pH before it is combined with the Na_2-EDTA. Like the Na_2-EDTA, the Tris should also be prepared as a concentrated stock solution (e.g., 0.1 M). To make 100 mL of Tris stock, dissolve 1.21 g of Tris in about 70 mL of water, adjust it to pH 7.6 with HCl, and then bring it to 100 mL final volume.

Step 4. Use the $C_1V_1 = C_2V_2$ equation twice, once for each solute, to calculate how much of each stock will be required to make a solution of TE with the proper concentration of both solutes.

To calculate how much 0.1 M stock of Tris is required:

$$C_1V_1 = C_2V_2$$

$$(0.1 \text{ M})(?) = (0.010 \text{ M})(100 \text{ mL})$$

$$? = \textbf{10 mL}$$

(Note that the units must be consistent on both sides of the equation, so 10 mM was converted to 0.010 M.)

To calculate how much 0.5 M Na_2-EDTA stock is required:

$$C_1V_1 = C_2V_2$$

$$(0.5 \text{ M})(?) = (0.001 \text{ M})(100 \text{ mL})$$

$$? = 0.2 \text{ mL} = 200 \text{ μL}$$

Step 5. Combine 10 mL of Tris stock and 200 μL of Na_2-EDTA stock, and bring the solution to the final volume, 100 mL, with water.

Step 6. Check the pH.

23.4 PREPARATION OF SOLUTIONS IN THE LABORATORY

23.4.1 GENERAL CONSIDERATIONS

Box 23.5 outlines a general procedure for making a solution in the laboratory and summarizes some of the issues discussed so far in these two chapters on solutions.

23.4.2 ASSURING THE QUALITY OF A SOLUTION

23.4.2.1 Documentation

All steps of solution preparation require documentation, including procedure(s) followed, the details of the

---

BOX 23.5   A GENERAL PROCEDURE TO MAKE A SOLUTION IN THE LABORATORY

1. *Read and understand the procedure; identify and collect materials and equipment needed.* Be sure equipment is properly maintained and calibrated.
2. *Determine the amount of each solute or stock solution that is required.*
3. *Weigh the amount of each solute required, or measure the volume of each stock solution.*
4. *Dissolve the weighed compounds, or mix the stock solutions, in less than the desired final volume of solvent.*
5. *Bring the solution to the correct pH, if necessary.*
6. *Bring the solution to volume.* Check the pH, if necessary.

Notes:

- *Water.* Unless instructed otherwise, use the highest quality (most purified) water available to make solutions. Water purification is discussed in Chapter 24.
- *Grades of Chemicals.* Be sure that the purity and form of chemicals match your requirements. Using a lower grade may adversely affect the reproducibility of results. See Table 23.3.
- *Contamination.* Do not return chemicals to stock containers so as not to contaminate the stocks of chemicals. In addition, avoid putting spatulas into stock containers, although it may be acceptable to use a clean spatula. To protect yourself and to avoid contaminating solutions, never touch any chemicals or the interior of glassware containers.
- *Labeling.* Label the container of every solution. Include the following on the label: the name and/or formula of the solution, the concentration of the solution, the solvent used, the date prepared, your name or initials, and safety information such as whether the solution is toxic, radioactive, acidic, basic, and so on. Additional information, such as the lot number of the solution, may be required.
- *Concentrated and Dilute Acids and Bases.* The term *concentrated* refers to acids and bases in water at the concentration that is their maximum solubility at room temperature (Table 23.4). For example, concentrated hydrochloric acid has a molarity of 11.6 and is 36% acid. In contrast, concentrated nitric acid is 15.7 M and 72% acid. Use proper safety practices as described in Table 9.5. When working with concentrated acids, put the acid into water and not vice versa to avoid hazardous splashing and overheating. Do not neutralize a strong acid with a strong base because this creates an exothermic reaction that can be intense enough to melt a container.
- *Hygroscopic Compounds.* **Hygroscopic compounds** *absorb moisture from the air.* A compound that has absorbed an unknown amount of water from the air may not be acceptable for use. Hygroscopic compounds, therefore, should be kept in tightly closed containers and should be exposed to air as little as possible. It may be advisable to purchase such compounds in small amounts so that once their container is opened, they are rapidly used and are not stored. These compounds may be stored in desiccators to help protect them from moisture. In some laboratories, it is common practice to record the date a bottle of a chemical is first opened. A chemical that has been opened and stored too long may need to be discarded.

(Continued)

- **Promoting Dissolution.** Some chemicals do not go into solution readily, and mild heat or a slightly acidic or alkaline environment might be helpful. (See, for example, the procedure for preparing TE buffer in Section 23.3.3.) If using heat, be aware of temperature effects on pH. Also, some substances are degraded by heat, such as some vitamins and antibiotics. If heat or other special methods are required, be sure this is included in any written standard procedure.
- **Sterilization.** It is often good practice to sterilize solutions by autoclaving or ultrafiltration to prevent bacterial and fungal growth. (These practices are discussed in Chapter 24.) Note, however, that some solution components are sensitive to heat.

TABLE 23.3

Grades of Chemical Purity

Chemicals are manufactured with varying degrees of purity. More purified preparations are usually more expensive.

1. *Reagent (ACS)-* or *Analytical Reagent (ACS)*-grade chemicals are certified to contain impurities below the specifications of the American Chemical Society Committee on Analytical Reagents. If the bottle was not contaminated by previous use, these chemicals are of high purity and are usually acceptable for making solutions.
2. *Reagent* grade. If the American Chemical Society has not established specifications for a particular chemical, the manufacturer may do so and label the chemical reagent grade. The manufacturer must then list the maximum allowable impurities for the chemical. These reagents are also usually acceptable for most laboratory uses.
3. *USP and NF* grades. The U.S. Pharmacopeia (USP) and the National Formulary (NF) jointly publish a book of pharmacopeial standards for chemical and biological drug substances, dosage forms, compounded preparations, excipients, medical devices, and dietary supplements. USP- and NF-grade chemicals meet or exceed the standards in their joint publication. Such chemicals are often acceptable for general laboratory use although you might want to review their specifications to make sure they meet the requirements of a particular laboratory procedure.
4. *CP* means chemically pure and is usually similar to, or slightly less pure than, reagent grade. **Purified** means that the manufacturer has removed many contaminants. Read the manufacturer's specifications to determine whether these chemicals are acceptable for your purposes.
5. *Laboratory grade*. These chemicals are commonly used in educational settings, but would not typically be used where consistent and traceable chemicals are required.
6. *Commercial-* and *Technical*-grade chemicals are intended for industrial use. *Practical* grade is intended for use in organic syntheses. In general, commercial-, technical-, and practical-grade chemicals are not used for laboratory solutions.
7. *Spectroscopic*-grade chemicals are intended for use in spectrophotometry because they have very little absorbance in the UV or IR range.
8. *HPLC*-grade reagents are organic solvents that are very pure and are intended for use as the mobile phase in HPLC. (See Chapter 34.)
9. *Primary Standard*-grade chemicals are the most highly purified chemicals and are intended to be used as standards.
10. *Others.* Chemical manufacturers often have their own grades, such as "molecular biology grade that has been analyzed for nucleases and proteases." Read the manufacturer's catalog to determine their specifications.

TABLE 23.4

Acids and Bases

Acid or Base	Formula	FW	Moles per Liter	Density (g/mL)	% Solution (w/w)
Acetic acid, glacial	CH_3COOH	60.1	17.4	1.05	99.5
Hydrochloric acid, concentrated	HCl	36.5	11.6	1.18	36
Nitric acid, concentrated	HNO_3	63.0	16.0	1.42	71
Sulfuric acid, concentrated	H_2SO_4	98.1	18.0	1.84	96
Sodium hydroxide, pellets	NaOH	40.0	Solid	—	—
Sodium hydroxide, diluted	—	—	6.1	1.22	20

Source: For more information on acids and bases, see *The Merck Index: An Encyclopedia of Chemicals, Drugs, and Biologicals.* 15th ed. Royal Society of Chemistry Publishing, 2013.

preparation (e.g., weights, volumes, and pH), problems or deviations from the procedure, components of the solution including the name of the manufacturer(s), manufacturer's lot numbers, who prepared the solution, date of preparation, handling after preparation, maintenance and calibration of equipment used, and safety considerations. It is always good practice to check and record the pH of a biological solution. Thorough documentation is important for reproducibility, even in laboratories that do not adhere to a formal quality system.

Figure 23.4 is an example of a form used in a biotechnology company that has a centralized solution-making facility. The side of the form in Figure 23.4a is completed when a solution is requested by a laboratory scientist, and the side in Figure 23.4b is filled in by the person who prepares the solution. In many companies, such information is entered directly into a computer system, but paper can also be used. Note the comprehensive documentation that is associated with this solution.

(a) MEDIA PREPARATION FORM (*SECTION A*)

ITEM NUMBER	LOT NUMBER

A-1.

1. NAME OF MEDIA/SOLUTION (*and pH @ Temp.*): _____

2. Requested By _____ Room No. _____ Dept. No. _____

3. Project # (*if applicable*)_____ 4. PIP # (*if applicable*)_____

5. Date and Time Ordered_____ Date and time Required _____

NOTE: Please allow a minimum of 24 hours for buffers and solutions, and 48 hours for agar-based plates and New Formulations.

A-2. TOTAL QUANTITY REQUIRED: _____

A-3. Check Product Type:

1. MEDIA PREP PRODUCT ☐   2. PRODUCT ☐   3. SHOP ORDER COMPONENT ☐

4. NEW FORMULATION ☐ *Check with Media Prep if unsure if request is a New Formulation*

   1) If a MEDIA PREP PRODUCT, write BPCS ITEM NUMBER in the box above. Media Prep will fill in LOT NUMBER.

   2) If a PRODUCT, write Production BPCS ITEM NUMBER and SHOP ORDER (LOT) NUMBER in the boxes above. Attach (staple) a copy of the CURRENT Procedure to the Media Preparation Form.

   3) If PRODUCTION SHOP ORDER COMPONENT, list PRODUCT ITEM # and SHOP ORDER # below.

      ITEM #: _____   SHOP ORDER #: _____

   4) If a NEW FORMULATION, identify the source of the new formulation. e.g., Notebook # and Page #, Journal article (Journal name, vol., article, and page), Reference Book (Title and page), etc. Attach a copy to this form.

      SOURCE: _____

      Complete the following sections of the Bill Of Materials on the reverse side (*SECTION B*) for NEW FORMULATIONS: Component, Component Item No., and Final Conc. for Media/Solution Components. If BPCS Item No.'s are unavailable, write vendor name (after the component) in Component Column, and vendor Item No. in the Item Column.

A-4. Check items to be included in prep   A-5. DISTRIBUTION

A-4	A-5	
☐ Autoclave	# of Plates:	Vol. Liquid/Per Vol. Container:
☐ Filter Sterilize	# of Flasks:	Vol. Liquid/Per Vol. Container:
☐ Filter (to remove extraneous debris)	# of Bottles:	Vol. Liquid/Per Vol. Container:
WATER SOURCE:	# of Tubes:	Vol. Liquid/Per Vol. Container:
☐ Deionized (DI.)	# of Slants:	Vol. Liquid/Per Vol. Container:
☐ NANOpure	# of Carboys:	Vol. Liquid/Per Vol. Container:
☐ Other (Describe):	Other:	

A-6. SPECIAL INSTRUCTIONS/COMMENTS: _____

_____

_____

A-7. DELIVERY INSTRUCTIONS: _____

☐ Check this box if you wish to receive a copy of the completed Media Preparation Form.

FIGURE 23.4 Media preparation form. (a) This part of the form is completed by the person requesting a solution. (b) This information is entered into the computer system when the solution is made.

(Continued)

(b) MEDIA PREPARATION FORM (*SECTION B*)   page ___ of ___
(Complete all sections. Write a line or dash in sections that do not apply.)

ITEM NUMBER	LOT NUMBER

NAME OF MEDIA/SOLUTION: _____

TARGET pH @ Temp. _____ (1 X pH @ Temp.) _____

CONDUCTIVITY RANGE/(Dilution): _____

SOP NUMBER FOLLOWED: _____

BILL OF MATERIALS   QUANTITY PREPARED: _____

Component (vendor)	Item Number	Lot Number	Final Conc.	Amount per Liter	Total Amount	Equipment	Amount Added

WATER *(circle source and write location)* Deionized/NANOpure/Ultrafiltered_____

EQUIPMENT AND MEASUREMENTS:

Balance(s) *ID #:* (1) _____ (2) _____ (3) _____

pH Meter *ID #:*_____ Conductivity Meter *ID #:*_____

Mixing Vessel: _____ Dispensing Equipment: _____

Filtered with: _____

Filter Sterilized with: _____

Autoclave(s) *ID #, cycle time, and run #:* _____

CALIBRATE pH ELECTRODE

Standard Buffers:_____ Slope/Efficiency @ Temp: _____

Initial pH: _____ pH adjusted with: _____

Final pH @ Temp: _____ 1 X pH @ Temp:_____

CONDUCTIVITY/Dilution: _____

DEVIATIONS FROM PROCEDURE *(list components, critical steps, etc., that differ from the procedure)*_____

_____

COMMENTS:_____

_____

_____

Location _____ Hours_____ Prepared by_____ Date _____

FIGURE 23.4 (*Continued*) Media preparation form. (a) This part of the form is completed by the person requesting a solution. (b) This information is entered into the computer system when the solution is made.

23.4.2.2 Traceability

In production facilities, it is essential that every component used in producing a product be traceable. If a solution is used in production, documentation that identifies each component of that solution must be connected to the product. This system ensures that if a problem arises with a product, every component of every solution that went into that product can be identified. In a research setting, it is also essential to be able to identify every reagent used in an experiment to ensure reproducibility.

23.4.2.3 Standard Procedures

It is critical that solutions are prepared consistently to ensure reproducibility in research laboratories, and to reduce variability in production settings. Note that there is often some ambiguity in published solution recipes from the scientific literature, and also sometimes in technical manuals. For example, the temperature of the solution may be unstated, or it may be unclear as to whether a specified pH refers to a particular solute, or whether it is the pH of the final solution. Standard procedures, therefore, play a key role in ensuring consistency and clarity in solution preparation. Even in laboratories that do not adhere to a particular quality system, it is good practice to develop standard procedures for preparation of solutions that are routinely made. For solutions that are not made routinely, be certain to record the details of preparation, the source of the solution "recipe," or the purpose of its components. Chapter 25 explores considerations that researchers use when developing solution "recipes" for particular applications.

Standard procedures should include all details, such as:

- full information about every chemical in the solution sufficient that the same chemical could be ordered again;
- details on calculations;
- what source of water should be used;
- details on if and how solutions are brought to proper pH;

- whether a final adjustment of pH should be performed after the solution is brought to volume;
- whether the solution should be sterilized, and if so, how;
- how the solution should be stored and how long it is stable;
- if a solution is used routinely, the acceptable range for the solution's pH and conductivity;
- instruments involved in the solution's preparation (including calibration instructions); and
- safety concerns.

When preparing a solution according to a standard procedure, follow the directions exactly to avoid irreproducibility. Modifications may have unintended and unpredictable consequences.

Examples of industry standard procedures to make solutions are shown in Figure 23.5.

(a)   1.0 FORMULATION   DNA BUFFER   DAB-00S

Component	Item Number	FW or Conc.	Amount/L	Final Concentration
2 MTris-HCl, pH 7.3	TRIS-73S	2 M	5 mL	10 mM
5 MNaCl	NACL-05S	5 M	2 mL	10 mM
0.25 MEDTA, pH 8.0	EDTA-25S	0.25M	4 mL	1 mM
D.I. H$_2$O		—	Q.S.	—

Target pH: 7.50 ± 0.10 @ 25°C
Conductivity Range: 55–75 μS/cm @ 25°C, 1:40 dilution (2.5 mL/97.5 mL D.I. H$_2$O)

2.0 SAFETY PRECAUTIONS

    2.1 Wear safety glasses, labcoat, and disposable laboratory gloves at all times during the preparation.

3.0 PROCEDURE

    3.1 Read referenced SOPs.

    3.2 Use only preapproved components as specified by the item number listed in the Formulation.

    3.3 Add components to the mixing vessel in the order listed above.

    3.4 Q.S. to final volume with D.I. H$_2$O.

    3.5 Allow to mix for minimum of 10 minutes on a stir plate or with a motorized lab stirrer (e.g., Lightning Mixer).

    3.6 Check final pH with a *calibrated* pH electrode.

    3.7 Filter sterilize (if requested) or autoclave (if requested).

    3.8 Store at room temperature (20–25°C).

FIGURE 23.5 Procedures to make solutions. (a) Procedure to make DNA buffer. This procedure uses the strategy of preparing concentrated stocks and diluting them to make the desired solution. QS is the same as "bring to volume." (b) Procedure to make 2 M Tris–HCl, pH 7.3. This is one of the components of the DNA buffer.

(b) 1.0 FORMULATION 2 M TRIS-HCL, pH 7.3

Component	Item Number	FW or Conc.	Amount/L	Final Concentration
Trizma HCl	014426	157.6	274.0 g	1.74 M
Trizma Base	014393	121.1	32.0 g	0.26 M
D.I. H$_2$O		—	Q.S.	—

Target pH: 7.30–7.35 @ 25°C
Conductivity Range: 150–230 μS/cm @ 25°C, 1:1000 dilution (100 μL/100 mL D.I. H$_2$O)

2.0 SAFETY PRECAUTIONS

  2.1 Wear Safety Glasses, Labcoat, and disposable laboratory gloves at all times during the preparation.

3.0 PROCEDURE

Note: Quantities of Tris HCl and Tris Base are obtained from the Trizma Mixing Table (see Sigma Trizma Technical Bulletin No. 106B). Values given in the table are specific for a 0.05 M solution. Multiply values by 40 to determine quantities of Tris HCl and Tris base needed for a 2 M solution. Increasing the total Tris concentration from 0.05 M to 0.5 M will increase the pH by about 0.05 pH units, and increasing the total Tris concentration to 2 M generally causes an increase of about 0.05 to 0.10 pH units.

Note: For Tris buffers, *as temperature increases pH decreases.* As a general rule for Tris buffers: *pH changes an average of 0.03 pH units per °C below 25°, and pH changes an average of 0.025 pH units per °C above 25°C.*

  3.1 Read referenced SOPs.

  3.2 Use only preapproved components as specified by the item number listed in the Formulation.

  3.3 Start with D.I. H$_2$O at approximately 60% of the total volume to be prepared. Add components in the order listed above, and allow to mix until dissolved. *Note:* Mixing of Tris components in water results in an endothermic reaction, and the temperature of the solution cools significantly.

  3.4 Bring volume up to approximately 90% of the final volume, allow to mix and to warm up to near room temperature (20–25°C).

  3.5 Check pH with a *calibrated* pH electrode near room temperature (20–25°C).

  3.6 If the pH is within ± 0.05 units @ 25°C, Q.S. to final volume with D.I. H$_2$O and allow to mix for a minimum of 5 minutes.

    3.6.1 If pH is not within acceptable limits, adjust carefully with concentrated HCl until pH is within acceptable limits. Q.S. to final volume with D.I. H$_2$O and allow to mix for a minimum of 5 minutes.

  3.7 Check final pH with a *calibrated* pH electrode.

  3.8 Check pH of a 0.05 M sample (2.5 mL of Tris buffer diluted to 100 mL).

  3.9 Check conductivity of the 2 M solution @ 1:1000 dilution.

  3.10 Filter to remove extraneous particles and dirt.

  3.11 Store at room temperature (20–25°C).

  3.12 *Note:* For large volume preparations (100 liters or more), heat approximately 30 liters of D.I. H$_2$O per 100 liters to approximately 80–90°C. Weigh out Tris powders and add to mix tank. Add approximately 10 liters of D.I. H$_2$O to the Tris powders. Add the 30 liters of heated D.I. H$_2$O. Bring to 90% of the total volume to be prepared. Perform steps 3.3 through 3.9 above.

FIGURE 23.5 (*Continued*) Media preparation form. (a) This part of the form is completed by the person requesting a solution. (b) This information is entered into the computer system when the solution is made.

23.4.2.4 Instrumentation

Balances, pH meters, and other instruments used in the preparation of a solution must be properly maintained and calibrated, and maintenance and calibration records must be kept for each instrument. When a solution is made, the instruments used are recorded. Note that the form in Figure 23.4b has blanks that relate to instrumentation. For example, the form has a space to record the slope/efficiency of the pH meter that was used. (See Chapter 20, Section 20.4.1.2 for an explanation of the significance of this value.)

23.4.2.5 Stability and Expiration Date

If possible, every solution should be labeled with an expiration date based on its stability during storage. Where stability information is not available, the solution should still be labeled with the preparation date. Manufacturers may provide storage and stability information for chemicals they sell.

23.4.2.6 Using Conductivity for Quality Control

A solution that is made consistently should always have about the same conductivity (conductivity is introduced in Chapter 20, Section 20.7) because it should have the same constituents. Therefore, based on experience, it is possible to specify the range within which the conductivity should be for a particular solution. The conductivity can then be measured each time the solution is prepared for quality-control purposes. An unexpected conductivity reading is a warning that there is a problem. The use of conductivity for quality control is included in the procedures in Figure 23.5.

Practice Problems

1. Suppose you start with 5X solution A and you want a final volume of 10 μL of 1X solution A. How much solution A do you need?
2. How would you mix 75 mL of 95% ethanol if you have 250 mL of 100% ethanol?
3. How would you mix 25 mL of 35% acetic acid from 25% acetic acid?
4. If you have 65 mL of 0.3 M Na_2SO_4, how much 0.1 mM solution can you make?
5. a. How would you prepare a 0.1 M stock solution of Tris buffer, pH 7.9? The FW of Tris is 121.1.
   b. How would you use this Tris stock to prepare a 10 mM solution of Tris, pH 7.9?
6. Using the information in Table 23.1, suggest a buffer to maintain a system at pH 8.5.
7. Calculate the pH of a Tris buffer solution that is pH 7.5 at 25°C and then is brought to a temperature of 65°C.
8. Write a procedure to first prepare stock solutions of Tris, Mg acetate, and NaCl, as shown in the table below, and then to use these stock solutions to prepare breaking buffer.

Stock Solutions for Breaking Buffer		
1 M Tris, pH 7.6 at 4°C	Tris	FW = 121.1
1 M Mg acetate	Magnesium acetate·$4H_2O$	FW = 214.40
1 M NaCl	NaCl	FW = 58.44

Breaking Buffer
0.2 M Tris, pH 7.6 at 4°C
0.2 M NaCl
0.01 M Mg acetate
0.01 M β-mercaptoethanol
5% (v/v) Glycerol

Notes:

i. The recipe specifies that the Tris buffer should be pH 7.6 at 4°C; therefore, bring it to the proper pH at this specified temperature.

ii. β-Mercaptoethanol comes as a liquid. It can be weighed out in grams, but it is easier to measure β-mercaptoethanol with a pipette. The density of β-mercaptoethanol is 1.1 g/mL, FW = 78.13. This means that 1 mL of β-mercaptoethanol weighs 1.1 g.

iii. β-Mercaptoethanol is extremely volatile, so there are safety precautions involved in its use. Refer to its Safety Data Sheet to find out how to use this chemical safely. It is a good policy to check safety precautions whenever working with a new chemical. (See Chapter 9.)

iv. Magnesium acetate may be purchased as the anhydrous or tetrahydrate form. The formula weight given in the problem is for the tetrahydrate form.

9. Deoxynucleotide triphosphates (dNTPs) are the building blocks of DNA. There are four deoxynucleotide triphosphates: dATP, dTTP, dGTP, and dCTP. Suppose that you are going to repeatedly perform a procedure that calls for 10 μL of a solution that contains a mixture of:

1.25 mM each of dATP, dTTP, dGTP, and dCTP

You do not have any deoxynucleotide triphosphates in your laboratory, so you are going to buy them. You look in a catalog and find the following entry:

DESCRIPTION: Deoxynucleotide triphosphates are provided at a concentration of 100 mM in water.

Product	Amount in Vial	Catalog #
dATP	40 μmoles	XXXA
dCTP	40 μmoles	XXXC
dGTP	40 μmoles	XXXG
dTTP	40 μmoles	XXXT

a. Observe that the four nucleotides are purchased separately and will then need to be combined into one solution. Decide how many vials of each nucleotide to purchase. Assume you will make enough of the solution to perform the procedure 100 times.

b. Outline how you would prepare the nucleotide solution.

10. Explain the concept of traceability as illustrated by the form in Figure 23.4.

11. Explain how the buffer in Figure 23.5b in Section 23.4.2.3 is brought to the proper pH.

24 Solutions: Associated Procedures and Information

24.1 WATER PURIFICATION

24.1.1 WATER AND ITS CONTAMINANTS

24.1.1.1 Introduction

We have seen in the previous two chapters that water, which is critical for life, is the primary component of most biological solutions. The quality of water used to prepare solutions is of utmost importance. For example, cells in culture will die if extremely small amounts of contaminants are present. Minor levels of impurities in water can introduce major errors into analytical procedures. Impurities in the water used to make a pharmaceutical product can have deadly consequences for a patient. We must therefore consider how purified water is obtained in a biotechnology facility.

24.1.1.2 Water Is Not Pure

Water is an excellent solvent, and it readily dissolves contaminants from a wide variety of sources (Table 24.1). In the environment, water receives contaminants from industry, municipalities, and agriculture. Water dissolves minerals from rocks and soil that make it "hard." A variety of organic materials from dead plants, animals, and microorganisms are naturally present in water. Gases from the air dissolve in water. Suspended particles enter water from silt, pipe scale, dust, and other sources. In the laboratory, contaminants may leach into water from glass, plastic, and metal containers.

If water is not sterilized, unwanted microorganisms will readily grow in it and may release toxic bacterial by-products termed *endotoxins* and *pyrogens*. **Pyrogens** *are derived from lipopolysaccharides that are found in the cell walls of gram-negative bacteria. Pyrogen levels are measured in* **endotoxin units (EU/mL).** Pyrogens were thought to be toxic, so they are often called *endotoxins*. In fact, pyrogens are not directly toxic; rather, they induce a dangerous immune response in mammals. This immune response leads to fever (hence the name *pyrogen*, "heat producing"), shock, and cardiovascular failure. Pyrogens induce adverse effects when present at minute levels; therefore, their elimination is of critical importance in the manufacture of drugs for human and animal use. Even sterile water can contain pyrogens. The term *endotoxin* may be used synonymously with *pyrogen*, or it is sometimes used more broadly to refer to any contaminating by-product of bacteria.

Removing these many and varied contaminants from water is an important task in almost any laboratory and is a major operation and expense in bioprocessing industries. Laboratories and industrial facilities use a variety of carefully engineered processes to meet their requirements for purified water.

24.1.2 WHAT IS PURE WATER?

There actually is no such thing as "pure water." Although a variety of effective methods are used to

TABLE 24.1

Contaminants of Water

Types of Contaminants

1. *Dissolved Inorganics*. This term refers to matter, not derived from plants, animals, or microorganisms, that usually dissociates in water to form ions. Calcium and magnesium, the minerals that make water "hard," are examples of dissolved inorganics, as are the heavy metals cadmium and mercury. Silica and metals can leach from glass and metal containers.

2. *Dissolved Organics*. Organic compounds contain carbon and hydrogen. Organic materials in water may be the result of natural vegetative decay processes, and they may also be human-made substances such as pesticides.

3. *Suspended Particles*. This category includes diverse materials, such as silt, dust, undissolved rock particles, and metal scale from pipes.

4. *Dissolved Gases*. Water can dissolve naturally occurring gases, such as carbon dioxide and oxygen, as well as gases from industrial processes, such as nitric and sulfuric oxides.

5. *Microorganisms*. This category includes bacteria, viruses, and other small agents.

6. *Pyrogens/Endotoxins*. See text for an explanation of these terms.

7. *Nucleases*. These are enzymes that break apart nucleic acids (DNA and RNA). They are of concern in facilities where molecular biology techniques are performed.

Related Terminology

1. *Total Organic Carbon (TOC)*. A measure of the concentration of organic contaminants in water.

2. *Total Solids (TS)*. A measure of the concentration of both the dissolved (TDS) and suspended solids (TSS) in a solution. Total dissolved solids include both organics (such as pyrogens and tannin) and inorganics (such as ions and silica). Total suspended solids include organics (such as algae and bacteria) and inorganics (such as silt). It is common to approximate TDS by measuring the water's conductivity, although conductivity is actually a measure only of conductive species.

remove contaminants from water, regardless of how stringently it is purified, it is never totally free of contaminants. Professional societies therefore publish standards that contain specifications for the relative purity of water to be used in certain applications.

Water purity standards historically were often linked to the method by which the water was purified; distillation in particular was required to produce high-quality grades of water. The requirement to use a particular purification method limited the use of modern technologies that are often more efficient and cost-effective than older methods. Common water purity standards therefore no longer require particular methods of purification. Current standards do require that certain key contaminants be measured, and that their levels do not exceed specified limits.

ASTM International establishes specifications for reagent (laboratory) water that are widely used as the basis for defining water quality. ASTM standards identify four types of laboratory water. Type I is the highest class of purity. Type I water is used for most analytical procedures (e.g., trace metal analysis) because it has very low levels of contaminants. Many biotechnology procedures require Type I water, such as tissue culture, immunological assays, preparation of standard solutions, molecular biology procedures, and reconstitution of lyophilized (freeze-dried) materials.

Type II water may have somewhat higher levels of impurities than Type I, but it is still suitable for much routine laboratory work, such as microbiology procedures and some assays. Type III water is acceptable for some applications, such as preparing microscope slides to examine tissue. Type III water can also be used to feed systems that will produce Type I and Type II water.

ASTM International further classifies water according to grade. There are three grades, A, B, and C, that can be applied to the three types of water. These grades vary in their permissible levels of bacterial contaminants and endotoxin levels. ASTM International specifications are shown in Tables 24.2 and 24.3.

There are applications where water that exceeds Type I standards is required. Analytical tests using high-performance liquid chromatography (HPLC), for example, are extremely sensitive and can detect organic compounds in samples at levels as low as a few parts per billion. Type I water may have more impurities than this, and therefore, it is not acceptable for the most sensitive HPLC work. Cell culture techniques may require water with lower pyrogen levels and fewer contaminants than are specified by Type I standards. There are water purification systems available that meet these stringent requirements, but they

TABLE 24.2

ASTM International Requirements for Purity of Purified Water

Water Type	I	II	III
Minimum resistivity at 25°C (MΩ-cm)	18.0	1.00	4.0
Total organic carbon, max (ppb)	50	50	200
Sodium, max (ppb)	1	5	10
Chlorides, max (ppb)	1	5	10
Total silica, max (ppb)	3	3	500
Water Type	**A**	**B**	**C**
Max bacteria count (CFU/100 mL)	1	10	1,000
Max endotoxin (endotoxin units/mL)	0.03	0.25	NA

TABLE 24.3

ASTM International Water: Methods of Production

Type I Purify to a conductivity of 20 μS/cm or less at 25°C by distillation followed by polishing with a mixed bed of ion exchange materials and a 0.2 μm membrane filter. Type I reagent water may be produced with alternate technologies as long as the specifications are met and the water is shown to be appropriate for its application.

Type II Purify by distillation to a conductivity of less than 1.0 μS/cm at 25°C. Ion exchange, distillation, or reverse osmosis and organic absorption may be required prior to distillation if the purity required cannot be obtained in a single distillation. Type II reagent water may be produced with alternate technologies as long as the specifications are met and the water is shown to be appropriate for its application.

Type III Purify by distillation, ion exchange, continuous electrodeionization, reverse osmosis, or a combination of these, followed by polishing with a 0.45 μm membrane filter. Type III reagent water may be produced with alternate technologies as long as the specifications are met and the water is shown to be appropriate for its application.

must properly be operated and maintained to deliver water of the highest quality.

The College of American Pathologists (CAP) and the Clinical and Laboratory Standards Institute (CLSI) establish specifications for water used in clinical laboratories. Their standards describe clinical laboratory reagent water (CLRW) that is analogous to ASTM International's Type I and Type II water. CLSI also identifies types of water for other purposes, such as autoclaving and washing laboratory materials.

The Japanese, European Union, and United States Pharmacopeias contain specifications for water used in manufacturing and testing pharmaceutical products. These pharmacopeia standards distinguish between purified water (PW), sterile purified water, and water for injection (WFI). **Purified water** *is produced using a suitable purification process from source water that complies with EPA (Environmental Protection Agency) drinking water regulations, or comparable regulations in the EU and Japan. PW cannot be used for manufacturing drugs that will be* injected, but can be used for nonsterile drugs. PW is also used for laboratory tests and assays, unless otherwise stated, and can be sterilized and used for sterile, noninjected drugs. PW can be made by any method, but must meet certain purity standards. WFI *is defined as water purified by distillation or a purification process that is equivalent or superior to distillation in the removal of chemicals and microorganisms.* WFI must meet the same requirements as purified water or sterilized purified water and, in addition, must meet the requirements for reduced bacterial and endotoxin contaminants. Table 24.4 summarizes European Union and United States Pharmacopeia standards for water.

24.1.3 How Is the Purity of Water Evaluated?

24.1.3.1 Conductivity and Resistivity

Purified water must be monitored to be certain that it meets its quality requirements. If you examine Tables 24.2–24.4, you will see that conductivity, and

TABLE 24.4
Pharmacopeia Requirements for Purity of Purified Water

Property	European Union Pharmacopoeia	United States Pharmacopeia
Nitrates	< 0.2 ppm	—
Heavy metals	< 0.1 ppm	—
TOC	< 500 µg/L	< 500 µg/L
Conductivity	< 4.3 µS/cm at 20°C	< 1.3 µS/cm at 25°C
Bacteria (guideline)	< 100 CFU/mL	< 100 CFU/mL
Additional requirements for WFI	Conductivity: < 1.1 µS/cm at 20°C Bacteria: < 10 CFU/100 mL Endotoxins: < 0.25 international units/mL	Bacteria: < 10 CFU/100 mL Endotoxins: < 0.25 endotoxin units/mL

its inverse, resistivity, are used as a primary basis for defining water purity types. (See Chapter 20 for information about these measurements.) The more ions present in water, the better it conducts current. Conversely, the fewer the ions present, the more resistive the water to current flow. Highly purified water has almost no ions and, therefore, is a very poor conductor of current.

Resistivity is measured in units of megohm-centimeters (MΩ-cm). The theoretical maximum value for the resistivity of water is 18.2–18.3 MΩ-cm at 25°C; the closer the resistivity of purified water to this value, the better its quality. The resistivity (or conductivity) of water that has undergone a purification process is therefore measured as an indication of the water's purity. There are meters placed in water systems that can measure conductivity and/or resistivity. By convention, conductivity is usually measured and reported for values above 1 µS/cm at 25°C, and resistivity for conductivity values below 1 µS/cm at 25°C.

Nonionic contaminants do not affect resistivity/conductivity, so even if the resistance of water is ≥ 17.0 MΩ-cm, it may contain contaminants that must be monitored in other ways.

24.1.3.2 Bacterial Counts

Bacterial counts are used to monitor levels of microorganisms in water. A particular volume of the water is filtered through a membrane that traps bacteria on its surface. The filter is incubated on nutrient medium. *The number of colonies growing on the medium is counted to give the number of* **colony-forming units (CFUs)** *per volume of water.* The various standards for water quality specify a maximum number of colony-forming units that are acceptable.

24.1.3.3 Pyrogens

Most analytical and research laboratories do not monitor pyrogen contamination, but this assay is essential in the pharmaceutical industry and may be important in cell culture facilities. The ***Limulus* amebocyte lysate (LAL) test** *is the common, traditional method used to monitor pyrogens in a water sample.* The LAL test requires an extract of blood from the horseshoe crab, *Limulus polyphemus*. Different dilutions of the water to be tested are mixed with the *Limulus* blood extract, and any pyrogens present cause the extract to clot. The clotting results can be converted to **endotoxin units per milliliter (EU/mL)**. Newer assays exist that do not rely on horseshoe crabs, whose populations can adversely be affected by the demand for their blood.

24.1.3.4 Total Organic Carbon

Total organic carbon (TOC) is another parameter that is used to evaluate the quality of purified water. Organic contaminants can interfere with analyses and can indicate that bacteria have contaminated the water treatment system. The unit of measure of TOC is parts per million (ppm), or parts per billion (ppb), or µg/L (1 µg/L = 1 ppb). At the beginning of a water treatment system, the TOC levels are usually in the 2–5 ppm range. After treatment, highly purified water might have TOC levels 1,000 times better, in the 1–5 ppb range. There are TOC-measuring instruments in water purification systems that are used to monitor and report this parameter.

24.1.3.5 pH

Ultrapure water that is exposed to air reacts with carbon dioxide, resulting in a pH of about 5.7. If it is in a covered container, pure water will have a pH of about 6.0. In some situations, the pH of the water is important and is monitored, but its value is not specified for Type I and Type II water.

24.1.3.6 Other

Other water quality parameters, such as particulate and silicate levels, can be tested where necessary. Whether various tests are performed, and if so, how often, depends on the needs of the laboratory or industrial facility.

24.1.4 METHODS OF WATER PURIFICATION

24.1.4.1 Overview

Source or **feed water** *is the water that is to be purified.* Most laboratories and biotechnology facilities begin with municipal (tap) water that is partially purified to make it safe for drinking. Such water has reduced levels of microorganisms and other contaminants, but contains high levels of chlorine and varying levels of many other contaminants. If you look at the shower head or faucets in your home, you may see a buildup of mineral contamination from hard water. These minerals are not harmful to you, but they must be removed before the water can be used for any laboratory or production purpose. The quality of municipal water varies greatly depending on geographic location, the season, treatment methods used, and other factors. One of the oldest biotechnology industries, the brewing industry, has a tradition of choosing manufacturing sites based on the quality of local water available.

Some water sources contain contaminants that are difficult to remove, even with the best treatment methods. Because source water is so variable, and because each application has different purity requirements, water purification systems are tailored to meet the needs of a particular laboratory or industrial facility. There are a variety of methods of purifying water that vary in their costliness and ability to remove different kinds of contaminants. These are summarized in Table 24.5 and introduced in more detail below. Because each purification method removes some, but not all, types of contaminants, it is common for water purification systems to use more than one purification method.

24.1.4.2 Distillation

Distillation was invented hundreds of years ago and is still sometimes used for preparing laboratory water. **Distilled water** *is prepared by heating water until it vaporizes* (Figure 24.1). The water vapor rises until it reaches a **condenser** *where cooling water lowers the temperature of the vapor and it condenses back to the liquid form.* The condensed liquid flows into a collection container. Most contaminants remain behind in the original vessel when the water is vaporized. Ionized solids, organic contaminants that boil at a temperature higher than water, pyrogens, and microorganisms are all removed from distilled water. A few contaminants, however, cannot be removed from water by distillation. These include dissolved gases such as carbon dioxide, chlorine, and ammonia; and organic contaminants that volatilize at a low temperature.

Distillation is a well-established method of water purification that was the traditional "gold standard." It removes a wider variety of contaminants than other methods. (See Table 24.5.) Distillation, however, is

---

TABLE 24.5

A Comparison of Water Purification Methods

	Distillation	Deionization	Electrodeionization	Filtration	Reverse Osmosis	Adsorption	Ultrafiltration	UV Oxidation
Dissolved ionized solids	E/G	E	E	P	G	P	P	P
Dissolved organics	G	P	P	P	G	E	P	E
Dissolved gases	G	E	E	P	P	G	P	P
Particulates	E	P	P	E	E	P	E	P
Bacteria	E	P	P	E	E	P	E	E
Pyrogens	E	P	P	P	E	G	E	P
Nucleases	P	P	P	P	P	G	G	P

Source: Based primarily on information from: *Thermo Scientific Waterbook*, Thermo Scientific. 2016. (Used with permission from Thermo Fisher Scientific, the copyright owner.)

E = excellent; G = good; P = poor.

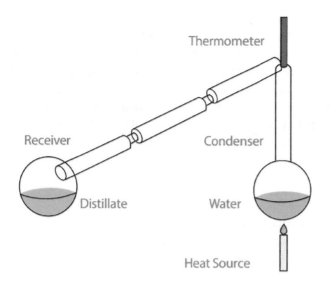

FIGURE 24.1 Distillation. (Used with permission from Thermo Fisher Scientific, the copyright owner.)

expensive compared with other methods because of the substantial energy input required to heat the water. Distillation is slow, and therefore, distilled water is generally prepared in advance and stored for later use. This storage of the distillate can be problematic because plasticizers and other substances will leach out of plastic storage containers and re-contaminate the water. In addition, bacteria grow in water that has been standing. Stills require frequent maintenance to remove scale buildup when the source water is hard. For all these reasons, distillation is being replaced in many facilities by newer technologies, as has been acknowledged by changes to water purity standards. Nonetheless, you may still see distillation being used, for example, in schools where the system will be shut down during the summer.

24.1.4.3 Ion Exchange

Ion exchange *removes ionic contaminants from water.* During ion exchange, water flows through cartridges packed with bead-shaped resins, called *ion exchange resins.* As the water flows past the beads, ions in the water are exchanged for ions that are bound to the beads. There are two types of resins. **Cationic resins** *bind and exchange positive ions.* **Anionic resins** *bind and exchange negative ions.*

Water softeners (such as many people have in their homes) are a type of ion exchanger. **Water softeners** *exchange positive ions that make water hard, particularly calcium and magnesium, with sodium ions.* Water softeners thus remove calcium and magnesium from water and add sodium to it. Water softeners use

cationic beads that initially have Na⁺ ions loosely bound to them. As water flows past these beads, positive ionic contaminants in the water exchange places with the Na⁺ ions. The contaminants remain linked to the beads while the sodium ions enter the water stream. Water softening may be used as a preliminary step in water purification, but it is not adequate for laboratory water purification because it leaves sodium ions and all negative ions in the water.

Deionization *is an ion exchange process used in the laboratory to remove "all" ionic contaminants from water* (Figure 24.2). Deionization involves both cation and anion exchange beads that are housed inside a cartridge. The anionic resin beads initially have loosely bound hydroxyl ions, whereas the cationic beads have hydrogen ions loosely bound to their surfaces. As water passes by the anion exchange beads, negatively charged ions in the solution take the place of the OH^- ions on the beads. The OH^- ions enter the water stream, but the contaminants remain bound to the resin beads. Positive ionic contaminants in the water similarly exchange places with H^+ ions that are attached to the cationic beads. The hydrogen ions enter the water; the positive ionic contaminants remain bound to the beads. The H^+ ions and the OH^- ions that

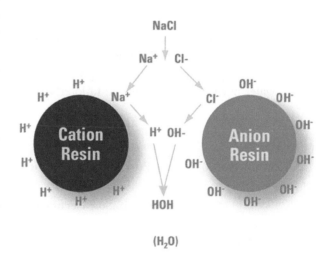

FIGURE 24.2 Deionization. This illustrates what happens if there is NaCl in the water stream. As the water passes between the beads, Na⁺ ions change places with H⁺ ions on the cationic beads, and Cl⁻ ions exchange with the OH⁻ ions on the anionic beads. The result is that H⁺ ions enter into the water stream where they combine with OH⁻ ions to form water molecules. The NaCl is removed from the water because it remains bound to the beads. An analogous process occurs for other ionic contaminants in the water stream. (Used with permission from Thermo Fisher Scientific, the copyright owner.)

enter the water stream combine with one another to form more water molecules.

After a period of use, cation and anion contaminants from the water will have replaced most of the active hydrogen and hydroxyl sites in the resins and the resistivity reading of the system will decline (or the conductivity will increase). At this point, the cartridges will need to be replaced or regenerated. Alternatively, **electrodeionization, EDI,** *is a newer variant of deionization that uses electrical power to continuously split water molecules, eliminating the need for regenerating the resin.* EDI has the advantage that it is continuous and therefore provides more consistent water purity than conventional methods. It is, however, currently a more expensive option than standard deionization units.

Deionized water is purer than tap or softened water because the ionic contaminants have been removed, but it is generally not acceptable for making laboratory solutions because it may contain organic contaminants, pyrogens, and microorganisms.

24.1.4.4 Carbon Adsorption

Carbon adsorption *is an effective method for removing dissolved organic compounds from water.* Carbon adsorption uses **activated carbon,** *which is traditionally made by charring wood by-products at high temperatures and then "activating" it by exposure to high-temperature steam.* There are also synthetic forms of activated carbon for water treatment systems. Natural activated carbon can release some ionic contaminants into the water being treated; the synthetic versions are less likely to release ionic contaminants.

Activated carbon has an extremely porous, honeycomb-like structure. As water passes through activated carbon, chlorine and organic impurities adsorb (stick) to the carbon and are thus removed (Figure 24.3). Activated carbon is sometimes used before deionization to protect the resins from contamination by organic materials and chlorine. Activated carbon can also be used in the same cartridge as deionization resin, effectively removing organic and ionic contaminants in one cartridge.

24.1.4.5 Filtration Methods

Filtration *removes particles as the water passes through the pores or spaces of a filter.* There are five types of filters used in water treatment:

1. **Depth filters** *are made of sand or matted fibers and retain particles through their entire depth by entrapping them* (Figure 24.4a). Depth filters are useful for removing relatively large particles, greater than about 10 μm, from the water stream. (Pollen and some yeast cells are examples of substances in this size range.) Depth filters may be used at the beginning of a purification system as an inexpensive method to remove large particles and debris that might otherwise clog and foul the downstream elements of the system.

2. **Microfiltration membrane filters (also called screen filters or microporous membranes)** *are made by carefully polymerizing cellulose esters or other materials so that a membrane with a specific pore size is formed* (Figure 24.4b). Water treatment filters often have pore sizes on the order of 0.20 μm because bacteria are too large to penetrate such pores. Microfiltration membrane filters are useful for removing bacteria and particles above a specified pore size, but they are not useful for removing smaller, dissolved molecules.

3. **Ultrafiltration membranes** *will prevent the passage of dissolved molecules, including most organics* (Figure 24.4c). Materials down to 1,000 D in molecular weight (smaller than most proteins) can be separated from water using ultrafilters. Small, dissolved inorganic molecules can pass through ultrafilters. Two important applications of ultrafilters are the removal of viruses and pyrogens from water.

4. **Reverse-osmosis (RO) membranes** *are more restrictive than ultrafiltration membranes* (Figure 24.4d). Materials down to 300 D in molecular weight can be removed from water using RO, and so this method is effective in removing viruses, bacteria, and pyrogens. Moreover, RO membranes also reject ions and very small dissolved particles, such as sugars.

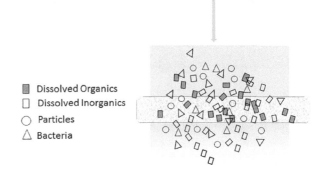

■ Dissolved Organics
□ Dissolved Inorganics
○ Particles
△ Bacteria

FIGURE 24.3 Activated carbon. As water flows through the porous activated carbon, chlorine and dissolved organics are selectively adsorbed by the carbon.

In reverse osmosis, water under pressure flows through multiple layers of thin RO membrane. The membrane allows water to pass through, but rejects a large proportion (on the order of 90%–99%) of all types of impurities (particles, pyrogens, microorganisms, and dissolved organic and inorganic materials). The permeate (water that has passed through the membrane) will contain very low levels of contaminants that are able to pass through even this restrictive membrane, but most types of contaminants are greatly reduced. Because RO membranes are very restrictive, the flow rate through them is slow. This process is called "reverse osmosis" because water is forced away from salts.

Reverse osmosis is similar to ultrafiltration in that both involve membranes that reject small materials. The major differences between ultrafiltration and RO include the following: (1) RO can retain very small solutes, including salt ions, that can pass through ultrafiltration membranes. (2) Higher pressures are used in RO to force water through the membrane. (3) An ultrafiltration membrane retains particles based almost solely on their size. RO membranes retain materials based both on their size and on ionic charge.

RO is a relatively inexpensive process and is often used for pretreatment in Type I and Type II water purification systems. When used for pretreatment of tap water, RO removes about 90% of impurities. This level of purification is often sufficient to make Type III and Type IV water, and so RO systems may be stand-alone in some settings.

5. **Nanofiltration (NF) membranes** *are in between ultrafilters and reverse-osmosis membranes.* RO can remove the smallest solute molecules, in the range of 0.0001 micrometer in diameter and smaller; nanofiltration removes molecules in the 0.001 micrometer range. NF is essentially a lower-pressure version of reverse osmosis that is used for applications where the utmost purity of product water is not required. Like RO, NF is also capable of removing bacteria, viruses, and other organics. Since NF requires lower pressure than does RO, energy costs are lower than those for a comparable RO treatment system.

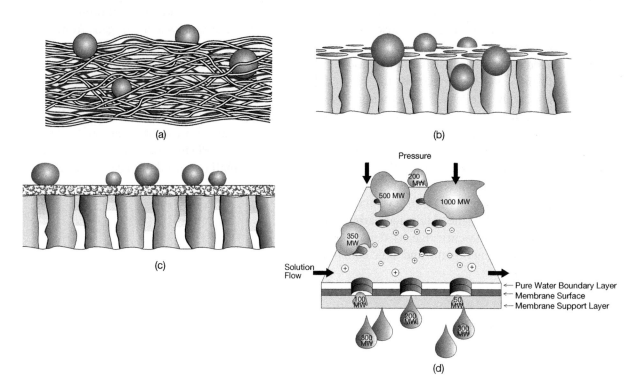

FIGURE 24.4 Filtration methods. (a) Depth filter. (b) Membrane filter with pores of a particular size. (c) Ultrafilter used to retain molecules that are above a particular molecular weight. (d) Reverse osmosis. (a–c reproduced with permission from Merck KGaA, Darmstadt, Germany, and/or its affiliates. d is courtesy of Osmonics.)

24.1.4.6 Ultraviolet Oxidation and Sterilization

Ultraviolet (UV) oxidation *removes organic contaminants from water.* Water is passed continuously (for example, for 30 minutes) across the light path of an ultraviolet lamp that emits light at a wavelength of 185 nm. The organic compounds in the water oxidize to simple compounds such as CO_2. With UV oxidation, TOC can be reduced to very low levels, as low as 5 ppb.

UV light at a wavelength of 254 nm will kill bacteria and is therefore used to sterilize water. A Type I water purification unit will typically have a dual-wavelength (185 and 254 nm) lamp, while a Type II water system might have only a 254 nm lamp. The combination of UV oxidation with an ultrafilter can remove pyrogens and nucleases from Type I water, which is important in biotechnology settings.

24.1.4.7 Water Purification Systems

Most biotechnology and pharmaceutical facilities use water that is purified by a combination of methods (Figure 24.5). The choice of methods and system design will depend on the applications for which the water is to be used and the quality of source water available. For example, a water purification system to prepare Type I water for a molecular biology laboratory might include the following steps:

1. Tap (feed) water may or may not be pretreated with a water softening unit.
2. The feed water passes through a reverse-osmosis unit that removes about 90% of contaminants. The RO water is stored in a tank.
3. The water then moves across a dual-wavelength UV light that both kills microorganisms and oxidizes organics.
4. The water then moves into a cartridge that combines deionization beads with activated carbon to remove nearly all ions and further reduce the level of organics.

5. Finally, just before use, the water runs through an ultrafilter that has a pore size small enough to retain residual pyrogens, nucleases, and bacteria.

24.1.4.8 Handling of Reagent Water

Highly purified water is an aggressive solvent. It will readily leach contaminants from any vessel in which it is stored and will also dissolve CO_2 from the air. Surprisingly, bacteria can multiply in purified water during storage. Type I water, therefore, cannot be stored and must be used immediately after it is produced. For this reason, water purification systems have a storage tank that is located before the final purification step(s). When purified water is required, the user first flushes the system for a minute or so, discarding the rinse water. As water is flushed through the system, the resistivity increases until it reaches an acceptable level. A value of 17.0 MΩ-cm is considered acceptable for Type I water in many laboratories. When the resistivity reading is suitably high, the user withdraws water for use. Most high-purity water systems store water only for a short time (e.g., a day) and then recirculate it back again through the purification system.

Note also that Type I water is highly reactive with metals and can cause pitting and damage to metal equipment. Therefore, Type I water should be avoided in certain laboratory equipment, such as glassware washers, water baths, and autoclaves. When in doubt, check with the instrument manufacturer.

24.1.4.9 A Note about Terminology

There are a variety of methods used to purify water. Distillation, however, is the traditional method, and it is common for people to speak of highly purified water as being "distilled" (DI) whether it is purified by RO, ultrafiltration, or actually by distillation. Note, however, that the term "DI" often means "deionized."

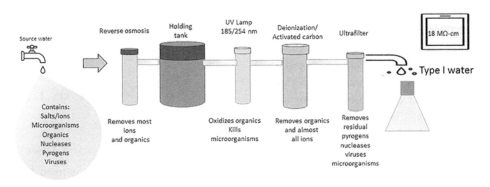

FIGURE 24.5 A multistep water purification system.

Deionized water is not the same as "distilled" water, nor is it the same as Type I water. We therefore advocate avoiding the term "DI" altogether. In this book, the term *purified water* is used to mean water that has been distilled, or that has been run through a system as shown in Figure 24.5. Some laboratory professionals prefer to name water by its level of purity, for example, ASTM Type I or Type II. Because the terminology of water purification is used inconsistently, and because in a given facility there may be more than one source of water, standard procedures should explicitly state the source of water required.

24.1.5 OPERATION AND MAINTENANCE OF WATER PURIFICATION SYSTEMS

24.1.5.1 Maintenance

Water purification systems need to be operated and maintained properly to ensure high-quality water. The maintenance required and its frequency will vary depending on the type of laboratory, the contaminants present in the source water, and any regulations that are applicable. Some general considerations include the following:

- **Distilled water apparatus.** A distilled water apparatus requires frequent, sometimes daily, cleaning; otherwise, it can become contaminated by microorganisms and their by-products. Distilled water should be prepared freshly each day because it can absorb contaminants from storage vessels and microorganisms can grow in it.
- **Ion exchange resins.** Conventional ion exchange resins must be regenerated once they have exchanged all their Na^+, H^+, or OH^- ions for contaminants. Regeneration is sometimes performed on-site, and sometimes cartridges are sent off-site for maintenance. Microorganisms can attach to deionization beads, thereby adding contaminants to the water. Deionization systems are therefore periodically sanitized.
- **Filters.** Filters must be tested when installed to be certain they have no small holes and are effective. They must be periodically flushed and sanitized to remove contaminants from their surfaces. Filters must be replaced periodically because they can develop defects over time and bacteria can actually grow through their pores.
- **Activated carbon.** Activated carbon may become colonized by bacteria and may become clogged by particulates. Activated

carbon beds must therefore be washed to remove particulates. Activated carbon must also be periodically recharged or replaced as active sites are filled by contaminants.
- **UV lamps.** As UV lamps age, their light output decreases. Bulbs must therefore be replaced about once a year.

24.1.5.2 Documentation

As with any other aspect of a quality-control program, the operation, maintenance, and monitoring of water purification systems requires planning and documentation. This includes developing standard procedures, training operators, planning routine maintenance and monitoring, establishing logbooks, developing written procedures to deal with problems, and documenting corrective actions taken.

Example Problem 24.1

a. Ultrapure, sterile water is stored in a very clean, sterile glass bottle for 4 days. Speculate as to the quality of this water after storage.
b. Ultrapure, sterile water is stored in a very clean, sterile plastic bottle for 4 days. Speculate as to the quality of this water after storage.

Answer

a, b. Although the bottles are clean and the conditions are sterile, ultrapure water is an aggressive solvent and will leach heavy metals (which are minor contaminants of glass) from the glass container into the water. Ultrapure water will similarly leach organic impurities from the plastic and will become contaminated. Ultrapure water cannot be stored for four days without degrading.

24.2 GLASS AND PLASTIC LABWARE

24.2.1 GLASS LABWARE

24.2.1.1 Types of Glass and Their Characteristics

Glass and plastic labware is used for measuring volumes, preparing and storing solutions, housing cells and microorganisms, and so on. There are a variety of glass and plastic items available for laboratory use, so these items can be matched to their application.

Various grades of glass differing in strength, heat resistance, and other factors are used to make beakers, flasks, graduated cylinders, pipettes, and other labware. Glass is relatively inert when in contact with chemicals and is therefore used to make containers for chemicals. Glass, however, may adsorb metal ions and proteins from solutions and is not recommended for the storage of metal standards. Also, trace amounts of contaminants, such as sodium, barium, and aluminum can leach out of glass containers. In normal laboratory work, these trace contaminants are not significant. However, in some analytical and pharmaceutical applications, they are problematic.

Borosilicate glass *is used to make general-purpose laboratory glassware that is strong, is resistant to heat and cold shock, and does not contain discernable contamination by heavy metals.* Borosilicate glass is sold, for example, under the trade names of Pyrex (Corning Glass Works) and Kimax (Kimble Chase Life Sciences and Research Products). Corex (Corning Glass Works) is a type of glass that is resistant to physical stress and is therefore used to make centrifuge tubes. Type I glass refers to highly resistant borosilicate glass used in the manufacture of ampules, vials, and other containers for pharmaceutical applications. (See also the case study later in this chapter, "Sometimes the Container is as Important as the Contents.") Additional information about types of glass is shown in Table 24.6.

24.2.1.2 Cleaning Glassware: Overview

Some glassware is disposable, but much glassware is reusable and must be cleaned properly after use. Although "washing the dishes" may not appear to be an interesting aspect of biotechnology, the individuals in a laboratory who are responsible for cleaning labware play a key role in ensuring quality results. Dirty labware may lead to invalid chemical analyses, alter experimental results, inhibit the growth of cells, or add contaminants to a product. Determining how to properly clean labware and equipment, and proving that cleaning methods are effective are critical concerns of scientists and engineers in the biotechnology/pharmaceutical industry. The FDA frequently cites companies for improper cleaning of equipment and surfaces.

There are many procedures to clean glassware after use in the laboratory. Which procedure is best depends on the items to be cleaned, the soils contaminating the labware, and the application for which they will be used. For example, analytical methods may be extremely sensitive to minute ionic contaminants, while cell culture requires sterile labware. Proteinaceous deposits, organic residues, metals, greases, and microorganisms are examples of contaminants that may need to be removed from glassware.

Note that safety precautions are required when washing labware that was used for hazardous materials, or if the cleaning agents used are dangerous.

TABLE 24.6
Common Types of Glass

Type	Qualities	Purpose
Borosilicate; for example, Pyrex and Kimax	High resistance to heat and cold shock. Resistant to most chemicals (but not certain strong acids). Has a small amount of added boron. Low in metal contaminants.	General-purpose labware
Corex	Aluminosilicate glass. High resistance to pressure and scratching.	Centrifugation
High silica (quartz)	Greater than 96% silicate. Transparent. Excellent optical properties. Very resistant to temperature changes.	Optical devices, mirrors, cuvettes, volumetric glassware
Low actinic[a]	Red-tinted to reduce light exposure of contents.	Used to make vessels to contain light-sensitive compounds
Flint	Soda–lime glass containing oxides of sodium, silicon, and calcium. Poor resistance to temperature changes and high temperature. Poor chemical resistance. Inexpensive.	Disposable glassware, such as pipettes
Optical	Soda–lime, lead, and borosilicate.	Used in prisms, lenses, and mirrors

[a] Actinic light is defined as light capable of producing a photochemical reaction. This is usually near-ultraviolet or blue visible light with wavelengths between 290 and 450 nm.

(Information in this table is from multiple sources.)

In general, glassware washing has five steps:

1. *Pre-rinse.* A pre-rinse or soak immediately after use prevents contaminants from drying onto labware.
2. *Contaminant removal.* Contaminants are typically removed using detergents or solvents in conjunction with a physical method such as scrubbing or exposure to jets of water.
3. *Rinse.* All traces of detergent and cleaning solvents are rinsed away.
4. *Final rinses.* Any residues from the rinse in Step 3 are removed.
5. *Drying.*

24.2.1.3 Detergents and Cleaning Agents

There are various types of detergents used for cleaning labware that vary in their pH, additives, and other qualities. It is necessary to choose the correct detergent and to dilute and use it properly. The choice of detergent is based on what is to be cleaned, what contaminants must be removed, and the physical method of cleaning that will be used. Table 24.7 lists some general guidelines to help determine which detergent to use.

Chromic acid washes were often routinely used in the past to remove proteinaceous and lipid contaminants from glassware. The use of chromic acid, however, has disadvantages. It is corrosive and hazardous, so extreme caution is necessary to avoid injury when using such washes. Chromic acid contains chromium, a heavy metal, which requires costly disposal in a special hazardous waste site. Because of the hazards and costs associated with chromic acid baths, they are now less common, though still sometimes used. Proper use of detergents, milder acids, such as 1 M HCl and 1 M HNO_3, and organic solvents usually eliminate the need for chromic acid washes.

24.2.1.4 Physical Cleaning Methods

Detergents are used in conjunction with physical methods of cleaning. Depending on the situation, a variety of physical methods are used to remove contaminants from labware. These include the following:

1. **Soaking.** This simply involves submerging the items to be cleaned in a cleaning solution. The items are sometimes rinsed and used immediately, but further cleaning is usually required.
2. **Manual cleaning.**
   - Items are washed by hand using a cloth, sponge, or brush followed by thorough rinsing.
   - Manual washing is not suitable for cleaning large numbers of items.
   - Gloves and eye protection should be used as described by the detergent manufacturer, and if labware was used for hazardous substances.
   - Centrifuge glassware and tubes, metal equipment, and most plasticware should not be cleaned with metal brushes or other abrasives. Scratches on centrifuge tubes can lead to material fatigue and breakage when the tube experiences the force of centrifugation.
   - Solvents such as acetone, alcohols, and methylene chloride can be used with glassware to remove greases and oils that are not removed by laboratory detergents.

TABLE 24.7

General Guidelines for Choosing a Detergent

1. *Alkaline Cleaners.* A broad range of organic and inorganic contaminants are readily removed from glassware by mildly alkaline, general-purpose cleaners.
2. *Acid Cleaners.* Metallic and inorganic residues are often solubilized and removed by acid cleaners.
3. *Protease Enzyme Cleaners.* Proteinaceous residues are effectively digested by protease enzyme cleaners.
4. *Cleaners for Radioactive Residues.* Radioactive residues are often removed by detergents with high chelating (binding) capacity. Detergents are available that are specifically designed to remove radioactive contaminants.
5. *Dishwashers.* Laboratory dishwasher detergents should be low foaming.
6. *Manual Cleaning.* For manual, ultrasonic, and soak cleaning, a mildly alkaline (pH 8–10), foaming detergent will often work well.
7. *Instructions.* For machine washing, prepare and use a detergent according to the manufacturer's instructions. It is important to use the right amount and type of detergent to match the machine's requirements.
8. *Tissue Culture.* Use detergents specifically formulated for cleaning items used in tissue culture.

3. **Machine washing.**
   - Many biotechnology facilities wash labware in commercial glassware washers. Machines can save time when a large number of items must be washed and can provide a higher level of consistency and effectiveness than manual washing.
   - Glassware washers have a single chamber that is sequentially filled and emptied through its cycle. A typical cycle consists of a pre-wash with room temperature water, a wash cycle of 2–15 minutes in a detergent solution at 60°C–90°C, and multiple rinses. The labware may be dried in the washer or in a separate drying oven.
   - Some machine washers use HEPA-filtered air to dry glassware. This helps prevent airborne particles from settling in and on glassware. (See Section 10.2.2.1, for information about HEPA filters.)
   - Items should be loaded so that their open ends face the spray nozzles.
   - Items should be loaded in a way that does not block the spray arms.
   - Narrow neck flasks are cleaned most effectively using spray spindles designed to fit inside the flasks.
   - Difficult-to-clean articles should be placed in the center of the rack, preferably with the spray nozzles pointing directly into them.
   - Small items should be placed in baskets.
   - Graduated cylinders should be placed at an angle so that their base does not trap a large amount of dirty wash water, thus contaminating the rinse water.
   - Commercial labware washers typically use deionized water as final rinses to avoid contaminating glassware with tap water impurities. Some washers include a built-in water softening system.
   - The machine's chamber must be kept clean. A periodic (monthly or quarterly) acid wash cycle with an empty chamber is recommended to remove scale buildup that can clog nozzles and deposit white calcium scale.

4. **Automatic siphon pipette washing.** Reusable glass pipettes are usually washed in long cylinders designed for this purpose. The cylinders are connected to a water supply so that the pipettes can alternately be filled with cleaning solution and drained. General directions include the following:
   - Presoak the pipettes by complete immersion in the washing solution immediately after use.
   - Place detergent in the bottom of the washer.
   - Place the pipettes in a holder in the washer.
   - Attach the washer to the water supply and run the water so that the pipettes are filled and drained completely.
   - Continue to wash until all the detergent has washed and drained through the pipettes.
   - Use a final rinse as appropriate.

5. **Ultrasonic washing.** Ultrasonic washers expose items to high-pitched sound waves that effectively penetrate and clean crevices, narrow spaces, and other difficult-to-reach surfaces. Metal and glass are usually safely washed in these devices, but some types of plastic are weakened by ultrasonic cleaning. General directions include the following:
   - Dilute the cleaning solution and add to the tank, run the machine several minutes to degas the solution, and allow the heater to come to temperature.
   - Place items in racks or baskets in the machine.
   - Align irregularly shaped items so that the long axis faces the transducer that generates sound waves.
   - Immerse the articles for 2–10 minutes with ultrasonic unit turned on.

6. **Clean-in-place (CIP).** Large, nonmovable items, such as fermentation vessels or pipes, must be cleaned in place. These items are often designed so that cleaning agents can be pumped into and circulated through the device.

24.2.1.5 Rinsing

Rinsing is a key step in washing labware. Standard washing procedures often specify that items should be rinsed three times with water. For machine washing, this means three rinse cycles are used. For manual washing, the item is filled and emptied at least three times. Tap water may be used in certain applications for initial rinses, but it should be followed by several rinses in purer water. Deionized water heated to 60°C to 70°C is an effective rinsing agent. The type of water for the final rinse depends on the application for which the item will be used.

24.2.1.6 Drying and Final Steps

Labware should not be dried by wiping the interior with any type of towel or tissue because they can contaminate the item with fibrous and chemical residues. Glassware can be allowed to air-dry in a washer, or on racks in the laboratory, or can be more rapidly dried in heated drying ovens at temperatures below 140°C. Note, however, that volumetric glassware should not be heated because heat causes expansion and contraction of the glass that may alter its calibration.

After washing, glassware should be visually inspected to be sure it is clean and not cracked. Cracked glassware should be repaired or discarded. A simple, common test of the cleanliness of glassware is that it wets uniformly with distilled water. Grease, oils, and other contaminants cause water to bead, or flow unevenly across the glass surface. Uneven wetting of the surface of volumetric glassware will alter the volumes measured and may distort the meniscus. It is good practice to cover cleaned and dried glassware so that it does not collect dust during storage.

24.2.1.7 Quality Control of Labware Washing

Each facility must have procedures for washing items according to the applications for which those items will be used. Laboratories that meet CGMP and ISO 9000 requirements must demonstrate that their cleaning procedures are, in fact, effective. Determining the effectiveness of a cleaning procedure can be difficult, and there is no standard method for doing so. Some general considerations for assuring the quality of washing operations include the following:

- Machine washers should be monitored to see that they maintain the proper temperature throughout their cycle and have no blocked spindles, and that the quality of the incoming water is adequate.
- Some models of machine washers helpfully provide automatic documentation of washing cycles.
- The effectiveness of a cleaning method can be evaluated by intentionally soiling items with substances that are visible or otherwise detectable. If residues are detected after cleaning, the washing method needs improvement.
- In pharmaceutical facilities and other sites where consistently and stringently clean labware is required, testing methods may be very sophisticated. Washed items may be periodically swabbed, and the swabs tested by

HPLC, spectrophotometry, or other methods that detect contaminants.
- As with any other aspect of a quality-control program, the operation, maintenance, and monitoring of washing systems requires planning and documentation. Standard procedures, operator training, plans for routine maintenance and monitoring, logbooks, written procedures to deal with problems, and documentation of corrective actions taken are components of a quality program.

24.2.2 PLASTIC LABWARE

24.2.1.1 Types of Plastics and Their Characteristics

Much common labware is made of plastic, not glass. Storage vials, microfuge tubes (tubes used in small, tabletop centrifuges), plates to hold cultured cells, and pipette tips are just a few examples of plasticware. Plastics are less breakable than glass, and they come in a variety of types that vary in their properties and applications. Some types of plastics are more resistant than others to heat, acids, organic solvents, and the forces in a centrifuge. Some plastics can be microwaved, while others are affected by this radiation. Some plastics are porous and stored liquids may evaporate from them. Some plastics are transparent, and others are not. Table 24.8 briefly summarizes the qualities of a few common types of plastics and their uses.

Plastics have the disadvantage that they are often incompatible with certain chemicals and may darken, become brittle, crack, or dissolve when exposed to them. There are a number of different types of plastic, each with its own chemical compatibilities. Figure 24.6 shows an abbreviated chemical compatibility chart of the type that can be consulted before using a particular type of plastic with a particular chemical. A more complete version can be found in manufacturers' catalogs, for example in the "Nalgene Labware Catalog," Nalgene Company (www.nalgenelabware.com/tech-data/chemical/index.asp). Note also that plastics vary widely in their compatibility with high and low temperatures and their compatibility with various sterilization methods.

24.2.1.2 Leachates from Plastics

In the early part of the twenty-first century, the general public became aware of the fact that chemical substances leach out of plastics during ordinary use

TABLE 24.8

Common Plastics and Their Qualities

Type	Qualities	Purpose
Polyethylene	Biologically and chemically inert, resistant to solvents including dilute acids. Resistant to breakage. Strong oxidizing agents will eventually make polyethylene brittle. Not resistant to autoclaving or dry heat.	General-purpose; used for wide variety of labware
Polypropylene	Translucent, can be autoclaved, resistant to solvents. More susceptible than polyethylene to strong oxidizers. Brittle at 0°C. Biologically inert. Thin-wall products permeable to CO_2 and O_2.	General-purpose
Polystyrene	Relatively inexpensive. Rigid, clear, brittle. Biologically inert. Sensitive to organic solvents. Used where transparency is an advantage. Melts in autoclave. Permeable to CO_2. Can be sterilized by ethylene oxide gas or gamma irradiation and is often sold pre-sterilized by one of these methods.	Widely used for disposable plasticware and tissue culture products.
Polyvinyl chloride (PVC)	Can be made soft and pliable. Often contains plasticizers that make it unsuitable for medical applications.	Often used for laboratory tubing.
Polycarbonate	Transparent and strong. Sensitive to bases, concentrated acids, and some solvents. Temperature-resistant.	Used for centrifuge-ware due to strength and also for other general labware
Teflon™ (Chemours Company)	Highly resistant to chemicals and temperature extremes. Inert; comparatively costly. Can be sterilized repeatedly by all methods except gamma radiation.	Wide variety of uses including fittings in instruments, stir bars, stopcocks, tubing, bottle cap liners, water distribution systems, and centrifuge tubes
Nylon	Strong, resistant to abrasion, resistant to organic solvents. Poor resistance to acids, oxidizing agents, and certain salts.	Widely used for membranes and filters in biology

Source: Based primarily on information from Thermo Fisher Catalog, https://www.thermofisher.com/us/en/home/life-science/lab-plasticware-supplies/plastic-material-selection.html.

(a) **Chemical Resistance Classification**

	ETFE	FLPE	HDPE	LDPE	PC	PETG	FEP/PFA	PMP	PP/PPCO	TPE**
Acids, dilute or weak	E	E	E	E	E	G	E	E	E	G
Acids *strong and concentrated	E	G	G	G	N	N	E	E	G	F
Alcohols, aliphatic	E	E	E	E	G	G	E	E	E	E
Aldehydes	E	G	G	G	F	G	E	G	G	G
Bases/Alkali	E	F	E	E	N	N	E	E	E	F
Esters	G	G	G	G	N	F	E	E	G	N
Hydrocarbons, aliphatic	E	E	G	F	G	G	E	G	G	E
Hydrocarbons, aromatic	G	E	N	N	N	N	E	N	N	N
Hydrocarbons, halogenated	G	G	N	N	N	N	E	N	N	F
Ketones, aromatic	G	G	N	N	N	N	E	F	N	N
Oxidizing Agents, strong	E	F	F	F	F	F	E	G	F	N

*Except for oxidizing acids; for oxidizing acids, see "Oxidizing Agents, strong."
** TPE gaskets.

(b) **Resin Codes**

ETFE	Tefzel† ETFE (ethylene-tetrafluoroethylene)
FEP	Teflon† FEP (ethylene-tetrafluoroethylene)
FLPE	fluorinated high-density polyethylene
HDPE	high-density polyethylene
LDPE	low-density polyethylene
PC	polycarbonate
PETG	polyethylene terephthalate copolyester
PFA	Teflon† PFA (perflouroalkoxy)
PMP	polymethylpentene ("TPX")
PP	polypropylene
PPCO	polypropylene copolymer
TPE	thermoplastic elastomer

† Or equivalent
Tefzel and Teflon are registered trademarks of DuPont.

FIGURE 24.6 Summary of resistance of various types of plastics to common chemicals. (a) Table of resistance. (b) Codes used to name plastics. (From "Thermo Scientific Nalgene Bottles and Carboys Technical Brochure," Thermo Scientific Company. 2012. Used with permission from Thermo Fisher Scientific, the copyright owner.) E = excellent. Thirty days of constant exposure causes no damage. G = good. Little or no damage after 30 days of exposure to the reagent. F = fair. Some effect after 7 days of constant exposure to the reagent. Depending on the plastic, the effect may be crazing, cracking, loss of strength, or discoloration. N = not recommended. Immediate damage may occur. Depending on the plastic, the effect may be crazing, cracking, loss of strength, deformation, permeation loss, dissolution, or discoloration.

and can be ingested, inhaled, or otherwise taken up by humans. These leachable chemicals are widely present in everything from baby bottles to cosmetics. Some of these chemicals are endocrine disrupters, which means they affect or mimic the effects of hormones. It is postulated that endocrine disrupters might have adverse effects such as altering fetal development, causing early onset of puberty, increasing the risks of breast cancer and obesity, and triggering behavioral and developmental problems in children. The significance of endocrine disrupters in humans is unclear, but in 2012, in response to public concerns, the FDA banned the use of one plastic additive, bisphenol A (BPA), from baby bottles and children's cups. The issue of the impact of plastic leachates on humans, and also on the environment, is far from resolved and continues to be a source of discussion and contention.

The fact that substances can leach from plastics has been noted in the scientific community for much longer. The FDA has been regulating leachates from plastic food packaging and plastic drug storage vials for many years. Researchers have found that substances leached from plastics during filtration and other procedures may have effects on experimental systems. (See the Bibliography for this unit for examples.) Whether the effects of plastic leachates are significant depends on many factors relating to the plastics, and the conditions of the experiment or system. For example, heat and long centrifugation times promote leaching from some types of plastics, but not others. There is no single recommendation to avoid problems with plastic leachates, other than to be aware of the potential problem. It is good practice to document the source of all plastic consumables to assist with troubleshooting if a problem is detected. (See, for example, the case study in Section 4.5.1, relating

---

Case Study: Sometimes the Container is as Important as the Contents

Most of us do not spend much time pondering the properties of different types of glass, other than to decide whether or not our casserole dish is safe for baking our dinner tonight. It therefore may come as a surprise to learn that there are scientists and engineers who devote their professional life to studying glass. In 2020, Corning, a major manufacturer of glass products in the United States, received a $204,000,000 grant from the United States government to expand manufacturing of a new kind of glass vial to use with COVID-19 vaccines. The Corning vial uses Valor glass, a product that emerged from Corning's high-tech materials science laboratories.

Before considering the properties of Valor glass, let's think about why vaccines and other drug products are packaged, shipped, and stored in glass vials. The disadvantage to glass is obvious; it is breakable. The first COVID-19 vaccines not only needed to be transported great distances, but also needed to be frozen to extremely low temperatures and then thawed – a process that stresses glass and can cause cracks and breaks. Obviously, broken vaccine vials, or worse still, vials containing undetected glass shards, pose dangers to patients and result in loss of product. Nonetheless, glass is the industry standard for vaccine vials. Why not plastic, which does not break? If you consider Table 24.8, you will see that plastics each have disadvantages. Chemical substances can extract out of plastics into their contents; this is a major problem for a container that holds an injectable pharmaceutical product. Another problem is that gases can sometimes slowly penetrate plastics, resulting in either the loss of moisture from their contents, or the introduction of oxygen and carbon dioxide from the air. Thus, one of the less glamorous, yet critical, tasks during the pandemic was quickly obtaining billions of glass vials to hold vaccines. This is the situation in which Corning was able to obtain a grant of hundreds of millions of dollars to manufacture a new kind of glass vial.

What is different about Valor glass? Conventional pharmaceutical vials are made of borosilicate glass, which gets its strength from boron. Glass for vials begins as tubes of glass that are extruded in ovens that reach thousands of degrees in temperature. Sometimes these high temperatures can evaporate the boron, eventually leading to small fragments of glass that flake off, contaminating the contents of the vial. Researchers came up with a solution to this problem, that is, substituting aluminum oxide for boron, along with adding an external coating to reduce friction between vials when they are filled with drug. The result is a new kind of super-strong glass that they named "Valor glass."

(Continued)

Case Study (*Continued*): Sometimes the Container is as Important as the Contents

The development of a new pharmaceutical product, such as Valor glass vials, does not end with its creation by scientists. Any new pharmaceutical product needs to obtain FDA approval (more about the FDA and its approval process in Unit X of this textbook). While most of us probably did not notice this event, the FDA approved Valor glass for use in pharmaceutical vials in 2019, an event that did create a stir in the pharmaceutical news media.

While we, as biotechnologists, probably are more fascinated with the mRNA in the new COVID-19 vaccines than we are with their vials, without the manufacture of tens of billions of glass vials, this life-saving drug would never have made it into the arms of people around the world.

to a problem with plastic storage tubes.) It is also advisable to use plastic consumable products, where possible, from a manufacturer who provides information about the additives they use, their consistency in manufacturing, and their quality-control practices.

24.2.1.3 Cleaning Plasticware

Much of the plasticware used in laboratories is disposable, but some plasticware is reusable, such as plastic beakers and graduated cylinders. The general directions already discussed for cleaning glassware apply to plasticware as well. There are, however, some additional factors to consider in washing plasticware, as summarized in Table 24.9.

24.3 STERILIZATION OF SOLUTIONS

Laboratory solutions are sometimes sterilized to protect them from degradation by microorganisms. There are two methods commonly used to sterilize solutions in a biotechnology laboratory: **autoclaving** and **filtration.** An **autoclave** *is a pressure cooker; materials to be sterilized are heated by steam under pressure.* The pressure in an autoclave is usually held at 15–20 lb/in^2 (psi) above normal pressure. This moist heat is very effective at destroying microorganisms; when water is present, bacteria are killed at lower temperatures than when heat alone is used. Sterilization with an autoclave is relatively fast and efficient and is the method of choice in many applications. Figure 24.7 shows the basic features of an autoclave, and Table 24.10 summarizes general autoclaving guidelines.

TABLE 24.9
Washing Plastic Labware

1. *Avoid alkaline detergents, especially when cleaning polycarbonate items.* Manufacturers make detergents that have a neutral pH and are recommended for plasticware.
2. *Do not use abrasive cleaners or scouring pads on any plasticware.*
3. *Periodically disassemble and clean spigots and threads on bottles, carboys, and closures to prevent salt buildup, which can cause leakage.*
4. *LDPE and PETG (see Figure 24.6b for resin codes) cannot be cleaned in commercial glassware washers due to temperature limitations, unless the washer allows suitable temperature control.*
5. *Repeated washing in a dishwasher can weaken polycarbonate labware.* Polycarbonate labware that has been exposed to high stress, such as centrifugation or use under vacuum, should be washed by hand using a mild, neutral pH, non-abrasive detergent.
6. *Keep machine washer cycle time to a minimum and remove plasticware as soon as it is cool.*
7. *Cover metal machine washer spindles with soft tubing or other materials to avoid abrasion.*
8. *Plasticware should be held in place during machine washing.*
9. *For high-sensitivity analytical work, soaking plasticware for up to eight hours in 1 N HCl will help remove trace metals.* Trace organics can be removed with alcohol cleaners.
10. *Minimize cleaning with organic solvents as some plastics are sensitive to these solvents.*
11. *Some types of plastic cannot be autoclaved.* In addition, plastics that can be autoclaved may require longer cycle times than glass because the heat transfer properties of plastic and glass are different. Cycle lengths for plastics need to be determined empirically (by experimentation) for each liquid and container.
12. *Some types of plastics discolor when washed in a commercial washer, but may still be suitable for use.*

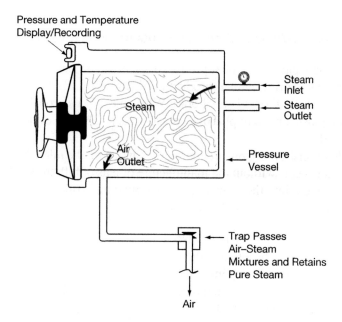

FIGURE 24.7 The basic features of an autoclave. Items to be sterilized are placed in the chamber, and the chamber door is tightly sealed. Air is purged from the chamber at the beginning of the sterilization cycle. After the air is removed, steam under pressure is introduced into the chamber and the temperature rises to the desired level. The autoclave is maintained under pressure at a high temperature for a predetermined time. The steam is then exhausted from the chamber, and the sterile materials are removed.

TABLE 24.10

Autoclaving Guidelines

1. ***Steam sterilization requires that the steam directly contact every surface and article being sterilized.*** For this reason:
   a. Do not overcrowd an autoclave. Allow spaces between items so that steam can contact all surfaces. When sterilizing instruments or fabrics, the steam must penetrate to, and contact each and every surface and fiber.
   b. Anhydrous materials such as oils, greases, and powders resist contact with water and therefore cannot be sterilized in an autoclave.
   c. Air can impede the contact of steam with a material. Autoclaves therefore evacuate air before introducing steam into the sterilization chamber. It is possible for air to be trapped in a material being autoclaved. For example, if a tube or needle is plugged, then air cannot escape, steam cannot enter, and that tube or needle will not be sterilized.
   d. Aqueous solutions do not require direct steam contact, but they do require adequate heating.
2. ***Containers should always be loosely capped.*** This ensures that steam can penetrate to all surfaces. Loose caps also allow pressure to equilibrate inside and outside the container. Containers can shatter if the pressure is not equilibrated. Flasks are often covered by cotton plugs or loosely twisted caps.
3. ***It is important that the autoclave be held at the proper temperature for the proper amount of time.*** The minimum temperature at which sterilization occurs is 121°C (250°F). Too little time at this temperature leads to failure of sterilization; too much time can degrade media containing carbohydrates, agar, and other sensitive materials. It is standard practice to hold the autoclave at high temperature for 20 minutes. Depending on what is being autoclaved, however, 20 minutes may not be adequate. For example, plastics are poor conductors of heat. It may take much longer than 20 minutes to sterilize a 1 L solution in a plastic container. The minimum time required for sterilization of materials in Pyrex vessels similarly varies depending on the container. Containers larger than 1 L are likely to require more than 20 minutes sterilization time. It is best to autoclave materials that are similar together, so that the optimal sterilization conditions will be the same for all items in the autoclave.
4. ***The quality of the water used to make steam is important.*** Volatile impurities in the water will be carried into all autoclaved items and solutions. It is common to use water that was deionized or treated by reverse osmosis.

(Continued)

TABLE 24.10 (*Continued*)

Autoclaving Guidelines

5. *Use a "slow exhaust" setting when autoclaving solutions.* Solutions do not boil during autoclaving even though the temperature in the autoclave is above the boiling point of water. This is because the pressure of the steam prevents boiling. After sterilization, however, if the steam is allowed to exit the chamber quickly, then solutions will boil violently out of their containers. When solutions are autoclaved, the autoclave should therefore be set to exhaust the steam slowly.

6. *Note that even when steam is released slowly, solutions will lose about 3%–5% of their liquid due to evaporation.* People sometimes add an extra 5% of purified water to each solution being autoclaved to compensate for this loss.

7. *Some materials should be autoclaved in separate containers because they will alter one another in some way.* For example, phosphates and glucose can interact with one another when subjected to heat so that the solution is degraded.

8. *Containers with liquids should not be filled more than 75% full to allow for fluid expansion and to prevent overflow.*

9. *Autoclaves attain very high temperatures and pressures and must be safely operated.* Safe operation requires that the instrument be properly maintained, and the operators understand how to use it safely. Note that older autoclaves sometimes lack the safety features of newer models.

After items have been subjected to autoclaving, it is important to know whether all microorganisms were indeed killed. The best way to test whether sterilization has occurred during autoclaving is to use spore strips. **Spores** *are dormant microorganisms.* Autoclaving should involve sufficient heat and sufficient time to kill spores. **Spore strips** *are dried pieces of paper to which large numbers of nonpathogenic, highly heat-resistant bacterial spores, such as those of* Geobacillus stearothermophilus, *are adhered.* Spore strips are placed in the autoclave in the area that is most inaccessible to steam, such as in a flask or bottle. After autoclaving, the strips are transferred to growth medium, being careful not to contaminate them with organisms from the air or laboratory. If sterilization in the autoclave was successful, the spores will not grow. It is important when performing spore tests to use a positive control and a negative control. The positive control is a spore strip that was not autoclaved and should grow if the spores are viable and the medium is properly prepared. The negative control is to place strips without spores in growth medium. No growth should occur in the negative control unless there is contamination.

Although spore strips are recommended for testing autoclave function, they require time to incubate, so they do not provide instantaneous information regarding the effectiveness of an autoclave cycle. There are convenient products used to identify materials that have been through a cycle in an autoclave. For example, autoclave tape can be wrapped on objects and changes colors when subjected to pressurized steam. Such devices are typically used in research laboratories to show that an item went through a sterilization cycle. Note, however, that autoclave tape and other such materials may change colors before all microorganisms are killed.

Some solution components are heat sensitive and must be sterilized by methods that do not involve heat. Heat-sensitive substances include proteins, vitamins, antibiotics, animal serum, and volatile chemicals. These materials are usually prepared as concentrated stock solutions and are then sterilized by filtration through a sterile filter with pores that are less than $0.2\,\mu m$ in diameter. The filter-sterilized components are then added to the rest of the solution using care to maintain the sterility of the mixture. Filtration through membrane filters is effective for removing microorganisms and is standard practice in many laboratories. Note that viruses will penetrate microfilters. In situations where viruses and pyrogens must be removed, as in the preparation of pharmaceuticals, ultrafilters can be used.

As with any other aspect of a quality program, the operation, maintenance, and monitoring of sterilization systems requires planning and documentation. Most modern autoclaves automatically record information for every cycle such as date, temperature reached, pressure, and time. Other aspects of quality, such as logbooks and operator training, are also essential.

Practice Problems

1. Pyrogens are negatively charged molecules that range in molecular weight from about 1,000 D to more than 10,000 D. What type of filtration method can remove pyrogens?

2. Viruses are about the same size as the protein albumin and are slightly bigger than pyrogens. What type of filtration method will remove them from a solution?

3. Is table salt, NaCl, removed from a solution that is passed through an ultrafilter? Is it removed by deionization?

4. Many water purification systems recirculate already purified water back through the purification steps. Why?

5. The term *log reduction value* (*LRV*) describes mathematically the ability of an ultrafiltration membrane to remove pyrogens (or other contaminants) from water. The higher the LRV, the greater the rejection of pyrogens by the membrane. The equation for LRV is:

$$\text{log reduction value} = \log\left(\frac{\text{feed water pyrogen concentration}}{\text{product water concentration}}\right)$$

For example, a filter was challenged by running water through it containing pyrogens at a concentration of 10,715 EU/mL. The endotoxin concentration in the product water was 0.0023 EU/mL. What was the LRV?

$$\text{log reduction value} = \log\left(\frac{10,715 \text{ EU/mL}}{0.0023 \text{ EU/mL}}\right) \approx 6.7$$

In another challenge, the same membrane was used to filter water containing pyrogens at a level of 4.5 EU/mL. After filtration, the water had less than 0.001 EU/mL. What was the LRV?

Questions for Discussion

1. You are placed in charge of purchasing a water purification system for a laboratory. For each of the following types of laboratories, what would be important considerations to discuss with manufacturers? What standards would your purified water need to meet? What purification methods do you think would be most appropriate for your water?
   a. You work in an analytical testing laboratory where you test air and water samples for pollutants.
   b. You work in a biotechnology research laboratory on a variety of genetic engineering projects.

2. A biotechnology company has successfully been purifying its water for several years. Workers suddenly discover that one of their products is no longer effective, and they trace the problem to the water used in production. Discuss factors that might have led to their sudden difficulties with water quality. Suggest changes they could make in their operating procedures to help avoid similar crises in the future.

3. If you have a water purification system in your laboratory, find out how it is designed and what routine maintenance it requires.

4. Write an SOP to clean glassware in your laboratory manually. If you do not work in a laboratory, write an SOP to clean glassware in a research laboratory where proteins are purified and analyzed.

5. If there is a machine washer in your laboratory, read its manual. Describe potential problems that can arise and how these problems can be recognized.

6. If you have a dishwasher at home, examine the quality of its performance. Devise a method to evaluate how effectively your dishes are cleaned.

25 Laboratory Solutions to Support the Activity of Biological Macromolecules

25.1 INTRODUCTION

Chapters 22–24 discussed the principles, calculations, and practical details regarding the preparation of laboratory solutions. For example, Section 23.3.2 explained how to calculate the required amount of each component of SM buffer. We have not, however, addressed the question of *why* SM buffer has the various components that it does. This chapter explores the purposes of the components used in solutions in biology laboratories.

Biologists work in the laboratory with materials derived from organisms. These materials include proteins, DNA, RNA, and cells. The activity of all these biological materials is intimately associated with their structure; in fact, the relationship between structure and function is one of the most important themes in biology. For example, the function of enzymes is to catalyze reactions. For catalysis to occur, the enzyme must recognize and bind the reactants involved. Recognition and binding occur because the enzyme's structure complements that of the reactants (Figure 25.1). Enzymes do not

function if their normal structure is degraded. Thus, to protect their function it is critical that laboratory solutions have components that provide conditions that preserve the structural integrity of biological molecules.

There are many different solutions used in the biology laboratory. The components of these solutions differ from one another because the requirements of various biological systems vary. For example, a solution to support the structure and activity of an enzyme is likely to differ from a solution for manipulating DNA. A solution that is used to store a material is likely to differ from a solution for isolating that material.

This chapter begins by briefly describing the structure and function of proteins. Then, the types of laboratory solutions that are used with proteins are considered. The second part of this chapter briefly discusses the structure and function of nucleic acids (DNA and RNA) followed by an exploration of how nucleic acids are manipulated and maintained in biological solutions. Chapter 31 considers some of the considerations involved in maintaining living, intact cells.

DOI: 10.1201/9780429282799-31

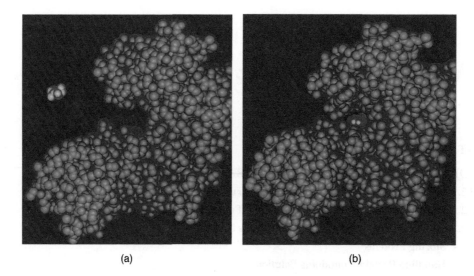

(a)   (b)

FIGURE 25.1   Example of the relationship between protein structure and function. An enzyme and its substrate. The small substrate molecule (red) (a) binds within a groove in the larger enzyme that is (b) complementary in shape and ionic properties. Many proteins bind other molecules at sites whose shapes are complementary.

25.2 WORKING WITH PROTEINS IN LABORATORY SOLUTIONS

25.2.1 An Overview of Protein Function and Structure

25.2.1.1 Protein Function and Amino Acids: The Subunits of Proteins

Proteins perform a myriad of essential functions in cells (Table 25.1). This functional diversity is possible because proteins are structurally diverse. As with the English language, where an immense number of words

with varied meaning are formed using only 26 letters, so diverse proteins are formed using primarily 20 different amino acid subunits. Each of the 20 amino acids has the same core structure consisting of an **amino group**, NH_2, and a **carboxyl group**, $COOH$, attached to a central carbon atom (Figure 25.2a). What makes each amino acid distinct is the presence of a characteristic **side chain**, or **R group**, which has a particular structure and chemical properties. The diversity and versatility of proteins is due to the wide variety of characteristics conferred on proteins by the various amino acid R groups.

A protein is formed by amino acid subunits linked together in a chain. *The bonds that connect the amino acids to one another are* **covalent peptide bonds** (Figure 25.2b). Every protein has a different sequence of amino acids that determines the final structure and function of the protein (Figure 25.2c).

25.2.1.2 The Four Levels of Organization of Protein Structure

Every protein has a distinct, complex three-dimensional structure. To describe these varied protein structures, people speak of four levels of organization: primary, secondary, tertiary, and quaternary (Figure 25.3). Secondary, tertiary, and quaternary structure are sometimes collectively referred to as *higher-order* structure.

1. **Primary structure** *refers to the linear sequence of amino acids that comprise the protein.*
2. **Secondary structure** *refers to regularly repeating patterns of twists or kinks of the*

TABLE 25.1

Protein Functions

Function	Examples
Metabolism	*Enzymes* catalyze reactions
Cellular control	*Hormones* affect metabolism of target cells
Structure	*Keratin* forms the structure of hair and nails
Storage	*Ovalbumin* (egg white) stores nutrients
Transport	*Hemoglobin* transports O_2 and CO_2
Mobility	*Actin* and *myosin* cause muscle contraction
Defense	*Antibodies* destroy invading microorganisms
Gene regulation	*Repressor proteins* in bacteria "turn off" genes
Recognition	*Receptors* on cell surfaces recognize specific hormones

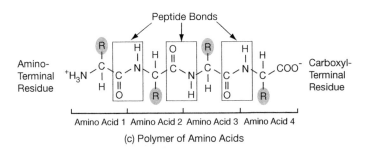

(a) Core Amino Acid Structure

(b) Peptide Bond

(c) Polymer of Amino Acids

FIGURE 25.2 Protein structure. (a) The core structure of an amino acid consists of an amino and a carboxyl group attached to a central carbon atom. Each amino acid has a different attached R group. (b) Proteins are polymers of amino acids joined by peptide bonds. A peptide bond is a covalent bond formed when the carboxyl group of one amino acid joins with the amino group of a second amino acid. (c) Every protein has a different sequence of amino acids that determines the structure and function of the protein.

amino acid chain. Two common types of secondary structure are called the **α-helix** and **β-pleated sheet**. *Regions of proteins that do not have regularly repeating structures are often said to have a* **random coil** *secondary structure, although their folding patterns are not truly random.*

Secondary structure is held together by weak, noncovalent, molecular interactions, called *hydrogen bonds*, between nonadjacent amino acids. **Hydrogen bonds** *form when a hydrogen atom that is bonded to an electronegative atom (such as O or N) is also partially bonded to another atom (usually also O or N). Thus, a hydrogen bond forms because a hydrogen atom is shared by two other atoms.*

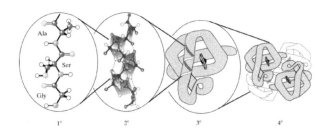

FIGURE 25.3 The four levels of protein structural organization: Primary (1°) structure is the sequence of amino acids, represented here by three-letter codes. Certain proteins or sections of proteins form regularly repeating patterns, called secondary (2°) structure. Many proteins require complex tertiary (3°) folding to provide the right structure and conformation for recognition, binding, and solubility. Quaternary (4°) structure occurs when folded chains associate with one another to form complexes.

3. **Tertiary structure** *refers to the three-dimensional globular structure formed by bending and twisting of the protein.* Globular structures include regions of α-helices, β-pleated sheets, and other secondary structures that are folded together in a way that is characteristic of that protein. Tertiary structure is generally stabilized by multiple weak, noncovalent interactions. These include the following:

- **Hydrogen bonds** that form when a hydrogen atom is shared by two other atoms.
- **Electrostatic interactions** that occur between charged amino acid side chains. **Electrostatic interactions** *are attractions between positive and negative sites on macromolecules.* These weak interactions bring together amino acids and stabilize protein folding (Figure 25.4).
- **Hydrophobic interactions** *that are due to the tendency of the aqueous environment of the cell to exclude hydrophobic amino acids.* Proteins spontaneously fold in such a way that hydrophobic amino acids fold into the interior of the protein, protected from the aqueous environment. In contrast, hydrophilic amino acids tend to be found on the protein's surface.

In addition to these weak, noncovalent interactions, some proteins have covalent **disulfide bonds** that stabilize their three-dimensional structure. **Disulfide bonds** *form when sulfurs from two cysteine molecules (cysteine is an amino acid) bind to one another with the loss of two hydrogens* (Figure 25.5). Disulfide bond formation is an example of oxidation. **Oxidation** *is defined as the removal of electrons from a substance.* In biological systems, oxidation often involves the removal of an electron accompanied by a proton (i.e., a hydrogen atom).

4. **Quaternary structure** *refers to the complex formed when two or more protein chains associate with one another.* Some proteins have separately synthesized subunit chains that join together to form a functional protein complex.

The covalent bonds that link amino acids into their primary chain structure are much stronger than the noncovalent interactions that stabilize higher-order structure. Relatively little energy (e.g., in the form of heat) is needed to disrupt the weaker, noncovalent bonds. Higher-order structures persist because a large number of weak interactions stabilize them.

The "weakness" of the noncovalent interactions that stabilize higher-order protein structure confers two important characteristics on proteins. First, proteins are flexible, and this flexibility is critical to their biological function. For example, enzymes change shape in order to bind to their substrates. Second, the normal folding and shape of proteins are readily disrupted by changes in temperature, pH, and other conditions. There are many circumstances in the laboratory, therefore, where the primary structure of the amino acid chain remains intact, but the complex shape of the protein is lost.

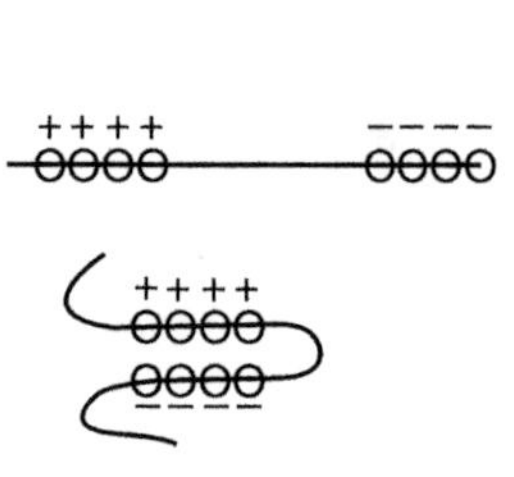

Folding Due to
Electrostatic Attraction

FIGURE 25.4 Electrostatic interactions and protein folding. Some R groups are positively charged, and some are negatively charged. Positively charged amino acids attract negatively charged amino acids and *vice versa*. Interactions between positively and negatively charged amino acids are called electrostatic interactions. Electrostatic interactions between charged R groups contribute to tertiary folding of the protein.

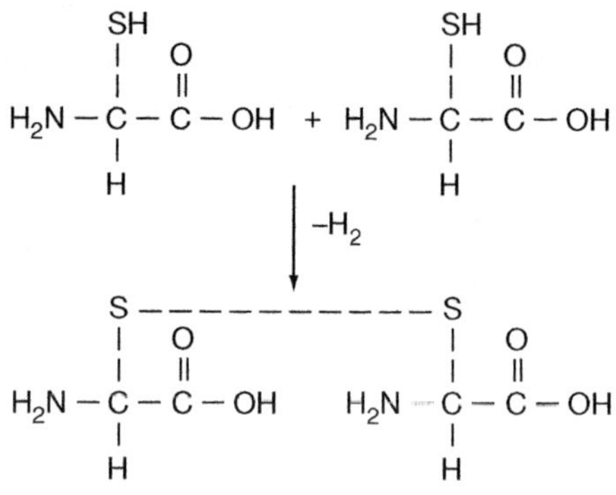

FIGURE 25.5 Disulfide bonds. Disulfide bonds contribute to tertiary folding. A disulfide bond is formed between two cysteine molecules.

25.2.2 How Proteins Lose Their Structure and Function

Laboratory solutions for proteins usually must provide suitable conditions so that the proteins' normal structures, and therefore their normal activity, are maintained. To understand how the structure of proteins is protected in laboratory solutions, it is necessary to understand how that structure can be destroyed. There are various ways by which proteins lose their normal structure and therefore their ability to function normally:

1. **Proteins can *denature* or unfold so that their three-dimensional structure is altered but their primary structure remains intact**. Because the amino acid side chain interactions that stabilize protein folding are relatively weak, environmental factors including high temperature, low or high pH, and high ionic strength can cause denaturation. The changes that occur in eggs when they are hard-boiled is a familiar example of protein denaturation. When boiled, the translucent egg protein, albumin, is denatured and becomes hard egg white.

    There are many laboratory situations where denaturation occurs reversibly under one set of circumstances and proteins regain their normal conformation when suitable conditions are restored. Denaturation can also be irreversible, as when eggs are hard-boiled.

2. **The primary structure of proteins can be broken apart by enzymes, called *proteases*, that digest the covalent peptide bonds between amino acids**. *This digestive process is called* proteolysis. Cells contain proteases sequestered in membrane-bound sacs called *lysosomes*. When cells are disrupted, the lysosomes break and release their proteases. In the laboratory, it is therefore necessary to minimize the activities of cellular proteases to protect proteins from proteolysis. Methods used to minimize proteolysis include low temperature, short working times, and addition of chemicals that inhibit proteases. Microbes also release proteases, and therefore antimicrobial agents are often added to solutions to protect proteins.

3. **Sulfur groups on cysteines may undergo oxidation to form disulfide bonds that are not normally present**. Extra disulfide bonds can form when proteins are removed from their normal environment and are exposed to the oxidizing conditions of the air. Because disulfide bonds affect protein folding, additional disulfide bonds may produce undesirable changes in protein tertiary structure. Disulfide bond formation is an oxidation reaction; therefore, antioxidizing chemicals, called reducing agents, are often added to protein solutions to prevent unwanted disulfide bonding.

4. **Proteins can aggregate with one another, leading to precipitation from solution**. Each protein requires specific conditions in order to remain in solution.

5. **Proteins can readily adsorb onto (stick to) surfaces**. This is a somewhat different concern than the first four issues because the proteins do not actually lose their normal structure, but their function is nonetheless lost. Surface charges on proteins make them "sticky," and many proteins will adsorb onto almost any surface. Glassware and plasticware provide large surfaces that can adsorb proteins from solution, thus reducing protein activity or the amount of protein recovered during isolation. Adsorption can be a significant problem in experiments involving low levels of protein. Similarly, in experiments that involve dosing a test animal with a particular amount of a protein substance, it is problematic if an unknown percentage of the dosing compound is lost to adsorption before administration to the animal. As we saw in Chapter 24, there are a variety of glassware and plasticware types used in laboratories. Some proteins adsorb significantly more to one type of labware than another, so the amount of adsorption to a particular surface needs to be determined for each protein. In addition, when protein solutions are filtered (as is sometimes done to sterilize the solution), proteins may adsorb to the filters and be lost. Whenever possible, proteins should be prepared at a relatively high concentration (e.g., 1 mg/mL) in buffer. High concentration stabilizes the protein's structure and helps minimize loss due to the protein adsorbing onto the surface of its storage container. When small amounts of proteins must be stored, analysts sometimes add an inert carrier protein, usually bovine serum albumin (BSA). The carrier protein sticks to and covers up the sticky sites on the glass or plastic container so that most of the protein of interest remains in solution. (For a detailed

discussion of adsorption to labware, see, for example, Goebel-Stengel, Miriam, et. al. "The Importance of Using Optimal Plasticware and Glassware in Studies Involving Peptides." *Analytical Biochemistry*, vol. 414, no. 1, 2011, pp. 38-46. doi:10.1016/j.ab.2011.02.009.)

Note that there are situations where protein adsorption can be exploited in the laboratory. For example, western blotting is a technique used to identify proteins based on their interactions with specific antibodies. Western blotting requires immobilizing proteins on a solid surface so that they are not washed away during rinses. Protein immobilization is accomplished by adsorbing the proteins to a nylon or plastic membrane. (See Chapter 29 for a discussion of western blotting.)

25.2.3 THE COMPONENTS OF LABORATORY SOLUTIONS FOR PROTEINS

25.2.3.1 Introduction

Laboratory solutions for proteins must provide them with suitable conditions. What constitutes "suitable conditions," however, varies greatly. Because proteins differ from one another, the conditions for manipulating one protein may not work for another. Even when considering a single protein, the optimal conditions differ depending on the situation. For example, extracting proteins from intact tissue requires disrupting the cells and may require extracting the proteins from membranes and/or tight associations with other molecules. Special *lysis buffers* are used for this type of cellular disruption. (Lysis is discussed in more detail later in this chapter.) The solutions that are used when one wants a protein to function properly in a test tube are different than the lysis buffers used to disrupt cells. Yet a different set of conditions is required for storing a protein. Thus, the solution requirements for each protein and each task must be determined individually, often by experimentation.

Understanding biological solutions is complicated by the fact that chemical agents that are used in solutions may perform more than one role depending on the situation. Moreover, the same agent may be desirable in one situation and problematic in another. For example, detergents play a familiar, useful role as cleaning agents. Detergents can also act to solubilize or denature proteins. Detergents are sometimes intentionally added to solutions because their effects are useful. At other times, detergents are scrupulously removed from solutions because their effects are detrimental.

TABLE 25.2
Classes of Agents Used in Laboratory Solutions for Proteins

Buffers	Precipitants
Salts	Reducing agents
Cofactors	Metal chelators
Detergents	Antimicrobial agents
Organic solvents	Protease inhibitors
Denaturants and solubilizing agents	Storage stabilizers

Although laboratory solutions for proteins vary significantly depending on the circumstances, there are classes of agents commonly found in protein solutions. These classes of agents are summarized in Table 25.2, and general principles regarding these agents will be discussed.

25.2.3.2 Proteins and pH: The Importance of Buffers

Buffers, as discussed in Chapter 23, are important agents in most biological solutions because pH must be controlled. Proteins that reside in the cell's cytoplasm normally have many charged amino acid side groups that interact with one another and affect the structure of the protein. The charges on these side groups are affected by the pH of their solution. The solution pH, therefore, affects the three-dimensional structure of proteins, their activity, and their solubility. There is no universal rule as to what pH is optimum, although pH extremes generally denature proteins. The pH of a laboratory solution usually should be maintained as closely as possible to the pH found in the protein's native environment, which is typically somewhere between pH 6 and 8.

25.2.3.3 Salts

Salts *are compounds made up of positive and negative ions.* Biological systems frequently require salts, and many laboratory solutions contain salts at a specific concentration.[1] **Ionic strength** *is a measure of the charges from all ions in an aqueous solution.* The general importance of ionic strength to biological molecules in laboratory solutions can be described as follows. Due to electrostatic interactions, two macromolecules in solution tend to attract one another if they have opposite charges, tend to repulse one another if they have the same charges, and have neither interaction if the molecules are

---

[1] It is not surprising that salts play many roles in biological systems; after all, life evolved in the sea. This ancestry is reflected in the fact that salts perform a number of essential roles in organisms and salt levels are rigorously controlled in living systems.

not charged. Ions in a solution also interact with macromolecules based on charge. High concentrations of ions in a solution will tend to counteract or shield opposite macromolecular charges.

In protein solutions, therefore, salts can affect the electrostatic interactions between the side chains of amino acids. At low ionic strengths, the ions in solution have little effect on these interactions (Figure 25.6a). An increased ionic strength tends to stabilize protein structures by shielding unpaired charged groups on protein surfaces. At still higher ionic strengths, many proteins are denatured, presumably because normal sites of electrostatic interaction between charged groups on amino acids are neutralized by the salt ions (Figure 25.6b).

The ionic strength of a solution affects both the three-dimensional structure of proteins and their solubility. Most cytoplasmic proteins are soluble at salt concentrations of about 0.15–0.20 M, but often precipitate at higher or lower salt concentrations. This phenomenon is frequently exploited in the laboratory to selectively isolate proteins from other solution components. (See Chapter 34). Salt is added to the solution until the proteins of interest precipitate. The precipitated proteins can then be retrieved. Salt-precipitated proteins usually regain their normal structure and solubility when the salt is removed. High salt concentrations retard microbial action; therefore, proteins are sometimes stored as salt precipitates.

The salt concentration in a solution can also be manipulated to control the binding of proteins to a solid matrix. Certain solid materials are designed so that they bind proteins when particular conditions of salt concentration and pH are met. Under these conditions, a desired protein may selectively bind to the solid matrix, leaving contaminants behind in the solution. The isolated protein is then recovered by changing the salt concentration so that the protein leaves the matrix and rinses into fresh buffer solution. This is a process performed in certain types of chromatography, as will be explored in Chapter 34.

The exact effects of ionic strength on a protein depend on the protein, the type of salts involved, and the pH of the solution. Recipes for protein solutions therefore must specify the pH, the type of salts, and their concentrations. Table 25.3 summarizes the important roles that salts play in protein solutions.

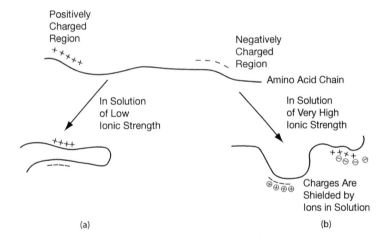

FIGURE 25.6 The effect of ionic strength on macromolecular surface charges. (a) A solution of low ionic strength. (b) A solution of high ionic strength.

TABLE 25.3

Significant Roles of Salts in Protein Solutions

1. *Salts affect the three-dimensional structure of proteins and may be manipulated to stabilize structure, or to intentionally denature a protein.*
2. *Salts affect the solubility of proteins and may be manipulated to keep proteins in solution, or to cause them to precipitate from solution.*
3. *Salts may be used to control the binding of a protein to a solid matrix.*
4. *Salts may be used to control the binding of proteins to other macromolecules* (as also discussed in Section 25.3.3.3).
5. *Salts may be added as cofactors for enzymatic reactions*, as discussed in the next section.

25.2.3.4 Cofactors

Cofactors are *chemical substances that many enzymes require for activity.* Cofactors include metal ions, such as Fe^{++}, Mg^{++}, and Zn^{++}. *Cofactors may also be complex organic molecules called* **coenzymes**, such as **nicotinamide adenine dinucleotide (NAD)** and **coenzyme A.** Cofactors are sometimes added to, or removed from, protein solutions. For example, Mg^{++} is a cofactor for nucleases, enzymes that digest DNA and RNA. In some situations, the salt, magnesium sulfate, is added to solutions to provide Mg^{++}, while in other applications, it is carefully avoided. Magnesium is also a necessary cofactor for the polymerase enzyme that amplifies DNA in the polymerase chain reaction (PCR, Chapter 27). If there is too little magnesium in the solution mixture for a PCR, then the yield of product is reduced. If there is too much magnesium in the reaction mixture, then there can be undesired, nonspecific background amplification.

25.2.3.5 Detergents

Many types of detergents are used in the laboratory. **Detergents** *are a class of chemicals that have both a hydrophobic "tail" and a hydrophilic "head."* Detergents therefore have two natures: They can be both water-soluble and lipid-soluble. *Some detergents have hydrophilic portions that are ionized in solution (***ionic detergents***), whereas others have hydrophilic sections that are not ionized in solution (***nonionic detergents***).* **SDS (sodium dodecyl sulfate)** is an example of an ionic detergent; **Triton X-100** is a nonionic detergent.

Detergents interact with and modify the structure of proteins. Detergents are sometimes intentionally added to protein solutions for varied reasons (in addition to their familiar role as cleaning agents):

- *Detergents may be used to solubilize proteins that are associated with lipid membranes.* Proteins that are associated with lipid membranes tend to be hydrophobic and difficult to bring into solution. Because detergents both are soluble in water and have hydrophobic properties, they can interact with and increase the water solubility of hydrophobic proteins. Nonionic detergents are generally preferred for solubilizing membrane-bound proteins because they tend not to be denaturing.
- *Detergents are sometimes added to solutions to intentionally denature proteins.* Ionic detergents tend to effectively denature proteins. Denaturants are discussed below.

- *Detergents may be added in low levels to solutions to help prevent protein aggregation and prevent proteins from adsorbing onto surfaces, such as glass.*

Some cautions regarding detergents are as follows:

- Some detergents, such as Triton X-100, absorb UV light and can interfere with spectrophotometric assays.
- Commercial detergents intended for cleaning may contain contaminating substances that affect proteins. To avoid such contaminants, manufacturers purify detergents that are intended for protein work.
- Detergents denature proteins, and therefore, detergent residues from cleaning glassware and equipment must be thoroughly removed.

25.2.3.6 Organic Solvents

Organic solvents, such as **ethanol** and **acetone**, are sometimes added to biological solutions. Organic solvents may be used to solubilize membrane proteins and occasionally to cause them to precipitate from a solution as part of a purification strategy. Organic solvents, however, cause most proteins to denature irreversibly and should generally be avoided when protein activity must be retained.

25.2.3.7 Denaturants, Solubilizing Agents, and Precipitants

Although we have been primarily considering how solutions protect the normal structure and function of proteins, there are situations where it is desirable to denature the higher-order structure of proteins. For example, polyacrylamide gel electrophoresis (PAGE; see Chapter 34) is a method used to separate proteins from one another by forcing them to migrate through a gel-like matrix in the presence of an electrical current. Before electrophoresis, the proteins may be intentionally denatured with the detergent SDS so that their three-dimensional structures do not affect their movement through the gel. Ionic detergent is preferred for this task because it is a more effective denaturing agent than nonionic detergent. (In addition to denaturing proteins, SDS also confers a uniform negative charge on all the proteins so that differences in their charges do not affect their migration in the gel. When SDS is used with PAGE, proteins separate from one another based only on differences in their sizes.) Proteins do not need to maintain their normal activity during this type of electrophoresis, so denaturation is not a problem.

Urea at high concentrations (4–8 M) and **guanidinium salts** are other effective denaturants. Urea is frequently used when proteins are separated from one another by a method called *isoelectric focusing*. Like SDS–PAGE, isoelectric focusing requires that the proteins be completely unfolded. SDS, however, is not used in isoelectric focusing because SDS confers a charge on the proteins that interferes with the isoelectric focusing method.

Urea, SDS, and guanidinium salts not only denature proteins, but they also cause them to be soluble in aqueous solutions. Some proteins are not normally soluble in laboratory solutions and are therefore difficult to manipulate in the laboratory. For example, bacteria may produce valuable recombinant proteins, but sequester them in insoluble inclusion bodies. Guanidinium hydrochloride is sometimes used to solubilize and recover the desired recombinant proteins from the bacterial inclusion bodies.

We have just seen that some agents that denature proteins cause the proteins to be soluble in aqueous solutions. There are also agents that simultaneously denature and cause proteins to precipitate out of solution. **TCA (trichloroacetic acid)** and the organic solvents ethanol, acetone, methanol, and chloroform are examples of agents that both denature and precipitate proteins. Such precipitation can be exploited in the laboratory to separate large macromolecules (including proteins, DNA, and RNA) from smaller molecules such as salts, amino acids, and nucleotides. Simultaneous denaturation and precipitation is useful, for example, when a protein is to be analyzed for its amino acid content. In this case, maintaining the protein's activity and three-dimensional structure is unnecessary.

There are situations where it is desirable to precipitate a protein, but the protein will later be required to exhibit normal activity. In such cases, precipitants such as **ammonium sulfate** and **PEG (polyethylene glycol)** are used. These agents cause proteins to precipitate without irreversibly denaturing them. Ammonium sulfate is an inexpensive salt and is commonly used to separate proteins from one another during protein purification procedures. As noted earlier, salt-precipitated proteins generally recover their normal activity when the salt is removed.

25.2.3.8 Reducing Agents

Recall that when proteins are exposed to the oxidizing conditions of the air, sulfur groups on cysteines may undergo oxidation to form unwanted disulfide bonds. Laboratory workers sometimes add **reducing agents** to solutions to simulate the reduced intracellular environment and prevent unwanted disulfide bond formation. **β-Mercaptoethanol** and **DTT (dithiothreitol)** at a final concentration of 1–5 mM are commonly used for this purpose. These agents should be added right before use because they are not stable in solution.

25.2.3.9 Metals and Chelating Agents

Metal ions, such as Ca^{++} and Mg^{++}, are frequently released when cells are disrupted and are often present as contaminants in reagents and water. Metals can accelerate the formation of undesired disulfide bonds and can act as cofactors for proteases. **Chelating agents**, therefore, are often added to laboratory solutions *to bind and remove metal ions from solution*. The most common chelator is **EDTA (ethylenediaminetetraacetic acid)**.

25.2.3.10 Antimicrobial Agents

Antimicrobial agents are used to kill microbes and to prevent proteolysis caused by microbes. Common examples are **sodium azide** (NaN_3) at a final concentration of 0.02–0.05% (w/v) and **thimerosal** at a final concentration of 0.01% (w/v). Sodium azide is toxic. It may affect the activity of some proteins.

25.2.3.11 Protease Inhibitors

Proteases are enzymes that break the covalent bonds holding proteins together. Proteases thus destroy the primary structure of proteins. Proteases are released into the cell extract when cells are disrupted. Until the proteins of interest are purified from the extract, protease inhibitors may be necessary to prevent protein degradation. EDTA is considered a protease inhibitor because it removes metal ion cofactors from solution. **PMSF (phenylmethanesulfonyl fluoride)** and **pepstatin** are examples of agents used to inhibit specific classes of proteases. Protease activity is also reduced by keeping protein solutions cold and working quickly to isolate proteins of interest from the cell extract.

25.2.4 Storing Proteins

25.2.4.1 Low Temperature as a Method to Protect Proteins during Storage

Proteins are seldom stored at room temperature because they are rapidly degraded, often as a result of microbial activity. Rather, cold temperatures are used during storage of proteins (and other biological materials) to reduce microbial growth and slow the degradation of proteins by proteolytic enzymes and spontaneous breakdown. Temperatures that are commonly used for storing proteins are summarized in Table 25.4.

TABLE 25.4

Different Degrees of Cold for Protein Storage

Temperature and Conditions	Maximum Recommended Time	Comments	Number of Times Sample May Be Removed from Storage
4°C (ice bath)	Hours	Used on the lab bench while performing procedures involving enzymes and other proteins	Many
4°C (refrigerator)	1 month	Requires sterile conditions or addition of antimicrobial agent	Many
−20°C (with 25%–50% glycerol or ethylene glycol)	1 year	Usually requires sterile conditions or addition of antimicrobial agent	Many
Frozen at −70°C to −80°C or in liquid nitrogen	Years	Does not require sterile conditions or addition of antimicrobial agent	Once; repeated freeze–thaw cycles generally degrade proteins
Lyophilized (usually also frozen)	Years	Does not require sterile conditions or addition of antimicrobial agent	Once; it is impractical to lyophilize a sample multiple times

Primary Source: "Protein Stability and Storage," Thermo Scientific. 2009. http://tools.thermofisher.com/content/sfs/brochures/TR0043-Protein-storage.pdf.

Laboratory refrigerators are set to 4°C. Water and biological materials do not freeze at this temperature, and some biological activities continue, but at a slowed rate. This condition is adequate for the storage of many proteins, particularly for short periods. Proteins are sometimes stored refrigerated in the form of stable salt precipitates.

Freezing is commonly used for long-term storage. Although water freezes at 0°C, this temperature does not prevent biological and chemical activity. Laboratory freezers therefore go down to −20°C for routine storage. Even −20°C, however, is not sufficiently cold to freeze concentrated solutions of biological molecules. Moreover, degradative enzymes that may be present in a biological sample can slowly destroy proteins and other cellular materials stored at −20°C. Freezers that attain temperatures in the range of −70°C to −80°C are therefore preferred for storing substances extracted from cells.

Although freezing is a good way to store laboratory materials, potentially destructive physical events occur during the freezing and thawing processes. Different components in a solution can freeze and thaw at different rates allowing materials to be exposed to extremes of pH and salt concentration. Another problem is that ice crystals form during freezing and melt during thawing. Ice crystals can damage biological substances. It is important to avoid multiple freeze–thaw cycles in order to limit damage related to ice crystals. To avoid multiple freeze–thaw cycles, *solutions containing proteins (or other biological substances) are divided into small volumes that are frozen in individual containers, called* **aliquots**. Only as many aliquots as are needed at a given time are thawed; any unused, thawed material is discarded. This technique ensures that a substance is thawed only one time. (It is critical that all aliquots are labeled carefully to avoid later confusion.)

Biological materials should be stored in freezers that are not "frost-free" to avoid repeated freeze–thaw cycles. Note that laboratory freezers are not usually "frost-free," whereas those used at home are. Therefore, laboratory freezers are seldom purchased from an ordinary appliance store. It is possible to purchase insulated containers to help prevent thawing if an ordinary frost-free freezer must be used.

25.2.4.2 Additives Used to Protect Proteins during Storage

Various additives are used to extend the storage life of proteins. Care must be taken, however, since these additives contaminate the sample. Check that no deleterious interaction with the protein of interest will occur when using additives. Additives include the following:

- **Carrier proteins** (e.g., 0.1%–1% **bovine serum albumin**, BSA) are sometimes added to stabilize dilute protein solutions and avoid the adsorption of scarce protein to the surface of a storage container.
- **Cryoprotectants**, used at concentrations from 10% to 50%, stabilize protein solutions during freezing and thawing. **Glycerol** and **ethylene glycol** are common cryoprotectants that keep solutions from actually freezing, prevent ice crystal formation, and improve protein stability. It is common to receive enzymes from suppliers that have been stabilized with glycerol. Note that glycerol may affect subsequent steps (e.g., chromatography or restriction enzyme digests) when the protein is removed from storage.
- **Protease inhibitors**, as previously discussed.
- **Antimicrobial agents**, as previously discussed.
- **Reducing agents**, as previously discussed.
- **Nonionic detergents** at low concentrations (e.g., 0.15%) may be added to prevent protein aggregation and adsorption onto the surface of the storage container.
- **Proprietary products** are sold by manufacturers to help stabilize proteins during storage.

25.2.4.3 Freeze-Drying to Store Proteins

Freeze-drying, also known as **lyophilization**, *is a process in which a material is first frozen and then dried under a vacuum.* Freeze-drying is familiar outside the laboratory as a method of preserving foods for astronauts and hikers. Freeze-drying is used in the laboratory to preserve proteins and other materials (e.g., antibodies,

enzymes, and oligonucleotide primers for PCR). It is common to receive materials from manufacturers that have been lyophilized for storage and shipment.

The purpose of freeze-drying is to halt biological activity and degradation by removing all water. Vacuum is used during the freeze-drying process to cause ice to change directly from solid to vapor without passing through a liquid phase. Removing water from a substance in this way leaves the basic structure and composition of the substance intact, while inhibiting enzymatic and microbial degradation. Freeze-dried substances in tightly sealed vials can be stably stored for long periods of time.

When it is time to use a freeze-dried material, water or buffer is added and the original properties of the material are restored quickly. (If the material does not go into solution easily, it may have denatured during the freeze-drying process.) Manufacturers who make freeze-dried products usually provide guidance for their rehydration. If not used immediately, rehydrated materials are usually aliquoted and frozen. Note that some freeze-dried products are lightweight and can easily be lost unless care is exercised when opening their vial and handling the product.

25.2.5 Handling Protein-Containing Solutions

Some considerations in handling solutions containing proteins are summarized in Table 25.5.

25.2.6 Summary of Protein Solution Components

Table 25.6 summarizes the components of protein solutions in tabular form.

TABLE 25.5

General Rules for Protein Handling to Reduce Degradation

1. *Wear gloves when handling protein solutions to avoid introducing contaminating proteases from your hands.*
2. *Mix protein solutions gently since some proteins are denatured by vigorous shaking.*
3. *Use only clean glassware.* All glassware and plasticware should be well washed and rinsed. Residual detergents or metal ions may have deleterious effects on proteins. A final rinse with EDTA-containing solution is sometimes used to eliminate metal ions.
4. *Use sterile buffers, solutions, and glassware to reduce bacterial contamination when working with, or storing proteins for long times.* Consider the use of antimicrobial agents.
5. *In general, keep protein-containing solutions cold in the laboratory.*
6. *Use only high-purity water to prepare buffers and other protein-containing solutions.*
7. *Be aware of potential protein loss due to adsorption to labware surfaces.*

TABLE 25.6

Summary of Common Components of Solutions Used to Maintain Protein Structure and Function

Type of Agent	Function	Examples
Buffering agents	Maintain pH	Tris, phosphate, acetate buffers
Salts	Control ionic strength, cofactors	NaCl, MgCl$_2$
Antimicrobial agents	Prevent microbial contamination	Na-azide
Reducing agents	Prevent unwanted disulfide bond formation	β-Mercaptoethanol, DTT
Protease inhibitors	Prevent proteolysis	EDTA, PMSF, pepstatin
Chelators	Remove metals from solution	EDTA, EGTA[a]
Detergents	Solubilize membrane proteins	Triton X-100, SDS
	Prevent aggregation and adsorption	
Denaturants	Denature proteins	SDS, urea, guanidinium salts
• Solubilizing denaturants		
• Precipitating denaturants		Acetone, phenol/ chloroform
Precipitants	Precipitate proteins	Ammonium sulfate, PEG, TCA
Cofactors	Required for enzyme activity	Mg^{++}, Fe^{++}, CoA
Organic solvents	Sometimes used to solubilize membrane proteins and occasionally to precipitate proteins during isolation	Ethanol, acetone
Storage stabilizers	Protect proteins during freezing	Glycerol

[a] **EGTA** is similar to EDTA, but it has a higher binding capacity for calcium than for magnesium. This chelator is commonly used when the regulation of calcium is desired.

Practice Problems: Proteins

1. *EcoR1* is an enzyme that is used in the laboratory to cleave DNA at specific sites. The first recipe below is for the storage buffer used by the manufacturer to store and ship *EcoR1* enzyme. The second recipe is for the buffer used when the enzyme cleaves DNA – when maximal activity is required. When the enzyme is used in the laboratory, a very small amount of the enzyme (in storage buffer) is removed and diluted in fresh activity buffer.

   a. Suggest what the purpose of each solution component might be.

   b. Explain the differences between the solution for storage and the solution for activity. Why are the two solutions not identical?

Storage Buffer (Store at −20°C)	**Activity Buffer**
10 mM Tris–HCl, pH 7.4	90 mM Tris–HCl, pH 7.5
400 mM NaCl	50 mM NaCl
0.1 mM EDTA	10 mM MgCl$_2$
1 mM DTT	
0.15% Triton X-100	
0.5 mg/mL BSA	
50% Glycerol	

2. Examine the following recipes that come from protein manuals. Suggest a function for the various components.

---

a. Sample Buffer Used When Loading Proteins into Electrophoresis Gel

60 mM Tris–HCl (pH 6.8)

2% SDS

14.4 mM β-mercaptoethanol

0.1% Bromophenol blue (dye that moves ahead of proteins, used to decide when to halt electrophoresis)

25% Glycerol (increases the solution density causing the proteins to sink into the gel)

b. Restriction Enzyme Buffer (Low Salt)

10 mM Tris–HCl (pH 7.5)

10 mM $MgCl_2$

1 mM DTT

---

3. Based on the answers to questions 1 and 2, what are the common components of protein solutions? What is the purpose of these components?

25.3 WORKING WITH NUCLEIC ACIDS IN LABORATORY SOLUTIONS

25.3.1 An Overview of Nucleic Acid Structure and Function

25.3.1.1 DNA Structure

DNA (deoxyribonucleic acid) is a biomolecule that comprises genes. **Genes** *encode the amino acid sequences for proteins.* The progress in understanding genes is leading to numerous biotechnology products and major research advances. At the same time, a flood of techniques have been and are being developed for working with nucleic acids in the laboratory. As with proteins, it is essential to understand the physical and chemical nature of nucleic acids when manipulating them in laboratory solutions.

DNA is a linear polymer of **deoxyribonucleotide** ("nucleotide") subunits. **Nucleotides** *are composed of a phosphate group, a five-carbon sugar (deoxyribose), and one of four nitrogenous bases (adenine, guanine, cytosine, or thymine)* (Figure 25.7). **DNA** *consists of nucleotides connected into strands by covalent **phosphodiester bonds***. The bonds link the phosphate group from one nucleotide to the sugar of another. The result is a "backbone" of alternating phosphates and sugars with the nitrogenous bases sticking out to the side.

The phosphate groups on each nucleotide readily give up H^+ ions; in fact, DNA is called an *organic acid.* Thus, in solution at neutral pH, negatively charged oxygens are exposed along the backbone of DNA strands. Nucleic acids dissolve readily in aqueous solutions because of the negatively charged backbone. The bases themselves are hydrophobic, but they face inward and are protected from the aqueous cellular environment in double-stranded DNA.

Chromosomal DNA exists in the cell as a *double helix* of two strands of DNA as shown in Figure 25.7. The bases of the two strands pair with one another so that a cytosine on one strand is always across from a guanine on the opposite strand, and an adenine is always across from a thymine. Cytosine and guanine are said to be *complementary*; adenine and thymine are also *complementary*. Double-stranded DNA is like a ladder with the bases forming the rungs on the inside and the phosphate–sugar backbone forming the outside rails. Imagine that the ladder is taken and twisted – this twist causes DNA to be called a double *helix*.

Complementary pairs of bases are held together by hydrogen bonds. Recall that hydrogen bonds are relatively weak compared to covalent bonds. Therefore, the phosphodiester bonds that link the nucleotide backbone are much stronger than the bonds holding the two strands together. This means that there are many conditions in the laboratory where the two DNA strands separate from one another, yet the strands themselves remain intact.

25.3.1.2 RNA Structure and Function

RNA (ribonucleic acid) *is a polymer of nucleotides similar to DNA except that the sugar is ribose and the nitrogenous base, uracil, is present in place of thymine.* There are various types of RNA with different structures and functions, including messenger RNA (mRNA), transfer RNA (tRNA), and ribosomal RNA (rRNA). **Messenger RNA** *carries information about the amino acid sequence of a protein from the nucleus to the cytoplasm.* **Transfer RNA** *shuttles amino acids to the ribosomes during protein synthesis.* **Ribosomal RNA** *is part of the structure of the ribosomes.* Other types of RNA play various roles in controlling gene expression.

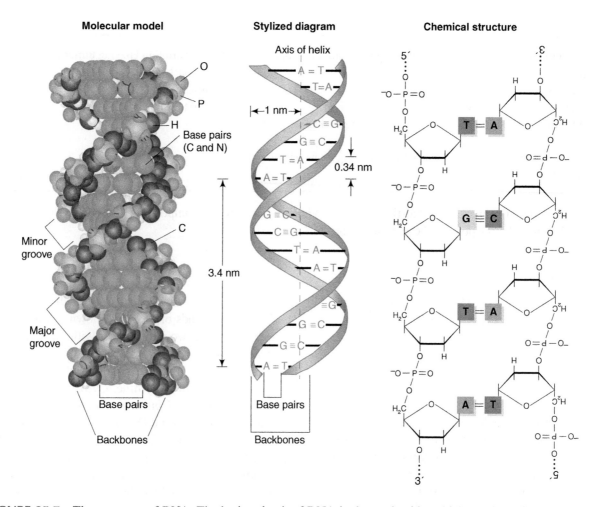

FIGURE 25.7 The structure of DNA. The basic subunit of DNA is the nucleotide, which consists of a sugar, a phosphate group, and a nitrogenous base. The subunits of DNA are connected by covalent phosphodiester bonds between the phosphate group on one nucleotide and the sugar on the next. In solution, the phosphate groups lose a H^+ and so have a negative charge. Chromosomal DNA is a double helix consisting of two strands of DNA held together by hydrogen bonds between complementary bases. Observe that three hydrogen bonds stabilize each G–C linkage, but only two hydrogen bonds stabilize each A–T linkage.

RNA is single-stranded and is shorter than chromosomal DNA. Although RNA is single-stranded, complementary bases *within* an RNA strand sometimes pair with one another. These bonds and other weak interactions cause RNA to fold into various complex conformations (Figure 25.8). As with proteins, the varied conformations of RNA molecules are important in controlling their functions.

25.3.2 LOSS OF NUCLEIC ACID STRUCTURE

Like proteins, nucleic acids can lose their normal structure in several ways:

1. **DNA and RNA can be denatured.** When speaking of DNA, **denaturation** *means that the two complementary strands separate*

from one another. The relatively weak hydrogen bonds that hold together the two strands can be disrupted by changes in pH (pH greater than 10.5 will denature DNA), high temperature, or addition of organic compounds such as **urea** and **formamide**. Low salt concentrations also promote denaturation. Denaturation of DNA is reversible under suitable conditions; **renaturation** *occurs when complementary base sequences reestablish hydrogen bonds.*

The base composition of double-stranded DNA determines *the temperature at which it denatures, its* **melting temperature** or T_m. Three hydrogen bonds form between guanine and cytosine base pairs, while only two form between adenine and thymine base

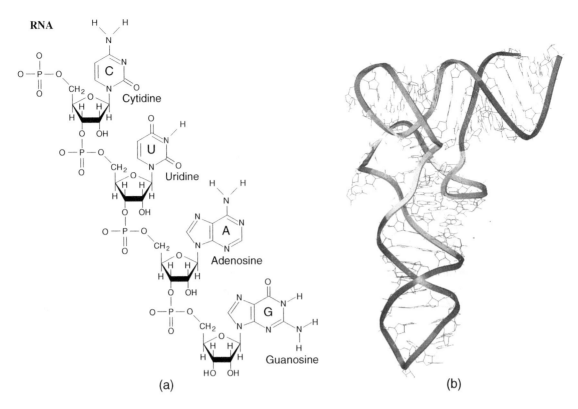

FIGURE 25.8 The structure of RNA. (a) Linear polymer of RNA. (b) Example of RNA with secondary structure (folding) due to multiple weak molecular interactions.

pairs. Therefore, the total bond strength between G–C pairs is stronger than between A–Ts. A DNA molecule that is comparatively rich in G–Cs denatures at a higher temperature than the one that is rich in A–Ts. At pH 7.0 and room temperature, DNA solutions are highly viscous. When DNA is denatured, it becomes noticeably less viscous, and it absorbs more UV light at a wavelength of 260 nm.

Local denaturation of DNA is an essential step in transcription (when RNA is copied from template DNA) and replication (when a cell divides and makes a copy of its DNA for the daughter cell). There are situations in the laboratory where denaturation is similarly desirable, and other situations where denaturation should be avoided.

When speaking of RNA, **denaturation** *means that the normal folding of the molecule is lost.*

2. **The covalent bonds that connect nucleotides into DNA and RNA strands can be broken by enzymes called *nucleases*.** Nucleases are found on human skin, in cells, and in cellular extracts.

DNA-degrading nucleases are relatively fragile, require Mg++ as a cofactor, and are commonly destroyed during routine DNA isolation procedures. In contrast, RNA nucleases are ubiquitous, difficult to destroy, and generally do not require metal ion cofactors to be active. The RNA nuclease, **RNase A**, can even survive periods of boiling or autoclaving. RNA nucleases frequently contaminate glassware and other laboratory items and make RNA difficult to manipulate in the laboratory. Agents and practices that help reduce the loss of RNA are shown in Table 25.7.

3. **Chromosomal DNA is a long, fragile molecule that is easily sheared into shorter lengths.** Vigorous vortexing, stirring, or passing a DNA solution through a small orifice, such as a hypodermic needle, will shear DNA.

TABLE 25.7

Commonly Cited Guidelines for Preventing the Loss of RNA

1. *Ribonucleases are released into solution when cells are disrupted.* Enzymes are proteins; therefore, strong protein-denaturing agents are often used to destroy these endogenous RNases when cells are disrupted for RNA isolation. Protein-denaturing agents used in RNA solutions include **6 M urea**, **SDS**, and **guanidinium salts** (**guanidine thiocyanate** and **guanidine hydrochloride**), sometimes in conjunction with a reducing agent such as β-mercaptoethanol. Tissue may also be rapidly frozen to inactivate endogenous ribonucleases. However, freezing temperatures disrupt cells and can cause the release of ribonucleases that become active when the material is thawed.

2. *People's hands are a major source of RNase contamination and gloves should be worn when working with RNA.* Once gloves have come in contact with a surface that was touched by skin (e.g., a pen, notebook, laboratory bench, and pipette), the gloves should be changed.

3. *Sterile technique should always be used when working with RNA because microorganisms on airborne dust particles are another major source of ribonuclease contamination.*

4. *Disposable sterile plasticware is seldom contaminated with RNases.* However, it is possible to treat plasticware to ensure the absence of RNases. Different types of plastics require different treatments. (For example, a table with methods for treating various plastics can be found in *Thermo Scientific Nalgene Bottles and Carboys Technical Brochure*, Thermo Scientific. 2012. http://tools.thermofisher.com/content/sfs/brochures/D01705.pdf.)

   Once packages of plasticware are opened, the contents must be protected from dust and contaminants.

5. *Non-disposable glassware can be baked at 300 °C for four hours to inactivate RNases.*

6. *The active site of RNase A contains a histidine amino acid; this active site can be destroyed by modifying the histidine.* The chemical ***DEPC (diethyl pyrocarbonate)*** in low concentrations can inactivate the histidine-binding site. DEPC is therefore frequently used to treat equipment and solutions that will be used for RNA work. Some points regarding the use of DEPC are as follows:

   a. DEPC is thought to be carcinogenic and should be used in a hood and handled while wearing gloves and suitable lab attire. See Chapter 9, p. 191 for more information about handling DEPC safely.

   b. DEPC can inhibit enzymatic reactions and can interact with nucleic acids. After DEPC inactivates RNases, it therefore needs to be completely degraded, usually by autoclaving, which breaks the DEPC down to carbon dioxide and ethanol. Recommendations for autoclaving time vary between 15 minutes per liter to 1 hour.

   c. To treat solutions, add 0.05%–0.1% DEPC to the solution in a glass container and mix well. The solution can be incubated overnight at room temperature or two hours at 37°C and then is autoclaved.

   d. Do not use DEPC to treat solutions with Tris buffer because Tris reacts with DEPC.

   e. DEPC will dissolve some plastic pipettes and so glass pipettes should be used. DEPC has been reported to be incompatible with polycarbonate and polystyrene containers.

   f. Glassware can be soaked in DEPC to inactivate RNases. Recommendations for soaking time vary from 30 to 60 minutes to overnight. The glassware is then autoclaved for 30 minutes to inactivate the DEPC.

   g. DEPC can be used to treat water by mixing 0.1% DEPC with the water and allowing the mixture to sit at least 6 hours at room temperature. However, many water purification systems effectively remove RNases and the use of DEPC for such purified water is not required. Check with the manufacturer of your water treatment system.

7. *It is useful to have a separate laboratory space and separate equipment for working with RNA.* This is particularly important if DNA work is also performed in the laboratory because it is common to intentionally add RNase to DNA solutions to destroy unwanted, contaminating RNA. Once RNase is present in the laboratory, it is difficult to eliminate from pipettes, centrifuges, pH electrodes, and other devices.

8. *Products (such as water, tips, tubes, buffers, enzymes, and bovine serum albumin) are all potential sources of RNase contamination.* Manufacturers produce many products that have been treated to remove ribonucleases and certify them to be RNase-free.

9. *Enzymes that are ribonuclease inhibitors can be purchased from molecular biology suppliers.* These inhibitors bind to and inactivate some classes of RNase.

10. *Clean surfaces.* Commercial products are available to remove RNases from benchtops and other exposed surfaces.

(Continued)

TABLE 25.7 (*Continued*)

Commonly Cited Guidelines for Preventing the Loss of RNA

11. *Other proprietary products.* Manufacturers make proprietary products for use when working with RNA. These include products to detect RNases and products that take the place of DEPC.
12. *Cold.* Cold temperatures can protect RNA, as described in the case study "Cold Chain Logistics and mRNA Vaccines" below.

Primary sources of information for this table are "Technical Bulletin 159: Working with RNA," Thermo Fisher. www.thermofisher.com/us/en/home/life-science/dna-rna-purification-analysis/rna-extraction/working-with-rna.html.

"*Diethyl pyrocarbonate:* Product Information," Sigma-Aldrich. www.sigmaaldrich.com/content/dam/sigma-aldrich/docs/Sigma/Product_Information_Sheet/d5758pis.pdf.

Case Study: Cold Chain Logistics and mRNA Vaccines

The COVID-19 pandemic has introduced the general public to a number of topics that were previously only on the minds of technical professionals. In the fall of 2020, the term "cold chain logistics" began to appear in news reports and economic analyses. At the same time, orders rose for specialized – expensive – deep-freeze units that can maintain temperatures of −80°C. One newspaper headline said, "Covid-19 Vaccine Race Turns Deep Freezers into a Hot Commodity" (Hopkins, Jared. "Covid-19 Vaccine Race Turns Deep Freezers into a Hot Commodity." *WSJ*, 4 September 2020, www.wsj.com/articles/covid-19-vaccine-race-turns-deep-freezers-into-a-hot-commodity-11599217201.) What is this about? The answer is that the first two vaccines against the virus that causes COVID-19 are mRNA vaccines (as introduced in Section 1.2.4 in this text). These two vaccines involve injecting mRNA, encapsulated in a lipid shell, into recipients. The recipients' cells use the mRNA as a template to produce a viral protein, which initiates an immune response. Other, more conventional, vaccine strategies use viral proteins, viral DNA, or inactivated viruses to initiate an immune response. mRNA vaccines have significant advantages over these other forms of vaccine in their speed of development and ease of manufacturing.

However, as biotechnologists have known for decades, because of its chemical structure, RNA is a particularly fragile biological molecule. In the body, mRNA performs its work and then rapidly degrades. In contrast, genomic DNA exists stably in the nucleus for the life of the cell. RNA is particularly sensitive to degradation by heat. Manufacturers of one of the mRNA vaccines therefore use deep cold, around −80°C, to protect their fragile mRNA vaccine products – this is about the temperature of the South Pole on a winter day. Such low temperatures can be achieved, but require expensive freezers that are seldom present in hospitals or clinics. Therefore, the challenge is to manufacture, distribute, and rapidly dispense billions of doses of vaccine to sites around the world, all at ultra-low temperatures much colder than a normal freezer can provide. Hospitals and public health facilities are investing in new freezers to maintain the fragile, valuable products. One company, Pfizer, has devised specialized containers, of the size of a suitcase, that can be filled with dry ice to attain temporary cold temperatures during shipping, and that contain temperature sensors for continuous monitoring. But someone has to make all this dry ice and a shortage had to be prevented. Moreover, dry ice emits carbon dioxide gas when it melts, making it hazardous on long airplane flights. Shipping companies must use sophisticated methods to handle dry ice, track the temperature of the vaccines at all times, and verify the location of their airplanes and trucks to ensure rapid delivery at a suitable temperature. In yet another offshoot of this problem, there were not enough pharmaceutical-grade glass vials that can handle these low temperatures. At least one glass manufacturer was reported to have built new facilities to manufacture appropriate vials. Even with all these efforts, mRNA vaccines may be impractical in some regions. In some countries, any type of medical product requiring even ordinary refrigeration is impractical. Thus, issues of working with fragile RNA molecules, as we introduce in this chapter, have become an international priority that raises complex logistic and social issues.

25.3.3 Laboratory Solutions for Nucleic Acids

25.3.3.1 Introduction

The structure and function of nucleic acids is not as variable as is that of proteins. For example, there are only four different nucleotides comprising DNA, but there are 20 different amino acids commonly found in proteins. There are, however, a multitude of different tasks performed in bioscience laboratories that involve DNA, and there are numerous techniques used to manipulate nucleic acids; therefore, there are many solutions used for DNA work. A summary of the basic classes of agents found in DNA solutions is provided in Table 25.8, and then these agents are discussed in more detail below.

25.3.3.2 Nucleic Acids and pH

As with proteins, many solutions that are used with DNA and RNA are buffered to help preserve the normal structure and function of the nucleic acids.

There are times when pH is manipulated to affect the structure of nucleic acids. For example:

1. **Manipulation of pH is a technique used for controlling DNA denaturation and renaturation in certain procedures.**
2. **Treatment with mild acids, such as 5% trichloroacetic acid (TCA) at or below room temperature, will precipitate nucleic acids.**
3. **Prolonged exposure to dilute acid at or above room temperature, or briefer exposure to more concentrated acid (e.g., 15 minutes in 1 M HCl at 100°C) will cause depurination of DNA and RNA (removal of most of the purine bases, adenine and guanine).** The pyrimidines (cytosine, uracil, and thymine) in these nucleic acids can also be removed by exposure to more concentrated acids or by longer periods of exposure.
4. **Strong acid treatments will break the phosphodiester bonds that join nucleotides.**
5. **Scientists can exploit the fact that DNA and RNA react differently under mildly alkaline conditions.** For example, 0.3 M KOH at 37°C for 1 hour will cleave the phosphodiester bonds in RNA, but the phosphodiester bonds of DNA will remain intact under these conditions.

25.3.3.3 Salts

Like proteins, nucleic acids are sensitive to the ionic strength of their solution. Consider the following three important cases:

1. **Denaturation of double-stranded DNA is controlled in the laboratory by the ionic strength of the solution in conjunction with its temperature and pH.** Consider DNA dissolved in a solution of low ionic strength (Figure 25.9a). The negatively charged phosphate groups cause the two DNA strands to repel one another. In contrast, in a solution of moderate ionic strength, such as 0.4 M NaCl, Na$^+$ ions are attracted to negatively charged phosphate groups. The negative charges of the phosphate groups are thus neutralized, so the two DNA strands do not repel one another (Figure 25.9b). Double-stranded DNA, therefore, is less stable and is more easily denatured in a solution with 0.01 M salt than it is in a solution of 0.4 M salt. As a result, DNA denatures at a lower temperature in 0.01 M NaCl than it does in 0.4 M NaCl. In solutions with salt concentrations between 0.01 M NaCl and 0.4 M NaCl, the likelihood of denaturation is intermediate.

2. **Hybridization** *(binding)* **of single-stranded DNA with short strands of complementary DNA or RNA is affected by the ionic strength of the solution. Stringency** *refers to the reaction conditions used when single-stranded, complementary nucleic acids are allowed to hybridize.* At high stringency, binding occurs only between strands with perfect complementarity. (Perfect complementarity means that every guanine is base-paired with a cytosine, and every adenine is base-paired with a thymine.) At lower stringency, there can be some mismatch of bases across the strands and hybridization still occurs. There are situations in the laboratory where high stringency is required and other situations where lower stringency is desirable.

TABLE 25.8

Classes of Agents Used in Laboratory Solutions for Nucleic Acids

Buffers and agents that affect pH	Nuclease inhibitors
Salts	Metal chelators
Organic solvents	Coprecipitants
Nucleases	Antimicrobial agents

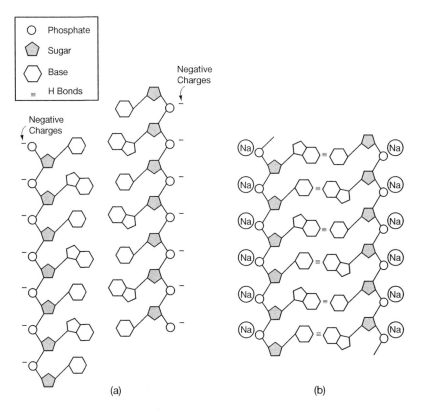

FIGURE 25.9 The effect of ionic strength on nucleic acid denaturation and renaturation. (a) In a solution of low ionic strength, the two DNA strands are negatively charged and repel one another, making denaturation more likely. (b) In a solution of moderate ionic strength, positive ions shield negatively charged phosphates and complementary base-pairing will occur.

Strands come together more easily when the salt concentration is high and the temperature is relatively low. Under these conditions, hybridization can occur even if there are some base pair mismatches; the conditions have lower stringency (Figure 25.10). In contrast, when the temperature is higher and the salt concentration is lowered, the conditions are more stringent.

3. **The ionic strength of the solution affects interactions between proteins and DNA.** Protein–DNA interactions are of critical importance in the cell. For example, such interactions are involved in transcription and replication. Electrostatic interactions between negatively charged phosphates on DNA and positively charged amino acids promote protein–DNA binding. However, electrostatic interactions are not specific. This means that electrostatic interactions do not promote the ability of an enzyme to find and bind to a *specific sequence* of DNA. Electrostatic interactions are affected by the concentration of salt in the solution. As salt increases, the strength of electrostatic interactions decreases. For example:

- **Chromatin** *is the native form of DNA, which is composed of DNA bound to many different proteins.* In the laboratory, 0.01 M NaCl chromatin is stable, but in 1 M NaCl, it dissociates almost completely to yield free DNA molecules and proteins. This effect is thought to be due to the loss of electrostatic interactions that normally stabilize chromatin. When NaCl is added, the salt ions shield the various charged sites and disrupt the electrostatic bonding.
- *Restriction enzymes that cut DNA at specific sites require the correct concentration of salts to cut DNA selectively.* At lower salt concentrations, electrostatic interactions increase, allowing the restriction enzyme to nonspecifically bind to and cut the DNA. A restriction enzyme is said to exhibit **star activity** *when it "mistakenly" cleaves DNA sequences that are not its proper specific target.* Star activity is an example of nonspecific protein–DNA interaction. Reduced salt concentration is one of several factors that promote star activity.

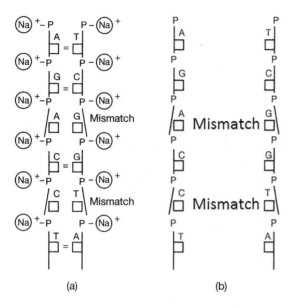

FIGURE 25.10 The effect of ionic strength on stringency. (a) At low temperatures and high ionic strength, some base pair mismatches can be tolerated. In this drawing, many Na⁺ ions are present (high ionic strength). These ions shield the negative charges on the phosphate groups, so the two DNA strands do not tend to repel one another. This makes the conditions less stringent, and so bonds form between matched base pairs. (b) At higher temperatures and lower ionic strength, mismatches are not tolerated for hybridization. The mismatched segments will remain separate.

Specific interactions between proteins and particular base sequences of DNA are not electrostatic. Unlike electrostatic interactions, specific protein–DNA interactions are frequently stable in higher salt concentrations.

25.3.3.4 Nucleases and Nuclease Inhibitors

When working with proteins in laboratory solutions, it is common (and sensible) to avoid conditions that will damage proteins. In contrast, when working with nucleic acids, the first agents used are those that denature, precipitate, and destroy proteins. This is because nucleases (which are proteins) are released when cells are disrupted. These endogenous nucleases must be destroyed, or the nucleic acids will be degraded. **Proteinase K** is an enzyme that is used to help degrade nucleases and histones (proteins associated with DNA in the chromosome). Proteinase K is active over a wide range of pH, salt, detergent, and temperature conditions. Chelators are also used as nuclease inhibitors because many nucleases require a metal ion cofactor for activity. Chelators bind and remove metal ions from solutions.

25.3.3.5 Nucleases and Chelators

While nucleases are often avoided, they can be harnessed to manipulate nucleic acids in important ways. For example, *DNA may be intentionally cleaved using nucleases called* **restriction endonucleases**. Restriction endonucleases recognize and cleave DNA phosphodiester bonds at specific sites. Another example is the use of the enzyme **RNase A** when isolating DNA to intentionally degrade contaminating RNA.

Most nucleases require Mg^{++} as a cofactor. The levels of chelators and metal ions, therefore, are controlled in nucleic acid solutions. For example, the chelator, EDTA is typically added to nucleic acid solutions during storage to reduce unwanted nuclease activity. Mg^{++} is added when nuclease activity is desired.

25.3.3.6 Organic Solvents

Organic solvents have commonly been used to isolate DNA and RNA from other components of cellular extracts. A standard extraction procedure involves addition of the organic solvents, **phenol** and **chloroform**, which simultaneously disrupt cell membranes and denature proteins (including endogenous nucleases). Phenol and chloroform are not water-soluble, so the mixture is centrifuged to separate it into an organic phenol/chloroform phase and an aqueous phase. Denatured proteins move into the organic phase or the interface between the two phases. The nucleic acids remain in the aqueous phase. A low concentration of **isoamyl alcohol** is usually added to aid in clean separation of the organic and aqueous phases. Phenol is relatively unstable in storage, so **8-hydroxyquinoline** is usually added to phenol as an antioxidant. Note that phenol is hazardous and must be handled with great care (see Chapter 9, p. 200). There are now commonly used commercial products that isolate nucleic acids without the use of these organic solvents.

25.3.3.7 Precipitants

Ethanol plays an important role in working with nucleic acids because it precipitates DNA and RNA, physically isolating them from a solution containing other components. Nucleic acids do not lose their structural or functional integrity when isolated with phenol/chloroform and/or ethanol.

Isopropanol is sometimes substituted for ethanol when the final solution volume must be minimized because less isopropanol is needed than ethanol to precipitate DNA.

Salt is added when nucleic acids are ethanol-precipitated because relatively high concentrations of monovalent cations (0.1–0.5 M) promote aggregation and precipitation of nucleic acid molecules. Various salts are used in different situations; however, sodium acetate is most common.

Case Study: Isolating Plasmids: The Roles of Solution Components

Consider as an example the solutions used in a common procedure: the isolation of plasmids from bacteria. **Plasmids** are small circular molecules of DNA found in bacteria that exist separate from the bacterial chromosome. In nature, plasmids sometimes carry genes that are responsible for the spread of antibiotic resistance among bacteria.

In the laboratory, scientists can modify plasmids and use them to carry genes of interest into host bacteria. Once inside the bacteria, the plasmids multiply as the bacteria divide. The plasmids can be isolated from the host bacteria using various techniques, such as the "plasmid minipreparation procedure." The conventional minipreparation procedure takes advantage of the small, coiled nature of the plasmids to separate them from the larger chromosomal DNA. The minipreparation procedure will be discussed. As we go through each step, we will explain the roles various solution components are thought to play in the separation.

Minipreparation Procedure to Isolate Plasmids from Host Bacteria

The Solutions

LB (Luria broth) (1 L)	**GTE buffer**
10 g Tryptone	25 mM Tris–HCl, pH 8.0
5 g Yeast extract	10 mM EDTA
5 g NaCl	50 mM glucose
Sterilize by autoclaving	
Potassium acetate/acetic acid	**SDS/NaOH**
60 mL of 5 M potassium acetate	1% SDS
11.5 mL of glacial acetic acid	0.2 N NaOH
28.5 mL of H_2O	
TE buffer	
10 mM Tris–HCl, pH 7.5–8.0	
1 mM EDTA	

The Procedure

1. *Grow a plasmid-containing bacterial culture overnight in LB medium.* LB medium contains nutrients for bacterial growth.
2. *Remove 1.5 mL of the culture and spin down the cells in a microfuge (a small centrifuge).*
3. *Discard the supernatant (liquid) and completely resuspend the cells in 100 μL of ice-cold GTE buffer.* The buffer is kept cold to inhibit nucleases released during the procedure. GTE buffer contains Tris to maintain the proper pH (pH 8.0), EDTA to bind divalent cations (Mg^{++} and Ca^{++}) in the lipid bilayer (thus weakening the cell membrane), and glucose, which may act as an osmotic support (osmosis is considered in Chapter 31) or may help to prevent cells from clumping.
4. *Add 200 μL of SDS/NaOH and gently mix.* The bacterial cells are lysed by the SDS, which solubilizes the membrane lipids and cellular proteins. SDS also denatures cellular proteins. The NaOH denatures both the chromosomal and plasmid DNA. The strands of the circular plasmid DNA remain intertwined. The suspension is mixed gently to avoid shearing the chromosomal DNA.

(Continued)

Case Study (*Continued*): Isolating Plasmids: The Roles of Solution Components

5. *Set on ice for 10 minutes.* This "resting step" allows time for the SDS/NaOH to contact all the cells and the processes in step 4 to go to completion.

6. *Add 150 µL of cold potassium acetate/glacial acetic acid and mix gently.* A white precipitate will appear. The acetic acid neutralizes the sodium hydroxide added in step 4, allowing the DNA to renature. The potassium acetate causes the SDS to precipitate, along with associated proteins and lipids. The large chromosomal DNA strands renature partially into a tangled web that precipitates along with the SDS complex. In contrast, the small, coiled renatured plasmids remain suspended. In this step, therefore, they are separated from chromosomal DNA, many proteins, and the detergent.

7. *Centrifuge the mixture; save the supernatant.* The discarded precipitate (the material at the bottom of the centrifuge tube) contains cellular material, linear DNA, protein, and detergent, whereas the supernatant (the liquid in the top of the centrifuge tube) contains the suspended plasmids.

8. *Remove 400 µL of the supernatant, add an equal volume of isopropanol, and mix.* Isopropanol, similar to ethanol, precipitates DNA. Isopropanol is used here because a smaller volume of isopropanol is required than ethanol. The alcohol quickly precipitates DNA and more slowly precipitates proteins. This step is therefore done quickly.

9. *Centrifuge and save the pellet.* The alcohol-precipitated plasmid DNA is in the pellet.

10. *Wash the pellet in 200 µL of 100% ethanol.* Although isopropanol precipitated the DNA in the preceding step, an ethanol wash helps remove extra salts and other unwanted contamination.

11. *Centrifuge and save the pellet.*

12. *Resuspend the pellet in 15 µL of Tris/EDTA (TE) buffer.* TE is often used for short-term storage of DNA. Tris maintains the pH, and EDTA chelates divalent cations to inhibit nuclease activity.

(This case study is based primarily on information in Micklos, David, and Greg Freyer. *DNA Science: A First Course, Second Edition.* Cold Spring Harbor Laboratory Press, 2003.)

25.3.3.8 Coprecipitants

Alcohol precipitation is ineffective if only small amounts of nucleic acids are present in a solution. Inert coprecipitants, such as **yeast tRNA** or **glycogen**, are therefore sometimes added to solutions to help precipitate nucleic acids.

25.3.3.9 Antimicrobial Agents

Antimicrobial agents are often added to DNA solutions to inhibit the growth of microorganisms during storage.

25.3.4 STORING DNA AND RNA

DNA is relatively stable as long as nucleases are absent. Some sources recommend storing genomic DNA, plasmids, and various small DNA molecules at 4°C for short periods. For long-term storage, DNA may be more stable if stored frozen. EDTA is often added to DNA storage solutions because it chelates magnesium ions and inhibits endonucleases.

RNA requires special care during storage because ribonucleases are ubiquitous and difficult to destroy. Recommendations include storing RNA for short periods at −20°C in RNase-free H_2O (with 0.1 mM EDTA) or TE buffer (10 mM Tris; 1 mM EDTA). The EDTA must be RNase-free. For longer-term storage (up to a year), storage in a −80°C freezer is recommended. Another suggestion is to store RNA as a precipitate in a NH_4OAc/ethanol mixture at −80°C. RNA stored in this way should be aliquoted to avoid multiple freeze–thaw cycles and to reduce the risk of introducing RNases into the stock.

25.3.5 SUMMARY OF NUCLEIC ACID SOLUTION COMPONENTS

Table 25.9 summarizes a number of solution components in terms of their functions in laboratory procedures.

TABLE 25.9
Typical Components of Nucleic Acid Solutions

Procedure	Components of Solutions
Isolation on and Purification of DNA from Cells	
Before DNA can be studied, sequenced, or recombined, it must be isolated and purified.	
Degrade the cell wall	The enzyme lysozyme (when working with bacteria)
Remove membranes	Detergents: SDS, sarkosyl
Remove proteins and RNA	Enzymes: Proteinase K, RNase A
Denature and remove proteins	Phenol/chloroform extraction
Preferentially precipitate DNA	Ethanol in high salt
	Isopropanol with salt
Concentrate DNA in aqueous solution	*sec*-Butanol
Purify specific types of nucleic acid	CsCl density gradient
	Commercial spin columns, glass beads
	Ion exchange resins
Preferentially degrade chromosomal DNA to isolate plasmids	NaOH+SDS
Nucleic Acid Storage Buffers	
Protects Nucleic Acids During Storage.	
Maintain proper pH	Buffers, commonly Tris
Chelate metals (inhibit nucleases)	EDTA
Maintain ionic strength	NaCl
Enzymatic Digestion and Modification	
Enzymes are used to digest, recombine, and modify DNA. These enzymes require proper solution conditions.	
Maintain proper pH	Buffers, commonly Tris
Prevent the loss of enzyme due to adsorption to container	Inert carriers, commonly BSA
Enzyme cofactors	Divalent cations: Mg^{++}, Mn^{++}, Fe^{++}
Stabilize protein and lower freezing temperature	Glycerol
Prevent aggregation and sticking	Low levels of detergents such as Triton X-100
Inhibit proteases	EDTA
Electrophoresis	
DNA molecules can be separated by size during migration in an electrical field.	
Gel matrix	Agarose, acrylamide
Running buffers	Tris–borate–EDTA, Tris–acetate–EDTA
Denaturing gels (lower the melting temperature of the double-stranded DNA)	Formamide, urea
Tracking dyes to visualize the migration of DNA	Xylene cyanol, bromophenol blue
Dyes to detect the presence of DNA	Ethidium bromide, methylene blue
Transformation/Transfection of Purified DNA into Cells	
Introduction of foreign DNA into a cell. In bacteria, it is necessary to prepare the cell; in eukaryotic cells, it is necessary to prepare the DNA.	
Preparation of bacterial cells to be "competent"	$CaCl_2$, RbCl, DMSO
Uptake of precipitated DNA by eukaryotic cells	$CaPO_4$ precipitation of DNA
Fusion of eukaryotic cells with other membranes	Liposomes, lipid-encased DNA
Permeabilization of eukaryotic membranes	DEAE-dextran, DMSO, glycerol
Hybridization	
Process in which denatured DNA binds (hybridizes) with complementary strands of DNA or RNA.	
Separation of DNA strands	NaOH, heat
Hybridization solution components:	
"Blocking agents" to fill nonspecific binding sites	BSA, nonfat dry milk, casein, gelatin
"Crowding agents" – inert compounds that effectively reduce the volume of the solution	Denhardt's reagent: Ficoll, PVP, PEG
Nonspecific nucleic acid to reduce background by binding nonspecific sites	Salmon sperm DNA, tRNA

Example Problem 25.1

Southern blotting and probe hybridization is a method in which DNA fragments separated from one another by electrophoresis are adhered to a plastic or nylon membrane and are then exposed to a single-stranded nucleic acid probe. The probe will bind to any single-stranded fragments adhered to the membrane that have a nucleotide sequence complementary to the probe. Before the DNA is adhered to the membrane, it is soaked in a solution containing NaOH. What is the function of the NaOH in this soaking solution?

Answer

In order for the probe to bind to DNA fragments, the DNA fragments (and the probe) must be single-stranded. The complementary bases will then be available for hybridization. The NaOH is added to the solution to denature the double-stranded DNA.

25.4 CASE STUDIES TO ILLUSTRATE CONCEPTS IN THIS CHAPTER

Let's consider several case studies that illustrate the ideas introduced in this chapter.

Case Study: The Commonplace Ingredients in Two State-of-the-Art Biological Solutions

Early in 2021, two COVID-19 vaccines catapulted into the pandemic scene, both of them based on the biological molecule, mRNA. Using mRNA in a vaccine is a concept that had been in development for decades, but these two COVID-19 vaccines were the first widespread application of the method. One of the vaccines is manufactured by the company Moderna, and the other by the company Pfizer–BioNTech.

The mRNA in the vaccines codes for a protein that is part of the SARS-CoV-2 virus and that the virus requires for its invasion of human cells. After an mRNA vaccine is injected into a person, the recipient's cells synthesize the viral protein encoded by the mRNA. The cells take the newly synthesized viral protein and place it on the membrane that surrounds each cell. The recipient's immune system detects the viral protein and recognizes that the protein does not belong in the body. The immune system, over a period of days, builds an immune response against the viral protein and thus protects the vaccinated person against the illness, COVID-19.

The ingredients in the two mRNA vaccines are publicly available and are summarized in Table 25.10. Observe that these vaccines are biological solutions, as we have defined the term in this unit. They are aqueous solutions that contain and support the activity of a biological molecule, in this case, mRNA. The critical ingredient in each vaccine solution is mRNA. But mRNA is actually just a small part of the product. In the Moderna vaccine, the amount of mRNA is 100 μg, and in the Pfizer–BioNTech vaccine, it is just 30 μg.

All the rest of the vaccine ingredients are there to contain, stabilize, and support the activity of the mRNA. When we are talking about pharmaceutical products, these extra ingredients have a special name; they are called **excipients**. So, what are the excipients in the two vaccines? The first is a category that we have not yet talked about in this chapter. This category is lipids, which are fatty molecules. Each vaccine contains four types of lipids. The purpose of the lipids is to encapsulate and protect the mRNA as it journeys from a syringe into human cells. Another ingredient in the mRNA vaccines is sugar. Sugar plays various roles in biological solutions, some of which will be discussed in Chapter 31. Sugar is probably present in the vaccines to protect the mRNA from damage during freezing.

(Continued)

**Case Study (*Continued*): The Commonplace Ingredients
in Two State-of-the-Art Biological Solutions**

A third and major component of these vaccines is a buffer, which, as we have seen throughout this unit, is the base ingredient in most common biological solutions. Pfizer–BioNTech uses PBS, an extremely common buffer introduced in Chapter 23. In the vaccine, PBS is composed of sodium phosphate dihydrate, potassium chloride, monobasic potassium phosphate, and sodium chloride. The Moderna vaccine uses the other common biological buffer introduced in Chapter 23, Tris. In addition, Moderna uses a second buffer, sodium acetate combined with acetic acid. Thus, a major part of both vaccines is a very commonplace ingredient, that is, a conventional buffer. Of course, we must remember that all the vaccine ingredients are dissolved in water that was purified to pharmaceutical specifications, as discussed in Chapter 24. It is thus evident that the creators of these vaccines successfully introduced a new and exciting method of vaccine production based on an encapsulated biological molecule, mRNA. But they also built on existing methods for creating buffered biological solutions – methods that have been developed and optimized over many years in thousands of laboratories around the world.

TABLE 25.10

Ingredients of the Moderna and Pfizer–BioNTech mRNA Vaccines

	Moderna	Pfizer–BioNTech
mRNA (Active Ingredient)	100 μg	30 μg
Lipids (To Form the Capsule Around mRNA)		
ALC-0315	No	Yes
ALC-0159 (a polyethylene glycol derivative)	No	Yes
DMPC	No	Yes
Cholesterol	Yes	Yes
SM-102	Yes	No
PEG2000-DMG (a polyethylene glycol derivative)	Yes	No
DSPC	Yes	No
Buffer Ingredients		
Potassium chloride	No	Yes
Monobasic potassium phosphate	No	Yes
Sodium chloride	No	Yes
Dibasic sodium phosphate dihydrate	No	Yes
Tromethamine (Tris)	Yes	No
Tromethamine hydrochloride (Tris)	Yes	No
Acetic acid	Yes	No
Sodium acetate	Yes	No
Sugar		
Sucrose	Yes	Yes
Final Volume in Pharmaceutical-Grade Water (WFI)	0.5 mL	0.3 mL

Primary Sources:

Food and Drug Administration. Moderna COVID-19 Vaccine EUA Letter of Authorization. 25 February 2021. www.fda.gov/media/144636/download.

Food and Drug Administration. Pfizer–BioNTech COVID-19 Vaccine EUA Letter of Authorization. 25 February 2021. www.fda.gov/media/144412/download.

Neubert, Jonas. "Personal Blog: Exploring the Supply Chain of the Pfizer/BioNTech and Moderna COVID-19 Vaccines." 10 January 2021. blog.jonasneubert.com/2021/01/10/exploring-the-supply-chain-of-the-pfizer-biontech-and-moderna-covid-19-vaccines/.

Case Study: Optimizing and Avoiding Problems with Solutions

The early phases of research often include finding optimal solution components for a particular application, sometimes by trying various combinations of agents until the best results are obtained. It is essential to understand the roles of the various components when optimizing solution recipes.

A study that looked at the effects of different laboratory solutions on the results of an experiment was reported by Favre and Rudin. (Favre, Nicolas, and Werner Rudin. "Salt-Dependent Performance Variation of DNA Polymerases in Co-Amplification PCR." *BioTechniques*, vol. 21, no. 1, 1996, pp. 28–30. doi:10.2144/96211bm04.) These authors were using a technique called reverse transcriptase PCR to amplify and detect specific mRNAs. The scientists were attempting to amplify two different mRNAs simultaneously in the same reaction tube. They tested reverse transcriptase enzymes from four manufacturers, in each case using the buffer solution provided by the manufacturers. They found that the enzymes/buffers from all four manufacturers were effective when the two different mRNAs were amplified individually in separate tubes. When they attempted to amplify both types of mRNA together in the same tube, however, only one manufacturer's kit was effective. Further study showed that the effective kit contained buffer with 200 mM Tris–HCl, whereas the other manufacturers provided buffer with 100 mM Tris–HCl. They concluded that the buffer concentration was critical in the amplification and detection of two mRNAs simultaneously.

Case Study: The Importance of Lysis Buffer Components

Another study that investigated the effects of buffer components was reported by Ignatoski and Verderame. (Ignatoski, Kathleen M. Woods, and Michael F. Verderame. "Lysis Buffer Composition Dramatically Affects Extraction of Phosphotyrosine-Containing Proteins." *BioTechniques*, vol. 20, no. 5, 1996, pp. 794–96. doi:10.2144/96205bm13.) In order to understand this example, it is necessary to understand the role of lysis buffers. All cells are surrounded by a cell membrane that maintains the structural integrity of the cell and also controls the interactions of the cell with its external environment. Cell membranes are composed of **phospholipids** with embedded proteins and glycoproteins. **Phospholipids** *have a negatively charged, hydrophilic phosphate group and "tails" that are hydrophobic hydrocarbon chains.* The hydrophilic "heads" face out to the cytoplasm or the exterior of the cell, whereas the hydrophobic tails face one another.

Cell membranes must be broken (lysed) to get access to the molecules inside. *Solutions whose primary function is to lyse these structures are called* **lysis buffers**. Membranes are usually lysed by detergents that, like the membranes, have both hydrophilic and hydrophobic portions. Mild detergents are preferred, such as Triton X-100 or sarkosyl (N-lauroylsarcosine).

Ignatoski and Verderame were comparing proteins in cultured cells that were transformed with an oncogene (a cancer-causing gene) with proteins in cultured cells that were not transformed. The cells were broken open with lysis buffer, releasing their proteins into solution. The released solubilized proteins were analyzed. The scientists discovered that the components of the lysis buffer had a significant effect on which proteins they detected and therefore, their conclusions regarding differences between transformed and non-transformed cells.

Ignatoski and Verderame compared the effects of three different lysis buffers. All three buffers included protease inhibitors and also the ingredients shown in Table 25.11.

The authors observed that the three lysis buffers appear to "be solubilizing different subsets of cellular proteins," and they postulated that:

1. **The type and concentration of detergent could have an effect on which proteins are detected.** Ionic detergents, such as Na^+ deoxycholate, disrupt the cell membrane more completely than do nonionic detergents, such as NP-40 or Triton X-100. Thus, more proteins might have been released into solution when ionic detergents were used.

(Continued)

Case Study (*Continued*): The Importance of Lysis Buffer Components

2. **The ionic strength of the buffer plays a critical role in protein solubilization.** Proteins need to be soluble in aqueous solutions to be detected. High salt concentrations might disrupt interactions of proteins with insoluble cellular components, thereby promoting protein solubility. The three buffers contained different salts, which may have caused the proteins solubilized in each buffer to be different.

3. **pH can affect detergent solubility, especially of Na⁺ deoxycholate, thereby changing which proteins are solubilized.** The pH of Buffer C was different from the other two buffers; therefore, the proteins that were detected may have been affected.

4. **The buffer effectiveness may have been affected by the presence of phosphate.** Buffer B was the only buffer containing phosphate, and this buffer resulted in the solubilization of a different subset of proteins than the other buffers.

The authors concluded that the lysis buffer components can play a role in affecting experimental results and therefore "several different lysis buffers . . . should be tested to obtain optimal experimental conditions."

It is clear from this example that experimental reproducibility requires careful attention to the solutions that are used. If various researchers use subtly different lysis buffers, they might reach dramatically different conclusions about the proteins involved in cancer. Similarly, if an individual researcher does not carefully control the composition of various batches of his or her buffers, then experimental results might be frustratingly variable.

TABLE 25.11

Lysis Buffer Ingredients

Lysis Buffer A	Lysis Buffer B	Lysis Buffer C
50 mM Tris–HCl, pH 7.5	10 mM Tris–HCl, pH 7.4	30 mM Tris–HCl pH 6.8
150 mM NaCl	50 mM NaCl	150 mM NaCl
1% Nonidet P-40 (detergent)	50 mM NaF	1% NP-40 (detergent)
0.25% Na⁺ deoxycholate (detergent)	1% Triton X-100 (detergent)	0.5% Na⁺ deoxycholate (detergent)
10 mg/mL BSA	5 mM EDTA	0.1% SDS (detergent)
	150 mM Na_3VO_4 (inhibits phosphatase enzymes)	
	30 mM $Na_4P_2O_7$	

As the preceding examples illustrate, the composition of solutions has a critical effect on results. Once solution recipes have been developed for a particular task, it is therefore essential to follow these recipes exactly as written (or to control and document alterations in solution components carefully).

In a production setting, optimizing a solution requires consideration not only of the effectiveness of the solution, but also of the cost and availability of the components. Large quantities of raw materials will be needed when the product goes into mass production. Therefore, raw materials should be chosen that are as inexpensive as possible and yet meet the requirements of the process. It is necessary to find suppliers of the solution components who can provide sufficient raw materials of consistently high quality. In a production setting, once the solution recipes have been established and suppliers have been found, it is particularly important not to modify any aspect of the solution without careful consideration and control of the changes.

Practice Problems: DNA

1. Suggest the possible purpose of the components of the following commonly used solutions.
   a. **TBE buffer – electrophoresis buffer**
      0.089 M Tris base (pH 8.3)
      0.089 M boric acid
      0.002 M EDTA
   b. **TE buffer – storage**
      10 mM Tris–HCl (pH 8.0)
      1 mM EDTA
   c. **Hybridization buffer**
      40 mM PIPES buffer (pH 6.4)
      1 mM EDTA
      0.4 M NaCl
      80% Formamide.
2. Examine the recipe for hybridization buffer, shown above. Comment on the salt concentration in terms of stringency.
3. (Modified from Freifelder, David. *Principles of Physical Chemistry with Applications to the Biological Sciences*. 2nd ed. Jones and Bartlett, 1985.) The enzyme RNA polymerase binds to DNA and controls the synthesis of a complementary strand of RNA from a DNA template. The binding of RNA polymerase to DNA in cells is specific in that only a particular required sequence of DNA is copied into RNA. It has been observed that in a test tube in **0.2 M NaCl**, four RNA polymerase molecules bind to a particular DNA molecule and synthesize the same, specific RNA molecule that is found in living organisms. In contrast, in a test tube in **0.01 M NaCl**, about 50 RNA polymerase molecules bind to the DNA and a great many different RNA molecules are synthesized.
   a. What does this tell us about the interactions between DNA and RNA polymerase?
   b. What are the implications of these observations for working with RNA polymerase in the laboratory?
4. **PCR, the polymerase chain reaction,** *is a method used to copy a specific region of DNA many thousands of times* (see Chapter 27). PCR requires **primers** *that are short, single strands of DNA*. The primers are complementary to regions flanking the DNA region of interest. There are three phases to the PCR cycle:
        Phase 1: The two strands of the parent DNA molecule that includes the sequence of interest are separated by heating, typically to 92°C.
        Phase 2: The solution is quickly cooled, typically to 72°C, and the single-stranded DNA hybridizes with the primers.
        Phase 3: Once the primers have annealed, the thermostable enzyme, *Taq 1* DNA polymerase, is able to copy the DNA region of interest, beginning from the location where the primers have bound.
        The PCR cycle is repeated many times, greatly amplifying the number of copies of the DNA of interest.
   a. Examine the following recipe for PCR buffer. Discuss the salt concentration in terms of DNA denaturation and hybridization of the primer to the DNA.

PCR buffer	
Component	**Final Concentration**
KCl	500 mM
Tris–HCl (pH 8.0)	100 mM
$MgCl_2$.	20 mM

   b. Explain the role of the $MgCl_2$ in the reaction mix.
   c. *Taq 1* polymerase is functional at high temperatures. Why is this type of DNA polymerase essential for the PCR method to work?
5. What do you think would happen to a bacterial culture with the addition of dishwashing detergent? Consider the membranes, proteins, and nucleic acids.
6. You have a sample of DNA (10 μg in 50 μL) that is stored in TE buffer (10 mM Tris, pH 7.6, and 1 mM EDTA). Upon addition of 100 μL of ethanol, no precipitate is seen. Why not? What do you do to recover the DNA?

Questions for Discussion

1. Suppose you are working in a university research laboratory and are taking over a project that was initiated by another investigator. The project involves isolating DNA from a particular type of fungus. One of the problems the previous investigator had begun to explore, but had not resolved, was that the DNA isolated was inconsistent in its qualities. The DNA sometimes seemed to be fragmented, sometimes more DNA was isolated than other times, and sometimes, when the DNA was digested with restriction endonucleases, the digestion did not appear to go to completion, even when the enzymes were known to be effective. Discuss the components of the various solutions involved that might be affecting the results. How could you investigate the various components of the solutions in a systematic, well-documented way?

2. Suppose you are now working in a production facility producing the DNA described in Discussion Question 1. Discuss the particular concerns you would have in this situation.

CHAPTER APPENDIX: ALPHABETICAL LISTING OF COMMON COMPONENTS OF BIOLOGICAL SOLUTIONS

Acrylamide. Monomers that polymerize to form a polyacrylamide matrix used in gel electrophoresis (PAGE).

bis-Acrylamide. Cross-linker added to acrylamide monomers before polymerization. Controls the porosity of the resulting gel.

Agarose. Matrix composed of complex carbohydrates derived from seaweed, used for gel electrophoresis.

Ammonium acetate. Salt added to DNA in solution to aid in ethanol precipitation. At a concentration of 2 M, it will preferentially facilitate ethanol precipitation of larger DNA and allow nucleotides to remain in solution.

Ammonium persulfate. A free radical generator added to unpolymerized acrylamide to stimulate polymerization.

Ampicillin. Antibiotic that inhibits bacterial cell wall synthesis.

Avidin. Egg white protein that has a high affinity for biotin; may be conjugated with an enzyme or other compound as part of a detection system.

BCIP (5-bromo-4-chloro-3-indolyl phosphate *p*-toluidine salt). A substrate for the enzyme alkaline phosphatase. In conjunction with NBT produces a blue color indicating the presence of this enzyme.

Biotin. Vitamin that has a high affinity for avidin. Biotin can be conjugated to various molecules, then bound to and detected by an indicator-linked avidin molecule.

Bromophenol blue. Dye used to visualize sample movement in gel electrophoresis.

BSA (bovine serum albumin). A protein often used as a carrier to help protect proteins; also used to block nonspecific binding sites.

sec-Butanol. Water-insoluble alcohol used to reduce the water in aqueous DNA solutions, thus concentrating DNA into a smaller volume.

CAA (casamino acids). Digested protein added to bacterial growth media as a source of amino acids.

CaCl$_2$ (calcium chloride). Salt used to increase the permeability of cellular membranes. Bacteria incubated in CaCl$_2$ at cold temperatures become competent to take up DNA.

Calcium phosphate. Salt that forms a precipitate with DNA. The complex is introduced into cells in culture.

Chloramphenicol. Antibiotic used to amplify plasmids in bacteria by restricting protein synthesis but not nucleic acid synthesis. Also the substrate for CAT (chloramphenicol acetyltransferase) enzyme assay.

Coomassie (Brilliant) Blue stain. Blue stain that binds to proteins. Used to visualize protein bands in gels after electrophoresis and in protein assays.

CsCl (cesium chloride). Salt used to form density gradients during ultracentrifugation to separate DNA by buoyant density.

CTAB (cetyltrimethylammonium bromide). Nonionic detergent. Precipitates DNA at salt concentrations below 0.5 M; at high salt concentrations, complexes with polysaccharides and protein.

DEAE-dextran (diethylaminoethyl-dextran). Used to permeabilize mammalian cells to facilitate DNA uptake.

Denhardt's reagent. Solution used in hybridization reactions as a buffer and as a blocking agent. Contains BSA, Ficoll, and PVP.

DEPC (diethyl pyrocarbonate). RNase inhibitor. Added to water, solutions, and glassware to inhibit the action of nonspecific RNases.

Detergents. Water-soluble compounds that have a hydrophobic tail. Detergents have multiple uses, including solubilization of membranes and membrane proteins, cell lysis, denaturing proteins, wetting surfaces, and emulsification of chemicals.

Dextran sulfate. Used as a "crowding agent" to effectively increase the concentration of nucleic acids in a hybridization mixture.

DMSO (dimethyl sulfoxide). Compound used to solubilize chemicals and to permeabilize cells.

DNase. Any member of a class of naturally occurring enzymes that digest DNA.

dNTP (deoxynucleotide triphosphate). Designates any of the four deoxynucleotides or a mixture of the four deoxynucleotides that comprise nucleic acids.

DTT (dithiothreitol). A reducing agent added to protein solutions to inhibit the formation of unwanted disulfide bonds.

EDTA (ethylenediaminetetraacetic acid). Chelator of divalent ions. Added to DNA solutions to reduce the activity of nucleases and modifying enzymes by binding to and removing cofactors.

EGTA (ethylene glycol-bis[β-aminoethyl ether]-N,N,N′,N′-tetraacetic acid). Chelator of divalent ions; binds preferentially to Ca^{++}.

Ethanol. An alcohol that dehydrates and precipitates DNA when in a high-salt buffer.

Ether. Volatile organic solvent; used to remove residual phenol from aqueous solutions.

Ethidium bromide. Dye that binds to nucleic acids and fluoresces when exposed to UV light. Used to visualize and quantify nucleic acids.

Ficoll. Inert polymer that acts as a "blocking agent" and/or "crowding agent" when included in a hybridization solution. A component of Denhardt's reagent. Crowding agents are relatively inert substances that, by increasing the concentration of macromolecules in a solution, can alter the properties of proteins and nucleic acids.

Formamide. Chemical added to destabilize nucleic acid duplexes.

Glycerol. Viscous, syrupy liquid added to solutions to reduce their freezing temperature, allowing solutions to be very cold without freezing. Increases the permeability of membranes and helps to stabilize proteins.

Glycogen. Inert carrier added to help ethanol precipitate low concentrations of DNA.

Guanidine isothiocyanate (GITC) also **guanidine thiocyanate (GTC).** Salt that denatures proteins; added to preparations when isolating RNA to denature and remove contaminating RNases and to free RNA from proteins.

HEPES. Commercially available nontoxic buffer often used in cell culture.

8-Hydroxyquinoline. Antioxidant, added to redistilled phenol to prevent unwanted oxidation products.

Isoamyl alcohol. Added to phenol/chloroform during extraction of protein from DNA to reduce interface foaming.

Isopropanol. Alcohol that will precipitate DNA when in a high-salt solution.

KCl (potassium chloride). Commonly used salt.

LiCl (lithium chloride). Salt which at 4 M will preferentially precipitate high molecular weight RNA.

β-Mercaptoethanol (also 2-mercaptoethanol). Reducing agent; added to protein solutions to restrict the formation of unwanted disulfide bonds. Can be used to help inactivate RNases.

MgCl₂ (magnesium chloride). Source of Mg^{++}, essential cofactor of many enzymes that act on nucleic acids.

NaCl (sodium chloride). Table salt, most common salt used to increase the ionic strength of a solution.

NaOH (sodium hydroxide). Strong base; at 0.2 M will denature DNA.

NBT (nitro blue tetrazolium). Used with BCIP as a color indicator for detection of alkaline phosphatase activity.

Nitrocellulose. Membrane; used to immobilize nucleic acids during a Southern blot.

PBS (phosphate-buffered saline). Isotonic buffer used for intact cells.

PCA (perchloric acid). Acid used to denature and precipitate nucleic acids and proteins.

PEG (polyethylene glycol). Waxy polymer available in varying sizes used to precipitate virus particles and protein. Added to hybridization solutions to reduce the liquid concentration in the solution.

Phenol. Water-insoluble organic solvent used as a protein denaturant. When added to an aqueous solution of DNA and protein, the protein is denatured and separates into the phenol phase and interface, and the nucleic acids are left in the aqueous phase.

Phenol/chloroform. Addition of chloroform to phenol increases the phenol solubility of denatured proteins.

Phenol/chloroform/isoamyl alcohol. Mixture of solvents used in DNA isolation to denature and remove proteins from the solution. Isoamyl alcohol is added to improve phase separation.

PMSF (phenylmethylsulfonyl fluoride). A protease inhibitor; added to crude cell extracts to reduce degradation of protein by proteases.

Potassium acetate. Salt commonly added with ethanol to precipitate DNA.

Protease inhibitors. Compounds that inhibit the activity of unwanted proteases.

PVP (polyvinylpyrrolidone). Inert polymer added to Denhardt's solution as a "crowding agent."

RNase. Any one of a class of enzymes that degrade RNA.

SDS (sodium dodecyl sulfate; also known as **lauryl sulfate, sodium salt).** Ionic detergent that is effective as a protein denaturant. Commonly used to denature proteins before separation by electrophoresis.

Sodium azide (NaN₃). Antimicrobial agent added to some solutions to reduce contamination and degradation by microorganisms.

Sorbitol. Inert sugar sometimes used as an osmotic support to protect cells after their cell wall is removed.

Spermine. A naturally occurring compound that occurs in all eukaryotes that is sometimes used in the laboratory to help precipitate DNA.

SSC (standard saline citrate). Buffer used in Southern blotting.

Streptomycin sulfate. Antibiotic that inhibits protein synthesis in bacteria.

TAE (Tris–acetate–EDTA buffer). Buffer for gel electrophoresis of nucleic acids.

TBE (Tris–borate–EDTA buffer). Buffer for gel electrophoresis of nucleic acids.

TCA (trichloroacetic acid). Protein and nucleic acid denaturant and precipitant.

TEMED (N,N,N′,N′-tetramethylethylenediamine). Stimulates polymerization of polyacrylamide.

Tetracycline. Antibiotic that inhibits protein synthesis in bacteria.

Tris (tris[hydroxymethyl]aminomethane). Commercial buffer used commonly in molecular biology.

Triton X-100. Nonionic detergent that solubilizes cell membranes.

Urea. RNA denaturant; destabilizes DNA by reducing the temperature at which the DNA double helix denatures; protein denaturant.

X-gal (5-bromo-4-chloro-3-indolyl β-D-galactopyranoside). Color indicator for the presence of the enzyme β-galactosidase.

Xylene cyanol. Tracking dye used in nucleic acid electrophoresis.

UNIT VII

Quality Assays and Tests

Chapters in this Unit

◆ Chapter 26: Introduction to Quality Laboratory Tests and Assays

◆ Chapter 27: Achieving Reproducible Results with Polymerase Chain Reaction Assays

◆ Chapter 28: Measurements Involving Light – Part B: Assays

◆ Chapter 29: Achieving Reproducible Immunoassay Results

Quality assays provide data that are reproducible and can be trusted when making decisions. This unit is intended to help you develop the understanding necessary to avoid stumbling into avoidable pitfalls, and to effectively troubleshoot problems that do arise when performing laboratory tests and assays.

DOI: 10.1201/9780429282799-32

Case Study: Lab Mistake Set Back Sanofi–GSK Shot by Months

This was the alarming headline from the Wall Street Journal, December 12, 2020, describing a laboratory error that delayed the hopes of a vaccine to prevent COVID-19 manufactured by Sanofi–GSK by four or five months. "It's a sad setback," reported Thomas Triomphe, Sanofi's executive vice president for vaccines. Mr. Triomphe explained that the problem occurred because of faulty reagents that were used when the potency of their antigen was measured. The antigen is a protein from the virus causing COVID-19 that triggers immunity when injected into a person. Because of the faulty reagents, the doses that were supplied for initial clinical trials in volunteers were incorrect. Therefore, the company was required to halt their clinical trials and reformulate their vaccine. This unfortunate turn of events occurred at a time when the world was waiting for billions of doses of vaccines to end the hardship caused by an international pandemic. This mishap is a sobering reminder of the critical importance of laboratory reagents; errors in the production of reagents can have serious, even life-threatening consequences.

The Sanofi–GSK story both highlights the vital importance of reagents, and addresses the importance of **assays**, *which are measurements of a property of a sample.* The antigen potency tests performed by Sanofi–GSK are a type of assay. A measurement of the amount of cholesterol in a blood sample, the lead level in paint chips, or the activity of a DNA-cutting enzyme are similar examples of assays. *The property or entity that is measured in an assay is the* **analyte.** In the Sanofi–GSK situation, the analyte is the viral antigen.

Chapter 26 introduces general principles, terminology, and quality considerations relating to assays and tests.

Chapters 27 and 29 include discussions of two categories of assay of particular interest to biotechnologists. The first category is the polymerase chain reaction (PCR) that is used to amplify a sequence of DNA (Chapter 27). The second category is **immunoassays,** *which take advantage of antibodies to perform assays of proteins* (Chapter 29). Both PCR-based assays and immunoassays depend on phenomena that occur naturally in cells: PCR depends on the fact that during cell division, DNA primers *specifically recognize* and *bind* to sequences of chromosomal DNA. Immunoassays depend on the fact that antibodies *specifically recognize* and *bind* to agents that are foreign to the organism.

As we will see in this unit, both PCR and immunoassays can be extremely specific. In nature, DNA primers find and bind to exact target sequences of DNA amidst billions of base pairs. Antibodies detect and bind to a specific target amidst vast numbers of other molecules. This specificity is an essential feature of the assays. Just as specific target recognition and binding give these assays the advantage of *specificity*, so they also make them very *sensitive*. Both PCR

and immunoassays can be used to detect specific molecules at extremely low concentrations in a sample.

As we will also discuss in this unit, both PCR and immunoassays can have poor reproducibility, if analysts do not control for factors that lead to spurious results. Contamination, poor raw materials, faulty reagent preparation, lack of suitable controls, and other mistakes can lead to misleading, inaccurate, and unreproducible results in both types of assays. The analysts at Sanofi–GSK can attest to the importance of these issues.

Chapter 28 focuses on a third category of assays of importance in biotechnology settings (and many other settings as well), that is, spectrophotometric assays. Chapter 28 continues our discussion of light measurements that began in Chapter 21. In this unit, we explore how measuring light absorbance can be used to obtain information about samples.

BIBLIOGRAPHY FOR UNIT VII

METHOD VALIDATION

Various organizations provide information relating to the validation of assays, analytical procedures, methods, and tests. The International Conference on Harmonisation of Technical Requirements for Registration of Pharmaceuticals for Human Use (ICH) has been helpful in harmonizing definitions and in determining the basic requirements for validation in the pharmaceutical area.

A few key documents and interesting articles are listed here:

Ermer, Joachim, and Nethercote, Phil W. *Method Validation in Pharmaceutical Analysis: A Guide to Best Practice.* 2nd ed. Wiley-VCH, 2015.

Food and Drug Administration. "Bioanalytical Method Validation Guidance for Industry." 2018. https://www.fda.gov/media/70858/download

International Conference on Harmonisation of Technical Requirements for Registration of Pharmaceuticals for Human Use "ICH Q2(R1): Validation of Analytical Procedures: Text and Methodology." 2005. https://www.fda.gov/regulatory-information/search-fda-guidance-documents/q2-r1-validation-analytical-procedures-text-and-methodology.

Kanarek, Alex D. "Method Validation Guidelines." *BioPharm International*, September 15, 2005. (A good summary of validation principles.)

The United States Pharmacopeial Convention, Inc. "Chapter <1225> Validation of Compendial Methods" *United States Pharmacopeia—National Formulary (USP–NF)*. Rockville, MD. Updated and published annually.

IMMUNOLOGICAL ASSAYS

Baker, Monya. "Reproducibility Crisis: Blame It on the Antibodies." *Nature*, vol. 521, no. 7552, 2015, pp. 274–76. doi:10.1038/521274a.

Biocompare. "Video Series: Reproducibility Issues in Life Science Research" "What Researchers Need to Know About Antibody Manufacturing." https://www.biocompare.com/reproducibility/ (A series with various dates.)

Bordeaux, Jennifer, et al. "Antibody Validation." *BioTechniques*, vol. 48, no. 3, 2010, pp. 197–209. doi:10.2144/000113382.

Bradbury, Andrew and Plückthun, Andreas. "Standardize Antibodies Used in Research." *Nature*, vol. 518, no. 7537, 2015. doi:10.1038/518027a.

Buchwalow, Igor, et al. "Non-Specific Binding of Antibodies in Immunohistochemistry: Fallacies and Facts." *Scientific Reports*, vol. 1, no. 1, p. 28, 2011. doi:10.1038/srep00028.

Cox, K.L., Devanarayan, V., et al. Immunoassay Methods. 2012. In: Sittampalam, G.S., Grossman, A., et al., editors. *Assay Guidance Manual* [Internet]. Bethesda (MD): Eli Lilly & Company and the National Center for Advancing Translational Sciences [Updated 2019 Jul 8].

Eaton, Samantha L., et al. "Total Protein Analysis as a Reliable Loading Control for Quantitative Fluorescent Western Blotting." *PLoS One*, vol. 8, no. 8, 2013, p. e72457. doi:10.1371/journal.pone.0072457.

Fosang, Amanda J., and Colbran, Roger J. "Transparency Is the Key to Quality." *Journal of Biological Chemistry*, vol. 290, no. 50, 2015, pp. 29692–94. doi:10.1074/jbc.e115.000002.

Goodman, Simon. L. "The Antibody Horror Show: An Introductory Guide for the Perplexed." *New Biotechnology*, vol. 45, 2018, pp. 9–13. doi:10.1016/j.nbt.2018.01.006.

Freedman, Leonard P., et al. "[Letter to the Editor] The Need for Improved Education and Training in Research Antibody Usage and Validation Practices." *BioTechniques*, vol. 61, no. 1, 2016, pp. 16–18. doi:10.2144/000114431.

Ghosh, Rajeshwary, et al. "The Necessity of and Strategies for Improving Confidence in the Accuracy of Western Blots." *Expert Review of Proteomics*, vol. 11, no. 5, 2014, pp. 549–60. doi:10.1586/14789450.2014.939635.

Gilda, Jennifer E., et al. "Western Blotting Inaccuracies with Unverified Antibodies: Need for a Western Blotting Minimal Reporting Standard (WBMRS)." *PLoS One*, vol. 10, no. 8, 2015, p. e0135392. doi:10.1371/journal.pone.0135392.

Janes, Kevin A. "An Analysis of Critical Factors for Quantitative Immunoblotting." *Science Signaling*, vol. 8, no. 371, 2015, p. rs2. doi:10.1126/scisignal.2005966.

Moritz, Christian P. "Tubulin or Not Tubulin: Heading Toward Total Protein Staining as Loading Control in Western Blots." *Proteomics*, vol. 17, no. 20, 2017, p. 1600189. doi:10.1002/pmic.201600189.

Prassas, Ioannis, et al. "False Biomarker Discovery Due to Reactivity of a Commercial ELISA for CUZD1 with Cancer Antigen CA125." *Clinical Chemistry*, vol. 60, no. 2, 2014, pp. 381–88. doi: 10.1373/clinchem.2013.215236.

Roncador, Giovanna, et al. "The European Antibody Network's Practical Guide to Finding and Validating Suitable Antibodies for Research." *MAbs*, vol. 8, no. 1, 2015, pp. 27–36. doi:10.1080/19420862.2015.1100787.

Uhlen, Mathias, et al. "A Proposal for Validation of Antibodies." *Nature Methods*, vol. 13, no. 10, 2016, pp. 823–27. doi:10.1038/nmeth.3995. (Discusses collaborative work to identify problems with antibodies and effective methods of antibody validation.)

Weller, Michael G. "Ten Basic Rules of Antibody Validation." *Analytical Chemistry Insights*, vol. 13, 2018, pp. 1–5. doi:10.1177/1177390118757462.

Antibody and reagent manufacturers provide a number of high-quality, free, online resources. A few of these are listed here.

Abcam. "Immunohistochemistry Application Guide" 2019. https://docs.abcam.com/pdf/kits/immunohistochemistry-ihc-application-guide.pdf.

Abcam. "Protocols Book" 2010. https://www.abcam.com/protocols/limited-edition-protocols-book (Includes protocols for a variety of immunological assays.)

Abcam. "Recommended Controls for Western Blot" no date given. https://www.abcam.com/protocols/recommended-controls-for-western-blot.

Abcam. "Western Blot Protocol" 2019. https://www.abcam.com/protocols/general-western-blot-protocol.

Abcam. "A Guide to Antibody Validation" 2020. https://www.abcam.com/primary-antibodies/a-guide-to-antibody-validation.

Abcam. "Antibody Storage Guide." 2020. https://www.abcam.com/protocols/antibody-storage-guide.

AbClonal. "Technical Frequently Asked Questions" https://abclonal.com/technical-faqs/.

Bio-Rad. "Can We Trust Western Blots?" Webinar. Aldrin Gomes, Presenter. https://www.bio-rad.com/en-us/life-science-research/support/webinars/can-we-trust-western-blots?ID=1516764668641. (Discusses several issues relating to obtaining reproducible western blots.)

Cell Signaling Technology, Inc. "A Guide to Successful Immunofluorescence" 2017. http://media.cellsignal.com/www/pdfs/resources/product-literature/application-if-brochure.pdf.

Hashemi, S.M. "Protocol for the Preparation and Fluorescent IHC Staining of Paraffin-embedded Tissue Sections" R&D Systems, 2011. https://faqs.tips/post/protocol-for-the-preparation-and-fluorescent-ihc-staining-of-paraffin-embedded-tissue-sections.html.

Leica Biosystems. Geoffrey Rolls. "An Introduction to Specimen Preparation" No date given. https://www.leica-microsystems.com/science-lab/how-to-prepare-your-specimen-for-immunofluorescence-microscopy/. (A well-illustrated introductory article.)

Li-Cor, Inc. "Decoding JBC Publication Guidelines" 2019. https://www.licor.com/bio/guide/westerns/decoding_guidelines.

Li-Cor, Inc. "Protocol" Determining the Linear Range for Quantitative Western Blot Detection" 2017. https://www.licor.com/documents/dxrkfkwdd2ypgrcrnpx8z8l4hhia22p2.

Li-Cor, Inc. "Good Westerns Gone Bad" 2019. https://www.licor.com/documents/dq6jb8sgnkiwlas0g99b5hq0tsuvcyyz.

Li-Cor, Inc. "Technical Note: Western Blot Normalization Handbook" 2017. https://www.licor.com/documents/gtk1a4mrphretj1gvmzld0f920g7h8sh.

Li-Cor, Inc. "Quantitative Western Blots. Is Your Question Quantitative or Qualitative?" 2019. https://www.licor.com/bio/applications/quantitative-western-blots/.

Li-Cor, Inc. "How to Validate Antibodies" 2020. https://www.licor.com/bio/guide/westerns/validate_antibodies.

ThermoFisher Scientific. "Antibody Production and Purification Technical Handbook" 2010. http://tools.thermofisher.com/content/sfs/brochures/1601975-Antibody-Production-Purification-Guide.pdf.

ThermoFisher Scientific. "Overview of Immunohistochemistry (IHC)" 2019. https://www.thermofisher.com/us/en/home/life-science/protein-biology/protein-biology-learning-center/protein-biology-resource-library/pierce-protein-methods/overview-immunohistochemistry.html.

ThermoFisher Scientific. "Western Blotting Handbook and Troubleshooting Guide" 2014. https://tools.thermofisher.com/content/sfs/brochures/1602761-Western-Blotting-Handbook.pdf. (Includes helpful trouble-shooting guide and information that is helpful in understanding how various reagents and materials work.)

ThermoFisher Scientific. "Antibody Bootcamp: Rising to the Fitness Challenge." *Nature*, vol. 560, no. 7719, 2018. http://assets.thermofisher.com/TFS-Assets/BID/Reference-Materials/antibody-reproducibility-validation-specificity-tests-nature-articles.pdf. (Advertising supplement with a number of useful articles on this topic.)

COLLABORATIONS/PROJECTS THAT AIM TO IMPROVE THE REPRODUCIBILITY OF IMMUNOASSAYS AND OTHER ASSAY METHODS

Antibodypedia (www.antibodypedia.com). This is a publicly available database that allows researchers to compare validation data for different antibodies. As of 2021, >4.3 million antibodies had been reviewed from 92 providers. These represent the protein products of 94% of human genes. More than two million validation experiments were used to obtain data. This allows researchers to search for antibodies against a particular target.

The Antibody Registry (https://antibodyregistry.org/). This is a useful resource for determining whether an antibody has been validated for specific uses and for getting technical data sheets about antibodies. This website uses the RRIDs, as created in the Resource Identification Portal.

Antibody Resource page (https://www.antibodyresource.com/). This tool allows users to search multiple antibody companies at once.

Biocompare Antibody Search Tool (https://www.biocompare.com/Antibodies/). This tool lets researchers easily search over 3 million antibodies from hundreds of antibody suppliers.

Human Protein Atlas (https://www.proteinatlas.org/). This is an international effort to use antibodies to map all the human proteins in tissues and cells. More than 20,000 antibodies from more than 50 providers are being studied, and validation is essential for the success of this project.

The Resource Identification Portal (https://scicrunch.org/resources). One of the major issues in reproducibility is that it is necessary to know exactly what reagents were used in previous work. The Resource Identification Portal was created in support of the Resource Identification Initiative, which aims to promote research resource identification, discovery, and reuse. The portal offers a central location for obtaining and exploring **Research Resource Identifiers (RRIDs)** – persistent and unique identifiers for referencing a research resource.

26 Introduction to Quality Laboratory Tests and Assays

26.1 INTRODUCTION

The products of a laboratory are data, information, and knowledge. Assays are one of the primary means by which laboratory analysts obtain data or information. We broadly define an **assay** *as any test used to analyze a characteristic of a sample.* As is true for measurements, a "good" assay might be defined as one that provides data that can be trusted when making decisions or reaching conclusions. This chapter introduces considerations that are important in obtaining "good," reliable results from assays.

There are a vast number of assays that are performed in biotechnology laboratories. Research scientists use assays to look for the effects of treatments on their experimental subjects. Pharmaceutical analysts test samples from animal and human subjects to determine what happens to a drug in the body. Development scientists in a biotechnology company might use an assay to determine whether bacteria that were transformed with a gene of interest are making the protein for which it codes. Forensic scientists use DNA "fingerprinting," an assay method that gives information about the identity of a person. Quality-control analysts might use a series of assays to evaluate a product to see whether it meets its specifications. The series might include separate tests for bacterial, viral, DNA, and protein contaminants. Clinicians use diagnostic assays to determine if a patient has a particular condition.

Some assays are **quantitative** *and provide numerical data* (e.g., a measure of the concentration or potency of a substance). Other assays provide *non-numerical* **qualitative** *information* (e.g., whether or not binding of two molecules occurs, whether a substance is present, whether a suspect might have been at a crime scene, the chemical identities of the components of a mixture, and whether or not a person is infected with a pathogen).

Performing assays is such an integral part of all laboratory work that it is not surprising to find that there are many relevant and interrelated terms; Table 26.1 discusses this terminology. Table 26.2 provides examples to illustrate the wide diversity of assay methods used in biotechnology laboratories. The case study "Diagnostic Assay Validation in the Time of COVID-19, Part I" explores some of the considerations that are important when choosing a particular assay method.

DOI: 10.1201/9780429282799-33

TABLE 26.1
Terminology Relating to Assays

Accuracy. *The closeness of a test result to the true or accepted value.*

Analyte. *A substance whose presence and/or level is evaluated using an assay.*

Analytical Method. *A laboratory method used to determine features of a sample.*

Artifact. *A distortion or error.* For example, in electron microscopy an artifact might be a substance that appears to be a component of a cell but, in fact, was accidentally created by the analyst during the preparation of the tissue sample.

Assay (noun). *A test used to analyze a characteristic of a sample, such as its composition, purity, or activity.* The terms *assay*, *test*, and *method* are often used interchangeably, although the term *assay* is not generally applied to a test of an instrument's performance.

Assay (verb). *To determine.* For example, "The technician *assayed* the protein concentration in a sample."

Bioanalytical Method. *A term that commonly refers to any test of a biological material.*

Bioassay or **Biological Assay.** *1. Analysis to quantify the biological activity(ies) of one or more of a sample's components by determining its capacity for producing an expected biological activity. 2. Any assay that involves cells, tissues, or organisms as test subjects.* Bioassays are used for many purposes (for example, testing the potency of a drug).

Cell-Based Assay. *Commonly describes any assay that involves living cells as the test subjects, a type of bioassay.* Cell-based assays are used, for example, to test a material for viral contaminants, and to look for a cellular response to a drug product.

Characterization (in the context of drugs). *A process in which a molecular entity's physical, chemical, and functional properties are determined using specific assays.* The qualities of the entity are then defined in terms of the results of those assays.

Compendium. *A concise, detailed collection of information about an extensive topic, usually in book form.*

Diagnostic Assay. *An assay used in a clinical setting to determine if a patient has a particular condition.*

Limit of Detection (LOD). *The lowest concentration of the material of interest that can be detected by the method.*

Limit of Quantitation (LOQ). *The lowest concentration of the material of interest that the method can quantify with acceptable accuracy and precision.*

Linearity. *The ability of a method to give test results that are directly proportional to the concentration of the material of interest (within a given concentration range).*

Matrix. *The physical material in which a sample or analyte is located, or from which it must be isolated, for example, blood, soil, or fiber.*

Method. *Detailed description of the means of performing an assay.* This term is often used as a synonym for *procedure, analytical procedure, assay,* or *test.*

Method Validation (Assay Validation). *A process used to demonstrate the ability of an assay method to reliably obtain a desired result.* The method validation is a formal process required by regulatory authorities in some situations.

Precision. *The degree of agreement between individual test results when a procedure is applied over and over again to portions of the same sample.*

Preparative Method (in contrast to an analytical method). *A method that produces a material or product for further use (perhaps for commercial sale, or perhaps for further experimentation).* Chromatography, for example, is used to separate the components of a sample from one another. If the components are separated in order to purify an enzyme, then chromatography is being used as a preparative method. Chromatography is also frequently used to identify the components present in a mixture, in which case chromatography is used as an analytical method.

Protocol. *1. Synonym for* procedure *or* method. *2. The formal design or action plan of a research study.*

Qualitative Analysis. *A test to determine the nature of the component(s) of a sample.*

Quantitative Analysis. *A test of how much of a particular analyte(s) is in a sample.*

Range. *The range of concentrations, from the lowest to the highest, that a method can measure with acceptable results.*

Repeatability. *The precision of measurements made under uniform conditions.*

Reproducibility (sometimes called ruggedness). *The precision of measurements made under nonuniform conditions, such as in two different laboratories.*

Resolution. *The smallest difference between two entities that can be separated and detected.*

Robustness. *The capacity of a method to remain unaffected by variations in method parameters.*

Sample. *A subset of the whole that represents the whole. (For example, a patient's blood sample represents all the blood in the patient's body.)*

(Continued)

TABLE 26.1 (*Continued*)
Terminology Relating to Assays

Specificity (sometimes called *selectivity*). *A measure of the extent to which a method can unequivocally determine the presence of a particular compound in a sample in the presence of other materials that may be expected to be present.*

System Suitability (according to FDA). *"Determination of instrument performance (e.g., sensitivity and chromatographic retention) by analysis of a reference standard prior to running the analytical batch."*

Test (according to AOAC International). *"A technical operation that consists of the determination of one or more characteristics...of a given product, material, equipment, organism, physical phenomenon, process, or service according to a specified procedure."* The result of a test is normally recorded in a document called a test report or certificate. Note that a test may or may not involve a sample, but an assay, as we are using the term, does involve a sample.

TABLE 26.2
Examples of Assay Methods and Associated Instrumentation Common in Biotechnology Laboratories

I. **Chromatographic Methods.** *A large class of methods used to separate the components of a mixture from one another and identify them.* Chromatography is used, for example, to determine the identity, quantity, and purity of a material, and during stability testing to look for degradation products. (Chromatography is explained in more detail in Chapter 34.)

II. **Electrophoretic Methods.** *Like chromatographic methods, electrophoretic methods separate and identify sample components. Separation is based on the components' relative ability to migrate when placed in an electrical field.* Electrophoretic methods are widely used in the analysis of proteins and DNA, for example, to confirm the identity of a DNA fragment, to look for protein contaminants in a product, and to roughly quantify a protein or DNA product. (Electrophoresis is explained in more detail in Chapter 34.)

    A. **Agarose Gel Electrophoresis.** *A method that separates sample components as they migrate through a gel matrix composed of a highly purified form of agar; commonly used to separate DNA fragments on the basis of base pair length.*

    B. **Polyacrylamide Gel Electrophoresis (PAGE).** *A method used to determine the molecular weights of molecules.* A common adaptation of PAGE, sodium dodecyl sulfate-PAGE (SDS–PAGE) is used to separate proteins on the basis of their molecular weights. Protein samples are treated with the negatively charged detergent, SDS. SDS denatures the proteins and confers a negative charge on them so that they migrate in the electrical field toward the positively charged end of the gel. Smaller proteins migrate through the matrix more quickly than larger ones. After separation, the proteins are stained for visualization, or are detected by western blotting, as described in Chapter 29.

    C. **Isoelectric Focusing (IEF).** *A method capable of detecting a single-charge difference between protein molecules; molecules are separated as they run through a pH gradient in an electrical field.* IEF detects post-translational modifications of proteins that result in net charge differences. IEF is used, for example, to evaluate protein identity and purity, and to look for protein changes during stability testing.

    D. **Capillary Electrophoresis.** *A miniaturized instrumental version of electrophoresis that is amenable to automation, requires very little sample, and provides quantitative information.* Capillary electrophoresis is widely used for analyzing food samples, forensic samples, oligonucleotides, proteins, peptides, and DNA. It is used in genome sequencing. It is also used in the pharmaceutical industry to help determine the identity, purity, and structure of small molecules and protein-based drugs.

    E. **Two-Dimensional Gel Electrophoresis.** *A type of high-resolution gel electrophoresis in which proteins are first separated by isoelectric focusing and then are separated based on differences in their molecular weight.* This method is important in proteomics because it can resolve hundreds of proteins on a single gel and can also be used to detect post-translational modifications.

(Continued)

TABLE 26.2 (*Continued*)

Examples of Assay Methods and Associated Instrumentation Common in Biotechnology Laboratories

III. **Immunological Methods.** *Methods that depend on the specific interaction of an antigen and its antibody.* Most commonly, the analyte is a protein or a protein derivative. These methods are used for many purposes, for example, testing the identity of a biopharmaceutical product, testing a material for the presence of specific contaminants, studying the localization of a cellular protein, and, in a clinical setting, determining whether a patient has been exposed to, or is infected with, a particular pathogen (e.g., the HIV virus or the SARS-CoV-2 virus). (Immunological methods are explained in Chapter 29.)

IV. **Polymerase Chain Reaction (PCR) Methods.** *Methods that greatly amplify the amount of DNA with a specific, desired sequence.* PCR has a vast number of assay applications, such as testing drug products for specific contaminants, testing food to see if it has been genetically modified, and human identity testing. (PCR is explored in Chapter 27.)

  A. **Endpoint PCR (Conventional PCR).** The original version of PCR in which the amplification is allowed to go to completion and then the amplified product is visualized using gel electrophoresis.

  B. **Quantitative Real-Time PCR (qPCR).** A version of PCR that quantifies the amount of DNA present at the same time as it is amplified.

  C. **Reverse Transcriptase PCR (RT-PCR).** A version of PCR that is used for the detection and quantitation of specific RNA sequences. **Reverse transcriptase** *is an enzyme that converts RNA into DNA. The resulting DNA is called* **cDNA**.

   A note about terminology: There is confusion in the acronyms for PCR because RT may stand for "real-time" or "reverse transcriptase." Real-time PCR is generally (though not always) used synonymously with quantitative PCR. When reading the literature, be sure you understand how the author is using the abbreviation "RT."

V. **Ultraviolet and Visible Light Spectrophotometric Methods.** *Methods that rely on the interaction of the sample with light.* A wide variety of spectrophotometric methods are used to measure analyte quantity, activity, identity, and purity in samples. (These assays are discussed in Chapter 28.)

VI. **Mass Spectrometry (Mass Spec).** *A powerful instrumental analytical technique used to identify and measure a wide variety of biological and chemical compounds based on differences in both their mass and charge.* Mass Spec can be used, for example, to identify what compounds are present in a sample and to determine the structure of a protein.

VII. **Methods Relating to Microbes/Viruses**

  A. **Microbial Testing, Bioburden.** *Methods of testing the microbial population associated with an unsterilized product or component; relates to the cleanliness of the production environment and the handling of the product.* Bioburden testing is required in GMP-compliant facilities.

  B. **Microbiological Testing, Sterility.** *Methods that test for the presence of bacterial and fungal contaminants in sterilized preparations.* The U.S. Pharmacopeia sterility test is a common method in which the test sample is incubated in nutrient-rich growth medium for a specified number of days and the medium is monitored daily for turbidity, an indicator of microbial growth.

  C. ***Limulus* Amebocyte Lysate (LAL) Test.** *A sensitive test for the presence of endotoxin contaminants that is based on the ability of endotoxin to cause a coagulation reaction in a reagent purified from the blood of the horseshoe crab.* Endotoxins (also called pyrogens) are bacterial by-products that elicit a dangerous immune response in mammals when present at very low levels.

  D. **USP Rabbit Pyrogen Test.** *Sample is injected into rabbits that are monitored to see if the test mixture elicits a fever response.*

  E. **Viral Contaminants Testing.** *Methods of testing for the presence of viral contaminants.*

   i. Infectivity Assays: *The test sample is inoculated onto susceptible cultured cells.* The susceptible cells are incubated and observed over a period of time (e.g., 28 days). If virus is present, the cells become infected and damaged, with visible morphological changes.

   ii. Transmission Electron Microscopy: Viral particles are very small, but they can be visualized using the high magnifications achievable with a transmission electron microscope. Electron microscopy is used, for example, to look for endogenous viruses in host cells used for biopharmaceutical production.

Many of the concepts and terms that were discussed in previous chapters also apply to assays. Chapters 13 and 14 include math tools that can be used for analyzing the results of assays. Chapter 15 discusses related topics including accuracy, precision, and error.

Chapters 27–29 continue the discussion that begins in this chapter by delving into the polymerase chain reaction (PCR), spectrophotometric, and immunological assays, respectively. These three classes of assay are of particular importance to biotechnologists.

Case Study: Diagnostic Assay Validation in the Time of COVID-19, Part I

Diagnostic assays *are tests that are used to determine if an individual has, or previously suffered from, a particular disease or disorder.* With the COVID-19 pandemic, diagnostic testing has become a major public health necessity and a topic of great interest to the general public. There are two general categories of COVID-19 diagnostic testing. The first category includes various tests that look for SARS-CoV-2 viral material in a sample swabbed from a person's respiratory tract. The purpose of this category of tests is to determine if a person is currently infected with the SARS-CoV-2 virus, the causative agent of COVID-19. It is important to find infected people because they must be isolated from others and treated. Also, people who have been in contact with an infected individual must be quarantined to prevent further spread of the virus. The second category of COVID-19 diagnostic tests includes methods that look for antibodies that a person has made against the SARS-CoV-2 virus. The presence of antibodies indicates that a person was infected in the past or has been infected for a long enough period of time to generate antibodies. When antibodies are the analyte of interest, a sample of blood is taken from the subject, because antibodies circulate in the blood.

Let's consider which of the assay methods listed in Table 26.2 are best adapted for COVID-19 diagnostics. It is always necessary to choose an assay method that matches the type of analyte, is accurate, and is efficient. Viruses consist of nucleic acid, either DNA or RNA, surrounded by proteins. The SARS-CoV-2 virus is an RNA virus (Figure 26.1). Therefore, when the purpose of the assay is to look for current infection by the SARS-CoV-2 virus, the analyte must be either the virus's RNA or a protein from the virus. Observe in Table 26.2 that PCR methods are commonly used when the analyte is nucleic acid. Reverse transcriptase PCR can be used when the analyte is RNA, as is the case with the SARS-CoV-2 virus. At the time of writing, the majority of COVID-19 infection assays look for the virus's RNA using reverse transcriptase-qPCR (RT-qPCR). A PCR method is required for diagnostics because PCR greatly amplifies

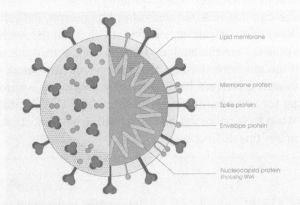

FIGURE 26.1 SARS-CoV-2 virus structure. The viral genome is surrounded by a lipid membrane and various proteins that perform tasks relating to viral entry into cells and replication. The viral RNA is most often the target of assays that detect infection with the virus, but one or more of the viral proteins, or a segment of a viral protein, can also be the analyte in a diagnostic assay. (Image courtesy of Abcam.)

(Continued)

Case Study (*Continued*): Diagnostic Assay Validation in the Time of COVID-19, Part I

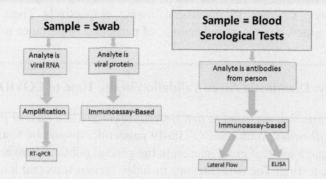

FIGURE 26.2 Two categories of COVID-19 testing. The first category of tests look for a viral component, either viral RNA or protein. When RNA is the target, PCR (or other similar amplification-based) assays are used. At the time of writing, reverse transcriptase-qPCR is the most common. RT-qPCR begins by converting viral RNA into DNA, followed by qPCR. When proteins are the assay target, as is the case when either viral protein or patient antibodies are the analyte, then immunoassay methods are used. At the time of writing, two different forms of immunoassay are commonly used for COVID-19 antibody testing: lateral flow assays that are fast and convenient, and ELISAs that can be optimized to be very sensitive and accurate. (ELISAs will be discussed in Chapter 29.)

the amount of nucleic acid present in a sample. Amplification is necessary because the amount of viral RNA present in a patient sample can be very low and it is mixed with nucleic acids from the patient and with nucleic acids from any bacteria or other organisms present in the patient's respiratory tract. PCR greatly amplifies only the unique SARS-CoV-2 viral nucleic acid, thus generating enough viral material to be detectable. Reverse transcriptase-qPCR is very sensitive, can provide accurate results, and is efficiently run inside a microprocessor-controlled instrument. Moreover, qPCR instruments can run multiple samples simultaneously. For all these reasons, RT-qPCR has been the method of choice for diagnosing current infection with SARS-CoV-2. We will note that it is also possible to assay for one of the viral proteins in order to determine if a person is infected with the virus. Referring to Table 26.2, you can see that immunological methods are useful when the analyte is a protein, and, in fact, there are immunological diagnostic assays that detect viral protein in a swab from the respiratory tract.

Now let's consider those assays that look for antibodies in a person, indicating previous infection with the SARS-CoV-2 virus. When antibodies are the analyte, PCR is not the method of choice because antibodies are proteins. When proteins are the analyte, immunological methods are commonly used. There are a number of COVID-19 immunological assays that can specifically detect only antibodies against the SARS-CoV-2 virus, even though a person's blood will have antibodies against a great many other substances. Thus, when looking for current infection with the SARS-CoV-2 virus, PCR-based methods are most commonly used. But when looking for antibodies signaling past infection, immunological assays are chosen. Figure 26.2 summarizes this distinction.

26.2 THE COMPONENTS OF AN ASSAY

26.2.1 OVERVIEW

Figure 26.3 provides an overview of an assay. Assays are performed on samples where a **sample** *is a part of the whole that represents the whole.* The samples in an assay undergo some procedure (method) in which the property of interest becomes detectable. Often a

laboratory instrument is involved in testing the samples. Information from the samples may be compared to information from a reference standard or reference material that is well characterized with respect to the property of interest. The data collected in the assay are analyzed, displayed, and/or stored. Statistical methods are often used in the analysis of data from assays.

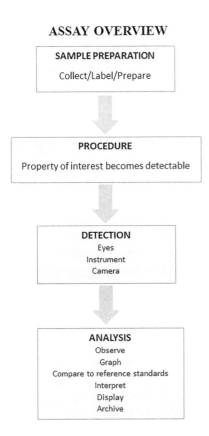

ASSAY OVERVIEW

SAMPLE PREPARATION

Collect/Label/Prepare

PROCEDURE

Property of interest becomes detectable

DETECTION

Eyes
Instrument
Camera

ANALYSIS

Observe
Graph
Compare to reference standards
Interpret
Display
Archive

FIGURE 26.3 Overview of an assay. Assays are performed on samples that undergo some process in which a property of the sample becomes detectable and can be evaluated. The data collected from the sample(s) are analyzed, displayed, and archived.

26.2.2 THE SAMPLE

26.2.2.1 Obtaining a Representative Sample

The samples that are analyzed in a laboratory assay or test represent the "whole." The "whole" is the object, batch, material, area, or population of individuals under investigation. The "whole" might be, for example, a batch of a biopharmaceutical product, the blood in an experimental animal, all patients with a disease, or a field of genetically modified corn.

It may be straightforward to obtain a sample that truly represents the whole. A blood sample from a patient is generally acknowledged to be representative of the whole of that person's blood. Devising a sampling strategy can be challenging, however, if the "whole" is heterogeneous. A classic problem in biological experimentation is how to sample an experimental agricultural plot. A field is likely to be heterogeneous with respect to its topography, soil moisture, soil type, sunlight, insects, and a host of other factors. Suppose

that scientists who are performing field tests of a new strain of corn go to the field and collect a sample consisting of plants located near their parked car. These corn plants probably experienced different conditions than plants elsewhere in the field and so are not a representative sample of the entire corn crop. Similarly, a fermenter is heterogeneous over time. As a fermentation process proceeds, changes occur in the number of cells present, the chemical composition of the broth, oxygen availability, pH, and so on. Obtaining a representative sample requires understanding the heterogeneity of the system of interest and then devising a sampling strategy.

Another issue in sampling is that humans usually have preferences when sampling. Even when people try not to be biased, they unknowingly have preferences. For example, when people try to list a random series of numbers with no inclination for particular numbers, there still are patterns in their choices because of subconscious preferences.

Statisticians may be consulted to develop a sampling strategy when large and complex studies are performed, or when a sampling process is being developed for a new product. In a smaller or simpler study, certain basic sampling guidelines are often employed. Box 26.1 describes some common methods used to help obtain a representative sample.

26.2.2.2 Sample Preparation

Biological samples frequently require extensive preparation before they can be analyzed. These preparations may greatly affect the results of a measurement or an assay. For example, in molecular biology, extracted materials from disrupted cells might be treated with harsh chemicals, centrifuged, stored, and otherwise manipulated. There is an example in Section 25.4 of this text of a study in which investigators were comparing cells that contained an oncogene (a cancer-causing gene) with cells that did not have the oncogene. The scientists found that their results varied depending on how the cells were prepared. The effects of sample preparation thus had the potential to obscure the effects due to the oncogene.

It is not uncommon during sample preparation that certain components of a sample will be selectively lost while others are retained. Similarly, biological structures may be altered and biological activities, such as enzymatic activity, may be lost. All such alterations in the samples are causes of systematic error. Sample preparation is thus a major issue in the analysis of biological samples. It is often necessary to perform

---

BOX 26.1   OBTAINING A REPRESENTATIVE SAMPLE

1. *When selecting a sample of individuals from a population:*
   a. *Devise a numbering system to assign an identification number to all the potential members of the sample.* For example, if a sample is to be 40 students chosen at random from all the students in a college, then the students' college ID numbers can be used.
   b. *Throw a die or pick numbers from a bowl to choose the individuals who will actually be in the sample.*
   c. *As an alternative to picking numbers from a bowl or throwing a die, use a computer to generate random numbers.* Computers can generate numbers with no preferences for certain numbers or patterns and are particularly helpful when a sample is large.
2. *When selecting a sample of an area:*
   a. *Draw or map the area.*
   b. *Place a grid with horizontal and vertical lines over the map.*
   c. *Number the boxes formed by the grid.*
   d. *Use a random number generator to choose grid boxes.* Sample the areas under the chosen grid boxes.
3. *When sampling biological solutions in the laboratory:*
   a. *Be certain liquids are well mixed before withdrawing a sample.*
   b. *When a liquid is removed from a freezer, be sure all the ice is melted before removing a sample.* Mix the liquid. Different components of the liquid may thaw at different rates.
   c. *When taking a sample from a suspension, such as a bacterial culture, make sure the suspension is evenly distributed.*
   d. *When sampling a powder, either mix it or take a random part of the solid.*
   e. *If a material is known to be heterogeneous, it may be advisable to take a series of samples from different parts of the material and combine them into one sample. This is called a* **composite sample**.

---

preliminary investigations of the effects of sample preparation before conducting assays or experiments. A few basic strategies to deal with sample preparation problems are summarized in Box 26.2.

26.2.3 THE METHOD

Assay methods are often invented by research scientists who need to assess some feature of their research subjects. Scientists publish their methods in scientific research literature, and then other people adapt the methods for various purposes. Sometimes research and development scientists in a biotechnology company invent new methods or modify existing methods as commercial products. In these cases, the companies sell kits with the materials needed to perform the method along with instructions to do it. There are, for example, biotechnology companies that sell kits for detecting DNA and for quantifying proteins in samples.

The methods by which assays are performed sometimes must be standardized, among individuals, laboratories, and nations. In certain disciplines, there are organizations that make recommendations for standard methods. A **standard method** *is a document established by consensus and approved by a recognized body that establishes rules or guidelines to make a procedure consistent among various people.* Table 26.3 lists a few of the organizations involved in establishing standard methods that are of interest to biotechnologists. Often standard methods for particular applications are collected and published in a compendium. The **U.S. Pharmacopeia (USP)** *is a compendium that contains hundreds of accepted methods for tests commonly performed in the pharmaceutical industry.* The USP methods have been validated (a term defined later in this chapter), and the Food and Drug Administration (FDA) accepts these assays for use in the pharmaceutical industry. Another compendium is the **Official Methods of**

BOX 26.2   AVOIDING SYSTEMATIC ERRORS IN SAMPLE PREPARATION

1. *Prepare all the samples, controls, and standards in a study identically.*
   a. If there are samples from experimental and control groups, be certain that they are treated in the same way.
   b. Create written procedures for sample handling that are followed by all analysts.
   c. Devise methods to ensure that all reagents, materials, and equipment used in sample preparation are consistent.
2. *Test for the loss of specific components.* This can often be accomplished by adding a known amount of the material of interest to a sample. (This is called *spiking* the sample.) The sample is then taken through the sample preparation steps. The amount of the material recovered after sample preparation is compared to the amount initially added.
3. *Test different sample preparation procedures.* Experiment with multiple sample preparation methods to determine their effects on the results. While time-consuming, this step can be critical for obtaining high-quality results.

TABLE 26.3

A Selected List of Agencies and Organizations that Are Involved in the Standardization of Measurements and Assays

ANSI (American National Standards Institute)
Administrator and coordinator of the US private sector voluntary standardization system. ANSI does not itself develop American National Standards; rather, it facilitates their development by establishing consensus among qualified groups. ANSI is the US representative to ISO. Examples of services provided by ANSI are:

- Promoting the use of US standards internationally.
- Playing an active role in the governance of ISO.
- Promoting the use of standards, including ISO 9000 in the United States.

AOAC International (Formerly the Association of Official Analytical Chemists)
An independent association of scientists devoted to promoting methods validation and quality measurements in the analytical sciences. Examples of services provided by AOAC are:

- Developing validated standard methods of analysis in microbiology and chemistry.
- Publishing validated standard methods in the compendium, *Official Methods of Analysis of AOAC International.*

ASTM International (Formerly American Society for Testing and Materials)
Coordinates efforts by manufacturers, consumers, and representatives of government and academia to develop consensus standards for materials, products, systems, and services. Examples of services provided by ASTM are:

- Developing and publishing technical standards (e.g., those covering the requirements for volume-measuring glassware). More than 12,000 ASTM standards are published each year.
- Providing technical publications and training courses.

CLSI (Clinical and Laboratory Standards Institute) (Formerly NCCLS)
An organization that promotes voluntary consensus standards relating to laboratory procedures, methods, and protocols applicable to clinical laboratories. (Clinical laboratories perform assays on patient samples.)

(Continued)

TABLE 26.3 (*Continued*)

A Selected List of Agencies and Organizations that Are Involved in the Standardization of Measurements and Assays

ISO (International Organization for Standardization)

A worldwide federation of national standards bodies from more than 160 countries that promotes the development of standards and related activities in the world with a view to facilitating the international exchange of goods and services. Examples of services provided by ISO are:

- Publishing and updating the SI (metric) system of units. The SI system is covered by a series of 14 international standards.
- Developing by consensus international standards, including the ISO 9000 quality series.

NIST (National Institute for Standards and Technology) (Formerly the National Bureau of Standards)

A US federal agency that works with industry and government to advance measurement science and develop standards. Examples of services provided by NIST are:

- Providing calibration services and primary standards for mass, temperature, humidity, fluid flow, and other physical properties.
- Providing standard reference materials to be used in biological, chemical, and physical assays.
- Performing research and development of new standards, including standards for biotechnology.

USP (United States Pharmacopeia)

An organization that promotes public health by establishing and disseminating officially recognized standards relating to medicines and other healthcare technologies. Examples of services provided by USP are:

- Developing validated standard methods of analysis for pharmaceutical and health-related products.
- Publishing validated standard methods in the compendium *the United States Pharmacopeia*.
- Distributing reference materials to be used with the methods outlined in *the United States Pharmacopeia*.

Analysis of AOAC International, *which contains microbiological and chemical analysis methods.* Table 26.4 lists a few major compendia used in bioscience laboratories.

Biotechnology companies that make products must have assays to test their final products, incoming raw materials, and in-process samples (samples taken while production is ongoing). Sometimes research and development scientists create new assays for testing new products. This might be the case in a biopharmaceutical company that is developing a novel therapeutic. More commonly, development scientists modify and optimize existing methods, including those from compendia. The creation and optimization of test methods is an important part of product development. As the product and the processes by which a product is made are refined, so too are the methods to support it.

26.2.4 REFERENCE MATERIALS

Many assays and tests require *chemical or biological substances whose compositions are established and that are used for comparison.* These substances are called **reference materials**, **reference standards**, **controls**, or simply, **standards.** For example, when a sample is tested using a chromatography or electrophoresis method, the result is a profile or "fingerprint" that varies depending on the composition of the sample. Different materials result in different profiles. The profiles of samples can be matched to the profiles of standards of known composition. If the standard and the sample profiles match, then they may be the same material. (Note, however, that two or more different materials sometimes have very similar profiles.)

Obtaining and characterizing reference materials is critical in the development of many assay methods. Standards sometimes have to be prepared in the laboratory that is developing the assay, in which case considerable effort will be required to establish and document their composition, stability, and other qualities. In many cases, reference standards of acceptable quality can be purchased. The National Institute of Standards and Technology (NIST) maintains more than 1,300 different materials, called **standard reference materials (SRMs)**, whose compositions are

TABLE 26.4

Examples of Methods Compendia

Bacteriological Analytical Manual

FDA and AOAC International

Contains methods for the microbiological analysis of foods. Currently available online at https://www.fda.gov/food/laboratory-methods-food/bacteriological-analytical-manual-bam.

British Pharmacopeia, BP

British Pharmacopoeia Commission Secretariat

Contains validated standard methods of analysis for pharmaceutical and health-related products; accepted in the United Kingdom.

European Pharmacopeia, EP

European Directorate for the Quality of Medicines – Council of Europe

Contains validated standard methods of analysis for pharmaceutical and health-related products; accepted in Europe outside the United Kingdom.

Official Methods of Analysis of AOAC International

AOAC International

Contains a variety of microbiological and chemical analysis methods.

Standard Methods for the Examination of Water and Wastewater

A joint publication of the American Public Health Association, the American Water Works Association, and the Water Environment Federation.

Contains current practices for the analysis of water.

The United States Pharmacopeia/National Formulary, USP-NF

United States Pharmacopeia

Contains validated standard methods of analysis for pharmaceutical and health-related products; accepted by the U.S. Food and Drug Administration.

---

determined as exactly as possible with modern methods. A wide variety of SRMs for chemistry, physics, and manufacturing are available. A biotechnology subdivision at NIST sells SRMs for biotechnology applications, such as DNA standards for quality control in forensic DNA testing. The U.S. Pharmacopeia organization also sells well-characterized, certified reference standards that the FDA accepts for use by pharmaceutical manufacturers.

Chemical and biological reference materials are available from commercial suppliers, as well as agencies such as NIST and USP. Commercially manufactured reference materials often have associated documentation that shows how their properties were determined, and the certainty that the manufacturer has in their values. Table 26.5 shows the documentation for a reference standard containing a mix of minerals. Note the care that was used to determine the levels of each mineral and the description of the variability associated with the measurements. A well-characterized standard like this mineral mixture might be used, for example, in a situation where regulatory decisions will be made based on test results.

Biologists commonly use products of biological derivation that are difficult and expensive to characterize as fully as the mineral mixture described in Table 26.5. Biological materials that are reasonably well characterized are used to determine the response of a method or an instrument to the substance of interest.

Regardless of the source of reference materials, they must be carefully and securely stored to avoid loss, degradation, or damage. Well-characterized reference materials are a valuable resource in a laboratory.

26.2.5 DOCUMENTATION

Issues relating to documentation of assays are shown in Box 26.3. This box is derived from the FDA's Good Manufacturing Practices regulations. It describes the items that should be recorded when conducting tests on samples in quality-control laboratories in pharmaceutical companies. Although the information in this box is intended for pharmaceutical analysts, it is relevant in any setting where assays are performed.

TABLE 26.5

Specifications for a Mineral Mixture

Property	Units	Consensus Value[a]	Standard Deviation[b]
Alkalinity	mg/L	93.56	4.35
Calcium	mg/L	27.21	2.2
Chloride	mg/L	91.74	4.09
Conductivity	µmho/cm	661	15.6
Fluoride	mg/L	2.39	0.18
Magnesium	mg/L	12.18	0.66
Potassium	mg/L	9.08	0.68
Sodium	mg/L	83.28	4.14

[a] The amount of each mineral in the mixture was rigorously analyzed in 230 laboratories, and the mean of the results is reported as the "Consensus Value."

[b] The Standard Deviation column indicates the variability in the data from the various laboratories.

**BOX 26.3   RULES FOR LABORATORY RECORDS QUOTED
FROM GMP REGULATIONS, 21 CFR 211.194**

a. Laboratory records shall include complete data derived from all tests as follows:
  1. A description of the sample received for testing with identification of source (that is, the location from which the sample was taken), quantity, lot number or any other distinctive code, the date sample was taken, and the date sample was received for testing. [Note also that the person taking the sample should usually be identified.]
  2. A statement of each method used in the testing of the sample. The statement shall indicate the location of data that establish that the methods used in the testing of the sample meet proper standards of accuracy and reliability.
  3. A statement of the weight or measure of sample used for each test, where appropriate.
  4. A complete record of all data secured in the course of each test, including all graphs, charts, and spectra from laboratory instrumentation, properly identified to show the particular component tested.
  5. A record of all calculations performed in connection with the test, including units of measure, [and] conversion factors.
  6. A statement of the results of tests and how the results compare with established standards.
  7. The initials or signature of the person who performs the test and the date(s) the tests were performed.
  8. a. The initials or signature of a second person showing that the original records have been reviewed for accuracy, completeness, and compliance with established standards.
     b. Complete records shall be maintained of any modification of an established method employed in testing. Such records shall include the reason for the modification and data to verify that the modification produced results that are at least as accurate and reliable for the material being tested as the established method.
     c. Complete records shall be maintained of any testing and standardization of laboratory reference standards, reagents, and standard solutions.
     d. Complete records shall be maintained of the periodic calibration of laboratory instruments, apparatus, gauges, and recording devices.

26.3 AVOIDING ERRORS IN ASSAYS AND TESTS

26.3.1 METHODS ARE BASED ON ASSUMPTIONS

Assay methods are based on certain assumptions and have potential inaccuracies and uncertainties. Consider, for example, a conventional test used to determine if a substance is contaminated by bacteria. Bacteria are too small to see with our eyes and are usually too numerous to count individually. It is necessary, therefore, to use an indirect assay to visualize and count bacteria in a substance. One such method is the spread plate method:

1. Obtain a sample from the substance to be tested.
2. Prepare an appropriate serial dilution of this sample.
3. Apply 0.1 mL of the diluted sample onto petri plates that contain sterile bacterial nutrient medium and distribute the sample evenly over the plate.
4. Incubate the plates overnight to allow bacterial growth.
5. If bacteria were present in the sample, visible bacterial colonies will grow on the plates; count these colonies. Each colony is called a *colony-forming unit (CFU)*.
6. The number of bacteria in the original substance is calculated based on the assumptions that:
   a. Each visible colony is derived from a single bacterium that was present in the sample.
   b. Each living bacterium in the diluted sample forms a colony on the plate.

This counting method is indirect; the number of colonies on the plate is used as an indicator of the number of bacteria in the original substance. Although the spread plate counting method is simple and widely accepted, there are cautions in its use including:

- The bacteria (if present) in the original substance must be uniformly distributed when the sample is taken so that the sample represents the whole substance.
- The method assumes that dead or non-reproductive bacteria are unimportant since these will not be counted.
- The method is inaccurate if adjacent bacteria merge to form a single colony.

- The method is inaccurate if too few or too many bacteria are plated.
- The method is inaccurate if the plates are incubated too briefly or too long.

The spread plate method is widely used, and its limitations are well understood in the scientific community. When using less-common assay methods and new methods, it is important for the analyst to evaluate the method's limitations and understand its applications.

26.3.2 THE USE OF POSITIVE AND NEGATIVE CONTROLS

Positive and negative controls are critical components of most assays used in biotechnology laboratories. For example, if one were using the plate count method to determine the number of bacteria in a sample, it would be good practice to remove 0.1 mL of the diluting solution (as used in Step 2) and apply it to another petri plate that contains nutrient medium. The diluting solution should be sterile, as should be the petri plate and the nutrient medium. Therefore, no colonies should grow on this plate. This is a *negative control*. If any colonies appear on the negative control plate, it means that the diluting material or the plates were contaminated with bacteria to begin with, or that contaminants were introduced accidentally sometime during the procedure. In the event of contamination, none of the results of the test are valid and it must be repeated with fresh, uncontaminated materials. A positive control would be to plate 0.1 mL of a substance known to contain bacteria; bacteria should grow. If they do not, the growth medium or conditions are not correct, and none of the results of the test are valid.

Let's further consider positive and negative controls by discussing another example, this one involving antibodies. *Methods involving antibodies are called* **immunochemical methods**. Antibodies have evolved over millennia to find, recognize, and bind specific invading substances (e.g., viruses and bacteria) from amidst a myriad of molecules. (See Chapter 1, Section 1.2.3, for an explanation of antigens and antibodies.) Different antibodies bind to different antigen targets. Analysts take advantage of the remarkable specificity of antibodies to target and visualize a substance of interest. In one type of immunochemical method, antibodies are applied to cellular material that has been prepared for electron microscopy to find and identify a subcellular structure. Suppose we are using this method to see whether a particular tissue contains

Sample is positive	Sample is negative
True positive result	False positive result
False negative result	True negative result

FIGURE 26.4 True *versus* false test results. In a situation where a sample is either positive or negative for an analyte, there are four possible test results. A sample that is really positive for the analyte can have either a true-positive or a false-negative result when tested. A sample that is really negative for the analyte might have a true-negative or a false-positive result when tested.

a hormone receptor of interest. There are four possible outcomes when the tissue is treated (Figure 26.4):

1. The tissue might contain the receptor and might form a visible deposit when treated by immunological methods. This is called a **true-positive result** *because the result is correct and it is positive for the receptor.*

2. The tissue might contain the receptor, but because of some problem in the procedure, might fail to form a visible deposit when treated. This is called a **false-negative result.** *The result is negative, and it is incorrect.*

3. The tissue might not contain the receptor and not form a visible deposit when treated. This is called a **true-negative result**. *The result is negative for the receptor, and it is correct.*

4. The tissue might not contain the receptor, but it might form a visible deposit because of nonselective binding or another problem in the procedure. This is called a **false-positive result**. *The result is positive, and it is incorrect.*

In order to help avoid false results, positive and negative controls can be used. A **negative control** *is tissue that is like the test sample, but is known not to contain the target of interest and therefore should not form a visible deposit* (Figure 26.5). The negative control tissue is treated exactly the same way as the other samples. If the negative control tissue does show a positive result, then we know there was too much antibody or there was another problem in the procedure and so we would not trust the results on the samples. The routine use of negative controls helps avoid false-positive results in assays.

A **positive control** *is tissue known to contain the target of interest and therefore should form a visible deposit if the reagents and procedures used were good.* Thus, if the positive control fails to form a visible deposit, then we know there was a problem in the

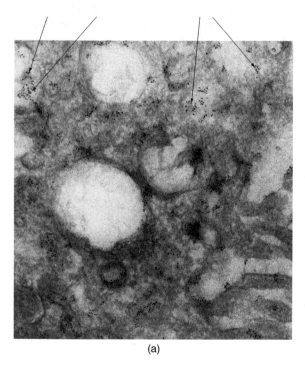

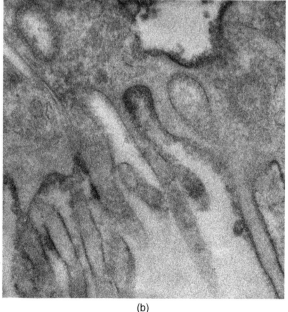

(a)   (b)

FIGURE 26.5 Using a negative control in an immunochemical assay method. (a) The small, round, dark deposits in this electron micrograph (arrows) represent antibodies bound to the specific receptor protein of interest. (b) A negative control. This is tissue known not to have the receptor of interest. The negative control tissue is processed alongside the experimental samples and is treated exactly the same as experimental tissue. No antibody deposits appear in this tissue.

Quality Laboratory Tests and Assays

Positive control	Helps guard against false negative results
Negative control	Helps guard against false positive results

FIGURE 26.6 The importance of positive and negative controls.

procedure. The routine use of positive controls helps avoid false-negative results in assays (Figure 26.6).

In addition to the preceding possibilities, the results of an assay might be ambiguous. For example, in the immunochemical assay example, the deposit might be difficult to see, the tissue may appear to be labeled in some areas, but not in other similar areas, various tissue samples might unexpectedly differ, and so on. Ambiguous results are frustrating and typically mean that the experimental test system has not been optimized adequately. More preliminary testing may reveal conditions that can be slightly altered to give better results.

The use of positive and negative controls is applicable in almost all assay situations, not only immunochemical and microbiological assays. For example, positive and negative controls are used in clinical testing of patient samples. Consider a blood test used to see if a patient is infected with a certain parasite. A *positive control* would be blood that is known to test positive for the parasite. The positive control is run alongside the patient's sample and helps detect false-negative results. (A *false negative* occurs when a person who is infected shows up as uninfected.) A negative control (i.e., blood that is known not to react with the test) is used to avoid a false-positive result. (A false positive occurs when a person who is uninfected appears to be infected.)

A note about terminology: When an assay or a test is performed, there are often more materials tested than just the "experimental" samples. There are also the positive and negative controls and the reference materials. All of these are often referred to as *samples*. A negative control, for example, may be called a *negative control sample*. The samples that are the actual subject of the analysis may be called the *unknowns*, the *unknown samples*, the *experimental samples*, or the *test samples*. It is essential that all the *samples* are processed and analyzed in the same way, whatever terminology we use to name them.

In summary, when performing tests and assays on biological samples, it is important to be aware of the assumptions, limitations, and potential errors inherent in those tests. Most biological tests and assays are

complex and so require development (discussed in the next section) and practice to get the best results. It is also vital to use positive and negative controls whenever possible to detect problems in the procedure.

26.3.3 REPLICATES

Replicates, similar to positive and negative controls, help analysts ensure trustworthy assay results. **Replicates** *involve repeating an assay more than once in order to assess the variability associated with the system being studied.*

There are differences in the way the term "replicate" is used in the literature. In this situation, we consider a replicate to be a repeat of an assay that involves the same sample and the same assay procedure (Figure 26.7). For example, suppose a given study involves taking a blood sample from experimental and control

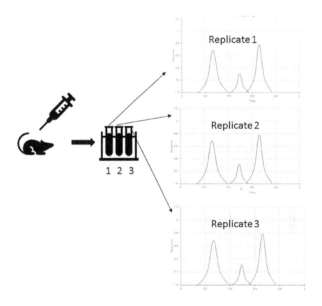

FIGURE 26.7 An example where an assay is replicated three times. A blood sample from an experimental mouse is subdivided, and each portion is prepared according to the same procedure. The prepared sample is then analyzed using an instrument that detects the components of the sample. These replicates help assess the variability in the procedure and the operation of the instrument. Observe that this type of replication process does not assess the variability that might exist from one mouse to another, only the variability in the assay procedure.[1]

---

[1] The term "technical replicate" is used for the type of replicate described here. In the performance of experiments, it is also important to consider "biological replicates." In Figure 26.7, biological replicates would involve performing the assay on multiple mice to look for variability between experimental subjects.

animals, processing the blood in a certain way, and running the prepared samples through an instrument that provides a readout of a particular analyte in the blood. It is possible to subdivide the blood sample from each animal and run it more than once through the entire assay procedure; each of these runs is a "replicate." Three replicates are commonly recommended although different situations call for different numbers of replicates. (In complex experimental studies, statisticians may be consulted to determine the proper number of replicates.) Replicates are expected to provide roughly the same assay results, but some variability is expected in sensitive assays. How much variability is acceptable needs to be determined for each situation. In a quantitative assay, if the variability among the replicates is acceptable, the results might be averaged. Sometimes the use of replicates reveals unexpected variability that is due to an error, such as a malfunctioning instrument. In such cases, the use of replicates alerts the analyst that there is a problem that must be corrected.

26.4 METHOD DEVELOPMENT AND VALIDATION

26.4.1 METHOD DEVELOPMENT

Many methods can be found in compendia, or can be purchased in kit form from manufacturers, or can be found in manuals and scientific literature. But sometimes new methods need to be developed for a particular application. New method development is often complex and requires experimentation and optimization of many factors.

Finding the basic concept for an assay is the first step. The concept must involve some sort of interaction or effect that can be detected and that is relevant to the property being analyzed. For example, a scientist interested in studying the functions of a hormone receptor might devise an immunochemical method to detect the receptor. A scientist interested in studying chemotherapeutic agents might devise a cell-based assay that detects biochemical changes in cultured cells exposed to the drugs.

Scientists experiment with a variety of factors during assay development to optimize the assay's performance and to eliminate interferences and ambiguous results. Development scientists must decide, for example, what chemical tag to use to label antibodies when optimizing an immunochemical assay. Scientists will likely experiment with more than one type of label to see which is most efficient. Assays that monitor the activity of enzymes require optimizing the temperature, the buffer, and the concentrations of substrates and cofactors. The polymerase chain reaction, which is discussed in more detail in Chapter 27, requires an instrument that cycles through a series of temperatures in a controlled fashion. The temperatures and their duration must be optimized. Many assays require an instrument for detection and/or to separate the components of a mixture. Instrumental methods always require optimization of the equipment parameters. Assays require positive and negative controls that must be optimized. During method development, the conditions of the method, the materials required, the controls, and the steps to perform the assay are all established.

Data analysis is another aspect of assays that needs to be developed. Data analysis may involve calculations, comparisons with standards, and comparisons with published data. Computers are often used to help eliminate artifacts, annotate data to make it easier for a human viewer to interpret, store data, and report data. Computers make it easier for an operator to work with data, but computer software must be extensively tested to ensure its validity.

The case study below provides an example of method development. (For another example, see the case study in Chapter 3, "Crime Scene Investigation: A Behind-the-Scenes Story," in Section 3.1.4.)

Case Study: Assay Development in a Biotechnology Company

BellBrook Labs, a biotechnology company formerly located in Madison, Wisconsin, created assay systems for sale to pharmaceutical companies where the assays are used for high-throughput screening (HTS). HTS assays are used to identify potential drugs by looking for compounds that inhibit the pathological effects of an aberrant protein associated with a specific disease. Pharmaceutical companies have "libraries" with hundreds of thousands of compounds that are possible drug candidates. The limiting factor for the pharmaceutical companies is finding reliable, rapid methods of screening all their compounds to find the few that might have a therapeutic effect – which is where BellBrook Labs found its business niche.

(Continued)

Case Study (*Continued*): Assay Development in a Biotechnology Company

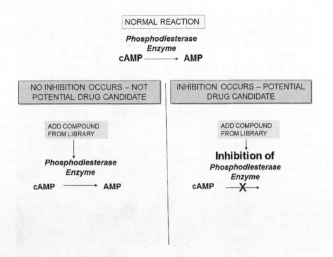

FIGURE 26.8   An HTS assay for compounds that inhibit phosphodiesterase enzymes.

One of BellBrook's patented discoveries was an assay system that targets phosphodiesterase proteins. Phosphodiesterases are enzymes that convert the molecule cyclic AMP (cAMP) into AMP. Although phosphodiesterases are normal enzymes, in certain situations they behave aberrantly and become associated with many diseases, including rheumatoid arthritis, diabetes, and erectile dysfunction. BellBrook scientists therefore developed an assay system that detects the conversion of cAMP to AMP as a measure of the ability of test compounds to inhibit the phosphodiesterase enzymes. The assay depends on an antibody that BellBrook manufactures. The antibody is exquisitely sensitive and can distinguish cAMP from AMP. It binds AMP, which is the product of the reaction of interest. BellBrook used a fluorescence polarization instrument to detect the AMP–antibody complex. Pharmaceutical companies can use this assay system to test their thousands of potential drug compounds by adding them to mixtures containing the phosphodiesterase enzymes and cAMP, along with various ions and other molecules that are required by the reaction. After a set length of time, they place the reaction mixtures in a fluorescence polarization instrument to detect whether cAMP has been converted to AMP. If cAMP is converted to AMP, then the potential drug candidate had no effect. If the conversion of cAMP to AMP is inhibited by a particular compound, then that compound can further be explored as a potential drug to treat various diseases (Figure 26.8).

Before the BellBrook assay system was ready for sale, months of experimentation and optimization were required. This task was given to Dr. Rebecca Josvai, an R&D scientist at BellBrook Labs, and her colleagues. Dr. Josvai's tasks included:

- testing the various components of the reaction mixture at many concentrations;
- testing different preparations of antibody to see which is the most sensitive and specific;
- testing for potential causes of erroneous results, for example, spontaneous conversion of cAMP to AMP in the absence of enzyme;
- optimizing the conditions for the reaction, such as temperature and time;
- optimizing settings of the fluorescence polarization instrument;
- testing the assay using molecules known to be inhibitory (positive controls) and molecules known not to be inhibitory (negative controls) to be sure the assay always performs as expected;
- scaling up the assay so that thousands of compounds can be rapidly tested robotically; and
- determining what controls should be used every time the assay is run in order to detect potential errors.

(*Continued*)

Case Study (*Continued*): Assay Development in a Biotechnology Company

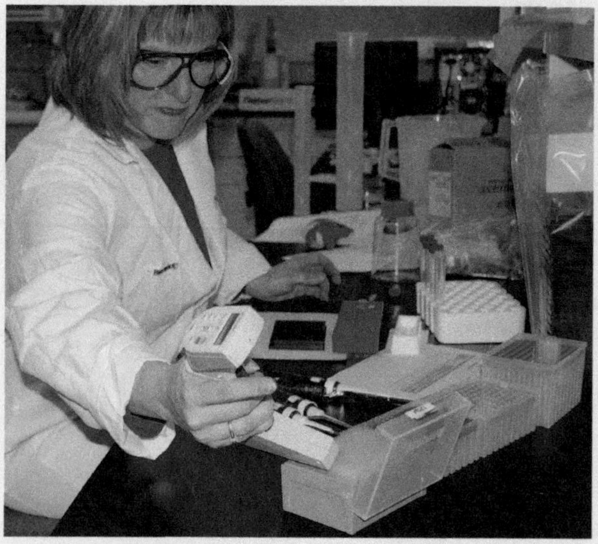

FIGURE 26.9 Dr. Josvai setting up multiple reactions in 384-well plates.

Dr. Josvai is shown in Figure 26.9 setting up a series of reactions to determine the effect of enzyme concentration. She uses a 384-well plate; each well is like a tiny test tube. To avoid errors, Dr. Josvai follows a spreadsheet that she previously prepared that tells her how much of each reaction component to place in each well.

After the reactions have occurred, Dr. Josvai places the 384-well plate in the fluorescence polarization instrument, which collects data from all 384 wells and sends it to a computer. Once the assay has been successfully optimized, it will be tested using a small robotic apparatus that automatically dispenses set volumes of liquids into the wells of plates. In a pharmaceutical company, robotic apparatuses are used to screen many thousands of compounds far faster than a human could do so (see Figure 18.17).

The information from each of Dr. Josvai's experiments is saved on a computer spreadsheet and documented in a laboratory notebook, which is signed and dated daily, and is countersigned by a witness at regular intervals.

The optimization of this assay illustrates some of the many efforts required for assay development. It is important to predict (and prevent) potential pitfalls of the assay. All aspects of assay development must be documented, including the conditions of each assay test, the results of various experiments, the exact materials used, suppliers of reagents, and instrument parameters. Observe that each of Dr. Josvai's experiments is actually 384 experiments, each conducted in its own tiny well. With so much data coming from a single experiment, a computer is required to store, process, and display the information. For BellBrook Labs, all this experimentation and optimization is necessary to create a reliable product that provides value to its customers.

26.4.2 Method Validation

26.4.2.1 Overview

Method validation is a formal process that occurs after method development, to verify the ability of the method to reliably obtain a desired result. As explained by John Butler, from the National Institute of Standards and Technology, "*Validation* refers to the process of demonstrating that a laboratory procedure is robust, reliable, and reproducible in the hands of the personnel performing the test in that laboratory. A *robust method* is one in which successful results are obtained a high percentage of the time and few, if any, samples need to be repeated. A *reliable method* refers to one in which the obtained results are accurate and correctly reflect the sample being tested. A *reproducible method* means that the same or very similar results are obtained each time a sample is tested."

Good Manufacturing Practices require formal validation of test methods used in the pharmaceutical industry. Although the process of method validation discussed in this chapter is based primarily on requirements in the pharmaceutical industry, the more general principles of validating a method are relevant in any laboratory. In research environments, applying these principles helps ensure reproducible experimental results.

A variety of parameters are tested during validation; the particular parameters depend on the method and the organization involved. For example, the International Council for Harmonisation (ICH) and the United States Pharmacopeia (USP) provide guidelines for method validation for the pharmaceutical industry. Their guidelines are similar, but not identical. Generally, the tests performed during validation fall into one of the following categories that are described in more detail below: (1) **accuracy**, (2) **precision**, (3) **limit of detection**, (4) **limit of quantitation**, (5) **specificity/selectivity**, (6) **linearity**, (7) **range**, (8) **robustness**, (9) **ruggedness**, and (10) **system suitability**.

26.4.2.2 Accuracy

It seems obvious that an assay or test method should give the "right" answer. **Accuracy** *is the closeness of a test result to the true or accepted value.* Accuracy is sometimes evaluated by testing a reference standard whose true or expected value for the test is known. Accuracy is also sometimes evaluated by comparing the results of the method being validated with another method that was previously validated and is considered to be correct. Accuracy can be expressed as the difference between the result obtained for the test and the known or accepted value expected.

In the pharmaceutical industry, it is common to test accuracy by adding known amounts of the analyte to samples – called *spiking* the samples – followed by a determination of the percent recovery of the added material. If the accuracy of the assay is "perfect," the assay will detect 100% of the spiked material. The ICH recommends collecting accuracy data from a minimum of nine determinations over a minimum of three concentration levels covering the specified range (e.g., three concentrations, three replicates each).

26.4.2.3 Precision and Reproducibility

Precision *is the degree of agreement between individual test results when the procedure is applied over and over again to portions of the same sample.* There are different types of precision. Measurements repeated in succession on the same day tend to be relatively consistent. In contrast, measurements performed on different days, by different people, and using different materials and equipment tend to be far more variable. **Repeatability** *is the precision of measurements made under uniform conditions.* **Reproducibility** *(sometimes called ruggedness) is the precision of measurements made under nonuniform conditions, such as in two different laboratories.* Repeatability and reproducibility are therefore two practical extremes of precision.

26.4.2.4 Limit of Detection

Limit of detection (LOD) *is the lowest concentration of the material of interest that can be detected* by the method. (The term **sensitivity or analytical sensitivity** sometimes is used synonymously with limit of detection.) Every method has a limit below which it cannot detect the analyte of interest. Analytical methods, therefore, can never prove that a particular substance is not present in a sample; rather, they show that it is not present at a concentration above a certain detection limit. For some situations, the limit of detection of a test is very important. For example, an analyst who develops a test for the purpose of detecting viral contamination in a drug product wants to be sure of the minimum number of viral particles that must be present to give a positive result. In this case, the limit of detection of the test must be validated. In other situations, this parameter may not require validation. For example, an assay that is designed solely to confirm the identity of a drug substance assumes that the substance being analyzed constitutes the bulk of the material present. In this case, the limit of detection is unlikely to require validation.

Noise in an analytical instrument (explained in Section 16.3.2.3) often determines the limit of detection and the limit of quantitation for an assay. Therefore, signal-to-noise ratio is often determined as part of a validation procedure.

26.4.2.5 Limit of Quantitation

Limit of quantitation (LOQ) *is the lowest concentration of the material of interest that the method can quantify with acceptable accuracy and precision.* The limit of quantitation is usually a higher concentration than the limit of detection. As for the limit of detection, the limit of quantitation may or may not require validation, depending on the nature of the test.

26.4.2.6 Specificity/Selectivity

Specificity *is a measure of the extent to which a method can unequivocally determine the presence of* *a particular compound in a sample in the presence of other materials that may be expected to be present.* Other materials might include impurities, degradation products, and matrix. A very specific test will only give a positive result to the compound of interest. A less specific test might erroneously give a positive result when a substance similar to the compound of interest is present. Note that some sources call this parameter **selectivity** (although other sources note slight distinctions in the two terms).

Testing specificity might involve adding the substance of interest to different relevant matrices (e.g., blood, soil, and buffer) and adding possible interfering agents (e.g., metabolites, detergents, and decomposition products) and seeing if the same results are obtained with and without the potentially interfering substances. Validating specificity is critical in a method such as DNA fingerprinting that must identify a single individual in a sample that may include many impurities.

26.4.2.7 Linearity and Range

Linearity *is the ability of a method to give test results that are directly proportional to the concentration of the material of interest (within a given concentration range).* **Range** *is the range of concentrations, from the lowest to the highest, that a method can measure with acceptable results.* Tests and assays have a particular, limited range in which they are linear. (This idea is discussed in more detail in Chapter 28.) When an assay or a test is developed, it is customary to determine the linear range. Validating linearity is important for many quantitative methods.

26.4.2.8 Robustness/Ruggedness

Robustness (sometimes used synonymously with ruggedness) *is the capacity of a method to remain unaffected by small, but deliberate variations in method parameters and provides an indication of its reliability during normal usage.* For example, chromatographic analytical methods usually require a particular buffer pH. A chromatographic method that has good specificity in a range from pH 6.5 to 6.9 is more robust (relative to pH) than a method that requires a pH between 7.2 and 7.3. Robustness is important because there are often subtle and not so subtle differences between laboratories and individuals performing a particular method.

Example 26.1

A chromatographic method of analyzing tricyclic antidepressants was evaluated for its robustness relative to pH. The amount of time it took for a particular drug to run through the column was checked at pH 6.9, 7.0, and 7.1. The results were as follows:

pH	Running Time
6.9	19.28
7.0	21.34
7.1	23.48

The percent difference between the running times for this drug at pH 6.9 and 7.0 is 9.65% and between pH 7.0 and 7.1 is 10.03%. This test would likely be considered to have poor robustness relative to pH. This is because it is difficult in the laboratory to absolutely ensure that a solution has a pH within 0.1 of a pH unit; it is not unusual for a solution to vary 0.1 pH units under normal laboratory conditions. In chromatography, 10% differences in running time could adversely affect the results of the test by causing sample components to shift positions and interfere with one another.

(This example is derived from "Contemporary Issues in Regulatory Compliance" by Michael E. Swartz, presented in a workshop by Waters Corporation on October 10, 1996.)

26.4.2.9 System Suitability

A method can enter into routine use after it has been developed and validated. System suitability tests may be thought of as a way to periodically verify that the method is still effective for routine use. The FDA defines **system suitability** as: *"Determination of instrument performance (e.g., sensitivity and chromatographic retention) by analysis of a reference standard prior to running the analytical batch."* System suitability requires that all the components of a system be tested together. If a method involves chromatography, for example, then the instrument, analyst, and reagents must all be tested together to be sure the whole system functions properly. As defined by the FDA, system suitability testing requires a characterized reference standard that contains the substances present in the actual samples along with expected impurities. The reference standard is run before the actual samples. Parameters, such as resolution between peaks and reproducibility, are determined for the reference standard and compared against specifications previously established for the method. Standards of known absorbance can similarly be used to periodically test a spectrophotometric method.

26.4.2.10 Validation Parameters Required

It is not always necessary to validate every analytical performance parameter. The type of assay method and its intended use determine which parameters need to be investigated. For example, assays that test the main ingredient in a drug product do not require validating the limit of detection or quantitation. In contrast, assays that measure the level of an impurity require validating the limit of quantitation, linearity, and range.

It is also common to distinguish between full validation and partial validation. **Full validation** is necessary when "inventing" a new assay and when developing methods to test a new drug entity. **Partial validation** is used for methods that have already been validated. Partial validation would be appropriate, for example, when analysts in a laboratory begin using an assay that was successfully used elsewhere, or begin using an assay that is from the USP. In a research laboratory, partial validation might be used when introducing an assay from a commercial kit or adopted from a manual. Partial validation may also be appropriate when a new instrument is purchased for an existing assay, or a new type of sample is to be tested with an existing method. Partial validation might also be referred to as "performance verification."

Forensic scientists distinguish between developmental and internal validations where **developmental validation** *involves new technologies.* **Internal validation**, *on the other hand, involves verifying that established procedures examined previously during developmental validation will work effectively in one's own laboratory.* Developmental validation is typically performed by commercial manufacturers of assay kits and large labs, such as the FBI Laboratory, while internal validation is the primary form of validation performed in smaller local and state forensic DNA laboratories.

26.4.2.11 Diagnostic Assay Validation

Two additional parameters must be validated for diagnostic assays. These are described in the case study Diagnostic Assay Validation in the Time of COVID-19, Part II.

Case Study: Diagnostic Assay Validation in the Time of COVID-19, Part II

Validation is essential for diagnostic tests, as it is for all assays. But for diagnostic tests, validation has two key parameters not yet discussed in this chapter: **diagnostic sensitivity** and **diagnostic specificity**. Although the terms "sensitivity" and "specificity" have already been introduced, they have a specific meaning in the context of diagnostic assays.

- **Diagnostic sensitivity (diagnostic SN)** *is defined as the percent of known affected individuals that test positive. It is an indication of false-negative results.* For example, if all people who are infected by SARS-CoV-2 have a positive result when tested for the virus, then the diagnostic SN is 100%. But if 5% of the affected people have a false-negative test result, then the diagnostic SN is 95%.
- **Diagnostic specificity (diagnostic SP)** *is defined as the percent of known uninfected individuals that test negative. It is an indication of false-positive results.* For example, if all unaffected people test negative, then the diagnostic SP is 100%. But if, for example, 10% of the unaffected people have a false-positive test result, then the diagnostic SP is 90%.

Figure 26.10 is a mnemonic to help remember the distinction between the two terms.

To validate diagnostic sensitivity and diagnostic specificity, it is necessary to have samples that are known to be positive for the condition, and samples that are known to be negative. These controls, or known samples, are run through the test, and the numbers of true-positive, true-negative, false-positive, and false-negative results are tallied. Diagnostic sensitivity and specificity can then be calculated, as illustrated in the example in Figure 26.11.

(Continued)

Case Study (*Continued*): Diagnostic Assay Validation in the Time of COVID-19, Part II

Diagnostic SN ⟶ False Negative
Diagnostic SP ⟶ False Positive

FIGURE 26.10 Diagnostic sensitivity indicates the percent of false-negative test results, while diagnostic specificity indicates the percent of false-positive results.

Diagnostic sensitivity is calculated as True positive/(True positive + False negative)

Diagnostic specificity is True negative/(True negative + False positive)

	Samples Known to Be Infected (600)	Samples Known *Not* to Be Infected (1400)
Positive Test Result	570 (True positive)	46 (False positive)
Negative Test Result	30 (False negative)	1354 (True negative)
	Diagnostic SN	Diagnostic SP

This example

$$\text{Diagnostic sensitivity} = \frac{TP}{TP+FN} = \frac{570}{600} = 95\%$$

$$\text{Diagnostic specificity} = \frac{TN}{TN+FP} = \frac{1354}{1400} \approx 96.7\%$$

FIGURE 26.11 Validation calculations for diagnostic sensitivity and specificity. The numbers in this example are from an article by R. H. Jacobson, New York State College of Veterinary Medicine, Cornell University, Ithaca, New York, USA. https://www.fws.gov/aah/PDF/PrinValDiagAssayforInfDis%20-%20JACOBSON.pdf.

There are additional parameters that must be validated for diagnostic assays in addition to diagnostic sensitivity and specificity. It is generally important to validate the limit of detection, much as one would for any analytical assay. For example, for a COVID-19 assay that detects virus, it is important to know how much virus must be present in a sample to give a positive result. Also, there is a concept much like analytical specificity (as specificity was defined earlier in this chapter) that is evaluated for diagnostic assays. **Cross-reactivity** *is the tendency of the assay to give a positive result in the presence of other similar analytes.* For example, in the case of COVID-19, if an assay detects other coronaviruses in addition to SARS-CoV-2 (such as a virus that causes the common cold), it is called cross-reactivity. Cross-reactivity can cause false-positive results. When a diagnostic assay is validated, researchers try to determine if there are likely cross-reactive agents that might impact the accuracy of the test.

Understanding diagnostic specificity and sensitivity is particularly relevant in the days of the COVID-19 pandemic. We will consider why this is so using serological assays (in use prior to vaccinations) as an example. It was important to know what percent of the population had been infected by this virus and had recovered. *The standard way to test for a past infection is to look for antibodies in a person's blood; such an assay is termed a* **serological assay**. Knowing who has antibodies against the virus is important because:

- If a number of people have antibodies and were not aware of being sick, then we know that in many people the illness is mild. This would be good news because it indicates that infection is not always serious. But if people can be infected without knowing it, they might unknowingly spread the illness to other people. Knowing if many people have asymptomatic infection is important for determining prevention strategies.
- Knowing who has antibodies against the virus can help epidemiologists track the disease's spread and predict where it will occur next.

(Continued)

Case Study (*Continued*): Diagnostic Assay Validation in the Time of COVID-19, Part II

- At the individual level, if a person knows he or she has antibodies, it might indicate that the person is now protected against the disease. This individual might decide to participate in higher-risk activities, such as visiting a friend, or returning to a job with frequent public contact. In the winter of 2021, no one knew with certainty if antibodies against the virus that causes COVID-19 will confer lasting immunity, but it is known that often antibodies do protect against disease.

The final bulleted point, that is, using a serological assay to guide the decisions of an individual, is problematic. As we already noted, a quality assay is one that can be trusted when making decisions. In the case of the serological assays available in 2020, many of the assays are probably not trustworthy for this type of decision, even if their diagnostic sensitivity and specificity are 95%. (Some of the commercial assays do achieve 95% sensitivity and specificity, but some do not.) Why is this so? It seems that if an assay provides true results at a 95% rate, and if that assay provides a positive result for an individual, then the person should be 95% sure that he or she does have antibodies. But this is not the case. We will show why with an example:

- Suppose a serological test for antibodies against the COVID-19 virus has a diagnostic SP of 95%.
- Suppose that the percent of people in the population who really do have antibodies against this virus is 5%.
- 100 people are tested; 5 out of the 100 (5%) are expected to have a positive test result because they do have antibodies.
- But, with a diagnostic specificity of 95%, 5% of people tested are expected to have a *false*-positive result – that would be 5 people.
- Thus, there might be an equal number of false-positive and true-positive test results in this population. Given a 50–50 chance that the result is correct, an individual cannot trust the assay results to make decisions!

A few notes about this analysis: First, as the percent of people with antibodies in the population increases, a positive result becomes more and more likely to be correct. Second, as the assays approach 100% diagnostic specificity, which some tests do, the results become more and more likely to be correct. However, a number of diagnostic tests have about 95% diagnostic specificity, despite being optimized and properly developed. Third, note that an assay that is not useful to guide an individual's decision-making is still usable at the population level. Epidemiologists can adjust rates for the entire population to compensate for false positives. The adjusted result can then be used to guide policy recommendations.

As a final note, once people are vaccinated, serological assays are not useful in determining if a person was ever infected because vaccination causes the person to develop antibodies. However, serological assays are still important during a pandemic to monitor the antibodies generated by vaccination and to determine what happens to antibody levels over time.

Examples of calculations exploring diagnostic specificity and sensitivity are provided in Practice Problems at the end of this chapter.

26.5 SUMMARY OF GENERAL PRINCIPLES THAT HELP ENSURE TRUSTWORTHY ASSAY RESULTS

Data, information, and knowledge are the products generated in laboratories. In order to have quality laboratory products, it is essential to have quality assays, that is, assays that can be trusted when making a decision. For example, in a clinical laboratory a quality assay result is one that can be trusted when making a decision about how to treat a patient. In a quality-control laboratory, a good assay can be trusted to make a decision about whether or not to release a product to consumers. In a research laboratory, investigators require data they can trust when drawing conclusions, writing scientific papers, applying for grants, and designing further studies. This chapter introduced the general features of assays and the means by which analysts create, develop, validate, optimize, and perform assays to get quality results.

Assay creation and development may or may not occur in the laboratory where an assay is ultimately used. For example, manufacturers have developed and marketed convenient kits to assay the amount of protein in samples; these kits are used in laboratories around the world for a wide variety of purposes. In other cases, novel assays are created and developed by researchers to use in their own laboratories. Validation is part of the development process in which a series of parameters are tested to learn how an assay performs, the limits of the assay, and how to best achieve accurate and reproducible results.

It is necessary to optimize assays for each application. For example, as we will see in Chapter 28, spectrophotometric assays require optimization of the wavelength chosen for analysis. Immunological assays, which will be described in Chapter 29, require optimization of many factors relating to the antibodies, buffers, and sample preparation.

The use of proper controls is also essential in ensuring quality assay results. Negative controls are samples that should not provide a positive result, and positive controls are samples that should provide a positive result. Negative controls guard against false-positive results; positive controls guard against false-negative results. Devising good control samples is part of developing and optimizing an assay method. Consider, for example, the situation introduced in Chapter 5 where we saw that the first assay kit developed in the United States to detect infection by the SARS-CoV-2 virus was defective. In that case, the negative control samples appeared to have viral RNA when they should not have had any virus. The fact that the test kit was defective was a major setback, but the negative control did exactly what it was supposed to do – it alerted investigators to contamination in the

environment in which the kits were manufactured. Knowing that there was a contaminant allowed the developers to fix the problem and subsequently create a good test kit. Without the negative control, it is conceivable that the contamination problem would have taken longer to detect with more severe consequences.

So far, we have defined a quality assay as one that can be trusted when making decisions. We can refine our definition at this point and say that a quality assay is one that can be trusted when making decisions – with an understanding of how much uncertainty there may be in the result. Consider the validation of the COVID-19 serological assays discussed in the case study "Diagnostic Assay Validation in the Time of COVID-19, Part II." We saw in that case study that diagnostic assay kits usually generate a certain number of false-positive and false-negative results. The number of false positives and negatives may be very low, but it is seldom possible to create a perfectly accurate diagnostic test. Validation allows analysts to estimate the percent of time the assay will provide an erroneous result. As we saw in that case study, if a serological COVID-19 test has a 95% diagnostic specificity, it would be unwise to trust a positive antibody result if an individual is deciding whether to visit friend's home. However, a test with a 95% diagnostic specificity may be trustworthy enough for epidemiologists to estimate the antibody level in a population. The epidemiologists can take the assay's uncertainty into account as they draw conclusions about the population. Practice Problems 7–11 also address the concept of uncertainty in assay results

The following three chapters in this unit explore in some depth important classes of assays in biotechnology: PCR, spectrophotometric, and immunological assays. While we cannot explore every class of assay in depth in this textbook, the general principles explored in this unit apply to a great many laboratory situations.

Practice Problems

1. A company produces a product that is packaged into vials that are then loaded onto trays. A technician tests the first vial in each tray as part of a quality-control program. Is this a good method of obtaining a random sample?
2. A research laboratory is developing a chromatographic method to check blood samples for the presence of a particular hormone. The scientists take blood and add to it some of the hormone of interest and a second hormone that is structurally similar to see if the method can distinguish between the

two. Which of the following characteristics of the assay are they checking? (1) accuracy, (2) precision, (3) limit of detection, (4) limit of quantitation, (5) specificity (selectivity), (6) linearity, (7) range, (8) robustness.

3. A research team is developing a method of DNA fingerprinting. The method involves cleaving DNA samples with restriction enzymes that recognize and cut at specific base sequences. The resulting fragments of DNA are separated from one another using electrophoresis and are made visible with dye. DNA from different sources forms different patterns when treated with this method.

   a. DNA restriction enzymes are sensitive to the level of salt in their buffer. The researchers therefore test the method using three buffers with different salt concentrations. Which of the characteristics of an assay listed in Problem 2 are they checking?

   b. If too little DNA is used in this method, then the pattern is not visible. If too much DNA is used, then the pattern is diffuse and is not useful. The researchers therefore evaluated the method to discover the minimum and maximum amounts of DNA that can be used. Which of the characteristics of an assay listed in Problem 2 are they checking?

4. The term *sensitivity* with reference to a biological assay is related to how low a level of target the assay can detect. The more sensitive the assay, the lower the level that can be detected. Which of the following statements is true?

   a. There is a risk associated with optimizing an assay to increase its sensitivity. This risk is that the number of false-negative results will increase.

   b. There is a risk associated with optimizing an assay to increase its sensitivity. This risk is that the number of false-positive results will increase.

5. Multiple choice: In an immunological assay, nonspecific binding tends to cause which of the following:

   a. False-negative results
   b. False-positive results
   c. The positive control comes up negative
   d. The negative control comes up positive
   e. b and d
   f. a and c

6. The directions for a home pregnancy test kit include the following instructions: "A pinkish-purplish line will form in the control (upper) window to tell you the test is working correctly. A positive result is two lines, one in each of the two windows (the upper and lower). A negative pregnancy test is a single line in the upper (control) window and no line in the lower window." Explain the upper window; why does a line form there? What is the significance of this line?

   Practice Problems 7–11 relate to a situation where the interpretation of the results of an assay can be difficult. In this case, the data relate to an imaginary assay for an imaginary disease, called philolaxis disease.

   The serum level of the imaginary protein, protein PHX, is elevated in patients suffering from philolaxis disease. The level of this protein varies from individual to individual and is approximately normally distributed. (You may want to review the section on the normal distribution in Chapter 14.) Figure 26.12a shows the distribution of the level of this protein in healthy patients; Figure 26.12b shows the distribution in ill patients. The distributions for healthy and ill patients are superimposed in Figure 26.12c. The level of protein PHX is used diagnostically to test whether a person is sick with philolaxis disease. You can see that there is variability in the serum levels of PHX for both affected and unaffected individuals and there is overlap in the two distributions. For this reason, there can be difficulty in classifying some patients as having the disease or not, based on this assay, and this can lead to a certain amount of uncertainty. It is important to understand why there is uncertainty when interpreting test results.

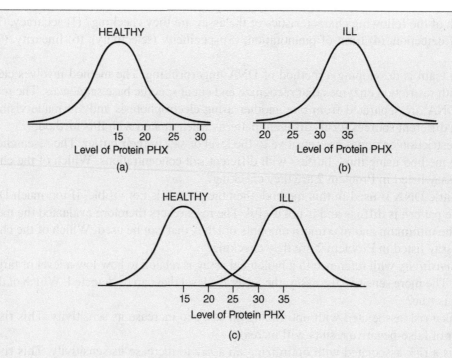

FIGURE 26.12A–C Data for Practice Problem 7.

7. a. Does a person with a serum level of 10 of protein PHX have philolaxis disease?
   b. Does a person with a level of 20 have philolaxis disease?
   c. Does a person with a level of 25 have philolaxis disease?
   d. Does a person with a level of 40 have philolaxis disease?
8. When clinicians test for philolaxis disease based on serum levels of protein PHX, they select a cutoff score. (See illustration 26.12d.) If a patient's level of protein PHX is above the cutoff score, the result is considered to be positive for philolaxis disease. If the patient's level is below the cutoff score, the result is considered to be negative for the disease.

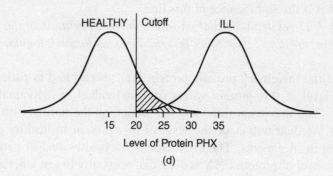

FIGURE 26.12D Data for Practice Problem 8.

   a. Multiple choice: If the clinicians decide that the cutoff score should be 20, about what percent of the time can they expect to get a false-positive result?
      (i) 0% **(ii)** 15% **(iii)** 50% **(iv)** 100%
   b. If the clinicians decide that the cutoff score should be 20, about what percent of the time can they expect to get a false-negative result?
      (i) 0% **(ii)** 15% **(iii)** 50% **(iv)** 100%

9. Multiple choice:

   a. If the clinicians decide that the cutoff score should be 25 (see illustration 26.12e), about what percent of the time can they expect a false-positive result?

       (i) 0% **(ii)** 2.5% **(iii)** 50% **(iv)** 100%

   b. If the clinicians decide that the cutoff score should be 25, about what percent of the time can they expect a false-negative result?

       (i) 0% **(ii)** 2.5% **(iii)** 50% **(iv)** 100%

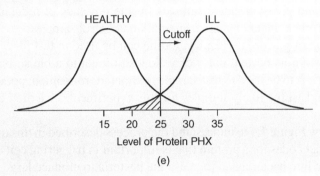

FIGURE 26.12E   Data for Practice Problem 9.

10. Multiple choice:

   a. If the clinicians decide that the cutoff score should be 30 (see illustration 26.12f), about what percent of the time can they expect to get a false-positive result?

       (i) 0% **(ii)** 15% **(iii)** 50% **(iv)** 100%

   b. If the clinicians decide that the cutoff score should be 30, about what percent of the time can they expect to get a false-negative result?

       (i) 0% **(ii)** 15% **(iii)** 50% **(iv)** 100%

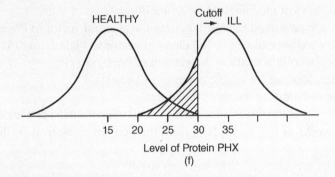

FIGURE 26.12F   Data for Practice Problem 10.

11. Consider HIV testing (for AIDS). If a person is told they do not have the virus, when in fact they do, they might unknowingly infect other people and also would not receive treatment. It is therefore common practice to use two tests for AIDS. The first test has a high rate of false positives and a low rate of false negatives. This test is relatively inexpensive and is used to screen for all samples that might be positive. All blood that tests positive with the first method is retested with a second assay that is more expensive but is better able to distinguish a true-positive result. Note that often in the laboratory there is more than one method available to test for a certain material or phenomenon. Each test is likely to have its own strengths and weaknesses, but together, the tests provide more reliable information than any test by itself. For this reason, when possible, it is good practice to use more than one test to confirm critical results.

In the case of philolaxis disease, if clinicians want the lowest possible level of false negatives, what cutoff score should be chosen?

12. Benedict's test is used to determine if a sample contains reducing sugars. (Reducing sugars are those that contain –CHO or –CO groups. Glucose is an example of a reducing sugar; sucrose is a nonreducing sugar.) Benedict's test is performed by adding Benedict's reagent to the sample and heating the mixture. Benedict's reagent turns from blue to green in the presence of small amounts of reducing sugars and turns reddish-orange in the presence of an abundance of reducing sugars. Suppose this test is to be used to test apple, orange, and pineapple juice samples for the presence of reducing sugars. List the likely samples for this experiment, including experimental and control samples.

13. Starch is a coiled chain of glucose molecules. Iodine can be used to test for the presence of starch in a sample. Iodine turns from its normal color of yellowish-brown to a bluish-black color when starch is present. Suppose iodine is to be used to test samples extracted from onion, potato, and green peppers for the presence of starch. List the likely samples for this experiment, including experimental and control samples.

14. You may want to review Figure 1.4 to understand the process described in this question. Researchers isolate the human gene that codes for a protein that is important in triggering cell division. The researchers want to insert the gene into bacteria and then use the bacteria to produce large quantities of the protein. These are the tasks they perform:
    - Isolate the DNA of interest.
    - Insert this DNA into a plasmid vector. The vector also contains a gene for resistance to the antibiotic ampicillin.
    - Cause bacteria to take up the plasmid vector. (The process in which the bacteria take up the plasmid is called *transformation*.)
    - Prepare the following plates on which to culture bacteria:
      Plate A: Contains nutrient medium and ampicillin.
      Plate B: Contains nutrient medium and no ampicillin.
      Plate C: Contains nutrient medium and ampicillin.
      Plate D: Contains nutrient medium and no ampicillin.
    - Add bacteria that are presumed to have taken up the plasmid vector to Plate A and Plate B.
    - Add bacteria that were not exposed to the plasmid vector to Plate C and Plate D.
    - Incubate the plates overnight to allow the growth of bacteria.
    - Examine the plates, looking for visible colonies of bacteria.

    Answer the following questions about this procedure:
    a. If the procedure works as the researchers plan, then on which plate(s) do they expect colonies to grow?
    b. If the procedure works as the researchers plan, then on which plate(s) do they expect colonies will not grow?
    c. Which of the plates are controls? Why are they used?

    Note: In the following questions, we ask how many people would be expected to have a positive or negative test for antibodies against a novel virus that causes a pandemic for which no vaccine is yet available. It is acceptable to come up with answers that are not whole numbers; for example, 4.5 people might have a particular test result.
    Remember that all the answers here are estimates.

15. Suppose:
    For a novel virus, the true prevalence of anti-virus antibodies in the populations is 5%.
    This test's diagnostic specificity (diagnostic SP) is 95% and its diagnostic sensitivity (diagnostic SN) is 90%, then:
    a. What is the test's rate of false positives? What is the test's rate of true negatives?
    b. What is the test's rate of false negatives? What is the test's rate of true positives?

    c. In a population of 200 people, about how many really have antibodies?

    d. In a population of 200 people, about how many really do not have antibodies?

    e. In a population of 200 people, how many of the people with antibodies will be expected to have a positive result and how many with antibodies will be expected to have a negative result with this test?

    f. In a population of 200 people, how many of the people without antibodies will be expected to have a positive result and how many without antibodies will be expected to have a negative result?

    g. In total, how many of these people will be expected to have a positive result with this test?

    h. Of the total positive results, what percent is correct? What percent is incorrect?

16. Suppose:

    *The true prevalence of anti-novel virus antibodies in the populations is **10%**.*

    *This test's diagnostic specificity is **90%** and its diagnostic sensitivity is **90%**, then:*

    a. What is the test's rate of false positives?

    b. What is the test's rate of false negatives?

    c. In a population of 100 people, about how many really have antibodies?

    d. In a population of 100 people, about how many really do not have antibodies?

    e. In a population of 100 people, how many of the people with antibodies will be expected to have a positive result and how many with antibodies will be expected to have a negative result with this test?

    f. In a population of 100 people, how many of the people without antibodies will be expected to have a positive result and how many without antibodies will be expected to have a negative result?

    g. In total, how many of these people will be expected to have a positive result with this test?

    h. Of the total positive results, what percent is correct? What percent is incorrect?

17. Suppose:

    *The true prevalence of anti-novel virus antibodies in the populations is **10%**.*

    *This test's diagnostic specificity is **95%** and its diagnostic sensitivity is **90%**, then:*

    a. What is the test's rate of false positives?

    b. What is the test's rate of false negatives?

    c. In a population of 200 people, about how many really have antibodies?

    d. In a population of 200 people, about how many really do not have antibodies?

    e. In a population of 200 people, how many of the people with antibodies will be expected to have a positive result and how many with antibodies will be expected to have a negative result with this test?

    f. In a population of 200 people, how many of the people without antibodies will be expected to have a positive result and how many without antibodies will be expected to have a negative result?

    g. In total, how many of these people will be expected to have a positive result with this test?

    h. Of the total positive results, what percent is correct? What percent is incorrect?

18. Suppose:

    *The true prevalence of anti-novel virus antibodies in the populations is **20%**.*

    *This test's diagnostic specificity is **90%** and its diagnostic sensitivity is **90%**, then:*

    a. What is the test's rate of false positives?

    b. What is the test's rate of false negatives?

    c. In a population of 100 people, about how many really have antibodies?

    d. In a population of 100 people, about how many really do not have antibodies?

    e. In a population of 100 people, how many of the people with antibodies will be expected to have a positive result and how many with antibodies will be expected to have a negative result with this test?

    f. In a population of 100 people, how many of the people without antibodies will be expected to have a positive result and how many without antibodies will be expected to have a negative result?

    g. In total, how many of these people will be expected to have a positive result with this test?

    h. Of the total positive results, what percent is correct? What percent is incorrect?

19. a. What happens to the percent of positive test results that are correct when the diagnostic specificity increases?
    b. Using a particular test kit, what happens to the percent of positive test results that are correct as the incidence of antibodies increases in the population?
20. Why do we care about the percent of erroneous test results when talking about serology tests that detect antibodies against an agent that causes a disease?

27 Achieving Reproducible Results with Polymerase Chain Reaction Assays

27.1 INTRODUCTION

This chapter discusses PCR-based assays. **PCR**, which stands for **polymerase chain reaction**, *allows analysts to select a specific target sequence of DNA that is present in low amounts and create millions of copies of that sequence.* The PCR method was introduced by Kary Mullis in 1983. In the ensuing years, PCR has widely been adopted to become one of the most important techniques in molecular biology. PCR can be used as a preparative tool to generate significant amounts of a specific fragment of DNA to use for another purpose, for example, for cloning or for transfecting cells. In this chapter, however, we are interested in PCR as it is used in assays, that is, to obtain information about a sample.

PCR is about amplification, that is, expanding the amount of DNA present in a test tube. As an analogy, imagine a piece of paper with the information of interest written on it. If that piece of paper is run through a photocopier, there will be two pieces of paper with the same information. If those two pieces of paper are run through the photocopier again, there will be four pieces of paper, all with the information of interest

written on them. Repeating the cycle again yields eight pieces of paper; repeating the cycle yet again provides sixteen papers. If this process is repeated 30 times, there will be more than one billion pieces of paper, all with the same information written on them! PCR is like this, but in PCR, DNA is amplified, not pieces of paper. The result of a single PCR run is millions or even billions of copies of a piece of DNA of interest. This type of amplification, where the amount of product doubles at every cycle, is called "exponential."

PCR is used in assays to see if a target DNA sequence, usually a particular gene, is present in a sample. Suppose, for example, a patient is suspected to be suffering from infection by a particular pathogenic bacterium. A PCR diagnostic test may be used to confirm the diagnosis. For a PCR assay to be useful for this purpose, there must be a gene or DNA sequence, called the *target*, which is found only in that pathogenic bacterium. A PCR diagnostic test can then be used to determine whether this target sequence is present in a sample (e.g., blood or sputum) taken from the patient. DNA is extracted from the sample and is subjected to PCR. If the target gene is present, it is

DOI: 10.1201/9780429282799-34

amplified to create many copies; this amplified DNA can then be detected. If the target gene is present, the test is positive. If the target gene is not detected, the test is negative. The reason we need PCR is that, without amplification, not enough of the target gene from the bacterium would be present to be detectable, even if the patient is infected by this organism. PCR methods are usually much faster than other methods of pathogen diagnosis, such as growing up the infectious agent in culture and then identifying it using biochemical methods. PCR assays in the clinic can therefore help a patient receive proper treatment sooner than other diagnostic methods.

Another example is the use of PCR in food safety testing to look for pathogen contaminants in meat, produce, and dairy products. Suppose, for example, there is an outbreak of food poisoning thought to be associated with bagged spinach. Spinach samples from bags suspected to be involved can be rapidly tested for various possible pathogens using PCR. PCR assays can therefore allow many spinach samples to be rapidly screened before the food is distributed to consumers. As in medical diagnosis, PCR's speed and sensitivity are great advantages in this situation where waiting a few days for answers could result in deaths.

PCR is a vital part of modern methods of DNA fingerprinting where it allows investigators to prepare DNA fingerprints from evidence containing only trace amounts of DNA. Residual DNA from a hair root or small drops of blood at a crime scene can yield enough DNA for a fingerprint when PCR is used for amplification. If the fingerprint matches a suspect, then it provides evidence that the suspect was at the crime scene.

PCR is also a critically important tool in many research laboratories. PCR is used in studies of evolution, development, cancer, and many other research areas. For example, PCR methods can be used to compare the expression of genes in normal and cancerous cells.

27.2 HOW DOES PCR WORK?

27.2.1 AMPLIFICATION

PCR creates duplicate copies of a sequence of DNA by mimicking the cell's natural ability to replicate (copy) DNA. DNA **replication** *is an enzymatic process in which a cell that is getting ready to divide duplicates its DNA to make two identical copies, one of which will be passed along to each daughter cell.*

To understand replication (and thus PCR), recall that DNA is composed of strands consisting of a sequence of four nucleotide building blocks: *A, T, G,* and *C.* (Refer to Chapter 1 and particularly Figure 1.1 for more information about DNA structure.) Double-stranded DNA consists of two strands of DNA. An *A* on one strand always pairs with a *T* on the opposite strand, and a *G* always pairs with a *C.* We say that *A* is "complementary" to *T,* and *G* is "complementary" to *C.* Because an *A* must always be across from a *T,* and a *G* across from a *C,* the sequence of one strand specifies the sequence of the opposite strand. During replication, the two original strands of DNA separate from one another. Each strand then serves as the **template**, or *guide,* for the manufacture of a new, complementary strand (Figure 27.1). Replication requires an enzyme, DNA polymerase, which performs the work of linking nucleotide subunits one after another to form the new DNA strands.

In PCR, the analyst simulates DNA replication by combining the required ingredients in a test tube and

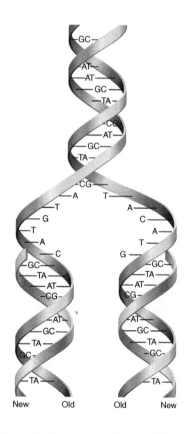

FIGURE 27.1 Replication duplicates DNA when a cell divides. At the beginning of replication, there is one double-stranded DNA molecule. The strands separate from one another, and each strand serves as the template to make a new complementary strand. After replication, there are two identical double-stranded DNA molecules.

providing suitable conditions for a polymerase enzyme to synthesize new strands of DNA. The ingredients of a reaction mixture for PCR are as follows:

- The four nucleotide subunits: *A, T, G*, and *C*.
- Template (sample) DNA. The sample DNA may or may not contain the target sequence.
- Two DNA primers, to be discussed below.
- A DNA polymerase enzyme. In PCR, the enzyme must be heat-stable for reasons that will become clear shortly. *Taq* DNA polymerase, an enzyme isolated from bacteria that live in hot springs, is commonly used for this purpose.
- Buffer and cofactor(s) to provide suitable conditions for the polymerase enzyme.

Although the PCR method simulates normal DNA replication, there are some differences between PCR, and replication as it occurs naturally in a cell. One of these differences is that in PCR, only a short section of the DNA template is copied (typically less than 10,000 base pairs). The section of DNA that is replicated is the target sequence (Figure 27.2).

How is it that in PCR the DNA polymerase enzyme only duplicates this target region rather than all of the DNA present? The answer lies in an important characteristic of DNA polymerases. These polymerase enzymes require a short section of double-stranded DNA to initiate, or *prime* synthesis. A **primer** *is a short piece of single-stranded DNA (or RNA) that is complementary to one end of a section of the DNA that is to be replicated.* The primer binds to single-stranded DNA creating a short double-stranded section. The DNA polymerase enzyme finds this double-stranded section and begins there to synthesize a new DNA strand. So, in PCR, there are primers that recognize and bracket the target region.

The first step in PCR is to separate the two strands of double-stranded DNA. This is achieved by using a high temperature. Once the strands are separated, the primers can find their targets and bind (Figure 27.3). Observe in this figure that there is a short primer that is complementary to a sequence at the beginning of the target region on one strand of the DNA. There is a second short primer that is complementary to a sequence at the end of the target region on the opposite strand of DNA. Observe how the primers thus recognize and bind to the ends of the target sequence so that they bracket the target portion of the DNA template that will be amplified.

The analyst controls the starting and ending points for DNA replication in PCR by the choice of specific primers. In order to design these primers, the analyst must know the DNA sequence of the target DNA, or must at least know the DNA sequence of the ends of the target region. PCR assays therefore rely on our knowledge of the DNA sequences of specific organisms and viruses – knowledge that is increasing at an exponential rate.

A second important difference between PCR and natural DNA replication is that, in nature, replication occurs once when a cell divides. In the laboratory during PCR, the target DNA is duplicated over and over again to generate millions of copies. This is accomplished by providing temperature conditions that promote repeated cycles of replication, After combining the components of the reaction mixture together in a tube, the analyst puts the tube in a device, called a **thermocycler**, *which subjects the mixture to a series of repeated temperature cycles.* As shown in Figure 27.3, at the beginning of each

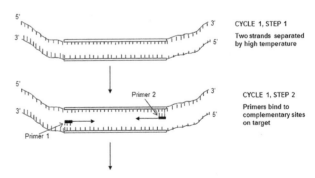

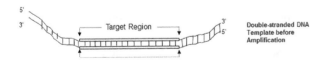

FIGURE 27.2 A simplified sketch of double-stranded DNA before PCR amplification. The template DNA is isolated from the sample. This template DNA is actually much longer than is drawn here and extends out in both directions. In PCR, however, we are interested in only a short stretch of the template DNA, the target region. This target region might be, for example, a stretch of DNA that is characteristic of a pathogen or a stretch of DNA that is used to help distinguish one individual from another.

FIGURE 27.3 The beginning of PCR. To begin a PCR, the two strands of DNA are separated from one another using high temperature. Once they are separated, the two primers find their complementary (target) regions of DNA. One primer binds to one strand of DNA, and the other primer to the other strand. (See text for more details.)

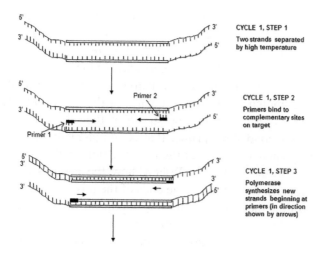

CYCLE 1, STEP 1

Two strands separated by high temperature

CYCLE 1, STEP 2

Primers bind to complementary sites on target

CYCLE 1, STEP 3

Polymerase synthesizes new strands beginning at primers (in direction shown by arrows)

FIGURE 27.4 One cycle of the polymerase chain reaction. In the first step of PCR, the two strands of the DNA template are separated by high temperature. Next, the temperature is lowered so that the primers complementary to the ends of the target sequence can bind to the template. The primers bracket the ends of the target region of DNA. In the final step of each cycle of PCR, a DNA polymerase enzyme builds new double-stranded DNA molecules by adding nucleotides complementary to the target sequence, beginning with each primer and moving only in one direction, as shown by the arrows. This process is repeated for 20–40 cycles; assuming 100% efficiency, the number of copies of the target sequence doubles after every cycle.

cycle, step 1, the template DNA is separated into its two strands. This is accomplished by subjecting the DNA to elevated temperatures, usually between 92°C and 94°C. Next, step 2, the temperature is brought down to a temperature in the range of 37°C–72°C. At this lower temperature, the primers associate with (anneal to) complementary regions of the target DNA (if the target is present). In the final step, step 3, usually at 72°C, *Taq* DNA polymerase copies each strand starting from the primer on that strand. To do this, it takes the needed nucleotide from the reaction mixture and attaches it to the growing strand (Figure 27.4). Because the *Taq* polymerase comes from a bacterium that lives in hot springs, the enzyme is heat-stable at the temperatures used in PCR.

The three steps of each cycle of PCR, as shown in Figure 27.4, result in two double-stranded DNA target molecules where originally there was only one. The process is then repeated, resulting in four target molecules, and again to give eight, and so on. Figure 27.5 illustrates three cycles of PCR, resulting in eight copies of the original target sequence. Usually, the process is continued for 25–40 cycles until millions of copies of the target sequence have been created. Note that if

the target sequence is not present in the DNA template, then the primers do not bind the template and no amplification occurs.

27.2.2 Detection of the PCR Product

The tube containing the reaction mixture is removed from the thermocycler after it has completed the proper number of cycles. If the amplification was successful, then there is a product, called the **amplicon**, *consisting of millions of copies of the target sequence.* The amplicon must then be detected. The conventional method of detection is to analyze a portion of the tube's contents using agarose gel electrophoresis followed by staining with the dye, ethidium bromide. Agarose gel electrophoresis separates DNA fragments based on differences in their sizes, with smaller fragments moving farther through an agarose gel support than larger fragments. Ethidium bromide allows DNA to be visualized and photographed because it fluoresces when excited with UV light. PCR-amplified DNA is thus detectable as a band on an agarose gel, as shown in Figure 27.6.

A molecular weight marker that contains DNA fragments of known sizes is run alongside the PCR samples during electrophoresis. The different fragments in the molecular weight marker move different distances through the gel. The size of the expected amplicon is known based on the length of the sequence bracketed by the primers, as shown in Figure 27.3. If there is a band of DNA visible on the gel in the sample lane, then its mobility through the gel is compared to the mobility of the molecular weight markers. This allows the analyst to estimate the size of the amplicon in the sample. The sample must produce an amplicon of the expected size to be considered a positive result. If a band appears on a gel, but it is not of the size expected for the amplicon, then the result is negative. (Proper controls, discussed below, must also be included in the gel.)

Figure 27.7 summarizes the steps in a PCR-based assay.

Observe that as we have described PCR so far, it is a qualitative assay. This type of PCR provides a yes or no answer; was the target DNA present in the sample? If the target was present (and if the assay worked properly), then a band is present on the gel. If the target was not present in the sample, then no band appears. Qualitative PCR is suitable to answer many important questions. However, there are times when a quantitative version of PCR is desirable, as will be explained later in this chapter.

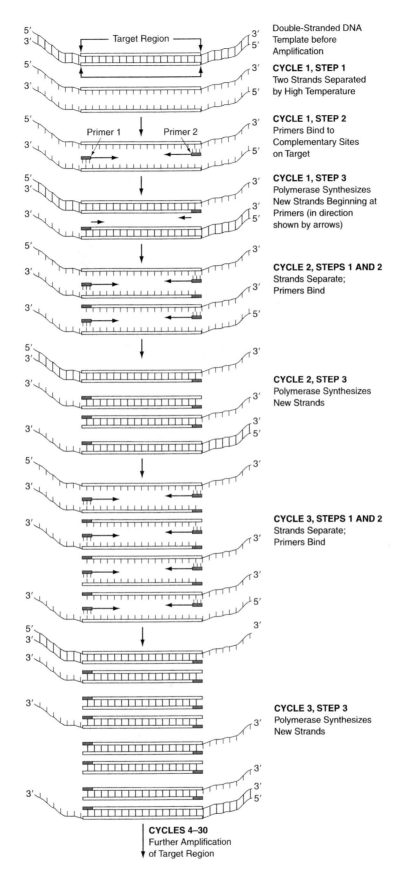

Double-Stranded DNA Template before Amplification

CYCLE 1, STEP 1
Two Strands Separated by High Temperature

CYCLE 1, STEP 2
Primers Bind to Complementary Sites on Target

CYCLE 1, STEP 3
Polymerase Synthesizes New Strands Beginning at Primers (in direction shown by arrows)

CYCLE 2, STEPS 1 AND 2
Strands Separate; Primers Bind

CYCLE 2, STEP 3
Polymerase Synthesizes New Strands

CYCLE 3, STEPS 1 AND 2
Strands Separate; Primers Bind

CYCLE 3, STEP 3
Polymerase Synthesizes New Strands

CYCLES 4–30
Further Amplification of Target Region

FIGURE 27.5   PCR involves many cycles of replication. The process illustrated in Figure 27.4 is repeated over and over again in PCR to result in millions of copies of the target sequence. While this diagram may seem intimidating, observe that in PCR the same three steps are repeated over and over again. Eventually, the primers and nucleotides are all consumed and the amplification ceases.

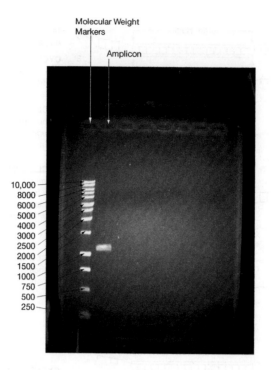

FIGURE 27.6 Detecting a PCR product. The bright bands on the gel represent DNA fragments. Lane 1 (on the left) contains a molecular weight marker consisting of a number of DNA fragments whose sizes are shown on the far left. The higher the band is on the gel, the larger the fragment. The band observable in Lane 2 is a PCR product, the amplicon. In this experiment, the amplicon was expected to be 1,100 base pairs (bp) in size. The amplicon migrated a little less distance in the gel than the 1,000 bp fragment in the molecular weight marker lane, which is consistent with the amplicon being 1,100 bp in size.

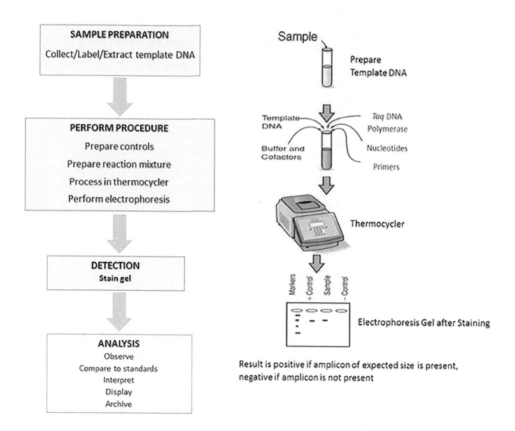

FIGURE 27.7 Overview of a PCR-based assay method involving electrophoresis for detection.

27.2.3 Exponential Amplification

In PCR, the number of copies of target DNA ideally doubles with each cycle. This means that the relationship between the cycle number and the number of copies of target DNA present is an exponential relationship, similar to the ones introduced in Chapter 13. In Chapter 13, we saw that the number of bacteria in a culture doubles at regular intervals; here, the number of DNA molecules doubles at each cycle. Assuming that a PCR is perfectly efficient, Equation 27.1 is used to calculate how many copies of the target sequence will be present after a certain number of cycles:

$$N = 2^t(N_0) \qquad (27.1)$$

where:
    N is the number of copies of the target sequence after amplification
    t is the number of cycles
    N_0 is the number of copies of the target sequence initially present in the reaction mixture.

Observe that this equation is basically the same as the equation for bacterial growth shown in Chapter 13 because this is the same type of relationship. In Chapter 13, the equation for bacterial growth included a term, "t," for the number of generations that had elapsed. For PCR, the analogous term is still "t," but in PCR, "t" is the number of cycles that have occurred. For bacterial growth, there was a term "N_0," the number of bacteria present initially. For PCR, N_0 is the number of copies, or the amount, of the target sequence originally present in the sample. Thus, amplification in PCR and bacterial reproduction are mathematically analogous.

As is the case for bacterial growth, exponential DNA amplification cannot continue forever. There is eventually a plateau. The plateau is due to a number of factors influencing the reaction, most obviously the depletion of reactants. It is common for PCRs to begin to plateau after about 40 cycles.

Table 27.1 and Figure 27.8 illustrate the PCR relationship between copy number and cycles that have occurred when there are initially 1,000 copies of the target present. This table and figure show early cycles, before the reaction approaches the plateau.

TABLE 27.1

The Relationship between the Cycle Number and Copies of the Target DNA Present ($N_0 = 100$)

Cycle Number	Copies of the Target DNA Present
0	100
1	200
2	400
3	800
4	1,600
5	3,200
6	6,400
7	12,800
8	25,600
9	51,200
10	102,400
11	204,800
12	409,600
13	819,200
14	1,638,400
15	3,276,800
16	6,553,600
17	13,107,200

Example Problem 27.1

In a forensic sample, 300 copies of a target sequence are present. PCR is performed for 30 cycles. How many copies of the target sequence are theoretically expected at the end of PCR amplification?

Answer

Apply Equation 27.1.

$$N = 2^t(N_0)$$

$$N = 2^{30}(300 \text{ copies}) \approx 3.22 \approx 10^{11} \text{ copies}$$

You can see that PCR can possibly produce billions of copies of a target DNA sequence.

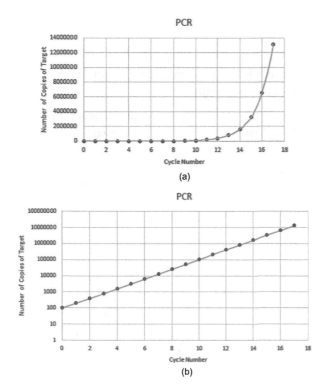

(a)

(b)

FIGURE 27.8 The relationship between the cycle number and DNA copies is exponential, although there is an eventual plateau. In this example, there were 100 copies of the target initially, $N_0 = 100$. (a) The relationship between the cycle number and the number of copies of the target DNA plotted on a linear scale. (b) The same relationship plotted on a semi-logarithmic plot. Observe the similarity between these graphs and those of the bacterial growth in Figures 13.15 and 13.16.

27.3 QUANTITATIVE PCR

Quantitative PCR (qPCR) *involves monitoring DNA amplification as it occurs.* For this reason, another name for qPCR is "real-time" PCR because the PCR amplification products are observed in "real time." qPCR provides quantitative information that can be important in many situations. For example, if PCR is being used to diagnose certain viral diseases, it is important to know whether a patient's sample contains a lot of virus, just a little virus, or none that is detectable.

In order to monitor a PCR amplification as it occurs, there must be a way to make the amplified DNA visible. It would be impractical to run an electrophoresis gel after each cycle of PCR, so another means of detecting DNA is required. The solution is to use fluorescent dyes that bind to the DNA that is produced during the amplification and make the DNA visible to a fluorimeter. (See Section 28.7.4 for a description of fluorimeters.) The more DNA that is produced, the

more dye-binding that occurs, and the more fluorescence that appears. qPCR thus requires more complex instrumentation than conventional PCR. Both types of PCR take place inside a thermocycler that controls the temperature of the reaction mix. qPCR additionally requires a light source to excite the fluorescent dyes (causing them to emit light) and a fluorimeter to detect and quantify their fluorescence (Figure 27.9).

Figure 27.10 shows what happens over time inside the test tube in a PCR assay. In the early cycles of PCR, the amount of DNA present is doubling, but it

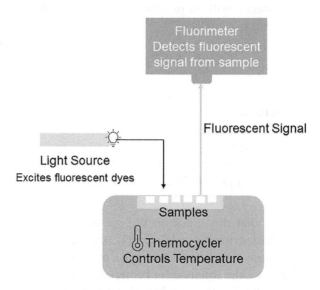

FIGURE 27.9 qPCR instrumentation. qPCR requires a thermocycler, as does conventional PCR, and also requires a light source and a fluorimeter. As the amplification proceeds, more and more DNA product is present. The DNA binds to a fluorescent dye that is energized by the light source, thus emitting fluorescent light that is detected by the fluorimeter.

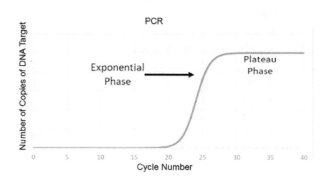

FIGURE 27.10 The time course of PCR. Initially, the DNA present in the tube is not detectable. After a certain number of cycles, so much DNA is present that it becomes visible and the exponential amplification of PCR is evident. Eventually, the reaction mixture is depleted and the system plateaus.

is not detectable. At a certain point, there is enough DNA present to be detected. After this point, the exponential increase in DNA is evident. Eventually, all the primers and nucleotides in the reaction mixture are depleted and there is no more increase in the DNA product; this is the plateau region in the graph. When conventional (qualitative) PCR is used, what we see on the electrophoresis gel is the DNA present at the end of the process, in the plateau region. Conventional PCR, using gel electrophoresis for detection, is therefore often called "endpoint" PCR.

To understand how qPCR is used to obtain quantitative information, consider what happens when two samples are compared. One sample begins with a lot of target DNA, the other sample has little. As you can see in Figure 27.11, the sample that begins with more DNA, shown in orange, reaches a detectable level of fluorescence before the other sample, shown in blue. The **threshold cycle (C_t)** *is the cycle where the fluorescence crosses a threshold and is bright enough to reliably detect.* As shown in Figure 27.11, the orange sample has a C_t value of about 17 and the blue sample a C_t value of about 22.5. The more starting DNA that is present in a sample, the lower the C_t value. Thus, **C_t values** *provide a measure of the amount of target DNA initially present in a sample.* In this particular example, we do not know in absolute terms how much DNA was initially present in either tube; we only know that the sample shown in orange began with more DNA

than the sample shown in blue. In order to determine the starting amount of target DNA in the samples in absolute terms, it is necessary to use standards with known amounts of target DNA. The C_t values for the standards are then used to create a standard curve.

Observe in Figure 27.11 that both the orange and blue samples reach the same endpoint, or plateau at the end of the run, even though the orange sample began with more target DNA.

27.4 REVERSE TRANSCRIPTASE PCR

So far, we have discussed situations where the starting sample template is DNA that is amplified via PCR. However, there are many situations where the starting template material is RNA. For example, if a researcher is interested in comparing the levels of gene expression between two or more treatments, then mRNA is of interest. This is because gene expression begins with the transcription of DNA into mRNA. Therefore, a higher level of expression of a particular gene results in a higher level of the mRNA for that gene. Another situation where RNA is the beginning template is in the diagnosis of diseases caused by viruses that use RNA as their genetic material. If one is using PCR to look for these viruses, RNA is the starting material. When RNA is the starting nucleic acid, it is necessary to convert it to DNA before performing the amplification. This is achieved by adding an enzyme, called reverse transcriptase, to the sample. The enzyme converts the RNA into DNA, which is then called complementary DNA (cDNA). The term RT-PCR is often used when a PCR procedure begins with the conversion of RNA into cDNA. If this step is followed by quantitative PCR, then the term RT-qPCR is applied. Figure 27.12 diagrams the process of reverse transcription.

27.5 OBTAINING TRUSTWORTHY RESULTS FROM A PCR ASSAY

27.5.1 PCR Has Potential Pitfalls

PCR can seem deceptively simple to perform. An analyst isolates DNA from a sample following a standard method, purchases quality-assured reagents from commercial suppliers, pipettes each reagent into a tube, and places the tube in the thermocycler or a qPCR instrument. The analyst turns on the device and returns a few hours later to remove a tube laden with amplified DNA, or to view a graph showing the amplification over time and the C_t value. In reality, similar to all assays, PCR is not so simple, and there are

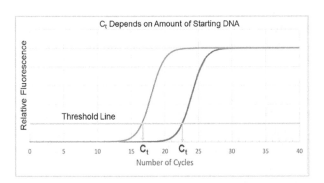

FIGURE 27.11 qPCR run of two samples. Two samples both containing the target DNA of interest are shown. The sample drawn in orange begins with a higher level of the target DNA than the sample shown in blue. Both attain the same number of DNA copies at the end of PCR, but the orange sample does so more quickly. The C_t value is the cycle number where the amount of DNA, and hence the fluorescence, crosses a threshold and can be reliably detected. The C_t value for the orange sample is lower than for the blue sample. Thus, the C_t value provides quantitative information about the amount of target DNA initially present in a sample.

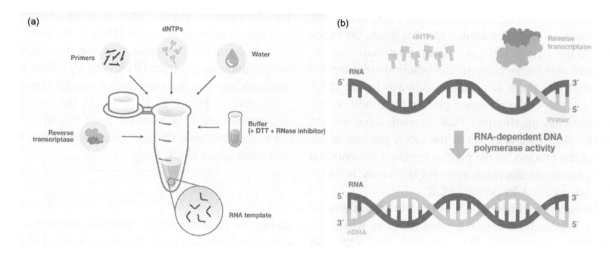

FIGURE 27.12 A simplified view of reverse transcription. (a) Reverse transcription begins with strands of RNA that is to be turned into cDNA. The RNA is placed into a tube along with primer that recognizes the RNA of interest, reverse transcriptase enzyme, nucleotides, and buffer. (b) The enzyme uses the RNA as a template to create a complementary strand of cDNA. The result is a double-stranded hybrid molecule with RNA (black in this diagram) and DNA (orange). Depending on the situation, the hybrid may undergo further processing to become double-stranded DNA. (Used with permission from Thermo Fisher Scientific, the copyright owner.)

many pitfalls that can lead to deceptive results, even in a routine, optimized PCR-based assay. It is therefore important to understand this method in order to get trustworthy results.

A PCR assay is used to answer a number of questions, such as: was a particular suspect present at a crime scene? Is a patient infected with a particular pathogen? Is a particular microorganism present in a batch of cheese? Does one sample contain more target DNA than another sample? In conventional PCR, if a band is visible on a gel, one might consider this to be a positive result. If no band is visible, this might be called a negative result. But how trustworthy is the result? Is it possible that a visible band is a false positive? Is it possible that no band appears, but this is a false-negative result? In fact, both false-positive and false-negative results can occur in both conventional and qPCR assays. In order to assure that a result is trustworthy, it is necessary to have controls. In this context, the word *control* takes on two slightly different meanings. First, there are positive and negative controls that are used with PCR, in the sense that was discussed on in Section 26.3.2. Second, the assay must be "controlled" in the sense that the analyst uses proper techniques to avoid introducing contaminants and errors into the system. We will consider both types of *control* in more detail.

27.5.2 SAMPLE PREPARATION

As with all assays, PCR assays require that the sample be chosen correctly and prepared properly. Sample selection and preparation vary greatly depending on the purpose of the assay. At a crime scene, for example, the selection of sample may involve searching the scene for evidence from which DNA can be isolated. In food safety testing, a subset of food items must be chosen in some reasonable way, since obviously all the food from a given farm, store, or batch cannot be tested. Improper sample selection and preparation can result in erroneous results.

Once the sample is chosen, template DNA must be prepared from it. The preparation method is likely to involve separating the DNA from the matrix in which it is found and removing contaminants that might interfere with the analysis. There are many methods of isolating template DNA for PCR. In general, sample preparation begins by lysing the cells containing the template DNA. Sometimes this involves mechanical grinding or shearing. Detergents are often added to open the cell and nuclear membranes. Enzymes may be used to break open cells and also to destroy proteins from the sample. Proteinase K, which breaks the bonds holding amino acids together in proteins, is commonly used for this purpose. Sometimes the template DNA is extracted with organic solvents followed by precipitation with ethanol. Alternatively, molecular biology companies sell proprietary products that conveniently extract and clean up DNA for PCR.

Some DNA preparation methods extensively purify the template away from all contaminants in the sample, but the trend is toward rapid methods that often provide minimal DNA cleanup. This is because the

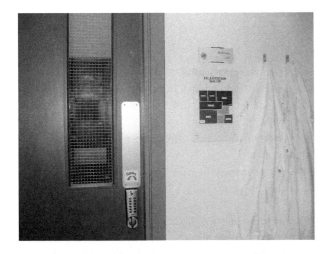

FIGURE 27.13 Use of facility design to help avoid contamination in PCR. It is not always possible to have a separate PCR facility, but a facility such as this can reduce contamination. This is a picture of the entrance to a separate PCR suite located inside of a contract research laboratory that performs preclinical drug testing. (See Chapter 36 for more information on this type of testing.) There is a lock on the door to restrict access to only individuals trained in PCR. The locking system will not allow a person to enter the suite more than once a day to prevent an analyst from introducing DNA contamination from previous assays on their hands or body. People change their lab coat before entering the facility and then move in one direction through the PCR suite to avoid contaminating samples with already-amplified DNA. Once inside the suite, there are separate rooms for setting up the PCR, running the reaction, and performing electrophoresis. Doors only open in one direction so a person cannot take materials backward through the facility. Personnel change their lab coats every time they move to a different room.

PCR is usually sensitive to the target DNA, even if it is present in the midst of other cellular constituents and nontarget DNA sequences. Note, however, that there can be inhibitory substances present that may bind required cofactors or otherwise reduce the activity of the polymerase enzyme. Inhibitors can come from many sources including the matrix from which the DNA was isolated. Foods, for example, have many potential inhibitory agents including proteases, nucleases, collagen, and fatty acids. If inhibitors are found to be a problem in a particular PCR method, then it is important to add steps to purify the template away from the inhibitory substances. It may require a period of experimentation to find a method that effectively prepares template DNA from a specific source. If DNA is not properly extracted, the assay might give false-negative or ambiguous results.

27.5.3 Avoiding Contamination

PCR is extremely sensitive – this is one of its major advantages. Even minuscule amounts of DNA can be amplified so that they are detectable after PCR. This sensitivity, however, also makes PCR extremely vulnerable to contamination. Unsuspecting investigators have thought that they were investigating DNA from experimental samples, when they were, in fact, investigating their own DNA that fell into the test tube in minute amounts, or other contaminating DNA from a pipette or another object. PCR requires carefully avoiding contamination to avoid false positives. Sometimes contaminating DNA does not cause problems because it does not contain the sequence recognized by the primers and so is not amplified. Other times contaminating DNA can totally invalidate an assay. For example, DNA typing in forensic laboratories involves primers that recognize human DNA. Therefore, if a few skin cells from the analyst fall into the reaction mixture, then the analyst will be fingerprinting himself or herself – not a person present at the crime scene.

The amplicon from previous assays is an important potential source of DNA contamination; this is called cross-contamination. If even the smallest amount of previously amplified DNA finds its way into a new reaction mixture, it will serve as template in the subsequent reaction, thus leading to a false-positive result. The tubes containing amplified product must therefore be treated carefully. Simply opening these tubes is problematic because it creates aerosols that may contain millions of copies of the target DNA. These copies may float in the air, fall on the counter, be picked up by an analyst's sleeve, contaminate an analyst's glove, or fall on a piece of equipment. It is essential to use proper practices to avoid cross-contamination between samples.

Ideally, there should be distinct, dedicated rooms or areas in a PCR laboratory (Figure 27.13). One is a space for preparing the sample, including extracting the DNA. This space may also be used for preparing buffers and other reagents. A second space should be used for combining the ingredients of the reaction mixture. A third space is set aside for the thermocycler. A fourth space is used to open the tubes after amplification and for performing detection by electrophoresis or another method. In an ideal situation, people and materials would always move in one direction, from the sample preparation area to the detection area. Before moving to the next area, personnel would change lab coats and gloves. Personnel would ideally

---

TABLE 27.2

Avoiding Contamination When Performing PCR Assays

- Separate preamplification and post-amplification steps by using different laboratory areas.
- Maintain separated, dedicated equipment (e.g., pipettes and microcentrifuge) and supplies (e.g., microcentrifuge tubes and pipette tips) for specimen preparation, assay setup, and handling amplified nucleic acids.
- Keep the thermocycler and electrophoresis equipment in the post-amplification space.
- Do not bring items from the post-amplification area back into the preamplification area, including pens and notebooks.
- Label items to indicate if they belong to the pre- or post-amplification area.
- Consider using a dedicated PCR workstation to isolate a small area for a specific PCR-related task. PCR workstations are enclosed cabinets that protect a work area from drafts. They usually contain germicidal UV lights that can be turned on to help destroy DNA contaminants.
- Prepare and store PCR reagents separately from other reagents and use them only for PCR.
- Aliquot PCR reagents and throw away any unused portions after an aliquot is opened.
- Maintain dedicated micropipettes that are only used for preparing PCR mixtures. These micropipettes should be used with DNase/RNase-free tips that have aerosol barriers (as shown in Figure 18.14). Change aerosol barrier pipette tips between all manual liquid transfers.
- Work surfaces, micropipettes, and centrifuges should be routinely cleaned and decontaminated with cleaning products (e.g., DNA/RNA remover, ethanol, 10% bleach) to minimize the risk of nucleic acid contamination.
- Wear a clean lab coat and disposable gloves (not previously worn) when setting up assays.
- Change gloves between samples and whenever contamination is suspected.
- Keep reagent and reaction tubes capped or covered as much as possible.
- Always check the expiration date on reagents and kits prior to use. Do not use expired reagent. Do not substitute or mix reagent from different kit lots or from different manufacturers.
- Use DNase/RNase-free disposable plasticware and pipettes reserved for DNA/RNA work to prevent cross-contamination with DNases/RNases from shared equipment.

---

begin in the morning in the preparation area and end the day in the detection area. The purpose of this one-way traffic is to prevent the product of a previous PCR assay from entering the area where new reactions are set up.

It is not always possible to have separate rooms for PCR. Table 27.2 lists practices that can help prevent cross-contamination in any laboratory setting.

27.5.4 POSITIVE AND NEGATIVE CONTROLS

There are various controls that are used to help recognize false-positive and false-negative PCR results. One way to check for contamination is to routinely run negative controls alongside the test samples. The simplest PCR negative control is a tube that has all the components of the reaction mixture except one (usually template DNA). This tube should not have an amplicon after the PCR. If, however, contaminants enter the PCR (e.g., from the analyst's skin, aerosols in the air, contaminated pipettes, or contaminated reagents), then an electrophoresis band(s) or trace on the qPCR graph may appear in this control lane, alerting the analyst to the presence of contamination.

DNA from a different, related organism can also be used as a negative control. In food testing, for example, a negative control might be template DNA from a non-pathogenic organism that is related to the pathogen of interest. If the primers are properly designed, the DNA from the related nonpathogenic organism will not be amplified. This negative control therefore tests for the specificity of the primers. It also tests for unoptimized replication conditions (e.g., temperatures that are too low) that allow the primers to bind nonspecifically to the "wrong" DNA sequence.

A false-negative result can occur in PCR for a variety of reasons. Inhibitors, for example, can cause false-negative results. False-negative results can also occur because one or more components of the reaction mix were improperly prepared, or because the reaction conditions were suboptimal. It is therefore common to use an external positive control. An external positive control is a sample that is known to contain the target DNA. The positive control is amplified in a separate tube alongside the test samples. The positive control DNA should produce an amplicon. If it does not, inhibitors might be present, one or more reagents might be improperly prepared, the cycling conditions might be suboptimal, or another problem may have occurred. Figure 27.14 illustrates the use of positive and negative controls.

Some analysts use internal amplification controls as an alternative to an external positive control. In conventional PCR, an internal amplification control is template DNA that is similar to the target of interest, but a section of its DNA has been deleted. This type of control is added into the reaction mixture in the same tube as the test sample. In a properly functioning assay, the internal control target DNA will bind to the primers, just as the target of interest does. The internal control DNA will also be amplified, just as the target is amplified. But the internal control template will produce a smaller-sized amplicon because it is missing part of the DNA sequence of interest. The smaller-sized amplicon should always appear, even if the sample is negative for the target DNA. Its presence proves that the reagents worked properly and that inhibitors did not prevent amplification. If the assay is positive for the target DNA, then two amplicons will be present that are easily distinguished. If neither amplicon is present, then the assay result cannot be trusted.

As a further control, one must always run a molecular weight marker alongside the samples and controls, assuming identification of the amplicon is based on its electrophoretic mobility. If a band is present, but is of the wrong size, then the assay is negative, and the band is attributed to nonspecific primer binding. The consistent use of a molecular weight marker when using detection by electrophoresis is one of the guards against false positives. The molecular weight markers may also be considered to be standards, in the sense discussed in Section 26.2.4. Table 27.3 summarizes these controls and standards.

27.5.5 Design and Optimization of a PCR Assay

The method development phase for a PCR assay involves a number of activities:

- It is necessary to find a unique sequence or partial sequence for the target DNA in order to design the primers. The *specificity* of the assay depends on the specificity of the primers; if they bind to DNA other than the sequence of interest, then the assay will not be *specific*.
- It is necessary to optimize the sample selection and preparation of template DNA. This step will vary greatly depending on the nature of the sample and the purpose of the assay.
- It is necessary to optimize the reaction components and conditions. Magnesium, for example, is a required cofactor for the polymerase enzyme. If too little magnesium is present, then less amplicon may form, and the assay will lack *sensitivity*. If, however, too much magnesium is present, it may promote nonspecific binding, thus reducing the *specificity* of the assay.
- The cycling conditions must be optimized. If the temperature of the primer extension phase is too low, nonspecific binding may occur, again reducing assay *specificity*. If the temperature of the primer extension phase is too high, the *sensitivity* of the assay may be impaired.
- The controls for the assay must be selected. A no-template negative control is easy to prepare; positive controls are sometimes more difficult to obtain. An internal amplification control may be created.

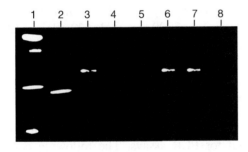

FIGURE 27.14 Positive and negative controls in PCR. The gel is from a clinical test where PCR is used diagnostically to detect a genetic disease. **Lane 1 contains a molecular weight marker** where each of the bright bands is a DNA fragment of a different, known length. **Lanes 2 and 3 are positive controls**. Lane 2 is amplified DNA from a patient known to have the disease of interest. Lane 3 is amplified DNA from a healthy person. Observe that the DNA band in Lane 2 has moved farther than the one in Lane 3. Clinicians know where the bands in healthy people and affected people should occur relative to the reference markers in Lane 1. If the expected results for healthy and affected people had not appeared, then the analysts would know there was a problem in the procedure. **Lanes 4, 5, and 8 are negative controls**. Each was lacking some necessary component of the reaction mixture and therefore should have no bright bands at all. The presence of any bands in any of these negative control lanes would indicate that contamination had occurred. As expected, the negative control lanes are dark. **Lanes 6 and 7** are patient samples. Both patients test as healthy since their DNA band is at the same level in the gel as that of a healthy control.

TABLE 27.3

Common Controls and Standards in PCR

Control or Standard	Type	Purpose
No template in reaction mixture	Negative control	Detects contaminating DNA if present; *avoids false positives*
Template DNA from related organism	Negative control	Detects nonspecific primer binding; *avoids false positives*
Purified template known to contain the target sequence	External positive control	Detects inhibition of the reaction, failure to add all reaction components, problems with reaction mixture components, problems with cycling conditions; *avoids false negatives*
Purified template containing a shortened target sequence	Internal positive control	Detects inhibition of the reaction, failure to add all reaction components, problems with reaction mixture components, problems with cycling conditions; *avoids false negatives*
Molecular weight markers	Standards	Analyzes the size of amplicon; *avoids false positives*

Like all assays, sophisticated PCR-based assays will not produce trustworthy results unless the analyst understands their pitfalls and carefully provides the proper controls.

27.5.6 MINIMAL INFORMATION FOR PUBLICATION OF QUANTITATIVE REAL-TIME PCR EXPERIMENTS

Thorough documentation and reporting is a critical aspect of ensuring that results obtained in one experiment are reproducible in other experiments, either in the same laboratory or in different laboratories. Insufficient reporting and documentation have been identified as a particular problem with qPCR. In response to this problem, a team of scientists introduced "The Minimum Information for Publication of Quantitative Real-Time PCR Experiments (MIQE) Guidelines." The purpose of the MIQE Guidelines is to help ensure the integrity of the scientific literature, promote consistency between laboratories,

and increase experimental transparency. The MIQE Guidelines describe the minimum information necessary for evaluating qPCR experiments. For more information about the reporting guidelines and an analysis of potential problems with qPCR that affect reproducibility, consult the complete article. (Bustin, Stephen A. et al., "The MIQE Guidelines: *M*inimum *I*nformation for Publication of *Q*uantitative Real-Time PCR *E*xperiments." *Clinical Chemistry*, Vol. 55, no. 4, 2009, pp. 611–622, doi.org/10.1373/clinchem.2008.112797.)

27.5.7 VALIDATION

As described earlier in this chapter, validation is essential for PCR assays. The following case study "Diagnostic Assay Validation in the Time of COVID-19, Part III" introduces some of the key steps in validating a qPCR diagnostic assay.

Case Study: Diagnostic Assay Validation in the Time of COVID-19, Part III

The marketing of diagnostic assays is regulated by the Food and Drug Administration (FDA) because diagnostic assays are medical products. A poorly designed and manufactured diagnostic assay might provide untrustworthy results that endanger the health of patients. Therefore, the FDA requires optimization and validation of diagnostic test kits before they are released to the public. The COVID-19 pandemic is a special situation where rapid deployment of diagnostic tests was of utmost importance. In response, in the early months of the pandemic, the FDA developed an abbreviated, emergency validation protocol for COVID-19 diagnostic kits. This protocol was intended to provide as much assurance as possible of the assay's trustworthiness, and be relatively rapidly completed. In this emergency protocol, the FDA specifies which parameters are required for validating RT-qPCR assays, and which are required for serological assays. In this case study, we will consider only the validation of RT-qPCR assays.

(Continued)

Case Study (*Continued*): Diagnostic Assay Validation in the Time of COVID-19, Part III

The validation parameters that must be tested for PCR-based assays are as follows:

1. ***Limit of detection (LOD).*** Recall that the LOD is the lowest concentration of the material of interest, in this case viral RNA, that can be detected. The LOD is determined by taking a sample that is known to contain a certain amount of inactivated virus or a certain amount of viral RNA, and serially diluting it to progressively lower concentrations. FDA specifies that the LOD is the lowest concentration at which 19 out of 20 replicate samples known to contain virus give a positive result.

2. ***Clinical evaluation.*** This is the term FDA uses for validating diagnostic sensitivity and diagnostic specificity as defined in Chapter 26. Recall that these parameters are evaluated using samples known to be positive, and other samples known to be negative for the analyte of interest. Early in the pandemic, it was difficult to obtain samples from patients that were known to be positive for the SARS-CoV-2 virus. Thus, early diagnostic tests were often validated using "contrived" clinical specimens. These are made by spiking (adding) viral RNA or inactivated virus into leftover clinical samples taken from the respiratory tract of people who were known not to be infected by SARS-CoV-2. By late spring 2020, many positive patient samples were available. (Whether a sample is positive or negative is determined using a previously approved assay.) At the time of writing, both patient samples and contrived specimens are used for validation studies.

3. ***Inclusivity.*** This is a validation parameter that is specific to PCR assays, although in a broad way it might be considered a form of specificity testing. This parameter refers to the primers used for amplification. The FDA requires that the primers recognize every known sequence of SARS-CoV-2 RNA. As viruses reproduce, small variations in their sequence arise by mutation. Thus, there is not a single, universal RNA sequence for SARS-CoV-2. The primers used in amplification for a diagnostic test must recognize SARS-CoV-2 viral RNA, even in the presence of subtle sequence differences. Tests for inclusivity are performed using computer software designed for this purpose.

4. ***Cross-reactivity.*** Cross-reactivity testing is a type of specificity testing in which one tests whether the amplification method recognizes other viruses or common respiratory flora in addition to the SARS-CoV-2 virus. FDA recommends performing this testing with samples of virus and bacteria that might be expected to occur in samples from the respiratory tract. If the assay method performs properly, it will be specific and will not give a positive result when any agent other than SARS-CoV-2 is present.

To make this validation protocol clearer, we will consider the validation of an assay kit that received authorization for emergency use from the FDA in the spring of 2020. The assay uses real-time reverse transcription PCR to detect SARS-CoV-2 viral RNA. As is true of all such assay kits, the kit comes with primers that detect a specific part of the viral genome. The kit also comes with probes that give rise to fluorescent light and enable the amplification to be followed in real time. The validation results for this kit are as follows:

1. ***Limit of detection.*** To validate the LOD, the manufacturer of the PCR kit purchased viral SARS-CoV-2 RNA from ATCC (a respected supplier of cells and other reagents located in Manassas, VA). Scientists took a known amount of this purchased viral RNA and spiked it into samples obtained from the upper respiratory tract of patients. (No information about the patients is provided, but we must assume they were not infected with SARS-CoV-2.) The spiked samples were then serially diluted, and the dilutions were tested using the PCR kit. Each dilution was tested with either 21 or 24 replicates. A portion of the data provided is shown in Table 27.4. Based on these data, the LOD of the COVID-19 assay kit is 5 RNA copies per reaction (1.25 copies/μL). This is the lowest dilution where more than 95% of the replicates test positive for viral RNA.

(*Continued*)

Case Study (*Continued*): Diagnostic Assay Validation in the Time of COVID-19, Part III

TABLE 27.4

LOD Results of a COVID-19 Assay Kit

Copies per Reaction	Copies per µL	Positive Replicates	Average C_t
40	10	21/21	33.2
30	7.5	24/24	33.8
20	5	24/24	34.4
10	2.5	24/24	35.3
5	**1.25**	**21/21**	**36.4**
2.5	0.63	19/21	37.6

2. *Inclusivity.* The primers in this kit recognize a portion of the SARS-CoV-2 genome that is highly conserved (seldom undergoes mutation). Scientists used a computer program to compare their primers to all publicly available SARS-CoV-2 sequences and found that their primers would recognize every known viral strain.

3. *Cross-reactivity.* As with inclusivity, this validation parameter was evaluated solely based on a computer program. The computer tested the genetic sequence of 31 pathogens that might be expected in patient samples. According to their analysis, none of these would react with their primers and provide a false-positive result. Given more time, it would be advisable to evaluate cross-reactivity by actually running the assay with a panel of pathogens that might realistically be expected to occur in patient samples. Using a computer program to evaluate this parameter was presumably sufficient, given the need to rapidly validate and bring to market COVID-19 assays.

4. *Clinical evaluation.* Validating diagnostic specificity and sensitivity requires obtaining specimens from patients known to be infected, and samples from patients known not to be infected. In this case, scientists used samples that had been tested using another, previously approved PCR test kit. The kit they chose was developed by the Centers for Disease Control (CDC). They obtained 63 samples that tested positive with the CDC kit, and 67 that were negative. The scientists compared the results from their new kit to the results from the CDC kit. A portion of their test data is shown in Table 27.5. Based on these data, they estimate their diagnostic specificity to be roughly 98% and their diagnostic sensitivity to be roughly 93%. These values are acceptable to the FDA.

TABLE 27.5

Clinical Evaluation of a New PCR Test Kit

	Samples "Known" to be Infected Based on CDC Test Kit (63)	Samples "Known" Not to be Infected Based on CDC Test Kit (67)
Positive test result with the new kit	62	5
Negative test result with the new kit	1	62
	Estimated diagnostic sensitivity 62/63 ≈ 98%	Estimated diagnostic specificity 62/67 ≈ 93%

Performing clinical evaluation using samples that were identified as positive or negative using the CDC test kit is acceptable at the time of writing, but it has the problem that it assumes the CDC test kit provides perfect results. This may or may not be a reasonable assumption. (For a good analysis of limitations in RT-qPCR testing for COVID-19, see Woloshin, Steven, Patel, Neeraj, and Kesselheim, Aaron S. "False Negative Tests for SARS-CoV-2 Infection—Challenges and Implications." *The New England Journal of Medicine*, vol. 383, no. 6, p. e38, 2020. DOI: 10.1056/NEJMp2015897.)

Practice Problems

1. PCR is a reaction that occurs in a test tube and is catalyzed by an enzyme. What are the components that need to be added to the test tube to make a PCR mixture?

2. Transcribing RNA into cDNA is also a reaction that occurs in a test tube and is catalyzed by an enzyme. What are the components that need to be added to the test tube to make a reverse transcriptase reaction mixture?

3. If you begin a PCR with 1,000 copies of a target DNA sequence, then after one cycle of PCR you expect to have 2,000 copies (under ideal conditions). After two cycles of PCR you expect to have 4,000 copies of target. How many copies of target DNA do you expect after 5 cycles of PCR, assuming ideal conditions?

4. If you begin PCR with 25 pg of DNA, how much DNA do you expect after 6 cycles, assuming ideal conditions?

5. The gel in Figure 27.15 is from an assay where PCR is being used to look for a bacterial pathogen in food samples. Lane 1 contains a molecular weight marker where each of the bright bands is a DNA fragment of a different, known length. Lane 2 is a positive control where a food sample was intentionally spiked with DNA from the bacterial pathogen. Lane 3 is a food sample known not to have contamination. Lane 4 is a no-template control. Lanes 5–8 are grocery store foods being tested for contamination by the pathogen. Discuss these results.

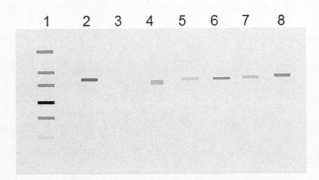

FIGURE 27.15   Result of PCR assay of bacterial pathogens in foods.

6. A diagnostic test is being developed for a (hypothetical) viral disease caused by the (hypothetical) virus SARS-CoV-3. (While there is no such virus at this time, scientists predict that there will be future pandemics caused by as yet unknown coronaviruses.) A clinical evaluation study is performed as part of method validation. The data are shown in the following table. Based on these data, what is the diagnostic sensitivity of the new assay? What is the diagnostic specificity of the new assay?

Clinical Evaluation of a New PCR Test Kit		
	Samples "Known" to be Infected (73)	Samples "Known" Not to be Infected (124)
Positive test result with the new kit	68	12
Negative test result with the new kit	5	112

7. Another test kit is also being developed to test for SARS-CoV-3. A clinical evaluation study is performed as part of method validation. The data are shown in the following table. Based on these data, what is the diagnostic sensitivity of the new assay? What is the diagnostic specificity of the new assay?

Clinical Evaluation of a New PCR Test Kit

	Samples "Known" to be Infected (100)	Samples "Known" Not to be Infected (100)
Positive test result with the new kit	93	9
Negative test result with the new kit	7	91

8. During optimization studies of the new test kit, researchers try raising the magnesium level in the reaction mixture. Do you think this will increase or decrease the diagnostic sensitivity of the test? Do you think this will increase or decrease the diagnostic specificity of the test?

9. During optimization studies of the new test kit, researchers try raising the temperature of the primer extension step. Do you think this will increase or decrease the diagnostic sensitivity of the test? Do you think this will increase or decrease the diagnostic specificity of the test?

10. Suppose the scientists developing this new PCR test want to be as certain as they can be that it will detect every case of infection with the SARS-CoV-3 virus. In this situation, would they prefer to have the utmost diagnostic sensitivity or the utmost diagnostic specificity possible?

11. The test kit for the SARS-CoV-3 virus tests for the presence of viral spike protein mRNA in patient samples. The test kit is supplied with a vial of SARS-CoV-3 virus spike protein mRNA. What is the purpose of this mRNA? How is it used when the test is performed?

28 Measurements Involving Light – Part B: Assays

28.1 INTRODUCTION

Chapter 21 described the nature of light and the design, operation, and performance verification of spectrophotometers that measure light. This chapter explains how measurements involving light and spectrophotometers are used in assays to obtain information about biological samples.

Spectrophotometers can be used to help answer essential questions in the laboratory, including: (1) What is the identity or nature of the component(s) of a sample? These are qualitative assays (Figure 28.1a). (2) How much of an analyte is present in a sample? These are quantitative assays (Figure 28.1b).

There are many spectrophotometric assay methods. For example, there are spectrophotometric methods

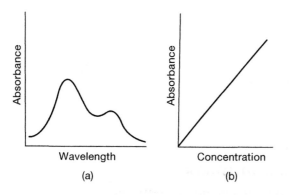

FIGURE 28.1 Qualitative versus quantitative spectrophotometry. (a) Qualitative applications use absorbance spectra to give information about the component(s) of a sample. (Recall that an absorbance spectrum is the absorbance measured for a sample [Y-axis] over a range of different wavelengths [X-axis]). (b) Quantitative applications determine the concentration of a substance in a sample. Concentration (or amount) is on the X-axis, and, as for qualitative applications, absorbance is on the Y-axis. This type of analysis is performed at a specific wavelength. Selection of the proper wavelength for quantitative analysis of a particular analyte is described later in this chapter. (See Section 13.1.3 for an introduction to standard curves.)

to confirm the identity of drug products (a qualitative application), to measure the levels of water pollutants (quantitative), to determine the levels of proteins in a sample (quantitative), and to study the activity of enzymes (quantitative). These spectrophotometric analytical methods must be developed, optimized, and validated. Note that ensuring that an assay method works effectively is a separate task from verifying that the instrument involved is properly functioning.

This chapter begins by explaining the general principles of qualitative and quantitative analyses by spectrophotometry. The chapter then looks at the development, optimization, and validation of spectrophotometric methods. In addition, there are three sections that briefly introduce associated spectrophotometric and fluorescent methods.

28.2 QUALITATIVE APPLICATIONS OF SPECTROPHOTOMETRY

Qualitative applications of spectrophotometry use the spectral features of a sample to obtain information about the nature of the sample's component(s) (Figure 28.1a). It is sometimes possible to use an absorbance spectrum like a "fingerprint" to identify an unknown substance. The infrared (IR) spectra of organic compounds are complex and form particularly distinctive

"fingerprints." Organic chemists therefore frequently use IR spectra to identify compounds.

Ultraviolet (UV) and visible (Vis) spectra are less complex and distinctive than IR spectra, so they are used less commonly for identification of unknown substances. There are, however, situations where information about a sample can be obtained from its UV–Vis spectrum. The U.S. Pharmacopeia includes tests of drug identity in which an absorbance spectrum is used to help confirm the identity of a drug or other compounds. In these identity tests, the spectrum of the test sample is compared with the spectrum of a pure reference material. If the sample and reference have identical spectra, then it is likely they are the same compound. UV–Vis identity tests are simple and rapid, and the equipment to prepare an absorbance spectrum is widely available.

Cyanocobalamin, a form of vitamin B12, provides an example of a U.S. Pharmacopeia identity assay. To perform this assay, a UV–Vis spectrum of the sample being tested is prepared. The test sample must have three distinct peaks at 278, 361, and 550 nm, respectively (Figure 28.2a). The most intense peak must be at 361 nm. (The 550 nm peak is responsible for the vitamin's red color.) Additionally, calculations are made of the ratio of the absorbances of the peaks at 361 and 278 nm, and the ratio of the absorbances of the peaks

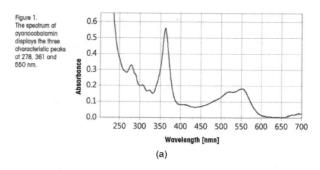

Figure 1.
The spectrum of cyanocobalamin displays the three characteristic peaks at 278, 361 and 550 nm.

(a)

	Acceptance criteria (USP 35-NF 30)	Sample
Peak 1	278 ±1	278.4
Peak 2	361 ±1	361.0
Peak 3	550 ±2	549.6
Ratio A361/A278	1.70–1.90	1.788
Ratio A361/A550	3.15–3.40	3.212

Table 1.
Acceptance criteria and measured results of a cyanocobalamin sample.

(b)

FIGURE 28.2 Analysis of vitamin B12 using qualitative spectrophotometry based on the method in the USP 35–NF-30. (a) A sample spectrum for vitamin B12 demonstrating the peaks at 278, 361, and 550 nm. (b) In this table, the acceptance criteria are shown in the second column. The data for the sample spectrum in (a) are in the final column of this table. This sample meets all the criteria for vitamin B12. (Courtesy of METTLER TOLEDO.)

at 361 and 550 nm. These calculated ratios must be within a certain range (Figure 28.2b). If the expected three peaks are present, and if the ratios of the peak absorbances are within the acceptable ranges, then the assay result is consistent with cyanocobalamin. Otherwise, the test is negative for this vitamin.

28.3 INTRODUCTION TO QUANTITATION WITH SPECTROPHOTOMETRY: STANDARD CURVES AND BEER'S LAW

28.3.1 CONSTRUCTING A STANDARD CURVE

Quantitative applications of spectrophotometry have the purpose of determining the concentration (or amount) of an analyte in a sample. Quantitative analysis with a spectrophotometer typically involves a "standard curve." A **standard curve (calibration curve)** *for spectrophotometric analysis is a graph of analyte concentration (X-axis) versus absorbance (Y-axis).*

To construct a calibration curve, standards are prepared with known concentrations of analyte. The absorbances of the standards are determined at a specified wavelength, and the results are graphed. Given a standard curve, it is possible to determine the concentration of an analyte in a test sample based on the sample's absorbance. The following example shows how this is done.

Example 28.1

a. Construct a standard curve for red food coloring.
b. Determine the concentration of red food coloring in a sample with an absorbance of 0.50.
   Step 1. Prepare a series of standards of known concentration by diluting a stock solution with purified water.
   Step 2: Prepare one tube that has no dye. This is the blank. (In this case, the blank is simply purified water.)
   Step 3. Place the blank in the spectrophotometer at the specified wavelength, and adjust the spectrophotometer to 0 absorbance.
   Step 4. Read the absorbance of each standard at the specified wavelength. Example results are shown in the following table.

Step 5. Plot the data on a graph with standard concentration on the X-axis and absorbance on the Y-axis, as shown in the following graph.
Step 6. Draw a best-fit line to connect the points. (See Chapter 13 for an introduction to best-fit lines and the Appendix of this chapter for a discussion of statistical methods to determine a best-fit line.)
Step 7. Read the absorbance of the test sample at the specified wavelength.
Step 8. Determine the concentration of the test sample based on the standard curve.

Example Data

Concentration of Standard	Absorbance
0.0 ppm	0.00
2.0 ppm	0.14
4.0 ppm	0.28
6.0 ppm	0.43
8.0 ppm	0.57
10.0 ppm	0.69
12.0 ppm	0.85
Test sample	0.50

---

a. The standard curve is:

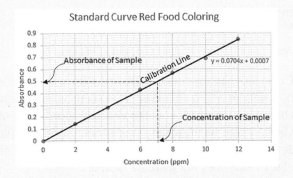

Standard Curve Red Food Coloring

b. Based on the standard curve, the concentration that corresponds to an absorbance of 0.5 is about **7.1 ppm.**

28.3.2 THE EQUATION FOR THE CALIBRATION LINE: BEER'S LAW

Information about how an analyte interacts with light can be obtained from the line on a calibration curve.

This section explores the features of that line in more detail.

The line on a calibration curve has an equation, as does every line. Recall from Chapter 13 that the general equation for a line is:

$$Y = mX + a$$

where:
    m = the slope
    a = the Y-intercept.

For the line on a calibration plot, the Y-axis is absorbance and the X-axis is concentration (or amount). Therefore, in the equation for the line on a spectrophotometer calibration curve, we can substitute A (for absorbance) and C (for concentration) as follows:

$$A = mC + a$$

For a calibration line, the Y-intercept is ideally zero. (This is because a Y-intercept of zero means that both the absorbance and analyte concentration are zero. A blank, containing no analyte, is used routinely to set the instrument to zero absorbance.) Thus, the equation for the line on a calibration curve is:

$$A = mC + 0 \text{ or simply}$$

$$A = mC$$

In words:

Absorbance = (slope of calibration line) · (Concentration of analyte)

Consider the slope of a calibration line. If the slope is relatively steep, it means that there is a dramatic change in absorbance as the concentration increases. On the other hand, if the slope is not very steep, then as the concentration of the analyte increases, the absorbance does not increase dramatically.

What determines the steepness of the slope of a standard curve? One factor is the nature of the analyte. Recall from Chapter 21 that different compounds have differing patterns of absorbance of light at different wavelengths. (For compounds that absorb light in the visible range, this pattern relates to the color of the compound.) For example, observe in Figure 28.3a that Compound A absorbs more light at 550 nm than does Compound B. (Assume that the concentrations of both the compounds were equal and that the

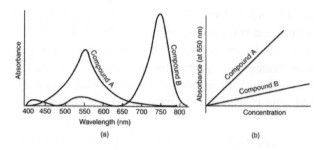

FIGURE 28.3 The slope of the calibration line. (a) Compound A absorbs more light at 550 nm than does Compound B. (b) The calibration line for Compound A is steeper than for Compound B at 550 nm.

conditions were the same when the spectra were constructed.) If we plot a standard curve for Compounds A and B at 550 nm, as is shown in Figure 28.3b, the line for Compound A will have a steeper slope than for Compound B. This is because Compound A has a greater inherent tendency to absorb light at 550 nm than does Compound B. Thus, at a given wavelength, the slope of a calibration line will vary from one compound to another.

A second factor that affects the slope of the calibration line is the path length. If the light has to pass through a longer path, then there is more absorbing material present, and more of the light will be absorbed. Thus, the two main factors that affect the slope are the following:

1. The tendency of the compound of interest to absorb light at the wavelength used.
2. The path length.

The inherent tendency of a material to absorb light at a particular wavelength is called its **absorptivity**. The greater the absorptivity of a material, the more it absorbs light at that wavelength. The absorptivity of a specific compound at a particular wavelength is a constant. (Note, however, that changing the pH, salt concentration, or other features of a solution can sometimes alter the absorptivity constant of a substance.)

The value of the absorptivity constant of a particular compound at a specific wavelength has many names, including the **absorptivity constant** (abbreviated with a Greek alpha, α) and the **absorption coefficient.** It is sometimes termed the **extinction coefficient** because it is an indication of the tendency of a compound to "extinguish" or absorb light. The absorptivity constant has units that vary depending on the units in which the concentration of the analyte is expressed.

The units for expressing concentration include moles per liter, milligrams per milliliter, and parts per million. If concentration is expressed in moles per liter, then the absorptivity constant is variously termed: **the molar absorptivity constant, ϵ (the Greek letter epsilon), the molar extinction coefficient,** or **the molar absorption coefficient.** Because the terminology varies, it is necessary to note the units in your own calculations and to be aware of the units in the calculations of others.

Returning to the equation for the calibration line, we can rewrite the equation to include the path length and the analyte's absorptivity, both of which contribute to the slope. The equation thus becomes:

$$A = mC$$

$$A = (\alpha b)C$$

where:
A = the absorbance
α = the absorptivity constant for that compound at that wavelength
b = the path length
C = the concentration
$(\alpha b) = m =$ the slope.

This equation, which shows the relationship between absorbance, concentration, absorptivity, and path length, is famous. It forms the basis for quantitative analysis by absorption spectrophotometry. Its discovery is variously attributed to Beer, Lambert, and Bouguer, but it is generally referred to simply as Beer's law. **Beer's law** *states that the amount of light emerging from a sample is reduced by three things:*

1. The concentration of absorbing substance in the sample (C in the equation).
2. The distance the light travels through the sample (path length or b).
3. The probability that a photon of a particular wavelength will be absorbed by the specific material in the sample (the absorptivity or α).

The following example problems illustrate some ways in which Beer's law can be used for quantitative analysis to determine the concentration of an analyte in a sample.

Example Problem 28.1

a. Using Beer's law, what is the concentration of the analyte in a sample whose molar absorptivity constant is 15,000 L/mole-cm and whose absorbance is 1.30 AU in a 1 cm cuvette?
b. Suppose that the absorbance of the preceding analyte is measured in a 1.25 cm cuvette. Will its absorbance be less than, more than, or equal to 1.30?
c. Will changing from a 1 cm cuvette to a 1.25 cm cuvette affect the molar absorptivity constant?

Answer

a. Substituting into the equation for Beer's law:

$$A = \alpha bC$$

$$1.30 = \frac{(15{,}000\ \text{L})(1\ \cancel{\text{cm}})C}{\text{mole} - \cancel{\text{cm}}}$$

The cm cancels:

$$1.30 = \frac{(15{,}000\ \text{L})C}{\text{mole}}$$

Solving for the concentration:

$$C = \frac{1.30}{\dfrac{(15{,}000\ \text{L})}{\text{mole}}} \approx 8.67 \times 10^{-5}\ \frac{\text{mole}}{\text{L}}$$

b. This change in cuvette will affect the path length. Because the path length is longer, the absorbance will be greater than 1.30.
c. The absorptivity constant is an intrinsic property of the analyte and is unaffected by the cuvette.

Example Problem 28.2

DNA polymerase is an enzyme that participates in the assembly of DNA strands from building blocks of nucleotides. The absorptivity constant of this enzyme at 280 nm (under specified pH conditions) is 0.85 mL/(mg)cm. (Worthington, Von, ed. *Worthington Enzyme Manual: Enzymes and Related Biochemicals.* Freehold, NJ: Worthington Biochemical Corp., 1993.) If a solution of DNA polymerase has an absorbance of 0.60 at 280 nm (under the proper conditions), what is its concentration?

Answer

Substituting into the equation for Beer's law:

$$A = \alpha bC$$

$$0.60 = \frac{0.85 \text{ mL}(1 \text{ cm})(C)}{\text{cm(mg)}}$$

$$C = \frac{0.60}{\dfrac{0.85 \text{ mL}}{\text{mg}}}$$

$$\approx 0.71 \frac{\text{mg}}{\text{mL}}$$

28.3.3 CALCULATING THE ABSORPTIVITY CONSTANT FROM A STANDARD CURVE

The absorptivity constant is an important property of an analyte. It is a useful skill to be able to calculate and report the absorptivity constant for a particular compound of interest based on data from your own spectrophotometer and in your own laboratory. This section shows two strategies to determine the absorptivity constant for a particular compound at a specific wavelength. The first strategy requires measuring the absorbance of a single standard and then calculating the absorptivity constant based on that single measurement. The second strategy is more reliable and involves constructing a standard curve and calculating the absorptivity constant based on its slope.

Strategy 1: Calculating the Absorptivity Constant Based on a Single Standard

Beer's equation can be rearranged as follows:

$$\text{Absorptivity constant} = \frac{\text{Absorbance}}{(\text{path length})(\text{concentration})}$$

If the absorbance and concentration of a single standard are known, these values can be entered into the preceding equation to determine the absorptivity constant. This method is illustrated in the following example.

Example 28.2

A standard containing 75 ppm of Compound Y is placed in a 1 cm cuvette. The absorbance of the standard is 1.20 at 450 nm. Assuming that there is a linear relationship between the concentration of Compound Y and the absorbance, and assuming that the standard was diluted properly, what is the absorptivity constant for this compound at 450 nm?

Substituting into the equation for Beer's law:

$$\text{absorptivity constant} = \frac{\text{absorbance}}{(\text{concentration})(\text{path length})}$$

$$\alpha = \frac{1.20}{75 \text{ ppm} (1 \text{ cm})}$$

$$= 0.016 \text{ / ppm} \cdot \text{cm}$$

The absorptivity constant at 450 nm, based on this single standard, therefore, is 0.016/ppm-cm.

This first strategy will give a value for the absorptivity constant. However, it is not the best method to use because it is based on only one standard. If that single standard is diluted incorrectly or if for some reason the absorbance value is slightly off what it should be, then the absorptivity constant calculated will also be inaccurate.

Strategy 2: Calculating the Absorptivity Constant from a Standard Curve

A better way to calculate the absorptivity constant for a particular compound at a specified wavelength is to base it on the absorbance of a series of standards. This is readily accomplished using a standard curve. Recall that the slope of the calibration line is *the absorptivity constant* (a) multiplied by *the path length*. Further, recall that the equation for the slope of a line is:

$$m = \frac{Y_2 - Y_1}{X_2 - X_1}$$

where:

    X_1 and X_2 are the X-coordinates for any two points on the line and

    Y_1 and Y_2 are the corresponding Y-coordinates for the same two points.

Determining the absorptivity constant based on the slope of a standard curve is illustrated in the following example and is summarized in Box 28.1.

Example 28.3

The calibration curve at 550 nm for Compound Q is shown. What is the absorptivity constant for this compound at 550 nm? (Assume the path length is 1 cm.)

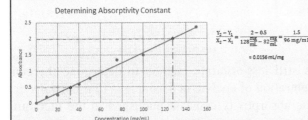

Determining Absorptivity Constant

$$\frac{Y_2 - Y_1}{X_2 - X_1} = \frac{2 - 0.5}{128\frac{mg}{mL} - 32\frac{mg}{mL}} = \frac{1.5}{96\,mg/mL}$$

$$= 0.0156\,mL/mg$$

Step 1. Calculate the slope of the calibration line. This can be done "by hand" as described in Section 13.1, and as

---

shown in the graph. Alternatively, graphing software can be used to determine the slope of the line. For this example, graphing software shows the slope to be 0.0157, which is slightly different – and more accurate – than the result obtained using the "by hand" method. Note, however, that the slope has units, which the software program does not show. In this case, the slope is 0.0156 mL/mg.

Step 2. The path length is 1 cm.

Step 3. From Beer's law, the slope of the calibration line is:

slope = (path length)(absorptivity constant)

$$\text{absorptivity constant} = \frac{\text{slope}}{\text{path length}}$$

In this example:

$$\text{absorptivity constant} = \frac{\textbf{0.0156 mL}}{\textbf{(1 cm)(mg)}}$$

28.3.4 Variations on a Theme

28.3.4.1 Overview

Now that we have explored the Beer's law equation, we will consider more examples of how it is used to determine the concentration (or amount) of an analyte in a sample. In general, the best way to determine the

---

BOX 28.1   DETERMINATION OF THE ABSORPTIVITY CONSTANT FROM A CALIBRATION CURVE

1. *Prepare a calibration line with concentration on the X-axis and absorbance on the Y-axis.*
2. *Calculate the slope of the calibration line using the equation:*

$$m = \frac{Y_2 - Y_1}{X_2 - X_1}$$

3. *Determine the path length for the instrument.* (The path length is generally 1 cm, assuming a standard type of cuvette and holder is used.)
4. *Solve the equation:*

$$\text{absorptivity constant} = \frac{\text{slope}}{\text{path length}}$$

concentration of an analyte in a sample is to construct a calibration curve based on a series of standards. There are, however, alternatives to constructing a calibration curve each time a sample is analyzed. These variations can be used successfully, but it is imperative to be aware of (and avoid) their potential pitfalls.

28.3.4.2 Variation 1: Determining Concentration Using a Calibration Curve from a Previous Analysis

In laboratories where a particular quantitative analysis is done routinely, it may be determined that a calibration curve needs to be produced and checked only once in a while. The same calibration curve is then used regularly for all samples. This method is accurate only if the following assumptions are true:

- **The instrument must remain calibrated and in good working order over time.** For example, if the spectrophotometer becomes unaligned so the wavelength is not the same as it was when the curve was first constructed, then using the standard curve will lead to inaccurate results.
- **The solvents and reagents must be consistent.** Reagents may change with storage. If the batch of reagents used when making the standard curve varies from a batch used for samples at a later time, then the results may be inaccurate.
- **Assay conditions, such as incubation time and temperature, must be consistent.**
- **A proper blank must be used when constructing the standard curve and when measuring the absorbances of the samples.**

Note that it is good practice to regularly check the absorbance of a control sample with a known analyte concentration. If the control has an unexpected reading, then the analyst is alerted that there may be a problem.

28.3.4.3 Variation 2: Determining Concentration Based on the Absorptivity Constant

A second alternative is to determine the absorptivity constant as shown in Box 28.1. Then, for subsequent samples the concentration of analyte in a sample can be determined based on Beer's law:

$$A = \alpha b C$$

Therefore,

$$C = A / \alpha b$$

This method was illustrated in Example Problems 28.1 and 28.2. This method is accurate only if the following assumptions are true:

- **The relationship between absorbance and concentration must be linear.**
- **The instrument must remain calibrated and in good working order over time.**
- **The solvents and reagents must be consistent.**
- **Assay conditions, such as incubation time and temperature, must be consistent.**
- **A proper blank must be used.**
- **The sample's concentration must be in the linear range of the assay (discussed in more detail later).**

28.3.4.4 Variation 3: Determining Concentration Based on an Absorptivity Constant from the Literature

A still less-accurate method to determine the concentration of analyte in a sample is to use a value for the absorptivity constant published in the literature. Absorptivity constants for various compounds at various wavelengths are reported in articles, catalogs, and other technical literature. In principle, spectrophotometers can be calibrated so that if the absorbance of the same sample is measured in various instruments, the same absorbance values will be obtained. In practice, two instruments often do not read identical absorbance values for a given sample. Moreover, the conditions when spectrophotometric measurements are made (such as solutions used, temperature, and pH) are likely to vary from one laboratory to another. An absorptivity constant derived in one laboratory is therefore seldom exactly reproducible in other circumstances. Nonetheless, there are situations where constants from the literature can be used to give estimates of concentrations. For example, if an analyst is comparing two methods of purifying a protein to see which is more efficient, it is reasonable to estimate and compare the amount of protein in the two preparations based on an absorptivity constant from the literature. In contrast, if the analyst wants to know how much protein is actually in each of the preparations, then a standard curve for that protein must be prepared by the analyst.

Example Problem 28.3

Suppose a manufacturer provides an absorptivity constant for an enzyme.

a. What features of your spectrophotometer must match those of the manufacturer in order for you to get the same absorptivity constant for this enzyme?
b. If your instrument does not match the company's instrument, is the absorptivity constant listed in the catalog useful information?

Answer

a. To reproduce an absorptivity constant obtained on one instrument using another instrument:

Both instruments must be calibrated for wavelength and photometric accuracy. In addition, the degree of "monochromaticity" of light exiting the monochromator must be the same. (The more monochromatic the light, the higher the absorptivity constant will be for a given compound.)
b. An absorptivity constant from the literature can be useful as an approximation. In addition, if the enzyme is measured consistently on the same instrument, then it can be compared in a relative way from batch to batch or from assay to assay.

28.3.4.5 Variation 4: Determining Concentration Using a Single Standard Rather than a Series of Standards

It is possible to use a single standard each time samples are analyzed. Then, the amount of analyte in the sample is proportional to the amount in the standard, as illustrated in the following example.

Example 28.4

A standard is prepared with 10 mg/mL of analyte.

The absorbance of the standard = 1.6.
The sample's absorbance is 0.8.
What is the concentration of the analyte in the sample?

$$\frac{10 \text{ mg / mL}}{1.6} = \frac{?}{0.8}$$

$$? = \textbf{5 mg / mL}$$

This method is accurate only if the following assumptions are true:

- **The single standard must be prepared accurately.** Manufacturers sometimes provide carefully tested standards for a particular method.
- **The relationship between analyte concentration and absorbance must be linear.**
- **The instrument must be set to zero absorbance using a properly made blank containing no analyte.**
- **The sample and the standard must have absorbances in the linear range of the assay.**

Example Problem 28.4

A standard containing 20 mg/mL of Compound Z is placed in a 1 cm cuvette. The absorbance of the standard is 1.20 at 600 nm. A sample containing Compound Z has an absorbance of 0.50. Assuming that there is a linear relationship between the concentration of Compound Z and the absorbance, and assuming that the standard was diluted properly, what is the concentration of Compound Z in the sample?

Answer

Let's think about this in two ways.

STRATEGY 1: USING PROPORTIONS

This is a proportional relationship; therefore,

$$\frac{20 \text{ mg / mL}}{1.2} = \frac{?}{0.5}$$

$$? \approx 8.33 \text{ mg / mL}$$

STRATEGY 2: BASED ON THE ABSORPTIVITY CONSTANT

Calculate the absorptivity constant based on this one standard. Substituting into the equation for Beer's law:

$$A = \alpha b C$$

$$1.20 = \alpha (1 \text{ cm})(20 \text{ mg / mL})$$

$$\alpha = \frac{1.20}{\dfrac{20 \text{ mg(cm)}}{\text{mL}}}$$

The absorptivity constant, based on this single standard, is:

$$\alpha = 0.06 \text{ mL / mg(cm)}$$

From Beer's law, therefore, if the absorbance of the sample is 0.50, its concentration is:

$$A = \alpha b C$$

$$0.50 = \frac{0.06 \text{ mL} (1 \text{ cm}) C}{\text{mg(cm)}}$$

$$C = \frac{0.50}{\dfrac{0.06 \text{ mL}}{\text{mg}}}$$

$$C \approx 8.33 \text{ mg / mL}$$

28.3.5 DEVIATIONS FROM BEER'S LAW

We have now seen how Beer's law, which states that the absorbance of a compound is directly proportional to its concentration, is the basis for quantitation using UV–Vis methods. In many systems, Beer's law holds true and it is therefore a very useful equation. There are situations, however, where real systems deviate from Beer's law. The causes of deviation from Beer's law include the following:

1. **At high absorbance levels, stray light causes spectrophotometers to have a nonlinear response.**
2. **At very low levels of absorbance, a spectrophotometer may be inaccurate.**
3. **There may be one or more components in the sample that interfere with the measurement of absorbance of a particular analyte.** We will discuss interference in a later section of this chapter.

Consider what occurs when absorbance is very high or low. Observe that the calibration line drawn in Figure 28.4 has two thresholds. At very low absorbance levels, the calibration curve deviates from linearity. At high absorbance levels, the instrument is not able to measure the small amount of light passing through the sample accurately. Thus, at high and low concentrations of analyte, the relationship between absorbance and concentration is not linear; therefore, Beer's law does not apply.

All quantitative spectrophotometric assays have a range of concentration within which the values obtained are reliable. Above this concentration range, the absorbance readings are too high to be useful, and below this range, the absorbance values are too low. It is essential that the standards and samples all have concentrations such that their absorbance falls in the range where the values obtained are accurate. Thus, a very important cause of deviation from Beer's law is failure to stay within the linear range of the assay. Note also that all quantitative analytical methods have a range in which their results are acceptably accurate and reproducible.

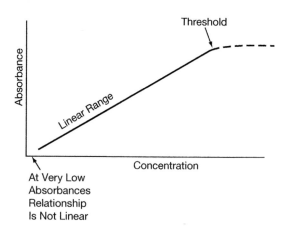

FIGURE 28.4 The relationship between absorbance and concentration is linear at intermediate concentrations of the analyte, but not at high or low concentrations.

Example Problem 28.5

Suppose you have prepared a standard curve as shown in the following graph. What is the concentration of the analyte in a sample whose absorbance is 1.80?

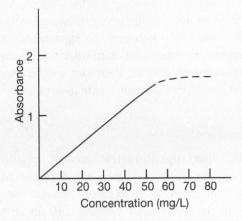

Answer

The absorbance of the sample is too high; it does not fall in the linear range of the standard curve. All we can tell is that the concentration of the analyte is above about 60 mg/mL. To determine the concentration of the analyte, it is necessary to dilute some of the sample and repeat the assay.

Example Problem 28.6

Suppose you repeat the assay as discussed in the preceding example problem, this time taking 1 mL of the sample and diluting it with 9 mL of purified water. Now, the absorbance of the sample is 1.0. What is the concentration of the analyte in this sample?

Answer

Reading from the standard curve, the concentration of the sample is about 40 mg/L. However, the sample was diluted 10X. Therefore, multiply by ten, which gives an answer of **400 mg/L.**

Although there is a lower limit of concentration below which an analyte cannot be detected, spectrophotometry is generally a method used to evaluate analytes that are present at very low levels in a sample. You can easily demonstrate that this is so by taking food coloring and trying to measure its absorbance. You will find that the absorbance of undiluted food coloring is far too intense to measure in a spectrophotometer. In fact, food coloring must be diluted a great deal before its absorbance falls into the linear range of a spectrophotometer. Samples often must be diluted before they can be analyzed spectrophotometrically, as is illustrated in the example problem below involving NADH.

Example Problem 28.7

Assume that a spectrophotometer is able to read accurately in the range from 0.1 to 1.8 AU. The molar absorptivity constant for NADH is 15,000 L/mole-cm at 260 nm. Using Beer's law, calculate the concentration range of NADH that can be quantitated accurately at this wavelength based on the limits of the spectrophotometer.

Answer

This involves the calculation of the molar concentrations that will produce absorbances of 0.1 and 1.8. From Beer's law:

Substituting 0.1 and 1.8 into the equation (and assuming a 1.0 cm path length):

$$C = \frac{A}{\alpha b}$$

$$C = \frac{0.1}{\dfrac{(15{,}000\text{ L})(1.0\text{ cm})}{(\text{cm})\text{mole}}} \approx 6.7 \times 10^{-6}\text{ mole / L}$$

$$C = \frac{1.8}{\dfrac{(15{,}000\text{ L})(1.0\text{ cm})}{(\text{cm})\text{mole}}} \approx 120 \times 10^{-6}\text{ mole / L}$$

The range of NADH concentrations that can be detected at this wavelength with this spectrophotometer is therefore from **6.7 × 10⁻⁶ to 120 × 10⁻⁶ mole/L**. These are dilute solutions of NADH.

28.4 MORE ABOUT QUANTITATIVE ASSAYS WITH SPECTROPHOTOMETRY

28.4.1 COLORIMETRIC ASSAYS

A visible spectrophotometer can only be used to analyze a material that absorbs visible light. Biological materials by themselves are usually colorless (i.e., they do not absorb visible light). **Colorimetric assays** are used to analyze materials that are naturally colorless. A **colorimetric assay** *is one in which a colorless substance of interest is exposed to another compound and/or to conditions that cause it to become colored:*

$$\text{Substance without color} + \text{Reagent(s)} \xrightarrow[\text{Conditions}]{\text{Proper}} \text{Product with Color}$$

For example, there are several commonly used colorimetric assays to measure the concentration of proteins in a sample. One such assay, the biuret method involves dissolving copper sulfate in an alkaline solution and adding it to the protein sample. Complexes form between the copper ions and nitrogen atoms in the proteins. These complexes produce a purple color that is measured in a spectrophotometer at 550 nm. The more protein present, the more intense the purple color.

When performing a colorimetric assay, the standards, samples, and the blank should be handled identically. For example, if the samples are subjected to heating, cooling, the addition of various reagents, or other treatments, then the standards and the blank should also be treated in these ways. Note also that when performing colorimetric assays, if a sample is too concentrated to be in the linear range of the instrument, then it must be diluted *before* the reagent(s) is added. Diluting the final reaction mixture will not produce accurate results.

28.4.2 TURBIDIMETRY AND THE ANALYSIS OF BACTERIAL SUSPENSIONS

Turbid solutions *contain small suspended particles that both absorb and scatter light.* Scattered light is generally deflected away from the detector, so it appears to have been absorbed when, in reality, it was not. Turbidity can also cause an apparent shift in absorbance peaks because shorter wavelengths are scattered more readily than longer ones. How much a beam is attenuated by scattering depends on the optics of the instrument, the orientation of the cuvette, and the uniformity of the suspension. Any inconsistencies in a turbid system, therefore, will give inconsistent absorbance

readings. Turbid samples should normally be avoided in spectrophotometry. Suspended materials can be removed by centrifugation or filtration of the sample.

A useful exception to the rule of avoiding turbid samples is in evaluating the concentration of microorganisms in a sample. The greater the concentration of bacteria in a sample, the more they scatter light, and the higher the apparent absorbance of the solution. Although absorbance is not actually being measured, within limits, the relationship between the apparent absorbance reading and the concentration of microorganisms is linear. Apparent absorbance, therefore, can be used as an indirect measure of cell number in a suspension.

28.4.3 KINETIC ASSAYS

Kinetic spectrophotometric assays *measure the changes over time in concentration of reactants or products in a chemical reaction.* Kinetic assays are useful in the analysis of enzymes. In an enzymatic reaction, one or more substrates are acted upon by the enzyme and converted to product(s):

$$\text{Substrate} \xrightarrow{\text{Enzyme}} \text{Product(s)}$$

As enzymatically catalyzed reactions proceed, the substrate(s) is consumed, and product(s) appears. In an enzyme assay, therefore, either the appearance of product or the disappearance of substrate over time is monitored.

Although enzyme assays are a class of quantitative assay, the quantitation of enzymes is different from that of other proteins. Proteins other than enzymes are typically measured in terms of their concentration, for example, in terms of milligrams per milliliter. Enzymes can be measured not only in terms of their concentration, but also in units of activity. Activity is a measure of the amount of substrate that is converted to product by the enzyme in a specified amount of time under certain conditions. Thus, time is an important factor in enzyme assays. Figure 28.5 shows a plot of a kinetic assay.

28.5 DEVELOPING AND VALIDATING SPECTROPHOTOMETRIC METHODS

28.5.1 DEVELOPING EFFECTIVE METHODS FOR SPECTROPHOTOMETRY

28.5.1.1 Overview

Spectrophotometric methods delineate the steps for analyzing a sample using a spectrophotometer. For example, a colorimetric method might involve

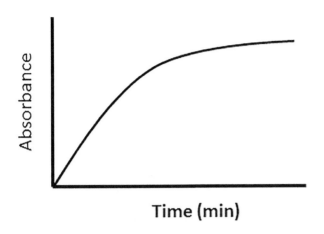

Time (min)

FIGURE 28.5 An example of a kinetic assay. Ingested alcohol is rapidly distributed throughout the bloodstream. The removal of alcohol from the body is a time-dependent, enzymatically catalyzed process that occurs in the liver. Alcohol dehydrogenase is the enzyme responsible for the initial metabolism of alcohol. The enzymatic reaction catalyzed by alcohol dehydrogenase results in the formation of NADH, which absorbs light at 340 nm. This graph shows the results of an assay for alcohol dehydrogenase in which the activity of alcohol dehydrogenase is monitored by following the change in absorbance at 340 nm over time.

(1) adding a reagent to samples, standards, and the blank, (2) heating the mixtures at a certain temperature for a certain time, (3) cooling them for a set period, (4) reading their absorbances, (5) constructing a standard curve, and (6) determining the concentration of analyte in the samples based on the standard curve. A spectrophotometric identification method might involve scanning a sample over a range of wavelengths and comparing peaks in the spectrum to those of a standard. There are thousands of published spectrophotometric analytical methods available. Sources of methods include laboratory manuals, research articles, the U.S. Pharmacopeia, and the manuals of AOAC and ASTM International.

Spectrophotometric methods are widely used partly because spectrophotometers are present in most laboratories, can be relatively inexpensive instruments, and are relatively easy to operate. Another important advantage to spectrophotometric methods is that an individual substance in a mixture can be analyzed without separating the components of the sample from one another. For quantitative analysis, the presence of multiple substances in a sample is not a problem as long as a wavelength exists where only the material of interest absorbs light. Spectral analysis of samples with multiple components may be possible if features of their spectra can be distinguished.

Developing and optimizing a spectrophotometric method requires finding a basis for the method (e.g., a colorimetric reaction) that is specific to the material of interest, finding the optimum wavelength for analysis (for quantitative assays), determining the type of sample for which the method is appropriate, determining the concentrations of sample for which the assay is accurate, discovering the characteristics of the sample that might interfere with the accuracy of the method, and determining what instrument performance features are necessary. Some of these factors are discussed in more detail in the upcoming sections. We will explore methods that are commonly used in biotechnology laboratories to assay proteins and nucleic acids to illustrate some of the issues that arise when using spectrophotometric assay methods.

28.5.1.2 The Sample and Interferences

The sample is a key part of a spectrophotometric method. As already discussed, each method is accurate only within a certain range of analyte concentrations. The analyte concentration in the sample, therefore, must match the requirements of the assay. Another important requirement is that the sample does not contain substances that interfere with the assay. An **interfering substance,** *in its broadest sense, may be considered as any material in the sample that leads to an inaccurate result in the analysis.*

Interferences may cause absorbance readings to be incorrectly high or low. In absorption spectrophotometry, interferences are often compounds that absorb light at the same wavelength as the analyte and therefore result in an absorbance value that is too high. Impurities that reduce (quench) light readings are often encountered in fluorescence assays (discussed later).

Figure 28.6 illustrates the effect of an interfering substance in a sample. The interfering substance in this example is a detergent, Triton X-100, which is sometimes used in biological solutions to solubilize membrane proteins, and to prevent protein aggregation. Detergents are also commonly used as cleaning agents. Both proteins and Triton X-100 absorb light at 280 nm. Residual detergent, therefore, must be removed from protein preparations before they are analyzed by spectrophotometry at 280 nm.

Colorimetric reactions are frequently used to visualize an analyte in the presence of other substances. For example, consider a sample containing a mixture of proteins, nucleic acids, and other cellular components. To analyze only the proteins, reagents can be added to the sample that selectively react with the proteins

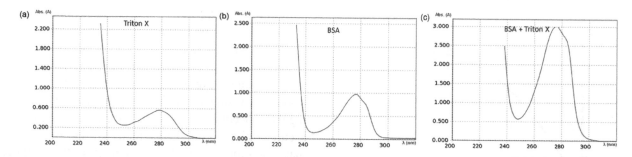

FIGURE 28.6 Triton X-100 as an interference in the analysis of protein. (a) The absorbance spectrum of Triton X. (b) The absorbance spectrum of 1 mg/mL of the protein bovine serum albumin (BSA). (c) The absorbance spectrum of the same sample of BSA as shown in **b**, but with a small amount of Triton X-100 added. Although the spectrum for BSA appears similar to that in (b), the absorbance at 280 nm is significantly raised due to the contamination of Triton X.

to form a colored product. There are situations, however, where a color-forming reagent(s) reacts with a substance in the sample in addition to the analyte. A substance that also reacts in an undesired fashion with colorimetric reagents is an interference. Table 28.1 summarizes practical points relating to interfering substances in colorimetric analyses.

It is generally not possible to compensate for the effect of an interfering substance by using a blank. A blank contains the solvent and reagents that are intentionally added to the sample. Interfering materials, by their nature, are variable and unpredictable substances in the sample that cannot be intentionally included in the blank.

The methods used to deal with interferences vary depending on the situation. In some cases, it is possible to find an analytical wavelength where the analyte absorbs light and the interfering substance(s) does not. In other situations where the nature of an interfering substance is known, it is possible to analyze more than one compound: the analyte and the interference(s). Because each component individually obeys Beer's law, and because both compounds are known, the concentration of each can be determined. A simple

example of this approach will be discussed later in the section on UV methods. An alternative spectrophotometric method sometimes exists that is insensitive to the interference(s) present in a sample. In other situations, where the nature of an interfering substance is unknown and a suitable analytical wavelength cannot be found, the sample must be purified to eliminate the interference.

28.5.1.3 Choosing the Proper Wavelength for Quantitative Analysis

Quantitative analyses are performed at a specific wavelength. Method development therefore requires finding the optimal wavelength based on the absorbance spectra for standards containing the analyte and based on representative samples.

The most important requirement when choosing the analytical wavelength is that the substance of interest absorbs light at that wavelength. The more strongly the analyte absorbs light at the chosen wavelength, the better the assay method will be able to detect low levels of compound. A second factor to consider is the nature of the absorbance peak. It is desirable that the peak of absorbance is not too narrow (i.e., the natural

---

TABLE 28.1

Symptoms That an Interfering Substance is Present in a Colorimetric Assay

1. *Interference is sometimes detectable if the color formed by a sample mixture is not of the same hue as that of the standards.*
2. *Interference is indicated if a precipitate forms or turbidity occurs when the sample and reagent are mixed together, but not when the standards and reagent are mixed together.* (This behavior is occasionally due to a concentration of analyte in the sample that is too high. In this case, simply diluting the sample will eliminate the problem.)
3. *If color does not form at the same rate or with the same stability in a sample as in the standards, an interfering substance(s) may be present in the sample.*

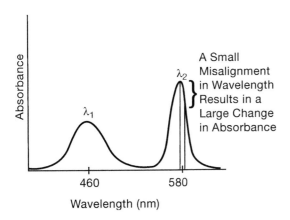

FIGURE 28.7 Determining the optimal wavelength for quantitative analysis. The wavelength chosen must be one at which the analyte absorbs light. There is a strong absorbance peak at 580 nm; however, because the peak is sharp, a small misalignment in the wavelength adjustment will cause a major change in absorbance. Unless very low levels of analyte must be measured, 460 nm therefore appears to be a better choice.

bandwidth of the peak should be fairly wide). If the peak is narrow, then any small error in the wavelength setting of the spectrophotometer will result in a large change in absorbance. For example, observe in Figure 28.7 that at λ_2 this compound strongly absorbs light, but the peak is very sharp, so a small error in the wavelength setting will result in a large change in the absorbance measured. At λ_1 there is a peak with a broader natural bandwidth that is probably the best choice of wavelength for analysis of this compound. If very low levels of analyte must be measured, then the maximum sensitivity is required and λ_2 might be used for analysis.

The presence of interfering substances in the sample is a complicating factor in choosing an analytical wavelength. If interfering substances are present, it may be possible to choose a wavelength for analysis at which the analyte absorbs less light, but the interfering substances are not absorptive.

Thus, the factors that are important in choosing the optimal wavelength are the following:

1. **The analyte should have a peak of absorbance at the wavelength chosen.**
2. **The absorbance peak should ideally be broad.**
3. **Interfering substances should not be present that absorb light at the chosen wavelength.**

Example Problem 28.8

Based on the following absorbance spectrum of Compound A, what is the optimal wavelength for performing quantitative analysis of Compound A?

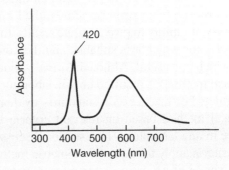

Answer

The wavelength of 600 nm appears to be a good choice because the compound absorbs light at that wavelength and the peak is not as "sharp" as it is at 420 nm.

Example Problem 28.9

Based on the following absorbance spectra, what is the optimal wavelength for performing quantitative analysis of Compound A in samples?

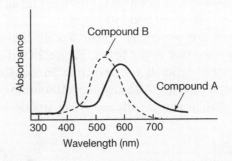

Answer

There is an absorbing substance, Compound B, that absorbs light at 600 nm. The wavelength of 420 nm appears to be a better choice in this situation because Compound A absorbs light at this wavelength and there is minimal interference.

28.5.2 METHOD VALIDATION

Validation of an analytical method provides documented evidence that the method will perform as is required. The criteria used to validate methods include linearity and range, limit of detection and limit of quantitation, specificity/selectivity, ruggedness, accuracy, and precision (see Chapter 26 for an introduction to these criteria). These criteria address the limitations of an assay, its ability to give accurate results, the types of samples for which it is suitable, and the conditions under which it is useful. Although formal validation of analytical methods is required only in laboratories that are regulated or meet certain standards, evaluating an analytical method based on relevant criteria is good practice in any laboratory. This section discusses how these criteria apply to spectrophotometric methods.

The **range of an assay** *is the interval between the upper and lower levels of analyte that can be measured accurately and with precision.* Factors that determine the range of an assay include the absorptivity of the analyte or the intensity of color produced in a colorimetric reaction, the capabilities of the spectrophotometer being used, and the choice of wavelength. In addition, the conditions of the assay may affect the range. For example, the intensity of color of some solutions is pH dependent.

The range of a spectrophotometric assay is usually considered to be the range in which the relationship between absorbance and concentration is linear. Figure 28.8 shows a standard curve for a spectrophotometric assay. At concentrations of analyte above about 130 μg/mL, the relationship between concentration and absorbance is not linear. The linear range for this assay is from about 5 to 130 μg/mL.

There are situations where a calibration curve can be used that is not linear. Figure 28.9 illustrates a nonlinear calibration curve from an assay used to quantitate proteins. Although the relationship between concentration and absorbance is not strictly linear at

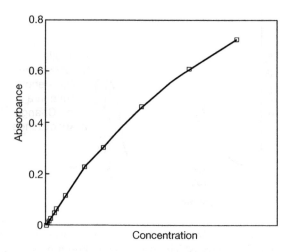

FIGURE 28.9 The standard curve from a Lowry's assay for protein. The assay gave linear results at the lower concentrations, but it was not linear at higher concentrations.

higher concentrations, this assay is still useful given a standard curve with sufficient points to clearly show the relationship.[1] (Remember that Beer's law does not apply if the relationship between absorbance and concentration is nonlinear.)

Sensitivity is the ratio of the change in the instrument response to a corresponding change in the stimulus. For a spectrophotometer, this is the ratio of the change in absorbance to a corresponding change in analyte concentration:

$$\text{Sensitivity} = \frac{\text{change in absorbance}}{\text{change in analyte concentration}}$$

This concept is illustrated in Figure 28.3. Compound A had more inherent tendency to absorb light at 550 nm than Compound B. At 550 nm, therefore, a spectrophotometer is more sensitive to Compound A than to Compound B. Thus, in spectrophotometry, the sensitivity of a method is affected largely by the absorptivity of the analyte or by the intensity of color developed in a colorimetric assay.

The **limit of detection** of an assay *is the lowest concentration of analyte in a sample that can be detected, but not necessarily quantified.* **The limit of quantitation** *is the lowest concentration of analyte that can reliably be quantified.* The limits of detection and quantitation of an assay are affected by the sensitivity of the method. The more sensitive the method, the lower the level of analyte that can be detected.

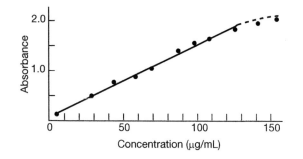

FIGURE 28.8 Linear range of an assay. This assay is useful in the linear range from about 5 to 130 μg/mL.

---

[1] Many spectrophotometers are equipped with software that automatically draws the best-fit line connecting the points of a standard curve. If an assay is known to be nonlinear, these automatic line fitting programs should not be used.

In addition, the limits are affected by noise. As the concentration of the analyte approaches zero, the signal becomes buried in the noise, and the level of analyte falls below the detection limit of the instrument. Thus, if a method needs to be able to detect trace levels of analyte, the method should be very sensitive and the noise in the spectrophotometer should be as low as possible. (Refer to Section 16.3.2.3 where issues relating to noise are introduced.)

Specificity (also called selectivity) *is the ability of an assay to distinguish the compound of interest in the presence of other materials.* Selectivity in spectrophotometry, therefore, relates to the ability of a method to measure an analyte accurately in the presence of potentially interfering substances in the sample.

Ruggedness *is the degree of reproducibility of test results obtained by the analysis of the same samples under a variety of normal test conditions.* A rugged method gives accurate results over a range of conditions. Many spectrophotometric assays are affected by solution conditions, such as pH, temperature, and salt concentration. This is because the properties and structures of biological molecules change as their solution conditions change. As the molecules change form, so their absorptivity may also change. Colorimetric reactions similarly may be affected by solution conditions; therefore, these conditions need to be controlled in spectrophotometric methods.

Accuracy is the closeness of a test result to the "true" value. One of the most important and basic ways to ensure accuracy in a quantitative spectrophotometric assay is to prepare a standard curve each time an assay is performed, using carefully prepared standards.

Precision and **long-term reproducibility** are often the most important aspects of an assay. For example, suppose spectrophotometry is being used to monitor the amount of protein present in a preparation as the protein is purified. Protein purification has multiple steps that may take place over a number of days. An assay method must give consistent results; otherwise, it is useless to monitor the purification.

28.6 BIOLOGICAL SPECTROPHOTOMETRIC ASSAYS

28.6.1 COLORIMETRIC PROTEIN ASSAYS

Colorimetric assays of proteins are an important application of spectrophotometry in biology. Most proteins do not naturally absorb light in the visible range. Colorimetric methods, therefore, have been devised in which proteins are reacted with certain dyes with the development of an intense color. Protein assays illustrate some of the issues that arise in using colorimetric methods.

Four common colorimetric protein assay methods are summarized in Table 28.2. The methods vary in features such as their range and their tendency to be affected by interfering substances. It is necessary, therefore, to choose the protein assay method that best suits your samples. It is important that the range of analyte concentration that can be measured by the assay matches the range of concentrations expected in the samples. Thus, if low levels of protein are to be measured, the Lowry or Bradford methods are preferred over the biuret method. Note that kits are available commercially for performing protein assays that can provide convenience and consistency.

Protein assays are often not linear. As shown in Figure 28.9, these assays can still be used, but a standard curve is required to establish the relationship between absorbance and concentration. Colorimetric protein assays are all affected to some extent by interfering substances. As can be seen in Table 28.2, interferences are particularly problematic for the Lowry assay.

The Lowry and Bradford methods both develop more intense color with some proteins than with others. To get accurate results with these methods, therefore, it is necessary to construct a standard curve for each protein of interest. For example, if the protein to be measured is DNA polymerase, then the standard curve should be constructed with DNA polymerase to get the best accuracy. In practice, it is common to construct a standard curve for a protein assay using bovine serum albumin (BSA) because the protein of interest is often expensive or difficult to obtain in large quantities, whereas BSA is readily available and is relatively inexpensive. Using BSA as the standard will give results that are not accurate. However, in many situations, using BSA to make the standard curve will give a useful estimate of protein concentration. For example, proteins are purified in a series of steps. After each step, the amount of protein present is measured. If this measurement is consistently made with reference to BSA, then the relative effectiveness of each purification step can be determined. In contrast, if an analyst wants to know the absolute amount of a specific protein in a preparation, then a standard curve using that protein is required.

TABLE 28.2
Colorimetric Protein Concentration Methods

Method	Principle	Approximate Range	Linearity	Interfering Substances	Comments
Biuret	Peptides react with Cu^2 in alkaline solution to yield a purple complex that has an absorption maximum at around 540 nm.	500–8,000 µg/mL	Not linear at all concentrations	Some interfering substances, including Tris buffer.	The biuret method is less susceptible to protein-to-protein variation because it measures the peptide bonds in a protein. Its major disadvantage is its relatively low sensitivity.
Lowry	Color results from the reaction of Folin's phenol reagent with certain amino acids in proteins.	1–300 µg/mL	May be nonlinear above about 40 µg/mL	Many compounds interfere including phenols, glycine, ammonium sulfate, and Tris buffer.	A sensitive method that is not completely specific to proteins; other substances can react with the reagent and interfere. The amount of color development depends on the percentage of tyrosine and tryptophan amino acids in the protein. Therefore, some proteins react more intensely than others.
Bradford	Color results from a reaction under acidic conditions with Coomassie Brilliant Blue G-250 reagent. Reagent changes from red to blue when the dye binds protein.	25–1,400 µg/mL	May be nonlinear at higher concentrations	Relatively few interfering substances, but some detergents interfere.	A specific and sensitive assay that is available in kit form and is relatively easy to perform.
BCA (Smith assay; bicinchoninic acid)	This method combines the biuret reaction with the highly sensitive and selective colorimetric detection of the cuprous cation using a reagent containing bicinchoninic acid.	20–2,000 µg/mL	Good linearity	A number of interfering substances, including EGTA, lipids, and phenol red.	BCA is compatible with samples that contain up to 5% surfactants (detergents). Compared to most dye-binding methods, the BCA assays are less affected by protein compositional differences, providing greater protein-to-protein uniformity. Readily available in kit form.

Source: Information in this table is from Copeland, Robert A. *Methods for Protein Analysis.* Chapman and Hall, 1994, and from the literature of manufacturers who provide kits to facilitate performing these assays.

Example Problem 28.10

A Lowry assay was performed. A standard curve was constructed as shown below using the protein bovine serum albumin. Three samples of another protein, Protein X, were then analyzed. The absorbances of the samples were the following:

Sample A	Absorbance = 0.10
Sample B	Absorbance = 0.58
Sample C	Absorbance = 1.3

a. What is the concentration of protein in the three samples based on the standard curve?
b. Is it reasonable to use BSA to prepare the standard curve when the protein of interest is Protein X?
c. Would it affect the reproducibility of the assay if a new batch of Folin's reagent were used? (Folin's phenol reagent is the color-producing reagent in the Lowry assay.)
d. Would it affect the reproducibility of the assay if the bulb on the spectrophotometer became less bright between assays?

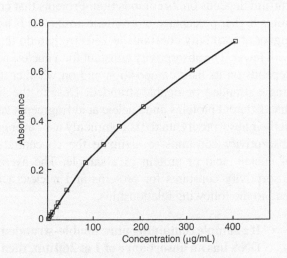

Answer

a. Sample A has a concentration of about 40 μg/mL. Sample B has a concentration of 300 μg/mL.

The absorbance of Sample C exceeds the range of the standard curve. In this case, it is necessary to repeat the assay, diluting the sample before the color-developing reagents are added.
b. Because the Lowry method primarily detects tyrosine and tryptophan amino acids, and because the percent tyrosine and tryptophan vary from protein to protein, it is not desirable to prepare a standard curve with a protein different than the one being evaluated. This practice is necessary, however, in situations where sufficient amounts of purified protein of interest to make a standard curve are unavailable or excessively costly.
c. As long as a new standard curve is prepared each time the assay is performed, and both batches of reagent are properly made, there should be no problem. (The results may not be accurate if a standard curve is not prepared each time the assay is run.)
d. Changes in the bulb will not affect the reproducibility of the assay as long as a new standard curve is prepared each time the assay is performed.

28.6.2 Analysis of DNA, RNA, and Proteins

28.6.2.1 The UV Absorbance Spectra of Nucleic Acids and Proteins

Although most biological molecules do not intrinsically absorb light in the visible range, they do absorb ultraviolet light. Biologists take advantage of UV absorbance to quickly estimate the concentration of DNA, RNA, and proteins in a sample. These methods require a spectrophotometer that works in both the visible and UV ranges. Points of general importance regarding spectrophotometric methods are illustrated by these applications.

Proteins have two UV absorbance peaks: one between 215 and 230 nm, where peptide bonds absorb, and another at about 280 nm due to light absorption by aromatic amino acids (tyrosine, tryptophan, and phenylalanine). DNA and RNA have an absorbance maximum at approximately 260 nm and an absorbance

TABLE 28.3

Wavelengths that Are Relevant to the Measurement of Nucleic Acids and Proteins

Wavelength	Significance	Comments
215–230 nm	Minimum absorbance for nucleic acids. Peptide bonds in proteins absorb light.	Measurements are generally not performed in this range because commonly used buffers and solvents, such as Tris, also absorb light here.
260 nm	Nucleic acids have maximum absorbance.	Purines' absorbance maximum is slightly below 260; pyrimidines' maximum is slightly above 260. Purines have a higher molar absorptivity than pyrimidines. The absorbance maximum and absorptivity of a segment of DNA, therefore, depend on its base composition.
270 nm	Phenol absorbs strongly.	Phenol may be a contaminant in nucleic acid preparations.
280 nm	Aromatic amino acids absorb light.	Nucleic acids also have some absorbance at this wavelength.
320 nm	Neither proteins nor nucleic acids absorb light of this wavelength.	Used for background correction because neither nucleic acids nor proteins absorb at this wavelength.

minimum at about 230 nm. Certain subunits of nucleic acids (purines) have an absorbance maximum slightly below 260 nm, whereas others (pyrimidines) have a maximum slightly above 260 nm. Therefore, although it is common to say that the absorbance peak of nucleic acids is 260 nm, in reality, the absorbance maxima of different fragments of DNA vary somewhat depending on their subunit composition. Table 28.3 summarizes the UV wavelengths relevant to the measurement of nucleic acids and proteins.

Figure 28.10 shows the absorbance spectra for DNA and proteins. Note that although proteins have lower absorbance at the absorbance peak of nucleic acids, 260 nm, both proteins and nucleic acids absorb light at 280 nm. Therefore, if nucleic acids and proteins are mixed in the same sample, their spectra interfere with one another.

28.6.2.2 Concentration Measurements of Nucleic Acids and Proteins

Consider two UV methods of determining concentration in a sample containing only proteins or only nucleic acids. The first method involves constructing a standard curve; the second is a "shortcut" method based on absorptivity constants from the literature.

It is possible to determine the concentration of nucleic acids or proteins based on their absorbance at a wavelength of 260 or 280 nm, respectively. A calibration curve using standards of known concentration can be constructed. For accurate results, the standard curve should be prepared using the protein of interest or DNA that is similar to that in the sample being measured. The linear range for DNA is reported to be

from about 5 to 100 μg/mL. Depending on the protein, UV analysis of proteins at 280 nm has a linear range from about 0.1 to 5 mg/mL.

Biologists commonly use a "shortcut" to provide a rough estimate of the concentration of nucleic acid or protein in a sample based on the sample's absorbance at 260 or 280 nm. This shortcut method uses absorptivity constants. Recall that given an absorptivity constant, it is possible to bypass the preparation of a standard curve by applying Beer's law (see Section 28.3.4.3).

The absorptivity constant for a particular protein at 280 nm depends on its composition. Proteins that contain a higher percentage of aromatic amino acids have higher absorptivity constants at 280 nm than do those with fewer. The absorptivity constant for a nucleic acid depends on its base composition and on whether it is single-stranded or double-stranded. Despite the fact that different proteins and nucleic acid fragments vary in their absorptivity, analysts commonly use "average" absorptivity constants to estimate the concentration of nucleic acid or protein in a sample. The average absorptivity constants for proteins and nucleic acids lead to the following relationships:

- **If a sample containing pure double-stranded DNA has an absorbance of 1 at 260 nm, then it contains approximately 50 μg/mL of double-stranded DNA.** Figure 28.11 is a standard curve for double-stranded DNA. This standard curve is consistent with the estimate that a DNA sample with an absorbance of 1 contains about 50 μg/mL of DNA.

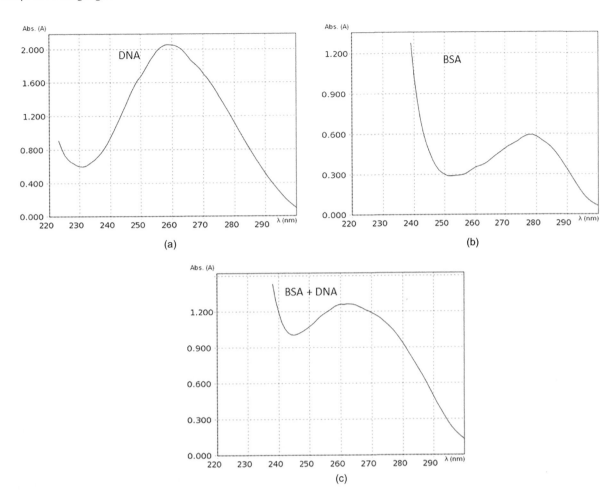

FIGURE 28.10 Absorbance spectra for DNA and protein. (a) The absorbance spectrum for DNA, 0.1 mg/mL. (b) The absorbance spectrum for protein (BSA), 1 mg/mL. (c) The absorbance spectrum for a mixture of DNA and protein, DNA 0.05 mg/L–BSA 0.5 mg/mL. Distinct peaks for DNA and protein cannot be resolved.

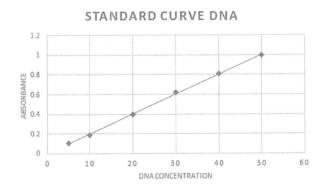

FIGURE 28.11 A standard curve for DNA at 260 nm. This standard curve is consistent with the literature value that an absorbance of 1 corresponds to roughly 50 μg/mL of double-stranded DNA. (The DNA is Sigma Catalog #D6898, DNA sodium salt from herring testes, Type XIV.)

- **If a sample containing pure single-stranded DNA has an absorbance of 1 at 260 nm, then it contains approximately 33 μg/mL of DNA.**
- **If a sample containing pure RNA has an absorbance of 1 at 260 nm, then it contains approximately 40 μg/mL of RNA.**
- **Values for proteins vary. A very rough rule is that if a sample containing pure protein has an absorbance of 1 at 280 nm, then it contains approximately 1 mg/mL of protein.** For example, 1 mg/mL of bovine serum albumin is reported to have an A_{280} value of 0.7. In contrast, antibodies (which are a type of protein) at a concentration of 1 mg/mL are reported to have an A_{280} between 1.35 and 1.2. (Values from Harlow, E., and Lane, D. *Antibodies: A Laboratory Manual.* Academic Press, 1988, p. 673.)

Example Problem 28.11

A laboratory scientist isolated DNA from bacterial cells and was interested in estimating the amount of DNA present in the preparation. The preparation had a volume of 2 mL. The scientist removed 50 µL from the preparation, added 450 µL of buffer, and then read the absorbance of the dilution at 260 nm. The absorbance was 0.65. Assuming the sample was pure, about how much double-stranded DNA was present in the original 2,000 µL preparation?

Answer

A proportion equation can be set up based on the relationship that a sample containing 50 µg/mL of pure double-stranded DNA has an absorbance of 1 at 260 nm:

$$\frac{1}{\frac{50\ \mu g}{mL}} = \frac{0.65}{?}$$

$$? = 32.5\ \mu g/mL$$

Since the preparation was diluted, the concentration was:

$$32.5\ \mu g/mL \times 10 = 325\ \mu g/mL$$

To estimate the amount of DNA in the original preparation, note that there were 2 mL of isolated product. The concentration of DNA in that preparation was about 325 µg/mL. The 2 mL of the original preparation, therefore, had about:

$$2\ mL \times 325\ \mu g/mL$$

$$= \textbf{650 µg double-stranded DNA}$$

28.6.2.3 Estimation of the Purity of a Nucleic Acid Preparation

Analysts often use UV spectrophotometry to roughly estimate the purity of a solution of nucleic acids. This method involves measuring the absorbance of the solution at two wavelengths, usually 260 nm and 280 nm,

and calculating the ratio of the absorbance at 260 nm to the absorbance at 280 nm. They then consider that:

- **An A_{260}/A_{280} ratio of 2.0 is characteristic of pure RNA.**
- **An A_{260}/A_{280} ratio of 1.8 is characteristic of pure DNA.**
- **An A_{260}/A_{280} ratio of about 0.6 is characteristic of pure protein.**

A ratio of 1.8–2.0 is therefore desired when purifying nucleic acids. (Note that this method does not actually distinguish DNA and RNA from one another.) A ratio less than 1.7 means there is probably a contaminant in the solution, typically either protein or phenol (which is a chemical sometimes used in the isolation of DNA). These calculations are performed so routinely that many spectrophotometers can perform them automatically and display the resulting values.

It is important, however, to be aware that the 260/280 ratios are only approximations and sometimes may be deceptively inaccurate. (Practice Problem 24 at the end of this chapter contains student data obtained when testing the effects of impurities on the 260/280 ratios.) The reasons for inaccuracy include the following:

1. The A_{260}/A_{280} method is based on the spectral characteristics of "average" proteins and nucleic acids. In reality, proteins and nucleic acids vary from one another, so their spectra may vary from one another.
2. These UV methods assume that the spectrophotometer is accurately calibrated. If the spectrophotometer is displaced by as little as 1 nm, the values may be significantly affected. (See, for example, Manchester, Keith L. "Value of A260/A280 Ratios for Measurement of Purity of Nucleic Acids." *BioTechniques*, vol. 19, no. 2, 1995, pp. 208–10.)
3. These UV methods are affected by the pH and ionic strength (salt concentration) of the buffer, resulting in variability. (See, for example, Wilfinger, William W. "Effect of pH and Ionic Strength on the Spectrophotometric Assessment of Nucleic Acid Purity." *BioTechniques* vol. 22, no. 3, 1997, pp. 474–80.)

These cautions apply not only to UV analysis of nucleic acids and proteins, but to other spectrophotometric methods. It is always important to be aware of potential inaccuracy when using absorptivity constants reported in the literature. The type of spectrophotometer used and its calibration will affect results in almost any method based on absorptivity constants. It is also important to consider matrix effects (i.e., effects due to the solvent and other components in the sample).

28.6.2.4 Multicomponent Analysis; the Warburg–Christian Assay Method

(Multicomponent analysis is a more complex application of spectrophotometry, and some readers may prefer to skip this section.)

It is sometimes desirable to estimate the concentrations of protein and of nucleic acids in a sample that contains both. The basic premise of an analysis such as this, a **multicomponent analysis,** is that each substance in the mixture individually obeys Beer's law; therefore, the absorbances of two or more components in a mixture add together. For example, if nucleic acids and proteins are mixed in a sample, then the total absorbances at 280 nm and 260 nm are:

$$A_{280} = A_{280} \text{ proteins} + A_{280} \text{ nucleic acids}$$

$$A_{260} = A_{260} \text{ proteins} + A_{260} \text{ nucleic acids}$$

To relate the absorbances at 260 and 280 nm to the concentrations of proteins and nucleic acids, these equations can be rewritten in terms of absorptivity constants for proteins and nucleic acids at each wavelength. Warburg and Christian performed these calculations using absorptivity constants for a yeast protein and for yeast RNA and published the results in 1941. (Warburg, Otto, and Christian, Walter. "Isolierung und Kristallisation des Gärungsferments Enolase." *Biochem. Z.* vol. 310, 1941, pp. 384–421.)

They derived these two equations:

$$[\text{nucleic acid}] \approx 62.9 \, A_{260} - 36.0 \, A_{280} \text{ (in units of } \mu g/mL)$$

$$[\text{protein}] \approx 1.55 \, A_{280} - 0.757 \, A_{260} \text{ (in units of mg/mL)}$$

where the brackets indicate "concentration."

To determine the concentrations of proteins and nucleic acids in a mixture, one measures the absorbances at 260 and 280 nm and solves the preceding equations. The example problem below illustrates this method.

Example Problem 28.12

A technician wants to get a rough estimate of the concentration of proteins and nucleic acids in a cell preparation. The absorbance of the preparation at 260 nm is 0.940 and at 280 nm is 0.820. Based on the Warburg–Christian equations, what is the approximate concentration of nucleic acids and proteins in the preparation?

Answer

Substituting into the Warburg–Christian equations:

$$[\text{nucleic acid}] \approx 62.9(0.940) - 36.0(0.820) \approx 29.6 \, \mu g/mL$$

$$[\text{protein}] \approx 1.55(0.820) - 0.757(0.940) \approx 0.559 \, mg/mL$$

Thus, based on the Warburg–Christian equations, the concentration of nucleic acids in the preparation is estimated to be **29.6 μg/mL** and of proteins to be **0.559 mg/mL**.

The Warburg–Christian equations are approximations based on absorptivity constants derived for the yeast protein, enolase, and yeast RNA. These equations, therefore, cannot be expected to give accurate results when applied to other proteins and nucleic acids. It is possible, however, to generalize the logic used by Warburg and Christian by substituting into the equations the absorptivity constants determined for your analytes at specific wavelengths. If correct absorptivity constants and optimal wavelengths are used, it is possible to quantitate multiple components in a sample, even if their spectra overlap to some degree.

Modern spectrophotometric software is available that can automatically perform multicomponent analysis using simultaneous equations analogous to those shown above. For example, drug products may contain more than one component, such as a tablet that contains both paracetamol and aspirin. Manufacturers need a way to test the concentrations of both components in their products; multicomponent spectrophotometry can be used for this purpose. (See, for example, Lawson-Wood, Kathryn and Robertson, Ian,

Pharmaceutical Assay and Multicomponent Analysis using the LAMBDA 365 UV/Vis Spectrophotometer, PerkinElmer Application Note, 2016, https://www.perkinelmer.com/lab-solutions/resources/docs/

APP-Pharmaceutical-Assay-and-Multicomponent-Analysis-using-the-LAMBDA-365-012622_01.pdf.)

Box 28.2 summarizes the various UV methods described in this section.

BOX 28.2   APPROXIMATING THE CONCENTRATION AND PURITY OF DNA, RNA, OR PROTEIN IN A SAMPLE

1. **Concentration of double-stranded DNA $\approx$ 50 µg/mL × the absorbance at 260 nm**
2. **Concentration of single-stranded DNA $\approx$ 33 µg/mL × the absorbance at 260 nm**
3. **Concentration of RNA $\approx$ 40 µg/mL × the absorbance at 260 nm**

   Turbidity sometimes causes an apparent increase in the absorbance of a sample, leading to incorrect readings. To compensate for slight turbidity, a background correction can be used. Proteins and nucleic acids do not absorb at 320 nm. If a sample absorbs at 320 nm, the absorbance is due to turbidity. The absorbance at 320 nm can be subtracted from the readings at 260 and 280 nm:

4. **Concentration of double-stranded DNA $\approx$ 50 µg/mL $(A_{260} - A_{320})$**
5. **Concentration of single-stranded DNA $\approx$ 33 µg/mL $(A_{260} - A_{320})$**
6. **Concentration of RNA $\approx$ 40 µg/mL $(A_{260} - A_{320})$**
7. **Concentration of protein $\approx$ 1 mg/mL × the absorbance at 280 nm**
8. **Concentration of protein $\approx$ 15 mg/mL × the absorbance at 215 nm**

   Notes:
   - Values for proteins vary. The rule that 1 mg/mL of protein has an absorbance of 1 may or may not apply to a particular protein.
   - Tris and other common solvents also absorb light at 215 nm. For this reason, 280 nm is more commonly used for protein measurements.

9. *Purity, estimations (these are very rough estimates)*

   An A_{260}/A_{280} **ratio of 2.0 is characteristic of pure RNA.**

   An A_{260}/A_{280} **of 1.8 is characteristic of pure DNA.**

   An A_{260}/A_{280} **of 0.6 is characteristic of pure protein.**

   Notes:
   - A ratio of 1.8–2.0 is desired when purifying nucleic acids.
   - This method is not very sensitive to low levels of protein in a DNA sample.
   - This method does not distinguish between DNA and RNA.
   - A ratio of less than 1.7 means there is probably a contaminant in the solution, usually either protein or phenol (which is sometimes used during the extraction of DNA from cells).

WARBURG–CHRISTIAN EQUATIONS FOR SOLUTIONS CONTAINING A MIXTURE OF NUCLEIC ACIDS AND PROTEINS

10. [nucleic acid] $\approx 62.9\,A_{260} - 36.0\,A_{280}$ (in units of µg/mL)

11. [protein] $= 1.55\,A_{280} - 0.757\,A_{260}$ (in units of mg/mL)

Example Problem 28.13

An analyst in our laboratory was attempting to check the absorbance spectrum of a solvent to see whether it was suitable for use in the UV range. She filled a cuvette with the solvent and prepared an absorbance spectrum, which is shown in the graph. The experienced analyst recognized that this spectrum did not look "right." How might she troubleshoot this possible problem?

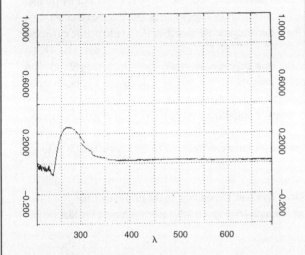

Answer

Various strategies are possible for troubleshooting this problem. It turns out that the cuvette in this case was dirty. The peak at 280 nm was due to protein that was left in the cuvette by another user and had dried onto the cuvette walls. This is unfortunately a common problem when users are not careful with cuvettes. The analyst soaked the cuvette for a couple of hours in purified water and rinsed it thoroughly. She then obtained a spectrum with no peak at 280 nm, indicating that the solvent has little absorption in the UV range (and that soaking and rinsing the cuvette with purified water cleaned it effectively in this case).

28.6.3 AVOIDING ERRORS IN SPECTROPHOTOMETRIC ANALYSES

The accuracy of spectrophotometric assays is affected by systematic errors. For example, in colorimetric assays, how long color is allowed to develop may affect the results. Color may be slow to develop in samples of low concentration and fast where concentration is high. Inconsistency in timing or a poor choice of development time may therefore lead to inaccurate results. Errors may occur if too little color-developing reagent is added to samples so that the assay underestimates the amount of analyte present in concentrated samples. Using a blank that contains a different solvent than the samples will lead to error. Many (but not all) errors can be avoided by preparing a standard curve each time an assay is performed rather than relying on an absorptivity constant from previous assays, a single standard, or a standard curve prepared at a different time.

Table 28.4 summarizes practical considerations relating to obtaining good results with spectrophotometric methods.

Example Problem 28.14

Suppose you are performing a colorimetric assay that includes a boiling step. Five minutes of boiling is necessary for complete color development. You become distracted and accidentally stop the boiling at only 3 minutes. How will this affect your results?

Answer

Assuming you prepare a standard curve each time you do the assay, and assuming the reagent still acts in such a way that color is proportional to the amount of analyte, then the standards should be affected by the shortened boiling time in the same way as the samples. The results, therefore, will still be accurate. You may observe, however, that the range for the assay is different than usual. Lower concentrations of analyte that are normally in the range of the assay may, in this case, not cause a detectable color change. (When you are familiar with an assay, any change in its performance should alert you to the fact that there is a problem.) In addition, note that if you do not prepare a standard curve each time the assay is performed, this type of mistake will lead to undetected errors in the results.

TABLE 28.4

Considerations in Obtaining Accurate Spectrophotometric Measurements

Factors Relating to the Instrument and Its Operation

1. *Always use a well-maintained and calibrated instrument.*
2. *Use the correct wavelength.* Assays are optimized at a particular wavelength, and irreproducibility or inaccuracy may occur at the improper wavelength.
3. *Allow the spectrophotometer light source to warm up, as directed by the manufacturer.* In most laboratories, the lamps are turned off when the instrument is not in use because lamps have a finite life span and are relatively expensive to replace. Note that the UV source may take longer to warm up than the visible light source.
4. *Use the proper cuvettes.*
   a. For high-accuracy work, use high-quality cuvettes that are clean and unscratched, and whose sides are exactly parallel.
   b. Use quartz cuvettes for work in the UV range. (There are certain brands of disposable plastic cuvettes that are intended for work in the UV range and may be suitable for some analyses. Be sure that, if you use plastic cuvettes, you are using the right ones for UV work.)
   c. Use matched cuvettes for the sample and the blank, or use the same cuvette sequentially for both the sample and blank.
   d. Do not use different cuvettes for each standard when making a standard curve; use a single cuvette and begin with the least concentrated standard. Rinse the cuvette thoroughly between each standard.
5. *Make certain cuvettes and glassware are clean.* In addition, drops or liquid smeared on the outside of the cuvette can cause erratic readings.
6. *Avoid fogging.* When the air is humid and the sample or blank is cold, moisture can condense on the cuvette leading to fogging.

Factors Relating to the Assay Method and the Sample

7. *Avoid samples and standards whose absorbance is very high or very low.* Depending on the instrument, the recommended range of absorbance may be about 0.5–1.8, but research-level spectrophotometers have a broader range in which they are accurate and precise.
8. *Be aware of sample characteristics that affect absorbance.*
   a. The color and the absorbance of some samples vary with the pH and ionic strength of the solution; in such cases, make sure the pH and ionic strength of standards and samples are the same.
   b. Sample absorbance may vary with temperature; in such cases, water-jacketed accessories are available for some spectrophotometers to hold the temperature constant during measurement. Otherwise, use a water bath or some other device to control the temperature of samples and standards when they are not in the spectrophotometer.
   c. In principle, the absorptivity of an analyte should be constant regardless of its concentration. In reality, some analytes change absorptivity as their concentration increases. In these cases, it is difficult to get a linear response.
   d. The ultraviolet absorption of a compound may vary in different solvents. Be sure, therefore, to record the solvent used and, for aqueous solvents, the pH. When following established methods, use the specified solvent.
9. *Be aware of time as a factor in an analysis.* The color intensity of samples reacted with color-developing reagents may increase or decrease over time, or the color may shift in absorbance peak. Time is a particularly important factor in assays involving enzymatic reactions.
10. *Avoid deteriorated reagents and standards.* Many color-developing reagents deteriorate over time. It may be helpful to check reagents regularly by measuring their absorbance versus a blank of purified water. Initial deterioration of reagents can often be recognized by changes in the reagent's absorbance. Discard reagents that become turbid or cloudy, that change colors, or that contain precipitate.
11. *Be aware that some samples are naturally fluorescent.* There are materials that fluoresce when irradiated with certain wavelengths of light. This results in erroneous transmittance values if the fluorescence falls on the detector. Fluorescent samples can be evaluated if a filter is available that blocks light of the fluorescent wavelength while allowing transmitted light to pass. The filter is placed between the sample and the detector.
12. *Avoid turbidity (except in some special cases where measurements of turbidity provide useful information).* Turbidity is a common source of error (e.g., environmental water samples are often turbid). Particles in suspension affect the transmittance of light and may cause it to appear that there is more absorbance than there really is. If the sample is not homogeneous, turbidity will result in inconsistent and incorrect results. Suspended material should generally be removed from samples prior to spectrophotometry by centrifugation or filtration. Buffers used for spectrophotometry may also require filtering.
13. *Use spectroscopy-grade solvents.*
14. *Avoid substances that interfere with a particular assay.*

Example Problem 28.15

You are preparing a standard curve for an assay that is routinely performed in your laboratory. This assay is known to have a linear relationship between absorbance and concentration in the range in which you are working. You obtain the data shown in the table below. Graph these data. What do you think about this standard curve? What should you do?

Standard Concentration (mg/mL)	Absorbance
0	0.24
2	0.36
4	0.55
6	0.69
8	0.85
10	1.01
12	1.3

Answer

The standard curve is shown below. The equation for the best-fit line is shown on the graph. The points are all close to the best-fit line, and superficially, this might seem to be a good standard curve. But observe on the graph that the line does not have a Y-intercept of zero. In theory, the Y-intercept for a spectrophotometric calibration curve (for the types of routine assays considered in this chapter) should be zero. This is because we always use a blank to set the instrument to zero absorbance when there is no analyte present. Therefore, zero concentration should correspond to zero absorbance. In practice, the Y-intercept might deviate a little bit from zero, but it should not deviate much. The Y-intercept on this graph is 0.2021, which is substantially different than zero. The fact that the Y-intercept is offset from zero by more than 0.2 absorbance units indicates that there was a problem somewhere in the production of this standard curve. For example, perhaps the blank was inserted improperly into the sample holder, or perhaps the blank was not prepared properly. A standard curve with a Y-intercept this high should alert the analyst that something is wrong that needs to be investigated. This standard curve should not be used to evaluate samples.

You might wonder how much the Y-intercept can deviate from zero and still be acceptable. There is no simple answer to this question. Limits would need to be determined in an individual laboratory for the particular assays in use. What is always true is that analysts must be attentive and must investigate unexpected variation in assays.

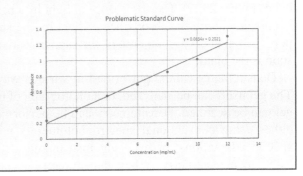

28.7 ASSOCIATED SPECTROPHOTOMETRIC AND FLUORESCENCE METHODS

28.7.1 INTRODUCTION

There are instruments in addition to spectrophotometers that measure interactions between light and a sample to obtain information about that sample. Some of these instruments allow samples to be studied that are not liquids in a cuvette. Other instruments are based on principles that are related to, but are not the same as, absorption spectrophotometry. There are also modified spectrophotometers that find a variety of uses in the laboratory. These instruments and associated technologies expand the applications that can be performed using light measurements. This section introduces three examples of light-measuring instrumentation that are of particular importance in the biotechnology laboratory.

28.7.2 GEL SCANNING/DENSITOMETRY

Electrophoresis is a widely used method in which mixtures of proteins or nucleic acids are separated from one another by allowing them to migrate through a gel matrix in the presence of an electrical field. The result of electrophoresis is that macromolecules of different sizes are separated into individual bands in the gel. The separated bands are invisible until they are stained with some type of dye. After staining, the band pattern on the gel, or a photograph of the gel, can be examined and used to obtain information about the sample.

Sometimes only the pattern of the bands in a gel is analyzed; other times, analysts are interested not only in the pattern, but also in quantifying the amount of nucleic acid or protein in each band. **Densitometry** *allows investigators to determine the amount of material in a band based on the intensity of its stain.* Densitometry uses light to measure how much material is present in a specific place on a gel or other matrices. When light is shined on a band in a gel, the amount of light absorbed is related to how much stain is present, which, in turn, depends on how much material is in the band (Figure 28.12). The absorbance of light by a band is compared with the absorbance of light by the support (gel) alone. Thus, an area of the support without stain is the blank.

Densitometry can be performed in various ways. The gel itself can be analyzed, or a photograph of the gel can be scanned. Densitometry can be performed either using conventional spectrophotometers with special adaptations, or by using instruments dedicated to densitometry. When densitometry is performed with a conventional spectrophotometer, a special holder is required to place the gel into the sample chamber. The holder must be moveable so that each band can be exposed to the light beam, one at a time. Optical arrangements are needed to limit the light beam to a specific band, and a mechanism must be provided to move the gel in the sample chamber.

28.7.3 Spectrophotometers or Photometers as Detectors for Chromatography Instruments

There are many instances in biotechnology laboratories where the components of a sample mixture are separated from one another before analysis. Chromatography is a commonly used separation method. During chromatography, samples plus solvents flow through a matrix held in cylindrical columns or on flat plates. Depending on their chemical and physical nature, some components of a mixture move more quickly through the column or across the plate than others, so the components are separated from one another by their rate of motion.

As drops of sample and solvent leave a chromatography column (are **eluted** from the column), they can be routed to flow through a detector. There are many types of detectors available, but spectrophotometers and photometers are common. A spectrophotometer or photometer used for chromatography does not involve cuvettes. Rather, there is a **flowcell** *through which the sample continuously flows.* As the sample moves through the flowcell, a continuous absorbance reading is produced (Figure 28.13). Because most biological materials absorb light in the UV range, these spectrophotometric detectors are frequently set to a UV wavelength (e.g., 280 nm).

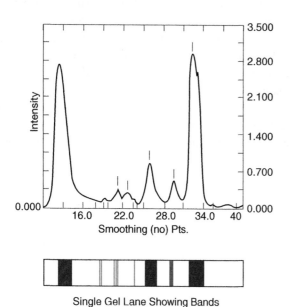

FIGURE 28.12 Densitometry of the bands on an electrophoresis gel. The peaks represent the amount of material in each band. (The Y-axis represents absorbance.) (Graph courtesy of Beckman Coulter Life Sciences.)

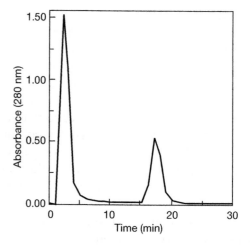

FIGURE 28.13 A spectrophotometer as a detector in chromatography. The absorbance of a sample was monitored as it eluted from a chromatography column. Time is on the X-axis, and absorbance on the Y-axis. The peaks represent the components of the sample that absorb light at 280 nm. We can see that one component of the sample elutes after about 4 minutes and another component elutes at about 17 minutes.

Chromatography instruments may use a photometer that can measure light absorbance at only a single wavelength, or a few discrete wavelengths. Variable wavelength detectors have a monochromator and can be adjusted to any wavelength in a certain range. In some cases, spectrophotometers with photodiode array detectors (PDAs) are used in conjunction with a chromatography instrument. The advantage of a PDA is that the absorbance spectrum of each material can be instantaneously generated as it elutes from the column. The resulting spectra are useful in identifying the separated components of the sample.

28.7.4 FLUORESCENCE SPECTROSCOPY

So far, we have discussed primarily the absorption of light by a sample. **Fluorescence** is another type of interaction that can give useful information about a sample. In **fluorescence,** *light interacts with matter in such a way that photons excite electrons to a higher-energy state. The electrons almost immediately drop back to their original energy level with the reemission of light.* The fluorescence of clothing or posters when exposed to "black lights" are familiar examples of this phenomenon.

Fluorophores *are molecules that fluoresce when exposed to light.* Most biological molecules are not naturally fluorescent. Fluorescent compounds, therefore, are used to label or tag desired biological structures. For example, ethidium bromide is a fluorophore that is hydrophobic, planar, and positively charged. Because of these properties, ethidium bromide spontaneously inserts (intercalates) into DNA molecules. One of the most common molecular biology techniques is to soak an electrophoresis gel containing fragments of DNA in a solution containing ethidium bromide. The ethidium bromide intercalates into the DNA. Ethidium bromide complexed with DNA fluoresces when exposed to light in the UV-B range. Therefore, ethidium bromide lights up the bands of DNA when the gel is placed on a UV light source.

Fluorophores can be used to tag antibodies that bind to their target biological structures with high affinity. (This application of fluorophores will be discussed in Chapter 29.) Fluorescent tags have also been used to label cancer cells, genes, proteins coded by specific genes, intracellular proteins, pathogens, and a host of other materials of biological importance.

The basic design of a fluorimeter (fluorescence spectrometer) is shown in Figure 28.14. Similar to a spectrophotometer, the fluorimeter has a light source that emits polychromatic light. The light enters a monochromator that functions to isolate light of a narrow range of wavelengths. In fluorescence spectroscopy, *the light that shines on the sample is called the* **excitation beam.** The wavelength of the excitation beam must be selected by the operator because different fluorophores are excited by different wavelengths of light. The excitation beam shines on the sample. Fluorophores in the sample are excited by the light and emit fluorescence. At this point, a fluorescent spectrophotometer differs from a UV–Vis spectrophotometer. The emitted light passes through a second monochromator that is adjusted by the operator to isolate the light that was emitted by the fluorophore. A photomultiplier tube detects the light emitted from the sample. Table 28.5 summarizes issues relating to fluorescence measurements.

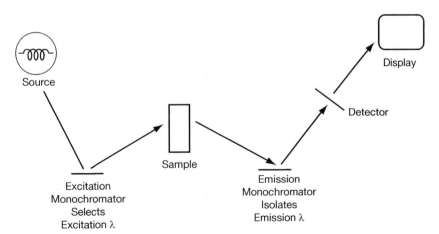

FIGURE 28.14 A fluorescence spectrometer. The excitation beam is directed at the sample. Fluorophores in the sample are excited by the excitation beam and emit fluorescent light, which passes through a second monochromator. The second monochromator isolates light of a wavelength matching that emitted by the fluorophore. A photomultiplier tube detects the light emitted by the sample.

TABLE 28.5

Issues Relating to Fluorescence Measurements

1. **Quenching** *occurs when an interference in the sample prevents the analyte from fluorescing.* This is a difficult problem to detect and often shows up as less sensitivity in an assay than is expected. Some fluorescent systems are quenched by oxygen from the air. Metal ions, even in low concentrations, as are present in some cleaning agents, may cause quenching.

2. *Fluorescence is much more sensitive to changes in the environment than is absorption.* Fluorescence increases as temperature decreases. On warm days, therefore, there may be noticeably less fluorescence than on cool days. Standards and samples should be at the same temperature. Some instruments control temperature to eliminate this variable.

3. *pH may affect the fluorescence of a sample.*

4. *There are fluorescent additives in plastics.* These additives may leach into samples that are stored in plastic containers.

5. *Most detergents used to wash glassware are strongly fluorescent.* Check that the detergent used in the laboratory does not contribute to fluorescence at the wavelength being measured.

6. *Stopcock grease and some filter papers contain fluorescent contaminants.*

7. *Microorganisms contaminating a solution may contribute fluorescence or may scatter emitted light.*

8. *The strong light sources used to excite the sample may bleach the fluorescent compounds.* Bleaching may be reduced by the choice of a different excitation wavelength, or by reducing the intensity of the excitation beam.

28.8 KEY POINTS ABOUT SPECTROPHOTOMETRIC ASSAYS

This chapter explores a common category of assay, that is, those that involve measuring the interaction of light with a sample. These assays typically involve a spectrophotometer, although, as introduced at the end of the chapter, there are other instruments that also measure light.

Spectrophotometers have detectors that measure the amount of light transmitted through a liquid sample that may contain analyte, divided by the amount of light transmitted through a blank with no analyte. This transmittance value is converted to absorbance, which is a measure of the amount of light absorbed by the sample. In most common assays, there is a linear relationship between the concentration of analyte in a sample, and its absorbance of light. This relationship is summarized by Beer's law:

absorbance =

absorptivity constant (path length) (concentration of analyte)

where:

The absorptivity constant describes the tendency of an analyte to absorb light of a specific wavelength, under specific conditions

Path length is the distance the light travels through the sample, which is determined by the dimensions of the sample holder (cuvette).

Some spectrophotometric assays are qualitative, such as assays that test the identity of a drug preparation. Qualitative analysis with a spectrophotometer is based on the spectral features of the sample, that is, the peaks and valleys that occur when the sample's absorbance is graphed versus wavelength. More commonly, spectrophotometers are used for quantitative analysis. Quantitative analysis involves measuring the absorbance of the sample at a specific wavelength, and comparing that absorbance to standards with known concentrations of the analyte. Biotechnologists, for example, routinely use spectrophotometry to determine the amount or concentration of proteins in samples. This can be accomplished by adding a reagent to the samples that chemically reacts with proteins to develop a particular color. The more protein present, the more color that appears. The spectrophotometer can quantify the amount of color that arises in the samples and standards. This type of assay is called "colorimetric" because it involves the measurement of a colored product.

Some spectrophotometers (not all) can be used to measure the absorbance of UV light by a sample. Biotechnologists often use UV spectrophotometry for measuring the amount and purity of proteins and nucleic acids.

Spectrophotometric assays are of interest both because they are commonly used and also because they illustrate concepts that are common to other categories of assays. One of these concepts is the idea of linearity; spectrophotometric assays generally provide linear results in a certain range of analyte concentration. At concentrations above or below this range, the assay no longer provides results that are proportional to the concentration of analyte. Beer's law summarizes this linear response. Many other quantitative assays also provide linear results, but only within a certain range of concentration.

Spectrophotometric assays also illustrate the importance of assay validation or verification, that is, understanding how the assay performs. With spectrophotometry, avoiding systematic errors involves understanding the limitations of the assay, its linearity and range, and materials that interfere with the assay. Many spectrophotometric assays are well characterized, and many are conveniently performed using commercial kits. Nonetheless, the analyst must still understand the assay well enough to select the right assay, use proper controls, follow the procedure properly, detect problems, and troubleshoot any problems that arise.

Practice Problems

1. a. Title the X- and Y-axes for each of the two graphs.
   b. Which one is called an absorbance spectrum? Which one is called a standard curve?
   c. Which is associated with quantitative analysis? Which is associated with qualitative analysis?

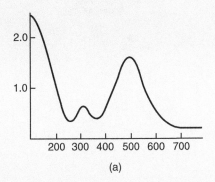

(a)

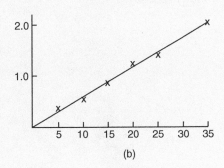

(b)

2. Label each of the following as a qualitative assay or a quantitative assay.
   a. Using an assay to determine the concentration of DNA in a sample.
   b. Using an infrared spectrum to identify an organic compound.
   c. Testing the amount of alcohol in a driver's blood.
   d. Determining the levels of lead in a child's blood.
   e. Determining whether a blood sample contains amphetamines.
   f. Determining whether or not a compound is oxygenated.
   g. Measuring the amount of product formed in an enzymatic reaction.
3. The following are absorption spectra for three compounds A, B, and C, respectively. Assume all three compounds were present at the same concentration when the spectra were plotted and that the analysis conditions were the same.
   a. What color is each compound?
   b. Which of these compounds will have the highest absorptivity constant at a wavelength of 540 nm? Explain.
   c. Which of these will be detectable in the lowest concentration at 540 nm? Explain.

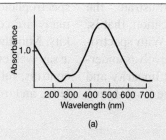

(a)

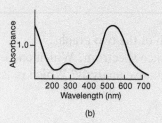

(b)

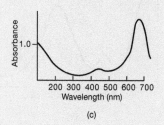

(c)

4. What is the linear range for the following assays?

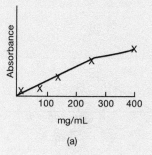

(a)

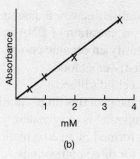

(b)

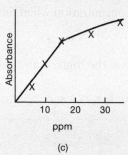

(c)

5. a. Prepare a standard curve from the data below.
   b. Is this graph linear? If all or part of the graph is linear, what is the equation for the line?
   c. If all or part of the graph is linear, determine the value of the absorptivity constant based on the line. Remember to include the units. (Assume the path length is 1 cm.)
   d. Does this graph show a threshold? If so, where? If so, why is there a threshold?

Standard #	Concentration (mg/L)	Absorbance
1	2	0.22
2	4	0.45
3	10	1.11
4	12	1.34
5	20	2.23
6	30	2.21
7	40	2.22
8	50	2.21

6. A certain compound is detected at 520 nm. A series of dilutions were prepared, and the data below were collected.
   a. Plot the data.
   b. Calculate the absorptivity constant based on the slope of the line. (Assume a path length of 1 cm.)
   c. Determine the amount of compound in the samples.

Standard #	Concentration (in mM)	Absorbance
1	0	0.0
2	80	0.2
3	160	0.4
4	240	0.8
5	320	1.0

Sample	Absorbance
A	0.18
B	0.31
C	0.96
D	1.0

7. a. Prepare a standard curve from the data below.
   b. Determine the absorptivity constant. (Assume the path length is 1 cm.)
   c. What is the concentration of an unknown which is diluted fivefold and has an absorbance of 0.30?

Standard #	Concentration (in mM)	Absorbance
1	1	0.22
2	2	0.46
3	5	1.08
4	6	1.34
5	10	2.18

For problems 8–18, assume Beer's law applies.

8. A certain sample has an absorbance of 1.6 when measured in a 1 cm cuvette.
   a. If the same sample's absorbance is measured in a 0.5 cm cuvette, will its absorbance be greater than, less than, or equal to 1.6?
   b. What is the absorbance of this sample in the 0.5 cm cuvette?

9. A 25 ppm solution of a certain compound has an absorbance of 0.87. What do we expect will be the absorbance of a 40 ppm solution? Explain.

10. A 10 mg/mL solution of a compound has a transmittance of 0.680.
    a. What is the absorbance of this solution?
    b. What will be the absorbance of a 16 mg/mL solution of this compound?
    c. What will be the transmittance of a 16 mg/mL solution of this compound?

11. A solution contains 30 mg/mL of a light-absorbing substance. In a 1 cm cuvette, its transmittance is 75% at a certain wavelength.
    a. What will be the transmittance of 60 mg/mL of the same substance?
    b. What will be the absorbance of 60 mg/mL of the same substance?
    c. What will be the transmittance of 90 mg/mL of the same substance?
    d. What will be the absorbance of 90 mg/mL of the same substance?

12. A standard solution of bovine serum albumin containing 1.0 mg/mL has an absorbance of 0.65 at 280 nm.
    a. Based on the BSA result, what is the protein concentration in a partially purified protein preparation if the absorbance of the preparation is 0.15 at 280 nm?
    b. Suggest two possible sources of error in this analysis.

13. The molar absorptivity constant of ATP is 15,400 L/mole-cm at 260 nm. (Value from Segel, I.H. *Biochemical Calculations*. 2nd ed. New York: John Wiley and Sons, 1976.) A solution of purified ATP has an absorbance of 1.6 in a 1 cm cuvette. Based on Beer's law, what is its concentration?

14. The molar absorptivity constant for NADH is 15,000 L/mole-cm at 260 nm. (Value from Segel, I.H. *Biochemical Calculations*. 2nd ed. New York: John Wiley and Sons, 1976.) A sample of purified NADH has an absorbance of 0.98 in a 1.2 cm cuvette. Based on Beer's law, what is its concentration?

15. The molar absorptivity constant for NADH is 15,000 L/mole-cm at 260 nm and is 6,220 L/mole-cm at 340 nm. The molar extinction coefficient of ATP is 15,400 L/mole-cm at 260 nm and zero at 340 nm. (Values from Segel, I.H. *Biochemical Calculations*. 2nd ed. New York: John Wiley and Sons, 1976.) How could you determine the concentration of NADH in a sample that contains ATP as well as NADH?

16. Calculate the range of molar concentrations that can be used to produce absorbance readings between 0.1 and 1.8 if a 1 cm cuvette is used and the compound has a molar absorptivity of 10,000 L/mole-cm.

17. a. A sample of purified double-stranded DNA has an absorbance of 4 at 260 nm. The technician therefore dilutes it, mixing 1 mL of the DNA solution with 9.0 mL of buffer. This time the A_{260} reading is 1.25. What is the approximate concentration of the DNA in the original sample based on the equations in Box 28.2?
    b. A sample of purified RNA has an absorbance at 260 nm of 2.4. The technician therefore dilutes it by mixing 1 mL of the RNA solution with 4.0 mL of buffer. This time the A_{260} reading is 0.63. Approximately how much RNA was in the original sample?

18. a. A sample of reasonably pure double-stranded DNA has an A_{260} of 0.85. What is the approximate concentration of DNA in the solution?
    b. A sample of reasonably pure RNA has an A_{260} of 3.8. Because a spectrophotometer does not accurately read absorbances that are so high, the analyst diluted the sample by adding 1 part sample to 9 parts buffer. The absorbance was then 0.69. What is the concentration of RNA in the solution?

19. Suppose a spectrophotometer has not been recently calibrated and its photometric accuracy is incorrect.
    a. Will the results of a quantitative assay be inaccurate if a standard curve is used?
    b. Will the results of a quantitative assay be inaccurate if a single standard is used?
    c. Will the results of a quantitative assay be inaccurate if the concentration of sample is calculated based on an absorptivity constant from the literature?

20. Suppose in a colorimetric assay, depending on the concentration of analyte in the sample, full color development requires anywhere from 3 to 5 minutes. You do not realize that there is a timing effect and allow the reaction to proceed for only 2 minutes. Will this cause an error in your results?

21. Suppose you are using a colorimetric assay that involves a color-developing reagent that deteriorates over time. You accidentally use an old bottle of the reagent that is less effective than it should be in causing a color change. What effect will this have on your results?

22. You are performing a spectrophotometer assay in which you use a buffer to prepare the standards and for the blank; however, the samples are centrifuged blood. Is this a potential source of error?

23. (Note that this problem is based on data obtained in our laboratory during a study of UV estimation methods for nucleic acids and proteins.)

   An analyst was evaluating the methods summarized in Box 28.2 to establish whether they would be useful in her investigations. She prepared three standards: (1) a standard containing pure double-stranded DNA, (2) a standard containing pure bovine serum albumin, and (3) a standard containing both double-stranded DNA and BSA mixed together. She measured the absorbance of the three standards at 260 and 280 nm and prepared an absorbance spectrum for all three standards. She then applied the equations shown in Box 28.2 to her data. Because she had prepared the standards from pure DNA and protein, she knew what the results of the calculations should have been. The steps she performed and the data she obtained are shown:

   Step 1. The technician dissolved 0.06 mg of DNA in 1 mL of a low-salt buffer. (The low-salt buffer was used as the blank also.)

   The A_{280} for the DNA was 0.260.
   The A_{260} for the DNA was 0.497.

   Step 2. She dissolved 0.5 mg of BSA in 1 mL of a low-salt buffer.

   The A_{280} for the protein was 0.276.
   The A_{260} for the protein was 0.176.

   Step 3. She prepared a solution containing 0.06 mg/mL DNA and 0.5 mg/mL BSA in a low-salt buffer.

   The A_{280} for the mixture was 0.547.
   The A_{260} for the mixture was 0.685.

   a. Estimate the concentration of DNA in the DNA standard based on its UV absorbance and on Equation 1 in Box 28.2. The concentration of DNA in the standard was 0.06 mg/mL (based on its weight). Comment on the accuracy of the UV estimation method.

   b. Estimate the concentration of protein based on its UV absorbance. Use the reported value that 1 mg/mL of BSA has an absorbance of 0.7 at 280 nm. The concentration of protein in the standard was 0.5 mg/mL (based on its weight). Comment on the accuracy of the UV estimation method.

   c. Check whether the absorbances at 260 and 280 nm were additive as predicted by the equations:

   $$A_{260} = A_{260} \text{ proteins} + A_{260} \text{ nucleic acids}$$

   $$A_{280} = A_{280} \text{ proteins} + A_{280} \text{ nucleic acids}$$

   Substitute into the first equation the A_{280} reading from the pure BSA sample and the A_{280} reading from the DNA sample. Substitute into the second equation the A_{260} readings for both pure samples. Comment as to whether the absorbances were additive.

   d. Calculate the A_{260}/A_{280} ratios for all three standards. Comment as to whether the ratios are as expected.

   e. (Optional) Substitute the A_{260} and A_{280} values obtained for the mixture into the Warburg–Christian equations, 10 and 11 in Box 28.2. Comment as to the accuracy of the Warburg–Christian equations.

   f. Comment on all the preceding results.

24. Ten students investigated the effect of protein contaminants on A_{260}/A_{280} ratios to see if this ratio provides a useful estimate of protein contamination. DNA and protein were dissolved in water to make stock solutions. DNA was at a concentration of 100 μg/mL, and protein was at a concentration of 1 mg/mL. The stock solutions were diluted as shown in the table below. Observe that the tubes contained progressively higher and higher amounts of contaminating protein while the amount of DNA was kept constant. One student's data for the absorbances at 260 and 280 nm are shown in the table.

   a. Fill in the table with the A_{260}/A_{280} ratios for this student's data.

   b. Comment on the use of this ratio as an indication of protein contamination based on this student's data. Note that other students obtained similar results.

Student Data for the Effect of Protein "Contamination" on A_{260}/A_{280} Ratios									
Tube #	µg DNA	µL DNA Stock (100 µg/mL)	µg Protein	µL Protein Stock (1 mg/mL)	µL Water	% DNA (Approximate)	A_{260}	A_{280}	A_{260}/A_{280} (Approximate)
1	50	500	0	0	500	100%	1.072	0.605	
2	50	500	5	5	495	91%	1.074	0.607	
3	50	500	20	20	480	71%	1.078	0.606	
4	50	500	50	50	450	50%	1.073	0.610	
5	50	500	150	150	350	25%	1.244	0.769	
6	50	500	250	250	250	17%	1.147	0.709	
7	50	500	350	350	150	12.5%	1.190	0.766	

CHAPTER APPENDIX: UNDERSTANDING THE LINE OF BEST FIT

OVERVIEW

A **standard curve** *displays the relationship between the concentration (or amount) of an analyte present and an instrument's response to that analyte.* Figure 28.15 shows a standard curve involving a spectrophotometer. Recall that the variable that is controlled by the investigator (in this example, the concentration of compound) is termed the *independent variable* and is plotted on the X-axis. The variable that varies in response to the independent variable (in this example, the absorbance) is the *dependent variable* and is plotted on the Y-axis. When a standard curve is prepared, we expect that the response of the instrument will be entirely determined by the concentration of analyte in the standard. If this expectation is met (and if the relationship between concentration and

instrument response is linear), then all the points will lie exactly on a line. However, there are sometimes slight errors or inconsistencies when a standard curve is constructed. These small errors cause the points to not lie exactly on a line, as shown in Figure 28.15.

The points in Figure 28.15 can easily be represented with a line, although the points do not all fall exactly on the line. In a situation like this, where a series of points approximate a line, we can use a ruler to draw "by eye" a line that seems to best describe the points. Drawing a line "by eye," however, is not the most accurate method, nor does it give consistent results (Figure 28.16). It is more accurate to use a statistical method to calculate the equation for the line that best describes the relationship between the two variables. *The statistical method used to determine the equation for a line is called* **the method of least squares.**

The method of least squares calculates the equation for a line such that the total deviation of all the points from the line is minimized (Figure 28.17). *The equation for the line calculated using the method of least squares is called the* **regression equation.** *The line itself is called the* **best-fit line.**

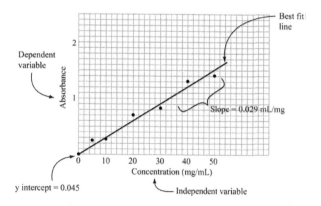

FIGURE 28.15 The relationship between the concentration of analyte and the response of an instrument, a standard curve. Due to slight inconsistencies, the points do not all lie exactly on a line. Also note that we expect the Y-intercept to be zero. It is close to zero, but, as is often observed in practice, the Y-intercept is not exactly zero.

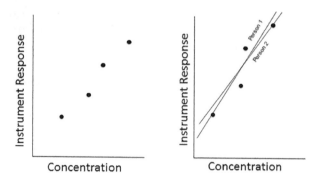

FIGURE 28.16 Drawing a best-fit line "by eye." Two people connect the concentration points into a best-fit line by eye with slightly different results.

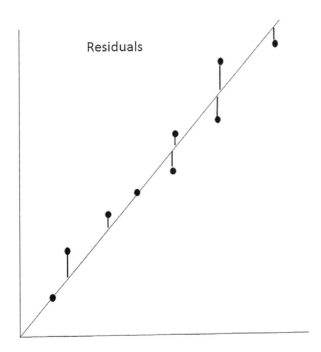

FIGURE 28.17 Residuals. The distance of a given point to the line is called the "residual." In this example, there are nine points, each with a residual distance to the line. The best-fit line minimizes the sum of the residuals of all the points. (Technically, the method minimizes the sum of the squared residuals.)

Recall that the equation for a line has the form:

$$Y = slope(X) + Y\ intercept\ or$$

$$Y = mX + a$$

where:
  m = the slope
  a = the Y-intercept
  X = the independent variable
  Y = the dependent variable.

The method of least squares calculates: (1) the slope and (2) the Y-intercept of the line that best fits the points. The method is based on the following assumptions:

- **The values of the independent variable, X, are controlled by the investigator.** In these spectrophotometric examples, the independent variable is the concentration or amount of analyte in each standard.
- **The dependent variable, Y, must be approximately normally distributed.** (For example, if the same sample of compound is measured many times with a very sensitive spectrophotometer, the absorbance values will vary slightly from reading to reading due to random error. These readings are expected to be normally distributed.)
- **If there is a relationship between the X and Y variables, it is linear.**

When the method of least squares is applied to the data graphed in Figure 28.15, the slope for the linear regression line is calculated to be 0.0285 mL/mg. The Y-intercept is calculated to be 0.045. Therefore, the equation for the line best fitting the points in Figure 28.15 is:

$$Y = \left(0.0285\frac{mL}{mg}\right)X + 0.0425$$

The slope and the Y-intercept can have units, and the units should be recorded. In spectrophotometry, the Y-intercept has no units because absorbance does not have units.

Most modern spectrophotometers automatically calculate the equation for the line of best fit when a standard curve is prepared. Computers and some calculators can also facilitate the determination of the equation for a line of best fit. An example of a standard curve graphed with Excel (a commonly used software program) is shown in Table 28.6 and Figure 28.18. In Excel, the best-fit line is termed a "trend line."

Note in Figure 28.18 that the Y-intercept is not exactly zero, although it is close to zero. In theory, the Y-intercept in a spectrophotometric assay should always be zero because a blank with no analyte is used to set the instrument to zero absorbance. In practice, the Y-intercept of the best-fit line often is not exactly zero, due to slight errors. The software for many spectrophotometers can be "told" either to force a calibration line to have a zero Y-intercept, or not. If the deviation from zero is minor, this is not usually a major concern

TABLE 28.6

Data for Determining the Line of Best Fit

Analyte Concentration (mg/mL)	Absorbance
0	0.01
5	0.05
10	0.09
20	0.24
30	0.31
40	0.47
50	0.51

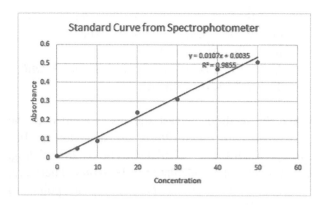

FIGURE 28.18   A standard curve as plotted by the software program Excel.

as long as you are consistent in how you use the software. But be alert to a situation where Y-intercept of a spectrophotometric calibration line deviates strongly from zero. This can indicate a problem with the assay that should be investigated.

Note in Figure 28.18 that Excel also provided a value called R^2, "R-squared." R-squared is a statistical measure of how closely the data fit the calculated best-fit line. (It is also known as the coefficient of determination.) The R-squared value is defined as the percentage of the response of the Y variable that is explained by the linear model[2]. Since it is a percentage, R-squared can vary between 0 and 1 or between 0% and 100%. The closer the R-squared is to 1 (or 100%), the better. In our example, the change in absorbance, that is, the Y value, is almost entirely explained by the changes in analyte concentration, the X value. Hence, the data fit the linear model quite well and there is a high value for R-squared, which is 0.9855. But R-squared is not exactly 1, so there is some variation in the Y values that is explained by other factors. Perhaps there was a small amount of pipetting error, or some fluctuation within the spectrophotometer, or some random error. The closer the R-squared is to 1, the more confident we can be that our linear plot correctly explains the variation in absorbance values.

---

[2] For an excellent and a more detailed explanation of the meaning of R^2, see *R-squared intuition,* Khan Academy, https://www.khanacademy.org/math/ap-statistics/bivariate-data-ap/assessing-fit-least-squares-regression/a/r-squared-intuition.

29 Achieving Reproducible Immunoassay Results

DOI: 10.1201/9780429282799-36

29.1 INTRODUCTION

29.1.1 What Is an Immunoassay?

Immunoassays *are a category of assays that rely on antibodies for interaction with an analyte.* Antibodies have exquisitely been optimized over thousands of years of evolution to "find" a substance even when it is present in only picogram amounts. This ability makes antibodies a powerful laboratory tool for detecting analytes in samples.

Consider some examples. Strep throat is an infection caused by the bacterium *Streptococcus pyogenes.* Strep throat infections usually resolve without problem. Occasionally, however, strep infection can lead to serious complications, such as rheumatic fever, which can damage the heart. Complications can be avoided by treating strep throat patients with antibiotics. However, most sore throats are caused by viruses, not bacteria, and patients with viral infections should not be treated with antibiotics (which only kill bacteria, not viruses). It is therefore important to determine whether a patient has strep throat before beginning antibiotic treatment. An immunoassay can be performed on mucus from the patient's throat to detect whether *Streptococcus pyogenes* is present. The strep test is only one example of a diagnostic immunoassay. In Chapter 26, we introduced immunoassays that are used to detect antibodies against the virus that causes COVID-19. Every day, thousands of medical conditions are diagnosed with immunoassays in clinical laboratories. There are even diagnostic immunoassays performed at home – home pregnancy tests are an example.

Just as immunoassays are widely used in testing laboratories, so they are also commonly used in biological research laboratories. Consider, for example, scientists who discover a protein receptor on the surface of cells that might be involved in controlling whether the cell divides. Such a receptor could play a role in cancer and might even be a target for a new drug. The scientists are interested in determining where this receptor is found. Is it found on cells in the breast, colon, or kidney? Is it found in areas where cells are actively dividing? Is it more common in cancer cells than in normal cells? Immunoassays provide a powerful means of studying these types of questions.

A search for "immunoassays" on PubMed (a database of biomedical literature) yielded half a million citations. In 2020, researchers reportedly spent $3.6 billion on antibodies for immunoassays (Grand View Research, Research Antibodies Market Size & Share Report, 2021-2028, https://www.grandviewresearch.

com/industry-analysis/research-antibodies-market). However, the power and widespread use of immunoassays makes them a source of irreproducibility, as discussed in Chapter 5 (Section 5.1). We saw in that chapter that researchers studying estrogen receptor β relied on antibodies that did not select for the proper target, resulting in 20 years of confusing research results. The purpose of this chapter is to introduce the concepts that analysts must understand in order to avoid error and obtain trustworthy, reproducible immunoassay results. We begin by introducing the basic principles and procedures of immunoassays, assuming that readers are unfamiliar with these techniques. This introduction omits most procedural details, but some points relating to the technique are provided in the appendices to this chapter. (More information about specific techniques can be found in the resources in the Unit Bibliography.) After the basic principles of immunoassays are introduced, we discuss some of the important strategies that are used to improve the reliability of immunoassays.

29.1.2 The Basic Concept of an Immunoassay

All assays begin with a sample that might (or might not) contain an analyte of interest. When discussing immunoassays, *the analyte is also called the* **target** or **antigen**. In any assay, the sample is subjected to a procedure that results in a detectable product if the analyte is present. In immunoassays, the target, if present, interacts with an antibody that specifically recognizes and binds to it, forming an antibody–target complex. This complex is initially invisible. Therefore, a **reporter molecule** *is necessary to make the complex visible to the eye and/or a particular sensor or instrument.* To illustrate, consider the example where scientists are investigating a receptor protein located on the surface of cells. The target of the immunoassay (the analyte) is the receptor protein. An antibody is manufactured that recognizes this receptor protein. (The receptor protein is also called an "antigen" because it is the target of an antibody.) After the antibody against the target is manufactured, it is *linked* or **conjugated** to a reporter, for example, a compound that emits fluorescent light (Figure 29.1a). The scientists can then use this reporter-linked antibody as a probe to see if the receptor is present, and if so, where it is located (Figure 29.1b). If the receptor protein is present, it will interact with the antibody, resulting in a target–antibody–reporter complex that is detectable by the light emitted by the reporter (Figure 29.1c).

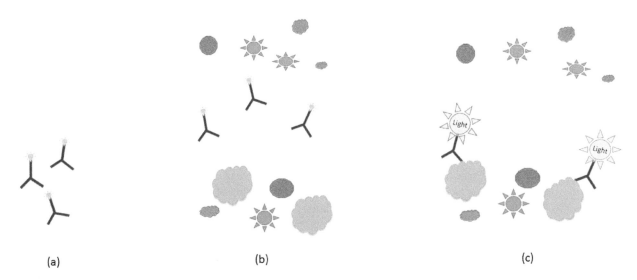

(a)   (b)   (c)

FIGURE 29.1 The basic concept of an immunoassay. (a) The antibody, purple, is manufactured to recognize the antigen (target) of interest and is conjugated with a reporter, yellow. (b) The antibody–reporter is incubated with a sample that contains a target analyte, green, and also other nontarget molecules. (c) The antibody–reporter binds the analyte, but not other substances in the sample. The resulting target–antibody–reporter complex is visualized by the signal (light) emitted by the reporter.

29.1.3 Terminology

There are various terms relating to immunoassays, and there is sometimes inconsistency in terminology. We therefore provide Table 29.1 to define terms as they will be used in this chapter.

29.2 FEATURES COMMON TO ALL IMMUNOASSAYS

29.2.1 Antibodies: Structure and Function

All immunoassays involve antibodies, so it is important to understand their characteristics. Antibodies were

---

TABLE 29.1

Vocabulary Relating to Immunoassays

Affinity. A measure of the attraction of one biological molecule toward another molecule. Antibodies and antigens have a high affinity for one another, and therefore, an antibody will recognize and bind to a specific antigen, even if that antigen is present in low concentration amidst many other molecules.

Analyte. The substance that is the subject of an assay. When discussing immunoassays, the term "target" or "antigen" is often used interchangeably with the term "analyte."

Antibody. A protein that circulates in the blood, produced in response to a specific substance that the body recognizes as being foreign, and that aids in protecting the body from the foreign substance.

Antigen. Any substance that the body recognizes as being foreign and in response to which the body produces antibodies.

Antigenicity. The capacity of an antigen to be recognized and bound by an antibody.

Autofluorescence. The situation where materials that are naturally present in a sample emit fluorescence in the absence of any reporter fluorophore.

Background or noise. In immunoassays, anything detected that is not due to an interaction of the antibody with the target of interest.

Cell lysate/lysis. "Lysis" occurs when a population of cells is broken open, releasing their cellular contents into a buffer solution. The resulting fluid containing the contents of the lysed cells is the "cell lysate."

Chromogenic assays ("Chromo-"; Greek for "color"). Assays involving a reporter that produces a colored deposit.

Cross-reactivity. The situation when an antibody recognizes and binds to a protein other than its intended target.

Endogenous. Anything produced or synthesized within an organism, as contrasted with a substance added to it.

Endogenous enzymes. Enzymes that are present in the sample being studied, as contrasted with enzymes that are added as part of an immunoassay procedure.

(Continued)

TABLE 29.1 (*Continued*)

Vocabulary Relating to Immunoassays

Enzyme immunoassays (EIAs). Any immunoassay in which the reporter is an enzyme.

Epitope. A molecular region on the surface of an antigen that elicits the production of an antibody, and that binds to the corresponding antibody.

　Conformational epitope. Discontinuous amino acids on an antigen that, due to their shape, come together to form an antibody-binding surface with a specific three-dimensional structure.

　Linear epitope. A series of specific amino acids that are adjacent to one another and form the antibody-binding region.

Immunochemistry. We use this term to refer to immunoassays on proteins or protein derivatives that have been isolated and bound to an artificial substrate, such as a thin membrane, or the plastic in a multiwell plate.

Immunocytochemistry (ICC). Immunoassays on cultured cells that are adhered to tissue culture plates, slides, multiwell plates, or other transparent surfaces. ICC assays typically involve fluorescent reporters and so are sometimes called "IF" assays.

Immunofluorescence (IF). Any immunoassay in which the reporter is a fluorescent compound.

Immunogen. The purified protein, synthetic protein, or peptide that is injected into an animal to stimulate an immune response and the production of antibodies. This term is often used synonymously with the word "antigen."

Immunohistochemistry (IHC). Immunoassays on tissue that has been preserved and thinly sliced to reveal its internal architecture.

Isotype. The class of an antibody, either IgA, IgD, IgE, IgM, or IgG. The five types of antibody play somewhat different roles in the body, although all are involved in the body's defense. Ig stands for "immunoglobulin," another name for "antibody."

Matrix. The sample environment in which the target is found. In IHC and ICC, the matrix is tissue or cells. In immunochemistry, the matrix is an artificial substrate, such as the bottom of the wells in a plastic multiwell plate. Matrix can also refer to the type of sample in which the analyte is found, for example, blood.

Monoclonal antibodies. A population of antibodies that all respond to a single epitope, traditionally all produced by a single clone of cells.

Multiwell plate. A covered plastic, transparent plate typically with 96 or 384 sample wells arranged in a rectangular matrix. Each well on the plate is similar to a tiny test tube for a different sample.

Nonspecific binding. Any situation where there is antibody binding to a site other than the target, for example, where an antibody sticks to the plastic of a multiwell plate instead of to its target.

Paratope. The portion of an antibody that binds to an epitope.

Polyclonal antibodies. A heterogeneous mixture of antibodies that respond to more than one epitope from a single antigen.

Recombinant antibodies (rAbs). Antibodies constructed using recombinant DNA technologies. These antibodies are often considered to be a form of monoclonal antibody, and some authors call them "recombinant monoclonal antibodies."

Reporter. A substance conjugated (linked) to an antibody that enables the antibody to be visualized by eye, microscopically, on film, with a camera's sensor, or with a specific instrument.

Sensitivity. The lowest concentration of an analyte that can be detected in a given assay.

Serum. The watery part of blood that does not clot and that contains proteins not involved in clotting, including electrolytes, antibodies, and hormones.

Specificity and selectivity. There is inconsistency in the way these words are used in the immunoassay literature. Specificity refers to an antibody's ability to recognize and bind to the intended target. Some authors use the term "selectivity" to mean the antibody's preference to bind its target protein in the presence of other sample proteins. With this usage of the terms, an antibody might be *specific* in that it does recognize its intended target, yet it might not be *selective* in that it cross-reacts with unintended targets.

Substrate. (1) The surface on which an immunochemical assay is performed, for example, the bottoms of the wells of a multiwell plate or a flat membrane. (2) A chemical(s) that is added to an enzymatic system in order to generate a detectable product.

Target. The antigen of interest. The analyte.

Validation of antibodies. The process of ensuring that an antibody works properly in a particular application. In the research literature, the term does not mean that any aspect of an immunoassay is "validated" to work in a clinical, diagnostic setting, as the FDA might use the term.

previously introduced on in Chapter 1 (Section 1.2.3) where the use of antibodies as therapeutics was discussed. Figure 1.10 shows the overall structure of an antibody with its variable and constant regions.

Recall from Chapter 1 that the **variable regions** *are the locations on the antibody where recognition of, and binding to a target occurs.* The portion of the antibody that recognizes and binds antigens is abbreviated "Fab," which stands for "antigen-binding fragment." An individual has millions of antibodies with different variable regions to recognize different potential invading substances. Each of these different variable regions has a different amino acid sequence. Each antibody also has a **constant region**. The constant region does not participate in recognition and binding to an antigen.

There are five different types or "classes" of antibody called: IgA, IgD, IgE, IgM, and IgG. The five types of antibody play somewhat different roles in the body, although all are involved in the body's defense. Ig stands for "immunoglobulin," another name for "antibody." The **constant region of an antibody** *determines the class of the antibody and does not vary within a given class.* For example, in an individual, all the IgA antibodies have the same amino acid sequence comprising their constant region. All the IgG antibodies have the same amino acid sequence comprising their constant region and that sequence is somewhat different from the sequence in the constant region of IgA antibodies. The constant portion of an antibody is abbreviated "Fc," which you can think of as standing for "constant fragment." In this chapter, we only discuss IgG antibodies because they are the type most commonly used in immunoassays.

Antibodies function by recognizing and binding to **antigens**. Antigens are entities that elicit the production of antibodies. Although an antigen can be any three-dimensional surface that binds to an antibody, in most biotechnology applications, antigens are proteins. Therefore, for the purpose of discussing immunoassays, we will assume that antigens are proteins or protein derivatives. If the target is a virus, bacterium, spore, or some other large structure, antibodies recognize a protein on its surface.

Recall also from Chapter 1 that antibodies recognize and bind to only a portion of an antigen. *The portion of the antigen recognized by an antibody is called an* **epitope** (Figure 29.2). Antibodies recognize specific protein epitopes based on the epitope's sequence of amino acids and/or its physical structure.

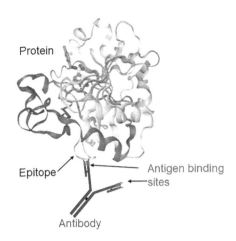

FIGURE 29.2 An epitope is only a portion of an antigenic protein. In this illustration, the antibody recognizes a particular sequence of amino acids that are part of a protein and are displayed when the protein has its normal conformation. (Not drawn to scale.)

Antibodies for immunoassays have high affinity for (attraction to) their antigens and can find and bind to their antigen even when the antigen is present in low concentrations amidst many other molecules. Antigen–antibody interaction involves multiple weak chemical interactions, including electrostatic interactions, hydrogen bonds, hydrophobic interactions, and Van der Waals forces. (Van der Waals' forces are short-range, weak, transient interactions between molecules.) As discussed previously in Chapter 25, such weak interactions between molecules can be influenced by the pH and salt concentration in solutions used to work with antibodies. Similarly, later in this chapter, we will discuss assays that involve immobilizing antigens and antibodies on various types of surfaces, for example, certain plastics. The interaction of antigens and antibodies with these surfaces is also influenced by pH and salt concentration. Therefore, solutions for immunoassays must be carefully prepared with optimized pH and salt concentrations.

29.2.2 ANTIBODIES: MANUFACTURING FOR IMMUNOASSAYS

As noted in Chapter 1, conventional manufacturing of antibodies begins with injecting the antigen of interest into a rabbit, mouse, or other animal. After the animal makes antibodies that recognize the antigen, it is possible to process the animal's serum into polyclonal or monoclonal antibodies (as explained in section 1.2.3).

A **polyclonal antibody preparation** *is a heterogeneous mixture of antibodies that a host animal produces in response to the antigen.* A polyclonal antibody preparation can contain antibodies that recognize different epitopes on a single antigen. In contrast, as discussed in Chapter 1, monoclonal antibodies are isolated from a single B cell clone and therefore all the antibodies in the preparation are identical to one another. In a monoclonal antibody preparation, all the antibodies recognize the same epitope on an antigen.

Techniques have been developed in recent years to manufacture **recombinant antibodies (rAbs)** *using recombinant DNA technologies.* **Recombinant antibodies** are often considered to be a type of monoclonal antibody, although they are not produced in the same way as Köhler and Milstein described (see Chapter 1). Recombinant methods allow manufacturers to make fully humanized antibody drugs that patients' immune systems do not recognize as being foreign. Recombinant antibodies are also valuable in the laboratory for immunoassays because they are reproducible. Monoclonal antibodies made using hybridoma cells are more consistent than polyclonal antibodies, but, over time, the cells can become genetically unstable, leading to the loss of their antibody product. In contrast, once the sequence of an antibody is known, genetic engineering tools can always be used to create rAbs consistently and in quantity. Recombinant antibodies are increasingly being used for immunoassays as they become available.

29.2.3 Antibodies: Proper Handling

Antibodies are critical, expensive reagents that must be handled properly. Antibodies inside a human, or other animal, exist in the blood in a very stable environment where temperature and pH are constant and stabilizing macromolecules are present. When antibodies are purified away from an animal, this stable environment is lost, and antibodies become subject to degradation.

The basic principle of protecting antibodies during storage is to keep them cold and avoid freezing–thawing–refreezing. To achieve this, antibodies are usually aliquoted upon arrival in the laboratory. **Aliquoting** *means to subdivide into small units.* When a vial of antibodies is received, the contents are aliquoted into multiple small tubes and are stored in a freezer and thawed only once, with any remainder kept at 4°C or discarded. When an analyst begins a new assay, only the aliquot(s) required is removed from storage. Aliquoting reagents avoids repeated freeze–thaw cycles and also avoids contamination that might be introduced by pipetting multiple times from a single vial.

Another important practice is to store antibodies at as high a concentration as possible, generally >1 mg/mL. Storage at high concentration helps stabilize antibody structure and reduces loss due to adsorption to the storage vial. Antibodies should not be diluted until an assay is about to be performed.

With proper storage and handling, most antibodies remain active for months or years. Box 29.7 in Appendix I of this chapter provides general guidance relating to aliquoting, storing, and diluting antibodies. Most manufacturers provide specific information about how to handle and store their antibody products, and the manufacturer's directions should be followed.

29.2.4 Antibodies: Conjugated to Reporters

29.2.4.1 Radioisotopes

The antibodies used in immunoassays are conjugated with reporter molecules that cause a detectable signal to appear. Figure 29.1 illustrates a situation where light is the signal that is detected after binding, but there are other types of signal as well. Rosalyn Sussman Yalow and Solomon Berson invented immunoassays in the 1950s, and Yalow received a Nobel Prize for this work in 1977. These innovative scientists used radioisotopes as the reporter. The signal emitted by radioisotopes, radioactivity, can be detected with specialized instruments and photographic film. However, radioisotopes require special handling and disposal due to associated health hazards. They are also expensive and have a limited shelf life. Other reporters are therefore usually preferred now.

29.2.4.2 Enzymes

Enzymes are frequently used as reporters (Figure 29.3). *Immunoassays that use enzymes as reporters are termed* **enzyme immunoassays (EIAs)**.

Enzymes catalyze reactions that turn substrates into products:

$$\text{Substrate} \xrightarrow{\text{Enzyme}} \text{Product(s)}$$

When enzymes are used as reporters, the enzyme is conjugated (attached) to the Fc portion of the antibody. When an assay is performed, the antibody–conjugate is allowed to bind to its target. Subsequently, a substrate(s) for the linked enzyme is introduced. The enzyme catalyzes a reaction converting the substrate(s) into a product that is detectable by eye, or with a microscope, or other instrument. The enzyme horseradish peroxidase (HRP) is an example of a

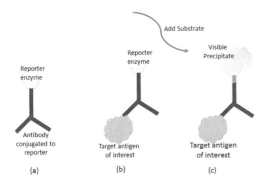

(a)   (b)   (c)

FIGURE 29.3 Enzymes as reporters. (a) An antibody that recognizes a target is conjugated to an enzyme reporter, shown in gray. (b) The antibody is allowed to find and bind its target. (c) A substrate solution is introduced, resulting in a product that forms a detectable precipitate wherever the antibody–target is present.

common EIA reporter. Its substrate, 3,3′-diaminobenzidine (DAB), forms a brown precipitate in the presence of the enzyme:

$$\text{DAB} \xrightarrow{\text{Enzyme (HRP)}} \text{Brown Precipitate} \bullet$$

An assay involving this type of reporter is termed a **chromogenic assay**, *because the signal is a colored precipitate.* ("Chromo-" means "color.")

29.2.4.3 Fluorophores (Light)

Fluorescence *occurs when certain molecules absorb light of one wavelength and then relax to a lower-energy state with the emission of light of a second wavelength* (Figure 29.4). *Molecules with this property are called* **fluorophores**, and they can be used as reporters. Fluorophore-labeled structures appear colored; hence, the signal from this type of reporter is colored light. There are different fluorophores that absorb and emit light of different colors.

When fluorophores are used, an instrument, commonly a fluorescent microscope, provides light of the wavelength that is absorbed by the specific fluorophore(s) in use. Fluorescent signal (light) can be recorded on photosensitive film, but is now more likely to be recorded using the sensor on a digital camera.

An advantage to fluorophore reporters is that it is possible to use more than one fluorophore, each of which fluoresces with a different color, in order to detect

(a)   (b)   (c)

FIGURE 29.4 Fluorescent reporters. (a) Fluorescence occurs when the electrons in the atoms of a molecule absorb light of one wavelength, in this illustration, 488 nm. (b) Electrons are excited to a higher-energy state by the absorption of light. (c) The electrons then return to a lower-energy level with the emission of light of a second wavelength, in this case, 507 nm. The emitted light provides a detectable signal.

more than one target at the same time. A disadvantage to the use of fluorophore reporters is the phenomenon of autofluorescence. **Autofluorescence** *occurs when materials naturally present in the samples emit fluorescence in the absence of any reporter fluorophore.* Autofluorescence causes light that appears to be from the reporter, but is not.

29.2.4.4 Chemiluminescent Compounds

Chemiluminescence, CL, *similar to fluorescence, occurs when a substance is excited to a higher-energy state and then relaxes to a lower-energy state with the release of light.* The difference between fluorescence and chemiluminescence is that, in the latter, the substance is excited to a higher-energy state with a chemical reaction, not with light absorption.

Horseradish peroxidase and alkaline phosphatase are enzymes that can catalyze a chemiluminescent reaction and are commonly used as chemiluminescent reporters. When a chemiluminescent reporter is used, the reporter-conjugated antibody is allowed to bind its target and then a substrate for chemiluminescence is added. The enzyme catalyzes a reaction that causes the substrate to be excited to a high-energy intermediate. The intermediate then relaxes to a low-energy state with the release of light (Figure 29.5).

Luminol is one of the substrates commonly used for chemiluminescence. Luminol, in the presence of HRP and hydrogen peroxide, forms an intermediate product named 3-aminophthalate. As the energy in 3-aminophthalate decays, it returns to a low-energy level with the release of light:

$$\text{Luminol} + \text{H}_2\text{O}_2 \xrightarrow{\text{Enzyme (HRP)}} \text{3-aminophthalate } \textit{excited} \longrightarrow \text{3-aminophthalate } \textit{normal energy state} + \textbf{LIGHT}$$

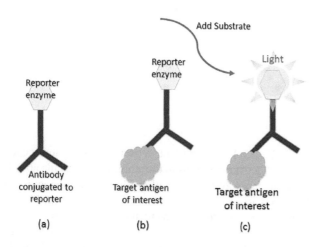

FIGURE 29.5 Chemiluminescence. (a) An antibody is conjugated to an enzyme reporter such as HRP. (b) The antibody–reporter recognizes and binds its target. (c) A substrate is introduced resulting in a product that emits light wherever the antibody–target is present.

Observe that HRP and alkaline phosphatase are enzymes that are commonly used as reporters in both chromogenic and chemiluminescent assays. For a chromogenic immunoassay, a substrate is added that forms a colored precipitate. For a chemiluminescent assay, a different substrate, such as luminol, is added, resulting in light emission.

Both chemiluminescent and fluorescent reporters provide light as the signal, but chemiluminescence has an advantage in that almost no molecules are inherently chemiluminescent. However, chemiluminescence has the disadvantage that the signal fades with time as substrate is depleted. Therefore, with chemiluminescence, assay results are greatly affected by the amount of substrate added and the timing of photographing a resulting image.

Case Study: Chemiluminescence and Forensics

If you have watched any crime scene investigation dramas, you are probably familiar with chemiluminescence. Forensic analysts use a chemiluminescent compound, luminol, to detect blood in a crime scene. Luminol and hydrogen peroxide react in the presence of iron in blood. The iron acts as a catalyst (as do enzymes) that causes a chemiluminescent reaction to occur, leading to the emission of light. This light is bright enough to be seen and photographed by the investigator (Figure 29.6).

$$\text{Luminol} + H_2O_2 \xrightarrow{\text{Iron from Blood}} \text{3-aminophthalate } \textit{excited} \longrightarrow \text{3-aminophthalate } \textit{normal energy state} + \textbf{LIGHT}$$

FIGURE 29.6 The use of luminol shows that an attempt was made to clean blood from the floor. (Image credit: Minnesota Bureau of Criminal Apprehension.)

29.2.5 DIRECT VERSUS INDIRECT LABELING

29.2.5.1 Indirect Assays Improve the Sensitivity of an Immunoassay

Regardless of the type of reporter used, the detection method for immunoassays can be either direct or indirect. In the **direct method**, *the antibody that is linked to a reporter binds its target, and this target–antibody–reporter complex is visualized.* This is the method that is illustrated in Figures 29.1, 29.3, and 29.5.

The direct method sometimes works well, but in many situations the target is present in low amounts and

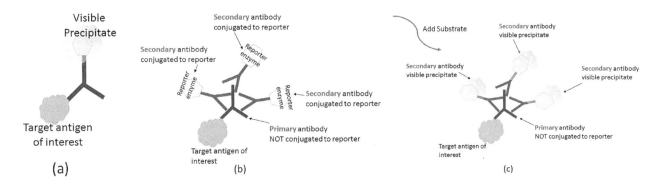

FIGURE 29.7 Direct versus indirect labeling. (a) In direct labeling, the primary antibody is conjugated to a reporter, finds its target, and is visualized. (b) Indirect labeling increases the sensitivity of an assay. The antibody that finds the target, the primary antibody, is not labeled with a reporter. A secondary antibody against the primary antibody is labeled. More than one secondary antibody can bind each primary, thus amplifying the signal from the target. (c) When substrate is added, more visible precipitate is formed with indirect labeling than with direct labeling.

the resulting signal is not strong enough for detection. It is possible to amplify the signal from the antibody–target complex by using a **secondary antibody** *that recognizes and binds the original antibody* (Figure 29.7). In this situation, the **primary antibody**, *the one that recognizes the target*, is *not* bound to a reporter. It is the *secondary antibody* that is linked to the reporter. Several molecules of secondary antibody can bind each primary antibody–target complex, thus amplifying the signal and increasing the sensitivity of the assay. Thus, in direct immunoassays the reporter is conjugated to the *primary antibody*, and in indirect assays the reporter is conjugated to the *secondary antibody*.

29.2.5.2 The Antibodies Used for Indirect Assays

Let's consider how secondary antibodies are manufactured. Secondary antibodies are generally polyclonal. They are produced by inoculating an animal with antibodies from the animal species used to make the primary antibody. For example, suppose the primary antibody in an assay is a monoclonal antibody made in mice. This primary antibody is the one that will recognize and bind the target analyte. To make secondary antibodies, an animal, such as a rabbit, is injected with mouse antibodies (Figure 29.8a). The immunized rabbit will make polyclonal antibodies that recognize the constant, Fc, region of all mouse antibody molecules (Figure 29.8b). These polyclonal antibodies can be isolated from the rabbit's serum and conjugated to a reporter (Figure 29.8c). These rabbit-derived secondary antibodies will recognize and bind to the constant region of any mouse antibody molecule (Figure 29.8d).

Observe in the example in Figure 29.8 that the rabbit-derived secondary antibodies will recognize the constant region of mouse antibody molecules regardless of the structure of the antigen-binding (Fab) regions. This means that analysts can use the same secondary antibody in a range of experiments. Suppose, for example, that a researcher is investigating three different receptor proteins on the surface of human cells. These experiments will require three different primary antibodies to recognize the three different receptors. But, if all three primary antibodies

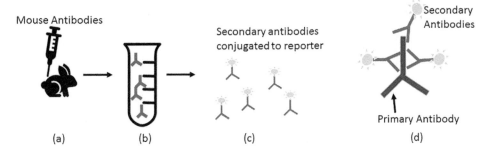

FIGURE 29.8 An example of manufacturing secondary antibodies for indirect assays. (a) The primary antibody in this example is a mouse monoclonal. Therefore, a rabbit is immunized with mouse antibodies. (b) The rabbit makes polyclonal antibodies (red) against mouse antibodies, which are isolated from the rabbit's blood. (c) The rabbit polyclonal antibodies are conjugated to a reporter to create secondary antibodies. (d) These secondary antibodies will recognize, and bind to, the Fc region of any mouse antibodies. More than one secondary antibody can bind to each mouse antibody molecule.

are produced in the same species of animal, let's say mice, it would be sufficient to obtain one secondary antibody that was created to recognize mouse antibody molecules.

We used an example involving mouse- and rabbit-derived antibodies in Figure 29.8, but other combinations of two animal species are also used to make primary and secondary antibodies. Whichever species are used, it is important to confirm that the primary and secondary antibodies perform properly together. There are many immunoassay kits commercially available that conveniently contain a primary antibody and a complementary secondary antibody. The antibodies in these kits have been optimized together. Diagnostic assays generally rely on FDA-approved test kits that have been carefully optimized and validated to work in diagnostic settings. In research situations, there may not be convenient test kits available. In such situations, analysts must obtain a primary antibody that is specific to their target, select an appropriate secondary antibody, and test the pair together.

There are advantages and disadvantages to both direct and indirect immunoassays. The direct method is faster because it requires fewer steps. However, indirect assays have the advantage of being sensitive to lower levels of target than direct assays.

29.2.6 Antibodies: Naming and Catalog Listings

Let's consider how antibodies are named and how you might look for antibodies for a particular purpose in a catalog. When naming primary antibodies, the name of the antigen must be included. For example, if the target of interest is the human epidermal growth factor receptor (EGFR) protein, and the primary antibody is a mouse monoclonal, then the antibody could be named: *mouse anti-human EGFR (monoclonal).*

In the example in Figure 29.8, where a secondary antibody was created by inoculating rabbits with mouse IgG, the resulting antibody would be named: "rabbit anti-mouse" or "rabbit α-mouse." As another example, suppose the primary antibody in an assay

is a recombinant human antibody. A secondary antibody made by inoculating goats with human IgG could be used; this secondary would be called "goat anti-human."

Figure 29.9 shows a catalog listing for a primary antibody that recognizes the protein estrogen receptor alpha. If you were to use this antibody, there is important information here that must be documented to ensure reproducibility. In this figure, the various columns are interpreted as follows:

- Column 1, **Product**, states the antigen, estrogen receptor alpha (ESR1), recognized by this antibody. The antibody is monoclonal, and OTI1B1 is its specific clone name. The clone name is critical identification information to document if you use this product because this manufacturer (and others) provides dozens of antibodies, all of which are intended to recognize estrogen receptor alpha. The manufacturer further states that this antibody was referenced in four articles.
- Column 2, **Figures**, shows that the manufacturer tested the antibody to verify its effectiveness.
- Column 3, **Target**, indicates that this particular monoclonal antibody recognizes estrogen receptor alpha in humans. The manufacturer sells other antibodies that recognize the same antigen in multiple species, or only in certain other species.
- Column 4, **Details**, states that this product was derived from a mouse B cell and is monoclonal.
- Column 5, **Application**, lists the types of immunoassay in which this antibody was tested and found to work effectively. (Two of these, western blot (WB) and immunohistochemistry (IHC), are applications that will be discussed later in this chapter.) When selecting an antibody, you will want to choose one that was tested in the type of immunoassay you plan to perform.

Product	Figures ⇕	Target	Details	Application	Price (USD)
OriGene ESR1 Monoclonal Antibody (OTI1B1), TrueMAB™		Human	Mouse Monoclonal	WB IHC (P)	426.00
🔖 4 References	10 figures				Cat # TA503759 100 µL

FIGURE 29.9 Excerpt for a primary antibody from an antibody catalog. See text for explanation. (Used with permission from Thermo Fisher Scientific, the copyright owner.)

| Invitrogen | | Goat | Mouse | Alexa Fluor® Plus | WB ICC/IF | 329.00 |
| Goat anti-Mouse IgG (H+L) Highly Cross-Adsorbed Secondary Antibody, Alexa Fluor Plus 555 | 13 figures | | | 555 | | Cat # A32727 1 mg |

FIGURE 29.10   Excerpt for a secondary antibody from an antibody catalog. See text for explanation. (Used with permission from Thermo Fisher Scientific, the copyright owner.)

- Column 6, **Price**, provides ordering details including the manufacturer's catalog number. Similar to the clone name, it is essential to document the name of the manufacturer and the manufacturer's catalog number.

Figure 29.10 shows a catalog listing for a secondary antibody that recognizes mouse primary antibodies. If you were to use this antibody, there is important information here that must be documented to ensure reproducibility. In this figure, the various columns are interpreted as follows:

- Column 1, **Product**, describes the product. It is a secondary antibody made in goats that recognizes the heavy and light chains of mouse IgG molecules. It is conjugated to a reporter, Alexa Fluor Plus 555, which is a fluorophore.
- Column 2, **Figures**, shows that the manufacturer tested the antibody to verify its effectiveness.
- Column 3, **Host**, reiterates that the antibody was made in goats.
- Column 4, **Target Species**, reiterates that the antibody recognizes mouse IgG.
- Column 5, **Conjugate**, reiterates that the antibody is conjugated to a reporter, Alexa Fluor Plus 555.
- Column 6, **Application**, lists the types of immunoassay in which this antibody was tested and found to work effectively. Western blotting (WB), immunocytochemistry (ICC), and immunofluorescence (IF) are applications that will be discussed later in this chapter.
- Column 7, **Price**, provides ordering details including the manufacturer's catalog number, which must always be documented.

29.2.7 THE CONCEPT OF SIGNAL-TO-NOISE RATIO IN IMMUNOASSAYS

The concept of signal-to-noise ratio, which was introduced in Chapter 15, applies to immunoassays and is of critical importance. In an immunoassay, signal emanates from a reporter conjugated to an antibody that has recognized and bound a target of interest. If a fluorescent or chemiluminescent reporter is used, then the signal consists of light. If a chromogenic reporter is used, then signal is a colored deposit that is present wherever the target is located. Noise, which is also termed "background" or "background noise," occurs whenever something is detected that is not related to the reporter–antibody–target complex. Background can obscure signal from the true target and make the results of an assay difficult to interpret (Figure 29.11). Background can even appear to be coming from the target, leading to a false-positive result. In all assays, it is essential to reduce noise as much as possible while increasing signal.

There are varied causes of background in immunoassays such as:

- *Autofluorescence*. This is when some material present in a sample is naturally fluorescent. Light from such a material is a potential cause of background when using fluorophore reporters.
- ***Endogenous enzymes. Endogenous enzymes** are those present in a sample, as contrasted with enzymes that are added as reporters.* To understand why endogenous enzymes can be problematic, recall that chromogenic and chemiluminescent detection methods involve an enzymatic reporter system. Substrate for the enzyme is added to visualize where antibody has bound its target. If, however, the reporter enzyme is also naturally present in the tissue or cells being studied (is endogenous), then a deposit will form, or light will be created, when substrate for the enzyme is added – whether or not the antibody finds its target. Some tissues are more likely than others to have certain endogenous enzymes. For example, when using horseradish peroxidase (HRP) as a reporter, nonspecific background may occur due to endogenous peroxidase activity in tissues such as kidney and liver, and samples containing red blood cells. When

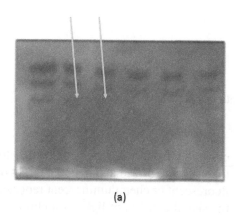
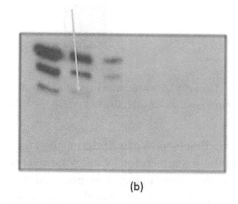

(a)   (b)

FIGURE 29.11   An example of noise on a western blot. Western blots are a type of immunochemical assay that will be discussed in more detail later in this chapter. Western blots are performed on a flat membrane (sheet) made of special plastics or a cellulose-based material called nitrocellulose. The membrane should be as light as possible after the immunoassay steps are performed, except where there are bands of dark color. These bands are the signal from the target proteins. (a) A blot with excessive noise. The deep gray hue covering the entire membrane is background noise. The arrows point to areas that might, or might not, be faint bands from target proteins. The overall gray background makes it impossible to determine whether or not the faint areas are signal from the target. This background might have been due to incomplete washing steps, excessive antibody incubations, nonspecific binding of antibodies to the membrane, and various sorts of contamination. (b) A blot with less background noise. In this case, the arrow points to a band that is faint, but can be distinguished from the background. (Used with permission from Thermo Fisher Scientific, the copyright owner.)

assaying those types of tissue samples, HRP reporter systems should be avoided.

- *Cross-reactivity.* **Cross-reactivity** *is where an antibody recognizes an epitope that is not its desired target.* For example, an antibody manufactured to recognize a receptor protein of interest might cross-react with a similar receptor protein that is not the intended target. In many situations, an antibody that exhibits cross-reactivity is unsuitable for use. Cross-reactivity will be discussed further in the section of this chapter on assay validation.
- *Contamination.* Incomplete washing, degraded reagents, dirty incubation trays, and other such problems can lead to high background noise.
- *Excessive incubation times.*
- *Nonspecific binding.* **Nonspecific binding** *occurs whenever an antibody binds to something other than its target.* For example, certain immunoassays are performed on a surface or membrane that is designed to readily bind proteins. However, the primary and secondary antibodies (which are themselves proteins) may stick nonspecifically to these surfaces, resulting in undesired background.

Nonspecific binding is often a problem that must be considered in immunoassays. To help prevent nonspecific binding, **blocking buffers** *are used that contain* **blocking agents** *to occupy sites where the target is not present* (Figure 29.12). **Blocking agents** *are usually a protein, the choice of which depends on the assay.* A good blocking agent prevents nonspecific binding to the substrate and does not react with the primary antibody, secondary antibody, or target protein. The use of a blocking buffer should reduce noise and therefore increase the sensitivity and reproducibility of the assay. There are certain blocking agents that are inexpensive and popular. However, selecting a blocking agent should be done thoughtfully, as is discussed in Table 29.3 in Appendix I of this chapter.

Regardless of the cause, it is vital to recognize background noise and not interpret it as a positive result. Controls are essential for this purpose and will be discussed later in this chapter. Method optimization is another defense against background. It is often

a.   b.

FIGURE 29.12   Blocking agent used to block unoccupied sites on a membrane. (a) The antigen of interest is shown in green. Without blocking, there are unoccupied sites on the membrane surface where antibodies might stick nonspecifically. (b) Blocking agent, purple, is added to occupy sites on the membrane, so the antibodies cannot stick to these sites.

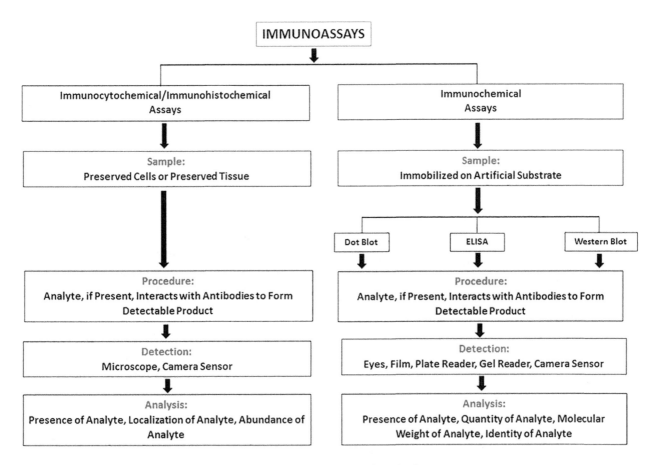

FIGURE 29.13 An overview of the immunoassays to be discussed in this chapter.

possible to optimize the procedure of an immunoassay to reduce background and raise signal. Optimization will also be discussed in a later section of this chapter.

29.2.8 OVERVIEW OF THE STEPS IN AN IMMUNOASSAY PROCEDURE

All assays begin with a sample. Many types of biological samples are analyzed using immunoassays. In the medical clinic, assays might involve serum, urine, or mucus samples from a patient. In a research setting, tissue from animals, or cultured cells might be analyzed. The methods used to prepare these samples will vary depending on the sample and the type of immunoassay involved. Regardless of their origin or the type of assay, all samples must be prepared properly because the methods used to prepare samples can greatly impact the assay results.

After the sample is prepared, it is subjected to a series of incubations and washes that allow the immunoassay reagents to interact with the sample. The details of each immunoassay method vary, but the

general steps common to various types of immunoassays are shown in the flowchart in Box 29.1.

29.2.9 CATEGORIZING IMMUNOASSAYS

There are dozens of varieties of immunoassays involving various combinations of samples and procedures, although the flow chart in Box 29.1 is generally applicable to all these varieties. These varieties can be organized into two broad categories. The first category includes immunoassays that we term "immunochemical." In **immunochemical assays,** *the targets are removed from their normal cellular site and are analyzed on an artificial surface, also called a "substrate."* In order to look for a protein target using an immunochemical assay, it is necessary to extract the proteins from the cells and solubilize them into a buffer. This is so that the protein constituents can be immobilized on a substrate. Protein extraction requires **lysing**, *breaking open*, the cells' membranes. This is usually accomplished with lysis buffers. Lysis will be discussed in more detail later in this chapter.

Immunochemical assays are a broad category with many variations on the basic idea of immobilizing the components of the sample on an artificial substrate. We will introduce three common variants:

- **Dot blots,** which are performed using a membrane to immobilize the samples.
- **ELISAs,** where the samples are immobilized in the bottoms of the wells of multiwell plates.
- **Western blots,** which involve an electrophoresis step before immobilizing the samples on a membrane.

The other broad category of immunoassays includes immunocytochemical (ICC) and immunohistochemical (IHC) assays. **ICC** and **IHC** *assays are performed on cells or tissues, respectively, and are used to detect a target within its cellular location.* When immunochemical assays are performed, cellular structure is intentionally destroyed in order to release protein targets into a solution so that they can be bound to the artificial substrate. In contrast, with ICC and IHC,

the goal is to preserve the cellular structure as much as possible in order to visualize a target in its cellular location. We will introduce the overall principles of ICC and IHC after discussing immunochemical assays.

Figure 29.13 summarizes the types of assays we will introduce.

29.3 DOT BLOTS

The **dot blot method** *is a relatively simple immunoassay used to detect if a target protein is present in a sample (i.e., a cell lysate) and possibly to roughly estimate the concentration of the target, if it is present.* Dot blots have various applications, for example, to optimize antibody concentrations when an assay is being developed.

Dot blots immobilize the assay components on thin membranes (sheets) typically made of nitrocellulose or polyvinylidene difluoride (PVDF). The steps of a dot blot procedure are shown in Box 29.2; these parallel the steps shown in Box 29.1.

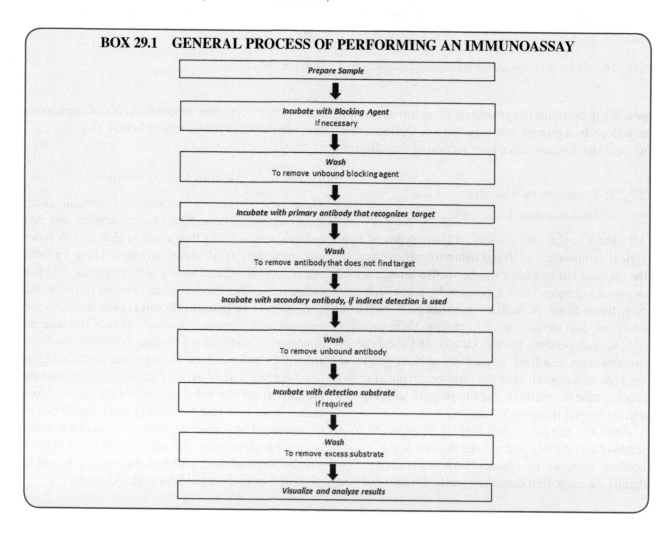

BOX 29.1   GENERAL PROCESS OF PERFORMING AN IMMUNOASSAY

Prepare Sample

↓

Incubate with Blocking Agent
If necessary

↓

Wash
To remove unbound blocking agent

↓

Incubate with primary antibody that recognizes target

↓

Wash
To remove antibody that does not find target

↓

Incubate with secondary antibody, if indirect detection is used

↓

Wash
To remove unbound antibody

↓

Incubate with detection substrate
If required

↓

Wash
To remove excess substrate

↓

Visualize and analyze results

BOX 29.2   A CONVENTIONAL DOT BLOT PROCEDURE

Prepare test samples, standards, and controls.

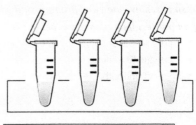

Use a pencil to lightly draw a grid on the dot blot membrane to indicate where samples will be applied.

Using a narrow-mouth pipette tip, spot each sample onto the membrane at the center of each gridded area.
- A typical volume for a spot is 2–10 µL.
- Minimize the area that the solution penetrates by applying it slowly.
- Choose a sample volume that will not spread between spots on the grid and will not exceed the binding capacity of the membrane.
- Sample should wick into the membrane and not spread across the membrane.

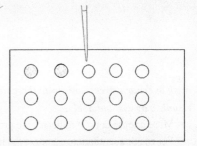

Let the membrane dry.
- Block the membrane by immersing in enough blocking solution to cover the membrane as it lays in a tray.
- Incubate 30 minutes to 1 hour.
- Pour the blocking solution into a waste container.
- Do not allow the membrane to dry for the rest of the procedure.

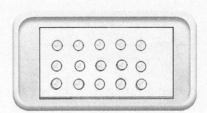

Wash off unbound blocking buffer.
- Cover the membrane with wash buffer and incubate 5 minutes.
- Pour wash buffer into waste container.
- Repeat three times with wash buffer.

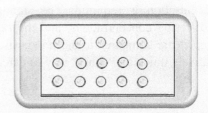

Dilute primary antibody.
- Apply primary antibody evenly over the membrane.
- Incubate with primary antibody 1 hour at room temperature.
- Pour off antibody.

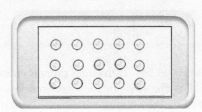

(Continued)

BOX 29.2 (*Continued*) A CONVENTIONAL DOT BLOT PROCEDURE

Wash to remove unbound primary antibody three times with washing buffer, as above.

If an indirect immunoassay is being used, dilute secondary antibody.
- Apply secondary antibody–reporter evenly over membrane.
- Incubate 30 minutes at room temperature.
- Pour off antibody.

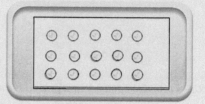

Wash to remove unbound secondary antibody three times with washing buffer, as above.
Incubate with substrate for the reporter.

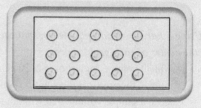

Wash to remove excess substrate three times with washing buffer, as above.
Visualize by eye and analyze.
- More intense deposit indicates the presence of more target protein.
- For rough quantitation, compare the intensity of colored deposit in samples to the intensity of deposit in positive standards of known concentration.

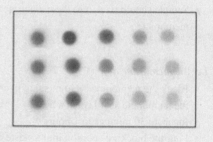

An example of using a dot blot to estimate protein concentration in samples is shown in Figure 29.14. A standard containing the protein of interest at a known concentration is required. The standard is diluted to differing concentrations, and the results from the samples are compared to the standards. This can be a relatively quick and simple way to determine whether a protein is present in samples, and its approximate concentration.

Dot blots can be used to *detect* an analyte and to *estimate the quantity* of an analyte. Dot blots provide less information than ELISAs and western blots, which will be discussed next. Dot blots are therefore useful when a relatively fast and simple assay is required, but other immunoassay methods are more informative.

Diluted standards   **Samples**

FIGURE 29.14 Estimating protein concentration with a dot blot. The first six spots are from a standard that was diluted to known concentrations. The last three spots are from the sample, and these can be compared to the standards to roughly estimate protein concentration in the sample.

29.4 ELISAs

29.4.1 ELISA PROCEDURES

ELISAs (enzyme-linked immunosorbent assays) are a type of immunochemical assay with many applications in testing and research laboratories. ELISAs can be performed both qualitatively and quantitatively. They are very sensitive; the typical detection range for an ELISA is reported to be 0.1–1 fmole or 0.01–0.1 ng. ELISAs also have the advantage of being relatively easy to perform, although, as will be discussed later, it is important to be aware of potential sources of erroneous results.

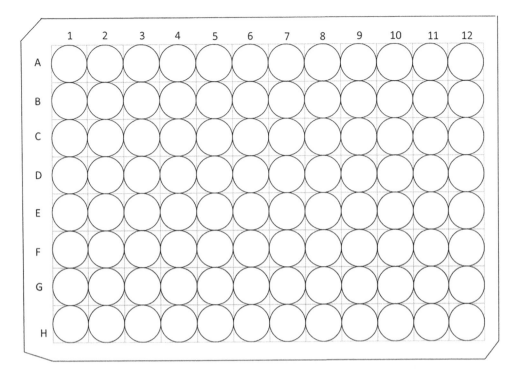

FIGURE 29.15 A plastic 96-well plate. Using a plate like this, 96 different samples can be tested at once. Each well has a label. For example, the first row is labeled A1–A12. The second row has wells B1–B12, and so on.

The term "ELISA" implies an immunoassay that is performed in multiwell plastic plates, allowing multiple samples to be assayed simultaneously. Each well on a plate is similar to a tiny test tube for a different sample. A common type of multiwell plate has 96 wells (Figure 29.15). The plastic of the plate provides a surface, or substrate, for the interaction between antibody and target. The plastic used to manufacture the plate (usually polystyrene or polyvinyl chloride) is designed so that proteins adsorb (stick) to it tightly. The "stickiness" of the plastic is important because it makes it possible to immobilize antigens and antibodies of interest.

The bottoms of the wells of the multiwell plates used for ELISAs are transparent. The use of these plates gives ELISAs an important advantage over dot blots; the results of an ELISA can be read in a plate reader. A **plate reader** *is a type of spectrophotometer that is configured so that it can read the absorbance of light in each well of a multiwell plate.* This means that quantitative colorimetric methods, such as those described in Chapter 28, can be applied to ELISAs. This makes ELISAs particularly valuable for quantitation.

There are various formats for an ELISA. We will just consider one format, the sandwich method of ELISA, in order to illustrate the general principles of this type of assay. In the **sandwich method**, *there are two antibodies that recognize the target.* The first is a **capture antibody** that is initially adsorbed to the bottoms of the wells. The samples are then added to the wells, and the analyte of interest (if present) *binds to the capture antibody.* Other proteins in the sample do not bind to the capture antibody, and so they are subsequently washed away in wash buffer. *Another antibody that recognizes the target is then applied,* **the primary antibody (also called the detection antibody).** The analyte (if present) is now *sandwiched* between the capture antibody and the primary antibody. If direct detection is used, the primary antibody is conjugated to the reporter. If indirect detection is used, a secondary antibody–reporter is then added.

Box 29.3 illustrates the sandwich format with direct detection. In this box, the bottoms of two wells are illustrated. A negative control sample that is known not to contain the analyte of interest is applied to well A1. A positive control sample known to contain a high concentration of the analyte of interest is applied to well A2. The result is that there is no color in well A1 at the end of the assay and there is intense color in well A2.

The overall procedure for an ELISA shown in Box 29.3 is much like that described in Boxes 29.1 and 29.2, but the use of a capture antibody is a special feature of this ELISA format.

BOX 29.3   ELISA SANDWICH ASSAY WITH DIRECT DETECTION

Prepare samples, standards, and controls.

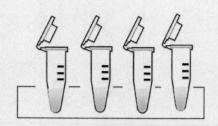

Pipette capture antibody solution into each well of a 96-well plate.

- Capture antibody sticks to bottom of wells.
- Two wells are illustrated here, A1 and A2.

Well A1   Well A2

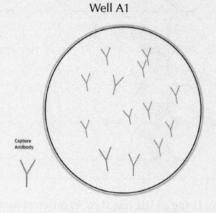

Capture Antibody

Wash with washing buffer to remove unbound capture antibody.
Pipette blocking buffer into wells.

- Blocking buffer contains blocking agent that covers unoccupied sites on well surfaces.
- Incubate.
- Blocking agent is shown in purple.

Well A1   Well A2

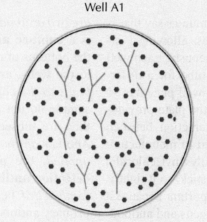

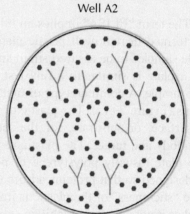

Wash with washing buffer to remove unbound blocking agent.
Pipette samples, standards, and controls into individual wells.

- Incubate.
- If the target is present, it will be bound by capture antibody.
- Analyte is shown in green.

Sample added to well A1: control with no analyte.

Sample added to well A2: has analyte.

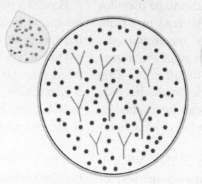

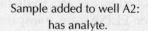

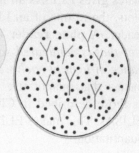

(Continued)

Analyte binds to capture antibody in the well.

Well A1 has no analyte, so no binding occurred.

Well A2 has analyte, so binding occurred.

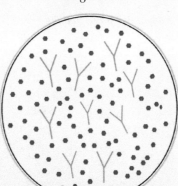
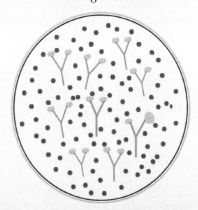

Wash with washing buffer to remove unbound sample.
Dilute primary antibody.

- The primary antibody is conjugated to an enzymatic reporter that will generate a colored deposit after substrate is added.

Pipette primary antibody into wells.

- Antibody recognizes and binds analyte, if present.
- Analyte molecules (if present) are now sandwiched between capture and primary antibodies.

Well A1

Well A2

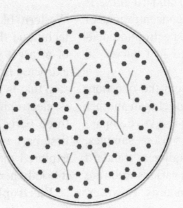
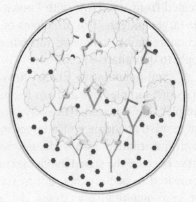

Wash to remove unbound primary antibody.
Incubate with substrate for the reporter enzyme.

Well A1

Well A2 – Color appears in wells with analyte.

(*Continued*)

BOX 29.3 (*Continued*)   ELISA SANDWICH ASSAY WITH DIRECT DETECTION

Wash to remove excess substrate.

The more the target in a well, the more the product that forms and the more intense the color that appears in the well.

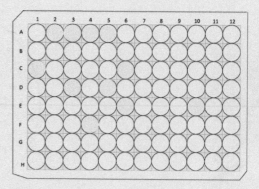

Stop the reaction while it is in the linear range where the amount of color is proportional to the amount of target. Dilute sulfuric acid is an example of a substance used for this purpose.

Put plate into a plate reader that detects and quantifies how much color is present in each well.

- The more the analyte that is present in a well, the more the color (signal) that will be present. As the signal increases, the absorbance measured by the plate reader increases.

(Photo is provided courtesy of Molecular Devices, LLC.)

29.4.2 QUANTITATION WITH ELISAS

ELISAs can be qualitative, quantitative, or semi-quantitative. A qualitative ELISA answers the question: Is a target present in a sample? A quantitative ELISA answers the question: What is the quantity of a target in a sample? A standard curve is used for quantitative ELISAs (Figure 29.16). The standard curve is created from standards with known concentrations of the target antigen. The principles of spectrophotometric quantitation that were introduced in Chapter 28 apply to quantitative ELISAs.

A semi-quantitative ELISA answers the question: Does one sample have more or less target than another sample? Semi-quantitative assays are used to compare the relative abundance of a target in various samples. Semi-quantitative assays do not require a standard curve because the signal intensities in various wells are compared to one another, not to standards. However, for semi-quantitative analyses, the linear range of the assay must be known, and the signal from all samples must be in the linear range. This concept was discussed in Section 28.3.5.

Table 29.5 in Appendix I of this chapter provides additional procedural considerations relating to assuring trustworthy ELISA results.

29.5 WESTERN BLOTS

29.5.1 BASIC WESTERN BLOT PROCEDURE

Western blotting is an important immunochemical technique that is widely used to detect the presence of a target protein in a complex mixture of proteins extracted from cells. Western blotting involves a major step that is not present in ELISAs or dot blotting, that is, electrophoresis. The western blotting procedure has three parts, as illustrated in Figure 29.17, where the first part is electrophoresis. Electrophoresis will be explored more fully in Chapter 34, but we introduce it briefly here.

Electrophoresis *is a method used to separate different molecules in a mixture from one another, using electrical current.* Electrophoresis is used in western blotting to separate proteins from one another; in other situations, electrophoresis is used to separate differently

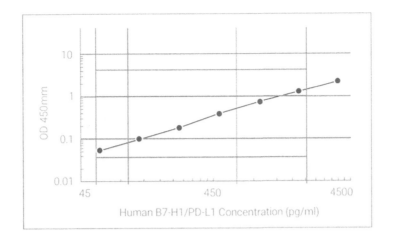

Standard #	8	7	6	5	4	3	2	1
Standard concentration (pg/ml)	0	62.5	125	250	500	1000	2000	4000
OD$_{450}$	0.003	0.055	0.101	0.183	0.388	0.744	1.272	2.194

FIGURE 29.16 A standard curve for an ELISA. Observe that this standard curve is analogous to those in Chapter 28. It is somewhat different, however, in that the *log* of the absorbance is plotted on the Y-axis and the *log* of concentration is on the X-axis. Taking the log of both the X- and Y-axes still results in a linear plot of concentration *versus* absorbance, but has the advantage of "spreading out" the graph in the lower concentrations, thus allowing the analyst to detect nonlinearity at very low concentrations. (Image created by PeproTech, Inc. ELISA Products Booklet, p. 4.)

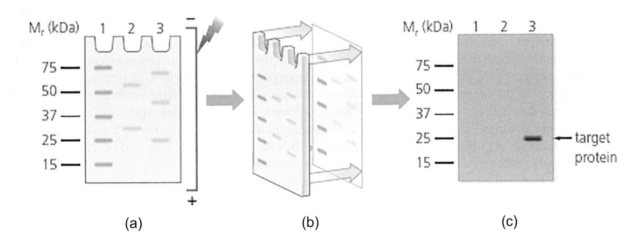

(a) (b) (c)

FIGURE 29.17 The three parts of a western blotting procedure. (a) Electrophoresis. The components of the samples are separated using electrophoresis. The first lane contains a standard consisting of a mixture of proteins of known molecular weights. The molecular weights of the proteins in the bands in the samples, Lanes 2 and 3, can be estimated by comparison with the standards in Lane 1. There are two protein bands in Lane 2 with approximate molecular weights of 60,000 and 30,000 D. There are three bands in Lane 3 with approximate molecular weights of 70,000, 45,000, and 25,000 D. (b) Blotting. The proteins in the gel are transferred, "blotted," onto a membrane. (c) Immunoassay. *The result of the immunoassay is the* **immunoblot**. In this example, one protein is detected using immunoassay. By comparison with molecular weight markers in Lane 1, the target protein in Lane 3 has a molecular weight of roughly 25,000 D. (Courtesy of Bio-Rad Laboratories, Inc., © 2021.)

sized DNA fragments. To begin electrophoresis, samples, standards, and controls, each with the same amount of protein, are carefully pipetted into the wells of a gel. If you have never seen an electrophoresis gel, imagine a slab of translucent Jell-O with shallow depressions, the "wells," at one end; (Figure 29.17). After the samples are loaded into the wells, the gel is placed inside an apparatus that applies current to it. Molecules in the sample migrate through the gel under the influence of the electrical current. In the most common form of protein electrophoresis, SDS–PAGE, proteins in the sample separate from one another based solely on differences in their sizes. After separation, the current is turned off and the gel is removed and placed in a staining solution that reveals the bands of separated proteins (Figure 29.17a). (Alternatively, there is a stain-free method that allows visualization of protein-containing bands by exposing the gel to a UV light.) The first lane (column) in this illustrated gel contains a standard consisting of a mixture of proteins with known molecular weights. The standard is used to estimate the molecular weights of the proteins in the samples. Estimation is performed by comparing the distance migrated by each band in the samples, to the distances migrated by the proteins in the standard mixture, as shown in the illustration.

SDS–PAGE effectively *separates* the proteins in samples from one another and allows estimation of their molecular weights. However, electrophoresis alone does not *identify* any of the separated proteins. Many proteins have roughly the same molecular weight, and it is not possible to be certain that a particular band seen on a gel contains a target of interest. If an analyst wants to confirm that a particular band on a gel does indeed contain a particular analyte, something else is required – for example, an immunoassay. The most obvious approach to performing an immunoassay would be to add to the gel a reporter-conjugated antibody specific to the target protein to see which bands, if any, are detected. However, this strategy will not work because the three-dimensional, semi-liquid gel is not suitable for an immunoassay. Instead, the proteins must be transferred to, and immobilized on,

the surface of a two-dimensional membrane, similar to that used for a dot blot (Figure 29.17b). *The process of transferring the proteins from the gel onto the membrane is called* **blotting** *or* **immunoblotting**; blotting is the second part of a western blot procedure. Once the proteins are immobilized on the membrane, an immunoassay can be performed (Figure 29.17c). The immunoassay is the third part of western blotting. This third stage is also called "immunodetection."

The word "blotting" in the term western blotting refers to the process of transferring proteins from the electrophoresis gel onto the surface of the membrane. The word "western" in western blotting is a play on words: "Southern" blotting is a method used to detect DNA sequences and was named after its inventor, Ed Southern. There is also "northern" blotting, which is used to detect RNA fragments.

Electrophoresis is the key feature that distinguishes western blotting from other immunoassay methods. Since the proteins in western blotting are commonly separated from one another based on differences in their molecular weights, western blots can be used to estimate or confirm the molecular weight of a target protein, as illustrated in Figure 29.17c. Western blots thus provide a piece of valuable information that dot blots and ELISAs cannot provide, but western blots take longer to prepare and are limited in the number of samples that can be run simultaneously on a single electrophoresis gel. Western blots are often used to provide semi-quantitative information, but, as we will see later, semi-quantitative western blotting is not as straightforward as quantitation with an ELISA.

Box 29.4 provides a generalized procedure for western blotting.

BOX 29.4 GENERALIZED WESTERN BLOT PROCEDURE

Perform a protein assay on all samples, standards, and controls to determine their total protein concentration. This step is necessary because it is important to load the same amount of protein into each lane of the electrophoresis gel.

- For SDS–PAGE, use detergent to denature proteins in samples, standards, and controls so that their varying shapes and charges do not affect how they migrate through the gel.
- Detergent confers uniform negative charge on all proteins, so migration through the gel is not affected by charge differences.

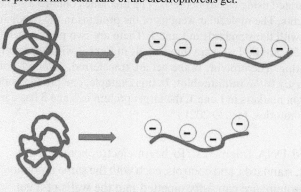

(Continued)

Prepare gel.
- Pipette molecular weight standards into Lane 1.
- Pipette the samples and controls into other lanes.
- Load the same amount of protein into each well.

Perform electrophoresis to separate components of samples, standards, and controls.
- Denatured proteins separate based on size, with larger proteins migrating less distance and smaller proteins migrating farther.

Transfer proteins to a membrane.
- A specialized blotting apparatus is usually used for this purpose.

Determine locations of proteins and their molecular weights using a total protein staining method (unless using stain-free technology).
- Photograph result.
- Remove stain.

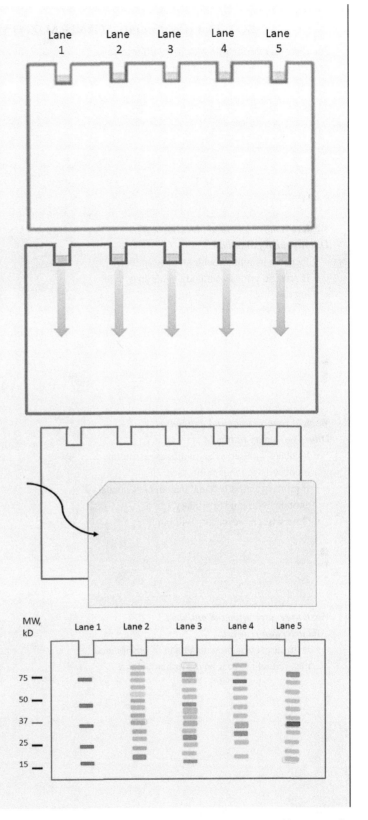

(*Continued*)

BOX 29.4 (*Continued*)   GENERALIZED WESTERN BLOT PROCEDURE

Incubate membrane with blocking buffer to cover nonspecific sites on the membrane.

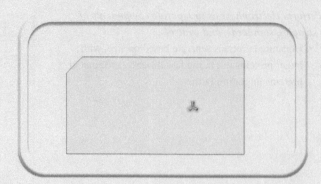

Wash to remove unbound blocking buffer.
Dilute primary antibody.
- Incubate membrane with primary antibody.
- If present, primary antibody binds analyte of interest.

Wash to remove unbound antibody.
Dilute secondary antibody.
- Incubate with secondary antibody, assuming indirect detection is being used.
- If primary antibody found the target, secondary antibody will bind to primary.
- Otherwise, no secondary will bind.

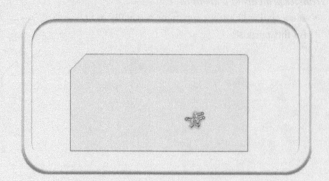

Wash to remove unbound antibody.
Add detection reagents.
- If the target was present, a band of precipitated material will appear on the immunoblot.

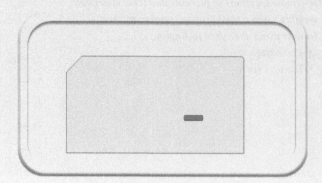

Wash to remove detection reagents.
Analyze the result.
- In this example, one band, the target analyte protein, is present in the sample that was loaded into Lane 4.
- The molecular weight of this band can be determined by comparison with the molecular weight standards in Lane 1, based on the photo taken earlier.

29.5.2 Detection and Analysis Methods for Western Blotting

After a western blot assay is performed, the analyst will know which sample(s) (if any) contained the analyte of interest, and the approximate molecular weight of the analyte. This analysis is greatly enhanced by the use of image analysis instruments and software (Figure 29.18). Image analysis systems perform these tasks:

- They *acquire* and *import* images of the results into a computer using image acquisition tools.
- They use software to *analyze* the images.
- They provide a *report*.

Software for western blotting can use the images of an electrophoresis gel and immunoblot to accurately:

- Determine and *subtract background noise*.
- *Determine the molecular weights* of proteins by comparison with standards.
- *Compare the amounts of protein* in different bands.

29.5.3 Quantitation with Western Blots

Quantitation of Western blots ... can be fraught with traps for the unwary investigator...

29.5.3.1 Overview

We have so far discussed only qualitative western blotting, which is used to determine the presence or absence of a protein of interest, and to provide information about a target's approximate molecular weight. **Semi-quantitative western blots** *are used to make quantitative comparisons between treatments.* Semi-quantitative analysis is based on the assumption that a more darkly stained band on a western blot contains more protein than one that is lighter. Semi-quantitative western blots have been used for thousands of critical research studies in biology. However, there are difficulties in achieving trustworthy, reproducible data from such studies. Difficulties with semi-quantitative western blots have been identified as a significant contributor to the "reproducibility crisis." Let's consider a hypothetical example. Suppose researchers are interested in determining if treating cells with experimental drugs causes an increase in the expression of a target protein. In such a study, the band intensities from untreated cells are compared to the band intensities from cells treated with each experimental drug (Figure 29.19). This is semi-quantitative analysis because the researchers are comparing the relative amounts of proteins in different samples. In this hypothetical example, immunoassay results show that the protein of interest was clearly detected in Lanes 3–5. A faint band is visible in the untreated cells in Lane 2. The researchers would like to be able to say that treatments with drugs induce cells to make more of the protein of interest and Drug C is the most effective. However, this type of semi-quantitative analysis has pitfalls. If proper procedures are not used, it is possible that the drugs had little or no effect, or that Drug A or B might have had more of an effect than Drug C. The following sections will consider why this is so.

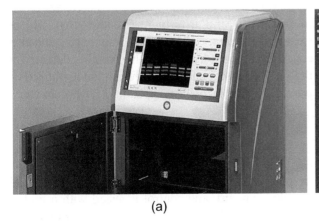

(a)

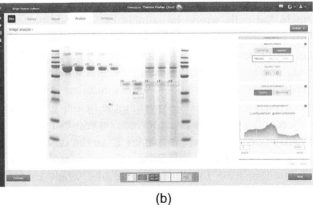

(b)

FIGURE 29.18 Image analysis systems. (a) An image analysis device used to capture an image of a western blot or an electrophoresis gel. Devices similar to this can detect the signal from chromogenic, chemiluminescent, and fluorescent reporters. (b) Once an image is captured, software is used to analyze the results, in this case, the bands on a gel. (Used with permission from Thermo Fisher Scientific, the copyright owner.)

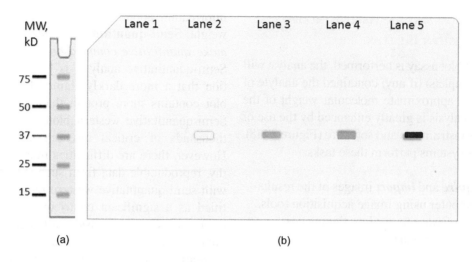

FIGURE 29.19   A hypothetical experiment. An experiment is performed in which cells are tested with three experimental drugs to see if any of them cause the cells to make more of a particular protein of interest. The protein of interest is known to have a molecular weight of 38,000 kDa. (a) The stained molecular weight markers. (b) The western blot. Lane 1: The molecular weight marker lane, shown with stain on the left. Lane 2: Extracted proteins from cells that were not treated with any drug. Lane 3: Extracted proteins from cells that were treated with experimental Drug A. Lane 4: Extracted proteins from cells that were treated with experimental Drug B. Lane 5: Extracted proteins from cells that were treated with experimental Drug C.

29.5.3.2 Sample Preparation

Sample preparation is the first step in western blotting, and it is the first step where issues can arise that may impact the results. Sample preparation begins with lysing (breaking open) the cells being studied. **Lysis buffer** *is the essential reagent that is used to extract and solubilize the proteins that are in the cells.* Lysis buffers have various components that help to break apart cell membranes while protecting proteins from degradation. Once the proteins have been solubilized into the lysis buffer, the extract is centrifuged to get rid of undissolved, unwanted cellular material, which moves to the bottom of the centrifuge tube and is discarded.

Ideally, all the proteins of interest that are present in the cells would be soluble in the lysis buffer. Unfortunately, the situation is not always ideal and not all proteins are readily solubilized in a particular lysis buffer. Lysis buffers with different ingredients at different concentrations will solubilize different assortments of proteins. An example of the effect of different lysis buffers on experimental results has previously been explored in Chapter 25 in the case study "The Importance of Lysis Buffer Components." Other researchers have found similar results when studying western blotting. For example, a study showed that the commonly used western blotting lysis buffer, RIPA, efficiently solubilized many cytoplasmic proteins; however, tubulin and intermediate filament proteins were substantially lost due to insolubility in RIPA. After studying not only RIPA but also several

other lysis buffers, the author concluded: "These results collectively showed that lysis buffer composition substantially affects the results of quantitative immunoblotting." (Janes, Kevin A., "An Analysis of Critical Factors for Quantitative Immunoblotting." *Science Signaling*, vol. 8, no. 371, p. rs2. doi: 10.1126/scisignal.2005966.)

Consider the hypothetical research study shown in Figure 29.19. Such a study may or may not have been influenced by the choice of lysis buffer. For example, if the lysis buffer only partially extracted the protein of interest, there may have been more of the protein of interest present in the untreated cells and in those treated with Drugs A and B than was detected in this immunoblot. Similarly, in real studies that rely on western blot data, the choice of lysis buffer constituents could profoundly affect the proteins detected in the final immunoblot. More information about lysis is provided in Table 29.4 in Appendix I of this chapter.

29.5.3.3 Controlling for the Amount of Protein in Each Lane

After the cell lysis and centrifugation steps, and prior to performing a western blot immunoassay, each sample, standard, and control is tested with a protein assay to determine its total protein concentration. The same amount of protein is then loaded into every lane. Hence, in theory, if one were to quantify the amount of total protein in each lane of a western blot, the result should be as shown in Figure 29.20a where the

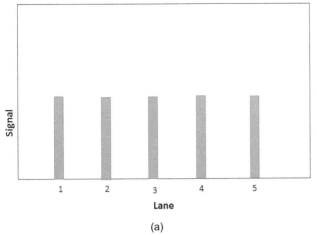
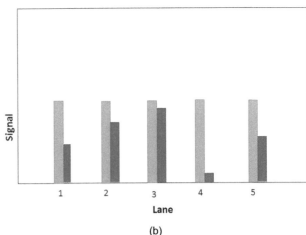

FIGURE 29.20 Ideal semi-quantitative western blotting. (a) Each lane contains a different sample. The total protein, blue bars, in each lane is the same. (b) The relative amount of the protein of interest, red bars, varies between samples.

blue bars represent the amount of total protein in each lane. In this ideal situation, if the signal intensity from the target protein varies, as shown with the red bars in Figure 29.20b, it provides evidence that the relative amount of target protein is varying between the samples. For example, the relative amount of the target protein, relative to total protein, is higher in the sample in Lane 3 compared to that in Lane 4. If different conditions are being tested, for example, protein expression in cells treated with different drugs, these results suggest that the drugs do, indeed, cause differences in response.

A difficulty with semi-quantitative western blotting analysis is that it is difficult to ensure that the total protein in every lane is, indeed, the same. Even if the analyst properly performs a protein assay, has perfect pipetting technique, and loads exactly the same amount of protein into every well, the transfer efficiency of protein from the gel to the blot will likely vary among different proteins. Small proteins (<10 kDa) may not be retained by the membrane, large proteins (>140 kDa) may not be transferred to the membrane, and varying gel concentrations may affect transfer efficiency. (Ghosh, Rajeshwary, et al. "The Necessity of and Strategies for Improving Confidence in the Accuracy of Western Blots." *Expert Review of Proteomics*, vol. 11, no. 5, 2014, pp. 549–60. doi: 10.1586/14789450.2014.939635.)

An additional problem is that it is possible for the membrane to become oversaturated if there is too much protein in a band (Figure 29.21). If any band is oversaturated, then some of the protein in that band will be lost.

When the amount of protein transferred from a gel to a membrane varies from lane to lane, the result is similar to that shown in Figure 29.22. In this situation,

FIGURE 29.21 Membrane saturation. In this example, more protein is present in the electrophoresis gel than can be transferred to a membrane. The excess protein will wash away, and the amount of protein in the band will be underestimated. To avoid this problem, the sample should have been diluted so that less protein was loaded into that lane. (Based on information from Li-Cor, Inc. "Quantitative Western Blots. Is Your Question Quantitative or Qualitative?" 2019. https://www.licor.com/bio/applications/quantitative-western-blots/.)

the signal from the target protein in Lane 4 is lower than in Lane 3, but the amount of total protein in Lane 4 is also lower, making interpretation of the result difficult. Consider again the hypothetical study illustrated in Figure 29.19. If the same amount of protein was not successfully transferred from each lane onto the membrane, then the intensity of band staining does not tell us whether one or another drug is more effective. It is even possible that the drugs had little or no effect.

There is a solution to the problem shown in Figure 29.22; it is called "normalization." The goal of **normalization** *is to correct for differences in the amounts of total protein in different lanes.* Normalization allows analysts to detect the variation among the blue bars in Figure 29.22 and mathematically adjust for these differences.

There are two approaches to normalization: using a "housekeeping protein" (HKP) and "total protein" (TPN) methods. The use of housekeeping proteins has traditionally been the most common method of

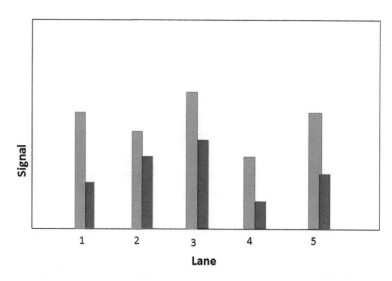

FIGURE 29.22 Non-ideal western blot result. The amount of total protein in each lane (blue bars) varies, making comparisons of the target protein levels (red bars) difficult. There may be less target protein in Lane 4 than in Lane 3 simply because less protein was successfully transferred to the membrane.

normalization. However, the total protein method is now preferred by many analysts and we will consider it first.

The TPN method of normalization is logical; in this method, one measures the amount of total protein in each lane on the blot. While the idea is simple, in practice it can be challenging, and so various approaches have been devised to determine the amount of protein in a lane. The most common methods rely on stains that are applied to the membrane before or after immunodetection. These stains are chosen because they bind evenly to all proteins (Figure 29.23). Ponceau S is a commonly used, rapid,

and economical TPN stain. Manufacturers produce other proprietary TPN stains that also bind evenly to all proteins. A western blot membrane can be briefly incubated in a TPN stain and then be photographed, or scanned using an image analyzer designed for this application. The resulting image will show each protein band, and the intensity of the bands' staining will be related to the amount of protein present. After imaging, the TPN stain is removed by washing in PBS or wash buffer. The membrane is then processed by immunoassay, and another image is captured that shows only the target protein in each lane where it is present. There are now two images: the TPN image

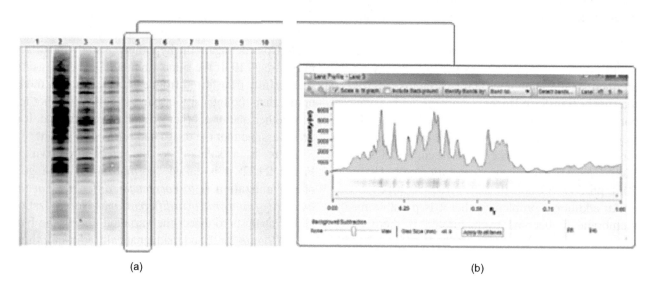

(a)　　　　　　　　　　　　　　　　　(b)

FIGURE 29.23 TPN staining. (a) An image of a blot that has been stained with a proprietary total protein stain. (b) Image analysis software is being used to determine the amount of protein in each band in Lane 5 of the gel. The heights of the green peaks are related to the amount of protein in each band. (Courtesy of Bio-Rad Laboratories, Inc., © 2021.)

and the immunoassay image. These two images can be processed with image analysis software that uses the intensity of TPN staining to evaluate the amount of total protein in every lane. The software automatically compensates for differences in total protein loading across lanes and provides a readout of the relative amounts of target in each lane.

TPN staining methods are effective and commonly used, but analysts must avoid problems that can cause untrustworthy results. For example, Ponceau S staining decreases quickly over time, and the image must be captured while the stain is intense. Ponceau S needs to be applied carefully to assure even staining and excess stain needs to be carefully washed away to avoid uneven background. The proprietary TPN products may improve the reliability of the TPN method.

Housekeeping proteins *are those that are required in all cells for basic cellular function and are therefore always present.* The housekeeping protein method of normalization is different from the TPN method because it only assesses the presence of a single protein (in addition to the target) in each lane. The HKP method involves performing an immunoassay for a housekeeping protein, in addition to the immunoassay for the target protein. This means that two primary antibodies are required: one for the analyte of interest, and one for the housekeeping protein. Actin is an example of a housekeeping protein that is part of the cytoskeleton and contributes to the shape and motility of cells. Actin is commonly used in western blotting when the housekeeping method of normalization is applied. Tubulin is another ubiquitous protein commonly used for this purpose. If the intensity of staining for the housekeeping protein is consistent, then it is assumed that the amount of total protein in the lanes is also consistent. If there is variation in the intensity of the housekeeping protein from lane to lane, this variation can be analyzed.

The HKP method is illustrated in Figure 29.24. In this example, tubulin is the housekeeping protein used for normalization. The target protein is Lnk, which is a protein involved in the immune system. The question that is being studied is whether Lnk increases over time if cells are treated with TNF (a protein associated with inflammation). Lanes 3–5 do show increasing intensity of Lnk staining, which is consistent with TNF affecting the expression of this protein. The staining of tubulin, shown in the bottom row, appears consistent over time, indicating that the differences in Lnk are real and not a result of differences in the amount of total protein in each lane. Image analysis software could be used to analyze this image more accurately

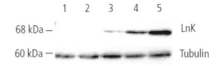

FIGURE 29.24 An example of the housekeeping protein method of normalization. Top row: The western blot was probed first with anti-Lnk antibody showing an increase in stain intensity over time after treatment with TNF. Lane 3 is 1 hour after TNF treatment, Lane 4 is 2 hours after treatment, and Lane 5 is 3 hours after treatment. Bottom row: The results of assay with anti-tubulin antibody to confirm that the same amount of protein was loaded in each lane. Lanes 1 and 2 are negative controls that are not expected to express Lnk. (Courtesy of Bio-Rad Laboratories, Inc., © 2021.)

than is possible by eye, but the image demonstrates the use of the HKP method.

While the housekeeping protein method of normalization has widely been used, its basic assumption is no longer considered to be valid. The expression of housekeeping proteins, which was formerly thought to be consistent, in fact, can vary with different treatments. (Eaton, Samantha L., et al. "Total Protein Analysis as a Reliable Loading Control for Quantitative Fluorescent Western Blotting." *PLoS One*, vol. 8, no. 8, 2013, p. e72457. doi: 10.1371/journal.pone.0072457.)

Therefore, researchers using the HKP method are being asked by journals and reviewers to prove that the expression of the housekeeping protein does not change in the course of their experiments. For example, if analysts were submitting Figure 29.24 as part of a paper for publication, they would need to demonstrate that the expression of tubulin does not change when cells are treated with TNF. This requirement would add complexity to their experiments. An additional problem with this method is that housekeeping genes are expressed at a high level in cells, while the target protein may be expressed at only a low level. Therefore, the amount of protein required per lane to detect the protein of interest often results in a signal that causes the housekeeping gene's band to be oversaturated (see Figure 29.21).

29.5.3.4 Linearity

For semi-quantitative applications, analysts would like to be able to say that if one band on a western blot has twice as much signal as another, then the sample from which it was taken contained twice as much target protein. This statement is true only if the bands' intensities are in the linear range of detection. The linear range of detection is the range of sample loading that

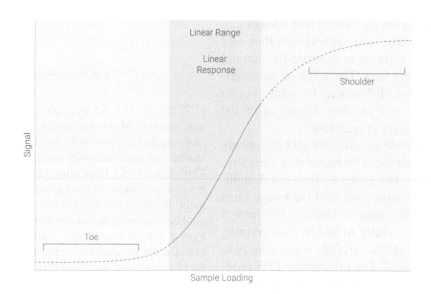

FIGURE 29.25   Linearity in western blotting. If too much sample ("shoulder region") or too little sample ("toe region") is loaded onto a gel for western blotting, then signal will not be directly proportional to the amount of target present. (Copyright Li-Cor, Inc., used with permission.)

produces a linear relationship between the amount of target loaded onto the membrane, and the signal intensity recorded by a detector. As we have previously discussed, all quantitative assays are only linear in a certain range of signal. This is true of immunoassays, as illustrated in Figure 29.25.

Let's consider some of the reasons why western blot results might not be linear. One cause of nonlinearity was previously mentioned; that is, the membrane can be oversaturated if too much protein is present (see Figure 29.21). In this situation, the relationship between sample loading and signal falls into the shoulder region of the graph in Figure 29.25. In the shoulder region, increases in the amount of target have little effect on the signal. Conversely, if too little sample is loaded and a band is faint, it may not be detectable above noise. This moves the relationship into the toe region of the graph.

The type of reporter used will affect the range in which the relationship between sample loading and signal is linear. Some reporters provide more intense signal than others. If a reporter provides a more intense signal, then lower levels of target will fall into the linear range of the assay. Chemiluminescence is frequently used for semi-quantitative western blotting because it is very sensitive and allows lower levels of target to be in the linear detection range. However, chemiluminescence is based on an enzymatic reaction that occurs over time. Therefore, the timing of image capture can affect the signal that is detected. Fluorescence is also commonly used as a reporter for

western blot immunoassays and is not as sensitive to timing. Fluorescence, however, suffers from the potential for autofluorescence, which causes background noise. Noise makes it difficult to distinguish faint bands and limits the linear range in the toe region.

The method used to detect signal from the reporter also affects the linear range. In the past, photographic film was used to capture western blot images when chemiluminescent and fluorescent reporters were involved. However, the dynamic range of film (the range of signal intensities that can be detected) is so limited that film is now less often used for this purpose. Digital imaging instruments that detect and capture fluorescent and chemiluminescent images have taken the place of film. Digital imaging devices have a much wider linear range than film, but there are still limitations in the range in which they can detect signal in a linear fashion. (This is the same idea as discussed for spectrophotometers, in Section 28.3.5.)

Because various factors affect linearity, analysts must experiment to determine the relationship between how much protein they load onto a gel, and the resulting intensity of target bands. To do this, it is common practice to make a series of dilutions of the target protein from the highest concentration expected in real samples to the lowest. Analysts then run these dilutions on a gel, blot the gel, and perform an immunoassay for their target protein. Based on the results, they determine the range in which sample loading and signal intensity are linear. An example is shown in Figure 29.26.

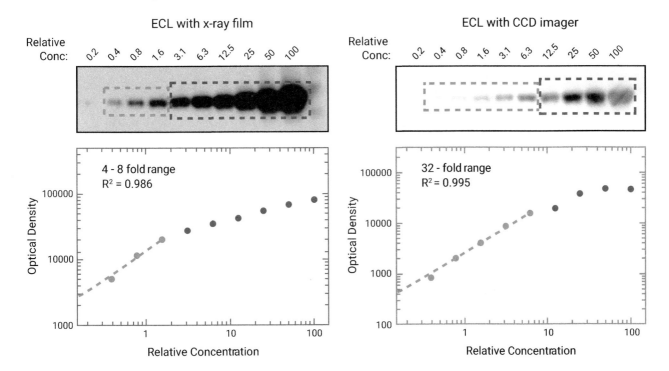

FIGURE 29.26 Determining the linear range of a western blot assay. The image on the left shows differing dilutions of target protein as captured on photographic film. At the top is an image of the film. Only the three most dilute samples create signal that increases in a linear fashion, as is graphed below the image using image analysis software. The right side shows the results of the same analysis, but using a digital camera to capture the image. In this case, the first five dilutions show a linear response between sample loading and signal intensity. This comparison demonstrates that the digital camera has a wider linear range of detection than film, but even the camera becomes saturated when signal is very high. (Copyright Li-Cor, Inc., used with permission.)

Appendix II, *Journals Provide Guidance about Good Practices for Semi-quantitative Western Blotting*, provides additional information about ensuring reproducibility when performing semi-quantitative western blots.

29.6 IMMUNOCYTOCHEMICAL ASSAYS AND IMMUNOHISTOCHEMICAL ASSAYS

29.6.1 INTRODUCTION

The immunoassays we have discussed so far require that cells and tissues be destroyed, and their protein molecules are extracted and suspended in lysis buffers. **Immunocytochemistry (ICC)** and **immunohistochemistry (IHC)** *are quite different in that they combine immunoassays with microscopy to visualize targets in their cellular location.* ICC and IHC provide less quantitative information than western blots or ELISAs, but there are many applications where visualizing the cellular/tissue location of a target is essential. As the Human Protein Atlas states:

Spatial partitioning of biological processes is a phenomenon fundamental to life. At the cellular level, proteins function to catalyze, conduct and control most processes at specific times and locations. Hence, the activity of how cells generate and maintain their spatial organization is central to understanding the mechanisms of the living cell. Protein function is predominantly determined by its subcellular localization, due to that cellular compartments, such as organelles, offer environments with different physiological conditions, constituents and interaction partners.

(The Human Protein Atlas, https:// v16.proteinatlas.org/humancell)

Immunocytochemistry *refers to immunoassays performed on cells that were grown in cell cultures* (Figure 29.27). **Immunohistochemistry** *refers to immunoassays performed on tissue samples*, for example, kidney tissue from a laboratory mouse, or a human biopsy sample (Figure 29.28). **Tissue** *consists of interconnected cells that are organized to perform a specific role in the organism.* For example, muscle

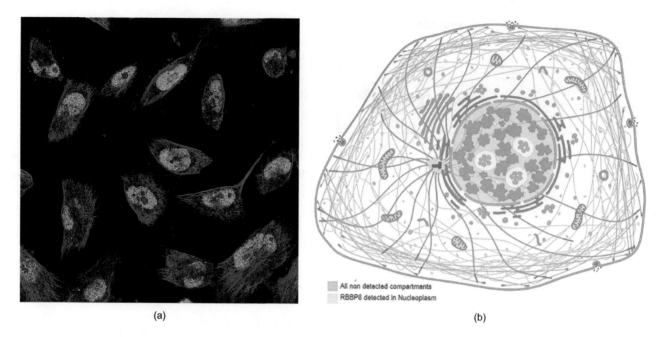

(a)   (b)

FIGURE 29.27 Example of ICC in a study of a gene involved in cancer. The gene RBBP8, when mutated, is involved in retinoblastoma, a rare form of eye cancer found in young children. (a) ICC image of cells tested by immunoassay for the protein encoded by the gene, RBBP8. A green fluorescent reporter stains RBBP8 protein and is localized inside the nuclei. A red fluorescent reporter stains tubulin to show the overall shape of the cells. (b) Based on multiple images, the localization of RBBP8 is summarized: Gray represents parts of the cell where there is no RBBP8, and green shows its presence in the nucleoplasm (the interior of the nucleus). (Image credit: Human Protein Atlas, CC by 3.0 US. Images available from https://v17.proteinatlas.org/ENSG00000101773-RBBP8/cell#human.)

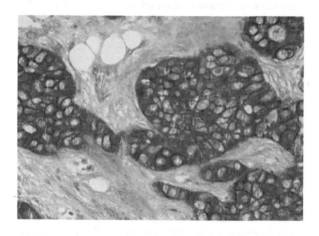

FIGURE 29.28 Example of the use of IHC in diagnosis of breast cancer. As discussed in Section 1.2.3, patients whose cancer cells overexpress the Her2 receptor protein can be treated with the drug Herceptin. Patient biopsies may therefore be tested by immunoassay to see if the Her2 receptor is overexpressed. This is a slide of cancer tissue that overexpresses the Her2 receptor. The brown color is the result of a chromogenic reporter and indicates abundant Her2 receptor protein. (Photo courtesy of Abcam, Inc.)

tissue is composed of interconnected cells with structures related to motility, while the cells in nerve tissue have a structure and function that allows them to convey information.

IHC and ICC have many applications in disease diagnosis, drug development, and biological research. A diagnostic application of IHC is illustrated in Figure 29.28. Pathologists routinely use IHC to look for protein tumor markers in biopsy tissue to diagnose whether a tumor is benign or malignant, and its stage, and to identify the cell type and origin of a cancer. IHC and ICC are used during drug development to detect changes in protein expression that occur as a result of drug treatment. In a research laboratory, ICC and IHC immunoassays are used in the study of development and regeneration, the control of cell division, the functions of different genes, and so on.

29.6.2 Sample Preparation for Immunocytochemistry and Immunohistochemistry

29.6.2.1 Fixation

Sample preparation is an important part of any assay, and the methods by which samples are prepared always have the potential to skew assay results. Sample preparation for ICC and IHC is particularly complex because it is difficult to preserve the structure of cells and tissues. As soon as tissue is excised, or cells are removed from their culture medium, cellular structures begin to break down due to enzymes found within the cells, and degradation by bacteria. *The process of halting degradation and preserving cellular structure is called* **fixation**. Ideally, fixation preserves "life-like" biological structures by quickly inhibiting degradative enzymes and stabilizing the structural components of cells. In fact, while useful fixation methods have been developed, tissue and cell structures are altered in various ways by the methods used to prepare the samples. Microscopic images are valuable, but imperfect, windows into the intricate and dynamic structures that support life.

Chemicals, including formaldehyde and paraformaldehyde, are most often used to preserve cells and tissues for IC and IHC, although there are also methods involving rapid freezing that are sometimes effective. Fixation with any method may denature protein targets. It is important to realize that some primary antibodies recognize denatured proteins, while others do not. Therefore, the choice of primary antibody must be made with an understanding of the effects of a particular fixation procedure. Table 29.6 in Appendix I of this chapter has more details about chemical fixation.

29.6.2.2 Blocking

ICC and IHC are performed in a cellular matrix where the immunoassay antibodies may recognize and weakly bind to nontarget proteins that have regions similar to the target epitope. Antibody binding to nontarget proteins results in background that can mask the signal from the target protein and lead to ambiguous results. Blocking buffers are therefore often used prior to antibody incubation, as previously described for dot blots and ELISAs. However, blocking agents can interfere with the target antigen and may therefore reduce the signal from the target. While blocking is common when performing IHC or ICC, it may not always improve the results. (Buchwalow, Igor, et al. "Non-Specific Binding of Antibodies in Immunohistochemistry: Fallacies and Facts." *Scientific Reports*, vol. 1, no. 1, p. 28, 2011. doi: 10.1038/srep00028.)

29.6.3 More about Immunocytochemistry

To begin ICC, cultured cells are seeded on a support material compatible with microscopy, such as glass-bottom cell culture dishes, glass coverslips, and commercially available multiwell chambers mounted on glass slides. (See Chapters 30 and 31 for information about cell culture.) When the cells have grown to a suitable density, they are incubated in a fixative and then washed. Following fixation, the cells are permeabilized if the target protein is inside the cells. (Some fixation procedures simultaneously fix and permeabilize the cells.) **Permeabilization** *involves modifying the cell membranes to allow the immunoassay reagents to enter.* Permeabilization is not necessary if the target epitope is located on the outer surface of the cells. Detergents are used for permeabilization because they disrupt lipid structures and create pores in the plasma membrane. Table 29.7 in Appendix I of this chapter provides more details about permeabilization.

The immunoassay procedure follows permeabilization and blocking. Both direct and indirect immunocytochemical assays are common. Fluorophores are most often used as the reporters for ICC, resulting in colored light emanating from the target. This light can be visualized by viewing the cells with a fluorescent microscope. More than one fluorophore can be used at the same time, allowing more than one protein to be visualized. Fluorescent small molecule dyes are also used to stain cellular structures in addition to the target, as shown in Figure 29.29. For example, a dye called DAPI binds to DNA and glows with a blue color, so the nuclei of cells can be visualized. The purpose of DAPI staining is to provide orientation by showing where the nuclei are located in a field of view. Box 29.5 illustrates a generalized procedure for ICC.

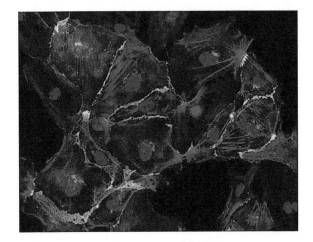

FIGURE 29.29 Immunofluorescent image. This image shows cultured cells treated with DAPI (blue) to stain the nuclei, and two antibodies. One of the antibodies labeled actin filaments (red), and the other antibody labeled β-catenin, a protein involved in a multitude of developmental and regulatory processes (green). (Illustration reproduced courtesy of Cell Signaling Technology, Inc. www.cellsignal.com.)

BOX 29.5   GENERALIZED IMMUNOCYTOCHEMISTRY PROCEDURE

Seed cells onto transparent support, such as a chamber
slide, shown here.

- Incubate in culture medium until cells reach desired
density.
- (Image used with permission, copyright 2012,
Corning, Inc.)

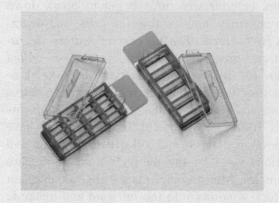

Wash to remove culture medium.
Incubate cells in a fixative.

- Common fixatives are toxic and should be used in a
fume hood.

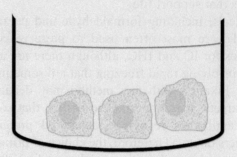

Wash to remove fixative.
*Move slide or dish into a humid light-tight box or covered dish to prevent drying and later to prevent exposure of
fluorophore to light.*
*If necessary, permeabilize with detergent or ice-cold
methanol.*

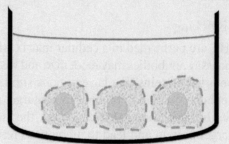

Wash to remove permeabilization agent.
Block, if necessary.
Wash to remove blocking agent.

- Dilute primary antibody.
- Incubate in primary antibody.

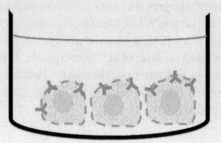

(Continued)

Wash to remove unbound antibody.
- Dilute secondary antibody, assuming indirect labeling is used.
- Incubate in secondary antibody.
- Secondary antibody is labeled with a fluorescent reporter.

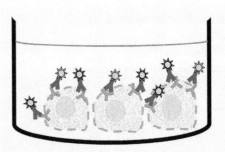

Wash to remove unbound antibody.
Incubate with DAPI to stain nuclei.

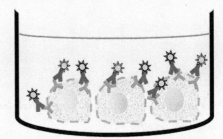

Pipette mounting medium over slides and place coverslips on top.
- Visualize cells with fluorescent microscope (assuming a fluorescent reporter was used) and photograph images for analysis.
- For long-term storage, store flat at 4°C in dark.
- (Image courtesy of Olympus Corporation.)

29.6.4 MORE ABOUT IMMUNOHISTOCHEMISTRY

Immunohistochemistry sample preparation differs from that for ICC. Immunocytochemistry begins with a thin, single layer of cells grown on a glass support that can be placed directly on a microscope after processing. IHC begins with blocks of tissue excised from an animal or patient. These tissue blocks cannot be placed directly under a microscope, but rather must be cut into thin slices to reveal their inner cellular structures. Sample preparation for IHC begins by incubating the blocks of tissue in a fixative immediately after excision. The fixed tissue is too soft to slice cleanly, so after fixation, it is embedded in a firm matrix; paraffin wax is most frequently used for this purpose. *The blocks of paraffin-embedded tissue can then be sectioned (sliced) using a specialized instrument called a* **microtome**. Paraffin is necessary to support the tissue during sectioning, but after sectioning, the paraffin must be removed because it prevents the immunoassay reagents from penetrating the sample. Deparaffinization has traditionally been done by dissolving the wax in xylene, but this solvent is flammable, toxic, and volatile. Xylene-free alternatives are now available from manufacturers.

Paraffin is not the only agent that can interfere with immunoassays. Fixation causes macromolecules in tissues to be cross-linked with one another. This cross-linking stabilizes their structures, but can also inhibit antibody access to targets. Therefore, tissue sections may require treatment to make the target's epitope accessible prior to immunoassay. *This process is called* **epitope or antigen retrieval**. Epitope/antigen retrieval is typically performed by heating or boiling the deparaffinized sections in buffer or by using enzymatic treatments. It may be necessary to experiment to determine whether antigen retrieval is necessary and, if so, how to best accomplish it. (Table 29.7 in Appendix I of this chapter provides more details about permeabilization.)

Box 29.6 illustrates a generalized procedure for IHC.

BOX 29.6   GENERALIZED IMMUNOHISTOCHEMISTRY PROCEDURE

After excision, incubate tissue blocks in a fixative.

- Use a fume hood.
- Neutral, buffered formalin is commonly used.

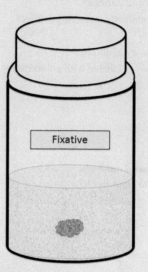

Fixative

Remove water from tissue by bathing in progressively more concentrated ethanol.

Infiltrate tissue with paraffin wax.

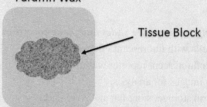

Paraffin Wax

Tissue Block

Slice block of tissue into thin sections using a microtome.

(Photo courtesy of Leica Biosystems, division of Leica Microsystems Inc., https://www.leicabiosystems.com/histology-equipment/microtomes/leica-rm-coolclamp/)

Mount thin sections onto slides.

(Photo courtesy of Leica Biosystems, division of Leica Microsystems Inc., https://www.leicabiosystems.com/knowledge-pathway/an-introduction-to-specimen-preparation/)

(Continued)

Remove paraffin and rehydrate sections.
Retrieve antigen, if necessary.
Block, if necessary.

Wash to remove excess blocking agent.
Dilute primary antibody.
- Incubate in primary antibody.

Wash to remove excess primary antibody.
Dilute secondary antibody if indirect labeling
is used.
- Incubate in secondary antibody.

Wash to remove unbound secondary antibody.
Incubate slides with substrate for reporter, if
necessary.

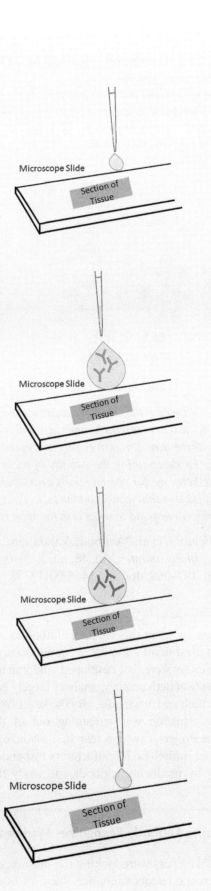

(Continued)

BOX 29.6 (*Continued*) GENERALIZED IMMUNOHISTOCHEMISTRY PROCEDURE

Wash to remove excess substrate.
Add mounting medium and seal slides with coverslips.
Visualize and photograph images for analysis.
(SP3 photo is used with permission from
 Thermo Fisher Scientific, the copyright
 owner.)

29.7 AVOIDING IRREPRODUCIBLE IMMUNOASSAY RESULTS

29.7.1 USING VALIDATED ANTIBODIES

Antibodies are among the most frequently used tools in basic science research and in clinical assays, but there are no universally accepted guidelines … for determining the validity of these reagents. Furthermore, for commercially available antibodies, it is clear that what is on the label does not necessarily correspond to what is in the tube.

(Bordeaux J. et al. "Antibody Validation."
Biotechniques, vol. 48, no. 3, 2010,
pp. 197–209. doi: 10.2144/000113382.)

29.7.1.1 The Problem

As we have seen so far in this chapter, antibodies are the key reagent used in all varieties of immunoassays. When immunoassays were first developed, and into the latter half of the twentieth century, analysts largely had to manufacture their own antibodies. By the late 1990s, new genetic information was streaming out of the Human Genome Project, cancer research, immunology studies, and other initiatives. Manufacturers responded enthusiastically by producing antibodies to study the

protein products of newly discovered genes so that, currently, there are tens of thousands of antibodies sold commercially. Accompanying reagents, buffers, and kits are also widely available, and analysts now are seldom required to manufacture their own immunoassay tools. This is a good situation, because as new genes are discovered, manufacturers provide antibodies to study them. This is a problematic situation because researchers often do not know how these antibodies were made, or whether they are of acceptable quality. In the past 10 or so years, published studies have demonstrated frustrating research failures relating to problems with antibodies. Two examples are provided in this textbook as case studies: One is shown below, and the other is on p. 105 in Chapter 5. Recall that the case study in Chapter 5 discussed IHC studies of estrogen receptor beta, ERβ, which was hoped to be a new target for breast cancer drug therapy. As described in that case study, after 20 years of intense research, ERβ was a source of confusion, not treatments. The authors attribute the problem to poorly functioning antibodies that did not, in fact, recognize ERβ. (Andersson, S., Sundberg, M., Pristovsek, N. et al. "Insufficient Antibody Validation Challenges Oestrogen Receptor Beta Research." *Nature Communications*, vol. 8, 2017, article 15840. doi: 10.1038/ncomms15840.) In fact, antibodies are

---

Case Study: Antibody Recognizes Marker for Ovarian Cancer Instead of Pancreatic Cancer

A team of researchers were looking for substances in the blood that could be assayed for the early detection of pancreatic adenocarcinoma. Such an assay would be analogous to the prostate-specific antigen (PSA) test that is widely used to screen men for prostate cancer. PSA is a protein secreted by the prostate.

(Continued)

Case Study (*Continued*): Antibody Recognizes Marker for Ovarian Cancer Instead of Pancreatic Cancer

Small amounts of PSA normally circulate in the blood, but the level of PSA is generally elevated in men with prostate cancer. Therefore, it is relatively straightforward to screen for prostate cancer by analyzing blood PSA levels (although elevated PSA levels can be due to causes other than cancer). Pancreatic cancer is difficult to detect in its early, treatable stages, and so a blood test for this cancer would be very valuable. A team of pancreatic cancer researchers used bioinformatics data to find proteins that are secreted, and are specific to the pancreas, with the idea that possibly one of these proteins could be assayed in blood to screen for pancreatic cancer. The researchers identified one protein that looked promising: CUZD1. They purchased a commercial ELISA test kit for CUZD1. However, after 2 years of work they discovered that the kit's antibody actually recognized a completely different antigen, that is, CA125, which is elevated in ovarian cancer but is not useful for detecting pancreatic cancer. During these 2 years, they spent roughly $100,000 in kits, reagents, and patient serum samples. They estimated that with the associated labor and other costs, their total expenditure was approximately $500,000. They state that "During this process, we also wasted thousands of highly valuable patient samples and raised false expectations due to the misleading results." (Prassas, Ioannis, et al. "False Biomarker Discovery Due to Reactivity of a Commercial ELISA for CUZD1 with Cancer Antigen CA125." *Clinical Chemistry*, vol. 60, no. 2, 2014, pp. 381–88. doi: 10.1373/clinchem.2013.215236.)

thought to be one of the most significant causes of the "reproducibility crisis" discussed in Chapter 5.

The scientific community has responded to the antibody problem with efforts to develop methods for **antibody validation**, that is, *ensuring that the performance of a specific antibody is suitable for its intended use*. Antibody validation determines whether an antibody has the following characteristics:

- *Specificity/selectivity* – the antibody should recognize its intended target and should not cross-react with molecules other than its intended target.
- *Sensitivity* – the antibody should recognize low levels of target protein amidst other molecules.
- *Reproducibility* – the antibody should function predictably over time and from lot to lot.

While antibody validation is of critical importance, it is also challenging. Some of the challenges are as follows:

- *There are tens of thousands of antibodies commercially available*. For example, there are more than 6,500 antibodies available that target EGFR, a receptor protein functioning in cell growth and division. From the manufacturer's perspective, it is expensive and time-consuming to validate an entire inventory containing thousands of antibodies. Adding further complication, more than one distributor may sell the same antibody under a different name. This makes it difficult for researchers to know what product to choose, and exactly what they are buying.

- *An antibody that is effective in one application might not work in another.* The major cause of this variable performance is that the target protein can have different forms in different situations. For example, the proteins in a western blot are denatured. In contrast, proteins in ICC are not intentionally denatured, but chemical fixatives might alter their structure. Some antibodies recognize their target protein when the protein is denatured, others only when it is intact, yet others might recognize either form. Some antibodies might not recognize a target unless it has normal post-translational modifications; other antibodies might. To add further complexity, different antibodies that recognize a single protein target can recognize different epitopes on that target, with differing functionality in immunoassays. This means that an antibody product that works for western blots might not work for other types of immunoassay, and *vice versa*. Therefore, every antibody needs to be validated for each application in which it is to be used, adding time and expense.

- **Irreproducibility between lots of the same antibody has been detected in various studies.** (See, for example, Bordeaux, Jennifer, et al.,

"Antibody Validation." *BioTechniques*, vol. 48, no. 3, 2010, pp. 197–209. doi: 10.2144/000113382, for a summary of these studies.) Inconsistency has been observed even with monoclonal antibodies, which are expected to be consistent from lot to lot.

- **Cross-reactivity is an important issue that is particularly problematic when the target is present in low levels**. In these situations, a procedure might be optimized so that the target is detected but at the cost of cross-reactivity. For example, longer incubation times or more concentrated antibody might ensure that a low abundance target is recognized. But these conditions also promote cross-reactivity. Therefore, an antibody might appear to be "good," and yet be unsuitable for a study where the target is present at low levels.

29.7.1.2 Finding Solutions

Scientific organizations, researchers, testing organizations, journal editors and reviewers, and antibody manufacturers have all collaborated to address the challenges of antibody validation. In 2016, key stakeholders met at the Asilomar Conference Grounds to participate in a workshop, "Antibody Validation: Standards, Policies, and Practices." This workshop resulted in recommendations for antibody validation process standards, proficiency testing, and user antibody rating services. These recommendations have guided much current thinking about antibodies. (See the Bibliography for this unit for more references about antibody validation.)

Antibody validation occurs during the process of developing a new antibody product, or testing existing inventory, and is primarily the responsibility of manufacturers. Once an antibody enters production, reproducibility requires testing each lot against the original lot that was validated during development. This is essential with polyclonal antibodies, but is also advisable with monoclonal antibodies to ensure that the clones making the antibodies have remained stable over time. Testing reproducibility between lots is also the manufacturer's task. However, it is advisable to check new lots of an antibody in your own laboratory to be sure that results are consistent from lot to lot.

A manufacturer or researcher is unlikely to use all the available validation methods, but combining several strategies is the most effective way to demonstrate that an antibody is suitable for use. We will introduce a few of the common methods used for antibody validation.

29.7.1.3 Methods of Antibody Validation

29.7.1.3.1 Method 1: Relative Expression in Different Cell Lines

Sometimes there are cell lines available that express a target of interest, and other cell lines that do not. When such cell lines are available, an antibody can be tested to see if it binds selectively in the cell lines with the target, and not in the other cell lines (Figure 29.30).

The example in Figure 29.31 involves the same general strategy as illustrated in Figure 29.30, but this time validates two antibodies for use in western blotting. In cell lines where the target is expected to be present, a band should appear at the proper molecular weight for that target. No bands should appear in cell lines that do not have that target. In Figure 29.31, a panel of lysates from various cell lines was prepared and western blot assays were performed with the two antibodies being validated. Only one of the two antibodies passed this test.

29.7.1.3.2 Method 2: Knockout and Knockdown Cell Lines

This method involves using cells grown in culture and knocking out the expression of the target gene. One method often used to produce "knockouts" is CRISPR-Cas9 (see Section 2.1.2.3 in Chapter 2). If an antibody is specific to the target, it will bind in the original cells, but will not bind in the lysates from knockout cells. The advantage of this strategy is that it confirms that an antibody binds to the protein that is the product of a gene of interest, and not to any other proteins in the cells. A similar method is to use RNAi (see Section 2.1.2.2 in Chapter 2) to knockdown or reduce the expression of the gene of interest (Figure 29.32). A limitation of these methods is that genes that code for proteins vital to the life of a cell cannot be knocked out or severely reduced.

29.7.1.3.3 Method 3: Cell Treatments

It is sometimes possible to treat cells with a chemical or other treatment that is known to affect the expression of a target molecule, perhaps eliminating its expression, or increasing it. In these cases, the antibody should not bind any target if the protein expression was eliminated by the treatment. If the treatment increases the expression of the target, then enhanced antibody binding should be observed. A limitation of this method is that it is only useful if a target is well understood, which is not always the case.

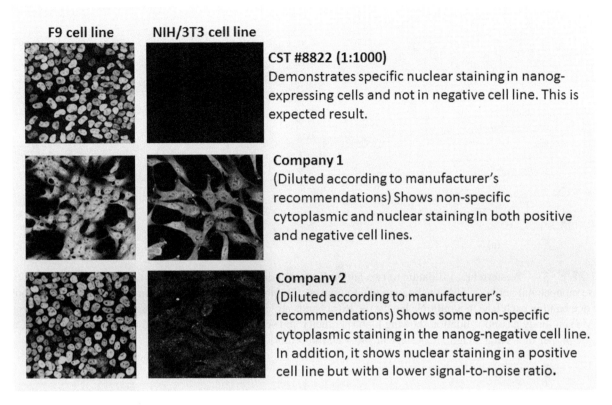

FIGURE 29.30 Antibody validation using different cell lines. In this validation example, antibodies to the NANOG protein from three companies were compared in two different cell lines, F9 and NIH/3T3. NANOG is a protein associated with differentiation of pluripotent cells. In the top row, the antibody from the first manufacturer detected its target in F9 cells, but not in NIH/3T3 cells; this is the expected result. So, the first manufacturer's antibody passes this test. In the middle row, antibody from another manufacturer was tested. The NIH/3T3 cells exhibit nonspecific binding in both their nuclei and cytoplasm. The use of this manufacturer's antibody would result in false-positive assay results and confusion as to where in the cell the target is localized. The antibody from the third manufacturer, shown in the bottom row, also binds to NIH/3T3 cells. Although the nonspecific binding is less in the bottom row than in the middle row, this off-target label would still cause problems in immunoassays. Thus, in this example, only one out of three antibodies (the top row) "passes" the validation. (Illustration reproduced courtesy of Cell Signaling Technology, Inc. www.cellsignal.com.)

29.7.1.3.4 Method 4: Independent Antibodies

Sometimes there are two or more antibodies raised against the same target, each of which recognizes a different epitope. Similarly, there are sometimes polyclonal and monoclonal antibodies against the same target. The function of antibodies that are expected to recognize the same target can be compared in a validation study and should be comparable.

29.7.1.3.5 Method 5: Peptide Array

A **peptide array** *is a glass or plastic chip onto which a collection of peptides has been deposited.* An antibody to be validated can be applied to the chip and should bind only to the peptide of interest. This can be a rapid screening process and requires little antibody sample volume. A limitation of this method is that the peptides on the chip do not have post-translational modifications

or normal conformations. Therefore, some antibodies will fail this validation test, but would work for studies where the target is in its natural conformation with normal post-translational modifications.

29.7.1.3.6 Method 6: Complementary Strategies (Also Called Orthogonal Strategies)

This term is used to mean that the performance of an antibody undergoing validation is correlated with a method that does not involve antibodies at all. For example, if a protein is expressed in cells, then both the protein and its mRNA should be present. Assuming this is the case, then an antibody should detect a target protein when the mRNA for that protein is present. The antibody should not detect anything when the mRNA for the target protein is absent. There are methods that enable analysts to test for the presence or

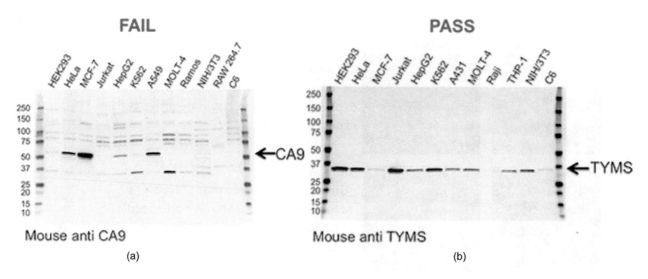

FIGURE 29.31 Western blot validation of two antibodies using different cell lines. (a) A carbonic anhydrase IX (CA9) mouse monoclonal antibody failed validation due to nonspecific binding and a low signal-to-noise ratio. There should be only one band in each lane, not multiple bands; the extra bands are due to nonspecific binding. (b) A thymidylate synthase (TYMS) antibody passed validation showing high specificity and sensitivity. Only one band is present in each lane, and it is at the proper molecular weight. (Courtesy of Bio-Rad Laboratories, Inc., © 2021.)

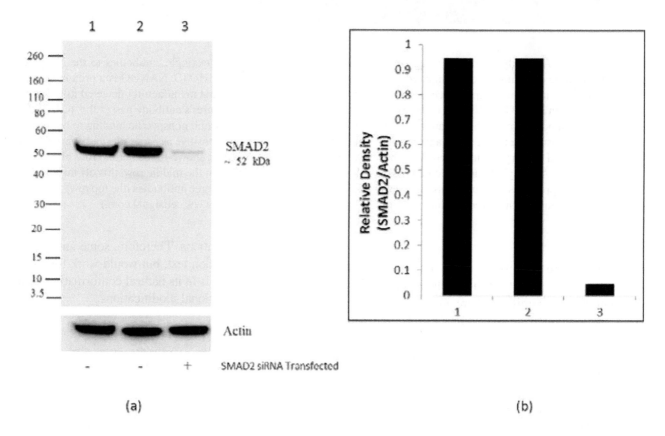

FIGURE 29.32 Knockdown method of antibody validation. (a) The expression of the protein called "SMAD2" was reduced in HeLa cells using an siRNA method. A western blot was then performed to validate an anti-SMAD2 antibody. The samples in Lanes 1 and 2 are lysates from control cells where the knockdown method was not used and the sample in Lane 3 is a lysate from the knockdown cells. Observe that Lanes 1 and 2 have strong bands, indicating that the antibody found and bound to SMAD2. In Lane 3, only a faint band was observed. In this semi-quantitative assay, the housekeeping gene, actin, was used for normalization. (b) The western blot was analyzed with image analysis software, and the results are shown in the graph. The intensity of each band was normalized by the software using the relative intensities of the actin bands. (Used with permission from Thermo Fisher Scientific, the copyright owner.)

FIGURE 29.33 Manufacturer's catalog listing for validated antibodies. These antibodies were validated using knockout cell lines. ICC or western blot data are provided. More information about each antibody and its validation is available in separate data sheets. (Photo courtesy of Abcam, Inc.)

absence of specific mRNAs, but obviously, this validation method involves time and resources.

Figure 29.33 shows a catalog listing for antibodies that were validated by the manufacturer and that have been reviewed by some users. While this information does not guarantee that a particular antibody will work for a specific project, it does provide assurance that the antibody was tested and found to be effective for some applications.

29.7.2 Optimizing the Immunoassay Procedure

29.7.2.1 Overview

Using validated antibodies is critical, but antibody validation alone is not enough to ensure trustworthy immunoassay results. Optimization is also critical. While validation is primarily the responsibility of manufacturers, optimization is performed by the analyst. **Optimization** *involves experimentation to determine the most effective assay parameters, such as antibody concentrations and incubation times.* Note that for clinical purposes (such as testing for strep), kits are often available that have extensively been tested. The performance of these kits still needs

to be verified in the purchasing laboratory, but minimal optimization should be required. In contrast, in a research setting, each step of the immunoassay procedure may need to be optimized for each application. In the next sections, we will introduce the following aspects of optimization:

- *Determining the optimal dilutions for the primary and secondary antibodies.*
- *Determining optimal sample preparation methods and the amount of sample to be assayed.*
- *Optimizing the detection system, including selecting a reporter and determining optimal concentration of detection reagents.*
- *Optimizing incubation times and wash steps.*
- *Optimizing reagents including lysis buffer, washing buffer, and blocking agent.*

29.7.2.2 Determining the Optimal Dilutions for the Primary and Secondary Antibodies

One of the key steps in optimization is determining the correct dilutions of primary and secondary antibodies. If the antibodies are diluted too much, then the target may not be visualized, even if it is present. If the antibodies are not diluted enough, most antibodies will exhibit cross-reactivity, meaning they will bind to molecules that are not their target (Figure 29.34). Cross-reactivity results in background that can cause erroneous or ambiguous results. The optimal antibody dilutions will provide a high signal-to-noise ratio. Most manufacturers provide a starting point that tells the user how much to dilute the antibody. However, this should be tested for each application. Figure 29.35

Primary Antibody
1:500
Secondary Antibody
1:5,000

Primary Antibody
1:5,000
Secondary Antibody
1:50,000

FIGURE 29.34 In this optimization experiment, better results were obtained when primary and secondary antibodies were more dilute. (Used with permission from Thermo Fisher Scientific, the copyright owner.)

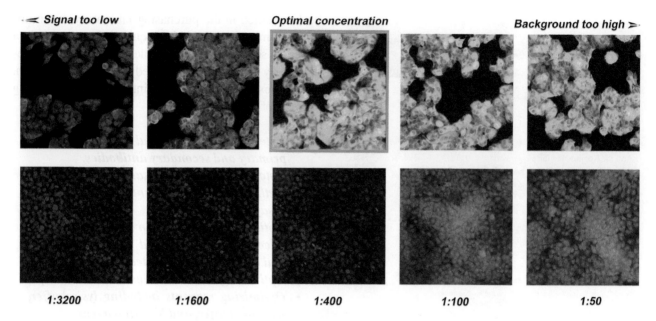

FIGURE 29.35 Optimization of the dilution of an antibody for IF analysis: The upper row of images is from a cell line known to have the target of interest (a protein named MUC1). The lower row of images is from a cell line known not to have the target of interest. Moving from left to right, the antibody becomes more concentrated. The antibody is conjugated to a green fluorescent reporter, so the target is expected to be green. The red color is due to a stain, propidium iodide, that stains the nuclei of living cells. Propidium iodide is used for orientation. Observe in the upper row that the green color of the target becomes more intense as the concentration of antibody increases; however, there is little improvement in the 1:100 or 1:50 dilutions. Observe in the lower row that as the concentration of the primary antibody increases, that is, the antibody is diluted less and less, the negative cell line begins to exhibit some green fluorescence. This is background noise and is undesirable. At 1:400, the target in the positive cell line is clearly stained and there is little green stain in the negative control: This is the antibody concentration where the signal-to-noise ratio is optimized. (Illustration reproduced courtesy of Cell Signaling Technology, Inc., www.cellsignal.com.)

is an example of checking, or "titrating" a primary antibody to the proper concentration. Figure 29.36 also illustrates the principle of diluting antibodies in such a way as to optimize signal-to-noise ratio.

Figure 29.37 illustrates using a "checkerboard" strategy to optimize the concentrations of primary and secondary antibodies for an ELISA. The same strategy can be used with dot blots. Dot blots are useful for optimizing western blots because they are relatively fast.

As is evident from the examples above, optimization of antibody dilutions is critical to improving the signal-to-noise ratio and avoiding false-positive, false-negative, and ambiguous results.

29.7.2.3 Determining Optimal Sample Preparation Methods and the Amount of Sample to Be Assayed

Sample preparation is always important, as previously discussed for western blots. For ICC and IHC, sample handling before fixation, fixation time, the fixative used, antigen retrieval/permeabilization, detection system, and other parameters can all affect the antigenicity of the target (Figure 29.38). The manufacturer

of a particular antibody might provide guidance as to the best fixative if they have validated the antibody for use in ICC and IHC.

29.7.2.4 Optimizing the Detection System

As we have seen, enzymatic, fluorescent, and chemiluminescent detection systems are all widely used. Each needs to be optimized. For example, with chemiluminescence, the secondary antibody is conjugated to an enzyme that reacts with a substrate to generate light. This light is not permanent, as is, for example, a chromogenic deposit. If there is too much secondary antibody, and therefore too much enzyme, then the light that is generated flashes and then rapidly decreases, making it difficult to capture a good image. This can result in variability among different experiments. A chemiluminescent assay must therefore be optimized so that the chemiluminescent signal lasts several hours, thus permitting consistent imaging. Similarly, the concentration of substrate for enzymatic systems must be optimized, along with the incubation time in substrate. Fluorescent reporters require ensuring that autofluorescence is not a problem in a given type of sample.

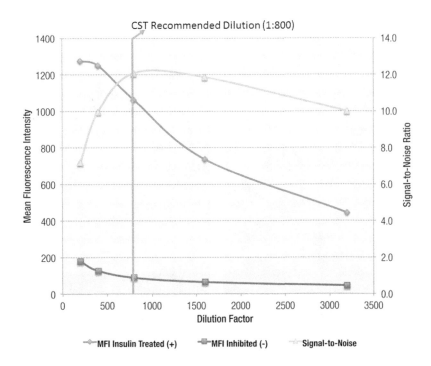

FIGURE 29.36 Optimization of primary antibody dilution to obtain the best signal-to-noise ratio. In this example, cells were treated with agents that either enhanced the expression of the protein target (blue line), or inhibited its expression (red line). The X-axis is dilution, meaning that the primary antibody is progressively more dilute moving from left to right. Observe that even though there is more fluorescence in the positive cells (blue line) at higher concentration, there is also some nonspecific signal in the negative cells (red line) at higher concentrations. The green line graphs the signal-to-noise ratio at each dilution. The optimal concentration has the highest signal-to-noise ratio and is calculated to be 1:800. (Illustration reproduced courtesy of Cell Signaling Technology, Inc., www.cellsignal.com.)

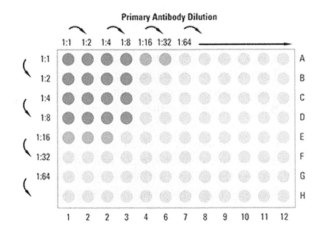

FIGURE 29.37 Optimizing the dilutions of primary and secondary antibodies. Optimizing primary and secondary antibody dilutions for ELISA using a checkerboard pattern. The amount of sample is held constant. The ideal concentration of primary and secondary antibodies will result in clearly detectable color. In most cases, you will want to use the least amount of antibody that gives a strong signal, so based on this evidence, you would use a 1:8 dilution of both primary and secondary antibodies. Subsequent optimization tests can be performed with differing amounts of sample and other reagents. (Used with permission from Thermo Fisher Scientific, the copyright owner.)

29.7.2.5 Optimizing Incubation Times

Just as the antibodies' dilutions impact the signal-to-noise ratio, so also do the incubation times. If the incubation period is too short, the sensitivity of the assay might be reduced, but if incubations are too long, background may appear.

29.7.2.6 Optimizing Reagents

Reagents used for immunoassays include lysis buffers, washing buffers, detection reagent buffers, antibody dilution buffers, and blocking buffers. The composition and pH of all these reagents need to be optimized. There are standard procedures that provide a starting point for preparing these reagents, but they may need optimization for a particular application. For example, the composition of lysis buffer needs to be optimized for different sample types, as previously discussed in this chapter.

29.7.3 USING SUITABLE CONTROLS

29.7.3.1 Overview

Let's assume that validated antibodies have been obtained, and an immunoassay has been optimized. There is yet another requirement to assure trustworthy

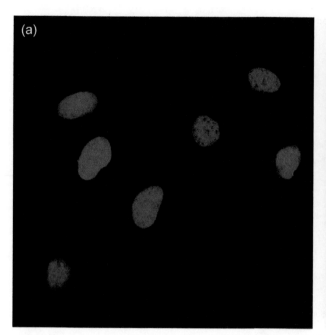
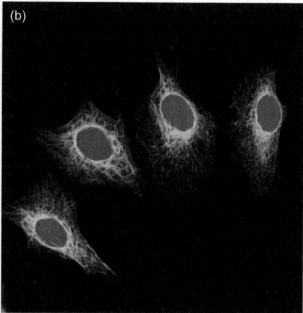

FIGURE 29.38 The effect of different fixatives on results. An antibody against keratin was tested in HeLa cells. The antibody was conjugated to a green fluorescent reporter. Propidium iodide was used as a stain to show the nuclei. When formaldehyde was used as the fixative (a), the target was not detected by the antibody. When methanol was used (b), the target was detected. In this situation, methanol fixation was clearly preferable; however, there are other situations where formaldehyde might provide better results. (Illustration reproduced courtesy of Cell Signaling Technology, Inc., www. cellsignal.com.)

results, that is, the inclusion of controls with each run of the assay. The word "control" has several meanings, and so let's consider what it means in the context of immunoassays.

Controls *in an assay are used to detect, and thereby help guard against, false-positive, false-negative, and ambiguous results.* Controls are often termed "control samples" or even just "samples." But controls are not used for the purpose of looking for an analyte; they are used to evaluate whether an assay is working properly. The "test" samples, also called "experimental" samples, are the samples where the analyst is testing for a target analyte. In most situations, the test samples and control samples are processed and run alongside one another each time an immunoassay is performed.

There are two basic types of controls for immunoassays:

1. **A negative control** *should not provide any signal.* A negative control helps avoid false-positive results. Consider how this works. Suppose that test sample A contains a target of interest and test sample B does not. Suppose further that the blocking agent used in the assay was made incorrectly and the blocking agent is ineffective. In this case, the primary

antibody binds nonspecifically to nontarget molecules. In this situation, both test samples A and B will appear to contain the target. This is a *true*-positive result for test sample A, but it is a *false*-positive result for test sample B. The analyst might not realize the problem and might report erroneous results for test sample B. Now, assume that the analyst includes a negative control alongside test samples A and B. In this situation of ineffective blocking agent, the negative control sample will provide signal and the analyst will know that something is wrong. Once the problem is recognized and blocking is optimized, the negative control sample will not generate any signal, nor will test sample B. The analyst will recognize the absence of signal from test sample B as a true-negative result.

2. **A positive control** *should provide signal.* A positive control helps avoid false-negative results. Consider how this works. Suppose again that test sample A contains the target of interest and test sample B does not. Further suppose that a reagent, such as the primary antibody, was stored improperly and has become degraded. In this case, both samples

A and B will give negative results in the assay. For sample B, this is a *true*-negative result because it does not contain the analyte. But for sample A, this is a *false*-negative result because it does contain the analyte. The analyst might not realize the problem and might report erroneous results for test sample A. However, a positive control can help avoid this mistake. A properly prepared positive control will provide signal that is clearly distinguishable from background noise – if the assay is working properly. In this example, the positive control would have no signal because the primary antibody was ineffective. In this case, the analyst will know that the assay did not work, and will not report erroneous results.

Ideally, every immunoassay run should include at least one positive control sample and one negative control sample, although this is not always practical. There are also situations where more than two controls are used, particularly during runs where an assay is being optimized.

There are variations in the approaches to creating positive and negative controls, depending on the sample type, the immunoassay method, the availability of materials, and the factors that might cause error in a particular situation. For example, it is important to have a control that detects autofluorescence when fluorophore reporters are used, but autofluorescence is not a problem when chromogenic reporters are involved.

29.7.3.2 Negative Controls

Some examples of negative controls for immunoassays include the following:

- **Negative control – blank (ELISA).** The results of quantitative ELISAs are usually analyzed with a plate reader, which is a type of spectrophotometer. As we saw in Chapter 21, spectrophotometers require a blank. The blank can be considered a type of negative control. *No sample at all is applied to the blank well(s)*, but all other steps of the assay are performed, that is, antibody incubations, blocking, washes, etc. The blank controls for any contribution of the plate itself, or residual reagents, to the measured absorbance. Multiwell plates for use in ELISA are designed to readily transmit light, and therefore, the absorbance of light by the blank should be very low. High absorbance in a blank well

may indicate ineffective wash steps leading to contamination, or it may indicate a problem with one or more reagents. Too much substrate for an enzyme reporter can also cause absorbance in the blank well(s). Further investigation of the plate-washing procedure, buffers, substrate, and instrumentation should be conducted if the blank absorbance is high.

- **Negative control – a sample known not to contain the analyte of interest.** In ELISAs or western blots, nonspecific binding of primary or secondary antibodies to the multiwell plate or membrane can lead to noise that can be mistakenly interpreted as arising from the analyte. Similarly, in IC and IHC, antibodies might bind to nontarget molecules. Blocking agents are an important means of avoiding nonspecific binding, but a negative control is also required to be sure the blocking was effective. This negative control differs from a blank in that a "sample" is applied, but this is *a control sample that is known not to contain the analyte of interest*. All steps (including blocking) are performed on this negative control sample. Any apparent signal from this control is a cause for concern and should be investigated. To illustrate, suppose that investigators are testing various types of cells to see if they harbor a pathogen. In this experiment, the cells are lysed, releasing their contents into a buffer solution. Samples from the lysates from each type of cells are analyzed with an ELISA. Ideally, a negative control would be available consisting of a lysate from cells known not to harbor the pathogen. If there is apparent signal in the well(s) containing the negative control, it might be due to nonselective binding of antibodies to the wells, or possibly due to cross-reaction of the primary antibody with something in the cells other than the pathogen of interest.
- **Negative control – knockout cell line.** Knockout cell lines can be used as negative controls, in addition to their value during antibody validation. Since these cells do not express the target protein, they make an excellent negative control.
- **Negative control – blocking peptide.** It is important in all immunoassays to ensure that the primary antibody is specific to its intended target. Some antibody manufacturers offer a **blocking peptide** *that specifically*

masks the target protein. This blocking peptide is used to create a negative control sample. To prepare this negative control, a sample is incubated with the blocking peptide. The blocking peptide masks the target protein, and so, when this control is run through an immunoassay, it should not have any signal – assuming the primary antibody is specific to its target. If there is signal from this control, it indicates that the primary antibody recognized something in the sample other than the target protein. Purchasing the blocking peptide from the manufacturer adds cost, but is one method to help assure that a primary antibody is specific.

- **Negative control – no primary antibody.** To prepare this control, *the primary antibody is replaced with the buffer* used to dilute the antibodies. All other steps of the assay are performed as usual, and the sample is applied. There should be no signal from this control sample; if there is, it is cause for concern. If there is signal, and if an indirect assay is being used, it is possible that the secondary antibody is nonspecifically binding to some component of the sample, or to the surface on which the assay is being conducted. This type of negative control is sometimes called a "reagent control." This is a relatively simple control to include and can be used routinely without adding considerable time or expense.

- **Negative control – no secondary antibody.** To prepare this control, *the secondary antibody is replaced with the buffer* used to dilute the antibodies. All other steps of the assay are performed as usual, and the sample is applied. In an indirect assay, if there is no secondary antibody, there should be no signal because the reporter is not present. Any signal in this control indicates a problem that needs to be resolved, for example, autofluorescence if a fluorophore reporter is being used. This is also sometimes called a "reagent control." This is also a relatively simple control to include.

- **Negative control – sample material not incubated with immunoassay reagents.** When using fluorescence, it is important to have a negative control consisting of *a sample that is not incubated with the immunoassay reagents.* This control is used to look for autofluorescence, which is potentially a cause of high background and false-positive results. In IHC, tissue fixation may induce autofluorescence, particularly when using formaldehyde or paraformaldehyde. Autofluorescence may also be caused by the presence of fluorescent compounds that naturally occur in tissue. If autofluorescence is a problem, there are non-aldehyde fixatives that may be preferable, and agents that might block autofluorescence. Using freezing rather than chemical fixation may also reduce this problem. If autofluorescence cannot be blocked, a chromogenic detection system may be required.

- **Negative control – a sample incubated with substrate only.** Recall that it is possible for tissues and cells to have endogenous enzyme activity. This can be a problem in IHC. It is possible to check for endogenous enzyme activity by *incubating a control sample with substrate,* but no other immunoassay reagents. If this control sample changes color, then endogenous enzyme is probably responsible. This effect can sometimes be blocked by incubating the sample with a special blocking agent. For example, a 10- to 15-minute incubation in 0.3% hydrogen peroxide is usually sufficient blocking when the reporter is HRP.

29.7.3.3 Positive Controls

- **Positive control – sample known to contain the target.** A positive result from this control sample indicates that the assay is working. There are various strategies for obtaining a positive control. For immunochemical assays, analysts might use a purified preparation of the antigen. It may also be possible to purchase or generate cell lysates that contain a target protein. For immunohistochemistry and immunocytochemistry, cells or tissues known to have the target analyte are used. These cells or tissues are fixed and treated along with the test samples to create positive controls.

- **Positive control – spike.** A spike control *is a type of positive control in which a known amount of the target protein is added to the sample matrix.* For example, if one is testing blood serum samples, then the sample matrix is serum. It is possible to obtain serum without the target protein and add to it a known amount of the target protein. The immunoassay is then performed, and the amount of target protein is determined. The performance of the assays is calculated based on the percent recovery of spiked material. A spiked serum control can also be compared to another positive control where the target protein is diluted in buffer instead of serum. The two positive controls should provide roughly the same amount of signal. This comparison tests whether the matrix (the serum in this example) in some way interferes with the assay.

- **Positive control – IgG.** This type of positive control tests the activity of a secondary antibody and the reporter system. For example, in an ELISA, a well can be coated with IgG from the same species as the one in which the primary antibody was made. This IgG contains a mixture of antibody proteins isolated from the blood of an animal. For example, if the primary antibody is a mouse monoclonal, then a well might be coated with a preparation of IgG molecules isolated from mouse serum. The secondary antibody should recognize this mouse IgG, and so there should be signal in this well. If there is little or no signal in this well, it indicates a problem with the reporter, or the secondary antibody itself.
- **Positive reagent control (IHC) – ubiquitous target.** Tissue is incubated with an antibody that detects a target that is found in all cells. This control ensures that the methodology and reagents are working as usual.

Not every type of control will be required in every situation, and determining which controls are necessary is part of optimizing and developing a particular immunoassay for a particular type of sample.

29.7.3.4 Replicates

While replicates are not "controls," as we have defined them, it is good practice to include replicates of controls, standards, and samples. Three replicates are commonly recommended although different situations call for different numbers of replicates. The replicates are expected to roughly have the same results; how much variability is normal needs to be determined for each situation. If the variability among the replicates is greater than the variability allowed, then an investigation is required. Variability among replicates might be due to inconsistent pipetting, inaccurate dilution calculations, using the wrong buffer, and other systematic errors.

29.7.4 Providing Sufficient Documentation

Comprehensive documentation is another aspect of ensuring that you or another analyst can reproduce immunoassay results. This begins with thoroughly documenting information about the antibodies used. For example, the *Journal of Neuroscience Research* requires the following information about each antibody used when articles including immunoassay data are submitted for publication:

- The name of the antibody.
- The structure of the immunogen (antigen) against which the animal was immunized. (Note that a vague reference to a part of the molecule is not acceptable.)
- The manufacturer, catalog and/or lot number, and the RRID. (See the Bibliography for this unit for an explanation of RRID.)
- The species in which the antibody was made, and whether it is a monoclonal or polyclonal antibody.
- The concentration at which the antibody was used.

A team of scientists published recommendations to enhance transparency in the reporting (documentation) of western blot data. The team calls their recommendations the "Western Blotting Minimal Reporting Standard (WBMRS)." The WBMRS, excerpted in Table 29.2, identifies and clarifies a number of factors that can influence western blot results, and the details that must be documented to ensure that the work can be reproduced by others. Observe that extensive documentation is required to ensure transparency. Although these standards were developed for reporting western blotting assay results, they are largely applicable to other immunoassays as well.

29.7.5 Summary of Good Practices for Immunoassays

While manufacturers have much responsibility for antibody quality, the final user has a critical role in ensuring quality immunoassay results. The following are the recommendations to keep in mind:

1. *Be aware that the quality of your results is directly affected by the antibodies that you use and that using the wrong antibodies will jeopardize the results.* Do not purchase antibodies based solely on cost or easy availability.

TABLE 29.2

Recommended Minimal Reporting Standards

Item that Must be Documented	Reason
1. *The primary and secondary antibodies used, the catalog number, company purchased from, and lot number if it is a polyclonal antibody.*	Different antibodies to the same protein can give different results. Polyclonal antibodies can vary from lot to lot.
2. *Molecular mass of the band of interest should be shown on the blot.*	Antibodies might recognize bands that are not the proper molecular mass of the target protein. In fact, some antibodies that are known to only recognize a different molecular weight protein are sold by some companies.
3. *The amount of total protein loaded onto the gel and the method used to determine the protein concentration should be stated.*	Use of too much protein results in inaccurate quantitation.
4. *Type, amount, and extent of use of blocking reagent.*	The type and amount of blocking reagents can significantly affect the number of nonspecific bands detected by an antibody.
5. *Washing solution used, how often, and for how long.*	The buffer used and amount of washing affect the background membrane staining and sometimes the number of nonspecific bands detected by an antibody.
6. *Amount and incubation time of primary antibody. Primary antibody buffer.*	The proper primary antibody amount and buffer are critical for good western blots.
7. *Amount and incubation time of secondary antibodies. Secondary antibody buffer.*	The secondary antibody amount and buffer are important for good western blots.
8. *How the image was collected.*	If X-ray film was used, then the source of X-ray film is needed. X-ray film has lower intensity than commercial analyzers.
9. *The reagent used for detection and how long the membrane was incubated in the reagent, including the company that manufactured the reagent.*	Different chemiluminescent reagents have distinct properties. Different fluorescent secondary antibodies also sometimes have dissimilar properties.
10. *The software (including the company that created the program and version number) used for data analysis. Basic information about how the quantitation was carried out.*	If quantitation is incorrectly done, then no matter how well the western blot was carried out, the data from the western blot will be inaccurate.

Source: Excerpted from Gilda, Jennifer E., et al. "Western Blotting Inaccuracies with Unverified Antibodies: Need for a Western Blotting Minimal Reporting Standard (WBMRS)." *PLOS ONE*, vol. 10, no. 8, 2015, p. e0135392. doi: 10.1371/journal.pone.0135392.

2. *Investigate your target protein, using a literature search and finding information from other analysts.* Evaluate your planned assays. Decide whether to use a polyclonal or monoclonal antibody to recognize the target. It is common to use a monoclonal antibody to maximize the specificity of an assay because monoclonal antibodies recognize only one epitope. However, a polyclonal antibody preparation might work better in some situations. If the target is slightly altered, for example, by mutation or a conformational change, polyclonal antibodies are more likely to recognize it, while monoclonal antibodies are less likely to do so. If you decide to use a monoclonal antibody, determine the desired features of the epitope. For example, membrane proteins often have a domain that extends into the cytoplasm, and another domain that is on the cell surface. It may be possible to purchase antibodies against either of the domains, and one or the other might work better for a given application.

3. *Once you have determined the characteristics a new antibody should possess, research antibody providers.* The resources listed in

the Bibliography for this unit are a starting point; other online resources also exist. Check manufacturer's catalogs and study the data sheets for antibodies that you are considering. Make sure the antibody has been validated in the type of application you will be using. For example, if you will be performing western blots, be sure the antibody is validated for that purpose. A thoroughly studied antibody is best. You can also look at the citations provided by the manufacturer to see if others have successfully used a particular antibody in an application similar to yours.

4. *Do not expect to purchase a single antibody to use in varied applications.*

5. *If it is anticipated that an antibody will be used extensively over many years, a monoclonal antibody might provide more reproducible results than a polyclonal antibody.*

A recombinant antibody, if available, might provide even more reproducibility over time.

6. *Handle antibodies carefully.* More information about antibodies is in Box 29.7 in Appendix I of this chapter.

7. *Be sure your primary and secondary antibodies are compatible with one another.*

8. *Optimize your assay.*

9. *Prepare all buffers and reagents carefully and store them according to manufacturer's recommendations.*

10. *Check dilution calculations to be sure they are correct.*

11. *Include controls in every assay run.*

12. *Prepare a written protocol before you begin to work and follow it carefully.* It may be helpful to check off each step as it is performed.

13. *Document thoroughly, as previously described.*

Practice Problems

1. The antibody you receive from a manufacturer has a concentration of 1 mg/mL. It is aliquoted into 20 μL aliquots in the freezer. You want to use it at a concentration of 1 μg/mL, and you need 5 mL of diluted antibody.
   a. How much antibody and how much diluent are required?
   b. How many aliquots should be thawed?
   c. What should you do with any unused, thawed antibody?

2. The optimal dilutions of antibodies are best determined by first selecting a fixed incubation time and then making a series of dilutions and testing them to see which dilution works the best. This is sometimes called a "titration experiment." For example, if an antibody product data sheet suggests using a 1:1,000 dilution of their product for western blotting, you might experiment with dilutions of:
   A. 1/500
   B. 1/1,000
   C. 1/2,000
   D. 1/4,000
   Suggest a strategy to use serial dilutions of the antibody received from the manufacturer so that you will have at least 2 mL of each of these four dilutions for your titration experiment. Antibodies are expensive, so suggest a strategy that minimizes waste.

3. You are planning to titrate the primary antibody, and you need 1 mL for each run. You want the following dilutions:
   A. 1/100
   B. 1/1,000
   C. 1/5,000
   D. 1/25,000
   Suggest a strategy to use serial dilutions to make each of the above dilutions.

4. An ELISA is being used to determine the level of parathyroid hormone (PTH) in human patient blood samples. There is a disorder where a person's parathyroid glands produce too much PTH, potentially causing debilitating effects such as bone loss and kidney malfunction. This disorder can be treated surgically by removing one or more of the parathyroid glands. It is critical that a person be diagnosed

correctly; a person whose hormone levels are normal should not be mistakenly subjected to surgery, and a person whose hormone levels are abnormal should not be left untreated. The table below shows a diagram of a 96-well ELISA plate that is set up to conduct a PTH ELISA of samples from three patients:

- Column 1 and column 2 contain the standards that are used to create the standard curve. Each standard was prepared in replicate (twice). Thus, the standards in column 1 and column 2 are of the same concentration.
- Columns 3 and 4 contain +(positive) control samples in replicate.
- Columns 5 and 6 contain patient samples, also in replicate.

SCENARIO 1

Practice Problem 4, Scenario 1. Arrangement of Standards, Samples, and Controls in a 96-Well Plate

	1	2	3	4	5	6	7	8	9	10	11	12
A	Standard 0 pg/mL	Standard 0 pg/mL	+ Control 1 30 pg/mL	+ Control 1 30 pg/mL	Patient 1	Patient 1 Replicate						
B	Standard 20 pg/mL	Standard 20 pg/mL	+ Control 2 60 pg/mL	+ Control 2 60 pg/mL	Patient 2	Patient 2 Replicate						
C	Standard 40 pg/mL	Standard 40 pg/mL	+ Control 3 100 pg/mL	+ Control 3 100 pg/mL	Patient 3	Patient 3 Replicate						
D	Standard 80 pg/mL	Standard 80 pg/mL										
E	Standard 100 pg/mL	Standard 100 pg/mL										
F	Standard 200 pg/mL	Standard 200 pg/mL										
G												
H												

The ELISA is performed, and the plate is put into the plate reader. The measured absorbance value for each well is shown here:

Practice Problem 4, Scenario 1. Plate Reader Values for the Standards, Samples, and Controls in the 96-Well Plate

	1	2	3	4	5	6	7	8	9	10	11	12
A	0.004	0.001	0.239	0.242	0.550	0.546						
B	0.161	0.153	0.481	0.479	1.344	1.400						
C	0.310	0.311	0.801	0.799	0.371	0.357						
D	0.623	0.626										
E	0.785	0.783										
F	1.569	1.566										
G												
H												

Your task is to analyze these results. To do so, perform the following tasks:

Step 1. *Construct a standard curve based on the values for the standards:*

- *Average the two plate reader values for each standard.*
- *Prepare a graph with the value of the standard on the X-axis and the average plate reader measurements on the Y-axis.*

- *Do the points appear to lie on a single line? If the points appear to form a line, connect them into a line. This is the standard curve.*
- *Be sure to label the axes and give the graph a title.*

Step 2. ***Evaluate the results for the positive and negative controls:***

- *Use the replicate standard with 0 pg/mL of PTH as the negative control. Average the replicate readings.*
- *Average the replicate values for the positive controls.*
- *Use the standard curve to determine the concentration of PTH in the + controls.*
- *The values for the controls should make sense; that is, the plate reader value for the 0 pg/mL should be zero. Note, however, that there is some noise in readings and the value may be close to, but not exactly, zero. Similarly, the values for the three + controls must lie in a particular range. Assume that makers of this ELISA kit specify that the value for the negative control must be in the range of +1 to −3 pg/mL. The value for the first + control must be in the range of 27–33 pg/mL, the second + control must be in the range of 57–63 pg/mL, and the third + control must be in the range of 97–103 pg/mL.*
- *If the positive and negative controls do not lie in the specified range, then DO NOT PROCEED with patient sample analysis. Consult your supervisor (or instructor).*

Step 3. ***If the + and − controls values are in the specified ranges, then evaluate the patient samples:***

- *Average the replicate values.*
- *Use the standard curve to determine the PTH value in each patient sample.*
- *Assume that it has been determined, based on previous analysis of many patients, that the normal range for PTH values in this ELISA should be between 23 and 90 pg/mL. Based on this information, evaluate whether each patient's values are in the normal range.*

SCENARIO 2

The tables below provide another example with more patient samples. Analyze these results as previously described.

Practice Problem 4, Scenario 2. Arrangement of Standards, Samples, and Controls in a 96-Well Plate

	1	2	3	4	5	6	7	8	9	10	11	12
A	Standard 0 pg/mL	Standard 0 pg/mL	+ Control 1 30 pg/mL	+ Control 1 30 pg/mL	Patient 4	Patient 4 Replicate						
B	Standard 30 pg/mL	Standard 30 pg/mL	+ Control 2 60 pg/mL	+ Control 2 60 pg/mL	Patient 5	Patient 5 Replicate						
C	Standard 50 pg/mL	Standard 50 pg/mL	+ Control 3 100 pg/mL	+ Control 3 100 pg/mL	Patient 6	Patient 6 Replicate						
D	Standard 80 pg/mL	Standard 80 pg/mL										
E	Standard 100 pg/mL	Standard 100 pg/mL										
F	Standard 200 pg/mL	Standard 200 pg/mL										
G												
H												

Practice Problem 4, Scenario 2. Plate Reader Values for the Standards, Samples, and Controls in the 96-Well Plate That Is Illustrated Above

	1	2	3	4	5	6	7	8	9	10	11	12
A	0.002	0.004	0.988	0.978	0.601	0.603						
B	0.901	0.906	1.505	1.500	1.450	1.440						
C	1.503	1.499	1.861	1.782	2.103	2.098						
D	2.403	2.402										
E	2.891	2.789										
F	2.931	3.102										
G												
H												

5. This is a protocol from a manufacturer (Leica) where three human protein targets, A, B, and C, are to be assayed simultaneously using ICC with immunofluorescent reporters. For such an experiment, there must be three different primary antibodies and three different secondary antibodies. Each of the three different secondary antibodies is conjugated to a different fluorescent reporter. Protein A is visualized with a reporter that produces green light, Protein B is visualized with a reporter that produces yellow light, and Protein C with a reporter that produces red light. The table below shows the species of animal in which the primary antibodies are manufactured.

   a. Fill in the row to name the secondary antibodies required. The secondary antibodies are all made in goat.

Target Protein	Protein A	Protein B	Protein C
Target species	Human	Human	Human
Primary antibody	Anti-Protein A	Anti-Protein B	Anti-Protein C
Species in which primary antibody was made	Mouse	Rabbit	Rat
Fill in the name of the secondary antibody			
Fluorochrome excitation/emission	490/525	556/573	650/665

   b. Referring to Table 21.1, what should be the approximate wavelength of light produced by each of the three reporters?

   c. In the column for Protein A, you see the numbers 490/525 for "Fluorochrome excitation/emission." What do these numbers mean?

6. You are going to follow the protocol below to perform a western blot. (Note that this is just an excerpt from a real protocol.) When you received the primary antibody, the data sheet said there was 0.2 mg in the vial at a concentration of 1.0 mg/mL. You aliquoted the contents into ten storage vials. Before use, you need to dilute the primary antibody.

   a. What is the diluent?

   b. How much diluent do you need? How much of the primary antibody do you need?

   c. How many aliquots do you need to remove from the freezer?

Protocol:

1. Following SDS–PAGE, transfer proteins onto blotting membrane according to the manufacturer's instructions.

2. Place blot into blocking solution for 2 hours at room temperature.

3. Rinse the blot briefly with wash buffer, and then add 10 mL of primary antibody diluted in the wash buffer to a concentration of 5 µg/mL.
4. Wash the blot extensively in wash buffer (3×10 minutes) with gentle agitation.
5. Add 10 mL of appropriate enzyme-conjugated secondary antibody diluted in wash buffer, and incubate for 1 hour at RT with gentle agitation.
6. Wash the membrane with gentle agitation.
7. Add appropriate enzyme substrate solution, and incubate as recommended by the manufacturer to visualize protein bands.

CHAPTER APPENDIX 29.1: TECHNICAL INFORMATION REGARDING IMMUNOASSAYS

This appendix contains additional technical information regarding immunoassays. Box 29.7 describes general methods for aliquoting, storing, handling, and diluting antibodies. Table 29.3 provides guidelines relating to blocking buffer. Table 29.4 provides

information relating to lysis buffers. Table 29.5 has considerations relating to ELISA procedures. Table 29.6 has information about chemical fixatives used for IHC. Table 29.7 has general information about permeabilization and antigen retrieval.

Technical literature from antibody manufacturers is an excellent source of information about procedural details. Multiple resources are suggested in the Unit Introduction.

BOX 29.7 ALIQUOTING, STORING, HANDLING, AND DILUTING ANTIBODIES

1. *If the antibody arrives frozen, thaw it, and place it on ice*. If it is liquid, immediately place it on ice. If necessary, storage at 4°C upon receipt of the antibody and before aliquoting is generally acceptable for 1–2 weeks unless the manufacturer's data sheet indicates otherwise.
2. *Gently vortex the thawed antibody solution and spin briefly (e.g., 10 or 20 seconds) at 5,000–10,000 ×g to collect all liquid into the bottom of the tube.*
3. *Check the total volume of the antibody solution and determine how many aliquots you can make.*
   a. Aliquots should be no smaller than 10 µL because the smaller the aliquot, the more the concentration is affected by evaporation and adsorption of the antibody onto the surface of the vial.
   b. The volume of each aliquot depends on how the antibody will be used.
   c. Typically, aliquots between 10 and 20 µL are appropriate. Use larger 20 µL aliquots for antibodies that are used with relatively low dilution, e.g., 1:100, and smaller volume aliquots for antibodies that will be highly diluted, e.g., 1:5,000.
4. *Transfer the desired volume into low-protein-binding microcentrifuge tubes.*
5. *Label the tubes and place in labeled storage box.*
6. *Consult the manufacturer's instructions for storage temperature.*
   a. Most sources recommend that antibodies are stored in –70°C or –20°C freezers that are not frost-free. (Frost-free freezers cycle between freezing and thawing, which should be avoided.)
   b. However, manufacturers may specify that enzyme-conjugated antibodies should not be frozen at all and should instead be stored at 4°C.
   c. Some protocols suggest adding the cryoprotectant, glycerol, to a final concentration of 50% to prevent freeze–thaw damage; glycerol will lower the freezing point to below –20°C. Storing solutions containing glycerol at –80°C is not advised since this is below the freezing point of glycerol. Be aware that glycerol can be contaminated with bacteria. If adding glycerol or any cryoprotectant, be certain it is sterile.

(Continued)

BOX 29.7 (*Continued*)   ALIQUOTING, STORING, HANDLING, AND DILUTING ANTIBODIES

7. *Antibodies conjugated to reporters, particularly fluorescent reporters, should be stored in dark tubes or wrapped in foil to avoid light exposure.*
8. *Be certain that tubes and boxes are completely labeled, including manufacturer's unique identifying information. Do not confuse tubes that come from different lots; different lots of antibody might be different, even if their catalog number is the same.*
9. *Save any accompanying literature or instructions in their proper place.*
10. *Sodium azide can be added to antibody preparations at a final concentration of 0.02% (w/v) to reduce microbial growth.*
    a. Many antibodies will already contain this preservative when shipped from the manufacturer, so be sure to check the manufacturer's data sheet before adding it.
    b. Note that sodium azide is toxic and should not be used if the antibodies are to be used with live cells or animals.
    c. Sodium azide interferes with some reporter enzyme assays.
    d. It is possible to remove sodium azide using dialysis or gel filtration (techniques that are discussed later in this text). The molecular weight of IgG is 150,000 daltons; the molecular weight of sodium azide is 65 daltons. This difference in size allows them to be separated.
11. *When you thaw an antibody, be certain to write the thaw date on the tube. Do not refreeze. Once thawed, antibodies should be kept cold and should be used as soon as possible.* Most sources recommend using thawed, diluted antibodies within one day. Antibodies may be stable longer at refrigerator temperature if they have not been diluted.
12. *Antibodies almost always must be diluted before use.*
    a. Manufacturers provide information to guide the dilution of the antibody, although it is often necessary to experiment to get the optimal dilution.
    b. The optimal working dilution of an antibody provides good sensitivity with a high signal-to-noise ratio. If the antibody is used at a concentration that is too high, it may cause high background and it is also wasteful. If the antibody is diluted too much, there will be a loss in sensitivity.
    c. The components of dilution buffer may need to be optimized. Manufacturers will usually provide information as to how to make dilution buffers.

TABLE 29.3

Considerations Relating to Blocking Buffer

1. *Basic principle of selecting a good blocking agent:* It is critical that the primary and secondary antibodies do not recognize the blocking agent and do not bind to it. If they do, then the blocking agent will be detected – exactly what we want to avoid.
2. *Common blocking agents:* **Dry milk** was one of the first blocking agents introduced and is still sometimes used because it is readily available and inexpensive. **Bovine serum albumin** (BSA) is another common, inexpensive blocking agent. However, some commercially available secondary antibodies, such as anti-bovine, anti-goat, and anti-sheep will react strongly with BSA or dry milk. Therefore, do not use BSA or dry milk for blocking when the secondary antibody was made to recognize cow, goat, or sheep IgG. Milk and BSA also might be unsuitable for assays where human serum samples are tested because humans sometimes develop antibodies against these products. In such cases, antibodies in patient serum can react with the blocking agent.
3. *Normal serum (5% v/v) from the host species of the labeled secondary antibody is a good blocking agent because the secondary antibody should not react with it.* For example, if the secondary antibody is a polyclonal antibody made in rabbits, then serum from rabbits would be a potential blocking agent. This is because rabbit antibodies should not recognize rabbit proteins; rabbit antibodies should only recognize "foreign" proteins.

(Continued)

TABLE 29.3 (*Continued*)
Considerations Relating to Blocking Buffer

4. *Never use a blocking agent derived from the host species of the* **primary** *antibody*. This is because a polyclonal secondary antibody will recognize proteins from the host species of the primary antibody. For example, suppose the primary antibody was made in a goat. The secondary antibody will typically be polyclonal and will recognize many goat proteins, including but not limited to the primary antibody. In this situation, consider what would happen if the blocking agent were derived from goat protein: The secondary antibody might recognize the blocking agent, causing nonspecific binding – again, exactly what we want to avoid!

5. *It is often necessary to experiment with several blocking agents and incubation conditions to find the one that reduces noise, but does not obscure the epitope of the target*. The concentration of the blocking agent and its incubation time need to be optimized to reduce noise as much as possible without affecting the signal from the target. Too much blocking agent can mask antibody–antigen interactions, while too little blocking agent will result in excessive background. Some sources warn that blocking overnight can result in poor immunoblots.

6. *Gentle, nonionic detergents in blocking buffer may help remove nonspecifically bound substances and enhance the effect of the blocking agent*.

7. *Manufacturers sell proprietary blocking buffers (where the exact ingredients are not disclosed)*. Ideally, manufacturers of such buffers guarantee that they are consistent from lot to lot. If you prepare your own blocking agents, be consistent in their preparation.

TABLE 29.4
Considerations Relating to Lysis Buffers

There are various lysis buffer formulations to match different types of samples. The usual ingredients include the following:

1. **Buffer.** Lysis buffers begin with an agent to buffer the pH. Tris buffer is very commonly used for this purpose.

2. **Detergents.**
   a. Most lysis buffers contain one or more detergents. (Detergents were introduced in Section 25.2.3.5.) Ionic detergents, such as sodium dodecyl sulfate (SDS), are strong detergents. Nonionic detergents, such as Triton X-100, are milder. When the target protein is found in the cytoplasm of cells and is not integrated into any membranes, a relatively mild lysis buffer is sufficient to extract the protein. However, if the target protein is inside a membrane-bound compartment, like the nucleus, or is bound to a membrane, like the plasma membrane, a stronger lysis buffer may be required.
   b. It is important to match the primary antibody and the lysis procedure. Some primary antibodies do not recognize their target protein when it is denatured. Harsh detergents can denature proteins, and so, if lysis requires an ionic detergent, the primary antibody must be able to recognize the denatured target protein. Antibody manufacturers should note on the antibody data sheet whether the antibody will recognize a denatured target.

3. **Salt.** The varied roles of salt ions in solutions were discussed in Chapter 25. Salt, often in the form of NaCl, is usually added to lysis buffers. Note that if the ion concentration is too high, the bands on a western blot may have a "smiley-face" appearance.

4. **Protease inhibitors.**
   a. Tissues and cells often contain large amounts of proteases, that is, enzymes that break down proteins. During lysis, these proteases are released. Therefore, protease inhibitors may be necessary to protect the target protein. Common protease inhibitors are PMSF (phenylmethylsulfonyl fluoride), aprotinin, leupeptin, pepstatin, and AEBSF-HCL (4-benzenesulfonyl fluoride hydrochloride). PMSF is highly effective and is the most popular choice for lysate preparation.
   b. Many proteases require a divalent metal ion to function, so a sequestering agent, such as EDTA, is also often used to inhibit protease activity.

5. **Reductant.** Reductants (such as dithiothreitol and beta-mercaptoethanol) may be added to stop the formation of unwanted disulfide bonds.

TABLE 29.5
Considerations Relating to ELISAs

1. *Avoid contaminating any components of the assay with proteases on skin or other laboratory contaminants.*
2. *Coating is the initial process of adsorbing antigen or antibody to the surface of the wells.*
   a. Proteins adsorb to plastic because of interactions between their amino acid side chains and the plastic surface. This process can be influenced by the incubation time, incubation temperature, the pH of the coating buffer, and the concentration of the coating agent.
   b. Typical coating conditions involve adding 50–100 μL of coating buffer containing antigen or antibody at a concentration of 1–10 μg/mL, and incubating overnight at 4°C or for 1–3 hours at 37°C.
   c. Coating requires optimization. If the concentration of antigen or antibody is too high, the wells may be oversaturated, which can inhibit antibody binding in later steps. Conversely, if the antigen or antibody concentration is too low, the assay sensitivity might be too low.
   d. Carbonate–bicarbonate buffer (0.2 M sodium carbonate–bicarbonate, pH 9.4) is commonly used. The high pH causes most proteins to have an overall negative charge which helps them adsorb to the plates; the plastics of the plates is positively charged. The high pH also helps to keep proteins solubilized in the buffer. There are other buffered solutions that are used, such as Tris-buffered saline (TBS) or phosphate-buffered saline (PBS) at physiological pH.
   e. The plates must remain moist throughout the incubation, and evaporation should be minimized.
3. *Use multiwell plates intended for ELISA.*
   a. These are usually flat-bottomed, polystyrene plates with high protein-binding capacity, batch-to-batch consistency, and good optical properties. However, if the reporter is fluorescent or chemiluminescent, then the base of the plate needs to be opaque – white and black plates are available.
   b. Avoid plates treated for tissue culture.
4. *The wells along the edges of a plate can experience different temperature and humidity conditions than the wells in the center of the plate.* Some analysts avoid using the wells along the edges of a plate. Alternatively, be certain that all reagents are equilibrated to the same temperature before use and seal the plate during incubation to avoid evaporation from the edges.
5. *When performing quantitative and semi-quantitative assays, be certain to work in the linear range of the assay.* It may be necessary to dilute samples and/or antibodies to be in this range.
6. *Use good pipetting technique to avoid imprecision and error.*
   a. Avoid splashing that can transfer materials from one well into another.
   b. Do not allow tips to touch the bottom or sides of the well to avoid removing bound sample.
   c. Avoid bubbles.
7. *Prepare the standards for a standard curve and controls in the same matrix as the sample whenever possible.* For example, if the target is a protein secreted from cells in culture, cell culture medium might be suitable for diluting standards and controls.
8. *Washing is critical to ensure that unbound materials are washed away at each step.*
   a. The two most commonly used wash buffers for ELISA are Tris-buffered saline and phosphate-buffered saline containing 0.05% (v/v) Tween 20. To wash a plate, the wells are filled and emptied either by aspiration or by inverting the plate and flicking off the solution. Generally, at least three washes of 5 minutes each are used after incubations. More washing, for example, six 5-minute washes, are recommended after incubation with the enzyme-conjugated antibody.
   b. Incomplete washing is a source of background noise; however, too much washing can reduce the sensitivity of the assay. Try adding more washes if background is high.
   c. When washing, fill each well with wash buffer at least as high as they were filled when coating.
9. *Analyze results as soon as the detection reaction is complete because signal can fade over time.* Be certain there are no fingerprints or liquid on the bottom of the plate before putting it into a plate reader.
10. *Use good labeling, documentation, and organizational practices to be certain every well receives the proper reagents and each step is clearly documented.*

Primary Source: Thermo Fisher Scientific technical literature, https://www.thermofisher.com/us/en/home/life-science/protein-biology/protein-biology-learning-center/protein-biology-resource-library/pierce-protein-methods/overview-elisa/factors-affecting-ELISA-signal.html.

TABLE 29.6
Considerations Relating to Fixatives

1. *Chemical fixation methods fall into one of the two classes: cross-linking reagents* (formaldehyde or paraformaldehyde) and *organic solvents* (methanol, ethanol, or acetone). Formaldehyde and paraformaldehyde fixatives cause cross-links to form between amino acids on adjacent proteins, thus stabilizing their structures. During fixation, proteins that are normally soluble are bound to insoluble proteins. Lipids, carbohydrates, and nucleic acids are trapped in a network of cross-linked proteins.

2. *Organic solvents preserve structures by precipitating proteins, thus stabilizing them.* Soluble molecules and lipids are poorly preserved with organic solvents, and so the cross-linking fixatives preserve cell structure better than organic solvents. However, organic solvents protect the antigenicity of some proteins better than the cross-linking preservatives. ("Protecting antigenicity" means that antibodies are able to find and bind their target proteins.)

3. *Formaldehyde and paraformaldehyde are toxic and should be used in a fume hood.*

4. *Antibody manufacturers may provide guidance on a fixation procedure that preserves the ability of their antibody to bind its target.* It may still be necessary to experiment with fixation reagents and procedures to optimize the preservation of various types of cells and tissues.

5. *A common fixation time for tissue is 18–24 hours; much shorter times are used for cells.*

6. *Under-fixation can result in no signal from the interior of tissue. Over-fixation can mask the target's epitope.*

7. *Samples with green fluorescent protein (which is commonly used for molecular biology studies) should not be fixed with organic solvents because they do not protect its structure and function.*

8. *Glutaraldehyde is a good fixative for the preservation of cellular structure and is commonly used for electron microscopy studies.* However, glutaraldehyde is strongly fluorescent and is seldom used for immunoassays involving light microscopy. If chromogenic reporters are used, glutaraldehyde may provide good structural details.

9. *Some target proteins are destroyed during routine fixation and paraffin embedding. As an alternative to chemical fixation, samples can be snap-frozen in liquid nitrogen.* Tissue preserved in this way can be sectioned on a special microtome, transferred to slides, and dried to preserve its morphology. Freezing does not preserve structure as well as chemical fixation, and sectioning frozen sections requires special equipment and skill. Chemical preservation methods are therefore more commonly used than freezing.

10. *Fixatives should be freshly diluted because older solutions may become fluorescent, contributing to background noise.*

TABLE 29.7
Considerations Relating to Permeabilization and Antigen Retrieval

As with fixation, the optimal permeabilization method varies depending on the target, the antibody, and the sample. Some general considerations are the following:

1. *Various detergents including Triton X-100, NP-40, saponin, and Tween 20 are used*; each one removes different molecules from cellular membranes to create variable sized pores.

2. *Triton X-100 will partially dissolve the nuclear membrane.* It is therefore a good detergent to use when access to nuclear targets is required.

3. *Triton X-100 can disrupt membrane proteins and should be avoided when the target is integrated in a membrane.*

4. *Tween 20 and saponin are mild detergents that create pores large enough for antibodies to pass through without totally dissolving membranes.* These detergents are suitable for immunoassays involving membrane-associated proteins, cytoplasmic proteins, and soluble nuclear proteins.

5. *A typical incubation time in detergent is 15–20 minutes, but the incubation time and detergent composition need to be optimized.*

6. *Ethanol or methanol are also sometimes used for permeabilization and can improve the signal obtained with certain targets, particularly those associated with organelles or the cytoskeleton.*

CHAPTER APPENDIX 29.2: JOURNALS PROVIDE GUIDANCE FOR REPORTING WESTERN BLOT RESULTS

The editors and reviewers of scientific journals have been participating in the conversation about reproducibility – or the lack thereof – and western blotting. For example, *The Journal of Biological Chemistry* (JBC) published the following statement that relates to semi-quantitative western blotting:

> *Quantitation of Western blots is not always required but it can be fraught with traps for the unwary investigator... It is not uncommon to "correct" Western blot signals for protein loading by normalizing to a second Western blot for a housekeeper protein, e.g. -actin, -tubulin... The problem with this approach is that a linear relationship between signal intensity and the mass or volume of sample loaded must be confirmed for every antigen. This is further complicated by the fact that some detection methods, in particular enhanced chemiluminescence using x-ray film, have a very restricted linear range, and careful attention to the experimental conditions is necessary to ensure linearity. It is typically better to normalize Western blots using total protein loading as the denominator... To avoid potential pitfalls and with a focus on improving transparency, the JBC strongly recommends that authors describe their methods used to quantify signal intensity, how the linearity of signal intensity*

with antigen loading was established, and how protein loading was normalized between lanes. We prefer that signal intensities are normalized to total protein by staining membranes with Coomassie Blue, Ponceau S, or other protein stains, and we strongly caution against the use of housekeeping proteins for normalization, unless there is a clear demonstration that expression of the housekeeping protein is unaffected by the experimental treatments.

(Fosang, Amanda J., and Roger J. Colbran. "Transparency Is the Key to Quality." *Journal of Biological Chemistry*, vol. 290, no. 50, 2015, pp. 29692–94. doi: 10.1074/jbc.e115.000002.)

These JBC guidelines relate to methods of normalization and ensuring linearity. There is another area of concern relating to western blots that JBC discusses in their guidelines to authors, that is, how results are displayed to ensure transparency. Transparent reporting is necessary to allow other researchers to understand the published data and reproduce (or fail to reproduce) studies involving western blotting. The interpretation of results from western blots can be impacted by the way that bands are printed, the contrast used in the images, and cutting and pasting together sections of images that alter the orientation of bands.

JBC Associate Editor Roger Colbran provides the following guidelines to help ensure transparency in reporting western blotting data:

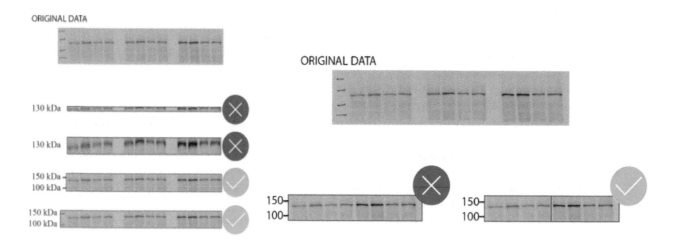

FIGURE 29.39 Examples of proper and improper methods of presenting western blot data. Items with a red X lose context and can be misinterpreted. Items with a green check mark are acceptable. (a) Methods to ensure that information about the molecular weight of bands is not obscured. (b) Where images are cut and pasted together, the splice must be clearly displayed. (Image courtesy of Elsevier.)

- "Blots should show full tonal range. A loss of tonal range is a loss of data." Raising contrast too much reduces the tonal range.
- "Crop immunoblots in a way that retains information about antigen size and antibody specificity."
- "Include positions of molecular weight markers above and below the band(s) of interest."

- "Avoid assembling figures of blots by splicing lanes from different sections of a gel. If blots must be spliced, borders must be clearly marked and explained in the figure legend."

Figure 29.39 shows images from JBC that demonstrate how western blot data should be presented to provide better transparency.

UNIT VIII

Cell Culture and Reproducibility

Chapters in This Unit

✦ Chapter 30: Introduction to Quality Practices for Cell Culture
✦ Chapter 31: Culture Media for Intact Cells

Every cell is a triumph of natural selection, and we're made of trillions of cells. Within us, is a little universe.

Carl Sagan

We saw in Chapter 1 that the dynamic "modern biotechnology" industry largely emerged from discoveries that made it possible to manufacture products – such as insulin and monoclonal antibodies – in cellular "factories." Biotechnology is thus rooted in an understanding of cell biology and the techniques required to work with cells.

The term **cell culture** *refers to maintaining any type of living cells in plates, flasks, vials, bioreactors, fermenters, or other vessels outside an organism.* (The term **tissue culture** *refers specifically to the in vitro propagation of cells derived from tissue of higher organisms.*) Cell culture is not a new discipline. In the late 1800s, Sydney Ringer was trying to understand the mysterious factors that caused a heart to beat. He invented a solution, called Ringer's solution, that allowed the heart of a frog to beat outside the body for a time so that he could study these factors. Over the years, other scientists have learned how to keep many kinds of isolated cells alive in dishes and flasks so that they could probe fundamental questions about life, for example: How do cells communicate? How do disease agents damage cells? What are the mechanisms by which drugs act on cells? What causes cells to become malignant? Cultured cells have long been important research tools for biologists.

DOI: 10.1201/9780429282799-37

Case Study: Cancer Research Depends on Cultured Cells

Cultured cells have been used in cancer research since the early 1900s, when successful tissue culture techniques were first developed. Since then, these techniques have advanced tremendously, allowing scientists to study both immortal cancer cell lines and shorter-lived cancer cells taken directly from tumors. Tissue culture provides a means to study the factors that make cells cancerous, as well as those that can be used to fight cancer. Novel cancer treatment drugs are frequently discovered from their ability to kill cancer cells in a dish. Further testing generally includes toxicity studies in cultured cells, avoiding safety issues in human testing and reducing the need to use experimental animal models.

Cultured cancer cells have long been used in basic research to determine why these cells behave as they do. In particular, studies of how cancer cells spread have yielded fresh insights into cancer **metastases** (*tumors that form in distant parts of the body from the original tumor*; these are the leading cause of death from cancer). In recent years, new techniques have been used to create three-dimensional cancer cell cultures, essentially tumors in a dish. These cultures could theoretically be used to develop personalized cancer treatments, by culturing the tumor cells of individual patients and studying their behavior and response to cancer treatments.

As we also saw in Chapter 2, new methods, such as those involving stem cells and gene editing, have led to many advances in regenerative medicine. Regenerative medicine uses cells themselves as therapies. CAR-T cells are already saving the lives of cancer patients. Hundreds of clinical trials are in progress using stem cells to regenerate tissue damaged by injury and disease. Thus, cultured cells have three main applications in biotechnology, as summarized in Figure 1.

It is beyond the scope of this text to introduce basic cell biology, or to comprehensively cover methods of cell culture. We therefore provide some useful resources in the Bibliography below. What we will discuss relates to the primary theme of this text; that is, we will introduce the fundamental principles and practices that must be understood in order to achieve reproducible, quality results in cell culture.

In Chapter 5, we noted that problems in cell culture practices have been identified as having a significant role in the "irreproducibility crisis." In 2005, an international task force published a "Guidance on Good Cell Culture Practice (GCCP)." (See the Bibliography below for the complete reference.) The 2005 GCCP document predates much of the discussion about irreproducibility, yet it anticipates concerns that have arisen about the reproducibility of cell culture work. To address concerns about cell culture, the GCCP document introduces a framework for quality based on six principles. Paraphrased, they are as follows:

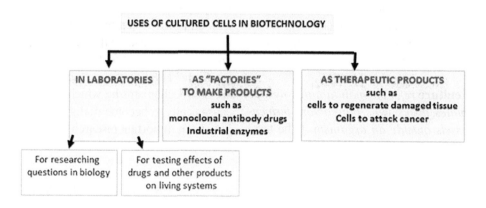

FIGURE 1   Three major uses of cultured cells in biotechnology. Cultured cells are used as laboratory tools to study fundamental questions and to test products. They can be used as "factories" to make products, particularly proteins. Cultured cells can also be used as therapies in regenerative medicine and cancer treatments.

Principle 1. It is necessary to understand the cell culture system in use and the factors that can affect it.

Principle 2. It is necessary to assure the quality of all materials and methods used in order to achieve valid and reproducible results.

Principle 3. It is necessary to maintain documentation sufficient to track the materials and methods used in order to permit the repetition of the work, and to enable the target audience to understand and evaluate the work.

Principle 4. It is necessary to establish and maintain adequate measures to protect individuals and the environment from any potential hazards.

Principle 5. It is necessary to comply with relevant laws and regulations, and with ethical principles.

Principle 6. It is necessary to provide relevant and adequate education and training for all personnel, to promote high-quality work and safety.

We discuss Principles 3, 4, and 5 in other chapters of this textbook, particularly Chapters 6, 10, and 36, respectively. In this unit, we address primarily Principle 2.

Chapter 30 discusses issues relating to the cells themselves, their authentication, and their handling.

Chapter 31 delves into the culture media that are essential to support the growth and function of cells. Note that Chapter 31 continues our discussion of biological solutions that began in Unit VI.

BIBLIOGRAPHY

GENERAL PRINCIPLES AND METHODS OF CELL CULTURE

ATCC has very helpful guides for a number of cell culture applications, from animal cell culture to stem cell culture to microbiology. The guides can be found on their website: https://www.atcc.org/en/Guides.aspx

Coecke, Sandra et al. "Guidance on Good Cell Culture Practice: A Report of the Second ECVAM Task Force on Good Cell Culture Practice." *ATLA*, vol. 33, no. 3, 2005, pp. 261–287. http://www.esactuk.org.uk/GCCPReportfinalvers.pdf

The Culture Collections of Public Health England have an excellent manual on Fundamental Techniques in Cell Culture, and a DVD Guide to Successful Cell Culture. These can be found at: https://www.phe-culturecollections.org.uk/promotions/labhandbook.aspx

https://www.phe-culturecollections.org.uk/promotions/cellculturedvd.aspx

Freshney, R. Ian. *Culture of Animal Cells: A Manual of Basic Technique and Specialized Applications,* 7th ed. Wiley-Blackwell, 2016. (Freshney's text is widely considered to be the most comprehensive and authoritative guide to animal cell culture.)

Geraghty, R. J., et al. "Guidelines for the Use of Cell Lines in Biomedical Research." *British Journal of Cancer*, vol. 111, no. 6, 2014, pp. 1021–46. doi:10.1038/bjc.2014.166. (Good review article by many of the prominent cell culture experts.)

Updates to the Guidance on Good Cell Culture Practice:

Pamies, David, et al. "Advanced Good Cell Culture Practice for Human Primary, Stem Cell-Derived and Organoid Models as well as Microphysiological Systems." *ALTEX*, vol. 35, no.3, 2018, pp. 353–378. doi:10.14573/altex.1710081

Ryan, John. *A Guide to Understanding and Managing Cell Culture Contamination.* Available from the Corning Life Science website, www.corning.com. (The 2002 version was authored by John Ryan. No author is provided for the 2017 version, but it is largely based on Ryan's work. This is an excellent and clearly written introduction.)

Sigma Aldrich. *Fundamental Techniques in Cell Culture: Laboratory Handbook*, 3rd edition. 2016. https://www.sigmaaldrich.com/content/dam/sigma-aldrich/docs/Sigma-Aldrich/General_Information/1/fundamental-techniques-in-cell-culture.pdf

Thermo Fisher Scientific. "Introduction to Cell Culture." https://www.thermofisher.com/us/en/home/references/gibco-cell-culture-basics/introduction-to-cell-culture.html

Yao, Tatsuma, and Asayama, Yuta. "Animal-Cell Culture Media: History, Characteristics, and Current Issues." *Reproductive Medicine and Biology*, vol. 16, no. 2, 2017, pp. 99–117. doi:10.1002/rmb2.12024.

AUTHENTICATION OF CELL LINES

Almeida, Jamie L., et al. "Standards for Cell Line Authentication and Beyond." *PLoS Biology*, vol. 14, no. 6, 2016, p. e1002476. doi:10.1371/journal.pbio.1002476.

Biocompare. "The importance of Cell Line Authentication, BioCompare Reproducibility: Issues in Life Science Research." (A documentary.) https://www.biocompare.com/Reproducibility/Cell-Line-Authentication/

Dunham, J.H. and Guthmiller, P. "Doing Good Science: Authenticating Cell Line Identity." https://www.promega.com/resources/pubhub/cell-line-authentication-with-strs-2012-update/ Updated 2012.

International Cell Line Authentication Committee (ICLAC). "Guide to Human Cell Line Authentication." 9 January 2014. https://iclac.org/resources/human-cell-line-authentication/

Reid, Y., et al. "Authentication of Human Cell Lines by STR DNA Profiling Analysis." In: Sittampalam, G.S., et al. editors. *Assay Guidance Manual* [Internet]. Eli Lilly &

Company and the National Center for Advancing Translational Sciences, 2004. Available from: https://www.ncbi.nlm.nih.gov/books/NBK144066/.

MYCOPLASMA

Journal Current Protocols. "Ask the Experts: How Mycoplasma Contamination Affects Results and Reproducibility." (Webinar.) March 19, 2020. https://www.workcast. com/register?cpak=9956658816964103. (A good overview of many issues relating to Mycoplasma contamination presented by experts in the field.)

Nikfarjam, L. and Farzaneh, P. "Prevention and detection of Mycoplasma contamination in cell culture." *Cell J.*, vol. 13, no. 4, 2012, pp. 203–212.

30 Introduction to Quality Practices for Cell Culture

30.1 INTRODUCTION

30.1.1 OVERVIEW

Many applications in biotechnology require cultured, living cells, that is, cells maintained inside plates, flasks, vials, dishes, bioreactors, or fermenters. Molecular biologists, for example, routinely grow bacterial cells for genetic transformation procedures. Researchers use cultured mammalian cells to study questions, for example: How do cells communicate? How do disease agents damage cells? What are the mechanisms by which drugs act on cells? What causes cells to become malignant? Analysts use cultured cells in pharmaceutical testing, for example, to study the effects and possible toxicity of new drugs, cosmetics, and chemicals. At a large scale, genetically modified cells are used to manufacture biopharmaceuticals and industrial enzymes. With the recent advances in stem cell technologies and regenerative medicine, the potential to use human and animal tissues grown outside the body for therapeutic purposes is moving from concept into reality.

In all these cell culture applications, the cells need to survive, grow, and reproduce while in an artificial, aqueous, "culture" environment. Additionally, humans must be able to evaluate the cells in various ways to be certain the cells are doing what we want them to do, and that they meet our quality requirements. Cell culturists must therefore have skills to accomplish the following:

- Providing a suitable environment for the cells that includes all the cells' environmental and nutritional requirements.
- Assessing cell growth and reproduction.

DOI: 10.1201/9780429282799-38

- Avoiding contamination of their cultures with external organisms from the environment.
- Characterizing the cells.

This chapter focuses on the last three bulleted requirements, and Chapter 31 discusses methods of providing a suitable aqueous environment for cells.

30.1.2 SIMILARITIES AND DIFFERENCES BETWEEN PROKARYOTES AND EUKARYOTES

There are two broad categories of cells that are cultured in biotechnology: prokaryotes and eukaryotes. Both prokaryotes and eukaryotes have these features: a plasma membrane that separates the cell's interior from the outside environment; DNA, the genetic information of the cell; ribosomes, to synthesize proteins; and a cytoplasm within which is found all the other cellular components. In **eukaryotic cells,** *the DNA is enclosed inside a membrane; i.e., there is a nucleus.* In contrast, in **prokaryotic cells,** *the DNA is not enclosed by a membrane, nor are there other membrane-enclosed organelles.* Bacteria are the most familiar type of prokaryote. The category of eukaryotes includes animal cells, plants cells, fungi, and protists (generally small, single-celled organisms). Both prokaryotic and eukaryotic cells are used for research and testing, and as "factories" to make products. The eukaryotic cells that are used as "factories" are usually (though not always) derived from mammals, such as hamsters or humans. Therefore, in this chapter, we limit our discussion of eukaryotic cells to those of mammalian origin. Cells that are used for therapeutic purposes generally are derived from a human source.

Bacterial cells are unicellular; each cell normally survives alone in the environment. Also, bacterial cells evolved to function in many different environments where they often experience fluctuations of temperature, nutrient availability, moisture, and so on. In contrast, mammalian cells evolved to survive in a multicellular body in a relatively constant environment. Mammalian cells in the body are bathed by nutrients; oxygen is brought to them via the bloodstream; waste products are removed; and temperature and pH are constant. These differences between the normal environments of bacterial and mammalian cells lead to some important differences in their requirements when they are cultured. The physical and nutritional demands of mammalian cells are far more stringent than those of bacteria. (There are bacteria that are difficult to maintain in culture, but they are not ordinarily

used in biotechnology applications.) Bacterial cells, for example, can often remain viable for days when stored in a refrigerator or incubator, or at room temperature. Mammalian cultured cells are much less tolerant. Mammalian cells must either be carefully stored – suspended in a frozen state in liquid nitrogen – or they must be growing and dividing in a controlled culture environment. If mammalian cells are not maintained in the exponential phase of growth (described below), then they can lose viability; there can be changes in their genetic material; and their morphology and metabolism can change. This means that mammalian cells in culture must be provided with fresh nutrients and living space on a regular basis, typically once every few days or even every day. Moreover, mammalian cells must be housed in a cell culture incubator, a specialized chamber that provides them with a consistent and suitable temperature, pH, and humidity level. Thus, because of the significant differences in the biology of prokaryotic and eukaryotic cells, there are important differences in how they are cultured in the laboratory. Therefore, in certain sections of this unit, we consider bacteria and mammalian cells separately.

30.1.3 BASIC PRINCIPLES OF MAMMALIAN CELL CULTURE

Cultured mammalian cells are derived from such sources as monkeys, mice, hamsters, and humans. Human cells can be obtained from biopsy specimens or from blood. **Primary cultures** *are those derived directly from tissue removed from a human or other animals.* The tissue is cut into small fragments and/or broken apart with enzymes and placed in sterile culture medium in a culture plate. The resulting cells are then coaxed into growing and dividing in this artificial environment. Primary cells are particularly valuable research tools because they retain the structural and functional characteristics of their tissue of origin. For example, the normal liver has a variety of important metabolic roles in the body and primary cells from the liver can reasonably be expected to retain specific liver enzymes. Primary cells, however, can only divide a limited number of times. As they grow and divide in culture, the cells' structural and morphological characteristics change, and they eventually die.

Some cells can be **transformed;** *these cells undergo genetic changes that change their growth properties.* Transformed cells acquire the ability to grow and divide indefinitely in culture. Transformation occasionally occurs spontaneously, but is more commonly induced by exposing cells to certain chemicals, viruses, or radiation. A **continuous cell line** *consists of cells from a single source that have been transformed and can be maintained indefinitely in culture.* Continuous cell lines are used in research and production, even though transformed cells have lost many of the morphological and functional properties of normal cells. Chinese hamster ovary (CHO) cells – the most common type of cells used for producing biopharmaceutical products – are a transformed, continuous cell line derived from the ovary cells of a hamster.

Mammalian cells are often **anchorage-dependent,** *which means that they require an appropriate solid surface on which to adhere.* This makes sense because mammalian cells normally grow inside the body where they are anchored in a particular location in a tissue. Anchorage-dependent cells normally grow as a **monolayer of adherent cells** *that are attached to the surface of their vessel.* (Monolayer means the cells do not grow on top of one another.) Primary cultures derived from solid tissue typically grow as a monolayer. Given suitable conditions, anchorage-dependent cells will multiply in culture until they reach **confluence**, that is, *completely cover the surface of their vessel.* Figure 30.1 shows cells that are about 90% confluent.

Anchorage-dependent cells are grown in vessels (usually made of disposable plastic) that are chemically treated to promote cell adherence. Tissue culture dishes (the term "tissue culture" refers to cells from "higher" organisms, not bacterial cells) look much like bacterial culture plates (Petri dishes), but anchorage-dependent cells will not adhere to, and grow in, untreated Petri dishes. Bacteria can be cultured in

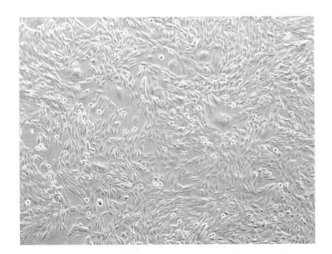

FIGURE 30.1 A plate of CHO cells that is about 90% confluent. This image was taken with the 10X objective in place. The cells were evenly dispersed on this plate and most fields of view looked similar.

tissue culture dishes, but because of the special interior coating, these dishes are much more expensive than bacteria Petri plates. In laboratories where both bacterial and tissue culture are practiced, it is important to check the labels and use treated dishes for eukaryotic cells and Petri dishes for bacterial cells.

In addition to dishes, there are a variety of other culture containers developed for mammalian cell lines; examples are shown in Figure 30.2. Tissue culture vessels are available with a variety of coatings to enhance adherence of specific cell types.

There are also mammalian cell types that grow **in suspension** *as single cells or as small clumps of cells floating in their culture medium.* Cells derived from blood, for example, tend to grow in suspension. The detailed methods of handling suspended cells and anchorage-dependent cells are somewhat different, but the same general principles apply to both.

We previously stated that, once removed from storage in liquid nitrogen, cultured mammalian cells must be maintained in the exponential phase of growth. Cell culturists must therefore provide the conditions required by exponentially growing cells. As cells grow and divide, they deplete the nutrients in their medium. Also, cell culture medium is buffered, but the buffering capacity can be exceeded if cells remain in the same medium for too long due to accumulation of acidic wastes. Therefore, to keep cells in the exponential phase of growth, *spent culture medium must be replaced with fresh medium;* this is called **"feeding the cells."** Feeding cells is a simple process that involves removing the old medium and replacing it with fresh medium. Many common cell types are fed every two or three days, although they may need to be fed sooner if the cells are approaching confluence or a pH change is observed.

If treated properly, cells will divide and multiply in culture and will eventually run out of space in their culture vessel. They must then be **subcultured** or **passaged,** *that is, divided by dilution and transferred into more plates with fresh culture medium* (Figure 30.3). Suspended cells can simply be diluted with culture medium and placed into more culture vessels. Adherent cells must be gently removed from the surface of their culture plate and then be *divided into two or more fresh vessels;* this is called **"splitting the cells."** Figure 30.4 summarizes some of the preceding discussion about deriving and working with a cultured cell line.

As you can see from the preceding discussion, cultured mammalian cells are manipulated in a variety of ways to keep them growing and dividing. They are placed in culture medium, put into and taken out of a cell culture incubator, visualized microscopically, fed, counted, passaged,

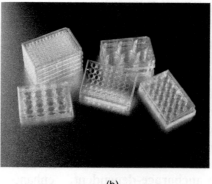

(a) (b) (c)

FIGURE 30.2 Examples of mammalian cell culture containers for laboratories. (a) Stackable plastic flask. (Credit:Jens Goepfert/Shutterstock.com) (b) Microplates with 6–96 wells. Six wells are shown here; the layout of a 96-well plate is shown in Figure 29.15. (Credit: Jens Goepfert/Shutterstock.com) (c) Roller bottles. To maximize the number of cells that can be grown in a single vessel, roller bottles of various sizes can be inoculated with cells in a relatively small volume of medium. The bottles are then incubated on their sides in an apparatus that slowly rotates the bottles to bathe the entire interior surface of the bottle with medium. This creates a much larger surface area for growth than using only the bottom of a vessel. ((b) and (c) used with permission, copyright 2012 Corning, Inc.)

and split. Continuous cell lines are divided into vials, frozen, and stored in liquid nitrogen. Subsequently, vials are removed from liquid nitrogen storage; the cells are thawed and placed in culture medium, and the process begins again (Figure 30.5). All these manipulations provide opportunities for undesired contamination to occur. Therefore, understanding contamination is essential for anyone working with cultured cells and we will consider this in detail later in this chapter.

30.1.4 Growth and Reproduction of Cells

30.1.4.1 Bacterial Cells in Liquid Culture

Cells grow and reproduce in culture; this is true of both mammalian and bacterial cells. Let's first consider bacterial cells. A bacterial cell divides to become two daughter cells, these two daughter cells divide to become four cells, and so on. This is called exponential growth and is discussed mathematically in Section 13.2.1.

As cultured cells grow and divide, they metabolize the nutrients in their medium and produce waste. If nothing is added to the culture after it is inoculated and nothing is removed from it, then four stages of growth will occur in a bacterial cell culture (Figure 30.6). In this figure, the $\log_{10}$ of cell concentration has been plotted on the y-axis with time on the x-axis. Observe the following four phases:

- **Lag phase.** *Immediately after the bacteria are inoculated into fresh medium, there is a phase with little cell division while the cells adapt.*

- **Exponential phase.** *A period of rapid growth follows the lag phase.* During this phase, cells are utilizing nutrients and excreting waste at maximum rates. This phase appears linear on the graph because cells *are doubling at a fixed interval, called the* **population doubling time**. You can determine the population doubling time by noting the amount of time it takes for a culture to double during the exponential phase (Figures 30.7 and 30.8).

- **Stationary phase.** *Growth levels off as a result of nutrient depletion and a buildup of waste in the medium.* Cells remain metabolically active, but they double at an increasingly longer time interval. Eventually, no net cell division occurs. On the growth curve graph, this phase appears as a plateau.

- **Death phase.** *Eventually, the accumulation of waste and depletion of nutrients leads to a phase where more cells die than reproduce, so the population declines.* This phase is not shown in Figures 30.6 and 30.7 because the cells will experience neither high enough levels of waste, nor low enough concentrations of nutrients, to enter this phase during normal culturing activities.

Let's consider how we might create a graph like the one in Figure 30.6 for a given bacterial culture. Clearly, it would require some method of counting the number of cells that are present at various time points. But bacterial cells are very small and difficult to see. They can be observed with a good light

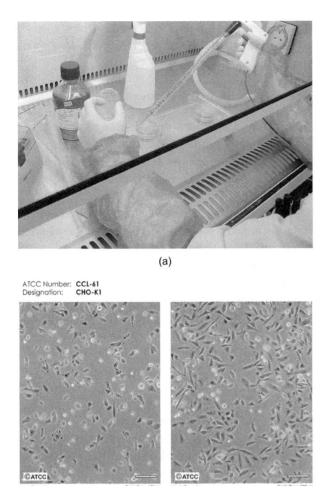

(a)

ATCC Number: **CCL-61**
Designation: **CHO-K1**

Low Density   Scale Bar = 100µm   High Density   Scale Bar = 100µm

(b)

FIGURE 30.3 Subculturing cells. (a) A technologist is shown pipetting fresh culture medium from the bottle on the left into small culture plates containing mammalian cells. The technologist is using aseptic procedures inside a laminar flow hood that helps protect the cultures from airborne contaminants. The technologist is wearing disposable sleeve protectors and gloves to further protect the cells from contaminants on his skin and lab coat. (b) Cultured mammalian cells as seen under a microscope. The left side of the image shows the cells at low density and the right side at higher density. (Images courtesy of ATCC.)

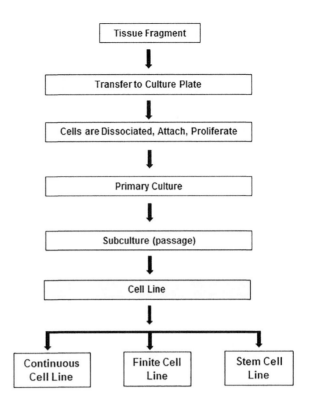

FIGURE 30.4 Mammalian cell culture flowchart. Mammalian cell culture begins with a fragment of tissue, usually from an animal or human biopsy. The fragment is transferred to a plastic plate, dissociated, and individual cells adhere to the plate surface. This is the primary culture that is expected to maintain the key characteristics of the original tissue and is likely to contain heterogeneous populations of cells. Under appropriate conditions, the cells grow and multiply to become a cell line with reduced heterogeneity. Eventually, the cells outgrow their plate and require passage into new culture vessels. Most cell lines are finite and will not divide indefinitely. But some cell lines are able to be subcultured indefinitely, forming a continuous cell line. Stem cell lines, as introduced in Chapter 2, are a special type of cell line with particular culture requirements. Stem cells are able to differentiate into multiple cell types. (Figure is adapted from Coecke, Sandra et al. "Guidance on Good Cell Culture Practice: A Report of the Second ECVAM Task Force on Good Cell Culture Practice." *ATLA,* vol. 33, no. 3, 2005, pp. 261–287.)

microscope, especially if the culture is diluted and special staining techniques are used to increase the cells' contrast. Nonetheless, it would be time-consuming and difficult to actually count the number of bacterial cells in a culture using a microscope. There are therefore two indirect methods used commonly to count the number of bacteria in a culture. One method is to determine the number of colony-forming units per milliliter (CFU/mL) of culture medium. This method is described in Section 26.3.1. The other

approach is to use a spectrophotometric method. The spectrophotometric method is based on the fact that the more bacteria present in a culture, the more turbid the culture will be. In a spectrophotometer, sample turbidity causes light to be scattered away from the detector and thus increases absorbance readings. The more bacteria present in a sample, the higher its apparent absorbance. The steps involved in counting

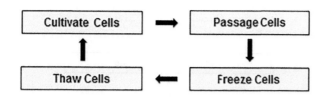

FIGURE 30.5 Culturing continuous cell lines. Cells are stored frozen in vials in liquid nitrogen and thawed as needed.

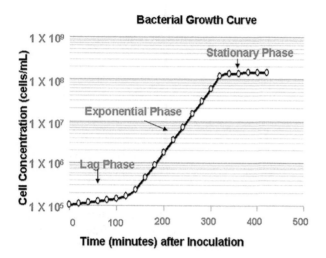

FIGURE 30.6 A typical bacterial growth curve. This graph was prepared on **semilog paper**, *a type of graph paper that substitutes for calculating logs*. The divisions on the y-axis are initially widely spaced and then become narrower as one moves toward powers of 10. This graph is on <u>semi</u>log paper where only the y-axis has logarithmic spacing; there are normal, linear subdivisions on the x-axis. The graph thus actually portrays the $\log_{10}$ (cell concentration) *versus* time. Exponential phase is the part of the growth curve that appears linear on this semilog plot.

bacterial cells and preparing a growth curve using a spectrophotometric method thus are as follows:

Step 1. Inoculate fresh nutrient medium with bacteria.

Step 2. Incubate the culture at a suitable temperature.

Step 3. Remove samples of culture at fixed time intervals.

Step 4. Measure sample absorbances.

Step 5. Convert absorbance values to cell concentration, based on a known constant that relates turbidity to cell concentration.

Step 6. Graph $\log_{10}$ (cell concentration) versus time.

Once a growth curve is prepared, it is possible to determine the doubling time for the cells, as shown in Figure 30.7.

30.1.4.2 Mammalian Cells

Mammalian cells follow the same stages of growth as bacterial cells (Figure 30.8). Mammalian cells experience a lag phase when they are first seeded into a new vessel with fresh medium. During this time, the cells are metabolically active, but the culture does not experience a net increase in cell number. Once the cells are accustomed to their new conditions, they begin to divide and soon enter a period of rapid, exponential growth. The cells will remain in the exponential stage for a number of generations, as long as they are properly maintained.

The population doubling time can be determined from the growth curve as shown in Figure 30.8. When working with mammalian cell lines, it is important to know the population doubling time so that you can predict when cells will be ready for a given manipulation. There are also many situations when culturists need a reasonable count of cell concentration. If, for example, the cells are to be used in an assay, the assay method will specify how many cells are required. In this case, it is necessary to count the cells before using them. Another time culturists count cells is when expanding

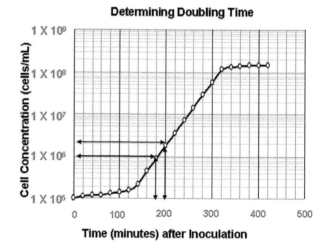

FIGURE 30.7 Determination of population doubling time. Two points are marked, each of them within the exponential phase, such that the cell concentration of one point is double that of the other point. The points chosen in this example correspond to cell concentrations of 1×10^6 and 2×10^6 cells/mL. On the x-axis, this corresponds to 20 minutes. Therefore, the population doubling time for this culture is 20 minutes.

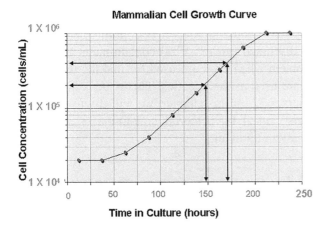

FIGURE 30.8 Mammalian cell growth curve and determination of doubling time. Like bacterial cells, these cells also go through lag phase, exponential phase, stationary phase, and then death phase (not shown). The cell concentration in this example doubles in about 24 hours from 2×10^5 to 4×10^5 cells/mL.

a culture to use the cells for production. Production technicians want to seed (put) the right number of cells into the production vessel. This is because mammalian cells are "social"; if too few cells are present, they do not thrive. If, however, too many cells are present, they use up the nutrient medium too quickly.

The method of preparing a mammalian growth curve follows the same general steps as for preparing a bacterial growth curve. As is the case for bacterial cells, an individual mammalian cell is too small to see with the naked eye, and tens of thousands of cells are likely to be present in each milliliter of growth medium. Mammalian cells, however, are larger than bacterial cells, and they can, with reasonable care, be diluted and counted using a microscope. It is not necessary to resort to indirect enumeration methods (using a spectrophotometer or colony counting method) as is commonly done with bacteria. Rather, it is common to count these cells using a **hemocytometer,** *a special glass microscope slide designed for this purpose.* The process and calculations involved in counting cells using a hemocytometer are found in the Appendix to this chapter. There are also automated cell counting instruments that facilitate cell counting and are commonly used in larger laboratories and industrial settings.

Compare Figures 30.7 and 30.8, the growth curves for a bacterial culture and a mammalian cell culture. Observe that the units on the x-axis of the bacterial growth curve are minutes, but the units on the x-axis of the mammalian cell growth curve are hours. This is because, in general, bacterial cells reproduce much more quickly than mammalian cells. For example,

the bacterium *Escherichia coli* (*E. coli*) can double its population every 20 minutes. In contrast, mammalian cell lines, such as the commonly used Chinese hamster ovary (CHO) cell line, require nearly a day to double in population. A practical result of this difference in growth rates is that mammalian cell cultures can be overrun by contaminating cells much more easily than bacterial cultures. This has many implications for handling cells. For example, with careful technique, it is possible to successfully manipulate bacterial cells on an open laboratory bench (meaning there is no enclosure or hood; this applies only to nonpathogenic species/strains). In contrast, any one of the vast number of contaminants floating around in the air can readily take over a mammalian cell culture. For this reason, mammalian cell culture is performed in special enclosures that are designed to exclude microbes. These manipulations will be described in more detail later in this chapter.

30.2 REPRODUCIBILITY REQUIRES AUTHENTICATION

30.2.1 OVERVIEW

As is true of all laboratory work, it is important to thoughtfully develop practices to assure good-quality, reproducible results. As we saw in Chapter 5, irreproducible research results are a matter of great concern to the biological sciences community and problems relating to cultured cells have been identified as one of the major culprits. We will consider why this is so in this section.

Authentication of cells *is the sum of the processes of testing the cells to be sure they are free of contamination by other cell lines, microbes, or other unwanted substances, and that they have the identity they are supposed to have.* Authentication of research reagents – including but not limited to cell lines – is an important part of ensuring research reproducibility.[1] Misidentified cell lines, and those with undetected contamination are a significant cause of irreproducibility. **Misidentified cells** *are those that are thought to be of one type, for example, kidney cells, but are really something else, for example cervical cancer cells.* Research that is performed with contaminated or misidentified cell lines is likely to be irreproducible and adds confusion to the scientific literature. We will therefore consider cell culture contamination and misidentification in more detail.

---

[1] The process of authenticating (validating) antibodies was discussed in Chapter 29.

30.2.2 Types of Contaminants: Chemical

There are chemical contaminants that might affect a cell culture. These include but are not limited to:

- Metal ions and other impurities from culture media and storage vessels.
- Leachates from plastics. (See Section 24.2.1.2 for more information.)
- Toxic substances caused by interactions between media components and light.
- Deposits and residues, such as detergents and disinfectants, on glassware and instruments.
- Impurities in the CO_2 gas used to maintain the pH in the incubator. (See Section 31.4.2.3 for more information.)
- Endotoxins. Endotoxins are by-products of certain bacteria that can cause a dangerous immune response in mammals. Endotoxins can adsorb to labware, thus potentially contaminating a culture. Endotoxins can also be found in water and culture media. It is possible to purchase certified nonpyrogenic (endotoxin-free) media and labware to reduce problems from these potential endotoxin sources.

As a general principle, scrupulous laboratory techniques, such as careful labware washing, using high-purity Type I water from a well-maintained water treatment system, and careful attention to the selection of labware and reagents will help avoid many chemical contaminants. (See Section 24.1.2 for an explanation of Type I water.)

30.2.3 Types of Contaminants: Biological

Cell culture media provide a hospitable environment for a number of unwanted intruders including:

- **Bacteria**
- **Mycoplasma** (which are a type of bacterium that is especially problematic in cell culture)
- **Fungi** (molds and yeasts, which are eukaryotic microorganisms)
- **Viruses**
- **Protozoans** (which are single-celled eukaryotic organisms; some are parasitic, and others are not)
- **Invertebrates** (which includes mites, insects, and spiders, common co-inhabitants of our homes and workplaces)
- **Cross-contaminants** (which lead to misidentification of cell lines).

Protozoans and invertebrates are possible, but less important contaminants of cell cultures and will not be discussed further. Viral contaminants are a major problem when cells are used for making therapeutics or are used as therapeutics. This issue is discussed in Chapter 37. In this chapter, we will discuss the problems caused by bacterial contaminants, fungi, and cross-contaminants.

30.2.4 Bacterial Contaminants in Mammalian Cell Culture

Bacterial cells are ubiquitous in the environment and on people's bodies. If a single bacterial cell falls into a culture of mammalian cells, that one bacterial cell can rapidly divide and take over the culture. Suppose, for example, that you transfer 10 mL of mammalian cell culture at a concentration of 10^4 cells/mL into a new flask. You are thus starting with a total of 10^5 mammalian cells. Your technique is nearly perfect – but you somehow also transfer a single bacterial cell. Just *one*. Now, assume that your bacterium is a strain that doubles every 20 minutes, while your mammalian cell line has a population doubling time of 18 hours. After 18 hours, your animal cell culture ideally will have doubled, leaving you with 2×10^5 cells. But what about that single bacterium? In the worst-case scenario, your contaminating bacterial population will have gone through 54 generation times. With infinite exposure to nutrients, oxygen, etc., you could now have 2^{54} bacteria present – or about 2×10^{16} cells! In reality, your bacterial population would have been limited by exhaustion of nutrients and buildup of waste. Still, the end result is a bacterial culture contaminated with mammalian cells – not the desired result at all. Figure 30.9 shows a mammalian cell culture plate that has been contaminated in just this way. This plate is worse than useless; it also has the potential to spread contamination to other plates.

It may be surprising to know that this is not the worst contamination scenario. Yes, this culture is worthless and must be discarded, but at least the problem is obvious. Unfortunately, contamination in cell culture, particularly mammalian cell culture, is often subtle and may go undetected for a long time. In a worst case, researchers have devoted years of work to their research, published what they thought were meaningful results, and later discovered that their cultures were in some way contaminated, thus making their work of dubious value. In a production setting, any products made with contaminated cells must be discarded. Consider how difficult and expensive this

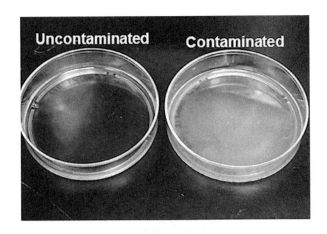

FIGURE 30.9 Cell culture contamination. The culture medium in an uncontaminated plate, left, is clear. An obviously contaminated cell culture plate on the right is cloudy due to the presence of many bacteria.

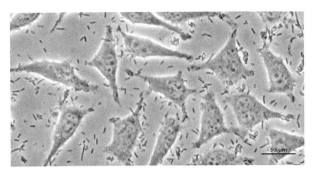

FIGURE 30.10 A plate of eukaryotic cells with significant bacterial contamination. The small, numerous, black, rod shaped objects are bacteria. (Image: Eppendorf AG, used with permission.)

would be if the product had already been distributed for use. Contamination is thus a major concern in cell culture, and every cell culturist must devote considerable thought and effort to avoiding it.

Bacterial contamination can often be detected macroscopically because it causes the medium to appear cloudy, as shown in Figure 30.9. Another symptom of bacterial contamination is a substantial drop in the pH of the medium caused by metabolic by-products of bacterial growth. The dye, phenol red, is frequently included in cell culture media as a pH indicator. Phenol red turns yellow if the culture medium becomes acidic, and so the color of the medium should be monitored as a method of detecting bacterial contamination.

Bacteria are much smaller than eukaryotic cells. They can sometimes be observed microscopically and appear as dark, rod-shaped structures, spheres, or spirals. Phase contrast microscopy is the best way to see bacteria, and a magnification of at least 100× is required. Sometimes bacteria can be observed microscopically to actively move around a culture plate, since many types are motile. Small dark moving objects are likely to be bacteria, but can also be harmless precipitates. If the number of such particles increases overnight, bacteria are likely the culprits. Figure 30.10 shows a picture taken with a microscope of a plate that is obviously contaminated by rod-shaped bacteria.

30.2.5 MYCOPLASMA

30.2.5.1 What is *Mycoplasma* Contamination?

Mycoplasma are a type of simple bacterium that lacks a cell wall and is small enough to pass through regular cell culture sterilization filters. (See Figure 10.13

for a size comparison.) There are more than 100 different species of *Mycoplasma*, but only five commonly contaminate cell cultures. These bacteria can be either intracellular (inside cells) or extracellular (outside cells) but in either case, they are not visible using conventional light microscopy, and they generally do not outgrow their host cells. This means that the cultures they contaminate do not show typical signs of bacterial contamination, such as pH changes, or turbidity in the medium. *Mycoplasma* are resilient to changes in pressure, temperature, and osmolarity and can survive in liquid nitrogen, which is the medium used to store mammalian cell cultures. Therefore, if a contaminated tube breaks open in a liquid nitrogen tank, *Mycoplasma* can contaminate all the cells stored in the tank. It is estimated that as many as 10%–20% of all cell lines have "silent" *Mycoplasma* infections. Some studies have found as many as 65% of cultures in particular settings to be contaminated. Figure 30.11 shows a cell with many *Mycoplasma* contaminants.

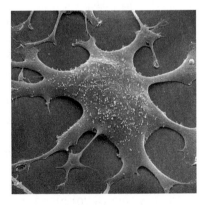

FIGURE 30.11 Scanning electron micrograph of *Mycoplasma* attached to a human fibroblast cell. In this image, the *Mycoplasma* are the small white objects sitting on the large cell. (Image credit: David M. Phillips/Science Source Images.)

This image was taken with a scanning electron microscope that provides much higher magnification than an ordinary light microscope.

Mycoplasma infections are of great concern in a laboratory or production facility because *Mycoplasma* can have wide-ranging effects on the expression of hundreds of genes in their host cells. Unusual proteins and metabolic products are often present in *Mycoplasma*-contaminated cultures, invalidating research results and possibly leading to retraction of published papers. For example, it has been suggested that some published studies where scientists intended to identify eukaryotic nucleases involved in the process of **apoptosis** (*programmed cell death*) have mistakenly identified *Mycoplasma* proteins instead. (Paddenberg R. et al. "Internucleosomal DNA fragmentation in cultured cells under conditions reported to induce apoptosis may be caused by mycoplasmal endonucleases. *European Journal of Cell Biology*," vol. 71, no.1, 1996, pp. 105–119.)

Mycoplasma is also a potential human pathogen, although a majority of people have *Mycoplasma* cells in their bodies without noticeable effects. Because *Mycoplasma* contamination is a serious problem in cell culture research, many journals are starting to require that authors demonstrate that they have tested their cultures and found them to be uncontaminated. In a production setting, *Mycoplasma* contamination can adversely affect the production of a desired protein and can have a significant economic impact. A facility that manufactures biopharmaceutical products in compliance with the Food and Drug Administration (FDA) regulations can be closed down for months to clean out bioreactors and decontaminate surfaces if *Mycoplasma* is found. (Ferreira, Meghaan M., "Cell Line Validation in Biomanufacturing," *Genetic Engineering and Biotechnology News*, vol. 38, no. 13, 2018.)

Mycoplasmal contamination can be a significant problem in the production of vaccines that originate in cultured cells. For this reason, many countries require that all vaccines be tested for the presence of *Mycoplasma*. One advantage of the current mRNA-based vaccines for COVID-19 is that they do not require cell cultures in their manufacture, eliminating this potential source of vaccine contamination.

Many new *Mycoplasma* infections are the result of contamination with previously infected cultures. Thus, if one cell line in a facility is found to be infected with *Mycoplasma*, it is imperative to check every culture. To avoid contamination, it is essential that different cell cultures be handled separately, with independent bottles of culture medium and sterile equipment. Regular *Mycoplasma* testing should be part of any tissue culture operation. Any new cell cultures brought into the laboratory should be quarantined and treated as though infected until proven otherwise. It is often recommended that research laboratories not accept "gifts" of cells from other laboratories because they might be contaminated. Table 30.1 outlines procedures that will help avoid introducing *Mycoplasma* into a culture.

TABLE 30.1

Avoiding *Mycoplasma* Contamination in a Laboratory

1. *Quarantine and test all new cell cultures.* Ideally maintain a designated incubator for this purpose. If possible, keep quarantined cells in a separate room.

2. *Perform regular **Mycoplasma** testing of all cultures in active use.* Methods of *Mycoplasma* testing are discussed below.

3. *Observe all routine aseptic technique strategies, wear clean personal protective equipment, and clean surfaces regularly.* Chapter 10 introduced methods of working in a biological safety cabinet, and basic principles of aseptic cell culture work are described later in this chapter.

4. *Work with only one culture in the hood at a time.* Wipe the hood surfaces with alcohol between cultures.

5. *Have each biological safety cabinet regularly certified by a qualified individual.* Cell culturists rely heavily on biological safety cabinets to exclude contaminants. These cabinets must have proper airflow to work correctly. (See also Section 10.2.2.)

6. *Limit talking when working in a cell culture cabinet.* Humans often carry *M. orale*, *M. fermentans*, and *M. hominis* in the mouth and throat. These species of *Mycoplasmas* account for more than half of all mycoplasma infections in cell cultures. (Nikfarjam L, Farzaneh P. "Prevention and detection of Mycoplasma contamination in cell culture." *Cell J.*, vol. 13, no. 4, 2012, pp. 203–212.) Talking both can impair the normal airflow in a cell culture cabinet and can release *Mycoplasma* into the air.

7. *Have a procedure for coughing and sneezing.* If you must cough or sneeze while performing cell culture work, push back from and turn away from the hood, cough or sneeze into the inside of your lab coat or into a tissue, change gloves, and be sure to frequently launder your lab coat.

30.2.5.2 Testing for *Mycoplasma*

In the United States, FDA regulations require that cells that are used for production of therapeutic products are *Mycoplasma*-tested regularly. Unfortunately, many research laboratories do not follow this practice, possibly because of cost or inconvenience. However, it is important to conduct routine *Mycoplasma* testing in all cell culture facilities because this contaminant is ubiquitous.

Mycoplasma testing can be accomplished in various ways, and there are now methods that are relatively fast, inexpensive, and effective. Two common strategies are as follows:

- ***Enzymatic-based assays.*** Relatively inexpensive and convenient enzymatic kits are commercially available for *Mycoplasma* detection. For example, there are kits that detect enzymes that are specifically released when *Mycoplasma* cells are lysed. To detect these enzymes, a sample of supernatant is removed from the culture and reagents are added to the sample. If *Mycoplasma* is present, its enzymes react with the reagents from the kit causing light to be emitted. The light is detected by an instrument called a **luminometer,** *which is similar to a spectrophotometer, but measures light emitted by a sample instead of light absorbed by a sample.* Results are available within 20 minutes, making this a fast and convenient assay method.
- ***PCR-based assay.*** PCR methods are somewhat more demanding and time-consuming than enzymatic assays, but are more sensitive. PCR-based assays can theoretically detect as few as 10 CFU/mL (a better sensitivity than is usually specified for enzymatic methods). PCR-based assays are recommended by global regulatory agencies for testing cells being used as therapeutics. Convenient PCR-based kits are available from manufacturers that detect multiple *Mycoplasma* species. (PCR-based assays were introduced in Chapter 27.)

30.2.5.3 If *Mycoplasma* is Found

Eliminating *Mycoplasma* from a contaminated culture can be difficult. There are antibiotics and other drugs that can be used to selectively kill *Mycoplasma* and are available from commercial suppliers. However, it can take weeks to clear a culture of contamination and during this time contamination of more cultures can occur. Therefore, one might choose to rescue an infected rare or special culture, such as a patient cell line that cannot be reproduced. However, if the culture is not essential, it is recommended to bleach (kill) and discard it. In addition to destroying or treating affected cultures, it is necessary to test all the other cultures in the facility if contamination is found in one. Reagents, such as culture media, may also need to be discarded to eliminate a possible source of the bacteria. Surfaces, water baths, and equipment should be cleaned. Clearly, finding *Mycoplasma* contamination can be stressful, but it is better to find it than to be unaware of its destructive effects.

30.2.6 Fungal Contaminants

Fungi are a diverse group of eukaryotic organisms that decompose and absorb organic matter. Mushrooms are a familiar type of fungus, but these are not cell culture contaminants. Yeasts are another type of fungus, which are valuable in biotechnology for their roles in food production. For example, yeasts are involved in the production of bread and wine. Yeasts, however, can also be unwelcome visitors in cell cultures. Contamination by yeasts can sometimes be detected by a cloudiness in the cell culture medium. At high magnification, yeasts can be visualized microscopically as ovoid, bright particles. They may float singly or may form chains and branches. Figure 30.12 shows mammalian cells surrounded by yeast particles.

Molds are another type of fungus. These have a filamentous structure and may be familiar to those of us who do not always consume the contents of our refrigerators promptly. The spores of molds are ubiquitous in the air, and they can fall into cell culture plates if careful aseptic technique is not followed. Mold spores can grow into colonies that may be visible floating on the surface of cell culture plates. The colonies may be gray, white, or greenish in color. Eventually, the medium may become cloudy and the pH of the medium may increase, causing the phenol red indicator to appear pink (Figure 30.13). Microscopically, it is possible to see the filamentous form of the mold colonies (Figure 30.14).

30.2.7 Cross-Contamination and Misidentified Cell Lines

Mycoplasma contamination is one of the major causes of irreproducibility in research that uses cultured cells. Misidentified cell lines are another critically important cause of irreproducibility. Misidentified cells are thought to be of one type, but are really something else. The primary way that cell lines become misidentified is through

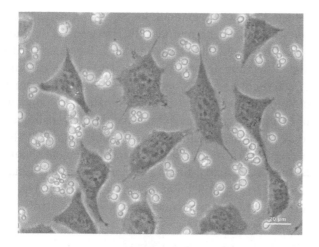

FIGURE 30.12 Microscopic examination of a cell culture contaminated by yeast. The yeast cells are the spherical, translucent objects. Observe that they are larger than bacterial contaminants. (Image: Eppendorf AG, used with permission.)

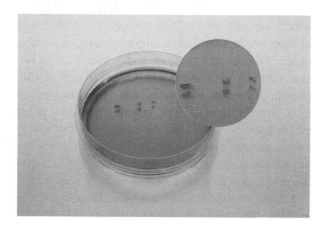

FIGURE 30.13 A cell culture plate with clearly evident mold contamination. (Image: Eppendorf AG, used with permission.)

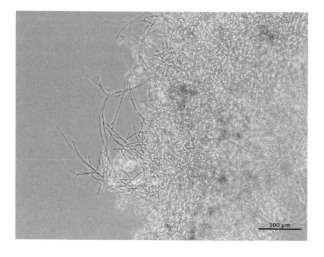

FIGURE 30.14 Microscopic view of mold contaminating a cell culture. (Image: Eppendorf AG, used with permission.)

cross-contamination, although erroneous labeling and errors in tissue identification by pathologists may account for some misidentifications. **Cross-contamination** *is where cells from one mammalian cell culture accidentally enter another mammalian cell culture.*

How does cross-contamination happen? Imagine that at some point in a cell culturist's work, one foreign cell is introduced into a previously pure cell culture. Perhaps the culturist makes this error by sharing a bottle of nutrient medium between two different cell line cultures, and a cell from one culture falls into the medium and is transferred to the other. If the contaminating cell divides more rapidly than the original cells, it can eventually dominate and change the identity of the culture. This problem can go undetected because many cell lines look the same microscopically and have similar patterns of growth. When those contaminated cells are shared with other labs – a common practice in the scientific community – that simple mistake escalates quickly.

Undetected cross-contamination of research cultures has led to years of research work being lost because researchers thought they were studying one type of cell, but in reality, were studying something else. In a production environment, cross-contaminated cells must be discarded at great expense. Another problem is that cross-contaminated cells can create unknown hazards for workers who are unaware of their exposure to contaminating organisms. This is a particular concern when there is cross-contamination with cells of human or primate origin because these might harbor pathogenic viruses (e.g., HIV or SARS).

A study in 2017 attempted to quantify the magnitude of the misidentification problem. The authors concluded: "We found 32,755 articles reporting on research with misidentified cells, in turn cited by an estimated half a million other papers." (Horbach, Serge P. J. M., and Halffman. Willem. "The Ghosts of HeLa: How Cell Line Misidentification Contaminates the Scientific Literature." *PLoS ONE*, vol. 12, no. 10, 2017, p. e0186281. doi:10.1371/journal.pone.0186281.) Others have tried to assess the financial impact of the problem. One expert estimated that $3.5 billion was spent on research that used just two of the many contaminated cell lines. (Neimark, Jill. "Line of Attack." *Science*, vol. 347, no. 6225, 2015, pp. 938–40. doi:10.1126/science.347.6225.938.) Obviously, misidentification is not a rare event and it has a great impact on the reliability and cost of biological research.

Misidentification is such a widespread problem in mammalian cell culture that the scientific community established the International Cell Line Authentication Committee (ICLAC) in 2012. The ICLAC website points out that every cell line has a story – the donor and the scientists who established it are part of that story. For some cell lines, we lose track of the story; the cell line becomes misidentified. ICLAC tracks cell lines that have been misidentified. In 2020, their registry listed more than 500 cell lines that were thought to be of one type, but were actually something else. The ICLAC website provides several case studies to illustrate this problem, one of which is summarized in the following case study, The "Notorious Chang Liver."

Case Study: The "Notorious Chang Liver"

The following case is modified from the ICLAC website (https://iclac.org/case-studies/chang-liver/).

In the early 1950s, Dr. R. Shihman Chang established a cell line derived from liver cells biopsied from a patient. He deposited these cells with ATCC for distribution to other scientists who needed a continuous liver cell line for research or testing. (ATCC is a global biological materials resource and standards organization that plays a major role in preserving and distributing biological resources, including cell lines.) In 1967, another researcher found evidence that the Chang liver cells that were being distributed were, in fact, HeLa cells. HeLa is an extremely aggressive and fast-growing cancer cell line that was isolated from a patient with cervical cancer. Advances in methods for identifying cell lines confirmed that the Chang "Liver" cell line is indeed HeLa; it consists of cervical, not liver cells. ATCC now labels this cell line as "HeLa [Chang Liver]" and warns users of the cross-contamination. Despite this warning, an ICLAC study found that between the years 2000 and the end of 2016, there were 352 articles published that inappropriately used the Chang Liver cells as a liver model. This is problematic; liver and cervical cells are not the same. For example, the liver is the primary organ where toxic materials that enter the body are detoxified by a group of liver-specific enzymes. Cervical cells do not perform this function. If a researcher thinks their research is looking at effects of liver enzymes, when in fact they are using cervical cells that do not make these enzymes, then their research is invalid.

30.2.8 Avoiding Cross-Contamination

There are a number of steps cell culturists can take to avoid cross-contamination and the use of misidentified cells. The most important is to only obtain cell lines from reliable sources that authenticate their cell lines and provide information about those cells. ATCC and other reputable suppliers test cells and provide authentication information for them. It is risky to accept cells directly from colleagues in other laboratories. If one does so, then they should be quarantined and tested before use.

Once a cell line has been introduced into a facility, it should be provided with its own bottles of culture medium and other reagents. Similarly, cell culturists should only handle one cell type at a time to avoid accidentally transferring a few cells from one culture to another. Keep in mind that it is possible for a single cell of one type (for example, a HeLa cell) to completely take over a culture of another type. Unfortunately, this event might go completely undetected unless the cells are routinely tested.

30.2.9 Methods of Testing Cultured Cells

30.2.9.1 Overview

Some journals have responded to the misidentification problem by requiring authors to submit data that show they have authenticated their cell lines. Beginning in 2016, the National Institutes of Health (NIH) Office of Extramural Research (OER) requires grant applicants to authenticate their cell lines (*Guidance: Rigor and Reproducibility in Grant Applications.* https://grants.nih.gov/policy/reproducibility/guidance.htm). Similarly, in 2015, the highly regarded family of journals, *Nature,* announced that they will require authors to confirm that they are not working on cells known to have been misidentified, and to provide details about the source and testing of their cell lines. At least 30 other journals have introduced similar requirements for authors who want to publish their work. Despite the fact that authenticating cell lines is recognized to be of critical importance, in 2017, ATCC reported that only one-third of laboratories test their cells for identity. Presumably, that percent is increasing as journals and grant funding agencies adopt more stringent policies and as awareness of the problem is growing.

30.2.9.2 Identity Testing of Human Cell Lines

One of the most common and widely accepted methods for authenticating cell lines is the use of short tandem repeat (STR) profiling (a form of DNA fingerprinting). STR profiling looks at regions in the genome where there are repeats of a short sequence of DNA, for example, GATA-GATA-GATA-... The number of these repeats varies from one individual to the next. In Chapter 2, we discussed how STR profiling is used in forensics, to see if a suspect was at a crime scene, and in ancestry testing, to construct family trees and to look for relatives. Similarly, STR profiles can be used to identify cell lines because each cell line has a characteristic STR profile, just as an individual person has his or her own profile.

There are regions in mammalian genomes where a short sequence of DNA is repeated over and over, for example, GATAGATAGATA. In this example, the repeated sequence is GATA and it is repeated three times. GATA is an example of a "short tandem repeat (STR)." Each cell has two copies of every chromosome (one inherited from the female parent and the other from the male parent), so there are two copies of each STR. For example, for the STR, GATA, an individual might have three repeats on the paternal chromosome and seven repeats on the maternal chromosome. In this situation, because the individual inherited two different versions of the repeat, we say the individual is heterozygous for that STR (Figure 30.15a). An individual can also be homozygous for a particular STR, in which case the number of repeats is the same on both chromosomes (Figure 30.15b).

There are many STRs in the genome, but the ones that are of use for identity testing are **polymorphic**, *which means that the number of repeats varies from one individual to the next.* For example, one individual might have GATA repeated four times on one chromosome and seven times on the other, while another individual might have GATA repeated eight times on one chromosome and six times on the other. Although STRs can be polymorphic, it is still possible for two people, or two mammalian cell lines, to share the same number of repeats for a given STR. Therefore, using only one STR, it is not possible to distinguish one person, or one cell line, from another. However, scientists

have calculated that if one tests at least eight different STRs, then there is a vanishingly low probability that two people, or two cell lines, will have the same profile of repeats; thus, the profile for a cell line based on eight STRs identifies that cell line. Note, however, that if two cell lines were created from one individual, for example, from a human liver biopsy and blood sample, then the profiles of both cell lines will be the same. As an aside, in forensic testing of people, more than eight STRs are used to be even more certain that a given profile matches only one person in the entire world (unless that person has an identical twin).

An example of an STR profile for a particular cell line is shown in tabular form in Table 30.2. The names of the STRs that are commonly used to authenticate cell lines are shown in the left column. The right column shows the number of repeats for each of the STRs for this particular cell line. For example, for an STR named TH01, this cell line is heterozygous with six repeats on one chromosome and nine on the other. Nine STRs were used to create this profile, plus amelogenin. Amelogenin is not an STR; it is a gene used to determine if the source of cells came from a male or a female source. This cell line came from a female; females are XX, and males are XY.

How do analysts determine the number of STRs at each locus? The answer is that they use a PCR-based method combined with capillary electrophoresis. (The basic method of PCR was introduced in Section 27.2, and capillary electrophoresis is a separation method that will be introduced in Section 34.3.4) In the particular version of PCR used for identity testing, the primers are labeled with a fluorescent tag. PCR amplifies each STR fragment until there are hundreds of thousands of each one present. Capillary electrophoresis is then used to separate the fragments from one another based on their size. For example, an STR that is repeated seven times is larger than the one that is repeated three times. The smaller fragment runs

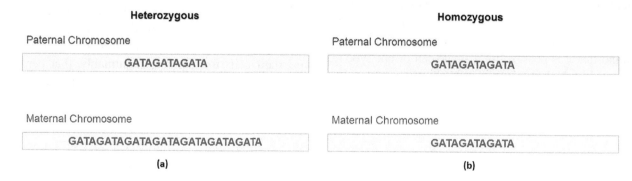

FIGURE 30.15 An STR. This particular STR has a repeat sequence of GATA. (a) This individual is heterozygous for this STR and inherited a different number of repeats from each parent. (b) This individual is homozygous for this STR.

TABLE 30.2

Example of an STR Profile for a Cell Line

STR Locus	Alleles (An Allele is a Version of a Gene Sequence)
TH01	6, 9
D21S11	29, 30
D5S818	11, 12
D13S317	11, 12
D7S820	10, 10
D16S539	11, 12
CSF1PO	11, 12
Amelogenin	X, X
vWA	17, 19
TPOX	8, 9

through the electrophoresis capillary more quickly than the larger fragment. In order to visualize the fragments, a laser light excites the fluorescent tags that were on the primers. A detector at the end of the electrophoresis capillary "sees" each fragment as it elutes out of the capillary. The result is called an **electropherogram** (Figures 30.16 and 30.17).

Figure 30.17 shows the electropherogram for the cell line shown in Table 30.2.

Figure 30.18 shows an example of a cell line that is contaminated with HeLa cells.

30.2.9.3 Authentication of Non-human Cell Lines

The situation for non-human cell lines is different than for human-derived lines. For human cell lines, convenient kits, reagents, databases, and commercial services exist to expedite STR cell line identity testing.

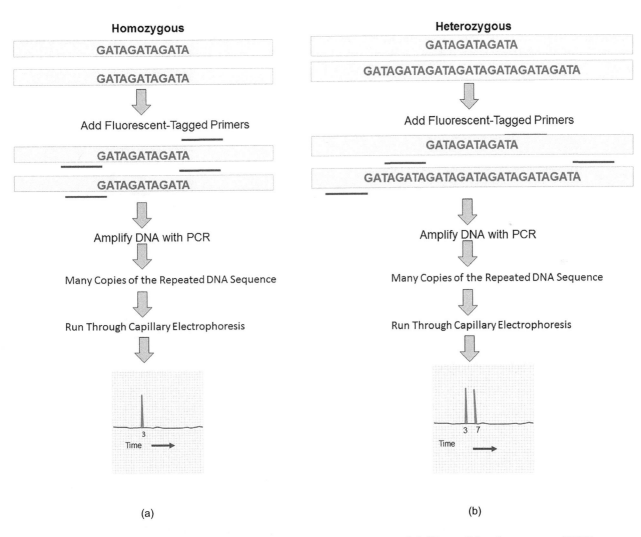

(a)   (b)

FIGURE 30.16 The process of cell line STR testing; one STR is illustrated. PCR amplifies the amount of DNA present. The primers that are used are labeled with a fluorescent tag. Capillary electrophoresis separates the fragments based on their size and generates an electropherogram. (a) A single peak occurs if the cell line is homozygous for this STR. (b) Two peaks occur if the cell line is heterozygous for this STR.

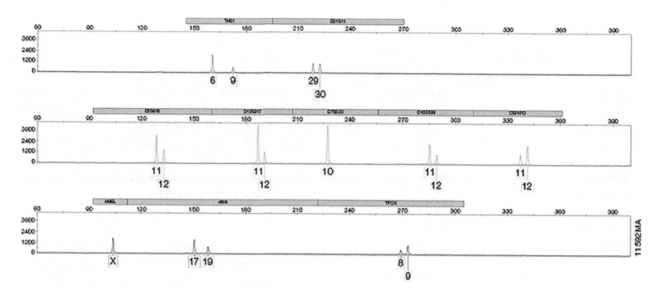

FIGURE 30.17 The electropherogram for the cell line shown in Table 30.2. There are nine STRs shown here, plus amelogenin. Three different strips are present because three different fluorescent tags were used for the PCR amplification. For example, TH01 is present in the top strip on the left. It is heterozygous with a fragment that is repeated six times and another that is repeated nine times. The cell line is homozygous for the STR named D7S820 and has ten repeats on both chromosomes. D7S820 is in the middle of the green strip. (Image reproduced with permission from Promega Corporation.)

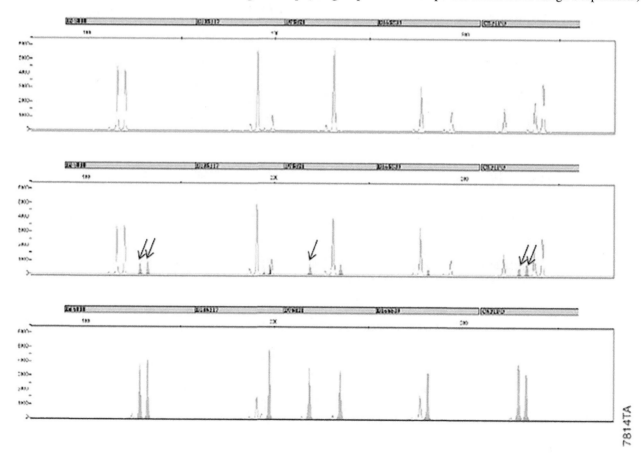

FIGURE 30.18 An STR profile that detected cell line contamination. The top strip is an STR profile from a cell line called HEK293. The middle strip is an STR profile from an HEK293 cell line that is contaminated with HeLa cells. Observe that there are additional peaks present, indicated by arrows. The bottom strip is an STR profile from HeLa cells, indicating the origin of the extra peaks. For simplicity, only five loci are shown. (Image reproduced with permission from Promega Corporation.)

At this time, none of this convenience exists for identity testing of non-human cell lines, although a database with mouse cell line STR profile information is available. There are, however, some methods that are commonly used to help authenticate non-human cell lines. Note that some of these same methods are also used to monitor human-derived cell lines, in addition to STR profiling. STR profiles are ideal for identifying a cell line, but not for monitoring changes in the characteristics of the cells with time. These additional authentication methods include the following:

1. **Phenotypic methods**. Phenotypic methods look at physical traits of a cell line. For example, it is routine in cell culture laboratories to examine the cells' appearance (morphology) nearly every day microscopically. When cells are contaminated or stressed, their morphology may change. As culturists work with a cell line over time, they become familiar with the expected morphology of the cells and can sometimes visually detect a problem. It is good practice to periodically photograph the cells so as to be able to go back later and look for morphological changes. The growth curve is another phenotypic trait that is frequently evaluated. A change in the growth properties of a cell line can be an indication of a problem. Unfortunately, a number of cell culture problems (such as *Mycoplasma* contamination) are not visible microscopically, so these phenotypic methods are necessary, but not sufficient.

2. **Karyotyping**. Karyotyping is a technique used to microscopically examine the number and appearance of the chromosomes from cells. Different species have different karyotypes, so this is a method that can be used for species identification. Karyotyping is also useful to detect gross genetic changes that sometimes occur in a cell line over time. A genetic change in a cell line that is dramatic enough to affect the chromosomes' appearance will likely have a profound effect on the properties and behavior of the cells.

3. **Isoenzyme analysis**. Isoenzymes *are enzymes that differ in amino acid sequence, but catalyze the same reaction.* Different species have different isoenzymes. These differences can be detected using a particular type of electrophoresis called isoelectric focusing. Convenient kits are available commercially to facilitate this type of analysis. (Electrophoresis is explained in Section 34.3.) An example is shown in Figure 30.19.

4. **"DNA barcoding."** This is a method of species identification based on sequencing a short stretch of DNA from a specific gene. The gene that is usually chosen is the cytochrome *c* oxidase I (COI or COX1) gene, found in mitochondrial DNA. An extensive database exists for species identification based on the DNA sequence for this gene. However, like isoenzyme analysis, DNA barcoding is only used to test the species of animal from which a cell line was derived. It is not nearly as discriminatory as STR profiling, which can identify a cell line from an individual person or animal.

5. **Genome sequencing**. As genome sequencing becomes less expensive and more available, it is becoming possible to sequence the DNA of a cell line to uniquely identify it. Genome sequencing opens the possibility of not only identifying individual cell lines, but also detecting small (and large) genetic changes that may occur with increasing numbers of passages. Cells are living systems and can change over time. Many cell lines are decades old. Over time, and with increasing numbers of passages,

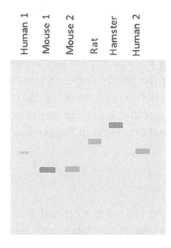

FIGURE 30.19 Simulated isoenzyme analysis result. Cells from cell lines representing different species were lysed and tested for the enzyme peptidase B. (Several other enzymes are also commonly tested to provide additional information but are not shown.) The simulated result shown here is based on data from: Steube, Klaus G., et al. "Isoenzyme Analysis as a Rapid Method for the Examination of the Species Identity of Cell Cultures." *In Vitro Cellular & Developmental Biology - Animal*, vol. 31, no. 2, 1995, pp. 115–19. doi:10.1007/bf02633971.

cells can undergo mutations and chromosomal alterations. These genetic alterations can influence all the properties of the cells, including their growth, metabolism, response to drugs and agents in the environment, and other aspects of their behavior. An additional advantage to sequencing is that it can be used to detect contaminating organisms in a culture. It is not currently common practice to routinely sequence cultured cells in research laboratories, but this may change in coming years.

30.2.9.4 General Recommendations

Cell line authentication is not a one-time process. It is often recommended that researchers perform authentication tests when they first receive a cell line, or derive it, to establish its identity. Cells should be tested again before they are frozen for storage, and periodically during the time they are actively growing, for example, every two months. Cells should be tested again at the end of a research study and before publication.

In a production environment, where the cells are used as "factories" to manufacture a product, extensive testing occurs when the cells are first genetically modified and designed for their purpose. Testing is repeated when cells are removed from cryogenic storage in order to begin production of a new batch of product. This testing minimally involves verifying the identity of the cells and ensuring they have not become contaminated. Testing is repeated yet again when the cells have finished making product to be certain that they have not become genetically altered or contaminated during the production process. Any alterations or contamination would mandate an investigation of the product, because a change in the cells might lead to an undesired alteration in the product they produce.

30.2.10 Documentation

Documentation is always a critical aspect of a quality system. As is true of any laboratory work, cell culturists must keep track of all materials and methods used. It is also essential to keep track of a cell line's "story" which can get complex as cells are used, frozen for storage, thawed, passaged, and shared with others.

Many cell culturists can rely on obtaining cells from repositories (such as ATCC) in which case the repository provides information about the cells' derivation and authentication. This information should be recorded, and care should be taken that the proper identification is always associated with every plate containing those cells.

Some cell culturists must derive their own cell lines, for example, if they are studying cells from patients with a particular disorder. Documentation requirements are particularly extensive when cell lines are initially derived because the cells may later be used for purposes not initially intended, such as for testing drugs, or producing therapeutics. Table 30.3 outlines some of the considerations for documentation in cell culture work. (For more comprehensive guidance on good documentation for cell culture, we suggest consulting Coecke, Sandra et al. "Guidance on Good Cell Culture Practice: A Report of the Second ECVAM Task Force on Good Cell Culture Practice." *ATLA*, vol. 33, no. 3, 2005, pp. 261–287.)

Note that when storing cells in liquid nitrogen, it is essential to use labels or markers that are suitable for cryogenic storage. Conventional labels will detach from vials, potentially causing serious misidentification problems. Many types of lab markers will similarly not persist in liquid nitrogen. Manufacturers sell products designed for labeling cryogen vials.

30.3 A BRIEF INTRODUCTION TO ASEPTIC TECHNIQUE

30.3.1 Overview

Now that we have outlined the scope of the contamination problem, let's consider some solutions. **Aseptic technique** *is a system of laboratory practices that minimize the risk of biological contamination in cultured cell systems.* The term aseptic technique is frequently used interchangeably with ***sterile technique,*** although sterile technique usually refers only to tissue culture practices (not bacterial culture).

Aseptic technique has two goals: first, preventing contaminants in the environment from entering a culture, and second, preventing cultured cells from causing adverse effects to humans in the laboratory or escaping to the external environment. This chapter concentrates on the protection of cultures, although the same practices are essential to protect humans from hazardous culture materials. (See also Chapter 10.)

Contaminating organisms that might enter a culture include bacteria, yeast, molds, and viruses, all of which are ubiquitous in the air, on surfaces and objects, on skin, and in a person's breath. The most essential part of aseptic technique is awareness – an understanding that we are surrounded by contaminants with the potential to destroy our cultures.

Table 30.4 summarizes key principles that guide aseptic technique. These principles are the same

TABLE 30.3

Documentation in Cell Culture Work

1. *Standard documentation, such as*:
   - Dates
   - Personnel
   - Objective of the work
   - Materials and equipment used, including calibration information for measuring equipment
   - Raw data
   - Quality-control tests
   - Procedures/SOPs used and any deviations.

2. *Cell culture information, such as*:
   - Culture media used, including supplements, supplier information, methods of preparation, and methods of sterilizing and testing for sterility
   - Culture vessels used, including any surface coatings, source, catalog information
   - Information about passaging and splitting cells
   - Morphology observed, growth characteristics
   - For storage, the passage number and identity of stored cells, cryoprotectant used, number of cells/vials, position in liquid nitrogen tank, viability after thaw, date, and operator
   - Dates and methods of monitoring and testing laminar flow cabinets
   - Monitoring information for incubators, refrigerators, freezers, liquid nitrogen tanks
   - Cell line authentication/testing
   - *Mycoplasma* testing.

3. *When deriving cell lines*:
   - Date
   - Personnel involved
   - Ethical issues, such as informed consent and privacy concerns for human donors
   - Hazards associated with handling the cells and tissues
   - Species/strain for animal sources
   - Source of subject, such as supplier for animals
   - Sex
   - Age
   - Health information (may be extensive for human cells)
   - Organ/tissue and method of obtaining
   - Cell types isolated/method of isolation
   - Pathogen testing
   - Methods used.

whether one is working with bacterial or mammalian cells. The specific practices for handling mammalian cells are, however, more stringent. As previously described, a single contaminating bacterial cell can quickly take over a culture of much slower-growing mammalian cells. Bacterial and fungal cultures are less susceptible to major contamination by a few stray organisms because they are typically inoculated with millions of rapidly dividing cells. This does not mean that one should be careless when handling bacteria, but it does mean that nonpathogenic bacteria can frequently be handled on an open laboratory bench if proper aseptic technique is used. In contrast, it is routine to use protective enclosures, such

as laminar flow cabinets, when manipulating mammalian cells and other slow-growing nonpathogenic cultures. Laminar flow cabinets were introduced in Section 10.2.2.1.

Developing the skills for successful aseptic manipulation of cultured cells is usually a matter of observation and practice. In many laboratories, new personnel learn and practice these techniques using a nonessential (and nonpathogenic) cell sample. The ability to repeatedly manipulate these cells without introducing external contamination usually indicates good aseptic technique. Keep in mind that while some types of contamination are easy to detect, others, such as *Mycoplasma,* may not be immediately apparent. Because contaminated

TABLE 30.4

General Principles of Aseptic Technique for Bacterial and Mammalian Cell Work

1. *First and foremost, remember that the air is filled with contaminating dust, microbes, viruses, spores, molds, etc*. Protect plates, culture medium, and equipment as much as possible from contact with air.

2. *Human skin and breath are rich sources of contaminating microorganisms*. Proper aseptic technique protects cultures from human contact and aerosols.

3. *An object, surface, or solution is sterile only if it contains no living organisms*. If any organisms are present, the item is nonsterile.

4. *Objects or solutions are not sterile unless they have been treated to eliminate microorganisms (e.g., autoclaved, heat-sterilized, and/or irradiated)*. All items used in the culture of living cells must be initially sterilized (e.g., growth media, test tubes, culture vessels, and pipettes). Some items are sterilized by the user, commonly by autoclaving or filtration. Alternatively, disposable sterile supplies and culture media may be purchased from suppliers.

5. *A sterile object, surface, or solution that comes in contact with a nonsterile item is no longer sterile*. Accidentally contaminated items must be replaced with sterile items before continuing a procedure.

6. *Air is not sterile unless it is sterilized in a closed container*. Whenever a container or object is open to the air, contaminating substances can fall into it. When materials (e.g., growth media, bacteria, and mammalian cells) are transferred from one location to another, special practices are used to minimize exposure to air and to avoid contact with nonsterile surfaces and objects.

7. *Equipment and surfaces must be cleaned on a routine basis*. This includes wiping laboratory benches each day, or after each use, with a suitable disinfecting solution; reducing dust; periodically cleaning water baths and adding clean water; cleaning all parts of an incubator; and other normal cleaning practices.

8. *Proper laboratory attire is required to protect cultures and workers*. For nonpathogenic bacterial culture, a normal laboratory coat, safety glasses, and gloves are routinely worn. A separate lab coat should be reserved for mammalian culture and should only be worn in the tissue culture space.

cultures are almost always useless, every step of every manipulation must be performed aseptically. It only takes one slip to destroy the purity of a culture. For this reason, one of the "wisdoms" heard in cell culture laboratories is, "When in doubt, throw it out." If you suspect that any object or sample may no longer be sterile, usually due to contact with a nonsterile item or surface, the object should be discarded immediately and replaced with a sterile substitute.

One of the easiest items to contaminate is the tip of sterile pipettes, which can easily contact the bench surface or nonsterile items during manipulations. It is helpful to think of pipettes, inoculating loops, and other sterile items as extensions of your hands and become conscious of their position at all times. Scientists who have excellent aseptic technique cannot always avoid contaminating sterile objects, but they are keenly aware of nonsterile contact and can therefore take steps to minimize any resulting contamination.

30.3.2 Aseptic Technique for Handling Bacterial Cultures

Bacterial work is often conducted on an open laboratory bench using specific techniques. Flame, normally supplied by a Bunsen burner, is used routinely when working on an open laboratory bench with bacteria.

Flame in microbial cell culture plays several roles:

- A flame can be used to kill the microbes on a device. For example, a metal loop can be used to transfer microbes from one place to another. The loop is first sterilized by flame.
- A flame can be used to create air currents that move upward and help keep dust and particles from falling onto a work surface.
- The tops of bottles are passed for a second or two through a flame to create air currents that keep microbes from falling into the bottles when they are opened and to keep microbes on the lips of the bottle from falling into the bottle.

Specific practices for manipulating bacterial cultures on an open laboratory bench are summarized in Box 30.1. Examples of such manipulations include inoculating cells into culture medium, transferring culture medium from a stock bottle into a plate or flask, diluting cells that have outgrown their container, and filter-sterilizing the medium.

BOX 30.1 SPECIFIC ASEPTIC PRACTICES FOR WORKING WITH BACTERIA ON AN OPEN LABORATORY BENCH

1. *All standard microbiological practices must be applied. (See Table 10.3.)*
2. *Prepare in advance so that all materials are available and cultures are exposed to the environment for the shortest possible time.*
3. *Avoid contact between any sterile object and any nonsterile item.*
4. *Prepare a clean, clear area in which to work. Swab the surface with 70% alcohol.*
5. *Place all required items in the work area, arranged so that they are accessible and do not require reaching over other items.*
6. *Set up one or more disposal containers with disinfectant for pipettes and any other contaminated items.*
7. *Always have additional disinfectant available in the work area in case of spills.*
8. *Work alongside a lit Bunsen burner, because the flame is thought to create an updraft that helps protect the culture from airborne contaminants.*
9. *Bacteria are generally transferred from one place to another with inoculating loops and needles that are sterilized by holding in a flame until they glow bright red; they are then cooled for 5–10 seconds.*
10. *Pipettes are often used for transferring liquids, including media that contain cells.*
    a. Non-disposable glass pipettes are placed in metal cans tip first and sterilized by autoclaving prior to use.
    b. Disposable sterile pipettes are available either in individual wrappers or in bags with multiple pipettes.
    c. Always open pipette wrappers or bags at the end away from the tip.
    d. When using any pipette, handle only the top of the pipette that never comes in contact with solutions.
    e. Open multi-pipette containers only as needed, and avoid touching the unused pipettes that remain in the can or bag.
    f. If the tip of the pipette touches any surface, discard it and its contents.
    g. Briefly pass the tip of a glass (not plastic) pipette through a flame for 2 or 3 seconds. This will not sterilize the tip, but is thought to prevent dust from landing on the tip and entering the culture. Be sure the tip is cool before using the pipette.
    h. Always use a fresh, sterile pipette, and never put the same pipette twice into a culture or into medium.
11. *When opening a bottle, flask, or plate:*
    a. Avoid putting the cap or lid down on a surface.
    b. Hold a cap in the crook of your left hand (if right-handed); do not touch the inside of the cap.
    c. Hold open tubes, bottles, and flasks at a slight angle away from your face, to minimize the chance that airborne particles will fall in.
    d. Flame the neck of a bottle briefly after opening it by passing it through a flame for 2 or 3 seconds.
    e. After use, immediately close the tops of bottles, flasks, tubes, or boxes of sterile pipette tips.
12. *Bacteria are frequently grown in glass or plastic Petri dishes or plates filled with a semisolid medium such as nutrient agar to support bacterial growth.*
    a. These dishes have loose-fitting lids that allow gas circulation, but prevent contaminants from entering the dish.

(Continued)

BOX 30.1 (*Continued*)   SPECIFIC ASEPTIC PRACTICES FOR WORKING WITH BACTERIA ON AN OPEN LABORATORY BENCH

b. When opening Petri dishes, hold the lid partway over the bottom of the plate with the open end away from your face, to protect the culture from airborne particles.

c. Always label the base of the dish with the cell information, medium, and date of culture. (Lids can sometimes be switched.) This should be done before cultures are added to the plates, using an indelible marker.

13. *Avoid getting drops of liquid onto the neck of a container underneath a cap that is to be closed.*

14. *Do not pass your arms, hands, or sleeves over any open plate, flask, or bottle, to prevent contamination.*

15. *When finished, clean the work area and swab with 70% alcohol. Be sure to wash your hands thoroughly.*

16. *Perform daily quality-control checks.*

a. Check cultures for contamination.

b. Check reagents and growth medium for sterility.

17. *Dispose of all contaminated waste according to proper procedure. (See Section 10.4.)*

30.3.3 Aseptic Technique: Mammalian Cell Culture Practices

30.3.3.1 Overview

The aseptic principles outlined in Box 30.1 apply to both bacterial and mammalian cell culture. The main difference between them is that mammalian tissue culture is almost always conducted within an enclosure rather than on an open bench. As discussed above, cells derived from mammalian tissue generally have significantly slower growth rates than bacteria and thus are more susceptible to rapid overgrowth by microbial contaminants. Mammalian cell culturists use a particular kind of laminar flow cabinet, called a biological safety cabinet (BSC) (also termed a cell culture hood) for cell culture work to minimize the likelihood of contamination. When used properly, BSCs provide effective containment and a sterile air supply to the working surface.

Typically, mammalian cell cultures are buffered with a bicarbonate/CO_2 buffering system. (See Section 31.4.2.3 for more information.) This buffering system requires that mammalian cell cultures be maintained in a CO_2-rich environment, usually an incubator that maintains a 5%–10% CO_2 atmosphere at 37°C. Because of the need for CO_2 gas exchange, tissue culture containers cannot be tightly sealed during incubation. (However, to maintain proper buffering, caps and lids should be tightened when containers are removed from their incubators.) Relatively high humidity must be maintained within CO_2 incubators to prevent cell culture media from evaporating, so most incubators have a bottom pan that is partially filled with water. This water can be a fertile medium for the growth of microorganisms, so it is essential that the pan be emptied, cleaned, and disinfected regularly, even if chemicals to inhibit microbial growth are added.

30.3.3.2 Proper Use of Class IIB Biosafety Cabinets

Class II cabinets are appropriate for both microbiology and tissue culture work, although the same cabinet should not be used for both. They provide effective containment and a sterile air supply to the working surface. These cabinets need to be inspected and tested periodically to ensure that the filter is in good condition and the airflow is adequate. A certified technician must decontaminate and replace the filters at regular intervals.

Many BSCs contain a UV germicidal lamp inside the work area. **Germicidal** *refers to methods that are capable of killing bacteria or other microorganisms.* When getting ready to use a BSC, first wipe down all the surfaces with an appropriate disinfectant. Turn on the UV lamp for a minimum of 15 minutes before starting work. Because of the radiation, the lamp must be turned off when the cabinet is in use. Serious eye damage and burns to the hands can result if the light is accidentally left on during cabinet use. While the UV light is not turned on all the time, some experts recommend leaving the air circulation system on constantly. Other procedures specify turning the airflow off at night and turning it back on at least 5 minutes before use to purge the hood of unfiltered air. Once the hood is purged of unsterile air and the UV light has been on for 15 minutes, the light is turned off and all the needed materials are placed in the hood. Only items that are needed in the workspace should be placed in the cabinet. Figure 30.20 shows a recommended layout for materials within a BSC.

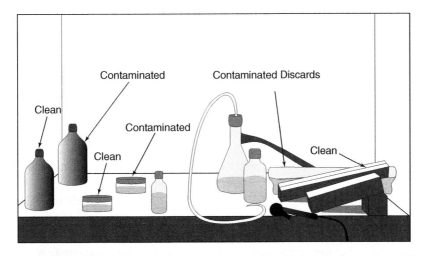

FIGURE 30.20   Suggested layout for work materials in a BSC. Note that all materials should be placed 4–6 inches back from the front airfoil. Materials should be arranged so that contaminated items do not pass over clean items.

Work in a BSC requires special technique. Plan all operations in a BSC so as to minimize disruptions in the airflow through the cabinet. Move your hands and other objects slowly into and out of the cabinet interior. Strong air currents created outside the cabinet by people moving rapidly, or laboratory doors opening and closing, may also disrupt the airflow pattern and reduce the effectiveness of a BSC. Box 30.2 provides guidelines for the proper use of Class II BSCs.

BOX 30.2   SAFE AND EFFECTIVE USE OF CLASS II BIOSAFETY CABINETS

Class II BSCs are designed to provide protection for workers, the work product, and the environment, when users observe basic guidelines for working in the cabinet.

GENERAL GUIDELINES

1. Understand how the cabinet works. (See Section 10.2.2.)
2. Do not disrupt the protective airflow pattern of the BSC.
3. Plan your work.
4. Minimize the storage of materials in and around the BSC.

CABINET SETUP GUIDELINES

1. Turn on the germicidal lamp if present; wait at least 15 minutes before proceeding with any activities in the cabinet. Turn off the lamp before beginning work inside the cabinet.
2. Turn on the airflow in the BSC at least 5 minutes before starting to place objects in the cabinet (if the BSC is turned off when not in use). Newer model BSCs are specifically designed for continuous operation, and turning off the BSC airflow at any time may adversely affect room air balance.
3. Be sure that all vents and airfoils are clean and unblocked.
4. Before starting to work, wash hands and lower arms thoroughly using a germicidal soap.
5. Wipe the work surface in the cabinet with 70% alcohol or other suitable disinfectants.
6. Assemble all the materials you will need for your work, and wipe each item with 70% alcohol before placing it inside the cabinet.
7. Do not overload the cabinet.
8. Wait another 5 minutes before beginning to work in the cabinet.

(Continued)

BOX 30.2 (*Continued*)   SAFE AND EFFECTIVE USE OF CLASS II BIOSAFETY CABINETS

CABINET USE GUIDELINES

1. Keep all materials at least 4 inches inside the cabinet face, and work with biological materials as deep within the cabinet as feasible.
2. Move arms slowly when removing or introducing new items into the BSC.
3. Do not place any objects over the front air intake vents or blocking the rear exhaust vents.
4. Segregate contaminated and clean items; work from clean areas toward contaminated areas (Figure 30.20).
5. Place a pan with disinfectant and a sharps container (if necessary) inside the BSC for pipette discard. To avoid disrupting the BSC airflow and introducing contaminants to the environment, do not use vertical pipette wash canisters on the floor outside the cabinet.
6. Place any equipment that may create air turbulence (such as a microcentrifuge) at the back of the cabinet (but not blocking the rear vents).
7. Clean any spills within the cabinet immediately; wait at least 10 minutes before resuming work.
8. When finished, remove all equipment and materials and wipe all interior surfaces of the cabinet with disinfectant.
9. Remove PPE and wash hands thoroughly before leaving the laboratory.

Note that flame is not recommended when working with mammalian cells in a cell culture hood. This is because the airflow in a cell culture hood is carefully controlled to prevent airborne contamination; heat disrupts the desired airflow. Heat can also damage the special filters used in laminar flow cabinets. Note also that flame in a cell culture hood is difficult to see and has resulted in serious burns to people who did not realize a burner was in use (and did not expect to find a flame in a laminar flow cabinet).

30.3.3.3 Checking Mammalian Cells

It is important to routinely check that your cell cultures are doing well. You do not want to invest time and expensive reagents on contaminated or otherwise unhealthy cells. There are several characteristics that you can monitor including the following:

- **Cell density and growth rate**. Each healthy cell line has a predictable growth rate and population doubling time. You can periodically count the cells to assess their growth rate. It is also sometimes possible to follow the culture microscopically, roughly assessing each day what percentage of the culture dish the cells have covered.
- **Appearance of the growth medium**. Contamination of monolayer cell cultures can be assessed in part by observing the appearance of the medium floating above the monolayer. If the medium looks turbid, there is quite possibly microbial contamination of the

cell culture. Cell death also indicates possible contamination.

- **Cell morphology**. Cell lines have a characteristic morphology. Once you are familiar with your specific cell line, you will know what cellular morphology to expect. If the cells exhibit an unusual morphology, you may suspect that they either are contaminated with another cell line or are not growing in optimal conditions. Cellular morphologies are generally classified as "epithelial-like," "lymphoblast-like," or "fibroblast-like."
    - *Epithelial-like cells* generally grow as an attached monolayer with a flat, polygonal shape, sometimes called "cobblestone" in appearance. CHO cells are epithelial. These are shown in Figure 30.3.
    - *Fibroblast-like cells* tend to be elongated and attached to the substrate, and they often grow in patterns that resemble swirls. NIH3T3 is a fibroblast cell line that is often used in research laboratories (Figure 30.21a).
    - *Lymphoblast-like cells* are generally round and grow in suspension rather than attached to a substrate (Figure 30.21b).
- **Uniformity of the culture**. If cells are contaminated by bacteria, yeast, or molds, you may be able to directly observe the contaminating organisms using high-power microscopy. This will generally not occur unless cultures are highly contaminated. If the cell

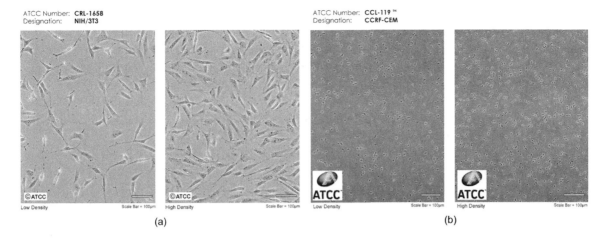

ATCC Number: **CRL-1658**
Designation: **NIH/3T3**

ATCC Number: **CCL-119** ™
Designation: **CCRF-CEM**

Low Density   Scale Bar = 100μm   High Density   Scale Bar = 100μm   Low Density   Scale Bar = 100μm   High Density   Scale Bar = 100μm

(a)   (b)

FIGURE 30.21 Morphology of mammalian cells in culture. (a) Fibroblast-like cells are elongated and grow attached to the substrate. (b) Lymphoblast-like cells are spherical and usually grow in suspension without attaching to the substrate. (Photos courtesy of ATCC.)

line harbors a low level of contamination, it is likely that you will miss this contamination using microscopy because of the small size and small number of offending organisms.

- **Attachment to substrate**. Monolayer cultures should be well attached to the substrate. A large percentage of unattached cells could indicate a problem. If cells appear to be "rounding up" or looking like they are getting ready to detach, they may be unhealthy or overcrowded.

- **Other signs of poor health**. Sometimes cell cultures are not harboring any contamination, yet are growing suboptimally. Poor growth can be caused by many factors, including problems with medium components, pH, temperature, or aeration.

Practice Problems

1. Go to the database of Cross-contaminated or Misidentified Cell Lines, https://standards.atcc.org/ kwspub/home/the_international_cell_line_authentication_committee-iclac_/Database_of_Cross_ Contaminated_or_Misidentified_Cell_Lines.pdf. For each of the following cell lines, determine if they are noted as being misidentified or cross-contaminated in the database. If so, what are they claimed to be? What are they actually?
   a. NIH3T3 cells
   b. BRCA5
   c. CHO-CD36
2. The doubling time of the cells in a culture is 35 minutes. Are these most likely prokaryotic or eukaryotic cells?
3. One of the guidelines for working with cultured cells in a biological safety cabinet is "Place a pan with disinfectant and a sharps container (if necessary) inside the BSC for pipette discard; to avoid disrupting the BSC airflow and introducing contaminants to the environment, do not use vertical pipette wash canisters on the floor outside the cabinet." Explain the rationale for this guideline. (The recommended placement of a pan for discarded pipettes is shown in Figure 30.20.)
4. A colleague gives you a vial of cells labeled human bone marrow stem cells. You plan to use these in studies of the biology of bone marrow stem cells. As part of routine testing, you karyotype the cells and find they have 40 chromosomes. What does this mean? What should you do?
5. Vero E6 is a cell line derived from the kidney cells of an African green monkey. This cell line is commonly used for viral research because a number of viruses replicate effectively in these cells. Their normal doubling time is roughly 22 hours. You are working in a laboratory that is studying antiviral compounds to treat patients infected by the Sars-CoV-2 virus, and you obtain Vero E6 cells for your

experiments. You routinely count the cells in your cultures every day. You plan to expose cells to virus on Thursday. Monday morning when you arrive at work you count the cells in your culture and observe 6.4×10^4 cells/mL. Tuesday morning when you arrive you count the cells in your culture and observe 9.0×10^4 cells/mL. Wednesday morning there are 1.1×10^5 cells/mL. What does this tell you?

6. MRC-5 (PD 30) is a cell line derived from human lung tissue. These cells are used in the production of certain vaccines, including those for rabies and hepatitis. The STR profile for this cell line is:

STR-PCR Data: Amelogenin: X,Y
CSF1PO: 11,12
D13S317: 11,14
D16S539: 9,11
D5S818: 11,12
D7S820: 10,11
THO1: 8
TPOX: 8
vWA: 15

You obtain a culture of these cells from a colleague and have them authenticated by STR analysis. A portion of the profile you receive is shown in Figure 30.22. Is this profile consistent with MRC-5 (PD 30) cells?

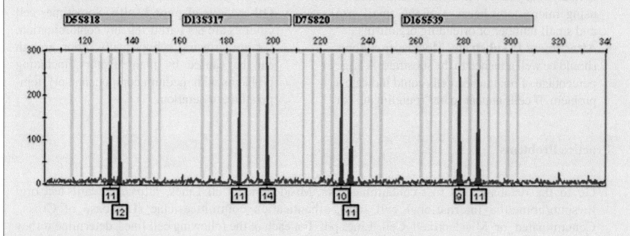

FIGURE 30.22   STR profile used for authentication. (Photo courtesy of ATCC.)

7. In your own words, summarize the requirements for ensuring that cells used for biotechnology applications are suitable for their purpose.

CHAPTER APPENDIX

Using a Hemocytometer to Count Mammalian Cells

Mammalian cells are typically counted in a special counting chamber. A **Neubauer hemocytometer**, shown in Figure 30.23, *is a special, modified microscope slide that is used to determine the concentration of cells in a suspension of cells.* A hemocytometer contains two identical wells, or chambers, into which a small volume of the cell suspension is pipetted (Figure 30.23a). Each chamber is

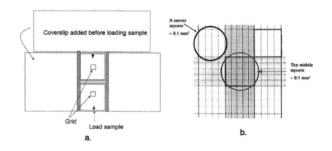

FIGURE 30.23   The hemocytometer is used for counting cells. (a) Top view of hemocytometer showing two identical chambers. (b) The grid as it appears under a microscope.

manufactured to hold an exactly specified volume of liquid. When you view the hemocytometer under a microscope, you see that the bottom surface of each well is marked into a grid pattern (Figure 30.23b). The grid consists of nine larger squares that are further subdivided into smaller squares. The four corner large squares are subdivided into 16 smaller squares. Each of the nine larger squares has dimensions of $1 \, mm \times 1 \, mm \times 0.1 \, mm$. Therefore, the volume of fluid over each of these large squares is $0.1 \, mm^3$. (Volume = length × width × height.) The units of mm^3 are not very helpful to us, so they are converted to units of mL: $0.1 \, mm^3 = 10^{-4} \, mL$.

The process of counting cells with a hemocytometer is as follows:

Step 1. Remove a sample of the cell suspension and dilute it with trypan blue.

Some of the cell culture suspension is removed and is diluted with the dye, trypan blue, and possibly also with sterile broth. Trypan blue is used so that living cells can be distinguished from dead ones. Living cells exclude the dye, but it readily passes through the membrane of dead cells. Thus, living cells are clear, while dead cells are blue.

Step 2. Apply the mixture to the hemocytometer.

A coverslip is placed on top of the hemocytometer slide. The cell suspension is mixed

well and then is carefully applied under the coverslip into both wells of the hemocytometer in such a way that each chamber is exactly filled.

Step 3. Count the cells.

Using a microscope, all the cells located over the four large corner squares and the center square in one of the chambers are counted. (See Figure 30.23.) A separate tally is kept of the viable and non-viable cells. Sometimes the cells in both chambers are counted, either to make sure that the counts are comparable, or to count more cells. (Some analysts try to count enough squares to get at least 100 cells total; others aim for at least 200 cells total.)

Step 4. Calculate percent viability and the concentration of viable cells.

The percent of viable cells is calculated. This value can be revealing. If there is a high concentration of dead cells, then the culture may not be healthy and should not be used for further experiments (although it may be diluted and scaled back up for future use).

Once the percent of viable cells is known, you can determine the concentration of viable cells in your culture. These calculations are discussed in Equation 30.1.

EQUATION 30.1 CALCULATION OF THE NUMBER OF CELLS/ML BASED ON HEMOCYTOMETER COUNTS

$$\frac{\text{Number of viable cells}}{\text{mL}} = \frac{\left(\# \text{ of viable cells in 5 large squares}\right) \times \text{reciprocal of dilution} \times 10^4}{5}$$

Consider the parts of this equation:

- The number of viable cells per mL is what we want to know. The number of viable cells in 5 large squares divided by 5 gives the average number of cells per large square. Note that if you count more or fewer squares, then you need to adjust the equation. For example, if you count only the four corner squares and not the center square, then you divide the total number of viable cells by four instead of five to get the average number of cells per large square.

- The dilution must be considered when calculating the number of cells per milliliter. In this situation, we use the reciprocal of the dilution (expressed as a fraction), since we are calculating back from the concentration of cells in a diluted sample to the concentration of cells in the original solution.

- Finally, each of the nine larger squares has a volume of $10^{-4} \, mL$, but we want to know the number of cells per 1 mL. Therefore, we multiply by 10^4 to calculate how many cells there are in 1 mL.

Example Problem 30.1

- 200 μL of a cell suspension is removed from a plate of cells.
- 300 μL of diluent is added to the suspension.
- 500 μL of trypan blue is added.
- The total volume is thus 1,000 μL.
- The total number of viable cells in the four corner plus center squares is 245.
- The total number of non-viable cells in these squares is 55.

What is the number of viable cells/mL?

Answer:

CALCULATIONS

1. The total number of cells counted = **245 + 55 = 300.**

2. The percent viable cells = **245/300 ≈ 81.7%**
3. The average number of viable cells per square = **245/5 = 49.**
4. The dilution is
   200 μL/ (200 μL + 300 μL + 500 μL) =
   200 μL /1,000 μL = 1/5.
5. The reciprocal of the dilution is **5/1 or 5.**
6. Using Equation 30.1, the concentration of viable cells in the original dish is estimated to be:

$$\frac{245}{5} \times 5 \times 10^4 = \frac{2,450,000 \text{ cells}}{\text{mL}}$$

31 Culture Media for Intact Cells

31.1 INTRODUCTION

This chapter extends our discussion of biological solutions that began in Unit VI to include solutions *that support living cells;* these solutions are called **nutrient media, culture media, growth media, cell culture media,** or simply **media.** Like the biological macromolecules DNA, RNA, and proteins, discussed in Chapter 25, cultured cells require a buffered aqueous environment that contains specific dissolved solutes in proper concentrations. Working with living cells, however, introduces some issues that we have not yet addressed. Perhaps the most obvious is that cells require nutrients if they are maintained in culture for more than a few hours. Another feature of intact cells is that they are surrounded by a cell membrane that is permeable to the flow of water and some, but not all, solutes. Intact cells therefore require an osmotically (defined below) balanced environment. Living cells also require a culture environment that is sterile, except for the cells of interest. (See Chapter 30 for a discussion of aseptic technique.)

This chapter discusses culture media for maintaining intact bacterial and mammalian cells. We first discuss osmotic considerations and isotonic solutions that are used for short-term (minutes to hours) maintenance of cells. The next section discusses media that support the nutritional and physical requirements for growth and reproduction of microbial cells, particularly bacteria. The final section in this chapter discusses the media requirements for cells derived from animals, particularly those of mammalian origin. (Plant tissue culture is outside the scope of this discussion.)

31.2 ISOTONIC SOLUTIONS

Intact cells contain water, salts, and various biological molecules enclosed by a cell membrane. The presence of a cell membrane adds an important dimension to the preparation of media. The cell membrane is **semipermeable,** *which means that it allows water and some small molecules to flow through unimpeded, whereas other molecules are restricted.* Cells exist in an aqueous environment, so water is constantly moving back and forth across the cell membrane. The net amount of water that flows into a cell must equal the net amount of water that leaves. If more water enters a cell than exits, then the cell swells, and may burst. Conversely, if more water

DOI: 10.1201/9780429282799-39

exits the cell than enters, then the cell shrinks. A cell is said to be in **osmotic equilibrium** *when the net rates of water flow into and out of the cell are equal.*

The flow of water across the cell membrane is controlled by the concentrations of solutes inside and outside the cell. The net flow of water across the cell membrane is from the side that has the lower solute concentration to the side that has the higher solute concentration. If there is a higher solute concentration outside the cell than inside, then more water flows out of the cell than enters it. Conversely, when there is a lower solute concentration outside the cell, then more water flows into the cell. **Osmosis** *is the net movement of water through a semipermeable membrane from a region of lesser solute concentration to a region of greater solute concentration.* **Osmotic pressure** *is the amount of pressure that needs to be exerted to halt the water's movement.*

The principle of osmotic pressure is illustrated in Figure 31.1. In Figure 31.1a, there is initially a lower solute concentration on the left side of the U-tube than on the right. The two sides are separated by a semipermeable membrane that allows water to flow freely, but is impermeable to the solute particles. The semipermeable membrane in this illustration is analogous to a cell membrane. On average, more water will flow from the left side into the right side than vice versa. The level of water will therefore rise on the right side (Figure 31.1b). Eventually, equilibrium will be achieved such that the gravitational pressure on the water equals the osmotic pressure.

Osmotic pressure is usually expressed in terms of the concentration of particles in a solution in units of either **milliosmolality** *or* **milliosmolarity** *where:*

Milliosmolality *is milliosmoles/kg of solvent*
Milliosmolarity *is milliosmoles/L of solution.*

Consider these two concentration expressions:

Numerators: The units in both numerators are milliosmoles (mOsm). **Osmoles** *are determined by the number of moles of particles in a solution. Accordingly,* **milliosmoles** *are determined by the number of millimoles of particles in a solution.*

Glucose, for example, does not dissociate when dissolved in water. Therefore, 1 millimole of glucose provides 1 millimole of solute particles in a solution.

For glucose: *1 millimole = 1 milliosmole*

In contrast, salts such as NaCl and KCl dissociate in water. One mole of NaCl dissociates to form one mole of Na^+ ions and one mole of Cl^- ions to make two moles of osmotically active solute particles.

Therefore, for these salts: *1 millimole = 2 milliosmole*

Denominators: Observe that the units in the denominator are either *kg,* for osmolality, or *L* for osmolarity. The volume of a solution may vary if the temperature changes, but the weight of a solution does not vary with temperature. The measurement of osmotic pressure often involves changing the temperature

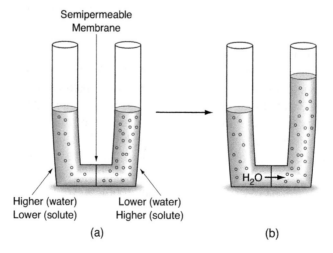

Semipermeable
Membrane

Higher (water)   Lower (water)
Lower (solute)   Higher (solute)

(a)   (b)

FIGURE 31.1 Osmotic pressure. The net flow of water is from the left side, where the solute concentration is lower, to the right side where the solute concentration is higher. The water level rises on the right side and eventually equilibrates when the osmotic pressure equals the force of gravity.

of the solution. Therefore, osmolality, which has units of kg, is more commonly used than osmolarity, which has units of L.

The most common method of measuring the osmotic pressure of a solution is with a **freezing point depression osmometer.** This device is based on the fact that particles dissolved in water depress the temperature at which the water freezes; the more the particles that are dissolved, the more the freezing point is depressed. A sample of the solution is frozen in the osmometer, and the temperature at which it freezes is measured. This is compared to standards of known osmolality to determine the osmolality of the sample.

The reason we care about osmotic pressure and its measurement is because cultured cells need to be surrounded by solutions that have about the same solute concentration as the interior of the cells to prevent them from swelling or shrinking. An **isotonic solution** *is one that has the same solute concentration as the interior of the cell* (Figure 31.2). **Physiological solutions** *are isotonic solutions that support intact cells in the laboratory.* Salts are commonly used in physiological solutions to maintain the proper osmotic equilibrium for cells. According to the literature from ATCC (an organization that maintains and distributes cultured cells), the useful range of osmolality of cell culture media for mammalian cells is 260–320 mOsm/kg. Human blood plasma is about 290 mOsm/kg. Culture media used for growing human stem cells have a relatively tight tolerance for osmolality and should be 290–300 mOsm/kg. Expressed in units of molarity, isotonic solutions for mammalian cells should be about

150 mM salt (assuming a salt that dissociates to form two ions in solution). Non-mammalian animal cell types, such as those from insects, vary in their osmotic requirements from about 155 mOsm/kg, to as much as 375 mOsm/kg.

The optimal osmolality for bacterial culture is variable since bacteria live in many different environments. Bacteria from the ocean, for example, have evolved in a different osmotic environment than those in fresh water. Bacteria often live in changeable environments (e.g., soil) and therefore are often more tolerant of variation in osmolality than mammalian cells. Nonetheless, extremes of osmolality do affect bacteria. Bacterial cells (unlike mammalian cells) are surrounded by a rigid cell wall. If there is a higher solute concentration outside a bacterial cell than inside, then water flows out of the cell and the cell membrane and contents shrink away from the cell wall. This condition inhibits bacterial growth and reproduction. For this reason, since ancient times, foods have been preserved by the addition of high levels of salts and sugars.

The simplest solution commonly used to maintain osmotic equilibrium is **standard (physiological) saline,** *which is 0.9% NaCl.* Most physiological solutions also have a buffering component in addition to salt. Table 31.1 shows examples of **balanced salt solutions** and **buffered salt solutions** *that are intended to maintain cells for short periods* (minutes to hours). Observe that in Chapter 25 we discussed the varied roles of salts in solutions used when working with proteins and nucleic acids. Here is yet another vital role for salts in biological solutions, that is, maintaining the proper osmotic strength of the aqueous environment.

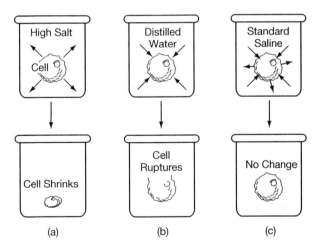

FIGURE 31.2 The flow of water across the cell membrane. (a) If there is a higher solute concentration outside the cell than inside, then water flows out of the cell and the cell shrinks. (b) If there is a lower solute concentration outside the cell, water flows into the cell, and it may burst. (c) Standard saline is an isotonic solution that is in equilibrium with the interior of the cell.

TABLE 31.1
Examples of Balanced Salt Solutions

Physiological saline (mammalian cells)	
NaCl	0.9% w/v
Phosphate-buffered saline	
NaCl	7.20 g/L
Na_2HPO_4 (anhydrous)	1.48 g/L
KH_2PO_4	0.43 g/L
Tris-buffered saline	
NaCl	0.9%
Tris, pH 7.2	10–50 mM
Ringer's solution (mammalian cells)	
$CaCl_2$	0.25 g/L
KCl	0.42 g/L
NaCl	9.00 g/L
Dulbecco's phosphate-buffered saline (DPBS)	
$CaCl_2$	0.10 g/L
$MgCl_2 \cdot 6H_2O$	0.10 g/L
KCl	0.20 g/L
KH_2PO_4 (anhydrous)	0.20 g/L
NaCl	8.00 g/L
Na_2HPO_4 (anhydrous)	1.150 g/L

For longer culture, cells require a medium that not only is osmotically balanced, but also contains nutrients and various other substances to support growth and reproduction. Cell culture media are formulated to provide the proper osmotic environment for cells while also providing these nutrients and growth-promoting constituents.

31.3　BACTERIAL CULTURE MEDIA

31.3.1　BACKGROUND

Bacterial growth media are aqueous mixtures of nutrients used to support the growth and reproduction of bacteria. The use of bacterial growth media is traced back to the 1800s when scientists, including Louis Pasteur and Robert Koch, conducted pioneering studies of the role of bacteria in causing disease. Early microbiologists grew bacteria on gelatin substrates in crude broths with nutrients derived from extracts of plant and animal tissues (e.g., beef brain and heart). Gelatin was not an ideal substrate because many bacteria digested the gelatin and it liquefied above room temperature. One of the advances that emerged from Koch's laboratory was the discovery that agar, a substance derived from seaweed, could be used to make a solid substrate on which bacteria readily grow. The use of agar was suggested by Angelina Hesse, the wife of

a physician who was working in Robert Koch's laboratory. Hesse was assisting her husband in the laboratory as an illustrator and laboratory technician. Hesse was aware of the use of agar in the East Indies as a cooking ingredient to help harden foods. She suggested using agar to harden the culture media in the laboratory – an idea that Koch and his colleagues tried with great success. The new agar medium proved to be far superior to extracts mixed with gelatin and enabled Koch and other scientists to isolate colonies of specific disease-causing organisms.

In the early 1900s, microbiologists discovered that the components of bacterial culture media can be prepared in large batches and then dehydrated, a form in which they are conveniently stored until needed. Commercial vendors began to manufacture dehydrated ingredients and combinations of ingredients for microbial media. Microbiologists learned to combine these readily available ingredients in different ways to best promote the growth of certain types of microorganisms, eventually creating hundreds of different media formulations optimized for different applications. Preparing culture media for bacteria today is often a straightforward procedure in which specified amounts of purchased, dehydrated components are dissolved in water, according to a recipe tailored for a specific bacterium and application. The resulting mixture is sterilized, usually by autoclaving. Compendia and handbooks are available with standard formulations and methods for food microbiology, medical microbiology, and environmental microbiology. (Examples of compendia are provided in this unit's Bibliography). Although the culture media used today are easier to prepare and more versatile than they were in the nineteenth century, extracts from plant and animal tissue, and agar from seaweed, are still staples of the microbiologist's pantry.

31.3.2　TYPES OF BACTERIAL CULTURE MEDIA

31.3.2.1　Liquid versus Solid

Bacterial culture media can be prepared to be liquid, solid, or semisolid. **Liquid culture media,** called **bacterial broths,** *are aqueous-based mixtures of nutrients that do not contain a hardening agent.* Bacteria grow suspended throughout a liquid broth. Broths are generally used for propagating large numbers of bacteria, either in the laboratory or for industrial-scale fermentation.

Solid and **semisolid media** *are aqueous-based mixtures of nutrients that contain agar as a hardening agent to provide a solid substrate for bacterial*

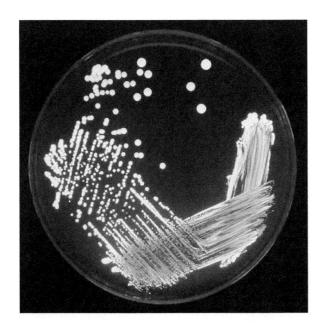

FIGURE 31.3 *Legionella sp.* colonies growing on the surface of an agar plate. Nutrient agar provides nutrients and a surface for the propagation and isolation of individual colonies. Each round white object at the top of this agar plate is a colony consisting of millions of bacterial cells all probably derived from a single parent cell. *Legionella* is a type of bacterium that caused an outbreak of pneumonia during the 1976 convention of the American Legion, sickening at least 180 people and killing 29. (Photo Courtesy of James Gathany, Centers for Disease Control and Prevention Public Health Image Library.)

growth. Bacteria can be grown as distinct colonies on top of solid media (Figure 31.3), or sometimes within the medium. Solid media are used to observe bacterial colony appearance, for isolating colonies derived from a single organism, and for storage. Semisolid media may be used for determining bacterial motility and for promoting growth of bacteria in the absence of oxygen.

31.3.2.2 Chemically Defined versus Chemically Undefined

Both liquid and solid culture media may be chemically defined or undefined. A **chemically defined medium** *is one in which all the ingredients and their quantities are known.* These media typically contain purified organic salts and simple organic compounds, such as glucose and amino acids. Defined media may also include trace elements, nucleotides, and vitamins.

Minimal media *are a type of defined medium that contain only the minimum nutrients required for a particular microbe to survive and reproduce.* Table 31.2 shows an example of a formulation for a defined, minimal medium that is used to culture naturally occurring

Escherichia coli. E. coli is a common bacterium, normally found in the intestine of humans and other animals, that has the ability to synthesize all its nutrients from relatively simple precursors.

A **complex or undefined medium** *is one in which the exact chemical constitution of the medium is not known.* Complex media are usually derived from materials of biological origin such as blood, milk, yeast, beef, or vegetables. These sources supply a mixture of amino acids and low molecular weight peptides, carbohydrates, vitamins, minerals, and trace metals.

Defined media have the advantage that they are consistent from batch to batch and do not contribute unknown impurities to experiments or production systems. Defined media, however, are often expensive because their components are purified. Complex media have the advantages that they usually provide all the growth factors that are required by bacteria, they tend to be less expensive than defined media, and they are broadly useful for culturing a variety of bacteria. Most medically important bacteria are cultured in complex media containing blood, serum, and tissue extracts. The major disadvantages to complex media are that they introduce lot-to-lot variability and unknown components into the system.

31.3.2.3 Selective and Differential Media

Selective media *inhibit the reproduction of unwanted organisms and/or encourage the reproduction of specific organisms.* For example, if a bacterium is resistant to a certain antibiotic, then that antibiotic can be added to the medium in order to prevent other cells, which are not resistant, from growing. Selective media are useful for the isolation of specific microorganisms from mixed populations.

Differential media *are designed to reveal differences among microorganisms or groups of microorganisms that are growing on the same medium.* Differential media usually contain a chemical that is utilized or altered by some microorganisms and not by others. When particular bacteria grow on differential medium, they cause a visible change, such as the production of a colored product. While selective media are used to allow the growth of only selected microorganisms, differential media allow the growth of multiple types, but cause them to have distinguishing characteristics. A growth medium can be both selective and differential. Selective and differential media may be either undefined or defined.

TABLE 31.2

Minimal Agar, Davis

Ingredient	Amount (per Liter) (g)	Function of the Ingredient
Agar	15.0	Hardens medium
K_2HPO_4	7.0	Buffers pH; provides P and K
KH_2PO_4	2.0	Buffers pH; provides P and K
$(NH_4)_2SO_4$	1.0	Buffers pH; provides N and S
Glucose	1.0	Provides C and energy source
Sodium citrate	0.5	Buffers pH
$MgSO_4 \cdot 7H_2O$	0.1	Provides S and Mg^{++}
		Magnesium is a cofactor for many metabolic reactions.

pH will be $7.0 + 0.2$ at 25°C.

Preparation:
1. Add components to cold purified water.
2. Bring to volume of 1 L.
3. Mix thoroughly.
4. Gently heat and bring to boiling.
5. Distribute into tubes or flasks.
6. Autoclave 15 minutes at 15 psi pressure at 121°C.
7. Pour into sterile Petri dishes, or leave in tubes.
8. Store at refrigerator temperature protected from direct light.

Source: Atlas, Ronald M. *Handbook of Microbiological Media.* 4th ed., CRC Press, 2010.

31.3.3 INGREDIENTS IN BACTERIAL CULTURE MEDIA

31.3.3.1 Agar

Agar, *which is derived from seaweed, is the most common hardening agent for bacterial media.* Bacteria do not digest agar, and it is relatively inexpensive. Agar does not melt until its temperature is about 84°C, but once melted it does not solidify again until its temperature drops to about 38°C. This is important because agar remains solid at most practical temperatures and, after melting, it can be poured into dishes or tubes at a convenient temperature. The typical concentration of agar is about 15.0 g/L (1%–2%). Lower concentrations of 7.5–10.0 g/L are used in soft or semisolid media.

31.3.3.2 Formulations that Reflect the Composition of Bacteria

Media formulations must supply the nutrients that bacteria need to grow and reproduce. Bacteria are about 70% water, so purified water (Chapter 24) is the primary ingredient in culture media. On a dry weight basis, bacteria are about half carbon, with lesser amounts of oxygen, nitrogen, hydrogen, and phosphorus, and 1% or less of sulfur, potassium, sodium, calcium, magnesium, chloride, and iron (Table 31.3). Growth media for bacteria provide nutrients in roughly the same percentages. Growth media additionally provide an energy source to the cells (unless they are able to use photosynthesis).

Once nutrients are inside a cell, they are chemically modified by metabolic processes to meet the nutritional needs of the cell. Some, but not all, bacteria are able to survive on a simple growth medium because they are able to synthesize everything they need from one or two carbon sources, a few salts, and a nitrogen source.

31.3.3.3 Sources of Nutrients

Meat extracts have long been used in complex (undefined) media to supply nutrients including carbon, nitrogen, vitamins, and trace minerals. Yeast extracts are another common constituent of complex media. **Yeast extract** *is the water-soluble portion of yeast cells that have been allowed to die so that the yeasts' digestive enzymes break down their proteins into simpler compounds while preserving the vitamins from the yeast.* Yeast extract is a rich source of amino acids and vitamins. It is typically used at a concentration of 0.3%–0.5% and has a pH of about 6.6 in media.

Peptones *are hydrolyzed (cleaved) proteins formed by enzymatic digestion or acid hydrolysis of natural substances including milk, meats, and vegetables.* Many complex media contain peptones as the main source of nitrogen-containing compounds (amino acids, peptides, and proteins). Peptones also provide

TABLE 31.3

Nutrient Requirements of Bacteria

Element	% Of Dry Weight	Function	Chemical Form Added to Culture Media
Carbon	≈ 50	Main building material of cells	**Organic;** simple sugars (e.g., glucose, acetate, or pyruvate), extracts (such as peptone, tryptone, and yeast extract) **Inorganic;** carbon dioxide (CO_2) or hydrogen carbonate salts (HCO_3^-)[a]
Oxygen	≈ 20	Constituent of water; electron acceptor in aerobic respiration	Water
Nitrogen	≈ 14	Constituent of amino acids, nucleic acids, and coenzymes	**Organic;** amino acids, nitrogenous bases, peptones **Inorganic;** NH_4Cl, $(NH_4)_2SO_4$, KNO_3, and for dinitrogen fixers, N_2
Hydrogen	≈ 8	Constituent of organic compounds and water	Present in most chemicals added to medium
Phosphorus	≈ 3	Constituent of nucleic acids and phospholipids	KH_2PO_4, Na_2HPO_4[a]
Sulfur	≈ 1	Constituent of certain amino acids and several coenzymes	Na_2SO_4, H_2S
Potassium	≈ 1	Main cellular inorganic cation and cofactor for certain enzymes	KCl, K_2HPO_4[a]
Magnesium	≈ 0.5	Inorganic cellular cation, cofactor for certain enzymes	$MgCl_2$, $MgSO_4$
Calcium	≈ 0.5	Inorganic cellular cation, cofactor for certain enzymes	$CaCl_2$, $Ca(HCO_3)_2$[a]
Iron	≈ 0.2	Component of cytochromes and certain non-heme iron proteins and a cofactor for certain enzymes	$FeCl_3$, $Fe(NH_4)(SO_4)_2$, Fe-chelates
Trace elements		Required in trace amounts for a variety of functions, for example, as cofactors for enzymes	$CoCl_2$, $ZnCl_2$, Na_2MoO_4, $CuCl_2$, $MnSO_4$, $NiCl_2$, Na_2SeO_4
Organic growth factors			Vitamins, amino acids, purines, pyrimidines

Sources: Todar, K. "Nutrition and Growth of Bacteria" in *Todar's On-Line Textbook of Bacteriology.* http://textbookofbacteriology.net/nutgro.html.

Sigma-Aldrich, "Microbiology Introduction." https://www.sigmaaldrich.com/technical-documents/articles/microbiology/microbiology-introduction.html.

[a] Also act as buffers.

vitamins, minerals, and carbohydrates. Casein, the principal protein in milk, is a common protein substrate for forming peptones, and the term *tryptone* has traditionally meant a digest of casein. Many microbiologists are now trying to avoid animal-derived substances in growth media, so vegetable sources, such as soybean meal, may be substituted to make "tryptones." Bacto Peptone, Bacto Tryptone, and Soytone are examples of the more than 50 commercial peptones available.

The enzymes used in the past for digestion of the substrate were derived from animal sources, particularly pork pancreas. Digestive enzymes for making peptones now are often manufactured by genetically modified organisms.

Defined media do not contain such substances as meat extracts or peptones. Nitrogen is commonly added to defined media in chemical form as nitrate, ammonium, or as purified amino acids. Glucose is a common carbon and energy source found in both undefined and defined media. (Dextrose is the name given to glucose produced from corn. Chemically, they are the same and you may see dextrose in a recipe for bacterial medium.) Other carbon/energy sources in both defined and undefined media include acetate, glycerol, certain lipids, and proteins. Sodium and potassium phosphates are used in defined and undefined media to provide phosphorus and to serve as buffers. Defined media contain sulfur in the form of inorganic salts of sulfate, hydrogen sulfide, sulfur granules, or thiosulfate, or in the form of organic compounds such as cysteine and methionine (both of which are amino acids).

31.3.3.4 Trace Elements

Trace elements are those substances required by cells in small amounts. Magnesium, calcium, and iron may be placed in this category, although these elements are required in higher amounts than substances such as cobalt. Trace elements are required for the activity of a number of enzymes. Iron is an essential component of cytochromes, which are molecules required for producing energy in cells. Defined media include trace elements supplied as mineral salts (e.g., $CaCl_2$ and $CuSO_4$). Complex media may rely on extracts and peptones to supply trace elements, but complex media may also be supplemented with mineral salts. Some trace elements are required in such small amounts that they are present naturally in sufficient amounts in agar and other media components.

31.3.3.5 Growth Factors

Naturally occurring *E. coli* can grow on simple, minimal media because they have biosynthetic pathways to convert glucose, nitrogen, sulfur, and other simple building blocks into all their required cellular molecules (e.g., proteins, vitamins, and nucleic acids). Many bacteria, however, cannot synthesize all their own molecular components. Those components that cannot be synthesized by bacterial cells and must be obtained from the environment are broadly termed *growth factors*. Vitamins, amino acids, purines, and pyrimidines (required for the synthesis of DNA and RNA) frequently fall into this category. All these components are added to defined media in purified forms (hence adding expense).

Growth factors in undefined media may be supplied by peptones and various extracts and/or may be added in purified form directly to the medium. Alternatively, serum is sometimes added to bacterial media to provide growth factors, proteins, vitamins, carbohydrates, lipids, amino acids, minerals, and trace elements. The exact composition of sera is unknown and varies from lot to lot, although lot-to-lot consistency has improved in recent years. Serum is discussed further later in this chapter when its use in mammalian cell culture is discussed.

31.3.3.6 Supplements

Bacterial culture media may contain other substances in addition to those listed above that are loosely classified as *supplements*. Examples include selective dyes, pH indicators, antibiotics, and chelating agents that bind to metals.

Molecular biologists frequently supplement their bacterial culture media with antibiotics. This is because antibiotic resistance is often used to select for those bacteria that have been transformed with a gene of interest. Molecular biologists use plasmids to carry a gene of interest into bacteria. The plasmids are engineered to carry both the gene of interest and also a gene that makes bacteria resistant to a particular antibiotic. Bacteria that are transformed by taking up the plasmid therefore have an antibiotic resistance gene and will grow on a culture medium that contains the antibiotic. Non-transformed bacteria that did not take up the plasmid will not grow when exposed to the antibiotic. Bacterial strains that carry plasmids with antibiotic selection markers should be cultured in the presence of the selective agent.

Note that antibiotics are often sensitive to light and should be protected from it. They are also sensitive to heat and therefore are sterilized by filtration (Chapter 32). Agar-containing media should be cooled to 50°C or lower before adding antibiotics. ATCC recommends that sterile stock solutions of antibiotics be stored in a −20°C freezer in the dark. ATCC further recommends that bacterial plates containing antibiotics should be stored in the dark no longer than three months at 4°C due to degradation of antibiotics.

31.3.4 Preparing Bacterial Culture Media

Like other biological solutions, bacterial growth media are made by combining the proper amounts of ingredients in the proper final volume of water so that each ingredient is present at the correct final concentration. Box 31.1 describes some common practices for making culture media. Box 31.2 and Figure 31.4 show common practices for pouring agar plates. Refer also to Chapter 30 for information about aseptic technique.

31.3.5 LB Agar/Broth

Escherichia coli bacteria are commonly used in the laboratory for recombinant DNA work. Naturally occurring *E. coli* can grow on minimal medium, but the strains used for recombinant DNA work have been mutated so that they require a complex growth medium. Their stringent nutrient requirements help ensure that laboratory *E. coli* strains cannot "escape" into the general environment. A complex medium is also useful for recombinant DNA work because cells grow more slowly and/or produce less DNA and protein in minimal medium than they do in complex medium. LB broth/agar is the complex bacterial medium most commonly used by molecular biologists for culturing *E. coli* in genetic transformation procedures. Lysogeny broth (LB) was created in the 1950s by Giuseppe Bertani, who was studying viruses that infect bacteria. (Lysogeny is a condition in which the viruses survive within a host bacterium but do not reproduce, nor do they burst the bacterial cell.)

BOX 31.1   COMMON PROCEDURES AND TIPS FOR BACTERIAL CULTURE MEDIA

STORAGE OF DEHYDRATED MEDIA

Dehydrated media are hygroscopic (absorb water) and are sensitive to heat, light, and extreme fluctuations in temperature.

1. *Store dehydrated media according to the manufacturer's directions, usually below 25°C in a dry area, away from direct sunlight and heat sources.* Some media are stored refrigerated.
2. *Record the date of receipt of each container of media on its label and the date that the container is first opened.*
3. *Check the expiration date; some media have longer shelf lives than others.* Discard after expiration date.
4. *Use media in the order of receipt; finish a bottle before opening a fresh one.*
5. *After use, tightly close the bottle.* Whenever possible avoid repeated opening and closing of media bottles since this will cause the media to degrade more quickly.
6. *Discard medium if it is not free-flowing, if its color has changed, or if it appears abnormal.*

PREPARATION OF MEDIA

1. *Water for reconstituting dehydrated microbial media should be freshly prepared by distillation, deionization, or reverse osmosis.* It is recommended that the conductivity be below 15 μS/cm and preferably under 5 μS/cm. Water should be stored in containers made from inert materials. Storage in polythene containers and flint glass is not recommended.
2. *Weigh out the required amount of dehydrated medium, or the various components of the medium; place in a clean, dry flask that is 2–3 times larger than the final volume desired.* Avoid inhaling the powder or prolonged skin contact.
3. *Add half the appropriate amount of water to the flask, swirl to mix. Pour the rest of the water down the sides of the flask to wash any powder adhering to the sides of the flask.* (Dried powder that adheres above the liquid level may not be sterilized during autoclaving and may cause contamination.)
   a. Most broths are clear at this point and do not require boiling to dissolve the components.
   b. If required by a procedure or manufacturer's directions, bring the mixture to the proper pH. Some procedures call for adjusting the pH before bringing the mixture to final volume, while others do so afterward.
4. *Agar: When agar is used in the medium, most manufacturers recommend melting the agar before autoclaving using a microwave, hot plate, boiling water, or steam bath. Heat the medium to boiling with frequent stirring or swirling to prevent overheating. Caution is required as agar media may boil over unexpectedly.* The purpose of this step is to ensure that the powdered agar completely dissolves and is uniformly distributed throughout the medium. This reduces the risk of contamination that may occur if any dry powder remains above the level of the liquid.

(Continued)

**BOX 31.1 (*Continued*)   COMMON PROCEDURES AND
TIPS FOR BACTERIAL CULTURE MEDIA**

a. Some procedures do not call for melting the agar before autoclaving since agar will dissolve in the autoclave. Be sure the agar is completely dissolved and uniformly mixed after autoclaving, if following this practice.

b. Culture media with a pH below 6.0 will hydrolyze agar and reduce its gelling properties. If using an acidic medium, avoid re-melting the agar once it has hardened.

c. Procedures may call for agar to be autoclaved separately from the other components, especially when making minimal media, to avoid formation of a precipitate. In these procedures, the other ingredients, such as salts, are prepared in a concentrated form and autoclaved. Glucose is prepared separately as a concentrate and is sterilized. Magnesium sulfate also is often prepared separately from other components. After sterilization, the agar, salts, and glucose (and other heat-labile ingredients, such as antibiotics) are combined aseptically.

5. ***When glucose is required, add it to water slowly with mixing; otherwise, the glucose will clump and resist dissolving.***

a. Glucose will turn brown if autoclaved with other ingredients, so some procedures call for glucose to be filter-sterilized.

b. Some procedures call for glucose to be autoclaved separately as a concentrated stock solution and added to the rest of the ingredients after autoclaving.

c. Some sources recommend that glucose not be autoclaved at a pH above 7.0 and that it not be autoclaved in the presence of amino acids.

6. ***Autoclave the solution(s) if required. (Refer also to Chapter 24 for more information on autoclaving.)***

a. A basic principle to keep in mind is that heat degrades media components, and therefore, heating should be minimized while still sufficient to eliminate microbial contaminants.

b. Use a calibrated, well-maintained autoclave to ensure that the temperature and pressure in the chamber are as expected.

c. *Loosely cap* flasks or tubes with cotton plugs, plastic foam plugs, or suitable plastic or metal caps.

d. Place tubes in racks.

e. Be sure flasks are less than two-thirds full.

f. Do not overload the autoclave and spread the flasks out uniformly in the chamber to optimize heat penetration, reduce cooling times, and ensure that each flask receives the same heat and pressure conditions.

g. Sterilize for 15–20 minutes at 15 psi, 121°C for quantities of liquid media up to one liter. If larger volumes are sterilized in one container, especially if the medium was not hot when placed in the autoclave, then a longer period should be employed. Note, however, that longer sterilization times might lead to additional degradation of media components. Do not use a higher temperature (unless directed to do so by a standard procedure) as this may degrade some components. (Recommendations for sterilization times, temperatures, and pressures are typically included in media recipes and/or in information from the manufacturers of dehydrated media.)

h. Avoid prolonged heating or over-sterilization, which may adversely affect some medium components and may cause agar to begin to precipitate. Repeated melting of solidified agar or long holding of melted agar at high temperature may also cause a precipitate to form.

i. Use a slow exhaust setting on the autoclave to release the pressure slowly at the end of sterilization; otherwise, the media will boil over, blowing the plugs from the tubes or flasks. The pressure, however, should drop rapidly enough to avoid excessive exposure to heat after the sterilization period. Around 10 minutes is recommended as the time to reach atmospheric pressure.

j. Remove media from the autoclave as soon as sterilization is complete.

k. Tighten caps once the media is cool (< 40°C).

l. Note that microwave ovens do not adequately sterilize culture media and are not used for this purpose.

(*Continued*)

7. Add supplemental, heat-sensitive ingredients just before use.

   a. *If agar is present, cool the medium to 50 to 55°C before adding other components.* (A flask containing liquid at 50°C feels hot, but can be held continuously with bare hands.) Broths can be cooled to room temperature before adding supplements.

   b. *Sterilize antibiotics by filtration through a filter unit with a pore size of 0.1 or 0.2 m.* If antibiotics are not used right after preparation, distribute them into aliquots and store in the dark at –20°C. (See also Practice Problem 4.)

   c. Aseptically add supplements to the medium. Swirl to mix; avoid bubbles.

8. ***Aseptically dispense sterile media into sterile tubes, flasks, or Petri dishes.***

9. ***Store prepared media at the temperature indicated in the product description.*** The length of time that prepared media can be stored varies depending on the type. Some manufacturers recommend storing simple broths and agars no longer than 6 months. Selective media may degrade more quickly.

10. ***For quality control, test the medium.***

    a. Test a portion of prepared tubes or plates by incubating them at an appropriate temperature (e.g., 30°C–37°C, or 20°C–25°C) for 2–5 days to check for sterility; do not reuse these tubes or plates.

    b. Test the prepared medium for growth performance by inoculating it with a stock culture.

    c. Periodically test stored medium to see that it continues to support growth as it did when freshly prepared and that it is not contaminated.

    d. Visually inspect prepared media before use. Determine that color, clarity, pH, and other characteristics of the culture medium are typical as listed in the product description.

    The information in this table can be found in multiple sources including:

    - Difco Laboratories. *The Difco and BBL Manual of Microbiological Culture Media.* 2nd edition. 2009.

      https://www.trios.cz/wp-content/uploads/sites/149/2016/08/DIFCO-A-BBL-MANUAL-2.pdf

    - Thermo Scientific. *Culture Media Preparation: Smart Notes.* https://www.thermofisher.com/content/dam/LifeTech/Thermo-Scientific/MBD/Marketing-Materials/General/Culture-Media-Preparation-Guide-EN.pdf

---

BOX 31.2   POURING NUTRIENT AGAR PLATES USING STANDARD ASEPTIC TECHNIQUES

1. ***This procedure uses standard aseptic technique. (See Section 30.3.)***
2. ***Prepare and autoclave nutrient agar; cool to a temperature of 50°C–55°C.*** Swirl gently to mix, avoiding air bubbles.
3. ***Label the bottom of sterile 100 mm plastic culture plates.***
4. ***Spread the plates out on lab bench; turn on Bunsen burner.***
5. ***Open the flask containing the nutrient agar, and pass the top through the Bunsen burner flame.***
6. ***Pour plates:***

   a. Carefully lift the lid of each culture plate; do not set the lid on the lab bench.

   b. Quickly pour in agar until the bottom plate is about 1/3 full (1/4 inch of agar). Immediately replace the lid after pouring the medium. Avoid bubbles. Swirl agar flask periodically to ensure homogeneity.

   c. Repeat for each plate, occasionally flaming the top of the flask.

   d. Hold the flask slightly tilted to prevent airborne contaminants from falling into the flask.

   e. Work quickly, but carefully, so that the agar does not harden before all the plates are poured.

7. ***Let plates cool with the lids on to solidify agar.*** Freshly poured plates are wet, which can allow applied bacteria to spread where they should not and can interfere with other applications. Many sources advocate allowing the plates to dry inverted at room temperature for 12–48 hours. They can also be dried in an inverted position in a 37°C incubator for 0.5–3 hours.

(Continued)

**BOX 31.2 (*Continued*)   POURING NUTRIENT AGAR PLATES
USING STANDARD ASEPTIC TECHNIQUES**

8. ***For storage, return dried plates to their plastic sleeves and turn them upside down to prevent
    condensate from pooling on the surface. Store at 4°C.***

   a. Plates that were stored in the cold should be warmed to room temperature for a few hours before
      use to avoid condensate.

   b. Store plates that contain antibiotics (which are light sensitive) in a dark room or wrapped in alu-
      minum foil at 4°C for not more than three months. Plates should be inverted.

   c. If plates are to be stored for more than 30 days, wrap in plastic to avoid moisture loss.

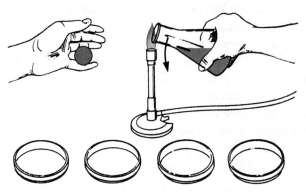

(a)   Remove the stopper, and flame the mouth of
        the flask.

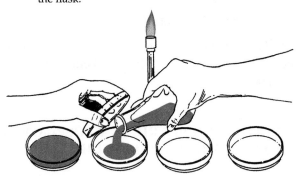

(b)   Remove the cover from one dish, and pour
        nutrient agar into the dish bottom.

FIGURE 31.4   Pouring agar plates using standard open bench aseptic techniques. (a) The cap is removed from the nutrient agar flask, and the mouth is passed briefly across the flame. If all of the nutrient agar will be poured into plates, the cap
can be set aside. If the flask will need to be closed again, then the cap is held in the left hand. (b) The cover is removed from
each dish, one at a time, and agar is poured into the plate. Observe how the principles of aseptic technique are applied.
The flask is flamed to create an updraft away from the lip. The plates are kept closed until needed, and the lid of the plates
is held above the open plate to protect it from the air. The cap is not placed down on the lab bench, because the bench is
assumed to be contaminated.

There are several different versions of LB broth
that vary in the amount of sodium chloride they contain, thus providing different osmotic conditions. The
concentration of salt can affect the yield of plasmid
obtained when the culture is being used for the purpose of isolating plasmids from transformed bacteria.
The low-salt formulations, Lennox and Luria, are useful for cultures requiring salt-sensitive antibiotics.

- LB-Miller (10 g/L NaCl)
- LB-Lennox (5 g/L NaCl)
- LB-Luria (0.5 g/L NaCl).

Box 31.3 provides an example of an LB formulation
that is used for making agar plates.

BOX 31.3   LB AGAR

Ingredient	Amount	Function of Component
Agar	15.0 g	Hardens medium
Tryptone	10.0 g	Supplies varied nutrients, including nitrogen, sulfur, carbon, and vitamins
Yeast extract	5.0 g	Supplies varied nutrients, including nitrogen, sulfur, and carbon, also vitamins and trace elements
Sodium chloride	10.0 g	Supplies sodium ions for membrane transport and osmotic balance
Water	Bring to 1 L	

Directions:

1. Mix ingredients in purified water.
2. Some procedures call for adjusting the pH of LB to 7.5 or 8 with sodium hydroxide. A disadvantage to using sodium hydroxide for this purpose is that the pH will not be buffered in the medium, which means that metabolic by-products from the bacteria will change the medium's pH during growth. A less commonly used option is to adjust the pH with 5–10 mM Tris buffer that is at the desired pH.
3. Sterilize mixture by autoclaving at 121°C for 15 minutes.
4. Cool medium to about 55°C before adding antibiotics and before pouring plates. A water bath can be used to bring the medium to the correct temperature. (The bottom of the flask will be warm to the touch, but will not burn the hand at this temperature.)
5. Add antibiotics, if required, from sterile stock solutions. (See also Practice Problem 4.)
6. Pour plates as directed in Box 31.2.

31.4 MEDIA FOR CULTURED MAMMALIAN CELLS

31.4.1 GENERAL REQUIREMENTS FOR MAMMALIAN MEDIA

Mammalian cell culture media must be carefully formulated and prepared. Some of the same ideas discussed for bacterial media also apply to mammalian culture media, but mammalian cells generally have more complex and stringent growth requirements than bacteria. Mammalian cells normally exist inside the body in contact with other cells where they are continuously bathed by blood with a complex, high-protein nutrient mixture. It is challenging to simulate this natural environment in culture vessels or bioreactors. The water used for mammalian cell culture must be of very high purity; the cells' gaseous environment must be tightly controlled; pH and osmolality must be maintained in narrow ranges; the culture medium must provide a complex mix of nutrients. Mammalian cell culture is particularly threatened by contamination because common contaminating organisms, such as bacteria and molds, multiply much more rapidly than mammalian cells and can quickly take over a cell culture. For all these reasons, performing mammalian cell culture requires training and practice.

Different cell types have different requirements for their growth media. The early cell culture technologists had to develop formulations to support the growth of their cells, but now there are existing recipes for media for routine culture of many cell types. Organizations that supply cells (e.g., ATCC) provide information on the cell culture medium recommended for culturing those cells. It is also possible to find information on cell culture media for specific cell lines in the scientific literature describing the origin of those cell lines.

Cell culture technologists in some facilities mix culture media themselves from the individual components, but the complexity of the formulations makes this practice relatively unusual in smaller-scale operations. High-quality, premixed mammalian cell culture media are conveniently (if expensively) available from commercial suppliers.

Biotechnology companies may perform cell culture at a very large scale using large volumes of cells to manufacture a product of interest. The use of cultured cells for production introduces some special concerns. The culture medium must not only provide for healthy cell growth and reproduction, but must also provide the building blocks (e.g., amino acids and sugars) for the synthesis of the product. The particular composition of the medium will therefore have a major impact on the yield of product made by those cells. Moreover, the constituents of the cell culture medium must be removed from the final product; this affects the downstream purification process. (Methods of product purification are introduced in Chapter 34.) The product may be used therapeutically in humans; this means it must be free of harmful contaminants that might be introduced from the medium. For all these reasons, finding a suitable culture medium for production is an important part of the development of a product that is manufactured by cells. Research and development scientists in industrial settings commonly perform extensive experimentation to optimize the culture medium formulation so that it provides the best yield of purified product most cost-effectively.

31.4.2 BASAL CELL CULTURE MEDIA

31.4.2.1 A Family of Basal Media

Mammalian cell culture is a more recent technology than bacterial culture. Many of the cell culture media and methods used today were developed in the mid-twentieth century by pioneer scientists including Harry Eagle, Renato Dulbecco, and Richard Ham. Eagle systematically investigated the nutritional requirements of cultured cells. Eagle's studies led to the formulation of a defined **basal liquid medium** *that contains a mixture of nutrients dissolved in a buffered, isotonic salt solution.*

Eagle's original basal medium formulation (BME) was designed to grow mouse fibroblasts and HeLa cells (a cell line from a human cervical carcinoma). This formulation was later modified in various ways to suit a wide range of cells, thus creating a "family" of basal media.

Some common basal media, such as minimal essential medium (MEM) and Dulbecco's modified Eagle's medium (DMEM), are relatively simple. Others are enriched with more constituents, such as additional vitamins and amino acids, trace elements, lipids, and nucleic acids. Ham's F-12 is an example of a richer medium. These richer media were developed to support specific cell types and also to provide the base for serum-free formulations (discussed in more detail below). The richer basal media are often termed *complex* because they have additional ingredients. Note that the word *complex* has a different meaning in bacterial culture, where it means that the exact components of the medium are not known. Some examples of common basal media and their characteristics are shown in Table 31.4.

Eagle's basal medium and others like it are incomplete; they are designed to be supplemented with animal blood serum. **Blood serum** *is the liquid component of blood from which blood cells and most clotting factors have been removed.* Serum provides a rich and complex source of proteins, polypeptides, amino acids, growth factors, lipids, carbohydrates, salts, hormones, and vitamins. Serum also helps to buffer the culture medium, inactivates proteolytic enzymes, increases medium viscosity (which reduces shear stress during pipetting or stirring), and conditions the growth surface of the culture vessel so that cells adhere better. This means that, when complete, these traditional (classic) mammalian cell culture media are undefined. Serum is discussed in more detail in a later section of this chapter.

31.4.2.2 Components of Basal Cell Culture Media

Table 31.5 shows the ingredients of Dulbecco's modified Eagle's medium (DMEM), an example of a classic basal medium formulation. (Formulations for other common media can be found in the references cited in this unit's Bibliography.) DMEM contains a defined mixture of many nutrients including:

* inorganic salts
* amino acids
* vitamins and
* glucose (as a source of energy and carbon).

A **balanced salt solution (BSS)** *is a mixture of inorganic salts in specific concentrations.* As discussed earlier in this chapter, these solutions are isotonic. Balanced salt solutions may be used by themselves to support the survival of mammalian cells for a few hours or so, but lack the nutrients required for longer periods of survival and growth. Cell culture media therefore contain a balanced salt solution plus nutrients.

Salts play multiple roles in mammalian cell culture media. They help maintain osmotic equilibrium. They help regulate the transport of materials across the cell membrane; sodium, potassium, and calcium are particularly important in membrane transport. Salts function as cofactors for enzymes. Salts also play a role in the attachment of adherent cells to the surface of their plate or flask. In situations where culturists want to grow cells in suspension, rather than adhered to the surface of their vessel, the calcium and magnesium levels are reduced in the medium.

There are a variety of salt solution recipes used in different media. Earle's and Hank's are two specific BSS formulations that are sometimes used to supply salts in basal cell culture media. Both Earle's and Hank's contain phosphate and bicarbonate for buffering

TABLE 31.4

Examples of Basal Cell Culture Media

Name	Features
Eagle's minimal essential medium (MEM)	A commonly used modification of Eagle's original formulation. Contains a higher concentration of amino acids than BME; supports cell growth for several days.
Dulbecco's modified Eagle's medium (DMEM)	A commonly used modification of Eagle's original formulation. Contains four times the concentration of vitamins and twice the concentration of amino acids as BME. Contains twice as much HCO_3 and CO_2 as BME to improve buffering capacity. Contains iron. Energy sources are optimized for protein production and nucleic acid metabolism. Originally contained 1 g/L of glucose. Available now with high glucose (4.5 mg/L) and low glucose (1 g/L).
RPMI-1640 (Roswell Park Memorial Institute)	An enriched formulation originally derived for human leukemia cells; now used for many mammalian cell lines.
McCoy's 5A	Developed in 1959 by McCoy and co-workers for cultivation of cells from a liver tumor. Subsequently modified to create a medium that supports the growth of cells derived from a wide variety of tissues.
Ham's F-10 and F-12	Ham's nutrient mixtures were originally developed to support the growth of specific cell types including Chinese hamster ovary and HeLa cells. Both mixtures were formulated for use with or without serum supplementation or with low serum concentration, depending on the cell type being cultured.
Dulbecco's modified Eagle's medium/Ham's F-12 Nutrient	During the past decades, researchers have learned to culture a variety of cell lines in medium that contains reduced serum or no serum. These media are supplemented with nutrients, growth factors, and hormones (e.g., insulin, transferrin, and epidermal growth factor). A 1:1 mixture of DMEM and Ham's F-12 Nutrient Mixture often provides the base for these low-serum and serum-free media. HEPES, an organic buffer, is included at a final concentration of 15 mM to compensate for the loss of buffering capacity that results from eliminating serum.
Iscove's Modified Dulbecco's Medium (IMDM)	A modification of DMEM containing additional amino acids and vitamins, selenium, sodium pyruvate, HEPES, and potassium nitrate instead of ferric nitrate. Has been useful in growing hybridomas and has been used as the base for serum-free formulations (discussed later in this chapter).

Primary Source: ATCC. *Animal Cell Culture Guide: Tips and Techniques for Continuous Cell Lines.* 2014. https://www.atcc. org/~/media/PDFs/Culture%20Guides/AnimCellCulture_Guide.ashx.

purposes (discussed below) and also contain calcium, magnesium, potassium, sodium, and phosphate. They differ in their concentrations of these salts. Hank's buffered salt solution contains a substantially lower concentration of sodium bicarbonate than Earle's BSS. Manufacturers will sometimes provide basal media with a choice of BSS, for example, Eagle's MEM with either Earle's salts or Hank's salts.

Amino acids are supplied to cultured cells for the synthesis of proteins and to supply energy. There are two types of amino acids: essential and nonessential. **Essential amino acids** *are those that cells cannot synthesize, plus cysteine and tyrosine.* Culture media must supply essential amino acids. The essential amino acids are L-arginine, L-cysteine, L-glutamine, L-histidine, L-isoleucine, L-leucine, L-lysine, L-methionine, L-phenylalanine, L-threonine, L-tryptophan, L-tyrosine, and L-valine.

L-Proline is required by Chinese hamster ovary cells, which are important in biopharmaceutical manufacturing. Nonessential amino acids may be added to culture media, either because a particular cell type cannot make them, or to enrich the medium to improve the growth of the cells.

Minimal basal media contain the B vitamins (folic acid, biotin, and pantothenate) plus choline, folic acid, inositol, and nicotinamide. Other vitamins are supplied by serum. Richer medium formulations supply other vitamins in addition to these, such as vitamin C (ascorbic acid) and vitamin E (alpha-tocopherol).

Most media formulations use glucose to supply energy and as a source of carbon. A few media substitute galactose or another sugar for the glucose. Glucose was originally thought to be the only energy source in basal media, but studies have shown that cultured cells also use amino acids for energy. The amino acid

TABLE 31.5

Components of DMEM

Inorganic salts (mg/L)	
NaCl	6,400.00
KCl	400.00
$CaCl_2 \cdot 2H_2O$	264.90
$MgSO_4 \cdot 7H_2O$	200.00
$NaH_2PO_4 \cdot 2H_2O$	140.00
$NaHCO_3$	3,700.00
Amino acids (mg/L)	
Arginine·HCl	84.00
Cystine·2Na	56.78
Glutamine	584.60
Glycine	30.00
Histidine·HCl·H_2O	42.00
Isoleucine	104.80
Leucine	104.80
Lysine·HCl	146.20
Methionine	30.00
Phenylalanine	66.00
Threonine	95.20
Tryptophan	16.00
Tyrosine·2Na	89.46
Valine	93.60
Trace elements (mg/L)	
$Fe(NO_3)_3 \cdot 9H_2O$	0.10
Vitamins (mg/L)	
Choline·Cl	4.00
Folic acid	4.00
Inositol	7.00
Nicotinamide	4.00
Pantothenate·Ca	4.00
Pyridoxal·HCl	4.00
Riboflavin	0.40
Thiamine·HCl	4.00
Other components (mg/L)	
Phenol red	10.00
Glucose	4,500.00
Pyruvate·Na	110.00
CO_2 (gas phase %)	10

L-glutamine is especially important for this purpose and is included at a high concentration relative to other amino acids.

Some basal culture media, including DMEM, contain sodium pyruvate. Pyruvate is part of one of the chemical pathways that cells use to create energy. It is added to media as an energy source and to provide a carbon skeleton for the synthesis of other molecules. Its addition helps the growth of some cell types, and it is particularly important when serum concentration is reduced in the medium.

There are richer basal medium formulations that provide additional nutrients, such as additional amino acids, proteins, vitamins, fatty acids, and lipids. These formulations are used for specific cell types and applications, and as a base for serum-free media.

31.4.2.3 Sodium Bicarbonate/ CO_2 Buffering System

Mammalian cells require a pH-controlled environment. Many cell types are maintained at a pH of 7.2–7.4, but some cells prefer a pH closer to 7.0 (e.g., Chinese hamster ovary cells) and others a somewhat higher pH of 7.4–7.7 (e.g., fibroblasts). Cells produce lactic acid during metabolism, and so a buffer is required to maintain the pH of their medium at a constant value.

Bicarbonate/CO_2 is the most common buffering system for mammalian cell culture because it is relatively inexpensive and nontoxic, and bicarbonate has a nutritional value. The bicarbonate/CO_2 buffer system functions both to bring the pH of the medium into a specified range and also to resist a change in pH if acids or bases are added to the medium. The bicarbonate/CO_2 buffer system requires addition of both carbon dioxide and bicarbonate to the culture medium. Sodium bicarbonate, $NaHCO_3$, is directly dissolved in the culture medium. Carbon dioxide, a gas, is most commonly provided by a CO_2 tank that is attached to the cell culture incubator and is adjusted with a regulator to deliver a steady flow of 5%–10% CO_2 in air. The caps for culture vessels are designed in such a way that contaminants are excluded, but gases can exchange with the culture medium, thus providing both carbon dioxide and oxygen. Sometimes, depending on the cell type, the culture medium, and the cell density, sufficient CO_2 is generated by the cells during metabolism and an external CO_2 tank is not required.

To understand this buffer system, consider first what happens when sodium bicarbonate is added to the culture medium. The sodium bicarbonate dissociates to form HCO_3^-, which undergoes these reactions:

$$HCO_3^- + H^+ \rightarrow H_2CO_3 \rightarrow CO_2 + H_2O$$

If low concentrations of bicarbonate alone are added to the culture medium, the HCO_3 will tend to remove hydrogen ions, leaving the solution somewhat basic. Next, consider what happens when carbon dioxide is added to the medium. Carbon dioxide from the air reacts with the water in the culture medium to form H_2CO_3, carbonic acid. The carbonic acid, in turn, ionizes:

$$H_2O + CO_2 \rightarrow H_2CO_3 \rightarrow H^+ + HCO_3^-$$

Culture Media for Intact Cells

The net result of increasing the level of carbon dioxide is to add hydrogen ions, which makes the solution somewhat acidic.

In cell culture, both CO_2 and bicarbonate are added to the medium together. In this case, equilibrium is reached between the reactions; the exact pH at which equilibrium occurs depends on the percent CO_2 and the concentration of bicarbonate provided. For example, for DMEM, 44 mM bicarbonate and 10% carbon dioxide are typically used. This concentration of bicarbonate by itself would bring the medium to a pH of about 7.8. The CO_2 alone would bring the medium to a pH of about 4.4. When both are present, the medium equilibrates at about pH 7.2.

This system is buffered in that if moderate amounts of either an acid or a base are added to the medium, the system shifts back to the desired equilibrium pH. If H^+ ions are added to the medium, then:

$$H^+ + HCO_3^- \rightarrow H_2CO_3 \rightarrow CO_2 + H_2O$$

If a base is added to the medium, then H^+ ions are removed from the system. In this case, more carbon dioxide reacts with water to produce more hydrogen ions:

$$H_2O + CO_2 \rightarrow H_2CO_3 \rightarrow H^+ + HCO_3^-$$

The system can thus compensate for small additions of acid or base and return the pH to the proper value.

There are limitations to the CO_2/bicarbonate buffering system. The medium may become basic very quickly if removed from the CO_2 incubator. An organic buffer, usually HEPES, is sometimes added to the medium (in addition to the CO_2/bicarbonate buffer) at concentrations of 10–20 mM to compensate for this problem, or to control the pH for cells that are very sensitive to pH change. HEPES can also be used alone to control pH. HEPES provides good control of pH, but it is more expensive than a bicarbonate buffer system and may be toxic to cells when present at higher concentrations. Note also that if HEPES is added to the medium, the equivalent concentration of NaCl is usually omitted to control the osmolality of the solution.

Most cell culture media include the pH indicator, phenol red, which allows the culturist to quickly judge the metabolic condition of a culture by the color of the medium. Below pH 7.0 the indicator is orange, becoming yellow at about pH 6.5. At pH 7.2–7.4 it is red, and above 7.5 it is reddish blue. Cells produce lactic acid during metabolism. When the amount of lactic acid rises above the buffering capacity of the medium, the HCO_3^- is depleted and the pH drops. The pH indicator turns orange, indicating that it is time to feed the cells with fresh medium.

31.4.2.4 Water

Water is the primary ingredient in cell culture media, and it must be highly purified and endotoxin-free. Culturists working in laboratories that require only small volumes of medium often purchase water purified for cell culture. Laboratories and production facilities where large volumes of culture media are required usually have on-site water purification systems. The water should minimally meet Type I water purity standards, as described in Chapter 24.

31.4.3 SUPPLEMENTS TO BASAL CULTURE MEDIA

31.4.3.1 L-Glutamine and Bicarbonate

Mammalian cell culture supplements *can be defined as materials that are sometimes added to basal media after the medium is prepared and/or that are added in varying concentrations.* L-Glutamine and glucose are common constituents of cell culture media that are often added as supplements right before use. L-Glutamine is an essential amino acid that is unstable in liquid solutions at temperatures of 4°C or higher. Over time, it dissociates to form toxic ammonium. Therefore, glutamine is prepared as a sterile, concentrated 100× stock, aliquoted, and stored frozen so that it does not degrade. L-Glutamine is typically used at a final concentration of 0.1–0.6 g/L. Cell culture media should be changed frequently to ensure availability of this amino acid and to avoid toxicity.

L-Alanyl-L-glutamine is sometimes used as a substitute for L-glutamine because it is stable over time. It is a dipeptide, that is, two amino acids chemically bonded together. Cells can retrieve L-glutamine by cleaving the peptide bond.

Sodium bicarbonate is usually included in liquid basal media preparations, but is often left out of dry preparations because it can liberate CO_2 during storage. Sodium bicarbonate at a concentration of 0.4–4.0 g/L is usually added to powders at the time of reconstitution and preparation.

31.4.3.2 Antibiotics

Some cell culture technologists add antibiotics to cell culture media in low concentrations to help avoid contamination, although most cell culture technologists avoid antibiotics whenever possible. The use of antibiotics can cover up poor sterile technique and can lead to low-level bacterial contamination that goes undetected. Antibiotic use can also encourage the growth of antibiotic-resistant bacteria. The use of antibiotics can therefore result in spurious results in cell culture experiments and is avoided in industry.

It is most common to add antibiotics for short periods when performing primary culture, or when establishing cultures of valuable cells. If antibiotics must be used, broad-spectrum antibiotics (e.g., ampicillin, gentamicin, kanamycin, and neomycin) are sometimes preferred because they are toxic to both gram-positive and gram-negative bacteria. Combinations of more specific antibiotics, such as penicillin and streptomycin, are also used.

Mycoplasma is a type of very small difficult-to-detect bacterium and was discussed in Chapter 30. In humans and animals, *Mycoplasma* can cause disease. These bacteria are troublesome contaminants in cell culture because they are difficult to detect, difficult to eradicate, and can subtly alter the properties of infected cells. Anti-mycoplasma agents include gentamicin and kanamycin. These are more commonly used with human-derived cells where *Mycoplasma* contamination can be a significant problem. Antifungal agents are also sometimes added, but should be used sparingly.

The use of antibiotics is not recommended for "curing" contaminated cultures. Unless contaminated cultures are irreplaceable, they should be discarded and an intensive decontamination program should be initiated. Note that Chapter 37 will discuss the use of master and working cell banks in industry. If a working cell bank becomes contaminated, a new vial from the protected master cell bank can be retrieved, thus avoiding the disastrous loss of a valuable cell line.

31.4.3.3 Other Supplements

There are a number of substances with specialized biological functions that are added to culture media for special purposes and for specific cell types. For example, MEM is a simple medium that is often supplemented with a solution of nonessential amino acids. This supplement can be prepared or purchased separately as a sterile stock (often 10 mM; 100X) that is aseptically added to the medium for a final concentration of 0.1 mM each. MEM can also be purchased with nonessential amino acids already added.

Selection agents are supplements used when cells are genetically transfected to cause the death of untransfected cells, as described above in the section on antibiotics in bacterial media. When cells are genetically transformed with a gene of interest, they are also transformed with a gene that enables them to detoxify a toxic agent. The toxic agent is added to the culture medium so that only genetically transformed cells survive.

31.4.3.4 Serum

Serum has traditionally been the most important supplement used for mammalian cell culture. It is also the most problematic for many current applications. Early cell culture technologists found that serum greatly enhances the growth of cultured cells. Serum contains a rich and complex array of many micronutrients and growth factors that promote cell reproduction. Serum also increases the buffering capacity of the medium and helps protect cells against mechanical damage, which can occur when cells are stirred or scraped from the surface of a plate.

The blood of fetal calves (fetal bovine serum, FBS) is a common source of serum because it is rich in embryonic growth factors. This blood is harvested when pregnant cows are slaughtered for food production. Other less-expensive sources of serum include calves and horses. Serum is processed from its animal source, sterilized by filtration, dispensed into aliquots, frozen, and stored at 20°C until use. When needed, serum is thawed and added aseptically to the culture medium, usually at a concentration of 5%–20% (v/v).

Procedures from different sources vary in their approach to adding serum (and other supplements). For example, when 10% serum is specified, some procedures call for adding 100 mL of serum to 1,000 mL of prepared medium. The result is that the concentration of serum is actually about 9.1% and the concentration of other components in the medium is also slightly altered from their reported values. The reason to prepare the medium this way is to avoid extra manipulation, for example, by removing 900 mL of a previously prepared medium and adding 100 mL of serum. Each manipulation provides an opportunity for contaminants to enter the medium. Many commonly used continuous cell lines are tolerant of variation in the exact concentration of the constituents of their medium, so a serum level that ranges between 9% and 10% will support their growth. This practice is problematic, however, when concentrations are communicated between technologists. Also, some cells are exacting in their growth requirements (e.g., primary cells) and concentrations of media constituents must be maintained in a narrow range.

Some procedures call for heat-inactivation of serum by incubating it at 56°C for 30 minutes. It is then dispensed into aliquots and frozen. Heat-inactivation of serum is thought to aid cell culture by inactivating complement (a group of blood proteins associated with the immune system). Heat-inactivation is useful for assays or procedures where complement is not desired (e.g., when cells will be used to prepare or assay viruses). Heat has also been used in the past to destroy

Mycoplasma, but this is no longer necessary because most serum suppliers use 0.1 μm filters to remove this small bacterial contaminant. It is advisable to avoid heat-inactivation unless it is required because heat can reduce or destroy growth factors, thus reducing the potency of the serum. Boxes 31.4 and 31.5 outline procedures for thawing and inactivating serum.

31.4.4 PREPARING STANDARD MEDIA

31.4.4.1 Purchasing Media

Premixed, quality-controlled mammalian cell culture media are available from commercial suppliers. This section discusses the use of standard commercially prepared cell culture media.

Media are supplied from manufacturers in three forms:

- 1X sterile liquids
- 10X sterile concentrates and
- powdered, dehydrated media.

The most convenient and expensive media are the 1X liquid media; these do not require the addition of water.

The 10X concentrates must be diluted by the user with sterile, highly purified water. Powdered media are less expensive than liquid, but require dissolution and sterilization by the user.

Most manufacturers provide standard media in several forms to provide flexibility for the user. It is possible, for example, to purchase DMEM with either 1.0 or 4.5 g/L of glucose. Some basal media can be purchased with either Hank's BSS or Earle's BSS. Sodium pyruvate and L-glutamine might be included in a medium mixture or might be left out.

The details of preparing media vary somewhat depending on which type is chosen. Some supplements are added before the medium is sterilized, while others are sterilized separately and are added aseptically to the rest of the medium.

31.4.4.2 Sterilizing Mammalian Cell Culture Media

Liquid 1X and 10X media are sterile when received from the manufacturer, but powdered media must be dissolved and sterilized by the user. In contrast to most bacterial media, mammalian cell culture media usually contain components that are heat sensitive and

---

BOX 31.4   PROCEDURE TO THAW SERUM

Serum can be stored for at least 2 years at −20°C or −70°C with little deterioration in growth-promoting activity.

1. *Remove serum from freezer, and refrigerate overnight at 2°C–8°C.*
2. *Preheat a water bath to 37°C.*
3. *Transfer the serum bottles to the 37°C water bath.*
4. *Agitate the bottles from time to time in order to mix the solutes that tend to concentrate at the bottom of the bottle.*
   a. Do not keep the serum at 37°C any longer than necessary to completely thaw it. It is not recommended to thaw serum at a higher temperature.
   b. Thawing serum in a bath above 40°C without mixing may lead to the formation of a precipitate inside the bottle.
   c. Do not subject serum to repeated freezing and thawing.

Note: According to ATCC, serum may be cloudy after thawing. This is because serum may contain small amounts of fibrinogen, a blood protein that may be converted to insoluble fibrin. ATCC has tested serum after this has happened, and their studies indicate that the serum is still suitable as a supplement for cell culture media. If the presence of this flocculent material is a concern, it can be removed by sterile filtration through a 0.45 μm filter. A precipitate can also form in serum that is incubated at 37°C for prolonged periods of time. The precipitate may include crystals of calcium phosphate. The formation of a calcium phosphate precipitate does not alter the performance of the serum as a supplement for cell culture.

Source: ATCC. *ATCC Animal Cell Culture Guide: Tips and Techniques for Continuous Cell Lines.*
https://www.atcc.org/~/media/PDFs/Culture%20Guides/AnimCellCulture_Guide.ashx2014.

BOX 31.5   PROCEDURE TO HEAT-INACTIVATE SERUM

1. *Preheat water bath to 56°C.* Ensure that there is sufficient water to immerse the bottle above the level of serum, but not up to the neck of the bottle.
2. *Mix thawed serum by gentle inversion, and place serum bottle in the 56°C water bath.*
3. *After the temperature of the water bath reaches 56°C again, continue to heat for an additional 30 minutes.* Mix gently every 5 minutes to insure uniform heating.
4. *Remove serum from water bath, and cool quickly because slow cooling can sometimes reverse the inactivation of complement activity.* ATCC recommends storing serum at –70°C if possible, otherwise at –20°C.

Source: ATCC. ATCC Animal Cell Culture Guide: Tips and Techniques for Continuous Cell Lines. https://www.atcc.org/~/media/PDFs/Culture%20Guides/AnimCellCulture_Guide.ashx2014.

cannot be autoclaved. Mammalian culture media are therefore sterilized by filtration. Filtration removes microorganisms because they are too large to pass through the pores of the filter. A pore size of $0.2\,\mu m$ is traditionally used because fungi and most bacteria are removed by this size pore. *Mycoplasma,* however, require $0.1\,\mu m$ filters for removal.

For filtering volumes in the milliliter range, a small filter can be attached to a syringe; pressure is applied to the syringe plunger to force the liquid through the filter. (See Figure 32.5) Syringe filtration is often used to sterilize concentrated supplements that are added to media in small volumes.

Laboratory-scale filtration of cell culture medium is commonly performed by placing a filtration unit on top of a sterile bottle and *applying vacuum to the underside of the filter unit; this is called* **negative pressure filtration**. (See Figure 32.4) Negative pressure filtration is commonly used for volumes up to a few liters. Alternatively, a *liquid culture medium can be forced through a filter with a pump that applies pressure to the top of the unit; this is called* **positive pressure filtration.** Positive pressure filtration is usually preferred when larger volumes of medium are filtered.

Table 31.6 summarizes general considerations relating to the preparation of mammalian cell culture media. Box 31.6 is an example of a procedure to prepare cell culture medium from a 1X liquid; Box 31.7 is an example of a procedure to prepare medium from a 10X liquid concentrate; and Box 31.8 is an example of a procedure to prepare medium from a commercially available powder mix.

31.4.5 SERUM-FREE MEDIA, ANIMAL PRODUCT-FREE MEDIA, PROTEIN-FREE MEDIA, AND DEFINED MEDIA

The use of serum in cell culture is problematic, despite its many benefits. Serum is expensive. Serum introduces lot-to-lot variability into the cell culture process because the individual animals from which it is taken vary. It is common for technologists in cell culture facilities to test several lots of sera for compatibility with a particular cell line and buy entire lots of the best sera. Even when this practice is used, the lot is eventually gone, and a different lot of serum must be introduced. This variability is a problem both in production settings and in research studies where it introduces unknown variables into the experiments.

Two additional serum-related problems arise when mammalian cells are used for biopharmaceutical production. First, serum introduces impurities that need to be removed from products during downstream processing, thus adding complexity and cost. Second, serum may be contaminated with pathogens (e.g., viruses and prions) that are difficult to remove even with the best modern methods. These pathogens may or may not be a risk to the cells themselves, but can contaminate a product made by the cells. In the 1980s and 1990s, some people in Europe died of a new variant of the neurological disease, Creutzfeldt–Jakob disease, thought to be transmitted from beef infected with a bovine pathogen. In response, in 1993 the FDA recommended that pharmaceutical manufacturers not use bovine-derived materials from cattle that resided in, or originated from, countries where this bovine disease had been diagnosed. This concern about bovine pathogens has now been extended to other animal-derived materials; if cows harbor pathogens that threaten humans, so can other animals.

TABLE 31.6

General Considerations Relating to Mammalian Cell Culture Media

1. *Water for cell culture can be sterilized by autoclaving.* Use glass or plastic bottles designed for autoclaving. Loosen the caps, and include 10% extra volume to allow for evaporation. Caps should not be tightened until bottles have cooled; otherwise, the vacuum resulting from cooling can cause breakage.

2. *Filtration (0.1–0.2 μm pore size) is used to sterilize heat-labile substances; most mammalian cell culture media contain such ingredients.* To filter protein supplements (e.g., hormones and growth factors), use only filters that are specified to be low protein binding.

3. *Media containing HEPES, riboflavin, and tryptophan can be photoactivated by normal fluorescent lighting.* This is problematic because photoactivation produces toxic hydrogen peroxide and free radicals. Short-term exposure is not normally a problem, but media should not be stored in lighted walk-in cold rooms or refrigerators with glass doors that allow room light to reach the media. The addition of sodium pyruvate to media is said to reduce or eliminate this problem. (For more information, see, for example, Lonza FAQs. https://knowledge.lonza.com/faq?id=789&search=stable+cell+lines)

4. *Common additions to media*:
   a. 1X liquid media are typically provided by the manufacturer without serum or L-glutamine.
   b. 10X concentrated liquid media are typically provided by the manufacturer without serum, L-glutamine, or sodium bicarbonate.
   c. Powdered media are typically provided by the manufacturer without serum or sodium bicarbonate.
   d. These substances are added to the medium just before use. (See Practice Problems 6 through 9.)

5. *Sodium bicarbonate may be added to media either in solid form prior to sterilization (e.g., for powdered media) or as a 7.5% sterile solution added after sterilization.* Follow the manufacturer's directions for the recommended amounts.

6. *L-Glutamine is an essential amino acid, required by virtually all mammalian and invertebrate cell lines, that is somewhat unstable in liquid media.*
   a. L-Glutamine is usually omitted from commercial liquid media and must be aseptically added from a sterile, concentrated stock solution prior to use.
   b. Additional L-glutamine can be added to media to extend the medium's shelf life.
   c. Liquid L-glutamine stock can be purchased from most commercial vendors of cell culture supplies.
   d. L-Glutamine concentrations for mammalian cells vary from 0.68 mM for Medium 199 to 4 mM for DMEM. Typical concentrations for cells in biomanufacturing are 2–4 mM.
   e. Dipeptides containing L-glutamine are a stable alternative to L-glutamine itself. Dipeptides are two amino acids linked by a chemical bond. Cells can retrieve L-glutamine by cleaving the peptide bond.

7. *Pyruvate is sometimes added to culture media. It is an intermediate in a metabolic pathway that provides energy to cells.*
   a. Pyruvate can pass readily into or out of cells, and its addition to culture media provides both an energy source and a carbon skeleton for the synthesis of other molecules.
   b. Pyruvate addition may be advantageous when maintaining certain specialized cells, when maintaining cells at low density, and when the serum concentration is reduced in the medium. It may also help reduce fluorescent light-induced phototoxicity.
   c. Pyruvate is usually added to a final concentration of 0.1 mM and is commercially available as a 10 mM (100X) stock solution.

8. *Antibiotics.*
   a. Antibiotics should be avoided for routine culture work because they can mask subtle bacterial or fungal contamination and may affect the growth of sensitive cells.
   b. Antibiotics are sometimes added for short periods to primary cultures or as a safeguard while expanding valuable cultures to produce cell banks.
   c. Typical concentrations in medium are 50–100 units penicillin G, 50 to 100 μg/mL of gentamicin sulfate, or 2.5 μg/mL of amphotericin B.

9. *HEPES is an effective organic buffer that is commonly used for cell culture.* However, it can be toxic to sensitive cells and has also been shown to increase the sensitivity of media to fluorescent light.

(Continued)

TABLE 31.6 (*Continued*)

General Considerations Relating to Mammalian Cell Culture Media

10. *Concentrated media are manufactured at a low pH because their components are less likely to precipitate when the pH is acidic.* After dilution, the medium might be too acidic or possibly too basic (depending on the formulation) and must be brought to the correct pH. Powdered media are usually brought to a pH 0.1 or 0.2 pH units below the desired final pH because their pH rises slightly during filtration.

11. *Adjusting the pH of media.* Sterile NaOH and HCl are used to bring media to the correct pH. They can be sterilized by filtration through a 0.2 μm filter; check that the filter membranes are compatible with these substances before use.

12. *Sterilized, complete media are commonly stored at 4°C in the dark if not used immediately.*
    a. Serum is stored separately and added to otherwise complete medium just before use.
    b. Consult the manufacturer to determine how long a medium can be stored.
    c. Do not freeze complete media because freezing causes some of the growth factors and/or vitamins to precipitate out of solution. These components can be difficult to get back into solution after thawing.

13. *Commercially prepared media are isotonic.* It is advisable to check the osmolality of the medium if it is supplemented with extra salt solutions or large volumes of buffering substances. Some drugs and hormones used to supplement media are initially dissolved in an acidic or basic solution, and the medium is brought to the correct pH after their addition. These acids and bases can significantly alter the osmolality of the final medium. It is therefore important to be conscious of osmotic effects when adding supplements to a medium (e.g., when adding a drug substance to test its effect or adding a supplement to enhance growth).

14. *Since the osmolality of culture media will rise as a result of evaporation, CO_2 incubators must be well humidified in order to prevent evaporation.*

15. *Media that contain Earle's salts are intended to be used with 5% carbon dioxide.*
    a. This is typically accomplished in an incubator provided with CO_2 gas and culture vessels that allow gas exchange.
    b. Alternatively, the vessels may be gassed with 5% CO_2 after filling.
    c. Bicarbonate levels of 2.2 g/L are typical with Earle's salts.

16. *Hank's salts have a lower bicarbonate level (0.35 g/L) and less buffering capacity than Earle's salts and are designed to be used without the addition of carbon dioxide to the gas phase.* A culture vessel that does not allow gas exchange should be used.

17. *Be aware that media formulations vary among suppliers, even for media with similar or identical names.* Read descriptions, formulations, and labels carefully to ensure that you use the appropriate medium for your cells and to ensure consistency.

Primary source: ATCC. ATCC Animal Cell Culture Guide: Tips and Techniques for Continuous Cell Lines. https://www.atcc.org/~/media/PDFs/Culture%20Guides/AnimCellCulture_Guide.ashx2014.

BOX 31.6   AN EXAMPLE OF A PROCEDURE FOR PREPARING A CELL CULTURE MEDIUM FROM 1X LIQUID

Use disposable, sterile cell culture plasticware or well-cleaned, sterilized glassware dedicated to cell culture.

1. *Check that serum, L-glutamine stock solution, and any other required supplements are ready.*
2. *Perform calculations (see Practice Problems for examples) to determine the amounts required for serum, glutamine, and any other additions required.*
3. *Thaw serum, glutamine, and any other frozen supplements by placing in a 37°C water bath.*
4. *Prepare laminar flow cabinet.*
5. *Swab the outside of the bottle of medium and any supplement tubes with 70% alcohol; place in laminar flow cabinet.*
6. *Using aseptic technique, add the correct amount of supplements to medium.*

(Continued)

7. ***Check that the pH equilibrates at the proper temperature.*** If it does not, alter the CO_2 concentration or adjust the pH of the medium.

8. ***Label the bottles with the name of medium, supplements, your name, and date.***

9. ***The complete medium is ready to use and has a limited shelf life.*** If stored, place in the dark at the temperature recommended by the manufacturer.

10. ***Quality control***:

    a. Incubate one or more bottles under cell culture conditions for three days to check for visible contamination (e.g., cloudiness, precipitate, flocculent matter, and change in pH). It is advisable to check both the basal medium and the complete medium with serum. Alternatively, remove 1 mL aliquots from each bottle of medium and transfer to 12-well culture plates. Incubate.

    b. Record how and when all glassware and equipment were sterilized, the lot numbers of filters used, all media component lot numbers and vendors, and how and when media were sterilized. Every bottle of culture medium should be individually labeled.

    c. If a separate microbiology area is available, it is possible to check the medium for contamination by applying some to a bacterial growth plate and incubating the plate; nothing should grow.

BOX 31.7   A PROCEDURE FOR PREPARING 1 L OF MEDIUM FROM A 10X LIQUID CONCENTRATE

Use disposable, sterile cell culture plasticware or well-cleaned, sterilized glassware dedicated to cell culture.

1. ***Check for precipitate.*** Precipitates will normally redissolve with dilution. If, however, a precipitate has formed due to degradation of a component of the medium, then the quality of the medium may be reduced, and it should be tested for its ability to support cell growth.

2. ***Check that sterile, highly purified water, serum, sterile 7.5% bicarbonate stock solution, glutamine stock solution, and any other required supplements are ready.***

3. ***Perform calculations (see Practice Problems for examples) to determine the amounts required for water, serum, glutamine, and any other additions required.***

4. ***Prepare laminar flow cabinet.***

5. ***Thaw serum, glutamine, bicarbonate, and any other frozen supplements in a 37°C water bath.***

6. ***Swab the outside of the bottle of medium concentrate and any tubes containing supplements with 70% alcohol; place in laminar flow cabinet.***

7. ***To make 1 L, using aseptic technique, measure 700–750 mL of sterile, highly purified water into a suitable sterile container.***

8. ***Using aseptic technique, gently stir and add 100 mL of 10X concentrate.***

9. ***Using aseptic technique, add the correct amount of 7.5% sodium bicarbonate based on manufacturer's instructions.***

10. ***Using aseptic technique, add sterile L-glutamine and other supplements.***

11. ***Adjust the pH with sterile NaOH or HCl.***

12. Add sterile, highly purified water to 1 L.

13. ***The complete medium is ready for use and has a limited shelf life.*** If stored, place in the dark at the temperature recommended by the manufacturer.

14. ***If medium is not used immediately, wait until the time of use and then add serum to the desired final concentration.***

    a. Store the basal nutrient medium and serum individually. Prepare the complete medium at the time of use and only in the volume necessary.

    b. Add serum at 1 to 20% v/v to otherwise complete medium.

15. ***Quality control, as in Box 31.6.***

BOX 31.8   A PROCEDURE FOR PREPARING 1 L OF MEDIUM FROM POWDER

- Use only highly purified water. Use disposable, sterile cell culture plasticware or well-cleaned, sterilized glassware dedicated to cell culture.
- Powdered media are extremely hygroscopic and must be protected from atmospheric moisture.
- The preparation of medium in concentrated form is not recommended because some of the amino acids have limited solubility and may precipitate in concentrated solutions to form insoluble salt complexes.
- Filter-sterilize the medium immediately after mixing to avoid microbial growth.

1. *Select a container as close in size to the final volume as possible. Measure out 90% of the final volume of water required.* Some sources recommend using sterile water and autoclaved containers at this point, although the medium will be filter-sterilized later.
2. *Slowly add the powdered medium to the water with gentle stirring.*
   a. Rinse the original package with a small amount of water to remove all traces of powder.
   b. Cover and stir with a sterile stir bar until dissolved.
   c. Do not use heat.
3. **Optional:** *Add the amount of HEPES that yields a concentration of 15 mM in the final volume of medium.* Omit this step if the powdered medium is already formulated with HEPES or if HEPES is not desired.
4. *Add the amount of sodium bicarbonate recommended by the supplier for use in a CO_2-controlled atmosphere.*
5. *Add NaOH or HCl with gentle stirring to adjust the pH.* Adjust the medium to 0.1 to 0.3 pH units below the desired final pH because the pH may rise this much with filter-sterilization.
6. *Bring to the final volume with highly purified water. (Some sources use sterile water.)*
7. *Sterilize the medium in a laminar flow hood by filtration through a 0.1 or 0.2 μm filter into sterile media bottles.*
8. *Cap tightly with sterile closures, and store in the dark at the temperature recommended for the product.*
9. *At the time of use, remove appropriate amount of medium using aseptic technique.*
   a. Add glutamine from a sterile stock solution to give a final concentration of 2 mM.
   b. Add serum to the desired final concentration at the time of use.
10. *Quality control, as in Box 31.6.*

Primary source: Sato, J.D. and Kan, M. "Media for Culture of Mammalian Cells." In Bonifacino, Juan S. et al. (Eds.) *Current Protocols in Cell Biology.* John Wiley and Sons, 2001. doi: 10.1002/0471143030. cb0102s00.

Serum therefore raises a variety of scientific concerns:

- Expense
- Lot-to-lot variability
- Introduction of unknown substances into the culture
- Product purification issues
- Pathogen contamination
- Regulatory concerns.

The first three of these concerns affect those who use cells either for research or for production purposes.

The latter issues are of particular concern when cells are used for biopharmaceutical manufacturing.

In response to these concerns about serum, researchers have been trying to determine the identities and functions of the many components of serum in order to develop serum-free media (SFM). It has been a challenging task to develop SFM that nourish cultured cells as effectively as serum-supplemented media. The main ingredients of serum are well known (e.g., the proteins albumin and transferrin), but serum contains numerous, diverse substances, many of them active at very low concentrations.

Serum-free media therefore require a great many additions, including:

- **Lipids**
- **Vitamins**
- **Essential trace metals** (including iron, zinc, copper, selenium, manganese, molybdenum, aluminum, silver, and nickel)
- **Attachment factors** (aid cells in attaching to substrates)
- **Cytokines** (small messenger proteins released by blood cells that are involved in communication between cells)
- **Hormones** (e.g., insulin, growth hormone, and steroids) and
- **Growth factors** (chemicals that play numerous roles in the promotion of cell growth and cell maintenance).

The first approach to creating SFM was to supplement enriched basal media with animal-derived protein hydrolysates (peptones) and/or with purified proteins from animal or human sources. Although some of the concerns relating to serum are alleviated by the use of these first-generation serum-free formulations, their high protein content can cause problems with downstream purification. Also, these formulations contain substances derived from animal sources and so raise the same concerns about animal pathogens as does serum. Furthermore, the use of hydrolysates introduces variability and undefined substances into the medium. Bovine serum albumin, BSA, for example, is an important serum protein that is often included in SFM. It is isolated from blood and usually contains a variable mixture of chemicals including fatty acids and other proteins.

As a next step in medium development, scientists substituted yeast and plant-based hydrolysates (e.g., from soy, wheat, and cottonseed) for animal-based hydrolysates. These substitutes have been suitable for many cell lines, including CHO and other production cell lines. Biopharmaceutical products made with these plant-based media have reached the market. Although these media do not contain materials directly isolated from animals, they nonetheless are undefined and may vary from lot to lot.

A third step in medium development is the creation of completely defined SFM in which all the medium requirements of the cells are provided in purified form. The result of these developments is that there is now an array of media optimized for various applications (Figure 31.5) that have one or more of the following characteristics:

- **Low serum.** Enriched formulations designed primarily to save money by reducing, but not eliminating, serum. (See Practice Problem 11.)
- **Serum-free.** Enriched formulations that eliminate the need for serum supplementation, but may contain animal-derived constituents and may be undefined.
- **Animal-derived component-free media (ADCF), also called xeno-free media.** Enriched formulations that do not require serum and whose constituents are not derived from animal sources. These media address safety concerns associated with animal-derived raw materials, but they may provide an uncharacterized mixture of peptides and trace elements derived from plant sources and may introduce lot-to-lot variability.
- **Protein-free media.** Formulations that do not contain proteins (e.g., BSA, insulin, and attachment factors). These media are a step toward completely defined media, but may contain both animal-derived components (depending on the manufacturer) and protein hydrolysates.
- **Defined media.** Contain only known, purified constituents, such as proteins made by recombinant DNA methods. May or may not be ADCF.

These newer cell culture media have many applications. In industry, new, optimized media formulations have

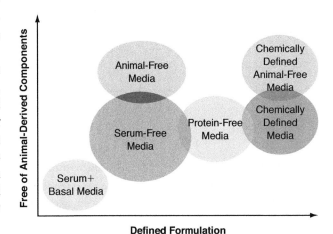

FIGURE 31.5 Trends in cell culture media. The trend in cell culture media is toward more defined media and media whose components are not isolated from animals. (Reprinted with permission from Hodge, Geoffrey. "Media Development for Mammalian Cell Culture." *Biopharm International*, May 1, 2005.)

contributed to increased productivity of biopharmaceutical production systems and have alleviated many of the concerns about animal pathogens. Scientists have benefited from the extensive research into serum and its properties because it has led to a better understanding of the many, and often subtle, factors that affect cell growth and division. There are, however, costs associated with the use of newer cell culture media. Unlike classic formulations (e.g., DMEM) that are readily adaptable to a variety of cell types and applications, the new formulations are tailored for specific cell types and specific applications. A medium that works for one cell type may not work for another. They can therefore be expensive and difficult to develop. Moreover, animal product-free, defined media require the addition of a number of highly purified constituents, each of which may be costly. Entirely avoiding animal sources in the production of these purified constituents may require producing dozens of substances in genetically modified organisms. Much remains to be learned before it becomes routine to use entirely defined, animal product-free, serum-free media for cell culture.

31.4.6 Cell Culturing Continues to Evolve

Mammalian cell culture is a rapidly evolving field. For example, biopharmaceutical companies continuously improve media formulations to provide higher product yields with more consistency and fewer animal-derived components. The introduction of stem cell culture is another example of how cell culture changes – in this case quite quickly. In 1998, James Thomson et al. published a landmark paper describing methods for culturing human embryonic stem cells. The ability to culture stem cells and control their differentiation (when a cell becomes a specific cell type,

such as nerve or muscle) was immediately recognized to have tremendous potential for use in regenerative medicine. This advance triggered a great deal of excitement, although there was also resistance to the use of embryonic cells. In 2007, a paper published by Shinya Yamanaka et al. introduced human induced pluripotent stem (iPS) cells. iPS cells are adult cells that have been genetically reprogrammed to be like embryonic stem cells. This is accomplished by forcing the cells to express genes that make them like embryonic stem cells. iPS cells alleviate concerns about using embryonic cells and open new possibilities for therapeutic and research use. But, not surprisingly, stem cells introduced new challenges for cell culturists.

One difference between working with these stem cells and other mammalian cells is that they must be maintained in an undifferentiated state until needed. Both embryonic and iPS cells have a natural tendency to differentiate. Keeping them in an undifferentiated state and then coaxing them into differentiating into a desired type requires special cell culture techniques. Early stem cell culture methods required the use of a **feeder layer**, *which is a layer of non-dividing cells seeded onto the culture plates in advance.* The stem cells were then introduced on top of the feeder layer. The feeder layer cells conditioned the culture medium by leaking growth factors and nutrients that the stem cells required. The feeder layer also provided a matrix to which the stem cells readily adhered. This system is not suitable for cells that will be used in clinical applications, and so more recent media formulations do not require feeder layers and are xeno-free. For example, TeSR™-E8™ is an animal component-free culture medium for stem cells that contains defined components added to a DMEM/F-12 basal medium. Cell media formulations will continue to evolve as different cell types come into use.

TABLE 31.7
Antibiotics

Antibiotic	Stock Solution Concentration	Recommended Working Concentration (µg/mL)	Amount Stock Required (per L)
Ampicillin[a] (sodium salt)	50 mg/mL in water	100	2 mL
Chloramphenicol	34 mg/mL in ethanol	170	___
Kanamycin	10 mg/mL in water	50	___
Streptomycin	10 mg/mL in water	50	___
Tetracycline HCl	5 mg/mL in ethanol	50	___

These values are for practice only. Consult a specific procedure for recommended stock and working concentrations.

[a] Ampicillin is commonly used as a selective agent, but note that beta-lactam antibiotics (including ampicillin) are not very stable and will slowly degrade when dissolved, even when frozen at 20°C. It is therefore best to reconstitute ampicillin shortly before use. The sodium salt is more soluble than ampicillin alone and is therefore used for culture media.

Practice Problems

1. What is the milliosmolarity of 0.9% NaCl w/v? (FW = 58.44.)
2. A kilogram of a particular solution contains 5% glucose monohydrate (w/w) (MW = 198) and 20 mmol of NaCl.
   a. How many milliosmoles are in the solution?
   b. What is the milliosmolality of the solution?
3. This is the recipe for a defined bacterial medium, as given in a manual of bacteriology.
   M63 medium, 5X
   10 g $(NH_4)_2SO_4$
   68 g KH_2PO_4
   2.5 mg $FeSO_4 \cdot 7H_2O$
   Adjust to pH 7 with KOH
   Directions: Dilute concentrated medium to 1X with sterile water. The following sterile solutions should be included, per liter:
   1 mL of 1 M $MgSO_4 \cdot 7H_2O$
   10 mL of 20% carbon source (glucose or glycerol)
   0.1 mL of 0.5% vitamin B1 (thiamine)
   a. How much concentrate should be used to make a 1X medium if 1 L is required?
   b. What is the concentration of glucose in the final solution?
      What is the concentration of $MgSO_4 \cdot 7H_2O$ in the final solution?
   c. What is the concentration of thiamine in the final solution?
4. Many molecular biology procedures require the addition of antibiotics to nutrient agar plates to select for bacteria that have taken up an antibiotic resistance gene. Antibiotics are not sterilized by autoclaving. They are prepared as concentrated stock solutions, filter-sterilized, and added to the nutrient agar medium after it has cooled to about 55°C. Table 31.7, p. 860, shows typical values for antibiotic concentrations in stock solutions and for final concentrations of antibiotics in the nutrient agar. Fill in the final column of the table with the volume of stock solution needed to make 1 L of nutrient agar. The first answer is filled in as an example.

Acetate differential agar (per liter)	
Sodium acetate	2.0 g
Magnesium sulfate	0.1 g
Sodium chloride	5.0 g
Monoammonium phosphate	1.0 g
Dipotassium phosphate	1.0 g
Bromothymol blue	0.08 g
Agar	20.0 g

5. The recipe for acetate differential agar from the website of BD Diagnostic Systems (http://www.bd.com/ds/) is shown below. This is a bacterial medium that is used to differentiate between *Shigella* and *E. coli* bacteria. *E. coli* can grow on this medium, but *Shigella* cannot. (Many species of *Shigella* cause a severe intestinal illness in humans called shigellosis, which causes high fever and acute diarrhea. Most species of *E. coli* are normal inhabitants of the healthy intestine in humans and other animals.)
   a. What is the source of carbon in this medium?
   b. What is the source of nitrogen in this medium?
   c. What is the likely purpose of the NaCl?
   d. What is the likely purpose of the phosphate compounds?
   e. Given that this is a differential medium, what is the likely purpose of the bromothymol blue?
   f. What is the purpose of the agar? What is the percent concentration of the agar?

6. A total volume of 500 mL of culture medium for mammalian cell culture is desired, including 20% serum. The culture medium is provided as a 1X liquid and has all required components, except serum. What volumes of medium and serum are required?

7. Suppose that the medium from question 6 does not contain glutamine and so it must be added. The glutamine is stored as a concentrated stock solution that is 100X. What volumes of medium, serum, and glutamine are required?

8. One liter of mammalian cell culture medium is to be prepared from 10X concentrate. Glutamine is available in a 200 mM stock solution. Undiluted serum is also available. Sodium bicarbonate (NaHCO$_3$) is available as a sterile 7.5% solution (which is the same as 0.89 M). You want the final concentration of the medium to be 1X, the glutamine to be 2 mM, the serum to be 10%, and the sodium bicarbonate to be 26 mM. How much is required of each of the following?
   a. 10X concentrate
   b. Glutamine stock
   c. Serum stock
   d. Sodium bicarbonate stock
   e. Sterile, highly purified water

9. Different mammalian cell culture media require different concentrations of sodium bicarbonate. Table 31.8 (shown below) shows the levels of sodium bicarbonate sometimes recommended for different media. Fill in the blanks in the table; the first row has been completed as an example.

10. Different mammalian cell culture media require different concentrations of glutamine. Table 31.9 (shown below) shows the levels of glutamine recommended by a manufacturer for different media. Fill in the blanks in the table; the first row has been completed as an example. The L-glutamine is assumed to be in a 200 mM stock solution. Its FW is 146.15.

11. Assume that serum costs $250 for a 500 mL bottle and culture medium costs $15 for a 500 mL bottle.
   a. How much would it cost to make 500 mL of cell culture medium that includes 10% serum?
   b. How much would it cost to make 500 mL of cell culture medium that includes 2% serum?
   For simplicity, assume in both cases that the entire bottle of medium is used, but only as much serum as is needed is removed.

12. List similarities and differences between mammalian cell culture media and bacterial media.

13. Use a web browser to find the patent application, US9284371B2. (See Chapter 3 for more information about patents and patent applications.)
   a. What is this patent about?
   b. What optimized characteristics of the cell culture process, in addition to pH, are described in the Claims?
   c. Why is it important to optimize these characteristics of the cell culture process?

TABLE 31.8
Recommended Additions of Sodium Bicarbonate

Medium	mL of NaHCO₃ Required, 7.5% Stock (per L)	g of Solid NaHCO₃ Required (per L)	Final Concentration of NaHCO₃ (mg/L)
DMEM	49.3	3.70	3,700
DMEM /Ham's F-12	32.5	——	——
Ham's F-12	——	——	1,176
MEM Earle's salts	——	2.20	——
MEM Hank's salts	4.7	——	——
RPMI-1640	——	——	2,000
McCoy's 5A	29.3	——	——
MEM alpha	——	——	2,200

These values are for practice only. Consult a specific procedure for recommended stock and working concentrations.

TABLE 31.9
Recommended Additions of L-Glutamine

Medium	mg/L in Final Medium	mM in Final Medium	mL/L of Stock Required
AMEM	292.3	2.0	10
BME	——	2.0	——
DMEM	——	——	20
F-12K	——	2.0	——
Ham's F-10	——	——	5
Ham's F-12	146.2	——	——
Iscove's MDM	584.6	——	——
EMEM	——	——	10
RPMI-1640	——	2.05	10.25

These values are for practice only. Consult a specific procedure for recommended stock and working concentrations.

UNIT IX

Basic Separation Methods

Chapters in this Unit

- ✦ Chapter 32: Introduction to Filtration
- ✦ Chapter 33: Introduction to Centrifugation
- ✦ Chapter 34: Introduction to Bioseparations

Humans have been separating substances from one another for a long time. As long as 2500 years ago, the Greek physician, Hippocrates, administered willow tea to ease the pain of childbirth, thus extracting aspirin from the willow bark and making it available to his patients. As long ago as this morning, many of us consumed coffee that involved the separation of the solid coffee grounds from a liquid, and the extraction of desirable coffee flavors and caffeine into our mugs. Separation techniques have come a long way since Hippocrates, and the biotechnology industry utilizes methods far more sophisticated than a coffee filter. Yet, as we will see in this unit, the principle of extracting a substance into a liquid, or the idea of filtering materials on the basis of size, is still quite relevant.

DOI: 10.1201/9780429282799-40

Case Study: Separation Techniques in the Real World

Biological separation techniques are essential in biotechnology, whether they are used for studying biological molecules of interest, or for purifying commercial products from a complex solution. However, separation methods are used for a variety of purposes outside the laboratory or production facility. If you have ever cooked spaghetti, you are familiar with the filtering techniques used to separate the pasta from the cooking water. You instinctively know that the holes in the strainer must be smaller than the pasta itself. If you've ever given a blood sample for testing, one of the first things that happens to the sample is centrifugation to separate the heavy blood cells from the plasma liquid for analysis. The brewing of tea or coffee is a chemical extraction process to solubilize flavor chemicals from the solid biological materials. If you drink decaffeinated coffee, that caffeine was removed from the coffee either by extraction with an organic chemical, or by a simple form of adsorption chromatography. And, since you have probably lived in the time of the COVID-19 pandemic, you have likely worn a face mask to separate yourself from virus aerosols, and to avoid spreading virus if you become infected. Cloth face masks function as depth filters – depth filters have been used in the laboratory for decades. In the face of COVID-19, many people who previously had little interest in home economics learned to sew face masks and became conversant in the weaves of various fabrics that affect their ability to reject viruses. In this unit, you will learn more about separation principles as they are applied in biotechnology facilities.

This unit introduces the fundamental principles of the most common bioseparation methods used in biotechnology. Bioseparation techniques are used whenever a biological material is purified for further study and analysis, as when gene segments are purified for later sequencing. Separation techniques are also essential whenever biological products are isolated and purified for commercial sale in large-scale, production settings.

The basic idea of bioseparations is that differences between substances can be exploited to isolate materials from one another. Consider, for example, the production of insulin to treat diabetic individuals. Insulin can be obtained from *E. coli* bacteria that have been genetically modified to produce insulin. But before the insulin can be injected into a patient, it must be isolated from the cells, and a series of bioseparation steps must be performed that sequentially remove impurities (such as cellular debris, other proteins, lipids, viruses, and microbial contaminants). The techniques used to extract, isolate, and purify the insulin are called *bioseparation methods.*

Chapter 32 discusses filtration, a familiar separation method that exploits size differences between substances to separate them from one another. A coffee filter separates the grounds from the coffee because the grounds are too large to penetrate the pores of the filter. In the example of insulin, filtration can similarly be used to trap and remove solid debris derived from the cells that make the insulin. Sophisticated filters with extremely fine pores can be used to trap and

remove viral contaminants. Chapter 32 explores filtration and demonstrates its use in various biotechnology applications.

Chapter 33 discusses centrifugation techniques, applications, and important safety issues. Centrifugation is another separation method that may be familiar to you. Centrifugation, like filtration, exploits size differences between substances to separate them from one another. Filtration and centrifugation are important methods, but are not sufficient to separate all biological materials of interest from one another.

Chapter 34 introduces two more essential families of bioseparation methods, electrophoresis and chromatography. Electrophoresis exploits size differences (and charge differences) between molecules to separate them from one another under the influence of an electrical field. The separation of DNA fragments of different sizes from one another using agarose gel electrophoresis is a common example of an electrophoretic separation.

Chromatography is a very important family of separation methods. Chromatography is particularly versatile because it exploits varied molecular properties to separate materials from one another. These properties include size, charge, solubility, and the affinity of some molecules for other molecules. Chromatography is also versatile in that it can be used to separate the components of very small-volume samples for analytical purposes, and to separate the components of very large volumes of material for manufacturing commercial products. Chromatography thus plays a vital role in both research and production environments.

In order to completely isolate and purify a particular biomolecule (e.g., insulin), it is necessary to perform a series of separation techniques in succession. Each separation technique removes certain contaminants from the product of interest. One of the challenges in biotechnology is the development of bioseparation strategies, consisting of a series of separation techniques, which effectively isolate and purify biological substances of interest. Chapter 34 includes a broad overview of this challenging task.

BIBLIOGRAPHY

There are many references available that provide detailed information about general principles of separation and about specific bioseparation techniques, such as centrifugation. The list below provides starting materials for a more in-depth study of purification methods.

GENERAL BIOSEPARATIONS

Although each of these references mentions proteins in their titles, many of the techniques and approaches described could be applied to other types of biomolecules as well.

Cutler, Paul. *Protein Purification Protocols (Methods in Molecular Biology)*. 2nd ed. Humana Press, 2003.

Rosenberg, Ian. *Protein Analysis and Purification: Benchtop Techniques*. 2nd ed. Birkhäuser, 2004.

Roe, Simon. *Protein Purification Techniques: A Practical Approach*. 2nd ed. Oxford University Press, 2001.

CENTRIFUGATION

Ford, T.C. and Graham, J.M. *An Introduction to Centrifugation*. Bios Publishers, 1991. A readable basic primer on centrifuges.

Graham, John. *Biological Centrifugation (The Basics)*. Garland Science, 2001. (Overview of basic principles.)

Rickwood, David, and Graham, John. *Biological Centrifugation*. Springer Verlag, 2001.

Useful technical brochures and documents can be obtained from centrifuge manufacturers and supply companies using the Internet. Examples are as follows:

Beckman Coulter, Inc. https://www.beckman.com/resources/fundamentals/principles-of-centrifugation. Beckman Coulter, Inc., also provides a rotor calculation resource at https://www.beckman.com/centrifuges/rotors/calculator. This can be used to calculate RPM, RCF (average and maximum), and k factor, based on the available information about any rotor. They also have a linked resource for run-time conversions with k factors.

Cole-Parmer Instrument Company. "Basics of Centrifugation." https://www.coleparmer.com/tech-article/basics-of-centrifugation.

Eppendorf. "A Short History of Centrifugation." 2018. https://handling-solutions.eppendorf.com/sample-handling/centrifugation/this-and-that/detailview/news/a-short-history-of-centrifugation/

SPECIALIZED BIOSEPARATION TECHNIQUES

Many techniques such as electrophoresis and chromatography are sufficiently specialized to be described in separate volumes. Below are some suggested references for detailed information beyond that provided by the general references above. In addition, the manufacturers and suppliers of materials for these techniques are usually excellent sources of information.

Dong, Michael W. *HPLC and UHPLC for Practicing Scientists*. Wiley-Interscience, 2019.

Miller, James M. *Chromatography: Concepts and Contrasts*. 2nd ed. Wiley-Interscience, 2004.

Rathor, Anurag S., and Velayudhan, Ajoy, eds. *Scale-Up and Optimization in Preparative Chromatography: Principles and Biopharmaceutical Applications*. CRC Press, 2002.

Westermeier, Reiner. *Electrophoresis in Practice: A Guide to Methods and Applications of DNA and Protein Separations*. 4th ed. WileyVCH, 2005.

WORKING WITH MONOCLONAL ANTIBODY PRODUCTS

Liu, Hui F., et al. "Recovery and Purification Process Development for Monoclonal Antibody Production." *MAbs*, vol. 2, no. 5, 2010, pp. 480–99. doi:10.4161/mabs.2.5.12645.

Wang, Xin, et al. "Molecular and Functional Analysis of Monoclonal Antibodies in Support of Biologics Development." *Protein & Cell*, vol. 9, no. 1, 2017, pp. 74–85. doi:10.1007/s13238-017-0447-x.

32 Introduction to Filtration

32.1 INTRODUCTION

32.1.1 THE BASIC PRINCIPLES OF FILTRATION

Filtration was introduced in Chapter 24 when we talked about preparing high-quality water for use in the laboratory or in production. In this chapter, we continue the discussion of filtration, but focus on other applications of filters in addition to water purification.

Filtration is a common separation method based on a simple principle: Particles smaller than a certain size pass through a porous filter material; particles larger than a certain size are trapped by the filter. Anyone who has strained spaghetti through a colander, made coffee with coffee filters, or played with a sieve at the beach has used filters. Gases, such as air, can be filtered as well as liquids. The air filters in cars and furnaces are commonplace examples.

The fluid and particles that pass through a filter are called the **filtrate** *or* **permeate.** *The materials trapped by the filter are sometimes called the* **retentate**. We are sometimes interested in the filtrate, as is the case with a coffee filter. We are interested other times in the material retained on the top of the filter, as is the case with spaghetti. We occasionally want to collect both the retained particles and the filtrate.

Filtration is also common in nature. Water is cleared of particulates as it passes through sandy soil to the groundwater. The kidneys are effective filtration devices that allow small unwanted metabolites to pass from blood into the urine, while retaining the relatively large blood cells.

Just as filtration plays various roles in the home and in nature, so it has a long history in the laboratory. Filtration was traditionally accomplished by folding a filter paper and placing it in a funnel for support (Figure 32.1). Liquids poured into the funnel would drip through the filter, and solids would remain on the surface. To speed up the process, analysts could add a vacuum pump to the flask. These simple filter paper systems are still used in laboratories to remove coarse particles from a solution.

Although the principle of filtration is simple, the study of filtration is complicated by its wide range of applications and the many types of filtration devices available. Filtration is used from the smallest scale in the laboratory, where samples of only a few microliters may be processed, to major industrial processes involving thousands of liters.

Depending on the application, the sample type, and the scale, filtration systems can vary greatly. Regardless of the simplicity or complexity of a filtration system, however, the following four components are present:

DOI: 10.1201/9780429282799-41

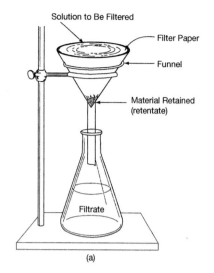

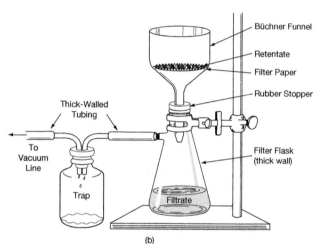

FIGURE 32.1 Laboratory filtration systems. (a) The simplest laboratory filtration system consists of a folded piece of filter paper supported in a conical funnel. Liquid moves through the filter due to the force of gravity. (b) A vacuum can be used to facilitate the movement of fluid through the filter. In this example, a Büchner-style funnel is used to support the filter paper, which lies flat within the funnel. Note that it is good practice to place a trap bottle between the vacuum flask and the source of vacuum to trap materials that may be pulled into the tubing. An alternative approach is to use an in-line disk filter to protect the vacuum source.

1. **The filter itself.**
2. **A support for the filter (like the funnels in Figure 32.1).**
3. **A vessel to receive the filtrate.**
4. **A driving force (such as gravity or vacuum) that drives the movement of fluids and particles through the filter.**

These four components are present both in a simple coffee maker and in a complex process filtration system in a pharmaceutical company.

32.1.2 OVERVIEW OF ISSUES IN FILTRATION

There are several issues that affect filtration and must be considered in designing or selecting a filtration system. The most obvious is **clogging**. Filters clog if they are covered by large particles, or by aggregations of smaller particles. Oils, lipids, and fats can form a film on the surface of a filter, which prevents filtration. In some situations, clogging is "cured" simply by replacing the clogged filter with a new one. In more sophisticated systems, particularly in industrial settings, clogging is reduced by moving the liquid to be filtered across the filter surface. Such systems are discussed later in this chapter.

Some substances, such as proteins, bind to certain filter materials. *When a component binds to the surface of a medium, such as a filter, it is called* **adsorption.** Adsorption tends to block the pores of a membrane, resulting in a lower rate of filtration. Adsorption also leads to loss of the adsorbing component in the sample. Adsorption has some chemical selectivity; that is, some macromolecules adsorb to certain surfaces more readily than others. Also, some materials used in the manufacture of filters readily adsorb various macromolecules, whereas others minimize binding.

Adsorption is different from absorption. **Absorption** *is when liquids are taken up into the entire depth of a material, as when water is absorbed by a sponge.* There is no chemical selectivity associated with absorption, but there is with adsorption. Both adsorption and absorption can occur in filtration.

Extractability is another issue in filtration. **Extractables** *are compounds from a filter that leach out and enter the sample being filtered.* (This issue was introduced in Chapter 24, Section 24.2.1.2. For our purposes, leaching and extraction are synonymous, although some sources distinguish the two.) Fibers from a filter may also enter the sample being filtered. The contamination of a sample by materials from the filter can be a serious problem in some applications. For example, in cell culture, extractable substances have been shown to inhibit the growth of cells.

In the pharmaceutical industry, filters have many applications, such as removing bacteria and viruses from products, and purifying water. It is essential that these filtration processes do not introduce impurities into the drugs. Another application of filtration is to remove particulates from samples before they are injected into instruments for analysis. Many types of instruments used in analyzing samples detect minute amounts of compounds, so impurities from filters must be meticulously avoided.

32.2 TYPES OF FILTRATION AND FILTERS

32.2.1 OVERVIEW

Recall from Chapter 24 that there are different classifications of filters based largely on the size of particles they retain (Figure 32.2).

1. **Macrofilters or general filters (depth filters)** *are used for the separation of particles on the order of 10 μm or larger.* A coffee filter and a common laboratory paper filter are examples of macrofilters.
2. **Microfilters** *are typically used to separate particles whose sizes range from about 0.1 μm to 10 μm.* Bacteria and whole cells fall in this size range.
3. **Ultrafilters** *are used to separate macromolecules on the basis of their molecular weight.* For example, large proteins can be separated from smaller ones using ultrafiltration. Ultrafilters allow such substances as small sugars, salts, amino acids, and inorganic acids to pass through.

4. **Nanofilters** *are used to remove dissolved molecules in the 0.001 micrometer range, which is smaller than those removed by ultrafiltration.* Nanofilters have applications in water treatment and industrial processing. In particular, nanofiltration has become one of the main methods used to remove viral contaminants from biopharmaceutical products.
5. **Reverse osmosis (RO) filters** *are more restrictive than nano- and ultrafiltration membranes and can remove substances down to 300 D in molecular weight.* These filters are effective in removing viruses, bacteria, and pyrogens. Moreover, RO filters also reject ions and very small dissolved particles, such as sugars. These filters are particularly important in laboratory treatment systems that make ultrapure water.

Note that the filters that are used for micro-, ultra-, and nanofiltration, and those used for RO are often called **membranes.**

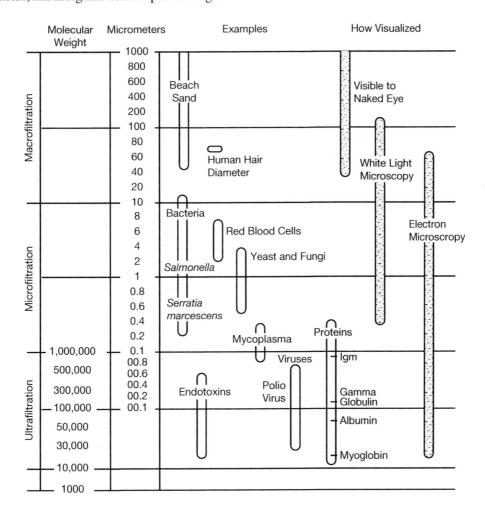

FIGURE 32.2 Filtration and size.

32.2.2 More about Macrofiltration

Macrofiltration filters are inexpensive devices used to remove relatively large particles, above about 10 μm, from a liquid. These filters are often used as "prefilters" to remove large particulates from a solution before it is filtered with a microfilter or ultrafilter. The prefilter prevents clogging of the more expensive, finer filter.

Macrofiltration filters are made of materials such as paper, woven polymer fibers, stainless steel mesh, or sand. These filters do not have pores or holes of a particular size; rather, they consist of a convoluted granular or fibrous matrix. Particles are trapped both on the filter surface and within the sinuous matrix. Because these filters trap particles throughout their depth, they are also called **depth filters**.

Macrofilters for the laboratory are usually made of cellulose (paper) or of glass or woven polymer fibers. In practice, all of these may be referred to as *filter paper,* even if they are not actually paper. Glass-fiber filters consist of long strands of borosilicate glass that are formed under high heat conditions. In contrast to filters made of paper, filters made of glass- and polymer-fibers allow faster flow rates, are stable under a wide range of temperatures, and are compatible with corrosive chemicals.

Manufacturers make a variety of paper filters and polymer- and glass-fiber filters. These filters vary in the density of their mesh and therefore in the sizes of particles they tend to trap. Filters that retain smaller particles tend to have slower flow rates than those that trap only larger particles. Different types of filter papers are graded by the amount of ash residue they leave when burned. Low ash level is important in some applications in analytical chemistry where a sample solution is filtered and then the filter paper is burned to leave behind only the sample. The ash level is also a measure of the purity of the fibers used to make the filter: the lower the ash level, the higher the purity. The term **quantitative grade** *refers to filter papers that leave little ash when burned and are intended for use in certain chemical analyses.* **Qualitative grade** *papers leave significant amounts of ash.*

Some filter papers are termed "hardened." **Hardened filter papers** *are treated to be stronger and better able to withstand vacuum filtration than unhardened papers.*

32.2.3 More about Microfiltration

32.2.3.1 Microfilters

Microfilters have many common applications in biotechnology. Microfiltration typically separates particles in the range of about 0.1–10 μm from a liquid or gas medium. **Microfiltration membranes** *are manufactured to have a particular pore size.* Particles larger than the rated size are retained on the surface of the membrane, while smaller particles pass through. When a manufacturer specifies the **pore size (absolute)**, *they mean that 100% of particles above that size will be retained by the membrane under specified conditions.* **Pore size (nominal)** *means that particles of that size will be retained with an efficiency below 100% (typically 90%–98%).* The methods of rating the nominal pore size can vary between manufacturers. In practice, pore size ratings refer to the size of particles retained, not to the actual physical dimensions of the pores.

Microfiltration membranes can be manufactured from a variety of plastic polymers, cellulose derivatives, metals, and ceramic materials, as shown in Table 32.1. While the first factor to consider in choosing a microfilter is pore size, different types of filter membranes differ in important ways. Recall from Chapter 24 that different plastic polymer types have different properties. In Chapter 24, we discussed the importance of these properties in terms of labware. The same properties are relevant when considering filtration. These properties include the following:

1. **Resistance to organic solvents**. Some membranes are dissolved by organic solvents, and others are not. When organic solvents are being filtered, it is necessary to choose a membrane made of a solvent-resistant material.
2. **Adsorption (binding) properties**. Some membrane materials readily adsorb single-stranded DNA, RNA, and proteins. There are applications in molecular biology where this binding is desirable because it immobilizes the macromolecules. For filtration, this binding is usually undesirable because the macromolecules are effectively "lost" from the sample.
3. **Surface smoothness**. Some membranes, called *screen membranes,* have a smooth surface with regularly spaced and evenly sized pores. Other membranes are more irregular in their pore structure. Smooth membranes are useful, for example, when the particles trapped on a membrane are to be viewed with a microscope.
4. **Extractables**. Some membranes have extremely low levels of extractables, whereas with others, there is extraction of materials from the membrane into the sample. Extraction is almost always undesirable.

TABLE 32.1

Examples of Materials Commonly Used to Make Microfilter Membranes

Material	Features	Examples of Applications
Cellulose acetate	Hydrophilic	General filtration
	Very low aqueous extractability	Sterilizing cell culture media
	Very low adsorption	General sterilization
	Not resistant to most organic solvents	
	Resistant to heat	
Nitrocellulose	Hydrophilic	General filtration
	Fast flow rates	Sterility testing
	Readily adsorbs nucleic acids and proteins	As a support for Southern blotting (where a DNA fragment is identified)
Polyethersulfone (PES) and polysulfone	Hydrophilic	Sterilization of culture media and other solutions
	Exceptional temperature and pH resistance	
	Low extractability	
	Low protein adsorption	
	Wide chemical compatibility	
	Autoclavable	
Nylon	Hydrophilic	Filtering HPLC samples
	Strong	
	Fast aqueous flow rates	
	Readily adsorbs proteins	
	Compatible with many solvents used in HPLC	
Polytetrafluoroethylene (PTFE)	Hydrophobic	Ideal for filtering gases, air
	Inert to most chemically aggressive solvents	Filtering organic solvents
	Resistant to acids and bases	
	Expensive	
Polyvinylidene difluoride (PVDF)	Naturally hydrophobic	Filtering HPLC samples
	Highly resistant to solvents	
	Very low adsorption	
Polypropylene	Hydrophilic	Filtering HPLC samples
	Low fiber release	
	Resistant to many solvents	

5. **Wetting properties**. Most membranes are hydrophilic, but some are hydrophobic. Almost all biological samples are aqueous and are filtered through hydrophilic membranes (which are readily wetted by water). Gases and organic solvents are almost always filtered with hydrophobic membranes because they must remain dry so that air or gas can pass freely through the filter. To assure that dryness, filters for gases are made using hydrophobic membranes, usually PTFE.

6. **Other characteristics**. These include the size of the filtration area with respect to sample size, strength of the filter, its resistance to heat, the rate at which fluids flow through it, and whether it can be autoclaved.

Examples of key properties of some of the most common microfiltration membrane materials are shown in Table 32.1. While there are a variety of factors to consider when choosing a filter, the choice is simplified for routine applications because manufacturers provide filtration units tailored specifically for particular purposes.

32.2.3.2 Applications of Microfiltration

One of the most important applications of microfiltration, both in the laboratory and in industry, is to remove contaminating bacteria, yeast, and fungi from solutions. Other methods of sterilization require heat or chemicals that are destructive to certain materials. For example, filtration is often used to sterilize cell culture media because these media contain heat-sensitive solutes, such as vitamins and antibiotics.

Various pore sizes of microfiltration membrane will retain various sized microorganisms:

- **0.10 μm,** recommended by some manufacturers to remove *Mycoplasma*, a very small type of bacterium that can contaminate cell cultures.
- **0.22 μm,** standard pore size for removing *Escherichia coli* and other bacteria.
- **0.65 μm,** used to remove fungi and yeast.
- **0.45–0.80 μm,** used for general particle removal.
- **1.0, or 2.5, or 5.0 μm,** for "coarse" particles.

Microfiltration is used both in the laboratory and in industry to sterilize both liquids and also gases, such as air and carbon dioxide. In many bacterial fermentations, air is supplied continuously to the fermentation vessel to provide agitation and oxygen. Hydrophobic microfilters are placed in the air stream to remove contaminating particles and microorganisms from the air before it enters the fermentation vessel. Filters may also be attached to supply lines for carbon dioxide and air running to animal cell culture vessels. Filters are used not only to protect the contents of fermentation vessels from contamination by outside air, but also to protect the facility from the vessel contents.

Microbiologists use microfiltration membranes when they monitor the levels of coliform bacteria in drinking water. A sample of water is filtered through a smooth membrane that retains bacteria on its surface. The membrane is then incubated on a petri dish with nutrient medium. Nutrients pass through the filter allowing the bacteria to grow into colonies. The colonies are counted to provide an estimate of the quantity of bacteria in the water supply. In an analogous way, particulates in air can be assessed by filtering large volumes of air through a membrane filter. The captured particulates can be analyzed, for example, by microscopic examination.

32.2.3.3 HEPA Filters

High-efficiency particulate air (HEPA) *filters are used to remove particulates, including microorganisms, from air.* HEPA filters are manufactured to retain particles as small as 0.3 μm. HEPA filters are depth filters made of glass microfibers that are formed into a flat sheet. The sheets are then pleated to increase their overall surface area. The pleated sheets are separated and supported by aluminum baffles. A diagram of a HEPA filter is shown in Figure 10.6 in Section 10.2.2.1.

HEPA filters have many applications both in the laboratory and in industry. HEPA filters are used in laboratory biological safety hoods to protect products from contamination and/or personnel from exposure to hazardous substances. In animal care facilities, HEPA filters are used on the tops of animal cages to protect valuable laboratory animals from infection with microorganisms. In industry, HEPA filters may be used to filter the air in entire rooms to protect products from contaminants.

32.2.4 More About Ultrafiltration

32.2.4.1 Ultrafilters

Ultrafilters are membranes that separate sample components on the basis of their molecular weight. Membranes used for ultrafiltration can separate particles with molecular weights ranging from about 1,000 to 1,000,000.

The term *molecular weight cutoff* (MWCO) is used to describe the sizes of particles separated by an ultrafilter. The **molecular weight cutoff** *is the lowest molecular weight solute that is 90% to 95% retained by the membrane.* MWCO values are not absolute because the degree to which a particular solute is retained by an ultrafiltration membrane is not entirely dependent on its molecular weight. The shape of the solute, its association with water, and its charge also affect its permeability through an ultrafiltration membrane. For example, a membrane is less likely to retain a linear molecule than a coiled, spherical molecule of the same molecular weight. In addition, the nature of the solvent, its pH, ionic strength, and temperature all affect the movement of solutes through membranes. By convention, if a membrane is rated to have a MWCO of 10,000, this means that the membrane will retain at least 90% of globular-shaped molecules whose molecular weight is 10,000 or greater.

When selecting an ultrafiltration membrane, many of the same considerations previously discussed come into play. The most common materials for these membranes are PES (polyethersulfone), polysulfone, and cellulose acetate. The MWCO you choose should generally be 3–5 times smaller than the molecular weight of the molecules to be retained on the membrane. If flow rate is a main consideration, a membrane with a higher MWCO will maximize flow rate in the filtration system, while a lower MWCO will maximize retention. Relative retention rates for specific molecules will also depend on factors such as molecular shape, pH, polarity, and other molecules present.

32.2.4.2 Applications of Ultrafiltration

The applications of ultrafiltration can be classified as either fractionation, concentration, or desalting. **Fractionation** *is the separation of larger particles from smaller ones.* For example, proteins that are significantly different in size can be separated from one another by ultrafiltration. A protein product that is made by cells in culture can be separated from other components of the cell culture medium in this way. Intact strands of DNA can similarly be separated from nucleotides.

In **concentration,** *solvent is forced through a filter, while solute is retained. The initial volume of the sample is thus reduced, and the high molecular weight species are concentrated above the filter.* For example, gel electrophoresis is used to separate and visualize proteins. Before electrophoresis, the proteins must be concentrated because only a very small volume can be applied to the gel. Ultrafiltration can be used for this purpose.

In **desalting,** *low molecular weight salt ions are removed from a sample solution.* Ultrafiltration is a simple method to remove salts because they readily penetrate the membranes, leaving the solutes of interest on the membrane surface.

32.2.5 Dialysis

Dialysis uses membranes such as those used for ultrafiltration. *Dialysis is based on differences in the concentrations of solutes between one side of the membrane and the other.* Solute molecules that are small enough to pass through the pores of the membrane will diffuse from the side with a higher concentration to the side with a lower concentration. The distinctive feature of dialysis is that differences in solute concentration provide the "driving force"; dialysis does not require pumps or a vacuum to force materials through the pores of the membrane.

Dialysis membranes are typically made of a thin mesh of cross-linked cellulose esters or PVDF (polyvinylidene fluoride). Dialysis membrane can be purchased as a roll of tubing, which is cut to length and washed thoroughly before use. As in ultrafiltration, dialysis membranes are chosen on the basis of their MWCO, which in most cases will be two to three times smaller than the molecules to be retained by the membrane.

Desalting is an example of the use of dialysis in the laboratory. During the process of purifying a protein, it is common to cause the protein to precipitate from solution by adding high concentrations of salt. Subsequent steps in the protein purification process require that the salt be removed; this is called *desalting.* The salts can be removed from the protein by placing the sample in a bag made of dialysis membrane (Figure 32.3a). The dialysis bag containing the sample is sealed at both ends and is suspended in a large volume of water or buffer solution. Thus, the concentration of salts is much higher inside the bag than outside. The relatively large protein molecules cannot penetrate the pores of the dialysis membrane and so remain inside the bag, but small molecules, including salt, readily move through the membrane. Over time, low molecular weight salt molecules inside the bag diffuse out to the water or buffer solution. The concentration of salt inside the bag and outside the bag eventually equalizes, and the system reaches equilibrium. The

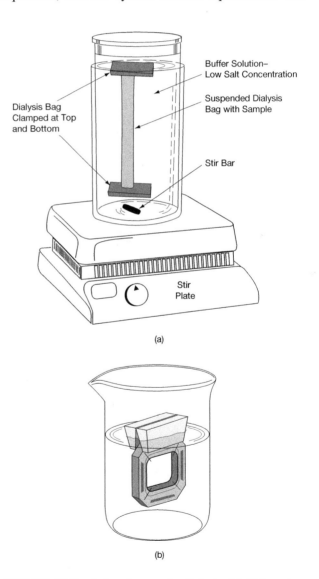

FIGURE 32.3 Laboratory-scale dialysis. (a) Using standard tubing. (b) Using a dialysis cassette.

salt molecules that were originally inside the dialysis bag are now distributed throughout both the large volume of buffer (or water) and the dialysis bag. The salts are not completely removed from the sample, but their concentration is much reduced. In order to further reduce the concentration of salt in the sample, the dialysis bag can be moved into fresh water or buffer solution and the process can be repeated.

Dialysis is relatively inexpensive (compared with ultrafiltration), simple, and gentle. Because dialysis relies on passive diffusion, however, it is a slow process. A number of special devices are available from manufacturers to make dialysis more efficient and convenient. For example, dialysis cassettes are sealed containers that eliminate potential sample leakage, as shown in Figure 32.3b. The sample is introduced into the cassette through a gasket using a needle and syringe. The cassette is then treated in the same manner as described for the tubing. Cassettes provide a greater surface area than plain tubing and can

therefore shorten dialysis time. They can also provide better recovery rates for small-volume samples.

32.3 FILTRATION SYSTEMS

32.3.1 SMALL-SCALE LABORATORY FILTRATION SYSTEMS

The principles of filtration are the same, whether the sample is 10 µL in the laboratory or 10,000 L in industry, but the design of filtration systems depends on the scale involved. The filter's size and shape, the support of the filter, the type of force used to move fluids through the filter, and the vessels involved can vary greatly depending on the scale.

Filtration requires a force to cause materials to flow through the filter. As illustrated in Figure 32.1, it is conventional to use vacuum filtration in the laboratory. Figure 32.4 shows laboratory vacuum filtration systems that take advantage of membrane filters.

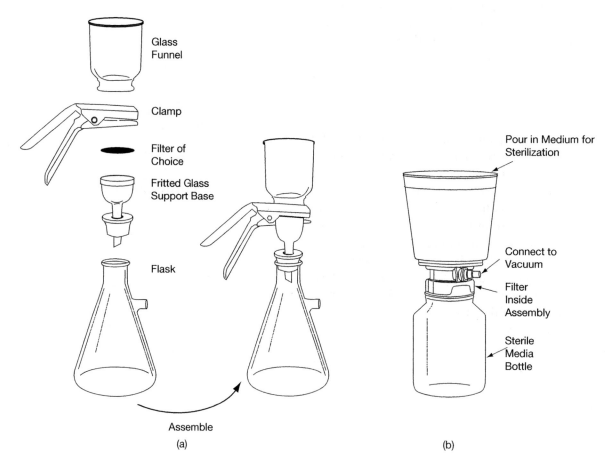

(a)   (b)

FIGURE 32.4 Laboratory filtration systems. (a) A membrane filter is mounted in a holder that clamps to the top of a flask. A vacuum is used to pull the material through the filter. The clamping structure allows the membrane to be firmly sealed in the reusable glass and steel supports. (b) A sterile filtration system used to conveniently sterilize cell culture media.

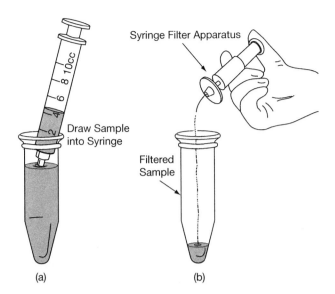

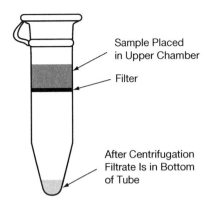

FIGURE 32.5 Syringe filter device. (a) The sample is drawn into the syringe. (b) The filter is attached to the syringe. Sample is forced through the filter by depressing the plunger.

FIGURE 32.6 A simple spin filtration system. Sample is loaded into the top of the tube. Sample is forced through the filter by the force of centrifugation.

Another laboratory method to gently force samples through a filter is to use a syringe. The sample is loaded into the syringe, and the disk-shaped filter unit is mounted on the end. The sample fluid is forced through the filter by depressing the plunger (Figure 32.5). Syringe filtration units contain microfilters of varying pore sizes. Sterile syringe filter units are available and are popular for sterilizing small volumes of sample, such as milliliter solutions of an antibiotic for cell culture. Very small syringe filter units are used for removing particulates from microliter volume samples prior to HPLC.

Spin filtration is a type of laboratory filtration system that uses a centrifuge to speed up filtration. Spin filtration units contain ultrafilters, or sometimes microfilters, that are housed within a centrifuge tube. The sample is placed in the tube on top of the filter, and the unit is spun in a centrifuge. During centrifugation, the liquid and smaller particles are forced through the filter and are captured in the bottom of the centrifuge tube. Larger molecules remain behind on the surface of the filter (Figure 32.6). Spin filters can be used to fractionate, concentrate, and desalt samples. They are often used in the molecular biology laboratory for small-volume samples of proteins, nucleic acids, antibodies, and viruses.

Ultrafiltration plates, shown in Figure 32.7, are ideal when high sample throughput is required. These are typically found in 96- and 384-well configurations. In most cases, individual membranes are sealed into each well to prevent cross-contamination. Samples are added to the wells, and filtration can be accomplished by either centrifugation or application of a vacuum. These plates are well suited for laboratory automation systems.

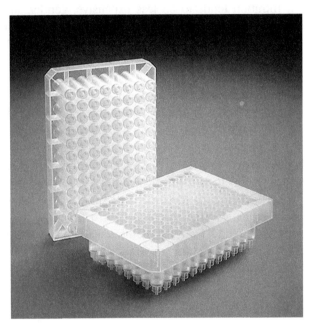

FIGURE 32.7 Multiwell filtration plates. Note the sample collection tubes under the filters. (Reproduced with permission from Merck KGaA, Darmstadt, Germany, and/or its affiliates.)

32.3.2 LARGE-SCALE FILTRATION SYSTEMS

Membrane filtration is widely used in the biotechnology industry, in both upstream and downstream applications, for collecting and purifying products. **Process filtration** *involves large volumes of liquids, as are found in biotechnology production facilities, pharmaceutical companies, and food production facilities.* Ultrafiltration is frequently used for sterilization of large-scale biological samples, and to separate cell products from media and cell contaminants.

Upstream filters are used to sterilize the culture medium components entering a bioreactor, including the air supply. The effective life span of the sterilization filters is increased with the use of prefilters that remove larger contaminants from the incoming air and liquids. The products of the bioreactor then proceed to downstream processing, the steps that purify a product made by cells.

Many biopharmaceuticals, for example, erythropoietin and tissue plasminogen activator, are made by large-scale cell culture, and the desired product is secreted by the cells into the surrounding liquid medium. Therefore, the first step in downstream processing is to separate the cells and cell debris from the liquid containing the product. *Separating cells and debris from a product is called* **clarification**. Clarification is often a costly and challenging process involving thousands of liters of material and a fragile protein product. Either filtration or centrifugation can be used for clarification, but filtration tends to be less expensive, gentler, and more convenient. There is a trend in industrial biotechnology toward using depth filtration to separate cells from their products. Clarification can be accomplished relatively quickly using cellulose-derived depth filters, which have a high capacity. Disposable filter units are increasingly common, eliminating the need for costly cleaning, sterilization, and validation procedures.

Filtration systems for industry must be designed to handle large volumes in a reasonably short time. Large-scale systems have sophisticated designs to maximize the surface area for filtration. The simplest way to do this is by using large sheets of filter membrane. More complex systems form the membranes into tubes, spirals, or pleats to maximize surface area. For example, **hollow-fiber ultrafilters** *are cylindrical cartridges*

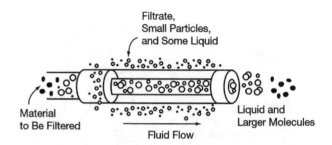

FIGURE 32.8 Hollow-fiber ultrafilter system. Sample flows through the lumen of the tubular filter. Small molecules pass through the filter.

packed with ultrafiltration membranes formed into hollow fibers. The liquid to be filtered flows through the lumen of the fibers. As molecules pass through the fiber core, substances smaller than the MWCO penetrate the membrane, whereas those that are larger are concentrated in the center (Figure 32.8).

Membrane clogging is a major problem in industrial-volume applications. One method to reduce clogging is to use **tangential flow,** or **cross-flow filtration,** *where the fluid to be filtered flows over the surface of the filter as well as through the filter.* The sweeping motion of the fluid clears the surface of the membrane, thus reducing clogging. Tangential flow filtration is mainly used for downstream ultrafiltration applications. This type of flow is illustrated in Figures 32.8 and 32.9.

We end this chapter by noting that filtration is a good introduction to bioseparations because the basic concept of filtration is easy to visualize. Filtration separates particles and molecules from one another based

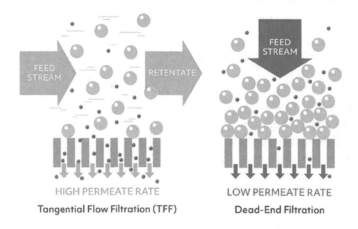

FIGURE 32.9 Comparison of tangential flow (cross-flow) filtration and standard filtration. A standard filter configuration is shown on the right. Molecules that do not pass through the pores of the filter eventually clog it. The diagram on the left shows the flow of sample in a tangential flow system. The "feed stream" is the material to be filtered that consists of a liquid containing molecules of various sizes. As the liquid flows through the system, small molecules and liquid are able to move through the filter while large molecules are retained. The pores of the filter are not clogged because the larger molecules keep moving through the system. (From Ruiz-García, Alejandro, Melián-Martel, N., and Nuez, Ignacio (2017). Short Review on Predicting Fouling in RO Desalination. *Membranes* vol. 7, 2017, p. 62. doi:10.3390/membranes7040062. CC BY 4.0.)

on differences in their sizes, or molecular weights. (However, we also note that adsorption and other, sometimes subtle, factors can be involved in filtration processes.) As we will see in Chapters 33 and 34, size differences between particles and molecules are also exploited in centrifugation, electrophoresis, and some forms of chromatography – all of these are important methods in biotechnology.

Filtration is a good example of a technology that is ancient, yet is constantly undergoing improvements that expand its applications and utility. Advances in materials science have led to filter materials that can effectively separate out molecules as small as viruses or even salt ions. Other engineering advances have led to filtration systems that facilitate efficient, large-scale purification processes. It is likely that you will encounter some form of filtration when working in almost any biotechnology setting.

Practice Problems

1. Order the following items in terms of size, from the smallest to the largest:
   Pollen grains (about 30 µm), albumin (a protein), sand grains, polio virus, red blood cells, *Serratia marcescens* (a type of bacterium), NaCl ions.
2. For each of the following separations, state whether it will involve macrofiltration, microfiltration, or ultrafiltration.
   a. Purifying antibodies from a liquid medium.
   b. Removing viruses from a vaccine.
   c. Removing salts from a solution also containing DNA.
   d. Sieving large particulates from water before it is treated in a sewage treatment plant.
   e. Sterilizing cell culture medium by removing bacteria.
   f. Removing pyrogens (fever-causing agents) from a drug product.
   g. Harvesting mammalian cells from a fermenter.
   h. Removing *Mycoplasma* from bovine serum. (Serum is sometimes added to cell culture media.)
3. The specifications for three filters are shown. Match the filters with the applications that follow:

Ready Separation Hollow-Fiber Filtration System

The "Ready Separation" System is a compact hollow-fiber filtration system capable of processing from 5 to 100 L. The system is sterilizable and can be purchased with microfiltration membranes from 0.1 to 0.80 µm or with ultrafiltration membranes from 300 to 500,000 MWCO.

Filter Type XYZ

The XYZ unit contains a hydrophobic, solvent-resistant, 0.2 µm membrane designed for use as a sterilizing filter for gases and liquids. The unit is a pyrogen-free, sterile, single-use device.

Filter Type ABC

The ABC filtration unit is a low protein binding, sterile filter for aqueous, proteinaceous substances. It is intended for applications where minimal sample loss is desired. It is a 0.22 µm filter, single-use product for use with syringes.

a. A fermentation process is being designed in which microorganisms produce an antibiotic that will be isolated from the broth. The microorganisms generate carbon dioxide, which is vented from the fermenter via a plastic tube. On the end of the tube is a filtration unit that keeps organisms from the outside air from contaminating the fermenter. Which filtration unit would be used?

b. In the laboratory, an antibiotic solution needs to be added to cell culture plates. It is heat-sensitive and is therefore sterilized by filtration. Which filter would be used?

c. An organic solvent is to be used with HPLC. It needs to be filtered to remove particulates. Which filter would be used?

d. A biotechnology company uses a large-scale cell culture process to produce a valuable protein that is being tested as a drug. The protein is secreted by the cells into the culture medium. Which type of filter might be used in the process of isolating the protein product from the cell culture medium?

33 Introduction to Centrifugation

A brief note about safety: Centrifuges pose a variety of hazards to laboratory personnel. Moreover, although centrifuges and their components appear solid and sturdy, they are, in fact, expensive devices that can easily be damaged by mishandling. Safe and proper handling of centrifuges is discussed throughout this chapter.

33.1 INTRODUCTION TO CENTRIFUGATION: PRINCIPLES AND INSTRUMENTATION

33.1.1 Basic Principles

33.1.1.1 Sedimentation

In 1869, Friedrich Miescher used a crude centrifuge to isolate a substance from cells that he termed "nuclein." We now know this substance to be DNA. This was a landmark discovery in the history of biotechnology that was made possible by centrifugation. Centrifuges continue to be an important tool for biotechnologists.

Centrifugation *is a separation method that is used routinely in all types of biotechnology facilities to separate and isolate particles in a solution.* To begin thinking about centrifugation, consider a cylinder filled with a slurry of gravel, sand, and water (Figure 33.1). We would expect that, over time, gravity would cause the particles to **sediment** *(i.e., to move through the liquid and settle to the bottom of the cylinder).* The larger gravel particles will sediment more quickly than the smaller sand particles so that there will eventually be a bottom layer of mostly gravel, a layer of mostly sand, and relatively clear liquid in the upper part of the cylinder. Thus, a separation, or fractionation process

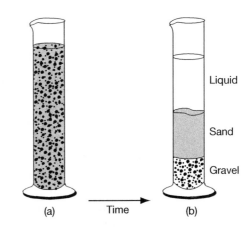

FIGURE 33.1 Sedimentation due to gravity. (a) A slurry consisting of gravel, sand, and water is placed in a cylinder. (b) The particles sediment according to their size and density under the influence of gravity.

DOI: 10.1201/9780429282799-42

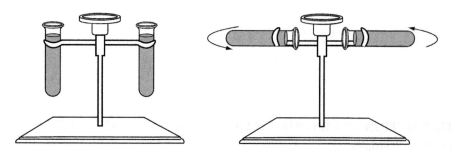

FIGURE 33.2   A simple centrifuge. Tubes containing a sample are rapidly spun about a central shaft, creating a force many times that of gravity.

will occur in which the particles separate from the liquid medium and from one another due to the force of gravity.

Biologists seldom need to separate gravel and sand from one another; rather, they are interested in such materials as cells, organelles, bacteria, and viruses. These biological materials, called *particles* in the context of centrifugation, are much smaller than sand and would take a long time to sediment due to the force of gravity alone. A **centrifuge** *is a piece of equipment that accelerates the rate of sedimentation by rapidly spinning the samples, thus creating a force many times that of gravity.*

A simplified centrifuge is illustrated in Figure 33.2. There is a central drive shaft that rotates during centrifugation. A **rotor** *sits on top of the drive shaft and holds tubes, bottles, or other sample containers.* As the drive shaft rotates, the sample containers spin rapidly, creating the force that facilitates the sedimentation of particles.

At the end of a common type of centrifugation run, the solution has separated into two phases: the **pellet** *at the bottom of the tube with sedimented particles,*

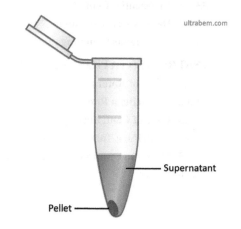

FIGURE 33.3   The pellet and supernatant after a centrifugation run. (Image credit: Ultrabem, CC0, via Wikimedia Commons.)

and the liquid **supernatant** *at the top of the tube* (**Figure 33.3**).

The case study shown below, "The Use of Centrifugation in the Plasmid Miniprep Procedure," illustrates how a centrifuge is used in a common separation procedure.

Case Study: The Use of Centrifugation in the Plasmid Miniprep Procedure

Plasmids are small circular molecules of DNA that exist separately from the bacterial chromosome. Plasmids are a vital tool in molecular biology because they can be used to transfer genes of interest into cells. (See Section 1.1.3.1 for more information.) We previously considered the plasmid miniprep procedure to isolate plasmids from bacteria in the case study on Chapter 25, pp. 639–640. This miniprep procedure is a type of separation technique in which one entity, the plasmids, is separated and purified out of cells. In the case study in Chapter 25, we were concerned with the roles of the various solution components involved in this separation method. Now, we extend our discussion of this method by talking about the role of centrifugation.

The plasmid miniprep procedure is a small-scale separation where the volumes are on the order of a milliliter. Therefore, a small centrifuge that holds small tubes is required; this type of centrifuge is often termed a "microcentrifuge" (Figure 33.4). The use of a microcentrifuge in the plasmid miniprep procedure is diagrammed in Figure 33.5.

The plasmid miniprep procedure begins with a culture of bacterial cells that contain the plasmid of interest. In Step 2, 1.5 mL of the bacterial culture is placed in a tube and spun in a microcentrifuge. The purpose

(Continued)

Case Study (*Continued*): The Use of Centrifugation in the Plasmid Miniprep Procedure

of this centrifugation step is to separate the bacterial cells from the culture medium. The supernatant is discarded, and the pelleted cells are retained.

In Step 3, the cells are resuspended in buffer and then the cells are lysed (broken open) in a series of steps (Steps 3–6). Observe that at the end of Step 6, the bacterial chromosomal DNA has precipitated out, but the plasmids are still suspended in the lysis buffer. This is another point where a centrifuge is used. The mixture is spun again in the microcentrifuge. This time the pellet is discarded because it contains the unwanted chromosomal DNA, along with other bacterial cellular materials, proteins, and detergent. The supernatant containing the plasmids is transferred to a fresh tube. Isopropanol is added to that tube (Step 8). The isopropanol, together with salt that is present in the solution, causes the plasmids' DNA to precipitate. At this point, the small plasmid particles are floating around in the solution. To obtain them, the solution is once again centrifuged. Now we are at Step 9 in the procedure. The supernatant is discarded, and the pellet is saved. Step 10 is a wash of the pellet with alcohol. The plasmids are once again floating around, suspended in the alcohol. So, there is one more centrifugation step, Step 11, where the pellet containing the plasmids is saved and the supernatant is discarded. The plasmids are now successfully separated from the bacterial cells and are ready for use in further procedures. (Note that the diagram in Figure 33.5 does not show the final wash step.) Thus, a centrifuge was an essential tool in the plasmid miniprep procedure. This is only one example of how centrifuges are used; there are many other applications, at both large and small scales.

FIGURE 33.4   A microcentrifuge. (Image courtesy of Gilson, Inc.)

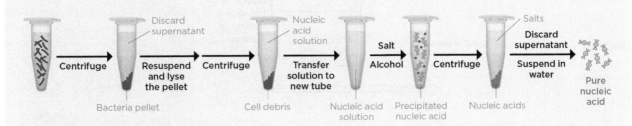

FIGURE 33.5   Schematic overview of the plasmid miniprep procedure highlighting the use of a centrifuge. (Image courtesy of Beckman Coulter Life Sciences.)

33.1.1.2 Force in a Centrifuge

A centrifuge is a device that applies force to a sample. To appreciate the nature of the force in a centrifuge, imagine that you tie a stone to the end of a string and whirl it rapidly in a circle above your head. You will feel a pull on your hand as the stone rotates. Whenever an object is forced to move in a circular path (as in a centrifuge), a force is generated. Now, consider what would happen if you whirled the stone faster – you would feel more pull. Similarly, the faster a sample is spun in a centrifuge, the more force is experienced by that sample. The speed of rotation in a centrifuge is expressed as **revolutions per minute (RPM).**

In addition to the speed of rotation, there is another (somewhat less obvious) factor that affects how much force is experienced by a sample in a centrifuge. If you whirl a stone tied to a long string, there will be more force on the stone than if you use a shorter string. In a centrifuge, the further a particle is from the center of rotation, the more force the particle experiences. *The distance from the center of rotation to the*

material of interest is called **the radius of rotation (r)** (Figure 33.6). In a centrifuge, the radius of rotation varies depending on the particular equipment being used.

There are thus two factors that determine the force experienced by a particle in a centrifuge:

1. **The speed of rotation, expressed as RPM.**
2. **The distance of the particle from the center of rotation, r.**

The force acting on samples in a centrifuge is expressed as the **relative centrifugal field (RCF).** The relative centrifugal field has units that are multiples of the earth's gravitational field (g). The force in a centrifuge might be expressed as, for example, 10,000 ×g or 10,000 RCF. Both these expressions mean that the force in the centrifuge is 10,000 times greater than the normal force of gravity.

It is possible to calculate the relative centrifugal field in a centrifuge given the speed of rotation and the radius of rotation. The equation for the relative centrifugal field is:

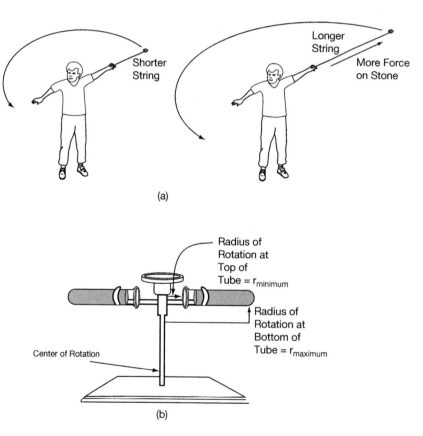

(a)

(b)

FIGURE 33.6 The radius of rotation. (a) The longer a string being whirled, the greater the radius of rotation and the more the force that is experienced by the stone. (b) In a centrifuge, the further a particle is from the center of rotation, the more the force it experiences. The radius of rotation depends on the rotor. You can usually find values for the maximum and minimum radii in the rotor's manual.

$$RCF = 11.2 \times r \left(\frac{RPM}{1,000} \right)^2$$

where:

RCF = the relative centrifugal field in units of ×g

RPM = the speed of rotation in revolutions per minute

r = the radius of rotation in centimeters.

You can find the radius of rotation for a particular rotor in its instruction manual.

This equation can be rearranged as:

$$RCF = 1.12 \times 10^{-5} (RPM)^2 \, r$$

Observe that two centrifuges may be spinning at the same rate, yet they might each be subjecting the samples to a different force. This is because the radius of rotation varies from one centrifuge/rotor configuration to another. For this reason, it is more useful to report to others the relative centrifugal field when a sample is centrifuged, rather than simply the speed of rotation.

Example Problem 33.1

Your centrifuge equipment has a maximum radius of rotation of 9.2 cm. If you spin a sample at a speed of 15,000 RPM, what is the maximum relative centrifugal field?

Answer

Substitute the values into the RCF equation:

$$RCF = 11.2 \times 9.2 \left(\frac{15,000}{1,000} \right)^2$$

$$RCF = 11.2 \times 9.2 \, (225) = 23,184 \times g$$

The relative centrifugal field is **23,184 ×g.**

Substituting into the rearranged form of the equation gives the same answer:

$$RCF = 1.12 \times 10^{-5} (RPM)^2 (r)$$

$$RCF = 1.12 \times 10^{-5} (15,000)^2 (9.2) = \mathbf{23,184 \times g}$$

Example Problem 33.2

A colleague tells you that a sample was centrifuged for a certain time at a speed of 15,000 RPM and that the force of centrifugation was 16,000 ×g. Your centrifuge equipment has a radius of rotation of 9.2 cm. What speed will you use to achieve a force of 16,000 ×g?

Answer

The unknown in this case is the speed of rotation. It is possible to algebraically rearrange the RCF equation to solve for RPM as follows:

$$RCF = 11.2 \times r \left(\frac{RPM}{1,000} \right)^2$$

$$\frac{RCF}{11.2 \, r} = \left(\frac{RPM}{1,000} \right)^2$$

$$\sqrt{\frac{RCF}{11.2 \, r}} = \left(\frac{RPM}{1,000} \right)$$

$$1,000 \sqrt{\frac{RCF}{11.2 \, r}} = RPM$$

Then, substituting into this equation gives:

$$1,000 \sqrt{\frac{16,000}{11.2 \, (9.2)}} = RPM$$

$$\mathbf{12,461 \approx RPM}$$

Thus, in order to achieve a force of 16,000 RCF, you will need to use a speed of 12,461 RPM. Observe that to obtain a force of 16,000 ×g, your centrifuge must be run at a different speed than your colleague's centrifuge. **It is the force in the centrifuge that ultimately matters, not the speed of rotation.**

Example 33.1

A colleague tells you that a sample was centrifuged at 30,000 ×g for 15 minutes. Your rotor can achieve a maximum RCF of only 20,000 ×g. What can you do?

It may be possible to compensate for the reduced force in your rotor by using a longer centrifugation run (but see the cautions at the end of this example). The time required in your rotor can be calculated using the following equation:

$$t_a = \frac{t_s \times RCF_s}{RCF_a}$$

where:

t_a = the time required in alternate rotor
t_s = the time specified in the procedure
RCF_a = the RCF of alternate rotor
RCF_s = the RCF specified in the procedure.

Thus,

$$t_a = \frac{15\ min \times 30,000 \times g}{20,000 \times g}$$

$t_a = 22.5\ min$

Spinning the sample for 22.5 minutes at 20,000 ×g will duplicate the required conditions. Note, however, that changing the time or the force of centrifugation may affect the results. Harder pelleting (more force) may damage sensitive biological materials. Longer spin times may lead to deterioration of some samples, particularly if the centrifuge is not refrigerated. Density gradient separations (discussed in a later section) are more complex than simple sedimentation and may not work if the time or the RCF is changed. For all these reasons, before changing centrifugation conditions, check that the changes will have no adverse effects.

33.1.1.3 Factors That Determine the Rate of Sedimentation of a Particle

It is reasonable that the greater the force during centrifugation, the more quickly the particles sediment. Another factor that affects the sedimentation rate is the viscosity of the liquid medium through which the particles are moving. The less viscous the medium, the faster the rate of sedimentation.

The rate of sedimentation of a particle also depends on its own characteristics. Larger particles sediment faster than smaller ones (as was also shown earlier in the example of sand and gravel sedimenting in a cylinder). It is also the case that the sedimentation rate of a particle depends on its density relative to the density of the surrounding liquid medium. Recall that density is the mass per unit volume. A stone, for example, is denser than a beach ball. If a stone and a beach ball are thrown into a pond, the stone will sink because it is denser than the water, but the ball will float. The density of particles similarly affects their movement in a liquid medium during centrifugation.

If a particle is denser than the liquid medium, then the particle sediments (sinks) during centrifugation. If the densities of the particle and the liquid medium are equal, the particle does not move relative to the medium – no matter how strong the relative centrifugal force is. Rather, the particle remains stationary in the medium. There are also situations where the medium is denser than a particle (as water is denser than a beach ball). In these situations, a particle can actually "float" in a centrifuge tube, or move in the opposite direction of the force of centrifugation. Thus,

1. **If the density of a particle and the liquid medium are equal, the particle does not move.**
2. **If the density of a particle is greater than the density of the medium, the particle sediments (moves away from the center of rotation).**
3. **If the density of a particle is less than the density of the medium, the particle moves toward the center of rotation (floats "upward" in the tube).**

For biologists, much of the time, the liquid medium in centrifugation is water, or a water-based buffer. The density of most biological materials, such as cells, organelles, and macromolecules, is greater than the density of the medium. Thus, in a simple separation in a water-based medium, most biological particles move toward the bottom of the tube. Lipids and fats, however, which are less dense than water, are found at the top of the tube after centrifugation. (Later in this chapter, we will consider what happens when the medium in centrifugation is not water.)

In summary, the factors that influence the rate of sedimentation of a particular particle include the following:

1. **The relative centrifugal field.** The higher the relative centrifugal field, the faster the rate of sedimentation.
2. **The viscosity of the medium.** The less viscous the medium through which the particles must move, the faster the rate of sedimentation.
3. **The size of the particle.** The larger the particle, the faster its rate of sedimentation.
4. **The difference in density between the particle and the medium.**[1]

33.1.1.4 Sedimentation Coefficients

You may see the term **sedimentation coefficient,** or **S value.** The sedimentation coefficient *is a value for a particular type of particle that describes its rate of sedimentation in a centrifuge.* The larger the *S* value, the faster the particle will sediment. The sedimentation coefficient is a physical constant that refers to a particular biological particle. Various proteins, different organelles, and nucleic acids each have their own sedimentation coefficient. Sedimentation coefficients have been measured for a variety of biological

TABLE 33.1

Examples of Sedimentation Coefficients

Biological Particle	Sedimentation Coefficient[a], S[b]
Soluble Proteins	
Ribonuclease (MW = 13,683)	1.6
Ovalbumin (MW = 45,000)	3.6
Hemoglobin (MW = 68,000)	4.3
Urease (MW = 480,000)	18.6
Other Materials	
E. coli rRNA	20
Calf liver DNA	20
Ribosomal subunits	40
Ribosomes	80
Tobacco mosaic virus	200
Bacteriophage T2	1,000
Mitochondria	15,000–70,000

[a] The sedimentation coefficient depends on the liquid medium and the temperature. Standard conditions for these coefficients are in water at 20°C.

[b] Sedimentation coefficients are expressed in Svedberg units, S, where one Svedberg unit = 10^{-13} seconds.

particles; some examples are shown in Table 33.1. Observe in this table that larger particles generally have larger sedimentation coefficients.

33.1.1.5 Applications of Centrifugation

Centrifugation has a long history in the biology research laboratory, where it is used in the separation, isolation, purification, and analysis of biological materials. Centrifugation also has a long history of use in clinical laboratories where it is used to separate the components of blood from one another. In the biotechnology industry, centrifugation may be used, for example, for the separation of cells from broth after fermentation.

Centrifugation applications may be broadly classified as either preparative or analytical. **Preparative centrifugation** *provides separated materials for later use.* Examples of applications of preparative centrifugation are in Table 33.2.

Analytical centrifugation *is used in specialized research settings to determine the molecular weight, purity, shape, and other physical characteristics of proteins and other macromolecules.* Analytical centrifugation involves specially designed, computer-controlled centrifuges that allow the operator to monitor the movement of particles. Analytical centrifugation is

---

[1] You may come across the following equation, Stokes equation, which combines these various factors as follows:

$$v = \frac{d^2 \left(\rho_p - \rho_m \right) g}{18 \, \mu}$$

where:

v = the velocity of sedimentation
d = the diameter of the particle
ρ_p = the density of the particle
ρ_m = the density of the medium through which the particle is sedimenting
g = the relative centrifugal force
μ = the viscosity of the medium.

Although this equation may seem formidable because it has several unfamiliar symbols, it is possible to obtain information from it fairly easily. Observe that the viscosity of the medium is in the denominator – the more viscous the medium, the slower the velocity of sedimentation. The g force is in the numerator – the higher the g force, the faster the sedimentation velocity. The difference in density between the particle and the medium is also in the equation. When a particle and the medium are of the same density, $(\rho_p - \rho_m) = 0$, which means the sedimentation velocity is zero – the particle does not move. If the particle is denser than the medium, the particle moves with a certain velocity toward the bottom of the tube. It is also possible for the particle to be less dense than the medium. In this case, the velocity is a negative number – that means the particle moves upward in the tube.

TABLE 33.2

Examples of Applications of Preparative Centrifugation

- *Separation of intact, single-cell suspensions such as bacteria, viruses, and blood cells from a liquid medium*
- *Separation of cellular organelles, such as mitochondria and ribosomes from disrupted cells*
- *Separation of biological macromolecules, including DNA, RNA, and protein from a solution* (The case study "**The Use of Centrifugation in the Plasmid Miniprep Procedure**" on pp. 882–883 falls into this category.)
- *Separation of plasma (the fluid portion of blood) from blood cells*
- *Separation of immiscible liquids from one another*
- *Separation of cells from broth after fermentation.*

not as routinely used as preparative centrifugation, but it has important applications in the biopharmaceutical industry. For example, therapeutic antibodies are an important class of biopharmaceutical drug product. One problem, however, is that these antibodies can form aggregates (clumps) that can be dangerous to patients. Analytical centrifugation has been used to study the formation and features of these aggregates for the purpose of reducing or eliminating them from drug preparations. Analytical centrifugation is an interesting application of centrifugation, but because it is not as routinely used as preparative centrifugation, it is not considered further in this text.

33.1.2 Design of a Centrifuge: Overview

The typical components of a centrifuge are shown in Figure 33.7. The motor, which turns the drive shaft, is located at the base of the unit. A rotor sits on top of the drive shaft and holds the sample containers. With the exception of older low-speed instruments, centrifuges have a protective enclosure so that the rotor sits in a chamber. In modern centrifuges, the chamber has a cover that is usually locked when the centrifuge is spinning. In higher-speed centrifuges, the enclosure and cover are constructed with reinforced materials which, in the event of a serious mishap, can withstand the tremendous force of a rotor that flies off the drive shaft. Older high-speed centrifuges may lack the protective safety features of newer models. Most centrifuges also have a braking device to slow the rotor, a microprocessor control device to set the speed, and a timer.

There are many styles and models of centrifuges, so it is convenient to classify them into types. The two main physical types are floor models and benchtop instruments. While benchtop centrifuges are convenient, floor models frequently offer higher sample capacities and sometimes higher speeds as well.

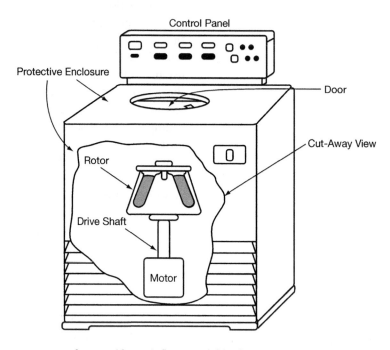

FIGURE 33.7   Basic components of a centrifuge. A floor model is shown.

TABLE 33.3
General Suitability of Different Centrifuges for Centrifugal Applications

Characteristics	Type			
	Low Speed	High Speed	Ultracentrifuge	Microfuge
Maximum speed (RPM)	10,000	30,000	150,000	15,000
Maximum force (×g)	8,000	100,000	1,000,000	30,000
Refrigeration	Sometimes	Yes	Yes	Sometimes
Vacuum	No	Sometimes	Yes	No
Examples of applications	Harvest intact plant and animal cells. Separate blood components. Harvest larger organelles (such as nuclei). Separate immiscible liquids.	Purify viruses. Isolate mitochondria, lysosomes, and other moderate-size organelles.	Purify small organelles such as ribosomal subunits. Purify dissolved DNA, RNA, and proteins.	Purify precipitated DNA. Perform small-volume separations.

A simple classification by speed is summarized in Table 33.3 and in the following list:

- **Low-speed centrifuges** typically attain rotation speeds less than 10,000 RPM and forces less than 8,000 ×g.
- **High-speed centrifuges** typically attain rotation speeds up to about 30,000 RPM and generate forces up to about 100,000 ×g.
- **Ultraspeed centrifuges**, or **ultracentrifuges**, typically attain rotation speeds up to about 150,000 RPM and generate forces up to 1,000,000 ×g.

Heat is generated as a rotor turns. High-speed centrifuges and ultracentrifuges, therefore, are usually refrigerated. Refrigeration is preferred for most biological samples to protect them from adverse effects due to heat, although intact cultured cells are usually centrifuged gently at room temperature. Some benchtop centrifuges can be installed in a cold room to protect samples during longer runs.

Ultracentrifuges have vacuum pumps that evacuate air from the chamber. This is necessary because the friction of air rubbing against the spinning rotor creates drag and heat. Vacuum is commonly achieved using a system with two pumps. The first pump is a "roughing pump" that draws air out of the system beginning at atmospheric pressure. A diffusion pump takes over when some vacuum has been established. Diffusion pumps generate heat and so require a cooling system.

Centrifuges can be classified on the basis of speed and also by their capacity or by the sizes and types of sample containers they can spin. For example, **microfuges** or **microcentrifuges** *are small centrifuges that are intended for small-volume samples in the microliter to 1 or 2 mL range.* Microfuges are useful in molecular biology laboratories for conveniently pelleting small-volume samples. (See Figure 33.4) At the other end of the spectrum, there are industrial, large-capacity centrifuges that are used in production facilities (e.g., to separate large volumes of cells from nutrient broth).

33.2 MODES OF CENTRIFUGE OPERATION

33.2.1 DIFFERENTIAL CENTRIFUGATION

Preparative centrifugation can be performed in several different modes. Of these, *differential centrifugation*, also called *differential pelleting*, is the simplest and is common in the biology laboratory. **Differential centrifugation** *separates a sample into two phases: a pellet consisting of sedimented materials, and a supernatant consisting of liquid and unsedimented particles,* (refer to Figure 33.3). A given type of particle may move into the pellet, remain in the supernatant, or be found in both phases. Which sample components move into the pellet, and which remain in the supernatant depends on the nature of the components, on the force, and on the duration of centrifugation. With this mode of centrifugation, all the particles of a greater density than the medium will eventually sediment, so this method is time dependent. The case study "The Use of Centrifugation in the Plasmid Miniprep Procedure" is an example of differential centrifugation.

In differential centrifugation, a sample is poured into a centrifuge tube or bottle. The sample is centrifuged for a certain time with a particular relative centrifugal force. As the sample spins, larger particles

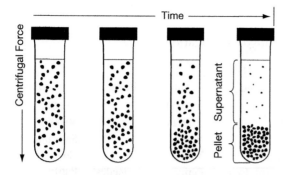

FIGURE 33.8 Differential centrifugation. A suspension of particles in a liquid medium is initially uniformly distributed in the centrifuge tube. During centrifugation, particles sediment into a pellet at differing rates depending mainly on their size.

sediment faster than smaller ones. By using a suitable combination of relative centrifugal force and time, a pellet enriched in a particular component can be obtained (Figure 33.8).

After centrifugation, the supernatant can readily be decanted from the tube. It is possible to continue by centrifuging the supernatant again at a higher force, thus obtaining a second pellet. This process can be repeated, resulting in a series of pellets containing progressively smaller particles. For example, a common application of differential centrifugation is the isolation of cellular organelles from a cellular homogenate. (A cellular homogenate is produced by disrupting cells and releasing their organelles, membranes, and soluble macromolecules into a liquid medium.) Nuclei are the largest organelle. It is possible to obtain a pellet enriched in nuclei by centrifuging a cellular homogenate for 10 minutes at 600 $\times$g. Other organelles tend not to move into the pellet given this relatively low centrifugal field and short run time. The supernatant is then removed and spun at a higher force. This process is continued until the smallest subcellular particles have been pelleted (Figure 33.9).

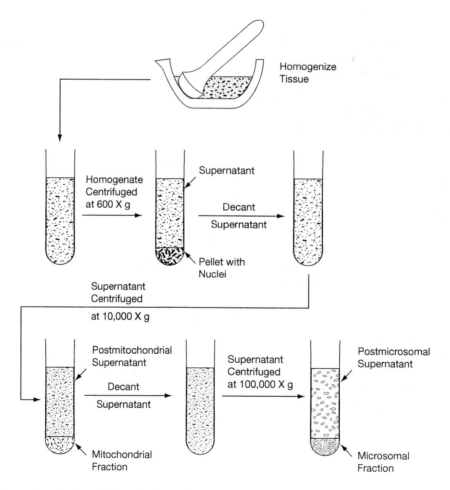

FIGURE 33.9 Example of using differential centrifugation to isolate cellular organelles. Various organelles are separated from one another as the sample is centrifuged at progressively higher forces. The final microsomal fraction includes fragments of the endoplasmic reticulum and attached ribosomes.

Although differential centrifugation is simple and widely used, it has disadvantages. When a sample mixture is initially poured into a centrifuge tube, the various sample components are dispersed throughout the tube. Thus, each pellet consists of sedimented components contaminated with whatever unsedimented particles were at the bottom of the tube initially. Moreover, the yield of smaller particles is lowered by their loss in early pellets. Differential centrifugation may also require starting and stopping the centrifuge repeatedly, which is time-consuming for the operator. Density gradient centrifugation is another mode of centrifugation that reduces or eliminates these problems.

33.2.2 DENSITY CENTRIFUGATION

33.2.2.1 The Formation of Density Gradients

A **density gradient** *is a column of fluid that increases in density from the top of the centrifuge tube to the bottom.* **Density gradient centrifugation** *includes techniques for separating molecules based on their rate of sedimentation or their buoyancy in a density gradient.* As different particles move through the density gradient, they separate into discrete bands (Figure 33.10).

Density gradients can be formed in three ways. Some high molecular weight solutes, such as cesium chloride (CsCl), form gradients spontaneously during centrifugation. With these media, the sample and the gradient material are mixed together and centrifuged. As centrifugation proceeds, a gradient forms.

Many density gradient media do not spontaneously form gradients, so the operator prepares the gradient before the separation is begun. A preformed gradient may either be a step gradient or a continuous gradient. A **step gradient** *is formed in layers with the densest layer on the bottom. Each layer has a different density and is clearly demarcated from the layers above and below.* A step gradient is prepared from separate solutions of the gradient medium, each of which has a different concentration and therefore a different density. The different solutions are carefully layered one on top of the other using a pipette or syringe (Figure 33.11a). The sample is loaded on top of the preformed gradient.

In **continuous gradients**, *there is a smooth increase in density from top to bottom with no sharp boundaries between layers.* Preformed continuous gradients are often produced by continuously mixing together two solutions of different density while simultaneously delivering the mixture to the centrifuge tube. This requires a special apparatus called a **gradient maker** (Figure 33.11b). Alternatively, if a gradient maker is unavailable, a preformed step gradient will diffuse into a continuous gradient over time.

There is a wide variety of materials used to make density gradients. Sucrose and Ficoll™ are examples of media often used for preformed gradients.

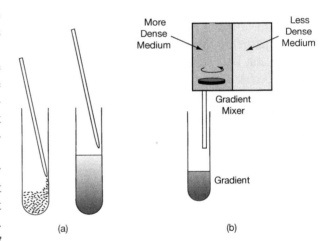

FIGURE 33.11 Preparing density gradients. (a) Layering a step gradient. (b) Preparing a continuous gradient using a gradient maker. A high-density (high-concentration) solution of the gradient medium is placed in one of the chambers of the gradient maker; this chamber contains a stir bar for mixing the solution. A lower-density solution is placed in the other chamber. A narrow tube connecting the two chambers is opened. At the same time, liquid is allowed to begin flowing out of the high-density chamber into the centrifuge tube. Only dense solution initially reaches the centrifuge tube. As time goes on, more and more of the less dense solution flows into the mixing chamber, progressively diluting the medium. As a result, the medium flowing into the centrifuge tube slowly decreases in density (concentration), thus forming a continuous density gradient in the centrifuge tube.

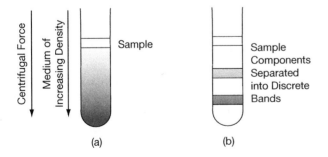

FIGURE 33.10 Density gradient centrifugation. (a) A sample is layered on top of a preformed density gradient. (b) During centrifugation, sample components separate into discrete bands as they move through the gradient.

Percoll™, a colloidal silica, is commonly used for separation of cells and organelles. Cesium chloride and other cesium salts are used in DNA separations. The medium chosen must be compatible with the sample, must have a sufficient range of densities to separate all the components of the sample, must not interfere with later steps in the procedure, must not be corrosive to the particular rotor being used, and so on. Because the requirements for the medium vary from application to application, different density gradient media have been developed and are available from manufacturers.

33.2.2.2 Rate Zonal Centrifugation

There are two methods of density centrifugation: rate zonal and isopycnic (also known as buoyant density centrifugation). **Rate zonal (size) centrifugation** *is a time-dependent method in which particles are separated as they sediment through a density gradient.* In the rate zonal method, the sample solution is layered on top of a preformed, usually continuous, density gradient. During centrifugation, the particles sediment through the gradient. The rate of movement of the particles depends primarily on their size, less so on their density, so different particles move at different rates. As the particles move at different speeds through the gradient, they separate into bands, or zones. (See Figure 33.10.)

In rate zonal centrifugation, the particles are denser than the densest part of the gradient. This means that if centrifugation is continued too long, the particles will all move to the bottom of the tube, mixing again into a pellet. Rate zonal centrifugation, therefore, is time dependent and is halted once the sample components have separated into discrete bands.

33.2.2.3 Isopycnic Centrifugation

Isopycnic centrifugation (also called buoyant density centrifugation) *is a method in which particles are separated based on their density alone.* As a simple example, suppose that it is necessary to separate two types of particles, one of which has a density of 1.50 g/mL and the other a density of 1.35 g/mL. This can be accomplished by layering the sample on top of a liquid medium that has a density of, for example, 1.40 g/mL. After centrifugation, the less-dense particles will form a floating band, whereas the denser particles will be in a pellet in the bottom of the tube.

In the previous simple example, there is not actually a density gradient; rather, there is a single density interface, called a *density barrier.* More complex isopycnic separations involve a density gradient. The gradient

may be a preformed step gradient, a preformed continuous gradient, or a gradient that forms spontaneously during centrifugation. However the gradient is formed, the density of the medium at the bottom of the gradient must be greater than the density of any of the particles to be separated.

During isopycnic centrifugation, particles move away from the center of rotation if they are denser than the medium in which they are located. Particles move toward the center of rotation if they are less dense than the medium at that point. Equilibrium is eventually reached when all the particles are banded into the portion of the density gradient where their density equals that of the medium around them.

Although the size of a particle does affect its *rate* of migration, buoyant density separations are ultimately based on density, not size. Isopycnic centrifugation thus differs from rate zonal and differential centrifugation where particle size affects the final separation.[2]

Observe that in isopycnic centrifugation, the sample can be loaded anywhere in the tube. If the sample is placed on top of the gradient, the various particles will migrate down the tube to reach their equilibrium positions. If the sample is loaded in the bottom of the tube with the density gradient on top, the sample components will float up to reach their equilibrium positions. The sample can also be homogeneously mixed throughout a medium that forms a spontaneous gradient. During centrifugation, the gradient forms and particles move till they are in equilibrium with the surrounding medium (Figure 33.12). Isopycnic centrifugation is therefore not time dependent.

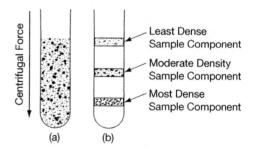

FIGURE 33.12 Isopycnic centrifugation. (a) A uniform mixture of gradient medium and sample is placed in a tube. (b) During centrifugation, particles separate into bands such that at equilibrium, their density equals that of the medium.

---

[2] There are many techniques that can separate materials on the basis of size, including filtration, rate zonal and differential centrifugation, electrophoresis, and chromatography. Isopycnic centrifugation is the only common method that separates materials based solely on differences in their densities.

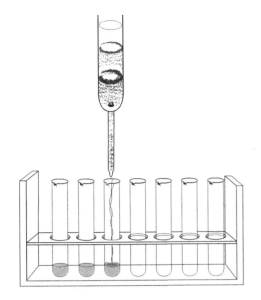

FIGURE 33.13 Collecting the fractions after density gradient centrifugation. Collecting fractions from the bottom of a tube.

After a density gradient separation is completed, there are various methods to collect the banded sample components. Figure 33.13 shows a technique in which the bottom of the tube is punctured and drops are collected in successive test tubes. Another method of retrieving the sample is to puncture the side of the tube with a needle at the site of a specific band. For this purpose, density gradient tubes are generally made of disposable polypropylene or other plastics.

33.2.3 Continuous Centrifugation

Continuous centrifugation is a mode of centrifugation intended for large-volume samples, such as when cells must be separated from a large-volume fermentation broth. In continuous centrifugation, the sample is continuously fed into a rotor that is rotating at its operating speed. Particles sediment in the rotor, while the supernatant is conveyed out of the rotor continuously. Special rotors, centrifuge equipment, and methods are required for continuous centrifugation.

The various modes of preparative centrifugation are summarized in Figure 33.14.

33.3 INSTRUMENTATION: ROTORS, TUBES, AND ADAPTERS

33.3.1 Types of Rotors

33.3.1.1 Horizontal Rotors

Samples are centrifuged inside centrifuge tubes (or bottles) that are placed in compartments in rotors. Individuals who use centrifuges must be able to choose rotors, tubes, and accessories for each application. The user must also be able to properly clean, sterilize (when appropriate), and store these centrifuge components. These practical aspects of centrifuge operation are discussed in this section.

There are three commonly used styles of rotor:

- **Horizontal, also called swinging bucket**
- **Fixed angle**
- **Vertical.**

Examples of these three rotor types are shown in Figure 33.15.

Because of differences in the way samples experience the force of centrifugation in these rotor types, the rotors vary in their relative suitability for specific centrifugal applications, as shown in Table 33.4.

The horizontal (swinging bucket) rotor allows tubes to swing up into a horizontal position during centrifugation (Figure 33.16). As centrifugation proceeds, particles move toward the bottom of the tube.

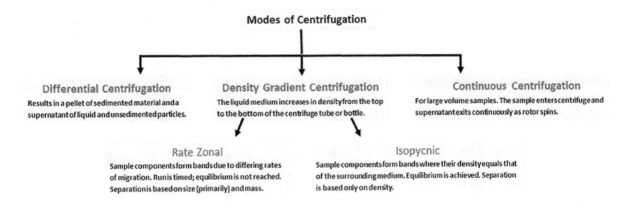

FIGURE 33.14 Modes of preparative centrifugation.

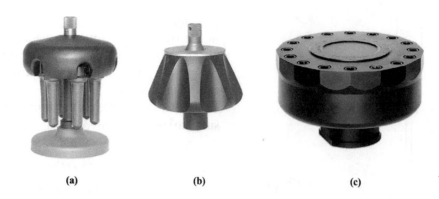

FIGURE 33.15 Common rotor types for centrifugation. (a) Horizontal (swinging bucket) rotor. (b) Fixed angle rotor. (c) Vertical rotor. (Images courtesy of Beckman Coulter Life Sciences.)

TABLE 33.4

General Suitability of Rotors for Different Centrifugal Applications

Applications

Rotor Type	Differential Centrifugation	Rate Zonal Centrifugation	Isopycnic Centrifugation
Horizontal (swinging bucket)	Slow	Good	Good for cells and organelles
Fixed angle	Excellent	Limited	Good for macromolecules Not as good for cells and organelles
Vertical	Not suitable	Excellent	Good

Particles beginning at the top of the tube move the entire length of the tube to form a pellet. This maximizes the path length through which the particles move. (As we will see, the path length is shorter in other types of rotors.) The long path length results in good separations of similar particles by rate zonal centrifugation. Swinging bucket rotors are also useful for differential centrifugation because they give well-formed pellets. A disadvantage to swinging bucket rotors is that, because of their design, they must usually be run at slower speeds than fixed angle or vertical rotors. Because the path length is longer and the speed is slower in horizontal rotors than in other rotors, separations take longer.

Observe in Figure 33.16a that the radius of rotation (r) for a particle varies depending on the particle's location in the tube. A particle at the top of the tube, closest to the center of rotation, is at the point where the radius of rotation is shortest, called r_{min} (minimum radius). In the middle of the tube, the radius is called r_{ave} (average radius), and at the bottom of the tube, the radius is called r_{max} (maximum radius). When calculating the g force experienced by a particle, it is necessary to consider its position in the tube. If the position is not known, use the average radius.

33.3.1.2 Fixed Angle Rotors

Fixed angle rotors *hold centrifuge tubes or bottles in compartments at a particular angle, usually between 15° and 40°* (Figure 33.17). The force of centrifugation causes particles to move outward toward the sides of the tubes. During differential pelleting, particles accumulate at the outer tube wall and then slide downward until they pellet along the side at the bottom of the tube (Figure 33.17b). The path length traveled by particles across the tube in a fixed angle rotor is relatively short. Fixed angle rotors, therefore, result in rapid pelleting (compared to horizontal rotors). A disadvantage to fixed angle rotors for pelleting is that fragile particles, such as intact cells, may be damaged as they are forced against the walls of the tube. Horizontal rotors are gentler.

As in a swinging bucket rotor, the radius of rotation varies for a particle depending on its location in the tube (Figure 33.17a). At the top of the tube, the radius is r_{min}; at the bottom of the tube, the radius is r_{max}. Particles at the bottom of a centrifuge tube in a fixed angle rotor therefore experience a higher g force than particles at the top of the tube.

Fixed angle rotors are used for density gradient separations as well as for differential pelleting; see Figure 33.17c–e. Note that the density gradient formed in

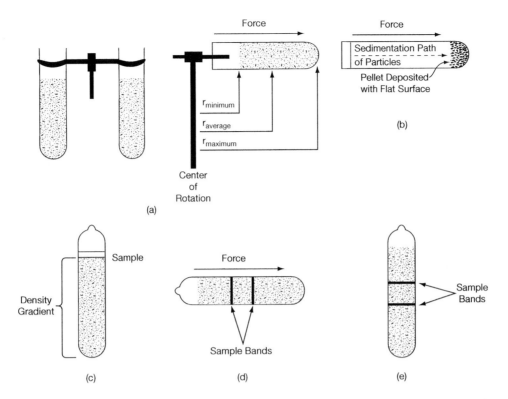

FIGURE 33.16 Swinging bucket (horizontal) rotor. (a) The rotor buckets swing outward to a horizontal position during centrifugation. Particles experience increasingly greater force as they move away from the center of rotation. (b) Differential centrifugation in a horizontal rotor. (c–e) Density gradient separation. (c) A sample is layered on top of a density gradient. (d) Bands form during centrifugation. (e) After centrifugation, the tube returns to a vertical position; the bands remain in their same orientation.

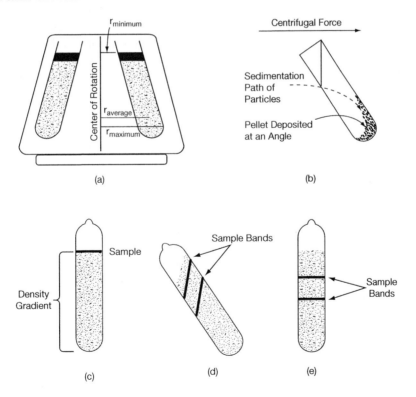

FIGURE 33.17 Fixed angle rotor. (a) Fixed angle rotors hold the tubes at an angle relative to the axis of rotation. The radius of rotation varies depending on the location of a particle in the tube. (b) Differential centrifugation in a fixed angle rotor. (c–e) Density gradient separation in a fixed angle rotor. (c) A sample is layered on top of a density gradient. (d) Particles form bands as centrifugation proceeds. (e) After centrifugation, when the tube is carefully removed from the rotor, the contents of the tube reorient.

a fixed angle rotor reorients horizontally when centrifugation ends and the tubes are removed from the rotor.

33.3.1.3 Near Vertical and Vertical-Tube Rotors

Near vertical-tube rotors *are a modification of fixed angle rotors that have a shallow angle of about 8°–10°.* These rotors are recommended for isopycnic separations where short run times are desired. Vertical rotors hold tubes straight up and down, resulting in a short path length and fast separation (Figure 33.18).

33.3.1.4 *K* Factors

You may come across the term *k factor* (clearing factor) in reference to a rotor, particularly when working with ultracentrifugation. The **k factor** *provides an estimate of the time required to pellet a particle of a known sedimentation coefficient.* The *k* factor varies for different rotors, depending on their minimum and maximum radii and on the speed of rotation. (Manufacturers typically report the *k* factor at the maximum speed that the rotor can attain.) The smaller the *k* factor for a rotor, the faster the separations using that rotor. One use of *k* factor, therefore, is in comparing the efficiency of various rotors.

The *k* factor is calculated by the manufacturer so that it is possible to approximate a particle's sedimentation time, in hours, using the following equation:

$$t = k / S$$

where:
  t = the predicted time (in hours) to move particles from the top of a tube, at r_{min}, to the bottom of the tube, at r_{max}
  k = the clearing factor
  S = the sedimentation coefficient in Svedberg units.

Note: This formula assumes that the liquid medium is water at 20°C. If the medium is denser or more viscous than water, the sedimentation time will be longer.

Example 33.2

A manufacturer reports that the *k* factor for a rotor is 100. How long will it take ribosomes to sediment through water to the bottom of a tube?

Ribosomes have a sedimentation coefficient of 80 (Table 33.1). Therefore,

$$t = \frac{100}{80} = 1.25 \text{ hours}$$

33.3.2 BALANCING A ROTOR

One of the most important tasks performed by the operator of a centrifuge is making sure that the rotor is used properly and that the tubes or bottles are balanced in the rotor. The purpose of balancing the load is to ensure that the rotor spins evenly on the drive shaft. This is essential for minimizing rotor stress and preventing damage to the centrifuge, rotor, and tubes. In worst-case accidents, improperly used, unbalanced, or damaged rotors have detached from the spindle and flown out of the centrifuge chamber at high velocity, wreaking havoc along their path (Figure 33.19). Fortunately, modern centrifuges are well armored, which helps prevent rotors from exiting the centrifuge in an accident. However, if a rotor detaches from the spindle at high

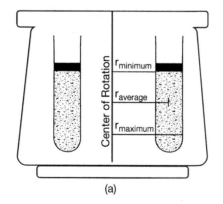
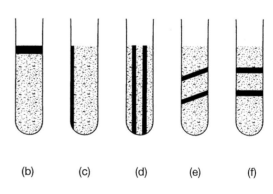

(a)   (b)   (c)   (d)   (e)   (f)

FIGURE 33.18 Vertical rotor. (a) Tubes are vertical in the rotor compartments. (b–f) Density gradient separation in a vertical rotor. (b) A sample is layered on top of a density gradient. (c) When centrifugation begins, the tube contents reorient under the influence of the relative centrifugal field. (d) The sample sediments into vertical bands. (e and f) When the rotor comes to a stop and the tubes are carefully removed from the rotor, the tube contents reorient forming horizontal bands.

FIGURE 33.19 Scene following an accident in which an ultracentrifuge rotor failed during operation. The subsequent "explosion" destroyed the centrifuge, ruined a nearby refrigerator and freezer, and made holes in the walls and ceiling. The centrifuge itself was propelled sideways, and cabinets containing hundreds of chemicals were damaged. Fortunately, cabinet doors prevented chemical containers from falling to the floor and breaking. A shock wave from the explosion shattered all four windows in the room and destroyed the control system for an incubator. Luckily, no one was injured. The cause of the accident was the use of a model of rotor that was not approved by the manufacturer for the particular model of centrifuge. (Courtesy of Cornell University, Environment, Health and Safety, Ithaca, NY.)

speed, even if it does not exit the centrifuge chamber, it will crash into the centrifuge chamber walls with great force, destroying the centrifuge chamber and smashing the rotor into small fragments. The damage to the centrifuge may be irreparable and will at best be extremely costly. A centrifuge accident will also spread the sample throughout the laboratory, usually as an aerosol – a serious problem if the sample is hazardous.

A rotor is balanced by placing samples and their containers symmetrically in the rotor (Figure 33.20).

Tubes or bottles across from one another should be of nearly equal weight. The manufacturer specifies how close in weight the tubes that are opposite from one another must be. For example, for a low-speed centrifuge the manufacturer might specify that the opposing tubes should be balanced so that there is no more than 0.5 g difference in their weight. For an ultracentrifuge, the manufacturer might specify that opposing tubes must be within 0.1 g of one another. Always adhere to the manufacturer's instructions for balancing tubes. Also note that for ultracentrifuges, tubes or bottles that are across from one another must

contain the same density profile as well as match in weight. Guidelines regarding proper balancing of a rotor and operation of a centrifuge are in Box 33.1.

33.3.3 CARE OF CENTRIFUGE ROTORS

33.3.3.1 Overview

Although rotors appear to be massive and sturdy, they are subjected to tremendous forces during centrifugation and can easily be damaged. A damaged or improperly used rotor can fail during centrifugation with catastrophic results. Always read the instructions that come with a rotor. General guidelines for care and cleaning that are relevant to most brands of rotor are discussed in this section and are summarized in Box 33.2.

Recall that the force generated in centrifugation is:

$$RCF = 11.2 \times r \left(\frac{RPM}{1,000} \right)^2$$

This equation tells us that the RCF generated is proportional to the square of the speed. For example, if the RPM is doubled, then the rotor is subjected to four times the original force. The stress on ultracentrifuge rotors, therefore, is considerably greater than that on lower-speed rotors. Small imperfections in an ultracentrifuge rotor can cause the rotor to fail when it is centrifuged at high speeds. For this reason, rotors, particularly those for the ultracentrifuge, are made to exacting specifications and must be correctly handled and maintained.

Each rotor is specified by the manufacturer for a maximum allowable speed above which it should never be run. Most high-speed and ultracentrifuge rotors have a protective mechanism that prevents them from being rotated at excessive speeds. This mechanism often involves an **overspeed disk** *that has black and white alternating sectors, or lines, and is affixed to the bottom of the rotor.* The number of sectors on an overspeed disk varies depending on the maximum allowable speed for the rotor. In the centrifuge, there is a photoelectric device that detects the lines on the overspeed disk during centrifugation. If the lines pass by the photoelectric device faster than an allowable limit, then the centrifuge automatically decelerates.

33.3.3.2 Chemical Corrosion and Metal Fatigue

Chemical corrosion *is a chemical reaction that causes a metal surface to become rusted or pitted.* Common salt solutions, such as NaCl, are particularly likely to cause corrosion and should be removed thoroughly if spilled on a rotor. Other corrosive agents

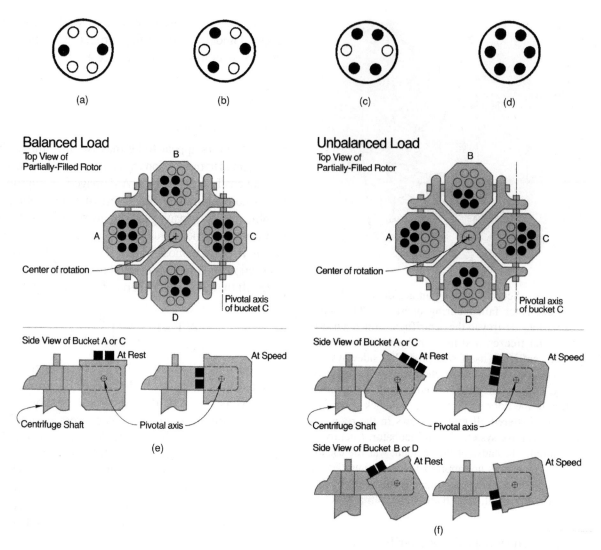

FIGURE 33.20 Balancing a rotor load. (a–d) Rotors containing two, three, four, and six tubes, respectively. Opposing tubes have the same weight (and density profile for ultracentrifuges). (e and f) The situation is more complex in buckets holding multiple tubes. Each bucket in this rotor holds ten tubes. (e) The opposing bucket sets A–C and B–D are loaded with an equal number of tubes, each filled with an equal weight of sample, and are balanced across the center of rotation. Each bucket is also balanced with respect to its pivotal axis. (f) This rotor is not balanced. None of the four buckets is balanced with respect to its pivotal axis. During centrifugation, buckets A and C will not reach a horizontal position, whereas buckets B and D will pivot past horizontal. Note also that the tubes in B and D are not arranged symmetrically across the center of rotation. (Reprinted with permission from Beckman Coulter Life Sciences.)

include strong acids and bases, alkaline laboratory detergents, and salts of heavy metals such as cesium, lead, silver, or mercury. Moisture that stands in the rotor compartments can also cause corrosion over time (Figure 33.21).

Even in the absence of chemical corrosion, metals fatigue, or become weakened, due to the repeated stress of centrifugation. This damage occurs inside the rotor, so it is not visible. Ultracentrifuge and high-speed rotors must therefore be derated and/or retired as time goes by and as they are used. *Derating means to run a rotor at a speed less than its originally rated maximum speed.* **Retiring** *a rotor means to discard*

it or return it to the manufacturer. The manufacturer specifies the conditions under which a rotor should be derated or retired. Because a rotor has a finite lifetime, it is essential to document its use, typically in a rotor log. Newer centrifuges may automatically record this information.

Low-speed rotors can be made of a variety of materials including aluminum, bronze, steel, and even high-impact plastic. High-speed rotors are frequently made of aluminum, which is light, strong, and relatively inexpensive. Aluminum rotors, however, are especially susceptible to corrosion. For ultracentrifugation, more expensive titanium rotors are preferred.

BOX 33.1 OPERATING A CENTRIFUGE

Be certain to consult the manufacturer's instructions before using any centrifuge. There are many different models of centrifuge with differing features, programmability, and safety mechanisms. It is essential that you understand the features of your own instrument. Guidelines that are generally applicable are shown in this box.

1. *For low- and high-speed centrifuges, opposing tubes can be balanced according to weight without considering density.*
    a. The simplest way to balance tubes or bottles is to place them on the opposite sides of a two-pan balance and bring them to equal weight.
    b. Remember to include caps and seals when balancing tubes.
    c. The manufacturer specifies how close in weight the tubes must be. For example, for a low-speed centrifuge the manufacturer may specify that the tubes should be balanced so that there is no more than 0.5 g difference in their weight.
    d. Opposing tubes may both contain sample, or one of the tubes may be a "blank" containing only a liquid, such as water.
2. *In an ultracentrifuge, the tubes that are across from one another must be of the same type with respect to density, as well as be of the same weight.*
    a. Fill opposing tubes with materials of the same density profile and make certain that their weights also match.
    b. Opposing tubes may both contain sample, or one of the tubes may be a "blank" containing only a liquid medium, or a density gradient to match the sample tubes.
3. *Tubes that are not directly across from one another can be of different weights and/or have different density profiles.*
4. *Do not exceed the maximum weight in any sample compartment, as specified by the rotor manufacturer.* Be sure to include the weight of any seals, plugs, or caps.
5. *Make sure the centrifuge chamber and the spindle are clean and dry before attaching a rotor.*
6. *Make sure the rotor is properly attached and tightened onto the spindle.* Newer instruments may use easy push button attachment mechanisms, but older systems require the user to carefully attach the rotor to the spindle with the proper amount of tightening.
7. *Make sure the rotor cover is present and is properly tightened.*
8. *All buckets must be in place in a swinging bucket rotor, even if some contain no samples.* If the buckets are numbered and have caps, they must be run in the correct positions with their caps.
9. *Do not operate the centrifuge at the critical speed.* At the **critical speed**, *any slight imbalance in the rotor will cause vibration.* You will feel and hear this vibration as the centrifuge passes through the critical speed.
10. *Pay attention to the brake setting.* The brake slows the rotor after a run is completed. On many centrifuges, it is possible to select a brake setting. If braking is too abrupt, some types of pellets will become resuspended.
11. *When using stainless steel tubes, certain plastic adapters, and high-density solutions, the centrifuge cannot be run at its maximum speed.* Be sure to consult the manufacturer's manual to determine permissible values of RPM.

Titanium is much more resistant to corrosion than aluminum, so titanium rotors are less likely to fail when repeatedly centrifuged. There are also newer-style, lightweight rotors that are not made of metal, but of a composite material that is composed of spun or woven carbon fibers embedded in epoxy resin. Carbon composite rotors have various advantages. Composite materials tend to be resistant to chemical corrosion and to fatigue. They accelerate and decelerate faster, resulting in shorter run times. Heavier metallic rotors are difficult to move in and out of centrifuges and can cause back injuries; lighter composite rotors are easier to handle. Note that titanium and carbon composite rotors usually have some aluminum or iron components, such as tube compartments, that are susceptible to chemical damage.

BOX 33.2   SUMMARY OF GUIDELINES FOR CLEANING, MAINTAINING, AND HANDLING ROTORS, TUBES, AND ADAPTERS

1. *Regularly clean rotor, adapters, and O-rings.*
   a. There are different materials used to construct rotors; consult the manufacturer to determine how to properly clean your rotor(s). The following are general recommendations.
   b. Use a mild, non-alkaline detergent and warm water.
   c. Thoroughly remove all disinfectants and detergents using purified water.
   d. With swinging bucket rotors, clean only the buckets themselves. Do not immerse the rotor portion because its pins cannot readily be dried and will rust if wetted.
   e. Air-dry and store rotors upside down to allow liquids to drain. Make sure rotors remain dry during storage. Do not store rotors in a cold room due to the humidity. If a chilled rotor is needed, cool immediately before use.
   f. Disinfect and sterilize rotors according to manufacturer's instructions.
   g. Do not wash in a dishwasher.
   h. Do not soak in detergent.

2. *Protect rotors from corrosive agents, including moisture, chemicals, alkaline solutions, and chloride salts.* Aluminum is particularly sensitive to corrosion. The cover and the compartments of titanium rotors may be aluminum; therefore, they are more vulnerable than the rest of the rotor.
   a. Immediately wash the rotor if it is exposed to corrosive materials.
   b. Visually inspect the rotor at least monthly, including inside the cavities and buckets. Look for hairline cracks, dents, white spots, white powder deposits, and pitting. If damage is detected, do not use the rotor without consulting a service technician.

3. *Do not scratch a rotor.*
   a. Use soft bristle brushes for cleaning.
   b. Do not place metal forceps or other implements into rotor compartments.
   c. Do not use metal implements to remove stuck tubes or to open tight lids, except as provided by the manufacturer.

4. *Check rotor O-rings before every run.* Many rotors have O-rings that maintain a seal during a run. These O-rings and the surfaces they contact must be kept clean and in good condition. Some manufacturers recommend lightly greasing the O-rings with silicone vacuum grease before reassembly. Check O-rings for cracks, tears, or an uneven surface; replace if necessary.

5. *Consult the ultracentrifuge manufacturer before using a rotor made by a different manufacturer.*

6. *Find out from the manufacturer how long, or for how many runs, a rotor is guaranteed.* Derate or retire a rotor as directed by the manufacturer.

7. *Never exceed the maximum speed of a rotor.*
   a. Never remove overspeed protective devices.
   b. Pay attention to the density of solutions in ultracentrifugation. It may be necessary to run the centrifuge more slowly than usual if a high-density solution is used.
   c. Uncapped tubes, partially filled tubes, and heavy tubes and/or samples may require that the rotor be used at speeds less than its maximum speed.

8. *Clean centrifuge tubes and bottles according to the manufacturer's directions.* General guidelines include the following:
   a. Reusable tubes, bottles, and adapters should be cleaned by hand with a mild detergent using soft, non-scratching brushes. Commercial dishwashers are likely to be too harsh.
   b. Bottles and tubes should not be dried in an oven; rather, they should be air-dried. Stainless steel tubes must be thoroughly dried.
   c. Some tubes and bottles can be autoclaved; however, note that this will reduce their life span.

9. <u>*Never*</u> *use a cracked tube or bottle or one that has become yellowed or brittle with age.*

10. *Make sure that tubes and bottles used for centrifugation are chemically compatible with the sample.*

(Continued)

11. ***Always make sure that the outer surfaces of sample tubes are completely dry.*** This not only avoids rotor corrosion, but also can reduce the possibility of tubes becoming stuck in the rotor.

12. ***Do not spin tubes or bottles at speeds above those for which they are rated.*** It may be necessary to reduce the speed when using plastic adapters or metal tubes.

13. ***Fill tubes and bottles to the height specified by the manufacturer.*** Some tubes must be entirely filled to prevent collapse during centrifuging while others may be filled to varying levels.

14. ***Use adapters to ensure a proper fit in the rotor compartments.*** Tubes or bottles that do not fit snugly may shift, resulting in breakage or a poorly formed pellet. Adapters hold tubes and bottles securely.

15. ***Maintain rotor and centrifuge logs.*** Some newer instruments log rotor and centrifuge usage automatically, but older models do not. Also document cleaning and decontamination.

FIGURE 33.21 Example of a corroded centrifuge rotor. (Image courtesy of Eppendorf AG.)

33.3.3.3 Situations That Require Using Rotors at Speeds Less than Their Maximum Rated Speed

A rotor is sometimes derated due to its age or the number of times it has been spun. There are other situations that also call for reducing the speed of rotation to avoid overstressing the rotor. One of these situations relates to the density of the samples being centrifuged. Every rotor is designed to centrifuge materials whose densities are below a certain maximum cutoff. This maximum density is often 1.2 g/mL. Some fixed angle and vertical-tube rotors are designed to hold a density of 1.7 g/mL, which is the maximum density of CsCl solutions used for DNA separations.

It is sometimes necessary to run a sample that exceeds the density limit for the rotor. In these cases, the rotor cannot be run at its maximum rated speed. The rotor manual will usually provide information about centrifuging samples whose density exceeds the rotor design limits. If the manual does not have this information, then there is an equation to estimate the maximum safe speed when centrifuging dense solutions:

$$\text{Derated speed} = \text{Maximum rotor speed in RPM}\sqrt{\frac{RD}{SD}}$$

where:

RD = the design limit of the rotor, as specified by the manufacturer

SD = the sample density.

Example Problem 33.3

A rotor is specified to have a maximum speed of 25,000 RPM and is designed for densities less than or equal to 1.2 g/mL. A density gradient is being centrifuged with a maximum density at the bottom of the tube of 1.5 g/mL. What is the maximum speed at which this gradient can be centrifuged?

Answer

$$\text{Derated speed} = 25{,}000 \text{ RPM}\sqrt{\frac{1.2}{1.5}} \approx 22{,}360 \text{ RPM}$$

So, in this situation, the maximum speed is **22,360 RPM**.

It is possible that the combination of rotor speed and temperature can be such that a gradient material precipitates out of solution. CsCl gradients, in particular, can form crystals in the bottom of the centrifuge tube at low temperatures. These crystals have a high density (4 g/mL) that will produce stress that far exceeds the design limits of most rotors. Consult the manufacturer regarding the use of CsCl gradients in a particular type of rotor.

There are other situations, in addition to the spinning of high-density samples, that require that a rotor be used at speeds less than its normal maximum speed. For example, samples or tubes are sometimes used that exceed the maximum weight for which the rotor was designed. The rotor must then be used at speeds less than its maximum speed, according to the manufacturer's directions.

33.3.3.4 Rotor Cleaning

Rotors must be kept clean. Rotors that receive heavy use should be cleaned on a routine basis, and every rotor should be thoroughly washed in the event of a spill. Detergents used to clean rotors must be mild and non-alkaline. Brushes used for cleaning must be soft so as not to scratch the rotor. Manufacturers sell mild detergents and brushes for use with rotors. Detergents must be thoroughly rinsed from rotors with purified water. Manufacturers recommend air-drying rotors and storing them upside down in a dry environment with their lids removed to avoid having moisture collect in the compartments.

Some rotors can be safely autoclaved, although repeated autoclaving may weaken ultracentrifuge rotors. Seventy percent alcohol is compatible with all rotor materials and therefore can be used for disinfection. Some manufacturers recommend sterilizing rotors with ethylene oxide, 2% glutaraldehyde, or UV light. Rotors contaminated with pathogenic or radioactive materials should be cleaned with agents that are compatible with the rotor. Note that alkaline detergents designed to remove radioactivity are not compatible with most rotors.

Rotors generally have O-rings and gaskets to ensure tight seals. These O-rings and gaskets can typically be lubricated with a thin layer of silicone vacuum grease. Worn or cracked O-rings and gaskets should be replaced.

Aluminum rotors should frequently be inspected for signs of corrosion, including rough spots, pitting, white powder deposits (which may be aluminum oxide), and heavy discoloration. If problems are observed, consult the manufacturer before using the rotor. Titanium and carbon composite rotors are much more corrosion resistant and require less-frequent inspection.

33.3.4 CENTRIFUGE TUBES, BOTTLES, AND ADAPTERS

33.3.4.1 General Considerations

Centrifuge tubes are containers for smaller-volume samples; centrifuge bottles are containers for larger-volume samples. Centrifuge tubes and bottles are available in varying styles and sizes and made of different materials. Some tubes and bottles are used with caps or seals; others are centrifuged open. Some tubes can be washed and used repeatedly, but others must be disposed of after one use. Because there are many types of sample containers for centrifugation, it is necessary to select those that match your samples and rotors.

Capped centrifuge tubes should always be used if a sample is potentially hazardous or volatile. Although screw caps are simple to use and appear to be leakproof when handled on the laboratory bench, they distort under the pressure of centrifugation. This distortion can allow samples to leak from the tubes. For hazardous samples, manufacturers make capping systems that compress onto and seal tubes and bottles firmly.

At low and moderate RCFs, most tubes and bottles can be centrifuged when only partially filled. At higher RCFs, only thick-walled tubes can be run partially filled; thin-walled tubes collapse during centrifugation.

A single rotor can cost thousands of dollars. To maximize their versatility, manufacturers design rotors so that they can flexibly accommodate different styles and sizes of tubes and bottles. This is accomplished with **adapters,** *plastic or rubber inserts that reduce the size of the sample compartment so that smaller tubes can be run in the same rotor.* There are many types of adapters (Figure 33.22). Some adapters

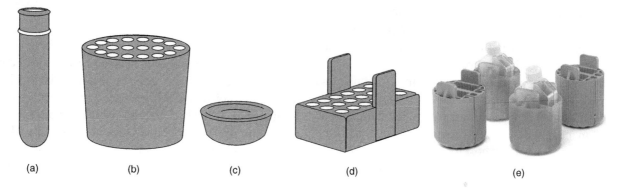

(a)   (b)   (c)   (d)   (e)

FIGURE 33.22 Adapters. Adapters allow varying sizes and shapes of tubes to be used in a single rotor. (a) An adapter that lines the rotor compartments and holds one tube. (b) An adapter that holds 18 tubes at once. (c) A cushion to support a glass or plastic bottle in a low-speed centrifuge. (d) An adapter to hold 18 tubes in a horizontal low-speed rotor. (e) Adapters to allow cell culture flasks to be spun in a centrifuge. Centrifugation is routinely used in cell culture when cells must be separated from their culture medium. (Image courtesy of Beckman Coulter Life Sciences.)

are tube-shaped and line the rotor compartment. Other adapters support the tops of tubes, and still others the bottoms. There are specialized adapters to hold 96-well plates and cell culture flasks. It is important to purchase the correct configurations of tubes, caps, and adapters for each rotor and each type of sample that is to be used in that rotor. It is essential that sample containers fit snugly in the rotor compartments. Be aware that it may be necessary to run a rotor at speeds less than its maximum speed when using adapters.

33.3.4.2 Materials Used to Make Centrifuge Tubes and Bottles

Materials used to make centrifuge tubes and bottles include glass, stainless steel, and various types of plastics (Table 33.5). These materials must be able to withstand the high forces of centrifugation. Depending on its composition and its design, every tube and bottle is rated for a maximum centrifugal force.

Plastics are relatively light and strong, so they are the preferred material for making tubes and bottles that will experience high centrifugal forces; however, plastics have limitations. Most plastic centrifuge tubes and bottles will soften and deform if they are centrifuged at temperatures above 25°C. Completely filled, thick-walled polyallomer (a type of plastic) tubes can be used at higher temperatures, but lower speeds and shorter run times are preferred. Also, as we have seen in previous chapters, different plastics have differing chemical compatibilities. Some plastics are incompatible with various solvents (e.g., acetone and alcohols) and with acids and bases. Most plastics will soften or dissolve when exposed to phenol. To determine whether a particular tube material is compatible with your sample, consult the chemical compatibility chart from the manufacturer. Even if the sample and centrifuge tube material appear to be compatible based on the chart, it may still be a good idea to perform a test run.

Stainless steel tubes and bottles can be used safely at high temperatures and are relatively inert, but stainless steel is heavy and limits the speed of centrifugation.

TABLE 33.5

Materials Used to Make Centrifuge Tubes and Bottles

1. *Glass. Glass is largely inert, which makes it chemically compatible with almost all samples.* Also, glass can be autoclaved repeatedly. Normal soda glass is unable to withstand forces much above $3,000 \times g$, whereas borosilicate glass is able to withstand at least $10,000–12,000 \times g$. The disadvantage to glass is the possibility of breakage. Glass therefore cannot be used at high RCFs. It is best to support glass tubes using a rubber pad or adapter, thus avoiding pressure between glass and metal. It is always good practice to do a test run with glass tubes or bottles before running them with potentially hazardous materials.

2. *Stainless steel. Stainless steel is resistant to organic solvents and to heat.* Stainless steel tubes can be reused many times as long as they are undamaged and uncorroded. These tubes are strong enough to be run partially filled. The disadvantage to metal tubes is that they are heavy and so require that a rotor be spun at speeds less than its maximum speed.

PLASTICS

3. *Polycarbonate. Polycarbonate is strong, transparent, and autoclavable.* Polycarbonate can be constructed with thick walls so that tubes can be only partially filled and do not collapse when centrifuged. Polycarbonate is commonly used for high-speed and ultracentrifuges. Polycarbonate, however, is not resistant to many solvents, including phenol and ethanol, both of which are commonly used in biology laboratories. Polycarbonate is also attacked by alkaline solutions, including many common laboratory detergents. Polycarbonate cannot withstand repeated autoclaving without some degradation. Polycarbonate can become brittle and should be visually inspected before use. Discard cracked or brittle tubes.

4. *Polysulfone. This plastic has many of the advantages of polycarbonate, but it is also resistant to alkaline solutions.* It is, however, attacked by phenol.

5. *Polypropylene and polyallomer. Polyallomer is commonly used for high-speed and ultracentrifuge tubes.* Both polyallomer and polypropylene can be autoclaved. Both can be made into thin-walled tubes that can be punctured readily to remove bands from gradients. Both plastics tend to be resistant to most organic solvents. Polyallomer is recommended for separating DNA in CsCl gradients because DNA does not adhere to the tube walls.

6. *Cellulose esters. These tubes are transparent, easy to pierce, and strong, but they cannot be autoclaved and are attacked by strong acids and bases and by some organic solvents.*

7. *Other plastics. Many companies make specialized plastic tubes and bottles with various properties and under their own trade names.* Check with the manufacturer before using these with acids, bases, and solvents, and before autoclaving. Also note the maximum speed at which they can be used.

Example Problem 33.4

A page from a rotor catalog is shown. Answer the following questions about the rotor described.

  a. What is the maximum speed that this rotor can withstand?
  b. What is the maximum force that this rotor can withstand?
  c. How many different types of tubes can be used with this rotor?
  d. What is the maximum volume that can be centrifuged in this rotor?
  e. Which types of tubes can withstand being centrifuged at 80,000 RPM?
  f. What is the function of the adapters? Which tubes require adapters?
  g. What happens to the maximum speed that can be used with this rotor if adapters are used?
  h. What is the maximum speed at which a density gradient with a maximum density of 1.4 g/mL can be run in this rotor?
  i. Is this rotor for a low-speed, high-speed, or ultracentrifuge?

FA-80 ROTOR

Description

25° fixed angle, titanium, 10 compartments, 3.5 mL maximum volume/compartment
Maximum speed 80,000 RPM, k factor 25.8
Designed for samples with a density less than or equal to 1.2 g/mL

	RCF (×g)	Radius (cm)
Maximum	460,000	6.55
Average	350,000	5.00
Minimum	250,000	3.41

Tubes

Capacity (mL)	Description	Maximum Speed	Cap Assembly	Adapters
3.5	Polyallomer, thin-walled tube	80,000	Multipiece	None
3.5	Polycarbonate, thick-walled tube	45,000	Screw cap	None
3.5	Polyallomer, thin-walled tube	80,000	Crimp seal	None
2.0	Polycarbonate, thick-walled tube	55,000	None	1 per compartment required
2.0	Cellulose acetate, butyrate, thick-walled tube	40,000	None	1 per compartment required
0.4	Cellulose acetate, butyrate, thick-walled tube	40,000	None	1 per compartment required

Answer

  a. The maximum speed for the rotor is 80,000 RPM.
  b. The maximum RCF generated by this rotor is 460,000 ×g at r_{max}.
  c. The manufacturer lists six types of tubes for this rotor.
  d. There are 10 compartments, each of which can hold a maximum of 3.5 mL, so the maximum total volume is 35 mL.
  e. Only the polyallomer tubes can be used at the rotor's maximum speed.
  f. Adapters allow tubes smaller than the sample compartment to be used. The tubes that hold smaller volumes require adapters.
  g. The use of adapters reduces the maximum speed at which the centrifuge can be run.

---

h. The rotor is designed for a density ≤ 1.2 g/mL. The square root equation can be used to determine the maximum rate given the higher density as follows:

$$\text{Derated speed} = 80{,}000 \text{ RPM} \sqrt{\frac{1.2}{1.4}} \approx 74{,}066 \text{ RPM} = \text{the maximum speed at this density.}$$

i. This is an ultracentrifuge rotor.

33.3.5 CENTRIFUGE MAINTENANCE AND TROUBLESHOOTING

33.3.5.1 Maintenance and Performance Verification

The details of operation vary from centrifuge to centrifuge, so it is necessary to read the manufacturer's instructions. It is especially important to read the manual when operating an ultracentrifuge because these instruments can easily be damaged with dangerous and costly results.

Centrifuges require periodic maintenance and performance verification that are usually performed by trained service technicians. The technicians make sure that the RPM values that are displayed correspond to the actual speed of the rotor, that the temperature in the chamber is maintained properly, and that timers and other controls work correctly. The vacuum and refrigeration systems must also be checked. Older centrifuges may have brushes that require replacement. All service and maintenance procedures should be documented.

33.3.5.2 Troubleshooting Common Problems

Some common centrifugation problems are described in Box 33.3.

BOX 33.3   TROUBLESHOOTING TIPS

Symptom 1: *VIBRATION DURING OPERATION*
   Vibration is due to rotor imbalance

1. Turn off the centrifuge.
2. Check that all containers are properly balanced and arranged properly.
3. Check that cushions, adapters, and caps match in all sample compartments.
4. Check that the centrifuge is on a level surface.
5. Increase speed gradually.

Symptom 2: *TUBE BREAKAGE*

1. Check that the type of tube in use is rated for the speed and force being used.
2. Reinspect all tubes and discard those with any cracks, pitting, or yellowing.
3. Check that the tubes are properly balanced.
4. Make sure that the adapters and cushions match the type of tubes.
5. Make sure the contents of the tubes are compatible with the materials used in constructing the tubes.
6. If using partially filled tubes, check that sufficient volume is being used to meet the manufacturer's specifications.

Symptom 3: *FINE GRAY OR WHITE POWDER OR DUST IN CHAMBER*
   This is caused by particles of glass sandblasting the inside of the centrifuge

1. Clean centrifuge thoroughly.
2. Run the centrifuge without samples several times and clean between each operation.

33.3.6 SAFETY

Various safety issues relating to rotors and centrifuge tubes have already been addressed in this chapter. Another risk that is less dramatic than a catapulting rotor, but is equally dangerous, relates to the centrifugation of hazardous materials such as viruses, radioactive substances, microbes, and carcinogenic compounds. Protection from hazardous materials is of great importance in biological centrifugation. In a worst case, a centrifuge tube that breaks during centrifugation can disperse hazardous aerosols throughout the laboratory and throughout the air-handling system in a building. There are more subtle concerns as well. A poorly sealed centrifuge tube cap can allow aerosols to escape. Screw caps on centrifuge tubes or bottles are likely to distort under the high force of centrifugation, allowing liquids and aerosols to be released. Simply opening a tube, particularly a tube with a "snap cap," creates aerosols. Pouring and decanting fluids from centrifuge tubes and bottles creates aerosols. Spills can occur while introducing samples into tubes or removing samples from tubes.

Because centrifugation provides many opportunities for contamination, it is necessary to use special precautions when spinning pathogenic or toxic materials. Biological safety cabinets can be used when filling and emptying centrifuge tubes or bottles. It is always important to be certain that tubes and bottles are compatible with samples and be certain not to exceed the force they can withstand. These practices are critical when spinning hazardous materials because tube breakage is more dangerous. There are centrifuges, rotors, and centrifuge tubes that are specifically designed for hazardous materials. Some rotors, for example, are designed to safely contain spilled materials in the event of tube breakage.

Radioactive materials can readily contaminate rotors and centrifuges. The commonly used radioisotope, ^{32}P, tends to bind tightly to metal surfaces, making it difficult to remove. This problem is exacerbated by the fact that alkaline detergents intended to remove radioactive contaminants are damaging to rotors.

Failure to follow proper practices when operating centrifuges can endanger laboratory staff, result in damage to the centrifuge or its accessories, and cause contamination of the centrifuge and the laboratory. Some safe handling guidelines are summarized in Box 33.4.

BOX 33.4   GUIDELINES FOR SAFE HANDLING OF CENTRIFUGES AND THEIR ACCESSORIES

1. *Carefully read the manufacturer's instructions.* Pay attention to warnings provided. Note limitations of rotors. Find out how to use tubes and adapters properly. Careful attention to the manufacturer's directions is especially important when operating high-speed and ultracentrifuges.
2. *Use special precautions when centrifuging pathogenic, toxic, or radioactive materials.*
   a. Make sure that tubes are properly capped and sealed. Screw caps are insufficient under the force of centrifugation.
   b. Avoid caps that snap open.
   c. Open tubes containing dangerous materials in a suitable hood or other protected enclosure.
   d. Pour and decant hazardous samples in a hood or protected enclosure.
   e. Clean spills promptly.
   f. Use special containment instrumentation as appropriate. Sealable tubes and rotors are available to contain hazardous materials.
   g. Some centrifuges are designed specifically for use with hazardous samples. Consider investing in such a device.
3. *Avoid touching any rotors with bare hands.* Assume that rotors and the insides of centrifuges are contaminated, but at the same time, try to avoid contamination. Assume that gloves are contaminated after touching a rotor and remove them before touching anything else. Never place bare fingers in any rotor compartment, both because of the possibility of contaminants, and because there may be broken glass.
4. *Protect service technicians from pathogens and radioactivity by careful cleaning of equipment before any repairs are performed.* Document the decontamination process.

(Continued)

5. ***Protect yourself from the spinning rotor.*** Although many centrifuges have locking covers that prevent access to a spinning rotor, some centrifuges lack this safety feature. There are also modes of centrifuge operation where the user must have access to the rotor as it spins.

   a. Do not touch a rotor that is in motion and do not slow a rotor with your hand.

   b. If you must work with a spinning rotor, then:
   - remove jewelry, such as necklaces and bracelets;
   - remove neckties and scarves;
   - tie back long hair; and
   - roll up and secure shirt sleeves.

6. ***Do not centrifuge large volumes of explosive or flammable materials, such as ethyl alcohol, chloroform, and toluene.*** (Note that in practice, small volumes of these materials are often centrifuged in low-speed centrifuges.)

7. ***Make sure the contents of a rotor are properly balanced before centrifugation.***

8. ***Always remain at the centrifuge as it begins to accelerate.*** An unbalanced rotor or other problems will become apparent when the centrifuge reaches critical speed.

Practice Problems

1. A portion of a table from a manufacturer's rotor manual is shown. Calculate the RCFs to fill in the blanks. Refer to the diagram to determine the values for r_{max}, r_{min}, and $r_{average}$ (Figure 33.23).

	RCFs Generated		
Speed (in RPM)	r_{max}	r_{min}	$r_{average}$
20,000	40,800	17,200	29,000
30,000			
40,000			

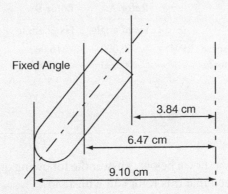

Fixed Angle

3.84 cm

6.47 cm

9.10 cm

FIGURE 33.23   Diagram for Practice Problem 1.

2. The following calculation is incorrect. Why? (Figure 33.24)
    Speed is 45,000 RPM.

$$RCF = 11.2 \times r \left(\frac{RPM}{1,000} \right)^2$$

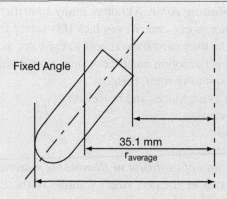

Fixed Angle

35.1 mm
$r_{average}$

FIGURE 33.24   Diagram for Practice Problem 2.

$$RCF = 11.2 \times 35.1 \left(\frac{45,000}{1,000} \right)^2 = 796,068$$

Incorrect answer!

3. A journal article includes the statement below. If you have a similar rotor with an r_{ave} of 2.6 cm, how many RPM will you need to use to duplicate the force used by these authors?

   The extract was centrifuged in a fixed angle rotor at 140,000 ×g for three hours.

4. Specifications for three rotors are shown. Answer the following questions about these rotors.
   a. Which rotor would be best for separating large volumes of bacterial cells after fermentation? Explain.
   b. Which would be used to isolate small, subcellular organelles and for buoyant density separations of macromolecules such as DNA and RNA? Explain.
   c. Which rotor would be useful for centrifuging 1 mL volumes of precipitated DNA in a molecular biology laboratory? Explain.

Type	Rotor A Fixed angle	Rotor B Fixed angle	Rotor C Fixed angle
Maximum speed (RPM)	60,000	16,000	10,000
Maximum force (×g)	400,000	40,000	6,000
Capacity (#×mL)	6×25	6×500	12×2

5. A portion from a rotor catalog is shown below. Answer the following questions about the rotor described.
   a. What is the maximum speed that this rotor can withstand?
   b. What is the maximum force that can be generated with this rotor?
   c. What is the maximum volume that can be centrifuged at one time in this rotor?
   d. What is the maximum number of tubes that can be centrifuged at once in this rotor? What is the maximum volume they can contain?
   e. Is this rotor for a low-speed, high-speed, or ultracentrifuge?

FA-13
Description

28° fixed angle, aluminum, six compartments, 315 mL maximum volume/compartment

Maximum speed 13,000 RPM, k factor 20

Designed for samples with a density less than or equal to 1.2 g/mL

Maximum RCF=27,500, maximum radius=14.60 cm

Tubes

Capacity (mL)	Description	Maximum Speed	Cap Assembly	Adapters
315	Stainless steel bottle	13,000	Stainless steel	None
280	Polysulfone bottle	13,000	Polypropylene seal	None
150	Polycarbonate, thick-walled	13,000	None	1 per compartment
30	Borosilicate glass tube	10,000	None	1 per compartment, each adapter holds 3 tubes
4	Polypropylene tube	13,000	Polypropylene seal	1 per compartment, each adapter holds 12 tubes

34 Introduction to Bioseparations

34.1 CHAPTER OVERVIEW

34.1.1 THE GOALS OF BIOSEPARATIONS

This unit explores a fundamental task of biotechnologists, that is, the separation of biomolecules from one another. The ability to isolate specific biomolecules is essential in both research and production settings. For hundreds of years, biologists have purified biomolecules in order to investigate their properties – and this is still an important task in many research laboratories. Research into the chemical properties of proteins and other biomolecules can provide knowledge of cellular processes and might, for example, lead to the discovery of new targets for drugs. In a production setting, safe and reliable purification processes are required to provide sufficient quantities of a product

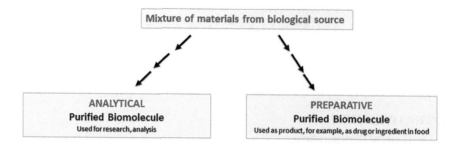

FIGURE 34.1   The two basic goals of separation methods. The goal of a bioseparation may be primarily analytical or preparative.

for large-scale testing, or for commercial sale. As biotechnologists, we are interested in both the research applications of bioseparation methods, and also their use in producing products of value to people.

When a separation procedure is performed to obtain information, we say it is an **analytical separation procedure**. For analytical procedures, aliquots of the sample are evaluated and are generally discarded (or stored away) after analysis is complete. When a separation procedure is performed to get a product, we say it is a **preparative separation procedure** (Figure 34.1).

In previous chapters of this unit, we introduced two commonly used, basic separation methods: centrifugation and filtration. Filtration is primarily a preparative method. For example, filtration is routinely used as a step in the preparation of purified water for further use. Centrifugation is also generally a preparative method, although, as noted in Chapter 33, there are analytical centrifuges for specialized applications.

34.1.2 Other Bioseparation Methods Explored in this Chapter

In this chapter, we will introduce two more separation methods: extraction and precipitation. Extraction and precipitation are traditional separation methods that have been used for decades. These methods are relatively simple in that they do not require complex instrumentation. We will then introduce two more separation methods: electrophoresis and chromatography. Electrophoresis and chromatography are both sophisticated instrumental methods that are actually "families" of techniques. It is beyond the scope of this textbook to cover every detail of the electrophoresis and chromatography "families," but we will introduce the basic principles of these important tools, and some of the common ways in which they are applied. Figure 34.2 summarizes the separation methods explored in this unit.

Electrophoretic separations generally involve only small volumes of substances, often less than a milliliter. Thus, while electrophoresis is occasionally used preparatively to prepare a small volume of a substance for further use, electrophoresis is most commonly an analytical separation method.

Chromatographic techniques are versatile and can be used at varying scales, from very low volumes to large volumes for preparative purposes in an industrial setting. Chromatography is thus sometimes used for analytical purposes and other times for preparative purposes. Chromatography techniques play a vital role in bioseparation strategies for most important biotechnology products.

34.1.3 Multistep Bioseparations

Biotechnology products usually must be isolated from a biological source, such as cultured cells, and must be purified away from contaminants. **Biomolecules** *are compounds that are produced by some type of biological source, such as a plant, animal, microorganism, or cultured cell.* **Purification or isolation** *is the separation of a specific biomolecule of interest from impurities in a manner that provides a useful end product.*

In order to purify a biological product completely, it is necessary to devise a purification strategy that involves a series of separation steps, as shown in Figure 34.3. Each separation step removes certain impurities from the product of interest, resulting in a progressively purer preparation. Assuming that bacterial or mammalian host cells are involved in producing the product, these impurities include proteins and nucleic acids that come from the host cells, cell culture media components, and viruses. At each step, impurities are removed, and the material of interest becomes purer. However, at each step, some of the desired product is unavoidably lost. For this reason, every purification step must be chosen carefully.

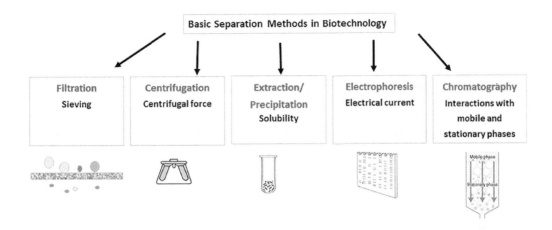

FIGURE 34.2   Summary of basic biotechnology separation methods that are explored in this unit.

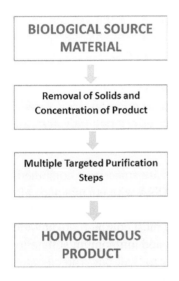

FIGURE 34.3   A simplified flowchart for a bioseparation strategy. Bioseparations begin with a source material (e.g., cultured cells or plants) that contains the product of interest. A bioseparation strategy must remove every impurity (or nearly every impurity) from the source material, leaving only the product of interest. The first step removes cells and solid debris so that the desired product is in a somewhat concentrated form. Subsequent steps remove various impurities eventually resulting in a purified, homogeneous product of interest.

34.1.4 Assays

In this chapter, we must also revisit assays, a topic that was introduced in Unit VII. Assays are critical in bioseparations because they allow us to monitor the progress of a separation procedure. At each separation step, we need to know the relative amounts of product versus impurities, but how do we obtain this information? The answer is that we must use assays that measure the concentration or amount of the product of interest, and assays that measure the concentration or amount of impurities. There are many different types of assays that are used to monitor bioseparation procedures and to test the purity of a resulting product. Some examples of common assays used in bioseparations are introduced in this chapter.

34.2 NON-INSTRUMENTAL METHODS OF SEPARATION

34.2.1 Extraction

There is a variety of relatively simple, non-instrumental separation methods available. Although these methods seldom remove all impurities from a product, they have the advantages that they are generally inexpensive and can be performed with large volumes of unpurified material. Many of these non-instrumental methods exploit the solubility characteristics of the desired product. The most common methods are classified as either extractions or precipitations.

Extraction methods *are based on the fact that a specific molecule will exhibit higher solubility in one medium compared with another.* For example, if you need to isolate a **hydrophobic** *(water-fearing)* compound from other cellular substances, you could add an organic solvent, such as ethyl acetate, to the aqueous mixture of unpurified cellular materials. By mixing the aqueous mixture derived from cells with ethyl acetate and shaking, the hydrophobic compound of interest will be extracted into the organic solvent (along with other hydrophobic impurities). The aqueous and organic phases will then separate from one another. Centrifugation is often used to accelerate this separation, but extraction can work without such an instrument. After the organic and aqueous phases have

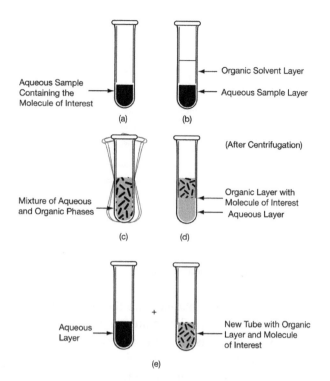

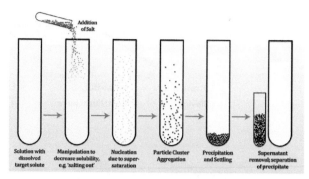

FIGURE 34.5 Concentration of a target molecule by salt precipitation. An appropriate salt is added to a solution containing a target molecule, in sufficient quantity to render the target insoluble. The target aggregates and then precipitates so that the pellet contains the target, which can then be dissolved in a smaller volume of solvent. (Copyright © Andrei Georgescu under Creative Commons 3.0 license.)

- *salt concentration*
- *pH*
- *temperature*
- *presence of alcohols, such as ethanol*
- *presence of organic polymers, such as polyethylene glycol (PEG).*

These factors are sometimes combined, as in the precipitation of DNA with ethanol and salt. (Examples of these techniques are also discussed in Chapter 25.)

One of the most common precipitation methods for proteins is to add ammonium sulfate to a cell homogenate in order to "salt out" (precipitate) the protein of interest. Increasing amounts of the ammonium salt are added to the protein solution. After each addition, the homogenate is centrifuged, and the supernatant is assayed for the protein of interest. When the appropriate concentration of salt is reached, the desired protein will ideally be (almost) entirely precipitated and can be collected for the next purification step. Figure 34.5 demonstrates the process of salting out a target molecule from a solution.

FIGURE 34.4 An example of a simple organic extraction. (a) Start with an aqueous mixture of cellular components (homogenate) containing the product of interest. (b) Add an equal volume of ethyl acetate, an organic solvent. (c) Mix the two liquids thoroughly. (d) Centrifuge the mixture to separate the organic and aqueous phases. (e) Collect the organic layer, which contains the product of interest. Note that in this example, the organic layer is of lower density than the aqueous layer. (Relative densities of extraction layers can be determined by consulting: Merck. *Merck Index.* 15th ed., Royal Society of Chemistry Publishing, 2013.)

separated, you can collect the organic phase containing virtually all of the compound of interest, while removing and discarding water-soluble impurities in the aqueous phase (Figure 34.4). Application of this method requires knowledge of the relative solubility of the molecule of interest in immiscible liquids.

34.2.2 Precipitation

Precipitation methods *are based on differences between molecules in their tendency to precipitate in particular liquids.* Manipulation of specific characteristics of the liquid medium of the sample can result in precipitation of certain molecules, which can then be separated from the liquid. In some cases, the desired product is precipitated; in other strategies, impurities are precipitated and thereby removed. The characteristics of the liquid that are most commonly adjusted are as follows:

34.3 THE FAMILY OF ELECTROPHORETIC METHODS

34.3.1 Overview of Electrophoretic Methods

Electrophoresis *is the separation of charged molecules in an electrical field.* Biotechnologists often use electrophoresis to separate different proteins from one another and to separate DNA fragments of different sizes. By performing such separations, it is possible to determine such molecular features as:

- *the length of DNA fragments, in base pairs*
- *the molecular weights of specific proteins*
- *the purity of an isolated protein*
- *the identity of an isolated protein*
- *impurities that might be present in a protein solution.*

Here we will briefly discuss three of the most common members of the electrophoresis family: agarose gel electrophoresis, polyacrylamide gel electrophoresis (PAGE), and capillary electrophoresis. Agarose gel electrophoresis is most commonly used to separate DNA fragments on the basis of their base pair lengths. PAGE is most commonly used to separate proteins on the basis of molecular weight. Capillary electrophoresis has previously been mentioned in this text because it is used in forensic DNA profiling, but it can be used for the analysis of a wide variety of protein and DNA samples.

34.3.2 Agarose Gel Electrophoresis

Many molecular biology procedures involve cutting DNA into fragments that can be separated from one another using agarose gel electrophoresis. **Agarose** *is a natural polysaccharide derived from agar, a substance found in some seaweeds.* When agarose powder is heated at low concentrations (0.8% is typical) in an appropriate buffer, the agarose dissolves into a viscous solution that gels into a semisolid (imagine a slab of Jell-O®) when cooled. The cooled gel is chemically stable, with a relatively high gel strength for thin sheets of agarose. Because agarose gels generally cannot support their own weights, electrophoresis is performed by laying the gel flat on top of a support in a plexiglass box, termed a horizontal gel box (Figure 34.6).

All DNA molecules have a uniform negative charge because of the regular occurrence of charged phosphate groups along the two DNA strands. When DNA samples are loaded into slots (wells) at the top of an agarose gel and the gel is exposed to a directional electric field, all DNA fragments will migrate through the gel in the direction of the positive electrode. Because the agarose gel acts as a molecular sieve, smaller DNA fragments will move to the positively charged end of the gel faster than larger fragments, which become entangled in the pores of the gel. DNA fragments migrate through the gel at a rate that is roughly related to the inverse log of the number of base pairs in the fragment. The result is that when the electric field is turned off, there are separated bands in the gel. Each band contains DNA fragments of the same base pair

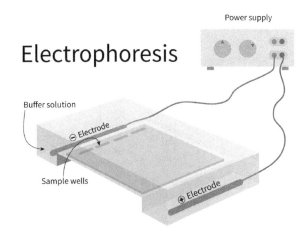

FIGURE 34.6 Agarose gel electrophoresis. The agarose gel is shown laying on top of a support in the horizontal gel box. Each sample was loaded into one of the slots (wells) at the top of the gel. A power supply (not shown) produced an electrical field that caused sample components to migrate through the gel at different rates, depending on their sizes. Observe that the DNA samples moved from the negative pole toward the positive pole. The separated components are visualized with ultraviolet light or with dyes. (Credit: M.PATTHAWEE/Shutterstock.com)

length. The separated bands of DNA are then most commonly stained with ethidium bromide, which allows the bands to be visualized under an ultraviolet light. If a standard consisting of DNA fragments of known base pair lengths is applied to the same gel alongside the samples, a reasonable approximation of base pair length for DNA bands in the samples can be made. Figure 34.6 illustrates agarose gel electrophoresis. (Note that ethidium bromide is a known mutagen, whose handling is discussed in Chapter 9.)

34.3.3 Polyacrylamide Gel Electrophoresis (PAGE)

Proteins can be separated using **polyacrylamide gels,** *which are made from acrylamide* (a neurotoxin – see Chapter 9) *and bisacrylamide polymerized in the presence of an appropriate initiator (ammonium persulfate) and the catalyst (N,N,N',N'-tetramethylethylenediamine, commonly called TEMED).* The pore size of a polyacrylamide gel can be adjusted by combining different amounts of acrylamide and bisacrylamide. Polyacrylamide gels are stronger than agarose gels of comparable thickness, so they can be used for vertical gel electrophoresis (PAGE) (Figure 34.7).

Unlike DNA molecules, proteins are not uniformly charged. At a given pH, proteins will have both positively and negatively charged amino acids in differing ratios. Therefore, PAGE can separate proteins on the

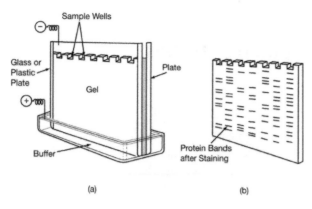

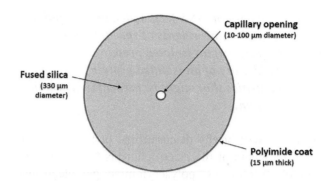

FIGURE 34.8 Cross section of capillary used for electrophoresis. Polyimide is a high-performance plastic.

FIGURE 34.7 Polyacrylamide gel electrophoresis (PAGE). (a) The gel is poured vertically between two glass plates. Sample is applied to wells at the top of the gel. (b) Protein bands are separated on the basis of their relative molecular weights and visualized with stains. *Note:* This is an example of SDS–PAGE.

basis of differences in their charges. A more common adaptation of PAGE, however, is SDS–PAGE. **Sodium dodecyl sulfate-PAGE (SDS–PAGE)** *is a technique that separates proteins on a polyacrylamide gel on the basis of molecular weight.* In this technique, protein samples are treated with the negatively charged detergent, SDS. SDS binds to proteins at a ratio of approximately one SDS molecule for every two amino acids in the protein. The resulting negative charge conferred by the SDS overshadows any native protein charge so that all proteins will migrate toward the positive electrode in an electric field, as shown in Figure 34.7. In this case, sieving of the proteins occurs in the gel according to the molecular weights of the proteins. Smaller proteins migrate more quickly than larger ones. The detergent also denatures the proteins, minimizing the differential effect of protein shape on migration rate. Higher concentrations of acrylamide and higher levels of cross-linking provide a gel with smaller pore sizes. Adjusting pore size controls the effective range of protein sizes that are separated on an individual gel. SDS–PAGE is frequently used both for the determination of the molecular weight of proteins and also for the determination of the purity of protein preparations, as will be discussed later in this chapter.

34.3.4 Capillary Electrophoresis

34.3.4.1 Overview

Agarose and polyacrylamide gel electrophoresis are both categorized as forms of "slab gel" electrophoresis because samples migrate through flat "slabs" of a gel-like substance. In contrast, **capillary electrophoresis**

(CE) *separates molecules as they move through the inside of very narrow tubes called* **capillary tubes.** The inner diameter of the capillary tubes is usually on the order of 10–100 μm (Figure 34.8). To put this in perspective, the thickness of a human hair is also around 100 μm. These fine capillary tubes are constructed of fused silica, which is basically a type of glass, and are coated with a thin protective layer of plastic. The length of a typical capillary tube is 30–60 cm. During a separation, samples in buffer flow through the capillary tubes under the influence of an electrical field. Hence, because an electrical field is involved, CE is categorized as a form of electrophoresis.

Capillary electrophoresis involves much higher voltages than slab gel electrophoresis, and therefore, CE separations are faster. Another advantage of capillary electrophoresis is that it requires very small sample volumes, in the nanoliter range. This is desirable where limited amounts of sample are available. CE is automated and therefore can efficiently run more samples in a given amount of time than typical slab gel electrophoresis methods. CE also provides more accurate quantitative results than slab gel electrophoresis. For all these reasons, CE has become the electrophoretic method of choice for many applications, including DNA identity testing, DNA sequencing, and certain biopharmaceutical quality-control tests. However, CE equipment is more expensive than slab gel setups and so there are situations in research and education laboratories where slab gel electrophoresis is still used as a cost-effective option.

34.3.4.2 Typical Configuration and Operation of CE

The basic instrumentation required for CE is relatively simple and is illustrated in Figure 34.9. The sample is introduced into one end of the capillary tube which is then placed between two buffer reservoirs. Electrodes

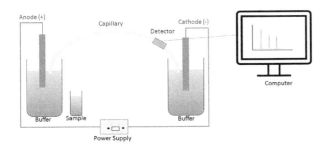

FIGURE 34.9 Diagram of a typical CE configuration. See text for explanation.

that are connected to a power supply are placed into the two reservoirs. A detector is present at one end of the capillary tube. In a conventional CE separation, when voltage is applied, the buffer inside the capillary flows from the anode (the positive pole) toward the cathode (the negative pole), bringing along with it the sample components. The components of the sample separate from one another as they flow through the capillary. A detector is used to detect each component as it reaches the end of the capillary tube. In applications involving DNA that has been amplified by PCR (such as STR identity testing; see below), fluorescently labeled primers cause the DNA fragments to fluoresce. In these applications, a fluorescence detector is used to monitor the separation. Other times a detector that measures the absorbance of light by each component is used instead (as described in Chapter 21). The output from the detector goes to a computer equipped with software to analyze the results. The data are displayed as an **electropherogram**, *where each component is displayed as a peak.*

Figure 34.9 illustrates a configuration where sample and buffer flow from the anode to the cathode. It is also possible to run CE in the opposite direction so that flow is from the cathode to the anode. In the latter case, the detector is placed near the anode.

Separation in CE is typically based on differences between molecules in their charges and/or their sizes and shapes. When CE is used for identity testing and gene sequencing, size is the basis for separation. CE can efficiently separate DNA fragments that vary in size by only a single base pair. As with agarose gel electrophoresis, smaller DNA fragments move through a CE column more quickly than larger fragments and the direction of flow is from the negative electrode to positive electrode.

34.3.4.3 CE in the Analysis of DNA

We introduced an important application of capillary electrophoresis previously in this text, that is, for identity testing. CE is the main method by which short tandem repeats (STRs) of DNA are separated from one another and visualized. Recall that STRs are places in the genome where the number of repeat units differs from one individual to another. STRs are therefore used to create DNA "fingerprints" that can identify an individual. Figure 3.3 shows an STR "fingerprint" as might be used in forensic testing where DNA from a suspect is tested to see if it matches DNA from a crime scene. Figures 30.16–30.18 show how the same method can be used to test the identity of cell lines to make sure they are not misidentified.

Another application of CE is in gene sequencing, that is, determining the order of nucleotides in a DNA strand. In the 1970s and 1980s, individual genes were laboriously sequenced, a task that required months of painstaking work. At that time, large slab gels were involved in gene sequencing. Each slab gel had to be carefully prepared, was only used one time, and was manually loaded with samples. Separations with these gels were slow, on the order of eight or more hours. The idea of sequencing an entire genome was unimaginable using the methods available at that time. The introduction of capillary electrophoresis in the late 1980s significantly facilitated gene sequencing. This is because capillary electrophoresis is readily automated; multiple samples can be run simultaneously; each separation is relatively fast; and the results can be quickly analyzed and visualized on a computer.

34.3.4.4 Obtaining Information from an Electropherogram

Observe the electropherogram in Figure 34.10. The electropherogram is a recording with time on the X-axis and detector reading on the Y-axis. In this illustration, the sample contained positively charged sodium ions that moved most quickly through the capillary, and so appeared first in the electropherogram. The negatively charged ions moved more slowly through the capillary and appeared later, with the larger negative ion appearing last.

When a UV or UV/Vis detector is used, the Y-axis of an electropherogram is absorbance. The Y-axis is an indication of how much substance passed by the detector at a given time. The X-axis is time elapsed. The result is a series of peaks. Each peak represents a different substance that was present in the original mixture within the sample. Some peaks are larger than others and enclose more area. The area under a peak is proportional to how much of a substance is present; the more area under the peak, the more of that substance that is present. Therefore, a CE run separates the components of a sample, assists with their identification (if suitable standards are available), and provides quantitative information about how much of each component

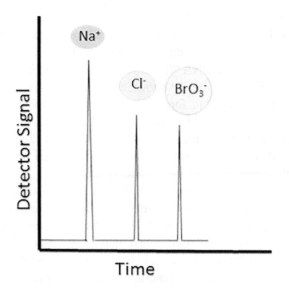

FIGURE 34.10 A simple separation based on charge and size using CE. Sample components separate in the column with positively charged ions and small ions moving most quickly. The heights of the peaks are related to the amount of that component present in the original sample.

is present. Note also that the peaks illustrated in Figure 34.10 are quite "sharp," that is, not broad in their shapes. This is a desirable feature of CE; it typically produces sharp peaks. When peaks are very sharp, it is possible to distinguish two different peaks even if they are quite close to one another on the X-axis.

34.3.4.5 CE in Biopharmaceutical QC Testing

CE has become important in biopharmaceutical quality-control laboratories where it is used to obtain information about products, such as monoclonal antibodies. Using CE, it is relatively efficient to see if a protein product has the correct identity, purity, and homogeneity. Identity testing asks the question, is the protein what it is supposed to be? A sample of the product being tested can be run alongside a reference standard of known identity. The test sample should move through the capillary with the same mobility as the reference. Purity testing asks the question, is the protein free of impurities? CE can be used to look for peaks that are due to impurities. Homogeneity testing asks the question, is the product free of alternative chemical structures? The last question relates to a common problem in biopharmaceutical production. When cells are used as "factories" to manufacture protein products, the proteins can have slight alterations that might affect their function as a drug. For example, the glycosylation patterns of the product may vary slightly. (Glycosylation is illustrated in Figure 1.8. It is the addition of complex, branched carbohydrates onto

a protein molecule.) *The varying forms of the same basic protein are called* **isoforms**. Figure 34.11 shows how CE can be used to detect isoforms of monoclonal antibody products.

Safety Note: Electrophoresis requires the same or even greater safety precautions as any other use of electricity. During electrophoresis, an electric current is being deliberately applied to a highly conductive liquid. Electrophoresis generates heat, which may result in the evaporation of the system buffer and may lead to short circuits, equipment damage, or fires. Gel boxes and CE devices are designed with safety features that should not be bypassed. Additional safety suggestions are provided in Chapter 8.

34.4 THE FAMILY OF CHROMATOGRAPHIC METHODS

34.4.1 Overview

The term *chromatography* describes the most widely used family of methods applied to the purification of biomolecules. Some type of chromatography is used in most biotechnology settings, for either preparative or analytical purposes (or both). *Chromatography techniques are based on the differential interaction of molecules between a stationary and a mobile phase.* As suggested by their names, the **stationary phase** *is an immobile material* and the **mobile phase** *is a liquid or gas that moves past the stationary phase.* When performing chromatography, the sample, which contains a mixture of molecules, is introduced into the flowing mobile phase. Sample molecules move along with the mobile phase past the stationary phase material (Figure 34.12). Different molecules have different properties that cause them to be more or less attracted to the mobile phase or the stationary phase. For example, one molecule might, under certain circumstances, tightly bind the stationary phase, while other molecules with different properties remain in the mobile phase and flow right through the chromatography system. Yet another molecule might be weakly attracted by the stationary phase and would therefore be slightly slowed in its flow. Different molecules thus move past the stationary phase at different rates and hence are separated from one another.

One of the advantages of chromatography is the versatility of techniques available. There are many different mobile and stationary phase materials that are used, and they are used in many different combinations. The most common type of chromatography in biotechnology settings involves some

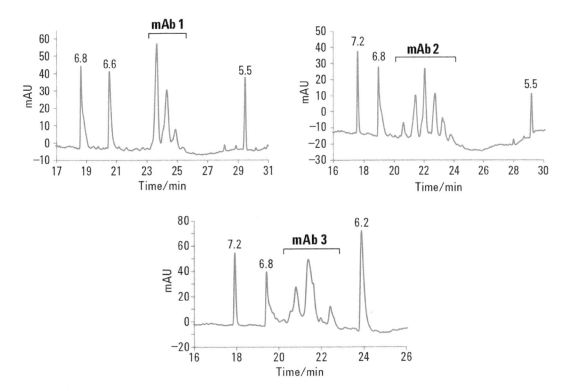

FIGURE 34.11 CE analysis of three different monoclonal antibodies. The three different monoclonal antibodies were purchased from biotechnology companies and were analyzed using CE. The peaks labeled 5.5, 6.6, 6.8, and 7.2 are the peaks of standards added when the samples were run. Monoclonal antibody 1 (mAb1) had three peaks indicating three isoforms. mAb2 had six isoforms, and mAb3 had what appears to be three isoforms, although the peaks are not ideally delimited using this CE protocol. (Wenz, Christian. Agilent Technologies, Inc. *Monoclonal Antibody Charge Heterogeneity Analysis by Capillary Isoelectric Focusing on the Agilent 7100 Capillary Electrophoresis System: Application note.* 2017. © Agilent Technologies, Inc. Reproduced with permission, Courtesy of Agilent Technologies, Inc.)

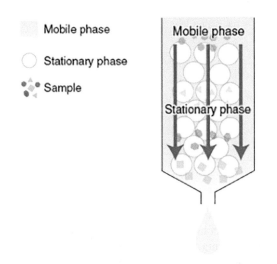

FIGURE 34.12 The basic concept of chromatography involves the relative interactions of sample components with a stationary phase and a mobile phase. (Image Courtesy of MBL International.)

type of solid material as the stationary phase, and some type of liquid as the mobile phase. When the *mobile phase is a liquid*, the method is termed **liquid**

chromatography, LC. There are also chromatography instruments that use a *gas as the mobile phase*, in which case the method is termed **gas chromatography, GC.** A gas mobile phase may be paired with a liquid stationary phase (sometimes abbreviated as GLC, gas–liquid chromatography). GC does have some applications in biotechnology, but is less common than LC and will not be discussed further in this text.

A derivation of LC and GC is **supercritical fluid chromatography (SFC)**, a *high-pressure chromatography system where carbon dioxide is usually used as the majority of the mobile phase.* (A **supercritical fluid** is *one that shows properties of both a gas and a liquid under specific conditions of temperature and pressure.*) SFC has the advantages of high speed, high specificity, and significantly lower amounts of solvents compared to traditional high-pressure LC. It is used in the pharmaceutical industry for preparatory purposes, because it can separate very closely related chemical structures. The technique is also suitable for analytical work. However, LC is the more common form of chromatography in most biotechnology settings and is the type of chromatography we will discuss here.

Chromatography is varied not only in terms of the selection of stationary and mobile phases, but also in terms of the type of instrumentation required. Early chromatography methods used planar (flat) stationary phases, such as paper, that require little specialized equipment and are actually non-instrumental methods. In paper chromatography, small spots of a sample are applied to special chromatography paper and allowed to dry. The end of the paper is then dipped into the mobile phase, which wicks through the paper, carrying along the various components of the sample. Paper chromatography, while an excellent introduction to the principle of chromatography, is not suitable for use in any practical research or production setting. **Thin-layer chromatography (TLC)** is another planar technique that does not require expensive equipment or materials and that is still sometimes used in professional settings for analysis of a sample. In TLC, *a thin layer of stationary phase material, usually silica, is spread on a glass or plastic plate, and the liquid mobile phase passes through the stationary phase by either capillary action or gravity.* Flexible plastic backing materials can provide custom sizes and ease of handling. Figure 34.13 illustrates planar chromatography. Observe that no special instrumentation or electricity is required.

Nowadays, the use of planar chromatography is uncommon, and most chromatography applications involve columns. For biopharmaceutical manufacturing, column chromatography is the only type of chromatography that is capable of operating at a commercial scale. **Column chromatography** *is performed with the stationary phase packed into a cylindrical container, the column.* The mobile phase, or **eluent**, passes through the column, propelled either by gravity or mechanically with a pump. **Elution** *occurs as the molecules in the sample distribute themselves between the two phases according to their affinities (attractions) for each phase* (Figure 34.14a). As the liquid flows through the column, different sample components elute from the column at different times. A detector, commonly a type of spectrophotometer, is used to "see" each component as it exits the column.

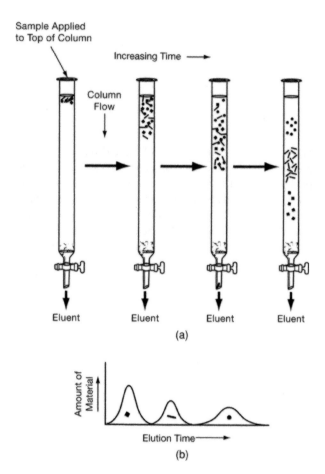

(a)

(b)

FIGURE 34.14 Elution of a sample mixture from a chromatographic column. (a) Separation of sample components. A sample containing three different molecules is loaded onto a column. The sample components are carried by the mobile phase and travel through the column at different speeds depending on the nature of their interaction with the stationary phase. (b) Diagram of the resulting chromatogram. The three molecules are graphed according to their elution times from the column. The Y-axis is a measure of the amount of material seen by the detector, and the X-axis is elution time. (In other cases, the X-axis can be the fraction number.)

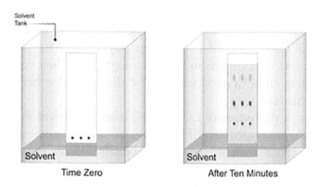

FIGURE 34.13 Planar chromatography. The sample is applied to a paper or thin-layer chromatography sheet and placed in a tank with a mobile phase, the solvent. The mobile phase moves up the paper or sheet via capillary action. Sample components that "prefer" the mobile phase move more quickly, while those that "prefer" the stationary phase move more slowly. The separated materials can be stained with dyes for visualization. (Image courtesy of Waters Corporation. Used with permission.)

The result of a chromatography run is thus a recording that shows the components of the sample exiting the column over time. *The recording is called a* **chromatogram** (Figure 34.14b). (Observe that a chromatogram is of the same form as an electropherogram.) When chromatography is performed for analytical purposes, analysts may only be interested in the chromatogram and they might discard the liquid that exits the column. But, if chromatography is performed for preparative purposes, then *the liquid that exits the column is collected into a series of tubes or other containers*, each of which is then called a **fraction.** Thus, if the chromatography run is being performed for a preparative purpose, the end result will include both the chromatogram and the series of fractions.

Example Problem 34.1

a. In the example shown in Figure 34.14, what would the chromatogram look like if the sample component illustrated as a bar had a higher affinity for the mobile phase than the other two components?

b. In the example shown in Figure 34.14, what would the chromatogram look like if the sample component illustrated as a bar had a higher affinity for the stationary phase than the other two components?

Answer

A column chromatography system can be as simple as the one shown in Figure 34.14 where gravity alone

moves the mobile phase through the column. More often, the liquid phase is driven through the column by the force generated by a mechanical pump operated at low, medium, or high pressure (Figure 34.15). Pumps are useful because they can control the flow of the mobile phase through the column and can accelerate the separation. In the system illustrated, the sample is injected into the mobile phase through an injection port. A detection system that "sees" molecules as they exit is located at the other end of the column. The detector is connected to a data recording device, usually a computer with software to analyze the data. Purified sample components can be recovered after the detector by collecting sequential fractions of the mobile phase. The various parts of the chromatography system are connected to one another with chemically inert tubing. Thus, for many chromatography systems, two of the essential skills for a chromatographer are an understanding of basic plumbing and the ability to use a wrench to maintain proper connections between system components.

For column chromatography, the stationary phase consists of small particles or resins (sometimes termed media) that are held in a column. These particles can be composed of a variety of different materials: silica, agarose, sepharose, acrylamide, etc., depending on the application. These particles can be physically or chemically modified to bind or repel particular molecules within complex mixtures.

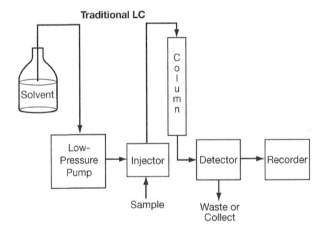

FIGURE 34.15 A traditional LC system. The components of a typical system include a reservoir from which the mobile phase flows (labeled solvent in this diagram), a pump, an injector to introduce the sample into the column, a column filled with stationary phase, a detector, and a recording device (usually a computer). Sometimes the eluted mobile phase is waste and is discarded. Other times, when the system is used for preparative purposes, fractions are collected.

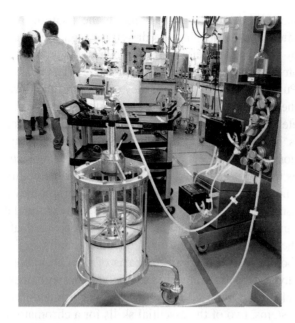

FIGURE 34.16 A moderate size column in a production facility. The column is in the foreground of the photo. The stationary phase is the white substance in the bottom of the column. The column is connected to other parts of the chromatography system with plastic tubing.

In many situations, the analysts who perform chromatography prepare their own columns by obtaining a glass or plastic column of a suitable size and filling it with a desired stationary phase material. This is called "packing the column." However, columns that are packed by the user are generally not used with high-pressure pumps.

For analytical work, smaller columns that require only a small volume of stationary phase are preferred. Separations with a small column can be relatively rapid and do not consume large amounts of sample or solvent. However, for preparative chromatography, it is common to use larger columns that can handle larger volumes of sample. With proper method development, large columns can provide very high product capacities together with effective removal of impurities.

This accounts for the widespread use of chromatographic techniques in production settings. Capacity increases with column volume (up to a point), although this increased capacity comes at an increased monetary cost. Figure 34.16 shows a moderate size column in its site in a production facility.

34.4.2 Modes of Column Chromatography

34.4.2.1 Overview

We previously said that in chromatography different components of a mixture interact in different ways with the mobile and stationary phases. This is because molecules differ in various physical properties, such as their size, shape, charge, solubility, and their affinity for other molecules. Chromatographic techniques can separate molecules based on differences between them in any of these molecular properties. Table 34.1 lists some of the molecular properties that are most commonly used as the basis for chromatographic separations. In this section, we will describe methods that are based on these properties with an emphasis on those that are often used for the separation and purification of protein biopharmaceuticals.

Partition chromatography refers to a group of methods *in which a specific molecule will exhibit higher solubility in one chromatographic phase compared to another and distribute itself accordingly.* This is the same principle as was previously described for extraction techniques. Partition chromatography can be distinguished from **adsorption chromatography**, *where specific molecules differ in their tendency to adsorb to a solid stationary phase.* The term **adsorb** means *to stick to a surface.* Adsorption techniques are usually used to separate larger molecules such as proteins, whereas partition chromatography is best suited for relatively small molecules. Ion exchange chromatography, hydrophobic interaction chromatography, and affinity chromatography, all of which will

---

TABLE 34.1

Molecular Characteristics Commonly Used as a Basis for Chromatography

Molecular Property	Chromatography Method
Polarity, or relative water solubility	Partition chromatography (reverse and normal phases)
Charge	Ion exchange chromatography (adsorption method)
Hydrophobic properties	Hydrophobic interaction chromatography (adsorption method)
Specific molecular binding properties	Affinity chromatography (adsorption method)
Molecular weight (size and shape)	Gel permeation chromatography

be discussed later, are categorized as forms of adsorption chromatography. In these adsorption methods, a component of the sample either sticks to the column matrix, or does not. If it does stick, it can be removed and collected later by changing the mobile phase.

34.4.2.2 Partitioning by Solubility: Normal Phase and Reverse Phase

The earliest partition chromatography techniques used a polar, hydrophilic (water-loving) stationary phase, and a mobile phase that was relatively nonpolar (hydrophobic). This is often referred to as *normal phase* chromatography. However, most modern partition methods use *reverse phase*, where the stationary phase is nonpolar and the mobile phase is polar relative to the stationary phase. Remember that molecules that are polar "like," or partition into, water more than molecules that are nonpolar. Figure 34.17 illustrates how partition chromatography separates molecules based on differences in their polarity. Observe that in normal phase chromatography, where the stationary phase is hydrophilic, the more polar components in a sample tend to move more slowly through the column because they "like" the stationary, hydrophilic phase more than the mobile phase. But in reverse phase chromatography, more polar components of a sample tend to move more quickly through the column because they partition into the more polar mobile phase.

It is relatively straightforward to imagine how a mobile phase can be chosen that has a desired level of hydrophobicity. Water is the most hydrophilic mobile phase we might imagine. If water is mixed with varying levels of nonpolar organic solvents, such as acetonitrile or methanol, it becomes less hydrophilic. The higher the percent of organic solvent in the mixture, the less hydrophilic it is. But how can the stationary phase be manufactured to be more or less hydrophilic? The answer lies in the way the stationary phase is made. In partition chromatography, solid stationary phases are usually formed from small microporous silica particles, as shown in Figure 34.18. These particles are nonreactive and uniform in their physical characteristics and provide a large surface area for potential interaction with molecules in the mobile phase. For reverse phase partition applications, the particles are modified by binding them with a hydrophobic liquid as shown in the diagram. Columns packed with these modified particles are then paired with a mobile phase that has the desired mix of water and organic solvent. Partition chromatography can be optimized to separate and quantitate a wide variety of molecules by choosing a column with a particular liquid bonded to the stationary silica particles, and by selecting a particular concentration of organic solvent.

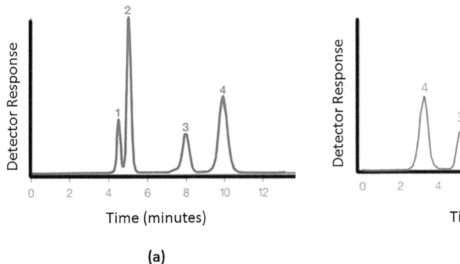

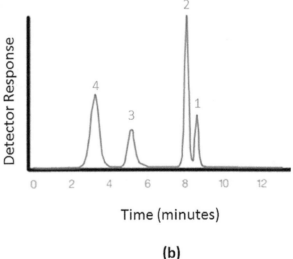

FIGURE 34.17 Comparison of elution patterns in (a) normal phase and (b) reverse phase chromatography. Molecule 1 is the most hydrophobic (nonpolar) of the molecules; Molecule 4 is the most polar; the relative polarities are 4>3>2>1. In normal phase chromatography, the stationary phase is hydrophilic. Therefore, in normal phase chromatography, component 4, the component that "likes" water the most, is retained longer on the column and elutes last, after 10 minutes. In contrast, in reverse phase chromatography, the mobile phase is hydrophilic. Therefore, component 4, which is the most polar, moves most quickly through the column and elutes after only 3 minutes.

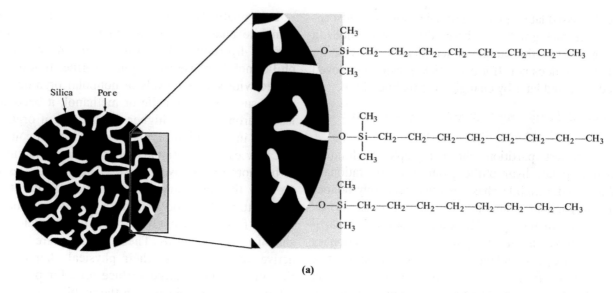

(a)

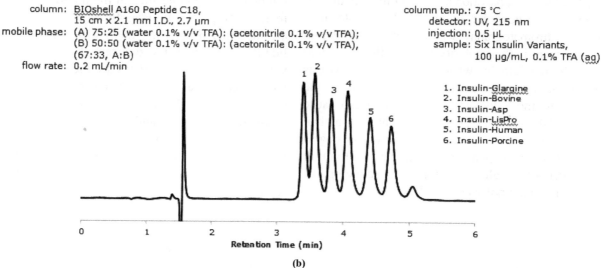

column: BIOshell A160 Peptide C18,
15 cm × 2.1 mm I.D., 2.7 µm
mobile phase: (A) 75:25 (water 0.1% v/v TFA): (acetonitrile 0.1% v/v TFA);
(B) 50:50 (water 0.1% v/v TFA): (acetonitrile 0.1% v/v TFA),
(67:33, A:B)
flow rate: 0.2 mL/min

column temp.: 75 °C
detector: UV, 215 nm
injection: 0.5 µL
sample: Six Insulin Variants,
100 µg/mL, 0.1% TFA (aq)

1. Insulin-Glargine
2. Insulin-Bovine
3. Insulin-Asp
4. Insulin-LisPro
5. Insulin-Human
6. Insulin-Porcine

Retention Time (min)

(b)

FIGURE 34.18   Reverse phase chromatography. (a) Microporous beads of silica are used as a base for the stationary phase. The beads are typically 1–10 µm in diameter with pores ~0.01 µm wide (for separating small molecules). The surfaces of the beads and pores are coated with the chemically bonded stationary phase. In this case, the chemical group is termed a C8 hydrocarbon. This chemical group is considered to be a liquid. (b) Chromatogram of a reverse phase separation of different but closely related forms of natural and recombinant insulin. This separation demonstrates a high degree of selectivity. In other words, the chromatography method can separate forms of insulin that are quite similar to one another. The printout at the top records all of the instrument parameters. (This chromatogram was obtained using high-performance liquid chromatography, a form of chromatography that is described later in this chapter.) (Reproduced with permission from Merck KGaA, Darmstadt, Germany, and/or its affiliates.)

34.4.2.3 Separation by Size

Gel permeation chromatography (GPC) *is a technique for separating molecules based on differences in their sizes/molecular weights using a column filled with porous gel particles.* GPC is also known as size-exclusion chromatography or gel filtration chromatography. If you refer to Table 34.1, you will see that gel permeation chromatography is categorized neither as a partition method, nor as an adsorption method.

In GPC, the mechanism of separation is based on molecular sieving rather than chemical interaction between the sample and the stationary phase.

The gel particles used for GPC are usually highly cross-linked polysaccharides (one typical material is Sephadex) with relatively large pore sizes, compared to other chromatographic media. In a gel permeation system, small molecules enter the pores of the stationary phase and are retarded on the column. Larger

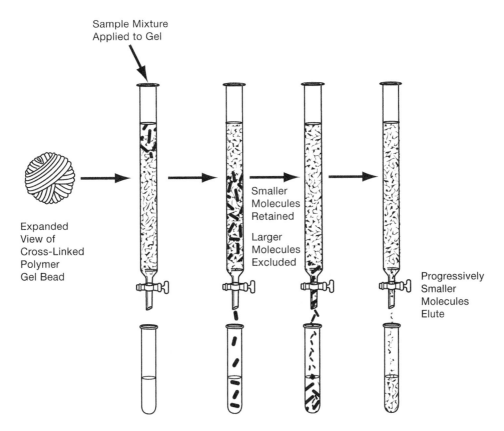

FIGURE 34.19   Diagram of gel permeation chromatography.

molecules are excluded from the pores and elute from the column relatively quickly (Figure 34.19). In other words, the larger the molecule, the faster it passes through the column. Thus, although both slab gel electrophoresis and GPC are based on sieving, they have the opposite pattern of separation. Like electrophoresis, GPC can be used to determine the approximate molecular weight of sample molecules. Gel permeation is a more difficult method to scale up than other modes of chromatography and therefore is commonly used as an analytical method or for small-scale preparative separations.

34.4.2.4 Separation by Charge

Ion exchange chromatography (IEC) *is an adsorption chromatography technique that separates biomolecules based on their net molecular charges. Inert stationary phase particles are coated with either positively or negatively charged molecules to create media called* **ion exchangers.** If the stationary phase particles are coated with positively charged molecules, then they will adsorb negatively charged molecules (anions) that flow by in the mobile phase. In this case, the column is termed an **anion exchange column.** While anions in the sample are adsorbed by the exchanger,

impurities that do not bind to the column are washed away with the mobile phase. The anions from the sample can then be released from the stationary phase and recovered from the column by changing the pH and/or salt concentration of the mobile phase. In a similar fashion, the stationary phase may be prepared by binding negatively charged molecules to the stationary phase particles. In this case, the column will adsorb positively charged molecules (cations) that flow by in the mobile phase and the column is termed a **cation exchanger** (Figure 34.20).

Most IEC applications require **gradient elution,** *in which the concentration of components in the mobile phase changes during the separation process to progressively elute more components from the stationary phase.* For example, a common type of gradient separation is one in which the concentration of salt in the mobile phase increases over the course of the chromatography run. For example, the mobile phase can smoothly transition from 0.2 M NaCl to 0.5 M NaCl, which is accomplished using a mechanical mixer that slowly adjusts the salt concentration. Alternatively, this can be accomplished in steps, for example, by beginning with a mobile phase that contains 0.2 M NaCl, then switching to a mobile phase with 0.3 M NaCl, and then

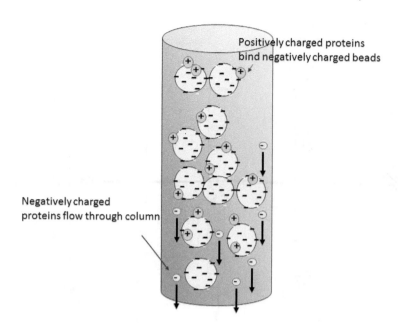

FIGURE 34.20 A cation exchange column. The beads that comprise the stationary phase (gray circles) are coated with a material that has a negative charge. Proteins with a net positive charge (green circles) adsorb to the beads, while proteins with a net negative charge flow through the column with the mobile phase. The proteins that have adsorbed to the column can subsequently be eluted by changing the chemical composition of the mobile phase, such as by increasing its salt concentration. This result is shown in Figure 34.21.

switching to a mobile phase of 0.5 M NaCl. Gradients allow separation of molecules with a wider range of properties and more quickly than is possible when the mobile phase does not vary in the concentration of its components. A chromatogram showing the result of a gradient IEC run is shown in Figure 34.21.

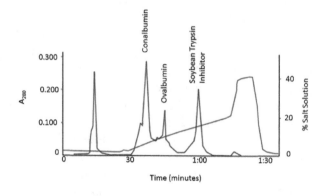

Time (minutes)

FIGURE 34.21 The results of an IEC run. This chromatogram shows elution of proteins (blue trace) as the concentration of salt is slowly increased (red trace). Proteins with only a few charged groups elute at lower salt concentrations, whereas proteins with many charged groups elute later at higher salt concentrations. This is because proteins with only a few charged groups bind less tightly to the stationary phase than proteins with more charged groups. (Image ©2020 Cytiva. Reproduced with permission from the owner.)

Ion exchange chromatography is commonly used in protein purification strategies. With careful optimization of the mobile phase conditions and selection of ion exchange matrices, this technique can provide reproducible, high-yield results. IEC is popular for biopharmaceutical applications because of the relative ease of scaling up the method for manufacturing purposes.

34.4.2.5 Separation by Affinity

Affinity chromatography *is used to purify biomolecules by exploiting their individual binding properties.* Inert column packing material is coated with a **ligand,** *a molecule that binds specifically to another molecule of interest.* Examples of familiar ligand/biomolecule pairings include enzyme/substrate and antibody/antigen. When the sample is passed through an affinity column, only the desired product binds to its ligand, while impurities are eluted with the mobile phase. The desired product can then be released (eluted) from the column by changing the mobile phase composition (Figure 34.22). If a suitable ligand exists, this is a very effective method of purifying a desired product away from impurities. A wide variety of ligand combinations are available.

Protein A affinity chromatography *is a form of affinity chromatography based on the specific and*

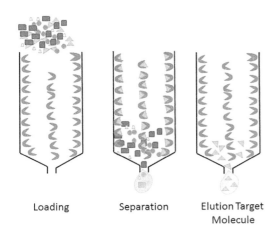

Loading   Separation   Elution Target
                                     Molecule

FIGURE 34.22 Affinity chromatography can be highly specific for target molecules. When the appropriate ligand is used, impurities can be washed through the column, leaving behind only the molecule of interest. In this cartoon, the target molecule is shown as yellow triangles. The target initially is retained by its ligand on the column's stationary phase, while other substances wash through the column. Finally, the isolated target molecules, the yellow triangles, are eluted by changing the conditions in the mobile phase.

reversible binding of antibody molecules to an immobilized protein A ligand. Protein A is a 56 kDa protein derived from the bacterium *Staphylococcus aureus*. Protein A binds the antibody molecules of many mammalian species. Protein A affinity chromatography is widely used as an early step in the purification of monoclonal antibodies (mAbs) because it reduces the overall number of processing steps necessary to produce a highly purified drug. Protein A affinity chromatography increases the yield and decreases the processing time required to obtain mAbs suitable for administration to a patient. Considering that mAbs are currently the largest class of therapeutic proteins, as described in Chapter 1, it is difficult to overestimate the importance of this method in biopharmaceutical manufacturing. In fact, some have argued that the commercial success of mAbs is not only related to their mechanisms of action, but is also due to the availability of Protein A for their purification. (See the case study, *A Purification Strategy for Monoclonal Antibody Drugs* later in this chapter.)

34.4.2.6 Separation by Hydrophobicity

Hydrophobic interaction chromatography (HIC) *separates proteins according to differences in their surface hydrophobicity*, taking advantage of a reversible interaction between these proteins and the hydrophobic surface of a column medium. Hydrophobic ("water-hating") molecules are relatively insoluble in water as contrasted with hydrophilic ("water-loving") molecules. Stationary phase beads can be coated with hydrophobic molecules that will interact with proteins having a hydrophobic surface. The addition of certain salts to the sample can enhance these interactions and promote binding (adsorption) to the column matrix. To selectively remove the bound hydrophobic proteins from the beads, the salt concentration is lowered gradually, and the sample components elute in order of hydrophobicity.

In theory, HIC is closely related to reverse phase chromatography (RPC, described previously) since both are based on interactions between hydrophobic patches on the surface of biomolecules, and the hydrophobic surfaces of the chromatography medium. In practice, the techniques are different. The surface of the beads in an RPC stationary phase is more hydrophobic than that of HIC media, which leads to stronger binding in RPC. Therefore, with RPC, in order to successfully elute the bound sample molecules, the binding must be reversed with organic solvents (alcohols, acetonitrile, etc.). These solvents can alter the structure of proteins (denature them); therefore, RPC is not ideal for protein separations. HIC offers an alternative by providing a less denaturing, milder, aqueous environment. In HIC, elution of the sample of interest is accomplished by changing the concentration of salt in the mobile phase, which is less likely to adversely affect the structure of the proteins in a sample.

Figure 34.23 illustrates how proteins with different degrees of surface hydrophobicity can be separated. The interaction between hydrophobic proteins and HIC resin is influenced significantly by the presence of certain salts in the mobile phase buffer. A high salt concentration enhances the interaction of the proteins with the stationary phase, while lowering the salt concentration weakens these interactions. In this example, the salt concentration in the mobile phase is initially relatively high. Therefore, all four proteins initially interact with the hydrophobic surface of the HIC medium and thus do not move through the column. Next, the ionic strength of the mobile phase buffer is gradually reduced so that the proteins interact less with the stationary phase and begin to elute from the column. The protein with the lowest degree of hydrophobicity is eluted first. The most hydrophobic protein elutes last, requiring a greater reduction in salt concentration to reverse its interaction with the stationary phase.

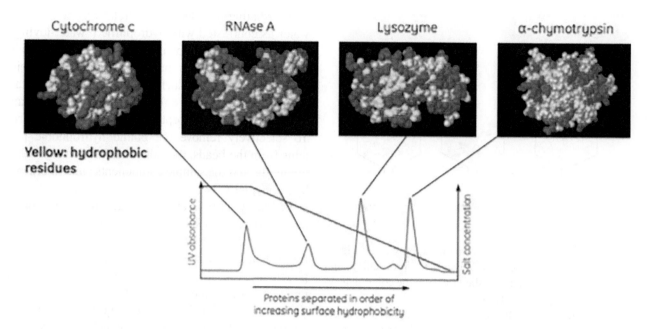

FIGURE 34.23 Separation of proteins with hydrophobic interaction chromatography. Proteins are separated on Phenyl Sepharose according to differences in their surface hydrophobicity. (Yellow indicates hydrophobic, and red hydrophilic amino acid residues.) Observe that the salt concentration (red trace) is initially relatively high, allowing all four proteins to interact with the stationary phase. Then, the salt concentration in the mobile phase is gradually decreased and the proteins elute, one at a time. Cytochrome c elutes first because it has the most hydrophilic groups relative to its hydrophobic groups. (Note that this is the opposite of IEC where elution is accomplished with *increasing* salt concentration.) (Image ©2020 Cytiva. Reproduced with permission from the owner.)

34.4.3 COLUMN CHROMATOGRAPHY SYSTEMS

34.4.3.1 High-Performance Liquid Chromatography

Column chromatography systems are available with differing characteristics. There are systems with a range of capacities that operate with a range of pump pressures and mobile phase flow rates. The terminology associated with different chromatography systems can be confusing since the acronyms for different systems are often used interchangeably.

High-performance liquid chromatography (HPLC) *is a set of chromatographic techniques that involve specially packed columns and high-pressure pumps.* In principle, HPLC could be adapted to any of the chromatographic principles shown in Table 34.1 but in practice, reverse phase chromatography is the most commonly used method.

An equipment setup for HPLC is shown in Figure 34.24. While the solvent delivery system is more complex than a low-pressure LC system, as shown in Figure 34.15, the same principles and components apply. Solvent from one or more reservoirs move to a solvent delivery system, which includes a high-pressure pump. The sample is injected through a port leading to a **guard column** (not shown in the figure), *which is an inexpensive, disposable chromatography column that protects the expensive HPLC column from impurities in the sample.* The guard column is placed immediately after the sample injector port and contains a relatively coarse packing material that retains particulate and sometimes specific chemical impurities to prevent fouling of the separation column. After passing through the guard column, the sample proceeds through the HPLC separation column. Because the solvent flows under high pressure, HPLC components are connected with very fine-bore steel tubing and special high-pressure fittings.

HPLC columns have stationary phases made of smaller and harder particles than lower-pressure methods because smaller particles provide the highest separation efficiencies. HPLC columns cannot be packed effectively without special equipment; therefore, pre-packed columns are usually purchased from various manufacturers. High-pressure pumps are required to move the mobile phase through the tight packing of the stationary phase.

Traditionally, HPLC is used for analytical purposes, to characterize and quantify the components of a sample, and to determine purity. For this reason,

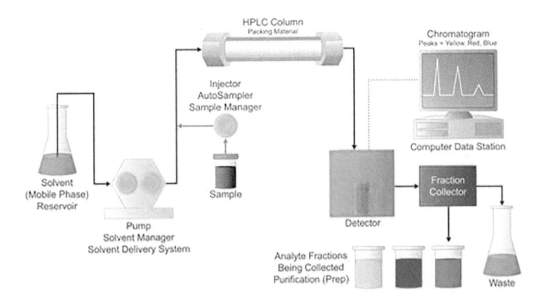

FIGURE 34.24 A typical HPLC system. See text for details. (Image courtesy of Waters Corporation. Used with permission.)

HPLC systems are extremely important in research and quality-control (QC) laboratories. (For an example, see Figure 34.18b.) Contract testing companies might have entire rooms filled with dozens of HPLC instruments.

34.4.3.2 HPLC vs. Fast Protein Liquid Chromatography

Fast protein liquid chromatography (FPLC, also known as **fast performance liquid chromatography**) *is the term that is often used for column chromatography systems that are used to purify larger biomolecules, such as proteins.* Although FPLC and HPLC systems have similar components, FPLC systems are used for preparative applications. FPLC therefore involves much larger columns and can handle much more material than HPLC. FPLC systems are most often operated at lower pressure than HPLC systems. In contrast to the analytical purposes of HPLC, the aim of the FPLC user is usually to obtain as much purified product as possible, so these systems are more commonly found in production facilities or pilot plants. FPLC can be readily scaled to process anywhere from milligrams of mixtures, using columns with a total volume of 5 mL or less, to industrial production of kilograms of purified protein, using columns with volumes of many liters.

In contrast to HPLC, the pressure used for FPLC is relatively low, but the flow rate is relatively high. Since stability at high pressure is not as important for FPLC as it is for HPLC, different column materials can be used. While HPLC uses silica for the stationary phase (as shown in Figure 34.18a), FPLC can use a variety of materials, including agarose and acrylamide. In traditional HPLC, pressure-resistant stainless steel is necessary for column hardware, but transparent, biocompatible glass columns can be used for FPLC (as shown in Figure 34.16). Mobile phase solvents for HPLC must be of the highest quality, generally designated "HPLC grade" by vendors, to avoid introduction of impurities into the system. FPLC systems routinely use buffers as the mobile phase, and these buffers can be prepared properly by the user.

Table 34.2 summarizes the features of FPLC, as contrasted with HPLC. Figure 34.25 shows a comparison of the types of chromatograms obtained with HPLC and FPLC. Observe that the peaks obtained with HPLC are sharper and more clearly defined than the peaks obtained with FPLC. This makes HPLC an excellent analytical technique. With FPLC, what matters is that the fractions containing the product of interest can be obtained with minimal contamination by other materials.

34.5 DEVELOPING BIOSEPARATION STRATEGIES TO PURIFY BIOMOLECULES

34.5.1 BIOSEPARATION STRATEGIES BEGIN WITH A SOURCE

Now that we have introduced a variety of separation techniques, we will see how these techniques are combined to purify a desired product. We will primarily

TABLE 34.2

Comparison of the Features of FPLC and HPLC

	FPLC	HPLC
Goal	Purification	Analysis
Samples	Large biomolecules	Up to ~3,000 daltons
Sample volume	100 μL to many liters	< 100 μL
Pressure	Lower pressure, max 40 bar	Higher pressure, 1,500 bar
Eluent	Buffer	Solvents
Material hardware	Biocompatible (often glass)	Stainless steel
Column matrix (resin)	Agarose/polymer, big particle sizes, higher flow rates	Silica beads, pressure stable, small particle sizes, lower flow rates
Methods	Size exclusion, ion exchange, affinity, or hydrophobic interaction	Reverse phase, normal phase

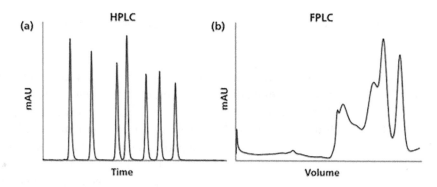

FIGURE 34.25 Comparison of HPLC and FPLC chromatograms. (a) A typical HPLC run showing retention time on the X-axis. Observe that the peaks are sharp and clearly separated from one another. (b) A typical FPLC run with volume (fraction number) on the X-axis. (Reprinted with permission from LC/GC. LC/GC is a copyrighted publication of MJH Life Sciences, LLC. All rights reserved.)

discuss these strategies in the context of a facility producing a product, rather than in the context of a research setting. However, the same general principles apply whether a biomolecule is being separated for use as a product, or for an analytical purpose.

The beginning of a bioseparation strategy is a source that contains the biomolecule of interest. There is a variety of biological sources that are commonly used to obtain biomolecules. Natural sources such as plants, animal tissues, blood, and other biological fluids have long been used to obtain biomolecules for research, or as products. For example, insulin to treat diabetic patients was originally isolated from the pancreas of animals grown by the meat industry. These natural sources, however, usually provide complex and relatively uncharacterized molecular mixtures, which require complex purification strategies to isolate the biomolecule of interest. Moreover, the amount of a product that can be obtained is often limited in natural sources. In some cases, as when a biomolecule

is found in human tissue, or an endangered plant, the natural source cannot be used for mass production.

Due to the complexity of isolating pure products from natural sources and the limited amount of the desired substance that may be available, biotechnologists most often use as "factories" cultured animal cells or cultured microorganisms grown using fermentation technology, as was introduced in Chapter 1. In most cases, these cultured cells are genetically engineered to produce the molecule of interest, as is the case with insulin. Humanized monoclonal antibody drugs are another example of an important biotechnology product. These drugs are usually manufactured by genetically modified cultured cells and cannot be isolated from any natural source.

In general, the biological source of a biomolecule product should be chosen or engineered to produce the greatest amount possible of usable biomolecule, as economically as possible. Table 34.3 summarizes some of the major variables in selecting a source for a biomolecule.

TABLE 34.3

Considerations in Choosing a Biological Source for Molecules

These are some of the major considerations when choosing a source for a product of interest:

- *The availability of the source*
- *The total amount of product present in the source*
- *The volume of material that will need to be processed to obtain the product of interest*
- *The stability of the product generated by the source*
- *The amount of product that can be isolated from the source relative to the amount of impurities that will need to be removed*
- *The nature of the impurities present in the source*
- *Any potential pathogenicity of the source*
- *The cost of culture media and equipment required to grow source cells*

In an industrial setting, production methods are categorized as upstream or downstream processes. **Upstream processes,** such as fermentation or cell culture, *are the source of the starting material of interest.* **Downstream processes** *are the separation procedures that result in a purified product.* Whereas downstream purification processes represent much of the cost of production, they are based on the choices made in the upstream system. Studies show that changes in upstream sources of biomolecules (e.g., switching from an animal source to a genetically engineered microorganism) can have a much greater effect on final production costs than changes in downstream processing. However, costs must be balanced with other significant factors, as we will discuss in this part of the chapter.

34.5.2 Purifying the Biomolecule of Interest

34.5.2.1 Setting Goals

Strategies to purify a specific biomolecule involve a series of steps. There is no "standard" series of specific steps that can be used to purify all biomolecules. Proteins, the cellular products most likely to be purified in biotechnology facilities, differ greatly in their molecular properties. The series of steps used to purify Protein X, therefore, is unlikely to be effective for Protein Y, unless the two proteins are closely related in structure and source. Also, many separation techniques can be performed under almost infinitely variable conditions (such as pH, temperature, and ionic strength). The conditions under which a technique is performed must therefore be individually optimized for every product. Moreover, the optimal strategy to purify a particular material of interest will depend to some extent on the source of starting material. This is because the best strategy to purify a product depends partly on the nature of the

biomolecule being purified, and partly on the nature of the impurities that must be removed. For example, the strategies used to purify insulin differ depending on whether its source is animal pancreas or genetically engineered *E. coli* cells.

However, even though there are no universal procedures applicable to every biomolecule, there are general requirements for the design of an appropriate bioseparation strategy. These requirements include the following:

- *Selecting the best possible source for the product;*
- *Understanding and prioritizing the goals of the purification process;*
- *Developing a specific assay for the product of interest and its impurities; and*
- *Developing an optimal sequence of methods to:*
  a. *release the product from its source in a soluble form and*
  b. *separate the product from impurities.*

Once the source is known, a first step in developing a bioseparation strategy is thus to determine the goals of the process. Questions to ask include the following: How pure does the biomolecule need to be for our purpose? How much of the biomolecule is required? How important is low cost? Figure 34.26 summarizes the three primary goals of a purification strategy: optimizing purity, optimizing yield, and reducing costs. These three goals are illustrated as a triangle to emphasize that all purification strategies require compromises. For example, the highest biomolecule purities are usually obtained only at high cost and with reduced yield. Yield is intimately related to cost-effectiveness and maintaining biomolecule quality. Optimizing either

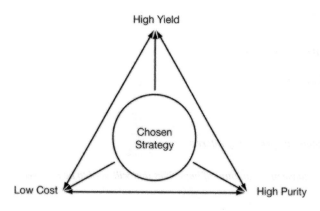

FIGURE 34.26 The primary goals for bioseparation strategies. All bioseparation strategies require prioritizing these goals.

yield or purity will usually raise the cost of a purification process.

Each of these goals varies in importance compared with the others, depending on the nature of the final biomolecule desired. For example, in a research setting, high purity may be essential for studies of the biomolecule's properties and cost may be less of a consideration because only small amounts of material will be processed. In a pharmaceutical production setting, a drug product requires the utmost purity, even if the cost is high. In contrast, a commercial product not intended for use as a drug may be acceptable in a less purified, less expensive form. In production facilities, the cost of materials and equipment is a major concern when designing a large-scale purification strategy. A purification method that works well and is cost-effective in the research laboratory may not be practical at large scale.

Additional considerations that must be considered when designing a bioseparation strategy include the following:

- **Safety requirements** (for example, some cultured cells and microorganisms are potentially pathogenic and will require special protections for personnel)
- **Simplicity of the procedures** (more complex procedures can lead to more difficulties when put into routine use)
- **Reliability of the methods**
- **Volumes of source material and unpurified product that can be handled with available equipment**
- **Speed of method development and scaling up in order to publish research results or get a product to market**
- **Regulatory concerns.**

Purification strategies for regulated commercial products must meet federal and state regulations and receive approval from appropriate agencies (Chapter 36). Once a product and its preparation procedure have been approved, the procedures for product purification must be rigorously followed and documented. In order to change these procedures, the company may again be required to prove the safety and/or effectiveness of the final product to regulatory agencies, which is a process that can require years of additional testing. There is considerable incentive, therefore, to develop an optimal purification strategy early in the process of product development.

Any product that is injected into humans as a drug must meet especially strict quality specifications for potency, homogeneity, and purity. It must be sterile and free of any contaminating substances, especially impurities such as proteins or nucleic acids that could induce adverse immune responses. Viruses must be removed. A series of assays will be required to test all these parameters.

34.5.2.2 Principles that Guide Development of a Purification Strategy

Because there are many purification techniques available, it is necessary to choose the techniques to be used for a particular product, and the order in which those techniques are performed. The simplest starting point for developing a bioseparation strategy is to follow the procedure for a similar molecule, although this approach varies widely in its success.

Despite the variabilities between one bioseparation procedure and another, there are certain principles that commonly guide the development of a bioseparation strategy. All bioseparations start with a mixture of materials from the biological source. One of the first priorities is to remove any live organisms from the mixture. Any known biohazardous impurities should be removed as early in the separation process as possible. Early steps will generally require the highest **capacity,** *which is the volume of material that can be processed simultaneously by a technique.* Figure 34.27 illustrates differences in column capacities. Capacity also refers to the amount of the product of interest that can be separated by the technique. The first definition is usually more significant early in the separation process, when volumes may be high (such as a supernatant from a 1,000 L fermentation vessel). Because it is usually expensive to process large volumes, volume reduction is an early concern. Typically, the simplest techniques are used early in the separation process, when the source mixture is complex. These early steps in a purification strategy usually take advantage of any

FIGURE 34.27 Preparative HPLC columns of varying capacities. Column capacity is related to the length and diameter of the column. (Used with permission from Thermo Fisher Scientific, the copyright owner.)

major differences between the desired product and the expected impurities. For example, if the product is a large molecule contaminated by many low molecular weight materials, an early separation step based on size would be appropriate.

Later steps in a purification process usually involve techniques that remove impurities similar in nature to the product itself. Later steps are often more expensive and have a lower capacity than earlier steps. Analysts use the terms **low resolution** and **high resolution** to distinguish different types of separation techniques. **Resolution,** or **resolving power,** *is the relative ability of a technique to distinguish between the product of interest and its impurities.* Earlier steps in the purification process are usually of low resolution; later ones are of higher resolution. **Selectivity** is a related term. The **selectivity** of a technique *is its ability to separate a specific component from a heterogeneous mixture, based on molecular properties.* The resolving power of a method is directly dependent on its selectivity. **Low-resolution purification methods** *are those that generally have high capacities and can be performed quickly, but also have relatively low selectivity.* Low-resolution methods eliminate impurities that are very different from the product of interest, but not those that have similar molecular properties. Low-resolution methods provide an efficient starting point for a bioseparation strategy, when the molecule of interest represents only a small portion of source material. In contrast, **high-resolution techniques** *are very selective and can remove impurities similar to the biomolecule of interest.*

Another basic principle that guides the development of a bioseparation strategy is to never purify the product more than is necessary for its application. Each purification step results in the loss of some of

the product of interest. Therefore, the highest product yields will be obtained with the fewest steps. Four to six sequential methods typically are optimal to purify and concentrate biomolecules.

Yet another basic principle is that techniques should be coordinated so that the end product of one step provides a suitable substrate for the next. For example, after a centrifugation step, a pelleted product should be dissolved in a volume of a buffer that is compatible with the next step.

Figure 34.28 is a flowchart of a generalized bioseparation strategy based on these general principles. Observe that the flowchart begins with two branches. This is because sometimes the biomolecule of interest is sequestered inside of the host cells and so the cells must be broken apart to release the biomolecule. In other cases, host cells can be genetically modified so that they secrete the desired product into the culture medium; this is a desirable situation since the biomolecule is readily accessible. The various steps shown in this flowchart will be described further in the sections below.

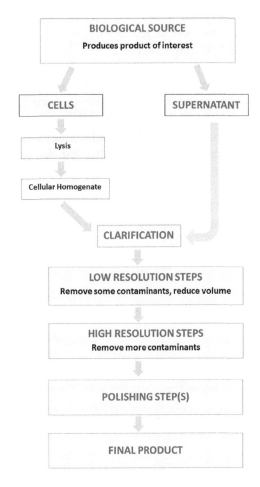

FIGURE 34.28 A generalized flowchart for a bioseparation strategy. See text for explanation.

34.5.2.3 Preparing a Cell Homogenate and Clarification

As shown in Figure 34.28, bioseparation strategies begin by releasing the substance of interest from cells and solubilizing it into an aqueous medium. In the case of secreted molecules, liquid supernatants can simply be separated from solid matter (cells) by centrifugation or filtration. In cases where the product is found inside the cell, the starting point is the preparation of a cell homogenate. *A* **cell homogenate** *is a suspension of cell contents in liquid, produced by disrupting the outer cell membrane, the cell wall if present, and some of the internal structure of the cell.* **Lysis** *is the breaking open of the cells.* Cell homogenates generally contain intact organelles, such as mitochondria; cytoplasmic constituents; membrane fragments; proteins; sugars; and nucleic acids, all suspended in an aqueous medium. Many techniques can be used to lyse cells and prepare a cellular homogenate (Figure 34.29). Cells with relatively fragile membranes, such as liver cells, require less strenuous methods than plant cells, which have a tough outer cell wall. Most lysis techniques can be adjusted to be more or less disruptive, as required by the type of starting material.

There are many factors to consider when preparing a cell homogenate. For example, the biomolecule of interest may be floating in the cytoplasm or it may be associated with cellular organelles. Depending on the situation, different buffers and techniques are used for homogenization. Consider also that the interior of an intact cell is buffered and protected from oxidizing

agents that can destroy molecules, or reduce their biological activity. Solutions used for homogenization, therefore, usually protect the biomolecule of interest by including antioxidizing agents, and buffers to control the pH. It is important to remember that mechanical disruption methods, such as grinding or sonication (the use of high-pitched sound), can generate considerable amounts of heat within the sample. These procedures should be performed under chilled conditions when possible. Cell homogenization is performed as gently as possible, to avoid loss of product, while still releasing as much of the biomolecule of interest as possible.

After cell lysis, the resulting homogenate is clarified. **Clarification** *is the removal of unwanted solid matter, usually by centrifugation or filtration.* If the product of interest is located within a cellular organelle, such as mitochondria, differential centrifugation can be performed to isolate the organelle of interest (described in Chapter 33).

34.5.2.4 After Clarification: Molecular Properties That Are Exploited for Bioseparations

The steps after clarification progressively remove impurities from the product of interest. Many bioseparation strategies begin with extractions and/or precipitations, techniques that we have categorized as non-instrumental. These low-resolution techniques are relatively inexpensive, have a high capacity, and are used both to remove some impurities from the product of interest and also to reduce the volume of material present. High-resolution methods, most often involving chromatography, are generally applied after low-resolution techniques have removed the bulk of impurities that are dissimilar to the product of interest. These methods have higher costs and lower capacities than low-resolution methods.

Most purification techniques select one or sometimes two molecular properties as the basis for separation. There are various molecular properties that are exploited for separations. These include the following:

- *Size or molecular weight.* (For example, the primary basis for all conventional filtration separations is the size of the materials to be separated. Centrifugation, electrophoresis, and chromatography also can separate materials from one another based on differences in their sizes, or molecular weights.)

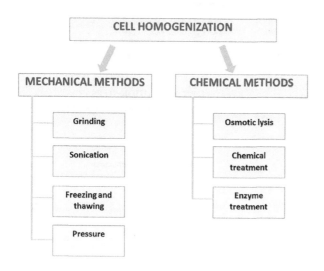

FIGURE 34.29 Common cell homogenization methods. The cell homogenization method used depends on the intracellular location of the molecule of interest, the cell type involved, and other factors.

- ***Molecular charge.*** (For example, ion exchange chromatography separations depend on differences in charge between different molecules.)
- ***Relative solubility of the molecule in water or other solvents.*** (Organic extractions, described earlier in this chapter, are an example.)
- ***Relative solubility of the molecule in the presence of salts.*** (Salting out, described earlier in this chapter, is an example.)
- ***Affinity.*** (Affinity chromatography is a commercially important example.)

By using a sequence of steps that separate molecules based on three or more different properties, it is generally possible to purify any biomolecule (Figure 34.30).

34.5.3 COMMON PROBLEMS ENCOUNTERED WITH PURIFICATION STRATEGIES

The most common problem encountered during a bioseparation procedure is the apparent loss of product during a specific separation step. If this occurs during initial strategy development, it suggests that an inappropriate separation method was chosen. For example, if you are performing an affinity chromatography technique and your product will not elute from the column, you may have inadvertently chosen a column matrix that binds irreversibly with the product of interest. During development, it is therefore important to use only an aliquot of a larger sample so that you can return to the results of a previous step and try another strategy.

Product is sometimes unexpectedly lost during a separation step that was previously proven to be effective. In these cases, there is usually a technical difficulty that can be identified. Note that these same difficulties can occur during strategy development, leading to the incorrect assumption that a method is not effective for purifying your product. Whenever large amounts of product disappear during a separation step, consider whether the factors outlined in Table 34.4 may be involved. Note that some of the problems indicated in Table 34.4 are the result of technical difficulties (e.g., using the wrong buffer) and others are the relatively unpredictable result of source contamination (e.g., the presence of enzyme inhibitors). In the latter case, either the interfering substance must be removed, or an alternative source of product must be used. To summarize the most basic precautions against losing product during a bioseparation strategy:

- ***In general, the fewer the purification steps used, the higher the ultimate product yield.***
- ***With many biomolecules, stability is a major concern. Maintain a constant temperature (usually chilled) and pH, and proceed from one step to the next as quickly as feasible.***
- ***Never discard any fraction during a separation procedure. Save everything until the end of the process, when the product has been recovered.***

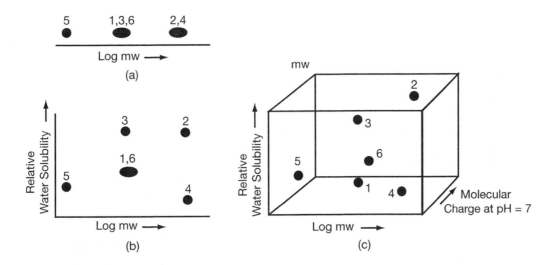

FIGURE 34.30 Isolation of specific molecules by their characteristics. This separation starts with a mixture of six molecules. (a) Separation on the basis of molecular weight gives clear separation of only Molecule 5. (b) Separation by both molecular weight and relative water solubility purifies all of the components except Molecules 1 and 6. (c) Adding a third separation parameter, charge, allows all components of the mixture to be purified.

Case Study: A Purification Strategy for Monoclonal Antibody Drugs

Monoclonal antibodies are among the most commercially important products of biotechnology, as was previously explored in Chapter 1. These protein biopharmaceuticals are used in the treatment of a variety of disorders, such as cancer, arthritis, and inflammatory diseases. The use of monoclonal antibodies to treat COVID-19 patients, or to prevent disease in people at high risk of infection, is a recent addition to this list of monoclonal antibody applications. At the time of writing, monoclonal antibody therapy is being used to treat COVID-19. Clinical trials show that Regeneron's mAb treatment, a combination of two antibodies called casirivimab and imdevimab, reduces COVID-19-related hospitalization or deaths in high-risk patients by about 70%. Monoclonal antibodies used in this way can complement the use of vaccines to curtail new infections.

As is true of all biopharmaceuticals, monoclonal antibody drugs must be highly purified products. The designs of their purification strategies thus prioritize the removal of virtually all impurities that come from the host cells or from processing steps, and the removal of any viruses present. In the case of drugs to treat COVID-19, speed to market is also an important goal. Meeting the latter goal has been facilitated in companies that have already introduced similar drugs, such as monoclonal antibodies to treat Ebola. A generalized flowchart for a monoclonal antibody purification strategy is shown in Figure 34.31. Observe that after clarification this strategy goes immediately to a high-resolution chromatographic method, that is, protein A affinity. It is very expensive to start with a protein A affinity separation. The cost is justified because protein A affinity is an efficient method of achieving purity in one step. Observe also that two steps are used to reduce or eliminate the serious risk caused by viral impurities.

TABLE 34.4

Factors that May Result in Product Loss during Purification Procedures

There are many factors that can contribute to a major loss of product during a bioseparation procedure. Some of these are relatively easy to identify (at least in retrospect), and others are difficult to predict. Common problems and examples include:

Problems leading to product loss

- *Discarding the fraction that contains the product of interest* (e.g., discarding a supernatant containing product while keeping a pellet of debris)
- *Performing separations using poor-quality reagents* (e.g., using an ion exchange buffer that was improperly prepared)
- *Poor product stability* (e.g., there may be enzymatic degradation caused by impurities in the mixture, or a product may be inherently unstable)
- *Unexpected precipitation of product* (e.g., product may be suspended in an inadequate amount of liquid)

Problems with the product assay

- *Performing an inappropriate product assay* (lacking adequate selectivity or specificity)
- *Using an incorrect buffer for the product* (e.g., certain salts may interfere with enzyme assays)
- *Presence of activity inhibitors* (if the presence of product is measured by its activity)
- *Loss of a required cofactor* (e.g., an enzyme may require the presence of magnesium or other ions for activity)

34.5.4 LARGE-SCALE OPERATIONS

All purification strategies are developed at the laboratory scale before moving to a production setting, and so the ability to convert a strategy to a large-scale operation is essential. **Scaling up** *is the process of converting a small-scale laboratory procedure to* *one that will be appropriate for large-scale product purification.* When testing purification strategies for commercial purposes, the research and development team selects and optimizes methods on a small scale. Development scientists then need to increase the scale of operations and test the robustness of the strategy.

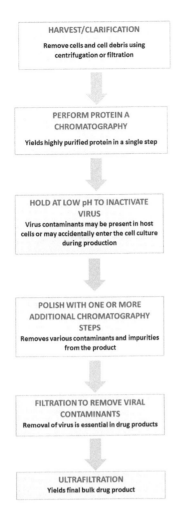

FIGURE 34.31 Flowchart of a strategy to purify a monoclonal antibody drug product. (Based on information in Liu, Hui F., et al. "Recovery and Purification Process Development for Monoclonal Antibody Production." *MAbs*, vol. 2, no. 5, 2010, pp. 480–99. doi:10.4161/mabs.2.5.12645.)

Scale-up must be considered from the start of product development. The working conditions (e.g., pH, temperature, and ion concentrations) must be achievable under large-scale conditions. Analysts need to consider the possibility of variations in biological source materials and must design techniques that will be relatively insensitive to the most likely variations. Some laboratory purification procedures may not be economically sound or physically possible in an industrial setting. For example, using sonication to lyse cells may work well in a laboratory, but it becomes impractical when hundreds of liters of cellular material is present. Breaking cells open with grinding or high-pressure techniques is more practical on a large scale.

Reproducibility of separation techniques in a production setting is essential. Having a simple assay method and documentation process for each purification step will contribute to cost-effectiveness and quality control. Of course, safety issues must be considered during scale-up. Strategies must be designed to eliminate any potential biohazards early in the process, and appropriate containment facilities and procedures are essential.

In reality, many potentially useful biotechnology products never reach the consumer market due to the high costs of downstream processes. The scale of required equipment and labor costs are major considerations. The ability to automate bioseparation procedures can increase cost-effectiveness to some extent, but not all procedures are easily automated. The ability to reduce source volumes quickly, apply less expensive, low-resolution methods, and achieve high product yields are key factors in reducing purification costs for biotechnology products.

34.6 ASSAYS ARE REQUIRED TO MONITOR A PURIFICATION PROCEDURE

34.6.1 OVERVIEW

This section discusses how analysts use assays to monitor a purification procedure. This discussion focuses on proteins because of their importance in a wide variety of biotechnology settings.

Assays are of vital importance in monitoring the progress and success of a bioseparation strategy. Assays are used initially as researchers optimize the separation strategy. Later, technicians rely on assays to routinely monitor the separation process. As discussed in Chapter 26, accuracy, precision, and sensitivity are key requirements for the assays used in a bioseparation situation.

34.6.2 ELECTROPHORESIS AND MASS SPECTROSCOPY

Electrophoresis plays an important role in bioseparations because it can be used to analyze and verify purification results. We have previously mentioned the importance of capillary electrophoresis in biopharmaceutical quality-control laboratories to analyze product purity. SDS–PAGE is another electrophoresis method that can be used to routinely check product purity after each step in a purification strategy. Samples are removed after each step and are analyzed by electrophoresis (Figure 34.32). Initially, many bands are present on the gel because many impurities are present. Each sequential step in a purification procedure ideally reduces impurities, and so fewer bands should appear on an electrophoresis gel. When samples of the final product are analyzed

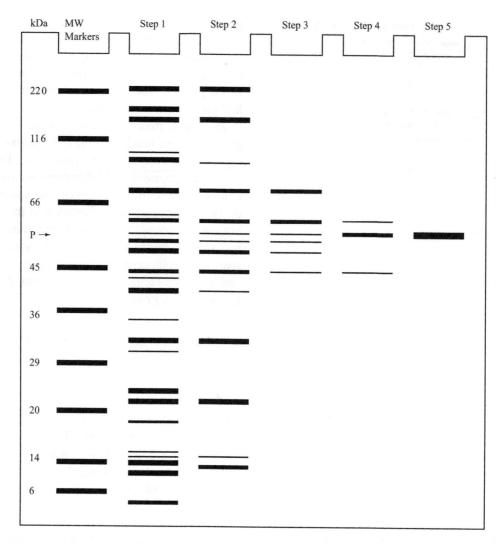

FIGURE 34.32 Illustration of SDS–PAGE analysis of sequential purification of a protein. Lane 1 on the left shows the bands from a protein molecular weight marker standard, used to estimate the size of protein bands in the samples. In this case, the protein of interest (the band labeled P) is known to be approximately 59 kDa (kilodaltons). The next lane is the sample from Step 1, which is cell homogenization. This sample contains the protein of interest and all protein impurities initially present. Step 2 in this purification scheme is an ammonium sulfate precipitation. Some impurities are removed in this low-resolution technique. Gel permeation chromatography in Step 3 removes proteins with a molecular weight significantly different from the desired product. Step 4, partition chromatography, eliminates most other proteins from the sample. The final step in this particular strategy is affinity chromatography, which is highly specific for the protein of interest and results in a single protein band.

by electrophoresis, there should be only one band. The presence of a single band on an SDS–PAGE gel is considered one measure of protein purity. However, the protein band may still contain impurities of the same molecular weight and properties as the protein of interest, and so at least one additional method must be used to confirm the purity of the product.

Mass spectroscopy (MS) is a specialized technique frequently used for identifying and quantitating molecules within a sample. MS *measures the mass-to-ionic charge ratios of molecules.* MS is a powerful analytical tool that allows the user to identify

impurities within samples and is sometimes coupled with LC (as LC–MS) as a detection system. MS is far more sensitive and specific than a UV detector, but at a much higher cost and complexity. MS is an important technique, but describing it in any detail is outside the scope of this text.

34.6.3 Assays for Specific Contaminants

In a production setting, there are likely to be multiple assays for specific contaminants, such as endotoxins, viruses, bacteria, host cell proteins, and host cell

DNA. For example, specific PCR assays, as introduced in Chapter 27, can be used to look for specific viral or bacterial contaminants. As another example, there are infectivity assays that detect the presence of viruses by their ability to infect cultured cells. These types of assays for specific contaminants may or may not be used in a research setting.

34.6.4 TOTAL PROTEIN ASSAYS, SPECIFIC ACTIVITY, AND YIELD

In order to monitor the purification of a desired protein product, it is necessary to know both how much of the desired protein is present, and also how much contaminating protein is present. When proteins are purified, the amount of total protein (desired protein plus contaminating protein) can be measured after every step. In a purification procedure, the method used to measure total protein must be reliable, fast, and sensitive to low protein concentrations, and relatively insensitive to potential interfering agents (e.g., buffer components or nucleic acids). Common spectrophotometric assay methods for total protein determination were discussed in Section 28.6.1. To ensure consistency, a single protein assay method should be chosen and then used throughout the purification process.

During a multistep purification process, the protein of interest is successively isolated from multiple impurities. After each separation step, it is common practice to set aside an aliquot of the product mixture for analysis. This aliquot is used to assay the amount of the desired protein present and the amount of total protein present. Based on these two measurements, it is possible to calculate the specific activity of the protein of interest, and the product yield after each step in the purification process. **Specific activity** *is the amount (or units) of the protein of interest, divided by the total amount of protein in a sample.* As protein purification proceeds, the product of interest should become purer at each step, so the specific activity of the preparation should increase. **Yield** *is the amount of the product of interest.* A 100% yield would be a situation where every last bit of the product of interest that was present in the source material is isolated and retained. In practice, each purification step results in loss of some of the product of interest and so it is not possible to achieve a 100% yield. **Percent yield, or percent recovery**, *is the percent of the total amount of the product of interest that was initially present that is actually retrieved after purification.* In a yield analysis, all starting materials must be accounted for because there cannot be additional product generated by the purification process. If yield is calculated to be greater than 100%, there must be a mistake. Yield analysis is thus part of quality-control testing performed during production.

The results of this process are illustrated in Figure 34.33; each purification step results both in the removal of impurities and in some unavoidable loss of the biomolecule of interest. As protein purification proceeds, the specific activity therefore increases, but the total amount of the desired product (calculated as the percent yield or recovery) decreases.

34.6.5 SPECIFIC ACTIVITY AND YIELD DETERMINATIONS FOR ENZYMES

Biologically active molecules are generally assayed by their activity. In this section, we will use enzymes as an example. Enzymes are an important class of proteins that are used in many processes in the food, agricultural, cosmetic, and pharmaceutical industries. In some cases, they can be used as therapeutic proteins for disease treatment. For example, as described in Table 1.1, Section 1.2.1, Kanuma® and Elelyso® are enzyme replacement drugs that are used for the treatment of enzyme deficiency in lysosomal storage disorders such as lysosomal acid lipase deficiency and Gaucher disease, respectively.

Enzymes are generally assayed by their ability to catalyze a reaction. Enzymatic reactions involve the conversion of a substrate(s) into a product(s):

$$ \text{Substrate} \xrightarrow{\text{Enzyme}} \text{Products} $$

Enzyme assays generally involve the measurement of either the disappearance of substrate or the appearance of product. Enzyme activity is typically expressed as units, based on the rate of the reaction the enzyme catalyzes. There are two standard expressions of enzyme activity. The **international unit (IU) of enzyme activity** *is defined as the amount of enzyme necessary to catalyze transformation of 1.0 μmol of substrate to product per minute under optimal measurement conditions.* (Conditions, such as pH and temperature, may affect the activity of an enzyme and so must be optimized.) The **SI unit of enzyme activity** *is defined as the amount of enzyme necessary to catalyze transformation of 1.0 mol of substrate to product per second under optimal measurement conditions (such as optimal pH and temperature). This is defined as a* **katal** *or* **kat** *unit.* The ability to measure enzyme activity is dependent on knowing the ideal reaction conditions.

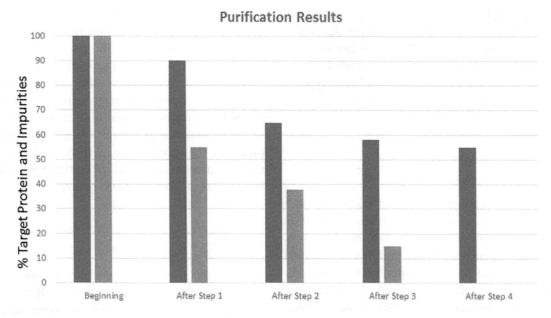

FIGURE 34.33 A step-by-step strategy that purifies a protein. This graph illustrates a protein purification strategy with four steps, labeled on the X-axis. The blue colored bars represent the amount of target protein present, expressed as a percent, and the orange bars are the impurities present. Initially, 100% of the target protein and its impurities are present. Observe that after the first purification step, some of the target protein is lost. The blue bar is lower, indicating that about 90% of the original amount of target protein is present. By the end of Step 4, less than 60% of the desired protein is present. Thus, the yield of target protein decreases. This seems undesirable, but observe the orange bars that represent the impurities. The impurities are steadily reduced so that by the end of Step 4, none can be detected. This bioseparation strategy is successful because the desired protein has been separated from the impurities.

Example Problem 34.2

An assay for enzyme Q measures the disappearance of substrate. In 15 minutes, 12 mmol of substrate is converted to product. What is the activity of the enzyme preparation in international units?

Answer

12 mmol = 12,000 μmol. The rate of conversion of substrate to product is 12,000 μmol in 15 minutes = 800 μmol/minute. The preparation therefore contains **800 IU** of enzyme.

Some enzymes, such as proteases, which digest substrates of variable or unknown molecular weight, cannot be quantitated in these standard units. For these proteins, units are defined according to the assay used, providing a relative measure of enzyme activity. Information about specific activity is useful when purchasing enzymes or purified proteins. You will frequently be offered various grades of product, based on the purity of protein present. This will allow you to choose among purity levels depending on your

application. In general, less purified grades are less expensive because less effort is required, and greater product yield is possible.

Example Problem 34.3

You need to purchase enough Enzyme Z to perform 100 reactions. You determine that this will require approximately 8,000 U of enzyme. The catalog from your favorite enzyme supplier offers you the following choices:

Enzyme Z

One unit of enzyme activity catalyzes the conversion of 1 μmol of A to B in 1 minute at 25°C at pH 7.5.

Grade 1	From aardvark liver	100 mg	$50
	Activity: 10,000 U/g	500 mg	$200
		1 g	$350
Grade 2	From aardvark liver	100 mg	$200
	Activity: 50,000 U/g	500 mg	$700
		1 g	$1,200
Grade 3	From aardvark liver 1 g	1 g	$30
	Activity: 2,000 U/g	5 g	$130
		10 g	$250

a. What enzyme grade would be your best choice if your main priority is to find the least expensive option?

b. What would be your best choice if you need the purest enzyme grade possible?

Answer

First, you want to consider the specific activities of each enzyme grade and determine what quantity you need to purchase:

Grade 1: At a specific activity of 10,000 U/g, you will need 0.8 g to provide the 8,000 activity units you need. This will require eight times 100 mg for a total of $400, or 1 g for $350.

Grade 2: At 50,000 U/g, you will need 160 mg, or 2 times 100 mg for $400.

Grade 3: At 2,000 U/g, you will need 4 g of protein, at four times 1 g for $120, or 5 g at $130.

   a. The least expensive option is Grade 3, which is the least pure.

   b. The highest specific activity, and therefore the most concentrated activity, is found in Grade 2.

34.6.6 ASSAYS FOR PROTEINS OTHER THAN ENZYMES

Most of the therapeutic, recombinant protein products that are currently on the market, and many that are in development, are not enzymes. (See Chapter 1 for more information about recombinant products.) For these proteins, many of which are signaling molecules or molecules that bind to signaling molecules, an appropriate activity assay will vary according to their particular biological activity. The concept of a signaling molecule is illustrated in Figure 34.34.

Signaling molecules have various effects on cells. Some signaling molecules cause target cells to proliferate, while others cause target cells to die. Yet other signaling molecules may cause target cells to differentiate, or migrate. Sometimes the protein product of interest is not itself a signaling molecule, but rather is a molecule that interacts with a signaling molecule to exert an effect. (See, for example, Figures 1.12 and 1.13.) When the product of interest is a signaling molecule, or is a protein that interacts with a signaling molecule, it is necessary to develop an assay that measures the protein's effect on living cells under standardized conditions. Assays that require living cells are called "bioassays" or "cell-based assays." Assay validation, as introduced in Chapter 26, can be particularly complex for bioassays because the assay's performance will be impacted by all the possible variables in a cell culture. These variables include type of cell, cell density, age of cells, passage number, nutritional state, culture medium components, and, of course, the technique of the cell culturist.

Specific activity for bioassays is defined as units/mg, as it is for enzymes, but a unit of activity is defined in a different way. The biological effect of the molecule can be reported as the **EC50, effective concentration.** *This is the concentration of the protein of interest that induces a response that is halfway between the baseline and the maximum response in*

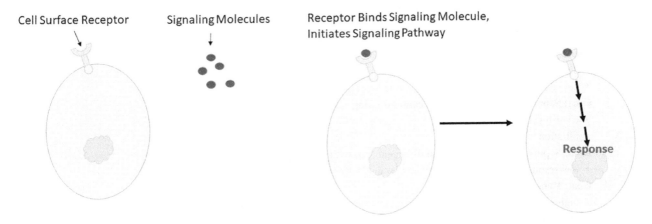

FIGURE 34.34 The concept of cell signaling. Cells have receptors on their surfaces that recognize and bind specific signaling molecules that are secreted by other cells. When the receptor binds to its signaling molecule, a pathway is triggered inside the cell that ultimately causes the cell to respond in some way.

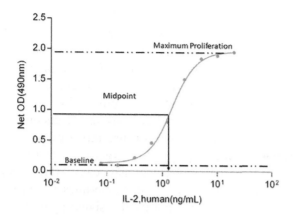

FIGURE 34.35 Example of a spectrophotometric proliferation assay. The X-axis plots the amount of a signaling molecule of interest that was added to different cell cultures. The Y-axis, optical density (OD), tells us how many cells were present in the culture; the higher the OD, the more the cells proliferated (reproduced). As the dose of the signaling molecule increased, the number of cells present also increased, until a plateau was reached. The EC50 value is then calculated by finding the signaling molecule concentration halfway between the baseline and maximum response. In this case, the EC50 is about 2 ng/mL of signaling molecule. Since one unit of activity is defined as the amount of this particular signaling molecule that has 50% of the maximal proliferative effect per milliliter, 2 ng equals one unit of this molecule. (Adapted from STEMCELL Technologies, Inc. "Biological Activity of Cytokines: Specific Activity (units/mg) vs. International Units." Tech Tips and Protocols, 2020. https://www.stemcell.com/biological-activity-of-cytokines-specific-activity-units-mg-vs-international-units.html.)

the specific assay (50%). For example, Figure 34.35 shows the effect of a particular signaling molecule that causes cells to proliferate. One unit of activity can then be defined as the amount per mL of this particular signaling molecule that has 50% of the maximal proliferative effect.

Consider as another example the monoclonal antibody drug for macular degeneration that is illustrated in Figure 1.13. This drug acts by binding to the signaling molecule VEGF. By binding to VEGF, the drug inhibits the proliferation of cells in the blood vessels of the eye. A bioassay for this drug needs to be able to measure how effectively the product inhibits the effect of VEGF. This type of assay is called a "neutralization assay" because it measures the amount of inhibition of VEGF's effect.

Assay development is thus a critical – and often challenging – part of designing a product purification operation.

34.7 SUMMARY

In this unit, we have seen how bioseparation methods take advantage of differences in the size, charge, solubility, and affinities of different biomolecules. Table 34.5 summarizes the overall process of bioseparation that we discussed in this chapter.

This chapter also provides a segue into the final unit of this text, which talks about regulatory affairs. Regulatory affairs relate to commercially important

---

TABLE 34.5
The Overall Process of a Bioseparation Strategy

The sequential stages of a purification process should follow a logical pattern. Although there are exceptions, the series of steps selected will address the following goals. Examples of appropriate techniques are provided for each goal. Assays are used at every step to monitor yield and purification progress.

1. **Goal: Solubilization of the product of interest**
   - disruption of cells, and/or
   - collection of supernatants containing soluble product
2. **Goal: Separation of product from contaminating solid material (clarification)**
   - centrifugation
   - filtration
3. **Goal: Separation of product from divergent impurities**
   - low-resolution purification techniques, such as salt fractionation and extractions
4. **Goal: Removal of all impurities**
   - high-resolution purification techniques, such as ion exchange chromatography
   - ultrafiltration to remove viruses
5. **Goal: Product polishing, concentration, and preparation for end use**
   - dialysis to reduce final volume
   - gel filtration to remove salts
   - ultrafiltration to reduce volume and remove any remaining impurities

products, notably (though not exclusively) medical-related products, and the interaction of government agencies with those involved in the manufacture and distribution of those products. Downstream processing, which was introduced in this chapter, is a key part of manufacturing commercially important biotechnology products.

Thus, we began this unit with the mention of a physician extracting a useful drug 2,500 years ago. Technology has advanced a long way since then, and separation methods have become far more sophisticated. However, the basic concept of taking a biological source and extracting a useful product from it is still a fundamental part of the modern biotechnology world.

Practice Problems

1. You have received a vial of enzyme that is labeled as containing 4 kat (SI unit) of enzyme. How many IUs of enzyme are in the vial?
2. You are presented with the following data from a series of purification steps for the (imaginary) enzyme "comatase." Fill in the blanks in the table.

Purification of Comatase from *E. coli*

Purification Step	Volume (mL)	Total Protein (mg)	Total Activity (IU)	Specific Activity (IU/mg)	Yield (%)
I. Homogenization	100,000	12,350	9,000	____	100
II. Dialysis	5,000	10,233	8,289	0.81	92.1
III. Organic extraction	80	3,860	____	1.65	____
IV. Ion exchange chromatography	10	1,140	5,625	____	62.5
V. PAGE	2	386	4,688	12.15	____

3. Given the purification scheme data shown in Problem 2, which purification step gave the greatest:
   a. relative increase in product purity?
   b. loss of product?
   c. absolute reduction in product volume (the greatest decrease in actual volume)?
   d. relative reduction in product volume (the greatest % decrease from the previous step)?
4. Suppose you are given the assignment of isolating an (imaginary) enzyme, "enzylase," which is secreted into the fermentation broth of the (imaginary) bacterium *B. techi*. The enzyme has an approximate molecular weight of 60,000. You grow 3 L of *B. techi*, which you then spin down in a centrifuge. You discard the broth and weigh the cells. You break open the cells using a grinding method and centrifuge the resulting paste. You discard the cellular debris and assay the supernatant for the enzylase. You are disappointed to find no activity. You test your assay with "store-bought" enzylase and find that the detection assay works well. What important mistake did you make?
5. You are interested in an enzyme produced by a particular bacterium. You grow 10 L of the microorganism in a fermenter. You centrifuge the 10 L to obtain a pellet of cells and set aside the supernatant. You resuspend the cell pellet in 20 mL of buffer, break open the cells by sonication, and remove the cellular debris by centrifugation. The enzyme of interest is in the intracellular supernatant. (Assume there is still 20 mL of supernatant.) The supernatant is called the crude extract. You remove 0.5 mL of the crude extract and perform an enzyme assay. You find there are 500 units of enzyme activity in the 0.5 mL sample of crude extract. You take another 0.5 mL sample of crude extract and perform a protein assay. You find there is 25 mg of protein present in the 0.5 mL of crude extract.
   a. Draw a flowchart of this procedure.
   b. What is the specific activity of enzyme in the crude extract?
   c. How many units of enzyme activity were in the original 10 L, assuming perfect efficiency and no loss at each step?

6. Choose the most probable answer to the following questions:
  a. During purification of a protein, the specific activity should:
     i. Increase as the purification proceeds.
     ii. Decrease as the purification proceeds.
     iii. Remain the same throughout the purification.
  b. During purification of a protein, the amount of total protein present should:
     i. Increase as the purification proceeds.
     ii. Decrease as the purification proceeds.
     iii. Remain the same throughout the purification.
  c. During purification of a protein, the yield is likely to
     i. Increase as the purification proceeds.
     ii. Decrease as the purification proceeds.
     iii. Remain the same throughout the purification.

7. Before developing a strategy to purify a protein, it is important to learn about the biochemical characteristics of that protein. List at least three features of the protein that are important to know when developing a strategy to purify a protein.

8. The enzyme, β-galactosidase, was purified from a culture of *E. coli*. The purification steps were as follows:

Step 1. Ten liters of cells was grown in a fermenter.

Step 2. The 10 L of cells plus broth were centrifuged to separate the cells from the broth. The result was a pellet containing the *E. coli* cells.

Step 3. The cells were suspended in buffer and were broken apart by sonication (a method that uses high-pitched sound). The resulting cell suspension was centrifuged again. The pellet and the supernatant were assayed for β-galactosidase enzymatic activity. β-Galactosidase was detected only in the supernatant. This supernatant is called a crude extract.

Step 4. The supernatant was treated with salt to precipitate the β-galactosidase. The precipitated enzyme was collected by centrifugation and was resuspended in buffer. The resulting β-galactosidase preparation was assayed for β-galactosidase enzymatic activity and for total protein.

Step 5. Salt was removed from the β-galactosidase solution using dialysis. The resulting β-galactosidase preparation was assayed for β-galactosidase activity and for total protein.

Step 6. The resulting solution, after dialysis, was passed through a chromatography column. This step separated the β-galactosidase from a number of impurities. The resulting β-galactosidase preparation was assayed for β-galactosidase activity and for total protein.

Observe that the β-galactosidase was extracted from *E. coli* cells and was then purified using a series of steps. After each step, the activity of the β-galactosidase and the amount of total protein were tested. It would be possible to continue with further purification steps at this point, or to stop, depending on the final purity required. The results are shown in the following table.
  a. Fill in the blanks in the table.
  b. What happened to yield as the purification steps proceeded?
  c. What happened to specific activity as the purification proceeded?

Purification of β-Galactosidase from *E. Coli*

Purification Step	Total Protein (mg)	Total Activity (IU)	Specific Activity (IU/mg)	Yield (%)
Crude extract	2,140	1,360	0.64	100
Salt precipitation	760	1,350	___	99
Dialysis	740	___	1.8	___
Chromatography	390	1,240	___	___

9. Imagine that you are responsible for following the protocol for purifying your company's product. The maximum possible yield is calculated to be 10 g/L; however, after performing the separation you calculate the purified total sample yield as 22 g/L. Discuss how your team might handle this situation.

10. Three different proteins are present in a mixture. Protein A has a molecular weight of 100 kD, Protein B has molecular weight of 30 kD, and Protein C has a molecular weight of 200 kD. The three proteins are separated from one another using GPC, and the following chromatogram is obtained. Which peak is Antibody A? Which peak is Antibody B? Which peak is Antibody C?

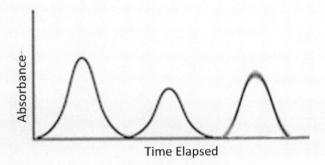

11. Convert the EC50 result shown in Figure 34.35 into specific activity.

UNIT X

Biotechnology and Regulatory Affairs

Chapters in this Unit

- ✦ Chapter 35: Biotechnology and the Regulation of Medical and Food Products
- ✦ Chapter 36: The Lifecycles and Regulation of Pharmaceutical Products
- ✦ Chapter 37: The Lifecycles and Regulation of Biopharmaceutical and Regenerative Medicine Products
- ✦ Chapter 38: Quality Systems in a Regulated Production Facility

---

Biotechnology is about the transformation of knowledge into products. These products can be of tremendous benefit to people, but only if they are properly conceived, developed, and manufactured. This unit is about the interaction between the government – which seeks to protect the consumer from dangerous and ineffective products – and the biotechnology industry. This interaction, which we term "regulatory affairs," is complex and continuously evolving.

DOI: 10.1201/9780429282799-44

Case Study: Regulation Saves Lives

In 1960, the United States avoided the most infamous drug-related tragedy of the twentieth century, that is, the use of the sedative, thalidomide. In 1957, a West German company began over-the-counter sales of thalidomide, a sedative and anti-anxiety drug. Thalidomide caused few side effects in the user and was considered a safe alternative to addictive barbiturates. By 1960, the drug was available in 46 countries. At that point, thalidomide had been found to be effective against morning sickness in pregnant women, accounting for substantial sales.

In 1960, the manufacturer applied to the Food and Drug Administration (FDA) for approval to market the drug in the United States. The application went to Dr. Frances Oldham Kelsey, who was newly in charge of reviewing new drug applications. Dr. Kelsey became concerned about the lack of adequate human testing, although the company had performed tests that technically met the legal requirements at that time. In particular, the drug had not been tested for fetal toxicity, which was not formally required at that time.

Rather than approve the application, Dr. Kelsey requested additional safety information from the drug company. Despite objections from the company and some officials within the FDA, she held her ground and continued to challenge the application. In doing so, she prevented a major tragedy from enveloping the United States.

By 1961, thalidomide was shown to be responsible for significant birth defects, in particular, phocomelia. This is a previously rare condition where limbs are severely deformed or missing. In some cases, feet and hands are connected directly with the torso. By 1962, when thalidomide sales were banned worldwide, ~10,000 babies had been born with thalidomide-induced phocomelia; only ~50% survived. This does not count the eye, ear, heart, and other birth defects associated with the drug (Figure 1).

Due to Dr. Kelsey's diligence, only 17 cases of phocomelia were identified in the United States. In 1962, the Kefauver-Harris Amendments to the Federal Food, Drug, and Cosmetic Act were passed by Congress, incorporating more stringent guidelines for new drug testing, and requiring drug companies to demonstrate both safety and effectiveness of their products. Soon after, President John F. Kennedy awarded Dr. Kelsey the President's Award for Distinguished Federal Civilian Service. (Figure 34.3.) Dr. Kelsey went on to have a long and successful career in the FDA.

The regulatory approval of new drugs is currently a stringent process that takes many years of research and clinical studies. The lessons of the drug thalidomide inform many aspects of this process.

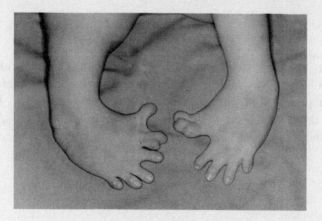

FIGURE 1   Congenital malformation of the feet caused by thalidomide consumption by pregnant mother. (Uploaded to Flickr by Otis Historical Archives National Museum of Health and Medicine. NCP14053, CC BY 2.0.)

The interaction of government with the biotechnology industry leads to a distinctive culture and work environment. Therefore, it is important for biotechnology professionals to have a basic understanding of regulatory affairs. This unit provides a general introduction to the history, vocabulary, and principles of regulation. We place particular emphasis on biopharmaceutical products because of their importance to biotechnologists.

Chapter 35 begins with a history of medical product regulation in the United States and provides background on the relationship between the biotechnology industry and regulatory agencies.

Chapter 36 introduces the regulatory lifecycle of pharmaceutical products in general.

Chapter 37 focuses on the regulation and lifecycles of biopharmaceutical products, including those that fall into the category of regenerative medicine. Biopharmaceutical therapeutics are generally produced in living systems, or are derived from living systems.

Chapter 38 discusses how the principles of quality systems and regulations are applied in production settings.

BIBLIOGRAPHY FOR UNIT X

For an engaging introduction to quality and regulatory affairs, we highly recommend the online resource: *Making the Call: Quality in Biomanufacturing* (www.franklinbiologics.org/). *Making the Call* is a "choose-your-own-adventure"-style, interactive video series. The videos tell the story of a fictional biopharmaceutical company, Franklin Biologics, and an imaginary drug, Squabanin. In this interactive series, viewers play the roles of various employees in the company as they strive to manufacture a life-saving drug (Figure 2).

There is a vast, fluid literature in the area of regulatory affairs, particularly as it relates to pharmaceuticals. Since the technology, regulations, and standards that impact biotechnology are changing quickly, the easiest way to enter the literature is through the Internet. Some useful sites are included in these references.

THE FDA WEBSITE IS VAST WITH A NUMBER OF USEFUL RESOURCES. EXAMPLES INCLUDE THE FOLLOWING:

Center for Biologics Evaluation and Research. "Current Good Tissue Practice (CGTP) and Additional Requirements for Manufacturers of Human Cells,

FIGURE 2 The quality-control laboratory in the video series *Making the Call: Quality in Biomanufacturing.*

Tissues, and Cellular and Tissue-Based Products (HCT/Ps)." *U.S. Food and Drug Administration*, 17 May 2019, www.fda.gov/regulatory-information/search-fda-guidance-documents/current-good-tissue-practice-cgtp-and-additional-requirements-manufacturers-human-cells-tissues-and.

Center for Biologics Evaluation and Research. "Development & Approval Process (CBER)." *U.S. Food and Drug Administration*, 27 January 2021, www.fda.gov/vaccines-blood-biologics/development-approval-process-cber. (Links to information about various approval pathways for medical products.)

Center for Devices and Radiological Health. "PMA Approvals." *U.S. Food and Drug Administration*, 23 August 2018, www.fda.gov/medical-devices/device-approvals-denials-and-clearances/pma-approvals. (Has information about how FDA regulates medical devices.)

Center for Devices and Radiological Health. "Premarket Approval (PMA)." *U.S. Food and Drug Administration*, 16 May 2019, www.fda.gov/medical-devices/premarket-submissions/premarket-approval-pma. (Has information about how FDA regulates medical devices.)

Center for Drug Evaluation and Research. "Data Integrity and Compliance With Drug CGMP. Questions and Answers Guidance for Industry." *U.S. Food and Drug Administration*, 13 December 2018, www.fda.gov/regulatory-information/search-fda-guidance-documents/data-integrity-and-compliance-drug-cgmp-questions-and-answers-guidance-industry.

Center for Drug Evaluation and Research. "Investigating Out-of-Specification Test Results for Pharmaceutical Production." *U.S. Food and Drug Administration*, 24 August 2018, www.fda.gov/regulatory-information/search-fda-guidance-documents/investigating-out-specification-test-results-pharmaceutical-production.

Center for Drug Evaluation and Research. "Pharmaceutical Quality for the 21st Century A Risk-Based Approach Progress Report." *U.S. Food and Drug Administration*, 27 November 2015, www.fda.gov/about-fda/center-drug-evaluation-and-research-cder/pharmaceutical-quality-21st-century-risk-based-approach-progress-report.

Center for Drug Evaluation and Research. "Process Validation: General Principles and Practices." *U.S. Food and Drug Administration*, 24 August 2018, www.fda.gov/regulatory-information/search-fda-guidance-documents/process-validation-general-principles-and-practices.

Center for Drug Evaluation and Research. "Q2B Validation of Analytical Procedures: Methodology." *U.S. Food and Drug Administration*, 14 April 2020, www.fda.gov/regulatory-information/search-fda-guidance-documents/q2b-validation-analytical-procedures-methodology. (This guideline provides the basis for method validation in the US pharmaceutical and biopharmaceutical industries.)

Center for Drug Evaluation and Research. "Quality Systems Approach to Pharmaceutical Current Good Manufacturing Practice Regulations." *U.S. Food and*

Drug Administration, 16 April 2020, www.fda.gov/regulatory-information/search-fda-guidance-documents/quality-systems-approach-pharmaceutical-current-good-manufacturing-practice-regulations. (This guidance document is intended to help manufacturers that are implementing modern quality systems and risk management approaches to meet the requirements of the Agency's CGMPs.)

Office of the Commissioner. "Part 11, Electronic Records; Electronic Signatures - Scope and Application." *U.S. Food and Drug Administration,* 24 August 2018, www.fda.gov/regulatory-information/search-fda-guidance-documents/part-11-electronic-records-electronic-signatures-scope-and-application.

Office of the Commissioner. "The FDA History Vault." *U.S. Food and Drug Administration,* 1 February 2018, www.fda.gov/about-fda/virtual-fda-history/fda-history-vault.

Office of the Commissioner. "Protecting and Advancing Women's Health: The FDA Office of Women's Health." *U.S. Food and Drug Administration,* www.fda.gov/files/science%20&%20research/published/10--Years-and-Beyond--FDA-Office-of-Women's-Health.pdf

ICH GUIDELINES

There are a number of ICH documents that are relevant to biotechnology (www.ich.org). All of them are periodically updated. ICH guidelines are divided into four major topic categories and are coded according to those topics (i.e., all "Quality" topics start with "Q").

Q – Quality. Topics relating to chemical and pharmaceutical quality assurance. Examples: Q1 Stability Testing, Q3 Impurity Testing, and Q9 Quality Risk Management.

S – Safety. Topics relating to preclinical studies. Examples: S6 Preclinical Safety Evaluation of Biotechnology-Derived Pharmaceuticals, S1 Carcinogenicity Testing, and S2 Genotoxicity Testing.

E – Efficacy. Topics relating to clinical studies in human subjects. Examples: E4 Dose Response Studies and E6 Good Clinical Practices.

M – Multidisciplinary. Topics that do not fit into one of the above categories. Examples: M1 Medical Terminology and M2 Electronic Standards for Transmission of Regulatory Information.

The following ICH guidelines are significant in outlining a risk-based, lifecycle approach to process validation:

**ICH Q8 (Pharmaceutical Development),
ICH Q9 (Quality Risk Management), and
ICH Q10 (Pharmaceutical Quality System).**

RELEVANT PORTIONS OF THE UNITED STATES CODE OF FEDERAL REGULATIONS

The Code of Federal Regulations is available at: https://www.govinfo.gov/app/collection/cfr.

21 CFR 11. Federal government rules that regulate electronic (computer) records, signatures, and other forms of electronic documentation.

21 CFR 58. Federal government rules that regulate preclinical laboratory studies of pharmaceutical products.

21 CFR 210. Federal government rules that define Current Good Manufacturing Practices for manufacturing, processing, packing, and holding drugs in general.

21 CFR 211. Federal government rules that define Current Good Manufacturing Practices for finished pharmaceuticals.

21 CFR 820. Federal government rules with which medical device manufacturers must comply, called "Quality System Regulation" (QSR).

OTHER

Betts, Roy. "The Food Safety Modernization Act- Global Impacts for Microbiological Food Safety." *Culture.* Vol. 36. no. 1, 2016, pp. 1–6. http://assets.thermofisher.com/TFS-Assets/MBD/Reference-Materials/Culture-36-1-FSMA-Microbiological-Impacts-LT2215A.pdf.

Blum, Deborah. *The Poison Squad: One Chemist's Single-Minded Crusade for Food Safety at the Turn of the Twentieth Century.* Penguin Books, 2018.

Chapman, Kenneth G. "A History of Validation in the United States: Part 1." *Pharmaceutical Technology,* October 1991, pp. 82–96. (Historical presentation; includes discussion of some issues and controversies relating to validation.)

Dougan, Stewart. "The Product is the Process - The Master Cell Bank I." caribbeanbiopharma.com/the-process-is-the-product-the-master-cell-bank–1.

Dumont, Jennifer, et al. "Human Cell Lines for Biopharmaceutical Manufacturing: History, Status, and Future Perspectives." *Critical Reviews in Biotechnology,* vol. 36, no. 6, 2015, pp. 1110–22. doi: 10.3109/07388551.2015.1084266.

Gorsky, Igor. "A Brief History Of Parenteral Process Validation — How We Got Here." *Pharmaceutical online Newsletter,* 7 July 2020. www.bioprocessonline.com/doc/a-brief-history-of-parenteral-process-validation-how-we-got-here-0001

Guo, J., et al. "PHP7 Recent Developments in Pompe Disease Therapy." *Value in Health,* vol. 15, no. 4, 2012, p. A14. doi:10.1016/j.jval.2012.03.084.

Harris, Richard. "A Search for New Ways to Pay for Drugs that Cost a Mint." Aired on *National Public Radio.* 14 November 2018. www.npr.org/sections/health-shots/2018/11/14/665782026/a-search-for-new-ways-to-pay-for-drugs-that-cost-a-mint.

Hoinowski, Alex M., Motola, Sol, Davis, Richard J., and McArdle, James V. "Investigation of Out-of-Specification Results." *Pharmaceutical Technology,* January 2002, pp. 40–50. pdfs.semanticscholar.org/de74/3346ee0e1a6008125c8b2b2a240d5f4ef270.pdf

Immel, Barbara. "A Brief History of the GMPs: The Power of Storytelling." *Immel Resources, LLC.* 2001–2004. biomanufacturing.org/uploads/files/30542959636 2804820-brief-history-of-gmps.pdf. (A good article about the history of drug regulation in the United States.)

Justia US Law. "United States v. Barr Laboratories, Inc., 812 F. Supp.458 (D.N.J. 1993)." 2020. law. justia.com/cases/federal/district-courts/FSupp/ 812/458/1762275/. (Justia is an organization whose "mission is to advance the availability of legal resources for the benefit of society." They provide a detailed analysis of the Barr decision.)

Kingham, Richard, et al. "Key Regulatory Guidelines for the Development of Biologics in the United States and Europe." In Wei Wang and Manmohan Singh (Eds.), *Biological Drug Products: Development and Strategies* (1st ed.). John Wiley & Sons, Inc. 2014, pp. 75–109.

The National Institutes of Health. "NIH Guidelines for Research Involving Recombinant or Synthetic Nucleic Acid Molecules (NIH Guidelines)." *Department of Health and Human Services.* April 2019. osp.od.nih. gov/wp-content/uploads/NIH_Guidelines.pdf.

The Pew Charitable Trust. "FDA's Framework for Regulating Regenerative Medicine Will Improve Oversight." October 2019. www.pewtrusts.org/-/ media/assets/2019/10/fdasframeworkforregulatin- gregenerativemedicine_v2.pdf.

Plavsic, Mark. "An Integrated Approach to Ensure the Viral Safety of Biotherapeutics. *BioPharm International* Vol. 29, no. 5, 2016, pp. 40–45.

Roy, Michael J. *Biotechnology Operations: Principles and Practices.* CRC Press. 2011. (A textbook that discusses quality and regulatory concerns in biotechnology companies.)

Tetzlaff, Ronald F., Shepherd, Richard E., and LeBlanc, Armand J. "The Validation Story: Perspectives on the Systematic GMP Inspection Approach and Validation Development." *Pharmaceutical Technology,* March 1993, pp. 100–16. (Historical presentation of issues leading to validation.)

Van Norman, Gail A. "Drugs, Devices, and the FDA: Part 2: An Overview of Approval Processes: FDA Approval of Medical Devices." *JACC: Basic to Translational Science.* Vol. 1, no. 4, June 2016, pp. 277–287.

Wang, Jun, and Chow, Shein-Chung. "On the Regulatory Approval Pathway of Biosimilar Products." *Pharmaceuticals*, Vol. 5, no. 4, 2012, pp. 353–68. doi:10.3390/ ph5040353.

Wazade MB, Walde, SR and Ittadwar, AM. "An Overview of Pharmaceutical Process Validation and Process Control Variables of Tablets Manufacturing Processes in Industry." *Int J Pharm Res Sci.* Vol. 3. no. 9, 2012, pp. 3007–3022.

HOMEPAGES FOR AGENCIES AND ORGANIZATIONS

The American Association for Laboratory Accreditation: http://www.a2la.org

The Environmental Protection Agency: http://www.epa.gov

The Food and Drug Administration: http://www.fda.gov

The International Conference on Harmonisation: http:// www.ich.org

The United States Department of Agriculture: http://www. usda.gov

The United States Pharmacopeia: http://www.usp.org

35 Biotechnology and the Regulation of Medical and Food Products

35.1 THE CONTINUING EVOLUTION OF DRUG AND FOOD REGULATION IN THE UNITED STATES

35.1.1 INTRODUCTION

Unit I provided an overview of the science that led to biotechnology products such as anticancer therapeutics, vaccines, gene therapies, medical diagnostic tools, and genetically modified food crops. In this unit, we turn to some of the regulatory issues relating to these types of products.

Medical products and foods are stringently regulated by the government, and these regulations have a profound effect on the biotechnology workplace. This chapter explores issues relating to government regulation of medical and food-related products, the relationship between the government and the biotechnology and pharmaceutical industries, and the evolution of that relationship. For simplicity, this chapter examines regulation primarily in the United States.

The regulated characteristics of products relate to safety, honesty (as in labeling), and, particularly for medical products, effectiveness, and reliability. A drug product, for example, must be correctly labeled, and neither a food nor a drug product should be contaminated with harmful microorganisms. The use of a mislabeled drug, or consumption of a contaminated food, can cause death. The individual consumer cannot determine whether a drug is properly labeled, or whether meat purchased in the supermarket has a suitably low level of microbial contamination. The public therefore looks to the government to enact and enforce laws that provide protection from unsafe or ineffective food and drugs.

Although we take for granted that the government regulates the production of drugs and foods, this was not always the case, as is illustrated dramatically in the history of food and drug regulation in the United States. The present system of regulation evolved over the years in response to often tragic events and abuses. Familiarity with these historic events is essential to understanding the regulatory landscape that exists today.

35.1.2 THE STORIES

The quality of drugs and foods in the United States was virtually unregulated until the early 1900s (Figure 35.1). A major industry emerged at that time to process foods for urban consumers. This burgeoning food industry was frequently filthy and poorly managed. There were various efforts to regulate food processing; however, no legislation was passed until 1906. At that time, Upton Sinclair published the novel, *The Jungle*, intended to publicize the difficulties facing immigrant workers in the Chicago stockyards. Sinclair vividly described the filthy conditions and alarming practices in the food industry. People were so outraged by Sinclair's descriptions that Congress passed a law to help regulate the production of food and, at the same time, drugs. This law was the original **Food and Drugs Act**, *which authorized regulations to ensure that manufacturers did not sell adulterated (contaminated) or misleadingly labeled food and drug products. A separate law enforcement agency was formed in 1927 to enforce legislation relating to food and drugs.* It was first known

as the Food, Drug, and Insecticide Administration, and then, in 1930, was named the **Food and Drug Administration (FDA)**.

The 1906 Food and Drugs Act failed to deal with the safety or effectiveness of drugs, so in 1933, FDA advisors recommended a complete revision of this inadequate law. A bill was introduced into the Senate, launching a 5-year legislative battle between those who wanted drug-law reform and the industry that vigorously fought its passage. During these years, an exhibition of dangerous foods, medicines, medical devices, and cosmetics was prepared to illustrate the shortcomings of the 1906 law. The exhibit showed, for example, an eyelash dye that had blinded women (Figure 35.2), lotions and creams that caused mercury poisoning, hair dyes that caused lead poisoning, and a weight loss drug that increased metabolic rate to such an extent that some users died. The FDA was powerless to seize these dangerous products because the 1906 act did not mandate drug safety. First Lady Eleanor Roosevelt borrowed the exhibit and invited congressional wives to the White House to view it. She and the congressional wives were among the most vigorous supporters of drug reform throughout the 5-year effort.

Eventually, a dramatic tragedy facilitated the passage of stronger laws. An antibiotic, sulfanilamide, was introduced in the 1930s. The drug was a major medical advance in the treatment of infectious diseases. However, sulfanilamide is relatively insoluble

FIGURE 35.1 Drugs of dubious value and safety were widely sold in the early 1900s. Many drug products contained alcohol, opium, and morphine, without any labeling to that effect. (Image credit: U.S. Food and Drug Administration History Office.)

FIGURE 35.2 Hazardous products persisted despite the 1906 FD&C Act. (Image credit: U.S. Food and Drug Administration History Office.)

and therefore was only available in a pill form that was difficult to administer to small children. In 1937, a chemist at the S. E. Massengill Company dissolved a batch of sulfanilamide in the toxic industrial solvent, diethylene glycol, in an effort to obtain a soluble form of the drug. The company marketed the resulting "elixir," resulting in at least 358 poisonings and 107 deaths, mostly children.

In response to mounting reports of deaths due to the sulfanilamide preparation, the FDA launched the first major recall of a drug product. The agency had the authority to seize the product, not because it caused deaths, but because it was inaccurately labeled an "elixir" when it did not contain alcohol. Most of the FDA's staff of inspectors and chemists, along with many local law enforcement officials, traveled throughout the country looking for the toxic sulfanilamide preparation. James Harvey Young tells the following story about the sulfanilamide recall:

"In Atlanta, an elderly FDA inspector, who was also ill, drove through the rain into the north Georgia mountains to a drugstore that could not be reached by phone . . . A pint bottle of the "elixir" had been sent to the druggist, and the inspector was charged with bringing it back. The druggist had the bottle but four ounces of its red contents were missing, all prescribed by a doctor for one patient, name unknown. When the physician returned from his rounds, he said that he kept no records but believed he had prescribed the medicine for a woman named Lula Rakes. He did not know where she lived, and her surname was not uncommon in the region. The inspector, shouting an inquiry to the deaf druggist, attracted the attention of a bystander who volunteered the guess that Lula lived eight miles over the ridge in Happy Hollow. The inspector drove over the dark mountain road, only to find Lula's home abandoned. A neighbor said that the Rake's family had moved one valley farther on. Driving onward the inspector at last found the house, and Lula was there. Busy with the moving, she had taken only a few doses of her medicine, but she could not recall what she had done with the bottle. An hour of conversation and exploring finally led to the medicine, with many other articles in a paper sack under the bed. The inspector began his weary way home.

(Young, James H. "Sulfanilamide and Diethylene Glycol," in *Chemistry and Modern Society:* Historical Essays in Honor of Aaron J. Ihde, edited by Parascandola, John and Whorton, James C. pp. 105–25. American Chemical Society, 1983.)

In other less-fortunate cases, inspectors were not able to find the drug before it was consumed. There were instances of druggists and doctors who falsified their records or lied when asked about the drug. Under the 1906 drug law, a company had no obligation to prove that their drug was safe, so Massengill was not held responsible for the deaths. The company was, however, fined a small amount for labeling the drug an "elixir."

In response to the sulfanilamide tragedy, drug reform became popular and resistance to regulation was overcome. A revised Food, Drug, and Cosmetic Act (FD&C Act) was passed in 1938 that contained the critical provision requiring that a manufacturer prove the safety of new drugs with animal and clinical studies, thus legislating the overall process of drug testing that will be described in Chapter 36.

In 1941, there were nearly 300 injuries or deaths from sulfathiazole tablets tainted with the sedative phenobarbital. This prompted the FDA to drastically tighten their control over the manufacturing of drugs.

In 1955, a manufacturer did not properly inactivate the virus used to make polio vaccine and 51 people contracted paralytic polio; 10 people died. This incident, and others like it, led to increased FDA factory inspections and increased testing of the safety of products before their release to the public.

One of the great successes of the FDA occurred in the early 1960s. At that time, the drug thalidomide, produced by a German firm, was commonly used in Europe for insomnia and for nausea in pregnant women. The German manufacturer wanted to license and distribute thalidomide in the lucrative US market. The American pharmaceutical firm, William S. Merrell, submitted a New Drug Application (NDA) to the FDA seeking approval of thalidomide in the United States. This application was assigned for review to the agency's newest medical officer, Dr. Frances Oldham Kelsey, who had joined the staff earlier that year. Dr. Kelsey did not think the NDA provided sufficient evidence of the drug's safety and refused to approve the application, thus blocking the legal entry of the drug into the United States (Figure 35.3). Thalidomide was later found to cause profound birth defects; thousands of malformed children were born in Europe, many of them with no arms or legs. News of the thalidomide tragedy allowed two legislators, Kefauver and Harris, to push stricter drug regulations through Congress that

FIGURE 35.3 Kelsey receives award for protecting US consumers from thalidomide. Kelsey received the President's Distinguished Federal Civilian Service Award in 1962 from President John F. Kennedy, the highest civilian honor available to government employees. (Image credit: U.S. Food and Drug Administration History Office.)

included, for example, provisions controlling testing of drugs in humans.

In 1963, the first set of Good Manufacturing Practice (GMP) regulations were published to detail how manufacturers should produce safe and effective drugs. A number of injuries and deaths in the 1960s and 1970s caused by contaminated products led to revised GMPs in 1978. These regulations included requirements that manufacturers maintain standard operating procedures, use validated systems (validation is discussed in Chapter 38), and retain extensive documentation. Other medical products in addition to drugs, such as heart pacemakers, were also included in these regulations.

In the 1970s, inspections of pharmaceutical animal testing laboratories revealed that toxicology studies

supporting new drug applications were poorly conceived and improperly conducted. These inadequacies led to the Good Laboratory Practice[1] regulations of 1976 whose goal is to assure the quality of data submitted to FDA in support of the safety of new products.

In 1989, a scandal relating to the manufacture of generic drugs rocked the industry. FDA investigators discovered that some generic drug manufacturers were guilty of illegal practices, fraud, and noncompliance with GMPs leading to defective drug products. These deficiencies led to increased inspections and FDA oversight of the generics industry.

35.1.3 THE HISTORY OF BIOTECHNOLOGY PRODUCT REGULATION

As we saw in Unit I, new biotechnologies have, in the past decades, dramatically reshaped the modern drug and medical product industries. In 1982, insulin produced by recombinant DNA methods became the first modern biotechnology product to be approved for market and sale. In 1986, the first monoclonal antibody product was approved, and in 1987, the first recombinant DNA product produced in a mammalian cell culture line was approved. Deciding how to regulate these new biotechnology drugs was a challenge to the regulatory system. As we saw in Chapter 1, many people, scientists and the public alike, saw the immense promise of biopharmaceutical products, but were concerned that they might pose unknown risks. This is a common theme with most new technologies; their potential benefits must be balanced against their potential risks.

Biopharmaceuticals like insulin are made using a novel method, but the products themselves are very similar to "traditional" products. For example, prior to 1982 insulin was extracted from animal organs. Genetic engineering allowed companies to produce insulin in genetically modified cells. The product is insulin in both cases, but the manufacturing process is different. The FDA had a regulatory system in place

---

[1] *Good Laboratory Practices (GLPs),* when used in reference to pharmaceuticals, relate to drug testing that is performed before a drug is allowed to be tested in humans. Much of this early drug testing involves experimentation with animals. The Environmental Protection Agency, EPA, also mandates "good laboratory practices." EPA's Good Laboratory Practices guide investigators who are studying the health and environmental effects of agrochemicals. The GLPs from the FDA and the EPA are not identical, but they do serve the same overall purpose of ensuring trustworthy test results. The term *good laboratory practices (glps)* is sometimes used much more broadly to refer to good practices and procedures in any type of laboratory.

by the 1970s for regulating products isolated from animal tissue. Regulators therefore had to decide whether biotechnology products should be regulated differently than other products (because their method of production is novel) or the same as traditional products (because the products themselves are similar). This is sometimes phrased as the *product versus process* controversy. At the heart of this debate was uncertainty as to whether the nature of biotechnology processes makes products derived by these methods inherently different from, and potentially riskier than, traditional products.

In the early 1970s, the use of immortal mammalian cell lines in pharmaceutical manufacturing was forbidden. This was because immortal cell lines are similar to cancer cells in that they are not subject to normal controls on growth, and they repeatedly divide. It was feared that the use of immortal cell lines might introduce a "factor" into the products that would be carcinogenic in patients. Furthermore, it was feared that mammalian cells might harbor pathogenic viruses that could contaminate products. By the mid-1970s, it became clear that immortal cell lines would be necessary for producing some biotechnology products that could not be made properly by bacteria. It was therefore imperative that the FDA resolve the issue of whether products made by new biotechnology methods were inherently more risky than traditional products, and whether they would require additional, or different regulation.

A number of scientific advances in the 1970s led to the eventual acceptance of products made using modern biotechnology methods. Experiments were performed that demonstrated that genetically manipulated organisms could be handled safely. Effective methods of inactivating viruses were developed. The use of continuous cell lines was first tested in other countries without adverse effects, and studies of cancer did not reveal a carcinogenic "factor" in continuous cell lines.

In 1974, the **National Institutes of Health (NIH)**, *a federal agency responsible for funding and overseeing research*, assembled an expert group to explore the safety of recombinant DNA technology. This group, **the Recombinant DNA Advisory Committee (RAC)**, reviewed and interpreted what was then known about the safety and risk of genetic manipulation methods. The result of their work was a set of guidelines for conducting experiments involving recombinant DNA. These guidelines, called **Guidelines for Research Involving Recombinant DNA Molecules (NIH Guidelines)**, were first published in 1976, and

they have since been revised several times. The NIH Guidelines cover such topics as methods of assessing the risk of a recombinant DNA experiment, methods of classifying organisms based on their risk, and methods of containing organisms so that they cannot escape and harm workers or reach the environment. Subsequent revisions of the Guidelines also discuss methods of working on a large scale with recombinant organisms (e.g., in a production facility), performing human gene therapy, and working with plants and animals. These revised guidelines outline basic practices that are applied to work with recombinant DNA in research and in the pharmaceutical industry in the United States, and sometimes in other countries. The FDA enforces the practices established in the NIH Guidelines when appropriate. (Safety issues related to recombinant DNA research are also discussed in Chapter 10.)

As a result of these developments throughout the 1970s, the FDA decided that its existing regulatory framework for complex molecular products could be extended to include products made by recombinant DNA methods. This means that (1) no additional laws or regulations were created to regulate biotechnology-derived products, (2) biotechnology companies manufacturing products for medical use must comply with CGMPs, GLPs, and other related regulations, and (3) biotechnology products must undergo the same approval processes as other drug products.

35.1.4 International Harmonization Efforts

An important worldwide trend has been the increasing internationalization of commerce. As companies extend their sales to markets outside their native country, they sometimes encounter difficulties meeting the varied regulatory requirements of different nations. There have therefore been important efforts to internationalize both standards and regulations. ISO standards, as introduced in Chapter 4, are an example. ISO standards are intended to increase consistency internationally in product quality and in production practices.

An important component of the global harmonization effort is the work of **the International Council on Harmonisation of Technical Requirements for Registration of Pharmaceuticals for Human Use (ICH)**. **ICH** *is an organization that brings together the regulatory authorities of the European Union, Japan, and the United States and experts from the pharmaceutical industry in the three regions to discuss scientific and technical aspects of pharmaceutical product* *regulation*. The organization makes recommendations on ways to achieve greater international harmonization in the interpretation and application of technical guidelines and requirements for medical product registration. Harmonization facilitates the mutual acceptance of data by regulatory bodies internationally, prevents unnecessary duplication of **clinical trials** (*studies involving human volunteers*), and reduces unnecessary animal testing without compromising safety and effectiveness.

ICH has established a number of technical guidelines, many of which are relevant to biotechnology companies that make biomedical products. The ICH E6 guidance document on Good Clinical Practices is widely used as a guide when performing clinical trials. Other important ICH guidelines cover such areas as: validation of bioanalytical procedures, viral safety evaluation of products, ensuring genetic stability of biotechnology products, gene therapy, and drug interaction studies. ICH guidelines are steadily becoming more important in guiding industry practice.

35.1.5 MEDICAL DEVICES

While much of this textbook focuses on pharmaceutical and biopharmaceutical products, it is important to also consider medical device regulation in the United States. **Medical devices**, according to the FDA, *are products, such as tongue depressors and pacemakers, which have a medical purpose, but are not drugs or biologics.* This category includes a broad array of items, such as stethoscopes, surgical instruments, pacemakers, patient ventilators, and diagnostic kits. Many items that are medical devices are not what we would call biotechnology products. Some medical devices, however, do fall under the biotechnology umbrella, particularly tests that diagnose disease, and associated reagents and cultured cells. For example, biotechnology techniques (PCR and immunoassays) are used in the tests that diagnose COVID-19 illness, or determine if a person has antibodies against the SARS-CoV-2 virus. Kits to perform these diagnostic tests are regulated by the FDA as medical devices. Similarly, cells that are grown in culture in order to perform viral or genetic testing are considered to be medical devices. The term **in vitro diagnostic (IVD) product** *is used by the FDA to refer to such items.*

Medical devices were not a major consideration when the 1906 Food and Drugs Act and the 1938 FD&C Act were passed to regulate food, drugs, and cosmetics. In the 1970s, a number of women using the Dalkon Shield, a contraceptive device, were seriously injured, drawing attention to the need to strengthen the regulation of medical devices. In 1976, the Medical Device Amendments to the FD&C Act were passed to expand the Act to include promoting the safety and effectiveness of medical devices. New medical devices introduced after May 1976 were required to obtain FDA approval before they could be marketed. The amendments also allowed the FDA to ban devices that were not safe or effective, and required medical device companies to follow Good Manufacturing Practices (GMPs), register their establishments, and list devices with the FDA. They were also required to report any adverse events (AEs) to the FDA.

The Medical Device Amendments created a three-class, risk-based classification system for medical devices with different regulatory pathways for each class:

- *Class I medical devices are those that pose low risk to consumers if improperly made*, e.g., bandages, canes, and tongue depressors. Many IVDs fall into this category.
- *Class II medical devices are those that could pose moderate risk to consumers*, e.g., acupuncture needles and surgical drapes. Some IVDs fall into this category.
- *Class III medical devices are those that could pose high risk to consumers*, e.g., heart valves and pacemakers. As of May 2020, COVID-19 diagnostic tests were thought likely to be classified as Class III medical devices because the consequences of a false result can have severe impacts on the community as a whole. (Congressional Research Service. "Development and Regulation of Domestic Diagnostic Testing for Novel Coronavirus (COVID-19): Frequently Asked Questions March 9, 2020." crsreports.congress.gov/product/pdf/R/R46261.) However, COVID-19 tests are, at the time of writing, being allowed to bypass some of the stringent regulatory requirements of Class III devices on an emergency basis.

Subsequent medical device acts in 1990 (the Safe Medical Devices Act) and in 1996 (the Food and Drug Administration Modernization Act) further strengthened the FDA's enforcement of medical devices and clarified regulatory requirements. Changes also were introduced to help harmonize the US medical device regulations with international standards, particularly ISO 13485, an international standard that supports medical device manufacturers in designing and maintaining quality systems. Consistent with the goal of harmonizing the US regulations with international quality standards, the 1996 act used the term "Quality

System Regulation" (QSR) to identify the regulations relating to medical devices.

At present, medical device manufacturers are required to develop a quality system commensurate with the type of medical device(s) they make, the risk represented by the device(s), the complexity of the device(s), and the complexity of the manufacturing facility. According to this paradigm, the quality system and regulatory oversight required for facilities making tongue depressors is quite different than that required for those making heart pacemakers. The medical device regulations also focus on building quality into each product from its inception and allow the FDA to look into records from research and development.

35.1.6 Risk-Based Quality Systems

We have seen from a historical perspective how and why the development and production of medical products came to be tightly controlled through laws and government oversight. The overall regulatory process that has evolved over the years is now well established and accepted. Even so, the relationship between the government and the pharmaceutical/biopharmaceutical and medical device industries is not a static one; it continues to evolve as the world changes. In September 2002, the FDA rolled out a new initiative called *Pharmaceutical CGMPs for the 21st Century: A Risk-Based Approach*. The motivation behind this initiative was different than others coming from the FDA; it was not a reaction to a tragedy or a scandal. Rather, the new initiative was partly a response to an industry climate in which scientific discoveries and technological advances are being made daily but are slow to be utilized by the pharmaceutical industry. The FDA was therefore looking for a regulatory system that protects patients and yet encourages innovation. The initiative was also a response to the fact that the FDA was finding itself stretched with fewer resources to regulate increasing numbers of companies and products. The FDA was seeking more efficient, cost-effective methods of inspection and oversight. The new initiative was also partly in response to political pressures to lower the cost of drugs. The FDA reasoned that the cost of drugs can be reduced if companies are encouraged to use modern, efficient production methods. The FDA was also trying to align its regulatory practices within its various divisions and with international regulatory bodies.

As the title of the initiative suggests, managing risk is a key to the new initiative. **Risk** *is defined as the combination of the probability of the occurrence of harm and the severity of that harm.* The International Conference on Harmonisation, ICH, says:

The two primary principles of risk management are:

- The evaluation of the risk to quality should be based on scientific knowledge and ultimately linked to the protection of the patient; and
- The level of effort, formality and documentation of the quality risk management process should be commensurate with the level of risk.

In a risk-based quality system, a company must identify potential risks, analyze the probability of the risk and the severity of harm, and then devise a plan to control the risk, focusing on the most critical issues. They must also evaluate whether the plan worked, and if not, modify it. A major goal of risk management is to focus time and attention on the things that matter, and not waste resources on things that do not. For a pharmaceutical company, this means identifying the most critical attributes of their products (e.g., the absence of contamination and proper potency), identifying the aspects of production that affect the critical attributes, and controlling those aspects.

In the risk-based regulatory model, the FDA continues to provide regulatory oversight to the pharmaceutical industry. But, in a concept paper introducing the initiative, the FDA stated:

FDA resources are used most effectively and efficiently to address the most significant health risks . . . In order to provide the most effective public health protection, FDA must match its level of effort against the magnitude of risk. Resource limitations prevent uniformly intensive coverage of all pharmaceutical products and production . . . . The intensity of FDA oversight needed will be related to several factors including the degrees of a manufacturer's product and process understanding and the robustness of the quality system controlling their process. For example, change to complex products (e.g., proteins . . .) made with complex manufacturing processes may need more oversight.

While the FDA's 2002 initiative resulted in significant changes, the idea of basing decisions on risk is not new. As previously mentioned, medical device regulation took a risk-based approach dating back to the 1970s. One of the influential events in the biotechnology product versus process discussion was a policy statement issued in 1992 by President George H.W. Bush, "Four Principles of Regulatory Review for Biotechnology" (Office of Science and Technology Policy, Executive Office of the President. "Exercise of Federal Oversight Within

Scope of Statutory Authority: Planned Introductions of Biotechnology Products Into the Environment." 1992). This statement set forth four broad principles of regulation for biotechnology. The first of these principles is:

Federal government regulatory oversight should focus on the characteristics and risks of the biotechnology product—not the process by which it is created. Products developed through biotechnology do not per se pose risks to human health and the environment; risk depends instead on the characteristics and use of individual products. Biotechnology products that pose little or no risk should not be subject to unnecessary regulatory review during testing and commercialization. This allows agencies to concentrate resources in areas that may pose substantial risks and leaves relatively unfettered the development of biotechnology products posing little or no risk.

35.2 THE REGULATORY PROCESS

35.2.1 CONGRESS PASSES LAWS

We have seen that the development and production of certain products, including drugs and foods, are regulated by the government. In this section, we introduce the mechanism by which regulations are created.

The regulation of food and medical products has often been driven by a tragedy, a problem, or an advance in science and technology. The public and their elected officials respond by enacting laws that are intended to reduce risks while maximizing benefits. A government agency is empowered to interpret and enforce laws through a system of regulations (Figure 35.4). For example, in the United States, the **Food and Drug Administration** *is the regulatory agency responsible for ensuring the safety, effectiveness, and reliability of medical products, foods, and cosmetics as set forth in various federal legislative acts, most notably the Food, Drug, and Cosmetic Act of 1938.* A brief quote from that act is shown in Figure 35.5. Note in this Act the focus on "filth" and "insanitary conditions." This harkens back to Upton Sinclair's reports of abysmal conditions in the food industry. Regulations change as the legislature, regulatory agencies, the public, and the industry respond to events and to one another.

REGULATION OF PHARMACEUTICAL PRODUCTS

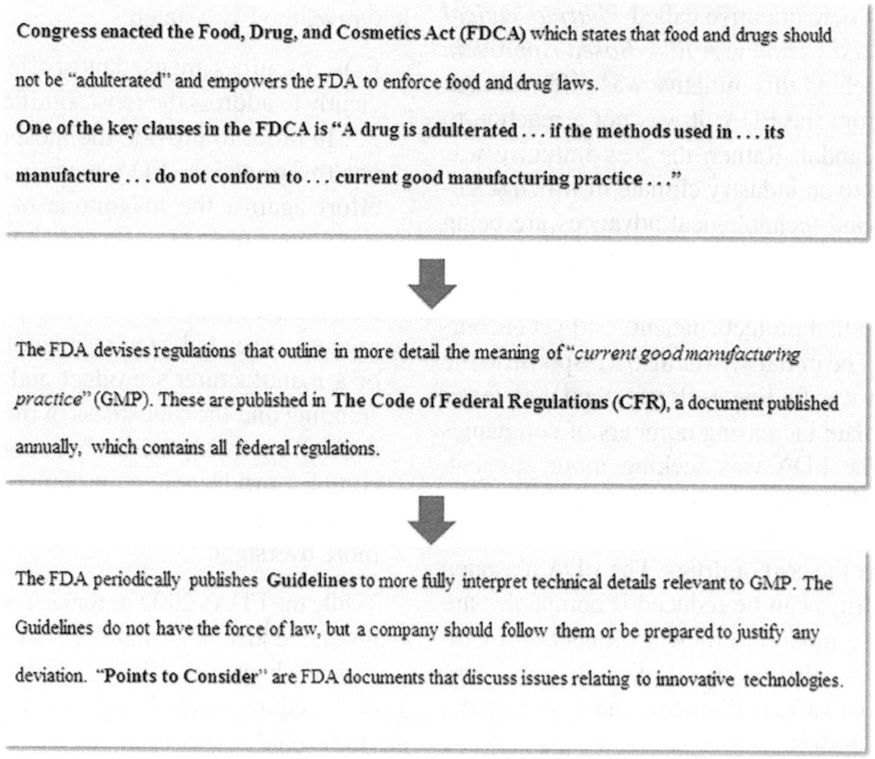

Congress enacted the Food, Drug, and Cosmetics Act (FDCA) which states that food and drugs should not be "adulterated" and empowers the FDA to enforce food and drug laws.

One of the key clauses in the FDCA is "A drug is adulterated . . . if the methods used in . . . its manufacture . . . do not conform to . . . current good manufacturing practice . . ."

The FDA devises regulations that outline in more detail the meaning of "*current good manufacturing practice*" (GMP). These are published in **The Code of Federal Regulations (CFR)**, a document published annually, which contains all federal regulations.

The FDA periodically publishes **Guidelines** to more fully interpret technical details relevant to GMP. The Guidelines do not have the force of law, but a company should follow them or be prepared to justify any deviation. "**Points to Consider**" are FDA documents that discuss issues relating to innovative technologies.

FIGURE 35.4 An example of regulation. Congress passed an act and authorized a federal agency to interpret and enforce it. The agency devises regulations and enforcement strategies to bring about compliance with the intent of Congress.

Federal Food, Drug, and Cosmetic Act Adulterated Drugs 501 [351] *A drug or device shall be deemed adulterated—(a) (1) if it consists in whole or in part of any filthy . . . substance; or (2) (A) if it has been prepared, packed, or held under insanitary conditions whereby it may have been contaminated with filth . . . or (B) if it is a drug and the methods used in, or the facilities or controls used for, its manufacture, processing, packing, or holding do not conform to or are not operated . . . in conformity with current good manufacturing practice to assure that such drug meets the requirements of the Act as to safety and has the identity and strength, and meets the quality and purity characteristics, which it purports or is represented to possess . . .*

FIGURE 35.5 A brief, but important excerpt from the FD&C Act of 1938. This act forbids adulteration of food and drugs and authorizes the regulation of food and drug quality in the United States.

35.2.2 THE UNITED STATES CODE OF FEDERAL REGULATIONS

The regulations that affect the medical products industry are found in the **Code of Federal Regulations** (CFR), *which is a codification of the rules of the United States federal government.* The CFR contains the complete and official text of the regulations that are enforced by federal agencies. Individuals working in regulated segments of the biotechnology industry must familiarize themselves with the relevant sections and adhere to them in their work. Box 35.1 explains how the regulations are organized and how to access specific regulations.

35.3 THE ORGANIZATION OF THE FDA

The FDA regulates the development, manufacture, marketing, and labeling of prescription and nonprescription drugs, blood products, vaccines, tissues for transplantation, medical devices, radiological products (including cellular telephones), animal drugs and feed, tobacco, and cosmetics. The FDA also regulates the safety and labeling of all foods except for meat and poultry, which are regulated by the U.S. Department of Agriculture. The FDA is headed by the Commissioner of Food and Drugs, who is appointed by the President of the United States and is confirmed by the U.S. Senate. The FDA is organized into Centers that are summarized in Table 35.1.

For historical reasons, there are two related, but separate regulatory systems within the FDA for pharmaceuticals and biopharmaceuticals. The FDA distinguishes between so-called drugs and biologics.

The FD&C Act defines **drugs** as *"articles intended for use in the diagnosis, cure, mitigation, treatment, or prevention of disease in man or other animals."* Drugs *are regulated by* **the Center for Drug Evaluation and Research, CDER**, one of the program centers in the FDA. Small, chemically synthesized therapeutic products are likely to be regulated as drugs. Many common prescription and over-the-counter items are drugs.

According to the FDA, "**Biologics**, *in contrast to drugs that are chemically synthesized, are derived from living sources (such as humans, animals, and microorganisms).* Most biologics are complex mixtures that are not easily identified or characterized, and many biologics are manufactured using biotechnology." The products discussed in Chapters 1 and 2 are mostly considered to be biologics. Biologics are regulated by the **Center for Biologics Evaluation and Research, CBER**. Substances that are presently regulated as biologics include but are not limited to:

- **Monoclonal antibodies** for therapeutic use.
- **Proteins** intended for therapeutic use, including cytokines (e.g., interferons), enzymes (e.g., thrombolytics), and other novel proteins. This category includes therapeutic proteins derived from plants, animals, or microorganisms, and recombinant versions of these products.
- **Antitoxins, antivenins, and venoms**.
- **Blood, blood components, and plasma-derived products**.
- **Vaccines**.
- **Human tissue for transplantation**.

BOX 35.1   THE ORGANIZATION OF THE CODE OF FEDERAL REGULATIONS

The Code of Federal Regulations (CFR) contains the complete and official text of the regulations that are enforced by US federal agencies. The CFR is organized as follows:

- The CFR is divided into 50 **titles** that represent broad areas subject to federal regulations.
- Each title is divided into **chapters** that are assigned to various agencies issuing regulations pertaining to that broad subject area.
- Each chapter is divided into **parts** covering specific regulatory areas.
- Large parts may be subdivided into **subparts**.
- Each part or subpart is then divided into **sections** – the basic unit of the CFR.
- Sometimes sections are subdivided further into paragraphs or subsections. Citations pertaining to specific information in the CFR will usually be provided at the section level.

An example of a typical CFR citation is *21 CFR 211.67(a)*. To interpret this:

- The number 21 is the CFR **title**. The broad subject area is "Food and Drugs."
- The number 211 is the **part**. Part 211 is entitled "Current Good Manufacturing Practice for Finished Pharmaceuticals."
- The number 67 refers to a particular **section**. This section is about "Equipment cleaning and maintenance."
- The (a) is the first paragraph or subsection within the section.

The areas of the CFR of primary interest to biotechnologists are Part 58, pertaining to GLPs; Parts 210 and 211, pertaining to pharmaceuticals; Part 820, pertaining to medical devices; Part 110 relating to the food industry; and Part 606 relating to the blood products industry.

There are several ways to access portions of the CFR using the Internet. The easiest way is to type the citation in the format shown in the example above into a search engine. The relevant portion should appear. It may also work to type in the name of the section or topic of interest, for example, "CFR equipment cleaning".

TABLE 35.1

Program Centers In the FDA

CDER, Center for Drug Evaluation and Research

Regulates prescription and over-the-counter drugs (including small molecules, synthesized compounds, and antibiotics).

CBER, Center for Biologics Evaluation and Research

Regulates biologics (e.g., vaccines, biopharmaceuticals, cellular products, blood and blood products, and allergens extracted from natural sources).

CDRH, Center for Devices and Radiological Health

Regulates medical devices (e.g., ultrasonic cleaners for medical instruments, diagnostic kits used in medical laboratories, and ventricular bypass devices) and many, but not all, *in vitro* diagnostic kits.

CFSAN, Center for Food Safety and Applied Nutrition

Promotes and protects the public health and economic interest by ensuring that foods are safe, nutritious, and honestly labeled.

NCTR, National Center for Tobacco Products

A center established in 2009 to implement the Family Smoking Prevention and Tobacco Control Act.

CVM, Center for Veterinary Medicine

Regulates the manufacture and distribution of food additives and drugs that will be given to animals.

- **Allergenic extracts** used for the diagnosis and treatment of allergic diseases and allergen patch tests.
- **Cellular products**, including products composed of human, bacterial, or animal cells (such as pancreatic islet cells for transplantation and stem cells).
- **Products used to diagnose, treat, and monitor persons with HIV and AIDS.**
- **Gene therapy products**.

Marketing new drugs requires FDA approval of a New Drug Application, NDA, or a New Animal Drug Application (NADA), or, for generic drugs, an Abbreviated New Drug Application (ANDA). These applications are reviewed by CDER. Marketing new biologics requires the company to file a Biologics License Application (BLA), which must be approved by CBER reviewers.

Medical devices are largely regulated by **the Center for Devices and Radiological Health (CDRH)**. There are a number of biotechnology companies that make diagnostic tools for detecting disease in humans; their products are regulated as devices.

There are also combination products. A **combination product** *is a product composed of any combination of a drug and a device; a biological product and a device; a drug and a biological product; or a drug, device, and a biological product.* For example, as discussed in Chapter 2, sometimes cells are grown on an engineered scaffold as part of a regenerative medicine therapeutic. The cells are a biological product, and the scaffold is a device; hence, this is a combination product. Prefilled delivery systems, such as an insulin injector pen, or an inhaler to treat asthma, are also combination products. There is an Office of Combination Products at FDA that classifies products and assigns them to an FDA center for review and regulation. Assignment to a center with primary jurisdiction for premarket review and post-market regulation, or a lead center, is based on a determination of the "primary mode of action" (PMOA) of the combination product. The PMOA is the part of the product that provides the most important therapeutic action of the combination. For example, if the PMOA of a device–drug combination product is attributable to the drug, then CDER would have primary jurisdiction for the combination product.

While the FDA is responsible for oversight of medical products, foods, and cosmetics in the United States, there are FDA counterparts in other countries, such as Health Canada in Canada and the Pharmaceutical and Food Safety Bureau in Japan. The European Medicines Agency (EMA) is a decentralized agency of the European Union that is the European Union's counterpart to the FDA.

35.4 THE REGULATION OF FOOD AND AGRICULTURE IN THE UNITED STATES

35.4.1 FOOD SAFETY

Ensuring safe food has been recognized as a major regulatory issue for more than 100 years, as evidenced by the passage of the original FD&C Act in 1906. Despite this, food-related illness and death are still common. The Centers for Disease Control and Prevention estimates that each year in the United States there are 48 million cases of foodborne illness, causing 3,000 deaths.

Major issues relating to food safety include contamination by pathogens, such as bacteria and parasites; possible adverse effects of intentional food additives, such as coloring agents and sweeteners; adverse effects of unintentional additives, such as pesticide residues and metals; and allergens (e.g., peanuts and peanut derivatives) that can cause serious risk to affected individuals.

In the past, food safety efforts mainly consisted of a system of inspectors who inspected slaughtered livestock in an effort to detect diseased animals and the presence of chemical residues. This testing, however, did not prevent contamination from occurring, nor did it detect all contamination that had occurred. The federal government, therefore, mandated that meat and poultry producers implement quality systems based on the same principles as ISO 9000 and GMP, that is, that quality cannot be inspected into the final product. Quality must instead be "built into" the product throughout its processing. In 2011, the **Food Safety Modernization Act (FSMA)** was signed into law, *giving the FDA greater power to monitor prevention of food safety problems instead of focusing on reaction to illnesses.* FSMA requires all food facilities to have a written food safety plan (called a HARPC; Hazard Analysis and Risk-Based Preventive Controls) based on an analysis of all hazards and designating prevention or minimization methods for each of these hazards. For example, hazards include the presence of pathogens, antibiotics, pesticide and hormone residues, additives, and foreign matter. To ensure appropriate compliance to the new standard, the FDA requires the food safety plan be created and implemented by a preventive controls-qualified individual (defined by FSMA). Particular emphasis is placed on monitoring food throughout the supply chain so that hazards may be mitigated. Efforts begin with farmers and other suppliers and extend through all activities involved in bringing food to the consumer.

35.4.2 INTRODUCTION TO REGULATORY AGENCIES AND RELEVANT LEGISLATION

Food safety, as relates to microbial contaminants, is clearly a major concern. But, while consumers need food to be safe and uncontaminated, they also want food to be abundant, nutritious, diverse, and economical. These latter objectives are often the result of innovations, such as pesticides that make food more abundant and less expensive, or food additives, such as nonnutritive (artificial) sweeteners, which make it more diverse. Biotechnology methods can also lead to innovations in food products. However, innovations sometimes lead to concerns about the safety of food, either to the consumer or to the environment. This has been the case for pesticides, nonnutritive sweeteners, and biotechnology-derived products. In the United States, there are several government agencies that divide responsibility for dealing with the various regulatory issues that arise relating to agriculture and agricultural innovations.

As we have already mentioned, the FDA has a major role in ensuring food safety. The FDA's regulatory authority for food, drugs, and cosmetics comes from the Food and Drugs Act of 1906 and the FD&C Act of 1938, as well as from other acts that relate to such issues as nutritional labeling and infant formula. Like the centers that regulate medical products, the FDA's **Center for Food Safety and Applied Nutrition (CFSAN)** includes chemists, nutritionists, microbiologists, and toxicologists so that its work has a scientific basis. The FDA has the primary responsibility for regulating food additives and new food products, except meats and poultry. As is the case for drugs, new food additives must be tested for safety and then be approved by the FDA before they are marketed; however, there are exceptions for materials that are generally recognized to be safe. The FDA is also responsible for ensuring that food products are produced properly and are not adulterated.

In addition to the FDA, **the United States Department of Agriculture (USDA), the Environmental Protection Agency (EPA), and the Centers for Disease Control and Prevention (CDC)** are also empowered to play roles in the regulation of food and agriculture (Tables 35.2 and 35.3). The USDA is responsible for the safety of meat, poultry, and eggs. The USDA also publishes voluntary standards for grading foods according to their quality. The EPA regulates pesticides used in agriculture and substances released to the environment. The CDC is responsible for disease-related surveillance and controls.

TABLE 35.2

Primary Federal Agencies Involved in the Regulation of Food and Agriculture in the United States

The United States Department of Agriculture (USDA)

The USDA enforces requirements for purity and quality of meat, poultry, and eggs, and it is involved in nutrition research and public education. The USDA regulates genetically engineered food plants through its division of Animal and Plant Health Inspection Service (APHIS).

Food and Drug Administration (FDA)

The FDA's Center for Food Safety and Applied Nutrition (CFSAN) is responsible for ensuring the safety and purity of all foods sold in interstate commerce except for meat, poultry, and eggs. (Foods that are sold only within a state are regulated by that state.) The FDA also regulates the composition, quality, and safety of food and color additives.

Environmental Protection Agency (EPA)

The EPA regulates pesticides, sets tolerance limits for pesticide residues in foods (which the FDA enforces), publishes directives for the safe use of pesticides, and establishes quality standards for drinking water. The EPA is also involved in the regulation of environmental releases of genetically modified organisms.

Centers for Disease Control and Prevention (CDC)

The CDC works with local, state, and federal partners to investigate outbreaks, and to implement systems to better detect, stop, and prevent them. The agency uses data to evaluate and revise foodborne disease prevention strategies and policies.

TABLE 35.3

Major Federal Legislative Acts Relating to the Regulation of Food and Agriculture

Food, Drug, and Cosmetic Act (FD&C Act) 1938

Gives the FDA the authority to regulate the safety and wholesomeness of foods and food additives.

Federal Meat Inspection Act and the Poultry Products Inspection Act 1957

Authorizes the USDA to ensure that meat and poultry products are safe, wholesome, and accurately labeled.

The Federal Insecticide, Fungicide, and Rodenticide Act (FIFRA) 1947

Makes EPA responsible for regulating the distribution, sale, use, and testing of pesticides; later determined to include pesticides that are the result of genetic modifications to organisms.

The Federal Plant Pest Act (FPPA) 1957

Regulates the introduction or release to the environment of "plant pests"; that is, any organisms, materials, or infectious substances that can cause damage to any plants or to products of plants. As such, the FPPA encompasses genetically engineered organisms if they might be plant pests. This act gives APHIS (part of USDA) the authority to regulate such introductions.

Toxic Substances Control Act (TSCA) 1976

The TSCA is intended to control chemicals that may pose a threat to human health or to the environment. The act requires reporting and record-keeping by chemical manufacturers and sets requirements for notifying the EPA when new chemicals are introduced. The provisions of this act primarily affect chemical manufacturers and their R&D laboratories; however, the act also authorizes the EPA to review new chemicals before they are introduced. Under this provision, the EPA has established a program to review microbial products that result from the introduction of genes from one type of bacterium into another type. EPA's rationale for this program is that microorganisms engineered to contain genes from another organism are new "chemicals." This is relevant, for example, if genetically engineered microorganisms are introduced into fields to control pests, or if they are introduced into the environment for purposes of cleaning contaminated soil or water.

Food Allergen Labeling and Consumer Protection Act (FALCPA) 2004

An amendment to the Federal Food, Drug, and Cosmetic Act that requires labeling of a food that contains an ingredient that is, or contains protein from, a "major food allergen."

The Food Safety Modernization Act (FSMA) 2011

Gives the FDA greater power to monitor prevention of food safety problems instead of focusing on reaction to illnesses. See text for more information.

35.4.3 RECOMBINANT DNA METHODS AND THE FOOD INDUSTRY

35.4.3.1 Safety

The first recombinant DNA food ingredient was approved by the FDA in 1990. It is an enzyme, chymosin (also called rennin), that is produced in *E. coli* transformed with the bovine chymosin gene. Rennin is the main enzyme used to make milk clot during the production of cheese. The traditional source of rennin is rennet, which is isolated from the fourth stomach of unweaned calves, which are killed for veal. The availability and price of rennet fluctuates depending on the availability of veal.

The FDA's concerns regarding the approval of rennin made by modified bacteria were the same as when a drug product made by recombinant DNA methods is introduced to supplement or replace a drug isolated from "natural" sources. The first concern is to establish

that the recombinant product is identical to the natural product, or that any differences have no effect on its safety or effectiveness. To demonstrate this for chymosin, the company performed a variety of identity tests both on the gene itself and on the resulting protein. They also performed activity assays that showed that the recombinant chymosin clotted milk. A second aspect of approving the recombinant DNA-derived chymosin was ensuring that the preparation was safe. As with recombinant drug products, this involved demonstrating that the processing steps removed contaminants, such as DNA and protein derived from the host bacterial cells. Researchers also demonstrated that the bacterial cells used in production were themselves nonpathogenic. Once the company producing the chymosin had demonstrated its equivalence to traditional rennin and its safety, the new version was approved.

The FDA's general policy is currently that if chemical, microbiological, and molecular biology studies

TABLE 35.4

External Forces That Impact the Regulatory Environment

- *Scientific discovery*. New discoveries and their applications result in new regulatory and potentially new ethical challenges (e.g., stem cell technologies as discussed in Chapter 2).
- *Patient advocacy groups*. Patients (and their caregivers) are often involved in the development of new medical products and often serve on ethics boards.
- *Policy groups*. For example, PhRMA is an organization representing the leading biopharmaceutical research companies in the United States. It is involved in the development of public policy relating to pharmaceutical development.
- *Globalization*. We live in a global environment. Changes in policy in one country may influence policy in another country. In this chapter, we introduced the ICH, an organization that helps to coordinate these changes and accelerate harmonization in the regulatory sphere.
- *Pressure for accelerated treatments*. There is pressure for treatments to be brought to market faster with limited resources. Part of the response to this pressure has been an emphasis on risk evaluation in the regulatory process, as discussed in this chapter. As we will discuss in Chapter 36, accelerated regulatory processes have been developed to respond to urgent needs, as, for example, the COVID-19 pandemic.
- *Public interest groups*. Consumers are increasingly engaged in regulatory affairs, perhaps as a result of the increase in readily available information through the Internet.

show that a food or additive is the same as, or substantially equivalent to, an accepted food-use product, then minimal extra testing or regulatory surveillance is required. A special review of a genetically engineered food product is required only when special safety issues exist. For example, a special review is required if a biotechnology product can cause allergic reactions, or has a different nutritional value than a traditional product.

35.4.3.2 Environmental Release of Genetically Modified Organisms

A number of transgenic plants have entered the market or have been approved for field testing. There are transgenic plants that can tolerate herbicides, resist insects or viruses, or produce modified fruits and flowers. For example, insect-resistant corn plants have been created by inserting into them a bacterial gene that codes for an insecticidal protein, as discussed in Chapter 1. There are also genetically modified bacteria that have applications in agriculture (e.g., to provide nitrogen to plants). Genetically modified microorganisms are similarly being investigated to clean polluted water and soil.

There are concerns that there will be adverse effects if organisms genetically modified by the methods of biotechnology are released into the environment (as occurs when a modified crop is grown in a field or when modified microorganisms are applied to the soil). For example, it is postulated that transgenic plants could cause the development of new, hardy weed species, or contribute to the loss of species diversity. It is feared that insects and other pests will become increasingly resistant to control agents because of the use of genetically modified crops. The regulation of field testing and planting transgenic crops is the responsibility of APHIS and the EPA. Their task has been difficult and particularly controversial. Regulatory agencies will presumably be able to establish increasingly effective policies as they gain more experience with these genetically modified organisms in the environment.

35.5 EVOLUTION OF THE REGULATORY ENVIRONMENT

The regulatory environment is always changing in response to unfolding events and external forces, as we have seen in the stories and discussions in this chapter; see also Table 35.4 and Figure 35.6.

The case study "Regulations Must Continue to Evolve" provides an example of the interaction between external events and regulatory change.

Case Study: Regulations Must Continue to Evolve

Cannabis (marijuana) and products derived from cannabis, once associated with flower children and black light posters, have entered mainstream America. As of August 2018, cannabis derivatives are legally available in 33 states to treat medical conditions; in some states, recreational marijuana is also legal.

(Continued)

Case Study (*Continued*): Regulations Must Continue to Evolve

Millions of people are producing, ingesting, and applying cannabinoid derivatives. Cannabis, however, is presently trapped in a regulatory quagmire: At the federal level, cannabis is a Schedule 1 drug, which means that it is a criminal act to "manufacture, distribute, or dispense, or possess with intent to manufacture, distribute, or dispense [it]." This classification has a serious consequence for the consumer; even if cannabis is legal in their state, the federal government is not regulating the quality and safety of these products. Various states have tried to fill the void with their own regulations, but there is inconsistency around the country. This is a common pattern in drug regulation; a change occurs, controversy ensues, and the government struggles to respond. Eventually, the regulatory system adjusts to the new situation.

What are major safety issues associated with cannabis products? Consider that cannabis is grown as a crop; it is not produced in a manufacturing facility. Growers therefore face the same problems as those growing food crops, including contamination of the plants with pesticides and mycotoxins. **Mycotoxins** *are metabolites produced by some molds* that readily colonize crops and can cause disease and death to humans and animals ingesting them. Mycotoxins are some of the most potent cancer-causing agents known. The FDA strictly regulates the levels of mycotoxins and pesticides in foods and animal feeds, so most consumers are never aware of these risks. But cannabis is not regulated by the Food and Drug Administration (FDA) and consumers are not protected in the same way as we are when purchasing food products.

Even though, at the time of writing, the regulatory status of cannabinoids is not sorted out, consumers, producers, scientists, equipment manufacturers, testing companies, and accreditation agencies have found that cannabis provides opportunity. In addition to cannabis growers and producers, there are dozens of companies making products to serve the industry. Scientists and analysts are developing assay methods to test cannabis products for pesticides, mycotoxins, and other contaminants while companies that make analytical equipment are launching new products for this burgeoning analytical industry.

Assaying cannabis for contaminants turns out to be difficult because the plant naturally contains over 400 chemical compounds that interfere with analysis. New testing methods to overcome the problems are being introduced; an example of a result of a new assay method is shown in Figure 35.7.

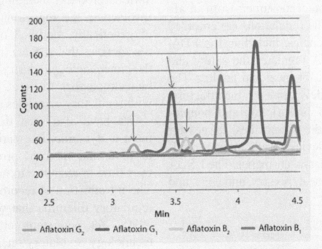

FIGURE 35.7 An analytical method to detect aflatoxin contamination in cannabis products. Aflatoxins are dangerous mycotoxins produced by certain fungal species. It is necessary to have an assay method that reliably detects them. This figure illustrates a method that uses analytical instrumentation in such a way that the different types of aflatoxins appear as peaks on a printout. (From Shimelis, Olga I., et. al. "Quick, Sensitive LC/MS/MS Analysis of Aflatoxins in Cannabis." *Reporter US*, vol. 34, no. 2, 2016. Reproduced with permission from Merck KGaA, Darmstadt, Germany, and/or its affiliates.)

(Continued)

Case Study: (*Continued*) Regulations Must Continue to Evolve

How can cannabis consumers protect themselves from dangerous contaminants? The most obvious way is to ask their retailer for a certificate of analysis (COA) showing that a product has been tested for contaminants and found to be safe. But it turns out that while this is a good idea, it is not enough – there must be confidence that the testing was performed properly. There is evidence that some testing laboratories tweak their data so that producers can say that their product is safe, when perhaps it is not. This is where laboratory certification, for example, in compliance with the ISO standard 17025, can be valuable. Certified laboratories must meet quality requirements and pass inspections. Several organizations and companies provide assistance to help cannabis testing laboratories become ISO 17025 compliant.

Cannabis is a rapidly changing industry, and by the time you read this, changes in the regulatory situation will almost certainly have occurred. You will be able to look back and see how the regulatory process evolved in response to a new situation. In the meantime, savvy consumers may develop a new awareness of the critical, and often overlooked, role of laboratory analysts in protecting us from harmful products.

Primary Sources

Goldman, Stephen and Nolte, Andrea. "Mycotoxins." *Marijuana Venture: The Journal of Professional Cannabis Growers and Retailers*, 27 December 2017. www.marijuanaventure.com/mycotoxins/.

Borchardt, Debra. "Cannabis Lab Testing Is the Industry's Dirty Little Secret." *Forbes*, 5 April 2017. www.forbes.com/sites/debraborchardt/2017/04/05/cannabis-lab-testing-is-the-industrys-dirty-little-secret/#25d5d2xx1220.

L'Heureux, Megan. "Testing for Pesticides and Mycotoxins in Cannabis: How to Meet Regulatory Requirements." *Cannabis Science and Technology*, 6 August 2018. www.cannabissciencetech.com/cannabis-voices/testing-pesticides-and-mycotoxins-cannabis-how-meet-regulatory-requirements.

Zimmer, Katrina. "Hazy Regulations." *The Scientist*, March 2020, pp. 52–55.

35.6 SUMMARY

In this chapter, we have seen that regulations have evolved over the years to protect consumers from unsafe and ineffective products. The regulations are enforced by various government agencies, including the FDA, USDA, and EPA. The FDA regulates the safety and labeling of all foods except for meat and poultry. The FDA also regulates the development, manufacture, marketing, and labeling of prescription and nonprescription drugs, blood products, vaccines, tissues for transplantation, medical devices, radiological products, animal drugs and feed, and cosmetics. The USDA enforces requirements for meat, poultry, and eggs, and

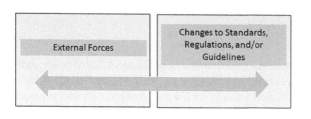

FIGURE 35.6 The regulatory environment evolves. External forces in science and society create the need for changes to standards, regulations, and/or guidelines. In turn, regulations, standards, and guidelines impact the biotechnology industry and the general public.

also plays a key role in regulating genetically modified plants through its APHIS division. The EPA regulates various products including pesticides, and notably is involved in the regulation of environmental releases of genetically modified organisms. Regulations and the agencies that enforce them have a major impact on the biotechnology industry. It is therefore important for biotechnologists to be at least minimally conversant with regulatory principles and issues.

We have also seen that the regulatory environment is dynamic. Regulations and standards relating to food and medical products continuously evolve over time in response to new advances and problems. The introduction of recombinant DNA drugs led to a regulatory dilemma that was mainly resolved by the 1990s. Yet, new stem cell technologies brought biotechnology regulatory questions into the next millennium. Similarly, the legalization of cannabis products in the United States led to a regulatory conundrum that is not resolved at the time of writing. Beginning in 2019, a worldwide pandemic led to an urgent need for rapid deployment of test kits, vaccine development, and novel treatments, resulting in a greatly accelerated response from the FDA. No doubt there will be regulatory issues in the news that are relevant to you whenever you are reading this text.

Practice Problems

1. Match the following:
   a. CVM, Center for Veterinary Medicine
   b. CDER, the Center for Drug Evaluation and Research
   c. Food, Drug, and Cosmetic Act, FD&C Act
   d. CBER, Center for Biologics Evaluation and Research
   e. Center for Food Safety and Applied Nutrition (CFSAN)
      1. Promotes and protects public health by ensuring that foods are safe, nutritious, and honestly labeled.
      2. Regulates biologics (e.g., vaccines, blood and blood products, many biopharmaceuticals, and allergens extracted from natural sources).
      3. Regulates prescription and over-the-counter drugs (including small molecules, synthesized compounds, antibiotics, and some biopharmaceuticals).
      4. Law intended to ensure that manufacturers do not sell adulterated (contaminated) or misleadingly labeled food and drug products.
      5. Regulates the manufacture and distribution of food additives and drugs that will be given to animals.
2. What is ICH? What is its role?
3. Review the ICH website: https://www.ich.org/. The guidelines issued by ICH are divided into Q, S, E, and M.
   a. What do these letters stand for?
   b. There is a guideline on stability testing, which determines how the quality of a drug product varies with time under the influence of a variety of environmental factors such as temperature, humidity, and light. Is this guideline in the Q, S, E or M series?
   c. There is a guideline that contains medical terminology. Is this guideline in the Q, S, E or M series?
   d. There is a guideline to assess whether a new drug might induce cancer in patients. Is this guideline in the Q, S, E or M series?
   e. There are guidelines that relate to clinical trials that assess the effectiveness of new treatments. Are these guidelines in the Q, S, E or M series?
4. In the United States, what agency is responsible for "ensuring the safety, effectiveness, and reliability of medical products, foods, and cosmetics?"
5. What is the difference between CBER and CDER?
6. How are biologics different from drugs?
7. Use a web search engine to find the Code of Federal Regulations for the Current Good Manufacturing Practice for Finished Pharmaceuticals, 21 CFR 211. What regulation could be cited if:
   a. Expiration dating is not labeled on a product.
   b. The production area has lightbulbs that are burned out and not replaced.
   c. There are not enough samples reserved for analytical testing of a product.
      (Hint, look at the table of contents.)
8. The Food Safety Modernization Act (FSMA) was signed into law in 2011.
   a. Give two examples of hazards present in foods.
   b. Why is it important to monitor food throughout the entire food chain?

36 The Lifecycles and Regulation of Pharmaceutical Products

36.1 INTRODUCTION: WHAT ARE PHARMACEUTICALS AND BIOPHARMACEUTICALS?

This chapter continues the discussion of product lifecycles that was introduced in Chapter 3, and regulatory affairs, as introduced in Chapter 35. This chapter discusses pharmaceutical and biopharmaceutical products in general, while Chapter 37 focuses on biopharmaceuticals. Pharmaceuticals/biopharmaceuticals are of significant medical value, have a major worldwide economic impact, provide interesting and challenging career opportunities, and are associated with major advances in the biological sciences.

To begin with some terminology, in the regulatory literature, **pharmaceutical products** *are chemical agents with therapeutic activity in the body that are used to treat, correct, or prevent the symptoms of illnesses, injuries, and disorders in humans and other animals.* The familiar word, *drug*, is generally used synonymously with *pharmaceutical* although drugs include not only therapeutic compounds but also agents that are used in the body for nontherapeutic (e.g., "recreational") or harmful purposes.

Biopharmaceutical is often used (as it was in Chapter 1) to refer to a category of pharmaceutical products that are manufactured by genetically modified organisms (GMOs). These products are generally proteins. The term "biopharmaceutical" is also used more broadly, for example, to refer to any drug manufactured by living cells or organisms (whether or not recombinant DNA techniques are involved), and also to include whole cells or tissues used therapeutically.

Biopharmaceutical can also refer to any drug that is a large biological molecule; these are usually proteins, but can be RNA or DNA (e.g., DNA for gene therapy). As biotechnologists, we are interested in biopharmaceutical agents according to any of these definitions.

It is helpful to contrast large biological molecule drugs (i.e., biopharmaceuticals) with small-molecule drugs in order to better understand the characteristics of biopharmaceuticals (however one defines the term). To give a sense of the relative sizes involved, aspirin is an example of a small-molecule drug with a molecular weight less than 200. Rituxan is an example of a large biological molecule drug (a monoclonal antibody) with a molecular weight of about 145,000. Insulin, discussed in Chapter 1, is one of the smallest protein drugs with a formula weight of about 6,000; see Figure 36.1.

Small-molecule drugs are often chemically synthesized in factories. Large biological molecule drugs cannot at this time be efficiently chemically synthesized, with a few exceptions (e.g., DNA used for gene therapy). Large biological molecule drugs must usually be isolated from organisms or manufactured in carefully nurtured living systems (cells, plants, or animals) and/or enzyme-based systems. Large biological molecule drugs are more difficult, complex, and expensive to produce than small-molecule agents. Moreover, the exact structure and function of a biological molecule is likely to be dependent on the particular system in which it is produced, which means that ensuring consistency and potency from batch to batch is challenging. Biological molecules are destroyed by heat, so ensuring their sterility poses another technical

DOI: 10.1201/9780429282799-46

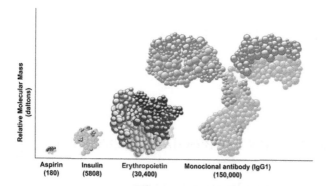

FIGURE 36.1 A small-molecule drug compared to several protein drugs. The approximate molecular weight of each drug molecule is indicated in the figure. Compared to insulin, or any other protein drug, aspirin is small. Aspirin's formula is $C_9H_8O_4$, meaning it has 21 atoms. Each sphere in the protein drugs represents an amino acid, composed of at least ten atoms. Insulin is a relatively small protein; erythropoietin is closer to average for a human protein. (From Mellstedt, H. "Clinical Considerations for Biosimilar Antibodies." *EJC Supplements*, vol. 11, no. 3, 2013, pp. 1–11. doi: 10.1016/S1359-6349(13)70001-6.)

challenge. The heat sensitivity of biological molecules also has implications relating to their stability during shipping and storage.

Although they are difficult and expensive to manufacture, protein therapeutics have the important advantage that they have a relevant, known activity in the body. Proteins are molecules that perform the work of cells. Insulin and growth hormone, for example, are proteins whose work is to carry messages between cells. It is therefore likely that a protein drug administered to play its normal role in the body (e.g., insulin administered to diabetics) will have a positive effect. In contrast, tens of thousands of small chemicals must be screened to find the one with a desired effect.

Small and large molecules tend to pose different concerns when administered to a patient. Proteins have the problem of being immunogenic. **Immunogenicity** *is the ability of an agent to stimulate an immune reaction.* Immunogenicity can be a problem in drug development in two ways. First, if a therapeutic agent is immunogenic, a patient may have a dangerous adverse (e.g., anaphylactic) reaction to it. Second, a patient may raise antibodies against the protein drug. The antibodies can bind to and inhibit the activity of the drug, thereby rendering it useless. A slightly different issue is that large biological molecules may not be toxic in the sense that a foreign chemical is toxic, but once administered to patients, they can be dangerous if their activity is exaggerated by the immune system. Organizations that develop biological drugs must evaluate their immunogenicity in tests on animals and human volunteers.

Small molecules, in contrast to large ones, are less likely to induce immunological effects. It is not unusual, however, for a small chemical agent to have off-target effects (i.e., it binds to unintended sites in cells with an undesirable effect). As the body breaks down small-molecule drugs, metabolites are generated that often have unintended consequences.

Yet another difference between small- and large-molecule drugs is that biological molecules are almost always destroyed in the gut by digestive enzymes, so they cannot be taken orally. This makes administering these drugs to a patient more difficult than administering small chemical drugs, which are typically taken by mouth. Some of the differences between small-molecule and protein pharmaceuticals are summarized in Table 36.1.

Despite the important distinctions between small pharmaceutical agents and large biopharmaceutical agents, the basic outline of their regulatory lifecycles is the same. For that reason, Chapter 36 is a discussion

TABLE 36.1
Comparison of Small Chemical and Protein Pharmaceuticals

Small-Molecule Therapeutic Agents	Large-Molecule, Protein Therapeutic Agents
Generally chemically synthesized	Produced by cells
Relatively inexpensive to manufacture	Expensive to manufacture
Relatively homogeneous	Production system may result in heterogeneous molecules
Relatively easily characterized	Complex and expensive to characterize
Typically enter cells and act on intracellular targets	Often act on cell surface receptors
Have less selectivity in their actions, therefore more side effects	Selective action with fewer side effects
Usually administered orally	Are destroyed in digestive tract; cannot be administered orally
Immunogenicity usually not a significant problem	Immunogenicity is a major consideration

of the lifecycle of pharmaceutical products in general. Chapter 37 turns to topics specific to the development stage of large biological molecule drugs and cell-based therapeutics that are manufactured using biotechnology methodologies.

36.2 THE PHARMACEUTICAL LIFECYCLE

36.2.1 Lifecycle Overview: An Expensive Process Punctuated by Important Milestones

The lifecycle of modern pharmaceuticals is a long one. Pharmaceuticals begin in research laboratories from which emerges a candidate drug substance, also called the **active pharmaceutical ingredient (API)**, *a compound that is thought to have therapeutic activity in the body*. The potential drug then enters a prolonged development period. The goal of the drug development process is to turn this promising drug candidate into a product for manufacture and sale. While the goal is straightforward, drug development is complex. It is dangerous to administer a poorly designed or improperly manufactured drug to a patient. Potential drugs therefore go through years of testing to ensure their safety and efficacy. The manufacturing processes for pharmaceuticals are similarly honed over years to ensure that they produce a consistent, safe, uncontaminated product.

The government regulates the development and then the manufacturing of pharmaceuticals to ensure that they are thoroughly tested, and properly manufactured. This oversight is provided by the Food and Drug Administration (FDA) in the United States and similar agencies in other countries. The FDA exerts its authority partly by controlling "gates" at key milestones in the pharmaceutical lifecycle. The interaction of regulatory agencies with the pharmaceutical product lifecycle contributes to their safety – and also poses challenges for companies that develop them.

Figure 36.2 is an outline of the lifecycle of a pharmaceutical product showing its important milestones. The research and discovery stage results in the identification of a promising possible drug agent, the first milestone. The development phase begins in laboratories and animal testing facilities. Early development is called the **preclinical** *stage because it comes before testing in humans*. If the candidate drug performs well in the preclinical testing, then the company can file with the FDA an **Investigational New Drug Application (IND)** *which requests permission to begin testing the substance in human subjects*. If the FDA grants the application, then another important milestone is passed and the product enters the **clinical development stage**, *where the product is tested in humans*. If the drug is successful in clinical testing, then the company can submit to the FDA a **New Drug Application (NDA)** *which requests permission to begin marketing the drug*. If the FDA approves the product, then a major milestone is passed, and the post-approval stage of the product's lifecycle begins. After approval, the company can market the product. Post-approval, the FDA will continue to interact with the company. The FDA periodically inspects the facilities of pharmaceutical manufacturers to be sure they are producing quality products in accordance with regulations. With additional post-approval studies, the company may be able to apply for label expansion, which means the drug can be used to treat more than one disorder.

The company or institution that is developing the drug is the **drug sponsor,** but the sponsoring organization is unlikely to perform all the tasks of development. Various development functions are often performed by **contract research organizations (CROs),** *which are*

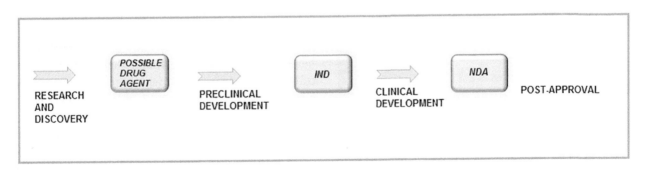

FIGURE 36.2 Overview of the lifecycle of a pharmaceutical product. In the United States, the FDA controls "gates" at critical milestones in the lifecycle. Similar agencies in other countries provide oversight of drug development and manufacture.

outside organizations paid to perform particular tasks.
It is common, for example, for the sponsor to contract
with a testing company that has expertise in animal
testing, or human testing. Smaller biotechnology companies often contract out development tasks ranging
from laboratory testing, through animal and human
testing, and into manufacturing. Even extremely large
pharmaceutical companies often contract with testing
companies for specialized services.

Drug development is costly. There is some debate
as to the actual cost of developing a new drug, but
estimates commonly are $1 billion and higher (e.g.,
Wouters, Olivier J., et al. "Estimated Research and
Development Investment Needed to Bring a New
Medicine to Market, 2009-2018." *JAMA*, vol. 323,
no. 9, 2020, p. 844. doi:10.1001/jama.2020.1166). It is
common to require 10–15 years from the time research
begins to the time a new pharmaceutical product
receives approval from the FDA. A major reason why
drug development is so costly is that the vast majority
of potential drugs fail (Figure 36.3). Although some
analysts dispute that drug companies spend as much as

a billion dollars to create a new drug, by any estimate
the cost of drug development is substantial in terms of
time and dollars spent.

36.2.2 THE BEGINNING OF THE LIFECYCLE: RESEARCH AND DISCOVERY

There are many ways by which therapeutic compounds
are discovered and/or designed; just a few examples
will be described here. (See also Chapters 1 and 2.)
One method, used since ancient times, is to search
for naturally occurring substances, particularly from
plants, which have potency for treating illness and
injury. The word *drug* is derived from an old Dutch
word *droog*, which means *dry*, since in the past, most
drugs were made from dried plants. Scientists now
are often able to identify and manufacture the active
ingredient in a traditional drug. Aspirin is a derivative of a natural product first isolated from willow bark
and used as a remedy more than a thousand years ago.
The active agent in remdesivir, an antiviral drug that
is being used to treat COVID-19, was first discovered
in a small marine creature, the sea sponge. The sea
sponges use this agent to protect themselves against
viruses and bacteria in the ocean. The search continues
today for therapeutic agents that can be harvested from
natural sources and are present in traditional remedies.
There are biotechnology companies that screen for
therapeutic compounds from such interesting sources
as poisonous fish, snakes, and insects; rainforest plants
and animals; and marine organisms.

Modern approaches to drug discovery are usually
based on a more scientific understanding of a disease
or injury process than was possible for our ancestors.
Sometimes medical researchers discover that a disease
occurs when a protein is missing in the patient (e.g.,
diabetes and hemophilia). In these cases, it is straightforward to imagine treating the disease by supplying
patients with the missing protein, either isolated from
natural sources or manufactured in genetically modified cells.

Researchers who study the roles of proteins in
health and disease have found that a protein often
exerts its effect in a cell, whatever that effect might
be, by binding to another molecule. In a normal cell,
this protein binding promotes the cell's health. In a
damaged cell (e.g., a cancer or an infected cell), the
binding of one protein to another molecule may trigger a harmful effect. Scientists reason that it might
be possible to ameliorate a disease if the binding
of involved proteins can be blocked or modified in
afflicted cells. Herceptin, discussed in Chapter 1, is

FIGURE 36.3 Failure contributes to the high cost of
drugs. Developing new drugs is a huge gamble with high
stakes. The vast majority of possible drug candidates fail at
some point in the drug development lifecycle, as illustrated
in this cartoon from the cover of *Science*. (Illustration by
Stephen Wagner. From *Science*, vol. 309, no. 721, 2005.
Reprinted with permission from AAAS.)

an example of a monoclonal antibody drug that was designed to block a disease process by binding to an involved protein.

Statins are a class of small-molecule drug that prevent disease by targeting, binding, and thereby blocking an enzyme. Medical researchers found that high levels of cholesterol in the blood are correlated with an elevated risk of heart disease. They reasoned that if they could find a drug to lower cholesterol, they could reduce the incidence of heart disease. Statins were discovered by an intensive search of thousands of fungal microorganisms for naturally occurring compounds that bind and inhibit an enzyme necessary for cholesterol production. Chemical modifications of the fungal compounds resulted in statin drugs that are used to lower cholesterol levels in millions of people. Statins are therefore drugs that are based on a scientific understanding of a protein's function, combined with the technique of searching natural sources for a useful compound.

Pharmaceutical companies have a sophisticated strategy for searching for new small-molecule drugs that have a desired effect in treating a particular disorder. **High-throughput screening (HTS)** *enables scientists to simultaneously test thousands of small chemical compounds for efficacy in a particular disease model.* First, researchers discover a specific protein that triggers detrimental effects in a disease; this is called the "target." Then researchers use HTS to search for a compound that binds and modifies the target with a desired effect. Pharmaceutical companies own "libraries" that often contain more than a million distinct chemical compounds. In high-throughput screening, scientists test all of these compounds (or a subset) in an automated laboratory test to see if the compounds interact with the target. HTS methods can typically test in excess of 100,000 compounds in a day, enabling a company to screen its entire library against a single target in less than 2 weeks. In order to do so, however, the company must have an assay that has been proven to reproducibly and accurately detect the interaction of the compound with the target. Developing a suitable assay can take months. (See Chapter 26 for a discussion of assay development.) The outcome of HTS is the identification of one or more "hits" – molecules that have activity against the desired target.

To be a useful drug, an agent must be sufficiently stable in the body to allow it to reach the target site in sufficient quantity, and for a long enough time to have a biologically relevant effect. The drug must be safe and not have side effects that would prevent its use for the intended disease. It also must have properties such that it can be manufactured and formulated in a practical manner. Often promising molecules identified in early research studies, such as HTS, have activity, but lack these additional requirements. Potential drug molecules therefore go through a process where they are assessed for activity, ability to bind only to the target of interest (selectivity), chemical attributes, solubility, and so on. The original "hit" molecules are then chemically modified and re-assayed. The relationships between various chemical modifications and the activity of the molecules in the assay are assessed. If several molecules are found that interact with the target, and if they share some common chemical features, then a new chemical(s) with those specific features might be synthesized. This iterative process leads to a better understanding of the molecules and the slow optimization of a particular candidate with desirable drug properties that can be further assessed in preclinical testing (Figure 36.4).

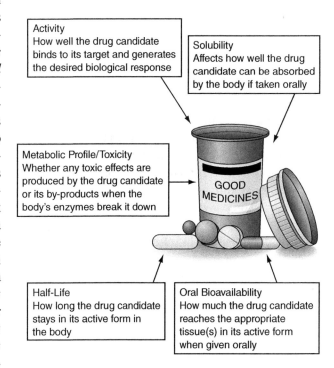

FIGURE 36.4 A good drug. This cartoon from a publication by the U.S. National Institutes of Health shows that a good drug combines a number of attributes, including the best possible activity, solubility, bioavailability, half-life (the time in the body required for the concentration of the drug to decrease to half of its starting dose), and metabolic profile. Attempting to improve one of these factors often affects other factors. For example, if scientists structurally alter a lead compound to improve its activity, they may also decrease its solubility or shorten its half-life. The final result must always be the best possible compromise.

Case Study: The Discovery of Gleevec

The first commercial success of high-throughput screening (HTS) is often considered to be the small-molecule drug Gleevec, which is used to treat chronic myeloid leukemia (CML). The roots of Gleevec go back to basic research performed in the 1970s by scientists who were studying viruses that cause leukemia in animals. Viruses are not known to play a role in causing most common human cancers, but researchers hoped to learn about the fundamental mechanisms of cancer initiation by studying animal models. These animal studies led to the important insight that viruses that cause cancer do so by "taking over" certain normal genes in the animals' cells. The viruses cause the genes to produce too much of their protein product which, in turn, eventually causes the cells to divide malignantly. The gene that is overexpressed in a particular mouse leukemia is called abl.

Meanwhile, in the 1960s and 1970s, other researchers discovered that a type of human cancer, CML, occurs when pieces of two different chromosomes break off and each piece reattaches to the opposite chromosome. This rearrangement fuses part of a specific gene from chromosome 22 (the bcr gene) with part of another gene from chromosome 9 (the abl gene). The resulting chromosome 22 is termed the Philadelphia chromosome (Figure 36.5). The activity of the protein normally encoded by the abl gene is tightly controlled and limited by the cell. In contrast, the protein produced by the fused bcr–abl gene is overexpressed, eventually causing the cell to divide malignantly.

The work on animal viral cancer and CML merged in 1982 when researchers discovered that abl, the gene that is overexpressed in mouse leukemia, has a counterpart in humans – the same gene that is overexpressed in CML. In mouse cancer, the gene is overexpressed because of a viral infection; in human cancer, the gene is overexpressed because of the breakage and fusion of the chromosomes. The cause is different, but the malignant result is the same.

The abl gene in both mice and humans encodes an enzyme made in blood cells. This enzyme is a key component of a normal signaling pathway that tells a blood cell it is time to divide. When the abl gene is overexpressed, it makes too much of the encoded enzyme, which turns on the signaling pathway and tells the cell to divide – over and over again – resulting in cancer.

Research in the 1980s showed that the abl enzyme must modify another protein, its substrate, in order to trigger the signaling pathway that tells the cell to divide. This knowledge set the stage for a search for a drug compound that would target the abl enzyme and block it from modifying its substrate.

Pharmaceutical companies initially were unenthused about searching for a CML drug because this is a relatively rare cancer and a drug to treat it seemed unlikely to be profitable. However, a large pharmaceutical company, Novartis, was performing high-throughput screening assays looking for small chemical compounds that would block other signaling proteins that are important in common cancers. They found a drug compound, ST1571, that binds to the abl enzyme (Figure 36.6). Novartis scientists gave some of the

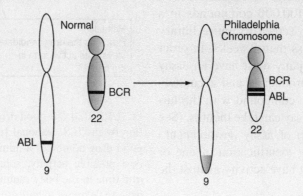

FIGURE 36.5 The Philadelphia chromosome. A rearrangement of genetic material between chromosomes 9 and 22 fuses part of a specific gene from chromosome 22 (the *bcr* gene) with part of another gene from chromosome 9 (the *abl* gene). The resulting chromosome 22 is termed the *Philadelphia chromosome*. The expression of the protein normally encoded by the *abl* gene is tightly controlled by the cell, but the protein produced by the fused *bcr–abl* gene is overexpressed.

(Continued)

Case Study (*Continued*): The Discovery of Gleevec

compound to Dr. Brian J. Druker, a university researcher. Druker tested the compound in cells and animals and found that it was indeed effective in blocking the activity of the CML abl enzyme. Druker's success in early testing led to an exploratory clinical trial with a small number of CML patients, which demonstrated that the compound had an effect in treating cancer in humans. Based on this early promising work, Novartis was convinced to continue to develop ST1571. The drug performed well in further testing, and it has few side effects compared to traditional chemotherapeutic agents. ST1571 was approved for patient use in 2001 in the United States where it is marketed as Gleevec.

Gleevec illustrates some common themes of drug discovery. First, the discovery of Gleevec had its roots in years of earlier basic research conducted by scientists in many laboratories, mostly in universities. Second, Gleevec illustrates the value of studying animal models to understand the basic principles of health and disease. The work on animal models first showed how genetic changes lead to cancer. Third, the discovery and development of Gleevec required academic researchers, medical researchers, scientists at Novartis, patient advocacy groups (who, at a certain critical juncture, encouraged Novartis to continue the development of Gleevec), and businesspeople, all of whom played a role in the discovery and development of the drug. The development of drugs and other medical products is usually so challenging that the efforts of many people and organizations are required. Fourth, the development of most biotechnology products is influenced by the potential of the product to be commercially successful. This has led to efforts by governments and organizations to find strategies to promote the development of drugs that are unlikely to be highly profitable, but are necessary to treat life-threatening diseases. *Drugs that are used to treat rare disorders, defined as affecting fewer than 200,000 Americans*, are called **orphan drugs**, and the markets are correspondingly small. The Orphan Drug Act, which was enacted in 1983 in the United States, provides tax relief and some marketing exclusivity for companies that develop drugs for less-common diseases.

Primary Sources

Waalen, Jill. "Gleevec's Glory Days." *Howard Hughes Medical Institute Bulletin*, vol. 14, no. 5. December 2001, pp. 10–15. https://www.hhmi.org/sites/default/files/Bulletin/2001/December/dec2001_fulltext.pdf.

Savage, David G., and Karen H. Antman. "Imatinib Mesylate — A New Oral Targeted Therapy." *New England Journal of Medicine*, vol. 346, no. 9, 2002, pp. 683–93. doi:10.1056/nejmra013339.

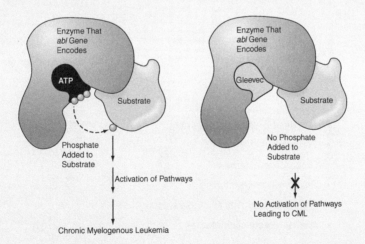

FIGURE 36.6 How Gleevec works. On the left, the enzyme encoded by the *abl* gene is shown bound to the molecule adenosine triphosphate (ATP). The enzyme transfers a phosphate from the ATP molecule to the substrate. The addition of the phosphate to the substrate leads to the activation of pathways that cause too many white blood cells to be made in the bone marrow, causing CML. Gleevec takes the place of ATP in the pocket of the enzyme. When Gleevec is present, phosphate is not transferred to the substrate and subsequent events that cause CML do not occur. (Modified from information in Goldman, John M., and Junia V. Melo. "Targeting the BCR-ABL Tyrosine Kinase in Chronic Myeloid Leukemia." *New England Journal of Medicine*, vol. 344, no. 14, 2001, pp. 1084–86. doi: 10.1056/nejm200104053441409.)

36.2.3 Preclinical Development

36.2.3.1 The Objectives of Preclinical Development

Whatever the means of its discovery, the identification of a candidate drug signals the beginning of the complex development process (Figure 36.7). The tasks of preclinical development include the following:

- Determining the physical and chemical properties of the candidate drug compound.
- Testing the candidate drug in vitro.
- Determining how to formulate the candidate drug for administration to test subjects and patients.
- Developing manufacturing methods for the candidate drug.
- Testing the candidate drug in cultured cells.
- Testing the candidate drug in animals to study its effects in the body and its safety.
- Developing analytical assays to test for the presence of the drug compound, its metabolites, activity, and other features.
- Securing intellectual property protection for the potential product, its uses, and its manufacture.

36.2.3.2 Product Development

Product development *is the process in which R&D scientists explore and optimize the features of a potential product.* This includes **product characterization**, *the process of studying a therapeutic compound to learn its structure, its interactions with other molecules, and its function.*

Any drug product needs to balance a number of key characteristics. Among these is the requirement that the drug can be formulated in such a way that it can be administered to a patient. A drug **formulation** *is the form in which a drug is administered to patients including its chemical components, stabilizers, the system of administration (e.g., pill or injection), and the mechanism by which the drug is targeted to its site of action.* Developing the formulation is part of product development that begins in the preclinical stage and may continue to be refined until the product is approved for sale. Formulation for large biological molecules means preparing the compound in such a way that it can be injected or otherwise introduced to the body. Small-molecule drugs usually must be soluble or must be chemically modified so that they are soluble.

During product development, studies will be conducted to determine the **stability** of the formulation, *its capacity to remain within its specifications over*

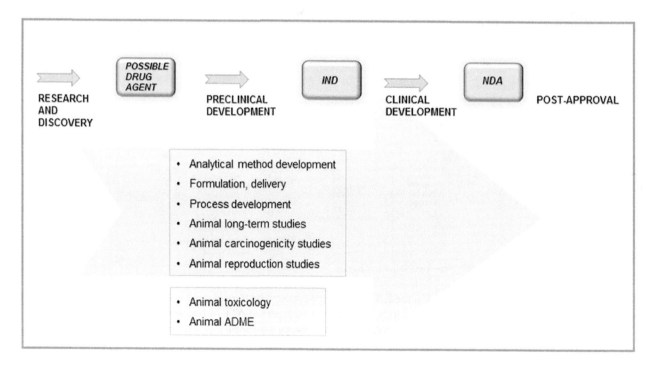

FIGURE 36.7 Preclinical development. A number of activities occur during the preclinical stage to develop the drug and its formulation and test its safety and effects *in vitro* and in animals. Some of these activities are shown here. Animal ADME studies are described below.

time. Laboratory teams test the product's potency, activity, pH, clarity, color, and particulate content over time. Specific analytical instruments may be used to look for breakdown products. It is common with drugs and other products to intentionally subject them to stresses (e.g., high temperature, freezing, exposure to light, and air) in the laboratory and check the product for changes. These studies accelerate the degradation of the drug compound and are used to investigate the degradation pathways and products of degradation. Stability testing is required to determine how expiration dates will be set for a product and to identify methods for its storage, shipping, and handling.

36.2.3.3 Process Development

Process development *is the development of methods for producing and purifying the drug substance.* The production process must result in a consistent and pure pharmaceutical substance. The process must be as cost-effective as possible. If the manufacturing process is too complex, then mistakes can occur that might harm patients, or might add expense and difficulties to production. Process development is ongoing. It begins in the preclinical stage before animal tests are initiated because there must be a source of drug to administer to the animals. The drug will initially be made in milligram amounts, but as development progresses, the process will continue to be refined and scaled up.

36.2.3.4 Preclinical Testing

A major part of the drug development process is to test the interactions of the drug with the body. The questions studied are as follows:

- What toxic effects, if any, does the drug have, and under what conditions?
- Where does the drug go in the body; how is it transported; does it accumulate in tissue?
- How does the drug interact with its target in the body?
- How does the body affect and modify the drug?

In vitro tests are those that are performed using biochemical assays, cultured cells (cells grown outside the body in dishes), *and other laboratory systems outside animals or humans.* There are many preclinical tests that are performed in vitro. For example, drugs that are ingested must be able to pass through the membranes of the cells lining the digestive tract; otherwise,

they will not be taken up by the body. It is possible to assay a compound's ability to do so in vitro by adding the substance to cultured gut cells and testing the ability of the compound to penetrate the cells. Another example is the use of bacteria or cultured cells to test if a compound is **mutagenic**, *that is, whether the compound causes a change in the genetic material of a cell after exposure.*

Animals are likely to have already played a role in the discovery and early testing of the activity of a potential drug and its interaction with the target in the body. If a candidate drug shows promise, and if it can be consistently manufactured, purified, and formulated in small amounts, then it advances to preclinical tests in animals.

Animal testing plays a key role in assessing the safety of a potential drug and its effects in the body before the agent is ever tested in humans. Animals do not respond exactly as people do to all compounds, but if a drug is not safe for animals, its development will stop – some drug candidates fail at this stage. In the future, it is likely that computer models, better in vitro assays, and new technologies, such as "organ or body on a chip" (see Section 2.15), will reduce our dependence on animal testing.

Animal testing is used largely to look at toxicity, what the drug does to the body. Experimenters administer varying doses of the drug to animals and look for toxic effects. This includes examining all the major organs (heart, lungs, brain, liver, kidney, and digestive system) and other parts of the body that may be relevant (e.g., skin for a drug delivered by a patch). At least two animal species are tested, one of which is almost always a rodent.

Experimenters also use animals to study what the body does to the drug. **Pharmacokinetics** *is the study of how a drug is changed in the body.* Experimenters administer the drug to the animals and test their urine, blood, and feces. They look at:

- **Absorption.** *This is the process by which a drug is transferred from its site of entry in the body to the body fluids, particularly blood.* For drugs that are taken orally, absorption requires that the drug can pass through the membranes of the cells that line the digestive tract. Drugs that are poorly absorbed when taken orally must be administered in another way.
- **Distribution.** *This relates to how the compound spreads through the body to various tissues and organs.*

- **Metabolism**. *This is the process by which compounds are broken down in the body by enzymes.* Small-molecule drugs are primarily metabolized in the liver by enzymes named cytochrome P450 enzymes. Metabolism converts the original molecule to new molecules called *metabolites*. The metabolites may be inert, in which case metabolism reduces the effects of the drug. Metabolites may also be active in the body, with either beneficial or harmful effects.
- **Excretion**. This is the elimination of compounds and their metabolites from the body, usually in either urine or feces.

The term **ADME** (*absorption, distribution, metabolism, excretion*) is often used in reference to these animal studies.

The conduct of preclinical animal testing was not scrutinized by the FDA prior to the mid-1970s. In 1975, however, FDA inspections of several pharmaceutical testing laboratories revealed poorly conceived and carelessly executed experiments, inaccurate record-keeping, poorly maintained animal facilities, and a variety of other problems. These deficiencies led the FDA to institute the **Good Laboratory Practice (GLP) regulations**, *which govern preclinical animal studies of pharmaceutical products.* GLPs require that testing laboratories follow written protocols and standard operating procedures (SOPs), have adequate facilities and equipment, provide proper animal care, properly record data, have well-trained and competent personnel, and conduct high-quality, valid toxicity tests. GLPs are also followed when conducting other nonclinical activities in addition to animal studies.

36.2.3.5 The Investigational New Drug Application

If the drug candidate is promising in preclinical testing, then the company *compiles data from preclinical studies and submits a plan for human subject testing to the Food and Drug Administration in the form of an* **Investigational New Drug Application (IND)**. According to the FDA website (www.fda.gov/drugs/types-applications/investigational-new-drug-ind-application), the IND contains information in three broad areas:

- **Information from animal studies**. Preclinical data are provided for FDA reviewers to assess whether the product is reasonably safe for initial testing in humans. Any previous

experience with the drug in humans (e.g., use in other countries) is also included.
- **Information relating to the composition and manufacture of the drug**. This includes the composition of the drug, the manufacturer, the drug's stability, and controls used when manufacturing the drug. Reviewers assess this information to ascertain whether the company can adequately produce and supply consistent batches of the drug.
- **Investigational plan**. Detailed protocols for the proposed clinical studies are submitted so that reviewers can assess whether the initial trials will expose human subjects to unnecessary risks. Reviewers also consider whether the trial design is likely to provide necessary information about drug function and safety. The study protocols include the number of human subjects to be studied, the criteria by which subjects will be chosen, the tests that will be performed on the subjects, and statistical methods by which data will be analyzed. Information must also be provided on the qualifications of clinical investigators who oversee the administration of the experimental compound. Finally, commitments are required to obtain informed consent from the research subjects, to obtain review of the study by an institutional review board (discussed below), and to adhere to the investigational new drug regulations.

The sponsor must wait 30 calendar days after submitting the IND before initiating any clinical trials. During this time, the FDA has the opportunity to review the IND to help assure that research subjects will not be subjected to unreasonable risk. This IND review serves as one of FDA's "gates." The application becomes effective, and testing in human volunteers can proceed, if the FDA does not disapprove the IND within 30 days.

36.2.4 CLINICAL DEVELOPMENT

36.2.4.1 Ethics

Ultimately, the only way to know if a new compound will be safe and effective in humans is to test it in humans. Clinical trials are the part of development where the safety and efficacy of the product are tested in human volunteers. Clinical trials are subject to stringent regulation to protect the safety and rights of these volunteers. The principles of conduct of clinical trials are set forth in internationally recognized documents, including notably the Declaration of Helsinki, first published by the World Medical Association in 1964. More recently, the International Conference on Harmonisation (http://www.ich.org) published the *E6 Good Clinical Practice* document that "is an international ethical and scientific quality standard for designing, conducting, recording, and reporting trials that involve the participation of human subjects. Compliance with this standard provides public assurance that the rights, safety, and well-being of trial subjects are protected, consistent with the principles that have their origin in the Declaration of Helsinki, and that the clinical trial data are credible." The principles contained in these and similar documents have been translated into regulations enforced by different agencies in different countries. In the United States, **Good Clinical Practices (GCPs)** *are regulations that protect the rights and safety of human subjects and ensure the scientific quality of clinical trials of drug safety and efficacy.*

GCPs cover many aspects of clinical trials. They require that clinical trials are designed in a scientifically rigorous fashion, that solid criteria are used to include or exclude potential volunteers, that statistical methods of data analysis are scientifically supported, that valid control groups are used, and so on. GCPs further require that participants be informed about the potential benefits and risks they will experience, that they understand the clinical trial, that their consent is given freely without coercion, and that their privacy is protected.

The company or institution that will perform the studies must also get approval from their institutional review board (IRB). The **IRB** *is a group composed of medical, scientific, and nonscientific community members who are responsible for protecting the rights, safety, and well-being of human subjects in clinical trials.* The IRB reviews the proposed trials, ensures that the method of obtaining informed consent is effective, and that relevant documentation is in order.

Specialized personnel play a key role during clinical trials, providing assurance that the trials adhere to written protocols. Later clinical trials typically involve patients and doctors at multiple sites (often international). It is necessary to assure that the selection of patients, administration of the test compounds, follow-up, and measurements of drug activity and safety are all performed in the same way, and that the interpretation of the results is consistent at all sites.

36.2.4.2 Phase I, II, and III Clinical Trials

Clinical trials are conducted in stages, each of which must be successful before continuing to the next phase (Figure 36.8). A successful product will often take on the order of 5 years to move through all the phases of clinical trials. **Phase I clinical trials** *are the first introduction of the proposed drug into humans.* Phase I trials normally include 20–80 healthy volunteers, but sometimes a small number of patients with the illness to be treated are tested at this point. (Phase I trials of toxic drugs, such as anticancer agents, involve small numbers of patients.) Phase I trials are conducted in dedicated medical facilities where subjects are monitored during the experiment. The primary purpose of phase I trials is to evaluate the safety of the agent. During phase I trials, the drug's pharmacokinetic and pharmacologic properties in healthy volunteers are also examined.

The first tests in humans usually consist of a single dose that is significantly lower (on a weight basis) than the dose at which adverse effects were observed in animals. This is followed by studies in which the dose is escalated to determine the maximum tolerated dose. Unpleasant side effects are common in phase I trials because these studies are intended to evaluate the maximum tolerated dose. If the maximum tolerated dose is below the expected therapeutic dose, then the drug fails and will not be further developed.

If a drug meets the safety requirements of phase I, then it enters phase II clinical trials. **Phase II trials** *are performed on a small number (usually 100–300) of patients with the condition the drug is expected to treat.* This phase is used to demonstrate clear drug activity and tolerance of the new compound in patients with mild to moderate disease or conditions. Patients are randomly assigned to receive either the new drug or a placebo (the formulation without the active drug ingredient). The new drug may also be compared to the best existing treatment rather than a placebo. Safety continues to be evaluated in phase II, and a dosage regimen for phase III trials is established. If the drug is successful in phase II, then it progresses to a broader **phase III trial** *involving on the order of 1,000 or more patients.* Phase III is critical to establish that a drug has the desired effects. During this phase, the safety of the drug continues to be evaluated, adverse reactions are monitored, an analysis of the drug's risks *versus* its benefits is performed, drug interactions are explored, and other data are collected. Phase III trials require a rigorous

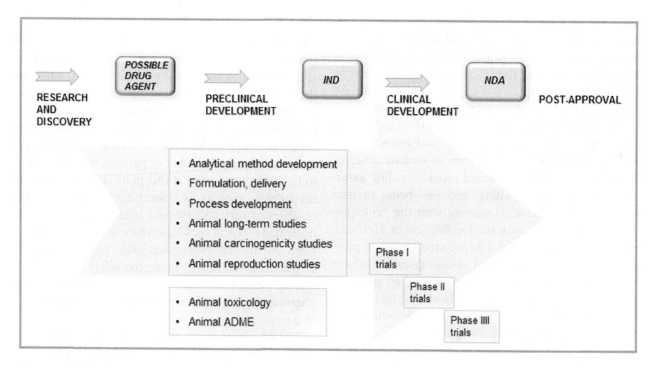

FIGURE 36.8 Clinical development. Clinical development is divided into three phases, I, II, and III, each of which is completed before the next begins. Phase I clinical trials are the first introduction of the proposed drug into humans and are generally designed to test the safety of the drug in a small group of healthy volunteers. Phase II trials test the effects of the drug on a small number of patients with the condition the drug is expected to treat. If the drug is successful in phase II, then it progresses to a broader phase III trial in order to establish that the drug has the desired effects.

statistical demonstration of clinical safety and patient benefit to ensure approval of the therapeutic.

Phase II and III clinical studies are normally, though not always, **double-blinded**, *meaning that neither the volunteers nor the clinicians know who is receiving the actual drug and who is receiving the placebo or standard treatment.* Double-blinded studies avoid results that are due to the psychological expectations of the participants and limit the potential for the investigators to bias the results, intentionally or unintentionally.

36.2.4.3 Other Activities

Even though many drugs fail clinical trials and are abandoned, the sponsor must still invest in a number of other activities during the years in which clinical trials are running. Process developers must work on the manufacturing and purification processes to produce the drug. Manufacturing teams must provide small amounts of the drug for clinical trials. Quality-control analysts must test these clinical trial materials to ensure their quality. If the product is successful, a team must learn how to scale up the production from levels measured in milligrams or grams to possibly hundreds of kilograms.

A manufacturing facility may need to be built or modified, and the production of the drug will need to be transferred from a pilot plant to the manufacturing facility. Laboratory analysts continue to work on developing analytical methods to test the substance. Manufacturing processes must be validated (see Chapter 38), and equipment must be qualified to prove that it works consistently to always deliver a quality product and give reliable results. Documentation systems are optimized throughout the development stages. Longer-term animal studies continue throughout the clinical phases to look at effects on reproduction and offspring, long-term toxicity, and the induction of cancer.

As the product enters phase III clinical trials, its characteristics must be "locked-in" so that the material that is to be sold for patients is demonstrably the same as the material that was tested in the phase III trials. This means that the product must be well characterized so that the company can prove that it is always making the same substance without variation. As will be discussed in Chapter 37, product characterization, which is usually relatively straightforward for small-molecule drugs, poses more challenges for large biological molecules.

36.2.4.4 The New Drug Application

If a drug passes all three phases of clinical testing, then the company may submit a New Drug Application (NDA) to the FDA that provides convincing evidence, based on its animal and human testing, that the new product is safe, reliable, and effective. The company must also demonstrate that it can reproducibly manufacture the therapeutic to a scale that supports broad usage by patients. This application is a critical gate controlled by the FDA. Observe that the FDA does not actually test each drug product itself; rather, FDA expert reviewers examine all the test results and information submitted by the company to determine whether a product is acceptable. An application for a new drug can contain hundreds or even thousands of pages of information and normally requires more than a year to review. However, the FDA has programs to speed up the availability of drugs that treat serious diseases, especially when the drugs are the first available treatment or if the drug has advantages over existing treatments. The FDA developed four distinct and successful approaches to making such drugs available as rapidly as possible:

- **Priority Review** *means the FDA aims to take action within 6 months of receiving the NDA.*
- **Breakthrough Therapy** *is a designation for drugs that may demonstrate substantial improvement over available therapies and for which review is expedited.*
- **Accelerated Approval** *is allowed for drugs that treat serious conditions and fill an unmet medical need.*
- **Fast Track** *is a process designed to facilitate the development and review of drugs that treat serious conditions and fill an unmet medical need.* Fast-tracking was initiated in the late 1990s, partially in response to the AIDS crisis where patients and their supporters advocated for quicker access to desperately needed medicines. Fast-tracking does not bypass any of the normal FDA "gates," but it can allow a sponsor to pass more quickly through them.

If the FDA review team decides that sufficient evidence supporting a new drug is provided in the NDA, then the new product will be approved or licensed by FDA and can be manufactured for commercial sale. This is a major accomplishment and is the beginning of the period where the company can profit from its product.

We will note that there is one more regulatory approach for medical products:

- **Emergency Use Authorization (EUA)** *is a special track used only in major emergencies to facilitate the use of unapproved products that diagnose, treat, or prevent serious or life-threatening diseases or conditions if there are no approved, available, or adequate alternatives.* This situation is rare but, during the COVID-19 crisis, the FDA declared a public health emergency and began to authorize the use of diagnostic assays, the conduct of vaccine clinical trials, the administration of new vaccines, and the use of certain drugs relating to COVID-19 with greatly expedited review. In such situations, the FDA must balance its primary responsibility for ensuring the safety and efficacy of medical products with the urgent need to quickly bring new products to market.

36.2.5 POST-APPROVAL

Once approved, the drug can be manufactured, marketed, and sold (Figure 36.9). Approved pharmaceutical products must be manufactured in compliance with Current Good Manufacturing Practices (CGMPs) in the United States and similar regulations in other countries.

Post-market surveillance is used to monitor the long-term safety of the product, product defects, and adverse reactions reported by patients and doctors. Adverse patient reactions must be reported to the FDA and in extreme cases have caused drugs to be withdrawn from the market. There may continue to be monitoring of patients from earlier clinical trials to look for long-term effects and adverse reactions, even after the drug is approved.

A company might want to expand the use of a drug to new applications (e.g., different types of cancer). In this case, the previous safety studies are usually still applicable, but new clinical studies are required to test the efficacy of the drug for the new purpose. If successful, these new studies allow the drug to be labeled for conditions in addition to the one for which the drug was initially approved.

If the manufacturing process for a drug changes substantially, the FDA may require new clinical trials to demonstrate that the product is still safe and effective. This is particularly true for large-molecule biological products. These products are so complex that they are

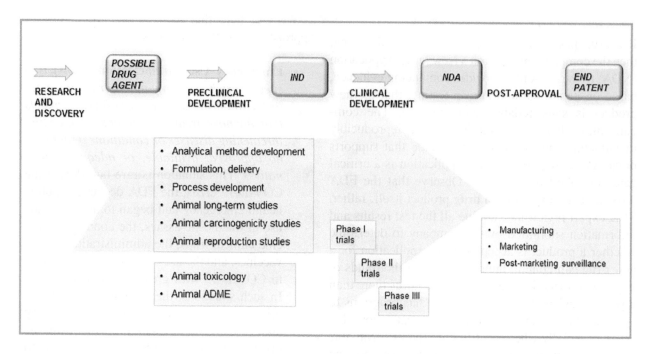

FIGURE 36.9 Post-approval. The drug can be manufactured, marketed, and sold after it is approved. It is still subject to FDA oversight and may require additional clinical testing and development.

difficult to completely characterize. Furthermore, the properties of a biopharmaceutical are tied to the production process used to make it. Significant changes to the manufacturing process can therefore result in a material not identical to that used in phase III clinical trials, and laboratory tests of the product might not detect the difference.

The patent protection of a drug, as described in Chapter 3, ends after a certain date at which time other companies may introduce generic versions of the drug. The pathway for introducing small-molecule generic drugs is reasonably straightforward, since these molecules are well characterized, and a generics manufacturer can usually demonstrate that their product is chemically identical to the original drug.

The generics pathway is not as simple for biopharmaceuticals as it is for small-molecule drugs. The exact nature of a biopharmaceutical depends on the specific cells or organisms in which it was produced, the specific genetic construct used (if it involves recombinant DNA methods), and the methods by which those cells or organisms were grown. These key aspects of the production process are trade secrets protected by the company that originally developed the therapeutic. Thus, it is extremely difficult (or impossible) to exactly recreate a biopharmaceutical product. Patents on the first biopharmaceuticals began to run out in the early

part of this century, and conflicts quickly arose over how generic biopharmaceuticals should be regulated. Some manufacturers of original biopharmaceuticals argued that generic biopharmaceuticals should not be permitted at all, while other manufacturers were anxious to make them. In 2009, the US Congress passed the "Biologics Price Competition and Innovation Act" that created a regulatory pathway for generic biopharmaceutical approvals. The purpose of this act was to increase access to biopharmaceuticals by allowing a generic version, thus broadening their accessibility to patients, and reducing healthcare costs. Along the way, *generic biopharmaceuticals got a new name to reflect the fact that they cannot be identical to the original products; they are now called* **biosimilars**. In order for a biosimilar to be approved, its sponsor must prove that the biosimilar is highly similar to, though not necessarily identical to, the previously approved "reference" biological drug. The new biosimilar drug must also have no meaningful differences in safety or effectiveness when it is administered to patients. This requires extensive testing of the biopharmaceutical product itself and usually some form of testing in humans, though not usually three full phases of clinical testing. There is thus a regulatory pathway for biosimilars that is abbreviated relative to the pathway for the original products. Eventually, the first biosimilar made it down the new abbreviated regulatory pathway and was approved in 2015. By March of 2020, there were 25 more with others in the pipeline.

36.2.6 COMPLEXITIES: BALANCING RISKS AND BENEFITS

Clinical testing generally proceeds without harm to volunteers, but it is not without risk. For example, in March 2006, six healthy volunteers were injected with a new type of monoclonal antibody that was different in its mechanism of action than previous monoclonal antibody therapeutics. This particular antibody triggered a reaction in all six individuals, causing their immune systems to violently overreact and resulting in serious injury to them. The case study below, "Death in Gene Therapy Study Raises Questions about Informed Consent," addresses another clinical trial issue, that of informed consent.

The process for developing a drug poses financial risks as well as health risks to volunteers. The drug development process is lengthy, and the pharmaceutical industry generally reports that only one out of every ten drugs that enters clinical testing will actually result in a marketed product. The low success rate of drugs coupled with the extensive resources required for their development contribute to the high costs of drugs.

The advantage to the complex, costly, and lengthy drug development process is that it does generally protect patients from overtly harmful new drugs. Before animal and human testing of new drug products was required by law, patients often died from dangerous drugs. The system, however, is not perfect. Clinical trials sometimes do not detect adverse effects that only show up after long-term administration of a drug, or adverse effects that harm only a small percent of patients. This is the case, for example, with Gleevec (see the Case Study on pp. 976–977). A study in 2006

Case Study: Death in Gene Therapy Study Raises Questions about Informed Consent

Eight years after the death of Jesse Gelsinger (see Section 2.1.2.1), another death occurred during a US gene therapy trial. In February 2007, Jolee Mohr, 36, was enrolled in a phase I/II study of a gene therapy protocol to potentially treat rheumatoid arthritis, an inflammatory disease of the joints. The gene therapy agent, tgAAC94, consisted of an adeno-associated virus (AAV) designed to deliver a gene that blocks tumor necrosis factor, a substance associated with joint inflammation. AAV is used as a vector in a number of gene therapy protocols, because the virus is only weakly immunogenic and does not cause human disease. The tgAAC94 trial involved the injection of the agent directly into the affected joints; in Mohr's case, the right knee. The study included 120 volunteers, with no previous adverse effects. Mohr, however, became ill within 24 hours of her second injection on July 2, 2007, and died 22 days later of massive internal bleeding and multiple organ failure. The gene therapy trial was suspended by Seattle-based Targeted Genetics Corporation, and the FDA was notified.

Mohr's death led to a government investigation by the National Institutes of Health Recombinant DNA Advisory Committee, which determined that Mohr died from complications of a systemic fungal infection. The committee found no evidence to indicate that her death was a direct result of the gene therapy, although they could not rule out the possibility of an indirect effect. They suggested that a major risk factor for Mohr may have been long-term drug treatment with the monoclonal antibody drug, Humira, which blocks TNF production systemically by suppressing the immune system. Immunosuppression may have been a key factor in Mohr's development of the opportunistic fungal infection. Since there was no evidence of direct harm from the gene therapy agent, the FDA approved the resumption of the direct tgAAC94 clinical trial in November 2007.[*]

It is customary in gene therapy trials for patients to continue taking medications that have been effective in treating their disorder. Humira is a fully human monoclonal antibody, the first to be approved for human use (in 2002). Abbott Laboratories received the 2007 Galen Prize for Best Biotechnology Product

---

[*] As of 2020, no gene therapy treatments for RA have been approved. RA does, however, continue to be an area of active research, and a number of potential treatments are being explored.

(Continued)

**Case Study (*Continued*): Death in Gene Therapy Study
Raises Questions about Informed Consent**

for Humira, which is currently a well-accepted, first-line treatment for severe rheumatoid arthritis and several other autoimmune conditions. An increased risk of infection is a well-known, potential side effect of this medication.

While the gene therapy trial may not have been directly responsible for Mohr's death, it has raised discussion of important bioethical issues that are central to all clinical trials. Many people have questioned whether Mohr, a young woman with a 5-year-old daughter, was a good candidate for a clinical safety study, especially since her arthritis had responded well to Humira treatment. In the cases of Mohr and Gelsinger, there has been much debate about the issue of informed consent. Mohr's husband maintains that he and his wife did not realize that the study had no intended benefit for patients, and that Jolee would not have volunteered had she been aware of this. In turn, Targeted Genetics Corporation and Mohr's doctor point out that they provided the Mohrs with an IRB-approved informed consent document and they had answered the family's questions fully. Mohr clearly signed the consent form, but some have questioned whether a patient without a medical background could properly comprehend a 15-page form or ask all of the appropriate questions, even when the form is written in relatively simple language.

These are questions that do not have easy answers. They emphasize the fact that pharmaceutical companies that wish to market new treatments must deal with human issues that extend far beyond drug development and production.

indicated that in a small percent of patients the drug appears to cause severe heart disease with long-term use. However, about 90% of CML patients treated with Gleevec survive 5 years or longer, whereas before the drug was approved in 1990, average survival was less than 5 years. The benefit of Gleevec to CML patients thus outweighs its risks, and patients are likely to continue to take the drug.

In a more sensational incident, the drug Vioxx, an anti-inflammatory drug used by roughly 80 million patients to treat pain and arthritis, was found to increase the risk of heart attack and stroke. Studies of Vioxx indicate that it increases the risk of heart disease and stroke, but only in a small percentage of people using it, and only after 18 months of taking the drug daily. This means that the adverse effect was unlikely to turn up in early drug testing. (See, for example, Bresalier, R. S., et al. "Cardiovascular Events Associated with Rofecoxib in a Colorectal Adenoma Chemoprevention Trial." *New England Journal of Medicine*, vol. 352, no. 11, 2005, p. 5. doi:10.1056/NEJMoa050493.)

The Vioxx case immediately led to hundreds of lawsuits and withdrawal of Vioxx from the market. Both Vioxx and Gleevec elevate heart disease risk with long-term use. CML patients, however, are at a very high risk of dying from cancer and so the benefit of the drug is thought to outweigh its risk, whereas Vioxx was used to treat nonfatal, though debilitating, diseases. The Vioxx incident was particularly complex and litigious because there is a question as to whether the drug companies knowingly withheld information pointing to the drug's risk. The FDA has also been criticized in the Vioxx case for failing to protect patients from serious risk.

These incidents point to the complexities, risks, and benefits of drug development. It is hoped that as new technologies are created (e.g., methods to use predictive genetic information from patients, computer models of drug interactions), fewer people will suffer unpredicted effects from drugs, and the drug development process will become more reliable.

Practice Problems

1. Matching. Match each of the activities below with the lifecycle phase that best fits it.
   i. Research/discovery
   ii. Preclinical development
   iii. Clinical development
   iv. Post-approval
      a. Determining whether a drug candidate is effective by evaluating its effect in several thousand human patients.
      b. Screening 100,000 small chemical compounds to find any that bind to a receptor found on the surface of pancreatic cancer cells.
      c. Performing tests in rats to see how their bodies metabolize a possible drug compound.
      d. Performing tests where a possible drug compound is added to plates containing cultured cells and determining if the compound is toxic to the cells.
      e. Using cultured intestinal cells to determine whether a compound has the ability to cross the cell membrane of the intestine.
      f. Devising and creating the genetic construct used for a monoclonal antibody product.
      g. Performing surveillance when a drug is administered to millions of patients to see if it causes rare adverse effects.
      h. Developing an assay to detect the drug compound in samples from blood and urine of animal and human subjects.
2. What is a "biopharmaceutical?"
3. What are the differences between "large-molecule" and "small-molecule" therapeutics?
4. Why is it difficult to ensure batch consistency and potency in biopharmaceutical batches?
5. True or False:
   a. Activity is how well the drug candidate binds to the cell wall. T/F
   b. How long the drug candidate stays in its active form in the body is known as its half-life. T/F
   c. The metabolic profile in drug development means determining if a volunteer has diabetes prior to entering into a clinical trial. T/F
6. What does it mean when a drug is "fast-tracked?"
7. What is an orphan disease?
8. Figure 36.7 shows the development timeline, from research and discovery to post-approval. During the preclinical phase, animal studies are conducted. Why is animal testing performed?
9. Figure 36.7 shows that after an IND is submitted, clinical development can be conducted. After NDA approval, post-approval studies are conducted. What is the benefit of post-approval studies?
10. What are the primary phases of clinical trials? What happens in each of the phases?
11. What is pharmacokinetics? Why are ADME studies conducted?
12. Provide two reasons for the high cost of drugs.
13. What is the difference between a generic and a biosimilar?
14. Examine Figure 36.7. Discuss how the process outlined in this figure relates to the historical incidents described in Chapter 34. This is a discussion question with various answers.
15. a. What is the role of Congress in ensuring the safety of food and medical products?
    b. What is the role of the FDA in ensuring the safety of food and medical products?
    c. What is the role of a pharmaceutical/biopharmaceutical company in ensuring the safety of food and medical products?
    d. What is the role of the consumer in ensuring the safety of food and medical products?
16. Use a web browser to find the following section of the code of federal regulations: 21CFR 211.22
    a. Explain the regulation in your own words.
    b. Is it all right for a company to have the production team take all responsibility for testing all incoming raw materials and all finished products?

Question for Discussion

This question is included to provide subjects for thought, discussion, and further research. It does not have a single answer, and no answer key is provided.

During World War II, Nazi physicians conducted brutal "experiments" on human adults, children, and infants in concentration camps. The victims of these "experiments" were frozen to death, mutilated and wounded, sickened, and often suffered greatly before their deaths. The Nuremberg Code is a set of ethical principles for **human experimentation** that was a response to these atrocities. The World Medical Association "Declaration of Helsinki: Recommendations Guiding Medical Doctors in Biomedical Research Involving Human Subjects," which expands on the Nuremberg Code, is a "statement of ethical principles to provide guidance to physicians and other participants in medical research involving human subjects" (https://www.wma.net).

The General Principles from the 2013 version of the Declaration of Helsinki are quoted below. (Some paragraphs have been omitted for the purposes of this question.)

1. Discuss how these basic principles apply to a company that is conducting clinical trials of a new drug substance. How can a research team ensure that its testing is ethical and conforms to the basic principles?
2. In 2008 the US FDA abandoned any mention of the Treaty of Helsinki and instead ruled that clinical trial compliance with the Good Clinical Practices, as developed by the International Conference on Harmonisation, is sufficient. This controversial decision was due primarily to the principles outlined in paragraphs 33 and 34, as shown below. Paragraphs 33 and 34 are sometimes thought to be of particular benefit to study subjects in developing countries. Research FDA's decision on the Internet, and decide whether you agree or disagree with the FDA's decision. Explain your position.

GENERAL PRINCIPLES

Paragraph 3. The Declaration of Geneva of the WMA binds the physician with the words, "The health of my patient will be my first consideration," and the International Code of Medical Ethics declares that, "A physician shall act in the patient's best interest when providing medical care."

Paragraph 4. It is the duty of the physician to promote and safeguard the health, well-being and rights of patients, including those who are involved in medical research. The physician's knowledge and conscience are dedicated to the fulfilment of this duty.

Paragraph 5. Medical progress is based on research that ultimately must include studies involving human subjects.

Paragraph 6. The primary purpose of medical research involving human subjects is to understand the causes, development and effects of diseases and improve preventive, diagnostic and therapeutic interventions (methods, procedures and treatments). Even the best proven interventions must be evaluated continually through research for their safety, effectiveness, efficiency, accessibility and quality.

Paragraph 7. Medical research is subject to ethical standards that promote and ensure respect for all human subjects and protect their health and rights.

Paragraph 8. While the primary purpose of medical research is to generate new knowledge, this goal can never take precedence over the rights and interests of individual research subjects.

Paragraph 9. It is the duty of physicians who are involved in medical research to protect the life, health, dignity, integrity, right to self-determination, privacy, and confidentiality of personal information of research subjects. The responsibility for the protection of research subjects must always rest with the physician or other health care professionals and never with the research subjects, even though they have given consent.

Paragraph 10. Physicians must consider the ethical, legal and regulatory norms and standards for research involving human subjects in their own countries as well as applicable international norms and standards. No national or international ethical, legal or regulatory requirement should reduce or eliminate any of the protections for research subjects set forth in this Declaration.

Paragraph 11. Medical research should be conducted in a manner that minimises possible harm to the environment.

Paragraph 12. Medical research involving human subjects must be conducted only by individuals with the appropriate ethics and scientific education, training and qualifications. Research on patients or healthy volunteers requires the supervision of a competent and appropriately qualified physician or other health care professional.

Paragraph 13. Groups that are underrepresented in medical research should be provided appropriate access to participation in research.

Paragraph 14. Physicians who combine medical research with medical care should involve their patients in research only to the extent that this is justified by its potential preventive, diagnostic or therapeutic value and if the physician has good reason to believe that participation in the research study will not adversely affect the health of the patients who serve as research subjects.

Paragraph 15. Appropriate compensation and treatment for subjects who are harmed as a result of participating in research must be ensured.

Paragraph 33. The benefits, risks, burdens and effectiveness of a new intervention must be tested against those of the best proven intervention(s), except in the following circumstances:

Where no proven intervention exists, the use of placebo, or no intervention, is acceptable; or

Where for compelling and scientifically sound methodological reasons the use of any intervention less effective than the best proven one, the use of placebo, or no intervention is necessary to determine the efficacy or safety of an intervention and the patients who receive any intervention less effective than the best proven one, placebo, or no intervention will not be subject to additional risks of serious or irreversible harm as a result of not receiving the best proven intervention.

Extreme care must be taken to avoid abuse of this option.

Paragraph 34. In advance of a clinical trial, sponsors, researchers and host country governments should make provisions for post-trial access for all participants who still need an intervention identified as beneficial in the trial. This information must also be disclosed to participants during the informed consent process.

37 The Lifecycles and Regulation of Biopharmaceutical and Regenerative Medicine Products

37.1 OVERVIEW

As we discussed in the beginning of this textbook, the biotechnology industry creates an array of products to diagnose, treat, and prevent diseases and disorders. Chapter 1 discussed the introduction of recombinant DNA technology, which led to dramatic improvements in the manufacturing of protein-based biopharmaceutical products. Chapter 2 introduced regenerative medicine, a category that includes stem cell and other cell-based therapies, gene therapies, and tissues for transplantation. From a regulatory standpoint, these biotechnology products are generally termed **biologics, or biological products**, *that is, a therapeutic that is produced from or by living organisms or contains components of living organisms.* Biological products often represent the cutting edge of biomedical research and may offer the most effective means to treat medical conditions that presently have no other treatments available.

The principles and processes that guide the regulation of these innovative biotechnology products are generally the same as those described in Chapter 36. Their development proceeds together with regulatory checkpoints where the sponsor/manufacturer must submit an application to the FDA. The FDA reviews the application, and, if approved, the product is allowed to move to the next stage. Although this overall regulatory process (pathway) is the same for drugs and biologics, there are ways in which the development of a biologic differs from the development of a small-molecule drug. This chapter begins with a discussion of some of the issues specific to biopharmaceutical development when cells are used as "factories" to produce therapeutic protein products. This topic is divided into two parts:

- **Product development**. *This includes tasks relating to understanding the biopharmaceutical molecule itself.*
- **Process development**. *This includes tasks relating to the development of methods for manufacturing and purifying the product.*

The latter part of this chapter introduces some of the issues and regulatory requirements relating to products that fall into the category of regenerative medicine.

DOI: 10.1201/9780429282799-47

37.2 PRODUCT DEVELOPMENT FOR PROTEIN BIOPHARMACEUTICALS

Protein biopharmaceuticals interact with the body in varied and complex ways. For example, insulin (which was among the first recombinant DNA products created) is a hormone that orchestrates the regulation of the blood's glucose level. Insulin, and other protein biopharmaceuticals, have complex functions, and correspondingly complex structures. The structural features of a protein biopharmaceutical determine how it will function in a patient. Every protein has a specific shape, a specific distribution of positive and negative charges on its surface, and sometimes a particular glycosylation pattern; these features allow the protein to do its specific work in the body. *The essential characteristics of a protein that enable it to function as a safe and effective drug are called* **critical quality attributes (CQAs)**. Understanding the structure of protein biopharmaceuticals and their CQAs is a multifaceted task that begins during the discovery and preclinical phases of drug development, and continues throughout development. This task requires many laboratory studies; Table 37.1 summarizes some of these studies. As you can see in this table, it is important to determine and document the structural features of the protein, and also other characteristics that might affect its performance, such as purity and stability.

Proteins are very sensitive to their environment. Proteins may degrade over time due to chemical interactions (e.g., amino acid interactions with oxygen) and physical changes (e.g., unfolding and aggregating). Since the proper function of a protein requires that it retain the correct structure, it is important that protein drugs are formulated in such a way that they are as stable as possible. During biopharmaceutical development, laboratory teams will research the most stable formulation for the drug (e.g., pH, buffer constituents) and how it should be handled, transported, and stored. Analysts will also perform stability testing where they will evaluate the drug's potency and general characteristics over time, and under various conditions. Analytical instruments, such as high-performance liquid chromatography instruments, as described in Chapter 34, are often used during stability testing to look for specific protein breakdown products that are associated with product degradation.

TABLE 37.1

Characterizing a Biopharmaceutical Protein Product

Identity Testing
- Determining and documenting the amino acid composition of the protein.
- Determining and documenting the amino acid sequence (order of the amino acids) for key regions of the protein.
- Determining and documenting the size of the protein.
- Determining and documenting the conformation (shape) of the protein.
- Determining and documenting any protein modifications, such as glycosylation.
- Determining and documenting how the protein behaves in electrophoresis, HPLC, and in other test systems.

Purity Testing
- Determining impurities from all sources that are associated with the product.
- Determining breakdown products that may occur.

Potency Testing
- Determining the therapeutic action of the protein.
- Finding an assay(s) to measure that activity.

Stability Testing
- Determining the stability of the product during storage and setting expiration dates.
- Determining the effects of heat, light, time, and other factors on the drug.
- Identifying breakdown products and how to avoid them.
- Determining proper storage conditions.

Example Problem 37.1

In various places in this text, we have talked about the new mRNA vaccines that have been developed to protect people from COVID-19. (See, for example, Section 1.2.4.) Recall that these mRNA vaccines code for a protein that is part of the SARS-CoV-2 virus and that the virus requires for its invasion of human cells. After an mRNA vaccine is injected into a person, the recipient's cells synthesize the viral protein encoded by the mRNA, thus "teaching" the recipient's immune system to recognize and destroy the SAR-CoV-2 virus. Recall also that the mRNA in the new vaccines is encapsulated in lipid nanoparticles. Think about what would be the CQAs for this vaccine – what are the critical features the vaccine must possess to be safe and effective? After you have made your list, compare it with the list below.

Answer

On November 30, 2020, Moderna, one of the companies making an mRNA vaccine, submitted an application for authorization to market their vaccine in Europe. In March 2021, the European Medicines Agency published an assessment report that listed the proposed CQAs for the Moderna mRNA vaccine.

We list those CQAs here in two categories, those relating to the identity of the key components of the vaccine and those relating to contamination:

Identity	Absence of Contaminants
mRNA identity	Purity and product-related impurities
Total RNA content	Lipid impurities
% RNA encapsulation	Bacterial endotoxin
Lipid identity	Sterility
Lipid content	
Mean particle size and polydispersity	
pH	
Osmolality	
Particulate matter	
Container content	

Source: European Medicines Agency. "Assessment report COVID-19: Vaccine Moderna." *European Medicines Agency*, 11 March 2021, www.ema.europa.eu/en/documents/assessment-report/covid-19-vaccine-moderna-epar-public-assessment-report_en.pdf.

37.3 PROCESS DEVELOPMENT

37.3.1 CREATION OF MASTER AND WORKING CELL BANKS

Biopharmaceutical protein products begin with the identification of a protein that is of therapeutic importance. Typically, the gene coding for the protein of interest (GOI) is obtained and inserted into a vector, which is then introduced into host cells (e.g., bacteria, yeast, and mammalian cells). The host cells will manufacture the protein, which is then isolated from the cells, purified, and formulated into a final product. This production system, which is quite unlike that used when small molecules are synthesized, is of special interest to us as biotechnologists.

The creation of the genetic construct that includes the GOI and the choice of a host cell system must be accomplished early in the discovery and development process. The successful creation of a system to manufacture the protein of interest requires understanding the gene and the protein for which it codes, knowledge that is acquired during the discovery phase of the product lifecycle. This early work is performed by scientists and technicians who are well versed in molecular biology.

Biotechnology production systems depend on cultured host cells that have been genetically modified to produce a desired product. (See the case study "The Process Defines the Product," later in this chapter.) When cells are continuously cultured over a period of time, mutations or other alterations in their characteristics can occur. Changes in the host cells can alter the product they produce – which is never desirable. Because of the complexity of biopharmaceutical molecules, a change in the final product might be subtle and difficult for analysts to recognize, yet might harm patients. It is therefore essential that there are methods to protect the stability of the host cell expression system. Early in the history of biopharmaceuticals, scientists and regulatory agencies decided that the creation of a master cell bank (MCB) is a key component in maintaining a stable production system when genetically engineered cells are involved. The **master cell bank** *is the source of cells used for production.* For bacterial production systems, the master cell bank is derived from a single, original colony of bacteria that was transformed with the genetic construct. If the production system uses mammalian or insect cells, then the master cell bank is derived from a single cell transfected with the genetic construct. In either case, the resulting cell(s) are allowed to divide a number of times and are then distributed into small aliquots (portions), each of which is placed in a vial and stored cryogenically

(in cold "suspended animation") to ensure its stability. *Each time a production run begins, an ampule from the master cell bank is thawed and allowed to multiply to form the* **working cell bank** (Figure 37.1).

The creation of the MCB is an important task of R&D scientists and technicians. They must document the history of the host cells including the details of their origin: the species, identity, age, and sex of the donor, if the cells are of animal origin; and the medical history of the cells' donor, if they are of human origin. The MCB is prepared using stringent contamination control procedures in specialized facilities by individuals highly skilled in cell culture.

The MCB is the source of all production cells, and hence, is extremely valuable. It is a critically important trade secret. The MCB is protected by the use of security measures, documentation, and alarm systems in the event of freezer malfunction (Figure 37.2). It is also common to store aliquots in more than one freezer and at more than one site.

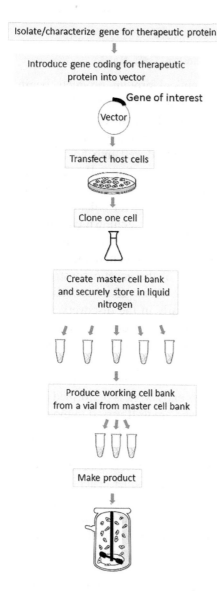

FIGURE 37.1 Generalized biopharmaceutical process development (mammalian cells). The gene coding for the therapeutic protein (GOI) is introduced into host cells. A master cell bank is derived from a host cell and stored under carefully controlled conditions. The master cell bank will be the source of all product and is therefore extremely valuable. A working cell bank is derived from a vial of the master cell bank to begin each new batch. There is extensive testing at each step to completely characterize and document the gene of interest, the vector, the host cell, the master cell bank, and the working cell bank.

FIGURE 37.2 Storage of a master cell bank. Special freezers containing liquid nitrogen provide an extremely low temperature of about −196°C. Each metal rack contains boxes full of cell aliquots. Note also that the technician is wearing special attire to help protect the cells from contamination. (Photo courtesy of Dan Felkner, WiCell.)

Case Study: The Process Defines the Product

The phrase "the process defines the product" is frequently used in reference to protein biopharmaceuticals. This is because biopharmaceutical production is entirely dependent on living, dynamic, cellular systems. Even slight alterations in the cells or their environment, such as a change in media formulation, or a mutation in the cells themselves, can profoundly affect the final structure of the protein product, and therefore how it functions in a patient. Small-molecule drugs are different in that small changes in the chemical processes may have no effect on the final product, or it may be relatively easy to detect changes in the final chemical structure of the drug. To make this distinction between small-molecule drugs and biopharmaceuticals clearer, let's consider the following case.

Pompe disease is a rare genetic disorder in which a baby is born lacking normal genes to synthesize the enzyme acid alpha-glucosidase (GAA). GAA is an enzyme that normally converts glycogen present inside of lysosomes into glucose. A deficiency in GAA results in lysosomal accumulation of glycogen, primarily (though not exclusively) in muscle tissues, which ultimately leads to severe muscle damage. In some cases, babies born with this disorder die before their first year. In other cases, affected individuals live into adulthood, but eventually must use a wheelchair and ultimately experience respiratory failure. Before 2006, this disease was untreatable, despite many efforts to find therapies. In 2006, the biopharmaceutical, Myozyme®, was approved by the FDA. Myozyme® is a protein biopharmaceutical, alglucosidase alfa, which is infused into patients and does the work of the enzyme GAA. Initially, Genzyme manufactured this biopharmaceutical in a 160 L bioreactor. However, there turned out to be a high demand for the drug, so the company built a new facility with a 4,000 L bioreactor. The same cell line and procedures were used in both facilities, but the larger bioreactor generated a protein with different carbohydrate side chains than the smaller bioreactor. Although scientists at Genzyme were experienced and knowledgeable, this difference in the protein product was unexpected and unexplained. The FDA determined that because of the side chain differences, the products made in the two bioreactors were different from one another. Genzyme was therefore required to perform new (expensive) clinical trials on patients with different forms of Pompe disease to prove the safety and efficacy of the biopharmaceutical produced by the new larger bioreactor. Genzyme then had to apply for a license in order to market the "new" product. Although the new product was proven safe and effective, the company also was required to use a different name and marketing materials for the two products from the differently sized bioreactors; the second is called Lumizyme®. This is an interesting example of how a seemingly simple change in a biomanufacturing process can have major effects on the resulting product and its regulation. Hence, the phrase, "the process defines the product." As discussed in Chapter 35, this is the key issue that arose when the industry and regulators were considering the introduction of "generic" biopharmaceuticals.

37.3.2 CONTAMINANTS AND IMPURITIES

Contaminants and impurities *are undesired substances that must be removed from products before they can be used.* The term "impurity" is used to refer to expected substances, such as proteins from the host cells, that must be removed. The term "contaminants" refers to unexpected materials, such as viral particles that somehow enter into the process. We will use the term "contaminant" to refer to both types of undesired materials.

The cell lines that are used to manufacture biopharmaceuticals are far less likely to carry harmful contaminants and impurities than tissues from humans or animals, but there are certain types of substances that are of concern when genetically engineered cells are used. Viruses are of particular concern when animal cells are used in production. Endotoxins can be a problem when bacterial cells are used (Figure 37.3). In addition, the host cells have their own proteins and DNA that must not be allowed to carry over into the product.

The raw materials used to grow the cells are another potential source of contamination. The culture medium used to nourish and sustain the cells may introduce viruses, prions, unwanted proteins, and other contaminants. Raw materials, particularly the culture medium, must be tested for these

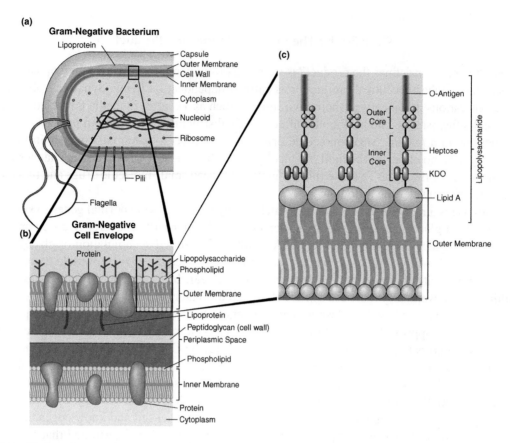

FIGURE 37.3 Endotoxins are dangerous – what are they? Endotoxins are contaminants derived from the breakdown of a certain type of bacterium (gram negative). Exposure to endotoxins can cause a severe reaction in humans and other animals, including fever, septic shock, and even death. Endotoxins are difficult to destroy and are a serious concern in the production of drugs and other medical products. Bacterial cells (unlike our cells) are surrounded by a cell wall. Endotoxins are a component of the outer part of the cell wall of gram-negative bacteria. (a) Half of a bacterium is shown for orientation. (b) A magnified view of a bacterial cell wall and its surrounding membranes. (c) On the outside of the outer membrane of the cell wall, there is a layer of lipid with associated proteins and sugars. Endotoxins are the lipid and sugar part of the outer membrane, i.e., the lipopolysaccharides. (Illustrations from earlier version of an online microbiology text from the University of Texas.)

contaminants. A section in Chapter 31 discussed the ongoing efforts to reduce the chance of introducing contaminants via the culture media.

Adventitious contaminants *come from outside the cell culture system, for example, from air, manufacturing operators, or nonsterile equipment.* These contaminants are introduced when Current Good Manufacturing Practices (CGMPs) are not followed. (CGMPs are discussed in more detail in Chapter 38.) The various types of contaminants that are of importance for biopharmaceuticals are summarized in Table 37.2.

TABLE 37.2

Contaminants of Concern in Products Made Using Biotechnology Methods

- **Adventitious Contaminants**. *Contaminants that come from outside the system that are introduced accidentally and sporadically during processing through a breakdown in good manufacturing practices.*
- **Endogenous Contaminants**. *Contaminants that are naturally found in the host cells.* For example, most mammalian host cell lines contain viruses that must be removed from any biopharmaceutical products produced by those cells.
- **Bacteria**. *Contaminants that are a concern in any pharmaceutical process.* Bacteria can contaminate a product and harm the patient. They can also infect and destroy the host cells.

(Continued)

TABLE 37.2 (*Continued*)

Contaminants of Concern in Products Made Using Biotechnology Methods

- **Mycoplasma**. *A type of very small, difficult-to-detect bacterium. Mycoplasma sp.* can cause disease in humans and animals. They are troublesome contaminants in cell culture because they are difficult to detect, are difficult to eradicate, and can subtly alter the properties of infected cells. (See also Chapter 30 for a discussion of mycoplasma detection.)
- **Viruses**. *Contaminants that can both affect host cells and possibly infect the patient.* Viruses can incorporate themselves into the genome of the host cells and can remain dormant for long periods of time. Mammalian cells used to produce protein products contain endogenous viruses.
- **Pyrogens/Endotoxins**. *Contaminants that are by-products of gram-negative bacteria.* Pyrogens cause fever and elicit a dangerous inflammatory response in mammals; they may remain in a preparation even when all bacteria have been killed.
- **Nucleic Acids (DNA and RNA) from the Host Cells**. There is concern that nucleic acids from the host cells might enter a patient's cells and have a harmful effect.
- **Proteins from the Medium and the Host Cells**. The biopharmaceutical protein of interest is not the only protein in the production system. The culture medium may contain proteins, and the host cells also have proteins. These extraneous proteins must be separated from the product. Residual proteins left from the production system may cause allergenic responses in patients. They also might have enzymatic effects, or might affect the regulation of genes in patients.
- **Yeast, Fungi, and Parasites**. These harmful agents might be accidentally introduced into the product from the growth medium, air, workers, and other sources.
- **Prions**. *A type of infectious agent, thought to be composed of misfolded proteins, which cause a variety of neurodegenerative diseases including scrapie, kuru, mad cow disease, and Creutzfeldt–Jakob disease.* Prions are a particular concern with any agent that is isolated from an animal source, including fetal bovine serum, which in the past was commonly added to cell culture medium to promote healthy cell growth. Animal-derived agents, such as serum, are now avoided whenever possible.
- **Residuals from the Cells' Growth Medium**. The host cells require a rich growth medium that contains many substances that should not be carried over into the final product, for example, insulin and albumin.
- **Residuals from Processing**. *Substances that contaminate the product during the purification and other processing steps.* All pharmaceuticals are processed in various ways to ensure their purity and to formulate them so they can be administered to patients. Residual materials from processing steps (e.g., chromatography solvents) must be removed.
- **Extractables from Plastic Components and Final Containers**. *Substances that leach out of plastics.* (See the case study "Detective Work Reveals the Culprit in Drug Mystery" below.)

Case Study: Detective Work Reveals the Culprit in Drug Mystery

Pure red cell aplasia (PRCA) is a severe and rare form of anemia. Beginning in 1998, doctors in France, Canada, the United Kingdom, and Spain began to notice an increase in PRCA among patients receiving the drug Eprex®, which is used to treat anemia due to renal failure. Eprex® is a biopharmaceutical product, erythropoietin, which is manufactured by Johnson & Johnson Company using genetically modified Chinese hamster ovary (CHO) cells. Erythropoietin is a glycoprotein normally produced in the kidney, which stimulates the production and maturation of red blood cells in the bone marrow. Although the incidence of PRCA was always low (about 1 case for every 5,000 patient-years of exposure), the increased rate of occurrence of the disease triggered a massive investigation by Johnson & Johnson. More than 100 laboratory scientists, epidemiologists, immunologists, clinicians, and quality-assurance personnel studied the problem over a period of 4 years. Recall that immunogenicity is a particular problem with protein therapeutics. According to Fred Bader, a vice president at Johnson & Johnson, nearly all therapeutic proteins elicit antibodies in some fraction of patients receiving the drug. The frequency ranges from 1 case per million doses to over 50% of patients treated. In some cases, these

(*Continued*)

Case Study (*Continued*): Detective Work Reveals the Culprit in Drug Mystery

antibodies have no clinical effect, but in other cases, they inhibit the therapeutic effect of the protein. Patients who were administered Eprex® and developed PRCA were found to have mounted an immune response against the recombinant erythropoietin drug. The antibodies produced by these patients not only neutralized the drug, but they also attacked their body's natural erythropoietin, thus shutting down their bone marrow's production of red blood cells and causing PRCA. Investigators therefore focused on factors that might trigger this type of immune response in patients receiving Eprex®. A scientific report published in 2006 showed that mice exposed only to Eprex® did not develop antibodies against it. Investigators therefore looked for adjuvants, agents which might trigger the mice to mount an immune response against the drug. Adjuvants are any substances distinct from the protein of interest (in this case, Eprex®) that trigger the immune system to respond to the protein of interest. The administration of various adjuvants combined with Eprex® resulted in an immune response in the mice that in some cases mimicked the PRCA seen in humans. Investigators therefore suspected an unknown adjuvant was causing problems in patients.

There are many agents that can act as adjuvants and were suspects in this case. These include other proteins from the host cells, extraneous chemicals from processing steps, trace minerals, DNA and RNA from the host cells, and other agents in the final formulation. Scientists focused on the most likely adjuvants that might be present in the drug preparation including CHO host cell proteins and polysorbate 80, which was added to the drug formulation to improve its physical properties. They also investigated the possible effects of plasticizers that might have leached from the rubber stoppers of the syringes used to administer Eprex®. Their investigations eventually identified leachates from the syringes as the culprits that triggered the serious immune response in human patients. The company switched to coated rubber stoppers to eliminate leaching, and this apparently solved the PRCA problem.

In an interesting twist to the case, some evidence implicates polysorbate 80 as a contributor to the leachate problem. Prior to 1998, human serum albumin (HAS), a blood product, was used to stabilize the recombinant erythropoietin drug. In 1998, regulatory authorities in the European Union directed Johnson & Johnson to remove human serum albumin from Eprex in order to eliminate the possibility of pathogen contaminants from a human blood product. The company replaced HSA with polysorbate 80 for their European product. Shortly thereafter, the problems with PRCA were reported across Europe, suggesting that polysorbate 80 contributed to the leachate problem. Johnson & Johnson points to this experience as a reason to require new clinical trials whenever any new process to produce biopharmaceuticals is introduced. As noted above, "the process defines the product."

Sources: Ryan, Mary H., et al. "An in Vivo Model to Assess Factors That May Stimulate the Generation of an Immune Reaction to Erythropoietin." *International Immunopharmacology*, vol. 6, no. 4, 2006, pp. 647–55. doi:10.1016/j.intimp.2005.10.001.

McCormick, Douglas. "Small Changes, Big Effects in Biological Manufacturing." *Pharmaceutical Technology*, November 2004. alfresco-static-files.s3.amazonaws.com/alfresco_images/pharma/2014/08/22/9fc2462d-9e9f-47c3-a2ea-fe97493c9ef5/article-132352.pdf.

Regulations constantly evolve in response to new issues, and there is now an FDA guidance for how sponsors should assess the immunogenicity of therapeutic proteins: Center for Drug Evaluation and Research. "Immunogenicity Assessment for Therapeutic Protein Products." *U.S. Food and Drug Administration*, 22 February 2018, www.fda.gov/regulatory-information/search-fda-guidance-documents/immunogenicity-assessment-therapeutic-protein-products.

37.3.3 TESTING THE CELL BANKS

The FDA requires significant testing to "establish all significant properties of the cells and the stability of these properties throughout the manufacturing process." These tests fall largely into three major categories:

- **Tests of the identity of the cells and the genetic construct.** Identity tests characterize both the host cells and the genetic construct that was transferred into them.
- **Purity tests** look for bacterial, fungal, viral, and sometimes prion contaminants.

- *Stability tests* demonstrate that the cells and their genetic constructs do not mutate or change with time. Stability testing requires thoroughly characterizing the genetic construct and the MCB so that tests can periodically be performed that will be able to detect any alterations in the expression system.

The master cell bank cells are initially tested during development. As each production run begins, the working cell bank used is again tested to see that it has maintained the proper characteristics. After every production run, the end-of-production cells are tested to be sure that they have not changed and have not been confused or contaminated with other cells.

Tables 37.3–37.5 are compiled from the catalogs of several companies that provide cell bank testing services. These tables are included to demonstrate the extensive, varied, and sophisticated laboratory activities required to characterize the cell banks. Observe that the MCB is most extensively tested. The working cell bank is subjected to limited testing, mainly to demonstrate that no adventitious agents have contaminated the system, and that the cells are the right ones. End-of-production cells are tested more thoroughly to ensure that they have not become contaminated (e.g., by the culture medium or careless handling), to check the stability of the cells and the genetic construct, and to ensure that viruses that may have been latent and undetected in the MCB have not been induced during culture. Viruses, if not detected and not removed from the final product, could cause harm to the patient.

TABLE 37.3
Testing a Master Cell Bank

Type of Assay	Assay	Description
Purity Testing		
Presence of microbial contaminants	General sterility	Tests for the presence of bacterial and fungal contaminants by incubating cell bank material in conditions favorable for contaminant growth.
	Mycoplasma sp.	Tests for the presence of *Mycoplasma* sp.
Presence of adventitious viruses	In vitro viral assay	Material from the master cell bank is added to indicator cells in culture dishes. Indicator cells are checked for evidence of viral infection after a specified period of time.
	In vivo viral assay	Animals are injected with material from the master cell bank and observed for clinical signs of viral infection and for serum antibodies against viral agents.
Presence of endogenous retroviruses	PERT	Retroviruses produce a characteristic enzyme, reverse transcriptase (RT); the presence of RT is the basis for many retrovirus assays. (RT enables retroviruses to use RNA as a template for the synthesis of a complementary DNA [cDNA] strand.) PERT assays amplify and detect cDNA which is produced only in the presence of RT.
	Transmission electron microscopy	Virus particles are visualized in cells using transmission electron microscopy.
Presence of specific viruses	Bovine and porcine viruses	Bovine viruses (e.g., from bovine serum and serum supplements) and porcine viruses (e.g., from the use of the enzyme trypsin) can contaminate the cell line.
	Human virus PCR panel	When human cells are used for production, testing is performed to detect HIV, Epstein–Barr, and other human pathogens. PCR amplification of specific viral nucleic acids is used.
	Rodent virus panel	When rodent cells are used for production, specific tests are done for rodent viruses.
Identification/Stability		
Characterization of cells	Isoenzymes	Isoenzymes are enzymes that differ from one another in a physical property, but still catalyze the same reactions. When isoenzymes are analyzed by electrophoresis, different cell lines produce different banding patterns that can be used for identification.
	DNA fingerprinting	DNA fingerprinting produces a banding pattern unique to each cell line.
	Karyology	Karyotyping images the morphology of the cells' chromosomes. Cell lines have characteristic karyotypes that should remain stable over time.

(Continued)

TABLE 37.3 (*Continued*)

Testing a Master Cell Bank

Type of Assay	Assay	Description
Characterization of genetic construct	Copy number	Determines the number of vectors containing the gene of interest that have integrated into the host cells.
	Restriction digest pattern of construct	Characterizes the genetic insert by cleaving the DNA construct with enzymes. The resulting DNA fragments are separated from one another using gel electrophoresis. The fragments are visualized, resulting in a pattern characteristic of the genetic construct.
	DNA sequencing	The DNA sequence of the protein-coding region of the genetic construct is determined.
Other		
Tumorigenicity (the ability of cells to cause tumors)	In vitro soft agarose	Many cell lines are able to live and replicate in soft agarose due to a decreased requirement for cell-to-cell and cell-substrate adhesion. Cells that can grow in soft agar or agarose share some qualities with cancer cells.

TABLE 37.4

Testing a Working Cell Bank

Type of Assay	Assay
Presence of microbial contaminants	General sterility
	Tests for *Mycoplasma* sp.
Presence of adventitious viruses	In vitro viral assay
	In vivo viral assay
Identification/Stability	Isoenzymes

TABLE 37.5

Testing End-of-Production Cells

Type of Assay	Assay
Purity Testing	
Presence of microbial contaminants	General sterility
	Tests for *Mycoplasma* sp.
Presence of adventitious viruses	In vitro viral assay
	In vivo viral assay
Presence of endogenous retroviruses	PERT
	Transmission electron microscopy
Identification/Stability	
Characterization of cells	Isoenzymes
Characterization of genetic construct	DNA fingerprinting
	Karyology
	Copy number
	Restriction digest pattern of construct

37.3.4 OPTIMIZATION OF PRODUCTION METHODS

The manufacture of a biopharmaceutical product is divided into two parts: "upstream" and "downstream" processing. **Upstream processing** *is the process of growing the host cells and their production of the desired product.* **Downstream processing** *involves isolating and purifying the product from the cells.* The optimization of upstream and downstream processing begins in the preclinical period of the product lifecycle and continues to be refined throughout development.

Fermentation and cell culture (upstream processing) are complex processes in which cells are grown under controlled conditions. Mammalian cells normally exist in the body where they are bathed by a complex, high nutrient mixture supplied by the bloodstream. This complex natural environment is absent in a manufacturing environment where cells must be supplied with defined (as much as possible) nutrients in the form of growth medium (see also Chapter 31). Cells vary in their nutrient requirements, depending on their type, and what product they are expressing. Optimizing the medium for each production system is therefore an important part of process development. Other aspects of the upstream process must also be optimized, such as how to aerate the medium and how to remove cellular wastes. During this optimization process, scientists must be careful to not only maximize cell growth, but also maximize gene expression – production of product is ultimately what matters. Upstream processing commonly has two phases: a growth phase during which the conditions are optimized to promote cell growth and reproduction, followed by a phase where conditions are optimized for protein production.

Downstream processing separates the protein of interest from all other substances, including the cells in which it was produced, the growth medium, and contaminants from any source. Proteins vary greatly from one another, and cells used for production also vary; therefore, every downstream process is different and must be optimized. Strategies to isolate a particular biopharmaceutical generally involve a series of steps, each of which employs a different separation method to remove different contaminants. (The fundamental principles of isolating and purifying biological products are covered in more detail in Chapter 34.)

Upstream and downstream processing have a somewhat different meaning when transgenic plants and animals are used for production. Upstream processing becomes growing the plants or animals in a controlled environment. Downstream processing might mean isolating a product from milk, or harvesting a crop and then purifying a component from part of the plants. Once the product is released from its initial source, the principles of downstream processing are similar, regardless of the method of upstream processing.

37.3.5 REMOVING CONTAMINANTS

One of the key concerns during process development is to ensure that contaminants are not present in the final product. Biotechnology products are proteins that cannot be sterilized by heat because they would be destroyed. Development scientists must therefore find other ways to remove infectious contaminants. In addition, as with all pharmaceutical production, routine steps to prevent contaminating the product during production are required. These routine safeguards include, for example, the use of special clean rooms with low levels of airborne particulates, and the sterilization of processing equipment.

One of the first tasks of the process development team is to identify all of the contaminants that might be present in the host cells and the growth medium and to develop methods of purification to remove those contaminants. As the process is developed, the team must perform **clearance studies** *in which a contaminant is intentionally added to the production system and the ability of the process to remove it is measured.* It is, for example, common to perform viral clearance studies in which viruses are intentionally added to the crude product and their removal by subsequent purification steps is measured. It is recommended that a series of clearance studies are performed using different types and sizes of viruses, DNA, and any additives that might be detrimental and whose removal must be documented.

All these development tasks relating to contamination require assays to tell if contaminants are present. There are a number of assays used to test host cells, raw materials, in-process samples, and end products for contamination. Each assay is usually specific for a certain type of contaminant. For example, one series of assays is used to look for DNA, RNA, and protein contaminants from the host cells. Other series of tests are used to look for viral, bacterial, and *Mycoplasma sp.* contaminants. Some of these assays are described in Table 37.3.

37.3.6 SUMMARY: PROTEIN BIOPHARMACEUTICALS

Figure 37.4 summarizes the lifecycle of biopharmaceutical products. Successful biopharmaceuticals have the same overall lifecycle as other pharmaceutical products, beginning with their birth in research laboratories, proceeding through preclinical development, submission of an IND, clinical development, and an application to the regulatory authorities which, if successful, is followed by approval and manufacturing. There are, however, differences in the details of the regulatory pathways, some of which are summarized in Table 37.6.

Although the regulatory pathways of small-molecule drugs and biopharmaceuticals are generally comparable, there are significant biological differences between them. First, protein biopharmaceuticals are a different sort of molecule than small-molecule drugs. Biopharmaceutical drugs are large complex molecules whose three-dimensional structure is critical to their activity. As shown in Table 37.1, characterizing biopharmaceuticals is multifaceted and laboratory-intensive. Assay methods need to be developed, validated, and used to determine the protein's structure, activity, stability, and breakdown products that appear if the protein degrades during storage and handling.

Not only are biopharmaceuticals themselves distinct from small-molecule drugs, but their production systems are also different. Small-molecule drugs are typically chemically synthesized. Protein biopharmaceuticals are often made by genetically modified living cells that express a protein of interest. Master cell banks are created for protein biopharmaceuticals to help ensure the stability of the production system and hence the consistency of the drug product. These cell

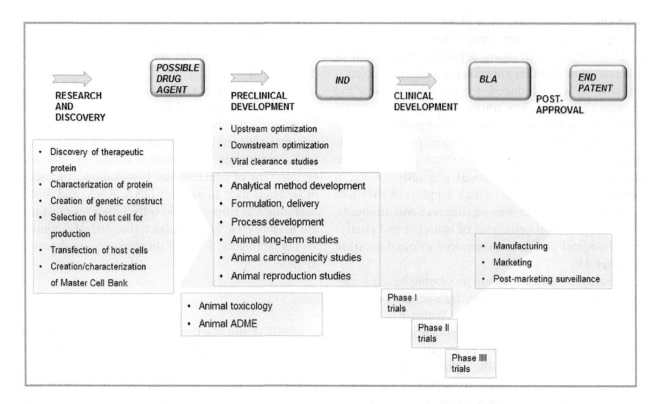

FIGURE 37.4   The lifecycle of protein biopharmaceuticals. The overall lifecycle of protein biopharmaceuticals is similar to that of other drugs, but there are additional requirements when using recombinant host cells as a manufacturing system, some of which are shown in this diagram. Observe that most biopharmaceuticals (e.g., insulin, human growth hormone, and monoclonal antibodies) go through a pathway that requires a Biologics License Application (BLA) rather than an NDA (as described in Chapter 36). The overall principles of regulation are the same in the two pathways, but the details and the historical events that led to the regulations are different.

TABLE 37.6
Regulatory Pathways for Drugs and Biologics

	Drugs	**Biologics**
FDA center oversight	Center for Drug Evaluation and Research (CDER)	Center for Biologics Evaluation and Research (CBER)
Applicable regulations	21 CFR Part 210 (Current Good Manufacturing Practice in Manufacturing Processing, Packing, or Holding of Drugs) 21 CFR Part 211 (Current Good Manufacturing Practice for Finished Pharmaceuticals) 21 CFR Part 11 (Electronic Records)	21 CFR Part 600 (Biological Products: General) 21 CFR Part 11 (Electronic Records)
Required Submissions to FDA		
Application submitted to FDA requesting permission to begin clinical trials	IND	IND
Comprehensive document that the manufacturer (sponsor) must submit to FDA with evidence supporting safety and efficacy. FDA then decides whether or not to approve the drug or biologic for marketing and sale.	NDA	BLA

banks are painstakingly created, tested, characterized, and stored.

Using cells as a production system requires careful management of the upstream processing conditions, such that the cells survive, thrive, and manufacture product efficiently. Optimizing these conditions is another challenging technical task.

The methods used to purify biopharmaceuticals are also distinctive because the biopharmaceutical agent is initially mixed in a broth that contains varied contaminants from the host cells. Avoiding adventitious contaminants requires strict adherence to practices that isolate and protect the cells. Removing endogenous contaminants requires developing practical, effective downstream purification methods. These are further challenges for biotechnologists.

The biopharmaceutical industry has grown steadily as the many technical issues associated with these drugs have become better understood. This has resulted not only in increasing numbers of valuable products, but also in a deepening of our understanding of cellular processes.

37.4 REGULATORY ISSUES RELATING TO HUMAN CELL AND TISSUE THERAPIES

37.4.1 Overview

So far in this chapter we have considered the regulatory and production issues for biotechnology therapeutics that consist of a molecule, typically a protein or protein derivative. In Chapter 2, we introduced another category of biological product, that is, a therapeutic consisting of living cells or tissues. This latter category includes, for example, stem cell treatments, immune cells used in cancer therapy, and tissues for implantation that are grown on engineered scaffolds. The FDA refers to these products as *human cells, tissues, and cellular and tissue-based products*, HCT/Ps. They define this category as: "articles containing or consisting of human cells or tissue that are intended for implantation, transplantation, infusion, or transfer into a human recipient."

A living, cell-based product is different from a protein biopharmaceutical (e.g., insulin) in some important ways. One significant feature of HCT/Ps is that there is a donor (person) involved who provides cells/tissues. *Sometimes the patient is also the donor and the cells/tissues are taken from the patient, processed in some way, and re-implanted back into the original patient.* In this situation, the cells or tissues are called **autologous**. In other situations, the donor is a different person from the patient; *cells/tissues from a donor other than the patient are called* **allogeneic**. For example, stem cells may be autologous or allogeneic. It is possible to create autologous induced pluripotent stem cells by using a cell from the patient (see Section 2.1.4). In contrast, embryonic stem cells used as therapeutics are always allogeneic. The manufacture of CAR-T cells, as discussed in Chapter 2, presently involves only autologous cells, although in the future scientists would like to be able to use allogeneic, "off-the-shelf," cells instead. When allogeneic material is used, it is imperative that no pathogens are introduced from the donor into the patient. From a regulatory perspective, minimizing the risk of implanting contaminated cells/tissues into a patient means that manufacturers must have strict criteria for selecting and testing donors, and for handling donor materials.

Another feature of HCT/Ps is that, in most cases, they contain living cells and therefore cannot be sterilized to remove pathogens. It is therefore essential to use rigorous sterile technique when conducting all procedures involving HCT/Ps. It is also necessary to avoid cross-contaminating one type of cell with another, or cells from one donor with another. Thus, from a regulatory perspective, the manufacturing practices that are used for HCT/Ps must be designed carefully to avoid contamination, and cross-contamination.

A few HCT/Ps, such as bone marrow transplants, have been used for decades, and their regulation is well understood. Others, such as stem cell therapies, are relatively new biotechnology products with less well-trodden regulatory pathways. With these new products, FDA has been trying to balance its essential role in ensuring safety and effectiveness with the societal goal of encouraging innovative new treatments. This balancing act has not always gone smoothly, as illustrated by the case study Stem Cell Therapies and the Regulatory Process: A Conundrum in Chapter 2 on pp. 45–46.

At the time of writing, relatively few HCT/Ps have been approved by the FDA. Those that have can be found on the FDA's website (https://www.fda.gov/vaccines-blood-biologics/cellular-gene-therapy-products/approved-cellular-and-gene-therapy-products). There are, however, hundreds of potential regenerative medicine products that are in clinical trials, and it is anticipated that at least some of these will ultimately be approved. Meanwhile, hundreds of clinics are marketing unapproved stem cell therapies as cures for a variety of ailments, and thousands of patients are buying into the promise of stem cell cures. In an attempt to clarify and control the regulatory situation, FDA issued four guidance documents in 2017 relating to cell- and tissue-based therapies. In these documents, the FDA laid out risk-based regulatory pathways for cell- and tissue-based products. Those products that pose more risk are regulated more stringently. (Recall that we introduced the concept of risk-based regulation in Chapter 35 when medical device regulations were discussed.) Using this paradigm for HCT/Ps, the FDA created three tiers:

Lowest tier: Products Posing the Least Risk to Patients. This category includes cells transplanted as part of fertility treatments between people who are intimate partners (e.g., semen). It also includes cells or tissues that are removed from a patient in a surgical procedure and then re-implanted into the same patient in the same procedure. For example, sometimes adipose tissue from one part of the body is used for cosmetic reconstruction in another part of the body. This tier of products is generally exempted from FDA regulation; this is called the "same surgical procedure" exemption.

Middle tier: Products that Pose Moderate Risk to Patients. These HCT/Ps are regulated by FDA with a particular emphasis on avoiding infection and contamination. The regulations that control these products are covered in a specific area of the code of federal regulations, 21 CFR, Parts 1270 and 1271. These regulations include a subpart that introduces the concept of Current Good Tissue Practices (CGTPs). We will describe CGTPs later in this chapter. Other key sections of 21 CFR, Parts 1270 and 1271 relate to donor screening, testing, and suitability determination. There are a few other requirements (such as a requirement to register one's establishment with the FDA), but notably there is no requirement that these products go through stringent preclinical and clinical testing to prove safety and efficacy. Rather, these products only need to comply with the requirements of 21 CFR, Parts 1270 and 1271.

Highest tier: Products that Pose the Highest Risk. According to the FDA, HCT/Ps that do not meet the requirements for either the lowest or middle tiers are regulated as biologics, drugs, devices, or combination products (depending on their exact nature). The products in this highest tier must therefore go through a pathway involving preclinical testing; applying for an IND; performing clinical trials; evaluating the results; and if the results are favorable, applying to the FDA for permission to market the new product. This highest regulatory tier includes, for example, human cells used in therapies involving the transfer of genetic material (e.g., cell nuclei, mitochondrial genetic material in an egg cell, and genetic material contained in a genetic vector). HCT/Ps regulated as drugs, devices, and/or biological products must be manufactured in accordance with both CGTP and CGMP requirements. The CGTP regulations supplement and do not replace the CGMP and quality system regulations applicable to drugs, devices, and biological products. Going through the full approval process is costly and time-consuming, so most companies would prefer to be regulated only by the requirements outlined in 21 CFR, Parts 1270 and 1271. However, the assignment of a product into a particular regulatory tier depends on the characteristics of the product, not the manufacturer's preference.[1]

Figure 37.5 is a flowchart that describes how the FDA assigns products to one of the three tiers. To understand this flowchart, it is necessary to understand some terms. **Homologous use** *means that the cells or tissues that are taken from a donor are used to perform the same function in the donor and recipient.* For example, skin cells taken from a donor must be used to repair skin in the recipient to meet this requirement. **Minimal manipulation** *means that the cells or tissues must not be processed in some way that changes their essential characteristics.* For example, centrifuging blood to obtain a higher concentration of stem cells is considered to be minimal manipulation because the stem cells' innate characteristics are not altered. However, culturing the stem cells and causing them to enter a particular developmental pathway is not "minimal" manipulation. **Systemic effects** *are defined as effects occurring in tissues distant from the site of the implanted HCT/P.* For example, drugs and biopharmaceuticals nearly always have a systemic effect because they are swallowed or injected at one site, and exert

---

[1] There is an abbreviated pathway for some regenerative medicine products termed the Regenerative Medicine Advanced Therapy (RMAT) designation. For more information on this pathway, consult the FDA website, www.fda.gov/vaccines-blood-biologics/cellular-gene-therapy-products/regenerative-medicine-advanced-therapy-designation.

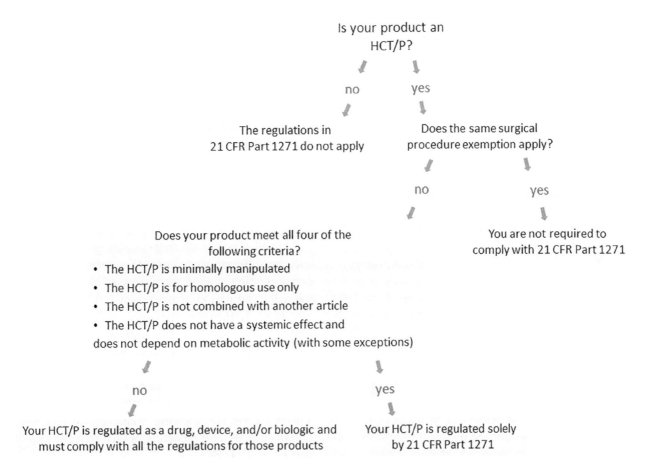

FIGURE 37.5 Flowchart to determine regulatory tier of an HCT/P. (Modified from: Center for Biologics Evaluation and Research. "Regulatory Considerations for Human Cells, Tissues, and Cellular and Tissue-Based Products: Minimal Manipulation and Homologous Use." *U.S. Food and Drug Administration*, 21 July 2020, www.fda.gov/regulatory-information/search-fda-guidance-documents/regulatory-considerations-human-cells-tissues-and-cellular-and-tissue-based-products-minimal.)

their effects elsewhere in the body. Observe that if a product does not meet all four requirements of "minimal manipulation," "homologous use," "not combined with another product," and having "no systemic effect," then preclinical and clinical testing, followed by FDA review and approval, are required.

37.4.2 CURRENT GOOD TISSUE PRACTICES (CGTPS)

Current Good Tissue Practices are introduced and outlined in 21 CFR 1271. They are found in Subpart D – Current Good Tissue Practice. This subpart begins with the statement: "You must recover, process, store, label, package, and distribute HCT/Ps, and screen and test cell and tissue donors, in a way that prevents the introduction, transmission, or spread of communicable diseases." The CGTPs are also described in a guidance document from the FDA, "Guidance for Industry Current Good Tissue Practice (CGTP) and Additional

Requirements for Manufacturers of Human Cells, Tissues, and Cellular and Tissue-Based Products (HCT/Ps)." (See the Bibliography for this unit for the complete reference.) Certain diseases, such as those caused by the human immunodeficiency virus, and the hepatitis B and C viruses, are known to be transmissible through the implantation or transfer of contaminated HCT/Ps. One major strategy to prevent such transmission is donor testing/screening to avoid using cells from an infected donor. However, the FDA points out that donor screening and testing, although crucial, are not sufficient to prevent the transmission of disease by HCT/Ps. In addition to controls on donors, each step in the manufacturing process needs to be appropriately controlled as described in the CGTPs. Anyone who has worked with cell culture, as described in Chapter 30, will be keenly aware that it is easy to introduce contaminants into a culture; it is often difficult to detect contaminants; and it is usually difficult or impossible to remove introduced contaminants. The CGTPs therefore discuss steps to ensure that:

- *Cells and tissues are handled appropriately.*
- *Work areas and surfaces are properly cleaned and disinfected.*
- *Errors in labeling are avoided.*
- *Donor testing records are not confused.*
- *Personnel have requisite skills to perform their jobs.*
- *Cross-contamination of one HCT/P with another is avoided.*
- *Facilities are suitable for HCT/P manufacturing.*
- *Record-keeping (documentation) is thorough.*

Consider, for example, record-keeping. The FDA guidance points out that "A single donor may be the source of a large number of HCT/Ps. It may be discovered, long after the donation and transplantations have been completed, that, due to an error in processing, the donor tissue was infected and capable of spreading communicable disease. Although it might be too late to prevent infections in the recipients, it would not be too late for the recipient to obtain treatment and take steps to avoid infecting others, such as close family members. Unless adequate records were maintained, and maintained for the period of time throughout which infections may be identified, it would be impossible to identify the recipients potentially infected by the donor's HCT/Ps. This would be a critical breakdown in the prevention of disease transmission." As another example, consider personnel. Personnel performing investigations of complaints or adverse reactions related to a possible communicable disease must have the training and experience to review and interpret clinical records, including pathology reports, laboratory results, and medical/surgical interventions.

FDA also points out that a single processing error, such as an improper practice that permitted bacterial contamination of all tissue processed at a location during a limited period of time, may have wide-ranging effects. It is therefore essential to track and report adverse events such as the transmission of communicable disease to recipients of donor tissue. Without such tracking, the common cause of seemingly isolated incidents would never come to light. Affected HCT/Ps would continue to place patients at risk. Therefore, HCT/P tracking, maintenance and retention of records, and reporting of adverse reactions are part of the CGTPs.

Practice Problems

1. List some similarities and some differences between the drugs Herceptin (see Chapter 1, pp. 20–21) and Gleevec (see Chapter 36, pp. 976–977).
2. Examine Tables 37.4 and 37.5.
    a. Based on these tables, list concerns relating to the use of cells for production.
    b. Why are end-of-production cells tested?
3. Which of the following tasks are required when developing and manufacturing any drug, and which are specific to biopharmaceutical products?
    a. Genetically engineering a plasmid to carry a gene of interest into host cells.
    b. Optimizing the cells' growth medium to maximize protein expression.
    c. Avoiding bacterial contamination of the product.
    d. Complying with FDA regulations during manufacturing.
    e. Complying with FDA regulations during clinical trials.
    f. Creating a master cell bank.
    g. Removing host cell proteins during purification.
4. Discuss the role of laboratory scientists, analysts, and technicians in the biotechnology industry. Where possible, include examples from Chapters 1 to 3. (The case studies provide examples that you might include in your answer.)
5. In the testing of a master cell bank and working cell bank, why is the testing of adventitious viruses critical?

6. Match the following:

a.	General sterility	1.	Material from the master cell bank is added to indicator cells in culture dishes. Indicator cells are checked for evidence of viral infection after a specified period of time.
b.	In vivo viral assay	2.	Determines the number of vectors containing the gene of interest that have integrated into the host cells.
c.	Copy number	3.	When human cells are used for production, testing is performed to detect HIV, Epstein–Barr, and other human pathogens. PCR amplification of specific viral nucleic acids is used.
d.	In vitro viral assay	4.	Animals are injected with material from the master cell bank and observed for clinical signs of viral infection and for serum antibodies against viral agents.
e.	Human virus PCR panel	5.	Tests for the presence of bacterial and fungal contaminants by incubating cell bank material in conditions favorable for contaminant growth.

7. What is the difference between *upstream processing* and *downstream processing*?
8. Why must the nutrient medium be optimized for each production system?
9. Proteins are similar to one another, and there is not much variation in cells used for production; therefore, every downstream process is similar and need not be optimized. True or False?

38 Quality Systems in a Regulated Production Facility

38.1 INTRODUCTION

This chapter provides a brief overview of some of the issues involved in manufacturing quality biotechnology products. Before discussing how companies make sophisticated products (such as recombinant proteins or transgenic plants), however, let's look at a product with which most of us are more familiar.

In Chapter 4, we introduced a team that is trying to establish a company to manufacture and sell chocolate chip cookies based on a favorite family recipe. Suppose the team has now completed their initial R&D phase, having experimented with recipes and ingredients, and having tested the quality of the cookies on a small sample of friends. As they scale up for commercial production, they will need to identify the features that made their cookies successful, and formalize these features into written specifications. They will need to set up their manufacturing facility and be certain that the processes for making cookies, which were effective in a standard size kitchen, will also work in a larger, commercial kitchen. They will need a program to ensure that their facilities and equipment

remain clean and operate properly. They will need written recipes and assistant bakers to follow the recipes. They must find ways to ensure that the cookies always bake at exactly the right temperature, for exactly the right length of time. They will need methods to ensure that all the incoming raw materials are of acceptable quality, and they will need to have suitable places to store ingredients, so they remain fresh. They will need to consider packaging, labeling, and shipping of the completed cookies. Whereas personally sampling the cookies was probably adequate quality control in the early stages of the venture, the team will need a more formal quality-control program once they begin commercial production. Furthermore, they will have to comply with all regulations relating to food processing, general safety regulations, and environmental regulations, and they might also voluntarily comply with quality standards for cookie makers. As you can see, manufacturing quality chocolate chip cookies is a complex goal with many components, all of which need to be planned and coordinated.

The manufacturing issues in a biotechnology company resemble those facing the industrious bakers. A biotechnology company must also ensure that adequate resources are available, including facilities, equipment, personnel, and raw materials. Just as the bakers must control the temperature of the ovens and the length of baking time, so must a biotechnology company control its processes that turn raw materials into products. Products – whether they are cookies or monoclonal antibodies – must be properly labeled, must be tested for final quality, and must be shipped to their destination. Skilled personnel must be available to carry out all the necessary tasks. Regulations and standards must be met. Although the basic issues are similar in the cookie kitchen and the biotechnology company, there are differences as well. A biotechnology company is likely to produce many different products and at a much larger scale than would the new bakery. Biotechnology products are likely to be far more technically complicated than chocolate chip cookies. The quality requirements in a biotechnology production facility are therefore even more complex and varied than the issues facing the bakery.

38.2 ISSUES RELATING TO RESOURCES

38.2.1 Facilities and Equipment

A production facility must be able to support the manufacture of products. Consider, for example, a facility in which a drug is manufactured. The building layout should be organized so that processing steps flow in an orderly fashion from one place to another. Raw materials that have been tested and accepted for use should be stored in a space separate from unapproved materials to avoid confusing the two. Materials that are nonsterile must be physically separated from sterile ones. The paths of finished products ideally should not cross the paths of raw materials because of the possibility of confusing them, or contaminating the finished product. The facility must have controls for environmental factors, such as humidity, temperature, dust, and particulates in the air. The facility must be sufficiently large to accommodate all equipment and personnel safely.

A manufacturing facility must be both well designed and maintained properly. This involves a housekeeping program to keep the facility clean, a pest and rodent control program, and environmental monitoring and control. The equipment and instruments in the facility likewise must be suitable for their purposes and must be properly maintained. An important aspect of equipment and facility monitoring is ensuring that all measuring devices operate properly. Measuring instruments are scattered throughout any facility. For example, thermometers are associated with freezers, refrigerators, sterilizing devices, incubators, and rooms. These thermometers must be functioning properly; otherwise, there is no assurance that materials are being held at the proper temperatures.

Example 38.1

Consider the development of a system to regulate production parameters.

It is determined during the development of a product that the temperature at which the product is produced affects its quality. Temperature, therefore, must be monitored and controlled in the production area of the facility. The program for the control of temperature involves the following aspects:

1. During the development of the product, laboratory tests are performed to determine the range of temperatures that is acceptable for production. Based on these tests, it is determined that the temperature of the processing area must remain between 19°C and 22°C.
2. The temperature in the facility is controlled by the heating and air conditioning system.
3. A thermostat is installed in the production area. The thermostat is designed so that it:
   - monitors the temperature on the production floor.
   - automatically prints out a continuous temperature recording.
   - connects to the heating and air conditioning system and turns them on or off as needed.
4. A program of scheduled maintenance is implemented to check all parts of the system regularly, including the thermostat and the heating and air conditioning equipment. Every week the thermostat is checked, and its readings verified to be accurate with a thermometer known to be correct. The heating and air conditioning systems

have monthly maintenance routines. The monitoring program is explained in documents. Records are maintained each time equipment is inspected, adjusted, or repaired.

5. If the temperature is about to go out of range in the production area, an alarm is triggered. There is a document that

explains to personnel what actions to take if the alarm sounds, how to record their activities and observations, and how to follow-up after the problem is resolved. This follow-up should describe what to do with the product that was in process when the alarm sounded.

Case Study: Example Relating to Facilities and Equipment Management

Water is one of the most important raw materials in biotechnology production facilities. This case study describes a situation where a company failed to observe basic quality practices, resulting in contamination of their water. This negligence resulted in patient illness. (The company was subsequently able to correct their deficiencies.)

In 1995, an epidemiologist working at a hospital in Milwaukee noticed that 12 hospital patients, all of whom were on ventilators, developed *Pseudomonas cepacia* infections. The hospital laboratory was able to find the bacterium responsible for the infection in the patients' sputum and in bottles of "Fresh Moment" mouthwash. Although this bacterium is seldom harmful to healthy people, *P. cepacia* can be dangerous to people who are ill. The FDA traced the manufacture of the mouthwash to a particular company, which was subsequently inspected by an FDA investigator. The investigator found five problems thought to have caused the contamination in the mouthwash:

- The company's purified water system had not been properly cleaned and tested since earlier in the year. The company's president told FDA officials that the company skipped the scheduled maintenance because it could not afford it.
- The reverse osmosis membranes of the water system had not been changed for more than a year, although this should be done every 4 months, according to the company's written procedure, to prevent microbial buildup. (Water purification systems are discussed in Chapter 24.)
- Employees failed to challenge the system with *P. cepacia* when they were testing the system. The challenge test would have involved intentionally adding the bacterium to water to see if their purification system could remove it.
- Employees did not use the appropriate hose clamps on equipment. The clamps they used appeared to be the type of hose clamps that are used in the automotive trades, and they were rusty, dirty, and discolored.
- Employees left doors open during production, allowing dust from outside to enter the manufacturing area.

(This account is excerpted from a report in the FDA Consumer, October 1996.)

38.2.2 Cleanrooms in Biotechnology

A **cleanroom** *is an area with a temperature-, pressure-, and humidity-controlled environment in which the levels of contaminants, including dust, aerosols, vapors, and microorganisms, are significantly reduced.* The purpose of most cleanrooms is to protect a product during testing or manufacturing. Cleanrooms vary in size from small enclosures inside a research laboratory, to factory floor-size manufacturing areas. They are used in the testing and manufacture of pharmaceutical/biopharmaceutical products in order to limit the presence of microorganisms (i.e., bacteria, viruses, fungi, and spores of any type) that might contaminate the drug product. Cleanrooms are essential in other types of manufacturing as well, for example, electronics, where the slightest particle can destroy the function of miniaturized products such as microprocessors. Cleanrooms are also used for the production and testing of medical devices and foods.

Cleanrooms are subject to standards that specify maximum particulate contamination levels. ISO 14644-1 is the global standard relating to cleanrooms and must be adhered to in all ISO-certified organizations that use this type of facility. Current Good Manufacturing Practice (CGMP) regulations (21 CFR Parts 210 and 211) also require that companies that produce pharmaceuticals and biological products using aseptic processing provide cleanroom conditions to protect their products from contamination. In **aseptic processing,** *the drug or biological product and its container are sterilized separately, frequently by different methods, and then packaged together under aseptic conditions to create a sterile final product.*

ISO standards designate cleanrooms as ISO Class 1 (cleanest) through 9, depending on the levels of particulates present (Table 38.1). The ISO Class number represents the logarithm of the number of particles ≥ 0.1 µm in size allowed per cubic meter. This means, for example, that an ISO Class 6 facility will have a maximum of 10^6 or 1,000,000 particles per m³. To put this in perspective, a Class 9 cleanroom, the least demanding category, has one-third the level of particulate contamination as an average hospital operating room. The cleanliness of a Class 1 facility cannot be found in any natural environment on Earth.

The FDA recommends that product handling take place in an ISO Class 5 or better cleanroom and that areas adjacent to the processing line be at least ISO Class 7. CGMPs require a rigorous monitoring program to document that both overall particle and microbiological levels remain below the specified limits. For certification, a cleanroom must pass at least five tests, which may include particle count, temperature, humidity, airflow velocity, air pressure, and filter leakage. Once certified, a cleanroom is retested at least every 2 years to confirm the maintenance of appropriate conditions.

Cleanrooms achieve particulate and microbiological control through careful design, operation, and human practices. A sample structure for a cleanroom facility is shown in Figure 38.1, although there are many possible designs for specific applications. The cleanroom is completely sealed off from the outside environment, which allows the operators to control air velocity and pressure, temperature, and humidity. A laminar airflow design is generally used (see Section 10.2.2.1), moving air from filtered ceiling ducts to a recirculating floor exhaust system as shown in Figure 38.1. HEPA and ultra-low penetration air (ULPA) filters are used to remove 99.999% of the airborne particles. Air is exchanged at 10–20 times the rate found in the average office. Manufacturing cleanrooms are operated at positive pressure (which slightly forces air out of the cleanroom), to prevent particle flow into the clean facility in case of air leaks. However, for safety reasons, any cleanrooms used for the handling of biohazards or toxic chemicals are operated under slight negative pressure, to avoid the potential escape of any hazardous materials.

TABLE 38.1

ISO 14644-1:2015 Classification for Cleanrooms

ISO Class Number	Maximum Concentration Limits (Particles/m³ of Air) for Particles Equal to or Greater than the Sizes Shown					
	>0.1 µm	>0.2 µm	>0.3 µm	>0.5 µm	>1 µm	>5 µm
ISO 1	10	2				
ISO 2	100	24	10	4		
ISO 3	1,000	237	102	35		
ISO 4	10,000	2,370	1,020	352	83	
ISO 5	100,000	23,700	10,200	3,520	832	
ISO 6	1,000,000	237,000	102,000	35,200	8,320	293
ISO 7				352,000	83,200	2,930
ISO 8				3,520,000	832,000	29,300
ISO 9				35,200,000	8,320,000	293,000

Extremely Clean ↑ *Clean* (left axis label)

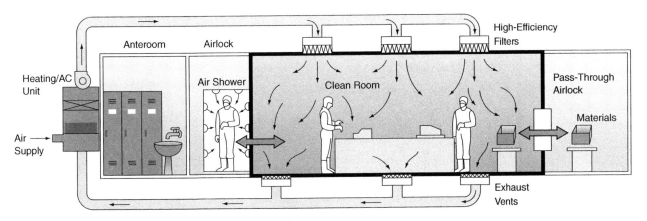

FIGURE 38.1 A simplified cleanroom design. This is a laminar design facility with an airlock shown on the left side for people to enter and exit the cleanroom. Sterile materials enter through a pass-through airlock on the right side of the figure. The exhaust and pressure-regulating systems are not shown.

Humans are the greatest source of contamination in the cleanroom environment, contributing approximately 75% of all particles found in a cleanroom. Under normal circumstances, humans generate particles of dead skin, hair, saliva droplets from coughing or talking, microorganisms, and even clothing lint. Chemically, humans can introduce skin oils, perspiration, and contaminants from beauty and hygiene products into the cleanroom environment. It is estimated that a person sitting at a desk sheds about 100,000 particles per minute and walking at a moderate pace generates 5,000,000–10,000,000 particles per minute! Even slight body movement can shed 4,000 bacteria into the air per minute.

Because humans are such a significant source of contamination, cleanroom operators must follow strict practices. For instance, before entering a cleanroom, operators must change shoes or put on protective shoe covers to avoid tracking contaminants into the facility. Cleanrooms that maintain high levels of protection have much more complex gowning requirements for operators. Intel, a company that manufactures computer components and other electronic products, has a set of 43 specific instructions that must be followed when entering or leaving their cleanrooms. Cleanrooms that are maintained at very low levels of contamination have a dedicated anteroom outside the main space of the cleanroom where users change clothes and store belongings. The anteroom is connected to the main room through an **airlock** (*a small room with interlocked doors between areas with different cleanliness standards*), which may also contain an air shower (shown in Figure 38.1). Everyone who enters such a cleanroom must wear some type of **bunny suit,** *an outfit of special protective clothing designed to prevent human introduction of particulates into the facility.* The items generally included in these suits are shown in Figure 38.2.

Once inside the cleanroom, humans must follow careful practices to avoid introducing contamination. Proper use of a cleanroom requires training and understanding of the many possible sources of contamination. All activities must be carried out using aseptic techniques in order to maintain product sterility. Air turbulence can cause particles to circulate, so movements within a cleanroom must be uniform, slow, and kept to a minimum. The relatively low humidity conditions in most facilities are conducive to electrostatic discharges, so anti-static mats, clothing, and materials within the facility are essential. All items brought into a cleanroom must be either specially designed to minimize particle shedding, or carefully decontaminated before entering. Pencils, ordinary paper, and tissues all generate particles and are therefore not brought into the cleanroom. Watches, jewelry, and keys shed trapped skin particles from their wearers and therefore are removed before donning a bunny suit. Even clicking pens generate particles when used; all writing must be done with one-piece ballpoint pens.

Given that humans are frequently the most disruptive element in a cleanroom, the current trend is toward isolating products and materials that require extreme cleanliness from all human operators. This generally involves the use within a cleanroom of **isolators** *which utilize barrier/isolation technology, where materials occupy an environment that is physically separated from human users.* FDA guidelines recommend that isolators maintain the air quality of at least an ISO 5 cleanroom. Depending on the characteristics of airflow within the isolator, these systems can be designed to prevent contamination of materials and products, or

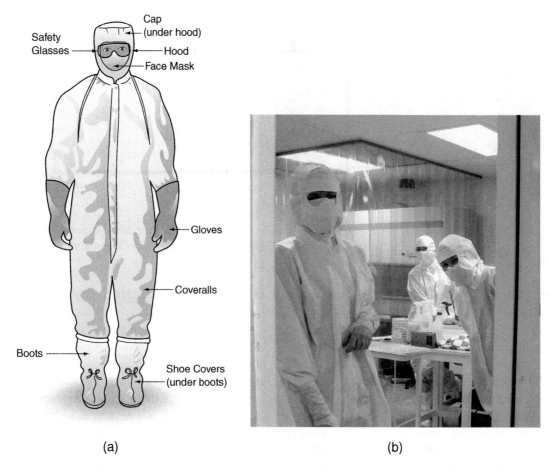

FIGURE 38.2 Examples of protective clothing used in cleanrooms. (a) The components of a bunny suit. (b) Bunny suits in use in a small production facility where master cell banks are prepared.

to protect humans from hazardous materials. A Class III biological safety cabinet is an example of the latter type of isolation system. (See Section 10.2.2.2) Isolators have a relatively low air volume compared to full-size cleanrooms and are particularly well suited for aseptic operations. While they do not eliminate the need for a surrounding cleanroom, they make it possible for the adjacent larger areas to meet somewhat less stringent contamination standards. The future of barrier/isolation technology may ultimately lead to isolator systems that can be used effectively without cleanrooms, as well as the development of automated systems to reduce contamination introduced by human operators. As with much laboratory technology, the equipment is only as effective as the user, as discussed in the case study below.

Case Study: Isolator Technology is Only as Effective as the Operator

A case of lab-acquired SARS virus occurred in 2003 in a BSL-4 facility in Taiwan. A worker was handling cultures of the SARS virus in a closed glove box appropriate for the facility. As the worker was finishing, he noticed a waste spill in an inaccessible area of the enclosure. Instead of the normal decontamination procedure, he sprayed the spill with alcohol and waited for 10–15 minutes before opening the box to transport materials. Alcohol is not effective for inactivating this virus. When the worker developed SARS, it was discovered that he had just attended a large scientific conference, accidentally exposing dozens of individuals to the virus. Luckily, no additional illnesses caused by the SARS virus were found, but many people were quarantined as a result. Knowing the characteristics and hazards of the organisms you handle in the laboratory can help prevent similar occurrences.

38.2.3 Handling Raw Materials

Raw materials (e.g., the chemicals, empty vials, and other components that go into making a product) are resources. Unlike facilities and major equipment, raw materials flow into the company, are used and transformed into products, and flow out of the company in another form. Systems must be in place to receive materials, document their arrival, and ensure that the materials received are the ones ordered. Once received, the materials must be quarantined until they have been tested and approved by the quality unit. This testing confirms that the material is what it is supposed to be, and that it is properly labeled. In a company that makes regulated products, raw materials are often chemically tested in the laboratory to be certain of their identity. Some of the principles that guide handling raw materials are summarized in Table 38.2.

TABLE 38.2

Handling Raw Materials

1. *It is essential not to confuse or mislabel items.* Consider, for example, the serious consequences that could occur if a component of a drug product is accidentally mislabeled.

2. *Raw materials intended for production should not be released to production until they have been tested and found to conform to their specifications.* To ensure that materials are not used prematurely, they are quarantined. Materials are not removed from quarantine until they have been approved for use by personnel from the quality unit.

3. *Storage conditions should take into consideration hazards associated with the material, its perishability, and its temperature and humidity requirements.* A system should be in place so that materials in storage can readily be located when needed.

4. *Traceability of materials must be assured.* Ensuring traceability requires, for example, that all incoming and outgoing materials are recorded, records of raw material components that go into each product are maintained, and items are properly labeled.

Case Study: Example of Warning Letters Relating to Resources and Facilities

The Federal Food and Drug Administration has inspectors who periodically inspect pharmaceutical facilities and biotechnology companies that make regulated products. If the inspectors observe violations of Good Manufacturing Practices, they note the violations on forms, called "483s," and in official warning letters sent to the company. Warning letters are considered a public record. Every year, a summary report of the 483s are published on the FDA website. These *Inspection Observation* data sets can be used by companies producing regulated products to understand the level of enforcement by the FDA and to identify areas of high risk in their companies.

The following are excerpts from a warning letter arising from inspections in 2018 that address issues relating to facility design, contamination, equipment maintenance, and raw materials testing. Observe how carefully inspectors checked the facilities and the details of their findings.

Warning Letter Excerpts:

Dear Mr....

The U.S. Food and Drug Administration (FDA) inspected your drug manufacturing facility [dates and addresses provided]. This warning letter summarizes significant violations of current good manufacturing practice (CGMP) regulations for finished pharmaceuticals. See 21 CFR, parts 210 and 211...Because your methods, facilities, or controls for manufacturing, processing, packing, or holding do not conform to CGMP, your drug products are adulterated within the meaning of section 501(a)(2)(B) of the Federal Food, Drug, and Cosmetic Act (FD&C Act) 21 U.S.C. 351(a)(2)(B)...

2. Your firm failed to test samples of each component for identity and conformity with all appropriate written specifications for purity, strength, and quality. Your firm also failed to validate and establish the reliability of your component supplier's test analyses at appropriate intervals (21 CFR 211.84(d)(1) and (2))...

(Continued)

Case Study: (*Continued*) Example of Warning Letters Relating to Resources and Facilities

Your firm... uses glycerin as an ingredient in ... drug products. Your firm failed to analyze lots of glycerin raw material from your supplier for the presence of diethylene glycol (DEG) and ethylene glycol (EG) prior to releasing it for use in drug product manufacturing. DEG contamination in glycerin has resulted in various lethal poisoning incidents in humans worldwide.

3. Your firm failed to use equipment in the manufacture, processing, packing, or holding of drug products that is of appropriate design, adequate size, and suitably located to facilitate operations for its intended use and for its cleaning and maintenance (21 CFR 211.63)...You failed to adequately design, qualify, and control your ... system used to manufacture drug products. A [piece of equipment] at your facility is ... open to the outdoor environment where it is exposed to vermin, animal waste, and various contaminants...

4. You did not follow your procedures to operate a sanitary.... water system. Further, your procedures to operate your.... water system were insufficient to ensure system control. You were unable to provide chemical and microbiological test results that demonstrate your firm can effectively design, maintain, sanitize, monitor, and control your.... water system to ensure that it consistently produces water that meets the [requirements for pharmaceutical grade water], USP monograph specifications and appropriate microbial limits....You also did not provide a comprehensive review of system design or a water system validation plan...

Until you correct all violations completely and we confirm your compliance with CGMP, FDA may withhold approval of any new applications or supplements listing your firm as a drug manufacturer... After you receive this letter, respond to this office in writing within 15 working days. Specify what you have done since our inspection to correct your violations and to prevent their recurrence. If you cannot complete corrective actions within 15 working days, state your reasons for delay and your schedule for completion.

Sincerely...

38.3 SPECIFICATIONS

Specifications *are descriptions that define and characterize properties that a product must possess based on its intended use.* There are specifications for raw materials and for products. Specifications are described in this quote from an FDA document (*Guideline on General Principles of Process Validation,* Food and Drug Administration, 1987).

During the research and development...phase, the desired product should be carefully defined in terms of its characteristics, such as physical, chemical, electrical and performance characteristics...It is important to translate the product characteristics into specifications as a basis for description and control of the product...For example...chemical characteristics would include raw material formulation ...The product's end use should be a determining factor in the development of product (and component) characteristics and specifications.

We briefly considered differences between road salt, table salt, and laboratory salt in Chapter 4. Table 38.3 continues this example by showing how specifications are used to clarify and define the characteristics of a product. The manufacturers' specifications for three salt products are shown. These specifications illustrate several points about specifications:

1. **Manufacturer's specifications describe properties that are important for that product based on its intended use.** Important specifications for all three salt products include their identity (NaCl, FW 58.44) and their purity. The specifications for road salt, however, mention anticaking agents; table salt specifications mention moisture content; and salt for laboratory use specifies maximum allowable levels for a number of trace contaminants. All three products are salt, all are called "sodium chloride," yet because they have different uses, they have specifications for different properties. All pertinent aspects of the product that impact its performance, reliability, effectiveness, safety, and stability should be considered when establishing specifications.

2. **The specifications for the same property may vary depending on the intended use.** Although all three salts are NaCl, the salt intended for use in the analytical laboratory

TABLE 38.3

Manufacturer's Specifications for Road Salt, Table Salt, and Salt Intended for Laboratory Use

	Road Salt (NaCl, FW 58.44)	Table Salt (NaCl, FW 58.44)	Analytical-Grade Salt (For Laboratory Use) (NaCl, FW 58.44)
Chemical Purity	Minimum 95% NaCl	Minimum 97% NaCl	Minimum 99% NaCl
Color	Clear to white, yellow, red, or black	Clear to white	Clear to white
Maximum allowed contaminants	Not specified	As < 0.5 ppm Cu < 2.0 ppm Pb < 2.0 ppm Cd < 0.5 ppm Hg < 0.10 ppm	Al < 0.0005% As < 0.0001% Ba < 0.0005% Ca < 0.002% Cd < 0.0005% Co < 0.0005% Cr < 0.0005% Cu < 0.0005% Fe < 0.0001% And so on
Physical requirements	90% of crystals between 2.36 and 12.5 mm	90% of crystals between 0.3 and 1.4 mm	95% of crystals between 0.18 and 0.3 mm
Allowed additives	Anticaking agents at 5–100 ppm Sodium ferrocyanide Ferric ferrocyanide	Coating agents, hydrophobic agents	Not allowed
Moisture	2%–3%	≤3%	Not specified

is required to be much purer than the other two products. It would be wasteful to spread ultrapurified salt on roads, and it would be unacceptable to use road salt in the laboratory. None of these products is "better" than the other; they are simply intended for different purposes.

3. **Specifications always are associated with analytical methods.** In order to tell whether a particular batch of salt meets its specifications, it is necessary to have a method to measure its purity, moisture content, and so on. In many cases, the evaluation of a product to see if it meets its specifications requires a sophisticated test. For example, the chocolate chip manufacturers can relatively easily count the number of chocolate chips in a cookie sample, but it requires an analytical method to analyze the percentage cocoa content of those chips. Analytical methods used to evaluate specifications must be proven to work properly.

4. **Specifications are written as a range of permissible values because there is inevitably some variation from product to product.** Acceptable ranges or limits should be established for each specified characteristic and should be expressed in readily measurable terms. (Note that the salt specifications express contaminants in terms of limits, that is, the maximum allowable contaminant levels. Crystal size is specified as a range of acceptable values.) The validity of the specification ranges or limits must be verified through testing the product. In addition, as mentioned earlier, the acceptable range may vary for the same compound depending on its intended use.

The preceding examples are of specifications for a final product. Specifications are also required for raw materials, equipment, and processes. For example, the quality of chocolate chips (a raw material) used to make cookies is important in determining the quality of the cookies. The requirements for the chips, therefore, must be specified. Processes also have endpoints that can be described and specified.

Establishing specifications is a key element in a quality program. Specifications document the product quality that must be achieved and are a contract between the producer and the consumer. The term **establishing specifications** *means:*

- *Defining the characteristics of the product, material, or process.*

- *Documenting those specifications.*
- *Ensuring that the specifications are met.*

Establishing specifications occurs during the development phase of a product. During development, the specifications may evolve and change. Once large-scale production begins, specifications should remain fixed.

Deciding what properties should be specified for a product, raw material, or process is not a simple task. Establishing specifications requires knowledge of how the product, material, or process will be used and the properties that will make it suitable for that use. It is also complex to define ranges for specifications. If the range for a specification is set too tightly (i.e., if the acceptable range of values is very narrow), then some adequate product or materials might be rejected. On the other hand, if the range of values for the specifications is too broad, then the quality of the product is not protected.

Because the setting of specifications is both a key component of a quality program and a difficult task, FDA scrutinizes specifications for products it regulates. FDA will not accept specifications if they are not complete, if they are unsuitable for the product, if their range is too broad, if they are unsubstantiated by testing, or if suitable analytical methods to test them are not available.

Example 38.2

Write specifications for an imaginary enzyme, called enzymase, which is used to destroy all enzymes, other than itself, in a reaction mixture.

Various features of the enzyme are studied and characterized so they can be specified. For example, the intended purpose of enzymase is to destroy other enzymes. An assay for enzyme destruction, therefore, is developed, and the activity of the enzyme preparation is specified using this assay. The enzyme must be purified before sale, and R&D researchers found that if the enzyme is at least 60% pure or better, then it functions well in various applications. Purity therefore needs to be specified. Stability is determined and specified. The enzyme is determined to require Mg^{++} as a cofactor, which is specified. The resulting specifications for enzymase are shown in Figure 38.3.

SPECIFICATIONS FOR ENZYMASE

MOLECULAR WEIGHT: 68kD

COFACTORS: Mg^{++}

INHIBITION: 50% inhibition by greater than 100 mM NaCl and chelators of Mg^{++}

INACTIVATION: Inactivated by heating to 70°C or by the addition of EDTA

SOURCE: Isolated from *B. subtilis* recombinant clone

PURITY: 60–80% pure with 125–130 units/mg dry weight

STABILITY: Supplied as a lyophilized powder and stable at least 2 years when stored at 2–8°C.

Reconstitute with 10 mM phosphate buffer at pH 5.8.

ASSAY: Enzymase is assayed by the modified enzymase assay at pH 5.8. Enzymase is incubated with purified enzyme for 60 minutes at 37°C. The extent of digestion is determined by the colorimetric ninhydrin method. The amino acids released are expressed as micromoles of leucine per milligram dry weight of enzyme. One unit is equal to one micromole of leucine released in 60 minutes.

FIGURE 38.3   Specifications for the imaginary enzyme, enzymase.

38.4 PROCESSES

Products are made by processes. A **process** is *a set of interrelated resources and activities that transform inputs into outputs.* **Inputs** *are raw materials.* **Outputs** *are products or intermediates that lead to products.* **Resources** *include personnel, facilities, equipment, and procedures.* **Activities** *involve one or more people who use the raw materials, instruments, equipment, and procedures to make a product or intermediate.* In order to obtain a quality product, it is essential that processes be carefully designed and developed by the R&D unit. Production operators in a biotechnology company participate in setting up, monitoring, and controlling the processes.

To illustrate a biotechnology process, consider microbial fermentation. Microbial fermentation is a process in which large numbers of microorganisms are grown in nutrient medium contained in large vats, called *fermenters.* During the fermentation process, the microorganisms produce a product, such as an enzyme or antibiotic, which can be isolated and purified. The *raw materials* in a fermentation process include bacteria and the components of the nutrient medium. The *activities* involved in the fermentation process include preparing the medium, adding the medium and bacteria to the fermenters, monitoring the condition of the microorganisms, and maintaining proper conditions in the fermenters. The *resources* involved in the fermentation process include raw materials, fermenters, instruments to monitor conditions in the fermenters, personnel to perform the required activities, and procedures for the operators to follow.

The design of a process begins during the research and development phase and is refined as the product enters production. Every process must be designed so that the inputs are efficiently converted into the desired product. An important aspect of designing a process is finding ways to monitor the process as it occurs, and finding ways to control the process so no problems arise. To effectively design, monitor, and control a process, it is necessary to understand it. For example, if there are minor contaminants in the raw materials, will this affect the process? Is the process very sensitive to a particular factor, such as temperature or pH? Do steps of the process need to be completed within a certain time limit? *The factors that are found potentially to affect the process adversely are factors that must be monitored and controlled.* These factors are sometimes called **critical control points**.

Monitoring and controlling a process may occur in various ways. *Sometimes samples are removed while the process is in progress and are analyzed for a critical characteristic.* Such testing is called **in-process testing**. Specifications for the in-process test results are established. If samples do not meet their specifications, then operators may follow a procedure to correct the problem, or they may initiate an investigation.

Sometimes processes are automatically monitored and controlled. For example, temperature and pH probes are routinely inserted into fermenters during production runs. If the temperature or pH exceeds certain limits, then operators may perform corrective actions, or a computer-controlled system may make adjustments. The *Pharmaceutical CGMPs for the 21st Century: A Risk-Based Approach* (2004), issued by the FDA, provides an overview of the current approach to risk management in pharmaceutical development. The included provisions encourage companies to use modern, automated process control methods. **Process analytical technology, PAT,** *is computer- based and relies on sophisticated detection methods that very quickly monitor the nature of a product and its intermediates and adjust manufacturing conditions as needed.* Modern PAT technology is being implemented, for example, to detect microbial contaminants in a drug batch during the manufacturing process, and to monitor levels of methionine and glucose (required nutrients) during fermentation. Improvements in real-time process monitoring and control are expected to improve the quality of products and the efficiency of production.

Table 38.4 summarizes some of the steps in designing a process.

Example 38.3

a. Discuss the development of a fermentation process to produce the hypothetical enzyme, enzymase.

1. The output of the process is enzymase.
2. An essential early step is to develop a method to analyze the amount of enzymase present. A spectrophotometric assay is developed that is used to measure the amount of enzymase present in samples of the broth.
3. The R&D team performs laboratory experiments to see which bacterial strain makes enzymase most efficiently; the medium formulation that results in the best enzyme yield; the temperature, pH, and

---

TABLE 38.4

Designing a Process

1. *The purpose of the process must be defined; that is, the desired output must be determined.*
2. *An endpoint(s) that demonstrates that the process has been performed satisfactorily must be defined.* A range of accepted values for the endpoint must be established; that is, specifications for the process must be written.
3. *A method to measure the desired endpoint(s) is required.*
4. *Raw materials and their specifications must be established.*
5. *The steps in the process must be determined, usually by experimentation.*
6. *The process must be scaled up for production.*
7. *An analysis of potential problems must be performed.* This may involve listing steps in the process where things can go wrong. These steps are sometimes called *critical points* or *control points*.
8. *Experiments must be performed to determine how the process must operate at each critical point in order to make a quality product.* For example, it might be established that the temperature during incubation must be between 20°C and 25°C.
9. *Methods to monitor the process must be developed.*
10. *Methods to control the process must be developed.*
11. *Effective record-keeping procedures must be developed.*
12. *All SOPs required for the process must be written and approved.*

---

aeration conditions under which the bacteria are most productive; and so on. These early experiments provide the information necessary to design the process.

4. The R&D team establishes the fermenter operating parameters, the nutrient medium formulation, and the process endpoint specifications. The endpoint specifications may include, for example, measurements of the number of bacterial cells per milliliter and the amount of enzymase produced per gram of bacteria.

(Note that the specifications for the endpoint for the fermentation process are not the same as they are for the final product. This is because the enzymase produced by the bacteria must go through a series of isolation and purification steps before it is ready for packaging and sale. These isolation and purification steps are together referred to as **downstream processing**. Thus, the fermentation process produces an intermediate product with its own specifications.)

5. The preparation of standard operating procedures (SOPs) begins in R&D and continues as the process is scaled up. Once the process is in place and production of enzymase begins, changes to the process are tightly controlled.

b. What processing controls are required for the fermentation of enzymase?

1. Preliminary experiments indicate that the hypothetical process is sensitive to pH, temperature, and oxygen levels in the nutrient medium. Early experiments are therefore performed to find the acceptable range for each of these three parameters, and to find means to keep them in the correct range. For example, it is found that the optimal pH for enzymase production is 6.6–7.0 and that the pH tends to drop slowly as the fermentation proceeds, eventually reaching pH 5.1. A basic solution must be slowly added to the broth to maintain an acceptable pH during the fermentation.

2. Based on these experiments, the R&D team establishes control

limits, expressed as a range, for temperature, pH, and oxygen level. Methods of maintaining each parameter in the desired range are devised. When the process is scaled up, these control ranges are confirmed to be achievable.

3. Methods of monitoring the three critical parameters are devised. Each parameter is monitored by a separate sterilized probe inserted into the media. The probe output is connected to a computer that displays and stores the values continuously. If any of the three parameters is too high or too low, then a message is displayed and the medium is adjusted: For pH, base is added; for temperature, cooling is initiated or turned off; and for oxygen, aeration is increased or decreased. These activities are documented throughout the process.

38.5 VALIDATION OF PROCESSES AND ASSOCIATED EQUIPMENT

38.5.1 INTRODUCTION TO VALIDATION

In a CGMP facility, after processes are developed, they must be **validated**. The goal of **validation** is to ensure that product quality is built into a product as it is produced. **Process validation** *is defined by FDA as "establishing documented evidence which provides a high degree of assurance that a specific process will consistently produce a product meeting its predetermined specifications and quality attributes."*

Validation in the pharmaceutical industry emerged from problems in the 1960s and 1970s. During this period, there were a number of cases of septicemia in patients caused by intravenous (IV) fluids that were contaminated by bacteria. The contamination was caused by serious problems in companies manufacturing medical products for IV use. The companies had processes in place that made products, and they had programs in place to test samples of their products; however, because it is not possible to test every single product, they still had contamination problems that led to patient deaths.

FDA's present philosophy regarding validation is easily understood in light of the septicemia history. FDA states that quality, safety, and effectiveness are designed and built into the product. Each step in the production process must be controlled to ensure that the finished product meets all quality and design specifications. Process validation is the method by which companies demonstrate that their activities, procedures, and processes consistently produce a quality result. For example, process validation of a sterilization process might involve extensive testing of the effectiveness of the process under varying conditions, when different materials are sterilized, with several operators, and with different contaminants. During this testing, the effectiveness of contaminant removal would be measured, the temperature and pressure at all locations in the sterilizer would be measured and documented, and any potential difficulties would be identified. Validation thus demonstrates that the process itself is effective. Validation and final product testing are recognized as two separate, complementary, and necessary parts of ensuring quality.

In addition to process validation, there is also validation of major equipment and of tests and assays. (Assay validation is discussed in Chapter 26.) Examples of what might be validated are shown in Table 38.5.

38.5.2 WHEN IS VALIDATION PERFORMED?

Planning for validation occurs throughout the development of the product. The actual validation process is usually performed before large-scale production and marketing of a product begins. Revalidation is

TABLE 38.5

Items That Are Validated

1. *Processes such as*:
   o Cleaning and sterilization
   o Fermentation
   o Purification, packaging, and labeling of a product
   o Chromatographic methods used to purify products
   o Mixing and blending
2. *Major equipment such as:*
   o Heating, ventilation, and air conditioning systems
   o Sterilization equipment
3. *Assays such as:*
   o Assays to analyze incoming raw materials
   o Assays to test whether product meets its specifications
   o Assays that check environmental conditions, such as microbial contamination

required whenever there are changes in raw materials, facilities, equipment, processes, or packaging that could affect the performance of the product.

Validation is important both in "traditional" pharmaceutical manufacturing and in the production of medical products using biotechnology methods. FDA's *Guidance for Industry, Process Validation: General Principles and Practices* (Center for Drug Evaluation and Research. "Process Validation: General Principles and Practices." *U.S. Food and Drug Administration*, August 24, 2018, www.fda.gov/regulatory-information/search-fda-guidance-documents/process-validation-general-principles-and-practices) is a general guide that is applicable to most manufacturing situations. There are also specific, detailed guidelines directly applicable to the validation of biotechnology products for medical use. For example, the "Final Guideline on Quality of Biotechnological Products: Analysis of the Expression Construct in Cells Used for Production of r-DNA Derived Protein Products" discusses the validation of processes involving recombinant DNA proteins. (Center for Drug Evaluation and Research. "Q5B Quality of Biotechnological Products: Analysis of the Expression Construct in Cells Used for Production of r-DNA Derived Protein Products." *U.S. Food and Drug Administration*, April 14, 2020, www.fda.gov/regulatory-information/search-fda-guidance-documents/q5b-quality-biotechnological-products-analysis-expression-construct-cells-used-production-r-dna.) Guidelines and resources related to validation are continuously being written and revised. The FDA website is an excellent source of current information.

Validation is a major undertaking that is expensive, time-consuming, and requires extensive planning and knowledge of the system being validated. The advantage to validation is that it helps to assure consistent product quality, greater customer satisfaction, and fewer costly product recalls. Although we are discussing validation primarily from the perspective of the medical products industry, the concept of validation is generally applicable. The cost of a defective batch of product or a product recall is very high in any industry. Ensuring that all production processes function properly to make a quality product makes good business sense.

38.5.3 PROCESS VALIDATION PLANNING

Every process is different. Validation begins, therefore, with an understanding of the process to be validated, how the process works, and what can go

---

TABLE 38.6

The Components of a Validation Protocol

1. *A description of the process to be validated.*
2. *An explanation of the features of the process that are to be evaluated.*
3. *A description of the equipment, raw materials, and intermediate products that are to be evaluated.*
4. *An explanation of what samples are to be collected and how they are to be selected.*
5. *Procedures for the tests and assays to be conducted on those samples.*
6. *An explanation of how the results of the assays are to be analyzed (including statistical methods).*
7. *A statement as to how many times each test is to be repeated.*
8. *SOPs describing how to run the process.*

---

wrong. Information required for validation is gathered throughout the development phase. The actual validation of a process does not occur until the process is in place, its specifications have been established, most of the factors that can adversely affect the process have been determined, and methods of monitoring and controlling the process have been identified.

Validation requires a **validation protocol,** *which is a document that details how the validation tests will be conducted.* The components of a validation protocol are summarized in Table 38.6.

Example 38.4

Consider the fermentation process to make the hypothetical enzymase. During R&D, the specifications for enzymase were established, an assay for its activity was developed, the process for its production was established, and critical parameters – such as temperature, pH, and aeration – were identified. The requirements for the fermenter equipment used to make enzymase (e.g., the size of the fermenter required, the presence of ports for insertion of probes, and how the fermenter would be cleaned) were similarly determined during development. If necessary, a suitable fermenter was built. SOPs to perform the fermentation-related activities were written. The process is then ready for validation.

38.5.4 Activities Involved in Validation

At the beginning of validation, measuring instruments must be calibrated to ensure that their readings are trustworthy. For example, a pH probe used to monitor the pH of a fermentation broth must be calibrated to ensure that it registers pH correctly. Instruments may be rechecked periodically during validation to ensure that their readings remain reliable. When validation is completed, the calibration of each instrument is rechecked to ensure that it is still correct. If any instrument has drifted out of calibration, then validation of the affected step(s) may need to be repeated.

Processes almost always involve equipment. In order for a process to proceed correctly, the equipment must be of good quality, properly installed, regularly maintained, and correctly operated. Equipment must therefore be validated or **qualified**, that is, *checked to ensure that it will function reliably under all the conditions that may occur during production.* Equipment qualification may be performed separately from process validation, but it is also a requirement for process validation.

Equipment qualification is often broken into steps. *First, the equipment is checked to be sure that it meets its design and purchase specifications and is properly installed; this is called* **installation qualification**. Installation qualification includes, for example, checking that instruction manuals, schematic diagrams, and spare parts lists are present; checking that all parts of the device are installed; checking that the materials used in construction were those specified; and making sure that fittings, attachments, cables, plumbing, and wiring are properly connected.

After installation, *the equipment can be tested to see that it performs within acceptable limits, which is called* **operational qualification**. For example, an autoclave might be tested to see that it reaches the proper temperature, plus or minus certain limits, in a set period of time; that it reaches the proper pressure, plus or minus certain limits; and so on. The penetration of steam to all parts of the chamber, the temperature of the autoclave in all areas of the chamber, the pressure achieved at various settings, and so on, would all be tested as part of the operational qualification of an autoclave.

Once all measuring instruments are calibrated, and all equipment is validated, process validation can be performed. The actual validation of the process will involve assessing the process under all the conditions that can be expected to occur during production. This includes running the process with all the raw materials that will be used, and performing all the involved activities according to their SOPs. Testing includes checking the process endpoint(s) under these conditions, and establishing that the process consistently meets its specifications.

Process validation also involves challenging the system with unusual circumstances. FDA speaks of the "worst-case" situation(s) that might be encountered during production. For example, a sterilization process might be challenged by placing large numbers of an especially heat-resistant, spore-forming bacterium in the corner of the autoclave known to be least accessible to steam. The effectiveness of bacterial killing under these "worst-case" conditions must meet the specifications for the process.

After all these validation activities have been performed, the data that were collected must be analyzed, as described in the validation protocol, and a report must be prepared. Successful validation demonstrates that a process is effective and reliable.

Example 38.5

Discuss briefly how the steps of validation – calibration, equipment qualification, process validation, challenge, and data analysis – apply to the fermentation process used to produce enzymase.

Validation protocol purpose. Does the fermentation process for producing crude enzymase reliably and repeatedly produce a crude enzyme that meets all its specifications?

1. **Calibration.** All the measuring instruments must be calibrated to ensure that their readings are correct. This includes, for example, the pH, temperature, and oxygen probes, and the instruments that monitor agitation rates and fluid flows.
2. **Fermenter equipment qualification.** The fermenter is a major piece of equipment that requires qualification before the fermentation process is validated. This involves:
   a. **Installation qualification.** Establishing that the fermenter and its ancillary equipment are properly installed and meet the specifications set by the design team.

For example, the following items would be inspected and their presence documented:

- The vessel where the bacteria and broth are contained must have features such as smooth walls, proper fittings and seals, and a pressure relief valve.
- The various ports from which samples are withdrawn, and into which nutrient medium can be added, must be machined so that contaminants cannot enter the system.
- Heating and cooling systems must be in place; their utility requirements must be provided; they must be connected to electricity and water, as required; and so on.
- Pumps must be present, have the proper ratings, be connected to power, and so on.
- All instruction manuals and schematic diagrams must be present, and spare parts lists must be available.

b. **Operational qualification.** Establishing that the fermenter operates as it should. For example, the following items would be tested:

- The heating and cooling systems function as specified.
- The air and water flow rates are acceptable.
- All switches function correctly.
- Alarms sound when they should.
- A critical feature of a fermenter is that it can be cleaned and sterilized. This requirement is typically tested by intentionally contaminating the fermenter with a test organism, which is a temperature-resistant bacterium. The fermenter is then run through its decontamination cycle and is checked for any bacteria remaining.
- Any computer associated with the fermentation process must be qualified. It is common to use computers to monitor and control fermenters. For example, a computer might be connected to the pH probe. The computer would continuously record the pH in the fermenter, would display the pH, might control a pump that feeds base into the fermenter as necessary, and might trigger an alarm if the pH exceeds preset limits.

3. **Process validation.** Bacteria and nutrient medium are introduced to the system, and the growth of bacteria and production of enzymase are tested. The fermentation process should be tested under varying conditions as might occur during production. Features to check include the following:

- The ability of the system to maintain the proper temperature, pH, and aeration.
- The absence of contamination in the system.
- Bacterial growth.
- The production of crude enzymase that meets specifications.
- **Worst-case challenge.** A challenge to the system might include, for example, allowing the fermentation process to run extra days to see whether slow-growing contaminants appear in the broth.
- **Analysis of data and preparation of report.** The information gathered during the validation is analyzed, summarized, and documented in a report.

Case Study: Example of Warning Letters Relating to Validation

The Federal Food and Drug Administration has inspectors who periodically inspect pharmaceutical facilities and biotechnology companies that make regulated products. If the inspectors observe violations of Good Manufacturing Practices, they note the violations on forms, called "483s," and in official warning letters sent to the company. The following are excerpts from two warning letters arising from inspections in 2018 that address issues relating to validation. Observe how carefully inspectors checked the facilities and the details of their findings.

Warning Letters Excerpts

Dear Mr…

Following our review of [your facility] we identified [multiple] concerns that are also included below.

2. [You failed] to validate and approve a process according to established procedures where the results of processing cannot be fully verified by subsequent inspection and tests. The validation activities and results must be documented, including the date and signature of the individual(s) approving the validation. [21 CFR 1271.230(a)]. For example, you failed to adequately validate the … manufacturing process, which cannot be fully verified, to ensure that your process removes contamination that is present, and to ensure that processing prevents the introduction, transmission, or spread of communicable disease through the use of the HCT/P [human cell and tissue products]. From October 2016 to December 2017, your firm processed HCT/Ps into …products and distributed those products without a validated process.

Dear Mr…

Production Process Validation

You did not validate the processes used to manufacture … your OTC drug products. You did not perform process qualification studies and lacked an ongoing program for monitoring process control to ensure stable manufacturing operations and consistent drug quality.

Inadequate Control of Water System

You failed to validate your … water system. Your firm has not demonstrated that you could effectively control, maintain, sanitize, and monitor the system so it consistently produces pharmaceutical grade water that, at a minimum, meets the USP monograph for purified water and appropriately stringent microbiological limits. You use water from this unvalidated system as a component in your drugs. You must design your water system to reproducibly yield suitable water for use in production operations.

38.5.5 UNPLANNED OCCURRENCES

We have so far considered many ways in which problems are avoided by careful planning, design, monitoring, and validation. Even in the best managed facility, however, there are *unexpected occurrences, called* **deviations**. Every company must be prepared for deviations and have a plan to deal with them. This plan includes having forms for documenting the deviation and a review process to decide what action to take. A supervisor and members of the quality unit normally review the problem and decide what, if any, corrective action to take. The corrective action and any associated investigation is documented. (See Section 4.5.3, CAPA.)

When a product or a raw material is out of specification, it is called a **nonconformance.**

Nonconformances are also documented. The material is quarantined. Problems with a product often require an investigation to determine the source of the problem and the extent of damage. Problems may have many sources including raw material deviations, human errors, equipment malfunctions, and other unforeseen occurrences. Damage may be minor, or may affect an entire batch of product. After investigation, the quality unit personnel decide what to do with the nonconforming material. If it is a final product, then it may need to be destroyed or reprocessed. A raw material may be returned to the vendor. As always, the investigation activities and finding should be documented and follow-up should occur.

38.6 THE QUALITY-CONTROL LABORATORY

38.6.1 INTRODUCTION

The quality-control (QC) laboratory plays a key role throughout production in any facility. QC analysts play a key role in testing:

- The production environment, including the air, and cleanliness of surfaces and equipment.
- Incoming raw materials.
- In-process samples.
- Final product before release.

These analysts therefore play an essential role in maintaining quality throughout the organization.

38.6.2 OUT-OF-SPECIFICATION RESULTS AND THE *BARR* DECISION

Suppose an analyst works in a QC laboratory, runs a test on a final product, and obtains a result that does not meet the specifications for that product. A drug product, for example, is specified to have a potency that is between 2,000 and 2,500 units/mg, but when potency tested gives a result of 3,200 units/mg. This result is called an **out-of-specification (OOS or out-of-spec) result**, *that is, a test result that is outside the range required for the product.* What happens now? There are two possible causes for this out-of-spec result: (1) This batch of drug product really is too potent, or (2) there was an error in the laboratory analysis. These two possibilities have very different implications for the company. The drug cannot be released for sale if it is really too potent, but if there was an error in the laboratory analysis, then the drug may actually be fine.

Companies must have procedures to handle out-of-spec results. An analyst's first instinct is often to immediately retest the drug product to see if the same result occurs again. If the result of the retest falls within the specified range, then the analyst might conclude that the original result was due to an unknown mistake, "throw away" the original result, and record the second result. In fact, in some companies analysts have been directed to retest samples over and over again until a passing result is obtained, or to average the results of more than one test to obtain a passing result. The FDA is very concerned with how laboratories handle OOS results in the pharmaceutical industry. If one QC test shows that a product fails to meet its specifications, but a later result shows it passes, how

does the company know which result was correct? If laboratory analysts are making mistakes that are just "thrown away," how can the company improve the quality of its testing? Simply repeating a test when the results are not expected is not acceptable in a GMP-compliant organization.

The practice of repeated retesting when OOS results were obtained (along with some other dubious practices) caused one company, Barr Laboratories, to undergo a costly and famous (within the pharmaceutical industry) courtroom battle. FDA investigators inspected Barr's manufacturing plants in 1989 and 1991 and recorded numerous criticisms, including some relating to how Barr handled OOS results. Barr QC analysts would, for example, average OOS values with in-spec values to get a passing value. They would discard raw data and perform multiple retests of samples with no defined endpoint (other than a passing value). They did not perform adequate investigations of failures (see Chapter 4 for a discussion of failure investigations), and they did not adequately validate their testing methods. FDA requested a court injunction requiring Barr to suspend, recall, or revamp numerous products. Barr filed a countersuit alleging that the FDA was applying unfair rules to them. The ensuing courtroom battle generated thousands of pages of testimony, hundreds of exhibits, and numerous lengthy declarations. Judge Alfred Wolin, who presided in the case, was responsible for sorting through the conflicting claims of the various experts called in by both sides. The principles that FDA now enforces relating to OOS results are largely based on Judge Wolin's legal rulings in 1993.

Judge Wolin ruled against Barr Laboratories in a number of instances and demanded recall of certain products, although he stopped short of shutting down their manufacturing operations. Judge Wolin acknowledged in his ruling that the CGMP regulations "create ambiguities" and "the regulations themselves, whose broad and sometimes vague instructions allow conflicting, but plausible, views of the precise requirements, transform what might be a routine evaluation into an arduous task." Judge Wolin ruled that how a company deals with OOS results is often not clear-cut because there are ambiguities and differences from one situation to another.

Judge Wolin did provide a number of specific rulings for the industry, despite the ambiguities relating to OOS results. An important part of his ruling was that a company could not simply retest a sample when an OOS result was obtained. He ruled that if an analyst obtains an OOS result, then the analyst must perform the following steps:

1. The analyst must report the result to a supervisor.
2. The analyst and supervisor must conduct an informal lab inspection, including discussing the testing procedure, reviewing all calculations, examining the instruments used, and reviewing the notebooks with the OOS result.
3. The analyst and supervisor must document the investigation results, recording any errors that might have been uncovered and any conclusions that were reached.

This informal investigation may reveal a clear laboratory error, for example, a calculation that was performed incorrectly. If an obvious error is found, then the initial result of the test can be invalidated (not used) and the sample can be retested.

A rigorous formal investigation must be performed if no obvious laboratory error is uncovered in the informal investigation. A formal investigation requires that the quality-assurance (QA) unit become involved and that the investigation team follow a previously written standard operating procedure. During the investigation, some limited retesting may occur, not to approve the product, but to determine the cause of the OOS result. If an error is found and the root cause can be documented, then the original test result can be invalidated and the sample can be retested. At this point, corrective actions should be taken to avoid further problems of the same nature.

If no laboratory error can be identified, then a more extensive investigation should ensue that will involve looking for errors in the manufacturing process. A report must be written at the end of the investigation that identifies the most probable cause of the OOS result, and what actions are to be taken, and by whom. The report must note if a defect in a batch is discovered, and if other batches might be affected. If a product has already been released to the public that is found to not conform to its specifications, then the FDA must be alerted within 72 hours of the discovery. The company must also take preventive actions to prevent recurrences of the problem in addition to taking corrective actions to save the batch, if possible, or to prevent the release of defective product. A guidance document published by the FDA in 2006 that includes recommendations for laboratory workers is summarized in Table 38.7. (Center for Drug Evaluation and Research. "Investigating Out-of-Specification Test Results for Pharmaceutical Production." *U.S. Food and Drug Administration*, August 24, 2018, www.fda.gov/regulatory-information/search-fda-guidance-documents/investigating-out-specification-test-results-pharmaceutical-production.)

While the *Barr* case relates to the pharmaceutical industry, there are lessons for any laboratory where samples are tested – which is to say, all laboratories. Most notably, *Barr* highlights the need for well-run quality systems in laboratories. The company spent

---

TABLE 38.7

Avoiding Laboratory Error

- Analysts should verify that all documentation relating to a product is in order and matches the label before performing a test on it.
- Analysts should verify that the proper test method is being used on a sample.
- Analysts should be certain that all instruments are properly calibrated before beginning the test and that any standards used to test the instruments perform as expected.
- If any standard does not give the expected result, then the test should be suspended and any data already acquired should be investigated.
- If an analyst makes an obvious error, such as spilling some of the sample, that error should be documented and the test begun again.
- An analyst should never finish a test where an error was known to have occurred.
- No samples should be discarded before the results of a test are checked. If the results are out of spec, then the samples must be saved for investigation.

The supervisor has further responsibilities including:

- Ensuring that there is a system to manage samples.
- Ensuring that all analysts have the required training to do their job.
- Notifying QA when necessary.
- Ensuring that all SOPs are available and up to date.
- Ensuring that proper documentation occurs during an investigation.

huge sums in court costs and recalled products largely as a result of a poorly managed laboratory quality system. (See also the costly consequences of poor-quality laboratory systems in the case study "OOS Results" below.) This incident also made clear some of the difficulties in designing a process that works for all situations. In the end, producing quality products relies on sound scientific judgment on the part of analysts and supervisors, coupled with a company's commitment to releasing only high-quality products.

Case Study: OOS Results

The Federal Food and Drug Administration inspectors periodically inspect pharmaceutical facilities. If the inspectors observe violations of Good Manufacturing Practices, they initially report them on Form 483. If the company does not fix the problems, the FDA sends a warning letter to the company. If the company does not correct the violations noted in the warning letter, then serious action follows. This case study reports on a situation where the FDA imposed very serious sanctions on a company that was not in compliance with CGMPs.

In 2002, Schering-Plough Corporation and two corporate officers signed a consent decree agreeing to take drastic measures to ensure that drug products manufactured at the company's New Jersey and Puerto Rico plants complied with GMP requirements. The company agreed to disgorge profits of $500 million. (Disgorgement of profits is a sanction based on the premise that an individual or corporation is not entitled to profits gained by illegal means.) They also agreed to pay up to $175 million and disgorge additional profits if the company failed to meet certain time frames. The company further reimbursed the government about $500,000 in inspection costs. They agreed to submit comprehensive facility work plans to FDA, have trained FDA-approved personnel at each facility to provide full-time oversight of all operations, and have consultants to conduct annual inspections of the facilities for 3 years. For at least 5 years, the company agreed to conduct regular audits of its operations and make reports to FDA regarding compliance. The company further agreed to periodic FDA inspections. If the company failed to comply with the consent decree, they could be held in contempt of court and could face additional penalties, and their plants could be shut down. This consent decree came after several inspections beginning in 1998 found numerous violations of GMP regulations that the company failed to adequately address. Some of these violations relate to OOS results and to investigations following OOS results. Consider the following short excerpts from a Form 483 issued on June 5, 2002. (This case is also discussed in Practice Problem 1.)

"After obtaining an out-of-specification assay result during the validation [of] the K-Dur extended release coating process, you reanalyzed the same sample preparation. When a second OOS result was obtained, you retested the sample using two analysts, this time obtaining two OOS results. These preparations were then further agitated and reanalyzed in duplicate by each analyst obtaining four OOS results for a total of eight OOS results. You then resampled the batch, tested the sample, and obtained passing results, this result was used as the official result to release the batch for further processing....The investigation states 'In a conversation with Technical Services [Personnel], an error in the original sampling process was made. The original results will be voided and resample results will be used as official for this test.' You could not provide a description of the sampling error, the specific procedure that was not followed, or any corrective action to prevent a recurrence....

Preliminary variance report 20010219 was initiated to document a low net yield and an interruption during the processing of...coated pellets...As part of the investigation, a special request was issued for analytical testing. However, the preliminary report does not list the results but instead states 'pending analytical test results.' However, the Recommendation for Material Disposition states 'Based on the investigation and special request results, approval of the lot is recommended' even though the person writing the report did not have the results of the testing yet.

After obtaining an Out of Specification test result of 121.6% for blend uniformity during the performance qualification of [a product] your firm reanalyzed the same sample preparation. The retest result of 121.9% confirmed the original result. A second sample...was tested obtaining a result of 122.0%. However this value was invalidated due to chromatography problems. [A] second analysis of the second sample obtained a value of 100%, this value was accepted without question..."

Practice Problems

1. In the case study "Example of Warning Letters Relating to Resources and Facilities," we discussed violations of Good Manufacturing Practices, or 483s. Every year, a summary report of the 483s issued is published on the FDA website. Access the FDA website (https://www.fda.gov/inspections-compliance-enforcement-and-criminal-investigations/inspection-references/inspection-observations) and locate the Inspectional Observations data sets. The Inspectional Observation data sets are organized by year on the FDA website. Download the Inspectional Observations data set labeled 2019.

   a. How many 483s were issued ("Sum Product Area 483s from System") between October 1, 2018, and September 30, 2019?

   b. Of the 483s issued between October 1, 2018, and September 30, 2019, how many were issued for Drugs? Devices? Biologics?

   c. Of the 483s issued for Drugs, there were 215 citations for violations of 21 CFR 211.22(d). What does 21 CFR 211.22(d) state? Why is this important in analytical testing?

   d. Of the 483s issued for Drugs, there were 53 citations for 21 CFR 211.25(a). 21 CFR 211.25 is related to personnel qualifications. In your own words, why are personnel qualifications important in GMP operations?

   e. Of the 483s issued for Biologics, there were 15 citations for 21 CFR 606.160(b). 21 CFR 606.160 addresses records. What records are discussed in this section? Why are these records important?

2. According to Table 38.1, what kind of room would be required to obtain a maximum concentration of $n = 100$ for particles > 0.1 µm?

3. What are types of resources used in the production of biopharmaceutical products?

4. Specifications *"are descriptions that define and characterize properties that a product must possess based on its intended use."* Which of the following are examples of specifications?

   a. $pH = 7.1–7.2$

   b. Color = aqua blue

   c. Bead size = 20–50

   d. Polypropylene plastic, pharmaceutical grade

5. With process analytical technology (PAT), processes are monitored in real time. What purpose does process analytical technology serve?

6. Validation is critical in the development of pharmaceutical products. What is the goal of validation?

7. What are the advantages and challenges of validation?

8. In the Barr decision, what should an analyst do when an OOS result is obtained?

9. True/False. To avoid a laboratory error:

   a. If any standard does not give the expected result, then the test should be suspended, and any data already acquired should be discarded.

   b. Analysts should verify that the proper test method is being used on a sample.

   c. Samples can be discarded before the results of a test are checked, as long as it has been documented in the sample use log.

10. The fines levied against Schering-Plough in 2002 were among the most severe in the history of FDA enforcement. Explain in your own words what the company did wrong with regard to handling OOS results. Why did the FDA take these errors so seriously?

Questions for Discussion

1. How could the Inspectional Observational data sets (as discussed in Practice Problem 1) be used by companies producing regulated products?

2. Look up a more recent Inspectional Observation data set and comment on the problems that are currently of most concern in the industry.

3. Warning letters are posted on the FDA website and are considered public record. You have been given the responsibility to write a response letter to the FDA. Research what should be included in a response letter and review the warning letter on p. 1028. Discuss what would you include in your response letter to the FDA.

4. Consider the warning letters on p. 1025 relating to validation.
   a. Why does the FDA emphasize validation?
   b. What situations do you think would require a company to revalidate a process?
   c. Procedures for cleaning equipment and work areas are extremely important in a pharmaceutical company. Describe how you think the company that received a warning letter regarding cleaning validation should have validated their procedures for cleaning equipment.

5. Consider the case study on p. 1011 relating to the contaminated mouthwash. Discuss how this company might restructure to improve the quality of its operations. For example, how might they prevent mistakes such as open doors and improper hose clamps?

Appendix: Answers to Practice Problems

CHAPTER 1

1. Cloning refers to the production of one or more copies of an original item. In Figure 1.4, it is a genetic sequence, or gene of interest, that is copied. Observe that the original transformed bacteria are placed in a fermenter where they reproduce many times to form many new bacteria. As the bacteria reproduce, so do the plasmids within them. Reproduction results in many new plasmids and many new copies of the original genetic sequence of interest.

2. See Glossary.

3. Remember that transfection refers to the introduction of "foreign" DNA into eukaryotic cells and transformation refers to the introduction of "foreign" DNA into prokaryotic cells.

   a. Producing crop plants that are resistant to the herbicide glyphosate involves **transfecting** the cells with a gene that confers resistance.

   b. Growing cultured human skin cells to use for tissue engineering requires growing large numbers of cells. This process does not require the introduction of new genetic material (although one can imagine scenarios in which transfection might be used to improve the procedure).

   c. Making wine does not require the introduction of DNA into cells.

   d. Bacteria do not naturally make human proteins and therefore must be **transformed** with the gene in order to produce human epidermal growth factor.

   e. Harvesting stem cells from an embryo does not require transfection or transformation.

4. **a** and **b**. Antibodies normally made by a person are proteins that generally recognize and bind foreign agents (e.g., bacteria and viruses). This binding signals other cells in the body to engulf and destroy the invading agent. Monoclonal antibody drugs, like naturally occurring antibodies, recognize and bind specific targets. You can see in Table 1.2 that the targets of established mAb drugs, however, are usually not foreign agents, such as bacteria or viruses. The targets rather are proteins normally found on the surfaces of cells that for some reason have become involved in a disease state. The binding of the mAb blocks the activity of the target protein. As shown in Table 1.2, many of the targets of currently available mAb drugs are related to cancer. Some are related to autoimmune diseases (such as Remicade®) and a few to various other disorders (such as Xolair®). Since Winter 2020 the FDA has issued Emergency Use Authorizations for multiple mAb drugs for the prevention of severe symptoms of COVID-19 in immunocompromised individuals and others who have been exposed to the causative virus or manifest early symptoms of the disease.

5.
   a. Synagis® *recognizes* and *binds* the F protein on the surface of the viral particles. In RSV infection, the F protein binds to surface proteins on lung cells. This binding leads to the internalization of the virus into lung cells, hence, infection. When the Synagis® antibody binds the F protein, the virus is *blocked* from binding to surface proteins on the lung cells and therefore is not internalized.

   b. Conventional vaccines stimulate the body to form *its own antibodies* against the infectious agent. Synagis® is itself an antibody and does not stimulate the patient to form his/her own antibodies against RSV.

6. Biopharmaceuticals are usually proteins. Like the proteins that we consume in food, biopharmaceuticals would be digested if they were taken orally. (Researchers are looking for ways to get around this problem to make oral administration of biopharmaceuticals possible.)

7. Traditional vaccines use a weakened or killed pathogen. These pathogens need to be grown up in mass and then processed to make vaccine. Growing and processing enough pathogen to make billions of vaccine doses is time-consuming. Protein-based vaccines that are manufactured using genetically modified cells also require time for cell growth and

protein purification. Synthesizing mRNA is faster because it does not require culturing pathogens or growing cells.

8.

   a. This is a protein-based vaccine.

   b.

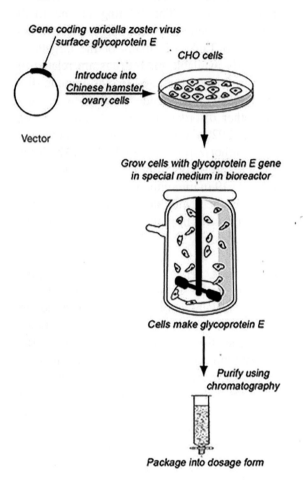

Gene coding varicella zoster virus surface glycoprotein E

CHO cells

Introduce into Chinese hamster ovary cells

Vector

Grow cells with glycoprotein E gene in special medium in bioreactor

Cells make glycoprotein E

Purify using chromatography

Package into dosage form

CHAPTER 2

1. Bone marrow stem cells were targeted in the X-SCID experiment. Bone marrow stem cells divide to make blood cells for the lifetime of a person. The cells that were targeted in Ashanti De Silva were not stem cells.

2. a, b, d involve the use of cells as factories. c. Apligraf®, the cells themselves are the major portion of the product.

3.

   a. CD33, CD34, and CD133 are proteins found on the surface of blood cells.

   b. CD33 is the target of a monoclonal antibody drug, Mylotarg®. This is because CD33 is found on cancerous blood cells.

   c. CD133 and CD34 are used to identify hematopoietic stem cells.

4. Hematopoietic stem cells differentiate into all types of blood cells including white blood cells, red blood cells, and platelets.

5.

   a. Parkinson's: cells in the brain that produce the neurotransmitter, dopamine.

   b. Diabetes: beta cells in the pancreas.

   c. Amyotrophic lateral sclerosis: motor neurons.

6. Mutations in the SARS-CoV-2 virus can affect the efficacy of any vaccine. In a worst-case scenario, the virus might mutate quickly and so any vaccine would have efficacy for only a limited period of time. We see this phenomenon with the influenza virus where people must receive a flu vaccine every year to keep up their immunity. Mutations might also make the virus more or less infectious or pathogenic (able to cause disease). At the time of writing, this has been the case with the delta variant of SARS-CoV-2. Studying mutations also has been important in allowing scientists to track the spread of the virus. For example, scientists found that the genome of the virus that caused the epidemic in New York City in spring 2020 was much like that of the virus found in Europe, while the genome of the virus causing the disease outbreak on the West Coast of the United States was similar to that in China. This provides information about how the virus entered and spread through the United States.

7. Answers may vary. Some main ideas:

   • The major difference is that the process that was used to create Dolly is intended to generate a whole, adult animal that is a clone of a single parent. Therapeutic cloning is intended to generate stem cells.

   • Both processes involve the introduction of genetic information from only one parent organism into an enucleated egg cell. (Sexual reproduction is not involved.) The egg cell divides to form an embryo.

   • For therapeutic cloning, stem cells are harvested from the blastocyst stage of the embryo. In whole animal cloning, the embryo is allowed to develop.

   • To create Dolly, the egg was implanted into a surrogate mother. For therapeutic cloning, the egg is cultured in the laboratory until the stem cells are harvested.

CHAPTER 3

1. There may be more than one possible match to some of these items because there are differences between companies in the division of tasks between functional units. Reasonable answers include:

   Research and development: a–c, f, h, k
   Production: e, i, j
   Quality control: g, m
   Quality assurance: d, l

2. You may think of many ideas, for example, your description of an ideal location, availability of wireless Internet, and access to transportation. Be sure to think about the laboratory needs of your company, since start-up companies usually are heavily involved in the testing and development of their products. The majority of your space may initially be devoted to the laboratory. A good laboratory space will have ample space for equipment and personnel, ample electrical power, access to water, good ventilation, laboratory benches, and so on.

3.

   a. This position is primarily R&D although it does include some QA/QC responsibilities. It is common in smaller, start-up companies for personnel to have multiple responsibilities.

   b. The company has created pancreatic cell lines using iPS techniques. These cells can be used to test prospective diabetes drugs to see if they have a positive biological effect. The cells can similarly be used to see if drug candidates are toxic or have adverse effects.

   c. No, this company appears to be involved in the research and development that comes before any testing of a drug in humans.

4.

   a and b. EP3472352A1. **Inventors**: Andrew Hutchinson, John V.W. Reynders, Guillermo Del Angel, Nina Jain, Christen D. Forbes, Xiao-Qin Ren, Barbara Burton. **Assignee**: Alexion Pharmaceuticals, Inc.

   US5972346A. **Inventors:** Pierre Hauser, Nathlie Marie-Josephe Claude Garcon, Pierre Desmons. **Assignee:** GlaxoSmithKline Biologicals SA.

   US9284371B2. **Inventors:** Itzcoatl A. Pla, Joseph G. Matuck, John C. Fann, Christof Schulz, Nichole A. Roy, David F. Bruton, James McIntire, Yu-Hsiang D. Chang, Thomas Seewoester. **Assignee:** AbbVie, Inc.

   The inventors are the individuals who actually created the invention. The assignees are the entities that have property rights to the patent and can license/profit from it. If researchers create an invention when working for a company or institution, the company/institution normally owns the invention. However, the inventors must be listed on the patent. Note that the assignee can change if the patent is transferred to another entity, but the inventorship, once correctly stated, cannot change.

   c. EP3472352A1: This invention relates to a diagnostic method to detect mutations in a particular gene, the LIPA gene. The application also details a series of specific nucleotides that are sometimes mutated, leading to changes in the amino acid sequence of the protein. The LIPA gene codes for an enzyme that is found in lysosomes and breaks down lipids. When the LIPA gene is mutated and its protein product does not form properly, a serious disease can occur.

   US5972346AL: This invention describes the components used in a vaccine to immunize people against the viral pathogen that causes hepatitis B. Note that the key component of this formulation is a protein, the hepatitis B surface antigen, that elicits an immune response in people administered the vaccine.

   US9284371B2: This invention relates to improved methods for producing a recombinant monoclonal antibody using cultured cells (as described in Chapter 1 in this textbook). The inventors describe improved culture media and culture conditions that increase the production of the desired protein.

   These three inventions thus include a diagnostic method, a method of producing a therapeutic vaccine, and methods to better culture cells.

   d. Figure 2.17 shows a report for a patient whose DNA was screened to look for mutations in genes that are sometimes mutated, leading to breast cancer. These are termed "pathogenic mutations." Similarly, the invention in this patent describes a group of pathogenic mutations that are associated with disease, in this case, LAL-D.

e. Adalimumab is another name for the drug, Humira®. See Table 1.2 in this text for more information.

f. When an individual is exposed to a foreign protein, in the form of either an invading pathogen or a vaccine, the immune system responds with the formation of antibodies to neutralize the pathogen and with immune cells (T cells) that assist in the destruction of the pathogen and of infected cells. The immune system "remembers" the invader and so, if a person is later exposed to that pathogen, the immune system is able to quickly neutralize the threat. This memory is key to developing an effective vaccine. The same strategy is being used to vaccinate people against COVID-19. With the SARS-CoV-2 virus, a protein on the surface of the virus, the spike protein, is a typical vaccine target. The spike protein allows the virus to invade cells. The figure in the Unit I Introduction shows the location of the spike protein on the surface of this virus. If people are vaccinated either with the spike protein, or with DNA or mRNA encoding the spike protein, they will make antibodies against the protein that will protect them in case of exposure to the actual virus.

CHAPTER 4

1. There are a variety of good business reasons to comply with ISO 9000 standards. Companies voluntarily comply with standards to help ensure that they have a quality product. The requirements of ISO 9000 guide the company in setting up organizational systems that promote quality. Companies also comply with these standards to demonstrate to their customers and other interested parties that their product is of good quality. Remember that there are costs associated with a poor-quality product ranging from loss of customers to lawsuits and legal proceedings.

2. This is a discussion question with various answers. It is particularly important to note that a change in the manufacturing process for a biopharmaceutical might change the resulting drug product. In the event of such a change, the resulting substance is not the one tested and approved by regulatory authorities for use in patients. An unapproved drug cannot be distributed for use in patients. Moreover, because biopharmaceuticals are complex molecules, such changes might be subtle and might not be detected through normal laboratory tests.

3. You can think of various answers to this question. Standard procedures help ensure that results from different individuals, and from the same individual over time, can be compared to one another. Standard procedures that are written by a knowledgeable individual ensure that everyone performs routine tasks correctly. Standard procedures help avoid breakage of equipment due to mistakes. Standard procedures are used to teach individuals how to do the work of that laboratory. When researchers do not follow consistent, correct procedures, their results may be incorrect or inconsistent. At best, this results in wasted time and supplies. At worst, the researchers may reach erroneous conclusions and publish erroneous results.

4.

a. A product must be tested to see if it meets its specifications before it can be released for sale to the public. This testing is performed by the QC unit. If a test of a product is performed and the product fails to meet the requirements of that test, the result is termed **OOS**. An OOS result means either that the product is defective in some way, or that the laboratory test was invalid for some reason. When an OOS result is obtained, there must be an investigation to find its cause. In this particular case study, 70% of all OOS results turned out to be problems with the testing procedure, and not the product. The erroneous results were invalidated. This means there were problems in the testing laboratory that were costing the company time and money and leading to unpredictability in how long it would take the laboratory to complete its work before releasing products.

b. In all the situations in the table, the analysts failed to follow the correct procedure to perform a test. Their failures to perform the tests correctly led to erroneous results. Initially, they tried retraining each analyst in how to perform the tests.

c. The root cause is the underlying cause of problems. The superficial cause of problems in this case study is that the laboratory analysts did not follow the procedures for performing laboratory tests. The root cause analysis looks at why the analysts failed to follow the procedures. The four hypotheses examined were as follows: The analysts were not sufficiently trained. A second hypothesis was that the procedures were poorly written and therefore the analysts were not able to follow them properly. A third possibility was that the analysts lacked support and supervision. A fourth possibility was that the analysts were rushing through the tests, presumably because they were overworked.

d. This is not explicitly stated, but we can infer that the root cause investigation showed that the analysts did have deficiencies. Some of the analysts were inexperienced, and some were not completely trained on all the methods used in the laboratory. The analysts also were not aware of, or were not convinced of, the importance of complying strictly with the procedures as written and were deviating from them. It appears therefore that the analysts were not adequately trained and that there was insufficient supervision.

e. To prevent further problems, the company provided more training to the analysts not only in how to perform the test procedures, but also in the importance of adhering to the procedures. Group training was particularly effective. The company also provided more supervision to employees in the form of a QA person who supported the laboratory supervisor.

5.
a. The company apparently made several errors. First, they did not perform a sufficiently rigorous investigation of a problem. Second, they did not initiate aggressive *corrective actions*. Third, they did not take any *preventive actions*. If you read the text carefully, you will see that the problem is that product that was placed by QA in quarantine and was

not supposed to be shipped was, in fact, shipped to five hospitals. This is a serious error in that these hospitals received "nonconforming" material, material that had not been shown to meet its requirements. There was apparently a form filed relating to a CAPA investigation of this incident(s). The investigation report was incomplete, however, since it lacked "the dates of these serious occurrences, the employees involved, the number of instances that product was actually either removed from quarantine or overridden in the computer system (SAP). The CAPA also did not list the number of units that were actually shipped, or the number of hospitals that actually received nonconforming product."

b. This is a discussion question with various answers. As in the case study "Analyst Errors," training of personnel might be part of a prevention plan. Other ideas may occur to you, such as increased security limiting access to the quarantine area, better computer systems to catch mistakes, and a system involving a supervisor's review.

c. The consequences so far include the shipment of potentially faulty products, which could conceivably adversely affect patients, lead to lawsuits, and damage the company's reputation. This warning letter, like all FDA warning letters, is publicly posted and, therefore, might negatively impact the company's reputation. The company could lose business and suffer financially. The consequences to a company that consistently fails to comply with FDA can be severe. These consequences include seizing the product from the company, preventing the company from marketing and selling the product, and fining the company.

6. a. 4, b. 3, c. 2, d. 1
7. Root causes are the "real" or underlying causes of a problem. If the root cause is not corrected, the problem might happen again.
8. Corrective action means to fix problems that have already occurred by finding and correcting the root cause(s). Preventive action involves looking for problems that have not already occurred and preventing them.

9. Many answers are possible. For example:
   - Training and technique, where one person performs a task differently than others.
   - Processing variability, such as temperature and length of time product is held at particular step.
   - Equipment, such as an item that functions differently than expected.
   - Raw materials – incoming raw materials may vary.

CHAPTER 5

1.
   a. This mistake will be evident as soon as you take a bite of cake.
   b. This cake will probably be inedible and likely will be discarded.
2. a, b, c. Unfortunately, this type of mistake might go unnoticed. If it does go unnoticed, it is likely to affect your experimental results. This sort of mistake could make your work irreproducible – by you or any other researcher. Like the cake in Problem 1, this solution ought to be discarded, but you may not realize it. It is essential to put controls in place to prevent such mistakes. For example, you might have a colleague check your calculations. You might test your completed solution using some type of instrument. You will certainly want to prepare to make the solution by writing out a procedure in advance, or by following a previously written procedure. As you read further in this text, you will learn more about how to avoid this type of mistake.
3. **a, b, c.** The issue here is the same as in Problem 2; your experiments are likely to be irreproducible. It is essential that all measuring equipment, such as scales, be properly maintained. Quality systems that relate to laboratory work always include requirements for proper maintenance and operation of measuring devices. One simple method to be sure a scale is correctly measuring weight is to place an item of known weight on the scale and see if the result is correct. Items of known weight are called "standards," and they are available for purchase from laboratory products suppliers. Unit V in this text provides extensive information about how to ensure that measuring devices operate properly.

CHAPTER 6

1.

FORMULATION OF XYZ COMPOUND

Clean Gene, Inc.

3550 Anderson St.
MADISON, WI 54909
Revision 01
Master Batch Record # 133
Approved by _____*Aaron Reid*_____   _____*Anna Gold*_____   _____*Sam Rothstein*_____

        Date 2/14/08 _____   Date 2/14/98 _____   Date 2/16/08 _____

Issued by: _____*Erin Jane*_____   **Date** 12/3/08 _____   **Lot #** 15.987

Product Name: **Very Good Product**

Strength: __10 Units/mL__   **Vial Size** __10 mL__   **Batch quantity:** __350 L__

Reference: Refer to separate Formulation SOP Q75 for quantities of each component.
        Refer to separate instrument/equipment SOP 76 for ID information

---

NOTE: PRODUCT IS TO BE STIRRED CONTINUOUSLY DURING COMPOUNDING AND FILLING

A.   COLLECTION OF WATER FOR INJECTION (WFI), USP.   **wrong temperature**

A1.   Collect approximately 370 *L* of WFI, USP, in a clean, calibrated vessel and cool to 24°C–28°C.
Vessel # ___7___   Amount collected ___365ℓ___   Initial Temperature (23°C)   Time ___8:00___ am/pm
Final Temperature ___26°C___   Time ___08:15___ am/pm
               ↑ **neither is**   _____*EJ* / *JM*_____
               circled   12/10/08 12/10/08

A2.   Close the water for injection valves.   _____*EJ* / *JM*_____
                                         12/10/08 12/10/08

A3.   Remove about 20*L* of the cooled WFI from step 1 and place in a clean, calibrated vessel.
     (This water will be used to bring the final formulation to the proper volume.)

Vessel # _____2_____   _____*EJ* / *JM*_____
                                         12/10/08 12/10/08

A4.   Remove about 5*L* of the cooled WFI from step 1 into a clean, calibrated vessel. (This water will be used to prepare
     the solutions used to adjust the pH.)

B.   PREPARATION OF pH ADJUSTING SOLUTIONS

B1.   Collect 750 *mL* cool WFI from the vessel in step A4 and place into a 1000 *mL* volumetric flask and add
     100 *g* of NaOH (Sodium Hydroxide # 875) USP and dissolve.   **wrong amount**

Amount of WFI collected ___750mℓ___   Amount NaOH added (115g)   (Lot # _____)   ← **missing**

Time step completed ___09:15___ am/pm   **no initials—why did**
              ↑ **not circled**   (___*JM*___)   **supervisor initial this?**
                               12/10/08

B2.   Using a water bath containing cold WFI, cool the solution prepared in B1 to 25°C ± 5°C.

Final Temperature (31°C) ← **wrong temperature**   _____*EJ* / *JM*_____
                                         12/10/08 12/10/08

B3.   Bring the solution to 1000 *mL* with cool WFI from the vessel in step A4.

Approximate volume of WFI added (300mℓ)   Time completed (09:00) (am/pm)
                                         _____*EJ* / *JM*_____
             ↑   ↑   12/10/08 12/10/08

     and so on   **wrong**   **time does not**
                volume   **make sense**

---

2. Some ideas: Commitment and directive documents state the intentions of the organization and direct personnel in how to accomplish those objectives. These documents "say what the organization does." Individuals are required to adhere to these written documents. Individuals provide evidence that they have done so by completing data collection documents, such as forms and laboratory notebooks. Data collection documents are the means by which the individual "says what he/she has done." A documentation system provides accountability on the part of the organization and the individuals within it.

3. You can imagine many documents that might be required. A few examples include:

- A batch record might be used to direct production, based on the entrepreneur's original recipe(s) and processes. This document would detail raw materials and equipment, how personnel must prepare, package, and store the cookies, and might provide blanks for them to fill in during the process.

- Standard operating procedures for operating ovens, checking their temperatures, cleaning the bakery, and other tasks would likely be prepared.

- Documents might describe requirements for incoming raw materials.

- Documents might be developed for shipping and receiving, indicating what materials come into the bakery, what products are shipped, and to where they are shipped.

- Other documents might be involved in proving to sanitation inspectors that the bakery operates according to local regulations.

- Documents relating to personnel would be required, such as training records, and time sheets.

- You can imagine that accounting and other business documents (outside the scope of this text) would be required.

4. Some ideas for your discussion:

- First warning letter, point 1. The first company did not establish written procedures to guide certain critical tasks performed by the quality-control unit.

- First warning letter, point 2. The company did not have a separate master manufacturing record (MMR) for each formulation (form of the product) that they manufacture. (An MMR is what we call a master batch record in this text.) The manufacturing record guides operators as they make products. It is possible to imagine that operators might make mistakes because they did not have specific instructions for each product. In a worst case, mistakes could result in a person being harmed by an improperly manufactured product.

- First warning letter, point 3. When a batch of product is to be made, the MMR is printed out; this is the batch record that moves along with the product. As operators make the batch, they fill in blanks on the batch record to record all the information about that batch. Apparently, the batch records in this company did not provide all the required information for each batch. Records that are made as a product is being manufactured are critical parts of documenting the quality of those products.

- Second warning letter: The blood bank appears to have "lost" all the records pertaining to what happened to their blood products over a period of time. Losing records is always unacceptable. This is why companies are required to prove that their electronic documentation systems are secure and reliable *before* putting them into routine use, as discussed previously in this chapter. This company's "loss" of its records could have severe consequences if there is a problem with any blood product that the company had shipped. They will have difficulty investigating the problem or issuing a recall.

CHAPTER 7

1. Assuming that substitution of a less hazardous solvent or decreasing the volume of the toxic solvent has been considered, you can reduce risk by:

- Preparing detailed SOPs for procedures involving the solvent.

- Ensuring that all personnel in the laboratory have received proper training in hazards associated with the solvent and proper techniques for its handling.

- Ensuring that all work with the solvents is carried out in a safe location (in this case, a fume hood).
- No one should ever work alone, but being particularly clear that this hazardous material should not be used when no one else is present.

2. The key emergency procedures are those relating to evacuation. This situation also demonstrates the importance of not working alone, remaining aware of the locations of colleagues, knowledge of evacuation procedures, and a safety-first attitude.

3. There is direct responsibility at four levels here:
   - Students, by using an inappropriate SOP for a large-scale operation, being unaware of potential hazards, and not informing the supervisor of scaling up the operation.
   - Supervisor, by being unaware of the specific lab activities of the graduate students, not providing an appropriate SOP for a scaled up reaction (assuming that this can be done relatively safely), not providing a proper safety training program, and/or not ensuring that the students fully participated in the training program.
   - Institution, by not providing a proper safety training program and/or not ensuring that the students fully participated in the program, and not ensuring that the supervisor was maintaining a "safety-first" culture in the laboratory.
   - Granting agency, by not ensuring that hazardous laboratory operations were performed safely in the laboratory.

4. The researcher erred by working alone, not wearing a lab coat, and wearing flammable clothing.

   The institution did not ensure that existing safety policies were being enforced.

CHAPTER 8

1. You should replace the cylinder cap, mark the cylinder as potentially damaged, and return it to the supplier. Do not attempt to use a lever to force the valve open.

2. Canvas shoes can absorb hazardous chemicals or become contaminated with biological materials. They do not provide a solid protective barrier for the feet.

3. Disposable latex gloves are too thin to provide adequate insulation from cryogenic materials. Thick gloves would be more appropriate.

4. A minimum of 15 minutes.

CHAPTER 9

1. The TWA means that the 50 ppm level is a time-weighted average over 8 hours. Actual air concentrations may vary from this average value.

2.
   a.
      - Keeping the floor dry.
      - Being careful walking around water sources.
      - Using a secondary container to transport the solvent bottle.
      - Storing the solvent in a non-glass container.

   b.
      - Notify other personnel of the problem.
      - Consider the toxic properties of the spilled liquid and order an evacuation if appropriate.
      - Be certain that any personal contamination with the solvent is cleaned up first.
      - If safe, proceed with proper spill cleanup techniques using a spill kit.

3. No! The air concentrations of a chemical that can be detected by the human nose are unrelated to toxic concentrations.

4.
   a. NaCl (least toxic; no TLV reported because inhalation is not a hazard); ethyl alcohol, acetone, phenol (most toxic), based on TLV values.
   b. The STEL value is the limit to which a person can be exposed for only 15 continuous minutes, up to four times during an 8-hour day. It is therefore a higher limit than the TLV value, which is the limit to which a person can be exposed 8 hours/ day, 40 hours/week.
   c. Acetone. Extinguish using water spray, dry chemical, carbon dioxide, or alcohol-resistant foam.

5. Some of the potential combinations here are dangerously incompatible. It is always essential to consult the SDS and any manufacturer's recommendations when storing unfamiliar chemicals. Note that this is only excerpted information, and there may be additional considerations in an actual laboratory.

   a. Nitric acid is a strong oxidizer and should be stored separately.

   b. Sodium chlorate is a strong oxidizer and should be stored separately.

   c. 2-Amino-2-(hydroxymethyl)-1,3-propanediol (Tris base), agarose, nicotinic acid (niacin), and D(+)-galactose are all non-reactive and relatively nonhazardous solids and can be stored together.

   d. Ethyl acetate and 2-propanol are both flammable liquid solvents and can be stored together.

   e. Sodium hydroxide is a strong base and should be stored separately.

   f. Acrylamide is a neurotoxin and should be stored separately.

   g. Sodium is a flammable solid and should be stored separately.

CHAPTER 10

1. Soaps and detergents have both a hydrophobic portion and a hydrophilic portion. The hydrophobic portion mixes with fats, while the hydrophilic portion mixes with water. Therefore, soaps and detergents can dissolve fats and allow them to be washed away with water. The SARS-CoV-2 virus is surrounded by a lipid membrane. This membrane is attacked or dissolved by soap, breaking open the virus and thus damaging it. The water can then wash away the viral residue. As an analogy, imagine trying to wash a butter dish. Butter is a type of fat, and it does not mix with water. If you try to wash the dish with water alone, the dish stays oily. But soap or detergent will dissolve the butter and allow it to be washed away with rinse water.

2.

   a. Some basic precautions:
      - Do not use glass containers for biohazards.
      - Use a sealed plastic container.
      - Use a secondary container for the flask.
      - Try to avoid using liquid cultures of pathogens.
      - Wear a lab coat to protect clothes.

   b. This individual's first priority is personal safety (assuming that no one else is involved in the accident). He should remove all contaminated items of clothing and immediately leave the room where the accident occurred. The doors should be closed, and no one should reenter the room for at least 10 minutes, to allow any aerosols to settle. The scientist should then wash any skin areas that have been exposed to the pathogen. He should use an antiseptic soap and vigorously scrub these areas without abrading the skin for at least 10 minutes. Following the emergency plan for his institution, he or a colleague should notify the safety office, or set up the materials to decontaminate the laboratory. Further medical treatment and/or surveillance may be necessary depending on the nature of the pathogen. Cleanup should follow the procedures outlined in Table 10.22.

3. Approximately 864 particles remain in air.

4. *Similarities:* Both refer to the importance of washing hands, decontaminating work surfaces, wearing PPE, avoiding mouth pipetting, and proper disposal of waste materials. *Differences:* Universal Precautions are OSHA regulations. They were developed for handling human blood and blood-related products. Universal Precautions are more detailed, but all of the requirements of each are compatible.

5. a, b, d. Organisms are classified as BSL-1. c. HeLa is classified as a BSL-2 cell line. **e.** SARS-CoV-2 is presently classified at BSL-3. It is possible that, with a vaccine and treatments, this virus might be reclassified. f. Once isolated, DNA is not pathogenic and is not classified in this way, although it would still be good laboratory practice to avoid contact with it. In this situation, your primary concern is likely to be to avoid contaminating the DNA sample. g. All work with human blood is classified as BSL-2. Therefore, while it used to be common for biology classes to involve a genuine blood typing activity, this is no longer done on an open lab bench in a school setting.

6.

BSL-2 Laboratory

BIOHAZARD

All Personal Protective Equipment shall be removed prior to leaving this work area. Eating, drinking, smoking applying cosmetics or lip balm and handling contacts lenses area prohibited in this work area.

Name of infectious agent(s):

Rabies virus

Transmission route(s):

Direct skin, eye, mucosal membrane, or inhalation of aerosols, ingestion

Special requirements for entering this area:

Authorized individuals only; BSL2 training required. PPE required.
Rabies vaccine offered.

Emergency contact
P.I. Name: C.J. Moore   Office: (409) 344-4515   Home: (409) 334-5603

EH&S office (M-F 8-4:30) 438-8325 (after hours) 911.

*Biosafety Level 2 work involving agents that may be associated with potential hazard to personnel or the environment.

Illinois State University
Environmental Health & Safety
Date 4/19/21

---

7. Step 3 requires opening both the tube with bacterial cells and the tube with DNA. Simply opening tubes has the potential to aerosolize a small portion of the tubes' contents. Of course, the tubes should be tightly capped when flicking to avoid spillage or aerosols.

Step 4, centrifugation, is always a major source of aerosols unless special containment equipment is available. Even if tubes appear to be closed, aerosols are likely to escape during centrifugation because the force of centrifugation can deform the plastic. If pathogenic materials are being used, this step requires special containment, as discussed in the text.

Step 8 requires opening the tube to add nutrient medium. This can create small amounts of aerosols. The nutrient medium is unlikely to be hazardous, but the bacterial cells might be. Shaking the tubes containing cells can create aerosols if the tubes are not completely closed. (Unlike centrifugation, tightly capping the tubes should be sufficient to keep aerosols from escaping while shaking.)

Step 9 requires opening the tube and pipetting the cells; both can create aerosols.

Step 10 will generate aerosols, and these can contain bacteria.

Step 11 can potentially create a microbial aerosol.

8. Various ideas are reasonable. Some topics to consider:
  - **Documentation.**
    Standard operating procedures can direct individuals in proper practices that avoid contamination.

    Rigorous recording of all activities can help identify the causes of problems and exposures.

    Chain of custody forms and documents to track samples can ensure that dangerous materials are not misplaced.

    Proper labeling can help ensure that pathogens are not accidentally handled improperly.
  - **Training.** All personnel must be adequately trained, and their mastery of safety practices should be tested. No untrained personnel should have access to hazardous materials.
  - **Regular inspections and audits** can ensure that all safety practices are followed.
  - **Regular medical surveillance of staff** can help identify exposures to pathogens.
  - **Access and security.** Access to the facility should be controlled by a security system.
  - **Individual responsibility.** No quality system will work unless individuals take responsibility for compliance. Senior personnel are responsible for setting a good example and emphasizing the necessity for compliance and the dangers of becoming complacent over time.

CHAPTER 11

MANIPULATION PRACTICE PROBLEMS: EXPONENTS AND SCIENTIFIC NOTATION (FROM P. 264)

1.
  a. $2^2 = 4$
  b. $3^3 = 27$
  c. $2^{-2} = 1/4 = 0.25$
  d. $3^{-3} = 1/27 \approx 0.037$
  e. $10^2 = 100$
  f. $10^4 = 10,000$
  g. $10^{-2} = 1/100 = 0.01$
  h. $10^{-4} = 1/10,000 = 0.0001$
  i. $5^0 = 1$
2.
  a. $2^2 \times 3^3 = 4 \times 27 = 108$
  b. $(14^3)(3^6) = (2,744)(729) = 2,000,376$

  c. $5^5 - 2^3 = (3,125) - (8) = 3,117$
  d. $\dfrac{5^7}{8^4} = \dfrac{78,125}{4,096} \approx 19.07$
  e. $(6^{-2})(3^2) = (1/36)(9) = 9/36 = 0.25$
  f. $(-0.4)^3 + (9.6)^2 = (-0.064) + (92.16) \approx 92.1$
  g. $a^2 \times a^3 = a^5$
  h. $\dfrac{c^3}{c^{-6}} = c^9$
  i. $(3^4)^2 = 3^8 = 6,561$
  j. $(c^{-3})^{-5} = c^{15}$
  k. $\dfrac{13}{(43)^2 + (13)^3} = \dfrac{13}{1,849 + 2,197}$

  $$= \dfrac{13}{4,046} \approx 0.0032$$
  l. $\dfrac{10^2}{10^3} = \dfrac{1}{10} = 10^{-1}$

3. *Larger numbers are **underlined***:
  a. 5×10^{-3}cm, $\underline{500 \times 10^{-1}}$cm
  b. $300 \times 10^{-3}\,\mu L$, $\underline{3,000 \times 10^{-2}}\,\mu L$
  c. 3.200×10^{-6}m, $\underline{3,200 \times 10^{-4}}$m
  d. $\underline{0.001 \times 10^1}$cm, 1×10^{-3}cm
  e. $\underline{0.008 \times 10^{-3}}$L, 0.0008×10^{-4}L

4.
  a. $54.0 = 5.40 \times 10^1$
  b. $4567 = 4.567 \times 10^3$
  c. $0.345000 = 3.45000 \times 10^{-1}$
  d. $10000000 = 1 \times 10^7$
  e. $0.009078 = 9.078 \times 10^{-3}$
  f. $540 = 5.40 \times 10^2$
  g. $0.003040 = 3.040 \times 10^{-3}$
  h. $200567987 = 2.00567987 \times 10^8$

5.
  a. $12.3 \times 10^3 = 12,300$
  b. $4.56 \times 10^4 = 45,600$
  c. $4.456 \times 10^{-5} = 0.00004456$
  d. $2.300 \times 10^{-3} = 0.002300$

e. $0.56 \times 10^6 = 560{,}000$

f. $0.45 \times 10^{-2} = 0.0045$

6.

a.

$$\frac{\left(4.725 \times 10^8\right)\left(0.0200\right)}{\left(3700\right)\left(0.770\right)} =$$

$$\frac{\left(4.725 \times 10^8\right)\left(2.00 \times 10^{-2}\right)}{\left(3.7 \times 10^3\right)\left(7.7 \times 10^{-1}\right)} =$$

$$\frac{9.45 \times 10^6}{28.49 \times 10^2} \approx 3.32 \times 10^3$$

b.

$$\frac{\left(1.93 \times 10^3\right)\left(4.22 \times 10^{-2}\right)}{\left(8.8 \times 10^8\right)\left(6.0 \times 10^{-6}\right)} =$$

$$\frac{8.1446 \times 10}{52.8 \times 10^2} = 0.154 \times 10^{-1} \approx 1.5 \times 10^{-2}$$

c. $(4.5 \times 10^3) + (2.7 \times 10^{-2}) = 4{,}500 + 0.027 = 4{,}500.027 = 4.5 \times 10^3$

d. $(35.6 \times 10^4) - (54.6 \times 10^6) = (0.356 \times 10^6) - (54.6 \times 10^6) = -54.244 \times 10^6 \approx -5.42 \times 10^7$

e. $(5.4 \times 10^{24}) + (3.4 \times 10^{26}) = (0.054 \times 10^{26}) + (3.4 \times 10^{26}) = 3.454 \times 10^{26} \approx 3.5 \times 10^{26}$

f. $(5.7 \times 10^{-3}) - (3.4 \times 10^{-6}) = (5{,}700 \times 10^{-6}) - (3.4 \times 10^{-6}) = 5{,}696.6 \times 10^{-6} \approx 5.7 \times 10^{-3}$

7.

a. $0.0050 \times 10^{-4} = 0.050 \times 10^{-5} = 0.50 \times 10^{-6} = 5.0 \times 10^{-7}$

b. $15.0 \times 10^{-3} = 1.50 \times 10^{-2} = 0.150 \times 10^{-1} = 0.00150 \times 10^1$

c. $5.45 \times 10^{-3} = 54.5 \times 10^{-4}$

d. $100.00 \times 10^1 = 1.0000 \times 10^3$

e. $6.78 \times 10^2 = 0.678 \times 10^3$

f. $54.6 \times 10^2 = 0.00546 \times 10^6$

g. $45.6 \times 10^8 = 4{,}560 \times 10^6$

h. $4.5 \times 10^{-3} = 450 \times 10^{-5}$

i. $356.98 \times 10^{-3} = 0.035698 \times 10^1$

j. $0.0098 \times 10^{-2} = 0.98 \times 10^{-4}$

MANIPULATION PRACTICE PROBLEMS: LOGARITHMS (FROM P. 268)

1.

a. 2

b. 3

2.

a. $\log 100 = 2$

b. $\log 10{,}000 = 4$

c. $\log 1{,}000{,}000 = 6$

d. $\log 0.0001 = -4$

e. $\log 0.001 = -3$

3.

a. $\log 7$ between 0 and 1

b. $\log 65.9$ between 1 and 2

c. $\log 89.0$ between 1 and 2

d. $\log 0.45$ between 0 and -1

e. $\log 0.0078$ between -2 and -3

4.

a. 4.18

b. 2.54

c. -2.0

d. -4.53

e. 3.082

f. -0.462

5.

a. antilog $4.8990 \approx 7.925 \times 10^4$

b. antilog $3.9900 \approx 9.772 \times 10^3$

c. antilog $-0.5600 \approx 0.2754$

d. antilog $-0.0089 \approx 0.9797$

e. antilog $9.8999 \approx 7.941 \times 10^9$

f. antilog $1.0000 = 10.00$

g. antilog $8.9000 \approx 7.943 \times 10^8$

6.

a. 0.35

b. 1.35

c. 2.35

d. 6.49

7.

a. antilog $-4.56 \approx 2.75 \times 10^{-5}\,M$

b. antilog $-5.67 \approx 2.14 \times 10^{-6}\,M$

c. antilog $-7.00 = 1.00 \times 10^{-7}\,M$

d. antilog $-1.09 \approx 8.13 \times 10^{-2}\,M$

e. antilog $-10.1 \approx 7.94 \times 10^{-11}\,M$

MANIPULATION PRACTICE PROBLEMS: MEASUREMENTS (FROM P. 272)

1. Larger:

a. $1\,\mu m$

b. $1\,cm$

c. $1{,}000\,mm$

d. $10\,m$

e. $1{,}000\,g = 1\,kg$

f. $1{,}000\,\mu m$

g. $1\,mg$

h. $1\,nm$

i. $10\,nm$

j. $1\,g$

k. $1,000\,\mu L = 1\,mL$
l. $1\,m$
m. $1\,L$
n. $500\,cm$
o. $1\,pL = 0.001\,nL$
p. $0.1\,\mu m$
q. $100\,cm$
r. $0.0001\,m$
2. $100\,g = 0.1\,kg$
3. $12\,oz \approx 355\,mL$

Manipulation Practice Problems: Equations (from p. 274)

1.
   a. A is equal to the value obtained if C is doubled.
   b. C is equal to the value obtained if A divided by D.
   c. Y is equal to the value obtained if X is multiplied by 2 and then 1 is added.
2.
   a. RCF increases if RPM increases.
   b. RCF decreases if r decreases.
   c. 11.18, 1,000
3.
   a. $A = 2C \qquad 4 = 2(2)$
   b. $C = \dfrac{A}{D} \qquad 12 = \dfrac{36}{3}$
   c. $Y = 2(x) + 1 \qquad 15 = 2(7) + 1$
4.
   a. $3X = 15 \qquad X = \dfrac{15}{3} = 5$
   b. $X = 25(5{-}4) \qquad X = 25(1) = 25$
   c. $-X = 3X - 1\,mg$

   $$1\,mg = 4X$$

   $$X = \frac{1}{4}\,mg$$

   d. $5X = 3X - 5\,mL + 34\,mL$

   $$2X = (34 - 5)\,mL$$

   $$2X = 29\,mL$$

   $$X = 14.5\,mL$$

   e. $\dfrac{X}{2} = 25(3)\,cm^2$

   $$\frac{X}{2} = 75\,cm^2$$

$$X = 2(75)\ cm^2$$

$$X = 150\,cm^2$$

f. $X = 3.0\,cm\,(2.0\,mg/mL)\,(2.0)$

$$X = 3.0\,cm\,(4.0\,mg\,/\,mL)$$

$$X = \frac{12.0\ cm\ mg}{mL}$$

g. $X = \dfrac{25\,mg}{mL}(4.0\ mL)\dfrac{3.0\,oz}{mg} = \dfrac{300\,mg\,mL\,oz}{mg\ mL}$

   $$= 300\ oz$$

5.
   a. $798\ lb^2$
   b. $\dfrac{15.2\ g}{3.1\ g} \approx 4.9$
   c. $\dfrac{25.2\ cm \times 34.5\ cm}{3.00} \approx 290\ cm^2$
   d. $\dfrac{5\ mL \times 3\ mL \times 2\ cm}{2\ cm} = 15\ mL^2$

CHAPTER 12

Manipulation Practice Problems: Proportions (from pp. 278)

1.
   a. $\dfrac{?}{5} = \dfrac{2}{10} \qquad ? = \dfrac{2 \times 5}{10} = 1$
   b. $\dfrac{?}{1\ mL} = \dfrac{10\ cm}{5\ mL}$

   $$? = \frac{10\ cm\,(1\ mL)}{5\ mL} = \frac{10\ cm}{5} = 2\ cm$$

   c. $\dfrac{0.5\ mg}{10\ mL} = \dfrac{30\ mg}{?}$

   $$? = \frac{(30\ mg)(10\ mL)}{0.5\ mg} = 600\ mL$$

   d. $\dfrac{50}{?} = \dfrac{100}{100} \qquad ? = \dfrac{(50)100}{100} = 50$

   e. $\dfrac{?}{30\ in^2} = \dfrac{15\ lb}{100\ in^2}$

   $$? = \frac{15\ lb\,(30\ in^2)}{100\ in^2} = 4.5\ lb$$

f. $\dfrac{?}{15} = \dfrac{30}{90}$ $? = \dfrac{30(15)}{90} = 5$

g. $\dfrac{100}{10} = \dfrac{50}{?}$ $? = \dfrac{50(10)}{100} = 5$

2.

a. **?** = time to drive to Denver and back

$$\dfrac{?}{2 \text{ way}} = \dfrac{50 \text{ min}}{1 \text{ way}}$$

$$? = \dfrac{(50 \text{ minutes})(2 \text{ way})}{1 \text{ way}} \quad ? = 100 \text{ minutes}$$

b. **?** = **amount** of baking soda for 33 loaves

$$\dfrac{1 \text{ t}}{1 \text{ loaf}} = \dfrac{?}{33 \text{ loaf}}$$

$$? = \dfrac{1 \text{ t}(33 \text{ loaf})}{1 \text{ loaf}} \quad ? = 33 \text{ t}$$

c. **?** = cost of 100 magazines

$$\dfrac{1.50}{1 \text{ magazine}} = \dfrac{?}{100 \text{ magazine}}$$

$$? = \dfrac{1.50(100 \text{ magazine})}{\text{magazine}} \quad ? = 150.00$$

d. **?** = margarine for 5 batches

$$\dfrac{\frac{1}{4}\text{cup}}{1 \text{ batch}} = \dfrac{?}{5 \text{ batch}}$$

$$? = \dfrac{\frac{1}{4}\text{cup}(5 \text{ batch})}{1 \text{ batch}} \quad ? = 1.25 \text{ cups}$$

e. **?** = baking powder for a half-batch

$$\dfrac{1/8 \text{ t}}{1 \text{ batch}} = \dfrac{?}{1/2 \text{ batch}}$$

$$? = \dfrac{1/8 \text{ t}(1/2 \text{ batch})}{1 \text{ batch}} = (1/8 \text{ t})(1/2) \quad ? = \dfrac{1}{16}\text{t}$$

f. **?** = oz in 4.5 bags of chips

$$\dfrac{1 \text{ bag}}{3 \text{ oz}} = \dfrac{4.5 \text{ bag}}{?}$$

$$\dfrac{(3 \text{ oz})(4.5 \text{ bag})}{1 \text{ bag}} = ? = 13.5 \text{ oz}$$

APPLICATION PRACTICE PROBLEMS:
PROPORTIONS (FROM PP. 278)

1. $\dfrac{10^2 \text{ cell}}{10^{-2} \text{ mL}} = \dfrac{?}{1 \text{ mL}}$ $? = 10^4 \text{cell}$

   (10^2 cell in 10^{-2} mL = 10^4 cell/mL)

2. $\dfrac{10 \text{ mL}}{1 \text{ tube}} = \dfrac{?}{37 \text{ tubes}}$ $? = 370 \text{ mL}$

3. $\dfrac{5 \times 10^1 \text{ paramecia}}{20 \text{ mL}} = \dfrac{?}{10^5 \text{ mL}}$

   $? = 2.5 \times 10^5$ paramecia

4. 5×10^4 g = 50 kg

$$\dfrac{315 \text{ larvae}}{1 \times 10^{-1} \text{ kg}} = \dfrac{?}{5 \times 10^1 \text{ kg}}$$

$$? = 1.6 \times 10^5 \text{ larvae}$$

5. $\dfrac{1 \times 10^9 \text{ bacteria}}{1 \text{ mL}} = \dfrac{?}{1,000 \times 10^3 \text{ mL}}$

   $? = 1 \times 10^{15}$ bacteria

6. $\dfrac{30 \text{ g}}{1,000 \text{ mL}} = \dfrac{?}{250 \text{ mL}}$ $? = 7.5 \text{ g}$

7. $\dfrac{100 \text{ mL ethanol}}{1,000 \text{ mL}} = \dfrac{?}{10^5 \text{ mL}}$

   $? = 10,000 \text{ mL ethanol} = 10 \text{ L}$

8. $10^4 \mu\text{L} = 10 \text{ mL}$

$$\dfrac{50 \text{ mg}}{1 \text{ mL}} = \dfrac{?}{10 \text{ mL}} \quad ? = 500 \text{ mg}$$

9. $\dfrac{60 \text{ amino acids}}{\text{minutes}} = \dfrac{?}{10 \text{ minutes}}$

   $? = 600$ amino acids

   1,000 amino acids − 600 amino acids
   = 400 amino acids left after 10 minutes

10.

	For 100 mL (g)	For 1,500 mL (g)
NaCl	20.0	300
Na azide	0.001	0.015
Mg sulfate	1.0	15
Tris	15.0	225

11.

a. $\dfrac{12\text{ g}}{1\text{ mol}} = \dfrac{?}{2\text{ mol}}$ $? = 24\text{ g}$

b. $\dfrac{12\text{ g}}{1\text{ mol}} = \dfrac{?}{0.5\text{ mol}}$ $? = 6\text{ g}$

12.

a. $\dfrac{1\text{ mol}}{58.5\text{ g}} = \dfrac{1.5\text{ mol}}{?}$ $? = 87.8\text{ g}$

   Alternatively, 58.5 g/mol × 1.5 mol = 87.8 g

b. $\dfrac{1\text{ mol}}{58.5\text{ g}} = \dfrac{0.75\text{ mol}}{?}$ $? = 43.9\text{ g}$

MANIPULATION PRACTICE PROBLEMS: PERCENTS (FROM P. 282)

1.

a. 98/100 = 98%

b. $110/10{,}004 \approx \dfrac{1}{100} = 1\%$

c. $\dfrac{3}{15} = \dfrac{1}{5} = \dfrac{20}{100} = 20\%$

d. $\dfrac{45}{45{,}002} \approx \dfrac{1}{1{,}000} = 0.1\%$

2.

a. $34\% = \dfrac{34}{100} = 0.34$

b. $89.5\% = \dfrac{89.5}{100} = 0.895$

c. $100\% = \dfrac{100}{100} = 1$

d. $250\% = \dfrac{250}{100} = \dfrac{2.5}{1} = 2.5$

e. $0.45\% = \dfrac{0.45}{100} = 0.0045$

f. $0.001\% = \dfrac{0.001}{100} = 0.00001$

3.

a. 15% of 450 = 0.15 × 450 = 67.5
b. 25% of 700 = 0.25 × 700 = 175
c. 0.01% of 1,000 = 0.0001 × 1,000 = 0.1
d. 10% of 100 = 0.1 × 100 = 10
e. 12% of 500 = 0.12 × 500 = 60
f. 150% of 1,000 = 1.5 × 1,000 = 1,500

4.

a. $\dfrac{15}{45} = 0.33 = 33\%$

b. $\dfrac{2}{2} = 1 = 100\%$

c. $\dfrac{10}{100} = 0.1 = 10\%$

d. $\dfrac{1}{100} = 0.01 = 1\%$

e. $\dfrac{1}{1{,}000} = 0.001 = 0.1\%$

f. $\dfrac{6}{40} = 0.15 = 15\%$

g. $\dfrac{0.1}{0.5} = 0.2 = 20\%$

h. $\dfrac{0.003}{89} \approx 0.0000337 \approx 0.0034\%$

i. $\dfrac{5}{10} = 0.5 = 50\%$

j. 0.05 = 5%
k. 0.0034 = 0.34%
l. 0.25 = 25%
m. 0.01 = 1%
n. 0.10 = 10%
o. 0.0001 = 0.01%
p. 0.0078 = 0.78%
q. 0.50 = 50%

5.

a. 20% of 100 = 20
b. 20% of 1,000 = 200
c. 20% of 10,000 = 2,000
d. 10% of 567 = 56.7
e. 15% of 1,000 = 150
f. 50% of 950 = 475
g. 5% of 100 = 5
h. 1% of 876 = 8.76
i. 30% of 900 = 270

6.

a. $\dfrac{1\text{ part}}{100\text{ parts}} = 1\%$

b. $\dfrac{3\text{ parts}}{50\text{ parts}} = 6\%$

c. $\dfrac{15\text{ parts}}{(15+45)\text{ parts}} = \dfrac{15\text{ parts}}{60\text{ parts}} = 25\%$

d. $\dfrac{0.05\text{ parts}}{1\text{ part}} = 0.05 = 5\%$

e. $\dfrac{1\text{ part}}{25\text{ parts}} = 0.04 = 4\%$

f. $\dfrac{2.35\text{ parts}}{(2.35+6.50)\text{ parts}} = \dfrac{2.35\text{ parts}}{8.85\text{ parts}}$

$\approx 0.266 = 26.6\%$

1. $\dfrac{25}{55}$ work   20 + hours ≈ 0.4545 ≈ 45%

$\dfrac{30}{55}$ work   19 − hours ≈ 0.545 ≈ 55%

2. 50% = 0.5 of total; total = 100 mL
   100 × 0.5 = 50 mL ethanol
3. 30% solution of ethylene glycol
   30% of total (250 mL)
   0.30 × 250 = 75 mL ethylene glycol
   Measure out the ethylene glycol, and add water until the volume is 250 mL.
4. 10% acetonitrile for 100 mL: 10 mL
   25% MeOH (methanol):   25 mL
   Measure out acetonitrile and MeOH, combine, and add water until the volume is 100 mL.
5. $\dfrac{5 \text{ mL}}{100 \text{ mL}}$ = 0.05 = 5% propanol solution
6. $\dfrac{15 \text{ mL EtOH}}{700 \text{ mL}}$ = 0.021 = 2.1% ethanol solution
7. $\dfrac{10 \text{ μL}}{1 \text{ mL}} = \dfrac{10 \text{ μL}}{1{,}000 \text{ μL}} = \dfrac{1}{100}$ = 0.01
   = 1% methanol solution
8. $\dfrac{15 \text{ mL acetone}}{1 \text{ L}} = \dfrac{15 \text{ mL}}{1{,}000 \text{ mL}}$
   = 0.015 or 1.5% acetone solution
9. 1,000 insects, 5% survive
   1,000 × 0.05 = 50 insects survive first winter
   50 insects × 100 = 5,000 insects after first summer
   5,000 × 0.05 = 250 insects after second winter
10. 250,000 people × 0.15 = 37,500 people die
11. Since thymine is 24%, adenine must also be 24% for a total of 48%. This leaves 52% that must be split equally between guanine and cytosine, at 26% each.
12. Of the subjects fed regular chips, 88 experienced gastrointestinal problems compared with 79 subjects fed Olestra chips.

1. Olive oil
2. $\dfrac{57.9 \text{ g}}{3.00 \text{ cm}^3} = \dfrac{?}{1 \text{ cm}^3}$? = $\dfrac{19.3 \text{ g}}{\text{cm}^3}$

3. Density = 1.26 g/mL

$\dfrac{1.26 \text{ g}}{1 \text{ mL}} = \dfrac{20.0 \text{ g}}{?}$? ≈ 15.9 mL

1.
a. 3.00 ft to cm (proportion method):

$$\dfrac{?}{3.00 \text{ ft}} = \dfrac{12 \text{ in}}{1 \text{ ft}}$$

$? = \dfrac{(12 \text{ in})(3.00 \text{ ft})}{1 \text{ ft}}$? = 36 in

$$\dfrac{?}{36 \text{ in}} = \dfrac{2.54 \text{ cm}}{1 \text{ in}}$$

$? = \dfrac{(2.54 \text{ cm})(36 \text{ in})}{1 \text{ in}}$? ≈ 91.4 cm

3.00 ft to cm (conversion factor method, also called unit cancelling method)

$$3.00 \text{ ft} \times \dfrac{12 \text{ in}}{1 \text{ ft}} \times \dfrac{2.54 \text{ cm}}{1 \text{ in}} ≈ 91.4 \text{ cm}$$

b. 100 mg to g (proportion method):

$$\dfrac{?}{100 \text{ mg}} = \dfrac{1 \text{ g}}{1{,}000 \text{ mg}}$$

$? = \dfrac{(1 \text{ g})(100 \text{ mg})}{1{,}000 \text{ mg}} = 0.1 \text{ g}$

100 mg to g (conversion factor method)

$$100 \text{ mg} \times \dfrac{1 \text{ g}}{1{,}000 \text{ mg}} = 0.1 \text{ g}$$

c. 12.0 in to mi (proportion method):

$$\dfrac{?}{12.0 \text{ in}} = \dfrac{1 \text{ ft}}{12.0 \text{ in}}$$

$? = \dfrac{(1 \text{ ft})(12.0 \text{ in})}{12.0 \text{ in}} = 1 \text{ ft}$

$$\dfrac{?}{1 \text{ ft}} = \dfrac{1 \text{ mi}}{5{,}280 \text{ ft}}$$

$? = \dfrac{(1 \text{ mi})(1 \text{ ft})}{5{,}280 \text{ ft}} = \dfrac{1 \text{ mi}}{5280} ≈ 0.000189 \text{ mi}$

$? = 1.89 \times 10^{-4}$ mile

12.0 in to mi (conversion factor method)

$$12.0 \text{ in} \times \frac{1 \text{ ft}}{12.0 \text{ in}} \times \frac{1 \text{ mi}}{5,280 \text{ ft}} \approx 1.89 \times 10^{-4} \text{ mi}$$

d. 100 in to km (proportion method):

$$\frac{?}{100 \text{ in}} = \frac{2.54 \text{ cm}}{1 \text{ in}}$$

$$? = \frac{(2.54 \text{ cm})(100 \text{ in})}{1 \text{ in}} \qquad ? = 254 \text{ cm}$$

$$\frac{?}{254 \text{ cm}} = \frac{1 \text{ m}}{100 \text{ cm}}$$

$$? = \frac{(254 \text{ cm})(1 \text{ m})}{100 \text{ cm}} \qquad ? = 2.54 \text{ m}$$

$$\frac{?}{2.54 \text{ m}} = \frac{1 \text{ km}}{1,000 \text{ m}}$$

$$? = \frac{(2.54 \text{ m})(1 \text{ km})}{1,000 \text{ m}} \qquad ? = 2.54 \times 10^{-3} \text{ km}$$

100 in to km (conversion factor method):

$$100 \text{ in} \times \frac{2.54 \text{ cm}}{1 \text{ in}} \times \frac{1 \text{ m}}{100 \text{ cm}} \times \frac{1 \text{ km}}{1,000 \text{ m}}$$

$$= 2.54 \times 10^{-3} \text{ km}$$

e. 10.0555 lb to oz (proportion method):

$$\frac{?}{10.0555 \text{ lb}} = \frac{16 \text{ oz}}{1 \text{ lb}}$$

$$? = \frac{(16 \text{ oz})(10.0555 \text{ lb})}{1 \text{ lb}} \qquad ? = 160.888 \text{ oz}$$

10.555 lb to oz (conversion factor method):

$$10.0555 \text{ lb} \times \frac{16 \text{ oz}}{1 \text{ lb}} = 160.888 \text{ oz}$$

f. 18.989 lb to g (proportion method):

$$\frac{?}{18.989 \text{ lb}} = \frac{453.6 \text{ g}}{1 \text{ lb}}$$

$$? = \frac{(18.989 \text{ lb})(453.6 \text{ g})}{1 \text{ lb}} \qquad ? \approx 8,613.4 \text{ g}$$

18.989 lb to g (conversion factor method):

$$18.989 \text{ lb} \times \frac{453.6 \text{ g}}{1 \text{ lb}} \approx 8,613.4 \text{ g}$$

g. 13 mi to km (proportion method):

$$\frac{?}{13 \text{ mi}} = \frac{1.609 \text{ km}}{1 \text{ mi}}$$

$$? = \frac{(1.609 \text{ km})(13 \text{ mi})}{1 \text{ mi}} \qquad ? \approx 21 \text{ km}$$

13 mi to km (conversion factor method):

$$13 \text{ mi} \times \frac{1.609 \text{ km}}{1 \text{ mi}} \approx 21 \text{ km}$$

h. 150 mL to L (proportion method):

$$\frac{?}{150 \text{ mL}} = \frac{1 \text{ L}}{1,000 \text{ mL}}$$

$$? = \frac{(150 \text{ mL})(1 \text{ L})}{1,000 \text{ mL}} \qquad ? = 0.150 \text{ L}$$

150 mL to L (conversion factor method):

$$150 \text{ mL} \times \frac{1 \text{ L}}{1,000 \text{ mL}} = 0.150 \text{ L}$$

i. 56.7009 cm to nm (proportion method):

$$\frac{?}{56.7009 \text{ cm}} = \frac{1 \text{ m}}{100 \text{ cm}}$$

$$? = 0.567009 \text{ m}$$

$$\frac{?}{0.567009 \text{ m}} = \frac{10^9 \text{ nm}}{1 \text{ m}}$$

$? = 5.67009 \times 10^8 \text{ nm}$

56.7009 cm to nm (conversion factor method):

$$56.7009 \text{ cm} \times \frac{1 \text{ m}}{10^2 \text{ cm}} \times \frac{10^9 \text{ cm}}{1 \text{ m}}$$

$$= 5.67009 \times 10^8 \text{ nm}$$

j. 500 nm to μm (proportion method):

$$\frac{?}{500 \text{ nm}} = \frac{1 \text{ μm}}{1,000 \text{ nm}}$$

$$? = \frac{(1 \text{ μm})(500 \text{ nm})}{1,000 \text{ nm}} \qquad ? = 0.500 \text{ μm}$$

500 nm to µm (conversion factor method):

$$500 \text{ nm} \times \frac{1 \text{ m}}{10^9 \text{ nm}} \times \frac{10^6 \text{ µm}}{1 \text{ m}} = 0.500 \text{ µm}$$

k.   10.0 nm to in (proportion method):

$$\frac{?}{10 \text{ nm}} = \frac{1 \text{ cm}}{10^7 \text{ nm}} \qquad ? = 1 \times 10^{-6} \text{ cm}$$

$$\frac{?}{1 \times 10^{-6} \text{ cm}} = \frac{1 \text{ in}}{2.54 \text{ cm}} \qquad ? = 3.94 \times 10^{-7} \text{ in}$$

10.0 nm to in (conversion factor method):

$$10.0 \text{ nm} \times \frac{1 \text{ m}}{10^9 \text{ nm}} \times \frac{100 \text{ cm}}{1 \text{ m}} \times \frac{1 \text{ in}}{2.54 \text{ cm}}$$

$$= 3.94 \times 10^{-7} \text{ in}$$

2. 10 km to mi (proportion method):

$$\frac{?}{10 \text{ km}} = \frac{1 \text{ mi}}{1.609 \text{ km}}$$

$$? = \frac{(1 \text{ mi})(10 \text{ km})}{1.609 \text{ km}} \qquad ? \approx 6.2 \text{ mi}$$

10 km to mi (conversion factor method):

$$10 \text{ km} \times \frac{1 \text{ mi}}{1.609 \text{ km}} \approx 6.2 \text{ mi}$$

3. $26.2 \text{ mi} \times \dfrac{1.609 \text{ km}}{1 \text{ mi}} \approx 42.2 \text{ km}$

4. 5 ft 4 in to m (proportion method):

$$\frac{?}{5 \text{ ft}} = \frac{12 \text{ in}}{\text{ft}} \qquad ? = \frac{(12 \text{ in})(5 \text{ ft})}{\text{ft}}$$

$$? = 60 \text{ in, plus } 4 \text{ in} = 64 \text{ in}$$

$$\frac{?}{64 \text{ in}} = \frac{1 \text{ m}}{39.37 \text{ in}}$$

$$? = \frac{(1 \text{ m})(64 \text{ in})}{39.37 \text{ in}} \qquad ? \approx 1.6 \text{ m}$$

5 ft 4 in to m (conversion factor method):

$$5 \text{ ft} \times \frac{12 \text{ in}}{1 \text{ ft}} = 60 \text{ in}$$

$$60 \text{ in} + 4 \text{ in} = 64 \text{ in}$$

$$64 \text{ in} \times \frac{1 \text{ m}}{39.37 \text{ in}} \approx 1.6 \text{ m}$$

5. 45 mi into km (proportion method):

$$\frac{?}{45 \text{ mi}} = \frac{1.609 \text{ km}}{1 \text{ mi}}$$

$$? = \frac{(1.609 \text{ km})(45 \text{ mi})}{1 \text{ mi}} \qquad ? \approx 72 \text{ km}$$

45 mi into km (conversion factor method):

$$45 \text{ mi} \times \frac{1 \text{ km}}{0.6214 \text{ mi}} \approx 72 \text{ km}$$

6. 55 mph to km/h (proportion method):

$$1 \text{ mph} = 1.609 \text{ kph}$$

$$\frac{1 \text{ mph}}{1.609 \text{ kph}} = \frac{55 \text{ mph}}{?}$$

$$? \approx 88.5 \text{ kph}$$

55 mph to km/h (conversion factor method):

$$55 \text{ mph} \times \frac{1 \text{ kph}}{0.6214 \text{ mph}} \approx 88.5 \text{ kph}$$

7. 3.0 ton to kg (proportion method):

$$\frac{?}{3.0 \text{ ton}} = \frac{2{,}000 \text{ lb}}{\text{ton}}$$

$$? = \frac{(2{,}000 \text{ lb})(3.0 \text{ ton})}{\text{ton}} \qquad ? = 6{,}000 \text{ lb}$$

$$\frac{?}{6{,}000 \text{ lb}} = \frac{1 \text{ kg}}{2.205 \text{ lb}}$$

$$? = \frac{(1 \text{ kg})(6{,}000 \text{ lb})}{2.205 \text{ lb}}$$

$? \approx 2{,}721.1 \text{ kg (round to } 2{,}700 \text{ kg)}$
3.0 ton to kg (conversion factor method):

$$3.0 \text{ ton} \times \frac{2{,}000 \text{ lb}}{1 \text{ ton}} \times \frac{1 \text{ kg}}{2.205 \text{ lb}} \approx 2721.1 \text{ kg}$$

(round to 2,700 kg)

8.

a. $2.50 for 12 oz (proportion method):

$$\frac{?}{12 \text{ oz}} = \frac{1 \text{ lb}}{16 \text{ oz}}$$

$$? = \frac{(1 \text{ lb})(12 \text{ oz})}{16 \text{ oz}} \qquad ? = 0.75 \text{ lb}$$

$$\frac{?}{0.75 \text{ lb}} = \frac{453.6 \text{ g}}{1 \text{ lb}}$$

$$? = \frac{(453.6 \text{ g})(0.75 \text{ lb})}{1 \text{ lb}} \qquad ? = 340.2 \text{ g}$$

$$340.2 \text{ g} / \$2.50 \approx 136.1 \text{ g} / \$1.00$$

b. $3.67 for 250 g (proportion method):

$$\frac{250 \text{ g}}{3.67} \approx 68.12 \text{ g} / 1.00\$$$

c. $4.50 for 0.300 kg:

$$\frac{?}{0.300 \text{ kg}} = \frac{1,000 \text{ g}}{\text{kg}}$$

$$? = \frac{(1,000 \text{ g})(0.300 \text{ lb})}{\text{kg}} \qquad ? = 300 \text{ g}$$

$$300 \text{ g} / \$4.50 \approx 66.7 \text{ g} / \$1.00$$

d. $2.35 for 0.75 lb:

$$\frac{?}{0.75 \text{ lb}} = \frac{453.6 \text{ g}}{\text{lb}}$$

$$? = \frac{(453.6 \text{ g})(0.75 \text{ lb})}{\text{lb}} \qquad ? = 340.2 \text{ g}$$

$$340.2 \text{ g} / \$2.35 \approx 144.8 \text{ g} / \$1.00 = \text{best value}$$

9. (Proportion method)

$$\frac{?}{1 \text{ week}} = \frac{7 \text{ days}}{1 \text{ week}} \quad ? = 7 \text{ days}$$

$$\frac{?}{7 \text{ days}} = \frac{24 \text{ hours}}{1 \text{ day}} \qquad ? = 24 \times 7 = 168 \text{ hours}$$

$$\frac{?}{168 \text{ hours}} = \frac{60 \text{ minutes}}{1 \text{ hours}}$$

$$? = 168 \times 60 = 10,080 \text{ minutes}$$

$$\frac{?}{10,080 \text{ minutes}} = \frac{60 \text{ s}}{1 \text{ minutes}}$$

$$? = 10,080 \times 60 = 604,800 \text{ seconds}$$

(Conversion factor method)

$$1 \text{ week} \times \frac{7 \text{ days}}{1 \text{ week}} = 7 \text{ days}$$

$$7 \text{ days} \times \frac{24 \text{ hours}}{1 \text{ day}} = 168 \text{ hours}$$

$$168 \text{ hours} \times \frac{60 \text{ minutes}}{1 \text{ hours}} = 10,080 \text{ minutes}$$

$$10,080 \text{ minutes} \times \frac{60 \text{ seconds}}{\text{minutes}} = 604,800 \text{ seconds}$$

10. 1/52 year = 1 week = 168 hours = 10,080 minutes = 604,800 seconds

11. 1 year = 52 weeks = 364 days = 524,160 minutes = 31,449,600 seconds

12. 1/5,280 mi = 1/3 yds = 1 ft = 12 in or 0.000189 mi ≈ 0.333 yds ≈ 1 ft ≈ 12 in

13. 0.0156 gal ≈ 0.125 pt = 2 oz

14. 1 km = 10^3 m = 10^5 cm = 10^6 mm = 10^9 μm = 10^{12} nm

15. 10^{-12} km = 10^{-9} m = 10^{-7} cm = 10^{-6} mm = 10^{-3} μm = 1 nm

16. 2.5×10^{-5} km = 2.5×10^{-2} m = 2.5 cm = 25 mm = 2.5×10^4 μm = 2.5×10^7 nm

17.
   a. 6.25 mm = 6.25×10^3 μm
   b. 0.00896 m = 8.96 mm
   c. 9,876,000 nm = 9.876 mm

18. 3.0×10^{-3} km = 3.0 m = 3.0×10^2 cm = 1.2×10^2 in

19. 5 kg = 5×10^3 g = 5×10^6 mg = 5×10^9 μg

20. 8.9×10^{-6} kg = 8.9×10^{-3} g = 8.9 mg = 8.9×10^3 μg

21. 2×10^{-17} kg = 2×10^{-14} g = 2×10^{-11} mg = 2×10^{-8} μg

22.
   a. 0.8657 g = 8.657×10^{-1} g = 8.657×10^2 mg
   b. 526 kg = 5.26×10^2 kg = 5.26×10^8 mg
   c. 63 g = 6.3×10^1 g = 6.3×10^7 μg
   d. 2.63×10^{-6} μg = 2.63×10^{-15} kg

23. 1 Ci = 3.7×10^{10} dps = 2.2×10^{12} dpm

24. 1 Ci = 1,000 mCi = 10^6 μCi = 3.7×10^{10} dps

25. 2.7×10^{-6} Ci = 2.7×10^{-3} mCi = 2.7 μCi = 10^5 dps

26. 1×10^{-4} Ci = 0.1 mCi = 100 μCi = 3.7×10^6 dps ≈ 2.2×10^8 dpm

27. 1 Ci = 10^3 mCi = 10^6 μCi = 3.7×10^{10} dps = 3.7×10^{10} Bq

28. 2.5×10^{-4} Ci $= 2.5 \times 10^{-1}$ mCi $= 250$ μCi $=$ 9.25×10^{6} dps $= 9.25 \times 10^{6}$ Bq

APPLICATION PRACTICE PROBLEMS:
UNIT CONVERSIONS (FROM P. 291)

1. If the medium requires 5 g/1 L, then 125 g of glucose is required for 25 L. The technician added:

$$\frac{?}{0.24 \text{ lb}} = \frac{453.6 \text{ g}}{1 \text{ lb}}$$

$$? = \frac{(453.6 \text{ g})(0.24 \text{ lb})}{\text{lb}} \qquad ? \approx 108.86 \text{ g}$$

The technician, therefore, added too little glucose.

2.

	For 1 L (g)	For 1 mL (mg)
NaCl	20.00	20
Na azide	0.001	0.001
Mg sulfate	1.000	1.000
Tris	15.00	15.00

3. 680 mg $= 0.680$ g
   So there is 0.680 g enzyme/1 g powder.

$$\text{need } \frac{10.0 \text{ oz}}{100.0 \text{ L}} = \frac{?}{500.0 \text{ L}} \qquad ? = 50.0 \text{ oz}$$

$$50.0 \text{ oz} \times \frac{1 \text{ lb}}{16 \text{ oz}} \times \frac{453.6 \text{ g}}{1 \text{ lb}} = 1417.5 \text{ g}$$

$$\frac{0.680 \text{ g enzyme}}{1 \text{ g powder}} = \frac{1417.5 \text{ g enzyme}}{?}$$

$? \approx 2{,}084.56$ g powder $\rightarrow$ Need 2.08 kg powder

MANIPULATION/APPLICATION PRACTICE
PROBLEMS: CONCENTRATION (FROM P. 292)

1. $$\frac{3 \text{ g}}{250 \text{ mL}} = \frac{?}{1{,}000 \text{ mL}}$$

$$? = \frac{(3 \text{ g})(1{,}000 \text{ mL})}{250 \text{ mL}} \qquad ? = 12 \text{ g}$$

2. $$\frac{25 \text{ g}}{1{,}000 \text{ mL}} = \frac{?}{100 \text{ mL}}$$

$$? = \frac{25 \text{ g} \times 100 \text{ mL}}{1{,}000 \text{ mL}} \qquad ? = 2.5 \text{ g}$$

3. $\dfrac{1 \text{ mg}}{1 \text{ mL}} = \dfrac{1 \text{ g}}{1 \text{ L}}$ so $\dfrac{1 \text{ g}}{1 \text{ L}} = \dfrac{?}{15 \text{ L}}$ $\quad ? = 15$ g

4. $$\frac{0.005 \text{ g Tris base}}{\text{L}} = \frac{?}{10^{-3} \text{L}}$$

$$? = \frac{(0.005 \text{ g})10^{-3} \text{L}}{\text{L}}$$

$? = 0.005 \text{ g} \times 10^{-3} \text{g} \qquad ? = 5 \times 10^{-6} \text{g}$

5. $$\frac{300 \text{ ng dioxin}}{100 \text{ g diapers}} = \frac{?}{1{,}000 \text{ g}}$$

$? = 3{,}000$ ng $= 3$ μg dioxin

6. $$\frac{?}{5 \text{ mL}} = \frac{5 \times 10^{-3} \text{ moles}}{1{,}000 \text{ mL}}$$

$$? = \frac{(5 \times 10^{-3} \text{ moles})5 \text{ mL}}{10^{3} \text{ mL}}$$

$? = 25 \times 10^{-6}$ moles $= 2.5 \times 10^{-5}$ moles

7. $(1 \text{ L} = 10^{6} \text{ μL})$

$$\frac{?}{1 \text{ μL}} = \frac{0.1 \text{ moles}}{10^{6} \text{ μL}} \qquad ? = \frac{(0.1 \text{ moles}) \text{μL}}{10^{6} \text{ μL}}$$

$? = 0.1 \times 10^{-6} = 1 \times 10^{-7}$ moles

8. $(1 \text{ L} = 10^{3} \text{ mL})$ $\qquad \dfrac{?}{78 \text{ mL}} = \dfrac{10^{-2} \text{ g}}{10^{3} \text{ mL}}$

$$? = \frac{(10^{-2} \text{ g})78 \text{ mL}}{10^{3} \text{ mL}}$$

$? = 78 \times 10^{-5}$ g $= 7.8 \times 10^{-4}$ g

9. $\dfrac{500 \text{ U}}{\text{mg}} = \dfrac{?}{3 \text{ mg}}$ $\quad ? = \dfrac{(500 \text{ U}) 3 \text{ mg}}{\text{mg}}$

$? = 1{,}500$ U

10. $$\frac{2{,}500 \text{ U}}{\text{mg}} = \frac{?}{5 \text{ mg}}$$

$$? = \frac{(2{,}500 \text{ U}) 5 \text{ mg}}{1 \text{ mg}} \qquad ? = 1.25 \times 10^{4} \text{U}$$

11. $\dfrac{?}{10^3 \text{ mL}} = \dfrac{3 \text{ g}}{\text{mL}}$

$? = \dfrac{(3 \text{ g}) \times 10^3 \text{ mL}}{\text{mL}}$ $? = 3 \times 10^3 \text{g}$

12. $\dfrac{5 \text{ µg}}{\text{L}} = \dfrac{5 \text{ µg}}{10^3 \text{ mL}}$

a. $\dfrac{5 \text{ µg}}{10^3 \text{ mL}} = \dfrac{?}{50 \text{ mL}}$

$? = \dfrac{(5 \text{ µg}) \, 50 \text{ mL}}{10^3 \text{ mL}}$

$? = \dfrac{250 \text{ µg}}{10^3} = 0.25 \text{ µg}$

b. $\dfrac{5 \text{ µg}}{10^3 \text{ mL}} = \dfrac{?}{500 \text{ mL}}$

$? = \dfrac{(5 \text{ µg}) \, 500 \text{ mL}}{10^3 \text{ mL}}$

$? = \dfrac{2500 \text{ µg}}{10^3} = 2.5 \text{ µg}$

c. $\dfrac{5 \text{ µg}}{10^3 \text{ mL}} = \dfrac{?}{100 \text{ mL}}$

$? = \dfrac{(5 \text{ µg}) \, 100 \text{ mL}}{10^3 \text{ mL}}$

$? = \dfrac{500 \text{ µg}}{10^3} = 0.5 \text{ µg}$

d. $\dfrac{5 \text{ µg}}{10^6 \text{ µL}} = \dfrac{?}{100 \text{ µL}}$

$? = \dfrac{(5 \text{ µg}) 100 \text{ µL}}{10^6 \text{ µL}} = \dfrac{500 \text{ µg}}{10^6} = 5 \times 10^{-4} \text{ µg}$

13.

a. $\dfrac{0.5 \text{ mg}}{1 \text{ mL}} = \dfrac{?}{5 \text{ mL}}$

$? = \dfrac{(0.5 \text{ mg}) \, 5 \text{ mL}}{1 \text{ mL}}$ $? = 2.5 \text{ mg}$

b. $\dfrac{0.5 \text{ mg}}{1 \text{ mL}} = \dfrac{?}{0.5 \text{ mL}}$

$? = \dfrac{(0.5 \text{ mg}) \, 0.5 \text{ mL}}{1 \text{ mL}}$ $? = 0.25 \text{ mg}$

c. $1 \text{ mL} = 10^3 \text{µL}$ $\dfrac{0.5 \text{ mg}}{10^3 \text{ µL}} = \dfrac{?}{100 \text{ µL}}$

$? = \dfrac{(0.5 \text{ mg}) \, 100 \text{ µL}}{10^3 \text{ µL}}$ $? = 0.05 \text{ mg}$

d. $1,000 \text{ µL} = 1 \text{ mL}$

$\dfrac{0.5 \text{ mg}}{\text{mL}} = \dfrac{?}{\text{mL}}$ $? = 0.5 \text{ mg}$

14. $\dfrac{100 \text{ molecules}}{10^9 \text{ molecules}} = \dfrac{1 \times 10^2}{10^9} = \dfrac{1}{10^7} < \dfrac{1}{10^6}$

Cannot be detected

15. $\dfrac{1 \text{ g}}{10^6 \text{kg}} = \dfrac{1 \text{ g}}{10^9 \text{ g}} \text{(more pure)}$

$\dfrac{10^{-2} \text{ mg}}{10^{-3} \text{ kg}} = \dfrac{10^{-5} \text{ g}}{\text{g}} = \dfrac{1 \text{ g}}{10^5 \text{ g}}$

MANIPULATION PRACTICE PROBLEMS: DILUTIONS (PART A) (FROM PP. 295)

1.
   a. 1 part orange juice in 4 parts total = 1/4
   b. 1 part orange juice to 3 parts water = 1:3
   c. 1:3 or 1/4
2.
   a. 1 mL/10 mL
   b. 1 mL/11 mL
   c. 3 mL/30 mL
   d. 3 mL/30 mL
   e. 0.5 mL/11.5 mL
3.
   a. 1/10
   b. 1/11
   c. 1/4
   d. 1/2
   e. 1/5
4.
   a. 1:9
   b. 1:10
   c. 1:9
   d. 1:9
   e. 1:22
5. 0.5 mL blood
   1.0 mL H$_2$O
   + 3.0 mL reagent
   4.5 mL
   0.5/4.5 dilution = 1/9

MANIPULATION PRACTICE PROBLEMS: DILUTIONS (PART B) (FROM PP. 297)

1. $\dfrac{?}{10\ mL} = \dfrac{1}{10}$ $? = \dfrac{(1)10\ mL}{10}$

   $? = 1\ mL$ blood
   $1\ mL + 9\ mL$ will give 1/10 dilution

2. $\dfrac{?}{250\ mL} = \dfrac{1}{300}$ $? = \dfrac{1(250\ mL)}{300}$

   $? \approx 0.833\ mL$ blood
   $0.833\ mL + 249.167\ mL = 250\ mL$

3. $\dfrac{?}{1\ mL} = \dfrac{1}{50}$ $? = \dfrac{1(1\ mL)}{50}$

   $? = 0.02\ mL$ blood
   $0.02\ mL + 0.98\ mL = 1\ mL$

4. $\dfrac{?}{1,000\ \mu L} = \dfrac{1}{100}$ $? = \dfrac{1(1,000\ \mu L)}{100}$

   $? = 10\ \mu L$ food coloring
   $10\ \mu L + 990\ \mu L = 1,000\ \mu L$

5. $\dfrac{?}{23\ mL} = \dfrac{3}{5}$ $? = \dfrac{3(23\ mL)}{5}$

   $? = 13.8\ mL$ of Q
   $13.8\ mL + 9.2\ mL = 23\ mL$

6. $\dfrac{?}{10\ mL} = \dfrac{1}{10}$ $? = \dfrac{1(10\ mL)}{10}$

   $? = 1\ mL$ stock
   $1\ mL + 9\ mL = 10\ mL$

7. $\dfrac{?}{15\ mL} = \dfrac{1}{5}$ $? = \dfrac{1(15\ mL)}{5}$

   $? = 3\ mL$ stock
   $3\ mL + 12\ mL = 15\ mL$

8. $\dfrac{?}{10^3\ \mu L} = \dfrac{1}{100}$ $? = \dfrac{1(10^3\ \mu L)}{100}$

   $? = 10\ \mu L$ stock
   $10\ \mu L + 990\ \mu L = 1,000\ \mu L$

9. $0.01 = 1/100$

   $$\dfrac{?}{50\ mL} = \dfrac{1}{100}$$

   $? = 0.5\ mL$ buffer
   $0.5\ mL + 49.5\ mL$

MANIPULATION PRACTICE PROBLEMS: DILUTIONS (PART C) (PP. 299)

1. $\dfrac{(50\%)\,1}{40} = 1.25\%$

2. $\dfrac{10\ mg}{mL} \times \dfrac{1}{10} = \dfrac{10\ mg}{10\ mL} = 1\ mg/mL$

3. $1:1 = \dfrac{1}{2}$ dilution

   so : $\dfrac{10\ mg}{mL} \times \dfrac{1}{2} = \dfrac{10\ mg}{2\ mL} = \dfrac{5\ mg}{mL}$

4. $1\ mL + 4\ mL = 1/5$ dilution.
   $1\ mL$ can make $5\ mL$ diluted solution.

5. $1,000\ mL \times \dfrac{1}{100} = \dfrac{1,000\ mL}{100} = 10\ mL$ original

   or $\dfrac{?}{1,000\ mL} = \dfrac{1}{100}$ $? = 10\ mL$

MANIPULATION PRACTICE PROBLEMS: DILUTIONS (PART D*) (P. 302)

1. $\dfrac{1}{10}$ dilution;

   • take $0.1\ mL$ food coloring $+ 0.9\ mL$ diluent
   • $\dfrac{1}{10} \times \dfrac{1}{25} = \dfrac{1}{250}$ take $1\ mL$ of above $+ 24\ mL$ diluent
   • $\dfrac{1}{250} \times \dfrac{1}{4} = \dfrac{1}{1,000}$ take $1\ mL$ of above $+ 3\ mL$ diluent

2.
   a.
   $\dfrac{1}{10^6}$ dilution – use three dilutions of 1/100, 1/100, 1/100
   • $0.1\ mL$ of culture $+ 9.9\ mL$ diluent $= 1/100$
   • take $0.1\ mL$ of above $+ 9.9\ mL = 1/100 \times 1/100 = 1/10^4$
   • take $0.1\ mL$ of above $+ 9.9\ mL = 1/100 \times 1/10^4 = 1/10^6$

   b. $\dfrac{1}{10} \times \dfrac{1}{10} \times \dfrac{1}{10} \times \dfrac{1}{10} \times \dfrac{1}{10} \times \dfrac{1}{10}$
   • $0.1\ mL$ culture $+ 0.9\ mL$ diluent $= 1/10$
   • take $0.1\ mL$ of above $+ 0.9\ mL = 1/10 \times 1/10 = 1/10^2$
   • take $0.1\ mL$ of above $+ 0.9\ mL = 1/10 \times 1/10^2 = 1/10^3$
   • take $0.1\ mL$ of above $+ 0.9\ mL = 1/10 \times 1/10^3 = 1/10^4$
   • take $0.1\ mL$ of above $+ 0.9\ mL = 1/10 \times 1/10^4 = 1/10^5$

---

[1] Note: For dilution problems, there are often several workable strategies. Only one strategy is shown in the answer key.

- take 0.1 mL of above + 0.9 mL = 1/10 × 1/10⁵ = 1/10⁶

3.
- 1 mL blood + 4 mL diluent = 1/5
- take 1 mL of above + 9 mL = 1/10 × 1/5 = 1/50
- take 1 mL of above + 4 mL = 1/5 × 1/50 = 1/250

Application Practice Problems:
Dilutions (from pp. 303)

1. Each tube requires 5 U.
   The enzyme concentration is:

$$\frac{1{,}000\ U}{1\ mL} = \frac{1{,}000\ U}{1{,}000\ \mu L} = \frac{1\ U}{1\ \mu L}$$

The 20 μL contains 20 units of enzyme. You can, therefore, make four tubes with 5 U per tube.

2. Each tube requires 5 × 0.01 U = 0.05 U.
   The enzyme concentration is 1 U/μL.

$$\frac{?}{0.05\ U} = \frac{1\ \mu L}{1\ U}$$

? = 0.05 μL = amount needed, but this volume is too low to measure.

Therefore, make 1/100 dilution, 10 μL enzyme + 990 μL = 1/100. After dilution, there are:

$$\frac{1\ U}{1\ \mu L} \times \frac{1}{100} = \frac{1\ U}{100\ \mu L} = \frac{10^{-2}\ U}{1\ \mu L}$$

Need 0.05 U/tube, so use 5 μL of dilution for each tube to 5 mL total volume.

3. Dilute $\dfrac{1}{500{,}000} = \dfrac{1}{50} \times \dfrac{1}{100} \times \dfrac{1}{100}$
   - 0.1 mL antibody + 4.9 mL diluent = 1/50
   - take 0.1 mL of above + 9.9 mL → 1/100 × 1/50 = 1/5,000
   - take 0.1 mL of above + 9.9 mL → 1/100 × 1/5,000 = 1/500,000

4. Have $\dfrac{1 \times 10^9\ cells}{1\ mL}$ want $\dfrac{2 \times 10^3\ cells}{1\ mL}$

$$= \frac{2 \times 10^2\ cells}{0.1\ mL}$$

$$\frac{1 \times 10^9}{2 \times 10^3} = \frac{1}{2} \times 10^6\ \text{or}\ 5 \times 10^5\ \text{dilution}$$

$$\frac{1}{100} \times \frac{1}{100} \times \frac{1}{50} = \frac{1}{5 \times 10^5}$$

- 0.1 mL culture + 9.9 mL diluent = 1/100 → (1/100) × (1 × 10⁹) = 10⁷ cells/mL
- 0.1 mL of above + 9.9 mL → 1/100 × 1/100 → (1/10⁴) × (1 × 10⁹) → 10⁵ cells/mL
- 0.1 mL of above + 4.9 mL → 1/50 × 1/10⁴

$$\to \frac{1}{\left(5 \times 10^5\right)} \times \left(1 \times 10^9\right) \to 2 \times 10^3\ \text{cells} / \text{mL}$$

$$= 200\ \text{cells} / 0.1\ \text{mL}$$

5. The first plate has too many colonies to count, and the third plate has fewer colonies than are desirable for a plate count. The middle plate has 21 colonies, so it is the one we will use for calculations. The concentration of cells added to this plate was about 21 cells/0.1 mL = 210 cells/mL. The dilution tube from which these cells were drawn had been diluted 1/10⁷. There were, therefore, originally about 210 × 10⁷ = 2.10 × 10⁹ bacterial cells/mL in the original broth.

6. The first plate can be used for calculations. There are about 43 colonies on the first plate, so the concentration of bacteria applied to that plate was about 430 cells/mL. The dilution tube from which these cells were drawn had been diluted 1/10⁶. There were, therefore, about 430 × 10⁶ = 4.30 × 10⁸ cells/mL in the original broth.

7. 10⁷ cells/mL; dilute to $\dfrac{100\ cells}{0.1\ mL}$
   = 1,000 cells/mL final concentration

$$\frac{10^7}{10^3} = 1 \times 10^4 = \text{dilution needed}$$

- 0.1 mL culture + 9.9 diluent = 1/100 = 1/10² → 10⁵ colonies/mL
- 0.1 mL culture + 9.9 diluent = 1/100 = 1/10² → 10³ colonies/mL

8. $\dfrac{1}{100}$ dilution had $\dfrac{50\ mg}{mL}$

$$\frac{50\ mg}{1\ mL} \times 100 = \frac{5{,}000\ mg}{1\ mL} = \frac{5\ g}{1\ mL}$$

= concentration of protein in undiluted sample.

$$\frac{5\ g}{1\ mL} = \frac{?}{100\ mL} \qquad ? = 500\ g$$

9. $\frac{1}{50}$ dilution had

$$\frac{87 \text{ mg}}{1 \text{ mL}} \times 50 = \frac{4,350 \text{ mg}}{1 \text{ mL}} = \frac{4.35 \text{ g}}{1 \text{ mL}}$$

= concentration of protein in undiluted sample.

10. The sample was diluted 1/5 and then 5/25 → (1/5) (5/25) = 5/125 total dilution. There was 3 mg of protein in 5 mL = 0.6 mg protein/mL. Multiplying times the dilution: 0.6 mg/mL × 125/5 = 15 mg/mL.

11. The sample was diluted 1/4 and then 10/30 → 10/120 total dilution.

There was 10 mg protein/5 mL = 2 mg/mL of protein in the solution. Multiplying times the dilution: 2 mg/mL × 120/10 = 24 mg/mL in original sample.

CHAPTER 13

MANIPULATION PRACTICE PROBLEMS:
GRAPHING (FROM PP. 309)

1.
   a. The amount of product increases in a linear fashion over time.
   b. The amount of product decreases in a non-linear fashion over time.
   c. The reaction rate is constant, regardless of the amount of compound X present.

2.
   a. (7,5)
   b. (−4, −4)
   c. (−5,5)
   d. (7, −4)

3. Change in X = −11; change in Y = −9; change in X = −1; change in Y = 9.

4.

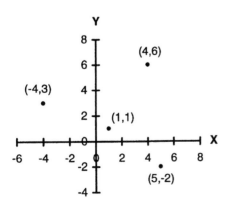

5. 12 <u>61</u>
   15 76
   <u>20</u> 101

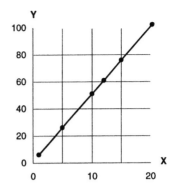

6. A = 3Q − 4

Q	A
−1	−7
0	−4
1	−1
2	2
3	5

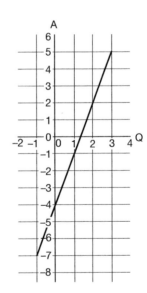

7.
   a. Slope 3, intercept 2
   b. Slope 0.2, intercept −1
   c. Slope 0.005, intercept 0

8.
   a. Slope 1, intercept (0,2)
   b. Slope −1, intercept (0,7.5)
   c. Slope 1.25, intercept (0,4)
   d. Slope −1, intercept (0,5)

9.
  a. Not linear
  b. linear
  c. linear
  d. not linear
10. All will form a line.
11.
  a, b.
   i. Slope = 10 cm/min, Y-intercept = (0 minutes, 1 cm)
   ii. Slope = −1, Y-intercept = (0,3)
   iii. Slope = 7 mg/cm, Y-intercept = (0 cm, 12 mg)

  c.

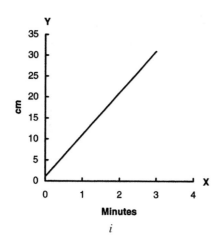

i

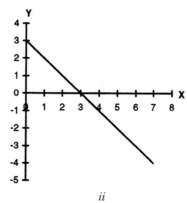

ii

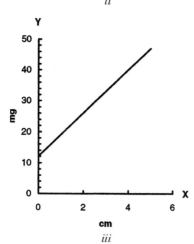

iii

12.

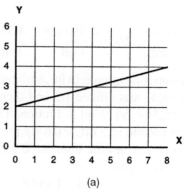

(a)

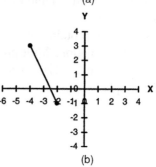

(b)

13.
  a. $Y = -1X + 2$
  b. $Y = 2.25X - 9$
  c. $Y = 3$
  d. $Y = -1X + 5.5$

14. **Line A:** $Y = \dfrac{3}{4} X$

   Line B: $Y = 2X + 2$

   Line C: $Y = \dfrac{1}{2} X + 2$

15. $Y = 1 X + 0$

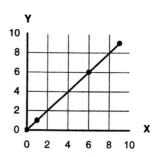

16.
  a. 9/5 and 32
  b. 32
  c. 9/5
  d.

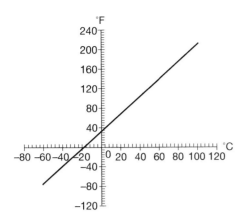

MANIPULATION AND APPLICATION PRACTICE PROBLEMS: QUANTITATIVE ANALYSIS (FROM P. 318)

1.
a.

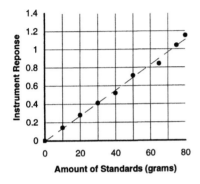

b. A 0.78 instrument response corresponds to 57 g. The sample was diluted 100×, so the amount in the undiluted sample was: 57 g × 100 = 5,700 g.

2. Stock solution at 100 mg/mL. First, decide what volume to make. For example, a strategy to make 10 mL of each standard:

Standard (mg/mL)	Dilution	Stock (mL)	Diluent (mL)
1	1/100	0.1	9.9
5	5/100	0.5	9.5
15	15/100	1.5	8.5
25	25/100	2.5	7.5
50	50/100	5.0	5.0
75	75/100	7.5	2.5
100		10.0	0

APPLICATION PRACTICE PROBLEMS: GRAPHING (FROM PP. 322)

1.
a. Yes. The graph shows a linear relationship between cancer incidence and exposure level, even with the lowest exposures to the agent.

b. No, the graph has a threshold. At low levels of exposure, below the threshold, no change in cancer incidence occurs.

c. Possible explanations include the following: (1) Individuals are able to detoxify and handle low levels of the agent, and (2) multiple receptors must be activated before an effect occurs.

d. A background level of cancer is found in the population even in the absence of exposure to the agent.

e. Regulatory agencies need to know whether low level exposure is safe in order to decide what levels (if any) of compound can be allowed. For example, this is important to know when deciding whether a particular pesticide can be used on crops, and if so, what levels are "safe."

2.
a. Graph **ii** indicates a relationship between the mass of the parent and the offspring where a higher parental mass is associated with a higher offspring mass.

b. Environmental factors may play a major role in determining mass. One experiment would be to take genetically identical clones of the plant and measure their mass under different, controlled environmental conditions. There are other experiments you might devise as well.

3.
a.

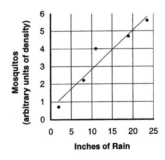

i

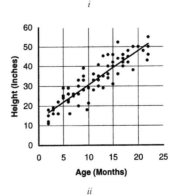

ii

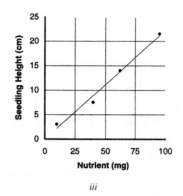

iii

b.
  i.   Slope = 0.22 mosquitos/in rain
  ii. Slope = 1.8 in/month
  iii. Slope = 0.24 cm/mg

c.
  i.   Y-intercept = (0 in rain, 0.6 mosquitos)
  ii. Y-intercept = (0 months, 13 in)
  iii. Y-intercept = (0 mg, 0 cm)

d.
  i.   Y = (0.22 mosquitos/in rain) X + 0.6 mosquitos
  ii. Y = (1.8 in/month) X + 13 in
  iii. Y = (0.24 cm/mg) X + 0 cm

e. 1.7 mosquitos (approximate)

f. Y = (0.22 mosquitos/in) (5 in) + 0.6 mosquitos = 1.7 mosquitos

g. Y = (1.8 in/month) (20 months) + 13 in = 49 in

h. Y = (0.24 cm/mg)(50 mg) + 0 cm = 12 cm

4.
  a. 80
  b. 80
  c. The slope is close to 1, so the test scores were roughly equal.
  d. The relationship is roughly linear, which means that there is a general relationship between the two scores. Students who did poorly on the midterm tended to do poorly on the final. Midterm scores appear therefore to be fairly predictive, although the next year's class could differ.

Application Practice Problems:
Exponential Equations (from pp. 332)

1.
  a.

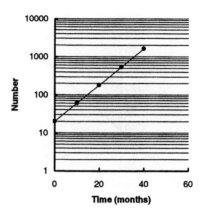

  b. Y = 3^x (20), where x = the number of generations

2. N = $(1/2)^t(N_0)$ N_0 = 300 μCi   half-life = 14 days
  a. t = 28 days/14 days = 2

$$N_2 = \left(\frac{1}{2}\right)^2 (300 \ \mu Ci) = 75 \ \mu Ci$$

  b. t = 140 days/14 days = 10

$$N_{10} = \left(\frac{1}{2}\right)^{10} (300 \ \mu Ci) = 0.293 \ \mu Ci$$

  c. t = 200 days/14 days = 14.3

$$N_{14.3} = \left(\frac{1}{2}\right)^{14.3} (300 \ \mu Ci) = 1.49 \times 10^{-2} \ \mu Ci$$

  d. t = 365 days/14 days = 26.1

$$N_{26.1} = \left(\frac{1}{2}\right)^{26.1} (300 \ \mu Ci) = 4.17 \times 10^{-6} \ \mu Ci$$

3.
  a. $N = \left(\frac{1}{2}\right)^{0.38} (600 \ \mu Ci) = 461 \ \mu Ci$

  b. $N = \left(\frac{1}{2}\right)^{0.77} (600 \ \mu Ci) = 352 \ \mu Ci$

  c. $N = \left(\frac{1}{2}\right)^{1.2} (600 \ \mu Ci) = 261 \ \mu Ci$

4.
  a. 10^6 bacteria
  b. 10^3 bacteria
  c. Using semilogarithmic paper results in a straight line on the graph that is easier to work with than an exponential graph. It also is used because of the wide range of values needed from 0 to 10^6.

5.

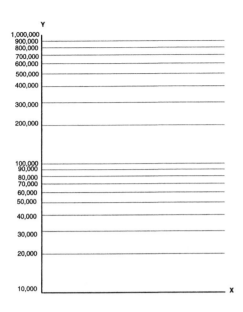

CHAPTER 14

1.
  a. Sample: The ten vials. Population: All the vials in the batch.
  b. Sample: The blood sample tested. Population: All the blood in that individual.
  c. The sample is the first graders who were tested. The population depends on the purpose of the study. If the study was intended to test the performance of all first graders in Franklin Elementary School, then the entire population was studied. If the sample was meant to represent all first graders in a city or in a larger group, then the population is that larger group.

2.
  a. $\bar{x} = 10.000$ g SD ≈ 0.001 g CV $\approx 0.012\%$
  b. $\bar{x} = 84\%$ SD $\approx 9.49\%$ CV $\approx 11.29\%$
  c. $\bar{x} = 235$ seconds   SD ≈ 33.56 seconds CV $\approx 14.28\%$
  d. $\bar{x} = 237$ seconds   SD ≈ 12.26 seconds CV $\approx 5.18\%$
     The means for the two methods are similar; however, the second method is more consistent. Based on this information, the second method seems better.

3.
  a. n = 33 $\bar{x} \approx 136.76$ mg

     SD ≈ 32.90 mg   CV $\approx 24.06\%$

  b. n = 25 $\bar{x} \approx 8.15$ g

     SD ≈ 1.47 g   CV $\approx 18.09\%$

  c. n = 19 $\bar{x} \approx 1,112.32$ mL

     SD ≈ 71.55 mL   CV $\approx 6.43\%$

4.
  a. n= 20 $\bar{x} \approx 33.05$ blossoms
     Median = 32.5 blossoms
     Range = 21–44 = 23 blossoms
     SD ≈ 5.57 blossoms
5. n = 12 $\bar{x} \approx 30.39$ kg   median = 28.6 kg
     Range = 18.2–52.2 = 34 kg   SD ≈ 9.90 kg
6. $\dfrac{\Sigma X}{30} = 75\%$ $\Sigma X = 2,250\%$

   2,250% + 100% = new total

   $$\dfrac{2,350\%}{31} \approx 75.8\%$$

7.
  a. Mean = 4, mode = 4
  b. Mean is approximately 5; mode is approximately 4.
  c. Mean is approximately 3; mode is approximately 4.
8.
  a. Normal
  b. Bimodal
  c. Skewed
9. **b** is less dispersed
10. **a** is less dispersed
11. About the same
12. **a** is less dispersed
13.
  a. n = 45 $\bar{x} \approx 9.55$ activity units
     SD ≈ 3.60 activity units
     Range = 0.4–15.1 activity units
     = 14.7 activity units
     Median = 10.1 activity units
  b.

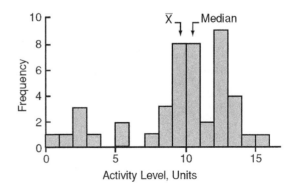

Frequency Table

Interval	Frequency
0–0.9	1
1.0–1.9	1
2.0–2.9	3
3.0–3.9	1
4.0–4.9	0
5.0–5.9	2
6.0–6.9	0
7.0–7.9	1
8.0–8.9	3
9.0–9.9	8
10.0–10.9	8
11.0–11.9	2
12.0–12.9	9
13.0–13.9	4
14.0–14.9	1
15.0–15.9	1

  c. Based on just this information, we cannot tell whether the DNA fragment was taken up or not because we do not know what level of enzyme activity exists in cells that have not been treated with the fragment.

14.

  a. $n = 20$ $\bar{x} \approx 1.29$ activity units
SD ≈ 0.93 activity units

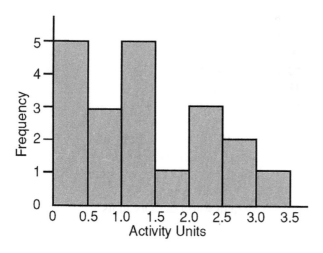

  b.

Interval	Frequency
0–0.4	5
0.5–0.9	3
1.0–1.4	5
1.5–1.9	1
2.0–2.4	3
2.5–2.9	2
3.0–3.4	1

  c. There is variation both in the group treated with the DNA fragment and in the group not treated. Using this assay, there is a low level of activity exhibited even in cells not treated with the DNA fragment; however, the mean enzyme activity of the two groups is different. There is also the suggestion graphically that the clones that were treated with the DNA fell into two categories: (1) those with higher levels of activity and (2) those with lower levels, similar to clones that were not treated with the DNA. Based on these observations, it appears that some of the treated cells took up the DNA and some did not. This hypothesis could be investigated by further study.

15. $n = 100$ $\bar{x} \approx 152.2$ cm   range $= 103$–197 cm
   $= 94$ cm

Frequency distribution table with data divided into 11 intervals:

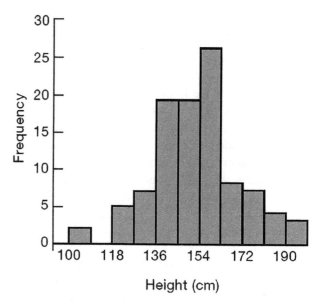

Interval (cm)	Frequency
100–108	2
109–117	0
118–126	5
127–135	7
136–144	19
145–153	19
154–162	26
163–171	8
172–180	7
181–189	4
190–198	3

16.

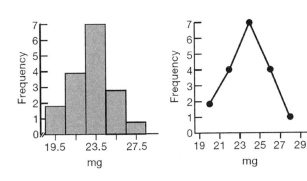

Interval (mg)	Frequency
19–20	2
21–22	4
23–24	7
25–26	3
27–28	1

17.
  a. Very certain. Common sense is sufficient to know this.
  b. Very certain. Common sense is sufficient to know this.
  c. Based on the standard deviation of the sample, we would expect 68% of customers to be in this range; however, a sample of five is too small to draw firm conclusions.
  d. Not very certain. A sample size of five is too small to draw firm conclusions.

18.
  a. 2.5%
  b. 16%

19.
  a. 99%
  b. 2.5%
  c. 68%
  d. 0.5%

20.
  a. The average appears to be in the mid-twenties and hovers at around ± 5; therefore, 18.1 mg appears a little low.
  b. Mean ≈ 27.16 mg, SD ≈ 3.87 mg. The mean −2 SD is 19.4; therefore, 18.1 mg is outside the range of 2 SD and probably should be investigated.

21.
  a. The values appear to hover around 65 leaves give or take about 10. It is difficult to tell whether 79 is a cause for concern.
  b. Mean ≈ 64.7 leaves, SD ≈ 9.7 leaves. The mean + 2 SD = 84.1; therefore, 79 does not appear to be unreasonable.

22. On several occasions, such as on March 22, the points lie outside the warning range; however, because later points are within the expected range, this is probably due to normal variation. The point on April 15 is out of the control range, and therefore, the process should be stopped and checked for problems. The points between May 31 and June 23 display an upward trend. A trend in one direction or another is also a cause for concern.

23.
  a. Mean = 5.99 pH units, SD ≈ 0.35 pH units
  b.

c. The pH values tend to decline with time.

d.

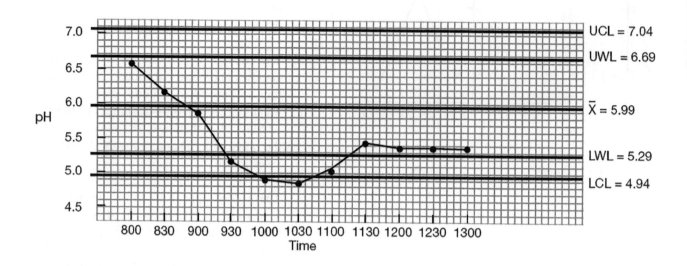

e. This process tended downward from the beginning, eventually reaching the lower control limit.

CHAPTER 15

1.

a. To calibrate and check the accuracy of a balance.

b. Recalibration involves comparison with a more accurate standard, or with one that has been stored and protected, rather than put into routine use.

   Recalibration is necessary because working standards may become damaged, dirty, dusty, or may otherwise be altered with time and use.

2. Readings will be too high. For example, a 200 g object will be read as slightly more than 200 g.

3. **a.** and **b.** We would expect all the readings made with this balance to be incorrect, but their consistency is unlikely to be affected by the improper calibration (adjustment) of the instrument. The accuracy of the instrument, therefore, would be adversely affected, but not its precision.

4. See Answer #3.

5. A pH meter needs to stabilize to give accurate results; therefore, this technician will sometimes have inaccurate results. In addition, precision requires that the operator perform a task in a consistent fashion;

therefore, the precision achieved by this technician will be poor. Failure to use an instrument properly, in this case, failure to allow for stabilization, is a systematic error. It would also be reasonable to label this as a gross error.

6. Precision. See text for explanation.

7. The standard deviation is about 0.15 mg/mL; therefore, it is reasonable to expect the next measurement to be 5.00 give or take about 0.15 mg/mL.

8. The most accurate measurements have the mean value at the "true value" (d). The most precise measurements have the narrowest curve with the smallest standard deviation (c).

9.

	Lab 1 (mg/mL)	Lab 2 (mg/mL)	Lab 3 (mg/mL)	Lab 4 (mg/mL)
Mean	37.5	≈ 38.1	≈ 38.0	≈ 38.4
Standard deviation	≈ 0.32	≈ 0.22	≈ 1.06	≈ 0.69

Laboratories 2 and 3 have good accuracy, but Laboratory 2 has the best precision.

10. Mean ≈ 37.5 mg/mL   SD ≈ 0.46 mg/mL

   The results from Laboratory 2 are more accurate and precise when the assays are completed at one time rather than over four months. (Many factors might lead to variability over time. Reagents or samples might slowly degrade, instruments may lose calibration or be replaced, personnel might change, and so on.)

11.

	Juan	Chris	Ilana	Mel
Mean	5.0002 g	5.0021 g	4.9999 g	5.0247 g
SD	≈ 0.0015 g	≈ 0.0001 g	≈ 0.0001 g	≈ 0.0516
% Error	−0.004%	−0.042%	0.002%	−0.494%

12. Mean ≈ 3.13   SD ≈ 0.019

The number of binding sites for an antibody is a whole number, in this case, presumably 3. The % error (a measure of accuracy) is −4.33%. The SD of the assay (a measure of precision) is ± 0.019 sites.

13.
  a. Subject 1: SD ≈ 0.15 mg/L
     Subject 2: SD ≈ 0.14 mg/L
     Subject 3: SD ≈ 0.10 mg/L
  b. Pooled mean ≈ 96.6 mg/L

14.
  a. $\bar{x}$ ≈ 7.16   SD ≈ 0.03
  b. Given the consistently high values and the fact that the instrument used to test the buffer was known to be properly working, we can be reasonably confident that the pH of this lot of buffer is about 7.16. This lot of buffer would be rejected.

15.
  a.

	Sample 1 (µg/m³)	Sample 2 (µg/m³)	Sample 3 (µg/m³)
$\bar{x}$	1.3	≈ 2.2	1.6
SD	0.10	≈ 0.12	0.10

  b. Pooled sample mean ≈ 1.7 µg/m³
     SD ≈ 0.42 µg/m³
  c. As we might expect, the standard deviation is higher for the pooled data because we would expect more variation among samples taken on different occasions than when an individual sample is tested three times.

16.
  a.

	Control Sample 1 (10 µg/L)	Control Sample 2 (100 µg/L)
$\bar{x}$	10 µg/L	≈ 101 µg/L
SD	≈ 1.6 µg/L	≈ 2.93 µg/L
RSD	≈ 16.3%	≈ 2.91%
% Error	0%	≈ -1.00%

  b. The accuracy and the precision of the assay may change at two different concentrations of the analyte.
  c. These analyses evaluate and document the precision and accuracy of the method. Because the assay components may change over time, this verification may be required numerous times. All such evaluations should be documented in an established way, such as with a form that is stored properly.
  d. The % error is an indication of the accuracy.

17.
  a. 56   2 significant figures
  b. 62   2 significant figures
  c. 35.9865   6 significant figures
  d. 8.25   3 significant figures
  e. 28.4 mL   3 significant figures
  f. 28.44 mL   4 significant figures

18.
  a. All three zeros are ambiguous.
  b. All six zeros are ambiguous.
  c. 0.00677
  d. 134,908,098, all zeros are significant.

19. **a.** 5 **b.** 3 **c.** 5 **d.** 3
20. **a.** 0.0 **b.** 1.0 **c.** 0.6 **d.** 0.2
21. lot a. rounded value = 10 mg/vial; yes, meets specification.

lot b. rounded value=10 mg/vial; yes, meets specification.

lot c. rounded value=8 mg/vial; no, does not meet specification.

lot d. rounded value=9 mg/vial; no, does not meet specification.

22. lot a. rounded value=0.03%; no, does not meet specification.

lot b. rounded value=0.02%; yes, meets specification.

lot c. rounded value=0.03%; no, does not meet specification.

lot d. rounded value=0.02%; yes, does meet specification.

23. **a, b.** The RSD is about 1.552% which, when rounded, is 2%. This is higher than the 1% specification; however, this does not necessarily indicate any problem with the spectrophotometer. Biological samples are often not homogeneous, and they are often not stable. With biological materials, therefore, variability is more likely to be due to the sample than to the instrument.

CHAPTER 16

1.
   a. Voltage source – batteries
      Resistance – bulb
   b. Voltage source – power company (wall outlet)
      Resistance – motor in the machine
   c. Voltage source – power company via power supply
      Resistance – the cells and the buffer in which they are suspended

2. The total standby current is 47.5 W.
   (47.5 W) (24 hours/day) (365 days)= 416,100 W-hours
   416,100 W-hours/1,000=416.1 kW-hours
   (416.1 kW-hours) ($0.09/kW-hr)≈$37.45 cost for standby current per year
   Multiplied by 50,000, it comes to about $1.87 million.

3. $V = I R$
   0.120 V=0.012 A (?)
   ?=10 ohms

4. $P = V I$
   ?=(120 V) (20 A)   ?=2,400 W

5. $V = I R$
   6 V=(0.400 A)(?)   ?=15 ohms

6. $P = V I$ maximum power=0.400 A (500 V) = 200 W

7. The total carrying capacity of the line in the laboratory is:

$$P = I V = (20 \text{ A})(120 \text{ V}) = 2,400 \text{ W}$$

The total power consumed by the fifteen units if they are run at their maximum power is:

$$200 \text{ W}(15) = 3,000 \text{ W}$$

The circuit, therefore, cannot handle all the units if they all are using 200 W.

8.

$P = V I$	$V = I R$
500 W=120 V (?)	120 V=4.2 A (?)
?≈4.2 A	?≈28.6 ohms

CHAPTER 17

1. Weigh out as close to 15 mg as possible, and then calculate the amount of water required to dilute the enzyme to a concentration of 15 mg/mL.

2.

a. ≈ 101.7 mL	**b.** ≈ 1.43 mL
c. ≈ 31.134 L	**d.** ≈ 10.14 mL

3. Assuming the technician has been careful to avoid temperature effects and drafts, the preparation is probably losing moisture because it was not dry to begin with.

4. Balance (b) has a greater response for a given weight, so it is more sensitive.

5. The mean value=9.999983333 g. The absolute error is 0.00002 g.
   The SD=0.000116905. Rounded, this is 0.0001. The balance meets its performance specifications.

6.
   a. The user first calibrates the balance and then checks the weight of a second NIST-traceable standard whose weight is close to that of the sample to be weighed. This verification is performed once a day when the balance is in use.

b. Form QF 15.3.6.3 is used to document the result of the performance verification. This includes the date, time, and operator name.

7. All subsequent readings with that balance will be a little too high. Dropping the standard is a gross error. The fact that subsequent readings will be a bit high is a systematic error.

8. **Midpoint**
Full scale = 500.001 g
Weight sum = 500.003 g
Difference = 0.002 g
Linearity error = 0.001 g
25%
Full scale = 500.001 g
Weight sum = 500.001 g
No error
75%
Full scale = 500.001 × 3 = 1,500.003 g
Weight sum = 1,499.996 g
Difference = 0.007 g
Linearity error = 0.00175 g

9.
a. Because 1,000 cm³ of air is displaced, the weight of air displaced is:

$$\left(1{,}000\,cm^3\right)\left(1.2\,mg\,/\,cm^3\right) = 1{,}200\,mg = 1.20\,g.$$

b. The weight of a sample varies depending on its location – such as whether it is in air or a vacuum. Its mass does not change.

CHAPTER 18

1. These answers ignore the final, estimated figure.
   a. 4.2 mL b. 2.4 mL c. 5.2 mL d. 6.4 mL
2.
   a. A volumetric flask or a graduated cylinder.
   b. A graduated cylinder
   c. A pipette or a micropipette
   d. A micropipette
   e. A micropipette
   f. A micropipette

3. There are many possible answers. Errors in use of a graduated cylinder or volumetric flask include the use of dirty glassware, improper reading of the meniscus, and use of a TC flask where TD is appropriate and vice versa. For micropipette errors, see Boxes 18.3 and 18.4 and Table 18.3.

4.
a. Graduated cylinder, ± 0.6 mL; volumetric flask Class A ± 0.08 mL, Class B ± 0.16 mL.
b. Graduated cylinder, ± 0.6 mL.
c. A serological pipette is one possibility with a tolerance of ± 0.02 mL. Other answers are possible.

   d–f From Table 18.5, permissible error.

	From Gilson Data (μL)	From ISO 8655 (μL)
d.	± 0.8	± 0.8
e.	± 0.10 or 0.075	± 0.20 or 0.12
f.	± 4	± 8.0

5. Brand B has the best precision and Brand C has the best accuracy, but they are basically similar and either brand is probably acceptable based on these specifications.

6.
a. ? = (0.9982 mg/μL) × 100 μL = 99.82 mg
b. The mean volume for the water in μL is 99.40 mg/0.9982 mg/μL ≈ 99.58 μL.

$$\%\ error\ =\ \frac{99.58\ \mu L - 100.00\ \mu L}{100\ \mu L} \times 100\%$$

$$=\ -0.42\%$$

c. Yes, it is within the specification for accuracy.

7.

<table>
<tr><td colspan="6">

Clean Gene, Inc.

VERIFICATION OF PERFORMANCE OF MICROPIPETTOR REPORT
FORM 232

CALIBRATION TECHNICIAN _9.E.9._

CALIBRATION DATE _9/22/19_   NEW DUE DATE FOR NEXT CALIBRATION _3/22/2020_

MICROPIPETTOR ID NUMBER _2127_   MANUFACTURER _Finestt Pipettes Inc._

MODEL NUMBER _micro B_   LOCATION _Lab #27_   TEMPERATURE _20°C_

RANGE _100–1000µL_   PRIMARY USER _R.E.9._

SUMMARY:

PREVERIFICATION CLEANING AND ADJUSTMENTS _Changed Seals & O-rings_

_____ PASS/FAIL _pass_

STATUS _Returned to lab for use_

Volume 1
</td></tr>
</table>

NOMINAL VOLUME (µL)	TUBE NUMBER	INITIAL WEIGHT OF TUBE	WEIGHT AFTER H_2O DISPENSED	NET WEIGHT OF DISPENSED H_2O	H_2O VOLUME
500	1	1.9355 g	2.4352 g	0.4997	500.60 µL
500	2	1.9877 g	2.4876 g	0.4999	500.80 µL
500	3	1.9787 g	2.4789 g	0.5002	501.10 µL
500	4	1.9850 g	2.4873 g	0.5023	503.20 µL
500	5	1.9755 g	2.4763 g	0.5008	501.70 µL
500	6	1.9387 g	2.4387 g	0.5000	500.90 µL

Mean Water Volume ($\bar{x}$) 501.38 µL % INACCURACY 0.28%   SD 0.97 µL   CV 0.19%

CHAPTER 19

1. $94°C \approx 201.2°F$
   $37°C \approx 98.6°F$
   $72°C \approx 161.6°F$
2. $81°C = 177.8°F$
3. $103.2°F = 39.6°C$

CHAPTER 20

1. $pH = -\log [H+]$
   $9 = -\log [H+]$
   $[H^+] = 10^{-9}$ mol/L
2. $pH\ 7.3 = -\log [H^+]$
   $= 5.0 \times 10^{-8}$ mol/L
3. $[H^+] = 1.2 \times 10^{-3}$ mol/L
   $pH = -\log [1.2 \times 10^{-3}] \approx 2.9$
4. $pH = -\log [0.01] = 2$
5. $[1 \times 10^{-4}] [H^+] = 1 \times 10^{-14}$
   $[H^+] = 1 \times 10^{-10}$
   $pH = 10$

6.
   a. 59.16 mV/pH unit
   b. 59.16 mV
   c. 54.2 mV/pH unit
   d. 54.2 mV

7.
   a. The pH goes down as the bacteria actively metabolize nutrients and their population grows.
   b. pH probes for fermentation are difficult to produce. Difficulties include probe sterilization, which usually involves exposure to pressurized steam. Probes are immersed in a turbid solution that contains numerous materials that can contaminate the electrode surface. They must be reliable because it is difficult to substitute a new probe during a fermentation run.
   c. The pH of a culture is an important indicator of its condition and therefore must be monitored. It may be necessary to adjust the pH

during a fermentation run. An unexpected value may indicate a problem. It is important to document that the pH values were those expected during a "good" run. The print recording provides a record of the pH.

CHAPTER 21

1. $100\,\mu m = 100 \times 10^3\,nm = 100,000\,nm$
   $30\,cm = 30 \times 10^7\,nm = 300,000,000\,nm$

2. X-rays have the most energy; green light the least.

3. The spectra are probably of the same compound. The heights of the peaks vary depending on the concentration of compound present.

4. Yes.

5. From Table 21.3:
   a. Safranin O red absorbs light at 525 nm and appears red.
   b. Brilliant green absorbs light at 620 nm and appears greenish-blue.
   c. Methyl orange absorbs light at 465 nm and appears orange.

6.
   a. α-Carotene absorbs and makes available to the plant the energy of light in the violet-blue region of the spectrum, between about 420 and 480 nm.
   b. A yellow-orange color.

7.
   a. Orange. **b.** Greenish blue.
   c. Blue-green algae.

8.
   a.

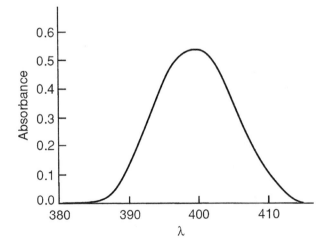

b. Yellow
c. Bandwidth about 15 nm

9.
   a.

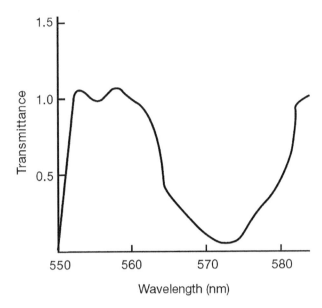

b. Violet to red-violet

10. $0.876 = 87.6\%$ T, $0.776 = 77.6\%$ T, $0.45 = 45\%$ T, $1.00 = 100\%$ T

11.

	A	**T (%)**	**A**
t = 0.876	≈ 0.057	T = 25	≈ 0.60
t = 0.776	≈ 0.110	T = 15	≈ 0.82
t = 0.45	≈ 0.35	T = 95	≈ 0.02
t = 1.00	≈ 0.00	T = 45	≈ 0.35

12.

Absorbance	**Transmittance (t)**	**% Transmittance (T)**
0.01	0.98	98
0.25	0.56	56
2.0	0.01	1.0

13. $1.24 \approx 0.058\,t$, $0.95 \approx 0.11\,t$, $1.10 \approx 0.079\,t$, $2.25 \approx 0.00562\,t$

14.
   a. Instrument B.
   b. Instrument B (less stray light).

c. Instrument B has gel scanning accessories.
d. Instrument A would be expected to be less costly and is probably suitable for routine work in many laboratories.
e. Instrument B (smaller spectral slit width).
f. Instrument B because it has better resolution.

15. Path length, cuvettes, photometer accuracy, wavelength accuracy.

16. Less than 1.95.

CHAPTER 22

PRACTICE PROBLEMS: WEIGHT PER VOLUME (FROM P. 561)

1. $\dfrac{0.1\ g}{1\ mL} = \dfrac{?}{100\ mL}$

$$? = 10\ g$$

Dissolve 10 g of $AgNO_3$ in purified water, and bring to 100 mL total solution in a graduated cylinder or volumetric flask.

2. $\dfrac{2\ mg}{1\ mL} = \dfrac{?}{50\ mL}$ $? = 100\ mg$

3. $\dfrac{100\ \mu g}{1\ mL} = \dfrac{?}{50\ mL}$

$$? = 5,000\ \mu g = 0.005\ g$$

Dissolve 0.005 g of proteinase K in purified water, and bring to a volume of 50 mL.

PRACTICE PROBLEMS: MOLARITY (FROM P. 566)

1. If you have a 3 L of a solution of potassium chloride at a concentration of 2 M, what is the solute? <u>Potassium chloride.</u> What is the solvent? <u>Water (purified).</u> What is the volume of the solution? <u>3 L.</u>
   Express 2 M as a fraction.

$$\frac{2\ mol\ of\ potassium\ chloride}{liter\ of\ solution}$$

2. Atomic weight of K = 39.10; atomic weight of Cl = 35.45. The FW of KCl, therefore, is: 39.10 + 35.45 = 74.55
   74.55 g/mol × 1 mol/L × 0.250 L ≈ 18.64 g of solute is required.

3. FW of KH_2PO_4 = 136.09.
   136.09 g/mol × 0.3 mol/L × 10 L = 408.27 g = solute required.

Dissolve the solute in less than 10 L of purified water and BTV.

4. FW of K_2HPO_4 = 174.18.
   174.18 g/mol × 0.1 M × 0.450 L ≈ 7.84 g = solute required.
   Dissolve the solute in less than 450 mL and BTV.

5. If 1 L of 1 M Tris buffer requires 121.10 g of solute, then 600 mL of 1 M solute requires:

$$\frac{?}{600\ mL} = \frac{121.10\ g}{1,000\ mL}$$

$$? = 72.66\ g$$

If 600 mL of 1 M Tris buffer requires 72.66 g of solute, then 0.4 M requires:

$$\frac{?}{0.4\ M} = \frac{72.66\ g}{1\ M}$$

$? \approx 29.06\ g$ = solute required.

6. You can use the anhydrous form, but you will need to add less than 25 g. The molecular weight of the hydrated form is 270.3, and the molecular weight of the anhydrous form is 162.2. To calculate how much of the anhydrous form to use, set up a proportion:

$$\frac{25\ g}{270.3\ g} = \frac{?}{162.2\ g}$$

$$? \approx 15.0\ g$$

Use 15.0 g of the anhydrous form instead of 25 g of the hydrated form.

7. 150 mg = 0.150 g. Using a proportion:

$$\frac{1\ mole}{230.11\ g} = \frac{?}{0.150\ g}$$

$? \approx 6.52 \times 10^{-4}$ moles = 652 μmoles

PRACTICE PROBLEMS: PERCENTS (FROM P. 569)

1. 35 mL × 0.95 = 33.25 mL
   Place 33.25 mL of 100% ethanol in a graduated cylinder and BTV 35 mL.

2. 200 g × 0.75 = 150 g

$$200\ g - 150\ g = 50\ g$$

Combine 150 g of resin and 50 g of acetone.

3. $600 \text{ mL} \times \dfrac{15 \text{ g}}{100 \text{ mL}} = 90 \text{ g}$

Dissolve 90 g of NaCl in water and BTV of 600 mL.

4. $\dfrac{50 \text{ g}}{500 \text{ mL}} = \dfrac{?}{100 \text{ mL}}$

$? = 10$ g. **a.** So this is a 10% solution. **b.** w/v.

5. 100 mg = 0.1 g and 1 L = 1,000 mL

$$\dfrac{0.1 \text{ g}}{1,000 \text{ mL}} = \dfrac{?}{100 \text{ mL}}$$

$? = 0.01$ g. **a.** So this is a 0.01% solution. **b.** w/v.

6. 25% = 25 g/100 mL of NaCl = 250 g/1,000 mL
   1 M NaCl = 58.44 g/1,000 mL

Using a proportion, if 58.44 μg/L is 1 M, then 250 g/L is 4.28 M.

$$\dfrac{58.44 \text{ g}/\text{L}}{1 \text{ M}} = \dfrac{250 \text{ g}/\text{L}}{?} \qquad ? \approx 4.28 \text{ M}$$

A second strategy:
Using the equation in the chapter summary:
M = w/v % × 10/FW, you get
M = 25 × 10/58.44 ≈ 4.28 M So, 25% NaCl is the same as a 4.28 M solution.

PRACTICE PROBLEMS: PARTS (FROM PP. 570)

1. 1 part + 3 parts = 4 parts total

$$\dfrac{1}{4} = \dfrac{?}{10 \text{ mL}} \qquad ? = 2.5 \text{ mL}$$

$$\dfrac{3}{4} = \dfrac{?}{10 \text{ mL}} \qquad ? = 7.5 \text{ mL}$$

Combine 2.5 mL of salt solution with 7.5 mL of water.

2. 1 part + 3.5 parts + 0.6 parts = 5.1 parts total

$$\dfrac{1}{5.1} = \dfrac{?}{200 \text{ mL}} \qquad ? \approx 39.2 \text{ mL chloroform}$$

$$\dfrac{3.5}{5.1} = \dfrac{?}{200 \text{ mL}} \qquad ? \approx 137.3 \text{ mL phenol}$$

$$\dfrac{0.6}{5.1} = \dfrac{?}{200 \text{ mL}} \qquad ? \approx 23.5 \text{ mL isoamyl alcohol}$$

Combine the preceding volumes of each component.

3. 5 parts + 3 parts + 0.1 parts = 8.1 parts total

$$\dfrac{5}{8.1} = \dfrac{?}{45 \text{ μL}} \qquad ? \approx 27.8 \text{ μL Solution A}$$

$$\dfrac{3}{8.1} = \dfrac{?}{45 \text{ μL}} \qquad ? \approx 16.7 \text{ μL Solution B}$$

$$\dfrac{0.1}{8.1} = \dfrac{?}{45 \text{ μL}} \qquad ? \approx 0.56 \text{ μL Solution C}$$

Note that it would be very difficult to accurately pipette 0.56 μL of a solution. You would probably mix a larger volume of the mixture and remove 45 μL from it.

4.

a. $3 \text{ ppm} = \dfrac{3 \text{ g}}{1 \times 10^6 \text{ g}} = \dfrac{3 \text{ g}}{1 \times 10^6 \text{ mL}}$

$$= \dfrac{3,000 \text{ mg}}{1 \times 10^6 \text{ mL}} = \dfrac{3 \times 10^{-3} \text{ mg}}{\text{mL}}$$

Another way to think about this is to recall that:

$$1 \text{ ppm in water} = \dfrac{1 \text{ μg}}{\text{mL}}$$

$$\text{so } 3 \text{ ppm} = \dfrac{3 \text{ μg}}{\text{mL}} = \dfrac{3 \times 10^{-3} \text{ mg}}{\text{mL}}$$

b. 3 ppm = 3 mg/L

5. $10 \text{ ppb} = \dfrac{10 \text{ g}}{1 \times 10^9 \text{ g}} = \dfrac{10 \text{ g}}{1 \times 10^9 \text{ mL}}$

$$= \dfrac{10 \times 10^3 \text{ mg}}{1 \times 10^9 \text{ mL}} = \dfrac{10 \times 10^{-3} \text{ mg}}{1000 \text{ mL}} = \dfrac{0.01 \text{ mg}}{\text{L}}$$

6. $100 \text{ ppm} = \dfrac{100 \text{ g}}{1 \times 10^6 \text{ mL}} = \dfrac{0.1 \text{ g}}{\text{L}}$

You could therefore dissolve 0.1 g of cadmium in water and BTV 1 L.

CHAPTER 23

1. $C_1 V_1 = C_2 V_2$
   (5X) (?) = (1X) (10 μL)
   ? = 2 μL
   You need 2 μL of solution A.

2. $C_1 V_1 = C_2 V_2$
   (100%) (?) = (95%) (75 mL)
   ? = 71.25 mL

Take 71.25 mL of 100% ethanol and BTV 75 mL. (The number 250 in the problem is extraneous.)

3. You cannot make a more concentrated solution from a less concentrated one.

4. $C_1 V_1 = C_2 V_2$
$(0.3 \text{ M}) (65 \text{ mL}) = (0.0001 \text{ M}) (?)$
$? = 195{,}000 \text{ mL} = 195 \text{ L}$
You can make quite a lot – 195 L!

5.
a. Decide how much stock solution to make. For example, 1 L.

    Using the equation given previously:

$$(121.1 \text{ g/mol}) (0.1 \text{ mol/L}) (1 \text{ L}) = 12.11 \text{ g}$$

    Weigh out 12.11 g of Tris, and dissolve in about 900 mL of water. Bring the solution to the proper pH using HCl. Then, bring the solution to 1 L. (Assume the solution is to be prepared and used at room temperature, since no temperature was specified.)

b. Decide how much solution to make, for example, 100 mL.

$$C_1 V_1 = C_2 V_2$$

$$(0.1 \text{ M}) (?) = (0.01 \text{ M})(100 \text{ mL}) \qquad ? = 10 \text{ mL}$$

    So, use 10 mL of Tris stock and BTV 100 mL. The pH may change slightly because of dilution, but should be acceptable for most applications. Record the final pH.

6. The pK_a of the buffer should be between 7.5 and 9.5. Any of the buffers in that range (e.g., HEPES or Tris) are acceptable on the basis of pK_a.

7. The pH of a Tris solution will decrease by 0.028 pH units for every degree increase in temperature. In this example, the temperature increases 40°C. The pH therefore decreases 1.12 pH units, to about 6.4.

8. Stocks:
- **1 M Tris (pH 7.6 at 4°C).**
  To make 1 L of stock, dissolve 121.1 g of Tris base in water. After the Tris is dissolved, be sure its temperature is reduced to 4°C by submerging it in an ice bath or placing it in the refrigerator. While the Tris is at this temperature, bring the pH to 7.6 with concentrated HCl. (Tris buffer will change pH as the temperature changes. As the temperature rises, the pH falls. The buffer will therefore require less HCl at 4°C than at room temperature.) Then BTV 1 L.
- **1 M Mg acetate.** To make 1 L of stock, dissolve 214.40 g of magnesium acetate (tetrahydrate) in water. Bring to volume.
- **1 M NaCl.** To make 1 L of stock, dissolve 58.44 g in water. Bring to volume.

Breaking Buffer to make 200 mL:

0.2 M Tris	**40 mL** of 1 M Tris (pH 7.6 at 4°C)
0.2 M NaCl	**40 mL** 1 M NaCl
0.01 M Mg acetate	**2 mL** 1 M Mg acetate
0.01 M β-mercaptoethanol	**142 μL** (see note)
5% Glycerol	**10 mL** glycerol

Bring to volume of 200 mL with water

Note: You want 200 mL of a 0.01 M solution of β-mercaptoethanol, so you need 0.156 g.

You can use a proportion to figure out how many milliliters will contain 0.156 g:

$$\frac{1.100 \text{ g}}{1 \text{ mL}} = \frac{0.156 \text{ g}}{?}$$

$? \approx 0.142 \text{ mL} = 142 \text{ μL}$
You therefore need 142 μL of β-mercaptoethanol.

---

9. **Step 1.** Each time a reaction is performed, 10 μL of the dNTP mixture is needed. You want to perform the reaction 100 times; therefore, you will need $10 \text{ μL} \times 100 = 1{,}000 \text{ μL}$.

    Step 2. Considering only a single dinucleotide, for example, how much of the dATP stock do you need? This is a $C_1 V_1 = C_2 V_2$ problem.

$$C_1 V_1 = C_2 V_2$$

$$(100 \text{ mM})(?) = (1.25 \text{ mM})(1{,}000 \text{ μL})$$

You need 12.5 μL of the dATP. You will similarly need 12.5 μL of the other three dNTPs.

    Step 3. Combining 12.5 μL of each dNTP gives a volume of 50 μL. The remaining solution will be water:

12.5 μL 100 mM dATP
12.5 μL 100 mM dCTP
12.5 μL 100 mM dGTP

12.5 μL 100 mM dTTP
950.0 μL water
1,000 μL total solution

Note that it is difficult to bring a solution to a volume of 1,000 μL. In this case, it is correct to calculate that you need 950 μL of water.

Step 4. How many vials of each dNTP will you need to purchase? The volume of each vial is not listed in the catalog. Rather, it tells you that the concentration of dNTP in each vial is 100 mM and there is 40 μmol (0.040 millimoles) in each vial. So what is the volume per vial?

A concentration of 100 mM means that there is 100 millimoles of the dNTP per liter. So, what volume contains 0.040 mmoles?

$$\frac{100 \text{ mmol}}{1,000 \text{ mL}} = \frac{0.040 \text{ mmol}}{?}$$

$$? = 0.400 \text{ mL} = 400 \text{ μL}$$

Each vial contains 400 μL of a given dNTP at a concentration of 100 mM. (This is a good time to review the difference between the words "amount" and "concentration.") One vial has more than enough dNTP to make the solution desired; purchase one vial of each dNTP.

10. Traceability in this context means that every component of each solution can be identified, located, and reordered when necessary. Three key pieces of information required for traceability are recorded on the form in Figure 23.4a: the vendor, the item number (catalog #), and the lot number. This is recorded for each component of each solution at the time of preparation, along with amounts, concentration, and presumably equipment used for measuring the components. Observe that the water used to make the solution is a critical component and its information is also recorded.

11. The procedure in Figure 23.5b (to make 2 M Tris-HCl at pH 7.3) uses the method of combining solutions of Tris-HCl and Tris base to obtain a desired pH (as described in Box 23.3 of this chapter). Note that in Step 3.6 the preparer is directed to check the pH and to adjust it if it is not close enough to the target pH. However, the procedure specifies that this is done with concentrated HCl, so we can assume that if the pH is too low, it is not acceptable to add NaOH.

CHAPTER 24

1. Pyrogens are removed by ultrafiltration and reverse osmosis (also by anion exchange resins).
2. Viruses are removed by ultrafilters and reverse osmosis.
3. Table salt is smaller (FW 58.44) than the cutoff for ultrafilters and so passes through them. NaCl is ionized in solution and therefore is removed by deionization.
4. Purified water is unstable. It absorbs gases from the air and leaches materials from the walls of the vessels in which it is contained. Microbes can grow in it. Continuous repurification is therefore used for Type I water.

5. $\log \text{ reduction value} = \log\left(\dfrac{4.5 \text{ EU} / \text{mL}}{0.001 \text{ EU} / \text{mL}}\right)$

$$= \log 4,500 \approx 3.7$$

CHAPTER 25

PRACTICE PROBLEMS: PROTEINS (FROM P. 630)

1.
   a. Storage Buffer
   - Low temperature. Limits bacterial growth and degradation of product; inhibits enzyme activity.
   - Tris-HCl. Buffer, maintains pH.
   - NaCl. Salt, maintains ionic strength at moderate level.
   - EDTA. Chelates Mg; inhibits enzyme activity during storage.
   - DTT. Reducing agent, prevents unwanted disulfide bond formation.
   - Triton X-100. Detergent, reduces adsorption of proteins to tube walls.
   - BSA. Added protein, reduces loss of protein due to adsorption and protease activity.
   - Glycerol. Prevents freezing of protein; stabilizes enzyme at low temperature.
      Activity Buffer
   - Tris. Buffer, maintains proper pH for activity.
   - NaCl. Salt, maintains ionic strength. Note that concentration of salt is significantly different than concentration used for storage.
   - $MgCl_2$. Salt, required cofactor for enzyme.

b. The two buffers have different functions. Storage requires protection of enzyme from degradation and adsorption. Activity is not required during storage. Activity buffer optimizes activity.

2.

a. Sample Buffer Used When Loading Proteins into Electrophoresis Gel
- Tris. Buffer, maintains pH.
- SDS. Detergent, denatures proteins; confers consistent charge on proteins.
- β-Mercaptoethanol. Reducing agent, inhibits disulfide bond formation.
- Bromophenol blue. Dye, for visualization.
- Glycerol. Causes proteins to sink into gel wells.

b. Restriction Enzyme Buffer (Low Salt)
- Tris. Buffer, maintains pH.
- $MgCl_2$. Salt, maintains ionic strength; provides Mg ions as cofactors.
- DTT. Reducing agent, protects against oxidation.

3. See Table 25.6.

PRACTICE PROBLEMS: DNA (FROM P. 646)

1.

a. TBE
- Tris base. Buffer, maintains pH.
- Boric acid. Bring Tris to proper pH.
- EDTA. Chelates Mg^{++}; inhibits nucleases.

b. TE Buffer
- Tris base. Buffer, maintains pH.
- EDTA. Chelates Mg^{++}; inhibits nucleases.

c. Hybridization Buffer
- PIPES. Buffer, maintains pH.
- EDTA. Chelates Mg^{++}; inhibits nucleases.
- NaCl. Salt, maintains ionic strength.
- Formamide. Lowers the melting temperature for DNA.

2. In 0.4 M salt, DNA tends to form duplexes readily; stringency is relatively low. (0.4 M is a relatively high concentration of salt.)

3.

a. One explanation is that DNA and RNA polymerase can bind one another *both* by nonspecific electrostatic interactions and by another kind of interaction that is specific to a particular site on the DNA. The electrostatic interactions decrease with increasing salt concentration.

b. This means that when working with RNA polymerase, the salt concentration will have an effect on the products made in the laboratory. One would choose a particular salt concentration based on the specificity required.

4.

a. The buffer has a relatively high concentration of salt that facilitates the hybridization of DNA to primer and the synthesis of new strands at 72°C. The melting temperature is raised because of the high salt concentration.

b. Mg^{++} is required by the *Taq 1* polymerase for its activity.

c. If *Taq 1* were sensitive to high temperature, it would be destroyed every time the temperature is raised to denature the DNA strands. In that case, more polymerase would have to be added after each cycle.

5. Most household detergents and shampoos contain SDS, which solubilizes membranes, denatures proteins, and releases DNA into solution.

6. DNA precipitation requires the presence of higher salt concentrations – bring the solution to 0.25 M Na acetate.

CHAPTER 26

1. It depends. If the first vial in each tray tends to be different in some way from the other vials, then this selection method would introduce a systematic error.

2. Specificity (selectivity).

3. a. Robustness. b. Limit of detection and range.

4. b. The risk is that the number of false-positive results will increase.

5. e. Nonspecific binding tends to cause false-positive results, and the negative control comes up positive.

6. This is a positive control and should come up positive, whether or not the person is pregnant. If it does not, then there is likely a problem in the reagents or in the way the test performed. The test kit instructions inform the user that if the positive control line did not appear, then the test did not work properly.

7.

a. No, 10 is not within the range for ill patients.

b. No.

c. Maybe. 25 is in the range of both healthy and ill patients.

d. Yes.

8. **a, b.** Any patient with a score of 20 or lower is healthy and therefore should have a negative result. Therefore, there is a 0% level of false negatives. However, some patients with scores above 20 are healthy and some are ill. Some people who are healthy, therefore, will get a false-positive result if a cutoff score of 20 is chosen. The false-positive rate will be about 15%.

9. **a, b.** Some people with a score of 25 are healthy, and some are ill. There will therefore be both false-positive and false-negative results. The rates of each will be equal, at about 2.5%.

10. **a, b.** Anyone with a score of 30 or higher is ill; therefore, the level of false positives will be 0. Some people, however, with a score of less than 30 are sick. There will therefore be about 15% false negatives.

11. A cutoff score of 20 will detect all cases.

12.

a. Apple juice (experimental sample)

b. Orange juice (experimental sample)

c. Pineapple juice (experimental sample)

d. A solution of glucose (positive control)

e. A solution of sucrose (negative control)

f. Pure water would be another negative control, probably in addition to a sucrose solution

13.

a. Onion extract (experimental sample)

b. Potato extract (experimental sample)

c. Green pepper extract (experimental sample)

d. Starch solution (positive control)

e. Water (negative control)

f. Glucose solution (negative control)

14. **a–c.** There should be colonies on Plate A derived from bacteria that successfully took up the ampicillin-containing plasmid. These are the bacteria that would be used to produce the protein product of interest.

Plate B is a control plate that confirms that the cells used for transformation were viable to begin with. Why is this important? Suppose, for example, that no colonies appear on Plate A. There are many explanations for this result. Perhaps none of the bacteria successfully took up the plasmid. Perhaps all the bacteria were killed by the transformation procedure. Perhaps there was a problem with the plasmid. Perhaps the bacteria were not viable to begin with. Plate B allows the investigators to be sure that at least the bacterial cells were viable to begin with. Note also that if the procedure works properly, then Plate B should have far more colonies than Plate A because transformation is a relatively rare event.

Plate C is a control plate. The non-transformed cells should not grow on this plate. If they do, then there is likely a problem with the activity of the ampicillin used to make the plates. (Or perhaps the cells already had antibiotic resistance.) If colonies appear on this plate, then the procedure is likely invalid, regardless of whether or not there are colonies on Plate A.

Plate D is a control to prove that the non-transformed cells are viable. There should be abundant growth on this plate. If not, perhaps the nutrient medium was improperly prepared, or the cells were not viable to begin with.

Plates B and D may be called "positive controls" in that the researchers expect bacteria to grow on them. Plate C may be considered a "negative control" in that no growth is expected on this plate.

15. *The prevalence of anti-COVID-19 antibodies in the populations is **5%**.*

*This test's diagnostic specificity is **95%** and its diagnostic sensitivity is **90%**, then:*

a. What is the test's rate of false positives? **5%**. What is the test's rate of true negatives? **95%**.

b. What is the test's rate of false negatives? **10%**. What is the test's rate of true positives? **90%**.

c. In a population of 200 people, about how many really have antibodies?

200 people (5% prevalence) → 200 (0.05) = **10** people with antibodies

d. In a population of 200 people, about how many really do not have antibodies?

200 − 10 = **190** without antibodies

e. In a population of 200 people, how many of the people *with* antibodies will be expected to have a positive result and how many with antibodies will be expected to have a negative result with this test?

The rate of false negatives is **10%**, and the rate of true positives is 90%.

10 people with antibodies×0.9=**9** people will test true positive

10×0.1=**1** will test false negative

f. In a population of 200 people, how many of the people *without* antibodies will be expected to have a positive result and how many without antibodies will be expected to have a negative result?

The rate of false positives is 5%, and the rate of true negatives is 95%.

190 people without antibodies×0.05=**9**.5 people will test false positive

190×0.95=**180.5** will test true negative

g. In total, how many of these people will be expected to have a positive result with this test?

9 true positive+9.5 false positive=**18.5** people will test positive

h. Of the total positive results, what percent is correct? What percent is incorrect?

9 true positive/18.5 total positive≈0.49 → **49%** correct positive results and **51%** incorrect

16. *The prevalence of anti-COVID-19 antibodies in the populations is **10%**.*

*This test's diagnostic specificity is **90%** and its diagnostic sensitivity is **90%**, then:*

a. What is the test's rate of false positives? **10%**

b. What is the test's rate of false negatives? **10%**

c. In a population of 100 people, about how many really have antibodies? **10**

d. In a population of 100 people, about how many really do not have antibodies? **90**

e. In a population of 100 people, how many of the people with antibodies will be expected to have a positive result and how many with antibodies will be expected to have a negative result with this test?

10 people with antibodies×0.9=**9** people will test true positive

10×0.1=**1** will test false negative

f. In a population of 100 people, how many of the people without antibodies will be expected to have a positive result and how many without antibodies will be expected to have a negative result?

90 people without antibodies×0.1=**9** people will test false positive

90×0.9=**81** will test true negative

g. In total, how many of these people will be expected to have a positive result with this test? **18**

h. Of the total positive results, what percent is correct? What percent is incorrect?

9/18=0.5 → **50%** correct and **50%** incorrect

17. *The prevalence of anti-COVID-19 antibodies in the populations is **10%**.*

*This test's diagnostic specificity is **95%** and its diagnostic sensitivity is **90%**, then:*

a. What is the test's rate of false positives? **5%**

b. What is the test's rate of false negatives? **10%**

c. In a population of 200 people, about how many really have antibodies? **20**

d. In a population of 200 people, about how many really do not have antibodies? **180**

e. In a population of 200 people, how many of the people with antibodies will be expected to have a positive result and how many with antibodies will be expected to have a negative result with this test?

20 people with antibodies×0.9=**18** people will test true positive

20×0.1=**2** will test false negative

f. In a population of 200 people, how many of the people without antibodies will be expected to have a positive result and how many without antibodies will be expected to have a negative result?

180 people without antibodies×0.05=**9** people will test false positive

180×0.95=**171** will test true negative

g. In total, how many of these people will be expected to have a positive result with this test? **27**

h. Of the total positive results, what percent is correct? What percent is incorrect?

18/27≈0.67 → **67%** correct and **33%** incorrect

18. *The prevalence of anti-COVID-19 antibodies in the populations is **20%**.*

*This test's diagnostic specificity is **90%** and its diagnostic sensitivity is **90%**, then:*

a. What is the test's rate of false positives? **10%**

b. What is the test's rate of false negatives? **10%**

c. In a population of 100 people, about how many really have antibodies? **20**

d. In a population of 100 people, about how many really do not have antibodies? **80**

e. In a population of 100 people, how many of the people with antibodies will be expected to have a positive result and how many with antibodies will be expected to have a negative result with this test?

20 people with antibodies$\times 0.9 = $**18** people will test true positive

$20 \times 0.1 = $**2** will test false negative

f. In a population of 100 people, how many of the people without antibodies will be expected to have a positive result and how many without antibodies will be expected to have a negative result?

80 people without antibodies$\times 0.1 = $**8** people will test false positive

$80 \times 0.9 = $**72** will test true negative

g. In total, how many of these people will be expected to have a positive result with this test? **26**

h. Of the total positive results, what percent is correct? What percent is incorrect?

$18/26 \approx 0.69 \rightarrow$ **69%** correct and **31%** incorrect

19.

a. What happens to the percent of positive test results that are correct when the diagnostic specificity increases? The percent of correct positive results increases as diagnostic specificity increases.

b. Using a particular test kit, what happens to the percent of positive test results that are correct as the incidence of antibodies increases in the population? The percent of correct positive results increases as the incidence of antibodies in the population increases.

20. Diagnostic tests are used to make decisions. For example, an individual might want to use a COVID-19 serology test to decide whether it is safe to attend an event in which crowds are expected, or whether it is safe to attend in-person classes. We are not yet certain that having antibodies confers immunity to COVID-19, but antibodies do confer immunity to other diseases. So, a person might make decisions based on what they think is their immune status. If a test provides mostly false results, the individual could be seriously misled. On a population level, epidemiologists can use test data to trace infection, to see if the rate of infection is subsiding or increasing, and to make decisions based on this information. At the population level, it is possible to adjust for false positives if one has an idea of their rate. Thus, a test can be useful at the population level, but wrong (and therefore dangerous) at the individual level.

CHAPTER 27

1. The basic components of PCR mixture are: the enzyme, usually Taq DNA polymerase; primers; dNTPs (nucleotides); buffer with proper cofactors; and DNA template.

2. The basic components of a reverse transcriptase reaction mixture are: the enzyme, reverse transcriptase; primers; dNTPs; buffer with proper components; and RNA template.

3. Beginning with 1,000 copies of a target DNA sequence, after 5 cycles of PCR, assuming ideal conditions, you expect:

$$N = 2^t (N_0)$$

$$N = 2^5 (1,000 \text{ copies}) = 32,000 \text{ copies}$$

This can also be solved intuitively: After 1 cycle there will be 2,000 copies, after 2 cycles 4,000 copies, after 3 cycles 8,000 copies, after 4 cycles 16,000 copies, and after 5 cycles 32,000 copies.

4. Begin PCR with 25 pg of DNA; after 6 cycles:

$$N = 2^t (N_0)$$

$$N = 2^6 (25 \text{ pg}) = 1,600 \text{ pg}$$

5. There is a problem with this result. The positive control in Lane 2 is fine as is the negative food sample control in Lane 3. But the no-template control lane, 4, should not have a band. A band in the no-template control lane indicates probable contamination, possibly where DNA from a previous assay somehow entered the tubes in this assay. Thus, the positive results in Lanes 5–8 are meaningless. It is likely that

the entire area needs to be carefully cleaned to remove any possible contaminants, new reagents need to be prepared, equipment needs to be cleaned, and the efficacy of the decontamination process needs to be verified.

6. Clinical Evaluation of a New PCR Test Kit

	Samples "Known" to be Infected (73)	Samples "Known" Not to be Infected Based (124)
Positive test result with the new kit	68	12
Negative test result with the new kit	5	112
	Diagnostic sensitivity 68/73 ≈ 93%	Diagnostic specificity 112/124 ≈ 90%

7. Clinical Evaluation of a New PCR Test Kit

	Samples "Known" to be Infected (100)	Samples "Known" Not to be Infected Based (100)
Positive test result with the new kit	93	9
Negative test result with the new kit	7	91
	Diagnostic sensitivity = 93/100 = 93%	Diagnostic specificity = 91/100 = 91%

8. Raising the magnesium level in the reaction mixture is likely to increase the diagnostic sensitivity of the test and decrease the diagnostic specificity of the test.

9. Raising the temperature of the primer extension step is likely to decrease the diagnostic sensitivity of the test and increase its specificity.

10. They want to optimize the diagnostic sensitivity to avoid false-negative results.

11. A reverse transcriptase reaction is part of the assay. The mRNA is a positive control to confirm that both the reverse transcriptase reaction and the PCR amplification work properly. When an assay is performed, the mRNA would be processed along with the patient samples (and probably a negative control). If this mRNA positive control does not come up positive, it would indicate a problem in the assay system and none of the results would be valid.

CHAPTER 28

1.
a, b. A is an absorbance spectrum. X-axis, wavelength; Y-axis, absorbance. B is a standard curve. X-axis, concentration (or amount of analyte); Y-axis, absorbance.
   c. An absorbance spectrum is associated with qualitative analysis. A standard curve is associated with quantitative analysis.

2. a. quantitative   b. qualitative
   c. quantitative   d. quantitative
   e. qualitative   f. qualitative
   g. quantitative

3.
   a. Compound A is orange. Compound B is red. Compound C is green.
   b. Compound B has the greatest absorptivity constant at 540 nm.
   c. Compound B will be detectable at 540 nm at a lower concentration than the other compounds because it has the largest signal at that wavelength.

4. Graph A: The linear range is about 0–250 mg/mL.

    Graph B: The linear range is at least about 0–3.5 mM.

    Graph C: The linear range is about 0–15 ppm.

5.
   a.

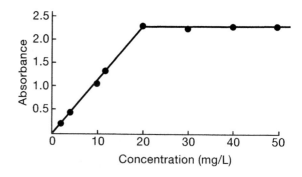

b. The graph is linear from 2 to 20 mg/L concentration.

$$\text{Slope} = m = \frac{Y_2 - Y_1}{X_2 - X_1} \approx \frac{1.1}{\dfrac{10 \text{ mg}}{L}} = \frac{0.11 \text{ L}}{mg}$$

$$A = \frac{0.11 \text{ L}}{mg} C + O$$

c. Absorptivity constant $= \alpha = \dfrac{\text{slope}}{\text{path length}}$

$$= \frac{0.11 \text{ L}}{(1 \text{ cm}) \text{ mg}}$$

d. This graph has an upper threshold where the graph plateaus. This is an indication of the limitation of the spectrophotometer at low light transmittances due to stray light.

6.

a.

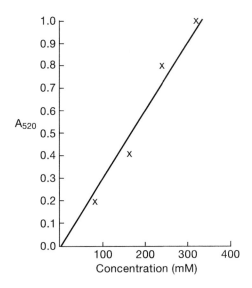

b. $\alpha = \dfrac{\text{slope}}{\text{path length}}$

Slope $\approx 0.3/100 \text{ mM} = 0.003/\text{mM}$

$$\alpha = \frac{0.003}{(\text{mM}) (1 \text{ cm})}$$

c. $A = \alpha \, b \, C$

$$C = \frac{A \text{ (mM)}}{0.003}$$

	Absorbance	≈ Concentration (in mM)
Sample A	0.18	60
Sample B	0.31	103.3
Sample C	0.96	320
Sample D	1.0	333.3

7.

a.

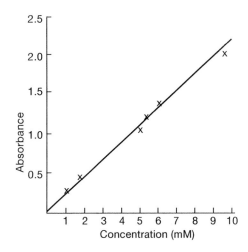

b. $\alpha = \dfrac{\text{slope}}{\text{path length}}$

$$\text{slope} \approx \frac{(1.0 - 0.5)}{(4.5 - 2.1)\text{mM}}$$

$$\alpha \approx \frac{0.21}{\text{mM (cm)}}$$

c. $C = \dfrac{0.30 \text{ (cm)(mM)}}{0.21 \text{ (1 cm)}}$

$$C \approx 1.43 \text{ mM diluted}$$
$$C \approx 1.43 \text{ mM} \times 5$$
$$= 7.15 \text{ mM undiluted}$$

8.

a. When the path length decreases the absorbance decreases.

b. 0.8

9. $\dfrac{25 \text{ ppm}}{0.87 \text{ AU}} = \dfrac{40 \text{ ppm}}{?}$ $? \approx 1.39 \text{ AU}$

The absorbance is a linear function of concentration, and this is a proportion problem.

10. $t = 0.680$

$$A = -\log(t)$$

a. $A = -\log (0.680) \approx 0.167$

b. Because there is a linear relationship between absorbance and concentration:

$$\frac{A}{16 \text{ mg/mL}} = \frac{0.167}{10 \text{ mg/mL}}$$

$$A \approx 0.267$$

c. $A = -\log (t)$

$$0.267 = -\log(t)$$

$$t \approx 0.541$$

11. $A = -\log (0.75) \approx 0.12$

a, b. $\dfrac{A}{60 \text{ mg/mL}} = \dfrac{0.12}{30 \text{ mg/mL}}$

Absorbance $= 0.24$ transmittance ≈ 0.58

c, d. $\dfrac{A}{90 \text{ mg/mL}} = \dfrac{0.12}{30 \text{ mg/mL}}$

Absorbance $= 0.36$ transmittance ≈ 0.44

12. $A_{280} = 0.65$ of 1 mg/mL

a. $\dfrac{?}{0.15} = \dfrac{1 \text{ mg/mL}}{0.65}$

$$? \approx 0.23 \text{ mg/mL}$$

b. The proteins in the preparation may not have the same aromatic amino acid composition as the BSA. There might be nucleic acids in the partially purified preparation that interfere at 280 nm. There are other possible answers as well.

13. $\alpha_{ATP} = 15{,}400$ L/mol-cm at 260 nm

$$A = \alpha b C$$

$$1.6 = \frac{15{,}400 \text{ L } (1 \text{ cm}) \text{ C}}{(\text{mole} - \text{cm})}$$

$$C = \frac{1.6 \text{ mole}}{15{,}400 \text{ L}} \approx 1.04 \times 10^{-4} \text{ mole/L}$$

14. $\alpha_{NADH} = \dfrac{15{,}000 \text{ L}}{\text{mole} - \text{cm}}$ at 260 nm

$$A = \alpha b C$$

$$0.98 = \frac{15{,}000 \text{ L}}{\text{mole} - \text{cm}} (1.2 \text{ cm}) \text{ C}$$

$$C = \frac{0.98 \text{ mole} - \text{cm}}{15{,}000 \ (1.2) \text{ cm} - \text{L}}$$

$$C \approx 5.44 \times 10^{-5} \text{ mole/L}$$

15. Although the highest absorbance peak for NADH is at 260 nm, it may be practical to use the **second** peak at 340 nm to determine the concentration because there is no interference from ATP at this wavelength.

16. $A = \alpha b C$

For lower limit : $0.1 = \dfrac{10{,}000 \text{ L } (1 \text{ cm}) \ (C)}{\text{mole } (\text{cm})}$

$$C = \frac{0.1 \text{ mole}}{10{,}000 \text{ L}}$$

$$C = 1 \times 10^{-5} \text{ mole/L}$$

For upper limit: $1.8 = \dfrac{10{,}000 \text{ L } (1 \text{ cm}) \ (C)}{\text{mole} - \text{cm}}$

$$C = \frac{1.8 \text{ mole}}{10{,}000 \text{ L}}$$

$$C = 1.8 \times 10^{-4} \text{ mole/L}$$

The range of concentrations that can be measured for this compound, therefore, is from 1.8×10^{-4} mol/L to 1×10^{-5} mol/L.

17.

a. $C = 50 \ \mu g/mL \ (1.25) = 62.5 \ \mu g/mL$
Because the solution was diluted, however, it is necessary to multiply the answer by 10, so 625 μg/mL = estimated concentration of DNA in the original sample

b. $C = (0.63) \ (40 \ \mu g/mL) = 25.2 \ \mu g/mL$
$25.2 \ \mu g/mL \times 5 = 126 \ \mu g/mL =$ estimated concentration of RNA in the original sample

18.

a. Substituting into Equation 1 from Box 28.2:

$$\text{Concentration} = \left(50 \ \mu g / mL\right)(0.85)$$

$$= 42.5 \ \mu g / mL$$

b. Substituting into Equation 3 from Box 28.2:
$(40\,\mu g/mL)\,(0.69)=27.6\,\mu g/mL$
$27.6\,\mu g/mL \times 10 = 276\,\mu g/mL$

19.

a. No. The results of an assay using a standard curve should be accurate because the values determined are compared with those on the standard curve.

b. If the single standard was properly prepared and the instrument was properly adjusted with a blank, and if the standard and sample are in the linear range of the assay, then the results should be accurate.

c. Yes, the results will be inaccurate because the value for the absorptivity constant was presumably based on a correctly calibrated instrument and therefore will not be relevant to this spectrophotometer.

20. This problem may be evident when you examine your standard curve if it causes the results to be nonlinear or low. If you prepare a standard curve alongside your samples, therefore, it may be evident that there is a problem.

21. Assuming you prepare a standard curve each time you do the assay, and assuming the reagent still acts in such a way that color is proportional to the amount of analyte, then the standards should be affected by the reagent in the same way as the samples. The results, therefore, will still be accurate. You may observe, however, that the range for the assay is different than usual. Low concentrations of analyte that are normally in the range of the assay may not cause a detectable color change.

22. Yes, this is a potential source of error. If there are substances in the blood that affect absorbance at the analytical wavelength, then the buffer blank will not compensate and the standard curve will not be correct for the samples.

23.

a. Concentration of double-stranded DNA $\approx$ $50\,\mu g/mL$ $(A_{260}) \approx 50\,\mu g/mL$ $(0.497) \approx$ $24.9\,\mu g/mL$

Thus, based on what she weighed, she had a solution of $60\,\mu g/mL$ DNA; based on the absorbance, the solution contained about $25\,\mu g/mL$ DNA. This is not particularly "close"; the UV method estimated

the concentration of DNA at less than half of what it should have been.

b. Concentration of protein $\approx 0.276/0.7 \approx$ $0.394\,mg/mL$

Thus, based on weight, she had a $0.50\,mg/mL$ solution of BSA. Based on its absorbance, she had a solution with about $0.40\,mg/mL$ BSA. The UV method thus gave a closer estimate for the protein concentration than for the DNA concentration.

c. $A_{260}=0.176+0.497=0.673$
$A_{280}=0.276+0.260=0.536$
The A_{260} value for the mixture was 0.685, and the A_{280} value for the mixture was 0.547.

The values for the mixture are close to the summed values for the pure samples. This means that the absorbances at both wavelengths were additive, as predicted.

d. For pure DNA: $A_{260}/A_{280}=0.497/0.260 \approx 1.91$
For pure BSA: $A_{260}/A_{280}=0.176/0.276 \approx$ 0.638
For the mixture: $A_{260}/A_{280}=0.685/0.547 \approx$ 1.25

All three ratios are consistent with the predicted ratios.

e. [Nucleic acid] $\approx 62.9(0.685) - 36.0\,(0.547)$
$\approx 23.4\,\mu g/mL$
[Protein] $\approx 1.55\,(0.547) - 0.757\,(0.685)$
$\approx 0.329\,mg/mL$

Thus, based on what she weighed, the DNA should have been $60\,\mu g/mL$ rather than $23.4\,\mu g/mL$. The protein should have been $0.5\,mg/mL$ rather than $0.329\,mg/mL$.

f. The UV methods did not give results that exactly matched the expected results for the standards. Many possible explanations of the differences are discussed in the text. Whether or not the estimates provided by the UV methods are "close enough" depends on the situation and the decisions that will be made based on the results. For example, if several DNA purification methods are being compared, then these UV estimates would be adequate to compare the efficacy of the purification methods. If a pharmaceutical company is trying to determine whether a recombinant protein is pure enough for drug use, however, these UV methods are not sufficiently accurate.

24.

a.

Tube #	µg DNA	µL DNA stock (100 µg/mL)	µg Protein	µL Protein stock (1 mg/mL)	µL Water	% DNA (approximate) (%)	A_{260}	A_{280}	A_{260}/A_{280} (approximate)
1	50	500	0	0	500	100	1.072	0.605	1.77
2	50	500	5	5	495	91	1.074	0.607	1.77
3	50	500	20	20	480	71	1.078	0.606	1.79
4	50	500	50	50	450	50	1.073	0.610	1.76
5	50	500	150	150	350	25	1.244	0.769	1.62
6	50	500	250	250	250	17	1.147	0.709	1.62
7	50	500	350	350	150	12.5	1.190	0.766	1.55

b. These data suggest that the A_{260}/A_{280} ratio of a DNA solution is not very sensitive to protein "contamination."

CHAPTER 29

1.
   a. Take 5 µL of antibody stock and add to 5 mL of dilution buffer. (Technically, you would add 4.995 mL of dilution buffer, but that is close enough to 5 mL to ignore the slight difference.) (Dilutions are discussed in Section 12.5.) **b.** One aliquot is needed. **c.** The antibody should be stable in the refrigerator for a period of time, as long as it is not diluted. The manufacturer may indicate how long the antibody is stable at refrigerator temperature. Discard unused, diluted antibody at the end of this time period.

2. **(More than one strategy may be correct.)**
   A. 1/500 (8 µL stock + 4 mL dilution buffer, technically 3,992 µL, but that is close enough to 4 mL that most people would not bother to measure 3,992 µL.)
   B. 1/1,000 (2,000 µL A + 2,000 µL dilution buffer)
   C. 1/2,000 (1,500 µL B + 1,500 µL dilution buffer)
   D. 1/4,000 (1,000 µL C + 1,000 µL dilution buffer)

   (To understand this strategy, consider dilution A. We want this to be a 1/500 dilution. The volume needs to be at least 2 mL and will need to be more than that because we will use some of it to make the next dilution. So, suppose you decide that you want to make 4 mL

of a 1/500 dilution. We saw in Chapter 12 that proportions can be used to calculate how to make such a dilution:

$$\frac{1}{500} = \frac{?}{4,000\,\mu L} \qquad ? = 8\,\mu L, \text{ so combine}$$

8 µL of the antibody product supplied by the manufacturer with 4 mL (4,000 µL) of dilution buffer. This is the dilution in tube A.

The second dilution is to be 1/1,000. If the 1/500 diluted antibody is diluted by 1/2, then the final dilution will be 1/1,000. Suppose we again decide to make 4 mL of this dilution:

$$\frac{1}{2} = \frac{?}{4,000\,\mu L} \qquad ? = 2,000\,\mu L, \text{ so combine}$$

2,000 µL of the dilution in tube A with 2,000 µL of dilution buffer. This is dilution B.

The third dilution is to be 1/2,000. If the 1/1,000 diluted antibody is diluted by 1/2, the final dilution will be 1/2,000. Suppose you decide to make 3 mL of this dilution:

$$\frac{1}{2} = \frac{?}{3,000\,\mu L} \qquad ? = 1,500\,\mu L, \text{ so combine}$$

1,500 µL of the dilution in tube B with 1,500 µL of dilution buffer. This is dilution C.

Finally, you need a dilution of 1/4,000. If the 1/2,000 diluted antibody is diluted by 1/2, the final dilution will be 1/4,000. Suppose you decide to make 2 mL of this dilution:

$$\frac{1}{2} = \frac{?}{2,000\,\mu L} \qquad ? = 1,000\,\mu L, \text{ so combine}$$

1,000 µL of the dilution in tube C with 1,000 µL of dilution buffer. This is dilution D.)

3. **(More than one strategy may be correct.)**
   A. 1/100 (20 µL stock received from manufacturer + 1,880 µL dilution buffer)

B. 1/1,000 (200 μL A + 1,800 μL dilution buffer)

C. 1/5,000 (400 μL B + 1,600 μL dilution buffer)

D. 1/25,000 (200 μL C + 800 μL dilution buffer)

4. Scenario 1

It is possible to determine the concentration of PTH in the controls and the patient samples in two ways:

The concentrations can be read from the graph, as is illustrated below in the Figure PTH Scenario 1. A line is drawn from the sample's average plate reader value on the Y-axis to

	1	2	Average of Values in Columns 1 and 2. These Values are Used to Construct the Standard Curve	Concentration in Standards (pg/mL)	3	4	Average of Values in Columns 3 and 4	5	6	Average of Values in Columns 5 and 6
A	0.004	0.001	0.003	0	0.239	0.242	0.241	0.550	0.546	0.548
B	0.161	0.153	0.157	20	0.481	0.479	0.480	1.344	1.400	1.372
C	0.310	0.311	0.311	40	0.801	0.799	0.800	0.371	0.357	0.364
D	0.623	0.626	0.625	80						
E	0.785	0.783	0.784	100						
F	1.569	1.566	1.568	200						

Scenario 1. Averaged Values for the Replicates. The averages of columns 1 and 2 are used to construct the standard curve shown below. The averages of columns 3 and 4 are the controls used to evaluate whether the assay was working properly. The averages of columns 5 and 6 are the patient samples.

The standard curve for this scenario is shown below The points all lie close to a best-fit line. The line is straight and does not plateau. The line runs through zero. So, the points appear to form a reasonable standard curve.

the best-fit line. For example, for the second +control the averaged value is 0.480 AU. Then another line is drawn from the intersection with the best-fit line to the X-axis, as shown. The result is 62 pg/mL. For the 100 pg/mL control, a similar procedure yields a value of 102 pg/mL. Alternatively, the equation for the trend line on the graph can be calculated or read from Excel and the values for Y can be plugged into the equation. One can then solve for X, which is the concentration of PTH. It is sometimes difficult to read values precisely off a graph, so the algebraic method may be more accurate.

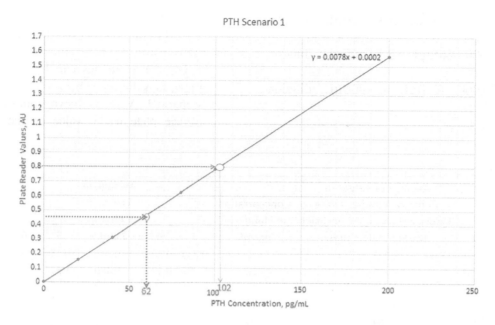

Excel was used to graph the best-fit line for the standard curve data. (Before graphing, the values of the two replicates were averaged.) The best-fit line was then used to find the values for the positive controls.

Algebraic Method:

The equation for the best-fit line from Excel is $Y = 0.0078X + 0.0002$.

Plugging in the average plate reader value for the 30 pg/mL + control, that is 0.241 AU, yields:

$$0.241 = 0.0078X + 0.0002$$

$$X \approx 30.87 \text{ pg/mL}$$

Plugging in the average plate reader value for the 60 pg/mL + control, that is 0.480 AU, yields:

$$0.480 = 0.0078X + 0.0002$$

$$X \approx 61.51 \text{ pg/mL}$$

Plugging in the average plate reader value for the 100 pg/mL + control, that is 0.800 AU, yields:

$$0.800 = 0.0078X + 0.0002$$

$$X \approx 102.53 \text{ pg/mL}$$

Analysis of controls: The values for the positive controls were calculated using both a graphical and algebraic method. The results of the two methods are consistent. All three + controls are within the specified range required. Therefore, it is possible to continue with analysis of the patient samples.

The replicates are averaged, and the PTH concentration is calculated, in this case, using the algebraic method:

Patient 1: $0.548 = 0.0078X + 0.0002$
  $X \approx 70.23 \text{ pg/mL}$

Patient 2: $1.372 = 0.0078X + 0.0002$
  $X \approx 175.87 \text{ pg/mL}$

Patient 3: $0.364 = 0.0078X + 0.0002$
  $X \approx 46.64 \text{ pg/mL}$

Based on these results, the PTH level of Patient 2 appears to be abnormal. Further medical follow-up would likely be recommended for this patient. The other two patient PTH levels are in the normal range.

Scenario 2

1	2	Average of Values in Columns 1 and 2. These Values Are Used to Construct the Standard Curve.	3	4	Average of Values in Columns 3 and 4	5	6	Average of Values in Columns 5 and 6
0.002	0.004	**0.003**	0.988	0.978	**0.983**	0.601	0.603	**0.602**
0.901	0.906	**0.904**	1.505	1.500	**1.503**	1.450	1.440	**1.445**
1.503	1.499	**1.501**	1.861	1.782	**1.822**	2.103	2.098	**2.101**
2.403	2.402	**2.403**						
2.891	2.789	**2.840**						
2.931	3.102	**3.017**						

Scenario 2. Averaged Values for the Replicates Using the Plate Reader Measurements. The averages of columns 1 and 2 are used to construct the standard curve shown in the figure below. The averages of columns 3 and 4 are the controls used to evaluate whether the assay was working properly. The averages of columns 5 and 6 are the patient samples.

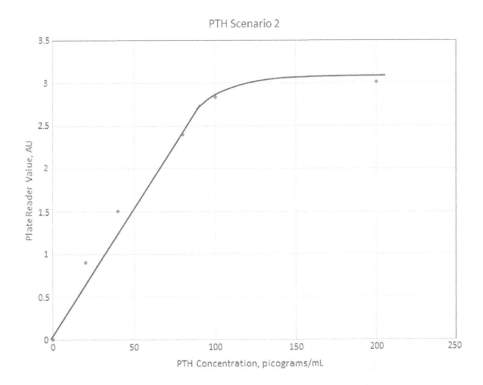

Excel was used to graph the standard curve data. (Before graphing, the values of each of the two replicates were averaged.)

The standard curve plateaus at higher PTH concentrations. It is easy to see that the third positive control, 100 pg/mL, appears to have too little PTH, based on the standard curve. There are various reasons why the standard curve might plateau, and the positive control might have the wrong result, but, whatever the reason, the results of this assay are not valid. We know that if the standard curve and the controls do not provide the right results, then the assay is invalid. Do not bother to evaluate the patient samples. In a workplace, the analyst might consult with a supervisor.

Note: *There are situations where assays are known to be inaccurate at higher and/ or lower levels of analyzed material. In these cases, the reporting range of the standard curve might be reduced to exclude the areas of inaccuracy. The positive controls would be used to establish the accurate reporting range. In such cases, patient samples that yield values outside the reporting range would need to be retested or would need to be tested using another method.*

5.
   a. See table:

6. **a, b, c.** You need $10 \text{ mL} \times 5 \mu\text{g/mL} = 50 \mu\text{g}$. The antibody is present at a concentration of 1.0mg/mL (which is the same as $1 \mu\text{g/}\mu\text{L}$), so you need **50 μL** of primary antibody stock. You will dilute it into **10 mL wash buffer**. Each aliquot has 20 μL, so you need to thaw **three aliquots.**

CHAPTER 30

1.
   a. NIH3T3 cells are not in the database. They are not known to be misidentified.
   b. BRCA5 cells were thought to be breast carcinoma cells, but are actually from a cervical adenocarcinoma.
   c. CHO-CD36 cells are not in the database. They are not known to be misidentified.
2. This is a rapid doubling time and is almost certainly a prokaryotic culture.
3. Each time you move your hands and arms into and out of the cabinet, the airflow is disrupted and there is a chance for contaminants to enter the cabinet. Also, used pipettes might contain living cells that are potentially pathogenic. Therefore, used pipettes are disinfected in the cabinet before removal.

Target Protein	Protein A	Protein B	Protein C
Target species	Human	Human	Human
Primary antibody	Anti-Protein A	Anti-Protein B	Anti-Protein C
Species in which primary antibody was made	Mouse	Rabbit	Rat
Name of the secondary antibody	Goat anti-mouse	Goat anti-rabbit	Goat anti-rat
Fluorochrome excitation/emission	490/525	556/573	650/665

   b. Green reporter should emit light in the 505–555 range, yellow 575–600, and red 650–780.
   c. 490 is the wavelength of light absorbed by the reporter, and 525 is the wavelength emitted, which is green light.

   Similarly, 556 is the wavelength of light absorbed by the second reporter and 573 is the wavelength of light it emits. This is on the edge of yellow and might appear yellowish-green to our eyes. 650 is the wavelength of light absorbed by the third reporter, and 665 is the wavelength of light it emits, which is red.

4. Humans have 46 chromosomes in their cells. These cells are therefore not human and have, apparently, been cross-contaminated. Mice have 40 chromosomes, and possibly, the cells in the vial are actually from a mouse cell line. Assuming your karyotyping was done properly, these cells should be discarded and your colleague should be notified.
5. The cells are dividing, but not at the predicted rate. (With a 22-hour doubling time, there should be about 2.6×10^5 cells/mL in the culture on Wednesday.) There are many possible explanations that should be investigated, such as a problem in the nutrient medium and a

problem with the incubator conditions. Also, it is important to be certain that you actually do have Vero E6 cells. In any event, the issue should be resolved before beginning critical experiments.

6. Yes.

7. Various answers are possible, but should minimally mention testing to be sure the cell line is not misidentified and the cells are not contaminated, particularly with *Mycoplasma sp.* Some answers might go further and include, for example, aseptic practices that reduce the chance of contaminating a culture.

CHAPTER 31

1. A 0.9% solution contains 0.9 g of solute/100 mL or 9 g/L. The molecular weight of NaCl is 58.44, so a 1 M solution contains 58.44 g/L.

$$\frac{58.44 \text{ g}/1{,}000 \text{ mL}}{1 \text{ M}} = \frac{9 \text{ g}/1{,}000 \text{ mL}}{?}$$

$$? \approx 0.154 \text{ M} = 154 \text{ mM}$$

NaCl dissociates in water to form 2 ions, both of which have an osmotic effect. Therefore, multiply by 2 to get milliosmoles: $2 \times 154 = 308$. So 0.9% NaCl is about 308 mOsm/L.

2.

a. 5% (w/w) = 50 g solute/kg.

The molecular weight of glucose monohydrate is 198.

$$\frac{198 \text{ g}}{1 \text{ mol}} = \frac{50 \text{ g}}{?} \qquad ? \approx 0.252 \text{ mole} = 252 \text{ mmol}$$

Glucose does not dissociate in solution, so 252 mmol of glucose = 252 mOsm.

NaCl dissociates in water into sodium and chloride ions, both of which have osmotic effects. Therefore, 20 mmol of NaCl = 40 mOsm.

Total: 252 mOsm + 40 mOsm = 292 mOsm.

b. 292 mOsm/kg.

3. The $C_1 V_1 = C_2 V_2$ applies since concentrated stocks are being used.

a. Concentrate:

$$C_1 V_1 = C_2 V_2$$

$$5X(?) = 1X(1{,}000 \text{ mL})$$

$? = 200$ mL of 5X concentrate is required

b. Concentration of glucose in final solution:

$$C_1 V_1 = C_2 V_2$$

$$20\%(10 \text{ mL}) = ?(1{,}000 \text{ mL}) ? = 0.2\%$$

c. Concentration of MgSO4·7H2O in final solution:

$$C_1 V_1 = C_2 V_2$$

$$1 \text{M} (1 \text{ mL}) = ?(1{,}000 \text{ mL})$$

$$? = 1 \text{mM}$$

d. Concentration of thiamine in final solution:

$$C_1 V_1 = C_2 V_2$$

$$0.5\% (0.1 \text{ mL}) = (?)(1{,}000 \text{ mL})$$

$$? = 0.00005\%$$

4. Table 31.7

Antibiotic	Amount Stock Required (per L)
Ampicillin (sodium salt)	2 mL
Chloramphenicol	5 mL
Kanamycin	5 mL
Streptomycin	5 mL
Tetracycline HCl	10 mL

5.

a. Sodium acetate is the carbon source.

b. Monoammonium phosphate is the nitrogen source.

c. NaCl is used to maintain osmotic balance.

d. Phosphates provide potassium, nitrogen, and phosphorus, and buffer the medium.

e. This is a selective medium, so the likely purpose of the bromothymol blue is to change color to indicate the selective growth of one of the types of bacteria.

(*E. coli* can utilize acetate as a carbon source, but *Shigella* cannot. *E. coli* produce alkaline by-products when they use acetate for growth. Bromothymol blue is a pH indicator that turns blue at higher values of pH, but is yellowish at lower values of pH. A blue color in this selective medium therefore is indicative of *E. coli* growth.)

f. The purpose of agar is to produce a solid surface on which bacteria can grow. The agar is present at a concentration of 2%.

6. 20% of 500 mL = 100 mL, so 100 mL of serum is required. 400 mL of medium is required to bring the volume to 500 mL total.

7. The $C_1V_1 = C_2V_2$ applies since concentrated stocks are being used.

$$C_1V_1 = C_2V_2$$

$$100 \times (?) = 1 \times (500\,mL) \qquad ? = 5\,mL$$

100 mL of serum
5 mL of glutamine
<u>395 mL of medium</u>
500 mL

8. The $C_1V_1 = C_2V_2$ applies since concentrated stocks are being used.

a. 10X cell culture medium concentrate required:

$$C_1V_1 = C_2V_2$$

$$10 \times (?) = 1 \times (1,000\,mL) \qquad ? = 100\,mL$$

b. Glutamine stock required:

$$C_1V_1 = C_2V_2$$

$$200\,mM \quad (?) = 2\,mM\,(1,000\,mL) \qquad ? = 10\,mL$$

c. Serum required:

$$1,000\,mL(0.10) = 100\,mL$$

or

$$C_1V_1 = C_2V_2$$

$$100\%(?) = 10\%(1,000\,mL) \qquad ? = 100\,mL$$

d. NaHCO₃ required:

$$C_1V_1 = C_2V_2$$

$$0.89\,M\,(?) = 0.026\,M\,(1,000\,mL) \qquad ? \approx 29.21\,mL$$

e. Water required = 1,000 − 239.21 mL = 760.79 mL

9. You may have slightly different answers due to rounding.

10. You may have slightly different answers due to rounding.

TABLE 31.8

Recommended Additions of Sodium Bicarbonate

Medium	mL of NaHCO₃ Required, 7.5% Stock (per L)	g Of Solid NaHCO₃ Required (per L)	Final Concentration of NaHCO₃ (mg/L)
DME	49.3	3.70	3,700
DME/Ham's F-12	32.5	<u>2.438</u>	<u>2,438</u>
Ham's F-12	15.7	<u>1.176</u>	1,176
MEM Earle's salts	<u>29.3</u>	2.20	<u>2,200</u>
MEM Hank's salts	4.7	<u>0.35</u>	<u>350</u>
RPMI-1640	<u>26.7</u>	<u>2.00</u>	2,000
McCoy's 5A	29.3	<u>2.20</u>	<u>2,200</u>
MEM alpha	<u>29.3</u>	<u>2.20</u>	2,200

TABLE 31.9

Recommended Additions of L-Glutamine

Medium	mg/L in Final Medium	mM in Final Medium	mL/L of Stock Required
AMEM	292.3	2.0	10
BME	292.3	2.0	10
DME	584.6	4.0	20
F-12K	292.3	2.0	10
Ham's F-10	146.2	1.0	5
Ham's F-12	146.2	1.0	5
Iscove's MDM	584.6	4.0	20
EMEM	292.3	2.0	10
RPMI-1640	299.6	2.05	10.25

11.
   a. With 10% serum cost is: $15 + cost of 50 mL of serum, which is $25. So total = $40.
   b. With 2% serum cost is: $15 + cost of 10 mL serum, which is $5. So total cost = $20.

   Reducing the use of serum results in substantial cost savings (50% in this case).

12. Key similarities
   - Both are aqueous solutions with solutes that support the growth and reproduction of living cells.
   - Both must be sterilized and must be handled using aseptic technique.
   - Both are osmotically balanced.
   - Both include one or more sources of energy, carbon, vitamins, and trace elements to support the growth and reproduction of cells.
   - Both may be defined or undefined.
   - Both may or may not contain hydrolysates and extracts from yeasts, plants, and animals.

Key differences

- Mammalian cell culture tends to be more exacting and more sensitive to contamination.
- Mammalian culture media are usually more complex than bacterial.
- Agar is used when culturing bacteria, not mammalian cells.
- Bacterial media are usually sterilized by autoclaving; mammalian cell media by filtration.

13.
   a. The invention describes improved methods for producing a recombinant monoclonal antibody product (specifically adalimumab, Humira®, used to treat arthritis and other inflammatory diseases) in mammalian cell culture.
   b. The invention describes optimal conditions for pH, temperature, dissolved oxygen level, supplementation with glucose and other nutrients, and maximum osmolarity. Note that these factors are not held constant over the course of the cell culture process, but rather are manipulated over time.
   c. The goal of optimizing all these factors is to increase the amount of drug product that can be isolated from the cells. In order to do so, the cells must grow and reproduce. However, the cells must also be coaxed into directing much of their nutritional resources into the production of drug product, rather than only making more cells. It often requires extensive experimentation to find the optimal combination of nutrients, pH, temperature, and other factors to drive the cells toward the desired goal of manufacturing abundant biopharmaceutical product.

CHAPTER 32

1. NaCl ions, albumin, polio virus, *Serratia marcescens,* red blood cells, pollen grains, sand grains
2. a–c. Ultrafiltration   d. macrofiltration
   e. microfiltration   f. ultrafiltration
   g. microfiltration   h. microfiltration

3. **a.** XYZ   **b.** ABC
   c. XYZ   **d.** "Ready Separation"

CHAPTER 33

1. $r_{max} = 9.10\,cm$, $r_{min} = 3.84\,cm$, $r_{average} = 6.47\,cm$

Speed (in RPM)	Approximate RCF Generated		
	r_{max}	r_{min}	$r_{average}$
20,000	40,800	17,200	29,000
30,000	91,700	38,700	65,200
40,000	163,100	68,800	115,900

2. The radius must be in centimeters to use this equation.

3. $1{,}000\sqrt{\dfrac{RCF}{11.2\ r}} = RPM$

$$= 1{,}000\sqrt{\dfrac{140{,}000}{11.2\ (2.6)}}$$

$$\approx 69{,}338\ RPM$$

4.
   a. For separating large volumes of cells from fermentation, a high-capacity rotor is more important than a rotor that can spin at high RCFs. Rotor B is therefore preferred.
   b. The separation of very small particles requires a rotor, like rotor A, which is capable of running at high RCFs.
   c. A microfuge is used for small-volume samples like this, rotor C.

5.
   a. The maximum speed for this rotor is 13,000 RPM.
   b. The maximum force that can be generated with this rotor is $27{,}500 \times g$.
   c. The maximum volume that can be centrifuged in this rotor is $6 \times (315\,mL) = 1{,}890\,mL$.
   d. This rotor can hold 72 tubes each with a volume of 4 mL= 288 mL total.
   e. High speed.

CHAPTER 34

1. 1 kat transforms 1 mol substrate to product in 1 second, so 4 kat transforms 4 mol substrate to product in one second and 240 mol to product in one minute. 240 mol$=2.4 \times 10^8\,\mu mol$. So 4 kat$=\mathbf{2.4 \times 10^8\,IU}$.

2. ***Purification of Comatase from E. coli***

Purification Step	Volume (mL)	Total Protein (mg)	Total Activity (IU)	Specific Activity (IU/mg)	Yield (%)
I. Homogenization	100,000	12,350	9,000	0.73	100
II. Dialysis	5,000	10,233	8,289	0.81	92.1
III. Organic extraction	80	3,860	6,369	1.65	70.7
IV. Ion exchange chromatography	10	1,140	5,625	4.93	62.5
V. PAGE	2	386	4,688	12.15	52.1

3. In evaluating the relative merits of specific purification steps, consider the main purpose of each step. In the example given, Step II provided a major reduction in product volume, but little purification (a resulting specific activity of 0.81 vs. 0.73 for the starting material).

a. Step IV provides the greatest relative increase in product purity, with approximately a threefold increase in the specific activity of the product. Step V gave a 2.5-fold purification.

b. The loss of product material in each step can be calculated by dividing the total product activity for each step by the activity present in the preceding step. Step III, therefore, gave the greatest loss of product activity:

(6,369 IU/8,289 IU) (100%) = 76.8% product recovery in this step.

c. Absolute reduction in product volume can be determined by looking at the final product volume for each step. In this example, Step II gave the greatest volume reduction, from 100,000 to 5,000 mL. In a production setting, this volume reduction would significantly ease the cost of operations.

d. Relative reduction in product volume is calculated in a similar manner to 3b. The volume after each step is divided by the volume at the beginning of the step. Step III, therefore, gave the greatest *relative* reduction in volume size:

(80 mL/5,000 mL) (100%) = 1.6%. This means that after organic extraction there was only 1.6% of the material before the extraction. (It might seem that dialysis should be the step with the greatest relative reduction in volume. But the relative volume reduction by dialysis is 5000 mL/100,000 mL = 0.05 × (100%) = 5%. Thus, 5% of the starting volume remained after dialysis. This is a larger remaining percent than after organic extraction.)

4. The enzyme was secreted into the broth, which was discarded after the first centrifugation step. It is good practice to keep all supernatants and pellets until you are certain you have the product of interest.

5.

a.

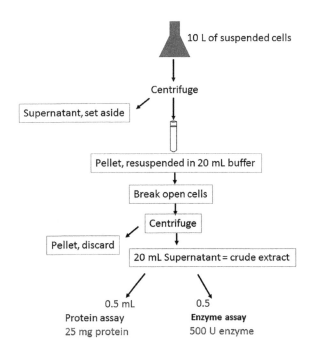

b. 20 U/mg protein.

c. If there are 500 units in 0.5 mL of the supernatant, then there are 20,000 units in the entire 20 mL.

6.

a. **i.** Specific activity should increase.

b. **ii.** Amount of total protein should decrease.

c. **ii.** Yield will decrease.

7. Some possible answers:
- Whether it is found inside cells or is secreted.
- If it is intracellular, its intracellular location.
- Molecular weight.
- Solubility characteristics in different solvents.
- Molecular charge.
- Stability at room temperature.

8.

a. Specific activity equals the amount of activity of the protein of interest divided by the total amount of protein present. Yield is the percent of the amount of starting activity that is still present after each purification step. In this example, 1,360 IU represents 100% yield.

Purification of β-Galactosidase from E. coli

	Total Protein	Total Activity	Specific Activity	Yield
Purification step	(mg)	(IU)	(IU/mg)	(%)
Crude extract	2,140	1,360	0.64	100
Salt precipitation	760	1,350	<u>1.78</u>	99
Dialysis	740	<u>1,332</u>	1.80	<u>98</u>
Chromatography	390	1,240	<u>3.18</u>	<u>91</u>

b. As we would predict, some of the β-galactosidase was lost at each step and the percent of total activity remaining decreased as the purification proceeded.

c. As the purification proceeded, the β-galactosidase became purer; therefore, the specific activity value increased.

9. This is a discussion question with many possible answers. The first thing to recognize is that the final yield cannot be higher than the maximum calculated yield, so there is a mistake somewhere. In consultation with a supervisor and colleagues, you would likely go back over the yield calculations. Calculations are an obvious place to look for a mistake. If the calculations are correct, the assay used to monitor the product purification should be checked. Possibly a buffer, temperature, pH, or some other condition was wrong when determining how much product was initially present or when calculating the amount of protein at the end of the purification. Troubleshooting might require preparing new reagents, recalibrating instruments, and performing assays again. All troubleshooting steps would need to be documented and justified.

10.

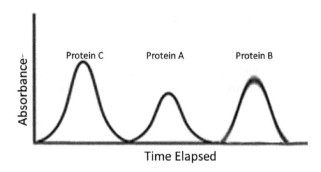

11. From the graph, it is determined that the EC50 is about 2 ng/mL, as indicated in the figure caption.

One unit of activity is defined as the amount of this particular signaling molecule that has 50% of the maximal proliferative effect per milliliter. Therefore, one unit of this signaling molecule is 2 ng. To calculate specific activity, we must know how many units there are per mg, using proportions.

A mg contains 1,000,000 ng.

$$\frac{1 \text{ unit}}{2 \text{ ng}} = \frac{?}{1,000,000 \text{ ng}}$$

? = 500,000, so the SA = **500,000 U/mg.**

CHAPTER 35

1. **a.** 5   **b.** 3   **c.** 4   **d.** 2   **e.** 1

2. ICH stands for the International Conference for Harmonisation of Technical Requirements for Registration of Pharmaceuticals for Human Use. ICH is an organization that brings together the regulatory authorities of the European Union, Japan, and the United States and experts from the pharmaceutical industry in the three regions to discuss scientific and technical aspects of pharmaceutical regulation. This work helps ensure that the products developed and manufactured in one country meet the regulatory requirements of other countries.

3.
   a. Q: Quality, S: Safety, E: Efficacy, and M: Multidisciplinary.

   b–e. Stability studies are in the Q series. Medical terminology is in the M series. Carcinogenicity risk is in the S series. Clinical trials are in the E series.

4. The Food and Drug Administration (FDA).

5. CBER is responsible for the oversight of biologics, whereas CDER is responsible for the oversight of prescription and over-the-counter drugs.

6. In general, biologics are derived from living sources (such as humans, animals, and microorganisms), whereas drugs are chemically synthesized.

7.
   a. 21 CFR Part 211.137 Expiration dating
   b. 21 CFR Part 211.44 Lighting
   c. 21 CFR Part 211.170 Reserve samples

8.
   a. Hazards may include the presence of pathogens, antibiotics, pesticide and hormone residues, additives, and foreign matter.
   b. Hazards can be introduced at any step in the food chain. By monitoring food throughout the supply chain, hazards can be mitigated.

CHAPTER 36

1. **a.** iii **b. i** **c.** ii **d.** i **e.** i **f.** i **g.** iv **h.** i. For d, you might also say that this is performed in the preclinical phase.

2. A biopharmaceutical refers to a category of pharmaceutical products that are manufactured by genetically modified organisms (GMOs). These products are generally proteins. The term "biopharmaceutical" is also used more broadly to refer to any drug manufactured by living cells or organisms (whether or not recombinant DNA techniques are involved), and also to include whole cells or tissues used therapeutically. Biopharmaceutical can also refer to any drug that is a large biological molecule; these are usually proteins, but can be RNA or DNA (e.g., DNA for gene therapy).

3. Small-molecule drugs have a smaller molecular weight and are often chemically synthesized. An example is aspirin (MW<200). A large-molecule drug, however, cannot at this time be efficiently chemically synthesized, with a few exceptions (e.g., DNA used for gene therapy and certain RNA products). Large biological molecule drugs must be isolated from organisms or manufactured in carefully nurtured living systems (cells, plants, or animals). Large biological molecule drugs are also more difficult, complex, and expensive to produce than small-molecule agents because living systems are required for their production. An example of a large-molecule drug is Rituxan (MW>145,000).

4. There is inherent variability between one biological system and another. For example, cultured cells can differ from one another if their culture conditions vary slightly. If the temperature, pH, or composition of their nutrient medium changes, this might alter the characteristics of any product they make.

5.
   a. False. Activity is how well the drug candidate binds to its target and generates the desired biological response.
   b. True, half-life is a measure of the time a drug remains in its active form in the body.
   c. False. The metabolic profile's purpose is to determine whether any toxic effects are produced by the drug candidate or its by-products when the body's enzymes break it down.

6. When a drug is fast-tracked, it undergoes the "process designed to facilitate the development and review of drugs that treat serious conditions and fill an unmet medical need." Examples of drugs that may be fast-tracked include those that treat: HIV/AIDS, Alzheimer's, heart failure, cancer, epilepsy, depression, diabetes, and orphan diseases.

7. An orphan disease is defined as a "condition that affects fewer than 200,000 people nationwide" (e.g., cystic fibrosis, Lou Gehrig's disease, and Tourette's syndrome).

8. Animal testing is performed to evaluate the safety of a potential drug and its effects in the body before the agent is tested in humans. At least two animal species are tested.

9. Post-approval studies allow companies to gather more data on a drug, e.g., drug–drug interactions. Post-approval studies are important for detecting side effects that are not evident when smaller populations are tested in clinical trials, or side effects that occur with long-term usage. Depending on post-approval studies, the company may choose to apply for label expansion, including other indications for which the drug may be used.

10. Phase I trials normally include a small number of healthy volunteers, but sometimes a small number of patients with the illness to be treated are tested at this point. The primary purpose of Phase I trials is to evaluate the safety of the agent. The drug's pharmacokinetic and pharmacologic properties in healthy volunteers are also examined. Phase II trials

are performed on a small number of patients with the condition the drug is expected to treat. This phase is used to demonstrate clear drug activity and tolerance of the new compound in patients with mild to moderate disease or conditions. Safety continues to be evaluated in Phase II, and a dosage regimen for Phase III trials is established. Phase III trials are performed on a large number of patients with the condition the drug is expected to treat. This phase is used to demonstrate that a drug has the desired effects. Also, during this phase, the safety of the drug continues to be evaluated, adverse reactions are monitored, an analysis of the drug's risks versus its benefits is performed, drug interactions are explored, and other data are collected.

11. Pharmacokinetics is the study of how a drug is changed in the body. ADME studies are conducted to evaluate how the drug is absorbed (A), distributed (D), metabolized (M), and excreted (E) from the body.

12. The drug development process is lengthy, and the pharmaceutical industry generally reports that only 1 out of every 10 drugs that enter clinical testing will actually result in a marketed product. The low success rate of drugs coupled with the extensive resources required for their development contribute to the high costs of drugs.

13. A generic is "chemically identical to the original drug," whereas a biosimilar is "highly similar to, though not necessarily identical to, the previously approved 'reference' biological drug."

14. Some points to include in your discussion: In order to have a regulatory process, such as the one outlined in Figure 36.7, the government had to accept responsibility for protecting the consumer. This occurred in 1906 with the passage of the first Food, Drug, and Cosmetic Act. You might also mention the beginnings of the FDA in 1927, because the process outlined in Figure 36.7 includes critical gates controlled by the FDA. Animal and human testing are major parts of the development process shown in Figure 36.7. The 1938 Food, Drug, and Cosmetic Act that was enacted in response to the sulfanilamide incident was of critical importance. This Act mandated that new drugs be tested for safety in animal and clinical studies. Other

incidents led to tighter control over the animal and clinical testing stages. In particular, the thalidomide tragedy led to tighter control of human testing, and inspections of animal testing facilities in the 1970s led to tighter control of animal testing.

15.

a. Congress enacts laws to protect consumers. These laws mandate overall processes for controlling the development and production of these products and give enforcement powers to government agencies.

b. The laws passed by Congress are broadly written. The FDA provides specifics in the form of the GLPs, GCPs, CGMPs, and also various guidance documents. The FDA controls critical gates during development through an approval process; see Figure 35.9. The FDA also oversees and inspects the manufacture of products.

c. Companies must comply with numerous requirements, including those outlined in GLPs, GCPs, and CGMPs. Companies must comply with directives from the FDA. Companies must design their products to be safe and effective and must put in place systems to ensure the quality of their products.

d. Consumers must take responsibility for their own health. This includes complying with the guidance shown on product labels (e.g., taking the proper dose of a drug) and storing and preparing food properly.

16.

a. This section is shown below. It outlines the responsibilities and authorities of the quality-control (QC) unit in a pharmaceutical/biopharmaceutical company. Use your own words to describe its provisions.

b. A separate QC unit is required by law. This means that the production unit cannot take final responsibility for these tests.

   211.22 Responsibilities of quality control unit.

a. *There shall be a quality control unit that shall have the responsibility and authority to approve or reject all components, drug product containers, closures, in-process materials, packaging material, labeling, and drug products, and the authority to review production records to assure that no errors have occurred or, if errors have*

*occurred, that they have been fully inves-
tigated. The quality control unit shall be
responsible for approving or rejecting
drug products manufactured, processed,
packed, or held under contract by another
company.*

b. *Adequate laboratory facilities for the testing and approval (or rejection) of components, drug product containers, closures,
packaging materials, in-process materials, and drug products shall be available
to the quality control unit.*

c. *The quality control unit shall have the
responsibility for approving or rejecting
all procedures or specifications impact-
ing on the identity, strength, quality, and
purity of the drug product.*

d. *The responsibilities and procedures applicable to the quality control unit shall be in
writing; such written procedures shall be
followed.*

CHAPTER 37

1. **Similarities include**:
   - Both drugs treat cancer by slowing uncon-
     trolled cell division.
   - Both drugs recognize and bind to a spe-
     cific target in the body that is associated
     with cancer cells.
   - The development of both drugs relied
     heavily on basic research into the mecha-
     nisms of cancer and cell division.
   - **Differences include**:
   - Gleevec is a small-molecule drug; Her-
     ceptin is a large-molecule (antibody) drug.
   - Gleevec treats CML; Herceptin treats
     breast cancer.
   - Gleevec targets the site where ATP binds
     to an enzyme; Herceptin targets receptors
     on the surface of breast cells.

   You can probably think of other similarities and differences as well.

2.
   a. Key concerns include:
      - Cells might be contaminated with
        agents that could cause disease in
        patients; these agents might be bacte-
        ria, fungi, viruses, or prions.
      - Host cells might mutate (change genet-
        ically), thereby possibly altering the
        product.

   - The genetic construct inserted into the
     cells might change or be lost over time,
     thereby changing the product.
   - Host cells might induce cancer in
     patients.

   b. End of production cells are tested to see if
      changes in the cells or the genetic construct
      have occurred over time. Any changes in
      the cells or the genetic construct could
      mean that the biopharmaceutical product
      is altered. The cells are also tested for con-
      taminating agents.

3. a, b, f, g relate to using genetically modified
   cells to make a product. These tasks therefore
   relate specifically to biopharmaceutical production. The other tasks are common to all drugs.

4. There is no single answer to this question.
   Be sure that your answer discusses the roles
   of various types of laboratory personnel. Key
   points include:
   - The role of laboratory personnel in per-
     forming basic research that elucidates
     the workings of cells and organisms and
     in uncovering the basic mechanisms of
     health and disease.

     Example: The scientific work of Erwin
     Chargaff, Rosalind Franklin, Maurice
     Wilkins, James Watson, and Francis Crick
     that elucidated the structure of DNA.
   - The role of laboratory personnel in the dis-
     covery of potential targets for drugs and in
     the discovery of new drug candidates.

     Example: The discovery of the Her2
     gene and its role in some breast cancers.
   - The role of laboratory personnel in the
     development of potential products.

     Example: Dawn Rabbach's work to
     develop a DNA fingerprinting method.

     **Biopharmaceutical development pro-
     vides many examples including:**
   - Testing drugs in animals and humans.
   - Optimizing methods for producing drugs
     in cells.
   - Optimizing methods of purifying drugs
     produced by cells.
   - Developing methods to test the activity,
     potency, and purity of drugs, raw materi-
     als, and in-process samples.
   - Scaling up production.
   - The role of quality-control analysts who
     test environmental samples, raw materials,
     in-process samples, and final products.

5. Testing is critical because adventitious viruses can cause illness in patients.

6. **a.** 5 **b.** 4 **c.** 2 **d.** 1 **e.** 3

7. Upstream processing is the process of growing cells and their production of the desired product. Downstream processing is the process of isolating and purifying the product from the cells.

8. Cells vary in their nutrient requirements depending on the cell type and the product they are expressing. Therefore, different culture media are required for efficient production of different products.

9. False. Proteins vary greatly from one another, and cells used for production also vary; therefore, every downstream process is different and must be optimized.

CHAPTER 38

1.

a. There were a total of n = 4,849 483s issued between October 1, 2018, and September 30, 2019.

b. The number = 779 for Drugs (≈16%), n = 822 for Devices (≈17%), and n = 116 for Biologics (≈2%).

c. 21 CFR 211.22(d) states: "The responsibilities and procedures applicable to the quality control unit shall be in writing; such written procedures shall be followed." This is a relatively simple statement, but it has a major impact: There is no leeway for deviation from established procedures in a QC unit. It is important for procedures to be documented and followed to ensure product quality. For example, if procedures are not documented or are not followed, it increases the risk of invalid analytical testing, which, in a worst case, could result in unsafe product being released to consumers.

d. Personnel qualifications are important in GMP operations to help ensure that employees have the appropriate education, training, and experience to perform their jobs properly. Any GMP training should be conducted by qualified individuals and with sufficient frequency (21 CFR 211.25a). Additionally, individuals responsible for supervising the manufacture, processing, packing, or holding of a drug needs to have the appropriate education, training, and experience (21 CFR 211.25b). There also needs to be sufficient qualified personnel to perform the manufacture, processing, packaging, or holding of a drug (21 CFR 211.25c). If personnel are not prepared to do their jobs, products that go to consumers might be unsafe or ineffective.

e. As discussed in Chapter 6, records are of vital importance. This section of the CFR lists key records relating to blood and blood products. A number of specific records required are listed; many of them relate to donors. (See Section 37.4.2 for more information about donors.)

  – Records are vital to ensure traceability. Suppose, for example, a lot of a blood product is found to be contaminated with a viral pathogen. It is imperative to find and recall all of that product and to find any patients who were administered contaminated products.

  – Records document that procedures for handling, storing, processing, and shipping the blood products are performed in such a way as to protect the product.

  – Quality-control records are retained to ensure that the in-process and final materials met specification prior to release of the product. This will vary for each product, but may include for example, pH measurements and HPLC recordings.

  – There are many other records listed, for example, records of facility inspections, visitor logs, and computer backup logs. This is for the general operation of the facility.

  – Observe that this is a comparatively long regulation with many specific details. This attests to the importance of documentation.

2. ISO 2.

3. Facilities, equipment, cleanrooms, raw materials, and skilled personnel.

4. All of them.

5. Process analytical technology (PAT) serves to improve the quality of products and the efficiency of production.

6. The goal of validation is to ensure that product quality is built into a product as it is produced.

7. Validation is a major undertaking that is expensive, is time-consuming, and requires extensive planning and knowledge of the system being validated. The advantage to validation is that it helps assure consistent product quality, greater customer satisfaction, and fewer costly product recalls.

8. Step 1. The analyst must report the result to a supervisor.

Step 2. The analyst and supervisor must conduct an informal lab inspection including discussing the testing procedure, reviewing all calculations, examining the instruments used, and reviewing the notebooks with the OOS result.

Step 3. The analyst and supervisor must document the investigation results, recording any errors that might have been uncovered and any conclusions that were reached.

Based on this initial investigation, it may be determined that the analyst made an error, and the sample should be retested. However, if no laboratory error is found, then further investigation must be conducted to determine if the test substance is unsuitable for its purpose.

9.
   a. False. If any standard does not give the expected result, then the test should be suspended and any data already acquired should be investigated.

   b. True.

   c. False. No samples should be discarded before the results of a test are checked. If the results are out of spec, then the samples must be saved for investigation.

10. In the K-Dur extended release coating process example, the analysts essentially tried to "test the sample into compliance." They kept testing it over and over again until they eventually obtained a passing result. They then disregarded the previous eight OOS results. There was no documented basis for disregarding the OOS results. Furthermore, no corrective action was taken. They essentially did the same thing in the blend uniformity test, rejecting without basis results that they did not like, and retesting until they obtained a result they did like. They did not document a clear reason for rejecting the OOS results in either situation, and they should have conducted a formal investigation. Remember that in cases like this, the product may, in fact, be defective and should not have been released for patient use.

Acronyms

ACGIH	American Conference of Governmental Industrial Hygienists	**CLSI**	Clinical and Laboratory Standards Institute
ACS	American Chemical Society	**CNS**	Central nervous system
ACT	Adoptive cell therapy	**COVID-19**	Coronavirus disease 2019
ADME	Absorption, distribution, metabolism, and excretion	**CPU**	Central processing unit
		CRO	Contract research organization
ANSI	American National Standards Institute	**DIN**	Deutsche Industrial Norms
APHIS	Animal and Plant Health Inspection Service	**DNA**	Deoxyribonucleic acid
		DOI	Digital object identification
API	Active pharmaceutical ingredient	**DOT**	Department of Transportation
ASTM	Formerly American Society for Testing and Materials. Now ASTM International	**DP**	Drug product
		DS	Drug substance
ATC	Automatic temperature compensating (probe)	**ELN**	Electronic laboratory notebook
		EMEA	European Agency for the Evaluation of Medicinal Products
ATCC	American Type Culture Collection	**EPA**	Environmental Protection Agency
AUP	Acceptable use policy	**EtBr**	Ethidium bromide
BLA	Biologics License Application	**FAQ**	Frequently asked question
BLAST	Basic local alignment search tool	**FDA**	Food and Drug Administration
BSE	Bovine spongiform encephalopathy	**FDCA**	Food, Drug, and Cosmetic Act
BSL	Biosafety level	**FIFRA**	Federal Insecticide, Fungicide, and Rodenticide Act
BTV	Bring to volume		
CAPA	Corrective and preventive actions	**FPLC**	Fast performance liquid chromatography
CAR-T	Chimeric antigen receptor T cell	**GB**	Gigabyte
CAS	Chemical Abstracts Service	**GC, GLC**	Gas–liquid chromatography
CBER	Center for Biologics Evaluation and Research	**GCP**	Good Clinical Practices
		GFI	Ground fault interrupt
CDC	Centers for Disease Control and Prevention	**GHz**	Gigahertz
		GLP	Good Laboratory Practices
CDER	Center for Drug Evaluation and Research	**GMLP**	Good Microbiological Laboratory Practices
CDRH	Center for Devices and Radiological Health		
		GMO	Genetically modified organism
CE	Capillary electrophoresis	**GPC**	Gel permeation chromatography
CFR	Code of Federal Regulations	**GWAS**	Genome-wide association study
CFSAN	Center for Food Safety and Applied Nutrition	**HACCP**	Hazard analysis critical control points
		HBV	Hepatitis B virus
CGLP	Current good laboratory practices	**HCS**	Federal Hazard Communication Standard
CGMP	Current Good Manufacturing Practices		
CGPM	Conférence Générale des Poids et Mesures/General Conference on Weights and Measures	**HEPA**	High-efficiency particulate air (filter)
		HGP	Human Genome Project
		HHS	U.S. Department of Health and Human Services
CHO	Chinese hamster ovary		
CHP	Chemical Hygiene Plan	**HIC**	Hydrophobic interaction chromatography
CIP	Clean-in-place		
CIPM	Comité International des Poids et Mesures/International Committee of Weights and Measures	**HIV**	Human immunodeficiency virus
		HPLC	High-performance liquid chromatography

HTS	High-throughput screening	**OLAW**	Office of Laboratory Animal Welfare
IACUC	Institutional Animal Care and Use Committee	**OMIM**	Online Mendelian Inheritance in Man
		OOS	Out of specification
IARC	International Agency for Research on Cancer	**OSHA**	Occupational Safety and Health Administration
ICH	International Conference on the Harmonisation of Technical Requirements for the Registration of Pharmaceuticals for Human Use	**PAGE**	Polyacrylamide gel electrophoresis
		PDA	Personal digital assistant
		PDB	Protein Data Bank
		PDF	Portable document format
IDLH	Immediately dangerous to life and health	**PEL**	Permissible exposure limit
ILAR	Institute for Laboratory Animal Research	**PHMSA**	Pipeline and Hazardous Materials Safety Administration
IND	Investigational New Drug Application		
IP	Intellectual property	**PIN**	Personal identification number
iPS	Induced pluripotent stem (cells)	**PK**	Pharmacokinetics
IRB	Institutional review board	**PMT**	Photomultiplier tube
ISE	Ion-selective electrode	**PPE**	Personal protective equipment
ISO	International Organization for Standardization	**PPM**	Parts per million
		QA	Quality assurance
IU	International unit of enzyme activity	**QC**	Quality control
JPEG	Joint Photographic Experts Group	**QS**	Quantum sufficit
JSA	Job Safety Analysis	**R&D**	Research and development
LAI	Laboratory-acquired infection	**RAC**	Recombinant DNA Advisory Committee
LAL	*Limulus* amebocyte lysate	**RCF**	Relative centrifugal field
LAN	Local area network	**RCRA**	Resource Conservation and Recovery Act of 1976
LC	Liquid chromatography		
LCD	Liquid crystal display		
LC$_{Lo}$	Lethal dose – low	**rDNA**	Recombinant DNA
LD$_{50}$	Lethal dose, 50%	**RFLP**	Restriction fragment length polymorphism
LED	Light-emitting diode		
LIMS	Laboratory information management system	**RNA**	Ribonucleic acid
		RNAi	RNA interference
mAb or Mab	Monoclonal antibody	**RO**	Reverse osmosis
MCB	Master cell bank	**RPC**	Reverse phase chromatography
MHz	Megahertz	**RPM**	Revolutions per minute
MPEG	Moving Picture Experts Group	**RSI**	Repetitive stress injury
mRNA	Messenger ribonucleic acid	**SARS-CoV-2**	Severe acute respiratory syndrome coronavirus 2
MSC	Mesenchymal (stromal) stem cell		
MWCO	Molecular weight cutoff	**SAS**	Statistical Analysis System
NCBI	National Center for Biotechnology Information	**SDS**	Safety Data Sheet
		SDS–PAGE	Sodium dodecyl sulfate–polyacrylamide gel electrophoresis
NCCLS	National Committee for Clinical Laboratory Standards		
		SFM	Serum-free medium
NCI	National Cancer Institute	**SI**	Système International d'Unités
NDA	New Drug Application	**SNP**	Single nucleotide polymorphism
NFPA	National Fire Protection Association	**SOP**	Standard operating procedure
NIH	National Institutes of Health	**SQL**	Structured query language
NIOSH	National Institute for Occupational Safety and Health	**SRM**	Standard reference material
		STEL	Short-term exposure limit
		STR	Short tandem repeat
OD	Optical density	**TB**	Terabyte
OIML	Organisation Internationale de Métrologie Légale/International Organization of Legal Metrology	**TEMED**	N,N,N′,N′-tetramethylethylenediamine
		TLC	Thin-layer chromatography
		TLV	Threshold limit value

TLV-C	Threshold limit value – ceiling	**URL**	Uniform resource locator
TLV-STEL	Threshold limit value – short-term exposure limit	**USB**	Universal serial bus
		USDA	United States Department of Agriculture
TLV-TWA	Threshold limit value – time-weighted average	**USP**	United States Pharmacopeia
		UV	Ultraviolet
tRNA	Transfer RNA	**VIS**	Visible
TSCA	Toxic Substances Control Act	**VPN**	Virtual private network
TWA	Time-weighted average	**WCB**	Working cell bank
UL	Underwriters Laboratories	**WFI**	Water for injection
UPS	Uninterruptible power supply	**Wi-Fi**	Wireless fidelity

Glossary Terms

% Recovery See **yield**.

Abscissa In graphing, the X-coordinate of a point; the distance of a point along the X-axis.

Absolute error The difference between the true value and the measured value. The plus (+) or minus (−) sign indicates whether the true value is above or below the measured value. *Absolute error = true value − measured value*.

Absolute zero The temperature at which thermal energy is virtually nonexistent; defined as 0 K or −273.15°C.

Absorbance (A) also called **optical density** A measure of the amount of light absorbed by a sample defined as: $A = \log_{10} (1/t) = -\log_{10} t$, where t is transmittance.

Absorbance scale accuracy See **photometric accuracy.**

Absorbance spectrum The plot or graphic representation of absorbance of a particular sample when exposed to light of various wavelengths, typically plotted with wavelength on the X-axis versus absorbance on the Y-axis.

Absorption Process in which one substance is taken up by another, as when a cotton ball absorbs water or liquid is taken up into the depth of a filtering material.

Absorption (in spectrophotometry) The loss of light of specific wavelengths as the light passes through a material and is converted to heat energy. *Note*: Spectrophotometers actually detect and measure transmittance, not absorbance. Absorbance is a calculated value.

Absorption (in the context of drug testing) A process by which a drug substance moves from the site of administration to the blood.

Absorptivity The inherent tendency of a material to absorb light of a certain wavelength. (Note that absorbance is a measured value that depends on the instrument used to measure it, whereas absorptivity is an intrinsic property of a material.) See also **Beer's law.**

Absorptivity constant (α) A value that indicates how much light is absorbed by a particular substance at a particular wavelength under specific conditions (such as temperature and solvent). It is sometimes calculated as: A/bC, where $A = absorbance$, $b = path\ length$, and $C = concentration$, and α has units that vary depending on the units of concentration and path length. See also **Beer's law.**

Absorptivity constant, molar (ε) The absorptivity constant when the concentration of analyte is expressed in the unit of moles per liter.

Acceptance (control chart usage) A decision that a process is operating in control. See also **control chart limits**.

Acceptance criteria (according to GMP) The minimum specifications for a product and the criteria for accepting or rejecting that product, together with a plan for sampling it.

Accepted reference value[1] "A value that serves as an agreed-upon reference for comparison, and which is derived as: (1) a theoretical or established value, based on scientific principles, (2) an assigned value, based on experimental work of some national or international organization [e.g., see definition of **standard reference material (SRM)**], or (3) a consensus or certified value, based on collaborative or experimental work under the auspices of a scientific or engineering group."

Accession number A unique identifier for a DNA sequence, supplied by GenBank and designed to facilitate database searches with that sequence.

Accuracy Closeness of agreement between a measurement or test result and the true value or the accepted reference value for that measurement or test. *Note:* The term accuracy, when applied to a set of observed values, is affected both by random error and by systematic error (bias). Because random components and bias components cannot be completely separated in routine use, the reported "accuracy" must be interpreted as a combination of these two elements. The use of the terms "imprecision" to describe random errors and "bias" to describe systematic errors will emphasize these distinct elements of variation. When assessing the systematic error (bias) of test methods or operations, the use of the term "bias" will avoid confusion.

Acid A compound that dissociates in water to release hydrogen ions and reacts with base to form

---

[1] Based on definitions in ASTM Standard E 456–06 "Standard Terminology Relating to Quality and Statistics."

neutral salts and water; can be corrosive and reactive.

Acid–base error A cause of inaccuracy when using pH indicator dyes. Indicators are themselves acids or bases and when added to unbuffered or weakly buffered solutions will cause a change in the pH of a solution.

Acidic solution An aqueous solution with a pH less than 7.

Acrylamide A small neurotoxic monomer that can be polymerized to relatively nontoxic polyacrylamide; see **polyacrylamide gels.**

Activated carbon Material that adsorbs and removes organic contaminants from water. Derived from wood and other sources that are charred at a high temperature to convert them to carbon. The carbon is "activated" by oxidation from exposure to high-temperature steam.

Active pharmaceutical ingredient (API) The substance or mixture of substances used in the manufacture of a drug product that has effect in the patient.

Activity (in pH measurement) The effective concentration of a solute that accounts for interactions among solutes, temperature, and other effects.

Acute Having a rapid onset.

Acute exposure A short-term contact or single dose of a substance or chemical.

Acute toxicity The harmful effects of any single dose of or short-term exposure to a chemical or other substances.

Adapter (in centrifugation) An insert placed in a rotor compartment that permits the use of a smaller-sized tube than could otherwise be accommodated.

Adenine (A) One of the four types of nucleotide subunit that comprise DNA and RNA.

Adherent cells (in mammalian cell culture) Cells that grow in a single layer attached to the surface of their culture vessel.

ADME (absorption, distribution, metabolism, and excretion) Testing performed during the development of a drug product that evaluates where the substance goes in the body, how it is metabolically altered, and how it is excreted or broken down.

Adoptive cell therapy Cell transfer therapy in which therapeutic cells are administered to a patient.

Adsorption A process in which a material sticks to the surface of a container, membrane, bead, filter membrane, or other solids.

Adsorption chromatography A chromatographic technique for separating molecules based on their relative affinities for a solid stationary phase and a gas or liquid mobile phase.

Adulterated (according to the FDCA) A food or drug that is produced by methods that do not conform to Current Good Manufacturing Practices, or that is made under unsanitary conditions, or that contains unacceptable contaminants.

Adventitious contaminant A contaminant that is accidentally and sporadically introduced from a source connected to the production system, for example, from a contaminated piece of equipment or from a person.

Adverse events Undesired effects or toxicity due to exposure to a pharmaceutical or medical product.

Aerosol Very small particles suspended in the air.

Affinity The ability of a material to bind specifically to other biomolecules.

Affinity chromatography A chromatographic technique for separating molecules by their ability to bind specifically to other molecules that are incorporated onto the surface of a stationary phase.

Agar A hardening agent derived from seaweed; used to make a solid nutrient substrate for bacterial growth.

Agarose A natural polysaccharide derived from agar, a substance found in some seaweeds.

Air displacement micropipette A device for measuring microliter volumes; designed so that there is an air cushion between the micropipette and the sample.

Air-purifying respirator A **respirator** that filters room air through canisters of various materials that remove specific contaminants.

Airborne particle retention A measure of the efficiency of an air filtration system.

Airfoils (in fume hoods) Grates located at the bottom and sides of the hood sash that help reduce air turbulence at the face opening of the hood.

Airlock A small room with interlocked doors between areas with different cleanliness standards.

Alcohol error (in pH measurement) A cause of inaccuracy when using pH indicator dyes. Alcohols may cause indicators to be a different color than they appear to be in aqueous solutions.

Aliquot The portions that result from subdividing a homogenous solution or substance into smaller units.

Alkaline error (in pH measurement) An error in the response of a pH-measuring electrode in which the electrode responds to Na and K. Alkaline error occurs when the pH values are above 9 or 10.

Alkalinity A measure of the capacity of water to accept H ions, that is, its acid-neutralizing ability. Carbonate, bicarbonate, and hydroxide ions are common contributors to alkalinity.

Allergen A substance that produces an allergic response in some individuals.

Allergy A reaction by the body's immune system to exposure to a specific chemical.

Alpha helix See **secondary structure**.

Alternating current (AC) Electrical current that cycles between flowing first in one direction and then in the other direction.

Ambient temperature The average temperature of the surrounding air that contacts the instrument or system being studied, often room temperature.

American Association for Accreditation of Laboratory Animal Care (AAALAC) An independent peer-review organization that ensures that companies, universities, hospitals, government agencies, and other research institutions surpass minimal animal care standards.

American Chemical Society (ACS) Professional organization that provides various services including setting standards for purity of chemical reagents.

American Conference of Governmental Industrial Hygienists (ACGIH) An organization of governmental, academic, and industrial professionals who develop and publish recommended threshold exposure limits for chemical and physical agents.

American National Standards Institute (ANSI) A national organization that sets standards related to safety and safety design. The US representative to ISO.

American Type Culture Collection (ATCC) A global, nonprofit organization that provides and distributes biological resources, particularly cells and tissues.

Amino acids A class of naturally occurring molecules that are the building blocks of proteins. Every amino acid contains a carbon atom bonded to an amino group (NH_2), a carboxyl group (COOH), a hydrogen atom, and a side chain. Each amino acid has a different side chain that gives it distinctive chemical properties.

Amino group Part of the core structure of amino acids, NH_3.

Ammeter A device to measure electrical current.

Amount How much of a substance is present. For example, 2 g or 4 cups.

Amperes (A or amp) The unit by which current is expressed. When 6.25×10^{18} electrons pass a point in the path of electricity's flow every second, the path is said to be carrying a current of 1 A.

Amplification The boosting of a weak signal.

Amplifier A device that boosts the voltage or current from a detector in proportion to the size of the original signal.

Analog "Smoothly changing." Analog measurement values are continuous as, for example, displayed by a meter with a needle that can point to any value on a scale.

Analog-to-digital converter (A/D converter) A device that converts an analog signal to a digital signal.

Analysis (in biology) The study of the specific chemical properties of a molecule.

Analyte A substance of interest whose presence, properties, and/or level is evaluated in a sample using an instrument, assay, or test.

Analytical balance An instrument that can accurately determine the weight of a sample to at least the nearest 0.0001 g.

Analytical method A test used to analyze, identify, or characterize a mixture, compound, chemical, or unknown material.

Analytical ultracentrifuge An ultracentrifuge that is designed to provide information about the sedimentation properties of particles.

Analytical wavelength (in spectrophotometry) The wavelength at which absorbance measurements are made in a particular assay.

Anaphylactic shock A sudden life-threatening reaction to allergen exposure.

Angstrom (Å) A unit of length that is 10^{-10} m, named after the Swedish physicist A.J. Ångström. A human hair has a diameter of about 500,000 Å.

Animal and Plant Health Inspection Service (APHIS) Division of the U.S. Department of Agriculture that oversees protection of animal and plant resources, including overseeing transgenic organisms.

Anion A negatively charged ion.

Anion exchange chromatography Ion exchange chromatography method where the stationary phase particles are coated with positively charged molecules to adsorb negatively charged molecules (anions) that flow by in the mobile phase.

Anion exchange resin A solid matrix material that has negatively charged ions available for exchange with negatively charged ions in a solution.

Anisotropic membrane A membrane in which the pore openings are larger on one side than on the other.

Annotation, genome (according to the Human Genome Project glossary) "Adding pertinent information such as gene coded for, amino acid sequence, or other commentary to the database entry of raw sequence of DNA bases."

Anode A positive electrode.

Anodized surface Thin, protective coating of aluminum oxide deposited electrochemically on aluminum rotors to help protect them from corrosion.

Antibody Protein made by immune system cells that recognizes and binds to substances invading the body and that aids in their destruction.

Antigens Substances that trigger the production of antibodies.

Antilogarithm (antilog) The number corresponding to a given logarithm. For example, $100 = 10^2$. The log of 100 is 2. The antilog of 2 is 100.

AOAC International An independent association of scientists devoted to promoting methods validation and quality measurements in the analytical sciences.

Artifact A distortion or error in the data. For example, in electron microscopy an artifact might be a substance that appears to be a component of a cell, but, in fact, was accidentally created by the stains used to prepare the sample for visualization.

Aseptic processing A process in which drug or biological products and their containers are sterilized separately, frequently by different methods, and then packaged together under aseptic conditions to create a sterile final product.

Aseptic technique A system of laboratory practices that minimize the risk of biological contamination.

Ash content The percent ash residue remaining when a paper filter is burned.

Asphyxiant A gaseous compound or vapor that can cause unconsciousness or death due to lack of oxygen.

Asphyxiation Interruption of normal breathing caused by lack of oxygen or breathing high concentrations of carbon dioxide or other gases.

Assay A test of a sample or system. An assay might measure a characteristic of a sample (e.g., identity, purity, and activity), a biological response, or an interaction between molecules. The terms "assay," "test," and "method" are often used interchangeably, although the term "assay" is not generally applied to a test of an instrument's performance and generally refers to an analysis of a sample. "Assay" can also be used as a verb, meaning "to determine"; for example, "the technician assayed the sample for protein."

ASTM International (formerly American Society for Testing and Materials) An organization that prepares and distributes standards to promote consistent procedures for measurement.

Atmospheric pressure Pressure due to the weight of the air that comprises the atmosphere and presses down on every object on earth.

Atomic absorption spectrophotometry A spectrophotometric technique based on the absorption of radiant energy by atoms; used to measure the concentration of metals.

Atomic weight See **gram atomic weight**.

Attenuated vaccine A vaccine prepared from live bacteria or viruses that have been weakened so that they elicit an immune response in the recipient, but do not cause disease.

Attenuator An electronic component that reduces the level of the signal from an instrument.

Audit trail (electronic) A secure, computer-generated, time and date-stamped record that allows the reconstruction of a course of events relating to the creation, modification, and deletion of an electronic record.

Authentication Unambiguous personal identification for data access.

Autoclave A laboratory pressure cooker that sterilizes materials with pressurized steam.

Automatic temperature compensating (ATC) probe A probe placed alongside pH electrodes that automatically reports the sample temperature to the meter.

Auxiliary scales (in thermometry) Extra scale markings at zero degrees and 100°C to assist in calibration and verification of the performance of the thermometer.

Background (in spectrophotometry) Light absorbance caused by anything other than the analyte.

Backup server A network computer that can automatically back up data from individual workstations.

Bacterial broth An aqueous mixture of nutrients to support the growth and reproduction of bacterial cells; prepared without a hardening agent.

Baffles (in fume hoods) Adjustable panels that direct the airflow within a fume hood.

Balance An instrument used to measure the weight of a sample by comparing the effect of gravity on the sample to the effect of gravity on objects of known mass.

Balanced salt solution (BSS) An isotonic mixture of inorganic salts in specific concentrations.

Band pass See **spectral bandwidth**.

Basal liquid medium (mammalian cell culture) Growth medium that contains a defined mixture of nutrients dissolved in a buffered physiological saline solution.

Base (for exponents) A number that is raised to a power. In the expression 10^3, the base is 10 and the exponent is 3.

Base (in the context of DNA and RNA) The molecules that distinguish the nucleotides that comprise DNA and RNA. In DNA there are four types of bases: adenine, cytosine, thymine, and guanine. RNA has uracil instead of thymine.

Base (in the context of pH) 1. A chemical that causes H^+ ions to be removed when dissolved in aqueous solutions. 2. A chemical that releases OH^- ions when dissolved in water.

Base pair (in the context of nucleic acids) Two complementary bases that lie across from each other on opposite strands of DNA. Adenine always pairs with thymine, and guanine with cytosine.

Baseline A line that shows an instrument's response in the absence of analyte; a reference point.

Basic properties (in the SI system) The fundamental measured properties for which the SI system defines units. These properties are length, mass, time, electrical current, thermodynamic temperature, luminous intensity, and amount of substance.

Basic research Research studies that are performed in order to understand nature.

Basic solution (in the context of pH) An aqueous solution with a pH greater than 7.

Batch (according to GMP) "[A] specific quantity of a drug or other material that is intended to have uniform character and quality, within specified limits, and is produced according to a single manufacturing order during the same cycle of manufacture."

Batch record An exact copy of a **master batch record**, but with an assigned lot number. The batch record is used to direct the manufacture of a product and is the document in which formulation and manufacturing activities are recorded.

Battery A device that harnesses the potential of electrochemical reactions to do work.

Beam See **lever**.

Beer's law (also Beer–Lambert or Beer–Bouguer law) A rule that states that the absorbance of a homogeneous sample is directly proportional to both the concentration (C) of the absorbing substance and the thickness of the sample in the optical path (b).

Beta-pleated sheet See **secondary structure**.

Bias (relating to measurements) "A systematic error that contributes to the difference between a population mean of the measurements or test results and an accepted or reference value."[1]

Bimetallic expansion thermometers Type of thermometer made of two different metals, each of which expands and contracts to a different extent as the temperature changes.

Bimodal distribution A frequency distribution with two peaks.

Bioaerosol An aerosol which includes biologically active materials.

Bioanalytical method Sometimes loosely refers to any test of a biological material. The FDA defines the term more narrowly (in "Guidance for Industry, Bioanalytical Method Validation") as a method used for the "quantitative determination of drugs and/or metabolites in biological matrices such as blood, serum, plasma, or urine . . . tissue and skin samples" taken from animal and human subjects.

Bioassay Commonly describes any assay that involves cells, tissues, or organisms as test subjects. Bioassays are used for many purposes, for example, in testing the potency of a drug.

---

[1] From ASTM E 131–05 "Standard Terminology Relating to Molecular Spectroscopy."

Bioburden The number of contaminating microbes on a material before that material is sterilized.

Biohazard A potentially dangerous biological organism or material.

Bioinformatics The field in biology that uses computers to analyze molecular data and address biological questions involving large amounts of information.

Biologic According to FDA, antitoxins, antivenins, and venoms; blood, blood components, plasma-derived products; childhood vaccines, including any future AIDS vaccines; human tissue for transplantation; allergenic extracts used for the diagnosis and treatment of allergic diseases and allergen patch tests; cellular products, including products composed of human, bacterial, or animal cells (such as pancreatic islet cells for transplantation); gene therapy products.

Biological activity According to FDA's "Biotechnology Inspection Guide," the level of activity or potency of a product as determined by tests in animals, in cells in culture, or in an in vitro biochemical assay.

Biologics License Application (BLA) A document submitted by a manufacturer to FDA containing information regarding manufacturing methods, testing results, and clinical trial data for a new biologic product. The application must be approved before the product can be marketed.

Biological macromolecule A large and complex molecule that has biological function (e.g., a hemoglobin molecule or a strand of RNA).

Biological safety cabinet Enclosure designed for the containment of biological hazards.

Biological solution A laboratory solution that supports the structure and/or function of biological molecules, intact cells, or microorganisms in culture.

Biometrics A method of verifying an individual's identity based on the measurement of physical features or repeatable actions that are unique to that person (e.g., fingerprint and retinal scan). A signature can be considered to be a biometric method.

Biomolecules Compounds produced by some type of biological source, such as a plant, animal, microorganism, or cultured cell.

Biopharmaceutical (1) A drug product that is manufactured using genetically modified organisms as a production system. (2) Most broadly, any drug manufactured by living cells or organisms (whether or not recombinant DNA techniques are involved); whole cells or tissues used therapeutically; any drug that is a large biological molecule – these are usually proteins, but can be RNA or DNA (e.g., DNA for gene therapy).

Bioreactor A specialized growth chamber used for producing a product in cells (usually refers to mammalian cells) in which conditions of temperature, nutrient level, aeration, pH, and mixing are controlled.

Biosafety level (BSL) Defined by NIH as the combinations of laboratory facilities, equipment, and practices that protect the laboratory, the public, and the environment from potentially hazardous organisms.

Biosafety level 1 (BSL-1) The lowest biosafety level; used when working with well-characterized strains of living microorganisms that are not known to cause disease in healthy adult humans.

Biosafety level 2 (BSL-2) The biosafety level used when working with agents that may cause human disease and therefore pose a hazard to personnel.

Biosafety levels 3 and 4 (BSL-3 and BSL-4) Biosafety levels generally associated with dangerous agents that are highly infectious.

Bioseparation methods Separation techniques that are used to extract, isolate, and purify specific biological products, or biomolecules (such as a particular protein).

Blank (relating to spectrophotometry) A reference that contains no analyte, but does contain the solvent (for a liquid sample) and any reagents that are intentionally added to the sample. The blank is held in a cuvette that is identical to that used for the sample, or the blank is alternately placed in the same cuvette as the sample. A spectrophotometer compares the interaction of light with the sample and with the blank in order to establish the absorbance due to the analyte.

BLAST (basic local alignment search tool) A pattern recognition tool used to search for similarities between nucleotide sequences or protein sequences.

Blood As defined by OSHA, refers to human blood, blood components, and any products made from human blood.

Blood serum The liquid component of blood from which blood cells have been removed.

Bloodborne pathogens As defined by OSHA, pathogenic microorganisms that are present in human blood and can cause disease in humans. These pathogens include, but are not limited to, hepatitis B virus (HBV) and human immunodeficiency virus (HIV).

"Blow-out" pipette A type of pipette that is calibrated so that the last drop is to be forcibly expelled from the tip of the pipette.

Blue litmus A pH indicator dye that changes from blue to red, denoting a change from alkaline to acid.

Boiling point The temperature at which a substance in the liquid phase transforms to the gaseous phase (under specified conditions of pressure).

Borosilicate glass A strong, temperature-resistant glass that does not contain discernable contamination by heavy metals and is used to manufacture general-purpose laboratory glassware.

Brain–computer interface (BCI) A computer-based system that acquires signals from the brain, interprets them, and translates them into commands that are sent to a device that carries out a desired action. BCIs can be used to assist people with paralysis, blindness, and other disabilities.

Breakthrough rate (related to gloves) A measurement of the time required for a chemical that is spilled on the outside of a glove to be detected on the inside of the glove.

Brilliant yellow A pH indicator dye that is yellow at pH 6.7 and changes to red at pH 7.9.

Bromocresol green A pH indicator dye that changes from yellow to blue at pH 4.0 to 5.4.

Bromocresol purple A pH indicator dye that changes from yellow at pH 5.2 to purple at pH 6.8.

Brushes (in centrifugation) A component of a centrifuge motor that conducts electrical current. Because brushes require maintenance, brushless motors are found in newer centrifuges.

BSE (bovine spongiform encephalopathy, also known as "mad cow disease") A fatal, neurodegenerative disease of cattle caused by an infectious prion agent. The disease is thought to be transmissible to humans and other animals.

BTV (bring to volume) The procedure in which solvent is added to solute (typically in a volumetric flask or graduated cylinder) until the total volume of the solution is exactly the final volume desired.

Buffer A substance or combination of substances that, when in aqueous solution, resists a change in H concentration even if acids or bases are added.

Buffered salt solution A solution that is intended to maintain living cells for short periods (minutes to hours) in an isotonic, pH-balanced environment.

Bulb (of a liquid-in-glass thermometer) A thin glass container at the bottom of a thermometer that is a reservoir for the liquid.

Bunny suit Special protective clothing designed to prevent human introduction of particulates into cleanroom facilities.

Buoyancy (principle of) Any object will experience a loss in weight equal to the weight of the medium it displaces.

Buoyancy error The discrepancy between mass and weight. The displacement of air by an object results in a slight buoyancy effect: Air slightly supports the object. The more air an object displaces, the more buoyant it is and the less it appears to weigh on a balance. Standards for balance calibration are typically made of metal with a density of 8.0 g/cm^3. In contrast, water has a density of only 1.0 g/cm^3. Because water is less dense than calibration standards, a 1 kg mass of water takes up more space and displaces more air than a 1 kg mass of metal. Because balances are calibrated with metal weights, there is a slight buoyancy error when less dense objects are weighed.

Buoyant density gradient centrifugation See **isopycnic centrifugation**.

Burette (also spelled "buret") A long graduated tube with a stopcock at one end that is used to accurately dispense known volumes.

Bypass fume hood Fume hood that has an opening at the top of the hood behind the sash, allowing air to enter the hood and bypass the working face, restricting face velocity when the sash is lowered.

Calcium A metallic element often found in water, typically as dissolved calcium carbonate; causes water hardness.

Calibration (1) To adjust a measuring system to bring it in accordance with external values. (2) A process that establishes, under specified conditions, the relationship between values indicated by a measuring instrument or measuring system and values of a trustworthy standard. Calibration permits the estimation of the uncertainty of the measuring instrument, or measuring system. The result of a calibration is recorded in a document.

Calibration (of an ion-sensitive electrode) The use of standards with known concentrations of the ion of interest to determine the relationship between the voltage response of the electrodes and the ion concentration in the sample.

Calibration (of a pH meter) The use of standards of known pH to determine the relationship between the voltage response of the electrodes and the pH of the sample.

Calibration (of a spectrophotometer) (1) The use of standards with known amounts of analyte to determine the relationship between light absorbance and analyte concentration. (2) Bringing the transmittance and wavelength values of a spectrophotometer into accordance with externally accepted values.

Calibration (of a thermometer) The process by which the display of a thermometer is associated with reference temperature values.

Calibration (of a volume-measuring device) (1) Placement of capacity lines, graduations, or other markings on the device so that they correctly indicate volume. (2) Adjustment of a dispensing device so that it dispenses accurate volumes.

Calibration curve See **standard curve**.

Calibration standard (for a balance) Objects whose masses are established and documented.

Calomel A paste consisting of mercury metal and mercurous chloride, Hg/Hg_2Cl_2.

Calomel electrode A type of reference electrode containing calomel. (This type of electrode is no longer commonly available due to hazards of mercury.)

Cancer A disease that is characterized by the uncontrolled growth of cells in different body organs.

Capacitor A component of electronic circuits that stores electrical energy.

Capacity The relative volume of sample that can be processed simultaneously by a technique; also refers to the amount of the product of interest that can be separated by the technique.

Capacity (for a balance) Maximum load that can be weighed on a particular balance as specified by the manufacturer.

Capacity line A line marked on an item of glassware to indicate the volume if the item is filled to that mark.

Capillary electrophoresis (CE) A technique that separates molecules as they move through the inside of very narrow tubes called capillary tubes.

Carbon adsorption A method of removing dissolved organic contaminants from water by using activated carbon.

Carboxyl group Part of the core structure of amino acids, COOH.

Carcinogen A compound that is capable or suspected of causing cancer in humans or animals.

Carpal tunnel syndrome A painful medical condition resulting from repetitive wrist and hand movements; caused by pressure on nerves due to deep tissue swelling in the wrist; an example of a repetitive stress injury.

Carrier An infected individual who is capable of spreading the infecting agent to other hosts; this individual may or may not show symptoms of disease.

Carryover Material from one sample that is carried to and contaminates another sample.

CAR T cell therapy A form of adoptive cell therapy.

Cathode A negative electrode.

Cation A positively charged ion.

Cation exchange chromatography Ion exchange chromatography method where the stationary phase particles are coated with negatively charged molecules to adsorb positively charged molecules (cations) that flow by in the mobile phase.

Cation exchange resin A solid matrix material that has positively charged ions available for exchange with positively charged ions in a solution.

Ceiling limit The air concentration of a chemical that is not to be exceeded at any time.

Cell constant (K) (for conductivity measurements) The ratio of the distance between the electrodes to the area of the electrodes.

Cell culture The process in which living cells derived from the tissue of multicellular animals are maintained in nutrient medium inside petri dishes, flasks, or other vessels.

Cell homogenate A suspension of cell contents in liquid, produced by disrupting the outer cell membrane and wall (if present) and some of the interior structure of the cell.

Cell line Cells in culture derived from a common ancestor cell that have acquired the ability to multiply indefinitely.

Cell-based assay Any assay that involves living cells as the test subjects, a type of bioassay. Cell-based assays are used, for example, to test a material for viral contaminants, to look for a response to a drug product, and to look for toxic effects of a compound.

Cell therapy Clinical methods that treat disease or injury with the use of whole cells or tissues.

Celsius scale (C) (also previously called centigrade scale) A temperature scale defined so that 0°C is the temperature at which pure water freezes and 100°C is the temperature at which water boils.

Center for Biologics Evaluation and Research (CBER) An FDA division that regulates biologics for human use.

Center for Devices and Radiological Health (CDRH) An FDA division that regulates medical devices and many, but not all, in vitro diagnostic kits.

Center for Drug Evaluation and Research (CDER) An FDA division that regulates drug products (including many, but not all, biotechnology products) such as insulin, heparin, aspirin, and erythropoietin.

Center for Food Safety and Applied Nutrition (CFSAN) An FDA division that promotes and protects the public's health and economic interest by ensuring that food is safe, nutritious, and honestly labeled.

Centers for Disease Control and Prevention (CDC) An agency of the Department of Health and Human Services whose mission is "to promote health and quality of life by preventing and controlling disease, injury, and disability."

Centi- (c) A prefix meaning 1/100.

Centimeter (cm) A unit of length that is 1/100 of a meter.

Central nervous system (CNS) The biological system that includes the brain, spinal cord, and system of neurons, the target of neurotoxins.

Centrifugal force The force that pulls a particle away from the center of rotation, in classical mechanics.

Centrifuge An instrument that generates centrifugal force, commonly used to help separate particles in a liquid medium from one another and from the liquid.

Certification (according to ISO) "Certification refers to the issuing of written assurance (the certificate) by an independent, external body that has audited an organization's management system and verified that it conforms to the requirements specified in the standard. Registration means that the auditing body then records the

certification in its client register. For practical purposes, the difference between the two terms is not significant and both are acceptable for general use."

Certified reference material Any reference material issued with documentation. See standard reference material.

Chain of custody Refers to the controlled and documented sequence of handling, storage, and disposition of a sample.

Change of state indicators Products that change color or form when exposed to heat.

Characterization (in the context of drugs) Process in which a molecular entity's physical, chemical, and functional properties are determined using specific assays. The qualities of the entity are then defined in terms of the results of those assays.

Checkpoint inhibitor A type of drug used to treat cancer that blocks proteins called checkpoints. Checkpoints are proteins made by some types of immune system cells, such as T cells, and some cancer cells. These checkpoints normally help keep immune responses from being too strong, but they can also keep T cells from killing cancer cells. When these checkpoints are blocked by checkpoint inhibitor drugs, T cells can kill cancer cells more effectively.

Chelator An agent that binds to and removes metal ions from solution.

Chemical Abstracts Service (CAS) number Identification number assigned to every commercial chemical compound.

Chemical corrosion A chemical reaction that causes a metal surface to become rusted or pitted.

Chemical fume hood A well-ventilated, enclosed, chemical- and fire-resistant work area which provides users access from one side.

Chemical Hygiene Plan (CHP) A written manual which outlines specific information and procedures necessary to protect workers from hazardous chemicals.

Chemical spill kits Preassembled materials for controlling and cleaning up small- to medium-sized laboratory spills.

Chemically defined growth medium An aqueous medium to support the growth and reproduction of cells that contains only known ingredients in known quantities.

Chinese hamster ovary (CHO) cells A cell line derived from the ovary of a Chinese hamster that is commonly used for biopharmaceutical production.

Chip A thin piece of silicon that contains all the components of an electrical circuit.

Chromatin The native form of DNA that is composed of DNA bound to several types of proteins.

Chromatography A group of bioseparation techniques based on the differential interaction of molecules between a stationary phase and a mobile phase. Because molecules differ in their relative attraction to the mobile phase, they will move past the stationary phase at differing rates.

Chromatophore, chromophore An atom or group of atoms or electrons in a molecule that absorb light.

Chromosomes Long DNA macromolecules that contain genes, regulatory sequences, and stretches of DNA of unknown function.

Chronic exposure Long-term, continuous or intermittent contact.

Chronic health hazard Biological harm resulting from long-term contact, or continuous or intermittent exposure to a hazard.

Circuit A complete path for current flow.

Circuit board A card on which components are mounted and connected to one another to form a functional unit.

Circuit breaker A safety device that limits the current flowing in a circuit.

Clarification (in bioseparations) The removal of unwanted solid matter after a bioseparation procedure, usually by centrifugation or filtration.

Clarification (in cell culture) The removal of cells and cellular debris from a culture medium.

Clean bench A type of laminar flow cabinet designed to provide a sterile work surface, but not worker protection.

Clean-in-place (CIP) The process of cleaning large, nonmovable items (e.g., fermentation vessels) in their place. CIP often involves pumping and circulating cleaning agents through the items.

Cleanroom An area with a temperature-, pressure-, and humidity-controlled environment in which the levels of particulate contaminants, including dust, aerosols, vapors, and microorganisms, are significantly reduced.

Clearance studies (1) Studies of a process in which a contaminant (e.g., a virus) is intentionally added to a system in order to test the ability of the process to remove it. (2) In pharmacokinetics, a candidate drug substance is administered to an organism and its clearance from the body is evaluated.

Clearing factor (k) (in centrifugation) A measure of the time required to sediment a particle in a particular rotor and under specified conditions.

Clearing time (t) (in centrifugation) The time required to sediment a particle. $t = k/S$, where t = time in hours to move particles from the top of tube to the bottom, k = clearing factor (provided by the manufacturer), and S = the sedimentation coefficient (in svedberg units, S).

Clinical and Laboratory Standards Institute (CLSI) (formerly NCCLS) An organization that promotes voluntary consensus standards for clinical laboratory testing of patient samples.

Clinical development The stages of drug development in which a candidate drug is tested in human volunteers.

Clinical laboratory Per OSHA, a workplace where diagnostic or other screening procedures are performed on blood or other potentially infectious materials.

Clinical trials See **clinical development**.

Clone A copy. (1) An exact copy of a DNA segment produced using recombinant DNA technology. (2) One or more cells derived from a single ancestral cell. (3) One or more organisms derived by asexual reproduction that is/are genetically identical (or nearly identical) to a parent.

Cloning Commonly refers to the production of one or more animals that have identical genetic information.

Closed system (in computers) A computer system in which access is controlled by the people who are responsible for the content of the system's records. For example, a system of computers that is only accessible to the individuals who work in a company is a closed system.

Code of Federal Regulations (CFR) A numerical system for the classification and identification of all federal regulations; all legally established federal regulations have a CFR number.

Coefficient (as relates to scientific notation) The first part of a number expressed in scientific notation. For example, the number 235 in scientific notation is expressed as 2.35×10^2, where "2.35" is the coefficient.

Coefficient of variation (CV) (relative standard deviation [RSD]) A measure that expresses the standard deviation in terms of the mean.

Coenzyme A complex organic molecule required by an enzyme for activity.

Cofactor A chemical substance required by an enzyme for activity. Cofactors include salt ions, such as Fe^{++}, Mg^{++}, and Zn^{++}.

Colony-forming unit (CFU) A unit that measures the number of bacteria in a sample. Measuring CFUs requires incubating a sample and counting the resulting colonies; each colony is assumed to have been derived from one bacterium.

Colorimeter An instrument used to measure the interaction of visible light with a sample.

Colorimetry Technique for measuring color.

Column chromatography Chromatography performed with the stationary phase packed into a cylindrical container that varies in width and length.

Combination electrode A measuring electrode and reference electrode that are combined into one housing.

Combustible A substance that can vigorously and rapidly burn under most conditions.

Comité International des Poids et Mesures/ International Committee of Weights and Measures (CIPM) An international body of eminent scientists who act together as authorities on matters related to measurement science and measurement standards.

Common logarithm (also log or log 10) The common log of a number is the power to which 10 must be raised to give that number. For example, $1{,}000 = 10^3$, so the log of 1,000 is 3. The log of 5 is approximately 0.6990, which means that $10^{0.6990} \approx 5$.

Common name The brand name, trade name, or name in common use for a chemical compound or mixture.

Compendium (of standard methods) A published collection of accepted methods in a particular field.

Complete immersion thermometer A liquid expansion thermometer that is designed to indicate temperature correctly when the entire thermometer is exposed to the temperature being measured. Compare with **partial immersion thermometer** and **total immersion thermometer**. (Based on definition in ASTM Standard E 344-07 "Terminology Relating to Thermometry and Hydrometry.")

Complete medium (mammalian cell culture) A medium that has all its components, including supplements and components that are added just before use.

Complex growth medium (1) Bacterial cell culture: An aqueous medium to support the growth and reproduction of cells that contains ingredients whose exact composition is unknown (e.g., contains extracts from plant and animal tissues). (2) Mammalian cell culture: Loosely refers to a medium that contains more ingredients than basal media.

Compound A substance composed of atoms of two or more elements that are bonded together.

Concentration A ratio where the numerator is the amount of a material of interest and the denominator is the volume (or sometimes mass) of the entire solution or mixture. For example, if 1 g of table salt is dissolved in water so that the total volume is 1 L, the concentration is 1 g/L. See also **ratio.**

Conception (in the context of patents) The formation, in the mind of the inventor, of the complete invention (as defined in the patent claim).

Condenser (in distillation) The site where cooling water lowers the temperature of water vapor, causing it to condense back to a liquid form.

Conductance A measurement of conductivity; the reciprocal of the resistance in ohms when measured in a $1\,cm^3$ cube of liquid at a specific temperature. Conductance = 1/resistance. The units are: 1/ohm = 1 mho = 1 siemens (S). 1 micromho (mho) = 1 mho/1,000,000.

Conductivity The inherent ability of a material to conduct electrical current. Used in water treatment to determine the level of ionized impurities present.

Conductivity meter A measuring instrument that measures the conductivity of a solution.

Conductor A material that offers little resistance to the flow of electricity.

Conférence Générale des Poids et Mesures/General Conference on Weights and Measures (CGPM) A body consisting of representatives of the governments that have subscribed to the Convention du Mètre.

Confidence interval A range of values (calculated based on the mean and standard deviation of a sample) which is expected to include the population mean with a stated level of confidence.

Congo red pH indicator dye that changes from red at pH 3.0 to blue at pH 5.0.

Constant Number in a particular equation that always has the same value.

Constant air volume (CAV) fume hood A fume hood that has a constant airflow through the exhaust duct; in these hoods, raising and lowering the sash changes the face velocity at the sash opening.

Containment The control of hazards and reduction of risk by isolation of the organism from the worker; see **primary containment** and **secondary containment**.

Contaminated Per OSHA, refers to the presence or the reasonably anticipated presence of blood or other potentially infectious materials on an item or surface.

Continuous density gradient A density gradient where there is a smooth increase in density from top to bottom with no sharp boundaries between layers.

Continuous flow centrifugation A mode of centrifugation (requiring special equipment) in which sample flows continuously into and out of the spinning rotor.

Contract research organization (CRO) A company contracted by a sponsor to perform preclinical or clinical drug testing.

Contraction chamber (of a liquid-in-glass thermometer) An enlargement of the capillary bore that holds some of the liquid volume, thereby allowing the overall length of the thermometer to be reduced.

Control (1) *Evaluation:* An evaluation to check, test, or verify; an item used to evaluate or verify a process, method, or experiment. (2) *Authority:* The act of guiding, directing, or managing. (3) *Stability:* A state in which the variability in a process is attributable only to chance (i.e., to the normal variability inherent in the process). (Based on ASTM Standard E 456-06 "Standard Terminology Relating to Quality and Statistics.")

Control chart A graphical display of test results together with limits in which the values are expected to lie if the process is in control. Control charts are useful in distinguishing between the normal variability inherent in a process and variability due to a problem or unusual occurrence.

Control chart limits Boundary lines on a control chart that define the limits of acceptable values based on standard deviations as determined in preliminary studies. The *lower warning limit* (LWL) is typically the mean minus two standard deviations, and the *upper warning limit* (UWL) is the mean plus two standard deviations. The warning limits define the range in which 95.5% of all values should lie. The *lower control limit* (LCL) is typically the mean minus three SD, and the *upper control limit* (UCL) is the mean plus 3 SD. These control limits define the range in which 99.7% of all points should lie.

Convention du Mètre A meeting that established a permanent organizational structure for member governments to act together on matters relating to units of measurement.

Coordinates The address for a point on a graph; values that describe the distances horizontally and vertically for the location of a point.

Corex A type of glass that is resistant to physical stress and is often used to make centrifuge tubes.

Corrective and preventive actions (CAPA) The processes by which a company responds to problems and failures. Corrective action means to fix problems that have already occurred and may happen again. Preventive action involves looking for problems that have not yet occurred and preventing them.

Corrosive A substance that will cause tissue damage or destruction at the site of contact.

Counterbalance A weight used in a mechanical balance that compensates for the weight of the sample.

Covalent bond A strong molecular bond formed by the sharing of electrons between atoms.

Covalent peptide bond A bond that connects adjacent amino acids together, thus forming chains; these bonds are relatively strong and are not readily dissociated by environmental changes, such as raised temperature.

COVID-19 (coronavirus disease 2019) A potentially fatal viral disease caused by the SARS-CoV-2 virus that caused a worldwide pandemic beginning late in 2019.

Cresol red pH indicator dye that changes from yellow at pH 7.2 to red at pH 8.8.

Creutzfeldt–Jakob disease A fatal, neurodegenerative human disease thought to be caused by misfolded proteins, prions.

CRISPR A powerful gene editing technology based on a bacterial antiviral system.

Critical speed (in centrifugation) A low speed at which any slight rotor imbalance will cause the rotor to vibrate.

Cross-contamination A situation where cells from one culture accidentally enter another culture.

Cryogenic Extremely cold substances; usually at temperatures below −78°C.

Cubic centimeter (cc, cm^3) A unit of volume that is equal to 1 mL.

Culture media See **growth media**.

Current (I) (electrical) The flow of electrical charge. Current may involve the movement of electrons in a conductor or the flow of ions in a solution.

Current Good Manufacturing Practices (CGMP) The currently accepted minimum standards and requirements for the manufacture, testing, and packaging of pharmaceutical products. "Current" means that a practice may be enforced, even if it is not in the published GMP regulations, if the practice has become accepted by industry.

Cursor A pointer used to indicate your location and select options on a computer screen.

Cushion (in centrifugation) Rubber or plastic pads positioned at the bottom of the compartments in a rotor, or placed underneath adapters, which support tubes and minimize breakage by distributing force over a larger area.

Cuvette A sample "test tube" that is designed to fit a spectrophotometer and is made of an optically defined material that is transparent to light of specified wavelengths.

Cytoplasm The substance of a cell outside the nucleus that is a fluid mixture of water, proteins, lipids, carbohydrates, and salts.

Cytosine (C) One of the four types of nucleotide subunit that comprise DNA.

Cytotoxic Damaging to cells.

Dalton (D) A unit of mass nearly equal to the mass of a hydrogen atom, 1.0000 on the atomic mass scale. A kilodalton (kDa) is a unit of mass equal to 1,000 D.

Dampening device A balance component that slows the oscillations of the moving parts of a balance so that the weight value can be read more quickly.

Dark current A signal that arises in the photodetector electronic circuits when no light shines on its surface.

Data Observations of a variable (singular, **datum**). The unprocessed facts from which we derive information.

Data mining The process of "digging" through very large amounts of data to discover relationships and other information.

Database An organized collection of data that is accessed through **database management software**.

Database management (DBM) software The programs that allow the user to search for, sort, look for patterns, and report selected data within a **database**.

Decontamination The use of physical or chemical means to remove, inactivate, or destroy pathogens on a surface or item to the point where they are no longer capable of transmitting infectious particles and the surface or item is considered safe for handling, use, or disposal (OSHA).

Defined medium See **chemically defined growth medium**.

Degradation rate (related to gloves) A measurement of the tendency of a chemical to physically change the properties of a glove on contact.

Degree (in thermometry) An incremental value or division in a temperature scale. For example, the Celsius scale is divided so that there are 100° between the freezing point and the boiling point of water, whereas the Fahrenheit scale is divided into 180° between the same two reference points.

Deionization A process in which dissolved ions are removed from a solution by passing the solution through a cartridge containing ion exchange resins. The resins are composed of beads that exchange hydrogen ions for cations in the solution and hydroxyl ions for anions in the solution. The ionic impurities remain associated with the beads, whereas the hydrogen and hydroxyl ions combine to form water.

Denaturation (of DNA) Separation of the two complementary strands of DNA from one another.

Denaturation (of proteins) Unfolding of a protein so that its normal three-dimensional structure is lost, but its primary structure remains intact.

Denominator The number written below the line in a fraction.

Densitometry A method used to quantify the amount of material on a solid medium (e.g., photographic negative or electrophoresis gel) by measuring its absorbance of light.

Density Mass per unit volume of a substance under specified environmental conditions.

Density gradient centrifugation Techniques for separating molecules based on their rate of sedimentation or their buoyancy in a density gradient. See also **isopycnic centrifugation** and **rate zonal centrifugation**.

Deoxyribonucleic acid (DNA) A linear polymer consisting of four types of molecular subunits, called nucleotides, connected one after another into long strands. DNA provides the molecular basis for inheritance, which is the passing of traits from parent to offspring.

Department of Transportation (DOT) An agency that regulates the transportation of hazardous materials.

Dependent variable A variable whose value changes depending on the value of the independent variable. For example, if an experimenter measures the growth of seedlings under different light intensities, the growth of the seeds is the dependent variable, and the light intensity is the independent variable. See also **variable** and **independent variable**.

Depth filter A filter made of matted fibers or sand that retains particles through its entire depth by entrapment.

Derate (in centrifugation) A situation where a rotor must be run at a speed lower than its originally specified maximum speed.

Dermatitis Redness, inflammation, or irritation of the skin.

Descriptive statistics Statistical methods that are used to describe and summarize data.

Detection limit (of a detector) The minimum level of the material or property of interest that causes a detectable signal.

Detector (general) An electronic transducer that generates an electrical signal in response to a physical or chemical property of a sample.

Detector (in a spectrophotometer) A device used to measure the amount of light transmitted through a sample.

Detergent Substances that have both a hydrophobic "tail" and a hydrophilic "head." Detergents therefore have two natures; they can be both water-soluble and lipid-soluble.

Deuterium arc lamp Source that produces light in the UV region, from about 185 to 375 nm.

Deviation (1) (of a data point) The difference between a data point and the mean. (2) (in a quality context) An unexpected occurrence. (3) (of a measurement) "The difference between a measurement … and its stated value or intended level."

Dewar flask A heavy multiwalled evacuated metal or glass container, used to hold cryogenic liquids.

Dialysis A separation method based on differences in the concentrations of solutes between one side of a membrane and the other.

Differential centrifugation A mode of centrifugation in which the sample is separated into two phases: a pellet consisting of sedimented material and a supernatant.

Differential medium A type of culture medium designed to reveal differences among microorganisms or groups of microorganisms that are growing on the same substrate.

Differentiation The process in which an embryonic cell, which originally has the capacity to become any type of cell in the body, matures into a particular cell type with a specialized structure and function (e.g., muscle, nerve, and skin).

Diffraction Bending of light that occurs when light passes an obstacle or narrow aperture.

Diffraction grating A device consisting of a series of evenly spaced grooves on a surface that is used to separate polychromatic light into its component wavelengths.

Digital Discontinuous measurement values (in contrast to continuous, analog values) as, for example, stored and displayed by a computer.

Digital certificate A physical item such as an identification card or USB plug-in that confirms your identity to a computer.

Digital filter A device used in electronic balances to stabilize their readout in the presence of drafts or vibrations.

Digital microliter pipettor An instrument used to dispense volumes in the microliter range that can be adjusted to deliver different volumes.

Digital-to-analog converter A signal processing device that converts a digitized signal to an analog signal.

Diluent A substance used to dilute another. For example, when concentrated orange juice is diluted, the diluent is water.

Dilution Addition of one substance (often, but not always, water) to another to reduce the concentration of the original substance.

Dilution series A group of solutions that have the same components, but at different concentrations.

DIN (Deutsche Industrial Norms) A German agency that provides engineering and measurement standards.

Diode An electronic component that allows electricity to flow in only one direction.

Direct current (DC) Current that always flows in the same direction.

Directory A search tool with a collection of links organized by topic.

Disinfection Destruction of most, but not all, microorganisms by means of heat, chemicals, or ultraviolet light.

Dispersing element (in a spectrophotometer) A part of a monochromator that separates polychromatic light into its component wavelengths.

Dispersion The separation of light into its component wavelengths.

Display (in measurement) Devices, such as meters, strip chart recorders, and computer screens, that display information in a form that is interpretable to a human or a computer.

Dissolved inorganics Water contaminants, not derived from plants, animals, or microorganisms, that usually dissociate in water to form ions.

Dissolved organics Water contaminants broadly defined to contain carbon and hydrogen. Organic materials in water may be the result of natural vegetative decay processes or may be human-made substances such as pesticides.

Distillation A water purification process in which impurities are removed by heating the water until it vaporizes. The water vapor is then cooled to a liquid and collected, leaving nonvolatile impurities behind.

Distribution The pattern of variation for a given variable.

Disulfide bond A covalent bond that forms in proteins when sulfurs from two cysteine molecules bind to one another with the loss of two hydrogens.

DNA fingerprinting A technique for distinguishing individuals based on differences in their DNA sequences.

DNA vaccines Vaccines that are made from vectors that have been genetically engineered to include the DNA coding for one or two specific proteins from the infectious agent. When injected into the cells of a person (or other animals), the DNA is expressed, leading to the synthesis of proteins from the infectious agent. The recipient's immune system responds to the new proteins by mounting a protective immune response.

DNase Deoxyribonuclease, a class of enzyme that breaks down the nucleotides in DNA.

Documentation Written records that guide activities and that record what has been done.

Dot diagram (plot) Simple graphical technique to represent a data set in which each datum is represented as a dot along an axis of values.

Double-beam spectrophotometer A type of spectrophotometer in which the sample and blank are placed simultaneously in the instrument so the absorbance of the sample can be continuously and automatically compared to that of the blank.

Double junction An electrode junction configuration used when the sample is incompatible with the reference electrode filling solution. A double junction separates the reference electrode filling solution from the sample.

Downstream (in filtration) The side of a filter facing the filtrate.

Downstream processes The separation procedures that result in a purified product.

Downstream processing The stages of processing, including isolation and purification, of a desired product that takes place after the product is made by a fermentation or cell culture process (upstream processing).

dpK$_a$/dt The change in the pH of a buffer in pH units with the change in its temperature in degree Celsius. The larger this value, the more the buffer will change pH with each degree of temperature change. Negative values for dpK$_a$/dt mean that there is a decrease in pH with an increase in temperature, and vice versa.

Driver (in computers) Software that allows the computer to control a peripheral device.

Drug Generally used synonymously with "pharmaceutical" although "drugs" include not only therapeutic compounds, but also agents that are used in the body for nontherapeutic (e.g., "recreational") or harmful purposes.

Drug discovery Methods for identifying new therapeutic agents.

Drug product (DP) A finished dosage form (e.g., tablet or vial) that contains a drug substance.

Drug substance (DS) The active ingredient in a drug product. See also active pharmaceutical ingredient.

Dry ice Frozen carbon dioxide in solid form.

Dynamic range (of a detector) The range of sample concentrations that can be accurately measured by the detector.

Efficacy (in the context of drugs) The ability of a drug to control or cure an illness or injury.

Electrical ground A conducting material that provides a pathway for current to the earth.

Electrical potential The potential energy of charges that are separated from one another and attract or repulse one another; measured in the unit of volts. Also commonly termed *voltage* and *electromotive force.*

Electrical shock The sudden stimulation of the body by electricity, when the body becomes part of an electrical circuit.

Electrochemical reaction Chemical reaction that occurs when metals are in contact with electrolyte solutions.

Electrode A metal and an electrolyte solution that participate in an electrochemical reaction.

Electrolyte A substance, such as an acid, base, or salt, that releases ions when dissolved in water.

Electrolyte solution A solution containing ions that conducts electrical current.

Electromagnetic radiation A form of energy that travels through space at high speeds. Electromagnetic radiation is classified into types based on wavelength including gamma rays, X-rays, ultraviolet (UV) light, visible (Vis) light, infrared (IR) light, microwaves, and radio waves.

Electromagnetic spectrum The range of all types of electromagnetic radiation from radiation with the longest wavelengths to those with the shortest.

Electromotive force (EMF) See **electrical potential**.

Electronic balance An instrument that determines the weight of a sample by comparing the effect of the sample on a load cell with the effect of standards.

Electronic laboratory notebook (ELN) A tablet computer with software that allows a computer to take the place of conventional paper laboratory notebooks.

Electronic records Text, graphics, data, audio, and pictorial information that is created, modified, maintained, archived, retrieved, or distributed by a computer system.

Electronic signature A computer equivalent to a handwritten signature. In its simplest form, can be a combination of a user ID plus password. It may also include identification based on biometric characteristics.

Electronics Instruments with components such as transistors, integrated circuits, and microprocessors that amplify, generate, or process electrical signals.

Electrophoresis A class of techniques in which molecules are separated from one another based on differences in their mobility when placed in a gel matrix and subjected to an electrical field.

Electroporator An instrument used to introduce drugs, genetic material, and other molecules into living cells suspended in an aqueous medium, by exposing the cells to a pulse of high voltage.

Electrostatic interactions Attractions between positive and negative sites on macromolecules. These weak interactions bring together amino acids and stabilize protein folding.

Eluent The mobile phase in chromatography.

Elution The passage of molecules through a chromatographic column, as the molecules in the sample distribute themselves between the two phases according to their affinities.

Elutriation A specialized centrifugation method in which the sedimentation of particles in a centrifugal field is opposed by the flow of liquid pumped toward the center of rotation. The flow rate and the centrifugation speed can be adjusted in such a way as to wash the cells out with the flowing liquid. This method minimizes the forces experienced by the particles and is useful for isolating fragile, living cells.

Embryotoxin A substance that is harmful to the developing fetus while showing little effect on the mother.

Emergent stem correction (thermometry) A method that corrects for the error introduced when a liquid-in-glass thermometer is immersed to a different depth than that at which it was calibrated.

Emission wavelength In fluorescence, the wavelength of light emitted by the sample as it fluoresces.

Encryption Creating encoded files for storage and transmission of data.

Encryption software Programs that apply mathematical keys to encode (sender) and decode (recipient) encrypted e-mails and files.

Endotoxin Materials derived from lipopolysaccharides that are released from the breakdown of gram-negative bacteria and that trigger a dangerous immune response and induce fever in animals. Can negatively affect the growth of cultured cells.

Endotoxin units/mL A measure of the concentration of pyrogens in water.

Energy The ability to do work.

Engineering controls As defined by OSHA, biohazard controls (such as sharps disposal containers) that isolate or remove the bloodborne pathogen hazards from the workplace.

Entrance slit width (in spectrophotometry) The size of the slit through which light enters the monochromator after being emitted by the source.

Entry routes A method of entry into the body, such as the mouth, lungs, or absorption through the skin.

Environmental Protection Agency (EPA) A U.S. government agency that, among other responsibilities, is involved in the regulation of environmental releases of genetically modified organisms.

Epitope The part of an antigen that an antibody recognizes and binds.

Equation A description of a relationship between two or more entities that uses mathematical symbols.

Equipment validation See validation

Equivalent weight (in reference to acids and bases) For an acid, 1 equivalent is equal to the number of grams of that acid that produces 1 mole of H^+ ions. For a base, 1 equivalent is equal to the number of grams of that base that supplies 1 mole of OH^-.

Equivalent weight (in reference to the concentration of solutes in body fluids) The number of grams of the solute that will produce 1 mole of ionic charge in solution.

Ergonomics The study of the effects of environmental factors on worker health and comfort, and the design of environments to increase worker health and productivity.

Error (1) The difference between a measured value and the "true" value (see also **percent error** and **absolute error**). (2) The cause of variability in measurements.

Ethidium bromide (EtBr) A mutagenic dye that reversibly intercalates into DNA molecules, allowing them to be visualized under ultraviolet light.

Ethylene oxide Reactive cyclic ether gas used for sterilization.

Etiological agent An organism that causes a specific disease in an infected host.

Eukaryotic organism Organisms whose cells have nuclei; plants, animals, and yeast are eukaryotic.

European Agency for the Evaluation of Medicinal Products (EMEA) An agency established by the European Union to coordinate scientific resources in member nations, in order to evaluate and supervise medicinal products for human and veterinary use.

Excitation wavelength In fluorescence, the wavelength of light that is absorbed by the molecule of interest and causes the compound to emit fluorescence.

Exit slit (in spectrophotometry) The slit through which light exits the monochromator.

Exit slit width (in spectrophotometry) The size of the slit through which light emerges from the monochromator.

Exome The sections of the genome that code for proteins.

Exothermic A chemical reaction that releases heat.

Expansion chamber (in thermometry) An enlargement of the capillary bore at the top of the thermometer to prevent buildup of excessive pressure.

Expansion thermometers Thermometers that rely on the expansion and contraction of a material in response to temperature.

Explosion A sudden release of large amounts of energy and gas within a confined area.

Explosive A Substance that is capable of rapid combustion, causing sudden release of heat, gas, and pressure.

Exponent A number used to show that a value (the base) should be multiplied by itself a certain number of times. The expression 10^3 means: $10 \times 10 \times 10$, which equals 1,000. The base is 10, and the exponent is 3.

Exponential equation A relationship between two or more entities whose equation includes a variable that is an exponent, for example, $y = 2^x$.

Exponential notation See **scientific notation**.

Exposure incident Defined by OSHA as a specific eye, mouth, mucous membrane, nonintact skin, or parenteral contact with blood or other potentially infectious materials that results from the performance of an employee's duties.

Expression (in the context of genetics) The process in which a cell makes the protein product encoded by a specific gene.

Expression system A host organism that has taken up a vector containing a gene of interest and that produces the protein encoded by the gene.

Extinction coefficient See **absorptivity constant**.

Extractables Contaminants that are leached into water from the materials used to construct filters, storage vessels, filters, tubing, and other items.

Extraction methods Bioseparation techniques based on the fact that molecules differ from one another in their solubility in various liquids.

Face velocity (refers to fume hoods) The rate of airflow into the entrance of the hood, measured in **linear feet per minute (fpm)**.

Fahrenheit scale (F) A temperature scale that sets the freezing point of water at 32° and the boiling point at 212°.

Fast performance liquid chromatography (FPLC) Column chromatography systems used to purify larger biomolecules, such as proteins.

FDA Guidance Documents Documents issued by FDA that contain information about new technologies and discuss concerns that should be addressed by pharmaceutical and related companies.

Federal Hazard Communication Standard (HazCom 2012) A federal standard which focuses on the availability of information concerning employee hazard exposure and applicable safety measures.

Federal Register A daily publication from the U.S. Government Printing Office that contains major revisions to regulations and announcements of proposed new regulations. It also contains announcements of meetings (e.g., FDA or EPA), seminars, and guidance documents.

Feed water The source water that enters a treatment process.

Fermentation (in the context of biotechnology) A process in which a product is produced by the mass culture of cells (usually refers to microorganisms) under controlled conditions.

Fermenter A vessel in which cells (usually microorganisms) are grown under controlled conditions of temperature, nutrient levels, aeration, pH, and mixing.

Fibrous junction A type of pH reference electrode junction composed of a fibrous material such as quartz or asbestos commonly used as a general-purpose junction.

Filling hole An opening in a reference electrode by which filling solution is introduced into the electrode.

Filling solution A solution of defined composition within an electrode. The filling solution sealed inside a pH-measuring electrode is called *internal filling solution,* and it normally consists of a buffered chloride solution. The solution that surrounds the reference electrode metal element is called *reference filling solution.* This electrolyte solution provides contact between the reference electrode metal element and the sample.

Filter A device used to separate components of samples on the basis of size; particles smaller than a certain size pass through a porous filter material; particles larger than a certain size are trapped by the filter.

Filter (in spectrometry) Material that blocks the passage of radiant energy in a particular manner with respect to wavelength.

Filtering (in electronics) Signal processing in which unwanted noise is removed from the signal generated by a detector.

Filter, neutral (in spectrometry) A filter that attenuates radiant energy by the same factor for all wavelengths within a certain spectral region.

Filtrate The fluid and any associated particles that have passed through a filter.

Filtration A separation method in which particles are separated from a liquid or gas by passage through a porous material.

Fire A chemical chain reaction between fuel and oxygen, which requires heat or another ignition source.

Fire triangle The combination of heat, fuel, and oxygen required to start a **fire.**

Fixed angle rotor A rotor that holds tubes at a fixed angle.

Fixed reference points (in thermometry) Physical systems whose temperatures are fixed by some physical process and hence are universal and repeatable, such as phase transitions.

Fixed resistor A resistor that provides a single, set level of resistance in a circuit.

Flammable Any substance that will ignite and burn readily in air.

Flash point The minimum temperature at which a compound gives off sufficient vapors to be ignited.

Flow rate (in filtration) The rate of flow of a gas or liquid through a filter at a given pressure and temperature. The flow rate determines the volume that can be filtered in a given amount of time.

Fluorescence Light emitted by an atom or molecule after it absorbs light with a shorter wavelength.

Fluorescence spectrometer An instrument used to analyze a sample based on its fluorescence.

Fluorophore A molecule that fluoresces when exposed to light.

Food and Drug Administration (FDA) The federal agency empowered by Congress to regulate the production of cosmetics, electronic products (e.g., X-ray machines), food, medical devices, diagnostic devices, and pharmaceuticals for human and veterinary use.

Food, Drug, and Cosmetic Act (FDCA) An act passed by the U.S. Congress in 1938 that requires that food and drugs not be "adulterated" and empowers the FDA to enforce food and drug laws.

Formula weight (FW) The weight, in grams, of one mole of a given compound. The FW is calculated by adding the atomic weights of the atoms that compose the compound.

Formulation The form in which a drug is administered to patients including its chemical components, the system of administration (e.g., pill or injection), and the mechanism by which the drug is targeted to its site of action.

Fouling When contaminants, such as packed bacteria, form a crust on a filter, membrane, or other surfaces, thus blocking further flow.

Freezing point The temperature at which a substance goes from the liquid phase to the solid phase.

Freezing point depression A lowering of the freezing point of a liquid due to the addition of impurities.

Freezing point depression osmometer A common device for measuring the osmotic pressure of a solution based on the fact that dissolved particles depress the temperature at which the water freezes. The temperature at which a sample freezes is compared to standards of known osmolality.

Frequency (in statistics) The number of times a particular value or range of values is observed in a data set.

Frequency (in electricity) The rate at which alternating current cycles back and forth, measured in cycles/sec or Hertz (Hz).

Frequency (of electromagnetic radiation) The number of waves of electromagnetic radiation that pass a given point per second, expressed in the unit of Hertz (Hz).

Frequency distribution A listing of the number of times each value for a variable occurs.

Frequency histogram A graphical display of data in which the frequency of each value or range of values is plotted as a bar.

Frequency polygon A graphical display of data in which the frequency of each value or range of values is plotted as a point.

Fritted junction (in pH measurement) A type of relatively slow-flowing pH reference electrode junction composed of a ceramic material consisting of many small particles pressed closely together.

Fume hood See **chemical fume hood**.

Functional units (in electronics) Combinations of electronic components connected to one another in various configurations that perform a particular function.

Fuse A safety device that limits the current flowing in a circuit.

Gain The degree to which an input voltage is amplified to become an output voltage.

Galvanometer An instrument that measures small electrical currents.

Gas chromatography (GC) Chromatography method in which the mobile phase is a gas.

Gas–liquid chromatography (GLC) Chromatography method in which a gas mobile phase is paired with a liquid stationary phase.

Gel filtration See **gel permeation chromatography**.

Gel permeation chromatography (GPC) A chromatographic technique for separating molecules by relative size and molecular weight using a column filled with porous gel particles.

GenBank An online database, maintained by NCBI, that contains virtually all reported DNA sequences.

Gene A sequence of DNA that occupies a specific location on a chromosome and codes for a particular characteristic in an organism.

Gene editing A technology for making targeted changes to DNA nucleotide sequences in cells.

Gene therapy Replacing a gene that is missing or correcting the function of a faulty gene in order to cure an illness.

Genetic engineering Methods that allow a gene from one organism to be transferred to another. Can also apply to methods used to remove a native gene from an organism.

Genetically modified organism (GMO) An organism into which genetic information from another organism has been transferred.

Genome-wide association study (GWAS) Studies that look at genomic markers in large numbers of people in order to identify variations associated with complex disorders.

Genomics Study of the genetic makeup of organisms including the base sequence and map of their DNA.

Germicidal Methods that are capable of killing bacteria or other microorganisms.

GFI (ground fault interrupt) circuits Safety circuits designed to shut off electric flow into the circuit if an unintentional grounding is detected; these are usually installed around sinks and other water sources.

Giga- The prefix for one billion, 10^9.

Gigabyte Approximately 1 billion bytes.

Gigahertz (GHz) One billion electrical cycles (**hertz**) per second; used to measure clock speed.

Glass electrode See **pH-measuring electrode**.

Glove box A gas-tight cabinet designed to provide total containment for extremely hazardous biological agents.

Glycoprotein A protein that has one or more attached sugars.

Glycosylation A common, important type of protein modification in which sometimes complex, branched carbohydrates are attached to a protein after it is assembled from amino acid building blocks.

Good Clinical Practices (GCP) The procedures and practices that govern the performance of clinical trials of potential pharmaceutical products.

Good Laboratory Practices (GLP) (1) FDA: The procedures and practices that must be followed when performing laboratory studies in animals in order to investigate the safety and toxicological effects of new drugs. (2) EPA: The procedures and practices that must be followed when investigating the effects of pesticides and other agrochemicals. (3) With small letters (i.e., good laboratory practices [glp]): A general term used to refer to all quality practices in any laboratory.

Good Manufacturing Practices (GMPs) See **Current Good Manufacturing Practices**.

Good Microbiological Laboratory Practices (GMLP) The basic practices that should be used when working with all microbiological organisms.

Gradient elution (in chromatography) Elution method in which the concentration of components in the mobile phase changes during the separation process to progressively elute more components from the stationary phase.

Gradient maker An apparatus used to make a continuous density gradient.

Graduated cylinders Cylindrical vessels calibrated to deliver various volumes.

Graduations Lines marked on glassware, plasticware, and pipettes that indicate volume.

Gram (g) The basic metric unit for mass.

Gram atomic weight The weight, in grams, of 6.02×10^{23} (Avogadro's number) atoms of a given element. For example, 1 mole of element carbon weighs 12.0 g.

Gravimetric method A method that involves the use of a balance; gravimetric methods are used to calibrate volume-measuring devices.

Gravimetric volume testing Determination of the "true" volume of a sample based on its weight. The volume of the liquid is calculated from its weight and its density at a particular temperature. This method takes advantage of the high accuracy and precision attainable with modern balances.

Gravity (g) The unit of measure for the rate of acceleration of gravity. The earth's normal gravity is defined as 1 g (or 1 g force).

Ground wire A separate wire attached to the metal frame of an appliance and connected to the earth through the third prong of the plug. Protects an operator in the event of a short circuit.

Growth factors (1) *Bacterial cell culture*: Often refers to components that cannot be synthesized by bacterial cells and must be obtained from the environment. (2) *Mammalian cell culture*: Refers to a molecular entity released by cells that causes other cells to proliferate.

Growth media Solutions that support the survival and growth of cells in culture. These solutions include nutrients, vitamins, salts, and other required substances.

Guanine (G) One of the four types of nucleotide subunit that comprise DNA and RNA.

Guard column An inexpensive, disposable chromatography column that protects the expensive HPLC column from contaminants in the sample.

Guidelines As defined by the FDA, documents produced by the FDA that establish principles and practices. Guidelines are not legal requirements; rather, they are practices that are acceptable to and recommended by the FDA.

Guidelines for Research Involving Recombinant DNA Molecules As defined by NIH, an influential document written by the Recombinant DNA Advisory Committee that covers topics such as methods of assessing the risk of a recombinant DNA experiment, methods of classifying organisms based on risk, and methods of containing recombinant organisms both on an experimental scale and in production settings.

GXP A term that includes CGMP, GLP, and GCP.

Half-life The time it takes for half of the initial number of radioactive atoms in a given sample to undergo radioactive decay.

Hardness (in reference to water) An indication of the concentration of calcium and magnesium salts in water.

Hardware The solid objects needed for computer functions.

Hazard The equipment, chemicals, and conditions that have a potential to cause harm.

Hazard Analysis and Critical Control Points (HACCP) A quality program implemented in the food industry that emphasizes reducing hazards throughout the production, slaughter, processing, and distribution of foods.

Hazard diamond system A system developed by the **National Fire Protection Association** to rate chemicals according to their flammability, reactivity, and general health hazards.

Hazardous Materials Identification System (HMIS) An alternate to the hazard diamond system that includes long-term chemical effects.

Heat The energy associated with the disordered motion of molecules in solids, liquids, and gases; thermal energy.

Heating The transfer of energy from the object with more random internal energy to an object with less.

HEPA filter A high-efficiency filter for particulate matter in air.

Hepatitis B A bloodborne infection that attacks the liver and causes inflammation.

Hepatitis B virus (HBV) The etiological agent for human hepatitis B.

Hertz (Hz) The unit of frequency of alternating current. 1 Hz = 1 cycle per second.

High-performance liquid chromatography (HPLC) An instrumental technique used to separate, analyze, and sometimes collect the components of a mixture.

High-resolution purification methods Techniques that have relatively high selectivities, but relatively low capacities; see also **low-resolution purification methods**.

High-speed centrifuge A centrifuge which spins samples at rates up to about 30,000 RPM and generates forces up to about $100,000 \times g$.

High-throughput screening (HTS) An automated method used by the pharmaceutical industry to quickly test thousands of compounds for possible therapeutic effects.

Hollow-fiber ultrafilter A cylindrical cartridge packed with ultrafiltration membranes formed into hollow, tubular fibers.

Horizontal laminar flow hood See **clean bench**.

Horizontal rotor (also called swinging bucket rotor) A rotor in which tubes or sample bottles swing into a horizontal position when centrifugal force is applied.

Human genome The entire sequence of nucleotide building blocks that make up the DNA in humans; often compared to a "blueprint" that contains the instructions for building a human being.

Human Genome Project A major government coordinated project that mapped and sequenced the human genome.

Human immunodeficiency virus (HIV) The etiological agent for acquired immunodeficiency syndrome.

Hybrid system (in the context of documentation) A system that uses both electronic and paper records. For example, a laboratory instrument might be attached to a computer that retrieves and processes data from the instrument, and then prints out a result that is signed and dated.

Hybridization The binding of single-stranded DNA or RNA to strands of complementary DNA or RNA.

Hybridoma An immortalized cell line that secretes monoclonal antibodies; results from the fusion of an antibody-producing cell with a cultured myeloma (tumor) cell.

Hydrates Compounds that contain chemically bound water. The weight of the bound water is included in the FW of hydrates. For example, calcium chloride can be purchased either as an anhydrous form with no bound water ($CaCl_2$, FW 111.0), or as a dihydrate ($CaCl_2 \cdot 2H_2O$, FW 147.0).

Hydrogen bond A type of weak chemical bond formed when a hydrogen atom bonded to an electronegative atom (e.g., F, O, or N) is shared by another electronegative atom.

Hydrogen ion H^+.

Hydrophilic "Liking" water; a substance that readily dissolves in water.

Hydrophilic (in filtration) A filter's ability to wet with water.

Hydrophobic Molecules that are relatively insoluble in aqueous liquids; water-hating. *In filtration*: A filter's resistance to wetting with water.

Hydrophobic interaction chromatography (HIC) A chromatographic technique for separating molecules based on their **hydrophobic** properties.

Hydroxide ion OH⁻.

Hygroscopic compound A compound that absorbs moisture from the air.

Hyperventilation An increased respiration rate that can induce dizziness.

Hypoallergenic A product label indicating that the product is less likely to trigger allergic reactions than similar products; note that this does not mean the product is nonallergenic.

Ice point check (in thermometry) A method of verifying the performance of a thermometer by checking that it reads 0°C (or 32°F) when it is in an ice bath.

Immediate danger to life and health (IDLH) Designation that indicates environmental conditions requiring respirator use and maximum personal protective equipment.

Immersion (of a liquid-in-glass thermometer) The depth to which a liquid-in-glass thermometer is immersed in the material whose temperature is to be measured.

Immersion line (in thermometry) A line etched onto the stem of a liquid-in-glass thermometer to show how far it should be immersed in the material whose temperature is to be measured.

Immunoassays A group of test methods based on the interaction between antibodies and the antigens to which they attach. Immunoassays can be used to detect the presence of proteins and to quantify how much is present.

Immunotherapy Any therapy that uses substances to stimulate or suppress the immune system to help the body fight cancer, infection, and other diseases.

Impervious Preventing passage of an organism or material.

Implosion The violent collapsing of a vessel with an internal pressure lower than the outside atmosphere.

In vitro "In glass"; refers to processes occurring in a test tube or other laboratory media.

In vivo "In life"; refers to processes occurring in a living organism.

Incident light Light that strikes, or shines on, a substance.

Incompatible (in chemical safety) Refers to combinations of chemicals that will react and cause hazardous conditions.

Independence (in sampling) Sampling in such a way that the choice of one member of a sample does not influence the choice of another.

Independent variable A variable whose value is controlled by the experimenter. See also **variable** and **dependent variable**.

Indicator electrode An electrode that responds selectively to a specific type of ion or molecule in solution.

Induced pluripotent stem (iPS) cells Adult cells treated to look and act like embryonic stem cells.

Infection Invasion and multiplication of microorganisms in tissue; may be associated with health effects.

Infectious The ability of an organism to spread to and invade another host organism.

Infectivity assay A method of testing for pathogens in which susceptible cells are grown with the substance being tested to see if the cells become infected.

Inferential statistics Statistical methods that are used to reach conclusions about a population based on a sample.

Inflammable Another term for **flammable** materials.

Information Processed data that have been collected, manipulated, and organized into an understandable form.

Infrared radiation The region of the electromagnetic spectrum from 780 to 2,500 nm.

Infrared thermometer A thermometer that senses the infrared energy emitted by an object and converts the infrared energy into an electrical signal that, in turn, is converted to a temperature reading.

Infringement (in the context of patents) A situation where one party holds a patent on an invention and another party uses that invention without permission.

Ingestion Entering the body through the mouth and digestive tract.

Inhalation Entering the body through the respiratory system and lungs.

Inorganic In water treatment, matter not derived from plants, animals, or microorganisms and that usually dissociates in water to form ions.

Input connector A connection that enables an instrument to receive electrical signals from another device.

Installation qualification (IQ) A set of activities designed to determine if a piece of equipment is installed correctly.

Institute of Laboratory Animal Resources (ILAR) An agency founded under the National Research Council to compile and disseminate information about laboratory animals and their care, and to promote humane care of these animals.

Institutional Animal Care and Use Committee (IACUC) The committee that reviews an institution's programs for humane care and use of animals, inspects institutional animal facilities, and reviews and approves all protocols using live vertebrate animals, among other duties.

Institutional Biosafety Committee (IBC) An on-site committee that reviews and provides oversight for almost all experiments involving recombinant or synthetic DNA molecules.

Institutional review board (IRB) A panel composed of medical, scientific, and nonscientific community members who are responsible for protecting the rights, safety, and well-being of human experimental subjects.

Instrument response time The time required for an indicating device to undergo a particular displacement following a change in the property being measured.

Insulator (in electricity) Materials in which the outer electrons of atoms are not free to move so electricity does not flow readily.

Integrated circuit (IC) A small electronic circuit with multiple components (such as resistors, capacitors, and diodes) built on a chip.

Integrator A signal processing mode in which the area under a peak is calculated.

Intellectual property (IP) A type of property that encompasses creations of the mind and intellect.

Interface Site where a measuring instrument contacts the material being measured.

Interface (in computers) The part of the system software that the user sees and interacts with.

Interference (in the context of an assay) Any substance in a sample that leads to an incorrect result in an analysis.

International Agency for Research on Cancer (IARC) An agency that determines the relative cancer hazard of materials.

International Conference on the Harmonisation of Technical Requirements for the Registration of Pharmaceuticals for Human Use (ICH) An organization that brings together the regulatory authorities of Europe, Japan, and the United States and experts from the pharmaceutical industry to discuss scientific and technical aspects of pharmaceutical product regulation with the purpose of harmonizing pharmaceutical regulatory requirements internationally.

International Organization of Legal Metrology (OIML) A worldwide intergovernmental organization whose aim is to harmonize regulations and controls relating to metrology in its member nations.

International prototype standard Formerly, the internationally accepted unit of mass as embodied in a platinum–iridium cylinder housed in France. The mass of this cylinder is, by definition, exactly 1 kg.

International Temperature Scale of 1990 (ITS-90) An internationally accepted temperature scale established in 1990 by the 18th General Conference on Weights and Measures. The scale uses reference temperatures based on the freezing, boiling, and triple points of various materials including water, hydrogen, and several metals.

Invention The discovery or creation of a new material (e.g., a drug, a protein made by a genetically modified organism), a new process (e.g., a new method of inserting genetic information into cells), a new use for an existing material, or any improvement of any of these.

Investigational New Drug Application (IND) An application filed with the Food and Drug Administration in which a sponsoring organization requests permission to launch clinical trials of an experimental drug.

Ion An atom or group of atoms with a net charge as a result of having lost or gained electrons. See **anion** and **cation**.

Ion exchange The process in which ions in solution are exchanged with similarly charged ions associated with ion exchange resins.

Ion exchange chromatography A chromatographic technique for separating molecules by molecular charge; based on ionic stationary phases and manipulation of salt concentrations and pH in the mobile phase.

Ion-selective electrode (ISE) A class of indicator electrode that responds to a specific ion in solution.

Ionic detergent A **detergent** whose hydrophilic portion is ionized in solution.

Ionic strength A measure of the charges from ions in an aqueous solution.

Irritant A substance that will cause irritation to the skin, eyes, or respiratory system.

ISO A network of the national standards institutes of more than 160 countries, on the basis of one member per country, with a Central Secretariat in Geneva, Switzerland, that coordinates the system.

ISO 9000 A set of internationally accepted standards that outline a system for quality management aimed at ensuring the quality of a product or service. ISO 9000 currently includes three quality standards: ISO 9000:2015 (fundamentals and vocabulary), ISO 9001:2015 (requirements that an organization needs to fulfill if it is to achieve consistent products and services that meet customer expectations), and ISO 9004:2018 (additional guidelines).

Isoelectric point The pH value at which a protein exhibits an overall neutral charge.

Isolation See **purification**.

Isopycnic centrifugation A method in which particles are separated based on their density alone; also called buoyant density centrifugation.

Isotonic solution A solution that has the same solute concentration as the interior of the cell.

Isotropic membrane A membrane in which the pore size is the same on both sides of the membrane.

IU (international unit of enzyme activity) The amount of enzyme necessary to catalyze transformation of $1.0\,\mu mol$ of substrate to product per minute under optimal measurement conditions.

Jack (in the context of electricity) A receptacle for a plug. The plug is inserted into the jack to complete a circuit.

Job safety analysis (JSA) A detailed analysis of each step in a procedure, identifying potential hazards and outlining accident prevention strategies.

Junction A part of a reference electrode that allows filling solution to flow into the solution whose pH is being measured. Also called a **salt bridge**.

Kat The designation for the **SI unit of enzyme activity**.

Kelvin scale A temperature scale based on the Celsius degree with 100 units between the freezing point and boiling point of water; however, the Kelvin scale sets the zero point as absolute zero, rather than the freezing point of water. $0°C = 273.15\,K$. (Units are in kelvin, K.)

Kilo- (k) A prefix meaning 1,000.

Kilogram (kg) A unit of mass equal to $1,000\,g$.

Kilometer (km) A unit of length equal to $1,000\,m$.

Kinetic spectrophotometric assay An assay that measures the changes over time in concentration of reactants or products in a chemical reaction.

Knife edge In mechanical balances, a support on which the beam rests that allows it to swing freely.

Laboratory-acquired infection (LAI) Infections that can be traced directly to laboratory organisms handled or contacted by the infected individuals.

Laboratory informatics The application of computers to collect, analyze, and manage laboratory data and information.

Laboratory information management system (LIMS) Computerized systems that provide integrated information and resource management for a variety of laboratory activities.

Lachrymator A substance that is an eye irritant and stimulates tear formation.

Lambda (λ) An older term sometimes used to mean $1\,\mu L$.

Laminar flow cabinet A cabinet with a sterile, directed airflow.

LAN (local area network) An intranet, generally operating within a building, school, or company. LANs are private and usually limited to a 5-mile radius.

Large-scale bioproduction Production of biological materials in quantities greater than 10 L.

Latency period Time from the first exposure to a toxic agent to the time when biological effects can be detected.

Latex A natural rubber product that is commonly found in gloves and many pieces of laboratory equipment and household items.

LC50 (lethal concentration, 50%) Concentration of a compound in air that will kill 50% of test animals.

LCLo (lethal concentration low) The lowest concentration recorded to cause lethality in test animals.

LD50 (lethal dose, 50%) Amount of a toxic compound given in a single dose that will cause death in 50% of test animals.

LDLo (lethal dose – low) The lowest single dosage known to cause lethality in test animals.

Leach (in the context of water purity) A process in which water dissolves substances from materials over which it flows. For example, minerals are leached as water flows over rocks.

Least squares method A statistical method used to calculate the equation for the line of best fit for a series of points.

Leveling Adjusting the balance to a level, horizontal position.

Leveling bubble A bubble that is used to determine whether the balance is level or not. When the balance is level, the bubble is centered in the window.

Leveling screws The screws used to adjust the balance so that it is level.

Lever (or beam) A rigid bar used in mechanical balances on which the sample to be weighed is balanced against objects of known mass.

L-Glutamine An essential amino acid required by cultured mammalian and insect cells.

License (in the context of patents) An agreement by a patent holder that it will not enforce the right of exclusion against the licensee (the party wishing to use the patented invention).

Light Electromagnetic radiation with wavelengths from 180 to 2,500 nm; includes the ultraviolet, visible, and infrared regions of the electromagnetic spectrum.

Light-emitting diode (LED) A diode that emits light when current flows through it.

Light scattering An interaction of light with small particles in which the light is bent away from its initial path.

Light source (in spectrophotometry) The bulb or lamp that emits the light that shines on the sample.

Limit of detection (LOD) The lowest level of the material of interest that a method or instrument can detect.

Limit of quantitation (LOQ) The lowest level of the material of interest that a method or instrument can quantify with acceptable accuracy and precision.

Limulus amebocyte lysate (LAL) test A test used to quantify the concentration of pyrogens in a sample that requires an extract of blood from the horseshoe crab, *Limulus polyphemus*. Pyrogens initiate clotting in the presence of this blood extract.

Line of best fit A line connecting a series of data points on a graph in such a way that the points are collectively as close as possible to the line.

Linear dispersion of a monochromator A measure of the ability of the monochromator to separate light into its component wavelengths.

Linear feet per minute (fpm) (in fume hoods) Measurement of **face velocity**.

Linear relationship A relationship between two properties such that when they are plotted on a graph the points form a straight line.

Linearity (1) The characteristic of a direct reading device. If a device is linear, calibration at two points (e.g., 0 and full scale) calibrates the device. (Two points determine a straight line.) (2) A measure of how well an instrument follows an ideal, linear relationship.

Linearity (for a balance) The ability of a balance to give readings that are directly proportional to the weight of the sample over its entire weighing range.

Linearity check (in pH measurement) A pH meter system is normally calibrated at two values of pH (e.g., pH 7.00 and pH 10.00). A linearity check tests whether the meter's response is linear through its entire range (e.g., also at pH 4.00).

Lipopolysaccharide (in reference to water quality) Molecules containing carbohydrates and lipids found in the outer wall of gram-negative bacteria. See **pyrogen.**

Liquid chromatography (LC) A general term for any chromatographic technique where the mobile phase is liquid.

Liquid crystal display (LCD) A digital device commonly used to display individual measurement values.

Liquid expansion thermometer See **liquid-in-glass thermometer.**

Liquid nitrogen The liquid form of nitrogen at −198°C, supplied in large compressed gas cylinders.

Liquid-in-glass thermometer A thermometer in which liquid, usually mercury or alcohol, expands or contracts with temperature changes and moves up or down the bore within a glass tube.

Liter (L) A metric system unit of volume. 1 L = 1 dm³. A liter is slightly more than a quart. See also **volume**.

Load (for a balance) The item(s) exerting force on the balance.

Load (in electricity). The electrical demand of a device expressed as power (watts), current (amps), or resistance (ohms).

Load cell A device that uses an electromagnetic force to compensate for, or counterbalance, the force of an object on the weighing pan.

Log, logarithm (log 10) See **common logarithm**.

Log-in Entering a name and password into the computer for identification.

Log-to-linear conversion Processing in which a signal that has a logarithmic relationship to the property being measured is converted to a signal with a linear relationship to the property being measured.

Lot defined by GMP: "[A] batch or a specific identified portion of a batch, having uniform character and quality within specified limits…"

Lot number (also, control number or batch number) An identifying number that is used to distinguish one lot from another.

Low-resolution purification methods Bioseparation techniques that generally have high capacities and can be performed quickly, but exhibit relatively low selectivity; see also **high-resolution purification methods**.

Low-speed centrifuge A centrifuge that spins samples at rates less than about 10,000 RPM.

Lyophilization Freeze-drying; a preservation method commonly used for proteins, which involves rapid freezing and drying under vacuum.

Lyse To break open cells and release their contents.

Lysis buffer A solution whose primary function is to lyse the cell membrane and/or cell wall.

Macrofiltration The removal of relatively large particles, above about 10 μm, from a liquid by passage through porous materials including paper, glass fibers, and cloth.

Magnetic dampening A device that helps stabilize the weighing pan in mechanical balances when samples are placed on it or weights are added to the beam, thus reducing the time required to reach a stable reading.

Manual dispenser (for reagent bottle) A device that dispenses set volumes from a bottle.

Mass The amount of matter in an object, expressed in units of grams. Mass is not affected by the location of the object or the effect of gravity.

Mass spectroscopy (MS) A specialized technique used for identifying and quantitating molecules within a sample, by measuring the mass-to-ionic charge ratios of molecules.

Master batch record Original step-by-step instructions that detail how to formulate or produce a product, including raw materials required, processing steps, controls, and required testing.

Master cell bank (MCB) Stored cells that are the source of cells used in a process. The stored cells are usually derived from a single cell, are fully characterized, and are aliquoted and stored cryogenically to assure genetic stability. The MCB is the source for the working cell bank.

Matrix The physical material in which a sample or analyte is located or from which it must be isolated, for example, blood, soil, or fiber.

Maximum tare The maximum container weight that can be automatically subtracted from the weight of a sample.

Mean The average. The sum of all values divided by the number of values.

Measurand A quantity that is measured (e.g., the volume of a particular sample under specified conditions, such as temperature and pressure).

Measurement Quantitative observation; numerical description.

Measurement system A related group of units, such as inches, feet, and miles. See also **United States Customary System, metric system, and SI system**.

Measures of central tendency Measures of the values about which a data set is centered (e.g., the mean, median, and mode).

Measures of dispersion Measures of the variability in a set of numerical data (e.g., range, variance, and standard deviation).

Measuring pipette A type of pipette calibrated with a series of graduation lines to allow the measurement of more than one volume.

Mechanical balance (laboratory) A balance that uses a lever and that does not generate an electrical signal in response to a sample. The object to be weighed is placed on a pan attached to the lever and is balanced against standards of known mass.

Median A statistic that is the middle value of a data set or the number that is greater than or equal to 50% of the values and less than or equal to 50% of the values.

Medical device Medical items that are not pharmaceuticals and are used in the diagnosis or treatment of disease or injury. Examples include sterilizers, test kits, cell counters, and pacemakers.

Megabyte (MB) Approximately 1 million bytes, 2^{20} bytes.

Megohm-cm A unit used to measure the resistivity of water. The fewer the dissolved ions in water, the higher its resistivity. The theoretical maximum ionic purity for water is 18.3 megohm-cm at 25°C.

Melting point The temperature at which a substance transforms from the solid phase to the liquid phase.

Melting temperature (for DNA), Tm The temperature at which DNA denatures. The bonds between guanines and cytosines are stronger than those between adenines and thymines; therefore, a DNA molecule that is comparatively rich in G–Cs denatures at a higher temperature than one that is rich in A–Ts.

Membrane filter A filter where particles are retained primarily on the surface of the membrane and in which there are pores or channels of defined size.

Meniscus Greek for "crescent moon." The surface of liquids in narrow spaces forms a curve known as a meniscus. The bottom of the meniscus is used as the point of reference in calibrating and using volumetric labware.

Mercury thermometer A liquid expansion thermometer containing liquid mercury. These thermometers are seldom in use now because of the hazards associated with mercury.

Messenger ribonucleic acid (mRNA) A class of RNA molecule that acts as an intermediary by transferring information from DNA in the nucleus to ribosomes in the cytoplasm of the cell. This information dictates the amino acid sequence of a protein to be manufactured by the cell.

Metadata Information that describes the content and context of the data. They help to reconstruct the original raw data. For example, a digital camera produces both a picture and also metadata that includes the camera shutter speed, f-stop, and other camera settings when the photo was taken.

Metallic resistance thermometer Thermometers having metal wires whose resistance increases as the temperature increases.

Meter (m) The basic metric unit for length.

Method The means of performing an analysis. A method describes the steps necessary to perform an analysis and related details, such as how the sample should be obtained and prepared, the reagents that are required, the setup and use of instruments, comparisons with reference materials, and calculations.

Method validation See **validation of an analytical method**.

Methyl red pH indicator dye that changes from red at pH 4.2 to yellow at pH 6.2.

Metric system A measurement system used in most laboratories and in much of the world whose basic units include meters, grams, and liters. These basic units are modified by the addition of prefixes that designate powers of 10.

Metrologist A person who studies and works with measurements.

Metrology The study of measurements.

Micro- (μ) A prefix meaning 1/1,000,000 or 10^{-6}.

Microampere (A) One millionth of an ampere.

Microarray A tool used by scientists to study thousands of genetic sequences simultaneously.

Microbalance An extremely sensitive balance that can accurately weigh samples to the nearest 0.000001 g.

Microchip Computer chip; an integrated circuit.

Microelectronics Electronics with very small components arranged into circuits.

Microfiltration The filtration of particles whose sizes are in the range from about 0.01 to 10 μm using membrane filters.

Microfiltration membrane filter Plastic polymeric filter that prevents the passage of microorganisms and particles physically larger than the filter's pore size.

Microfuge A small centrifuge intended for small-volume samples in the microliter to 1 or 2 mL range.

Microgram (μg) A unit of mass that is 1/1,000,000 g, or 10^{-6} g.

Microliter pipette or micropipette A term used in this book to refer to various styles of device that measure volumes in the 1 to 1,000 μL range.

Microliter (µL) A unit of volume that is 1/1,000,000 L, or 10^{-6} L.

Micrometer (µm) A unit of length that is 1/1,000,000 m or 10^{-6} m.

Micromolar (µM) A concentration expression that is 1/1,000,000 M. For example, 1 M NaCl is 58.44 g of NaCl in 1 L total volume of solution and 1 µM NaCl is 0.00005844 g of NaCl in 1 L of solution.

Micromole (µmol) An amount that is 1/1,000,000 of a mole. For example, 1 mole of NaCl is 58.44 g and 1 µmol NaCl is 0.00005844 g of NaCl.

Micron A micrometer.

Microorganism A microscopic organism; includes bacteria, fungi, and viruses.

Micropipette A term used in this book to refer to an instrument that measures volumes typically in the 1–1,000 µL range. This term is not generally used to refer to a simple hollow tube device, but rather refers to a more complex instrument.

Microprocessor An integrated circuit that can store and process information.

Mil (in gloves) A measurement of glove thickness, where 1 mil = 0.001 in.

Milli- (m) A prefix meaning 1/1,000 or 10^{-3}.

Milliampere (mA) One thousandth of an ampere.

Milligram (mg) A unit of mass that is 1/1,000 g.

Milligrams per cubic meter, mg/m³, of air A measure of air concentrations of chemicals.

Milliliter (mL) A unit of volume that is 1/1,000 L.

Millimeter (mm) A unit of length that is 1/1,000 m.

Millimolar (mM) A concentration expression that is 1/1,000 molar. For example, 1 M NaCl is 58.44 g of NaCl in 1 L total volume of solution and 1 mM NaCl is 0.05844 g in 1 L total volume.

Millimole (mmol) An amount that is 1/1,000 of a mole. For example, 1 mole of NaCl is 58.44 g and 1 mmole of NaCl is 0.05844 g.

Milliosmolality A concentration expression that is milliosmoles/kg of solvent.

Millivolt (mV) One thousandth of a volt.

Minimal medium Growth medium that contains only the minimum nutrients required for a particular microbe to survive and reproduce.

Mixed bed ion exchanger A design in which both cation and anion exchange resins are housed in the same cartridge.

Mobile phase (in chromatography) The liquid or gas that moves past the **stationary phase** in a chromatography system.

Mode The value that is most frequently observed in a set of data.

Mohr pipette A type of serological pipette that is calibrated so that the liquid in the tip is not part of the measurement.

Molality (m) An expression of concentration of a solute in a solution that is the number of moles of solute per kilogram of solvent.

Molarity (M) An expression of concentration of a solute in a solution that is the number of moles of solute dissolved in 1 L of total solution.

Mold A filamentous form of fungus and common contaminant of cell cultures.

Mole A mole of any element contains 6.02×10^{23} (Avogadro's number) atoms. Because some atoms are heavier than others, a mole of one element weighs a different amount than a mole of another element. A mole of a compound contains 6.02×10^{23} molecules of that compound.

Molecular sieving See gel permeation chromatography.

Molecular weight (MW) (of a compound) See **formula weight**.

Molecular weight cutoff (MWCO) In ultrafiltration, the lowest molecular weight solute that is retained by the membrane.

Monochromatic Light that is of one wavelength. In practice, "monochromatic" light is composed of a narrow range of wavelengths.

Monochromator A device used to separate polychromatic light into its component wavelengths and to select light of a certain wavelength (or a narrow range of wavelengths).

Monoclonal antibodies (Mabs) Exceptionally homogeneous populations of antibodies directed against a specific target and produced by cells that are derived from a single antibody-producing cell.

mRNA vaccine A type of vaccine to prevent disease caused by a pathogen, first launched widely in 2020 to combat COVID-19. This type of vaccine contains mRNA that codes for a protein that is part of the pathogen. Once inoculated, the recipient's body uses the mRNA to guide production of the pathogen protein, which then initiates an immune response against the pathogen.

MT Empty; used to mark empty gas cylinders.

Multicomponent spectrophotometric analysis Methods that allow simultaneous quantitation of more than one analyte in a sample.

Mutagen A substance that can cause changes in DNA, resulting in genetic alterations.

Mycoplasma Simple bacteria that lack a cell wall and are small enough to pass through regular cell culture sterilization filters.

National Center for Biotechnology Information (NCBI) A subdivision of NIH that acts as a public resource for molecular data and other biomedical information.

National Committee for Clinical Laboratory Standards (NCCLS) See **Clinical and Laboratory Standards Institute**.

National Fire Protection Association (NFPA) An organization that developed the visual labeling and rating system for heath, flammability, reactivity, and related hazards.

National Institute for Occupational Safety and Health (NIOSH) Public health service that tests and recommends chemical exposure limits.

National Institute of Standards and Technology (NIST) The national standard laboratory for the United States, formerly known as NBS, the National Bureau of Standards. NIST is a federal agency that works with industry and government to advance measurement science and develop standards.

National Institutes of Health (NIH) A part of the U.S. Department of Health and Human Services that is the primary federal agency for conducting and supporting medical research.

Natural bandwidth (in spectrophotometry) The width of the peak generated for a specific material when its absorbance is plotted versus wavelength, an intrinsic characteristic of the substance. Natural bandwidth is measured as the width of the peak at half its height.

Natural log (ln) Logarithms whose base is the number ≈ 2.7183, called "e".

Near vertical tube rotor A modification of a fixed angle rotor having a shallow angle of about $8°–10°$.

Neurotoxin A compound that can cause damage to the **central nervous system**.

Neutral litmus pH indicator dye that is red in acid conditions and blue in alkaline conditions.

Neutral position (in ergonomics) The body alignment that uses the least muscular energy and provides the best possible blood circulation.

Neutral red pH indicator dye that changes from red at pH 6.8 to orange at pH 8.0.

Neutral solution An aqueous solution with an equal number of hydrogen and hydroxide ions and a pH of 7.

Neutralize To chemically react acids and bases to form neutral salts and water; to bring the pH of a material or mixture to 7.0.

New Drug Application (NDA) An application filed with the Food and Drug Administration requesting permission for marketing and sale of a new drug product.

Noise (electrical) Unwanted electrical interference that creates a signal unrelated to the property being measured. Electrical noise may have two components:

Short term Random, rapid spikes in the electronics of an instrument.

Long term, or drift A relatively long-term increase or decrease in readings due to changes in the instrument and electronics.

Nominal pore size (in filtration) Rated pore size; a large percent (often 99.9%) of particles above the nominal pore size should be retained by the filter.

Nominal volume The desired volume for which a pipetting device is set.

Nomogram, nomograph A chart that can be used to determine either RCF, rpm, or rotor radius values (rarely), if the other two values are known.

Non-bypass fume hood A fume hood design where air enters the hood only at the bottom of the sash.

Nonconformance Situation in which a product or raw material does not meet its specifications.

Nondisclosure agreement A contract in which the parties promise to protect the confidentiality of secret information that is disclosed during employment or another type of business transaction.

Nonelectrolyte A substance that does not ionize when dissolved in solution.

Nonfiber-releasing filter A filter that is treated in such a way that it will not release fibers into the filtrate.

Nonflammable Not easily ignited and burned.

Nonionic detergent A **detergent** whose **hydrophilic** portion is not ionized in solution.

Normal distribution The frequency distribution of data that have a bell shape when graphed. The peak of a perfect normal distribution is the mean, median, and mode, and values are equally spread out on either side of that central high point.

Normality An expression of the concentration of a solute in a solution that is the number of equivalent weights of a solute per liter of solution. See **equivalent weight**.

Nuclease Any member of a class of enzymes that break the covalent bonds of DNA and RNA.

Nucleotide A subunit of RNA and DNA that consists of a phosphate group, a five-carbon sugar (ribose in RNA or deoxyribose in DNA), and one of four nitrogenous bases (adenine, guanine, cytosine, or uracil in RNA; adenine, guanine, cytosine, or thymine in DNA).

Numerator The number written above the line in a fraction.

Occupational Exposure to Hazardous Chemicals in Laboratories Standards (29 CFR Part 1910.1450) A set of federal standards which adapt and expand the **HCS** to apply to academic, industrial, and clinical laboratories.

Occupational exposure As defined by OSHA, reasonably anticipated skin, eye, mucous membrane, or parenteral contact with blood or other potentially infectious materials that may result from the performance of an employee's duties.

Occupational Safety and Health Administration (OSHA) The main federal agency responsible for monitoring workplace safety.

Off-center errors Differences in indicated weight when a sample is shifted to various positions on the weighing area of the weighing pan.

Ohm (Ω) Unit of resistance. One ohm is the value of resistance through which 1 V maintains a current of 1 A.

Ohm's law An equation that relates voltage, current, and resistance in a circuit: Voltage = (current) (resistance).

Ohmmeter An instrument used to measure electrical resistance in a circuit.

Open system (in the context of computers) A computer system that is not controlled by the persons who are responsible for the content of the system. For example, if a contract laboratory sends data to a company via the Internet, the system is open. Additional security must be in place for open systems as compared to closed systems.

Operational qualification (OQ) A set of activities designed to establish that a piece of equipment performs within acceptable limits.

Optical density (OD) See **absorbance**.

Order of magnitude One order of magnitude is 10^1. For example, 10^2 is said to be "two orders of magnitude" less than 10^4.

Ordinate In graphing, the Y-coordinate for a point; the distance of a point along the Y-axis.

Organelle A structurally distinct component of a cell that performs a particular job in the cell (e.g., ribosomes are an organelle that produces proteins).

Organic A broad category that refers to materials containing carbon and hydrogen.

Origin The intersection of the X- and Y-axes on a two-dimensional graph whose coordinates are (0,0).

O-ring (for a micropipette) Rubber ring used as a seal to prevent the leakage of air.

Orphan Drug Act Enacted in 1983 in the United States, provides tax relief and some marketing exclusivity for companies that develop drugs for less-common diseases.

Osmolality The number of osmoles of solute per kg of water.

Osmolarity The number of osmoles of solute per liter of solution.

Osmole An osmole of any substance is equal to 1 g molecular weight of that substance divided by the number of particles formed when the substance dissolves. For example, KCl ionizes when dissolved to form one K^+ and one Cl^- ion; therefore, 1 osmole of potassium chloride is equal to its molecular weight, 74.6 g, divided by 2 = 37.3 g. For a substance that does not ionize when dissolved, 1 mole = 1 osmole.

Osmosis The net movement of water through a semipermeable membrane from a region of lesser solute concentration to a region of greater solute concentration.

Osmotic equilibrium A condition where the rate of water flow into and out of a cell is equal.

Osmotic pressure The amount of pressure that would need to be exerted to halt water's movement by osmosis.

Osmotic support The cells of plants, yeasts, and microbes are sensitive to changes in osmotic pressure when their cell walls are removed. Materials such as sucrose, sorbitol, mannitol, and glucose may be added to solutions containing these cells to protect the cell's integrity when the cell wall is removed.

Other potentially infectious materials OSHA-defined term referring to (1) human body fluids such as semen, vaginal secretions, cerebrospinal fluid, synovial fluid, pleural fluid, pericardial fluid, peritoneal fluid, amniotic fluid, saliva in dental procedures, any body fluid that is visibly contaminated with blood, and all body fluids in situations where it is difficult or impossible to differentiate between body fluids. (2) Any unfixed tissue or organ (other than intact skin) from a human (living or dead). (3) HIV-containing cell or tissue cultures, organ cultures, and HIV- or HBV-containing culture medium or other solutions; and blood. Organs, or other tissues from experimental animals infected with HIV or HBV.

Outlier A data point that lies far outside the range of all the other data points.

Out-of-specification (OOS) result A finding that indicates a product fails to meet its requirements.

Output connector A connection that enables an instrument to send an electrical signal to another device.

Overspeed disc (in centrifugation) A striped black disc attached to the bottom of rotors which, in conjunction with a photoelectric device in the centrifuge, prevents a rotor from being spun at rates above its maximum safe speed.

Oxidation The removal of electrons from a substance.

Oxidizing agent A substance that gains electrons in a reaction or can oxidize another substance.

Ozone sterilization A method of killing bacteria by exposing them to ozone.

PAGE (polyacrylamide gel electrophoresis) A protein separation technique that involves applying an electric field to a **polyacrylamide gel**.

Parallel circuit A circuit in which the components are connected so that there are two or more paths for current.

Parameter A numerical statement about a population, comparable to a sample statistic (e.g., the true mean and standard deviation of a population are parameters).

Parametric statistical methods Statistical methods that assume the variables of interest are normally distributed in the population.

Parenteral OSHA defines as referring to piercing mucous membranes or the skin barrier through events such as needle sticks, human bites, cuts, and abrasions.

Partial immersion thermometer A liquid expansion thermometer designed to indicate temperature correctly when the bulb and a specified part of the stem are exposed to the temperature being measured. Compare with **complete immersion thermometer** and **total immersion thermometer**. (Based on definition in ASTM Standard E 344-07 "Terminology Relating to Thermometry and Hydrometry.")

Particle retention A term used in reference to depth filters to indicate the smallest particle size (in micrometers) that is retained by the filter. Particle retention ratings are nominal (i.e., a small percent of particles of the rated size will penetrate the filter).

Particulate A general term used to refer to a solid particle large enough to be removed by filtration.

Partition chromatography A chromatographic technique for separating molecules between two liquid phases based on their relative solubility in each phase.

Parts per billion (ppb) An expression of concentration of solute in a solution that is the number of parts of solute per billion parts of solution. Any units may be used, but must be the same for the solute and total solution.

Parts per million (ppm) An expression of concentration of solute in a solution that is the number of parts of solute per 1 million parts of total solution. Any units may be used, but must be the same for the solute and total solution.

Pasteur pipette A type of pipette used to transfer liquids from one place to another, but not to measure volume.

Patent A type of intellectual property protection that is an agreement between the government, represented by the Patent Office, and an inventor whereby the government gives the inventor the right to exclude others from using the invention in certain ways.

Path length The distance light passes through the sample; typically measured as the length of the cuvette in centimeters or millimeters.

Pathogen Any biological organism that can cause disease.

Pathogenicity The relative capability of an organism to cause disease in humans or other living organisms.

Pellet Components of a sample that have settled to the bottom of a container after centrifugation.

Pen drive See **flash drive**.

Peptones Hydrolyzed (cleaved) proteins formed by enzymatic digestion or acid hydrolysis of natural substances including milk, meats, and vegetables.

Percent (%) A fraction whose denominator is 100.

Percent Error $\dfrac{\text{True value} - \text{measured value}}{\text{true value}} (100\%)$

where the true value may be the value of an accepted reference material.

Percent solution An expression of the concentration of solute in a solution where the numerator is the amount of solute (in grams or milliliters) and the denominator is 100 units (usually milliliters) of total solution. There are three types of percent expressions that vary in their units:

Weight per volume percent, w/v The grams of solute per 100 mL of solution.

Volume percent, v/v The milliliters of solute per 100 mL of solution.

Weight percent, w/w The grams of solute per 100 g of solution.

Performance qualification (PQ) A set of activities that evaluate the performance of a piece of equipment under the conditions of actual use.

Performance verification A process of checking that an instrument is performing properly.

Permeate The fluid and particles that pass through a membrane filter.

Permeation rate (in gloves) A measurement of the tendency of a chemical to penetrate glove material.

Permissible exposure limit (PEL) Limit set by OSHA for the allowable concentration of a substance in air.

Peroxide former A chemical that produces peroxides or hydroperoxides with age or air contact.

Personal containment Standard practices used to reduce the spread of the microorganism; procedures used for handling and disposal of the hazard along with the practices of proper laboratory hygiene.

Personal protective equipment (PPE) Specialized clothing or equipment worn for protection against a hazard.

pH A measure representing the relative acidity or alkalinity of a solution; expressed as the negative log of the H^+ concentration when concentration is expressed in moles per liter; $pH = -\log [H^+]$.

pH indicator dyes Dyes whose color is pH dependent. Indicators can be directly dissolved in a solution or can be impregnated into strips of paper that are then dipped into the solution to be tested.

pH-measuring electrode An electrode whose voltage depends on the concentration of H^+ ions in the solution in which it is immersed, also called a **glass electrode**.

pH meter A term that is commonly used to refer to a specialized voltage meter and its accompanying electrodes used to measure pH.

Pharmaceuticals Chemical agents with therapeutic activity in the body that are used to treat, correct, or prevent the symptoms of illnesses, injuries, and disorders in humans and other animals.

Pharmacogenetics/pharmacogenomics The use of genetic tests to decide if a certain drug will be safe and effective for a particular person.

Pharmacokinetics (PK) The study of how the body acts on a drug; the absorption, distribution, metabolism, and excretion of the drug and its metabolites in the body over time.

Phase I clinical trials Studies of candidate pharmaceuticals typically performed on healthy human volunteers that focus on the safety of potential products.

Phase II clinical trials Studies of potential pharmaceuticals typically performed on a small number of diseased patients to determine the drug's clinical efficacy, dose, and potential side effects.

Phase III clinical trials Studies of potential pharmaceuticals performed on a population of several hundred to several thousand patients to establish therapeutic efficacy, side effects, longer-term safety, and recommended dose levels.

Phase transition Condition where a liquid turns to a solid or a vapor, or a vapor turns to liquid.

Phenol red pH indicator dye that changes from yellow at pH 6.8 to red at pH 8.2.

Phenolphthalein pH indicator dye that changes from colorless at pH 8.0 to red at pH 10.0.

Philadelphia chromosome A chromosomal rearrangement in which pieces of two different chromosomes break off and each piece reattaches to the opposite chromosome. This rearrangement fuses part of a specific gene from chromosome 22 (the bcr gene) with part of another gene from chromosome 9 (the abl gene). The resulting chromosome 22 is termed the Philadelphia chromosome and is associated with a type of leukemia.

Phosphodiester bond A covalent bond that connects the nucleotides that compose DNA or RNA. A phosphodiester bond is formed when the 5'-phosphate group of one nucleoside is joined to the 3'-hydroxyl group of the next nucleoside through a phosphate group bridge.

Photodetector A device that responds to light by producing an electrical signal that is proportional to the amount of incident light.

Photodiode array detector (PDA) A type of detector that can simultaneously determine the absorbance of a sample at a wide range of wavelengths.

Photoemissive surface A surface coated with a material that gives off electrons when bombarded by light.

Photometer An instrument that measures light absorbance, consisting of a light source, a filter to select the wavelength range, a sample holder, a detector, and a readout device.

Photometric accuracy (also called absorbance scale accuracy) The extent to which a measured transmittance (or absorbance) value agrees with the nationally or internationally accepted value.

Photometric linearity The ability of a spectrophotometer to give a linear relationship between the intensity of light hitting the detector and the detector response.

Photomultiplier tube (PMT) A common type of detector used in spectrophotometers that consists of a series of metal plates coated with a thin layer of a photoemissive material. Light transmitted through the sample "knocks" electrons from the surface, ultimately resulting in an electrical signal.

Photons Packages of energy of electromagnetic radiation.

Physical containment Containment strategies that include laboratory design and the physical barriers that workers use.

Physiological saline 0.9% NaCl.

Physiological solution An isotonic solution that is used to support intact cells or microorganisms in the laboratory.

Pilot plant A moderate size production facility where development of efficient production methods occurs. Used to produce clinical material until a larger facility is needed for commercial production.

Pipeline and Hazardous Materials Safety Administration (PHMSA) The agency within the Department of Transportation responsible for regulating and monitoring all transport of hazardous materials within the United States.

Pipette (also spelled "pipet") Hollow tube that allows liquids to be drawn in and dispensed from one end; generally used to measure volumes in the 0.1–25 mL range.

Pipette aid A device used to draw liquid into and expel it from pipettes.

pKa The pH at which a buffer stabilizes and is least sensitive to additions of acids or bases.

Plasmid A circular molecule of DNA found most often in bacteria, which exists separately from the bacterial chromosome and can replicate independently. Plasmids are often used as vectors to transport genetic information into a cell.

Platinum resistance thermometer A metallic resistance thermometer based on the fact that the resistance in a platinum wire increases as the temperature increases.

Plunger A part of a manual micropipette that is compressed by the operator as liquids are taken up and expelled.

Poise In a mechanical balance, a sliding weight mounted on a lever used to balance against a sample.

Polarity The characteristic of having a positive or negative charge.

Polarization layer (in filtration) A buildup of retained particles or solutes on a membrane.

Polarized An electrical device that has negative and positive identifications on its terminals. When connecting such devices to a source of voltage, the negative terminal of the device is connected to the negative terminal of the voltage source.

Polarized plug Plugs with different sized prongs to ensure that the plug is oriented correctly in the outlet.

Polished water Refers to high-quality water after it has gone through a final phase of treatment that removes "all" impurities.

Polyacrylamide gel A separation gel made from polymerized acrylamide and bisacrylamide in the presence of an appropriate initiator (ammonium persulfate) and catalyst (N,N,N′,N′-tetramethylethylenediamine, also called TEMED).

Polyacrylamide gel electrophoresis See **PAGE**.

Polychromatic light A combination of light of many wavelengths.

Polyclonal antibodies Antibodies derived from different B cell lines that are produced in response to a particular antigen, but potentially recognize different epitopes.

Polymodal distribution A frequency distribution with more than two peaks.

Population A group of events, objects, or individuals where each member of the group has some unifying characteristic(s). Examples of populations are all of a person's red blood cells and all the enzyme molecules in a test tube.

Pore size, absolute (in filtration) The size of particles that are retained with 100% efficiency by a microfilter under specified conditions.

Pore size, nominal (in filtration) The size of particles that are retained with an efficiency below 100% (typically 90%–98%) by a microfilter under specified conditions.

Porosity The percentage of a filter area that is porous.

Positive displacement micropipette A volume-measuring device, such as a syringe, where the sample comes in contact with the plunger and the walls of the pipetting instrument.

Positively charged A material in which electrons are depleted.

Potable water Water that is suitable for drinking.

Potential energy Energy is the ability to do work; potential energy is stored energy.

Potentiometer A type of variable resistor that can be adjusted to control the voltage or current in a circuit.

Power (P) The rate of using energy measured in the unit of watts: Power = (volts) (current).

Power supply A device that produces varying levels of AC and DC voltage, converts AC current to DC, and regulates the voltage level in the circuit.

Precipitation methods Bioseparation techniques that are based on differences between molecules in their tendency to precipitate from various liquids.

Precision (1) The consistency of a series of measurements or tests obtained under stipulated conditions. (See also **Reproducibility** and **Repeatability**.) (2) The fineness of increments of a measuring device; the smaller the increments, the better the precision.

Precision (for a balance) A measure of the consistency of a series of repeated weighings of the same object.

Precision medicine Strategies that tailor a medical treatment to the characteristics, particularly the genetic traits, of an individual.

Preclinical development The stages of drug development preceding the first trials in humans. Encompasses laboratory (in vitro) and animal testing.

Predicate Rules (in the context of 21 CFR Part 11) The GMP, GCP, GLP, and other regulations (as contrasted with the 21 CFR Part 11 regulations).

Prefilter A filter placed upstream from a membrane filter or from an instrument to protect the membrane or instrument from particulates that would cause clogging.

Preparative centrifugation Centrifugation for the purpose of obtaining material for further use.

Preparative method (in contrast to an analytical method) Method that produces a material or product for further use (perhaps for commercial use, or perhaps for further experimentation). Chromatography, for example, is used to separate the components of a sample from one another. If the components are separated in order to purify an enzyme product that will later be sold commercially, then chromatography is being used as a preparative method. Chromatography is also frequently used to help identify the components present in a mixture, in which case, chromatography is used as an analytical method.

Pretreatment (of water) Initial steps in a water treatment process that remove a large portion of contaminants and prolong the life of filters and cartridges used in later, more expensive steps.

Preventive maintenance A program of scheduled inspections of laboratory instruments and equipment that leads to minor adjustments or repairs and ensures that items are functioning properly.

Primary chemical container Containers supplied by the manufacturer; these should immediately be labeled with a date and username when they arrive in the laboratory.

Primary containment Protection of the worker and laboratory environment through the use of good microbiological technique and safety equipment.

Primary culture (mammalian cell culture) Cultured cells that are derived directly from excised tissue.

Primary standard In the United States, the physical items housed at NIST to which measurements are referenced (traced).

Primary structure (proteins) The linear sequence of amino acids that comprise a protein and are connected by covalent peptide bonds.

Printed circuit board A thin piece of insulating material onto which copper wires are printed using a chemical process. Electronic components are connected to one another on the board.

Prion A type of infectious agent, thought to be composed of misfolded proteins, which causes a variety of neurodegenerative diseases including BSE in cattle and **Creutzfeldt–Jakob disease** in humans.

Probability The likelihood of a particular outcome.

Probability distribution A theoretical distribution of values based on the calculated probabilities of values occurring.

Probe (temperature) A generic term for many types of temperature sensor.

Procedure Specified way to perform an activity.

Process Set of interrelated resources (such as personnel, facilities, equipment, techniques, and methods) that transforms inputs into outputs.

Process development The process in which R&D scientists develop the methods by which a product will be made, especially through increasingly larger scales of production.

Process filtration Filtration of large volumes of liquids, as are found in biotechnology production facilities, pharmaceutical companies, and food production facilities.

Process validation Activities that prove that a manufacturing process meets its requirements so that the final product will do what it is supposed to do safely and effectively. See **validation**.

Product Results of activities or processes. A product may be a service, a processed material, or knowledge.

Product development The process in which R&D scientists explore and optimize the features of a potential product.

Product water Water produced by a purification process.

Program A series of instructions, written in one of many computer languages, that allows the computer to perform a specific function.

Prokaryotic organism Organisms whose cells have only a few organelles and lack nuclei; includes bacteria and cyanobacteria.

Proportion Two ratios that have the same value, but different numbers. For example: $1/2 = 5/10$.

Proportionality constant A value that expresses the relationship between the numerators and the denominators in a proportional relationship.

Protease An enzyme which breaks peptide bonds, thus degrading proteins.

Protease inhibitor An agent that inhibits protease activity.

Protein Diverse biological molecules composed of amino acid subunits that control the structure, function, and regulation of cells.

Protein assay A test of the amount or concentration of protein in a sample.

Protein error A cause of error when using pH indicator dyes, where proteins react with indicators and affect their colors.

Proteolysis The breaking apart of the primary structure of a protein by the action of enzymes (proteases) that digest the peptide bonds connecting the amino acids.

Proteomics The study of how proteins interact with one another and work in living systems.

Protocol A step-by-step outline that tells an operator how to perform an experiment that is intended to answer a question.

PubMed A database of more than thirty million scientific reference articles.

Purification The separation of a specific material of interest from contaminants, in a manner that provides a useful end product.

Purified water (USP) Water that meets USP requirements and is used in a number of pharmaceutical and cosmetic applications. **Purified water** is defined by its chemical specifications and may be produced by a variety of techniques.

Pyrogen Materials derived from lipopolysaccharides that are released from gram-negative bacteria and that trigger a dangerous immune response and induce fever, hence their name pyrogen (heat producing). The term *endotoxin* is sometimes used as a synonym because these substances are toxic to animals.

Pyrophoric Chemicals that will ignite on contact with air.

QS Latin abbreviation, *quantum sufficit*, "as much as sufficient." See **BTV, bring to volume**.

Qualification process Process of demonstrating whether an entity is capable of fulfilling specified requirements.

Qualitative analysis The analysis of the identity of substance(s) present in a sample.

Qualitative filter paper A paper filter that leaves an appreciable percent of ash when burned.

Quality According to ISO, all the features of a product or service that are required by the customer or user.

Quality assurance (QA) An organizational unit in a company that provides confidence that product quality requirements are fulfilled, in part through the effective deployment and management of documentation.

Quality control (QC) A department responsible for monitoring processes and performing laboratory testing to ensure that products are of suitable quality.

Quality system The organization, structure, responsibilities, procedures, processes, and resources for ensuring the quality of a product or service.

Quantitation limit The lowest amount of analyte in a sample which can be quantitatively determined with suitable precision and accuracy.

Quantitative analysis The measurement of the concentration or amount of a substance of interest in a sample.

Quantitative filter paper A paper filter that leaves little ash when burned.

Quaternary protein structure A protein complex formed when two or more protein chains associate with one another.

Query (in databases) A text-based set of criteria used to search for and extract a desired subset of data from a database and then present it in a specific format.

R group (of proteins) Amino acids are distinguished from one another by a side chain, or *R group*, which has a particular structure and chemical properties.

Radiant energy Energy of electromagnetic waves.

Radius of rotation The distance from the center of rotation to the material being centrifuged.

Random error Error of unknown origin that causes a loss of precision.

Random phenomenon A phenomenon where the outcome of a single repetition is uncertain, but the results of many repetitions have a known, predictable pattern. For example, the outcome of a single flip of a coin might be a head or a tail; the outcome of many flips is predicted to be half heads and half tails.

Random sample A sample drawn in such a way that every member of a population has an equal chance of being chosen.

Random variability (random error) The situation where a group of observations or measurements vary by chance in such a way that they are sometimes a bit higher than expected, sometimes a bit lower, but overall average the "correct" or expected value. See also the contrasting term, **systematic variability**.

Range (1) A range of values, from the lowest to the highest, that a method or instrument can measure with acceptable results. (2) A statistical measure that is the difference between the lowest and the highest values in a set of data.

Rankine temperature scale A temperature scale that places 0 at absolute zero (like the Kelvin scale) and whose units,° R, are of the same size as a Fahrenheit degree.

Rate zonal density gradient centrifugation A method of density gradient separation in which particles move through a gradient at different rates based on their size and density.

Ratio The relationship between two quantities, for example, "25 miles per gallon" and "10 mg of NaCl per L."

Raw data The first record of an original observation. Depending on the situation, raw data may be recorded with pen by the operator, may be a paper output from an instrument, or may be recorded directly into a computer medium.

Reactivity The tendency of a chemical to undergo chemical reactions.

Readability (for a balance) The value of the smallest unit of weight that can be read. This may include the estimation of some fraction of a scale division or, in the case of a digital display, will represent the minimum value of the least significant digit. (According to ASTM document E 319-85.)

Readout See **display**.

Reagent-grade water Water suitable to be a solvent for laboratory reagents or for use in analytical or biological procedures.

Receptacle The half of a connector that is mounted on a wall, instrument panel, or other support and into which electrical devices are plugged.

Reciprocal A number related to another so that when multiplied together their product equals 1. To calculate the reciprocal of a number, divide 1 by that number. For example, the reciprocal of 5 is: 1/5 or 0.2.

Recirculation (of water) A process in which purified water is continuously recirculated and repurified. Recirculation helps prevent microbial growth and leaching of materials into the water from storage containers.

Recombinant DNA (rDNA) DNA that contains sequences of DNA from different sources that were brought together by techniques of molecular biology and not by traditional breeding methods.

Recombinant DNA Advisory Committee (RAC) An expert committee assembled by the National Institutes of Health that has reviewed and interpreted what is known about the risks associated with genetic manipulation methods and has published guidelines for safely working with recombinant DNA.

Rectification Conversion of alternating current into direct current.

Red litmus pH indicator dye that changes from red to blue, denoting a change from acid to alkaline.

Reducing agent A substance that donates electrons in a chemical reaction.

Reducing agents (in biological solutions) A chemical added to a solution to simulate the reduced intracellular environment and prevent unwanted oxidation reactions.

Reduction to practice (in the context of patents) Constructing a prototype of an invention or performing a method or process (as described in the patent claim).

Reference (in spectrophotometry) See blank.

Reference electrode An electrode that maintains a stable voltage for comparison with the pH-measuring electrode or an ion-sensitive electrode.

Reference junction (in temperature measurement) The cold junction in a thermocouple circuit that is held at a stable known temperature.

Reference material A material or substance one or more properties of which are sufficiently well established to be used for the calibration of an apparatus, for the assessment of a measuring method, or for assigning values to materials.

Reference point A temperature at which the reading of a thermometer is calibrated or verified.

Reference standards (for weights) Individual or combinations of weights whose mass values and uncertainties are known sufficiently well to allow them to be used to calibrate other weights, objects, or balances.

Reflection (of light) An interaction of light with matter in which the light strikes a surface, causing a change in the light's direction.

Regeneration (of ion exchange resins) A process in which the capacity of used ion exchange resins is restored. An acid rinse is used to restore cation resins, and a sodium hydroxide rinse is used to restore anion resins.

Regenerative medicine A broad area of biotechnology that encompasses strategies to restore normal function to tissues and organs that have been damaged due to injury, genetic problems, aging, or disease.

Regulated waste Liquid or semi-liquid blood or other potentially infectious materials; any contaminated items that would release blood or other potentially infectious materials in a liquid or semi-liquid state if compressed; any items that are caked with dried blood or other potentially infectious materials and are capable of releasing these materials during handling; contaminated sharps; and pathological and microbiological wastes containing blood or other potentially infectious materials (OSHA).

Regulations Requirements imposed by the government and having the force of law.

Regulator (in gas cylinders) A device which attaches to the valve of a compressed gas cylinder and decreases and modulates the pressure of the gas leaving the cylinder.

Regulatory affairs unit A functional unit in a company whose personnel interpret the rules and guidelines of regulatory agencies and ensure that the company complies with these requirements.

Regulatory agency A government body that is responsible for enforcing and interpreting legislative acts. The FDA, for example, is a regulatory agency responsible for ensuring the safety, effectiveness, and reliability of medical products, foods, and cosmetics as set forth in federal legislation.

Regulatory DNA Segments of DNA that act as "switches" to control which genes are turned on at a given time in a given cell.

Regulatory submissions Documents that are completed to meet the requirements of a government regulatory agency.

Relative centrifugal field (RCF) The ratio of a centrifugal field to the earth's force of gravity; the unit of RCF is in $\times$ g. The relative centrifugal field developed in a centrifuge depends on the speed of rotation and the distance to the center of rotation according to the formula: $RCF = 11.2 \times r \ (RPM/1,000)^2$, where RCF is the relative centrifugal field, RPM is revolutions per minute, and r is the radius of rotation in cm.

Relative standard deviation See **coefficient of variation**.

Renaturation (of DNA) Process in which complementary base pairs reestablish hydrogen bonds, bringing together two strands of DNA.

Repeatability Precision under repeatability conditions.

Repeatability conditions Conditions where independent test results are obtained with the same test method on identical test items in the same laboratory by the same operator using the same equipment within short intervals of time.

Repeatability standard deviation The standard deviation of test results obtained under repeatability conditions.

Repetitive stress injury (RSI) Physical injury caused by performing constant identical motions over long periods of time.

Representative sample Per GMP: "[A] sample that consists of a number of units that are drawn based on rational criteria such as random sampling and intended to assure that the sample accurately portrays the material being sampled."

Reproducibility Precision under reproducibility conditions.

Reproducibility conditions Conditions where test results are obtained with the same test method on identical test items in different laboratories with different operators using different equipment.

Reproducibility standard deviation The standard deviation of test results obtained under reproducibility conditions.

Research and development (R&D) The organizational unit in a company that finds ideas for products, performs research and testing to see if the ideas are feasible, and develops promising ideas into actual products.

Residue analysis Method in which particles of interest are separated and retained on a filter for further analysis.

Resin (ion exchange) Beads of synthetic material that have an affinity for certain ions.

Resistance (R, in the context of electricity) A measure of the difficulty electrons encounter when moving through a material, measured in units of ohms (Ω) or megohms (MΩ), where 1 megohm = 1,000,000 Ω.

Resistant (in safety) Indicates a relative inability to react with another material.

Resistivity 1/conductivity. The unit is ohm-cm.

Resistor An electronic component that resists the flow of current.

Resolution The smallest detectable increment of measurement.

Resolution (in bioseparations) The relative ability of a technique to distinguish between the product of interest and its contaminants. The resolving power of a method is directly dependent on its **selectivity**.

Resolution (in spectrophotometry) The separation (in nanometers) of two absorbance peaks that can just be distinguished.

Resource Conservation and Recovery Act of 1976 (RCRA) Legislation which provides a system for tracking hazardous waste, including toxic or reactive chemicals, from generation to disposal.

Respirator Breathing devices designed to reduce airborne hazards by manipulating the quality of the air supply.

Restriction endonucleases Enzymes that cleave DNA at sites with specific base pair sequences.

Restriction fragment length polymorphism (RFLP) analysis A method of analyzing DNA that can be used to identify individuals based on slight differences in the sequence of their DNA in specific regions of the genome. The method is also used to look for mutations associated with specific genetic diseases. The term *polymorphism* refers to the slight differences that exist between individuals in base pair sequences. The method involves incubating the test subject's DNA with restriction enzymes that recognize and cut the DNA at specific sequences. The resulting DNA fragments are separated from one another by electrophoresis and stained. The DNA fragments form a pattern of bands that differs among individuals depending on their sequence of DNA in the analyzed region.

Retentate The materials trapped by a filter.

Retiring (a rotor) Taking a rotor out of use due to age or amount of use it has received.

Reverse osmosis A process in which liquid is forced through a semipermeable membrane, leaving behind impurities; can remove very low molecular weight materials, including salts, from a liquid.

Reverse-osmosis membranes Very restrictive filters that prevent the passage or materials as small as dissolved ions.

Reversed phase (RP, in chromatography) Any chromatography method where the stationary phase is nonpolar and the mobile phase is polar relative to the stationary phase.

Revolutions per minute (RPM) A measure of the speed of rotation in a centrifuge.

Rheostat A type of variable resistor that can be adjusted to control the voltage and/or current in a circuit.

Ribonucleic acid (RNA) A single-stranded molecule comprised of nucleic acids that is found in the nucleus and cytoplasm of cells where it performs various roles, particularly those relating to protein synthesis. See also **mRNA** and **RNAi**.

Ribosomes The cellular organelle responsible for manufacturing proteins.

Rise (in graphing) An expression that describes the amount by which the Y-coordinate changes on a two-dimensional graph.

Risk The probability that a **hazard** will cause harm.

Risk assessment Estimation of the potential for human injury or property damage from an activity.

RNAi, RNA interference Short stretches of RNA molecules that can turn off the activity of specific genes.

RNase, Ribonuclease A class of enzymes that catalyze the cleavage of nucleotides in RNA.

Robustness (in the context of an assay method) The capacity of a method to give acceptable results when there are deliberate variations in method parameters.

Root-mean-square (RMS) noise A statistical calculation that "averages" the noise present over a period of time; the lower the value, the less noise is present.

Rotor The device that rotates in the centrifuge and holds the sample tubes or other containers.

Ruggedness Similar to robustness. The ruggedness of an analytical procedure is its ability to tolerate small variations in procedural conditions, which may include variation in volumes, temperatures, concentrations, pH, and instrument settings, without affecting the analytical result. It provides an indication of the applicability of the method in a variety of laboratory conditions.

Run (in graphing) An expression that describes the amount by which the X-coordinate changes on a two-dimensional graph.

Safety The elimination of hazards and risk, to the extent possible.

Safety Data Sheet (SDS) An OSHA-required technical document provided by chemical suppliers, describing the specific properties of a chemical.

Safety rules Procedures which are designed to prevent accidents, by controlling the risk of hazards in situations where the hazards cannot be eliminated entirely.

Salt A compound formed by replacing hydrogen in an acid by a metal (or a radical that acts like a metal). (When an acid and base combine, their ionic components dissociate. In solution, the H^+ and OH^- ions combine to form water. The other two ions combine to create a salt. For example, the salt NaCl is formed by the combination of solutions of NaOH and HCl.)

Salt bridge See **junction**.

Salt error A cause or error when using pH indicator dyes. Salts at concentrations above about 0.2 M can affect the color of pH indicators.

Sample A subset of the whole that represents the whole (e.g., a blood sample represents all the blood in an individual).

Sample chamber The location in which the sample is placed.

Sample statistic A numerical statement about a sample (e.g., the sample mean is a statistic).

SARS-CoV-2 The virus that causes COVID-19. It is a type of coronavirus, named for the crown-like protein spikes on their surfaces. There are many types of coronavirus that cause respiratory diseases in humans and other animals.

SAS (Statistical Analysis System) A complex software package designed for statistical analysis.

Scale Deposits of calcium carbonate that, for example, coat the inside surface of boilers or the surfaces of membranes.

Scale (of a thermometer) Graduations that indicate degrees, fractions of degrees, or multiples of degrees.

Scale-up (1) The process of converting a small-scale laboratory procedure to one that will be appropriate for large-scale product purification. (2) The transition between producing small amounts of a product to the manufacture of large quantities for clinical testing and ultimately commercial sale.

Scanning (in spectrophotometry) Process of determining the absorbance of a sample at a series of wavelengths.

Scatter plot A plot where all data points are graphed, but not grouped or connected.

Scattering (of light) Redirection of light in many directions by small particles.

Scientific notation (also called exponential notation) The use of exponents to simplify handling numbers that are very large or very small. For example, 4,500 in scientific notation is 4.5×10^3.

SDS (sodium dodecyl sulfate) A negatively charged detergent.

SDS–PAGE (sodium dodecyl sulfate-PAGE) A technique that separates proteins on the basis of molecular size, using **polyacrylamide gel electrophoresis**.

Search engine A program that finds web pages, creates topic indices for these pages, and then identifies the ones that match specific search criteria.

Secondary containment Protection of the environment outside the laboratory by the use of good laboratory design and safe practices.

Secondary standard A standard whose value is based on comparison with a primary standard.

Secondary structure (of proteins) Regularly repeating patterns of twists or kinks of an amino acid chain held together by hydrogen bonds. Two common types of secondary structure are the alpha-helix and beta-pleated sheet. Regions of proteins without regularly repeating structures are said to have a "random coil" secondary structure.

Sedimentation The settling out of particles from a liquid suspension.

Sedimentation coefficient (S) A measure of the sedimentation velocity of a particle. In practice, the sedimentation coefficient is a function of the size of a particle; larger particles have larger sedimentation coefficients.

Selective medium A type of growth medium for cells that inhibits the reproduction of unwanted organisms and/or encourages the reproduction of specific organisms.

Selectivity The ability of a technique to separate a specific component from a heterogeneous mixture based on molecular properties; see **specificity**.

Self-contained breathing apparatus A **respirator** that contains its own air supply and is appropriate for situations where the user is exposed to highly toxic gases.

Semiconductor A material that is more resistant to electron flow than a conductor, but is less resistant than an insulator. Semiconductors are important in the manufacture of electronic devices.

Semilog paper A type of graph paper that has a log scale on one axis and a linear scale on the other axis.

Semipermeable membrane A membrane that allows water and some small molecules to flow through unimpeded, whereas other molecules are restricted; the membrane that surrounds every living cell is semipermeable.

Sensitivity (for a laboratory balance) The smallest value of weight that will cause a change in the response of the balance that can be observed by the operator.

Sensitivity (of a detector or instrument) Response per amount of sample.

Sensitizer A substance or agent that may trigger an allergy directly, or cause an individual to develop an allergic reaction to an accompanying chemical.

Sensor See **transducer**.

Sephadex A **gel permeation** medium manufactured by GE Healthcare Life Sciences.

Serial dilutions Dilutions made in series (each one derived from the one before) and that all have the same dilution. For example, a series of 1/10 dilutions of an original sample would be: 1, 1/10, 1/100, 1/1,000, etc.

Series circuit A circuit configuration in which the components are connected in a "string," end to end, so that current can only flow in one pathway.

Serological assay A blood test that looks for antibodies associated with a particular disease.

Serological pipette Term that usually refers to a calibrated glass or plastic pipette that measures in the 0.1–25 mL range. These pipettes are calibrated so that the last drop is in the tip. This drop needs to be "blown out" to deliver the full volume of the pipette.

Serum (mammalian cell culture) The liquid component of blood from which blood cells and most clotting factors have been removed.

Serum-free medium (mammalian cell culture) (SFM) A culture medium that does not contain serum. These media are supplemented with proteins, growth factors, vitamins, hormones, and other constituents to take the place of the serum.

Sharps A term to describe laboratory items, such as razor blades and needles, that can cause cuts and lacerations.

Short circuit An unintended path for current flow that bypasses the resistance or load.

Short tandem repeat (STR) A defined region of DNA containing multiple copies of short sequences of bases (1 to 5 base pairs long) that are repeated a number of times; the number of repeats varies among individuals.

Short-term exposure limit (STEL) See **TLV-STEL**.

SI system (Système International d'Unités) A standardized system of units of measurement, adopted in 1960 by a number of international organizations, that is derived from the metric system.

SI unit of enzyme activity The amount of enzyme necessary to catalyze transformation of 1.0 mol of substrate to product per second under optimal measurement conditions; expressed in units of kat.

Siemens, S A unit of conductance.

Signal (in electricity) An electrical change that conveys information. The signal may be a change in voltage, current, or resistance.

Signal processing unit (signal processor) A device that modifies an electrical signal, for example, to amplify it.

Signal-to-noise ratio The instrument response due to the sample divided by the electronic noise in the system.

Significant figure A digit in a number that is a reliable indicator of value.

Silicon An abundant element that is used in the manufacture of electronic components.

Silver/silver chloride (Ag/AgCl) electrode A type of reference electrode that contains a strip of silver coated with silver chloride and immersed in an electrolyte solution of KCl and silver chloride.

Single-beam, double-pan balance A simple balance that compares the weight of the sample to the weight of a standard that is placed on a pan across a beam.

Single nucleotide polymorphisms (SNPs) Places in DNA where a single nucleotide differs from person to person.

Size-exclusion chromatography See **gel permeation chromatography**.

Skewed distribution A distribution in which the values tend to be clustered either above or below the mean.

Sleeve junction A type of reference electrode junction made by placing a hole in the side of the electrode housing and covering the hole with a glass or plastic sleeve. The sleeve junction is relatively fast flowing and is unlikely to become clogged.

Slope (of a line) Given a straight line plotted on a graph, the slope of the line is how steeply the line rises or falls. It is the ratio of vertical change to horizontal change between any two points on the line. Slope can be calculated by choosing any two points on the line and dividing the change in their Y-values by the change in their X-values.

Smoke generators Small tubes of chemicals, frequently including titanium tetrachloride, which generate highly visible white smoke from a chemical reaction; used to check the general function of fume hoods.

Softened water Water that has cations replaced with sodium ions.

Solid or semisolid medium (bacterial culture) Aqueous-based mixtures of nutrients that contain agar as a hardening agent to provide a solid substrate on which or in which microbes may be cultured.

Solid state An electrical circuit that uses semiconductor diodes and transistors instead of vacuum tubes. (The term "solid" is used because signal flows through a solid semiconductor material instead of through a vacuum.)

Solid-state electrode A newer type of electrode that relies on a small electronic chip to detect ions.

Solute A substance that is dissolved in some other material. For example, if table salt is dissolved in water, salt is the solute and water is the solvent.

Solution A homogeneous mixture in which one or more substances is (are) dissolved in another.

Solvent A substance that dissolves another. For example, if salt is dissolved in water, then water is the solvent.

Solvent cutoff (in spectrophotometry) The wavelength at which a particular solvent absorbs a significant amount of light. Solvent absorbance interferes with the analysis of the analyte.

Somatic cell nuclear transfer See **therapeutic cloning**.

Sonication device (sonicator) Laboratory equipment that disrupts cells with high-frequency sound waves.

Source (in spectrophotometry) The lamp or bulb used to provide light in a spectrophotometer.

Specific activity The amount (or units) of the protein of interest, divided by the total amount of protein in a sample.

Specification The defined limits within which physical, chemical, biological, and microbiological test results for a product should lie to ensure its quality.

Specificity (also called selectivity) A measure of the extent to which a method or instrument can determine the presence of a particular compound in a sample without interference from other materials present. Typically, these might include impurities, degradants, and matrix.

Spectral bandwidth (in spectrophotometry) Spectral bandwidth is a measure of the range of wavelengths emerging from the monochromator when a particular wavelength is selected.

Spectral slit width The physical width of the monochromator exit slit multiplied by the linear dispersion of the diffraction grating.

Spectrometer Any instrument used to measure the interaction of electromagnetic radiation with matter.

Spectrophotometer An instrument that measures the effect of a sample on the incident light beam. The instrument consists of a source of light, entrance slit, monochromator, exit slit, sample holder, detector, and readout device.

Spectroscopy The study and measurement of interactions of electromagnetic radiation with matter.

Spectrum Electromagnetic radiation separated or distinguished according to wavelength.

Spirit thermometer A liquid expansion thermometer containing an alcohol-based liquid.

Sponsor Individual, company, institution, or organization that initiates or pays for testing a potential drug product and submits regulatory documents to the FDA. The sponsor may or may not perform drug development activities itself.

Spontaneous density gradient A density gradient made of a medium that forms a gradient spontaneously under the influence of a centrifugal field.

Spore Dormant microorganism that is resistant to heat.

Spore strips Dried pieces of paper to which large numbers of nonpathogenic spores are adhered; used to test the success of sterilization.

SQL (Structured Query Language) A computer language specifically designed to find and retrieve data from databases.

Stability (in safety) The chemical characteristic of remaining unchanged over time.

Stability (in the context of pharmaceuticals) The capacity of a product to remain within its specifications over time.

Stability testing According to FDA, "The purpose of stability testing is to provide evidence on how the quality of a drug substance or drug product varies with time under the influence of a variety of environmental factors, such as temperature, humidity, and light, and to establish a retest period for the drug substance or a shelf life for the drug product and recommended storage conditions."

Standard Operating principle or requirement.

Standard (in spectrophotometry) A mixture including the analyte of interest dissolved in solvent and used to determine the relationship between absorbance and concentration for that analyte.

Standard (relating to measurement) (1) Broadly, any concept, method, or object that has been established by authority, custom, or agreement to serve as a model in the measurement of any property. (2) A physical object, the properties of which are known with sufficient accuracy to be used to evaluate another item; a physical embodiment of a unit. For example, a metal object whose mass is accurately known can be used by comparison to determine the mass of a sample. (3) In chemical or biological measurements, a standard often describes a substance or a solution that is used to establish the response of an instrument or an assay method to the analyte. (4) A document established by consensus and approved by a recognized body that establishes rules or guidelines to make a procedure consistent among various people.

Standard curve (also called a calibration curve) A graph that shows the relationship between a response (e.g., of an instrument) and a property that the experimenter controls (e.g., the concentration of a substance).

Standard curve (in spectrophotometry) A graph that shows the relationship between absorbance (on the Y-axis) and the amount or concentration of a substance (on the X-axis).

Standard deviation A measure of the dispersion of observed values or results expressed as the positive square root of the variance. A statistical measure of variability.

Standard method A technique to perform a measurement or an assay that is specified by an external organization to ensure consistency in a field. For example, ASTM specifies standard methods to calibrate volumetric glassware; the U.S. Pharmacopeia specifies methods to perform tests of pharmaceutical products.

Standard microbiological practices The basic practices that should be used when working with all microbiological organisms.

Standard operating procedure (SOP) A set of instructions for performing a routine method, manufacturing operation, administrative process, or maintenance operation.

Standard reference material (SRM) A well-characterized material available from NIST produced in quantity and certified for one or more physical or chemical properties.

Standard weight Any weight whose mass is known with a given uncertainty.

Star activity When a restriction enzyme "mistakenly" cleaves DNA at sequences that are not its proper target.

Stationary phase (in chromatography) The immobile matrix in a chromatography system.

Statistical process control The desired situation in which a manufacturing or measurement process behaves as expected. A certain amount of variability in the process is expected, but when the process is in control, the variability does not exceed previously determined limits.

Stem (of a liquid-in-glass thermometer) A glass capillary tube through which the mercury or organic liquid moves as temperature changes.

Stem cells, adult Undifferentiated cells found among differentiated cells in a tissue or organ that can differentiate when needed to form the specialized cell types found in that tissue.

Stem cells, embryonic Cells from an embryo that have the potential to differentiate into any cell in the body.

Stem enlargement (of a liquid-in-glass thermometer) A thickening of the stem that assists in the proper placement of the thermometer in a device (e.g., in an oven).

Step gradient A density gradient in which there are layers of medium with different densities, each of which is clearly demarcated from the layer above and below.

Sterile technique See **aseptic technique**.

Sterilization Destruction of all life forms on an object.

Sterilize Per OSHA, using a physical or chemical procedure to destroy all microbial life, including highly resistant bacterial spores.

Sterilizing filter A nonfiber-releasing filter that produces a filtrate containing no demonstrable microorganisms when tested as specified in the United States Pharmacopeia.

Stock solution A concentrated solution that is diluted to a working concentration.

Stray light Radiation that reaches the detector without interacting with the sample.

Stringency The reaction conditions used when single-stranded complementary nucleic acids are allowed to hybridize. At high stringency, binding occurs only between strands with perfect complementarity, whereas at lower stringency, some mismatches of base pairs are tolerated.

Strip chart recorder An analog display device that records the output of a detector continuously over time.

Strong acid A chemical that completely dissociates in water to release hydrogen ions.

Strong base A chemical that completely dissociates in water to release hydroxide ions.

Strong electrolyte A substance that ionizes completely in solution.

Sublimate Change directly from a solid to a gas form (as in dry ice).

Sum of squares For a set of data, the sum of the squared differences between each data point and the mean.

Supercritical fluid chromatography (SFC) A normal phase, high-pressure chromatography system where carbon dioxide is usually used as the majority of the mobile phase.

Superheated Refers to liquids that have been heated past their boiling point without the release of the gaseous phase; superheated liquids can boil over, sometimes violently, if jarred.

Supernatant The liquid medium above a pellet after centrifugation.

Supplements (in cell culture) Loosely defined as materials that are sometimes added to basal media after the medium is prepared and/or that are added in varying concentrations.

Surface filter See **membrane filter**.

Surge protector A multioutlet power regulator between the computer and the main power outlet, designed to smooth out momentary electrical spikes and fluctuations.

Suspended cells (mammalian cell culture) Cells that grow suspended in culture medium, either as individual cells or as small clumps of cells.

Suspended solids Undissolved solids that can be removed by filtration.

Svedberg, T A pioneer in centrifugation who worked on the mathematics, methods, and instrumentation for centrifugation.

Swinging bucket rotor See **horizontal rotor**.

Switch A device that controls current flow in a circuit.

Synthetic DNA Short DNA molecules that have been chemically synthesized rather than derived from living organisms.

System suitability A part of method validation based on the concept that the instruments used, the sample, and procedure must together provide acceptable results.

Systematic error An error that causes measurement results to be biased. See **systematic variability**.

Systematic variability (systematic error) The situation where a group of observations or measurements vary in such a way that they tend to be higher or lower than the true value or the expected value; this is also called bias. The cause of systematic error may be known or unknown. See also the contrasting term, **random variability**.

Table (in databases) A set of related information.

Tangential flow filtration A filtration mode used mainly in industry where the fluid to be filtered flows across the filter. The sweeping motion of the fluid clears the surface of the membrane, thus reducing clogging.

Tare A feature that allows a balance to automatically subtract the weight of the weighing vessel from the total weight of the sample plus container.

Target organ The body part or organ most likely to be affected by exposure to a chemical or hazard.

Technology The application of knowledge to make products useful to humans.

TEMED (N,N,N′,N′-tetramethylethylenediamine) A chemical catalyst for the polymerization of **acrylamide**.

Temperature A measure of the average energy of the randomly moving molecules that make up a substance. The tendency of a substance to lose or gain heat.

Temperature scale A scale derived by choosing two or more fixed reference points.

Teratogen, teratogenic A compound that can cause defects in a fetus when administered to the mother.

Tertiary structure The three-dimensional globular structure formed by bending and twisting of a protein.

Test According to ISO Guide 17025: A technical operation that consists of the determination of one or more characteristics of performance of a given product, material, equipment, organism, physical phenomenon, process, or service according to a specified procedure. The result of a test is normally recorded in a document sometimes called a test report or a test certificate. Note that a "test" may or may not involve a "sample."

Testing laboratory A place where analysts test samples.

Therapeutic cloning A technology, not yet accomplished in humans, in which stem cells would be created by removing the genetic material from the cell of a patient and transferring it to an enucleated human egg from a donor woman. The resulting embryo would be allowed to divide a few times, and the embryonic stem cells harvested to treat the patient.

Thermal conductivity The ability of a material to conduct heat.

Thermal equilibrium The condition where two objects are at the same temperature, so that there is no net transfer of internal energy from one to the other.

Thermal expansion An increase in the size of a material due to an increase in temperature.

Thermal expansion coefficient The amount that a particular material expands with a given rise in temperature.

Thermistor A thermometer consisting of a semiconductor material that has a large change in resistance when exposed to a small change in temperature.

Thermocouple A temperature-sensing device that consists of two dissimilar metals joined together. The thermocouple has a voltage output proportional to the difference in temperature between the junction whose temperature is being measured and the reference junction.

Thermocycler An instrument that holds multiple tubes and that repeatedly heats and cools the tubes to specified temperatures.

Thin-layer chromatography (TLC) A chromatographic technique where a thin layer of stationary phase material is spread on a glass plate, and the mobile phase passes through the stationary phase by either capillary action or gravity.

Threshold (for a graph) A change in a relationship plotted on a graph.

Throughput (in filtration) The length of time a liquid can flow through a filter before the membrane clogs.

Thymine (T) One of the four types of nucleotide subunit that compose DNA.

Tissue culture The in vitro propagation of cells derived from tissue of higher organisms.

Tissue engineering A type of regenerative medicine that typically involves the implantation of materials that combine living cells with engineered structural materials to restore, maintain, or improve damaged tissues or whole organs.

TLV (threshold limit value) The air concentration of a chemical that will not pose a health threat to most normal healthy workers; determined by the ACGIH.

TLV-C (threshold value limit – ceiling) The maximum allowable concentration of a material in air; concentrations should never exceed this value.

TLV-STEL (threshold limit value – short-term exposure limit) The concentration of a toxic material in air that limits exposure of a worker to 15 minutes; determined by the ACGIH.

TLV-TWA (threshold limit value – time-weighted average) The acceptable air concentration of a substance averaged over an 8-hour day; determined by the ACGIH.

To contain (TC) A method of calibrating glassware so that it contains the specified amount when exactly filled to the capacity line. The device will not deliver that amount if the liquid is poured out because some of the liquid will adhere to the sides of the container.

To deliver (TD) A method of calibrating glassware so that it delivers the specified amount when poured.

Tolerance The amount of error allowed in the calibration of a measuring item. For example, the tolerance for a "500 g" Class 1 mass standard is 1.2 mg, which means the standard must have a true mass between 500.0012 g and 499.9988 g.

Total bacteria count An estimation of the total number of bacteria in a solution based on an assay that involves incubating a sample and counting the number of colony-forming units (CFUs).

Total dissolved solids (TDS) A value representing all the solids dissolved in a solution.

Total immersion thermometer A liquid expansion thermometer designed to indicate temperature correctly when that portion of the thermometer containing the liquid is exposed to the temperature being measured. Compare with **complete immersion thermometer** and **partial immersion thermometer**.

Total organic carbon (TOC) A measure of the concentration of organic contaminants in water.

Total solids (TS) In water treatment, a measure of the concentration of both the dissolved (TDS) and suspended solids (TSS) in a solution.

Total squared deviation See **sum of squares**.

Toxic Poisonous; a substance's ability to cause harm to biological organisms or tissue or to cause adverse health effects.

Toxic materials Substances that are poisonous.

Toxic Substances Control Act Authorizes EPA to, among other things, review new chemicals before they are introduced into commerce, including the examination of microorganisms that have been genetically modified.

Toxicology The study of poisonous materials and their effects on living organisms and tissue; the study of adverse effects of a drug and its metabolites on the body.

Trace analysis Analysis of a material that is present in very low (e.g., ppm or ppb) concentrations.

Traceability The ability to trace the history, application, or location of an entity … by means of recorded identifications. The term *traceability* may have one of three main meanings:

1. in a product . . . sense it may relate to the origin of materials and parts, the product processing history, the distribution and location of the product after delivery;
2. in a calibration sense, it relates measuring equipment to national or international standards, basic physical constants or properties, or reference materials;
3. in a data collection sense, it relates calculations and data generated throughout the quality loop . . . sometimes back to the requirements for quality . . . for an entity.

Traceability (with reference to solutions) The process by which it is ensured that every component of a product can be identified and documented.

Tracking error (in spectrophotometry) A problem that can occur when an absorbance spectrum is scanned too quickly, resulting in absorbance peaks that are slightly shifted from their true locations.

Trade secrets Private information or physical materials that give a competitive advantage to the owner.

Transcription The cellular process in which information encoded in a DNA sequence is used to synthesize a corresponding mRNA molecule.

Transducer A device that senses one form of energy and converts it to another form.

Transfection Introduction of foreign DNA into host cells; usually refers to eukaryotic host cells.

Transformation (1) A process in which cells undergo genetic changes that alter the normal mechanisms controlling cell growth and reproduction. (2) The introduction of foreign genetic information into bacteria.

Transformer (in electronics) A device used to vary the input voltage entering an instrument.

Transgenic A plant or animal that is genetically modified by the introduction of foreign DNA using the techniques of biotechnology.

Transistor A semiconductor component used primarily for amplification and sometimes to switch a circuit on or off.

Translation (in the context of cell biology) The cellular process in which the information encoded in mRNA is used to manufacture a protein.

Translational medicine Applied medical research as contrasted with basic research.

Transmittance (t) The ratio of the amount of light transmitted through the sample to that transmitted through the blank.

Transmitted light Light that passes through an object.

Trend line In scatter plots, a statistically derived line that reflects the general trend in the data values, if one exists.

Triple point of water The temperature and pressure at which solid, liquid, and gas phases of a given substance are simultaneously present in equilibrium. The triple point for water is 0.01°C.

Tris (tris(hydroxymethyl)aminomethane) One of the most common buffers in biotechnology laboratories. Tris buffers over the normal biological range (pH 7–9), is nontoxic to cells, and is relatively inexpensive.

Tris base Unconjugated **Tris**; has a basic pH when dissolved in water.

Tris–HCl Tris buffer that is conjugated to HCl.

tRNA A small type of RNA that transfers a specific amino acid to a growing polypeptide chain at the ribosome during protein synthesis.

True value (for a measurement) The actual value for a measurement that would be obtained in the absence of any error.

Trueness The closeness of agreement between the population mean of the measurements or test results and the accepted reference value.

Tungsten filament A thin metal wire that emits light when heated and provides visible light in visible spectrophotometry.

Turbid solution One that contains numerous small particles that both absorb and scatter light.

TWA (time-weighted average) The concentration of a substance in air that is allowed when averaged over an eight-hour day; during the day, the actual concentrations will be higher and lower than the daily average concentration.

Two-point calibration If the response of a device is linear, calibration at two points (e.g., 0 and full scale) calibrates the device.

Two-stage gas regulator A gas regulator with a pair of valves that (1) greatly reduce the pressure of gas leaving the cylinder and (2) provide the operator with the ability to fine-tune the gas pressure reaching its destination.

U.S. National Prototype Standards (for mass) Platinum–iridium standards with mass values traceable to the International definition of a kilogram.

Ultrafiltration A membrane separation technique that is used to separate macromolecules on the basis of their molecular weight.

Ultrafiltration membrane A filter with pores small enough to remove large molecules. These membranes are rated in terms of their molecular weight cutoff.

Ultramicrobalance An extremely sensitive analytical balance that can accurately weigh samples to the nearest 0.0000001 g.

Ultrapure water Type I or better water.

Ultrasonic washing A method of cleaning items by exposing them to high-pitch sound waves that effectively penetrate and clean crevices, narrow spaces, and other difficult-to-reach spaces.

Ultraspeed centrifuge, or ultracentrifuge A centrifuge that rotates at speeds up to about 120,000 RPM and can generate forces up to 700,000 g.

Ultraviolet (UV) light A form of nonionizing radiation which makes up the light spectrum between visible light and X-rays at 180–380 nm.

Uncertainty An indication of the variability associated with a measured value that takes into account two major components of error: (1) bias, and (2) the random error attributed to the imprecision of the measurement process. Quantitative measurements of uncertainty generally require descriptive statements of explanation because of differing traditions of usage and because of differing circumstances.

Undefined growth medium See **complex growth medium**.

Underwriters Laboratories (UL) An organization which has developed codes for safe electrical devices.

Unit (of measure) A precisely defined amount of a property.

United States Customary System (USCS) The measurement system common in the United States that includes miles, pounds, gallons, inches, and feet.

United States Department of Agriculture (USDA) A US government agency that enforces requirements for purity and quality of meat, poultry, and eggs, and that is involved in nutrition research and education. The USDA is one of the government agencies that play a role in the regulation of genetically engineered food plants.

United States Pharmacopeia (USP) (1) An organization that promotes public health by establishing and disseminating officially recognized standards for the use of medicines and other healthcare technologies. (2) The compendium containing drug descriptions, specifications, and standard test methods for such parameters as drug identity, strength, quality, and purity. This compendium is recognized as a legal authority by the FDA.

Universal absorbent Polypropylene, expanded silicates, or other materials that can absorb virtually any liquid safely, including some corrosives; can be purchased in loose or pillow form.

Universal precautions An approach to infection control, where all human blood and human body fluids are treated as if known to be infectious for **HIV, HBV,** and other **bloodborne pathogens** (OSHA).

Upstream (in filtration) The side of a filter facing the incoming liquid or gas.

Upstream processes Biological processes, such as fermentation or cell culture, that produce the biomolecule of interest.

Upstream processing In biopharmaceutical manufacturing, the process of growing cells and their production of a desired product.

Uracil (U) One of the four types of nucleotide subunit that comprise RNA.

USB (universal serial bus) A universal type of port that provides high-speed, stable connections between up to 127 devices from a single port.

UV oxidation A process in which ultraviolet light breaks down organic impurities.

VAC (in electricity) An abbreviation for voltage when the current is alternating.

Validation A process or a set of activities that ensure that an individual piece of equipment, a process, or an analytical method reliably and effectively performs the function for which it is intended.

> **Process validation** Activities that prove that a manufacturing process meets its requirements so that the final product will do what it is supposed to do safely and effectively.

> **Equipment validation This type of validation has three phases:**

>> **Installation qualification (IQ)** A set of activities designed to determine if a piece of equipment is installed correctly.

>> **Operational qualification (OQ)** A set of activities designed to establish that a piece of equipment performs within acceptable limits.

>> **Performance qualification (PQ)** A set of activities that evaluates the performance of a piece of equipment under the conditions of actual use.

> **Validation (of an analytical procedure)** A process used by the scientific community to evaluate a test method or assay. This involves determining the ability of the method to reliably obtain a desired result, the conditions under which such results can be obtained, and the limitations of the method.

Value (in math) An assigned or calculated numerical quantity.

Variable (1) A property or a characteristic that can have various values (in contrast to a constant that always has the same value). See also **dependent variable** and **independent variable**. (2) A quality of a population that can be measured. For example, for the population of all six-year-old children, one could measure height, eye color, or favorite book. Variables may be classified as follows: (1) **Discrete variables** are those that can be counted, such as litter size or the number of bacterial colonies on a petri dish. (2) **Continuous variables** can be measured and can be whole numbers or any fraction of a whole number, such as weight or temperature. (3) **Qualitative variables** are attributes that are not numeric, such as color or flavor.

Variable air volume (VAV) fume hoods Fume hoods that maintain a relatively constant **face velocity** by changing the amount of air exhausted from the hood.

Variable resistor (in electronics) A device that can be adjusted to provide differing amounts of resistance to electron flow. (See also **rheostat** and **potentiometer**.)

Variance A measure of the squared dispersion of observed values or measurements expressed as a function of the sum of the squared deviations from the population mean or sample average.

VDC (in electricity) An abbreviation for voltage when the current is direct.

Vector An entity, such as a plasmid or a modified virus, into which a DNA fragment of interest is integrated, and that carries the DNA of interest into a host cell.

Velometer, Velocity meter An instrument used to measure the **face velocity** of a fume hood.

Verification Confirmation by examination and provision of objective evidence that specified requirements have been fulfilled. (1) In connection with the management of measuring equipment, verification provides a means of checking that the deviations between values indicated by a measuring instrument and corresponding known values are consistently smaller than the limits of permissible error defined in a standard regulation or specification relevant to the management of the measuring equipment. (2) In connection with measurement traceability, the meaning of the terms *verification* and *verified* is that of a simplified calibration, giving evidence that specified metrological requirements including the compliance with given limits of error are met.

Vertical rotor A rotor in which tubes are held upright in the sample compartments.

Viable organism A living organism that can reproduce.

Virulence The capacity of a biological organism or material to cause harm.

Virus Infectious particle containing either DNA or RNA surrounded by a protein coat.

Visible radiation The region of the electromagnetic spectrum from about 380–780 nm. Light of different wavelengths in this range is perceived as different colors.

Volatile Refers to chemicals that evaporate quickly at room temperature.

Voltage The potential energy of charges that are separated from one another and attract or repulse one another.

Voltmeter An instrument used to measure voltage.

Volts The unit of potential difference between two points in an electrical circuit.

Volume The amount of space a substance occupies, defined in the SI system as length×length×length = meters^3. More commonly, the liter (dm^3) is used as the basic unit of volume. See also **liter**.

Volumetric (transfer) pipette A pipette made of borosilicate glass and calibrated "to deliver" a single volume when filled to its capacity line at 20°C.

Volumetric glassware A term used generally to refer to accurately calibrated glassware intended for applications where high-accuracy volume measurements are required.

Waste Any laboratory material which has completed its original purpose and is being disposed of permanently.

Water aspirator A laboratory setup that creates a vacuum through a side arm to a faucet with flowing water.

Water for injection (WFI) Water used to make drugs for injection. The USP specifies the quality of WFI.

Water softener An ion exchange device that exchanges positive ions that make water hard, primarily calcium and magnesium, with sodium ions.

Watts (W) (in electricity) The unit of power.

Wavelength The distance from the crest of one wave to the crest of the next wave.

Wavelength accuracy The agreement between the wavelength the operator selects and the actual wavelength that exits the monochromator and shines on the sample.

Weak acid A chemical that partially dissociates in water with the release of hydrogen ions.

Weak base A chemical that partially dissociates in water with the release of hydroxide ions.

Weak electrolyte A substance that partially ionizes when dissolved.

Weak molecular interactions Attractive interactions between atoms that are more easily disrupted than covalent bonds.

Weigh boat A plastic or metal container designed to hold a sample for weighing.

Weighing paper Glassine coated paper used to hold small samples for weighing.

Weight The force of gravity on an object.

Wi-Fi (wireless fidelity) A high-speed wireless networking standard conforming to IEEE 802.11 standards.

Workers' compensation A no-fault state insurance system designed to pay for the medical expenses of workers who are injured on the job, or develop work-related medical problems.

Working cell bank (WCB) Cells derived from the master cell bank that are expanded and used for the production of a product.

Working standard A physical standard that is used to make measurements in the laboratory and that is calibrated against a primary or secondary standard.

X-axis The main horizontal line on a graph.

"X Solution" A stock solution where X means how many times more concentrated the stock is than normal. A 10 X solution is ten times more concentrated than the solution is normally prepared.

Xenotransplantation Transplantation of organs, tissues, or cells from one species to another.

Y-axis The main vertical line on a graph.

Y-intercept The point at which a line passes through the Y-axis, where $X = 0$, on a two-dimensional graph.

Yeast extract The water-soluble portion of yeast cells that have been allowed to die so that the yeasts' digestive enzymes break down their proteins into simpler compounds while preserving the vitamins from the yeast.

Yield The **% recovery**, or percent of the starting amount of the product of interest that can be recovered in purified form using a specific strategy.

Z factor A conversion factor that incorporates the buoyancy correction for air at a particular temperature and barometric pressure.

Zonal rotor Rotors that accommodate large-volume samples by containing the sample in a large cylindrical cavity rather than in individual tubes or bottles. Zonal rotors are used for both rate zonal and buoyant density gradient separations, but are seldom used for pelleting.

Zoonotic diseases, zoonoses Diseases that can be passed between different species.

Index

Note: **Bold** page numbers refer to tables and *italic* page numbers refer to figures.

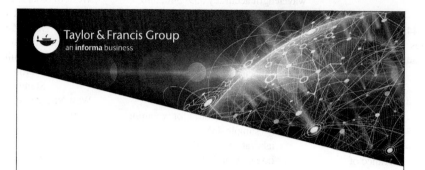